# Oxford Users' Guide to Mathematics

**Oxford Users' Guide to Mathematics**

**E. Zeidler**
*with* W. Hackbusch and H.R. Schwarz

KLF
QA
40
T4813
2004
WEB

# Oxford Users' Guide to Mathematics

*Edited by*

**Eberhard Zeidler**

Max Planck Institute for Mathematics in the Sciences
Leipzig

*Translated and typeset by*

**Bruce Hunt**

OXFORD
UNIVERSITY PRESS

Great Clarendon Street, Oxford OX2 6DP

Oxford University Press is a department of the University of Oxford.
It furthers the University's objective of excellence in research, scholarship,
and education by publishing worldwide in

Oxford New York

Auckland Bangkok Buenos Aires Cape Town Chennai
Dar es Salaam Delhi Hong Kong Istanbul Karachi Kolkata
Kuala Lumpur Madrid Melbourne Mexico City Mumbai
Nairobi São Paulo Shanghai Taipei Tokyo Toronto

Oxford is a registered trade mark of Oxford University Press
in the UK and in certain other countries

Published in the United States
by Oxford University Press Inc., New York

Translated by Bruce Hunt

English translation © Oxford University Press 2004

The moral rights of the author have been asserted

Database right Oxford University Press (maker)

Title of the original: Teubner-Taschenbuch der Mathematik, Vol 1

© 1996 B. G. Teubner, Stuttgart and Leipzig

Translation arranged with the approval of the Publisher B. G. Teubner
from the original German edition

All rights reserved. No part of this publication may be reproduced,
stored in a retrieval system, or transmitted, in any form or by any means,
without the prior permission in writing of Oxford University Press,
or as expressly permitted by law, or under terms agreed with the appropriate
reprographics rights organization. Enquiries concerning reproduction
outside the scope of the above should be sent to the Rights Department,
Oxford University Press, at the address above

You must not circulate this book in any other binding or cover
and you must impose this same condition on any acquirer

A catalogue record for this title is available from the British Library

Library of Congress Cataloging in Publication Data
(Data available)

ISBN 0 19 850763 1

10 9 8 7 6 5 4 3 2 1

Typeset in LaTeX by Bruce Hunt
Printed in Hong Kong
on acid-free paper by Nordica Printing Co., Ltd.

# *Foreword*

We live in an age in which mathematics plays a more and more important role, to the extent that it is hard to think of an aspect of human life to which it either has not provided, or does not have the potential to provide, crucial insights. Mathematics is the language in which quantitative models of the world around us are described. As subjects become more understood, they become more mathematical. A good example is medecine, where the Radon transform is what makes X-ray tomography work, where statistics form the basis of evaluating the success or failure of treatments, and where mathematical models of organs such as the heart, of tumour growth, and of nerve impulses are of key importance.

To apply mathematics in a different area requires of the mathematician the ability and willingness to learn about that area and understand its own language, and it requires of the specialist in that area a similar ability and willingness to learn the language of the appropriate parts of mathematics. Mathematics has its own internal language barriers too, so that to move from one part of the subject to another can require a major effort.

It is to these needs of furnishing basic understanding and overcoming language barriers that the *Oxford Users' Guide to Mathematics* responds. In editing it, Eberhard Zeidler has given us a remarkable panoramic overview of mathematics, ranging from elementary facts and concepts, to advanced and sophisticated techniques, lucidly and economically explained. The outcome of many years of work, it will provide readers of diverse backgrounds with the fundamental concepts and language on which deeper understanding and significant applications can be based.

*Oxford, December 2003* *John Ball*

# *Preface*

In the past few years, mathematics has made enormous strides forward. An eminent factor for this progress has been the construction and application of ever stronger and faster computers. Moreover, the extremely complicated problems of modern technology which pose themselves to engineers and natural scientists require highly sophisticated mathematics, in which routine knowledge no longer suffices and the boundaries between pure and applied mathematics are starting to melt.

The *Oxford Users' Guide to Mathematics* responds to the high standards required by the growing influence of computer science within the mathematical sciences and the increasingly close relationships between mathematics and the natural and engineering sciences. It conveys a lively, modern picture of mathematics aimed at a wide readership, including:

- students of high schools and undergraduates,
- graduate students of mathematics,
- students of engineering, natural sciences, economy and other directions of study which require mathematical background,
- practitioners who work in these fields,
- teachers, both in schools and at universities.

The needs of as broad an audience as this will be taken into account in our presentation by the consideration of a wide range of aspects, starting from elementary facts all the way up to modern, highly sophisticated results and methods. In addition, the presentation is very broad in its consideration of very diverse areas of mathematical research. In this respect, the book is both vertically and horizontally quite deep. At the same time, we go to great pains to motivate the material completely and explain the basic ideas in depth, both aspects of which are emphasized more in the text than are technical details. Also, applications of the ideas and methods play an important role in the development.

There are many interludes on historical background of the results or more generally on the period in which the results were first obtained. In addition to these remarks throughout the text, there is, at the end of the volume, a detailed sketch of the history of mathematics. This exemplifies the point of view that mathematics is more than a dry collection of formulas, definitions, theorems and manipulations with symbols. Rather, mathematics is an integral part of our culture and a wonderful medium for thought and discovery, which makes it possible to make progress on frontiers like the modern theory of elementary particles and cosmology, areas which cannot be understood without a mathematical model, as they so lie so far from our usual realm of perception and understanding.

In the introductory Chapter 0 we collect basic mathematical notions and facts which are often required by students, scientists and other practitioners, in the form of a reference book. A student of medicine, for example, can find here an elementary introduction to the methods of mathematical statistics, which will hopeful be of use in the writing of a statistical part of a thesis. The following three chapters are devoted to the three basic pillars of mathematics:

- analysis,
- algebra, and
- geometry.

These chapters are followed by a chapter devoted to

- foundations of mathematics (logic and set theory),

which takes into account in particular the needs and difficulties of beginning students. The last three chapters are then devoted to the most important fields of applications of mathematics, namely

- theory of variations and optimization,
- stochastics (probability theory and mathematical statistics), and
- scientific computation.

The possibilities which modern supercomputers offer have radically changed scientific computation. Whether mathematician, engineer or natural scientist, the practitioner today is in a position to carry out extensive experiments on the computer which make it possible to collect experience from examples in hitherto underdeveloped areas of mathematics. In this way, completely new questions arise and give new impulses for the development of mathematical theories. The last chapter is the first appearance, in pocket book form, of the modern theory of scientific computation, which, as mentioned above, has revolutionized the engineering sciences.

The past decade has seen the appearance of software systems which make it possible to carry out many routine jobs in mathematics on a standard PC. This is mentioned at the corresponding places in the text, where these methods are motivated and described. The bibliography at the end of the book was very carefully put together and gives you an idea of where to turn should you have questions above and beyond what is directly treated in the text. The level of references varies from introductory texts to classics and goes on to advanced monographs, reflecting the frontiers of modern research.

This book has a long history. The *Pocketbook of Mathematics* by I. N. Bronstein and K. A. Semendjajew was originally translated from Russian into German by Viktor Ziegler. It appeared in 1958, published by the B. G. Teubner Verlag in Leipzig, and has become in the mean time a standard in the German language with 18 editions until 1978. Toward the end of the last century it was decided to bring this classic up to date, not only with respect to the presented material, but also with respect to the breadth and kind of presentation; this was carried out under the supervision of Eberhard Zeidler, who wrote all chapters except that on scientific computing, which was authored by Wolfgang Hackbusch and Hans Rudolf Schwarz. This appeared, again by Teubner-Verlag, in 1996. The work was so fundamentally different than its predecessors that Oxford University Press felt it would be worth translating it, and they assigned Bruce Hunt that job of doing so. The translator has done his best to keep the spirit of the book as in the original, at the same time including a series of corrections which had been reported by astute readers or which were spotted in the process of translating the volume. Furthermore the translator has gone to great pains to improve the graphical quality of the text, as this improves the ease with which the material can be absorbed by the reader.

**Acknowledgments:** In addition to the pure translation of the volume, it was agreed with the publisher to typeset the entire book from scratch; in particular this meant retyping all the formulas and tables, as well as redoing all the illustrations. For help with these aspects, as well as extensive proof reading of the translation, both the translator and the editor are indebted to Micaela Krieger-Hauwede (illustrations), Lars Uhlmann (equations and tables) and Kerstin Fölting (equations, tables and a meticulous proof-reading of the entire text). Without their help the translation would have taken much longer than it did. The editor likes to thank the translator for his excellent job.

Frankfurt/Main<br>Leipzig<br>Fall 2003

Bruce Hunt<br>Eberhard Zeidler

# *Contents*

# *Introduction*

> *The greatest mathematicians like Archimedes, Newton and Gauss have always been able to combine theory and applications into one.*
>
> *Felix Klein* (1849–1925)

Mathematics has more than 5000 years of history. It is the most powerful instrument of the human mind, able to precisely formulate laws of nature. In this way it is possible to dwell into the secrets of nature and into the incredible, unimaginable extension of the universe. Fundamental branches of mathematics are

- algebra,
- geometry, and
- analysis.

Algebra is concerned with, at least in it original form, the solution of equations. Cuneiform writing from the days of King Hammurapi (eighteenth century BC) document that the practical mathematical thinking of the Babylonians was strongly algebra-oriented. On the other hand, the mathematical thought of ancient Greece, whose crowning achievement was the appearance of Euclid's *The Elements* (around 300 BC), was strongly influenced by geometry. Analytical thinking, based on the notion of limit, was not systematically developed until the creation of calculus by Newton and Leibniz in the seventeenth century.

Important branches of applied mathematics are aptly described by the following indications:

- ordinary and partial differential equations (describing the change in time of systems of nature, engineering and society),
- the calculus of variations and optimization,
- scientific computing (the approximation and simulation of processes with more and more powerful computing machines).

Foundations of mathematics are concerned with

- mathematical logic, and
- set theory.

These two branches of mathematics did not exist until the nineteenth century. Mathematical logic investigates the possibilities, but also the limits of mathematical proofs.

Because of its by nature very formal development, it is well-equipped to describe processes in algorithms and on computers, which are free of subjectivity. Set theory is basically a powerful language for formulating mathematics. Instead of dealing in this book with the formal aspects of set theory, we put our emphasis on the liveliness and broad nature of mathematics, something which has fascinated mankind for centuries.

In modern mathematics there are opposing tendencies visible. On the one hand, we observe an increase in the degree of specialization. On the other hand, there are open questions coming from the theory of elementary particles, cosmology and modern technology which have such a high degree of complexity that they can only be approached through a synthesis of quite diverse areas of mathematics. This leads to a unification of mathematics and to an increasing elimination of the non-natural split between pure and applied mathematics.

The history of mathematics is full of the appearance of new ideas and methods. We can safely assume that this tendency with continue on into the future.

# *0. Important Formulas, Graphical Representations and Tables*

*Everything should be made as simple as possible, but not simpler.*
*Albert Einstein* (1879–1955)

## 0.1 Basic formulas of elementary mathematics

### 0.1.1 Mathematical constants

*Table 0.1. Some frequently used mathematical constants.*

| *Symbol* | *Approximation* | *Notation* |
|---|---|---|
| $\pi$ | 3.14 15 92 65 | Ludolf number pi |
| e | 2.71 82 81 83 | Euler[1] number e |
| C | 0.57 72 15 67 | Euler constant |
| ln 10 | 2.30 25 85 09 | natural logarithm of the number 10 |

A table of the most important scientific constants can be found at the end of this handbook.

**Factorial:** Often the symbol

$$\boxed{n! := 1 \cdot 2 \cdot 3 \cdot \ldots \cdot n}$$

is used for the shown product; this product is called *n-factorial.* Moreover, we define $0! := 1$.

*Example 1:* $1! = 1$, $2! = 1 \cdot 2$, $3! = 1 \cdot 2 \cdot 3 = 6$, $4! = 24$, $5! = 120$ and $6! = 720$.

In statistical physics, one requires the value of $n!$ for $n$ around $10^{23}$. For such astronomical numbers, one has the *Stirling formula*

$$\boxed{n! = \left(\frac{n}{\mathrm{e}}\right)^n \sqrt{2\pi n}} \tag{0.1}$$

as a good approximation (cf. 0.7.3.2).

[1]Leonhard Euler (1707–1783) was the most productive mathematician of all times. His collected works fill 72 volumes and more than 5000 additional letters. His monumental lifetime work has shaped much of the modern mathematical science. At the end of this handbook there is a table of the history of mathematics, which should help the reader to orient her- or himself in the history of mathematics and its greatest contributors.

**Infinite series for $\pi$ and e:** The precise value of $\pi$ is given as the value of the convergent Leibniz series

$$\frac{\pi}{4} = 1 - \frac{1}{3} + \frac{1}{5} - \frac{1}{7} + \ldots \tag{0.2}$$

Because of the alternating sign of the terms, the error of the truncated series is always given by the following term. Thus, the terms listed on the right-hand side of (0.2) give an approximation of $\pi$ for which the error is at most $1/9$. This series, however, is not used for practical computations of values for $\pi$ on computers, because it converges very slowly. Contemporary approximations of $\pi$ are accurate up to more than 2 billion decimal places (cf. the more detailed discussion of the number $\pi$ in 2.7.7). The value of e is the value of the following convergent series

$$\mathrm{e} = 2 + \frac{1}{2!} + \frac{1}{3!} + \frac{1}{4!} + \ldots$$

For large numbers $n$, for example, one has approximately

$$\mathrm{e} = \left(1 + \frac{1}{n}\right)^n. \tag{0.3}$$

More precisely, the right-hand side of (0.3) approaches for larger and larger values of $n$ the value of the number e. One also writes for this

$$\boxed{\mathrm{e} = \lim_{n\to\infty}\left(1 + \frac{1}{n}\right)^n.}$$

In words: the number e is the limit of the sequence of numbers $\left(1 + \frac{1}{n}\right)^n$, as $n$ approaches infinity. With the help of the number e one can define the most important function in mathematics:

$$\boxed{y = \mathrm{e}^x.} \tag{0.4}$$

This is the Euler e-function (exponential function, cf. 0.2.5). The inverse function of (0.4) is the natural logarithm

$$\boxed{x = \ln y}$$

(cf. 0.2.6). In particular for powers of 10 one gets

$$\boxed{\ln 10^x = x \cdot \ln 10 = x \cdot 2.30\,25\,85\,09.}$$

Here $x$ can be an arbitrary real number.

**Representations of $\pi$ and e through continued fractions:** For more detailed investigations of the structure of numbers, one uses representations in terms of *continued fractions* instead of decimal numbers (cf. 2.7.5). The representations of $\pi$ and e in terms of continued fractions are displayed in Table 2.7.

**The Euler constant C:** The precise value of C is given by the formula

$$\mathrm{C} = \lim_{n\to\infty}\left(1 + \frac{1}{2} + \frac{1}{3} + \ldots + \frac{1}{n} - \ln(n+1)\right) = -\int_0^\infty \mathrm{e}^{-t}\ln t\,\mathrm{d}t\,.$$

For large natural numbers $n$, one thus has the approximation formula

$$1 + \frac{1}{2} + \frac{1}{3} + \ldots + \frac{1}{n} = \ln(n+1) + \mathrm{C}\,.$$

The Euler constant C appears in a surprisingly large number of mathematical formulas (cf. 0.7).

## 0.1.2 Measuring angles

**Degrees:** In Figure 0.1, some of the most often used angles, measured in degrees, are shown. An angle of 90° is also called a *right angle.* In ancient Sumeria near the Euphrates and Tigris rivers, more than 4,000 years ago, a number system with the basis 60 (sexagesimal system) was used. One can trace back to this usage the fact that the numbers 12, 24, 60 and 360 are used in such an important way in our measurement of time and angles. In addition to the degree, other measures for angles used in, for example, astronomy are the following smaller measurements:

$$1' \quad \text{(arc minute)} = \frac{1^\circ}{60}\,,$$
$$1'' \quad \text{(arc second)} = \frac{1^\circ}{3\,600}\,.$$

90° 45° 60° 180° 270° −90° −45°

*Figure 0.1. Postive and negative angles.*

*Example 1* (Astronomy): The face of the sun is about 30′ (half a degree) in the sky.

Because of the motion of the earth and the sun, the stars in the sky change their positions. Half the maximal change per year is called a *parallax.* This is equal to the angle $\alpha$, which the star would appear to see between the earth and the sun when they are at maximal distance from each other (cf. Fig 0.2 and Table 0.2).

*Table 0.2. Parallax and distance.*

| ***Star*** | ***Parallax*** | ***Distance*** |
|---|---|---|
| Proxima Centauri (nearest star) | 0.765″ | 4.2 light years |
| Sirius (brightest star) | 0.371″ | 8.8 light years |

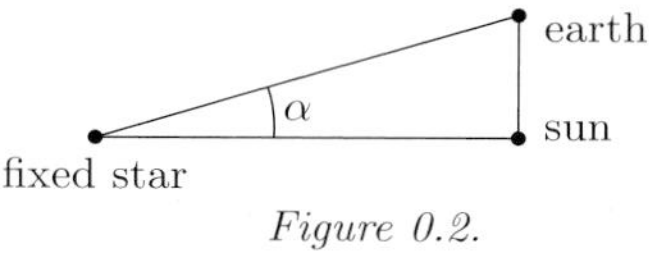


*Figure 0.2.*

A parallax of one arc second corresponds to a distance of 3.26 light years ($3.1 \cdot 10^{13}$ km). This distance is also referred to as a *parsec.*

**Radians:** A angle of $\alpha$ degrees ($\alpha^\circ$) corresponds to

$$\boxed{\alpha = 2\pi\left(\frac{\alpha^\circ}{360^\circ}\right)}$$

*radians.* Here $\alpha$ is the length of an arc on the unit circle which is cut out by the angle $\alpha^\circ$ (Figure 0.3). In Table 0.3 one finds often-used values for this measurement.

**Convention:** Unless stated otherwise, all angles in this book will be measured in radians.

*Table 0.3. Angles and radians.*

| ***Degrees*** | $1^\circ$ | $45^\circ$ | $60^\circ$ | $90^\circ$ | $120^\circ$ | $135^\circ$ | $180^\circ$ | $270^\circ$ | $360^\circ$ |
|---|---|---|---|---|---|---|---|---|---|
| ***Radians*** | $\frac{\pi}{180}$ | $\frac{\pi}{4}$ | $\frac{\pi}{3}$ | $\frac{\pi}{2}$ | $\frac{2\pi}{3}$ | $\frac{3\pi}{4}$ | $\pi$ | $\frac{3\pi}{2}$ | $2\pi$ |

$1' = \frac{\pi}{10\,800} = 0.000291, \quad 1'' = \frac{\pi}{648\,000} = 0.000005$

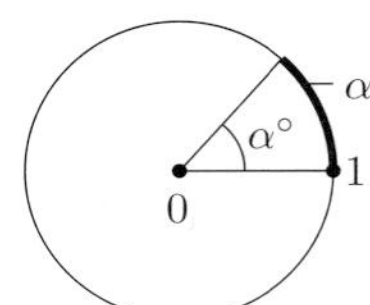


*Figure 0.3.*

**Sum of angles in a triangle:** In a triangle, the sum of the angles is always $\pi$, i.e.,

$$\boxed{\alpha + \beta + \gamma = \pi}$$

(cf. Figure 0.4).

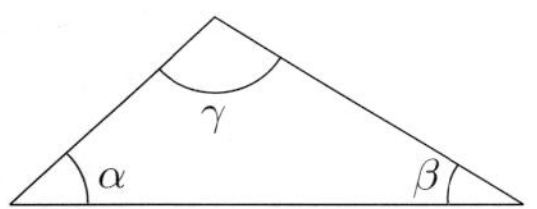


*Figure 0.4. Angles in a triangle.*

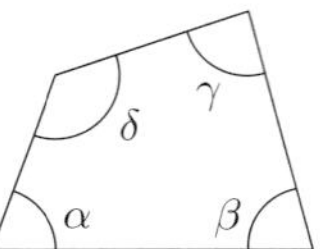


*Figure 0.5. Angles in a quadrangle.*

**Sum of angles in a quadrangle:** Since a rectangle can be decomposed into two triangles, the sum of angles must be $2\pi$, i.e.,

$$\boxed{\alpha + \beta + \gamma + \delta = 2\pi}$$

(cf. Figure 0.5).

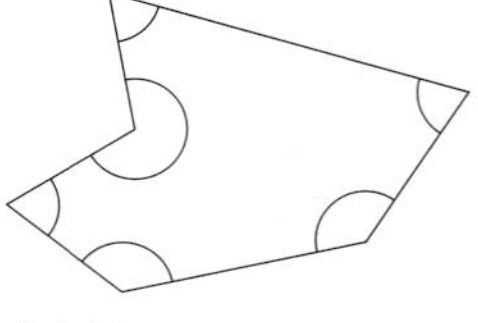

**(a)** pentagon **(b)** hexagon

*Figure 0.6. Pentagon and hexagon.*

**Sum of angles in an $n$-gon:** In general one has

$$\boxed{\text{Sum of the inside angles of a } n\text{-gon} = (n-2)\pi\,.}$$

*Example 2:* For a pentagon (5-gon) (resp. hexagon (6-gon)), the sum of the angels is $3\pi$ (resp. $4\pi$) (Figure 0.6).

### 0.1.3 Area and circumference of plane figures

In Table 0.4 the most important plane figures are illustrated. The meaning of the appearing trigonometric functions $\sin\alpha$ and $\cos\alpha$ is explained in detail in 0.2.8.

*Table 0.4. Surface area and circumference of several polygons.*

| ***Figure*** | ***Diagram*** | ***Area A*** | ***Circumference C*** |
|---|---|---|---|
| square | $a$, $a$ | $A = a^2$ ($a$ length of a side) | $C = 4a$ |
| rectangle | $b$, $a$ | $A = ab$ ($a, b$ lengths of the sides) | $C = 2a + 2b$ |
| parallelogram | $b$, $\gamma$, $h$, $a$ | $\boxed{A = ah = ab\sin\gamma}$ ($a$ length of the base, $b$ length of the side $h$ height) | $C = 2a + 2b$ |
| rhombus (equilateral parallelogram) | $a$, $\gamma$, $a$ | $A = a^2\sin\gamma$ | $C = 4a$ |
| trapezoid (quadrangle with two parallel sides) | $b$, $c$, $h$, $d$, $a$ | $A = \frac{1}{2}(a+b)h$ ($a, b$ length of the parallel sides, $h$ height) | $C = a + b + c + d$ |
| triangle | $b$, $\gamma$, $h$, $c$, $a$ | $\boxed{A = \frac{1}{2}ah = \frac{1}{2}ab\sin\gamma}$ ($a$ length of the base, $b, c$ length of the other sides, $h$ height, $s := C/2$) *Heronian formula for the area:* $\boxed{A = \sqrt{s(s-a)(s-b)(s-c)}}$ | $C = a + b + c$ |

*Table 0.4 (continued)*

| | | | |
|---|---|---|---|
| right triangle | 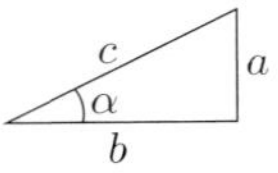 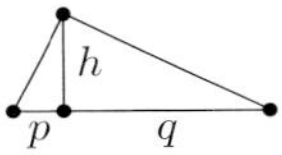 | $\boxed{A = \frac{1}{2}ab}$ <br> *relation between sides and angles:* <br> $\boxed{a = c\sin\alpha, \quad b = c\cos\alpha, \quad a = b\tan\alpha}$ <br> ($c$ hypotenuse [2], <br> $a$ opposite leg, <br> $b$ neighboring leg) <br> *Theorem of Pythagoras*[3]*:* <br> $\boxed{a^2 + b^2 = c^2}$ <br> *Euclidean relation for the height:* <br> $\boxed{h^2 = pq}$ <br> ($h$ height over the hypotenuse, <br> $p, q$ segment lengths) | $C = a + b + c$ |
| equilateral triangle | 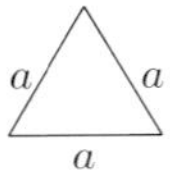 | $A = \frac{\sqrt{3}}{4}a^2$ | $C = 3a$ |
| circle | 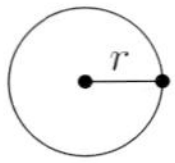 | $\boxed{A = \pi r^2}$ <br> ($r$ radius) | $\boxed{C = 2\pi r}$ |
| sector of a circle | 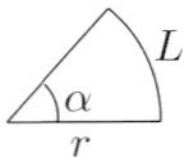 | $A = \frac{1}{2}\alpha r^2$ | $C = L + 2r$, <br> $L = \alpha r$ |
| annulus | 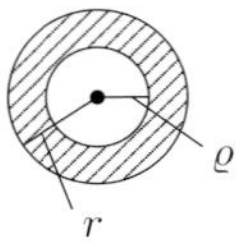 | $A = \pi(r^2 - \varrho^2)$ <br> ($r$ outer radius, <br> $\varrho$ inner radius) | $C = 2\pi(r + \varrho)$ |

[2]In a right triangle the side which is opposite the right angle is called the *hypotenuse*. The other sides are called *catheti* or simply *legs*.

[3]Pythagoras from Samos (at 500 BC) is considered to be the founder of the famous school of the Pythagoreans in ancient Greece. The theorem of Pythagoras however was know almost 1,000 years before that, by the Babylonians under the regent King Hammurapi (1728–1686 BC).

*Table 0.4 (continued)*

| | | | |
|---|---|---|---|
| parabola sector[4] | 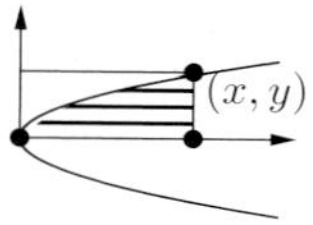 | $A = \frac{1}{3}xy$ | |
| hyperbola sector | 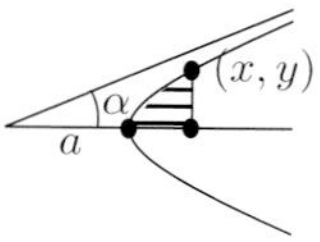 | $A = \frac{1}{2}\left(xy - ab \cdot \operatorname{arcosh}\frac{x}{a}\right)$ <br> $(b = a\tan\alpha)$ | |
| ellipse sector | 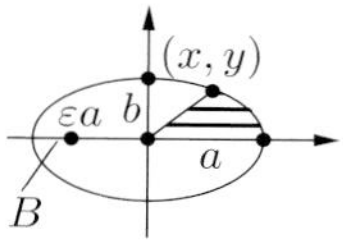 | $A = \frac{1}{2}ab \cdot \operatorname{arcosh}\frac{x}{a}$ | |
| ellipse | diagram as above, where $B$ is the focal point | $\boxed{A = \pi ab}$ <br> ($a, b$ lengths of the axi, $b < a$, $\varepsilon$ numerical eccentricity) | $\boxed{C = 4aE(\varepsilon)}$ <br> (cf. (0.5)) |

**The meaning of elliptic integrals for the calculation of the circumference of the ellipse:** The numerical eccentricity $\varepsilon$ of an ellipse is given by the formula

$$\boxed{\varepsilon = \sqrt{1 - \frac{b^2}{a^2}}.}$$

The geometric interpretation of $\varepsilon$ is to be found in the fact that the focal point of the ellipse has a distance from the center of the ellipse of $\varepsilon a$. For a circle, one has $\varepsilon = 0$. The larger the numerical eccentricity $\varepsilon$ is, the flatter the ellipse is.

It was already noticed in the eighteenth century that the *circumference of an ellipse can not* be calculated by elementary means. This circumference is given by $C = 4aE(\varepsilon)$, where we use the notation

$$\boxed{E(\varepsilon) := \int_0^{\pi/2} \sqrt{1 - \varepsilon^2 \sin^2\varphi}\,\mathrm{d}\varphi} \qquad (0.5)$$

for the *complete elliptic integral of the second kind* of Legendre. There are tabulated values for this integral (cf. 0.5.4). For an ellipse we always have that $0 \leq \varepsilon < 1$. As

[4]Parabola, hyperbola und ellipse will be considered in 0.1.7. The function arcosh will be introduced in 0.2.12.

approximations for all these values one has the series

$$E(\varepsilon) = 1 - \left(\frac{1}{2}\right)^2 \varepsilon^2 - \left(\frac{1\cdot 3}{2\cdot 4}\right)^2 \frac{\varepsilon^4}{3} - \left(\frac{1\cdot 3\cdot 5}{2\cdot 4\cdot 6}\right)^2 \frac{\varepsilon^6}{5} - \ldots$$
$$= 1 - \frac{\varepsilon^2}{4} - \frac{3\varepsilon^4}{64} - \frac{5\varepsilon^6}{256} - \ldots$$

The general theory of elliptic integrals was created in the nineteenth century (cf. 1.14.19).

**Regular $n$-gons:** A $n$-gon is said to be *regular*, if all the sides and angles are equal (Figure 0.7).

The distance from the center to one of the corners of the $n$-gon will be denoted by $r$. Then the geometry of a regular $n$-gon is determined by the following statements:

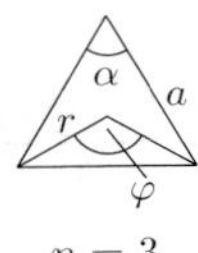

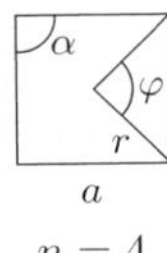

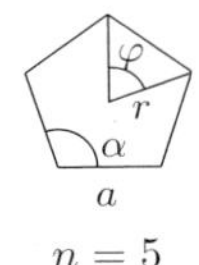

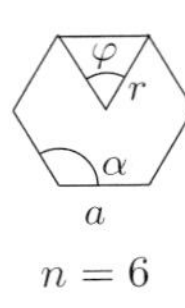


*Figure 0.7. Regular $n$-gons.*

| | |
|---|---|
| center angle | $\varphi = \frac{2\pi}{n}\,,$ |
| complementary angle | $\alpha = \pi - \varphi\,,$ |
| length of sides | $a = 2r\,\sin\frac{\varphi}{2}\,,$ |
| circumference | $C = na\,,$ |
| area | $A = \frac{1}{2}nr^2\sin\varphi\,.$ |

**Theorem of Gauss:** A $n$-gon with $n \leq 20$ can be constructed with the help of a ruler and compass, if and only if

$$n = 3\,,\ 4\,,\ 5\,,\ 6\,,\ 8\,,\ 10\,,\ 12\,,\ 15\,,\ 16\,,\ 17\,,\ 20\,.$$

In particular, such a construction is not possible for $n = 7\,,\ 9\,,\ 11\,,\ 13\,,\ 14\,,\ 18\,,\ 19$. This result is the consequence of Galois theory and will be considered in 2.6.6 in more detail.

## 0.1.4 Volume and surface area of solids

In Table 0.5 the most important three-dimensional figures are collected.

*Table 0.5. Volume and surface area of some solids.*

| *Solid* | *Diagram* | *Volume V* | *Surface area O* <br> *section area M* |
|---|---|---|---|
| cube | $a$ | $V = a^3$ <br> ($a$ length of sides) | $O = 6a^2$ |

Table 0.5 (continued)

| | | | |
|---|---|---|---|
| parallelepiped | 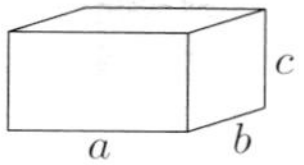 | $V = abc$ ($a, b, c$ lengths of sides) | $O = 2(ab+bc+ca)$ |
| ball | 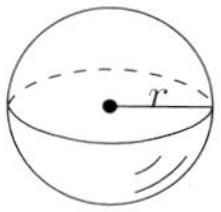 | $\boxed{V = \frac{4}{3}\pi r^3}$ ($r$ radius) | $\boxed{O = 4\pi r^2}$ |
| prism | 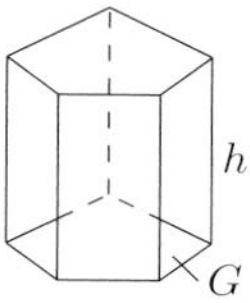 | $\boxed{V = Gh}$ ($G$ area of the base, $h$ height) | |
| cylinder | 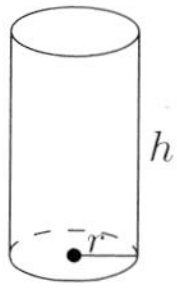 | $V = \pi r^2 h$ ($r$ radius, $h$ height) | $O = M + 2\pi r^2$, $M = 2\pi r h$ |
| solid annulus | 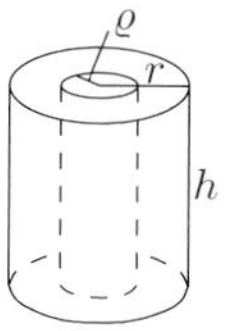 | $V = \pi h(r^2 - \varrho^2)$ ($r$ outer radius, $\varrho$ inner radius, $h$ height) | |
| pyramid | 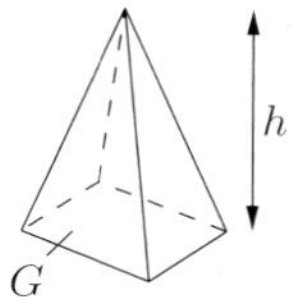 | $\boxed{V = \frac{1}{3}Gh}$ ($G$ area of the base, $h$ height) | |

*Table 0.5 (continued)*

| | | | |
|---|---|---|---|
| circular cone | 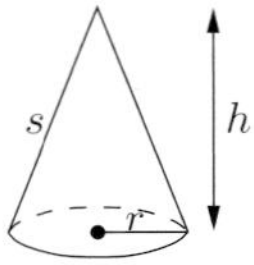 | $V = \frac{1}{3}\pi r^2 h$ ($r$ radius, $h$ height, $s$ length of a meridian) | $O = M + \pi r^2$, $M = \pi r s$ |
| capped pyramid | 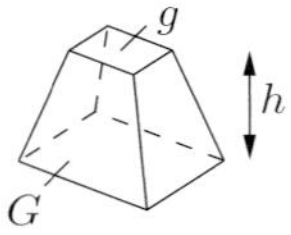 | $V = \frac{h}{3}(G + \sqrt{Gg} + g)$ ($G$ surface area of the base, $g$ area of the top) | |
| capped cone | 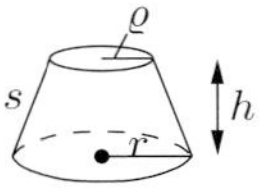 | $V = \frac{\pi h}{3}(r^2 + r\varrho + \varrho^2)$ ($r, \varrho$ radii, $h$ height, $s$ length of the side) | $O = M + \pi(r^2 + \varrho^2)$, $M = \pi s(r + \varrho)$ |
| obelisk | 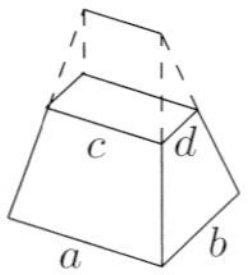 | $V = \frac{1}{6}(ab + (a + c)(b + d) + cd)$ ($a, b, c, d$ lengths of the sides) | |
| wedge (the sides are equilateral triangles) | 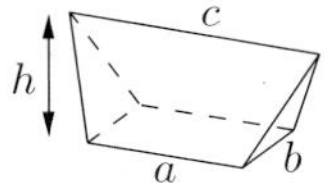 | $V = \frac{\pi}{6} bh(2a + c)$ ($a, b$ base side lengths, $c$ upper edge, $h$ height) | |
| section of a ball (bounded by a meridian) | 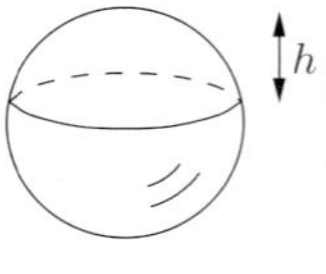 | $V = \frac{\pi}{3} h^2(3r - h)$ ($r$ radius of the ball, $h$ height) | $O = 2\pi rh$ (top part) |
| slice of a ball (bounded by two meridians) | 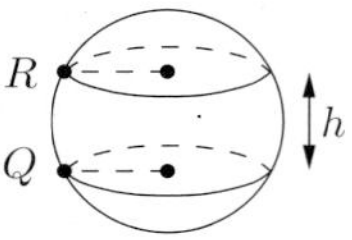 | $V = \frac{\pi h}{6}(3R^2 + 3\varrho^2 + h^2)$ ($r$ radius of the ball, $h$ height, $R$ and $\varrho$ radii of the meridians) | $O = 2\pi rh$ (middle layer) |

*Table 0.5 (continued)*

| | | | |
|---|---|---|---|
| torus | | $V = 2\pi r^2 \varrho$ ($r$ radius of the torus, $\varrho$ radius of the section) | $O = 4\pi^2 r \varrho$ |
| barrel (with circular section) | | $V = 0.0873\, h(2D + 2r)^2$ ($D$ diameter, $r$ radius at the top, $h$ height; the formula is an approximation) | |
| ellipsoid | | $V = \frac{4}{3}\pi abc$ ($a, b, c$ lengths of the axi, $c < b < a$) | see the formula of Legendre (L) for $O$ |

**The meaning of elliptic integrals for the calculation of the surface area of the ellipsoid:** The surface of an ellipsoid can not be calculated by elementary means. One requires again elliptic integrals. For this one has the formula of Legendre

$$\boxed{O = 2\pi c^2 + \frac{2\pi b}{\sqrt{a^2 - c^2}} \left( c^2 F(k, \varphi) + (a^2 - c^2) E(k, \varphi) \right)} \qquad \text{(L)}$$

with

$$k = \frac{a}{b} \frac{\sqrt{b^2 - c^2}}{\sqrt{a^2 - c^2}}, \qquad \varphi = \arcsin \frac{\sqrt{a^2 - c^2}}{a}.$$

The formulas for the elliptic integrals $E(k, \varphi)$ and $F(k, \varphi)$ can be found in 0.5.4

### 0.1.5 Volumes and surface areas of regular polyhedra

**Polyhedra:** A *polyhedron* is a solid which is bounded by elementary parts (plane figures).

The *regular polyhedra* (also called *Platonic solids*) have faces, all of which are congruent, regular $n$-gons of side length $a$, in which at all corners the same number of faces meet. There are precisely 5 regular polyhedra, which are listed in Table 0.6.

*Table 0.6. The five Platonic solids.*[5]

| ***Regular polyhedron*** | | ***Faces*** | ***Volume*** | ***Surface area*** |
|---|---|---|---|---|
| tetrahedron | 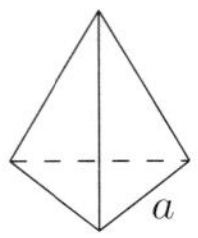 $a$ | 4 equilateral triangles | $\frac{\sqrt{2}}{12} \cdot a^3$ | $\sqrt{3}a^2$ |
| cube | 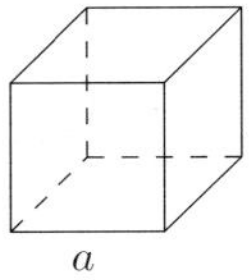 $a$ | 6 squares | $a^3$ | $6a^2$ |
| octahedron | 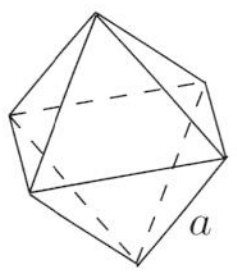 $a$ | 8 equilateral triangles | $\frac{\sqrt{2}}{3} \cdot a^3$ | $2\sqrt{3} \cdot a^2$ |
| dodecahedron | 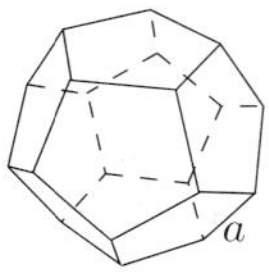 $a$ | 12 equilateral pentagons | $7.663 \cdot a^3$ | $20.646 \cdot a^2$ |
| icosahedron[6] | 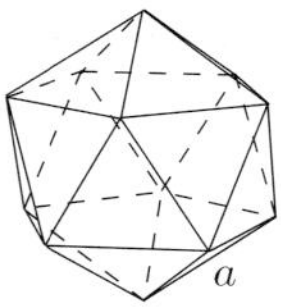 $a$ | 20 equilateral triangles | $2.182 \cdot a^3$ | $8.660 \cdot a^2$ |

**Euler's polyhedral formula:** The following relation is generally true for regular polyhedra:[7]

$$\boxed{\text{number of corners } c - \text{ number of edges } e + \text{number of faces } f = 2\,.}$$

[5] In this table, the common length of an edge is denoted by $a$. The fomulas for the volumes and areas of the dodecahedron and the icosahedron are approximations.

[6] The German mathematician Felix Klein wrote an famous book about the symmetries of the icosahedron and its relation to the equations of fifth degree, (cf. [22]).

[7] This formula is a special case of a general topological fact. Since the surfaces of the regular polyhedra are all homeomorph to the sphere, they have genus 0 and the Euler characteristic 2.

Table 0.7 verifies this formula.

*Table 0.7. The key numbers for the Platonic solids.*

| ***Regular polyhedron*** | $c$ | $e$ | $f$ | $c+e-f$ |
|---|---|---|---|---|
| tetrahedron | 4 | 6 | 4 | 2 |
| cube | 8 | 12 | 6 | 2 |
| octahedron | 6 | 12 | 8 | 2 |
| dodecahedron | 20 | 30 | 12 | 2 |
| icosahedron | 12 | 30 | 20 | 2 |

### 0.1.6 Volume and surface area of $n$-dimensional balls

The following formulas are necessary in statistical physics. In these formulas, $n$ is roughly of the size $10^{23}$. For such large values of $n$, one uses the Stirling formula for an approximation to the value of $n!$ (cf. (0.1)).

**Characterization of the solid ball by an inequality:** The $n$-dimensional ball $K_n(r)$ of radius $r$ with center at the origin is defined to be the set of all points $(x_1, \ldots, x_n)$ that satisfy the inequality

$$\boxed{x_1^2 + \ldots + x_n^2 \leq r^2.}$$

Here $x_1, \ldots, x_n$ are real numbers with $n \geq 2$. The boundary (surface) of this ball is formed by the set of all $(x_1, \ldots, x_n)$ which satisfy the inequality

$$\boxed{x_1^2 + \ldots + x_n^2 = r^2.}$$

For the volume $V_n$ and the surface area $O_n$ of $K_n(r)$ one has the following formulas of Jacobi:

$$\boxed{\begin{aligned} V_n &= \frac{\pi^{n/2} r^n}{\Gamma\left(\frac{n}{2}+1\right)}, \\ O_n &= \frac{2\pi^{n/2} r^{n-1}}{\Gamma\left(\frac{n}{2}\right)}. \end{aligned}}$$

The gamma function $\Gamma$ is considered in section 1.14.16. It satisfies the recursion formula

$$\Gamma(x+1) = x\Gamma(x) \quad \text{for all } x > 0$$

with $\Gamma(1) = 1$ and $\Gamma\left(\dfrac{1}{2}\right) = \sqrt{\pi}$. From this one gets for $m = 1, 2, \ldots$ the following formulas:

$$V_{2m} = \frac{\pi^m r^{2m}}{m!}, \qquad V_{2m+1} = \frac{2(2\pi)^m r^{2m+1}}{1 \cdot 3 \cdot 5 \cdot \ldots \cdot (2m+1)}$$

and

$$O_{2m} = \frac{2\pi^m r^{2m-1}}{(m-1)!}, \qquad O_{2m+1} = \frac{2^{2m+1} m! \pi^m r^{2m}}{(2m)!}.$$

*Example:* In the special case $n = 3$ and $m = 1$, one gets the well-known formulas

$$V_3 = \frac{4}{3}\pi r^3, \qquad O_3 = 4\pi r^2$$

for the volume $V_3$ and the surface area $O_3$ of the three-dimensional ball of radius $r$.

## 0.1.7 Basic formulas for analytic geometry in the plane

Analytic geometry describes geometric objects like lines, planes and conic sections by means of equations for the coordinates and investigates the geometric properties through manipulations with these inequalities. This process of increased use of arithmetic and algebra in geometry goes back to the philosopher, scientist and mathematician René Descartes (1596–1650), after whom the Cartesian coordinates have their name.

### 0.1.7.1 Lines

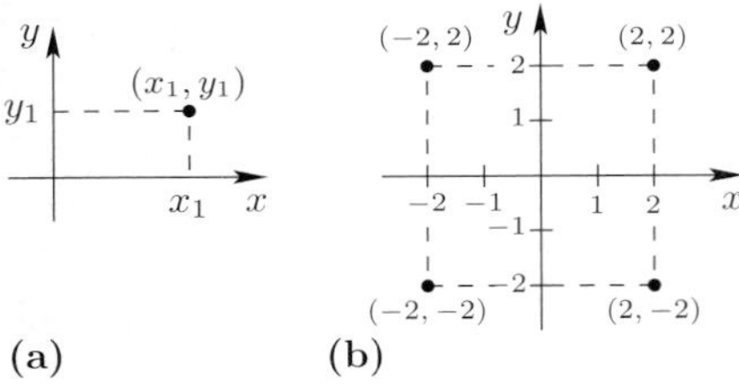


*Figure 0.8. Cartesian coordinates.*

All of the following formulas are in terms of a Cartesian coordinate system, in which the $y$-axis is perpendicular to the $x$-axis. The coordinates of a point $(x_1, y_1)$ are given as in Figure 0.8(a). The $x$ coordinate of a point left of the $y$-axis is negative, and the $y$ coordinate of a point underneath the $x$-axis is also negative.

*Example 1:* The points $(2,2)$, $(2,-2)$, $(-2,-2)$ and $(-2,2)$ are found in Figure 0.8(b).

**The distance $d$ of the two points $(x_1, y_1)$ and $(x_2, y_2)$:**

$$\boxed{d = \sqrt{(x_2 - x_1)^2 + (y_2 - y_1)^2}}$$

(Figure 0.9). This formula corresponds to the theorem of Pythagoras.

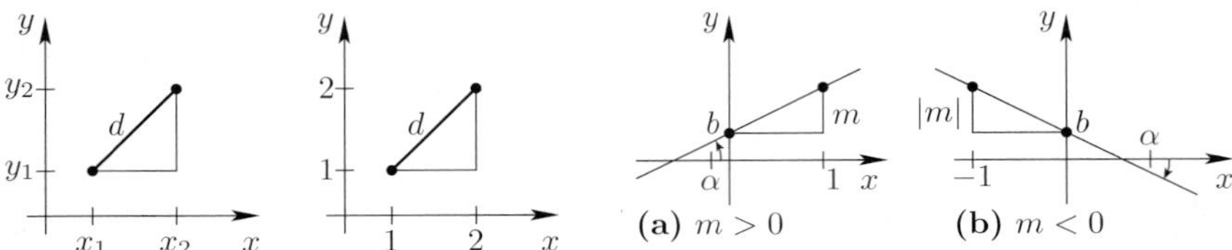


*Figure 0.9. The distance between two points.*

*Figure 0.10. The equation of a line.*

*Example 2:* The distance of the two points $(1, 1)$ and $(2, 2)$ is

$$d = \sqrt{(2-1)^2 + (2-1)^2} = \sqrt{2}\,.$$

**The equation of a line:**

$$\boxed{y = mx + b\,.} \tag{0.6}$$

Here $b$ is the intersection of the line with the $y$-axis (*$y$-intersect*), and the *slope* of the line is $m$ (Figure 0.10). For the *slope angle* $\alpha$ one has the relation

$$\boxed{\tan\alpha = m\,.}$$

(i) If one knows a point $(x_1\,,\ y_1)$ of the line and the slope $m$, then one gets the missing value of $b$ as $b = y_1 - mx_1$.

(ii) If one knows two points $(x_1\,,\ y_1)$ and $(x_2\,,\ y_2)$ on the line with $x_1 \neq x_2$, then:

$$\boxed{m = \frac{y_2 - y_1}{x_2 - x_1}\,, \qquad b = y_1 - mx_1\,.} \tag{0.7}$$

*Example 3:* The equation of the line through the two points $(1, 1)$ and $(3, 2)$ is

$$y = \frac{1}{2}x + \frac{1}{2}\,,$$

as by (0.7) we get $m = \dfrac{2-1}{3-1} = \dfrac{1}{2}$ and $b = 1 - \dfrac{1}{2} = \dfrac{1}{2}$ (Figure 0.11).

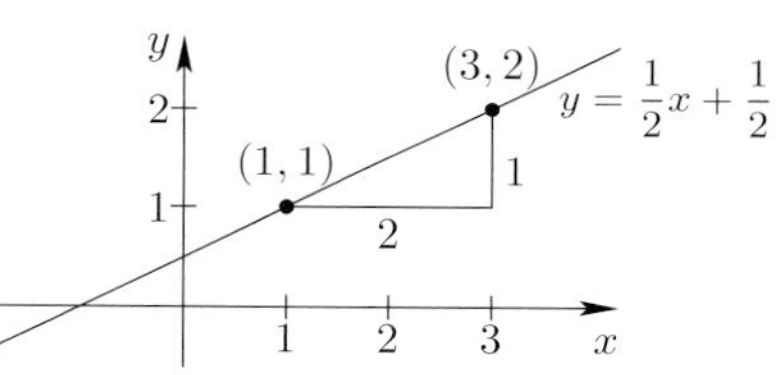


*Figure 0.11. The slope of a line.*

**Abscissa equation of a line:** If one divides the equation a line (0.6) by $b$ and sets $\dfrac{1}{a} := -\dfrac{m}{b}$, then one gets:

$$\boxed{\frac{x}{a} + \frac{y}{b} = 1\,.} \tag{0.8}$$

For $y = 0$ (resp. $x = 0$) one can read off from this that the line hits the $x$-axis at the point $(a, 0)$ (resp. the $y$-axis at the point $(0, b)$) (Figure 0.12(a)).

*Example 4:* If we divide the line equation

$$y = -8x + 4$$

by 4, it follows that $\dfrac{y}{4} = -2x + 1$ and consequently

$$2x + \frac{y}{4} = 1\,.$$

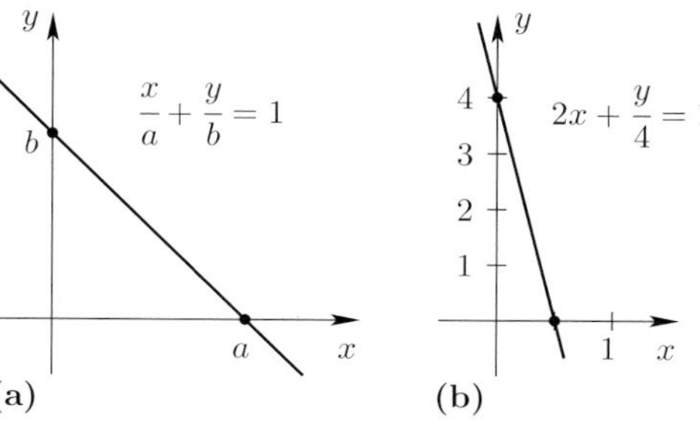


*Figure 0.12. The abscissa of a line.*

If we set $y = 0$, then we get $x = \dfrac{1}{2}$. Hence the line intersects the $x$-axis in the point $x = \dfrac{1}{2}$ (Figure 0.12(b)).

**Equation of the $y$-axis:**

$$\boxed{x = 0\,.}$$

This equation is not a special case of (0.6). It corresponds formally to a slope of $m = \infty$ (infinite slope).

**General equation of a line:** All lines are defined as the set of points satisfying the equation

$$\boxed{Ax + By + C = 0}$$

with real constants $A$, $B$ and $C$, which satisfy the condition $A^2 + B^2 \neq 0$.

*Example 5:* For $A = 1$, $B = C = 0$ one gets the equation $x = 0$ of the $y$-axis.

**Applications to linear algebra:** A series of problems in analytic geometry are most easily solved by using the language of vectors (linear algebra). This will be considered in section 3.3.

### 0.1.7.2 Circles

**The equation of a circle of radius $r$ with center at the point $(c, d)$:**

$$\boxed{(x-c)^2 + (y-d)^2 = r^2} \tag{0.9}$$

(Figure 0.13(a)).

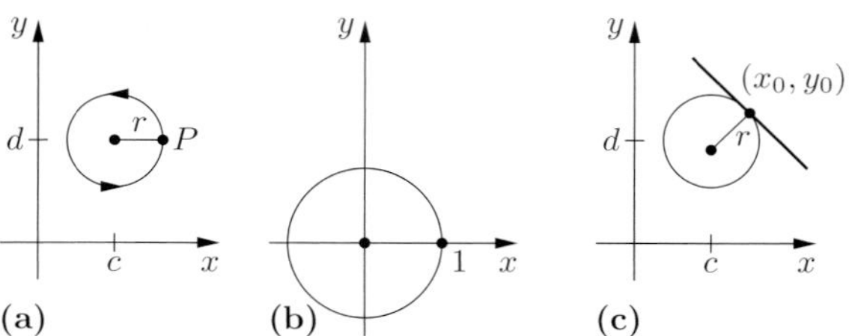


*Figure 0.13. Circles in the plane.*

*Example:* The equation of a circle of radius $r = 1$ with center at the origin $(0, 0)$ is (Figure 0.13(b)):

$$x^2 + y^2 = 1\,.$$

**Equation of the tangent to a circle:**

$$\boxed{(x-c)(x_0-c) + (y-d)(y_0-d) = r^2\,.}$$

This is the equation of the tangent to the circle (0.9) through the point $(x_0, y_0)$ (Figure 0.13(c)).

**Parameterization of the circle of radius $r$ with center at the point $(c, d)$:**

$$x = c + r\cos t\,, \qquad y = d + r\sin t\,, \qquad 0 \le t < 2\pi\,.$$

If one interprets $t$ as the time, then this starting point at $t = 0$ corresponds to the point $P$ in Figure 0.13(a). In the time given by parameters $t = 0$ to $t = 2\pi$, the circle is transversed exactly once counter-clockwise with constant speed (*mathematical positive direction*).

**Curvature $K$ of a circle of radius $R$:** By definition, one has

$$K = \frac{1}{R}\,.$$

### 0.1.7.3 Ellipse

**The equation of an ellipse with center at the origin:**

$$\frac{x^2}{a^2} + \frac{y^2}{b^2} = 1\,. \tag{0.10}$$

We assume $0 < b < a$. Then the ellipse lies symmetrically with respect to the origin. The length of the long (resp. short) axis of the ellipse is equal to $a$ (resp. $b$) (Figure 0.14(a)). One also introduces the following quantities:

$$\begin{aligned} \text{linear eccentricity} \quad & e = \sqrt{a^2 - b^2}\,, \\ \text{numerical eccentricity} \quad & \varepsilon = \frac{e}{a}\,, \\ \text{half-parameter} \quad & p = \frac{b^2}{a}\,. \end{aligned}$$

The two points $(\pm e, 0)$ are called the *focal points* $B_\pm$ of the ellipse (Figure 0.14(a)).

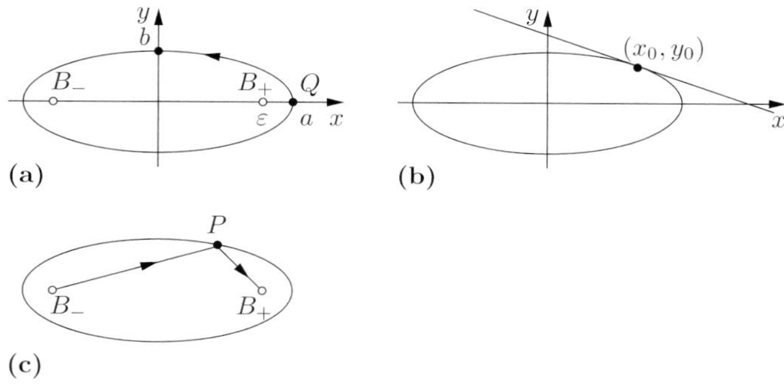


*Figure 0.14. The ellipse.*

**Equation of a tangent to the ellipse:**

$$\boxed{\frac{xx_0}{a^2} + \frac{yy_0}{b^2} = 1\,.}$$

This is the equation of the tangent to the ellipse (0.10) through the point $(x_0, y_0)$ (Figure 0.14(b)).

**Parameterization of an ellipse:**

$$\boxed{x = a\cos t\,, \qquad y = b\sin t\,, \qquad 0 \le t < 2\pi\,.}$$

When the parameter $t$ runs through the values from 0 to $2\pi$, the ellipse in (0.10) is run through once counter-clockwise. The starting value $t = 0$ corresponds to the point on the curve $Q$ (Figure 0.14(a)).

**Geometric characterization of an ellipse:** An ellipse is by definition the set of points $P$, whose sum of distances from two given points $B_-$ and $B_+$ is constant, equal to $2a$ (cf. Figure 0.14(c)).

These points are called the *focal points.*

**Construction:** To construct an ellipse, one fixes two points $B_-$ and $B_+$ which are to serve as focal points. Then one fixes the ends of a piece of string with a thumbtack to these focal points, and moves the pencil with the help of the string, keeping the string taut. The pencil then has drawn an ellipse (Figure 0.14(c)).

**Physical property of the focal points:** A light ray which is sent from one of the focal points $B_-$ and reflected on the ellipse, meets the other focal point $B_+$ (Figure 0.14(c)).

**Surface area and circumference of an ellipse:** See Table 0.4.

**The equation of an ellipse in polar coordinates, directrix property and curvature radii:** See section 0.1.7.6

#### 0.1.7.4 Hyperbola

**The equation of a hyperbola with center at the origin:**

$$\boxed{\frac{x^2}{a^2} - \frac{y^2}{b^2} = 1\,.} \qquad (0.11)$$

Here $a$ and $b$ are positive constants.

**Asymptotes of a hyperbola:** A hyperbola intersects the $x$-axis in the points $(\pm a, 0)$. The two lines

$$y = \pm\frac{b}{a}x$$

are called the *asymptotes* of the hyperbola. These lines approach the branches of the hyperbola as one moves out from the origin (Figure 0.15(b)).

**Focal points:** We define

$$\text{linear eccentricity} \quad e = \sqrt{a^2 + b^2}\,,$$
$$\text{numerical eccentricity} \quad \varepsilon = \frac{e}{a}\,,$$
$$\text{half-parameter} \quad p = \frac{b^2}{a}\,.$$

The two points $(\pm e, 0)$ are called the *focal points* $B_\pm$ of the hyperbola (Figure 0.15(a)).

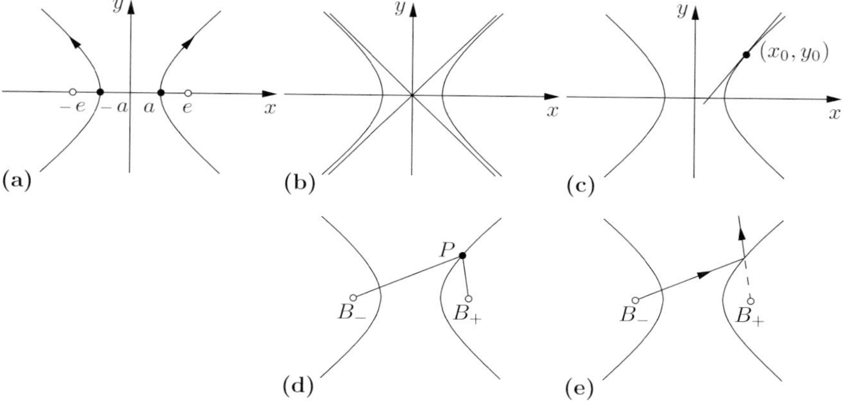


*Figure 0.15. Properties of the hyperbola.*

**Equation for the tangents to a hyperbola:**

$$\boxed{\frac{xx_0}{a^2} - \frac{yy_0}{b^2} = 1\,.}$$

This is the equation of the tangent to the hyperbola (0.11) through the point $(x_0, y_0)$ (Figure 0.15(c)).

**Parameterization of a hyperbola**[8]**:**

$$\boxed{x = a\cosh t\,, \qquad y = b\sinh t\,, \qquad -\infty < t < \infty\,.}$$

As the parameter $t$ runs through all real values, the right branch of the hyperbola in Figure 0.15(a) is run through once in the direction of the arrow in that picture. The initial point at $t = 0$ is the point $(a, 0)$ on the hyperbola. Similarly, the left hyperbola branch in Figure 0.15(a) is run through once by the parameterization

$$x = -a\cosh t\,, \qquad y = b\sinh t\,, \qquad -\infty < t < \infty\,.$$

**Geometric characterization of a hyperbola:** By definition, a hyperbola consists of all points $P$ whose difference of distances from two given points $B_-$ and $B_+$ is constant, equal to $2a$ (cf. Figure 0.15(d)). These points are again called the *focal points*.

[8] The hyperbolic functions $\cosh t$ and $\sinh t$ are treated in detail in 0.2.10 .

**Physical property of the focal points:** A light ray emerging from $B_-$ is reflected on the hyperbola in such a way that its backward extension passes through the other focal point $B_+$ (Figure 0.15(e)).

**Surface area of a hyperbola section:** See Table 0.4.

**Equation of hyperbolas in polar coordinates, directrix properties and curvature radii:** See section 0.1.7.6

#### 0.1.7.5 Parabola

**The equation of a parabola:**

$$\boxed{y^2 = 2px\,.} \tag{0.12}$$

Here $p$ is a positive constant (Figure 0.16). We define:

$$\text{linear eccentricity} \quad e = \frac{p}{2}\,,$$
$$\text{numerical eccentricity} \quad \varepsilon = 1\,.$$

The point $(e, 0)$ is called the *focal point* of the parabola (Figure 0.16(a)).

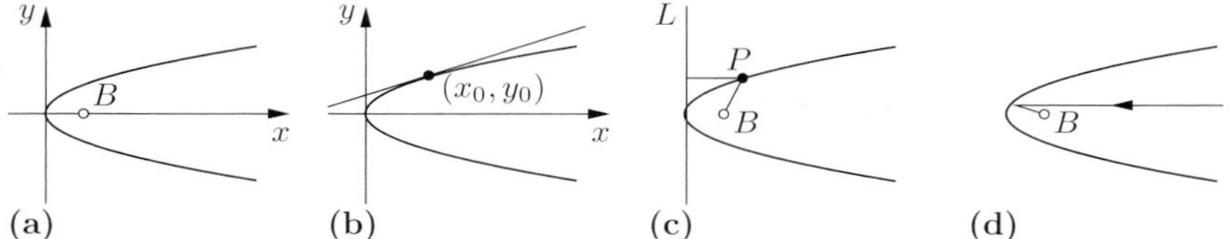


*Figure 0.16. Properties of the parabola.*

**The equation of a tangent to a parabola:**

$$\boxed{yy_0 = p(x + x_0)\,.}$$

This is the equation of the tangent to the parabola (0.12) through the point $(x_0, y_0)$ (Figure 0.16(b)).

**Geometric characterization of parabolas:** By definition, a parabola consists of all points $P$, whose distance from a fixed point $B$ (focal point) and a fixed line $L$ (directrix) is equal (Figure 0.16(c)).

**Physical property of the focal point (parabolic mirror):** A light ray, which is parallel to the $x$-axis and hits the parabola, is reflected in such a way that it passes through the focal point (Figure 0.16(d)).

**Surface area of a parabolic sector:** See Table 0.4.

**Equation of a parabola in polar coordinates and the curvature radii:** See section 0.1.7.6

### 0.1.7.6 Polar coordinates and conic sections

**Polar coordinates:** Instead of Cartesian coordinates, often *polar coordinates* are used, in order to take advantage of the symmetry of the equations in certain problems. The polar coordinates $(r, \varphi)$ of a point $P$ in the plane are given as in Figure 0.17 by the distance $r$ of the point $P$ from the origin $O$ and the angle $\varphi$ of the line segment $\overline{OP}$ with the $x$-axis. The following relation between the Cartesian coordinates $(x, y)$ and the polar coordinates $(r, \varphi)$ of a point $P$ hold:

$$\boxed{x = r\cos\varphi\,, \qquad y = r\sin\varphi\,, \qquad 0 \le \varphi < 2\pi\,.} \tag{0.13}$$

Moreover, one has

$$r = \sqrt{x^2 + y^2}\,, \qquad \tan\varphi = \frac{y}{x}\,.$$

**Conic sections:** By definition, a conic section is obtained by taking the section of a double circular cone with a plane (Figure 0.18). In this way, the following figures occur:

(i) *Regular conic sections:* Circle, ellipse, parabola or hyperbola.

(ii) *Degenerate conic sections:* two lines, one line or a point.

**Equation of regular conic sections in polar coordinates:**

$$\boxed{r = \frac{p}{1 - \varepsilon\cos\varphi}}$$

(cf. Table 0.8). The regular conic sections are characterized by the geometrical property, that they consists of all points $P$ for which the relation

$$\frac{r}{d} = \varepsilon$$

is constant, equal to $\varepsilon$, where $r$ is the distance from a fixed point $B$ (focal point) and $d$ is the distance from a fixed line $L$ (directrix).

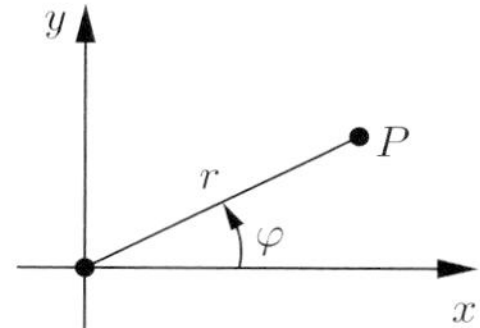


*Figure 0.17.*

*Figure 0.18.*

**Vertical circle and curvature radius:** In the apex $S$ of a regular conic sections, one can inscribe a circle in such a way, that it touches the conic section (i.e., has the same tangent as) at the point $S$. The radius of this vertical circle is called the *curvature radius* $R$ at the point $S$. The same construction is possible at an arbitrary point $P(x_0, y_0)$ of the conic section (cf. Table 0.9). The curvature $K$ at the point $P$ is given by definition by the formula

$$\boxed{K = \frac{1}{R_0}\,.}$$

*Table 0.8. Regular conic sections.*

| ***Conic section*** | ***Numerical eccentricity*** $\varepsilon$ | ***Linear eccentricity e*** | ***Half-parameter p*** | ***Directrix-property*** $\frac{r}{d} = \varepsilon$ |
|---|---|---|---|---|
| hyperbola[8] | $\varepsilon > 1$ | $e = \frac{\varepsilon p}{(1-\varepsilon)^2}$ | $p = \frac{b^2}{a}$ | 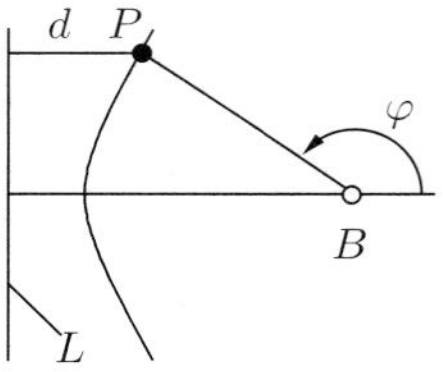 |
| parabola | $\varepsilon = 1$ | $e = \frac{p}{2}$ | | 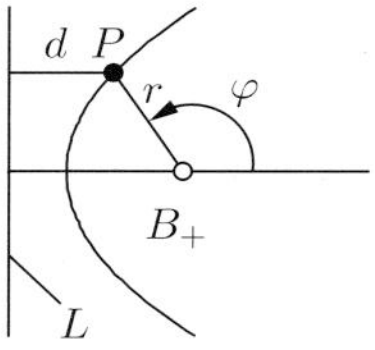 |
| ellipse | $0 \le \varepsilon < 1$ | $e = \frac{\varepsilon p}{1-\varepsilon^2}$ | $p = \frac{b^2}{a}$ | 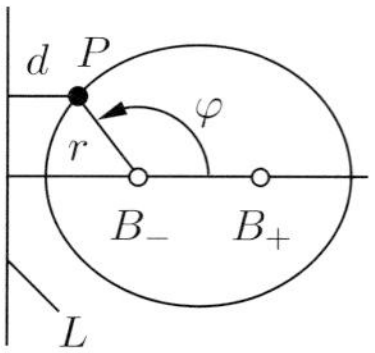 |
| circle | $\varepsilon = 0$ (limiting case $d = \infty$) | $e = 0$ | $p =$ radius $r$ | 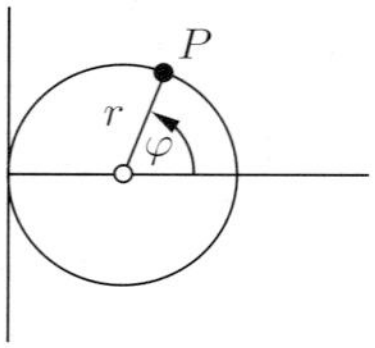 |

[8]Because of the inequalities $\varepsilon > 1$, $\varepsilon = 1$ and $\varepsilon < 1$, the Greek mathematician Appolonius of Perga (roughly 260–190 BC) introduced the nomenclature $\dot{\upsilon}\pi\varepsilon\rho\beta o\lambda\acute{\eta}$ (hyperbolé which means excess), $\pi\alpha\rho\alpha\beta o\lambda\acute{\eta}$ (parabolé which means equality) and $\tilde{\varepsilon}\lambda\lambda\varepsilon\iota\psi\iota\zeta$ (élleipsis which means deficiency).

*Table 0.9. Inscribed circles.*

| ***Conic*** | ***Equation*** | ***Curvature radius*** | ***Diagram*** |
|---|---|---|---|
| ellipse | $\frac{x^2}{a^2} + \frac{y^2}{b^2} = 1$ | $R_0 = a^2b^2\left(\frac{x_0^2}{a^4} + \frac{y_0^2}{b^4}\right)^{3/2}$, $R = \frac{b^2}{a} = p$ | |
| hyperbola | $\frac{x^2}{a^2} - \frac{y^2}{b^2} = 1$ | $R_0 = a^2b^2\left(\frac{x_0^2}{a^4} + \frac{y_0^2}{b^4}\right)^{3/2}$, $R = \frac{b^2}{a} = p$ | |
| parabola | $y^2 = 2px$ | $R_0 = \frac{(p + 2x_0)^{3/2}}{\sqrt{p}}$, $R = p$ | |

## 0.1.8 Basic formulas of analytic geometry of space

**Cartesian coordinates in space:** A spatial Cartesian coordinate system is given as shown in Figure 0.19 by three axi which are perpendicular to one another, which are denoted as the $x$-axis, $y$-axis and $z$-axis, and which are oriented in the same way as the thumb, the pointing finger and the middle finger of the right hand (right-handed system). The coordinates $(x_1, y_1, z_1)$ of a point are determined by perpendicular projection onto the axi.

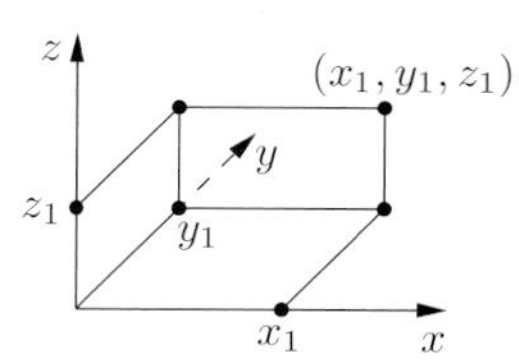


*Figure 0.19.*

**Equation of a line through the two points $(x_1, y_1, z_1)$ and $(x_2, y_2, z_2)$:**

$$\boxed{x = x_1 + t(x_2 - x_1)\,, \qquad y = y_1 + t(y_2 - y_1)\,, \qquad z = z_1 + t(z_2 - z_1)\,.}$$

The parameter $t$ runs through the real numbers and can be interpreted as the time (Figure 0.20(a)).

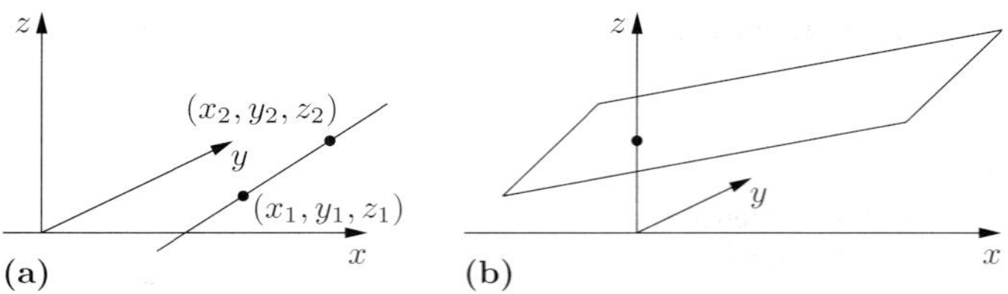


*Figure 0.20. Equations for lines and planes in three-space.*

**Distance $d$ between the two points $(x_1, y_1, z_1)$ and $(x_2, y_2, z_2)$:**

$$\boxed{d = \sqrt{(x_1 - x_2)^2 + (y_1 - y_2)^2 + (z_1 - z_2)^2}\,.}$$

**Equation of a plane:**

$$\boxed{Ax + By + Cz = D\,.}$$

The real constants $A, B$ and $C$ must fulfill the condition $A^2 + B^2 + C^2 \neq 0$ (Figure 0.20(b)).

**Applications of vector algebra to lines and planes in three-space:** See 3.3.

### 0.1.9 Powers, roots and logarithms

**Power laws:** For all positive real numbers $a, b$ and all real numbers $x, y$ one has:

$$\boxed{\begin{array}{lll} a^x a^y = a^{x+y}\,, & (a^x)^y = a^{xy}\,, & \\ (ab)^x = a^x b^x\,, & \left(\dfrac{a}{b}\right)^x = \dfrac{a^x}{b^x}\,, & a^{-x} = \dfrac{1}{a^x}\,. \end{array}}$$

It wasn't until after a long historical course of development that the notion of powers $a^x$ for *arbitrary real exponents* was realized (cf. 0.2.7).

**Important special cases:** For $n = 1, 2, \ldots$ one has:

1. $a^0 = 1\,, \quad a^1 = a, \quad a^2 = a \cdot a, \quad a^3 = a \cdot a \cdot a, \quad \ldots$
2. $a^n = a \cdot a \cdot \ldots \cdot a \quad$ ($n$ factors).
3. $a^{-1} = \dfrac{1}{a}\,, \quad a^{-2} = \dfrac{1}{a^2}\,, \quad \ldots, \quad a^{-n} = \dfrac{1}{a^n}.$
4. $a^{\frac{1}{2}} = \sqrt{a}\,, \quad a^{\frac{1}{3}} = \sqrt[3]{a}.$

**$n^{th}$ roots:** Let a positive real number $a$ be given. Then $x = a^{1/n}$ is the unique solution to the equation

$$\boxed{x^n = a\,, \qquad x \geq 0\,.}$$

In older literature the term $a^{1/n}$ is often denoted by $\sqrt[n]{a}$ ($n^{th}$ root). In manipulations with expressions involving such roots it is better to use the expression $a^{1/n}$, since then one can use the general rules for powers and is not restricted to 'rules for roots'.

*Example 1:* From $\left(a^{\frac{1}{n}}\right)^{\frac{1}{m}} = a^{\frac{1}{mn}}$ the root law $\sqrt[n]{\sqrt[m]{a}} = \sqrt[nm]{a}$ follows.

**Limit relation for general powers:** For $x = \frac{m}{n}$ with $m, n = 1, 2, \ldots$ the following relation holds:

$$a^x = \left(\sqrt[n]{a}\right)^m .$$

Moreover $a^{-x} = 1/a^x$. Hence the calculation of $a^x$ for arbitrary *rational* exponents $x$ can be *reduced* to the calculation of roots.

Now let an arbitrary real number $x$ be given. We choose a number sequence[9] $(x_k)$ of real numbers $x_k$ with

$$\lim_{k \to \infty} x_k = x .$$

Then we have

$$\boxed{\lim_{k \to \infty} a^{x_k} = a^x .}$$

This is an expression of the *continuity* of the exponential function (cf. 1.3.1.2). If one chooses in particular a sequence $x_k$ of rational numbers $x_k$, then the expressions $a^{x_k}$ can be expressed in terms of powers of roots, and $a^x$ is approximated for larger and larger $k$ more and more accurately.

*Example 2:* The approximate value of $\pi$ is given by $\pi = 3.14\ldots$ Therefore we have

$$a^{3.14} = a^{314/100} = \left(\sqrt[100]{a}\right)^{314}$$

is an approximation to the number $a^\pi$. Better and better approximations for $a^\pi$ can be obtained by incorporating more and more decimal places in the decimal representation $\pi = 3.14\,15\,92\ldots$

**The logarithm:** Let $a$ be a fixed, positive real number $a \neq 1$. For each given positive real number $y$ the equation

$$\boxed{y = a^x}$$

has a unique real solution $x$, which is denoted by

$$\boxed{x = \log_a y}$$

and is called the *logarithm of $y$ to base $a$.*[10]

**Laws for logarithms:** For all positive real numbers $c, d$ and all real numbers $x$ one has:

$$\boxed{\begin{array}{lll} \log_a(cd) = \log_a c + \log_a d\,, & & \log_a\left(\frac{c}{d}\right) = \log_a c - \log_a d\,, \\ \log_a c^x = x\log_a c\,, & \log_a a = 1\,, & \log_a 1 = 0\,. \end{array}}$$

From the relation $\log(cd) = \log c + \log d$ one sees that the logarithm has the fundamental property that multiplication of two numbers corresponds to the *addition* of the logarithms of those numbers.

[9] Limits of sequences of real numbers will be considered in 1.2.

[10] The word logarithm has a Greek root and means 'ratio number'.

**Historical remark:** In his monograph *Arithmetica integra* (Collected arithmetic), Michael Stifel noted in 1544 the the comparison of

$$\begin{array}{cccccc} 1 & a & a^2 & a^3 & a^4 & \ldots \\ 0 & 1 & 2 & 3 & 4 & \ldots \end{array}$$

allows the reduction of the multiplication of the numbers in the first row to the addition of the powers in the second row. This is precisely the basic idea of calculations with logarithms. Stifel remarks on this: "One could write an entire book on the properties of these wonderful numbers, but I have to be modest and close my eyes to this at this point." In the year 1614 the Scotch nobleman John Neper (or Napier) published the first incomplete tables of logarithms (with a base proportional to 1/e). These tables were improved bit for bit. After discussions with Henry Briggs, Neper agreed to use the basis 10 for all logarithms. In 1617 Briggs published a table of logarithms up to 14 decimal places (to base $a = 10$). The appearance of these tables was of great help to Kepler in the completion of his famous "Rudolfian tables" in 1624 (cf. 0.1.12). He propagated the advantages of this powerful new method of calculation with ardent zeal.

In our modern times with the widespread use of computers these tables are no longer of importance and represent a historical episode.

**Natural logarithms:** Logarithms $\log_{\mathrm{e}} y$ to base e are referred to as *natural logarithms* (logarithmus naturalis) and are denoted $\ln y$. If $a > 0$ is an arbitrary base, then one has the relation

$$\boxed{a^x = \mathrm{e}^{x \ln a}}$$

for all real numbers $x$. If one knows the natural logarithm, then one can find the logarithm to an arbitrary base by means of the formula

$$\boxed{\log_a y = \frac{\ln y}{\ln a}.}$$

*Example 3:* For $a = 10$ one has $\ln a = 2.302585\ldots$ and $\dfrac{1}{\ln a} = 0.434294\ldots$

In 1.12.1 we will give applications of the function $y = \mathrm{e}^x$ with the help of differential equations to radioactive decay and growth process. These examples show that the Euler number $\mathrm{e} = 2.718283\ldots$ is the natural base for the exponential function. The inverse of $y = \mathrm{e}^x$ gives $x = \ln y$. This motivates the nomenclature 'natural logarithm'.

## 0.1.10 Elementary algebraic formulas

### 0.1.10.1 The geometric and arithmetic series

**Summation symbol and product symbol:** We define

$$\sum_{k=0}^{n} a_k := a_0 + a_1 + a_2 + \ldots + a_n$$

and

$$\prod_{k=0}^{n} a_k := a_0 a_1 a_2 \ldots a_n \,.$$

**The finite geometric series:**

$$\boxed{a + aq + aq^2 + \ldots + aq^n = a\frac{1-q^{n+1}}{1-q}, \qquad n = 1, 2, \ldots} \tag{0.14}$$

This formula is valid for all real or complex numbers $a$ and $q$ with $q \neq 1$. The geometric series (0.14) is characterized by the fact that the *quotient* of two successive terms is a constant. With the help of the summation symbol one can write (0.14) in the form

$$\sum_{k=0}^{n} aq^k = a\frac{1-q^{n+1}}{1-q}, \qquad q \neq 1, \quad n = 1, 2, \ldots$$

*Example 1:* $1 + q + q^2 = \dfrac{1-q^3}{1-q} \qquad (q \neq 1)\,.$

**The arithmetic series:**

$$\boxed{a + (a+d) + (a+2d) + \ldots + (a+nd) = \frac{n+1}{2}(a + (a+dn))\,.} \tag{0.15}$$

The arithmetic series (0.15) is characterized by the property that the *difference* of two successive terms is a constant. In words:

> *The sum of an arithmetic series is equal to the sum of the first and the last term multiplied by half the total number of terms.*

With the help of the summation symbol, the formula (0.15) can be written:

$$\sum_{k=0}^{n}(a + kd) = \frac{n+1}{2}(a + (a+nd))\,.$$

Arithmetic series can be found in ancient texts of Babylonian and Egyptian times (around 2000 BC). Geometric series and the formula for the sum are found in Euclid's *Elements* (around 300 BC).

*Example 2:* It is reported that the teacher of the young Gauss (1777–1855) wanted a relaxing day by giving his students the assignment of adding the numbers 1 to 40. Just after assigning this, the little boy Gauss (who was to become one of the greatest mathematicians of all times) came to the teachers desk with his slate and the result of 820. It apparently was immediately clear to the youngster that instead of the original series $1 + 2 + \ldots + 40$ one should rather consider

$$\begin{array}{cccc} 1 & 2 & 3 & \ldots & 40 \\ 40 & 39 & 38 & \ldots & 1. \end{array}$$

Here we have 40 pairs of numbers (those in columns above) whose sum is 41. Consequently, the sum of the first series is half of the sum of these pairs, i.e., $20 \cdot 41 = 820$.

This is an example for a moment of inspiration in mathematics. A problem who initially seems to be quite complicated is reduced by some elegant trick to a different, easier problem which is quickly solved.

#### 0.1.10.2 Calculations with the summation and product symbols

**Summation symbol:** The following manipulations are often applied:

1. $\sum_{k=0}^{n} a_k = \sum_{j=0}^{n} a_j$, (change of summation index).

2. $\sum_{k=0}^{n} a_k = \sum_{j=N}^{n+N} a_{j-N}$, (shift of the summation index; $j = k + N$).

3. $\sum_{k=0}^{n} a_k + \sum_{k=0}^{n} b_k = \sum_{k=0}^{n} (a_k + b_k)$, (rule for addition).

4. $(\sum_{j=1}^{m} a_j)(\sum_{k=1}^{n} b_k) = \sum_{j=1}^{m}\sum_{k=1}^{n} a_j b_k$, (distributive law).

5. $\sum_{j=1}^{m}\sum_{k=1}^{n} a_{jk} = \sum_{k=1}^{n}\sum_{j=1}^{m} a_{jk}$, (commutative law).

**Product symbol:** Analogously to the summations symbol one has the following properties of the product symbol:

1. $\prod_{k=0}^{n} a_k = \prod_{j=0}^{n} a_j$ .

2. $\prod_{k=0}^{n} a_k = \prod_{j=N}^{n+N} a_{j-N}$ .

3. $\prod_{k=0}^{n} a_k \prod_{k=0}^{n} b_k = \prod_{k=0}^{n} a_k b_k$ .

4. $\prod_{j=1}^{m}\prod_{k=1}^{n} a_{jk} = \prod_{k=1}^{n}\prod_{j=1}^{m} a_{jk}$ .

#### 0.1.10.3 The binomial formula

**Three classical binomial formulas:**

| | |
|---|---|
| $(a+b)^2 = a^2 + 2ab + b^2$, | (first binomial formula), |
| $(a-b)^2 = a^2 - 2ab + b^2$, | (second binomial formula), |
| $(a-b)(a+b) = a^2 - b^2$, | (third binomial formula). |

These formulas are valid for all real or complex numbers $a$ and $b$. The second binomial formula is actually a consequence of the first, by replacing $b$ by $-b$.

**The general third binomial formula:** One has

$$\sum_{k=0}^{n} a^{n-k}b^k = a^n + a^{n-1}b + \ldots + ab^{n-1} + b^n = \frac{a^{n+1} - b^{n+1}}{a-b}$$

for all $n = 1, 2, \ldots$ and all real or complex numbers $a$ and $b$ with $a \neq b$.

**Binomial coefficients:** For all $k = 1, 2, \ldots$ and all real numbers $\alpha$ we set

$$\binom{\alpha}{k} := \frac{\alpha}{1} \cdot \frac{(\alpha-1)}{2} \cdot \frac{(\alpha-2)}{3} \cdot \ldots \cdot \frac{(\alpha-k+1)}{k} .$$

Furthermore let

$$\binom{\alpha}{0} := 1 .$$

*Example 1:* $\binom{3}{2} = \frac{3 \cdot 2}{1 \cdot 2} = 3$, $\binom{5}{3} = \frac{5 \cdot 4 \cdot 3}{1 \cdot 2 \cdot 3} = 10$.

**The general first binomial formula (binomial theorem):**

$$(a+b)^n = a^n + \binom{n}{1}a^{n-1}b + \binom{n}{2}a^{n-2}b^2 + \ldots + \binom{n}{n-1}ab^{n-1} + b^n . \tag{0.16}$$

This fundamental formula of elementary mathematics is valid for all $n = 1, 2, \ldots$ and all real or complex $a$ and $b$. With the help of the summation symbol, (0.16) can be written:

$$(a+b)^n = \sum_{k=0}^{n} \binom{n}{k} a^{n-k}b^k . \tag{0.17}$$

**The general second binomial formula:**

$$(a-b)^n = \sum_{k=0}^{n} \binom{n}{k} (-1)^k a^{n-k}b^k .$$

This formula follows immediately from (0.17) upon replacing $b$ by $-b$.

**Pascal triangle:** In Table 0.10 each coefficient is obtained as the sum of the two coefficients lying above the given one. This gives a convenient way to obtain the coefficients for the general binomial formulas.

*Example 2:*

$$\begin{aligned}(a+b)^3 &= a^3 + 3a^2b + 3ab^2 + b^3 ,\\ (a+b)^4 &= a^4 + 4a^3b + 6a^2b^2 + 4ab^3 + b^4 ,\\ (a+b)^5 &= a^5 + 5a^4b + 10a^3b^2 + 10a^2b^3 + 5ab^4 + b^5 .\end{aligned}$$

The Pascal triangle is named after Blaise Pascal (1623–1662), who at the age of 20 built the first addition machine. The modern computer language *Pascal* is named in his honor. One can also find the Pascal triangles for $n = 1, \ldots, 8$ in the Chinese monograph *The precious mirror of four elements* by Chu Shih–Chieh, written in 1303.

Table 0.10. Pascal's triangle.

| | Coefficients of the binomial formulas |
|---|---|
| $n = 0$ | 1 |
| $n = 1$ | 1   1 |
| $n = 2$ | 1   2   1 |
| $n = 3$ | 1   3   3   1 |
| $n = 4$ | 1   4   6   4   1 |
| $n = 5$ | 1   5   10   10   5   1 |

**Newton's binomial series for real exponents:** The 24-year old Isaac Newton (1643–1727) found by intuitive reasoning the general formula for the series:

$$(1+x)^\alpha = 1 + \binom{\alpha}{1}x + \binom{\alpha}{2}x^2 + \binom{\alpha}{3}x^3 + \ldots = \sum_{k=0}^{\infty}\binom{\alpha}{k}x^k\,. \tag{0.18}$$

For $\alpha = 1, 2, \ldots$, the infinite series (0.18) is actually *finite* an is nothing but the binomial formula.

**Theorem of Euler (1774):** The binomial series converges for all *real exponents* $\alpha$ and all complex numbers $x$ with $|x| < 1$.

It had been attempted for a long time to prove the convergence of this series. It wasn't until Euler was 67 that he succeeded, more than one hundred years after Newton's discovery of the series.

**The polynomial theorem:** This theorem generalizes the binomial theorem to more than two summands. Special cases are:

$$\begin{aligned}(a+b+c)^2 &= a^2+b^2+c^2+2ab+2ac+2bc\,,\\ (a+b+c)^3 &= a^3+b^3+c^3+3a^2b+3a^2c+3b^2c\\ &\quad + 6abc+3ab^2+3ac^2+3bc^2\,.\end{aligned}$$

The general form of this theorem for arbitrary real or complex non-vanishing numbers $a_1, \ldots, a_N$ and natural numbers $n = 1, 2, \ldots$ is:

$$(a_1 + a_2 + \ldots + a_N)^n = \sum_{m_1+\ldots+m_N=n} \frac{n!}{m_1!m_2!\cdots m_N!}\, a_1^{m_1}a_2^{m_2}\cdots a_N^{m_N}\,.$$

The summation here is over all $N$-tuples $(m_1, m_2, \ldots, m_N)$ of natural numbers running from 0 to $n$ and whose sum is $n$. Moreover $n! = 1\cdot 2\cdot\ldots\cdot n$.

**Properties of binomial coefficients:** For natural numbers $n, k$ with $0 \le k \le n$ and real or complex numbers $\alpha, \beta$ one has:

(i) *symmetry law*

$$\boxed{\binom{n}{k} = \binom{n}{n-k} = \frac{n!}{k!(n-k)!}\,.}$$

(ii) *addition law*[11]

$$\boxed{\binom{\alpha}{k} + \binom{\alpha}{k+1} = \binom{\alpha+1}{k+1}\,,} \tag{0.19}$$

$$\binom{\alpha}{0} + \binom{\alpha+1}{1} + \ldots + \binom{\alpha+k}{k} = \binom{\alpha+k+1}{k}\,,$$

$$\binom{\alpha}{0}\binom{\beta}{k} + \binom{\alpha}{1}\binom{\beta}{k-1} + \ldots + \binom{\alpha}{k}\binom{\beta}{0} = \binom{\alpha+\beta}{k}\,.$$

*Example 3:* If we set $\alpha = \beta = k = n$ in the last equation, then from the symmetry law we get the relation:

$$\binom{n}{0}^2 + \binom{n}{1}^2 + \ldots + \binom{n}{n}^2 = \binom{2n}{n}\,.$$

From the binomial theorem for $a = b = 1$ and $a = -b = 1$ we get:

$$\binom{n}{0} + \binom{n}{1} + \ldots + \binom{n}{n} = 2^n\,,$$

$$\binom{n}{0} - \binom{n}{1} + \binom{n}{2} - \ldots + (-1)^n\binom{n}{n} = 0\,.$$

### 0.1.10.4 Sums of powers and Bernoulli numbers

**Sums of natural numbers:**

$$\sum_{k=1}^{n} k = 1 + 2 + \ldots + n = \frac{n(n+1)}{2}\,,$$

$$\sum_{k=1}^{n} 2k = 2 + 4 + \ldots + 2n = n(n+1)\,,$$

$$\sum_{k=1}^{n} (2k-1) = 1 + 3 + \ldots + (2n-1) = n^2\,.$$

**Sums of squares:**

$$\boxed{\sum_{k=1}^{n} k^2 = 1^2 + 2^2 + \ldots + n^2 = \frac{n(n+1)(2n+1)}{6}\,,}$$

$$\sum_{k=1}^{n} (2k-1)^2 = 1^2 + 3^2 + \ldots + (2n-1)^2 = \frac{n(4n^2-1)}{3}\,.$$

[11]The Pascal triangle is based on the formula (0.19).

**Sums of third and fourth powers:**

$$\sum_{k=1}^{n} k^3 = 1^3 + 2^3 + \ldots + n^3 = \frac{n^2(n+1)^2}{4},$$

$$\sum_{k=1}^{n} k^4 = 1^4 + 2^4 + \ldots + n^4 = \frac{n(n+1)(2n+1)(3n^2+3n-1)}{30}.$$

**Bernoulli numbers:** Jacob Bernoulli (1645–1705) ran across these numbers as he attempted calculating an empirical formula for the sums of powers

$$S_n^p := 1^p + 2^p + \ldots + n^p$$

of natural numbers. He found for $n = 1, 2, \ldots$ and for the exponents $p = 1, 2, \ldots$ the general formula:

$$\boxed{S_n^p = \frac{1}{p+1} n^{p+1} + \frac{1}{2} n^p + \frac{B_2}{2}\binom{p}{1} n^{p-1} + \frac{B_3}{3}\binom{p}{2} n^{p-2} + \ldots + \frac{B_p}{p}\binom{p}{p-1} n\,.}$$

He also noticed that the sum of the coefficients always turns out to equal 1, i.e, we have

$$\frac{1}{p+1} + \frac{1}{2} + \frac{B_2}{2}\binom{p}{1} + \frac{B_3}{3}\binom{p}{2} + \ldots + \frac{B_p}{p}\binom{p}{p-1} = 1\,.$$

From this one gets for $p = 2, 3, \ldots$ successively the Bernoulli numbers $B_2, B_3, \ldots$ One also sets $B_0 := 1$ and $B_1 := -1/2$ (see Table 0.11). For odd numbers $n \geq 3$ one has $B_n = 0$. The recursion formula can also be written in the form

$$\boxed{\sum_{k=0}^{n} \binom{p+1}{k} B_k = 0\,.}$$

Symbolically, this equation is

$$(1+B)^{p+1} - B_{p+1} = 0\,,$$

if one agrees to replace $B^n$ by $B_n$ after multiplying out the expression on the left.

*Table 0.11. Bernoulli numbers $B_k$ ($B_3 = B_5 = B_7 = \ldots = 0$).*

| $k$ | $B_k$ | $k$ | $B_k$ | $k$ | $B_k$ | $k$ | $B_k$ |
|---|---|---|---|---|---|---|---|
| 0 | 1 | 4 | $-\frac{1}{30}$ | 10 | $\frac{5}{66}$ | 16 | $-\frac{3617}{510}$ |
| 1 | $-\frac{1}{2}$ | 6 | $\frac{1}{42}$ | 12 | $-\frac{691}{2730}$ | 18 | $\frac{43\,867}{798}$ |
| 2 | $\frac{1}{6}$ | 8 | $-\frac{1}{30}$ | 14 | $\frac{7}{6}$ | 20 | $-\frac{174\,611}{330}$ |

*Example:*

$$S_n^1 = \frac{1}{2}n^2 + \frac{1}{2}n\,,$$
$$S_n^2 = \frac{1}{3}n^3 + \frac{1}{2}n^2 + \frac{1}{6}n\,,$$
$$S_n^3 = \frac{1}{4}n^4 + \frac{1}{2}n^3 + \frac{1}{4}n^2\,,$$
$$S_n^4 = \frac{1}{5}n^5 + \frac{1}{2}n^4 + \frac{1}{3}n^3 - \frac{1}{30}n\,.$$

In addition one has:

$$\boxed{\frac{S_n^p}{p!} = \frac{B_0(n+1)^{p+1}}{0!(p+1)!} + \frac{B_1(n+1)^p}{1!p!} + \frac{B_2(n+1)^{p-1}}{2!(p-1)!} + \ldots + \frac{B_p(n+1)}{p!1!}\,.}$$

**Bernoulli numbers and infinite series:** For all complex numbers $x$ with $0 < |x| < 2\pi$, one has:

$$\boxed{\frac{x}{e^x - 1} = \frac{B_0}{0!} + \frac{B_1}{1!}x + \frac{B_2}{2!}x^2 + \ldots = \sum_{k=0}^{\infty} \frac{B_k}{k!}x^k\,.}$$

Bernoulli numbers also appear in the power series expansion of the functions

$$\tan x\,,\ \cot x\,,\ \tanh x\,,\ \coth x\,,\ \frac{1}{\sin x}\,,\ \frac{1}{\sinh x}\,,$$
$$\ln|\tan x|\,,\ \ln|\sin x|\,,\ \ln\cos x$$

(cf. 0.7.2).

Bernoulli numbers also play an important role in the summation of the inverses of powers of natural numbers. Euler discovered in 1734 the famous formula

$$\boxed{1 + \frac{1}{2^2} + \frac{1}{3^2} + \ldots = \sum_{n=1}^{\infty} \frac{1}{n^2} = \frac{\pi^2}{6}\,.}$$

More generally, Euler discovered for $k = 1, 2, \ldots$ the values[12]:

$$\boxed{1 + \frac{1}{2^{2k}} + \frac{1}{3^{2k}} + \ldots = \sum_{n=1}^{\infty} \frac{1}{n^{2k}} = \frac{(2\pi)^{2k}}{2(2k)!}|B_{2k}|\,.}$$

Even earlier, the brothers Johann and Jakob Bernoulli had tried for a long time to determine the value of these series.

[12]For this, Euler used the product formula

$$\sin \pi x = \pi x \prod_{m=1}^{\infty} \left(1 - \frac{x^2}{m^2}\right)\,,$$

which he had discovered and which holds for all complex numbers $x$; this is in fact a generalization of the fundamental theorem of algebra (cf. 2.1.6) to the sine function

#### 0.1.10.5 The Euler numbers

**Defining relations:** For all complex numbers $x$ with $|x| < \frac{\pi}{2}$, the infinite series

$$\boxed{\frac{1}{\cosh x} = 1 + \frac{E_1}{1!}x + \frac{E_2}{2!}x^2 + \ldots = \sum_{k=0}^{\infty} \frac{E_k}{k!}x^k}$$

converges. The coefficients $E_k$ which occur in this series are called *Euler numbers* (cf. Table 0.12). One has $E_0 = 1$ and for odd $n$, $E_n = 0$. The Euler numbers satisfy the symbolic equation

$$\boxed{(E+1)^n + (E-1)^n = 0\,, \qquad n = 1, 2, \ldots\,,}$$

in which one agrees to replace $E^n$ by $E_n$ after the multiplication has been carried out. This gives a convenient recursion formula for the $E_n$. The relation between the Euler and the Bernoulli numbers is, again in symbolic form, given by:

$$E_{2n} = \frac{4^{2n+1}}{2n+1}\left(B_n - \frac{1}{4}\right)^{2n+1}, \qquad n = 1, 2, \ldots$$

*Table 0.12. The Euler numbers $E_k$ ($E_1 = E_3 = E_5 = \ldots = 0$).*

| $k$ | $E_k$ | $k$ | $E_k$ | $k$ | $E_k$ |
|---|---|---|---|---|---|
| 0 | 1 | 6 | $-61$ | 12 | $2,702,765$ |
| 2 | $-1$ | 8 | $1,385$ | 14 | $-199,360,981$ |
| 4 | 5 | 10 | $-50,521$ | | |

**Euler numbers and infinite series:** The Euler numbers occur in the power series expansion of the functions

$$\frac{1}{\cosh x}, \quad \frac{1}{\cos x}$$

(cf. 0.7.2). For $k = 1, 2, \ldots$ one has in addition the formula

$$\boxed{1 - \frac{1}{3^{2k+1}} + \frac{1}{5^{2k+1}} - \ldots = \sum_{n=0}^{\infty} \frac{(-1)^n}{(2n+1)^{2k+1}} = \frac{\pi^{2k+1}}{2^{2k+2}(2k)!}|E_{2k}|\,.}$$

### 0.1.11 Important inequalities

The rules for manipulations with inequalities can be found in section 1.1.5

**The triangle inequality[13]:**

$$\boxed{\Big||z| - |w|\Big| \le |z - w| \le |z| + |w| \quad \text{for all } z, w \in \mathbb{C}\,.}$$

[13] The statement 'for all $a \in \mathbb{R}$' means that the formula is valid for all real numbers $a$. The statement 'for all $z \in \mathbb{C}$' means that the statement is valid for all complex numbers. Note that each real number is also a complex number. The absolute value $|z|$ of a real or complex number is introduced in 1.1.2.1.

In addition one has for $n$ complex summands $x_1, \ldots, x_n$ the triangle inequality

$$\boxed{\left|\sum_{k=1}^{n} x_k\right| \le \sum_{k=1}^{n} |x_k|\,.}$$

**The Bernoulli inequality:** For all real numbers $x \ge -1$ and $n = 1, 2, \ldots$ one has

$$\boxed{(1+x)^n \ge 1 + nx\,.}$$

**The binomial inequality:**

$$\boxed{|ab| \le \frac{1}{2}\left(a^2 + b^2\right) \qquad \text{for all } a, b \in \mathbb{R}\,.}$$

**The inequality for means:** For all positive real numbers $c$ and $d$ one has:

$$\boxed{\frac{2}{\dfrac{1}{c} + \dfrac{1}{d}} \le \sqrt{cd} \le \frac{c+d}{2} \le \sqrt{\frac{c^2 + d^2}{2}}\,.}$$

The means which appear here are called, from left to right, harmonic mean, geometric mean, arithmetic mean and quadratic mean. All these means lie in between the two values $\min\{c, d\}$ and $\max\{c, d\}$, which justifies the term mean.[14]

**Inequality for general means:** For positive real numbers $x_1, \ldots, x_n$ one has:

$$\boxed{\min\{x_1, \ldots, x_n\} \le h \le g \le m \le s \le \max\{x_1, \ldots, x_n\}\,.}$$

In this formula we have used the notations:

$$m := \frac{x_1 + x_2 + \ldots + x_n}{n} = \frac{1}{n}\sum_{k=1}^{n} x_k, \qquad \text{(arithmetic mean or mean value)},$$

$$g := (x_1 x_2 \ldots x_n)^{1/n} = \left(\prod_{k=1}^{n} x_k\right)^{1/n}, \qquad \text{(geometric mean)},$$

$$h := \frac{n}{\dfrac{1}{x_1} + \ldots + \dfrac{1}{x_n}}, \qquad \text{(harmonic mean)}$$

and

$$s := \left(\frac{1}{n}\sum_{k=1}^{n} x_k^2\right)^{1/2}, \qquad \text{(quadratic mean)}.$$

[14]The terms $\min\{c, d\}$ (resp. $\max\{c, d\}$) denote the smallest (resp. the largest) of the two numbers $c$ and $d$.

**The Young inequality:** One has

$$\boxed{|ab| \le \frac{|a|^p}{p} + \frac{|b|^q}{q} \quad \text{for all } a, b \in \mathbb{C}} \tag{0.20}$$

and all real exponents $p$ and $q$ which satisfy $p, q > 1$ and

$$\frac{1}{p} + \frac{1}{q} = 1\,.$$

In the special case $p = q = 2$ the Young inequality is nothing but the binomial inequality. If $n = 2, 3, \ldots$, then the general Young inequality is valid:

$$\boxed{\left|\prod_{k=1}^{n} x_k\right| \le \sum_{k=1}^{n} \frac{|x_k|^{p_k}}{p_k} \quad \text{for all } x_k \in \mathbb{C}} \tag{0.21}$$

and all real exponents $p_k > 1$ with $\displaystyle\sum_{k=1}^{n} \frac{1}{p_k} = 1\,.$

**The Schwarz inequality:**

$$\boxed{\left|\sum_{k=1}^{n} x_k y_k\right| \le \left(\sum_{k=1}^{n} |x_k|^2\right)^{1/2} \left(\sum_{k=1}^{n} |y_k|^2\right)^{1/2} \quad \text{for all } x_k, y_k \in \mathbb{C}\,.}$$

**The Hölder inequality**[15]**:** One has

$$\boxed{\big|(x|y)\big| \le \|x\|_p \|y\|_q \quad \text{for all } x, y \in \mathbb{C}^N}$$

and all real exponents $p, q > 1$ with $\dfrac{1}{p} + \dfrac{1}{q} = 1\,.$ The notations used are defined as follows:

$$(x|y) := \sum_{k=1}^{N} \overline{x}_k y_k \qquad \text{and} \qquad \|x\|_p := \left(\sum_{k=1}^{N} |x_k|^p\right)^{1/p}$$

as well as

$$\|x\|_\infty := \max_{1 \le k \le N} |x_k|\,.$$

The notation $\overline{x}_k$ denotes the complex conjugate number to $x_k$ (cf. 1.1.2).

**The Minkowski inequality:**

$$\boxed{\|x + y\|_p \le \|x\|_p + \|y\|_p \quad \text{for all } x, y \in \mathbb{C}^N\,,\ 1 \le p \le \infty\,.}$$

**Jensen's inequality:**

$$\boxed{\|x\|_p \le \|x\|_r \quad \text{for all } x \in \mathbb{C}^N\,,\ 0 < r < p \le \infty\,.}$$

[15] The statement 'for all $x \in \mathbb{C}^N$' means 'for all $N$-tuples $(x_1, \ldots, x_N)$ of complex numbers $x_k$'.

**Integral inequalities:** The following inequalities hold, provided the integral on the right hand side exists (and is therefore finite)[16]. In addition, the real coefficients $p, q > 1$ should satisfy the condition $\frac{1}{p} + \frac{1}{q} = 1$. Then:

(i) *triangle inequality*

$$\left| \int_G f \mathrm{d}x \right| \le \int_G |f(x)| \mathrm{d}x \,.$$

(ii) *Hölder inequality*

$$\left| \int_G f(x)g(x)\mathrm{d}x \right| \le \left( \int_G |f(x)|^p \mathrm{d}x \right)^{1/p} \left( \int_G |g(x)|^q \mathrm{d}x \right)^{1/q} .$$

In the special case $p = q = 2$ this reduces to the Schwarz inequality.

(iii) *Minkowski inequality* $(1 \le r < \infty)$

$$\left( \int_G |f(x) + g(x)|^r \mathrm{d}x \right)^{1/r} \le \left( \int_G |f(x)|^r \mathrm{d}x \right)^{1/r} + \left( \int_G |g(x)|^r \mathrm{d}x \right)^{1/r} .$$

(iv) *Jensen's inequality* $(0 < p < r < \infty)$

$$\left( \int_G (|f(x)|^p \mathrm{d}x \right)^{1/p} \le \left( \int_G |f(x)|^r \mathrm{d}x \right)^{1/r} .$$

**The Jensen convexity inequality:** Let $m = 1, 2, \ldots$ If the real valued function $F : \mathbb{R}^N \to \mathbb{R}$ is convex, then

$$F\left( \sum_{k=1}^m \lambda_k x_k \right) \le \sum_{k=1}^m \lambda_k F(x_k)$$

for all $x_k \in \mathbb{R}^N$ and all non-negative real coefficients $\lambda_k$ with $\sum_{k=1}^m \lambda_k = 1$ (cf. 1.4.5.5).

**The Jensen convexity inequality for integrals:**

$$F\left( \frac{\int_G p(x)g(x)\mathrm{d}x}{\int_G p(x)\mathrm{d}x} \right) \le \frac{\int_G p(x)F(g(x))\mathrm{d}x}{\int_G p(x)\mathrm{d}x} . \tag{0.22}$$

Here it is assumed that:

[16]These formulas hold under very general assumptions. One can use the classical one-dimensional integral (Riemann integral) $\int_G f \mathrm{d}x = \int_a^b f \mathrm{d}x$, the several variable classical integral or the modern Lebesgue integral. The values of the function $f(x)$ may be real or complex.

(i) The real valued function $F\colon \mathbb{R} \to \mathbb{R}$ is convex.

(ii) The function $p\colon G \to \mathbb{R}$ is non-negative and is integrable on the open set $G$ in $\mathbb{R}^N$ with $\int_G p\mathrm{d}x > 0$.

(iii) The function $g\colon G \to \mathbb{R}$ has the property that all integrals in (0.22) exist[17].

For example, one may choose $p(x) \equiv 1$.

**The fundamental convexity inequality:** Let $n = 1, 2, \ldots$ For all non-negative real numbers $x_k$ and $\lambda_k$ with $\lambda_1 + \lambda_2 + \ldots + \lambda_n = 1$ one has

$$\boxed{f^{-1}\left(\sum_{k=1}^{n} \lambda_k f(x_k)\right) \le g^{-1}\left(\sum_{k=1}^{n} \lambda_k g(x_k)\right),} \tag{0.23}$$

provided the following assumptions are fulfilled:

(i) The functions $f, g : [0,\infty[\to [0,\infty[$ are increasing and surjective. We denote by $f^{-1}, g^{-1} : [0,\infty[\to [0,\infty[$ the inverse functions to $f$ and $g$.

(ii) The composition $y = g\left(f^{-1}(x)\right)$ of functions is *convex* on the interval $[0,\infty[$.

Except for the triangle inequality one gets all the inequalities above from (0.23). The idea behind all of these is the fruitful notion of *convexity.*

*Example 1:* If we choose $f(x) := \ln x$ and $g(x) := x$, then we have $f^{-1}(x) = \mathrm{e}^x$ and $g^{-1}(x) = x$. From (0.23) we get the inequality for the weighted mean

$$\boxed{\prod_{k=1}^{n} x_k^{\lambda_k} \le \sum_{k=1}^{n} \lambda_k x_k,} \tag{0.24}$$

which is valid for all non-negative real numbers $x_k$ and $\lambda_k$ which satisfy $\sum\limits_{k=1}^{n} \lambda_k = 1$. This inequality is equivalent to the Young inequality (0.21).

In the special case $\lambda_k = 1/n$ for all $k$, the inequality (0.24) is just the inequality $g \le m$ between the geometric mean $g$ and the arithmetic mean $m$.

**The duality inequality:**

$$\boxed{(x|y) \le F(x) + F^*(y) \quad \text{for all } x, y \in \mathbb{R}^N.} \tag{0.25}$$

Here, the function $F : \mathbb{R}^N \to \mathbb{R}$ is given, and the *dual* function $F^* : \mathbb{R}^N \to \mathbb{R}$ is given by the relation

$$F^*(y) := \sup_{x\in\mathbb{R}^N} (x|y) - F(x).$$

[17] If $G := ]a,b[$ is an open bounded interval, then is it sufficient for example that $p$ and $g$ are continuous on $[a,b]$ (or more generally, almost everywhere continuous and bounded). In this case we have

$$\int_G \ldots \mathrm{d}x = \int_a^b \ldots \mathrm{d}x.$$

If $G$ is a bounded, open (non-empty) set in $\mathbb{R}^N$, then it is sufficient that $p$ and $g$ are continuous on the closure $\overline{G}$ (or more generally almost everywhere continuous and bounded).

*Example 2:* Let $N = 1, p > 1$ and $F(x) := \dfrac{|x|^p}{p}$ for all $x \in \mathbb{R}$. Then one has

$$F^*(y) = \frac{|y|^q}{q} \quad \text{for all } y \in \mathbb{R},$$

where $q$ is determined from the equation $\dfrac{1}{p} + \dfrac{1}{q} = 1$. In this special case, (0.25) is nothing but the Young inequality $xy \leq \dfrac{|x|^p}{p} + \dfrac{|y|^q}{q}$.

**Standard literature:** A large collection of further inequalities can be found in the standard references [19] and [15].

### 0.1.12 Application to the motion of the planets – a triumph of mathematics in space

*One can not have a pure understanding of what one has until one has a complete understanding of what others had before oneself.*

*Johann Wolfgang von Goethe* (1749–1832)

The results of the previous sections are correctly considered today to belong to elementary mathematics. Actually it was the result of centuries of toil and thought – always in interaction with the resolution of important questions put to man by nature – before these realizations, today considered to be elementary, could be attained. As an example of this we consider here in more detail planetary motion.

Conic sections were already investigated intensively in ancient times. To describe the location of the planets in the heavens, the ancient astronomers used the idea of Appolonius von Perga (roughly 260–190 BC) of *epicycles*. According to this theory, the planets move along a small circular orbit, which in turn moves along a larger circular orbit (cf. Figure 0.21(a)).

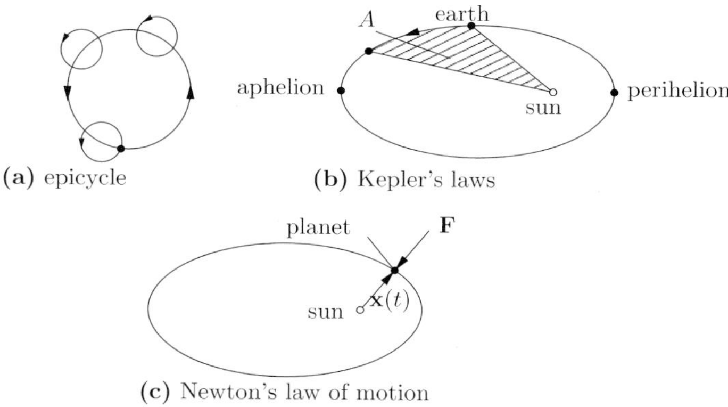


*Figure 0.21. Historical occurrences of conic sections.*

This theory gave a relatively accurate description of the apparent complicated annual motion of the planets in the sky. The theory of epicycles is a very vivid example for how the attempt to fit theory with observation can lead to a totally *wrong* model.

**Copernicus' view of the world:** In 1543, the year of death of Nicolaus Copernicus (born in 1473 in the old Polish Hansa city Toruń), his epochal work *De revolutionibus orbium coelestium* (On the motion of the heavenly orbits) appeared. In this work he broke with the tradition of the view of the world of the ancients, shaped by Ptolemy, according to which the earth was the center of the universe. On the contrary, Copernicus created the idea that the earth orbits the sun, while keeping the idea of *circular* orbits.

**The three Kepler laws:** Based on extensive observations of the Danish astronomer Tycho Brahe (1564–1601), Johannes Kepler (1571–1630, born in the city of Weil in Germany) found after extensive calculation the following *three laws for planetary motion* (Figure 0.21(b)):

1. *The planets move in elliptical orbits, with the sun at one of the focal points of the ellipse.*
2. *The motion sweeps out equal areas in equal times (denoted $A$ in Figure 0.21(b)).*
3. *The ratio of the square of the orbital period $T$ and the third power of the long axis $a$ of the ellipse is a constant for all planets*:

$$\boxed{\frac{T^2}{a^3} = \text{const}\,.}$$

The first two of the laws were published by Kepler in 1609 in his monograph *Astronomia nova* (New Astronomy). Ten years later the third law appeared in his thesis *Harmonices mundi* (World Harmonies)[18].

In 1624, Kepler finished the enormous work involved in completing the "Rudolfian tables", which the German Emperor Rudolf II had commissioned him with in 1601. These tables were used by astronomers for the next 200 years. With the help of these tables it was possible to precisely predict the motion of the planets and solar and lunar eclipses for all times past and future. In these days of computer computational power it is impossible to imagine what an achievement this was, particularly since for use in astronomy one needs very precise results, not just rough approximations. Kepler even had to work without tables for logarithms. The first table of logarithms was published by the Scotch nobleman Neper in 1614. Kepler immediately realized the computational power afforded by these mathematical tool, reducing multiplications to additions. In fact, Kepler's paper on this was of great help in spreading the popularity of logarithms.

**Newtonian mechanics:** Exactly one hundred years after the death of Copernicus, Isaac Newton – one of the true geniuses of human kind – was born in 1643 as the son of a leaseholder in a small village on the east coast of England. Lagrange wrote about him: "He is the luckiest of all; the system of the universe can only be discovered once". At the age of 26, Newton became Professor at the famous Trinity College in Cambridge (England). Already at the age of 23 he used the third of Kepler's laws to estimate the power of gravitational attraction and found that this must be proportional to the inverse of the square of the distance. In 1687 his famous book *Philosophiae naturalis principia mathematica* (Mathematical Principles of Science) appeared. In this book, he founded classical mechanics and derived and applied his famous *law of motion*

$$\boxed{\text{force} = \text{mass} \times \text{acceleration.}}$$

[18] Kepler discovered the third law on May 18, 1618, five days before the window incident in Prague, which began the thirty years' war.

At the same time he created the theory of differential and integral calculus. Newton's law written in modern notion is the differential equation for the motion of the planets

$$\boxed{m\mathbf{x}''(t) = \mathbf{F}(\mathbf{x}(t))\,.} \tag{0.26}$$

The vector $\mathbf{x}(t)$ describes the position of the planets[19] at the time $t$ (Figure 0.21(c)). The second derivative with respect to time, $\mathbf{x}''(t)$, corresponds to the vector of acceleration of the planet at time $t$, and the positive constant $m$ is the mass of the planet. The gravitational attraction of the sun according to Newton has the form

$$\boxed{\mathbf{F}(\mathbf{x}) = -\frac{\mathrm{G}mM}{|\mathbf{x}|^2}\mathbf{e}}$$

with the unit vector

$$\mathbf{e} = \frac{\mathbf{x}}{|\mathbf{x}|}\,.$$

The negative sign of $\mathbf{F}$ corresponds to the fact that gravitational force points in the direction $-\mathbf{x}(t)$, that is, from the planet toward the sun. Furthermore, $M$ denotes the mass of the sun, G is a universal constant of natural, called the *gravitational constant*:

$$\mathrm{G} = 6.6726 \cdot 10^{-11}\,\mathrm{m}^3\mathrm{kg}^{-1}\mathrm{s}^{-2}\,.$$

Newton found solutions of (0.26) which are ellipses

$$r = \frac{p}{1 - \varepsilon\cos\varphi}$$

(in polar coordinates) with the numerical eccentricity $\varepsilon$ and the half-parameter $p$ determined according to the following equations:

$$\varepsilon = \sqrt{1 + \frac{2ED^2}{\mathrm{G}^2m^3M^2}}\,, \qquad p = \frac{D^2}{\mathrm{G}^2m^3M^2}\,.$$

The energy $E$ and the angular momentum $D$ are determined from the position and the velocity of the planet at some fixed time. The orbital motion $\varphi = \varphi(t)$ is obtained by solving the equation

$$t = \frac{m}{D}\int_0^{\varphi} r^2(\varphi)\mathrm{d}\varphi$$

for the angle $\varphi$.

**Gauss rediscovers Ceres:** In the new years night of 1801 a tiny star of the magnitude 8 was discovered at the observatory in Palermo, which moved relatively quickly and then vanished again. This amounted to an incredible challenge for the astronomers of the day. Only 9 degrees of the orbit were known. The methods used up until then for celestial calculations failed. The 24-year old Gauss however succeeded in surmounting the difficulties of mastering an equation of the eighth degree, by developing totally new methods, which he published in 1809 in his work *Theoria motus corporum coelestium in sectionibus conicis Solem ambientium*[20].

[19] Vector calculus will be described in detail in 1.8.

[20] A translation of this title is *The theory of the motion of the planets, which move in conic sections around the sun.*

According to Gauss' calculations, Ceres could be observed again in the new years night 1802. Ceres was the first of the asteroids to be observed. It is estimated that there are approximately 50,000 such asteroids moving in a belt between Mars and Jupiter, whose total mass is just a few thousandths that of the earth. The diameter of Ceres is 768 km. It is the largest known asteroid.

**The discovery of Neptune:** During a night in March, 1781, Wilhelm Herschel discovered a new planet, which was later named Uranus and whose orbital period around the sun is 84 years (cf. Table 0.13). Two young astronomers, John Adams (1819–1892) in Cambridge and Jean Leverrier (1811–1877) in Paris, determined independently of each other the orbit of Uranus and concluded from the observed perturbation in Uranus' orbit the existence of a new planet, which according to Leverrier had been observed by Gottfried Galle in 1846 at the Berlin Observatory and received the name Neptune. This was a triumph of Newtonian mechanics and at the same time one of practical calculations in the theory of celestial motions.

From the observed perturbations in the motion of Neptune one later concluded the existence of a further, tiny planet very far from the sun, which was discovered in 1930 and was named Pluto after the Roman God of the underworld (cf. Table 0.13).

*Table 0.13. A model of the solar system scaled to* $1\,\mathrm{m} \mathrel{\hat{=}} 10^6\,\mathrm{km}$.

| *Planet* | *Distance from the sun* | *Orbital period* | *Numerical orbital eccentricity* $\varepsilon$ | *Planet's diameter* | *Comparative size* |
|---|---|---|---|---|---|
| Sun | – | – | – | 1.4 m | – |
| Mercury | 58 m | 88 days | 0.206 | 5 mm | pea |
| Venus | 108 m | 255 days | 0.007 | 12 mm | cherry |
| Earth | 149 m | 1 year | 0.017 | 13 mm | cherry |
| Mars | 229 m | 2 years | 0.093 | 7 mm | pea |
| Jupiter | 778 m | 12 years | 0.048 | 143 mm | coconut |
| Saturn | 1400 m | 30 years | 0.056 | 121 mm | coconut |
| Uranus | 2900 m | 84 years | 0.047 | 50 mm | apple |
| Neptune | 4500 m | 165 years | 0.009 | 53 mm | apple |
| Pluto | 5900 m | 249 years | 0.249 | 10 mm | cherry |

**The perihelion motion of Mercury:** The calculation of the orbits of the planets is quite complicated by virtue of the fact that not only the gravitational force of the sun, but also of the other planets must be accounted for. This in done in the context of mathematical *perturbation theory*, which in general considers the behavior of solutions under small perturbations of the (coefficients of the) equations. In spite of incredibly precise calculations, the orbit of the planet nearest to the sun, Mercury, had a rotation of the long axi of the ellipse describing its motion by 43 arc seconds a century, which was inexplicable. This discrepancy wasn't explained until the advent of Einstein's general theory of relativity in 1916.

**The background microwave radiation of the big bang:** There is a solution to the equations of the general theory of relativity which describes an expanding universe. The starting point of this expansion is referred to as the *big bang*. In 1965 the American physicists Penzias and Wilson at the Bell Laboratory in New Jersey discovered an extremely weak (microwave), completely isotropic (the same in all directions) radiation, which is now viewed to be a relict of and experimental evidence for the big bang from 15 billion years ago. This was a scientific sensation. Both scientist were awarded the Nobel prize for this discovery. Since the radiation can be viewed as a photon gas at the temperature of 3 degrees Kelvin (above absolute zero), one also speaks of the 3K radiation. The complete isotropy of this radiation on the other hand was for some time quite difficult to explain; it is an apparent contradiction to the formation of galaxies in the universe. In 1992, the satellite COBE., designed by George Smoot, after extensive preparations over several years, finally observed a detailed anisotropy in the background microwave radiation. This gives us a view back in time at the distribution of matter in the universe at the very young age of 300,000 years after the big bang and makes the formation of galaxies at about 10 billion years ago understandable.[21].

**Astrophysics, differential equations, numerics, fast computers and the death of the sun:** Our source of life, the sun, formed together with the planets about 5 billion years ago by attraction and compression of dark matter. Modern mathematics is in a position to describe the life and death of the sun. One uses a model for the sun which consists of a complicated system of differential equations, the derivation of which was the work of decades of astronomers. It is impossible to give exact solutions to this complicated system of differential equations. However, modern methods in numerics provides effective ways of calculating approximations to solutions with the computational power of supercomputers. The chair of Roland Bulirsch at the Technical University in Munich has carried through these calculations. This has been made vividly imaginable by motion pictures describing the solutions found in this way; these show how the sun at an age of about 11 billion years will start to expand to the orbit of Venus, at which time all life on the planet Earth will long have ceased to exist from the incredible heat caused by this expansion. Somewhat later the sun will start to collapse and will become a brown dwarf from which no more light can escape.

## 0.2 Elementary functions and their graphical representation

**Basic idea:** A real-valued function[22]

$$\boxed{y = f(x)}$$

assigns, in a unique fashion, a real number $y$ to the real number $x$. One must differentiate in thought between the function $f$ as an assignment and the value $f(x)$ of the function at the number $x$.

**(i)** The set of all $x$ for which the assignment is defined is called the *domain* $D(f)$ of the function $f$.

**(ii)** The set of image points $y$ for all $x \in D(f)$[23] is called the *range* $R(f)$ of the function

[21] The fascinating story of modern cosmology and of the COBE-project is described in the book [28].

[22] Real-valued functions are special *maps*. The definition and properties of general maps are discussed in 4.3.3. For simplicity, real-valued functions are also briefly referred to as real functions.

[23] The symbol $x \in D(f)$ indicates that $x$ is an element of the set $D(f)$.

$f$.

**(iii)** The set of all point pairs $(x, f(x))$ is called the *graph* $G(f)$ of the function $f$.

Functions can be defined by a *table of values* or by a *graphical representation.*

*Example:* For the function $y = 2x + 1$, the table of values is

| $x$ | 0 | 1 | 2 | 3 | 4 |
|---|---|---|---|---|---|
| $y$ | 1 | 3 | 5 | 7 | 9 |

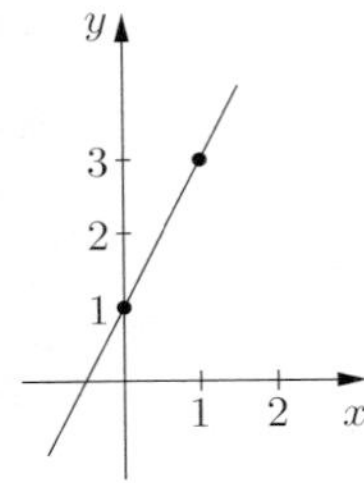


*Figure 0.22.*

The graphical representation of $y = 2x + 1$, the graph of $f$ is the plane of points $(x, y)$, is the line through the two points $(0, 1)$ and $(1, 3)$.

**Increasing and decreasing functions:** A function $f$ is said to be (strictly) *increasing* if

$$\boxed{x < u \text{ implies } f(x) < f(u)\,.} \tag{0.27}$$

A function $f$ is said to be *non-decreasing*, *decreasing* or *non-increasing*, if in (0.27) the symbol '$f(x) < f(u)$' is replaced by, in order

$$f(x) \le f(u)\,, \qquad f(x) > f(u)\,, \qquad f(x) \ge f(u)$$

(see Table 0.14).

*Table 0.14. Properties of functions.*

| ***Increasing*** | ***Non-decreasing*** | ***Decreasing*** | ***Non-increasing*** |
|---|---|---|---|
| | | | |
| ***Even*** | ***Odd*** | ***Periodic*** | |
| | | | |

**Basic idea of the inverse function:** We consider the function

$$\boxed{y = x^2\,, \qquad x \ge 0\,.} \tag{0.28}$$

The equation (0.28) has for each $y \ge 0$ exactly one solution $x \ge 0$, which one denotes by $\sqrt{y}$:

$$\boxed{x = \sqrt{y}\,.}$$

Exchanging formally $x$ with $y$, we get the *square root* function

$$\boxed{y = \sqrt{x}\,.} \tag{0.29}$$

The graph of the inverse function (0.29) is obtained from the graph of the original function (0.28) by reflecting the graph on the diagonal (Figure 0.23).

This construction can be carried out for arbitrary continuous, increasing functions (cf. 1.4.4). As we will see in the next sections, one get in this manner many important functions (for example $y = \ln x$, $y = \arcsin x$, $y = \arccos x$ etc.).

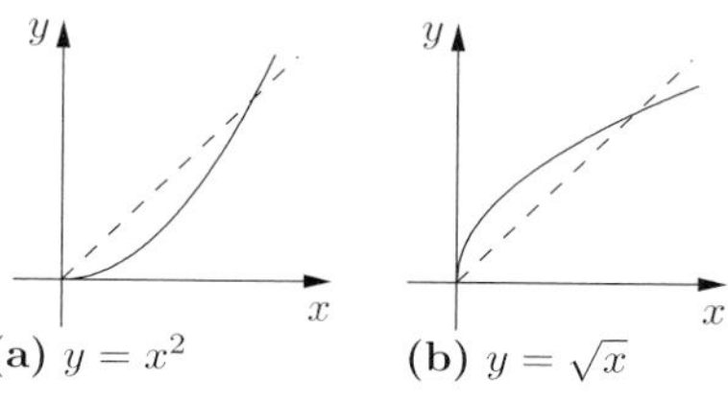


*Figure 0.23. Power functions.*

**Graphical representation of functions with Mathematica:** The software package Mathematica contains a built-in series of important mathematical functions. These can be displayed by tables of values or by plotting the graphs.

## 0.2.1 Transformation of functions

It suffices to know certain standard forms of functions. From these one can get graphical representations of other interesting functions by the processes of translation, dilation and reflection.

**Translation:** The graph of the function

$$\boxed{y = f(x-a) + b}$$

is obtained from the graph of $y = f(x)$ by the translation in which each point $(x, y)$ is shifted to $(x + a, y + b)$.

*Example 1:* The graph of $y = (x-1)^2 + 1$ is obtained from the graph of $y = x^2$ by the translation in which the point $(0,0)$ is translated to the point $(1,1)$ (Figure 0.24).

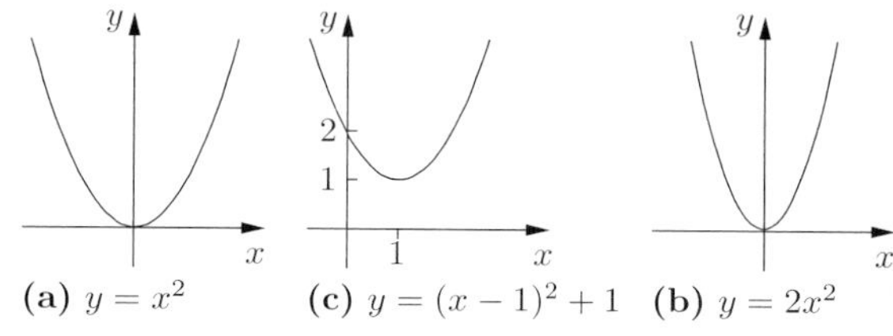


*Figure 0.24. Translation and dilation of a graph.*

**Dilation along axi:** The graph of the function

$$\boxed{y = b\,f\Big(\frac{x}{a}\Big)}$$

with fixed $a > 0$ and $b > 0$ is obtained from the graph of $y = f(x)$ by stretching the $x$-axis by a factor of $a$ and stretching the $y$-axis by a factor of $b$.

*Example 2:* From $y = x^2$ one gets $y = 2x^2$ by stretching the $y$-axis by a factor of 2 (Figure 0.24).

*Example 3:* From $y = \sin x$ one gets $y = \sin 2x$ by dilating the $x$-axis by a factor of $\frac{1}{2}$ (Figure 0.25).

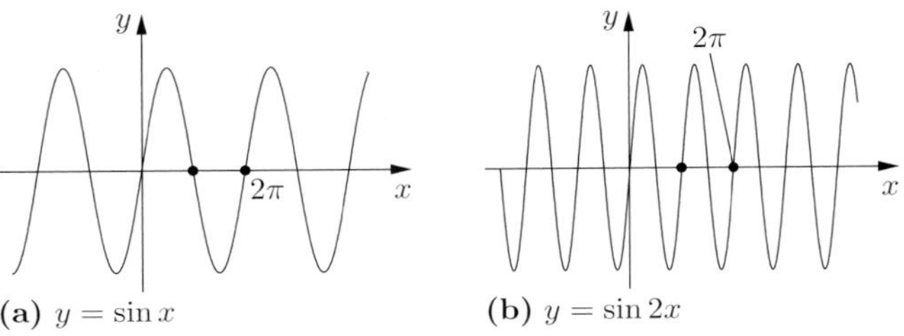


*Figure 0.25. Sinusoidal waves.*

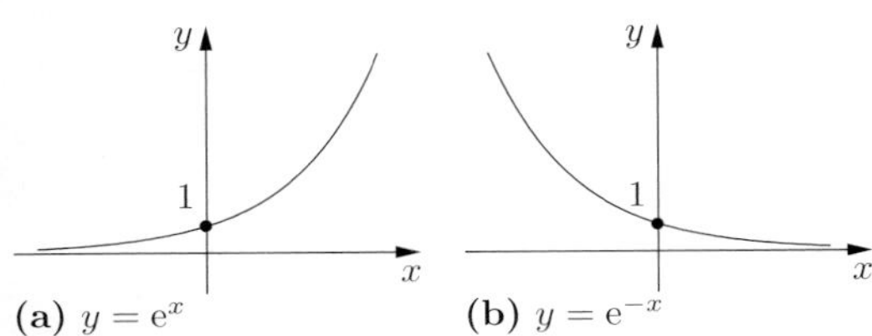


*Figure 0.26. Exponential functions.*

**Reflection:** The graphs of

$$\boxed{y = f(-x)} \quad \text{resp.} \quad \boxed{y = -f(x)}$$

are obtained from the graph of $y = f(x)$ by reflection on the $y$-axis (resp. on the $x$-axis).

*Example 4:* The graph of $y = e^{-x}$ results by reflecting the graph of $y = e^x$ on the $y$-axis (Figure 0.26).

**Even and odd functions:** A function $y = f(x)$ is said to be *even* (resp. odd), if

$$f(-x) = f(x) \qquad (\text{resp. } f(-x) = -f(x))$$

for all $x \in D(f)$ (Table 0.14).

The graph of an even (resp. odd) function is invariant under reflection of the $x$-axis (resp. reflection of both axi) on the origin.

*Example 5:* The function $y = x^2$ is even, while $y = x^3$ is odd.

**Periodic functions:** The function $f$ has by definition a *period* $p$, if

$$\boxed{f(x+p) = f(x) \quad \text{for all } x \in \mathbb{R},}$$

i.e., if the relation is satisfied for all real numbers $x$. The graph of a periodic function is invariant under translations of the $x$-axis by $p$.

*Example 6:* The function $y = \sin x$ has a period of $2\pi$ (Figure 0.25).

## 0.2.2 Linear functions

The linear function

$$\boxed{y = mx + b}$$

has a graph which is a line with slope $m$ and which has $y$-intercept $b$ (see Figure 0.10 in 0.1.7.1).

### 0.2.3 Quadratic functions

The simplest quadratic function

$$\boxed{y = ax^2} \tag{0.30}$$

for $a \neq 0$ has a graph which is a parabola (Figure 0.27). A general quadratic function

$$\boxed{y = ax^2 + 2bx + c} \tag{0.31}$$

can be put in the form

$$y = a\left(x + \frac{b}{a}\right)^2 - \frac{D}{a} \tag{0.32}$$

with the discriminant $D := b^2 - ac$ by means of *quadratic completion.* Thus (0.31) results from (0.30) by a translation which moves the apex $(0,0)$ to $\left(-\frac{b}{a}, -\frac{D}{a}\right)$.

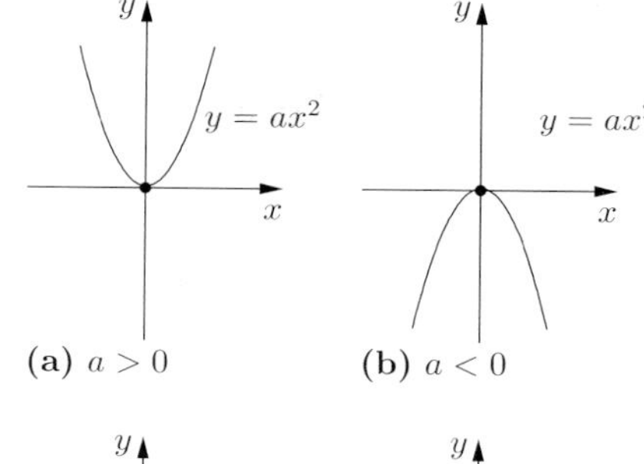


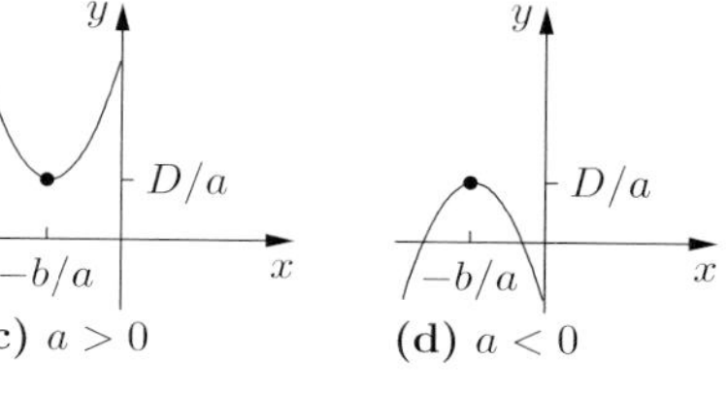


*Figure 0.27.*

**Quadratic equations:** The equation

$$\boxed{ax^2 + 2bx + c = 0}$$

has for real coefficients $a, b$ and $c$ with $a > 0$ the solutions

$$\boxed{x_\pm = \frac{-b \pm \sqrt{D}}{a} = \frac{-b \pm \sqrt{b^2 - ac}}{a}.}$$

*Case 1:* $D > 0$. There are two different real zeros $x_+$ and $x_-$, which correspond to two different points of intersection of the parabola (0.31) with the $x$-axis (Figure 0.28(a)).

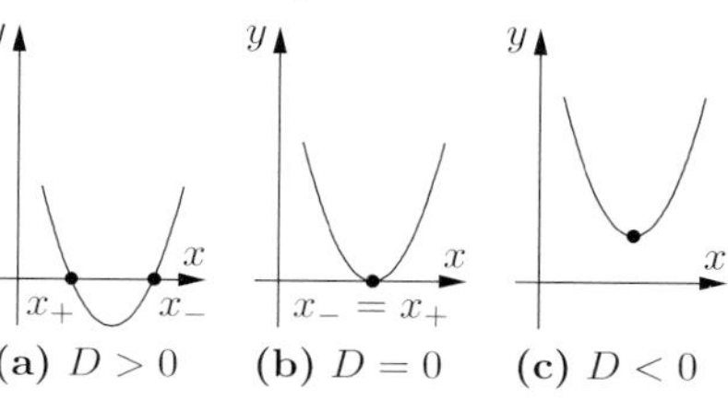


*Figure 0.28.*

*Case 2:* $D = 0$. There is one real zero $x_+ = x_-$. The parabola (0.31) is tangent to the $x$-axis (Figure 0.28(b)).

*Case 3:* $D < 0$. There are two *complex* zeros

$$x_\pm = \frac{-b \pm \mathrm{i}\sqrt{-D}}{a} = \frac{-b \pm \mathrm{i}\sqrt{ac - b^2}}{a},$$

where i is the imaginary unit with $\mathrm{i}^2 = -1$ (cf. 1.1.2). In this case the $x$-axis is not intersected by the (real) parabola (0.31) (Figure 0.28(c)).

*Example 1:* The equation $x^2 - 6x + 8 = 0$ has the two zeros

$$x_\pm = 3 \pm \sqrt{3^2 - 8} = 3 \pm 1,$$

that is $x_+ = 4$ and $x_- = 2$.

*Example 2:* The equation $x^2 - 2x + 1 = 0$ has the zero

$$x_\pm = 1 \pm \sqrt{1 - 1} = 1.$$

*Example 3:* For $x^2 + 2x + 2 = 0$ we get the zeros

$$x_\pm = -1 \pm \sqrt{1 - 2} = -1 \pm \mathrm{i}.$$

### 0.2.4 The power function

*Table 0.15. The power function $y = ax^n$.*

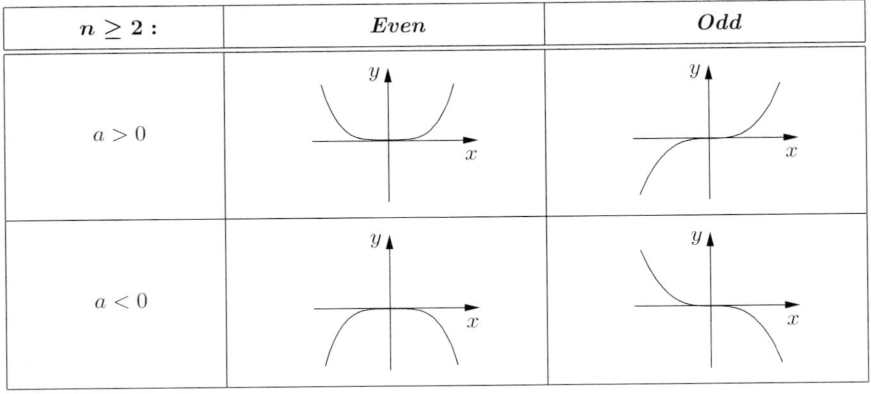

| $n \geq 2$ : | *Even* | *Odd* |
|---|---|---|
| $a > 0$ | | |
| $a < 0$ | | |

Let $n = 2, 3, \ldots$ The function

$$\boxed{y = ax^n}$$

for even $n$ is shaped similarly as $y = ax^2$ and for odd $n$ similarly as $y = ax^3$ (Table 0.15).

### 0.2.5 The Euler e-function

*The shortest path between two real points is through the complex domain.*
*Jacques Hadamard* (1865–1963)

In order to recognize deep connections among different parts of mathematics, it is important to consider the functions $e^x$, $\sin x$ and $\cos x$ also for complex arguments $x$.

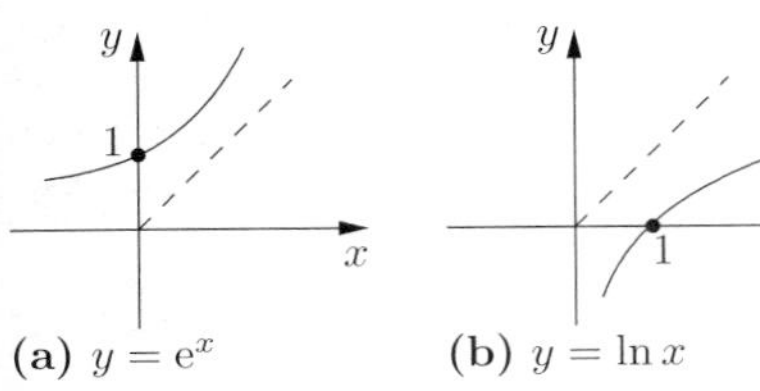


**(a)** $y = e^x$ **(b)** $y = \ln x$

*Figure 0.29.*

Complex numbers of the form $x = a + bi$ with real numbers $a$ and $b$ are discussed in detail in 1.1.2. One just has to note that the imaginary unit i satisfies the relation

$$\boxed{i^2 = -1.}$$

Every real number is at the same time a complex number.

**Definition:** For all complex numbers $x$, the infinite series[24]

$$\boxed{e^x := 1 + x + \frac{x^2}{2!} + \frac{x^3}{3!} + \ldots = \sum_{k=0}^{\infty} \frac{x^k}{k!}} \tag{0.33}$$

converges.

In this way the exponential function $y = e^x$ is defined for all complex arguments $x$,

[24] Infinite series are considered in detail in section 1.10.

which turns out to be the most important single function in all of mathematics. For real $x$ this function was introduced by Newton at the age of 33 in 1676 (Figure 0.29(a)).

**Addition theorem:** For all complex numbers $x$ and $z$ one has the fundamental formula:

$$\boxed{e^{x+z} = e^x e^z\,.}$$

Euler made the very surprising discovery about 75 years after Newton that the e-function and the trigonometric functions (for complex arguments) are closely related (see the Euler formula (0.35) in 0.2.8.). Therefore one refers to the exponential function $y = e^x$ as the Euler e-function. For $x = 1$ we get

$$e = 1 + 1 + \frac{1}{2!} + \frac{1}{3!} + \ldots$$

In addition the Euler limit formula holds[25]

$$\boxed{e^x = \lim_{n\to\infty}\left(1 + \frac{x}{n}\right)^n}$$

for all real numbers $x$. One has $e = 2.71828183$.

**Increasing property:** The function $y = e^x$ is strictly increasing and continuous for all real arguments.

**Behavior at infinity[26]:**

$$\boxed{\lim_{x\to+\infty} e^x = +\infty\,, \qquad \lim_{x\to-\infty} e^x = 0\,.}$$

For negative arguments of large absolute value the graph of $y = e^x$ approaches the $x$-axis asymptotically (Figure 0.29(a)). The limiting relation

$$\boxed{\lim_{x\to+\infty} \frac{e^x}{x^n} = +\infty\,, \qquad n = 1, 2, \ldots\,,}$$

states that the exponential function for large arguments *grows faster than every power function.*

**The complexity of computer algorithms:** If a computer algorithm depends on a natural number $N$ (if, for example, $N$ is the number of equations) and the needed computation time behaves like $e^N$, then the computation time explodes for large $N$, making the algorithm practically useless for large $N$. Investigations of this kind are done in the context of the modern *complexity theory*. Especially many algorithms used in computer algebra have a high complexity.

**Derivative:** The function $y = e^x$ is infinitely often differentiable for all real or complex number $x$, and the derivative is[27]

$$\boxed{\frac{de^x}{dx} = e^x\,.}$$

**Periodicity in the complex domain:** The Euler e-function has the complex period of $2\pi i$, that is, for all complex numbers $x$ one has:

$$\boxed{e^{x+2\pi i} = e^x\,.}$$

---

[25] Limits of sequences of numbers are introduced in 1.2.

[26] Limits of functions are investigated in 1.3.

[27] The notion of derivative of real or complex functions, one of the most fundamental notions of analysis, is found in 1.4.1 (resp. 1.14.3).

If one restricts oneself to real arguments $x$, then this periodicity is invisible (see Figure 0.29(a)).

**Non-vanishing of the e-function:** For all complex numbers $x$, we have $\mathrm{e}^x \neq 0$.[28]

### 0.2.6 The logarithm

**The inverse of the e-function:** Since the e-function is strictly increasing and continuous for all real arguments, the equation

$$\boxed{y = \mathrm{e}^x}$$

has a unique real number $x$ as solution for all $y > 0$, which is denoted

$$\boxed{x = \ln y}$$

and is called the *natural logarithm* (logarithmus naturalis). Formally exchanging $x$ and $y$, we get the function

$$\boxed{y = \ln x\,,}$$

which is the inverse function of the function $y = \mathrm{e}^x$. The graph of $y = \ln x$ is obtained from the graph of $y = \mathrm{e}^x$ by reflection on the diagonal (Figure 0.29(b)).

From the addition theorem $\mathrm{e}^{u+v} = \mathrm{e}^u \mathrm{e}^v$ the fundamental property of the logarithm follows:[29]

$$\boxed{\ln(xy) = \ln x + \ln y}$$

for all positive real numbers $x$ and $y$.

**Logarithm laws:** See section 0.1.9.

**Limit relations:**

$$\boxed{\lim_{x \to +0} \ln x = -\infty\,, \qquad \lim_{x \to +\infty} \ln x = +\infty\,.}$$

For every real number $\alpha > 0$ one has

$$\boxed{\lim_{x \to +0} x^\alpha \ln x = 0\,.}$$

It follows that the function $y = \ln x$ approaches minus infinity extremely slowly near $x = 0$.

**Derivative:** For all real numbers $x > 0$ one has

$$\boxed{\frac{\mathrm{d} \ln x}{\mathrm{d}x} = \frac{1}{x}\,.}$$

[28] More precisely, the map $x \longmapsto \mathrm{e}^x$ is a surjective map from the complex plane $\mathbb{C}$ onto $\mathbb{C} \setminus \{0\}$.

[29] If we set $x := \mathrm{e}^u$ and $y := \mathrm{e}^v$, then we get $xy = \mathrm{e}^{u+v}$. This yields $u = \ln x$, $v = \ln y$ and $u + v = \ln(xy)$.

### 0.2.7 The general exponential function

**Definition:** For every positive real numbers $a$ and every real number $x$ we set

$$\boxed{a^x := e^{x \ln a}\,.}$$

In this way the general exponential function $a^x$ is reduced to the e-function (Figure 0.30).

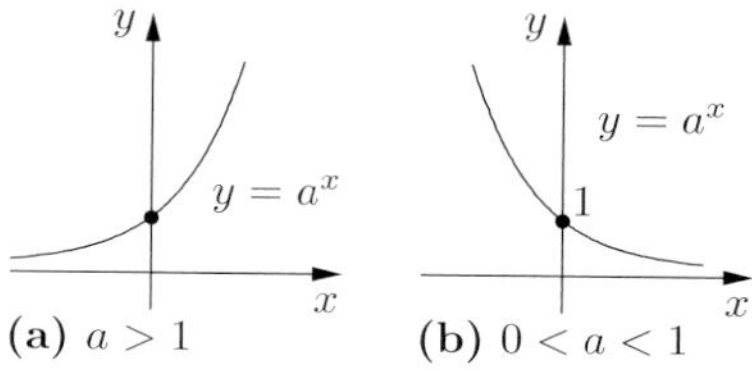


*Figure 0.30. The general exponential.*

**Power laws:** See section 0.1.9.

**General logarithm:** Let $a$ be a fixed positive real number $a \neq 1$. For every positive real number $y$, the equation

$$y = a^x$$

has a unique real solution $x$, which we denote by $x = \log_a y$. Formally exchanging $x$ and $y$, we get the inverse function to $y = a^x$:

$$y = \log_a x.$$

For this one has the relation

$$\boxed{\log_a y = \frac{\ln y}{\ln a}}$$

(cf. 0.1.9). One has $\ln a > 0$ for $a > 1$ and $\ln a < 0$ for $0 < a < 1$.

**Two important functional equations:** Let $a > 0$.

(i) The only continuous function[30] $f : \mathbb{R} \longrightarrow \mathbb{R}$, which satisfies the relation

$$\boxed{f(x+y) = f(x)f(y) \quad \text{for all } x, y \in \mathbb{R}}$$

together with the normalization $f(1) = a$, is the exponential function $f(x) = a^x$.

(ii) The only continuous function $g : \,]0, \infty[\, \longrightarrow \mathbb{R}$, which satisfies the condition

$$\boxed{g(xy) = g(x) + g(y) \quad \text{for all } x, y \in \,]0, \infty[}$$

together with the normalization $g(a) = 1$, is the logarithm $g(x) = \log_a x$.

Both of these statements show that the exponential and logarithm are very naturally and useful functions and that the mathematicians of the past certainly would have had to run across these functions sooner or later.

### 0.2.8 Sine and cosine

**Analytical definition:** From a modern perspective it is convenient to define the two functions $y = \sin x$ and $y = \cos x$ by means of their infinite series expansions

$$\boxed{\begin{aligned} \sin x &= x - \frac{x^3}{3!} + \frac{x^5}{5!} - \ldots = \sum_{k=0}^{\infty} (-1)^k \frac{x^{2k+1}}{(2k+1)!}\,, \\ \cos x &= 1 - \frac{x^2}{2!} + \frac{x^4}{4!} - \ldots = \sum_{k=0}^{\infty} (-1)^k \frac{x^{2k}}{(2k)!}\,. \end{aligned}} \tag{0.34}$$

[30]The notion of continuity will be introduced in 1.3.1.2.

These two series converge for all complex numbers[31] $x$.

**The Euler formula (1749):** For all complex numbers $x$ the following fundamental formula is valid:

$$\boxed{e^{\pm ix} = \cos x \pm i \sin x\,.} \tag{0.35}$$

This formula dominates the entire theory of trigonometric functions. The realtion (0.35) follows immediately from the power series expansions (0.33) and (0.34) for $e^{ix}$, $\cos x$ and $\sin x$, when one takes note of the fact that $i^2 = -1$. In 1.3.3 one can find important applications of this formula to the theory of vibrations. From (0.35) we get

$$\boxed{\sin x = \frac{e^{ix} - e^{-ix}}{2i}\,, \qquad \cos x = \frac{e^{ix} + e^{-ix}}{2}\,.} \tag{0.36}$$

These formulas, together with the addition theorem $e^{u+v} = e^u e^v$, easily yield the following fundamental addition theorems for sine and cosine.

**Addition theorems:** For all complex numbers $x$ and $y$ one has:

$$\boxed{\begin{aligned} \sin(x \pm y) &= \sin x \cos y \pm \cos x \sin y\,, \\ \cos(x \pm y) &= \cos x \cos y \mp \sin x \sin y\,. \end{aligned}} \tag{0.37}$$

**Evenness and oddness:** For all complex numbers $x$ one has:

$$\boxed{\sin(-x) = -\sin x\,, \qquad \cos(-x) = \cos x\,.}$$

**Geometric interpretation on a right triangle:** We consider a right triangle with an angle $x$ measured in radians (cf. 0.1.2). Then $\sin x$ and $\cos x$ are given as the ratios of the sides as shown in Table 0.16.

*Table 0.16. Interpretation of trigonometric functions in terms of a right triangle.*

| ***Right triangle*** | ***Sine*** | ***Cosine*** |
|---|---|---|
| 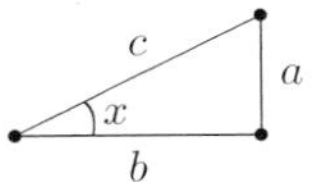  | $\sin x = \dfrac{a}{c}$ | $\cos x = \dfrac{b}{c}$ |
| $0 < x < \dfrac{\pi}{2}$ | (length of opposite side $a$ divided by the hypotenuse $c$) | (length of adjacent side $b$ divided by the hypotenuse $c$) |

[31] Compare the remarks made at the beginning of 0.2.5 about complex numbers.
The symbol '$\sin x$' is read 'sine of $x$' and the symbol '$\cos x$' is read 'cosine of $x$'. The latin word sinus means bulge. In older literature one also uses the functions

*secant:* $\sec x := \dfrac{1}{\cos x}$, *cosecant:* $\operatorname{cosec} x := \dfrac{1}{\sin x}$.

**Geometric interpretation on the unit circle:** Using the unit circle, the quantities $\sin x$ and $\cos x$ are just the lengths of the segments shown in Figure 0.31(a)-(d). From this one see immediately that $\sin x$ and $\cos x$ have the same values after a rotation of $2\pi$. This is the geometric interpretation of the $2\pi$-periodicity of the $\sin x$ and $\cos x$:

$$\boxed{\sin(x + 2\pi) = \sin x\,, \qquad \cos(x + 2\pi) = \cos x\,.} \tag{0.38}$$

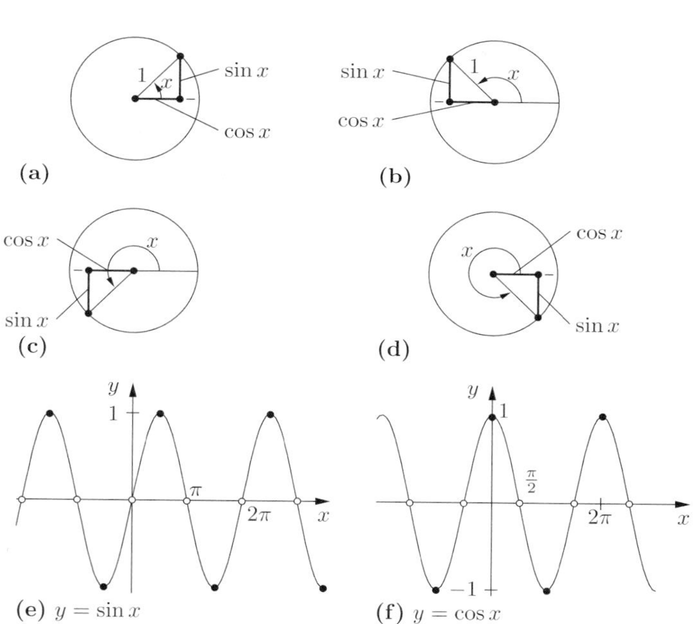


*Figure 0.31. Trigonometric functions and the unit circle.*

These relations hold for all complex arguments $x$. Looking at the unit circle again, one sees the following symmetries

$$\boxed{\sin(\pi - x) = \sin x\,, \qquad \cos(\pi - x) = -\cos x} \tag{0.39}$$

for $0 \le x \le \pi/2$. In fact these relations hold for all complex numbers $x$. Finally one gets from Figure 0.31(a) and the theorem of Pythagoras the relation

$$\boxed{\cos^2 x + \sin^2 x = 1\,,} \tag{0.40}$$

which holds not only for real angles $x$, but also for all complex arguments $x$. In the same way we get from the theorem of Pythogoras the values for $\sin x$ and $\cos x$ listed in Table 0.17 (cf. 3.2.1.2).

The validity of (0.38), (0.39) and (0.40) for all complex numbers follows easily from the addition theorem (0.37) and the relations $\sin 0 = \sin 2\pi = 0$ and $\cos 0 = \cos 2\pi = 1$.

*Table 0.17. Exact values of the sine and cosine functions for important angles.*

| $x$ | $0$ | $\frac{\pi}{6}$ | $\frac{\pi}{4}$ | $\frac{\pi}{3}$ | $\frac{\pi}{2}$ | $\frac{2\pi}{3}$ | $\frac{3\pi}{4}$ | $\frac{5\pi}{6}$ | $\pi$ | ***(radians)*** |
|---|---|---|---|---|---|---|---|---|---|---|
| | $0$ | $30°$ | $45°$ | $60°$ | $90°$ | $120°$ | $135°$ | $150°$ | $180°$ | ***(degrees)*** |
| $\sin x$ | $0$ | $\frac{1}{2}$ | $\frac{\sqrt{2}}{2}$ | $\frac{\sqrt{3}}{2}$ | $1$ | $\frac{\sqrt{3}}{2}$ | $\frac{\sqrt{2}}{2}$ | $\frac{1}{2}$ | $0$ | |
| $\cos x$ | $1$ | $\frac{\sqrt{3}}{2}$ | $\frac{\sqrt{2}}{2}$ | $\frac{1}{2}$ | $0$ | $-\frac{1}{2}$ | $-\frac{\sqrt{2}}{2}$ | $-\frac{\sqrt{3}}{2}$ | $-1$ | |

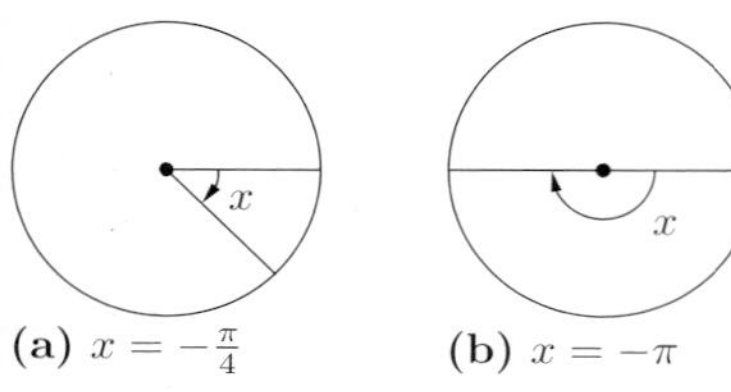


*Figure 0.32. Negative angles.*

**Negative angles:** Applying the geometric interpretation of the unit circle one gets the graphical representation of the functions $y = \sin x$ and $y = \cos x$ as shown in Figure 0.31(e),(f). Here negative angles $x < 0$ were introduced as in Figure 0.32, by measuring positive angles in counter-clockwise direction (*positive mathematical direction*) and negative angles in clockwise direction (*negative mathematical direction*).

**Zeros:** From Figure 0.31(e),(f) it follows that:

(i) The function $y = \sin x$ has zeros at the points $x = k\pi$, where $k$ is an arbitrary integer; in other words the set of zeros is given by $x = 0\,,\ \pm\pi\,,\ \pm 2\pi\,,\ \ldots$

(ii) The function $y = \cos x$ has zeros at the points $x = k\pi + \frac{\pi}{2}$, where $k$ is an arbitrary integer.

(iii) Both functions $y = \sin x$ and $y = \cos x$ have only real zeros in the complex plane. These zeros are those described in (i) and (ii).

**The law of translation:** It is suffient to know the values of $\sin x$ for all angles $x$ with $0 \le x \le \frac{\pi}{2}$. All other values can be obtained by the following formulas, which in turn are consequences of the addition theorems:

$$\sin\left(\frac{\pi}{2}+x\right) = \cos x\,, \qquad \sin(\pi + x) = -\sin x\,, \qquad \sin\left(\frac{3\pi}{2}+x\right) = -\cos x\,,$$
$$\cos\left(\frac{\pi}{2}+x\right) = -\sin x\,, \qquad \cos(\pi + x) = -\cos x\,, \qquad \cos\left(\frac{3\pi}{2}+x\right) = \sin x\,.$$

**De Moivre's formula for multiples of a given angle**[32]**:** Let $n = 2, 3, \ldots$ Then for

[32] This formula, found by de Moivre (1667–1754), inspired Euler to the discovery of his famous formula

$$e^{ix} = \cos x + i\sin x\,.$$

Today it is more convenient to work the other way around: de Moivre's formula (0.41) is a consequence of the Euler formula, using

$$\cos nx + i\sin nx = e^{inx} = \left(e^{ix}\right)^n = (\cos x + i\sin x)^n$$

and the binomial formula (cf. 0.1.10.3).

all complex numbers $x$, one has

$$\cos nx + \mathrm{i}\sin nx = \sum_{k=0}^{n} \mathrm{i}^k \binom{n}{k} \cos^{n-k} x \sin^k x\,. \tag{0.41}$$

Seperating here the real and imaginary part of the complex numbers, one gets

$$\cos nx = \cos^n x - \binom{n}{2}\cos^{n-2} x \sin^2 x + \binom{n}{4}\cos^{n-4} x \sin^4 x - \ldots \tag{0.42}$$

$$\sin nx = \binom{n}{1}\cos^{n-1} x \sin x - \binom{n}{3}\cos^{n-3} x \sin^3 x + \binom{n}{5}\cos^{n-5} x \sin^5 x - \ldots$$

For $n = 2, 3, 4$ we get the following special cases:

$$\begin{aligned} \sin 2x &= 2\sin x\cos x\,, & \cos 2x &= \cos^2 x - \sin^2 x\,,\\ \sin 3x &= 3\sin x - 4\sin^3 x\,, & \cos 3x &= 4\cos^3 x - 3\cos x\,,\\ \sin 4x &= 8\cos^3 x\sin x - 4\cos x\sin x\,, & \cos 4x &= 8\cos^4 x - 8\cos^2 x + 1\,. \end{aligned}$$

**The formula for half-angles:** For all complex numbers $x$ one has:

$$\sin^2\frac{x}{2} = \frac{1}{2}(1-\cos x)\,, \qquad \cos^2\frac{x}{2} = \frac{1}{2}(1+\cos x)\,,$$

$$\sin\frac{x}{2} = \begin{cases} \sqrt{\frac{1}{2}(1-\cos x)}\,, & 0 \le x \le \pi\,,\\ -\sqrt{\frac{1}{2}(1-\cos x)}\,, & \pi \le x \le 2\pi\,, \end{cases}$$

$$\cos\frac{x}{2} = \begin{cases} \sqrt{\frac{1}{2}(1+\cos x)}\,, & -\pi \le x \le \pi\,,\\ -\sqrt{\frac{1}{2}(1+\cos x)}\,, & \pi \le x \le 3\pi\,. \end{cases}$$

**Formulas for sums:** For all complex numbers $x$ and $y$ one has:

$$\begin{aligned} \sin x \pm \sin y &= 2\sin\frac{x\pm y}{2}\cos\frac{x\mp y}{2}\,,\\ \cos x + \cos y &= 2\cos\frac{x+y}{2}\cos\frac{x-y}{2}\,,\\ \cos x - \cos y &= 2\sin\frac{x+y}{2}\sin\frac{y-x}{2}\,,\\ \cos x \pm \sin x &= \sqrt{2}\sin\left(\frac{\pi}{4}\pm x\right). \end{aligned}$$

**Formulas for products of two factors:**

$$\begin{aligned} \sin x\sin y &= \frac{1}{2}\left(\cos(x-y) - \cos(x+y)\right),\\ \cos x\cos y &= \frac{1}{2}\left(\cos(x-y) + \cos(x+y)\right),\\ \sin x\cos y &= \frac{1}{2}\left(\sin(x-y) + \sin(x+y)\right). \end{aligned}$$

**Formulas for products of three factors:**

$$\sin x \sin y \sin z = \frac{1}{4}\left(\sin(x+y-z) + \sin(y+z-x) + \sin(z+x-y) - \sin(x+y+z)\right),$$

$$\sin x \cos y \cos z = \frac{1}{4}\left(\sin(x+y-z) - \sin(y+z-x) + \sin(z+x-y) + \sin(x+y+z)\right),$$

$$\sin x \sin y \cos z = \frac{1}{4}\left(-\cos(x+y-z) + \cos(y+z-x) + \cos(z+x-y) - \cos(x+y+z)\right),$$

$$\cos x \cos y \cos z = \frac{1}{4}\left(\cos(x+y-z) + \cos(y+z-x) + \cos(z+x-y) + \cos(x+y+z)\right).$$

**Formulas for powers:**

$$\sin^2 x = \frac{1}{2}(1-\cos 2x), \qquad \cos^2 x = \frac{1}{2}(1+\cos 2x),$$

$$\sin^3 x = \frac{1}{4}(3\sin x - \sin 3x), \qquad \cos^3 x = \frac{1}{4}(3\cos x + \cos 3x),$$

$$\sin^4 x = \frac{1}{8}(\cos 4x - 4\cos 2x + 3), \qquad \cos^4 x = \frac{1}{4}(\cos 4x + 4\cos 2x + 3).$$

More general formulas for $\sin^n x$ and $\cos^n x$ follow from de Moivre's formula (0.42).

**Addition theorems for three summands:**

$$\sin(x+y+z) = \sin x \cos y \cos z + \cos x \sin y \cos z + \cos x \cos y \sin z - \sin x \sin y \sin z,$$

$$\cos(x+y+z) = \cos x \cos y \cos z - \sin x \sin y \cos z - \sin x \cos y \sin z - \cos x \sin y \sin z.$$

All of these formulas are verified by expressing $\cos x$ and $\sin x$ as linear combinations of $e^{\pm ix}$ according to (0.36). Then it only remains to verify some elementary algebraic identities. One can also apply the addition theorem (0.37).

**The Euler product formula**[33]**:** For all complex numbers one has:

$$\boxed{\sin \pi x = \pi x \prod_{k=1}^{\infty}\left(1 - \frac{x^2}{k^2}\right).}$$

One can read off of this formula immediately exactly where the sine has zeros: $\sin \pi x$ has zeros at $x = 0, \pm 1, \pm 2, \ldots$ These zeros are in addition *simple* (see 1.14.6.3).

**Partial fraction decomposition:** For all complex numbers $x$ different from $0, \pm 1, \pm 2$, ... one has

$$\boxed{\frac{\cos \pi x}{\sin \pi x} = \frac{1}{x} + \sum_{k=1}^{\infty}\left(\frac{1}{x-k} + \frac{1}{x+k}\right).}$$

[33]Infinite products are considered in 1.10.6.

**Derivatives:** For all complex numbers $x$ one has:

$$\frac{\mathrm{d}\sin x}{\mathrm{d}x} = \cos x\,, \qquad \frac{\mathrm{d}\cos x}{\mathrm{d}x} = -\sin x\,.$$

**Parametrization of the unit circle with the aid of trigonometric functions:** See section 0.1.7.2.

**Applications of trigonometric functions in plane trigonometry (land surveying) and spherical trigonometry (navigation and air traffic):** See section 3.2.

**Historical remarks:** Ever since ancient times, the developement of trigonometry has been inseparably connected with technological developments in surveying and navigation, construction and use of calenders and the science of astronomy. Trigonometry had a heyday in the hands of the arabians in the 8th century. In 1260 the book *Treatise on the complete quadrilaterial* was written by at-Tusi, the most important islamic mathematician in the area of trigonometry. This book was the starting point of an independent branch of mathematics concerned with trigonometry. The most important European mathematician of the fifteenth century was Regiomontanus (1436–1476), whose name was in reality Johannes Müller. His most important work[34] *De triangulis omnimodis libri quinque* didn't appear until 1533, long after his death. This treatise contains a complete presentation of plane and spherical trigonometry, and founded the modern branch of mathematics referred to as trigonometry. Unfortunately all formulas in that book were expressed awkwardly in words.[35] Since Regiomontanus didn't have decimal numbers at his disposal[36], he used in the sense of Table 0.16 the formula

$$a = c\sin x \qquad \text{with} \quad c = 10,000,000\,.$$

His values for $a$ correspond to an accuracy of 7 decimal places for $\sin x$. Euler (1707–1783) was the first to use $c = 1$.

At the end of the sixteenth century, Vieta (1540–1603) calculated, in his monograph *Canon*, a table of trigonometric functions, which proceeds from arc minute to arc minute.

Just like tables for logarithms, tables for values of trigonometric functions are obsolete in the day of computers.

## 0.2.9 Tangent and cotangent

**Analytic definition:** For all complex numbers $x$ not equal to one of the values $\frac{\pi}{2} + k\pi$ with $k \in \mathbb{Z}$, we set[37]

$$\tan x := \frac{\sin x}{\cos x}\,.$$

We further define for all complex numbers $x$ not equal to one of the values $k\pi$ with $k \in \mathbb{Z}$ the function

$$\cot x := \frac{\cos x}{\sin x}\,.$$

[34] Translated into English the title means "Five books about all kinds of triangles".

[35] The use of formulas goes back to Vieta *In artem analyticam isagoge*, which appeared in 1591.

[36] Decimal numbers were introduced in 1585 by Stevin in his book *La disme* (The decimal system). This lead to the unification of measurements in continental Europe, based on the decimal system.

[37] We denote by the symbol $\mathbb{Z}$ the set of all integers $k = 0, \pm 1, \pm 2, \ldots$

**Translation property:** For all complex numbers $x$ with $x \neq k\pi$, $k \in \mathbb{Z}$, one has:

$$\boxed{\cot x = \tan\left(\frac{\pi}{2} - x\right).}$$

Because of this, all properties of the function cotangent follow directly from those of tangent.

**Geometric interpretation in a right triangle:** We consider a right triangle with the angle $x$ measured in radians (cf. 0.1.2). Then the values of $\tan x$ and $\cot x$ are given by the ratios of sides as shown in Table 0.18.

*Table 0.18. Interpretation of trigonometric functions in terms of a right triangle.*

| ***Right triangle*** | ***Tangent*** | ***Cotangent*** |
|---|---|---|
| 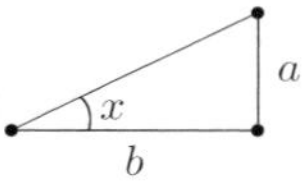  $0 < x < \frac{\pi}{2}$ | $\tan x = \frac{a}{b}$ (length of opposite side $a$ divided by length of adjacent side $b$) | $\cot x = \frac{b}{a}$ (length of adjacent side $b$ divided by length of opposite side $a$) |

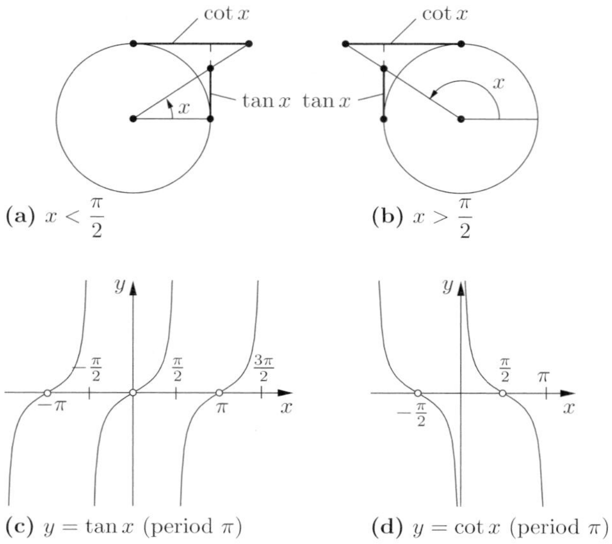


*Figure 0.33. Geometrical interpretation of the tangent and cotangent functions.*

**Geometric interpretation on the unit circle:** Using the unit circle, the values of $\tan x$ and $\cot x$ are the lengths of the segments shown in Figure 0.33(a),(b). One gets from this the special values listed in Table 0.19.

*Table 0.19. Exact values of tan and cot for important angles.*

| $x$ | $0$ | $\frac{\pi}{6}$ | $\frac{\pi}{4}$ | $\frac{\pi}{3}$ | $\frac{\pi}{2}$ | $\frac{2\pi}{3}$ | $\frac{3\pi}{4}$ | $\frac{5\pi}{6}$ | $\pi$ | ***(radians)*** |
|---|---|---|---|---|---|---|---|---|---|---|
| | $0$ | $30°$ | $45°$ | $60°$ | $90°$ | $120°$ | $135°$ | $150°$ | $180°$ | ***(degrees)*** |
| $\tan x$ | $0$ | $\frac{\sqrt{3}}{3}$ | $1$ | $\sqrt{3}$ | – | $-\sqrt{3}$ | $-1$ | $-\frac{\sqrt{3}}{3}$ | $0$ | |
| $\cot x$ | – | $\sqrt{3}$ | $1$ | $\frac{\sqrt{3}}{3}$ | $0$ | $-\frac{\sqrt{3}}{3}$ | $-1$ | $-\sqrt{3}$ | – | |

**Zeros and poles:** The function $y = \tan x$ has for complex arguments $x$ exactly the zeros $k\pi$ with $k \in \mathbb{Z}$ and precisely the poles $k\pi + \frac{\pi}{2}$ with $k \in \mathbb{Z}$. All of these zeros and poles are simple (Figure 0.33(c)).

The function $\cot x$ has for complex arguments $x$ exactly the poles $k\pi$ with $k \in \mathbb{Z}$ and exactly the zeros $k\pi + \dfrac{\pi}{2}$ with $k \in \mathbb{Z}$. Again, all of these zeros and poles are simple[38] (Figure 0.33(d)).

**Partial fraction decomposition:** For all complex numbers $x$ with $x \notin \mathbb{Z}$ one has:

$$\boxed{\cot \pi x = \frac{1}{x} + \sum_{k=1}^{\infty}\left(\frac{1}{x-k} + \frac{1}{x+k}\right).}$$

**Derivative:** For all complex numbers $x$ with $x \neq \frac{\pi}{2} + k\pi$ and $k \in \mathbb{Z}$ one has:

$$\boxed{\frac{\mathrm{d}\tan x}{\mathrm{d}x} = \frac{1}{\cos^2 x}.}$$

For all complex numbers $x$ with $x \neq k\pi$ and $k \in \mathbb{Z}$ one has:

$$\boxed{\frac{\mathrm{d}\cot x}{\mathrm{d}x} = -\frac{1}{\sin^2 x}.}$$

**Power series:** For all complex numbers $x$ with $|x| < \frac{\pi}{2}$ one has:

$$\tan x = x + \frac{x^3}{3} + \frac{2x^5}{15} + \frac{17x^7}{315} + \ldots = \sum_{k=1}^{\infty} 4^k(4^k - 1)\frac{|B_{2k}|x^{2k-1}}{(2k)!}.$$

For all complex numbers $x$ with $0 < |x| < \pi$ one has:

$$\begin{aligned}\cot x &= \frac{1}{x} - \frac{x}{3} - \frac{x^3}{45} - \frac{2x^5}{945} - \ldots \\ &= \frac{1}{x} - \sum_{k=1}^{\infty} \frac{4^k|B_{2k}|x^{2k-1}}{(2k)!}.\end{aligned}$$

Here $B_{2k}$ denote the Bernoulli numbers.

**Convention:** The following formulas hold for all complex arguments $x$ and $y$ with the exception of those arguments for which the function has a pole.

[38]The notion of simple zero or pole is defined in 1.14.6.3

**Periodicity:**

$$\boxed{\tan(x+\pi) = \tan x\,, \qquad \cot(x+\pi) = \cot x\,.}$$

**Oddness:**

$$\boxed{\tan(-x) = -\tan x\,, \qquad \cot(-x) = -\cot x\,.}$$

**Addition theorems:**

$$\boxed{\tan(x \pm y) = \frac{\tan x \pm \tan y}{1 \mp \tan x \tan y}\,, \qquad \cot(x \pm y) = \frac{\cot x \cot y \mp 1}{\cot y \pm \cot x}\,,}$$

$$\tan\left(\frac{\pi}{2} \pm x\right) = \mp \cot x\,, \qquad \tan(\pi \pm x) = \pm \tan x\,, \qquad \tan\left(\frac{3\pi}{2} \pm x\right) = \mp \cot x\,,$$
$$\cot\left(\frac{\pi}{2} \pm x\right) = \mp \tan x\,, \qquad \cot(\pi \pm x) = \pm \cot x\,, \qquad \cot\left(\frac{3\pi}{2} \pm x\right) = \mp \tan x\,.$$

**Multiples of arguments:**

$$\tan 2x = \frac{2\tan x}{1-\tan^2 x} = \frac{2}{\cot x - \tan x}\,, \qquad \cot 2x = \frac{\cot^2 x - 1}{2\cot x} = \frac{\cot x - \tan x}{2}\,,$$
$$\tan 3x = \frac{3\tan x - \tan^3 x}{1 - 3\tan^2 x}\,, \qquad \cot 3x = \frac{\cot^3 x - 3\cot x}{3\cot^2 x - 1}\,,$$
$$\tan 4x = \frac{4\tan x - 4\tan^3 x}{1 - 6\tan^2 x + \tan^4 x}\,, \qquad \cot 4x = \frac{\cot^4 x - 6\cot^2 x + 1}{4\cot^3 x - 4\cot x}\,.$$

**Half-arguments:**

$$\tan\frac{x}{2} = \frac{\sin x}{1+\cos x} = \frac{1-\cos x}{\sin x}\,,$$
$$\cot\frac{x}{2} = \frac{\sin x}{1-\cos x} = \frac{1+\cos x}{\sin x}\,.$$

**Sums:**

$$\tan x \pm \tan y = \frac{\sin(x \pm y)}{\cos x \cos y}\,, \qquad \cot x \pm \cot y = \pm\frac{\sin(x \pm y)}{\sin x \sin y}\,,$$
$$\tan x + \cot y = \frac{\cos(x-y)}{\cos x \sin y}\,, \qquad \cot x - \tan y = \frac{\cos(x+y)}{\sin x \cos y}\,.$$

**Products:**

$$\tan x \tan y = \frac{\tan x + \tan y}{\cot x + \cot y} = -\frac{\tan x - \tan y}{\cot x - \cot y}\,,$$
$$\cot x \cot y = \frac{\cot x + \cot y}{\tan x + \tan y} = -\frac{\cot x - \cot y}{\tan x - \tan y}\,,$$
$$\tan x \cot y = \frac{\tan x + \cot y}{\cot x + \tan y} = -\frac{\tan x - \cot y}{\cot x - \tan y}\,.$$

**Squares:**

| | | | | |
|---|---|---|---|---|
| $\sin^2 x$ | – | $1-\cos^2 x$ | $\dfrac{\tan^2 x}{1+\tan^2 x}$ | $\dfrac{1}{1+\cot^2 x}$ |
| $\cos^2 x$ | $1-\sin^2 x$ | – | $\dfrac{1}{1+\tan^2 x}$ | $\dfrac{\cot^2 x}{1+\cot^2 x}$ |
| $\tan^2 x$ | $\dfrac{\sin^2 x}{1-\sin^2 x}$ | $\dfrac{1-\cos^2 x}{\cos^2 x}$ | – | $\dfrac{1}{\cot^2 x}$ |
| $\cot^2 x$ | $\dfrac{1-\sin^2 x}{\sin^2 x}$ | $\dfrac{\cos^2 x}{1-\cos^2 x}$ | $\dfrac{1}{\tan^2 x}$ | – |

## 0.2.10 The hyperbolic functions $\sinh x$ and $\cosh x$

**Sinus hyperbolicus and cosinus hyperbolicus (hyperbolic sine and cosine):** For all complex numbers $x$ we define the functions

$$\boxed{\sinh x := \frac{e^x - e^{-x}}{2}, \quad \cosh x := \frac{e^x + e^{-x}}{2}.}$$

The function 'sinh' is read 'sinch', 'cosh' is read 'cosh'. For real arguments $x$ the graph is drawn in Figure 0.34.

**Relation to the trigonometric functions:** For all complex numbers $x$ one has:

$$\boxed{\sinh \mathrm{i}x = \mathrm{i}\sin x, \quad \cosh \mathrm{i}x = \cos x.}$$

Because of this relation every formula about the trigonometric functions sine and cosine gives rise to a formula about the hyperbolic functions cosh and sinh. For example $\cos^2 \mathrm{i}x + \sin^2 \mathrm{i}x = 1$ for all complex numbers $x$ implies the following formula:

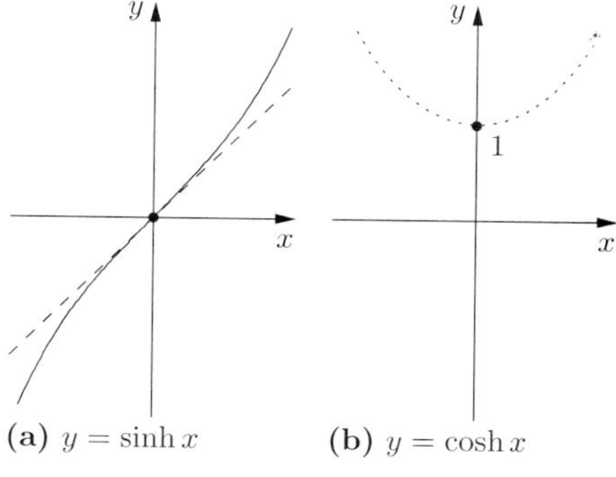


(a) $y = \sinh x$ (b) $y = \cosh x$

*Figure 0.34. Hyperbolic functions.*

$$\boxed{\cosh^2 x - \sinh^2 x = 1.}$$

The terminology hyperbolic function arises from the fact that these functions $x = a\cosh t$, $y = b\sinh t$, $t \in \mathbb{R}$ are the parameterization of a hyperbola (cf. 0.1.7.4).

The following formulas hold for all complex numbers $x$ and $y$.

**Evenness and oddness:**

$$\sinh(-x) = -\sinh x, \quad \cosh(-x) = \cosh x.$$

**Periodicity in the complex domain:**

$$\sinh(x + 2\pi\mathrm{i}) = \sinh x\,, \qquad \cosh(x + 2\pi\mathrm{i}) = \cosh x\,.$$

**Power series**:

$$\boxed{\sinh x = x + \frac{x^3}{3!} + \frac{x^5}{5!} + \frac{x^7}{7!} + \ldots\,, \qquad \cosh x = 1 + \frac{x^2}{2!} + \frac{x^4}{4!} + \frac{x^6}{6!} + \ldots}$$

**Derivative:**

$$\boxed{\frac{\mathrm{d}\sinh x}{\mathrm{d}x} = \cosh x\,, \qquad \frac{\mathrm{d}\cosh x}{\mathrm{d}x} = \sinh x\,.}$$

**Addition theorems:**

$$\boxed{\begin{aligned}\sinh(x \pm y) &= \sinh x \cosh y \pm \cosh x \sinh y\,,\\ \cosh(x \pm y) &= \cosh x \cosh y \pm \sinh x \sinh y\,.\end{aligned}}$$

**Doubled arguments:**

$$\begin{aligned}\sinh 2x &= 2 \sinh x \cosh x\,,\\ \cosh 2x &= \sinh^2 x + \cosh^2 x\,.\end{aligned}$$

**Half-arguments:**

$$\begin{aligned}\sinh\frac{x}{2} &= \sqrt{\frac{1}{2}(\cosh x - 1)} &&\text{for} \quad x \geq 0\,,\\ \sinh\frac{x}{2} &= -\sqrt{\frac{1}{2}(\cosh x - 1)} &&\text{for} \quad x < 0\,,\\ \cosh\frac{x}{2} &= \sqrt{\frac{1}{2}(\cosh x + 1)} &&\text{for} \quad x \in \mathbb{R}\,.\end{aligned}$$

**Formula of de Moivre:**

$$\boxed{(\cosh x \pm \sinh x)^n = \cosh nx \pm \sinh nx\,, \qquad n = 1, 2, \ldots}$$

**Sums:**

$$\begin{aligned}\sinh x \pm \sinh y &= 2 \sinh\frac{1}{2}(x \pm y) \cosh\frac{1}{2}(x \mp y)\,,\\ \cosh x + \cosh y &= 2 \cosh\frac{1}{2}(x + y) \cosh\frac{1}{2}(x - y)\,,\\ \cosh x - \cosh y &= 2 \sinh\frac{1}{2}(x + y) \sinh\frac{1}{2}(x - y)\,.\end{aligned}$$

### 0.2.11 The hyperbolic functions $\tanh x$ and $\coth x$

**Tangens hyperbolicus and cotangens hyperbolicus (hyperbolic tangent and cotangent):** For all complex numbers $x \neq \left(k\pi + \frac{\pi}{2}\right)\mathrm{i}$ with $k \in \mathbb{Z}$ we define a function

$$\boxed{\tanh x := \frac{\sinh x}{\cosh x}\,.}$$

For all complex numbers $x \neq k\pi\mathrm{i}$ with $k \in \mathbb{Z}$ we define the function

$$\boxed{\coth x := \frac{\cosh x}{\sinh x}.}$$

The graphical representation of these two functions for real arguments $x$ is given in Figure 0.35.

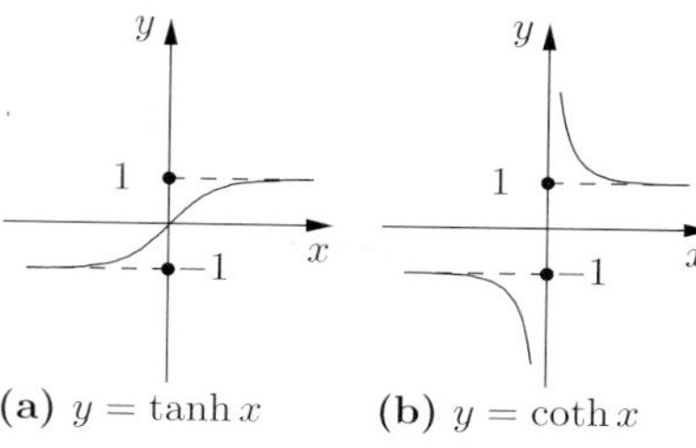


**(a)** $y = \tanh x$ **(b)** $y = \coth x$

*Figure 0.35. Hyperbolic functions.*

The following formulas hold for all complex arguments $x$ and $y$ for which the functions do not have poles[39]

**Relationship with the trigonometric functions:**

$$\boxed{\tanh x = -\mathrm{i} \tan \mathrm{i}x, \qquad \coth x = \mathrm{i} \cot \mathrm{i}x.}$$

*Table 0.20. Zeros and poles of the hyperbolic functions and trigonometric functions (all zeros and poles are simple).*

| ***Function*** | ***Period*** | ***Zeros*** ($k \in \mathbb{Z}$) | ***Poles*** ($k \in \mathbb{Z}$) | ***Parity*** |
|---|---|---|---|---|
| $\sinh x$ | $2\pi\mathrm{i}$ | $\pi k\mathrm{i}$ | – | odd |
| $\cosh x$ | $2\pi\mathrm{i}$ | $\left(\pi k + \frac{\pi}{2}\right)\mathrm{i}$ | – | even |
| $\tanh x$ | $\pi\mathrm{i}$ | $\pi k\mathrm{i}$ | $\left(\pi k + \frac{\pi}{2}\right)\mathrm{i}$ | odd |
| $\coth x$ | $\pi\mathrm{i}$ | $\left(\pi k + \frac{\pi}{2}\right)\mathrm{i}$ | $\pi k\mathrm{i}$ | odd |
| $\sin x$ | $2\pi$ | $\pi k$ | – | odd |
| $\cos x$ | $2\pi$ | $\pi k + \frac{\pi}{2}$ | – | even |
| $\tan x$ | $\pi$ | $\pi k$ | $\pi k + \frac{\pi}{2}$ | odd |
| $\cot x$ | $\pi$ | $\pi k + \frac{\pi}{2}$ | $\pi k$ | odd |

**Derivative:**

$$\boxed{\frac{\mathrm{d} \tanh x}{\mathrm{d}x} = \frac{1}{\cosh^2 x}, \qquad \frac{\mathrm{d} \coth x}{\mathrm{d}x} = -\frac{1}{\sinh^2 x}.}$$

[39] In older liturature also the following functions are used (hyperbolic secant and cosecant):

$$\operatorname{cosech} x, := \frac{1}{\sinh x} \quad \text{(cosecans hyperbolicus)},$$
$$\operatorname{sech} x, := \frac{1}{\cosh x} \quad \text{(secans hyperbolicus)}.$$

**Addition theorems:**

$$\tanh(x \pm y) = \frac{\tanh x \pm \tanh y}{1 \pm \tanh x \tanh y}, \qquad \coth(x \pm y) = \frac{1 \pm \coth x \coth y}{\coth x \pm \coth y}.$$

**Doubled arguments:**

$$\tanh 2x = \frac{2 \tanh x}{1 + \tanh^2 x}, \qquad \coth 2x = \frac{1 + \coth^2 x}{2 \coth x}.$$

**Half-arguments:**

$$\tanh \frac{x}{2} = \frac{\cosh x - 1}{\sinh x} = \frac{\sinh x}{\cosh x + 1},$$
$$\coth \frac{x}{2} = \frac{\sinh x}{\cosh x - 1} = \frac{\cosh x + 1}{\sinh x}.$$

**Sums:**

$$\tanh x \pm \tanh y = \frac{\sinh(x \pm y)}{\cosh x \cosh y}.$$

**Squares:**

| | | | | |
|---|---|---|---|---|
| $\sinh^2 x$ | – | $\cosh^2 x - 1$ | $\frac{\tanh^2 x}{1 - \tanh^2 x}$ | $\frac{1}{\coth^2 x - 1}$ |
| $\cosh^2 x$ | $\sinh^2 x + 1$ | – | $\frac{1}{1 - \tanh^2 x}$ | $\frac{\coth^2 x}{\coth^2 x - 1}$ |
| $\tanh^2 x$ | $\frac{\sinh^2 x}{\sinh^2 x + 1}$ | $\frac{\cosh^2 x - 1}{\cosh^2 x}$ | – | $\frac{1}{\coth^2 x}$ |
| $\coth^2 x$ | $\frac{\sinh^2 x + 1}{\sinh^2 x}$ | $\frac{\cosh^2 x}{\cosh^2 x - 1}$ | $\frac{1}{\tanh^2 x}$ | – |

**Power series expansion:** See section 0.7.2.

### 0.2.12 The inverse trigonometric functions

**The function arcsine**: The equation

$$y = \sin x, \qquad -\frac{\pi}{2} \leq x \leq \frac{\pi}{2},$$

has for every real number $y$ with $-1 \leq y \leq 1$ exactly one solution, which we denote by $x = \arcsin y$. Formally exchanging $x$ and $y$, we get the function

$$y = \arcsin x, \qquad -1 \leq x \leq 1.$$

The graph of this function is obtained from that of $y = \sin x$ by reflection on the diagonal[40] (see Table 0.21).

[40] In older literature the principal branch and other branches of the function $y = \arcsin x$ are used. This distinction can however lead to erroneous interpretation of (many-valued) formulas. In order to avoid that, we will use in this book only the one-to-one inverse function, which corresponds to the older principal branch (see Tables 0.21 and 0.22). The notation $y = \arcsin x$ means: $y$ is the size of the angle $y$ (measured in radians), whose sine has the value $x$ (latin: arcus cuius sinus est $x$). Instead of $\arcsin x$, $\arccos x$, $\arctan x$ and $\operatorname{arccot} x$ one speaks of the functions arcsine, arccosine, arctangent and arccotangent (of $x$).

*Table 0.21. Inverse trigonometric functions – graphs.*

| ***Original function*** | ***Inverse function*** |
|---|---|
| 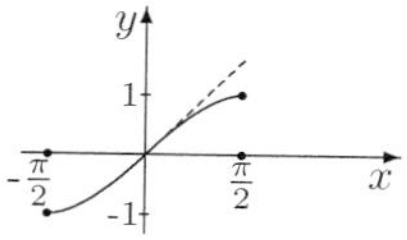 $y = \sin x$ | 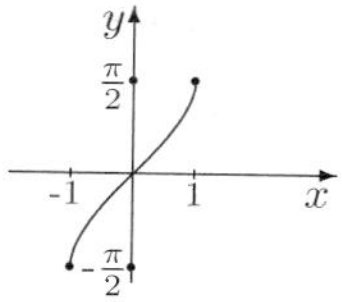 $y = \arcsin x$ |
| 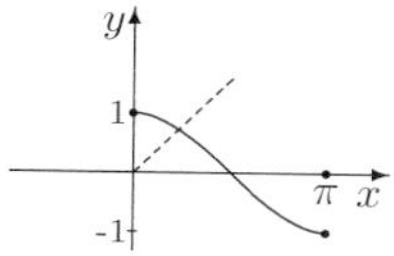 $y = \cos x$ | 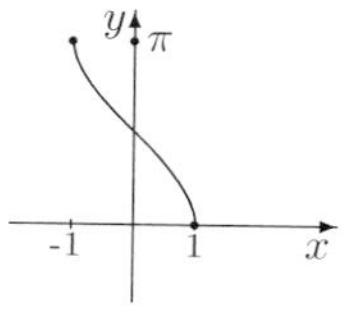 $y = \arccos x$ |
| 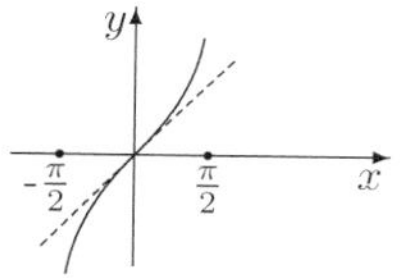 $y = \tan x$ | 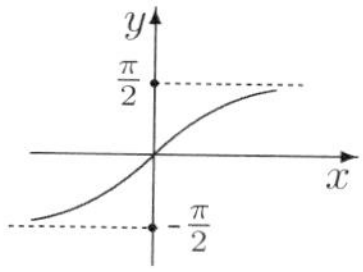 $y = \arctan x$ |
| 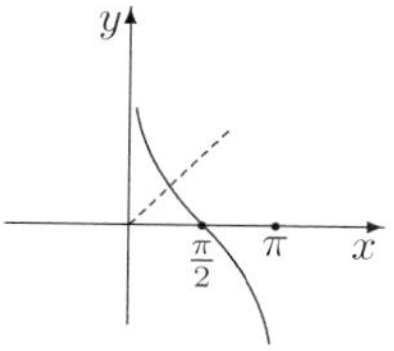 $y = \cot x$ | 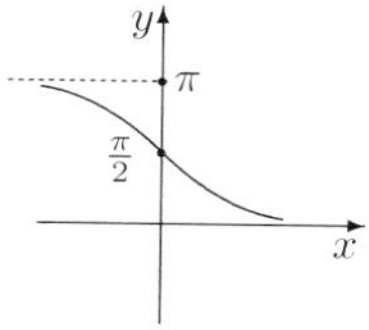 $y = \text{arccot} x$ |

*Table 0.22. Inverse trigonometric functions – formulas.*

| ***Equation*** | ***Bounds on y*** | ***Solutions x*** $(k \in \mathbb{Z})$ |
|---|---|---|
| $y = \sin x$ | $-1 \le y \le 1$ | $x = \arcsin y + 2k\pi\,, \quad x = \pi - \arcsin y + 2\pi k\,,$ |
| $y = \cos x$ | $-1 \le y \le 1$ | $x = \pm \arccos y + 2k\pi\,,$ |
| $y = \tan x$ | $-\infty < y < \infty$ | $x = \arctan y + k\pi\,,$ |
| $y = \cot x$ | $-\infty < y < \infty$ | $x = \text{arccot}\, y + k\pi\,.$ |

**Transformation formulas:** For all real numbers $x$ with $-1 < x < 1$ one has:

$$\arcsin x = -\arcsin(-x) = \frac{\pi}{2} - \arccos x = \arctan \frac{x}{\sqrt{1-x^2}}\,.$$

For all real numbers $x$ one has:

$$\arctan x = -\arctan(-x) = \frac{\pi}{2} - \operatorname{arccot} x = \arcsin \frac{x}{\sqrt{1+x^2}}\,.$$

**Derivative:** For all real numbers $x$ with $-1 < x < 1$ one has:

$$\boxed{\frac{\mathrm{d} \arcsin x}{\mathrm{d}x} = \frac{1}{\sqrt{1-x^2}}\,, \qquad \frac{\mathrm{d} \arccos x}{\mathrm{d}x} = -\frac{1}{\sqrt{1-x^2}}\,.}$$

For all real numbers $x$ one has:

$$\boxed{\frac{\mathrm{d} \arctan x}{\mathrm{d}x} = \frac{1}{1+x^2}\,, \qquad \frac{\mathrm{d} \operatorname{arccot} x}{\mathrm{d}x} = -\frac{1}{1+x^2}\,.}$$

**Power series:** See section 0.7.2.

### 0.2.13 The inverse hyperbolic functions

**Arcsinh:** The equation

$$y = \sinh x\,, \qquad -\infty < x < \infty\,,$$

has, for every real number $y$, exactly one solution, which is denoted by $x = \operatorname{arsinh} y$. Formally exchanging $x$ and $y$, we get the function

$$y = \operatorname{arsinh} x\,, \qquad -\infty < x < \infty\,.$$

The graph of this function is obtained from the graph of the function $y = \sinh x$ by a reflection on the diagonal[41] (see Table 0.23).

**Derivative:**

$$\begin{aligned}
\frac{\mathrm{d} \operatorname{arsinh} x}{\mathrm{d}x} &= \frac{1}{\sqrt{1+x^2}}\,, && -\infty < x < \infty\,,\\
\frac{\mathrm{d} \operatorname{arcosh} x}{\mathrm{d}x} &= \frac{1}{\sqrt{1-x^2}}\,, && x > 1\,,\\
\frac{\mathrm{d} \operatorname{artanh} x}{\mathrm{d}x} &= \frac{1}{1-x^2}\,, && |x| > 1\,,\\
\frac{\mathrm{d} \operatorname{arcoth} x}{\mathrm{d}x} &= \frac{1}{1-x^2}\,, && |x| < 1\,.
\end{aligned}$$

**Power series:** See section 0.7.2.

[41] The Latin names for the inverse hyperbolic functions are area sinus hyperbolicus, area cosinus hyperbolicus, area tangens hyperbolicus and area cotangens hyperbolicus (of $x$). The notation used here is from the fact that these functions give values which are the *arguments* of the hyperbolic functions.

*Table 0.23. Inverse hyperbolic functions – graphs.*

| ***Original function*** | ***Inverse function*** |
|---|---|
| 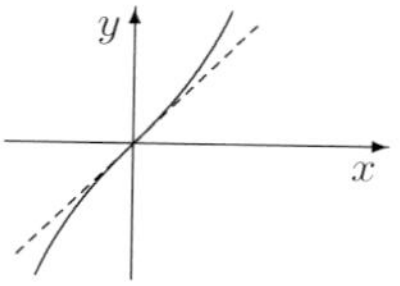 $y = \sinh x$ | 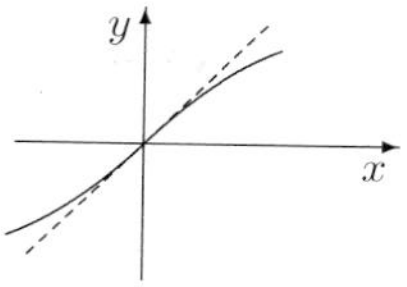 $y = \operatorname{arsinh} x$ |
| 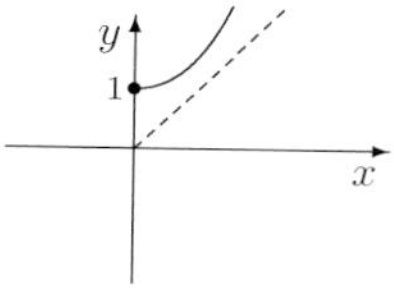 $y = \cosh x$ | 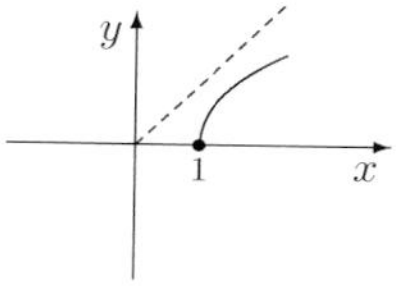 $y = \operatorname{arcosh} x$ |
| 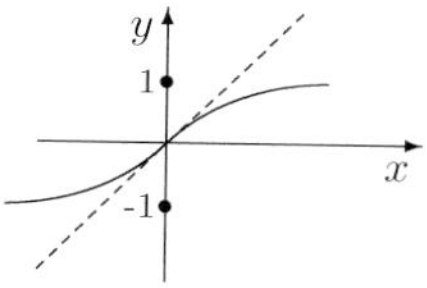 $y = \tanh x$ | 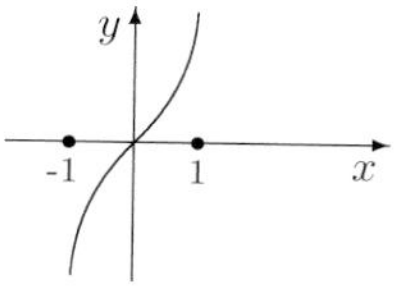 $y = \operatorname{artanh} x$ |
| 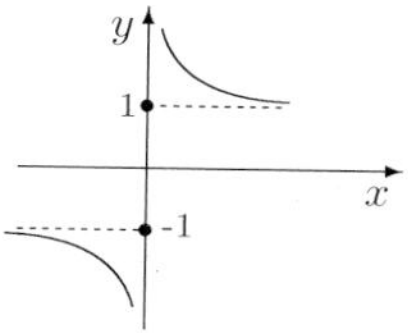 $y = \coth x$ | 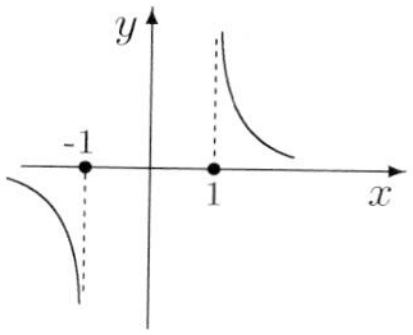 $y = \operatorname{arcoth} x$ |

*Table 0.24. Inverse hyperbolic functions – formulas.*

| ***Equation*** | ***Bounds on y*** | ***Solution x*** |
|---|---|---|
| $y = \sinh x$ | $-\infty < y < \infty$ | $x = \operatorname{arsinh} y = \ln\left(y + \sqrt{y^2+1}\right)$, |
| $y = \cosh x$ | $y \geq 1$ | $x = \pm\operatorname{arcosh} y = \pm\ln\left(y + \sqrt{y^2-1}\right)$, |
| $y = \tanh x$ | $-1 < y < 1$ | $x = \operatorname{artanh} y = \dfrac{1}{2}\ln\dfrac{1+y}{1-y}$, |
| $y = \coth x$ | $y > 1\,,\ y < -1$ | $x = \operatorname{arcoth} y = \dfrac{1}{2}\ln\dfrac{y+1}{y-1}$. |

**Transformation formulas:**

$$\operatorname{arsinh} x = (\operatorname{sgn} x)\operatorname{arcosh}\sqrt{1+x^2} = \operatorname{artanh}\frac{x}{\sqrt{1+x^2}}\,, \qquad -\infty < x < \infty\,,$$

$$\operatorname{arcosh} x = \operatorname{arsinh}\sqrt{x^2-1}\,, \qquad x \geq 1\,,$$

$$\operatorname{arcoth} x = \operatorname{artanh}\frac{1}{x}\,, \qquad -1 < x < 1\,.$$

### 0.2.14 Polynomials

A (real) *polynomial of degree n* is a function of the form

$$\boxed{y = a_n x^n + a_{n-1}x^{n-1} + \ldots + a_1 x + a_0\,.} \tag{0.43}$$

Here $n$ can take any of the values $n = 0, 1, 2, \ldots$, and all coefficients $a_k$ are real numbers with $a_n \neq 0$.

**Smoothness:** The function $y = f(x)$ in (0.43) is continuous and infinitely often differentiable in every point $x \in \mathbb{R}$. The first derivative is:

$$f'(x) = na_n x^{n-1} + (n-1)a_{n-1}x^{n-2} + \ldots + a_1\,.$$

**Behavior at infinity:** The function $y = f(x)$ in (0.43) behaves for $x \to \pm\infty$ in the same way as the function $y = ax^n$, i.e., for $n \geq 1$ one has[42]:

$$\lim_{x\to+\infty} f(x) = \begin{cases} +\infty & \text{for} \quad a_n > 0\,, \\ -\infty & \text{for} \quad a_n < 0\,, \end{cases}$$

$$\lim_{x\to-\infty} f(x) = \begin{cases} +\infty & \text{for} \quad a_n > 0 \quad \text{and } n \text{ even} \\ & \text{or for } a_n < 0 \text{ and } n \text{ odd}\,, \\ -\infty & \text{for} \quad a_n > 0 \quad \text{and } n \text{ odd} \\ & \text{or for } a_n < 0 \text{ and } n \text{ even}\,. \end{cases}$$

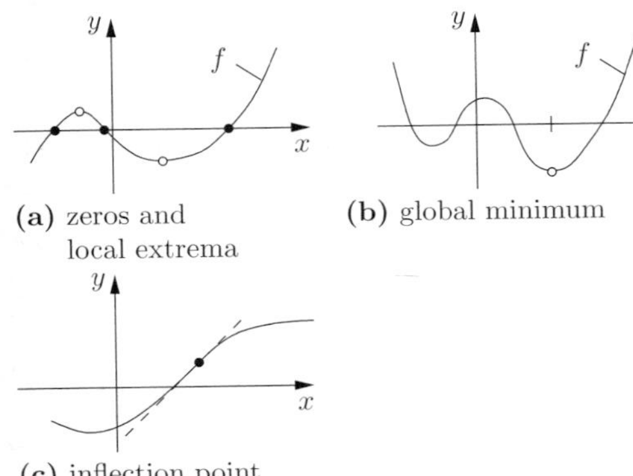


*Figure 0.36. Local properties of polynomials.*

**Zeros:** If $n$ is odd, then the graph of $y = f(x)$ intersects the $x$-axis at least once (Figure 0.36(a)). This point of intersection corresponds to a solution of the equation $f(x) = 0$.

**Global minimum:** If $n$ is even and $a_n > 0$, then $y = f(x)$ has a global minimum, i.e., there is a point $a$ with $f(a) \leq f(x)$ for all $x \in \mathbb{R}$ (Figure 0.36(b)).

If $n$ is even and $a_n < 0$, then $y = f(x)$ has a global maximum.

**Local extrema:** Let $n \geq 2$. Then the function $y = f(x)$ has

[42] The meaning of the limit symbol 'lim' will be explained in 1.3.1.1 .

at most $n-1$ local extrema, which are alternately local minima and local maxima.

**Inflection points:** Let $n \geq 3$. Then the graph of $y = f(x)$ has at most $n-2$ inflection points (Figure 0.36(c)).

## 0.2.15 Rational functions

### 0.2.15.1 Special rational functions

Let $b > 0$ be a fixed real number. The function

$$\boxed{y = \frac{b}{x}, \qquad x \in \mathbb{R}, \quad x \neq 0,}$$

represents a equilateral hyperboloid, which has the $x$- and the $y$-axis as asymptotes. The vertices are $S_\pm = \left(\pm\sqrt{b},\ \pm\sqrt{b}\right)$ (Figure 0.37).

**Behavior at infinity:** $\lim\limits_{x\to\pm\infty} \frac{b}{x} = 0$.

**Pole at the point $x = 0$:** $\lim\limits_{x\to\pm 0} \frac{b}{x} = \pm\infty$.

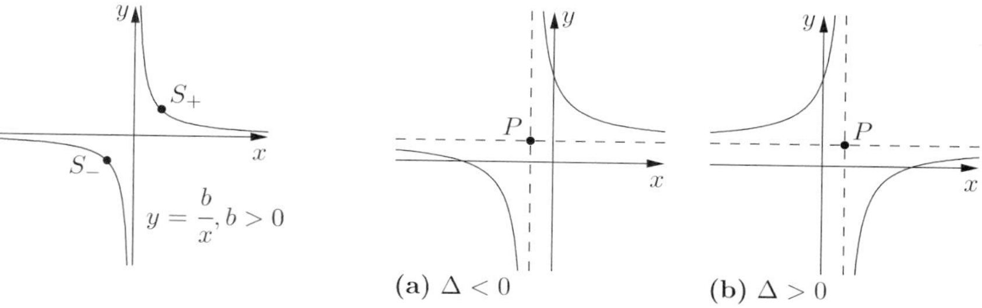


*Figure 0.37.*

*Figure 0.38.*

### 0.2.15.2 Rational function with linear numerators and denominators

Let the real numbers $a, b, c$ and $d$ be given with $c \neq 0$ and $\Delta := ad - bc \neq 0$. The function

$$\boxed{y = \frac{ax+b}{cx+d}, \qquad x \in \mathbb{R},\ x \neq -\frac{d}{c}} \tag{0.44}$$

is transformed by the change of coordinates $x = u - \frac{d}{c}$, $y = w + \frac{a}{c}$ to the simpler form

$$w = -\frac{\Delta}{c^2 u}.$$

Thus, the general equation (0.44) results for the normalized form $y = -\frac{\Delta}{c^2 x}$ by a simple change of coordinates, which maps the point $(0,0)$ to the point $P = \left(-\frac{d}{c}, \frac{a}{c}\right)$ (Figure 0.38).

### 0.2.15.3 Special rational function with a denominator of $n$th degree

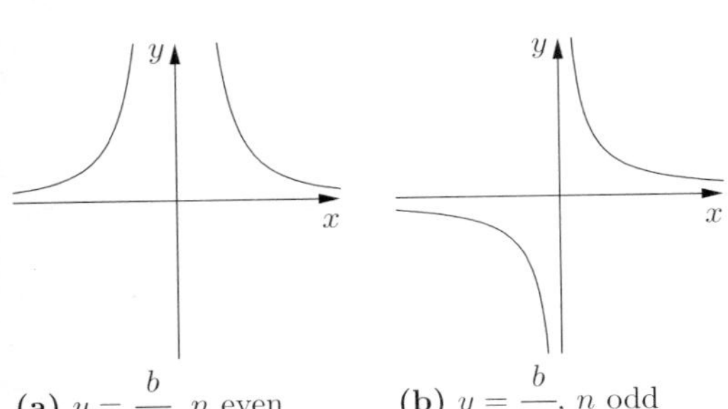


**(a)** $y = \dfrac{b}{x^n}$, $n$ even **(b)** $y = \dfrac{b}{x^n}$, $n$ odd

*Figure 0.39.*

Let $b > 0$ be given and $n = 1, 2, \ldots$ The function

$$y = \frac{b}{x^n}, \qquad x \in \mathbb{R}, \, x \neq 0,$$

is displayed in Figure 0.39.

#### 0.2.15.4 Rational functions with quadratic denominator

**Special case 1:** Let $d > 0$ be given. The functions

$$y = \frac{1}{x^2 + d^2}, \qquad x \in \mathbb{R},$$

and

$$y = \frac{x}{x^2 + d^2}, \qquad x \in \mathbb{R},$$

are pictured in Figure 0.40.

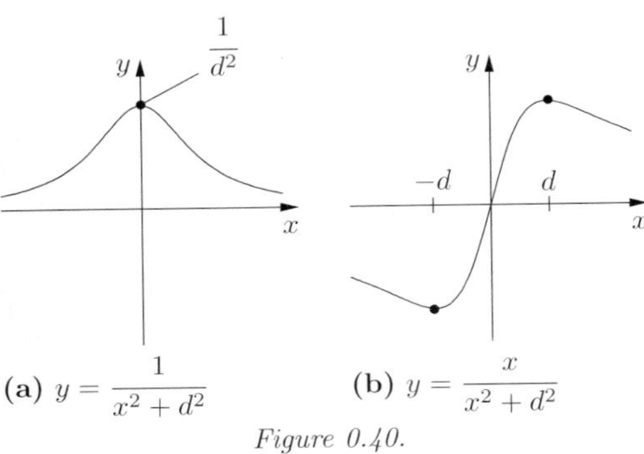


**(a)** $y = \dfrac{1}{x^2 + d^2}$ **(b)** $y = \dfrac{x}{x^2 + d^2}$

*Figure 0.40.*

**Special case 2:** Let two real numbers $x_\pm$ be given with $x_- < x_+$. The function $y = f(x)$ given by

$$y = \frac{1}{(x - x_+)(x - x_-)} \tag{0.45}$$

can be put in the form

$$y = \frac{1}{x_+ - x_-}\left(\frac{1}{x - x_+} - \frac{1}{x - x_-}\right).$$

This is a special case of the so-call partial fraction decomposition (cf. 2.1.7). One has:

$$\lim_{x \to x_+ \pm 0} f(x) = \pm\infty, \qquad \lim_{x \to x_- \pm 0} f(x) = \mp\infty, \qquad \lim_{x \to \pm\infty} f(x) = 0.$$

Thus the poles of the function are at the points $x_+$ and $x_-$ (see Figure 0.41).

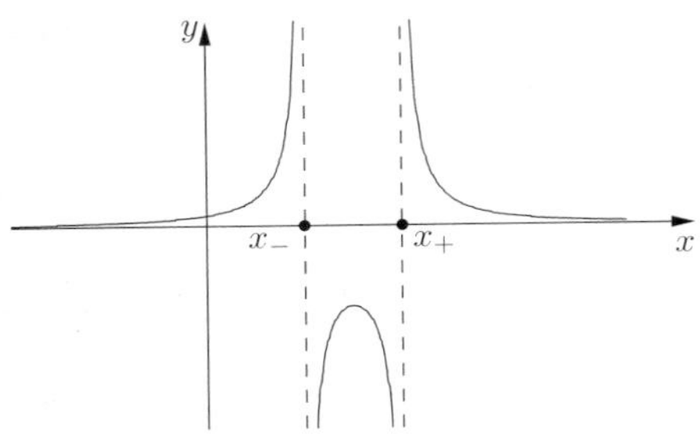


*Figure 0.41.*

**Special case 3:** The function

$$y = \frac{x - 1}{x^2 - 1} \tag{0.46}$$

is initially not defined at the point $x = 1$. However, if one uses the decomposition $x^2 - 1 = (x - 1)(x + 1)$, then we get

$$y = \frac{1}{x + 1}, \qquad x \in \mathbb{R}, \, x \neq -1.$$

One says that the function (0.46) has a *apparent singularity* at the point $x = 1$.

**General case:** Let real numbers $a, b, c$ and $d$ be given with $a^2 + b^2 \neq 0$. The behavior of the function $y = f(x)$ defined by

$$\boxed{y = \frac{ax + b}{x^2 + 2cx + d}} \tag{0.47}$$

depends in an essential way on the sign of the *discriminant* $D := c^2 - d$. Independently of this, one always has

$$\boxed{\lim_{x\to\pm\infty} f(x) = 0\,.}$$

*Case 1:* $D > 0$. Then one has

$$x^2 + 2cx + d = (x - x_+)(x - x_-)$$

with $x_\pm = -c \pm \sqrt{D}$. This yields the partial fraction decomposition

$$f(x) = \frac{A}{x - x_+} + \frac{B}{x - x_-}\,.$$

The constants $A$ and $B$ are determined by calculating the limits:

$$A = \lim_{x\to x_+}(x - x_+)f(x) = \frac{ax_+ + b}{x_+ - x_-}\,, \qquad B = \lim_{x\to x_-}(x - x_-)f(x) = \frac{ax_- + b}{x_- - x_+}\,.$$

There are poles at the points $x_\pm$.

*Case 2:* $D = 0$. In this case one has $x_+ = x_-$. We thus get

$$f(x) = \frac{ax + b}{(x - x_+)^2}\,.$$

This yields

$$\lim_{x\to x_+\pm 0} f(x) = \begin{cases} +\infty\,, & \text{if} \quad ax_+ + b > 0\,,\\ -\infty\,, & \text{if} \quad ax_+ + b < 0\,,\end{cases}$$

i.e., the point $x_+$ is a pole.

*Case 3:* $D < 0$. In this case we have $x^2 + 2cx + d > 0$ for all $x \in \mathbb{R}$. Consequently the function $y = f(x)$ in (0.47) is continuous and infinitely often differentiable for all points $x \in \mathbb{R}$, in other words, $f$ is smooth.

#### 0.2.15.5 The general rational function

A (real) *rational function* is a function $y = f(x)$ of the form

$$\boxed{y = \frac{a_n x^n + \ldots + a_1 x + a_0}{b_m x^m + \ldots + b_1 x + b_0}}\,,$$

where there are polynomials in both the numerator and the denominator (cf. 0.2.14).

**Behavior at infinity:** We set $c := a_n/b_m$. Then we have:

$$\boxed{\lim_{x\to\pm\infty} f(x) = \lim_{x\to\pm\infty} cx^{n-m}\,.}$$

From this we can discuss all possible cases.

*Case 1:* $c > 0$.

$$\lim_{x\to+\infty} f(x) = \begin{cases} c & \text{for } n = m\,, \\ +\infty & \text{for } n > m\,, \\ 0 & \text{for } n < m\,. \end{cases}$$

$$\lim_{x\to-\infty} f(x) = \begin{cases} c & \text{for } n = m\,, \\ +\infty & \text{for } n > m \text{ and } n - m \text{ even}\,, \\ -\infty & \text{for } n > m \text{ and } n - m \text{ odd}\,, \\ 0 & \text{for } n < m\,. \end{cases}$$

*Case 2:* $c < 0$. Here one must replace $\pm\infty$ by $\mp\infty$.

**Partial fraction decomposition:** The precise structure of rational functions is given by the partial fraction decomposition (cf. 2.1.7).

## 0.3 Mathematics and computers – a revolution in mathematics

*One can say that we live in the age of mathematics, and that our culture has been 'mathematized'. This is proved beyond a doubt by the widespread use of computers.*

*Arthur Jaffe*
*(Harvard University, Cambridge, USA)*

In solving mathematical problems one utilizes (at least) four important techniques:

(i) the use of numerical algorithms;

(ii) the algorithmic treatment of analytical, algebraic and geometric problems;

(iii) reference to tables and collections of formulas;

(iv) the graphical representation of situations.

Modern software programs can carry out all four of these effectively on computers:

(a) For the solution of standard problems of mathematics we suggest the system Mathematica.

(b) For more complicated problems of scientific calculations a combination of Maple and Matlab often leads to success.

(c) Many software packages also contain the program library Imsl math/stat/sfun library (International Mathematical and Statistical Library).

To solve a given mathematical problem, one should first check whether the problem is amenable to the procedures available in Mathamatica. This is the case for example for many of the problems considered in this book. Only after this has been checked with negative result should one resort to (b) or (c).

There is a long list of literature on this topic at the beginning of the bibliography. The handbooks listed there for using Mathematica are skillfully written and didacticly apt, addressed to a large audience. Experience shows that one requires a certain amount of time to get used to such programs, before they can be applied efficiently. It is worthwhile to invest this time; the gains are potentially great.

Modern software systems, which are continually being refined, are already able to do a great amount of the routine work for the user, freeing him for other activities. However, this can not replace a thorough occupation with the basics of mathematics. In this connection the following picture is helpful. At a construction site one sees today huge cranes, which do an enormous amount of work for humans. But still it is the humans which decide what is to be built and how the building should be designed. For this human qualities like phantasy and originality are required, something one can not expect (or want!) a machine to possess.

## 0.4 Tables of mathematical statistics and standard procedures for practitioners

The goal of this section is to give a large audience of potential readers an acquaintance with the basics and practical application of mathematical statistics. To meet this goal, we assume on the part of the reader almost nothing in the way of mathematical background. A discussion of the fundamentals of mathematical statistics can be found in 6.3.

**Mathematical statistics on the computer:** Elementary standard procedures can be done with Mathematica. More specialized statistical packages which are wide spread are **SPSS** and **SAS**.

### 0.4.1 The most important empirical data for sequences of measurements (trials)

Many measurements in technology, science or medicine have the characteristic property that the results of measurements vary from trial to trial. One says, that the measurements have a component of randomness. The quantity one wishes to measure, $X$, is called a *random variable.*

*Example 1:* The height $X$ of a person is random, i.e., $X$ is a random variable.

**Sequence of measurements:** If we measure a random quantity $X$, then we get measurements

$$\boxed{x_1, \ldots, x_n.}$$

*Example 2:* The Tables 0.25 and 0.26 show the result of measuring the height of 8 men in cm.

*Table 0.25*

| $x_1$ | $x_2$ | $x_3$ | $x_4$ | $x_5$ | $x_6$ | $x_7$ | $x_8$ | $\overline{x}$ | $\Delta x$ |
|---|---|---|---|---|---|---|---|---|---|
| 168 | 170 | 172 | 175 | 176 | 177 | 180 | 182 | 175 | 4,8 |

*Table 0.26*

| $x_1$ | $x_2$ | $x_3$ | $x_4$ | $x_5$ | $x_6$ | $x_7$ | $x_8$ | $\overline{x}$ | $\Delta x$ |
|---|---|---|---|---|---|---|---|---|---|
| 174 | 174 | 174 | 174 | 176 | 176 | 176 | 176 | 175 | 1,07 |

**Empirical mean and empirical standard deviation:** Two basic characteristics of a sequence of measurements $x_1, \ldots, x_n$ are the empirical mean

$$\boxed{\overline{x} := \frac{1}{n}\,(x_1 + x_2 + \ldots + x_n)}$$

and the empirical standard deviation $\Delta x$. The square of this quantity is given by the formula[43]

$$\boxed{(\Delta x)^2 := \frac{1}{n-1}\left[(x_1 - \overline{x})^2 + (x_2 - \overline{x})^2 + \ldots + (x_n - \overline{x})^2\right] .}$$

*Example 3:* For the values in Table 0.25 one gets

$$\overline{x} = \frac{1}{8}\big(168 + 170 + 172 + 175 + 176 + 177 + 180 + 182\big) = 175 .$$

One says that the average height of the men is 175 cm. The same average is obtained from the values in Table 0.26. A glance at the tables however shows that the variation in the values of Table 0.25 is much higher than that of Table 0.26. For Table 0.26 we get

$$\begin{aligned}(\Delta x)^2 &= \frac{1}{7}\left[(174-175)^2 + (174-175)^2 + \ldots + (176-175)^2\right]\\ &= \frac{1}{7}\big[1+1+1+1+1+1+1+1\big] = \frac{8}{7},\end{aligned}$$

in other words, $\Delta x = 1.07$. On the other hand, the values for Table 0.25 yield for the standard deviation from the equation

$$\begin{aligned}(\Delta x)^2 &= \frac{1}{7}\left[(168-175)^2 + (170-175)^2 + \ldots + (182-175)^2\right]\\ &= \frac{1}{7}\big[49+25+9+1+4+25+49\big] = 23\end{aligned}$$

the value $\Delta x = 4.8$.

**Rule of thumb:**

| The smaller the empirical standard deviation is, the smaller is the variation of the measurements from the mean $\overline{x}$. |
|---|

In the limiting case $\Delta x = 0$, all the measurements coincide with the mean $\overline{x}$.

**The distribution of the measurements – the histogram:** To get a general idea of the distribution of the measurements, one uses, especially for larger sets of measurements, a graphical representation called the *histogram*.

(i) One divides the measurements into several classes $K_1, K_2, \ldots, K_s$. These are neighboring intervals.

(ii) Let $m_r$ denote the number of measurements which belong to the class $K_r$.

(iii) If $n$ measurements have been made $x_1, \ldots, x_n$, then the quantity $\dfrac{m_r}{n}$ is called the relative frequency of the measurements with respect to the class $K_r$.

---

[43]The appearance of the denominator $n-1$ instead of that probably expected by many readers, namely $n$, can be justified by estimation theory. In fact the quantity $\Delta x$ is a *expectation faithful estimation* for the theoretical standard deviation $\Delta X$ of the random variable $X$ (cf. 6.3.2). For large $n$ the difference between $n$ and $n-1$ is negligible.

The quantity $(\Delta x)^2$ is called the *variance*.

(iv) One graphs the classes $K_j$, with a column of height $\frac{m_r}{n}$ over each $K_r$.

*Example 4:* In Table 0.27 the measurements for the heights of 100 men in centimeters are listed. The histogram constructed from these data is displayed in Figure 0.42.

*Table 0.27*

| ***Class*** $K_r$ | ***Measure-ments*** | ***Frequency*** $m_r$ | ***Relative frequency*** $\frac{m_r}{100}$ |
|---|---|---|---|
| $K_1$ | $150 \le x < 165$ | 2 | 0, 02 |
| $K_2$ | $165 \le x < 170$ | 18 | 0, 18 |
| $K_3$ | $170 \le x < 175$ | 30 | 0, 30 |
| $K_4$ | $175 \le x < 180$ | 32 | 0, 32 |
| $K_5$ | $180 \le x < 185$ | 16 | 0, 16 |
| $K_6$ | $185 \le x < 200$ | 2 | 0, 02 |

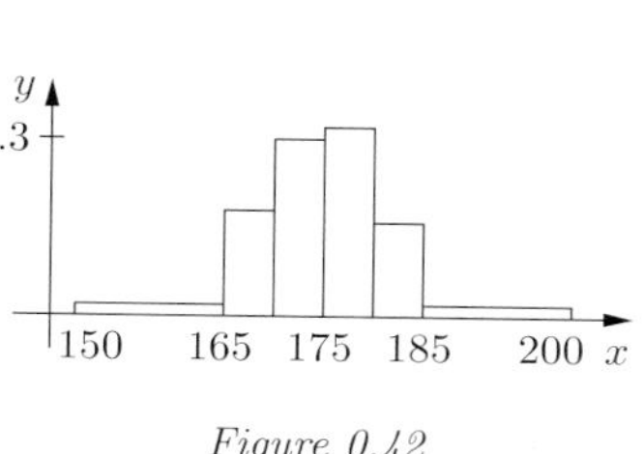


*Figure 0.42*

## 0.4.2 The theoretical distribution function

The sequences of trials of a random variable $X$ generally vary from trial to trial. For example, the measurements of heights of persons leads to different results if one measures all men in a house, a city or a state. The notion of theoretical distribution function is necessary in order to build up a theory of random variables.

**Definition:** The *theoretical distribution function* $\Phi$ of a random variable $X$ is defined by the following prescription:

$$\boxed{\Phi(x) := P(X < x)\ .}$$

This means that the value $\Phi(x)$ is equal to the *probability* that the random variable is less than the given number $x$.

**The normal distribution:** Many measured quantities follow a *normal* distribution. To explain this, we consider a Gauss bell curve

$$\boxed{\varphi(x) := \frac{1}{\sigma\sqrt{2\pi}} e^{-(x-\mu)^2/2\sigma^2}\ .} \tag{0.48}$$

Such a curve has a maximum at the point $x = \mu$. The smaller the positive value $\sigma$ is, the more it is more concentrated at the point $x = \mu$. One calls $\mu$ the *mean* and $\sigma$ the *standard deviation* of the normal distribution (Figure 0.43(a)).

> The area of the hatched surface in Figure 0.43(b) is equal to the probability that the random variable lies in the interval $[a, b]$.

The distribution function $\Phi$ of the normal distribution (0.48) is displayed in Figure 0.43(d). The value $\Phi(a)$ in Figure 0.43(d) is equal to the area of the surface under the bell curve, which lies to the left of $a$. The difference

$$\boxed{\Phi(b) - \Phi(a)}$$

is equal to the area of the hatched surface in Figure 0.43(b).

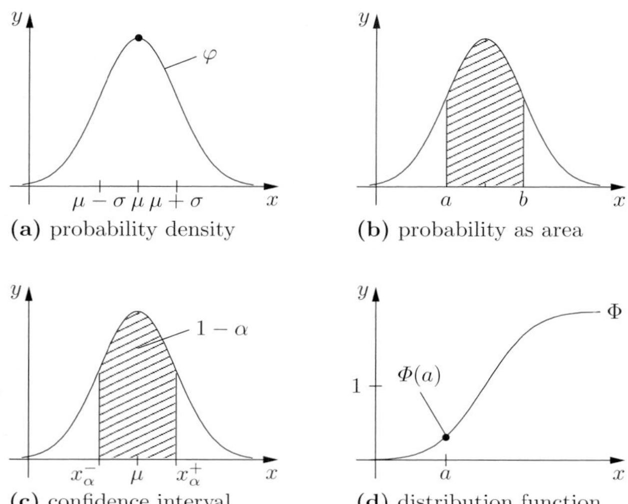

(a) probability density (b) probability as area

(c) confidence interval (d) distribution function

*Figure 0.43. Properties of the Gaussian normal distribution.*

**Confidence interval:** This notion is of central importance for mathematical statistics. The $\alpha$-confidence interval $[x_\alpha^-, x_\alpha^+]$ of the random variable $X$ is defined in such a way that the probability that all measurements of $X$ lie in the interval is $1-\alpha$, i.e., the probability that $x$ satisfies the inequality

$$x_\alpha^- \le x \le x_\alpha^+$$

is $1-\alpha$. In Figure 0.43(c) the endpoints of the interval $x_\alpha^+$ and $x_\alpha^-$ are chosen in such a way that they are symmetric around the mean $\mu$ and the area of the hatched surface is equal to $1-\alpha$. One has

$$x_\alpha^+ = \mu + \sigma z_\alpha\,, \qquad x_\alpha^- = \mu - \sigma z_\alpha\,.$$

The value of $z_\alpha$ for the important cases for many applications, namely $\alpha = 0.01\,,\ 0.05\,,\ 0.1$, are listed in Table 0.28.

*Table 0.28*

| $\alpha$ | 0.01 | 0.05 | 0.1 |
|---|---|---|---|
| $z_\alpha$ | 2.6 | 2.0 | 1.6 |

The random variable $X$ lies with the probability $1-\alpha$ in the $\alpha$-confidence interval.

*Example:* Let $\mu = 10$ and $\sigma = 2$. For $\alpha = 0.01$ we get

$$x_\alpha^+ = 10 + 2\cdot 2.6 = 15.2\,, \qquad x_\alpha^- = 10 - 2\cdot 2.6 = 4.8\,.$$

It follows that with a probability of $1-\alpha = 0.99$, the measured value $x$ lies between 4.8 and 15.2. Intuitively this means the following.

(a) If $n$ is a large number and we take a total of $n$ measurements of $X$, then there are approximately $(1-\alpha)n = 0.99n$ measured values in between 4.8 and 15.2.

(b) If we measure $X$ for example 1000 times, then approximately 990 of the values lie between 4.8 and 15.2.

### 0.4.3 Checking for a normal distribution

Many test procedures in applications are based on the assumption that a random variable $X$ follows a normal distribution. We describe here a simple graphical procedure to test *whether* $X$ is normally distributed.

(i) We draw a line in the $(z, y)$-coordinate plane, which contains the pairs of points $(z, \Phi(z))$, which we take from Table 0.29 (Figure 0.44). Note that the value on the $y$-axis in the present case has an irregular scale.

(ii) For given measured values $x_1, \ldots, x_n$ of $X$ we form the quantities

$$z_j := \frac{x_j - \overline{x}}{\Delta x}, \qquad j = 1, \ldots, n.$$

(iii) We calculate the numbers

$$\Phi_*(z_j) = \frac{1}{n}\,(\text{number of measured values } z_k \text{, which are smaller than } z_j)$$

and plot the points $\big(z_j, \Phi_*(z_j)\big)$ in Figure 0.44.

If these points lie approximately on the line drawn in (i), then $X$ is approximately normally distributed.

*Example:* The open circles in Figure 0.44 represent measurements which are approximately normally distributed.

*Table 0.29. Sample values of the normal distribution function.*

| $z$ | $-2.5$ | $-2$ | $-1.5$ | $-1$ | $-0.5$ | $0$ | $0.5$ | $1$ | $1.5$ | $2$ | $2.5$ |
|---|---|---|---|---|---|---|---|---|---|---|---|
| $\Phi(z)$ | 0.01 | 0.02 | 0.07 | 0.16 | 0.31 | 0.5 | 0.69 | 0.84 | 0.93 | 0.98 | 0.99 |

A more precise table of the values of $\Phi$ is given in Table 0.29. The diagram of Figure 0.44 can also be obtained as so-called probability paper.

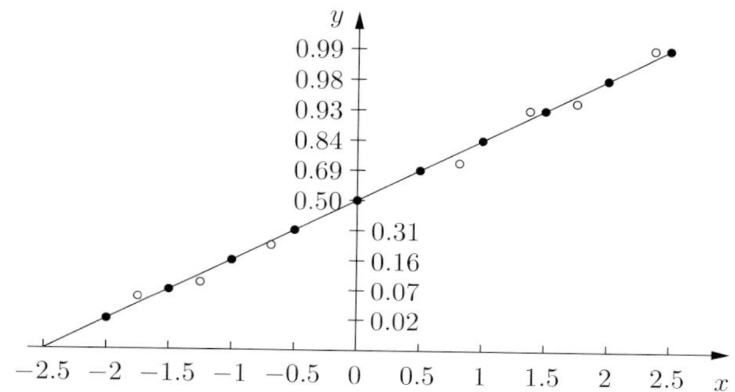


*Figure 0.44. Checking data for approximate normal distribution.*

**The $\chi^2$-fit test for normal distributions:** This test, which is much more significant than the intuitive method just explained with the probability paper, is described in 6.3.4.5.

### 0.4.4 The statistical evaluation of a sequence of measurements

We assume that $X$ is a normally distributed random variable, whose normal distribution (0.48) has mean $\mu$ and standard deviation $\sigma$.

**The confidence limit for the mean $\mu$:**

(i) We take $n$ measurements of the quantity $X$ and get the measurements $x_1, \ldots, x_n$.

(ii) We choose a small number $\alpha$ as the probability of being in error and determine the value of $t_{\alpha,m}$ with $m = n - 1$ from Table in 0.4.6.3.

Then the unknown mean $\mu$ for the normal distribution satisfies the inequality:

$$\boxed{\overline{x} - t_{\alpha,m}\frac{\Delta x}{\sqrt{n}} \le \mu \le \overline{x} + t_{\alpha,m}\frac{\Delta x}{\sqrt{n}}\,.}$$

This statement has a probability of $\alpha$ of being in error.

*Example 1:* For the measurements of heights listed in Table 0.25 one has $n = 8$, $\overline{x} = 175$, $\Delta x = 4.8$. If we choose $\alpha = 0.01$, then we get from 0.4.6.3 for $m = 7$ the value $t_{\alpha,m} = 3.5$. If we assume that the height is a normally distributed random variable, then we have, with error probability $\alpha = 0.01$, for the mean:

$$169 \le \mu \le 181\,.$$

**The confidence interval for the standard deviation $\sigma$:** With probability $\alpha$ of error, the standard deviation $\sigma$ satisfies the inequality:

$$\boxed{\frac{(n-1)\,(\Delta x)^2}{b} \le \sigma^2 \le \frac{(n-1)\,(\Delta x)^2}{a}\,.}$$

The values $a := \chi^2_{1-\alpha/2}$ and $b := \chi^2_{\alpha/2}$ are extracted from the table in 0.4.6.4 with $m = n - 1$ degrees of freedom.

*Example 2:* We consider again the height measurements listed in Table 0.25. For $\alpha = 0.01$ and $m = 7$ we get $a = 1.24$ and $b = 20.3$ from 0.4.6.4. Consequently we get with error probability $\alpha = 0.01$ the estimate

$$2.8 \le \sigma \le 11.40\,.$$

It is not surprising that these estimates are quite rough. This is because of the small number of measurements.

A more detailed justification is given in 6.3.3.

### 0.4.5 The statistical comparison of two sequences of measurements

Let two sequences of trials of random variables $X$ and $Y$ be given,

$$\boxed{x_1, \ldots, x_{n_1} \qquad \text{and} \qquad y_1, \ldots, y_{n_2}.} \tag{0.49}$$

Two basic questions are:

(i) Is there a dependence between the two sequences of measurements?

(ii) Is there a significant difference between the two random variables?

To investigate (i) one uses correlation coefficients. An answer to (ii) is provided by the $F$-test, the $t$-test and the Wilcoxon-test. This will be considered in the sequel.

#### 0.4.5.1 Empirical correlation coefficients

The empirical correlation coefficient of the two sequences of measurements (0.49) with $n_1 = n_2 = n$ is given by the number

$$\varrho = \frac{(x_1 - \overline{x})(y_1 - \overline{y}) + (x_2 - \overline{x})(y_2 - \overline{y}) + \ldots + (x_n - \overline{x})(y_n - \overline{y})}{(n-1)\Delta x \Delta y}.$$

One has $-1 \leq \varrho \leq 1$. For $\varrho = 0$ there is no dependency between the two sequences.

> The dependency between the two sequences is stronger the larger the quantity $\varrho^2$ is.

**Regression line:** If one plots the pairs $(x_j, y_j)$ of measurements in a Cartesian coordinate system, then the so-called *regression line*

$$\boxed{y = \overline{y} + \varrho \frac{\Delta y}{\Delta x}(x - \overline{x})}$$

is the line closest to the set of plotted points (Figure 0.45), i.e., this line is a solution of the minimum problem

$$\sum_{j=1}^{n} (y_j - a - bx_j)^2 \stackrel{!}{=} \min, \qquad a, b \text{ real},$$

and the minimal value is equal to $(\Delta y)^2 (1 - \varrho^2)$. The fit of the regression line on the measurements is thus optimal for $\varrho^2 = 1$.

*Table 0.30*

| $x_1$ | $x_2$ | $x_3$ | $x_4$ | $x_5$ | $x_6$ | $x_7$ | $x_8$ | $\overline{x}$ | $\Delta x$ |
|---|---|---|---|---|---|---|---|---|---|
| 168 | 170 | 172 | 175 | 176 | 177 | 180 | 182 | 175 | 5 |
| $y_1$ | $y_2$ | $y_3$ | $y_4$ | $y_5$ | $y_6$ | $y_7$ | $y_8$ | $\overline{y}$ | $\Delta y$ |
| 157 | 160 | 163 | 165 | 167 | 167 | 168 | 173 | 165 | 5 |

*Example:* For the two sequences of measurements listed in Table 0.30 one gets the correlation coefficient

$$\varrho = 0.96$$

with the regression line

$$y = \overline{y} + 0.96(x - \overline{x}). \qquad (0.50)$$

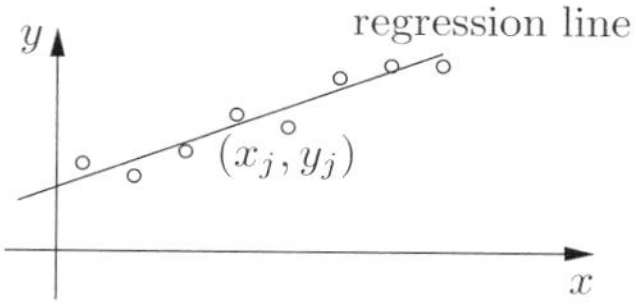

*Figure 0.45.*

Here there is a strong dependency between the two sequences of measurements. The measurements are approximated quite well by the regression line (0.50).

#### 0.4.5.2 The comparison of two means with the $t$-test

The $t$-test is often used in applications. With this test one can verify whether the means of two sequences of trials differ essentially from one another.

(i) We consider the two sequences $x_1, \ldots, x_{n_1}$ and $y_1, \ldots, y_{n_2}$ of two random variables $X$ and $Y$, which we assume are normally distributed.

In addition one assumes that the standard deviation of $X$ and $Y$ are the same. This assumption can be checked with the help of the $F$-test 0.4.5.3.

(ii) We calculate the number

$$t = \frac{\overline{x} - \overline{y}}{\sqrt{(n_1 - 1)(\Delta x)^2 + (n_2 - 1)(\Delta y)^2}} \sqrt{\frac{n_1 n_2 (n_1 + n_2 - 2)}{n_1 + n_2}} .$$

(iii) For a given $\alpha$ and $m = n_1 + n_2 - 1$ we determine the value $t_{\alpha,m}$ from the table in 0.4.6.3.

*Case 1:* Assume

$$\boxed{|t| > t_{\alpha,m} .}$$

In this case the means of $X$ and $Y$ differ, i.e., the differences between the measured empirical means $\overline{x}$ and $\overline{y}$ are not random, but have some explanation. One also says in this case that there is a significant difference between the random variables $X$ and $Y$.

*Case 2:* One has

$$\boxed{|t| < t_{\alpha,m} .}$$

One may assume that the means of $X$ and $Y$ do not differ significantly.

The statements both have a probability of error of $\alpha$. This means the following. If one applies this test in 100 different situations, then there is a probability that this will lead to a false conclusion in $100 \cdot \alpha$ cases.

*Example 1:* For $\alpha = 0.01$ there might be in 100 tests one case in which the test leads to a wrong conclusion.

*Example 2:* Two medicines $A$ and $B$ are being given to patients which have the same illness. The random variable is the number of days $X$ (resp. $Y$) for the medicine $A$ (resp. $B$) to be administered until the illness has been cured. Table 0.31 lists some measurements. For example, the mean duration until cure is 20 days under the use of medicine $A$.

*Table 0.31*

| | | | |
|---|---|---|---|
| medicine $A$ : | $\overline{x} = 20$ | $\Delta x = 5$ | $n_1 = 15$ patients |
| medicine $B$ : | $\overline{y} = 26$ | $\Delta y = 4$ | $n_2 = 15$ patients. |

We get

$$t = \frac{26 - 20}{\sqrt{14 \cdot 25 + 14 \cdot 16}} \cdot \sqrt{\frac{15 \cdot 15\,(30 - 2)}{15 + 15}} = 3.6 .$$

We find for the value of $\alpha = 0.01$ and $m = 15 + 15 - 1 = 29$ the value $t_{\alpha,m} = 2.8$ from the table in 0.4.6.3.

Because of $t > t_{\alpha,m}$ there is a significant difference between the two medicines, i.e., medicine $A$ is better than $B$.

#### 0.4.5.3 The *F*-test

This test verifies whether the standard deviations of two normally distributed random variables differ from one another.

(i) We consider the two sequences of measurements $x_1, \ldots, x_{n_1}$ and $y_1, \ldots, y_{n_2}$ of the two random variables $X$ and $Y$, both of which we assume follow a normal distribution.

(ii) We form the quotients

$$F := \begin{cases} \left(\dfrac{\Delta x}{\Delta y}\right)^2, & \text{if } \Delta x > \Delta y\,, \\ \left(\dfrac{\Delta y}{\Delta x}\right)^2, & \text{if } \Delta x \leq \Delta y\,. \end{cases}$$

(iii) We look up the bold-faced value $F_{0.01;m_1m_2}$ in 0.4.6.5 for $m_1 := n_1 - 1$ and $m_2 := n_2 - 1$.

*Case 1:* One has

$$\boxed{F > F_{0.01;m_1m_2}\,.}$$

In this case the standard deviations of $X$ and $Y$ are *not* the same, i.e., the difference between the measured empirical standard deviations $\Delta x$ and $\Delta y$ is not a random variation, but has some deeper meaning.

*Case 2:* One has

$$\boxed{F \leq F_{0.01;m_1m_2}\,.}$$

One may assume that the standard deviations of $X$ and $Y$ are essentially the same.

These statements both have a probability of error of 0.02. This means the following. Carrying out this test in 100 different situations, it is likely that in two of these situations the test leads to a wrong conclusion.

*Example:* We consider again the situation giving rise to the data of Table 0.31. One has $F = (\Delta x/\Delta y)^2 = 1.6$. From the table in 0.4.6.5 with $m_1 = m_2 = 14$ we find $F_{0.01;m_1m_2} = 3.7$. Because of $F < F_{0.01;m_1m_2}$ we can be assured that $X$ and $Y$ have the same standard deviation.

#### 0.4.5.4 The Wilcoxon-test

The $t$-test can only be applied to normally distributed quantities. The Wilcoxon-test is much more general and can be applied for example to check whether two sequences of trials come from random variables with different distributions, i.e., whether these quantities differ from each other in an essential way. This test is described in 6.3.4.5.

## 0.4.6 Tables of mathematical statistics

### 0.4.6.1 Interpolation of tables

**Linear interpolation:** Each table consists of entries and table values. In Table 0.32, $x$ denotes the entries and $f(x)$ the values.

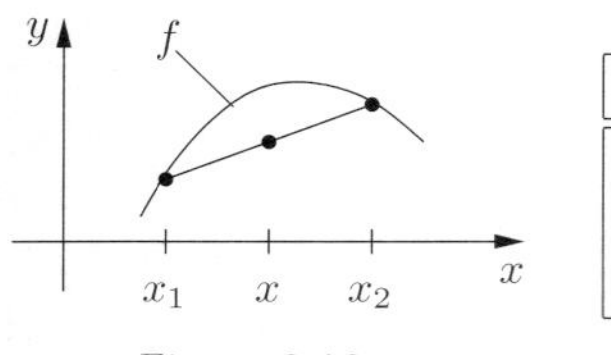


*Figure 0.46.*

Table 0.32

| $x$ | $f(x)$ |
|---|---|
| 1 | 0.52 |
| 2 | 0.60 |
| 3 | 0.64 |

**First basic problem: Interpolation of table values $f(x)$ for known entries $x$:** If the entry $x$ is not in the table, one can use the method of linear interpolation, which is illustrated in Figure 0.46. Here the graph of $y = f(x)$ is replaced by the secant between two of the pairs $(x_1, f(x_1))$ and $(x_2, f(x_2))$. The approximate value $f_*(x)$ for $f(x)$ is then derived from the linear interpolation formula:

$$\boxed{f_*(x) = f(x_1) + \frac{f(x_2) - f(x_1)}{x_2 - x_1}(x - x_1)\,.} \tag{0.51}$$

*Example 1:* Let $x = 1.5$. In Table 0.32 one finds the nearest entries

$$x_1 = 1 \qquad \text{and} \qquad x_2 = 2$$

with the values $f(x_1) = 0.52$ and $f(x_2) = 0.60$. From the interpolation formula (0.51) it follows

$$\begin{aligned} f_*(x) &= 0.52 + \frac{0.60 - 0.52}{1}(1.5 - 1) \\ &= 0.52 + 0.08 \cdot 0.5 = 0.56\,. \end{aligned}$$

**Second basic problem: interpolation of the entry $x$ for a known value $f(x)$:** To determine $x$ from $f(x)$, one uses the formula:

$$\boxed{x = x_1 + \frac{f(x) - f(x_1)}{f(x_2) - f(x_1)}(x_2 - x_1)\,.} \tag{0.52}$$

*Example 2:* Suppose that the value $f(x) = 0.62$ is given. The nearest table values in Table 0.32 are $f(x_1) = 0.60$ and $f(x_2) = 0.64$ with $x_1 = 2$ and $x_2 = 3$. Applying (0.52) we get

$$x = 2 + \frac{0.62 - 0.60}{0.64 - 0.60}(3 - 2) = 2 + \frac{0.02}{0.04} = 2.5\,.$$

**Higher precision with Mathematica:** Linear interpolation is a method of producing an approximation. For the needs of mathematical statistics this is quite sufficent. One should not be led to believe that more digits after the decimal point is an increase in accuracy in an endeavor like statistics which by its very nature is not precise.

In physics and technology, however, one often requires a higher precision. For this the method of quadratic interpolation is often applied. In the day of wide-spread computers one can use computer software programs to get very precise values for special functions (for example with Mathematica).

### 0.4.6.2 Normal distribution

*Table 0.33. The density function* $\varphi(z) = \frac{1}{\sqrt{2\pi}} \exp\left(-\frac{1}{2}z^2\right)$ *of the normalized, centered normal distribution.*

$\varphi(z)$

0 $z$

*Figure 0.47.*

| $z$ | 0 | 1 | 2 | 3 | 4 | 5 | 6 | 7 | 8 | 9 |
|---|---|---|---|---|---|---|---|---|---|---|
| 0.0 | $3989^{-4}$ | 3989 | 3989 | 3988 | 3986 | 3984 | 3982 | 3980 | 3977 | 3973 |
| 0.1 | $3970^{-4}$ | 3965 | 3961 | 3956 | 3951 | 3945 | 3939 | 3932 | 3925 | 3918 |
| 0.2 | $3910^{-4}$ | 3902 | 3894 | 3885 | 3876 | 3867 | 3857 | 3847 | 3836 | 3825 |
| 0.3 | $3814^{-4}$ | 3802 | 3790 | 3778 | 3765 | 3752 | 3739 | 3725 | 3712 | 3697 |
| 0.4 | $3683^{-4}$ | 3668 | 3653 | 3637 | 3621 | 3605 | 3589 | 3572 | 3555 | 3538 |
| 0.5 | $3521^{-4}$ | 3503 | 3485 | 3467 | 3448 | 3429 | 3410 | 3391 | 3372 | 3352 |
| 0.6 | $3332^{-4}$ | 3312 | 3292 | 3271 | 3251 | 3230 | 3209 | 3187 | 3166 | 3144 |
| 0.7 | $3123^{-4}$ | 3101 | 3079 | 3056 | 3034 | 3011 | 2989 | 2966 | 2943 | 2920 |
| 0.8 | $2897^{-4}$ | 2874 | 2850 | 2827 | 2803 | 2780 | 2756 | 2732 | 2709 | 2685 |
| 0.9 | $2661^{-4}$ | 2637 | 2613 | 2589 | 2565 | 2541 | 2516 | 2492 | 2468 | 2444 |
| 1.0 | $2420^{-4}$ | 2396 | 2371 | 2347 | 2323 | 2299 | 2275 | 2251 | 2227 | 2203 |
| 1.1 | $2179^{-4}$ | 2155 | 2131 | 2107 | 2083 | 2059 | 2036 | 2012 | 1989 | 1965 |
| 1.2 | $1942^{-4}$ | 1919 | 1895 | 1872 | 1849 | 1826 | 1804 | 1781 | 1758 | 1736 |
| 1.3 | $1714^{-4}$ | 1691 | 1669 | 1647 | 1626 | 1604 | 1582 | 1561 | 1539 | 1518 |
| 1.4 | $1497^{-4}$ | 1476 | 1456 | 1435 | 1415 | 1394 | 1374 | 1354 | 1334 | 1315 |
| 1.5 | $1295^{-4}$ | 1276 | 1257 | 1238 | 1219 | 1200 | 1182 | 1163 | 1145 | 1127 |
| 1.6 | $1109^{-4}$ | 1092 | 1074 | 1057 | 1040 | 1023 | 1006 | $9893^{-5}$ | 9728 | 9566 |
| 1.7 | $9405^{-5}$ | 9246 | 9089 | 8933 | 8780 | 8628 | 8478 | 8329 | 8183 | 8038 |
| 1.8 | $7895^{-5}$ | 7754 | 7614 | 7477 | 7341 | 7206 | 7074 | 6943 | 6814 | 6687 |
| 1.9 | $6562^{-5}$ | 6438 | 6316 | 6195 | 6077 | 5960 | 5844 | 5730 | 5618 | 5508 |
| 2.0 | $5399^{-5}$ | 5292 | 5186 | 5082 | 4980 | 4879 | 4780 | 4682 | 4586 | 4491 |
| 2.1 | $4398^{-5}$ | 4307 | 4217 | 4128 | 4041 | 3955 | 3871 | 3788 | 3706 | 3626 |
| 2.2 | $3547^{-5}$ | 3470 | 3394 | 3319 | 3246 | 3174 | 3103 | 3034 | 2965 | 2898 |
| 2.3 | $2833^{-5}$ | 2768 | 2705 | 2643 | 2582 | 2522 | 2463 | 2406 | 2349 | 2294 |
| 2.4 | $2239^{-5}$ | 2186 | 2134 | 2083 | 2033 | 1984 | 1936 | 1888 | 1842 | 1797 |
| 2.5 | $1753^{-5}$ | 1709 | 1667 | 1625 | 1585 | 1545 | 1506 | 1468 | 1431 | 1394 |
| 2.6 | $1358^{-5}$ | 1323 | 1289 | 1256 | 1223 | 1191 | 1160 | 1130 | 1100 | 1071 |
| 2.7 | $1042^{-5}$ | 1014 | $9871^{-6}$ | 9606 | 9347 | 9094 | 8846 | 8605 | 8370 | 8140 |
| 2.8 | $7915^{-6}$ | 7697 | 7483 | 7274 | 7071 | 6873 | 6679 | 6491 | 6307 | 6127 |
| 2.9 | $5953^{-6}$ | 5782 | 5616 | 5454 | 5296 | 5143 | 4993 | 4847 | 4705 | 4567 |
| 3.0 | $4432^{-6}$ | 4301 | 4173 | 4049 | 3928 | 3810 | 3695 | 3584 | 3475 | 3370 |
| 3.1 | $3267^{-6}$ | 3167 | 3070 | 2975 | 2884 | 2794 | 2707 | 2623 | 2541 | 2461 |
| 3.2 | $2384^{-6}$ | 2309 | 2236 | 2165 | 2096 | 2029 | 1964 | 1901 | 1840 | 1780 |
| 3.3 | $1723^{-6}$ | 1667 | 1612 | 1560 | 1508 | 1459 | 1411 | 1364 | 1319 | 1275 |
| 3.4 | $1232^{-6}$ | 1191 | 1151 | 1112 | 1075 | 1038 | 1003 | $9689^{-7}$ | 9358 | 9037 |
| 3.5 | $8727^{-7}$ | 8426 | 8135 | 7853 | 7581 | 7317 | 7061 | 6814 | 6575 | 6343 |
| 3.6 | $6119^{-7}$ | 5902 | 5693 | 5490 | 5294 | 5105 | 4921 | 4744 | 4573 | 4408 |
| 3.7 | $4248^{-7}$ | 4093 | 3944 | 3800 | 3661 | 3526 | 3396 | 3271 | 3149 | 3032 |
| 3.8 | $2919^{-7}$ | 2810 | 2705 | 2604 | 2506 | 2411 | 2320 | 2232 | 2147 | 2065 |
| 3.9 | $1987^{-7}$ | 1910 | 1837 | 1766 | 1698 | 1633 | 1569 | 1508 | 1449 | 1393 |
| 4.0 | $1338^{-7}$ | 1286 | 1235 | 1186 | 1140 | 1094 | 1051 | 1009 | $9687^{-8}$ | 9299 |
| 4.1 | $8926^{-8}$ | 8567 | 8222 | 7890 | 7570 | 7263 | 6967 | 6683 | 6410 | 6147 |
| 4.2 | $5894^{-8}$ | 5652 | 5418 | 5194 | 4979 | 4772 | 4573 | 4382 | 4199 | 4023 |
| 4.3 | $3854^{-8}$ | 3691 | 3535 | 3386 | 3242 | 3104 | 2972 | 2845 | 2723 | 2606 |
| 4.4 | $2494^{-8}$ | 2387 | 2284 | 2185 | 2090 | 1999 | 1912 | 1829 | 1749 | 1672 |
| 4.5 | $1598^{-8}$ | 1528 | 1461 | 1396 | 1334 | 1275 | 1218 | 1164 | 1112 | 1062 |
| 4.6 | $1014^{-8}$ | $9684^{-9}$ | 9248 | 8830 | 8430 | 8047 | 7681 | 7331 | 6996 | 6676 |
| 4.7 | $6370^{-9}$ | 6077 | 5797 | 5530 | 5274 | 5030 | 4796 | 4573 | 4360 | 4156 |
| 4.8 | $3961^{-9}$ | 3775 | 3598 | 3428 | 3267 | 3112 | 2965 | 2824 | 2960 | 2561 |
| 4.9 | $2439^{-9}$ | 2322 | 2211 | 2105 | 2003 | 1907 | 1814 | 1727 | 1643 | 1563 |

Remark: $3989^{-4}$ is to be understood to mean $3989 \cdot 10^{-4}$.

*Table 0.34. Probability integral* $\Phi_0(z) = \int\limits_0^z \varphi(x)\mathrm{d}x = \frac{1}{\sqrt{2\pi}} \int\limits_0^z \exp\left(-\frac{1}{2}x^2\right)\mathrm{d}x$ *of the normalized, centered normal distribution.*

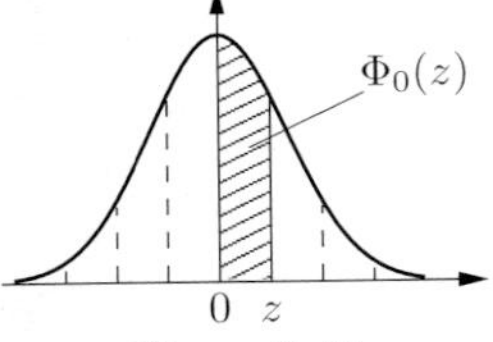

*Figure 0.48.*

The distribution function $\Phi(z) = \frac{1}{\sqrt{2\pi}} \int\limits_{-\infty}^z \exp\left(-\frac{1}{2}x^2\right)\mathrm{d}x$ is related to $\Phi_0(z)$ by the relation $\Phi(z) = \frac{1}{2} + \Phi_0(z)$; moreover, $\Phi_0(-z) = -\Phi_0(z)$.

| z | 0 | 1 | 2 | 3 | 4 | 5 | 6 | 7 | 8 | 9 |
|---|---|---|---|---|---|---|---|---|---|---|
| 0.0 | 0.0 000 | 040 | 080 | 120 | 160 | 199 | 239 | 279 | 319 | 359 |
| 0.1 | 398 | 438 | 478 | 517 | 557 | 596 | 636 | 675 | 714 | 753 |
| 0.2 | 793 | 832 | 871 | 910 | 948 | 987 | ·026 | ·064 | ·103 | ·141 |
| 0.3 | 0.1 179 | 217 | 255 | 293 | 331 | 368 | 406 | 443 | 480 | 517 |
| 0.4 | 554 | 591 | 628 | 664 | 700 | 736 | 772 | 808 | 844 | 879 |
| 0.5 | 915 | 950 | 985 | ·019 | ·054 | ·088 | ·123 | ·157 | ·190 | ·224 |
| 0.6 | 0.2 257 | 291 | 324 | 357 | 389 | 422 | 454 | 486 | 517 | 549 |
| 0.7 | 580 | 611 | 642 | 673 | 703 | 734 | 764 | 794 | 823 | 852 |
| 0.8 | 881 | 910 | 939 | 967 | 995 | ·023 | ·051 | ·078 | ·106 | ·133 |
| 0.9 | 0.3 159 | 186 | 212 | 238 | 264 | 289 | 315 | 340 | 365 | 389 |
| 1.0 | 413 | 438 | 461 | 485 | 508 | 531 | 554 | 577 | 599 | 621 |
| 1.1 | 643 | 665 | 686 | 708 | 729 | 749 | 770 | 790 | 810 | 830 |
| 1.2 | 849 | 869 | 888 | 907 | 925 | 944 | 962 | 980 | 997 | ·015 |
| 1.3 | 0.4 032 | 049 | 066 | 082 | 099 | 115 | 131 | 147 | 162 | 177 |
| 1.4 | 192 | 207 | 222 | 236 | 251 | 265 | 279 | 292 | 306 | 319 |
| 1.5 | 332 | 345 | 357 | 370 | 382 | 394 | 406 | 418 | 429 | 441 |
| 1.6 | 452 | 463 | 474 | 484 | 495 | 505 | 515 | 525 | 535 | 545 |
| 1.7 | 554 | 564 | 573 | 582 | 591 | 599 | 608 | 616 | 625 | 633 |
| 1.8 | 641 | 649 | 656 | 664 | 671 | 678 | 686 | 693 | 699 | 706 |
| 1.9 | 713 | 719 | 726 | 732 | 738 | 744 | 750 | 756 | 761 | 767 |
| 2.0 | 772 | 778 | 783 | 788 | 793 | 798 | 803 | 808 | 812 | 817 |
| 2.1 | 821 | 826 | 830 | 834 | 838 | 842 | 846 | 850 | 854 | 857 |
| 2.2 | 860<br>*966* | 864<br>*474* | 867<br>*906* | 871<br>*263* | 874<br>*545* | 877<br>*755* | 880<br>*894* | 883<br>*962* | 886<br>*962* | 889<br>*893* |
| 2.3 | 892<br>*759* | 895<br>*559* | 898<br>*296* | 900<br>*969* | 903<br>*581* | 906<br>*133* | 908<br>*625* | 911<br>*060* | 913<br>*437* | 915<br>*758* |
| 2.4 | 918<br>*025* | 920<br>*237* | 922<br>*397* | 924<br>*506* | 926<br>*564* | 928<br>*572* | 930<br>*531* | 932<br>*443* | 934<br>*309* | 936<br>*128* |
| 2.5 | 937<br>*903* | 939<br>*634* | 941<br>*323* | 942<br>*969* | 944<br>*574* | 946<br>*139* | 947<br>*664* | 949<br>*151* | 950<br>*600* | 952<br>*012* |
| 2.6 | 953<br>*388* | 954<br>*729* | 956<br>*035* | 957<br>*308* | 958<br>*547* | 959<br>*754* | 960<br>*930* | 962<br>*074* | 963<br>*189* | 964<br>*274* |
| 2.7 | 965<br>*330* | 966<br>*358* | 967<br>*359* | 968<br>*333* | 969<br>*280* | 970<br>*202* | 971<br>*099* | 971<br>*972* | 972<br>*821* | 973<br>*646* |
| 2.8 | 974<br>*449* | 975<br>*229* | 975<br>*988* | 976<br>*726* | 977<br>*443* | 978<br>*140* | 978<br>*818* | 979<br>*476* | 980<br>*116* | 980<br>*738* |
| 2.9 | 981<br>*342* | 981<br>*929* | 982<br>*498* | 983<br>*052* | 983<br>*589* | 984<br>*111* | 984<br>*618* | 985<br>*110* | 985<br>*588* | 986<br>*051* |

Remarks: 0.4 860 is to be interpreted here to mean 0.4 860 966.
*966*

A dot in front of an entry indicates a jump of one in decimal place.
For example in the line $z = 0.5$, the entry ·019 means .2 019.

*Table 0.34. (continued)*

| $z$ | 0 | 1 | 2 | 3 | 4 | 5 | 6 | 7 | 8 | 9 |
|---|---|---|---|---|---|---|---|---|---|---|
| 3.0 | 0.4 986<br>501 | 986<br>938 | 987<br>361 | 987<br>772 | 988<br>171 | 988<br>558 | 988<br>933 | 989<br>297 | 989<br>650 | 989<br>992 |
| 3.1 | 990<br>324 | 990<br>646 | 990<br>957 | 991<br>260 | 991<br>553 | 991<br>836 | 992<br>112 | 992<br>378 | 992<br>636 | 992<br>886 |
| 3.2 | 993<br>129 | 993<br>363 | 993<br>590 | 993<br>810 | 994<br>024 | 994<br>230 | 994<br>429 | 994<br>623 | 994<br>810 | 994<br>991 |
| 3.3 | 995<br>166 | 995<br>335 | 995<br>499 | 995<br>658 | 995<br>811 | 995<br>959 | 996<br>103 | 996<br>242 | 996<br>376 | 996<br>505 |
| 3.4 | 996<br>631 | 996<br>752 | 996<br>869 | 996<br>982 | 997<br>091 | 997<br>197 | 997<br>299 | 997<br>398 | 997<br>493 | 997<br>585 |
| 3.5 | 997<br>674 | 997<br>759 | 997<br>842 | 997<br>922 | 997<br>999 | 998<br>074 | 998<br>146 | 998<br>215 | 998<br>282 | 998<br>347 |
| 3.6 | 998<br>409 | 998<br>469 | 998<br>527 | 998<br>583 | 998<br>637 | 998<br>689 | 998<br>739 | 998<br>787 | 998<br>834 | 998<br>879 |
| 3.7 | 998<br>922 | 998<br>964 | 999<br>004 | 999<br>043 | 999<br>080 | 999<br>116 | 999<br>150 | 999<br>184 | 999<br>216 | 999<br>247 |
| 3.8 | 999<br>276 | 999<br>305 | 999<br>333 | 999<br>359 | 999<br>385 | 999<br>409 | 999<br>433 | 999<br>456 | 999<br>478 | 999<br>499 |
| 3.9 | 999<br>519 | 999<br>539 | 999<br>557 | 999<br>575 | 999<br>593 | 999<br>609 | 999<br>625 | 999<br>641 | 999<br>655 | 999<br>670 |
| 4.0 | 999<br>683 | 999<br>696 | 999<br>709 | 999<br>721 | 999<br>733 | 999<br>744 | 999<br>755 | 999<br>765 | 999<br>775 | 999<br>784 |
| 4.1 | 999<br>793 | 999<br>802 | 999<br>811 | 999<br>819 | 999<br>826 | 999<br>834 | 999<br>841 | 999<br>848 | 999<br>854 | 999<br>861 |
| 4.2 | 999<br>867 | 999<br>872 | 999<br>878 | 999<br>883 | 999<br>888 | 999<br>893 | 999<br>898 | 999<br>902 | 999<br>907 | 999<br>911 |
| 4.3 | 999<br>915 | 999<br>918 | 999<br>922 | 999<br>925 | 999<br>929 | 999<br>932 | 999<br>935 | 999<br>938 | 999<br>941 | 999<br>943 |
| 4.4 | 999<br>946 | 999<br>948 | 999<br>951 | 999<br>953 | 999<br>955 | 999<br>957 | 999<br>959 | 999<br>961 | 999<br>963 | 999<br>964 |
| 4.5 | 999<br>966 | 999<br>968 | 999<br>969 | 999<br>971 | 999<br>972 | 999<br>973 | 999<br>974 | 999<br>976 | 999<br>977 | 999<br>978 |
| 5.0 | 999<br>997 | | | | | | | | | |

### 0.4.6.3 Values $t_{\alpha, m}$ of the Student $t$-distribution

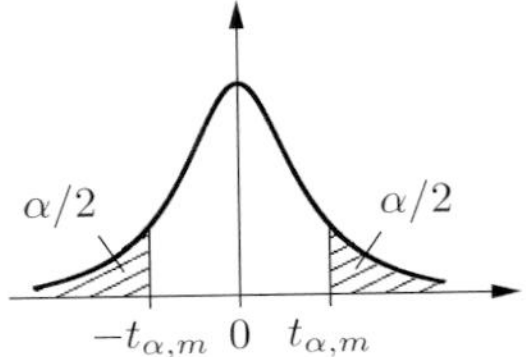


*Figure 0.49.*

| $m$ \ $\alpha$ | 0.10 | 0.05 | 0.025 | 0.020 | 0.010 | 0.005 | 0.003 | 0.002 | 0.001 |
|---|---|---|---|---|---|---|---|---|---|
| 1 | 6.314 | 12.706 | 25.452 | 31.821 | 63.657 | 127.3 | 212.2 | 318.3 | 636.6 |
| 2 | 2.920 | 4.303 | 6.205 | 6.965 | 9.925 | 14.089 | 18.216 | 22.327 | 31.600 |
| 3 | 2.353 | 3.182 | 4.177 | 4.541 | 5.841 | 7.453 | 8.891 | 10.214 | 12.922 |
| 4 | 2.132 | 2.776 | 3.495 | 3.747 | 4.604 | 5.597 | 6.435 | 7.173 | 8.610 |
| 5 | 2.015 | 2.571 | 3.163 | 3.365 | 4.032 | 4.773 | 5.376 | 5.893 | 6.869 |
| 6 | 1.943 | 2.447 | 2.969 | 3.143 | 3.707 | 4.317 | 4.800 | 5.208 | 5.959 |
| 7 | 1.895 | 2.365 | 2.841 | 2.998 | 3.499 | 4.029 | 4.442 | 4.785 | 5.408 |
| 8 | 1.860 | 2.306 | 2.752 | 2.896 | 3.355 | 3.833 | 4.199 | 4.501 | 5.041 |
| 9 | 1.833 | 2.262 | 2.685 | 2.821 | 3.250 | 3.690 | 4.024 | 4.297 | 4.781 |
| 10 | 1.812 | 2.228 | 2.634 | 2.764 | 3.169 | 3.581 | 3.892 | 4.144 | 4.587 |
| 12 | 1.782 | 2.179 | 2.560 | 2.681 | 3.055 | 3.428 | 3.706 | 3.930 | 4.318 |
| 14 | 1.761 | 2.145 | 2.510 | 2.624 | 2.977 | 3.326 | 3.583 | 3.787 | 4.140 |
| 16 | 1.746 | 2.120 | 2.473 | 2.583 | 2.921 | 3.252 | 3.494 | 3.686 | 4.015 |
| 18 | 1.734 | 2.101 | 2.445 | 2.552 | 2.878 | 3.193 | 3.428 | 3.610 | 3.922 |
| 20 | 1.725 | 2.086 | 2.423 | 2.528 | 2.845 | 3.153 | 3.376 | 3.552 | 3.849 |
| 22 | 1.717 | 2.074 | 2.405 | 2.508 | 2.819 | 3.119 | 3.335 | 3.505 | 3.792 |
| 24 | 1.711 | 2.064 | 2.391 | 2.492 | 2.797 | 3.092 | 3.302 | 3.467 | 3.745 |
| 26 | 1.706 | 2.056 | 2.379 | 2.479 | 2.779 | 3.067 | 3.274 | 3.435 | 3.704 |
| 28 | 1.701 | 2.048 | 2.369 | 2.467 | 2.763 | 3.047 | 3.250 | 3.408 | 3.674 |
| 30 | 1.697 | 2.042 | 2.360 | 2.457 | 2.750 | 3.030 | 3.230 | 3.386 | 3.646 |
| $\infty$ | 1.645 | 1.960 | 2.241 | 2.326 | 2.576 | 2.807 | 2.968 | 3.090 | 3.291 |

### 0.4.6.4 Values $\chi^2_\alpha$ of the $\chi^2$-distribution

Figure 0.50.

| Number $m$ of degrees of freedom | probability $\alpha$ | | | | | | | | | | | | | | | |
|---|---|---|---|---|---|---|---|---|---|---|---|---|---|---|---|---|
| | 0.99 | 0.98 | 0.95 | 0.90 | 0.80 | 0.70 | 0.50 | 0.30 | 0.20 | 0.10 | 0.05 | 0.02 | 0.01 | 0.005 | 0.002 | 0.001 |
| 1 | 0.000 16 | 0.000 6 | 0.003 9 | 0.016 | 0.064 | 0.148 | 0.455 | 1.07 | 1.64 | 2.7 | 3.8 | 5.4 | 6.6 | 7.9 | 9.5 | 10.83 |
| 2 | 0.020 | 0.040 | 0.103 | 0.211 | 0.446 | 0.713 | 1.386 | 2.41 | 3.22 | 4.6 | 6.0 | 7.8 | 9.2 | 10.6 | 12.4 | 13.8 |
| 3 | 0.115 | 0.185 | 0.352 | 0.584 | 1.005 | 1.424 | 2.366 | 3.67 | 4.64 | 6.3 | 7.8 | 9.8 | 11.3 | 12.8 | 14.8 | 16.3 |
| 4 | 0.30 | 0.43 | 0.71 | 1.06 | 1.65 | 2.19 | 3.36 | 4.9 | 6.0 | 7.8 | 9.5 | 11.7 | 13.3 | 14.9 | 16.9 | 18.5 |
| 5 | 0.55 | 0.75 | 1.14 | 1.61 | 2.34 | 3.00 | 4.35 | 6.1 | 7.3 | 9.2 | 11.1 | 13.4 | 15.1 | 16.8 | 18.9 | 20.5 |
| 6 | 0.87 | 1.13 | 1.63 | 2.20 | 3.07 | 3.83 | 5.35 | 7.2 | 8.6 | 10.6 | 12.6 | 15.0 | 16.8 | 18.5 | 20.7 | 22.5 |
| 7 | 1.24 | 1.56 | 2.17 | 2.83 | 3.82 | 4.67 | 6.35 | 8.4 | 9.8 | 12.0 | 14.1 | 16.6 | 18.5 | 20.3 | 22.6 | 24.3 |
| 8 | 1.65 | 2.03 | 2.73 | 3.49 | 4.59 | 5.53 | 7.34 | 9.5 | 11.0 | 13.4 | 15.5 | 18.2 | 20.1 | 22.0 | 24.3 | 26.1 |
| 9 | 2.09 | 2.53 | 3.32 | 4.17 | 5.38 | 6.39 | 8.34 | 10.7 | 12.2 | 14.7 | 16.9 | 19.7 | 21.7 | 23.6 | 26.1 | 27.9 |
| 10 | 2.56 | 3.06 | 3.94 | 4.86 | 6.18 | 7.27 | 9.34 | 11.8 | 13.4 | 16.0 | 18.3 | 21.2 | 23.2 | 25.2 | 27.7 | 29.6 |
| 11 | 3.1 | 3.6 | 4.6 | 5.6 | 7.0 | 8.1 | 10.3 | 12.9 | 14.6 | 17.3 | 19.7 | 22.6 | 24.7 | 26.8 | 29.4 | 31.3 |
| 12 | 3.6 | 4.2 | 5.2 | 6.3 | 7.8 | 9.0 | 11.3 | 14.0 | 15.8 | 18.5 | 21.0 | 24.1 | 26.2 | 28.3 | 30.9 | 32.9 |
| 13 | 4.1 | 4.8 | 5.9 | 7.0 | 8.6 | 9.9 | 12.3 | 15.1 | 17.0 | 19.8 | 22.4 | 25.5 | 27.7 | 29.8 | 32.5 | 34.5 |
| 14 | 4.7 | 5.4 | 6.6 | 7.8 | 9.5 | 10.8 | 13.3 | 16.2 | 18.2 | 21.1 | 23.7 | 26.9 | 29.1 | 31.3 | 34.0 | 36.1 |
| 15 | 5.2 | 6.0 | 7.3 | 8.5 | 10.3 | 11.7 | 14.3 | 17.3 | 19.3 | 22.3 | 25.0 | 28.3 | 30.6 | 32.8 | 35.6 | 37.7 |
| 16 | 5.8 | 6.6 | 8.0 | 9.3 | 11.2 | 12.6 | 15.3 | 18.4 | 20.5 | 23.5 | 26.3 | 29.6 | 32.0 | 34.3 | 37.1 | 39.3 |
| 17 | 6.4 | 7.3 | 8.7 | 10.1 | 12.0 | 13.5 | 16.3 | 19.5 | 21.6 | 24.8 | 27.6 | 31.0 | 33.4 | 35.7 | 38.6 | 40.8 |
| 18 | 7.0 | 7.9 | 9.4 | 10.9 | 12.9 | 14.4 | 17.3 | 20.6 | 22.8 | 26.0 | 28.9 | 32.3 | 34.8 | 37.2 | 40.1 | 42.3 |
| 19 | 7.6 | 8.6 | 10.1 | 11.7 | 13.7 | 15.4 | 18.3 | 21.7 | 23.9 | 27.2 | 30.1 | 33.7 | 36.2 | 38.6 | 41.6 | 43.8 |
| 20 | 8.3 | 9.2 | 10.9 | 12.4 | 14.6 | 16.3 | 19.3 | 22.8 | 25.0 | 28.4 | 31.4 | 35.0 | 37.6 | 40.0 | 43.0 | 45.3 |
| 21 | 8.9 | 9.9 | 11.6 | 13.2 | 15.4 | 17.2 | 20.3 | 23.9 | 26.2 | 29.6 | 32.7 | 36.3 | 38.9 | 41.4 | 44.5 | 46.8 |
| 22 | 9.5 | 10.6 | 12.3 | 14.0 | 16.3 | 18.1 | 21.3 | 24.9 | 27.3 | 30.8 | 33.9 | 37.7 | 40.3 | 42.8 | 45.9 | 48.3 |
| 23 | 10.2 | 11.3 | 13.1 | 14.8 | 17.2 | 19.0 | 22.3 | 26.0 | 28.4 | 32.0 | 35.2 | 39.0 | 41.6 | 44.2 | 47.3 | 49.7 |
| 24 | 10.9 | 12.0 | 13.8 | 15.7 | 18.1 | 19.9 | 23.3 | 27.1 | 29.6 | 33.2 | 36.4 | 40.3 | 43.0 | 45.6 | 48.7 | 51.2 |
| 25 | 11.5 | 12.7 | 14.6 | 16.5 | 18.9 | 20.9 | 24.3 | 28.2 | 30.7 | 34.4 | 37.7 | 41.6 | 44.3 | 46.9 | 50.1 | 52.6 |
| 26 | 12.2 | 13.4 | 15.4 | 17.3 | 19.8 | 21.8 | 25.3 | 29.2 | 31.8 | 35.6 | 38.9 | 42.9 | 45.6 | 48.3 | 51.6 | 54.1 |
| 27 | 12.9 | 14.1 | 16.2 | 18.1 | 20.7 | 22.7 | 26.3 | 30.3 | 32.9 | 36.7 | 40.1 | 44.1 | 47.0 | 49.6 | 52.9 | 55.5 |
| 28 | 13.6 | 14.8 | 16.9 | 18.9 | 21.6 | 23.6 | 27.3 | 31.4 | 34.0 | 37.9 | 41.3 | 45.4 | 48.3 | 51.0 | 54.4 | 56.9 |
| 29 | 14.3 | 15.6 | 17.7 | 19.8 | 22.5 | 24.6 | 28.3 | 32.5 | 35.1 | 39.1 | 42.6 | 46.7 | 49.6 | 52.3 | 55.7 | 58.3 |
| 30 | 15.0 | 16.3 | 18.5 | 20.6 | 23.4 | 25.5 | 29.3 | 33.5 | 36.3 | 40.3 | 43.8 | 48.0 | 50.9 | 53.7 | 57.1 | 59.7 |

### 0.4.6.5 Values $F_{0.05;m_1m_2}$ and values $F_{0.01;m_1m_2}$ (in boldface) of the F-distribution

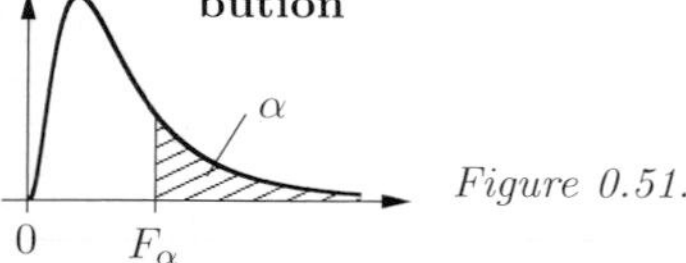


Figure 0.51.

| $m_2$ \ $m_1$ | 1 | 2 | 3 | 4 | 5 | 6 | 7 | 8 | 9 | 10 | 11 | 12 |
|---|---|---|---|---|---|---|---|---|---|---|---|---|
| 1 | 161<br>**4 052** | 200<br>**4 999** | 216<br>**5 403** | 225<br>**5 625** | 230<br>**5 764** | 234<br>**5 859** | 237<br>**5 928** | 239<br>**5 981** | 241<br>**6 022** | 242<br>**6 056** | 243<br>**6 083** | 244<br>**6 106** |
| 2 | 18.51<br>**98.50** | 19.00<br>**99.00** | 19.16<br>**99.17** | 19.25<br>**99.25** | 19.30<br>**99.30** | 19.33<br>**99.33** | 19.35<br>**99.36** | 19.37<br>**99.37** | 19.38<br>**99.39** | 19.39<br>**99.40** | 19.40<br>**99.41** | 19.41<br>**99.42** |
| 3 | 10.13<br>**34.12** | 9.55<br>**30.82** | 9.28<br>**29.46** | 9.12<br>**28.71** | 9.01<br>**28.24** | 8.94<br>**27.91** | 8.89<br>**27.67** | 8.85<br>**27.49** | 8.81<br>**27.34** | 8.79<br>**27.23** | 8.76<br>**27.13** | 8.74<br>**27.05** |
| 4 | 7.71<br>**21.20** | 6.94<br>**18.00** | 6.59<br>**16.69** | 6.39<br>**15.98** | 6.26<br>**15.52** | 6.16<br>**15.21** | 6.09<br>**14.98** | 6.04<br>**14.80** | 6.00<br>**14.66** | 5.96<br>**14.55** | 5.94<br>**14.45** | 5.91<br>**14.37** |
| 5 | 6.61<br>**16.26** | 5.79<br>**13.27** | 5.41<br>**12.06** | 5.19<br>**11.39** | 5.05<br>**10.97** | 4.95<br>**10.67** | 4.88<br>**10.46** | 4.82<br>**10.29** | 4.77<br>**10.16** | 4.74<br>**10.05** | 4.70<br>**9.96** | 4.68<br>**9.89** |
| 6 | 5.99<br>**13.74** | 5.14<br>**10.92** | 4.76<br>**9.78** | 4.53<br>**9.15** | 4.39<br>**8.75** | 4.28<br>**8.47** | 4.21<br>**8.26** | 4.15<br>**8.10** | 4.10<br>**7.98** | 4.06<br>**7.87** | 4.03<br>**7.79** | 4.00<br>**7.72** |
| 7 | 5.59<br>**12.25** | 4.74<br>**9.55** | 4.35<br>**8.45** | 4.12<br>**7.85** | 3.97<br>**7.46** | 3.87<br>**7.19** | 3.79<br>**7.00** | 3.73<br>**6.84** | 3.68<br>**6.72** | 3.64<br>**6.62** | 3.60<br>**6.54** | 3.57<br>**6.47** |
| 8 | 5.32<br>**11.26** | 4.46<br>**8.65** | 4.07<br>**7.59** | 3.84<br>**7.01** | 3.69<br>**6.63** | 3.58<br>**6.37** | 3.50<br>**6.18** | 3.44<br>**6.03** | 3.39<br>**5.91** | 3.35<br>**5.81** | 3.31<br>**5.73** | 3.28<br>**5.67** |
| 9 | 5.12<br>**10.56** | 4.26<br>**8.02** | 3.86<br>**6.99** | 3.63<br>**6.42** | 3.48<br>**6.06** | 3.37<br>**5.80** | 3.29<br>**5.61** | 3.23<br>**5.47** | 3.18<br>**5.35** | 3.14<br>**5.26** | 3.10<br>**5.18** | 3.07<br>**5.11** |
| 10 | 4.96<br>**10.04** | 4.10<br>**7.56** | 3.71<br>**6.55** | 3.48<br>**5.99** | 3.33<br>**5.64** | 3.22<br>**5.39** | 3.14<br>**5.20** | 3.07<br>**5.06** | 3.02<br>**4.94** | 2.98<br>**4.85** | 2.94<br>**4.77** | 2.91<br>**4.71** |
| 11 | 4.84<br>**9.65** | 3.98<br>**7.21** | 3.59<br>**6.22** | 3.36<br>**5.67** | 3.20<br>**5.32** | 3.09<br>**5.07** | 3.01<br>**4.89** | 2.95<br>**4.74** | 2.90<br>**4.63** | 2.85<br>**4.54** | 2.82<br>**4.46** | 2.79<br>**4.40** |
| 12 | 4.75<br>**9.33** | 3.89<br>**6.93** | 3.49<br>**5.95** | 3.26<br>**5.41** | 3.11<br>**5.06** | 3.00<br>**4.82** | 2.91<br>**4.64** | 2.85<br>**4.50** | 2.80<br>**4.39** | 2.75<br>**4.30** | 2.72<br>**4.22** | 2.69<br>**4.16** |
| 13 | 4.67<br>**9.07** | 3.81<br>**6.70** | 3.41<br>**5.74** | 3.18<br>**5.21** | 3.03<br>**4.86** | 2.92<br>**4.62** | 2.83<br>**4.44** | 2.77<br>**4.30** | 2.71<br>**4.19** | 2.67<br>**4.10** | 2.63<br>**4.02** | 2.60<br>**3.96** |
| 14 | 4.60<br>**8.86** | 3.74<br>**6.51** | 3.34<br>**5.56** | 3.11<br>**5.04** | 2.96<br>**4.70** | 2.85<br>**4.46** | 2.76<br>**4.28** | 2.70<br>**4.14** | 2.65<br>**4.03** | 2.60<br>**3.94** | 2.57<br>**3.86** | 2.53<br>**3.80** |
| 15 | 4.54<br>**8.68** | 3.68<br>**6.36** | 3.29<br>**5.42** | 3.06<br>**4.89** | 2.90<br>**4.56** | 2.79<br>**4.32** | 2.71<br>**4.14** | 2.64<br>**4.00** | 2.59<br>**3.89** | 2.54<br>**3.80** | 2.51<br>**3.73** | 2.48<br>**3.67** |
| 16 | 4.49<br>**8.53** | 3.63<br>**6.23** | 3.24<br>**5.29** | 3.01<br>**4.77** | 2.85<br>**4.44** | 2.74<br>**4.20** | 2.66<br>**4.03** | 2.59<br>**3.89** | 2.54<br>**3.78** | 2.49<br>**3.69** | 2.46<br>**3.62** | 2.42<br>**3.55** |
| 17 | 4.45<br>**8.40** | 3.59<br>**6.11** | 3.20<br>**5.18** | 2.96<br>**4.67** | 2.81<br>**4.34** | 2.70<br>**4.10** | 2.61<br>**3.93** | 2.55<br>**3.79** | 2.49<br>**3.68** | 2.45<br>**3.59** | 2.41<br>**3.52** | 2.38<br>**3.46** |
| 18 | 4.41<br>**8.29** | 3.55<br>**6.01** | 3.16<br>**5.09** | 2.93<br>**4.58** | 2.77<br>**4.25** | 2.66<br>**4.01** | 2.58<br>**3.84** | 2.51<br>**3.71** | 2.46<br>**3.60** | 2.41<br>**3.51** | 2.37<br>**3.43** | 2.34<br>**3.37** |
| 19 | 4.38<br>**8.18** | 3.52<br>**5.93** | 3.13<br>**5.01** | 2.90<br>**4.50** | 2.74<br>**4.17** | 2.63<br>**3.94** | 2.54<br>**3.77** | 2.48<br>**3.63** | 2.42<br>**3.52** | 2.38<br>**3.43** | 2.34<br>**3.36** | 2.31<br>**3.30** |
| 20 | 4.35<br>**8.10** | 3.49<br>**5.85** | 3.10<br>**4.94** | 2.87<br>**4.43** | 2.71<br>**4.10** | 2.60<br>**3.87** | 2.51<br>**3.70** | 2.45<br>**3.56** | 2.39<br>**3.46** | 2.35<br>**3.37** | 2.31<br>**3.29** | 2.28<br>**3.23** |
| 21 | 4.32<br>**8.02** | 3.47<br>**5.78** | 3.07<br>**4.87** | 2.84<br>**4.37** | 2.68<br>**4.04** | 2.57<br>**3.81** | 2.49<br>**3.64** | 2.42<br>**3.51** | 2.37<br>**3.40** | 2.32<br>**3.31** | 2.28<br>**3.24** | 2.25<br>**3.17** |
| 22 | 4.30<br>**7.95** | 3.44<br>**5.72** | 3.05<br>**4.82** | 2.82<br>**4.31** | 2.66<br>**3.99** | 2.55<br>**3.76** | 2.46<br>**3.59** | 2.40<br>**3.45** | 2.34<br>**3.35** | 2.30<br>**3.26** | 2.26<br>**3.18** | 2.23<br>**3.12** |
| 23 | 4.28<br>**7.88** | 3.42<br>**5.66** | 3.03<br>**4.76** | 2.80<br>**4.26** | 2.64<br>**3.94** | 2.53<br>**3.71** | 2.44<br>**3.54** | 2.37<br>**3.41** | 2.32<br>**3.30** | 2.27<br>**3.21** | 2.24<br>**3.14** | 2.20<br>**3.07** |

| $m_1$ | | | | | | | | | | | | |
|---|---|---|---|---|---|---|---|---|---|---|---|---|
| 14 | 16 | 20 | 24 | 30 | 40 | 50 | 75 | 100 | 200 | 500 | $\infty$ | $m_2$ |
| 245<br>**6 143** | 246<br>**6 169** | 248<br>**6 209** | 249<br>**6 235** | 250<br>**6 261** | 251<br>**6 287** | 252<br>**6 302** | 253<br>**6 323** | 253<br>**6 334** | 254<br>**6 352** | 254<br>**6 361** | 254<br>**6 366** | 1 |
| 19.42<br>**99.43** | 19.43<br>**99.44** | 19.44<br>**99.45** | 19.45<br>**99.46** | 19.46<br>**99.47** | 19.47<br>**99.47** | 19.48<br>**99.48** | 19.48<br>**99.49** | 19.49<br>**99.49** | 19.49<br>**99.49** | 19.50<br>**99.50** | 19.50<br>**99.50** | 2 |
| 8.71<br>**26.92** | 8.69<br>**26.83** | 8.66<br>**26.69** | 8.64<br>**26.60** | 8.62<br>**26.50** | 8.59<br>**26.41** | 8.58<br>**26.35** | 8.57<br>**26.27** | 8.55<br>**26.23** | 8.54<br>**26.18** | 8.53<br>**26.14** | 8.53<br>**26.12** | 3 |
| 5.87<br>**14.25** | 5.84<br>**14.15** | 5.80<br>**14.02** | 5.77<br>**13.93** | 5.75<br>**13.84** | 5.72<br>**13.74** | 5.70<br>**13.69** | 5.68<br>**13.61** | 5.66<br>**13.57** | 5.65<br>**13.52** | 5.64<br>**13.48** | 5.63<br>**13.46** | 4 |
| 4.64<br>**9.77** | 4.60<br>**9.68** | 4.56<br>**9.55** | 4.53<br>**9.47** | 4.50<br>**9.38** | 4.46<br>**9.29** | 4.44<br>**9.24** | 4.42<br>**9.17** | 4.41<br>**9.13** | 4.39<br>**9.08** | 4.37<br>**9.04** | 4.36<br>**9.02** | 5 |
| 3.96<br>**7.60** | 3.92<br>**7.52** | 3.87<br>**7.39** | 3.84<br>**7.31** | 3.81<br>**7.23** | 3.77<br>**7.14** | 3.75<br>**7.09** | 3.72<br>**7.02** | 3.71<br>**6.99** | 3.69<br>**6.93** | 3.68<br>**6.90** | 3.67<br>**6.88** | 6 |
| 3.53<br>**6.36** | 3.49<br>**6.27** | 3.44<br>**6.16** | 3.41<br>**6.07** | 3.38<br>**5.99** | 3.34<br>**5.91** | 3.32<br>**5.86** | 3.29<br>**5.78** | 3.27<br>**5.75** | 3.25<br>**5.70** | 3.24<br>**5.67** | 3.23<br>**5.65** | 7 |
| 3.24<br>**5.56** | 3.20<br>**5.48** | 3.15<br>**5.36** | 3.12<br>**5.28** | 3.08<br>**5.20** | 3.05<br>**5.12** | 3.02<br>**5.07** | 3.00<br>**5.00** | 2.97<br>**4.96** | 2.95<br>**4.91** | 2.94<br>**4.88** | 2.93<br>**4.86** | 8 |
| 3.03<br>**5.00** | 2.99<br>**4.92** | 2.93<br>**4.81** | 2.90<br>**4.73** | 2.86<br>**4.65** | 2.83<br>**4.57** | 2.80<br>**4.52** | 2.77<br>**4.45** | 2.76<br>**4.42** | 2.73<br>**4.36** | 2.72<br>**4.33** | 2.71<br>**4.31** | 9 |
| 2.86<br>**4.60** | 2.83<br>**4.52** | 2.77<br>**4.41** | 2.74<br>**4.33** | 2.70<br>**4.25** | 2.66<br>**4.17** | 2.64<br>**4.12** | 2.61<br>**4.05** | 2.59<br>**4.01** | 2.56<br>**3.96** | 2.55<br>**3.93** | 2.54<br>**3.91** | 10 |
| 2.74<br>**4.29** | 2.70<br>**4.21** | 2.65<br>**4.10** | 2.61<br>**4.02** | 2.57<br>**3.94** | 2.53<br>**3.86** | 2.51<br>**3.81** | 2.47<br>**3.74** | 2.46<br>**3.71** | 2.43<br>**3.66** | 2.42<br>**3.62** | 2.40<br>**3.60** | 11 |
| 2.64<br>**4.05** | 2.60<br>**3.97** | 2.54<br>**3.86** | 2.51<br>**3.78** | 2.47<br>**3.70** | 2.43<br>**3.62** | 2.40<br>**3.57** | 2.36<br>**3.49** | 2.35<br>**3.47** | 2.32<br>**3.41** | 2.31<br>**3.38** | 2.30<br>**3.36** | 12 |
| 2.55<br>**3.86** | 2.51<br>**3.78** | 2.46<br>**3.66** | 2.42<br>**3.59** | 2.38<br>**3.51** | 2.34<br>**3.43** | 2.31<br>**3.38** | 2.28<br>**3.30** | 2.26<br>**3.27** | 2.23<br>**3.22** | 2.22<br>**3.19** | 2.21<br>**3.17** | 13 |
| 2.48<br>**3.70** | 2.44<br>**3.62** | 2.39<br>**3.51** | 2.35<br>**3.43** | 2.31<br>**3.35** | 2.27<br>**3.27** | 2.24<br>**3.22** | 2.21<br>**3.14** | 2.19<br>**3.11** | 2.16<br>**3.06** | 2.14<br>**3.03** | 2.13<br>**3.00** | 14 |
| 2.42<br>**3.56** | 2.38<br>**3.49** | 2.33<br>**3.37** | 2.29<br>**3.29** | 2.25<br>**3.21** | 2.20<br>**3.13** | 2.18<br>**3.08** | 2.15<br>**3.00** | 2.12<br>**2.98** | 2.10<br>**2.92** | 2.08<br>**2.89** | 2.07<br>**2.87** | 15 |
| 2.37<br>**3.45** | 2.33<br>**3.37** | 2.28<br>**3.26** | 2.24<br>**3.18** | 2.19<br>**3.10** | 2.15<br>**3.02** | 2.12<br>**2.97** | 2.09<br>**2.86** | 2.07<br>**2.86** | 2.04<br>**2.81** | 2.02<br>**2.78** | 2.01<br>**2.75** | 16 |
| 2.33<br>**3.35** | 2.29<br>**3.27** | 2.23<br>**2.16** | 2.19<br>**3.08** | 2.15<br>**3.00** | 2.10<br>**2.92** | 2.08<br>**2.87** | 2.04<br>**2.79** | 2.02<br>**2.76** | 1.99<br>**2.71** | 1.97<br>**2.68** | 1.96<br>**2.65** | 17 |
| 2.29<br>**3.27** | 2.25<br>**3.19** | 2.19<br>**3.08** | 2.15<br>**3.00** | 2.11<br>**2.92** | 2.06<br>**2.84** | 2.04<br>**2.78** | 2.00<br>**2.71** | 1.98<br>**2.68** | 1.95<br>**2.62** | 1.93<br>**2.59** | 1.92<br>**2.57** | 18 |
| 2.26<br>**3.19** | 2.21<br>**3.12** | 2.15<br>**3.00** | 2.11<br>**2.92** | 2.07<br>**2.84** | 2.03<br>**2.76** | 2.00<br>**2.71** | 1.96<br>**2.63** | 1.94<br>**2.60** | 1.91<br>**2.55** | 1.90<br>**2.51** | 1.88<br>**2.49** | 19 |
| 2.22<br>**3.13** | 2.18<br>**3.05** | 2.12<br>**2.94** | 2.08<br>**2.86** | 2.04<br>**2.78** | 1.99<br>**2.69** | 1.97<br>**2.64** | 1.92<br>**2.56** | 1.91<br>**2.54** | 1.88<br>**2.48** | 1.86<br>**2.44** | 1.84<br>**2.42** | 20 |
| 2.20<br>**3.07** | 2.16<br>**2.99** | 2.10<br>**2.88** | 2.05<br>**2.80** | 2.01<br>**2.72** | 1.96<br>**2.64** | 1.94<br>**2.58** | 1.89<br>**2.51** | 1.88<br>**2.48** | 1.84<br>**2.42** | 1.82<br>**2.38** | 1.81<br>**2.36** | 21 |
| 2.17<br>**3.02** | 2.13<br>**2.94** | 2.07<br>**2.83** | 2.03<br>**2.75** | 1.98<br>**2.67** | 1.94<br>**2.58** | 1.91<br>**2.53** | 1.87<br>**2.46** | 1.85<br>**2.42** | 1.81<br>**2.36** | 1.80<br>**2.33** | 1.78<br>**2.31** | 22 |
| 2.15<br>**2.97** | 2.11<br>**2.89** | 2.05<br>**2.78** | 2.00<br>**2.70** | 1.96<br>**2.62** | 1.91<br>**2.54** | 1.88<br>**2.48** | 1.84<br>**2.41** | 1.82<br>**2.37** | 1.79<br>**2.32** | 1.77<br>**2.28** | 1.76<br>**2.26** | 23 |

| $m_2$ | $m_1$ | | | | | | | | | | | |
|---|---|---|---|---|---|---|---|---|---|---|---|---|
| | 1 | 2 | 3 | 4 | 5 | 6 | 7 | 8 | 9 | 10 | 11 | 12 |
| 24 | 4.26<br>**7.82** | 3.40<br>**5.61** | 3.01<br>**4.72** | 2.78<br>**4.22** | 2.62<br>**3.90** | 2.51<br>**3.67** | 2.42<br>**3.50** | 2.36<br>**3.36** | 2.30<br>**3.26** | 2.25<br>**3.17** | 2.22<br>**3.09** | 2.18<br>**3.03** |
| 25 | 4.24<br>**7.77** | 3.39<br>**5.57** | 2.99<br>**4.68** | 2.76<br>**4.18** | 2.60<br>**3.86** | 2.49<br>**3.63** | 2.40<br>**3.46** | 2.34<br>**3.32** | 2.28<br>**3.22** | 2.24<br>**3.13** | 2.20<br>**3.06** | 2.16<br>**2.99** |
| 26 | 4.23<br>**7.72** | 3.37<br>**5.53** | 2.98<br>**4.64** | 2.74<br>**4.14** | 2.59<br>**3.82** | 2.47<br>**3.59** | 2.39<br>**3.42** | 2.32<br>**3.29** | 2.27<br>**3.18** | 2.22<br>**3.09** | 2.18<br>**3.02** | 2.15<br>**2.96** |
| 27 | 4.21<br>**7.68** | 3.35<br>**5.49** | 2.96<br>**4.60** | 2.73<br>**4.11** | 2.57<br>**3.78** | 2.46<br>**3.56** | 2.37<br>**3.39** | 2.31<br>**3.26** | 2.25<br>**3.15** | 2.20<br>**3.06** | 2.16<br>**2.99** | 2.13<br>**2.93** |
| 28 | 4.20<br>**7.64** | 3.34<br>**5.45** | 2.95<br>**4.57** | 2.71<br>**4.07** | 2.56<br>**3.76** | 2.45<br>**3.53** | 2.36<br>**3.36** | 2.29<br>**3.23** | 2.24<br>**3.12** | 2.19<br>**3.03** | 2.15<br>**2.96** | 2.12<br>**2.90** |
| 29 | 4.18<br>**7.60** | 3.33<br>**5.42** | 2.93<br>**4.54** | 2.70<br>**4.04** | 2.55<br>**3.73** | 2.43<br>**3.50** | 2.35<br>**3.33** | 2.28<br>**3.20** | 2.22<br>**3.09** | 2.18<br>**3.00** | 2.14<br>**2.93** | 2.10<br>**2.87** |
| 30 | 4.17<br>**7.56** | 3.32<br>**5.39** | 2.92<br>**4.51** | 2.69<br>**4.02** | 2.53<br>**3.70** | 2.42<br>**3.47** | 2.33<br>**3.30** | 2.27<br>**3.17** | 2.21<br>**3.07** | 2.16<br>**2.98** | 2.13<br>**2.90** | 2.09<br>**2.84** |
| 32 | 4.15<br>**7.50** | 3.29<br>**5.34** | 2.90<br>**4.46** | 2.67<br>**3.97** | 2.51<br>**3.65** | 2.40<br>**3.43** | 2.31<br>**3.25** | 2.24<br>**3.13** | 2.19<br>**3.02** | 2.14<br>**2.93** | 2.10<br>**2.86** | 2.07<br>**2.80** |
| 34 | 4.13<br>**7.44** | 3.28<br>**5.29** | 2.88<br>**4.42** | 2.65<br>**3.93** | 2.49<br>**3.61** | 2.38<br>**3.39** | 2.29<br>**3.22** | 2.23<br>**3.09** | 2.17<br>**2.98** | 2.12<br>**2.89** | 2.08<br>**2.82** | 2.05<br>**2.76** |
| 36 | 4.11<br>**7.40** | 3.26<br>**5.25** | 2.87<br>**4.38** | 2.63<br>**3.89** | 2.48<br>**3.57** | 2.36<br>**3.35** | 2.28<br>**3.18** | 2.21<br>**3.05** | 2.15<br>**2.95** | 2.11<br>**2.86** | 2.07<br>**2.79** | 2.03<br>**2.72** |
| 38 | 4.10<br>**7.35** | 3.24<br>**5.21** | 2.85<br>**4.34** | 2.62<br>**3.86** | 2.46<br>**3.54** | 2.35<br>**3.32** | 2.26<br>**3.15** | 2.19<br>**3.02** | 2.14<br>**2.91** | 2.09<br>**2.82** | 2.05<br>**2.75** | 2.02<br>**2.69** |
| 40 | 4.08<br>**7.31** | 3.23<br>**5.18** | 2.84<br>**4.31** | 2.61<br>**3.83** | 2.45<br>**3.51** | 2.34<br>**3.29** | 2.25<br>**3.12** | 2.18<br>**2.99** | 2.12<br>**2.89** | 2.08<br>**2.80** | 2.04<br>**2.73** | 2.00<br>**2.66** |
| 42 | 4.07<br>**7.28** | 3.22<br>**5.15** | 2.83<br>**4.29** | 2.59<br>**3.80** | 2.44<br>**3.49** | 2.32<br>**3.27** | 2.24<br>**3.10** | 2.17<br>**2.97** | 2.11<br>**2.86** | 2.06<br>**2.78** | 2.03<br>**2.70** | 1.99<br>**2.64** |
| 44 | 4.06<br>**7.25** | 3.21<br>**5.12** | 2.82<br>**4.26** | 2.58<br>**3.78** | 2.43<br>**3.47** | 2.31<br>**3.24** | 2.23<br>**3.08** | 2.16<br>**2.95** | 2.10<br>**2.84** | 2.05<br>**2.75** | 2.01<br>**2.68** | 1.98<br>**2.62** |
| 46 | 4.05<br>**7.22** | 3.20<br>**5.10** | 2.81<br>**4.24** | 2.57<br>**3.76** | 2.42<br>**3.44** | 2.30<br>**3.22** | 2.22<br>**3.06** | 2.15<br>**2.93** | 2.09<br>**2.82** | 2.04<br>**2.73** | 2.00<br>**2.66** | 1.97<br>**2.60** |
| 48 | 4.04<br>**7.20** | 3.19<br>**5.08** | 2.80<br>**4.22** | 2.57<br>**3.74** | 2.41<br>**3.43** | 2.30<br>**3.20** | 2.21<br>**3.04** | 2.14<br>**2.91** | 2.08<br>**2.80** | 2.03<br>**2.72** | 1.99<br>**2.64** | 1.96<br>**2.58** |
| 50 | 4.03<br>**7.17** | 3.18<br>**5.06** | 2.79<br>**4.20** | 2.56<br>**3.72** | 2.40<br>**3.41** | 2.29<br>**3.19** | 2.20<br>**3.02** | 2.13<br>**2.89** | 2.07<br>**2.79** | 2.03<br>**2.70** | 1.99<br>**2.63** | 1.95<br>**2.56** |
| 55 | 4.02<br>**7.12** | 3.16<br>**5.01** | 2.78<br>**4.16** | 2.54<br>**3.68** | 2.38<br>**3.37** | 2.27<br>**3.15** | 2.18<br>**2.98** | 2.11<br>**2.85** | 2.06<br>**2.75** | 2.01<br>**2.66** | 1.97<br>**2.59** | 1.93<br>**2.53** |
| 60 | 4.00<br>**7.08** | 3.15<br>**4.98** | 2.76<br>**4.13** | 2.53<br>**3.65** | 2.37<br>**3.34** | 2.25<br>**3.12** | 2.17<br>**2.95** | 2.10<br>**2.82** | 2.04<br>**2.72** | 1.99<br>**2.63** | 1.95<br>**2.56** | 1.92<br>**2.50** |
| 65 | 3.99<br>**7.04** | 3.14<br>**4.95** | 2.75<br>**4.10** | 2.51<br>**3.62** | 2.36<br>**3.31** | 2.24<br>**3.09** | 2.15<br>**2.93** | 2.08<br>**2.80** | 2.03<br>**2.69** | 1.98<br>**2.61** | 1.94<br>**2.53** | 1.90<br>**2.47** |
| 70 | 3.98<br>**7.01** | 3.13<br>**4.92** | 2.74<br>**4.08** | 2.50<br>**3.60** | 2.35<br>**3.29** | 2.23<br>**3.07** | 2.14<br>**2.91** | 2.07<br>**2.78** | 2.02<br>**2.67** | 1.97<br>**2.59** | 1.93<br>**2.51** | 1.89<br>**2.45** |
| 80 | 3.96<br>**6.96** | 3.11<br>**4.88** | 2.72<br>**4.04** | 2.49<br>**3.56** | 2.33<br>**3.26** | 2.21<br>**3.04** | 2.13<br>**2.87** | 2.06<br>**2.74** | 2.00<br>**2.64** | 1.95<br>**2.55** | 1.91<br>**2.48** | 1.88<br>**2.42** |
| 100 | 3.94<br>**6.90** | 3.09<br>**4.82** | 2.70<br>**3.98** | 2.46<br>**3.51** | 2.31<br>**3.21** | 2.19<br>**2.99** | 2.10<br>**2.82** | 2.03<br>**2.69** | 1.97<br>**2.59** | 1.93<br>**2.50** | 1.89<br>**2.43** | 1.85<br>**2.37** |
| 125 | 3.92<br>**6.84** | 3.07<br>**4.78** | 2.68<br>**3.94** | 2.44<br>**3.47** | 2.29<br>**3.17** | 2.17<br>**2.95** | 2.08<br>**2.79** | 2.01<br>**2.66** | 1.96<br>**2.55** | 1.91<br>**2.50** | 1.87<br>**2.40** | 1.83<br>**2.33** |
| 150 | 3.90<br>**6.81** | 3.06<br>**4.75** | 2.66<br>**3.92** | 2.43<br>**3.45** | 2.27<br>**3.14** | 2.16<br>**2.92** | 2.07<br>**2.76** | 2.00<br>**2.63** | 1.94<br>**2.53** | 1.89<br>**2.44** | 1.85<br>**2.37** | 1.82<br>**2.31** |
| 200 | 3.89<br>**6.76** | 3.04<br>**4.71** | 2.65<br>**3.88** | 2.42<br>**3.41** | 2.26<br>**3.11** | 2.14<br>**2.89** | 2.06<br>**2.73** | 1.98<br>**2.60** | 1.93<br>**2.50** | 1.88<br>**2.41** | 1.84<br>**2.34** | 1.80<br>**2.27** |
| 400 | 3.86<br>**6.70** | 3.02<br>**4.66** | 2.62<br>**3.83** | 2.39<br>**3.36** | 2.23<br>**3.06** | 2.12<br>**2.85** | 2.03<br>**2.69** | 1.96<br>**2.55** | 1.90<br>**2.46** | 1.85<br>**2.37** | 1.81<br>**2.29** | 1.78<br>**2.23** |
| 1000 | 3.85<br>**6.66** | 3.00<br>**4.63** | 2.61<br>**3.80** | 2.38<br>**3.34** | 2.22<br>**3.04** | 2.11<br>**2.82** | 2.02<br>**2.66** | 1.95<br>**2.53** | 1.89<br>**2.43** | 1.84<br>**2.34** | 1.80<br>**2.27** | 1.76<br>**2.20** |
| $\infty$ | 3.84<br>**6.63** | 3.00<br>**4.61** | 2.60<br>**3.78** | 2.37<br>**3.32** | 2.21<br>**3.02** | 2.10<br>**2.80** | 2.01<br>**2.64** | 1.94<br>**2.51** | 1.88<br>**2.41** | 1.83<br>**2.32** | 1.79<br>**2.25** | 1.75<br>**2.18** |

| $m_1$ | | | | | | | | | | | | $m_2$ |
|---|---|---|---|---|---|---|---|---|---|---|---|---|
| 14 | 16 | 20 | 24 | 30 | 40 | 50 | 75 | 100 | 200 | 500 | $\infty$ | |
| 2.13 | 2.09 | 2.03 | 1.98 | 1.94 | 1.89 | 1.86 | 1.82 | 1.80 | 1.77 | 1.75 | 1.73 | 24 |
| **2.93** | **2.85** | **2.74** | **2.66** | **2.58** | **2.49** | **2.44** | **2.36** | **2.33** | **2.27** | **2.24** | **2.21** | |
| 2.11 | 2.07 | 2.01 | 1.96 | 1.92 | 1.87 | 1.84 | 1.80 | 1.78 | 1.75 | 1.73 | 1.71 | 25 |
| **2.89** | **2.81** | **2.70** | **2.62** | **2.54** | **2.45** | **2.40** | **2.32** | **2.29** | **2.23** | **2.19** | **2.17** | |
| 2.10 | 2.05 | 1.99 | 1.95 | 1.90 | 1.85 | 1.82 | 1.78 | 1.76 | 1.73 | 1.70 | 1.69 | 26 |
| **2.86** | **2.78** | **2.66** | **2.58** | **2.50** | **2.42** | **2.36** | **2.28** | **2.25** | **2.19** | **2.16** | **2.13** | |
| 2.08 | 2.04 | 1.97 | 1.93 | 1.88 | 1.84 | 1.81 | 1.76 | 1.74 | 1.71 | 1.68 | 1.67 | 27 |
| **2.82** | **2.75** | **2.63** | **2.55** | **2.47** | **2.38** | **2.33** | **2.25** | **2.22** | **2.16** | **2.12** | **2.10** | |
| 2.06 | 2.02 | 1.96 | 1.91 | 1.87 | 1.82 | 1.79 | 1.75 | 1.73 | 1.69 | 1.67 | 1.65 | 28 |
| **2.80** | **2.71** | **2.60** | **2.52** | **2.44** | **2.35** | **2.30** | **2.22** | **2.19** | **2.13** | **2.09** | **2.06** | |
| 2.05 | 2.01 | 1.94 | 1.90 | 1.85 | 1.80 | 1.77 | 1.73 | 1.71 | 1.67 | 1.65 | 1.64 | 29 |
| **2.77** | **2.69** | **2.57** | **2.49** | **2.41** | **2.33** | **2.27** | **2.19** | **2.16** | **2.10** | **2.06** | **2.03** | |
| 2.04 | 1.99 | 1.93 | 1.89 | 1.84 | 1.79 | 1.76 | 1.72 | 1.70 | 1.66 | 1.64 | 1.62 | 30 |
| **2.74** | **2.66** | **2.55** | **2.47** | **2.38** | **2.30** | **2.25** | **2.16** | **2.13** | **2.07** | **2.03** | **2.01** | |
| 2.01 | 1.97 | 1.91 | 1.86 | 1.82 | 1.77 | 1.74 | 1.69 | 1.67 | 1.63 | 1.61 | 1.59 | 32 |
| **2.70** | **2.62** | **2.50** | **2.42** | **2.34** | **2.25** | **2.20** | **2.12** | **2.08** | **2.02** | **1.98** | **1.96** | |
| 1.99 | 1.95 | 1.89 | 1.84 | 1.80 | 1.75 | 1.71 | 1.67 | 1.65 | 1.61 | 1.59 | 1.57 | 34 |
| **2.66** | **2.58** | **2.46** | **2.38** | **2.30** | **2.21** | **2.16** | **2.08** | **2.04** | **1.98** | **1.94** | **1.91** | |
| 1.98 | 1.93 | 1.87 | 1.82 | 1.78 | 1.73 | 1.69 | 1.65 | 1.62 | 1.59 | 1.56 | 1.55 | 36 |
| **2.62** | **2.54** | **2.43** | **2.35** | **2.26** | **2.17** | **2.12** | **2.04** | **2.00** | **1.94** | **1.90** | **1.87** | |
| 1.96 | 1.92 | 1.85 | 1.81 | 1.76 | 1.71 | 1.68 | 1.63 | 1.61 | 1.57 | 1.54 | 1.53 | 38 |
| **2.59** | **2.51** | **2.40** | **2.32** | **2.23** | **2.14** | **2.09** | **2.00** | **1.97** | **1.90** | **1.86** | **1.84** | |
| 1.95 | 1.90 | 1.84 | 1.79 | 1.74 | 1.69 | 1.66 | 1.61 | 1.59 | 1.55 | 1.53 | 1.51 | 40 |
| **2.56** | **2.48** | **2.37** | **2.29** | **2.20** | **2.11** | **2.06** | **1.97** | **1.94** | **1.87** | **1.83** | **1.80** | |
| 1.93 | 1.89 | 1.83 | 1.78 | 1.73 | 1.68 | 1.65 | 1.60 | 1.57 | 1.53 | 1.51 | 1.49 | 42 |
| **2.54** | **2.46** | **2.34** | **2.26** | **2.18** | **2.09** | **2.03** | **1.94** | **1.91** | **1.85** | **1.80** | **1.78** | |
| 1.92 | 1.88 | 1.81 | 1.77 | 1.72 | 1.67 | 1.63 | 1.58 | 1.56 | 1.52 | 1.49 | 1.48 | 44 |
| **2.52** | **2.44** | **2.32** | **2.24** | **2.15** | **2.06** | **2.01** | **1.92** | **1.89** | **1.82** | **1.78** | **1.75** | |
| 1.91 | 1.87 | 1.80 | 1.76 | 1.71 | 1.65 | 1.62 | 1.57 | 1.55 | 1.51 | 1.48 | 1.46 | 46 |
| **2.50** | **2.42** | **2.30** | **2.22** | **2.13** | **2.04** | **1.99** | **1.90** | **1.86** | **1.80** | **1.75** | **1.73** | |
| 1.90 | 1.86 | 1.79 | 1.75 | 1.70 | 1.64 | 1.61 | 1.56 | 1.54 | 1.49 | 1.47 | 1.45 | 48 |
| **2.48** | **2.40** | **2.28** | **2.20** | **2.12** | **2.03** | **1.97** | **1.88** | **1.84** | **1.78** | **1.73** | **1.70** | |
| 1.89 | 1.85 | 1.78 | 1.74 | 1.69 | 1.63 | 1.60 | 1.55 | 1.52 | 1.48 | 1.46 | 1.44 | 50 |
| **2.46** | **2.38** | **2.26** | **2.18** | **2.10** | **2.00** | **1.95** | **1.86** | **1.82** | **1.76** | **1.71** | **1.68** | |
| 1.88 | 1.83 | 1.76 | 1.72 | 1.67 | 1.61 | 1.58 | 1.52 | 1.50 | 1.46 | 1.43 | 1.41 | 55 |
| **2.43** | **2.34** | **2.23** | **2.15** | **2.06** | **1.96** | **1.91** | **1.82** | **1.78** | **1.71** | **1.67** | **1.64** | |
| 1.86 | 1.82 | 1.75 | 1.70 | 1.65 | 1.59 | 1.56 | 1.50 | 1.48 | 1.44 | 1.41 | 1.39 | 60 |
| **2.39** | **2.31** | **2.20** | **2.12** | **2.03** | **1.94** | **1.88** | **1.79** | **1.75** | **1.68** | **1.63** | **1.60** | |
| 1.85 | 1.80 | 1.73 | 1.69 | 1.63 | 1.58 | 1.54 | 1.49 | 1.46 | 1.42 | 1.39 | 1.37 | 65 |
| **2.37** | **2.29** | **2.18** | **2.09** | **2.00** | **1.90** | **1.85** | **1.76** | **1.72** | **1.65** | **1.60** | **1.56** | |
| 1.84 | 1.79 | 1.72 | 1.67 | 1.62 | 1.57 | 1.53 | 1.47 | 1.45 | 1.40 | 1.37 | 1.35 | 70 |
| **2.35** | **2.27** | **2.15** | **2.07** | **1.98** | **1.88** | **1.83** | **1.74** | **1.70** | **1.62** | **1.57** | **1.53** | |
| 1.82 | 1.77 | 1.70 | 1.65 | 1.60 | 1.54 | 1.51 | 1.45 | 1.43 | 1.38 | 1.35 | 1.32 | 80 |
| **2.31** | **2.23** | **2.12** | **2.03** | **1.94** | **1.85** | **1.79** | **1.70** | **1.66** | **1.58** | **1.53** | **1.49** | |
| 1.79 | 1.75 | 1.68 | 1.63 | 1.57 | 1.52 | 1.48 | 1.42 | 1.39 | 1.34 | 1.31 | 1.28 | 100 |
| **2.26** | **2.19** | **2.06** | **1.98** | **1.89** | **1.79** | **1.73** | **1.64** | **1.60** | **1.52** | **1.47** | **1.43** | |
| 1.77 | 1.72 | 1.65 | 1.60 | 1.55 | 1.49 | 1.45 | 1.39 | 1.36 | 1.31 | 1.27 | 1.25 | 125 |
| **2.23** | **2.15** | **2.03** | **1.94** | **1.85** | **1.75** | **1.69** | **1.59** | **1.55** | **1.47** | **1.41** | **1.37** | |
| 1.76 | 1.71 | 1.64 | 1.59 | 1.53 | 1.48 | 1.44 | 1.37 | 1.34 | 1.29 | 1.25 | 1.22 | 150 |
| **2.20** | **2.12** | **2.00** | **1.91** | **1.83** | **1.72** | **1.66** | **1.56** | **1.52** | **1.43** | **1.38** | **1.33** | |
| 1.74 | 1.69 | 1.62 | 1.57 | 1.52 | 1.46 | 1.41 | 1.35 | 1.32 | 1.26 | 1.22 | 1.19 | 200 |
| **2.17** | **2.09** | **1.97** | **1.88** | **1.79** | **1.69** | **1.63** | **1.53** | **1.48** | **1.39** | **1.33** | **1.28** | |
| 1.72 | 1.67 | 1.60 | 1.54 | 1.49 | 1.42 | 1.38 | 1.32 | 1.28 | 1.22 | 1.16 | 1.13 | 400 |
| **2.12** | **2.04** | **1.92** | **1.84** | **1.74** | **1.64** | **1.57** | **1.47** | **1.42** | **1.32** | **1.24** | **1.19** | |
| 1.70 | 1.65 | 1.58 | 1.53 | 1.47 | 1.41 | 1.36 | 1.30 | 1.26 | 1.19 | 1.13 | 1.08 | 1000 |
| **2.09** | **2.02** | **1.89** | **1.81** | **1.71** | **1.61** | **1.54** | **1.44** | **1.38** | **1.28** | **1.19** | **1.11** | |
| 1.69 | 1.64 | 1.57 | 1.52 | 1.46 | 1.39 | 1.35 | 1.28 | 1.24 | 1.17 | 1.11 | 1.00 | $\infty$ |
| **2.08** | **2.00** | **1.88** | **1.79** | **1.70** | **1.59** | **1.52** | **1.41** | **1.36** | **1.25** | **1.15** | **1.00** | |

#### 0.4.6.6 The Fischer Z-distribution

**Remark on the table:** The table contains the values of $z_0$, for which the probability: that the Fischer random variable $Z$ with $(r_1, r_2)$ degrees of freedom is not smaller than $z_0$; is equal to 0.01, in other words,

$$P(Z \geq z_0) = \int_{z_0}^{\infty} f(z)\,\mathrm{d}z = 0.01.$$

Here $f(z)$ is given by the formula

$$f(z) = \frac{2r_1^{\frac{r_1}{2}} r_2^{\frac{r_2}{2}}}{B\left(\frac{r_1}{2}, \frac{r_2}{2}\right)} \frac{\mathrm{e}^{r_1 z}}{(r_1 \mathrm{e}^{2z} + r_2)^{\frac{r_1+r_2}{2}}}.$$

| $r_2$ | $r_1$ | | | | | | | | | |
|---|---|---|---|---|---|---|---|---|---|---|
| | 1 | 2 | 3 | 4 | 5 | 6 | 8 | 12 | 24 | $\infty$ |
| 1 | 4.1535 | 4.2585 | 4.2974 | 4.3175 | 4.3297 | 4.3379 | 4.3482 | 4.3585 | 4.3689 | 4.3794 |
| 2 | 2.2950 | 2.2976 | 2.2984 | 2.2988 | 2.2991 | 2.2992 | 2.2994 | 2.2997 | 2.2999 | 2.3001 |
| 3 | 1.7649 | 1.7140 | 1.6915 | 1.6786 | 1.6703 | 1.6645 | 1.6569 | 1.6489 | 1.6404 | 1.6314 |
| 4 | 1.5270 | 1.4452 | 1.4075 | 1.3856 | 1.3711 | 1.3609 | 1.3473 | 1.3327 | 1.3170 | 1.3000 |
| 5 | 1.3943 | 1.2929 | 1.2449 | 1.2164 | 1.1974 | 1.1838 | 1.1656 | 1.1457 | 1.1239 | 1.0997 |
| 6 | 1.3103 | 1.1955 | 1.1401 | 1.1068 | 1.0843 | 1.0680 | 1.0460 | 1.0218 | 0.9948 | 0.9643 |
| 7 | 1.2526 | 1.1281 | 1.0682 | 1.0300 | 1.0048 | 0.9864 | 0.9614 | 0.9335 | 0.9020 | 0.8658 |
| 8 | 1.2106 | 1.0787 | 1.0135 | 0.9734 | 0.9459 | 0.9259 | 0.8983 | 0.8673 | 0.8319 | 0.7904 |
| 9 | 1.1786 | 1.0411 | 0.9724 | 0.9299 | 0.9006 | 0.8791 | 0.8494 | 0.8157 | 0.7769 | 0.7305 |
| 10 | 1.1535 | 1.0114 | 0.9399 | 0.8954 | 0.8646 | 0.8419 | 0.8104 | 0.7744 | 0.7324 | 0.6816 |
| 11 | 1.1333 | 0.9874 | 0.9136 | 0.8674 | 0.8354 | 0.8116 | 0.7785 | 0.7405 | 0.6958 | 0.6408 |
| 12 | 1.1166 | 0.9677 | 0.8919 | 0.8443 | 0.8111 | 0.7864 | 0.7520 | 0.7122 | 0.6649 | 0.6061 |
| 13 | 1.1027 | 0.9511 | 0.8737 | 0.8248 | 0.7907 | 0.7652 | 0.7295 | 0.6882 | 0.6386 | 0.5761 |
| 14 | 1.0909 | 0.9370 | 0.8581 | 0.8082 | 0.7732 | 0.7471 | 0.7103 | 0.6675 | 0.6159 | 0.5500 |
| 15 | 1.0807 | 0.9249 | 0.8448 | 0.7939 | 0.7582 | 0.7314 | 0.6937 | 0.6496 | 0.5961 | 0.5269 |
| 16 | 1.0719 | 0.9144 | 0.8331 | 0.7814 | 0.7450 | 0.7177 | 0.6791 | 0.6339 | 0.5786 | 0.5064 |
| 17 | 1.0641 | 0.9051 | 0.8229 | 0.7705 | 0.7335 | 0.7057 | 0.6663 | 0.6199 | 0.5630 | 0.4879 |
| 18 | 1.0572 | 0.8970 | 0.8138 | 0.7607 | 0.7232 | 0.6950 | 0.6549 | 0.6075 | 0.5491 | 0.4712 |
| 19 | 1.0511 | 0.8897 | 0.8057 | 0.7521 | 0.7140 | 0.6854 | 0.6447 | 0.5964 | 0.5366 | 0.4560 |
| 20 | 1.0457 | 0.8831 | 0.7985 | 0.7443 | 0.7058 | 0.6768 | 0.6355 | 0.5864 | 0.5253 | 0.4421 |
| 21 | 1.0408 | 0.8772 | 0.7920 | 0.7372 | 0.6984 | 0.6690 | 0.6272 | 0.5773 | 0.5150 | 0.4294 |
| 22 | 1.0363 | 0.8719 | 0.7860 | 0.7309 | 0.6916 | 0.6620 | 0.6196 | 0.5691 | 0.5056 | 0.4176 |
| 23 | 1.0322 | 0.8670 | 0.7806 | 0.7251 | 0.6855 | 0.6555 | 0.6127 | 0.5615 | 0.4969 | 0.4068 |
| 24 | 1.0285 | 0.8626 | 0.7757 | 0.7197 | 0.6799 | 0.6496 | 0.6064 | 0.5545 | 0.4890 | 0.3967 |
| 25 | 1.0251 | 0.8585 | 0.7712 | 0.7148 | 0.6747 | 0.6442 | 0.6006 | 0.5481 | 0.4816 | 0.3872 |
| 26 | 1.0220 | 0.8548 | 0.7670 | 0.7103 | 0.6699 | 0.6392 | 0.5952 | 0.5422 | 0.4748 | 0.3784 |
| 27 | 1.0191 | 0.8513 | 0.7631 | 0.7062 | 0.6655 | 0.6346 | 0.5902 | 0.5367 | 0.4685 | 0.3701 |
| 28 | 1.0164 | 0.8481 | 0.7595 | 0.7023 | 0.6614 | 0.6303 | 0.5856 | 0.5316 | 0.4626 | 0.3624 |
| 29 | 1.0139 | 0.8451 | 0.7562 | 0.6987 | 0.6576 | 0.6263 | 0.5813 | 0.5269 | 0.4570 | 0.3550 |
| 30 | 1.0116 | 0.8423 | 0.7531 | 0.6954 | 0.6540 | 0.6226 | 0.5773 | 0.5224 | 0.4519 | 0.3481 |
| 40 | 0.9949 | 0.8223 | 0.7307 | 0.6712 | 0.6283 | 0.5956 | 0.5481 | 0.4901 | 0.4138 | 0.2922 |
| 60 | 0.9784 | 0.8025 | 0.7086 | 0.6472 | 0.6028 | 0.5687 | 0.5189 | 0.4574 | 0.3746 | 0.2352 |
| 120 | 0.9622 | 0.7829 | 0.6867 | 0.6234 | 0.5774 | 0.5419 | 0.4897 | 0.4243 | 0.3339 | 0.1612 |
| $\infty$ | 0.9462 | 0.7636 | 0.6651 | 0.5999 | 0.5522 | 0.5152 | 0.4604 | 0.3908 | 0.2913 | 0.0000 |

### 0.4.6.7 Critical numbers for the Wilcoxon test

$\alpha = 0.05$

| | $n_2$ | | | | | | | | | | | |
|---|---|---|---|---|---|---|---|---|---|---|---|---|
| | 4 | 5 | 6 | 7 | 8 | 9 | 10 | 11 | 12 | 13 | 14 | $n_1$ |
| | – | – | – | – | 8.0 | 9.0 | 10.0 | 10.0 | 11.0 | 12.0 | 13.0 | 2 |
| | – | 7.5 | 8.0 | 9.5 | 10.0 | 11.5 | 12.0 | 13.5 | 14.0 | 15.5 | 16.0 | 3 |
| | 8.0 | 9.0 | 10.0 | 11.0 | 12.0 | 13.0 | 15.0 | 16.0 | 17.0 | 18.0 | 19.0 | 4 |
| | 9.0 | 10.5 | 12.0 | 12.5 | 14.0 | 15.5 | 17.0 | 18.5 | 19.0 | 20.5 | 22.0 | 5 |
| | | | 13.0 | 15.0 | 16.0 | 17.0 | 19.0 | 20.0 | 22.0 | 23.0 | 25.0 | 6 |
| 15 | 47.5 | | | 16.5 | 18.0 | 19.5 | 21.0 | 22.5 | 24.0 | 25.5 | 27.0 | 7 |
| 14 | 46.0 | 48.0 | | | 19.0 | 21.0 | 23.0 | 25.0 | 26.0 | 28.0 | 29.0 | 8 |
| 13 | 43.5 | 45.0 | 47.5 | | | 22.5 | 25.0 | 26.5 | 28.0 | 30.5 | 32.0 | 9 |
| 12 | 41.0 | 43.0 | 45.0 | 47.0 | | | 27.0 | 29.0 | 30.0 | 32.0 | 34.0 | 10 |
| 11 | 38.5 | 40.0 | 42.5 | 44.0 | 46.5 | | | 30.5 | 33.0 | 34.5 | 37.0 | 11 |
| 10 | 36.0 | 38.0 | 40.0 | 42.0 | 43.0 | 45.0 | | | 35.0 | 37.0 | 39.0 | 12 |
| 9 | 33.5 | 35.0 | 37.5 | 39.0 | 40.5 | 42.0 | 44.5 | | | 38.5 | 41.0 | 13 |
| 8 | 31.0 | 33.0 | 34.0 | 36.0 | 38.0 | 39.0 | 41.0 | 42.0 | | | 43.0 | 14 |
| 7 | 28.5 | 30.0 | 31.5 | 33.0 | 34.5 | 36.0 | 37.5 | 39.0 | 40.5 | | | |
| 6 | 26.0 | 27.0 | 29.0 | 30.0 | 32.0 | 33.0 | 34.0 | 36.0 | 37.0 | 38.0 | 39.0 | |
| 5 | 23.5 | 24.0 | 25.5 | 27.0 | 28.5 | 30.0 | 30.5 | 32.0 | 33.5 | 35.0 | 35.5 | |
| 4 | 20.0 | 21.0 | 23.0 | 24.0 | 25.0 | 26.0 | 27.0 | 28.0 | 29.0 | 30.0 | 32.0 | |
| 3 | 17.5 | 18.0 | 19.5 | 20.0 | 21.5 | 22.0 | 23.5 | 24.0 | 25.5 | 26.0 | 27.5 | |
| 2 | 14.0 | 15.0 | 15.0 | 16.0 | 17.0 | 18.0 | 18.0 | 19.0 | 20.0 | 21.0 | 22.0 | |
| $n_1$ | 15 | 16 | 17 | 18 | 19 | 20 | 21 | 22 | 23 | 24 | 25 | |
| | $n_2$ | | | | | | | | | | | |

$\alpha = 0.01$

| | $n_2$ | | | | | | | | | | | |
|---|---|---|---|---|---|---|---|---|---|---|---|---|
| | 4 | 5 | 6 | 7 | 8 | 9 | 10 | 11 | 12 | 13 | 14 | $n_1$ |
| | – | – | – | – | – | 13.5 | 15.0 | 16.5 | 17.0 | 18.5 | 20.0 | 3 |
| | – | – | 12.0 | 14.0 | 15.0 | 17.0 | 18.0 | 20.0 | 21.0 | 22.0 | 24.0 | 4 |
| | – | 12.5 | 14.0 | 15.5 | 18.0 | 19.5 | 21.0 | 22.5 | 24.0 | 25.5 | 28.0 | 5 |
| | | | 16.0 | 18.0 | 20.0 | 22.0 | 24.0 | 26.0 | 27.0 | 29.0 | 31.0 | 6 |
| 15 | 61.5 | | | 20.5 | 22.0 | 24.5 | 26.0 | 28.5 | 30.0 | 32.5 | 34.0 | 7 |
| 14 | 59.0 | 62.0 | | | 25.0 | 27.0 | 29.0 | 31.0 | 33.0 | 35.0 | 38.0 | 8 |
| 13 | 55.5 | 58.0 | 61.5 | | | 29.5 | 32.0 | 33.5 | 36.0 | 38.5 | 41.0 | 9 |
| 12 | 53.0 | 55.0 | 58.0 | 61.0 | | | 34.0 | 36.0 | 39.0 | 41.0 | 44.0 | 10 |
| 11 | 49.5 | 52.0 | 54.5 | 57.0 | 59.5 | | | 39.5 | 42.0 | 44.5 | 47.0 | 11 |
| 10 | 46.0 | 49.0 | 51.0 | 53.0 | 56.0 | 58.0 | | | 44.0 | 47.0 | 50.0 | 12 |
| 9 | 42.5 | 45.0 | 47.5 | 50.0 | 52.5 | 54.0 | 56.5 | | | 50.5 | 53.0 | 13 |
| 8 | 40.0 | 42.0 | 44.0 | 46.0 | 48.0 | 50.0 | 52.0 | 54.0 | | | 56.0 | 14 |
| 7 | 36.5 | 38.0 | 40.5 | 42.0 | 44.5 | 46.0 | 48.5 | 50.0 | 51.5 | | | |
| 6 | 33.0 | 35.0 | 36.0 | 38.0 | 40.0 | 42.0 | 44.0 | 45.0 | 47.0 | 49.0 | 51.0 | |
| 5 | 29.5 | 31.0 | 32.5 | 34.0 | 35.5 | 37.0 | 38.5 | 41.0 | 42.5 | 44.0 | 45.5 | |
| 4 | 25.0 | 27.0 | 28.0 | 30.0 | 31.0 | 32.0 | 34.0 | 35.0 | 37.0 | 38.0 | 40.0 | |
| 3 | 20.5 | 22.0 | 23.5 | 25.0 | 25.5 | 27.0 | 28.5 | 29.0 | 30.5 | 32.0 | 32.5 | |
| 2 | – | – | – | – | 19.0 | 20.0 | 21.0 | 22.0 | 23.0 | 24.0 | 25.0 | |
| $n_1$ | 15 | 16 | 17 | 18 | 19 | 20 | 21 | 22 | 23 | 24 | 25 | |
| | $n_2$ | | | | | | | | | | | |

### 0.4.6.8 The Kolmogorow–Smirnow λ-distribution

**Remark on the table:**

The tables on probability theory and mathematical statistics are taken in part from [17] and [27].

| $\lambda$ | $Q(\lambda)$ | $\lambda$ | $Q(\lambda)$ | $\lambda$ | $Q(\lambda)$ | $\lambda$ | $Q(\lambda)$ | $\lambda$ | $Q(\lambda)$ | $\lambda$ | $Q(\lambda)$ |
|---|---|---|---|---|---|---|---|---|---|---|---|
| 0.32 | 0.0000 | 0.66 | 0.2236 | 1.00 | 0.7300 | 1.34 | 0.9449 | 1.68 | 0.9929 | 2.00 | 0.9993 |
| 0.33 | 0.0001 | 0.67 | 0.2396 | 1.01 | 0.7406 | 1.35 | 0.9478 | 1.69 | 0.9934 | 2.01 | 0.9994 |
| 0.34 | 0.0002 | 0.68 | 0.2558 | 1.02 | 0.7508 | 1.36 | 0.9505 | 1.70 | 0.9938 | 2.02 | 0.9994 |
| 0.35 | 0.0003 | 0.69 | 0.2722 | 1.03 | 0.7608 | 1.37 | 0.9531 | 1.71 | 0.9942 | 2.03 | 0.9995 |
| 0.36 | 0.0005 | 0.70 | 0.2888 | 1.04 | 0.7704 | 1.38 | 0.9556 | 1.72 | 0.9946 | 2.04 | 0.9995 |
| 0.37 | 0.0008 | 0.71 | 0.3055 | 1.05 | 0.7798 | 1.39 | 0.9580 | 1.73 | 0.9950 | 2.05 | 0.9996 |
| 0.38 | 0.0013 | 0.72 | 0.3223 | 1.06 | 0.7889 | 1.40 | 0.9603 | 1.74 | 0.9953 | 2.06 | 0.9996 |
| 0.39 | 0.0019 | 0.73 | 0.3391 | 1.07 | 0.7976 | 1.41 | 0.9625 | 1.75 | 0.9956 | 2.07 | 0.9996 |
| 0.40 | 0.0028 | 0.74 | 0.3560 | 1.08 | 0.8061 | 1.42 | 0.9646 | 1.76 | 0.9959 | 2.08 | 0.9996 |
| 0.41 | 0.0040 | 0.75 | 0.3728 | 1.09 | 0.8143 | 1.43 | 0.9665 | 1.77 | 0.9962 | 2.09 | 0.9997 |
| 0.42 | 0.0055 | 0.76 | 0.3896 | 1.10 | 0.8223 | 1.44 | 0.9684 | 1.78 | 0.9965 | 2.10 | 0.9997 |
| 0.43 | 0.0074 | 0.77 | 0.4064 | 1.11 | 0.8299 | 1.45 | 0.9702 | 1.79 | 0.9967 | 2.11 | 0.9997 |
| 0.44 | 0.0097 | 0.78 | 0.4230 | 1.12 | 0.8374 | 1.46 | 0.9718 | 1.80 | 0.9969 | 2.12 | 0.9997 |
| 0.45 | 0.0126 | 0.79 | 0.4395 | 1.13 | 0.8445 | 1.47 | 0.9734 | 1.81 | 0.9971 | 2.13 | 0.9998 |
| 0.46 | 0.0160 | 0.80 | 0.4559 | 1.14 | 0.8514 | 1.48 | 0.9750 | 1.82 | 0.9973 | 2.14 | 0.9998 |
| 0.47 | 0.0200 | 0.81 | 0.4720 | 1.15 | 0.8580 | 1.49 | 0.9764 | 1.83 | 0.9975 | 2.15 | 0.9998 |
| 0.48 | 0.0247 | 0.82 | 0.4880 | 1.16 | 0.8644 | 1.50 | 0.9778 | 1.84 | 0.9977 | 2.16 | 0.9998 |
| 0.49 | 0.0300 | 0.83 | 0.5038 | 1.17 | 0.8706 | 1.51 | 0.9791 | 1.85 | 0.9979 | 2.17 | 0.9998 |
| 0.50 | 0.0361 | 0.84 | 0.5194 | 1.18 | 0.8765 | 1.52 | 0.9803 | 1.86 | 0.9980 | 2.18 | 0.9999 |
| 0.51 | 0.0428 | 0.85 | 0.5347 | 1.19 | 0.8823 | 1.53 | 0.9815 | 1.87 | 0.9981 | 2.19 | 0.9999 |
| 0.52 | 0.0503 | 0.86 | 0.5497 | 1.20 | 0.8877 | 1.54 | 0.9826 | 1.88 | 0.9983 | 2.20 | 0.9999 |
| 0.53 | 0.0585 | 0.87 | 0.5645 | 1.21 | 0.8930 | 1.55 | 0.9836 | 1.89 | 0.9984 | 2.21 | 0.9999 |
| 0.54 | 0.0675 | 0.88 | 0.5791 | 1.22 | 0.8981 | 1.56 | 0.9846 | 1.90 | 0.9985 | 2.22 | 0.9999 |
| 0.55 | 0.0772 | 0.89 | 0.5933 | 1.23 | 0.9030 | 1.57 | 0.9855 | 1.91 | 0.9986 | 2.23 | 0.9999 |
| 0.56 | 0.0876 | 0.90 | 0.6073 | 1.24 | 0.9076 | 1.58 | 0.9864 | 1.92 | 0.9987 | 2.24 | 0.9999 |
| 0.57 | 0.0987 | 0.91 | 0.6209 | 1.25 | 0.9121 | 1.59 | 0.9873 | 1.93 | 0.9988 | 2.25 | 0.9999 |
| 0.58 | 0.1104 | 0.92 | 0.6343 | 1.26 | 0.9164 | 1.60 | 0.9880 | 1.94 | 0.9989 | 2.26 | 0.9999 |
| 0.59 | 0.1228 | 0.93 | 0.6473 | 1.27 | 0.9206 | 1.61 | 0.9888 | 1.95 | 0.9990 | 2.27 | 0.9999 |
| 0.60 | 0.1357 | 0.94 | 0.6601 | 1.28 | 0.9245 | 1.62 | 0.9895 | 1.96 | 0.9991 | 2.28 | 0.9999 |
| 0.61 | 0.1492 | 0.95 | 0.6725 | 1.29 | 0.9283 | 1.63 | 0.9902 | 1.97 | 0.9991 | 2.29 | 0.9999 |
| 0.62 | 0.1632 | 0.96 | 0.6846 | 1.30 | 0.9319 | 1.64 | 0.9908 | 1.98 | 0.9992 | 2.30 | 0.9999 |
| 0.63 | 0.1778 | 0.97 | 0.6964 | 1.31 | 0.9354 | 1.65 | 0.9914 | 1.99 | 0.9993 | 2.31 | 1.0000 |
| 0.64 | 0.1927 | 0.98 | 0.7079 | 1.32 | 0.9387 | 1.66 | 0.9919 | | | | |
| 0.65 | 0.2080 | 0.99 | 0.7191 | 1.33 | 0.9418 | 1.67 | 0.9924 | | | | |

### 0.4.6.9 The Poisson distribution

$$P(X = r) = \frac{\lambda^r}{r!} e^{-\lambda}$$

| $r$ | $\lambda$ = 0.1 | 0.2 | 0.3 | 0.4 | 0.5 | 0.6 | 0.7 | 0.8 |
|---|---|---|---|---|---|---|---|---|
| 0 | 0.904 837 | 0.818 731 | 0.740 818 | 0.670 320 | 0.606 531 | 0.548 812 | 0.496 585 | 0.449 329 |
| 1 | 0.090 484 | 0.163 746 | 0.222 245 | 0.268 128 | 0.303 265 | 0.329 287 | 0.347 610 | 0.359 463 |
| 2 | 0.004 524 | 0.016 375 | 0.033 337 | 0.053 626 | 0.075 816 | 0.098 786 | 0.121 663 | 0.143 785 |
| 3 | 0.000 151 | 0.001 092 | 0.003 334 | 0.007 150 | 0.012 636 | 0.019 757 | 0.028 388 | 0.038 343 |
| 4 | 0.000 004 | 0.000 055 | 0.000 250 | 0.000 715 | 0.001 580 | 0.002 964 | 0.004 968 | 0.007 669 |
| 5 | – | 0.000 002 | 0.000 015 | 0.000 057 | 0.000 158 | 0.000 356 | 0.000 696 | 0.001 227 |
| 6 | – | – | 0.000 001 | 0.000 004 | 0.000 013 | 0.000 036 | 0.000 081 | 0.000 164 |
| 7 | – | – | – | – | 0.000 001 | 0.000 003 | 0.000 008 | 0.000 019 |
| 8 | – | – | – | – | – | – | 0.000 001 | 0.000 002 |

| $r$ | $\lambda$ = 0.9 | 1.0 | 1.5 | 2.0 | 2.5 | 3.0 | 3.5 | 4.0 |
|---|---|---|---|---|---|---|---|---|
| 0 | 0.406 570 | 0.367 879 | 0.223 130 | 0.135 335 | 0.082 085 | 0.049 787 | 0.030 197 | 0.018 316 |
| 1 | 0.365 913 | 0.367 879 | 0.334 695 | 0.270 671 | 0.205 212 | 0.149 361 | 0.105 691 | 0.073 263 |
| 2 | 0.164 661 | 0.183 940 | 0.251 021 | 0.270 671 | 0.256 516 | 0.224 042 | 0.184 959 | 0.146 525 |
| 3 | 0.049 398 | 0.061 313 | 0.125 510 | 0.180 447 | 0.213 763 | 0.224 042 | 0.215 785 | 0.195 367 |
| 4 | 0.011 115 | 0.015 328 | 0.047 067 | 0.090 224 | 0.133 602 | 0.168 031 | 0.188 812 | 0.195 367 |
| 5 | 0.002 001 | 0.003 066 | 0.014 120 | 0.036 089 | 0.066 801 | 0.100 819 | 0.132 169 | 0.156 293 |
| 6 | 0.000 300 | 0.000 511 | 0.003 530 | 0.012 030 | 0.027 834 | 0.050 409 | 0.077 098 | 0.104 196 |
| 7 | 0.000 039 | 0.000 073 | 0.000 756 | 0.003 437 | 0.009 941 | 0.021 604 | 0.038 549 | 0.059 540 |
| 8 | 0.000 004 | 0.000 009 | 0.000 142 | 0.000 859 | 0.003 106 | 0.008 102 | 0.016 865 | 0.029 770 |
| 9 | – | 0.000 001 | 0.000 024 | 0.000 191 | 0.000 863 | 0.002 701 | 0.006 559 | 0.013 231 |
| 10 | – | – | 0.000 004 | 0.000 038 | 0.000 216 | 0.000 810 | 0.002 296 | 0.005 292 |
| 11 | – | – | – | 0.000 007 | 0.000 049 | 0.000 221 | 0.000 730 | 0.001 925 |
| 12 | – | – | – | 0.000 001 | 0.000 010 | 0.000 055 | 0.000 213 | 0.000 642 |
| 13 | – | – | – | – | 0.000 002 | 0.000 013 | 0.000 057 | 0.000 197 |
| 14 | – | – | – | – | – | 0.000 003 | 0.000 014 | 0.000 056 |
| 15 | – | – | – | – | – | 0.000 001 | 0.000 003 | 0.000 015 |
| 16 | – | – | – | – | – | – | 0.000 001 | 0.000 004 |
| 17 | – | – | – | – | – | – | – | 0.000 001 |

| $r$ | $\lambda$ = 4.5 | 5.0 | 6.0 | 7.0 | 8.0 | 9.0 | 10.0 |
|---|---|---|---|---|---|---|---|
| 0 | 0.011 109 | 0.006 738 | 0.002 479 | 0.000 912 | 0.000 335 | 0.000 123 | 0.000 045 |
| 1 | 0.049 990 | 0.033 690 | 0.014 873 | 0.006 383 | 0.002 684 | 0.001 111 | 0.000 454 |
| 2 | 0.112 479 | 0.083 224 | 0.044 618 | 0.022 341 | 0.010 735 | 0.004 998 | 0.002 270 |
| 3 | 0.168 718 | 0.140 374 | 0.089 235 | 0.052 129 | 0.028 626 | 0.014 994 | 0.007 567 |
| 4 | 0.189 808 | 0.175 467 | 0.133 853 | 0.091 226 | 0.057 252 | 0.033 737 | 0.018 917 |
| 5 | 0.170 827 | 0.175 467 | 0.160 623 | 0.127 717 | 0.091 604 | 0.060 727 | 0.037 833 |
| 6 | 0.128 120 | 0.146 223 | 0.160 623 | 0.149 003 | 0.122 138 | 0.091 090 | 0.063 055 |
| 7 | 0.082 363 | 0.104 445 | 0.137 677 | 0.149 003 | 0.139 587 | 0.117 116 | 0.090 079 |
| 8 | 0.046 329 | 0.065 278 | 0.103 258 | 0.130 377 | 0.139 587 | 0.131 756 | 0.112 599 |
| 9 | 0.023 165 | 0.036 266 | 0.068 838 | 0.101 405 | 0.124 077 | 0.131 756 | 0.125 110 |
| 10 | 0.010 424 | 0.018 133 | 0.041 303 | 0.070 983 | 0.099 262 | 0.118 580 | 0.125 110 |
| 11 | 0.004 264 | 0.008 242 | 0.022 529 | 0.045 171 | 0.072 190 | 0.097 020 | 0.113 736 |
| 12 | 0.001 599 | 0.003 434 | 0.011 264 | 0.026 350 | 0.048 127 | 0.072 765 | 0.094 780 |
| 13 | 0.000 554 | 0.001 321 | 0.005 199 | 0.014 188 | 0.029 616 | 0.050 376 | 0.072 908 |
| 14 | 0.000 178 | 0.000 472 | 0.002 228 | 0.007 094 | 0.016 924 | 0.032 384 | 0.052 077 |
| 15 | 0.000 053 | 0.000 157 | 0.000 891 | 0.003 311 | 0.009 026 | 0.019 431 | 0.034 718 |
| 16 | 0.000 015 | 0.000 049 | 0.000 334 | 0.001 448 | 0.004 513 | 0.010 930 | 0.021 699 |
| 17 | 0.000 004 | 0.000 014 | 0.000 118 | 0.000 596 | 0.002 124 | 0.005 786 | 0.012 764 |
| 18 | 0.000 001 | 0.000 004 | 0.000 039 | 0.000 232 | 0.000 944 | 0.002 893 | 0.007 091 |
| 19 | – | 0.000 001 | 0.000 012 | 0.000 085 | 0.000 397 | 0.001 370 | 0.003 732 |
| 20 | – | – | 0.000 004 | 0.000 030 | 0.000 159 | 0.000 617 | 0.001 866 |
| 21 | – | – | 0.000 001 | 0.000 010 | 0.000 061 | 0.000 264 | 0.000 889 |
| 22 | – | – | – | 0.000 003 | 0.000 022 | 0.000 108 | 0.000 404 |
| 23 | – | – | – | 0.000 001 | 0.000 008 | 0.000 042 | 0.000 176 |
| 24 | – | – | – | – | 0.000 003 | 0.000 016 | 0.000 073 |
| 25 | – | – | – | – | 0.000 001 | 0.000 006 | 0.000 029 |
| 26 | – | – | – | – | – | 0.000 002 | 0.000 011 |
| 27 | – | – | – | – | – | 0.000 001 | 0.000 004 |
| 28 | – | – | – | – | – | – | 0.000 001 |
| 29 | – | – | – | – | – | – | 0.000 001 |

# 0.5 Tables of values of special functions

**Remark on the following tables:**
Some of these tables are taken from [21].

## 0.5.1 The gamma functions $\Gamma(x)$ and $1/\Gamma(x)$

**Remark on this table:** See also section 1.14.16.

| $x$ | $\Gamma(x)$ | $1/\Gamma(x)$ | $x$ | $\Gamma(x)$ | $1/\Gamma(x)$ | $x$ | $\Gamma(x)$ | $1/\Gamma(x)$ |
|---|---|---|---|---|---|---|---|---|
| 1.00 | 1.000 00 | 1.000 0 | 1.40 | 0.887 26 | 1.127 0 | 1.70 | 0.908 64 | 1.100 5 |
| 1.01 | 0.994 33 | 005 7 | 1.41 | 886 76 | 127 7 | 1.71 | 910 57 | 098 2 |
| 1.02 | 988 84 | 011 3 | 1.42 | 886 36 | 128 2 | 1.72 | 912 58 | 095 8 |
| 1.03 | 983 55 | 016 7 | 1.43 | 886 04 | 128 6 | 1.73 | 914 67 | 093 3 |
| 1.04 | 978 44 | 022 0 | 1.44 | 885 81 | 128 9 | 1.74 | 916 83 | 090 7 |
| 1.05 | 973 50 | 027 2 | 1.45 | 885 66 | 129 1 | 1.75 | 919 06 | 088 1 |
| 1.06 | 968 74 | 032 3 | 1.46 | 885 60 | 129 1 | 1.76 | 921 37 | 085 4 |
| 1.07 | 964 15 | 037 2 | 1.47 | 885 63 | 129 1 | 1.77 | 923 76 | 082 5 |
| 1.08 | 959 73 | 042 0 | 1.48 | 885 75 | 129 1 | 1.78 | 926 23 | 079 6 |
| 1.09 | 955 46 | 046 6 | 1.49 | 885 95 | 128 8 | 1.79 | 928 77 | 076 7 |
| 1.10 | 0.951 35 | 1.051 1 | 1.50 | 0.886 23 | 1.128 4 | 1.80 | 0.931 38 | 1.073 7 |
| 1.11 | 947 40 | 055 5 | 1.51 | 886 59 | 127 9 | 1.81 | 934 08 | 070 6 |
| 1.12 | 943 59 | 059 8 | 1.52 | 887 04 | 127 3 | 1.82 | 936 85 | 067 4 |
| 1.13 | 939 93 | 063 9 | 1.53 | 887 57 | 126 7 | 1.83 | 939 69 | 064 2 |
| 1.14 | 936 42 | 067 9 | 1.54 | 888 18 | 125 9 | 1.84 | 942 61 | 060 9 |
| 1.15 | 933 04 | 071 8 | 1.55 | 888 87 | 125 0 | 1.85 | 945 61 | 057 5 |
| 1.16 | 929 80 | 075 5 | 1.56 | 889 64 | 124 0 | 1.86 | 948 69 | 054 1 |
| 1.17 | 926 70 | 079 1 | 1.57 | 890 49 | 123 0 | 1.87 | 951 84 | 050 6 |
| 1.18 | 923 73 | 082 6 | 1.58 | 891 42 | 121 8 | 1.88 | 955 07 | 047 1 |
| 1.19 | 920 89 | 085 9 | 1.59 | 892 43 | 120 5 | 1.89 | 958 38 | 043 5 |
| 1.20 | 0.918 17 | 1.089 1 | 1.60 | 0.893 52 | 1.119 1 | 1.90 | 0.961 77 | 1.039 8 |
| 1.21 | 915 58 | 092 2 | 1.61 | 894 68 | 117 7 | 1.91 | 965 23 | 036 0 |
| 1.22 | 913 11 | 095 2 | 1.62 | 895 92 | 116 1 | 1.92 | 968 77 | 032 2 |
| 1.23 | 910 75 | 098 0 | 1.63 | 897 24 | 114 5 | 1.93 | 972 40 | 028 4 |
| 1.24 | 908 52 | 100 7 | 1.64 | 898 64 | 112 8 | 1.94 | 976 10 | 024 5 |
| 1.25 | 906 40 | 103 2 | 1.65 | 900 12 | 110 9 | 1.95 | 979 88 | 020 6 |
| 1.26 | 904 40 | 105 7 | 1.66 | 901 67 | 109 1 | 1.96 | 983 74 | 016 5 |
| 1.27 | 902 50 | 108 0 | 1.67 | 903 30 | 107 1 | 1.97 | 987 68 | 012 5 |
| 1.28 | 900 72 | 110 2 | 1.68 | 905 00 | 104 9 | 1.98 | 991 71 | 008 3 |
| 1.29 | 899 04 | 112 3 | 1.69 | 906 78 | 102 8 | 1.99 | 995 81 | 004 2 |
| 1.30 | 0.897 47 | 1.114 2 | | | | | | |
| 1.31 | 896 00 | 116 1 | | | | | | |
| 1.32 | 894 64 | 117 8 | | | | | | |
| 1.33 | 893 38 | 119 4 | | | | | | |
| 1.34 | 892 22 | 120 8 | | | | | | |
| 1.35 | 891 15 | 122 2 | | | | | | |
| 1.36 | 890 18 | 123 4 | | | | | | |
| 1.37 | 889 31 | 124 4 | | | | | | |
| 1.38 | 888 54 | 125 4 | | | | | | |
| 1.39 | 887 85 | 126 3 | | | | | | |

If $x$ is a natural number $n$ with $n \geq 1$, then

$$\Gamma(n) = (n-1)!,$$

so that, for example, $\Gamma(2) = 1$.
To calculate $\Gamma(x)$ for $x$ which is less than 1 but not an integer, one can use the formula

$$\Gamma(x) = \frac{\Gamma(x+1)}{x}.$$

If $x > 2$, then for the calculation the formula

$$\Gamma(x) = (x-1) \cdot \Gamma(x-1)$$

can be used.

*Examples:*

1. $\Gamma(-0.2) = \dfrac{\Gamma(0.8)}{-0.2} = -\dfrac{\Gamma(1.8)}{0.2 \cdot 0.8} = -\dfrac{0.931\,38}{0.16} = -5.821\,13.$
2. $\Gamma(3.2) = 2.2 \cdot \Gamma(2.2) = 2.2 \cdot 1.2 \cdot 2.\Gamma(1.2) = 2.2 \cdot 1.2 \cdot 0.918\,17 = 2.423\,97.$

## 0.5.2 Cylinder functions (also known as Bessel functions)

**Remark:** See also section 1.14.22.

| $x$ | $J_0(x)$ | $J_1(x)$ | $Y_0(x)$ | $Y_1(x)$ | $I_0(x)$ | $I_1(x)$ | $K_0(x)$ | $K_1(x)$ |
|---|---|---|---|---|---|---|---|---|
| 0.0 | +1.000 0 | +0.000 0 | $-\infty$ | $-\infty$ | 1.000 | 0.000 0 | $\infty$ | $\infty$ |
| 0.1 | +0.997 5 | +0.049 9 | -1.534 2 | -6.459 0 | 1.003 | 0.050 1 | 2.427 1 | 9. 853 8 |
| 0.2 | +0.990 0 | +0.099 5 | -1.081 1 | -3.323 8 | 1.010 | 0.100 5 | 1.752 7 | 4. 776 0 |
| 0.3 | +0.977 6 | +0.148 3 | -0.807 3 | -2.293 1 | 1.023 | 0.151 7 | 1.372 5 | 3. 056 0 |
| 0.4 | +0.960 4 | +0.196 0 | -0.606 0 | -1.780 9 | 1.040 | 0.204 0 | 1.114 5 | 2. 184 4 |
| 0.5 | +0.938 5 | +0.242 3 | -0.444 5 | -1.471 5 | 1.063 | 0.257 9 | 0.924 4 | 1. 656 4 |
| 0.6 | +0.912 0 | +0.286 7 | -0.308 5 | -1.260 4 | 1.092 | 0.313 7 | 0.777 5 | 1. 302 8 |
| 0.7 | +0.881 2 | +0.329 0 | -0.190 7 | -1.103 2 | 1.126 | 0.371 9 | 0.660 5 | 1. 050 3 |
| 0.8 | +0.846 3 | +0.368 8 | -0.086 8 | -0.978 1 | 1.167 | 0.432 9 | 0.565 3 | 0. 861 8 |
| 0.9 | +0.807 5 | +0.405 9 | +0.005 6 | -0.873 1 | 1.213 | 0.497 1 | 0.486 7 | 0. 716 5 |
| 1.0 | +0.765 2 | +0.440 1 | +0.088 3 | -0.781 2 | 1.266 | 0.565 2 | 0.421 0 | 0. 601 9 |
| 1.1 | +0.719 6 | +0.470 9 | +0.162 2 | -0.698 1 | 1.326 | 0.637 5 | 0.365 6 | 0. 509 8 |
| 1.2 | +0.671 1 | +0.498 3 | +0.228 1 | -0.621 1 | 1.394 | 0.714 7 | 0.318 5 | 0. 434 6 |
| 1.3 | +0.620 1 | +0.522 0 | +0.286 5 | -0.548 5 | 1.469 | 0.797 3 | 0.278 2 | 0. 372 5 |
| 1.4 | +0.566 9 | +0.541 9 | +0.337 9 | -0.479 1 | 1.553 | 0.886 1 | 0.243 7 | 0. 320 8 |
| 1.5 | +0.511 8 | +0.557 9 | +0.382 4 | -0.412 3 | 1.647 | 0.981 7 | 0.213 8 | 0. 277 4 |
| 1.6 | +0.455 4 | +0.569 9 | +0.420 4 | -0.347 6 | 1.750 | 1.085 | 0.188 0 | 0. 240 6 |
| 1.7 | +0.398 0 | +0.577 8 | +0.452 0 | -0.284 7 | 1.864 | 1.196 | 0.165 5 | 0. 209 4 |
| 1.8 | +0.340 0 | +0.581 5 | +0.477 4 | -0.223 7 | 1.990 | 1.317 | 0.145 9 | 0. 182 6 |
| 1.9 | +0.281 8 | +0.581 2 | +0.496 8 | -0.164 4 | 2.128 | 1.448 | 0.128 8 | 0. 159 7 |
| 2.0 | +0.223 9 | +0.576 7 | +0.510 4 | -0.107 0 | 2.280 | 1.591 | 0.113 9 | 0. 139 9 |
| 2.1 | +0.166 6 | +0.568 3 | +0.518 3 | -0.051 7 | 2.446 | 1.745 | 0.100 8 | 0. 122 7 |
| 2.2 | +0.110 4 | +0.556 0 | +0.520 8 | +0.001 5 | 2.629 | 1.914 | 0.089 27 | 0. 107 9 |
| 2.3 | +0.055 5 | +0.539 9 | +0.518 1 | +0.052 3 | 2.830 | 2.098 | 0.079 14 | 0. 094 98 |
| 2.4 | +0.002 5 | +0.520 2 | +0.510 4 | +0.100 5 | 3.049 | 2.298 | 0.070 22 | 0. 083 72 |
| 2.5 | -0.048 4 | +0.497 1 | +0.498 1 | +0.145 9 | 3.290 | 2.517 | 0.062 35 | 0. 073 89 |
| 2.6 | -0.096 8 | +0.470 8 | +0.481 3 | +0.188 4 | 3.553 | 2.755 | 0.055 40 | 0. 065 28 |
| 2.7 | -0.142 4 | +0.441 6 | +0.460 5 | +0.227 6 | 3.842 | 3.016 | 0.049 26 | 0. 057 74 |
| 2.8 | -0.185 0 | +0.409 7 | +0.435 9 | +0.263 5 | 4.157 | 3.301 | 0.043 82 | 0. 051 11 |
| 2.9 | -0.224 3 | +0.375 4 | +0.407 9 | +0.295 9 | 4.503 | 3.613 | 0.039 01 | 0. 045 29 |
| 3.0 | -0.260 1 | +0.339 1 | +0.376 9 | +0.324 7 | 4.881 | 3.953 | 0.034 74 | 0. 040 16 |
| 3.1 | -0.292 1 | +0.300 9 | +0.343 1 | +0.349 6 | 5.294 | 4.326 | 0.030 95 | 0. 035 63 |
| 3.2 | -0.320 2 | +0.261 3 | +0.307 0 | +0.370 7 | 5.747 | 4.734 | 0.027 59 | 0. 031 64 |
| 3.3 | -0.344 3 | +0.220 7 | +0.269 1 | +0.387 9 | 6.243 | 5.181 | 0.024 61 | 0. 028 12 |
| 3.4 | -0.364 3 | +0.179 2 | +0.229 6 | +0.401 0 | 6.785 | 5.670 | 0.021 96 | 0. 025 00 |
| 3.5 | -0.380 1 | +0.137 4 | +0.189 0 | +0.410 2 | 7.378 | 6.206 | 0.019 60 | 0. 022 24 |
| 3.6 | -0.391 8 | +0.095 5 | +0.147 7 | +0.415 4 | 8.028 | 6.793 | 0.017 50 | 0. 019 79 |
| 3.7 | -0.399 2 | +0.053 8 | +0.106 1 | +0.416 7 | 8.739 | 7.436 | 0.015 63 | 0. 017 63 |
| 3.8 | -0.402 6 | +0.012 8 | +0.064 5 | +0.414 1 | 9.517 | 8.140 | 0.013 97 | 0. 015 71 |
| 3.9 | -0.401 8 | -0.027 2 | +0.023 4 | +0.407 8 | 10.37 | 8.913 | 0.012 48 | 0. 014 00 |
| 4.0 | -0.397 1 | -0.066 0 | -0.016 9 | +0.397 9 | 11.30 | 9.759 | 0.011 16 | 0. 012 48 |
| 4.1 | -0.388 7 | -0.103 3 | -0.056 1 | +0.384 6 | 12.32 | 10.69 | 0.009 980 | 0. 011 14 |
| 4.2 | -0.376 6 | -0.138 6 | -0.093 8 | +0.368 0 | 13.44 | 11.71 | 0.008 927 | 0. 009 938 |
| 4.3 | -0.361 0 | -0.171 9 | -0.129 6 | +0.348 4 | 14.67 | 12.82 | 0.007 988 | 0. 008 872 |
| 4.4 | -0.342 3 | -0.202 8 | -0.163 3 | +0.326 0 | 16.01 | 14.05 | 0.007 149 | 0. 007 923 |
| 4.5 | -0.320 5 | -0.231 1 | -0.194 7 | +0.301 0 | 17.48 | 15.39 | 0.006 400 | 0. 007 078 |
| 4.6 | -0.296 1 | -0.256 6 | -0.223 5 | +0.273 7 | 19.09 | 16.86 | 0.005 730 | 0. 006 325 |
| 4.7 | -0.269 3 | -0.279 1 | -0.249 4 | +0.244 5 | 20.86 | 18.48 | 0.005 132 | 0. 005 654 |
| 4.8 | -0.240 4 | -0.298 5 | -0.272 3 | +0.213 6 | 22.79 | 20.25 | 0.004 597 | 0. 005 055 |
| 4.9 | -0.209 7 | -0.314 7 | -0.292 1 | +0.181 2 | 24.91 | 22.20 | 0.004 119 | 0. 004 521 |

| $x$ | $J_0(x)$ | $J_1(x)$ | $Y_0(x)$ | $Y_1(x)$ | $I_0(x)$ | $I_1(x)$ | $K_0(x)$ | $K_1(x)$ |
|---|---|---|---|---|---|---|---|---|
| 5.0 | −0.1776 | −0.3276 | −0.3085 | +0.1479 | 27.24 | 24.34 | $3691 \cdot 10^{-6}$ | $4045 \cdot 10^{-6}$ |
| 5.1 | −0.1443 | −0.3371 | −0.3216 | +0.1137 | 29.79 | 26.68 | $3308 \cdot 10^{-6}$ | $3619 \cdot 10^{-6}$ |
| 5.2 | −0.1103 | −0.3432 | −0.3313 | +0.0792 | 32.58 | 29.25 | $2966 \cdot 10^{-6}$ | $3239 \cdot 10^{-6}$ |
| 5.3 | −0.0758 | −0.3460 | −0.3374 | +0.0445 | 35.65 | 32.08 | $2659 \cdot 10^{-6}$ | $2900 \cdot 10^{-6}$ |
| 5.4 | −0.0412 | −0.3453 | −0.3402 | +0.0101 | 39.01 | 35.18 | $2385 \cdot 10^{-6}$ | $2597 \cdot 10^{-6}$ |
| 5.5 | −0.0068 | −0.3414 | −0.3395 | −0.0238 | 42.69 | 38.59 | $2139 \cdot 10^{-6}$ | $2326 \cdot 10^{-6}$ |
| 5.6 | +0.0270 | −0.3343 | −0.3354 | −0.0568 | 46.74 | 42.33 | $1918 \cdot 10^{-6}$ | $2083 \cdot 10^{-6}$ |
| 5.7 | +0.0599 | −0.3241 | −0.3282 | −0.0887 | 51.17 | 46.44 | $1721 \cdot 10^{-6}$ | $1866 \cdot 10^{-6}$ |
| 5.8 | +0.0917 | −0.3110 | −0.3177 | −0.1192 | 56.04 | 50.95 | $1544 \cdot 10^{-6}$ | $1673 \cdot 10^{-6}$ |
| 5.9 | +0.1220 | −0.2951 | −0.3044 | −0.1481 | 61.38 | 55.90 | $1386 \cdot 10^{-6}$ | $1499 \cdot 10^{-6}$ |
| 6.0 | +0.1506 | −0.2767 | −0.2882 | −0.1750 | 67.23 | 61.34 | $1244 \cdot 10^{-6}$ | $1344 \cdot 10^{-6}$ |
| 6.1 | +0.1773 | −0.2559 | −0.2694 | −0.1998 | 73.66 | 67.32 | $1117 \cdot 10^{-6}$ | $1205 \cdot 10^{-6}$ |
| 6.2 | +0.2017 | −0.2329 | −0.2483 | −0.2223 | 80.72 | 73.89 | $1003 \cdot 10^{-6}$ | $1081 \cdot 10^{-6}$ |
| 6.3 | +0.2238 | −0.2081 | −0.2251 | −0.2422 | 88.46 | 81.10 | $9001 \cdot 10^{-7}$ | $9691 \cdot 10^{-7}$ |
| 6.4 | +0.2433 | −0.1816 | −0.1999 | −0.2596 | 96.96 | 89.03 | $8083 \cdot 10^{-7}$ | $8693 \cdot 10^{-7}$ |
| 6.5 | +0.2601 | −0.1538 | −0.1732 | −0.2741 | 106.3 | 97.74 | $7259 \cdot 10^{-7}$ | $7799 \cdot 10^{-7}$ |
| 6.6 | +0.2740 | −0.1250 | −0.1452 | −0.2857 | 116.5 | 107.3 | $6520 \cdot 10^{-7}$ | $6998 \cdot 10^{-7}$ |
| 6.7 | +0.2851 | −0.0953 | −0.1162 | −0.2945 | 127.8 | 117.8 | $5857 \cdot 10^{-7}$ | $6280 \cdot 10^{-7}$ |
| 6.8 | +0.2931 | −0.0652 | −0.0864 | −0.3002 | 140.1 | 129.4 | $5262 \cdot 10^{-7}$ | $5636 \cdot 10^{-7}$ |
| 6.9 | +0.2981 | −0.0349 | −0.0563 | −0.3029 | 153.7 | 142.1 | $4728 \cdot 10^{-7}$ | $5059 \cdot 10^{-7}$ |
| 7.0 | +0.3001 | −0.0047 | −0.0259 | −0.3027 | 168.6 | 156.0 | $4248 \cdot 10^{-7}$ | $4542 \cdot 10^{-7}$ |
| 7.1 | +0.2991 | +0.0252 | +0.0042 | −0.2995 | 185.0 | 171.4 | $3817 \cdot 10^{-7}$ | $4078 \cdot 10^{-7}$ |
| 7.2 | +0.2951 | +0.0543 | +0.0339 | −0.2934 | 202.9 | 188.3 | $3431 \cdot 10^{-7}$ | $3662 \cdot 10^{-7}$ |
| 7.3 | +0.2882 | +0.0826 | +0.0628 | −0.2846 | 222.7 | 206.8 | $3084 \cdot 10^{-7}$ | $3288 \cdot 10^{-7}$ |
| 7.4 | +0.2786 | +0.1096 | +0.0907 | −0.2731 | 244.3 | 227.2 | $2772 \cdot 10^{-7}$ | $2953 \cdot 10^{-7}$ |
| 7.5 | +0.2663 | +0.1352 | +0.1173 | −0.2591 | 268.2 | 249.6 | $2492 \cdot 10^{-7}$ | $2653 \cdot 10^{-7}$ |
| 7.6 | +0.2516 | +0.1592 | +0.1424 | −0.2428 | 294.3 | 274.2 | $2240 \cdot 10^{-7}$ | $2383 \cdot 10^{-7}$ |
| 7.7 | +0.2346 | +0.1813 | +0.1658 | −0.2243 | 323.1 | 301.3 | $2014 \cdot 10^{-7}$ | $2141 \cdot 10^{-7}$ |
| 7.8 | +0.2154 | +0.2014 | +0.1872 | −0.2039 | 354.7 | 331.1 | $1811 \cdot 10^{-7}$ | $1924 \cdot 10^{-7}$ |
| 7.9 | +0.1944 | +0.2192 | +0.2065 | −0.1817 | 389.4 | 363.9 | $1629 \cdot 10^{-7}$ | $1729 \cdot 10^{-7}$ |
| 8.0 | +0.1717 | +0.2346 | +0.2235 | −0.1581 | 427.6 | 399.9 | $1465 \cdot 10^{-7}$ | $1554 \cdot 10^{-7}$ |
| 8.1 | +0.1475 | +0.2476 | +0.2381 | −0.1331 | 469.5 | 439.5 | $1317 \cdot 10^{-7}$ | $1396 \cdot 10^{-7}$ |
| 8.2 | +0.1222 | +0.2580 | +0.2501 | −0.1072 | 515.6 | 483.0 | $1185 \cdot 10^{-7}$ | $1255 \cdot 10^{-7}$ |
| 8.3 | +0.0960 | +0.2657 | +0.2595 | −0.0806 | 566.3 | 531.0 | $1066 \cdot 10^{-7}$ | $1128 \cdot 10^{-7}$ |
| 8.4 | +0.0692 | +0.2708 | +0.2662 | −0.0535 | 621.9 | 583.7 | $9588 \cdot 10^{-8}$ | $1014 \cdot 10^{-7}$ |
| 8.5 | +0.0419 | +0.2731 | +0.2702 | −0.0262 | 683.2 | 641.6 | $8626 \cdot 10^{-8}$ | $9120 \cdot 10^{-8}$ |
| 8.6 | +0.0146 | +0.2728 | +0.2715 | +0.0011 | 750.5 | 705.4 | $7761 \cdot 10^{-8}$ | $8200 \cdot 10^{-8}$ |
| 8.7 | −0.0125 | +0.2697 | +0.2700 | +0.0280 | 824.4 | 775.5 | $6983 \cdot 10^{-8}$ | $7374 \cdot 10^{-8}$ |
| 8.8 | −0.0392 | +0.2641 | +0.2659 | +0.0544 | 905.8 | 852.7 | $6283 \cdot 10^{-8}$ | $6631 \cdot 10^{-8}$ |
| 8.9 | −0.0653 | +0.2559 | +0.2592 | +0.0799 | 995.2 | 937.5 | $5654 \cdot 10^{-8}$ | $5964 \cdot 10^{-8}$ |
| 9.0 | −0.0903 | +0.2453 | +0.2499 | +0.1043 | 1094.0 | 1031.0 | $5088 \cdot 10^{-8}$ | $5364 \cdot 10^{-8}$ |
| 9.1 | −0.1142 | +0.2324 | +0.2383 | +0.1275 | 1202.0 | 1134.0 | $4579 \cdot 10^{-8}$ | $4825 \cdot 10^{-8}$ |
| 9.2 | −0.1367 | +0.2174 | +0.2245 | +0.1491 | 1321.0 | 1247.0 | $4121 \cdot 10^{-8}$ | $4340 \cdot 10^{-8}$ |
| 9.3 | −0.1577 | +0.2004 | +0.2086 | +0.1691 | 1451.0 | 1371.0 | $3710 \cdot 10^{-8}$ | $3904 \cdot 10^{-8}$ |
| 9.4 | −0.1768 | +0.1816 | +0.1907 | +0.1871 | 1595.0 | 1508.0 | $3339 \cdot 10^{-8}$ | $3512 \cdot 10^{-8}$ |
| 9.5 | −0.1939 | +0.1613 | +0.1712 | +0.2032 | 1753.0 | 1685.0 | $3006 \cdot 10^{-8}$ | $3160 \cdot 10^{-8}$ |
| 9.6 | −0.2090 | +0.1395 | +0.1502 | +0.2171 | 1927.0 | 1824.0 | $2706 \cdot 10^{-8}$ | $2843 \cdot 10^{-8}$ |
| 9.7 | −0.2218 | +0.1166 | +0.2179 | +0.2287 | 2119.0 | 2006.0 | $2436 \cdot 10^{-8}$ | $2559 \cdot 10^{-8}$ |
| 9.8 | −0.2323 | +0.0928 | +0.1045 | +0.2379 | 2329.0 | 2207.0 | $2193 \cdot 10^{-8}$ | $2302 \cdot 10^{-8}$ |
| 9.9 | −0.2403 | +0.0684 | +0.0804 | +0.2447 | 2561.0 | 2428.0 | $1975 \cdot 10^{-8}$ | $2072 \cdot 10^{-8}$ |
| 10.0 | −0.2459 | +0.0435 | +0.0557 | +0.2490 | 2816.0 | 2671.0 | $1778 \cdot 10^{-8}$ | $1865 \cdot 10^{-8}$ |

## Some values of Bessel functions of higher order $p$, for integral arguments

For $p = 0.5,\ 1.5$ and $2.5$ see the table *Spherical cylinder functions* below.

| $p$ | $J_p(1)$ | $J_p(2)$ | $J_p(3)$ | $J_p(4)$ | $J_p(5)$ |
|---|---|---|---|---|---|
| 0 | $+0.7652$ | $+0.2239$ | $-0.2601$ | $-0.3971$ | $-0.1776$ |
| 1.0 | $+0.4401$ | $+0.5767$ | $+0.3391$ | $-0.06604$ | $-0.3276$ |
| 2.0 | $+0.1149$ | $+0.3528$ | $+0.4861$ | $+0.3641$ | $+0.04657$ |
| 3.0 | $+0.01956$ | $+0.1289$ | $+0.3091$ | $+0.4302$ | $+0.3648$ |
| 3.5 | $+0.7186\cdot10^{-2}$ | $+0.06852$ | $+0.2101$ | $+0.3658$ | $+0.4100$ |
| 4.0 | $+0.2477\cdot10^{-2}$ | $+0.03400$ | $+0.1320$ | $+0.2811$ | $+0.3912$ |
| 4.5 | $+0.807\cdot10^{-3}$ | $+0.01589$ | $+0.07760$ | $+0.1993$ | $+0.3337$ |
| 5.0 | $+0.2498\cdot10^{-3}$ | $+0.7040\cdot10^{-2}$ | $+0.04303$ | $+0.1321$ | $+0.2611$ |
| 5.5 | $+0.74\cdot10^{-4}$ | $+0.2973\cdot10^{-2}$ | $+0.02266$ | $+0.08261$ | $+0.1906$ |
| 6.0 | $+0.2094\cdot10^{-4}$ | $+0.1202\cdot10^{-2}$ | $+0.01139$ | $+0.04909$ | $+0.1310$ |
| 6.5 | $+0.6\cdot10^{-5}$ | $+0.467\cdot10^{-3}$ | $+0.5493\cdot10^{-2}$ | $+0.02787$ | $+0.08558$ |
| 7.0 | $+0.1502\cdot10^{-5}$ | $+0.1749\cdot10^{-3}$ | $+0.2547\cdot10^{-2}$ | $+0.01518$ | $+0.05338$ |
| 8.0 | $+0.9422\cdot10^{-7}$ | $+0.2218\cdot10^{-4}$ | $+0.4934\cdot10^{-3}$ | $+0.4029\cdot10^{-2}$ | $+0.01841$ |
| 9.0 | $+0.5249\cdot10^{-8}$ | $+0.2492\cdot10^{-5}$ | $+0.8440\cdot10^{-4}$ | $+0.9386\cdot10^{-3}$ | $+0.5520\cdot10^{-2}$ |
| 10.0 | $+0.2631\cdot10^{-9}$ | $+0.2515\cdot10^{-6}$ | $+0.1293\cdot10^{-4}$ | $+0.1950\cdot10^{-3}$ | $+0.1468\cdot10^{-2}$ |

| $p$ | $J_p(6)$ | $J_p(7)$ | $J_p(8)$ | $J_p(9)$ | $J_p(10)$ |
|---|---|---|---|---|---|
| 0 | $+0.1506$ | $+0.3001$ | $+0.1717$ | $-0.09033$ | $-0.2459$ |
| 1.0 | $-0.2767$ | $-0.4683\cdot10^{-2}$ | $+0.2346$ | $+0.2453$ | $+0.04347$ |
| 2.0 | $-0.2429$ | $-0.3014$ | $-0.1130$ | $+0.1448$ | $+0.2546$ |
| 3.0 | $+0.1148$ | $-0.1676$ | $-0.2911$ | $-0.1809$ | $+0.05838$ |
| 3.5 | $+0.2671$ | $-0.3403\cdot10^{-2}$ | $-0.2326$ | $-0.2683$ | $-0.09965$ |
| 4.0 | $+0.3576$ | $+0.1578$ | $-0.1054$ | $-0.2655$ | $-0.2196$ |
| 4.5 | $+0.3846$ | $+0.2800$ | $+0.04712$ | $-0.1839$ | $-0.2664$ |
| 5.0 | $+0.3621$ | $+0.3479$ | $+0.1858$ | $-0.05504$ | $-0.2341$ |
| 5.5 | $+0.3098$ | $+0.3634$ | $+0.2856$ | $+0.08439$ | $-0.1401$ |
| 6.0 | $+0.2458$ | $+0.3392$ | $+0.3376$ | $+0.2043$ | $-0.01446$ |
| 6.5 | $+0.1833$ | $+0.2911$ | $+0.3456$ | $+0.2870$ | $+0.1123$ |
| 7.0 | $+0.1296$ | $+0.2336$ | $+0.3206$ | $+0.3275$ | $+0.2167$ |
| 8.0 | $+0.05653$ | $+0.1280$ | $+0.2235$ | $+0.3051$ | $+0.3179$ |
| 9.0 | $+0.02117$ | $+0.05892$ | $+0.1263$ | $+0.2149$ | $+0.2919$ |
| 10.0 | $+0.6964\cdot10^{-2}$ | $+0.02354$ | $+0.06077$ | $+0.1247$ | $+0.2075$ |

| $p$ | $J_p(11)$ | $J_p(12)$ | $J_p(13)$ | $J_p(14)$ | $J_p(15)$ |
|---|---|---|---|---|---|
| 0 | $-0.1712$ | $+0.04769$ | $+0.2069$ | $+0.1711$ | $-0.01422$ |
| 1.0 | $-0.1768$ | $-0.2234$ | $-0.07032$ | $+0.1334$ | $+0.2051$ |
| 2.0 | $+0.1390$ | $-0.08493$ | $-0.2177$ | $-0.1520$ | $+0.04157$ |
| 3.0 | $+0.2273$ | $+0.1951$ | $+0.3320\cdot10^{-2}$ | $-0.1768$ | $-0.1940$ |
| 3.5 | $+0.1294$ | $+0.2348$ | $+0.1407$ | $-0.06245$ | $-0.1991$ |
| 4.0 | $-0.01504$ | $+0.1825$ | $+0.2193$ | $+0.07624$ | $-0.1192$ |
| 4.5 | $-0.1519$ | $+0.06457$ | $+0.2134$ | $+0.1830$ | $+0.7984\cdot10^{-2}$ |
| 5.0 | $-0.2383$ | $-0.07347$ | $+0.1316$ | $+0.2204$ | $+0.1305$ |
| 5.5 | $-0.2538$ | $-0.1864$ | $+0.7055\cdot10^{-2}$ | $+0.1801$ | $+0.2039$ |
| 6.0 | $-0.2016$ | $-0.2437$ | $-0.1180$ | $+0.08117$ | $+0.2061$ |
| 6.5 | $-0.1018$ | $-0.2354$ | $-0.2075$ | $-0.04151$ | $+0.1415$ |
| 7.0 | $+0.01838$ | $-0.1703$ | $-0.2406$ | $-0.1508$ | $+0.03446$ |
| 8.0 | $+0.2250$ | $+0.04510$ | $-0.1410$ | $-0.2320$ | $-0.1740$ |
| 9.0 | $+0.3089$ | $+0.2304$ | $+0.06698$ | $-0.1143$ | $-0.2200$ |
| 10.0 | $+0.2804$ | $+0.3005$ | $+0.2338$ | $+0.08501$ | $-0.09007$ |

| $p$ | $J_p(16)$ | $J_p(17)$ | $J_p(18)$ | $J_p(19)$ | $J_p(20)$ |
|---|---|---|---|---|---|
| 0 | −0.174 9 | −0.169 9 | −0.013 36 | +0.146 6 | +0.167 0 |
| 1.0 | +0.090 40 | −0.097 67 | −0.188 0 | −0.105 7 | +0.066 83 |
| 2.0 | +0.186 2 | +0.158 4 | $-0.7533 \cdot 10^{-2}$ | −0.157 8 | −0.160 3 |
| 3.0 | −0.043 85 | +0.134 9 | +0.186 3 | +0.072 49 | −0.098 90 |
| 3.5 | −0.158 5 | +0.014 61 | +0.165 1 | +0.164 9 | +0.021 52 |
| 4.0 | −0.202 6 | −0.110 7 | +0.069 64 | +0.180 6 | +0.130 7 |
| 4.5 | −0.161 9 | −0.187 5 | −0.055 01 | +0.116 5 | +0.180 1 |
| 5.0 | −0.057 47 | −0.187 0 | −0.155 4 | $+0.3572 \cdot 10^{-2}$ | +0.151 2 |
| 5.5 | +0.067 43 | −0.113 9 | −0.192 6 | −0.109 7 | +0.059 53 |
| 6.0 | +0.166 7 | $+0.7153 \cdot 10^{-3}$ | −0.156 0 | −0.178 8 | −0.055 09 |
| 6.5 | +0.208 3 | +0.113 8 | −0.062 73 | −0.180 0 | −0.147 4 |
| 7.0 | +0.182 5 | +0.187 5 | +0.051 40 | −0.116 5 | −0.184 2 |
| 8.0 | $-0.7021 \cdot 10^{-2}$ | +0.153 7 | +0.195 9 | +0.092 94 | −0.073 87 |
| 9.0 | −0.189 5 | −0.042 86 | +0.122 8 | +0.194 7 | +0.125 1 |
| 10.0 | −0.206 2 | −0.199 1 | −0.073 17 | +0.091 55 | +0.186 5 |

## Spherical cylinder functions (Bessel functions) $J_{\pm(n+1/2)}$

| $x$ | $J_{1/2}$ | $J_{3/2}$ | $J_{5/2}$ | $J_{-1/2}$ | $J_{-3/2}$ | $J_{-5/2}$ |
|---|---|---|---|---|---|---|
| 0 | 0.000 0 | 0.000 0 | 0.000 0 | $+\infty$ | $-\infty$ | $+\infty$ |
| 1 | +0.671 4 | +0.240 3 | +0.049 5 | +0.431 1 | −1.102 5 | +2.876 4 |
| 2 | +0.513 0 | +0.491 3 | +0.223 9 | −0.234 8 | −0.395 6 | +0.828 2 |
| 3 | +0.065 0 | +0.477 7 | +0.412 7 | −0.456 0 | +0.087 0 | +0.369 0 |
| 4 | −0.301 9 | +0.185 3 | +0.440 9 | −0.260 8 | +0.367 1 | −0.014 6 |
| 5 | −0.342 2 | −0.169 7 | +0.240 4 | +0.101 2 | +0.321 9 | −0.294 4 |
| 6 | −0.091 0 | −0.327 9 | −0.073 0 | +0.312 8 | +0.038 9 | −0.332 2 |
| 7 | +0.198 1 | −0.199 1 | −0.283 4 | +0.227 4 | −0.230 6 | −0.128 5 |
| 8 | +0.279 1 | +0.075 9 | −0.250 6 | −0.041 0 | −0.274 0 | +0.143 8 |
| 9 | +0.109 6 | +0.254 5 | −0.024 8 | −0.242 3 | −0.082 7 | +0.269 9 |
| 10 | −0.137 3 | +0.198 0 | +0.196 7 | −0.211 7 | +0.158 4 | +0.164 2 |
| 11 | −0.240 6 | −0.022 9 | +0.234 3 | +0.001 1 | +0.240 5 | −0.066 6 |
| 12 | −0.123 6 | −0.204 7 | +0.072 4 | +0.194 4 | +0.107 4 | −0.221 2 |
| 13 | +0.093 0 | −0.193 7 | −0.137 7 | +0.200 8 | −0.108 4 | −0.175 8 |
| 14 | +0.211 2 | −0.014 1 | −0.214 3 | +0.029 2 | −0.213 3 | +0.016 6 |
| 15 | +0.134 0 | +0.165 4 | −0.100 9 | −0.156 5 | −0.123 5 | +0.181 2 |
| 16 | −0.057 4 | +0.187 4 | +0.092 6 | −0.191 0 | +0.069 4 | +0.178 0 |
| 17 | −0.186 0 | +0.042 3 | +0.193 5 | −0.053 2 | +0.189 2 | +0.019 9 |
| 18 | −0.141 2 | −0.132 0 | +0.119 2 | +0.124 2 | +0.134 3 | −0.146 6 |
| 19 | +0.027 4 | −0.179 5 | −0.055 8 | +0.181 0 | −0.037 0 | −0.175 1 |
| 20 | +0.162 9 | −0.064 7 | −0.172 6 | +0.072 8 | −0.166 5 | −0.047 8 |
| 21 | +0.145 7 | +0.102 3 | −0.131 1 | −0.095 4 | −0.141 1 | +0.115 5 |
| 22 | −0.001 5 | +0.170 0 | +0.024 7 | −0.170 1 | +0.009 2 | +0.168 8 |
| 23 | −0.140 8 | +0.082 5 | +0.151 6 | −0.088 6 | +0.144 6 | +0.069 8 |
| 24 | −0.147 5 | −0.075 2 | +0.138 1 | +0.069 1 | +0.144 6 | −0.087 2 |
| 25 | −0.021 1 | −0.159 0 | +0.002 0 | +0.158 2 | +0.014 8 | −0.159 9 |
| 26 | +0.119 3 | −0.096 6 | −0.130 5 | +0.101 2 | −0.123 2 | −0.087 0 |
| 27 | +0.146 9 | +0.050 3 | −0.141 3 | −0.044 9 | −0.145 2 | +0.061 0 |
| 28 | +0.040 8 | +0.146 6 | −0.025 1 | −0.145 1 | −0.035 7 | +0.149 0 |
| 29 | −0.098 3 | +0.107 4 | +0.109 4 | −0.110 8 | +0.102 1 | +0.100 3 |
| 30 | −0.143 9 | −0.027 3 | +0.141 2 | +0.022 5 | +0.143 2 | −0.036 8 |
| 31 | −0.057 9 | −0.133 0 | +0.045 0 | +0.131 1 | +0.053 7 | −0.136 3 |
| 32 | +0.077 8 | −0.115 2 | −0.088 6 | +0.117 7 | −0.081 4 | −0.110 0 |
| 33 | +0.138 9 | +0.006 1 | −0.138 3 | −0.001 8 | −0.138 8 | +0.014 5 |
| 34 | +0.072 4 | +0.118 2 | −0.062 0 | −0.116 1 | −0.069 0 | +0.122 2 |
| 35 | −0.057 8 | +0.120 2 | +0.068 0 | −0.121 9 | +0.061 2 | +0.116 6 |
| 36 | −0.131 9 | +0.013 4 | +0.133 0 | −0.017 0 | +0.132 4 | +0.006 0 |
| 37 | −0.084 4 | −0.102 7 | +0.076 1 | +0.100 4 | +0.081 7 | −0.107 0 |
| 38 | +0.038 4 | −0.122 6 | −0.048 0 | +0.123 6 | −0.041 6 | −0.120 3 |
| 39 | +0.123 1 | −0.030 9 | −0.125 5 | +0.034 1 | −0.124 0 | −0.024 5 |
| 40 | +0.094 0 | +0.086 5 | −0.087 5 | −0.084 1 | −0.091 9 | +0.091 0 |

**The $n^{th}$ zero of some Bessel functions**

| $n$ | $p = 0$ | $p = 1$ | $p = 2$ | $p = 3$ | $p = 4$ | $p = 5$ |
|---|---|---|---|---|---|---|
| 1 | 2.405 | 3.832 | 5.135 | 6.379 | 7.588 | 8.771 |
| 2 | 5.520 | 7.016 | 8.417 | 9.760 | 11.064 | 12.339 |
| 3 | 8.654 | 10.173 | 11.620 | 13.015 | 14.373 | 15.700 |
| 4 | 11.792 | 13.323 | 14.796 | 16.224 | 17.616 | 18.980 |
| 5 | 14.931 | 16.470 | 17.960 | 19.410 | 20.827 | 22.218 |
| 6 | 18.071 | 19.616 | 21.117 | 22.583 | 24.018 | 25.430 |
| 7 | 21.212 | 22.760 | 24.270 | 25.749 | 27.200 | 28.627 |
| 8 | 24.353 | 25.903 | 27.421 | 28.909 | 30.371 | 31.812 |
| 9 | 27.494 | 29.047 | 30.569 | 32.065 | 33.537 | 34.989 |

## 0.5.3 Spherical functions (Legendre polynomials)

**Remark:** See also section 1.13.2.13.

| $x = P_1(x)$ | $P_2(x)$ | $P_3(x)$ | $P_4(x)$ | $P_5(x)$ | $P_6(x)$ | $P_7(x)$ |
|---|---|---|---|---|---|---|
| 0.00 | –0. 500 0 | 0. 000 0 | 0. 375 0 | 0. 000 0 | –0. 312 5 | 0. 000 0 |
| 0.05 | –0. 496 2 | –0. 074 7 | 0. 365 7 | 0. 092 7 | –0. 296 2 | –0. 106 9 |
| 0.10 | –0. 485 0 | –0. 147 5 | 0. 337 9 | 0. 178 8 | –0. 248 8 | –0. 199 5 |
| 0.15 | –0. 466 2 | –0. 216 6 | 0. 292 8 | 0. 252 3 | –0. 174 6 | –0. 264 9 |
| 0.20 | –0. 440 0 | –0. 280 0 | 0. 232 0 | 0. 307 5 | –0. 080 6 | –0. 293 5 |
| 0.25 | –0. 406 2 | –0. 335 9 | 0. 157 7 | 0. 339 7 | +0. 024 3 | –0. 279 9 |
| 0.30 | –0. 365 0 | –0. 382 5 | +0. 072 9 | 0. 345 4 | 0. 129 2 | –0. 224 1 |
| 0.35 | –0. 316 2 | –0. 417 8 | –0. 018 7 | 0. 322 5 | 0. 222 5 | –0. 131 8 |
| 0.40 | –0. 260 0 | –0. 440 0 | –0. 113 0 | 0. 270 6 | 0. 292 6 | –0. 014 6 |
| 0.45 | –0. 196 2 | –0. 447 2 | –0. 205 0 | 0. 191 7 | 0. 329 0 | +0. 110 6 |
| 0.50 | –0. 125 0 | –0. 437 5 | –0. 289 1 | +0. 089 8 | 0. 323 2 | 0. 223 1 |
| 0.55 | –0. 046 2 | –0. 409 1 | –0. 359 0 | –0. 028 2 | 0. 270 8 | 0. 300 7 |
| 0.60 | +0. 040 0 | –0. 360 0 | –0. 408 0 | –0. 152 6 | 0. 172 1 | 0. 322 6 |
| 0.65 | 0. 133 8 | –0. 288 4 | –0. 428 4 | –0. 270 5 | +0. 034 7 | 0. 273 7 |
| 0.70 | 0. 235 0 | –0. 192 5 | –0. 412 1 | –0. 365 2 | –0. 125 3 | +0. 150 2 |
| 0.75 | 0. 343 8 | –0. 070 3 | –0. 350 1 | –0. 416 4 | –0. 280 8 | –0. 034 2 |
| 0.80 | 0. 460 0 | +0. 080 0 | –0. 233 0 | –0. 399 5 | –0. 391 8 | –0. 239 7 |
| 0.85 | 0. 583 8 | 0. 260 3 | –0. 050 6 | –0. 285 7 | –0. 403 0 | –0. 391 3 |
| 0.90 | 0. 715 0 | 0. 472 5 | +0. 207 9 | –0. 041 1 | –0. 241 2 | –0. 367 8 |
| 0.95 | 0. 853 8 | 0. 718 4 | 0. 554 1 | +0. 372 7 | +0. 187 5 | +0. 011 2 |

One has: $P_n(1) = 1$ for all $n = 1, 2, \ldots$

$$P_0(x) = 1, \qquad P_1(x) = x,$$

$$P_2(x) = \frac{1}{2}(3x^2 - 1),$$

$$P_3(x) = \frac{1}{2}(5x^3 - 3x),$$

$$P_4(x) = \frac{1}{8}(35x^4 - 30x^2 + 3).$$

$$P_5(x) = \frac{1}{8}(63x^5 - 70x^3 + 15x),$$

$$P_6(x) = \frac{1}{16}(231x^6 - 315x^4 + 105x^2 - 5),$$

$$P_7(x) = \frac{1}{16}(429x^7 - 693x^5 + 315x^3 - 35x),$$

## 0.5.4 Elliptic integrals

**Remark:** See also section 1.14.19.

**a) Elliptic integrals of the first kind $F(k, \varphi)$, $k = \sin\alpha$.**

| | $\alpha = 0°$ | 10° | 20° | 30° | 40° |
|---|---|---|---|---|---|
| $\varphi$=0° | 0.0000 | 0.0000 | 0.0000 | 0.0000 | 0.0000 |
| 10 | 0.1745 | 0.1746 | 0.1746 | 0.1748 | 0.1749 |
| 20 | 0.3491 | 0.3493 | 0.3499 | 0.3508 | 0.3520 |
| 30 | 0.5236 | 0.5243 | 0.5263 | 0.5294 | 0.5334 |
| 40 | 0.6981 | 0.6997 | 0.7043 | 0.7116 | 0.7213 |
| 50 | 0.8727 | 0.8756 | 0.8842 | 0.8982 | 0.9173 |
| 60 | 1.0472 | 1.0519 | 1.0660 | 1.0896 | 1.1226 |
| 70 | 1.2217 | 1.2286 | 1.2495 | 1.2853 | 1.3372 |
| 80 | 1.3963 | 1.4056 | 1.4344 | 1.4846 | 1.5597 |
| 90 | 1.5708 | 1.5828 | 1.6200 | 1.6858 | 1.7868 |

| | $\alpha = 50°$ | 60° | 70° | 80° | 90° |
|---|---|---|---|---|---|
| $\varphi$=0° | 0.0000 | 0.0000 | 0.0000 | 0.0000 | 0.0000 |
| 10 | 0.1751 | 0.1752 | 0.1753 | 0.1754 | 0.1754 |
| 20 | 0.3533 | 0.3545 | 0.3555 | 0.3561 | 0.3564 |
| 30 | 0.5379 | 0.5422 | 0.5459 | 0.5484 | 0.5493 |
| 40 | 0.7323 | 0.7436 | 0.7535 | 0.7604 | 0.7629 |
| 50 | 0.9401 | 0.9647 | 0.9876 | 1.0044 | 1.0107 |
| 60 | 1.1643 | 1.2126 | 1.2619 | 1.3014 | 1.3170 |
| 70 | 1.4068 | 1.4944 | 1.5959 | 1.6918 | 1.7354 |
| 80 | 1.6660 | 1.8125 | 2.0119 | 2.2653 | 2.4362 |
| 90 | 1.9356 | 2.1565 | 2.5046 | 3.1534 | $\infty$ |

**b) Elliptic integrals of the second kind $E(k, \varphi)$, $k = \sin\alpha$.**

| | $\alpha = 0°$ | 10° | 20° | 30° | 40° |
|---|---|---|---|---|---|
| $\varphi$=0° | 0.0000 | 0.0000 | 0.0000 | 0.0000 | 0.0000 |
| 10 | 0.1745 | 0.1745 | 0.1744 | 0.1743 | 0.1742 |
| 20 | 0.3491 | 0.3489 | 0.3483 | 0.3473 | 0.3462 |
| 30 | 0.5236 | 0.5229 | 0.5209 | 0.5179 | 0.5141 |
| 40 | 0.6981 | 0.3966 | 0.6921 | 0.6851 | 0.6763 |
| 50 | 0.8727 | 0.8698 | 0.8614 | 0.8483 | 0.8317 |
| 60 | 1.0472 | 1.0426 | 1.0290 | 1.0076 | 0.9801 |
| 70 | 1.2217 | 1.2149 | 1.1949 | 1.1632 | 1.1221 |
| 80 | 1.3963 | 1.3870 | 1.3597 | 1.3161 | 1.2590 |
| 90 | 1.5708 | 1.5589 | 1.5238 | 1.4675 | 1.3931 |

| | $\alpha = 50°$ | 60° | 70° | 80° | 90° |
|---|---|---|---|---|---|
| $\varphi$=0° | 0.0000 | 0.0000 | 0.0000 | 0.0000 | 0.0000 |
| 10 | 0.1740 | 0.1739 | 0.1738 | 1.1737 | 0.1736 |
| 20 | 0.3450 | 0.3438 | 0.3429 | 0.3422 | 0.3420 |
| 30 | 0.5100 | 0.5061 | 0.5029 | 0.5007 | 0.5000 |
| 40 | 0.6667 | 0.6575 | 0.6497 | 0.6446 | 0.6428 |
| 50 | 0.8134 | 0.7954 | 0.7801 | 0.7697 | 0.7660 |
| 60 | 0.9493 | 0.9184 | 0.8914 | 0.8728 | 0.8660 |
| 70 | 1.0750 | 1.0266 | 0.9830 | 0.9514 | 0.9397 |
| 80 | 1.1926 | 1.1225 | 1.0565 | 1.0054 | 0.9848 |
| 90 | 1.3055 | 1.2111 | 1.1184 | 1.0401 | 1.0000 |

**c) Complete elliptic integrals K and E, $k = \sin\alpha$; for $\alpha = 90°$, we set K$=\infty$, E$=1$.**

| $\alpha°$ | K | E | $\alpha°$ | K | E | $\alpha°$ | K | E |
|---|---|---|---|---|---|---|---|---|
| 0 | 1.5708 | 1.5708 | 30 | 1.6858 | 1.4675 | 60 | 2.1565 | 1.2111 |
| 1 | 1.5709 | 1.5707 | 31 | 1.6941 | 1.4608 | 61 | 2.1842 | 1.2015 |
| 2 | 1.5713 | 1.5703 | 32 | 1.7028 | 1.4539 | 62 | 2.2132 | 1.1920 |
| 3 | 1.5719 | 1.5697 | 33 | 1.7119 | 1.4469 | 63 | 2.2435 | 1.1826 |
| 4 | 1.5727 | 1.5689 | 34 | 1.7214 | 1.4397 | 64 | 2.2754 | 1.1732 |
| 5 | 1.5738 | 1.5678 | 35 | 1.7312 | 1.4323 | 65 | 2.3088 | 1.1638 |
| 6 | 1.5751 | 1.5665 | 36 | 1.7415 | 1.4248 | 66 | 2.3439 | 1.1545 |
| 7 | 1.5767 | 1.5649 | 37 | 1.7522 | 1.4171 | 67 | 2.3809 | 1.1453 |
| 8 | 1.5785 | 1.5632 | 38 | 1.7633 | 1.4092 | 68 | 2.4198 | 1.1362 |
| 9 | 1.5805 | 1.5611 | 39 | 1.7748 | 1.4013 | 69 | 2.4610 | 1.1272 |
| 10 | 1.5828 | 1.5589 | 40 | 1.7868 | 1.3931 | 70 | 2.5046 | 1.1184 |
| 11 | 1.5854 | 1.5564 | 41 | 1.7992 | 1.3849 | 71 | 2.5507 | 1.1096 |
| 12 | 1.5882 | 1.5537 | 42 | 1.8122 | 1.3765 | 72 | 2.5998 | 1.1011 |
| 13 | 1.5913 | 1.5507 | 43 | 1.8256 | 1.3680 | 73 | 2.6521 | 1.0927 |
| 14 | 1.5946 | 1.5476 | 44 | 1.8396 | 1.3594 | 74 | 2.7081 | 1.0844 |
| 15 | 1.5981 | 1.5442 | 45 | 1.8541 | 1.3506 | 75 | 2.7681 | 1.0764 |
| 16 | 1.6020 | 1.5405 | 46 | 1.8691 | 1.3418 | 76 | 2.8327 | 1.0686 |
| 17 | 1.6061 | 1.5367 | 47 | 1.8848 | 1.3329 | 77 | 2.9026 | 1.0611 |
| 18 | 1.6105 | 1.5326 | 48 | 1.9011 | 1.3238 | 78 | 2.9786 | 1.0538 |
| 19 | 1.6151 | 1.5283 | 49 | 1.9180 | 1.3147 | 79 | 3.0617 | 1.0468 |
| 20 | 1.6200 | 1.5238 | 50 | 1.9356 | 1.3055 | 80 | 3.1534 | 1.0401 |
| 21 | 1.6252 | 1.5191 | 51 | 1.9539 | 1.2963 | 81 | 3.2553 | 1.0338 |
| 22 | 1.6307 | 1.5141 | 52 | 1.9729 | 1.2870 | 82 | 3.3699 | 1.0278 |
| 23 | 1.6365 | 1.5090 | 53 | 1.9927 | 1.2776 | 83 | 3.5004 | 1.0223 |
| 24 | 1.6426 | 1.5037 | 54 | 2.0133 | 1.2681 | 84 | 3.6519 | 1.0172 |
| 25 | 1.6490 | 1.4981 | 55 | 2.0347 | 1.2587 | 85 | 3.8317 | 1.0127 |
| 26 | 1.6557 | 1.4924 | 56 | 2.0571 | 1.2492 | 86 | 4.0528 | 1.0086 |
| 27 | 1.6627 | 1.4864 | 57 | 2.0804 | 1.2397 | 87 | 4.3387 | 1.0053 |
| 28 | 1.6701 | 1.4803 | 58 | 2.1047 | 1.2301 | 88 | 4.7427 | 1.0026 |
| 29 | 1.6777 | 1.4740 | 59 | 2.1300 | 1.2206 | 89 | 5.4349 | 1.0008 |

$$F(k,\varphi) = \int_0^{\varphi} \frac{\mathrm{d}\psi}{\sqrt{1-k^2\sin^2\psi}} = \int_0^{\sin\varphi} \frac{\mathrm{d}x}{\sqrt{1-x^2}\sqrt{1-k^2x^2}},$$

$$E(k,\,\varphi) = \int_0^{\varphi} \sqrt{1-k^2\sin^2\psi}\,\mathrm{d}\psi = \int_0^{\sin\varphi} \sqrt{\frac{1-k^2x^2}{1-x^2}}\,\mathrm{d}x,$$

$$\mathrm{K} = F\Big(k,\,\frac{\pi}{2}\Big) = \int_0^{\pi/2} \frac{\mathrm{d}\psi}{\sqrt{1-k^2\sin^2\psi}} = \int_0^1 \frac{\mathrm{d}x}{\sqrt{1-x^2}\sqrt{1-k^2x^2}},$$

$$\mathrm{E} = E\Big(k,\,\frac{\pi}{2}\Big) = \int_0^{\pi/2} \sqrt{1-k^2\sin^2\psi}\,\mathrm{d}\psi = \int_0^1 \sqrt{\frac{1-k^2x^2}{1-x^2}}\,\mathrm{d}x.$$

$$4(n+1)^2 \int \mathrm{E}x^n\mathrm{d}x - (2n+3)(2n+5)\int \mathrm{E}x^{n+1}\mathrm{d}x$$
$$= (2n+3)^2 \int \mathrm{K}x^{n+1}\mathrm{d}x - 4(n+1)^2\int \mathrm{K}x^n\mathrm{d}x = 2x^{n+1}(\mathrm{E}-(2n+3)(1-x)\mathrm{K}).$$

### 0.5.5 Integral trigonometric and exponential functions

**Definition:** $\mathrm{Si}(x) = \int\limits_0^x \frac{\sin t}{t}\,\mathrm{d}t, \qquad \mathrm{si}(x) = \mathrm{Si}(x) - \frac{\pi}{2} = -\int\limits_x^\infty \frac{\sin t}{t}\,\mathrm{d}t,$

$$\mathrm{Ci}(x) = -\int\limits_x^\infty \frac{\cos t}{t}\,\mathrm{d}t, \qquad \mathrm{Ei}(x) = \int\limits_{-\infty}^x \frac{\mathrm{e}^t}{t}\,\mathrm{d}t,$$

$$\mathrm{li}(x) = \int\limits_0^x \frac{\mathrm{d}t}{\ln t}, \qquad \mathrm{li}(x) = \mathrm{Ei}(\ln x).$$

| $x$ | Si($x$) | Ci($x$) | Ei($x$) | $x$ | Si($x$) | Ci($x$) | Ei($x$) |
|---|---|---|---|---|---|---|---|
| 0.00 | 0.0000 | $-\infty$ | $-\infty$ | 0.40 | 0.3965 | −0.3788 | 0.1048 |
| 0.01 | 0.0100 | −4.0280 | −4.0179 | 0.41 | 0.4062 | −0.3561 | 0.1418 |
| 0.02 | 0.0200 | −3.3349 | −3.3147 | 0.42 | 0.4159 | −0.3341 | 0.1783 |
| 0.03 | 0.0300 | −2.9296 | −2.8991 | 0.43 | 0.4256 | −0.3126 | 0.2143 |
| 0.04 | 0.0400 | −2.6421 | −2.6013 | 0.44 | 0.4353 | −0.2918 | 0.2498 |
| 0.05 | 0.0500 | −2.4191 | −2.3679 | 0.45 | 0.4450 | −0.2715 | 0.2849 |
| 0.06 | 0.0600 | −2.2371 | −2.1753 | 0.46 | 0.4546 | −0.2517 | 0.3195 |
| 0.07 | 0.0700 | −2.0833 | −2.0108 | 0.47 | 0.4643 | −0.2325 | 0.3537 |
| 0.08 | 0.0800 | −1.9501 | −1.8669 | 0.48 | 0.4739 | −0.2138 | 0.3876 |
| 0.09 | 0.0900 | −1.8328 | −1.7387 | 0.49 | 0.4835 | −0.1956 | 0.4211 |
| 0.10 | 0.0999 | −1.7279 | −1.6228 | 0.50 | 0.4931 | −0.1778 | 0.4542 |
| 0.11 | 0.1099 | −1.6331 | −1.5170 | 0.51 | 0.5027 | −0.1605 | 0.4870 |
| 0.12 | 0.1199 | −1.5466 | −1.4193 | 0.52 | 0.5123 | −0.1436 | 0.5195 |
| 0.13 | 0.1299 | −1.4672 | −1.3287 | 0.53 | 0.5218 | −0.1271 | 0.5517 |
| 0.14 | 0.1399 | −1.3938 | −1.2438 | 0.54 | 0.5313 | −0.1110 | 0.5836 |
| 0.15 | 0.1498 | −1.3255 | −1.1641 | 0.55 | 0.5408 | −0.0953 | 0.6153 |
| 0.16 | 0.1598 | −1.2618 | −1.0887 | 0.56 | 0.5503 | −0.0800 | 0.6467 |
| 0.17 | 0.1697 | −1.2020 | −1.0172 | 0.57 | 0.5598 | −0.0650 | 0.6778 |
| 0.18 | 0.1797 | −1.1457 | −0.9491 | 0.58 | 0.5693 | −0.0504 | 0.7087 |
| 0.19 | 0.1896 | −1.0925 | −0.8841 | 0.59 | 0.5787 | −0.0362 | 0.7394 |
| 0.20 | 0.1996 | −1.0422 | −0.8218 | 0.60 | 0.5881 | −0.0223 | 0.7699 |
| 0.21 | 0.2095 | −0.9944 | −0.7619 | 0.61 | 0.5975 | −0.0087 | 0.8002 |
| 0.22 | 0.2194 | −0.9490 | −0.7042 | 0.62 | 0.6069 | +0.0046 | 0.8302 |
| 0.23 | 0.2293 | −0.9057 | −0.6485 | 0.63 | 0.6163 | 0.0176 | 0.8601 |
| 0.24 | 0.2392 | −0.8643 | −0.5947 | 0.64 | 0.6256 | 0.0303 | 0.8898 |
| 0.25 | 0.2491 | −0.8247 | −0.5425 | 0.65 | 0.6349 | 0.0427 | 0.9194 |
| 0.26 | 0.2590 | −0.7867 | −0.4919 | 0.66 | 0.6442 | 0.0548 | 0.9488 |
| 0.27 | 0.2689 | −0.7503 | −0.4427 | 0.67 | 0.6535 | 0.0666 | 0.9780 |
| 0.28 | 0.2788 | −0.7153 | −0.3949 | 0.68 | 0.6628 | 0.0782 | 1.0071 |
| 0.29 | 0.2886 | −0.6816 | −0.3482 | 0.69 | 0.6720 | 0.0895 | 1.0361 |
| 0.30 | 0.2985 | −0.6492 | −0.3027 | 0.70 | 0.6812 | 0.1005 | 1.0649 |
| 0.31 | 0.3083 | −0.6179 | −0.2582 | 0.71 | 0.6904 | 0.1113 | 1.0936 |
| 0.32 | 0.3182 | −0.5877 | −0.2147 | 0.72 | 0.6996 | 0.1219 | 1.1222 |
| 0.33 | 0.3280 | −0.5585 | −0.1721 | 0.73 | 0.7087 | 0.1322 | 1.1507 |
| 0.34 | 0.3378 | −0.5304 | −0.1304 | 0.74 | 0.7179 | 0.1423 | 1.1791 |
| 0.35 | 0.3476 | −0.5031 | −0.0894 | 0.75 | 0.7270 | 0.1522 | 1.2073 |
| 0.36 | 0.3574 | −0.4767 | −0.0493 | 0.76 | 0.7360 | 0.1618 | 1.2355 |
| 0.37 | 0.3672 | −0.4511 | −0.0098 | 0.77 | 0.7451 | 0.1712 | 1.2636 |
| 0.38 | 0.3770 | −0.4263 | +0.0290 | 0.78 | 0.7541 | 0.1805 | 1.2916 |
| 0.39 | 0.3867 | −0.4022 | 0.0672 | 0.79 | 0.7631 | 0.1895 | 1.3195 |

| $x$ | $\mathbf{Si}(x)$ | $\mathbf{Ci}(x)$ | $\mathbf{Ei}(x)$ | $x$ | $\mathbf{Si}(x)$ | $\mathbf{Ci}(x)$ | $\mathbf{Ei}(x)$ |
|---|---|---|---|---|---|---|---|
| 0.80 | 0.772 1 | 0.198 3 | 1.347 4 | 2.6 | 1.800 4 | 0.253 3 | 7.576 1 |
| 0.81 | 0.781 1 | 0.206 9 | 1.375 2 | 2.7 | 1.818 2 | 0.220 1 | 8.110 3 |
| 0.82 | 0.790 0 | 0.215 3 | 1.402 9 | 2.8 | 1.832 1 | 0.186 5 | 8.679 3 |
| 0.83 | 0.798 9 | 0.223 5 | 1.430 6 | 2.9 | 2.842 2 | 0.152 9 | 9.286 0 |
| 0.84 | 0.807 8 | 0.231 6 | 1.458 2 | 3.0 | 1.848 7 | 0.119 6 | 9.933 8 |
| 0.85 | 0.816 6 | 0.239 4 | 1.485 7 | 3.1 | 1.851 7 | 0.086 99 | 10.626 3 |
| 0.86 | 0.825 4 | 0.247 1 | 1.513 2 | 3.2 | 1.851 4 | 0.055 26 | 11.367 3 |
| 0.87 | 0.834 2 | 0.254 6 | 1.540 7 | 3.3 | 1.848 1 | +0.024 68 | 12.161 0 |
| 0.88 | 0.843 0 | 0.261 9 | 1.568 1 | 3.4 | 1.841 9 | −0.004 52 | 13.012 1 |
| 0.89 | 0.851 8 | 0.269 1 | 1.595 5 | 3.5 | 1.833 1 | −0.032 13 | 13.925 4 |
| 0.90 | 0.860 5 | 0.276 1 | 1.622 8 | 3.6 | 1.821 9 | −0.057 97 | 14.906 3 |
| 0.91 | 0.869 2 | 0.282 9 | 1.650 1 | 3.7 | 1.808 6 | −0.081 9 | 15.960 6 |
| 0.92 | 0.877 8 | 0.289 6 | 1.677 4 | 3.8 | 1.793 4 | −0.103 8 | 17.094 8 |
| 0.93 | 0.886 5 | 0.296 1 | 1.704 7 | 3.9 | 1.776 5 | −0.123 5 | 18.315 7 |
| 0.94 | 0.895 1 | 0.302 4 | 1.731 9 | 4.0 | 1.758 2 | −0.141 0 | 19.630 9 |
| 0.95 | 0.903 6 | 0.308 6 | 1.759 1 | 4.1 | 1.738 7 | −0.156 2 | 21.048 5 |
| 0.96 | 0.912 2 | 0.314 7 | 1.786 4 | 4.2 | 1.718 4 | −0.169 0 | 22.577 4 |
| 0.97 | 0.920 7 | 0.320 6 | 1.813 6 | 4.3 | 1.697 3 | −0.179 5 | 24.227 4 |
| 0.98 | 0.929 2 | 0.326 3 | 1.840 7 | 4.4 | 1.675 8 | −0.187 7 | 26.009 0 |
| 0.99 | 0.937 7 | 0.331 9 | 1.867 9 | 4.5 | 1.654 1 | −0.193 5 | 27.933 7 |
| 1.0 | 0.946 1 | 0.337 4 | 1.895 1 | 4.6 | 1.632 5 | −0.197 0 | 30.014 1 |
| 1.1 | 1.028 7 | 0.384 9 | 2.167 4 | 4.7 | 1.611 0 | −0.198 4 | 32.263 9 |
| 1.2 | 1.108 0 | 0.420 5 | 2.442 1 | 4.8 | 1.590 0 | −0.197 6 | 34.697 9 |
| 1.3 | 1.184 0 | 0.445 7 | 2.721 4 | 4.9 | 1.569 6 | −0.194 8 | 37.332 5 |
| 1.4 | 1.256 2 | 0.462 0 | 3.007 2 | 5.0 | 1.549 9 | −0.190 0 | 40.185 3 |
| 1.5 | 1.324 7 | 0.470 4 | 3.301 3 | 6 | 1.424 7 | −0.068 1 | 85.989 8 |
| 1.6 | 1.389 2 | 0.471 7 | 3.605 3 | 7 | 1.454 6 | +0.076 7 | 191.505 |
| 1.7 | 1.449 6 | 0.467 0 | 3.921 0 | 8 | 1.574 2 | +0.122 4 | 440.380 |
| 1.8 | 1.505 8 | 0.456 8 | 4.249 9 | 9 | 1.665 0 | +0.055 35 | 1 037.88 |
| 1.9 | 1.557 8 | 0.441 9 | 4.593 7 | 10 | 1.658 3 | −0.045 46 | 2 492.23 |
| 2.0 | 1.605 4 | 0.423 0 | 4.954 2 | 11 | 1.578 3 | −0.089 56 | 6 071.41 |
| 2.1 | 1.648 7 | 0.400 5 | 5.333 2 | 12 | 1.505 0 | −0.049 78 | 14 959.5 |
| 2.2 | 1.687 6 | 0.375 1 | 5.732 6 | 13 | 1.499 4 | +0.026 76 | 37 197.7 |
| 2.3 | 1.722 2 | 0.347 2 | 6.154 4 | 14 | 1.556 2 | +0.069 40 | 93 192.5 |
| 2.4 | 1.752 5 | 0.317 3 | 6.600 7 | 15 | 1.618 2 | +0.046 28 | 234 956.0 |
| 2.5 | 1.778 5 | 0.285 9 | 7.073 8 | | | | |

| $x$ | $\mathbf{Si}(x)$ | $\mathbf{Ci}(x)$ | $x$ | $\mathbf{Si}(x)$ | $\mathbf{Ci}(x)$ |
|---|---|---|---|---|---|
| 20 | 1.548 2 | +0.044 42 | 120 | 1.564 0 | +0.004 78 |
| 25 | 1.531 5 | −0.006 85 | 140 | 1.572 2 | +0.007 01 |
| 30 | 1.566 8 | −0.033 03 | 160 | 1.576 9 | +0.001 41 |
| 35 | 1.596 9 | −0.011 48 | 180 | 1.574 1 | −0.004 43 |
| 40 | 1.587 0 | +0.019 02 | 200 | 1.568 4 | −0.004 38 |
| 45 | 1.558 7 | +0.018 63 | 300 | 1.570 9 | −0.003 33 |
| 50 | 1.551 6 | −0.005 63 | 400 | 1.572 1 | −0.002 12 |
| 55 | 1.570 7 | −0.018 17 | 500 | 1.572 6 | −0.000 93 |
| 60 | 1.586 7 | −0.004 81 | 600 | 1.572 5 | +0.000 08 |
| 65 | 1.579 2 | +0.012 85 | 700 | 1.572 0 | +0.000 78 |
| 70 | 1.561 6 | +0.010 92 | 800 | 1.571 4 | +0.001 12 |
| 80 | 1.572 3 | −0.012 40 | $10^3$ | 1.570 2 | +0.000 83 |
| 90 | 1.575 7 | +0.009 99 | $10^4$ | 1.570 9 | −0.000 03 |
| 100 | 1.562 2 | −0.005 15 | $10^5$ | 1.570 8 | +0.000 00 |
| 110 | 1.579 9 | −0.000 32 | $\infty$ | $\pi/2$ | +0.000 00 |

### 0.5.6 Fresnel integrals

**Remark:** See also section 0.10.1.

| $x$ | $C(x)$ | $S(x)$ | $x$ | $C(x)$ | $S(x)$ | $x$ | $C(x)$ | $S(x)$ |
|---|---|---|---|---|---|---|---|---|
| 0.0 | 0 | 0 | 8.5 | 0.6129 | 0.5755 | 21.0 | 0.5738 | 0.5459 |
| 0.1 | 0.2521 | 0.0084 | 9.0 | 0.5608 | 0.6172 | 21.5 | 0.5423 | 0.5748 |
| 0.2 | 0.3554 | 0.0238 | 9.5 | 0.4969 | 0.6286 | 22.0 | 0.5012 | 0.5849 |
| 0.3 | 0.4331 | 0.0434 | 10.0 | 0.4370 | 0.6084 | 22.5 | 0.4607 | 0.5742 |
| 0.4 | 0.4966 | 0.0665 | 10.5 | 0.3951 | 0.5632 | 23.0 | 0.4307 | 0.5458 |
| 0.5 | 0.5502 | 0.0924 | 11.0 | 0.3804 | 0.5048 | 23.5 | 0.4181 | 0.5068 |
| 0.6 | 0.5962 | 0.1205 | 11.5 | 0.3951 | 0.4478 | 24.0 | 0.4256 | 0.4670 |
| 0.7 | 0.6356 | 0.1504 | 12.0 | 0.4346 | 0.4058 | 24.5 | 0.4511 | 0.4361 |
| 0.8 | 0.6693 | 0.1818 | 12.5 | 0.4881 | 0.3882 | 25.0 | 0.4879 | 0.4212 |
| 0.9 | 0.6979 | 0.2143 | 13.0 | 0.5425 | 0.3983 | 25.5 | 0.5269 | 0.4258 |
| 1.0 | 0.7217 | 0.2476 | 13.5 | 0.5846 | 0.4325 | 26.0 | 0.5586 | 0.4483 |
| 1.5 | 0.7791 | 0.4155 | 14.0 | 0.6047 | 0.4818 | 26.5 | 0.5755 | 0.4829 |
| 2.0 | 0.7533 | 0.5628 | 14.5 | 0.5989 | 0.5337 | 27.0 | 0.5738 | 0.5211 |
| 2.5 | 0.6710 | 0.6658 | 15.0 | 0.5693 | 0.5758 | 27.5 | 0.5541 | 0.5534 |
| 3.0 | 0.5610 | 0.7117 | 15.5 | 0.5240 | 0.5982 | 28.0 | 0.5217 | 0.5721 |
| 3.5 | 0.4520 | 0.7002 | 16.0 | 0.4743 | 0.5961 | 28.5 | 0.4846 | 0.5731 |
| 4.0 | 0.3682 | 0.6421 | 16.5 | 0.4323 | 0.5709 | 29.0 | 0.4518 | 0.5562 |
| 4.5 | 0.3252 | 0.5565 | 17.0 | 0.4080 | 0.5293 | 29.5 | 0.4314 | 0.5260 |
| 5.0 | 0.3285 | 0.4659 | 17.5 | 0.4066 | 0.4818 | 30.0 | 0.4279 | 0.4900 |
| 5.5 | 0.3724 | 0.3918 | 18.0 | 0.4278 | 0.4400 | 30.5 | 0.4420 | 0.4570 |
| 6.0 | 0.4433 | 0.3499 | 18.5 | 0.4660 | 0.4139 | 31.0 | 0.4700 | 0.4350 |
| 6.5 | 0.5222 | 0.3471 | 19.0 | 0.5113 | 0.4093 | 31.5 | 0.5048 | 0.4291 |
| 7.0 | 0.5901 | 0.3812 | 19.5 | 0.5528 | 0.4269 | 32.0 | 0.5379 | 9.4406 |
| 7.5 | 0.6318 | 0.4415 | 20.0 | 0.5804 | 0.4616 | 32.5 | 0.5613 | 0.4663 |
| 8.0 | 0.6393 | 0.5120 | 20.5 | 0.5878 | 0.5049 | 33.0 | 0.5694 | 0.4999 |

### 0.5.7 The function $\int_0^x e^{t^2}\,dt$

| $x$ | 0 | 1 | 2 | 3 | 4 | 5 | 6 | 7 | 8 | 9 |
|---|---|---|---|---|---|---|---|---|---|---|
| 0.0 | 0.0000 | 0.0100 | 0.0200 | 0.0300 | 0.0400 | 0.0500 | 0.0601 | 0.0701 | 0.0802 | 0.0902 |
| 0.1 | 0.1003 | 0.1104 | 0.1206 | 0.1307 | 0.1409 | 0.1511 | 0.1614 | 0.1717 | 0.1820 | 0.1923 |
| 0.2 | 0.2027 | 0.2131 | 0.2236 | 0.2341 | 0.2447 | 0.2553 | 0.2660 | 0.2767 | 0.2875 | 0.2983 |
| 0.3 | 0.3092 | 0.3202 | 0.3313 | 0.3424 | 0.3536 | 0.3648 | 0.3762 | 0.3876 | 0.3991 | 0.4107 |
| 0.4 | 0.4224 | 0.4342 | 0.4461 | 0.4580 | 0.4701 | 0.4823 | 0.4946 | 0.5070 | 0.5196 | 0.5322 |
| 0.5 | 0.5450 | 0.5579 | 0.5709 | 0.5841 | 0.5974 | 0.6109 | 0.6245 | 0.6382 | 0.6522 | 0.6662 |
| 0.6 | 0.6805 | 0.6949 | 0.7095 | 0.7243 | 0.7393 | 0.7544 | 0.7698 | 0.7853 | 0.8011 | 0.8171 |
| 0.7 | 0.8333 | 0.8497 | 0.8664 | 0.8833 | 0.9005 | 0.9179 | 0.9356 | 0.9536 | 0.9718 | 0.9903 |
| 0.8 | 1.0091 | 1.0282 | 1.0477 | 1.0674 | 1.0875 | 1.1079 | 1.1287 | 1.1498 | 1.1713 | 1.1932 |
| 0.9 | 1.2155 | 1.2382 | 1.2613 | 1.2848 | 1.3088 | 1.3332 | 1.3581 | 1.3835 | 1.4093 | 1.4357 |
| 1.0 | 1.463 | 1.490 | 1.518 | 1.547 | 1.576 | 1.606 | 1.636 | 1.667 | 1.699 | 1.731 |
| 1.1 | 1.765 | 1.799 | 1.833 | 1.869 | 1.905 | 1.942 | 1.980 | 2.019 | 2.059 | 2.099 |
| 1.2 | 2.141 | 2.184 | 2.228 | 2.272 | 2.318 | 2.365 | 2.414 | 2.463 | 2.514 | 2.566 |
| 1.3 | 2.620 | 2.675 | 2.731 | 2.789 | 2.848 | 2.909 | 2.972 | 3.037 | 3.103 | 3.171 |
| 1.4 | 3.241 | 3.313 | 3.387 | 3.463 | 3.542 | 3.622 | 3.705 | 3.791 | 3.879 | 3.970 |
| 1.5 | 4.063 | 4.159 | 4.259 | 4.361 | 4.467 | 4.575 | 4.688 | 4.803 | 4.923 | 5.046 |
| 1.6 | 5.174 | 5.305 | 5.441 | 5.581 | 5.726 | 5.876 | 6.030 | 6.190 | 6.356 | 6.527 |
| 1.7 | 6.704 | 6.887 | 7.076 | 7.272 | 7.475 | 7.685 | 7.903 | 8.128 | 8.362 | 8.604 |
| 1.8 | 8.85 | 9.11 | 9.38 | 9.66 | 9.95 | 10.25 | 10.57 | 10.89 | 11.23 | 11.58 |
| 1.9 | 11.94 | 12.32 | 12.70 | 13.11 | 13.54 | 13.98 | 14.43 | 14.91 | 15.40 | 15.92 |

## 0.5.8 Changing from degrees to radians

Arclength of the unit circle

| *Angle* | *Arc* | *Angle* | *Arc* | *Angle* | *Arc* |
|---|---|---|---|---|---|
| 1″ | 0.000 005 | 1° | 0.017 453 | 31° | 0.541 052 |
| 2 | 0.000 010 | 2 | 0.034 907 | 32 | 0.558 505 |
| 3 | 0.000 015 | 3 | 0.052 360 | 33 | 0.575 959 |
| 4 | 0.000 019 | 4 | 0.069 813 | 34 | 0.593 412 |
| 5 | 0.000 024 | 5 | 0.087 266 | 35 | 0.610 865 |
| 6 | 0.000 029 | 6 | 0.104 720 | 36 | 0.628 319 |
| 7 | 0.000 034 | 7 | 0.122 173 | 37 | 0.645 772 |
| 8 | 0.000 039 | 8 | 0.139 626 | 38 | 0.663 225 |
| 9 | 0.000 044 | 9 | 0.157 080 | 39 | 0.680 678 |
| 10 | 0.000 048 | 10 | 0.174 533 | 40 | 0.698 132 |
| 20 | 0.000 097 | 11 | 0.191 986 | 45 | 0.785 398 |
| 30 | 0.000 145 | 12 | 0.209 440 | 50 | 0.872 665 |
| 40 | 0.000 194 | 13 | 0.226 893 | 55 | 0.959 931 |
| 50 | 0.000 242 | 14 | 0.244 346 | 60 | 1.047 198 |
| | | 15 | 0.261 799 | 65 | 1.134 464 |
| 1′ | 0.000 291 | 16 | 0.279 253 | 70 | 1.221 730 |
| 2 | 0.000 582 | 17 | 0.296 706 | 75 | 1.308 997 |
| 3 | 0.000 873 | 18 | 0.314 159 | 80 | 1.396 263 |
| 4 | 0.001 164 | 19 | 0.331 613 | 85 | 1.483 530 |
| 5 | 0.001 454 | 20 | 0.349 066 | 90 | 1.570 796 |
| 6 | 0.001 745 | 21 | 0.366 519 | 100 | 1.745 329 |
| 7 | 0.002 036 | 22 | 0.383 972 | 120 | 2.094 395 |
| 8 | 0.002 327 | 23 | 0.401 426 | 150 | 2.617 994 |
| 9 | 0.002 618 | 24 | 0.418 879 | 180 | 3.141 593 |
| 10 | 0.002 909 | 25 | 0.436 332 | 200 | 3.490 659 |
| 20 | 0.005 818 | 26 | 0.453 786 | 250 | 4.363 323 |
| 30 | 0.008 727 | 27 | 0.471 239 | 270 | 4.712 389 |
| 40 | 0.011 636 | 28 | 0.488 692 | 300 | 5.235 988 |
| 50 | 0.014 544 | 29 | 0.506 145 | 360 | 6.283 185 |
| | | 30 | 0.523 599 | 400 | 6.981 317 |

*Examples:*

1) 52° 37′ 23″

| | |
|---|---|
| 50° | = 0.872 665 |
| 2° | = 0.034 907 |
| 30′ | = 0.008 727 |
| 7′ | = 0.002 036 |
| 20″ | = 0.000 097 |
| 3″ | = 0.000 015 |
| | 0.918 447 |

52° 37′ 23″ = 0.91845 rad

2) 5.645 radians (arclength)

| | | |
|---|---|---|
| 5.235 988 | = 300° | |
| 0.409 012 | | |
| 0.401 426 | = 23° | |
| 0.007 586 | | |
| 0.005 818 | = | 20′ |
| 0.001 768 | | |
| 0.001 745 | = | 6′ |
| 0.000 023 | = | 5″ |

5.645 rad = 323° 26′ 5″

The radian is the plane angle for which the quotient of the length of the corresponding circular arc and its radius is equal to 1 (abbreviated rad).

## 0.6 Table of prime numbers $\leq$ 4000

The prime number *twins*, i.e., two consecutive odd numbers which are prime, are indicated by boldface (starting at 41–43). It is known that there are infinitely many such twins.

| 2 | 3 | 5 | 7 | 11 | 13 | 17 | 19 | 23 | 29 |
|---|---|---|---|---|---|---|---|---|---|
| 31 | 37 | **41** | **43** | 47 | 53 | **59** | **61** | 67 | **71** |
| **73** | 79 | 83 | 89 | 97 | **101** | **103** | **107** | **109** | 113 |
| 127 | 131 | **137** | **139** | **149** | **151** | 157 | 163 | 167 | 173 |
| 179 | 181 | **191** | **193** | **197** | **199** | 211 | 223 | **227** | **229** |
| 233 | 239 | 241 | 251 | 257 | 263 | **269** | **271** | 277 | **281** |
| **283** | 293 | 307 | **311** | **313** | 317 | 331 | 337 | **347** | **349** |
| 353 | 359 | 367 | 373 | 379 | 383 | 389 | 397 | 401 | 409 |
| **419** | **421** | **431** | **433** | 439 | 443 | 449 | 457 | **461** | **463** |
| 467 | 479 | 487 | 491 | 499 | 503 | 509 | **521** | **523** | 541 |
| 547 | 557 | 563 | 569 | 571 | 577 | 587 | 593 | **599** | **601** |
| 607 | 613 | **617** | **619** | 631 | **641** | **643** | 647 | 653 | 659 |
| 661 | 673 | 677 | 683 | 691 | 701 | 709 | 719 | 727 | 733 |
| 739 | 743 | 751 | 757 | 761 | 769 | 773 | 787 | 797 | **809** |
| **811** | **821** | **823** | **827** | **829** | 839 | 853 | **857** | **859** | 863 |
| 877 | **881** | **883** | 887 | 907 | 911 | 919 | 929 | 937 | 941 |
| 947 | 953 | 967 | 971 | 977 | 983 | 991 | 997 | 1009 | 1013 |
| **1019** | **1021** | **1031** | **1033** | 1039 | **1049** | **1051** | **1061** | **1063** | 1069 |
| 1087 | **1091** | **1093** | 1097 | 1103 | 1109 | 1117 | 1123 | 1129 | **1151** |
| **1153** | 1163 | 1171 | 1181 | 1187 | 1193 | 1201 | 1213 | 1217 | 1223 |
| **1229** | **1231** | 1237 | 1249 | 1259 | **1277** | **1279** | 1283 | **1289** | **1291** |
| 1297 | **1301** | **1303** | 1307 | **1319** | **1321** | 1327 | 1361 | 1367 | 1373 |
| 1381 | 1399 | 1409 | 1423 | **1427** | **1429** | 1433 | 1439 | 1447 | **1451** |
| **1453** | 1459 | 1471 | **1481** | **1483** | **1487** | **1489** | 1493 | 1499 | 1511 |
| 1523 | 1531 | 1543 | 1549 | 1553 | 1559 | 1567 | 1571 | 1579 | 1583 |
| 1597 | 1601 | **1607** | **1609** | 1613 | **1619** | **1621** | 1627 | 1637 | 1657 |
| 1663 | **1667** | **1669** | 1693 | **1697** | **1699** | 1709 | **1721** | **1723** | 1733 |
| 1741 | 1747 | 1753 | 1759 | 1777 | 1783 | **1787** | **1789** | 1801 | 1811 |
| 1823 | 1831 | 1847 | 1861 | 1867 | **1871** | **1873** | **1877** | **1879** | 1889 |
| 1901 | 1907 | 1913 | **1931** | **1933** | **1949** | **1951** | 1973 | 1979 | 1987 |
| 1993 | **1997** | **1999** | 2003 | 2011 | 2017 | **2027** | **2029** | 2039 | 2053 |
| 2063 | 2069 | **2081** | **2083** | **2087** | **2089** | 2099 | **2111** | **2113** | **2129** |
| **2131** | 2137 | **2141** | **2143** | 2153 | 2161 | 2179 | 2203 | 2207 | 2213 |
| 2221 | **2237** | **2239** | 2243 | 2251 | **2267** | **2269** | 2273 | 2281 | 2287 |
| 2293 | 2297 | **2309** | **2311** | 2333 | **2339** | **2341** | 2347 | 2351 | 2357 |
| 2371 | 2377 | **2381** | **2383** | 2389 | 2393 | 2399 | 2411 | 2417 | 2423 |
| 2437 | 2441 | 2447 | 2459 | 2467 | 2473 | 2477 | 2503 | 2521 | 2531 |
| 2539 | 2543 | **2549** | **2551** | 2557 | 2579 | **2591** | **2593** | 2609 | 2617 |
| 2621 | 2633 | 2647 | **2657** | **2659** | 2663 | 2671 | 2677 | 2683 | **2687** |
| **2689** | 2693 | 2699 | 2707 | **2711** | **2713** | 2719 | **2729** | **2731** | 2741 |
| 2749 | 2753 | 2767 | 2777 | **2789** | **2791** | 2797 | **2801** | **2803** | 2819 |
| 2833 | 2837 | 2843 | 2851 | 2857 | 2861 | 2879 | 2887 | 2897 | 2903 |
| 2909 | 2917 | 2927 | 2939 | 2953 | 2957 | 2963 | **2969** | **2971** | **2999** |
| **3001** | 3011 | 3019 | 3023 | 3037 | 3041 | 3049 | 3061 | 3067 | 3079 |
| 3083 | 3089 | 3109 | **3119** | **3121** | 3137 | 3163 | **3167** | **3169** | 3181 |
| 3187 | 3191 | 3203 | 3209 | 3217 | 3221 | 3229 | **3251** | **3253** | **3257** |
| **3259** | 3271 | **3299** | **3301** | 3307 | 3313 | 3319 | 3323 | **3329** | **3331** |
| 3343 | 3347 | **3359** | **3361** | **3371** | **3373** | **3389** | **3391** | 3407 | 3413 |
| 3433 | 3449 | 3457 | **3461** | **3463** | **3467** | **3469** | 3491 | 3499 | 3511 |
| 3517 | **3527** | **3529** | 3533 | **3539** | **3541** | 3547 | **3557** | **3559** | 3571 |
| **3581** | **3583** | 3593 | 3607 | 3613 | 3617 | 3623 | 3631 | 3637 | 3643 |
| 3659 | **3671** | **3673** | 3677 | 3691 | 3697 | 3701 | 3709 | 3719 | 3727 |
| 3733 | 3739 | 3761 | **3767** | **3769** | 3779 | 3793 | 3797 | 3803 | **3821** |
| **3823** | 3833 | 3847 | **3851** | **3853** | 3863 | 3877 | 3881 | 3889 | 3907 |
| 3911 | **3917** | **3919** | 3923 | **3929** | **3931** | 3943 | 3947 | 3967 | 3989 |

## 0.7 Formulas for series and products

For infinite series and infinite products the notion of convergence is fundamental (cf. 1.10.1 and 1.10.6).

### 0.7.1 Special series

One gets important series by inserting special values in the power series listed in 0.7.2 or in Fourier series listed in 0.7.4.

#### 0.7.1.1 The Leibniz series and related series

$$\boxed{1-\frac{1}{3}+\frac{1}{5}-\ldots=\sum_{n=0}^{\infty}\frac{(-1)^n}{2n+1}=\frac{\pi}{4}\qquad\text{(Leibniz, 1676)},}$$

$$1-\frac{1}{2}+\frac{1}{3}-\ldots=\sum_{n=1}^{\infty}\frac{(-1)^{n+1}}{n}=\ln 2\,,$$

$$\ln\left(1-\frac{1}{2^2}\right)+\ln\left(1-\frac{1}{3^2}\right)+\ldots=\sum_{k=2}^{\infty}\ln\left(1-\frac{1}{k^2}\right)=-\ln 2\,,$$

$$\boxed{2+\frac{1}{2!}+\frac{1}{3!}+\ldots=\sum_{n=0}^{\infty}\frac{1}{n!}=\mathrm{e}\qquad\text{(Euler number)},}$$

$$\frac{1}{2!}-\frac{1}{3!}+\frac{1}{4!}-\ldots=\sum_{n=2}^{\infty}\frac{(-1)^n}{n!}=\frac{1}{\mathrm{e}}\,,$$

$$\boxed{1+\frac{1}{2}+\frac{1}{4}+\frac{1}{8}+\ldots=\sum_{n=0}^{\infty}\frac{1}{2^n}=2\qquad\text{(geometric series)},}$$

$$1-\frac{1}{2}+\frac{1}{4}-\frac{1}{8}+\ldots=\sum_{n=0}^{\infty}\frac{(-1)^n}{2^n}=\frac{2}{3}\qquad\text{(alternating geometric series)},$$

$$\frac{1}{1\cdot 2}+\frac{1}{2\cdot 3}+\frac{1}{3\cdot 4}+\ldots=\sum_{n=1}^{\infty}\frac{1}{n(n+1)}=1\,,$$

$$\frac{1}{1\cdot 3}+\frac{1}{3\cdot 5}+\frac{1}{5\cdot 7}+\ldots=\sum_{n=1}^{\infty}\frac{1}{(2n-1)(2n+1)}=\frac{1}{2}\,,$$

$$\frac{1}{1\cdot 3}+\frac{1}{2\cdot 4}+\frac{1}{3\cdot 5}+\ldots=\sum_{n=2}^{\infty}\frac{1}{(n-1)(n+1)}=\frac{3}{4}\,,$$

$$\frac{1}{3\cdot 5}+\frac{1}{7\cdot 9}+\frac{1}{11\cdot 13}+\ldots=\sum_{n=1}^{\infty}\frac{1}{(4n-1)(4n+1)}=\frac{1}{2}-\frac{\pi}{8}\,,$$

$$\frac{1}{1\cdot 2\cdot 3}+\frac{1}{2\cdot 3\cdot 4}+\frac{1}{3\cdot 4\cdot 5}+\ldots=\sum_{n=1}^{\infty}\frac{1}{n(n+1)(n+2)}=\frac{1}{4}\,,$$

$$\frac{1}{1\cdot 2\cdot 3\cdots k}+\frac{1}{2\cdot 3\cdots (k+1)}+\ldots = \sum_{n=1}^{\infty}\frac{1}{n(n+1)\cdots(n+k-1)} = \frac{1}{(k-1)(k-1)!}\,, \qquad k=2,3,\ldots,$$

$$\sum_{n=p+1}^{\infty}\frac{1}{n^2-p^2} = \frac{1}{2p}\left(1+\frac{1}{2}+\ldots+\frac{1}{2p}\right), \qquad p=1,2,\ldots, \qquad \text{(Jakob Bernoulli, 1689).}$$

### 0.7.1.2 Special values of the Riemannian ζ-function and related series

The series

$$\zeta(s) = 1+\frac{1}{2^s}+\frac{1}{3^s}+\ldots = \sum_{n=1}^{\infty}\frac{1}{n^s}$$

converges for all real numbers $s>1$ and more generally for all complex numbers $s$ with $\operatorname{Re} s > 1$. This function is of fundamental importance in the mathematical discipline of number theory, in particular with the distribution of prime numbers (see section 2.7.3). It is called the *Riemann ζ-function* and was studied by Euler and particularly by Riemann in 1859.

**The formula of L. Euler (1734)[44] :**

$$\zeta(2k) = 1+\frac{1}{2^{2k}}+\frac{1}{3^{2k}}+\ldots = \frac{(2\pi)^{2k}}{2(2k)!}B_{2k}\,, \qquad k=1,2,\ldots$$

*Special cases:*

$$\zeta(2) = 1+\frac{1}{2^2}+\frac{1}{3^2}+\ldots = \frac{\pi^2}{6}\,,$$

$$\zeta(4) = \frac{\pi^4}{90}\,, \qquad \zeta(6) = \frac{\pi^6}{945}\,, \qquad \zeta(8) = \frac{\pi^8}{9\,450}\,,$$

$$1-\frac{1}{2^{2k}}+\frac{1}{3^{2k}}-\frac{1}{4^{2k}}+\ldots = \sum_{n=1}^{\infty}\frac{(-1)^{n+1}}{n^{2k}} = \frac{\pi^{2k}\left(2^{2k}-1\right)}{(2k)!}|B_{2k}|\,, \qquad k=1,2,\ldots$$

*Special cases:*

$$1-\frac{1}{2^2}+\frac{1}{3^2}-\frac{1}{4^2}+\ldots = \sum_{n=1}^{\infty}\frac{(-1)^{n+1}}{n^2} = \frac{\pi^2}{12}\,,$$

$$1-\frac{1}{2^4}+\frac{1}{3^4}-\frac{1}{4^4}+\ldots = \sum_{n=1}^{\infty}\frac{(-1)^{n+1}}{n^4} = \frac{7\pi^4}{720}\,,$$

[44]The Bernoulli numbers $B_k$ and the Euler numbers $E_k$ can be found in sections 0.1.10.4 and 0.1.10.5.

$$1+\frac{1}{3^{2k}}+\frac{1}{5^{2k}}+\ldots=\sum_{n=0}^{\infty}\frac{1}{(2n+1)^{2k}}=\frac{\pi^{2k}\left(2^{2k-1}\right)}{2(2k)!}\,|B_{2k}|\,,\quad k=1,2,\ldots$$

*Special cases:*

$$1+\frac{1}{3^2}+\frac{1}{5^2}+\ldots=\frac{\pi^2}{8},$$
$$1+\frac{1}{3^4}+\frac{1}{5^4}+\ldots=\frac{\pi^4}{96}.$$

$$1-\frac{1}{3^{2k+1}}+\frac{1}{5^{2k+1}}-\ldots=\sum_{n=0}^{\infty}\frac{(-1)^n}{(2n+1)^{2k+1}}=\frac{\pi^{2k+1}}{2^{2k+2}(2k)!}\,|E_{2k}|\,,\quad k=0,1,2,\ldots$$

*Special cases:* For $k=0$ one gets the Leibniz series $1-\frac{1}{3}+\frac{1}{5}-\ldots$; for $k=1$:

$$1-\frac{1}{3^3}+\frac{1}{5^3}-\ldots=\frac{\pi^3}{32}.$$

#### 0.7.1.3 The Euler–McLaurin summation formula

**The asymptotic formula of Euler (1734):**

$$\lim_{n\to\infty}\left(1+\frac{1}{2}+\frac{1}{3}+\ldots+\frac{1}{n}-\ln(n+1)\right)=\mathrm{C}\,. \tag{0.53}$$

The Euler constant C has the value $\mathrm{C}=0.57721\,56649\,01532\ldots$, which had already been calculated by Euler. The asympotic formula (0.53) is a special case of the Euler–McLaurin summation formula (0.54).

**Bernoulli polynomials:**

$$B_n(x):=\sum_{k=0}^{\infty}\binom{n}{k}B_k x^{n-k}\,.$$

**Modified Bernoulli polynomials:**[45]

$$C_n(x):=B_n\big(x-[x]\big)\,.$$

**The Euler–McLaurin summation formula:** For $n=1,2,\ldots$ one has

$$f(0)+f(1)+\ldots+f(n)=\int_0^n f(x)\,\mathrm{d}x+\frac{f(0)+f(n)}{2}+S_n \tag{0.54}$$

with[46]

$$S_n:=\frac{B_2}{2!}f'+\frac{B_4}{4!}f^{(3)}+\ldots+\frac{B_{2p}}{(2p)!}f^{(2p-1)}\Big|_1^n+R_p\,,\qquad p=2,3,\ldots,$$

[45] We denote by $[x]$ the largest integer $n$ smaller than or equal to $x$: (Gauss bracket). The function $C_n$ coincides in the interval $[0,1[$ with $B_n$ and is extended periodically with period 1.

[46] The symbol $g\big|_1^n$ means $g(n)-g(1)$.

and the remainder term

$$R_p = \frac{1}{(2p+1)!}\int_0^n f^{(2p+1)}(x)C_{2p+1}(x)\,dx\,.$$

Here it is assumed that the function $f\colon [0,n] \to \mathbb{R}$ is sufficiently smooth, i.e., has continuous derivatives up to order $2p+1$ on the interval $[0,n]$.

#### 0.7.1.4 Infinite partial fraction decomposition

The following series converge for all complex numbers $x$ with the exception of those values for which the denominator vanishes[47]:

$$\cot \pi x = \frac{1}{x} + \sum_{k=1}^{\infty}\left(\frac{1}{x-k} + \frac{1}{x+k}\right),$$

$$\tan \pi x = -\sum_{k=1}^{\infty}\frac{1}{x-\left(k-\frac{1}{2}\right)} + \frac{1}{x+\left(k-\frac{1}{2}\right)},$$

$$\frac{\pi}{\sin \pi x} = \frac{1}{x} + \sum_{k=1}^{\infty}\frac{(-1)^k 2x}{x^2-k^2},$$

$$\left(\frac{\pi}{\sin \pi x}\right)^2 = \sum_{k=-\infty}^{\infty}\frac{1}{(x-k)^2},$$

$$\left(\frac{\pi}{\cos \pi x}\right)^2 = \sum_{k=-\infty}^{\infty}\frac{1}{\left(x-k+\frac{1}{2}\right)^2}.$$

### 0.7.2 Power series

**Comments on the power series table:** The power series listed in the following table converge for all complex numbers $x$ for which the stated inequalities hold. The properties of power series will be considered in more detail in 1.10.3.

The given first term in the series may be used as an approximation for $|x|$ sufficiently small.

*Example:* One has

$$\sin x = x - \frac{x^3}{6} + \frac{x^5}{120} - \frac{x^7}{5\,040} + \ldots = \sum_{k=0}^{\infty}\frac{(-1)^k x^{2k+1}}{(2k+1)!}.$$

If $|x|$ is small, then, approximately

$$\sin x = x\,.$$

Successively improved approximations are obtained by

$$\sin x = x - \frac{x^3}{6}, \qquad \sin x = x - \frac{x^3}{6} + \frac{x^5}{120} \qquad \text{etc.}$$

[47]These series are special cases of the theorem of Mittag–Leffler (cf. 1.14.6.4).

A sum $\sum_{k=-\infty}^{\infty}\ldots$ stands for the sum of the two infinite sums: $\sum_{k=0}^{\infty}\ldots + \sum_{k=-\infty}^{-1}\ldots$

For the frequently appearing factorials one can use the following table:

| $n$ | 0 | 1 | 2 | 3 | 4 | 5 | 6 | 7 | 8 | 9 | 10 |
|---|---|---|---|---|---|---|---|---|---|---|---|
| $n!$ | 1 | 1 | 2 | 6 | 24 | 120 | 720 | 5040 | 40,320 | 362,880 | 3,628,800 |

In the expansions of

$$\frac{x}{e^x - 1}\,, \ \tan x\,, \ \cot x\,, \ \frac{1}{\sin x} \equiv \operatorname{cosec} x\,, \ \tanh x\,, \ \coth x\,, \ \frac{1}{\sinh x} \equiv \operatorname{cosec} x$$

and

$$\frac{1}{\cosh x} \equiv \operatorname{sech} x\,, \ \frac{1}{\cos x} \equiv \sec x\,, \ \ln \cos x\,, \ \ln|x| - \ln|\sin x|\,,$$

respectively, the Bernoulli numbers $B_k$ resp. Euler numbers $E_k$ appear (see sections 0.1.10.4 and 0.1.10.5).

| ***Function*** | ***Power series expansion*** | ***Domain of convergence ($x \in \mathbb{C}$)*** |
|---|---|---|
| | **geometric series** | |
| $\frac{1}{1-x}$ | $1 + x + x^2 + x^3 + \ldots = \sum_{k=0}^{\infty} x^k$ | $\|x\| < 1$ |
| $\frac{1}{1+x}$ | $1 - x + x^2 - x^3 + \ldots = \sum_{k=0}^{\infty} (-1)^k x^k$ | $\|x\| < 1$ |
| | **The binomial series of Newton** | |
| $(1+x)^\alpha$ | $1 + \binom{\alpha}{1} x + \binom{\alpha}{2} x^2 + \ldots = \sum_{k=0}^{\infty} \binom{\alpha}{k} x^k$ ($\alpha$ is an arbitrary real number[48]) | $\|x\| < 1$ ($x = \pm 1$, $\alpha > 0$) |
| $(a+x)^\alpha$ | $a^\alpha \left(1 + \frac{x}{a}\right)^\alpha = a^\alpha + \alpha a^{\alpha-1} x + a^{\alpha-2} \binom{\alpha}{2} x^2 + \ldots$ $= \sum_{k=0}^{\infty} a^{\alpha-k} \binom{\alpha}{k} x^k$ ($a$ is a positive real number) | $\|x\| < a$ ($x = \pm a$ for $\alpha > 0$) |
| $(a+x)^n$ | $a^n + \binom{n}{1} a^{n-1} x + \binom{n}{2} a^{n-2} x^2 + \ldots + \binom{n}{1} a x^{n-1} + x^n$ ($n = 1, 2, \ldots$ ; $a$ and $x$ are arbitrary complex numbers) | $\|x\| < \infty$ |
| $(a+x)^{-n}$ | $(a+x)^{-n} := \frac{1}{(a+x)^n}$ | |
| $(a+x)^{1/n}$ | $(a+x)^{1/n} := \sqrt[n]{a+x}$ | |
| $(a+x)^{-1/n}$ | $(a+x)^{-1/n} := \frac{1}{\sqrt[n]{a+x}}$ | |

[48]The generalized binomial coefficients are defined by $\binom{\alpha}{1} = \alpha$, $\binom{\alpha}{2} = \frac{\alpha(\alpha-1)}{1 \cdot 2}$, $\binom{\alpha}{3} = \frac{\alpha(\alpha-1)(\alpha-2)}{1 \cdot 2 \cdot 3}$ etc.

**Special cases of the binomial series for integral exponents**
($a$ a complex number with $a \neq 0$)

| | | |
|---|---|---|
| $\frac{1}{(a \pm x)^n}$ | $\frac{1}{a^n} \mp \frac{nx}{a^{n+1}} + \frac{n(n+1)x^2}{2a^{n+2}} \mp \dots$ <br> $= \frac{1}{a^n} + \sum_{k=1}^{\infty} \frac{n(n+1)\dots(n-k+1)}{k!a^{n+k}}(\mp x)^k$ | $\|x\| < \|a\|$ |
| $\frac{1}{a \pm x}$ | $\frac{1}{a} \mp \frac{x}{a^2} + \frac{x^2}{a^3} \mp \frac{x^3}{a^4} + \dots = \sum_{k=0}^{\infty} \frac{(\mp x)^k}{a^{k+1}}$ | $\|x\| < \|a\|$ |
| $(a \pm x)^2$ | $a^2 \pm 2ax + x^2$ | $\|x\| < \infty$ |
| $\frac{1}{(a \pm x)^2}$ | $\frac{1}{a^2} \mp \frac{2x}{a^3} + \frac{3x^2}{a^4} \mp \frac{4x^3}{a^5} + \dots = \sum_{k=0}^{\infty} \frac{(k+1)(\mp x)^k}{a^{2+k}}$ | $\|x\| < \|a\|$ |
| $(a \pm x)^3$ | $a^3 \pm 3a^2x + 3ax^2 \pm x^3$ | $\|x\| < \infty$ |
| $\frac{1}{(a \pm x)^3}$ | $\frac{1}{a^3} \mp \frac{3x}{a^4} + \frac{6x^2}{a^5} \mp \frac{10x^3}{a^6} + \dots$ <br> $= \sum_{k=0}^{\infty} \frac{(k+1)(k+2)(\mp x)^k}{2a^{3+k}}$ | $\|x\| < \|a\|$ |

**Special cases of the binomial series for rational exponents**
($b$ a positive real number)

| | | |
|---|---|---|
| $\sqrt{b \pm x}$ | $\sqrt{b} \pm \frac{x}{2\sqrt{b}} - \frac{x^2}{8b\sqrt{b}} \pm \frac{x^3}{16b^2\sqrt{b}} - \dots$ <br> $= \sqrt{b} \pm \frac{x}{2\sqrt{b}} + \sum_{k=2}^{\infty} \frac{1 \cdot 3 \cdot 5 \dots (2k-3)(-1)^{k+1}(\pm x)^k}{(2 \cdot 4 \cdot 6 \dots 2k)b^{k-1}\sqrt{b}}$ | $\|x\| < b$ |
| $\frac{1}{\sqrt{b \pm x}}$ | $\frac{1}{\sqrt{b}} \mp \frac{x}{2b\sqrt{b}} + \frac{3x^2}{8b^2\sqrt{b}} \mp \frac{15x^3}{48b^3\sqrt{b}} + \dots$ <br> $= \frac{1}{\sqrt{b}} \mp \frac{x}{2b\sqrt{b}} + \sum_{k=2}^{\infty} \frac{1 \cdot 3 \cdot 5 \dots (2k-1)(-1)^k(\pm x)^k}{(2 \cdot 4 \cdot 6 \dots 2k)b^k\sqrt{b}}$ | $\|x\| < b$ |
| $\sqrt[3]{b \pm x}$ | $\sqrt[3]{b} \pm \frac{x}{3\sqrt[3]{b^2}} - \frac{x^2}{9b\sqrt[3]{b^2}} \pm \frac{5x^3}{81b^2\sqrt[3]{b^2}} - \dots$ <br> $= \sqrt[3]{b} \pm \frac{x}{3\sqrt[3]{b^2}} + \sum_{k=2}^{\infty} \frac{2 \cdot 5 \cdot 8 \dots (3k-4)(-1)^{k+1}(\pm x)^k}{(3 \cdot 6 \cdot 9 \dots 3k)b^{k-1}\sqrt[3]{b^2}}$ | $\|x\| < b$ |
| $\frac{1}{\sqrt[3]{b \pm x}}$ | $\frac{1}{\sqrt[3]{b}} \mp \frac{x}{3b\sqrt[3]{b}} + \frac{2x^2}{9b^2\sqrt[3]{b}} \mp \frac{14x^3}{81b^3\sqrt[3]{b}} + \dots$ <br> $= \frac{1}{\sqrt[3]{b}} + \sum_{k=1}^{\infty} \frac{4 \cdot 7 \cdot 10 \dots (3k-2)(-1)^k(\pm x)^k}{(3 \cdot 6 \cdot 9 \dots 3k)b^k\sqrt[3]{b}}$ | $\|x\| < b$ |

| Hypergeometric series (generalized binomial series) of Gauss | | |
|---|---|---|
| $F(\alpha,\beta,\gamma,x)$ | $1+\frac{\alpha\beta}{\gamma}x+\frac{\alpha(\alpha+1)\beta(\beta+1)}{2\gamma(\gamma+1)}x^2+\ldots$ <br> $=1+\sum_{k=1}^{\infty}\frac{\alpha(\alpha+1)\ldots(\alpha+k-1)\beta(\beta+1)\ldots(\beta+k-1)}{k!\gamma(\gamma+1)\ldots(\gamma+k-1)}x^k$ | $\|x\|<1$ |

| Special cases of the hypergeometric series | | |
|---|---|---|
| $(1+x)^\alpha$ | $=F(-\alpha,1,1,-x)$ | |
| $\arcsin x$ | $=xF\left(\frac{1}{2},\frac{1}{2},\frac{3}{2},x^2\right)$ | |
| $\ln(1+x)$ | $=xF(1,1,2,-x)$ | |
| $\mathrm{e}^x$ | $=\lim_{\beta\to+\infty}F\left(1,\beta,1,\frac{x}{\beta}\right)$ | |
| $P_n(x)$ | $=F\left(n+1,-n,1,\frac{1-x}{2}\right),\qquad n=0,1,2,\ldots$ | |
| (Legendre polynomials, see page 123 below) | | |
| $Q_n(x)$ | $=\frac{\sqrt{\pi}\Gamma(n+1)}{2^{n+1}\Gamma\left(n+\frac{3}{2}\right)}\cdot\frac{1}{x^{n+1}}F\left(\frac{n+1}{2},\frac{n+2}{2},\frac{2n+3}{2},\frac{1}{x^2}\right)$ | $\|x\|>1$ |
| (Legendre functions, see page 123 below) | | |

| Exponential function | | |
|---|---|---|
| $\mathrm{e}^x$ | $1+x+\frac{x^2}{2!}+\frac{x^3}{3!}+\ldots=\sum_{k=0}^{\infty}\frac{x^k}{k!}$ | $\|x\|<\infty$ |
| $\mathrm{e}^{bx}$ | $1+bx+\frac{(bx)^2}{2!}+\frac{(bx)^3}{3!}+\ldots=\sum_{k=0}^{\infty}\frac{(bx)^k}{k!}$ <br> ($b$ is a complex number) | $\|x\|<\infty$ |
| $a^x$ | $a^x=\mathrm{e}^{bx}$ with $b=\ln a$ ($a$ real and positive) | |
| $\frac{x}{\mathrm{e}^x-1}$ | $1-\frac{x}{2}+\frac{x^2}{12}-\frac{x^4}{7\,200}+\ldots=\sum_{k=0}^{\infty}\frac{B_k}{k!}x^k$ | $\|x\|<2\pi$ |

| Trigonometric functions and hyperbolic functions | | |
|---|---|---|
| $\sin \mathrm{i}x = \mathrm{i}\sinh x\,,\ \cos \mathrm{i}x = \cosh x\,,\ \sinh \mathrm{i}x = \mathrm{i}\sin x\,,\ \cosh \mathrm{i}x = \cos x$ (for all complex numbers $x$) | | |
| $\sin x$ | $x - \frac{x^3}{3!} + \frac{x^5}{5!} - \ldots = \sum_{k=0}^{\infty}(-1)^k \frac{x^{2k+1}}{(2k+1)!}$ | $\lvert x\rvert < \infty$ |
| $\sinh x$ | $x + \frac{x^3}{3!} + \frac{x^5}{5!} + \ldots = \sum_{k=0}^{\infty} \frac{x^{2k+1}}{(2k+1)!}$ | $\lvert x\rvert < \infty$ |
| $\cos x$ | $1 - \frac{x^2}{2!} + \frac{x^4}{4!} - \ldots = \sum_{k=0}^{\infty}(-1)^k \frac{x^{2k}}{(2k)!}$ | $\lvert x\rvert < \infty$ |
| $\cosh x$ | $1 + \frac{x^2}{2!} + \frac{x^4}{4!} + \ldots = \sum_{k=0}^{\infty} \frac{x^{2k}}{(2k)!}$ | $\lvert x\rvert < \infty$ |
| $\tan x$ | $x + \frac{x^3}{3} + \frac{2x^5}{15} + \frac{17x^7}{315} + \ldots = \sum_{k=1}^{\infty} 4^k\left(4^k - 1\right)\frac{\lvert B_{2k}\rvert x^{2k-1}}{(2k)!}$ | $\lvert x\rvert < \frac{\pi}{2}$ |
| $\tanh x$ | $x - \frac{x^3}{3} + \frac{2x^5}{15} - \frac{17x^7}{315} + \ldots = \sum_{k=1}^{\infty} 4^k\left(4^k - 1\right)\frac{B_{2k} x^{2k-1}}{(2k)!}$ | $\lvert x\rvert < \frac{\pi}{2}$ |
| $\frac{1}{x} - \cot x$ | $\frac{x}{3} + \frac{x^3}{45} + \frac{2x^5}{945} + \frac{x^7}{4\,725} + \ldots = \sum_{k=1}^{\infty} \frac{4^k \lvert B_{2k}\rvert x^{2k-1}}{(2k)!}$ | $0 < \lvert x\rvert < \pi$ |
| $\coth x - \frac{1}{x}$ | $\frac{x}{3} - \frac{x^3}{45} + \frac{2x^5}{945} - \frac{x^7}{4\,725} + \ldots = \sum_{k=1}^{\infty} \frac{4^k B_{2k} x^{2k-1}}{(2k)!}$ | $0 < \lvert x\rvert < \pi$ |
| $\frac{1}{\cos x}$ | $1 + \frac{x^2}{2} + \frac{5x^4}{24} + \frac{61x^6}{720} + \ldots = \sum_{k=0}^{\infty} \frac{\lvert E_k\rvert x^k}{k!}$ | $\lvert x\rvert < \frac{\pi}{2}$ |
| $\frac{1}{\cosh x}$ | $1 - \frac{x^2}{2} + \frac{5x^4}{24} - \frac{61x^6}{720} + \ldots = \sum_{k=0}^{\infty} \frac{E_k x^k}{k!}$ | $\lvert x\rvert < \frac{\pi}{2}$ |
| $\frac{1}{\sin x} - \frac{1}{x}$ | $\frac{x}{6} + \frac{7x^3}{360} + \frac{31x^5}{15\,120} + \frac{127x^7}{604\,800} + \ldots$ $= \sum_{k=1}^{\infty} \frac{2\left(2^{2k-1} - 1\right)}{(2k)!}\lvert B_{2k}\rvert x^{2k-1}$ | $0 < \lvert x\rvert < \pi$ |
| $\frac{1}{x} - \frac{1}{\sinh x}$ | $\frac{x}{6} - \frac{7x^3}{360} + \frac{31x^5}{15\,120} - \frac{127x^7}{604\,800} + \ldots$ $= \sum_{k=1}^{\infty} \frac{2\left(2^{2k-1} - 1\right)}{(2k)!} B_{2k} x^{2k-1}$ | $0 < \lvert x\rvert < \pi$ |

| Inverse trigonometric functions and inverse hyperbolic functions | | |
|---|---|---|
| $\arctan x$ | $x - \frac{x^3}{3} + \frac{x^5}{5} - \ldots = \sum_{k=0}^{\infty}(-1)^k \frac{x^{2k+1}}{2k+1}$ | $\lvert x\rvert < 1$ (or $x = \pm 1$) |
| $\frac{\pi}{4} = \arctan 1$ | $1 - \frac{1}{3} + \frac{1}{5} - \ldots$ (Leibniz series) | |
| $\operatorname{artanh} x$ | $x + \frac{x^3}{3} + \frac{x^5}{5} + \ldots = \sum_{k=0}^{\infty} \frac{x^{2k+1}}{2k+1}$ | $\lvert x\rvert < 1$ |
| $\frac{\pi}{2} - \operatorname{arccot} x$ | $\frac{\pi}{2} - \operatorname{arccot} x = \arctan x$ | |
| $\arctan \frac{1}{x}$ | $\frac{\pi}{2} - x + \frac{x^3}{3} - \frac{x^5}{5} + \ldots$ | $0 < x < 1$ |
| $\arctan \frac{1}{x}$ | $-\frac{\pi}{2} - x + \frac{x^3}{3} - \frac{x^5}{5} + \ldots$ | $-1 < x < 0$ |
| $\operatorname{arcoth} \frac{1}{x}$ | $x + \frac{x^3}{3} + \frac{x^5}{5} + \ldots$ | $0 < \lvert x\rvert < 1$ |
| $\arcsin x$ | $x + \frac{x^3}{6} + \frac{3x^5}{40} + \frac{15x^7}{336} + \cdots$ $= x + \sum_{k=1}^{\infty} \frac{1\cdot 3\cdot 5\ldots(2k-1)x^{2k+1}}{2\cdot 4\cdot 6\cdots 2k(2k+1)}$ | $\lvert x\rvert < 1$ |
| $\frac{\pi}{2} - \arccos x$ | $\frac{\pi}{2} - \arccos x = \arcsin x$ | |
| $\operatorname{arsinh} x$ | $x - \frac{x^3}{6} + \frac{3x^5}{40} - \frac{15x^7}{336} + \ldots$ $= x + \sum_{k=1}^{\infty} \frac{1\cdot 3\cdot 5\cdots(2k-1)(-1)^k x^{2k+1}}{2\cdot 4\cdot 6\cdots(2k)(2k+1)}$ | $\lvert x\rvert < 1$ |

| Logarithmic functions | | |
|---|---|---|
| $\ln(1+x)$ | $x - \frac{x^2}{2} + \frac{x^3}{3} - \frac{x^4}{4} + \ldots = \sum_{k=1}^{\infty}(-1)^{k+1} \frac{x^k}{k}$ | $\lvert x\rvert < 1$ (and $x = 1$) |
| $\ln 2$ | $1 - \frac{1}{2} + \frac{1}{3} - \frac{1}{4} + \ldots$ | |
| $-\ln(1-x)$ | $x + \frac{x^2}{2} + \frac{x^3}{3} + \frac{x^4}{4} + \ldots = \sum_{k=1}^{\infty} \frac{x^k}{k}$ | $\lvert x\rvert < 1$ (and $x = -1$) |
| $\ln \frac{1+x}{1-x} = 2 \operatorname{artanh} x$ | $2x + \frac{2x^3}{3} + \frac{2x^5}{5} + \frac{2x^7}{7} + \ldots = \sum_{k=1}^{\infty} \frac{2x^{2k+1}}{2k+1}$ | $\lvert x\rvert < 1$ |

| | | |
|---|---|---|
| $\ln\lvert x\rvert$ $-\ln\lvert\sin x\rvert$ | $\frac{x^2}{6}+\frac{x^4}{180}+\frac{x^6}{2\,835}+\ldots=\sum_{k=1}^{\infty}\frac{2^{2k-1}}{k(2k)!}\lvert B_{2k}\rvert x^{2k}$ | $0<\lvert x\rvert<\pi$ |
| $-\ln\cos x$ | $\frac{x^2}{2}+\frac{x^4}{12}+\frac{x^6}{45}+\frac{17x^8}{2\,520}+\ldots$ $=\sum_{k=1}^{\infty}\frac{2^{2k-1}\left(4^k-1\right)}{k(2k)!}\lvert B_{2k}\rvert x^{2k}$ | $\lvert x\rvert<\frac{\pi}{2}$ |
| $\ln\lvert\tan x\rvert$ $-\ln\lvert x\rvert$ | $\frac{x^2}{3}+\frac{7x^4}{90}+\frac{62x^6}{2\,835}+\ldots$ $=\sum_{k=1}^{\infty}\frac{4^k\left(2^{2k-1}-1\right)}{k(2k)!}\lvert B_{2k}\rvert x^{2k}$ | $0<\lvert x\rvert<\frac{\pi}{2}$ |

| Complete elliptic integrals | | |
|---|---|---|
| $K(k)$ | $\int_0^{\pi/2}\frac{\mathrm{d}\varphi}{\sqrt{1-k^2\sin^2\varphi}}=\frac{\pi}{2}\left(1+\frac{k^2}{4}+\frac{9k^4}{64}+\ldots\right)$ $=\frac{\pi}{2}\left(1+\sum_{n=1}^{\infty}\left(\frac{1\cdot3\cdot5\cdots(2n-1)}{2\cdot4\cdot6\cdots2n}\right)^2k^{2n}\right)$ | $\lvert k\rvert<1$ |
| $E(k)$ | $\int_0^{\pi/2}\sqrt{1-k^2\sin^2\varphi}\,\mathrm{d}\varphi=\frac{\pi}{2}\left(1-\frac{k^2}{4}+\frac{9k^4}{192}-\ldots\right)$ $=\frac{\pi}{2}\left(1+\sum_{n=1}^{\infty}\left(\frac{1\cdot3\cdot5\cdots(2n-1)}{2\cdot4\cdot6\cdots2n}\right)^2\frac{(-1)^nk^{2n}}{2n-1}\right)$ | $\lvert k\rvert<1$ |

| **The Euler gamma function (generalized factorial)** | | $x\in\mathbb{C}$ |
|---|---|---|
| | $\Gamma(x+1)=x!\,,\quad\Gamma(x+1)=x\Gamma(x)$ | $x\neq0,-1,-2,\ldots$ |
| | $\Gamma(x)=\int_0^{\infty}\mathrm{e}^{-t}t^{x-1}\mathrm{d}t$ | $\mathrm{Re}\,x>0$ |
| $\ln\Gamma(x+1)$ | $-\mathrm{C}x+\frac{\zeta(2)x^2}{2}-\frac{\zeta(3)x^3}{3}+\ldots=-\mathrm{C}x+\sum_{k=2}^{\infty}(-1)^k\frac{\zeta(k)x^k}{k}$ $=\frac{1}{2}\ln\frac{\pi x}{\sin\pi x}-\frac{1}{2}\ln\frac{1+x}{1-x}+(1-\mathrm{C})x$ $+\sum_{k=1}^{\infty}\frac{\big(1-\zeta(2k+1)\big)x^{2k+1}}{2k+1}$ (Legendre series[49]) | $\lvert x\rvert<1$ |
| $\Gamma(x+1)$ | $\sqrt{\frac{\pi x}{\sin\pi x}\cdot\frac{1-x}{1+x}}\exp\left((1-\mathrm{C})x+\sum_{k=1}^{\infty}\frac{\big(1-\zeta(2k+1)\big)x^{2k+1}}{2k+1}\right)$ | $\lvert x\rvert<1$ |

[49]Here C denotes the Euler constant, and $\zeta$ is the Riemannian $\zeta$-function.

| The Euler beta function | | |
|---|---|---|
| $B(x,y)$ | $B(x,y) := \dfrac{\Gamma(x)\Gamma(y)}{\Gamma(x+y)}$ | $x, y \in \mathbb{C}$, $x, y, x+y \neq 0, -1, -2, \ldots$ |
| | $B(x,y) = \displaystyle\int_0^1 t^{x-1}(1-t)^{y-1}\mathrm{d}t$ | $x > 0$, $y > 0$ |

| Bessel functions (cylinder functions) | | |
|---|---|---|
| $J_p(x)$ | $\dfrac{x^p}{2^p\Gamma(p+1)}\left(1 - \dfrac{x^2}{4(p+1)} + \dfrac{x^4}{32(p+1)(p+2)} - \ldots\right)$ $= \displaystyle\sum_{k=0}^{\infty} \frac{(-1)^k}{k!\Gamma(p+k+1)}\left(\frac{x}{2}\right)^{2k+p}$ The parameter $p$ is real with $p \neq -1, -2, \ldots$ | $\|x\| < \infty$, $x \notin ]-\infty, 0]$ |
| $J_{-n}(x)$ | $J_{-n}(x) = (-1)^n J_n(x)$, $\quad n = 1, 2, \ldots$ | $\|x\| < \infty$ |

| Neumann functions | | |
|---|---|---|
| $N_p(x)$ | $N_p(x) := \dfrac{J_p(x)\cos p\pi - J_{-p}(x)}{\sin p\pi}$ The parameter $p$ is real with $p \neq 0, \pm 1, \pm 2, \ldots$ | $\|x\| < \infty$, $x \notin ]-\infty, 0]$ |
| $N_m(x)$ | $N_m(x) := \lim\limits_{p \to m} N_p(x) = \dfrac{1}{\pi}\left(\dfrac{\partial J_p(x)}{\partial p} - (-1)^m \dfrac{\partial J_{-p}(x)}{\partial p}\right)_{p=m}$ $m = 0, \pm 1, \pm 2, \ldots$ | $0 < \|x\| < \infty$ |

| Hankel functions | | |
|---|---|---|
| $H_p^{(s)}(x)$ | $H_p^{(1)}(x) := J_p(x) + \mathrm{i}N_p(x)$ $H_p^{(2)}(x) := J_p(x) - \mathrm{i}N_p(x)$ The parameter $p$ is real. | $\|x\| < \infty$, $x \notin ]-\infty, 0]$ |

| Bessel functions with imaginary argument | | |
|---|---|---|
| $I_p(x)$ | $I_p(x) := \dfrac{J_p(\mathrm{i}x)}{\mathrm{i}^n} = \displaystyle\sum_{k=0}^{\infty} \frac{1}{k!\Gamma(p+k+1)}\left(\frac{x}{2}\right)^{2k+p}$ The parameter $p$ is real. | $\|x\| < \infty$, $x \notin ]-\infty, 0]$ |

| MacDonald functions | | |
|---|---|---|
| $K_p(x)$ | $K_p(x) := \dfrac{\pi\left(I_{-p}(x) - I_p(x)\right)}{2\sin p\pi}$ The parameter $p$ is real with $p \neq 0, \pm 1, \pm 2, \ldots$ | $\|x\| < \infty$, $x \notin ]-\infty, 0]$ |
| $K_m(x)$ | $K_m(x) := \lim\limits_{p \to m} K_p(x) = \dfrac{(-1)^m}{2}\left(\dfrac{\partial I_{-p}(x)}{\partial p} - \dfrac{\partial I_p(x)}{\partial p}\right)_{p=m}$ $m = 0, \pm 1, \pm 2, \ldots$ | $0 < \|x\| < \infty$ |

| **Gaussian error integral** $\operatorname{erf} x := \frac{2}{\sqrt{\pi}} \int_0^x \mathrm{e}^{-t^2} \mathrm{d}t$ | | |
|---|---|---|
| $\operatorname{erf} x$ | $\frac{2}{\sqrt{\pi}} \left( x - \frac{x^3}{3} + \frac{x^5}{10} - \ldots \right) = \frac{2}{\sqrt{\pi}} \sum_{k=0}^{\infty} \frac{(-1)^k x^{2k+1}}{k!(2k+1)}$ | $\lvert x\rvert < \infty$ |

| **Integral sine** $\operatorname{Si}(x) = \int_0^x \frac{\sin t}{t} \mathrm{d}t = \frac{\pi}{2} - \int_x^{\infty} \frac{\sin t}{t} \mathrm{d}t$ | | |
|---|---|---|
| $\operatorname{Si}(x)$ | $x - \frac{x^3}{18} + \frac{x^5}{600} - \ldots = \sum_{k=0}^{\infty} \frac{(-1)^k x^{2k+1}}{(2k+1)!(2k+1)}$ | $\lvert x\rvert < \infty$ |

| **Integral cosine** $\operatorname{Ci}(x) := -\int_x^{\infty} \frac{\cos t}{t} \mathrm{d}t$ | | $0 < x < \infty$ |
|---|---|---|
| $\ln x - \operatorname{Ci}(x) + \mathrm{C}$ | $\frac{x^2}{4} - \frac{x^4}{96} + \ldots = \sum_{k=1}^{\infty} \frac{(-1)^{k+1} x^{2k}}{(2k)!2k}$ (C Euler constant[50]) | $\lvert x\rvert < \infty$ |

| **Integral exponential function**[51] $\operatorname{Ei}(x) := \mathrm{PV} \int_{-\infty}^{x} \frac{\mathrm{e}^t}{t} \mathrm{d}t$ | | $-\infty < x < \infty$, $x \neq 0$ |
|---|---|---|
| $\operatorname{Ei}(x) - \ln\lvert x\rvert - \mathrm{C}$ | $x + \frac{x^2}{4} + \frac{x^3}{18} + \ldots = \sum_{k=1}^{\infty} \frac{x^k}{k!k}$ | $\lvert x\rvert < \infty$ |

| **Logarithmic integral**[52] $\operatorname{li}(x) := \mathrm{PV} \int_0^x \frac{\mathrm{d}t}{\ln t}$ | | $0 < x < 1$, $x > 1$ |
|---|---|---|
| | $\operatorname{li}(x) = \operatorname{Ei}(\ln x)$ | |

[50] The function $\ln x - \operatorname{Ci}(x) + \mathrm{C}$ is initially only defined for real positive $x$. The power series converges for all complex numbers and represents the *analytically extended* function $\ln x - \operatorname{Ci}(x) + \mathrm{C}$ (cf. 1.14.15).

[51] The notation $\mathrm{PV} \int \ldots$ denotes the principal value of the integral, i.e.,

$$\operatorname{Ei}(x) = \lim_{\varepsilon \to +0} \left( \int_{-\infty}^{-\varepsilon} \frac{\mathrm{e}^t}{t} \mathrm{d}t + \int_{\varepsilon}^{x} \frac{\mathrm{e}^t}{t} \mathrm{d}t \right).$$

For $x < 0$ the principal value coincides with the usual integral for the case at hand.

[52] For $0 < x < 1$ one has $\operatorname{li} x := \int_0^x \frac{\mathrm{d}t}{\ln t}$. For $x > 1$ one has

$$\operatorname{li} x = \lim_{\varepsilon \to +0} \left( \int_0^{1-\varepsilon} \frac{\mathrm{d}t}{\ln t} + \int_{1+\varepsilon}^{x} \frac{\mathrm{d}t}{\ln t} \right).$$

| | **Legendre polynomials**[53] $\quad n=0,1,2,\ldots$ | |
|---|---|---|
| $P_n(x)$ | $\frac{1}{2^n n!}\frac{\mathrm{d}^n\left(x^2-1\right)^n}{\mathrm{d}x^n}=\frac{(2n)!}{2^n(n!)^2}\left(x^n-\frac{n(n-1)}{2(2n-1)}x^{n-2}+\frac{n(n-1)(n-2)(n-3)}{2\cdot 4\cdot(2n-1)(2n-3)}x^{n-4}-\ldots\right)$<br>(If $n$ is even (resp. odd), the the last term is $x^0$ (resp. $x$)).<br>*Orthogonality relations:*<br>$\int_{-1}^{1} P_n(x)P_m(x)\mathrm{d}x=\frac{2\delta_{nm}}{2n+1}\,,\qquad n,m=0,1,\ldots$<br>*Special cases:*<br>$P_0(x)=1\,,\quad P_1(x)=x\,,\quad P_2(x)=\frac{1}{2}\left(3x^2-1\right)\,,$<br>$P_3(x)=\frac{1}{2}\left(5x^3-3x\right)\,,\quad P_4(x)=\frac{1}{8}\left(35x^4-30x^2+3\right)$ | $\|x\|<\infty$ |
| $\frac{1}{\sqrt{1-2xz+z^2}}$ | $P_0(x)+P_1(x)z+P_2(x)z^2+\ldots=\sum_{k=0}^{\infty}P_k(x)z^n$ | $\|z\|<1$ |
| | **Legendre functions** $\quad n=0,1,2,\ldots$ | |
| $Q_n(x)$ | $\frac{1}{2}P_n(x)\ln\frac{1+x}{1-x}-\sum_{k=1}^{N(n)}\frac{2n-4k+3}{(2k-1)(n-k+1)}P_{n-2k+1}(x)$<br>$N(n):=\begin{cases}\frac{n}{2} & \text{for even } n\\ \frac{n+1}{2} & \text{for odd } n\end{cases}$ | $x\in\mathbb{C}\backslash[-1,1]$ |
| | **Laguerre polynomials** $\quad n=0,1,2,\ldots$ | |
| $L_n^{\alpha}(x)$ | $\frac{\mathrm{e}^x x^{-\alpha}}{n!}\frac{\mathrm{d}^n\left(\mathrm{e}^{-x}x^{n+\alpha}\right)}{\mathrm{d}x^n}=\sum_{k=0}^{n}\binom{n+\alpha}{n-k}\frac{(-1)^k}{k!}x^k$<br>*Orthogonality relations:*<br>$\int_0^{\infty} x^{\alpha}\mathrm{e}^{-x}L_n^{\alpha}(x)L_m^{\alpha}(x)\mathrm{d}x=\delta_{nm}\Gamma(1+\alpha)\binom{n+\alpha}{n}\,,$<br>$n,m=0,1,\ldots\,,\ \alpha>-1$<br>*Special cases:*<br>$L_0^{\alpha}(x)=1\,,\quad L_1^{\alpha}(x)=1-x+\alpha$ | $\|x\|<\infty$ |

[54]The deeper meaning of the Legendre, Hermite and Laguerre polynomials will become clear in the context of the theory of complete orthonormal systems in Hilbert spaces.

| **Laguerre functions**[55] $\mathscr{L}_n^\alpha(x) := c_n^\alpha \mathrm{e}^{x/2} x^{-\alpha/2} \dfrac{\mathrm{d}^n}{\mathrm{d}x^n}(\mathrm{e}^{-x} x^{n+\alpha}),\ n = 0,1,2,\ldots$ | $\lvert x\rvert < \infty$ |
|---|---|
| *Orthogonality relations:* $$\int_0^\infty \mathscr{L}_n^\alpha(x)\mathscr{L}_m^\alpha(x)\mathrm{d}x = \delta_{nm}\,, \quad n,m = 0,1,2,\ldots$$ | $\alpha > -1$ fixed |

| Hermite polynomials $n = 0,1,2,\ldots$ | | |
|---|---|---|
| $H_n(x)$ | $\alpha_n(-1)^n \mathrm{e}^{x^2} \dfrac{\mathrm{d}^n \mathrm{e}^{-x^2}}{\mathrm{d}x^n}\,, \qquad \alpha_n := 2^{-n/2}(n!)^{-1/2}\pi^{-1/4}$ <br> *Special cases:* <br> $H_0(x) = \alpha_0\,, \quad H_1(x) = 2\alpha_1 x\,, \quad H_2(x) = \alpha_2\left(4x^2 - 2\right)$ | $\lvert x\rvert < \infty$ |

| **Hermite functions** $\mathscr{H}_n(x) := H_n(x)\mathrm{e}^{-x^2/2}\,, \quad n = 0,1,2,\ldots$ | $\lvert x\rvert < \infty$ |
|---|---|
| *Orthogonality relations:* $$\int_{-\infty}^\infty \mathscr{H}_n(x)\mathscr{H}_m(x)\mathrm{d}x = \delta_{nm}\,, \quad n,m = 0,1,2,\ldots$$ | |

## 0.7.3 Asymptotic series

An asymptotic expansion of a function is a representation of a function for very large values of the argument.

### 0.7.3.1 Convergent expansions

| *Function* | *Infinite series* | *Domain of convergence* |
|---|---|---|
| $\ln x$ | $\dfrac{x-1}{x} + \dfrac{(x-1)^2}{2x^2} + \dfrac{(x-1)^3}{3x^3} + \ldots = \sum\limits_{k=1}^\infty \dfrac{(x-1)^k}{kx^k}$ | $x > \dfrac{1}{2}$ ($x$ real) |
| $\arctan x$ | $\dfrac{\pi}{2} - \dfrac{1}{x} + \dfrac{1}{3x^3} - \dfrac{1}{5x^5} + \ldots$ | $x > 1$ ($x$ real) |
| $\arctan x$ | $-\dfrac{\pi}{2} - \dfrac{1}{x} + \dfrac{1}{3x^3} - \dfrac{1}{5x^5} + \ldots$ | $x < -1$ ($x$ real) |
| $\ln 2x - \operatorname{arcosh} x$ | $\dfrac{1}{4x^2} + \dfrac{3}{32x^4} + \dfrac{15}{288x^6} + \ldots$ <br> $= \sum\limits_{k=1}^\infty \dfrac{1\cdot 3\cdot 5\cdots(2k-1)}{2\cdot 4\cdot 6\cdots 2k(2k)}\cdot\dfrac{1}{x^{2k}}$ | $x > 1$ ($x$ real) |
| $\operatorname{arcoth} x$ | $\dfrac{1}{x} + \dfrac{1}{3x^3} + \dfrac{1}{5x^5} + \ldots$ | $\lvert x\rvert > 1$ ($x$ complex) |

[55] See section 1.13.2.13; the coefficients $c_n^\alpha$ are chosen so that $\mathscr{L}_n^\alpha$ satisfy $(\mathscr{L}_n^\alpha, \mathscr{L}_n^\alpha) = 1$.

#### 0.7.3.2 Asymptotic equality

We use the notation

$$\boxed{f(x) \cong g(x)\,, \qquad x \to a\,,}$$

to indicate that $\lim\limits_{x \to a} \dfrac{f(x)}{g(x)} = 1$.

| |
|---|
| $\left(1 + \frac{1}{2} + \frac{1}{3} + \ldots + \frac{1}{n}\right) - \ln(n+1) \cong \mathrm{C}\,, \quad n \to \infty \quad$ (C the Euler constant), |
| $n! \cong \left(\frac{n}{\mathrm{e}}\right)^n \sqrt{2\pi n}\,, \quad n \to \infty \quad$ (Stirling 1730), <br> $\ln n! \cong \left(n + \frac{1}{2}\right) \ln n - n + \frac{1}{2} \ln \sqrt{2\pi}\,, \quad n \to \infty\,.$ |

#### 0.7.3.3 Asymptotic expansions in the sense of Poincaré

Following Poincaré (1854–1912) one writes

$$\boxed{f(x) \sim \sum_{k=1}^{\infty} \frac{a_k}{x^k}\,, \qquad x \to +\infty\,,} \tag{0.55}$$

to indicate the behavior

$$f(x) = \sum_{k=1}^{n} \frac{a_k}{x^k} + o\left(\frac{1}{x^n}\right)\,, \qquad x \to +\infty\,,$$

for all $n = 1, 2, \ldots$[56] This kind of series was met by Poincaré during his deep investigations of celestial mechanics at the end of the nineteenth century. He ran across divergent series of the form (0.55). At the same time he discovered that such series are nevertheless quite natural, since the expansion contains important information about the function $f$.

**Stirling's series for the gamma function:**

$$\boxed{\ln \Gamma(x+1) - \left(x + \frac{1}{2}\right) \ln x + x - \ln \sqrt{2\pi} \sim \sum_{k=1}^{\infty} \frac{B_{2k}}{(2k-1)2k} \cdot \frac{1}{x^{2k-1}}\,, \qquad x \to +\infty\,.}$$

Here $B_{2k}$ denote the Bernoulli numbers.

**Asymptotic expansion of the Euler integral:**

$$\boxed{\int_x^{\infty} t^{-1} \mathrm{e}^{x-t} \mathrm{d}t \sim \frac{1}{x} - \frac{1}{x^2} + \frac{2!}{x^3} - \frac{3!}{x^4} + \ldots\,, \qquad x \to +\infty\,.}$$

[56] The symbol $o(\ldots)$ will be explained in 1.3.1.4. Explicitly, one has

$$\lim_{x \to +\infty} x^n \left(f(x) - \sum_{k=1}^{n} \frac{a_k}{x^k}\right) = 0 \qquad \text{for all} \quad n = 1, 2, \ldots$$

**Asymptotic representation of the Bessel and Neumann functions:**

$$J_p(x) = \sqrt{\frac{2}{\pi x}}\left(\cos\left(x - \frac{p\pi}{2} - \frac{\pi}{4}\right)\right) + o\left(\frac{1}{\sqrt{x}}\right), \qquad x \to +\infty,$$

$$N_p(x) = \sqrt{\frac{2}{\pi x}}\left(\sin\left(x - \frac{p\pi}{2} - \frac{\pi}{4}\right)\right) + o\left(\frac{1}{\sqrt{x}}\right), \qquad x \to +\infty.$$

The parameter $p$ is real.

**The method of stationary phase:** One has

$$\int_{-\infty}^{\infty} A(x)\mathrm{e}^{\mathrm{i}\omega p(x)}\mathrm{d}x \sim \frac{b\mathrm{e}^{\mathrm{i}p(a)}}{\sqrt{\omega}}\sum_{k=0}^{\infty}\frac{A_k}{\omega^k}, \qquad \omega \to +\infty,$$

with $b := \sqrt{2\pi\mathrm{i}/p''(a)}$ $(\mathrm{Re}\, b > 0)$ and

$$A_k := \sum_{\substack{n-m=k \\ 2n \geq 3m \geq 0}} \frac{1}{\mathrm{i}^k 2^n n! m! p''(a)} \frac{\mathrm{d}^{n+1}}{\mathrm{d}x^{n+1}}(P^m f)(a)$$

as well as $P(x) := p(x) - p(a) - \frac{1}{2}(x-a)^2 p''(a)$. Here it is assumed that the following conditions are satisfied.

(i) The complex valued phase factor $p\colon \mathbb{R} \to \mathbb{C}$ is infinitely often differentiable. One has $\mathrm{Im}\, p(a) = 0$ and $p'(a) = 0$ with $p''(a) \neq 0$.

(ii) One has $p'(x) \neq 0$ for all real numbers $x \neq a$. The imaginary part $\mathrm{Im}\, p(x)$ is non-negative for all real numbers $x$.

(iii) The real function $A\colon \mathbb{R} \to \mathbb{R}$ describing the amplitude is infinitely often differentiable and vanishes outside of some bounded interval.

This theorem is important in classical optics (limiting behavior for large angular frequencies $\omega$ and hence for small wavelengths $\lambda$) as well as in the modern theory of Fourier integral operators.

### 0.7.4 Fourier series

Note: See section 1.11.2.

1. $y = x$ for $-\pi < x < \pi$; $\quad y = 2\left(\frac{\sin x}{1} - \frac{\sin 2x}{2} + \frac{\sin 3x}{3} - \dots\right)$

For the arguments $\pm k\pi$, the series sums up to 0 according to Dirichlet's theorem.

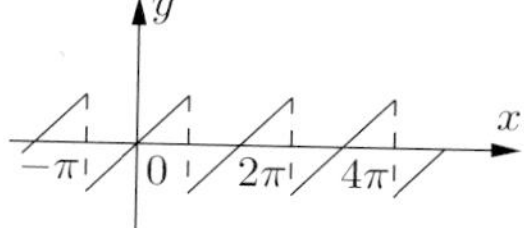


2. $y = |x|$ for $-\pi \leq x \leq \pi$; $\quad y = \frac{\pi}{2} - \frac{4}{\pi}\left(\cos x + \frac{\cos 3x}{3^2} + \frac{\cos 5x}{5^2} + \frac{\cos 7x}{7^2} + \dots\right)$

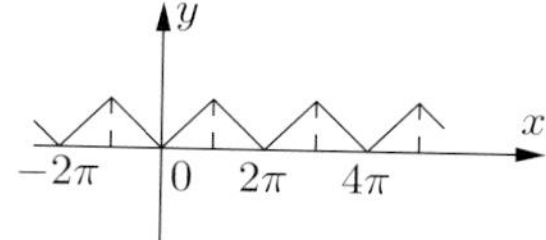


3. $y = x$ for $0 < x < 2\pi$; $\quad y = \pi - 2\left(\frac{\sin x}{1} + \frac{\sin 2x}{2} + \frac{\sin 3x}{3} + \dots\right)$

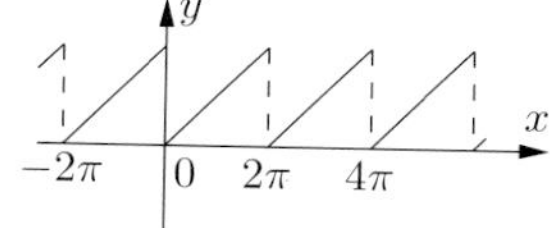


4.

$$y = \begin{Bmatrix} x & \text{for} & -\frac{\pi}{2} \leq x \leq \frac{\pi}{2} \\ \pi - x & \text{for} & \frac{\pi}{2} \leq x \leq \pi \\ -(\pi + x) & \text{for} & -\pi \leq x \leq -\frac{\pi}{2} \end{Bmatrix}; \quad y = \frac{4}{\pi}\left(\sin x - \frac{\sin 3x}{3^2} + \frac{\sin 5x}{5^2} - \dots\right)$$

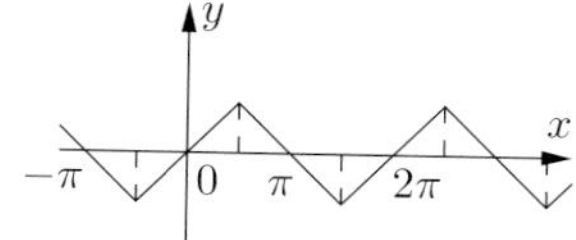


5. $y = \begin{Bmatrix} -a & \text{for} & -\pi < x < 0 \\ a & \text{for} & 0 < x < \pi \end{Bmatrix}; \quad y = \frac{4a}{\pi}\left(\sin x + \frac{\sin 3x}{3} + \frac{\sin 5x}{5} + \dots\right)$

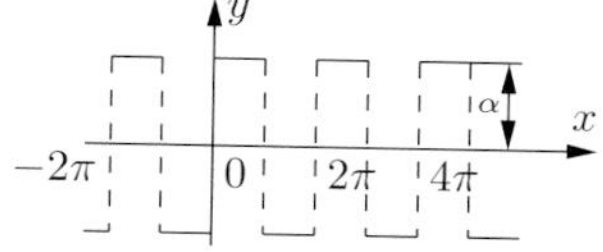

6. $y = \begin{cases} c_1 & \text{for} \quad -\pi < x < 0 \\ c_2 & \text{for} \quad 0 < x < \pi \end{cases};$

$$y = \frac{c_1 + c_2}{2} - 2\frac{c_1 - c_2}{\pi}\left(\sin x + \frac{\sin 3x}{3} + \frac{\sin 5x}{5} + \dots\right)$$

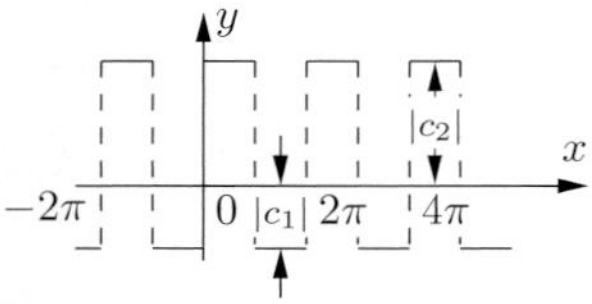

7.

$$y = \begin{cases} 0 & \text{for} \quad -\pi < x < -\pi + \alpha, \quad -\alpha < x < \alpha, \quad \pi - \alpha < x < \pi \\ a & \text{for} \quad \alpha < x < \pi - \alpha \\ -a & \text{for} \quad -\pi + \alpha < x < -\alpha \end{cases};$$

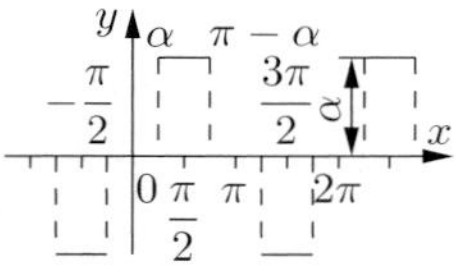

$0 < \alpha < \frac{\pi}{2}$

$$y = \frac{4a}{\pi}\left(\cos\alpha \sin x + \frac{1}{3}\cos 3\alpha \sin 3x + \frac{1}{5}\cos 5\alpha \sin 5x + \dots\right)$$

8.

$$y = \begin{cases} \frac{ax}{\alpha} & \text{for} \quad -\alpha \le x \le \alpha \\ a & \text{for} \quad \alpha \le x \le \pi - \alpha \\ -a & \text{for} \quad -\pi + a \le x \le -\alpha \\ \frac{a(\pi - x)}{\alpha} & \text{for} \quad \pi - \alpha \le x \le \pi \\ -\frac{a(x + \pi)}{\alpha} & \text{for} \quad -\pi \le x \le -\pi + \alpha \end{cases};$$

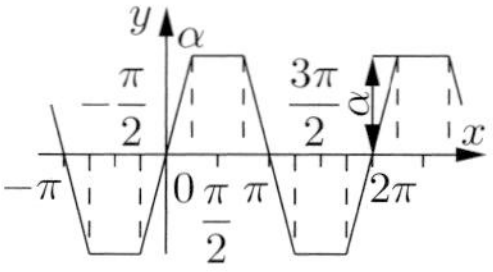

$$y = \frac{4a}{\pi\alpha}\left(\sin\alpha \sin x + \frac{1}{3^2}\sin 3\alpha \sin 3x + \frac{1}{5^2}\sin 5\alpha \sin 5x + \dots\right)$$

In particular, for $\alpha = \frac{\pi}{3}$:

$$y = \frac{6a\sqrt{3}}{\pi^2}\left(\sin x - \frac{1}{5^2}\sin 5x + \frac{1}{7^2}\sin 7x - \frac{1}{11^2}\sin 11x + \dots\right)$$

9. $y = x^2 \quad \text{for} \quad -\pi \le x \le \pi;$ $\qquad y = \frac{\pi^2}{3} - 4\left(\cos x - \frac{\cos 2x}{2^2} + \frac{\cos 3x}{3^2} - \dots\right)$

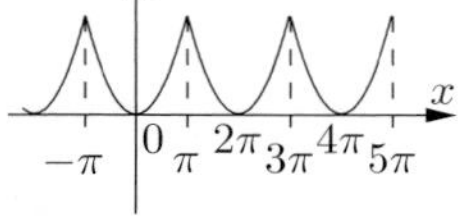

10. $y = \begin{cases} -x^2 & \text{for } \quad -\pi < x \le 0 \\ x^2 & \text{for } \quad 0 \le x < \pi \end{cases};$

$$y = 2\pi\left(\sin x - \frac{\sin 2x}{2} + \frac{\sin 3x}{3} - \dots\right) - \frac{8}{\pi}\left(\frac{\sin x}{1^3} + \frac{\sin 3x}{3^3} + \frac{\sin 5x}{5^3} + \dots\right)$$

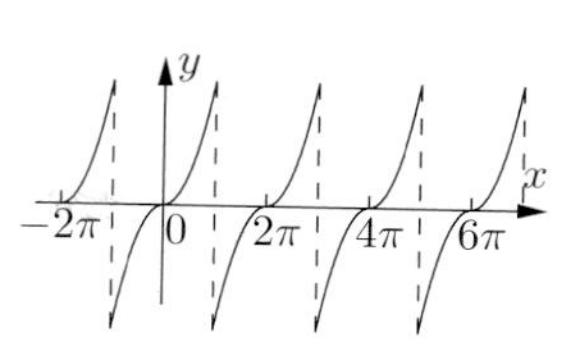


11. $y = x(\pi - x)$ for $0 \le x \le \pi$, an even extension in $(-\pi, 0)$;

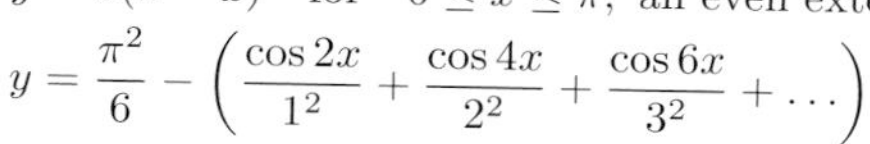

$$y = \frac{\pi^2}{6} - \left(\frac{\cos 2x}{1^2} + \frac{\cos 4x}{2^2} + \frac{\cos 6x}{3^2} + \dots\right)$$

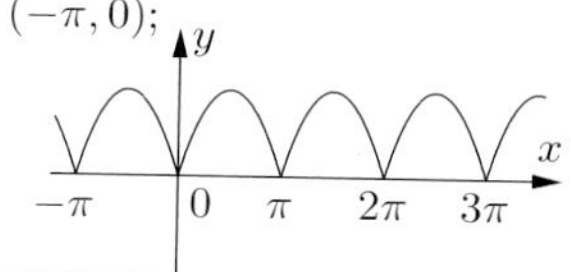


12. $y = x(\pi - x)$ for $0 \le x \le \pi$, an odd extension in $(-\pi, 0)$;

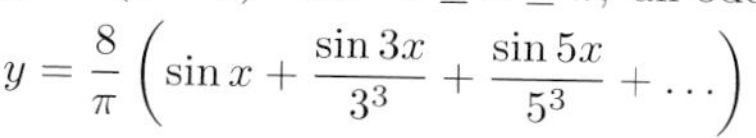

$$y = \frac{8}{\pi}\left(\sin x + \frac{\sin 3x}{3^3} + \frac{\sin 5x}{5^3} + \dots\right)$$

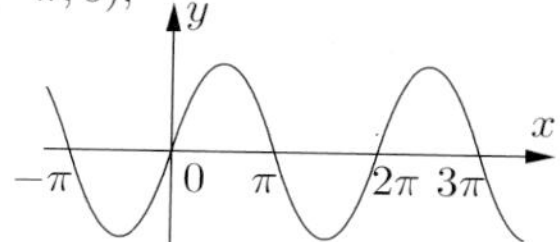


13. $y = Ax^2 + Bx + C$ for $-\pi < x < \pi$;

$$y = \frac{A\pi^2}{3} + C + 4A\sum_{k=1}^{\infty}(-1)^k\frac{\cos kx}{k^2} - 2B\sum_{k=1}^{\infty}(-1)^k\frac{\sin kx}{k}$$

14. $y = |\sin x|$ for $-\pi \le x \le \pi$;

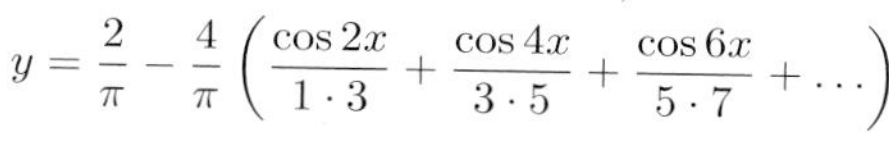

$$y = \frac{2}{\pi} - \frac{4}{\pi}\left(\frac{\cos 2x}{1\cdot 3} + \frac{\cos 4x}{3\cdot 5} + \frac{\cos 6x}{5\cdot 7} + \dots\right)$$

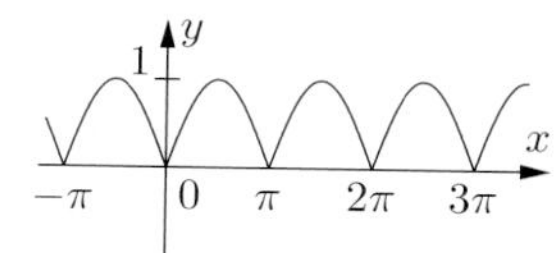


15. $y = \cos x$ for $0 < x < \pi$, an odd extension in $(-\pi, 0)$;

$$y = \frac{4}{\pi}\left(\frac{2\sin 2x}{1\cdot 3} + \frac{4\sin 4x}{3\cdot 5} + \frac{6\sin 6x}{5\cdot 7} + \dots\right)$$

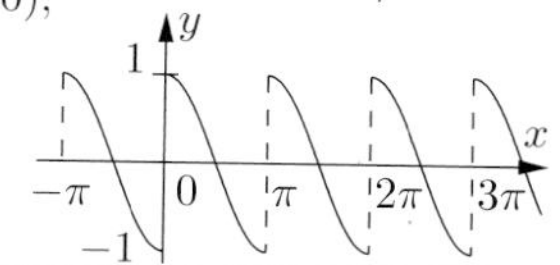


16. $y = \begin{cases} 0 & \text{for } \quad -\pi \le x \le 0 \\ \sin x & \text{for } \quad 0 \le x \le \pi \end{cases};$

$$y = \frac{1}{\pi} + \frac{1}{2}\sin x - \frac{2}{\pi}\left(\frac{\cos 2x}{1\cdot 3} + \frac{\cos 4x}{3\cdot 5} + \frac{\cos 6x}{5\cdot 7} + \dots\right)$$

17. $y = \cos ux$ for $-\pi \le x \le \pi$, $u$ arbitrary real, but not integer;

$$y = \frac{2u\sin u\pi}{\pi}\left(\frac{1}{2u^2} - \frac{\cos x}{u^2 - 1} + \frac{\cos 2x}{u^2 - 4} - \frac{\cos 3x}{u^2 - 9} + \dots\right)$$

18. $y = \sin ux$ for $-\pi < x < \pi$, $u$ arbitrary real, but not integer;

$$y = \frac{2\sin u\pi}{\pi}\left(-\frac{\sin x}{u^2 - 1} + \frac{2\sin 2x}{u^2 - 4} - \frac{3\sin 3x}{u^2 - 9} + \dots\right)$$

19. $y = x \cos x$ for $-\pi < x < \pi$;

$$y = -\frac{1}{2}\sin x + \frac{4\sin 2x}{2^2-1} - \frac{6\sin 3x}{3^2-1} + \frac{8\sin 4x}{4^2-1} - \dots$$

20. $y = x \sin x$ for $-\pi \le x \le \pi$;

$$y = 1 - \frac{1}{2}\cos x - 2\left(\frac{\cos 2x}{2^2-1} - \frac{\cos 3x}{3^2-1} + \frac{\cos 4x}{4^2-1} - \dots\right)$$

21. $y = \cosh ux$ for $-\pi \le x \le \pi$;

$$y = \frac{2u\sinh u\pi}{\pi}\left(\frac{1}{2u^2} - \frac{\cos x}{u^2+1^2} + \frac{\cos 2x}{u^2+2^2} - \frac{\cos 3x}{u^2+3^2} + \dots\right)$$

22. $y = \sinh ux$ for $-\pi < x < \pi$;

$$y = \frac{2\sinh u\pi}{\pi}\left(\frac{\sin x}{u^2+1^2} - \frac{2\sin 2x}{u^2+2^2} + \frac{3\sin 3x}{u^2+3^2} - \dots\right)$$

23. $y = \mathrm{e}^{ax}$ for $-\pi < x < \pi, \quad a \ne 0$;

$$y = \frac{2}{\pi}\sinh a\pi\left(\frac{1}{2a} + \sum_{k=1}^{\infty}\frac{(-1)^k}{a^2+k^2}(a\cos kx - k\sin kx)\right)$$

In the following examples the problem is not so much how to develop a given function into a Fourier series than the converse question: to which functions do certain simple trigonometric series converge?

24. $$\sum_{k=1}^{\infty}\frac{\cos kx}{k} = -\ln\left(2\sin\frac{x}{2}\right), \quad 0 < x < 2\pi$$

25. $$\sum_{k=1}^{\infty}\frac{\sin kx}{k} = \frac{\pi - x}{2}, \quad 0 < x < 2\pi$$

26. $$\sum_{k=1}^{\infty}\frac{\cos kx}{k^2} = \frac{3x^2 - 6\pi x + 2\pi^2}{12}, \quad 0 \le x \le 2\pi$$

27. $$\sum_{k=1}^{\infty}\frac{\sin kx}{k^2} = -\int_0^x \ln\left(2\sin\frac{z}{2}\right)\mathrm{d}z, \quad 0 \le x \le 2\pi$$

28. $$\sum_{k=1}^{\infty}\frac{\cos kx}{k^3} = \int_0^x \mathrm{d}z\int_0^z \ln\left(2\sin\frac{t}{2}\right)\mathrm{d}t + \sum_{k=1}^{\infty}\frac{1}{k^3}, \quad 0 \le x \le 2\pi$$

$$\left(\sum_{k=1}^{\infty}\frac{1}{k^3} = \frac{\pi^3}{25.79436\dots} = 1.202\,06\dots\right)$$

29. $$\sum_{k=1}^{\infty}\frac{\sin kx}{k^3} = \frac{x^3 - 3\pi x^2 + 2\pi^2 x}{12}, \quad 0 \le x \le 2\pi$$

30. $$\sum_{k=1}^{\infty}(-1)^{k+1}\frac{\cos kx}{k} = \ln\left(2\cos\frac{x}{2}\right), \quad -\pi < x < \pi$$

31. $\sum_{k=1}^{\infty}(-1)^{k+1}\frac{\sin kx}{k} = \frac{x}{2}, \quad -\pi < x < \pi$

32. $\sum_{k=1}^{\infty}(-1)^{k+1}\frac{\cos kx}{k^2} = \frac{\pi^2 - 3x^2}{12}, \quad -\pi \le x \le \pi$

33. $\sum_{k=1}^{\infty}(-1)^{k+1}\frac{\sin kx}{k^2} = \int_0^x \ln\left(2\cos\frac{z}{2}\right) dz, \quad -\pi \le x \le \pi$

34. $\sum_{k=1}^{\infty}(-1)^{k+1}\frac{\cos kx}{k^3} = \sum_{k=1}^{\infty}(-1)^{k+1}\cdot\frac{1}{k^3} - \int_0^x dz \int_0^z \ln\left(2\cos\frac{t}{2}\right) dt, \quad -\pi \le x \le \pi$

35. $\sum_{k=1}^{\infty}(-1)^{k+1}\frac{\sin kx}{k^3} = \frac{\pi^2 x - x^3}{12}, \quad -\pi \le x \le \pi$

36. $\sum_{k=0}^{\infty}\frac{\cos(2k+1)\,x}{2k+1} = -\frac{1}{2}\ln\left(\tan\frac{x}{2}\right), \quad 0 < x < \pi$

37. $\sum_{k=0}^{\infty}\frac{\sin(2k+1)\,x}{2k+1} = \frac{\pi}{4}, \quad 0 < x < \pi$

38. $\sum_{k=0}^{\infty}\frac{\cos(2k+1)\,x}{(2k+1)^2} = \frac{\pi^2 - 2\pi x}{8}, \quad 0 \le x \le \pi$

39. $\sum_{k=0}^{\infty}\frac{\sin(2k+1)\,x}{(2k+1)^2} = -\frac{1}{2}\int_0^x \ln\left(\tan\frac{z}{2}\right) dz, \quad 0 \le x \le \pi$

40. $\sum_{k=0}^{\infty}\frac{\cos(2k+1)\,x}{(2k+1)^3} = \frac{1}{2}\int_0^x dz \int_0^z \ln\left(\tan\frac{t}{2}\right) dt + \sum_{k=0}^{\infty}\frac{1}{(2k+1)^3}, \quad 0 \le x \le \pi$

41. $\sum_{k=0}^{\infty}\frac{\sin(2k+1)\,x}{(2k+1)^3} = \frac{\pi^2 x - \pi x^2}{8}, \quad 0 \le x \le \pi$

42. $\sum_{k=0}^{\infty}(-1)^k\frac{\cos(2k+1)\,x}{2k+1} = \frac{\pi}{4}, \quad -\frac{\pi}{2} < x < \frac{\pi}{2}$

43. $\sum_{k=0}^{\infty}(-1)^k\frac{\sin(2k+1)\,x}{2k+1} = -\frac{1}{2}\ln\left[\tan\left(\frac{\pi}{4} - \frac{x}{2}\right)\right], \quad -\frac{\pi}{2} < x < \frac{\pi}{2}$

44. $\sum_{k=0}^{\infty}(-1)^k\frac{\cos(2k+1)\,x}{(2k+1)^2} = -\frac{1}{2}\int_0^{\frac{\pi}{2} - x} \ln\left(\tan\frac{z}{2}\right) dz, \quad -\frac{\pi}{2} \le x \le \frac{\pi}{2}$

45. $\sum_{k=0}^{\infty}(-1)^k\frac{\sin(2k+1)\,x}{(2k+1)^2} = \frac{\pi x}{4}, \quad -\frac{\pi}{2} \le x \le \frac{\pi}{2}$

46. $\displaystyle\sum_{k=0}^{\infty}(-1)^k \frac{\cos(2k+1)\,x}{(2k+1)^3} = \frac{\pi^3 - 4\pi x^2}{32}, \qquad -\frac{\pi}{2} \le x \le \frac{\pi}{2}$

47. $\displaystyle\sum_{k=0}^{\infty}(-1)^k \frac{\sin(2k+1)\,x}{(2k+1)^3} = \frac{1}{2}\int\limits_0^{\frac{\pi}{2}-x} \mathrm{d}z \int\limits_0^{z} \ln\left(\tan\frac{t}{2}\right)\mathrm{d}t + \sum_{k=0}^{\infty}\frac{1}{(2k+1)^3}, \ -\frac{\pi}{2} \le x \le \frac{\pi}{2}$

### 0.7.5 Infinite products

The convergence of infinite products will be considered in section 1.10.6.

| *Function* | *Infinite product* | *Discoverer* | *Domain of convergence* ($x \in \mathbb{C}$) |
|---|---|---|---|
| $\sin \pi x$ | $\pi x \prod_{k=1}^{\infty}\left(1 - \frac{x^2}{k^2}\right)$ | (Euler 1734) | $\lvert x\rvert < \infty$ |
| $\frac{\pi}{2}$ | $\prod_{k=1}^{\infty}\frac{(2k)^2}{4k^2-1}$ | (Wallis 1655) | |
| $\Gamma(x+1)$ | $\prod_{n=1}^{\infty}\frac{\left(1+\frac{1}{n}\right)^x}{1+\frac{x}{n}}$ | (Euler) | $\lvert x\rvert < \infty$ <br> $(x \ne -1, -2, \ldots)$ |
| | $\lim\limits_{k\to\infty}\frac{k!k^x}{(x+1)(x+2)\ldots(x+k)}$ | (Gauss) | |
| | $\mathrm{e}^{-Cx}\prod_{n=1}^{\infty}\frac{\mathrm{e}^{x/n}}{1+\frac{x}{n}}$ | (Weierstrass; C Euler constant) | |
| $\zeta(x)$ | $\prod_p \left(1 - \frac{1}{p^x}\right)^{-1}$ | (Euler[57]) | $\lvert x\rvert > 1$ |

**Further examples:**

$$\prod_{k=2}^{\infty}\left(1 - \frac{1}{k^2}\right) = \frac{1}{2}, \qquad \prod_{k=0}^{\infty}\left(1 + x^{2^k}\right) = \frac{1}{1-x} \qquad (x \in \mathbb{C},\ \lvert x\rvert < 1),$$

$$\sqrt{\frac{1}{2}}\sqrt{\frac{1}{2}+\frac{1}{2}\sqrt{\frac{1}{2}}}\sqrt{\frac{1}{2}+\frac{1}{2}\sqrt{\frac{1}{2}+\frac{1}{2}\sqrt{\frac{1}{2}}}}\ldots = \frac{2}{\pi} \quad \text{(product of Vieta 1579)},$$

$$\prod_{k=1}^{\infty}\left(1 - \frac{1}{(2k)^2}\right) = \frac{2}{\pi}, \qquad \prod_{k=1}^{\infty}\left(1 - \frac{1}{(2k+1)^2}\right) = \frac{\pi}{4},$$

$$\left(\frac{2}{1}\right)\left(\frac{4}{3}\right)^{1/2}\left(\frac{6\cdot 8}{5\cdot 7}\right)^{1/4}\left(\frac{10\cdot 12\cdot 14\cdot 16}{9\cdot 11\cdot 13\cdot 15}\right)^{1/8}\ldots = \mathrm{e},$$

$$\prod_{k=1}^{\infty}\frac{\sqrt[k]{\mathrm{e}}}{1+\frac{1}{k}} = \mathrm{e}^{C} \qquad (\text{Euler constant } C = 0.577\,215\ldots).$$

[57] The product is to be taken over all prime numbers $p$, and $\zeta(s)$ denotes the Riemannian $\zeta$-Funktion.

# 0.8 Tables for differentiation of functions

## 0.8.1 Differentiation of elementary functions

*Table 0.35. First derivatives.*

| ***Function*** $f(x)$ | ***Derivative***[58] $f'(x)$ | ***Validity for real numbers***[59] | ***Validity for complex numbers***[59] |
|---|---|---|---|
| $C$ (constant) | $0$ | $x \in \mathbb{R}$ | $x \in \mathbb{C}$ |
| $x$ | $1$ | $x \in \mathbb{R}$ | $x \in \mathbb{C}$ |
| $x^2$ | $2x$ | $x \in \mathbb{R}$ | $x \in \mathbb{C}$ |
| $x^n \ (n = 1, 2, \ldots)$ | $nx^{n-1}$ | $x \in \mathbb{R}$ | $x \in \mathbb{C}$ |
| $\frac{1}{x}$ | $-\frac{1}{x^2}$ | $x \neq 0$ | $x \neq 0$ |
| $\frac{1}{x^n} \ (n = 1, 2, \ldots)$ | $-\frac{n}{x^{n+1}}$ | $x \neq 0$ | $x \neq 0$ |
| $x^\alpha = \mathrm{e}^{\alpha \cdot \ln x}$ ($\alpha$ real) | $\alpha x^{\alpha-1}$ | $x > 0$ | $x \neq 0, -\pi < \arg x < \pi$ |
| $\sqrt{x} = x^{\frac{1}{2}}$ | $\frac{1}{2\sqrt{x}}$ | $x > 0$ | $x \neq 0, -\pi < \arg x < \pi$ |
| $\sqrt[n]{x} = x^{\frac{1}{n}} \ (n = 2, 3, \ldots)$ | $\frac{\sqrt[n]{x}}{nx}$ | $x > 0$ | $x \neq 0, -\pi < \arg x < \pi$ |
| $\ln x$ | $\frac{1}{x}$ | $x > 0$ | $x \neq 0, -\pi < \arg x < \pi$ |
| $\log_a x = \frac{\ln x}{\ln a}$ $(a > 0,\ a \neq 1)$ | $\frac{1}{x \ln a}$ | $x > 0$ | $x \neq 0, -\pi < \arg x < \pi$ |
| $\mathrm{e}^x$ | $\mathrm{e}^x$ | $x \in \mathbb{R}$ | $x \in \mathbb{C}$ |
| $a^x = \mathrm{e}^{x \cdot \ln a}$ $(a > 0,\ a \neq 1)$ | $a^x \ln a$ | $x \in \mathbb{R}$ | $x \in \mathbb{C}$ |
| $\sin x$ | $\cos x$ | $x \in \mathbb{R}$ | $x \in \mathbb{C}$ |
| $\cos x$ | $-\sin x$ | $x \in \mathbb{R}$ | $x \in \mathbb{C}$ |
| $\sinh x$ | $\cosh x$ | $x \in \mathbb{R}$ | $x \in \mathbb{C}$ |
| $\cosh x$ | $\sinh x$ | $x \in \mathbb{R}$ | $x \in \mathbb{C}$ |
| $\tan x$ | $\frac{1}{\cos^2 x}$ | $x \neq k\pi + \frac{\pi}{2}, k \in \mathbb{Z}$ | $x \neq k\pi + \frac{\pi}{2},\ k \in \mathbb{Z}$ |

[58] Instead of $f'(x)$ one also writes $\frac{\mathrm{d}f(x)}{\mathrm{d}x}$ or $\frac{\mathrm{d}y}{\mathrm{d}x}$.

[59] $x \in \mathbb{R}$ (resp. $x \in \mathbb{C}$) means that the derivative for all real (resp. complex) numbers exists. The notation $k \in \mathbb{Z}$ stands for $k = 0, \pm 1, \pm 2, \ldots$

Table 0.35. (continued)

| | | | |
|---|---|---|---|
| $\cot x$ | $-\dfrac{1}{\sin^2 x}$ | $x \neq k\pi, k \in \mathbb{Z}$ | $x \neq k\pi,\ k \in \mathbb{Z}$ |
| $\tanh x$ | $\dfrac{1}{\cosh^2 x}$ | $x \in \mathbb{R}$ | $x \neq \mathrm{i}k\pi + \dfrac{\mathrm{i}\pi}{2},\ k \in \mathbb{Z}$ |
| $\coth x$ | $-\dfrac{1}{\sinh^2 x}$ | $x \in \mathbb{R}$ | $x \neq \mathrm{i}k\pi,\ k \in \mathbb{Z}$ |
| $\arcsin x$ | $\dfrac{1}{\sqrt{1-x^2}}$ | $-1 < x < 1$ | $\|x\| < 1$ |
| $\arccos x$ | $-\dfrac{1}{\sqrt{1-x^2}}$ | $-1 < x < 1$ | $\|x\| < 1$ |
| $\arctan x$ | $\dfrac{1}{1+x^2}$ | $x \in \mathbb{R}$ | $\|\mathrm{Im}\, x\| < 1$ |
| $\mathrm{arccot} x$ | $-\dfrac{1}{1+x^2}$ | $x \in \mathbb{R}$ | $\|\mathrm{Im}\, x\| < 1$ |
| $\mathrm{arsinh} x$ | $\dfrac{1}{\sqrt{1+x^2}}$ | $-1 < x < 1$ | $\|x\| < 1$ |
| $\mathrm{arcosh} x$ | $\dfrac{1}{\sqrt{x^2-1}}$ | $x > 1$ | $-\pi < \arg(x^2-1) < \pi$,<br>$x \neq \pm 1$ |
| $\mathrm{artanh} x$ | $\dfrac{1}{1-x^2}$ | $\|x\| < 1$ | $\|x\| < 1$ |
| $\mathrm{arcoth} x$ | $\dfrac{1}{1-x^2}$ | $\|x\| > 1$ | $\|x\| \in \mathbb{C} \backslash [-1, 1]$ |

Table 0.36. Higher derivatives.

| ***Function $f(x)$*** | ***$n^{th}$ derivative $f^{(n)}(x)$*** | ***Validity for real numbers*** | ***Validity for complex numbers*** |
|---|---|---|---|
| $x^m\ (m = 1, 2, \ldots)$ | $m(m-1)\ldots(m-n+1)x^{m-n}$<br>$(= 0 \quad \text{for} \quad n > m)$ | $x \in \mathbb{R}$ | $x \in \mathbb{C}$ |
| $x^\alpha\ (\alpha\ \text{ real})$ | $\alpha(\alpha-1)\ldots(\alpha-n+1)x^{\alpha-n}$ | $x > 0$ | $x \neq 0$,<br>$-\pi < \arg x < \pi$ |
| $\mathrm{e}^{ax}\ (a \in \mathbb{C})$ | $a^n \mathrm{e}^{ax}$ | $x \in \mathbb{R}$ | $x \in \mathbb{C}$ |
| $\sin bx$<br>$(b \in \mathbb{C})$ | $b^n \sin\left(bx + \dfrac{n\pi}{2}\right)$ | $x \in \mathbb{R}$ | $x \in \mathbb{C}$ |
| $\cos bx$<br>$(b \in \mathbb{C})$ | $b^n \cos\left(bx + \dfrac{n\pi}{2}\right)$ | $x \in \mathbb{R}$ | $x \in \mathbb{C}$ |
| $\sinh bx$ | $b^n \sinh bx$ for even $n$,<br>$b^n \cosh bx$ for odd $n$ | $x \in \mathbb{R}$ | $x \in \mathbb{C}$ |

Table 0.36. (continued)

| | | | |
|---|---|---|---|
| $\cosh bx$ | $b^n \cosh bx$ for even $n$, $b^n \sinh bx$ for odd $n$ | $x \in \mathbb{R}$ | $x \in \mathbb{C}$ |
| $a^{bx}$ $(a > 0, b \in \mathbb{C})$ | $(b \cdot \ln a)^n a^{bx}$ | $x \in \mathbb{R}$ | $x \neq 0,\ -\pi < \arg x < \pi$ |
| $\ln x$ | $(-1)^{n-1}\dfrac{(n-1)!}{x^n}$ | $x > 0$ | $x \neq 0,\ -\pi < \arg x < \pi$ |
| $\log_a x$ $(a > 0,\ a \neq 1)$ | $(-1)^{n-1}\dfrac{(n-1)!}{x^n \ln a}$ | $x > 0$ | $x \neq 0,\ -\pi < \arg x < \pi$ |

### 0.8.2 Rules for differentiation of functions of one variable

Table 0.37. Rules for differentiation.[60]

| ***Rule*** | ***Formula in Leibniz' notation*** |
|---|---|
| rule of sums | $\dfrac{\mathrm{d}(f+g)}{\mathrm{d}x} = \dfrac{\mathrm{d}f}{\mathrm{d}x} + \dfrac{\mathrm{d}g}{\mathrm{d}x}$ |
| multiplication by a constant | $\dfrac{\mathrm{d}(Cf)}{\mathrm{d}x} = C\dfrac{\mathrm{d}f}{\mathrm{d}x} \quad (C \text{ constant})$ |
| product rule | $\dfrac{\mathrm{d}(fg)}{\mathrm{d}x} = \dfrac{\mathrm{d}f}{\mathrm{d}x}g + f\dfrac{\mathrm{d}g}{\mathrm{d}x}$ |
| quotient rule | $\dfrac{\mathrm{d}\left(\dfrac{f}{g}\right)}{\mathrm{d}x} = \dfrac{\dfrac{\mathrm{d}f}{\mathrm{d}x}g - f\dfrac{\mathrm{d}g}{\mathrm{d}x}}{g^2}$ |
| chain rule | $\dfrac{\mathrm{d}y}{\mathrm{d}x} = \dfrac{\mathrm{d}y}{\mathrm{d}z}\dfrac{\mathrm{d}z}{\mathrm{d}x}$ |
| inverse function | $\dfrac{\mathrm{d}x}{\mathrm{d}y} = \dfrac{1}{\left(\dfrac{\mathrm{d}y}{\mathrm{d}x}\right)}$ |

**Applying the sum rule:** *Example 1:* Using Table 0.35 we get

$$(\mathrm{e}^x + \sin x)' = (\mathrm{e}^x)' + (\sin x)' = \mathrm{e}^x + \cos x\,, \quad x \in \mathbb{R}\,,$$
$$(x^2 + \sinh x)' = (x^2)' + (\sinh x)' = 2x + \cosh x\,, \quad x \in \mathbb{R}\,,$$
$$(\ln x + \cos x)' = (\ln x)' + (\cos x)' = \frac{1}{x} - \sin x\,, \quad x > 0\,.$$

**Applying the rule for multiplication by a constant:** *Example 2:*

$$(2\mathrm{e}^x)' = 2(\mathrm{e}^x)' = 2\mathrm{e}^x\,, \quad (3\sin x)' = 3(\sin x)' = 3\cos x\,, \quad x \in \mathbb{R}\,,$$
$$(3x^4 + 5)' = (3x^4)' + (5)' = 3 \cdot 4x^3 = 12x^3\,, \quad x \in \mathbb{R}\,.$$

**Applying the product rule:** *Example 3:*

$$(x\mathrm{e}^x)' = (x)'\mathrm{e}^x + x(\mathrm{e}^x)' = 1 \cdot \mathrm{e}^x + x\mathrm{e}^x = (1+x)\mathrm{e}^x\,, \quad x \in \mathbb{R}\,,$$
$$(x^2 \sin x)' = (x^2)' \sin x + x^2(\sin x)' = 2x \sin x + x^2 \cos x\,, \quad x \in \mathbb{R}\,,$$
$$(x \ln x)' = (x)' \ln x + x(\ln x)' = \ln x + 1\,, \quad x > 0\,.$$

[60] The precise assumptions for the validity of the formulas can be found in 1.4. These rules hold for functions of one real or complex variable. The Examples 1 to 6 remain valid for complex arguments $x$.

**Applying the quotient rule:** *Example 4:*

$$(\tan x)' = \left(\frac{\sin x}{\cos x}\right)' = \frac{(\sin x)'\cos x - \sin x(\cos x)'}{\cos^2 x}$$
$$= \frac{\cos^2 x + \sin^2 x}{\cos^2 x} = \frac{1}{\cos^2 x}\,.$$

This derivative exists for all $x$ for which the denominator $\cos x$ is non-vanishing, i.e., such that $x \neq k\pi + \frac{\pi}{2}$ with $k = 0, \pm 1, \pm 2, \ldots$

**Applying the chain rule:** *Example 5:* To differentiate

$$y = \sin 2x,$$

we write

$$y = \sin z\,, \qquad z = 2x\,.$$

The chain rule yields

$$y' = \frac{\mathrm{d}y}{\mathrm{d}x} = \frac{\mathrm{d}y}{\mathrm{d}z}\frac{\mathrm{d}z}{\mathrm{d}x} = (\cos z)\cdot 2 = 2\cos 2x\,.$$

*Example 6:* The differentiation of

$$y = \cos(3x^4 + 5)$$

is affected by setting $y = \cos z$, $z = 3x^4 + 5$ and calculating

$$y' = \frac{\mathrm{d}y}{\mathrm{d}x} = \frac{\mathrm{d}y}{\mathrm{d}z}\frac{\mathrm{d}z}{\mathrm{d}x} = (-\sin z)\cdot 12x^3 = -12x^3\sin(3x^4 + 5)\,.$$

**Applying the rule for the derivative of the inverse function:** Inverting the function

$$y\mathrm{e}^x\,, \qquad -\infty < x < \infty\,,$$

yields

$$x = \ln y\,, \qquad y > 0\,.$$

From this we get

$$\frac{\mathrm{d}\ln y}{\mathrm{d}y} = \frac{\mathrm{d}x}{\mathrm{d}y} = \frac{1}{\left(\dfrac{\mathrm{d}y}{\mathrm{d}x}\right)} = \frac{1}{\mathrm{e}^x} = \frac{1}{y}\,, \qquad y > 0\,.$$

### 0.8.3 Rules for differentiating functions of several variables

**Partial derivative:** If the function $f = f(x, w, \ldots)$ depends on $x$ and further variables $w, \ldots$, then the partial derivative

$$\boxed{\frac{\partial f}{\partial x}}$$

is formed by viewing $f$ as a function of $x$ alone, viewing the other variables as constants, and forming the derivative with respect to $x$.

*Example 1:* Let $f(x) = Cx$ with the constant $C$. Then one has

$$\frac{\mathrm{d}f(x)}{\mathrm{d}x} = C\,.$$

In the same way, we get for $f(x,u,v) = (\mathrm{e}^v \sin u)x$ the partial derivative

$$\frac{\partial f(x,u,v)}{\partial x} = \mathrm{e}^v \sin u\,.$$

This is because one views $u, v$ and hence also $C = \mathrm{e}^v \sin u$ as constants.

*Example 2:* Let $f(x) = \cos(3x^4 + C)$, where $C$ denotes a constant. By Example 6 in 0.8.2 one has

$$\frac{\mathrm{d}f(x)}{\mathrm{d}x} = -12x^3 \sin(3x^4 + C)\,.$$

In the function $f(x,u) = \cos(3x^4 + \mathrm{e}^u)$ we view $u$ and hence also $C = \mathrm{e}^u$ as constants and get

$$\frac{\partial f(x,u)}{\partial x} = -12x^3 \sin(3x^4 + \mathrm{e}^u)\,.$$

*Example 3:* For $f(x,y) := xy$ one gets

$$f_x(x,y) = \frac{\partial f(x,y)}{\partial x} = y\,, \qquad f_y(x,y) = \frac{\partial f(x,y)}{\partial y} = x\,.$$

*Example 4:* For the function $f(x,y) := \dfrac{x}{y} = xy^{-1}$ we get

$$f_x(x,y) = \frac{\partial f(x,y)}{\partial x} = y^{-1}\,, \qquad f_y(x,y) = \frac{\partial f(x,y)}{\partial y} = -xy^{-2}\,.$$

*Table 0.38. The chain rule.*[61]

$$f = f(x,y)\,, \quad f_x := \frac{\partial f}{\partial x}\,, \quad f_y := \frac{\partial f}{\partial y}$$

| ***Name*** | ***Formula*** |
|---|---|
| total differential | $\mathrm{d}f = f_x \mathrm{d}x + f_y \mathrm{d}y$ |
| chain rule | $\dfrac{\partial f}{\partial w} = f_x \dfrac{\partial x}{\partial w} + f_y \dfrac{\partial y}{\partial w}$ |

For the chain rule in Table 0.38 we view $x = x(w, \ldots)$ and $y = y(w, \ldots)$ as functions of $w$ and (perhaps) other variables. A similar rule holds for functions $f = f(x_1, \ldots, x_n)$. One has the *total differential*

$$\boxed{\mathrm{d}f = f_{x_1}\mathrm{d}x_1 + \ldots + f_{x_n}\mathrm{d}x_n}$$

which gives the following expression for the chain rule

$$\boxed{\frac{\partial f}{\partial w} = f_{x_1}\frac{\partial x_1}{\partial w} + \ldots + f_{x_n}\frac{\partial x_n}{\partial w}\,,}$$

[61] The precise assumptions for the validity of these rules can be found in 1.5 These rules are valid for functions of real or complex variables.

in case the functions $x_1, \ldots, x_n$ depend on $w$ and further variables. If $x_1, \ldots, x_n$ are only functions of $w$, then one uses the notation $\dfrac{\mathrm{d}}{\mathrm{d}w}$ instead of $\dfrac{\partial}{\partial w}$. This yields the special form of the chain rule

$$\boxed{\frac{\mathrm{d}f}{\mathrm{d}w} = f_{x_1}\frac{\mathrm{d}x_1}{\mathrm{d}w} + \ldots + f_{x_n}\frac{\mathrm{d}x_n}{\mathrm{d}w}\,.}$$

**Applications of the chain rule:** *Example 5:* We set $f(t) := x(t)y(t)$. From Example 3 we get the expression

$$\boxed{\mathrm{d}f = f_x\,\mathrm{d}x + f_y\,\mathrm{d}y = y\,\mathrm{d}x + x\,\mathrm{d}y}$$

for the total differential. From this we then get

$$f'(t) = \frac{\mathrm{d}f}{\mathrm{d}t} = y\frac{\mathrm{d}x}{\mathrm{d}t} + x\frac{\mathrm{d}y}{\mathrm{d}t} = y(t)x'(t) + x(t)y'(t)\,.$$

This is the *product rule*, which consequently may be viewed as a special case of the chain rule for functions of several variables.

*Example 6:* For the function

$$f(t) := \frac{x(t)}{y(t)}$$

we get from Example 4 the total differential

$$\boxed{\mathrm{d}f = f_x\,\mathrm{d}x + f_y\,\mathrm{d}y = y^{-1}\,\mathrm{d}x - xy^{-2}\,\mathrm{d}y}$$

and $$f'(t) = \frac{\mathrm{d}f}{\mathrm{d}t} = \frac{x'(t)}{y(t)} - \frac{x(t)y'(t)}{y(t)^2} = \frac{x'(t)y(t) - x(t)y'(t)}{y(t)^2}\,.$$

This is nothing but the *quotient rule.*

## 0.9 Tables of integrals

*Differentiation is handicraft — integration is an art.*

*Folklore*

**Differentiation and integration on the computer:** For this one can advantageously use the software system Mathematica.

### 0.9.1 Integration of elementary functions

The formula

$$\boxed{\int f(x)\,\mathrm{d}x = F(x)\,, \qquad x \in D\,,}$$

means

$$\boxed{F'(x) = f(x) \qquad \text{for all} \quad x \in D\,.}$$

Thus, the function $F$ has the property that the derivative of $F$ is the function $f$ on the set $D$. One refers to $F$ as a *primitive* or as an *indefinite integral* of $f$. In this sense, integration is the inverse process of differentiation.

(i) Real case: If $x$ is a real variable and $D$ denotes an interval, then one gets all possible indefinite integrals of $f$ on the set $D$ by adding to some fixed indefinite integral different constants. To express this fact, one writes

$$\int f(x)\,dx = F(x) + C\,, \qquad x \in D\,.$$

(ii) Complex case: Let $D$ be a domain in the complex plane. All statements above remain valid, if $C$ is now taken to be a *complex* constant.

*Table 0.39. Basic integrals.*

| ***Function $f(x)$*** | ***Indefinite integral***[62] $\int f(x)dx$ | ***Validity for real numbers***[63] | ***Validity for complex numbers***[63] |
|---|---|---|---|
| $C$ (constant) | $Cx$ | $x \in \mathbb{R}$ | $x \in \mathbb{C}$ |
| $x$ | $\frac{x^2}{2}$ | $x \in \mathbb{R}$ | $x \in \mathbb{C}$ |
| $x^n$ $(n = 1, 2, \ldots)$ | $\frac{x^{n+1}}{n+1}$ | $x \in \mathbb{R}$ | $x \in \mathbb{C}$ |
| $\frac{1}{x^n}$ $(n = 2, 3, \ldots)$ | $\frac{1}{(1-n)x^{n-1}}$ | $x \neq 0$ | $x \neq 0$ |
| $\frac{1}{x}$ | $\ln x$ | $x > 0$ | $x \neq 0, -\pi < \arg x < \pi$ |
| $\frac{1}{x}$ | $\ln \lvert x \rvert$ | $x \neq 0$ | |
| $x^\alpha$ ($\alpha$ real, $\alpha \neq -1$) | $\frac{x^{\alpha+1}}{\alpha+1}$ | $x > 0$ | $x \neq 0, -\pi < \arg x < \pi$ |
| $\sqrt{x} = x^{\frac{1}{2}}$ | $\frac{2}{3}x\sqrt{x}$ | $x > 0$ | $x \neq 0, -\pi < \arg x < \pi$ |
| $e^x$ | $e^x$ | $x \in \mathbb{R}$ | $x \in \mathbb{C}$ |
| $a^x$ $(a > 0,\ a \neq 1)$ | $\frac{a^x}{\ln a}$ | $x \in \mathbb{R}$ | $x \in \mathbb{C}$ |
| $\sin x$ | $-\cos x$ | $x \in \mathbb{R}$ | $x \in \mathbb{C}$ |
| $\cos x$ | $\sin x$ | $x \in \mathbb{R}$ | $x \in \mathbb{C}$ |
| $\tan x$ | $-\ln \lvert \cos x \rvert$ | $x \neq (2k+1)\frac{\pi}{2}$ $(k \in \mathbb{Z})$ | |
| $\cot x$ | $\ln \lvert \sin x \rvert$ | $x \neq k\pi \quad (k \in \mathbb{Z})$ | |
| $\frac{1}{\cos^2 x}$ | $\tan x$ | $x \neq (2k+1)\frac{\pi}{2}$ | $x \neq (2k+1)\frac{\pi}{2} \quad (k \in \mathbb{Z})$ |

[62] Only one indefinite integral is listed.

[63] $x \in \mathbb{R}$ (resp. $x \in \mathbb{C}$) means the corresponding formula is valid for all real (resp. complex) numbers. Moreover, $k \in \mathbb{Z}$ stands for $k = 0, \pm 1, \pm 2, \ldots$ For the functions $\ln x$, $\sqrt{x}$ and $x^{\alpha+1}$ we use the principal branch for the functions with complex arguments, which are obtained from the values for $x > 0$ by analytic continuation (cf. 1.14.15).

*Table 0.39. (continued)*

| | | | |
|---|---|---|---|
| $\dfrac{1}{\sin^2 x}$ | $-\cot x$ | $x \neq k\pi$ | $x \neq k\pi \quad (k \in \mathbb{Z})$ |
| $\sinh x$ | $\cosh x$ | $x \in \mathbb{R}$ | $x \in \mathbb{C}$ |
| $\cosh x$ | $\sinh x$ | $x \in \mathbb{R}$ | $x \in \mathbb{C}$ |
| $\tanh x$ | $\ln \cosh x$ | $x \in \mathbb{R}$ | |
| $\coth x$ | $\ln \lvert \sinh x \rvert$ | $x \neq 0$ | |
| $\dfrac{1}{\cosh^2 x}$ | $\tanh x$ | $x \in \mathbb{R}$ | $x \neq \mathrm{i}(2k+1)\dfrac{\pi}{2} \quad (k \in \mathbb{Z})$ |
| $\dfrac{1}{\sinh^2 x}$ | $-\coth x$ | $x \neq 0$ | $x \neq \mathrm{i}k\pi \quad (k \in \mathbb{Z})$ |
| $\dfrac{1}{a^2+x^2} \quad (a > 0)$ | $\dfrac{1}{a}\arctan\dfrac{x}{a}$ | $x \in \mathbb{R}$ | |
| $\dfrac{1}{a^2-x^2} \quad (a > 0)$ | $\dfrac{1}{2a}\ln\left\lvert\dfrac{a+x}{a-x}\right\rvert$ | $x \neq a$ | |
| $\dfrac{1}{\sqrt{a^2-x^2}} \quad (a > 0)$ | $\arcsin\dfrac{x}{a}$ | $\lvert x \rvert < a$ | |
| $\dfrac{1}{\sqrt{a^2+x^2}} \quad (a > 0)$ | $\operatorname{arsinh}\dfrac{x}{a}$ | $x \in \mathbb{R}$ | |
| $\dfrac{1}{\sqrt{x^2-a^2}} \quad (a > 0)$ | $\operatorname{arcosh}\dfrac{x}{a}$ | $\lvert x \rvert > a$ | |

## 0.9.2 Rules for integration

### 0.9.2.1 Indefinite integrals

The rules for calculation of definite integrals can be found in Table 0.41.

*Table 0.40. Rules for calculation of indefinite integrals.*[64]

| ***Name of rule*** | ***Formula*** |
|---|---|
| sum rule | $\int (u+v)\,\mathrm{d}x = \int u\,\mathrm{d}x + \int v\,\mathrm{d}x$ |
| constant multiples | $\int \alpha u\,\mathrm{d}x = \alpha \int u\,\mathrm{d}x \quad (\alpha \quad \text{constant})$ |
| integration by parts | $\int u'v\,\mathrm{d}x = uv - \int uv'\,\mathrm{d}x$ |
| substitution formula | $\int f(x)\,\mathrm{d}x = \int f(x(t))\dfrac{\mathrm{d}x}{\mathrm{d}t}\,\mathrm{d}t$ |
| logarithmic derivative | $\int \dfrac{f'(x)}{f(x)}\,\mathrm{d}x = \ln \lvert f(x) \rvert$ |

[64]The precise assumptions for the validity of these rules will be formulated in sections 1.6.4 and 1.6.5.

The substitution rule is often used in the mnemonic very convenient formulation[65]

$$\boxed{\int f(t(x))\,\mathrm{d}t(x) = \int f(t)\,\mathrm{d}t} \tag{0.56}$$

with $\mathrm{d}t(x) = t'(x)\,\mathrm{d}x$. In many cases (0.56) is more convenient than the formulation in Table 0.40, for which one must in addition assume that $x'(t) \neq 0$ in order to guarentee the existence of the inverse function $t = t(x)$.

In all cases in which (0.56) can be applied, the existence of the inverse function is *not* necessary.

**Examples of substitutions:** *Example 1:* We want to calculate the integral

$$J = \int \sin(2x+1)\,\mathrm{d}x\,.$$

To do this, we set $t := 2x + 1$. The inverse function is

$$x = \frac{1}{2}(t-1)\,.$$

It follows that $\dfrac{\mathrm{d}x}{\mathrm{d}t} = \dfrac{1}{2}$. The substitution formula in Table 0.40 yields

$$J = \int (\sin t)\frac{1}{2}\,\mathrm{d}t = -\frac{1}{2}\cos t = -\frac{1}{2}\cos(2x+1)\,.$$

Now we use the formula (0.56) to calculate $J$.

Because of $\dfrac{\mathrm{d}(2x+1)}{\mathrm{d}x} = 2$ one has $\mathrm{d}(2x+1) = 2\,\mathrm{d}x$. This yields

$$\begin{aligned}\int \sin(2x+1)\,\mathrm{d}x &= \int \frac{1}{2}\sin(2x+1)\,\mathrm{d}(2x+1) = \int \frac{1}{2}\sin t\,\mathrm{d}t\\ &= -\frac{1}{2}\cos t = -\frac{1}{2}\cos(2x+1)\,.\end{aligned}$$

After a certain amount of practice, one will note the following formula:

$$\boxed{\int \sin(2x+1)\,\mathrm{d}x = \int \frac{1}{2}\sin(2x+1)\,\mathrm{d}(2x+1) = -\frac{1}{2}\cos(2x+1)\,.}$$

Generally speaking, one should first check whether (0.56) can be applied. There are situations where (0.56) cannot be immediately applied (see Example 3 in 1.6.5).

*Example 2:* From $\dfrac{\mathrm{d}x^2}{\mathrm{d}x} = 2x$ it follows that

$$\int \mathrm{e}^{x^2} x\,\mathrm{d}x = \int \frac{1}{2}\mathrm{e}^{x^2}\,\mathrm{d}x^2 = \int \frac{1}{2}\mathrm{e}^t\,\mathrm{d}t = \frac{1}{2}\mathrm{e}^t = \frac{1}{2}\mathrm{e}^{x^2}\,.$$

[65]Exchanging the role of $x$ and $t$, one gets by the substitution rule in Table 0.40 the formula

$$\int f(t(x))t'(x)\,\mathrm{d}x = \int f(t)\,\mathrm{d}t\,,$$

which corresponds to (0.56).

With more experience you will just write

$$\boxed{\int \mathrm{e}^{x^2} x \, \mathrm{d}x = \int \frac{1}{2} \mathrm{e}^{x^2} \, \mathrm{d}x^2 = \frac{1}{2} \mathrm{e}^{x^2} .}$$

*Example 3:*

$$\int \frac{x \, \mathrm{d}x}{1+x^2} = \int \frac{\mathrm{d}x^2}{2(1+x^2)} = \frac{1}{2} \ln(1+x^2) .$$

More examples applying substitution can be found in 0.9.4. and 1.6.5.

**Examples of integration by parts:** *Example 4:* To calculate $\int x \sin x \, \mathrm{d}x$, we set

$$\begin{aligned} u' &= \sin x \, , & v &= x \, , \\ u &= -\cos x \, , & v' &= 1 \, . \end{aligned}$$

This leads to

$$\begin{aligned} \int x \sin x \, \mathrm{d}x &= \int u' v \, \mathrm{d}x = uv - \int u v' \, \mathrm{d}x \\ &= -x \cos x + \int \cos x \, \mathrm{d}x = -x \cos x + \sin x \, . \end{aligned}$$

*Example 5:* To calculate $\int \arctan x \, \mathrm{d}x$ we choose

$$\begin{aligned} u' &= 1 \, , & v &= \arctan x \, , \\ u &= x \, , & v' &= \frac{1}{1+x^2} \, . \end{aligned}$$

This yields

$$\begin{aligned} \int \arctan x \, \mathrm{d}x &= \int u' v \, \mathrm{d}x = uv - \int u v' \mathrm{d}x \\ &= x \arctan x - \int \frac{x \mathrm{d}x}{1+x^2} \\ &= x \arctan x - \frac{1}{2} \ln(1+x^2) \end{aligned}$$

by Example 3.

More examples of integration by parts can be found in 1.6.4.

### 0.9.2.2 Definite integrals

The most important rules are gathered in Table 0.41.

Table 0.41. Rules for calculation of definite integrals.[66]

| Name of rule | Formula |
|---|---|
| substitution formula | $\int_\alpha^\beta f(x(t))x'(t)\,\mathrm{d}t = \int_a^b f(x)\,\mathrm{d}x$ <br> $(x(\alpha) = a, \quad x(\beta) = b, \quad x'(t) > 0)$ |
| integration by parts | $\int_a^b u'v\,\mathrm{d}x = uv\Big\|_a^b - \int_a^b uv'\,\mathrm{d}x\,.$ |
| fundamental theorem of calculus due to Newton and Leibniz | $\int_a^b u'\mathrm{d}x = u\Big\|_a^b$ |

The fundamental theorem of calculus results from the formula for integration by parts in Table 0.41 by setting $v = 1$ there.

*Example:* $\int_a^b \sin x\,\mathrm{d}x = -\cos x\Big|_a^b = -\cos b + \cos a.$

This follows from the fact that for $u := -\cos x$ one has $u' = \sin x$.

More examples are to be found in 1.6.4.

### 0.9.2.3 Integrals of functions of several variables

The rules of Table 0.41 for one-dimensional integrals (integrals of functions of a single variable) correspond to similar formulas for integrals of functions of several variables (higher-dimensional integrals), which are gathered in Table 0.42.

Table 0.42. Formulae for integrals of functions of several variables.[67]

| Name of rule | Formula |
|---|---|
| substitution formula | $\int_{x(H)} f(x)\,\mathrm{d}x = \int_H f(x(t))\|\det x'(t)\|\,\mathrm{d}t$ |
| integration by parts | $\int_G (\partial_j u)v\,\mathrm{d}x = \int_{\partial G} uvn_j\,\mathrm{d}F - \int_G u\partial_j v\,\mathrm{d}x$ |
| theorem of Gauss | $\int_G \partial_j u\,\mathrm{d}x = \int_{\partial G} un_j\,\mathrm{d}F$ |
| theorem of Gauss–Stokes | $\boxed{\int_M \mathrm{d}\omega = \int_{\partial M} \omega}$ |
| theorem of Fubini (iterated integration) | $\int_{\mathbb{R}^2} f(x,y)\,\mathrm{d}x\mathrm{d}y = \int_{-\infty}^{\infty}\Big(\int_{-\infty}^{\infty} f(x,y)\,\mathrm{d}x\Big)\,\mathrm{d}y$ |

[66] The precise assumptions are found in 1.6. We set $f\Big|_a^b := f(b) - f(a)$.

[67] The precise assumptions and an explanation of the notations are in 1.7.

**Remarks:**

(i) The theorem of Gauss in Table 0.42 results form the formula for integration by parts, upon setting $v = 1$ there.

(ii) The theorem of Gauss–Stokes generalizes the fundamental theorem of calculus to manifolds (for example, curves, surfaces, domains, etc.).

(iii) In fact, both the formula for integration by parts and the theorem of Gauss are special cases of the theorem of Gauss–Stokes, which is one of the central theorems in all of mathematics (cf. 1.7.6).

Applications of these rules are contained in sections 1.7.1 ff.

### 0.9.3 Integration of rational functions

Every rational function can be uniquely written as a sum of so-called partial fractions

$$\boxed{\frac{A}{(x-a)^n}.} \tag{0.57}$$

Here $n = 1, 2, \ldots$, and $A$ and $a$ are real or complex numbers (cf. 2.1.7). The partial fractions (0.57) can be immediately integrated following the rules in Table 0.43.[68]

*Table 0.43. Integration of partial fractions.*

| | |
|---|---|
| $\displaystyle\int \frac{\mathrm{d}x}{(x-a)^n} = \frac{1}{(1-n)(x-a)^{n-1}}$ | $x \in \mathbb{R}, \quad n = 2, 3, \ldots, \quad a \in \mathbb{C}$ |
| $\displaystyle\int \frac{\mathrm{d}x}{x-a} = \ln\lvert x-a\rvert$ | $x \in \mathbb{R}, \quad x \neq a, \quad a \in \mathbb{R}$ |
| $\displaystyle\int \frac{\mathrm{d}x}{x-a} = \ln\lvert x-a\rvert + \mathrm{i}\arctan\frac{x-\alpha}{\beta}$ | $x \in \mathbb{R}, \quad a = \alpha + \mathrm{i}\beta, \quad \beta \neq 0$ |

*Example 1:* From

$$\frac{1}{x^2-1} = \frac{1}{2}\left(\frac{1}{x-1} - \frac{1}{x+1}\right)$$

it follows that

$$\int \frac{\mathrm{d}x}{x^2-1} = \frac{1}{2}\Big(\ln\lvert x-1\rvert - \ln\lvert x+1\rvert\Big) = \frac{1}{2}\ln\left\lvert\frac{x-1}{x+1}\right\rvert.$$

*Example 2:* Because of

$$\frac{1}{x^2+1} = \frac{1}{2\mathrm{i}}\left(\frac{1}{x-\mathrm{i}} - \frac{1}{x+\mathrm{i}}\right)$$

we get

$$\begin{aligned}\int \frac{\mathrm{d}x}{x^2+1} &= \frac{1}{2\mathrm{i}}\Big(\ln\lvert x-\mathrm{i}\rvert + \mathrm{i}\arctan x - \ln\lvert x+\mathrm{i}\rvert - \mathrm{i}\arctan(-x)\Big)\\ &= \arctan x\,, \qquad x \in \mathbb{R}.\end{aligned}$$

[68]The method described here is particularly easy to comprehend, because it uses the extension to complex numbers. Avoiding the complex numbers requires a delicate case-by-case study.

Note that $|x - \mathrm{i}| = |x + \mathrm{i}|$ and $\arctan(-x) = -\arctan x$ for $x \in \mathbb{R}$.

*Example 3:* According to (2.30) one has

$$f(x) := \frac{x}{(x-1)(x-2)^2} = \frac{1}{x-1} - \frac{1}{x-2} + \frac{2}{(x-2)^2}\,.$$

It follows from this that

$$\int f(x)\,\mathrm{d}x = \ln|x-1| - \ln|x-2| - \frac{2}{x-2}\,.$$

Arbitrary rational functions can be written as a sum of a polynomial and a fraction in reduced terms.

*Example 4:* $\dfrac{x^2}{1+x^2} = 1 - \dfrac{1}{1+x^2}$,

$$\int \frac{x^2\,\mathrm{d}x}{1+x^2} = \int \mathrm{d}x - \int \frac{\mathrm{d}x}{1+x^2} = x - \arctan x\,.$$

**Use of Mathematica:** In our age of computers one only attempts to calculate very easy expressions of the above kind by hand. Otherwise one applies one of the applicable computer algebra programs, for example Mathematica.

## 0.9.4 Important substitutions

We list some types of integrals which can be solved by means of some *universal* substitution. In particular cases, however, special substitutions may be more convenient in trimming down the necessary computations. Nowadays this work is done by computer algebra systems.

Very few integrals can be solved in closed form in terms of elementary functions.

**Polynomials of several variables:** A *polynomial* $P = P(x_1, \ldots, x_n)$ of the variables $x_1, \ldots, x_n$ is a finite sum of expressions of the form

$$a_{i_1 \ldots i_n} x_1^{\alpha_1} x_2^{\alpha_2} \cdots x_n^{\alpha_n}\,.$$

Here $a_{\ldots}$ denotes a complex number, and all exponents $\alpha_j$ are equal to one of the values $1, 2, \ldots$

**Rational functions of several variables:** A *rational function* $R = R(x_1, \ldots, x_n)$ of the variables $x_1, \ldots, x_n$ is an expression of the form

$$\boxed{R(x_1, \ldots, x_n) := \frac{P(x_1, \ldots, x_n)}{Q(x_1, \ldots, x_n)}}\,,$$

where $P$ and $Q$ are polynomials.

**Convention:** In what follows, $R$ will always denote a rational function.

**Type 1:** $\boxed{\int R(\sinh x, \cosh x, \tanh x, \coth x, \mathrm{e}^x)\,\mathrm{d}x\,.}$

*Solution:* One expresses $\sinh x$, etc. in terms of $\mathrm{e}^x$ and uses the substitution

$$\boxed{t = \mathrm{e}^x\,, \qquad \mathrm{d}t = t\,\mathrm{d}x\,.}$$

This yields a rational function in $t$, which can be decomposed into a partial fraction decomposition (cf. 0.9.3). Explicitly one has

$$\sinh x = \frac{1}{2}(\mathrm{e}^x - \mathrm{e}^{-x})\,, \quad \cosh x = \frac{1}{2}(\mathrm{e}^x + \mathrm{e}^{-x})\,,$$
$$\tanh x = \frac{\sinh x}{\cosh x}\,, \quad \coth x = \frac{\cosh x}{\sinh x}\,.$$

*Example 1:* $J := \int \frac{\mathrm{d}x}{2\cosh x} = \int \frac{\mathrm{d}x}{\mathrm{e}^x + \mathrm{e}^{-x}} = \int \frac{\mathrm{d}t}{t\left(t + \frac{1}{t}\right)}$

$$= \int \frac{\mathrm{d}t}{t^2 + 1} = \arctan t = \arctan \mathrm{e}^x\,.$$

*Example 2:* $J := \int 8\sinh^2 x\,\mathrm{d}x = \int 2(\mathrm{e}^{2x} - 2 + \mathrm{e}^{-2x})\,\mathrm{d}x$
$$= \mathrm{e}^{2x} - 4x - \mathrm{e}^{-2x}\,.$$

In this particular case the substitution $t = \mathrm{e}^x$ is unnecessary.

*Example 3:* For the calculation of $J := \int \sinh^n x \cosh x\,\mathrm{d}x$ it is useful to apply (0.56). This gives

$$J = \int \sinh^n x\,\mathrm{d}\sinh x = \frac{\sinh^{n+1} x}{n+1}\,, \qquad n = 1, 2, \ldots$$

This approach corresponds to the substitution $t = \sinh x$.

$$\textbf{Type 2:} \qquad \boxed{\int R(\sin x, \cos x, \tan x, \cot x)\,\mathrm{d}x\,.} \tag{0.58}$$

*Solution:* Express $\sin x$, etc., in terms of $\mathrm{e}^{\mathrm{i}x}$ and use the substitution

$$\boxed{t = \mathrm{e}^{\mathrm{i}x}\,, \qquad \mathrm{d}t = \mathrm{i}t\,\mathrm{d}x\,.}$$

This yields a rational function in $t$, which can again be decomposed into partial fractions (cf. 0.9.3). Explicitly one has

$$\sin x = \frac{1}{2\mathrm{i}}(\mathrm{e}^{\mathrm{i}x} - \mathrm{e}^{-\mathrm{i}x})\,, \quad \cos x = \frac{1}{2}(\mathrm{e}^{\mathrm{i}x} + \mathrm{e}^{-\mathrm{i}x})\,,$$
$$\tan x = \frac{\sin x}{\cos x}\,, \qquad \cot x = \frac{\cos x}{\sin x}\,.$$

Instead of this method, the substitution[69]

$$\boxed{t = \tan\frac{x}{2}\,, \qquad -\pi < x < \pi} \tag{0.59}$$

[69] In the words of M. Spivak, this is "undoubtedly the world's sneakiest substitution". Applying this substitution transforms any integral which involves only sin and cos, combined by addition, multiplication, and division, into the integral of a rational function.

always leads to a solution. One has

$$\cos x = \frac{1-t^2}{1+t^2}\,, \qquad \sin x = \frac{2t}{1+t^2}\,, \qquad \mathrm{d}x = \frac{2\mathrm{d}t}{1+t^2}\,.$$

*Example 4:* $J := \int 8\cos^2 x\mathrm{d}x = \int 2(\mathrm{e}^{2\mathrm{i}x} + 2 + \mathrm{e}^{-2\mathrm{i}x})\,\mathrm{d}x$

$$= \frac{1}{\mathrm{i}}(\mathrm{e}^{2\mathrm{i}x} - \mathrm{e}^{-2\mathrm{i}x}) + 4x = 2\sin 2x + 4x\,.$$

In the present case we can conclude more quickly

$$J = \int (4\cos 2x + 4)\,\mathrm{d}x = 2\sin 2x + 4x$$

because of $2\cos^2 x = \cos 2x + 1$.

*Example 5:* From (0.56) it follows that

$$\int \frac{\sin x\,\mathrm{d}x}{\cos^2 x} = \int \frac{-\mathrm{d}\cos x}{\cos^2 x} = -\frac{1}{\cos x}\,.$$

**Type 3:** $\boxed{\int R\left(x, \sqrt[n]{\frac{\alpha x+\beta}{\gamma x+\delta}}\right)\mathrm{d}x\,,}$ $\qquad \alpha\delta - \beta\gamma \neq 0\,, \quad n = 2, 3, \ldots$

*Solution:* Use the substitution[70]

$$\boxed{t = \sqrt[n]{\frac{\alpha x+\beta}{\gamma x+\delta}}\,, \quad x = \frac{\delta t^n - \beta}{\alpha - \gamma t^n}\,, \quad \mathrm{d}x = n(\alpha\delta - \beta\gamma)\frac{t^{n-1}\,\mathrm{d}t}{(\alpha - \gamma t^n)^2}\,.}$$

This reduces the integral to one of a rational function of $t$, for which the method of partial fraction decomposition can be applied (cf. 0.9.3).

*Example 6:* The substitution $t = \sqrt{x}$ yields $x = t^2$ and $\mathrm{d}x = 2t\,\mathrm{d}t$. Thus one gets

$$\int \frac{x-\sqrt{x}}{x+\sqrt{x}}\,\mathrm{d}x = \int \frac{t^2 - t}{1+t} 2\,\mathrm{d}t = 2\int\left(t - 2 + \frac{2}{t+1}\right)\mathrm{d}t$$
$$= 2\left(\frac{t^2}{2} - 2t + 2\ln|t+1|\right)$$
$$= 2\left(\frac{x}{2} - 2\sqrt{x} + 2\ln|1+\sqrt{x}|\right).$$

**Type 4:** $\boxed{\int R(x, \sqrt{\alpha x^2 + 2\beta x + \gamma})\,\mathrm{d}x\,.}$

Let $\alpha \neq 0$. With the help of the quadratic completion

$$\alpha x^2 + 2\beta x + \gamma = \alpha\left(x + \frac{\beta}{\alpha}\right)^2 - \frac{\beta^2}{\alpha} + \gamma,$$

this type of integral can be reduced to one of those listed in Table 0.44. It is also possible to apply the Euler substitution (cf. Table 0.45).

[70]If an integral contains roots of different degrees, it can be reduced to type 3 by passing to the smallest common multiple of the degrees of the roots. For example one has

$\sqrt[3]{x} + \sqrt[4]{x} = (\sqrt[12]{x})^4 + (\sqrt[12]{x})^3\,.$

*Table 0.44. Algebraic functions of degree 2.*

| ***Integral*** $(a > 0)$ | ***Substitution*** | ***Validity*** |
|---|---|---|
| $\int R(x, \sqrt{a^2 - (x+b)^2}\,\mathrm{d}x$ | $x + b = a \sin t$ | $-\dfrac{\pi}{2} < t < \dfrac{\pi}{2}$, $\mathrm{d}x = a \cos t \,\mathrm{d}t$ |
| $\int R(x, \sqrt{a^2 + (x+b)^2}\,\mathrm{d}x$ | $x + b = a \sinh t$ | $-\infty < t < \infty$, $\mathrm{d}x = a \cosh t \,\mathrm{d}t$ |
| $\int R(x, \sqrt{(x+b)^2 - a^2}\,\mathrm{d}x$ | $x + b = a \cosh t$ | $t > 0$, $\mathrm{d}x = a \sinh t \,\mathrm{d}t$ |

$$\cos^2 t + \sin^2 t = 1\,, \qquad \cosh^2 t - \sinh^2 t = 1.$$

*Example 7:*
$$\int \frac{\mathrm{d}x}{\sqrt{a^2 + x^2}} = \int \frac{a \cosh t \,\mathrm{d}t}{\sqrt{a^2 + a^2 \sinh^2 t}} = \int \frac{a \cosh t \,\mathrm{d}t}{a \cosh t} = \int \mathrm{d}t = t = \operatorname{arsinh} \frac{x}{a}\,.$$

*Table 0.45. The Euler substitutions for* $\int R(x, \sqrt{\alpha x^2 + 2\beta x + \gamma})\,\mathrm{d}x$.

| ***Case*** | ***Substitution*** |
|---|---|
| $\alpha > 0$ | $\sqrt{\alpha x^2 + 2\beta x + \gamma} = t - x\sqrt{\alpha}$ |
| $\gamma > 0$ | $\sqrt{\alpha x^2 + 2\beta x + \gamma} = tx + \sqrt{\gamma}$ |
| $\alpha x^2 + 2\beta x + \gamma = \alpha(x - x_1)(x - x_2)$<br>$x_1, x_2$ real, $x_1 \neq x_2$ | $\sqrt{\alpha x^2 + 2\beta x + \gamma} = t(x - x_1)$ |

**Type 5:** $\boxed{\int R(x, \sqrt{\alpha x^4 + \beta x^3 + \gamma x^2 + \delta x + \mu})\,\mathrm{d}x\,.}$

Here we assume that $\alpha \neq 0$ or $\alpha = 0$ and $\beta \neq 0$.

These so-called *elliptic integrals* can be solved by substitutions with elliptic functions in the same manner as in Table 0.44 (cf. 1.14.19).

**Type 6:** $\boxed{\int R(x, w(x))\,\mathrm{d}x\,.}$

Here $w = w(x)$ is an algebraic function, i.e., this function satisfies some equation of the form $P(x, w) = 0$, where $P$ is a polynomial in $x$ and $w$. These kinds of integrals are called *Abelian integrals.*

*Example 8:* For $w^2 - a^2 + x^2 = 0$ one has $w = \sqrt{a^2 - x^2}$.

The theory of Abelian integrals was developed in the nineteenth century by Abel, Riemann and Weierstrass and lead to profound discoveries in complex function theory and topology (Riemann surfaces) and in algebraic geometry (cf. 3.8.1).

**Type 7:** $\boxed{\int x^m(\alpha+\beta x^n)^k \,\mathrm{d}x\,.}$

These so-called binomial integrals can be integrated in elementary terms if and olny if one of the cases listed in Table 0.46 applies. The substitutions listed there lead to integrals of rational functions, which again may be solved by the method of partial fraction decomposition (cf. 0.9.3).

*Table 0.46. Binomial integrals* $\int x^m(\alpha+\beta x^n)^k\,\mathrm{d}x$ *($m, n, k$ rational).*

| ***Case*** | ***Substitution*** |
|---|---|
| $k \in \mathbb{Z}$[71] | $t = \sqrt[r]{x}$ ($r$ smallest common multiple of the denominators of $m$ and $n$) |
| $\dfrac{m+1}{n} \in \mathbb{Z}$ | $t = \sqrt[q]{\alpha+\beta x^n}$ ($q$ denominator of $k$) |
| $\dfrac{m+1}{n} + k \in \mathbb{Z}$ | $t = \sqrt[q]{\dfrac{\alpha+\beta x^n}{x^n}}$ |

### 0.9.5 Tables of indefinite integrals

**Comments on using these tables.**

1. For simplicity the constant of integration has been omitted. Here the term $\ln f(x)$ is understood as $\ln|f(x)|$.

2. If the principal function is presented by a power series, then there is no elementary representation for this function.

3. The symbol $*$ marks formulas which are valid also for functions of complex variables.

4. Notations: $\mathbb{N}$ denotes the set of natural numbers, $\mathbb{Z}$ the integers and $\mathbb{R}$ denotes the real numbers.

#### 0.9.5.1 Integrals of rational functions

We first assume that we are given a function $L$ with

$$\boxed{L = ax+b,\ a \neq 0.}$$

1.* $\displaystyle\int L^n \mathrm{d}x = \frac{1}{a(n+1)}L^{n+1} \qquad (n \in \mathbb{N},\quad n \neq 0).$

2. $\displaystyle\int L^n \mathrm{d}x = \frac{1}{a(n+1)}L^{n+1} \qquad \left(n \in \mathbb{Z};\quad n \neq 0,\ n \neq -1;\right.$ $\left.\text{if } n<0,\ x \neq -\frac{b}{a};\ \text{for } n=-1, \text{ see No. } 6\right).$

3. $\displaystyle\int L^s \mathrm{d}x = \frac{1}{a(s+1)}L^{s+1} \qquad (s \in \mathbb{R},\ s \neq 0,\ s \neq -1,\ L>0).$

[71]This means that $k$ is an integer.

4.* $\int x \cdot L^n \mathrm{d}x = \frac{1}{a^2(n+2)} L^{n+2} - \frac{b}{a^2(n+1)} L^{n+1} \quad (n \in \mathbb{N},\ n \neq 0).$

5. $\int x \cdot L^n \mathrm{d}x = \frac{1}{a^2(n+2)} L^{n+2} - \frac{b}{a^2(n+1)} L^{n+1} \quad \left(n \in \mathbb{Z},\ n \neq 0,\ n \neq -1,\ n \neq -2;\ \text{if } n < 0,\ x \neq -\frac{b}{a}\right).$

6. $\int \frac{\mathrm{d}x}{L} = \frac{1}{a} \ln L \quad \left(x \neq -\frac{b}{a}\right).$

7. $\int \frac{x\mathrm{d}x}{L} = \frac{x}{a} - \frac{b}{a^2} \ln L \quad \left(x \neq -\frac{b}{a}\right).$

8. $\int \frac{x\mathrm{d}x}{L^2} = \frac{b}{a^2 L} + \frac{1}{a^2} \ln L \quad \left(x \neq -\frac{b}{a}\right).$

9. $\int \frac{x\mathrm{d}x}{L^n} = \int x \cdot L^{-n} \mathrm{d}x$ (see No. 5).

10. $\int \frac{x^2\mathrm{d}x}{L} = \frac{1}{a^3} \left(\frac{1}{2} L^2 - 2bL + b^2 \ln L\right) \quad \left(x \neq -\frac{b}{a}\right).$

11. $\int \frac{x^2\mathrm{d}x}{L^2} = \frac{1}{a^3} \left(L - 2b \ln L - \frac{b^2}{L}\right) \quad \left(x \neq -\frac{b}{a}\right).$

12. $\int \frac{x^2\mathrm{d}x}{L^3} = \frac{1}{a^3} \left(\ln L + \frac{2b}{L} - \frac{b^2}{2L^2}\right) \quad \left(x \neq -\frac{b}{a}\right).$

13. $\int \frac{x^2\mathrm{d}x}{L^n} = \frac{1}{a^3} \left(\frac{-1}{(n-3)L^{n-3}} + \frac{2b}{(n-2)L^{n-2}} - \frac{b^2}{(n-1)L^{n-1}}\right) \quad \left(n \in \mathbb{N},\ n > 3,\ x \neq -\frac{b}{a}\right).$

14. $\int \frac{x^3\mathrm{d}x}{L} = \frac{1}{a^4} \left(\frac{L^3}{3} - \frac{3bL^2}{2} + 3b^2 L - b^3 \ln L\right) \quad \left(x \neq -\frac{b}{a}\right).$

15. $\int \frac{x^3\mathrm{d}x}{L^2} = \frac{1}{a^4} \left(\frac{L^2}{2} - 3bL + 3b^2 \ln L + \frac{b^3}{L}\right) \quad \left(x \neq -\frac{b}{a}\right).$

16. $\int \frac{x^3\mathrm{d}x}{L^3} = \frac{1}{a^4} \left(L - 3b \ln L - \frac{3b^2}{L} + \frac{b^3}{2L^2}\right) \quad \left(x \neq -\frac{b}{a}\right).$

17. $\int \frac{x^3\mathrm{d}x}{L^4} = \frac{1}{a^4} \left(\ln L + \frac{3b}{L} - \frac{3b^2}{2L^2} + \frac{b^3}{3L^3}\right) \quad \left(x \neq -\frac{b}{a}\right).$

$$18. \int \frac{x^3 \mathrm{d}x}{L^n} = \frac{1}{a^4}\left[\frac{-1}{(n-4)L^{n-4}} + \frac{3b}{(n-3)L^{n-3}} - \frac{3b^2}{(n-2)L^{n-2}} + \frac{b^3}{(n-1)L^{n-1}}\right]$$

$$\left(x \neq -\frac{b}{a},\ n \in \mathbb{N},\ n > 4\right).$$

$$19. \int \frac{\mathrm{d}x}{xL^n} = -\frac{1}{b^n}\left[\ln\frac{L}{x} - \sum_{i=1}^{n-1}\binom{n-1}{i}\frac{(-a)^i x^i}{iL^i}\right]$$

$$\left(b \neq 0,\ x \neq -\frac{b}{a},\ x \neq 0,\ n \in \mathbb{N},\ n > 0\right).$$

For $n = 1$ the sum is trivial (contains no terms).

$$20. \int \frac{\mathrm{d}x}{x^2 L} = -\frac{1}{bx} + \frac{a}{b^2}\ln\frac{L}{x} \qquad \left(x \neq -\frac{b}{a},\ x \neq 0\right).$$

$$21. \int \frac{\mathrm{d}x}{x^2 L^n} = -\frac{1}{b^{n+1}}\left[-\sum_{i=2}^{n}\binom{n}{i}\frac{(-a)^i x^{i-1}}{(i-1)L^{i-1}} + \frac{L}{x} - na\ln\frac{L}{x}\right]$$

$$\left(x \neq -\frac{b}{a},\ x \neq 0,\ n \in \mathbb{N},\ n > 1\right).$$

$$22. \int \frac{\mathrm{d}x}{x^3 L} = -\frac{1}{b^3}\left[a^2\ln\frac{L}{x} - \frac{2aL}{x} + \frac{L^2}{2x^2}\right] \qquad \left(x \neq -\frac{b}{a}\quad x \neq 0\right).$$

$$23. \int \frac{\mathrm{d}x}{x^3 L^2} = -\frac{1}{b^4}\left[3a^2\ln\frac{L}{x} + \frac{a^3 x}{L} + \frac{L^2}{2x^2} - \frac{3aL}{x}\right] \qquad \left(x \neq -\frac{b}{a}\quad x \neq 0\right).$$

$$24. \int \frac{\mathrm{d}x}{x^3 L^n} =$$

$$-\frac{1}{b^{n+2}}\left[-\sum_{i=3}^{n+1}\binom{n+1}{i}\frac{(-a)^i x^{i-2}}{(i-2)L^{i-2}} + \frac{a^2L^2}{2x^2} - \frac{(n+1)aL}{x} + \frac{n(n+1)a^2}{2}\ln\frac{L}{x}\right]$$

$$\left(x \neq -\frac{b}{a},\ x \neq 0,\ n \in \mathbb{N},\ n > 2\right).$$

**Remark:** Suppose we are given $\int x^m L^n \mathrm{d}x = \frac{1}{a^{m+1}}\int (L-b)^m L^n \mathrm{d}x$; if $n \in \mathbb{N},\ n \neq 0$, then $L^n$ on the left hand side is treated as a binomial representation (see section 2.2.2.1); if $m \in \mathbb{N},\ m \neq 0$, then $(L-b)^m$ on the right hand side is treated as a binomial representation; for $n \in \mathbb{N}$ and $m \in \mathbb{N}$ and $m < n$ the right hand side representation is preferable.

**Integrals containing two linear functions $ax + b$ and $cx + d$:** Here we make the following assumptions:

$$\boxed{L_1 = ax + b,\ L_2 = cx + d,\ D = bc - ad,\ a,\ c \neq 0,\ D \neq 0.}$$

If $D = 0$, there exists a number $s$ for which : $L_2 = s \cdot L_1$.

25. $\int \frac{L_1}{L_2} \mathrm{d}x = \frac{ax}{c} + \frac{D}{c^2} \ln L_2 \qquad \left(x \neq -\frac{b}{a},\, x \neq -\frac{d}{c}\right).$

26. $\int \frac{\mathrm{d}x}{L_1 L_2} = \frac{1}{D} + \ln \frac{L_2}{L_1} \qquad \left(x \neq -\frac{b}{a},\, x \neq -\frac{d}{c}\right).$

27. $\int \frac{x\mathrm{d}x}{L_1 L_2} = \frac{1}{D} \left(\frac{b}{a} \ln L_1 - \frac{d}{c} \ln L_2\right) \qquad \left(x \neq -\frac{b}{a},\, x \neq -\frac{d}{c}\right).$

28. $\int \frac{\mathrm{d}x}{L_1^2 L_2} = \frac{1}{D} \left(\frac{1}{L_1} + \frac{c}{D} \ln \frac{L_2}{L_1}\right) \qquad \left(x \neq -\frac{b}{a},\, x \neq -\frac{d}{c}\right).$

29. $\int \frac{x\mathrm{d}x}{L_1^2 L_2} = \frac{d}{D^2} \ln \frac{cL_1}{aL_2} - \frac{b}{aDL_1} \qquad \left(x \neq -\frac{b}{a},\, x \neq -\frac{d}{c}\right).$

30. $\int \frac{x^2\mathrm{d}x}{L_1^2 L_2} = \frac{b^2}{a^2 DL_1} + \frac{b(bc - 2ad)}{a^2 D^2} \ln \left(\frac{1}{a} L_1\right) + \frac{d^2}{cD^2} \ln \left(\frac{1}{c} L_2\right)$
$\left(x \neq -\frac{b}{a},\, x \neq -\frac{d}{c}\right).$

31. $\int \frac{\mathrm{d}x}{L_1^2 L_2^2} = \frac{-1}{D^2} \left(\frac{a}{L_1} + \frac{c}{L_2} - \frac{2ac}{D} \ln \frac{cL_1}{aL_2}\right) \qquad \left(x \neq -\frac{b}{a},\, x \neq -\frac{d}{c}\right).$

32. $\int \frac{x\mathrm{d}x}{L_1^2 L_2^2} = \frac{1}{D^2} \left(\frac{b}{L_1} + \frac{d}{L_2} - \frac{cb + ad}{D} \ln \frac{cL_1}{aL_2}\right) \qquad \left(x \neq -\frac{b}{a},\, x \neq -\frac{d}{c}\right).$

33. $\int \frac{x^2\mathrm{d}x}{L_1^2 L_2^2} = \frac{-1}{D^2} \left(\frac{b^2}{aL_1} + \frac{d^2}{cL_2} - \frac{2bd}{D} \ln \frac{cL_1}{aL_2}\right) \qquad \left(x \neq -\frac{b}{a},\, x \neq -\frac{d}{c}\right).$

**Integrals containing a quadratic function $ax^2 + bx + c$:**

$$\boxed{Q = ax^2 + bx + c,\; D = 4ac - b^2,\; a \neq 0,\; D \neq 0.}$$

For $D = 0$, $Q$ is a square of a linear function; if $Q$ is in the denominator of the fraction, then no zeroes of $Q$ are allowed to lie in the interval of integration.

34. $$\int \frac{\mathrm{d}x}{Q} = \begin{cases} \frac{2}{\sqrt{D}} \arctan \frac{2ax + b}{\sqrt{D}} & \text{(for } D > 0\text{)}, \\ -\frac{2}{\sqrt{-D}} \operatorname{artanh} \frac{2ax + b}{\sqrt{-D}} & \text{(for } D < 0 \text{ and } |2ax + b| < \sqrt{-D}\text{)}, \\ \frac{1}{\sqrt{-D}} \ln \frac{2ax + b - \sqrt{-D}}{2ax + b + \sqrt{-D}} & \text{(for } D < 0 \text{ and } |2ax + b| > \sqrt{-D}\text{)}. \end{cases}$$

35. $\int \frac{\mathrm{d}x}{Q^n} = \frac{2ax + b}{(n-1)DQ^{n-1}} + \frac{(2n-3)2a}{(n-1)D} \int \frac{\mathrm{d}x}{Q^{n-1}}.$

36. $$\int \frac{x\,\mathrm{d}x}{Q} = \frac{1}{2a}\ln Q - \frac{b}{2a}\int \frac{\mathrm{d}x}{Q} \qquad \text{(see No. 34).}$$

37. $$\int \frac{x\,\mathrm{d}x}{Q^n} = -\frac{bx+2c}{(n-1)DQ^{n-1}} - \frac{b(2n-3)}{(n-1)D}\int \frac{\mathrm{d}x}{Q^{n-1}}.$$

38. $$\int \frac{x^2\,\mathrm{d}x}{Q} = \frac{x}{a} - \frac{b}{2a^2}\ln Q + \frac{b^2-2ac}{2a^2}\int \frac{\mathrm{d}x}{Q} \qquad \text{(see No. 34).}$$

39. $$\int \frac{x^2\,\mathrm{d}x}{Q^n} = \frac{-x}{(2n-3)aQ^{n-1}} + \frac{c}{(2n-3)a}\int \frac{\mathrm{d}x}{Q^n} - \frac{(n-2)b}{(2n-3)a}\int \frac{x\,\mathrm{d}x}{Q^n}$$

(see No. 35 and 37).

40. $$\int \frac{x^m\,\mathrm{d}x}{Q^n} = -\frac{x^{m-1}}{(2n-m-1)aQ^{n-1}} + \frac{(m-1)c}{(2n-m-1)a}\int \frac{x^{m-2}\,\mathrm{d}x}{Q^n}$$
$$-\frac{(n-m)b}{(2n-m-1)a}\int \frac{x^{m-1}\,\mathrm{d}x}{Q^n} \qquad (m \neq 2n-1;\ \text{for } m = 2n-1 \text{ see No. 41}).$$

41. $$\int \frac{x^{2n-1}\,\mathrm{d}x}{Q^n} = \frac{1}{a}\int \frac{x^{2n-3}\,\mathrm{d}x}{Q^{n-1}} - \frac{c}{a}\int \frac{x^{2n-3}\,\mathrm{d}x}{Q^n} - \frac{b}{a}\int \frac{x^{2n-2}\,\mathrm{d}x}{Q^n}.$$

42. $$\int \frac{\mathrm{d}x}{xQ} = \frac{1}{2c}\ln \frac{x^2}{Q} - \frac{b}{2m}\int \frac{\mathrm{d}x}{Q} \qquad \text{(see No. 34).}$$

43. $$\int \frac{\mathrm{d}x}{xQ^n} = \frac{1}{2c(n-1)Q^{n-1}} - \frac{b}{2c}\int \frac{\mathrm{d}x}{Q^n} + \frac{1}{c}\int \frac{\mathrm{d}x}{xQ^{n-1}}.$$

44. $$\int \frac{\mathrm{d}x}{x^2Q} = \frac{b}{2c^2}\ln \frac{Q}{x^2} - \frac{1}{cx} + \left(\frac{b^2}{2c^2} - \frac{a}{c}\right)\int \frac{\mathrm{d}x}{Q} \qquad \text{(see No. 34).}$$

45. $$\int \frac{\mathrm{d}x}{x^mQ^n} = -\frac{1}{(m-1)cx^{m-1}Q^{n-1}} - \frac{(2n+m-3)a}{(m-1)c}\int \frac{\mathrm{d}x}{x^{m-2}Q^n}$$
$$-\frac{(n+m-2)b}{(m-1)c}\int \frac{\mathrm{d}x}{x^{m-1}Q^n} \qquad (m > 1).$$

46. $$\int \frac{\mathrm{d}x}{(fx+g)Q} = \frac{1}{2(cf^2-gbf+g^2a)}\left[f\ln \frac{(fx+g)^2}{Q}\right] + \frac{2ga-bf}{2(cf^2-gbf+g^2a)}\int \frac{\mathrm{d}x}{Q}$$

(see No. 34).

## Integrals containing the quadratic function $a^2 \pm x^2$:

$$Q = a^2 \pm x^2, \quad P = \begin{cases} \arctan\dfrac{x}{a} & \text{for the sign "+"}, \\ \operatorname{artanh}\dfrac{x}{a} = \dfrac{1}{2}\ln\dfrac{a+x}{a-x} & \text{for the sign "$-$" and } |x| < a, \\ \operatorname{arcoth}\dfrac{x}{a} = \dfrac{1}{2}\ln\dfrac{x+a}{x-a} & \text{for the sign "$-$" and } |x| > a. \end{cases}$$

In the case of a double sign in a formula the upper sign corresponds to $Q = a^2 + x^2$ and the lower one to $Q = a^2 - x^2$, $a > 0$.

47. $\displaystyle\int \frac{dx}{Q} = \frac{1}{a}P.$

48. $\displaystyle\int \frac{dx}{Q^2} = \frac{x}{2a^2Q} + \frac{1}{2a^3}P.$

49. $\displaystyle\int \frac{dx}{Q^3} = \frac{x}{4a^2Q^2} + \frac{3x}{8a^4Q} + \frac{3}{8a^5}P.$

50. $\displaystyle\int \frac{dx}{Q^{n+1}} = \frac{x}{2na^2Q^n} + \frac{2n-1}{2na^2}\int\frac{dx}{Q^n}.$

51. $\displaystyle\int \frac{x\,dx}{Q} = \pm\frac{1}{2}\ln Q.$

52. $\displaystyle\int \frac{x\,dx}{Q^2} = \mp\frac{1}{2Q}.$

53. $\displaystyle\int \frac{x\,dx}{Q^3} = \mp\frac{1}{4Q^2}.$

54. $\displaystyle\int \frac{x\,dx}{Q^{n+1}} = \mp\frac{1}{2nQ^n}.$

55. $\displaystyle\int \frac{x^2dx}{Q} = \pm x \mp aP.$

56. $\displaystyle\int \frac{x^2dx}{Q^2} = \mp\frac{x}{2Q} \pm \frac{1}{2a}P.$

57. $\displaystyle\int \frac{x^2dx}{Q^3} = \mp\frac{x}{4Q^2} \pm \frac{x}{8a^2Q} \pm \frac{1}{8a^3}P.$

58. $\displaystyle\int \frac{x^2dx}{Q^{n+1}} = \mp\frac{x}{2nQ^n} \pm \frac{1}{2n}\int\frac{dx}{Q^n}.$

59. $\displaystyle\int \frac{x^3dx}{Q} = \pm\frac{x^2}{2} - \frac{a^2}{2}\ln Q.$

60. $\displaystyle\int \frac{x^3dx}{Q^2} = \frac{a^2}{2Q} + \frac{1}{2}\ln Q.$

61. $\displaystyle\int \frac{x^3dx}{Q^3} = -\frac{1}{2Q} + \frac{a^2}{4Q^2}.$

62. $\displaystyle\int \frac{x^3dx}{Q^{n+1}} = -\frac{1}{2(n-1)Q^{n-1}} + \frac{a^2}{2nQ^n}.$

In numbers 50, 54 and 58 above we require $n \neq 0$; in 62, $n > 1$.

63. $\displaystyle\int \frac{dx}{xQ} = \frac{1}{2a^2}\ln\frac{x^2}{Q}.$

64. $\displaystyle\int \frac{dx}{xQ^2} = \frac{1}{2a^2Q} + \frac{1}{2a^4}\ln\frac{x^2}{Q}.$

65. $\displaystyle\int \frac{dx}{xQ^3} = \frac{1}{4a^2Q^2} + \frac{1}{2a^4Q} + \frac{1}{2a^6}\ln\frac{x^2}{Q}.$

66. $\displaystyle\int \frac{dx}{x^2Q} = -\frac{1}{a^2x} \mp \frac{1}{a^3}P.$

67. $\displaystyle\int \frac{\mathrm{d}x}{x^2Q^2} = -\frac{1}{a^4x} \mp \frac{x}{2a^4Q} \mp \frac{3}{2a^5}P.$

68. $\displaystyle\int \frac{\mathrm{d}x}{x^2Q^3} = -\frac{1}{a^6x} \mp \frac{x}{4a^4Q^2} \mp \frac{7x}{8a^6Q} \mp \frac{15}{8a^7}P.$

69. $\displaystyle\int \frac{\mathrm{d}x}{x^3Q} = -\frac{1}{2a^2x^2} \mp \frac{1}{2a^4}\ln\frac{x^2}{Q}.$

70. $\displaystyle\int \frac{\mathrm{d}x}{x^3Q^2} = -\frac{1}{2a^4x^2} \mp \frac{1}{2a^4Q} \mp \frac{1}{a^6}\ln\frac{x^2}{Q}.$

71. $\displaystyle\int \frac{\mathrm{d}x}{x^3Q^3} = -\frac{1}{2a^6x^2} \mp \frac{1}{a^6Q} \mp \frac{1}{4a^4Q^2} \mp \frac{3}{2a^8}\ln\frac{x^2}{Q}.$

72. $\displaystyle\int \frac{\mathrm{d}x}{(b+cx)Q} = \frac{1}{a^2c^2 \pm b^2}\left[c\ln(b+cx) - \frac{c}{2}\ln Q \pm \frac{b}{a}P\right].$

**Integrals containing a cubic function $a^3 \pm x^3$:**

$K = a^3 \pm x^3$; in the case of a "±", the upper sign means $K = a^3 + x^3$, the lower one means $K = a^3 - x^3$.

73. $\displaystyle\int \frac{\mathrm{d}x}{K} = \pm\frac{1}{6a^2}\ln\frac{(a \pm x)^2}{a^2 \mp ax + x^2} + \frac{1}{a^2\sqrt{3}}\arctan\frac{2x \mp a}{a\sqrt{3}}.$

74. $\displaystyle\int \frac{\mathrm{d}x}{K^2} = \frac{x}{3a^3K} + \frac{2}{3a^3}\int\frac{\mathrm{d}x}{K}$ (see No. 73).

75. $\displaystyle\int \frac{x\mathrm{d}x}{K} = \frac{1}{6a}\ln\frac{a^2 \mp ax + x^2}{(a \pm x)^2} \pm \frac{1}{a\sqrt{3}}\arctan\frac{2x \mp a}{a\sqrt{3}}.$

76. $\displaystyle\int \frac{x\mathrm{d}x}{K^2} = \frac{x^2}{3a^3K} + \frac{1}{3a^3}\int\frac{x\mathrm{d}x}{K}$ (see No. 75).

77. $\displaystyle\int \frac{x^2\mathrm{d}x}{K} = \pm\frac{1}{3}\ln K.$

78. $\displaystyle\int \frac{x^2\mathrm{d}x}{K^2} = \mp\frac{1}{3K}.$

79. $\displaystyle\int \frac{x^3\mathrm{d}x}{K} = \pm x \mp a^3\int\frac{\mathrm{d}x}{K}$ (see No. 73).

80. $\displaystyle\int \frac{x^3\mathrm{d}x}{K^2} = \mp\frac{x}{3K} \pm \frac{1}{3}\int\frac{\mathrm{d}x}{K}$ (see No. 73).

81. $\int \frac{\mathrm{d}x}{xK} = \frac{1}{3a^3} \ln \frac{x^3}{K}.$

82. $\int \frac{\mathrm{d}x}{xK^2} = \frac{1}{3a^3K} + \frac{1}{3a^6} \ln \frac{x^3}{K}.$

83. $\int \frac{\mathrm{d}x}{x^2K} = -\frac{1}{a^3x} \mp \frac{1}{a^3} \int \frac{x\mathrm{d}x}{K}$ (see No. 75).

84. $\int \frac{\mathrm{d}x}{x^2K^2} = -\frac{1}{a^6x} \mp \frac{x^2}{3a^6K} \mp \frac{4}{3a^6} \int \frac{x\mathrm{d}x}{K}$ (see No. 75).

85. $\int \frac{\mathrm{d}x}{x^3K} = -\frac{1}{2a^3x^2} \mp \frac{1}{a^3} \int \frac{\mathrm{d}x}{K}$ (see No. 73).

86. $\int \frac{\mathrm{d}x}{x^3K^2} = -\frac{1}{2a^6x^2} \mp \frac{x}{3a^6K} \mp \frac{5}{3a^6} \int \frac{\mathrm{d}x}{K}$ (see No. 73).

**Integrals containing the biquadratic (quartic) function $a^4 \pm x^4$:**

87. $$\int \frac{\mathrm{d}x}{a^4 + x^4} = \frac{1}{4a^3\sqrt{2}} \ln \frac{x^2 + ax\sqrt{2} + a^2}{x^2 - ax\sqrt{2} + a^2} + \frac{1}{2a^3\sqrt{2}} \left( \arctan\left(\frac{x\sqrt{2}}{a} + 1\right) + \arctan\left(\frac{x\sqrt{2}}{a} - 1\right) \right).$$

88. $\int \frac{x\mathrm{d}x}{a^4 + x^4} = \frac{1}{2a^2} \arctan \frac{x^2}{a^2}.$

89. $$\int \frac{x^2\mathrm{d}x}{a^4 + x^4} = -\frac{1}{4a\sqrt{2}} \ln \frac{x^2 + ax\sqrt{2} + a^2}{x^2 - ax\sqrt{2} + a^2} + \frac{1}{2a\sqrt{2}} \left( \arctan\left(\frac{x\sqrt{2}}{a} + 1\right) + \arctan\left(\frac{x\sqrt{2}}{a} - 1\right) \right).$$

90. $\int \frac{x^3\mathrm{d}x}{a^4 + x^4} = \frac{1}{4} \ln(a^4 + x^4).$

91. $\int \frac{\mathrm{d}x}{a^4 - x^4} = \frac{1}{4a^3} \ln \frac{a + x}{a - x} + \frac{1}{2a^3} \arctan \frac{x}{a}.$

92. $\int \frac{x\mathrm{d}x}{a^4 - x^4} = \frac{1}{4a^2} \ln \frac{a^2 + x^2}{a^2 - x^2}.$

93. $\int \frac{x^2 \mathrm{d}x}{a^4 - x^4} = \frac{1}{4a} \ln \frac{a+x}{a-x} - \frac{1}{2a} \arctan \frac{x}{a}.$

94. $\int \frac{x^3 \mathrm{d}x}{a^4 - x^4} = -\frac{1}{4} \ln(a^4 - x^4).$

**Special cases of integration by use of a partial fraction decomposition:**

95. $\int \frac{\mathrm{d}x}{(x+a)(x+b)(x+c)} = u \int \frac{\mathrm{d}x}{x+a} + v \int \frac{\mathrm{d}x}{x+b} + w \int \frac{\mathrm{d}x}{x+c},$

$$u = \frac{1}{(b-a)(c-a)}, \quad v = \frac{1}{(a-b)(c-b)}, \quad w = \frac{1}{(a-c)(b-c)},$$

$a,\ b,\ c,$ are pairwise different.

96. $\int \frac{\mathrm{d}x}{(x+a)(x+b)(x+c)(x+d)} = t \int \frac{\mathrm{d}x}{x+a} + u \int \frac{\mathrm{d}x}{x+b} + v \int \frac{\mathrm{d}x}{x+c} + w \int \frac{\mathrm{d}x}{x+d},$

$$t = \frac{1}{(b-a)(c-a)(d-a)}, \quad u = \frac{1}{(a-b)(c-b)(d-b)},$$
$$v = \frac{1}{(a-c)(b-c)(d-c)}, \quad w = \frac{1}{(a-d)(b-d)(c-d)},$$

$a,\ b,\ c,\ d,$ pairwise distinct.

97. $\int \frac{\mathrm{d}x}{(a+bx^2)(c+dx^2)} = \frac{1}{bc-ad} \left( \int \frac{b\,\mathrm{d}x}{a+bx^2} - \int \frac{d\,\mathrm{d}x}{c+dx^2} \right) \qquad (bc - ad \neq 0).$

98. $\int \frac{\mathrm{d}x}{(x^2+a)(x^2+b)(x^2+c)} = u \int \frac{\mathrm{d}x}{x^2+a} + v \int \frac{\mathrm{d}x}{x^2+b} + w \int \frac{\mathrm{d}x}{x^2+c},$

$u,\ v,\ w,\ a,\ b,\ c,$ see No. 95.

### 0.9.5.2 Integrals of irrational functions

**Integrals containing the square root $\sqrt{x}$ and the linear function $a^2 \pm b^2 x$:**

$$L = a^2 \pm b^2 x, \quad M = \begin{cases} \arctan \dfrac{b\sqrt{x}}{a} & \text{for the sign “ + ”}, \\ \dfrac{1}{2} \ln \dfrac{a + b\sqrt{x}}{a - b\sqrt{x}} & \text{for the sign “ − ”}. \end{cases}$$

In the case of a double sign in a formula the upper sign corresponds to $L = a^2 + b^2 x$ and the lower one corresponds to $L = a^2 - b^2 x$.

99. $\displaystyle\int \frac{\sqrt{x}\mathrm{d}x}{L} = \pm 2\frac{\sqrt{x}}{b^2} \mp \frac{2a}{b^3}M.$

100. $\displaystyle\int \frac{\sqrt{x^3}\mathrm{d}x}{L} = \pm\frac{2\sqrt{x^3}}{3b^2} - \frac{2a^2\sqrt{x}}{b^4} + \frac{2a^3}{b^5}M.$

101. $\displaystyle\int \frac{\sqrt{x}\mathrm{d}x}{L^2} = \mp\frac{\sqrt{x}}{b^2 L} \pm \frac{1}{ab^3}M.$

102. $\displaystyle\int \frac{\sqrt{x^3}\mathrm{d}x}{L^2} = \pm\frac{2\sqrt{x^3}}{b^2 L} + \frac{3a^2\sqrt{x}}{b^4 L} - \frac{3a}{b^5}M.$

103. $\displaystyle\int \frac{\mathrm{d}x}{L\sqrt{x}} = \frac{2}{ab}M.$

104. $\displaystyle\int \frac{\mathrm{d}x}{L\sqrt{x^3}} = -\frac{2}{a^2\sqrt{x}} \mp \frac{2b}{a^3}M.$

105. $\displaystyle\int \frac{\mathrm{d}x}{L^2\sqrt{x}} = \frac{\sqrt{x}}{a^2 L} + \frac{1}{a^3 b}M.$

106. $\displaystyle\int \frac{\mathrm{d}x}{L^2\sqrt{x^3}} = -\frac{2}{a^2 L\sqrt{x}} \mp \frac{3b^2\sqrt{x}}{a^4 L} \mp \frac{3b}{a^5}M.$

**Other integrals containing the square root $\sqrt{x}$:**

107. $\displaystyle\int \frac{\sqrt{x}\mathrm{d}x}{p^4 + x^2} = -\frac{1}{2p\sqrt{2}}\ln\frac{x + p\sqrt{2x} + p^2}{x - p\sqrt{2x} + p^2} + \frac{1}{p\sqrt{2}}\arctan\frac{p\sqrt{2x}}{p^2 - x}.$

108. $\displaystyle\int \frac{\mathrm{d}x}{(p^4 + x^2)\sqrt{x}} = \frac{1}{2p^3\sqrt{2}}\ln\frac{x + p\sqrt{2x} + p^2}{x - p\sqrt{2x} + p^2} + \frac{1}{p^3\sqrt{2}}\arctan\frac{p\sqrt{2x}}{p^2 - x}.$

109. $\displaystyle\int \frac{\sqrt{x}\mathrm{d}x}{p^4 - x^2} = \frac{1}{2p}\ln\frac{p + \sqrt{x}}{p - \sqrt{x}} - \frac{1}{p}\arctan\frac{\sqrt{x}}{p}.$

110. $\displaystyle\int \frac{\mathrm{d}x}{(p^4 - x^2)\sqrt{x}} = \frac{1}{2p^3}\ln\frac{p + \sqrt{x}}{p - \sqrt{x}} + \frac{1}{p^3}\arctan\frac{\sqrt{x}}{p}.$

**Integrals containing the square root function $\sqrt{ax + b}$:**

L=ax+b

111. $\displaystyle\int \sqrt{L}\mathrm{d}x = \frac{2}{3a}\sqrt{L^3}.$

112. $\int x\sqrt{L}\mathrm{d}x = \dfrac{2(3ax-2b)\sqrt{L^3}}{15a^2}.$

113. $\int x^2\sqrt{L}\mathrm{d}x = \dfrac{2(15a^2x^2-12abx+8b^2)\sqrt{L^3}}{105a^3}.$

114. $\int \dfrac{\mathrm{d}x}{\sqrt{L}} = \dfrac{2\sqrt{L}}{a}.$

115. $\int \dfrac{x\mathrm{d}x}{\sqrt{L}} = \dfrac{2(ax-2b)}{3a^2}\sqrt{L}.$

116. $\int \dfrac{x^2\mathrm{d}x}{\sqrt{L}} = \dfrac{2(3a^2x^2-4abx+8b^2)\sqrt{L}}{15a^3}.$

117. $\int \dfrac{\mathrm{d}x}{x\sqrt{L}} = \begin{cases} \dfrac{1}{\sqrt{b}}\ln\dfrac{\sqrt{L}-\sqrt{b}}{\sqrt{L}+\sqrt{b}} & \text{for } b>0, \\ \dfrac{2}{\sqrt{-b}}\arctan\sqrt{\dfrac{L}{-b}} & \text{for } b<0. \end{cases}$

118. $\int \dfrac{\sqrt{L}}{x}\mathrm{d}x = 2\sqrt{L} + b\int \dfrac{\mathrm{d}x}{x\sqrt{L}}$ (see No. 117).

119. $\int \dfrac{\mathrm{d}x}{x^2\sqrt{L}} = -\dfrac{\sqrt{L}}{bx} - \dfrac{a}{2b}\int \dfrac{\mathrm{d}x}{x\sqrt{L}}$ (see No. 117).

120. $\int \dfrac{\sqrt{L}}{x^2}\mathrm{d}x = -\dfrac{\sqrt{L}}{x} + \dfrac{a}{2}\int \dfrac{\mathrm{d}x}{x\sqrt{L}}$ (see No. 117).

121. $\int \dfrac{\mathrm{d}x}{x^n\sqrt{L}} = -\dfrac{\sqrt{L}}{(n-1)bx^{n-1}} - \dfrac{(2n-3)a}{(2n-2)b}\int \dfrac{\mathrm{d}x}{x^{n-1}\sqrt{L}}.$

122. $\int \sqrt{L^3}\mathrm{d}x = \dfrac{2\sqrt{L^5}}{5a}.$

123. $\int x\sqrt{L^3}\mathrm{d}x = \dfrac{2}{35a^2}(5\sqrt{L^7} - 7b\sqrt{L^5}).$

124. $\int x^2\sqrt{L^3}\mathrm{d}x = \dfrac{2}{a^3}\left(\dfrac{\sqrt{L^9}}{9} - \dfrac{2b\sqrt{L^7}}{7} + \dfrac{b^2\sqrt{L^5}}{5}\right).$

125. $\int \dfrac{\sqrt{L^3}}{x}\mathrm{d}x = \dfrac{2\sqrt{L^3}}{3} + 2b\sqrt{L} + b^2\int \dfrac{\mathrm{d}x}{x\sqrt{L}}$ (see No. 117).

126. $\int \dfrac{x\mathrm{d}x}{\sqrt{L^3}} = \dfrac{2}{a^2}\left(\sqrt{L} + \dfrac{b}{\sqrt{L}}\right).$

127. $\int \frac{x^2 \mathrm{d}x}{\sqrt{L^3}} = \frac{2}{a^3}\left(\frac{\sqrt{L^3}}{3} - 2b\sqrt{L} - \frac{b^2}{\sqrt{L}}\right).$

128. $\int \frac{\mathrm{d}x}{x\sqrt{L^3}} = \frac{2}{b\sqrt{L}} + \frac{1}{b}\int \frac{\mathrm{d}x}{x\sqrt{L}}$ (see No. 117).

129. $\int \frac{\mathrm{d}x}{x^2\sqrt{L^3}} = -\frac{1}{bx\sqrt{L}} - \frac{3a}{b^2\sqrt{L}} - \frac{3a}{2b^2}\int \frac{\mathrm{d}x}{x\sqrt{L}}$ (see No. 117).

130. $\int L^{\pm n/2}\mathrm{d}x = \frac{2L^{(2\pm n)/2}}{a(2\pm n)}.$

131. $\int xL^{\pm n/2}\mathrm{d}x = \frac{2}{a^2}\left(\frac{L^{(4\pm n)/2}}{4\pm n} - \frac{bL^{(2\pm n)/2}}{2\pm n}\right).$

132. $\int x^2L^{\pm n/2}\mathrm{d}x = \frac{2}{a^3}\left(\frac{L^{(6\pm n)/2}}{6\pm n} - \frac{2bL^{(4\pm n)/2}}{4\pm n} + \frac{b^2L^{(2\pm n)/2}}{2\pm n}\right).$

133. $\int \frac{L^{n/2}\mathrm{d}x}{x} = \frac{2L^{n/2}}{n} + b\int \frac{L^{(n-2)/2}}{x}\mathrm{d}x.$

134. $\int \frac{\mathrm{d}x}{xL^{n/2}} = \frac{2}{(n-2)bL^{(n-2)/2}} + \frac{1}{b}\int \frac{\mathrm{d}x}{xL^{(n-2)/2}}.$

135. $\int \frac{\mathrm{d}x}{x^2L^{n/2}} = -\frac{1}{bxL^{(n-2)/2}} - \frac{na}{2b}\int \frac{\mathrm{d}x}{xL^{n/2}}.$

**Integrals containing square root functions $\sqrt{ax+b}$ and $\sqrt{cx+d}$:**

$$\boxed{L_1 = ax+b,\quad L_2 = cx+d,\quad D = bc-ad,\quad D \neq 0.}$$

136. $\int \frac{\mathrm{d}x}{\sqrt{L_1L_2}} = \begin{cases} \frac{2\,\mathrm{sgn}(a)\,\mathrm{sgn}(L_1)}{\sqrt{-ac}}\arctan\sqrt{-\frac{cL_1}{aL_2}} & \text{for } ac<0,\ \mathrm{sgn}(L_1) = \frac{L_1}{|L_1|}, \\ \frac{2\,\mathrm{sgn}(a)\,\mathrm{sgn}(L_1)}{\sqrt{ac}}\,\mathrm{artanh}\sqrt{\frac{cL_1}{aL_2}} & \text{for } ac>0, \text{ and } |cL_1| < |aL_2|. \end{cases}$

137. $\int \frac{x\mathrm{d}x}{\sqrt{L_1L_2}} = \frac{\sqrt{L_1L_2}}{ac} - \frac{ad+bc}{2ac}\int \frac{\mathrm{d}x}{\sqrt{L_1L_2}}$ (see No. 136).

138. $\int \frac{\mathrm{d}x}{\sqrt{L_1}\sqrt{L_2^3}} = -\frac{2\sqrt{L_1}}{D\sqrt{L_2}}.$

139. $\int \frac{\mathrm{d}x}{L_2\sqrt{L_1}} = \begin{cases} \frac{2}{\sqrt{-Dc}}\arctan\frac{c\sqrt{L_1}}{\sqrt{-Dc}} & \text{for } Dc<0, \\ \frac{1}{\sqrt{Dc}}\ln\frac{c\sqrt{L_1}-\sqrt{Dc}}{c\sqrt{L_1}+\sqrt{Dc}} & \text{for } Dc>0. \end{cases}$

140. $$\int \sqrt{L_1 L_2}\mathrm{d}x = \frac{D + 2aL_2}{4ac}\sqrt{L_1 L_2} - \frac{D^2}{8ac}\int \frac{\mathrm{d}x}{\sqrt{L_1 L_2}}$$ (see No. 136).

141. $$\int \sqrt{\frac{L_2}{L_1}}\mathrm{d}x = \operatorname{sgn}(L_1)\left(\frac{1}{a}\sqrt{L_1 L_2} - \frac{D}{2a}\int \frac{\mathrm{d}x}{\sqrt{L_1 L_2}}\right)$$ (see No. 136).

142. $$\int \frac{\sqrt{L_1}\mathrm{d}x}{L_2} = \frac{2\sqrt{L_1}}{c} + \frac{D}{c}\int \frac{\mathrm{d}x}{L_2\sqrt{L_1}}$$ (see No. 139).

143. $$\int \frac{L_2^n \mathrm{d}x}{\sqrt{L_1}} = \frac{2}{(2n+1)a}\left(\sqrt{L_1}L_2^n - nD\int \frac{L_2^{n-1}\mathrm{d}x}{\sqrt{L_1}}\right).$$

144. $$\int \frac{\mathrm{d}x}{\sqrt{L_1}L_2^n} = -\frac{1}{(n-1)D}\left(\frac{\sqrt{L_1}}{L_2^{n-1}} + \left(n - \frac{3}{2}\right)a\int \frac{\mathrm{d}x}{\sqrt{L_1}L_2^{n-1}}\right).$$

145. $$\int \sqrt{L_1}L_2^n \mathrm{d}x = \frac{1}{(2n+3)c}\left(2\sqrt{L_1}L_2^{n+1} + D\int \frac{L_2^n \mathrm{d}x}{\sqrt{L_1}}\right)$$ (see No. 143).

146. $$\int \frac{\sqrt{L_1}\mathrm{d}x}{L_2^n} = \frac{1}{(n-1)c}\left(-\frac{\sqrt{L_1}}{L_2^{n-1}} + \frac{a}{2}\int \frac{\mathrm{d}x}{\sqrt{L_1}L_2^{n-1}}\right).$$

**Integrals containing the square root function $\sqrt{a^2 - x^2}$:**

$$\boxed{Q = a^2 - x^2.}$$

147. $$\int \sqrt{Q}\mathrm{d}x = \frac{1}{2}\left(x\sqrt{Q} + a^2 \arcsin\frac{x}{a}\right).$$

148. $$\int x\sqrt{Q}\mathrm{d}x = -\frac{1}{3}\sqrt{Q^3}.$$

149. $$\int x^2\sqrt{Q}\mathrm{d}x = -\frac{x}{4}\sqrt{Q^3} + \frac{a^2}{8}\left(x\sqrt{Q} + a^2 \arcsin\frac{x}{a}\right).$$

150. $$\int x^3\sqrt{Q}\mathrm{d}x = \frac{\sqrt{Q^5}}{5} - a^2\frac{\sqrt{Q^3}}{3}.$$

151. $$\int \frac{\sqrt{Q}}{x}\mathrm{d}x = \sqrt{Q} - a\ln\frac{a + \sqrt{Q}}{x}.$$

152. $$\int \frac{\sqrt{Q}}{x^2}\mathrm{d}x = -\frac{\sqrt{Q}}{x} - \arcsin\frac{x}{a}.$$

153. $$\int \frac{\sqrt{Q}}{x^3}\mathrm{d}x = -\frac{\sqrt{Q}}{2x^2} + \frac{1}{2a}\ln\frac{a + \sqrt{Q}}{x}.$$

154. $\displaystyle\int \frac{\mathrm{d}x}{\sqrt{Q}} = \arcsin\frac{x}{a}.$

155. $\displaystyle\int \frac{x\mathrm{d}x}{\sqrt{Q}} = -\sqrt{Q}.$

156. $\displaystyle\int \frac{x^2\mathrm{d}x}{\sqrt{Q}} = -\frac{x}{2}\sqrt{Q} + \frac{a^2}{2}\arcsin\frac{x}{a}.$

157. $\displaystyle\int \frac{x^3\mathrm{d}x}{\sqrt{Q}} = \frac{\sqrt{Q^3}}{3} - a^2\sqrt{Q}.$

158. $\displaystyle\int \frac{\mathrm{d}x}{x\sqrt{Q}} = -\frac{1}{a}\ln\frac{a+\sqrt{Q}}{x}.$

159. $\displaystyle\int \frac{\mathrm{d}x}{x^2\sqrt{Q}} = -\frac{\sqrt{Q}}{a^2x}.$

160. $\displaystyle\int \frac{\mathrm{d}x}{x^3\sqrt{Q}} = -\frac{\sqrt{Q}}{2a^2x^2} - \frac{1}{2a^3}\ln\frac{a+\sqrt{Q}}{x}.$

161. $\displaystyle\int \sqrt{Q^3}\mathrm{d}x = \frac{1}{4}\left(x\sqrt{Q^3} + \frac{3a^2x}{2}\sqrt{Q} + \frac{3a^4}{2}\arcsin\frac{x}{a}\right).$

162. $\displaystyle\int x\sqrt{Q^3}\mathrm{d}x = -\frac{1}{5}\sqrt{Q^5}.$

163. $\displaystyle\int x^2\sqrt{Q^3}\mathrm{d}x = -\frac{x\sqrt{Q^5}}{6} + \frac{a^2x\sqrt{Q^3}}{24} + \frac{a^4x\sqrt{Q}}{16} + \frac{a^6}{16}\arcsin\frac{x}{a}.$

164. $\displaystyle\int x^3\sqrt{Q^3}\mathrm{d}x = \frac{\sqrt{Q^7}}{7} - \frac{a^2\sqrt{Q^5}}{5}.$

165. $\displaystyle\int \frac{\sqrt{Q^3}}{x}\mathrm{d}x = \frac{\sqrt{Q^3}}{3} + a^2\sqrt{Q} - a^3\ln\frac{a+\sqrt{Q}}{x}.$

166. $\displaystyle\int \frac{\sqrt{Q^3}}{x^2}\mathrm{d}x = -\frac{\sqrt{Q^3}}{x} - \frac{3}{2}x\sqrt{Q} - \frac{3}{2}a^2\arcsin\frac{x}{a}.$

167. $\displaystyle\int \frac{\sqrt{Q^3}}{x^3}\mathrm{d}x = -\frac{\sqrt{Q^3}}{2x^2} - \frac{3\sqrt{Q}}{2} + \frac{3a}{2}\ln\frac{a+\sqrt{Q}}{x}.$

168. $\displaystyle\int \frac{\mathrm{d}x}{\sqrt{Q^3}} = \frac{x}{a^2\sqrt{Q}}.$

169. $\int \frac{x\mathrm{d}x}{\sqrt{Q^3}} = \frac{1}{\sqrt{Q}}.$

170. $\int \frac{x^2\mathrm{d}x}{\sqrt{Q^3}} = \frac{x}{\sqrt{Q}} - \arcsin\frac{x}{a}.$

171. $\int \frac{x^3\mathrm{d}x}{\sqrt{Q^3}} = \sqrt{Q} + \frac{a^2}{\sqrt{Q}}.$

172. $\int \frac{\mathrm{d}x}{x\sqrt{Q^3}} = \frac{1}{a^2\sqrt{Q}} - \frac{1}{a^3}\ln\frac{a+\sqrt{Q}}{x}.$

173. $\int \frac{\mathrm{d}x}{x^2\sqrt{Q^3}} = \frac{1}{a^4}\left(-\frac{\sqrt{Q}}{x} + \frac{x}{\sqrt{Q}}\right).$

174. $\int \frac{\mathrm{d}x}{x^3\sqrt{Q^3}} = -\frac{1}{2a^2x^2\sqrt{Q}} + \frac{3}{2a^4\sqrt{Q}} - \frac{3}{2a^5}\ln\frac{a+\sqrt{Q}}{x}.$

**Integrals containing the square root function $\sqrt{x^2+a^2}$:**

$$\boxed{Q = x^2 + a^2.}$$

175. $\int \sqrt{Q}\mathrm{d}x = \frac{1}{2}\left(x\sqrt{Q} + a^2\operatorname{arcsinh}\frac{x}{a}\right) = \frac{1}{2}[x\sqrt{Q} + a^2(\ln(x+\sqrt{Q}) - \ln a)].$

176. $\int x\sqrt{Q}\mathrm{d}x = \frac{1}{3}\sqrt{Q^3}.$

177. $\int x^2\sqrt{Q}\mathrm{d}x = \frac{x}{4}\sqrt{Q^3} - \frac{a^2}{8}\left(x\sqrt{Q} + a^2\operatorname{arsinh}\frac{x}{a}\right)$

$$= \frac{x}{4}\sqrt{Q^3} - \frac{a^2}{8}[x\sqrt{Q} + a^2(\ln(x+\sqrt{Q}) - \ln a)].$$

178. $\int x^3\sqrt{Q}\mathrm{d}x = \frac{\sqrt{Q^5}}{5} - \frac{a^2\sqrt{Q^3}}{3}.$

179. $\int \frac{\sqrt{Q}}{x}\mathrm{d}x = \sqrt{Q} - a\ln\frac{a+\sqrt{Q}}{x}.$

180. $\int \frac{\sqrt{Q}}{x^2}\mathrm{d}x = -\frac{\sqrt{Q}}{x} + \operatorname{arsinh}\frac{x}{a} = -\frac{\sqrt{Q}}{x} + \ln(x+\sqrt{Q}) - \ln a.$

181. $\int \frac{\sqrt{Q}}{x^3}\mathrm{d}x = -\frac{\sqrt{Q}}{2x^2} - \frac{1}{2a}\ln\frac{a+\sqrt{Q}}{x}.$

182. $\displaystyle\int \frac{\mathrm{d}x}{\sqrt{Q}} = \operatorname{arsinh}\frac{x}{a} = \ln(x + \sqrt{Q}) - \ln a.$

183. $\displaystyle\int \frac{x\mathrm{d}x}{\sqrt{Q}} = \sqrt{Q}.$

184. $\displaystyle\int \frac{x^2\mathrm{d}x}{\sqrt{Q}} = \frac{x}{2}\sqrt{Q} - \frac{a^2}{2}\operatorname{arsinh}\frac{x}{a} = \frac{x}{2}\sqrt{Q} - \frac{a^2}{2}(\ln(x + \sqrt{Q}) - \ln a).$

185. $\displaystyle\int \frac{x^3\mathrm{d}x}{\sqrt{Q}} = \frac{\sqrt{Q^3}}{3} - a^2\sqrt{Q}.$

186. $\displaystyle\int \frac{\mathrm{d}x}{x\sqrt{Q}} = -\frac{1}{a}\ln\frac{a + \sqrt{Q}}{x}.$

187. $\displaystyle\int \frac{\mathrm{d}x}{x^2\sqrt{Q}} = -\frac{\sqrt{Q}}{a^2x}.$

188. $\displaystyle\int \frac{\mathrm{d}x}{x^3\sqrt{Q}} = -\frac{\sqrt{Q}}{2a^2x^2} + \frac{1}{2a^3}\ln\frac{a + \sqrt{Q}}{x}.$

189. $$\int \sqrt{Q^3}\mathrm{d}x = \frac{1}{4}\left(x\sqrt{Q^3} + \frac{3a^2x}{2}\sqrt{Q} + \frac{3a^4}{2}\operatorname{arsinh}\frac{x}{a}\right)$$
$$= \frac{1}{4}\left(x\sqrt{Q^3} + \frac{3a^2x}{2}\sqrt{Q} + \frac{3a^4}{2}(\ln(x + \sqrt{Q}) - \ln a)\right).$$

190. $\displaystyle\int x\sqrt{Q^3}\mathrm{d}x = \frac{1}{5}\sqrt{Q^5}.$

191. $$\int x^2\sqrt{Q^3}\mathrm{d}x = \frac{x\sqrt{Q^5}}{6} - \frac{a^2x\sqrt{Q^3}}{24} - \frac{a^4x\sqrt{Q}}{16} - \frac{a^6}{16}\operatorname{arsinh}\frac{x}{a}$$
$$= \frac{x\sqrt{Q^5}}{6} - \frac{a^2x\sqrt{Q^3}}{24} - \frac{a^4x\sqrt{Q}}{16} - \frac{a^6}{16}(\ln(x + \sqrt{Q}) - \ln a).$$

192. $\displaystyle\int x^3\sqrt{Q^3}\mathrm{d}x = \frac{\sqrt{Q^7}}{7} - \frac{a^2\sqrt{Q^5}}{5}.$

193. $\displaystyle\int \frac{\sqrt{Q^3}}{x}\mathrm{d}x = \frac{\sqrt{Q^3}}{3} + a^2\sqrt{Q} - a^3\ln\frac{a + \sqrt{Q}}{x}.$

194. $$\int \frac{\sqrt{Q^3}}{x^2}\mathrm{d}x = -\frac{\sqrt{Q^3}}{x} + \frac{3}{2}x\sqrt{Q} + \frac{3}{2}a^2\operatorname{arsinh}\frac{x}{a}$$
$$= -\frac{\sqrt{Q^3}}{x} + \frac{3}{2}x\sqrt{Q} + \frac{3}{2}a^2(\ln(x + \sqrt{Q}) - \ln a).$$

195. $\displaystyle\int \frac{\sqrt{Q^3}}{x^3}\mathrm{d}x = -\frac{\sqrt{Q^3}}{2x^2} + \frac{3}{2}\sqrt{Q} - \frac{3}{2}a\ln\left(\frac{a+\sqrt{Q}}{x}\right).$

196. $\displaystyle\int \frac{\mathrm{d}x}{\sqrt{Q^3}} = \frac{x}{a^2\sqrt{Q}}.$

197. $\displaystyle\int \frac{x\mathrm{d}x}{\sqrt{Q^3}} = -\frac{1}{\sqrt{Q}}.$

198. $\displaystyle\int \frac{x^2\mathrm{d}x}{\sqrt{Q^3}} = -\frac{x}{\sqrt{Q}} + \mathrm{arsinh}\frac{x}{a} = -\frac{x}{\sqrt{Q}} + \ln(x+\sqrt{Q}) - \ln a.$

199. $\displaystyle\int \frac{x^3\mathrm{d}x}{\sqrt{Q^3}} = \sqrt{Q} + \frac{a^2}{\sqrt{Q}}.$

200. $\displaystyle\int \frac{\mathrm{d}x}{x\sqrt{Q^3}} = \frac{1}{a^2\sqrt{Q}} - \frac{1}{a^3}\ln\frac{a+\sqrt{Q}}{x}.$

201. $\displaystyle\int \frac{\mathrm{d}x}{x^2\sqrt{Q^3}} = -\frac{1}{a^4}\left(\frac{\sqrt{Q}}{x} + \frac{x}{\sqrt{Q}}\right).$

202. $\displaystyle\int \frac{\mathrm{d}x}{x^3\sqrt{Q^3}} = -\frac{1}{2a^2x^2\sqrt{Q}} - \frac{3}{2a^4\sqrt{Q}} + \frac{3}{2a^5}\ln\frac{a+\sqrt{Q}}{x}.$

**Integrals containing the square root function $\sqrt{x^2 - a^2}$:**

$$\boxed{Q = x^2 - a^2,\ x > a > 0.}$$

203. $\displaystyle\int \sqrt{Q}\,\mathrm{d}x = \frac{1}{2}\left(x\sqrt{Q} - a^2\mathrm{arcosh}\frac{x}{a}\right) = \frac{1}{2}[x\sqrt{Q} - a^2(\ln(x+\sqrt{Q}) - \ln a)].$

204. $\displaystyle\int x\sqrt{Q}\,\mathrm{d}x = \frac{1}{3}\sqrt{Q^3}.$

205. $\displaystyle\int x^2\sqrt{Q}\,\mathrm{d}x = \frac{x}{4}\sqrt{Q^3} + \frac{a^2}{8}\left(x\sqrt{Q} - a^2\mathrm{arcosh}\frac{x}{a}\right)$

$$= \frac{x}{4}\sqrt{Q^3} + \frac{a^2}{8}[x\sqrt{Q} - a^2(\ln(x+\sqrt{Q}) - \ln a)].$$

206. $\displaystyle\int x^3\sqrt{Q}\,\mathrm{d}x = \frac{\sqrt{Q^5}}{5} + \frac{a^2\sqrt{Q^3}}{3}.$

207. $\displaystyle\int \frac{\sqrt{Q}}{x}\mathrm{d}x = \sqrt{Q} - a\arccos\frac{a}{x} = \sqrt{Q} - a[\ln(x+\sqrt{Q}) - \ln a].$

208. $$\int \frac{\sqrt{Q}}{x^2}\mathrm{d}x = -\frac{\sqrt{Q}}{x} + \operatorname{arcosh}\frac{x}{a} = -\frac{\sqrt{Q}}{x} + \ln(x+\sqrt{Q}) - \ln a.$$

209. $$\int \frac{\sqrt{Q}}{x^3}\mathrm{d}x = -\frac{\sqrt{Q}}{2x^2} + \frac{1}{2a}\arccos\frac{a}{x} = -\frac{\sqrt{Q}}{2x^2} + \frac{1}{2a}[\ln(x+\sqrt{Q}) - \ln a].$$

210. $$\int \frac{\mathrm{d}x}{\sqrt{Q}} = \operatorname{arcosh}\frac{x}{a} = \ln(x+\sqrt{Q}) - \ln a.$$

211. $$\int \frac{x\,\mathrm{d}x}{\sqrt{Q}} = \sqrt{Q}.$$

212. $$\int \frac{x^2\mathrm{d}x}{\sqrt{Q}} = \frac{x}{2}\sqrt{Q} + \frac{a^2}{2}\operatorname{arcosh}\frac{x}{a} = \frac{x}{2}\sqrt{Q} + \frac{a^2}{2}[\ln(x+\sqrt{Q}) - \ln a].$$

213. $$\int \frac{x^3\mathrm{d}x}{\sqrt{Q}} = \frac{\sqrt{Q^3}}{3} + a^2\sqrt{Q}.$$

214. $$\int \frac{\mathrm{d}x}{x\sqrt{Q}} = \frac{1}{a}\arccos\frac{a}{x}.$$

215. $$\int \frac{\mathrm{d}x}{x^2\sqrt{Q}} = \frac{\sqrt{Q}}{a^2x}.$$

216. $$\int \frac{\mathrm{d}x}{x^3\sqrt{Q}} = \frac{\sqrt{Q}}{2a^2x^2} + \frac{1}{2a^3}\arccos\frac{a}{x}.$$

217. $$\int \sqrt{Q^3}\,\mathrm{d}x = \frac{1}{4}\left(x\sqrt{Q^3} - \frac{3a^2x}{2}\sqrt{Q} + \frac{3a^4}{2}\operatorname{arcosh}\frac{x}{a}\right)$$
$$= \frac{1}{4}\left(x\sqrt{Q^3} - \frac{3a^2x}{2}\sqrt{Q} + \frac{3a^4}{2}[\ln(x+\sqrt{Q}) - \ln a]\right).$$

218. $$\int x\sqrt{Q^3}\,\mathrm{d}x = \frac{1}{5}\sqrt{Q^5}.$$

219. $$\int x^2\sqrt{Q^3}\,\mathrm{d}x = \frac{x\sqrt{Q^5}}{6} + \frac{a^2x\sqrt{Q^3}}{24} - \frac{a^4x\sqrt{Q}}{16} + \frac{a^6}{16}\operatorname{arcosh}\frac{x}{a}$$
$$= \frac{x\sqrt{Q^5}}{6} + \frac{a^2x\sqrt{Q^3}}{24} - \frac{a^4x\sqrt{Q}}{16} + \frac{a^6}{16}[\ln(x+\sqrt{Q}) - \ln a].$$

220. $$\int x^3\sqrt{Q^3}\,\mathrm{d}x = \frac{\sqrt{Q^7}}{7} + \frac{a^2\sqrt{Q^5}}{5}.$$

221. $$\int \frac{\sqrt{Q^3}}{x}\mathrm{d}x = \frac{\sqrt{Q^3}}{3} - a^2\sqrt{Q} + a^3\arccos\frac{a}{x}.$$

222. $$\int \frac{\sqrt{Q^3}}{x^2}\mathrm{d}x = -\frac{\sqrt{Q^3}}{2} + \frac{3}{2}x\sqrt{Q} - \frac{3}{2}a^2 \operatorname{arcosh}\frac{x}{a}$$
$$= -\frac{\sqrt{Q^3}}{2} + \frac{3}{2}x\sqrt{Q} - \frac{3}{2}a^2 \left[\ln(x + \sqrt{Q}) - \ln a\right].$$

223. $$\int \frac{\sqrt{Q^3}}{x^3}\mathrm{d}x = -\frac{\sqrt{Q^3}}{2x^2} + \frac{3\sqrt{Q}}{2} - \frac{3}{2}a \arccos\frac{a}{x}.$$

224. $$\int \frac{\mathrm{d}x}{\sqrt{Q^3}} = -\frac{x}{a^2\sqrt{Q}}.$$

225. $$\int \frac{x\mathrm{d}x}{\sqrt{Q^3}} = -\frac{1}{\sqrt{Q}}.$$

226. $$\int \frac{x^2\mathrm{d}x}{\sqrt{Q^3}} = -\frac{x}{\sqrt{Q}} + \operatorname{arcosh}\frac{x}{a} = -\frac{x}{\sqrt{Q}} + \ln(x + \sqrt{Q}) - \ln a.$$

227. $$\int \frac{x^3\mathrm{d}x}{\sqrt{Q^3}} = \sqrt{Q} - \frac{a^2}{\sqrt{Q}}.$$

228. $$\int \frac{\mathrm{d}x}{x\sqrt{Q^3}} = -\frac{1}{a^2\sqrt{Q}} - \frac{\operatorname{sgn}(x)}{a^3}\arccos\frac{a}{x}; \qquad \begin{array}{ll} \operatorname{sgn}(x) = 1 & \text{for } x > 0, \\ \operatorname{sgn}(x) = -1 & \text{for } x < 0. \end{array}$$ [72]

229. $$\int \frac{\mathrm{d}x}{x^2\sqrt{Q^3}} = -\frac{1}{a^4}\left(\frac{\sqrt{Q}}{x} + \frac{x}{\sqrt{Q}}\right).$$

230. $$\int \frac{\mathrm{d}x}{x^3\sqrt{Q^3}} = \frac{1}{2a^2x^2\sqrt{Q}} - \frac{3}{2a^4\sqrt{Q}} - \frac{3}{2a^5}\arccos\frac{a}{x}.$$

**Integrals containing the square root function $\sqrt{ax^2 + bx + c}$:**

$$\boxed{Q = ax^2 + bx + c,\ D = 4ac - b^2,\ d = \frac{4a}{D}.}$$

231. $$\int \frac{\mathrm{d}x}{\sqrt{Q}} = \begin{cases} \frac{1}{\sqrt{a}}\ln(2\sqrt{aQ} + 2ax + b) + C & \text{for } a > 0, \\ \frac{1}{\sqrt{a}}\operatorname{arsinh}\frac{2ax + b}{\sqrt{D}} + C_1 & \text{for } a > 0,\ D > 0, \\ \frac{1}{\sqrt{a}}\ln(2ax + b) & \text{for } a > 0,\ D = 0, \\ -\frac{1}{\sqrt{-a}}\arcsin\frac{2ax + b}{\sqrt{-D}} & \text{for } a < 0,\ D < 0. \end{cases}$$

[72] This integral is valid also for $x < 0$ if $|x| > a$.

232. $\displaystyle\int \frac{dx}{Q\sqrt{Q}} = \frac{2(2ax+b)}{D\sqrt{Q}}.$

233. $\displaystyle\int \frac{dx}{Q^2\sqrt{Q}} = \frac{2(2ax+b)}{3D\sqrt{Q}}\left(\frac{1}{Q}+2d\right).$

234. $\displaystyle\int \frac{dx}{Q^{(2n+1)/2}} = \frac{2(2ax+b)}{(2n-1)DQ^{(2n-1)/2}} + \frac{2d(n-1)}{2n-1}\int \frac{dx}{Q^{(2n-1)/2}}.$

235. $\displaystyle\int \sqrt{Q}\,dx = \frac{(2ax+b)\sqrt{Q}}{4a} + \frac{1}{2d}\int \frac{dx}{\sqrt{Q}}$ (see No. 231).

236. $\displaystyle\int Q\sqrt{Q}\,dx = \frac{(2ax+b)\sqrt{Q}}{8a}\left(Q+\frac{3}{2d}\right) + \frac{3}{8d^2}\int \frac{dx}{\sqrt{Q}}$ (see No. 231).

237. $\displaystyle\int Q^2\sqrt{Q}\,dx = \frac{(2ax+b)\sqrt{Q}}{12a}\left(Q^2+\frac{5Q}{4d}+\frac{15}{8d^2}\right) + \frac{5}{16d^3}\int \frac{dx}{\sqrt{Q}}$ (see No. 231).

238. $\displaystyle\int Q^{(2n+1)/2}dx = \frac{(2ax+b)Q^{(2n+1)/2}}{4a(n+1)} + \frac{2n+1}{2d(n+1)}\int Q^{(2n-1)/2}\,dx.$

239. $\displaystyle\int \frac{x\,dx}{\sqrt{Q}} = \frac{\sqrt{Q}}{a} - \frac{b}{2a}\int \frac{dx}{\sqrt{Q}}$ (see No. 231).

240. $\displaystyle\int \frac{x\,dx}{Q\sqrt{Q}} = -\frac{2(bx+2c)}{D\sqrt{Q}}.$

241. $\displaystyle\int \frac{x\,dx}{Q^{(2n+1)/2}} = -\frac{1}{(2n-1)aQ^{(2n-1)/2}} - \frac{b}{2a}\int \frac{dx}{Q^{(2n+1)/2}}$ (see No. 234).

242. $\displaystyle\int \frac{x^2\,dx}{\sqrt{Q}} = \left(\frac{x}{2a}-\frac{3b}{4a^2}\right)\sqrt{Q} + \frac{3b^2-4ac}{8a^2}\int \frac{dx}{\sqrt{Q}}$ (see No. 231).

243. $\displaystyle\int \frac{x^2dx}{Q\sqrt{Q}} = \frac{(2b^2-4ac)x+2bc}{aD\sqrt{Q}} + \frac{1}{a}\int \frac{dx}{\sqrt{Q}}$ (see No. 231).

244. $\displaystyle\int x\sqrt{Q}\,dx = \frac{Q\sqrt{Q}}{3a} - \frac{b(2ax+b)}{8a^2}\sqrt{Q} - \frac{b}{4ad}\int \frac{dx}{\sqrt{Q}}$ (see No. 231).

245. $\displaystyle\int xQ\sqrt{Q}\,dx = \frac{Q^2\sqrt{Q}}{5a} - \frac{b}{2a}\int Q\sqrt{Q}\,dx$ (see No. 236).

246. $\displaystyle\int xQ^{(2n+1)/2}dx = \frac{Q^{(2n+3)/2}}{(2n+3)a} - \frac{b}{2a}\int Q^{(2n+1)/2}dx$ (see No. 238).

247. $\displaystyle\int x^2\sqrt{Q}\,\mathrm{d}x = \left(x - \frac{5b}{6a}\right)\frac{Q\sqrt{Q}}{4a} + \frac{5b^2-4ac}{16a^2}\int\sqrt{Q}\,\mathrm{d}x$ (see No. 235).

248. $$\int\frac{\mathrm{d}x}{x\sqrt{Q}} = \begin{cases} \dfrac{1}{\sqrt{c}}\ln\dfrac{-2\sqrt{cQ}+2c+bx}{2x} & \text{for } c>0, \\ -\dfrac{1}{\sqrt{c}}\operatorname{arsinh}\dfrac{bx+2c}{x\sqrt{D}} & \text{for } c>0,\ D>0, \\ -\dfrac{1}{\sqrt{c}}\ln\dfrac{bx+2c}{x} & \text{for } c>0,\ D=0, \\ \dfrac{1}{\sqrt{-c}}\arcsin\dfrac{bx+2c}{x\sqrt{-D}} & \text{for } c<0,\ D<0. \end{cases}$$

249. $\displaystyle\int\frac{\mathrm{d}x}{x^2\sqrt{Q}} = -\frac{\sqrt{Q}}{cx} - \frac{b}{2c}\int\frac{\mathrm{d}x}{x\sqrt{Q}}$ (see No. 248).

250. $\displaystyle\int\frac{\sqrt{Q}\,\mathrm{d}x}{x} = \sqrt{Q} + \frac{b}{2}\int\frac{\mathrm{d}x}{\sqrt{Q}} + c\int\frac{\mathrm{d}x}{x\sqrt{Q}}$ (see No. 231 and 248).

251. $\displaystyle\int\frac{\sqrt{Q}\,\mathrm{d}x}{x^2} = -\frac{\sqrt{Q}}{x} + a\int\frac{\mathrm{d}x}{\sqrt{Q}} + \frac{b}{2}\int\frac{\mathrm{d}x}{x\sqrt{Q}}$ (see No. 231 and 250).

252. $\displaystyle\int\frac{Q^{(2n+1)/2}}{x}\mathrm{d}x = \frac{Q^{(2n+1)/2}}{2n+1} + \frac{b}{2}\int Q^{(2n-1)/2}\,\mathrm{d}x + c\int\frac{Q^{(2n-1)/2}}{x}\mathrm{d}x$
(see No. 238 and 248).

253. $\displaystyle\int\frac{\mathrm{d}x}{x\sqrt{ax^2+bx}} = -\frac{2}{bx}\sqrt{ax^2+bx}.$

254. $\displaystyle\int\frac{\mathrm{d}x}{\sqrt{2ax-x^2}} = \arcsin\frac{x-a}{a}.$

255. $\displaystyle\int\frac{x\,\mathrm{d}x}{\sqrt{2ax-x^2}} = -\sqrt{2ax-x^2} + a\arcsin\frac{x-a}{a}.$

256. $\displaystyle\int\sqrt{2ax-x^2}\,\mathrm{d}x = \frac{x-a}{2}\sqrt{2ax-x^2} + \frac{a^2}{2}\arcsin\frac{x-a}{a}.$

**Integrals containing the square root of other expressions:**

257. $$\int\frac{\mathrm{d}x}{(ax^2+b)\sqrt{cx^2+d}} = \begin{cases} \dfrac{1}{\sqrt{b}\sqrt{ad-bc}}\arctan\dfrac{x\sqrt{ad-bc}}{\sqrt{b}\sqrt{cx^2+d}} & (ad-bc>0), \\ \dfrac{1}{2\sqrt{b}\sqrt{bc-ad}}\ln\dfrac{\sqrt{b}\sqrt{cx^2+d}+x\sqrt{bc-ad}}{\sqrt{b}\sqrt{cx^2+d}-x\sqrt{bc-ad}} & (ad-bc<0). \end{cases}$$

258. $\displaystyle\int \sqrt[n]{ax+b}\,\mathrm{d}x = \frac{n(ax+b)}{(n+1)a}\sqrt[n]{ax+b}.$

259. $\displaystyle\int \frac{\mathrm{d}x}{\sqrt[n]{ax+b}} = \frac{n(ax+b)}{(n-1)a}\,\frac{1}{\sqrt[n]{ax+b}}.$

260. $\displaystyle\int \frac{\mathrm{d}x}{x\sqrt{x^n+a^2}} = -\frac{2}{na}\ln\frac{a+\sqrt{x^n+a^2}}{\sqrt{x^n}}.$

261. $\displaystyle\int \frac{\mathrm{d}x}{x\sqrt{x^n-a^2}} = \frac{2}{na}\arccos\frac{a}{\sqrt{x^n}}.$

262. $\displaystyle\int \frac{\sqrt{x}\,\mathrm{d}x}{\sqrt{a^3-x^3}} = \frac{2}{3}\arcsin\sqrt{\left(\frac{x}{a}\right)^3}.$

**Recursive formulas for the integrals of special polynomials:**

263.* $\displaystyle\int x^m(ax^n+b)^k\,\mathrm{d}x$

$$= \frac{1}{m+nk+1}\left[x^{m+1}(ax^n+b)^k + nkb\int x^m(ax^n+b)^{k-1}\,\mathrm{d}x\right]$$

$$= \frac{1}{bn(k+1)}\left[-x^{m+1}(ax^n+b)^{k+1} + (m+n+nk+1)\int x^m(ax^n+b)^{k+1}\,\mathrm{d}x\right]$$

$$= \frac{1}{(m+1)b}\left[x^{m+1}(ax^n+b)^{k+1} - a(m+n+nk+1)\int x^{m+n}(ax^n+b)^k\mathrm{d}x\right]$$

$$= \frac{1}{a(m+nk+1)}\left[x^{m-n+1}(ax^n+b)^{k+1} - (m-n+1)b\int x^{m-n}(ax^n+b)^k\mathrm{d}x\right].$$

### 0.9.5.3 Integrals of trigonometric functions[72]

**Integrals which contain the function $\sin\alpha x$ ($\alpha$ a real parameter):**

264.* $\displaystyle\int \sin\alpha x\,\mathrm{d}x = -\frac{1}{\alpha}\cos\alpha x.$

265.* $\displaystyle\int \sin^2\alpha x\,\mathrm{d}x = \frac{1}{2}x - \frac{1}{4\alpha}\sin 2\alpha x.$

266.* $\displaystyle\int \sin^3\alpha x\,\mathrm{d}x = -\frac{1}{\alpha}\cos\alpha x + \frac{1}{3\alpha}\cos^3\alpha x.$

267.* $\displaystyle\int \sin^4\alpha x\,\mathrm{d}x = \frac{3}{8}x - \frac{1}{4\alpha}\sin 2\alpha x + \frac{1}{32\alpha}\sin 4\alpha x.$

[73] For integrals of functions which contain, in addition to $\sin x$ and $\cos x$, also hyperbolic functions and $\mathrm{e}^{ax}$, see No. 428 ff.

268.* $\int \sin^n \alpha x \, dx = -\frac{\sin^{n-1} \alpha x \cos \alpha x}{n\alpha} + \frac{n-1}{n} \int \sin^{n-2} \alpha x \, dx \qquad (n \text{ an integer} > 0).$

269.* $\int x \sin \alpha x \, dx = \frac{\sin \alpha x}{\alpha^2} - \frac{x \cos \alpha x}{\alpha}.$

270.* $\int x^2 \sin \alpha x \, dx = \frac{2x}{\alpha^2} \sin \alpha x - \left( \frac{x^2}{\alpha} - \frac{2}{\alpha^3} \right) \cos \alpha x.$

271.* $\int x^3 \sin \alpha x \, dx = \left( \frac{3x^2}{\alpha^2} - \frac{6}{\alpha^4} \right) \sin \alpha x - \left( \frac{x^3}{\alpha} - \frac{6x}{\alpha^3} \right) \cos \alpha x.$

272.* $\int x^n \sin \alpha x \, dx = -\frac{x^n}{\alpha} \cos \alpha x + \frac{n}{\alpha} \int x^{n-1} \cos \alpha x \, dx \qquad (n > 0).$

273.* $\int \frac{\sin \alpha x}{x} \, dx = \alpha x - \frac{(\alpha x)^3}{3 \cdot 3!} + \frac{(\alpha x)^5}{5 \cdot 5!} - \frac{(\alpha x)^7}{7 \cdot 7!} + \ldots$

The integral $\int_0^x \frac{\sin t}{t} \, dt$ is referred to as the integral sine, denoted $\mathrm{Si}(x)$ (see section 0.5.5).

$\mathrm{Si}(x) = x - \frac{x^3}{3 \cdot 3!} + \frac{x^5}{5 \cdot 5!} - \frac{x^7}{7 \cdot 7!} + \ldots$

274. $\int \frac{\sin \alpha x}{x^2} \, dx = -\frac{\sin \alpha x}{x} + \alpha \int \frac{\cos \alpha x \, dx}{x}$ (see No. 312).

275. $\int \frac{\sin \alpha x}{x^n} \, dx = -\frac{1}{n-1} \frac{\sin \alpha x}{x^{n-1}} + \frac{\alpha}{n-1} \int \frac{\cos \alpha x}{x^{n-1}} dx$ (see No. 314).

276. $\int \frac{dx}{\sin \alpha x} = \int \operatorname{cosec} \alpha x \, dx = \frac{1}{\alpha} \ln \tan \frac{\alpha x}{2} = \frac{1}{\alpha} \ln (\operatorname{cosec} \alpha x - \cot \alpha x).$

277. $\int \frac{dx}{\sin^2 \alpha x} = -\frac{1}{\alpha} \cot \alpha x.$

278. $\int \frac{dx}{\sin^3 \alpha x} = -\frac{\cos \alpha x}{2\alpha \sin^2 \alpha x} + \frac{1}{2\alpha} \ln \tan \frac{\alpha x}{2}.$

279. $\int \frac{dx}{\sin^n \alpha x} = -\frac{1}{\alpha(n-1)} \frac{\cos \alpha x}{\sin^{n-1} \alpha x} + \frac{n-2}{n-1} \int \frac{dx}{\sin^{n-2} \alpha x} \qquad (n > 1).$

280. $$\int \frac{x \, dx}{\sin \alpha x} = \frac{1}{\alpha^2} \left( \alpha x + \frac{(\alpha x)^3}{3 \cdot 3!} + \frac{7(\alpha x)^5}{3 \cdot 5 \cdot 5!} + \frac{31(\alpha x)^7}{3 \cdot 7 \cdot 7!} + \frac{127(\alpha x)^9}{3 \cdot 5 \cdot 9!} + \ldots + \frac{2(2^{2n-1} - 1)}{(2n+1)!} B_{2n} (\alpha x)^{2n+1} + \ldots \right).$$

$B_{2n}$ are the Bernoulli numbers (see section 0.1.10.4).

281. $\int \frac{x \, dx}{\sin^2 \alpha x} = -\frac{x}{\alpha} \cot \alpha x + \frac{1}{\alpha^2} \ln \sin \alpha x.$

282. $$\int \frac{x \, dx}{\sin^n \alpha x} = -\frac{x \cos \alpha x}{(n-1) \alpha \sin^{n-1} \alpha x} - \frac{1}{(n-1)(n-2) \alpha^2 \sin^{n-2} \alpha x} + \frac{n-2}{n-1} \int \frac{x \, dx}{\sin^{n-2} \alpha x} \qquad (n > 2).$$

283. $\displaystyle\int \frac{\mathrm{d}x}{1+\sin\alpha x} = -\frac{1}{\alpha}\tan\left(\frac{\pi}{4}-\frac{\alpha x}{2}\right).$

284. $\displaystyle\int \frac{\mathrm{d}x}{1-\sin\alpha x} = \frac{1}{\alpha}\tan\left(\frac{\pi}{4}+\frac{\alpha x}{2}\right).$

285. $\displaystyle\int \frac{x\,\mathrm{d}x}{1+\sin\alpha x} = -\frac{x}{\alpha}\tan\left(\frac{\pi}{4}-\frac{\alpha x}{2}\right)+\frac{2}{\alpha^2}\ln\cos\left(\frac{\pi}{4}-\frac{\alpha x}{2}\right).$

286. $\displaystyle\int \frac{x\,\mathrm{d}x}{1-\sin\alpha x} = \frac{x}{\alpha}\cot\left(\frac{\pi}{4}-\frac{\alpha x}{2}\right)+\frac{2}{\alpha^2}\ln\sin\left(\frac{\pi}{4}-\frac{\alpha x}{2}\right).$

287. $\displaystyle\int \frac{\sin\alpha x\,\mathrm{d}x}{1\pm\sin\alpha x} = \pm x+\frac{1}{\alpha}\tan\left(\frac{\pi}{4}\mp\frac{\alpha x}{2}\right).$

288. $\displaystyle\int \frac{\mathrm{d}x}{\sin\alpha x(1\pm\sin\alpha x)} = \frac{1}{\alpha}\tan\left(\frac{\pi}{4}\mp\frac{\alpha x}{2}\right)+\frac{1}{\alpha}\ln\tan\frac{\alpha x}{2}.$

289. $\displaystyle\int \frac{\mathrm{d}x}{(1+\sin\alpha x)^2} = -\frac{1}{2\alpha}\tan\left(\frac{\pi}{4}-\frac{\alpha x}{2}\right)-\frac{1}{6\alpha}\tan^3\left(\frac{\pi}{4}-\frac{\alpha x}{2}\right).$

290. $\displaystyle\int \frac{\mathrm{d}x}{(1-\sin\alpha x)^2} = \frac{1}{2\alpha}\cot\left(\frac{\pi}{4}-\frac{\alpha x}{2}\right)+\frac{1}{6\alpha}\cot^3\left(\frac{\pi}{4}-\frac{\alpha x}{2}\right).$

291. $\displaystyle\int \frac{\sin\alpha x\,\mathrm{d}x}{(1+\sin\alpha x)^2} = -\frac{1}{2\alpha}\tan\left(\frac{\pi}{4}-\frac{\alpha x}{2}\right)+\frac{1}{6\alpha}\tan^3\left(\frac{\pi}{4}-\frac{\alpha x}{2}\right).$

292. $\displaystyle\int \frac{\sin\alpha x\,\mathrm{d}x}{(1-\sin\alpha x)^2} = -\frac{1}{2\alpha}\cot\left(\frac{\pi}{4}-\frac{\alpha x}{2}\right)+\frac{1}{6\alpha}\cot^3\left(\frac{\pi}{4}-\frac{\alpha x}{2}\right).$

293. $\displaystyle\int \frac{\mathrm{d}x}{1+\sin^2\alpha x} = \frac{1}{2\sqrt{2}\alpha}\arcsin\left(\frac{3\sin^2\alpha x-1}{\sin^2\alpha x+1}\right).$

294. $\displaystyle\int \frac{\mathrm{d}x}{1-\sin^2\alpha x} = \int\frac{\mathrm{d}\,x}{\cos^2\alpha x} = \frac{1}{\alpha}\tan\alpha x.$

295.* $\displaystyle\int \sin\alpha x\,\sin\beta x\,\mathrm{d}x = \frac{\sin(\alpha-\beta)\,x}{2(\alpha-\beta)}-\frac{\sin(\alpha+\beta)\,x}{2(\alpha+\beta)}$

($|\alpha|\neq|\beta|$; for $|\alpha|=|\beta|$ see No. 265).

296. $$\int \frac{\mathrm{d}x}{\beta+\gamma\sin\alpha x} = \begin{cases} \dfrac{2}{\alpha\sqrt{\beta^2-\gamma^2}}\arctan\dfrac{\beta\tan\alpha x/2+\gamma}{\sqrt{\beta^2-\gamma^2}} & \text{for } \beta^2>\gamma^2,\\[2ex] \dfrac{1}{\alpha\sqrt{\gamma^2-\beta^2}}\ln\dfrac{\beta\tan\alpha x/2+\gamma-\sqrt{\gamma^2-\beta^2}}{\beta\tan\alpha x/2+\gamma+\sqrt{\gamma^2-\beta^2}} & \text{for } \beta^2<\gamma^2. \end{cases}$$

297. $\displaystyle\int \frac{\sin\alpha x\,\mathrm{d}x}{\beta+\gamma\sin\alpha x} = \frac{x}{\gamma}-\frac{\beta}{\gamma}\int\frac{\mathrm{d}x}{\beta+\gamma\sin\alpha x}$ (see No. 296).

298. $\displaystyle\int \frac{\mathrm{d}x}{\sin\alpha x(\beta+\gamma\sin\alpha x)} = \frac{1}{\alpha\beta}\ln\tan\frac{\alpha\beta}{2}-\frac{\gamma}{\beta}\int\frac{\mathrm{d}x}{\beta+\gamma\sin\alpha x}$ (see No. 296).

299. $\displaystyle\int \frac{\mathrm{d}x}{(\beta+\gamma\sin\alpha x)^2} = \frac{\gamma\cos\alpha x}{\alpha(\beta^2-\gamma^2)(\beta+\gamma\sin\alpha x)}+\frac{\beta}{\beta^2-\gamma^2}\int\frac{\mathrm{d}x}{\beta+\gamma\sin\alpha x}$
(see No. 296).

300. $\displaystyle\int \frac{\sin\alpha x\,\mathrm{d}x}{(\beta+\gamma\sin\alpha x)^2} = \frac{\beta\cos\alpha x}{\alpha(\gamma^2-\beta^2)(\beta+\gamma\sin\alpha x)}+\frac{\gamma}{\gamma^2-\beta^2}\int\frac{\mathrm{d}x}{\beta+\gamma\sin\alpha x}$
(see No. 296).

301. $$\int \frac{\mathrm{d}x}{\beta^2 + \gamma^2 \sin^2 \alpha x} = \frac{1}{\alpha\beta\sqrt{\beta^2+\gamma^2}} \arctan \frac{\sqrt{\beta^2+\gamma^2}\tan\alpha x}{\beta} \qquad (\beta > 0).$$

302. $$\int \frac{\mathrm{d}x}{\beta^2 - \gamma^2 \sin^2 \alpha x} = \begin{cases} \dfrac{1}{\alpha\beta\sqrt{\beta^2-\gamma^2}} \arctan \dfrac{\sqrt{\beta^2-\gamma^2}\tan\alpha x}{\beta}, & \beta^2 > \gamma^2, \beta > 0, \\ \dfrac{1}{2\alpha\beta\sqrt{\gamma^2-\beta^2}} \ln \dfrac{\sqrt{\gamma^2-\beta^2}\tan\alpha x + \beta}{\sqrt{\gamma^2-\beta^2}\tan\alpha x - \beta}, & \gamma^2 > \beta^2, \beta > 0. \end{cases}$$

**Integrals which contain the function $\cos \alpha x$:**

303.* $$\int \cos \alpha x \,\mathrm{d}x = \frac{1}{\alpha} \sin \alpha x.$$

304.* $$\int \cos^2 \alpha x \,\mathrm{d}x = \frac{1}{2}x + \frac{1}{4\alpha} \sin 2\alpha x.$$

305.* $$\int \cos^3 \alpha x \,\mathrm{d}x = \frac{1}{\alpha} \sin \alpha x - \frac{1}{3\alpha} \sin^3 \alpha x.$$

306.* $$\int \cos^4 \alpha x \,\mathrm{d}x = \frac{3}{8}x + \frac{1}{4\alpha} \sin 2\alpha x + \frac{1}{32\alpha} \sin 4\alpha x.$$

307.* $$\int \cos^n \alpha x \,\mathrm{d}x = \frac{\cos^{n-1} \alpha x \sin \alpha x}{n\alpha} + \frac{n-1}{n} \int \cos^{n-2} \alpha x \,\mathrm{d}x \qquad (n \in \mathbb{N}).$$

308.* $$\int x \cos \alpha x \,\mathrm{d}x = \frac{\cos \alpha x}{\alpha^2} + \frac{x \sin \alpha x}{\alpha}.$$

309.* $$\int x^2 \cos \alpha x \,\mathrm{d}x = \frac{2x}{\alpha^2} \cos \alpha x + \left(\frac{x^2}{\alpha} - \frac{2}{\alpha^3}\right) \sin \alpha x.$$

310.* $$\int x^3 \cos \alpha x \,\mathrm{d}x = \left(\frac{3x^2}{\alpha^2} - \frac{6}{\alpha^4}\right) \cos \alpha x + \left(\frac{x^3}{\alpha} - \frac{6x}{\alpha^3}\right) \sin \alpha x.$$

311.* $$\int x^n \cos \alpha x \,\mathrm{d}x = \frac{x^n \sin \alpha x}{\alpha} - \frac{n}{\alpha} \int x^{n-1} \sin \alpha x \,\mathrm{d}x \qquad (n \in \mathbb{N}).$$

312. $$\int \frac{\cos \alpha x}{x} \mathrm{d}x = \ln(\alpha x) - \frac{(\alpha x)^2}{2 \cdot 2!} + \frac{(\alpha x)^4}{4 \cdot 4!} - \frac{(\alpha x)^6}{6 \cdot 6!} + \ldots$$

The improper integral $\int_x^\infty \frac{\cos t}{t} \mathrm{d}t$ is called the *integral cosine*, denoted $\mathrm{Ci}(x)$ (see section 0.5.5).

$$\mathrm{Ci}(x) = \mathrm{C} + \ln x - \frac{x^2}{2 \cdot 2!} + \frac{x^4}{4 \cdot 4!} - \frac{x^6}{6 \cdot 6!} + \ldots$$

C denotes the Euler constant (see section 0.1.1).

313. $$\int \frac{\cos \alpha x}{x^2} \mathrm{d}x = -\frac{\cos \alpha x}{x} - \alpha \int \frac{\sin \alpha x \,\mathrm{d}x}{x} \qquad \text{(see No. 273).}$$

314. $$\int \frac{\cos \alpha x}{x^n} \mathrm{d}x = -\frac{\cos \alpha x}{(n-1)x^{n-1}} - \frac{\alpha}{n-1} \int \frac{\sin \alpha x \,\mathrm{d}x}{x^{n-1}} \qquad (n \neq 1, \quad \text{see No. 275}).$$

315. $$\int \frac{\mathrm{d}x}{\cos \alpha x} = \int \sec \alpha x \,\mathrm{d}x = \frac{1}{\alpha} \ln \tan\left(\frac{\alpha x}{2} + \frac{\pi}{4}\right) = \frac{1}{\alpha} \ln(\sec \alpha x + \tan \alpha x).$$

316. $\int \frac{dx}{\cos^2 \alpha x} = \frac{1}{\alpha} \tan \alpha x.$

317. $\int \frac{dx}{\cos^3 \alpha x} = \frac{\sin \alpha x}{2\alpha \cos^2 \alpha x} + \frac{1}{2\alpha} \ln \tan\left(\frac{\pi}{4} + \frac{\alpha x}{2}\right).$

318. $\int \frac{dx}{\cos^n \alpha x} = \frac{1}{\alpha(n-1)} \frac{\sin \alpha x}{\cos^{n-1} \alpha x} + \frac{n-2}{n-1} \int \frac{dx}{\cos^{n-2} \alpha x} \quad (n > 1).$

319. $$\int \frac{x\,dx}{\cos \alpha x} = \frac{1}{\alpha^2}\left(\frac{(\alpha x)^2}{2} + \frac{(\alpha x)^4}{4 \cdot 2!} + \frac{5(\alpha x)^6}{6 \cdot 4!} + \frac{61(\alpha x)^8}{8 \cdot 6!} + \frac{1\,385(\alpha x)^{10}}{10 \cdot 8!} + \ldots + \frac{E_{2n}(\alpha x)^{2n+2}}{(2n+2)(2n)!} + \ldots\right).$$

$E_{2n}$ are the Euler numbers (see section 0.1.10.5).

320. $\int \frac{x\,dx}{\cos^2 \alpha x} = \frac{x}{\alpha} \tan \alpha x + \frac{1}{\alpha^2} \ln \cos \alpha x.$

321. $$\int \frac{x\,dx}{\cos^n \alpha x} = \frac{x \sin \alpha x}{(n-1)\alpha \cos^{n-1} \alpha x} - \frac{1}{(n-1)(n-2)\alpha^2 \cos^{n-2} \alpha x} + \frac{n-2}{n-1} \int \frac{x\,dx}{\cos^{n-2} \alpha x} \quad (n > 2).$$

322. $\int \frac{dx}{1 + \cos \alpha x} = \frac{1}{\alpha} \cot\left(\frac{\alpha x}{2}\right).$

323. $\int \frac{dx}{1 - \cos \alpha x} = -\frac{1}{\alpha} \cot \frac{\alpha x}{2}.$

324. $\int \frac{x\,dx}{1 + \cos \alpha x} = \frac{x}{\alpha} \tan \frac{\alpha x}{2} + \frac{2}{\alpha^2} \ln \cos \frac{\alpha x}{2}.$

325. $\int \frac{x\,dx}{1 - \cos \alpha x} = -\frac{x}{\alpha} \cot \frac{\alpha x}{2} + \frac{2}{\alpha^2} \ln \sin \frac{\alpha x}{2}.$

326. $\int \frac{\cos \alpha x\,dx}{1 + \cos \alpha x} = x - \frac{1}{\alpha} \tan \frac{\alpha x}{2}.$

327. $\int \frac{\cos \alpha x\,dx}{1 - \cos \alpha x} = -x - \frac{1}{\alpha} \cot \frac{\alpha x}{2}.$

328. $\int \frac{dx}{\cos \alpha x(1 + \cos \alpha x)} = \frac{1}{\alpha} \ln \tan\left(\frac{\pi}{4} + \frac{\alpha x}{2}\right) - \frac{1}{\alpha} \tan \frac{\alpha x}{2}.$

329. $\int \frac{dx}{\cos \alpha x(1 - \cos \alpha x)} = \frac{1}{\alpha} \ln \tan\left(\frac{\pi}{4} + \frac{\alpha x}{2}\right) - \frac{1}{\alpha} \cot \frac{\alpha x}{2}.$

330. $\int \frac{dx}{(1 + \cos \alpha x)^2} = \frac{1}{2\alpha} \tan \frac{\alpha x}{2} + \frac{1}{6\alpha} \tan^3 \frac{\alpha x}{2}.$

331. $\int \frac{dx}{(1 - \cos \alpha x)^2} = -\frac{1}{2\alpha} \cot \frac{\alpha x}{2} - \frac{1}{6\alpha} \cot^3 \frac{\alpha x}{2}.$

332. $\int \frac{\cos \alpha x\,dx}{(1 + \cos \alpha x)^2} = \frac{1}{2\alpha} \tan \frac{\alpha x}{2} - \frac{1}{6\alpha} \tan^3 \frac{\alpha x}{2}.$

333. $\int \frac{\cos \alpha x\,dx}{(1 - \cos \alpha x)^2} = \frac{1}{2\alpha} \cot \frac{\alpha x}{2} - \frac{1}{6\alpha} \cot^3 \frac{\alpha x}{2}.$

334. $$\int \frac{dx}{1+\cos^2 \alpha x} = \frac{1}{2\sqrt{2}\alpha} \arcsin\left(\frac{1-3\cos^2 \alpha x}{1+\cos^2 \alpha x}\right).$$

335. $$\int \frac{dx}{1-\cos^2 \alpha x} = \frac{dx}{\sin^2 \alpha x} = -\frac{1}{\alpha}\cot \alpha x.$$

336.* $$\int \cos\alpha x \cos \beta x \, dx = \frac{\sin(\alpha-\beta)x}{2(\alpha-\beta)} + \frac{\sin(\alpha+\beta)x}{2(\alpha+\beta)} \quad (|\alpha| \neq |\beta|;$$
for $|\alpha| = |\beta|$ see No. 304).

337. $$\int \frac{dx}{\beta+\gamma\cos\alpha x} = \begin{cases} \dfrac{2}{\alpha\sqrt{\beta^2-\gamma^2}} \arctan \dfrac{(\beta-\gamma)\tan\alpha x/2}{\sqrt{\beta^2-\gamma^2}} & \text{for } \beta^2 > \gamma^2, \\ \dfrac{1}{\alpha\sqrt{\gamma^2-\beta^2}} \ln \dfrac{(\gamma-\beta)\tan \alpha x/2 + \sqrt{\gamma^2-\beta^2}}{(\gamma-\beta)\tan\alpha x/2 - \sqrt{\gamma^2-\beta^2}} & \text{for } \beta^2 < \gamma^2. \end{cases}$$

338. $$\int \frac{\cos\alpha x \, dx}{\beta+\gamma\cos\alpha x} = \frac{x}{\gamma} - \frac{\beta}{\gamma}\int \frac{dx}{\beta+\gamma\cos\alpha x} \qquad \text{(see No. 337).}$$

339. $$\int \frac{dx}{\cos\alpha x(\beta+\gamma\cos\alpha x)} = \frac{1}{\alpha\beta}\ln\tan\left(\frac{\alpha x}{2}+\frac{\pi}{4}\right) - \frac{\gamma}{\beta}\int\frac{dx}{\beta+\gamma\cos\alpha x}$$
(see No. 337).

340. $$\int \frac{dx}{(\beta+\gamma\cos\alpha x)^2} = \frac{\gamma\sin\alpha x}{\alpha(\gamma^2-\beta^2)(\beta+\gamma\cos\alpha x)} - \frac{\beta}{\gamma^2-\beta^2}\int\frac{dx}{\beta+\gamma\cos\alpha x}$$
(see No. 337).

341. $$\int \frac{\cos\alpha x\,dx}{(\beta+\gamma\cos\alpha x)^2} = \frac{\beta\sin\alpha x}{\alpha(\beta^2-\gamma^2)(\beta+\gamma\cos\alpha x)} - \frac{\gamma}{\beta^2-\gamma^2}\int\frac{dx}{\beta+\gamma\cos\alpha x}$$
(see No. 337).

342. $$\int \frac{dx}{\beta^2+\gamma^2\cos^2\alpha x} = \frac{1}{\alpha\beta\sqrt{\beta^2+\gamma^2}}\arctan\frac{\beta\tan\alpha x}{\sqrt{\beta^2+\gamma^2}} \quad (\beta > 0).$$

343. $$\int \frac{dx}{\beta^2-\gamma^2\cos^2\alpha x} = \begin{cases} \dfrac{1}{\alpha\beta\sqrt{\beta^2-\gamma^2}}\arctan\dfrac{\beta\tan\alpha x}{\sqrt{\beta^2-\gamma^2}}, & \beta^2>\gamma^2,\ \beta>0, \\ \dfrac{1}{2\alpha\beta\sqrt{\gamma^2-\beta^2}}\ln\dfrac{\beta\tan\alpha x - \sqrt{\gamma^2-\beta^2}}{\beta\tan\alpha x + \sqrt{\gamma^2-\beta^2}}, & \gamma^2>\beta^2,\ \beta>0. \end{cases}$$

**Integrals which contain the functions $\sin\alpha x$ and $\cos\alpha x$:**

344.* $$\int \sin\alpha x\cos\alpha x\,dx = \frac{1}{2\alpha}\sin^2\alpha x.$$

345.* $$\int \sin^2\alpha x\cos^2\alpha x\,dx = \frac{x}{8} - \frac{\sin 4\alpha x}{32\alpha}.$$

346.* $$\int \sin^n\alpha x\cos\alpha x\,dx = \frac{1}{\alpha(n+1)}\sin^{n+1}\alpha x \qquad (n \in \mathbb{N}, \text{ see No. 358}).$$

347.* $\int \sin \alpha x \cos^n \alpha x \,\mathrm{d}x = -\frac{1}{\alpha(n+1)} \cos^{n+1} \alpha x \quad (n \in \mathbb{N}, \text{ see No. 357}).$

348.* $$\int \sin^n \alpha x \cos^m \alpha x \,\mathrm{d}x = -\frac{\sin^{n-1} \alpha x \cos^{m-1} \alpha x}{\alpha(n+m)} + \frac{n-1}{n+m} \int \sin^{n-2} \alpha x \cos^m \alpha x \,\mathrm{d}x$$
$$= \frac{\sin^{n+1} \alpha x \cos^{m-1} \alpha x}{\alpha(n+m)} + \frac{m-1}{n+m} \int \sin^n \alpha x \cos^{m-2} \alpha x \,\mathrm{d}x$$
$(m, n \in \mathbb{N}; n > 0;$ see No. 359, No. 370, No. 381).

349. $\int \frac{\mathrm{d}x}{\sin \alpha x \cos \alpha x} = \frac{1}{\alpha} \ln \tan \alpha x.$

350. $\int \frac{\mathrm{d}x}{\sin^2 \alpha x \cos \alpha x} = \frac{1}{\alpha} \left[ \ln \tan \left( \frac{\pi}{4} + \frac{\alpha x}{2} \right) - \frac{1}{\sin \alpha x} \right].$

351. $\int \frac{\mathrm{d}x}{\sin \alpha x \cos^2 \alpha x} = \frac{1}{\alpha} \left( \ln \tan \frac{\alpha x}{2} + \frac{1}{\cos \alpha x} \right).$

352. $\int \frac{\mathrm{d}x}{\sin^3 \alpha x \cos \alpha x} = \frac{1}{\alpha} \left( \ln \tan \alpha x - \frac{1}{2 \sin^2 \alpha x} \right).$

353. $\int \frac{\mathrm{d}x}{\sin \alpha x \cos^3 \alpha x} = \frac{1}{\alpha} \left( \ln \tan \alpha x + \frac{1}{2 \cos^2 \alpha x} \right).$

354. $\int \frac{\mathrm{d}x}{\sin^2 \alpha x \cos^2 \alpha x} = -\frac{2}{\alpha} \cot 2\alpha x.$

355. $\int \frac{\mathrm{d}x}{\sin^2 \alpha x \cos^3 \alpha x} = \frac{1}{\alpha} \left[ \frac{\sin \alpha x}{2 \cos^2 \alpha x} - \frac{1}{\sin \alpha x} + \frac{3}{2} \ln \tan \left( \frac{\pi}{4} + \frac{\alpha x}{2} \right) \right].$

356. $\int \frac{\mathrm{d}x}{\sin^3 \alpha x \cos^2 \alpha x} = \frac{1}{\alpha} \left( \frac{1}{\cos \alpha x} - \frac{\cos \alpha x}{2 \sin^2 \alpha x} + \frac{3}{2} \ln \tan \frac{\alpha x}{2} \right).$

357. $\int \frac{\mathrm{d}x}{\sin \alpha x \cos^n \alpha x} = \frac{1}{\alpha(n-1) \cos^{n-1} \alpha x} + \int \frac{\mathrm{d}x}{\sin \alpha x \cos^{n-2} \alpha x}$
$(n \neq 1$, see No. 347, No. 351, No. 353).

358. $\int \frac{\mathrm{d}x}{\sin^n \alpha x \cos \alpha x} = -\frac{1}{\alpha(n-1) \sin^{n-1} \alpha x} + \int \frac{\mathrm{d}x}{\sin^{n-2} \alpha x \cos \alpha x}$
$(n \neq 1$, see No. 346, No. 350, No. 352).

359. $$\int \frac{\mathrm{d}x}{\sin^n \alpha x \cos^m \alpha x} = -\frac{1}{\alpha(n-1)} \frac{1}{\sin^{n-1} \alpha x \cos^{m-1} \alpha x}$$
$$+ \frac{n+m-2}{n-1} \int \frac{\mathrm{d}x}{\sin^{n-2} \alpha x \cos^m \alpha x}$$
$$= \frac{1}{\alpha(m-1)} \frac{1}{\sin^{n-1} \alpha x \cos^{m-1} \alpha x}$$
$$+ \frac{n+m-2}{m-1} \int \frac{\mathrm{d}x}{\sin^n \alpha x \cos^{m-2} \alpha x}$$
$(m, n \in \mathbb{N}; n > 0;$ see No. 348, No. 370, No. 381).

360. $\displaystyle\int \frac{\sin \alpha x \,\mathrm{d}x}{\cos^2 \alpha x} = \frac{1}{\alpha \cos \alpha x} = \frac{1}{\alpha} \sec \alpha x.$

361. $\displaystyle\int \frac{\sin \alpha x \,\mathrm{d}x}{\cos^3 \alpha x} = \frac{1}{2\alpha \cos^2 \alpha x} = \frac{1}{2\alpha} \tan^2 \alpha x + \frac{1}{2\alpha}.$

362. $\displaystyle\int \frac{\sin \alpha x \,\mathrm{d}x}{\cos^n \alpha x} = \frac{1}{\alpha(n-1) \cos^{n-1} \alpha x}.$

363. $\displaystyle\int \frac{\sin^2 \alpha x \,\mathrm{d}x}{\cos \alpha x} = -\frac{1}{\alpha} \sin \alpha x + \frac{1}{\alpha} \ln \tan \left(\frac{\pi}{4} + \frac{\alpha x}{2}\right).$

364. $\displaystyle\int \frac{\sin^2 \alpha x \,\mathrm{d}x}{\cos^3 \alpha x} = \frac{1}{\alpha} \left[\frac{\sin \alpha x}{2 \cos^2 \alpha x} - \frac{1}{2} \ln \tan \left(\frac{\pi}{4} + \frac{\alpha x}{2}\right)\right].$

365. $\displaystyle\int \frac{\sin^2 \alpha x \,\mathrm{d}x}{\cos^n \alpha x} = \frac{\sin \alpha x}{\alpha(n-1) \cos^{n-1} \alpha x} - \frac{1}{n-1} \int \frac{\mathrm{d}x}{\cos^{n-2} \alpha x}$

$(n \in \mathbb{N}, n > 1,$ see No. 315, No. 316, No. 318).

366. $\displaystyle\int \frac{\sin^3 \alpha x \,\mathrm{d}x}{\cos \alpha x} = -\frac{1}{\alpha} \left(\frac{\sin^2 \alpha x}{2} + \ln \cos \alpha x\right).$

367. $\displaystyle\int \frac{\sin^3 \alpha x \,\mathrm{d}x}{\cos^2 \alpha x} = \frac{1}{\alpha} \left(\cos \alpha x + \frac{1}{\cos \alpha x}\right).$

368. $\displaystyle\int \frac{\sin^3 \alpha x \,\mathrm{d}x}{\cos^n \alpha x} = \frac{1}{\alpha} \left[\frac{1}{(n-1) \cos^{n-1} \alpha x} - \frac{1}{(n-3) \cos^{n-3} \alpha x}\right] \qquad (n \in \mathbb{N},\ n > 3).$

369. $\displaystyle\int \frac{\sin^n \alpha x}{\cos \alpha x} \mathrm{d}x = -\frac{\sin^{n-1} \alpha x}{\alpha(n-1)} + \int \frac{\sin^{n-2} \alpha x \,\mathrm{d}x}{\cos \alpha x} \qquad (n \in \mathbb{N}, n > 1).$

370. $\displaystyle\int \frac{\sin^n \alpha x}{\cos^m \alpha x} \mathrm{d}x = \frac{\sin^{n+1} \alpha x}{\alpha(m-1) \cos^{m-1} \alpha x} - \frac{n-m+2}{m-1} \int \frac{\sin^n \alpha x}{\cos^{m-2} \alpha x} \mathrm{d}x$

$(m, n \in \mathbb{N};\ m > 1),$

$$= -\frac{\sin^{n-1} \alpha x}{\alpha(n-m) \cos^{m-1} \alpha x} + \frac{n-1}{n-m} \int \frac{\sin^{n-2} \alpha x \,\mathrm{d}x}{\cos^m \alpha x}$$

$(m \neq n,$ see No. 348, No. 359, No. 381),

$$= \frac{\sin^{n-1} \alpha x}{\alpha(m-1) \cos^{m-1} \alpha x} - \frac{n-1}{m-1} \int \frac{\sin^{n-2} \alpha x \,\mathrm{d}x}{\cos^{m-2} \alpha x}$$

$(m, n \in \mathbb{N};\ m > 1).$

371. $\displaystyle\int \frac{\cos \alpha x \,\mathrm{d}x}{\sin^2 \alpha x} = -\frac{1}{\alpha \sin \alpha x} = -\frac{1}{\alpha} \operatorname{cosec} \alpha x.$

372. $\displaystyle\int \frac{\cos \alpha x \,\mathrm{d}x}{\sin^3 \alpha x} = -\frac{1}{2\alpha \sin^2 \alpha x} = -\frac{\cot^2 \alpha x}{2\alpha} - \frac{1}{2\alpha}.$

373. $\displaystyle\int \frac{\cos \alpha x \,\mathrm{d}x}{\sin^n \alpha x} = -\frac{1}{\alpha(n-1) \sin^{n-1} \alpha x}.$

374. $\displaystyle\int \frac{\cos^2 \alpha x \,\mathrm{d}x}{\sin \alpha x} = \frac{1}{\alpha} \left(\cos \alpha x + \ln \tan \frac{\alpha x}{2}\right).$

375. $\int \frac{\cos^2 \alpha x \, dx}{\sin^3 \alpha x} = -\frac{1}{2\alpha}\left(\frac{\cos \alpha x}{\sin^2 \alpha x} + \ln \tan \frac{\alpha x}{2}\right).$

376. $\int \frac{\cos^2 \alpha x \, dx}{\sin^n \alpha x} = -\frac{1}{(n-1)}\left(\frac{\cos \alpha x}{\alpha \sin^{n-1} \alpha x} + \int \frac{dx}{\sin^{n-2} \alpha x}\right)$
$(n \in \mathbb{N},\ n > 1$, see No. 279).

377. $\int \frac{\cos^3 \alpha x \, dx}{\sin \alpha x} = \frac{1}{\alpha}\left(\frac{\cos^2 \alpha x}{2} + \ln \sin \alpha x\right).$

378. $\int \frac{\cos^3 \alpha x \, dx}{\sin^2 \alpha x} = -\frac{1}{\alpha}\left(\sin \alpha x + \frac{1}{\sin \alpha x}\right).$

379. $\int \frac{\cos^3 \alpha x \, dx}{\sin^n \alpha x} = \frac{1}{\alpha}\left[\frac{1}{(n-3)\sin^{n-3} \alpha x} - \frac{1}{(n-1)\sin^{n-1} \alpha x}\right] \quad (n \in \mathbb{N},\ n > 3).$

380. $\int \frac{\cos^n \alpha x}{\sin \alpha x} dx = \frac{\cos^{n-1} \alpha x}{\alpha(n-1)} + \int \frac{\cos^{n-2} \alpha x \, dx}{\sin \alpha x} \quad (n \neq 1).$

381. $\int \frac{\cos^n \alpha x \, dx}{\sin^m \alpha x} = -\frac{\cos^{n+1} \alpha x}{\alpha(m-1)\sin^{m-1} \alpha x} - \frac{n-m+2}{m-1}\int \frac{\cos^n \alpha x \, dx}{\sin^{m-2} \alpha x}$

$(m, n \in \mathbb{N};\ m > 1),$

$$= \frac{\cos^{n-1} \alpha x}{\alpha(n-m)\sin^{m-1} \alpha x} + \frac{n-1}{n-m}\int \frac{\cos^{n-2} \alpha x \, dx}{\sin^m \alpha x}$$

$(m \neq n,$ see No. 348, No. 359, No. 370),

$$= -\frac{\cos^{n-1} \alpha x}{\alpha(m-1)\sin^{m-1} \alpha x} - \frac{n-1}{m-1}\int \frac{\cos^{n-2} \alpha x \, dx}{\sin^{m-2} \alpha x}$$

$(m, n \in \mathbb{N};\ m > 1).$

382. $\int \frac{dx}{\sin \alpha x(1 \pm \cos \alpha x)} = \pm\frac{1}{2\alpha(1 \pm \cos \alpha x)} + \frac{1}{2\alpha} \ln \tan \frac{\alpha x}{2}.$

383. $\int \frac{dx}{\cos \alpha x(1 \pm \sin \alpha x)} = \mp\frac{1}{2\alpha(1 \pm \sin \alpha x)} + \frac{1}{2\alpha} \ln \tan \left(\frac{\pi}{4} + \frac{\alpha x}{2}\right).$

384. $\int \frac{\sin \alpha x \, dx}{\cos \alpha x(1 \pm \cos \alpha x)} = \frac{1}{\alpha} \ln \frac{1 \pm \cos \alpha x}{\cos \alpha x}.$

385. $\int \frac{\cos \alpha x \, dx}{\sin \alpha x(1 \pm \sin \alpha x)} = -\frac{1}{\alpha} \ln \frac{1 \pm \sin \alpha x}{\sin \alpha x}.$

386. $\int \frac{\sin \alpha x \, dx}{\cos \alpha x(1 \pm \sin \alpha x)} = \frac{1}{2\alpha(1 \pm \sin \alpha x)} \pm \frac{1}{2\alpha} \ln \tan \left(\frac{\pi}{4} + \frac{\alpha x}{2}\right).$

387. $\int \frac{\cos \alpha x \, dx}{\sin \alpha x(1 \pm \cos \alpha x)} = -\frac{1}{2\alpha(1 \pm \cos \alpha x)} \pm \frac{1}{2\alpha} \ln \tan \frac{\alpha x}{2}.$

388. $\int \frac{\sin \alpha x \, dx}{\sin \alpha x \pm \cos \alpha x} = \frac{x}{2} \mp \frac{1}{2\alpha} \ln(\sin \alpha x \pm \cos \alpha x).$

389. $\int \frac{\cos \alpha x \, dx}{\sin \alpha x \pm \cos \alpha x} = \pm\frac{x}{2} + \frac{1}{2\alpha} \ln(\sin \alpha x \pm \cos \alpha x).$

390. $$\int \frac{\mathrm{d}x}{\sin \alpha x \pm \cos \alpha x} = \frac{1}{\alpha\sqrt{2}} \ln \tan \left(\frac{\alpha x}{2} \pm \frac{\pi}{8}\right).$$

391. $$\int \frac{\mathrm{d}x}{1 + \cos \alpha x \pm \sin \alpha x} = \pm\frac{1}{\alpha} \ln \left(1 \pm \tan \frac{\alpha x}{2}\right).$$

392. $$\int \frac{\mathrm{d}x}{\beta \sin \alpha x + \gamma \cos \alpha x} = \frac{1}{\alpha\sqrt{\beta^2 + \gamma^2}} \ln \tan \frac{\alpha x + \phi}{2}$$

$$\text{with} \quad \sin \phi = \frac{\gamma}{\sqrt{\beta^2 + \gamma^2}} \quad \text{and} \quad \tan \phi = \frac{\gamma}{\beta}.$$

393. $$\int \frac{\sin \alpha x \,\mathrm{d}x}{\beta + \gamma \cos \alpha x} = -\frac{1}{\alpha\gamma} \ln(\beta + \gamma \cos \alpha x).$$

394. $$\int \frac{\cos \alpha x \,\mathrm{d}x}{\beta + \gamma \sin \alpha x} = \frac{1}{\alpha\gamma} \ln(\beta + \gamma \sin \alpha x).$$

395. $$\int \frac{\mathrm{d}x}{\beta + \gamma \cos \alpha x + \delta \sin \alpha x} = \int \frac{\mathrm{d}\left(x + \frac{\phi}{\alpha}\right)}{\beta + \sqrt{\gamma^2 + \delta^2}\,\sin(\alpha x + \phi)}$$

$$\text{with} \quad \sin \phi = \frac{\gamma}{\sqrt{\gamma^2 + \delta^2}} \quad \text{and} \quad \tan \phi = \frac{\gamma}{\delta} \qquad \text{(see No. 296).}$$

396. $$\int \frac{\mathrm{d}x}{\beta^2 \cos^2 \alpha x + \gamma^2 \sin^2 \alpha x} = \frac{1}{\alpha\beta\gamma} \arctan \left(\frac{\gamma}{\beta} \tan \alpha x\right).$$

397. $$\int \frac{\mathrm{d}x}{\beta^2 \cos^2 \alpha x - \gamma^2 \sin^2 \alpha x} = \frac{1}{2\alpha\beta\gamma} \ln \frac{\gamma \tan \alpha x + \beta}{\gamma \tan \alpha x - \beta}.$$

398. $$\int \sin \alpha x \cos \beta x \,\mathrm{d}x = -\frac{\cos(\alpha + \beta)\, x}{2(\alpha + \beta)} - \frac{\cos(\alpha - \beta)\, x}{2(\alpha - \beta)}$$

$$(\alpha^2 \neq \beta^2, \text{ for } \alpha = \beta \text{ see No. 344}).$$

**Integrals which contain the function $\tan \alpha x$:**

399. $$\int \tan \alpha x \,\mathrm{d}x = -\frac{1}{\alpha} \ln \cos \alpha x.$$

400. $$\int \tan^2 \alpha x \,\mathrm{d}x = \frac{\tan \alpha x}{\alpha} - x.$$

401. $$\int \tan^3 \alpha x \,\mathrm{d}x = \frac{1}{2\alpha} \tan^2 \alpha x + \frac{1}{\alpha} \ln \cos \alpha x.$$

402. $$\int \tan^n \alpha x \,\mathrm{d}x = \frac{1}{\alpha(n-1)} \tan^{n-1} \alpha x - \int \tan^{n-2} \alpha x \,\mathrm{d}x.$$

403. $$\int x \tan \alpha x \,\mathrm{d}x = \frac{\alpha x^3}{3} + \frac{\alpha^3 x^5}{15} + \frac{2\alpha^5 x^7}{105} + \frac{17\alpha^7 x^9}{2835} + \ldots + \frac{2^{2n}(2^{2n} - 1) B_{2n} \alpha^{2n-1} x^{2n+1}}{(2n+1)!} + \ldots$$

404. $$\int \frac{\tan \alpha x \, dx}{x} = \alpha x + \frac{(\alpha x)^3}{9} + \frac{2(\alpha x)^5}{75} + \frac{17(\alpha x)^7}{2205} + \dots + \frac{2^{2n}(2^{2n}-1)B_{2n}(\alpha x)^{2n-1}}{(2n-1)(2n)!} + \dots$$

$B_{2n}$ are the Bernoulli numbers (see 0.1.10.4).

405. $$\int \frac{\tan^n \alpha x}{\cos^2 \alpha x} dx = \frac{1}{\alpha(n+1)} \tan^{n+1} \alpha x \qquad (n \neq -1).$$

406. $$\int \frac{dx}{\tan \alpha x \pm 1} = \pm \frac{x}{2} + \frac{1}{2\alpha} \ln(\sin \alpha x \pm \cos \alpha x).$$

407. $$\int \frac{\tan \alpha x \, dx}{\tan \alpha x \pm 1} = \frac{x}{2} \mp \frac{1}{2\alpha} \ln(\sin \alpha x \pm \cos \alpha x).$$

**Integrals which contain the function $\cot \alpha x$:**

408. $$\int \cot \alpha x \, dx = \frac{1}{\alpha} \ln \sin \alpha x.$$

409. $$\int \cot^2 \alpha x \, dx = -\frac{\cot \alpha x}{\alpha} - x.$$

410. $$\int \cot^3 \alpha x \, dx = -\frac{1}{2\alpha} \cot^2 \alpha x - \frac{1}{\alpha} \ln \sin \alpha x.$$

411. $$\int \cot^n \alpha x \, dx = -\frac{1}{\alpha(n-1)} \cot^{n-1} \alpha x - \int \cot^{n-2} \alpha x \, dx \qquad (n \neq 1).$$

412. $$\int x \cot \alpha x \, dx = \frac{x}{\alpha} - \frac{\alpha x^3}{9} - \frac{\alpha^3 x^5}{225} - \dots - \frac{2^{2n} B_{2n} \alpha^{2n-1} x^{2n+1}}{(2n+1)!} \dots$$

$B_{2n}$ are the Bernoulli numbers (see section 0.1.10.4).

413. $$\int \frac{\cot \alpha x \, dx}{x} = -\frac{1}{\alpha x} - \frac{\alpha x}{3} - \frac{(\alpha x)^3}{135} - \frac{2(\alpha x)^5}{4725} - \dots - \frac{2^{2n} B_{2n} (\alpha x)^{2n-1}}{(2n-1)(2n)!} - \dots$$

$B_{2n}$ are the Bernoulli numbers (see section 0.1.10.4).

414. $$\int \frac{\cot^n \alpha x}{\sin^2 \alpha x} dx = -\frac{1}{\alpha(n+1)} \cot^{n+1} \alpha x \qquad (n \neq -1).$$

415. $$\int \frac{dx}{1 \pm \cot \alpha x} = \int \frac{\tan \alpha x \, dx}{\tan \alpha x \pm 1} \qquad \text{(see No. 407).}$$

### 0.9.5.4 Integrals which contain other transcendental functions

**Integrals which contain $e^{\alpha x}$:**

416.* $$\int e^{\alpha x} dx = \frac{1}{\alpha} e^{\alpha x}.$$

417.* $$\int x e^{\alpha x} dx = \frac{e^{\alpha x}}{\alpha^2}(\alpha x - 1).$$

418.* $$\int x^2 e^{\alpha x} dx = e^{\alpha x} \left( \frac{x^2}{\alpha} - \frac{2x}{\alpha^2} + \frac{2}{\alpha^3} \right).$$

419.* $\int x^n \mathrm{e}^{\alpha x}\mathrm{d}x = \frac{1}{\alpha}x^n\mathrm{e}^{\alpha x} - \frac{n}{\alpha}\int x^{n-1}\mathrm{e}^{\alpha x}\mathrm{d}x.$

420. $\int \frac{\mathrm{e}^{\alpha x}}{x}\mathrm{d}x = \ln x + \frac{\alpha x}{1\cdot 1!} + \frac{(\alpha x)^2}{2\cdot 2!} + \frac{(\alpha x)^3}{3\cdot 3!} + \dots$

The improper integral $\int\limits_{-\infty}^{x} \frac{\mathrm{e}^t}{t}\mathrm{d}t$ is called the *integral exponential function* and denoted Ei(x). For $x > 0$, this integral diverges at the point $t = 0$; Ei(x) is then the principal value of the improper integral (see sections 0.5.5 und 0.7.2).

$$\int\limits_{-\infty}^{x} \frac{\mathrm{e}^t}{t}\mathrm{d}t = \mathrm{C} + \ln x + \frac{x}{1\cdot 1!} + \frac{x^2}{2\cdot 2!} + \frac{x^3}{3\cdot 3!} + \dots + \frac{x^n}{n\cdot n!} + \dots$$

(C is the Euler constant, see 0.1.1).

421. $\int \frac{\mathrm{e}^{\alpha x}}{x^n}\mathrm{d}x = \frac{1}{n-1}\left(-\frac{\mathrm{e}^{\alpha x}}{x^{n-1}} + \alpha\int \frac{\mathrm{e}^{\alpha x}}{x^{n-1}}\mathrm{d}x\right) \qquad (n \in \mathbb{N},\ n > 1).$

422. $\int \frac{\mathrm{d}x}{1+e^{\alpha x}} = \frac{1}{\alpha}\ln\frac{\mathrm{e}^{\alpha x}}{1+\mathrm{e}^{\alpha x}}.$

423. $\int \frac{\mathrm{d}x}{\beta+\gamma\mathrm{e}^{\alpha x}} = \frac{x}{\beta} - \frac{1}{\alpha\beta}\ln(\beta+\gamma\mathrm{e}^{\alpha x}).$

424. $\int \frac{\mathrm{e}^{\alpha x}\mathrm{d}x}{\beta+\gamma\mathrm{e}^{\alpha x}} = \frac{1}{\alpha\gamma}\ln(\beta+\gamma\mathrm{e}^{\alpha x}).$

425. $\int \frac{\mathrm{d}x}{\beta\mathrm{e}^{\alpha x}+\gamma\mathrm{e}^{-\alpha x}} = \begin{cases} \frac{1}{\alpha\sqrt{\beta\gamma}}\arctan\left(\mathrm{e}^{\alpha x}\sqrt{\frac{\beta}{\gamma}}\right) & (\beta\gamma > 0), \\ \frac{1}{2\alpha\sqrt{-\beta\gamma}}\ln\frac{\gamma+\mathrm{e}^{\alpha x}\sqrt{-\beta\gamma}}{\gamma-\mathrm{e}^{\alpha x}\sqrt{-\beta\gamma}} & (\beta\gamma < 0). \end{cases}$

426. $\int \frac{x\mathrm{e}^{\alpha x}\mathrm{d}x}{(1+\alpha x)^2} = \frac{\mathrm{e}^{\alpha x}}{\alpha^2(1+\alpha x)}.$

427. $\int \mathrm{e}^{\alpha x}\ln x\,\mathrm{d}x = \frac{\mathrm{e}^{\alpha x}\ln x}{\alpha} - \frac{1}{\alpha}\int \frac{\mathrm{e}^{\alpha x}}{x}\mathrm{d}x$ (see No. 420).

428.* $\int \mathrm{e}^{\alpha x}\sin\beta x\,\mathrm{d}x = \frac{\mathrm{e}^{\alpha x}}{\alpha^2+\beta^2}(\alpha\sin\beta x - \beta\cos\beta x).$

429.* $\int \mathrm{e}^{\alpha x}\cos\beta x\,\mathrm{d}x = \frac{\mathrm{e}^{\alpha x}}{\alpha^2+\beta^2}(\alpha\cos\beta x + \beta\sin\beta x).$

430.* $\int \mathrm{e}^{\alpha x}\sin^n x\,\mathrm{d}x = \frac{\mathrm{e}^{\alpha x}\sin^{n-1}x}{\alpha^2+n^2}(\alpha\sin x - n\cos x) + \frac{n(n-1)}{\alpha^2+n^2}\int \mathrm{e}^{\alpha x}\sin^{n-2}x\,\mathrm{d}x$ (see No. 416 and 428).

431.* $\int \mathrm{e}^{\alpha x}\cos^n x\,\mathrm{d}x = \frac{\mathrm{e}^{\alpha x}\cos^{n-1}x}{\alpha^2+n^2}(\alpha\cos x + n\sin x) + \frac{n(n-1)}{\alpha^2+n^2}\int \mathrm{e}^{\alpha x}\cos^{n-2}x\,\mathrm{d}x$ (see No. 416 and 429).

432.* $\displaystyle\int x e^{\alpha x} \sin \beta x \, dx = \frac{x e^{\alpha x}}{\alpha^2+\beta^2}(\alpha \sin \beta x - \beta \cos \beta x) - \frac{e^{\alpha x}}{(\alpha^2+\beta^2)^2}[(\alpha^2-\beta^2)\sin \beta x - 2\alpha\beta \cos \beta x].$

433.* $\displaystyle\int x e^{\alpha x} \cos \beta x \, dx = \frac{x e^{\alpha x}}{\alpha^2+\beta^2}(\alpha \cos \beta x + \beta \sin \beta x) - \frac{e^{\alpha x}}{(\alpha^2+\beta^2)^2}[(\alpha^2-\beta^2)\cos \beta x + 2\alpha\beta \sin \beta x].$

**Integrals which contain $\ln x$:**

434. $\displaystyle\int \ln x \, dx = x(\ln x - 1).$

435. $\displaystyle\int (\ln x)^2 dx = x[(\ln x)^2 - 2\ln x + 2].$

436. $\displaystyle\int (\ln x)^3 dx = x[(\ln x)^3 - 3(\ln x)^2 + 6\ln x - 6].$

437. $\displaystyle\int (\ln x)^n dx = x(\ln x)^n - n\int (\ln x)^{n-1} dx \qquad (n \neq -1,\ n \in \mathbb{Z}).$

438. $\displaystyle\int \frac{dx}{\ln x} = \ln \ln x + \ln x + \frac{(\ln x)^2}{2 \cdot 2!} + \frac{(\ln x)^3}{3 \cdot 3!} + \dots$

The integral $\displaystyle\int_0^x \frac{dt}{\ln t}$ is called the *logarithmic integral*, denoted li(x). For $x > 1$ it diverges at the point $t = 1$. In this case the value of the function li(x) is understood to be the principal value of the improper integral (see 0.5.5 and 0.7.2).

439. $\displaystyle\int \frac{dx}{(\ln x)^n} = -\frac{x}{(n-1)(\ln x)^{n-1}} + \frac{1}{n-1}\int \frac{dx}{(\ln x)^{n-1}}$
$(n \in \mathbb{N},\ n > 1$, see No. 438).

440. $\displaystyle\int x^m \ln x \, dx = x^{m+1}\left[\frac{\ln x}{m+1} - \frac{1}{(m+1)^2}\right] \qquad (m \in \mathbb{N}$, see No. 443).

441. $\displaystyle\int x^m (\ln x)^n dx = \frac{x^{m+1}(\ln x)^n}{m+1} - \frac{n}{m+1}\int x^m (\ln x)^{n-1} dx$
$(m, n \in \mathbb{N}$, see No. 444, 446, and 450).

442. $\displaystyle\int \frac{(\ln x)^n}{x} dx = \frac{(\ln x)^{n+1}}{n+1}.$

443. $\displaystyle\int \frac{\ln x}{x^m} dx = -\frac{\ln x}{(m-1)x^{m-1}} - \frac{1}{(m-1)^2 x^{m-1}} \qquad (m \in \mathbb{N},\ m > 1).$

444. $\displaystyle\int \frac{(\ln x)^n}{x^m} dx = -\frac{(\ln x)^n}{(m-1)x^{m-1}} + \frac{n}{m-1}\int \frac{(\ln x)^{n-1}}{x^m} dx \qquad (m, n \in \mathbb{N},\ m > 1,$
see No. 441, 446).

445. $\displaystyle\int \frac{x^m dx}{\ln x} = \int \frac{e^{-y}}{y} dy$ with $y = -(m+1)\ln x$ (see No. 420).

446. $$\int \frac{x^m \mathrm{d}x}{(\ln x)^n} = -\frac{x^{m+1}}{(n-1)(\ln x)^{n-1}} + \frac{m+1}{n-1}\int \frac{x^m \mathrm{d}x}{(\ln x)^{n-1}}$$ ($m, n \in \mathbb{N}$, $n > 1$, see No. 441, 444).

447. $$\int \frac{\mathrm{d}x}{x \ln x} = \ln \ln x.$$

448. $$\int \frac{\mathrm{d}x}{x^n \ln x} = \ln \ln x - (n-1)\ln x + \frac{(n-1)^2(\ln x)^2}{2 \cdot 2!} - \frac{(n-1)^3(\ln x)^3}{3 \cdot 3!} + \ldots$$

449. $$\int \frac{\mathrm{d}x}{x(\ln x)^n} = \frac{-1}{(n-1)(\ln x)^{n-1}}$$ ($n \in \mathbb{N}$, $n > 1$).

450. $$\int \frac{\mathrm{d}x}{x^m(\ln x)^n} = \frac{-1}{x^{m-1}(n-1)(\ln x)^{n-1}} - \frac{m-1}{n-1}\int \frac{\mathrm{d}x}{x^m(\ln x)^{n-1}}$$ ($m, n \in \mathbb{N}$, $n > 1$, see No. 441, 444 and 446).

451. $$\int \ln \sin x \,\mathrm{d}x = x \ln x - x - \frac{x^3}{18} - \frac{x^5}{900} - \ldots - \frac{2^{2n-1} B_{2n} x^{2n+1}}{n(2n+1)!} - \ldots$$

452. $$\int \ln \cos x \,\mathrm{d}x = -\frac{x^3}{6} - \frac{x^5}{60} - \frac{x^7}{315} - \ldots - \frac{2^{2n-1}(2^{2n}-1)B_{2n}}{n(2n+1)!}x^{2n+1} - \ldots$$

453. $$\int \ln \tan x \,\mathrm{d}x = x \ln x - x + \frac{x^3}{9} + \frac{7x^5}{450} + \ldots + \frac{2^{2n}(2^{2n-1}-1)B_{2n}}{n(2n+1)!}x^{2n+1} + \ldots$$

$B_{2n}$ are the Bernoulli numbers (see section 0.1.10.4).

454. $$\int \sin \ln x \,\mathrm{d}x = \frac{x}{2}(\sin \ln x - \cos \ln x).$$

455. $$\int \cos \ln x \,\mathrm{d}x = \frac{x}{2}(\sin \ln x + \cos \ln x).$$

456. $$\int \mathrm{e}^{\alpha x} \ln x \,\mathrm{d}x = \frac{1}{\alpha}\mathrm{e}^{\alpha x} \ln x - \frac{1}{\alpha}\int \frac{\mathrm{e}^{\alpha x}}{x}\mathrm{d}x$$ (see No. 420).

**Integrals which contain hyperbolic functions:**

457.* $$\int \sinh \alpha x \,\mathrm{d}x = \frac{1}{\alpha}\cosh \alpha x.$$

458.* $$\int \cosh \alpha x \,\mathrm{d}x = \frac{1}{\alpha}\sinh \alpha x.$$

459.* $$\int \sinh^2 \alpha x \,\mathrm{d}x = \frac{1}{2\alpha}\sinh \alpha x \cosh \alpha x - \frac{1}{2}x.$$

460.* $$\int \cosh^2 \alpha x \,\mathrm{d}x = \frac{1}{2\alpha}\sinh \alpha x \cosh \alpha x + \frac{1}{2}x.$$

461. $$\int \sinh^n \alpha x \,\mathrm{d}x =$$
$$\begin{cases} \frac{1}{\alpha n}\sinh^{n-1}\alpha x \cosh \alpha x - \frac{n-1}{n}\int \sinh^{n-2}\alpha x \,\mathrm{d}x & (n \in \mathbb{N}, n > 0),^{79} \\ \frac{1}{\alpha(n+1)}\sinh^{n+1}\alpha x \cosh \alpha x - \frac{n+2}{n+1}\int \sinh^{n+2}\alpha x \,\mathrm{d}x & (n \in \mathbb{Z},\ n < -1). \end{cases}$$

[79]In this case the formula is also true for complex numbers $x$.

462. $\int \cosh^n \alpha x \, dx =$

$$\begin{cases} \frac{1}{\alpha n} \sinh \alpha x \cosh^{n-1} \alpha x + \frac{n-1}{n} \int \cosh^{n-2} \alpha x \, dx & (n \in \mathbb{N}, n > 0),^{79} \\ -\frac{1}{\alpha(n+1)} \sinh \alpha x \cosh^{n+1} \alpha x + \frac{n+2}{n+1} \int \cosh^{n+2} \alpha x \, dx & (n \in \mathbb{Z},\ n < -1). \end{cases}$$

463.* $\int \sinh \alpha x \sinh \beta x \, dx = \frac{1}{\alpha^2 - \beta^2}(\alpha \sinh \beta x \cosh \alpha x - \beta \cosh \beta x \sinh \alpha x), \quad \alpha^2 \neq \beta^2.$

464.* $\int \cosh \alpha x \cosh \beta x \, dx = \frac{1}{\alpha^2 - \beta^2}(\alpha \sinh \alpha x \cosh \beta x - \beta \sinh \beta x \cosh \alpha x), \quad \alpha^2 \neq \beta^2.$

465.* $\int \cosh \alpha x \sinh \beta x \, dx = \frac{1}{\alpha^2 - \beta^2}(\alpha \sinh \beta x \sinh \alpha x - \beta \cosh \beta x \cosh \alpha x), \quad \alpha^2 \neq \beta^2.$

466.* $\int \sinh \alpha x \sin \alpha x \, dx = \frac{1}{2\alpha}(\cosh \alpha x \sin \alpha x - \sinh \alpha x \cos \alpha x).$

467.* $\int \cosh \alpha x \cos \alpha x \, dx = \frac{1}{2\alpha}(\sinh \alpha x \cos \alpha x + \cosh \alpha x \sin \alpha x).$

468.* $\int \sinh \alpha x \cos \alpha x \, dx = \frac{1}{2\alpha}(\cosh \alpha x \cos \alpha x + \sinh \alpha x \sin \alpha x).$

469.* $\int \cosh \alpha x \sin \alpha x \, dx = \frac{1}{2\alpha}(\sinh \alpha x \sin \alpha x - \cosh \alpha x \cos \alpha x).$

470.* $\int \frac{dx}{\sinh \alpha x} = \frac{1}{\alpha} \ln \tanh \frac{\alpha x}{2}.$

471. $\int \frac{dx}{\cosh \alpha x} = \frac{2}{\alpha} \arctan e^{\alpha x}.$

472. $\int x \sinh \alpha x \, dx = \frac{1}{\alpha} x \cosh \alpha x - \frac{1}{\alpha^2} \sinh \alpha x.$

473.* $\int x \cosh \alpha x \, dx = \frac{1}{\alpha} x \sinh \alpha x - \frac{1}{\alpha^2} \cosh \alpha x.$

474. $\int \tanh \alpha x \, dx = \frac{1}{\alpha} \ln \cosh \alpha x.$

475. $\int \coth \alpha x \, dx = \frac{1}{\alpha} \ln \sinh \alpha x.$

476. $\int \tanh^2 \alpha x \, dx = x - \frac{\tanh \alpha x}{\alpha}.$

477. $\int \coth^2 \alpha x \, dx = x - \frac{\coth \alpha x}{\alpha}.$

**Integrals which contain inverse trigonometric functions:**

478. $\int \arcsin \frac{x}{\alpha} dx = x \arcsin \frac{x}{\alpha} + \sqrt{\alpha^2 - x^2} \qquad (|x| < |\alpha|).$

479. $\int x \arcsin \frac{x}{\alpha} \, dx = \left( \frac{x^2}{2} - \frac{\alpha^2}{4} \right) \arcsin \frac{x}{\alpha} + \frac{x}{4} \sqrt{\alpha^2 - x^2} \qquad (|x| < |\alpha|).$

480. $\int x^2 \arcsin \frac{x}{\alpha}\,\mathrm{d}x = \frac{x^3}{3} \arcsin \frac{x}{\alpha} + \frac{1}{9}(x^2 + 2\alpha^2)\sqrt{\alpha^2 - x^2} \qquad (|x| < |\alpha|).$

481. $\int \frac{\arcsin \frac{x}{\alpha}\,\mathrm{d}x}{x} = \frac{x}{\alpha} + \frac{1}{2\cdot 3\cdot 3}\frac{x^3}{\alpha^3} + \frac{1\cdot 3}{2\cdot 4\cdot 5\cdot 5}\frac{x^5}{\alpha^5} + \frac{1\cdot 3\cdot 5}{2\cdot 4\cdot 6\cdot 7\cdot 7}\frac{x^7}{\alpha^7} + \dots$

482. $\int \frac{\arcsin \frac{x}{\alpha}\,\mathrm{d}x}{x^2} = -\frac{1}{x}\arcsin \frac{x}{\alpha} - \frac{1}{\alpha}\ln \frac{\alpha + \sqrt{\alpha^2 - x^2}}{x} \qquad (|x| < |\alpha|).$

483. $\int \arccos \frac{x}{\alpha}\,\mathrm{d}x = x \arccos \frac{x}{\alpha} - \sqrt{\alpha^2 - x^2} \qquad (|x| < |\alpha|).$

484. $\int x \arccos \frac{x}{\alpha}\,\mathrm{d}x = \left(\frac{x^2}{2} - \frac{\alpha^2}{4}\right) \arccos \frac{x}{\alpha} - \frac{x}{4}\sqrt{\alpha^2 - x^2} \qquad (|x| < |\alpha|).$

485. $\int x^2 \arccos \frac{x}{\alpha}\,\mathrm{d}x = \frac{x^3}{3} \arccos \frac{x}{\alpha} - \frac{1}{9}(x^2 + 2\alpha^2)\sqrt{\alpha^2 - x^2} \qquad (|x| < |\alpha|).$

486. $\int \frac{\arccos \frac{x}{\alpha}\,\mathrm{d}x}{x} = \frac{\pi}{2}\ln x - \frac{x}{\alpha} - \frac{1}{2\cdot 3\cdot 3}\frac{x^3}{\alpha^3} - \frac{1\cdot 3}{2\cdot 4\cdot 5\cdot 5}\frac{x^5}{\alpha^5} - \frac{1\cdot 3\cdot 5}{2\cdot 4\cdot 6\cdot 7\cdot 7}\frac{x^7}{\alpha^7} - \dots$

487. $\int \frac{\arccos \frac{x}{\alpha}\,\mathrm{d}x}{x^2} = -\frac{1}{x}\arccos \frac{x}{\alpha} + \frac{1}{\alpha}\ln \frac{\alpha + \sqrt{\alpha^2 - x^2}}{x} \qquad (|x| < |\alpha|).$

488. $\int \arctan \frac{x}{\alpha}\,\mathrm{d}x = x \arctan \frac{x}{\alpha} - \frac{\alpha}{2}\ln(\alpha^2 + x^2).$

489. $\int x \arctan \frac{x}{\alpha}\,\mathrm{d}x = \frac{1}{2}(x^2 + \alpha^2) \arctan \frac{x}{\alpha} - \frac{\alpha x}{2}.$

490. $\int x^2 \arctan \frac{x}{\alpha}\,\mathrm{d}x = \frac{x^3}{3} \arctan \frac{x}{\alpha} - \frac{\alpha x^2}{6} + \frac{\alpha^3}{6}\ln(\alpha^2 + x^2).$

491. $\int x^n \arctan \frac{x}{\alpha}\,\mathrm{d}x = \frac{x^{n+1}}{n+1} \arctan \frac{x}{\alpha} - \frac{\alpha}{n+1}\int \frac{x^{n+1}\,\mathrm{d}x}{\alpha^2 + x^2} \qquad (n \in \mathbb{N}, \text{ see No. 494}).$

492. $\int \frac{\arctan \frac{x}{\alpha}\,\mathrm{d}x}{x} = \frac{x}{\alpha} - \frac{x^3}{3^2\alpha^3} + \frac{x^5}{5^2\alpha^5} - \frac{x^7}{7^2\alpha^7} + \dots \qquad (|x| < |\alpha|).$

493. $\int \frac{\arctan \frac{x}{\alpha}\,\mathrm{d}x}{x^2} = -\frac{1}{x}\arctan \frac{x}{\alpha} - \frac{1}{2\alpha}\ln \frac{\alpha^2 + x^2}{x^2}.$

494. $\int \frac{\arctan \frac{x}{\alpha}\,\mathrm{d}x}{x^n} = -\frac{1}{(n-1)x^{n-1}} \arctan \frac{x}{\alpha} + \frac{\alpha}{n-1}\int \frac{\mathrm{d}x}{x^{n-1}(\alpha^2 + x^2)}$
$(n \in \mathbb{N}, \text{ see No. 491}).$

495. $\int \operatorname{arccot} \frac{x}{\alpha}\,\mathrm{d}x = x \operatorname{arccot} \frac{x}{\alpha} + \frac{\alpha}{2}\ln(\alpha^2 + x^2).$

496. $\int x \operatorname{arccot} \frac{x}{\alpha}\,\mathrm{d}x = \frac{1}{2}(x^2 + \alpha^2) \operatorname{arccot} \frac{x}{\alpha} + \frac{\alpha x}{2}.$

497. $\int x^2 \operatorname{arccot} \frac{x}{\alpha}\,\mathrm{d}x = \frac{x^3}{\alpha} \operatorname{arccot} \frac{x}{\alpha} + \frac{\alpha x^2}{6} - \frac{\alpha^3}{6}\ln(\alpha^2 + x^2).$

498. $\displaystyle\int x^n \operatorname{arccot}\frac{x}{\alpha}\mathrm{d}x = \frac{x^{n+1}}{n+1}\operatorname{arccot}\frac{x}{\alpha} + \frac{\alpha}{n+1}\int\frac{x^{n+1}\mathrm{d}x}{\alpha^2+x^2}$ ($n \in \mathbb{N}$, see No. 501).

499. $\displaystyle\int\frac{\operatorname{arccot}\frac{x}{\alpha}\,\mathrm{d}x}{x} = \frac{\pi}{2}\ln x - \frac{x}{\alpha} + \frac{x^3}{3^2\alpha^3} - \frac{x^5}{5^2\alpha^5} + \frac{x^7}{7^2\alpha^7} - \ldots$

500. $\displaystyle\int\frac{\operatorname{arccot}\frac{x}{\alpha}\,\mathrm{d}x}{x^2} = -\frac{1}{x}\operatorname{arccot}\frac{x}{\alpha} + \frac{1}{2\alpha}\ln\frac{\alpha^2+x^2}{x^2}.$

501. $\displaystyle\int\frac{\operatorname{arccot}\frac{x}{\alpha}\,\mathrm{d}x}{x^n} = -\frac{1}{(n-1)x^{n-1}}\operatorname{arccot}\frac{x}{\alpha} - \frac{\alpha}{n-1}\int\frac{\mathrm{d}x}{x^{n-1}(\alpha^2+x^2)}$
($n \in \mathbb{N}$, see No. 498).

#### 0.9.5.4.1 Integrals which contain inverse hyperbolic functions:

502. $\displaystyle\int \operatorname{arsinh}\frac{x}{\alpha}\,\mathrm{d}x = x\operatorname{arsinh}\frac{x}{\alpha} - \sqrt{x^2+\alpha^2}.$

503. $\displaystyle\int \operatorname{arcosh}\frac{x}{\alpha}\,\mathrm{d}x = x\operatorname{arcosh}\frac{x}{\alpha} - \sqrt{x^2-\alpha^2}$ $(|\alpha| < |x|)$.

504. $\displaystyle\int \operatorname{artanh}\frac{x}{\alpha}\,\mathrm{d}x = x\operatorname{artanh}\frac{x}{\alpha} + \frac{\alpha}{2}\ln(\alpha^2-x^2)$ $(|x| < |\alpha|)$.

505. $\displaystyle\int \operatorname{arcoth}\frac{x}{\alpha}\,\mathrm{d}x = x\operatorname{arcoth}\frac{x}{\alpha} + \frac{\alpha}{2}\ln(x^2-\alpha^2)$ $(|\alpha| < [x|)$.

## 0.9.6 Tables of definite integrals

### 0.9.6.1 Integrals which contain exponential functions

We consider here integrals containing exponential functions combined with algebraic, trigonometric and logarithmic functions.

1. $\displaystyle\int_0^\infty x^n\,\mathrm{e}^{-\alpha x}\,\mathrm{d}x = \frac{\Gamma(n+1)}{\alpha^{n+1}}$ $(\alpha, n \in \mathbb{R}, \quad \alpha > 0, \quad n > -1)$.

   (Gamma function $\Gamma(n)$, see 0.5.1).

   For $n \in \mathbb{N}$ this integral evaluates to $\dfrac{n!}{\alpha^{n+1}}$.

2. $$\int_0^\infty x^n \mathrm{e}^{-\alpha x^2}\,\mathrm{d}x = \begin{cases} \dfrac{\Gamma\left(\dfrac{n+1}{2}\right)}{2\alpha^{\left(\frac{n+1}{2}\right)}} & (n, \alpha \in \mathbb{R}, \quad \alpha > 0, \quad n > -1), \\ \dfrac{1\cdot 3\ldots(2k-1)\sqrt{\pi}}{2^{k+1}\alpha^{k+1/2}} & (n = 2k, \quad k \in \mathbb{N}), \\ \dfrac{k!}{2\alpha^{k+1}} & (n = 2k+1, \quad k \in \mathbb{N}), \end{cases}$$

   (see No. 1).

3. $\int_0^\infty e^{-\alpha^2 x^2}\,dx = \frac{\sqrt{\pi}}{2\alpha}$ for $\alpha > 0$.

4. $\int_0^\infty x^2 e^{-\alpha^2 x^2}\,dx = \frac{\sqrt{\pi}}{4\alpha^3}$ for $\alpha > 0$.

5. $\int_0^\infty e^{-\alpha^2 x^2} \cos\beta x\,dx = \frac{\sqrt{\pi}}{2\alpha} e^{-\beta^2/4\alpha^2}$ for $\alpha > 0$.

6. $\int_0^\infty \frac{x\,dx}{e^x - 1} = \frac{\pi^2}{6}$.

7. $\int_0^\infty \frac{x\,dx}{e^x + 1} = \frac{\pi^2}{12}$.

8. $\int_0^\infty \frac{e^{-\alpha x}\sin x}{x}\,dx = \operatorname{arccot}\alpha = \arctan\frac{1}{\alpha}$ for $\alpha > 0$.

9. $\int_0^\infty e^{-x}\ln x\,dx = -C \approx -0.577\,2$.

10. $\int_0^\infty e^{-x^2}\ln x\,dx = \frac{1}{4}\Gamma'\left(\frac{1}{2}\right) = -\frac{\sqrt{\pi}}{4}(C + 2\ln 2)$.

11. $\int_0^\infty e^{-x^2}\ln^2 x\,dx = \frac{\sqrt{\pi}}{8}\left[(C + 2\ln 2)^2 + \frac{\pi^2}{2}\right]$.

In 9 – 11, C is the Euler constant (see 0.1.1).

12. $$\int_0^{\pi/2} \sin^{2a+1} x \cos^{2b+1} x\,dx = \begin{cases} \frac{\Gamma(a+1)\Gamma(b+1)}{2\Gamma(a+b+2)} = \frac{1}{2}B(a+1,\, b+1) & (a, b \in \mathbb{R}), \\ \frac{a!b!}{2(a+b+1)!} & (a, b \in \mathbb{N}). \end{cases}$$

$B(x,y) = \frac{\Gamma(x)\cdot\Gamma(y)}{\Gamma(x+y)}$ is the so-called Beta-function or, as it is also called, the Euler integral of the first kind, and $\Gamma(x)$ is the Gamma function, also called the Euler integral of the second kind (see No. 1).

13. $\int_{-\pi}^{\pi} \sin(mx)\sin(nx)\,dx = \delta_{m,n}\cdot\pi \quad (m, n \in \mathbb{N})$.[80]

14. $\int_{-\pi}^{\pi} \cos(mx)\sin(nx)\,dx = 0 \quad (m, n \in \mathbb{N})$.

[80] $\delta_{m,n} = 0$ for $m \neq n$, $\delta_{m,n} = 1$ for $m = n$, Kronecker symbol.

15. $\displaystyle\int_{-\pi}^{\pi} \cos(mx)\cos(nx)\,\mathrm{d}x = \delta_{m,n}\cdot\pi \quad (m,n\in\mathbb{N}).$[80]

16. $\displaystyle\int_0^\infty \frac{\sin\alpha x}{x}\,\mathrm{d}x = \begin{cases} \dfrac{\pi}{2} & \text{for} \quad \alpha > 0, \\ -\dfrac{\pi}{2} & \text{for} \quad \alpha < 0. \end{cases}$

17. $\displaystyle\int_0^\infty \frac{\sin\beta x}{x^s}\,\mathrm{d}x = \frac{\pi\beta^{s-1}}{2\Gamma(s)\sin s\pi/2}, \qquad 0 < s < 2.$

18. $\displaystyle\int_0^a \frac{\cos\alpha x\,\mathrm{d}x}{x} = \infty \quad (a\in\mathbb{R}).$

19. $\displaystyle\int_0^\infty \frac{\cos\beta x}{x^s}\,\mathrm{d}x = \frac{\pi\beta^{s-1}}{2\Gamma(s)\cos s\pi/2}, \qquad 0 < s < 1.$

20. $\displaystyle\int_0^\infty \frac{\tan\alpha x\,\mathrm{d}x}{x} = \begin{cases} \dfrac{\pi}{2} & \text{for} \quad \alpha > 0, \\ -\dfrac{\pi}{2} & \text{for} \quad \alpha < 0. \end{cases}$

21. $\displaystyle\int_0^\infty \frac{\cos\alpha x - \cos\beta x}{x}\,\mathrm{d}x = \ln\frac{\beta}{\alpha}.$

22. $\displaystyle\int_0^\infty \frac{\sin x\,\cos\alpha x}{x}\,\mathrm{d}x = \begin{cases} \dfrac{\pi}{2} & \text{for} \quad |\alpha| < 1, \\ \dfrac{\pi}{4} & \text{for} \quad |\alpha| = 1, \\ 0 & \text{for} \quad |\alpha| > 1. \end{cases}$

23. $\displaystyle\int_0^\infty \frac{\sin x}{\sqrt{x}}\,\mathrm{d}x = \int_0^\infty \frac{\cos x}{\sqrt{x}}\,\mathrm{d}x = \sqrt{\frac{\pi}{2}}.$

24. $\displaystyle\int_0^\infty \frac{x\,\sin\beta x}{\alpha^2+x^2}\,\mathrm{d}x = \operatorname{sgn}(\beta)\frac{\pi}{2}\mathrm{e}^{-|\alpha\beta|}, \qquad (\operatorname{sgn}(\beta) = -1 \text{ for } \beta < 0,\ \operatorname{sgn}(\beta) = 1 \text{ for } \beta > 0).$

25. $\displaystyle\int_0^\infty \frac{\cos\alpha x}{1+x^2}\,\mathrm{d}x = \frac{\pi}{2}\mathrm{e}^{-|\alpha|}.$

26. $\int_{0}^{\infty} \frac{\sin^2 \alpha x}{x^2}\, dx = \frac{\pi}{2}|\alpha|.$

27. $\int_{-\infty}^{+\infty} \sin(x^2)\, dx = \int_{-\infty}^{+\infty} \cos(x^2)\, dx = \sqrt{\frac{\pi}{2}}.$

28. $\int_{0}^{\pi/2} \frac{\sin x\, dx}{\sqrt{1 - a^2 \sin^2 x}} = \frac{1}{2a} \ln \frac{1+a}{1-a} \quad \text{for} \quad |a| < 1.$

29. $\int_{0}^{\pi/2} \frac{\cos x\, dx}{\sqrt{1 - a^2 \sin^2 x}} = \frac{1}{a} \arcsin a \quad \text{for} \quad |a| < 1.$

30. $\int_{0}^{\pi/2} \frac{\sin^2 x\, dx}{\sqrt{1 - a^2 \sin^2 x}} = \frac{1}{a^2}(\mathrm{K} - \mathrm{E}) \quad \text{for} \quad |a| < 1.$

31. $\int_{0}^{\pi/2} \frac{\cos^2 x\, dx}{\sqrt{1 - a^2 \sin^2 x}} = \frac{1}{a^2}\left[\mathrm{E} - (1 - a^2)\mathrm{K}\right] \quad \text{for} \quad |a| < 1.$

In 30 and 31, E and K are the complete elliptic integrals:

$$\mathrm{E} = E\left(a, \frac{\pi}{2}\right), \ \mathrm{K} = F\left(a, \frac{\pi}{2}\right) \quad \text{(see section 0.5.4).}$$

32. $\int_{0}^{\pi} \frac{\cos \alpha x\, dx}{1 - 2\beta \cos x + \beta^2} = \frac{\pi \beta^\alpha}{1 - \beta^2} \quad \alpha \in \mathbb{N}, \quad |\beta| < 1.$

#### 0.9.6.2 Integrals containing logarithmic functions

33. $\int_{0}^{1} \ln \ln x\, dx = -\mathrm{C} \approx -0.577\,2$, where C is the Euler constant (see 0.1.1).

34. $\int_{0}^{1} \frac{\ln x}{x - 1}\, dx = \frac{\pi^2}{6}.$

35. $\int_{0}^{1} \frac{\ln x}{x + 1}\, dx = -\frac{\pi^2}{12}.$

36. $\int_{0}^{1} \frac{\ln x}{x^2 - 1}\, dx = \frac{\pi^2}{8}.$

37. $\displaystyle\int_0^1 \frac{\ln(1+x)}{x^2+1}\,\mathrm{d}x = \frac{\pi}{8}\ln 2.$

38. $\displaystyle\int_0^1 \frac{(1-x^\alpha)(1-x^\beta)}{(1-x)\ln x}\,\mathrm{d}x = \ln\frac{\Gamma(\alpha+1)\Gamma(\beta+1)}{\Gamma(\alpha+\beta+1)},\ (\alpha > -1,\ \beta > -1,\ \alpha+\beta > -1).$

39. $\displaystyle\int_0^1 \ln\left(\frac{1}{x}\right)^\alpha \mathrm{d}x = \Gamma(\alpha+1)\quad(-1 < \alpha < \infty)$, where $\Gamma(x)$ is the Gamma function (see No. 1).

40. $\displaystyle\int_0^1 \frac{x^{\alpha-1}-x^{-\alpha}}{(1+x)\ln x}\,\mathrm{d}x = \ln\tan\frac{\alpha\pi}{2}\qquad(0 < \alpha < 1).$

41. $\displaystyle\int_0^{\pi/2} \ln\sin x\,\mathrm{d}x = \int_0^{\pi/2} \ln\cos x\,\mathrm{d}x = -\frac{\pi}{2}\ln 2.$

42. $\displaystyle\int_0^{\pi} x\ln\sin x\,\mathrm{d}x = -\frac{\pi^2\ln 2}{2}.$

43. $\displaystyle\int_0^{\pi/2} \sin x\ln\sin x\,\mathrm{d}x = \ln 2 - 1.$

44. $\displaystyle\int_0^{\infty} \frac{\sin x}{x}\ln x\,\mathrm{d}x = -\frac{\pi}{2}\mathrm{C}.$

45. $\displaystyle\int_0^{\infty} \frac{\sin x}{x}\ln^2 x\,\mathrm{d}x = \frac{\pi}{2}\mathrm{C}^2 + \frac{\pi^3}{24},$ where C is the Euler constant (see 0.1.1).

46. $\displaystyle\int_0^{\pi} \ln(\alpha\pm\beta\cos x)\,\mathrm{d}x = \pi\ln\frac{\alpha+\sqrt{\alpha^2-\beta^2}}{2}\qquad(\alpha\ge\beta).$

47. $\displaystyle\int_0^{\pi} \ln(\alpha^2-2\alpha\beta\cos x+\beta^2)\,\mathrm{d}x = \begin{cases} 2\pi\ln\alpha & (\alpha\ge\beta>0),\\ 2\pi\ln\beta & (\beta\ge\alpha>0).\end{cases}$

48. $\displaystyle\int_0^{\pi/2} \ln\tan x\,\mathrm{d}x = 0.$

49. $$\int_0^{\pi/4} \ln(1+\tan x)\,\mathrm{d}x = \frac{\pi}{8}\ln 2.$$

### 0.9.6.3 Integrals which contain algebraic functions

50. $$\int_0^1 x^a(1-x)^b\,\mathrm{d}x = 2\int_0^1 x^{2a+1}(1-x^2)^b\,\mathrm{d}x = \frac{\Gamma(a+1)\Gamma(b+1)}{\Gamma(a+b+2)} = B(a+1,b+1).$$

For $B(x,y)$ and $\Gamma(x)$, see No. 12.

51. $$\int_0^\infty \frac{\mathrm{d}x}{(1+x)x^\alpha} = \frac{\pi}{\sin\alpha\pi} \quad \text{for} \quad \alpha<1.$$

52. $$\int_0^\infty \frac{\mathrm{d}x}{(1-x)x^\alpha} = -\pi\cot\alpha\pi \quad \text{for} \quad \alpha<1.$$

53. $$\int_0^\infty \frac{x^{\alpha-1}}{1+x^\beta}\,\mathrm{d}x = \frac{\pi}{\beta\sin\dfrac{\alpha\pi}{\beta}} \quad \text{for} \quad 0<\alpha<\beta.$$

54. $$\int_0^1 \frac{\mathrm{d}x}{\sqrt{1-x^\alpha}} = \frac{\sqrt{\pi}\,\Gamma\left(\dfrac{1}{\alpha}\right)}{\alpha\Gamma\left(\dfrac{2+\alpha}{2\alpha}\right)}.$$

$\Gamma(x)$ is the Gamma function (see No. 1).

55. $$\int_0^1 \frac{\mathrm{d}x}{1+2x\cos\alpha+x^2} = \frac{\alpha}{2\sin\alpha} \qquad \left(0<\alpha<\frac{\pi}{2}\right).$$

56. $$\int_0^\infty \frac{\mathrm{d}x}{1+2x\cos\alpha+x^2} = \frac{\alpha}{\sin\alpha} \qquad \left(0<\alpha<\frac{\pi}{2}\right).$$

# 0.10 Tables on integral transformations

## 0.10.1 Fourier transformation

Legend of symbols occuring in the table:

C: Euler constant (C= 0.577 215 67 ...)

$$\Gamma(z) = \int_0^\infty \mathrm{e}^{-t} t^{z-1} \mathrm{d}t, \qquad \operatorname{Re} z > 0 \qquad \text{(Gamma function),}$$

$$J_\nu(z) = \sum_{n=0}^\infty \frac{(-1)^n (\frac{1}{2}z)^{\nu+2n}}{n!\,\Gamma(\nu+n+1)} \qquad \text{(Bessel functions),}$$

$$K_\nu(z) = \frac{1}{2}\pi\big(\sin(\pi\nu)\big)^{-1}[I_{-\nu}(z) - I_\nu(z)] \quad \text{with}$$

$$I_\nu(z) = \mathrm{e}^{-\frac{1}{2}\mathrm{i}\pi\nu} J_\nu(z\mathrm{e}^{\frac{1}{2}\mathrm{i}\pi}) \qquad \text{(modified Bessel functions),}$$

$$\left.\begin{aligned} C(x) &= \frac{1}{\sqrt{2\pi}} \int_0^x \frac{\cos t}{\sqrt{t}} \mathrm{d}t \\ S(x) &= \frac{1}{\sqrt{2\pi}} \int_0^x \frac{\sin t}{\sqrt{t}} \mathrm{d}t \end{aligned}\right\} \qquad \text{(Fresnel integrals),}$$

$$\left.\begin{aligned} \operatorname{Si}(x) &= \int_0^x \frac{\sin t}{t} \mathrm{d}t \\ \operatorname{si}(x) &= -\int_x^\infty \frac{\sin t}{t} \mathrm{d}t = \operatorname{Si}(x) - \frac{\pi}{2} \end{aligned}\right\} \qquad \text{(Elliptic sine),}$$

$$\operatorname{Ci}(x) = -\int_x^\infty \frac{\cos t}{t} \mathrm{d}t \qquad \text{(Elliptic cosine).}$$

Occasionally we use the notation $\exp(x)$ for $\mathrm{e}^x$. Furthermore, $[x]$ denotes the Gauss bracket, i.e., the largest integer $n$ for which $n \leq x$.

#### 0.10.1.1 Fourier cosine transform:

| $f(x)$ | $F(y) = \sqrt{\frac{2}{\pi}} \int_0^\infty f(x) \cos(xy) \mathrm{d}x$ |
|---|---|
| $1, \quad 0 < x < a$<br>$0, \quad x > a$ | $\sqrt{\frac{2}{\pi}} \frac{\sin(ay)}{y}$ |
| $x, \quad 0 < x < 1$<br>$2 - x, \quad 1 < x < 2$<br>$0, \quad x > 2$ | $4\sqrt{\frac{2}{\pi}} \left(\cos y \ \sin^2 \frac{y}{2}\right) y^{-2}$ |
| $0, \quad 0 < x < a$<br>$\frac{1}{x}, \quad x > a$ | $-\sqrt{\frac{2}{\pi}} \mathrm{Ci}(ay)$ |
| $\frac{1}{\sqrt{x}}$ | $\frac{1}{\sqrt{y}}$ |
| $\frac{1}{\sqrt{x}}, \quad 0 < x < a$<br>$0, \quad x > a$ | $\frac{2C(ay)}{\sqrt{y}}$ |
| $0, \quad 0 < x < a$<br>$\frac{1}{\sqrt{x}}, \quad x > a$ | $\frac{1 - 2C(ay)}{\sqrt{y}}$ |
| $(a + x)^{-1}, \quad a > 0$ | $\sqrt{\frac{2}{\pi}} \left[ - \mathrm{si}(ay) \sin(ay) - \mathrm{Ci}(ay) \cos(ay) \right]$ |
| $(a - x)^{-1}, \quad a > 0$ | $\sqrt{\frac{2}{\pi}} \left[ \cos(ay) \mathrm{Ci}(ay) + \sin(ay) \left( \frac{\pi}{2} + \mathrm{Si}(ay) \right) \right]$ |
| $(a^2 + x^2)^{-1}$ | $\sqrt{\frac{\pi}{2}} \frac{\mathrm{e}^{-ay}}{a}$ |
| $(a^2 - x^2)^{-1}$ | $\sqrt{\frac{\pi}{2}} \frac{\sin(ay)}{y}$ |
| $\frac{b}{b^2 + (a - x)^2} + \frac{b}{b^2 + (a + x)^2}$ | $\sqrt{2\pi} \ \mathrm{e}^{-by} \cos(ay)$ |
| $\frac{a + x}{b^2 + (a + x)^2} + \frac{a - x}{b^2 + (a - x)^2}$ | $\sqrt{2\pi} \ \mathrm{e}^{-by} \sin(ay)$ |
| $(a^2 + x^2)^{-\frac{1}{2}}$ | $\sqrt{\frac{2}{\pi}} \ K_0(ay)$ |

| $f(x)$ | $F(y) = \sqrt{\frac{2}{\pi}} \int_0^\infty f(x) \cos(xy) \mathrm{d}x$ |
|---|---|
| $(a^2 - x^2)^{-\frac{1}{2}}, \quad 0 < x < a$<br>$0, \quad x > a$ | $\sqrt{\frac{\pi}{2}}\, J_0(ay)$ |
| $x^{-\nu}, \quad 0 < \operatorname{Re}\nu < 1$ | $\sqrt{\frac{2}{\pi}} \sin\left(\frac{\pi\nu}{2}\right) \Gamma(1-\nu) y^{\nu-1}$ |
| $\mathrm{e}^{-ax}$ | $\sqrt{\frac{2}{\pi}}\, \frac{a}{a^2 + y^2}$ |
| $\frac{\mathrm{e}^{-bx} - \mathrm{e}^{-ax}}{x}$ | $\frac{1}{\sqrt{2\pi}} \ln\left(\frac{a^2 + y^2}{b^2 + y^2}\right)$ |
| $\sqrt{x}\, \mathrm{e}^{-ax}$ | $\frac{\sqrt{2}}{2} (a^2 + y^2)^{-\frac{3}{4}} \cos\left(\frac{3}{2} \arctan\left(\frac{y}{a}\right)\right)$ |
| $\frac{\mathrm{e}^{-ax}}{\sqrt{x}}$ | $\left(\frac{a + (a^2 + y^2)^{\frac{1}{2}}}{a^2 + y^2}\right)^{\frac{1}{2}}$ |
| $x^n \mathrm{e}^{-ax}$ | $\sqrt{\frac{2}{\pi}}\, n!\, a^{n+1} (a^2 + y^2)^{-(n+1)} \cdot \sum_{0 \le 2m \le n+1} (-1)^m \binom{n+1}{2m} \left(\frac{y}{a}\right)^{2m}$ |
| $x^{\nu-1} \mathrm{e}^{-ax}$ | $\sqrt{\frac{2}{\pi}}\, \Gamma(\nu) (a^2 + y^2)^{-\frac{\nu}{2}} \cos\left(\nu \arctan\left(\frac{y}{a}\right)\right)$ |
| $\frac{1}{x}\left(\frac{1}{2} - \frac{1}{x} + \frac{1}{\mathrm{e}^x - 1}\right)$ | $-\frac{1}{\sqrt{2\pi}} \ln(1 - \mathrm{e}^{-2\pi y})$ |
| $\mathrm{e}^{-ax^2}$ | $\frac{\sqrt{2}}{2}\, a^{-\frac{1}{2}} \exp\left(-\frac{y^2}{4a}\right)$ |
| $x^{-\frac{1}{2}} \exp\left(-\frac{a}{x}\right)$ | $\frac{1}{\sqrt{y}}\, \mathrm{e}^{-\sqrt{2ay}} \left(\cos\sqrt{2ay} - \sin\sqrt{2ay}\right)$ |
| $x^{-\frac{3}{2}} \exp\left(-\frac{a}{x}\right)$ | $\sqrt{\frac{2}{a}}\, \mathrm{e}^{-\sqrt{2ay}} \cos\sqrt{2ay}$ |
| $\ln x, \quad 0 < x < 1$<br>$0, \quad x > 1$ | $-\sqrt{\frac{2}{\pi}}\, \frac{\operatorname{Si}(y)}{y}$ |
| $\frac{\ln x}{\sqrt{x}}$ | $-\frac{1}{\sqrt{y}} \left(\mathrm{C} + \frac{\pi}{2} + \ln 4y\right)$ |
| $(x^2 - a^2)^{-1} \ln\left(\frac{x}{a}\right)$ | $\sqrt{\frac{\pi}{2}} \cdot \frac{1}{a} \Big(\sin(ay)\operatorname{Ci}(ay) - \cos(ay)\operatorname{si}(ay)\Big)$ |

| $f(x)$ | $F(y) = \sqrt{\frac{2}{\pi}} \int\limits_0^\infty f(x) \cos(xy) \mathrm{d}x$ |
|---|---|
| $(x^2 - a^2)^{-1} \ln(bx)$ | $\sqrt{\frac{\pi}{2}} \cdot \frac{1}{a} \Big( \sin(ay) \big[ \mathrm{Ci}(ay) - \ln(ab) \big] - \cos(ay) \mathrm{si}(ay) \Big)$ |
| $\frac{1}{x} \ln(1+x)$ | $\frac{1}{\sqrt{2\pi}} \left[ \left( \mathrm{Ci}\left(\frac{y}{2}\right) \right)^2 + \left( \mathrm{si}\left(\frac{y}{2}\right) \right)^2 \right]$ |
| $\ln \left\| \frac{a+x}{b-x} \right\|$ | $\sqrt{\frac{2}{\pi}} \cdot \frac{1}{y} \Big\{ \frac{\pi}{2} \big[ \cos(by) - \cos(ay) \big] + \cos(by) \mathrm{Si}(by)$ $+ \cos(ay) \mathrm{Si}(ay) - \sin(ay) \mathrm{Ci}(ay) - \sin(by) \mathrm{Ci}(by) \Big\}$ |
| $\mathrm{e}^{-ax} \ln x$ | $-\sqrt{\frac{2}{\pi}} \frac{1}{a^2 + y^2} \left[ a\mathrm{C} + \frac{a}{2} \ln(a^2 + y^2) + y \arctan\left(\frac{y}{a}\right) \right]$ |
| $\ln \left( \frac{a^2 + x^2}{b^2 + x^2} \right)$ | $\frac{\sqrt{2\pi}}{y} \left( \mathrm{e}^{-by} - \mathrm{e}^{-ay} \right)$ |
| $\ln \left\| \frac{a^2 + x^2}{b^2 - x^2} \right\|$ | $\frac{\sqrt{2\pi}}{y} \Big( \cos(by) - \mathrm{e}^{-ay} \Big)$ |
| $\frac{1}{x} \ln \left( \frac{a+x}{a-x} \right)^2$ | $-2\sqrt{2\pi}\, \mathrm{si}(ay)$ |
| $\frac{\ln(a^2 + x^2)}{\sqrt{a^2 + x^2}}$ | $-\sqrt{\frac{2}{\pi}} \left[ \left( \mathrm{C} + \ln\left(\frac{2y}{a}\right) \right) K_0(ay) \right]$ |
| $\ln \left( 1 + \frac{a^2}{x^2} \right)$ | $\sqrt{2\pi}\, \frac{1 - \mathrm{e}^{-ay}}{y}$ |
| $\ln \left\| 1 - \frac{a^2}{x^2} \right\|$ | $\sqrt{2\pi}\, \frac{1 - \cos(ay)}{y}$ |
| $\frac{\sin(ax)}{x}$ | $\sqrt{\frac{\pi}{2}}, \quad y < a$ <br> $\frac{1}{2}\sqrt{\frac{\pi}{2}}, \quad y = a$ <br> $0, \quad y > a$ |
| $\frac{x \sin(ax)}{x^2 + b^2}$ | $\sqrt{\frac{\pi}{2}}\, \mathrm{e}^{-ab} \cosh(by), \quad y < a$ <br> $-\sqrt{\frac{\pi}{2}}\, \mathrm{e}^{-by} \sinh(ab), \quad y > a$ |

| $f(x)$ | $F(y)=\sqrt{\frac{2}{\pi}}\int_0^\infty f(x)\cos(xy)\mathrm{d}x$ |
|---|---|
| $\dfrac{\sin(ax)}{x(x^2+b^2)}$ | $\sqrt{\frac{\pi}{2}}\,b^{-2}\big(1-\mathrm{e}^{-ab}\cosh(by)\big),\quad y<a$<br>$\sqrt{\frac{\pi}{2}}\,b^{-2}\mathrm{e}^{-by}\sinh(ab),\quad y>a$ |
| $\mathrm{e}^{-bx}\sin(ax)$ | $\dfrac{1}{\sqrt{2\pi}}\left[\dfrac{a+y}{b^2+(a+y)^2}+\dfrac{a-y}{b^2+(a-y)^2}\right]$ |
| $\dfrac{\mathrm{e}^{-x}\sin x}{x}$ | $\dfrac{1}{\sqrt{2\pi}}\arctan\left(\dfrac{2}{y^2}\right)$ |
| $\dfrac{\sin^2(ax)}{x}$ | $\dfrac{1}{2\sqrt{2\pi}}\ln\left\lvert 1-4\dfrac{a^2}{y^2}\right\rvert$ |
| $\dfrac{\sin(ax)\sin(bx)}{x}$ | $\dfrac{1}{\sqrt{2\pi}}\ln\left\lvert\dfrac{(a+b)^2-y^2}{(a-b)^2-y^2}\right\rvert$ |
| $\dfrac{\sin^2(ax)}{x^2}$ | $\sqrt{\frac{\pi}{2}}\left(a-\frac{1}{2}y\right),\quad y<2a$<br>$0,\quad y>2a$ |
| $\dfrac{\sin^3(ax)}{x^2}$ | $\dfrac{1}{4\sqrt{2\pi}}\Big\{(y+3a)\ln(y+3a)+(y-3a)\ln\lvert y-3a\rvert$<br>$-(y+a)\ln(y+a)-(y-a)\ln\lvert y-a\rvert\Big\}$ |
| $\dfrac{\sin^3(ax)}{x^3}$ | $\frac{1}{4}\sqrt{\frac{\pi}{2}}\,(3a^2-y^2),\quad 0<y<a$<br>$\frac{1}{2}\sqrt{\frac{\pi}{2}}\,y^2,\quad y=a$<br>$\frac{1}{8}\sqrt{\frac{\pi}{2}}\,(3a-y)^2,\quad a<y<3a$<br>$0,\quad y>3a$ |
| $\dfrac{1-\cos(ax)}{x}$ | $\dfrac{1}{\sqrt{2\pi}}\ln\left\lvert 1-\dfrac{a^2}{y^2}\right\rvert$ |
| $\dfrac{1-\cos(ax)}{x^2}$ | $\sqrt{\frac{\pi}{2}}\,(a-y),\quad y<a$<br>$0,\quad y>a$ |

| $f(x)$ | $F(y)=\sqrt{\frac{2}{\pi}}\int\limits_0^\infty f(x)\cos(xy)\mathrm{d}x$ |
|---|---|
| $\dfrac{\cos(ax)}{b^2+x^2}$ | $\sqrt{\frac{\pi}{2}}\,\dfrac{\mathrm{e}^{-ab}\cosh(by)}{b},\quad y<a$ <br> $\sqrt{\frac{\pi}{2}}\,\dfrac{\mathrm{e}^{-by}\cosh(ab)}{b},\quad y>a$ |
| $\mathrm{e}^{-bx}\cos(ax)$ | $\dfrac{b}{\sqrt{2\pi}}\left[\dfrac{1}{b^2+(a-y)^2}+\dfrac{1}{b^2+(a+y)^2}\right]$ |
| $\mathrm{e}^{-bx^2}\cos(ax)$ | $\dfrac{1}{\sqrt{2b}}\exp\left(-\frac{a^2+y^2}{4b}\right)\cosh\left(\dfrac{ay}{2b}\right)$ |
| $\dfrac{x}{b^2+x^2}\tan(ax)$ | $\sqrt{2\pi}\cosh(by)\left(1+\mathrm{e}^{2ab}\right)^{-1}$ |
| $\dfrac{x}{b^2+x^2}\cot(ax)$ | $\sqrt{2\pi}\cosh(by)\left(\mathrm{e}^{2ab}-1\right)^{-1}$ |
| $\sin(ax^2)$ | $\dfrac{1}{2\sqrt{a}}\left(\cos\left(\dfrac{y^2}{4a}\right)-\sin\left(\dfrac{y^2}{4a}\right)\right)$ |
| $\sin\left[a(1-x^2)\right]$ | $-\dfrac{1}{\sqrt{2a}}\cos\left(a+\dfrac{\pi}{4}+\dfrac{y^2}{4a}\right)$ |
| $\dfrac{\sin(ax^2)}{x^2}$ | $\sqrt{\frac{\pi}{2}}\,y\left[S\left(\dfrac{y^2}{4a}\right)-C\left(\dfrac{y^2}{4a}\right)\right]+\sqrt{2a}\,\sin\left(\dfrac{\pi}{4}+\dfrac{y^2}{4a}\right)$ |
| $\dfrac{\sin(ax^2)}{x}$ | $\sqrt{\frac{\pi}{2}}\left\{\dfrac{1}{2}-\left[C\left(\dfrac{y^2}{4a}\right)\right]^2-\left[S\left(\dfrac{y^2}{4a}\right)\right]^2\right\}$ |
| $\exp(-ax^2)\sin(bx^2)$ | $\dfrac{1}{\sqrt{2}}\left(a^2+b^2\right)^{-\frac{1}{4}}\exp\left(-\frac{1}{4}ay^2(a^2+b^2)^{-1}\right)$ <br> $\times\sin\left[\dfrac{1}{2}\arctan\left(\dfrac{b}{a}\right)-\dfrac{by^2}{4(a^2+b^2)}\right]$ |
| $\cos(ax^2)$ | $\dfrac{1}{2\sqrt{a}}\left[\cos\left(\dfrac{y^2}{4a}\right)+\sin\left(\dfrac{y^2}{4a}\right)\right]$ |
| $\cos\left[a(1-x^2)\right]$ | $\dfrac{1}{\sqrt{2a}}\sin\left(a+\dfrac{\pi}{4}+\dfrac{y^2}{4a}\right)$ |
| $\exp(-ax^2)\cos(bx^2)$ | $\dfrac{1}{\sqrt{2}}\left(a^2+b^2\right)^{-\frac{1}{4}}\exp\left(-\frac{1}{4}ay^2(a^2+y^2)^{-1}\right)$ <br> $\times\cos\left[\dfrac{by^2}{4(a^2+b^2)}-\dfrac{1}{2}\arctan\left(\dfrac{b}{a}\right)\right]$ |

| $f(x)$ | $F(y) = \sqrt{\frac{2}{\pi}} \int\limits_0^\infty f(x) \cos(xy) \mathrm{d}x$ |
|---|---|
| $\frac{1}{x} \sin\left(\frac{a}{x}\right)$ | $\sqrt{\frac{\pi}{2}}\, J_0(2\sqrt{ay})$ |
| $\frac{1}{\sqrt{x}} \sin\left(\frac{a}{x}\right)$ | $\frac{1}{2\sqrt{y}} \left[\sin(2\sqrt{ay}) + \cos(2\sqrt{ay}) - \mathrm{e}^{-2\sqrt{ay}}\right]$ |
| $\left(\frac{1}{\sqrt{x}}\right)^3 \sin\left(\frac{a}{x}\right)$ | $\frac{1}{2\sqrt{a}} \left[\sin(2\sqrt{ay}) + \cos(2\sqrt{ay}) + \mathrm{e}^{-2\sqrt{ay}}\right]$ |
| $\frac{1}{\sqrt{x}} \cos\left(\frac{a}{x}\right)$ | $\frac{1}{2\sqrt{y}} \left[\cos(2\sqrt{ay}) - \sin(2\sqrt{ay}) + \mathrm{e}^{-2\sqrt{ay}}\right]$ |
| $\left(\frac{1}{\sqrt{x}}\right)^3 \cos\left(\frac{a}{x}\right)$ | $\frac{1}{2\sqrt{a}} \left[\cos(2\sqrt{ay}) - \sin(2\sqrt{ay}) + \mathrm{e}^{-2\sqrt{ay}}\right]$ |
| $\frac{1}{\sqrt{x}} \sin(a\sqrt{x})$ | $\frac{2}{\sqrt{y}} \left[C\left(\frac{a^2}{4y}\right) \sin\left(\frac{a^2}{4y}\right) - S\left(\frac{a^2}{4y}\right) \cos\left(\frac{a^2}{4y}\right)\right]$ |
| $\exp(-bx) \sin(a\sqrt{x})$ | $\frac{a}{\sqrt{2}} (b^2 + a^2)^{\frac{3}{4}} \exp\left(-\frac{1}{4} a^2 b (b^2 + y^2)^{-1}\right) \times \cos\left[\frac{a^2 y}{4(b^2 + y^2)} - \frac{3}{2} \arctan\left(\frac{y}{b}\right)\right]$ |
| $\frac{\sin(a\sqrt{x})}{x}$ | $\sqrt{2\pi} \left[S\left(\frac{a^2}{4y}\right) + C\left(\frac{a^2}{4y}\right)\right]$ |
| $\frac{1}{\sqrt{x}} \cos(a\sqrt{x})$ | $\sqrt{\frac{2}{y}} \sin\left(\frac{\pi}{4} + \frac{a^2}{4y}\right)$ |
| $\frac{\exp(-ax)}{\sqrt{x}} \cos(b\sqrt{x})$ | $\sqrt{2} (a^2 + y^2)^{-\frac{1}{4}} \exp\left(-\frac{1}{4} a b^2 (a^2 + b^2)^{-1}\right) \times \cos\left[\frac{b^2 y}{4(a^2 + y^2)} - \frac{1}{2} \arctan\left(\frac{y}{a}\right)\right]$ |
| $\exp(-a\sqrt{x}) \cos(a\sqrt{x})$ | $a\sqrt{2}\, (2y)^{-\frac{3}{2}} \exp\left(-\frac{a^2}{2y}\right)$ |
| $\frac{\mathrm{e}^{-a\sqrt{x}}}{\sqrt{x}} [\cos(a\sqrt{x}) - \sin(a\sqrt{x})]$ | $\frac{1}{\sqrt{y}} \exp\left(-\frac{a^2}{2y}\right)$ |

### 0.10.1.2 Fourier sine transform:

| $f(x)$ | $F(y)=\sqrt{\frac{2}{\pi}}\int\limits_0^\infty f(x)\sin(xy)\mathrm{d}x$ |
|---|---|
| $1,\quad 0<x<a$<br>$0,\quad x>a$ | $\sqrt{\frac{2}{\pi}}\,\frac{1-\cos(ay)}{y}$ |
| $x,\quad 0<x<1$<br>$2-x,\quad 1<x<2$<br>$0,\quad x>2$ | $4\sqrt{\frac{2}{\pi}}\,y^{-2}\sin y\,\sin^2\left(\frac{y}{2}\right)$ |
| $\frac{1}{x}$ | $\sqrt{\frac{\pi}{2}}$ |
| $\frac{1}{x},\quad 0<x<a$<br>$0,\quad x>a$ | $\sqrt{\frac{2}{\pi}}\,\mathrm{Si}(ay)$ |
| $0,\quad 0<x<a$<br>$\frac{1}{x},\quad x>a$ | $-\sqrt{\frac{2}{\pi}}\,\mathrm{si}(ay)$ |
| $\frac{1}{\sqrt{x}}$ | $\frac{1}{\sqrt{y}}$ |
| $\frac{1}{\sqrt{x}},\quad 0<x<a$<br>$0,\quad x>a$ | $\frac{2S(ay)}{\sqrt{y}}$ |
| $0,\quad 0<x<a$<br>$\frac{1}{\sqrt{x}},\quad x>a$ | $\frac{1-2S(ay)}{\sqrt{y}}$ |
| $\left(\frac{1}{\sqrt{x}}\right)^3$ | $2\sqrt{y}$ |
| $(a+x)^{-1},\quad a>0$ | $\sqrt{\frac{2}{\pi}}\Big[\sin(ay)\mathrm{Ci}(ay)-\cos(ay)\mathrm{si}(ay)\Big]$ |
| $(a-x)^{-1},\quad a>0$ | $\sqrt{\frac{2}{\pi}}\left[\sin(ay)\mathrm{Ci}(ay)-\cos(ay)\left(\frac{\pi}{2}+\mathrm{Si}(ay)\right)\right]$ |
| $\frac{x}{a^2+x^2}$ | $\sqrt{\frac{\pi}{2}}\,\mathrm{e}^{-ay}$ |
| $(a^2-x^2)^{-1}$ | $\sqrt{\frac{2}{\pi}}\cdot\frac{1}{a}\Big[\sin(ay)\mathrm{Ci}(ay)-\cos(ay)\mathrm{Si}(ay)\Big]$ |

| $f(x)$ | $F(y)=\sqrt{\frac{2}{\pi}}\int\limits_0^\infty f(x)\sin(xy)\mathrm{d}x$ |
|---|---|
| $\frac{b}{b^2+(a-x)^2}-\frac{b}{b^2+(a+x)^2}$ | $\sqrt{2\pi}\ \mathrm{e}^{-by}\sin(ay)$ |
| $\frac{a+x}{b^2+(a+x)^2}-\frac{a-x}{b^2+(a-x)^2}$ | $\sqrt{2\pi}\ \mathrm{e}^{-by}\cos(ay)$ |
| $\frac{x}{a^2-x^2}$ | $-\sqrt{\frac{\pi}{2}}\cos(ay)$ |
| $\frac{1}{x(a^2-x^2)}$ | $\sqrt{\frac{\pi}{2}}\ \frac{1-\cos(ay)}{a^2}$ |
| $\frac{1}{x(a^2+x^2)}$ | $\sqrt{\frac{\pi}{2}}\ \frac{1-\mathrm{e}^{-ay}}{a^2}$ |
| $x^{-\nu}, \qquad 0<\operatorname{Re}\nu<2$ | $\sqrt{\frac{2}{\pi}}\cos\left(\frac{\pi\nu}{2}\right)\Gamma(1-\nu)y^{\nu-1}$ |
| $\mathrm{e}^{-ax}$ | $\sqrt{\frac{2}{\pi}}\ \frac{y}{a^2+y^2}$ |
| $\frac{\mathrm{e}^{-ax}}{x}$ | $\sqrt{\frac{2}{\pi}}\arctan\left(\frac{y}{a}\right)$ |
| $\frac{\mathrm{e}^{-ax}-\mathrm{e}^{-bx}}{x^2}$ | $\sqrt{\frac{2}{\pi}}\left[\frac{1}{2}y\ \ln\left(\frac{b^2+y^2}{a^2+y^2}\right)+b\ \arctan\left(\frac{y}{b}\right)-a\ \arctan\left(\frac{y}{a}\right)\right]$ |
| $\sqrt{x}\ \mathrm{e}^{-ax}$ | $\frac{\sqrt{2}}{2}(a^2+y^2)^{-\frac{3}{4}}\sin\left[\frac{3}{2}\arctan\left(\frac{y}{a}\right)\right]$ |
| $\frac{\mathrm{e}^{-ax}}{\sqrt{x}}$ | $\left(\frac{(a^2+y^2)^{\frac{1}{2}}-a}{a^2+y^2}\right)^{\frac{1}{2}}$ |
| $x^n\mathrm{e}^{-ax}$ | $\sqrt{\frac{2}{\pi}}\ n!\,a^{n+1}(a^2+y^2)^{-(n+1)}\cdot\sum\limits_{m=0}^{[\frac{1}{2}n]}(-1)^m\binom{n+1}{2m+1}\left(\frac{y}{a}\right)^{2m+1}$ |
| $x^{\nu-1}\mathrm{e}^{-ax}$ | $\sqrt{\frac{2}{\pi}}\ \Gamma(\nu)(a^2+y^2)^{-\frac{\nu}{2}}\sin\left(\nu\arctan\left(\frac{y}{a}\right)\right)$ |
| $\exp\left(-\frac{1}{2}x\right)\left(1-\mathrm{e}^{-x}\right)^{-1}$ | $-\frac{1}{\sqrt{2\pi}}\ \tanh(\pi y)$ |

| $f(x)$ | $F(y)=\sqrt{\frac{2}{\pi}}\int\limits_0^\infty f(x)\sin(xy)\mathrm{d}x$ |
|---|---|
| $x\mathrm{e}^{-ax^2}$ | $\sqrt{\frac{2}{a}}\,\frac{y}{4a}\,\exp\left(-\frac{y^2}{4a}\right)$ |
| $x^{-\frac{1}{2}}\exp\left(-\frac{a}{x}\right)$ | $\frac{1}{\sqrt{y}}\,\mathrm{e}^{-\sqrt{2ay}}\left[\cos\sqrt{2ay}+\sin\sqrt{2ay}\right]$ |
| $x^{-\frac{3}{2}}\exp\left(-\frac{a}{x}\right)$ | $\sqrt{\frac{2}{a}}\,\mathrm{e}^{-\sqrt{2ay}}\sin\sqrt{2ay}$ |
| $\ln x,\quad 0<x<1$<br>$0,\quad x>1$ | $\sqrt{\frac{2}{\pi}}\,\frac{\mathrm{Ci}(y)-\mathrm{C}-\ln y}{y}$ |
| $\frac{\ln x}{x}$ | $-\sqrt{\frac{\pi}{2}}\,(\mathrm{C}+\ln y)$ |
| $\frac{\ln x}{\sqrt{x}}$ | $\frac{1}{\sqrt{y}}\left[\frac{\pi}{2}-\mathrm{C}-\ln 4y\right]$ |
| $x(x^2-a^2)^{-1}\ln(bx)$ | $\sqrt{\frac{\pi}{2}}\left[\cos(ay)\left[\ln(ab)-\mathrm{Ci}(ay)\right]-\sin(ay)\mathrm{si}(ay)\right]$ |
| $x(x^2-a^2)^{-1}\ln\left(\frac{x}{a}\right)$ | $-\sqrt{\frac{\pi}{2}}\left[\cos(ay)\mathrm{Ci}(ay)+\sin(ay)\mathrm{si}(ay)\right]$ |
| $\mathrm{e}^{-ax}\ln x$ | $\sqrt{\frac{2}{\pi}}\,\frac{1}{a^2+y^2}\left[a\ \arctan\left(\frac{y}{a}\right)-\mathrm{C}y-\frac{1}{2}y\ \ln(a^2+y^2)\right]$ |
| $\ln\left\|\frac{a+x}{b-x}\right\|$ | $\sqrt{\frac{2}{\pi}}\cdot\frac{1}{y}\Big\{\ln\left(\frac{a}{b}\right)+\cos(by)\mathrm{Ci}(by)-\cos(ay)\mathrm{Ci}(ay)$<br>$+\sin(by)\mathrm{Si}(by)-\sin(ay)\mathrm{Si}(ay)$<br>$+\frac{\pi}{2}\left[\sin(by)+\sin(ay)\right]\Big\}$ |
| $\ln\left\|\frac{a+x}{a-x}\right\|$ | $\frac{\sqrt{2\pi}}{y}\ \sin(ay)$ |
| $\frac{1}{x^2}\ln\left(\frac{a+x}{a-x}\right)^2$ | $\frac{2\sqrt{2\pi}}{a}\left[1-\cos(ay)-ay\ \mathrm{si}(ay)\right]$ |
| $\ln\left(\frac{a^2+x^2+x}{a^2+x^2-x}\right)$ | $\frac{2\sqrt{2\pi}}{y}\ \exp\left(-y\sqrt{a^2-\frac{1}{4}}\right)\sin\left(\frac{y}{2}\right)$ |
| $\ln\left\|1-\frac{a^2}{x^2}\right\|$ | $\frac{2}{y}\sqrt{\frac{2}{\pi}}\left[\mathrm{C}+\ln(ay)-\cos(ay)\mathrm{Ci}(ay)-\sin(ay)\mathrm{Si}(ay)\right]$ |

| $f(x)$ | $F(y)=\sqrt{\frac{2}{\pi}}\int_0^\infty f(x)\sin(xy)\mathrm{d}x$ |
|---|---|
| $\ln\left(\frac{a^2+(b+x)^2}{a^2+(b-x)^2}\right)$ | $\frac{2\sqrt{2\pi}}{y}\ \mathrm{e}^{-ay}\sin(by)$ |
| $\frac{1}{x}\ln\lvert 1-a^2x^2\rvert$ | $-\sqrt{2\pi}\ \mathrm{Ci}\left(\frac{y}{a}\right)$ |
| $\frac{1}{x}\ln\left\lvert 1-\frac{a^2}{x^2}\right\rvert$ | $\sqrt{2\pi}\ [\mathrm{C}+\ln(ay)-\mathrm{Ci}(ay)]$ |
| $\frac{\sin(ax)}{x}$ | $\frac{1}{\sqrt{2\pi}}\ln\left\lvert\frac{y+a}{y-a}\right\rvert$ |
| $\frac{\sin(ax)}{x^2}$ | $\sqrt{\frac{\pi}{2}}\,y, \quad 0<y<a$<br>$\sqrt{\frac{\pi}{2}}\,a, \quad y>a$ |
| $\frac{\sin(\pi x)}{1-x^2}$ | $\sqrt{\frac{2}{\pi}}\sin y, \quad 0\le y\le\pi$<br>$0, \quad y\ge\pi$ |
| $\frac{\sin(ax)}{b^2+x^2}$ | $\sqrt{\frac{\pi}{2}}\frac{\mathrm{e}^{-ab}}{b}\sinh(by), \quad 0<y<a$<br>$\sqrt{\frac{\pi}{2}}\frac{\mathrm{e}^{-by}}{b}\sinh(ab), \quad y>a$ |
| $\mathrm{e}^{-bx}\sin(ax)$ | $\frac{1}{\sqrt{2\pi}}\,b\left[\frac{1}{b^2+(a-y)^2}-\frac{1}{b^2+(a+y)^2}\right]$ |
| $\frac{\mathrm{e}^{-bx}\sin(ax)}{x}$ | $\frac{1}{4}\sqrt{\frac{2}{\pi}}\ln\left(\frac{b^2+(y+a)^2}{b^2+(y-a)^2}\right)$ |
| $\mathrm{e}^{-bx^2}\sin(ax)$ | $\frac{1}{\sqrt{2b}}\exp\left(-\frac{1}{4}\frac{a^2+y^2}{b}\right)\sinh\left(\frac{ay}{2b}\right)$ |
| $\frac{\sin^2(ax)}{x}$ | $\frac{1}{4}\sqrt{2\pi}, \quad 0<y<2a$<br>$\frac{1}{8}\sqrt{2\pi}, \quad y=2a$<br>$0, \quad y>2a$ |

| $f(x)$ | $F(y)=\sqrt{\frac{2}{\pi}}\int\limits_0^\infty f(x)\sin(xy)\mathrm{d}x$ |
|---|---|
| $\frac{\sin(ax)\sin(bx)}{x}$ | $0, \quad 0<y<a-b$<br>$\frac{1}{4}\sqrt{2\pi}, \quad a-b<y<a+b$<br>$0, \quad y>a+b$ |
| $\frac{\sin^2(ax)}{x^2}$ | $\frac{1}{4}\sqrt{\frac{2}{\pi}}\left[(y+2a)\ln(y+2a)+(y-2a)\ln\lvert y-2a\rvert-\frac{1}{2}y\ln y\right]$ |
| $\frac{\sin^2(ax)}{x^3}$ | $\frac{1}{4}\sqrt{2\pi}\,y\left(2a-\frac{y}{2}\right), \quad 0<y<2a$<br>$\sqrt{\frac{\pi}{2}}\,a^2, \quad y>2a$ |
| $\frac{\cos(ax)}{x}$ | $0, \quad 0<y<a$<br>$\frac{1}{4}\sqrt{2\pi}, \quad y=a$<br>$\sqrt{\frac{\pi}{2}}, \quad y>a$ |
| $\frac{x\cos(ax)}{b^2+x^2}$ | $-\sqrt{\frac{\pi}{2}}\,\mathrm{e}^{-ab}\sinh(by), \quad 0<y<a$<br>$\sqrt{\frac{\pi}{2}}\,\mathrm{e}^{-by}\cosh(ab), \quad y>a$ |
| $\sin(ax^2)$ | $\frac{1}{\sqrt{a}}\left[\cos\left(\frac{y^2}{4a}\right)C\left(\frac{y^2}{4a}\right)+\sin\left(\frac{y^2}{4a}\right)S\left(\frac{y^2}{4a}\right)\right]$ |
| $\frac{\sin(ax^2)}{x}$ | $\sqrt{\frac{\pi}{2}}\left[C\left(\frac{y^2}{4a}\right)-S\left(\frac{y^2}{4a}\right)\right]$ |
| $\cos(ax^2)$ | $\frac{1}{\sqrt{a}}\left[\sin\left(\frac{y^2}{4a}\right)C\left(\frac{y^2}{4a}\right)-\cos\left(\frac{y^2}{4a}\right)S\left(\frac{y^2}{4a}\right)\right]$ |
| $\frac{\cos(ax^2)}{x}$ | $\sqrt{\frac{\pi}{2}}\left[C\left(\frac{y^2}{4a}\right)+S\left(\frac{y^2}{4a}\right)\right]$ |
| $\mathrm{e}^{-a\sqrt{x}}\sin(a\sqrt{x})$ | $\frac{a}{2y\sqrt{y}}\ \exp\left(-\frac{a^2}{2y}\right)$ |

### 0.10.1.3 Fourier transform

| $f(x)$ | $F(y) = \frac{1}{\sqrt{2\pi}} \int_{-\infty}^{\infty} f(x) e^{-ixy} dy$ |
|---|---|
| $\exp\left(-\frac{x^2}{2}\right)$ | $\exp\left(-\frac{y^2}{2}\right)$ |
| $\exp\left(-\frac{x^2}{4a}\right)$, $\operatorname{Re} a > 0$, $\operatorname{Re}\sqrt{a} > 0$ | $\sqrt{2a}\, e^{-ay^2}$ |
| $A$ for $a \le x \le b$<br>$0$ otherwise | $\frac{iA}{y\sqrt{2\pi}} \left(e^{-iby} - e^{-iay}\right)$ for $y \neq 0$ |
| $e^{-ax} \cos bx$ for $x \geq 0$<br>$0$ for $x < 0$<br>$(b \geq 0,\ a > 0)$ | $\frac{a + iy}{\sqrt{2\pi}\left((a+iy)^2 + b^2\right)}$ |
| $e^{-ax} e^{ibx}$ for $x \geq 0$<br>$0$ for $x < 0$<br>$(b \geq 0,\ a > 0)$ | $\frac{1}{\sqrt{2\pi}\left(a + i(y-b)\right)}$ |
| $\delta_\varepsilon(x) := \frac{\varepsilon}{\pi(x^2+\varepsilon^2)}$ $(\varepsilon > 0)$ | $\frac{1}{\sqrt{2\pi}} e^{-\varepsilon\vert x\vert}$ |
| $\delta$ (Dirac delta distribution) | $\frac{1}{\sqrt{2\pi}}$ (D) |
| $\frac{1}{\sqrt{\vert x\vert}}$ | $\frac{1}{\sqrt{\vert y\vert}}$ (D) |
| $\frac{\operatorname{sgn} x}{\sqrt{\vert x\vert}}$ | $-i\frac{\operatorname{sgn} y}{\sqrt{\vert y\vert}}$ (D) |

Formulas indicated by (D) are to be understood in the sense of distributions (generalized functions; see [212]).

Numerous other formulas can be obtained from the relation

$$\begin{aligned} 2F(y) &= \sqrt{\frac{2}{\pi}} \int_0^\infty \left(f(x) + f(-x)\right) \cos(xy)\, dx \\ &\quad -i\sqrt{\frac{2}{\pi}} \int_0^\infty \left(f(x) - f(-x)\right) \sin(xy)\, dx, \end{aligned}$$

using the previous tables for the Fourier cosine and Fourier sine transforms.

## 0.10.2 Laplace transformation

### 0.10.2.1 Table of the inverse transformations for functions whose Laplace transformation is a rational function

The table is ordered by the degree of the denominator. It is complete up to degree 3 and contains a few functions with denominators of higher degree.

| $\mathscr{L}\{f(t)\}$ | $f(t)$ |
|---|---|
| $\frac{1}{s}$ | $1$ |
| $\frac{1}{s+\alpha}$ | $e^{-\alpha t}$ |
| $\frac{1}{s^2}$ | $t$ |
| $\frac{1}{s(s+\alpha)}$ | $\frac{1}{\alpha}\left[1-e^{-\alpha t}\right]$ |
| $\frac{1}{(s+\alpha)(s+\beta)}$ | $\frac{1}{\beta-\alpha}\left[e^{-\alpha t}-e^{-\beta t}\right]$ |
| $\frac{s}{(s+\alpha)(s+\beta)}$ | $\frac{1}{\alpha-\beta}\left[\alpha\,e^{-\alpha t}-\beta\,e^{-\beta t}\right]$ |
| $\frac{1}{(s+\alpha)^2}$ | $t\,e^{-\alpha t}$ |
| $\frac{s}{(s+\alpha)^2}$ | $e^{-\alpha t}(1-\alpha t)$ |
| $\frac{1}{s^2-\alpha^2}$ | $\frac{1}{\alpha}\sinh(\alpha t)$ |
| $\frac{s}{s^2-\alpha^2}$ | $\cosh(\alpha t)$ |
| $\frac{1}{s^2+\alpha^2}$ | $\frac{1}{\alpha}\sin(\alpha t)$ |
| $\frac{s}{s^2+\alpha^2}$ | $\cos\alpha t$ |
| $\frac{1}{(s+\beta)^2+\alpha^2}$ | $\frac{1}{\alpha}\,e^{-\beta t}\sin\alpha t$ |
| $\frac{s}{(s+\beta)^2+\alpha^2}$ | $e^{-\beta t}\left[\cos\alpha t-\frac{\beta}{\alpha}\sin\alpha t\right]$ |

| $\mathscr{L}\{f(t)\}$ | $f(t)$ |
|---|---|
| $\frac{1}{s^3}$ | $\frac{1}{2}t^2$ |
| $\frac{1}{s^2(s+\alpha)}$ | $\frac{1}{\alpha^2}\left(\mathrm{e}^{-\alpha t}+\alpha t-1\right)$ |
| $\frac{1}{s(s+\alpha)(s+\beta)}$ | $\frac{1}{\alpha\beta(\alpha-\beta)}\left[(\alpha-\beta)+\beta\mathrm{e}^{-\alpha t}-\alpha\mathrm{e}^{-\beta t}\right]$ |
| $\frac{1}{s(s+\alpha)^2}$ | $\frac{1}{\alpha^2}\left[1-\mathrm{e}^{-\alpha t}-\alpha t\,\mathrm{e}^{-\alpha t}\right]$ |
| $\frac{1}{(s+\alpha)(s+\beta)(s+\gamma)}$ | $\frac{1}{(\alpha-\beta)(\beta-\gamma)(\gamma-\alpha)}$ $\times\left[(\gamma-\beta)\mathrm{e}^{-\alpha t}+(\alpha-\gamma)\mathrm{e}^{-\beta t}+(\beta-\alpha)\mathrm{e}^{-\gamma t}\right]$ |
| $\frac{s}{(s+\alpha)(s+\beta)(s+\gamma)}$ | $\frac{1}{(\alpha-\beta)(\beta-\gamma)(\gamma-\alpha)}$ $\times\left[\alpha(\beta-\gamma)\mathrm{e}^{-\alpha t}+\beta(\gamma-\alpha)\mathrm{e}^{-\beta t}+\gamma(\alpha-\beta)\mathrm{e}^{-\gamma t}\right]$ |
| $\frac{s^2}{(s+\alpha)(s+\beta)(s+\gamma)}$ | $\frac{1}{(\alpha-\beta)(\beta-\gamma)(\gamma-\alpha)}$ $\times\left[-\alpha^2(\beta-\gamma)\mathrm{e}^{-\alpha t}-\beta^2(\gamma-\alpha)\mathrm{e}^{-\beta t}-\gamma^2(\alpha-\beta)\mathrm{e}^{-\gamma t}\right]$ |
| $\frac{1}{(s+\alpha)(s+\beta)^2}$ | $\frac{1}{(\beta-\alpha)^2}\left[\mathrm{e}^{-\alpha t}-\mathrm{e}^{-\beta t}-(\beta-\alpha)t\,\mathrm{e}^{-\beta t}\right]$ |
| $\frac{s}{(s+\alpha)(s+\beta)^2}$ | $\frac{1}{(\beta-\alpha)^2}\left[-\alpha\mathrm{e}^{-\alpha t}+\left[\alpha+\beta t(\beta-\alpha)\right]\mathrm{e}^{-\beta t}\right]$ |
| $\frac{s^2}{(s+\alpha)(s+\beta)^2}$ | $\frac{1}{(\beta-\alpha)^2}\left[\alpha^2\mathrm{e}^{-\alpha t}+\beta\left[\beta-2\alpha-\beta^2 t+\alpha\beta t\right]\mathrm{e}^{-\beta t}\right]$ |
| $\frac{1}{(s+\alpha)^3}$ | $\frac{t^2}{2}\,\mathrm{e}^{-\alpha t}$ |
| $\frac{s}{(s+\alpha)^3}$ | $\mathrm{e}^{-\alpha t}\,t\left[1-\frac{\alpha}{2}t\right]$ |
| $\frac{s^2}{(s+\alpha)^3}$ | $\mathrm{e}^{-\alpha t}\left[1-2\alpha t+\frac{\alpha^2}{2}t^2\right]$ |
| $\frac{1}{s\left[(s+\beta)^2+\alpha^2\right]}$ | $\frac{1}{\alpha^2+\beta^2}\left[1-\mathrm{e}^{-\beta t}\left(\cos\alpha t+\frac{\beta}{\alpha}\sin\alpha t\right)\right]$ |

| $\mathscr{L}\{f(t)\}$ | $f(t)$ |
|---|---|
| $\dfrac{1}{s(s^2+\alpha^2)}$ | $\dfrac{1}{\alpha^2}(1-\cos\alpha t)$ |
| $\dfrac{1}{(s+\alpha)(s^2+\beta^2)}$ | $\dfrac{1}{\alpha^2+\beta^2}\left[\mathrm{e}^{-\alpha t}+\dfrac{\alpha}{\beta}\sin\beta t-\cos\beta t\right]$ |
| $\dfrac{s}{(s+\alpha)(s^2+\beta^2)}$ | $\dfrac{1}{\alpha^2+\beta^2}\left[-\alpha\,\mathrm{e}^{-\alpha t}+\alpha\cos\beta t+\beta\sin\beta t\right]$ |
| $\dfrac{s^2}{(s+\alpha)(s^2+\beta^2)}$ | $\dfrac{1}{\alpha^2+\beta^2}\left[\alpha^2\mathrm{e}^{-\alpha t}-\alpha\beta\sin\beta t+\beta^2\cos\beta t\right]$ |
| $\dfrac{1}{(s+\alpha)\left[(s+\beta)^2+\gamma^2\right]}$ | $\dfrac{1}{(\beta-\alpha)^2+\gamma^2}\left[\mathrm{e}^{-\alpha t}-\mathrm{e}^{-\beta t}\cos\gamma t+\dfrac{\alpha-\beta}{\gamma}\,\mathrm{e}^{-\beta t}\sin\gamma t\right]$ |
| $\dfrac{s}{(s+\alpha)\left[(s+\beta)^2+\gamma^2\right]}$ | $\dfrac{1}{(\beta-\alpha)^2+\gamma^2}\left[-\alpha\mathrm{e}^{-\alpha t}+\alpha\mathrm{e}^{-\beta t}\cos\gamma t-\dfrac{\alpha\beta-\beta^2-\gamma^2}{\gamma}\,\mathrm{e}^{-\beta t}\sin\gamma t\right]$ |
| $\dfrac{s^2}{(s+\alpha)\left[(s+\beta)^2+\gamma^2\right]}$ | $\dfrac{1}{(\beta-\alpha)^2+\gamma^2}\left[\alpha^2\mathrm{e}^{-\alpha t}+((\alpha-\beta)^2+\gamma^2-\alpha^2)\mathrm{e}^{-\beta t}\cos\gamma t-\left(\alpha\gamma+\beta\left(\gamma-\dfrac{(\alpha-\beta)\beta}{\gamma}\right)\right)\mathrm{e}^{-\beta t}\sin\gamma t\right]$ |
| $\dfrac{1}{s^4}$ | $\dfrac{1}{6}t^3$ |
| $\dfrac{1}{s^3(s+\alpha)}$ | $\dfrac{1}{\alpha^3}-\dfrac{1}{\alpha^2}t+\dfrac{1}{2\alpha}t^2-\dfrac{1}{\alpha^3}\,\mathrm{e}^{-\alpha t}$ |
| $\dfrac{1}{s^2(s+\alpha)(s+\beta)}$ | $-\dfrac{\alpha+\beta}{\alpha^2\beta^2}+\dfrac{1}{\alpha\beta}t+\dfrac{1}{\alpha^2(\beta-\alpha)}\,\mathrm{e}^{-\alpha t}+\dfrac{1}{\beta^2(\alpha-\beta)}\,\mathrm{e}^{-\beta t}$ |
| $\dfrac{1}{s^2(s+\alpha)^2}$ | $\dfrac{1}{\alpha^2}t(1+\mathrm{e}^{-\alpha t})+\dfrac{2}{\alpha^3}(\mathrm{e}^{-\alpha t}-1)$ |
| $\dfrac{1}{(s+\alpha)^2(s+\beta)^2}$ | $\dfrac{1}{(\alpha-\beta)^2}\left[\mathrm{e}^{-\alpha t}\left(t+\dfrac{2}{\alpha-\beta}\right)+\mathrm{e}^{-\beta t}\left(t-\dfrac{2}{\alpha-\beta}\right)\right]$ |
| $\dfrac{1}{(s+\alpha)^4}$ | $\dfrac{1}{6}t^3\mathrm{e}^{-\alpha t}$ |
| $\dfrac{s}{(s+\alpha)^4}$ | $\dfrac{1}{2}t^2\mathrm{e}^{-\alpha t}-\dfrac{\alpha}{6}t^3\mathrm{e}^{-\alpha t}$ |

| $\mathscr{L}\{f(t)\}$ | $f(t)$ |
|---|---|
| $\dfrac{1}{(s^2+\alpha^2)(s^2+\beta^2)}$ | $\dfrac{1}{\beta^2-\alpha^2}\left[\dfrac{1}{\alpha}\sin\alpha t-\dfrac{1}{\beta}\sin\beta t\right]$ |
| $\dfrac{s}{(s^2+\alpha^2)(s^2+\beta^2)}$ | $\dfrac{1}{\beta^2-\alpha^2}\left[\cos\alpha t-\cos\beta t\right]$ |
| $\dfrac{s^2}{(s^2+\alpha^2)(s^2+\beta^2)}$ | $\dfrac{1}{\beta^2-\alpha^2}\left[-\alpha\sin\alpha t+\beta\sin\beta t\right]$ |
| $\dfrac{s^3}{(s^2+\alpha^2)(s^2+\beta^2)}$ | $\dfrac{1}{\beta^2-\alpha^2}\left[-\alpha^2\cos\alpha t+\beta^2\cos\beta t\right]$ |
| $\dfrac{1}{(s^2-\alpha^2)(s^2-\beta^2)}$ | $\dfrac{1}{\beta^2-\alpha^2}\left[\dfrac{1}{\beta}\sinh\beta t-\dfrac{1}{\alpha}\sinh\alpha t\right]$ |
| $\dfrac{s}{(s^2-\alpha^2)(s^2-\beta^2)}$ | $\dfrac{1}{\beta^2-\alpha^2}\left[\cosh\beta t-\cosh\alpha t\right]$ |
| $\dfrac{s^2}{(s^2-\alpha^2)(s^2-\beta^2)}$ | $\dfrac{1}{\beta^2-\alpha^2}\left[\beta\sinh\beta t-\alpha\sinh\alpha t\right]$ |
| $\dfrac{s^3}{(s^2-\alpha^2)(s^2-\beta^2)}$ | $\dfrac{1}{\beta^2-\alpha^2}\left[\beta^2\cosh\beta t-\alpha^2\sinh\alpha t\right]$ |
| $\dfrac{1}{(s^2+\alpha^2)^2}$ | $\dfrac{1}{2\alpha^2}\left[\dfrac{1}{\alpha}\sin\alpha t-t\cos\alpha t\right]$ |
| $\dfrac{s}{(s^2+\alpha^2)^2}$ | $\dfrac{1}{2\alpha}\ t\sin\alpha t$ |
| $\dfrac{s^2}{(s^2+\alpha^2)^2}$ | $\dfrac{1}{2\alpha}\left[\sin\alpha t+\alpha t\cos\alpha t\right]$ |
| $\dfrac{s^3}{(s^2+\alpha^2)^2}$ | $\dfrac{1}{2}\left[2\cos\alpha t-\alpha t\sin\alpha t\right]$ |
| $\dfrac{1}{(s^2-\alpha^2)^2}$ | $\dfrac{1}{2\alpha^2}\left[t\cosh\alpha t-\dfrac{1}{\alpha}\sinh\alpha t\right]$ |
| $\dfrac{s}{(s^2-\alpha^2)^2}$ | $\dfrac{1}{2\alpha}\ t\sinh\alpha t$ |
| $\dfrac{s^2}{(s^2-\alpha^2)^2}$ | $\dfrac{1}{2\alpha}\left[\sinh\alpha t+\alpha t\cosh\alpha t\right]$ |
| $\dfrac{s^3}{(s^2-\alpha^2)^2}$ | $\dfrac{1}{2}\left[2\cosh\alpha t+\alpha t\sinh\alpha t\right]$ |

| $\mathscr{L}\{f(t)\}$ | $f(t)$ |
|---|---|
| $\dfrac{1}{s^2(s^2+\alpha^2)}$ | $\dfrac{1}{\alpha^2}\left[t-\dfrac{1}{\alpha}\sin\alpha t\right]$ |
| $\dfrac{1}{s^2(s^2-\alpha^2)}$ | $\dfrac{1}{\alpha^2}\left[\dfrac{1}{\alpha}\sinh\alpha t-t\right]$ |
| $\dfrac{1}{s^4+\alpha^4}$ | $\dfrac{1}{\sqrt{2}\,\alpha^3}\left[\cosh\dfrac{\alpha}{\sqrt{2}}t\ \sin\dfrac{\alpha}{\sqrt{2}}t-\sinh\dfrac{\alpha}{\sqrt{2}}t\ \cos\dfrac{\alpha}{\sqrt{2}}t\right]$ |
| $\dfrac{s}{s^4+\alpha^4}$ | $\dfrac{1}{\alpha^2}\sin\dfrac{\alpha}{\sqrt{2}}t\ \sinh\dfrac{\alpha}{\sqrt{2}}t$ |
| $\dfrac{s^2}{s^4+\alpha^4}$ | $\dfrac{1}{\sqrt{2}\,\alpha}\left[\cos\dfrac{\alpha}{\sqrt{2}}t\ \sinh\dfrac{\alpha}{\sqrt{2}}t+\sin\dfrac{\alpha}{\sqrt{2}}t\ \cosh\dfrac{\alpha}{\sqrt{2}}t\right]$ |
| $\dfrac{s^3}{s^4+\alpha^4}$ | $\cos\dfrac{\alpha}{\sqrt{2}}t\ \cosh\dfrac{\alpha}{\sqrt{2}}t$ |
| $\dfrac{1}{s^4-\alpha^4}$ | $\dfrac{1}{2\alpha^3}\left[\sinh\alpha t-\sin\alpha t\right]$ |
| $\dfrac{s}{s^4-\alpha^4}$ | $\dfrac{1}{2\alpha^2}\left[\cosh\alpha t-\cos\alpha t\right]$ |
| $\dfrac{s^2}{s^4-\alpha^4}$ | $\dfrac{1}{2\alpha}\left[\sinh\alpha t+\sin\alpha t\right]$ |
| $\dfrac{s^3}{s^4-\alpha^4}$ | $\dfrac{1}{2}\left[\cosh\alpha t+\cos\alpha t\right]$ |
| $\dfrac{1}{s^2(s^2+\alpha^2)}$ | $\dfrac{1}{\alpha^2}\left[t-\dfrac{1}{\alpha}\sin\alpha t\right]$ |
| $\dfrac{1}{s^2(s^2-\alpha^2)}$ | $\dfrac{1}{\alpha^2}\left[\dfrac{1}{\alpha}\sinh\alpha t-t\right]$ |
| $\dfrac{1}{s^n}$ | $\dfrac{1}{(n-1)!}\,t^{n-1}$ |
| $\dfrac{1}{(s+\alpha)^n}$ | $\dfrac{1}{(n-1)!}\,t^{n-1}\mathrm{e}^{-\alpha t}$ |
| $\dfrac{1}{s(s+\alpha)^n}$ | $\dfrac{1}{\alpha^n}\left[1-\sum_{k=0}^{n-1}\dfrac{(\alpha t)^k}{k!}\ \mathrm{e}^{-\alpha t}\right]$ |
| $\dfrac{1}{s(\alpha s+1)\dots(\alpha s+n)}$ | $\dfrac{1}{n!}\left(1-\exp\left(-\frac{t}{\alpha}\right)\right)^n$ |

### 0.10.2.2 The Laplace transform of a few non-rational functions

In what follows, $\gamma$ denotes the constant $\gamma = \mathrm{e}^{\mathrm{C}}$; C is referred to as the *Euler constant*, defined as follows (see also section 0.1.1)

$$\mathrm{C} = \lim_{n\to\infty}\left(\sum_{\nu=1}^{n}\frac{1}{\nu} - \ln n\right) = 0.577\,215\,67\ldots$$

| $\mathscr{L}\{f(t)\}$ | $f(t)$ |
|---|---|
| $\dfrac{\ln s}{s}$ | $-\ln \gamma t$ |
| $-\dfrac{\ln \gamma s}{s}$ | $\ln t$ |
| $-\sqrt{\dfrac{\pi}{s}}\ \ln 4\gamma s$ | $\dfrac{\ln t}{\sqrt{t}}$ |
| $\dfrac{\ln s}{s^{n+1}}$ | $\dfrac{t^n}{n!}\left[1 + \dfrac{1}{2} + \ldots + \dfrac{1}{n} - \ln \gamma t\right]$ |
| $\dfrac{1}{s^{n+1}}\left[\sum\limits_{\nu=1}^{n}\dfrac{1}{\nu} - \ln \gamma s\right]$ | $\dfrac{t^n}{n!}\ \ln t$ |
| $\dfrac{(\ln s)^2}{s}$ | $(\ln \gamma t)^2 - \dfrac{\pi^2}{6}$ |
| $\dfrac{(\ln \gamma s)^2}{s}$ | $(\ln t)^2 - \dfrac{\pi^2}{6}$ |
| $\dfrac{1}{s^\alpha \ln s}$ | $\displaystyle\int_\alpha^\infty \frac{t^{u-1}}{\Gamma(u)}\,\mathrm{d}u$ (also valid for $\alpha = 0$) |
| $\ln\left(\dfrac{s+\alpha}{s-\alpha}\right)$ | $\dfrac{2}{t}\sinh \alpha t$ |
| $\ln\left(\dfrac{s-\alpha}{s-\beta}\right)$ | $\dfrac{1}{t}(\mathrm{e}^{\beta t} - \mathrm{e}^{\alpha t})$ |
| $\ln\left(\dfrac{s^2+\alpha^2}{s^2+\beta^2}\right)$ | $\dfrac{2}{t}(\cos \beta t - \cos \alpha t)$ |
| $\dfrac{1}{\sqrt{s}}$ | $\dfrac{1}{\sqrt{\pi t}}$ |
| $\dfrac{1}{s\sqrt{s}}$ | $2\sqrt{\dfrac{t}{\pi}}$ |

| $\mathscr{L}\{f(t)\}$ | $f(t)$ |
|---|---|
| $\dfrac{s+\alpha}{s\sqrt{s}}$ | $\dfrac{1+2\alpha t}{\sqrt{\pi t}}$ |
| $\sqrt{s-\alpha}-\sqrt{s-\beta}$ | $\dfrac{1}{2t\sqrt{\pi t}}\left(\mathrm{e}^{\beta t}-\mathrm{e}^{\alpha t}\right)$ |
| $\sqrt{\sqrt{s^2+\alpha^2}-s}$ | $\dfrac{\sin\alpha t}{t\sqrt{2\pi t}}$ |
| $\sqrt{\dfrac{\sqrt{s^2+\alpha^2}-s}{s^2+\alpha^2}}$ | $\sqrt{\dfrac{2}{\pi t}}\ \sin\alpha t$ |
| $\sqrt{\dfrac{\sqrt{s^2+\alpha^2}+s}{s^2+\alpha^2}}$ | $\sqrt{\dfrac{2}{\pi t}}\ \cos\alpha t$ |
| $\sqrt{\dfrac{\sqrt{s^2-\alpha^2}-s}{s^2-\alpha^2}}$ | $\sqrt{\dfrac{2}{\pi t}}\ \sinh\alpha t$ |
| $\sqrt{\dfrac{\sqrt{s^2-\alpha^2}+s}{s^2-\alpha^2}}$ | $\sqrt{\dfrac{2}{\pi t}}\ \cosh\alpha t$ |
| $\dfrac{1}{\sqrt{s}}\sin\dfrac{\alpha}{s}$ | $\dfrac{\sinh\sqrt{2\alpha t}\ \sin\sqrt{2\alpha t}}{\sqrt{\pi t}}$ |
| $\dfrac{1}{s\sqrt{s}}\sin\dfrac{\alpha}{s}$ | $\dfrac{\cosh\sqrt{2\alpha t}\ \sin\sqrt{2\alpha t}}{\sqrt{\alpha\pi}}$ |
| $\dfrac{1}{\sqrt{s}}\cos\dfrac{\alpha}{s}$ | $\dfrac{\cosh\sqrt{2\alpha t}\ \cos\sqrt{2\alpha t}}{\sqrt{\pi t}}$ |
| $\dfrac{1}{s\sqrt{s}}\cos\dfrac{\alpha}{s}$ | $\dfrac{\sinh\sqrt{2\alpha t}\ \cos\sqrt{2\alpha t}}{\sqrt{\alpha\pi}}$ |
| $\dfrac{1}{\sqrt{s}}\sinh\dfrac{\alpha}{s}$ | $\dfrac{\cosh 2\sqrt{\alpha t}-\cos 2\sqrt{\alpha t}}{2\sqrt{\pi t}}$ |
| $\dfrac{1}{s\sqrt{s}}\sinh\dfrac{\alpha}{s}$ | $\dfrac{\sinh 2\sqrt{\alpha t}-\sin 2\sqrt{\alpha t}}{2\sqrt{\alpha\pi}}$ |
| $\dfrac{1}{\sqrt{s}}\cosh\dfrac{\alpha}{s}$ | $\dfrac{\cosh 2\sqrt{\alpha t}+\cos 2\sqrt{\alpha t}}{2\sqrt{\pi t}}$ |

| $\mathscr{L}\{f(t)\}$ | $f(t)$ |
|---|---|
| $\frac{1}{s\sqrt{s}}\cosh\frac{\alpha}{s}$ | $\frac{\sinh 2\sqrt{\alpha t}+\sin 2\sqrt{\alpha t}}{2\sqrt{\alpha\pi}}$ |
| $\frac{1}{s^z}, \qquad (\operatorname{Re} z>0)$ | $\frac{t^{z-1}}{\Gamma(z)}$ |
| $\frac{1}{\sqrt{s}}\exp\left(\frac{1}{s}\right)$ | $\frac{\cosh 2\sqrt{t}}{\sqrt{\pi t}}$ |
| $\frac{1}{s\sqrt{s}}\exp\left(\frac{1}{s}\right)$ | $\frac{\sinh 2\sqrt{t}}{\sqrt{\pi}}$ |
| $\arctan\frac{\alpha}{s}$ | $\frac{\sin\alpha t}{t}$ |
| $\frac{\sin\left(\beta+\arctan\frac{\alpha}{s}\right)}{\sqrt{s^2+\alpha^2}}$ | $\sin(\alpha t+\beta)$ |
| $\frac{\cos\left(\beta+\arctan\frac{\alpha}{s}\right)}{\sqrt{s^2+\alpha^2}}$ | $\cos(\alpha t+\beta)$ |

### 0.10.2.3 The Laplace transform of a few piecewise continuous functions

In what follows, the symbol $[t]$ denotes the largest number $n$ with $n \le t$ (the function [-] is called the *Gauss bracket*). Correspondingly, $f\big([t]\big) = f(n)$ for $n \le t < n+1$; $n = 0, 1, 2, \ldots$

| $\mathscr{L}\{f(t)\}$ | $f(t)$ |
|---|---|
| $\frac{1}{s(\mathrm{e}^s-1)}$ | $[t]$ |
| $\frac{1}{s(\mathrm{e}^{\alpha s}-1)}$ | $\left[\frac{t}{\alpha}\right]$ |
| $\frac{1}{(1-\mathrm{e}^{-s})s}$ | $[t]+1$ |
| $\frac{1}{(1-\mathrm{e}^{-\alpha s})s}$ | $\left[\frac{t}{\alpha}\right]+1$ |

| $\mathscr{L}\{\boldsymbol{f(t)}\}$ | $\boldsymbol{f(t)}$ |
|---|---|
| $\dfrac{1}{s(e^s-\alpha)} \qquad (\alpha \neq 1)$ | $\dfrac{\alpha^{[t]}-1}{\alpha-1}$ |
| $\dfrac{e^s-1}{s(e^s-\alpha)}$ | $\alpha^{[t]}$ |
| $\dfrac{e^s-1}{s(e^s-\alpha)^2}$ | $[t]\,\alpha^{[t]-1}$ |
| $\dfrac{e^s-1}{s(e^s-\alpha)^3}$ | $\dfrac{1}{2}[t]\,\big([t]-1\big)\alpha^{[t]-2}$ |
| $\dfrac{e^s-1}{s(e^s-\alpha)(e^s-\beta)}$ | $\dfrac{\alpha^{[t]}-\beta^{[t]}}{\alpha-\beta}$ |
| $\dfrac{e^s+1}{s(e^s-1)^2}$ | $[t]^2$ |
| $\dfrac{(e^s-1)(e^s+\alpha)}{s(e^s-\alpha)^3}$ | $[t]^2\alpha^{[t]-1}$ |
| $\dfrac{(e^s-1)\sin\beta}{s\big(e^{2s}-2e^s\cos\beta+1\big)}$ | $\sin\beta[t]$ |
| $\dfrac{(e^s-1)(e^s-\cos\beta)}{s\big(e^{2s}-2e^s\cos\beta+1\big)}$ | $\cos\beta[t]$ |
| $\dfrac{(e^s-1)\alpha\sin\beta}{s\big(e^{2s}-2\alpha e^s\cos\beta+\alpha^2\big)}$ | $\alpha^{[t]}\sin\beta[t]$ |
| $\dfrac{(e^s-1)(e^s-\alpha\cos\beta)}{s\big(e^{2s}-2\alpha e^s\cos\beta+\alpha^2\big)}$ | $\alpha^{[t]}\cos\beta[t]$ |
| $\dfrac{e^{-\alpha s}}{s}$ | $\begin{Bmatrix} 0 & \text{for} & 0<t<\alpha \\ 1 & \text{for} & \alpha<t \end{Bmatrix}$ <br>  |
| $\dfrac{1-e^{-\alpha s}}{s}$ | $\begin{Bmatrix} 1 & \text{for} & 0<t<\alpha \\ 0 & \text{for} & \alpha<t \end{Bmatrix}$ <br>  |
| $\dfrac{e^{-\alpha s}-e^{-\beta s}}{s}$ | $\begin{Bmatrix} 0 & \text{for} & 0<t<\alpha \\ 1 & \text{for} & \alpha<t<\beta \\ 0 & \text{for} & \beta<t \end{Bmatrix}$ <br>  |

| $\mathscr{L}\{f(t)\}$ | $f(t)$ | |
|---|---|---|
| $\dfrac{(1-\mathrm{e}^{-\alpha s})^2}{s}$ | $\begin{cases} 1 & \text{for} \quad 0<t<\alpha \\ -1 & \text{for} \quad \alpha<t<2\alpha \\ 0 & \text{for} \quad 2\alpha<t \end{cases}$ | 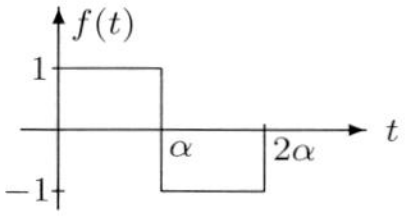  |
| $\dfrac{(\mathrm{e}^{-\alpha s}-\mathrm{e}^{-\beta s})^2}{s}$ | $\begin{cases} 0 & \text{for} \quad 0<t<2\alpha \\ 1 & \text{for} \quad 2\alpha<t<\alpha+\beta \\ -1 & \text{for} \quad \alpha+\beta<t<2\beta \\ 0 & \text{for} \quad 2\beta<t \end{cases}$ | 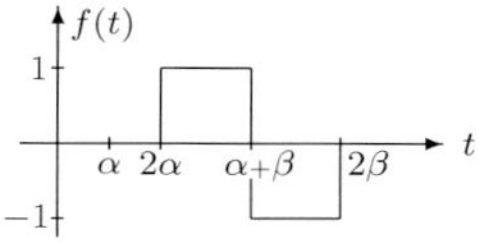  |
| $\dfrac{(1-\mathrm{e}^{-\alpha s})^2}{s^2}$ | $\begin{cases} t & \text{for} \quad 0<t<\alpha \\ 2\alpha-t & \text{for} \quad \alpha<t<2\alpha \\ 0 & \text{for} \quad 2\alpha<t \end{cases}$ | 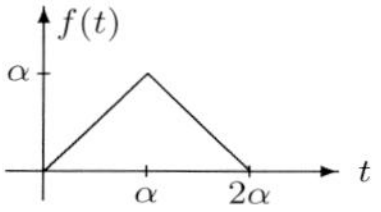  |
| $\dfrac{(\mathrm{e}^{-\alpha s}-\mathrm{e}^{-\beta s})^2}{s^2}$ | $\begin{cases} 0 & \text{for} \quad 0<t<2\alpha \\ t-2\alpha & \text{for} \quad 2\alpha<t<\alpha+\beta \\ 2\beta-t & \text{for} \quad \alpha+\beta<t<2\beta \\ 0 & \text{for} \quad 2\beta<t \end{cases}$ | 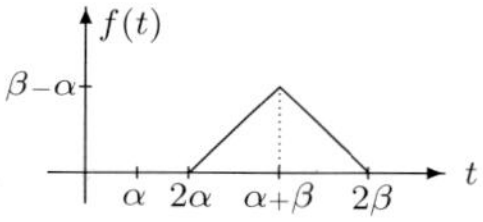  |
| $\dfrac{\beta \mathrm{e}^{-\alpha s}}{s(s+\beta)}$ | $\begin{cases} 0 & \text{for} \quad 0<t<\alpha \\ 1-\mathrm{e}^{-\beta(t-\alpha)} & \text{for} \quad \alpha<t \end{cases}$ | 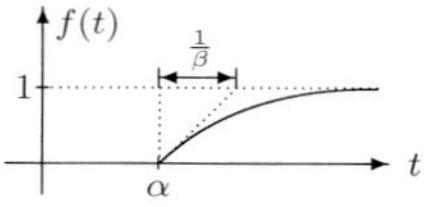  |
| $\dfrac{\mathrm{e}^{-\alpha s}}{s+\beta}$ | $\begin{cases} 0 & \text{for} \quad 0<t<\alpha \\ \mathrm{e}^{-\beta(t-\alpha)} & \text{for} \quad \alpha<t \end{cases}$ | 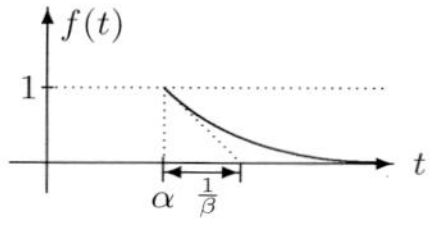  |
| $\dfrac{1-\mathrm{e}^{-\alpha s}}{s^2}$ | $\begin{cases} t & \text{for} \quad 0<t<\alpha \\ \alpha & \text{for} \quad \alpha<t \end{cases}$ | 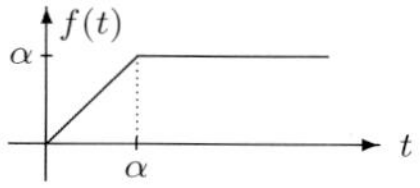  |
| $\dfrac{\mathrm{e}^{-\alpha s}-\mathrm{e}^{-\beta s}}{s^2}$ | $\begin{cases} 0 & \text{for} \quad 0<t<\alpha \\ t-\alpha & \text{for} \quad \alpha<t<\beta \\ \beta-\alpha & \text{for} \quad \beta<t \end{cases}$ | 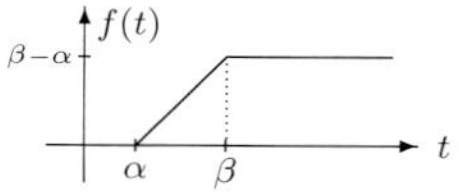  |
| $\dfrac{1}{s(1+\mathrm{e}^{-\alpha s})}$ | $\begin{cases} 1 \text{ for } 2n\alpha<t<(2n+1)\alpha \\ 0 \text{ for } (2n+1)\alpha<t<(2n+2)\alpha \end{cases}$ <br> $n=0,1,2,\ldots$ | 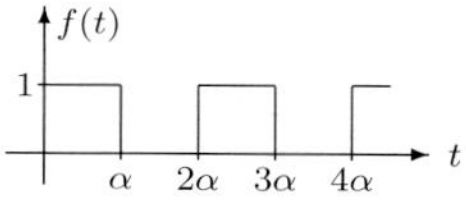  |

| $\mathscr{L}\{f(t)\}$ | $f(t)$ |
|---|---|
| $\dfrac{1}{s(1+e^{\alpha s})}$ | $\begin{Bmatrix} 0 \text{ for } 2n\alpha < t < (2n+1)\alpha \\ 1 \text{ for } (2n+1)\alpha < t < (2n+2)\alpha \end{Bmatrix}$ $n = 0, 1, 2, \dots$ 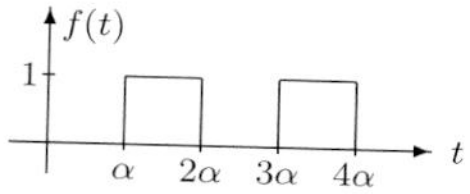  |
| $\dfrac{1-e^{-\alpha s}}{s(1+e^{-\alpha s})}$ | $\begin{Bmatrix} 1 \text{ for } 2n\alpha < t < (2n+1)\alpha \\ -1 \text{ for } (2n+1)\alpha < t < (2n+2)\alpha \end{Bmatrix}$ $n = 0, 1, 2, \dots$ 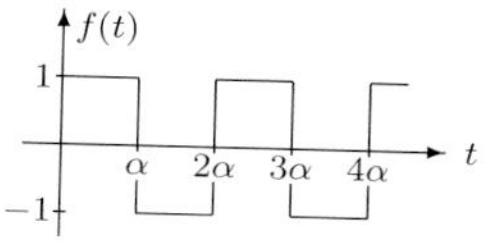  |
| $\dfrac{e^{-\alpha s}-1}{s(1+e^{-\alpha s})}$ | $\begin{Bmatrix} -1 \text{ for } 2n\alpha < t < (2n+1)\alpha \\ 1 \text{ for } (2n+1)\alpha < t < (2n+2)\alpha \end{Bmatrix}$ $n = 0, 1, 2, \dots$ 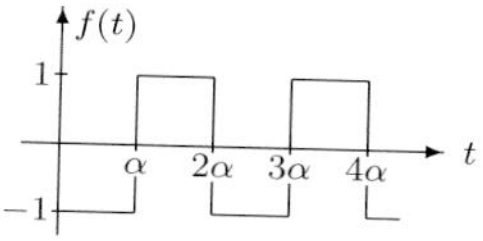  |
| $\dfrac{1-e^{-\alpha s}}{s(1+e^{\alpha s})}$ | $\begin{Bmatrix} 0 \text{ for } 0 < t < \alpha \\ 1 \text{ for } (2n+1)\alpha < t < (2n+2)\alpha \\ -1 \text{ for } (2n+2)\alpha < t < (2n+3)\alpha \end{Bmatrix}$ $n = 0, 1, 2, \dots$ 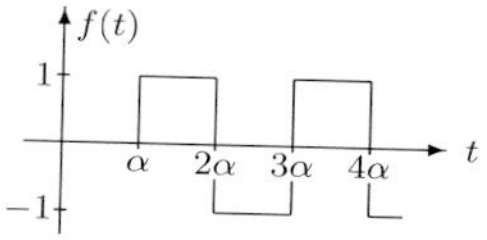  |
| $\dfrac{1-e^{-\alpha s}}{s(e^{\alpha s}+e^{-\alpha s})}$ | $\begin{Bmatrix} 0 \text{ for } 2n\alpha < t < (2n+1)\alpha \\ 1 \text{ for } (4n+1)\alpha < t < (4n+2)\alpha \\ -1 \text{ for } (4n+3)\alpha < t < (4n+4)\alpha \end{Bmatrix}$ $n = 0, 1, 2, \dots$ 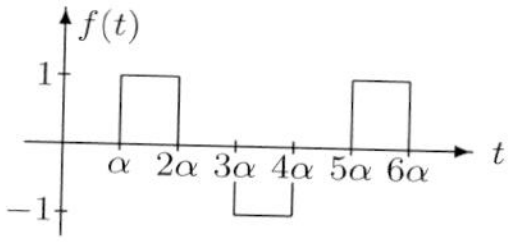  |
| $\dfrac{1-e^{-\frac{\alpha}{\nu}s}}{s(1-e^{-\alpha s})}$ | $\begin{Bmatrix} 1 \quad \text{for } n\alpha < t < \left(n+\dfrac{1}{\nu}\right)\alpha \\ 0 \quad \text{for } \left(n+\dfrac{1}{\nu}\right)\alpha < t < (n+1)\alpha \end{Bmatrix}$ $\nu > 1; \quad n = 0, 1, 2, \dots$ 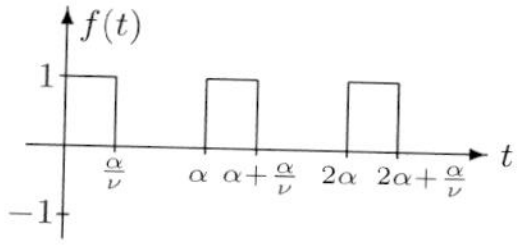  |
| $\dfrac{1-e^{-\alpha s}}{s^2(1+e^{-\alpha s})}$ | $\begin{Bmatrix} \dfrac{t}{\alpha} - 2n & \text{for } 2n\alpha < t < (2n+1)\alpha \\ -\dfrac{t}{\alpha} + 2(n+1) & \text{for } (2n+1)\alpha < t < (2n+2)\alpha \end{Bmatrix}$ $n = 0, 1, 2, \dots$ 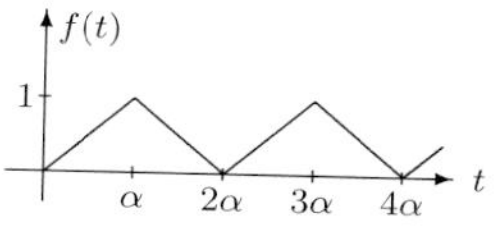  |
| $\dfrac{(1-e^{-\alpha s})^2}{\alpha s^2(1-e^{-4\alpha s})}$ | $\begin{Bmatrix} \dfrac{t}{\alpha} - 4n & \text{for} & 4n\alpha < t < (4n+1)\alpha \\ -\dfrac{t}{\alpha} + 4n + 2 & \text{for} & (4n+1)\alpha < t < (4n+2)\alpha \\ 0 & \text{for} & (4n+2)\alpha < t < (4n+4)\alpha \end{Bmatrix}$ $n = 0, 1, 2, \dots$ 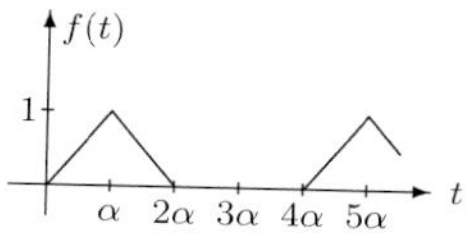  |

| $\mathscr{L}\{f(t)\}$ | $f(t)$ |
|---|---|
| $\dfrac{\alpha s + 1 - e^{\alpha s}}{\alpha s^2(1 - e^{\alpha s})}$ | $\dfrac{t}{\alpha} - n$ for $n\alpha < t < (n+1)\alpha$<br>$n = 0, 1, 2, \ldots$ 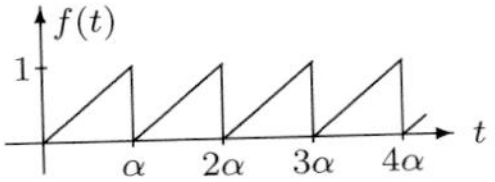 |
| $\dfrac{1 - (1 + \alpha s)e^{-\alpha s}}{\alpha s^2(1 - e^{-2\alpha s})}$ | $\left\{\begin{array}{l} \dfrac{t}{\alpha} - 2n \text{ for } 2n\alpha < t < (2n+1)\alpha \\ 0 \text{ for } (2n+1)\alpha < t < (2n+2)\alpha \end{array}\right\}$<br>$n = 0, 1, 2, \ldots$ 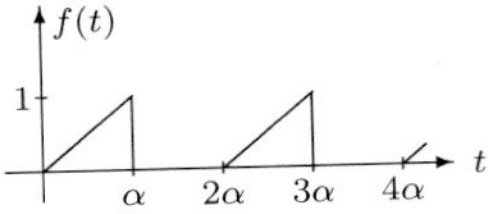 |
| $\dfrac{2 - \alpha s - (2 + \alpha s)e^{-\alpha s}}{\alpha s^2(1 - e^{-\alpha s})}$ | $\dfrac{2t}{\alpha} - (2n+1)$ for $n\alpha < t < (n+1)\alpha$<br>$n = 0, 1, 2, \ldots$ 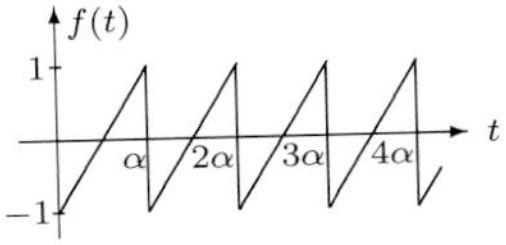 |
| $\dfrac{2(1 - e^{-\alpha s})}{\alpha s^2(1 + e^{-\alpha s})} - \dfrac{1}{s}$ | $\left\{\begin{array}{l} \dfrac{2t}{\alpha} - (4n+1) \text{ for } 2n\alpha < t < (2n+1)\alpha \\ -\dfrac{2t}{\alpha} + 4n + 3 \text{ for } (2n+1)\alpha < t < (2n+2)\alpha \end{array}\right\}$<br>$n = 0, 1, 2, \ldots$ 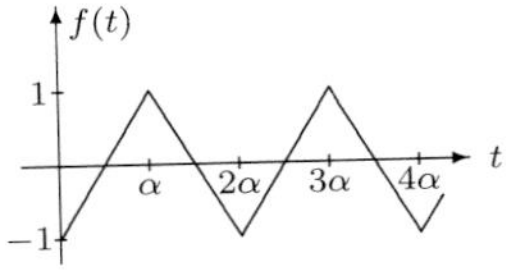 |
| $\dfrac{\nu(\nu - 1) + \nu e^{-\alpha s} - \nu^2 e^{-\frac{\alpha s}{\nu}}}{(\nu - 1)\alpha s^2(1 - e^{-\alpha s})}$ | $\left\{\begin{array}{l} \dfrac{\nu t}{\alpha} - n \\ \quad \text{for } n\alpha < t < \left(n + \dfrac{1}{\nu}\right)\alpha \\ -\dfrac{\nu}{\alpha(\nu - 1)}t + \dfrac{\nu(n+1)}{\nu - 1} \\ \quad \text{for } \left(n + \dfrac{1}{\nu}\right)\alpha < t < (n+1)\alpha \end{array}\right\}$<br>$\nu > 1; \quad n = 0, 1, 2, \ldots$ 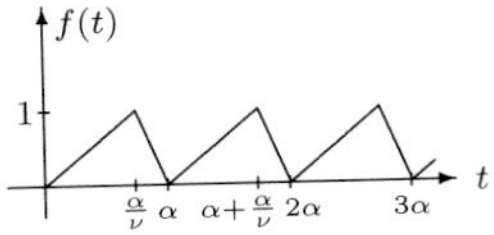 |
| $\dfrac{\nu - (\nu + \alpha s)\exp\left(-\frac{\alpha s}{\nu}\right)}{\alpha s^2(1 - e^{-\alpha s})}$ | $\left\{\begin{array}{l} \dfrac{\nu}{\alpha}t - \nu n \text{ for } n\alpha < t < \left(n + \dfrac{1}{\nu}\right)\alpha \\ 0 \text{ for } \left(n + \dfrac{1}{\nu}\right)\alpha < t < (n+1)\alpha \end{array}\right\}$<br>$\nu > 1; \quad n = 0, 1, 2, \ldots$ 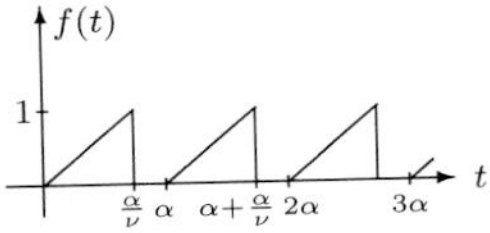 |

### 0.10.3 The $Z$-transformation

| $f_n$ | $\mathscr{Z}\{f_n\} = F(z)$ |
|---|---|
| $1$ | $\dfrac{z}{z-1}$ |
| $(-1)^n$ | $\dfrac{z}{z+1}$ |
| $n$ | $\dfrac{z}{(z-1)^2}$ |
| $n^2$ | $\dfrac{z(z+1)}{(z-1)^3}$ |
| $n^k$ | $\dfrac{N_k(z)}{(z-1)^{k+1}}$ [81] |
| $(-1)^n n^k$ | $\dfrac{(-1)^{k+1}N_k(-z)}{(z+1)^{k+1}}$ |
| $\binom{n}{m}; \quad n \geq m-1$ | $\dfrac{z}{(z-1)^{m+1}}$ |
| $(-1)^n \binom{n}{m}; \quad n \geq m-1$ | $\dfrac{(-1)^m z}{(z+1)^{m+1}}$ |
| $\binom{n+k}{m}; \quad k \leq m$ | $\dfrac{z^{k+1}}{(z+1)^{m+1}}$ |
| $a^n$ | $\dfrac{z}{z-a}$ |
| $a^{n-1}; \quad n \geq 1$ | $\dfrac{1}{z-a}$ |
| $(-1)^n a^n$ | $\dfrac{z}{z+a}$ |
| $1-a^n$ | $\dfrac{z(1-a)}{(z-1)(z-a)}$ |

[81] The polynomials $N_k(z)$ can be recursively calculated as follows:

$$N_1(z) = z; \quad N_{k+1}(z) = (k+1)zN_k(z) - (z^2 - z)\frac{\mathrm{d}}{\mathrm{d}z}N_k(z).$$

| $f_n$ | $\mathscr{Z}\{f_n\} = F(z)$ |
|---|---|
| $na^n$ | $\dfrac{za}{(z-a)^2}$ |
| $n^k a^n$ | $\dfrac{a^{k+1} N_k\left(\frac{z}{a}\right)}{(z-a)^{k+1}}$ |
| $\dbinom{n}{m} a^n; \quad n \geq m-1$ | $\dfrac{a^m z}{(z-a)^{m+1}}$ |
| $\dfrac{1}{n}; \quad n \geq 1$ | $\ln \dfrac{z}{z-1}$ |
| $\dfrac{(-1)^{n-1}}{n}; \quad n \geq 1$ | $\ln\left(1+\dfrac{1}{z}\right)$ |
| $\dfrac{a^{n-1}}{n}; \quad n \geq 1$ | $\dfrac{1}{a} \ln \dfrac{z}{z-a}$ |
| $\dfrac{a^n}{n!}$ | $\exp\left(\frac{a}{z}\right)$ |
| $\dfrac{n+1}{n!} a^n$ | $\left(1+\dfrac{a}{z}\right) \exp\left(\dfrac{a}{z}\right)$ |
| $\dfrac{(-1)^n}{(2n+1)!}$ | $\sqrt{z} \sin \dfrac{1}{\sqrt{z}}$ |
| $\dfrac{(-1)^n}{(2n)!}$ | $\cos \dfrac{1}{\sqrt{z}}$ |
| $\dfrac{1}{(2n+1)!}$ | $\sqrt{z} \sinh \dfrac{1}{\sqrt{z}}$ |
| $\dfrac{1}{(2n)!}$ | $\cosh \dfrac{1}{\sqrt{z}}$ |
| $\dfrac{a^n}{(2n+1)!}$ | $\sqrt{\dfrac{z}{a}} \sinh \sqrt{\dfrac{a}{z}}$ |
| $\dfrac{a^n}{(2n)!}$ | $\cosh \sqrt{\dfrac{a}{z}}$ |
| $\mathrm{e}^{\alpha n}$ | $\dfrac{z}{z-\mathrm{e}^{\alpha}}$ |

| $f_n$ | $\mathscr{Z}\{f_n\} = F(z)$ |
|---|---|
| $\sinh \alpha n$ | $\dfrac{z \sinh \alpha}{z^2 - 2z \cosh \alpha + 1}$ |
| $\cosh \alpha n$ | $\dfrac{z(z - \cosh \alpha)}{z^2 - 2z \cosh \alpha + 1}$ |
| $\sinh(\alpha n + \varphi)$ | $\dfrac{z\big(z \sinh \varphi + \sinh(\alpha - \varphi)\big)}{z^2 - 2z \cosh \alpha + 1}$ |
| $\cosh(\alpha n + \varphi)$ | $\dfrac{z\big(z \cosh \varphi - \cosh(\alpha - \varphi)\big)}{z^2 - 2z \cosh \alpha + 1}$ |
| $a^n \sinh \alpha n$ | $\dfrac{za \sinh \alpha}{z^2 - 2za \cosh \alpha + a^2}$ |
| $a^n \cosh \alpha n$ | $\dfrac{z(z - a \cosh \alpha)}{z^2 - 2za \cosh \alpha + a^2}$ |
| $n \sinh \alpha n$ | $\dfrac{z(z^2 - 1) \sinh \alpha}{(z^2 - 2z \cosh \alpha + 1)^2}$ |
| $n \cosh \alpha n$ | $\dfrac{z\big((z^2 + 1) \cosh \alpha - 2z\big)}{(z^2 - 2z \cosh \alpha + 1)^2}$ |
| $\sin \beta n$ | $\dfrac{z \sin \beta}{z^2 - 2z \cos \beta + 1}$ |
| $\cos \beta n$ | $\dfrac{z(z - \cos \beta)}{z^2 - 2z \cos \beta + 1}$ |
| $\sin(\beta n + \varphi)$ | $\dfrac{z\big(z \sin \varphi + \sin(\beta - \varphi)\big)}{z^2 - 2z \cos \beta + 1}$ |
| $\cos(\beta n + \varphi)$ | $\dfrac{z\big(z \cos \varphi - \cos(\beta - \varphi)\big)}{z^2 - 2z \cos \beta + 1}$ |
| $\mathrm{e}^{\alpha n} \sin \beta n$ | $\dfrac{z\, \mathrm{e}^{\alpha} \sin \beta}{z^2 - 2z\, \mathrm{e}^{\alpha} \cos \beta + \mathrm{e}^{2\alpha}}$ |
| $\mathrm{e}^{\alpha n} \cos \beta n$ | $\dfrac{z(z - \mathrm{e}^{\alpha} \cos \beta)}{z^2 - 2z\, \mathrm{e}^{\alpha} \cos \beta + \mathrm{e}^{2\alpha}}$ |

| $f_n$ | $\mathscr{Z}\{f_n\} = F(z)$ |
|---|---|
| $(-1)^n e^{\alpha n} \sin \beta n$ | $\dfrac{-z\, e^{\alpha} \sin \beta}{z^2 + 2e^{\alpha} \cos \beta + e^{2\alpha}}$ |
| $(-1)^n e^{\alpha n} \cos \beta n$ | $\dfrac{z(z + e^{\alpha} \cos \beta)}{z^2 + 2e^{\alpha} \cos \beta + e^{2\alpha}}$ |
| $n \sin \beta n$ | $\dfrac{z(z^2 - 1) \sin \beta}{(z^2 - 2z \cos \beta + 1)^2}$ |
| $n \cos \beta n$ | $\dfrac{z\big((z^2 + 1) \cos \beta - 2z\big)}{(z^2 - 2z \cos \beta + 1)^2}$ |
| $\dfrac{\cos \beta n}{n}; \quad n \geq 1$ | $\ln\left(\dfrac{z}{\sqrt{z^2 - 2z \cos \beta + 1}}\right)$ |
| $\dfrac{\sin \beta n}{n}; \quad n \geq 1$ | $\arctan\left(\dfrac{\sin \beta}{z - \cos \beta}\right)$ |
| $(-1)^{n-1} \dfrac{\cos \beta n}{n}; \quad n \geq 1$ | $\ln\left(\dfrac{\sqrt{z^2 + 2z \cos \beta + 1}}{z}\right)$ |
| $(-1)^{n-1} \dfrac{\sin \beta n}{n}; \quad n \geq 1$ | $\arctan\left(\dfrac{\sin \beta}{z + \cos \beta}\right)$ |
| $\dfrac{\cos \beta n}{n!}$ | $\cos \dfrac{\sin \beta}{z} \exp\left(\dfrac{\cos \beta}{z}\right)$ |
| $\dfrac{\sin \beta n}{n!}$ | $\sin \dfrac{\sin \beta}{z} \exp\left(\dfrac{\cos \beta}{z}\right)$ |

**Remark:** An inequality $n \geq \nu$ on the left hand side of the table indicates that in the construction of the $Z$-transform, the summation starts with $\nu$, for example, $\mathscr{Z}\{a^{n-1}\} = \sum_{n=1}^{\infty} a^{n-1} z^{-n}$.

# 1. Analysis

*Data aequatione quotcunque fluentes quantitae involvente fluxiones invernire et vice versa.*[1]

*Newton in a letter to Leibniz from 1676*

The most fundamental notion in analysis is that of a *limit*. Many of the important concepts in mathematics and physics can be defined in terms of limits, for example velocity, acceleration, work, energy, power, action, volume and surface of a body, length and curvature of a curve, curvature of a surface, etc. The heart of analysis is the *calculus*, which was independently discovered by Newton (1643–1727) and Leibniz (1646–1716). With a few rare exceptions this notion was *not* known in antiquity. Today analysis is one of the most important fundamental notions in the mathematical description of the natural sciences (see Figure 1.1).

However, analysis only develops its true capacity in interaction with other disciplines of the mathematical sciences, like algebra, number theory, geometry, stochastic and numerics.

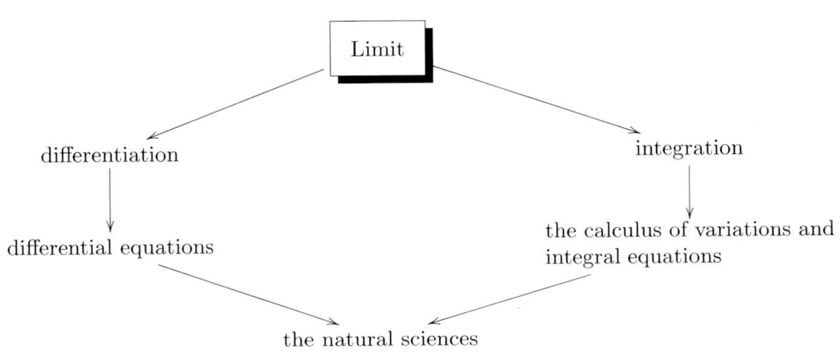


*Figure 1.1. The notion of a limit is central in mathematics.*

[1] A modern translation of this is: 'It is useful to differentiate functions and solve differential equations.' Actually, Newton encrypted this Latin sentence in the following anagram (letter riddle):

6a cc d ae 13e ff 7i 3l 9n 4o 4q rr 4s 9t 12v x,

which means that the letter 'a' occurs 6 times, etc. Newton's words 'fluentes' and 'fluxiones' correspond to the modern words 'function' and 'derivative'. It appears that a solution of this anagram would be as great a intellectual achievement as the discovery of calculus!

# 1.1 Elementary analysis

*Concepts without intuition are empty, intuition without concepts is blind*

*Immanuel Kant* (1724–1804)

## 1.1.1 Real numbers

**Intuitive introduction to the real numbers:**[2] Start with a line $G$ with two points, called 0 and 1, and marked as in Figure 1.2(a). Each point $a$ on $G$ corresponds to a *real number*, and in this manner one gets the *real number line*. For simplicity we use the same notation for a point $a$ of the line $G$ and the real number which corresponds to it. The segment from 0 to 1 is called the unit segment $E$ in $G$.

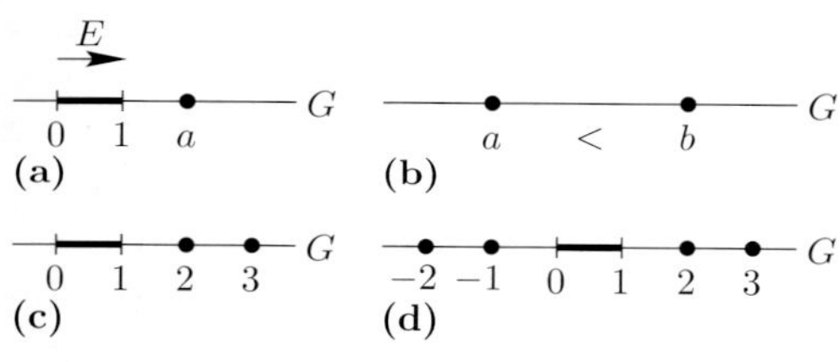


*Figure 1.2. The real number line.*

**Order:** For two arbitrary real numbers $a$ and $b$ we write the symbol

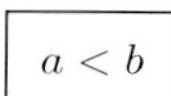

$$a < b$$

if and only if the point $a$ is strictly to the left of $b$, and we say in this case that $a$ is *less than* $b$ (Figure 1.2(b)). We write $a \leq b$ if $a < b$ or $a = b$ and say in this case that $a$ is less than or equal to $b$.

The real number $a$ is said to be positive (resp. negative, non-negative) if and only if $0 < a$ (resp. $a < 0$, $0 \leq a$).

**Evolution of the concept of number:** The concept of a number is actually one of the greatest achievements of abstraction ever made by the human mind. Instead of speaking of 'two trees', 'two stones', etc., one day the abstract notion of 'two' was applied. This would seem to have occured in the newer stone age (Neolithic period) roughly 10,000 years ago, when in Asia and Europe the ice age ended and humanoids began to settle. Cave paintings in France and Spain from about 15,000 years ago attest to the fact that the humanoids of this period already had a keen sense of forms.

About 3000 BC the first Sumarian settlements were established in Mesopotamia near the Tigres–Eufrates rivers (modern Iraq). The high level of mathematical achievments of the Babylonians and Assyrians goes back to those of the Sumarians. These used a number system with basis 60 (known as a sexagesimal system, 60 as opposed to our modern 10). The Babylonians adapted this system and added about 600 BC an empty position, which corresponded to our modern number 0.

#### 1.1.1.1 Natural numbers and integers

The numbers

$$0, \ 1, \ 2, \ 3, \ \ldots,$$

[2] A stringent definition of real numbers and a discussion of the difficulties involved in irrational numbers in the history of the subject can be found in 1.2.2. The very words used for some types of numbers such as 'irrational', 'imaginary' and 'transcendental' attest to the epistemological difficulties which had to be surmounted over the centuries.

which correspond to successive segments of unit length, are called *natural numbers*[3] (see Figure 1.2(c)) Moreover, we call the numbers

$$\ldots, -3,\ -2,\ -1,\ 0,\ 1,\ 2,\ 3,\ \ldots$$

*integers* (see Figure 1.2(d)). One can imagine our line $G$ as being a thermometer, in which case the negative numbers would correspond to temperatures below freezing (zero). Addition of two integers corresponds to the addition of the temperatures.

*Example 1:* The equation

$$\boxed{-3+5=2}$$

can be interpreted in the following manner: if the temperature in the early morning is three below zero and the temperature rises five degrees by noon, then we have at that time a temperature of two degrees.

The multiplication of integers is by means of the following rule:

$$\boxed{\begin{array}{l}\text{`positive times positive} = \text{negative times negative} = \text{positive'}\\ \text{`positive times negative} = \text{negative times positive} = \text{negative'.}\end{array}} \tag{1.1}$$

The division of integers is by means of the following rule:

$$\boxed{\begin{array}{l}\text{`positive divided by positive} = \text{negative divided by negative} = \text{positive'}\\ \text{`positive divided by negative} = \text{negative divided by positive} = \text{negative'.}\end{array}} \tag{1.2}$$

We write $+12$ instead of 12, etc.

*Example 2:*

$$3\cdot 4 = (+3)(+4) = +12 = 12,$$

$$(-3)(+4) = -12, \qquad (+3)(-4) = -12, \qquad (-3)(-4) = 12,$$

$$(-12)\div(+4) = -3, \quad (-12)\div(-4) = 3, \quad 12\div(-4) = -3.$$

From $3\cdot 4 = 12$ it follows that $12\div 4 = 3$. As expected, a comparison of the second and third line above shows that the inverse of multiplication remains correct for integers.

#### 1.1.1.2 Rational numbers

**Basic idea:** One runs into rational numbers (fractions) when one attempts to break the unit segment into parts.

*Example 1:* Let $n$ be a proper natural number. If we divide the unit segment $E$ into $n$ equal parts, we get points which we denote by

$$\frac{1}{n}, \frac{2}{n}, \frac{3}{n}, \ldots, \frac{n-1}{n}, \frac{n}{n} = 1.$$

In particular, for $n = 2$ (resp. $n = 4$) we get the numbers

$$\frac{1}{2}, \frac{2}{2} = 1 \qquad \text{or} \qquad \frac{1}{4}, \frac{2}{4}, \frac{3}{4}, \frac{4}{4} = 1$$

[3]Through the influence of set theory and computer sciences it has become common use to include the number 0 in the set of natural numbers. The positive natural numbers would then be called the *proper* natural numbers.

(see Figure 1.3(a)). In a fraction $\frac{m}{n}$, $m$ (resp. $n$) is called the *numerator* (resp. *denominator*).

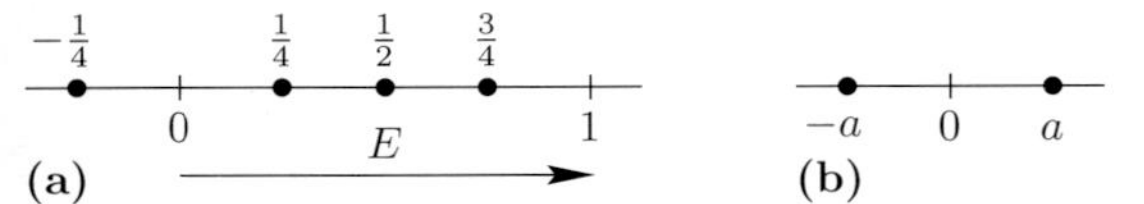


*Figure 1.3. Fractions on the real number line.*

**Reducing and expanding fractions:** According to Figure 1.3(a) one has

$$\frac{2}{4} = \frac{1}{2}.$$

This relation is a special case of the following general rule:

> A fraction is left unchanged upon multiplication of both denominator and numerator by the same natural number $n$ (resp. divison by $n$).

This process is called 'expanding the fraction' (resp. 'reducing the fraction').

*Example 2:* Expanding $\dfrac{2}{3}$ by 4 gives

$$\frac{2}{3} = \frac{2 \cdot 4}{3 \cdot 4} = \frac{8}{12}.$$

Simplifying $\dfrac{8}{12}$ by dividing by 4 yields $\dfrac{8}{12} = \dfrac{8 \div 4}{12 \div 4} = \dfrac{2}{3}$.

**Multiplication of fractions:** Fractions are multiplied with one another according to the following rule:

> 'Numerator times numerator and denominator times denominator'.

*Example 3:* $\dfrac{2}{3} \cdot \dfrac{3}{4} = \dfrac{2 \cdot 3}{3 \cdot 4} = \dfrac{6}{12}$.

**Division of fractions:** The *reciprocal* of a fraction is the fraction obtained by switching numerator and denominator. The rule for division is:

> Division of fractions is the multiplication by the reciprocal of the second fraction.

*Example 4:* The reciprocal of $\dfrac{3}{5}$ is $\dfrac{5}{3}$. Consequently

$$\frac{2}{7} \div \frac{3}{5} = \frac{2}{7} \cdot \frac{5}{3} = \frac{2 \cdot 5}{7 \cdot 3} = \frac{10}{21}.$$

**Addition of fractions:** Two fractions are added by first extending them so that they have a common denominator and then adding the numerators.

*Example 5:* $\dfrac{1}{2} + \dfrac{3}{4} = \dfrac{2}{4} + \dfrac{3}{4} = \dfrac{5}{4}$.

This procedure is the same thing as the general rule of 'cross-multiplication':

$$\boxed{\frac{a}{b} + \frac{c}{d} = \frac{ad + bc}{bd}.}$$

*Example 6*: $\frac{1}{2}+\frac{3}{4}=\frac{1\cdot 4+2\cdot 3}{2\cdot 4}=\frac{4+6}{8}=\frac{10}{8}$. Upon reduction one gets from this $\frac{5}{4}$.

**Negative fractions:** If $m$ and $n$ are proper natural numbers, then one gets the point $-\frac{m}{n}$ by *reflecting the point* $\frac{m}{n}$ *through the origin (point 0)*. The sign of a fraction of integers is determined according to the rule (1.2).

*Example 7:* $\frac{(-3)}{(-4)}=+\frac{3}{4}=\frac{3}{4}$, $\quad \frac{(-3)}{4}=-\frac{3}{4}$, $\quad \frac{3}{(-4)}=-\frac{3}{4}$.

The apparently arbitrary choice of signs in the rules (1.1) and (1.2) are actually determined by demanding that for real numbers 'simple' rules for calculations should hold. For example, demanding the truth of the associative law

$$a(b+c)=ab+ac,$$

one gets for the values $a=4,\ b=+3$ and $c=-3$

$$4(3+(-3))=4\cdot 0=0,$$

and

$$4(3+(-3))=4\cdot 3+4\cdot(-3)=12+4\cdot(-3)=0,$$

or in other words $4(-3)=-12$, coinciding with the value given by the rule (1.1).

**Definition:** All real numbers with a representation $\frac{a}{b}$ with integers $a$, $b$ are called *rational numbers.*

Those real numbers which are not rational are called *irrational.* For example, $\sqrt{2}$ is irrational; a classical proof of this known in antiquity will be given in section 4.2.1.

The following notations have become customary in the modern literature:

$\mathbb{N}$:= set of natural numbers,
$\mathbb{Z}$:= set of integers,
$\mathbb{Q}$:= set of rational numbers,
$\mathbb{R}$:= set of real numbers,
$\mathbb{C}$:= set of complex numbers.

**Powers:** For a real number $a\neq 0$ we set:

$$a^0:=1,\quad a^1:=a,\quad a^2:=a\cdot a,\quad a^3:=a\cdot a\cdot a,\quad \ldots,$$
$$a^{-1}:=\frac{1}{a},\quad a^{-2}:=\frac{1}{a^2},\quad a^{-3}:=\frac{1}{a^3},\quad \ldots$$

*Example 8:* $2^0=1,\quad 2^2=4,\quad 2^3=8,\quad 2^{-1}=\frac{1}{2},\quad 2^{-2}=\frac{1}{4}$.

#### 1.1.1.3 Decimal numbers

**Basic idea:** In our daily life the decimal system is used.[4] The number symbol 123 is actually an abbreviation for the number

$$1\cdot\mathbf{10}^2+2\cdot\mathbf{10}^1+3\cdot\mathbf{10}^0.$$

[4]The volume *La disme* (The decimal system) appeared in 1585 by Simon Stevin. After that time all measurements used on the European continent were unified to the decimal system.

In the same way the symbol 2.43 stands for the sum:

$$2 \cdot \mathbf{10}^0 + 4 \cdot \mathbf{10}^{-1} + 3 \cdot \mathbf{10}^{-2}. \tag{1.3}$$

Finally, a symbol like 2.43567... stands for the (unique) real number $x$ which satisfies the following *infinite* set of *inequalities*:

$$\begin{aligned} &2.4 &&\leq x < 2.4 + \mathbf{10}^{-1}, \\ &2.43 &&\leq x < 2.43 + \mathbf{10}^{-2}, \\ &2.435 &&\leq x < 2.435 + \mathbf{10}^{-3}, \\ &\ldots \end{aligned} \tag{1.4}$$

**Expansion in decimal fractions:** In what follows all $a_j$ are assumed to be integers with $a_j = 0,\ 1, \ldots, 9$ and $a_n \neq 0$. Moreover let $n = 0,\ 1,\ 2, \ldots$ be a natural, $m$ a proper natural number.

(i) The symbol

$$\boxed{a_n a_{n-1} \ldots a_0 . a_{-1} a_{-2} \ldots a_{-m}}$$

stands for the sum

$$a_n \cdot \mathbf{10}^n + a_{n-1} \cdot \mathbf{10}^{n-1} + \ldots + a_0 \cdot \mathbf{10}^0 + a_{-1} \cdot \mathbf{10}^{-1} + a_{-2} \cdot \mathbf{10}^{-2} + \ldots + a_{-m} \cdot \mathbf{10}^{-m}$$

in the same way as in (1.3).

(ii) The symbol $a_n a_{n-1} \ldots a_0 . a_{-1} a_{-2} \ldots$ stands for the (unique) real number $x$, which as in (1.4) satisfies the following infinite chain of inequalities:

$$\boxed{a_n \cdots a_0 . a_{-1} a_{-2} \ldots a_{-m} \leq x < a_n \cdots a_0 . a_{-1} a_{-2} \ldots a_{-m} + \mathbf{10}^{-m}, \ m = 1, 2, \ldots}$$

Each real number can be expanded in a unique fashion in such a decimal fraction.

**Theorem:** A real number is rational, if and only if the decimal fraction expansion is finite or periodic.

*Example 1:* The numbers $\frac{1}{4} = 0.25$ and $\frac{1}{3} = 0.333333\cdots$ are rational. On the other hand, the decimal expansion of $\sqrt{2}$:

$$\sqrt{2} = 1.414213562\cdots, \tag{1.5}$$

has no period.

**Rules for rounding real numbers:** The goal of the process of rounding is to pass from an infinite decimal expansion to a *finite* one, with as little error as possible.

*Example 2:*

(i) Rounding 2.3456... up gives 2.346.

(ii) Rounding 2.3454... down gives 2.345.

(iii) Rounding 2.3455... up gives 2.346.

(iv) Rounding 2.3465... down gives 2.346.

The error is in each case less that 0.0005.

The following *rule* is applied: If the last digit is $0-4$ (resp. $6-9$), the one rounds down (resp. up). If the last digit is 5, then one can round up or down, the decision being dictated by the stipulation that after rounding the last digit is *even*.[5] However, the usual procedure is to simply round 5 up.

**The integer part of a real number:** The notation $[a]$ (known as the *Gauss bracket*) for a real number $a$ denotes the integer part of $a$, which is by definition the smallest integer $g$ with $g \leq a$.

*Example 3:* $[2] = 2, \quad [1.99] = 1, \quad [-2.5] = -3.$

### 1.1.1.4 Binary numbers

One gets the system of *binary numbers* by replacing the **10** in the expansions above by **2**, and $a_j$ is chosen to be 0 or 1. An arbitrary real number can be written in a unique manner as a binary expansion in which only the coefficients 0 or 1 occur. Because of this property binary numbers are the system used by computers.

*Example:* In the binary system the symbol 1010.01 stands for the sum

$$1 \cdot \mathbf{2}^3 + 0 \cdot \mathbf{2}^2 + 1 \cdot \mathbf{2}^1 + 0 \cdot \mathbf{2}^0 + 0 \cdot \mathbf{2}^{-1} + 1 \cdot \mathbf{2}^{-2},$$

which in the decimal system is the number $8 + 2 + \frac{1}{4} = 10.25$.

**Other number systems:** By replacing the number **10** in 1.1.1.3 by any chosen fixed natural number $\boldsymbol{\beta} \geq 2$, one gets a number system with basis $\boldsymbol{\beta}$. For example the Sumerians in Mesopotamia used a sexagesimal system with basis 60 around 2000 BC. Our division of time into 60 minutes per hour and of the circle into 360 degrees goes back to the Sumerians.

The Mayas in Mexico and the Celts in Europe used a system with basis $\boldsymbol{\beta} = 20$. The ancient Egyptians used a decimal system with special symbols for each decimal unit. The number system of the Romans used the same principle. The Roman digits

M, D, C, L, X, V and I

stand for 1000, 500, 100, 50, 10, 5 and 1 in that order. The symbol MDCLXVII for example corresponds to the number 1667. Such a system is not adapted to doing complicated computations.

### 1.1.1.5 Intervals

Let $a$ and $b$ be real numbers with $a < b$. By definition, a compact *interval* (with endpoints $a$ and $b$) is the set (see Figure 1.4(a))

$$[a, b] := \{x \in \mathbb{R} \mid a \leq x \leq b\}.$$

In words, the interval $[a, b]$ consists of all real numbers $x$ in $\mathbb{R}$ such that $a \leq x \leq b$. Further we define[6] (see Figures 1.4(b)-(d))

$$\begin{aligned}
]a, b[ &:= \{x \in \mathbb{R} \mid a < x < b\}; && \text{(open interval)} \\
[a, b[ &:= \{x \in \mathbb{R} \mid a \leq x < b\}; && \text{(right half-open interval)} \\
]a, b] &:= \{x \in \mathbb{R} \mid a < x \leq b\}. && \text{(left half-open interval)}
\end{aligned}$$

[5]This statistical strategy for rounding has as consequence that after a long time of calculations with rounding, the error is smaller than by consequently rounding 5 up or down.

[6]Instead of $]a, b[$, $[a, b[$, $]a, b]$ one also writes $(a, b)$, $[a, b)$, $(a, b]$ in that order. The notation above has become customary in the modern literature to avoid confusion with the ordered pair of numbers $(a, b)$.

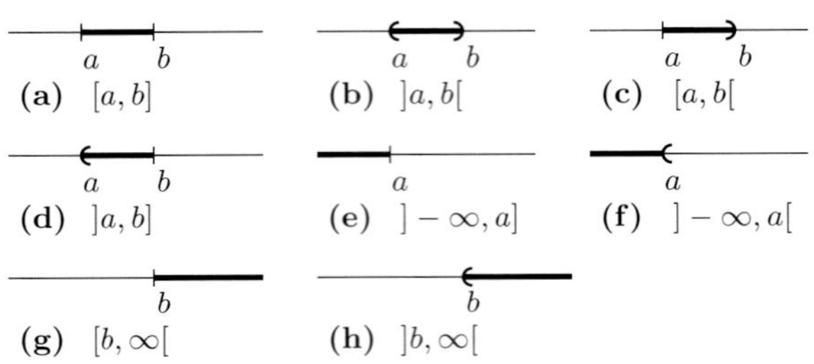


*Figure 1.4. Intervals on the real number line.*

Often one used the following infinite intervals (see Figures 1.4(e)-(h)):

$$]-\infty, a] := \{x \in \mathbb{R} \mid x \leq a\}, \qquad ]-\infty, a[ := \{x \in \mathbb{R} \mid x < a\},$$
$$[b, \infty[ := \{x \in \mathbb{R} \mid b \leq x\}, \qquad ]b, \infty[ := \{x \in \mathbb{R} \mid b < x\}.$$

The set $\mathbb{R}$ of real numbers is also denoted by $]-\infty, \infty[$.

### 1.1.2 Complex numbers

**Formal introduction of the complex numbers:**[7] There is no real number $x$ which satisfies the equation

$$x^2 = -1.$$

This is the way the Italian mathematician Raphael Bombelli introduce the symbol $\sqrt{-1}$ in the middle of the sixteenth century. Leonhard Euler (1707–1783) used the symbol i for this. The so-called *imaginary unit* satisfies the equation

$$\boxed{\mathrm{i}^2 = -1.} \tag{1.6}$$

Euler discovered the following basic relation, valid for all real numbers $x$, $y$:

$$\boxed{\mathrm{e}^{x+\mathrm{i}y} = \mathrm{e}^x(\cos y + \mathrm{i} \cdot \sin y),} \tag{1.7}$$

which yields an unexpected connection between the exponential function and the trigonometric functions.[8] This formula is constantly applied in the theory of oscillations (compare 1.1.3)

**Cartesian representation:** A *complex number* is a symbol of the form

$$\boxed{x + \mathrm{i}y}$$

[7]The rigorous introduction of the complex numbers as the algebra of ordered pairs $(x, y)$ is carried out in 2.5.3 in the context of field theory.

[8]In electro-technical literature one uses the symbol j instead of i, to avoid confusion with the notation for the current strength.

where $x$ and $y$ are real numbers. Real numbers correspond to the special case $y = 0$. For a long time complex numbers seemed to be rather mysterious. It was Gauss who gave them their right place in mathematics, by considering $x+\mathrm{i}y$ as point in the Cartesian plane; he showed that the calculation with complex numbers can be given a geometric interpretation (Figure 1.5). Today the complex numbers are ubiquitous in many parts of mathematics, physics and technology.

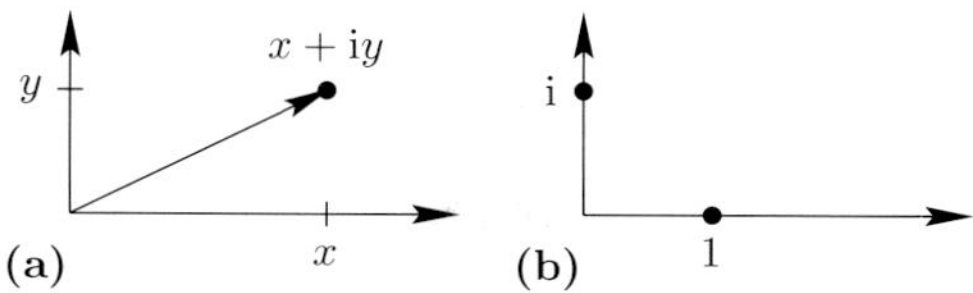


*Figure 1.5. Graphical representation of complex numbers.*

One calculates with complex numbers by applying the usual formulas to the numbers $x + \mathrm{i}y$, taking into consideration the fact that $\mathrm{i}^2 = -1$. In particular, one calls the number

$$\boxed{\overline{x + \mathrm{i}y} := x - \mathrm{i}y}$$

the *complex conjugate*[9] number to $x + \mathrm{i}y$.

*Example 1* (Addition): $(2 + 3\mathrm{i}) + (1 + 2\mathrm{i}) = 3 + 5\mathrm{i}$.

*Example 2* (Multiplication): $(2 + 3\mathrm{i})(1 + 2\mathrm{i}) = 2 + 3\mathrm{i} + 4\mathrm{i} + 6\mathrm{i}^2 = 2 + 7\mathrm{i} - 6 = -4 + 7\mathrm{i}$. For all real numbers $x$ and $y$, the relation

$$\boxed{(x + \mathrm{i}y)(x - \mathrm{i}y) = x^2 + y^2} \tag{1.8}$$

is valid.

*Example 3* (Division): From (1.8) it follows that

$$\frac{1 + 2\mathrm{i}}{3 + 2\mathrm{i}} = \frac{(1 + 2\mathrm{i})(3 - 2\mathrm{i})}{(3 + 2\mathrm{i})(3 - 2\mathrm{i})} = \frac{3 + 6\mathrm{i} - 2\mathrm{i} - 4\mathrm{i}^2}{9 + 4} = \frac{1}{13}(7 + 4\mathrm{i}).$$

This method of expanding by the complex conjugate of the denominator is applicable generally.

#### 1.1.2.1 Absolute value

By definition the absolute value of a complex number $z = x + \mathrm{i}y$ is defined by

$$\boxed{|z| := \sqrt{x^2 + y^2}.}$$

Geometrically this is the length of the vector defined by $z$ (see Figure 1.6(a)).

*Example 1:* For a real number $x$, one has

$$|x| := \begin{cases} x, & \text{if } x \geq 0, \\ -x, & \text{if } x < 0. \end{cases}$$

Moreover, $|\mathrm{i}| = 1$, $|1 + \mathrm{i}| = \sqrt{1^2 + 1^2} = \sqrt{2}$.

**Distance:** For two complex numbers $z$ and $w$, the *distance* between them is

$$|z - w|,$$

[9] *Complex* conjugate to avoid confusion with other (numerous) notions of conjugate objects in mathematics.

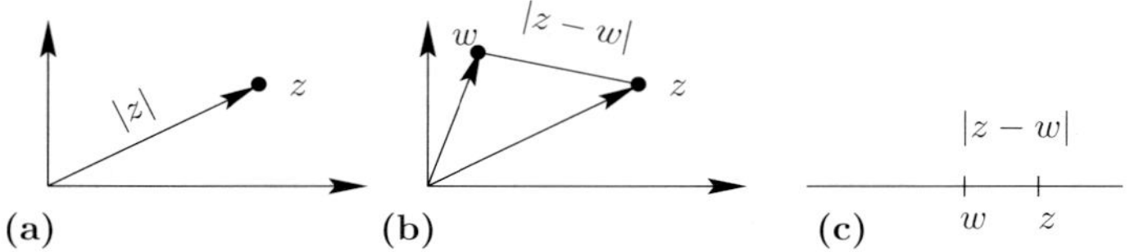


*Figure 1.6. Graphical representation of the distance between complex numbers.*

(see Figure 1.6(b),(c)). In particular, $|z|$ is the distance of $z$ from the origin.

**Triangle inequality:** For all complex numbers $z$ and $w$, we have the important *triangle inequality*:

$$\boxed{||z| - |w|| \leq |z \pm w| \leq |z| + |w|.} \tag{1.9}$$

In particular, this says that $|z+w| \leq |z| + |w|$, which means that the length of the vector $z+w$ is at most the sum of the lengths of the vectors $z$ and $w$ (Figure 1.8(b)). Moreover,

$$|zw| = |z||w|, \qquad \left|\frac{z}{w}\right| = \frac{|z|}{|w|}, \qquad |z| = |\overline{z}|,$$

where $w \neq 0$ must be assumed when $w$ occurs in the denominator.

**Complex numbers in polar coordinates:** If we use polar coordinates, then we have for the complex number $z = x + \mathrm{i}y$ the representation:

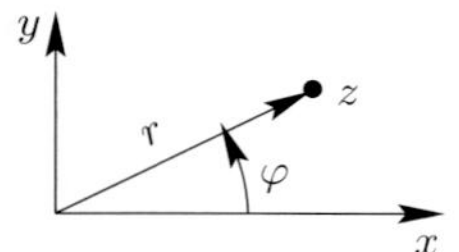


*Figure 1.7.*

$$\boxed{z = r(\cos\varphi + \mathrm{i}\cdot\sin\varphi), \qquad -\pi < \varphi \leq \pi, \quad r = |z|.}$$

Here $\varphi$ denotes the angle of this vector with the $x$-axis (Figure 1.7). The Euler formula (1.7) yields the elegant representation:

$$\boxed{z = r\mathrm{e}^{\mathrm{i}\varphi}, \qquad -\pi < \varphi \leq \pi, \quad r = |z|.}$$

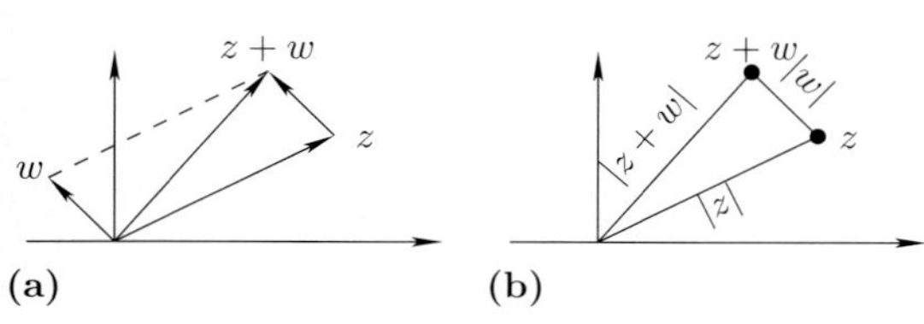


*Figure 1.8. The addition of complex numbers.*

One calls the angle $\arg z := \varphi$ the principal value of the *argument* of $z$. Making the assumption $-\pi < \varphi \leq \pi$ determines $\varphi$ through $z$ uniquely.

All angles $\psi$ with $z = r\mathrm{e}^{\mathrm{i}\psi}$ are called arguments of $z$. One has

$$\psi = \varphi + 2\pi k, \qquad k = 0, \pm 1, \pm 2, \ldots$$

*Example 2:* $\mathrm{i} = \mathrm{e}^{\mathrm{i}\pi/2}, \quad -1 = \mathrm{e}^{\mathrm{i}\pi}, \quad |\mathrm{i}| = |-1| = 1, \quad \arg \mathrm{i} = \frac{\pi}{2}, \quad \arg(-1) = \pi.$

### 1.1.2.2 Geometric interpretation of the operations with complex numbers

One has:

(i) The *addition* of two complex numbers $z$ and $w$ corresponds to the addition of the vectors they define (Figure 1.8(a)).
(ii) The *multiplication* of two complex numbers

$$\boxed{r\mathrm{e}^{\mathrm{i}\varphi} \cdot \varrho\mathrm{e}^{\mathrm{i}\psi} = r\varrho\mathrm{e}^{\mathrm{i}(\varphi+\psi)}}$$

corresponds to a dilated rotation, i.e., the lengths of the vectors are multiplied and the angles are added.
(iii) The *division* of two complex numbers

$$\boxed{\frac{r\mathrm{e}^{\mathrm{i}\varphi}}{\varrho\mathrm{e}^{\mathrm{i}\psi}} = \frac{r}{\varrho}\mathrm{e}^{\mathrm{i}(\varphi-\psi)}}$$

corresponds to a division of the lengths of the corresponding vectors and a subtraction of the angles.

(iv) *Reflection.* The transition from the complex number $z = x + \mathrm{i}y$ to its complex conjugate $\overline{z} = x - \mathrm{i}y$ is the geometric operation of reflecting $z$ on the real axis (Figure 1.9(a)). The transition from $z$ to $-z$ is a reflection through the origin (Figure 1.9(b)). The transition from $z$ to the reciprocal $(\overline{z})^{-1}$ of the conjugate number corresponds to a reflection on the unit circle, i.e., the image and inverse image points lie on the same line through the origin, and the product of their distances is equal to 1 (Figure 1.9(c)).

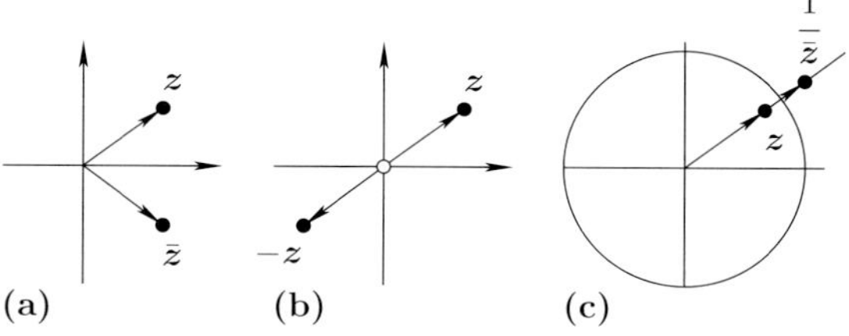


*Figure 1.9. Graphical representation of conjugate and inverse complex numbers.*

#### 1.1.2.3 Rules for arithmetic

**Addition and multiplication:** For all complex numbers $a,\ b,\ c$ one has

$$\boxed{\begin{array}{lll} a+(b+c)=(a+b)+c, & a(bc)=(ab)c, & \text{(associativity)},\\ a+b=b+a, \quad ab=ba, & & \text{(commutativity)},\\ a(b+c)=ab+ac, & & \text{(distributivity)}.\end{array}}$$

*Example 1:* $(a+b)^2 = (a+b)(a+b) = a^2 + ab + ba + b^2 = a^2 + 2ab + b^2$.

**Rules for signs:** For all complex numbers $a,\ b$ one has

$$\boxed{\begin{array}{lll} (-a)(-b)=ab, & (-a)\,b=-ab, & a(-b)=-ab,\\ -(-a)=a, & (-1)\,a=-a. & \end{array}}$$

*Example 2:* $(a-b)^2 = (a-b)(a-b) = a^2 - ab - ba + b^2 = a^2 - 2ab + b^2$.

*Example 3:* $(a+b)(a-b) = a^2 + ab - ba - b^2 = a^2 - b^2$.

**Arithmetic with fractions:** In what follows we assume that all complex numbers appearing in the denominators of fractions are non-zero. For all complex numbers $a$, $b$, $c$ and $d$, one has:

$$
\begin{array}{ll}
\frac{a}{b} = \frac{c}{d} \iff ad = bc, & \text{(equality)},^{10} \\
\frac{a}{b} \cdot \frac{c}{d} = \frac{ac}{bd}, & \text{(multiplication)}, \\
\frac{a}{b} \pm \frac{c}{d} = \frac{ad \pm bc}{bd}, & \text{(addition and subtraction)}, \\
\frac{\left(\frac{a}{b}\right)}{\left(\frac{c}{d}\right)} = \frac{ad}{bc}, & \text{(division)}.
\end{array}
$$

**Transition to complex conjugate numbers:** If $a$, $b$, $c$ and $d$ are arbitrary complex numbers with $d \neq 0$, then one has:

$$\overline{a \pm b} = \overline{a} \pm \overline{b}, \qquad \overline{ab} = \overline{a} \cdot \overline{b}, \qquad \overline{\left(\frac{c}{d}\right)} = \frac{\overline{c}}{\overline{d}}.$$

Let $z = x + \mathrm{i}y$. The component $x$ (resp. $y$) is called the *real part* (resp. *imaginary part*) of $z$. We write for this $x = \operatorname{Re} z$ (resp. $y = \operatorname{Im} z$). One has:

$$\operatorname{Re} z = \frac{1}{2}(z + \overline{z}), \qquad \operatorname{Im} z = \frac{1}{2\mathrm{i}}(z - \overline{z}).$$

### 1.1.2.4 Roots of complex numbers

Let the complex number $a = |a|\mathrm{e}^{\mathrm{i}\varphi}$ with $-\pi < \varphi \leq \pi$ and $a \neq 0$.

**Theorem:** For fixed $n = 2, 3, \ldots$ the *cyclotomic equation*

$$\boxed{x^n = a}$$

has precisely the solutions

$$\boxed{x = \sqrt[k]{|a|}\left(\cos\left(\frac{2\pi k + \varphi}{n}\right) + \mathrm{i} \cdot \sin\left(\frac{2\pi k + \varphi}{n}\right)\right), \qquad k = 0, 1, \ldots, n-1.}$$

These numbers, which we can also write in the form $x = \sqrt[k]{|a|}\mathrm{e}^{\mathrm{i}(2\pi k + \varphi)/n}$, $k = 0, \ldots, n-1$ are called the $n$th roots of the complex number $a$. These $n$th roots divide the circle of radius $\sqrt[k]{|a|}$ into $n$ equal parts.

*Example 1:* For $a = 1$ and $n = 2, 3, 4$, the $n$th roots of $a = 1$ (these are called *roots of unity*) are displayed in Figure 1.10

[10] In words: $\frac{a}{b} = \frac{c}{d}$ implies $ad = cb$ and vice versa.

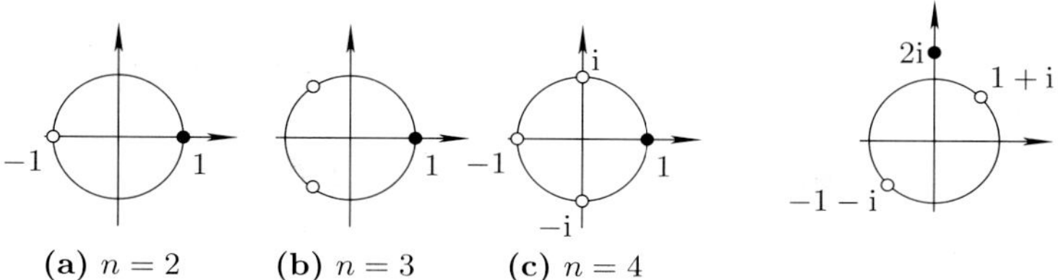


*Figure 1.10.* *Roots of unity.*

*Figure 1.11.*

*Example 2:* The 2nd roots of $2\mathrm{i} = 2\mathrm{e}^{\mathrm{i}\pi/2}$ are:

$$x = \sqrt{2}\left(\cos\frac{\pi}{4} + \mathrm{i}\cdot\sin\frac{\pi}{4}\right) = 1 + \mathrm{i},$$

$$x = \sqrt{2}\left(\cos\left(\pi + \frac{\pi}{4}\right) + \mathrm{i}\cdot\sin\left(\pi + \frac{\pi}{4}\right)\right) = -(1 + \mathrm{i})$$

(see Figure 1.11).

### 1.1.3 Applications to oscillations

Let the function $f$ of *period* $T > 0$ be given, meaning that

$$\boxed{f(t + T) = f(t) \qquad \text{for all } t \in \mathbb{R}.}$$

We also define

$$\nu := \frac{1}{T} \text{ (frequency)}, \qquad \omega := 2\pi\nu \text{ (angular frequency)}.$$

*Example 1* (sine): The function

$$\boxed{y := A \cdot \sin(\omega t + \alpha)}$$

describes an oscillation of the angular frequency $\omega$ and amplitude $A$ (Figure 1.12(a)). The number $\alpha$ is called the *phase displacement.*

*Example 2* (sinusoidal wave): Let $A > 0$ and $\omega > 0$ be given. The function

$$\boxed{y(x, t) = A \cdot \sin(\omega t + \alpha - kx)} \tag{1.10}$$

describes a wave of amplitude $A$ and wavelength $\lambda := 2\pi/k$, which spreads out from left to right (Figure 1.12(b)) with the so-called *phase velocity*

$$\boxed{c := \frac{\omega}{k}.}$$

The number $k$ is called the *wave number.* A point $(x, y(x, t))$ which moves in time according to the law $kx = \omega t + \alpha - \dfrac{\pi}{2}$ corresponds to wave crest of height $A$ moving from left to right with velocity $c$.

In physics and technology it is customary to denote such waves by the complex function

$$\boxed{Y(x,t) := C \cdot \mathrm{e}^{\mathrm{i}(\omega t - kx)}}$$

with the *complex amplitude* $C = A\mathrm{e}^{\mathrm{i}\alpha}$. The imaginary part of $Y(x,t)$ corresponds to $y(x,t)$ in (1.10).

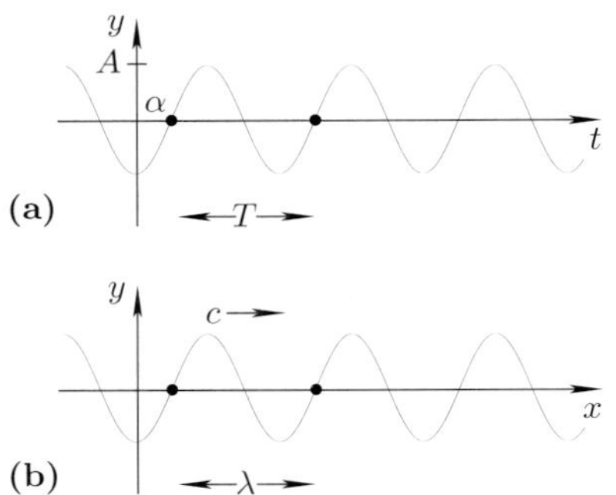


*Figure 1.12. Wave functions (oscillations).*

### 1.1.4 Calculations with equalities

**Operations on an equation:** Let $a$, $b$ and $c$ be arbitrary real (or complex) numbers. Calculations with equalities is dominated by the following rules:

| | | | |
|---|---|---|---|
| $a = b$ | $\Rightarrow$ | $a + c = b + c,$ | (addition), |
| $a = b$ | $\Rightarrow$ | $a - c = b - c,$ | (subtraction), |
| $a = b$ | $\Rightarrow$ | $ac = bc,$ | (multiplication), |
| $a = b,\ c \neq 0$ | $\Rightarrow$ | $\dfrac{a}{c} = \dfrac{b}{c},$ | (division), |
| $a = b,\ a \neq 0$ | $\Rightarrow$ | $\dfrac{1}{a} = \dfrac{1}{b},$ | (reciprocal). |

In words:

(i) One may add the same number to both sides of an equality, the result being again an equality.

(ii) One may multiply both sides of an equality by the same number.

(iii) One may divide both sides of an equality by the same (non-zero) number.

(iv) One may form the reciprocal of an equality.

In cases (iii) and (iv) one must always be careful to observe that

| Division by zero is not allowed! |
|---|

Heuristically an equation is like a scale in equilibrium. This equilibrium is not disturbed by doing the same thing on both sides at the same time.

**Operations on two equations:** For all real (or complex) numbers $a$, $b$, $c$ and $d$ one has:

$$
\boxed{
\begin{array}{lcll}
a=b,\ c=d & \Rightarrow & a+c=b+d, & \text{(addition of two equations)},\\
a=b,\ c=d & \Rightarrow & a-c=b-d, & \text{(subtraction of two equations)},\\
a=b,\ c=d & \Rightarrow & ac=bd, & \text{(multiplication of two equations)},\\
a=b,\ c=d,\ c\neq 0 & \Rightarrow & \dfrac{a}{c}=\dfrac{b}{d}, & \text{(division of two equations)}.
\end{array}
}
$$

**Solving equations:** A *solution* of the equation

$$\boxed{2x+3=7} \tag{1.11}$$

is a number which satisfies (1.11) when we replace $x$ by this number. One refers to $x$ as a *variable* or *indeterminant.*

*Example 1:* The equation (1.11) has a unique solution $x=2$.

*Proof:* First step: suppose that a number $x$ is a solution of (1.11). Subtracting 3 from the left and right hand side of (1.11), we get

$$2x=4.$$

Now dividing the left and right sides by 2 gives

$$x=2.$$

This shows that *if* (1.11) has a solution, then it must equal 2.

Second step: We now show that 2 actually *is* a solution of (1.11). This follows from the elementary equality $2\cdot 2+3=7$. □

The second step is also called the 'check'. Often a mathematical mistake occurs by confusing the first step with a complete proof (cf. 4.2.6.2).

*Example 2* (a system of linear equations): Let real (or complex) numbers $a$, $b$, $c$, $d$, $\alpha$, $\beta$ with $ad-bc\neq 0$ be given. The system of equations

$$\boxed{\begin{aligned} ax+by&=\alpha,\\ cx+dy&=\beta\end{aligned}} \tag{1.12}$$

then has a unique solution

$$x=\frac{\alpha d-\beta b}{ad-bc},\qquad y=\frac{a\beta-c\alpha}{ad-bc}. \tag{1.13}$$

*Proof:* First step: suppose the numbers $x$ and $y$ satisfy the equation (1.12). We multiply the first (resp. second) equation of (1.12) by $d$ (resp. $(-b)$). This yields

$$\begin{aligned} adx+bdy&=\alpha d,\\ -bcx-bdy&=-b\beta.\end{aligned}$$

Adding both of these equations then gives

$$(ad-bc)x=\alpha d-\beta b.$$

After dividing both sides by $ad - bc$ we get the expression of (1.13) for $x$.

Now multiply the first (resp. second) equation of (1.12) on both sides by $(-c)$ (resp. $a$). This gives

$$\begin{aligned} -cax - cby &= -c\alpha, \\ acx + ady &= a\beta. \end{aligned}$$

Adding these two equations then yields

$$(ad - bc)y = a\beta - c\alpha.$$

After dividing both sides by $ad - bc$ we get the expression (1.13) for $y$. This shows that any solution of (1.12) must have the form (1.13), if it exists. In particular, there is (if any) a *unique* solution.

Second step (check): we insert the values of (1.13) for $x$ and $y$ into the equations (1.12), and after an easy computation, find that these are in fact solutions. □

*Example 3* (quadratic equations): Let $b$ and $c$ be real numbers satisfying $b^2 - c > 0$. Then the quadratic equation

$$x^2 + 2bx + c = 0 \tag{1.14}$$

has exactly the two solutions

$$x = -b \pm \sqrt{b^2 - c}. \tag{1.15}$$

*Proof:* First step: If $x$ is a solution of (1.14), then add $b^2 - c$ to both sides of (1.14) to get

$$x^2 + 2bx + b^2 = b^2 - c.$$

This gives $(x + b)^2 = b^2 - c$, and from this we get $x + b = \pm\sqrt{b^2 - c}$. Adding $(-b)$ to both sides of this equation gives (1.15). This shows that all solutions of (1.14) are of the form (1.15), if they exist.

Second step (check): Insert the expression (1.15) for $x$ into the equation (1.14). An easy calculation shows that these values for $x$ are indeed solutions. □

### 1.1.5 Calculations with inequalities

**Manipulations with inequalities:** For arbitrary real $a$, $b$ and $c$, the following rules hold:

| | | | |
|---|---|---|---|
| $a \leq b$ | $\Rightarrow$ | $a + c \leq b + c,$ | (addition), |
| $a \leq b$ | $\Rightarrow$ | $a - c \leq b - c,$ | (subtraction), |
| $a \leq b,\ c \geq 0$ | $\Rightarrow$ | $ac \leq bc,$ | (multiplication), |
| $a \leq b,\ c < 0$ | $\Rightarrow$ | $bc \leq ac,$ | |
| $a \leq b,\ c > 0$ | $\Rightarrow$ | $\frac{a}{c} \leq \frac{b}{c},$ | (division), |
| $a \leq b,\ c < 0$ | $\Rightarrow$ | $\frac{b}{c} \leq \frac{a}{c},$ | |
| $0 < a \leq b$ | $\Rightarrow$ | $\frac{1}{b} \leq \frac{1}{a},$ | (reciprocal). |

This means: one may add or subtract the same number from both sides of an inequality, or multiply or divide both sides by a *positive* number, without changing the validity of the inequality. Multiplication or division of both sides by a negative number *turns the inequality symbol around.*

**Manipulations with two inequalities:** For all real numbers $a$, $b$, $c$ and $d$, one has

| | | | |
|---|---|---|---|
| $a \leq b,\ c \leq d$ | $\Rightarrow$ | $a + c \leq b + d,$ | (addition), |
| $a \leq b,\ 0 \leq c \leq d$ | $\Rightarrow$ | $ac \leq bd,$ | (multiplication). |

All these rules for manipulations with inequalities remain true, when the inequality $\leq$ is replaced throughout by strict inequality $<$.

*Example 1:* For all real numbers $a$ and $b$ we have the inequality

$$ab \leq \frac{1}{2}(a^2 + b^2).$$

*Proof:* From $0 \leq (a - b)^2$ one gets, applying the binomial formula

$$0 \leq a^2 - 2ab + b^2.$$

Adding $2ab$ to both sides gives $2ab \leq a^2 + b^2$. The claim follows from this, upon division of both sides by 2. □

*Example 2:* For all real numbers $a$ the following inequality is valid:

$$\frac{a^4}{1 + a^2} \leq a^4.$$

*Proof:* From $1 \leq 1 + a^2$ it follows that $\frac{1}{1 + a^2} \leq 1$ by the rule for reciprocals. Multiplying both sides by $a^4$ then yields the statement. □

*Example 3:* Let $a$ and $b$ be real numbers with $a \neq 0$. We wish to examine the linear inequality

$$\boxed{ax - b \geq 0} \tag{1.16}$$

for real $x$.

(i) For $a > 0$, (1.16) holds if and only if $x \geq \frac{b}{a}$.

(ii) For $a < 0$, (1.16) holds if and only if $x \leq \frac{b}{a}$.

*Example 4:* For given real numbers $a$, $b$ and $c$ with $a > 0$, we consider the quadratic inequality

$$\boxed{ax^2 + 2bx + c \geq 0} \tag{1.17}$$

with the so-called *discriminant* $D := b^2 - ac$.

(i) For $D \leq 0$ any real number $x$ is a solutions of (1.17).

(ii) For $D > 0$ the set of solution of (1.17) consists of all real numbers $x$ which satisfy

$$x \leq \frac{-b - \sqrt{D}}{a} \quad \text{or} \quad x \geq \frac{-b + \sqrt{D}}{a}.$$

A selection of important inequalities can be found in 0.1.11.

## 1.2 Limits of sequences

### 1.2.1 Basic ideas

**Sequences:** *Example 1:* We consider the sequence of real numbers $(a_n)$ with

$$a_n := \frac{1}{n}, \qquad n = 1, 2, \ldots$$

As $n$ grows in magnitude, the value of $a_n$ approaches 0 (see Table 1.1). To describe this behavior we write

$$\boxed{\lim_{n\to\infty} a_n = 0}$$

and say that the *limit* of the sequence $(a_n)$ is 0.

Table 1.1

| $n$ | 1 | 2 | 10 | 100 | 1000 | 10000 | ... |
|---|---|---|---|---|---|---|---|
| $a_n$ | 1 | 0.5 | 0.1 | 0.01 | 0.001 | 0.0001 | ... |

*Example 2:* The sequence $(b_n)$ with $b_n := \dfrac{n}{n+1}$ approaches for large $n$ the value 1. We again write $\lim\limits_{n\to\infty} b_n = 1$.

**Functions:** In many applications of mathematics in science, technology and economics, the notion of limits plays a particularly important role. The notion of the *limit of a function* is reduced to the notion of the limit of a sequence as above.

*Example 3:* Consider the function

$$f(x) := \begin{cases} x^2 & \text{for all real numbers } x \neq 0\,, \\ 1 & \text{for } x = 0 \end{cases}$$

(Figure 1.13).

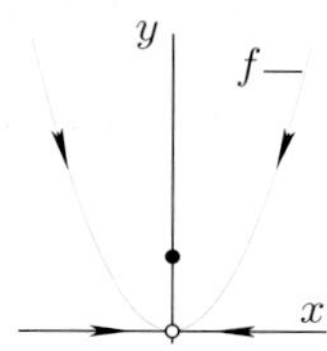


*Figure 1.13.*

We write

$$\boxed{\lim_{x\to a} f(x) = b,}$$

if and only if for every sequence $(a_n)$ with $a_n \neq a$ for all $n$, we have the following:

$$\boxed{\text{From } \lim_{n\to\infty} a_n = a \text{ it follows that } \lim_{n\to\infty} f(a_n) = b.}$$

For the function $f(x)$ in this example one has

$$\lim_{x\to 0} f(x) = 0, \tag{1.18}$$

since from $a_n \neq 0$ for all $n$ and $\lim\limits_{n\to\infty} a_n = 0$ it follows that $\lim\limits_{n\to\infty} f(a_n) = \lim\limits_{n\to\infty} a_n^2 = 0$.

The relation (1.18) corresponds to our intuitive impression: if the point $x$ approaches from the right (of the left) the point 0, then the corresponding values of the function approach 0. The value of $f$ at 0 is irrelevant to these considerations.

Since the limit of a sequence of rational numbers can be irrational, one needs a rigorous development of the theory of limits, arising from a rigorous introduction of the real numbers, which we describe in the following section.

### 1.2.2 The Hilbert axioms for the real numbers

Around 500 BC a member of the Pythagorean school in ancient Greece discovered that the length $d$ of the diagonals of a unit square is incommensurable with the length of the sides, i.e., the ratio of $d$ by the length of the sides is *not* a rational number. By the Pythagorean theorem it follows by Figure 1.14 that

$$d^2 = 1^2 + 1^2,$$

which implies $d = \sqrt{2}$. In modern terminology this citizen of ancient Greece discovered the *irrationality* of the number $\sqrt{2}$ (see section 4.2.1). This discovery destroyed the harmonic picture of the universe of the Pythagoreans and triggered a deep shock. According to legend, the discoverer of this fact was thrown by other members of the Pythagorean school into the ocean during a journey at sea.

The difficulties of irrational numbers were mastered by the – next to Archimedes (281-212 BC) the most important mathematician of antiquity – Eudoxus of Knidos (410-350 BC). It wasn't until 2000 years later that Dedekind returned to the ideas of Eudoxus in 1872, in order to create a mathematically rigorous definition of irrational numbers.

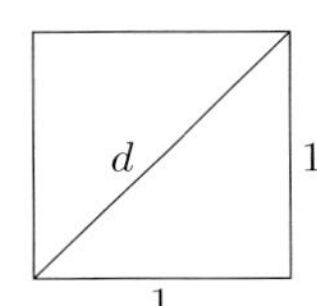


*Figure 1.14.*

Following the example of the *Elements* of Euclid (around 300 BC), mathematical theories are often constructed axiomatically, i.e., one builds the theory on some simple principles. The principles (laid down in a set of axioms) need not be proved. Usually the axioms are the result of a long and tedious mathematical examination of the situation. Starting from the axioms, one deduces through logical conclusions the entire theory.

#### 1.2.2.1 The axioms

We postulate the existence of a set $\mathbb{R}$, whose elements are called real numbers and which satisfy the following axioms (F), (O) and (C).

(F) *Field axioms.* The set $\mathbb{R}$ is a field with the neutral element 0 for addition and neutral element 1 for multiplication.

(O) *Ordering axiom.* For any two given real numbers $a$ and $b$, exactly one of the following three relations holds:

$$a < b, \quad a = b, \quad b < a, \qquad \text{(trichotomy)}.$$

For arbitrary real $a$, $b$ and $c$ the following hold:

(i) The relations $a < b$ and $b < c$ imply $a < c$, (transitivity),

(ii) The relation $a < b$ implies $a + c < b + c$, (monotony of addition),

(iii) The relations $a < b$ and $0 < c$ imply $ac < bc$, (monotony of multiplication).

A *Dedekind section* $(A, B)$ is an ordered pair of non-empty sets of real numbers such that any real number lies in one of them and $a \in A$ and $b \in B$ implies $a < b$.

(C) *Completeness axiom.* For any Dedekind section $(A, B)$ there is exactly one real number $\alpha$ with the property that

$$\boxed{a \le \alpha \le b \quad \text{for all } a \in A,\ b \in B.}$$

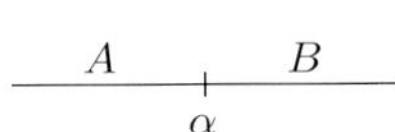


*Figure 1.15.*

Intuitively, (C) means that the real number line doesn't have any holes (Figure 1.15).

The axiom (F) means that the set of real numbers satisfies the following conditions.

**Definition of a field:** A set $K$ is called a *field*, if it satisfies the following.

*Addition.* Two arbitrary elements $a$ and $b$ in $K$ are assigned a unique third element of $K$, denoted $a+b$. For all $a$, $b$ and $c$ in $K$ we have the relations:

$$\boxed{\begin{array}{rl} (a+b)+c = a+(b+c), & \text{(associativity)}, \\ a+b = b+a, & \text{(commutativity)}. \end{array}}$$

There is a unique element in $K$, denoted 0, such that

$$\boxed{a+0=a}$$

for all $a \in K$. This element is called the *neutral element for addition.*

For every element $a \in K$ there is a unique element $b \in K$ such that

$$\boxed{a+b=0.}$$

This element is called the *additive inverse* of $a$. This element $b$ is also written $-a$.

*Multiplication.* Two arbitrary elements $a$ and $b$ in $K$ are assigned a unique third element in $K$, which we denote by $ab$ (or also $a \cdot b$). For all $a$, $b$ and $c$ in $K$ we have:

$$\boxed{\begin{array}{rl} (ab)c = a(bc), & \text{(associativity)}, \\ (ab) = (ba), & \text{(commutativity)}, \\ a(b+c) = ab+ac, & \text{(distributivity)}. \end{array}}$$

There is a unique element in $K$, denoted 1, with $1 \neq 0$ and

$$\boxed{a \cdot 1 = a}$$

for all $a \in K$. This element is called the *neutral element for multiplication.*

For every $a \in K$ with $a \neq 0$, there is a unique element $b \in K$ such that

$$\boxed{ab=1.}$$

This element is called the *multiplicative inverse* to $a$. In this case the element $b$ is also written $a^{-1}$.

The notion of a field is one of the most basic in all of mathematics. Many mathematical objects are fields (see 2.5.3). In general field theory the symbol $e$ is also used for the element 1.

**Consequences of the axioms:** All rules of arithmetic for the real numbers (rules for signs, fractions, equalities and inequalities) can be derived from these axioms.

The axioms (F) and (O) are also valid for the rational numbers; the axiom (C) is however false.

**Uniqueness:** If $\mathbb{R}$ and $\mathbb{R}'$ are two sets which satisfy all the axioms (F), (O) and (C), then the field $\mathbb{R}$ is isomorphic to $\mathbb{R}'$ by an isomorphism which respects the ordering axiom. This means: there is a bijective map $\varphi : \mathbb{R} \longrightarrow \mathbb{R}'$, such that for all $a,\ b \in \mathbb{R}$ the following conditions are satisfied:

(i) $\varphi(a+b) = \varphi(a) + \varphi(b)$.

(ii) $\varphi(ab) = \varphi(a)\varphi(b)$.

(iii) From $a \leq b$ it follows that $\varphi(a) \leq \varphi(b)$.

This implies that one can do calculations in $\mathbb{R}'$ just as in $\mathbb{R}$.

#### 1.2.2.2 The law of induction

Intuitively one gets the set of natural numbers $0, 1, 2 \ldots$, by continued addition $0,\ 0+1,\ 1+1$, etc. To arrive at a mathematically rigorous definition one must proceed along a different (apparently more complicated) route.

*Inductive sets:* A set $M$ of real numbers is called *inductive*, if it contains 0 and the implication $a \in M \Rightarrow a+1 \in M$ holds.

By definition the set of natural numbers consists of the intersection of all inductive sets. This means that $\mathbb{N}$ is the smallest inductive set.

**Law of induction:** Let $A$ be a set of natural numbers with the following properties:

(i) $0 \in A$,

(ii) $n \in A$ implies $n+1 \in A$.

Then $A = \mathbb{N}$.

*Proof:* The set $A$ is inductive. Since $\mathbb{N}$ is the smallest inductive set, $\mathbb{N} \subset A$. Since $A$ is a set of natural numbers, $A \subset \mathbb{N}$, and consequently $A = \mathbb{N}$.

Many proofs in mathematics are based on the law of induction. This will be considered in detail in 4.2.2.

#### 1.2.2.3 Supremum and infinimum

**Theorem:** The set $\mathbb{R}$ of real numbers has an Archimedian order, i.e., to every real number $x$ there is a real number $y$ with $x < y$.

**Bounds:** A set $M$ of real numbers is said to be *bounded above* (resp. *bounded below*), if and only if there is a real number $S$ such that

$$\boxed{x \leq S \text{ for all } x \in M \qquad (\text{resp. } S \leq x \text{ for all } x \in M).}$$

The number $S$ is called an *upper bound* (resp. *lower bound*) of the set $M$.

A set of real numbers is bounded if and only if it is bounded above *and* below.

**Supremum:** Every non-empty set $M$ of real numbers which is bounded above has a smallest upper bound. This bound is denoted

$$\boxed{\sup M.}$$

The supremum of $M$ is not necessarily contained in $M$.

For a non-empty set $M$ of real numbers which is unbounded above we set $\sup M := +\infty$.

*Example 1:* For $M := \{0, 1\}$ the set of upper bounds for $M$ consists of all real numbers $S$ with $S \geq 1$. Therefore $\sup M = 1$.

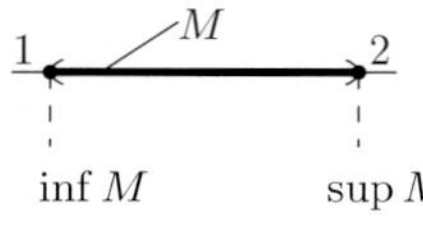


*Figure 1.16.*

*Example 2:* For the open interval $M := ]1, 2[$ the set of upper bounds consists of all real $S$ with $S \geq 2$. Thus $\sup M = 2$; in this case the supremum does not belong to $M$ (Figure 1.16).

**Infimum:** Any non-empty set $M$ of real numbers which is bounded below has a largest lower bound. This is denoted

$$\boxed{\inf M.}$$

The infinimum of $M$ does not necessarily belong to $M$.

For a non-empty set of real numbers $M$ which is unbounded below, we set $\inf M := -\infty$.

*Example 3*: For the set $M := \{0, 1\}$ the set of lower bounds consists of all real numbers $S$ with $S \leq 0$. Thus $\inf M = 0$.

*Example 4:* The set of lower bounds for the open interval $M := ]1, 2[$ consists of the set of real numbers $S$ with $S \leq 1$. Therefore $\inf M = 1$ (Figure 1.16). Here also the infinimum does not belong to $M$.

*Example 5:* $\inf \mathbb{R} = -\infty$, $\sup \mathbb{R} = +\infty$, $\inf \mathbb{N} = 0$ and $\sup \mathbb{N} = +\infty$.

### 1.2.3 Sequences of real numbers

In formulating the notion of limit, modern analysis uses the geometric language of neighborhoods. Doing things in this manner allows a very general formulation of the notion of limit in the context of topology (see [212]).

#### 1.2.3.1 Finite limits

**Neighborhoods:** An *$\varepsilon$-neighborhood* $U_\varepsilon(a)$ of a real number $a$ is the set of real numbers $x$ such that the distance from $x$ to $a$ is less than $\varepsilon$, i.e., in set-theoretic notation

$$U_\varepsilon(a) := \{x \in \mathbb{R} : |x - a| < \varepsilon\}$$

(Figure 1.17(a)). A set of real numbers $U(a)$ is a neighborhood of $a$, if it contains *some* $\varepsilon$-neighborhood of $a$:

$$\boxed{U_\varepsilon(a) \subseteq U(a)}$$

for some $\varepsilon > 0$.

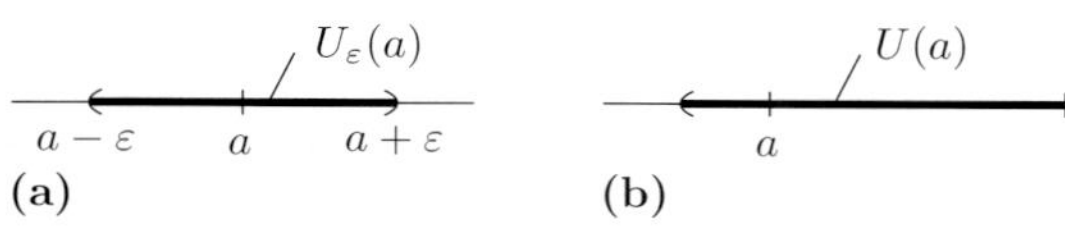


*Figure 1.17. Neighborhoods of a point a.*

If one uses the notion of interval, then $U_\varepsilon(a) = ]a - \varepsilon, a + \varepsilon[$. However, a neighborhood $U(a)$ as just defined need not be an interval, but rather must just contain some interval.

**The fundamental definition of limit:** Let $(a_n)$ be a sequence of real numbers.[11] We write

$$\boxed{\lim_{n \to \infty} a_n = a,} \tag{1.19}$$

[11]This means that for any natural number $n$ there is a real number $a_n$ assigned which belongs to the sequence.

if and only if any $\varepsilon$-neighborhood of the real number $a$ contains all but finitely many of the $a_n$. In this case we say that the sequence $(a_n)$ *converges* to the limit $a$.

In other words, the relation (1.19) holds if and only if for every real $\varepsilon > 0$ there is a natural number $n_0(\varepsilon)$ (depending on $\varepsilon$), such that

$$\boxed{|a_n - a| < \varepsilon \qquad \text{for all } n \geq n_0(\varepsilon).}$$

*Example 1:* We set $a_n := \frac{1}{n}$ for $n = 1, 2, \ldots$ Then we have

$$\lim_{n\to\infty} \frac{1}{n} = 0.$$

*Proof:* For every real $\varepsilon > 0$ there is a natural number $n_0(\varepsilon)$ with $n_0(\varepsilon) > \frac{1}{\varepsilon}$. Consequently,

$$|a_n| = \frac{1}{n} < \varepsilon \quad \text{for all } n \geq n_0(\varepsilon).$$

□

*Example 2:* For a constant sequence $a_n = a$ for all $n$, we get $\lim\limits_{n\to\infty} a_n = a$.

**Theorem:**

(i) A limit, if it exists, is unique.

(ii) This limit does not change if *finitely* many of the terms of the sequence are changed.

**Rules for manipulating limits:** For any two sequences $(a_n)$ and $(b_n)$, both of which converge to a finite limit, the following hold:

| | |
|---|---|
| $\lim\limits_{n\to\infty}(a_n + b_n) = \lim\limits_{n\to\infty} a_n + \lim\limits_{n\to\infty} b_n,$ | (rule of sums), |
| $\lim\limits_{n\to\infty}(a_n b_n) = \lim\limits_{n\to\infty} a_n \lim\limits_{n\to\infty} b_n,$ | (product rule), |
| $\lim\limits_{n\to\infty} \frac{a_n}{b_n} = \frac{\lim\limits_{n\to\infty} a_n}{\lim\limits_{n\to\infty} b_n},$ | (quotient rule[12]), |
| $\lim\limits_{n\to\infty} \lvert a_n\rvert = \lvert\lim\limits_{n\to\infty} a_n\rvert,$ | (absolute value rule), |
| from $a_n \leq b_n$ for all $n$ it follows $\lim\limits_{n\to\infty} a_n \leq \lim\limits_{n\to\infty} b_n,$ | (inequality rule). |

*Example 3* (products): $\lim\limits_{n\to\infty} \frac{1}{n^2} = \lim\limits_{n\to\infty} \frac{1}{n} \lim\limits_{n\to\infty} \frac{1}{n} = 0.$

*Example 4* (sums): $\lim\limits_{n\to\infty} \frac{n+1}{n} = \lim\limits_{n\to\infty} \left(1 + \frac{1}{n}\right) = \lim\limits_{n\to\infty} 1 + \lim\limits_{n\to\infty} \frac{1}{n} = 1.$

#### 1.2.3.2 Improper limits

**Neighborhoods:** We set

$$U_E(+\infty) :=]E, \infty[, \qquad U_E(-\infty) :=] - \infty, E[.$$

[12] Here one must in addition assume that $\lim_{n\to\infty} b_n \neq 0$.

A set $U(+\infty)$ of real numbers is called a *neighborhood of infinity*, if there is a real number $E$ such that

$$\boxed{U_E(+\infty) \subseteq U(+\infty).}$$

In the same way, $U(-\infty)$ is a set such that $U_E(-\infty) \subseteq U(-\infty)$ for some fixed real number $E$ (Figure 1.18).

**Definition:** Let $(a_n)$ be a sequence of real numbers. We write

$$\boxed{\lim_{n\to\infty} a_n = +\infty,} \tag{1.20}$$

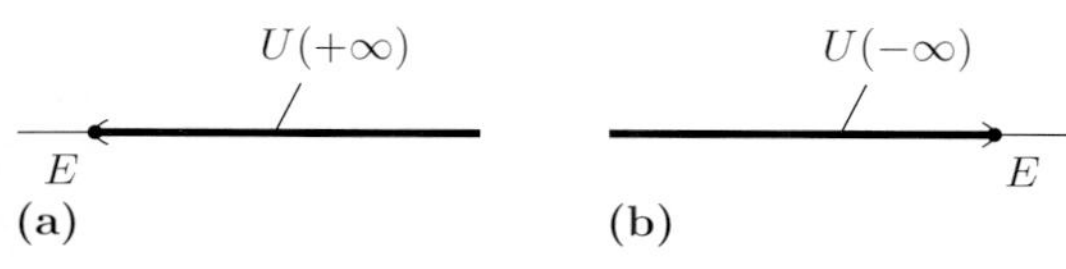


*Figure 1.18. Neighborhoods of $\pm$ infinity.*

if and only if all but finitely many $a_n$ lie in every neighborhood $U(+\infty)$.

In other words, the relation (1.20) holds if and only if for every real number $E$ we can find a natural number $n_0(E)$ such that

$$\boxed{a_n > E \qquad \text{for all } n \geq n_0(E).}$$

*Example 1:* $\lim\limits_{n\to\infty} n = +\infty$.

In the same way we write

$$\lim_{n\to\infty} a_n = -\infty,$$

if and only if all but finitely many of the $a_n$ lie in every neighborhood $U(-\infty)$.[13]

**Reflection principle:** One has $\lim\limits_{n\to\infty} a_n = -\infty$ if an only if $\lim\limits_{n\to\infty} (-a_n) = +\infty$.

*Example 2:* $\lim\limits_{n\to\infty} (-n) = -\infty$.

**Rules for manipulations with infinity:** Let $-\infty < a < \infty$. Then we have

Addition:

$$a + \infty = +\infty, \qquad +\infty + \infty = +\infty,$$
$$a - \infty = -\infty, \qquad -\infty - \infty = -\infty.$$

Multiplication:

$$a(\pm\infty) = \begin{cases} \pm\infty & \text{for } a > 0, \\ \mp\infty & \text{for } a < 0, \end{cases}$$

$$(+\infty)(+\infty) = +\infty, \quad (-\infty)(-\infty) = +\infty, \quad (+\infty)(-\infty) = -\infty.$$

Division:

$$\frac{a}{\pm\infty} = 0, \qquad \frac{\pm\infty}{a} = \begin{cases} \pm\infty & \text{for } a > 0, \\ \mp\infty & \text{for } a < 0. \end{cases}$$

[13]Let $\lim_{n\to\infty} a_n = a$. In older literature one speaks of convergence for $a \in \mathbb{R}$, but of determined divergence in case $a = \pm\infty$. In modern mathematics one has a general notion of convergence. In this sense $(a_n)$ converges for all values of $a$ (cf. Example 4 in 1.3.2.1). This modern point of view, which we adapt here, has the distinct advantage of avoiding the consideration of different cases, see 1.2.4.

For example, the symbolic notation '$a + \infty = +\infty$' means that from the relations

$$\lim_{n\to\infty} a_n = a \quad \text{and} \quad \lim_{n\to\infty} b_n = +\infty \tag{1.21}$$

we always get

$$\lim_{n\to\infty} (a_n + b_n) = +\infty$$

as a consequence. Similarly, the notation '$a(+\infty) = +\infty$ for $a > 0$' means that from (1.21) and $a > 0$, it follows that

$$\lim_{n\to\infty} a_n b_n = +\infty.$$

*Example 3:* $\lim_{n\to\infty} n^2 = \lim_{n\to\infty} n \lim_{n\to\infty} n = +\infty$.

**Rational expressions:** We set

$$a := \frac{\alpha_k n^k + \alpha_{k-1} n^{k-1} + \ldots + \alpha_0}{\beta_m n^m + \beta_{m-1} n^{m-1} + \ldots + \beta_0}, \qquad n = 1, 2, \ldots$$

for fixed $k, m = 0, 1, 2\ldots$ and fixed real numbers $\alpha_r, \beta_s$ with $\alpha_k \neq 0$ and $\beta_m \neq 0$. Then we have:

$$\lim_{n\to\infty} a_n = \begin{cases} \dfrac{\alpha_k}{\beta_m} & \text{for } k = m, \\ 0 & \text{for } k < m, \\ +\infty & \text{for } k > m \text{ and } \alpha_k/\beta_m > 0, \\ -\infty & \text{for } k > m \text{ and } \alpha_k/\beta_m < 0. \end{cases}$$

*Example 4:* $\lim_{n\to\infty} \dfrac{n^2+1}{n^3+1} = 0$.

**Indeterminant expressions:** In the case of

$$\boxed{+\infty - \infty, \ 0 \cdot (\pm\infty), \ \frac{0}{0}, \ \frac{\infty}{\infty}, \ 0^0, \ 0^\infty, \ \infty^0} \tag{1.22}$$

one must be extremely careful! There are no general rules for manipulating these expressions. In different cases one gets different results.

*Example 5* ($+\infty - \infty$):

$$\lim_{n\to\infty} (2n - n) = +\infty, \quad \lim_{n\to\infty} (n - 2n) = -\infty, \quad \lim_{n\to\infty} ((n+1) - n) = 1.$$

*Example 6* ($0 \cdot \infty$):

$$\lim_{n\to\infty} \left(\frac{1}{n} \cdot n\right) = 1, \quad \lim_{n\to\infty} \left(\frac{1}{n} \cdot n^2\right) = \lim_{n\to\infty} n = +\infty.$$

In certain case expressions as in (1.22) can be given meaning and calculated by means of l'Hospital' rule (cf. 1.3.1.3).

### 1.2.4 Criteria for convergence of sequences

**Basic idea:** *Example 1:* We consider the iteration procedure

$$a_{n+1} = \frac{a_n}{2} + \frac{1}{a_n}, \qquad n = 0, 1, 2, \ldots, \tag{1.23}$$

with a fixed initial value $a_0 := 2$. To calculate the limit of the sequence $(a_n)$, we assume the limit

$$\boxed{\lim_{n\to\infty} a_n = a} \tag{1.24}$$

exists with $a > 0$. From (1.23) we conclude

$$\lim_{n\to\infty} a_{n+1} = \lim_{n\to\infty} \left(\frac{a_n}{2} + \frac{1}{a_n}\right),$$

or $a = \dfrac{a}{2} + \dfrac{1}{a}$. This yields $2a^2 = a^2 + 2$, or $a^2 = 2$ and finally $a = \sqrt{2}$. Consequently we get

$$\lim_{n\to\infty} a_n = \sqrt{2}. \tag{1.25}$$

The following example contains a false conclusion.

*Example 2:* We consider the iteration process

$$a_{n+1} = -a_n, \qquad n = 0, 1, 2, \ldots \tag{1.26}$$

with initial value $a_0 := 1$. The same method yields

$$\lim_{n\to\infty} a_{n+1} = -\lim_{n\to\infty} a_n.$$

This implies $a = -a$, or $a = 0$, in other words we have $\lim\limits_{n\to\infty} a_n = 0$.

On the other hand from (1.26) we get immediately $a_n = (-1)^n$ for all $n$, and this sequence is not convergent. Where is the mistake? The answer is:

| This convenient method of computing the limit of an iteration process is only valid if the existence of the limit is insured. |
|---|

Because of this it is important to have some general criteria for checking when a sequence converges. Such criteria are discussed in section 1.2.4.1 and 1.2.4.3 below.

**Theorem:** The iteration scheme (1.23) converges, i.e., one has $\lim\limits_{n\to\infty} a_n = \sqrt{2}$.

*Sketch of the proof:* One shows

$$a_0 \geq a_1 \geq a_2 \geq \ldots \geq 1. \tag{1.27}$$

This means that the sequence $(a_n)$ is non-increasing and bounded below. The convergence criterion in 1.2.4.1 gives the existence of the limit (1.24), which proves the claim (1.25).[14]

**Bounded sequences:** A sequence of real numbers $(a_n)$ is said to be *bounded below* (resp. *bounded above*), if there is a real number $S$ such that

$$\boxed{a_n \leq S \qquad \text{for all } n}$$

(resp. $S \leq a_n$ for all $n$). A sequence is said to be bounded, if it is bounded above and below.

*Criterium for boundedness:* Every sequence of real numbers which converges to a finite limit, is bounded.

*Consequence:* An unbounded sequence of real numbers can not converge to a finite limit.

*Example 3:* The sequence $(n)$ of natural numbers is unbounded above. Consequently this sequence does not converge to a finite limit.

[14] A more detailed proof of (1.27) will be given below in 4.2.4 as an application of induction.

#### 1.2.4.1 Increasing and decreasing sequences

**Definition:** A sequence $(a_n)$ of real numbers is said to be *increasing* (res. *decreasing*), if

$$\boxed{n \le m \text{ implies } a_n \le a_m}$$

(resp. $n \le m$ implies $a_n \ge a_m$).

**Convergence criterion:** Every increasing sequence of real numbers $(a_n)$ converges to a finite or to an infinite limit.[15]

(i) If $(a_n)$ is bounded above, then $\lim\limits_{n\to\infty} a_n = a$ for some finite $a \in \mathbb{R}$.

(ii) If $(a_n)$ is unbounded above, then $\lim\limits_{n\to\infty} a_n = +\infty$.

Set $M := \{a_n \,|\, n \in \mathbb{N}\}$. Then one has $\lim\limits_{n\to\infty} a_n = \sup M$.

*Example:* We choose the sequence $a_n := 1 - \dfrac{1}{n}$. This sequence is decreasing and bounded above. Moreover, $\lim\limits_{n\to\infty} a_n = 1$.

#### 1.2.4.2 The Cauchy criterion for convergence

**Definition:** A sequence $(a_n)$ of real numbers is called a *Cauchy sequence*, if for every $\varepsilon > 0$ there is $n_0(\varepsilon) \in \mathbb{N}$ such that

$$\boxed{|a_n - a_m| < \varepsilon \qquad \text{for all } n, m \ge n_0(\varepsilon).}$$

**Cauchy criterion:** A sequence of real numbers is convergent if and only if it is a Cauchy sequence.

#### 1.2.4.3 Subsequences

**Subsequences:** Let $(a_n)$ be a sequence of real numbers. We choose indices $k_0 < k_1 < \cdots$ and set

$$\boxed{b_n := a_{k_n}, \qquad n = 0, 1, \ldots}$$

Then the sequence $(b_n)$ is called a *subsequence*[16] of $(a_n)$.

*Example 1:* Let $a_n := (-1)^n$. If we set $b_n := a_{2n}$, then $(b_n)$ is a subsequence of $(a_n)$. Explicitly, one has

$$a_0 = 1, \quad a_1 = -1, \quad a_2 = 1, \quad a_3 = -1, \quad \ldots,$$

$$b_0 = a_0 = 1, \quad b_1 = a_2 = 1, \ldots, \quad b_n = a_{2n} = 1, \quad \ldots$$

---

[15] In the same way, every decreasing sequence $(a_n)$ of real numbers converges to a finite or to an infinite limit.

(i) If $(a_n)$ is bounded below, then $\lim_{n\to\infty} a_n = a$ for some finite $a \in \mathbb{R}$.

(ii) If $(a_n)$ is unbounded below, then $\lim_{n\to\infty} a_n = -\infty$. If one sets $M := \{a_n : n \in \mathbb{N}\}$, then $\lim_{n\to\infty} a_n = \inf M$.

[16] It is often convenient to denote such a subsequence by $(a_{n'})$, which means we set $a_{1'} := b_1$, $a_{2'} := b_2$, etc.

**Accumulation point:** Let $-\infty \leq a \leq \infty$. Then $a$ is called an *accumulation point* of the sequence $(a_n)$, if there is a subsequence $(a_{n'})$ with

$$\boxed{\lim_{n\to\infty} a_{n'} = a.}$$

The set of all accumulation points of $(a_n)$ is called the *limit set* of $(a_n)$.

**Theorem of Bolzano–Weierstrass:** (i) Every sequence of real numbers has an accumulation point.

(ii) Every bounded sequence of real numbers has a real number which is an accumulation point.

**The limit superior:** Let $(a_n)$ be a sequence of real numbers. We set[17]

$$\boxed{\overline{\lim_{n\to\infty}}\, a_n := \text{largest accumulation point of } (a_n)}$$

and $\underline{\lim}_{n\to\infty}\, a_n :=$ smallest accumulation point of $(a_n)$.

**Subsequence criterion for convergence:** Let $-\infty \leq a \leq \infty$. For a sequence $(a_n)$ of real numbers, the following are equivalent:

(i) $\lim_{n\to\infty} a_n = a$, and

(ii)

$$\boxed{\overline{\lim_{n\to\infty}}\, a_n = \underline{\lim_{n\to\infty}}\, a_n = a.}$$

*Example 2:* Let $a_n := (-1)^n$. For the two subsequences $(a_{2n})$ and $(a_{2n+1})$, one has

$$\lim_{n\to\infty} a_{2n} = 1 \quad \text{and} \quad \lim_{n\to\infty} a_{2n+1} = -1.$$

There for $a = 1$ and $a = -1$ are accumulation points of $(a_n)$, and these are all the accumulation points. Consequently

$$\overline{\lim_{n\to\infty}}\, a_n = 1, \qquad \underline{\lim_{n\to\infty}}\, a_n = -1.$$

Since these values do not coincide, the sequence $(a_n)$ cannot be convergent.

*Example 3:* For $a_n := (-1)^n n$ one has $\lim_{n\to\infty} a_{2n} = +\infty$ and $\lim_{n\to\infty} a_{2n+1} = -\infty$. These are all the accumulation points of this sequence. Consequently we get

$$\overline{\lim_{n\to\infty}}\, a_n = +\infty, \qquad \underline{\lim_{n\to\infty}}\, a_n = -\infty.$$

As these two values again do not coincide, this sequence is divergent (i.e., not convergent).

**Special cases:** Let $(a_n)$ be a sequence of real numbers and let $-\infty \leq a \leq \infty$.

(i) If $\lim_{n\to\infty} a_n = a$, then $a$ is the only accumulation point of $(a_n)$, and every subsequence of $(a_n)$ converges to the same value $a$.

[17]This definition makes sense, since $(a_n)$ really does have a largest and smallest accumulation point (which may be $\pm\infty$). One calls $\overline{\lim}_{n\to\infty}\, a_n$ (resp. $\underline{\lim}_{n\to\infty}\, a_n$) the *limit superior* or upper limit (resp. *limit inferior* or lower limit) of the sequence $(a_n)$.

(ii) If a subsequence of a Cauchy sequence $(a_n)$ converges to $a \in \mathbb{R}$, then $a$ is the only accumulation point of $(a_n)$ and one has $\lim\limits_{n\to\infty} a_n = a$.

## 1.3 Limits of functions

### 1.3.1 Functions of a real variable

We consider functions $y = f(x)$ of a real variable $x$ with real values $f(x)$.

#### 1.3.1.1 Limits

**Definition:** Let $-\infty \leq a, b \leq \infty$. We write

$$\boxed{\lim_{x\to a} f(x) = b,}$$

if, for every sequence $(x_n)$ in the domain of $f$ with $x_n \neq a$ for every $n$, we have[18]

$$\boxed{\lim_{n\to\infty} x_n = a \qquad \text{implies} \qquad \lim_{n\to\infty} f(x_n) = b.}$$

In particular, we write

$$\lim_{x\to a+0} f(x) = b \qquad \text{resp.} \qquad \lim_{x\to a-0} f(x) = b,$$

if only sequences $(x_n)$ with $x_n > a$ for all $n$ (resp. $x_n < a$ for all $n$) are to be considered $(a \in \mathbb{R})$.

**Manipulations:** Since the notion of limit of a function is defined in terms of the limit of a sequence of numbers, the manipulations for the latter give manipulations for the limits here. In particular, for $-\infty \leq a \leq \infty$ one has

$$\boxed{\begin{aligned} &\lim_{x\to a}(f(x) + g(x)) = \lim_{x\to a} f(x) + \lim_{x\to a} g(x),\\ &\lim_{x\to a} f(x)g(x) = \lim_{x\to a} f(x) \lim_{x\to a} g(x),\\ &\lim_{x\to a} \frac{f(x)}{h(x)} = \frac{\lim\limits_{x\to a} f(x)}{\lim\limits_{x\to a} h(x)}. \end{aligned}}$$

Here the additional assumption is made that all the limits on the right hand side exist and are finite, and in the last expression, $\lim\limits_{x\to a} h(x) \neq 0$.

These manipulations remain correct for $x \longrightarrow a + 0$ and $x \longrightarrow a - 0$ for $a \in \mathbb{R}$.

*Example 1:* Let $f(x) := x$. Then for all $a \in \mathbb{R}$,

$$\lim_{x\to a} f(x) = a.$$

Indeed, $\lim\limits_{n\to\infty} x_n = a$ implies $\lim\limits_{n\to\infty} f(x_n) = a$.

[18] The function $f$ need not be defined at $a$ for these considerations. We only require that the domain of $f$ contains at least one sequence $(x_n)$ with the limit property listed above.

*Example 2:* Let $f(x) := x^2$. Then we have

$$\lim_{x \to a} x^2 = \lim_{x \to a} x \lim_{x \to a} x = a^2.$$

*Example 3:* We define

$$f(x) := \begin{cases} 1 & \text{for } x > a, \\ 2 & \text{for } x = a, \\ -1 & \text{for } x < a \end{cases}$$

(see Figure 1.19). Then we have

$$\lim_{x \to a+0} f(x) = 1, \qquad \lim_{x \to a-0} = -1.$$

One calls the limit $\lim_{x \to a+0} f(x)$ (resp. $\lim_{x \to a-0} f(x)$) the right-sided (resp. left-sided) limit of $f$ at $a$.

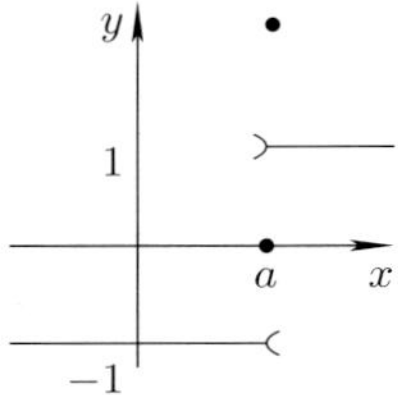


Figure 1.19.

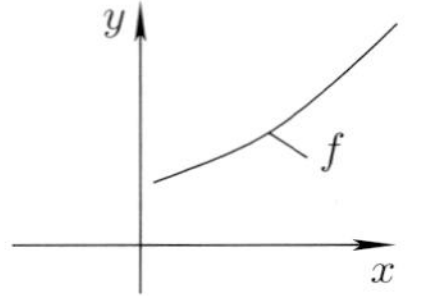

(a) continuous function (b) non-continuous function

Figure 1.20.

#### 1.3.1.2 Continuous functions

Intuitively the notion of continuous function is one which has no jumps (Figure 1.20).

**Definition:** Let $a \in M$. The function $f : M \subseteq \mathbb{R} \longrightarrow \mathbb{R}$ is said to be *continuous at a point* $a$, if for every neighborhood $U(f(a))$ of the image point $f(a)$ there is a neighborhood $U(a)$, such that[19]

$$\boxed{x \in U(a) \text{ and } x \in M \quad \text{imply} \quad f(x) \in U(f(a)).}$$

In other words, $f$ is continuous at $a$, if for every real $\varepsilon > 0$ there is a real number $\delta > 0$ such that

$$|f(x) - f(a)| < \varepsilon \qquad \text{for all } x \in M \quad \text{with } |x - a| < \delta.$$

**Limit criterion:** $f$ is continuous at a point $a$ if and only if[20]

$$\boxed{\lim_{x \to a} f(x) = f(a).}$$

[19] One also writes $f(U(a)) \subseteq U(f(a))$.

[20] This means: for every sequence $(x_n)$ in $M$ with $\lim_{n \to \infty} x_n = a$, one has $\lim_{n \to \infty} f(x_n) = f(a)$.

**Manipulations:** If $f, g : M \subseteq \mathbb{R} \longrightarrow \mathbb{R}$ are continuous at $a$, then

(i) The sum $f + g$ and the product $fg$ are continuous at $a$.

(ii) The quotient $\frac{f}{g}$ is continuous at $a$, if $g(a) \neq 0$.

We now consider the composition of two functions

$$\boxed{H(x) := F(f(x)).}$$

We also write $H = F \circ f$ for this.

**Continuity of composed functions:** The function $H$ is continuous at $a$ if $f$ is continuous at $a$ and $F$ is continuous at the point $f(a)$.

**Differentiability and continuity:** If the function $f : M \subseteq \mathbb{R} \longrightarrow \mathbb{R}$ is differentiable at $a$, then $f$ is continuous at $a$ (cf. 1.4.1).

*Example:* The function $y = \sin x$ is differentiable at every point $a \in \mathbb{R}$. It follows that

$$\boxed{\lim_{x \to a} \sin x = \sin a.}$$

Similar statements hold for $y = \cos x$, $y = \mathrm{e}^x$, $y = \cosh x$, $y = \sinh x$, $y = \arctan x$ and for every polynomial $y = a_0 + a_1 x + \ldots + a_n x^n$ with real coefficients $a_0, \ldots, a_n$.

The following theorems show that continuous functions have very pleasant properties. Let $-\infty < a < b < \infty$.

**Theorem of Weierstrass:** Every continuous function $f : [a, b] \longrightarrow \mathbb{R}$ has a minimum and a maximum.

More precisely this means that there are $\alpha, \beta \in [a, b]$ with

$$\boxed{f(\alpha) \leq f(x) \quad \text{for all } x \in [a, b]}$$

(minimum) and $f(x) \leq f(\beta)$ for all $x \in [a, b]$ (maximum) (see Figure 1.21).

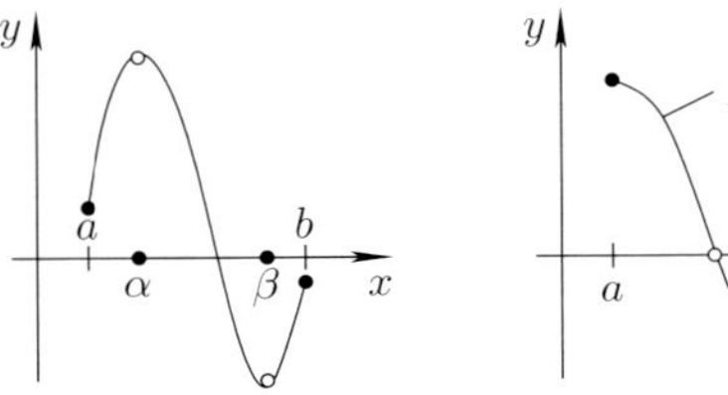


*Figure 1.21.* *Figure 1.22.*

**Theorem of Bolzano:** If the function $f : [a, b] \longrightarrow \mathbb{R}$ is continuous and $f(a)f(b) \leq 0$, then the equation

$$\boxed{f(x) = 0, \quad x \in [a, b]}$$

has a solution (Figure 1.22).

**Mean value theorem:** If the function $f : [a, b] \longrightarrow \mathbb{R}$ is continuous, then the equation

$$\boxed{f(x) = \gamma, \quad x \in [a, b]}$$

has a solution for all $\gamma$ with $\min\limits_{a \le x \le b} f(x) \le \gamma \le \max\limits_{a \le x \le b} f(x)$.

#### 1.3.1.3 L'Hospital's rule

This important rule allows the evaluation of expressions of the form $\frac{0}{0}$ and $\frac{\infty}{\infty}$. It is

$$\boxed{\lim_{x \to a} \frac{f(x)}{g(x)} = \lim_{x \to a} \frac{f'(x)}{g'(x)}.} \tag{1.28}$$

It is assumed that:

(i)There are limits $\lim\limits_{x \to a} f(x) = \lim\limits_{x \to a} g(x) = b$ with $b = 0$ or $b = \pm\infty$, and $-\infty \le a \le \infty$.

(ii) There is a neighborhood $U(a)$ such that the derivatives $f'(x)$ and $g'(x)$ exist for all $x \in U(a)$ and $x \ne a$.

(iii) One has $g'(x) \ne 0$ for all $x \in U(a)$, $x \ne a$.

(iv) The limit on the right hand side of (1.28) exists.[21]

*Example 1* $\left(\frac{0}{0}\right)$: One has $\lim\limits_{x \to 0} \sin x = \lim\limits_{x \to 0} x = 0$. From (1.28) it follows that

$$\lim_{x \to 0} \frac{\sin x}{x} = \lim_{x \to 0} \frac{\cos x}{1} = \cos 0 = 1,$$

because of the continuity of $\cos x$.

*Example 2* $\left(\frac{\infty}{\infty}\right)$:

$$\lim_{x \to +\infty} \frac{\ln x}{x} = \lim_{x \to +\infty} \frac{\frac{1}{x}}{1} = 0,$$

$$\lim_{x \to +\infty} \frac{\mathrm{e}^x}{x} = \lim_{x \to +\infty} \frac{\mathrm{e}^x}{1} = +\infty.$$

**Variants of l'Hospital's rule:** Sometimes one must apply l'Hospital's rule repeatedly before one gets a well-defined limit on the right-hand side.

$$\boxed{\lim_{x \to a} \frac{f(x)}{g(x)} = \lim_{x \to a} \frac{f'(x)}{g'(x)} = \ldots = \lim_{x \to a} \frac{f^{(n)}(x)}{g^{(n)}(x)}.}$$

*Example 3:* $\lim\limits_{x \to +\infty} \dfrac{\mathrm{e}^x}{x^2} = \lim\limits_{x \to +\infty} \dfrac{\mathrm{e}^x}{2x} = \lim\limits_{x \to +\infty} \dfrac{\mathrm{e}^x}{2} = +\infty.$

[21] A similar statement holds for $s \to a + 0$ (resp. $a \to a - 0$) with $a \in \mathbb{R}$. In this case one requires the assumptions (ii) and (iii) only for points $x \in U$ for which $x > a$ (resp. $x < a$).
The notion of derivative $f'(x)$ will be introduced in 1.4.1.

Expressions of the form $0 \cdot \infty$ are brought into the form $\frac{\infty}{\infty}$ to which l'Hospital's rule applies.

*Example 4:* $\lim\limits_{x\to+0} x\ln x = \lim\limits_{x\to+0} \dfrac{\ln x}{\frac{1}{x}} = \lim\limits_{x\to+0} \dfrac{\frac{1}{x}}{\left(-\frac{1}{x^2}\right)} = \lim\limits_{x\to+0}(-x) = 0.$

Expressions of the form $\infty - \infty$ are brought into the form $\infty \cdot a$ for some finite value $a$.

*Example 5:* $\lim\limits_{x\to+\infty} (\mathrm{e}^x - x) = \lim\limits_{x\to+\infty} \mathrm{e}^x \left(1 - \dfrac{x}{\mathrm{e}^x}\right) = \lim\limits_{x\to+\infty} \mathrm{e}^x \lim\limits_{x\to+\infty} \left(1 - \dfrac{x}{\mathrm{e}^x}\right) = \lim\limits_{x\to+\infty} \mathrm{e}^x$ $= +\infty$. This follows from Example 2, where one has

$$\lim_{x\to+\infty} \left(1 - \frac{x}{\mathrm{e}^x}\right) = 1.$$

The following formula is also very handy:

$$a^x = \mathrm{e}^{x\cdot\ln a}.$$

It can be applied to expressions of the form $0^0$, $\infty^0$ or $0^\infty$.

*Example 6* ($\infty^0$): From $x^{1/x} = \mathrm{e}^{\frac{\ln x}{x}}$ and Example 2, one gets

$$\lim_{x\to+\infty} x^{1/x} = \mathrm{e}^0 = 1.$$

#### 1.3.1.4 The order of magnitude of functions

For many considerations it is sufficient to have a good understanding of the qualitative behavior of functions. For this there are convenient symbols $O(g(x))$ and $o(g(x))$, due to Landau. Let $-\infty \le a \le \infty$.

**Definition** (Asymptotic equality): We write

$$\boxed{f(x) \cong g(x), \qquad x \to a\,,}$$

if and only if $\lim\limits_{x\to a} \dfrac{f(x)}{g(x)} = 1$.

*Example 1:* The equality $\lim\limits_{x\to 0} \dfrac{\sin x}{x} = 1$ implies $\sin x \cong x$, $x \to 0$.

**Definition:** We write

$$\boxed{f(x) = O(g(x)), \quad x \to a,} \tag{1.29}$$

if there is a neighborhood $U(a)$ of $a$ and a real number $K$ such that

$$\left|\frac{f(x)}{g(x)}\right| \le K \qquad \text{for all } x \in U(a) \qquad \text{with } x \ne a. \tag{1.29*}$$

**Theorem:** The relation (1.29) holds if the finite limit $\lim\limits_{x\to a} \dfrac{f(x)}{g(x)}$ exists.

*Example 2:* The equality $\lim\limits_{x\to+\infty} \dfrac{3x^2+1}{x^2} = 3$ implies $3x^2 + 1 = O(x^2)$, $x \to +\infty$.

**Definition:** We write[22]

$$\boxed{f(x) = o(g(x)), \qquad x \to a,}$$

[22]Let $a \in \mathbb{R}$. In the same way one introduces the symbols $f(x) \cong g(x)$, $f(x) = O(g(x))$ and $f(x) = o(g(x))$ for $x \to a + 0$ (resp. $x \to a - 0$). The inequality (1.29*) is in general true only for $x \in U(a)$ with $x > a$ (resp. $x < a$).

if $\lim\limits_{x\to a}\dfrac{f(x)}{g(x)}=0$.

*Example 3:* One has $x^n=o(x),\ x\to 0$ for $n=2,3,\ldots$

*Example 4:*

(i) $\dfrac{x^2}{x^2+2}\cong 1,\ x\to\infty$.

(ii) $\dfrac{1}{x^2+2}\cong\dfrac{1}{x^2},\ x\to+\infty$.

(iii) $\sin x=O(1)$ for $x\to a$ and all $a$ with $-\infty\le a\le\infty$.

(iv) $\ln x=o\left(\frac{1}{x}\right)$ for $x\to+0$ and $\ln x=o(x)$ for $x\to+\infty$.

(v) $x^n=o(\mathrm{e}^x)$ for $x\to+\infty$ and $n=1,2,\ldots$

The last statement (v) means that the function $y=\mathrm{e}^x$ grows faster than every power $x^n$ as $x\to+\infty$.

## 1.3.2 Metric spaces and point sets

**Motivation:** One of the characteristics of modern mathematics is the tendency to extend notions and methods to more and more abstract situations. This allows for solutions giving a great deal of insight of more and more complicated problems and for seeing the connections between apparently completely different and disjoint areas of study. This procedure also turns out too be highly economical, since it replaces the necessity of more and more different notions by their derivation from a few very basic ones.

To carry the notion of limits of functions of a single variable over to functions of several variables, it is advantageous to introduce metric spaces. The full power of the modern point of view becomes apparent in the study of functional analysis. This branch of mathematics was developed in the 20th century (cf. [212]).

### 1.3.2.1 The notion of distance and convergence

**Metric spaces:** In a metric space one has a notion of distance between two points. A non-empty set $X$ is called a *metric space*, if for every ordered pair of points, $(x,y)$ in $X$, there is assigned a real number $d(x,y)\ge 0$, such that for all $x,y,z\in X$, the following statements are true:

(i) $d(x,y)=0$ if and only if $x=y$,

(ii) $d(x,y)=d(y,x)$ (symmetry),

(iii) $d(x,z)\le d(x,y)+d(y,z)$ (triangle inequality).

The number $d(x,y)$ is called the *distance* between $x$ and $y$. By definition, the empty set is also a metric space.

**Theorem:** Every subset of a metric space is again a metric space with the same distance function.

**Limits:** Let $(x_n)$ be a sequence of points in a metric space $X$. We write

$$\boxed{\lim_{n\to\infty} x_n = x,}$$

if $\lim_{n\to\infty} d(x_n, x) = 0$, that is, if the distance between $x_n$ and $x$ approaches zero as $n \to \infty$.

**Uniqueness:** If a limit exists, then it is uniquely determined.

*Example 1:* The set $\mathbb{R}$ of real numbers is a metric space with distance function

$$d(x, y) := |x - y| \qquad \text{for all } x, y \in \mathbb{R}.$$

The notion of distance induced by this metric is the usual (naive) one (see 1.2.3.1).

*Example 2:* The set $\mathbb{R}^N$ is by definition the set of all $N$-tuples $x = (\xi_1, \ldots, \xi_N)$ of real numbers $\xi_j$. Let $y = (\eta_1, \ldots, \eta_N)$ be another element of $\mathbb{R}^N$, and set

$$\boxed{d(x, y) := \sqrt{\sum_{j=1}^{N} (\xi_j - \eta_j)^2}.}$$

This makes $\mathbb{R}^N$ a metric space. For $N = 1, 2, 3$ the induced notion of distance in $\mathbb{R}$, $\mathbb{R}^2$ and $\mathbb{R}^3$ coincides with the usual (naive) notion (see Figure 1.23).

Furthermore we define

$$\boxed{|x - y| := d(x, y)}$$

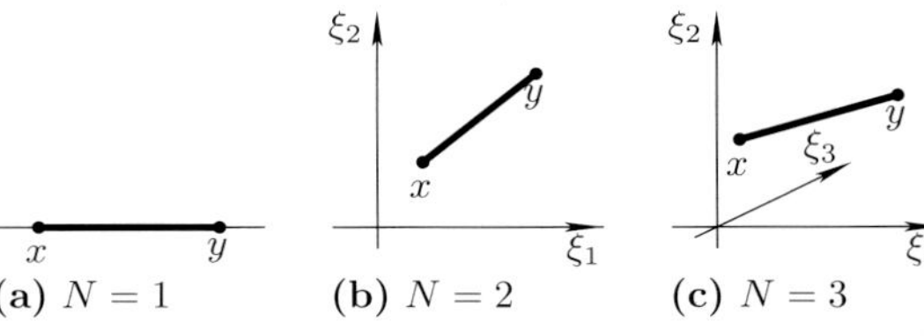


*Figure 1.23. Distances in $\mathbb{R}^n$.*

and let $|x| = \sqrt{\sum_{j=1}^{N} \xi_j^2}$ denote the *Euclidean norm* of $x$. Intuitively $|x|$ is the distance of $x$ to the origin.

Let $(x_n)$ be a sequence in $\mathbb{R}^N$ with components $x_n = (\xi_{1n}, \ldots, \xi_{Nn})$ and let $x = (\xi_1, \ldots, \xi_N)$ be as above. Then the convergence

$$\boxed{\lim_{n\to\infty} x_n = x}$$

in the metric space $\mathbb{R}^N$ is equivalent to the component-wise convergence

$$\boxed{\lim_{n\to\infty} \xi_{jn} = \xi_j \quad \text{for all } j = 1, \ldots, N.}$$

*Example 3:* In the special case $N = 2$, the convergence $\lim_{n\to\infty} x_n = x$ corresponds to the visible fact that the points $x_n$ get closer and closer to the point $x$ (see Figure 1.24).

*Example 4* (the unit circle): We consider the situation depicted in Figure 1.25. Each point $x$ of the real line $\mathbb{R}$ corresponds to a unique point $x_*$ on the unit circle of radius 1. The north pole $\mathsf{N}$ doesn't correspond to any point of $\mathbb{R}$. Usually one replaces the north pole by the points $+\infty$ and $-\infty$. We define

$$\overline{\mathbb{R}} := \mathbb{R} \cup \{+\infty, -\infty\}.$$

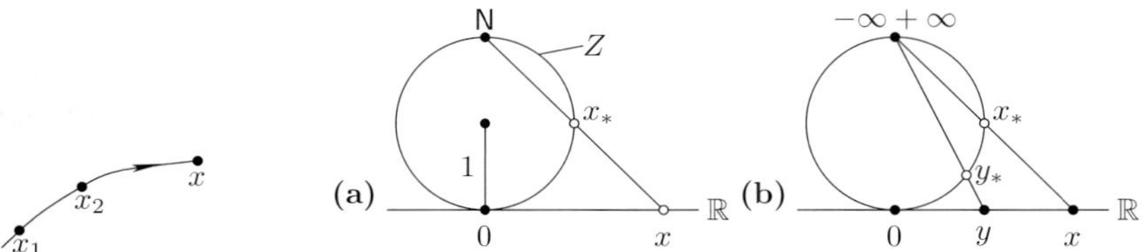


*Figure 1.24.*

*Figure 1.25. A distance function on the unit circle.*

The set $\overline{\mathbb{R}}$ becomes a metric space by setting

$$d(x, y) := \text{ arc length between } x_* \text{ and } y_* \text{ on the unit circle } Z.$$

We agree on the convention

$$d(-\infty, \infty) := 2\pi.$$

For example, we have $d(\pm\infty, 0) = \pi$. Let $(x_n)$ be a sequence of real numbers. The convergence

$$\lim_{n\to\infty} x_n = x$$

with $-\infty \le x \le +\infty$ in the sense of the metric $d$ on $\overline{\mathbb{R}}$ means that the corresponding points $(x_n)_*$ on the unit circle converge to the point $x_*$. This is equivalent to the classical notion of convergence (see 1.2.3).

In this manner, the classical notion of convergence to either finite or infinite values is given a uniform definition which derives from the metric notion of convergence in the metric space $Z$ of the unit circle.

#### 1.3.2.2 Special sets

Let $M$ be a subset of a metric space $X$.

**Bounded sets:** The non-empty set $M$ is said to be *bounded*, if there is a real number $R > 0$ such that

$$d(x, y) \le R \qquad \text{for all } x, y \in M.$$

The empty set is by definition bounded.

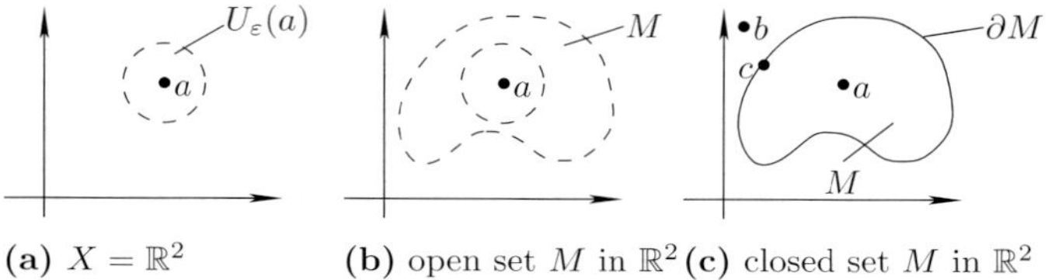


**(a)** $X = \mathbb{R}^2$ **(b)** open set $M$ in $\mathbb{R}^2$ **(c)** closed set $M$ in $\mathbb{R}^2$

*Figure 1.26. Open and closed sets in $\mathbb{R}^2$.*

**Neighborhoods:** Let $\varepsilon > 0$. We set

$$U_\varepsilon(a) := \{x \in X \mid d(a,x) < \varepsilon\},$$

and call it the *$\varepsilon$-neighborhood* of $a$. In other words, the $\varepsilon$-neighborhood $U_\varepsilon(a)$ of the point $a$ consists of all the points $x$ in the metric space $X$ whose distance to $a$ is $< \varepsilon$ (Figure 1.26).

A set $U(a)$ is called a neighborhood of $a$, if it contains some $\varepsilon$-neighborhood $U_\varepsilon(a)$.

**Open sets:** A set $M$ is said to be *open*, if for every $a \in M$ there is neighborhood $U(a)$ of $a$ contained in $M$, $U(a) \subseteq M$.

**Closed sets:** The set $M$ is said to be *closed*, if the complement $X - M$ is open.

**Interior and exterior:** The point $a \in X$ is said to be an *interior* (or inner) point of $M$, if there is a neighborhood $U(a)$ of $a$ contained in $M$, $U(a) \subseteq M$ (Figure 1.26(c)).

The point $b$ is said to be an *exterior* (or outer) point of $M$, if there is a neighborhood $U(b)$ which does *not* belong to $M$, $U(b) \subseteq X - M$.

The point $c$ is said to be a *boundary* point of $M$, if $c$ is neither an inner nor an exterior point of $M$ (Figure 1.26(c)).

The set of all interior (resp. exterior) points of $M$ is denoted $\operatorname{int} M$ (resp. $\operatorname{ext} M$).

**Boundary and closure:** The set $\partial M$ of all boundary points of $M$ is called the *boundary* of $M$ (Figure 1.26(c)). Furthermore, the set

$$\boxed{\overline{M} := M \cup \partial M}$$

is called the *closure* of $M$.

**Theorem:** (i) The interior $\operatorname{int} M$ of $M$ is the largest open set contained in $M$.

(ii) The closure $\overline{M}$ of $M$ is the smallest closed set containing $M$.

(iii) One has a decomposition into disjoint sets:

$$\boxed{X = \operatorname{int} M \cup \operatorname{ext} M \cup \partial M,}$$

which means that every point $x \in X$ belongs to *exactly* one of the sets $\operatorname{int} M, \operatorname{ext} M, \partial M$.

**Accumulation point:** A point $a \in X$ is called an *accumulation point* of $M$, if every neighborhood of $a$ contains a point of $M$ other than $a$ itself.

**Theorem of Bolzano–Weierstrass:** Every infinite unbounded set of $\mathbb{R}^N$ has an accumulation point.

#### 1.3.2.3 Compactness

The notion of compactness is among the most important in all of analysis.

A subset $M$ of a metric space is said to be *compact*, if every open cover of $M$ (collection of open sets whose union contains $M$) contains a *finite* sub-cover, i.e., there is a finite subset of that collection of open sets whose union still contains $M$.

A set is said to be *relatively compact*, if its closure is compact.

**Theorem:** (i) Every compact set is closed and bounded.

(ii) Every relatively compact set is bounded.

**Characterization in terms of convergent sequences:** Let $M$ be a subset of a metric space.

(i) $M$ is closed if and only if every convergent sequence $(x_n)$ in $M$ has a limit in $M$.

(ii) $M$ is relatively compact if and only if every sequence in $M$ has a convergent subsequence.

(iii) $M$ is compact if and only if every sequence in $M$ has a convergent subsequence whose limit belongs to $M$.

**Subsets of $\mathbb{R}^N$:** Let $M$ be a subset of $\mathbb{R}^N$. The following three statements are equivalent:

(i) $M$ is compact.

(ii) $M$ is closed and bounded.

(iii) Every sequence in $M$ has a convergent subsequence whose limit belongs to $M$.

Moreover, the following three statements are also equivalent:

(a) $M$ is relatively compact.

(b) $M$ is bounded.

(c) Every sequence in $M$ contains a convergent subsequence.

### 1.3.2.4 Connectedness

A subset $M$ of a metric space is said to be *arc-wise connected*, if there is a continuous curve in $M$ joining any two points $x, y \in M$[23] (see Figure 1.27).

**Domains:** A subset of a metric space is said to be a *domain*, if it is open, arc-wise connected and not empty.

**Simply connected sets:** A subset $M$ of a metric space is said to be *simply connected*, if it is arc-wise connected and every closed curve in $M$ can be retracted continuously to a point[24] (Figure 1.28).

### 1.3.2.5 Examples

*Example 1* ($X = \mathbb{R}$): Let $-\infty < a < b < +\infty$.

(i) The interval $[a, b]$ is a compact set in $\mathbb{R}$. It is also closed and bounded.

(ii) The interval $]a, b[$ is open and bounded.

(iii) A subset of $\mathbb{R}$ is arc-wise connected if and only if it is an interval.

(iv) Every real number is an accumulation point of the set of rational numbers.

---

[23]The notion of continuous curve is defined as follows: there exists a continuous map $\varphi : [0, 1] \to M$ with $\varphi(0) = x$ and $\varphi(1) = y$. The continuity of $\varphi$ means that

$$\lim_{n \to \infty} \varphi(t_n) = \varphi(t)$$

for every sequence $(t_n)$ in $[0, 1]$ for which $\lim_{n \to \infty} t_n = t$.

[24]This means that for each closed curve, i.e., continuous $\varphi : [0, 1] \to M$ with $\varphi(0) = \varphi(1)$, there is a continuous function $H = H(t, x)$ from $[0, 1] \times M$ into $M$, such that

$$H(0, x) = \varphi(x) \quad \text{and} \quad H(1, x) = x_0$$

for all $x \in M$, where $x_0$ is some fixed point in $M$.

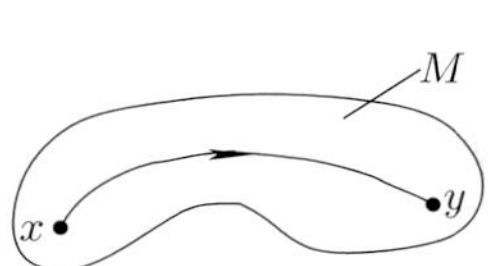


*Figure 1.27.* *Arc-wise connectedness.*

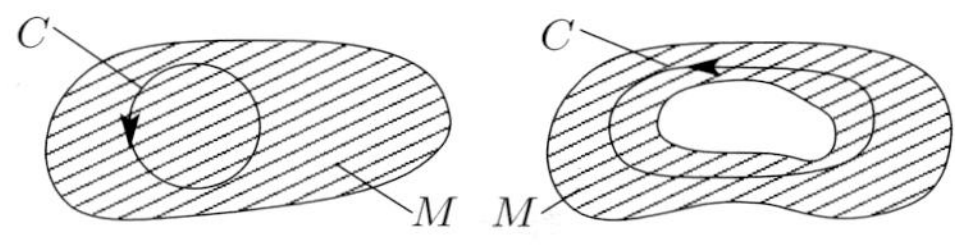


*Figure 1.28.* *The notion of simply connected and non-simply connected sets.*

(v) The half-open interval $[a, b[$ is neither open nor closed. It is however bounded and relatively compact.

*Example 2* ($X = \mathbb{R}^2$): Let $r > 0$. We set

$$M := \{(\xi_1, \xi_2) \in \mathbb{R}^2 \mid \xi_1^2 + \xi_2^2 < r^2\}.$$

Then $M$ is the interior of a circle of radius $r$ centered at the origin (Figure 1.29(a)). One may check that the boundary and closure are given as follows:

$$\partial M = \{(\xi_1, \xi_2) \in \mathbb{R}^2 \mid \xi_1^2 + \xi_2^2 = r^2\},$$

$$\overline{M} = \{(\xi_1, \xi_2) \in \mathbb{R}^2 \mid \xi_1^2 + \xi_2^2 \leq r^2\}$$

(see Figure 1.29(b)).

(i) The set $M$ is open, bounded, arc-wise connected, simply connected and relatively compact.

(ii) The set $M$ is a simply connected domain.

(iii) The set $M$ is neither closed nor compact.

(iv) The set $\overline{M}$ is closed, bounded, compact, arc-wise connected and simply connected.

(v) The boundary $\partial M$ is closed, bounded, compact and arc-wise connected, but not simply connected.

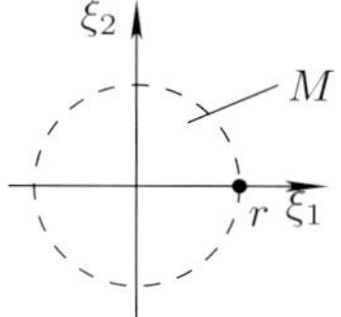


**(a)** open circle $M$

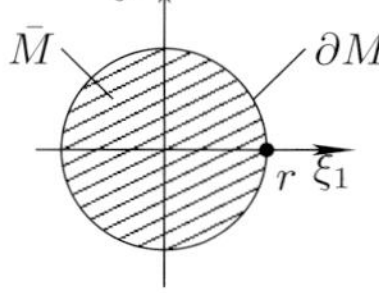


**(b)** $\bar{M} = M \cup \partial M$

*Figure 1.29.*

*Example 3* (unit circle): The set $\mathbb{R}$ is unbounded with respect to the classical distance function, hence not compact (cf. Example 1 in 1.3.2.1).

On the other hand, the metric space $\mathbb{R} \cup \{\pm\infty\}$ introduced in Example 4 of 1.3.2.1 is bounded and compact. This is the deeper reason for the fact that one can handle finite and infinite limits in a uniform manner.

The notions introduced above can be generalized to metric and topological spaces (see [212]).

### 1.3.3 Functions of several variables

Most of the functions which occur in applications depend on more than one variable, for example space and time coordinates. We abbreviate this by writing $y = f(x)$ with $x = (\xi_1, \ldots, \xi_N)$, where all $\xi_j$ are real variables.

### 1.3.3.1 Limits

Let $f : M \longrightarrow Y$ be a function[25] from a metric space $M$ to a metric space $Y$. We write

$$\boxed{\lim_{x \to a} f(x) = b,}$$

if and only if for every sequence $(x_n)$ in the domain of $f$ with $x_n \neq a$, and for all $n$, we have:[26]

$$\boxed{\lim_{n\to\infty} x_n = a \quad \text{implies} \quad \lim_{n\to\infty} f(x_n) = b.}$$

*Example 1:* For the function $f : \mathbb{R}^2 \longrightarrow \mathbb{R}$ given by $f(u, v) := u^2 + v^2$ we have

$$\lim_{(u,v)\to(a,b)} f(u, v) = a^2 + b^2.$$

Indeed, for an arbitrary sequence $(u_n, v_n)$ with $\lim\limits_{n\to\infty} (u_n, v_n) = (a, b)$ we have $\lim\limits_{n\to\infty} u_n = a$ and $\lim\limits_{n\to\infty} v_n = b$. Consequently,

$$\lim_{n\to\infty} (u_n^2 + v_n^2) = a^2 + b^2.$$

*Example 2:* For the function

$$f(u, v) := \begin{cases} \dfrac{u}{v} & \text{for } (u, v) \neq (0, 0), \\ 0 & \text{for } (u, v) = (0, 0) \end{cases}$$

the limit $\lim\limits_{(u,v)\to(0,0)} f(u, v)$ does not exist. This is because the sequence $(u_n, v_n) = \left(\dfrac{1}{n}, \dfrac{1}{n}\right)$ satisfies

$$\lim_{n\to\infty} f(u_n, v_n) = 1,$$

while for $(u_n, v_n) = \left(\dfrac{1}{n^2}, \dfrac{1}{n}\right)$ we get

$$\lim_{n\to\infty} f(u_n, v_n) = \lim_{n\to\infty} \frac{1}{n} = 0.$$

### 1.3.3.2 Continuity

**Definition:** A map $f : M \longrightarrow Y$ between two metric spaces $M$ and $Y$ is said to be *continuous at a point* $a$, if for every neighborhood of the image $U(f(a))$ there is a neighborhood of $a$ satisfying:

$$\boxed{f(U(a)) \subseteq U(f(a)).}$$

This means $x \in U(a)$ implies $f(x) \in U(f(a))$.

The function $f$ is called *continuous*, if it is continuous at every point $a \in M$.

---

[25] The definition and properties of general functions can be found in 4.3.3.

[26] The function $f$ need not be defined in the point $a$. We only require that the domain of $f$ contains *some* sequence $(x_n)$ with the limit property stated.

**Limit criterion:** For a function $f : M \longrightarrow Y$ and a point $a \in M$, the following three statements are equivalent:[27]

(i) $f$ is continuous at $a$.

(ii) $\boxed{\lim_{x \to a} f(x) = f(a).}$

(iii) For every $\varepsilon > 0$ there is a $\delta > 0$ such that $d(f(x), f(a)) < \varepsilon$ for all $x$ with $d(x, a) < \delta$.

**Theorem:** A function $f : M \longrightarrow Y$ is continuous if and only if the inverse images of open sets are open.

**Law of composition:** If $f : M \longrightarrow Y$ and $F : Y \longrightarrow Z$ are continuous, then the composed map

$$\boxed{F \circ f : M \to Z}$$

is also continuous. We have $(F \circ f)(x) := F(f(x))$.

**Manipulations:** If the functions $f, g : M \longrightarrow \mathbb{R}$ are continuous in a point $a$, then:

| | | |
|---|---|---|
| $f + g$ | is continuous in $a$, | (rule of sums), |
| $fg$ | is continuous in $a$, | (rule for products), |
| $\dfrac{f}{g}$ | is continuous in $a$, if $g(a) \neq 0$, | (rule for quotients). |

**Component rule:** Let $f(x) = (f_1(x), \ldots, f_k(x))$. Then the following two statements are equivalent:

(i) $f_j : M \longrightarrow \mathbb{R}$ is continuous at $a$ for every $j$.

(ii) $f : M \longrightarrow \mathbb{R}^k$ is continuous in $a$.

*Example:* Let $x = (\xi_1, \xi_2)$. Every polynomial

$$p(x) = \sum_{j,k=0}^{m} a_{jk} \xi_1^j \xi_2^k$$

with real coefficients $a_{jk}$ is continuous at every point $x \in \mathbb{R}^2$.

A similar statement holds also for polynomials in $N$ variables.

**Principle of invariance:** Let $f : M \longrightarrow Y$ be a continuous map between two metric spaces. Then

(i) $f$ maps compact sets to compact sets.

(ii) $f$ maps arc-wise connected sets to arc-wise connected sets.

**Theorem of Weierstrass:** A continuous function $f : M \longrightarrow \mathbb{R}$ from a non-empty compact subset $M$ of a metric space has a minimum and a maximum.

This is true in particular for non-empty, bounded and closed subsets of $\mathbb{R}^N$.

[27] The condition (ii) means that $\lim_{n \to \infty} f(x_n) = f(a)$ for every sequence $(n_n)$ in $M$ with $\lim_{x \to \infty} x_n = a$.

**Bolzano's theorem on zeros:** Let $f : M \longrightarrow \mathbb{R}$ be a continuous function on an arc-wise connected subset of a metric space. If there are two points $a, b \in M$ with $f(a)f(b) \le 0$, then the equation

$$\boxed{f(x) = 0, \qquad x \in M}$$

has a solution.

**Mean value theorem:** If $f : M \longrightarrow \mathbb{R}$ is continuous and $M$ is arc-wise connected, then the image $f(M)$ is an interval.

In the special case that $f(a) < f(b)$ for two points $a, b \in M$, the equation

$$\boxed{f(x) = \gamma, \qquad x \in M}$$

has a solution for every real number $\gamma$ for which $f(a) \le \gamma \le f(b)$.

## 1.4 Differentiation of functions of a real variable

### 1.4.1 The derivative

**Definition:** We consider a real function $y = f(x)$ of a real variable $x$, which is defined in a neighborhood of a point $p$. The *derivative* $f'(p)$ of $f$ at the point $p$ is defined as the finite limit

$$\boxed{f'(p) = \lim_{h \to 0} \frac{f(p+h) - f(p)}{h}.}$$

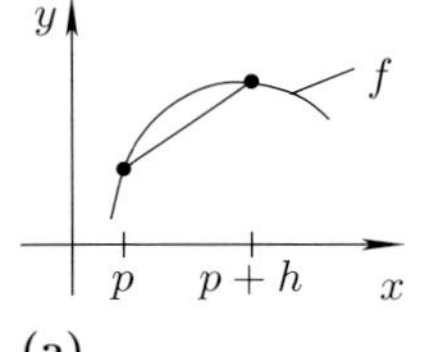


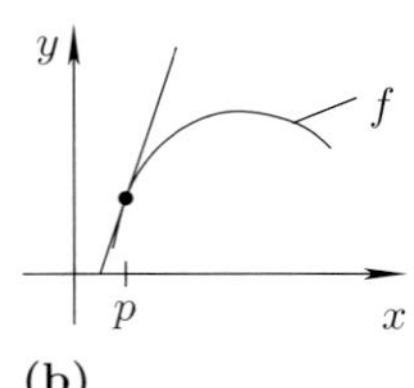


*Figure 1.30. The derivative.*

**Geometric interpretation:** The number

$$\frac{f(p+h) - f(p)}{h}$$

is the slope of the secant in Figure 1.30(a). For $h \to 0$ the secant intuitively approaches the tangent. Thus we define:

$$\boxed{f'(p) \text{ is the slope of the tangent of the graph of } f \text{ at the point } (p, f(p)).}$$

The corresponding *equation of the tangent* is then:

$$\boxed{y = f'(p)(x - p) + f(p).}$$

*Example 1:* For the function $f(x) := x^2$ we get

$$f'(p) = \lim_{h \to 0} \frac{(p+h)^2 - p^2}{h} = \lim_{h \to 0} \frac{2ph + h^2}{h} = \lim_{h \to 0} (2p + h) = 2p.$$

**Table of important derivatives:** See section 0.8.1.

**The notation of Leibniz:** Let $y = f(x)$. Instead of $f'(p)$ one also writes

$$f'(p) = \frac{\mathrm{d}f}{\mathrm{d}x}(p) \quad \text{or} \quad f'(p) = \frac{\mathrm{d}y}{\mathrm{d}x}(p).$$

If we set $\Delta f := f(x) - f(p)$, $\Delta x := x - p$ and $\Delta y = \Delta f$, then we have

$$\boxed{\frac{\mathrm{d}f}{\mathrm{d}x}(p) = \lim_{\Delta x \to 0} \frac{\Delta f}{\Delta x} = \lim_{\Delta x \to 0} \frac{\Delta y}{\Delta x}.}$$

This notation was introduced by Gottfried Wilhelm Leibniz (1646–1716) and has turned out to be extremely convenient, as many of the important rules of manipulations with derivatives follow just from the notation. This is a property which one expects from well-chosen mathematical notation.

**The relation between continuity and differentiability:** If $f$ is differentiable at a point $p$ (i.e., the derivative of $f$ exists at $p$), then it is also continuous there (one says 'then it is all the more continuous').

The converse statement is false. For example, the function $f(x) := |x|$ is continuous at $x = 0$, but it is not differentiable at that point (although it is differentiable at all other points), the reason being that the graph of this function has no tangent at $x = 0$ (Figure 1.31).

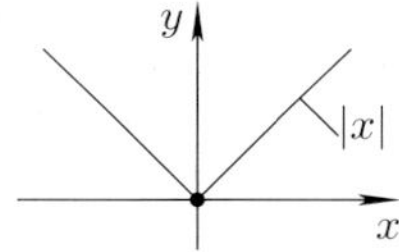


*Figure 1.31.*

**Higher derivatives:** If we set $g(x) := f'(x)$, then by definition

$$\boxed{f''(p) := g'(p).}$$

We also write $f^{(2)}(p)$ for this, or in Leibniz notation:

$$f''(p) = \frac{\mathrm{d}^2 f}{\mathrm{d}x^2}(p).$$

Similarly, we define $f^{(n)}(p)$ for $n = 2, 3, \ldots$.

*Example 2:* For $f(x) := x^2$ we have

$$f'(x) = 2x, \quad f''(x) = 2, \quad f'''(x) = 0, \quad f^{(n)}(x) = 0 \quad \text{for } n = 4, 5, \ldots$$

(see 0.8.1).

**Basic rules:** Suppose $f$ and $g$ are differentiable at a point $x$, and let $\alpha, \beta$ be real numbers. Then:

$$\boxed{\begin{array}{ll} (\alpha f + \beta g)'(x) = \alpha f'(x) + \beta g'(x), & \text{(rule of sums)}, \\ (fg)'(x) = f'(x)g(x) + f(x)g'(x), & \text{(product rule)}, \\ \left(\dfrac{f}{g}\right)'(x) = \dfrac{f'(x)g(x) - f(x)g'(x)}{g(x)^2}, & \text{(quotient rule)}. \end{array}}$$

In the case of the quotient rule one must of course assume that $g(x) \neq 0$.

Examples can be found in 0.8.2.

**The Leibniz product rule:** If $f$ and $g$ are $n$-times differentiable in a point $x$, then for $n = 1, 2, \ldots$ one has

$$\boxed{(fg)^{(n)}(x) = \sum_{k=0}^{n} \binom{n}{k} f^{(n-k)}(x) g^{(k)}(x).}$$

This rule of differentiation has a similarity with the binomial formula (see 0.1.10.3). In particular, for $n = 2$ one has

$$(fg)''(x) = f''(x)g(x) + 2f'(x)g'(x) + f(x)g''(x).$$

*Example 3:* We consider the function $h(x) := x \cdot \sin x$. If we set $f(x) := x$ and $g(x) := \sin x$, then $f'(x) = 1$, $f''(x) = 0$ and $g'(x) = \cos x$, $g''(x) = -\sin x$. Consequently,

$$h''(x) = 2\cos x - x \sin x.$$

**Functions of class $C[a,b]$:** Let $[a,b]$ be a compact interval. We denote the space of all continuous functions $f : [a,b] \longrightarrow \mathbb{R}$ by $C[a,b]$. Moreover we set[28]

$$||f|| := \max_{a \le x \le b} |f(x)|.$$

**Functions of class $C^k[a,b]$:** This class consists of all functions $f \in C[a,b]$, which have continuous derivatives $f'$, $f''$, $\ldots$, $f^{(k)}$ on the open interval $]a,b[$, each of which can be extended to a continuous function on $[a,b]$.

We define[29]

$$||f||_k := \sum_{j=0}^{k} \max_{a \le x \le b} |f^{(j)}(x)|.$$

**Type $C^k$:** We say that a function in a neighborhood of a point $p$ is *of type* $C^k$, if in an open neighborhood of $p$ it has $k$ derivatives which are continuous.

## 1.4.2 The chain rule

The fundamental *chain rule* is easiest to remember in the suggestive Leibniz notation:

$$\boxed{\frac{dy}{dx} = \frac{dy}{du}\frac{du}{dx}.} \qquad (1.30)$$

*Example 1:* In order to differentiate the function $y = f(x) = \sin x^2$, we write

$$y = \sin u, \quad u = x^2$$

and apply the chain rule. According to 0.8.1. we have

$$\frac{dy}{du} = \cos u, \qquad \frac{du}{dx} = 2x.$$

[28]With respect to this norm $||f||$, $C[a,b]$ is a Banach space, cf. [212].

[29]We set $f^{(0)}(x) := f(x)$. As above, $C^k[a,b]$ is a Banach space with respect to the norm $||f||_k$.

Consequently, it follows from (1.30) that

$$f'(x) = \frac{\mathrm{d}y}{\mathrm{d}x} = \frac{\mathrm{d}y}{\mathrm{d}u}\frac{\mathrm{d}u}{\mathrm{d}x} = 2x\cos u = 2x\cos x^2.$$

*Example 2:* Let $b > 0$. For the function $f(x) := b^x$, we have

$$f'(x) = b^x \ln b, \qquad x \in \mathbb{R}.$$

*Proof:* We have $f(x) = \mathrm{e}^{x\ln b}$, and set $y = \mathrm{e}^u$, $u = x\ln b$. By 0.8.1 we have

$$\frac{\mathrm{d}y}{\mathrm{d}u} = \mathrm{e}^u, \qquad \frac{\mathrm{d}u}{\mathrm{d}x} = \ln b.$$

Hence, it follows from (1.30) that

$$f'(x) = \frac{\mathrm{d}y}{\mathrm{d}x} = \frac{\mathrm{d}y}{\mathrm{d}u}\frac{\mathrm{d}u}{\mathrm{d}x} = \mathrm{e}^u \ln b = b^x \ln b. \qquad \square$$

The precise formulation of (1.30) is as follows.

**Theorem** (Chain rule): For a composed function $F(x) := g(f(x))$ the derivative at a point $p$ exists and is given by:

$$\boxed{F'(p) = g'(f(p))f'(p)}$$

under the following assumptions:

(i) The function $f : M \longrightarrow \mathbb{R}$ is defined in neighborhood $U(p)$ of $p$ and the derivative $f'(p)$ exists.

(ii) The function $g : N \longrightarrow \mathbb{R}$ is defined in a neighborhood $U(f(p))$ of $f(p)$ and the derivative $g'(f(p))$ exists.

**Barriers of thought:** The chain rule shows that precise mathematical formulations can be much more unwieldy than suggestive rules. This unfortunately often leads to barriers between mathematicians and physicists and engineers, which have to be overcome somehow. In fact it is a good idea to know both the suggestive and formal rules as well as the precise mathematical formulations, in order to on the one hand do computations with a minimum of work and on the other hand be aware of possible incorrect applications of formal rules.

### 1.4.3 Increasing and decreasing functions

**Criterion for increasing (or decreasing):** Let $-\infty \le a < b \le +\infty$, and let $f : ]a, b[\longrightarrow \mathbb{R}$ be differentiable.

(i) $f$ is non-decreasing (resp. non-increasing), if

$$\boxed{f'(x) \ge 0 \qquad \text{for all } x \in ]a, b[}$$

(resp. $f'(x) \le 0$ for all $x \in ]a, b[$).

(ii) If $f'(x) > 0$ for all $x \in ]a, b[$, then in fact $f$ is increasing[30] in $]a, b[$.

(iii) If $f'(x) < 0$ for all $x \in ]a, b[$, then in fact $f$ is decreasing in $]a, b[$.

[30] The definition of increasing and decreasing functions is in section 0.2, with pictures in Table 0.14

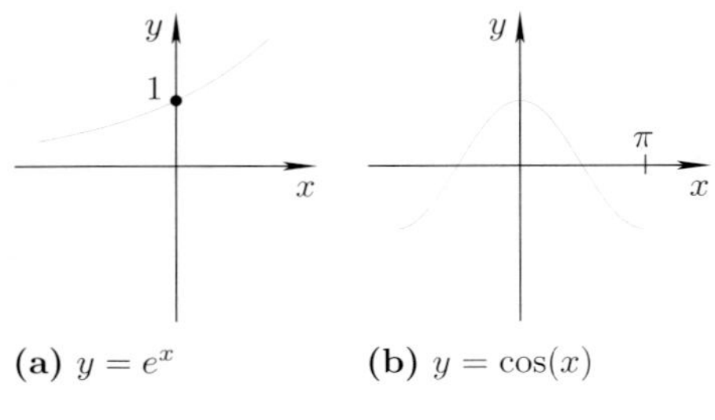


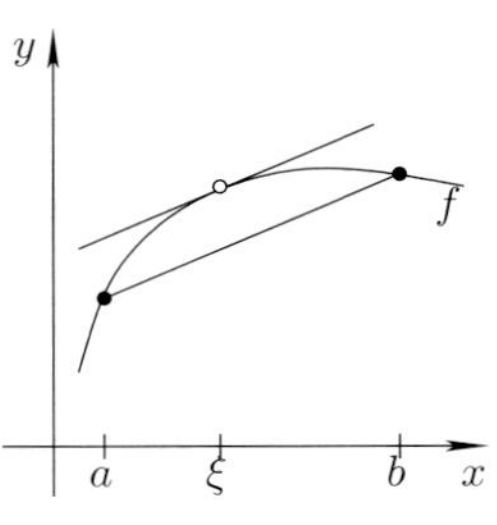


*Figure 1.32. Increasing and decreasing functions.*

*Figure 1.33. The mean value theorem.*

*Example 1:* Let $f(x) := \mathrm{e}^x$. It follows from $f'(x) = \mathrm{e}^x > 0$ for all $x \in \mathbb{R}$ that $f$ is an increasing function on $\mathbb{R}$ (Figure 1.32(a)).

*Example 2:* We take $f(x) := \cos x$. Since $f'(x) = -\sin x < 0$ for all $x \in ]0, \pi[$, if follows that $f$ is decreasing on this interval (Figure 1.32(b)).

**Mean value theorem:** Let $-\infty < a < b < +\infty$. If $f : [a, b] \longrightarrow \mathbb{R}$ is differentiable on the open interval $]a, b[$, then there is a number $\xi \in ]a, b[$ with

$$\boxed{\frac{f(b) - f(a)}{b - a} = f'(\xi).}$$

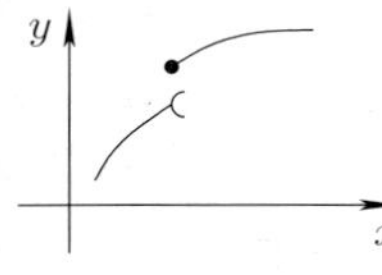


*Figure 1.34.*

Intuitively this means that in Figure 1.33 the secants have the same slope as the tangent for *some* point $\xi$.

**Theorem of Lebesgue:** Let $-\infty \le a < b \le +\infty$. For a strictly increasing function $f : ]a, b[ \longrightarrow \mathbb{R}$ one has:

(i) $f$ is continuous except for finitely many points, at which the left and right sided limits exist.

(ii) $f$ is differentiable almost everywhere[31] (Figure 1.34).

## 1.4.4 Inverse functions

Many important functions are the inverse functions of known functions (cf. (0.28)).

### 1.4.4.1 Local inverses

The rule for differentiating inverse functions is easiest to remember using the suggestive Leibniz notation:

$$\boxed{\frac{\mathrm{d}x}{\mathrm{d}y} = \frac{1}{\frac{\mathrm{d}y}{\mathrm{d}x}}.} \qquad (1.31)$$

*Example 1:* The inverse function to $y = x^2$ is

$$x = \sqrt{y}, \qquad y > 0.$$

[31] This means that there is a set $M$ of Lebesgue measure zero such that $f$ is differentiable for all points *not* in $M$ (see 1.7.2).

One has $\frac{\mathrm{d}y}{\mathrm{d}x} = 2x$. By (1.31) we have

$$\frac{\mathrm{d}\sqrt{y}}{\mathrm{d}y} = \frac{\mathrm{d}x}{\mathrm{d}y} = \frac{1}{\frac{\mathrm{d}y}{\mathrm{d}x}} = \frac{1}{2x} = \frac{1}{2\sqrt{y}}.$$

*Example 2:* For the function $f(x) := \sqrt{x}$ one has

$$f'(x) = \frac{1}{2\sqrt{x}}, \qquad x > 0.$$

This follows from Example 1, by exchanging $y$ and $x$ there.

*Example 3:* The inverse function to $y = \mathrm{e}^x$ is

$$x = \ln y, \quad y > 0$$

(cf. 0.2.6). One has $\frac{\mathrm{d}y}{\mathrm{d}x} = \mathrm{e}^x$. From (1.31) it follows that

$$\frac{\mathrm{d}\ln y}{\mathrm{d}y} = \frac{\mathrm{d}x}{\mathrm{d}y} = \frac{1}{\frac{\mathrm{d}y}{\mathrm{d}x}} = \frac{1}{\mathrm{e}^x} = \frac{1}{y}.$$

*Example 4:* For the function $f(x) := \ln x$ we get

$$f'(x) = \frac{1}{x}, \qquad x > 0.$$

This follows from Example 3, again by exchanging $x$ and $y$.

The precise formulation of (1.31) is as follows.

**Theorem on local inverse functions:** Assume the function $f : M \subseteq \mathbb{R} \longrightarrow \mathbb{R}$ is defined in a neighborhood $U(p)$ of a point $p$ and is differentiable at the point $p$ with $f'(p) \neq 0$. Then

(i) The inverse function $g$ to $f$ exists in a neighborhood of the point $f(p)$.[32]

(ii) The inverse function $g$ is differentiable at $f(p)$ and the derivative is given by

$$\boxed{g'(f(p)) = \frac{1}{f'(p)}.}$$

#### 1.4.4.2 The theorem on global inverses

In mathematics one carefully discriminates between

(a) local behavior (that is, behavior in the neighborhood of a point, or the behavior *in the small*) and

(b) global behavior (that is behavior *in the large*). Usually global results are much more difficult to prove than local results. A strong tool for deriving global results is topology (cf. [212]).

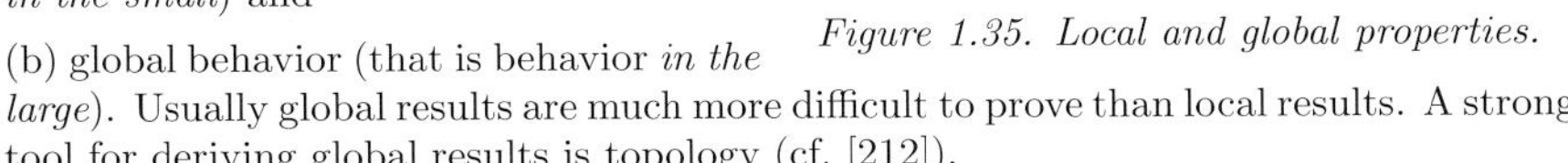


**(a)** local **(b)** global

*Figure 1.35. Local and global properties.*

[32]That means that the equation $y = f(x)$, $x \in U(p)$ can be uniquely inverted for $y \in U(f(p))$ and yields $x = g(y)$ (Figure 1.35(a)). We write $f^{-1}$ for $g$.

**Theorem:** Let $-\infty < a < b < \infty$, and suppose the function $f : [a, b] \longrightarrow \mathbb{R}$ is strictly increasing. Then the inverse function $f^{-1}$ exists,

$$f^{-1} : [f(a), f(b)] \to [a, b],$$

in other words, the equation

$$\boxed{f(x) = y, \qquad x \in [a, b]}$$

has for every $y \in [f(a), f(b)]$ a unique solution $x$, which one denotes by $x = f^{-1}(y)$ (Figure 1.35(b)).

If $f$ is continuous, then so is $f^{-1}$.

**Theorem on global inverse functions:** Let $-\infty < a < b < \infty$, and suppose $f : [a, b] \longrightarrow \mathbb{R}$ is a continuous function which is differentiable on the open interval $]a, b[$ with

$$\boxed{f'(x) > 0 \quad \text{for all } x \in ]a, b[.}$$

Then there exists a continuous inverse function $f^{-1} : [f(a), f(b)] \longrightarrow [a, b]$ with derivative

$$\boxed{(f^{-1})'(y) = \frac{1}{f'(x)}, \qquad y = f(x),}$$

for all $y \in ]f(a), f(b)[$.

### 1.4.5 Taylor's theorem and the local behavior of functions

The Taylor series of a function can be used to get many statements about the local behavior of a function $y = f(x)$ in the neighborhood of a point $p$.

#### 1.4.5.1 Basic ideas

To study the behavior of a function in the neighborhood of the point $x = 0$ we make the ansatz[33]

$$f(x) = a_0 + a_1 x + a_2 x^2 + \ldots$$

To determine the coefficients $a_0, a_1, \ldots$, we differentiate *formally*

$$\begin{aligned} f'(x) &= a_1 + 2a_2 x + 3a_3 x^2 + \ldots, \\ f''(x) &= 2a_2 + 2 \cdot 3a_3 x + \ldots, \\ f'''(x) &= 2 \cdot 3a_3 + \ldots \end{aligned}$$

At $x = 0$ we get the formal expansion

$$a_0 = f(0), \quad a_1 = f'(0), \quad a_2 = \frac{f''(0)}{2}, \quad a_3 = \frac{f'''(0)}{2 \cdot 3}, \quad \ldots$$

[33]This German word cannot be correctly translated, which is why it has become customary to use it in the mathematics and physics literature. It means we just try something out, in this case the particular way of writing the function locally, and see what this leads to.

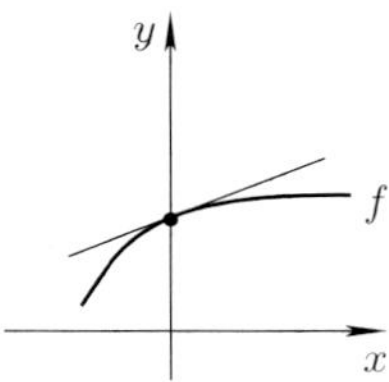


Figure 1.36.

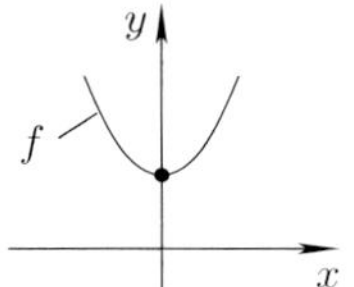


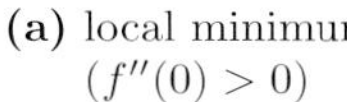


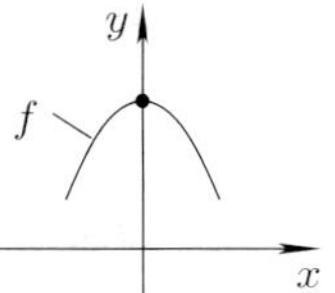


**(b)** local maximum
$(f''(0) < 0)$

*Figure 1.37. Local extrema: minima and maxima.*

This gives the following basic formula, known as *Taylor series*

$$f(x) = f(0) + f'(0)x + \frac{f''(0)}{2!}x^2 + \frac{f'''(0)}{3!}x^3 + \dots \tag{1.32}$$

**Local behavior at the point** $x = 0$: From (1.32) we can see the local behavior of the function $f$ at the point $x = 0$, which we now explain. For this we use the known behavior of the power function $y = x^n$ (see Table 0.15).

*Example 1* (tangent): The first approximation is $f(x) = f(0) + f'(0)x$, which tells us that locally the function is approximated by the line (Figure 1.36)

$$y = f(0) + f'(0)x.$$

This line is just the tangent to $f$ at the point $x = 0$.

*Example 2* (local minimum or maximum): Suppose that $f'(0) = 0$ and $f''(0) \neq 0$. Then near $x = 0$ the function $f$ behaves like (this is the first approximation in this case)

$$f(0) + \frac{f''(0)}{2}x^2.$$

For $f''(0) > 0$ (resp. $f''(0) < 0$) one therefore has at $x = 0$ a *local minimum* (resp. *a local maximum*) (Figure 1.37).

*Example 3* (horizontal inflection point): If $f'(0) = f''(0) = 0$ and $f'''(0) \neq 0$, then near $x = 0$ the function $f$ behaves like

$$f(0) + \frac{f'''(0)}{3!}x^3.$$

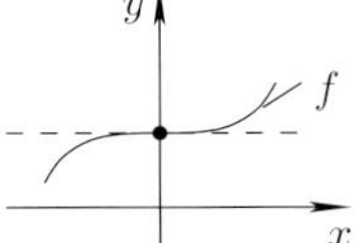


**(a)** horizontal inflection point $(f'''(0) > 0)$

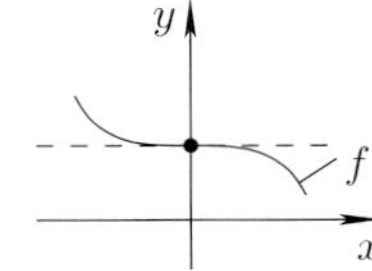


**(b)** horizontal inflection point $(f'''(0) < 0)$

*Figure 1.38. Local extrema: inflection points.*

This corresponds to a *horizontal inflection point* at $x = 0$ (Figure 1.38).

*Example 4* (local minimum): Suppose that $f^{(n)}(0) = 0$ for $n = 1, \dots, 125$ and $f^{(126)}(0) > 0$. Then locally near $x = 0$ the function behaves like

$$f(0) + ax^{126}$$

with $a := f^{(126)}(0)/126!$ The only important fact is that $x^{126}$ is an even power of $x$ and $a$ is positive. Because of these two facts, $f$ behaves locally like Figure 1.37(a), that is $f$ has a local minimum at $x = 0$. The difference is *quantitative*, not *qualitative*; looking at the function under a magnifying glass one would see that $f$ is much more flat than Figure 1.37(a).

This example shows the *universal applicability* of this method also in cases where many derivatives of the function vanish.

**Local curvature:** The function

$$\boxed{g(x) := f(x) - (f(0) + f'(0)x)}$$

describes the difference between $f$ and the tangent at the point $x = 0$. By (1.32) we have

$$g(x) = \frac{f''(0)}{2!}x^2 + \frac{f'''(0)}{3!}x^3 \ldots$$

Because of this, the derivatives $f''(0), f'''(0), \ldots$ of the function describe how the graph of $f$ looks near the tangent at $x = 0$. This gives us information on the *curvature* of (the graph of) $f$.

*Example 5* (local convexity and local concavity): Suppose now that $f''(0) \neq 0$. Then, near $x = 0$, $g$ behaves like

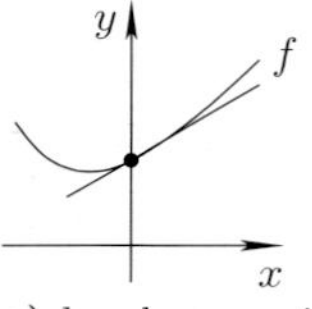


**a)** local convexity ($f''(0) > 0$)

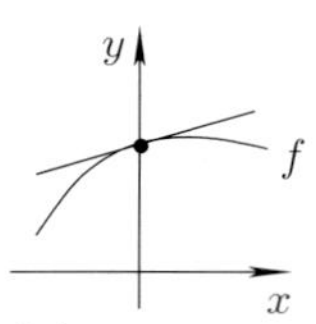


**(b)** local concavity ($f''(0) < 0$)

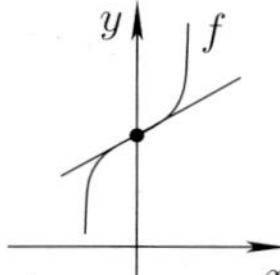


**(c)** inflection point ($f''(0)=0, f'''(0)>0$)

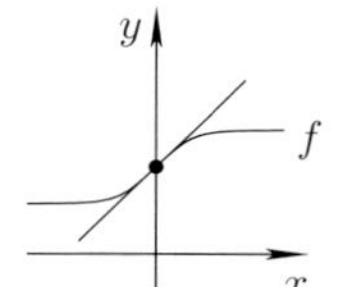


**(d)** inflection point ($f''(0)=0, f'''(0)<0$)

*Figure 1.39. Local curvature of functions.*

$$\frac{f''(0)}{2!}x^2.$$

From this one gets the results:

(i) For $f''(0) > 0$ the graph of $f$ lies locally near $x = 0$ above the tangent (*local convexity*, see Figure 1.39(a)).

(ii) For $f''(0) < 0$ the graph of $f$ lies locally near $x = 0$ under the tangent (*local concavity*, see Figure 1.39(b)).

*Example 6* (inflection point): Suppose $f''(0) = 0$ and $f'''(0) \neq 0$. Then $g$ is locally near $x = 0$ like

$$\frac{f'''(0)}{3!}x^3.$$

Thus the graph of $f$ lies locally at $x = 0$ on both sides of the tangent (Figure 1.39(c),(d)).

**L'Hospital's rule:** This rule described in 1.3.3 follows formally immediately from (1.32). To see this, we note

$$f(x) = f(0) + f'(0)x + \frac{f''(0)}{2!}x^2 + \frac{f'''(0)}{3!}x^3 + \ldots,$$

$$g(x) = g(0) + g'(0)x + \frac{g''(0)}{2!}x^2 + \frac{g'''(0)}{3!}x^3 + \ldots.$$

*Example 7* $\left(\frac{0}{0}\right)$: Suppose that $f(0) = g(0) = 0$ and $g'(0) \neq 0$. Then we get

$$\lim_{x\to 0} \frac{f(x)}{g(x)} = \lim_{x\to 0} \frac{x\left(f'(0) + \frac{f''(0)}{2}x + \ldots\right)}{x\left(g'(0) + \frac{g''(0)}{2}x + \ldots\right)} = \frac{f'(0)}{g'(0)}.$$

### 1.4.5.2 The remainder term

The formal considerations in 1.4.5.1 can be made rigorous by estimating the error in (1.32). This is done by the following formula

$$f(x) = f(p) + f'(p)(x-p) + \frac{f''(p)}{2!}(x-p)^2 + \ldots \\ \ldots + \frac{f^{(n)}(p)}{n!}(x-p)^n + R_{n+1}(x). \tag{1.33}$$

Here the remainder term $R_{n+1}(x)$ has the form

$$R_{n+1}(x) = \frac{f^{(n+1)}(p+\vartheta(x-p))}{(n+1)!}(x-p)^{n+1}, \qquad 0 < \vartheta < 1. \tag{1.34}$$

With the help of the summation symbol, (1.33) can be written

$$f(x) = \sum_{k=0}^{n} \frac{f^{(k)}(p)}{k!}(x-p)^k + R_{n+1}(x).$$

**Taylor's theorem:** Let $J$ be an open interval with $p \in J$ and suppose that $f : J \longrightarrow \mathbb{R}$ is $(n+1)$-times differentiable on $J$. Then for every $x \in J$ there is a number $\vartheta \in ]0,1[$, such that the representation (1.33) with the remainder term (1.34) is valid.

This is the most important theorem in local analysis.

**Application to infinite series:**[34] One has

$$f(x) = \sum_{k=0}^{\infty} \frac{f^{(k)}(p)}{k!}(x-p)^k, \tag{1.35}$$

if the following assumptions are fulfilled:

(i) The function $f : J \longrightarrow \mathbb{R}$ is infinitely often differentiable on the open interval $J$ and $p \in J$.

(ii) For fixed $x \in J$ and every $n = 1, 2, \ldots$ there are numbers $\alpha_n(x)$ such that we have the following *estimate*:

$$\left| \frac{f^{(n+1)}(p+\vartheta(x-p))}{(n+1)!}(x-p)^{n+1} \right| \le \alpha_n(x) \qquad \text{for all } \vartheta \in ]0,1[$$

with $\lim\limits_{n\to\infty} \alpha_n(x) = 0$.

*Example* (expansion of the sine function): Let $f(x) := \sin x$. Then

$$f'(x) = \cos x, \quad f''(x) = -\sin x, \quad f'''(x) = -\cos x, \quad f^{(4)}(x) = \sin x,$$

hence $f'(0) = 1$, $f''(0) = 0$, $f'''(0) = -1$, $f^{(4)}(0) = 0$ and so on. From (1.33) with $p = 0$ we get

$$\sin x = x - \frac{x^3}{3!} + \frac{x^5}{5!} - \ldots + \frac{(-1)^{n-1}x^{2n-1}}{(2n-1)!} + R_{2n}(x)$$

[34] Infinite series are considered in detail in 1.10.

for all $x \in \mathbb{R}$ and $n = 1, 2, \ldots$ with the *error estimate*:[35]

$$|R_{2n}(x)| = \left| \frac{f^{(2n)}(\vartheta x)}{(2n)!} x^{2n} \right| \le \frac{|x|^{2n}}{(2n)!}.$$

From $\lim_{n\to\infty} \frac{|x|^{2n}}{(2n)!} = 0$ we obtain from this

$$\sin x = x - \frac{x^3}{3!} + \ldots = \sum_{n=1}^{\infty} \frac{(-1)^{n-1} x^{2n-1}}{(2n-1)!} \qquad \text{for all } x \in \mathbb{R}.$$

**The integral remainder term:** If the function $f : J \longrightarrow \mathbb{R}$ is of type $C^{n+1}$ (see end of section 1.4.1) on the open interval $J$ with $p \in J$, then (1.33) holds for all $x \in J$, where the remainder term has the form:

$$R_{n+1}(x) = \left( \int_0^1 (1-t)^n f^{(n+1)}(p + (x-p)t)\mathrm{d}t \right) \frac{(x-p)^{n+1}}{n!}.$$

#### 1.4.5.3 Local extrema and critical points

**Definition:** A function $f : M \longrightarrow \mathbb{R}$ on a metric space $M$ is said to have a *local minimum* (resp. *local maximum*) at a point $a \in M$, if there is a neighborhood $U(a)$ with

$$f(a) \le f(x) \quad \text{for all } x \in U(a) \tag{1.36}$$

(resp. $f(x) \le f(a)$ for all $x \in U(a)$).

The function $f$ has a *local strict miminum*, if instead of (1.36) one has

$$f(a) < f(x) \quad \text{for all } x \in U(a) \quad \text{with } x \neq a$$

and analog for *local strict maximum*.

Local extrema (local extremal points) are by definition either local minima or local maxima (cf. Figure 1.40(a),(b)).

**Starting data:** We consider a function

$$f : ]a, b[ \longrightarrow \mathbb{R}$$

with $p \in ]a, b[$.

**Critical points:** A point $p$ is called a *critical point* of $f$, if the derivative $f'(p)$ exists and fulfills

$$f'(p) = 0.$$

[35]Note that $f^{(k)}(\vartheta x) = \pm \sin \vartheta x$, $\pm \cos \vartheta x$, $|\sin \vartheta x| \le 1$ and $|\cos \vartheta x| \le 1$ hold for all real $x$ and $\vartheta$.

Intuitively this means that the tangent at the point $p$ is horizontal.

**Horizontal inflection point:** This is a critical point of $f$ which is neither a local minimum nor local maximum (Fig 1.40).

**Necessary condition for a local extremum:** If the function $f$ has a local extremum at the point $p$ and the derivative $f'(p)$ exists, then $p$ is a critical point of $f$, i.e., $f'(p) = 0$.

**Sufficient conditions for a local extremum:** If $f$ is of type $C^{2n}$, $n \geq 1$ in a neighborhood of $p$ and

$$f'(p) = f''(p) = \ldots = f^{(2n-1)}(p) = 0,$$

and, moreover

$$\boxed{f^{(2n)}(p) > 0,}$$

(resp. $f^{(2n)}(p) < 0$), then $f$ has a local minimum (resp. maximum).

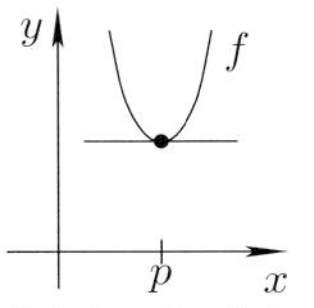

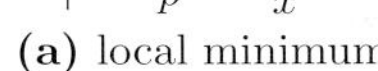
(a) local minimum

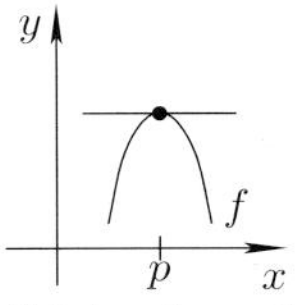

(b) local maximum

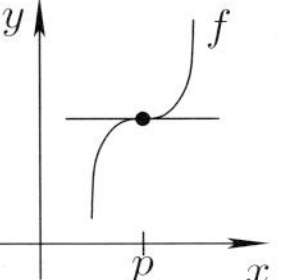

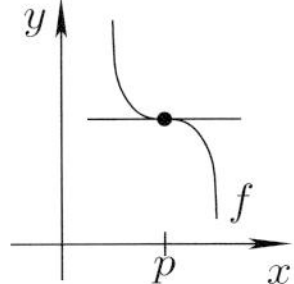

(c) horizontal inflection points

*Figure 1.40. Local extrema.*

**Sufficient condtions for a horizontal inflection point:** If $f$ is of type $C^{2n+1}$, $n \geq 1$ in an open neighborhood of $p$ and one has

$$f'(p) = f''(p) = \ldots = f^{(2n)}(p) = 0$$

as well as

$$\boxed{f^{(2n+1)}(p) \neq 0,}$$

then $f$ has a horizontal inflection point at $p$.

*Example 1:* For $f(x) := \cos x$ one has $f'(x) = -\sin x$ and $f''(x) = -\cos x$. This gives

$$f'(0) = 0 \quad \text{and} \quad f''(0) < 0.$$

Moreover $f$ has a local maximum at $x = 0$. From

$$\cos x \leq 1 \quad \text{for all } x \in \mathbb{R}$$

one gets that the function $y = \cos x$ even has a global maximum at $x = 0$ (Figure 1.41(a)).

(a) $y = \cos x$ (b) $y = x^3$

*Figure 1.41.*

*Example 2:* For $f(x) := x^3$ we get $f'(x) = 3x^2$, $f''(x) = 6x$ and $f'''(x) = 6$. Hence

$$f'(0) = f''(0) = 0 \quad \text{and} \quad f'''(0) \neq 0.$$

Consequently, $f$ has a horizontal inflection point at $x = 0$ (Figure 1.41(b)).

#### 1.4.5.4 Curvature

**The relative position of the graph to the tangent:** The function

$$g(x) := f(x) - f(p) - f'(p)(x - p)$$

describes the difference between the function $f$ and the tangent at $p$. We define:

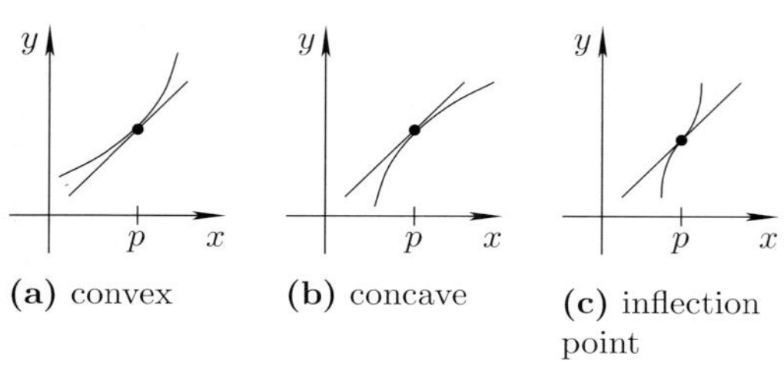


*Figure 1.42. Locally convex and concave functions.*

(i) The function $f$ is *locally convex* at $p$, if the function $g$ has a local minimum at $p$.

(ii) $f$ is *locally concave* at $p$, if $g$ has a local maximum at $p$.

(iii) $f$ has an *inflection point* at $p$, if $g$ has a horizontal inflection point.

In (i) (resp. (ii)) the graph of $g$ lies above (resp. below) the tangent at the point $p$.

In (iii) the graph of $f$ lies locally near $p$ on both sides of the tangent (Figure 1.42).

**Necessary conditions for an inflection point:** If $f$ is of type $C^2$ in a neighborhood of $p$ and if $f$ has an inflection point there, then

$$f''(p) = 0.$$

**Sufficient condition for an inflection point:** Suppose $f$ is of type $C^k$ in the neighborhood of a point $p$ and satisfies

$$f''(p) = f'''(p) = \ldots = f^{(k-1)}(p) = 0 \tag{1.37}$$

for odd $k \geq 3$, and moreover $f^{(k)}(p) \neq 0$. Then $f$ has an inflection point at $p$.

**Sufficient condition for local convexity:** Suppose $f$ is of type $C^k$ in a neighborhood of $p$. If one of the following conditions is satisfied, then $f$ is locally convex at $p$.

(i) $f''(p) > 0$ and $k = 2$.

(ii) $f^{(k)}(p) > 0$ and (1.37) for even $k \geq 4$.

**Sufficient conditions for local concavity:** Suppose $f$ is of type $C^k$ in a neighborhood of $p$. Then $f$ is locally concave if one of the following conditions is satisfied:

(i) $f''(p) < 0$ and $k = 2$.

(ii) $f^{(k)}(p) < 0$ and (1.37) for even $k \geq 4$.

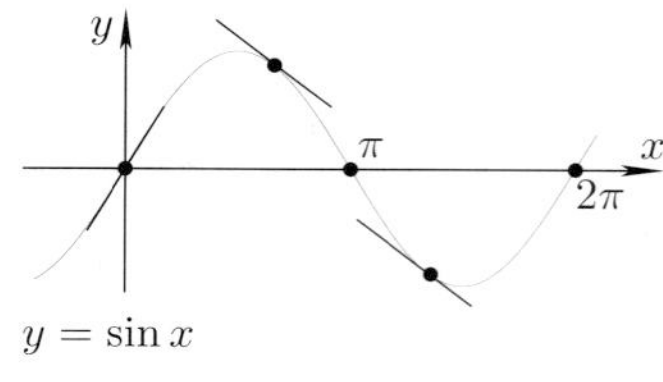


*Figure 1.43.*

*Example:* Let $f(x) := \sin x$. Then $f'(x) = \cos x$, $f''(x) = -\sin x$ and $f'''(x) = -\cos x$.

(i) Because of $f''(0) = 0$ and $f'''(0) \neq 0$, the point $x = 0$ is an inflection point.

(ii) For $x \in ]0, \pi[$ one has $f''(x) < 0$, hence $f$ is locally concave there.

(iii) For $x \in ]\pi, 2\pi[$ one has $f''(x) > 0$, hence $f$ is locally convex there.

(iv) Because of $f''(\pi) = 0$ and $f'''(\pi) \neq 0$, $x = \pi$ is an inflection point (Figure 1.43).

### 1.4.5.5 Convex functions

Convexity is the most simple kind of non-linearity. Often energy and negative entropy functions are convex. Moreover, convex functions play an important role in the calculus of variations and in optimization (see Chapter 5).

**Definition:** A set $M$ of a linear space is said to be *convex*, if we have the implication

$$\boxed{x, y \in M \Rightarrow tx + (1-t)y \in M \quad \text{for all} \quad t \in [0,1].}$$

Geometrically this means that the secant joining two points in $M$ also belongs to $M$ (Figure 1.44)

A function $f : M \longrightarrow \mathbb{R}$ is said to be *convex*, if the set $M$ is convex and

$$\boxed{f(tx + (1-t)y) \le tf(x) + (1-t)f(y)} \qquad (1.38)$$

for all $x, y \in M$ and all real $t \in ]0,1[$. If in (1.38) one has $<$ instead of $\le$, $f$ is said to be *strictly convex* (Figure 1.45).

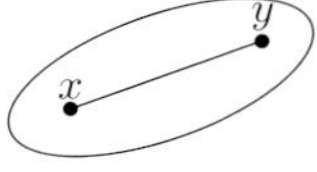


*Figure 1.44.*

A function $f : M \longrightarrow \mathbb{R}$ is said to be *concave* (resp. *strictly concave*), if $-f$ is convex (resp. strictly convex).

*Example:* The real function $f : M \longrightarrow \mathbb{R}$ from an interval $M$ is convex (resp. strictly convex), if and only if the secant joining two points of the graph of $f$ are above (resp. strictly above) the graph of $f$ (cf. Figure 1.45).

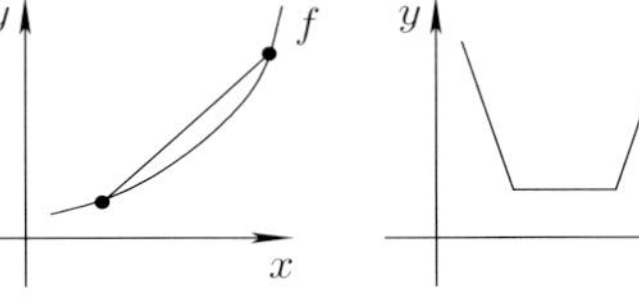


**(a)** strictly convex **(b)** convex

*Figure 1.45.*

**Criteria for convexity:** A function $f : J \longrightarrow \mathbb{R}$ on an open interval $J$ has the following properties.

(i) If $f$ is convex, then $f$ is continuous on $J$.

(ii) If $f$ is convex, then in every point $x \in J$ the right-sided derivative[36] $f_+(x)$ and the left-sided derivative $f_-(x)$ exist, and fulfill

$$f_-(x) \le f_+(x).$$

(iii) If the first derivative $f'$ of $f$ exists on $J$, then[37]

$$\boxed{f \text{ is (strictly) convex on } J \iff f' \text{ is (strictly) increasing on } J.}$$

(iv) If the second derivative $f''$ exists on $J$, then

$$\boxed{\begin{array}{l} f''(x) \ge 0 \text{ on } J \iff f \text{ is convex on } J, \\ f''(x) > 0 \text{ on } J \iff f \text{ is strictly convex on } J. \end{array}}$$

#### 1.4.5.6 Application to the analysis of graphs

In order to determine the qualitative behavior of the graph of a function $f : M \subseteq \mathbb{R} \longrightarrow \mathbb{R}$, one proceeds as follows:

(i) First determine the set of points where $f$ is non continuous.

(ii) Determine the behavior of $f$ near these points by calculating the one-sided derivatives, if possible.

---

[36] $f'_\pm(x) := \lim_{h\to\pm 0} \frac{f(x+h)-f(x)}{h}$.

[37] The symbol $A \Rightarrow B$ means that $A$ implies $B$, and $A \iff B$ means that both $A \Rightarrow B$ and $B \Rightarrow A$ hold.

(iii) Determine the behavior of $f$ 'at infinity' by calculating the limit $\lim_{x\to\pm\infty} f(x)$, if these exist.

(iv) Determine the zeros of $f$ by solving the equation $f(x) = 0$.

(v) Determine the critical points of $f$ by solving the equation $f'(x) = 0$.

(vi) Classify the types of critical points of $f$ (minima, maxima, inflection points, see 1.4.5.3).

(vii) Determine the domains where $f$ is increasing or decreasing by studying the sign of the first derivative $f'$ at different points (see 1.4.3).

(viii) Determine the curvature of $f$ by studying the sign of $f''(x)$ (convexity, concavity, see 1.4.5.4).

(ix) Finally determine the zeros of $f''(x)$ to see whether these are inflection points (see 1.4.5.4).

*Example:* We want to plot the graph of the function

$$f(x) := \begin{cases} \dfrac{x^2+1}{x^2-1} & \text{for } x \le 2, \\ \dfrac{5}{3} & \text{for } x > 2. \end{cases}$$

(i) From

$$\lim_{x\to 2\pm 0} f(x) = \frac{5}{3}$$

we see that the function $f$ is continuous at $x = 2$. Hence $f$ is continuous for all $x \neq \pm 1$ and for all $x \in \mathbb{R}$, $x \neq \pm 1$, $x \neq 2$ differentiable.

(ii) From

$$f(x) = \frac{2}{(x-1)(x+1)} + 1 \quad \text{for } x \le 2$$

we get

$$\lim_{x\to 1\pm 0} f(x) = \pm\infty, \qquad \lim_{x\to -1\pm 0} f(x) = \mp\infty.$$

(iii) We have $\lim_{x\to-\infty} f(x) = 1$ and $\lim_{x\to+\infty} f(x) = \frac{5}{3}$.

(iv) The equation $f(x) = 0$ has no solutions, so the graph of $f$ does not intersect the $x$-axis.

(v) Let $x \neq \pm 1$. Taking derivatives, we get

$$f'(x) = -\frac{4x}{(x^2-1)^2} \qquad \text{for } x < 2,$$

$$f''(x) = \frac{4(3x^2+1)}{(x^2-1)^3} \qquad \text{for } x < 2,$$

and $f'(x) = f''(x) = 0$ for $x > 2$. From

$$\lim_{x\to 2-0} f'(x) = -\frac{8}{9}, \qquad \lim_{x\to 2+0} f'(x) = 0$$

we obtain that for $x = 2$, there is no tangent to the graph. The equation

$$f'(x) = 0, \qquad x < 2$$

has precisely one solution, given by $x = 0$.

(vi) From $f'(0) = 0$ and $f''(0) < 0$ we see that at $x = 0$ there is a local minimum.

(vii) From

$$f'(x)\begin{cases} > 0 & \text{on } ]-\infty, -1[ \text{ and } ]-1, 0[, \\ < 0 & \text{on } ]0, 1[ \text{ and } ]1, 2[ \end{cases}$$

it follows: $f$ is strictly increasing in the intervals $]-\infty, -1[$ and $]-1, 0[$, and strictly decreasing in $]0, 1[$ and $]1, 2[$.

(viii) From

$$f''(x)\begin{cases} > 0 & \text{on } ]-\infty, -1[ \text{ and } ]1, 2[, \\ < 0 & \text{on } ]-1, 1[ \end{cases}$$

it follows that $f$ is strictly convex on the interval $]-\infty, -1[$ and $]1, 2[$, and strictly concave on $]-1, 1[$.

(ix) The equation $f''(x) = 0$, $x < 2$ has no solution. Hence for $x < 2$ there is no inflection point.

In conclusion, we see that the graph of $f$ has the form of Figure 1.46.

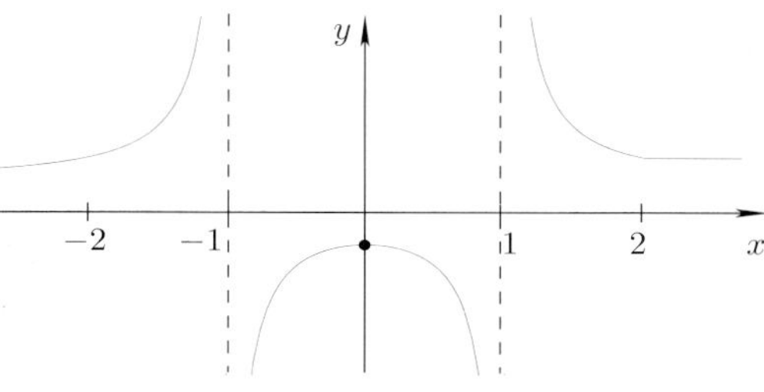


*Figure 1.46. The graph of the function $f$ above.*

### 1.4.6 Complex valued functions

We consider functions $f : M \subseteq \mathbb{R} \longrightarrow \mathbb{C}$, which are defined on an interval $M$ and take on complex values. We decompose $f$ into its real and imaginary parts,

$$\boxed{f(x) = \alpha(x) + \beta(x)\mathrm{i}} \qquad (1.39)$$

with $\alpha(x), \beta(x) \in \mathbb{R}$. The derivative is defined by the limit

$$f'(x) = \lim_{h \to 0} \frac{f(x+h) - f(x)}{h}$$

if this exists.[38]

**Theorem:** The derivative $f'(x)$ exists if and only if the derivatives $\alpha'(x)$ and $\beta'(x)$ exist. In this case one has

$$\boxed{f'(x) = \alpha'(x) + \beta'(x)\mathrm{i}.} \qquad (1.40)$$

*Example:* For $f(x) := \mathrm{e}^{\mathrm{i}x}$ one has

$$f'(x) = \mathrm{i}\mathrm{e}^{\mathrm{i}x}, \qquad x \in \mathbb{R}.$$

*Proof:* From the Euler formula $f(x) = \cos x + \mathrm{i}\sin x$ and (1.40) it follows that $f'(x) = -\sin x + \mathrm{i}\cos x = \mathrm{i}(\cos x + \mathrm{i}\sin x)$. □

[38]The meaning of the limit is the same as for real functions of a real variable, using convergence of sequences in the domain and sequences of complex numbers in the range (cf. 1.14.2). Explicitly, this means

$$f'(x) = \lim_{n \to \infty} \frac{f(x+h_n) - f(x)}{h_n}$$

for all sequences $(h_n)$ in $M$ with $\lim_{n\to\infty} h_n = 0$ and $h_n \neq 0$ for all $n$.

# 1.5 Derivatives of functions of several real variables

In this section we denote the points of $\mathbb{R}^N$ by $x = (x_1, \ldots, x_N)$, where all $x_j$ are real numbers. We write $y = f(x)$ instead of $y = f(x_1, \ldots, x_N)$.

## 1.5.1 Partial derivatives

**Basic idea:** For the function $f(u) := u^2C$ for some constant $C$ we have by 0.8.2 the derivative

$$\frac{\mathrm{d}f}{\mathrm{d}u} = 2uC. \tag{1.41}$$

Let

$$\boxed{f(u,v) := u^2v^3.}$$

If we view $v$ to be a constant and differentiate $f$ as a function of $u$ with respect to $u$, then we get as in (1.41) the so-called *partial derivative*

$$\boxed{\frac{\partial f}{\partial u} = 2uv^3} \tag{1.42}$$

with respect to the variable $u$. Of course we can do the same, considering $u$ as constant and differentiating $f$ as a function of $v$, we get

$$\frac{\partial f}{\partial v} = 3u^2v^2. \tag{1.43}$$

Summarizing:

> Partial derivatives are formed by considering a function of several variables just as a function of one of the variables; treating all other variables as a constant.

*Higher partial derivatives* are obtained in the same manner. For example, treating $u$ in (1.42) as a constant, one has:

$$\frac{\partial^2 f}{\partial v \partial u} = \frac{\partial}{\partial v}\left(\frac{\partial f}{\partial u}\right) = 6uv^2.$$

If we consider $v$ instead to be constant in (1.43), then we get

$$\frac{\partial^2 f}{\partial u \partial v} = \frac{\partial}{\partial u}\left(\frac{\partial f}{\partial v}\right) = 6uv^2.$$

We use the following notation:

$$\boxed{f_u := \frac{\partial f}{\partial u}, \quad f_v := \frac{\partial f}{\partial v}, \quad f_{uv} := (f_u)_v = \frac{\partial^2 f}{\partial v \partial u}, \quad f_{vu} := (f_v)_u = \frac{\partial^2 f}{\partial u \partial v}.}$$

For sufficiently smooth functions we have the convenient symmetry

$$\boxed{f_{uv} = f_{vu}}$$

(compare with the Schwarz' theorem (1.44)).

**Definition:** Let $f : M \subseteq \mathbb{R}^N \longrightarrow \mathbb{R}$ be a function and let $p$ be an inner point of $M$. If the limit

$$\boxed{\frac{\partial f}{\partial x_1}(p) := \lim_{h \to 0} \frac{f(p_1 + h, p_2, \ldots, p_N) - f(p_1, \ldots, p_N)}{h},}$$

exists, then we say that $f$ has a *partial derivative* with respect to $x_1$. Other partial derivatives are defined similarly.

The following terminology is often used in analysis.

**The class $C^k(G)$ of smooth functions:** Let $G$ be an open set of $\mathbb{R}^N$. $C^k(G)$ is the set of all functions $f : G \longrightarrow \mathbb{R}$, which have continuous partial derivatives up to order $k$. If $f \in C^k(G)$, we also say that $f$ is *of type* $C^k$.

**The class $C^k(\overline{G})$:** We let $\overline{G} = G \cup \partial G$ denote the closure of $G$ (see 1.3.2.2). The set $C^k(\overline{G})$ consists of all functions $f : \overline{G} \longrightarrow \mathbb{R}$ with $f \in C^k(G)$, for which all partial derivatives up to order $k$ can be continuously extended to the closure $\overline{G}$.[39]

**Schwarz' theorem:** If the function $f : M \subseteq \mathbb{R}^N \longrightarrow \mathbb{R}$ is of type $C^2$ in a open neighborhood of $p$, then

$$\boxed{\frac{\partial^2 f(p)}{\partial x_j \partial x_m} = \frac{\partial^2 f(p)}{\partial x_m \partial x_j}, \qquad j, m = 1, \ldots, N.} \tag{1.44}$$

More generally, if $f$ is of type $C^k$, $k \geq 2$ on some open neighborhood of $p$, then the order of taking partial derivatives up to order $k$ is irrelevant.

*Example 1:* For $f(u, v) = u^4 v^2$ one has $f_u = 4u^3 v^2$, $f_v = 2u^4 v$ and

$$f_{uv} = f_{vu} = 8u^3 v.$$

Moreover, $f_{uu} = 12u^2 v^2$ and

$$f_{uuv} = f_{uvu} = 24u^2 v.$$

**Notations:** To simplify our notations, we will write

$$\boxed{\partial_j f := \frac{\partial f}{\partial x_j}.}$$

*Example 2:* Equation (1.44) means that $\partial_j \partial_m f(p) = \partial_m \partial_j f(p)$.

### 1.5.2 The Fréchet derivative

**Basic idea:** We want to extend the notion of derivative to functions $f : M \subseteq \mathbb{R}^N \longrightarrow \mathbb{R}^K$ of several variables. The starting point is the relation

$$\boxed{f(p + h) - f(p) = f'(p)h + r(h)} \tag{1.45}$$

[39]Omitting the superscript $k$ or setting $k = 0$, $C(G)$ or $C^0(G)$ (resp. $C(\overline{G})$ or $C^0(\overline{G})$) denotes the space of all continuous functions $f : G \longrightarrow \mathbb{R}$ (resp. $f : \overline{G} \longrightarrow \mathbb{R}$). Moreover, $C^\infty(G)$ (resp. $C^\infty(\overline{G})$) consists of those functions which belong to $C^k(G)$ (resp. $C^k(\overline{G})$) for all $k$.

for all $h$ in a neighborhood $U(0)$ of the origin with

$$\boxed{\lim_{h\to 0}\frac{r(h)}{|h|}=0.} \tag{1.46}$$

The general philosophy of modern mathematics, hidden behind this definition, is:

$$\boxed{\text{differentiation means linearization.}} \tag{1.47}$$

**The one-dimensional classical special case:** Let $J$ be an interval with $p \in J$. The function $f : J \longrightarrow \mathbb{R}$ has a derivative $f'(p)$ if and only if the decomposition (1.45) with (1.46) holds.

*Proof:* If the classical derivative

$$f'(p)=\lim_{h\to 0}\frac{f(p+h)-f(p)}{h} \tag{1.48}$$

exists, we define

$$\varepsilon(h) := \frac{f(p+h)-f(p)}{h} - f'(p) \qquad \text{for } h \neq 0$$

and $\varepsilon(0) := 0$ as well as $r(h) := h\varepsilon(h)$. Then (1.48) implies (1.45) and (1.46). Conversely, if one has a decomposition as in (1.45) for some fixed $f'(p)$ for which (1.46) holds, then (1.48) follows. □

**The modern point of view:** The classical definition (1.48) is absolutely inconvenient to extend the notion of derivative to functions of several variables, since in that case the analog of $h$ is a vector $h \in \mathbb{R}^N$; it makes no sense to divide by such a vector. But the *decomposition formula* (1.45) always makes sense. This is why the modern theory of differentiation is based on (1.45) and the general strategy (1.47).[40]

**Differentiating functions from $\mathbb{R}^N$ to $\mathbb{R}^K$:** Suppose $M$ is a subset of $\mathbb{R}^N$ which contains a neighborhood of the point $p$. A map

$$\boxed{f : M \subseteq \mathbb{R}^N \to \mathbb{R}^K} \tag{1.49}$$

has the form $y = f(x)$ with columns[41]

$$x=\begin{pmatrix} x_1\\ \vdots \\ x_N\end{pmatrix},\quad f(x)=\begin{pmatrix} f_1(x)\\ \vdots \\ f_K(x)\end{pmatrix},\quad y=\begin{pmatrix} y_1\\ \vdots \\ y_K\end{pmatrix},\quad h=\begin{pmatrix} h_1\\ \vdots \\ h_N\end{pmatrix}.$$

[40] This elegant point of view can be immediately extended to operators acting in infinite-dimensional Hilbert and Banach spaces and is an important tool in non-linear functional analysis for solving non-linear differential and integral equations (cf. [212]). Briefly summarized, we can say:

> Modern differential calculus approximates non-linear operators by linear operators and allows one to apply the easy theory of linear algebra to the study of difficult non-linear problems.

[41] We use columns in order to be able to write (1.45) as a matrix equation; matrices and determinants are dealt with in section 2.1. Using transposed matrices we could also write $x = (x_1, \ldots, x_N)^{\mathsf{T}}$ and $f(x) = (f_1(x), \ldots, f_K(x))^{\mathsf{T}}$.

The Euclidean norm of $h$ is defined by the expression

$$|h| := \left( \sum_{j=1}^{N} h_j^2 \right)^{1/2}.$$

**Definition:** The map $f$ in (1.49) is in a point $p$ *Fréchet differentiable*, if there is a

$$\boxed{(K \times N)\text{-matrix } f'(p)}$$

such that a decomposition (1.45) with (1.46) holds.

In this case the matrix $f'(p)$ is called the Fréchet derivative of $f$ at the point $p$.

**Convention:** Instead of Fréchet derivative we will forthwith speak of the F-derivative.[42]

**Main theorem:** If the function $f$ in (1.49) is of type $C^1$ in a neighborhood of a point $p$, then the F-derivative $f'(p)$ exists, and satisfies $f'(p) = (\partial_j f_k(p))$. Explicitly this is the matrix

$$\boxed{f'(p) = \begin{pmatrix} \partial_1 f_1(p), & \partial_2 f_1(p), & \ldots, & \partial_N f_1(p) \\ \partial_1 f_2(p), & \partial_2 f_2(p), & \ldots, & \partial_N f_2(p) \\ \ldots & & & \\ \partial_1 f_K(p), & \partial_2 f_K(p), & \ldots, & \partial_N f_K(p) \end{pmatrix}}$$

of the first partial derivatives of the components $f_k$ of $f$. The matrix $f'(p)$ is also referred to as the *Jacobian* matrix of $f$ at the point $p$.

**Jacobian determinant:** Suppose that $N = K$. Then the determinant $\det f'(p)$ of the matrix $f'(p)$ is called the Jacobian (functional) determinant of $f$ and is written

$$\boxed{\frac{\partial(f_1, \ldots, f_N)}{\partial(x_1, \ldots, x_N)} := \det f'(p).}$$

*Example 1* ($K = 1$): For a real function $f : M \subseteq \mathbb{R}^N \longrightarrow \mathbb{R}$ with $N$ real variables one has

$$f'(p) = (\partial_1 f(p), \ldots, \partial_N f(p)).$$

If for example, $f(x) := x_1 \cos x_2$, then one has $\partial_1 f(x) = \cos x_2$, $\partial_2 f(x) = -x_1 \sin x_2$ , hence

$$f'(0,0) = (\partial_1 f(0,0), \partial_2 f(0,0)) = (1,0).$$

To connect with the idea of linearization we use the Taylor expansion

$$\cos h_2 = 1 - \frac{(h_2)^2}{2} + \ldots$$

For $p = (0,0)^\mathsf{T}$, $h = (h_1, h_2)^\mathsf{T}$ and small values $h_1$, $h_2$, we get from this

$$\boxed{f(p+h) - f(h) = h_1 + r(h),}$$

[42]This notion of derivative was introduced by the French mathematician René Maurice Fréchet (1878–1956) at the beginning of this century. Fréchet, who is also responsible for the theory of metric spaces, is together with David Hilbert one of the fathers of modern analytic thinking (see [212]).

where $r$ denotes terms of higher order. From $f'(p) = (1,0)$ one also has

$$f(p+h) - f(p) = f'(p)h + r(h) = (1,0)\begin{pmatrix} h_1 \\ h_2 \end{pmatrix} + r(h).$$

We can view $f'(p)h = h_1$ as a linear approximation of $f(h) = h_1 \cos h_2$ if $h_1$ and $h_2$ are sufficiently small.

*Example 2:* We set

$$x = \begin{pmatrix} x_1 \\ x_2 \end{pmatrix}, \qquad f(x) = \begin{pmatrix} f_1(x) \\ f_2(x) \end{pmatrix}.$$

Then we have

$$f'(p) = \begin{pmatrix} \partial_1 f_1(p) & \partial_2 f_1(p) \\ \partial_1 f_2(p) & \partial_2 f_2(p) \end{pmatrix}$$

and

$$\frac{\partial(f_1, f_2)}{\partial(x_1, x_2)} = \det\, f'(p) = \begin{vmatrix} \partial_1 f_1(p) & \partial_2 f_1(p) \\ \partial_1 f_2(p) & \partial_2 f_2(p) \end{vmatrix},$$

hence

$$\det\, f'(p) = \partial_1 f_1(p)\partial_2 f_2(p) - \partial_2 f_1(p)\partial_1 f_2(p).$$

In the special case that $f_1(x) = ax_1 + bx_2,\ f_2(x) = cx_1 + dx_2$, we have

$$f(x) = \begin{pmatrix} a & b \\ c & d \end{pmatrix}\begin{pmatrix} x_1 \\ x_2 \end{pmatrix}$$

and

$$f'(p) = \begin{pmatrix} a & b \\ c & d \end{pmatrix}, \qquad \det\, f'(p) = \begin{vmatrix} a & b \\ c & d \end{vmatrix} = ad - bc.$$

As to be expected, the linearization of a linear map is just the map itself, so $f'(p)x = f(x)$.

### 1.5.3 The chain rule

The important chain rule allows the differentiation of composed functions. In the spirit of the general linearizations strategy (1.47), this rule says

> The linearization of composed mappings is the composition of the linearizations of the individual maps. (1.50)

#### 1.5.3.1 Basic idea

Let

$$z = F(u, v), \qquad u = u(x), \quad v = v(x).$$

It is our goal to differentiate the composed function $z = F(u(x), v(x))$ by $x$. Following the Leibniz notation, the chain rule follows formally from the formula

$$\mathrm{d}F = F_u \mathrm{d}u + F_v \mathrm{d}v \tag{1.51}$$

by (formal) division

$$\boxed{\frac{\mathrm{d}F}{\mathrm{d}x} = F_u \frac{\mathrm{d}u}{\mathrm{d}x} + F_v \frac{\mathrm{d}v}{\mathrm{d}x}.} \tag{1.52}$$

If $u$ and $v$ depend on other variables than just $x$, that is

$$u = u(x, y, \ldots), \qquad v = v(x, y, \ldots),$$

then the usual derivative in (1.52) must be replaced by partial derivatives. This gives

$$\boxed{\frac{\partial F}{\partial x} = F_u \frac{\partial u}{\partial x} + F_v \frac{\partial v}{\partial x}.} \tag{1.53}$$

Replacing $x$ by $y$, we get

$$\frac{\partial F}{\partial y} = F_u \frac{\partial u}{\partial y} + F_v \frac{\partial v}{\partial y}.$$

If $F$ depends on further variables, i.e., $y = F(u, v, w, \ldots)$, the one uses the relation

$$\mathrm{d}F = F_u \mathrm{d}u + F_v \mathrm{d}v + F_w \mathrm{d}w + \ldots$$

and proceeds in the same manner.

*Example:* Let $F(u, v) := uv^2$ and $u = x^2$, $v = x$. We set

$$F(x) := F(u(x), v(x)) = x^4. \tag{1.54}$$

Using (1.52), we get

$$\frac{\mathrm{d}F}{\mathrm{d}x} = F_u \frac{\mathrm{d}u}{\mathrm{d}x} + F_v \frac{\mathrm{d}v}{\mathrm{d}x} = v^2(2x) + 2uv = 4x^3.$$

The same result follows directly from $F'(x) = 4x^3$.

**Precise notation:** The formula (1.52) is quite suggestive, but not completely precise; in fact, on the left hand side $F$ is a function of $x$, while on the right hand side it is a function of $u$ and $v$. If we are trying to be precise, then we should change the notations, for example setting

$$H(x) := F(u(x), v(x)).$$

Then the precise statement of the chain rule, which completely states all arguments, is

$$\boxed{H'(x) = F_u(u(x), v(x))u'(x) + F_v(u(x), v(x))v'(x).} \tag{1.55}$$

For

$$H(x, y) := F(u(x, y), v(x, y))$$

we get

$$H_x(x, y) = F_u(u(x, y), v(x, y))u_x(x, y) + F_v(u(x, y), v(x, y))v_y(x, y). \tag{1.56}$$

Since the formulas (1.55) and (1.56) are rather unwieldy compared with (1.52) and (1.53), they are not often used in calculations. However, for more theoretical purposes, like proving theorems, it can be essential to have the more precise notations as above.

**Physicists' notation from thermodynamics:** Let $E$ denote the energy of a system. Then the symbol

$$\boxed{\left(\frac{\partial E}{\partial V}\right)_T,}$$

means that one views $E = E(V, T)$ as a function of the volume $V$ and the temperature $T$, and forms the partial derivatives with respect to $V$. One the other hand,

$$\left(\frac{\partial E}{\partial p}\right)_V,$$

means that $E = E(p, V)$ is viewed as a function of the pressure $p$ and the volume $V$, and the partial derivatives are with respect to $p$. In this way the energy is denoted by a unified symbol, the notation makes it clear, which variables the function depends on, and one can use the advantages of the Leibniz notation (1.51) and (1.53).

### 1.5.3.2 Derivatives of composed functions

**Basic formulas:** We consider the composed function

$$\boxed{H(x) := F(f(x)).}$$

Explicitly this means

$$H_m(x) := F_m(f_1(x), \ldots, f_K(x)), \qquad m = 1, \ldots M,$$

with $x = (x_1, \ldots, x_N)$. Our goal is to derive the chain rule

$$\boxed{\frac{\partial H_m}{\partial x_n}(p) = \sum_{k=1}^{K} \frac{\partial F_m}{\partial f_k}(f(p)) \frac{\partial f_k}{\partial x_n}(p)} \tag{1.57}$$

for $m = 1, \ldots, M$ and $n = 1, \ldots, N$. Written as a matrix equation, (1.57) is

$$\boxed{H'(p) = F'(f(p))f'(p).} \tag{1.58}$$

Because $H = F \circ f$ this can also be written

$$\boxed{(F \circ f)'(p) = F'(f(p))f'(p),} \tag{1.59}$$

which is similar to the linearization (1.50).

A function is said to be locally at a point $p$ of type $C^k$, if the function is of type $C^k$ in a neighborhood of $p$.

**The chain rule:** The formulas (1.57)-(1.59) hold and the composed function $H = F \circ f$ is locally at $p$ of type $C^1$, provided the following assumptions are satisfied:

(i) The function $f : D(f) \subseteq \mathbb{R}^N \longrightarrow \mathbb{R}^K$ is locally at $p$ of type $C^1$.

(ii) The function $F : D(F) \subseteq \mathbb{R}^K \longrightarrow \mathbb{R}^M$ is locally at $f(p)$ of type $C^1$.

**The product formula for functional determinants:** In the case $M = K = N$, (1.58) leads to the determinant formula

$$\boxed{\det H'(p) = \det F'(f(p))\det f'(p).}$$

This is equivalent with the Jacobi product formula

$$\frac{\partial(H_1,\ldots,H_N)}{\partial(x_1,\ldots,x_N)}(p) = \frac{\partial(F_1,\ldots,F_N)}{\partial(f_1,\ldots,f_N)}(f(p))\frac{\partial(f_1,\ldots,f_N)}{\partial(x_1,\ldots,x_N)}(p).$$

## 1.5.4 Applications to the transformation of differential operators

Differential equations can often be simplified by passing to new coordinates. We illustrate this with the example of polar coordinates. The same considerations can be applied to arbitrary coordinate transformations.

**Polar coordinates:** Instead of Cartesian coordinates $x, y$, we introduce

$$\boxed{x = r\cos\varphi, \qquad y = r\sin\varphi, \qquad -\pi < \varphi \le \pi,} \tag{1.60}$$

which are called *polar coordinates* $r$, $\varphi$ (Figure 1.47). We set

$$\alpha := \arctan\frac{y}{x}, \qquad x \neq 0;$$

then the inverse of this coordinate transformation is

$$r = \sqrt{x^2+y^2}, \tag{1.61}$$

$$\varphi = \begin{cases} \alpha & \text{for } x > 0,\ y \in \mathbb{R}, \\ \pm\pi + \alpha & \text{for } x < 0,\ y \gtrless 0, \\ \pm\dfrac{\pi}{2} & \text{for } x = 0,\ y \gtrless 0, \\ \pi & \text{for } x < 0,\ y = 0. \end{cases}$$

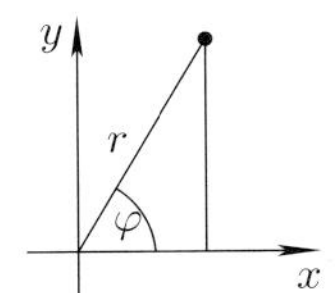


*Figure 1.47.*

**Transformation of a function to polar coordinates:** The transformation of a function $F = F(x, y)$ from Cartesian to polar coordinates $r$, $\varphi$ is affected by setting

$$\boxed{f(r,\varphi) := F(x(r,\varphi), y(r,\varphi)).} \tag{1.62}$$

**Transformation of the Laplace operator:** Suppose the function $F : \mathbb{R}^2 \longrightarrow \mathbb{R}$ is of type $C^2$. Then the usual Laplace operator

$$\boxed{\Delta F := F_{xx} + F_{yy}, \qquad (x,y) \in \mathbb{R}^2,} \tag{1.63}$$

is transformed to the expression

$$\boxed{\Delta f = f_{rr} + \frac{1}{r^2}f_{\varphi\varphi} + \frac{1}{r}f_r, \qquad r > 0.} \tag{1.64}$$

**Consequence:** The relation (1.64) implies immediately that the function

$$f(r) := \ln r, \qquad r > 0,$$

is a solution of the partial differential equation $\Delta f = 0$. Transforming back, we see that

$$F(x,y) = \ln\sqrt{x^2+y^2}, \qquad x^2+y^2 \neq 0,$$

is a solution of the Laplace equation $\Delta F = 0$, a fact which is much more difficult to see in these coordinates.

**Notations:** Often the symbol $F$ is used instead of $f$ in (1.64). Although this notation is inconsistent, it is very convenient in applications in physics and technology.

We now describe two methods for deriving the transformed equation (1.64).

**First method:** We start from the identity

$$\boxed{F(x,y) = f(r(x,y), \varphi(x,y)).}$$

Taking derivatives with respect to $x$ and $y$ with the help of the chain rule gives

$$\begin{aligned} F_x(x,y) &= f_r(r(x,y),\varphi(x,y))r_x(x,y) + f_\varphi(r(x,y),\varphi(x,y))\varphi_x(x,y), \\ F_y(x,y) &= f_r(r(x,y),\varphi(x,y))r_y(x,y) + f_\varphi(r(x,y),\varphi(x,y))\varphi_y(x,y). \end{aligned}$$

Taking derivatives a second time, now using the product rule and the chain rule, gives

$$\begin{aligned} F_{xx} &= (f_{rr}r_x + f_{r\varphi}\varphi_x)r_x + f_r r_{xx} + (f_{\varphi r}r_x + f_{\varphi\varphi}\varphi_x)\varphi_x + f_\varphi\varphi_{xx}, \\ F_{yy} &= (f_{rr}r_y + f_{r\varphi}\varphi_y)r_y + f_r r_{yy} + (f_{\varphi r}r_y + f_{\varphi\varphi}\varphi_y)\varphi_y + f_\varphi\varphi_{yy}. \end{aligned}$$

Thus, we get the result

$$\begin{aligned} F_{xx} + F_{yy} = \;& f_{rr}(r_x^2 + r_y^2) + f_{\varphi\varphi}(\varphi_x^2 + \varphi_y^2) + 2f_{r\varphi}(\varphi_x r_x + \varphi_y r_y) \\ & + f_r(r_{xx} + r_{yy}) + f_\varphi(\varphi_{xx} + \varphi_{yy}). \end{aligned} \tag{1.65}$$

First assume that $x \neq 0$. Then (1.61) implies

$$\begin{aligned} r_x &= \frac{x}{\sqrt{x^2+y^2}} = \cos\varphi, & r_y &= \frac{y}{\sqrt{x^2+y^2}} = \sin\varphi, \\ \varphi_x &= -\frac{y}{x^2+y^2} = -\frac{\sin\varphi}{r}, & \varphi_y &= \frac{x}{x^2+y^2} = \frac{\cos\varphi}{r}. \end{aligned}$$

Differentiating again with respect to $x$ and $y$ using the chain rule gives

$$\begin{aligned} r_{xx} &= (-\sin\varphi)\varphi_x = \frac{\sin^2\varphi}{r}, & r_{yy} &= (\cos\varphi)\varphi_y = \frac{\cos^2\varphi}{r}, \\ \varphi_{xx} &= \frac{2\cos\varphi\sin\varphi}{r^2} = -\varphi_{yy}. \end{aligned}$$

These relations, together with (1.65), yields the formula (1.64) for $x \neq 0$. The case of $x = 0$, $y \neq 0$ follows from (1.64) by taking the limit $x \to 0$.

**Second method:** We now use the identity

$$\boxed{f(r,\varphi) = F(x(r,\varphi), y(r,\varphi)).}$$

To simplify notations, we write $f$ instead of $F$. Taking derivatives with respect to $r$ and $\varphi$, using the chain rule gives

$$\begin{aligned} f_r &= f_x x_r + f_y y_r = f_x \cos\varphi + f_y \sin\varphi, \\ f_\varphi &= f_x x_\varphi + f_y y_\varphi = -f_x r \sin\varphi + f_y r \cos\varphi. \end{aligned}$$

Solving for $f_x$ and $f_y$ in this equation we obtain

$$f_x = Af_r + Bf_\varphi, \qquad f_y = Cf_r + Df_\varphi \tag{1.66}$$

with

$$A = \cos\varphi, \qquad B = -\frac{\sin\varphi}{r}, \qquad C = \sin\varphi, \qquad D = \frac{\cos\varphi}{r}.$$

We write $\partial_x = \dfrac{\partial}{\partial x}$, etc. Then (1.66) is equivalent to the following *key formula*

$$\boxed{\partial_x = A\partial_r + B\partial_\varphi, \qquad \partial_y = C\partial_r + D\partial_\varphi.}$$

From this we obtain

$$\begin{aligned} \partial_x^2 &= (A\partial_r + B\partial_\varphi)(A\partial_r + B\partial_\varphi) \\ &= A\partial_r(A\partial_r) + B\partial_\varphi(A\partial_r) + A\partial_r(B\partial_\varphi) + B\partial_\varphi(B\partial_\varphi). \end{aligned}$$

The product rule implies $\partial_r(A\partial_r) = (\partial_r A)\partial_r + A\partial_r^2$ etc. Thus,

$$\partial_x^2 = AA_r\partial_r + A^2\partial_r^2 + BA_\varphi\partial_r + 2AB\partial_\varphi\partial_r + AB_r\partial_\varphi + BB_\varphi\partial_\varphi + B^2\partial_\varphi^2.$$

Exchanging $A$ with $C$ and $B$ with $D$, one gets similarly

$$\partial_y^2 = CC_r\partial_r + C^2\partial_r^2 + DC_\varphi\partial_r + 2CD\partial_\varphi\partial_r + CD_r\partial_\varphi + DD_\varphi\partial_\varphi + D^2\partial_\varphi^2.$$

Because of the relations

$$\begin{aligned} &A_r = C_r = 0, \qquad A_\varphi = -\sin\varphi, \qquad C_\varphi = \cos\varphi, \\ &B_r = \frac{\sin\varphi}{r^2}, \qquad B_\varphi = -\frac{\cos\varphi}{r}, \qquad D_r = -\frac{\cos\varphi}{r^2}, \qquad D_\varphi = -\frac{\sin\varphi}{r} \end{aligned}$$

we finally get

$$\Delta = \partial_x^2 + \partial_y^2 = \partial_r^2 + \frac{1}{r^2}\partial_\varphi^2 + \frac{1}{r}\partial_r,$$

which is the transformed formula (1.64).

The second method does not use the inverse formula (1.61), which can be a great simplification in complicated calculations.

### 1.5.5 Application to the dependency of functions

**Definition:** Let $f_k : G \longrightarrow \mathbb{R}$ be a $C^1$-function, $k = 1, \ldots, K+1$, where $G$ is a non-empty open set of $\mathbb{R}^N$. We say that $f_{K+1}$ *depends* on $f_1, \ldots, f_K$, if and only if there is a $C^1$-function $F : \mathbb{R}^K \longrightarrow \mathbb{R}$ with

$$\boxed{f_{K+1}(x) = F(f_1(x), \ldots, f_K(x)) \qquad \text{for all } x \in G.}$$

**Theorem:** The dependency relation is satisfied, provided the rank of the two matrices[43]

$$(f_1'(x), \ldots, f_{K+1}'(x)) \quad \text{and} \quad (f_1'(x), \ldots, f_K'(x))$$

is constant, equal to $r$ for some $r$ with $1 \le r \le K$.

*Example:* Let $f_1(x) := \mathrm{e}^{x_1}$, $f_2(x) := \mathrm{e}^{x_2}$ and $f_3(x) := \mathrm{e}^{x_1+x_2}$. Then we have

$$f_1'(x) = \begin{pmatrix} \partial_1 f_1(x) \\ \partial_2 f_1(x) \end{pmatrix} = \begin{pmatrix} \mathrm{e}^{x_1} \\ 0 \end{pmatrix}, \qquad f_2'(x) = \begin{pmatrix} 0 \\ \mathrm{e}^{x_2} \end{pmatrix}.$$

Because of the relation $\det(f_1'(x), f_2'(x)) = \mathrm{e}^{x_1}\mathrm{e}^{x_2} \neq 0$, we get from this

$$\operatorname{rank}(f_1'(x), f_2'(x)) = \operatorname{rank}(f_1'(x), f_2'(x), f_3'(x)) = 2$$

for all $x = (x_1, x_2)$ in $\mathbb{R}^2$. Consequently, $f_3$ is dependent on $f_1$ and $f_2$ in $\mathbb{R}^2$. In fact, one has the explicit relation

$$f_3(x) = f_1(x) f_2(x).$$

## 1.5.6 The theorem on implicit functions

### 1.5.6.1 An equation with two real variables

We want to solve the equation

$$\boxed{F(x,y) = 0} \tag{1.67}$$

for $y$, where $x, y \in \mathbb{R}$ and $F(x,y) \in \mathbb{R}$. That is, we are looking for a function $y = y(x)$ with

$$F(x, y(x)) = 0.$$

We assume that we know some fixed solution of the equation, i.e.,

$$\boxed{F(q,p) = 0.} \tag{1.68}$$

Moreover, we require

$$\boxed{F_y(q,p) \neq 0.} \tag{1.69}$$

**Theorem on implicit functions:** If the function $F : D(F) \subseteq \mathbb{R}^2 \longrightarrow \mathbb{R}$ is of type $C^k$, $k \ge 1$ in some neighborhood of the point $(q,p)$, and the conditions (1.68) and (1.69) are both satisfied, then the equation (1.67) can be uniquely solved at the point $(q,p)$ for $y$[44] (Figure 1.48).

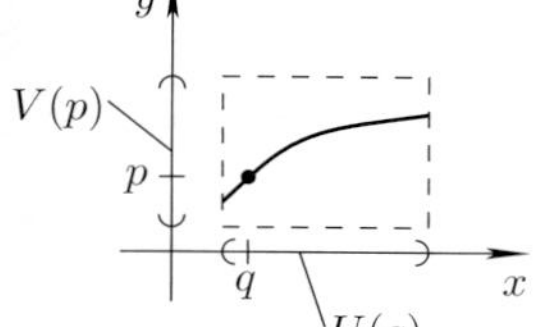


*Figure 1.48.*

The solution $y = y(x)$ is locally at $q$ of type $C^k$.

**The method of implicit differentiation:** To calculate the derivatives of the solution $y = y(x)$, we differentiate the equation

$$F(x, y(x)) = 0$$

with respect to $x$ and use the chain rule. This yields

$$F_x(x, y(x)) + F_y(x, y(x)) y'(x) = 0, \tag{1.70}$$

[43] We write $f_j'(x)$ as a column matrix.

[44] This means there are open neighborhoods $U(q)$ and $V(p)$, such that the equation (1.67) has a unique solution $y(x) \in V(p)$ for every $x \in U(q)$.

which in turn implies

$$\boxed{y'(x) = -F_y(x, y(x))^{-1}F_x(x, y(x)).} \tag{1.71}$$

The higher derivatives of $y$ are gotten by differentiating (1.71). However, it is more convenient to take the derivative of (1.70) with respect to $x$. This gives

$$F_{xx}(x, y(x)) + 2F_{xy}(x, y(x))y'(x) + F_{yy}(x, y(x))y'(x)^2 + F_y(x, y(x))y''(x) = 0.$$

From this one can solve for $y''(x)$. One proceeds similarly for higher derivatives.

**Approximation formula:** The Taylor expansion of $y$ for a solution of $F(x, y) = 0$ gives the approximation

$$\boxed{y = p + y'(q)(x - q) + \frac{y''(q)}{2!}(x - q)^2 + \ldots}$$

*Example:* Let $F(x, y) := \mathrm{e}^y \sin x - y$. Then $F(0, 0) = 0$ and $F_y(x, y) = \mathrm{e}^y \sin x - 1$, hence $F_y(0, 0) \neq 0$. Consequently, the equation

$$\mathrm{e}^y \sin x - y = 0 \tag{1.72}$$

can be uniquely solved near (0,0) for $y$. To get an approximation of the solution $y = y(x)$, we make the ansatz

$$y(x) = a + bx + cx^2 + \ldots$$

Because $y(0) = 0$, one has $a = 0$. Using the power series expansion of the exponential

$$\mathrm{e}^y = 1 + y + \ldots, \qquad \sin x = x - \frac{x^3}{3!} + \ldots, \tag{1.73}$$

we get from (1.72) and (1.73) the equation

$$x - bx + x^2(\ldots) + x^3(\ldots) + \ldots = 0.$$

Comparing coefficients yields $b = 1$, hence

$$y = x + \ldots$$

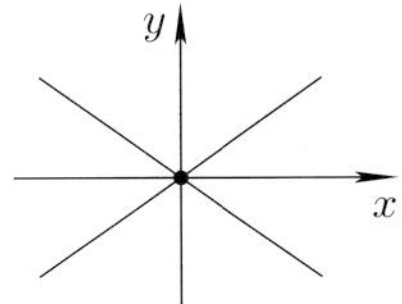


*Figure 1.49.*

**Bifurcation:** Let $F(x, y) := x^2 - y^2$. Then we have $F(0, 0) = 0$ and $F_y(0, 0) = 0$. Since (1.69) is not satisfied, the equation $F(x, y) = 0$ cannot be locally solved at $(0, 0)$ uniquely for $y$. In fact, the equation

$$\boxed{x^2 - y^2 = 0}$$

has two solutions, $y = \pm x$. Hence the point $(0, 0)$ is a point where the solutions *branch* (a *bifurcation point*)[45] (Figure 1.49).

[45] The general theory of bifurcations has many interesting applications in physics (cf. [212]).

#### 1.5.6.2 Systems of equations

The calculus of the F-derivative is flexible enough that it can be immediately generalized to apply to systems of non-linear equations

$$\boxed{F(x,y)=0} \tag{1.74}$$

with $x \in \mathbb{R}^N$, $y \in \mathbb{R}^M$ and $F(x,y) \in \mathbb{R}^M$. One only has to pay attention to the fact that $F_y(q,p)$ is a matrix and the decisive condition $F_y(q,p) \neq 0$ has to be replaced by

$$\boxed{\det F_y(q,p) \neq 0.} \tag{1.75}$$

Let $(q,p)$ be a solution of

$$\boxed{F(q,p)=0, \qquad q \in \mathbb{R}^N,\ p \in \mathbb{R}^M.} \tag{1.76}$$

**Theorem on implicit functions:** If the function $F : D(F) \subseteq \mathbb{R}^{N+M} \longrightarrow \mathbb{R}^M$ is of type $C^k$, $k \geq 1$ in a neighborhood of the point $(q,p)$, and if the two conditions (1.75) and (1.76) are satisfied, then the equation (1.74) has a unique solution $y$ at the point $(q,p)$.

The solution $y = y(x)$ is locally of type $C^k$ at the point $q$. The formula (1.71) for the F-derivative $y'(x)$ remains valid, now as a matrix equation.

**Explicit formulation:** The system (1.74) is explicitly

$$F_k(x_1,\ldots,x_N,y_1,\ldots,y_M)=0, \qquad k=1,\ldots,M, \tag{1.77}$$

and $F_y(x,y)$ is the matrix of the first partial derivatives of $F_k$ with respect to $y_m$. Replacing $y_m$ by $y_m(x_1,\ldots,x_n)$ in (1.77) and taking derivatives with respect to $x_n$, one has

$$\frac{\partial F_k}{\partial x_n}(x,y(x)) + \sum_{m=1}^{M} \frac{\partial F_k}{\partial y_m}(x,y(x))\frac{\partial y_m}{\partial x_n}(x) = 0, \qquad n=1,\ldots,N.$$

Solving this equation for $\partial y_m/\partial x_n$ yields the matrix equation (1.71).

### 1.5.7 Inverse mappings

#### 1.5.7.1 Homeomorphisms

**Definition:** Let $X$ and $Y$ be metric spaces (for example subsets of $\mathbb{R}^N$). A map $f : X \longrightarrow Y$ is called a *homeomorphism* if $f$ is bijective and both $f$ and $f^{-1}$ are continuous (see 4.3.3).

**Theorem on homeomorphisms:** A bijective continuous map $f : X \longrightarrow Y$ on a compact set is a homeomorphism.

This theorem generalizes the theorem on global inverse real functions on a compact interval (see 1.4.4.2).

#### 1.5.7.2 Local diffeomorphisms

**Definition:** Let $X$ and $Y$ be non-empty open sets of $\mathbb{R}^N$, $N \geq 1$. The map $f : X \longrightarrow Y$ is called a $C^k$-*diffeomorphism*, if $f$ is bijective and both $f$ and $f^{-1}$ are of type $C^k$.

**The main theorem on local diffeomorphisms:** Let $1 \leq k \leq \infty$. Assume the map $f : M \subseteq \mathbb{R}^N \longrightarrow \mathbb{R}^N$ is of type $C^k$ on an open neighborhood $V(p)$ of $p$, and that

$$\boxed{\det f'(p) \neq 0.}$$

Then $f$ is a local $C^k$-diffeomorphism[46] at the point $p$.

*Example:* We consider the map

$$\boxed{u = g(x, y), \qquad v = h(x, y)} \tag{1.78}$$

with $u_0 := g(x_0, y_0)$, $v_0 := h(x_0, y_0)$. The functions $g$ and $h$ are both assumed to be of type $C^k$, $1 \leq k \leq \infty$ in a neighborhood of the point $(x_0, y_0)$, and it is assumed that

$$\boxed{\begin{vmatrix} g_u(x_0, y_0) & g_v(x_0, y_0) \\ h_u(x_0, y_0) & h_v(x_0, y_0) \end{vmatrix} \neq 0.}$$

Then the map (1.78) is a local $C^k$-diffeomorphism at the point $(x_0, y_0)$.

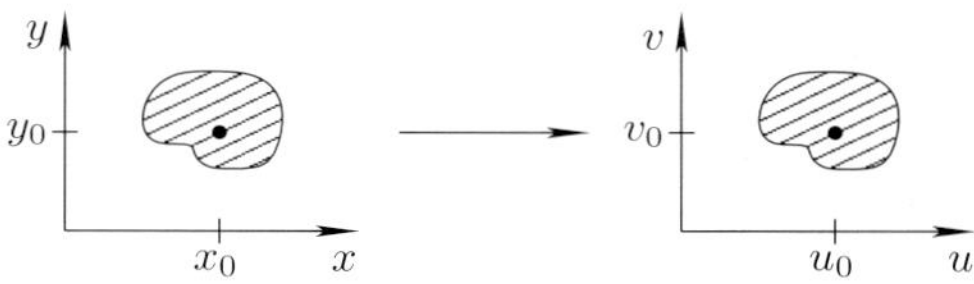


Figure 1.50. A local diffeomorphism.

This means that the map in (1.78), depicted in Figure 1.50, can be inverted in a neighborhood of the point $(u_0, v_0)$, and the inverse mapping

$$x = x(u, v), \qquad y = y(u, v)$$

in smooth in a neighborhood of $(u_0, v_0)$, that is, $x$, $y$ are of type $C^k$.

#### 1.5.7.3 Global diffeomorphisms

**The Theorem of Hadamard on global diffeomorphisms:** Let $1 \leq k \leq \infty$. Suppose a $C^k$-map $f : \mathbb{R}^N \longrightarrow \mathbb{R}^N$ satisfies the two conditions:

$$\left.\begin{array}{l} \lim_{|x| \to \infty} |f(x)| = +\infty, \text{ and} \\ \det f'(x) \neq 0 \end{array}\right\} \text{ for all } x \in \mathbb{R}^N. \tag{1.79}$$

Then $f$ is a $C^k$-diffeomorphism.[47]

*Example:* Let $N = 1$. For the function $f(x) := \sinh x$, $f'(x) = \cosh x > 0$ imply that (1.79) is satisfied. Thus $f : \mathbb{R} \longrightarrow \mathbb{R}$ is a $C^\infty$-diffeomorphism (Figure 1.51).

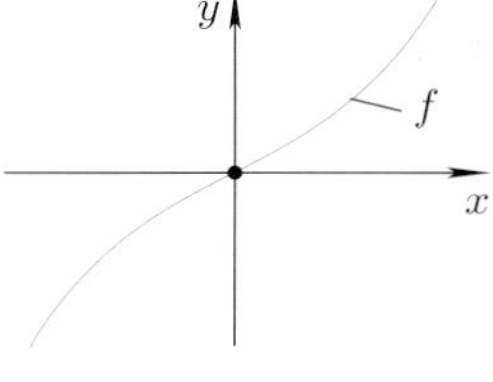


Figure 1.51.

[46] This means that $f$ is a $C^k$-diffeomorphism from some appropriately chosen open neighborhood $U(p)$ to an open neighborhood $U(f(p))$.

[47] In the special case where $N = 1$, we have $\det f'(x) = f'(x)$.

#### 1.5.7.4 Generic behavior of solutions

**Theorem:** Let $f : \mathbb{R}^N \longrightarrow \mathbb{R}^N$ be of type $C^1$ such that $\lim_{|x|\to\infty} |f(x)| = \infty$. Then there is an open and dense[48] set $D \subseteq \mathbb{R}^N$, such that the equation

$$\boxed{f(x) = y, \quad x \in \mathbb{R}^N} \tag{1.80}$$

has at most finitely many solutions for each $y \in D$.

One abbreviates this conclusion by saying: there are generically finitely many solutions. More precisely, one has the following clear situation.

(i) *Perturbations.* If a value $y_0 \in \mathbb{R}^N$ is given, then there is, in every neighborhood of $y_0$, a point $y \in \mathbb{R}^N$ for which the equation (1.80) has at most finitely many solutions, in other words, by perturbing $y_0$ slightly one can get favorable behavior of the solutions.

(ii) *Stability.* If the equation (1.80) has for some point $y_1 \in D$ at most finitely many solutions, then there is a neighborhood $U(y_1)$ in which (1.80) has only finitely many solutions for all $y \in U(y_1)$.

### 1.5.8 The $n^{th}$ variation and Taylor's theorem

**$n^{th}$ variation:** Let $f : U(p) \subseteq \mathbb{R}^N \longrightarrow \mathbb{R}$ be a function defined in a neighborhood of a point $p$. For $h \in \mathbb{R}^N$ we set

$$\varphi(t) := f(p + th),$$

where the real parameter $t$ is allowed to vary in a small neighborhood of $t = 0$. If the $n^{th}$ derivative $\varphi^{(n)}(0)$ exists, then the number

$$\boxed{\delta^n f(p; h) := \varphi^{(n)}(0)}$$

is called the $n^{th}$ *variation* of the function $f$ at the point $p$ in the direction of $h$.

**Directional derivatives:** For $n = 1$ we set $\delta f(p; h) := \delta^1 f(p; h)$ and call this expression the *directional derivative* of $f$ at the point $p$ in the direction of $h$. Explicitly, one has

$$\delta f(p; h) = \lim_{t \to 0} \frac{f(p + th) - f(p)}{t}.$$

**Theorem:** Let $n \geq 1$. If the function $f : U(p) \subseteq \mathbb{R}^N \longrightarrow \mathbb{R}$ is of type $C^n$ in an open neighborhood of the point $p$, then one has

$$\boxed{\delta f(p; h) = \sum_{k=1}^{N} h_k \frac{\partial f(p)}{\partial x_k}}$$

and

$$\boxed{\delta^r f(p; h) = \left( \sum_{k=1}^{N} h_k \frac{\partial}{\partial x_k} \right)^r f(p), \qquad r = 1, \ldots, n.}$$

[48] A set $D$ is *dense* in $\mathbb{R}^N$, if its closure in $\mathbb{R}^N$ is all of $\mathbb{R}^N$, i.e., $\overline{D} = \mathbb{R}^N$.

*Example:* For $N = n = 2$, we have

$$\left(h_1\frac{\partial}{\partial x_1} + h_2\frac{\partial}{\partial x_2}\right)^2 = h_1^2\frac{\partial^2}{\partial x_1^2} + 2h_1h_2\frac{\partial^2}{\partial x_1\partial x_2} + h_2^2\frac{\partial^2}{\partial x_2^2}.$$

If we use the more convenient notation $\partial_j := \partial/\partial x_j$, then we get from this

$$\begin{aligned} \delta f(p;h) &= h_1\partial_1 f(p) + h_2\partial_2 f(p), \\ \delta^2 f(p;h) &= h_1^2\partial_1^2 f(p) + 2h_1h_2\partial_1\partial_2 f(p) + h_2^2\partial_2^2 f(p). \end{aligned}$$

**The general form of Taylor's theorem:** Let $f : U \subseteq \mathbb{R}^N \longrightarrow \mathbb{R}$ be a function of type $C^{n+1}$ on the open convex set $U$. For all points $x, x+h \in U$, we have

$$\boxed{f(x+h) = f(x) + \sum_{k=1}^{n} \frac{\delta^k f(x;h)}{k!} + R_{n+1}}$$

with the remainder term

$$\boxed{R_{n+1} = \frac{\delta^{n+1} f(x+\vartheta h;h)}{(n+1)!},}$$

where the number $\vartheta$ depends on $x$ and satisfies $0 < \vartheta < 1$. In addition, one has

$$R_{n+1} = \int_0^1 \frac{(1-\tau)^n}{n!}\delta^{n+1} f(x+\tau h;h)\mathrm{d}\tau.$$

**Local behavior of functions:** One can use the Taylor expansion in an entirely similar manner as in 1.4.5 to study the local behavior of functions. Important results in this respect can be found in 5.4.1.

### 1.5.9 Applications to estimation of errors

Usually measurements in physics contain measurement errors. The theoretical error estimation is used to relate errors in the arguments of functions with those of the values.

**Functions of a real variable:** We consider the function

$$\boxed{y = f(x)}$$

and set

$$\begin{aligned} \Delta x &= \text{error of the argument } x, \\ \Delta f &= f(x+\Delta x) - f(x) = \text{error of the value of the function } f(x), \\ \frac{\Delta f}{f(x)} &= \text{relative error of the value } f(x). \end{aligned}$$

From Taylor's theorem it follows that

$$\Delta f = f'(x)\Delta x + \frac{f''(x+\vartheta\Delta x)}{2}(\Delta x)^2$$

with $0 < \vartheta < 1$. From this we get the *error estimate*:

$$|\Delta f - f'(x)\Delta x| \leq \frac{(\Delta x)^2}{2} \sup_{0<\eta<1} |f''(x+\eta\Delta x)|.$$

*Example 1:* For $f(x) := \sin x$ we have $f''(x) = -\sin x$, hence

$$|\Delta f - \Delta x \cdot \cos x| \leq \frac{(\Delta x)^2}{2}.$$

For example, when $\Delta x = 10^{-3}$ we have $(\Delta x)^2 = 10^{-6}$.
For sufficiently small errors $\Delta x$ one uses the general approximation formula

$$\Delta f = f'(x)\Delta x.$$

**Functions of several variables:** For the function $y = f(x_1, \ldots, x_N)$ we set

$$\Delta f := f(x_1 + \Delta x_1, \ldots, x_N + \Delta x_N) - f(x_1, \ldots, x_N).$$

The approximation formula here is

$$\Delta f = \sum_{k=1}^{N} \frac{\partial f(x)}{\partial x_j} \Delta x_j.$$

**Chain rule:** For a composed function $H(x) = F(f_1(x), \ldots, f_m(x))$ with $x = (x_1, \ldots, x_N)$ we have

$$\Delta H = \sum_{k=1}^{m} \frac{\partial F}{\partial f_k}(f_1(x), \ldots, f_m(x))\Delta f_k.$$

*Example 2* (summation rule): For $H(x) = \sum_{k=1}^{m} f_k(x)$ we get

$$\Delta H = \sum_{k=1}^{m} \Delta f_k,$$

that is, the absolute errors add up.

*Example 3* (product rule): For $H(x) = \prod_{k=1}^{m} f_k(x)$ one gets

$$\frac{\Delta H}{H(x)} = \sum_{k=1}^{m} \frac{\Delta f_k}{f_k(x)},$$

that is, the relative logarithmic errors add up.

*Example 4* (quotient rule): For $H(x) = \frac{f(x)}{g(x)}$ one has

$$\boxed{\frac{\Delta H}{H(x)} = \frac{\Delta f}{f(x)} - \frac{\Delta g}{g(x)},}$$

that is, the relative logarithmic errors are subtracted.

**The Gaussian law of propagation of errors:** We consider the function

$$\boxed{z = f(x, y).}$$

Suppose we are given measurements

$$x_1, \ldots, x_n \qquad \text{and} \qquad y_1, \ldots, y_m$$

of the $x$ and $y$. From this we get the values of $z$, $z_{jk} = f(x_j, y_k)$. By definition, the averages $\overline{x}, \overline{y}$ and the variances $\sigma_x^2, \sigma_y^2$ are given by

$$\overline{x} = \frac{1}{n}\sum_{j=1}^{n} x_j, \qquad \overline{y} = \frac{1}{m}\sum_{k=1}^{m} y_k,$$

$$\sigma_x^2 = \frac{1}{n-1}\sum_{j=1}^{n}(x_j - \overline{x})^2, \qquad \sigma_y^2 = \frac{1}{m-1}\sum_{k=1}^{m}(y_k - \overline{y})^2.$$

According to Gauss one has the following approximate relations for sufficiently large $n$ and $m$:

$$\boxed{\begin{aligned} \overline{z} &= f(\overline{x}, \overline{y}), \\ \sigma_z^2 &= f_x(\overline{x}, \overline{y})^2 \sigma_x^2 + f_y(\overline{x}, \overline{y})^2 \sigma_y^2. \end{aligned}}$$

This relation is called the *Gaussian law of error propagation.*

### 1.5.10 The Fréchet differential

*It is important to take note of the way notations can help with discoveries. In this way the work of the mind can be wonderfully reduced.*

*Gottfried Wilhelm Leibniz* (1646–1716)

**Leibniz differential calculus:** The notion of differential is of fundamental importance in modern analysis, geometry and mathematical physics. For Leibniz these were differentials $\mathrm{d}f$ of *infinitely small size*, which reflected his philosophical ideas about the smallest mental components of the world. The inconcise, but extremely convenient notion of the infinitely small can be found even today in the physics and technological literature. In order to give the convenient Leibniz notion a sound basis, one introduces the Fréchet differential $\mathrm{d}f(x)$. This was done at the beginning of this century by the French mathematician Maurice Fréchet (1878–1956). For a function

$$f : \mathbb{R}^N \longrightarrow \mathbb{R}$$

the *Fréchet differential* is a *linear map*

$$\mathrm{d}f(x) : \mathbb{R}^N \longrightarrow \mathbb{R}$$

which is to be defined, which assigns to every $h \in \mathbb{R}^N$ a real number $\mathrm{d}f(x)h$, and which in addition satisfies the linearity condition

$$\mathrm{d}f(x)(\alpha h + \beta k) = \alpha \mathrm{d}f(x)h + \beta \mathrm{d}f(x)k$$

for all $\alpha, \beta \in \mathbb{R}$ and all $h, k \in \mathbb{R}^N$ (cf. 1.5.10.2).

Differentials are linear maps.

**The differential calculus of Cartan:** The same notion of differential is the basis of the elegant *differential calculus of Cartan*, which the great French mathematician Élie Cartan (1869–1961) introduced at the end of the last century and provides a useful extension of the Leibniz calculus. The Cartan calculus is one of the most powerful and most often used instruments of modern mathematics and physics.[49]

**Advantages for practical calculations:** Both the Leibniz and the Cartan differential calculus have a decided practical advantage, in that one only needs to remember a few easy rules. The rest is done by the calculus itself. We emphasize this by first presenting the formal rules and only then worrying about the rigorous justification. For most practical calculations, one only needs to know the formal rules.

### 1.5.10.1 The formal Leibniz differential calculus

Let a function $y = f(x)$ with $x = (x_1, \ldots, x_N)$ be given. According to Leibniz, one does computations with differentials by applying the following rules:

| | | |
|---|---|---|
| (i) | $\mathrm{d}f = \sum_{j=1}^{N} \frac{\partial f}{\partial x_j} \mathrm{d}x_j,$ | (total differential), |
| (ii) | $\mathrm{d}(f+g) = \mathrm{d}f + \mathrm{d}g,$ | (rule of sums), |
| (iii) | $\mathrm{d}(fg) = (\mathrm{d}f)g + f\mathrm{d}g,$ | (product rule), |
| (iv) | $\mathrm{d}^2 x_j = 0,$ | (infinitely small). |

One has to keep in mind that the last rule (iv) only holds for the *arguments*.

This calculus has shown itself to be extremely flexible.

**Transformation rules for differentials:** This calculus is applied particularly often to transform functions to a new set of variables. If we have

$$x_j = x_j(u_1, \ldots, u_M), \qquad j = 1, \ldots, N,$$

then we get from the rule for the total differential the fundamental *transformation rule for differentials:*

$$\mathrm{d}x_j = \sum_{m=1}^{M} \frac{\partial x_j}{\partial u_m} \mathrm{d}u_m. \tag{1.81}$$

[49] Leibniz' notion of the infinitely small has been given a rigorous basis in *non-standard analysis*, by extending the real numbers by new quantities, which are called infinitely small or infinitely large numbers, and with which (in the context of an extended logic) one can calculate rigorously (cf. [53]).

This yields

$$\boxed{\mathrm{d}f = \sum_{j=1}^{N} \frac{\partial f}{\partial x_j}\mathrm{d}x_j = \sum_{j=1}^{N}\sum_{m=1}^{M} \frac{\partial f}{\partial x_j}\frac{\partial x_j}{\partial u_m}\mathrm{d}u_m.} \tag{1.82}$$

Comparing this with

$$\mathrm{d}f = \sum_{m=1}^{M} \frac{\partial f}{\partial u_m}\mathrm{d}u_m$$

we get

$$\frac{\partial f}{\partial u_m} = \sum_{j=1}^{N} \frac{\partial f}{\partial x_j}\frac{\partial x_j}{\partial u_m}.$$

The Leibniz calculus gives in this way the chain rule automatically.

*Example 5* (chain rule for higher derivatives): Let $h(x) := f(z)$, $z = g(x)$. As argument (dependent variable) we choose $x$, which by rule (iv) implies

$$\mathrm{d}^2 x = 0.$$

From the product rule we get

$$\begin{aligned} \mathrm{d}z &= g'\mathrm{d}x, \\ \mathrm{d}^2 z &= \mathrm{d}(\mathrm{d}z) = \mathrm{d}(g'\mathrm{d}x) = \mathrm{d}g'\mathrm{d}x + g'^2 x = \mathrm{d}g'\mathrm{d}x = g''\mathrm{d}x^2 \end{aligned}$$

and

$$\begin{aligned} \mathrm{d}f &= f'\mathrm{d}z, \\ \mathrm{d}^2 f &= \mathrm{d}(\mathrm{d}f) = (\mathrm{d}f')\mathrm{d}z + f'\mathrm{d}^2 z \\ &= f''\mathrm{d}z^2 + f'g''\mathrm{d}x^2 = (f''g'^2 + f'g'')\mathrm{d}x^2. \end{aligned}$$

It follows from this that

$$\boxed{\frac{\mathrm{d}^2 f(x)}{\mathrm{d}x^2} = f''(z)g'(x)^2 + f'(z)g''(x), \qquad z = g(x).} \tag{1.83}$$

*Rigorous proof of (1.83)*: Taking derivatives of $h(x) = f(g(x))$ using the chain rule yields

$$h'(x) = f'(g(x))g'(x).$$

Differentiating again and using the chain rule and the product rule, we get

$$h''(x) = f''(g(x))g'(x)^2 + f'(g(x))g''(x),$$

which is (1.83). □

For functions of several variables it is often advantageous to use differentials instead of partial derivatives. For this reason this calculus is popular in the physics and technological literature.

#### 1.5.10.2 Fréchet differentials and higher Fréchet derivatives

Fréchet differentials (F-differentials for short) can be defined rigorously by using *decompositions* of functions with appropriate remainder terms. For historical reasons one uses both F-differentials as well as F-derivatives. In fact, however, these two notions coincide. Consider a function

$$\boxed{f : U(x) \subseteq \mathbb{R}^N \to \mathbb{R},}$$

which is defined in a neighborhood of a point $x$. Also set

$$\partial_j f := \frac{\partial f}{\partial x_j}, \qquad \partial_j \partial_k f := \frac{\partial^2 f}{\partial x_j \partial x_k} \qquad \text{etc.}$$

**The F-differential $\mathrm{d}f(x)$:** By definition the function $f$ has a F-differential at a point $x$, if and only if the decomposition

$$\boxed{f(x+h) - f(x) = \mathrm{d}f(x)h + r(h)}$$

is valid for all $h \in \mathbb{R}^N$ in a neighborhood of the origin, where one assumes in addition the two conditions:

(i) $\mathrm{d}f(x) : \mathbb{R}^N \longrightarrow \mathbb{R}$ is a linear mapping.

(ii) The remainder term is $o$-small,[50] $r(h) = o(|h|), \ h \to 0$.

**The relation to the F-derivative:** One also calls $\mathrm{d}f(x)$ the F-derivative, denoted $f'(x)$. Moreover, we call

$$\boxed{\mathrm{d}f(x)h = f'(x)h, \qquad h \in \mathbb{R}^N,}$$

the value of the F-differential of the function $f$ at the point $x$ in the direction of $h$.

**Relation to the first variation:** If the F-differential $\mathrm{d}f(x)$ exists, then the first variation of $f$ at the point $x$ in the direction $h$ also exists, and

$$\delta f(x;h) = \mathrm{d}f(x)h, \quad h \in \mathbb{R}^N.$$

**Existence theorem:**[51] If the function $f$ is of type $C^1$ in an open neighborhood of a point $x$, then one has

$$\boxed{\mathrm{d}f(x)h = \sum_{j=1}^{N} \partial_j f(x) h_j.} \tag{1.84}$$

**The second F-differential $\mathrm{d}^2 f(x)$:** By definition, a function $f$ has at a point $x$ a second F-differential, if there is a decomposition of differentials

$$\mathrm{d}f(x+h)k - \mathrm{d}f(x)k = \mathrm{d}^2 f(x)(k,h) + r(h,k)$$

for all $h \in \mathbb{R}^N$ in a neighborhood of the origin and for all $k \in \mathbb{R}^N$, where we in addition assume:

---

[50] This means $\lim_{h \to 0} \frac{r(h)}{|h|} = 0$ with $|h| = \left( \sum_{j=1}^{N} h_j^2 \right)^{1/2}$. The reading is 'little o-small', as there is also a notion of 'big O-small'.

[51] For most practical purposes it suffices to know the formulas (1.84), (1.85) and (1.86). The more general definition here is given because it is easier to generalize to abstract operators, which is of utmost importance for modern theoretical and numerical treatment of non-linear differential and integral equations (cf. [212]).

(i) $\mathrm{d}^2 f(x) : \mathbb{R}^N \times \mathbb{R}^N \longrightarrow \mathbb{R}$ is a bilinear mapping.

(ii) The remainder term is $o$ of $h$, i.e., $\sup\limits_{|k|\le 1} |r(h,k)| = o(|h|),\ h \to 0$. To simplify notations, one writes for this also $\mathrm{d}^2 f(x)hk := \mathrm{d}^2 f(x)(h,k)$ and $\mathrm{d}^2 f(x)h^2 := \mathrm{d}^2 f(x)(h,h)$.

**The second F-derivative $f''(x)$:** One also denotes the differential $\mathrm{d}^2 f(x)$ as the second F-derivative $f''(x)$ of the function $f$ in the point $x$. Furthermore, the *value* of this differential of the function $f$ at the point $x$ with respect to the directions $h$ and $k$ is

$$\boxed{\mathrm{d}^2 f(x)hk = f''(x)hk, \qquad h,k \in \mathbb{R}^N.}$$

**Relation with the second variation:** If the second F-differential $\mathrm{d}^2 f(x)$ exists, then the second variation of $f$ at $x$ exists for every direction $h$ which satisfies

$$\delta^2 f(x;h) = \mathrm{d}^2 f(x)h^2, \qquad h \in \mathbb{R}^N.$$

Similarly one defines the $n^{th}$ F-differentials $\mathrm{d}^n f(x)$.

**Existence theorem:** Let $n \ge 2$. If $f$ of type $C^n$ in a neighborhood of a point $x$, then:

$$\boxed{\mathrm{d}^2 f(x)hk = \sum_{r,s=1}^{N} \partial_r \partial_s f(x) h_r h_s.} \tag{1.85}$$

More generally, one has

$$\boxed{\mathrm{d}^n f(x)h^{(1)} \ldots h^{(n)} = \sum_{r_1,\ldots,r_n=1}^{N} \partial_{r_1}\partial_{r_2}\ldots\partial_{r_n} f(x) h^{(1)}_{r_1} h^{(2)}_{r_2} \ldots h^{(n)}_{r_n}.} \tag{1.86}$$

In particular one has the symmetry of the second derivatives

$$\mathrm{d}^2 f(x)hk = \mathrm{d}^2 f(x)kh \qquad \text{for all}\ \ h,k \in \mathbb{R}^N.$$

Similarly, $\mathrm{d}^n f(x)h^{(1)} \cdots h^{(n)}$ is invariant under an arbitrary permutation of $h^{(1)}, \ldots, h^{(n)}$, $h^{(i)} \in \mathbb{R}^N$.

### 1.5.10.3 Rigorous justification of the Leibniz differential calculus

If one views the Leibniz differentials as F-differentials, then it is easy to give a rigorous justification to the Leibniz differential calculus.

**The Leibniz formula for the total differential:** If a function $f$ is of type $C^1$ in a neighborhood of a point $x$, then one has:[52]

$$\boxed{\mathrm{d}f(x) = \sum_{j=1}^{N} \partial_j f(x) \mathrm{d}x_j} \tag{1.87}$$

with

$$\boxed{\mathrm{d}x_j h = h_j \ \text{ for all }\ h \in \mathbb{R}^N.}$$

[52] We abbreviate notations slightly be writing $\mathrm{d}x_j$ instead of $\mathrm{d}x_j(x)$.

*Proof:* We set $f(x) := x_j$. Form (1.84) we get

$$\mathrm{d}x_j(x)h = \sum_{k=1}^{N} \partial_k f(x) h_k = h_j$$

where we have used $\partial_k f(x) = \dfrac{\partial x_j}{\partial x_k} = \delta_{kj}$. The statement (1.87) is then

$$\mathrm{d}f(x)h = \sum_{j=1}^{N} \partial_j f(x) \mathrm{d}x_j h = \sum_{j=1}^{N} \partial_j f(x) h_j.$$

But this is equivalent to (1.84). □

**The differential operator d:** We define

$$\boxed{\mathrm{d} := \sum_{j=1}^{N} \mathrm{d}x_j \partial_j.}$$

The relation (1.87) can then be written in the following way:

$$\boxed{\mathrm{d}f(x) = \mathrm{d} \otimes f(x),}$$

if we agree to let $\partial_j \otimes f(x) := \partial_j f(x)$.

**The Leibniz product formula:** Let functions $f, g : U(x) \subseteq \mathbb{R}^N \longrightarrow \mathbb{R}$ be of type $C^1$ in a neighborhood of $x$. Then

$$\boxed{\mathrm{d}(fg)(x) = g(x)\mathrm{d}f(x) + f(x)\mathrm{d}g(x).}$$

*Proof:* This follows from (1.87) and the product rule for derivatives: $\partial_j(fg) = g\partial_j f + f\partial_j g$.

**The Leibniz transformation formula:** We assume that

$$x_j = x_j(u_1, \ldots, u_M), \qquad j = 1, \ldots, N,$$

that is, the quantities $x_j$ depend on the variables $u_m$. Furthermore we set $F(u) := f(x(u))$. Then one has

$$\boxed{\mathrm{d}x_j(u) := \sum_{m=1}^{M} \frac{\partial x_j(u)}{\partial u_m} \mathrm{d}u_m} \tag{1.88}$$

and

$$\boxed{\mathrm{d}F(u) = \sum_{j=1}^{N} \frac{\partial f}{\partial x_j}(x(u)) \mathrm{d}x_j(u).} \tag{1.89}$$

This corresponds to (1.81) and (1.82).

*Proof:* The formula (1.88) follows from (1.87).

Applying (1.87) to the function $F$, it follows from the chain rule that

$$\mathrm{d}F(u) = \sum_{m=1}^{M} \frac{\partial F(u)}{\partial u_m}\mathrm{d}u_m = \sum_{m=1}^{M}\sum_{j=1}^{N} \frac{\partial f}{\partial x_j}(x(u))\frac{\partial x_j(u)}{\partial u_m}\mathrm{d}u_m.$$

This is (1.89). □

**Leibniz' formula for the second differential:** Let a function $f$ be of type $C^2$ in a neighborhood of $x$. Then

$$\boxed{\mathrm{d}^2 f(x) = \mathrm{d}\otimes\mathrm{d}f(x).} \tag{1.90}$$

More explicitly, this is

$$\boxed{\mathrm{d}^2 f(x) = \sum_{j,m=1}^{N} \partial_j\partial_m f(x)\mathrm{d}x_j \otimes \mathrm{d}x_m} \tag{1.91}$$

and

$$\boxed{\mathrm{d}^2 x_j(x) = 0.} \tag{1.92}$$

Here we are using the tensor product, i.e., the following relation

$$(\mathrm{d}x_r \otimes \mathrm{d}x_s)(h,k) := (\mathrm{d}x_r h)(\mathrm{d}x_s k) = h_r k_s. \tag{1.93}$$

*Proof:* The formula (1.91) follows from (1.85) together with (1.93). If we set $f(x) := x_j$, then the second partial derivatives of $f$ with respect to $x_1, x_2, \ldots$ all vanish identically. Thus (1.85) implies (1.92). □

**Comparison of Leibniz' and Cartan's differential calculus:** The tensor product $\otimes$ and the exterior (outer) product $\wedge$ are of particular importance in multilinear algebra (cf. 2.4.2).

(i) The Leibniz calculus is based on the operator d and the tensor product $\otimes$. One then has for example the product $\mathrm{d}^2$ as

$$\boxed{\mathrm{d}^2 = \mathrm{d}\otimes\mathrm{d}.}$$

(ii) The Cartan calculus is based on the operator d and the exterior product $\wedge$. In this case the operator $\mathrm{d}^2$ is given by

$$\boxed{\mathrm{d}^2 = \mathrm{d}\wedge\mathrm{d}.}$$

Instead of (1.93) one has in the Cartan calculus the following relation

$$(\mathrm{d}x_r \wedge \mathrm{d}x_s)(h,k) = (\mathrm{d}x_r h)(\mathrm{d}x_s k) - (\mathrm{d}x_r k)(\mathrm{d}x_s h) = h_r k_s - k_r h_s \tag{1.94}$$

for all $h, k \in \mathbb{R}^N$.

### 1.5.10.4 The formal Cartan differential calculus

In order to simplify notation, we now agree that if two identical indices appear in a formula, then one forms the sum over that index (Einstein summation convention). In this section, the index will run from 1 to $N$. For example, we have

$$a_j \mathrm{d}x_j = \sum_{j=1}^{N} a_j \mathrm{d}x_j .$$

**The product symbol $\wedge$:** The Cartan differential calculus follows from the Leibniz calculus, by inserting a $\wedge$; one must now worry about signs, as $\wedge$ is *anti-commutative*:

$$\boxed{\mathrm{d}x_j \wedge \mathrm{d}x_m = -\mathrm{d}x_m \wedge \mathrm{d}x_j .} \tag{1.95}$$

From (1.95) it follows also that $\mathrm{d}x_m \wedge \mathrm{d}x_m = -\mathrm{d}x_m \wedge \mathrm{d}x_m$, or in other words

$$\boxed{\mathrm{d}x_m \wedge \mathrm{d}x_m = 0.}$$

*Example 1:* $\mathrm{d}x_1 \wedge \mathrm{d}x_2 \wedge \mathrm{d}x_3 = -\mathrm{d}x_2 \wedge \mathrm{d}x_1 \wedge \mathrm{d}x_3 = \mathrm{d}x_2 \wedge \mathrm{d}x_3 \wedge \mathrm{d}x_1$.

*Example 2:* $\mathrm{d}x_1 \wedge \mathrm{d}x_1 \wedge \mathrm{d}x_2 = 0$.

**Rule for permutations:** The product

$$\mathrm{d}x_{j_1} \wedge \mathrm{d}x_{j_2} \wedge \ldots \wedge \mathrm{d}x_{j_r} \tag{1.96}$$

does not change upon an *even* permutation of the factors, but changes its sign under an *odd* permutation,[53] and vanishes if two factors coincide.

**Differential forms:** A *differential form* of degree $r$ is a linear combination of products of the form (1.96).

Functions are by definition differential forms of degree 0.

*Example 3:*

$$\begin{array}{ll} \omega = a_j \mathrm{d}x_j , & \text{(degree 1)}, \\ \omega = a_{jk} \mathrm{d}x_j \wedge \mathrm{d}x_k , & \text{(degree 2)}, \\ \omega = a_{jkm} \mathrm{d}x_j \wedge \mathrm{d}x_k \wedge \mathrm{d}x_m , & \text{(degree 3)}. \end{array}$$

The coefficients $a_j, a_{jk}$ and $a_{jkm}$ are functions of $x = (x_1, \ldots, x_N)$.

**The three basic rules:**

(i) *Addition:* Differential forms are added in the usual way and multiplied by functions in the usual way.

(ii) *Multiplication:* Differential forms are multiplied in the usual way with the operator $\wedge$, paying attention to (1.95).

(iii) *Differentiation:* For a function $f$ one has the Leibniz rule:

$$\boxed{\mathrm{d}f = (\partial_j f)\mathrm{d}x_j .} \tag{1.97}$$

For a form $\omega = a_{j_1 \cdots j_r} \mathrm{d}x_{j_1} \wedge \cdots \wedge \mathrm{d}x_{j_r}$ one has the Cartan rule:

$$\boxed{\mathrm{d}\omega = \mathrm{d}a_{j_1 \ldots j_r} \wedge \mathrm{d}x_{j_1} \wedge \ldots \wedge \mathrm{d}x_{j_r} .} \tag{1.98}$$

[53] See 2.1.1 for definitions about permutations.

These three basic rules completely determine the calculus with differential forms.

*Example 4:* For $a = a(x, y)$ and $b = b(x, y)$, one has

$$\mathrm{d}a = a_x \mathrm{d}x + a_y \mathrm{d}y, \qquad \mathrm{d}b = b_x \mathrm{d}x + b_y \mathrm{d}y.$$

*Example 5:* For $\omega = a\mathrm{d}x + b\mathrm{d}y$ one gets for the exterior derivative of $\omega$:

$$\begin{aligned} \mathrm{d}\omega &= \mathrm{d}a \wedge \mathrm{d}x + \mathrm{d}b \wedge \mathrm{d}y = (a_x \mathrm{d}x + a_y \mathrm{d}y) \wedge \mathrm{d}x + (b_x \mathrm{d}x + b_y \mathrm{d}y) \wedge \mathrm{d}y \\ &= (b_x - a_y)\mathrm{d}x \wedge \mathrm{d}y. \end{aligned}$$

Here the following facts were used:

$$\mathrm{d}x \wedge \mathrm{d}x = \mathrm{d}y \wedge \mathrm{d}y = 0 \quad \text{and} \quad \mathrm{d}x \wedge \mathrm{d}y = -\mathrm{d}y \wedge \mathrm{d}x.$$

*Example 6:* Let $c = c(x, y)$ be given. For $\omega = c\mathrm{d}x \wedge \mathrm{d}y$ we get

$$\mathrm{d}\omega = \mathrm{d}c \wedge \mathrm{d}x \wedge \mathrm{d}y = (c_x \mathrm{d}x + c_y \mathrm{d}y) \wedge \mathrm{d}x \wedge \mathrm{d}y = 0.$$

Here again one uses the fact that a $\wedge$-product with *two equal factors* vanishes identically.

*Example 7:* Let

$$\omega = a\mathrm{d}x + b\mathrm{d}y + c\mathrm{d}z,$$

where $a, b$ and $c$ depend on $x, y$ and $z$. Then we have[54]

$$\boxed{\mathrm{d}\omega = (b_x - a_y)\mathrm{d}x \wedge \mathrm{d}y + (c_y - b_z)\mathrm{d}y \wedge \mathrm{d}z + (a_z - c_x)\mathrm{d}z \wedge \mathrm{d}x.}$$

This follows from

$$\begin{aligned} \mathrm{d}\omega &= \mathrm{d}a \wedge \mathrm{d}x + \mathrm{d}b \wedge \mathrm{d}y + \mathrm{d}c \wedge \mathrm{d}z \\ &= (a_x \mathrm{d}x + a_y \mathrm{d}y + a_z \mathrm{d}z) \wedge \mathrm{d}x + (b_x \mathrm{d}x + b_y \mathrm{d}y + b_z \mathrm{d}z) \wedge \mathrm{d}y \\ &\quad + (c_x \mathrm{d}x + c_y \mathrm{d}y + c_z \mathrm{d}z) \wedge \mathrm{d}z. \end{aligned}$$

*Example 8:* For

$$\omega = a\mathrm{d}y \wedge \mathrm{d}z + b\mathrm{d}z \wedge \mathrm{d}x + c\mathrm{d}x \wedge \mathrm{d}y$$

we get

$$\boxed{\mathrm{d}\omega = (a_x + b_y + c_z)\mathrm{d}x \wedge \mathrm{d}y \wedge \mathrm{d}z.}$$

This follows from the computation

$$\begin{aligned} \mathrm{d}\omega &= (a_x \mathrm{d}x + a_y \mathrm{d}y + a_z \mathrm{d}z) \wedge \mathrm{d}y \wedge \mathrm{d}z + (b_x \mathrm{d}x + b_y \mathrm{d}y + b_z \mathrm{d}z) \wedge \mathrm{d}z \wedge \mathrm{d}x \\ &\quad + (c_x \mathrm{d}x + c_y \mathrm{d}y + c_z \mathrm{d}z) \wedge \mathrm{d}x \wedge \mathrm{d}y. \end{aligned}$$

**Transformation of differential forms to a new set of variables:** For this one uses the Leibniz rule. The number of old and new variables are immaterial for this.

*Example 9:* If we apply the variable transformation

$$x = x(t), \quad y = y(t), \quad z = z(t)$$

to

$$\omega = a\mathrm{d}x + b\mathrm{d}y + c\mathrm{d}z,$$

[54]Observe the great symmetry in all of these formulas. The summands are obtained by cyclic permutations of $a, b, c$ and $x, y, z$.

then we get $\mathrm{d}x = x'\mathrm{d}t$, etc. This yields

$$\boxed{\omega(ax' + by' + cz')\mathrm{d}t.}$$

*Example 10:* Applying the variable transformation

$$x = x(u, v), \quad y = y(u, v)$$

to

$$\omega = a\mathrm{d}x \wedge \mathrm{d}y$$

with $a = a(x, y)$ yields

$$\omega = a(x_u y_v - x_v y_u)\mathrm{d}u \wedge \mathrm{d}v.$$

This formula also follows from

$$\mathrm{d}x = x_u \mathrm{d}u + x_v \mathrm{d}v, \qquad \mathrm{d}y = y_u \mathrm{d}u + y_v \mathrm{d}v$$

together with the relation $\omega = (x_u \mathrm{d}u + x_v \mathrm{d}v) \wedge (y_u \mathrm{d}u + y_v \mathrm{d}v)$. With the help of the Jacobi determinant

$$\frac{\partial(x, y)}{\partial(u, v)} = \begin{vmatrix} x_u & x_v \\ y_u & y_v \end{vmatrix} = x_u y_v - x_v y_u$$

one can also write this in the form

$$\boxed{\omega = a\frac{\partial(x, y)}{\partial(u, v)}\mathrm{d}u \wedge \mathrm{d}v.} \tag{1.99}$$

*Example 11:* Consider the change of variables

$$x = x(u, v, w), \quad y = y(u, v, w), \quad z = z(u, v, w)$$

and apply this to

$$\omega = a\mathrm{d}x \wedge \mathrm{d}y \wedge \mathrm{d}z.$$

This gives the expression

$$\boxed{\omega = a\frac{\partial(x, y, z)}{\partial(u, v, w)}\mathrm{d}u \wedge \mathrm{d}v \wedge \mathrm{d}w} \tag{1.100}$$

with the functional determinant

$$\frac{\partial(x, y, z)}{\partial(u, v, w)} = \begin{vmatrix} x_u & x_v & x_w \\ y_u & y_v & y_w \\ z_u & z_v & z_w \end{vmatrix}$$

This follows from $\omega = a\mathrm{d}x \wedge \mathrm{d}y \wedge \mathrm{d}z$ and

$$\begin{aligned} \mathrm{d}x &= x_u \mathrm{d}u + x_v \mathrm{d}v + x_w \mathrm{d}w, \\ \mathrm{d}y &= y_u \mathrm{d}u + y_v \mathrm{d}v + y_w \mathrm{d}w, \\ \mathrm{d}z &= z_u \mathrm{d}u + z_v \mathrm{d}v + z_w \mathrm{d}w. \end{aligned}$$

*Example 12:* Via the change of variables

$$x = x(u, v), \quad y = y(u, v), \quad z = z(u, v)$$

we get from

$$\omega = a\mathrm{d}y \wedge \mathrm{d}z + b\mathrm{d}z \wedge \mathrm{d}x + c\mathrm{d}x \wedge \mathrm{d}y$$

the expression

$$\boxed{\omega = \left( a\frac{\partial(y,z)}{\partial(u,v)} + b\frac{\partial(z,x)}{\partial(u,v)} + c\frac{\partial(x,y)}{\partial(u,v)} \right) \mathrm{d}u \wedge \mathrm{d}v.}$$

**Manipulations:** Let $\omega, \mu$ and $\vartheta$ denote arbitrary differential forms of degree $\geq 0$. We set

$$a \wedge \omega := a\omega, \qquad \omega \wedge a := a\omega,$$

if $a$ is a function.

(i) *Associative law:*

$$\omega \wedge (\mu \wedge \vartheta) = (\omega \wedge \mu) \wedge \vartheta.$$

(ii) *Distributive law:*

$$\omega \wedge (\mu + \vartheta) = \omega \wedge \mu + \omega \wedge \vartheta.$$

(iii) *Supercommutativity:*

$$\boxed{\omega \wedge \mu = (-1)^{rs} \mu \wedge \omega \qquad (r \text{ degree of } \omega,\ s \text{ degree of } \mu).}$$

(iv) *Rule for products of differential forms:*

$$\boxed{\mathrm{d}(\omega \wedge \mu) = \mathrm{d}\omega \wedge \mu + (-1)^r \omega \wedge \mathrm{d}\mu.}$$

(v) *Poincaré lemma:* One always has $\mathrm{d}^2 = 0$, i.e.,

$$\boxed{\mathrm{d}(\mathrm{d}\omega) = 0.}$$

(vi) *Rule of exchange:* The operations of differentiation and of change of variables can be exchanged (these operations commute with each other).[55]

**Mnemonics:** The basic formula (1.98) for differentiation can be easily remembered, by writing $\mathrm{d} \wedge \omega$ instead of $\mathrm{d}\omega$. Then formally one has:

$$\begin{aligned} \mathrm{d} \wedge \omega &= \mathrm{d}x_j \partial_j \wedge a_{j_1 \ldots j_r} \mathrm{d}x_{j_1} \wedge \ldots \wedge \mathrm{d}x_{j_r} \\ &= \partial_j a_{j_1 \ldots j_r} \mathrm{d}x_j \wedge \mathrm{d}x_{j_1} \wedge \ldots \wedge \mathrm{d}x_{j_r} = \mathrm{d}a_{j_1 \ldots j_r} \wedge \mathrm{d}x_{j_1} \wedge \ldots \wedge \mathrm{d}x_{j_r}. \end{aligned}$$

Poincaré's rule $\mathrm{d}(\mathrm{d}\omega) = 0$ then also follows formally:

$$\begin{aligned} \mathrm{d} \wedge (\mathrm{d} \wedge \omega) &= (\mathrm{d} \wedge \mathrm{d}) \wedge \omega = (\partial_j \mathrm{d}x_j) \wedge (\partial_k \mathrm{d}x_k) \wedge \omega \\ &= \tfrac{1}{2}(\partial_j \partial_k - \partial_k \partial_j) \mathrm{d}x_j \wedge \mathrm{d}x_k \wedge \omega = 0 \end{aligned}$$

because of $\partial_j \partial_k - \partial_k \partial_j = 0$.

[55] This means that it is immaterial whether one first forms $\mathrm{d}\omega$ and then applies a coordinate change, or whether one first changes variables and then forms $\mathrm{d}\omega$ with respect to the new variables. It is this fact which makes the Cartan calculus so flexible.

### 1.5.10.5 Rigorous justification of the Cartan differential calculus and its applications

In order to make what we have discussed up to this point mathematically rigorous, it will suffice to understand the $\wedge$-product in the sense of multilinear algebra, which is given by (1.94). Then the differential formula (1.98) is a definition of $d\omega$, and the remaining statements can be verified by direct computation.

The Cartan differential calculus has the following *applications* (see [212]):

(i) Iterated integrals and integral along curves on $m$-dimensional surfaces (cf. 1.7.6).

(ii) The theorem of Stokes $\int_{\partial M} \omega = \int_M d\omega$, which generalizes the fundamental theorem of calculus to higher dimensions and which contains the classical integral theorems of Gauss, Green and Stokes as special cases (cf. 1.7.6.ff).

(iii) Poincaré's theorem on the solutions of $d\omega = \mu$ and applications of this in vector analysis (cf. 1.9.11).

(iv) The theorem of Cartan–Kähler on the solution of systems of differential forms

$$\omega_1 = 0, \ \omega_2 = 0, \ldots, \ \omega_k = 0,$$

which contain general systems of partial differential equations as special cases (cf. 1.13.5.4).

(v) Tensor analysis.

(vi) The special theory of relativity and electrodynamics.

(vii) Calculus on manifolds.

(viii) Thermodynamics.

(ix) Symplectic geometry, classical mechanics and classical statistical physics.

(x) Riemannian geometry, Einstein's general theory of relativity, cosmology, and the standard model in particle physics.

(xi) Lie groups and symmetry.

(xii) Differential topology and de Rham cohomology.

(xiii) Modern differential geometry, curvature of principal bundles, gauge theories in high energy physics, and string theory.

This list of applications shows that the Cartan differential calculus plays an important role in many areas of modern mathematics and physics.

## 1.6 Integration of functions of a real variable

Many methods for the explicit calculation of integrals and an extensive list of known integrals can be found in section 0.9

### 1.6.1 Basic ideas

The precise mathematical formulation of the following considerations can be found in 1.6.2 ff.

**The limit of a sum**: The integral

$$\int_a^b f(x)\mathrm{d}x$$

is equal to the area of the hatched region under the graph of $f$ in Figure 1.52(a). This area can be calculated by choosing an approximation by means of rectangles as in Figure 1.52(b) and then taking the limit as the rectangles get thinner and thinner. This means[56]

$$\int_a^b f(x)\mathrm{d}x = \lim_{n\to\infty} \sum_{k=1}^{n} f(x_k)\Delta x. \tag{1.101}$$

Here we divide the compact interval $[a, b]$ into $n$ equal parts. The division points are then given by

$$x_k = a + k\Delta x, \qquad k = 0, 1, 2, \ldots n,$$

with

$$\Delta x := \frac{b-a}{n}.$$

In particular $x_0 = a$ and $x_n = b$.

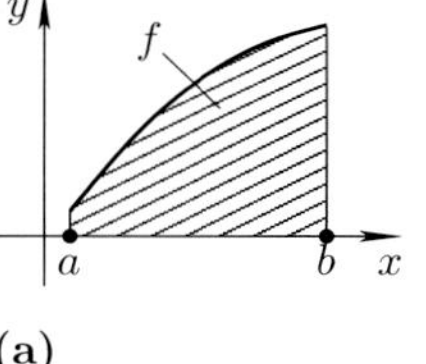


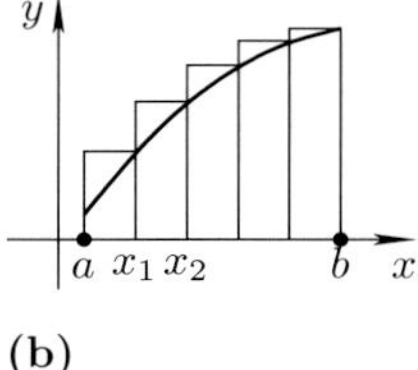


*Figure 1.52. Approximation of surface area.*

**Practical calculation of integrals:**
Newton and Leibniz discovered the fundamental formula

$$\int_a^b F'(x)\mathrm{d}x = F(b) - F(a), \tag{1.102}$$

which is called the fundamental theorem of calculus.[57] This formula shows that the integration can be viewed as the inverse of differentiation. Henceforth we will write

$$F(x)\Big|_a^b := F(b) - F(a).$$

[56]The area of a rectangle of width $\Delta x$ and height $f(x_k)$ is equal to $f(x_k)\Delta x$. Consequently, the expression

$$\sum_{k=1}^{n} f(x_k)\Delta x$$

is the sum of the areas of the individual rectangles in Figure 1.52(b).

[57]A formal motivation for the formula (1.102) is given by passing to the infinite limit:

$$\sum \frac{\Delta F}{\Delta x}\Delta x = \sum \Delta F = F(b) - F(a).$$

In Leibniz notation one gets the parallel formulas

$$\int_a^b \frac{\mathrm{d}F}{\mathrm{d}x}\mathrm{d}x = \int_a^b \mathrm{d}F$$

*Example 1:* Let $F(x) := x^2$. From $F'(x) = 2x$ we get

$$\int_a^b 2x\mathrm{d}x = x^2\Big|_a^b = b^2 - a^2.$$

*Example 2:* Let $F(x) := \sin x$. From $F'(x) = \cos x$ we get

$$\int_a^b \cos x\mathrm{d}x = \sin x\Big|_a^b = \sin b - \sin a.$$

**Primitive functions:** Let $J$ be an open interval. A function $F : J \longrightarrow \mathbb{R}$ is called an *primitive function* to $f$ on $J$, if

$$\boxed{F'(x) = f(x) \qquad \text{for all } \ x \in J.}$$

**Theorem:** If $F$ is a primitive function for $f$ on $J$, then all primitive functions for $f$ on $J$ are of the form

$$F + C,$$

where $C$ is an arbitrary real constant.

One also writes for this

$$\boxed{\int f(x)\mathrm{d}x = F(x) + C \quad \text{ on } \ J} \tag{1.104}$$

and calls the set of all primitive functions on the right-hand side in (1.104) the *indefinite integral* of $f$ on $J$. From (1.102) we get

$$\boxed{\int_a^b f(x)\mathrm{d}x = F\Big|_a^b.} \tag{1.105}$$

The calculation of integrals is thus reduced to the calculation of the primitive functions.

**Table of important primitive functions:** This table can be found in 0.9.1.

*Example 3:* Because of $(-\cos x)' = \sin x$ one has

$$\int \sin x\mathrm{d}x = -\cos x + C$$

and

$$\int_a^b \sin x\mathrm{d}x = -\cos x\Big|_a^b = \cos a - \cos b.$$

---

and

$$\int_a^b \mathrm{d}F = F(b) - F(a). \tag{1.103}$$

A rigorous justification of (1.103) is given in the context of general measure theory, which is also valid for a certain class of non-continuous functions $F$ (cf. [212]).

*Example 4:* Let $\alpha$ be a real number with $\alpha \neq 0$. From $(x^\alpha)' = \alpha x^{\alpha-1}$ it follows that

$$\int \alpha x^{\alpha-1} \mathrm{d}x = x^\alpha + C$$

and

$$\int_a^b \alpha x^{\alpha-1} \mathrm{d}x = x^\alpha \Big|_a^b = b^\alpha - a^\alpha .$$

**Integration of non-continuous functions:** A differentiable function is always continuous, hence only sufficiently smooth functions can be differentiated.

However, one can integrate a large class of non-continuous functions.

*Example 5:* We set

$$f(x) := \begin{cases} 1 & \text{for} \quad x < 2, \\ 3 & \text{for} \quad x > 2, \\ c & \text{for} \quad x = 2. \end{cases}$$

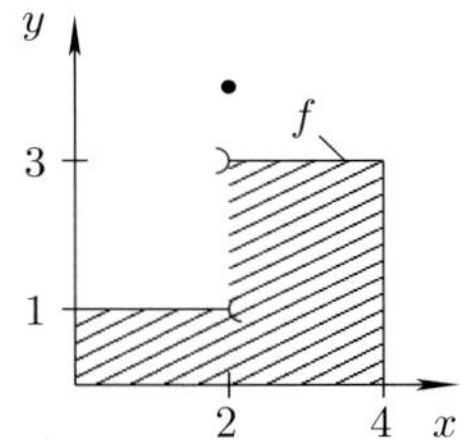


*Figure 1.53. Integration of discontinuous functions.*

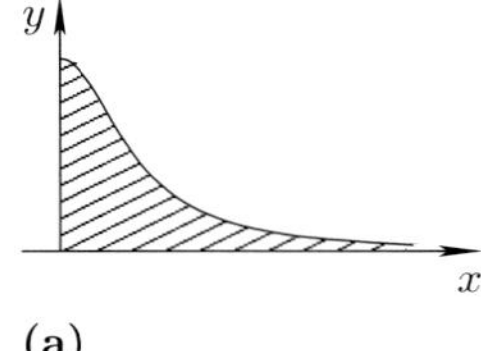


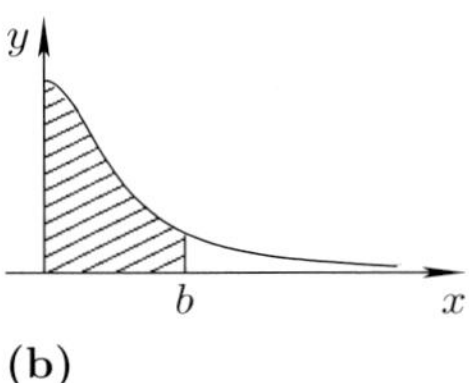


*Figure 1.54. Approximation of the integral on an unbounded interval.*

Because of the additivity of areas we expect according to Figure 1.53 the relation:

$$\int_0^4 f \mathrm{d}x = \int_0^2 f \mathrm{d}x + \int_2^4 f \mathrm{d}x = 2 \cdot 1 + 2 \cdot 3 = 8.$$

Here the discontinuity of $f$ at the point $x = 2$ is irrelevant. Intuitively the area of the region under the graph of the function $f$ in Figure 1.52(a) does not change when the value of $f$ is changed in finitely many points.

**Integration over unbounded intervals:** The integral $\int_0^\infty \frac{\mathrm{d}x}{1+x^2}$ corresponds intuitively to the area of the hatched region in Figure 1.54(a). This area can be calculated in the obvious way as the limit

$$\boxed{\int_0^\infty \frac{\mathrm{d}x}{1+x^2} = \lim_{b \to +\infty} \int_0^b \frac{\mathrm{d}x}{1+x^2} = \frac{\pi}{2}}$$

(Figure 1.54(b)). Here we have used the formula

$$\int_0^b \frac{\mathrm{d}x}{1+x^2} = \arctan x \Big|_0^b = \arctan b - \arctan 0 = \arctan b.$$

**Integration of unbounded functions:** The integral $\int_0^1 \frac{\mathrm{d}x}{\sqrt{x}}$ corresponds to the area of the hatched region in Figure 1.55(a).

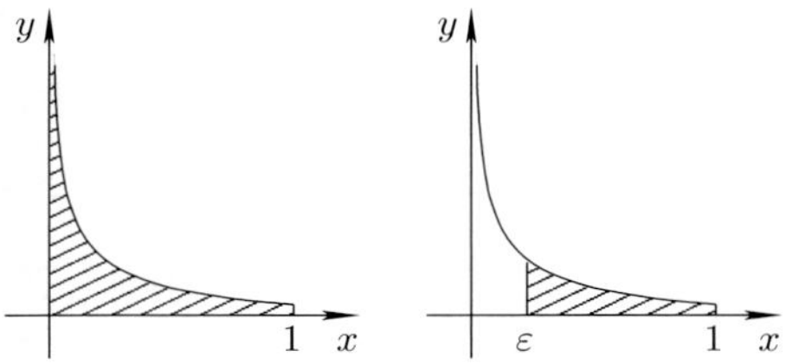

This area can be calculated by passing to the limit

$$\int_0^1 \frac{\mathrm{d}x}{\sqrt{x}} = \lim_{\varepsilon \to +0} \int_\varepsilon^1 \frac{\mathrm{d}x}{\sqrt{x}} = 2$$

(Figure 1.55(b)). Here we have used the formula

$$\int_\varepsilon^1 \frac{\mathrm{d}x}{\sqrt{x}} = 2\sqrt{x}\Big|_\varepsilon^1 = 2 - 2\sqrt{\varepsilon}.$$

*Figure 1.55. Integration of unbounded functions.*

**Measure and integral:** The great mathematician of ancient Greece, Archimedes (287-212 BC) calculated an approximation to the circumference of the unit circle by approximating the circle by a 96-gon. In this way he got the approximation 6.28 for the number $2\pi$. After him, many mathematicians and physicists worked on calculating 'measures' for sets (length of curves, areas of surfaces, volumes, masses, charges, etc.). At the beginning of this century, the French mathematician Henri Lebesgue (1875–1941) developed a general measure theory, which allows the assignment of measures to subsets of a given set, calculations which can be performed in a satisfactory manner; in particular, limits can be formed in this theory. In this way, Lebesgue solved completely the ancient problem of finding and calculating measures. The notion of the Lebesgue measure contains the so-called Lebesgue integral, of which the classical (Riemannian) integral (1.101) is a special case.

For reasons of didactics, this classical integral is still part of the syllabus in schools and colleges. But in modern mathematics and physics, one really needs the full power of the general notion of the Lebesgue integral (for example in probability theory, calculus of variations, theory of partial differential equations, quantum theory and so forth). The reason for the superiority of the modern Lebesgue integral is the basic *limit formula*

$$\lim_{n\to\infty} \int_a^b f_n(x)\mathrm{d}x = \int_a^b \lim_{n\to\infty} f_n(x)\mathrm{d}x, \tag{1.106}$$

which holds under very mild assumptions for the Lebesgue integral, but does not hold for the classical integral. It can happen that the integral on the left in (1.106) exists as a classical integral, while the limit function $\lim_{n\to\infty} f(x)$ is so terribly non-continuous that the right-hand side of (1.106) does not exist in the classical sense, but only in the sense of the Lebesgue integral.

In this Volume we consider only the classical notion of integral. The modern measure-theoretic notion is explained in [212]. The following statements about one-dimensional integrals $\int_a^b f(x)\mathrm{d}x$ can be immediately extended to integrals of several variables (cf. 1.7).

### 1.6.2 Existence of the integral

Let $-\infty < a < b < \infty$.

**First existence theorem:** If a function $f : [a, b] \longrightarrow \mathbb{C}$ is continuous, then the integral $\int_a^b f(x)\mathrm{d}x$ exists in the sense of (1.101).

**Sets of vanishing one-dimensional measure:** A subset $M$ of $\mathbb{R}$ is said to have one-dimensional *vanishing Lebesgue measure*, if for every real $\varepsilon > 0$ there is an at most countable set of intervals $J_1,\ J_2, \ldots$, which cover the set $M$ and whose total length is less than $\varepsilon$.

*Example 1:* Every set $M$ which consists of finitely or countably many real numbers, has zero one-dimensional Lebesgue measure.

Since the set $\mathbb{Q}$ of rational numbers is countable, it follows in particular that $\mathbb{Q}$ has zero Lebesgue measure.

**Functions which are continuous almost everywhere:** A function $f : [a, b] \longrightarrow \mathbb{R}$ is *almost everywhere continuous* (or *continuous almost everywhere*), if there is a set $M$ of zero Lebesgue measure, such that $f$ is continuous for all $x \in [a, b] - M$.

*Example 2:* The function pictured in Figure 1.56 contains finitely many points of discontinuity and is therefore almost everywhere continuous.

**Second existence theorem:** If a function $f : [a, b] \longrightarrow \mathbb{R}$ is bounded[58] and almost everywhere continuous, then the integral $\int_a^b f(x)\mathrm{d}x$ in the sense of (1.101) exists.

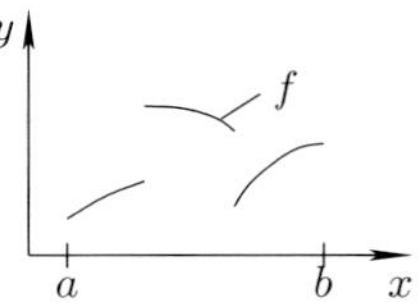


*Figure 1.56.*

**Complex-valued functions:** A complex-valued function $f : [a, b] \longrightarrow \mathbb{C}$ can be put in the form

$$\boxed{f(x) = \varphi(x) + \mathrm{i}\psi(x),}$$

where $\varphi(x)$ denotes the real and $\psi(x)$ the imaginary part of the complex number $f(x)$. The function $f$ is continuous in a point $x$ precisely when both $\varphi$ and $\psi$ are continuous in $x$.

Both of the existence theorems above continue to hold without change for complex-valued functions $f : [a, b] \longrightarrow \mathbb{C}$. In this case, $f$ is bounded and almost everywhere continuous if both $\varphi$ and $\psi$ enjoy these properties. For the integral one gets the following formula:

$$\boxed{\int_a^b f(x)\mathrm{d}x = \int_a^b \varphi(x)\mathrm{d}x + \mathrm{i}\int_a^b \psi(x)\mathrm{d}x.}$$

**Properties of the integral:** Let $-\infty < a < c < b < \infty$, suppose the functions $f, g : [a, b] \longrightarrow \mathbb{C}$ are bounded and almost everywhere continuous, and let $\alpha, \beta \in \mathbb{C}$.

(i) *Linearity:*

$$\int_a^b (\alpha f(x) + \beta g(x))\mathrm{d}x = \alpha \int_a^b f(x)\mathrm{d}x + \beta \int_a^b g(x)\mathrm{d}x.$$

(ii) *Triangle inequality:*

$$\boxed{\left|\int_a^b f(x)\mathrm{d}x\right| \le \int_a^b |f(x)|\mathrm{d}x \le (b-a) \sup_{a \le x \le b} |f(x)|.}$$

[58]The boundedness of $f$ means that $|f(x)| \le$ constant for all $x \in [a, b]$.

(iii) *Addition rule:*

$$\int_a^c f(x)\mathrm{d}x + \int_c^b f(x)\mathrm{d}x = \int_a^b f(x)\mathrm{d}x.$$

(iv) *Principle of invariance:* The integral $\int_a^b f(x)\mathrm{d}x$ does not change when the values of the function $f$ are changed on a set $M$ of zero Lebesgue measure.

(v) *Monotony:* If $f$ and $g$ are real-valued functions, then $f(x) \le g(x)$ for all $x \in [a,b]$ implies the inequality

$$\int_a^b f(x)\mathrm{d}x \le \int_a^b g(x)\mathrm{d}x.$$

**Mean value theorem for integrals:** One has the relation

$$\boxed{\int_a^b f(x)g(x)\mathrm{d}x = f(\xi)\int_a^b g(x)\mathrm{d}x}$$

for some appropriate $\xi \in [a,b]$, if the function $f : [a,b] \longrightarrow \mathbb{R}$ is continuous and $g : [a,b] \longrightarrow \mathbb{R}$ is non-negative, bounded and almost everywhere continuous.

*Example 3:* For the special case $g(x) \equiv 1$, we get

$$\int_a^b f(x)\mathrm{d}x = f(\xi)(b-a).$$

*Example 4:* If $f : [a,b] \longrightarrow \mathbb{R}$ is almost everywhere continuous and if $m \le f(x) \le M$ for all $x \in [a,b]$, then one has

$$\int_a^b m\mathrm{d}x \le \int_a^b f(x)\mathrm{d}x \le \int_a^b M\mathrm{d}x,$$

that is, $(b-a)m \le \int_a^b f(x)\mathrm{d}x \le (b-a)M$.

### 1.6.3 The fundamental theorem of calculus

**Fundamental theorem:** Let $-\infty < a < b < \infty$. For a function $F : [a,b] \longrightarrow \mathbb{C}$ of class[59] $C^1$, the following relation holds:

$$\boxed{\int_a^b F'(x)\mathrm{d}x = F(b) - F(a).}$$

*Example:* Because of $(\mathrm{e}^{\alpha x})' = \alpha\mathrm{e}^{\alpha x}$ for all $x \in \mathbb{R}$ ($\alpha$ a complex number), we have

$$\alpha\int_a^b \mathrm{e}^{\alpha x}\mathrm{d}x = \mathrm{e}^{\alpha x}\Big|_a^b = \mathrm{e}^{\alpha b} - \mathrm{e}^{\alpha a}.$$

[59] This means that both the real and the imaginary part of $F$ belong to $C^1[a,b]$.

In the following $f : [a,b] \longrightarrow \mathbb{C}$ denotes a continuous function.

**Differentiation with respect to the upper limit of integration:** We set

$$\boxed{F_0(x) := \int_a^x f(t)\mathrm{d}t, \qquad a \le x \le b;}$$

we then have

$$F'(x) = f(x) \qquad \text{for all } x \in ]a,b[ \tag{1.107}$$

with $F = F_0$.

**Existence of a primitive function:**

(i) The function $F_0 : [a,b] \longrightarrow \mathbb{C}$ is the unique solution in the class $C^1$ of the differential equation (1.107) with $F_0(a) = 0$. In particular, $F_0$ is an primitive function for $f$ on the interval $]a,b[$.

(ii) All $C^1$-solutions $F_0 : [a,b] \longrightarrow \mathbb{C}$ of (1.107) are obtained as

$$F_0(x) + C,$$

where $C$ is an arbitrary complex constant.

(iii) If $F : [a,b] \longrightarrow \mathbb{R}$ is a solution of (1.107) in the class $C^1$, then

$$\int_a^b f(x)\mathrm{d}x = F(b) - F(a).$$

## 1.6.4 Integration by parts

**Theorem:** Let $-\infty < a < b < \infty$. For the $C^1$-functions $u, v : [a,b] \longrightarrow \mathbb{C}$ one has:

$$\boxed{\int_a^b u'v\mathrm{d}x = uv\Big|_a^b - \int_a^b uv'\mathrm{d}x.} \tag{1.108}$$

*Proof:* From the fundamental theorem of calculus together with the product rule for differentiation we have:

$$\int_a^b (u'v + uv')\mathrm{d}x = \int_a^b (uv)'\mathrm{d}x = uv\Big|_a^b.$$

$\square$

*Example 1:* To calculate the integral

$$A := \int_1^2 2x \ln x\mathrm{d}x$$

we set

$$u' = 2x, \qquad v = \ln x,$$
$$u = x^2, \qquad v' = \frac{1}{x}.$$

From (1.108) we get

$$A = x^2 \ln x\Big|_1^2 - \int_1^2 x\mathrm{d}x = x^2 \ln x - \frac{x^2}{2}\Big|_1^2 = 4\ln 2 - \frac{3}{2}.$$

*Example 2:* To find the value of the integral

$$A = \int_a^b x \sin x \mathrm{d}x$$

we set

$$\begin{aligned} u' &= \sin x, & v &= x, \\ u &= -\cos x, & v' &= 1. \end{aligned}$$

According to (1.108), this means

$$A = -x\cos x\Big|_a^b + \int_a^b \cos x\mathrm{d}x = -x\cos x + \sin x\Big|_a^b.$$

*Example 3* (iterated partial integration): To calculate the integral

$$B = \int_a^b \frac{1}{2}x^2 \cos x\mathrm{d}x$$

we set

$$\begin{aligned} u' &= \cos x, & v &= \frac{1}{2}x^2, \\ u &= \sin x, & v' &= x. \end{aligned}$$

From (1.108) we get

$$B = \frac{1}{2}x^2 \sin x\Big|_a^b - \int_a^b x\sin x\mathrm{d}x.$$

The last integral is calculated as in Example 2 by using partial integration again.

**Indefinite integrals:** Under the same assumptions as for (1.108) we have:

$$\boxed{\int u'v\mathrm{d}x = uv - \int uv'\mathrm{d}x \quad \text{on} \quad ]a,b[.}$$

## 1.6.5 Substitution

**Basic idea:** We wish to transform the integral $\int_a^b f(x)\mathrm{d}x$ by making the substitution

$$\boxed{x = x(t).}$$

Using the Leibniz differential calculus, the formal rule

$$\mathrm{d}x = \frac{\mathrm{d}x}{\mathrm{d}t}\mathrm{d}t$$

yields the formula

$$\boxed{\int_a^b f(x)\mathrm{d}x = \int_\alpha^\beta f(x(t))\frac{\mathrm{d}x}{\mathrm{d}t}(t)\mathrm{d}t,} \tag{1.109}$$

which can be rigorously justified (see Figure 1.57).

**Theorem:** The formula (1.109) is valid under the following assumptions:

(a) The function $f : [a, b] \longrightarrow \mathbb{C}$ is bounded and almost everywhere continuous.

(b) The $C^1$-function $x : [\alpha, \beta] \longrightarrow \mathbb{R}$ satisfies the condition[60]

$$\boxed{x'(t) > 0 \text{ for all } t \in ]\alpha, \beta[} \tag{1.110}$$

and $x(\alpha) = a, \ x(\beta) = b$.

The important condition (1.110) guarantees that the function $x = x(t)$ is strictly increasing on $[\alpha, \beta]$, and hence the unique invertibility $t = t(x)$ for the transformation of variables. Without assuming (1.110), the formula (1.109) leads to completely wrong results.

*Example 1:* To calculate the integral

$$A = \int_a^b \mathrm{e}^{2x}\mathrm{d}x$$

we set $t = 2x$. This gives

$$x = \frac{t}{2}, \qquad \frac{\mathrm{d}x}{\mathrm{d}t} = \frac{1}{2}.$$

For $x = a, b$ this implies $t = 2a, 2b$, hence $\alpha = 2a$ and $\beta = 2b$. From (1.109) it then follows that

$$A = \int_\alpha^\beta \frac{1}{2}\mathrm{e}^t\mathrm{d}t = \frac{1}{2}\mathrm{e}^t\Big|_\alpha^\beta = \frac{1}{2}\mathrm{e}^{2x}\Big|_a^b = \frac{\mathrm{e}^{2b} - \mathrm{e}^{2a}}{2}.$$

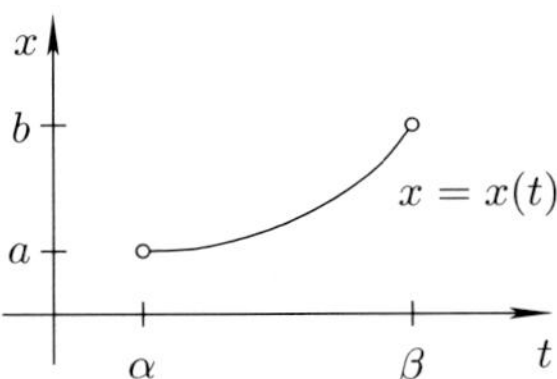


*Figure 1.57.*

**Substitution in indefinite integrals:** According to the Leibniz notation the formal rule for substitution in indefinite integrals is

$$\boxed{\int f(x)\mathrm{d}x = \int f(x(t))\frac{\mathrm{d}x}{\mathrm{d}t}\mathrm{d}t.} \tag{1.111}$$

One has to consider two cases:

(i) During the calculation, no inverse function is required.

[60]If $x'(t) < 0$ for all $t \in ]\alpha, \beta[$, then one must pass from $x(t)$ to $-x(t)$.

(ii) One requires an inverse function during the calculation.

In the non-critical case (i), one can always use (1.111). In the second case (ii) however, one can only use intervals on which the inverse function to $x = x(t)$ exists.

In case one is not sure, one should always carry out the check $F'(x) = f(x)$ *after* the calculation $\int f(x)\mathrm{d}x = F(x)$.

*Example 2:* To calculate the integral

$$A = \int \mathrm{e}^{x^2} 2x\mathrm{d}x$$

we set $t = x^2$. From $\dfrac{\mathrm{d}t}{\mathrm{d}x} = 2x$, we get

$$\boxed{2x\mathrm{d}x = \mathrm{d}t.}$$

This yields[61]

$$A = \int \mathrm{e}^t \mathrm{d}t = \mathrm{e}^t + C = \mathrm{e}^{x^2} + C \qquad \text{on } \mathbb{R}.$$

The check for this is $\left(\mathrm{e}^{x^2}\right)' = \mathrm{e}^{x^2} 2x$.

*Example 3:* To determine the integral

$$B = \int \frac{\mathrm{d}x}{\sqrt{1-x^2}},$$

we choose the substitution $x = \sin t$. Then $\dfrac{\mathrm{d}x}{\mathrm{d}t} = \cos t$, and we get

$$B = \int \frac{\cos t}{\sqrt{1-\sin^2 t}}\mathrm{d}t = \int \frac{\cos t}{\cos t}\mathrm{d}t = \int \mathrm{d}t = t + C = \arcsin x + C.$$

In these formal considerations we have used the inverse function $t = \arcsin x$, hence we must be careful in deducing on which interval the derived expression for $B$ is valid. We begin with the substitution

$$\boxed{x = \sin t, \qquad -\frac{\pi}{2} < t < \frac{\pi}{2}}$$

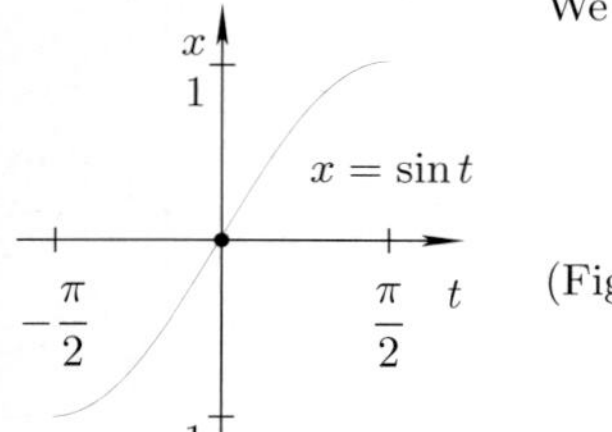

*Figure 1.58.*

(Figure 1.58). The corresponding inverse function is

$$t = \arcsin x, \qquad -1 < x < 1.$$

For all $t \in \left]-\frac{\pi}{2}, \frac{\pi}{2}\right[$ one has $\cos t > 0$, hence

$$\sqrt{1-\sin^2 t} = \sqrt{\cos^2 t} = \cos t.$$

---

[61] The following notation is particular helpful for the mnemonics:

$$\int \mathrm{e}^{x^2} 2x\mathrm{d}x = \int \mathrm{e}^{x^2} \mathrm{d}x^2 = \int \mathrm{e}^t \mathrm{d}t = \mathrm{e}^t + C = \mathrm{e}^{x^2} + C.$$

Once you are used to this you can use the even shorter version:

$$\int \mathrm{e}^{x^2} 2x\mathrm{d}x = \int \mathrm{e}^{x^2} \mathrm{d}x^2 = \mathrm{e}^{x^2} + C.$$

Thus, we get in sum

$$B=\int\frac{\mathrm{d}x}{\sqrt{1-x^2}}=\arcsin x+C,\quad -1<x<1.$$

**List of important substitutions:** This can be found in 0.9.4.

### 1.6.6 Integration on unbounded intervals

Integrals on unbounded intervals are calculated by first calculating integrals on bounded intervals and then passing to the limit where the length of the interval goes to infinity.[62]

Let $a\in\mathbb{R}$. Then one has:

$$\int_a^\infty f(x)\mathrm{d}x=\lim_{b\to+\infty}\int_a^b f(x)\mathrm{d}x, \tag{1.112}$$

$$\int_{-\infty}^a f(x)\mathrm{d}x=\lim_{b\to-\infty}\int_b^a f(x)\mathrm{d}x \tag{1.113}$$

and

$$\int_{-\infty}^\infty f(x)\mathrm{d}x=\int_{-\infty}^a f(x)\mathrm{d}x+\int_a^\infty f(x)\mathrm{d}x. \tag{1.114}$$

**Criterion for existence:** Assume that the function $f:J\longrightarrow\mathbb{C}$ is almost everywhere continuous and satisfies the estimate:

$$|f(x)|\le\frac{\text{const}}{(1+|x|)^\alpha}\qquad\text{for all}\ \ x\in J$$

and fixed $\alpha>1$. Then the following hold:

(i) If $J=[a,\infty[$ (resp. $J=]-\infty,a]$), then the finite limit in (1.112) (resp. in (1.113)) exists.

(ii) If $J=]-\infty,\infty[$, then the finite limits in (1.112) and (1.113) exist for all $a\in\mathbb{R}$, and the sum on the right-hand side of (1.114) is independent of the choice of $a$.

*Example:*

$$\int_0^\infty\frac{\mathrm{d}x}{1+x^2}=\lim_{b\to+\infty}\int_0^b\frac{\mathrm{d}x}{1+x^2}=\lim_{b\to+\infty}\arctan b=\frac{\pi}{2};$$

$$\int_{-\infty}^0\frac{\mathrm{d}x}{1+x^2}=\lim_{b\to-\infty}(-\arctan b)=\frac{\pi}{2};$$

$$\int_{-\infty}^\infty\frac{\mathrm{d}x}{1+x^2}=\int_{-\infty}^0\frac{\mathrm{d}x}{1+x^2}+\int_0^\infty\frac{\mathrm{d}x}{1+x^2}=\frac{\pi}{2}+\frac{\pi}{2}=\pi.$$

[62] In older literature the term 'improper integral' is used for this. This word is misleading, as also this kind of integral is a special case of the general notion of Lebesgue integral, and obeys also the general rules laid down by this theory. For the Lebesgue integral it is irrelevant whether the integrand or interval is unbounded.

### 1.6.7 Integration of unbounded functions

Let $-\infty < a < b < \infty$. The starting point is the relation between limits

$$\boxed{\int_a^b f(x)\mathrm{d}x = \lim_{\varepsilon\to+0} \int_{a-\varepsilon}^b f(x)\mathrm{d}x.} \tag{1.115}$$

**Criterion for existence:** Suppose the function $f : ]a, b] \longrightarrow \mathbb{C}$ is almost everywhere continuous and satisfies the estimate:

$$\boxed{|f(x)| \le \frac{\text{const}}{|x-a|^\alpha} \qquad \text{for all} \quad x \in ]a, b]}$$

for fixed $\alpha < 1$. Then the limit in (1.115) exists and is finite.

*Example:* Let $0 < \alpha < 1$. From

$$\int_\varepsilon^1 \frac{\mathrm{d}x}{x^\alpha} = \frac{x^{1-\alpha}}{1-\alpha}\bigg|_\varepsilon^1 = \frac{1}{1-\alpha}(1-\varepsilon^{1-\alpha})$$

it follows that

$$\int_0^1 \frac{\mathrm{d}x}{x^\alpha} = \lim_{\varepsilon\to+0} \int_\varepsilon^1 \frac{\mathrm{d}x}{x^\alpha} = \frac{1}{1-\alpha}.$$

In a similar manner one treats the case

$$\boxed{\int_a^b f(x)\mathrm{d}x = \lim_{\varepsilon\to+0} \int_a^{b-\varepsilon} f(x)\mathrm{d}x.} \tag{1.116}$$

**Criterion for existence:** Suppose the function $f : [a, b[ \longrightarrow \mathbb{C}$ is almost everywhere continuous and satisfies the estimate:

$$\boxed{|f(x)| \le \frac{\text{const}}{|x-b|^\alpha} \qquad \text{for all} \ \ x \in [a, b[}$$

and fixed $\alpha < 1$. Then the limit (1.116) exists and is finite.

### 1.6.8 The Cauchy principal value

Let $-\infty < a < c < b < \infty$. We define the *Cauchy principal value* by the formula

$$PV \int_a^b \frac{\mathrm{d}x}{x-c} = \lim_{\varepsilon\to+0} \left( \int_a^{c-\varepsilon} \frac{\mathrm{d}x}{x-c} + \int_{c+\varepsilon}^b \frac{\mathrm{d}x}{x-c} \right). \tag{1.117}$$

Let $\varepsilon > 0$ be sufficiently small. Because of the relations

$$\int_a^{c-\varepsilon} \frac{\mathrm{d}x}{x-c} = \ln|x-c|\bigg|_a^{c-\varepsilon} = \ln\varepsilon - \ln(c-a),$$

$$\int_{c+\varepsilon}^{b} \frac{\mathrm{d}x}{x-c} = \ln|x-c|\Big|_{c+\varepsilon}^{b} = \ln(b-c) - \ln\varepsilon,$$

we obtain

$$PV\int_{a}^{b} \frac{\mathrm{d}x}{x-c} = \ln(b-c) - \ln(c-a) = \ln\frac{b-c}{c-a}.$$

For $\varepsilon \to +0$ we have $\ln\varepsilon \to -\infty$, and hence because of the special choice of limits in the integrals of (1.117), the dangerous terms in $\ln\varepsilon$ cancel each other.

The integral $\int_a^b \frac{\mathrm{d}x}{x-c}$ does not exist in either the classical nor the Lebesgue sense. Therefore, the Cauchy principal value is a genuine extension of the notion of (Lebesgue) integral.

### 1.6.9 Application to arc length

**Arc length of a plane curve:** The *arc length* of a curve

$$x = x(t), \quad y = y(t), \quad a \le t \le b, \tag{1.118}$$

is defined to be the value of the integral

$$\boxed{s := \int_a^b \sqrt{x'(t)^2 + y'(t)^2}\,\mathrm{d}t.} \tag{1.119}$$

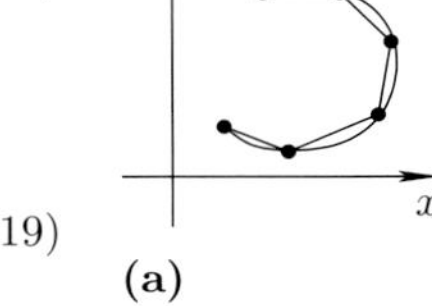


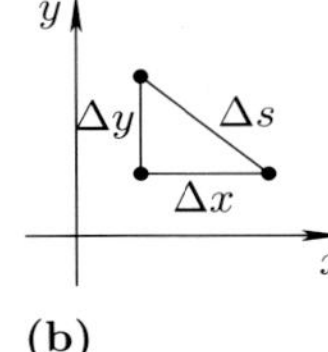


*Figure 1.59. Arc length.*

**Standard motivation:** Following the example of Archimedes (287–212 BC), we approximate the curve by an open polygon (Figure 1.59(a)).

The *theorem of Pythagoras* yields for the length of one secant of this polygon

$$(\Delta s)^2 = (\Delta x)^2 + (\Delta y)^2$$

(Figure 1.59(b)). From this it follows that

$$\frac{\Delta s}{\Delta t} = \sqrt{\left(\frac{\Delta x}{\Delta t}\right)^2 + \left(\frac{\Delta y}{\Delta t}\right)^2}. \tag{1.120}$$

The length of the secant is approximately equal to

$$s = \sum \Delta s = \sum \frac{\Delta s}{\Delta t}\Delta t. \tag{1.121}$$

Passing to the limit in which we let the length of the individual secants go to zero, the integral expression (1.119) is a continuous analog of (1.121).

**Refined motivation:** We suppose that the curve has an arc length and let $s(\tau)$ denote the arc length between the points on the curve defined by $t = a$ and $t = \tau$ (Figure 1.60(b) with $s(\tau) = m(\tau)$). From (1.120) the following differential equation follows for $\Delta t \to 0$:

$$\begin{aligned} &s'(\tau) = \sqrt{x'(\tau)^2 + y'(\tau)^2}, \qquad a \le \tau \le b, \\ &s(a) = 0, \end{aligned} \tag{1.122}$$

which has the unique solution

$$s(\tau) = \int_a^\tau \sqrt{x'(t)^2 + y'(t)^2}\, \mathrm{d}t$$

according to (1.107).

*Example:* For the arc length of the unit circle given by the curve

$$x = \cos t, \qquad y = \sin t, \qquad 0 \le t \le 2\pi,$$

we get

$$s = \int_0^{2\pi} \sqrt{x'(t)^2 + y'(t)^2}\, \mathrm{d}t = \int_0^{2\pi} \sqrt{\sin^2 t + \cos^2 t}\, \mathrm{d}t = \int_0^{2\pi} \mathrm{d}t = 2\pi.$$

### 1.6.10 A standard argument from physics

**The mass of a curve:** Let $\varrho = \varrho(s)$ be the density of mass of the curve (1.118) per unit of length. According to definition, the mass $m(\sigma)$ of part of this curve of the length $\sigma$ is given by

$$\boxed{m(\sigma) = \int_0^\sigma \varrho(s)\mathrm{d}s.} \tag{1.123}$$

If we relate this expression to the parameter $t$ of the curve, we get for the mass of the piece of curve from the points corresponding to the parameter values $t = a$ and $t = \tau$ the formula

$$m(s(\tau)) = \int_0^\tau \varrho(s(t))\frac{\mathrm{d}s}{\mathrm{d}t}\mathrm{d}t.$$

This yields

$$\boxed{m(s(\tau)) = \int_0^\tau \varrho(s(t))\sqrt{x'(t)^2 + y'(t)^2}\, \mathrm{d}t.} \tag{1.124}$$

**Standard motivation:** We subdivide the curve into small pieces with the mass $\Delta m$ and the arc length $\Delta s$ (Figure 1.60(a)). Then we have approximately

$$m = \sum \Delta m = \sum \frac{\Delta m}{\Delta s}\Delta s = \sum \varrho \Delta s = \sum \varrho \frac{\Delta s}{\Delta t}\Delta t. \tag{1.125}$$

If we let the pieces get smaller and smaller, the the formula (1.124) for $\Delta t \to 0$ is a continuous analog of (1.125).

This kind of considerations have been used in physics ever since the times of Newton, as a way to motivate formulas for physical quantities, which are defined in terms of integral expressions.

**Refined motivation:** We start with the mass function $m = m(s)$ and assume that the integral representation (1.123) with a continuous function $\varrho$ is valid. Differentiation of (1.123) at the point $\sigma = s$ gives

$$m'(s) = \varrho(s)$$

(cf. (1.107)). The function $\varrho$ is thus the derivative of the mass with respect to the arc length and is therefore called the *length density*. The formula (1.124) follows then from (1.123) with the help of the rule for substitution of integrals.

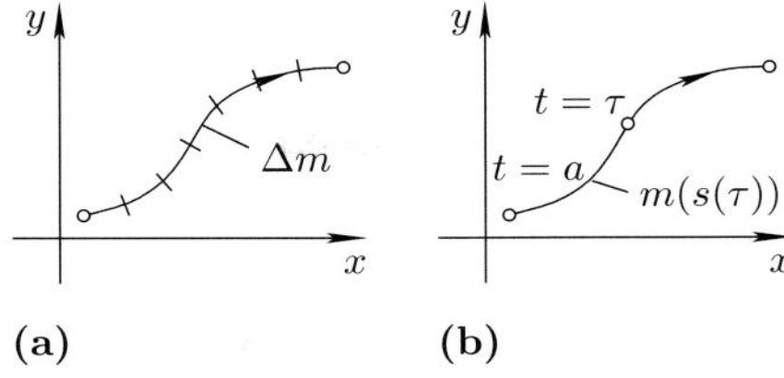


*Figure 1.60.*

## 1.7 Integration of functions of several real variables

The calculation of integrals is concerned with limits of finite sums, which occur for example in the computation of volumes, areas, length of curves, masses, charges, center of masses, moments of inertia or probabilities. In general one has

| differentiation = linearization of functions (or maps), integration = limits of sums of values of functions. |
|---|

The most important results in the theory of integration are sketched in the following keywords:

(i) principle of Cavalieri (the theorem of Fubini),

(ii) rule of substitution,

(iii) fundamental theorem of calculus (theorem of Gauss–Stokes),

(iv) integration by parts (a special case of (iii)).

The principle (i) allows the computation of iterated integrals (integrals of functions of several variables) to be reduced to that of one-dimensional integrals.

In older literature one uses in addition to integrals for volumes a series of further notions of integrals: curve integrals of the first and second kind, surface integrals of the first and second kind, etc. In passing to higher dimensions $n = 4, 5, \ldots$, which is necessary for example in the theory of general relativity or statistical mechanics, the situation gets even more confusing. Notations and notions of this kind have been poorly chosen and completely conceal the following, very simple principle:

| Integration over an area of integration $M$ of arbitrary dimension (domains, curves, surfaces, etc.) corresponds to the integration $\int_M \omega$ of differential forms $\omega$. |
|---|

If one takes this point of view, then there are only a few rules which one needs to remember, in order to get all the important rules in a form very convenient for mnemonics in the context of Cartan differential calculus.

### 1.7.1 Basic ideas

The following heuristical considerations will be rigorously justified later in 1.7.2.

**The mass of a rectangle:** Let $-\infty < a < b < \infty$ and $-\infty < c < d < \infty$. We consider the rectangle $R := \{(x,y) : a \le x \le b,\ c \le y \le d\}$, which is assumed to be covered by a mass density $\varrho$ (Figure 1.61(a)). To calculate the mass from $R$, we set

$$\Delta x := \frac{b-a}{n}, \qquad \Delta y := \frac{d-c}{n}, \qquad n = 1, 2, \ldots$$

and $x_j := a + j\Delta x$, $y_k := c + k\Delta y$ with $j, k = 0, \ldots, n$. We subdivide the rectangle $R$ in small sub-rectangles with the upper right corner $(x_j, y_k)$ and length of sides $\Delta x$, $\Delta y$. The mass of such a sub-rectangle is approximately equal to

$$\Delta m = \varrho(x_j, y_k)\Delta x \Delta y.$$

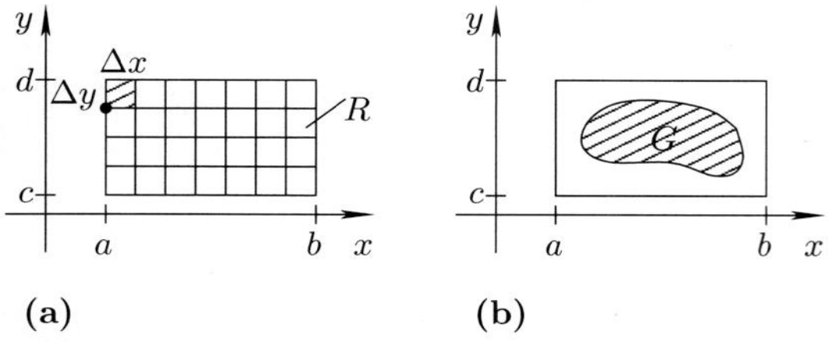

*Figure 1.61. Calculation of volumes.*

Therefore, it makes sense to defined $R$ by the limit relation

$$\boxed{\int_R \varrho(x,y)\mathrm{d}x\mathrm{d}y := \lim_{n\to\infty} \sum_{j,k=1}^{n} \varrho(x_j, y_k)\Delta x \Delta y.} \tag{1.126}$$

**Iterated integration (theorem of Fubini):** One can reduce the calculation of an integral over $R$ to that of one-dimensional integrals by means of the formula

$$\boxed{\int_R \varrho(x,y)\mathrm{d}x\mathrm{d}y = \int_c^d \left(\int_a^b \varrho(x,y)\mathrm{d}x\right)\mathrm{d}y = \int_a^b \left(\int_c^d \varrho(x,y)\mathrm{d}y\right)\mathrm{d}x,} \tag{1.127}$$

which is of great practical value.[63]

*Example 1:* $\displaystyle\int_R \mathrm{d}x\mathrm{d}y = \int_c^d \left(\int_a^b \mathrm{d}x\right)\mathrm{d}y = \int_c^d (b-a)\mathrm{d}y = (b-a)(d-c)$. This is the area of the rectangle $R$.

*Example 2:* From the relation

$$\int_a^b 2xy\mathrm{d}x = 2y\int_a^b x\mathrm{d}x = yx^2\big|_a^b = y(b^2 - a^2)$$

it follows that

$$\int_R 2xy\mathrm{d}x\mathrm{d}y = \int_c^d \left(\int_a^b 2xy\mathrm{d}x\right)\mathrm{d}y = \int_c^d y(b^2-a^2)\mathrm{d}y = \frac{1}{2}(d^2 - c^2)(b^2 - a^2).$$

[63]The formula (1.127) follows by taking the limit $n \longrightarrow \infty$ in the following commutation relation for finite sums:

$$\sum_{j,k=1}^{n} \varrho\Delta x\Delta y = \sum_{k=1}^{n}\left(\sum_{j=1}^{n} \varrho\Delta x\right)\Delta y = \sum_{j=1}^{n}\left(\sum_{k=1}^{n} \varrho\Delta y\right)\Delta x,$$

where $\varrho$ is written instead of (more correctly) $\varrho(x_j, y_k)$.

**The mass of a bounded domain:** To calculate the mass of a domain $D$ with the mass density $\varrho$, we choose a rectangle $R$, which contains $D$ and set

$$\boxed{\int_D \varrho(x,y)\mathrm{d}x\mathrm{d}y := \int_R \varrho_*(x,y)\mathrm{d}x\mathrm{d}y} \tag{1.128}$$

with

$$\varrho_*(x,y) := \begin{cases} \varrho(x,y) & \text{for } (x,y)\in D \\ 0 & \text{outside of } D.^{64} \end{cases}$$

These considerations can be generalized in a completely analogous manner to higher dimensions. Instead of rectangles one has in that case $n$-cubes, where $n$ is the dimension of the domain (cf. Figure 1.61(b)).

**The principle of Cavalieri:** We consider the situation illustrated in Figure 1.62. We have

$$\int_a^b \varrho_*(x,y)\mathrm{d}x = \int_{\alpha(y)}^{\beta(y)} \varrho_*(x,y)\mathrm{d}x = \int_{\alpha(y)}^{\beta(y)} \varrho(x,y)\mathrm{d}x.$$

Note that $\varrho_*(x,y)$ coincides with $\varrho(x,y)$ on the interval $[\alpha(y),\beta(y)]$ for fixed $y$, and it vanishes outside of this interval. From (1.127) and (1.128) it consequently follows that

$$\boxed{\int_D \varrho(x,y)\mathrm{d}x\mathrm{d}y = \int_c^d \left(\int_{\alpha(y)}^{\beta(y)} \varrho(x,y)\mathrm{d}x\right)\mathrm{d}y.}$$

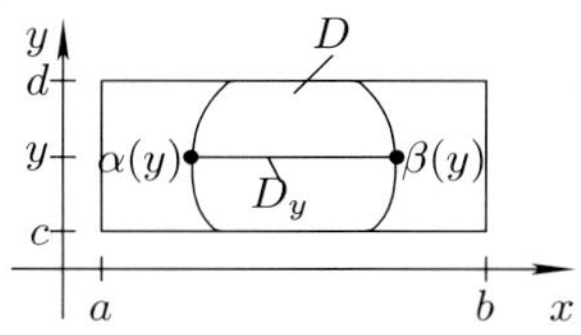


*Figure 1.62. A y-section.*

This formula can also be written in the abbreviated form

$$\boxed{\int_D \varrho(x,y)\mathrm{d}x\mathrm{d}y = \int_c^d \left(\int_{D_y} \varrho(x,y)\mathrm{d}x\right)\mathrm{d}y} \tag{1.129}$$

with the so-called $y$-section of $D$:

$$D_y = \{x\in\mathbb{R} \mid (x,y)\in D\}.$$

The formula (1.129) does not depend on the special form of the domain[65] $D$, shown in Figure 1.62 (see also Figure 1.63(a)).

[64] If $\varrho$ also takes negative values, then one may view $\varrho$ as a surface charge density, and $\int_D \varrho(x,y)\mathrm{d}x\mathrm{d}y$ is interpreted as the total charge in the domain $D$.

For $\varrho\equiv 1$ the integral $\int_D \mathrm{d}x\mathrm{d}y$ is just the area of the domain $D$.

[65] If we introduce the $x$-section of $D$ as $D_x := \{y\in\mathbb{R} \mid (x,y)\in D\}$, then similarly to (1.129) one has the formula

$$\int_D \varrho(x,y)\mathrm{d}x\mathrm{d}y = \int_a^b \left(\int_{D_x} \varrho(x,y)\mathrm{d}y\right)\mathrm{d}x$$

(cf. Figure 1.63(b)).

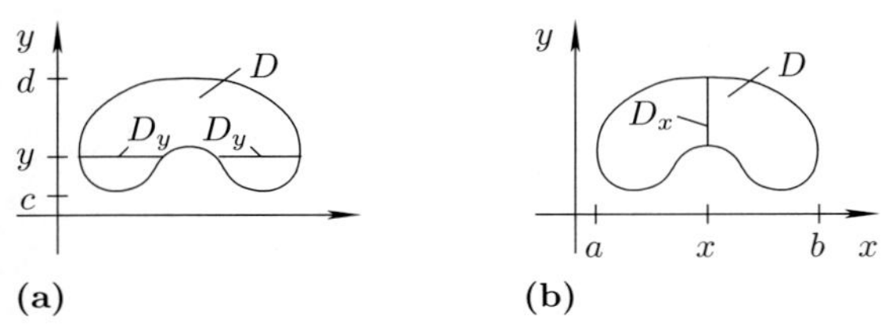


*Figure 1.63. y and x-sections of a domain D.*

The equation (1.129) can also be generalized to higher dimensions and corresponds to the general principle of the theory of integration, which was originally discovered in its easiest form before Newton and Leibniz by the student of Galileo, Bonaventura Cavalieri (1598–1647) in his main work *Geometria indivisibilius continuorum*, which appeared in 1653.

*Example 3* (volume of a cone over a circle): Let $D$ be a cone over a circle of radius $R$ and height $h$. For the volume of $D$ one has the formula

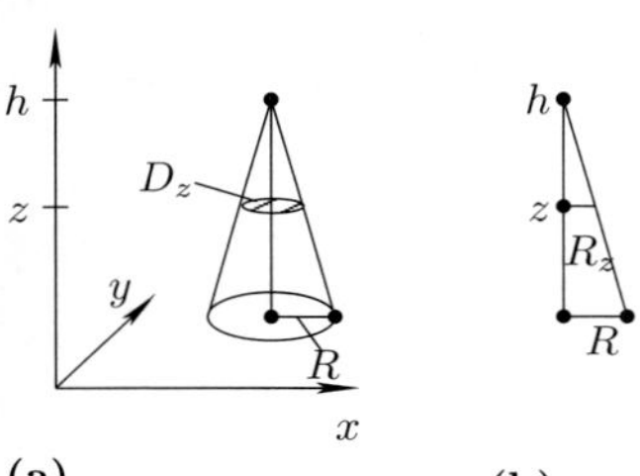


*Figure 1.64.*

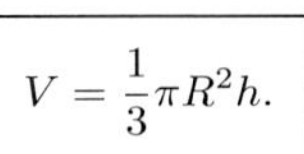

$$\boxed{V = \frac{1}{3}\pi R^2 h.}$$

To derive this formula, we use the principle of Cavalieri:

$$V = \int_D \mathrm{d}x\mathrm{d}y\mathrm{d}z = \int_0^h \left( \int_{D_z} \mathrm{d}x\mathrm{d}y \right) \mathrm{d}z.$$

The $z$-section $D_z$ is a disc of radius $R_z$ (Figure 1.64(a)). Hence one has

$$\int_{D_z} \mathrm{d}x\mathrm{d}y = \text{ surface area } A \text{ of a disc of radius } R_z, \text{ which is } A = \pi R_z^2$$

(cf. Example 4). From Figure 1.64(b) we see that

$$\frac{R_z}{R} = \frac{h-z}{h}.$$

Hence we get

$$V = \int_0^h \frac{\pi R^2}{h^2}(h-z)^2 \mathrm{d}z = -\frac{\pi R^2}{3h^2}(h-z)^3 \Big|_0^h = \frac{1}{3}\pi R^2 h.$$

**The substitution rule and Cartan differential calculus:**

We consider a map

$$\boxed{x = x(u,v) \qquad y = y(u,v),} \tag{1.130}$$

which maps the domain $D'$ in the $(u,v)$ plane to the domain $D$ in the $(x,y)$ plane (Figure 1.65).

The proper transformation law for the integral $\int_D \varrho(x,y)\mathrm{d}x\mathrm{d}y$ is derived in the following formal manner from Cartan's differential calculus. For this we write

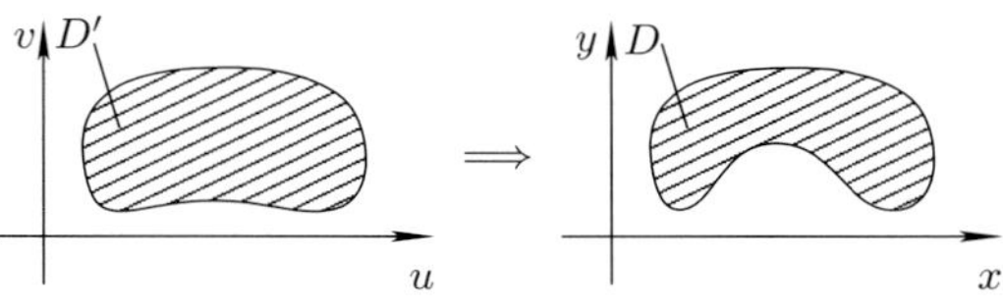


*Figure 1.65. The substitution rule.*

$$\int\limits_D \varrho(x,y)\mathrm{d}x\mathrm{d}y = \int\limits_D \omega$$

with

$$\omega = \varrho\mathrm{d}x \wedge \mathrm{d}y.$$

Using the transformation (1.130), we get

$$\omega = \varrho\frac{\partial(x,y)}{\partial(u,v)}\mathrm{d}u \wedge \mathrm{d}v$$

(cf. Example 10 in 1.5.10.4). In this manner we have derived the following fundamental formula for substitutions:

$$\boxed{\int\limits_D \varrho(x,y)\mathrm{d}x\mathrm{d}y = \int\limits_{D'} \varrho(x(u,v),y(u,v))\frac{\partial(x,y)}{\partial(u,v)}\mathrm{d}u\mathrm{d}v,} \tag{1.131}$$

which can be rigorously justified. Here one must assume that[66]

$$\frac{\partial(x,y)}{\partial(u,v)}(u,v) > 0 \qquad \text{for all } (u,v) \in D'.$$

**Application to polar coordinates:** The transformation

$$x = r\cos\varphi, \qquad y = r\sin\varphi, \qquad -\pi < \varphi \le \pi,$$

maps the usual Cartesian coordinates $(x,y)$ to polar coordinates $(r,\varphi)$ (Figure 1.66(a)). Then we have

$$\boxed{\int\limits_D \varrho(x,y)\mathrm{d}x\mathrm{d}y = \int\limits_{D'} \varrho r\,\mathrm{d}r\mathrm{d}\varphi.} \tag{1.132}$$

This follows from (1.131) with [67]

$$\frac{\partial(x,y)}{\partial(r,\varphi)} = \begin{vmatrix} x_r & x_\varphi \\ y_r & y_\varphi \end{vmatrix} = \begin{vmatrix} \cos\varphi & -r\sin\varphi \\ \sin\varphi & r\cos\varphi \end{vmatrix} = r(\cos^2\varphi + \sin^2\varphi) = r.$$

[66] It is allowed that $\frac{\partial(x,y)}{\partial(u,v)}$ vanishes at finitely many points.

[67] An intuitive motivation for (1.132) is given by dividing the domain into small pieces $\Delta F = r\Delta r\Delta\varphi$ and carrying out the limit $\Delta F \to 0$ in the sum (cf. Figure 1.66(b)):

$$\sum \varrho\Delta F = \sum \varrho r\Delta r\Delta\varphi.$$

*Example 4* (area inscribed in a circle): Let $D$ be the domain bounded by a circle of radius $R$. For the surface area $A$ of $D$ we get (Figure 1.67):

$$A = \int_D \mathrm{d}x\mathrm{d}y = \int_{D'} r\ \mathrm{d}r\mathrm{d}\varphi = \int_{r=0}^{R} \left( \int_{\varphi=-\pi}^{\pi} r\ \mathrm{d}\varphi \right) \mathrm{d}r = 2\pi \int_0^R r\ \mathrm{d}r = \pi R^2.$$

**The fundamental theorem of calculus and Cartan's differential calculus:** In the one-dimensional case the fundamental theorem of Newton and Leibniz is:

$$\int_a^b F'(x)\mathrm{d}x = F\Big|_a^b.$$

We write this formula in the form

$$\boxed{\int_D \mathrm{d}\omega = \int_{\partial D} \omega} \qquad (1.133)$$

with $\omega = F$ and $D = ]a, b[$. Note that $\mathrm{d}\omega = \mathrm{d}F = F'(x)\mathrm{d}x$.

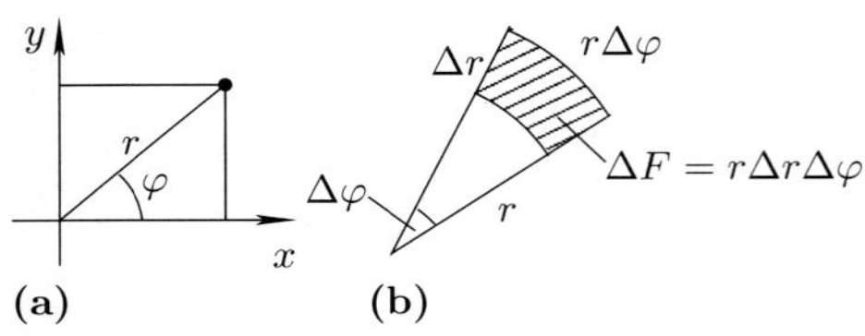


*Figure 1.66. Polar coordinates.*

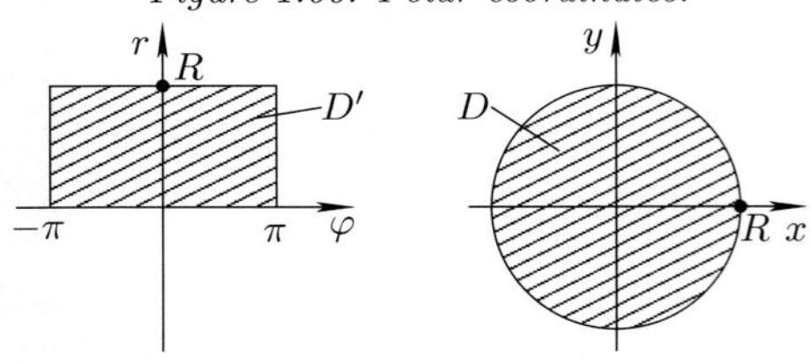


*Figure 1.67. The area inscribed by a circle.*

The beautiful elegance of the differential calculus of Cartan can be seen in the fact that (1.133) holds for domains, curves and surfaces of arbitrary dimensions (cf. 1.7.6).

Since (1.133) contains the classical theorems of Gauss and Stokes from field theory (vector analysis) as special cases, (1.133) is referred to as the *theorem of Gauss–Stokes* (or more briefly just the general theorem of Stokes).

*Example 5* (Theorem of Gauss in the plane): Let $D$ be a plane domain with boundary $\partial D$, which has a parameterization given by

$$x = x(t), \qquad y = y(t), \qquad \alpha \le t \le \beta,$$

and which as in Figure 1.68 is positively oriented in the mathematical sense. We choose the one-form

$$\omega = a\mathrm{d}x + b\mathrm{d}y.$$

Then we have

$$\mathrm{d}\omega = (b_x - a_y)\mathrm{d}x \wedge \mathrm{d}y$$

(cf. Example 5 in 1.5.10.4). The formula (1.133) is then

$$\int_D (b_x - a_y)\mathrm{d}x \wedge \mathrm{d}y = \int_{\partial D} a\mathrm{d}x + b\mathrm{d}y. \qquad (1.134)$$

The differential calculus of Cartan shows us at the same time how these integrals are to be calculated. In the integral to the left in (1.134) one simply replaces $\mathrm{d}x \wedge \mathrm{d}y$ by $\mathrm{d}x\mathrm{d}y$ (this rule being valid for domains of arbitrary dimension). In the integral to the right in (1.134) we relate $\omega$ to the parameterization of $\partial D$. This gives

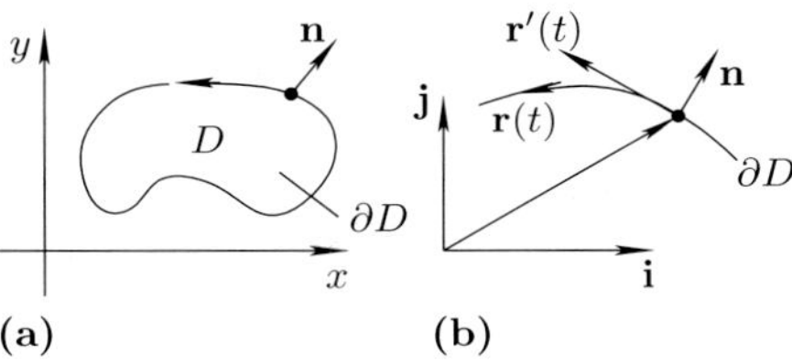


*Figure 1.68. An n-frame at a point on the boundary $\partial D$.*

$$\omega = \left(a\frac{\mathrm{d}x}{\mathrm{d}t} + b\frac{\mathrm{d}y}{\mathrm{d}t}\right)\mathrm{d}t$$

and

$$\int\limits_{\partial D} \omega = \int\limits_{\alpha}^{\beta} (ax' + by')\,\mathrm{d}t.$$

Thus (1.134) is written in classical notion,[68]

$$\boxed{\int\limits_{D} (b_x - a_y)\,\mathrm{d}x\mathrm{d}y = \int\limits_{\alpha}^{\beta} (ax' + by')\,\mathrm{d}t.} \tag{1.135}$$

This is the theorem of Gauss in the plane.

**Application to integration by parts:** The following relations hold:

$$\boxed{\begin{aligned} \int\limits_{D} u_x v\mathrm{d}x\mathrm{d}y &= \int\limits_{\partial D} uvn_x\mathrm{d}s - \int\limits_{D} uv_x\mathrm{d}x\mathrm{d}y, \\ \int\limits_{D} u_y v\mathrm{d}x\mathrm{d}y &= \int\limits_{\partial D} uvn_y\mathrm{d}s - \int\limits_{D} uv_y\mathrm{d}x\mathrm{d}y. \end{aligned}} \tag{1.136}$$

Here $\mathbf{n} = n_x\mathbf{i} + n_y\mathbf{j}$ is the outer normal vector at a boundary point and $s$ denotes the arc length of the (supposed sufficiently smooth) boundary curve. The formulas (1.136) generalize the one-dimensional formula

$$\int\limits_{\alpha}^{\beta} u'v\mathrm{d}x = uv\Big|_{\alpha}^{\beta} - \int\limits_{\alpha}^{\beta} uv'\mathrm{d}x.$$

We now wish to show that (1.136) follows from (1.135). For this, we view the parameter $t$ along the boundary curve $\partial D$ as the time. If a point moves along $\partial D$, then its equation of motion is

$$\mathbf{r}(t) = x(t)\mathbf{i} + y(t)\mathbf{j}$$

---

[68] Including also the arguments in the notation, this is more precisely the following formula:

$$\int\limits_{D} (b_x(x,y) - a_y(x,y))\,\mathrm{d}x\mathrm{d}y = \int\limits_{\alpha}^{\beta} \big(a(x(t),y(t))x'(t) + b(x(t),y(t))y'(t)\big)\,\mathrm{d}t.$$

with the velocity vector

$$\mathbf{r}'(t) = x'(t)\mathbf{i} + y'(t)\mathbf{j}.$$

The vector $\mathbf{N} = y'(t)\mathbf{i} - x'(t)\mathbf{j}$ is, because of the relation $\mathbf{r}'(t)\mathbf{N} = x'(t)y'(t) - y'(t)x'(t) = 0$, perpendicular to the tangent vector $\mathbf{r}'(t)$ and points to the outside of $D$. For the corresponding unit vector $\mathbf{n}$, we get the expression

$$\mathbf{n} = \frac{\mathbf{N}}{|\mathbf{N}|} = \frac{y'(t)\mathbf{i} - x'(t)\mathbf{j}}{\sqrt{x'(t)^2 + y'(t)^2}} = n_x\mathbf{i} + n_y\mathbf{j}$$

(cf. Figure 1.68(a)). Moreover one has $\dfrac{\mathrm{d}s}{\mathrm{d}t} = \sqrt{x'(t)^2 + y'(t)^2}$ (cf. 1.6.9). Hence

$$n_x \frac{\mathrm{d}s}{\mathrm{d}t} = y'(t).$$

If we set $b := uv$ and $a \equiv 0$ in (1.135), then we get

$$\int_D (uv)_x \mathrm{d}x\mathrm{d}y = \int_\alpha^\beta uvy'\mathrm{d}t = \int_\alpha^\beta uvn_x \frac{\mathrm{d}s}{\mathrm{d}t}\mathrm{d}t = \int_{\partial D} uvn_x \mathrm{d}s.$$

Because of the product rule $(uv)_x = u_x v + uv_x$ this is the first formula in (1.136). The second formula is obtained in exactly the same manner, this time by setting $a := uv$.

**Integration over unbounded domains:** As in the one-dimensional case the integral over an unbounded domain is derived by approximating the domain $D$ by bounded domains, and then passing to the limit in which the approximating domains grow in size beyond all bounds:

$$\boxed{\int_D f\mathrm{d}x\mathrm{d}y = \lim_{n\to\infty} \int_{D_n} f\mathrm{d}x\mathrm{d}y.}$$

The notations are chosen such that $D_1 \subseteq D_2 \subseteq \cdots$ and $D = \bigcup_{n=1}^\infty D_n$.

*Example 6:* We set $r := \sqrt{x^2 + y^2}$, and take the domain defined by

$$D := \{(x, y) \in \mathbb{R}^2 \,|\, 1 < r < \infty\},$$

the exterior of the unit circle. We approximate $D$ by annuli (Figure 1.69(a))

$$D_n = \{(x, y) \,|\, 1 < r < n\}.$$

For $\alpha > 2$ we have

$$\boxed{\int_D \frac{\mathrm{d}x\mathrm{d}y}{r^\alpha} = \lim_{n\to\infty} \int_{D_n} \frac{\mathrm{d}x\mathrm{d}y}{r^\alpha} = \frac{2\pi}{\alpha - 2}.}$$

Using polar coordinates, the above follows from the relation

$$\int_{D_n} \frac{\mathrm{d}x\mathrm{d}y}{r^\alpha} = \int_{r=1}^{n} \left( \int_{\varphi=0}^{2\pi} \frac{r\,\mathrm{d}r\mathrm{d}\varphi}{r^\alpha} \right) = 2\pi \int_1^n r^{1-\alpha}\mathrm{d}r = 2\pi \frac{r^{2-\alpha}}{2-\alpha}\bigg|_1^n = \frac{2\pi}{\alpha - 2}\left(1 - \frac{1}{n^{\alpha-2}}\right).$$

**Integration of unbounded functions:** Again, as in the one-dimensional case, one approximates the domain $D$ by domains on which the function *is* bounded.

*Example 7:* Let $D := \{(x, y) \in \mathbb{R}^2 \,|\, r \leq R\}$ be the interior of a circle of radius $R$. We approximate this ring by annuli (Figure 1.69(b))

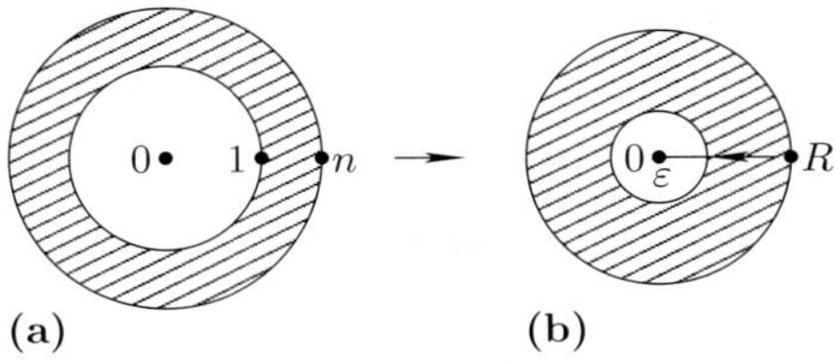


*Figure 1.69. Integrating unbounded functions (b) and over unbounded domains (a).*

$$D_\varepsilon := D \backslash U_\varepsilon(0),$$

where $U_\varepsilon(0) := \{(x, y) \in \mathbb{R}^2 \,|\, r < \varepsilon\}$ is the interior of a circle of radius $\varepsilon$. For $0 < \alpha < 2$ one has

$$\boxed{\int_D \frac{dxdy}{r^\alpha} = \lim_{\varepsilon \to 0} \int_{D_\varepsilon} \frac{dxdy}{r^\alpha} = \frac{2\pi R^{2-\alpha}}{2 - \alpha}.}$$

Again using polar coordinates, this follows from the relation

$$\int_{D_\varepsilon} \frac{dxdy}{r^\alpha} = \int_{r=\varepsilon}^{R} \left( \int_{\varphi=0}^{2\pi} \frac{r\,dr d\varphi}{r^\alpha} \right) = 2\pi \int_\varepsilon^R r^{1-\alpha} dr = \frac{2\pi}{2-\alpha}(R^{2-\alpha} - \varepsilon^{2-\alpha}).$$

### 1.7.2 Existence of the integral

Let $N$ be a natural number $\geq 1$. The points of $\mathbb{R}^N$ will be denoted $x = (x_1, \ldots, x_N)$. Furthermore, let $|x| := \sqrt{\left(\sum_{j=1}^{N} x_j^2\right)}$.

**The reduction principle:** We reduce the integration over subsets $D$ of $\mathbb{R}^N$ to the integration over all of $\mathbb{R}^N$ by the formula

$$\boxed{\int_D f(x) dx := \int_{\mathbb{R}^N} f_*(x) dx.}$$

Here we have set

$$f_*(x) := \begin{cases} f(x) & \text{on } D \\ 0 & \text{outside of } D. \end{cases}$$

The function $f_*$ is in general *non-continuous* at the boundary points of $D$ (jump to a zero value). Therefore we are naturally led to the integration of (sufficiently nice) non-continuous functions.

**Sets of vanishing $N$-dimensional measure:** A subset $D$ of $\mathbb{R}^N$ has by definition vanishing $N$-dimensional Lebesgue measure, when for every real number $\varepsilon > 0$ there is a set of at most countably many parallelepipeds $R_1, R_2, \ldots$ which cover $D$ and whose total measure is less than $\varepsilon$.[69]

---

[69] An $N$-dimensional *parallelepiped* is a set of the form

$$R := \{x \in \mathbb{R}^N \,|\, -\infty < a_j \leq x_j \leq b_j < \infty,\ j = 1, \ldots, N\}$$

The classical volume (measure) of $R$ is by definition given by the formula $\text{meas}(R) := (b_1 - a_1)(b_2 - a_2) \cdots (b_N - a_N)$.

*Example 1:* A set of finitely many or countably many points of $\mathbb{R}^N$ has vanishing $N$-dimensional Lebesgue measure.

*Example 2:* (i) Any reasonable (bounded or unbounded) curve in $\mathbb{R}^2$ has vanishing 2-dimensional Lebesgue measure.

(ii) Any reasonable (bounded or unbounded) surface in $\mathbb{R}^3$ has vanishing 3-dimensional Lebesgue measure.

(iii) Any reasonable (bounded or unbounded) subset of $\mathbb{R}^N$ with dimension $< N$ has vanishing $N$-dimensional Lebesgue measure.

**Almost everywhere valid properties:** A property holds *almost everywhere* on a subset $D$ of $\mathbb{R}^N$, if it is valid for all points of $D$ with possible exception of a set with vanishing $N$-dimensional Lebesgue measure.

*Example 3:* Almost all real numbers are irrational, since the exceptions to this, the set of rational numbers, is countable, hence has vanishing Lebesgue measure.

**Admissible domains of integration:** A subset $D$ of $\mathbb{R}^N$ is said to be *admissible*, if the boundary has vanishing $N$-dimensional Lebesgue measure.

**Admissible functions:** A real or complex valued function $f : D \subseteq \mathbb{R}^N \longrightarrow \mathbb{C}$ is said to be *admissible*, if $f$ is almost everywhere continuous on an admissible set $D$ and one of the following conditions is satisfied:

(i) Let $\alpha > N$. For all points $x \in D$ we have the following *estimate*:

$$\boxed{|f(x)| \leq \frac{\text{const}}{(1+|x|)^\alpha}.} \tag{1.137}$$

(ii) Let $0 < \beta < N$. There are at most finitely many points $p_1, \ldots, p_J$ and bounded neighborhoods $U(p_1), \ldots, U(p_J)$ in $\mathbb{R}^N$, such that for all $x \in U(p_j) \cap D$ with $x \neq p_j$ we have the following *estimate*:

$$\boxed{|f(x)| \leq \frac{\text{const}}{|x-p_j|^\beta}, \qquad j = 1, \ldots, J.} \tag{1.138}$$

Moreover the estimate (1.137) is satisfied for all points $x \in D$, which are outside of all the neighborhoods $U(p_j)$.

**Remarks:** (a) If a function $f$ is bounded on a bounded set $D$, that is $\sup\limits_{x \in D} |f(x)| < \infty$, then the condition (i) is automatically fulfilled.

(b) If the set $D$ is unbounded, then (i) means that $|f(x)| \to 0$ for $|x| \to \infty$ sufficient fast.

(c) In case (ii) the function $f$ may possibly have singularities in the points $p_1, \ldots, p_J$, where $|f(x)|$ may approach infinity for $x \to p_j$, but *not too fast.*[70]

**Theorem of existence:** For every admissible function $f : D \subseteq \mathbb{R}^N \longrightarrow \mathbb{C}$ the integral $\int_D f(x)\mathrm{d}x$ exists.

**Construction of the integral:** First let $N = 2$.

[70]The function $f$ need not be defined in the points $p_j$. For those points we set $f_*(p_j) := 0$, $j = 1, \ldots, J$.

(a) For a rectangle $R$ the limit[71]

$$\int_R f_*(x)\mathrm{d}x := \lim_{n\to\infty} \sum_{j,k=1}^{n} f_*(x_{1j}, x_{2k})\Delta x_1 \Delta x_2$$

exists.

(b) We choose a sequence $R_1 \subseteq R_2 \subseteq \cdots$ of rectangles with $\mathbb{R}^2 = \bigcup_{m=1}^{\infty} R_m$. Then the limit

$$\int_{\mathbb{R}^2} f_*(x)\mathrm{d}x := \lim_{m\to\infty} \int_{R_m} f_*(x)\mathrm{d}x$$

exists and is independent of the choice of rectangles.

(c) We set $\int_D f(x)\mathrm{d}x := \int_{\mathbb{R}^2} f_*(x)\mathrm{d}x$.

In the general case $N \geq 1$ one proceeds in the same manner, just replacing the rectangles $R_i$ by parallelepipeds.

For the empty set $D = \emptyset$, we define $\int_D f(x)\mathrm{d}x = 0$.

**Connection with the Lebesgue integral:** For admissible functions the value of the integral we have constructed coincides with the value of the general Lebesgue integral defined by the Lebesgue measure on $\mathbb{R}^N$.

**Standard examples:**

(i) The integral

$$\int_{\mathbb{R}^N} \mathrm{e}^{-|x|^2}\mathrm{d}x$$

exists. By Example 3 in 1.7.4, the value of this integral is $(\sqrt{\pi})^N$.

(ii) Let $D$ be an admissible bounded domain in $\mathbb{R}^3$. If the function $\varrho : D \longrightarrow \mathbb{R}$ is almost everywhere continuous and bounded, then the integral

$$U(p) = -\mathrm{G}\int_D \frac{\varrho(x)}{|x-p|}\mathrm{d}x$$

exists for all points $p \in \mathbb{R}^3$, and we set

$$F_j(p) := \mathrm{G}\int_D \frac{\varrho(x)(x_j - p_j)}{|x-p|^3}\mathrm{d}x, \qquad j = 1, 2, 3.$$

If we interpret $\varrho(x)$ as a mass density at the point $x$, then the function $U$ is the gravity potential, and the vector

$$\mathbf{F}(p) = F_1(p)\mathbf{i} + F_2(p)\mathbf{j} + F_3(p)\mathbf{k}$$

describes the gravitational pull (force) acting at the point $p$, which is generated by the mass distribution induced by $\varrho$. Moreover,

$$\mathbf{F}(p) = -\mathbf{grad}\, U(p) \qquad \text{for all } p \in \mathbb{R}^3.$$

The gravitational constant is denoted by G.

[71] We are using the notations introduced in (1.126).

**Measure of a set:** If $D$ is a bounded subset of $\mathbb{R}^N$, whose boundary $\partial D$ has vanishing $N$-dimensional Lebesgue measure, then the integral

$$\boxed{\operatorname{meas}(D) := \int_D \mathrm{d}x} \tag{1.139}$$

exists and is by definition the measure of the set $D$.

### 1.7.3 Calculations with integrals

Let $D$ and $D_n$ denote admissible sets in $\mathbb{R}^N$, and assume the functions $f, g : D \subseteq \mathbb{R}^N \longrightarrow \mathbb{C}$ are admissible; let $\alpha, \beta \in \mathbb{C}$. Then one has the following properties for manipulations with the integral.

(i) *Linearity:*

$$\int_D (\alpha f(x) + \beta g(x))\mathrm{d}x = \alpha \int_D f(x)\mathrm{d}x + \beta \int_D g(x)\mathrm{d}x.$$

(ii) *Triangle inequality:*[72]

$$\boxed{\left| \int_D f(x)\mathrm{d}x \right| \leq \int_D |f(x)|\mathrm{d}x.}$$

(iii) *Invariance principle:* The integral $\int_D f(x)\mathrm{d}x$ does not change when $f$ is changed on a set of vanishing $N$-dimensional Lebesgue measure.

(iv) *Additivity with respect to domains of integration:* One has

$$\int_D f(x)\mathrm{d}x = \int_{D_1} f(x)\mathrm{d}x + \int_{D_2} f(x)\mathrm{d}x,$$

if $D$ is the disjoint union of two domains $D_1$ and $D_2$.

(v) *Monotony:* If $f$ and $g$ are real functions, then the inequality $f(x) \leq g(x)$ for all $x \in D$ implies an inequality

$$\boxed{\int_D f(x)\mathrm{d}x \leq \int_D g(x)\mathrm{d}x.}$$

**Mean value theorem in integration:** One has

$$\boxed{\int_D f(x)g(x)\mathrm{d}x = f(\xi) \int_D g(x)\mathrm{d}x}$$

[72]If $D$ and $f$ are bounded, then one has in addition

$$\int_D |f(x)|\mathrm{d}x \leq \operatorname{meas}(D) \cdot \sup_{x \in D} |f(x)|.$$

for appropriate $\xi \in \overline{D}$, provided the function $f : \overline{D} \longrightarrow \mathbb{R}$ is continuous on a compact, arc connected set $\overline{D}$, and the non-negative function $g : D \longrightarrow \mathbb{R}$ is admissible.

**Convergence with respect to the domain of integration:** One has

$$\boxed{\int_D f(x)\mathrm{d}x = \lim_{n\to\infty} \int_{D_n} f(x)\mathrm{d}x}$$

provided $D = \bigcup_{n=1}^{\infty} D_n$ with $D_1 \subseteq D_2 \subseteq \cdots$ and the function $f : D \longrightarrow \mathbb{C}$ is admissible.

**Convergence with respect to the integrand:** One has

$$\boxed{\lim_{n\to\infty} \int_D f_n(x)\mathrm{d}x = \int_D \lim_{n\to\infty} f_n(x)\mathrm{d}x,}$$

provided the following assumptions are satisfied:

(i) All functions $f_n : D \longrightarrow \mathbb{C}$ are almost everywhere continuous.

(ii) There is an admissible function $h : D \longrightarrow \mathbb{R}$ with

$$|f_n(x)| \leq h(x) \qquad \text{for almost all } x \in D \text{ and all } n.$$

(iii) The limit $f(x) := \lim_{n\to\infty} f_n(x)$ exists for almost all $x \in D$ and the limit function $f$ is in $D$ almost everywhere continuous.[73]

### 1.7.4 The principle of Cavalieri (iterated integration)

Let $\mathbb{R}^N = \mathbb{R}^K \times \mathbb{R}^M$, that is, $\mathbb{R}^N = \{(y,z) \,|\, y \in \mathbb{R}^K,\ z \in \mathbb{R}^M\}$.

**Theorem of Fubini:** If the function $f : \mathbb{R}^N \longrightarrow \mathbb{C}$ is admissible, then one has the relation:

$$\boxed{\int_{\mathbb{R}^N} f(y,z)\mathrm{d}y\mathrm{d}z = \int_{\mathbb{R}^M} \left( \int_{\mathbb{R}^K} f(y,z)\mathrm{d}y \right) \mathrm{d}z = \int_{\mathbb{R}^K} \left( \int_{\mathbb{R}^M} f(y,z)\mathrm{d}z \right) \mathrm{d}y.} \tag{1.140}$$

*Example 1:* For $N = 2$ and $K = M = 1$ we get

$$\int_{\mathbb{R}^2} f(y,z)\,\mathrm{d}y\mathrm{d}z = \int_{-\infty}^{\infty} \left( \int_{-\infty}^{\infty} f(y,z)\mathrm{d}y \right) \mathrm{d}z.$$

In case $N = 3$, this implies that for variables $x, y, z \in \mathbb{R}$ we have the formula

$$\int_{\mathbb{R}^3} f(x,y,z)\mathrm{d}x\mathrm{d}y\mathrm{d}z = \int_{\mathbb{R}^2} \left( \int_{\mathbb{R}} f(x,y,z)\mathrm{d}x \right) \mathrm{d}y\mathrm{d}z = \int_{-\infty}^{\infty} \left( \int_{-\infty}^{\infty} \left( \int_{-\infty}^{\infty} f(x,y,z)\mathrm{d}x \right) \mathrm{d}y \right) \mathrm{d}z.$$

In a similar manner, the calculation of an integral over $\mathbb{R}^N$ is reduced to the successive calculation of one-dimensional integrals.

[73]In the points where the limit does not exist, the value of $f$ can be prescribed arbitrarily.

In case $f$ can be written as a product $f(y,z)=a(y)b(z)$, one has

$$\int_{\mathbb{R}^2} a(y)b(z)\mathrm{d}y\mathrm{d}z = \int_{-\infty}^{\infty} a(y)\mathrm{d}y \int_{-\infty}^{\infty} b(z)\mathrm{d}z.$$

A similar formula is also valid in $\mathbb{R}^N$.

*Example 2* (Gaussian distribution): One has

$$A := \int_{-\infty}^{\infty} \mathrm{e}^{-x^2}\mathrm{d}x = \sqrt{\pi}.$$

*Proof:* We use an elegant, classical trick. Iterated integration yields

$$B := \int_{\mathbb{R}^2} \mathrm{e}^{-x^2-y^2}\mathrm{d}x\mathrm{d}y = \int_{\mathbb{R}^2} \mathrm{e}^{-x^2}\mathrm{e}^{-y^2}\mathrm{d}x\mathrm{d}y = \int_{-\infty}^{\infty} \mathrm{e}^{-x^2}\mathrm{d}x \int_{-\infty}^{\infty} \mathrm{e}^{-y^2}\mathrm{d}y = A^2.$$

On the other hand, using polar coordinates, we get

$$\begin{aligned} B &= \int_{r=0}^{\infty}\left(\int_0^{2\pi} \mathrm{e}^{-r^2} r\,\mathrm{d}\varphi\right)\mathrm{d}r = 2\pi\int_0^{\infty} \mathrm{e}^{-r^2} r\,\mathrm{d}r = \lim_{r\to R} 2\pi \int_0^R \mathrm{e}^{-r^2} r\,\mathrm{d}r \\ &= \lim_{R\to\infty} -\pi\mathrm{e}^{-r^2}\Big|_0^R = \lim_{R\to\infty} \pi\left(1-\mathrm{e}^{-R^2}\right) = \pi. \end{aligned}$$

□

*Example 3:* One has

$$\int_{\mathbb{R}^N} \mathrm{e}^{-|x|^2}\mathrm{d}x = (\sqrt{\pi})^N.$$

*Proof:* For $N=3$ we have $\mathrm{e}^{-|x|^2} = \mathrm{e}^{-u^2-v^2-w^2}$ and hence

$$\int_{\mathbb{R}^3} \mathrm{e}^{-u^2}\mathrm{e}^{-v^2}\mathrm{e}^{-w^2}\mathrm{d}u\mathrm{d}v\mathrm{d}w = \left(\int_{-\infty}^{\infty} \mathrm{e}^{-u^2}\mathrm{d}u\right)\left(\int_{-\infty}^{\infty} \mathrm{e}^{-v^2}\mathrm{d}v\right)\left(\int_{-\infty}^{\infty} \mathrm{e}^{-w^2}\mathrm{d}w\right) = (\sqrt{\pi})^3.$$

A completely analogous conclusion holds for arbitrary $N$.

**Principle of Cavalieri:** If the function $f : D \subseteq \mathbb{R}^N \longrightarrow \mathbb{C}$ is admissible, then

$$\boxed{\int_D f(y,z)\mathrm{d}y\mathrm{d}z = \int_{\mathbb{R}^M}\left(\int_{D_z} f_*(y,z)\mathrm{d}y\right)\mathrm{d}z.} \tag{1.141}$$

In this formula, the $z$-section $D_z$ of $D$ is as defined above,

$$D_z := \{y \in \mathbb{R}^K : (y,z) \in D\}.$$

This principle follows from (1.140), by replacing the function $f$ in (1.140) by its trivial extension

$$f_*(y,z) := \begin{cases} f(y,z) & \text{on } D, \\ 0 & \text{otherwise.} \end{cases}$$

Applications of this principle can be found in 1.7.1.

### 1.7.5 Substitution

**Theorem:** Let $D'$ and $D$ be open subsets of $\mathbb{R}^N$. Then one has

$$\boxed{\int\limits_D f(x)\mathrm{d}x = \int\limits_{D'} f(x(u))|\mathrm{det}x'(u)|\mathrm{d}u,} \tag{1.142}$$

provided the function $f : D \longrightarrow \mathbb{C}$ is admissible and the map defined by $x = x(u)$ is a $C^1$-diffeomorphism of $D'$ onto $D$.

If we set $x = (x_, \ldots, x_N)$ and $u = (u_1, \ldots, u_N)$, then the determinant $\det x'(u)$ is the Jacobian functional determinant, i.e.,

$$\det x'(u) = \frac{\partial(x_1, \ldots, x_N)}{\partial(u_1, \ldots, u_N)}$$

(cf. 1.5.2).

**Application to polar coordinates, cylinder coordinates, spherical coordinates:** See section 1.7.9.

**Application to differential forms:** For $\omega = f(x)\mathrm{d}x_1 \wedge \cdots \wedge \mathrm{d}x_N$ we define

$$\boxed{\int\limits_D \omega := \int\limits_D f(x)\mathrm{d}x.}$$

Here the symbol $\int_D f(x)\mathrm{d}x$ stands for $\int_D f(x_1, \ldots, x_N)\mathrm{d}x_1 \cdots \mathrm{d}x_N$.

**Transformation principle:** Let $x = x(u)$ be a function as in (1.142). Then the integral $\int_D \omega$ is unchanged when $\omega$ is transformed to the new coordinates $u$, provided this transformation preserves the orientation, i.e., one has $\det x'(u) > 0$ on $D'$.

*Proof:* One has the formula

$$\omega = f(x(u))\mathrm{det}\, x'(u)\mathrm{d}u_1 \wedge \mathrm{d}u_2 \wedge \ldots \wedge \mathrm{d}u_N$$

(cf. Example 11 in 1.5.10.4). Thus, the rule for substitutions (1.142) yields

$$\int\limits_D f(x)\mathrm{d}x_1 \wedge \ldots \wedge \mathrm{d}x_N = \int\limits_H f(x(u))\mathrm{det}\, x'(u)\mathrm{d}u_1 \wedge \ldots \wedge \mathrm{d}u_N. \qquad \square$$

### 1.7.6 The fundamental theorem of calculus (theorem of Gauss–Stokes)

*One can ask what the deepest mathematical theorem is which unquestionably has a clearly defined physical interpretation. For me the natural candidate for such a theorem is the Theorem of Stokes.*

*René Thom (born 1923)*

The basic formula of the general theorem of Stokes[74] is:

$$\boxed{\int_M \mathrm{d}\omega = \int_{\partial M} \omega.} \qquad (1.143)$$

This exceptionally elegant formula is a generalization of the fundamental theorem of calculus of Newton and Leibniz, which is

$$\int_a^b F'(x)\mathrm{d}x = F\big|_a^b,$$

to higher dimensions.

**Theorem of Stokes:** Let $M$ be a $n$-dimensional real oriented compact manifold ($n \geq 1$) with the coherently oriented boundary $\partial M$, and let $\omega$ be a $(n-1)$-form on $\partial M$ which is in the class $C^1$. Then the formula (1.143) holds.

**Comments:** The mathematical objects occurring in this theorem will be introduced in detail in [212]. For the time being, we recommend that the reader consider the fundamental formula (1.143) in an intuitive manner without worrying about the precise formulation.

(i) It is useful to think of $M$ as a bounded curve, a bounded $m$-dimensional surface ($m = 2, 3, \ldots$) or the closure of a bounded open set in $\mathbb{R}^N$.

(ii) *Parameter:* If writing the integrals in local coordinates, then these latter should be arbitrary, for both $M$ and $\partial M$.

(ii) *Decomposition principle:* If it is not possible to describe $M$ (resp. $\partial M$) be a single set of local coordinates (parameters), then one decomposes $M$ (resp. $\partial M$) into a disjoint union of *finitely many* parts, each of which does have such a global parameter. Then the sum of the integrals of the parts is the sought for integral over $M$.

Then one has the convenient fact:

| Cartan's differential calculus works by itself. |
|---|

One only has to take care that the local coordinates on $M$ near a boundary point $P$ is *synchronized* with the local coordinate one has on the boundary $\partial M$ near $P$ (this is the notion of coherently oriented boundary). This will be intuitively explained in the following examples-[75]

### 1.7.6.1 Applications of the classical integral theorem of Gauss

We consider the 2-form

$$\omega = a\mathrm{d}y \wedge \mathrm{d}z + b\mathrm{d}z \wedge \mathrm{d}x + c\mathrm{d}x \wedge \mathrm{d}y.$$

According to Example 8 in 1.5.10.4 we have

$$\mathrm{d}\omega = (a_x + b_y + c_z)\mathrm{d}x \wedge \mathrm{d}y \wedge \mathrm{d}z.$$

[74] One usually refers to the Theorem of Gauss–Stokes just as *Stokes' Theorem.*

[75] The general synchronization principle in arbitrary dimensions can be found in [212].

In order for $\int_{\partial M} \omega$ to be well-defined, $\partial M$ must be two-dimensional. Hence we assume that $M$ is the closure of a bounded open (non-empty) set in $\mathbb{R}^3$ with a sufficiently smooth boundary $\partial M$, which has a parameterization

$$x = x(u,v), \qquad y = y(u,v), \qquad z = z(u,v)$$

If we relate $\omega$ to the parameters $u$ and $v$, then we get

$$\omega = \left( a\frac{\partial(y,z)}{\partial(u,v)} + b\frac{\partial(z,x)}{\partial(u,v)} + c\frac{\partial(x,y)}{\partial(u,v)} \right) \mathrm{d}u \wedge \mathrm{d}v$$

(cf. Example 12 in 1.5.10.4). The formula $\int_M \mathrm{d}\omega = \int_{\partial M} \omega$ then yields the classical theorem of Gauss for three-dimensional domains $M$:

$$\int\limits_M (a_x + b_y + c_z)\,\mathrm{d}x\mathrm{d}y\mathrm{d}z = \int\limits_{\partial M} \left( a\frac{\partial(y,z)}{\partial(u,v)} + b\frac{\partial(z,x)}{\partial(u,v)} + c\frac{\partial(x,y)}{\partial(u,v)} \right) \mathrm{d}u\mathrm{d}v. \qquad (1.144)$$

**Transition to vector notation:** We introduce the coordinate vector $\mathbf{r} := x\mathbf{i} + y\mathbf{j} + z\mathbf{k}$. Then the expression

$$\mathbf{r}_u(u,v) = x_u(u,v)\mathbf{i} + y_u(u,v)\mathbf{j} + z_u(u,v)\mathbf{k}$$

is a tangent vector to the coordinate line $y = \text{const}$ through the point $P(u,v)$ on $\partial M$ (Figure 1.70(b)). Similarly, $\mathbf{r}_v(u,v)$ is a tangent vector to the coordinate line $u = \text{const}$ through the point $P(u,v)$. The equation of the *tangent plane* at the point $P(u,v)$ is then

$$\mathbf{r} = \mathbf{r}(u,v) + p\mathbf{r}_u(u,v) + q\mathbf{r}_v(u,v),$$

where $p$, $q$ are real parameters and $\mathbf{r}(u,v)$ is the coordinate vector of the point $P(u,v)$. In order for the tangent vectors $\mathbf{r}_u(u,v)$ and $\mathbf{r}_v(u,v)$ to span a plane, we must assume that $\mathbf{r}_u(u,v) \times \mathbf{r}_v(u,v) \neq 0$, i.e., that these two vectors are not parallel or anti-parallel. The unit vector

$$\mathbf{n} := \frac{\mathbf{r}_u(u,v) \times \mathbf{r}_v(u,v)}{|\mathbf{r}_u(u,v) \times \mathbf{r}_v(u,v)|} \qquad (1.145)$$

is perpendicular to the tangent plane at the point $P(u,v)$ and is therefore a normal vector. Coherent orientation of $M$ and $\partial M$ means here that this vector points to the *outside* (Figure 1.70(a)).

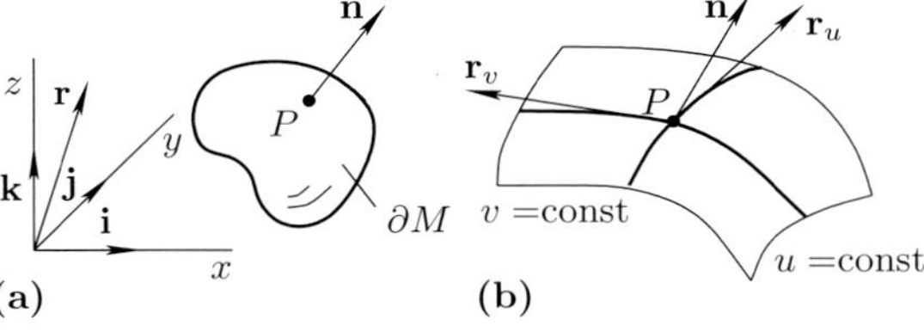


*Figure 1.70. Coordinates for the integral theorem of Gauss.*

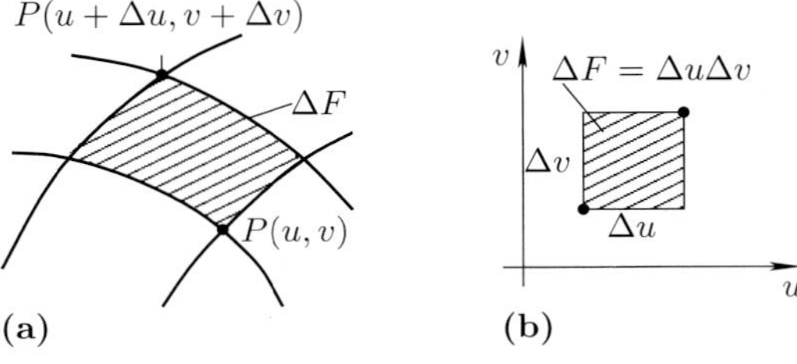


*Figure 1.71. Volume on a surface.*

If we introduce the vector field $\mathbf{J} := a\mathbf{i} + b\mathbf{j} + c\mathbf{k}$, then the theorem of Gauss for three-dimensional domains can be written in the classical form:

$$\boxed{\int\limits_M \operatorname{div} \mathbf{J} \mathrm{d}x = \int\limits_{\partial M} \mathbf{J}\mathbf{n}\, \mathrm{d}F.} \tag{1.146}$$

Here we have set

$$\mathrm{d}F = |\mathbf{r}_u \times \mathbf{r}_v| \mathrm{d}u\mathrm{d}v = \sqrt{\left(\frac{\partial(x,y)}{\partial(u,v)}\right)^2 + \left(\frac{\partial(y,z)}{\partial(u,v)}\right)^2 + \left(\frac{\partial(z,x)}{\partial(u,v)}\right)^2}\, \mathrm{d}u\mathrm{d}v.$$

Intuitively the surface element $\Delta F$ of $\partial M$ is approximately given by

$$\boxed{\Delta F = |\mathbf{r}_u(u,v) \times \mathbf{r}_v(u,v)| \Delta u \Delta v} \tag{1.147}$$

(Figure 1.71(a)).[76]

The physical interpretation of (1.146) can be found in 1.9.7. The integral $\int_{\partial M} g\mathrm{d}F$ is intuitively given by decomposing the surface $\partial M$ into small parts $\Delta F$ and refining the decomposition. To describe this we also use the abbreviated formula

$$\int\limits_{\partial M} g\mathrm{d}F = \lim_{\Delta F \to 0} \sum g\Delta F. \tag{1.148}$$

**Surfaces in three-dimensional space:** The formulas above for the tangent plane, the normal unit vector and the surface element are valid for arbitrary (sufficiently smooth) surfaces in $\mathbb{R}^3$.

#### 1.7.6.2 Applications to the classical integral theorem of Stokes

We consider the 1-form

$$\omega = a\mathrm{d}x + b\mathrm{d}y + c\mathrm{d}z.$$

According to Example 7 in 1.5.10.4 one has

$$\mathrm{d}\omega = (c_y - b_z)\mathrm{d}y \wedge \mathrm{d}z + (a_z - c_x)\mathrm{d}z \wedge \mathrm{d}x + (b_x - a_y)\mathrm{d}x \wedge \mathrm{d}y.$$

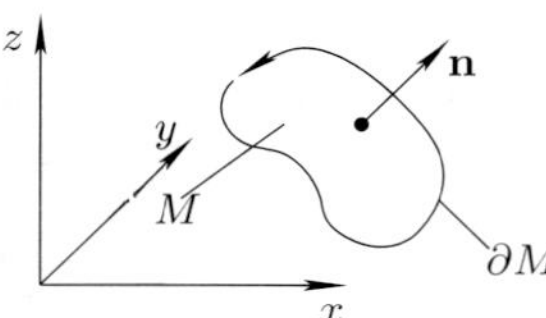


Figure 1.72.

In order for $\int_{\partial M} \omega$ to make sense, the boundary $\partial M$ must be one-dimensional, that is, it must be a curve, which bounds the surface $M$ (Figure 1.72). The surface $M$ has a parameterization

$$x = x(u,v), \qquad y = y(u,v), \qquad z = z(u,v),$$

and the parameterization of the boundary curve is:

$$x = x(t), \qquad y = y(t), \qquad z = z(t), \qquad \alpha \le t \le \beta.$$

As in Figure 1.72, a coherent orientation of $M$ and of $\partial M$ is given by the rule that the normal vector (1.145) together with the oriented curve $\partial M$ are oriented as the thumb and the first two fingers of the *right* hand (right-hand rule).

[76] In the special case of a plane, $x = u$, $y = v$ and $z = 0$, one has $\Delta F = \Delta u \Delta v$ (cf. Figure 1.71(b)).

If we relate $\omega$ (resp. $\mathrm{d}\omega$) to the corresponding parameters $t$ (resp. $(u,v)$), then we get

$$\begin{aligned}\omega &= \left(a\frac{\mathrm{d}x}{\mathrm{d}t} + b\frac{\mathrm{d}y}{\mathrm{d}t} + c\frac{\mathrm{d}z}{\mathrm{d}t}\right)\mathrm{d}t,\\ \mathrm{d}\omega &= \left((c_y - b_z)\frac{\partial(y,z)}{\partial(u,v)} + (a_z - c_x)\frac{\partial(z,x)}{\partial(u,v)} + (b_x - a_y)\frac{\partial(x,y)}{\partial(u,v)}\right)\mathrm{d}u\wedge\mathrm{d}v\end{aligned}$$

(cf. Example 12 in 1.5.10.4). The formula $\int_M \mathrm{d}\omega = \int_{\partial M}\omega$ then gives the theorem of Stokes for surfaces in $\mathbb{R}^3$ in its classical form:

$$\begin{aligned}&\int_M \left((c_y - b_z)\frac{\partial(y,z)}{\partial(u,v)} + (a_z - c_x)\frac{\partial(z,x)}{\partial(u,v)} + (b_x - a_y)\frac{\partial(x,y)}{\partial(u,v)}\right)\mathrm{d}u\mathrm{d}v\\ &= \int_{\partial M}(ax' + by' + cz')\mathrm{d}t.\end{aligned}$$

If we write $\mathbf{B} = a\mathbf{i} + b\mathbf{j} + c\mathbf{k}$, then we get

$$\boxed{\int_M (\mathbf{curl}\,\mathbf{B})\mathbf{n}\,\mathrm{d}F = \int_\alpha^\beta \mathbf{B}(\mathbf{r}(t))\mathbf{r}'(t)\,\mathrm{d}t.} \tag{1.149}$$

We will give a physical interpretation of this formula in 1.9.8.

#### 1.7.6.3 Applications to curve integrals

**The formula for the potential:** We consider a 0-form $\omega = U$ in $\mathbb{R}^3$ with the function $U = U(x,y,z)$. Then we have

$$\mathrm{d}U = U_x\mathrm{d}x + U_y\mathrm{d}y + U_z\mathrm{d}z.$$

We choose a curve $M$ with the parameterization

$$x = x(t), \qquad y = y(t), \qquad z = z(t), \qquad \alpha \le t \le \beta. \tag{1.150}$$

Transforming $U$ to the parameter $t$ yields

$$\mathrm{d}U = \left(U_x(P(t))\frac{\mathrm{d}x}{\mathrm{d}t} + U_y(P(t))\frac{\mathrm{d}y}{\mathrm{d}t} + U_z(P(t))\frac{\mathrm{d}z}{\mathrm{d}t}\right)\mathrm{d}t.$$

The theorem of Stokes gives us here the so-called potential formula:

$$\boxed{\int_M \mathrm{d}U = U(P) - U(Q).} \tag{1.151}$$

Here $Q$ is the starting point, $P$ is the end point of the curve $M$. Explicitly the potential formula is

$$\boxed{\int_\alpha^\beta (U_x x' + U_y y' + U_z z')\mathrm{d}t = U(P) - U(Q).}$$

Using the language of vector analysis, we have $\mathrm{d}U = \mathbf{grad}\, U\, \mathrm{d}\mathbf{r}$, and we get

$$\int_M \mathbf{grad}\, U\, \mathrm{d}\mathbf{r} = U(P) - U(Q)$$

and

$$\int_\alpha^\beta (\mathbf{grad}\, U)(\mathbf{r}(t))\mathbf{r}'(t)\mathrm{d}t = U(\mathbf{r}(\beta)) - U(\mathbf{r}(\alpha))$$

with $\mathbf{r}(t) = x(t)\mathbf{i} + y(t)\mathbf{j} + z(t)\mathbf{k}$. The physical interpretation of this formula can be found in 1.9.5.

**Integrals over 1-forms (curve integrals):** Let the 1-form

$$\omega = a\mathrm{d}x + b\mathrm{d}y + c\mathrm{d}z$$

be given and the curve $M$ with the parameterization (1.150). The integral

$$\int_M \omega = \int_\alpha^\beta (ax' + by' + cz')\mathrm{d}t$$

is called a *curve integral*.[77]

**Independence of the chosen path:** Let $D$ be a contractible[78] domain in $\mathbb{R}^3$, and let $M$ be a curve in $D$ with the $C^1$-parameterization (1.150). Furthermore let

$$\omega = a\mathrm{d}x + b\mathrm{d}y + c\mathrm{d}z$$

be a 1-form of class $C^2$, i.e., $a, b$ and $c$ are real $C^2$-functions on $D$. Then we have

(i) We assume that there exists a $C^1$-function $U : D \longrightarrow \mathbb{R}$ with

$$\boxed{\omega = \mathrm{d}U \qquad \text{on } D,} \tag{1.152}$$

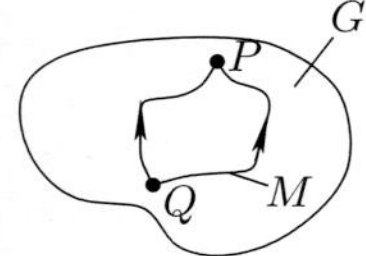


*Figure 1.73.*

i.e., the following hold:

$$a = U_x, \qquad b = U_y, \qquad c = U_z \qquad \text{on } D\,.$$

Then the integral $\int_M \omega$ is independent of the path, i.e., because of the potential formula $\int_M \mathrm{d}M = U(P) - U(Q)$ the integral only depends on the starting and end points of the curve $M$ (Figure 1.73).

(ii) The equation (1.152) has a unique $C^1$-solution $U$, if the integrability condition

$$\boxed{\mathrm{d}\omega = 0 \qquad \text{on } D}$$

[77] More precisely this formula is

$$\int_M \omega = \int_\alpha^\beta (a(P(t))x'(t) + b(P(t))y'(t) + c(P(t))z'(t))\mathrm{d}t$$

with $P(t) := (x(t), y(t), z(t))$.

[78] Intuitively this means that the domain $D$ can be continuously contracted to a point. The precise definition is given in 1.9.11.

is satisfied. According to 1.7.6.2 this is equivalent to the condition

$$c_y = b_z, \qquad a_z = c_x, \qquad b_x = c_y \qquad \text{on } D.$$

(iii) The equation (1.152) has a $C^1$-solution $U$, if and only if the integral $\int_M \omega$ is independent of the path of integration $M$ for every $C^1$-curve in $D$.

The statement (ii) is a special case of the Poincaré lemma (cf. 1.9.11). Moreover (iii) is a special case of the de Rham theorem. A deeper understanding of these results is possible in the context of differential topology (de Rham cohomology), cf. [212].

The physical interpretation of this result can be found in 1.9.5.

*Example:* We want to integrate the 1-form

$$\omega = x\mathrm{d}x + y\mathrm{d}y + z\mathrm{d}z$$

along the line $M: \ x = t, \ y = t, \ z = t$ with $0 \le t \le 1$. Then we have $\omega = tx'\mathrm{d}t + ty'\mathrm{d}t + tz'\,\mathrm{d}t = 3t\,\mathrm{d}t$, hence

$$\int_M \omega = \int_0^1 3t\,\mathrm{d}t = \frac{3}{2}t^2\Big|_0^1 = \frac{3}{2}.$$

Because of the relation $\mathrm{d}\omega = \mathrm{d}x \wedge \mathrm{d}x + \mathrm{d}y \wedge \mathrm{d}y + \mathrm{d}z \wedge \mathrm{d}z = 0$, this integral in $\mathbb{R}^3$ is independent of the path of integration. In fact, $\omega = \mathrm{d}U$ with $U = \frac{1}{2}(x^2 + y^2 + z^2)$. This yields

$$\int_M \omega = \int_M \mathrm{d}U = U(1,1,1) - U(0,0,0) = \frac{3}{2}.$$

**Properties of curve integrals:**

(i) *Addition of curves:*

$$\boxed{\int_A \omega + \int_B \omega = \int_{A+B} \omega.}$$

Here $A + B$ denotes the curve which arises when one first runs through $A$ and then through $B$ (Figure 1.74(a)).

(ii) *Reversing the orientation on curves:*

$$\boxed{\int_{-M} \omega = -\int_M \omega.}$$

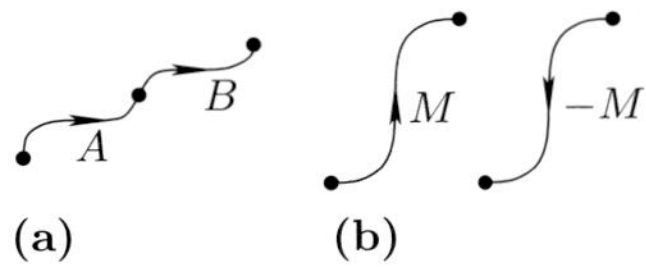


*Figure 1.74. Reversal of orientation.*

Here $-M$ denotes the curve, which is obtained from $M$ by reversing the orientation (Figure 1.74(b)).

All of the properties of contour integrals $\int_M \omega$ described in this section hold in a similar manner for curves $x_j = x_j(t), \ \alpha \le t \le \beta$ in $\mathbb{R}^N$, with $j = 1, \ldots, N$.

### 1.7.7 The Riemannian surface measure

According to Riemann (1826–1866) the knowledge of the arc length leads immediately to an expression for the surface area of a surface and the volume of a domain in curved coordinates.

Let $\mathscr{D}$ be a domain in $\mathbb{R}^m$. An $m$-dimensional surface is given in $\mathbb{R}^N$ by the parameterization

$$\boxed{x = x(u), \qquad u \in \mathscr{D},}$$

where $u = (u_1, \ldots, u_m)$ and $x = (x_1, \ldots, x_N)$ (cf. Figure 1.75 with $u_1 = u$, $u_2 = v$ and $x_1 = x$, $x_2 = y$, $x_3 = z$).

**Definition:** For a curve $x = x(t)$, $\alpha \leq t \leq \beta$ in $\mathbb{R}^N$, the *arc length* between the curve points with parameter values $t = \alpha$ and $t = \tau$ is given by the formula

$$\boxed{s(\tau) = \int\limits_\alpha^\tau \left( \sum_{j=1}^N x_j'(t)^2 \right)^{1/2} \mathrm{d}t.}$$

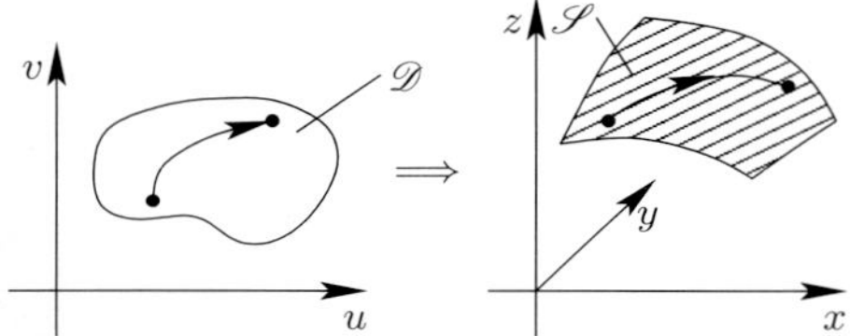


*Figure 1.75. Arc length in a domain.*

A motivation for this definition is given in 1.6.9.

**Theorem:** Every curve $u = u(t)$ in the parameter domain $\mathscr{D}$ corresponds to a curve

$$x = x(u(t))$$

on a surface element $\mathscr{S}$, whose arc length satisfies the differential equation:

$$\boxed{\left( \frac{\mathrm{d}s(t)}{\mathrm{d}t} \right)^2 = \sum_{j,k=1}^m g_{jk}(u(t)) \frac{\mathrm{d}u_j(t)}{\mathrm{d}t} \frac{\mathrm{d}u_k(t)}{\mathrm{d}t}} \tag{1.153}$$

One calls the terms

$$g_{jk}(u) := \sum_{n=1}^N \frac{\partial x_n(u)}{\partial u_j} \frac{\partial x_n(u)}{\partial u_k}$$

the components of the *metric tensor*. Instead of (1.153) one writes symbolically

$$\boxed{\mathrm{d}s^2 = g_{jk} \mathrm{d}u_j \mathrm{d}u_k.}$$

This corresponds to the approximation $(\Delta s)^2 = g_{jk} \Delta u_j \Delta u_k$.

*Proof:* We set $x_j(t) = x_j(u(t))$, we then have the relation

$$s'(t)^2 = \sum_{n=1}^N x_n'(t) x_n'(t),$$

and the chain rule yields $x_n'(t) = \sum\limits_{j=1}^m \frac{\partial x_n}{\partial u_j} \frac{\mathrm{d}u_j}{\mathrm{d}t}$. □

**Volume form:** We set $g := \det(g_{jk})$ and define the *volume form* $\mu$ of the surface $\mathscr{F}$ by:

$$\boxed{\mu := \sqrt{g}\, \mathrm{d}u_1 \wedge \ldots \wedge \mathrm{d}u_m.}$$

**Surface integral:** We define

$$\int_{\mathscr{F}} \varrho \mathrm{d}F := \int_{\mathscr{F}} \varrho\mu.$$

In classical notation this corresponds to the formula

$$\int_{\mathscr{F}} \varrho \mathrm{d}F = \int_{\mathscr{D}} \varrho\sqrt{g}\,\mathrm{d}u_1\mathrm{d}u_2\ldots\mathrm{d}u_m.$$

**Physical interpretation:** If we view $\varrho$ as a mass density (resp. charge density), then $\int_{\mathscr{F}} \varrho\mathrm{d}F$ is equal to the amount of mass (resp. charge) on $\mathscr{F}$. For $\varrho \equiv 1$, $\int_{\mathscr{F}} \varrho\mathrm{d}F$ is equal to the surface measure of $\mathscr{F}$.

**Application to surfaces in $\mathbb{R}^3$:** Let a surface $\mathscr{F}$ in $\mathbb{R}^3$ be given with parameterization

$$x = x(u,v), \qquad y = y(u,v), \qquad z = z(u,v), \qquad (u,v) \in \mathscr{D}$$

in Cartesian coordinates $x, y$ and $z$ (Figure 1.75). Then one has

$$\mathrm{d}s^2 = E\mathrm{d}u^2 + 2F\mathrm{d}u\mathrm{d}v + G\mathrm{d}v^2.$$

with

$$E := \mathbf{r}_u^2 = x_u^2 + y_u^2 + z_u^2, \quad G := \mathbf{r}_v^2 = x_v^2 + y_v^2 + z_v^2,$$
$$F := \mathbf{r}_u\mathbf{r}_v = x_u x_v + y_u y_v + z_u z_v,$$

hence $g = EG - F^2$. The volume form is given by $\mu = \sqrt{EG - F^2}\,\mathrm{d}u \wedge \mathrm{d}v$. The surface integral has the form

$$\int_{\mathscr{F}} \varrho\mathrm{d}F = \int_{\mathscr{F}} \varrho\mu = \int_{\mathscr{D}} \varrho\sqrt{EG - F^2}\mathrm{d}u\mathrm{d}v.$$

**Generalization to Riemannian manifolds:** See [212].

### 1.7.8 Integration by parts

Surface integrals play a central role in generalizing the classical formula of partial integration

$$\int_a^b uv'\mathrm{d}x = uv\Big|_a^b - \int_a^b u'v\mathrm{d}x$$

to higher dimensions. In place of the usual derivative one has partial derivatives, and the boundary term $uv|_a^b$ is replace by a boundary integral. This gives:

$$\int_D u\partial_j v\,\mathrm{d}x = \int_{\partial D} uvn_j\,\mathrm{d}F - \int_D v\partial_j u\,\mathrm{d}x. \tag{1.154}$$

**Theorem:** Let $D$ be a bounded, open, non-empty set in $\mathbb{R}^N$ with a piecewise smooth boundary[79] $\partial D$ and the outer normal vector $\mathbf{n} = (n_1, \ldots, n_N)$. Then for all $C^1$-functions $u, v : \overline{D} \longrightarrow \mathbb{C}$, the formula (1.154) for integration by parts holds.

**Comment:** The formula

$$\int_D \partial_j w \, \mathrm{d}x = \int_{\partial D} w n_j \, \mathrm{d}F \tag{1.155}$$

follows from the general theorem of Stokes $\int_D \mathrm{d}\omega = \int_{\partial D} \omega$, in case $\omega$ is an $(N-1)$-form. This follows in the same way that the integral theorem of Gauss in 1.7.6.1 followed from that theorem. The formula (1.154) follows immediately from (1.155), by setting $w = uv$ and applying the product rule $\partial_j(uv) = v\partial_j u + u\partial_j v$.

**Application to the formula of Green:** Let $\Delta u := u_{xx} + u_{yy} + u_{zz}$, i.e., $\Delta$ is the Laplace operator in $\mathbb{R}^3$. Then the following *Green's formula* holds:

$$\boxed{\int_D (v\Delta u - u\Delta v)\mathrm{d}x = \int_{\partial D} \left( v\frac{\partial u}{\partial n} - u\frac{\partial v}{\partial n} \right) \mathrm{d}F.}$$

Here $\frac{\partial u}{\partial n} = n_1 u_x + n_2 u_y + n_3 u_z$ denotes the normal outer derivative with normal outer vector $\mathbf{n} = n_1\mathbf{i} + n_2\mathbf{j} + n_3\mathbf{k}$ (Figure 1.70(a)).

*Proof:* We write $\mathrm{d}V$ for $\mathrm{d}x\mathrm{d}y\mathrm{d}z$. Integration by parts yields

$$\int_D u v_{xx} \, \mathrm{d}V = \int_{\partial D} u v_x n_1 \, \mathrm{d}F - \int_D u_x v_x \, \mathrm{d}V,$$

$$\int_D u_x v_x \, \mathrm{d}V = \int_{\partial D} u_x v n_1 \mathrm{d}F - \int_D u_{xx} v \, \mathrm{d}V.$$

Similar formulas hold for $y$ and $z$, hence summing the contributions leads to (1.154). □

The formula for integration by parts pays a fundamental role in the modern theory of partial differential equations, because it makes it possible to introduce the notion of *generalized derivative.* This is connected with the theory of distributions and Sobolev spaces. Distributions are objects which extend the classical notion of function and which have the exceptional property of being infinitely often differentiable, (see [212]).

## 1.7.9 Curvilinear coordinates

We denote by **i, j** and **k** the unit vectors in the directions of the coordinate axi of a Cartesian coordinate system $(x, y, z)$. Furthermore we set $\mathbf{r} = x\mathbf{i} + y\mathbf{j} + z\mathbf{k}$ (Figure 1.70(a)).

### 1.7.9.1 Polar coordinates

**Coordinate transformation** (Figure 1.76):

$$\boxed{x = r\cos\varphi, \qquad y = r\sin\varphi, \qquad -\pi < \varphi \le \pi, \quad r \ge 0.}$$

[79] This boundary is allowed to have reasonable corners and edges. The precise assumption here is that $\partial D \in C^{0,1}$ (cf. [212]).

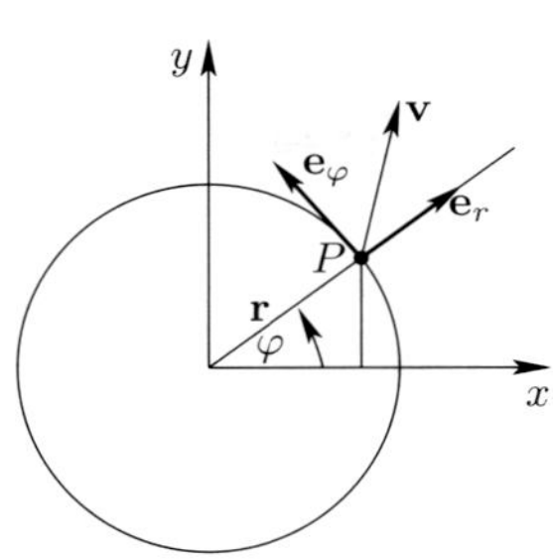


Figure 1.76. Polar coordinates.

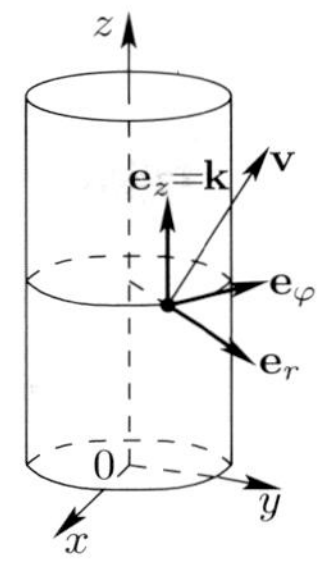


Figure 1.77. Cylinder coordinates.

**Natural basis vectors $\mathbf{e}_r,\ \mathbf{e}_\varphi$ at the point $P$:**

$$\begin{aligned}\mathbf{e}_r &= \mathbf{r}_r = x_r\mathbf{i} + y_r\mathbf{j} = \cos\varphi\mathbf{i} + \sin\varphi\mathbf{j},\\ \mathbf{e}_\varphi &= \mathbf{r}_\varphi = x_\varphi\mathbf{i} + y_\varphi\mathbf{j} = -r\sin\varphi\mathbf{i} + r\cos\varphi\mathbf{j}.\end{aligned}$$

### 1.7.9.2 Cylinder coordinates

**Coordinate transformation:**

$$\boxed{x = r\cos\varphi, \qquad y = r\sin\varphi, \qquad z = z, \qquad -\pi < \varphi \le \pi, \quad r \ge 0.}$$

**Natural basis vectors $\mathbf{e}_r,\ \mathbf{e}_\varphi,\ \mathbf{e}_z$ at a point $P$:**

$$\begin{aligned}\mathbf{e}_r &= \mathbf{r}_r = x_r\mathbf{i} + y_r\mathbf{j} = \cos\varphi\mathbf{i} + \sin\varphi\mathbf{j},\\ \mathbf{e}_\varphi &= \mathbf{r}_\varphi = x_\varphi\mathbf{i} + y_\varphi\mathbf{j} = -r\sin\varphi\mathbf{i} + r\cos\varphi\mathbf{j},\\ \mathbf{e}_z &= \mathbf{r}_z = \mathbf{k}.\end{aligned}$$

**Coordinate lines:**

| | | | | |
|---|---|---|---|---|
| (i) | $r =$ variable, | $\varphi =$ const, | $z =$ const : | ray perpendicular to the $z$-axis. |
| (ii) | $\varphi =$ variable, | $r =$ const, | $z =$ const : | great circle on the surface of the cylinder given by $r =$ const. |
| (iii) | $z =$ variable, | $r =$ const, | $\varphi =$ const : | line parallel to the $z$-axis. |

Through each point $P$, which is not the origin, there pass exactly three coordinate lines with the tangent vectors $\mathbf{e}_r,\ \mathbf{e}_\varphi,\ \mathbf{e}_z$, which are moreover perpendicular to one another (Figure 1.77).

**Decomposition of a vector $\mathbf{v}$ at a point $P$:**

$$\mathbf{v} = v_1\mathbf{e}_r + v_2\mathbf{e}_\varphi + v_3\mathbf{e}_z.$$

We call $v_1,\ v_2$ and $v_3$ the natural components of the vector $\mathbf{v}$ at the point $P$ in cylinder coordinates.

**Arc element:** $\mathrm{d}s^2 = \mathrm{d}r^2 + r^2\mathrm{d}\varphi^2 + \mathrm{d}z^2$.

**Volume form:** $\mu := r\,\mathrm{d}r \wedge \mathrm{d}\varphi \wedge \mathrm{d}z = \mathrm{d}x \wedge \mathrm{d}y \wedge \mathrm{d}z$.

**Volume integral:**

$$\boxed{\int \varrho\mu = \int \varrho r\,\mathrm{d}r\mathrm{d}\varphi\mathrm{d}z = \int \varrho\,\mathrm{d}x\mathrm{d}y\mathrm{d}z.}$$

This formula corresponds to the substitution rule.

**Cylinders of fixed radius:** For the cylinder given by $r = \text{const}$ one has for the arc length the formula

$$\mathrm{d}s^2 = r^2\mathrm{d}\varphi^2 + \mathrm{d}z^2$$

with the volume form $\mu = r\mathrm{d}\varphi \wedge \mathrm{d}z$ and the area integral

$$\boxed{\int \varrho\,\mathrm{d}F = \int \varrho\mu = \int \varrho r\,\mathrm{d}\varphi\mathrm{d}z.}$$

*Example:* The surface area of the cylinder of radius $r$ and height $h$ is determined by calculating the value of the integral:

$$\int_{z=0}^{h} \left( \int_{\varphi=-\pi}^{\pi} r\mathrm{d}\varphi \right) \mathrm{d}z = 2\pi r h.$$

### 1.7.9.3 Spherical coordinates

**Coordinate transformation:**

$$\boxed{x = r\cos\varphi\cos\theta, \qquad y = r\sin\varphi\cos\theta, \qquad z = r\sin\theta,}$$

with $-\pi < \varphi \leq \pi$, $-\dfrac{\pi}{2} \leq \theta \leq \dfrac{\pi}{2}$ and $r \geq 0$.

**Natural basis vectors $\mathbf{e}_\varphi$, $\mathbf{e}_\theta$, $\mathbf{e}_r$ at a point $P$:**

$$\begin{aligned}
\mathbf{e}_\varphi &= \mathbf{r}_\varphi = -r\sin\varphi\cos\theta\mathbf{i} + r\cos\varphi\cos\theta\mathbf{j}, \\
\mathbf{e}_\theta &= \mathbf{r}_\theta = -r\cos\varphi\sin\theta\mathbf{i} - r\sin\varphi\sin\theta\mathbf{j} + r\cos\theta\mathbf{k}, \\
\mathbf{e}_r &= \mathbf{r}_r = \cos\varphi\cos\theta\mathbf{i} + \sin\varphi\cos\theta\mathbf{j} + \sin\theta\mathbf{k}.
\end{aligned}$$

The surface $r = \text{const}$ is the surface $S_r$ of a sphere centered at the origin of radius $r$.

**Coordinate lines:**

| | | | | |
|---|---|---|---|---|
| (i) | $\varphi = \text{variable}$, | $r = \text{const}$, | $\theta = \text{const}$ : | great circle on $S_r$ of geographical latitude $\theta$. |
| (ii) | $\theta = \text{variable}$, | $r = \text{const}$, | $\varphi = \text{const}$ : | half a great circle on $S_r$ of geographical longitude $\varphi$. |
| (iii) | $r = \text{variable}$, | $\varphi = \text{const}$, | $\theta = \text{const}$ : | ray from the origin outwards. |

Through each point $P$, distinct from the origin, there are exactly three coordinate lines which pass through the point; these have tangent vectors $\mathbf{e}_\varphi$, $\mathbf{e}_\theta$, $\mathbf{e}_r$ which are perpendicular to one another (Figure 1.78).

**Decomposition of a vector v at the point $P$:**

$$\mathbf{v} = v_1\mathbf{e}_\varphi + v_2\mathbf{e}_\theta + v_3\mathbf{e}_r.$$

We call $v_1$, $v_2$ and $v_3$ the natural components of the vector $\mathbf{v}$ at the point $P$ in spherical coordinates.

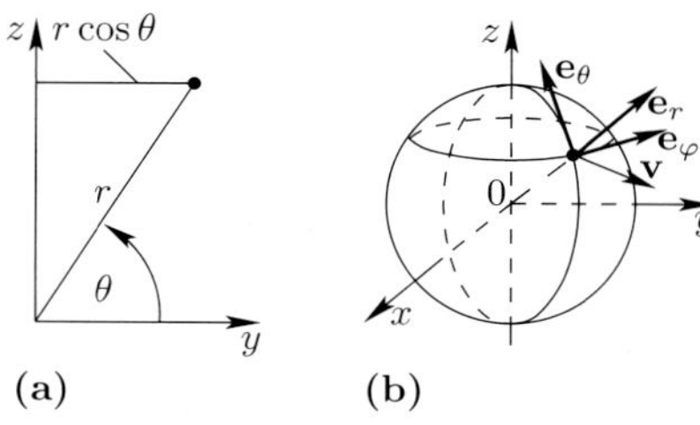


*Figure 1.78. Spherical coordinates.*

**Arc element:** $\mathrm{d}s^2 = r^2\cos^2\theta \mathrm{d}\varphi^2 + r^2\mathrm{d}\theta^2 + \mathrm{d}r^2$.

**Volume form:** $\mu := r^2\cos\theta \mathrm{d}\varphi \wedge \mathrm{d}\theta \wedge \mathrm{d}r = \mathrm{d}x \wedge \mathrm{d}y \wedge \mathrm{d}z$.

**Volume integral:**

$$\boxed{\int \varrho\mu = \int \varrho r^2 \cos\theta \, \mathrm{d}\varphi\mathrm{d}\theta\mathrm{d}r = \int \varrho \, \mathrm{d}x\mathrm{d}y\mathrm{d}z.}$$

This formula corresponds again to the rule of substitution.

**Spherical surface:** On $S_r$ the formula for the arc length is given by

$$\mathrm{d}s^2 = r^2\cos^2\theta \mathrm{d}\varphi^2 + r^2\mathrm{d}\theta^2,$$

with the volume form $\mu = r^2\cos\theta \mathrm{d}\varphi \wedge \mathrm{d}\theta$ and the surface integral

$$\boxed{\int \varrho \, \mathrm{d}F = \int \varrho\mu = \int \varrho r^2 \cos\theta \, \mathrm{d}\varphi\mathrm{d}\theta.}$$

*Example:* The surface area $F$ of a sphere of radius $r$ is determined as the value of the integral

$$F = \int_{\theta=-\pi/2}^{\pi/2} \int_{\varphi=-\pi}^{\pi} r^2 \cos\theta \, \mathrm{d}\varphi\mathrm{d}\theta = 2\pi r^2 \int_{-\pi/2}^{\pi/2} \cos\theta \, \mathrm{d}\theta = 4\pi r^2.$$

*Table 1.2. Curvilinear coordinates.*

| | ***Polar coordinates*** | ***Cylinder coordinates*** | ***Spherical coordinates*** |
|---|---|---|---|
| ***Length element*** $\mathbf{ds^2}$ | $\mathrm{d}r^2 + r^2\mathrm{d}\varphi^2$ | $\mathrm{d}r^2 + r^2\mathrm{d}\varphi^2 + \mathrm{d}z^2$ | $\mathrm{d}r^2 + r^2\mathrm{d}\theta^2 + r^2\cos^2\theta\mathrm{d}\varphi^2$ |
| ***Special volume element*** $\boldsymbol{v}$ | | $r\mathrm{d}\varphi\mathrm{d}z\mathrm{d}r$ | $r^2\cos\theta\mathrm{d}\varphi\mathrm{d}\theta\mathrm{d}r$ |
| ***Surface element*** | $r\mathrm{d}r\mathrm{d}\varphi$ (plane surface) | $r\mathrm{d}\varphi\mathrm{d}z$ (cylinder surface with the radius $r$) | $r^2\cos\theta\mathrm{d}\varphi\mathrm{d}\theta$ (sphere with the radius $r$) |

### 1.7.10 Applications to the center of mass and center of inertia

The formulas for mass, center of mass and center of inertia are found in Table 1.3.

*Example* (ball): We consider a solid ball of radius $R$ centered at the origin in a Cartesian coordinate system $(x, y, z)$ (Figure 1.79(a)).

*Volume:*

$$M = \int r^2 \cos\theta \mathrm{d}\theta \, \mathrm{d}\varphi \mathrm{d}r = \int_0^R r^2 \mathrm{d}r \int_{-\pi}^{\pi} \mathrm{d}\varphi \int_{-\pi/2}^{\pi/2} \cos\theta \, \mathrm{d}\theta = \frac{4\pi R^3}{3}.$$

We have used here spherical coordinates (cf. Table 1.2).

*Center of mass:* This is located at the center of the ball, i.e., at the origin.

*Table 1.3. Integral quantities in curvilinear coordinates.*

| | ***Mass M*** **(*ϱ density*)** | ***Vector of center of mass*** **(r = *x*i + *y*j + *z*k)** | ***Center of inertia with respect to the z-axis*** |
|---|---|---|---|
| ***Curve*** $\mathscr{C}$ | $M = \int_{\mathscr{C}} \varrho \, \mathrm{d}s$ (for $\varrho \equiv 1$: the curve length of $\mathscr{C}$) | $\mathbf{r}_{\mathrm{cm}} = \frac{1}{M} \int_{\mathscr{C}} \mathbf{r} \varrho \, \mathrm{d}s$ | $\Theta_z = \int_{\mathscr{C}} (x^2 + y^2) \varrho \mathrm{d}s$ |
| ***Surface*** $\mathscr{F}$ | $M = \int_{\mathscr{F}} \varrho \, \mathrm{d}F$ (for $\varrho \equiv 1$: the surface area of $\mathscr{F}$) | $\mathbf{r}_{\mathrm{cm}} = \frac{1}{M} \int_{\mathscr{F}} \mathbf{r} \varrho \, \mathrm{d}F$ | $\Theta_z = \int_{\mathscr{F}} (x^2 + y^2) \varrho \, \mathrm{d}F$ |
| ***Solid*** $\mathscr{G}$ **(d*V* = d*x*d*y*d*z*)** | $M = \int_{\mathscr{G}} \varrho \, \mathrm{d}V$ (for $\varrho \equiv 1$: the volume of $\mathscr{D}$) | $\mathbf{r}_{\mathrm{cm}} = \frac{1}{M} \int_{\mathscr{G}} \mathbf{r} \varrho \, \mathrm{d}V$ | $\Theta_z = \int_{\mathscr{G}} (x^2 + y^2) \varrho \, \mathrm{d}V$ |

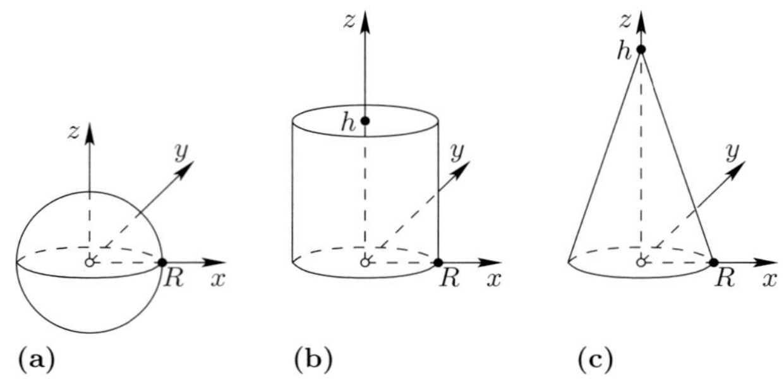


*Figure 1.79. Solid ball, cylinder and circular cone.*

*Center of inertia* (with respect to the $z$-axis):

$$\Theta_z = \int (r\cos\theta)^2 r^2 \cos\theta \, \mathrm{d}\theta \mathrm{d}\varphi \mathrm{d}r = \int_0^R r^4 \mathrm{d}r \int_{-\pi}^{\pi} \mathrm{d}\varphi \int_{-\pi/2}^{\pi/2} \cos^3\theta \, \mathrm{d}\theta = \frac{2}{5} R^2 M.$$

*Example 2* (spherical cylinder): We consider a spherical cylinder of radius $R$ and height $h$ (Figure 1.79(b)).

*Volume:*

$$M = \int r\, \mathrm{d}r\mathrm{d}\varphi\mathrm{d}z = \int_0^R r\, \mathrm{d}r \int_{-\pi}^{\pi} \mathrm{d}\varphi \int_0^h \mathrm{d}z = \pi R^2 h.$$

Here we are using the cylinder coordinates (cf. Table 1.2).

*Center of mass:* $z_{\mathrm{cm}} = \dfrac{h}{2}$, $x_{\mathrm{cm}} = y_{\mathrm{cm}} = 0$.

*Center of inertia with respect to the $z$-axis:*

$$\Theta_z = \int r^2 \cdot r\, \mathrm{d}\varphi\mathrm{d}r\mathrm{d}z = \int_0^R r^3 \mathrm{d}r \int_{-\pi}^{\pi} \mathrm{d}\varphi \int_0^h \mathrm{d}z = \frac{1}{2} R^2 M.$$

*Example 3* (circular cone): We consider a cone over a circle of radius $R$ and of height $h$ (Figure 1.79(c)).

*Volume:*

$$M = \int_{z=0}^{h} \left( \int_{D_z} \mathrm{d}x\mathrm{d}y\mathrm{d}z \right) = \int_0^h \pi R_z^2 \mathrm{d}z = \frac{1}{3} \pi R^2 h.$$

Here we apply the principle of Cavalieri. According to Example 3 in 1.7.1 one has $R_z = (h - z)R/h$.

*Center of mass:*

$$z_{\mathrm{cm}} = \frac{1}{M} \int_{z=0}^{h} \left( \int_{D_z} z\mathrm{d}x\mathrm{d}y \right) \mathrm{d}z = \frac{1}{M} \int_0^h z\pi R_z^2 \mathrm{d}z = \frac{h}{4}, \qquad x_{\mathrm{cm}} = y_{\mathrm{cm}} = 0.$$

*Center of inertia with respect to the $z$-axis:*

$$\begin{aligned} \Theta &= \int (x^2 + y^2)\mathrm{d}x\mathrm{d}y\mathrm{d}z = \int_0^h \left( \int_{D_z} r^3 \mathrm{d}r\mathrm{d}\varphi \right) \mathrm{d}z = \int_0^h \left( \int_0^{R_z} r^3 \mathrm{d}r \right) \left( \int_{-\pi}^{\pi} \mathrm{d}\varphi \right) \mathrm{d}z \\ &= \int_0^h \frac{\pi}{2} R_z^4\, \mathrm{d}z = \frac{3}{10} M R^2. \end{aligned}$$

**First Guldinian rule:** The volume of a body of rotation can be found by multiplying the area $F$ of a meridian section lying on the side of the axis of rotation $S$ with the length of the trajectory which the center of mass of $S$ follows during one revolution (cf. Figure 1.80).

**Second Guldinian rule:** The area of the surface of a body of rotation can be found by multiplying the length of the boundary $\partial S$ of the meridian section with the length of the trajectory transcribed by the center of mass of the boundary $\partial S$ upon one revolution.

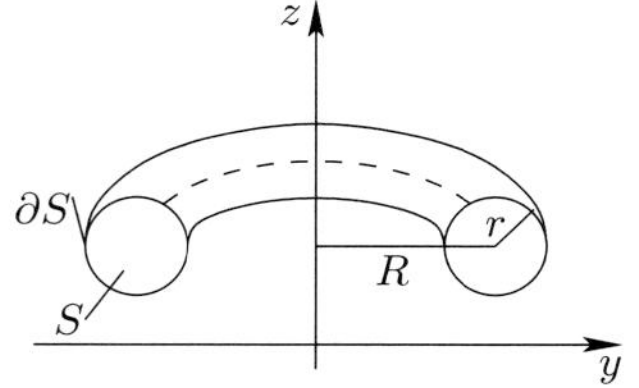


*Figure 1.80.*

*Example 4* (torus): The meridian section $S$ of a torus is a circle of radius $r$ with surface area $F = \pi r^2$ and the circumference $U = 2\pi r$ (Figure 1.80). The center of mass of $S$ and of $\partial S$ is the center of the circle, which has a distance of $R$ from the $z$-axis. According to the *Guldinian rule*, one has:

$$\text{volume of the torus} = 2\pi RF = 2\pi^2 Rr^2,$$
$$\text{surface area of the torus} = 2\pi RU = 4\pi^2 Rr.$$

### 1.7.11 Integrals depending on parameters

We consider the function

$$F(p) = \int_D f(x,p)\mathrm{d}x, \qquad p \in P,$$

where $D$ is an admissible subset of $\mathbb{R}^N$ and $P$ is an open set of $\mathbb{R}^M$. We call $p$ a parameter. Furthermore, let $\partial_j := \partial/\partial p_j$.

**Continuity:** The function $F : P \longrightarrow \mathbb{C}$ is continuous, if:

(i) The function $f(.,p) : D \longrightarrow \mathbb{C}$ is admissible and continuous for every parameter $p \in P$.

(ii) There is an admissible function $h : D \longrightarrow \mathbb{R}$ with

$$|f(x,p)| \leq h(x) \qquad \text{for all } x \in D,\ p \in P.$$

**Differentiability:** The function $F : P \longrightarrow \mathbb{C}$ is of type $C^1$ with

$$\partial_j F(p) = \int_D \partial_j f(x,p)\mathrm{d}x, \qquad j = 1, \ldots, N,$$

for all $p \in P$, if in addition to (i) and (ii), the following conditions are satisfied:

(a) The function $f(.,p) : D \longrightarrow \mathbb{C}$ is of type $C^1$ for every parameter $p \in P$.

(b) There is an admissible function $h_j : D \longrightarrow \mathbb{R}$ with

$$|\partial_j f(x,p)| \leq h_j(x) \qquad \text{for all } x \in D,\ p \in P \text{ and } j = 1, \ldots N.$$

In these definitions, $f(.,p)$ denotes the function which for fixed $p$ assigns to each point $x$ the value $f(x,p)$.

**Integration:** Let $P$ be an admissible subset of $\mathbb{R}^M$. Then the integral

$$\int_P F(p)\mathrm{d}p = \int_{D\times P} f(x,p)\,\mathrm{d}x\mathrm{d}p,$$

exists, if $f : P \times D \longrightarrow \mathbb{R}$ is admissible.

*Example:* Let $-\infty < a < b < \infty$ and $-\infty < c < d < \infty$. We set $Q := \{(x,y) \in \mathbb{R}^2 : a \leq x \leq b,\ c \leq y \leq d\}$ and choose $p := y$ as a parameter.

(i) If $f : Q \longrightarrow \mathbb{C}$ is continuous, then the function

$$F(p) := \int_a^b f(x,p)\mathrm{d}x$$

is continuous for every parameter $p \in [c,d]$. Moreover,

$$\int_c^d F(p)\,\mathrm{d}p = \int_Q f(x,p)\,\mathrm{d}x\mathrm{d}p.$$

(ii) If $f : Q \longrightarrow \mathbb{C}$ is of type $C^1$, then the same is true for $F$ on $[c,d]$, and we get the derivative $F'(p)$ by differentiating under the integral sign:

$$F'(p) = \int_a^b f_p(x,p)\mathrm{d}x \qquad \text{for all } p \in ]c,d[.$$

**Application to integral transformations:** See section 1.11.

# 1.8 Vector algebra

**Scalars, vectors and affine vectors:** Quantities whose values are given by real numbers are called *scalars* (for example, mass, charge, temperature, work, energy, power). On the other hand, quantities which have in addition to a scalar component a direction are called *vectors* (for example, speed, acceleration, force, electric or magnetic field force). In physics, the base (starting point) of the vector is often of importance, which leads to the notion of (based) vectors. If the starting point is irrelevant, then one is dealing with *affine vectors.*

**Definition of vector:** Let a point $O$ be given; a vector $\mathbf{F}$ starting at the point $O$ is an arrow whose origin is at the point $O$ (see Figure 1.81).

The length of the arrow is denoted by $|\mathbf{F}|$. If $P$ is the endpoint of the arrow, we also write $\mathbf{F}= \overrightarrow{OP}$. The vector $\mathbf{0} := \overrightarrow{OO}$ is called the *zero vector* at the point $O$. Vectors whose length is unity are called *unit vectors.*

**Physical interpretation:** One may view $\mathbf{F}$ as a force, which acts on the point $O$.

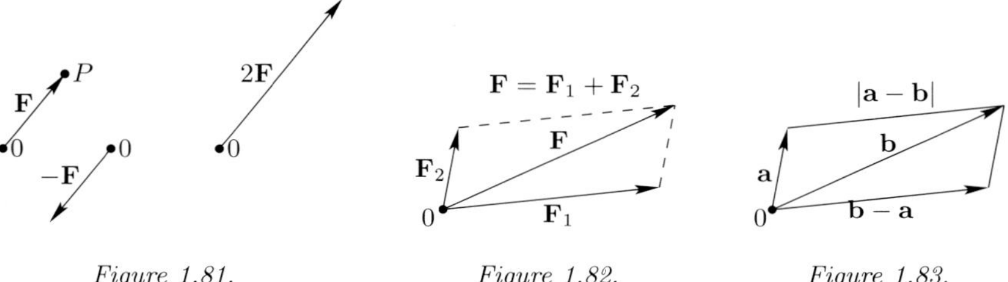


*Figure 1.81.* *Figure 1.82.* *Figure 1.83.*

## 1.8.1 Linear combinations of vectors

**Definition of scalar multiplication of a vector** (Figure 1.81): Let $\mathbf{F}$ be a vector based at the point $O$, and let $\alpha$ be a real number.

(i) $-\mathbf{F}$ is a vector with base point $O$, with the same length as $\mathbf{F}$, but with the opposite direction as $\mathbf{F}$.

(ii) For $\alpha > 0$, the vector $\alpha\mathbf{F}$ is a vector again based at $O$, but in the same direction as $\mathbf{F}$ and of length equal to $\alpha|\mathbf{F}|$.

(iii) For $\alpha < 0$, we set $\alpha\mathbf{F} := |\alpha|(-\mathbf{F})$, and for $\alpha = 0$, $\alpha\mathbf{F} := \mathbf{0}$.

**Definition of vector addition:** Let $\mathbf{F}_1$ and $\mathbf{F}_2$ be two vectors, both based at the point $O$; the *sum*

$$\boxed{\mathbf{F}_1 + \mathbf{F}_2}$$

is by definition a vector based at the point $O$, which is defined as the diagonal in the parallelogram formed by $\mathbf{F}_1$ and $\mathbf{F}_2$ as in Figure 1.82.

**Physical interpretation:** If $\mathbf{F}_1$ and $\mathbf{F}_2$ are two forces acting on the point $O$, then $\mathbf{F}_1 + \mathbf{F}_2$ is the resulting force acting there (parallelogram of forces).

The difference $\mathbf{b} - \mathbf{a}$ for two vectors $\mathbf{a}$ and $\mathbf{b}$ is defined by the formula $\mathbf{b} + (-\mathbf{a})$. One has $\mathbf{a} + (\mathbf{b} - \mathbf{a}) = \mathbf{b}$ (Figure 1.83).

**Basic laws:** Let $\mathbf{a}$, $\mathbf{b}$ and $\mathbf{c}$ be vectors, based at a point $O$, and let $\alpha$, $\beta$ be real numbers. Then one has:

$$\boxed{\begin{array}{lll} \mathbf{a}+\mathbf{b}=\mathbf{b}+\mathbf{a}, & (\mathbf{a}+\mathbf{b})+\mathbf{c}=\mathbf{a}+(\mathbf{b}+\mathbf{c}), & \\ \alpha(\beta\mathbf{a})=(\alpha\beta)\mathbf{a}, & (\alpha+\beta)\mathbf{a}=\alpha\mathbf{a}+\beta\mathbf{a}, & \alpha(\mathbf{a}+\mathbf{b})=\alpha\mathbf{a}+\alpha\mathbf{b}. \end{array}} \tag{1.156}$$

We call $\alpha\mathbf{a} + \beta\mathbf{b}$ a *linear combination* of the vectors $\mathbf{a}$ and $\mathbf{b}$. Moreover one has

$$\begin{aligned} &|\mathbf{a}| = 0 \quad \text{if and only if} \quad \mathbf{a} = \mathbf{0}; \\ &|\alpha\mathbf{a}| = |\alpha|\,|\mathbf{a}|; \\ &||\mathbf{a}| - |\mathbf{b}|| \le |\mathbf{a} \pm \mathbf{b}| \le |\mathbf{a}| + |\mathbf{b}|. \end{aligned} \tag{1.157}$$

We use the notation $V(O)$ to denote the set of all vectors based at the point $O$. Using the modern language of mathematics, (1.156) says that $V(O)$ is a real vector space (cf. 2.3.2). Equation (1.157) then implies that $V(O)$ is in fact a *normed* vector space (cf. [212]). The distance function

$$d(\mathbf{a}, \mathbf{b}) := |\mathbf{b} - \mathbf{a}| \tag{1.158}$$

in addition gives $V(O)$ the structure of a *metric* space. Here $|\mathbf{b} - \mathbf{a}|$ is the distance between the endpoints of the vectors $\mathbf{b}$ and $\mathbf{a}$ (Figure 1.83).

**Linear independence:** The vectors $\mathbf{F}_1, \ldots, \mathbf{F}_r$ in $V(O)$ are said to be *linearly independent*, if the equation

$$\alpha_1\mathbf{F}_1 + \ldots + \alpha_r\mathbf{F}_r = 0$$

with real numbers $\alpha_1, \ldots, \alpha_r$ implies that $\alpha_1 = \cdots = \alpha_r = 0$.

*Example:* (i) Two vectors in $V(O)$, both different from $\mathbf{0}$, are linearly independent if and only if they do not lie on a line, i.e., if they are not collinear.

(ii) Three vectors in $V(O)$, all different from the zero vector, are linearly independent, if and only if they do not lie in a plane, i.e., they are not coplanar.

### 1.8.2 Coordinate systems

The maximal number of linear independent vectors in $V(O)$ is equal to three. Because of this, one says that the linear space $V(O)$ has *dimension* equal to three.

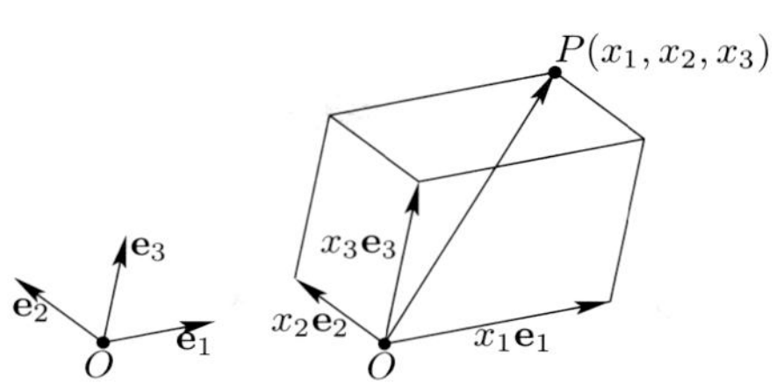


*Figure 1.84. A coordinate basis.*

**Basis:** If $\mathbf{e}_1,\ \mathbf{e}_2,\ \mathbf{e}_3$ are three linear independent vectors in $V(O)$, one can write an arbitrary vector $\mathbf{r}$ in $V(O)$ in a unique way as

$$\boxed{\mathbf{r} = x_1\mathbf{e}_1 + x_2\mathbf{e}_2 + x_3\mathbf{e}_3.}$$

The real numbers $x_1,\ x_2,\ x_3$ are called the components of $\mathbf{r}$ with respect to the basis $\mathbf{e}_1,\ \mathbf{e}_2,\ \mathbf{e}_3$. At the same time, $x_1,\ x_2,\ x_3$ are the coordinates of the endpoint $P$ of $\mathbf{r}$ (Figure 1.84).

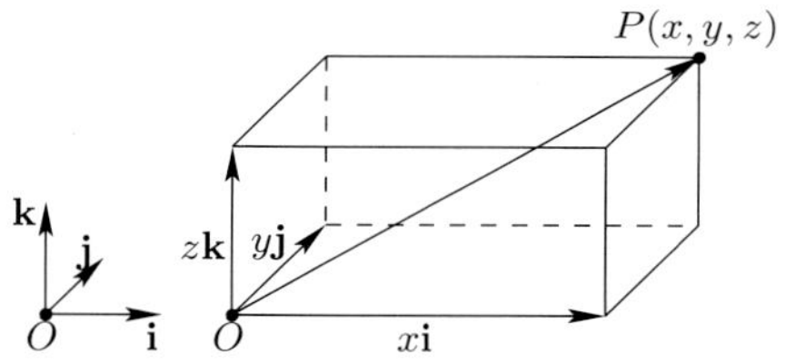


*Figure 1.85. Cartesian coordinates.*

A Cartesian coordinate system is given by three vectors of unit length $\mathbf{i},\ \mathbf{j},\ \mathbf{k}$ which are perpendicular to one another, and which are *right-handed* with respect to each other, i.e., they point (are oriented) like the thumb and the first two fingers of the right hand. Each vector $\mathbf{r}$ in $V(O)$ then has a unique representation of the form

$$\boxed{\mathbf{r} = x\mathbf{i} + y\mathbf{j} + z\mathbf{k},}$$

where $x,\ y$ and $z$ are called the Cartesian coordinates of the endpoint $P$ of $\mathbf{r}$ (Figure 1.85).

Moreover, one has

$$\boxed{|\mathbf{r}| := \sqrt{x^2 + y^2 + z^2}.}$$

The corresponding situation in the plane is shown in Figure 1.86.

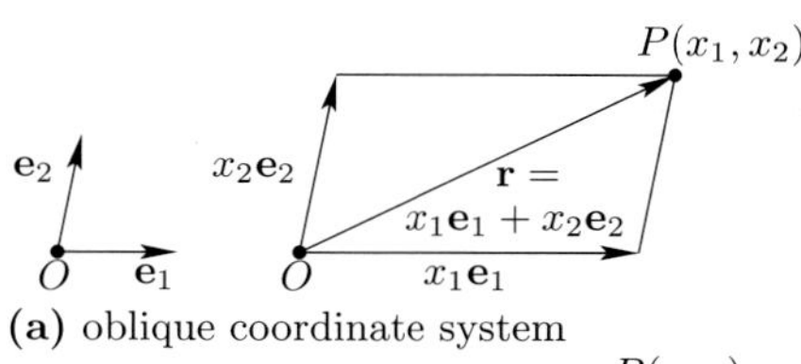


**(a)** oblique coordinate system

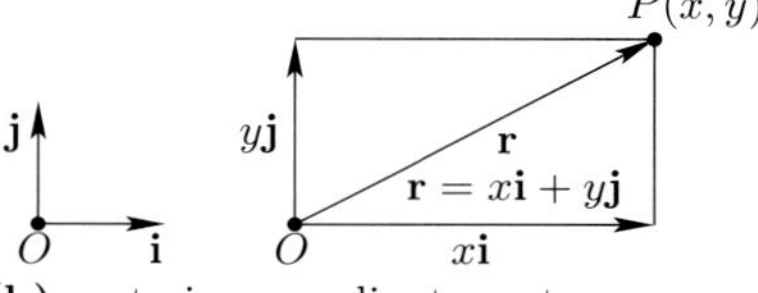


**(b)** cartesian coordinate system

*Figure 1.86. The length of vectors.*

**Free vectors:** Two vectors which are based at the same point or at two different points are said to be *equivalent*, if they have the same direction and the same length. We write for this

$$\mathbf{a} \sim \mathbf{c}$$

(Figure 1.87). Furthermore we denote by $[\mathbf{a}]$ the equivalence class of the vector $\mathbf{a}$, i.e., the set of all vectors $\mathbf{c}$ which are equivalent to $\mathbf{a}$. All elements in the equivalence class $[\mathbf{a}]$ are called *representatives* of $\mathbf{a}$. Each such equivalence class $[\mathbf{a}]$ is called a *free vector*.

**Geometric interpretation:** A free vector $[\mathbf{a}]$ represents a translation in space. Each representative $\mathbf{c} = \overrightarrow{QP}$ indicates that the point $Q$ is mapped to $P$ under the translation (Figure 1.87).

**Sum of free vectors:** We define

$$[\mathbf{a}] + [\mathbf{b}] := [\mathbf{a} + \mathbf{b}].$$

This means that free vectors are added component-wise, and this operation is independent of the choice of representative. Similarly, we define

$$\alpha[\mathbf{a}] := [\alpha\mathbf{a}].$$

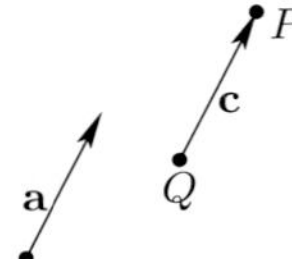


*Figure 1.87.*

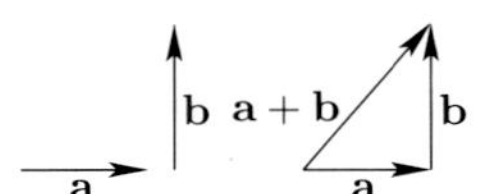


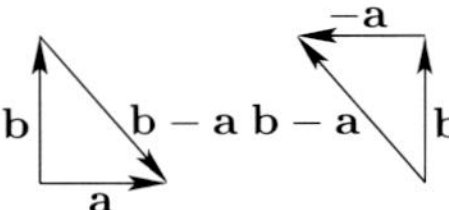


*Figure 1.88. The sum and difference of vectors.*

**Convention:** To simplify the notation, one writes $\mathbf{a} = \mathbf{c}$ instead of $\mathbf{a} \sim \mathbf{c}$ and $\mathbf{a} + \mathbf{b}$ (resp. $\alpha\mathbf{a}$) instead of $[\mathbf{a}] + [\mathbf{b}]$ (resp. $\alpha[\mathbf{a}]$).

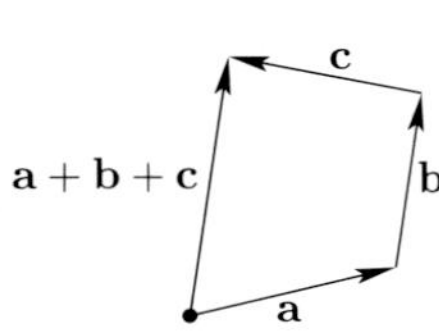


*Figure 1.89.*

This corresponds to the proceedure of working with (usual) vectors and viewing these as being equal if they have the same direction and the same length. Instead of free vectors we will usual just speak of vectors in the sequal.

*Example:* The sum $\mathbf{a} + \mathbf{b}$ of two vectors $\mathbf{a}$ and $\mathbf{b}$ is given by moving $\mathbf{a}$ and $\mathbf{b}$ by a parallel translation until they have the same base point, then adding them by the parallelogram method described above (Figure 1.88). In the same way one obtains $\mathbf{a} + \mathbf{b} + \mathbf{c}$ (Figure 1.89).

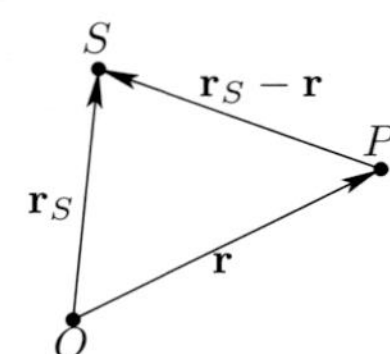


*Figure 1.90.*

**Gravitational attraction of the sun:** If the sun has a mass $M$ at the point $S$ in space and there is a planet with the mass $m$ at the point $P$, then the gravitational force of the sun acting on the planet is

$$\mathbf{F}(P) = \frac{GmM(\mathbf{r}_{\text{cm}} - \mathbf{r})}{|\mathbf{r}_{\text{cm}} - \mathbf{r}|^3} \tag{1.159}$$

with $\mathbf{r} := \overrightarrow{OP}$, $\mathbf{r}_{\text{cm}} := \overrightarrow{OS}$ and G denotes the gravitational constant (Figure 1.90).

## 1.8.3 Multiplication of vectors

**Definition of the scalar product:** The *scalar product* of two vectors $\mathbf{a}$ and $\mathbf{b}$, denoted $\mathbf{ab}$, is defined to be the number

$$\boxed{\mathbf{ab} := |\mathbf{a}|\,|\mathbf{b}|\cos\varphi,}$$

where $\varphi$ is the angle between the two vectors $\mathbf{a}$ and $\mathbf{b}$, which is to be chosen in such a way that $0 \leq \varphi \leq \pi$ (Figure 1.91).

**Orthogonality:** Two vectors $\mathbf{a}$ and $\mathbf{b}$ are said to be *orthogonal*,[80] if

$$\boxed{\mathbf{ab} = 0.}$$

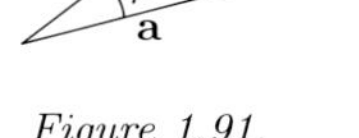


*Figure 1.91.*

If $\mathbf{a} \neq 0$ and $\mathbf{b} \neq 0$, then this condition is satisfied if and only if $\varphi = \pi/2$.

The zero vector $\mathbf{0}$ is by convention orthogonal to all vectors.

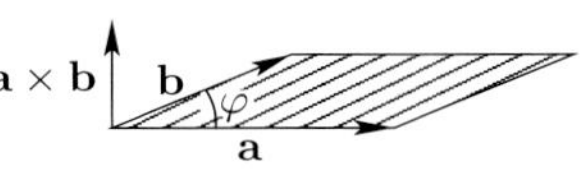


*Figure 1.92.*

**Definition of the vector product:** The *vector product* of two vectors $\mathbf{a}$ and $\mathbf{b}$, denoted $\mathbf{a} \times \mathbf{b}$, is defined to be a vector with the length

$$\boxed{|\mathbf{a}|\,|\mathbf{b}| \sin\varphi}$$

(this is just the surface area of the parallelogram spanned by $\mathbf{a}$ and $\mathbf{b}$), which is perpendicular to both $\mathbf{a}$ and $\mathbf{b}$, in such a way that the three vectors $\mathbf{a}$, $\mathbf{b}$ and $\mathbf{a} \times \mathbf{b}$, in that order, form a right-handed system, provided $\mathbf{a} \neq 0$ and $\mathbf{b} \neq 0$ (Figure 1.92).

**Laws:** For arbitrary vectors $\mathbf{a}$, $\mathbf{b}$ and $\mathbf{c}$ and arbitrary real numbers $\alpha$, we have the following laws:

$$\boxed{\begin{array}{ll} \mathbf{ab} = \mathbf{ba}, & \mathbf{a} \times \mathbf{b} = -(\mathbf{b} \times \mathbf{a}), \\ \alpha(\mathbf{ab}) = (\alpha\mathbf{a})\mathbf{b}, & \alpha(\mathbf{a} \times \mathbf{b}) = (\alpha\mathbf{a}) \times \mathbf{b}, \\ \mathbf{a}(\mathbf{b}+\mathbf{c}) = \mathbf{ab} + \mathbf{ac}, & \mathbf{a} \times (\mathbf{b}+\mathbf{c}) = (\mathbf{a} \times \mathbf{b}) + (\mathbf{a} \times \mathbf{c}), \\ \mathbf{a}^2 := \mathbf{aa} = |\mathbf{a}|^2, & \mathbf{a} \times \mathbf{a} = \mathbf{0}. \end{array}}$$

Moreover, the vector product has the following properties:

(i) The vector product vanishes, $\mathbf{a} \times \mathbf{b} = 0$, if and only if either one of the two factors is the zero vector, or $\mathbf{a}$ and $\mathbf{b}$ are parallel or anti-parallel.

(ii) One has $\mathbf{a} \times \mathbf{b} = 0$ if and only if $\mathbf{a}$ and $\mathbf{b}$ are linearly dependent.

(iii) The vector product is *not* commutative, i.e., in case $\mathbf{a} \times \mathbf{b} \neq 0$, then $\mathbf{a} \times \mathbf{b} \neq \mathbf{b} \times \mathbf{a}$.

**Products of several vectors:**

*Development law:*

$$\mathbf{a} \times (\mathbf{b} \times \mathbf{c}) = \mathbf{b}(\mathbf{ac}) - \mathbf{c}(\mathbf{ab}).$$

*Lagrange identity:*

$$(\mathbf{a} \times \mathbf{b})(\mathbf{c} \times \mathbf{d}) = (\mathbf{ac})(\mathbf{bd}) - (\mathbf{bc})(\mathbf{ad}).$$

**Triple product:** We define

$$(\mathbf{abc}) := (\mathbf{a} \times \mathbf{b})\mathbf{c}.$$

Under a permutation of $\mathbf{a}$, $\mathbf{b}$ and $\mathbf{c}$, the triple product $(\mathbf{abc})$ gets multiplied by the sign of the permutation, i.e., one has

$$(\mathbf{abc}) = (\mathbf{bca}) = (\mathbf{cab}) = -(\mathbf{acb}) = -(\mathbf{bac}) = -(\mathbf{acb}).$$

[80] One also says that $\mathbf{a}$ and $\mathbf{b}$ are *perpendicular* to one another.

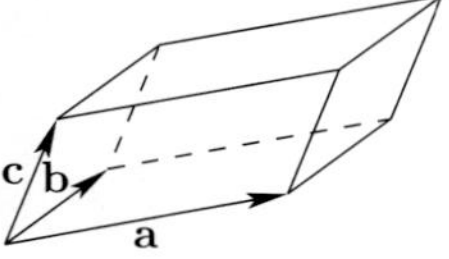


*Figure 1.93.*

Geometrically, the triple product **abc** is the volume of the parallelepiped which is spanned by **a**, **b** and **c** (Figure 1.93). Moreover, one has

$$(\mathbf{abc})(\mathbf{efg}) = \begin{vmatrix} \mathbf{ae} & \mathbf{af} & \mathbf{ag} \\ \mathbf{be} & \mathbf{bf} & \mathbf{bg} \\ \mathbf{ce} & \mathbf{cf} & \mathbf{cg} \end{vmatrix}.$$

The three vectors **a**, **b** and **c** are linearly independent, if and only if one of the following two conditions is satisfied:

(a) $(\mathbf{abc}) \neq 0$.

(b) $\begin{vmatrix} \mathbf{aa} & \mathbf{ab} & \mathbf{ac} \\ \mathbf{ba} & \mathbf{bb} & \mathbf{bc} \\ \mathbf{ca} & \mathbf{cb} & \mathbf{cc} \end{vmatrix} \neq 0$ (Gram determinant).

**Expressions in a Cartesian coordinate system:** In terms of the Cartesian coordinates in the representations

$$\mathbf{a} = a_1\mathbf{i} + a_2\mathbf{j} + a_3\mathbf{k}, \quad \mathbf{b} = b_1\mathbf{i} + b_2\mathbf{j} + b_3\mathbf{k}, \quad \mathbf{c} = c_1\mathbf{i} + c_2\mathbf{j} + c_3\mathbf{k},$$

the products above are expressed as follows:

$$\mathbf{ab} = a_1b_1 + a_2b_2 + a_3b_3,$$

$$\mathbf{a} \times \mathbf{b} = \begin{vmatrix} \mathbf{i} & \mathbf{j} & \mathbf{k} \\ a_1 & a_2 & a_3 \\ b_1 & b_2 & b_3 \end{vmatrix} = (a_2b_3 - a_3b_2)\mathbf{i} + (a_3b_1 - a_1b_3)\mathbf{j} + (a_1b_2 - a_2b_1)\mathbf{k},$$

$$(\mathbf{abc}) = \begin{vmatrix} a_1 & a_2 & a_3 \\ b_1 & b_2 & b_3 \\ c_1 & c_2 & c_3 \end{vmatrix}.$$

**Expressions in general coordinate systems:** Let $\mathbf{e}_1, \mathbf{e}_2$ and $\mathbf{e}_3$ be arbitrary linearly independent vectors. Then we call the set of three vectors

$$\mathbf{e}^1 := \frac{\mathbf{e}_2 \times \mathbf{e}_3}{(\mathbf{e}_1\mathbf{e}_2\mathbf{e}_3)}, \qquad \mathbf{e}^2 := \frac{\mathbf{e}_3 \times \mathbf{e}_1}{(\mathbf{e}_1\mathbf{e}_2\mathbf{e}_3)}, \qquad \mathbf{e}^3 := \frac{\mathbf{e}_1 \times \mathbf{e}_2}{(\mathbf{e}_1\mathbf{e}_2\mathbf{e}_3)}$$

the reciprocal basis to the basis $\mathbf{e}_1$, $\mathbf{e}_2, \mathbf{e}_3$. Every vector **a** has unique decompositions

$$\mathbf{a} = a^1\mathbf{e}_1 + a^2\mathbf{e}_2 + a^3\mathbf{e}_3 \quad \text{and} \quad \mathbf{a} = a_1\mathbf{e}^1 + a_2\mathbf{e}^2 + a_3\mathbf{e}^3.$$

We call $a^1$, $a^2$, $a^3$ the *contravariant* coordinates and $a_1$, $a_2$, $a_3$ the *covariant* coordinates of **a**. Then one has:

$$\mathbf{ab} = a_1b^1 + a_2b^2 + a_3b^3,$$

$$\mathbf{a} \times \mathbf{b} = (\mathbf{e}_1\mathbf{e}_2\mathbf{e}_3) \begin{vmatrix} \mathbf{e}^1 & \mathbf{e}^2 & \mathbf{e}^3 \\ a^1 & a^2 & a^3 \\ b^1 & b^2 & b^3 \end{vmatrix},$$

$$(\mathbf{abc}) = (\mathbf{e}_1\mathbf{e}_2\mathbf{e}_3) \begin{vmatrix} a^1 & a^2 & a^3 \\ b^1 & b^2 & b^3 \\ c^1 & c^2 & c^3 \end{vmatrix}.$$

In a Cartesian coordinate system one has the relations

$$\mathbf{i} = \mathbf{e}_1 = \mathbf{e}^1, \qquad \mathbf{j} = \mathbf{e}_2 = \mathbf{e}^2, \qquad \mathbf{k} = \mathbf{e}_3 = \mathbf{e}^3, \qquad a_j = a^j, \qquad j = 1, 2, 3.$$

In particular, in this case the contravariant and the covariant coordinates coincide.

**Applications of vector algebra in geometry:** See section 3.3.

# 1.9 Vector analysis and physical fields

*We need an analysis which is of geometric nature and describes physical situations as directly as algebra expresses quantities.*

*Gottfried Wilhelm Leibniz* (1646–1716)

The topic of vector analysis is the study of functions of vectors (vector-valued functions) with the help of calculus. It is one of the most fundamental mathematical instruments for the description of classical physical fields (hydrodynamics, elasticity, heat conduction and electrodynamics). The field theory in use today in modern physics (special and general theories of relativity, gauge field theory of elementary particles) are based on Cartesian differential calculus and tensor analysis, both of which contain classical vector analysis as special cases (cf. [212]).

**Characteristic invariance:** All of the operations with vectors described in the sequel are independent of the chosen coordinate system. This is the reason why vector analysis is such an important tool in the description of geometric properties and physical phenomena.

## 1.9.1 Velocity and acceleration

**Limits:** If $(\mathbf{a}_n)$ is a sequence of vectors and $\mathbf{a}$ is some fixed vector, where we assume that all the $\mathbf{a}_n$ as well as $\mathbf{a}$ are based at a point $O$, we write

$$\boxed{\mathbf{a} = \lim_{n\to\infty} \mathbf{a}_n,}$$

if and only if $\lim_{n\to\infty} |\mathbf{a}_n - \mathbf{a}| = 0$, i.e., the distance between the endpoints of $\mathbf{a}_n$ and $\mathbf{a}$ approaches 0 as $n \longrightarrow \infty$ (Figure 1.94).

With the help of this notion of limit, one can extend many properties which have been introduced for real-valued functions to the case of vector functions.

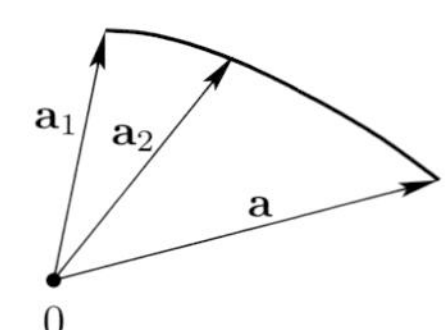

*Figure 1.94.*

**Trajectory:** We choose a fixed point $O$. Let $\mathbf{r}(t) := \overrightarrow{OP(t)}$. Then the equation

$$\boxed{\mathbf{r} = \mathbf{r}(t), \qquad \alpha \le t \le \beta,}$$

describes the motion of a particle whose coordinates at the time $t$ is $P(t)$.

**Continuity:** The vector function $\mathbf{r} = \mathbf{r}(t)$ is said to be *continuous*, if $\lim_{s\to t} \mathbf{r}(s) = \mathbf{r}(t)$.

**Velocity vector:** We define the *derivative* of the vector-valued function $\mathbf{r}(t)$ by the formula

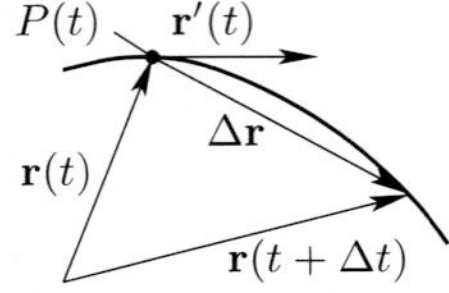


Figure 1.95.

$$\boxed{\mathbf{r}'(t) := \lim_{\Delta t \to 0} \frac{\Delta \mathbf{r}}{\Delta t}}$$

with $\Delta \mathbf{r} := \mathbf{r}(t + \Delta t) - \mathbf{r}(t)$ (Figure 1.95). The vector $\mathbf{r}'(t)$ has the direction of the tangent to the trajectory at the point $P(t)$ (in the direction of increasing values of $t$). In physics, $\mathbf{r}'(t)$ is referred to as the velocity vector at the point $P(t)$, and its length $|\mathbf{r}'(t)|$ is by definition the speed of the particle at the time $t$.

**Acceleration vector:** The second derivative

$$\boxed{\mathbf{r}''(t) := \lim_{\Delta t \to 0} \frac{\mathbf{r}'(t + \Delta t) - \mathbf{r}'(t)}{\Delta t}}$$

is by definition the *acceleration vector* of the particle at the time $t$, and $|\mathbf{r}''(t)|$ is the acceleration at time $t$.

**The fundamental Newton law of motion in classical mechanics:**

$$\boxed{m\mathbf{r}''(t) = \mathbf{F}(\mathbf{r}(t), t).} \tag{1.160}$$

Here, $m$ denotes the mass of the particle, and $\mathbf{F}(\mathbf{r}, t)$ is the force, which acts on the endpoint of the vector $\mathbf{r}$ at the time $t$. Written in words, the equation (1.160) says:

Force is equal to mass times acceleration.

This law holds even if the force depends on the velocity, i.e., $\mathbf{F} = \mathbf{F}(\mathbf{r}, \mathbf{r}', t)$.

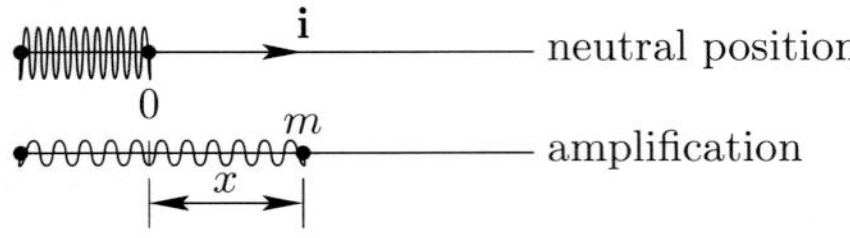


Figure 1.96. The harmonic oscillator.

*Example* (harmonic oscillator): The motion of the endpoint of a spring aligned on a line, of mass $m$, is given by $\mathbf{r}(t) = x(t)\mathbf{i}$ with a force acting in the other direction $\mathbf{F} := -kx\mathbf{i}$; here $k > 0$ is a constant describing the physical nature of the spring.[81] This yields the Newtonian law of motion

$$mx''(t)\mathbf{i} = -kx\mathbf{i}.$$

[81] This law of motion can be derived from the following general consideration: according to Taylor's theorem, one has for small values of $x$ (small amplitude) the approximation

$$\mathbf{F}(x) = \mathbf{F}(0) + x\mathbf{a} + x^2\mathbf{b} + x^3\mathbf{c} + \ldots$$

If there is no motion, then there should also be no force; in other words, $\mathbf{F}(0) = 0$. Moreover, we expect the symmetry $\mathbf{F}(-x) = -\mathbf{F}(x)$. From this is follows that $\mathbf{b} = 0$. Since the force is in the direction opposite to that of the motion, it must act in the direction of $-\mathbf{i}$. Thus $\mathbf{a} = -k\mathbf{i}$ and $\mathbf{c} = -l\mathbf{i}$ for some positive constants $k$ and $l$. Hence we get the law of motion

$$\boxed{\mathbf{F}(x) = -kx\mathbf{i} - lx^3\mathbf{i}}$$

for what is called the aharmonic oscillator. The harmonic oscillator is the special case when $l = 0$.

If we set $\omega^2 = k/m$, then we get the differential equation for the harmonic oscillator:

$$x'' + \omega^2 x = 0$$

(Figure 1.96). This differential equation will be solved in 1.11.1.2.

**Coordinate representation:** If we choose a Cartesian coordinate system with coordinates $(x, y, z)$ and which has the point $O$ as its origin, then for an arbitrary vector we have

$$\boxed{\mathbf{r}(t) = x(t)\mathbf{i} + y(t)\mathbf{j} + z(t)\mathbf{k}}$$

and

$$\boxed{\begin{aligned}\mathbf{r}^{(n)}(t) &= x^{(n)}(t)\mathbf{i} + y^{(n)}(t)\mathbf{j} + z^{(n)}(t)\mathbf{k},\\ |\mathbf{r}^{(n)}(t)| &= \sqrt{x^{(n)}(t)^2 + y^{(n)}(t)^2 + z^{(n)}(t)^2}.\end{aligned}}$$

In particular, $\mathbf{r}'(t) = \mathbf{r}^{(1)}(t)$, $\mathbf{r}''(t) = \mathbf{r}^{(2)}(t)$ etc. This means that the $n$th derivative $\mathbf{r}^{(n)}(t)$ exists precisely when the $n$th derivatives $x^{(n)}(t)$, $y^{(n)}(t)$ and $z^{(n)}(t)$ exist. Let $\mathbf{a}_n = \alpha_n\mathbf{i} + \beta_n\mathbf{j} + \gamma_n\mathbf{k}$ and $\mathbf{a} = \alpha\mathbf{i} + \beta\mathbf{j} + \gamma\mathbf{k}$. Then we have

$$\lim_{n\to\infty} \mathbf{a}_n = \mathbf{a},$$

if and only if, $\alpha_n \longrightarrow \alpha$, $\beta_n \longrightarrow \beta$ and $\gamma_n \longrightarrow \gamma$ for $n \longrightarrow \infty$ obtains (convergence of the coordinates, or as one says, coordinate-wise convergence).

**Smoothness:** The $C^k$-property of $\mathbf{r} = \mathbf{r}(t)$ is defined in the same way as for real function (cf. 1.4.1).

The function $\mathbf{r} = \mathbf{r}(t)$ is of type $C^k$ on an interval $[a, b]$, if and only if, all component functions $x = x(t)$, $y = y(t)$ and $z = z(t)$ are in the class $C^k$ on $[a, b]$ in some fixed Cartesian coordinate system.

**Taylor series:** If $\mathbf{r} = \mathbf{r}(t)$ is of type $C^{n+1}$ on an interval $[a, b]$, then we have

$$\boxed{\mathbf{r}(t+h) = \mathbf{r}(t) + h\mathbf{r}'(t) + \frac{h^2}{2}\mathbf{r}''(t) + \ldots + \frac{h^n}{n!}\mathbf{r}^{(n)}(t) + \mathbf{R}_{n+1}}$$

for all $t,\ t+h \in [a, b]$, with the error estimate

$$|\mathbf{R}_{n+1}| \le \frac{h^{n+1}}{(n+1)!} \sup_{s\in[a,b]} |\mathbf{r}^{(n+1)}(s)|.$$

## 1.9.2 Gradient, divergence and curl

As usual, $U_x$, $U_{xx}$, etc., denote partial derivatives.

**Gradient:** In a Cartesian coordinate system with origin at a point $O$ we consider the function

$$T = T(P)$$

with $P = (x, y, z)$, and define the *gradient* of $T$ in the point $P$ by the formula

$$\boxed{\mathbf{grad}\, T(P) = T_x(P)\mathbf{i} + T_y(P)\mathbf{j} + T_z(P)\mathbf{k}.}$$

Often one writes $T(\mathbf{r})$ instead of $P$, where $\mathbf{r}$ is the vector $\mathbf{r} = \overrightarrow{OP}$.

**Directional derivatives:** If $\mathbf{n}$ is a unit vector (i.e., a vector of length one), then the derivative of $T$ at the point $P$ is defined by means of the formula

$$\frac{\partial T(P)}{\partial \mathbf{n}} := \lim_{h \to 0} \frac{T(\mathbf{r} + h\mathbf{n}) - T(\mathbf{r})}{h}.$$

One has:

$$\boxed{\frac{\partial T(P)}{\partial \mathbf{n}} = \mathbf{n}(\mathbf{grad}\, T(P)),}$$

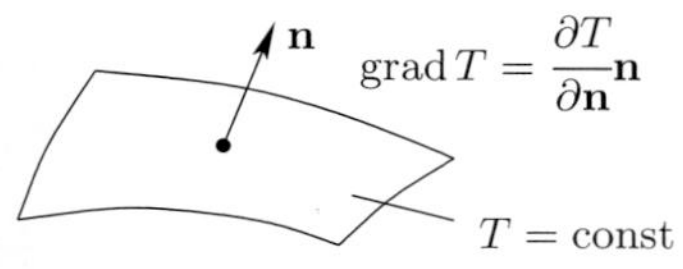


*Figure 1.97.*

if $T$ is of type $C^1$ in a neighborhood of the point $P$. If $\mathbf{n}$ denotes the normal vector to a surface, then $\partial T/\partial \mathbf{n}$ is called the *normal derivative.*

**Physical interpretation:** If $T(P)$ is the temperature at the point $P$, then one has:

(i) The vector $\mathbf{grad}\ T(P)$ is perpendicular to the surface of constant temperature $T = \text{const}$ and points in the direction of increasing temperature.[82]

(ii) The length $|\mathbf{grad}\ T(P)|$ of the gradient is equal to the normal derivative $\frac{\partial T(P)}{\partial \mathbf{n}}$ (see Figure 1.97).

The function $f(h) := T(\mathbf{r} + h\mathbf{n})$ describes the temperature on the line through the point $P$ in the direction of $\mathbf{n}$. One has

$$\frac{\partial T(P)}{\partial \mathbf{n}} = f'(0).$$

**Scalar fields:** Real-valued functions are also called *scalar functions* in physics (for example one speaks of a scalar temperature field).

**Divergence and curl:** Suppose we are given a *vector field*

$$\mathbf{F} = \mathbf{F}(P),$$

i.e., an assignment of a vector $\mathbf{F}(P)$ to each point $P$. For example $\mathbf{F}(P)$ might be a force acting at the point $P$. Instead of $\mathbf{F}(P)$ one also often writes $\mathbf{F}(\mathbf{r})$ with $\mathbf{r} = \overrightarrow{OP}$. In a Cartesian coordinate system we have the representation

$$\mathbf{F}(P) = a(P)\mathbf{i} + b(P)\mathbf{j} + c(P)\mathbf{k}.$$

We define the *divergence* of the vector field $\mathbf{F}$ at the point $P$ by

$$\boxed{\operatorname{div} \mathbf{F}(P) := a_x(P) + b_y(P) + c_z(P)}$$

and the *curl* of the vector field $\mathbf{F}$ by

$$\boxed{\mathbf{curl}\, \mathbf{F}(P) := (c_y - b_z)\mathbf{i} + (a_z - c_x)\mathbf{j} + (b_x - a_y)\mathbf{k},}$$

where $c_y$ stands for $c_y(P)$, etc.

[82]The surface $T = \text{const}$ is called a *level surface* of the function $T$.

**Vector gradient:** For a fixed vector $\mathbf{v}$ we define

$$\boxed{(\mathbf{v}\,\mathbf{grad})\mathbf{F}(P) := \lim_{h\to 0}\frac{\mathbf{F}(\mathbf{r}+h\mathbf{v})-\mathbf{F}(\mathbf{r})}{h},}$$

where $\mathbf{r} = \overrightarrow{OP}$. In particular, if $\mathbf{n}$ is a unit vector, then one calls

$$\boxed{\frac{\partial \mathbf{F}(P)}{\partial \mathbf{n}} := (\mathbf{n}\,\mathbf{grad})\mathbf{F}(P)}$$

the *derivative of the vector field* $\mathbf{F}$ *in the direction of* $\mathbf{n}$ at the point $P$. In a Cartesian coordinate system one has

$$(\mathbf{v}\,\mathbf{grad})\mathbf{F}(P) = (\mathbf{v}\,\mathbf{grad}\,a(P))\mathbf{i} + (\mathbf{v}\,\mathbf{grad}\,b(P))\mathbf{j} + (\mathbf{v}\,\mathbf{grad}\,c(P))\mathbf{k}.$$

**Laplace operator:** We define

$$\boxed{\Delta T(P) := \operatorname{div}\mathbf{grad}\,T(P)}$$

and

$$\boxed{\Delta \mathbf{F}(P) := \mathbf{grad}\operatorname{div}\mathbf{F}(P) - \mathbf{curl}\,\mathbf{curl}\,\mathbf{F}(P).}$$

In a Cartesian coordinate system one has:

$$\Delta T(P) = T_{xx}(P) + T_{yy}(P) + T_{zz}(P),$$
$$\Delta \mathbf{F}(P) = \mathbf{F}_{xx}(P) + \mathbf{F}_{yy}(P) + \mathbf{F}_{zz}(P).$$

From $\mathbf{F} = a\mathbf{i} + b\mathbf{j} + z\mathbf{k}$ it follows $\Delta\mathbf{F} = (\Delta a)\mathbf{i} + (\Delta b)\mathbf{j} + (\Delta c)\mathbf{k}$.

**Invariance:** The expressions $\mathbf{grad}\,T$, $\operatorname{div}\mathbf{F}$, $\mathbf{curl}\,\mathbf{F}$, $(\mathbf{v}\,\mathbf{grad})\mathbf{F}$, $\Delta T$ and $\Delta\mathbf{F}$ have a meaning which is independent of the chosen Cartesian coordinate system. In all Cartesian coordinate systems these expressions are given by the same formulas. The following definitions also have such an invariant meaning.

A vector field $\mathbf{F} = \mathbf{F}(P)$ on an open set $D$ of $\mathbb{R}^3$ is said to be of type $C^k$, if and only if all components $a$, $b$ and $c$ in any Cartesian coordinate system are of type $C^k$.

**Curved coordinates:** The formulas for $\mathbf{grad}\,T$, $\operatorname{div}\mathbf{F}$ and $\mathbf{curl}\,\mathbf{F}$ in cylinder and spherical coordinates can be found in Table 1.5. The corresponding formulas in arbitrary curved coordinates can be elegantly given using tensor analysis (cf. [212]).

**Physical interpretation:** Compare sections 1.9.3 – 1.9.10.

### 1.9.3 Applications to deformations

The operations $\operatorname{div}\mathbf{u}$ and $\mathbf{curl}\,\mathbf{u}$ on a vector field $\mathbf{u}$ play a fundamental role in the description of the behavior of deformations. The deformation of a elastic body under the influence of forces is given by

$$\boxed{\mathbf{y}(\mathbf{r}) = \mathbf{r} + \mathbf{u}(\mathbf{r})\,.}$$

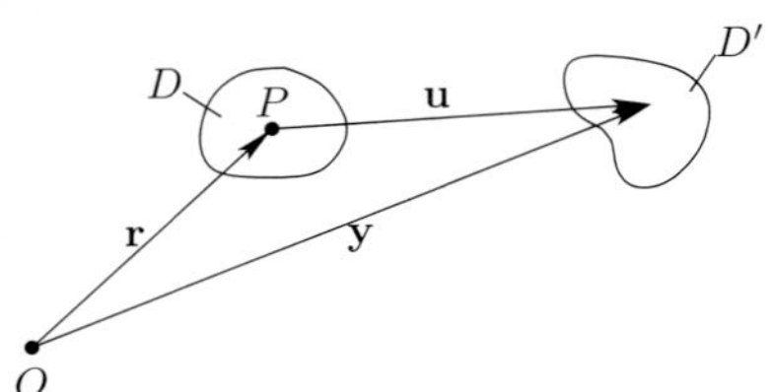


Figure 1.98.

Here $\mathbf{r} := \overrightarrow{OP}$ and $\mathbf{y}(\mathbf{r})$ are vectors based at the point $O$ (Figure 1.98). For simplification of the relations we identify the vector $\mathbf{r}$ with its endpoint $P$.

The endpoint of the vector $\mathbf{r}$ is transformed into the endpoint of $\mathbf{y}(\mathbf{r}) = \mathbf{r} + \mathbf{u}(\mathbf{r})$ under this deformation. We set

$$\boxed{\boldsymbol{\omega} := \frac{1}{2}\mathbf{curl}\ \mathbf{u}(P)\ .}$$

We denote the line through the point $O$ in the direction of $\boldsymbol{\omega}$ by $A$.

**Theorem:** Let $\mathbf{u} = \mathbf{u}(P)$ be a $C^1$-vector field in the domain $D \subset \mathbb{R}^3$. Then one has:

(i) A small increment of volume at the point $P$ is rotated in a first approximation by the angle $|\boldsymbol{\omega}|$ with respect to the axis $A$. In addition, it is rotated in the direction of the main axis, and there is a translation involved.

(ii) If the first partial derivatives of the coordinates of $\mathbf{u}$ are sufficiently small, then $\operatorname{div}\mathbf{u}(P)$ is in a first approximation equal to the relative change in volume of a small increment of volume at the point $P$.

This explains the names 'curl' and 'divergence' of the vector field: the curl describes the rotation[83] of the vector field, while the divergence describes for example a change of mass of a chemical substance during a chemical reaction (cf. 1.9.7).

We wish to discuss this theorem in more detail. The starting point of our considerations is the decomposition

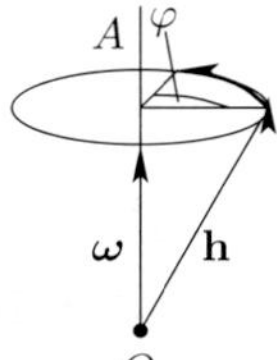


Figure 1.99.

$$\boxed{\mathbf{y}(\mathbf{r}+\mathbf{h}) = \mathbf{r} + \mathbf{u}(\mathbf{r}) + (\mathbf{h} + \boldsymbol{\omega}\times\mathbf{h}) + D(\mathbf{r})\mathbf{h} + \mathbf{R},}$$

with a remainder term $|\mathbf{R}| = o(|\mathbf{h}|)$ for $|\mathbf{h}| \longrightarrow 0$.

**Infinitesimal rotation:** Let $T(\boldsymbol{\omega})\mathbf{h}$ denote the vector based at $O$, which is the rotation of $\mathbf{h}$ with respect to the axis $A$ by an angle of $\varphi := |\boldsymbol{\omega}|$ (Figure 1.99). Then one has

$$T(\boldsymbol{\omega})\mathbf{h} = \mathbf{h} + \boldsymbol{\omega}\times\mathbf{h} + o(\varphi), \qquad \varphi \to 0.$$

For this reason one denotes the product $\boldsymbol{\omega}\times\mathbf{h}$ as the *infinitesimal rotation* of $\mathbf{h}$.

**Dilations:** There are three pairwise perpendicular unit vectors $\mathbf{b}_1, \mathbf{b}_2$ and $\mathbf{b}_3$ based at $P$ and positive real numbers $\lambda_1$, $\lambda_2$ and $\lambda_3$ such that:

$$D(\mathbf{r})\mathbf{h} = \sum_{j=1}^{3} \lambda_j(\mathbf{h}\mathbf{b}_j)\mathbf{b}_j.$$

Consequently one has $D(\mathbf{r})\mathbf{b}_j = \lambda_j\mathbf{b}_j$, $j = 1,2,3$. One refers to $\mathbf{b}_j$, $j = 1,2,3$ as the *principal axi* of the stretching at the point $P$. If we consider a Cartesian coordinate system, then $\lambda_1$, $\lambda_2$ and $\lambda_3$ are the three eigenvalues of the matrix $(d_{jk})$ whose entries are given by the relations

$$d_{jk} := \frac{1}{2}\left(\frac{\partial u_k}{\partial x_j} + \frac{\partial u_j}{\partial x_k}\right).$$

[83] In German, this is referred to as **rot u**, which is short for 'Rotation', the German word for rotation.

Here we have $\mathbf{u} = u_1\mathbf{i} + u_2\mathbf{j} + u_3\mathbf{k}$. Moreover, the columns of the matrix $(d_{jk})$ are the coordinates of the principal axi $\mathbf{b}_1$, $\mathbf{b}_2$, $\mathbf{b}_3$.

**Applications to the equations of hydrodynamics and elasticity:** See [212].

### 1.9.4 Calculus with the nabla operator

**The nabla operator:** Many formulas of vector analysis can be found in Table 1.4. These formulas can be verified by a series of straightforward calculations in Cartesian coordinates. However, the same results can be derived much more easily by applying calculus with the nabla operator.[84] For this one introduces the nabla operator as follows:

$$\nabla := \mathbf{i}\frac{\partial}{\partial x} + \mathbf{j}\frac{\partial}{\partial y} + \mathbf{k}\frac{\partial}{\partial z}.$$

Then one has:

$$\boxed{\mathbf{grad}\, T = \nabla T, \qquad \mathrm{div}\mathbf{F} = \nabla\mathbf{F}, \qquad \mathbf{curl}\,\mathbf{F} = \nabla \times \mathbf{F}.}$$

Another common notation for $\nabla$ is $\dfrac{\partial}{\partial \mathbf{r}}$.

**Rules for calculus with $\nabla$:**

(i) Write the expression you wish to calculate as a formal product with the help of $\nabla$.

(ii) Linear combinations are multiplied distributively:

$$\nabla(\alpha X + \beta Y) = \alpha\nabla X + \beta\nabla Y.$$

Here $X$ and $Y$ are function or vectors, and $\alpha$ and $\beta$ are real numbers.

(iii) A product of the form $\nabla(XY)$ is written as follows:

$$\nabla(XY) = \nabla(\overline{X}Y) + \nabla(X\overline{Y}). \tag{1.161}$$

For this it is irrelevant whether $X$ and $Y$ are vectors or functions. The bar indicates that the corresponding factor is differentiated.

(iv) Transform the expressions $\nabla(\overline{X}Y)$, $\nabla(X\overline{Y})$ strictly according to the rules of vector algebra in such a way that all expressions *without* a bar are to the *left* of $\nabla$ (and all objects with a bar are to the right). During these manipulations, treat $\nabla$ as a *vector*.

(v) Finally write the formal products involving $\nabla$ as expressions of vector analysis, for example

$$U(\nabla\overline{V}) = U(\mathbf{grad}\, V), \quad \mathbf{v} \times (\nabla \times \overline{\mathbf{w}}) = \mathbf{v} \times \mathbf{curl}\,\mathbf{w} \text{ etc.}$$

Here the bars are dropped.

This formal calculus takes into account that $\nabla$ is on the one hand a vector and on the other a differential operator. The formula (1.161) is just the product formula for differentiation. For three factors, use the rule

$$\nabla(XYZ) = \nabla(\overline{X}YZ) + \nabla(X\overline{Y}Z) + \nabla(XY\overline{Z}).$$

*Example 1:* $\mathbf{grad}(U + V) = \nabla(U + V) = \nabla U + \nabla V = \mathbf{grad}\, U + \mathbf{grad}\, V$.

[84] The name 'nabla' and the symbol $\nabla$ come from a Phoenician string instrument.

*Table 1.4. Rules of vector calculus*

***Gradient***

$$\mathbf{grad}\, c = 0, \quad \mathbf{grad}(cU) = c\,\mathbf{grad}\, U \qquad (c = \text{const}),$$
$$\mathbf{grad}(U+V) = \mathbf{grad}\, U + \mathbf{grad}\, V, \quad \mathbf{grad}(UV) = U\,\mathbf{grad}\, V + V\,\mathbf{grad}\, U,$$
$$\mathbf{grad}\,(\mathbf{vw}) = (\mathbf{v}\,\mathbf{grad}\,)\mathbf{w} + (\mathbf{w}\,\mathbf{grad})\mathbf{v} + \mathbf{v} \times \mathbf{curl}\,\mathbf{w} + \mathbf{w} \times \mathbf{curl}\,\mathbf{v},$$
$$\mathbf{grad}(\mathbf{cr}) = \mathbf{c} \qquad (\mathbf{c} = \text{const}),$$
$$\mathbf{grad}\, U(r) = U'(r)\frac{\mathbf{r}}{r} \qquad (\text{central field; } r = |\mathbf{r}|),$$
$$\mathbf{grad}\, F(U) = F'(U)\,\mathbf{grad}\, U,$$
$$\frac{\partial U}{\partial \mathbf{n}} = \mathbf{n}(\mathbf{grad}\, U) \qquad (\text{directional derivative in the direction of the unit vector } \mathbf{n}),$$
$$U(\mathbf{r}+\mathbf{a}) = U(\mathbf{r}) + \mathbf{a}(\mathbf{grad}\, U(r)) + \ldots \qquad (\text{Taylor expansion}).$$

***Divergence***

$$\operatorname{div}\mathbf{c} = 0, \quad \operatorname{div}(c\mathbf{v}) = c \operatorname{div}\mathbf{v} \qquad (c, \mathbf{c} = \text{const}),$$
$$\operatorname{div}(\mathbf{v}+\mathbf{w}) = \operatorname{div}\mathbf{v} + \operatorname{div}\mathbf{w}, \quad \operatorname{div}(U\mathbf{v}) = U \operatorname{div}\mathbf{v} + \mathbf{v}(\mathbf{grad}\, U),$$
$$\operatorname{div}(\mathbf{v} \times \mathbf{w}) = \mathbf{w}(\mathbf{curl}\,\mathbf{v}) - \mathbf{v}(\operatorname{curl}\mathbf{w}),$$
$$\operatorname{div}(U(r)\mathbf{r}) = 3U(r) + rU'(r) \qquad (\text{central field; } r = |\mathbf{r}|),$$
$$\operatorname{div}\mathbf{curl}\,\mathbf{v} = 0.$$

***Curl***

$$\mathbf{curl}\,\mathbf{c} = 0, \quad \mathbf{curl}(c\mathbf{v}) = c\,\mathbf{curl}\,\mathbf{v} \qquad (c, \mathbf{c} = \text{const}),$$
$$\mathbf{curl}(\mathbf{v}+\mathbf{w}) = \mathbf{curl}\,\mathbf{v} + \mathbf{curl}\,\mathbf{w}, \quad \mathbf{curl}(U\mathbf{v}) = U\,\mathbf{curl}\,\mathbf{v} + (\mathbf{grad}\, U) \times \mathbf{v},$$
$$\mathbf{curl}(\mathbf{v} \times \mathbf{w}) = (\mathbf{w}\,\mathbf{grad})\mathbf{v} - (\mathbf{v}\,\mathbf{grad})\mathbf{w} + \mathbf{v}\operatorname{div}\mathbf{w} - \mathbf{w}\operatorname{div}\mathbf{v},$$
$$\mathbf{curl}(\mathbf{c} \times \mathbf{r}) = 2\mathbf{c},$$
$$\mathbf{curl}\,\mathbf{grad}\,\mathbf{v} = 0,$$
$$\mathbf{curl}\,\mathbf{curl}\,\mathbf{v} = \mathbf{grad}\operatorname{div}\mathbf{v} - \Delta\mathbf{v}.$$

***Laplace operator***

$$\Delta U = \operatorname{div}\mathbf{grad}\, U,$$
$$\Delta\mathbf{v} = \mathbf{grad}\operatorname{div}\mathbf{v} - \mathbf{curl}\,\mathbf{curl}\,\mathbf{v}.$$

***Vector gradient***

$$2(\mathbf{v}\,\mathbf{grad})\mathbf{w} = \mathbf{curl}(\mathbf{w} \times \mathbf{v}) + \mathbf{grad}(\mathbf{vw}) + \mathbf{v}\operatorname{div}\mathbf{w} - \mathbf{w}\operatorname{div}\mathbf{v} - \mathbf{v} \times \mathbf{curl}\,\mathbf{w} - \mathbf{w} \times \mathbf{curl}\,\mathbf{v},$$
$$\mathbf{w}(\mathbf{r}+\mathbf{a}) = \mathbf{w}(\mathbf{r}) + (\mathbf{a}\,\mathbf{grad})\mathbf{w}(\mathbf{r}) + \ldots \qquad (\text{Taylor expansion}).$$

### 1.11.1 The Laplace transformation

An extensive table of Laplace transformations of various functions can be found in 0.10.2.

As the mathematician Doetsch noticed a few decades after Heaviside, one can justify Heaviside's calculus with the help of a transformation which goes back to Laplace (1749–1827). The basic formula is

$$F(s) := \int_0^\infty e^{-st} f(t)\,dt, \qquad s \in H_\gamma.$$

Here $H_\gamma := \{s \in \mathbb{C} \mid \operatorname{Re} s > \gamma\}$ denotes a half-plane in the complex plane (Figure 1.114). One calls $F$ the *Laplace transform* of $f$ and also writes $F = \mathscr{L}\{f\}$ for this.

**The class $K_\gamma$ of admissible functions:** Let $\gamma$ be a real number. By definition, $K_\gamma$ consists of all functions $f : [0, \infty[ \longrightarrow \mathbb{C}$ satisfying the weak growth condition

$$|f(t)| \leq \text{const}\, e^{\gamma t} \qquad \text{for all } t \geq 0.$$

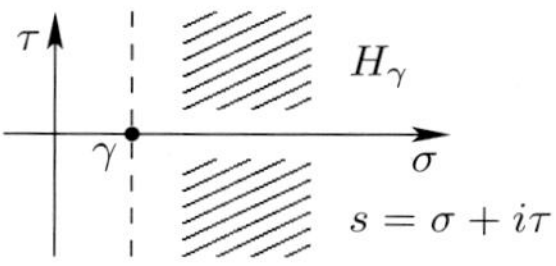


*Figure 1.114.*

**Theorem of existence:** For $f \in K_\gamma$, the Laplace transformation $F$ of $f$ exists and is holomorphic, i.e., infinitely often differentiable, on the half-strip $H_\gamma$. The derivatives are obtained by differentiating under the integral sign. For example one has:

$$F'(s) = \int_0^\infty e^{-st}(-tf(t))\,dt \qquad \text{for all } s \in H_\gamma.$$

**Theorem of uniqueness:** If two functions $f, g \in K_\gamma$ have the same Laplace transformations on $H_\gamma$, then $f = g$.

**Convolution:** We denote by $R$ the entirety of all continuous functions $f : [0, \infty[ \longrightarrow \mathbb{C}$. For $f, g \in R$, we define the *convolution* $f * g \in R$ by the formula

$$(f * g)(t) := \int_0^t f(\tau) g(t - \tau)\,d\tau \qquad \text{for all } t \geq 0.$$

For all $f, g, h \in R$ one has:[99]

(i) $f * g = g * f$, (commutativity),

(ii) $f * (g * h) = (f * g) * h$, (associativity),

(iii) $f * (g + h) = f * g + f * h$, (distributivity),

(iv) from $f * g = 0$ it follows that $f = 0$ or $g = 0$.

[99] The properties (i) to (iv) show that $R$ with respect to the 'multiplication' $*$ and the usual addition of functions forms a commutative ring without divisors of zero (an integral domain), cf. 2.5.2.

#### 1.11.1.1 The basic laws

**Law 1** (exponential function):

$$\mathscr{L}\left\{\frac{t^n}{n!}\mathrm{e}^{\alpha t}\right\} = \frac{1}{(s-\alpha)^{n+1}}, \qquad n = 0, 1, \ldots, s \in H_\sigma.$$

Here $\alpha$ denotes an arbitrary complex number whose real part is $\sigma$.

*Example 1:* $\mathscr{L}\{\mathrm{e}^{\alpha t}\} = \dfrac{1}{s-\alpha}, \qquad \mathscr{L}\{t\mathrm{e}^{\alpha t}\} = \dfrac{1}{(s-\alpha)^2}.$

**Law 2** (linearity): For $f, g \in K_\gamma$, and $a, b \in \mathbb{C}$, one has

$$\mathscr{L}\{af + bg\} = a\mathscr{L}\{f\} + b\mathscr{L}\{g\}.$$

**Law 3** (differentiation): Assume the function $f \in K_\gamma$ is of type $C^n$, $n \geq 1$. We set $F := \mathscr{L}\{f\}$. Then we have for all $s \in H_\gamma$:

$$\mathscr{L}\{f^{(n)}\}(s) := s^n F(s) - s^{n-1}f(0) - s^{n-2}f'(0) - \ldots - f^{(n-1)}(0).$$

*Example 2:* $\mathscr{L}\{f'\} = sF(s) - f(0), \qquad \mathscr{L}\{f''\} = s^2F(s) - sf(0) - f'(0).$

**Law 4** (convolution): For $f, g \in K_\gamma$, we have:

$$\mathscr{L}\{f * g\} = \mathscr{L}\{f\}\mathscr{L}\{g\}.$$

#### 1.11.1.2 Applications to differential equations

**The universal method:** The Laplace transformation turns out to be a universal tool, to solve in a very elegant way

> ordinary differential equations of arbitrary order with constant coefficients as well as systems of such equations.

Equations of this kind occur for example quite often in control engineering.

One uses the following solution steps:

(i) Transform the given differential equation (D) into an algebraic equation (A) with the help of the linearity and differentiation law for the Laplace transformation (laws 2 and 3).

(ii) The equation (A) is a linear equationor a system of linear equations and can be solved with the methods of linear algebra. This solution is a rational function and a partial fraction decomposition (cf. 2.1.7) of this rational function can be formed.

(iii) These partial fractions are individually transformed back with the help of law 1 (exponential function).

(iv) Inhomogeneous terms of the differential equation give rise to product terms in the image space, which can be transformed back by using the convolution (law 4).

To get the partial fraction decomposition, one needs to determine the zeros of the denominator, which correspond under the inverse transformation to the frequencies of the characteristic oscillations of the system.

*Example 1* (harmonic oscillator): The oscillation $x = f(t)$ of a spring at the time $t$ under the influence of an exterior force $\mathbf{f} = \mathbf{f}(t)$ is described by the differential equation

$$\boxed{\begin{aligned} &f'' + \omega^2 f = \mathbf{f}, \\ &f(0) = a, \quad f'(0) = b, \end{aligned}} \tag{1.188}$$

with $\omega > 0$ (cf. 1.9.1).

We set $F := \mathscr{L}\{f\}$ and $\mathbf{F} := \mathscr{L}\{\mathbf{f}\}$. From the first line of (1.188) it follows that

$$\mathscr{L}\{f''\} + \omega^2 \mathscr{L}\{f\} = \mathscr{L}\{\mathbf{f}\}$$

because of the linearity of the Laplace transformation (law 2). The law of differentiation (law 3) then yields

$$s^2 F - as - b + \omega^2 F = \mathbf{F}$$

which has the solution

$$F = \frac{as+b}{s^2+\omega^2} + \frac{\mathbf{F}}{s^2+\omega^2}.$$

The partial fraction decomposition of the this is

$$F = \frac{a}{2}\left(\frac{1}{s-\mathrm{i}\omega} + \frac{1}{s+\mathrm{i}\omega}\right) + \frac{b}{2\mathrm{i}\omega}\left(\frac{1}{s-\mathrm{i}\omega} - \frac{1}{s+\mathrm{i}\omega}\right) + \frac{\mathbf{F}}{2\mathrm{i}\omega}\left(\frac{1}{s-\mathrm{i}\omega} - \frac{1}{s+\mathrm{i}\omega}\right).$$

Applying now exponentiation (law 1) and convolution (law 4) leads to

$$f(t) = a\left(\frac{\mathrm{e}^{\mathrm{i}\omega t} + \mathrm{e}^{-\mathrm{i}\omega t}}{2}\right) + b\left(\frac{\mathrm{e}^{\mathrm{i}\omega t} - \mathrm{e}^{-\mathrm{i}\omega t}}{2\mathrm{i}\omega}\right) + \mathbf{F} * \left(\frac{\mathrm{e}^{\mathrm{i}\omega t} - \mathrm{e}^{-\mathrm{i}\omega t}}{2\mathrm{i}\omega}\right).$$

The Euler formula $\mathrm{e}^{\mathrm{i}\omega t} = \cos\omega t \pm \mathrm{i}\sin\omega t$ gives the solution:

$$\boxed{f(t) = f(0)\cos\omega t + \frac{f'(0)}{\omega}\sin\omega t + \frac{1}{\omega}\int_0^t (\sin\omega(t-\tau))\mathbf{f}(\tau)\mathrm{d}\tau.}$$

This representation of the solution displays for the engineer or the physicist how the individual quantities influence the system. For example, a pure cosine wave with the angular frequency $\omega$ is obtained for $f'(0) = 0$ and $\mathbf{f} \equiv 0$, i.e., when the system is a rest at $t = 0$ and there are no exterior forces.

In case $f(0) = f'(0) = 0$ (this means that in addition to being at rest at $t = 0$, the position of the spring is at the origin) the system is only influenced by the exterior force, and we get

$$\boxed{f(t) = \int_0^t G(t,\tau)\mathbf{f}(\tau)\mathrm{d}\tau.}$$

The function $G(t,\tau) := \dfrac{1}{\omega}\sin\omega(t-\tau)$ is called the *Green's function* of the harmonic oscillator.

*Example 2* (harmonic oscillator with friction):

$$\boxed{\begin{array}{l} f'' + 2f' + f = 0, \\ f(0) = 0, \qquad f'(0) = b. \end{array}}$$

Applying the Laplace transformation to this, we get

$$s^2F - b + 2sF + F = 0,$$

which means

$$F = \frac{b}{s^2 + 2s + 1} = \frac{b}{(s+1)^2}.$$

The inverse transformation (law 1) yields the solution

$$\boxed{f(t) = f'(0)te^{-t}.}$$

One has $\lim_{t \to +\infty} f(t) = 0$. This means that the system returns to a position at rest after sufficient time.

*Example 3* (electrical circuit): We consider an electrical circuit with a resistance $R$, a coil with the inductance $L$ and the potential difference $V = V(t)$ (Figure 1.115). The differential equation for the current $I(t)$ at the time $t$ is:

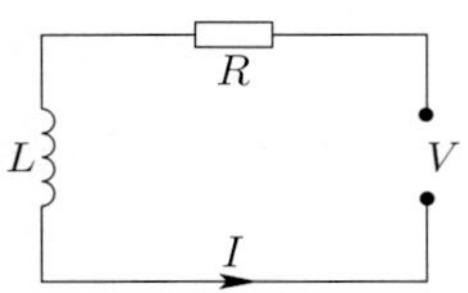


*Figure 1.115.*

$$\boxed{\begin{array}{rcl} LI' + RI & = & V, \\ I(0) & = & a. \end{array}}$$

We set $F := \mathscr{L}\{I\}$ and $K := \mathscr{L}\{V\}$. For simplicity we set $L = 1$. As in Example 1, we get

$$sF - I(0) + RF = K$$

with the solution

$$F = \frac{I(0)}{s+R} + K\left(\frac{1}{s+R}\right).$$

From laws 3 and 4 we get for the solution:

$$\boxed{I(t) = I(0)e^{-Rt} + \int_0^t e^{-R(t-\tau)}V(\tau)d\tau.}$$

From this one can see that the resistance $R > 0$ has a dampening effect.

*Example 4:* We consider the differential equation

$$\boxed{\begin{array}{rcll} f^{(n)} & = & g, & \\ f(0) & = & f'(0) = \ldots = f^{(n-1)}(0) = 0, & \qquad n = 1, 2, \ldots \end{array}}$$

Applying the Laplace transformation leads to

$$s^nF = G,$$

which means $F = G\left(\frac{1}{s^n}\right)$. Exponentiating (law 1) leads to $\mathscr{L}\left\{\frac{t^{n-1}}{(n-1)!}\right\} = \frac{1}{s^n}$. Convolution then gives the solution

$$f(t) = \int_0^t \frac{(t-\tau)^{n-1}}{(n-1)!} g(\tau)\mathrm{d}\tau.$$

In the special case where $n = 1$ this is $f(t) = \int_0^t g(\tau)\mathrm{d}\tau$.

*Example 5* (a system of differential equations):

$$f' + g' = 2k, \quad f' - g' = 2h,$$
$$f(0) = g(0) = 0.$$

Here the Laplace transformation leads to the *system of linear equations* (cf. 2.1.4.)

$$sF + sG = 2K, \quad sF - sG = 2H$$

with the solution

$$F = (K + H)\frac{1}{s}, \quad G = (K - H)\frac{1}{s}.$$

According to law 1 we have $\mathscr{L}\{1\} = \frac{1}{s}$. The inverse transformation using convolution yields $f = (k + h) * 1$ and $g = (k - h) * 1$. This means:

$$f(t) = \int_0^t (k(\tau) + h(\tau))\mathrm{d}\tau, \qquad g(t) = \int_0^t (k(\tau) - h(\tau))\mathrm{d}\tau.$$

### 1.11.1.3 Further rules

**Translation:** $\mathscr{L}\{f(t-b)\} = \mathrm{e}^{-bs}\mathscr{L}\{f(t)\}$ for $b \in \mathbb{R}$.

**Dampening:** $\mathscr{L}\{\mathrm{e}^{-\alpha t}f(t)\} = F(s+\alpha)$ for $\alpha \in \mathbb{C}$.

**Similarity:** $\mathscr{L}\{f(at)\} = \frac{1}{a}F\left(\frac{s}{a}\right)$ for $a > 0$.

**Multiplication:** $\mathscr{L}\{t^n f(t)\} = (-1)^n F^{(n)}(s)$ for $n = 1, 2, \ldots$

**Inverse transformation:** If $f \in K_\gamma$, then one has

$$f(t) = \frac{1}{2\pi}\int_{-\infty}^{\infty} \mathrm{e}^{(\sigma+\mathrm{i}\tau)t}F(\sigma + \mathrm{i}\tau)\mathrm{d}\tau \qquad \text{for all } t \geq 0,$$

where $\sigma$ denotes some fixed number with $\sigma > \gamma$ and $F$ denotes the Laplace transformation of $f$.

### 1.11.2 The Fourier transformation

Extensive tables of Fourier transforms of functions can be found in 0.10.1.

**Basic ideas:** The basic formula is

$$f(t) = \frac{1}{\sqrt{2\pi}} \int_{-\infty}^{\infty} F(\omega) \mathrm{e}^{\mathrm{i}\omega t} \mathrm{d}\omega \tag{1.189}$$

with the amplitude

$$F(\omega) = \frac{1}{\sqrt{2\pi}} \int_{-\infty}^{\infty} f(t) \mathrm{e}^{-\mathrm{i}\omega t} \mathrm{d}t. \tag{1.190}$$

We set $\mathscr{F}\{f\} := F$ and call $F$ the *Fourier transform* of $f$. Moreover, the set of all Fourier transforms form a space called the *Fourier space*. The basic property of the Fourier transformation is obtained upon differentiating (1.189) with respect to $t$:

$$f'(t) = \frac{1}{\sqrt{2\pi}} \int_{-\infty}^{\infty} \mathrm{i}\omega F(\omega) \mathrm{e}^{\mathrm{i}\omega t} \mathrm{d}\omega. \tag{1.191}$$

Hence the derivative $f'$ passes over to a multiplication $\mathrm{i}\omega F$ in the Fourier space.

**Physical interpretation:** Let $t$ be the time. The formula (1.189) displays the time-dependent process $f = f(t)$ as a continuous superposition of oscillations

$$F(\omega) \mathrm{e}^{\mathrm{i}\omega t}$$

of angular frequency $\omega$ and amplitude $F(\omega)$.

The influence of the angular frequency on the behavior of the function $f$ increases with the absolute value $|F(\omega)|$.

*Example 1* (rectangular momentum): The Fourier transform of the function

$$f(t) := \begin{cases} 1 & \text{for} \quad -a \leq t \leq a, \\ 0 & \text{otherwise} \end{cases}$$

is

$$F(\omega) = \frac{1}{\sqrt{2\pi}} \int_{-a}^{a} \mathrm{e}^{-\mathrm{i}\omega t} \mathrm{d}t = \begin{cases} \dfrac{2 \sin a\omega}{\omega\sqrt{2\pi}} & \text{for} \quad \omega \neq 0, \\[2ex] \dfrac{2a}{\sqrt{2\pi}} & \text{for} \quad \omega = 0. \end{cases}$$

*Example 2* (dampened oscillations): Let $\alpha$ and $\beta$ be positive numbers. The Fourier transform of the function

$$f(t) := \begin{cases} \mathrm{e}^{-\alpha t} \mathrm{e}^{\mathrm{i}\beta t} & \text{for} \quad t \geq 0, \\ 0 & \text{for} \quad t < 0 \end{cases}$$

is

$$F(\omega) = \frac{1}{\sqrt{2\pi}} \frac{1}{\alpha + \mathrm{i}(\omega - \beta)}$$

with

$$|F(\omega)| = \frac{1}{\sqrt{2\pi}\,(\alpha^2 + (\omega - \beta)^2)}.$$

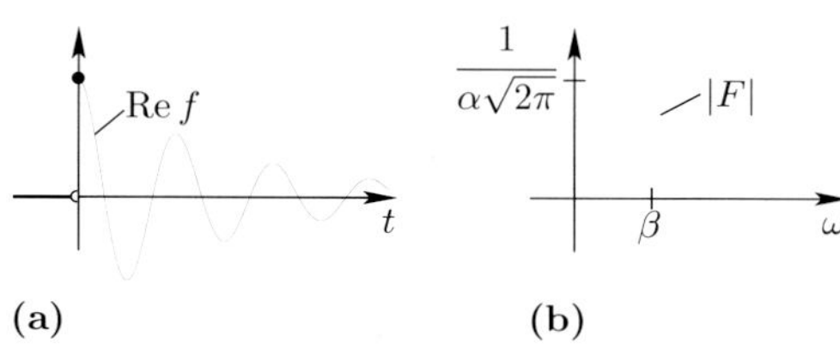


*Figure 1.116. Dampened oscillations.*

According to Figure 1.116, the absolute value of the amplitude $|F|$ has a maximum at the dominating frequency $\omega = \beta$. This maximum is steeper when the dampening is small, i.e., when $\alpha$ is small.

*Example 3* (Gaussian normal distribution): The non-normalized Gaussian distribution $f(t) := e^{-t^2/2}$ has the splendid property of coinciding with its Fourier transformation.

**The Dirac 'delta function', white noise and generalized functions:** Let $\varepsilon > 0$. The Fourier transform of the function

$$\delta_\varepsilon(t) := \frac{\varepsilon}{\pi(\varepsilon^2 + t^2)}$$

is

$$F_\varepsilon(\omega) = \frac{1}{\sqrt{2\pi}} e^{-\varepsilon|\omega|}$$

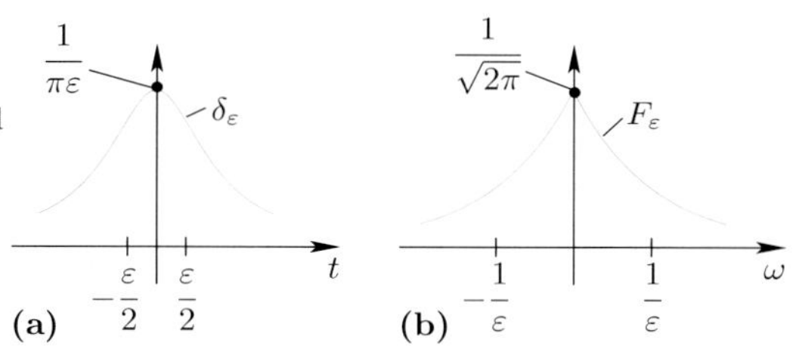


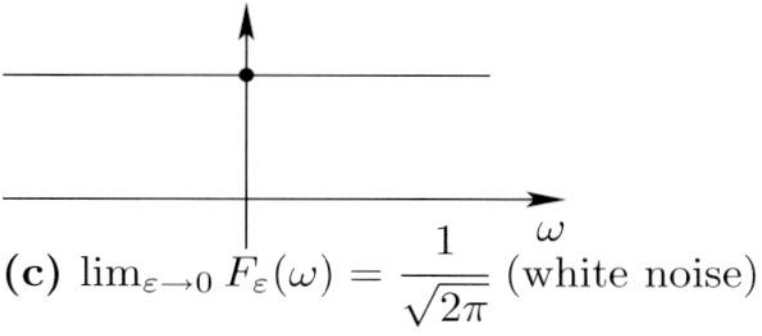


**(c)** $\lim_{\varepsilon\to 0} F_\varepsilon(\omega) = \frac{1}{\sqrt{2\pi}}$ (white noise)

*Figure 1.117. The Dirac delta function and white noise.*

(Figure 1.117). The following considerations are fundamental to the understanding of modern physical literature.

(i) *The limiting process $\varepsilon \longrightarrow 0$ in Fourier space.* We get

$$\boxed{\lim_{\varepsilon\to+0} F_\varepsilon(\omega) = \frac{1}{\sqrt{2\pi}} \qquad \text{for all } \omega \in \mathbb{R}.}$$

Hence the amplitude is constant for all frequencies $\omega$. One speaks of 'white noise'.

(ii) *Formal limiting process $\varepsilon \longrightarrow 0$ in the domain.* Physicists are particularly interested in whether the real process

$$\delta(t) := \lim_{\varepsilon\to 0} \delta_\varepsilon(t)$$

corresponds to white noise. Formally we get

$$\delta(t) := \begin{cases} +\infty & \text{for} \quad t = 0, \\ 0 & \text{for} \quad t \neq 0 \end{cases} \tag{1.192}$$

and

$$\delta(t) = \frac{1}{2\pi} \int_{-\infty}^{\infty} e^{i\omega t} d\omega. \tag{1.193}$$

Moreover, from $\int_{-\infty}^{\infty} \delta_\varepsilon(t)\mathrm{d}t = 1$ one gets formally the relation

$$\int_{-\infty}^{\infty} \delta(t)\mathrm{d}t = 1. \tag{1.194}$$

(iii) *Rigorous justification.* There is no *classical* function $y = \delta(t)$ with the properties (1.192) and (1.194). Also, the integral (1.193) diverges. In spite of this, physicists have used the Dirac delta function, introduced by the renowned physicist Paul Dirac since about 1930, with great success.

The history of mathematics shows that successful formal calculations can always be rigorously justified in an appropriate formulation. In the case at hand, this justification was delivered around 1950 by the French mathematician Laurent Schwartz in the context of his theory of distributions (generalized functions). These are mathematical objects which are infinitely differentiable and which can be manipulated more conveniently than functions. In place of the Dirac delta function one has the Schwartz delta distribution. This wonderful modern extension of the classical differential calculus of Newton and Leibniz can be found in 10.4. of section II.

**Fourier cosine and sine transformations:** For $\omega \in \mathbb{R}$ we define the Fourier cosine transformation by the formula

$$\boxed{F_c(\omega) := \sqrt{\frac{2}{\pi}} \int_0^{\infty} f(t) \cos \omega t \; \mathrm{d}t}$$

and similarly, the Fourier sine transformation is given by the formula

$$\boxed{F_s(\omega) := \sqrt{\frac{2}{\pi}} \int_0^{\infty} f(t) \sin \omega t \; \mathrm{d}t.}$$

We also write $\mathscr{F}_c\{f\}$ and $\mathscr{F}_s\{f\}$ for $F_c$ and $F_s$, respectively.

**Theorem of existence:** Let $f : \mathbb{R} \longrightarrow \mathbb{C}$ be almost everywhere continuous and suppose that

$$\int_{-\infty}^{\infty} |f(t)t^n|\mathrm{d}t < \infty$$

for fixed $n = 0, 1, \ldots$. Then we have:

(i) In case $n = 0$, $\mathscr{F}\{f\}, \mathscr{F}_c\{f\}$ and $\mathscr{F}_s\{f\}$ are continuous on $\mathbb{R}$ and

$$\boxed{2\mathscr{F}\{f\} = \mathscr{F}_c\{f(t) + f(-t)\} - \mathrm{i}\mathscr{F}_s\{f(t) - f(-t)\}.}$$

(ii) If $n \geq 1$, $\mathscr{F}\{f\}, \mathscr{F}_c\{f\}$ and $\mathscr{F}_s\{f\}$ are of type $C^n$ on $\mathbb{R}$. The derivatives are obtained by differentiating under the integral sign. For example, for all $\omega \in \mathbb{R}$ one has

$$F(\omega) = \frac{1}{\sqrt{2\pi}} \int_{-\infty}^{\infty} f(t)\mathrm{e}^{-\mathrm{i}\omega t}\mathrm{d}t,$$

$$F'(\omega) = \frac{1}{\sqrt{2\pi}} \int_{-\infty}^{\infty} f(t)(-\mathrm{i}t)\mathrm{e}^{-\mathrm{i}\omega t}\mathrm{d}t.$$

#### 1.11.2.1 The main theorem:

**The $\mathscr{L}_p$ spaces:** By definition, the space $\mathscr{L}_p$ consists of all functions $f : \mathbb{R} \longrightarrow \mathbb{C}$, which are almost everywhere continuous and for which the $p^{th}$ power of the absolute value of the function is integrable,

$$\int_{-\infty}^{\infty} |f(t)|^p \mathrm{d}t < \infty.$$

**The Schwartz space $\mathscr{S}$:** A function $f : \mathbb{R} \longrightarrow \mathbb{C}$ is contained in $\mathscr{S}$ by definition, if and only if $f$ is infinitely often differentiable and

$$\sup_{t\in\mathbb{R}} |t^k f^{(n)}(t)| < \infty$$

for all $k, n = 0, 1, \ldots$, i.e., the function $f$ and all of its derivatives approach zero rapidly as $t \longrightarrow \pm\infty$.

**The classical theorem of Dirichlet–Jordan:** Suppose a function $f \in \mathscr{L}_1$ has in addition the following properties:

(i) There are finitely many points $t_0 < t_1 < \ldots < t_m$, such that in each interval, where the real and imaginary part of $f$ is increasing or decreasing and continuous on each $]t_j, t_{j+1}[$.

(ii) In each of the points $t_j$, the left and right limit of $f$, $f(t_j \pm 0) := \lim_{\varepsilon\longrightarrow+0} f(t_j \pm \varepsilon)$ exists.

Then the Fourier transform of $f$, denoted $F$, exists, and for all $t \in \mathbb{R}$, one has

$$\boxed{\frac{f(t+0) + f(t-0)}{2} = \frac{1}{\sqrt{2\pi}} \int_{-\infty}^{\infty} F(\omega)\mathrm{e}^{\mathrm{i}\omega t}\mathrm{d}\omega.}$$

In the points where $f$ is continuous, the left side of the above expression is equal to $f(t)$.

**Corollary:** (i) The Fourier transformation (1.190) determines a bijective map $\mathscr{F} : \mathscr{S} \longrightarrow \mathscr{S}$, which assigns to a function $f$ its Fourier transform. The inverse transformation is given by the classical formula (1.189).

(ii) This transformation translates differentiation into multiplication and conversely. More precisely, one has for all $f \in \mathscr{S}$ and all $n = 1, 2, \ldots$ the relations

$$\boxed{\mathscr{F}\{f^{(n)}\}(\omega) = (\mathrm{i}\omega)^n F(\omega) \qquad \text{for all } \omega \in \mathbb{R}} \tag{1.195}$$

and

$$\boxed{\mathscr{F}\{(-\mathrm{i}t)^n f\}(\omega) = F^{(n)}(\omega) \qquad \text{for all } \omega \in \mathbb{R}} \tag{1.196}$$

hold. More generally, the formula (1.195) holds for each function $f : \mathbb{R} \longrightarrow \mathbb{C}$ of type $C^n$ such that $f, f', \ldots, f^{(n)} \in \mathscr{L}_1$. Moreover, the relation (1.196) holds under the weaker assumption that the two functions $f$ and $t^n f(t)$ belong to $\mathscr{L}_1$.

### 1.11.2.2 Rules for calculations

**Rule for differentiation and multiplication:** See (1.195) and (1.196).

**Linearity:** For all $f, g \in \mathscr{L}_1$ and $a, b \in \mathbb{C}$ one has:

$$\boxed{\mathscr{F}\{af + bg\} = a\mathscr{F}\{f\} + b\mathscr{F}\{g\}.}$$

**Translation:** Let $a, b$ and $c$ be real numbers with $a \neq 0$. For each function $f \in \mathscr{L}_1$ one has the relation:

$$\boxed{\mathscr{F}\{\mathrm{e}^{\mathrm{i}ct} f(at + b)\}(\omega) = \frac{1}{a}\mathrm{e}^{\mathrm{i}b(\omega - c)/a} F\left(\frac{\omega - c}{a}\right) \qquad \text{for all } \omega \in \mathbb{R}.}$$

**Convolution:** If both $f$ and $g$ belong to $\mathscr{L}_1$ *and* to $\mathscr{L}_2$, then one has

$$\boxed{\mathscr{F}\{f * g\} = \mathscr{F}\{f\}\mathscr{F}\{g\}}$$

with the convolution

$$(f * g)(t) := \int_{-\infty}^{\infty} f(\tau)g(t - \tau)\mathrm{d}\tau.$$

**The Parseval equality:** For all functions $f \in \mathscr{S}$ one has:

$$\boxed{\int_{-\infty}^{\infty} |f(t)|^2 \mathrm{d}t = \int_{-\infty}^{\infty} |F(\omega)|^2 \mathrm{d}\omega.}$$

Here $F$ denotes the Fourier transform of $f$.

**The connection between the Fourier and the Laplace transformations:** Let $\sigma$ be a real number. We set

$$f(t) := \begin{cases} \mathrm{e}^{-\sigma t}\sqrt{2\pi}\, g(t) & \text{for} \quad t \geq 0, \\ 0 & \text{for} \quad t < 0. \end{cases}$$

Moreover let $s := \sigma + \mathrm{i}\omega$. The Fourier transform of the functions of this special form is given by

$$F(s) = \int_0^{\infty} \mathrm{e}^{-\mathrm{i}\omega t}\mathrm{e}^{-\sigma t} g(t)\mathrm{d}t = \int_0^{\infty} \mathrm{e}^{-st} g(t)\mathrm{d}t.$$

This is the Laplace transform of $g$.

### 1.11.3 The $Z$-transformation

| An extensive table of $Z$-transforms of various functions can be found in 0.10.3. |
|---|

The $Z$-transformation can be viewed as a discrete version of the Laplace transformation. It is used to solve difference equations with constant coefficients.

We consider the sequence of complex numbers

$$f = (f_0, f_1, \ldots).$$

The basic formula is then

$$\boxed{F(z) := \sum_{n=0}^{\infty} \frac{f_n}{z^n}.}$$

One calls $F$ the $Z$-transform of $f$ and writes $F = \mathscr{Z}\{f\}$.

*Example:* Let $f := (1, 1, \ldots)$. The geometric series yields

$$F(z) = 1 + \frac{1}{z} + \frac{1}{z^2} + \ldots = \frac{z}{z-1}$$

for all $z \in \mathbb{C}$ with $|z| > 1$.

**The class $\mathscr{K}_\gamma$ of admissible sequences:** Let $\gamma \geq 0$. The class $\mathscr{K}_\gamma$ consists by definition of the set of all sequences $f$ which satisfy the property

$$\boxed{|f_n| \leq \text{const } \mathrm{e}^{\gamma n}, \qquad n = 0, 1, 2, \ldots}$$

We denote by $R_\gamma := \{z \in \mathbb{C} : |z| > \gamma\}$ the outside of a disc of radius $\gamma$ around the origin.

**Theorem of existence:** For $f \in \mathscr{K}_\gamma$, the $Z$-transform $F$ of $f$ exists and is holomorphic on $R_\gamma$.

**Theorem of uniqueness:** If the $Z$-transforms of two sequences $f, g \in \mathscr{K}_\gamma$ coincide on $R_\gamma$, then $f = g$.

**Inverse transformation:** If $f \in \mathscr{K}_\gamma$, one obtains $f$ from the $Z$-transform $F$ by the formula

$$f_n = \frac{1}{2\pi \mathrm{i}} \int_C F(z) z^{n-1} \mathrm{d}z, \qquad n = 0, 1, 2, \ldots$$

Here the integral is carried out over a circle $C := \{z \in \mathbb{C} : |z| = r\}$ with radius $r > \gamma$.

**Convolution:** For two sequences we define the convolution $f * g$ by the formula

$$(f * g)_n := \sum_{k=0}^{n} f_k g_{n-k}, \qquad n = 0, 1, 2, \ldots$$

One has $f * g = g * f$.

**Translation operator:** We define $Tf$ by the formula

$$(Tf)_n := f_{n+1}, \qquad n = 0, 1, 2, \ldots .$$

Then one has $(T^k f) = f_{n+k}$ for $n = 0, 1, 2, \ldots$ and $k = 0, \pm 1, \pm 2, \ldots$

### 1.11.3.1 The basic laws

Let $F$ be the $Z$-transform of $f$.

**First law** (linearity): For $f, g \in \mathscr{K}_\gamma$ and $a, b \in \mathbb{C}$ one has

$$\boxed{\mathscr{Z}\{af + bg\} = a\mathscr{Z}\{f\} + b\mathscr{Z}\{g\}.}$$

**Second law** (translation): For $k = 1, 2, \ldots$ one has:

$$\boxed{\mathscr{Z}\{T^k f\} = z^k F(z) - \sum_{j=0}^{k-1} f_j z^{k-j}}$$

and

$$\mathscr{Z}\{T^{-k} f\} = z^{-k} F(z).$$

*Example:*

$$\begin{aligned} \mathscr{Z}\{Tf\} &= zF(z) - f_0 z, \\ \mathscr{Z}\{T^2 f\} &= z^2 F(z) - f_0 z^2 - f_1 z. \end{aligned}$$

**Third law** (convolution): For $f, g \in \mathscr{K}_\gamma$ one has:

$$\boxed{\mathscr{Z}\{f * g\} = \mathscr{Z}\{f\}\mathscr{Z}\{g\}.}$$

**Fourth law** (Taylor expansion): If we set $G(\zeta) := F(1/\zeta)$, then we have

$$\boxed{f_n = \frac{G^{(n)}(0)}{n!}, \qquad n = 0, 1, 2, \ldots}$$

**Fifth law** (partial fraction decomposition): For $a \in \mathbb{C}$ one has:

$$\begin{aligned} F(z) &= \frac{1}{z-a}, & f &= (0, 1, a, a^2, a^3, \ldots), \\ F(z) &= \frac{1}{(z-a)^2}, & f &= (0, 0, 1, 2a, 3a^2, 4a^3, \ldots), \\ F(z) &= \frac{1}{(z-a)^3}, & f &= \left(0, 0, 0, 1, \binom{3}{2} a, \binom{4}{2} a^2, \ldots\right), \\ F(z) &= \frac{1}{(z-a)^4}, & f &= \left(0, 0, 0, 0, 1, \binom{4}{3} a, \binom{5}{3} a^2, \ldots\right). \end{aligned} \tag{1.197}$$

Applying the same law one gets the inverse transformation for $\dfrac{1}{(z-a)^5}, \dfrac{1}{(z-a)^6}, \ldots$

The inverse transformation of $\dfrac{z}{(z-a)^n}$ is obtained from the partial fraction decomposition:

$$F(z) = \frac{z}{(z-a)^n} = \frac{a}{(z-a)^n} + \frac{1}{(z-a)^{n-1}}, \qquad n = 2, 3, \ldots$$

*Proof of the fifth law:* The geometric series yields

$$\frac{1}{z-a} = \frac{1}{z}\left(\frac{1}{1-\frac{a}{z}}\right) = \frac{1}{z} + \frac{a}{z^2} + \frac{a^2}{z^3} + \ldots \tag{1.198}$$

From this the expression (1.197) for the $Z$-transform of $f = (0, 1, a, a^2, \ldots)$ follows. Differentiating (1.198) with respect to $z$ yields

$$\frac{1}{(z-a)^2} = \frac{1}{z^2} + \frac{2a}{z^3} + \frac{3a^2}{z^4} + \ldots$$

etc.

### 1.11.3.2 Applications to difference equations

**The universal method:** The $Z$-transform is a universal tool for solving equations of the form

$$\boxed{\begin{array}{ll} f_{n+k} + a_{k-1} f_{n+k-1} + \ldots + a_0 f_n = h_n, & n = 0, 1, \ldots, \\ f_r = \beta_r, \qquad r = 0, 1, , \ldots, k-1 & \text{(initial value).} \end{array}} \tag{1.199}$$

Here we are given complex numbers $\beta_0, \ldots, \beta_{k-1}$ and $h_0, h_1, \ldots$. The solution will be the complex numbers $f_k, f_{k+1}, \ldots$. If we set

$$\boxed{\Delta f_n := f_{n+1} - f_n,}$$

then we have

$$\Delta^2 f_n = \Delta(\Delta f_n) = \Delta(f_{n+1} - f_n) = f_{n+2} - f_{n+1} - (f_{n+1} - f_n) = f_{n+2} - 2f_{n+1} + f_n,$$

etc. Therefore (1.199) can be expressed as a linear combination of $f_n, \Delta f_n, \ldots, \Delta^k f_n$. This is why one refers to (1.199) as a difference equation of degree $k$ with constant complex coefficients $a_0, \ldots, a_{k-1}$.

One applies the following steps in the solution:

(i) By applying linearity and translation (laws 1 and 2), one gets an equation for the $Z$-transform $F$, which can be solved immediately and yields a rational function $F$.

(ii) Applying partial fraction decomposition and law 5, we get the solution $f$ of the original problem (1.199).

*Example:* The solution of the difference equation of second order

$$\boxed{\begin{array}{ll} f_{n+2} - 2f_{n+1} + f_n = h_n, & n = 0, 1, \ldots \\ f_0 = 0, \qquad f_1 = \beta & \end{array}} \tag{1.200}$$

is

$$\boxed{f_n = n\beta + \sum_{k=2}^{n} (k-1) h_{n-k}, \qquad n = 2, 3, \ldots} \tag{1.201}$$

In order to get this expression, we write (1.200) first in the form

$$T^2 f - 2Tf + f = h.$$

The law of translation (law 2) then gives us

$$\mathscr{Z}\{Tf\} = zF(z), \qquad \mathscr{Z}\{T^2 f\} = z^2 F(z) - \beta z.$$

From this we get in turn

$$(z^2 - 2z + 1)F = \beta z + H,$$

which implies

$$F(z) = \frac{\beta z}{(z-1)^2} + \frac{H}{(z-1)^2}.$$

The partial fraction decomposition which this leads to is

$$F(z) = \frac{\beta}{(z-1)^2} + \frac{\beta}{z-1} + \frac{H}{(z-1)^2}.$$

Law 5 then gives the inverse transformation

$$\frac{1}{z-1} \Rightarrow \varphi := (0, 1, 1, 1, \ldots), \qquad \frac{1}{(z-1)^2} \Rightarrow \psi := (0, 0, 1, 2, \ldots).$$

Applying finally law 3 we get

$$f = \beta\psi + \beta\varphi + \psi * h.$$

This is the solution given in (1.200).

#### 1.11.3.3 Some further rules for calculations

**Multiplication:** $\mathscr{Z}\{nf_n\} = -zF'(z)$.
**Similarity:** For every complex number $\alpha \neq 0$, we have:

$$\boxed{\mathscr{Z}\{\alpha^n f_n\} = F\left(\frac{z}{\alpha}\right).}$$

**Rule of differences:** For $k = 1, 2, \ldots$ and $F := \mathscr{Z}\{f\}$, one has

$$\boxed{\mathscr{Z}\{\Delta^k f\} = (z-1)^k F(z) - z\sum_{r=0}^{k-1}(z-1)^{k-r-1}\Delta^r f_0}$$

with $\Delta^r f_0 := f_0$ for $r = 0$.

**Rule of sums:** $\mathscr{Z}\left\{\sum_{k=0}^{n-1} f_k\right\} = \frac{F(z)}{z-1}$.

**Rule for residues:** If $F$ is the $Z$-transform of a rational function $f$ with poles $a_1, \ldots, a_J$, then we have:

$$\boxed{f_n = \sum_{j=1}^{J} \operatorname*{Res}_{a_j}\ (F(z)z^{n-1}), \qquad n = 0, 1, \ldots}$$

The residue of a function $g$ with a pole of order $m$ at a point $a$ is calculated by the following formula:

$$\operatorname*{Res}_{a}\ g(z) = \frac{1}{(m-1)!}\lim_{z\to a}\frac{\mathrm{d}^{m-1}}{\mathrm{d}z^{m-1}}(g(z)(z-a)^m).$$

## 1.12 Ordinary differential equations

*Differential equations form the basis for the scientific view of the world.*

*Vladimir Igorovich Arnol'd*

**Solving differential equations with Mathematica:** This software package is able to solve differential equations numerically and also to give the solution in closed form if this is possible.

A relatively complete list of ordinary and partial differential equations whose solution is known in closed form can be found in classics on the subject [113].

**Smoothness:** We say a function is smooth if it is in the class $C^\infty$, that is, if it is infinitely often differentiable.

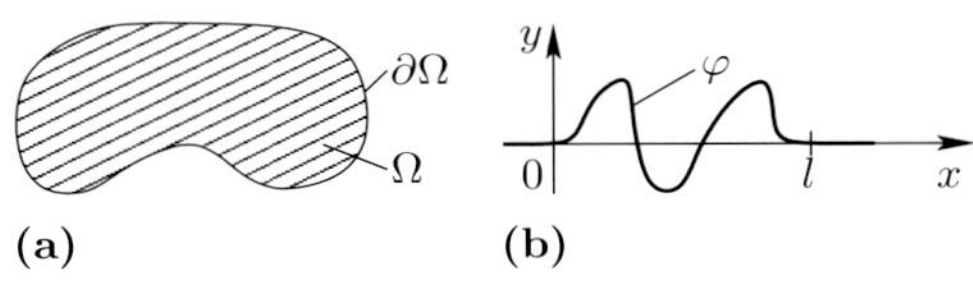


*Figure 1.118. Functions of class $C_0^\infty(\Omega)$.*

A domain $\Omega \subset \mathbb{R}^N$ will be called a domain with smooth boundary, if the boundary $\partial\Omega$ is smooth, that is, the domain $\Omega$ lies locally on one side of the boundary $\partial\Omega$, and this boundary can be described locally by a smooth function (Figure 1.118(a)). Domains with smooth boundaries have no corners.

The class of functions which are smooth in the domain $\Omega$ and which have carrier contained inside a compact subset of $\Omega$, i.e., which vanish outside a compact subset of $\Omega$, will be denoted $C_0^\infty(\Omega)$.

*Example:* The function $\varphi$ depicted in Figure 1.118(b) belongs to the class $C_0^\infty(0, l)$. This function is smooth and vanishes outside the interval $]0, l[$ as well as in a neighborhood of $x = 0$ and $x = l$.

### 1.12.1 Introductory examples

#### 1.12.1.1 Radioactive decay

Consider a radioactive substance (for example radium, which was discovered in 1898 by the husband and wife scientists Curie). Such a material has the property that certain of its atoms are continually decaying.

Let $N(t)$ denote the number of atoms at time $t$ which have not decayed. Then the following law holds:

$$\boxed{\begin{aligned} &N'(t) = -\alpha N(t), \\ &N(0) = N_0 \quad \text{(initial value).} \end{aligned}} \tag{1.202}$$

This equation contains one derivative of the sought-for function and is called a differential equation for that reason. The initial value describes the fact that at the beginning time $t = 0$ the number of atoms which have not yet decayed is equal to $N_0$. The positive constant $\alpha$ is the *constant of decay*.

**Existence and uniqueness result:** Problem (1.202) has a unique solution (Figure 1.119(a)):

$$\boxed{N(t) = N_0 e^{-\alpha t}, \quad t \in \mathbb{R}.} \tag{1.203}$$

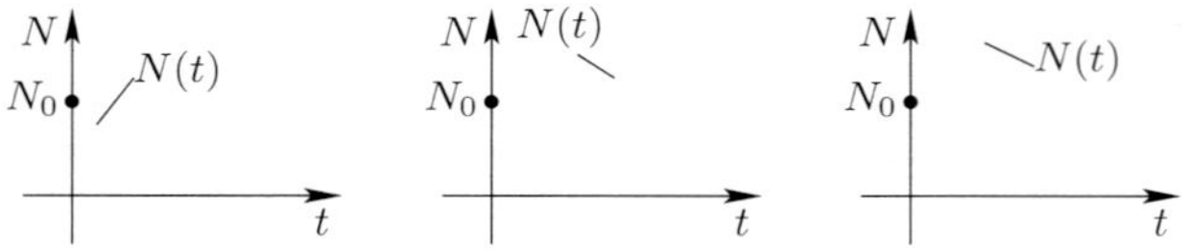

**(a)** radioactive decay **(b)** growth **(c)** decelerated growth

*Figure 1.119. Solutions of the differential equations $N'(t) = -\alpha N(t)$, $N'(t) = \alpha N(t)$ and $N'(t) = \alpha N(t) - \beta N(t)^2$.*

*Proof:* (i) (Existence). Differentiating (1.203) gives

$$N'(t) = -\alpha N_0 e^{-\alpha t} = -\alpha N(t).$$

Furthermore, $N(0) = N_0$.

(ii) (Uniqueness). The right-hand side of the differential equation $N' = -\alpha N$ is of type $C^1$ with respect to $N$. The global uniqueness theorem in 1.12.4.2 then gives the uniqueness of the solution. □

**General solution:** The general solution of the differential equation (1.202) is obtained by choosing $N_0$ arbitrarily. This insures (1.203) with the constant $N_0$.

The following examples can be treated in a similar manner.

**Motivation for the differential equation:** It is interesting that the differential equation (1.202) can be derived without knowing anything about the precise process of radioactive decay. For this consider the Taylor series

$$N(t + \Delta t) - N(t) = A\Delta t + B(\Delta t)^2 + \cdots . \tag{1.204}$$

Our assumption is that $A$ is proportional to the amount $N(t)$. Because of decay, we have $N(t + \Delta t) - N(t) < 0$ for $\Delta t > 0$. Therefore $A$ must be negative, and we set

$$A = -\alpha N(t). \tag{1.205}$$

From (1.204) we get

$$N'(t) = \lim_{\Delta t \to 0} \frac{N(t + \Delta t) - N(t)}{\Delta t} = A = -\alpha N(t).$$

**Well-posedness of the problem:** Small changes in the starting amount $N_0$ lead to small changes in the solutions.

To describe this in more detail, we introduce the norm

$$||N|| := \max_{0 \le t \le \mathscr{T}} |N(t)|.$$

For two solutions $N$ and $N_*$ of the differential equation (1.202) we then have

$$\boxed{||N - N_*|| \le |N(0) - N_*(0)|.}$$

**Stability:** The solution is asymptotically stable, meaning that it tends towards an equilibrium solution for large times. More precisely, we have:

$$\boxed{\lim_{t\to+\infty} N(t) = 0.}$$

This means that after sufficiently long times all atoms have decayed.

#### 1.12.1.2 The equation for growth

We now let $N(t)$ denote the number of a particular kind of pathogens at time $t$. We assume that the reproduction of this species obeys (1.204) with $A = \alpha N(t)$. From this we get the *equation for growth*

$$\boxed{\begin{aligned} &N'(t) = \alpha N(t), \\ &N(0) = N_0 \quad \text{(initial value)}. \end{aligned}} \tag{1.206}$$

**Existence and uniqueness:** The problem (1.206) has a unique solution (Figure 1.119(b)):

$$\boxed{N(t) = N_0 \mathrm{e}^{\alpha t}, \quad t \in \mathbb{R}.} \tag{1.207}$$

**The problem is ill-posed:** Small changes in the boundary condition $N_0$ grow in time to large differences in the solution:

$$\boxed{||N - N_*|| = \mathrm{e}^{\alpha \mathscr{T}} |N(0) - N_*(0)|.}$$

**Instability:**

$$\lim_{t\to+\infty} N(t) = +\infty.$$

> Processes with constant speed of growth grow in time beyond all bounds and lead to a catastrophe already after relatively short times.

#### 1.12.1.3 Impeded growth (logistic equation)

The equation

$$\boxed{\begin{aligned} &N'(t) = \alpha N(t) - \beta N(t)^2, \\ &N(0) = N_0 \quad \text{(initial value)} \end{aligned}} \tag{1.208}$$

with positive constants $\alpha$ and $\beta$ deviates from the growth equation (1.206) by the impedance term, which describes the difficulties of the population in finding sufficient food. The equation (1.208) is special case of a so-called Riccati differential equation (see 1.12.4.7).[100]

[100] The logistic equation (1.208) was suggested in 1838 as the equation for the growth of the human population on earth by the Belgian mathematician Verhulst.

**Rescaling:** We change the units for the number of particles $N$ and the time $t$, that is, we introduce new variables $\mathscr{N}$ and $\tau$ with

$$\boxed{N(t) = \gamma \mathscr{N}(\tau), \quad t = \delta\tau.}$$

We then get from (1.208) the equation

$$\begin{aligned}\frac{\mathrm{d}N}{\mathrm{d}t} &= \frac{\mathrm{d}(\gamma\mathscr{N})}{\mathrm{d}\tau}\frac{\mathrm{d}\tau}{\mathrm{d}t} = \gamma\mathscr{N}'(\tau)\frac{1}{\delta} \\ &= \alpha\gamma\mathscr{N}(\tau) - \beta\gamma^2\mathscr{N}(\tau)^2.\end{aligned}$$

If we choose $\delta := 1/\alpha$ and $\gamma := \alpha/\beta$, then we get the new equation

$$\boxed{\begin{aligned}&\mathscr{N}'(\tau) = \mathscr{N}(\tau) - \mathscr{N}(\tau)^2, \\ &\mathscr{N}(0) = \mathscr{N}_0 \quad \text{(initial condition).}\end{aligned}} \tag{1.209}$$

**Determination of the equilibrium states:** The time-independent solutions (*equilibrium states*) of (1.209) are given by

$$\mathscr{N}(\tau) \equiv 0 \quad \text{and} \quad \mathscr{N}(\tau) \equiv 1.$$

*Proof:* From $\mathscr{N}(\tau) = \text{const}$ and (1.209) it follows that $\mathscr{N}^2 - \mathscr{N} = 0$, which implies $\mathscr{N} = 0$ or $\mathscr{N} = 1$. □

**Existence and uniqueness:** Let $0 < \mathscr{N}_0 \leq 1$. Then the problem (1.209) has for all times $\tau$ the solution (see Figure 1.119(c))

$$\boxed{\mathscr{N}(\tau) = \frac{1}{1 + C\mathrm{e}^{-\tau}}}$$

with $C := (1 - \mathscr{N}_0)/\mathscr{N}_0$.

For $\mathscr{N}_0 = 0$ the problem (1.209) has for all times $\tau$ the unique solution $\mathscr{N}(\tau) \equiv 0$.

**Stability:** If $0 < \mathscr{N}_0 \leq 1$, the system develops for large times into the equilibrium solution $\mathscr{N} \equiv 1$, that is

$$\boxed{\lim_{\tau \to +\infty} \mathscr{N}(\tau) = 1.} \tag{1.210}$$

The equilibrium solution $\mathscr{N} \equiv 1$ is stable, that is small changes in the number of particles $\mathscr{N}$ at the initial time $\tau = 0$ leads by (1.210) after sufficient time to this equilibrium solution.

The equilibrium solution $\mathscr{N} \equiv 0$ on the other hand is instable. Small changes in the number of particles at the initial time $\tau = 0$ lead by (1.210) to drastic changes in the solution after relatively small times.

#### 1.12.1.4 Explosion in finite time (blowing up)

The differential equation

$$\boxed{N'(t) = 1 + N(t)^2, \quad N(0) = 0,} \tag{1.211}$$

has a unique solution in the interval $]-\pi/2, \pi/2[$

$$N(t) = \tan t$$

(see Figure 1.120).

We have

$$\lim_{t\to\frac{\pi}{2}-0} N(t) = +\infty.$$

The unusual thing here is that the solution becomes infinite in finite time. This is a model for a self-induction process, which is feared for example by engineers in chemical factories.

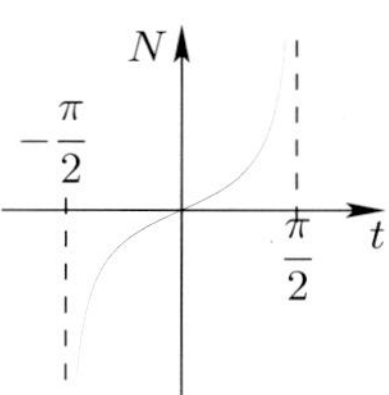


*Figure 1.120.*

#### 1.12.1.5 The harmonic oscillator and characteristic oscillations

**Simple spring:** We consider a point mass of mass $m$, which moves along the $x$-axis under the influence of a spring, which develops a force which is proportional to the distance from $x = 0$, $\mathbf{F}_0 := -kx\mathbf{i}$, with an additional external force $\mathbf{F}_1 := \mathscr{F}(t)\mathbf{i}$. The Newtonian law of force tells us that the force is equal to the mass times the acceleration, $m\mathbf{x}'' = \mathbf{F}_0 + \mathbf{F}_1$, where $\mathbf{x} = x\mathbf{i}$. This yields the differential equation

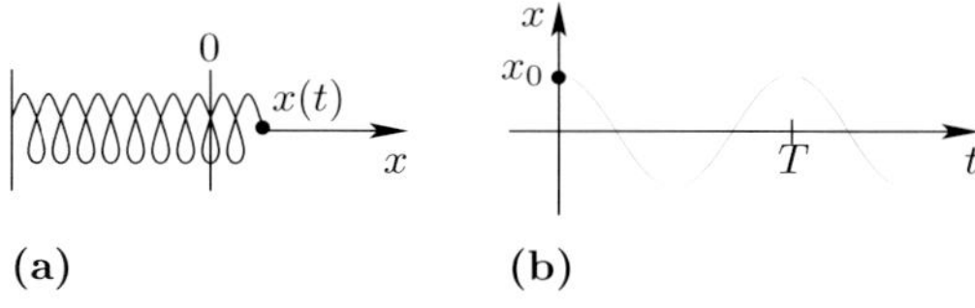


*Figure 1.121. Initial value problem for oscillations.*

$$\begin{aligned} x''(t) + \omega^2 x(t) &= F(t), \\ x(0) &= x_0 \quad \text{(initial position)}, \\ x'(0) &= v \quad \text{(initial velocity)}. \end{aligned} \tag{1.212}$$

Here $\omega := \sqrt{k/m}$ and $F := \mathscr{F}/m$. The force function $F : [0, \infty[\to \mathbb{R}$ is assumed to be continuous.

**Existence and uniqueness:** The problem has a unique solution for all times[101]

$$x(t) = x_0 \cos \omega t + \frac{v}{\omega} \sin \omega t + \int_0^t G(t,\tau) F(\tau) \mathrm{d}\tau \tag{1.213}$$

with the Greens function $G(t,\tau) := \frac{1}{\omega} \sin \omega (t - \tau)$.

**Characteristic oscillations:** If the external force vanishes, i.e., $F \equiv 0$, then one calls a solution of (1.213) a *characteristic oscillation* of the harmonic oscillator. This is the superposition of a sinusoidal wave and a cosinosidal wave with the frequency $\omega$ and the period

$$T = \frac{2\pi}{\omega}.$$

*Example:* Figure 1.121(b) shows the characteristic oscillation $x = x(t)$ which incurs, if the point mass on the $x$-axis is not central at $t = 0$ and is not in motion at this time, i.e., $x_0 \neq 0$ and $v = 0$.

[101] This solution can be calculated with the help of the Laplace transformation (see (1.188)).

**Well-posedness of the problem:** Small changes in the initial position $x_0$, the initial velocity $v$ and the external force $F$ lead to small changes in the motion. More precisely, for two solutions $x$ and $x_*$ of (1.212), we have the inequality

$$||x - x_*|| \leq |x(0) - x_*(0)| + \frac{1}{\omega}|x'(0) - x'_*(0)| + \frac{\mathscr{T}}{\omega} \max_{0 \leq t \leq \mathscr{T}} |F(t) - F_*(t)|$$

with

$$||x - x_*|| := \max_{0 \leq t \leq \mathscr{T}} |x(t) - x_*(t)|.$$

Here $[0, \mathscr{T}]$ is an arbitrary time interval.

**Eigenvalue problem:** The problem

$$\boxed{\begin{aligned} &-x''(t) = \lambda x(t), \\ &x(0) = x(l) = 0 \quad \text{(boundary condition)} \end{aligned}}$$

is called an eigenvalue problem. The number $l > 0$ is given. A non-trivial solution $x \not\equiv 0$ will be called an eigensolution $(x, \lambda)$. The corresponding number $\lambda$ is then called the eigenvalue.

**Theorem:** All eigensolutions are given by

$$\boxed{x(t) = C\sin(n\omega_0 t), \quad \lambda = n^2\omega_0^2, \quad \omega_0 = \frac{\pi}{l}, \quad n = 1, 2, \ldots}$$

Here $C$ is an arbitrary non-zero constant.

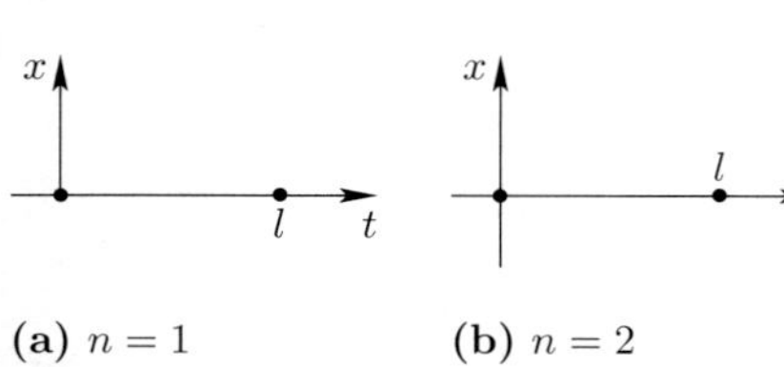


**(a)** $n = 1$ **(b)** $n = 2$

*Figure 1.122.*

*Proof:* We use the solution $x(t) = \frac{v}{\omega}\sin\omega t$ of (1.213) with $x_0 = 0$, $F \equiv 0$ and determine the frequency $\omega$ so that the mass point meets $x = 0$ at time $l$ (Figure 1.122). From $\sin(\omega l) = 0$ we get $\omega l = n\pi$ with $n = 1, 2, \ldots$. This yields $\omega = n\frac{\pi}{l} = n\omega_0$. Differentiating $x(t) = C\sin(n\omega_0 t)$ then gives

$$x''(t) = -\lambda x(t)$$

with $\lambda = n^2\omega_0^2$.

#### 1.12.1.6 Dangerous resonance effects

We consider the harmonic oscillator (1.212) with the periodic external force

$$\boxed{F(t) := \sin\alpha t.}$$

**Definition:** This external force is *in resonance* with the vibrations of the harmonic oscillator, if $\alpha = \omega$, that is, the frequency $\alpha$ of the external force is the same as the frequency of the characteristic oscillation.

In this case the external force actually amplifies the characteristic oscillations. This phenomenon is feared by engineers. For example, in the case of bridges, one has to take care that the vibrations generated by traffic are not in resonance with the characteristic oscillations of the bridge. The construction of earth quake resistant high rises is based on the fact that the resonance effects of the earth quake vibrations are avoided.

The following considerations show how resonance effects arise mathematically.

**The non-resonance case:** Let $\alpha \neq \omega$. Then the unique solution of (1.212) for all times $t$ with periodic external force $F(t) := \sin \alpha t$ is:

$$x(t) = x_0 \cos \omega t + \frac{v}{\omega} \sin \omega t + \frac{\sin \alpha t + \sin \omega t}{2(\alpha + \omega)\omega} - \frac{\sin \alpha t - \sin \omega t}{2(\alpha - \omega)\omega}. \tag{1.214}$$

This solution is bounded for all times.

**The resonance case:** Let $\alpha = \omega$. Then the unique solution of (1.212) for all times $t$ with external force $F(t) = \sin \omega t$ is:

$$\boxed{x(t) = x_0 \cos \omega t + \frac{v}{\omega} \sin \omega t + \frac{\sin \omega t}{2\omega^2} - \frac{t}{2\omega} \cos \omega t.} \tag{1.215}$$

The last term $t \cdot \cos \omega t$, which describes a vibration of the external force of frequency $\omega$, is dangerous, as it grows without bound as $t$ grows, which in real life can lead to the destruction of structures (Figure 1.123(a)).

The occurrence of the dangerous resonance term $t \cdot \cos \omega t$ is understandable, when one realizes that the resonance solution (1.215) can be derived from the non-resonance solution (1.214) by passing to the limit $\alpha \to \omega$.

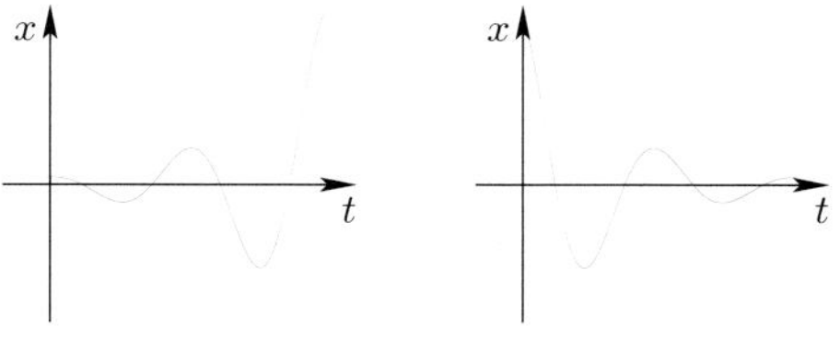


**(a)** resonance **(b)** damped oscillation

*Figure 1.123.*

### 1.12.1.7 Dampening

If there is an additional resistance force $\mathbf{F}_2 = -\gamma \mathbf{x}'$, $\gamma > 0$ acting on the mass point in 1.12.1.5 which is proportional to the speed of the mass point, then we get from the equation of motion $m\mathbf{x}'' = \mathbf{F}_0 + \mathbf{F}_2 = -k\mathbf{x} - \gamma \mathbf{x}'$ the differential equation

$$\boxed{\begin{array}{l} x''(t) + \omega^2 x(t) + 2\beta x'(t) = 0, \\ x(0) = x_0, \qquad x'(0) = v \end{array}} \tag{1.216}$$

with the positive constant $\beta := \gamma/2m$.

**An ansatz:** We make the ansatz

$$\boxed{x = \mathrm{e}^{\lambda t}.}$$

From (1.216) we get $(\lambda^2 + \omega^2 + 2\beta\lambda)\mathrm{e}^{\lambda t} = 0$, in other words,

$$\boxed{\lambda^2 + \omega^2 + 2\beta\lambda = 0}$$

with the solution $\lambda_\pm = -\beta \pm \mathrm{i}\sqrt{\omega^2 - \beta^2}$. If $C$ and $D$ are arbitrary constants, then the function

$$x = C\mathrm{e}^{\lambda_+ t} + D\mathrm{e}^{\lambda_- t}$$

is a solution of (1.216). The constants $C$ and $D$ are determined from the initial conditions. We also use the Euler formula $\mathrm{e}^{(a+\mathrm{i}b)t} = \mathrm{e}^{\alpha t}(\cos bt + \mathrm{i}\sin bt)$.

**Existence and uniqueness:** Let $0 < \beta < \omega$. Then the problem (1.216) has the unique solution for all times $t$[102]

$$\boxed{x = x_0\mathrm{e}^{-\beta t}\cos\omega_* t + \frac{v + \beta x_0}{\omega_*}\mathrm{e}^{-\beta t}\sin\omega_* t} \qquad (1.217)$$

with $\omega_* := \sqrt{\omega^2 - \beta^2}$. These are dampened vibrations.

*Example:* If the mass point is at rest at $t = 0$ with $x_0 \neq 0$ but $v = 0$, then one finds the dampened vibrations (1.217) in Figure 1.123(b).

#### 1.12.1.8 Chemical reactions and the inverse problem of chemical reaction kinetics

Let $m$ chemical substances $A_1, \ldots, A_m$ be given, together with a chemical reaction among them of the form

$$\sum_{j=1}^{m} \nu_j A_j = 0$$

with the so-called stochiometrical coefficients $\nu_j$. Moreover let $N_j$ be the number of molecules[103] of the substance $A_j$. We denote by

$$\boxed{c_j := \frac{N_j}{V}}$$

the density of the substance $A_j$. Here $V$ is the cumulative volume of the reaction.

*Example:* The reaction

$$2A_1 + A_2 \longrightarrow 2A_3$$

means that two molecules of substance $A_1$ combine with a single molecule $A_2$ to form two molecules of a third substance $A_3$. One can also write this as follows:

$$\nu_1 A_1 + \nu_2 A_2 + \nu_3 A_3 = 0$$

with $\nu_1 = -2$, $\nu_2 = -1$ and $\nu_3 = 2$. An example of this is the reaction

$$2\mathrm{H}_2 + \mathrm{O}_2 \longrightarrow 2\mathrm{H}_2\mathrm{O},$$

which describes for formulation of two molecules of water from two molecules of hydrogen and an oxygen molecule.

**The fundamental equation of chemical reaction kinetics:**

$$\boxed{\begin{aligned} &\frac{1}{\nu_j}\frac{\mathrm{d}c_j}{\mathrm{d}t} = kc_1^{n_1}c_2^{n_2}\cdots c_m^{n_m}, \quad j = 1, \ldots, m, \\ &c_j(0) = c_{j0} \quad \text{(initial conditions).} \end{aligned}} \qquad (1.218)$$

[102]This solution can be derived with the help of the Laplace transformation (see 1.11.1.2).

[103]The number of such in chemistry is measured in terms of moles, where one mole contains Avegadro's number $6.023 \cdot 10^{23}$ of molecules.

Here $k$ is the positive reaction speed constant, which depends on the pressure $p$ and the temperature $T$. The numbers $n_1$, $n_2, \ldots$ are called reaction orders. The unknowns are the dependence on time of the densities of the particles $c_j(t)$.

**Remark:** Chemical reactions usual will involve sub-reactions with their own products on the way of completing the reaction. In this way a huge number of systems like (1.218) are needed to describe the actual chemical changes that occur. In many cases one does not really know what these sub-reactions are nor does one know the reaction speed constant $k$ or the reaction orders $n_j$. This leads to the difficult problem of deducing the constants $k$ and $n_j$ from measurements of the $c_j(t)$, using (1.218). This is a so-called *inverse problem.*[104]

**Applications in biology:** Equations of the form (1.218) or variants thereof occur often in biology. In this case $N_j$ is the number of living beings of some kind (compare also the equation for growth (1.206) and the dampened equation of growth (1.208)).[105]

In the following sections we consider a series of fundamental phenomena in which ordinary and partial differential equations occur. The knowledge of these phenomena is quite helpful in understanding the theory of differential equations.

## 1.12.2 Basic notions

Many processes in nature and in technology are described by differential equations.

(i) Systems which have *finitely many degrees of freedom* correspond to ordinary differential equations (for example the motion of finitely many mass points in Newtonian mechanics).

(ii) Systems with *infinitely many degrees of freedom* correspond to partial differential equations (for example the motion of elastic bodies, liquids, gases, electromagnetic fields and quantum systems, the description of reaction or diffusion processes in biology and chemistry or the variation in time of our universe).

The basic equations of the different disciplines in physics are all differential equations. The starting point for many of them is the *Newton law of motion* for a particle (for example a planet or star) of mass $m$:

$$\boxed{m\mathbf{x}''(t) = \mathbf{F}(\mathbf{x}(t), t).} \tag{1.219}$$

This law is, expressed in words, the mass of the particle times its acceleration is equal to the force acting on it. One is looking for the trajectory of the particle,

$$\boxed{\mathbf{x} = \mathbf{x}(t),}$$

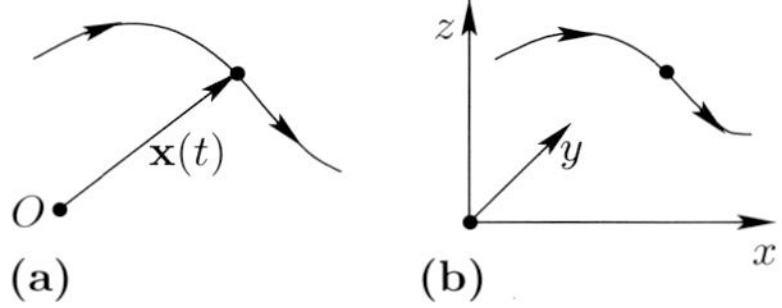


*Figure 1.124.*

which satisfies the equation (1.219) (Figure 1.124). It is typical for a differential equation that it contains, in addition to the function being sought for, also derivatives of the latter (hence of course the name 'differential equation').

---

[104] At the Konrad–Zuse–Zentrum in Berlin there are very effective computer programs for this kind of inverse problem which have been developed by Prof. Dr. Peter Deuflhard and his coworkers.

[105] In complicated cases more terms occur in (1.218).

**Ordinary differential equations:** If the function which is being sought depends only on a single real variable (for example on time $t$), then the differential equation is referred to as *ordinary*.

*Example 1:* In (1.219) one has an ordinary differential equation.

**Partial differential equations:** In physical field theories the quantities involved (for example the temperature or the strength of the electromagnetic field) depends on several variables (for example on time and position). The differential equation for these quantities then contains partial derivatives of the function being sought for, and is thus called a *partial differential equation.*

*Example 2:* The temperature field $T = T(x, y, z, t)$ of a body satisfies in many cases the *heat conduction equation*

$$\boxed{T_t - \kappa\Delta T = 0} \tag{1.220}$$

with $\Delta T := T_{xx} + T_{yy} + T_{zz}$. Here $T(x, y, z, t)$ denotes the temperature at the position $(x, y, z)$ at the time $t$. The material constant $\kappa$ characterizes the ability of heat conduction of the body.

#### 1.12.2.1 The fundamental 'infinitesimal' epistemological strategy in the natural sciences

The differential equation (1.219) describes the behavior of a trajectory at an 'infinitesimal level', which means roughly speaking that it is valid for *extremely small times t.*[106] This is part of an astounding epistemological phenomenon, which roughly speaking says the following.

> At the 'infinitesimal level' (i.e., for extremely small times and extremely small spatial distances), all processes in nature become very simple and can be expressed in terms of a few basic equations.

These basic equations then encode an incredible amount of information. It is the job of mathematics to decode this information, i.e., solve the differential equation for reasonable times and spatial distances.

With the creation of infinitesimal calculations, i.e., calculus, Newton and Leibniz have given us the key to a deeper understanding of phenomena occuring in the natural sciences. This achievement of the human mind can not be esteemed highly enough.

#### 1.12.2.2 The role of initial conditions

The differential equation of Newton (1.219) describes all possible motions of the particle of mass $m$. Astronomers, for example, are really interested in calculating the trajectories of heavenly bodies. In order to calculate these trajectories, one must introduce in addition to the equations of Newton the situation of the bodies in question at a given

[106] Ever since Newton (1643–1727) and Leibniz (1646–1716) one speaks of 'infinitesimal' or 'infinitely small' times and spatial distances. A precise mathematical interpretation of this notion can be obtained using modern non-standard analysis (cf. [53]). Traditionally in rigorous mathematics the notion of 'infinitely small' is not used, but rather replaced by the consideration of limits.

time $t_0$. More precisely, one has to consider the following problem.

$$\boxed{\begin{aligned} &m\mathbf{x}''(t) = \mathbf{F}(\mathbf{x}(t), t) && \text{(equation of motion)},\\ &\mathbf{x}(t_0) = \mathbf{x}_0 && \text{(initial position)},\\ &\mathbf{x}'(t_0) = \mathbf{v}_0 && \text{(initial velocity)}.\end{aligned}} \tag{1.221}$$

**Existence and uniqueness result:** We assume the following:

(i) At the initial time $t_0$ the position $\mathbf{x}_0$ and the velocity vector $\mathbf{v}_0$ of this point (the space object) are given.

(ii) The force field $\mathbf{F} = \mathbf{F}(\mathbf{x}, t)$ is sufficiently smooth (for example of type $C^1$) for all positions $\mathbf{x}$ and all times $t$ in a small neighborhood of the initial time $t_0$.

Then there is a spatial neighborhood $U(\mathbf{x}_0)$ and a time-interval $J(t_0)$ such that the problem (1.221) has exactly one solution which is a trajectory

$$\mathbf{x} = \mathbf{x}(t),$$

which remains in $U(\mathbf{x}_0)$ for all times $t \in J(t_0)$.[107]

This result surprisingly insures the existence of a trajectory only for sufficiently small times. But in general one can not expect more to hold. It is possible, that

$$\boxed{\lim_{t \to t_1} |\mathbf{x}(t)| = \infty,}$$

i.e., the force $\mathbf{F}$ is so strong, that the particle reaches 'infinity' in a finite time $t_1$.

*Example* (model problem): For the force $F(x) := 2mx(1 + x^2)$, the differential equation

$$\begin{aligned} &mx'' = F(x),\\ &x(0) = 0, \qquad x'(0) = 1\end{aligned}$$

has the unique solution

$$x(t) = \tan t, \qquad -\frac{\pi}{2} < t < \frac{\pi}{2}.$$

Here we have

$$\lim_{t \to \frac{\pi}{2} - 0} x(t) = +\infty.$$

To insure that a solution for all times exists, one uses the following general principle:

$$\boxed{\textit{A-priori} \text{ estimates insure global solutions.}}$$

This can be found in section 1.12.9.8.

### 1.12.2.3 The role of stability

The gravitational field of the sun has the form

$$\mathbf{F} = -\frac{GMm}{|\mathbf{x}|^3}\mathbf{x} \tag{1.222}$$

[107] This is a special case of the general existence and uniqueness theorem of Picard–Lindelöf (cf. section 1.12.4.1).

in which $M$ is the mass of the sun, $m$ is the mass of the massive particle or body in the gravitational field of the sun and G denotes the gravitational constant. At the center of the sun $\mathbf{x} = \omega$, this force field has a singularity.

**The famous problem of the stability of the solar system:** This problem is summed up in the following two questions.

(a) Are the trajectories of the planets stable, i.e., do they change their form in long times very little?

(b) Is it possible that a planet could collide into the sun or escape from the solar system altogether?

Many great mathematicians have worked on this problem ever since Lagrange (1736–1813). First it was attempted to express the trajectories of the planets in closed terms through 'elementary functions'. Toward the end of the nineteenth century, however, it was realized by Poincaré (1854–1912) that this is not possible, not even in principle. This led to two completely new directions of development in mathematics.

**(I) Abstract proofs of existence and topology:** Since it did not seem possible to write down explicitly solutions, it was attempted to at least prove their existence by indirect, abstract methods. This lead to the development of fixed-point theorems, which will be described in detail in [212]. One of the topological fundamental principles toward the existence of solutions of ordinary and partial differential equations is the famous *principle of Leray–Schrauder*, which originated in 1934 and is

| *A-priori* estimates insure the existence of solutions. |
|---|

**(II) Dynamical systems and topology:** Scientists and engineers are often interested not in the precise form of solutions, but only in the fundamental properties of their behavior (for example the existence of stable equilibrium positions of stable periodic vibrations or also the possible transition to chaos). This set of problems is the topic of investigation of the science of dynamical systems, which will be described in detail in [212].

Those branches of mathematics which are more interested in the *qualitative behavior* of objects is *topology.* Both topology and the theory of dynamical systems were initiated by the great French mathematician Poincaré in connection with his fundamental investigations of stability in celestial mechanics.[108] A readable account of his historical discoveries in this regard can be found in the book [217].

Some aspects of stability theory were developed in the middle of the nineteenth century by engineers. They were interested in constructing machines, buildings and bridges in such a way that they were stable and not easily destroyed by the elements (wind and storms).

Fundamental general mathematical results of stability theory were obtained by the Russian mathematician Liapunov in 1892. With this, stability theory came into being as a separate mathematical discipline, which is still today the subject of intensive research. For many complicated problems the stability properties of the solutions are still not

[108]In 1892 the first of a three-volume series *Les méthodes nouvelles de la mécanique céleste* appeared. With this set of volumes, Poincaré continued a tradition started by the volume *La mécanique analytique* by Lagrange (1788) and the five volumes entitled *La mécanique céleste* by Laplace in 1799.

noindent known today. Note:

> Mathematically correct solutions can be completely irrelevant in the real world, if they are instable and therefore cannot be realized in nature.

A similar statement is true for numerical procedures made for computers. Only *stable numerical procedures*, that is, ones which are robust with respect to rounding errors, are of use.

The problem of stability of the solar system is still unsolved today. In 1955, Kolmogorov and later also Arnol'd and Moser showed that the perturbation of quasiperiodic motions (as the motion of the solar system) are very sensitive to the kind of perturbation and can end in chaos (KAM-theory). A miniscule particle can possibly lead to a change in the total motion of the system. For this reason, the question of stability of the solar system will never be solved by theoretical considerations. Week-long computations with supercomputers (for example at the famous Massechusettes Institute of Technology (MIT) in Boston) have shown that the solar system will remain stable for at least the next one million years.

#### 1.12.2.4 The role of boundary conditions and the fundamental idea of Green's functions

In addition to initial conditions, we also have to consider *boundary conditions.*

*Example 3* (elastic rod): The motion (displacement) $y = y(x)$ of an elastic rod under the influence of an external force is described by the following mathematical problem:

$$\begin{array}{rcll} -\kappa y''(x) & = & f(x) & \text{(equilibrium of forces)}, \\ y(0) & = & y(l) = 0 & \text{(boundary condition)}. \end{array} \tag{1.223}$$

Here, $\left(\int_a^b f(x)\mathrm{d}x\right)\mathbf{j}$ denotes the force which acts in the interval $[a, b]$ on the rod in the $y$-direction, i.e., $f(x)$ is the density of the exterior force at the point $x$. The positive material constant $\kappa$ describes the elastic properties of the rod. The boundary condition describes the fact that the rod is spanned at the points $x = 0$ and $x = l$ (Figure 1.125(a)).

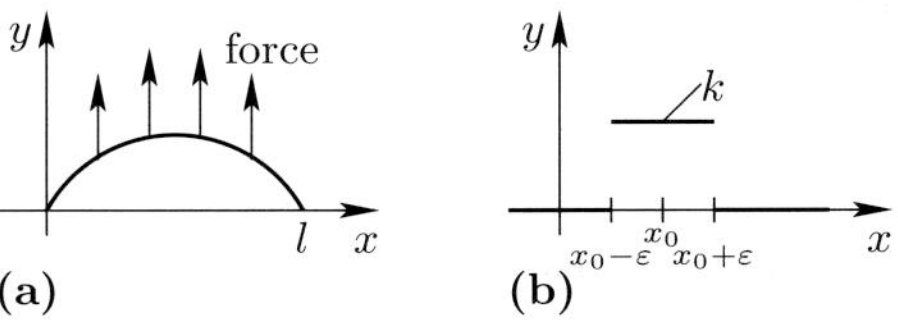


*Figure 1.125. The differential equation for an elastic rod.*

**Connection with the calculus of variations:**

> The appearance of boundary conditions is typical for problems which are related to variations.

For example, (1.223) results as the Euler–Lagrange equation from the principle of least action

$$\boxed{\begin{aligned}\int_0^l L(y(x), y'(x))\mathrm{d}x \stackrel{!}{=} \text{stationary},\\ y(0) = y(l) = 0\end{aligned}}$$

with the Lagrange function $L := \frac{\kappa}{2} y'^2 - fy$ (cf. 5.1.2). Here, one has:

$$\int_0^l L\mathrm{d}x = \text{the elastic energy of the rod minus the work done by the force.}$$

**Representation of solutions with the help of the Green's function:** The unique solution of (1.123) is given by the formula

$$\boxed{y(x) = \int_0^l G(x, \xi) f(\xi)\mathrm{d}\xi.} \tag{1.224}$$

Here, the *Green's function* of the problem (1.223) is

$$G(x, \xi) := \begin{cases} \dfrac{(l-\xi)x}{l\kappa} & \text{for } 0 \le x \le \xi \le l, \\ \dfrac{(l-\xi)\xi}{l\kappa} & \text{for } 0 \le \xi < x \le l. \end{cases}$$

**Physical interpretation of the Green's function:** We choose the force density

$$f_\varepsilon(x) = \begin{cases} \dfrac{1}{2\varepsilon} & \text{for } x_0 - \varepsilon \le x \le x_0 + \varepsilon \\ 0 & \text{otherwise,} \end{cases}$$

which for smaller and smaller $\varepsilon$ is more and more concentrated at the point $x_0$ and which corresponds to a total force of

$$\int_0^l f_\varepsilon(x)\mathrm{d}x = 1$$

(Figure 1.125(b)). The motion determined by this is denoted by $y_\varepsilon$. Then one has

$$\boxed{\lim_{\varepsilon \to 0} y_\varepsilon(x) = G(x, x_0).}$$

**The formal use of the Dirac delta function:** Often physicists write the formal expression

$$\delta(x - x_0) = \lim_{\varepsilon \to 0} f_\varepsilon(x) = \begin{cases} +\infty & \text{for } x = x_0, \\ 0 & \text{otherwise} \end{cases}$$

and say that $y(x) := G(x, x_0)$ is a solution of the initial value problem (1.223) for the point–density force $f(x) := \delta(x - x_0)$. The function $\delta(x - x_0)$ is called the *Dirac delta function.* In this formal sense, we have

$$\boxed{\begin{aligned} -\kappa G_{xx}(x, x_0) &= \delta(x - x_0) && \text{on } ]0, l[, \\ G(0, x_0) &= G(l, x_0) = 0 && \text{(boundary condition).} \end{aligned}} \tag{1.225}$$

**The precise mathematical formulation in the context of the theory of distributions:** The Green's function is a solution of the boundary value problem

$$\begin{array}{|rl|}\hline -\kappa G_{xx}(x,x_0) = \delta_{x_0} & \text{on } ]0,l[, \\ G(0,x_0) = G(l,x_0) = 0 & \text{(boundary condition).} \\ \hline \end{array} \tag{1.226}$$

Here, $\delta_{x_0}$ stands for the delta distribution, and the solution (1.226) is to be understood in the sense of distributions.[109]

Green's functions were introduced around 1830 by the English mathematician and physicist George Green (1793–1841). The general strategy is:

> The Green's function described physical effects which are generated by sharply concentrated exterior influences $\mathscr{E}$.
> The effect of general exterior influences is derived as the supposition of exterior influences with similar (sharp) forms as $\mathscr{E}$.

The method of Green's functions is used intensively in all areas of physics, since it allows a localization of physical effects and shows how general physical effects are constructed. In quantum field theory, for example, Green's functions are calculated with the help of Feynman integrals (path integrals).

The formula (1.224) for the solution represents the action of an arbitrary force as the superposition of individual forces

$$G(x,\xi)f(\xi)$$

which are localized at the point $\xi$.

#### 1.12.2.5 The role of boundary–initial conditions

In physical fields theories, one must prescribe the structure of the fields at the initial time $t_0$ and at the boundary of a domain. Often one sets $t_0 := 0$.

*Example* (heat conduction): In order to uniquely determine the distribution of temperature in a body, one needs to know the distribution at the initial time $t = 0$ and the distribution for all times $t \geq 0$ along the boundary. Therefore, one has to add to the heat conduction equation (1.220) the following conditions:

$$\begin{array}{|rll|}\hline T_t - \kappa\Delta T = 0, & P \in D,\ t \geq 0, & \\ T(P,0) = T_0(P), & P \in D, & \text{(initial temperature),} \\ T(P,t) = T_1(P,t), & P \in \partial D,\ t \geq 0 & \text{(boundary temperature).} \\ \hline \end{array} \tag{1.228}$$

[109] The theory of distributions, which was created around 1950 by the French mathematician Laurent Schwartz, is described in [212]. The equation (1.226) then means

$$-\kappa \int_0^l G(x,x_0)\varphi''(x)\mathrm{d}x = \delta_{x_0}(\varphi) = \varphi(x_0) \tag{1.227}$$

for all test functions $\varphi \in C_0^\infty(0,l)$. The equation (1.227) follows formally upon multiplying (1.225) by $\varphi$, partially integrating over $[0,l]$ twice and using

$$\int_0^l \delta(x-x_0)\varphi(x)\mathrm{d}x = \varphi(x_0).$$

Here we have set $P := (x, y, z)$.

**Theorems on existence and uniqueness:** If $D$ is a bounded domain with smooth boundary in $\mathbb{R}^3$, and if the prescribed initial temperature $T_0$ as well as the given boundary temperature $T_1$ are smooth, then the heat conduction equation (1.228) has a uniquely determined smooth solution $T$ in $D$ for all times $t \geq 0$.

#### 1.12.2.6 Well-posed problems

In order for a mathematical model in the form of a differential equation to be of use in the investigation of scientific phenomena, the differential equation must possess the following properties:

(i) There is a unique solution.

(ii) Small changes in the initial conditions lead to small changes in the solutions.

Problems for which these properties hold are referred to as *well-posed.* In each case, one must make more precise what is to be understood under 'small' changes.

In the case of time-dependent problems, one is in addition interested in the global stability:

(iii) The solution exists for all times $t \geq t_0$ and tends to an equilibrium position for $t \to +\infty$.

*Example 1:* The initial value problem

$$y' = -y, \qquad y(0) = \varepsilon \tag{1.229}$$

is well-posed for $\varepsilon = 0$, since (1.229) has for arbitrary $\varepsilon$ the unique solution

$$y(t) = \varepsilon \mathrm{e}^{-t}, \qquad t \in \mathbb{R}. \tag{1.230}$$

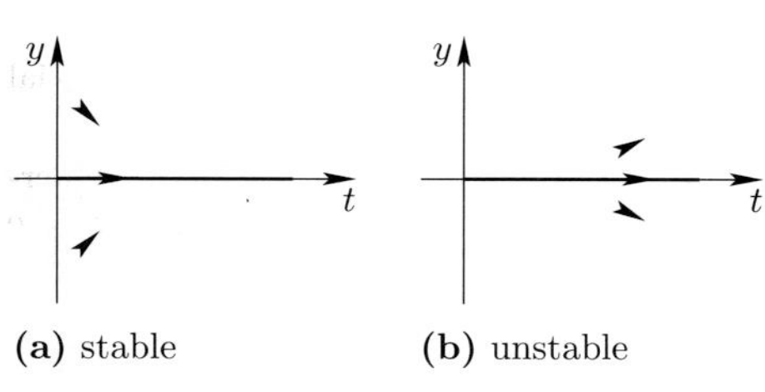


*Figure 1.126. Perturbations of solutions of differential equations.*

If $\varepsilon = 0$, we get the equilibrium solution $y(t) = 0$. For small perturbations $\varepsilon$, the solution (1.230) changes but slightly and because of

$$\lim_{t \to +\infty} \varepsilon \mathrm{e}^{-t} = 0$$

for large times $t$, it tends in that case to the equilibrium solution $y = 0$ (Figure 1.126(a)).

*Example 2:* The initial value problem

$$y' = y, \qquad y(0) = \varepsilon$$

is for $\varepsilon = 0$ not well-posed. Indeed, the uniquely determined solution

$$y(t) = \varepsilon \mathrm{e}^{t}$$

blows up for every, arbitrarily small, initial value $\varepsilon \neq 0$ (Figure 1.126(b)).

*Example 3* (ill-posed inverse problem): A satellite measures the gravitational field of the earth. From this, one would like to determine the density $\varrho$ of the earth; one is particularly interested in localizing oil fields.

This problem is ill-posed, i.e., the density $\varrho$ cannot be uniquely determined from the measurements.

#### 1.12.2.7 Reduction to integral equations

*Example 1:* The problem

$$y'(t) = g(t), \qquad y(0) = a,$$

has the unique solution

$$y(t) = a + \int_0^t g(\tau)\mathrm{d}\tau.$$

Therefore, one can reduce the more general problem

$$\boxed{y'(t) = f(t, y(t)), \qquad y(0) = a,}$$

to the equivalent problem

$$\boxed{y(t) = a + \int_0^t f(\tau, y(\tau))\mathrm{d}\tau.}$$

This equation contains an unknown function $y$ under the integral sign and is therefore referred to as an *integral equation.* This integral equation can be solved by means of the iteration process

$$\boxed{y_{n+1}(t) = a + \int_0^t f(\tau, y_n(\tau))\mathrm{d}\tau, \qquad y_0(t) \equiv a, \quad n = 0, 1, 2, \ldots.}$$

*Example 2:* The boundary value problem

$$\boxed{\begin{aligned} -\kappa y''(x) &= f(x, y(x)), \\ y(0) &= y(l) = 0, \end{aligned}}$$

can, because of the formula (1.224) for a solution, be reduced to the equivalent integral equation

$$y(x) = \int_0^l G(x, \xi) f(\xi, y(\xi))\mathrm{d}\xi$$

Integral equations will be dealt with systematically in [212]. In the classical theory of partial differential equations, one used to pass to equivalent integral equations by use of Green's functions. This method, however, turns out to be quite toilsome for more complicated problems, and in some cases yields no result at all. In the more modern, functional analytic theory of partial differential equations, which arose around 1935, partial differential equations are viewed from the beginning as equations for differential operators, without passing over to integral equations.

#### 1.12.2.8 The importance of integrability conditions

For functions of the class $C^2$, $u = u(x, y)$, we have the commutativity of partial differentiation:

$$\boxed{u_{xy} = u_{yx}.}$$

This relation plays an important role in many questions in the theory of partial differential equations.

*Example 1:* The equation

$$u_x = x, \qquad u_y = x$$

can not possess any solutions. Indeed, for a solution $u$, we would have $u_{xy} = u_{yx}$, which because of $u_{xy} = 0$ and $u_{yx} = 1$ cannot possibly be true.

*Example 2:* Let the $C^1$-functions $f = f(x,y)$, $g = g(x,y)$ be given, defined on a domain $D$ in $\mathbb{R}^2$. We consider the equation

$$\boxed{u_x(x,y) = f(x,y), \qquad u_y(x,y) = g(x,y) \quad \text{on } D.}$$

If a $C^2$-solution $u$ exists, then because of $u_{xy} = u_{yx}$, the so-called *integrability condition*

$$\boxed{f_y(x,y) = g_x(x,y) \quad \text{on } D}$$

must hold. This condition is sufficient for the existence of a solution, if $D$ is simply connected. The solution can then be determined as the curve integral

$$u(x,y) = \text{const} + \int_{(x_0,y_0)}^{(x,y)} f\mathrm{d}x + g\mathrm{d}y,$$

which is independent of the path of integration.

If one chooses $D$ as a small disk around a point, then $D$ is simply connected. Thus the validity of the integrability condition is sufficient for solving the initial problem $u_x = f$, $u_y = g$ locally.

It is a basic experience which has been made again and again, that necessary integrability conditions are also sufficient for the *local* solvability. In contrast, the *global* solvability depends in a decisive way on the structure (topology) of the domain.[110] Integrability conditions play an important role in many applications:

(i) The fundamental *theorema egregium* of Gauss represents an integrability condition for the derivative equations of a surface (cf. 3.6.3.3).

(ii) The two facts: every circulation-free force field (like for example the gravitational field) has a potential, and: the electromagnetic field is the derivative of a four-potential, are both consequences of integrability conditions.

(iii) Important relations in thermodynamics follow from the integrability conditions for Gibbs' equation, which is closely related to the first and second theorem of thermodynamics (cf. 1.13.1.10).

The appropriate set-up for the elegant treatment of integrability conditions is given by the Cartan–Kähler theorem on differential forms (cf. 1.13.5.4).

### 1.12.3 The classification of differential equations

**The order of a differential equation:** The highest derivative occuring in a differential equation is called the order.

*Example 1:* The Newtonian law of motion

$$m\mathbf{x}'' = \mathbf{F}$$

[110] A very deep result in this direction is the de Rham theorem (see [212]).

contains two time derivatives and is thus of order two.

The heat equation

$$T_t - \kappa(T_{xx} + T_{yy} + T_{zz}) = 0 \tag{1.231}$$

contains one time derivative and two derivatives in each of the space variables. This differential equation therefore has order two.

**Systems of differential equations:** The heat equation (1.231) is a differential equation for the temperature, viewed as a function of time. In case there are more than one equation or function in the differential equation, one speaks of systems of differential equations.

*Example 2:* If we use Cartesian coordinates, then for a vector of position we can write $\mathbf{x} := x\mathbf{i} + y\mathbf{j} + z\mathbf{k}$. Decomposing the force vector in the same way, $\mathbf{F} := X\mathbf{i} + Y\mathbf{j} + Z\mathbf{k}$, the Newtonian law of action of Example 1 can be written:

$$mx''(t) = X(P(t), t), \quad my''(t) = Y(P(t), t), \quad mz''(t) = Z(P(t), t)$$

with $P := (x, y, z)$. This is a system of differential equations of order two.

### 1.12.3.1 The principle of reduction

> Any differential equation and any system of differential equations can be reduced to an equivalent system of differential equations *of first order* by introducing appropriate variables.

*Example 1:* The differential equation of second order

$$y'' + y' + y = 0$$

is transformed into the equivalent system of differential equations of first order

$$\begin{aligned} &y' = p, \\ &p' + p + y = 0 \end{aligned}$$

by introducing the new variable $p := y'$. In a similar manner, the equation

$$y''' + y'' + y' + y = 0$$

is transformed into

$$\begin{aligned} &p = y', \quad q = p', \\ &q' + q + p + y = 0, \end{aligned}$$

by introducing $p := y'$ and $q := y''$.

*Example 2:* The Laplace equation

$$u_{xx} + u_{yy} = 0$$

is transformed into the equivalent system of first order

$$\begin{aligned} &p = u_x, \quad q = u_y, \\ &p_x + q_y = 0 \end{aligned}$$

by changing variables to $p := u_x$ and $q := u_y$.

#### 1.12.3.2 Linear differential equations and the principle of supposition

A linear differential equation for the unknown function $u$ has by definition the form

$$\boxed{Lu = f,} \tag{1.232}$$

where the right hand side $f$ also depends on the variable $u$ and the differential operator $L$ has the characteristic property:

$$\boxed{L(\alpha u + \beta v) = \alpha Lu + \beta Lv} \tag{1.233}$$

for all sufficiently smooth functions $u, v$ and all real numbers $\alpha, \beta$.

*Example 1:* The ordinary differential equation

$$u''(t) = f(t)$$

is linear. To see this, set

$$Lu := \frac{\mathrm{d}^2 u}{\mathrm{d}t^2}.$$

Often one writes simply $L := \frac{\mathrm{d}^2}{\mathrm{d}t^2}$. From the additivity of differentiation we get

$$L(\alpha u + \beta v) = (\alpha u + \beta v)'' = \alpha u'' + \beta v'' = \alpha Lu + \beta Lv.$$

Consequently, the linearity condition (1.233) is satisfied.

*Example 2:* The most general ordinary differential equation of $n$th order for the function $u = u(t)$ has the form

$$\boxed{a_0 u + a_1 u' + a_2 u'' + \ldots + a_n u^{(n)} = f,}$$

where all coefficients $a_i$ and $f$ are functions of the time $t$ and $a_n \neq 0$.

The most general linear differential equation is a linear combination of partial derivatives with coefficients, which are all functions of the same variables as the function which is the solution of the equation.

*Example 3:* Let $u = u(x, y)$. The differential equation

$$a u_{xx} + b u_{yy} = f \tag{1.234}$$

is linear in case $a = a(x, y)$, $b = b(x, y)$ and $f = f(x, y)$ are all functions of only the independent variables $x$ and $y$.

In case the right-hand side $f = f(x, y, u)$ also depends on the unknown $u$, then (1.234) is a non-linear differential equation.

**Homogenous equations:** A linear differential equation (1.232) is said to be *homogenous*, if $f \equiv 0$, otherwise it is called *inhomogenous*.

**The principle of superposition:** (i) For a *homogenous* differential equation, a linear combination $\alpha u + \beta v$ of solutions $u$ and $v$ is also a solution of the equation. (ii) For an *inhomogenous* equation one has the following important rule:

$$\boxed{\begin{array}{l}\text{The general solution of the inhomogenous equation}\\ = \text{a particular solution of the inhomogenous equation}\\ + \text{the general solution of the homogenous equation.}\end{array}} \tag{1.235}$$

*Example 4:* We consider the differential equation

$$\boxed{u' = 1 \quad \text{on } \mathbb{R}.}$$

A particular solution is $u = t$. The general solution of the homogenous equation $u' = 0$ is $u = \text{const}$. Hence, the general solution of the inhomogenous system is

$$u = t + \text{const}.$$

#### 1.12.3.3 Non-linear differential equations

> Non-linear differential equations describe processes with interactions.

Most processes in nature are processes with some kind of interaction. This explains the importance of non-linear differential equations in the natural sciences. An apparent exception to this rule is formed by the set of Maxwell equations for the electromagnetic field. However, these equations only describe a part of the total phenomena of the electromagnetic fields. The complete equations of *quantum electrodynamics* describe interactions between electromagnetic waves (photons), electrons and positrons. These equations are indeed non-linear.

> For non-linear differential equations the superposition principle is not valid.

*Example 1:* The Newton equations of motion for a planet in the gravitational field of the sun is

$$m\mathbf{x}''(t) = -\frac{GmM}{|\mathbf{x}(t)|^3}\mathbf{x}(t).$$

These equations (one for each coordinate of the vector $\mathbf{x}$) are non-linear and describe the gravitational interaction of the sun and the planets.

**Semilinear equations:** If $L$ is a linear differential operator of order $n$ as in (1.232), one calls an equation of the form

$$Lu = f(u)$$

*semilinear*; the right-hand side of $f$ depends on the sought-for function $u$ and its derivatives up to order $n - 1$.

*Example 2:* The equation $u_{xx} + u_{yy} = f(u, u_x, u_y, x, y)$ is semilinear.

**Quasilinear equations:** An equation which is linear in its highest derivative is called *quasilinear.*

*Example 3:* The equation $au_{xx} + bu_{yy} = f$ is quasilinear, if $a, b$ and $f$ depend (only) on $x, y, u, u_x$ and $u_y$.

#### 1.12.3.4 Stationary and non-stationary processes

A process which depends on the time $t$ is said to be *non-stationary*. *Stationary* process are by definition those which are independent of time.

> Stationary processes correspond to equilibrium configurations
> in nature, technology and economy.

#### 1.12.3.5 Equilibrium configurations

**Stable equilibrium configurations:** A configuration in equilibrium is said to be *stable*, if the system is unaffected by small perturbations, i.e., after a small perturbation out of the equilibrium position, the configuration returns after a finite time to the equilibrium.

| In our physical world, only stable equilibrium configurations occur. |
|---|

*Unstable* equilibrium configurations are those which leave equilibrium indefinitely after small perturbations.

**The principle of equilibrium:** In order to find the equilibrium solutions of a differential equation for a non-stationary process, one sets all time derivatives in the differential equation to zero and solves the resulting differential equation.

*Example 1:* Consider the non-stationary heat equation

$$\boxed{T_t - \kappa \Delta T = 0.} \tag{1.236}$$

We get the equilibrium solutions by searching for solutions $T = T(x, y, z)$ which are independent of time. Thus one has $T_t \equiv 0$. Thus we are looking for solutions to the stationary heat equation

$$\boxed{-\kappa \Delta T = 0.} \tag{1.237}$$

One expects that a non-stationary (time dependent) heat distribution tends to an equilibrium under mild assumptions at $t \to \infty$, i.e., certain solutions of (1.236) tend to a solution of (1.237) as $t \to \infty$.

This expectation can be rigorously justified for a general situation under appropriate assumptions.

We explain this in the following simple model problem.

*Example 2:* Let $a \neq 0$. In order to find the equilibrium solutions of the differential equation

$$\boxed{y'(t) = ay(t)}$$

we assume that the solution does not depend on time. Thus $y'(t) = 0$ and hence

$$\boxed{y(t) = 0 \quad \text{for all } t.}$$

According to Examples 1 and 2 in 1.12.2.6 this equilibrium solution for $a = -1$ is stable and for $a = 1$ it is unstable (Figure 1.126).

#### 1.12.3.6 The method of comparing coefficients – a general method of solution

*Example 1:* To solve the equation

$$\boxed{u'' = u + 1}$$

we use the Taylor expansion

$$u(t) = u(0) + u'(0)t + u''(0)\frac{t^2}{2} + u'''(0)\frac{t^3}{3!} + \dots .$$

If we know $u(0)$ and $u'(0)$, then all higher derivatives $u''(0), u'''(0), \dots$ can be calculated from the differential equation. At the same time, we must include the following initial value problem to get a unique solution

$$\boxed{u'' = u + 1, \quad u(0) = a, \quad u'(0) = b.}$$

Then we get

$$\begin{aligned} &u''(0) = u(0) + 1 = a + 1, \quad u'''(0) = u'(0) = b, \\ &u^{(2n)}(0) = a + 1, \quad u^{(2n+1)}(0) = b, \quad n = 1, 2, \dots \end{aligned}$$

> In the same way one can solve any ordinary differential equation from which the highest derivatives can be solved for.

The same method can also be applied to partial differential equations. In this case, however, a new effect enters the game, which has to do with the characteristics of the differential equation.

*Example 2:* Let the function $\varphi = \varphi(x)$ be given. We are looking for a function $u := u(x, y)$ which solves the following initial value problem:

$$\boxed{\begin{aligned} &u_y = u, \\ &u(x, 0) = \varphi(x) \quad \text{(initial value)}, \end{aligned}}$$

i.e., we prescribe the values of $u$ along the $x$-axis. To proceed, we take the Taylor expansion of the function at the origin,

$$u(x, y) = u(0, 0) + u_x(0, 0)x + u_y(0, 0)y + \dots .$$

For this method to yield a well-defined expression, we have to be able to determine all the partial derivatives of $u$ from the initial values and differential equation. In the case at hand this is in fact possible. The initial condition $u(x, 0) = \varphi(x)$ yields first all the partial derivatives with respect to $x$:

$$u(0, 0) = \varphi(0), \quad u_x(0, 0) = \varphi'(0), \quad u_{xx}(0, 0) = \varphi''(0) \quad \text{and so on.}$$

From the differential equation $u_y = u$ we get

$$u_y(0, 0) = u(0, 0) = \varphi(0).$$

All remaining derivatives can be determined by differentiating the differential equation:

$$u_{yx}(0, 0) = u_x(0, 0) \quad \text{and so on.}$$

*Example 3:* The situation is completely different for the initial value problem

$$\boxed{\begin{aligned} &u_y = u, \\ &u(0, y) = \psi(y) \quad \text{(initial value)}, \end{aligned}}$$

for which we prescribe the values of $u$ along the $y$-axis.

In this case we have no information whatsoever about the derivatives with respect to $x$. In fact, the differential equation and the initial value problems can contradict one another, so that *no solutions* at all exist. For example, a solution would have to fulfill $u_y(0,0) = \psi'(0)$ and at the same time $u_y(0,0) = u(0,0) = \psi(0)$, and hence

$$\boxed{\psi(0) = \psi'(0).}$$

This necessary condition for the existence of a solution is referred to as a *compatibility condition*; if this condition is not satisfied, then no solutions can exist.

*Example 4:* Let the line $\ell : y = \alpha x$ with $\alpha \neq 0$ be given. To solve the initial value condition

$$\boxed{\begin{array}{l} u_y = u, \\ u \text{ is known along } \ell \quad \text{(initial value)} \end{array}}$$

we choose $\ell$ as $\xi$-coordinate axis and introduce new coordinates $\xi, y$ (Figure 1.127(b)).

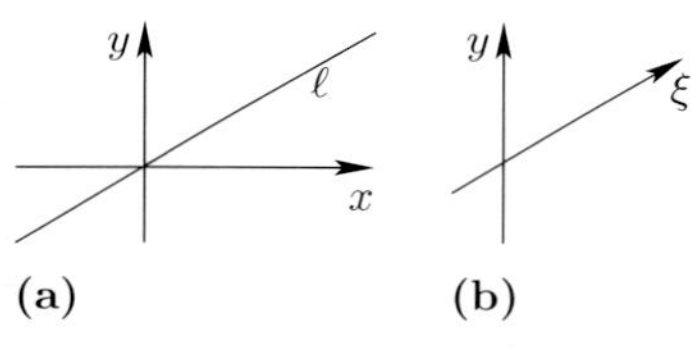


Figure 1.127.

If we write the function $u$ in these new coordinates, then we get the problem

$$\boxed{\begin{array}{l} u_y = u, \\ u(\xi, 0) = \varphi(\xi), \end{array}}$$

which can be solved in the same way as Example 2 above.[111]

**Characteristics:** Because of the course of Examples 2 to 4 above, one says that in these cases the $y$-axis is *characteristic* for the differential equation

$$u_y = u$$

while all other lines through the origin are not characteristic.

The general theory of characteristics and their physical interpretation will be considered in 1.13.3. The behavior of characteristics of a differential equation can at the same time be used to classify partial differential equations (elliptic, parabolic and hyperbolic, cf. 1.13.3.2).

Roughly speaking one has

> Characteristics correspond to the initial conditions of systems which do not determine solutions uniquely or which allow no solutions at all.

From a physical point of view characteristics are important because they describe wave fronts. The propagation of waves is the most important mechanism in nature for transporting energy.

*Example 5:* The equation

$$u(x,t) = \varphi(x - ct) \tag{1.238}$$

[111]The rigorous justification of the power series method described here can be carried out for ordinary and partial differential equations using the theorems of Cauchy and Cauchy–Kowalewskaja (cf. 1.12.9.3 and 1.13.5.1).

describes the propagation of a wave with speed $c$ from left to right (Figure 1.128).

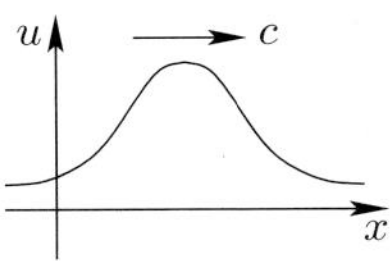


Figure 1.128.

(i) If we prescribe the values of $u$ at the initial time $t = 0$, we can determine the function $\varphi$ from the equation $u(x, 0) = \varphi(x)$ uniquely.

(ii) If on the other hand we prescribe $u$ along the line $x - ct =$ const $= a$, then only the value $\varphi(a)$ is determined, and the function $\varphi$ is arbitrary up to this value.

#### 1.12.3.7 Important information which can be derived from a differential equation without actually solving it

In many cases it is not possible to explicitly solve a differential equation. Because of this it is of great value to derive as much physically relevant information directly from a given differential equation. Among others, the following kinds of information can be obtained in this manner:

(i) conservation laws (such as conservation of energy);

(ii) equations for the wave fronts (characteristics), see 1.13.3.1;

(iii) maximum principles (cf. 1.13.4.2);

(iv) criteria for stability (cf. 1.12.7).

*Example* (conservation of energy): Let $x = x(t)$ be a solution of the differential equation

$$mx''(t) = -U'(x(t)).$$

If we set

$$E(t) := \frac{mx'(t)^2}{2} + U(x(t)),$$

then we have

$$\frac{\mathrm{d}E(t)}{\mathrm{d}t} = mx''(t)x'(t) + U'(x(t))x'(t) = 0,$$

in other words,

$$\boxed{E(t) = \text{const.}}$$

This is the law of conservation of energy.

In deriving this relation we only used the differential equation and not the form of a solution.

#### 1.12.3.8 Symmetry and conservation laws

*Example 1* (conservation of energy): We consider the Euler–Lagrange equation (see sections 5.1.1 and 14.5.2)

$$\boxed{\frac{\mathrm{d}}{\mathrm{d}t} L_{q'}(P(t)) - L_q(P(t)) = 0} \qquad (1.239)$$

for the Lagrange function $L$ and a function $P(t) := (q(t), q'(t))$. If we set

$$E(t) := q'(t)L_{q'}(P(t)) - L(P(t)),$$

then for every solution of (1.239) we have the following law of conservation of energy:

$$\boxed{E(t) = \text{const.}} \tag{1.240}$$

Conservation laws are of fundamental importance in nature and are responsible for the stability of different forms in the world around us. A world without conservation laws would be complete chaos.

What is the deeper reason for the occurrence of such conservation laws? One answer to this difficult question is the following:

> Symmetries in our world are responsible for the conservation laws we observe.

The rigorous mathematical formulation of this basic epistemological principle is the content of the famous theorem of Emmy Noether, proved in 1918. This theorem, which is of extreme importance for theoretical physics, will be considered in more detail in [212].

*Example 2:* The *conservation of energy* (1.240) is a special case of the Noether's theorem. The corresponding symmetry property derives from the fact that the Lagrange function $L$ does not depend on time $t$. Because of this, the equation (1.239) is invariant under translations in time. This means: if

$$q = q(t)$$

-s a solution of (1.239), then the function

$$q = q(t + t_0)$$

is also a solution, where $t_0$ denotes an arbitrary time constant.

**Definition:** One says a physical system is *invariant under time translations*, if the following statement holds: if a physical process $\mathscr{P}$ is possible, then also any process which is derived from $\mathscr{P}$ by adding a time constant is also possible.

> Systems which are invariant under time translations possess a conserved quantity, which is called *energy.*

*Example 3:* The gravitational field of our sun is independent of time. Therefore, all motions of the planets around the sun are invariant under time translations, thus any motion resulting from an existing motion by a translation in time is possible.

If the gravitational field of the sun were time dependent, then the choice of an initial value (time) would be of utmost importance in determining the motion of the planets. In this case conservation of energy for planetary motion would not hold (cf. 1.9.6).

> Systems which are invariant under a rotation possess a conserved quantity, which one refers to as *angular momentum.* Conservation of angular momentum means: if a process $\mathscr{P}$ is possible then any other motion which derives from $\mathscr{P}$ by a rotation is also possible.

*Example 4:* The gravitational field of our sun is rotationally symmetric. Thus angular momentum is a conserved quantity for our solar system.

> Systems which are invariant under translations possess a conserved quantity which is referred to as *momentum*.

*Example 5:* If one lets the location of the sun in the solar system be variable instead of being situated at the origin, our solar system is invariant under translations. Thus, momentum is a conserved quantity for the solar system. This is equivalent to the statement that the center of mass of the solar system moves in a straight line with constant velocity (cf. 1.9.6).

#### 1.12.3.9 Strategies for obtaining uniqueness results

For ordinary differential equations there is a very simple general uniqueness result (cf. 1.12.4.2). For partial differential equations, however, the situation is much more complicated. One has the following methods for checking uniqueness of solutions.

(i) The method of energy, which is based on the law of conservation of energy (cf. 1.13.4.1) and

(ii) maximum principles (cf. 1.13.4.2).

We explain the basic idea in two simple examples.

**Energy method:** *Example 1:* The initial value problem

$$mx'' = -x, \quad x(0) = a, \quad x'(0) = b \tag{1.241}$$

has at most one solution.

*Proof:* Suppose that there were two different solutions $x_1$ and $x_2$. Then as in all uniqueness results, we consider the difference

$$y(t) := x_1(t) - x_2(t).$$

We will be finished if we can show that $y(t) \equiv 0$.

To do this, we first note the equation for $y$ which results when we subtract (1.241) for $x = x_1$ and $x = x_2$:

$$my'' = -y, \quad y(0) = y'(0) = 0.$$

From conservation of energy as in 1.12.3.7 with $U = y^2/2$, we have

$$\frac{my'(t)^2}{2} + \frac{y(t)^2}{2} = \text{const} = E.$$

If we consider the initial time $t = 0$, then we get from the relations $y(0) = y'(0) = 0$ the relation $E = 0$. Hence

$$y(t) = 0.$$

The simple physical idea behind this proof is the intuitively clear fact:

> If a system in which conservation of energy holds is at rest at an initial time, then the system has no energy and remains at rest for all times.

**Maximum principle:** *Example 2:* Let $\Omega$ be a bounded domain in $\mathbb{R}^3$. Every solution $T$ of the stationary heat equation

$$T_{xx} + T_{yy} + T_{zz} = 0 \quad \text{on } \Omega, \tag{1.242}$$

which is of type $C^2$ on the closure $\overline{\Omega} = \Omega \cup \partial\Omega$, attains its maximum and minimum on the boundary.

In particular, if $T \equiv 0$ on $\partial\Omega$ for a solution of (1.242), then $T \equiv 0$ on $\overline{\Omega}$.

*Physical interpretation:* If the temperature at the boundary vanishes, then there can be no point $P$ in the interior such that $T(P) \neq 0$. Otherwise, the rate of change of temperature would lead to a time independent heat flow, which contradicts the time independent (stationary) property of the system.

*Example 3* (uniqueness): Let the function $T_0$ be given. The boundary value problem

$$\begin{aligned} &T_{xx} + T_{yy} + T_{zz} = 0 \qquad &&\text{on } \Omega, \\ &T = T_0 &&\text{on } \partial\Omega \end{aligned} \tag{1.243}$$

has at most a single solution $T$.

*Proof:* If $T_1$ and $T_2$ are different solutions, the difference $T := T_1 - T_2$ fulfills the equation (1.243) with $T_0 = 0$. Hence, by Example 2, $T \equiv 0$.

## 1.12.4 Elementary methods of solution

**Laplace transformation:** Every linear differential equation of arbitrary order with *constant coefficients* as well as every system of such equations can be solved with the help of the Laplace transformation (cf. 1.11.1.2).

**Quadratures:** By definition, a differential equation can be solved by *quadratures*, if the solution can be found by computing integrals. The following so-called elementary solution methods are of this type.[112]

### 1.12.4.1 The local existence and uniqueness theorem

$$\boxed{\begin{aligned} &x'(t) = f(t, x(t)), \\ &x(t_0) = x_0 \quad \text{(initial value).} \end{aligned}} \tag{1.244}$$

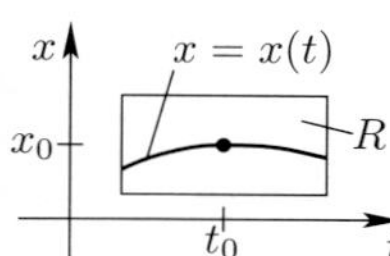


Figure 1.129.

**Definition:** Let a point $(t_0, x_0) \in \mathbb{R}^2$ be given. The initial value problem (1.244) is *locally uniquely solvable*, if and only if, there is a rectangle $R := \{(t_0, x) \in \mathbb{R}^2 : |t - t_0| \leq \alpha,\ |x - x_0| \leq \beta\}$, such that in $R$ a unique solution $x = x(t)$ of (1.244) exists (Figure 1.129).[113]

**The theorem of Picard (1890) and Lindelöf (1894):** If $f$ is of type $C^1$ in a neighborhood of a point $(t_0, x_0)$, then the initial value problem is locally uniquely solvable. This solution can be obtained by means of the iteration scheme[114]

$$\boxed{x_{n+1}(t) = x_0 + \int_{t_0}^{t} f(\tau, x_n(\tau))\mathrm{d}\tau, \qquad n = 0, 1, \ldots}$$

[112]The most general differential equation can not be solved by quadratures.

A general symmetry principle, through which many differential equations can be solved by quadratures, was discovered by the Norwegian mathematician Sophus Lie (1842–1899). This principle, which uses the theory of transformation groups, can be found in [212].

[113]This means there is a unique solution $x = x(t)$ of (1.244) with $|x(t) - x_0| \leq \beta$ for all times $t$ with $|t - t_0| \leq \alpha$.

[114]The proof of this relies on the Banach fixed-point theorem, and can be found in [212].

The zeroth approximation is the constant function $x_0(t) \equiv x_0$.

**Relaxing the assumptions:** It is sufficient that one of the following assumptions is satisfied:

(i) $f$ and the partial derivative $f_x$ are continuous in a neighborhood of the point $(t_0, x_0)$.

(ii) $f$ is continuous in a neighborhood $U$ of the point $(t_0, x_0)$ and Lipschitz continuous with respect to $x$, i.e., one has

$$|f(t,x) - f(t,y)| \leq \text{const}|x - y|$$

for all points $(t, x)$ and $(t, y)$ in $U$.

In fact, (i) is a special case of (ii).

**The theorem of Peano (1890):** If $f$ is continuous in a neighborhood $U$ of the point $(t_0, x_0)$, then the initial value problem (1.244) is locally solvable.

However, the uniqueness of such a solution can not be guaranteed.[115]

**Generalization to systems:** All the statements above remain valid, when (1.244) represents a system of differential equations. In this case $x = (x_1, \ldots, x_n)$ and $f = (f_1, \ldots, f_n)$. The equation (1.244) is in this case explicitly:[116]

$$\boxed{\begin{aligned} &x_j'(t) = f_j(t, x(t)), \quad j = 1, \ldots, n, \\ &x_j(t_0) = x_{j0}. \end{aligned}} \tag{1.245}$$

According to the reduction principle, an arbitrary system of arbitrary order can be reduced to one of the form (1.245) (cf. 1.12.3.1).

**Generalization to systems of differential equations with complex variables:** All statements remain valid, with appropriate modifications, if the $x_j$ are complex variables and the values $f_j(t, x)$ are complex.

**Global existence and uniqueness results:** In this respect, see 1.12.9.1.

### 1.12.4.2 The global theorem of uniqueness

**Theorem:** Suppose that $x = x(t)$, $t_1 < t < t_2$ is a solution of (1.244) such that every point $(t, x(t))$ has a neighborhood $U$ on which $f$ is of type $C^1$. Then the initial value[117] problem (1.244) has no further solution on the interval $]t_1, t_2[$ (Figure 1.130).

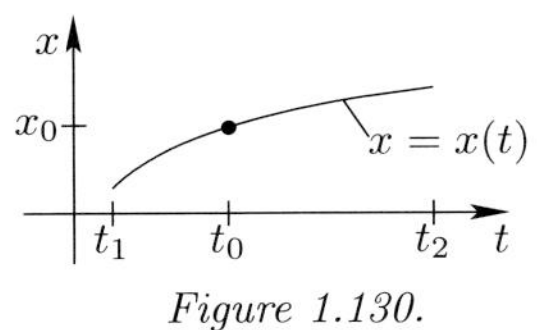


*Figure 1.130.*

*Proof idea:* A further solution would necessarily differ from the given one in some point, which is impossible by the Theorem of Picard–Lindelöf.

**Generalization:** A similar result holds for real and complex-valued systems.

---

[115] The theorem of Peano can be proved with the fixed point theorem of Schauder, which is based on the notion of compactness (cf. [212]).

[116] We denote in this case by $f_x$ the *matrix* $(\partial f_j / \partial x_k)$ of the first partial derivatives with respect to $x_1, \ldots, x_n$, and we set:

$$|x| := \left( \sum_{j=1}^{n} |x_j|^2 \right)^{\frac{1}{2}}.$$

[117] It is sufficient that one of the two following conditions is satisfied:

(i) $f$ and $f_x$ are continuous on $U$.

(ii) $f$ is continuous on $U$ and in addition Lipschitz continuous with respect to $x$.

#### 1.12.4.3 A general strategy for finding solutions

Physicists and engineers (as well as mathematicians in the seventeenth and eighteenth centuries) developed mnemonic and very simple formal methods for solving differential equations (cf. for example 1.12.4.4). If one has found a 'solution' with the help of one of these methods, then there are two important questions:

(a) Is this in fact a solution?

(b) Is this the only solution or are there more solutions, which the formal method does not find?

The answers are:

(a) A solution can be recognized by carrying out the differentiation occurring in the differential equation explicitly.

(b) Use the global uniqueness result of 1.12.4.2.

In the next section we will consider applications of this strategy.

#### 1.12.4.4 Separation of variables

$$\boxed{\begin{aligned} &\frac{\mathrm{d}x}{\mathrm{d}t} = f(t)g(x), \\ &x(t_0) = x_0 \quad \text{(initial value).} \end{aligned}} \tag{1.246}$$

**Formal method:** The differential calculus of Leibniz elegantly yields immediately

$$\frac{\mathrm{d}x}{g(x)} = f(t)\mathrm{d}t$$

and

$$\int \frac{\mathrm{d}x}{g(x)} = \int f(t)\mathrm{d}t.$$

If one wants to accommodate initial conditions, then one writes:

$$\boxed{\int_{x_0}^{x} \frac{\mathrm{d}x}{g(x)} = \int_{t_0}^{t} f(t)\mathrm{d}t.} \tag{1.247}$$

**Theorem:** If $f$ is continuous in a neighborhood of $t_0$ and $g$ is of type $C^1$ in a neighborhood of $x_0$ and $g(x_0) \neq 0$, then the initial value problem (1.246) is locally uniquely solvable. The solution can be obtained by solving equation (1.247) for $x$.

**Remark:** This theorem insures the solution for times which lie in a small neighborhood of the initial time. In general it is not possible to obtain more than this (cf. Example 2). However in concrete cases, one can use this method to obtain a candidate $x = x(t)$ for a solution, which may exist for longer times. It is advantageous at this point to use the strategy set out in 1.12.4.3.

*Example 1:* We consider the initial value problem

$$\boxed{\begin{aligned} &\frac{\mathrm{d}x}{\mathrm{d}t} = ax(t), \\ &x(0) = x_0 \quad \text{(initial value).} \end{aligned}} \tag{1.248}$$

Here $a$ is a real constant. The unique solution of (1.248) is

$$\boxed{x(t) = x_0 \mathrm{e}^{at}, \quad t \in \mathbb{R}.} \tag{1.249}$$

*Formal method:* We assume first that $x_0 > 0$. Then applying separation of variables yields

$$\int_{x_0}^{x} \frac{\mathrm{d}x}{x} = \int_0^t a\mathrm{d}t.$$

From this it follows that $\ln x - \ln x_0 = at$, hence $\ln \dfrac{x}{x_0} = at$, i.e., $\dfrac{x}{x_0} = \mathrm{e}^{at}$.

*Exact solution:* Differentiating (1.249) gives

$$x' = ax_0\mathrm{e}^{at} = ax,$$

i.e., the function $x$ in (1.249) is indeed a solution of (1.248).

Since the right side $f(x,t) := ax$ is of type $C^1$, the global uniqueness theorem 1.12.4.2 shows that there are no other solutions.

These considerations are valid for all $x_0 \in \mathbb{R}$, while for example the formal method fails for $x_0 = 0$ because of '$\ln 0 = -\infty$'.

*Example 2:*

$$\boxed{\begin{array}{l} \dfrac{\mathrm{d}x}{\mathrm{d}t} = \dfrac{(1+x^2)}{\varepsilon}, \\ x(0) = 0 \quad \text{(initial value).} \end{array}} \tag{1.250}$$

Here $\varepsilon > 0$ is a constant. The uniquely determined solution of (1.250) is:

$$\boxed{x(t) = \tan\frac{t}{\varepsilon}, \quad -\frac{\varepsilon\pi}{2} < t < \frac{\varepsilon\pi}{2}.}$$

Moreover we have $x(t) \to +\infty$ as $t \to \dfrac{\varepsilon\pi}{2} - 0$. The smaller $\varepsilon$ is, the faster the solution blows up (see Figure 1.131).

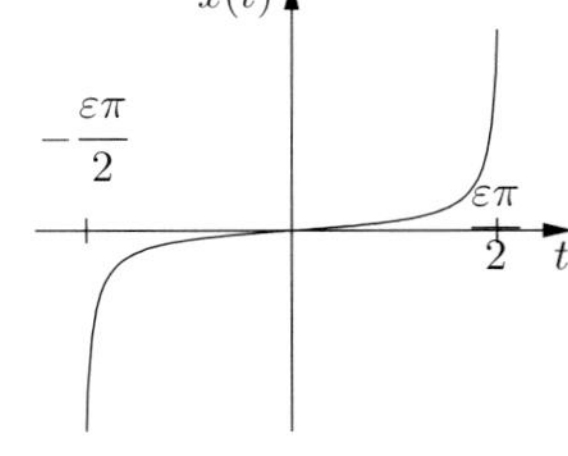


*Figure 1.131.*

*Formal method:* Separation of variables yields

$$\int_0^x \frac{\mathrm{d}x}{1+x^2} = \int_0^t \frac{\mathrm{d}t}{\varepsilon},$$

so that $\arctan x = t/\varepsilon$, i.e., $x = \tan(t/\varepsilon)$.

*Exact solution:* Conclude as we have in Example 1. □

#### 1.12.4.5 The linear differential equation and the propagator

$$\boxed{\begin{array}{rcl} x' &=& A(t)x + B(t), \\ x(0) &=& x_0 \quad \text{(initial value)} \end{array}} \tag{1.251}$$

**Theorem:** If $A, B : J \longrightarrow \mathbb{R}$ are continuous on an open interval $J$, then the initial value problem (1.251) has on $J$ the unique solution

$$\boxed{x(t) = P(t, t_0)x_0 + \int_{t_0}^{t} P(t, \tau)B(\tau)\mathrm{d}\tau} \tag{1.252}$$

with the so-called *propagator*

$$\boxed{P(t, \tau) := \exp\left(\int_{\tau}^{t} A(s)\mathrm{d}s\right).}$$

For the propagator we have

$$P_t(t, \tau) = A(t)P(t, \tau), \quad P(\tau, \tau) = 1, \tag{1.253}$$

and $P(t_3, t_1) = P(t_3, t_2)P(t_2, t_1)$ for $t_1 < t_2 < t_3$.

*Proof:* Differentiating (1.252) yields

$$x'(t) = P_t(t, t_0)x_0 + \int_{t_0}^{t} P_t(t, \tau)B(\tau)\mathrm{d}\tau + P(t, t)B(t).$$

Taking (1.253) into account, we obtain

$$x'(t) = A(t)x(t) + B(t).$$

The uniqueness follows from the global uniqueness result in section 1.12.4.2. □

The fundamental importance of the propagator in physical processes will be discussed in 1.12.6.1.

One is led to the solution (1.252) upon applying a method called variation of constants, which was invented by Lagrange (1736–1813) for treatment of problems of planetary motion.

**Variation of constants:** Step 1: *Solution of the homogenous problem.* If we set $B \equiv 0$, then we get from (1.251) by separation of variables the expression

$$\int_{x_0}^{x} \frac{\mathrm{d}x}{x} = \int_{t_0}^{t} A(s)\mathrm{d}s.$$

For $x_0 > 0$ this gives $\ln \dfrac{x}{x_0} = \displaystyle\int_{t_0}^{t} A(s)\mathrm{d}s$, hence

$$x = C \exp\left(\int_{t_0}^{t} A(s)\mathrm{d}s\right) \tag{1.254}$$

with the constant $C = x_0$.

Step 2: *Solution of the inhomogenous problem.* Lagrange's idea is that by introducing the perturbation $B$, the constant $C = C(t)$ becomes time dependent. Differentiation of (1.254) yields

$$x' = C' \exp\left(\int_{t_0}^{t} A(s)\mathrm{d}s\right) + Ax.$$

By comparison with the given differential equation $x' = Ax + B$, we get the differential equation

$$C'(t) = \exp\left(-\int_{t_0}^{t} A(s)\mathrm{d}s\right) B(t)$$

with the solution

$$C(t) = x_0 + \int_{t_0}^{t} \exp\left(-\int_{t_0}^{\tau} a(s)\mathrm{d}s\right) B(\tau)\mathrm{d}\tau.$$

This is (1.252).

**Application of the principle of superposition:** *Example 3:* A first guess leads to the special solution $x = 1$ of the differential equation

$$x' = x - 1. \tag{1.255}$$

The homogenous equation $x' = x$ has the general solution $x = \text{const} \cdot \mathrm{e}^t$ by (1.248). Therefore, by the superposition principle in 1.12.3.2 the differential equation (1.255) has the general solution

$$x = \text{const} \cdot \mathrm{e}^t + 1.$$

#### 1.12.4.6 The differential equation of Bernoulli

$$\boxed{x' = A(t)x + B(t)x^{\alpha}, \quad \alpha \neq 1.}$$

By the substitution $y = x^{1-\alpha}$, one gets from this the linear differential equation

$$y' = (1-\alpha)Ay + (1-\alpha)B.$$

This differential equation was first studied by Jakob Bernoulli (1654–1705).

#### 1.12.4.7 The Riccati differential equation and problems of control

$$\boxed{x' = A(t)x + B(t)x^2 + C(t).} \tag{1.256}$$

If one knows a special solution $x_*$, then the linear differential equation

$$-y' = (A + 2x_* B)y + B \tag{1.257}$$

is obtained by the substitution $x = x_* + \dfrac{1}{y}$.

*Example:* The inhomogenous logistic equation

$$x' = x - x^2 + 2$$

has the special solution $x = 2$. The corresponding differential equation (1.257) is $y' = 3y + 1$ and has the general solution $y = \dfrac{1}{3}\left(C\mathrm{e}^{3t} - 1\right)$. The general solution

$$x = 2 + \frac{3}{C\mathrm{e}^{3t} - 1}, \quad t \in \mathbb{R}$$

follows from this and (1.256), with the constant $C$.

**The double ratio:** If one knows three solutions $x_1$, $x_2$ and $x_3$ of (1.256), then one gets the general solution $x = x(t)$ of (1.256) from the condition that the *double ratio* of these four functions

$$\boxed{\frac{x(t) - x_2(t)}{x(t) - x_1(t)} : \frac{x_3(t) - x_2(t)}{x_3(t) - x_1(t)}}$$

is constant. This equation, which was studied by Jacopo Count Riccati (1676–1754) plays a central role today in (linear) control theory with quadratic cost function (cf. 5.3.2).

#### 1.12.4.8 The homogenous differential equation

$$\boxed{x' = F(x, t).}$$

If $F(\lambda x, \lambda t) = F(x, t)$ for all $\lambda \in \mathbb{R}$, then $F(x, t) = f\left(\frac{x}{t}\right)$, meaning that $F$ is in reality only a function of the ratio $\frac{x}{t}$. The substitution

$$y = \frac{x}{t}$$

then leads to the differential equation

$$\frac{\mathrm{d}y}{\mathrm{d}t} = \frac{f(y) - y}{t},$$

which can be solved via separation of variables.

*Example:* The differential equation

$$x' = \frac{x}{t} \tag{1.258}$$

transforms to the equation $y' = 0$ upon substituting $y = x/t$; this has the solution $y = \text{const}$. Therefore,

$$x = \text{const} \cdot t$$

is the general solution of (1.258).

#### 1.12.4.9 The exact differential equation

$$\boxed{\begin{aligned} &\frac{\mathrm{d}x}{\mathrm{d}t} = \frac{f(x,t)}{g(x,t)}, \\ &x(t_0) = x_0 \quad \text{(initial value).} \end{aligned}} \tag{1.259}$$

Let $g(x_0, t_0) \neq 0$.

**Definition** The differential equation (1.259) is said to be *exact*, if and only if the functions $f$ and $g$ are of type $C^1$ in a neighborhood $U$ of $(t_0, x_0)$ and on $U$ the integrability condition

$$\boxed{f_x(x, t) = -g_t(x, t)} \tag{1.260}$$

is satisfied.

**Theorem:** In case of exactness, the equation (1.259) is locally uniquely solvable. The solution is obtained from the equation

$$\boxed{\int_{x_0}^{x} g(\xi, t_0)\mathrm{d}\xi = \int_{t_0}^{t} f(x_0, \tau)\mathrm{d}\tau} \tag{1.261}$$

after solving for $x$.

This is a generalization of the method of separation of variables (cf. 1.12.4.4).

**Total differential:** The formula (1.261) for the solution is equivalent to the following procedure.

(i) Write the differential equation (1.259) in the form

$$g\mathrm{d}x - f\mathrm{d}t = 0.$$

(ii) Determine the function $F$ as the solution of the equation

$$\mathrm{d}F = g\mathrm{d}x - f\mathrm{d}t.$$

The condition for solvability $\mathrm{d}(\mathrm{d}F) = 0$ is satisfied because of $\mathrm{d}(\mathrm{d}F) = (g_t + f_x)\mathrm{d}t \wedge \mathrm{d}x$ and (1.260).

(iii) Solve the equation

$$F(x, t) = F(x_0, t_0)$$

for $x$ and get a solution $x = x(t)$ of (1.259). Explicitly, this means

$$F(x, t) = \int_{x_0}^{x} f(\xi, t_0)\mathrm{d}\xi - \int_{t_0}^{t} f(x_0, \tau)\mathrm{d}\tau + \text{const.}$$

(1.261) follows from this.

In simple cases one doesn't need to use (1.261), since the function $F$ can easily be guessed. The following example demonstrates this.

*Example:* The equation

$$\boxed{\frac{\mathrm{d}x}{\mathrm{d}t} = -\frac{3x^2t^2 + x}{2xt^3 + t}} \tag{1.262}$$

is written in the form

$$(2xt^3 + t)\mathrm{d}x + (3x^2t^2 + x)\mathrm{d}t = 0.$$

One easily guesses that the equation

$$\mathrm{d}F = F_x\mathrm{d}x + F_t\mathrm{d}t = (2xt^3 + t)\mathrm{d}x + (3x^2t^2 + x)\mathrm{d}t$$

has the solution $F(x, t) = x^2t^3 + xt$. The equation $F(x, t) = \text{const.}$, that is

$$\boxed{x^2t^3 + xt = \text{const.}}$$

describes a family of curves, which represent the general solution of (1.262).

#### 1.12.4.10 The Euler multiplier

If the differential equation (1.261) is not exact, then one can try, by multiplying denominator and numerator by $M(x, y)$, to get a new differential equation

$$\boxed{\frac{\mathrm{d}x}{\mathrm{d}t} = \frac{M(x,t)f(x,t)}{M(x,t)g(x,t)}}$$

which is exact. If the factor $M$ achieves this, then it is called an *Euler multiplier.*

*Example:* If we apply this to the equation

$$\boxed{\frac{\mathrm{d}x}{\mathrm{d}t} = -\frac{3x^4t^3 + x^3t}{2x^3t^4 + t^2x^2}}$$

with $M := 1/x^2t$, then we get the exact differential equation (1.262).

#### 1.12.4.11 Differential equations of higher order

**Type 1 (Energy trick):**

$$\boxed{x'' = f(x).} \tag{1.263}$$

Let $F(x) = \displaystyle\int f\mathrm{d}x$, that is, $F$ is a primitive function of $f$. The equation (1.263) is, for non-constant solutions, equivalent to the so-called *equation of energy conservation*

$$\boxed{\frac{x'^2}{2} - F(x) = \text{const.}} \tag{1.264}$$

Indeed, differentiation gives

$$\boxed{\frac{\mathrm{d}}{\mathrm{d}t}\left(\frac{x'^2}{2} - F(x)\right) = (x'' - f(x))\, x'.}$$

Accordingly, every solution of (1.263) is also a solution of (1.264). The converse holds for non-constant solutions.

An application of this to the determination of the cosmological limiting velocity of the earth will be considered in 1.12.5.1.

**Type 2 (Reduction to lower order):**

$$\boxed{x'' = f(x', t), \quad x(t_0) = x_0, \quad x'(t_0) = v.} \tag{1.265}$$

By the substitution $y = x'$, one obtains the equation of first order

$$y' = f(y, t), \quad y(t_0) = v$$

whose solution $y = y(t)$ yields the solution

$$x(t) = x_0 + \int_{t_0}^{t} y(\tau)\mathrm{d}\tau$$

of (1.265).

**Type 3 (The trick of the inverse function):**

$$\boxed{x'' = f(x, x'), \quad x(t_0) = x_0, \quad x'(t_0) = v.} \tag{1.266}$$

We use the formal Leibniz' calculus and set

$$p := \frac{\mathrm{d}x}{\mathrm{d}t}.$$

From (1.266) it then follows that

$$\frac{\mathrm{d}p}{\mathrm{d}t} = f(x, p).$$

The chain rule $\dfrac{\mathrm{d}p}{\mathrm{d}x} = \dfrac{\mathrm{d}p}{\mathrm{d}t}\dfrac{\mathrm{d}t}{\mathrm{d}x} = \dfrac{\mathrm{d}p}{\mathrm{d}t}\dfrac{1}{p}$ then implies

$$\frac{\mathrm{d}p}{\mathrm{d}x} = \frac{f(x,p)}{p}, \qquad p(x_0) = v.$$

From the solution $p = p(x)$ of this equation we obtain from $\dfrac{\mathrm{d}t}{\mathrm{d}x} = \dfrac{1}{p}$ the function

$$t(x) = t_0 + \int_{x_0}^{x} \frac{\mathrm{d}\xi}{p(\xi)}.$$

The inverse of $t = t(x)$ then is the sought-for solution of (1.266).

These formal considerations may be rigorously justified.

**Type 4 (Variation of constants):**

$$\boxed{a(t)x'' + b(t)x' + c(t)x = d(t).}$$

If one knows a special solution $x_*$ of the homogenous equation with $d \equiv 0$, then setting

$$x(t) = C(t)x_*(t)$$

leads the the linear differential equation of first order

$$ax_*y' + (2ax_*' + bx_*)y = d$$

with $C = y'$, i.e., $C = \displaystyle\int y\mathrm{d}t$.

**Type 5 (Euler–Lagrange equation):**

$$\boxed{\frac{\mathrm{d}}{\mathrm{d}t}L_{x'} - L_x = 0.}$$

Here $L = L(x, x', t)$. All differential equations which can be obtained from variational problems have this form. Special methods for solving this kind of equation can be found in section 5.1.1.

### 1.12.4.12 The geometric interpretation of differential equations of first order

Let the differential equation

$$\boxed{x' = f(t, x)} \qquad (1.267)$$

be given. This equation associates to each point $(t, x)$ a number $m := f(t, x)$.
In each point $(t, x)$ we draw a short segment through this point with a slope of $m$. In this way, a field of directions is formed (Figure 1.132(a)). The solutions of (1.267) are exactly the curves $x = x(t)$, which fit into this field of directions, i.e., the slope of the curve at the point $(x(t), t)$ is equal to $m = f(t, x(t))$ (Figure 1.132(b)).

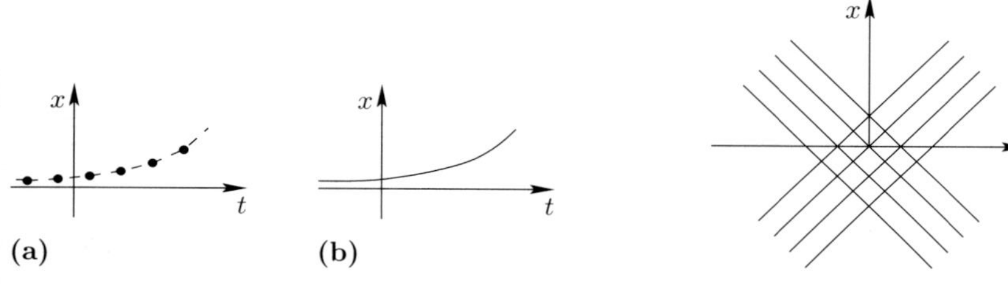


*Figure 1.132.* *Figure 1.133.*

**Branching of solutions (bifurcations):** In the case of an implicit differential equation

$$\boxed{F(t, x, x') = 0} \qquad (1.268)$$

it is possible that, for each point $(t, x)$, there are several directional elements $m$ which satisfy the equation $F(t, x, m) = 0$. Several different solutions can pass through such points.

*Example:* The differential equation

$$x'^2 = 1$$

has the two families of curves $x = \pm t + \text{const}$ as solution curves (Figure 1.133).

### 1.12.4.13 Envelopes and singular solutions

**Theorem:** If the differential equation (1.268) has a family of solutions with an envelope,[118] then the envelope is also a solution of the differential equation and is called a *singular solution.*

**Construction:** In case a singular solution of (1.268) exists, then it can be obtained by solving the system of equations

$$F(t, x, C) = 0, \quad F_{x'}(t, x, C) = 0 \qquad (1.269)$$

and eliminating the constant $C$.[119]

[118] This notion is defined in 3.7.1.

[119] We assume here that such a (local) solution is possible by the theorem on implicit functions.

*Example:* The Clairaut differential equation[120]

$$\boxed{x = tx' - \frac{1}{2}x'^2} \tag{1.270}$$

has the family of solutions consisting of the lines

$$\boxed{x = tC - \frac{1}{2}C^2} \tag{1.271}$$

with a constant $C$. Its envelope is obtained, according to 3.7.1, by differentiating (1.271) with respect to $C$, that is, using

$$0 = t - C$$

to eliminate $C$ from (1.271). This yields

$$\boxed{x = \frac{1}{2}t^2} \tag{1.272}$$

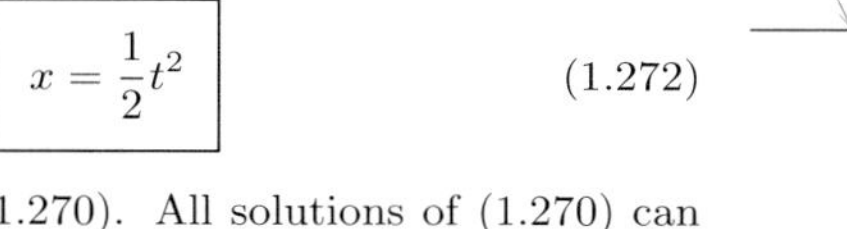

as singular solution of (1.270). All solutions of (1.270) can be obtained from the parabola (1.272) and its set of tangent lines (1.271) (Figure 1.134).

*Figure 1.134.*

The same result can be obtained by using the method of (1.269).

#### 1.12.4.14 The method of contact transformations of Legendre

**Basic idea:** The contact transformation of Legendre (1752–1833) is used, when a given differential equation

$$\boxed{f(t, x, x') = 0} \tag{1.273}$$

is complicated when expressed in terms of the derivative $x'$ of the sought-for function $x = x(t)$, but simple as a function of $x$. In this case the transformed differential equation

$$\boxed{F(\tau, \xi, \xi') = 0} \tag{1.274}$$

for a sought-for function $\xi = \xi(\tau)$ has a simple structure with respect to the derivative $\xi'$. The derivative $x'$ is transformed under the Legendre transformation to the dependent variable $\tau$.

**Definition:** The *Legendre transformation* $(t, x, x') \mapsto (\tau, \xi, \xi')$ is defined by the relations:

$$\boxed{\tau = x', \quad \xi = tx' - x, \quad \xi' = t.} \tag{1.275}$$

Here all variables are considered to be real-valued. The inverse transformation is given symmetrically by

$$\boxed{t = \xi', \quad x = \tau\xi' - \xi, \quad x' = \tau.} \tag{1.276}$$

[120]The French mathematician, physicist and astronomer Alexis Claude Clairaut (1713–1765) worked in Paris.

There is a transformation $F$ of $f = f(t, x, x')$, naturally associated to the relation (1.276):

$$\boxed{F(\tau, \xi, \xi') := f(\xi', \tau\xi' - \xi, \tau).}$$

**1.12.4.14.1 The fundamental invariance property of the Legendre transformation:**

> Solutions of differential equations are invariant under Legendre transformations.

**Theorem:** (i) If $x = x(t)$ is a solution of the differential equation (1.273), then the function $\xi = \xi(\tau)$ which is given by the parameterization

$$\tau = x'(t), \quad \xi = tx'(t) - x(t)$$

is a solution of the transformed equation (1.274).

(ii) If conversely $\xi = \xi(\tau)$ is a solution of the differential equation (1.274), then the function $x = x(t)$, given by the parameterization

$$\boxed{t = \xi'(\tau), \quad x = \tau\xi'(\tau) - \xi(\tau)} \tag{1.277}$$

is a solution of the original equation (1.273).

**1.12.4.14.2 Application to the differential equation of Clairaut:** The differential equation

$$\boxed{x - tx' = g(x')} \tag{1.278}$$

is transformed under the Legendre transformation (1.275) into the equation

$$\boxed{-\xi = g(\tau).}$$

This equation, which no longer contains derivatives, can be trivially solved. From the inverse transformation (1.277) we get the parameterization

$$t = -g'(\tau), \quad x = -\tau g'(\tau) + g(\tau)$$

which is a solution of (1.278). From this we obtain the family of solutions

$$\boxed{x = \tau t + g(\tau)}$$

of (1.278) with a parameter $\tau$. This is a collection of lines.

**1.12.4.14.3 Application to the Lagrangian differential equation:**

$$\boxed{a(x')t + b(x')x + c(x') = 0.} \tag{1.279}$$

The Legendre transformation (1.276) yields here the linear differential equation

$$\boxed{a(\tau)\xi' + b(\tau)(\tau\xi' - \xi) + c(\tau) = 0}$$

which is easily solved (cf. 1.12.4.5). The inverse transformation (1.277) yields the solution $x = x(t)$ of (1.279).

#### 1.12.4.14.4 The geometric interpretation of the Legendre transformation:

> The basic geometric idea of the Legendre transformation is that a curve is viewed not as a set of points, but rather as the envelope of its tangents.

The equation of these tangents is the equation of a curve in tangent coordinates (Figure 1.135). We want to give an analytic description of this idea.

Let the equation

$$\boxed{x = x(t)}$$

of a curve $C$ in point coordinates $(t, x)$ be given. The equation of the *tangent* to this curve in a fixed point $(t_*, x(t_*))$ of the curve $C$ is

$$\boxed{x = \tau t - \xi} \tag{1.280}$$

with the slope $\tau$ and the intersection $-\xi$ with the $x$-axis (Figure 1.135(a)). Hence we have:

$$\begin{aligned} \tau &= x'(t_*), \\ \xi &= t_* x'(t_*) - x(t_*). \end{aligned} \tag{1.281}$$

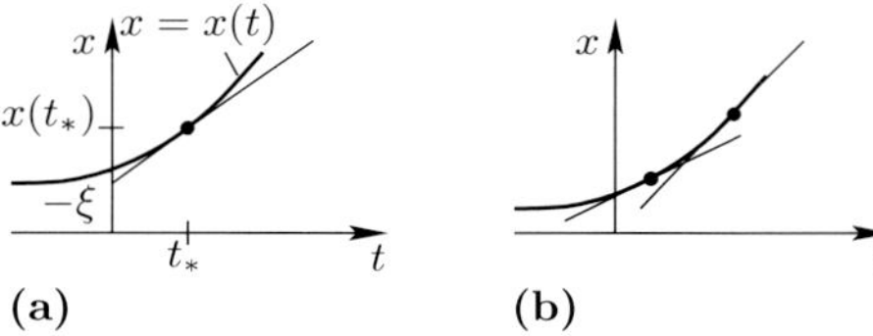


*Figure 1.135. The envelope of tangents on a curve.*

Each tangent is uniquely determined by its tangent coordinates $(\tau, \xi)$. The union of all of these tangents of $C$ is given by the equation

$$\boxed{\xi = \xi(\tau).} \tag{1.282}$$

(i) The equation (1.282) of the curve $C$ in tangent coordinates is obtained from (1.281) by eliminating the parameter $t_*$.

(ii) If conversely the equation (1.282) of $C$ in tangent coordinates is given, then one obtains according to (1.280) the collection of tangents of $C$ in the form

$$x = \tau t - \xi(\tau).$$

The envelope of this family of tangents is calculated by eliminating the parameter $\tau$ from the system

$$x = \tau t - \xi(\tau), \quad \xi'(\tau) - t = 0$$

(cf. 3.7.1). This yields the equation $x = x(t)$ of $C$.

The transition from point coordinates to tangent coordinates corresponds precisely to the Legendre transformation.

*Example:* The equation of the curve $x = \mathrm{e}^t$, $t \in \mathbb{R}$ is, in tangent coordinates, given by (Figure 1.136):

$$\xi = \tau \ln \tau - \tau, \quad \tau > 0.$$

The following fact is decisive:

> The Legendre transformation transforms the directional elements $(t, x, x')$ of curves into the directional elements $(\tau, \xi, \xi')$ of other curves (Figure 1.136).

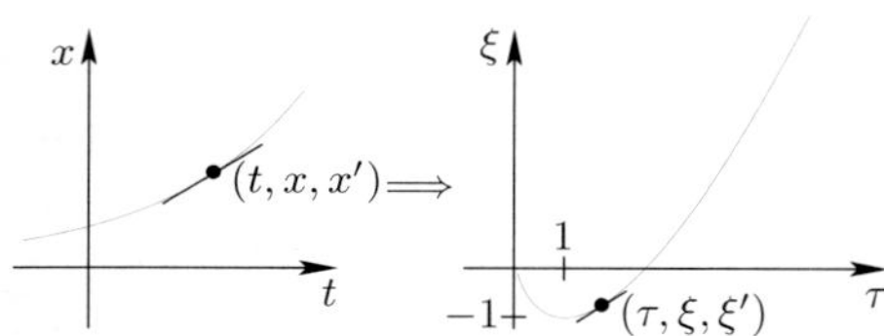


*Figure 1.136. The Legendre transformation.*

For this reason one refers to the Legendre transformation as a contact transformation (cf. 1.13.1.11).

The following considerations represent the analytic heart of the Legendre transformation and can be applied in a universal manner to arbitrary systems of ordinary and partial differential equations. This illustrates the elegance and flexibility of the differential calculus of Cartan.

**1.12.4.14.5 Differential forms and the product trick of Legendre:** If we set $\tau = x'$, then we can put the original differential equation

$$F(t, x, x') = 0 \tag{1.283}$$

in the equivalent form

$$\boxed{\begin{aligned} F(t, x, \tau) &= 0, \\ \mathrm{d}x - \tau \mathrm{d}t &= 0. \end{aligned}} \tag{1.284}$$

We will explain in (1.287) that the form (1.284) is more to the point for purely geometrical reasons than (1.283). The trick of Legendre is now to apply the product rule for differential forms[121]

$$\boxed{\mathrm{d}(\tau t) = \tau \mathrm{d}t + t\mathrm{d}\tau.} \tag{1.285}$$

In this way, the new equation

$$\begin{aligned} &F(t, x, \tau) = 0, \\ &\mathrm{d}(\tau t - x) - t\mathrm{d}\tau = 0 \end{aligned} \tag{1.286}$$

arises from equation (1.284). This equation is particularly simple, if we introduce the new variable

$$\boxed{\xi := \tau t - x.}$$

From (1.286) it then follows that $\mathrm{d}\xi - t\mathrm{d}\tau = 0$, that is $\xi' = t$. We hence obtain

$$\tau = x', \quad \xi = x't - x, \quad \xi' = t.$$

Thus the equation (1.286) is transformed into the form

$$\boxed{\begin{aligned} F(\xi', \tau t - \xi, \tau) &= 0, \\ \mathrm{d}\xi - t\mathrm{d}\tau &= 0 \end{aligned}}$$

which is equivalent to the transformed equation

$$G(\tau, \xi, \xi') = 0.$$

**The advantage of formulating differential equations in the language of the differential calculus of Cartan:**

*Example 1:* The differential equation

[121] This effective general trick is also the basis of the Legendre transformation in theoretical mechanics and thermodynamics (cf. 1.13.1.11).

$$\frac{\mathrm{d}x}{\mathrm{d}t} = \frac{x}{t} \tag{1.287}$$

is ill-posed, since it contains a singular point at $t = 0$. Moreover, (1.287) does not accurately reflect the geometric situation which is given by the family of directional elements as in Figure 1.137.

The curves $x = \text{const} \cdot t$ and $t = 0$ fit to this family of directional elements. However, the solution $t = 0$ does not appear in (1.287). The correct geometric formulation is based on obtaining solutions with the parameterization

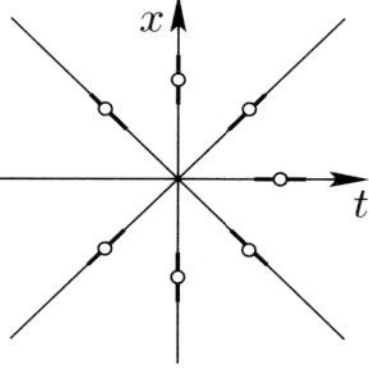


*Figure 1.137.*

$$x = x(p), \quad t = t(p),$$

and instead of (1.287), considering the equation

$$\frac{\mathrm{d}x(p)}{\mathrm{d}p} t(p) - x(p) \frac{\mathrm{d}t(p)}{\mathrm{d}p} = 0.$$

This can be written in the abbreviated form

$$t\mathrm{d}x - x\mathrm{d}t = 0.$$

This corresponds to the procedure in (1.284), in which we have eliminated the variable $\tau$ for simplicity.

The full power of this formulation becomes apparent when considering arbitrary systems of partial differential equations in the language of differential forms, in the form of the fundamental theorem of Cartan–Kähler (cf. 1.13.5.4).

**1.12.4.14.6 Application to differential equations of the second order:** In this case one uses the Legendre transformation

$$\boxed{\tau = x', \quad \xi = tx' - x, \quad \xi' = t, \quad \xi'' = \frac{1}{x''}}$$

with the inverse transformation

$$\boxed{t = \xi', \quad x = \tau\xi' - \xi, \quad x' = \tau, \quad x'' = \frac{1}{\xi''}.}$$

*Example 2:* Applying the Legendre transformation to

$$x''x' = 1 \tag{1.288}$$

yields

$$\xi'' = \tau$$

with the general solution

$$\xi = \frac{\tau^3}{6} + C\tau + D.$$

Because of $x = \tau\xi' - \xi$ and $t = \xi'$ we obtain from this the solution of (1.288) in the parameterized form

$$x = \frac{\tau^3}{3} - D, \quad t = \frac{\tau^2}{2} + C$$

with the parameter $\tau$ and the constants $C$ and $D$.

## 1.12.5 Applications

### 1.12.5.1 The escape velocity of the earth

The radial movement $r = r(t)$ of a rocket in the gravitational field of the earth, whose radius is taken to be $R$, is given by the Newtonian law of motion

$$\boxed{mr'' = -\frac{GMm}{r^2}, \quad r(0) = R, \quad r'(0) = v} \tag{1.289}$$

($m$ is the mass of the rocket, $M$ the mass of the earth, G is the gravitational constant). The law of conservation of energy (1.264) yields

$$r'^2 = \frac{2GM}{r} + \text{const.}$$

Taking the initial conditions into account, this means

$$r'^2 = \frac{2GM}{r} + \left(v^2 - \frac{2GM}{R}\right).$$

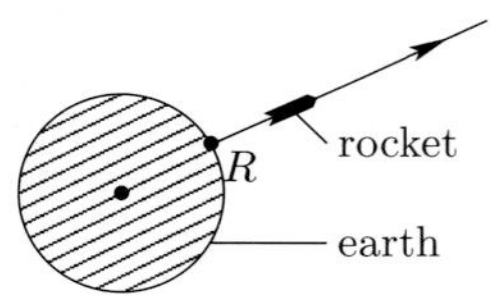


*Figure 1.138.*

We want to determine the starting velocity $v$ of the rocket which is necessary for the rocket to escape the gravitational field, i.e., such that $r'(t) > 0$ for all times $t$ (Figure 1.138). The smallest possible value for $v$ is obtained from the equation $v^2 - 2GM/R = 0$, that is

$$\boxed{v = \sqrt{\frac{2GM}{R}} = 11.2\,\text{km/s}.}$$

This is the sought for escape velocity of the earth. This starting velocity $v$ corresponds to the motion of the rocket given by

$$r(t) = \left(R^{3/2} + \frac{3}{2}\sqrt{2GM}\,t\right)^{2/3}.$$

### 1.12.5.2 The two body problem

The two body problem in celestial mechanics can be reduced to a one body problem for the relative motion of the two bodies with respect to one another, yielding the Kepler laws of motion.

**The Newtonian law of motion:** We study the motion

$$\mathbf{x}_1 = \mathbf{x}_1(t) \quad \text{and} \quad \mathbf{x}_2 = \mathbf{x}_2(t)$$

for two celestial bodies with the masses $m_1$ and $m_2$ and the total mass $m = m_1 + m_2$. For example, $m_1$ could be the mass of the sun and $m_2$ the mass of one of the planets (Figure 1.139). The laws of motion are

$$\boxed{\begin{array}{l} m_1\mathbf{x}_1'' = \mathbf{F}, \quad m_2\mathbf{x}_2'' = -\mathbf{F}, \\ \mathbf{x}_j(0) = \mathbf{x}_{j0}, \quad \mathbf{x}_j{}'(0) = \mathbf{v}_j, \quad j = 1,2 \quad \text{(initial value)} \end{array}} \tag{1.290}$$

with the Newtonian gravitational force

$$\mathbf{F} = \mathrm{G}\frac{m_1 m_2(\mathbf{x}_2 - \mathbf{x}_1)}{|\mathbf{x}_2 - \mathbf{x}_1|^3}$$

(G the gravitational constant). The appearance of the forces $\mathbf{F}$ and $-\mathbf{F}$ in (1.290) corresponds to the Newtonian law *actio = reactio.*

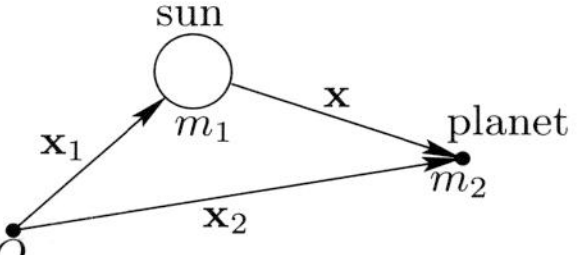


*Figure 1.139.*

**Separation of the motion of the center of mass:** For the center of mass

$$\mathbf{y} := \frac{1}{m}(m_1\mathbf{x}_1 + m_2\mathbf{x}_2)$$

of this system and the total momentum $\mathbf{P} = m\mathbf{y}'$ one gets from (1.290) the *conservation of the total momentum*

$$\mathbf{P}' = 0,$$

which means $m\mathbf{y}'' = 0$ with the solution

$$\mathbf{y}(t) = \mathbf{y}(0) + t\mathbf{y}'(0).$$

The center of mass hence moves on a straight line with constant velocity. Explicitly, one has

$$\mathbf{y}(0) = \frac{1}{m}(m_1\mathbf{x}_{10} + m_2\mathbf{x}_{20}), \quad \mathbf{y}'(0) = \frac{1}{m}(m_1\mathbf{v}_1 + m_2\mathbf{v}_2).$$

For the relative motion with respect to the center of mass

$$\mathbf{y}_j := \mathbf{x}_j - \mathbf{y}$$

we get the equations of motion

$$m_1\mathbf{y}_1'' = \mathbf{F}, \quad m_2\mathbf{y}_2'' = -\mathbf{F}, \quad m_1\mathbf{y}_1 + m_2\mathbf{y}_2 = 0.$$

In the case of the system consisting of the sun and a planet, the center of mass $\mathbf{y}$ lies inside the sun.

**The relative motion of the two celestial bodies with respect to each other:** For

$$\mathbf{x} := \mathbf{x}_2 - \mathbf{x}_1$$

one gets from $\mathbf{x} = \mathbf{y}_2 - \mathbf{y}_1$ the law of motion

$$\boxed{\begin{aligned} &m_2\mathbf{x}'' = -\frac{\mathrm{G}mm_2\mathbf{x}}{|\mathbf{x}|^3}, \\ &\mathbf{x}(0) = \mathbf{x}_{20} - \mathbf{x}_{10}, \quad \mathbf{x}'(0) = \mathbf{v}_2 - \mathbf{v}_1. \end{aligned}} \tag{1.291}$$

This is the *one body problem* for a body of mass $m_2$ in the gravitational field of a body of mass $m$.

If one knows the solution of (1.291), then one gets a solution of the original equation of motion (1.290) through

$$\mathbf{x}_1(t) = \mathbf{y}(t) - \frac{m_2}{m}\mathbf{x}(t), \quad \mathbf{x}_2(t) = \mathbf{y}(t) + \frac{m_1}{m}\mathbf{x}(t).$$

**Conservation laws:** From (1.291), the *law of conservation of energy and angular momentum*

$$\frac{m_2}{2}\mathbf{x}'(t)^2 + U(\mathbf{x}(t)) = \mathrm{const} = E, \tag{1.292}$$

$$m_2\mathbf{x}(t) \times \mathbf{x}'(t) = \text{const} = \mathbf{N} \tag{1.293}$$

follows, with

$$U(\mathbf{x}) = -\frac{Gmm_2}{|\mathbf{x}|}, \quad E = \frac{m_2}{2}\mathbf{x}'(0)^2 + U(\mathbf{x}(0)), \quad \mathbf{N} = m_2\mathbf{x}(0) \times \mathbf{x}'(0)$$

(cf. 1.9.6). We choose initial conditions such that $\mathbf{N} \neq 0$.

**Plane motion:** From the conservation of angular momentum (1.293), it follows that $\mathbf{x}(t)\mathbf{N} = 0$ for all times $t$. Therefore the motion is constrained in a plane perpendicular to the vector $\mathbf{N}$. We choose a Cartesian $(x, y, z)$-coordinate system with the $z$-axis in the direction of the vector $\mathbf{N}$. Then $\mathbf{x}(t)$ lies in the $(x, y)$-plane, i.e., one has

$$\mathbf{x}(t) = x(t)\mathbf{i} + y(t)\mathbf{j}.$$

**Polar coordinates** (Figure 1.140): We choose polar coordinates

$$x = r\cos\varphi, \quad y = r\sin\varphi$$

and introduce the unit vectors

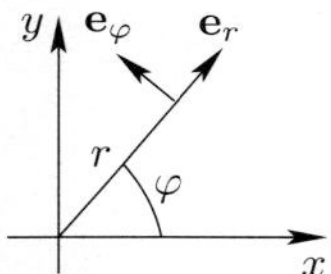


*Figure 1.140.*

$$\mathbf{e}_r := \cos\varphi\mathbf{i} + \sin\varphi\mathbf{j}, \quad \mathbf{e}_\varphi := -\sin\varphi\mathbf{i} + \cos\varphi\mathbf{j}.$$

The motion is then given by $\varphi = \varphi(t)$, $r = r(t)$. We choose in addition the $x$-axis in such a way that $\varphi(0) = 0$.

Differentiation with respect to the time $t$ yields

$$\mathbf{e}_r' = (-\sin\varphi\mathbf{i} + \cos\varphi\mathbf{j})\varphi' = \varphi'\mathbf{e}_\varphi.$$

From the equation for the trajectory

$$\boxed{\mathbf{x}(t) = r(t)\mathbf{e}_r(t)}$$

we get $\mathbf{x}' = r'\mathbf{e}_r + r\varphi'\mathbf{e}_\varphi$. Because of $\mathbf{e}_r\mathbf{e}_\varphi = 0$, one has $\mathbf{x}'^2 = (r')^2 + (r')^2(\varphi')^2$. Thus we get from (1.292) the system of equations

$$\boxed{\begin{aligned} m_2\left(r'^2 + r^2\varphi'^2 - \frac{2Gm}{r}\right) &= 2E, \\ m_2r^2\varphi' &= |\mathbf{N}|. \end{aligned}} \tag{1.294}$$

**Theorem:** The solution of (1.294) is

$$\boxed{r = \frac{p}{1 + \varepsilon\cos\varphi}.} \tag{1.295}$$

The motion in time is then given by

$$t = \frac{m_2}{|\mathbf{N}|}\int_0^\varphi r(\varphi)^2 \mathrm{d}\varphi, \tag{1.296}$$

with constants

$$p := \mathbf{N}^2/\alpha m_2, \quad \varepsilon := \sqrt{1 + 2E\mathbf{N}^2/m_2\alpha^2}, \quad \alpha := Gmm_2. \tag{1.297}$$

*Proof:* This is easily verified by straightforward differentiation.[122] □

We discuss the case $0 \leq \varepsilon < 1$.

**First Kepler law:** The planets move along ellipses, with the sun lying in one of the focal points (Figure 1.141).

*Proof:* In Cartesian coordinates $x, y$, the equation (1.295) turns into the equation

$$\frac{(x+\varepsilon a)^2}{a^2} + \frac{y^2}{b^2} = 1$$

for the trajectory. Here one has $a := p/(1-\varepsilon^2)$ and $b := p/\sqrt{1-\varepsilon^2}$. The sun is centered at the focal point $(0,0)$ of the ellipse. □

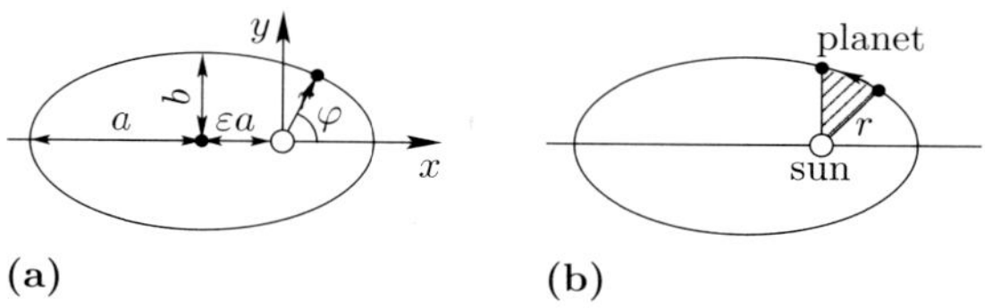


*Figure 1.141. The Kepler laws of plantary motion.*

**The second law of Kepler:** The orbits of the planets sweep out equal areas in equal times.

*Proof:* In the time interval $[s,t]$, the area

$$\frac{1}{2}\int_s^t r^2\varphi' \mathrm{d}t = \frac{(t-s)|\mathbf{N}|}{2m_2} \tag{1.298}$$

is swept out. □

**The third law of Kepler:** The ratio of the square of the period $T$ and the third power of the long semiaxis $a$ is a constant for all planets.

*Proof:* The surface area of the elliptic trajectory is equal to $\pi ab$. From (1.298) with $t = T$ and $s = 0$, we get

$$\frac{1}{2}\pi ab = \frac{T|\mathbf{N}|}{2m_2}.$$

From $a = p/(1-\varepsilon^2)$ and (1.297), we obtain

$$\frac{T^2}{a^3} = \frac{4\pi^2}{G(m_1+m_2)}. \tag{1.299}$$

Since the mass of the sun $m_1$ is very large compared with that of the planet $m_2$, we may disregard $m_2$ in (1.299) in a first approximation. □

Our considerations show that the third Kepler law is only an approximation. Kepler (1571–1630) derived his law through the study of extensive data from observations of the planets. The law he got empirically and the mathematical derivation which was later discovered by Newton (1643–1727) was a great achievement of the mathematics and physics of the day.

---

[122] One is lead to the formula (1.295) by using

$$\frac{\mathrm{d}\varphi}{\mathrm{d}r} = \frac{\varphi'(t)}{r'(t)} = F(r)$$

and integrating this equation by means of separation of variables. The function $F(r)$ is derived from the first equation in (1.294). Moreover, (1.296) follows from the second equation in (1.294). In 1.13.1.5 we will use the elegant method of Jacobi for calculating the solution.

## 1.12.6 Systems of linear differential equations and the propagator

For systems of linear differential equations with variable, continuous coefficients, there is a complete theory of solutions. In the case of constant coefficients, the Laplace transformation (cf. 1.11.1.2) is the universal method of solution.

### 1.12.6.1 Linear systems of first order

$$\boxed{\begin{aligned} \mathbf{x}' &= A(t)\mathbf{x} + B(t), \quad t \in J, \\ \mathbf{x}(t_0) &= a \quad \text{(initial value).} \end{aligned}} \tag{1.300}$$

Here $\mathbf{x} = (x_1, \ldots, x_n)^\mathsf{T}$ is a column matrix of complex numbers $x_j$. Moreover, $A(t)$ is a complex $(n \times n)$ matrix, and $B(t)$ is a complex matrix with $n$ columns. We let $J$ denote an open interval in $\mathbb{R}$ with $t_0 \in J$ (for example, $J = \mathbb{R}$). Written in components, (1.300) becomes:

$$x_j' = \sum_{k=1}^{n} a_{jk}(t)x_k + b_j(t), \quad j = 1, \ldots, n.$$

**Existence and uniqueness theorems:** If the components of $A$ and $B$ are continuous functions on $J$, then the initial value problem has for every $a \in \mathbb{C}^n$ exactly one solution $\mathbf{x} = \mathbf{x}(t)$ on $J$.

If $A$, $B$ and $a$ are real, then the solution is also real.

**Propagator:** The solution has the convenient representation

$$\boxed{\mathbf{x}(t) = P(t, t_0)a + \int_{t_0}^{t} P(t, \tau)B(\tau)\mathrm{d}t} \tag{1.301}$$

where $P$ is the so-called *propagator.*

**Constant coefficients:** If $A(t) = \text{const}$, then one has[123]

$$\boxed{P(t, \tau) = \mathrm{e}^{(t-\tau)A}.}$$

**The fundamental formula of Dyson:** In the general case the propagator possesses the convergent series representation

$$P(t, \tau) := I + \sum_{k=1}^{\infty} \int_{\tau}^{t} \int_{\tau}^{t_1} \cdots \int_{\tau}^{t_{k-1}} A(t_1)A(t_2)\ldots A(t_k)\mathrm{d}t_k \ldots \mathrm{d}t_2\mathrm{d}t_1.$$

If one introduces the time ordering operator $\mathscr{T}$ by means of the formula

$$\mathscr{T}\,(A(t)A(s)) := \begin{cases} A(t)A(s) & \text{for} \quad t \geq s, \\ A(s)A(t) & \text{for} \quad s \geq t, \end{cases}$$

[123] The series

$$\mathrm{e}^{(t-\tau)A} = I + (t-\tau)A + \frac{(t-\tau)^2}{2}A^2 + \frac{(t-\tau)^3}{3!}A^3 + \cdots$$

converges for all $t, \tau \in \mathbb{R}$ component-wise.

then one has

$$P(t,\tau) := I + \sum_{k=1}^{\infty} \frac{1}{k!} \int_\tau^t \int_\tau^t \cdots \int_\tau^t \mathscr{T}\left(A(t_1)A(t_2)\ldots A(t_k)\right) \mathrm{d}t_k \ldots \mathrm{d}t_2 \mathrm{d}t_1 .$$

A shorter and elegant way of writing this is as follows:

$$\boxed{P(t,\tau) = \mathscr{T}\left(\exp \int_\tau^t A(s)\mathrm{d}s\right).}$$

This formula of Dyson can be extended to operator equations in Banach spaces. It plays a key role in the construction of the $S$-matrix (scattering matrix) in quantum field theory. The $S$-matrix contains all information about scattering processes of elementary particles, which are conducted in modern accelerators.

**Remark:** If $B(0) \equiv 0$, then the propagator describes the solution $\mathbf{x}(t) = P(t, t_0)a$ of the homogenous equation. Formula (1.301) shows that one has the following situation:

| The propagator of the homogenous problem allows the construction of the general solution of the inhomogenous problem by means of superposition. |
|---|

This a fundamental principle in physics, which is valid in much more general situations than that of (1.300) and which for this reason can be applied to partial differential equations and operator equations in infinite-dimensional spaces.

**The propagator equation:** For arbitrary times $t$ and $t_1 \le t_2 \le t_3$ one has:

$$\boxed{P(t_1,t_3) = P(t_1,t_2)P(t_2,t_3)}$$

with $P(t,t) = I$.

The following classical considerations are appropriate for the special structure of (1.300) and, in contrast to the previous ones, cannot be generalized.

**Fundamental solution:** We consider a $(n \times n)$-matrix

$$\mathbf{X}(t) := (X_1(t), \ldots, X_n(t))$$

with the column vectors $X_j$, where the following hold:

(i) Every column $X_j$ is a solution of the homogenous equation $X_j' = AX_j$.

(ii) $\det \mathbf{X}(t_0) \neq 0$.[124]

Then $\mathbf{X} = \mathbf{X}(t)$ is called a *fundamental solution* of $\mathbf{x}' = A(t)\mathbf{x}$.

**Theorem:** For a fundamental solution the following hold:

(i) The general solution of $\mathbf{x}' = A(t)\mathbf{x}$ has the form

$$\mathbf{x}(t) = \sum_{j=1}^{n} C_j X_j(t) \tag{1.302}$$

with arbitrary constants $C_1, \ldots, C_n$.

[124] One calls $\det \mathbf{X}(t)$ the *Wronskian determinant.* From (ii) one has also $\det \mathbf{X}(t) \neq 0$ for all $t$.

(ii) The propagator has the form[125]

$$P(t,\tau) = \mathbf{X}(t)\mathbf{X}(\tau)^{-1}.$$

*Example:* The system

$$\begin{array}{ll} x_1' = x_2 + b_1(t), & x_2' = -x_1 + b_2(t), \\ x_1(t_0) = a_1, & x_2(t_0) = a_2 \end{array} \tag{1.303}$$

is, written in matrix notation, the matrix equation

$$\begin{pmatrix} x_1' \\ x_2' \end{pmatrix} = \begin{pmatrix} 0 & 1 \\ -1 & 0 \end{pmatrix} \begin{pmatrix} x_1 \\ x_2 \end{pmatrix} + \begin{pmatrix} b_1 \\ b_2 \end{pmatrix},$$

or more briefly $\mathbf{x}' = A\mathbf{x} + \mathbf{b}$. One recognizes immediately that $x_1 = \cos(t - t_0)$, $x_2 = -\sin(t - t_0)$ and $x_1 = \sin(t - t_0)$, $x_2 = \cos(t - t_0)$ are solutions of (1.303) with $b_1 \equiv 0$ and $b_2 \equiv 0$. These solutions form the columns of the fundamental solution

$$\mathbf{X}(t) = \begin{pmatrix} \cos(t - t_0) & \sin(t - t_0) \\ -\sin(t - t_0) & \cos(t - t_0) \end{pmatrix}.$$

Because of $\mathbf{X}(t_0) = I$ (identity matrix), we get from this the propagator

$$P(t, t_0) = \mathbf{X}(t).$$

The inhomogenous problem (1.303) has a solution

$$\mathbf{x}(t) = P(t, t_0)a + \int_{t_0}^{t} P(t,\tau)b(\tau)\mathrm{d}\tau.$$

For $t_0 = 0$ this corresponds to the explicit formula for the solution:

$$x_1(t) = a_1 \cos t + a_2 \sin t + \int_0^t \left(b_1(\tau)\cos(t-\tau) + b_2(\tau)\sin(t-\tau)\right)\mathrm{d}\tau,$$

$$x_2(t) = -a_1 \sin t + a_2 \cos t + \int_0^t \left(-b_1(\tau)\sin(t-\tau) + b_2(\tau)\cos(t-\tau)\right)\mathrm{d}\tau.$$

The same result would be obtained with the help of the Laplace transformation.

The formula

$$P(t,\tau) = P(t,s)P(s,\tau)$$

for the propagator is equivalent to the addition theorems for sine and cosine. Since (1.303) corresponds to the motion of a harmonic oscillator (cf. 1.12.6.2), this gives the addition theorem an immediate physical interpretation.

[125] The formula (1.301) for the solution can be obtained by setting

$$\mathbf{x}(t) = \sum_{j=1}^{n} C_j X_j(t).$$

This is the method of variation of constants of Lagrange (cf. 1.12.4.5).

#### 1.12.6.2 Linear differential equations of arbitrary order

$$y^{(n)} + a_{n-1}(t)y^{(n-1)}(t) + \ldots + a_0(t)y = f(t),$$
$$y(t_0) = \alpha_0, \quad y'(t_0) = \alpha_1, \ \ldots\ , y^{(n-1)}(t_0) = \alpha_{n-1}.$$

This problem can be reduced to a systems of linear differential equations of the first order of the form (1.300) by introducing new variables $x_1 = y,\ x_2 = y', \ldots, x_n = y^{(n-1)}$. Analogously, linear systems of arbitrary order are handled.

*Example:* The equation for the harmonic oscillator

$$y'' + y = f(t),$$
$$y(t_0) = a_1, \quad y'(t_0) = a_2,$$

is transformed into the form (1.303) with $b_1 = 0$ and $b_2 = f$ by making the substitution $x_1 = y,\ x_2 = y'$.

### 1.12.7 Stability

We consider the linear differential equation

$$\mathbf{x}' = A\mathbf{x} + \mathbf{b}(\mathbf{x}, t), \qquad \mathbf{x}(0) = \mathbf{a} \quad \text{(initial value)} \tag{1.304}$$

where $\mathbf{x} = (x_1, \ldots, x_n)^\mathsf{T}$, $A$ is a real, time independent $(n \times n)$-matrix and the components of $\mathbf{b} = (b_1, \ldots, b_n)^\mathsf{T}$ are real $C^1$-functions for all $\mathbf{x}$ in a neighborhood of the origin for all times $t \geq 0$. Moreover, let

$$\mathbf{b}(\mathbf{0}, t) \equiv \mathbf{0}.$$

Hence $\mathbf{x}(t) \equiv \mathbf{0}$ is a solution of (1.304) which corresponds to a point of equilibrium of the system. We also abbreviate this behavior by speaking of an equilibrium point at $\mathbf{x} = \mathbf{0}$.

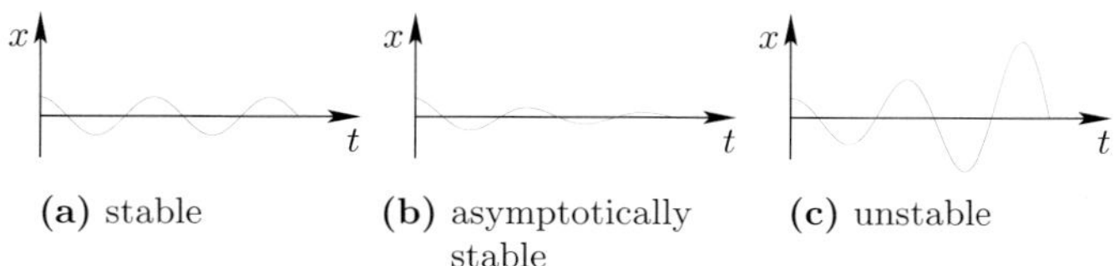


**(a)** stable **(b)** asymptotically stable **(c)** unstable

*Figure 1.142. Stability of solutions of a linear differential equation.*

**Definition:** (i) *Stability* (Figure 1.142(a)): The equilibrium point $\mathbf{x} = \mathbf{0}$ is *stable* if and only if for each $\varepsilon > 0$ there is a number $\delta > 0$ such that from

$$|t| < \delta,$$

the existence of a unique solution $\mathbf{x} = \mathbf{x}(t)$ of (1.304) follows, with

$$|\mathbf{x}(t)| < \varepsilon \quad \text{for all times} \quad t \geq 0.$$

This means that sufficiently small perturbations of the equilibrium configuration at $\mathbf{x} = \mathbf{0}$ remain small for all $t \geq 0$.

(ii) *Asymptotic stability* (Figure 1.142(b)): The equilibrium point $\mathbf{x} = \mathbf{0}$ is *asymptotically stable* if and only if it is stable and in addition there is a number $\delta_* > 0$ such that, for each solution with $|\mathbf{x}(0)| < \delta_*$, we have the limit relation

$$\lim_{t\to\infty} \mathbf{x}(t) = \mathbf{0}.$$

This means that small perturbations of the equilibrium configuration at time $t = 0$ return to their starting configuration after a sufficiently long time.

(iii) *Instability* (Figure 1.142(c)): The equilibrium point $\mathbf{x} = \mathbf{0}$ is *instable* if and only if it is not stable.

**The Theorem on stability by Liapunov (1892):** Suppose a perturbation $\mathbf{b}$ of the linear system $\mathbf{x}' = A\mathbf{x}$ with constant coefficients is sufficiently small, i.e., one has

$$\lim_{|\mathbf{x}|\to\mathbf{0}} \left( \sup_{t\geq 0} \frac{|\mathbf{b}(\mathbf{x},t)|}{|\mathbf{x}|} \right) = 0. \qquad (1.305)$$

Then one has:

(i) The equilibrium point $\mathbf{x} = \mathbf{0}$ is asymptotically stable, if all eigenvalues[126] $\lambda_1, \ldots, \lambda_m$ of the matrix $A$ are in the left half plane, i.e., have negative real part for all $j$ (Figure 1.143(a)).

(ii) The equilibrium point $\mathbf{x} = \mathbf{0}$ is unstable, if an eigenvalue of $A$ is in the right half plane, i.e., one has $\operatorname{Re}\lambda_j > 0$ for some $j$ (Figure 1.143(b)).

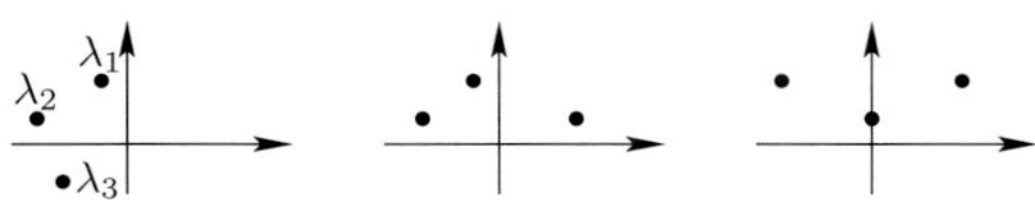


**(a)** asymptotically stable **(b)** unstable **(c)** critical

*Figure 1.143. Eigenvalues of linear differential operators.*

If an eigenvalue of $A$ is on the imaginary axis, then the *method of the center manifold* (see [212]) must be applied (displayed in Figure 1.143(c)).

To effectively apply the stability criterion of Liapunov to complicated problems in control theory, one requires a criterion for when the equation $\det(A - \lambda I) = 0$ has a zero in the left half plane, *without calculating the eigenvalues explicitly.* This problem, which was posed by Maxwell in 1868, was solved in 1875 by the English physicist Routh and independently by the German mathematician Hurwitz in 1895.

**The criterion of Routh–Hurwitz:** All zeros of the polynomial

$$\boxed{a_n\lambda^n + a_{n-1}\lambda^{n-1} + \ldots + a_1\lambda + a_0 = 0}$$

with real coefficients $a_j$ and fulfilling $a_n > 0$ lie in the left half plane, if and only if, all determinants

$$a_1, \quad \begin{vmatrix} a_1 & a_0 \\ a_3 & a_2 \end{vmatrix}, \quad \begin{vmatrix} a_1 & a_0 & 0 \\ a_3 & a_2 & a_1 \\ a_5 & a_4 & a_3 \end{vmatrix}, \ldots, \begin{vmatrix} a_1 & a_0 & 0 & 0\ldots 0 \\ a_3 & a_2 & a_1 & 0\ldots 0 \\ \ldots & & & \ddots \\ a_{2n-1} & a_{2n-2} & a_{2n-3} & \ldots a_n \end{vmatrix}$$

[126] The notions of eigenvalues and eigenvectors of matrices is introduced in section 2.2.1.

(with $a_m = 0$ for $m > n$) are positive.

**Applications:** The equilibrium point $x = 0$ of the differential equation

$$a_n x^{(n)} + a_{n-1} x^{(n-1)} + \ldots + a_1 x' + a_0 x = b(x, t)$$

with $b(0, t) \equiv 0$ is asymptotically stable, if the smallness criterion (1.305) and the criterion of Routh–Hurwitz are satisfied.

*Example 1:* The equilibrium point $x = 0$ of the differential equation

$$x'' + 2x' + x = x^n, \quad n = 1, 2, \ldots$$

is asymptotically stable.

*Proof:* One has

$$a_1 = 2 > 0, \quad \begin{vmatrix} a_1 & a_0 \\ a_3 & a_2 \end{vmatrix} = \begin{vmatrix} 2 & 1 \\ 0 & 1 \end{vmatrix} = 2 > 0.$$

□

**Generalization:** In order to investigate the stability of an arbitrary solution $y_*$ of the differential equation

$$y' = f(y, t),$$

one sets $y = y_* + x$. This yields a differential equation

$$x' = g(x, t)$$

with the equilibrium point $x = 0$. The stability of this point is by definition equivalent to the stability of $y_*$.

*Example 2:* The solution $y(t) \equiv 1$ of the differential equation

$$y'' + 2y' + y - 1 - (y - 1)^n = 0, \quad n = 2, 3, \ldots$$

is asymptotically stable.

*Proof:* If we set $y = x + 1$, then $x$ satisfies the differential equation of Example 1, and $x = 0$ is asymptotically stable.

## 1.12.8 Boundary value problems and Green's functions

In the following section the classical theory, which is based on the work of Sturm (1803–1855) and Liouville (1809–1882) and whose generalization to partial differential equations has played an important role in the development of analysis in this century, is described.[127]

### 1.12.8.1 The inhomogenous problem

We consider the boundary value problem

$$\boxed{\begin{aligned} -\left(p(x)y'\right)' + q(x)y &= f(x), \quad a \le x \le b, \\ y(a) = y(b) &= 0. \end{aligned}} \qquad (1.306)$$

[127] The relation to the theory of integral equations and to functional analysis (Hilbert–Schmidt–theory) can be found in [212]. The theory of singular boundary value problems of Hermann Weyl (1885–1955) and its modern extensions, which represent a gem of mathematics, are also described in [212].

The given real functions $p$ and $q$ are assumed to be smooth on the compact interval $[a, b]$ with $p(x) > 0$ on $[a, b]$. The given real-valued function $f$ is assumed to be continuous on $[a, b]$. We are looking for the real-valued function $y = y(x)$.

The problem (1.306) is said to *homogenous*, if and only if, $f(x) \equiv 0$.

**The Fredholm alternative:** (i) If the homogenous problem (1.306) has only the trivial solution $y \equiv 0$, then the inhomogenous problem (1.306) has for every $f$ exactly one solution. This solution can be represented by the integral

$$y(x) = \int_a^b G(x, \xi) f(\xi) \mathrm{d}\xi$$

with the continuous symmetric Green's function $G$, i.e., one has

$$G(x, \xi) = G(\xi, x) \quad \text{for all} \quad x, \xi \in [a, b].$$

(ii) If the homogenous problem (1.306) has a non-trivial solution $y_*$, then the inhomogenous problem (1.306) has a solution if and only if the solvability criterion

$$\boxed{\int_a^b y_*(x) f(x) \mathrm{d}x = 0}$$

is satisfied for the function $f$ occuring in the right-hand side of (1.306).

**Uniqueness conditions:** (a) If $y$ and $z$ are non-trivial solution of the equation $-(py')' + qy = 0$, and are not constant multiples of one another, then the case (i) occurs if and only if $y(a)z(b) - y(b)z(a) \neq 0$.

(b) The condition $\max\limits_{a \le x \le b} q(x) \ge 0$ is sufficient for the occurrence of case (i).

**Construction of the Green's function:** Suppose case (i) occurs. We choose functions $y_1$ and $y_2$ with

$$-(py_j')' + qy_j = 0 \quad \text{on} \quad [a, b], \quad j = 1, 2,$$

and the initial conditions

$$y_1(a) = 0, \quad y_1'(a) = 1 \quad \text{as well as} \quad y_2(b) = 0, \quad y_2'(b) = \frac{1}{p(b)y_1(b)}.$$

Then we have:

$$G(x, \xi) := \begin{cases} y_1(x)y_2(\xi) & \text{for} \quad a \le x \le \xi \le b, \\ y_2(x)y_1(\xi) & \text{for} \quad a \le \xi \le x \le b. \end{cases}$$

*Example 1:* The boundary value problem

$$-y'' = f(x) \quad \text{on} \quad [0, 1], \quad y(0) = y(1) = 0,$$

has for every continuous function $f : [0, 1] \longrightarrow \mathbb{R}$ the unique solution

$$y(x) = \int_0^1 G(x, \xi) f(\xi) \mathrm{d}\xi$$

with the Green's function

$$G(x, \xi) = \begin{cases} x(1 - \xi) & \text{for} \quad 0 \le x \le \xi \le 1, \\ \xi(1 - x) & \text{for} \quad 0 \le \xi \le x \le 1. \end{cases}$$

The uniqueness of the solution follows from (b) above.

**Nullstellensatz of Sturm:** Let $J$ be a finite or infinite interval. Then every non-trivial solution $y$ of the differential equation

$$-\left(p(x)y'\right)' + q(x)y = 0 \quad \text{on } J \tag{1.307}$$

has only simple zeros, and of these at most countably many, which moreover have no accumulation point at infinity.

**Sturm's theorem on separating zeros:** If $y$ is a solution of (1.307) and $z$ is a solution of

$$-\left(p(x)z'\right)' + q^*(x)z = 0 \quad \text{on } J \tag{1.308}$$

with $q^*(x) \le q(x)$ on $J$, then there is a zero of $y$ between any two zeros of $z$.

*Example 2:* Let $\gamma \in \mathbb{R}$. Then every solution $v = v(\xi)$ of the *Bessel differential equation*

$$\xi^2 v'' + \xi v' + (\xi^2 - \gamma^2)v = 0$$

on the interval $]0, \infty[$ has countably many zeros.

*Proof:* By making the substitution $x := \ln \xi$ and $y(x) := v(\mathrm{e}^x)$, we get the differential equation (1.307) with $q(x) := \mathrm{e}^{2x} - \gamma^2$. We set $q^*(x) := 1$ and choose a number $x_0$ such that

$$q^*(x) \le q(x) \qquad \text{for all} \qquad x_0 \le x.$$

The function $z := \sin x$ satisfies the differential equation (1.308) and has only countably many zeros in the interval $J := [x_0, \infty[$. Therefore the claim follows from the theorem of Sturm above. □

**Oscillations:** Every non-trivial solution $y$ of the differential equation

$$y'' + q(x)y = 0$$

has at most countably many zeros, provided the function $q$ is continuous in the interval $J := [a, \infty[$ and one of the two following conditions is satisfied:

(a) $q(x) \ge 0$ in $J$ and $\displaystyle\int_J q\,\mathrm{d}x = \infty$.

(b) $\displaystyle\int_J |q(x) - \alpha|\,\mathrm{d}x < \infty$ for some fixed number $\alpha > 0$.

In case (b) one can conclude in addition the boundedness of $y$ on the interval $J$.

#### 1.12.8.2 The corresponding variational problem

We set

$$F(y) := \int_a^b (py'^2 + qy^2 - 2fy)\,\mathrm{d}x.$$

We let $Y$ denote the set of all $C^2$-functions $y : [a, b] \longrightarrow \mathbb{R}$ which satisfy the boundary conditions $y(a) = y(b) = 0$.

**Theorem:** The variational problem

$$\boxed{F(y) :\overset{!}{=} \min., \quad y \in Y} \tag{1.309}$$

is equivalent to the original problem (1.306).

**The method of approximation of Ritz:** We choose functions $y_1, \ldots, y_n \in Y$ and consider the approximation problem

$$\boxed{F(c_1 y_1 + \ldots + c_n y_n) \overset{!}{=} \min., \quad c_1, \ldots, c_n \in \mathbb{R}} \tag{1.310}$$

instead of (1.309). This a problem of the form "minimize $G(c)$, $c \in \mathbb{R}^n$". The necessary condition for a solution of (1.310) is

$$\frac{\partial G(c)}{\partial c_j} = 0, \quad j = 1, \ldots, n.$$

This yields for $c$ the following system of linear equations

$$\boxed{Ac = b} \tag{1.311}$$

with $c = (c_1, \ldots, c_n)^{\mathrm{T}}$, $A = (a_{jk})$ and $b = (b_1, \ldots, b_n)^{\mathrm{T}}$, as well as

$$a_{jk} := \int_a^b (p y_j' y_k' + q y_j y_k) \mathrm{d}x, \quad b_k := \int_a^b f y_k \mathrm{d}x.$$

The approximate solution of (1.306) and (1.309) is then

$$\boxed{y = c_1 y_1 + \ldots + c_n y_n.} \tag{1.312}$$

**Ritz' procedure for the eigenvalue problem:** From the solutions of the matrix eigenvalue problem

$$\boxed{Ac = \lambda c}$$

one obtains according to (1.312) an approximation $\lambda$ and $y$ for the eigenvalues and eigenfunctions, respectively, of the following problem (1.313).

#### 1.12.8.3 The eigenvalue problem

Instead of (1.306), we now consider the similar boundary value problem

$$\boxed{\begin{aligned} -(p(x)y')' + q(x)y = \lambda y, \quad a \le x \le b,\\ y(a) = y(b) = 0. \end{aligned}} \tag{1.313}$$

A real number $\lambda$ is called an *eigenvalue* of (3.313), if there is a non-trivial solution $y$ of (1.313). The eigenvalue $\lambda$ is called *simple*, if all corresponding eigenfunctions are equal up to a constant multiple.

*Example:* The problem

$$-y'' = \lambda y, \quad 0 \le x \le \pi, \quad y(0) = y(\pi) = 0$$

has the eigenfunctions $y_n = \sin nx$ and the eigenvalues $\lambda_n = n^2$, $n = 1, 2, \ldots$

**Existence theorem:** (i) All eigenvalues of (1.313) form a sequence

$$\lambda_1 < \lambda_2 < \cdots \quad \text{with} \quad \lim_{n \to \infty} \lambda_n = +\infty.$$

(ii) These eigenvalues are all simple. The corresponding eigenfunctions $y_1, y_2, \ldots$ can be normalized in such a way that

$$\int_a^b y_j(x)y_k(x)\mathrm{d}x = \delta_{jk}, \quad j,k = 1,2,\ldots$$

(iii) The $n$th eigenfunction $y_n$ has exactly $n-1$ zeros in the interior of the interval $[a,b]$, and all of these zeros are simple.

(iv) For the smallest eigenvalue we have the estimate $\lambda_1 > \min_{a\le x\le b} q(x)$.

**The fundamental theorem on developing solutions:** (i) Every $C^1$-function $f : [a,b] \longrightarrow \mathbb{R}$ which satisfies the boundary conditions $f(a) = f(b) = 0$, can be described by an absolutely and uniformly convergent power series

$$\boxed{f(x) = \sum_{n=1}^{\infty} c_n y_n(x), \quad a \le x \le b,} \tag{1.314}$$

with the *generalized Fourier coefficients*

$$c_n := \int_a^b y_n(x)f(x)\mathrm{d}x.$$

(ii) If the function $f : [a,b] \longrightarrow \mathbb{R}$ is only assumed to be almost everywhere continuous and satisfies $\int_a^b f(x)^2\mathrm{d}x < \infty$, then the generalized Fourier series (1.314) converges in the sense of mean squared convergence, i.e., one has

$$\lim_{n\to\infty} \int_a^b \left( f(x) - \sum_{k=1}^{n} c_k y_k \right)^2 \mathrm{d}x = 0.$$

**Asymptotic behavior of eigensolutions:** For $n \longrightarrow \infty$ one has

$$\lambda_n = \frac{n^2\pi^2}{\varphi(b)^2} + O(1)$$

and

$$y_n(x) = \frac{\sqrt{2}}{\sqrt{\varphi(b)}\sqrt[4]{p(x)}} \sin\frac{n\pi\varphi(x)}{\varphi(b)} + O\left(\frac{1}{n}\right).$$

Here we have set

$$\varphi(x) := \int_a^x \sqrt{\frac{1}{p(\xi)}}\mathrm{d}\xi.$$

**The minimum principle:** Let

$$F(y) := \frac{1}{2}\int_a^b (py'^2 + qy^2)\mathrm{d}x, \quad (y|z) := \int_a^b yz\mathrm{d}x.$$

Moreover, let $Y$ be the set of all $C^2$-functions $y : [a,b] \longrightarrow \mathbb{R}$ with $y(a) = y(b) = 0$.

(i) For the first eigenfunction (the eigenfunction to the first eigenvalue) $y_1$, one has the extremal (minimization) problem:

$$F(y) \stackrel{!}{=} \min., \quad (y|y) = 1, \quad y \in Y.$$

(ii) The second eigenfunction $y_2$ is obtained from the extremal problem:

$$F(y) \stackrel{!}{=} \text{min.}, \quad (y|y) = 1, \quad (y|y_1) = 0, \quad y \in Y.$$

(iii) The $n$th eigenfunction $y_n$ is the solution of the extremal problem:

$$F(y) \stackrel{!}{=} \text{min.}, \quad (y|y) = 1, \quad (y|y_k) = 0, \quad k = 1, \ldots, n-1, \quad y \in Y.$$

For the $n$th eigenvalue one has $\lambda_n := F(y_n)$.

**The minimum–maximum principle of Courant:** The $n$th eigenvalue $\lambda_n$ is obtained directly from the equation

$$\boxed{\lambda_n = \max_{Y_n} \min_{y \in Y_n} F(y), \quad n = 2, 3, \ldots .} \tag{1.315}$$

This is to be understood as follows. We choose fixed functions $z_1, \ldots, z_{n-1}$ in $Y$. Then $Y_n$ consists of precisely those functions $y \in Y$ which satisfy

$$(y|y) = 1 \quad \text{and} \quad (y|z_k) = 0, \quad k = 1, 2, \ldots, n-1.$$

For each choice of $Y_n$ we calculate the minimal value $F(y)$ on $Y_n$. Then $\lambda_n$ is the maximum of all of these possible minimal values.

**Comparison theorem:** From $p(x) \leq p^*(x)$ and $q(x) \leq q^*(x)$ on $[a, b]$ it follows that

$$\boxed{\lambda_n \leq \lambda_n^*, \quad n = 1, 2, \ldots}$$

for the corresponding eigenvalues of the original problem (1.313).

This is an immediate consequence of (1.315).

## 1.12.9 General theory

### 1.12.9.1 The global existence and uniqueness theorem

Consider the problem

$$\boxed{\begin{aligned} x'(t) &= f(t, x(t)), \\ x(t_0) &= x_0 \quad \text{(initial value).} \end{aligned}} \tag{1.316}$$

Let $x = (x_1, \ldots, x_n)$ and $f = (f_1, \ldots, f_n)$. For this system of higher order we use the same notations as in (1.245).

**Theorem:** Suppose the function $f : U \subseteq \mathbb{R}^{n+1} \longrightarrow \mathbb{R}$ is of type $C^1$ on an open set $U$, and suppose the point $(x_0, t_0)$ belongs to $U$. Then the initial value problem (1.316) has a uniquely determined *maximal solution*

$$\boxed{x = x(t),} \tag{1.317}$$

i.e., this solution is defined from boundary to boundary of $U$ and can therefore not be further extended in $U$ (Figure 1.144).

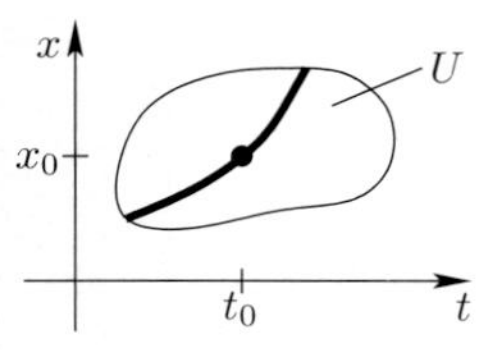

*Figure 1.144.*

**Corollary:** It is sufficient that one of the following conditions hold:

(i) $f$ and $f_x$ are continuous in $U$.

(ii) $f$ is continuous in $U$ and $f$ is locally Lipschitz continuous[128] in $U$ with respect to $x$.

#### 1.12.9.2 Smooth (differentiable) dependence on initial values and parameters

We denote the maximal solution (1.317) by

$$x = X(t; x_0, t_0; p), \tag{1.318}$$

where we admit the possibility that the right hand side $f = f(x, t, p)$ in (1.316) depends also on parameters $p = (p_1, \ldots, p_m)$ which vary in an open set $P$ of $\mathbb{R}^m$.

**Theorem:** If $f$ is of type $C^k$ with $k \geq 1$ in $U \times P$, then $X$ in (1.318) is also of type $C^k$ with respect to all variables $(t, x_0, t_0, p)$ in the domain of existence of the maximal solution.

#### 1.12.9.3 Power series and the Cauchy theorem

**Theorem of Cauchy:** If $f$ is analytic,[129] then also the maximal solution $x = x(t)$ of (1.316) is analytic.

The local power series expansion of $x = x(t)$ is obtained by the method of comparing coefficients.

*Example:* For the initial value problem

$$x' = x, \quad x(0) = 1$$

we get $x'(0) = 1$ and similarly $x^{(n)}(0) = 1$ for all $n$. From this we get the solution

$$\begin{aligned} x(t) &= x(0) + x'(0)t + x''(0)\frac{t^2}{2!} + \ldots \\ &= 1 + t + \frac{t^2}{2!} + \ldots = \mathrm{e}^t. \end{aligned}$$

**Remark:** The theorem of Cauchy remains valid for complex variables $t, x_1, \ldots, x_n$ and complex-valued functions $f_1, \ldots, f_n$.

#### 1.12.9.4 Integral equations:

We consider the integral equation

$$\boxed{x(t) \leq \alpha + \int_0^t f(s)x(s)\mathrm{d}s \quad \text{on} \quad J} \tag{1.319}$$

for the unknown $x$, where $J := [0, T]$.

[128] To every point in $U$ there is a neighborhood $V$, such that $|f(x,t) - f(y,t)| \leq \text{const}|x - y|$ for all $(t, y) \in V$.

[129] This means that for each point in $U$ there is a neighborhood in which $f$ can be expanded in an absolutely convergent (in all variables) power series.

**The Lemma of Gronwall (1918):** Suppose the function $x : J \longrightarrow \mathbb{R}$ is continuous and obeys the integral equation (1.319) with a real number $\alpha$ and a non-negative function $f : J \longrightarrow \mathbb{R}$. Then one has

$$\boxed{x(t) \le \alpha \mathrm{e}^{F(t)} \quad \text{on } J}$$

with $F(t) := \int_0^t f(s)\mathrm{d}s$.

### 1.12.9.5 Differential inequalities

We consider the following system of equations and inequalities:

$$\boxed{\begin{array}{ll} x'(t) \le f\left(x(t)\right) & \text{on } J, \\ y'(t) = f\left(y(t)\right) & \text{on } J, \\ x(0) \le y(0), & \end{array}} \tag{1.320}$$

where $J := [0, T]$ and $f : [0, \infty[ \longrightarrow [0, \infty[$ is a strictly increasing function of type $C^1$.

**Theorem:** If $x$ and $y$ are $C^1$-functions satisfying the relation (1.320) with $x(t) \ge 0$ on $J$, then one has

$$\boxed{x(t) \le y(t) \quad \text{on } J.}$$

**Corollary:** If the $C^1$-functions $x$ and $y$ satisfy the relations

$$\begin{array}{ll} x'(t) \ge f\left(x(t)\right) & \text{on } J, \\ y'(t) = f\left(y(t)\right) & \text{on } J, \\ 0 \le y(0) \le x(0) & \end{array}$$

and if moreover $x(t) \ge 0$ on $J$, then

$$0 \le y(t) \le x(t) \quad \text{on } J.$$

*Example:* Let $x = x(t)$ be a solution of the differential equation

$$x'(t) = F(x(t)), \quad x(0) = 0$$

with $F(x) \ge 1 + x^2$ for all $x \in \mathbb{R}$. Then one has

$$x(t) \ge \tan t, \quad 0 \le t \le \frac{\pi}{2},$$

and hence also $\lim_{t \to \frac{\pi}{2} - 0} x(t) = +\infty$.

*Proof:* We set $f(y) := 1 + y^2$. For $y(t) := \tan t$ we get

$$y'(t) = 1 + y^2.$$

The statement now follows from the corollary. □

#### 1.12.9.6 Blowing up of solutions in finite time

We consider the real system of first order

$$\boxed{\begin{aligned} x'(t) &= f\,(x(t), t)\,, \\ x(0) &= x_0 \end{aligned}} \tag{1.321}$$

where we have set $x = (x_1, \ldots, x_n)$, $f = (f_1, \ldots, f_n)$ and let in addition $\langle x|y\rangle = \sum_{j=1}^{n} x_j y_j$.

We make the following assumptions:

(A1) The function $f : \mathbb{R}^{n+1} \longrightarrow \mathbb{R}$ is of type $C^1$.

(A2) We have $\langle f(x,t)|x\rangle \geq 0$ for all $(x,t) \in \mathbb{R}^{n+1}$.

(A3) There are constants $b > 0$ and $\beta > 2$ such that $\langle f(x,t)|x\rangle \geq b|x|^{\beta}$ for all $(x,t) \in \mathbb{R}^{n+1}$ with $|x| \geq |x_0| > 0$.

**Theorem:** There is a number $T > 0$ with

$$\lim_{t \to T-0} |x(t)| = \infty,$$

i.e., the solution blows up in finite time.

#### 1.12.9.7 The existence of global solutions

The faster than linear growth of $f$ in (A3) is responsible for the blowing up of the solution. The situation is dramatically different if the growth is at most linear.

(A4) There are positive constants $c$ and $d$ with

$$|f(x,t)| \leq c|x| + d \quad \text{for all} \quad (x,t) \in \mathbb{R}^{n+1}.$$

**Theorem:** If the assumptions (A1) and (A4) are satisfied, then the initial value problem (1.321) has a unique solution which exists for all times $t$.

#### 1.12.9.8 The principle of *a-priori* estimates

We assume:

(A5) If a solution of the initial value problem (1.321) exists on an open interval $]t_0 - T, t_0 + T[$, then one has

$$\boxed{|x(t)| \leq C} \tag{1.322}$$

with a constant $C$, which may depend on $T$.

**Theorem:** Under the assumptions (A1) and (A5) the initial value problem (1.321) has a unique solution which exists for all times.

**Remark:** One calls an estimate like (1.322) an *a-priori estimate.* The theorem above is a special case of the following general principle in mathematics:[130]

> *A-priori* estimates guarentee the existence of solutions.

[130] A general statement in this direction is the Leray–Schauder principle (cf. [212]).

*Example:* The initial value problem

$$x' = \sin x, \quad x(0) = x_0$$

has for every $x_0 \in \mathbb{R}$ a unique solution, which exists for all times.

*Proof:* If $x = x(t)$ is a solution on $[-T, T]$, then one has

$$x(t) = x_0 + \int_{-T}^{T} \sin x(t) \mathrm{d}t.$$

Because of $|\sin x| \leq 1$ for all $x$, we obtain the following *a-priori* estimate:

$$|x(t)| \leq |x_0| + \int_{-T}^{T} \mathrm{d}t = |x_0| + 2T. \qquad \square$$

To obtain such *a-priori* estimates, one can utilize differential inequalities.

## 1.13 Partial differential equations

*Of all the mathematical disciplines, the theory of differential equations is the most important. All branches of physics pose problems which can be reduced to the integration of differential equations. More generally, the way of explaining all natural phenomena which depend on time is given by the theory of differential equations.*

*Sophus Lie (1894)*

In this section we consider elements of the theory of partial differential equations. The modern theory is based on the notion of generalized derivatives and the application of the theory of Sobolev spaces in the context of functional analysis. This latter topic will be considered in more detail in [212]. Since partial differential equations describe very diverse type of phenomena which occur in nature, it is not surprising that the theory is far from being finished. There are a series of basic and deep questions which have no satisfactory answer even today.

The basic ideas of this theory, which also occur in the theory of ordinary differential equations, can be found in 1.12.1.

**The great spectrum of types of solutions of partial differential equations:** Partial differential equations have as a rule classes of functions as solutions.

*Example 1:* Let $\Omega$ be a non-empty, open set in $\mathbb{R}^N$. The differential equation

$$u_{x_1}(x) = 0 \text{ on } \Omega$$

has as solutions precisely the set of functions which do not depend on $x_1$.

*Example 2:* The differential equation

$$u_{xy} = 0 \text{ on } \mathbb{R}^2$$

has as set of smooth solutions precisely the set of functions of the form

$$u(x, y) := f(x) + g(y),$$

where $f$ and $g$ are smooth.

For physical problems it is not finding the most general solution which is of interest, but rather, describing concrete processes. To do this, one adds to the differential equation certain constraints, which describe the physical system at a starting time (initial value) or along a boundary (boundary value).

Many phenomena of the theory of partial differential equations can be visualized by interpreting the equations physically. We shall pursue this method of inquiry systematically in this chapter.

## 1.13.1 Equations of first order of mathematical physics

### 1.13.1.1 Conservation laws and the method of characteristics

Consider the equation

$$\boxed{E_t + f(x,t)E_x = 0.} \tag{1.323}$$

where $x = (x_1, \ldots, x_n)$ and $f = (f_1, \ldots, f_n)$. Together with this linear homogenous partial differential equation of the first order for the sought-for function $E = E(x,t)$, we consider the system of ordinary differential equations of first order[131]

$$x' = f(x,t). \tag{1.334}$$

The solutions $x = x(t)$ of (1.324) are called the *characteristics* of (1.323).

Suppose the function $f : \Omega \subseteq \mathbb{R}^{n+1} \longrightarrow \mathbb{R}$ is smooth in a domain $\Omega$. A *conserved quantity* (also referred to as an *integral* of (1.324)) is a function $E$ which is constant along every solution of (1.324), i.e., $E$ is constant along all characteristics.

**Conserved quantities:** A smooth function $E$ is a solution of (1.323) if and only if it is a conserved quantity for the characteristics.

*Example 1:* We set $x = (y,z)$. The equation

$$E_t + zE_y - yE_z = 0 \tag{1.325}$$

has a smooth solution

$$E = g(y^2 + z^2) \tag{1.326}$$

with an arbitrary smooth function $g$. This is the most general smooth solution of (1.325).

*Proof:* The equation for the characteristics $y = y(t),\ z = z(t)$ is

$$\begin{aligned} y' &= z, \quad z' = -y, \\ y(0) &= y_0, \quad z(0) = z_0, \end{aligned} \tag{1.327}$$

with the solutions

$$y = y_0 \cos t + z_0 \sin t, \quad z = -y_0 \sin t + z_0 \cos t. \tag{1.328}$$

These are circles $y^2 + z^2 = y_0^2 + z_0^2$ of radius $\sqrt{y_0^2 + z_0^2}$. Hence (1.326) is the most general conserved quantity. □

[131] More explicitly one has

$$E_t + \sum_{j=1}^{n} f_j(x,t)E_{x_j} = 0$$

and

$$x_j'(t) = f_j(x(t),t), \quad j = 1, \ldots, n.$$

**The initial value problem:** In addition to (1.323) we now consider also an initial value problem:

$$\boxed{\begin{aligned} &E_t + f(x,t)E_x = 0, \\ &E(x,0) = E_0(x) \quad \text{(initial condition).} \end{aligned}} \tag{1.329}$$

**Theorem:** If the given function $E_0$ is smooth in a neighborhood of the point $x = p$, then the problem (1.329) has a unique solution in a sufficiently small neighborhood of $(p,0)$, and this solution is smooth.

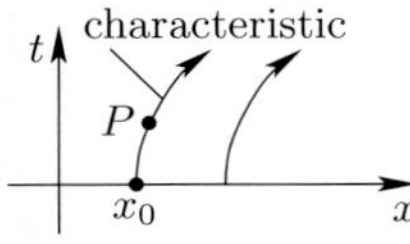


*Figure 1.145.*

If we vary $E_0$, then we get the general solution in a small neighborhood of the point $(p,0)$.

**Construction of the solution with the help of the method of characteristics:** Through each point $x = x_0,\ t = 0$, a characteristic passes, which we denote by

$$x = x(t, x_0) \tag{1.330}$$

(Figure 1.145). The solution $E$ of (1.329) must be constant along these characteristics, which means that we have

$$E(x(t,x_0),t) = E_0(x_0).$$

If we solve the equation (1.330) for $x_0$, we get $x_0 = x_0(x,t)$ and $E(x,t) = E_0(x_0(x,t))$. This is the sought-for solution.

*Example 2:* The initial value problem

$$\begin{aligned} &E_t + zE_y - yE_z = 0, \\ &E(y,z,0) = E_0(y,z) \quad \text{(initial condition)} \end{aligned} \tag{1.331}$$

has for every smooth function $E_0 : \mathbb{R}^2 \longrightarrow \mathbb{R}$ the unique solution

$$E(y,z,t) = E_0(y\cos t - z\sin t, z\cos t + y\sin t)$$

for all $x, y, t \in \mathbb{R}$.

*Proof:* According to Example 1, the characteristics are circles. If we solve the equation of the characteristics (1.328) for $y_0,\ z_0$, then we get

$$y_0 = y\cos t - z\sin t, \qquad z_0 = z\cos t + y\sin t.$$

The solution of (1.331) is obtained from $E(x,y,t) = E_0(y_0, z_0)$. □

**Historical note:** If one knows $n$ linearly independent[132] conserved quantities $E_1, \ldots, E_n$ of the equation $x' = f(x,t)$ for the characteristics, and if $C_1, \ldots, C_n$ are constants, then one obtains upon solving the equation

$$E_j(x,t) = C_j, \quad j = 1, \ldots, n,$$

locally the general solution $x = x(t;C)$ of $x' = f(x,t)$.

During the nineteenth century it was attempted to solve the three body problem in celestial mechanics in this manner. This problem is given by a system of differential equations of second order for the nine components of the trajectory. This is equivalent

[132] This means that $\det E'(x) \neq 0$ on $\Omega$ with $E'(x) = (\partial_k E/\partial x_j)$.

to a system of first order with 18 variables. Thus one requires 18 conserved quantities. The conservation of momentum (the motion of the center of mass), angular momentum and energy yield only ten scalar conserved quantities. In 1887 and 1889, respectively, Bruns and Poincaré showed that for a large class of functions there are no further integrals. Thus it was seen to be *impossible* to obtain a closed solution of the three body problem by using integrals of motion. The deeper reason for this failure is the fact that the three body problem can in fact be *chaotic*.

In treating the $n$-body problem with $n \geq 3$ one uses in our present age of satellites abstract existence and uniqueness results and effective numerical procedures which arise from these for calculating the trajectories.

#### 1.13.1.2 Conserved quantities, shock waves and the condition for entropy due to Lax

*Although differential equations which determine the motion of fluids have been written down, the integration of these equations has only been successful in the cases in which the differences in pressure are infinitely small.*

*Bernhard Riemann (1860)*[133]

In the dynamics of liquids one often encounters shock waves, which for example can be observed as the sonic boom of supersonic aircraft. Shock waves of this kind, which correspond to a point of non-continuity of the mass density $\rho$, make the treatment of the dynamics of fluids extremely difficult. The equation

$$\boxed{\begin{aligned} &\rho_t + f(\rho)_x = 0, \\ &\quad \rho(x,0) = \rho_0(x) \quad \text{(initial value)} \end{aligned}} \tag{1.332}$$

is the simplest mathematical model possible to describe the phenomena of shock waves. We assume the function $f : \mathbb{R} \longrightarrow \mathbb{R}$ is smooth.

*Example 1:* In the special case $f(\rho) = \rho^2/2$, we obtain the so-called *Burger's equation*

$$\rho_t + \rho\rho_x = 0, \quad \rho(x,0) = \rho_0. \tag{1.333}$$

**Physical interpretation:** We consider the distribution of mass along the $x$-axis; let $\rho(x,t)$ denote the density at the point $x$ and time $t$. We introduce the mass density current vector

$$\mathbf{J}(x,t) := f(\rho(x,t))\mathbf{i}$$

we may then write the equation (1.332) in the form $\rho_t + \mathbf{div}\,\mathbf{J} = 0$, i.e., the equation (1.332) describes the conservation of mass (cf. 1.9.7).

**Characteristics:** The lines

$$\boxed{x = v_0 t + x_0 \quad \text{with} \quad v_0 := f'(\rho_0(x_0))} \tag{1.334}$$

are called *characteristics*. One has:

> Every smooth solution $\rho$ of the equation of conservation (1.332) is constant along the characteristics.

[133] In his fundamental treatise *On the propagation of air waves of finite amplitude*, Riemann laid a cornerstone for the mathematics of the theory of fluids and non-linear hyperbolic differential equations, which describe non-linear wave processes. This paper, together with a commentary by Peter Lax, can be found in Riemann's collected works [182].

This may be given the following physical interpretation: a point of mass, which is at the point $x_0$ at time $t=0$, moves according to (1.334) with a constant velocity $v_0$. Collisions of such points of mass lead to discontinuities in $\rho$, which we refer to as shocks.

**Shock waves:** Let $f''(\rho) > 0$ for all $\rho \in \mathbb{R}$, i.e., the function $f'$ is assumed to be monotonously increasing. If $x_0 < x_1$ and

$$\boxed{\rho_0(x_0) > \rho_0(x_1),}$$

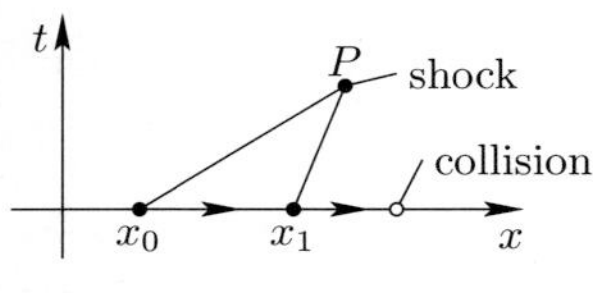


*Figure 1.146.*

then $v_0 > v_1$ in (1.334), and a point of mass which starts at point $x_0$ catches up with a point of mass which starts at $x_1$. In an $(x, t)$-diagram, the corresponding characteristics intersect each other at a point $P$ (Figure 1.146).

Since the density $\rho$ must be constant along the characteristic, $\rho$ is necessarily discontinuous at $P$. By definition, this means that at the point $P$ we have a shock wave.

**Solution of the initial value problem:** If the initial density $\rho_0$ is smooth, then we get a solution $\rho$ of the initial value problem (1.332) be setting

$$\rho(x, t) := \rho_0(x_0),$$

where $(x, t)$ and $x_0$ are connected by the equation (1.334).

This solution is unique there and also smooth, as long as the characteristics do not intersect in the $(x, t)$-plane.

> If $f''(\rho) > 0$ in $\mathbb{R}$, then the equation (1.332) for the conserved quantity has no smooth solution for all $\rho$ for all times $t \geq 0$, in spite of the smoothness of the initial function $\rho_0$.

The points of discontinuity develop through the shocks.

**Generalized solutions:** In order to study the behavior of the discontinuities precisely, we will refer to the function $\rho$ as a *generalized solution* of the equation (1.332), if

$$\boxed{\int_{\mathbb{R}^2_+} (\rho\varphi_t + f(\rho)\varphi_x)\mathrm{d}x\mathrm{d}t = 0} \tag{1.335}$$

for all test functions $\varphi \in C_0^\infty(\mathbb{R}^2_+)$.[134]

**The jump along a shock wave:** Let the characteristic

$$\mathscr{S} : x = v_0 t + x_0$$

be given, together with a generalized solution $\rho$ of (1.332), which is smooth except for jumps along the characteristic. The one-sided limit of $\rho$ from the left and right of the characteristic will be denoted by $\rho_+$ and $\rho_-$ (Figure 1.147). Then the following fundamental condition on the jump is satisfied:

[134] The function $\varphi$ is a *test function*, if it is smooth and vanishes outside of a compact set in $\mathbb{R}^2_+ := \{(x, t) \in \mathbb{R}^2 : t > 0\}$. The relation (1.335) follows by multiplying the equation (1.332) with $\varphi$ and applying partial integration.

$$\boxed{v_0 = \frac{f(\rho_+) - f(\rho_-)}{\rho_+ - \rho_-}.} \qquad (1.336)$$

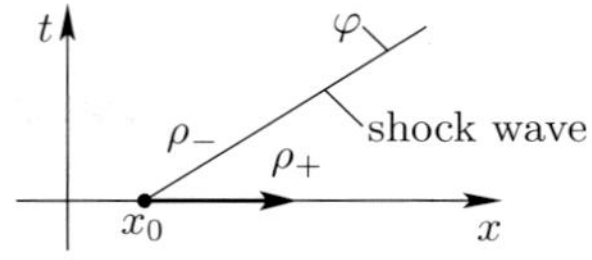


Figure 1.147.

Conditions on the location of the jumps were first introduced into the study of fluids by Riemann in 1860 and then a few years later in a more general formulation by Rankine and Hugoniot. The relation (1.336) relates the velocity $v_0$ of the shock wave to the jump of the density.

**The condition of Lax on the entropy (1957):** The jump condition of (1.336) is also valid for rarefications. However, these are eliminated from consideration by the so-called *entropy condition*

$$\boxed{f'(\rho_-) > v_0 > f'(\rho_+).} \qquad (1.337)$$

**Physical discussion:** We consider a metal cylinder with a moving piston, and two different fluids (or gases) on each side with different densities $\rho_-$ and $\rho_+$. The piston will then only move from left to right, if $\rho_- > \rho_+$. This is a compression process (Figure 1.148). Rarefaction processes with $\rho_- < \rho_+$, in which the density is smaller in front of the moving piston than behind it, are not observed in nature.

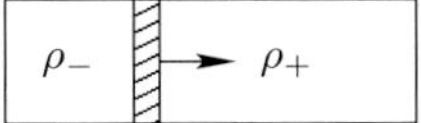


Figure 1.148.

| The second law of thermodynamics decides whether a process is possible in nature or not. |
|---|

Only those processes in closed systems which have a non-decreasing entropy are possible. The entropy condition (1.337) is a replacement for the second law of thermodynamics in the model (1.332).

**Applications to Burger's equation:** We consider the initial value problem (1.333).

*Example 2:* Suppose the initial density is given by the function:

$$\rho_0(x) := \begin{cases} 1 & \text{for} \quad x \le x_0 \\ 0 & \text{for} \quad x > x_0. \end{cases}$$

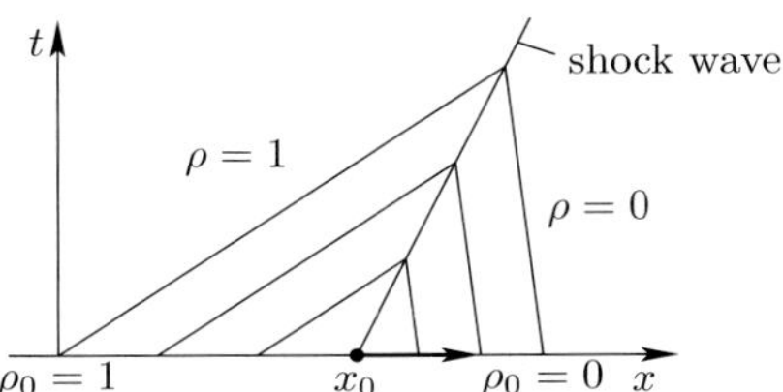


Figure 1.149.

Let $f(\rho) := \rho^2/2$ and $\rho_- = 1$ and $\rho_+ = 0$. The jump condition (1.336) yields the relation

$$v_0 = \frac{f(\rho_+) - f(\rho_-)}{\rho_+ - \rho_-} = \frac{1}{2}.$$

The shock wave thus propagates with the velocity $v_0 = 1/2$ from left to right. In front of the shock wave (resp. behind it), the density is $\rho_+ = 0$ (resp. $\rho_- = 1$). This is a compression process, for which from $f'(\rho) = \rho$ we see that the entropy condition

$$\rho_- > v_0 > \rho_+$$

is satisfied (Figure 1.149).

*Example 3:* If the initial density is

$$\rho_0(x) := \begin{cases} 0 & \text{for} \quad x \le x_0 \\ 1 & \text{for} \quad x > x_0, \end{cases} \tag{1.338}$$

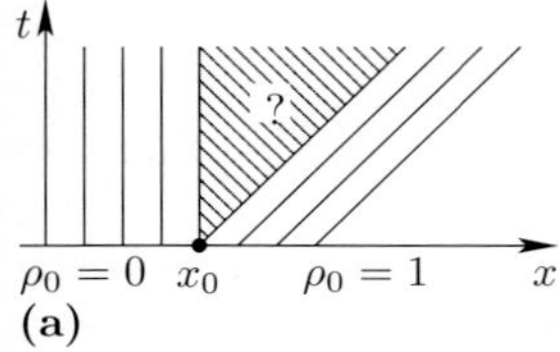


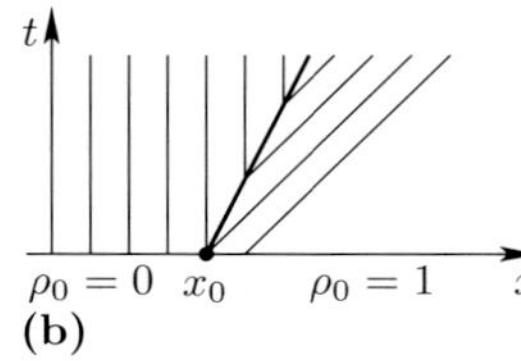


*Figure 1.150. Rarefication process for Burger's equation.*

then we observe the same shock wave as in Example 2. However the density in this case is $\rho_+ = 1$ (resp. $\rho_- = 0$) in front of the wave (resp. behind it). This is a physically impossible rarefication process, for which the entropy condition is violated (Figure 1.150(b)).

The characteristics of the initial conditions (1.338) are depicted in Figure 1.150(a). Here there is a hatched area which is not covered by the characteristics and in which the solution is indeterminate. There are many possibilities to fill this hole, so that generalized solutions occur. However, none of these solutions make sense physically.

#### 1.13.1.3 The Hamilton–Jacobi differential equations

Suppose the Hamilton function $H = H(q, \tau, p)$ is given. In addition to the *canonical equations*

$$\boxed{q' = H_p, \quad p' = -H_q} \tag{1.339}$$

for the sought-for trajectory $q = q(\tau)$, $p = p(\tau)$, we consider, following Jacobi, also the *Hamilton–Jacobi partial differential equation*

$$\boxed{S_\tau + H(q, \tau, S_q) = 0} \tag{1.340}$$

for the sought-for function $S = S(q, \tau)$. Here $q = (q_1, \ldots, q_n)$ and $p = (p_1, \ldots, p_n)$. If $H$ does not depend on $\tau$, then $H$ is a conserved quantity for (1.339). In classical mechanics $H$ is then the energy of the system.

The theory is particularly elegant if it is formulated in the language of symplectic geometry. To do this, one requires the canonical differential form

$$\boxed{\sigma := p \, \mathrm{d}q} \tag{1.341}$$

and the corresponding *symplectic form*[135]

$$\boxed{\omega = -\mathrm{d}\sigma.}$$

[135] In components one has

$$q_j' = H_{p_j}, \quad p_j' = -H_{q_j}.$$

Moreover, $S_q = (S_{q_1}, \ldots, S_{q_n})$ and

$$\sigma = \sum_{j=1}^{n} p_j \mathrm{d}q_j, \quad \omega = \sum_{j=1}^{n} \mathrm{d}q_j \wedge \mathrm{d}p_j.$$

**The fundamental duality between light rays and wave fronts:** In geometric optics the curves

$$\boxed{q = q(\tau)} \tag{1.342}$$

are referred to as *light rays*, where $q$ and $\tau$ denote space variables. The equation (1.340) is called the *eikonal equation*. The surfaces

$$\boxed{S = \text{const}} \tag{1.343}$$

correspond to the *wave fronts*, which are perpendicular to the light rays. If we calculate the integral

$$S(q,\tau) = \int_{(q_0,\tau_o)}^{(q,\tau)} (p(\sigma)q'(\sigma) - H(q(\sigma),\sigma,p(\sigma)))\, d\sigma \tag{1.344}$$

along one of the light rays $q = q(\sigma)$, $p = p(\sigma)$ which connect the points $(q_0, \tau_0)$ and $(q,\tau)$, then $S(q,\tau)$ is the time which the light ray requires to transverse the distance between these two points.

Since there is a close relation between the physics of light rays and of wave fronts, one expects also a close relation between the two equations (1.339) and (1.340). The following two famous theorems, due to Jacobi and Lagrange, verify that this is indeed the case.[136]

**The Hamiltonian analogy between mechanics and geometric optics:** It was the idea of the Irish mathematician and physicist William Rowan Hamilton (1805–1865) to apply the methods of geometric optics to the study of classical mechanics. In mechanics, $q = q(\tau)$ corresponds to the motion of a system of particles at time $\tau$. The integral (1.344) describes the action that is transported along the trajectory. Action is a fundamental physical quantity with the dimension of energy times time (cf. 5.1.3).

We denote by $Q = (Q_1, \ldots, Q_m)$ and $P = (P_1, \ldots, P_m)$ real parameters. The following result contains an important method for solving the equations of motion in celestial mechanics in more complicated cases. From the viewpoint of geometric optics, this theorem shows that systems of light rays can be obtained from systems of wave fronts.

**Theorem of Jacobi (1804–1851):** If one has a smooth solution $S = S(q,\tau,Q)$ of the Hamilton–Jacobi partial differential equation (1.340), then one obtains by means of

$$\boxed{-S_Q(q,\tau,Q) = P, \quad S_q(q,\tau,Q) = p} \tag{1.345}$$

a system of solutions[137]

$$q = q(\tau; Q, P), \quad p = p(\tau; Q, P)$$

of the canonical equation (1.339), which depends on $Q$ and $P$, that is, on $2m$ real parameters.

Application of this result will be considered in 1.13.1.4 and 1.13.1.5 below.

---

[136] The most general relation between arbitrary non-linear partial differential equations of the first order and systems of ordinary differential equations of the first order will be described in the theorem of Cauchy in section 1.13.5.2.

[137] The notation here is that $S_Q = \partial S/\partial Q$, etc. We are assuming that the equation $S_Q(q,\tau,Q) = P$ can be solved for $q$. This is locally the case if $\det S_{Qq}(q_0,\tau_0,Q) \neq 0$.

The basic idea of the following theorem is that the eikonal function of the wave fronts can be constructed out of the system of light rays. This goal cannot be achieved with an arbitrary system of rays, but rather it is required that the rays form a *Lagrangian manifold.*

We assume the Hamiltonian function $H$ is smooth.

**The theorem of Lagrange (1736–1813) and symplectic geometry:** Let a system of solutions

$$q = q(\tau, Q), \quad p = p(\tau, Q) \tag{1.346}$$

of the canonical differential equation (1.339) with $q_Q(\tau_0, Q_0) \neq 0$ be given. Then the equation $q = q(\tau, Q)$ can be solved in a neighborhood of $(\tau_0, Q_0)$ for the variable $Q$, which yields a relation $Q = Q(\tau, q)$.

The system (1.346) is now assumed to form a *Lagrangian manifold* in the neighborhood of $Q_0$ at time $\tau_0$, i.e., the symplectic form $\omega$ vanishes identically for $\tau_0$ along the solutions of the system.[138]

We consider the curve integral

$$\boxed{\mathscr{S}(Q,\tau) = \int_{(Q_0,\tau_0)}^{(Q,\tau)} (pq_\tau - H)\,\mathrm{d}t + pq_Q\,\mathrm{d}Q,}$$

where $q$ and $p$ are given by (1.346). This curve integral is independent of the path of integration and yields by means of

$$\boxed{S(q,\tau) := \mathscr{S}(Q(q,\tau),\tau)}$$

a solution $S$ of the Hamilton–Jacobi differential equation (1.340).

**Corollary:** The system of solutions (1.346) form for all times $\tau$ a Lagrangian manifold.

**The solution of the initial value problem:**

$$\boxed{\begin{aligned} S_\tau + H(q, S_q, \tau) &= 0, \\ S(q,0) &= 0 \quad \text{(initial value).}^{139} \end{aligned}} \tag{1.347}$$

---

[138] Because of $\omega = \sum_{i=1}^{n} \mathrm{d}q_i \wedge \mathrm{d}p_i$, this condition implies that

$$\sum_{j,k=1}^{n} [Q_j, Q_k]\,\mathrm{d}Q_j \wedge \mathrm{d}Q_k = 0,$$

hence

$$\boxed{[Q_j, Q_k](t_0, Q) = 0, \quad k, j = 1, \ldots, n,}$$

for all parameters $Q_i$. Here we are using the *brackets* which were introduced by Lagrange

$$[Q_j, Q_k] := \sum_{i=1}^{n} \frac{\partial q_i}{\partial Q_j}\frac{\partial p_i}{\partial Q_k} - \frac{\partial q_i}{\partial Q_k}\frac{\partial p_i}{\partial Q_j}.$$

Implicitly it was already clear to Lagrange that the notion of symplectic geometry is of great importance for the mathematical description of classical mechanics. However, this notion of geometry was not *explicitly* applied until around 1960, in order to understand deeper properties of many classical considerations, but also to get new insights. This will be described in more detail in 1.13.1.7 and more generally in [212]. The modern standard reference for symplectic geometry and its many applications is the text [156].

**Theorem:** We assume that the Hamilton function $H = H(q, \tau, p)$ is smooth in a neighborhood of the point $(q_0, 0, 0)$. Then the initial value problem (1.347) has a unique solution in a sufficiently small neighborhood of the point $(q_0, 0, 0)$, and this solution is smooth.

**Construction of the solution:** We solve the initial value problem

$$q' = H_p, \quad p' = -H_q, \quad q(\tau_0) = Q, \quad p(\tau_0) = 0$$

for the canonical equations. The corresponding system of solutions $q = q(\tau, Q)$, $p = p(\tau, Q)$ yields by the Theorem of Lagrange the solution $S$ of (1.347).

#### 1.13.1.4 Applications to geometric optics

The movement of the light ray $q = q(\tau)$ in the $(\tau, q)$-plane can be obtained by the *principle of Fermat* (1601–1665):

$$\boxed{\begin{gathered}\int_{\tau_0}^{\tau_1} \frac{n(q(\tau))}{c}\sqrt{1 + q'(\tau)^2}\,\mathrm{d}\tau \stackrel{!}{=} \min., \\ q(\tau_0) = q_0, \quad q(\tau_1) = q_1.\end{gathered}} \tag{1.348}$$

Here $n(q)$ is the indicatrix, i.e., the index of refraction at the point $q(\tau)$ ($c/n(q)$ is the speed of light in the medium), and $c$ is the speed of light in the vacuum. A light ray hence moves in such a way as to transverse the distance between two points in a minimum of time.

**The Euler–Lagrange equations:** If we introduce the Lagrangian $L(q, q', \tau) := \frac{n(q)}{c}\sqrt{1 + (q')^2}$, then every solution $q = q(\tau)$ of (1.348) satisfies the ordinary differential equation of second order

$$\frac{\mathrm{d}}{\mathrm{d}\tau} L_{q'} - L_q = 0,$$

that is

$$\boxed{\frac{\mathrm{d}}{\mathrm{d}\tau} \frac{nq'}{\sqrt{1 + q'^2}} = n_q\sqrt{1 + q'^2}.} \tag{1.349}$$

To simplify notations we choose the units of measurement so that $c = 1$.

**The Hamiltonian canonical equations:** The *Legendre transformation*

$$p = L_{q'}(q, q', \tau), \quad H = pq' - L$$

yields the Hamiltonian

$$H(q, p, \tau) = -\sqrt{n(q)^2 - p^2}.$$

The Hamiltonian canonical equations $q' = H_p$, $p' = -H_q$ are then:

$$\boxed{q' = \frac{p}{\sqrt{n^2 - p^2}}, \quad p' = \frac{n_q n}{\sqrt{n^2 - p^2}}.} \tag{1.350}$$

[139] The general initial value problem with the initial conditions $S(q, 0) = S_0(q)$ can be immediately reduced to the case considered in (1.347), by replacing $S$ by the difference $S - S_0$.

This a system of ordinary differential equations of the first order.

**The Hamilton–Jacobi differential equation:** The equation $S_\tau + H(q, S_q, \tau) = 0$ is

$$S_\tau - \sqrt{n^2 - S_q^2} = 0.$$

This corresponds to the eikonal equation

$$\boxed{S_\tau^2 + S_q^2 = n^2.} \tag{1.351}$$

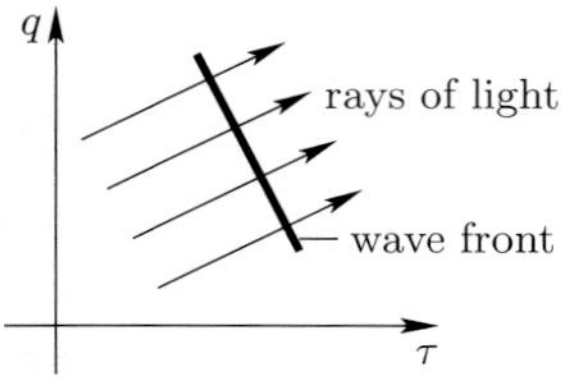

Figure 1.151.

**The method of solution of Jacobi:** We consider the special case $n \equiv 1$, which corresponds to propagation of light in a vacuum. Obviously

$$S = Q\tau + \sqrt{1 - Q^2}q$$

is a solution of (1.351), which depends on the parameter $Q$. According to (1.345) we get upon setting $-S_Q = P$, $p = S_q$ a system of solutions of the canonical equations

$$\frac{Qq}{\sqrt{1 - Q^2}} - \tau = P, \quad p = \sqrt{1 - Q^2},$$

which depends on the two constants $Q$ and $P$, hence is a general solution. This system is a set of lines $q = q(\tau)$ of light rays, which are perpendicular to the straight wave fronts $S = \text{const}$ (Figure 1.151).

### 1.13.1.5 Applications to the two body problem

**The Newtonian equation of motion:** According to 1.12.5.2, the two body problem (for example for the sun and one of the planets) leads to the equation

$$\boxed{m_2 \mathbf{q}'' = \mathbf{F}} \tag{1.352}$$

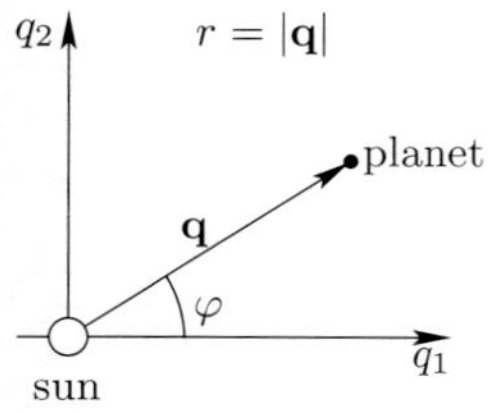


Figure 1.152.

for the plane motion $\mathbf{q} = \mathbf{q}(t)$, where the sun is taken to be at the origin (Figure 1.152). Here $m_1$ denotes the mass of the sun, $m_2$ the mass of the planet, $m = m_1 + m_2$ is the total mass, and G is the gravitational constant. The force is given by the relation

$$\mathbf{F} = -\mathbf{grad}\ U = -\frac{\alpha \mathbf{q}}{|\mathbf{q}|^3} \quad \text{with} \quad U := -\frac{\alpha}{|\mathbf{q}|}, \quad \alpha := \mathrm{G}m_2 m.$$

**The total energy $E$:** This is given as the sum of the kinetic and the potential energy. This gives

$$E = \frac{1}{2} m_2 (\mathbf{q}')^2 + U(\mathbf{q}).$$

**The Hamiltonian canonical equations:** We introduce the momentum $\mathbf{p} = m_2 \mathbf{q}'$ (mass times velocity). From the expression above for the energy $E$ we get the Hamiltonian

$$H = \frac{\mathbf{p}^2}{2m_2} + U(\mathbf{q}).$$

If we set $\mathbf{q} = q_1\mathbf{i} + q_2\mathbf{j}$ and $\mathbf{p} = p_1\mathbf{i} + p_2\mathbf{j}$, then the canonical equations $q_j' = H_{p_j}$, $p_j' = -H_{q_j}$ can be written in vector notation:

$$\boxed{\mathbf{q}' = \frac{\mathbf{p}}{m_2}, \qquad \mathbf{p}' = -\mathbf{grad}\, U.}$$

This equation is equivalent to the Newtonian equations of motion (1.352).

**The Hamilton–Jacobi equation:** For the sought-for function $S = S(q,t)$, the equation $S_t + H(q, S_q) = 0$ becomes explicitly

$$S_t + \frac{S_q^2}{2m_2} + U(\mathbf{q}) = 0$$

with $S_q = \mathbf{grad}\ S$. In order to solve this equation conveniently, it is important to pass to *polar coordinates* $r$, $\varphi$ at this point. This gives the equation

$$\boxed{S_t + \frac{1}{2m_2}\left(S_r^2 + \frac{S_\varphi^2}{r^2}\right) - \frac{\alpha}{r} = 0.} \tag{1.353}$$

**The method of solution of Jacobi:** We look for a two-parameter system of solutions $S = S(r, \varphi, t, Q_1, Q_2)$ of (1.353) with the parameters $Q_1$ and $Q_2$. Setting

$$S = -Q_1 t + Q_2\varphi + s(r)$$

yields with the help of (1.353) the ordinary differential equation

$$-Q_1 + \frac{1}{2m_2}\left(s'(r)^2 + \frac{Q_2^2}{r^2}\right) - \frac{\alpha}{r} = 0.$$

This means that $s'(r) = f(r)$, where

$$f(r) := \sqrt{2m_2\left(Q_1 + \frac{\alpha}{r}\right) - \frac{Q_2^2}{r^2}}.$$

Hence we obtain

$$s(r) = \int f(r)\mathrm{d}r.$$

According to (1.345), the equation for the trajectory is obtained from the equation $-S_{Q_j} = P_j$ with a constant $P_j$. This gives

$$P_1 = t - \int \frac{m_2\,\mathrm{d}r}{f(r)}, \qquad P_2 = -\varphi + \int \frac{Q_2\,\mathrm{d}r}{r^2 f(r)}.$$

For simplification we set $Q_1 = E$ and $Q_2 = N$. Integrating the second equation, we obtain

$$\varphi = \arccos \frac{\dfrac{N}{r} - \dfrac{m_2\alpha}{N}}{\sqrt{2m_2E + \dfrac{m_2^2\alpha^2}{N}}} + \text{const.}$$

We may choose const = 0. We have thus derived the equation

$$\boxed{r = \frac{p}{1 + \varepsilon \cos\varphi}} \tag{1.354}$$

for the orbit, where $p := N^2/m_2\alpha$ and $\varepsilon := \sqrt{1 + 2EN^2/m_2\alpha^2}$. The calculation of the energy and the angular momentum of this motion imply that the constant $E$ is the energy and the constant $N$ is the absolute value $|\mathbf{N}|$ of the vector of angular momentum $\mathbf{N}$.

The trajectories (1.354) are conic sections, which are ellipses for $0 < \varepsilon < 1$. The Kepler laws for the orbits of the planets can be derived from these solutions.

#### 1.13.1.6 The canonical transformation of Jacobi

**Canonical transformations:** A diffeomorphism

$$Q = Q(q,p,t), \quad P = P(q,p,t), \quad T = t,$$

is called a *canonical transformation* of the canonical equation

$$\boxed{q' = H_p, \quad p' = -H_q,} \tag{1.355}$$

if this transforms the equation into a new canonical equation

$$\boxed{Q' = \mathscr{H}_P, \quad P' = -\mathscr{H}_Q.} \tag{1.356}$$

The idea is that by choosing the canonical transformation judiciously, the solution of (1.355) can be reduced to the simpler problem (1.356). This is the most important method to solve complicated equations in celestial mechanics.

**The generating function of Jacobi:** Let a function $S = S(q,Q,t)$ be given. Using the relation

$$\boxed{\mathrm{d}S = p\,\mathrm{d}q - P\,\mathrm{d}Q + (\mathscr{H} - H)\mathrm{d}t,} \tag{1.357}$$

a canonical transformation is generated by setting

$$P = -S_Q(q,Q,t), \quad p = S_q(q,Q,t), \tag{1.358}$$

and

$$\mathscr{H} = S_t + H.$$

Here we are assuming that the equation $p = S_q(q,Q,t)$ can be solved for $Q$ by means of the theorem on implicit functions.

**Jacobi's method:** If we choose $S$ as a solution of the Hamilton–Jacobi differential equation $S_t + H = 0$, then $\mathscr{H} \equiv 0$. The transformed canonical equation (1.356) becomes trivial and has the solution $Q =$ const and $P =$ const. Therefore, (1.358) is the method of Jacobi given in (1.345).

**Symplectic transformations:** The transformation $Q = Q(q,p)$, $P = P(q,p)$ is now assumed to be *symplectic*, i.e., it satisfies the condition

$$\mathrm{d}(P\,\mathrm{d}Q) = \mathrm{d}(p\,\mathrm{d}q).$$

Then this transformation is canonical with $\mathscr{H} = H$.

*Proof:* From the relation $\mathrm{d}(p\,\mathrm{d}q - P\,\mathrm{d}Q) = 0$ we see that the equation

$$\mathrm{d}S = p\,\mathrm{d}q - P\,\mathrm{d}Q$$

has locally a solution $S$ by 1.9.11, which according to (1.357) generates a canonical transformation. □

#### 1.13.1.7 The hydrodynamic interpretation of Hamiltonian mechanics and symplectic geometry

> *The interaction between mathematics and physics has always played a pronounced role. The physicist, who only has a rudimentary grasp of mathematics, is at a great disadvantage. The mathematician, who has no interest in the physical applications, loses an opportunity for motivation and deeper insights.*
>
> *Martin Schechter,*
> *University of California*

A particularly intuitive and elegant interpretation of Hamiltonian mechanics is obtained by using a hydrodynamical picture in the $(q,p)$-phase space and the language of differential forms. A key role is played in this setup by the Hamiltonian function $H$ and the three differential forms

$$\boxed{\sigma := p\,\mathrm{d}q, \quad \omega := -\mathrm{d}\sigma, \quad \sigma - H\,\mathrm{d}t.}$$

The symplectic form $\omega$ is responsible for the ability of symplectic geometry for giving a closed mathematical description.

In what follows suppose that all functions and curves which occur are smooth. Moreover, only bounded domains with smooth boundaries shall be considered. Pathological curves and domains are excluded from consideration.

**Classical currents in $\mathbb{R}^3$**

**Integral curves:** Suppose we are given a velocity field $\mathbf{v} = \mathbf{v}(\mathbf{x}, t)$. The curves $\mathbf{x} = \mathbf{x}(t)$ which satisfy the differential equation

$$\boxed{\mathbf{x}'(t) = \mathbf{v}(\mathbf{x}(t), t), \quad \mathbf{x}(0) = \mathbf{x}_0}$$

are called *integral curves* or *flow lines* of the vector field. These curves describe the flow of fluid particles (Figure 1.153(a)). We set

$$F_t(\mathbf{x}_0) := \mathbf{x}(t),$$

i.e., $F_t$ associates to each point $P$ of the fluid the point $P_t$, which denotes the position of the particle at time $t$ which starts at $P$ at time $t = 0$. One calls $F_t$ the *flow operator* at time $t$.[140]

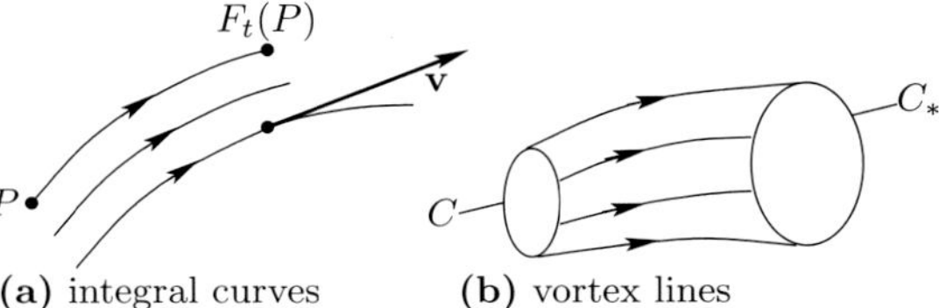


*Figure 1.153. Currents in $\mathbb{R}^3$.*

[140] The general theory of flows on manifolds will be considered in [212]. In his construction of the theory of Lie groups and algebras, Sophus Lie (1842–1899) used in an essential way the notion of flows (one-parameter subgroups). For ease of notation we identify the vector $\mathbf{x}$ with its endpoint in $P$.

**The theorem of flows:**

$$\boxed{\frac{\mathrm{d}}{\mathrm{d}t}\int\limits_{F_t(\Omega)} h(x,t)\,\mathrm{d}x = \int\limits_{F_t(\Omega)} (h_t + (\operatorname{div} h)\mathbf{v})(x,t)\,\mathrm{d}x.} \tag{1.359}$$

*Example:* If $h = \rho$ is the mass density, then the conservation of mass means that in (1.359) the integral on the left vanishes. If we contract the domain $\Omega$ to a point, then from the vanishing of the integral on the right-hand side yields the so-called *continuity equation*

$$\rho_t + \operatorname{div}(\rho\mathbf{v}) = 0.$$

**Vortex lines:** The curves $\mathbf{x} = \mathbf{x}(t)$ which satisfy

$$\boxed{\mathbf{x}'(t) = \frac{1}{2}(\mathbf{curl}\,\mathbf{v})(\mathbf{x}(t),t)}$$

are called *vortex lines.* The contour integral

$$\int\limits_C \mathbf{v}\,\mathrm{d}\mathbf{x}$$

along a closed curve $C$ is called the *circulation* of the velocity field along $C$. If the field $\mathbf{v}$ is vortex-free, i.e., if $\mathbf{curl}\,\mathbf{v} \equiv 0$, then the circulation along every closed curve vanishes. It then follows from Stokes theorem that

$$\int\limits_{\partial F} \mathbf{v}\,\mathrm{d}\mathbf{x} = \int\limits_F (\mathbf{curl}\,\mathbf{v})\mathbf{n}\,\mathrm{d}F = 0,$$

if $C$ is the boundary $\partial F$ of a surface $F$. In general the circulation is non-vanishing and yields a measure of the strength of the vortex in the fluid. There are two important conservation laws for the circulation along curves.

**The vortex theorem of Helmholtz (1821–1894):** One has the relation

$$\boxed{\int\limits_C \mathbf{v}\,\mathrm{d}\mathbf{x} = \int\limits_{C_*} \mathbf{v}\,\mathrm{d}\mathbf{x},}$$

if the curve $C_*$ is obtained from $C$ by parallel transport along vortex lines (Figure 1.153(b)).

**The vortex theorem of Kelvin (1824–1907):** In an ideal fluid one has the relation

$$\boxed{\int\limits_C \mathbf{v}\,\mathrm{d}\mathbf{x} = \int\limits_{F_t(C)} \mathbf{v}\,\mathrm{d}\mathbf{x}.}$$

Here $F_t(C)$ consists of those particles at time $t$, which belong to the closed curve $C$ at time $t = 0$. Hence the circulation along a closed curve, which consists of particles of an ideal fluid, remains constant in time.

**Ideal fluids:** As opposed with the vortex theorem of Helmholtz, the velocity field $\mathbf{v}$ in the theorem of Kelvin must be the solution of the *Euler equations of motion* for an ideal fluid. These equations are

$$\begin{aligned}
&\rho\mathbf{v}_t + \rho(\mathbf{v}\,\mathbf{grad})\mathbf{v} = -\rho\,\mathbf{grad}\,U - \mathbf{grad}\,p \quad \text{(equation of motion)},\\
&\rho_t + \operatorname{div}(\rho\mathbf{v}) = 0 \quad \text{(conservation of mass)},\\
&\rho(\mathbf{x},t) = f(p(\mathbf{x},t)) \quad \text{(pressure–density law } \rho = f(p)).
\end{aligned} \tag{1.360}$$

Here $\rho$ denotes the density, $p$ the pressure and $\mathbf{f} = -\mathbf{grad}\,U$ the force density.

**Incompressible fluids:** In the case of constant density $\rho = \text{const}$, one speaks of an *incompressible fluid.* It then follows from the continuity equation (conservation of mass) in (1.360) that we have the so-called *incompressibility condition*:

$$\operatorname{div}\mathbf{v} = 0.$$

**Conservation of volume:** In an incompressible fluid the flow is volume-preserving, i.e., the fluid particles that are in a domain $\Omega$ at time $t = 0$ are in $F_t(\Omega)$ at time $t$ and both of these domains have the same volume (Figure 1.154). Analytically this means

$$\int_{\Omega} \mathrm{d}x = \int_{F_t(\Omega)} \mathrm{d}x.$$

*Proof:* This follows from the flow equation (1.359) with $h \equiv 1$ and $\operatorname{div}\mathbf{v} = 0$. □

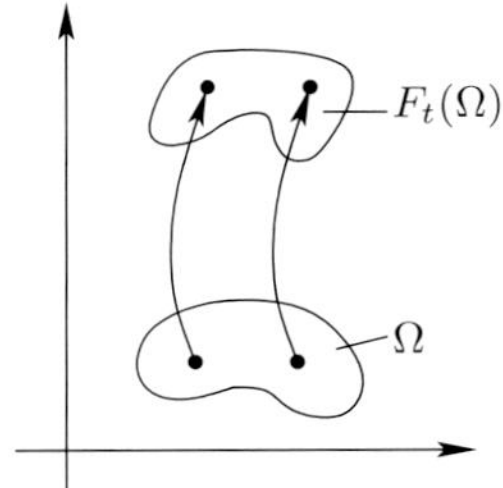

*Figure 1.154. Conservation of volume.*

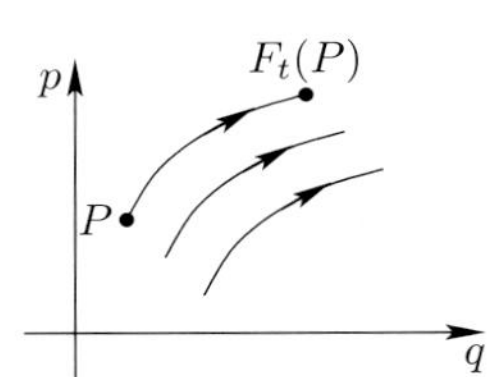


*Figure 1.155. Hamiltonian flows.*

### The Hamiltonian flow

**Phase space:** let $q = (q_1, \ldots, q_n)$ and $p = (p_1, \ldots, p_n)$, so that $(q,p) \in \mathbb{R}^{2n}$. We denote this $(p,q)$-space as the *phase space.* Moreover let $q = q(t)$ and $p = p(t)$ be solutions of the canonical equations

$$\begin{aligned}
&q'(t) = H_p(q(t),p(t)), \quad && p'(t) = -H_q(q(t),p(t)),\\
&q(0) = q_0, && p(0) = p_0.
\end{aligned}$$

We set

$$F_t(q_0, p_0) := (q(t), p(t)).$$

In this manner one obtains what is by definition called a *Hamiltonian flow* (Figure 1.155).

**Conservation of energy:** The Hamiltonian function is constant along the integral curves of the Hamiltonian flow.

**The theorem of Liouville (1809–1882):** The Hamiltonian flow is volume preserving.

**Remark:** This theorem states that

$$\int_{\Omega} \mathrm{d}q\,\mathrm{d}p = \int_{F_t(\Omega)} \mathrm{d}q\,\mathrm{d}p. \tag{1.361}$$

The differential form $\theta := \mathrm{d}q_1 \wedge \mathrm{d}q_2 \wedge \ldots \wedge \mathrm{d}q_n \wedge \mathrm{d}p_1 \wedge \ldots \wedge \mathrm{d}p_n$ is the *volume form* of the phase space. The important relation with the symplectic form $\omega$ is contained in the formula

$$\boxed{\theta = \alpha_n \omega \wedge \omega \wedge \ldots \wedge \omega}$$

with $n$ factors and a constant $\alpha_n$. The relation (1.361) then corresponds to the formula

$$\boxed{\int_{\Omega} \theta = \int_{F_t(\Omega)} \theta.}$$

**The generalized vortex theorem of Helmholtz and the invariant integral of Hilbert:** Let $C$ and $C_*$ be two closed curves, whose points are connected via integral curves of a Hamiltonian flow. Then one has

$$\boxed{\int_{C} p\,\mathrm{d}q - H\,\mathrm{d}t = \int_{C_*} p\,\mathrm{d}q - H\,\mathrm{d}t.}$$

This integral is called the invariant integral of Hilbert (or the *absolute integral invariant of Poincaré–Cartan*).

**The generalized vortex theorem of Kelvin:** If $C$ is a closed curve in phase space, then one has the relation

$$\boxed{\int_{C} p\,\mathrm{d}q = \int_{F_t(C)} p\,\mathrm{d}q.}$$

This integral is called the *relative integral invariant of Poincaré*.

**The parallel transport of curves and tangent vectors induced by the Hamiltonian flow** (Figure 1.156): Let the curve

$$C : q = q(\alpha), \quad p = p(\alpha)$$

in phase space be given, which passes through the point $P$ for the parameter value $\alpha = 0$. The Hamiltonian flow carries the point $P$ into the point

$$\boxed{P_t := F_t(P)}$$

and transports the curve $C$ into the curve $C_t$. Moreover, the tangent vector v to the curve $C$ at the point $P$ is carried into the tangent vector $v_t$ to the curve $C_t$ at the point $P_t$. If two curves $C$ and $C'$ both have the same tangent vector at the point $P$, then the image curves $C_t$ and $C'_t$ both have the same tangent vector $v_t$ at the point $P_t$. In this manner we obtain a transformation $v \mapsto v_t$. We write

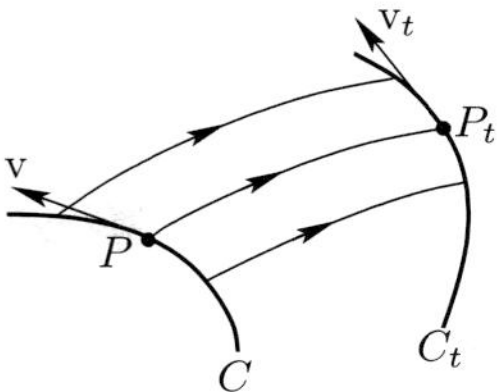


*Figure 1.156.*

$$\boxed{v_t = F_t'(P)v \quad \text{for all } v \in \mathbb{R}^{2n}}$$

to denote this transformation.[141] One calls $F_t'(P)$ the *linearization of the flow operator* at the point $P$. In fact, $F_t'(P)$ is just the Fréchet derivative of $F_t$ at the point $P$.

**The natural transformation of differential forms induced by the Hamiltonian flow:** Let $\mu$ be a 1-form. We define the 1-form $F_t^*\mu$ by means of the natural relation

$$\boxed{(F_t^*\mu)_P(v) := \mu_{P_t}(v_t) \quad \text{for all } v \in \mathbb{R}^{2n}.}$$

We call $F_t^*\mu$ the *pull-back* (with respect to the given flow) of the differential form $\mu$. Indeed, the values of $F_t^*\mu$ at the point $P$ only depend on the form $\mu$ at the point $P_t$ (Figure 1.156).

In the same way one defines the pull-back for arbitrary differential forms. For example, this is for a 2-form

$$(F_t^*\omega)_P(v,w) := \omega_{P_t}(v_t, w_t) \quad \text{for all } v,w \in \mathbb{R}^{2n}.$$

Analogously, the pull-back can be introduced when $F_t$ is replaced by some diffeomorphism $F$.

The pull-back is used to verify invariance properties of the differential form with respect to the flow in a very elegant manner.

**Compatibility of pull-back with the outer product:** For arbitrary differential forms $\mu$, $\nu$, one has[142]

$$\boxed{F_t^*(\mu \wedge \nu) = F_t^*\mu \wedge F_t^*\nu} \tag{1.362}$$

and

$$\int_\Omega F_t^*\mu = \int_{F_t(\Omega)} \mu. \tag{1.363}$$

This statement remain correct if $F_t$ is replaced by an arbitrary diffeomorphism.

---

[141] This is a kind of differentiation, since we have

$$v = \left( \frac{dq(0)}{d\alpha}, \frac{dp(0)}{d\alpha} \right)$$

and

$$v_t = \frac{d}{d\alpha} F_t(q(\alpha), p(\alpha)) \Big|_{\alpha=0}$$

[142] The general rules for this analysis can be found in [212].

**Symplectic transformations**

Let $F : \Omega \subseteq \mathbb{R}^{2n} \longrightarrow F(\Omega)$ be a diffeomorphism from a domain $\Omega$ of the phase space $\mathbb{R}^{2n}$, which we assume is of the form

$$F : P = P(q,p), \quad Q = Q(q,p).$$

We call $F$ a *symplectic transformation,* if the symplectic form $\omega$ remains invariant under $F$, i.e., if

$$\boxed{F^*\omega = \omega.}$$

Written in components, this equation becomes

$$\boxed{\sum_{i=1}^{n} \mathrm{d}Q_i \wedge \mathrm{d}P_i = \sum_{i=1}^{n} \mathrm{d}q_i \wedge \mathrm{d}p_i}\,.$$

**Theorem:** If $F$ is a symplectic transformation, then one has:

(i) $F^*\theta = \theta$.

(ii) $\displaystyle\int_\Omega \theta = \int_{F(\Omega)} \theta$ ($F$ is volume preserving).

(iii) For the canonical form $\sigma$ there is locally a function $S$ with

$$F^*\sigma - \sigma = \mathrm{d}S. \tag{1.364}$$

(iv) For a closed curve $C$ one has:

$$\int_C \sigma = \int_{F(C)} \sigma.$$

These basic statements can be obtained easily and elegantly from the differential calculus of Cartan.

*Proof:* (i): From (1.362) we have

$$F^*\theta = F^*(\omega \wedge \ldots \wedge \omega) = F^*\omega \wedge \ldots \wedge F^*\omega = \omega \wedge \ldots \wedge \omega = \theta.$$

(ii): The relation (1.363) yields

$$\int_\Omega \theta = \int_\Omega F^*\theta = \int_{F(\Omega)} \theta.$$

(iii): We have $\mathrm{d}(F^*\sigma - \sigma) = F^*\mathrm{d}\sigma - \mathrm{d}\sigma = -F^*\omega + \omega = 0$. Hence the equation (1.364) has according to the Poincaré lemma locally a unique solution $S$ (cf. 1.9.11).

(iv): One has $\displaystyle\int_C \mathrm{d}S = 0$. Thus the relation (1.363) yields

$$\int_C F^*\sigma = \int_{F(C)} \sigma.$$

Thus (iv) follows from (iii). □

**The main theorem of the Hamiltonian formulation of mechanics:**

> For every time $t$, the map $F_t$ generated by the Hamiltonian flow is *symplectic.*

Therefore, in the above theorem, $F$ may be replaced throughout by $F_t$.

**The canonical equations:** The velocity field v of the Hamiltonian flow satisfies the equation:[143]

$$\boxed{\mathrm{v} \lrcorner \omega = \mathrm{d}H.} \tag{1.365}$$

This is the most elegant formulation of the Hamiltonian canonical equations. The appearance of the symplectic form $\omega$ in these equations is the key for the applications of symplectic geometry in classical mechanics.

**Symplectic invariance of the canonical equations:** The canonical equations (1.365) are invariant under symplectic transformation, i.e., under a symplectic transformation, the canonical equations

$$q' = H_p, \quad p' = -H_q,$$

are transformed into new canonical equations

$$Q' = H_P, \quad P' = -H_Q.$$

*Proof of* (1.365): Let $q = q(t),\ p = p(t)$ be an integral curve of the Hamiltonian flow. Then for the velocity vector v at the time $t$ we have the relation

$$\mathrm{v} = (q'(t), p'(t)).$$

If we moreover set $\mathrm{w} = (a, b)$ with $a,\ b \in \mathbb{R}^n$, then equation (1.365) tells us that

$$\omega(\mathrm{v}, \mathrm{w}) = \mathrm{d}H(\mathrm{w}) \quad \text{for all} \quad \mathrm{w} \in \mathbb{R}^{2n}.$$

From $\omega = \sum_i \mathrm{d}q_i \wedge \mathrm{d}p_i$ and

$$(\mathrm{d}q_i \wedge \mathrm{d}p_i)(\mathrm{v}, \mathrm{w}) = \mathrm{d}q_i(\mathrm{v})\mathrm{d}p_i(\mathrm{w}) - \mathrm{d}q_i(\mathrm{w})\mathrm{d}p_i(\mathrm{v}) = q_i'(t)a_i - p_i'(t)b_i$$

we obtain

$$\sum_{i=1}^{n} q_i'(t)a_i - p_i'(t)b_i = \sum_{i=1}^{n} H_{q_i}a_i + H_{p_i}b_i$$

for all $a_i,\ b_i \in \mathbb{R}$. Upon comparing coefficients, this yields

$$q_i' = H_{p_i}, \quad p_i' = -H_{q_i}.$$

These are the canonical equations. □

**Lagrangian manifolds:** Let $D$ be an open set in $\mathbb{R}^n$. An $n$-dimensional surface

$$\mathscr{F} : q = q(C), \quad p = p(C), \quad C \in D,$$

[143] The symbol $\mathrm{v} \lrcorner \omega$ denotes the so-called *inner product* of v with $\omega$. This is a linear functional, defined by means of the relation

$$(\mathrm{v} \lrcorner \omega)(w) = \omega(\mathrm{v}, w) \quad \text{for all} \quad w \in \mathbb{R}^{2n}.$$

Instead of $\mathrm{v} \lrcorner \omega$ one also writes $i_{\mathrm{v}}(\omega)$.

which has a tangent plane at each point,[144] is called a *Lagrangian manifold*, if $\omega$ vanishes on $\mathscr{F}$, i.e., one has

$$\boxed{\omega_P(\mathrm{v}, \mathrm{w}) = 0, \quad P \in \mathscr{F},} \tag{1.366}$$

for all tangent vectors v and w of $\mathscr{F}$ in the point $P$.[145] Geometrically this means that each tangent space of $\mathscr{F}$ is isotropic with respect to the symplectic form induced by $\omega$ on it (cf. 3.9.8).

**Invariance:** Lagrangian manifolds are transformed to Lagrangian manifolds under symplectic transformations.

### 1.13.1.8 Poisson brackets and integrable systems

We consider the canonical equations

$$\boxed{p' = -H_q(q,p), \quad q' = H_p(q,p)} \tag{1.367}$$

for the sought-for motion $q = q(t)$, $p = p(t)$ with $q,\ p \in \mathbb{R}^n$. Our goal is to find conditions for the system (1.367) to have solutions, which after an appropriate change of coordinates are of the form

$$\boxed{\varphi_j(t) = \omega_j t + \text{const}, \quad j = 1, \ldots, n.} \tag{1.368}$$

Here $\varphi_j$ are angle coordinates of period $2\pi$, i.e., $(\varphi_1, \ldots, \varphi_n)$ and $(\varphi_1 + 2\pi, \ldots, \varphi_n + 2\pi)$ describe the same state of the system.

**Quasi-periodic motions:** In (1.368) each coordinate corresponds to a periodic motion with the angular frequency of $\omega_j$. As these frequencies $\omega_1, \ldots, \omega_n$ may very well be different, the motion as a whole is referred to as *quasi-periodic*. The set

$$T := \{\varphi \in \mathbb{R}^n \,|\, 0 \le \varphi_j \le 2\pi,\ j = 1, \ldots, 2\pi\}$$

is called an $n$-dimensional torus. In this description, the boundary points are identified, if the coordinates $\varphi_j$ coincide or if they differ by integral multiples of $2\pi$. The motion (1.368) takes place along an $n$-dimensional torus $T$.

*Example 1:* For $n = 2$ the Figure 1.157(a) pictures the situation. Here one identifies in a natural manner the points which lie on opposite sides of the rectangle. If these points of $T$ are glued together, then one gets the geometric torus $\mathscr{T}$ depicted in Figure 1.157(b).

(i) If the ratio $\omega_1/\omega_2$ is a *rational* number, then the trajectory $\varphi_1 = \omega_1 t + \text{const}$, $\varphi_2 = \omega_2 t + \text{const}$ consists of finitely many pieces, which return to the starting position. The corresponding curve on $\mathscr{T}$ is a closed curve, wrapping around $\mathscr{T}$ finitely many times before closing.

(ii) If $\omega_1/\omega_2$ is an *irrational* number, then the trajectory covers both $T$ and $\mathscr{T}$ densely without returning to its starting point (Fig 1.157(c)).

[144] This means $(q'(C), p'(C)) = n$ on $G$.

[145] If one uses the Lagrangian brackets introduced in 1.13.1.3, then (1.366) is equivalent to the equation

$$[C_j, C_k](P) = 0 \quad \text{for all} \quad P \in \mathscr{F} \quad \text{and all} \quad j, k.$$

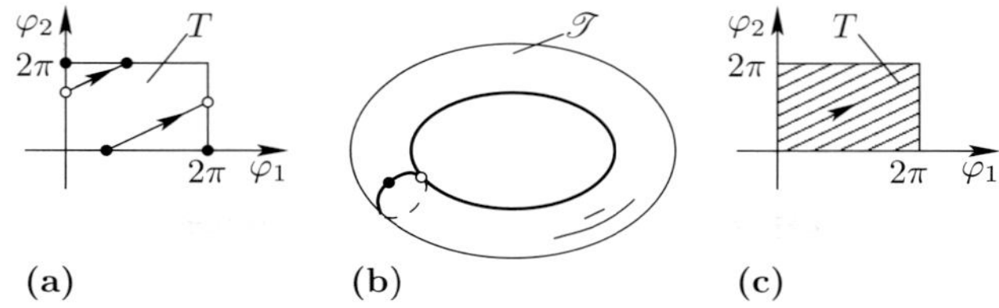


*Figure 1.157. Quasi-periodic motion in coordinates as in (1.368).*

**Poisson brackets:**[146] For two smooth functions $f = f(q,p)$ and $g = g(q,p)$ the *Poisson bracket* is defined by the formula

$$\{f,g\} := \sum_{j=1}^{n} \frac{\partial f}{\partial p_j}\frac{\partial g}{\partial q_j} - \frac{\partial f}{\partial q_j}\frac{\partial g}{\partial p_j}.$$

**Liouville's (1809–1882) theorem:** Let $n$ smooth conserved quantities $F_1, \ldots, F_n : \mathbb{R}^{2n} \longrightarrow \mathbb{R}$ of the canonical equations (1.367) be given with $F_1 = H$, which are *in involution*, i.e., for which

$$\{F_j, F_k\} \equiv 0 \quad \text{for} \quad j,k = 1,\ldots,n.$$

Moreover we assume that the set $M_\alpha$ of all points $(q,p) \in \mathbb{R}^{2n}$ with

$$F_j(q,p) = \alpha_j, \quad j = 1,\ldots,n,$$

for fixed $\alpha \in \mathbb{R}^n$ form a compact, connected $n$-dimensional manifold, i.e., the matrix $(\partial_k F_j)$ of first partial derivatives has rank $n$ at every point of $M_\alpha$.

Then $M_\alpha$ is diffeomorphic to a $n$-dimensional torus $T$, where the trajectories $q = q(t)$, $p = p(t)$ which are solutions of the canonical equations (1.367) represent the quasi-periodic motion (1.368) on $M_\alpha$.

**The foliation by invariant tori:** Suppose there is an open neighborhood $U$ of $\alpha$, so that a neighborhood of $M_\alpha$ in the phase space $\mathbb{R}^{2n}$ is diffeomorphic to the product

$$\boxed{T \times U.}$$

Here the set $T \times \{I\}$ with $I \in U$ is diffeomorphic to $M_I$. In particular, the parameter value $I = \alpha$ belongs to $M_\alpha$. Through this diffeomorphism

$$\varphi = \varphi(q,p), \quad I = I(q,p) \tag{1.369}$$

the original canonical equations (1.367) transform into the new canonical equations

$$\boxed{I' = -\mathscr{H}_\varphi(I) = 0, \quad \varphi' = \mathscr{H}_I(I).} \tag{1.370}$$

[146]One can construct a Poissonian mechanics parallel to Hamiltonian mechanics, which utilizes the Poisson brackets (cf. 5.1.3). Poissonian mechanics is based on the fact that the vector fields on a manifold form a Lie algebra, while Hamiltonian mechanics takes advantage of the fact that the cotangent bundle of a manifold has a natural symplectic structure (cf. [212]). Poissonian mechanics plays a key role in the quantization of classical mechanics as carried out by Heisenberg in 1924, which was the birth of quantum mechanics.

The solution this yields is

$$I_j = \text{const}, \quad \varphi_j = \omega_j t + \text{const}, \quad j = 1, \ldots, n. \tag{1.371}$$

Here $\omega_j := \partial H(I)/\partial I_j$. The variables $I_j$ are called action variables; together with the angle variables $\varphi_j$ they form the set of *action-angle variables* of the integrable Hamiltonian system.

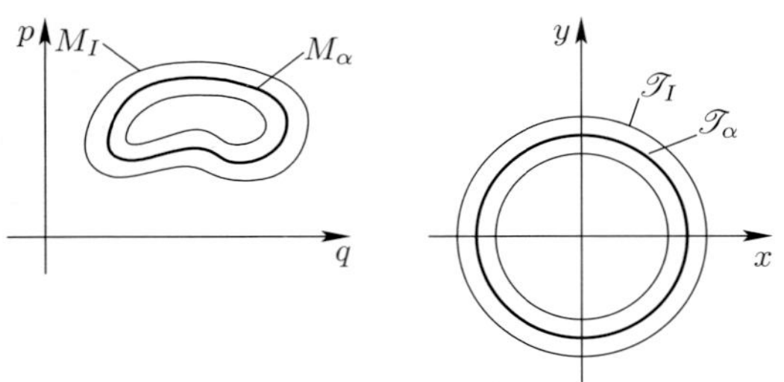


*Figure 1.158. Invariant tori and action-angle variables.*

The curve (1.371) corresponds to a motion on $M_I$. Here $M_I$ is the set of all points $(q,p) \in \mathbb{R}^{2n}$ in phase space which satisfy

$$F_j(q,p) = I_j, \quad j = 1, \ldots, n.$$

We refer to $M_I$ as an *invariant torus.* If $I$ is in a sufficiently small neighborhood of $\alpha$, then $M_I$ is obtained from $M_\alpha$ by a small deformation.

*Example 2:* In case $n = 1$, the situation is as in Figure 1.158. Here one has $T := \{\varphi \in \mathbb{R} : 0 \leq \varphi \leq 2\pi\}$, where the points $\varphi = 0$ and $\varphi = 2\pi$ are identified. The closed curves in $(q,p)$-phase space map to the system of circles

$$x = I\cos\varphi, \quad y = I\sin\varphi$$

of which the radii are $I \in U$. Thus $M_I$ corresponds to the circle $\mathscr{T}_I$ of radius $I$ (action variable).

### 1.13.1.9 Perturbations of integrable systems (KAM theory)

The decisive question here is: How does an integrable system behave under a small perturbation? The natural answer, that the situation is only slightly deformed, is unfortunately false. The reason is that resonances between the angular frequencies $\omega_1, \ldots, \omega_n$ can occur.

Instead of the integrable system (1.370) we consider the perturbed system

$$\boxed{I' = -H_\varphi(I,\varphi,\varepsilon), \quad \varphi' = H_I(I,\varphi,\varepsilon)} \tag{1.372}$$

with the perturbed Hamiltonian $H := \mathscr{H}(I) + \varepsilon\mathscr{H}_*(I,\varphi)$ for a small parameter $\varepsilon$. The Kolmogorov–Arnol'd–Moser theory (KAM theory), which was initiated by Kolmogorov in 1953 and extended significantly some years later by V. I. Arnol'd and J. Moser, is concerned with the behavior of the perturbed system (1.372).

**Definition:** An invariant torus $T \times \{I\}$ on which the system of solutions (1.371) lives, is called *resonant*, if and only if there are rational numbers $r_1, \ldots, r_n$, not all of which vanish, such that

$$r_1\omega_1 + \ldots + r_n\omega_n = 0.$$

The following results are formulated for *non-degenerate* systems, i.e., for which the determinant of the second partial derivatives $\det(\partial^2\mathscr{H}(I_0)/\partial I_j\partial I_k)$ of the unperturbed Hamiltonian in non-vanishing.

**Theorem:** *If the perturbation parameter $\varepsilon$ is sufficiently small, then most of the non-resonant tori of the unperturbed system remain only slightly deformed, and the qualitative behavior of the trajectories on these tori is unaffected by the perturbation.*

The intricacy of the situation lies in the fact that certain non-resonant tori *can* be destroyed under the perturbation. Moreover, in the unperturbed system the set of non-resonant and resonant tori is not clearly separated, i.e., in a arbitrarily small neighborhood of non-resonant tori there may be resonant tori.

> Under the smallest possible perturbation, the qualitative behavior of the trajectories on the invariant tori may change dramatically. It is even possible that chaotic motion, in a sense the opposite of integrable motion, may ensue.

**Application to the stability of the solar system:** If one in a first approximation neglects the gravitational force between the planets, as well as the force of the planets on the sun, then each planet moves in a periodic orbit around the sun (an ellipse with the sun at one of the focal points, the first Kepler law, cf. 1.12.5.2) with different periods (frequencies). This situation corresponds to a quasi-periodic motion. Once the interactions of the planets are included in the model, one obtains a perturbation of this quasi-periodic motion. According to the KAM theory it is in principle impossible to prove the stability of the solar system for all times, as this behavior depends decisively on the initial values, which can only be known up to a certain precision.

Many aspects of classical and modern celestial mechanics can be found in the encyclopedia volume [121] as well as in [122].

#### 1.13.1.10 Gibbs' equations in thermodynamics

**First law of thermodynamics:**

$$\boxed{E'(t) = Q'(t) + A'(t).} \tag{1.373}$$

**Second law of thermodynamics:**

$$\boxed{Q'(t) \leq T(t)S'(t).} \tag{1.374}$$

These equations describe the behavior of the temporal development of general thermodynamical systems. These are systems which consist of a large number of particles (for example molecules or photons). The quantities appearing in these equations are:

| | |
|---|---|
| $Q(t)$ | thermal energy, which is added to the system in the time interval $[0, t]$; |
| $A(t)$ | work done on the system in the time interval $[0, t]$; |
| $E(t), S(t), T(t)$ | inner energy, entropy, and the absolute temperature, respectively, of the system at the time $t$. |

If $Q'(t) = T(t)S'(t)$ for all times $t$, then the process is called *reversible*. Otherwise the process is called *irreversible*.

If the system is closed, then this means in particular that no thermal energy is added from outside the system, i.e., $Q(t) \equiv 0$. From the second law (1.374) it follows that in this case

$$S'(t) \geq 0.$$

Hence one has:

In a closed thermodynamic system, the entropy is non-decreasing.

In a closed system, a process is reversible if and only if $S'(t) = 0$, i.e., the entropy is *constant.*

The following equation characterizes an important class of thermodynamical systems.

**Gibbs' law:**

$$\mathrm{d}E = T\,\mathrm{d}S - p\,\mathrm{d}V + \sum_{j=1}^{r} \mu_j \mathrm{d}N_j. \tag{1.375}$$

This equation holds for thermodynamical systems, whose state can be characterized by the following parameters:

$T$: absolute temperature, $V$: volume, $N_j$: the number of particles in the $j^{th}$ substance.

The other quantities are then functions of these parameters:

$$\begin{aligned} &E = E(T,V,N) && \text{(inner energy)},\\ &S = S(T,V,N) && \text{(entropy)},\\ &p = p(T,V,N) && \text{(pressure)},\\ &\mu_j = \mu_j(T,V,N) && \text{(chemical potential of the } j^{th} \text{ substance)}. \end{aligned}$$

Here one has $N = (N_1, \ldots, N_r)$. The equation (1.375) is equivalent to the system of partial differential equations of the first order:

$$E_T = TS_T, \quad E_V = TS_V - p, \quad E_{N_j} = TS_{N_j} + \mu_j, \quad j = 1, \ldots, r.$$

A thermodynamical process is described in this situation by the equation

$$T = T(t), \quad V = V(t), \quad N = N(t), \quad t_0 \leq t \leq t_1. \tag{1.376}$$

This includes the functions

$$E(t) := E(\mathscr{P}(t)), \quad S(t) := S(\mathscr{P}(t)) \tag{1.377}$$

with $\mathscr{P}(t) := (T(t), V(t), N(t))$. Moreover we obtain $Q = Q(t)$ and $A = A(t)$ by integrating

$$\begin{aligned} Q'(t) &= T(t)S'(t),\\ A'(t) &= -p(\mathscr{P}(t))V'(t) + \sum_{j=1}^{r} \mu_j(\mathscr{P}(t))N_j'(t) \end{aligned} \tag{1.378}$$

with $Q(t_0) = A(t_0) = 0$.

**Theorem 1:** If one knows a solution of the basic Gibbs equation (1.375), then the thermodynamical process (1.376) – (1.378) satisfies the first and second law of thermodynamics. This process is reversible.

**The special case of a gas or liquid:** We consider the system which consists of $N$ molecules of some sort of particle with the molecular mass of $m$. Then $M = Nm$ is the total mass. Gibbs' equation in this case amounts to

$$\boxed{\mathrm{d}E = T\,\mathrm{d}S - p\,\mathrm{d}V + \mu\,\mathrm{d}N.} \tag{1.379}$$

We introduce the following quantities:

$$\rho := \frac{M}{V} \quad \textit{mass density,}$$

$$e := \frac{E}{M} \quad \textit{specific inherent energy,}$$

$$s := \frac{S}{M} \quad \textit{specific entropy.}$$

Then $e$, $s$, $p$ and $\mu$ are functions of $T$ and $\rho$. Moreover we define the *specific heat* by the relation $c(T,\rho) := e_T(T,\rho)$.

**Theorem 2:** For $T > 0$ and $\rho > 0$ suppose we are given two smooth functions

$$\boxed{p = p(T,\rho) \quad \text{and} \quad c = c(T,\rho),}$$

where the constraint $c_\rho = -p_{TT}T/\rho^2$ is satisfied. We also assume we are given values $e(T_0,\rho_0)$ and $s(T_0,\rho_0)$. Then the uniquely determined solution of the Gibbs' equations (1.379) is:

$$e(T,\rho) = e(T_0,\rho_0) + \int_{(T_0,\rho_0)}^{(T,\rho)} c\,\mathrm{d}T + \rho^{-2}(p - p_T T)\mathrm{d}\rho,$$

$$s(T,\rho) = s(T_0,\rho_0) + \int_{(T_0,\rho_0)}^{(T,\rho)} T^{-1}c\,\mathrm{d}T - \rho^{-2}p_T\,\mathrm{d}\rho,$$

$$\mu(T,\rho) = e(T,\rho) - Ts(T,\rho) + \frac{p(T,\rho)}{\rho}.$$

All of these curve integrals are independent of the path of integration.

**Remark:** The state condition $p = p(T,\rho)$ and the specific heat $c(T,\rho)$ have to be determined experimentally. Then all other thermodynamical quantities $e$, $s$ and $\mu$ are determined from these.

*Example 1:* For an ideal gas at room temperature $T$ one has:

$$p = r\rho T \quad \text{(equation of state)}, \qquad c = \frac{\alpha r}{2} \quad \text{(specific heat)}.$$

Here $r$ is called the gas constant, and $\alpha$ corresponds to the excited degrees of freedom (usually one has $\alpha = 3, 5, 6$, respectively, for gases consisting of one, resp. two resp. $n$ with $n \geq 3$ atoms). From this we obtain the relations

$$e = cT + \text{const}, \qquad s = c\ln(T\rho^{1-\gamma}) + \text{const},$$

$$\mu = e - Ts + \frac{p}{\rho} \qquad \text{(chemical potential)}$$

with $\gamma := 1 + r/c$.

**The Legendre transformation and thermodynamical potentials:** In thermodynamics one often requires a change of variables. This can be done quite elegantly with the help of Gibbs' law.

*Example 2:* Gibbs' equation

$$\mathrm{d}E = T\,\mathrm{d}S - p\,\mathrm{d}V + \mu\,\mathrm{d}N$$

shows that $S$, $V$ and $N$ are natural variables of the *inner energy*. From $E = E(S, V, N)$ it follows that

$$T = E_S, \quad p = -E_V, \quad \mu = E_N.$$

Because of $E_{SV} = E_{VS}$ etc. we get from this the *integrability conditions*

$$T_V(\mathscr{P}) = -p_S(\mathscr{P}), \quad p_N(\mathscr{P}) = -\mu_V(\mathscr{P}), \quad T_N(\mathscr{P}) = \mu_S(\mathscr{P}),$$

where we have set $\mathscr{P} := (S, V, N)$.

*Table 1.7. Important thermodynamical potentials.*

| ***Potential*** | ***Total differential*** | ***Natural variable*** | ***Interpretation of the derivative*** |
|---|---|---|---|
| inner energy $E$ | $\mathrm{d}E = T\,\mathrm{d}S - p\,\mathrm{d}V + \mu\,\mathrm{d}N$ | $E(S, V, N)$ | $E_S = T,\ E_V = -p,$ $E_N = \mu$ |
| free energy $F = E - TS$ | $\mathrm{d}F = -S\,\mathrm{d}T - p\,\mathrm{d}V + \mu\,\mathrm{d}N$ | $F(T, V, N)$ | $F_T = -S,\ F_V = -p,$ $F_N = \mu$ |
| entropy S | $T\,\mathrm{d}S = \mathrm{d}E + p\,\mathrm{d}V - \mu\,\mathrm{d}N$ | $S(E, V, N)$ | $TS_E = 1,\ TS_V = p,$ $TS_N = -\mu$ |
| enthalpy $H = E + pV$ | $\mathrm{d}H = T\,\mathrm{d}S - V\,\mathrm{d}p + \mu\,\mathrm{d}N$ | $H(S, p, N)$ | $H_S = T,\ H_p = -V,$ $H_N = \mu$ |
| free enthalpy $G = F + pV$ | $\mathrm{d}G = -S\,\mathrm{d}T - V\,\mathrm{d}p + \mu\,\mathrm{d}N$ | $G(T, p, N)$ | $G_T = -S,\ G_p = -V,$ $G_N = \mu$ |
| statistical potential $\Omega = F - \mu N$ | $\mathrm{d}\Omega = -S\,\mathrm{d}T - p\,\mathrm{d}V - N\,\mathrm{d}\mu$ | $\Omega(T, V, \mu)$ | $\Omega_T = -S,\ \Omega_V = -p,$ $\Omega_\mu = -N$ |

*Example 3:* The function $F := E - TS$ is called the *free energy*. Because of the relation $\mathrm{d}F = \mathrm{d}E - T\mathrm{d}S - S\mathrm{d}T$ one has

$$\mathrm{d}F = -S\,\mathrm{d}T - p\,\mathrm{d}V + \mu\,\mathrm{d}N.$$

Hence $T$, $V$ and $N$ are the most natural variables of $F$. From $F = F(T, V, N)$ we then get

$$S = -F_T, \quad p = -F_V, \quad \mu = F_N.$$

One calls $E$ and $F$ *thermodynamic potentials*, as one can obtain all other thermodynamical quantities from these by differentiation. Further thermodynamical potentials are listed in Table 1.7.

### 1.13.1.11 The contact transformations of Lie

*By pursuing Plücker's ideas about the changing of elements of space, I arrived in 1868 at the general notion of a contact transformation.*

*Sophus Lie* (1842–1899)

In mathematics one can often simplify problems by carrying out an appropriate transformation. For differential equations the contact transformations turn out to be the proper kind of transformation to achieve this. These are a generalization of the Legendre transformation, whose geometric interpretation was discussed in 1.12.1.14. It is important that:

> Under contact transformations solutions of differential equations are carried over to solutions of the transformed differential equations.

In addition to the traditional transformation of dependent and independent variables, in contact transformations the *derivatives* of the variables can be used as independent variables.

**Definition:** Let $x = (x_1, \ldots, x_n)$ and $p = (p_1, \ldots, p_n)$. Moreover let $X = (X_1, \ldots, X_n)$ and $P = (P_1, \ldots, P_n)$. A *contact transformation*

$$\boxed{X = X(x,u,p), \quad P = P(x,u,p), \quad U = U(x,u,p)} \tag{1.380}$$

is a diffeomorphism of an open set $G$ of $\mathbb{R}^{2n+1}$ to an open set $\Omega$ of $\mathbb{R}^{2n+1}$, such that the relation

$$\boxed{\mathrm{d}U - \sum_{j=1}^{n} P_j \mathrm{d}X_j = \rho(x,u,p)\left(\mathrm{d}u - \sum_{j=1}^{n} p_j \mathrm{d}x_j\right)} \tag{1.381}$$

is satisfied in $G$, where the smooth function $\rho$ is non-vanishing on $G$.

**Theorem:** If $u = u(x)$ is a solution of the differential equation

$$f(x,u,u') = 0 \tag{1.382}$$

with $u' = (u_{q_1}, \ldots, u_{q_n})$, then $U = U(X)$ is a solution of the differential equation

$$F(X,U,U') = 0,$$

which is obtained from (1.382) with the help of the contact transformation (1.380), where we have set

$$p_j = \frac{\partial u}{\partial x_j}, \quad j = 1, \ldots, n.$$

Then one obtains also $P_j = \partial U/\partial X_j$, $j = 1, \ldots, n$.

**The general Legendre transformation:**

$$\boxed{\begin{aligned} &U = \sum_{j=1}^{k} p_j x_j - u, \quad X_j = p_j, \quad P_j = x_j, \quad j = 1, \ldots, k, \\ &X_r = x_r, \quad P_r = p_r, \quad r = k+1, \ldots, n. \end{aligned}} \tag{1.383}$$

$k$ can take on any value $1 \leq k \leq n$. For $k = n$ the last row is vacant. From the product rule $\mathrm{d}(p_j x_j) = p_j \mathrm{d}x_j + x_j \mathrm{d}p_j$, (1.383) follows for $\rho = -1$. Thus (1.383) is a contact transformation.

*Example 1:* The Legendre transformation of thermodynamics is obtained from (1.383) for $k = 1$. For example in this case we have $E = u$ (inner energy) and $F = -U$ (free energy) (cf. 1.13.1.10).

*Example 2:* The Legendre transformation of mechanics is (1.383) with the Lagrange function $L = u$ and the Hamiltonian $H = U$ (cf. 5.1.3). In this case one has $x_j = q'_j$ (velocity coordinates).

## 1.13.2 Equations of mathematical physics of the second order

*The equations for the flow of heat as well as those for the oscillations of acoustic bodies and of fluids belong to an area of analysis which has recently been opened, and which is worth examining in the greatest detail.*

*Jean Baptiste Joseph Fourier,*
*Théorie analytique de la chaleur*[147]*, 1822*

### 1.13.2.1 The universal Fourier method

The basic idea of the Fourier method consists in representing the solutions of partial differential equations of second order in the form

$$u(x,t) = \sum_{k=0}^{\infty} a_k(x) b_k(t). \tag{1.384}$$

In important cases the term $a_k(x)b_k(t)$ corresponds to a characteristic oscillation of the physical system. The following general principle:

> The development in time of many physical systems is given as the superposition of characteristic states (for example, of characteristic oscillations).

is hidden behind (1.384).

This principle was used first by Daniel Bernoulli in 1730 to treat oscillations of rods and strings. The sound of every music instrument as well as of every singing voice is described by expressions of the form (1.384), where $a_k(x)b_k(t)$ represent the fundamental and higher tones, the intensity of which determines the quality of the sound. Interestingly, Euler did not believe the claim made by Daniel Bernoulli that one can obtain the general time development with the help of (1.384). One must remember that, at this time, there was no accepted general notion of a function and convergence of infinite series.

In his work *Analytic theory of heat*, which appeared in 1822, the method (1.384) of Fourier was developed as an important tool in mathematical physics. However, it wasn't until the beginning of the twentieth century that one obtained a deeper understanding of this method through application of methods of functional analysis. This will be discussed in more detail in [212].

[147] The analytic theory of heat.

#### 1.13.2.2 Applications to vibrating strings

$$\boxed{\begin{array}{lll} \frac{1}{c^2}u_{tt} - u_{xx} = 0, & 0 < x < L, t > 0 & \text{(differential equation)}, \\ u(0,t) = u(L,t) = 0, & t \geq 0 & \text{(boundary value)}, \\ u(x,0) = u_0(x), & 0 \leq x \leq L & \text{(initial position)}, \\ u_t(x,0) = u_1(x), & 0 \leq x \leq L & \text{(initial velocity)}. \end{array}} \tag{1.385}$$

This problem describes the motion of a string of the length $L$, which is fixed at the ends. The function $u$ has the interpretation: $u(x,t)$ = the displacement of the string at the point $x$ at time $t$ (Figure 1.159). The number $c$ corresponds to the velocity of propagation of the string waves.

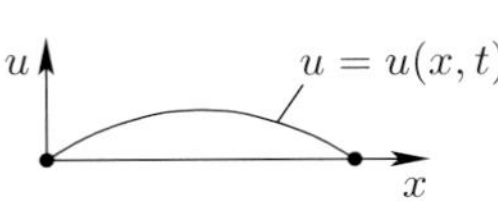


*Figure 1.159.*

In order to simplify the notions, we let $L = \pi$ and $c = 1$.

**Existence and uniqueness result:** Let $u_0$ and $u_1$ be a given smooth, odd function of period $2\pi$. Then the problem (1.385) has the unique solution

$$\boxed{u(x,t) = \sum_{k=1}^{\infty} (a_k \sin kt + b_k \cos kt) \sin kx.} \tag{1.386}$$

The notations are such that

$$u_0(x) = \sum_{k=1}^{\infty} b_k \sin kx, \quad u_1(x) = \sum_{k=1}^{\infty} ka_k \sin kx, \tag{1.387}$$

that is, the coefficients $b_k$ (resp. $ka_k$) are the Fourier coefficients of $u_0$ (resp. $u_1$). Explicitly this means

$$b_k = \frac{2}{\pi} \int_0^{\pi} u_0(x) \sin kx \, dx, \quad a_k = \frac{2}{k\pi} \int_0^{\pi} u_1(x) \sin kx \, dx.$$

**Physical interpretation:** The solution (1.386) corresponds to a superposition of characteristic oscillations

$$u(x,t) = \sin kt \sin kx \quad \text{and} \quad u(x,t) = \cos kt \sin kx$$

of the suspended string with the angular frequency $\omega = k$.

The following considerations are typical of applications of the Fourier method.

**Motivation for the given solution:** (i) We first seek *special solutions* of the original problem (1.385) in the form of a *product*:

$$\boxed{u(x,t) = \varphi(x)\psi(t).}$$

(ii) The *initial conditions* $u(0,t) = u(\pi,t) = 0$ can be satisfied by setting:

$$\varphi(0) = \varphi(\pi) = 0$$

(iii) *The $\lambda$-trick:* From the differential equation $u_{tt} - u_{xx} = 0$ we obtain

$$\varphi(x)\psi''(t) = \varphi''(x)\psi(t).$$

This equation can be satisfied by setting

$$\frac{\psi''(t)}{\psi(t)} = \frac{\varphi''(x)}{\varphi(x)} = \lambda$$

with an unknown real number $\lambda$. In this way we obtain the two equations

$$\boxed{\varphi''(x) = \lambda\varphi(x), \quad \varphi(0) = \varphi(\pi) = 0} \tag{1.388}$$

and

$$\psi''(t) = \lambda\psi(t). \tag{1.389}$$

One calls (1.388) the *boundary–eigenvalue problem* with the eigenvalue parameter $\lambda$.

(iv) *Non-trivial solutions* of (1.388) are

$$\boxed{\varphi(x) = \sin kx, \quad \lambda = -k^2, \qquad k = 1, 2, \ldots}$$

If we set $\lambda = -k^2$ in (1.389), then we obtain the solutions

$$\psi(t) = \sin kt, \quad \cos kt.$$

(v) *Superposition* of these special solutions yields

$$u(x,t) = \sum_{j=1}^{\infty}(a_k \sin kt + b_k \cos kt) \sin kx$$

with unknown coefficients $a_k$ and $b_k$. Differentiating with respect to $t$ yields

$$u_t(x,t) = \sum_{k=1}^{\infty}(ka_k \cos kt - kb_k \sin kt) \sin kx.$$

(vi) From the *initial conditions* $u(x,0) = u_0$ and $u_t(x,0) = u_1(x)$ we then get the equation (1.387) for the determination of the $a_k$ and $b_k$.

#### 1.13.2.3 Applications to a rod conducting heat

$$\boxed{\begin{array}{lll} T_t - \alpha T_{xx} = 0, & 0 < x < L, t > 0 & \text{(differential equation)},\\ T(0,t) = T(L,t) = 0, & t \geq 0 & \text{(boundary temperature)},\\ T(x,0) = T_0(x), & 0 \leq x \leq L & \text{(initial temperature)}. \end{array}} \tag{1.390}$$

This problem describes the distribution of the temperature in a rod of length $L$. The function $T$ has the interpretation: $T(x,t) =$ the temperature of the rod at the point $x$ at the time $t$. The positive number $\alpha$ is a material constant.

For simplification of the calculations set $L = \pi$ and $\alpha = 1$.

**Existence and uniqueness result:** Let a smooth, odd function $T_0$ of period $2\pi$ be given. Then the problem (1.390) has the unique solution:

$$\boxed{T(x,t) = \sum_{k=1}^{\infty} b_k \mathrm{e}^{-k^2 t} \sin kx.}$$

Here one has

$$T_0(x) = \sum_{j=1}^{\infty} b_k \sin kx,$$

i.e., the coefficients $b_k$ are the Fourier coefficients of $T_0$. Explicitly this means

$$b_k = \frac{2}{\pi} \int_0^{\pi} T_0(x) \sin kx \; dx.$$

This result is analogous to 1.13.2.2.

#### 1.13.2.4 The instantaneous heat equation

$$\begin{array}{lll} s\mu T_t - \kappa \Delta T = 0, & x \in \Omega, t > 0 & \text{(differential equation)}, \\ T(x,t) = T_0(x), & x \in \partial\Omega, t \geq 0 & \text{(boundary temperature)}, \\ T(x,0) = T_1(x), & x \in \Omega & \text{(initial temperature)}. \end{array} \tag{1.391}$$

This problem describes the distribution of temperature in a bounded domain $\Omega$ of $\mathbb{R}^3$ with smooth boundary $\partial\Omega$. The function $T$ has the meaning: $T(x,t) =$ the temperature at the point $x = (x_1, x_2, x_3)$ at the time $t$. The physical meaning of the constants $s$, $\mu$ and $\kappa$ can be found in (1.170). The operator

$$\Delta T := \sum_{j=1}^{3} \frac{\partial^2 T}{\partial x_j^2} \tag{1.392}$$

is called the *Laplace operator.*

**Existence and uniqueness theorem:** Let two smooth functions $T_0$ and $T_1$ be given. Then the problem (1.391) has a unique solution. This solution is smooth.

**Heat sources:** A similar result holds if one replaces the differential equation in (1.391) by

$$s\mu T_t - \kappa \Delta T = f, \qquad x \in \Omega, \quad t > 0 \tag{1.393}$$

with a smooth function $f = f(x,t)$. The function $f$ describes heat sources (cf. (1.170)).

**The initial value problem for all of space:**

$$\begin{array}{lll} T_t - \alpha \Delta T = 0, & x \in \mathbb{R}^3, t > 0 & \text{(differential equation)}, \\ T(x,0) = T_0(x), & x \in \mathbb{R}^3 & \text{(initial temperature)}. \end{array} \tag{1.394}$$

**Existence and uniqueness result:** Let a continuous, bounded function $T_0$ be given. Then the problem (1.394) has the unique solution

$$T(x,t) = \frac{1}{(4\pi\alpha t)^{3/2}} \int_{\mathbb{R}^3} \exp\left(\frac{-|x-y|^2}{4\alpha t}\right) T_0(y) \; dy \tag{1.395}$$

for all $x \in \mathbb{R}^3$ and $t > 0$.[148] Moreover, one has

$$\lim_{x \to +0} T(x,t) = T_0(x) \qquad \text{for all} \quad x \in \mathbb{R}^3.$$

**Remark:** The solution $T$ in (1.395) is smooth for all times $t > 0$, even though the initial temperature $T_0$ at time $t = 0$ is only continuous. This smoothing out effect if typical for all flow processes (like heat conduction and diffusion).

#### 1.13.2.5 The instantaneous diffusion equation

The equation (1.391) also describes diffusion processes. In that case $T$ is the density of the number of particles (number of molecules per volume). Similarly, (1.394) and (1.395) describe diffusion processes in $\mathbb{R}^3$.

*Example:* The initial density of particles $T_0$ in (1.394) is concentrated near the origin, i.e., one has

$$T_0(x) = \begin{cases} \dfrac{3N}{4\pi\varepsilon^3} & \text{for } |x| \leq \varepsilon, \\ 0 & \text{otherwise.} \end{cases}$$

The corresponds to precisely $N$ particles near the origin. It then follows from (1.395) after passing to the limit $\varepsilon \to 0$ that the solution is

$$\boxed{T(x,t) = \frac{N}{(4\pi\alpha t)^{3/2}} \exp\left(\frac{-|x|^2}{4}\right) \alpha t, \qquad t > 0,\, x \in \mathbb{R}^3,} \tag{1.396}$$

where $T$ denotes the density of the number of particles. The particles, which are initially concentrated near the origin, diffuse into the entire space. From a microscopic point of view this is a stochastic process for the Brownian motion of the particles (cf. 6.4.4).

#### 1.13.2.6 The stationary heat equation

If the temperature $T$ does not depend on the time $t$, then the stationary heat equation is obtained from the instantaneous heat equation (1.393), leading to

$$\boxed{-\kappa \Delta T = f, \quad x \in \Omega,} \tag{1.397}$$

which is also called the *Poisson equation*. The heat current density vector is given by

$$\mathbf{J} = -\kappa\, \mathbf{grad}\, T.$$

Here $\Omega$ is a bounded domain in $\mathbb{R}^3$ with smooth boundary $\partial\Omega$. In addition to the differential equation (1.397) one can consider three different kinds of boundary conditions.

(i) First boundary condition:

$$\boxed{T = T_0 \quad \text{on} \quad \partial\Omega.}$$

[148] In order to insure the uniqueness of the solution, one must require in addition

$$\sup_{x \in \mathbb{R}^3, 0 \leq t \leq \tau} |T(x,t)| < \infty$$

for all $\tau > 0$, i.e., the temperature must remain bounded in the time interval $[0, \tau]$.

(ii) Second boundary condition:

$$\boxed{\mathbf{Jn} = g \qquad \text{on} \quad \partial\Omega.}$$

Here $\mathbf{n}$ denotes the outer normal vector along the boundary $\partial\Omega$.

(iii) Third boundary condition:

$$\boxed{\mathbf{Jn} = hT + g \qquad \text{on} \quad \partial\Omega.}$$

Here we assume $h > 0$ on the boundary $\partial\Omega$. Moreover one has

$$\mathbf{Jn} \equiv -\kappa \frac{\partial T}{\partial n} \qquad \text{on} \quad \partial\Omega.$$

**Physical interpretation:** In the first boundary condition the boundary temperature is known, while in the latter two the outer normal component of the heat conduction density vector is known on the boundary $\partial\Omega$ (Figure 1.160).

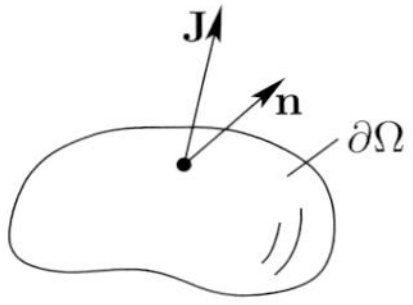

*Figure 1.160.*

**Existence and uniqueness result:** Let $F$, $g$ and $h$ be given smooth functions.

(i) The first and third boundary value problems for the Poisson equation (1.397) are uniquely solvable.

(ii) The second boundary value problem for the Poisson equation (1.397) is solvable if and only if

$$\int_{\Omega} f \, \mathrm{d}V = \int_{\partial\Omega} g \, \mathrm{d}F.$$

The solution $T$ is then unique up to an additive constant.

**Variational principles:** (i) Every smooth solution of the minimum problem

$$\begin{gathered} \int_{\Omega} \left( \frac{\kappa}{2} (\mathbf{grad}\, T)^2 - FT \right) \mathrm{d}x \stackrel{!}{=} \min., \\ T = T_0 \qquad \text{on} \quad \partial\Omega, \end{gathered} \tag{1.398}$$

is a solution of the first boundary value problem for the Poisson equation (1.397).

(ii) Every smooth solution of the minimum problem

$$\int_{\Omega} \left( \frac{\kappa}{2} (\mathbf{grad}\, T)^2 - fT \right) \mathrm{d}x + \int_{\partial\Omega} \left( \frac{1}{2} hT^2 + gT \right) \mathrm{d}F \stackrel{!}{=} \min.$$

is a solution of the third boundary value problem for the Poisson equation (1.397). In case $h \equiv 0$, the second boundary value problem also has a solution.

**The first boundary value problem for a ball $B_R$ of radius $R$:**

$$\boxed{\Delta T = 0 \qquad \text{on} \quad B_R, \qquad T = T_0 \quad \text{on} \quad \partial B_R.} \tag{1.399}$$

Let $B_R$ be an open ball in $\mathbb{R}^3$ of radius $R$ centered at the origin. If $T_0$ is continuous on the boundary $\partial B_R$, then the problem (1.399) has a unique solution

$$T(x) = \frac{1}{4\pi R} \int\limits_{\partial K_R} \frac{R^2 - |x|^2}{|x-y|^3} T_0(y) \, \mathrm{d}F_y \quad \text{for all } x \in B_R.$$

The function $T$ is continuous on the closed ball $\overline{B}_R$.

**Remark:** Even though the temperature of the boundary $T_0$ is only continuous, the temperature $T$ in the interior of the ball has derivatives of arbitrary order. The strong smoothing effect is typical for stationary processes.

### 1.13.2.7 Properties of harmonic functions

Let $\Omega$ be a domain in $\mathbb{R}^3$.

**Definition:** A function $T : \Omega \longrightarrow \mathbb{R}$ is called *harmonic*, if $\Delta T = 0$ on $\Omega$. Here $\Delta$ is the Laplace operator (see (1.392)).

We can interpret $T$ as a temperature distribution in $\Omega$ (without heat sources).

**Smoothness:** Every harmonic function $T : \Omega \longrightarrow \mathbb{R}$ is smooth.

**Lemma of Weyl:** If the functions $T_n : \Omega \longrightarrow \mathbb{R}$ are harmonic, and if

$$\lim_{n\to\infty} \int\limits_{\Omega} T_n \varphi \, \mathrm{d}x = \int\limits_{\Omega} T\varphi \, \mathrm{d}x \quad \text{for all } \varphi \in C_0^\infty(\Omega) \tag{1.400}$$

with a continuous function $T : \Omega \longrightarrow \mathbb{R}$, then $T$ is is harmonic.

The condition (1.400) is satisfied in particular if the sequence $(T_n)$ converges uniformly to $T$ on every compact subset of $\Omega$.

**Mean value property:** A continuous function $T : \Omega \longrightarrow \mathbb{R}$ is harmonic, if

$$T(x) = \frac{1}{4\pi R^2} \int\limits_{|x-y|=R} T(y) \, \mathrm{d}F$$

for all balls of radius $R$ in $\Omega$, for all $R$.

**Maximum principle:** A non-constant harmonic function $T : \Omega \longrightarrow \mathbb{R}$ has in $\Omega$ neither a minimum nor a maximum.

**Corollary 1:** Let $T : \overline{\Omega} \longrightarrow \mathbb{R}$ be a non-constant continuous function. If $T$ is harmonic in $\Omega$, then $T$ attains its minimum and its maximum on the boundary $\partial\Omega$.

*Physical motivation:* If there were a maximal temperature in $\Omega$, then this would lead to an instantaneous heat flow, which is in contradiction to the stationarity of the situation.

**Corollary 2:** Let $\Omega$ be a bounded domain with exterior $\Omega_* := \mathbb{R}^3 - \overline{\Omega}$. If $T : \overline{\Omega}_* \longrightarrow \mathbb{R}$ is continuous and harmonic on $\Omega_*$ with $\lim\limits_{|x|\to\infty} T(x) = 0$, then one has

$$|T(x)| \le \max_{y \in \partial\Omega} |T(y)| \quad \text{for all } x \in \overline{\Omega}.$$

**The Harnack inequality:** If $T$ is harmonic and non-negative on the ball $B_R := \{x \in \mathbb{R}^3 : |x| < R\}$, then one has the inequality

$$\frac{R(R-|x|)}{(R+|x|)^2} T(0) \le T(x) \le \frac{R(R+|x|)}{(R-|x|)^2} T(0) \quad \text{for all } x \in B_R.$$

#### 1.13.2.8 The wave equation

***The one-dimensional wave equation***

$$\boxed{\frac{1}{c^2}u_{tt} - u_{xx} = 0, \qquad x, t \in \mathbb{R}.} \tag{1.401}$$

We interpret $u = u(x,t)$ as the displacement of a vibrating, infinite string at the point $x$ and the time $t$. This equation is called the *one-dimensional wave equation.*

**Theorem:** The general smooth solution of (1.401) has the form

$$u(x,t) = f(x - ct) + g(x + ct)$$

with arbitrary smooth functions $f, g : \mathbb{R} \longrightarrow \mathbb{R}$.

**Physical interpretation:** The solution $u(x,t) = f(x - ct)$ corresponds to a wave, which propagates from left to right with the velocity $c$ and at the time $t = 0$ the form $u(x,0) = f(x)$ (Figure 1.161(a)). Similarly, $u(x,t) = g(x + ct)$ corresponds to a wave which propagates from right to left with the velocity $c$.

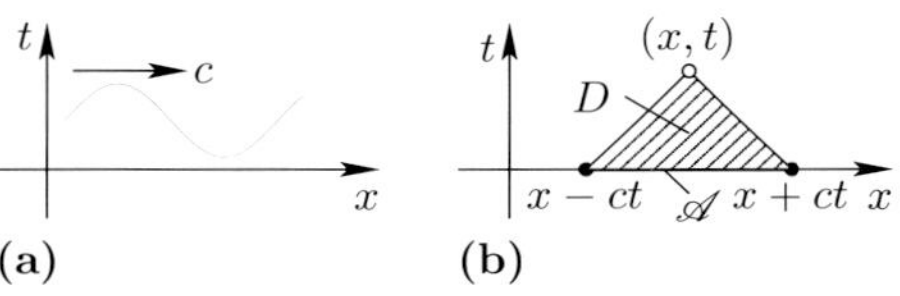


*Figure 1.161. The initial value problem for the wave equation.*

**The existence and uniqueness result for the initial value problem:** Let $u_0, u_1 : \mathbb{R} \longrightarrow \mathbb{R}$ and $f : \mathbb{R}^2 \longrightarrow \mathbb{R}$ be smooth functions. Then the problem

$$\boxed{\begin{aligned} &\frac{1}{c^2}u_{tt} - u_{xx} = f(x,t), \qquad x, t \in \mathbb{R}, \\ &u(x,0) = u_0(x), \quad u_t(x,0) = u_1(x), \quad x \in \mathbb{R}, \end{aligned}}$$

has the unique solution

$$\boxed{u(x,t) = \frac{1}{2}(u_0(x - ct) + u_0(x + ct)) + \frac{1}{2c}\int\limits_{\mathscr{A}} u_1(\xi)\,\mathrm{d}\xi + \frac{c}{2}\int\limits_D f\,\mathrm{d}x\,\mathrm{d}t.}$$

Here $\mathscr{A} := [x - ct, x + ct]$ and $D$ is the triangle which is pictured in Figure 1.161(b). The line $x = \pm ct + \text{const}$ are called *characteristics.* The sides of $D$ which start in the point $(x,t)$ are among those characteristics.

**Domain of dependence:** Let $f \equiv 0$. Then the solution $u$ at the point $x$ at time $t$ depends only on the initial values $u_0$ and $u_1$ on $\mathscr{A}$. Therefore one refers to $\mathscr{A}$ as the *domain of dependence* of the point $(x,t)$ (Figure 1.161(b)).

**Remark:** It is typical for wave processes as opposed to diffusion or stationary processes that no smoothing of the initial solution takes place.

***The two-dimensional wave equation***

**Existence and uniqueness result:** Let $u_0, u_1 : \mathbb{R}^2 \longrightarrow \mathbb{R}$ be smooth functions. Then the initial value problem

$$\boxed{\begin{aligned} &\frac{1}{c^2}u_{tt} - \Delta u = 0, \qquad x \in \mathbb{R},\ t > 0, \\ &u(x,0) = u_0(x), \quad u_t(x,0) = u_1(x), \quad x \in \mathbb{R}^2, \end{aligned}} \tag{1.402}$$

has the unique solution

$$u(x,t) = \frac{1}{2\pi c} \int\limits_{B_{ct}(x)} \frac{u_1(y)}{(c^2t^2 - |y-x|^2)^{1/2}}\, \mathrm{d}y + \frac{\partial}{\partial t}\left(\frac{1}{2\pi c} \int\limits_{B_{ct}(x)} \frac{u_0(y)}{(c^2t^2 - |y-x|^2)^{1/2}}\, \mathrm{d}y\right).$$

Here $B_{ct}(x)$ denotes a ball of radius $ct$ centered at $x$.

***The three-dimensional wave equation***

**Existence and uniqueness result:** If $u_0, u_1 : \mathbb{R}^3 \longrightarrow \mathbb{R}$ and $f : \mathbb{R}^4 \longrightarrow \mathbb{R}$ are smooth functions, then the initial value problem

$$\boxed{\begin{array}{l} \frac{1}{c^2}u_{tt} - \Delta u = f(x,t), \qquad x \in \mathbb{R}^3, \quad t > 0, \\ u(x,t) = u_0(x), \quad u_t(x,0) = u_1(x), \quad x \in \mathbb{R}^3 \end{array}} \tag{1.403}$$

has the unique solution

$$\boxed{u(x,t) = t\mathscr{M}_{ct}^x(u_1) + \frac{\partial}{\partial t}(t\mathscr{M}_{ct}^x(u_0)) + \frac{1}{4\pi} \int\limits_{B_{ct}(x)} \frac{f\left(t - \frac{|y-x|}{c}, y\right)}{|y-x|} \mathrm{d}y.}$$

Here $\mathscr{M}_r^x(u)$ denotes the mean value

$$\mathscr{M}_r^x(u) := \frac{1}{4\pi r^2} \int\limits_{\partial B_r(x)} u \, \mathrm{d}F.$$

In the above formula, $\partial B_r(x)$ denotes the boundary of a ball $B_r(x)$ of radius $r$ centered at $x$ (this is a sphere of radius $r$).

**Domain of dependence:** Let $f \equiv 0$. Then the solution $u$ at the point $x$ and time $t$ depends only on the values of $u_0$ and $u_1$ and the first derivatives of $u_0$ on the set $\mathscr{A} := \partial B_{ct}(x)$, which is therefore called the *domain of dependence* of $(x, t)$.

**Clear transmission of signals and the Huygens' principle in $\mathbb{R}^3$:** Explicitly, $\mathscr{A}$ consists of all points $y$ which satisfy

$$|y - x| = ct.$$

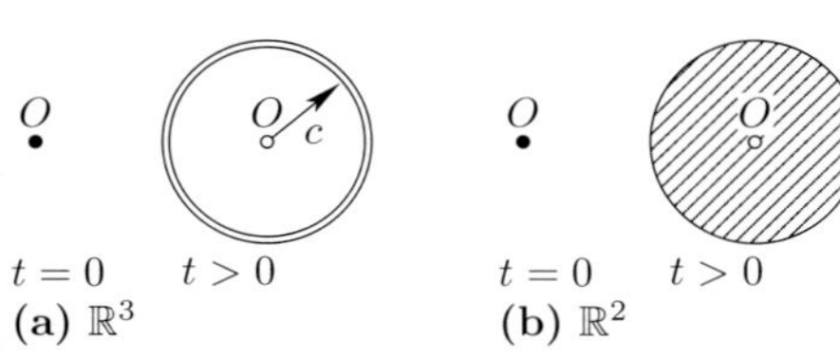


*Figure 1.162. Huygens' principle in $\mathbb{R}^2$ and $\mathbb{R}^3$.*

This corresponds to a clear transmission of signals with the velocity $c$. One also refers to the validity of the Huygens' principle in $\mathbb{R}^3$ instead of clear transmission of signals. If $u_0$ and $u_1$ are concentrated in a small neighborhood of the origin $x = 0$ at time $t = 0$, then the perturbation which these functions represent propagates with the velocity $c$ and is therefore at time $t$ concentrated in a small neighborhood of the surface $\partial B_{ct}(0)$ of the ball (Figure 1.162(a)).

**Non-validity of Huygens' principle in $\mathbb{R}^2$:** In the case of dimension two, the dependence of the solution at the point $x$ at time $t$ is given by $\mathscr{A} = B_{ct}(x)$. Therefore there is no clear transmission of signals. A small perturbation, which is concentrated at $x = 0$

at time $t = 0$ propagates to the entire disc $B_{ct}(0)$ (Figure 1.162(b)). To get an intuitive feeling for this situation, think of living in flatland. The non-validity of Huygens' principle in this two-dimensional world makes it impossible to listen to radio or watch TV: all signals arrive at your antenna completely distorted through the superposition (heterodyning) of the signals sent at different times.

#### 1.13.2.9 The Maxwell equations of electrodynamics

The initial value problem for the Maxwell equations consists in prescribing the electric and magnetic fields at the time $t = 0$. Moreover one has the electric charge density $\rho$ and the electric current density vector $\mathbf{j}$ defined for all times in all of space, where the equation of continuity

$$\rho_t + \operatorname{div} \mathbf{j} = 0$$

must be satisfied. If these quantities are smooth, then they determine unique electric and magnetic fields for all times in all of space. The explicit solution together with a detailed investigation of the Maxwell equations will be given in [212].

#### 1.13.2.10 Electrostatics and Green's functions

**The basic equations of electrostatics:**

$$\boxed{\begin{aligned} -\varepsilon_0 \Delta U &= \rho \quad \text{on } \Omega, \\ U &= U_0 \quad \text{on } \partial\Omega. \end{aligned}} \tag{1.404}$$

Let $\Omega$ be a bounded domain in $\mathbb{R}^3$ with smooth boundary. The electromagnetic potential $U$ is to be determined, with given boundary values $U_0$ and given external charge density $\rho$. In the special case where $U_0 \equiv 0$, the boundary $\partial\Omega$ consists of an electrically conducting material ($\varepsilon_0$ is the dielectric constant of the vacuum).

**Theorem 1:** If the functions $\rho : \overline{\Omega} \longrightarrow \mathbb{R}$ and $U_0 : \partial\Omega \longrightarrow \mathbb{R}$ are smooth, then the problem (1.404) has a unique solution. The corresponding electric field is $\mathbf{E} = -\mathbf{grad}\, U$.

**The Green's function $G$:**

$$\boxed{\begin{aligned} -\varepsilon_0 \Delta G(x,y) &= 0 \quad \text{on } \Omega,\ x \neq y, \\ G(x,y) &= 0 \quad \text{on } \partial\Omega, \\ G(x,y) &= \frac{1}{4\pi\varepsilon_0 |x-y|} + V(x). \end{aligned}} \tag{1.405}$$

We fix the point $y \in \Omega$. Suppose the function $V$ is smooth on $\overline{\Omega}$.

**Theorem 2:** (i) For every fixed point $y \in \Omega$ the problem (1.405) has a unique solution $G$.

(ii) One has $G(x,y) = G(y,x)$ for all $x, y \in \Omega$, that is, $G$ is a symmetric function.

(iii) The unique solution of (1.404) is obtained by the formula

$$\boxed{U(x) = \int\limits_{\Omega} G(x,y)\rho(y)\, \mathrm{d}y - \int\limits_{\partial\Omega} \frac{\partial G(x,y)}{\partial n_y} U_0(y)\, \mathrm{d}F_y.}$$

Here $\partial/\partial n_y$ denotes the outer normal derivative with respect to $y$.

**Physical interpretation:** The Green's function $x \mapsto G(x,y)$ corresponds to an electrostatic potential with a point charge of strength $Q = 1$ at the point $y$ in the domain $\Omega$, which is bounded by an electrical conductor. In the language of distributions, one has:

$$\begin{aligned} -\varepsilon_0 \Delta G(x,y) &= \delta_y \quad \text{on } \Omega, \\ G(x,y) &= 0 \quad \text{on } \partial\Omega. \end{aligned} \tag{1.406}$$

Here $\delta_y$ denotes the Dirac distribution (cf. [212]). The first line of (1.406) is equivalent to the relation

$$-\varepsilon_0 \int_\Omega G(x,y)\Delta\varphi(x)\, dx = \varphi(y) \quad \text{for all } \varphi \in C_0^\infty(\Omega).$$

*Example 1:* The Green's function for the ball $B_R := \{x \in \mathbb{R}^3 : |x| < R\}$ is

$$G(x,y) = \frac{1}{4\pi\varepsilon_0|x-y|} - \frac{R}{4\pi\varepsilon_0|y||x-y_*|} \qquad \text{for all } x, y \in B_R.$$

The point $y_* := \dfrac{R^2}{|y|^2}y$ is obtained from the point $y$ by reflecting on the sphere $\partial B_R$.

*Example 2:* The Green's function for the half-space $H_+ := \{x \in \mathbb{R}^3 : x_3 > 0\}$ has the form

$$G(x,y) = \frac{1}{4\pi\varepsilon_0|x-y|} - \frac{1}{4\pi\varepsilon_0|x-y_*|} \qquad \text{for all } x, y \in H_+.$$

The point $y_*$ is obtained from $y$ by reflecting on the plane $x_3 = 0$.

#### 1.13.2.11 The Schrödinger equation of quantum mechanics and the hydrogen atom

**Classical motion:** For a particle of mass $m$ in a force field $\mathbf{F} = -\mathbf{grad}\, U$ with a potential $U$ the Newton equations of motion are:

$$m\mathbf{x}'' = \mathbf{F}.$$

The energy of the system is given by

$$\boxed{E = \frac{\mathbf{p}^2}{2m} + U(x),} \tag{1.407}$$

where $\mathbf{p} = m\mathbf{x}'$ denotes the momentum.

**Quantized motion:** In quantum mechanics the motion of a particle is given by the *Schrödinger equation*

$$\boxed{\mathrm{i}\hbar\psi_t = -\frac{\hbar}{2m}\Delta\psi + U\psi} \tag{1.408}$$

(h is the Planck constant, and $\hbar = h/2\pi$). The complex valued wave function $\psi = \psi(x,t)$ of the particle satisfies the normalization:

$$\int_{\mathbb{R}^3} |\psi(x,t)|^2\, dx = 1.$$

The number

$$\int_\Omega |\psi(x,t)|^2 \, dx$$

is equal to the probability that the particle in contained in the domain $\Omega$ at the time $t$.

**Rule for quantization:** The Schrödinger equation (1.408), which was formulated by Schrödinger in 1926, is obtained from the classical formula (1.407) for the energy upon making the replacements:

$$E \Leftrightarrow \mathrm{i}\hbar \frac{\partial}{\partial \mathrm{t}}, \quad \mathbf{p} \Leftrightarrow \frac{\hbar}{\mathrm{i}} \mathbf{grad} \,.$$

Then $\mathbf{p}^2$ changes to $-\hbar \mathbf{grad}^2 = \hbar\Delta$.

**States with strict energy levels:** We call the differential operator

$$H := -\frac{\hbar}{2m}\Delta + U$$

the *Hamilton operator* of the quantum mechanical system. If the function $\varphi = \varphi(x)$ is an eigenfunction of $H$ with eigenvalue $E$, i.e., if

$$H\varphi = E\varphi,$$

then the function

$$\psi(x,t) = \mathrm{e}^{-\mathrm{i}tE/\hbar} \, \varphi(x)$$

is a solution of the Schrödinger equation (1.408). By definition, $\psi$ corresponds to the state of a particle with the energy $E$.

**The hydrogen atom:** The motion of an electron of mass $m$ and charge $e < 0$ around the nucleus of a hydrogen atom of charge $|e|$ corresponds to the potential

$$U(x) = -\frac{e^2}{4\pi\varepsilon_0},$$

(where $\varepsilon_0$ is the dielectrical constant of the vacuum). The corresponding Schrödinger equation has, in spherical coordinates, the solution

$$\boxed{\psi = \mathrm{e}^{-\mathrm{i}E_n t/\hbar} \, \frac{1}{r}\sqrt{\frac{2}{nr_0}} L_{n-l-1}^{2l+1}\left(\frac{2r}{nr_0}\right) Y_l^m(\varphi,\theta)} \tag{1.409}$$

with the so-called *quantum numbers* $n = 1, 2, \ldots$ and $l = 0, 1, 2, \ldots, n-1$ as well as $m = l, l-1, \ldots, -l$. The function $\psi$ in (1.409) corresponds to states of the electron with energy

$$\boxed{E_n = -\frac{\gamma}{n^2}.}$$

Here the quantity $\gamma$ is given by $\gamma := e^4 m/8\varepsilon_0^2 \mathrm{h}^2$. Moreover, $r_0 := 4\pi\varepsilon_0 \mathrm{h}^2/me^2 = 5 \cdot 10^{-11}\mathrm{m}$ is the Bohr radius of the atom. The definition of the special functions occuring in (1.409) is explained in section 1.13.2.13.

**Orthogonality:** Two functions $\psi$ and $\psi_*$ of the form (1.409) which belong to different quantum numbers are orthogonal, i.e., one has

$$\int_{\mathbb{R}^3} \overline{\psi(x,t)}\psi_*(x,t) \, dx = 0 \qquad \text{for all} \quad t \in \mathbb{R}.$$

In case $\psi = \psi_*$, the integral is equal to unity (=1).

**The spectrum of the hydrogen atom:** If an electron lying in the energy shell with energy $E_n$ jumps to the energy shell with the *lower* energy $E_k$, then a photon of energy $\Delta E = E_n - E_k$ is emitted with the frequency $\nu$ given by the following formula:

$$\boxed{\Delta E = \mathrm{h}\nu.}$$

A deeper understanding of quantum mechanics is only possible in the context of functional analysis. This is discussed in [212].

### 1.13.2.12 The harmonic oscillator in quantum mechanics and Planck's law of radiation

**Classical motion:** The equation

$$mx'' = -m\omega^2 x$$

corresponds to the oscillation of a point of mass $m$ on the $x$-axis with the energy

$$\boxed{E = \frac{p^2}{2m} + \frac{m\omega^2}{2}}$$

and momentum $p = mx'$.

**Quantized motion:** In quantum mechanics the motion of the particle is determined by the Schrödinger equation

$$\boxed{\mathrm{i}\hbar\psi_t = -\frac{\hbar}{2m}\psi_{xx} + \frac{m\omega^2}{2}\psi} \tag{1.410}$$

together with the normalization

$$\int_{\mathbb{R}} |\psi(x,t)|^2 \, \mathrm{d}x = 1.$$

The number

$$\int_a^b |\psi(x,t)|^2 \, \mathrm{d}x$$

is equal to the probability that the particle can be found in the interval $[a, b]$. The Schrödinger equation (1.410) has the solutions

$$\psi = \mathrm{e}^{-\mathrm{i}E_n t/\hbar} \, \frac{1}{x_0} H_n\left(\frac{x}{x_0}\right)$$

with $n = 0, 1, \ldots$ and $x_0 := \sqrt{\hbar/m\omega}$ (cf. 1.13.2.13). These are states of energy

$$\boxed{E_n = \hbar\omega\left(\mathrm{n} + \frac{1}{2}\right).} \tag{1.411}$$

For $\Delta E := E_{n+1} - E_n$ we get

$$\boxed{\Delta E = \hbar\omega.} \tag{1.412}$$

**The Planck law of radiation (1900):** The equation (1.412) contains the famous *Planck quantum formula*, which is the starting point of quantum mechanics and, as opposed to the unsuccessful attempts using classical mechanics, gives the correct radiation law. The energy $E$, which is emitted by a star of temperature $T$ with the surface area $F$ in the time interval $\Delta t$, is according to Planck, equal to

$$E = 2\pi \mathrm{h}\mathrm{c}^2 F \Delta t \int\limits_0^\infty \frac{\mathrm{d}\lambda}{\lambda^5(\mathrm{e}^{\mathrm{hc}/kT\lambda} - 1)}$$

where $\lambda$ is the wavelength of the light, h is the Planck constant, c is the speed of light and $k$ is the Boltzmann constant.

**The zero point energy of Heisenberg:** The formula (1.411) was obtained by Heisenberg in 1924 in the context of his matrix mechanics. In this manner Heisenberg created quantum mechanics. The most interesting point of (1.411) is the fact that the lowest state $n = 0$ has a non-vanishing energy $E_0 = \hbar\omega/2$. This leads to the fact that the lowest states of a quantum field with its infinitely many degrees of freedom has an 'infinitely large' energy. This behavior is one of the reasons for the unsurmountable difficulties involved in forming a mathematically rigorous quantum field theory.

#### 1.13.2.13 Special functions of quantum mechanics

**Orthonormal systems:** Let $X$ be a Hilbert space with scalar product $(u, v)$. Then the set of elements $u_0, u_1, \ldots$ form a complete *orthonormal system* in $X$, if

$$(u_k, u_m) = \delta_{km}$$

for all $k, m = 0, 1, 2, \ldots$ and every element $u \in X$ can be written in the form

$$u = \sum_{k=0}^{\infty} (u_k, u) u_k.$$

This means that

$$\lim_{n\to\infty} ||u - \sum_{k=0}^{n} (u_k, u) u_k|| = 0$$

with $||v|| := (v, v)^{1/2}$. In the following let $x \in \mathbb{R}$. Most of the functions discussed below have been introduced in section 0.7.2.

**Hermitian functions:** (see p. 124)

$$\mathscr{H}_n(x) := \frac{(-1)^n}{\sqrt{2^n n! \sqrt{\pi}}}\, \mathrm{e}^{x^2/2} \frac{\mathrm{d}^n}{\mathrm{d}x^n} \mathrm{e}^{-x^2}.$$

For $n = 0, 1, 2, \ldots$, these functions satisfy the differential equation

$$-y'' + x^2 y = (2n+1) y$$

and form a complete orthonormal system in the Hilbert space $L_2(-\infty,\infty)$ with the scalar product[149]

$$(u,v) := \int_{-\infty}^{\infty} u(x)v(x)\,\mathrm{d}x.$$

**Normalized Legendre polynomials:**

$$\mathscr{P}_n(x) := \sqrt{\frac{2n+1}{2^{2n+1}(n!)^2}}\frac{\mathrm{d}^n(1-x^2)^n}{\mathrm{d}x^n}.$$

For $n = 0, 1, \ldots$, these functions satisfy the differential equation

$$-((1-x^2)y')' = n(n+1)y$$

and form a complete orthonormal system in the Hilbert space $L_2(-1,1)$ with the scalar product

$$(u,v) = \int_{-1}^{1} u(x)v(x)\,\mathrm{d}x.$$

**Generalized Legendre polynomials:**

$$\mathscr{P}_l^k(x) := \sqrt{\frac{(l-k)!}{(l+k)!}}(1-x^2)^{l/2}\frac{\mathrm{d}^k\mathscr{P}_l(x)}{\mathrm{d}x^k}.$$

For $k = 0, 1, 2, \ldots$ and $l = k, k+1, k+2, \ldots$, these functions satisfy the differential equation

$$-((l-x^2)y')' + k^2(1-x^2)^{-1}y = l(l+1)y$$

and form a complete orthonormal system in the Hilbert space $L_2(-1,1)$.

**Normalized Laguerre functions:**

$$\mathscr{L}_n^\alpha(x) := c_n^\alpha \mathrm{e}^{x/2}x^{-\alpha/2}\frac{\mathrm{d}^n}{\mathrm{d}x^n}(\mathrm{e}^{-x}x^{n+\alpha}).$$

For a fixed $\alpha > -1$ and $n = 0, 1, \ldots$, these functions satisfy the differential equation

$$-4(xy')' + \left(x + \frac{\alpha^2}{x}\right)y = 2(2n+1+\alpha)y$$

and form a complete orthonormal system in the Hilbert space $L_2(0,\infty)$ with the scalar product

$$(u,v) := \int_{0}^{\infty} u(x)v(x)\,\mathrm{d}x.$$

The positive constant $c_n^\alpha$ can be chosen such that $(L_n^\alpha, L_n^\alpha) = 1$.

**Spherical functions:**

$$\mathscr{Y}_l^m(\varphi,\theta) := \frac{1}{\sqrt{2\pi}}\mathscr{P}_l^{|m|}(\sin\theta)\mathrm{e}^{\mathrm{i}m\varphi}.$$

[149] A function $u : \mathbb{R} \longrightarrow \mathbb{R}$ belongs to $L_2(-\infty,\infty)$ if and only if $(u,u) < \infty$, where the integral is to be understood in the sense of Lebesgue (cf. [212]). In particular the continuous or almost everywhere continuous functions $u : \mathbb{R} \longrightarrow \mathbb{R}$ belong to $L_2(-\infty,\infty)$ if and only if $(u,u) < \infty$. Similarly the spaces $L_2(-1,1)$ etc. can be defined.

Here $r$, $\varphi$ and $\theta$ are spherical coordinates (see section 1.7.9.3). For $l = 0, 1, \ldots$ and $m = l, l-1, \ldots, -l$, these functions form a complete orthonormal system in the Hilbert space $L_2(S^2)_{\mathbb{C}}$ consisting of the complex-valued functions on the unit sphere $S^2 := \{x \in \mathbb{R}^3 : |x| = 1\}$ with the scalar product

$$(u, v) := \int_{S^2} \overline{u(x)} v(x) \, \mathrm{d}F.$$

#### 1.13.2.14 Non-linear partial differential equations in the natural sciences

*The limit is the location of knowledge.*
*Paul Tillich*

Important processes in nature are described by complicated non-linear differential equations. Among these are the equations of hydrodynamics, gas dynamics, elasticity, chemical processes, general relativity (cosmology), quantum electrodynamics and gauge theories (the standard model in the theory of elementary particles). The non-linear terms which occur correspond to *interactions*.

In these problems the methods of classical mathematics fail. One requires modern functional analysis for their treatment. A central tool for this is the theory of Sobolev spaces. These are spaces of functions which are not smooth and which possess only generalized derivatives (in the sense of distributions). These Sobolev spaces are at the same time the appropriate tool in studying the convergence of modern numerical procedures.

These questions will be handled in detail in [212].

### 1.13.3 The role of characteristics

Important information of physical processes involving waves can be gleaned from the corresponding differential equations without actually solving them. There is a difference between weak discontinuities (jumps of higher derivatives) and strong discontinuities (jumps of the functions themselves). Weak discontinuities are connected with characteristics and yield for example the following important physical statements:

(i) Electromagnetic waves are transversal;
(ii) Sound waves are longitudinal;
(iii) Elastic waves can be both transversal and longitudinal.

Strong discontinuities correspond to shock waves in the dynamics of gases and the corresponding Rankine–Hugoniot conditions.

**The behavior of jumps:** Let the surface

$$\mathscr{F} : \psi(x) = 0$$

be given in $\mathbb{R}^{N+1}$ with $x = (x_1, \ldots, x_N)$. The normal vector $\mathbf{n}$ in the point $x$ is given by[150]

$$\mathbf{n} := \frac{\psi'(x)}{|\psi'(x)|}.$$

[150] Explicitly one has $\psi' = (\partial_1 \psi, \ldots, \partial_N \psi)$ and $\mathbf{n} = (n_1, \ldots, n_N)$ with $n_k = \dfrac{\partial_k \psi(x)}{\left(\sum_{j=1}^{N} |\partial_k \psi(x)|^2\right)^{1/2}}$.

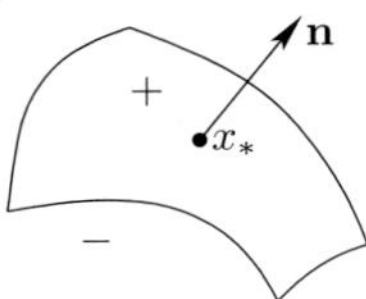


Figure 1.163.

We define the size of the jump by the relation

$$\boxed{[u](x) := u_+(x) - u_-(x)}$$

with $u_\pm(x) = \lim\limits_{h\to+0} u(x \pm h\mathbf{n})$ (Figure 1.163).

**Notations:** Let

$$x = (x_1, \ldots, x_N), \quad u = (u_1, \ldots, u_M)^\mathsf{T}, \quad \partial_j v := \frac{\partial v}{\partial x_j}, \quad b = (b_1, \ldots, b_M)^\mathsf{T}.$$

In order to have a convenient notation for higher partial derivatives, we introduce multi-indices $\alpha = (\alpha_1, \ldots, \alpha_N)$ as a tuple of natural numbers $\alpha_1, \ldots, \alpha_N$ and write

$$\partial^\alpha v := \partial_1^{\alpha_1} \partial_2^{\alpha_2} \ldots \partial_N^{\alpha_N} v = \frac{\partial^{|\alpha|} v}{\partial x_1^{\alpha_1} \ldots \partial x_N^{\alpha_N}}$$

with $|\alpha| := \alpha_1 + \cdots + \alpha_N$. Similarly, let

$$\lambda^\alpha := \lambda_1^{\alpha_1} \lambda_2^{\alpha_2} \ldots \lambda_N^{\alpha_N} \qquad \text{for all} \quad \lambda \in \mathbb{R}^N.$$

In particular, if $\alpha = (0, \ldots, 0)$, we set $\partial^\alpha v := v$ and $\lambda^\alpha := 1$.

### 1.13.3.1 Characteristics and the propagation of discontinuities

**Quasilinear systems:** We consider the initial value problem

$$\boxed{\begin{aligned} &\sum_{|\alpha|\le m} a_\alpha(x, \partial u) \partial^\alpha u = b(x, \partial u), \\ &\partial^\beta u = c_\beta \quad \text{on the surface } \mathscr{F} \text{ for all } \beta \text{ with } |\beta| \le m-1 \end{aligned}} \tag{1.413}$$

with smooth coefficient functions $a_\alpha$, $b$ and $c_\beta$. Each symbol $a_\alpha$ denotes a quadratic $(N \times N)$-matrix. In (1.413) the summation is over all derivatives of $u$ up to order $m$, where the zeroth derivative is understood to be the function $u$ itself. The coefficients $a_\alpha$ and $b$ are only supposed to contain derivatives of $u$ up to order $m-1$. A system of equations of this type is referred to as a *quasilinear system* of order $m$. If all the coefficient functions $a_\alpha$ and $b$ are independent of $u$, then this is in fact a linear system.

Let the surface $\mathscr{F}$ be defined by the equation

$$\psi(x) = 0 \tag{1.414}$$

with $\psi'(x) \neq 0$ in all points of $\mathscr{F}$.

**Symbol:** To the differential equation (1.413) we associate its *symbol*

$$\boxed{\mathscr{S}(x, u(x), \lambda) := \det\left(\sum_{|\alpha|=m} a_\alpha(x, \partial u(x)) \lambda^\alpha\right), \qquad \lambda \in \mathbb{R}^N.}$$

If the system (1.413) is linear, the symbol $\mathscr{S}(x, \lambda)$ does not depend on $u$.

The symbol $\mathscr{S}$ contains fundamental information about the system of solutions of (1.413).

**Characteristics:** The surface $\mathscr{F} : \psi(x) = 0$ is called a *characteristic*, if and only if the function $\psi$ satisfies the differential equation

$$\boxed{\mathscr{S}(x, u(x), \psi'(x)) = 0.}$$

A curve $x = x(\sigma)$ is called a *bicharacteristic* to the characteristic $\psi$, if[151]

$$\boxed{x'(\sigma) = \mathscr{S}_\lambda(x(\sigma), u(x(\sigma)), \psi'(x(\sigma))).}$$

**Physical interpretation:** For the Maxwell equations the bicharacteristics correspond to the light waves, and the characteristics are the wave fronts of these light waves (cf. 1.13.3.3).

The differential equation $\mathscr{S}(x, u(x), \psi'(x)) = 0$ is of the first order in $\psi$. According to a general result of Cauchy one can build the solution surfaces of partial differential equations of first order from curves, which in the case of the function $\psi$ correspond to the bicharacteristics (cf. 1.13.5.2).

**Theorem on discontinuities:** Let $u = u(x)$ be a solution of the quasilinear system (1.413), which, together with its derivatives of up to $(m-1)^{st}$ order, is continuous in a neighborhood of $\mathscr{F}$. The possible jumps of the $m^{th}$ derivatives of $u$ at $x$ obey the following:

(i) Kinematic compatibility condition:

$$\sum_{|\alpha|=m} a_\alpha(x, \partial u(x))[\partial^\alpha u](x) = 0.$$

(ii) Dynamical compatibility condition:

$$[\partial^\alpha u](x) = \psi'(x)^\alpha \rho \qquad \text{for all} \quad \alpha \quad \text{with} \quad |\alpha| = m.$$

Here $\rho$ is a fixed vector in $\mathbb{R}^M$. One has $\rho \neq 0$ if $(\psi, u)$ is characteristic in $x$, i.e., $\mathscr{S}(x, u(x), \psi'(x)) = 0$.[152]

**Weak correctness of the initial value problem:** For smooth functions $u, \psi$ the following statements are equivalent:

(i) All derivatives of $u$ up to order $m$ at $x$ are uniquely determined by the differential equation and the initial values of (1.413).

(ii) $(\psi, u)$ is not characteristic at $x$, i.e., $\mathscr{S}(x, u(x), \psi'(x)) \neq 0$.

### 1.13.3.2 Applications to the classification of differential equations

Let $u$ be a smooth solution of the quasilinear system (1.413). The following classification depends in general on $u$. For linear systems the described classification is, however, independent of $u$.

---

[151] Explicitly one has $x_k'(\sigma) = \dfrac{\partial \mathscr{S}}{\partial \lambda_k}(x(\sigma), u(x(\sigma)), \psi'(x(\sigma))), \quad k = 1, \ldots, N.$

[152] We set $\psi' = (\partial_1 \psi, \ldots, \partial_N \psi)$ and

$$\psi'(x)^\alpha = \partial_1^{\alpha_1}\psi(x)\partial_2^{\alpha_2}\psi(x)\ldots\partial_N^{\alpha_N}\psi(x).$$

We fix a point $x$ and consider the $\lambda$-polynomial

$$\boxed{\mathscr{P}(\lambda) := \mathscr{S}(x, u(x), A\lambda), \qquad \lambda \in \mathbb{R}^N,}$$

with a fixed invertible real $(N \times N)$-matrix $A$ and the column vector $\lambda$. The following conditions should be satisfied for a fixed choice of $A$ (for example, $A\lambda \equiv \lambda$).

**Definition:** (i) The quasilinear system (1.413) is called *elliptic* at the point $x$, if $\lambda = 0$ is the only zero of $\mathscr{P}$.

(ii) (1.413) is said to be *parabolic* at $x$, if the polynomial $\mathscr{P}$ is degenerate, i.e., it depends on fewer than $N$ variables.

(iii) (1.413) is said to be *strictly hyperbolic* at $x$, if the equation

$$\mathscr{P}(\lambda) = 0, \qquad \lambda \in \mathbb{R}^N,$$

has exactly $MN$ different real solutions $\lambda_N$ for every non-vanishing tuple $(\lambda_1, \ldots, \lambda_{N-1}) \in \mathbb{R}^{N-1}$.

**Qualitative behavior:** Roughly speaking one has the following facts:

(i) Elliptic problems correspond to *stationary processes* in nature. Here there are no characteristics. The solutions have no discontinuities.

(ii) Parabolic problems correspond to *flow processes* (for example heat conduction and diffusion). These processes have a smoothing effect in time.

(iii) Hyperbolic problems belong to *wave processes*. Here the propagation of discontinuities along wave fronts is an important mechanism for transporting physical effects in nature.

**Classification of equations of second order:** We consider the equation

$$\sum_{j,k=1}^{n} a_{jk}\partial_j\partial_k u + \sum_{j=1}^{n} a_j\partial_j u + au = f \tag{1.415}$$

with $\partial_j := \partial/\partial x_j$ and with a symmetric real matrix $A = (a_{jk})$. It is our goal to find a real function $u = u(x)$ which solves the equation. The coefficients $a_{jk}$, $a_j$ and $a$ are assumed to be real numbers. The corresponding *symbol* is obtained by

$$\mathscr{S}(\lambda) := \sum_{j,k=1}^{n} a_{jk}\lambda_j\lambda_k,$$

i.e., $\mathscr{S}(\lambda) = \lambda^{\mathsf{T}}A\lambda$. The equation for the *characteristics* is $\psi(x) = 0$. The function $\psi$ satisfies the equation $\mathscr{S}(\psi'(x)) = 0$, that is

$$\sum_{j,k=1}^{n} a_{jk}\partial_j\psi\partial_k\psi = 0.$$

The equation for the *bicharacteristics* $x = x(\sigma)$ is:

$$x'(\sigma) = 2A\psi'(x(\sigma)).$$

**Theorem:** (i) The equation (1.415) is elliptic if and only if the eigenvalues of $A$ are all positive (or all negative).

(ii) The equation (1.415) is parabolic if and only if at least one eigenvalue of $A$ vanishes.

(iii) The equation (1.415) is strictly hyperbolic if and only if an eigenvalue of $A$ is positive and all other eigenvalues are negative (or the other way around).

*Example 1:* The Laplace equation

$$\boxed{u_{xx} + u_{yy} = 0}$$

can be written in the form $\partial_1^2 u + \partial_2^2 u = 0$. The symbol is therefore

$$\mathscr{S}(\lambda) := \lambda_1^2 + \lambda_2^2, \qquad (\lambda_1, \lambda_2) \in \mathbb{R}^2.$$

From $\mathscr{S}(\lambda) = 0$ it follows that $\lambda = 0$. Therefore the Laplace equation is *elliptic.*

*Example 2:* The heat equation

$$\boxed{u_t - u_{xx} = 0}$$

has the symbol $\mathscr{S}(\lambda) = -\lambda_1^2$ and is hence degenerate, as it does not depend on $\lambda_2$. Consequently, the heat equation is *parabolic.*

The equation for the characteristics $\psi(x,t) = 0$ is determined by the solution of $\psi_x^2 = 0$. The family of solutions $\psi = t + \text{const}$ correspond to the lines

$$\boxed{t = \text{const}}$$

as characteristics (Figure 1.164(a)).

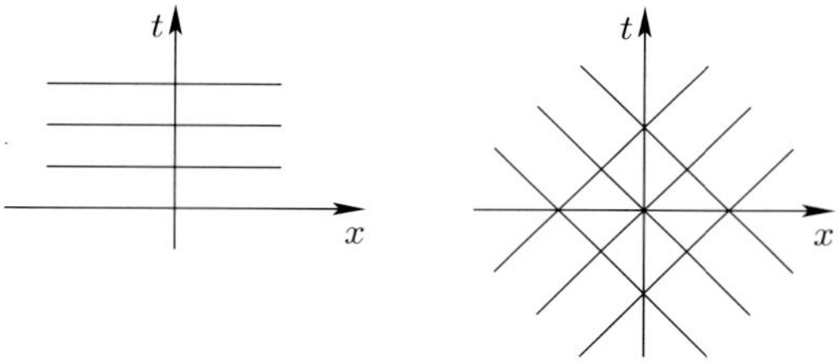

**(a)** $t = \text{const}$ **(b)** $x = \pm\, ct + \text{const}$

*Figure 1.164. Characteristics of the heat and vibrating string equations.*

*Example 3:* The equation of a vibrating string

$$\boxed{\frac{1}{c^2}u_{tt} - u_{xx} = 0}$$

has the symbol $\mathscr{S}(\lambda) := \frac{1}{c^2}\lambda_1^2 - \lambda_2^2$. For every real number $\lambda_1 \neq 0$ the equation $\mathscr{S}(\lambda) = 0$ has two real solutions $\lambda_2$. Hence the equation of the vibrating string is *strictly hyperbolic.* The equation for the characteristics is $\psi(x,t) = 0$. The function $\psi$ is obtained as the solution of the equation

$$\frac{1}{c^2}\psi_t^2 - \psi_x^2 = 0.$$

The family of solutions $\psi = \pm ct - x + \text{const}$ corresponds to the characteristics (Figure 1.164(b)):

$$\boxed{x = \pm ct + \text{const}.}$$

**Remark:** As the speed $c$ of propagation increases, these characteristics approach the characteristic $t = \text{const}$ of the heat equation. In fact, the initial perturbations of the heat equation are propagated with arbitrarily high speed. This fact is at contradiction with the Einstein principle, according to which physical effects can propagate with at most the speed of light. Therefore the heat equation must be modified in special relativity.

*Example 4:* Let $\Delta u := u_{xx} + u_{yy} + u_{zz}$.

(i) The Poisson equation

$$\Delta u = f$$

is *elliptic.*

(ii) The heat equation

$$u_t - \alpha \Delta u = f$$

is parabolic.

(iii) The wave equation

$$\frac{1}{c^2} u_{tt} - \Delta u = f$$

is *strictly hyperbolic.* The characteristics $\psi(x, y, z, t) = 0$ of the wave equation, which correspond to moving wave fronts, belong to solutions $\psi$ of the *eikonal equation*

$$\frac{1}{c^2}\psi_t^2 - (\mathbf{grad}\,\psi)^2 = 0.$$

In particular, for the characteristics of the form $\psi = ct - \varphi(x, y, z)$ one obtains the *bicharacteristics* $\mathbf{x} = \mathbf{x}(t)$ from the differential equation

$$\mathbf{x}'(t) = \mathbf{grad}\,\varphi(\mathbf{x}(t)).$$

These are curves which are perpendicular to the wave surfaces $\varphi(x, y, z) = \text{const}$. In the particular case when $\varphi(x) = \alpha x + \beta y + \gamma z + \delta$, the characteristics

$$\alpha x + \beta y + \gamma z + \delta = ct$$

correspond to planes which propagate at the speed $c$ in the normal direction. The bicharacteristics are lines, which are perpendicular to these planes.

### 1.13.3.3 Applications to electromagnetic waves

We consider the initial value problem for the Maxwell equations

$$\begin{aligned} &\mathbf{curl}\ \mathbf{E} = -\mathbf{B}_t, \quad \mathbf{curl}\ \mathbf{B} = \frac{1}{\mathrm{c}^2}\mathbf{E}_t, \\ &\mathbf{E} = \mathbf{E}_0 \quad \text{and} \quad \mathbf{B} = \mathbf{B}_0 \quad \text{at time} \quad t = 0 \end{aligned} \tag{1.416}$$

for the electric field vector $\mathbf{E}$ and the magnetic field vector $\mathbf{B}$ in a vacuum in the absence of electrical charges and currents. Here c denotes the speed of light in a vacuum. Suppose that $\operatorname{div}\mathbf{E}_0 = 0$ and $\operatorname{div}\mathbf{B}_0 = 0$. For the solutions of (1.416) this means automatically that $\operatorname{div}\mathbf{E} = \operatorname{div}\mathbf{B} = 0$ for all times $t$.

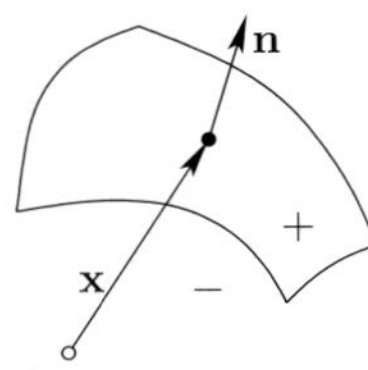

*Figure 1.165.*

**Characteristics:** The equation for the characteristics of (1.416) are $\psi = 0$. The function $\psi$ is a solution of the equation

$$\left(\frac{1}{\mathrm{c}^2}\psi_t^2 - (\mathbf{grad}\,\psi)^2\right)\psi_t^4 = 0.$$

We consider solutions of the form $\psi(\mathbf{x}, t) = \varphi(\mathbf{x}) - ct$ with $(\mathbf{grad}\,\varphi)^2 \equiv 1$. Then

$$\mathscr{F} : \varphi(\mathbf{x}) - \mathrm{c}t = 0$$

corresponds to the propagation of the wave front with speed c in the direction of the unit normal vector

$$\mathbf{n} = \mathbf{grad}\,\varphi(x)$$

(Figure 1.165). One has $\mathbf{x} = x_1\mathbf{i} + x_2\mathbf{j} + x_3\mathbf{k}$ and $\mathbf{n} = n_1\mathbf{i} + n_2\mathbf{j} + n_3\mathbf{k}$.

**Jump condition:** If the electromagnetic fields **E**, **B** are continuous along the wave front $\mathscr{F}$, then the possible jumps of the first derivatives of **E** and **B** at **x** at the time $t$ satisfy the relations[153]

$$\boxed{[\partial_k \mathbf{E}] = \mathbf{a}n_k, \quad [\partial_k \mathbf{B}] = \mathrm{c}^{-1}\mathbf{b}n_k, \qquad k = 1, 2, 3.}$$

Here **a** and **b** are vectors with $\mathbf{a}^2 + \mathbf{b}^2 \neq 0$ and

$$\mathbf{a} = \mathbf{b} \times \mathbf{n} \quad \text{and} \quad \mathbf{b} = \mathbf{n} \times \mathbf{a}.$$

Since **a** and **b** are perpendicular to the direction of propagation **n** of the wave front, we speak of *transversal waves.*

*Example:* If a radio station starts its daily program at time $t = 0$, then an electromagnetic wave front is generated, in such a way that **E** and **B** vanish in front of the wave front but are non-vanishing behind it.

### 1.13.3.4 Applications to elastic waves

We consider small deformations of elastic bodies, i.e., a point with position vector **x** passes during this deformation at time $t$ to a point with position vector

$$\mathbf{y} = \mathbf{x} + \mathbf{u}(\mathbf{x}, t).$$

The equations governing linear elasticity theory are:[154]

$$\boxed{\rho \mathbf{u}_{tt} = \kappa \Delta \mathbf{u} + (\lambda + \kappa)\,\mathbf{grad}\ \mathrm{div}\ \mathbf{u}.}$$

Here $\rho$ is the constant density of the body and $\kappa$ and $\lambda$ denote the Lamé material constants.

**Characteristics:** The solution $\psi$ of the equation

$$\left(\frac{1}{c_{\mathrm{tr}}^2}\psi_t^2 - (\mathbf{grad}\,\psi)^2\right)^2 \left(\frac{1}{c_{\mathrm{l}}^2}\psi_t^2 - (\mathbf{grad}\,\psi)^2\right) = 0 \tag{1.417}$$

with

$$c_{\mathrm{tr}} := \sqrt{\frac{\kappa}{\rho}}, \qquad c_{\mathrm{l}} := \sqrt{\frac{\lambda + 2\kappa}{\rho}}$$

yield the characteristics $\mathscr{F} : \psi(\mathbf{x}, t) = 0$.

**Transversal waves:** Let $\varphi$ be a function with $(\mathbf{grad}\,\varphi)^2 \equiv 1$. Then the function

$$\psi := c_{\mathrm{tr}}t - \varphi(\mathbf{x})$$

is a solution of (1.417), where $\mathscr{F}$ corresponds to a surface, which moves with the speed $c_{\mathrm{tr}}$ in the direction of the normal vector $\mathbf{n} := \mathbf{grad}\,\varphi(\mathbf{x})$ of the surface (Figure 1.165). The condition for discontinuities for the second derivatives of **u** at the point **x** of the wave front at time $t$ is:

$$\boxed{[\partial_j \partial_k \mathbf{u}] = \mathbf{a}n_j n_k, \qquad j, k = 1, 2, 3.}$$

[153] $[f] := f_+ - f_-$ with $f_\pm(\mathbf{x}, t) := \lim_{h \to +0} f(\mathbf{x} \pm h\mathbf{n}, t)$.

[154] The general equations of *non-linear* elasticity theory can be found in [212].

The vector $\mathbf{a} \neq 0$ is perpendicular to $\mathbf{n}$. We therefore speak of a transversal wave.

**Longitudinal waves:** If we replace $c_{\text{tr}}$ by $c_{\text{l}}$, then we get the same result, with the vector $\mathbf{a}$ now being parallel to $\mathbf{n}$, i.e., we now have a longitudinal wave.

#### 1.13.3.5 Applications to sound waves

The Euler equations for the motion of a *compressible fluid* (without inner friction) are:

$$\begin{array}{rl} \rho \mathbf{v}_t + \rho(\mathbf{v}\,\mathbf{grad})\mathbf{v} = \mathbf{f} - \mathbf{grad}\,p & \text{(equation of motion)}, \\ \rho_t + \operatorname{div}(\rho\mathbf{v}) = 0 & \text{(conservation of mass)}, \\ p = p(\rho) & \text{(adiabatic equation of state).}^{155} \end{array}$$

The notations are as follows: $\mathbf{v}(\mathbf{x}, t)$ is the velocity of the fluid particles at time $t$ at the point $\mathbf{x}$, $\rho(\mathbf{x}, t)$ is the density of fluid at the point $\mathbf{x}$ at time $t$, $\mathbf{f}$ is the density of exterior forces and $p$ is pressure.

Let $\psi_t(\mathbf{x}, t) < 0$. The equation

$$\mathscr{F} : \psi(\mathbf{x}, t) = 0$$

describes the motion of a surface in $\mathbb{R}^3$ with the speed $c$ in the direction of the unit normal vector $\mathbf{n}$ of the surface at the point $\mathbf{x}$ at time $t$. One has:

$$c = -\frac{\psi_t(\mathbf{x}, t)}{|\mathbf{grad}\ \psi|}, \qquad \mathbf{n} = \frac{\mathbf{grad}\ \psi}{|\mathbf{grad}\ \psi|}.$$

A fluid particle at the point $\mathbf{x}$ of the surface at time $t$ has the velocity vector $\mathbf{v}$ with the normal component $\mathbf{vn}$. For the relative velocity $c - \mathbf{vn}$ between the wave front and the particle we obtain

$$c - \mathbf{vn} = -\frac{1}{|\mathbf{grad}\ \psi|} D_t\psi$$

with $D_t\psi := \psi_t + \mathbf{v}\,\mathbf{grad}\,\psi$.

**Characteristic equation:**

$$(D_t\psi)^2 \left( \frac{1}{c_{\mathrm{S}}^2} (D_t\psi)^2 - (\mathbf{grad}\,\psi)^2 \right) = 0. \tag{1.418}$$

Here we have set

$$c_{\mathrm{S}} := \sqrt{p'(\rho)}.$$

**Sound waves:** If $\psi$ is a solution of the equation

$$\frac{1}{c_{\mathrm{S}}^2} (D_t\psi)^2 - (\mathbf{grad}\,\psi)^2 = 0,$$

then (1.418) is satisfied. For the relative velocity we obtain

$$c - \mathbf{vn} = c_{\mathrm{S}}.$$

[155] If one introduces the density $\rho$ and the specific entropy density $s$, then one obtains the Dirac density relation $p = p(\rho)$ from $p = p(\rho, s)$ with $s = \text{const}$ (adiabatic process). The following considerations remain valid for gases (cf. 1.13.3.6).

The quantity $c_S$ is called the *speed of sound.* The condition for discontinuities of the first derivatives of $\mathbf{v}$ and $\rho$ at the point $\mathbf{x}$ at time $t$ is

$$\boxed{[\partial_j \mathbf{v}] = \mathbf{a} n_j, \qquad [\partial_j \rho] = b n_j}$$

with $\mathbf{a} = b c_S \rho^{-1}\mathbf{n}$ and $\mathbf{a}^2 + b^2 \neq 0$. Since the jump vector $\mathbf{a}$ is parallel to the surface normal vector $\mathbf{n}$, we have a *longitudinal wave.*

#### 1.13.3.6 Shock waves in the dynamics of gases

The equations of motion for gases (without friction and without heat conduction) are:

$$\begin{array}{rl} (\rho\mathbf{v})_t + \operatorname{div}(\rho\mathbf{v}\otimes\mathbf{v} + pI) = \mathbf{f} & \text{(conservation of momentum),}^{156} \\ \rho_t + \operatorname{div}(\rho\mathbf{v}) = 0 & \text{(conservation of mass),} \\ \epsilon_t + \operatorname{div}(\epsilon\mathbf{v} + p\mathbf{v}) = \mathbf{f}v & \text{(conservation of energy),} \\ (\rho s)_t + \operatorname{div}(\rho s\mathbf{v}) \geq 0 & \text{(entropy equation).} \end{array} \tag{1.419}$$

In addition one has the thermodynamic relations

$$\begin{aligned} &p = p(\rho, T), \quad e = e(\rho, T), \quad s = s(\rho, T), \\ &de = T ds + p\frac{d\rho}{\rho^2} \qquad \text{(Gibbs equation).} \end{aligned}$$

In contrast to the case of liquids, in working with gases one must account for the effects of thermodynamics.

The notations are as follows: $\mathbf{v}$ is the velocity vector, $p$ the pressure, $T$ the absolute temperature, $\rho$ the density, $\mathbf{f}$ the density of exterior forces, $e$ the specific inner energy density (inner energy per unit of mass), and $s$ is the specific entropy density. Moreover,

$$\epsilon := \frac{1}{2}\rho\mathbf{v}^2 + \rho e$$

is the density of total energy.

---

[156] The tensor product $\mathbf{a}\otimes\mathbf{b}$ is identified here with a linear operator, which is defined by the relation

$$(\mathbf{a}\otimes\mathbf{b})\mathbf{c} := \mathbf{a}(\mathbf{bc}) \quad \text{for all vectors } \mathbf{c}.$$

In older literature one used the notation $(\mathbf{a}\circ\mathbf{b})\mathbf{c} = \mathbf{a}(\mathbf{bc})$ and spoke of the *dyadic product.* The meaning of $\operatorname{div}(\mathbf{a}\otimes\mathbf{b})$ is given by the Gaussian integral formula

$$\int_G \operatorname{div}(\mathbf{a}\otimes\mathbf{b})dx = \int_{\partial G} (\mathbf{a}\otimes\mathbf{b})\mathbf{n}dS,$$

where $\mathbf{n}$ is the outer normal vector. In a Cartesian coordinate system with basis vectors $\mathbf{e}_1$, $\mathbf{e}_2$ and $\mathbf{e}_3$, one has the following representation in components:

$$\operatorname{div}(\mathbf{a}\otimes\mathbf{b}) = \sum_{j=1}^{3} \operatorname{div}(a_j\mathbf{b})\mathbf{e}_j = \sum_{j,k=1}^{3} \partial_k(a_j b_k)\mathbf{e}_j.$$

Because of conservation of mass $\varrho_t + \operatorname{div}(\rho\mathbf{v}) = 0$ the conservation of momentum (1.419) is equivalent to the equation of motion

$$\boxed{\rho\mathbf{v}_t + \rho(\mathbf{v}\ \mathbf{grad})\mathbf{v} = \mathbf{f} - \mathbf{grad}\ p.}$$

*Example:* In the special case of an ideal gas, under the condition of temperatures which are not too low, one has:

$$p = r\rho T, \quad e = cT, \quad s = c\ln(T\rho^{1-\gamma}), \quad \gamma = 1 + r/c, \tag{1.420}$$

where $r$ is the gas constant and $c$ is the specific heat capacity.

**The Rankine–Hugoniot discontinuity condition:** Let a moving surface

$$\mathscr{F} : \varphi(\mathbf{x}) - t = 0 \tag{1.421}$$

be given, which is not necessarily a characteristic. For the possible jumps of the physical quantities at a point $\mathbf{x}$ on the surface at the time $t$, one has:[157]

$$\boxed{\begin{aligned} &-[\rho\mathbf{v}] + [\rho\mathbf{v} \otimes \mathbf{v} + pI]\varphi'(\mathbf{x}) = \mathbf{0}, \\ &-[\rho] + [\rho\mathbf{v}]\varphi'(\mathbf{x}) = 0, \\ &-[\epsilon] + [\epsilon\mathbf{v} + p\mathbf{v}]\varphi'(\mathbf{x}) = 0, \\ &-[\rho s] + [\rho s\mathbf{v}]\varphi'(\mathbf{x}) \geq 0. \end{aligned}} \tag{1.422}$$

If discontinuities of this type occur, one denotes $\mathscr{F}$ as a *shock wave.* Supersonic aircraft for example generate shock waves. The theoretical and numerical treatment of the equations of the dynamics of gases are often quite made complicated by the occurrence of shock waves. These mathematical problems are of great importance for the construction of modern aircraft with minimal fuel consumption.

**Shock waves for conservation laws:** We consider the equation

$$\boxed{\mu_t + \operatorname{div} \mathbf{j} = g.} \tag{1.423}$$

If the solutions to this equation, $\mu$ and $\mathbf{j}$, have jumps at a point $\mathbf{x}$ on the surface $\mathscr{F}$ in (1.421), then one has:

$$\boxed{-[\mu] + [\mathbf{j}]\varphi'(\mathbf{x}) = 0.} \tag{1.424}$$

In the derivation of (1.424) it is assumed that the equation (1.423) is satisfied in the sense of distributions (cf. [212]).

Since the basic equations (1.419) of the dynamics of gases are of the form of conservation laws, the relations (1.422) for the jumps follow from (1.424).

**Sound waves in a gas:** The propagation of sound is a process which proceeds so quickly that the individual units of volume cannot exchange any heat with one another. Therefore the specific entropy density $s$ remains constant. From (1.420) we get for $s = \text{const}$ the relation

$$T = \text{const} \cdot \rho^{\gamma - 1}, \quad e = cT$$

---

[157] We set $\varphi' = \mathbf{grad}\,\varphi$ and

$$[f] = f_+ - f_- \quad \text{with} \quad f_\pm := \lim_{h \to +0} f(\mathbf{x} \pm h\mathbf{n}).$$

Here $\mathbf{n} := \varphi'(\mathbf{x})/|\varphi'(\mathbf{x})|$ is the unit normal vector to the surface $\mathscr{F}$ at the point $\mathbf{x}$ (cf. Figure 1.165). Moreover one has

$$[\rho\mathbf{v} \otimes \mathbf{v} + pI]\varphi'(\mathbf{x}) = [\rho\mathbf{v}(\mathbf{v}\varphi'(\mathbf{x}))] + [p]\varphi'(\mathbf{x}).$$

and the adiabatic equation of state

$$\boxed{p = \text{const} \cdot \rho^\gamma.}$$

Similarly as in 1.13.3.5 the speed of sound $c_S$ is obtained from the relation $c_S = \sqrt{p'(\rho)}$, hence

$$\boxed{c_S = \sqrt{\gamma p/\rho}.}$$

For the mixture of gases which form our atmosphere, experiments yield the value $\gamma \sim 1.4$.

## 1.13.4 General principles for uniqueness

### 1.13.4.1 The energy method

This method can be applied to problems for which the energy is conserved. To explain the general ideal of this method, we consider the equation for a vibrating string

$$\boxed{\begin{array}{lll} \frac{1}{c^2}u_{tt} - u_{xx} = f(x,t), & 0 < x < L, t > 0, & \\ u(0,t) = a(t), \quad u(L,t) = b(t), & t \geq 0 & \text{(boundary condition)}, \\ u(x,0) = u_0(x), \quad u_t(x,0) = u_1(x), & 0 \leq x \leq L & \text{(initial condition)}. \end{array}} \tag{1.425}$$

Here $u(x,t)$ denotes the displacement of the string at the point $x$ at time $t$.

**Uniqueness result:** The problem (1.425) has at most one smooth solution $u$.

*Proof:* For simplification we set $c = 1$. If $v$ and $w$ are two solutions of (1.425), then we set

$$u := v - w.$$

We must show that $u \equiv 0$. The function $u$ satisfies the equation (1.425) with

$$f \equiv 0, \quad a \equiv 0, \quad b \equiv 0, \quad u_0 \equiv 0 \quad \text{and} \quad u_1 \equiv 0. \tag{1.426}$$

We consider the function

$$E(t) := \int_0^L \frac{1}{2}\left(u_t(x,t)^2 + u_x(x,t)^2\right) \mathrm{d}x.$$

This corresponds to the energy of the string at time $t$.

(i) One has $E'(t) = 0$ for all times $t \geq 0$. Indeed, partial integration yields

$$E'(t) = \int_0^L (u_t u_{tt} + u_x u_{tx}) \mathrm{d}x = \int_0^L u_t(u_{tt} - u_{xx}) \mathrm{d}x + u_x(x,t)u_t(x,t)\Big|_0^L = 0$$

because of (1.425) and (1.426).

(ii) $E(0) = 0$. This follows from (1.425) and (1.426).

(iii) From (i) and (ii) it follows that $E(t) = 0$ for all times $t \geq 0$. This in turn implies

$$u_t(x,t) = u_x(x,t) = 0 \quad \text{for all} \quad x \in [0, L] \quad \text{and} \quad t \geq 0,$$

hence $u(x,t) = \text{const}$. Because of the initial condition $u(x,0) \equiv 0$ we obtain the desired result that $u \equiv 0$. □

**Basic physical idea:** The method of proof above uses the fact that one has conservation of energy and the initial energy vanishes. It follows that the energy vanishes for all times, and the system must be at rest.

The same argument may be applied to dissipative processes, in which the energy is a non-increasing function in time.

#### 1.13.4.2 Maximum principles

The basic physical ideas of the maximum principle is that the temperature differential in a body generates a flow of heat in the direction of lower temperature.

**Non-stationary heat equation:**

$$\boxed{\begin{array}{lll} T_t - \alpha\Delta T = f(x,t), & x \in \Omega, t > 0, & \\ T(x,t) = r(x), & x \in \partial\Omega, t \geq 0 & \text{(boundary value)}, \\ T(x,0) = T_0(x), & x \in \overline{\Omega} & \text{(initial value)}. \end{array}} \tag{1.427}$$

We denote by $\Omega$ a bounded domain in $\mathbb{R}^N$ with smooth boundary for $N \geq 2$. Assume that the material constant $\alpha$ is positive. We set $D := \overline{\Omega} \times [0, t_0]$ for a fixed time $t_0 > 0$.

**Maximum principle:** If $T$ is a smooth solution of (1.427) with $f \leq 0$ on $D$, and if the temperature $T$ attains its maximum in an interior point of $D$, then $T$ is constant on $D$.

**Inequality relations:** For a smooth solution $T$ of (1.427) one has:

(i) From $f \geq 0$ on $D$, $r \geq 0$ on $\partial\Omega$ and $T_0 \geq 0$ on $\overline{\Omega}$ it follows that $T \geq 0$ on $D$.

(ii) From $f \equiv 0$, $r \equiv 0$ and $T_0 \equiv 0$ it follows that $T \equiv 0$.

**Uniqueness result:** The problem (1.427) has at most one smooth solution $T$.

*Proof:* If $v$ and $w$ are two solutions, then the difference $T := v - w$ satisfies the equation (1.427) with $r \equiv 0$ and $T_0 \equiv 0$. From (ii) it then follows that $T \equiv 0$, hence $v \equiv w$. □

**The stationary heat equation:**

$$\boxed{\begin{array}{lll} -\alpha\Delta T = f(x), & x \in \Omega, & \\ T(x) = r(x), & x \in \partial\Omega & \text{(boundary value)}. \end{array}} \tag{1.428}$$

**Maximum principle:** If $T$ is a smooth solution of (1.428) with $f \leq 0$ on $\Omega$ and if the temperature $T$ attains its maximum on $\overline{\Omega}$ at a point of $\Omega$, then $T$ is constant on $\overline{\Omega}$.

**Inequality relations:** For a smooth solution $T$ of (1.428) one has:

(i) From $f \geq 0$ on $\Omega$ and $r \geq 0$ on $\partial\Omega$ it follows that $T \geq 0$ on $\overline{\Omega}$.

(ii) From $f \equiv 0$ and $r \equiv 0$ it follows that $T \equiv 0$.

**Uniqueness result:** The problem (1.428) has at most one smooth solution $T$.

### 1.13.5 General existence results

In this section we consider important classical results on the existence of solutions of partial differential equations. Modern existence results can be found in [212].

#### 1.13.5.1 The theorem of Cauchy–Kovalevski

$$\boxed{\begin{aligned} &u_t(x,t) = f(x,t,u), \\ &u(x,t_0) = \varphi(x) \qquad \text{(initial value).} \end{aligned}} \tag{1.429}$$

Here we use notions $x = (x_1, \ldots, x_n)$, $u = (u_1, \ldots, u_m)$ and $f = (f_1, \ldots, f_m)$. All quantities $t$, $x_j$ and $f_k$ are assumed to be complex. The solutions are the complex-valued functions $u_k$. We will call a function analytic if it can be developed in an *absolutely* convergent power series with respect to all of its variables.

We assume that $f$ is analytic in a neighborhood of the point $(x_0, t_0, u_0)$. Moreover we assume that $\varphi$ is analytic in a neighborhood of the point $x_0$ with $\varphi(x_0) = u_0$.

**Theorem of Augustin Cauchy (1789–1855) and Sonya Kovalevski (1850–1891):** The initial value problem (1.429) has a unique solution $u$ in a neighborhood of the point $(x_0, t_0)$, and this solution is analytic. The power series expansion can be derived by means of comparing coefficients.

*Example:*

$$\boxed{u_t = u, \qquad u(x,0) = x.}$$

We set $P := (0,0)$. From the initial conditions it follows that $u(P) = 0$, $u_x(P) = 1$, $u_{xx}(P) = 0$ etc. The differential equations yields $u_t(P) = u(P) = 0$, $u_{tt}(P) = u_t(P) = 0$, $u_{tx}(P) = u_x(P) = 1$. In this way we obtain the following solution in a neighborhood of the point $P$:

$$\begin{aligned} u(x,t) &= u(P) + u_x(P)x + u_t(P)t + \frac{1}{2}(u_{xx}(P)x^2 + 2u_{tx}(P)xt + u_{tt}(P)t^2) + \ldots \\ &= x + xt + \ldots \end{aligned}$$

#### 1.13.5.2 The theorem of Cauchy for partial differential equations of the first order

$$\boxed{\begin{aligned} &F(x,S,S_x) = 0, \\ &S(x_0(\sigma)) = S_0(\sigma) \quad \text{on } U \text{ (initial condition for } S\text{)}, \\ &S_x(x_0(\sigma)) = p_0(\sigma) \quad \text{on } U \text{ (initial condition for } S_x\text{)}. \end{aligned}} \tag{1.430}$$

The sought-for solution of the equations is the real function $S = S(x)$ of the real variables $x = (x_1, \ldots, x_n)$. Moreover $p = (p_1, \ldots, p_n)$. We view $F = F(x,S,p)$ as a function of the variables $x$, $S$ and $p$. Let $\sigma = (\sigma_1, \ldots, \sigma_{n-1})$ be a tuple of $n-1$ real parameters, which vary in a neighborhood of the origin in $\mathbb{R}^{n-1}$.

*Example 1:* The Hamilton–Jacobi differential equation

$$S_t + H(q, S_q) = 0$$

is a special case of (1.430) with $x_1 = q$, $x_2 = t$.

**Assumptions:** Let the smooth functions

$$x = x_0(\sigma), \quad p = p_0(\sigma), \quad S = S_0(\sigma) \quad \text{on} \quad U$$

be given, where we assume the following *compatibility condition* to hold[158]

$$\boxed{S_0'(\sigma) = p_0(\sigma)x_0'(\sigma) \quad \text{on } U} \tag{1.431}$$

as well as the *regularity condition*

$$\det(x_0'(0), F_p(P)) \neq 0,$$

which should be satisfied with $P := (x_0(0), S_0(0), p_0(0))$.

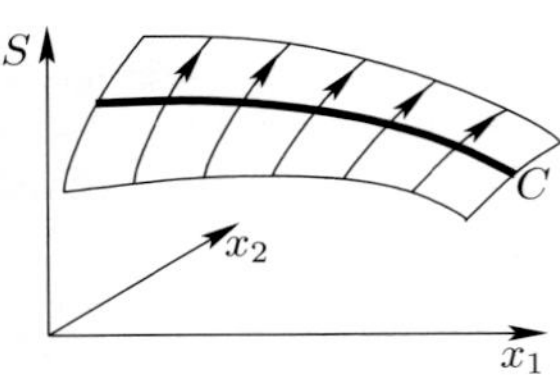

*Figure 1.166.*

**Geometric interpretation:** In the case $n = 2$ we are looking for a surface $S = S(x)$ passing through the curve

$$C : x = x(\sigma), \quad S = S_0(\sigma)$$

(Figure 1.166). For the construction of the solution, however, it turns out to be convenient to introduce in addition the quantity $p = S_x$. The chain rule then yields

$$S_0'(\sigma) = S_x(x_0(\sigma))x_0'(\sigma) = p_0(\sigma)x_0'(\sigma).$$

This is the compatibility condition (1.431).

**Theorem of Cauchy:** The initial value problem (1.430) has a unique solution $S = S(x)$ in a sufficiently small neighborhood of the point $x_0(0)$. This solution is smooth.

**Construction of the solution:** The surface $S = S(x)$ which is a solution is constructed as the union of curves

$$x = x(t;\sigma), \quad p = p(t;\sigma), \quad S = \mathscr{S}(t;\sigma)$$

with parameter $t$ and an additional parameter $\sigma$ (Figure 1.166). These curves satisfy the following system of ordinary differential equations, which one calls the *characteristic system* belonging to the partial differential equation (1.430):[159]

$$\boxed{\begin{array}{lll} x' = F_p, & S' = pF_p, & p' = -F_x - pF_S, \\ x(0) = x_0(\sigma), & S(0) = S_0(\sigma), & p(0) = p_0(\sigma). \end{array}} \tag{1.432}$$

The regularity condition is equivalent to the fact that one can solve the equation

$$\boxed{x = x(t;\sigma)}$$

[158] The classical texts on the subject are full of long and complicated formulas with lots of indices. Modern analysis prefers working with the notion of Fréchet derivative and can therefore be formulated much more briefly and elegantly.

The transition to components is provided by the relations

$$x'(\sigma) = \left(\frac{\partial x}{\partial \sigma_j}\right), \quad S_x = \left(\frac{\partial S}{\partial x_k}\right), \quad F_p = \left(\frac{\partial F}{\partial p_k}\right).$$

The compatibility conditions are then explicitly

$$\frac{\partial S_0}{\partial \sigma_j} = \sum_{k=1}^{n} p_{0k}\frac{\partial x_{0k}}{\partial \sigma_j}.$$

The determinant in the regularity condition contains $\dfrac{\partial x_0}{\partial \sigma_j}$ in the $j^{th}$ column and $F_p$ as the last column.

[159] To simplify notions we are writing here $x = x(t)$, $p = p(t)$ and $S = S(t)$ for $x = x(t;\sigma)$, $p = P(t;\sigma)$ and $S = \mathscr{S}(x,t)$.

in a neighborhood of the point $t = 0,\ \sigma = 0$ for $(t, \sigma)$. This yields a relation

$$t = t(x), \quad \sigma = \sigma(x).$$

From this we get the sought-for solution

$$\boxed{S(x) := \mathscr{S}(t(x), \sigma(x)).}$$

*Example 2:* For the Hamilton–Jacobi differential equation

$$S_t + H(q, t, S_q) = 0$$

the characteristic system (1.432) is, for $q = q(t),\ p = p(t)$:

$$q' = H_p(q, p, t), \quad p' = -H_q(q, p, t).$$

These are the *canonical Hamiltonian equations.* Moreover the equations

$$S' = pq' + P, \quad P' = -H_t(q, p, t)$$

belong for $S = S(t),\ P = P(t),\ q = q(t)$ and $p = p(t)$ to the characteristic system.

### 1.13.5.3 The theorem of Frobenius and integrability conditions

$$\boxed{\begin{array}{ll} \dfrac{\partial u}{\partial x_j}(x) = K_j(x, u(x)) & j = 1, \ldots, N, \\ u(a) = b & \text{(initial value).} \end{array}} \tag{1.433}$$

Given are the point $a \in \mathbb{R}^N$ and a real number $b$, as well as the smooth functions $K_j,\ j = 1, \ldots, N$ in a neighborhood of the point $(a, b)$ in $\mathbb{R}^{N+1}$. The sought-for solution is the real-valued function $u = u(x)$. The problem (1.433) is equivalent to the equation

$$\mathrm{d}u = K$$

with $K = \sum_{j=1}^{n} K_j(x, u)\mathrm{d}x_j$.

**Theorem of Georg Frobenius (1849–1917):** The initial value problem (1.433) has a unique solution in a sufficiently small neighborhood of the point $a$ if and only if the *integrability condition*

$$\boxed{\frac{\partial K_j(P)}{\partial x_m} + \frac{\partial K_j(P)}{\partial u}\frac{\partial u(x)}{\partial x_m} = \frac{\partial K_m(P)}{\partial x_j} + \frac{\partial K_m(P)}{\partial u}\frac{\partial u(x)}{\partial x_j}} \tag{1.434}$$

is satisfied for all $j, m = 1, \ldots, N$ with $P := (x, u)$ in a neighborhood of $(a, b)$.

**Remark:** The integrability conditions (1.434) follow immediately from the relations

$$\frac{\partial^2 u}{\partial x_j \partial x_m} = \frac{\partial^2 u}{\partial x_m \partial x_j}$$

and (1.433). A similar result holds for (1.433) with $u = (u_1, \ldots, u_M)$.

**Applications:** The theorem of Frobenius is an important tool in constructing surfaces (or more general manifolds).

(i) The proof of the main theorem in the theory of surfaces in 3.6.3.3 uses in an essential way the theorem of Frobenius. The integrability condition implies for example the famous *theorema egregium* of Gauss.

(ii) The construction of a Lie group from its Lie algebra is based on the theorem of Frobenius.

(iii) If $N = 3$ and the functions $K_j$ are independent of $u$, then (1.433) corresponds exactly to the vector equation

$$\mathbf{grad}\, u = \mathbf{F}, \quad u(a) = b.$$

Here one wants to determine the potential $-u$ of a given force field $\mathbf{F}$. The integrability condition in this case is

$$\mathbf{curl}\, \mathbf{F} = 0.$$

(iv) A general formulation of the theorem of Frobenius in the language of differential forms on manifolds can be found in [212].

#### 1.13.5.4 The theorem of Cartan–Kähler

**Basic ideas**

An arbitrary system of partial differential equations can be described as a system of equations for differential forms. The fundamental theorem of Cartan–Kähler guarantees for this kind of system unique solutions for regular initial value problems. For this it is assumed that the functions which occur in the formulation are analytic. This is a generalization of the theorem of Cauchy–Kovalevski. In fact, the theorem of Cartan–Kähler is a consequence of the the theorem of Cauchy–Kovalevski. The idea for proving this is to introduce appropriate local coordinates and solve for the first partial derivatives, leading to a system of *first order*, to which one can apply the theorem of Cauchy–Kovalevski.

*Example 1:* We consider the partial differential equation of first order

$$F(x, y, u, u_x, u_y) = 0. \tag{1.435}$$

If we set $p := u_x$ and $q := u_y$, then we get the equivalent system

$$\begin{aligned} F(x, y, u, p, q) &= 0, \\ \mathrm{d}u - p\mathrm{d}x - q\mathrm{d}y &= 0. \end{aligned} \tag{1.436}$$

This is a system of equations for differential forms, where we view functions as differential forms of the zeroth order.

*Example 2:* The partial differential equation of the second order

$$F(x, y, u, u_x, u_y, u_{xx}, u_{xy}, u_{yy}) = 0$$

is transformed into the equivalent system

$$\begin{aligned} F(x, y, u, p, q, a, b, c) &= 0, \\ \mathrm{d}u - p\mathrm{d}x - q\mathrm{d}y &= 0, \\ \mathrm{d}p - a\mathrm{d}x - b\mathrm{d}y &= 0, \\ \mathrm{d}q - b\mathrm{d}x - c\mathrm{d}y &= 0 \end{aligned}$$

through the substitutions $p := u_x$, $q := u_y$, $a := u_{xx}$, $b := u_{xy}$ and $c := u_{yy}$.

In a similar manner one can treat an arbitrary system of partial differential equations.

> The transition to differential forms greatly simplifies the treatment, since one can then apply the elegant differential calculus of Cartan.

The original theory, which originated in work of Riquier at the end of the nineteenth century, used differential equations and was very complicated and difficult to understand. Between 1904 and 1908 Élie Cartan discovered that the calculus of differential forms could advantageously be applied to this circle of problems. The final, very elegant formulation was provided by Erich Kähler in 1934.

**The procedure of forming the closure:** If a system

$$\omega_j = 0, \quad j = 1, \ldots, J$$

of differential forms is given, then the *closure* is formed by adding the Cartan derivatives

$$\mathrm{d}\omega_j = 0, \quad j = 1, \ldots, J.$$

In this way all dependencies between the partial derivatives of the coefficient functions are dealt with (integrability condition). Because of $\mathrm{dd}\omega \equiv 0$ (Poincaré Lemma), no new information is obtained by adjoining the Cartan derivatives once more. Therefore the process of forming the completion is finished after one step.

*Example 3:* The closure of the equation

$$a(x,y)\mathrm{d}x + b(x,y)\mathrm{d}y = 0$$

yields the relation $\mathrm{d}a \wedge \mathrm{d}x + \mathrm{d}b \wedge \mathrm{d}y = 0$, hence

$$\begin{aligned} &a\ \mathrm{d}x + b\ \mathrm{d}y = 0, \\ &(a_y - b_x)\ \mathrm{d}y \wedge \mathrm{d}x = 0. \end{aligned}$$

### Integral manifolds

The solutions of the systems of differential forms are called *integral manifolds.* The solutions can be found by trial and error, applying appropriate substitutions.

*Example 4:* We consider the closed system

$$\boxed{\begin{aligned} &y + f(x) = 0, \\ &\mathrm{d}y + f'(x)\mathrm{d}x = 0, \\ &\mathrm{d}x \wedge \mathrm{d}y = 0. \end{aligned}} \tag{1.437}$$

**Zero-dimensional integral manifolds:** A point $(x_0, y_0)$ is a solution of (1.437), if and only if $y_0 + f(x_0) = 0$.

**One-dimensional integral manifolds:** The curve

$$x = x(t), \quad y = y(t)$$

is a solution of (1.437), if

$$\begin{aligned} &y(t) + f(x(t)) = 0, \\ &[y'(t) + f'(x(t))x'(t)]\ \mathrm{d}t = 0, \\ &x'(t)y'(t)\ \mathrm{d}t \wedge \mathrm{d}t = 0. \end{aligned} \tag{1.438}$$

This is obtained by substituting $\mathrm{d}x = x'(t)\mathrm{d}t$ and $\mathrm{d}y = y'(t)\mathrm{d}t$ in (1.437). The system (1.438) is equivalent to the system

$$\begin{aligned} y(t) + f(x(t)) &= 0, \\ y'(t) + f'(x(t))x'(t) &= 0. \end{aligned}$$

The third equation of (1.438) is automatically satisfied because of $\mathrm{d}t \wedge \mathrm{d}t = 0$. In general one has:

> In searching for $r$-dimensional integral manifolds, it is sufficient to consider all differential forms up to order $r$.

**Two-dimensional integral manifolds:** The surface

$$x = x(t, s) \quad y = y(t, s)$$

is a solution of (1.437), if

$$\begin{aligned} &y(P) + f(x(P)) = 0, \\ &y_t(P)\ \mathrm{d}t + y_s(P)\ \mathrm{d}s + f'(x(P))[x_t(P)\ \mathrm{d}t + x_s(P)\ \mathrm{d}s] = 0, \\ &[x_t(P)y_s(P) - x_s(P)y_t(P)]\ \mathrm{d}t \wedge \mathrm{d}s = 0. \end{aligned} \tag{1.439}$$

Here we have set $P := (t, s)$. This follows upon making the substitution $\mathrm{d}x = x_t\mathrm{d}t + x_s\mathrm{d}s$ and $\mathrm{d}y = y_t\mathrm{d}t + y_s\mathrm{d}s$ from (1.437). The system (1.439) is equivalent to the following system:

$$\begin{aligned} &y(P) + f(x(P)) = 0, \\ &y_t(P) + f'(x(P))x_t(P) = 0, \\ &y_s(P) + f'(x(P))x_s(P) = 0, \\ &x_t(P)y_s(P) - x_s(P)y_t(P) = 0. \end{aligned}$$

Here the coefficients in (1.439) of $\mathrm{d}t$, $\mathrm{d}s$ and $\mathrm{d}t \wedge \mathrm{d}s$ have been set to zero.

**The pull-back $g^*\omega$ of a differential form $\omega$:** In order to conveniently formulate the notion of integral manifold above, we introduce the symbol $g^*\omega$. The equation

$$y = g(t), \quad t \in U,$$

with $y = (y_1, \ldots, y_n)$, $t = (t_1, \ldots, t_m)$ and an open set $U \subset \mathbb{R}^m$, describes an integral manifold given by the equation

$$\boxed{\omega = 0,}$$

if and only if

$$\boxed{g^*\omega(t) = 0, \quad t \in U.} \tag{1.440}$$

Here the pull-back $g^*\omega$ is obtained from $\omega$ upon transforming the $y$-coordinates to $t$-coordinates utilizing the substitution $y = g(t)$. If we denote by

$$e_1 := (1, 0, \ldots, 0), \quad e_2 := (0, 1, 0, \ldots, 0), \quad \ldots, \quad e_m := (0, \ldots, 0, 1)$$

the canonical basis of $\mathbb{R}^m$, then one has $\mathrm{d}t_j(e_k) = \delta_{jk}$, and equation (1.440) is equivalent to the relation

$$g^*\omega(t)(e_1, \ldots, e_m) = 0, \quad t \in U. \tag{1.441}$$

*Example 5:* Let the equation $\omega = 0$ be given in the form

$$\mathrm{d}y_1 \wedge \mathrm{d}y_2 = 0. \tag{1.442}$$

Moreover, set

$$y_j = g_j(t_1, t_2), \quad j = 1, 2. \tag{1.443}$$

We set $\partial_j := \partial/\partial t_j$. From $\mathrm{d}y_j = \partial_1 g_j \mathrm{d}t_1 + \partial_2 g_j \mathrm{d}t_2$, (1.440) corresponds here to the relation

$$(\partial_1 g_1 \partial_2 g_2 - \partial_2 g_1 \partial_1 g_2)\ \mathrm{d}t_1 \wedge \mathrm{d}t_2 = 0.$$

If we take account of $(\mathrm{d}t_1 \wedge \mathrm{d}t_2)(e_1, e_2) = \mathrm{d}t_1(e_1)\mathrm{d}t_2(e_2) - \mathrm{d}t_1(e_2)\mathrm{d}t_2(e_1) = 1$, then (1.441) is equivalent to

$$\partial_1 g_1 \partial_2 g_2 - \partial_2 g_1 \partial_1 g_2 = 0. \tag{1.444}$$

Hence the function (1.443) is a solution of (1.442) if and only if the equation (1.444) holds. This corresponds to the same method applied in Example 4.

Our goal is the formulation of global existence and uniqueness results. We first consider a local variant.

**The local existence and uniqueness theorem of Cartan–Kähler**

We study the initial value problem

$$\boxed{\begin{array}{l} \omega_k = 0, \quad k = 1, \ldots, K, \\ g(t_1, \ldots, t_p, 0) = a(t_1, \ldots, t_p) \text{ on } W_0 \quad \text{(initial value).} \end{array}} \tag{1.445}$$

The function $a$ is given. The object we are trying to determine is an integral manifold

$$\boxed{y = g(t_1, \ldots, t_{p+1}) \text{ on } W,}$$

where the coordinates $t_1, \ldots, t_{p+1}$ vary in an open neighborhood $W$ of $\mathbb{R}^{p+1}$. We denote by $W_0$ the set of all points of $W$ with $t_{p+1} = 0$. The fixed integer $p$ is assumed to fulfill the inequality $1 \le p < n$.

**Remark:** We consider the differential forms $\omega_j$ with respect to a fixed $y$-coordinate system, where $y = (y_1, \ldots, y_n)$ and $y \in \mathbb{R}^n$. The transformation

$$\boxed{y = \varphi(t), \quad t \in U}$$

introduces a new $t$-coordinate with $t = (t_1, \ldots, t_n)$ and $t \in \mathbb{R}^n$, where $U$ is an open neighborhood of the origin and $\varphi : U \longrightarrow V$ is a diffeomorphism of $U$ to an open neighborhood $V$ of the point $y_0$ in $\mathbb{R}^n$. We assume that $\varphi$ is analytic, i.e., that the components of $\varphi$ can be expanded in (absolutely convergent) power series. We denote by $e_1, \ldots, e_n$ the canonical basis in the $t$-coordinates, i.e., $e_1 = (1, 0, \ldots, 0)$, etc.

We assume that

$$a(t_1, \ldots, t_p) = \varphi(t_1, \ldots, t_p, 0, \ldots, 0)$$

on $W_0$.

**The dual polar space:** We choose a fixed point $t \in U$ and denote by $P(t)$ the *polar space*, the set of all vectors $v \in \mathbb{R}^n$ for which

$$\boxed{\omega_k^*(t)(e_1, \ldots, e_p, v) = 0, \quad k = 1, \ldots, K.}$$

Here $\omega_k^*(t)$ denotes the differential form $\omega_k$ after the transformation into the $t$-coordinates, i.e., $\omega_k^* := \varphi^*\omega_k$.

**Regularity of the initial value problem:** We speak of a *regular initial value problem,* if we have the following:

(a) There is a fixed number $r$, such that all $r$ vectors $e_{n-r+1}, \ldots, e_n$ form together with $e_1, \ldots, e_p$ a basis of the polar space $P(t)$ in each point $t \in \mathbb{R}^n$ in a sufficiently small neighborhood of $t = 0$. Here we require $1 \leq r < n - p$.

(b) The matrix whose entries are the first partial derivatives

$$\boxed{\frac{\partial}{\partial t_j} a^*\omega_k(t)(e_1, \ldots, e_p), \quad k = 1, \ldots, K, \ j = 1, \ldots, p}$$

has constant rank $n - p$ in a neighborhood of the point $(t_1, \ldots, t_p) = 0$ in $\mathbb{R}^p$.

Here $a^*\omega_k(t)$ is pulled back from $\omega_k$ by applying the transformation $y = a(t_1, \ldots, t_p)$ to the coordinates $t_1, \ldots, t_p$.

**Remark:** Condition (b) guarantees that the system (1.445) is regular along the $p$-dimensional initial surface

$$I_p : y = a(t_1, \ldots, t_p).$$

If condition (b) does not hold, then there is some singular behavior, which usually is connected with the fact that $I_p$ represents a characteristic.[160] In this case there can be infinitely many solution surfaces through $I_p$. The sought-for $(p+1)$-dimensional solution surface of the original system (1.445) is denoted by

$$I_{p+1} : y = a(t_1, \ldots, t_{p+1}).$$

For the uniqueness statement we need the $(n - r)$-dimensional surface

$$F : y = \varphi(t_1, \ldots, t_p, t_{p+1}, \ldots, t_{n-r}, 0, \ldots, 0),$$

where $(t_1, \ldots, t_{n-r})$ varies in neighborhood of the origin in $\mathbb{R}^{n-r}$. In order to pass from these considerations to the global formulation in the following sections, we introduce in addition the $n$-dimensional open neighborhood of the origin

$$M : y = \varphi(t_1, \ldots, t_n),$$

where $t$ varies in a neighborhood of the origin in $\mathbb{R}^n$. The situation is summarized by the following inclusions:

$$\boxed{I_p \subset I_{p+1} \subseteq F \subseteq M.} \tag{1.446}$$

**The local existence and uniqueness result:** We make the following assumptions:

(i) *Closedness:* The initial system (1.445) is closed, i.e., for every differential form $\omega_k$ in (1.445), also the Cartan derivative $d\omega_k$ belongs to (1.445).

(ii) *Analyticity:* The system (1.445) is analytic, i.e., the coefficients of the differential forms and the initial value function $a(.)$ can be developed in power series with real coefficients.[161]

[160] One can study the structure of characteristics of partial differential equations by analyzing non-regular initial value problems. In particular, one gets in this manner the characteristic system of Cauchy for partial differential equations of first order (cf. 1.13.5.2) and its generalization to partial differential equations of higher order and even of arbitrary systems. Details on this can be found for example in [141].

[161] The assumption of analyticity can be weakened to $C^\infty$-smoothness.

(iii) *Regularity:* The initial value problem is regular.

Then the initial value problem (1.445) has an analytic solution in a sufficiently small neighborhood of the origin in $\mathbb{R}^n$, in which the surface $F$ is contained.

**The global existence and uniqueness theorem of Cartan–Kähler**

We consider the system

$$\boxed{\omega_k = 0, \quad k = 1, \ldots, K} \tag{1.447}$$

of analytic differential forms of arbitrary order $\geq 0$ on a real analytic manifold $\mathscr{M}$ of dimension $n \geq 1$. The basic situation is now described by the inclusions

$$\boxed{\mathscr{I}_p \subset \mathscr{I}_{p+1} \subseteq \mathscr{F} \subseteq \mathscr{M}.} \tag{1.448}$$

**Global existence and uniqueness result:** We make the following assumptions:

(i) A $p$-dimensional submanifold $\mathscr{I}_p \subset \mathscr{M}$ is given with $1 \leq p < n$.

(ii) $\mathscr{I}_p$ is a regular integral manifold of the original system (1.447).

(iii) The polar space of $\mathscr{I}_p$ has in every point of $\mathscr{I}_p$ dimension $r+p$ with $1 \leq r < n-p$.

(iv) There is a submanifold $\mathscr{F}$ in $\mathscr{M}$ of dimension $n-r$ with $\mathscr{I}_p \subset \mathscr{F}$, where the tangent space of $\mathscr{F}$ is transversal to the polar space at every point of $\mathscr{I}_p$.

Then the system (1.447) has exactly one $(p+1)$-dimensional submanifold $\mathscr{I}_{p+1}$ of $\mathscr{M}$ which is an integral manifold with the property (1.448).

**Helpful remark for the reader:** Global statements are often formulated in modern mathematics in the elegant language of manifolds, which are presented in detail in [212]. However, the contents of the above theorem can be understood by every reader, without understanding that language. Just imagine that at every point of $\mathscr{M}$ one can introduce local $t$-coordinates, for which we have the situation of the local existence and uniqueness theorem for the problem (1.445). Locally, (1.448) is realized as the statement (1.446).

**Sketch of proof:** To obtain the global theorem, one first proves the local statement with the help of the theorem of Cauchy–Kovalevski and then extends this local solution with the help of *analytic continuation* to a global solution. This is where the assumptions on analyticity come in.

**Applications:** The theorem of Cauchy–Kovalevski has a large number of applications in differential geometry (the construction of manifolds with given properties) and in mathematical physics. We recommend the monograph [129] for details on this. The theorem of Frobenius for manifolds is a special case of the theorem of Cartan–Kähler. Applications of the theorem of Frobenius in thermodynamics will be discussed in [212].

**Differential ideals:**

Up to now we have considered finite systems

$$\omega_k = 0, \qquad k = 1, \ldots, K \tag{1.449}$$

of differential forms. Different choices of the forms $\omega_k$ can lead to equivalent systems. In order to eliminate this indeterminacy, one considers systems of the form

$$\boxed{\omega = 0, \quad \omega \in J.} \tag{1.450}$$

Here $J$ is a so-called *differential ideal*, i.e., one has

(i) $J$ is a real linear space of differential forms.

(ii) From $\omega \in J$ it follows that $\omega \wedge \mu \in J$ for every differential form $\mu$.

(iii) From $\omega \in J$ it follows that $\mathrm{d}\omega \in J$.

Every ideal of this kind possesses a finite basis $\omega_1, \ldots, \omega_K$.

Therefore (1.450) is equivalent to (1.449), where the concrete choice of basis of the ideal is irrelevant.

The procedure corresponds to a general strategy of modern algebraic geometry, which is to replace systems of equations by the systems of objects which are annihilated by an ideal.

## 1.14 Complex function theory

*The introduction of complex quantities into mathematics originates in and has its first applications in the theory of variables which are dependent upon one another via simple operations. Indeed, if one applies these dependencies in an extended sense by allowing the variables to be complex, then a hidden harmony and structure becomes apparent.*

*Bernhard Riemann, 1851*

*Riemann (1826–1866) is the man with glowing intuition. Through his overall genius he stands far above all his peers. Everywhere that his interest was aroused, he started the theory from scratch without worrying about tradition or constraints of existing systems. Weierstrass (1815–1897) was above all a logician; he works slowly, systematically, step by step. Where he works, he heads for a finished form of the theory.*

*Felix Klein* (1849–1925)

The development of the theory of functions of a complex variable took a rather winding path, as opposed to the modern theory of today with its extreme elegance, which belongs to the most beautiful and esthetically pleasing theories mathematics has to offer. This theory reaches into all branches of mathematics and physics. The formulation of modern quantum field theory for example is based in essence on the notion of a complex number.

The complex numbers were introduced by the Italian mathematician Bombelli in the middle of the sixteenth century, in order to solve equations of the third order. Euler (1707–1783) introduced instead of the number $\sqrt{-1}$ the symbol i and discovered the famous formula

$$\mathrm{e}^{x+\mathrm{i}y} = \mathrm{e}^x(\cos y + \mathrm{i}\sin y), \qquad x, y \in \mathbb{R},$$

which establishes a surprising and most important connection between trigonometric functions and the exponential function.

In his dissertation in 1799, Gauss provided the first (almost) complete proof of the *fundamental theorem of algebra*. For this proof he required complex numbers as a tool. Gauss eliminated the mysticism which surrounded complex numbers of the form $x + \mathrm{i}y$ up until that time, and showed that they may be interpreted as points $(x, y)$ in the complex (Gaussian) plane (cf. 1.1.2). There is much evidence that Gauss already knew many properties of the complex-valued functions at the beginning of the eighteenth century, in particular the relation to elliptic integrals. However, he never published any of this.

In his famous *Cours d'analyse* (course in analysis), Cauchy treated power series in 1821 and showed that series of this kind in the complex realm have a *circle of convergence.* In a fundamental piece of work in 1825, Cauchy considered contour integrals and discovered their *independence from the path of integration.* In this regard, he later developed a *calculus of residues* for the calculation of apparently complicated integrals.

In 1851, Riemann took a decisive step in the construction of a theory of complex-valued functions, when in his dissertation at Göttingen, with the title *Grundlagen für eine allgemeine Theorie der Funktionen einer veränderlichen komplexen Größe* (Foundations of a theory of functions of a complex variable), he founded the so-called geometric function theory, which uses *conformal maps* and which is distinguished by its intuitive appeal and the close proximity to physics.

Parallel to Riemann's work, Weierstrass developed rigorous analytic foundations for function theory based on power series. The work of both Riemann and Weierstrass was centered around the search for a deeper understanding of elliptic and more general Abelian integrals for algebraic functions. In this connection, completely new ideas are due to Riemann, out of which modern topology – the mathematics of qualitative behavior and form – sprouted.

In the last quarter of the nineteenth century Felix Klein and Henri Poincaré created the powerful structure of the theory of automorphic functions. This class of functions is a broad generalization of the periodic and doubly period (elliptic) functions and is closely related to Abelian integrals.

In 1907 Koebe and Poincaré proved independently the famous uniformization theorem, which represents one of the highlights of classical function theory and which completely clarifies the structure of Riemann surfaces. This uniformization theorem, which Poincaré sought for years to prove, is described in [212].

The first modern and complete presentation of classical function theory was given by Hermann Weyl in his book *Die Idee der Riemannschen Fläche* (The idea of Riemann surfaces), which is a pearl of the mathematical literature.[162]

Important new ideas in function theory were introduced in the fifties by the French mathematicians Jean Leray and Henri Cartan, who developed the notion and theory of sheaves, which is described in [212].

## 1.14.1 Basic ideas

**The local–global principle in analysis:**[163] The elegance of complex function theory is based on the following three basic facts:

(i) Every differentable complex-valued function on an open set can be developed locally in a power series, i.e., is analytic.

(ii) The contour integral of analytic functions in a simply connected domain is independent of the path of integration.

(iii) Every locally given function which is analytic in a neighborhood of a point can be uniquely extended to a global analytic function, provided one uses the notion of Riemann surface as the domain of definition of that extension. Hence one has:

| The local behavior of an analytic function uniquely determines its global behavior. | (1.451) |
|---|---|

[162] A new edition of this classic with commentaries has been published by Teubner-Verlag in 1996 (cf. [236]).

[163] The local-global principle in number theory can be found in 2.7.10.2 ($p$-adic numbers).

This kind of rigid behavior is in general not possessed by real-valued functions.

*Example 1:* The real-valued function $f(x) := x$ for $x \in [0, \varepsilon]$ can be extended in *infinitely many ways* to a differentiable function (Figure 1.167). The unique extension of this function, viewed as a complex-valued function, is the analytic function

$$f(z) = z, \quad z \in \mathbb{C}. \tag{1.452}$$

*Example 2:* The differentiable functions depicted in Figures 1.167(b),(c) can not be extended to analytic functions in the complex domain. This is because they locally coincide with the function (1.452), but *not* globally.

The principle (1.451) is quite important in physics. If one knows that a physical quantity is analytic, then one only needs to know its behavior in a small open set, in order to understand (or at least determine) its global behavior. This is the case for example for the elements of the $S$-matrix, which describe the scattering of elementary particles in modern particle accelerators. From this fact one gets the so-called *dispersion relations.*

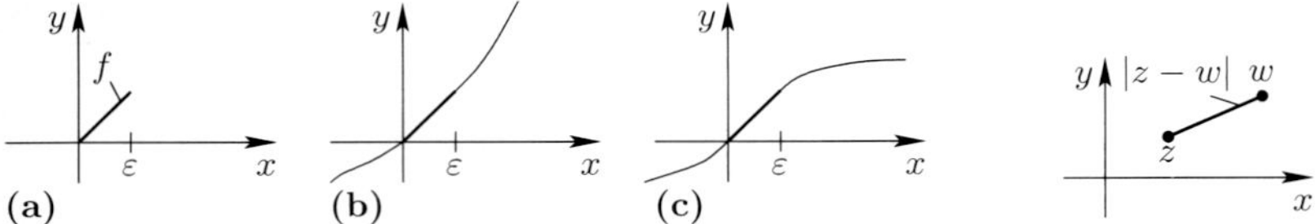


*Figure 1.167. Non-analytic functions in the plane.*

*Figure 1.168.*

## 1.14.2 Sequences of complex numbers

Every complex number $z$ can be written in the form

$$\boxed{z = x + \mathrm{i}y}$$

with real numbers $x$, $y$. One has

$$\boxed{\mathrm{i}^2 = -1.}$$

We write $\operatorname{Re} z := x$ (the *real part* of $z$) and $\operatorname{Im} z := y$ (the *imaginary part* of $z$). The set of complex numbers is denoted by $\mathbb{C}$. The rules for computations involving complex numbers can be found in 1.1.2.

**The metric of the complex plane $\mathbb{C}$:** If $z$ and $w$ are two complex numbers, then we define their *distance* by the formula

$$\boxed{d(z, w) := |z - w|.}$$

This is the classical notion of distance between two points in the plane (Figure 1.168). This gives $\mathbb{C}$ the structure of a metric space, and all of the general results for metric spaces (cf. 1.3.2) may be applied to $\mathbb{C}$.

**Convergence of complex sequences:** We say we have *convergence*, denoted

$$\boxed{\lim_{n\to\infty} z_n = z}$$

if and only if the sequence of complex numbers $(z_n)$ satisfies the relation

$$\boxed{\lim_{n\to\infty} |z_n - z| = 0.}$$

This is equivalent to

$$\lim_{n\to\infty} \operatorname{Re} z_n = \operatorname{Re} z \quad \text{and} \quad \lim_{n\to\infty} \operatorname{Im} z_n = \operatorname{Im} z.$$

*Example 3:* One has

$$\lim_{n\to\infty} \left(\frac{1}{n} + \frac{n}{n+1}\mathrm{i}\right) = \mathrm{i},$$

since $1/n \longrightarrow 0$ and $n/(n+1) \longrightarrow 1$ as $n \longrightarrow \infty$.

**Convergence of series with complex terms:** This convergence is defined by the relation

$$\sum_{k=0}^{\infty} a_k = \lim_{n\to\infty} \sum_{k=0}^{n} a_k.$$

Series of this kind were studied in section 1.10.

**Convergence of complex functions:** The limit

$$\lim_{z\to a} f(z) = b$$

is to be understood as meaning that for every sequence of complex numbers $(z_n)$ with $z_n \neq a$ for all $n$ and $\lim_{n\to\infty} z_n = a$ implies $\lim_{n\to\infty} f(z_n) = b$.

### 1.14.3 Differentiation

The connection between complex differentiation and the theory of partial differential equations is given by the fundamental Cauchy–Riemann differential equations.

**Definition:** Let a function $f : U \subseteq \mathbb{C} \longrightarrow \mathbb{C}$ be given, defined in a neighborhood of the point $z_0$. $f$ is said to be *complex differentiable* at $z_0$, if the limit

$$\boxed{f'(z_0) := \lim_{h\to 0} \frac{f(z_0 + h) - f(z_0)}{h}}$$

exists. The complex number $f'(z_0)$ is called the (complex) derivative of $f$ at the point $z_0$. We set

$$\frac{\mathrm{d}f(z_0)}{\mathrm{d}z} := f'(z_0).$$

*Example 1:* For $f(z) := z$ we get $f'(z) = 1$.

We set $z = x + \mathrm{i}y$ with $x, y \in \mathbb{R}$ and

$$f(z) = u(x, y) + \mathrm{i}v(x, y),$$

i.e., $u(x, y)$ (resp. $v(x, y)$) is the real (resp. imaginary) part of $f(z)$.

**The main theorem of Cauchy (1814) and Reimann (1851):** A complex-valued function $f : U \subseteq \mathbb{C} \longrightarrow \mathbb{C}$ is complex differentiable at $z_0 \in \mathbb{C}$ if and only if both $u$ and $v$ are Fréchet-differentiable at the point $(x_0, y_0)$ and the *Cauchy–Riemann differential equations*

$$\boxed{u_x = v_y, \qquad u_y = -v_x} \tag{1.453}$$

hold at the point $(x_0, y_0)$. In this case one has

$$f'(z_0) = u_x(x_0, y_0) + \mathrm{i}v_x(x_0, y_0).$$

**Holomorphic functions:** A function $f : U \subseteq \mathbb{C} \longrightarrow \mathbb{C}$ is said to be *holomorphic* on an open set $U$, if $f$ is complex differentiable in every point $z$ of $U$.

**Rules for differentiation:** The same rules as for real-valued differentiable functions, the rule of sums, the product and quotient rules as well as the chain rule, hold for complex derivatives (cf. 0.8.2). The rules for derivatives of inverse functions are considered in section 1.14.10 below.

**Power series:** A function

$$f(z) = a_0 + a_1(z - a) + a_2(z - a)^2 + a_3(z - a)^3 + \ldots$$

can be differentiated at every point of the interior of the *domain of convergence* (cf. 1.10.3). The derivative can be obtained by termwise differentiation of the power series, i.e., one has

$$f'(z) = a_1 + 2a_2(z - a) + 3a_3(z - a)^2 + \ldots .$$

*Example 2:* Let $f(z) := \mathrm{e}^z$. From the relation

$$f(z) = 1 + z + \frac{z^2}{2!} + \frac{z^3}{3!} + \ldots$$

it follows that

$$f'(z) = 1 + z + \frac{z^2}{2!} + \ldots = \mathrm{e}^z \qquad \text{for all } z \in \mathbb{C}.$$

**Table of derivatives:** For all functions which can be developed in power series, the real and complex derivatives coincide. A table of derivatives for important elementary functions can be found in 0.8.1.

**The differential operators $\partial_z$ and $\partial_{\overline{z}}$ of Poincaré:** If we set

$$\partial_z := \frac{1}{2}\left(\frac{\partial}{\partial x} - \mathrm{i}\frac{\partial}{\partial y}\right) \qquad \text{and} \qquad \partial_{\overline{z}} := \frac{1}{2}\left(\frac{\partial}{\partial x} + \mathrm{i}\frac{\partial}{\partial y}\right),$$

then the Cauchy–Riemann differential equations (1.453) can be written in the elegant form

$$\boxed{\partial_{\overline{z}} f(z_0) = 0 .}$$

Let $f : U \subseteq \mathbb{C} \longrightarrow \mathbb{C}$ be an arbitrary complex-valued function. We set $z = x + \mathrm{i}y$ and $\overline{z} = x - \mathrm{i}y$ as well as

$$\mathrm{d}z := \mathrm{d}x + \mathrm{i}\mathrm{d}y \quad \text{and} \quad \mathrm{d}\overline{z} := \mathrm{d}x - \mathrm{i}\mathrm{d}y.$$

If we write $f(x,y)$ instead of $f(z)$, then one has

$$\mathrm{d}f = u_x\mathrm{d}x + u_y\mathrm{d}y + \mathrm{i}(v_x\mathrm{d}x + v_y\mathrm{d}y).$$

From this it follows that

$$\boxed{\mathrm{d}f = \partial_z f\mathrm{d}z + \partial_{\overline{z}} f\mathrm{d}\overline{z}.}$$

If $f$ is complex differentiable at the point $z_0$, then one has

$$\mathrm{d}f = \partial_z f\mathrm{d}z.$$

### 1.14.4 Integration

The most important integration property of holomorphic functions is the independence of the contour integral $\int_C f(z)\mathrm{d}z$ from the path of integration on *simply connected domains* (integral theorem of Cauchy).

**Curves in the complex plane $\mathbb{C}$:** A curve $C$ in $\mathbb{C}$ is given by a function

$$\boxed{z = z(t),\ a \le t \le b} \qquad (1.454)$$

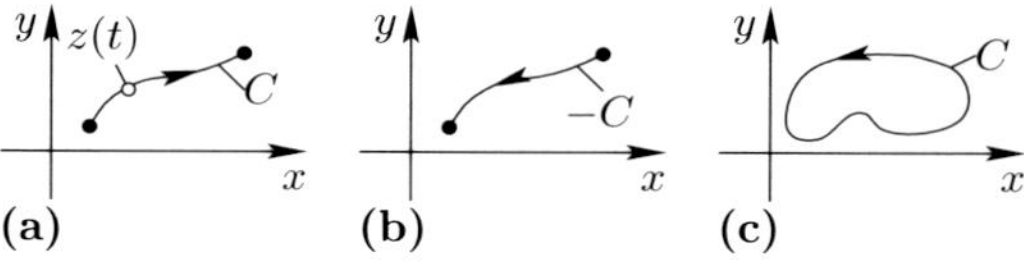


*Figure 1.169. Curves in the complex plane.*

(Figure 1.169(a)). Here $-\infty < a < b < \infty$. We set $z := x + \mathrm{i}y$, so that this corresponds to a real curve

$$x = x(t), \quad y = y(t), \quad a \le t \le b$$

in the plane $\mathbb{R}^2$. The curve $C$ is said to by of type $C^1$ if both functions $x = x(t)$ and $y = y(t)$ are of type $C^1$ on $[a,b]$.

**Jordan curves:** A curve $C$ is called a *Jordan curve*, if the map $A : t \mapsto z(t)$ is a homeomorphism on $[a,b]$.[164]

The curve $C$ in (1.454) is a *closed* curve, if $z(a) = z(b)$. A closed Jordan curve is the homeomorphic image of the circle $\{z \in \mathbb{C} : |z| = 1\}$ in the complex plane $\mathbb{C}$.

Jordan curves have regular behavior, i.e., there are no self-intersections as shown in Figure 1.169(c).

**Definition of curve integrals:** If a function $f : U \subseteq \mathbb{C} \longrightarrow \mathbb{C}$ is continuous on an open set $U$ and $C : z = z(t),\ a \le t \le b$ is a $C^1$-curve, then we define the *curve integral* by the formula

$$\boxed{\int_C f(z)\mathrm{d}z := \int_a^b f(z(t))z'(t)\mathrm{d}t.}$$

This definition is independent of the parameterization of the oriented curve $C$.[165] If the curve $C$ is closed, then one speaks of a *contour integral*.

[164] This means that $A$ is bijective and both $A$ and $A^{-1}$ are continuous. Because of the compactness of $[a,b]$ it is sufficient that $A$ is bijective and continuous.

[165] Here $C^1$-changes of parameter $t = t(\tau)$, $\alpha \le \tau \le \beta$ are allowed, where $t'(\tau) > 0$ for all $\tau$, i.e., $t = t(\tau)$ is strictly increasing.

**Triangle inequality:**

$$\left| \int_C f \, \mathrm{d}z \right| \le (\text{length of } C) \sup_{z \in C} |f(z)| .$$

**Change of orientation:** If we denote by $-C$ the curve which is obtained from $C$ by reversing the orientation, then one has (Figure 1.169(b)):

$$\int_{-C} f \, \mathrm{d}z = -\int_C f \, \mathrm{d}z .$$

**Main theorem of Cauchy (1825) and Morera (1886):** A continuous function $f : U \subseteq \mathbb{C} \longrightarrow \mathbb{C}$ on a *simply connected domain* $U$ is holomorphic if and only if the integral $\int_C f \mathrm{d}z$ is independent of the path of integration in $U$.[166]

*Example 1:* In Figure 1.170(a) we have $\int_C f \mathrm{d}z = \int_{C'} f \mathrm{d}z$. The fundamental notion of simply connected domains was introduced in section 1.3.2.4. Intuitively, simply connected domains have no *holes*.

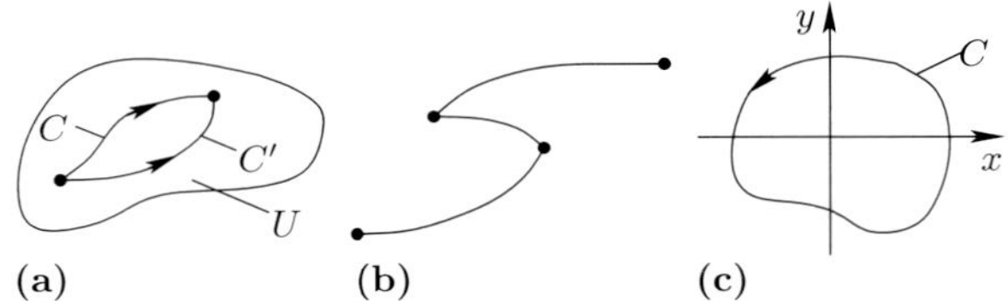


*Figure 1.170. Properties of contour integrals.*

**Corollary:** A continuous function $f : U \subseteq \mathbb{C} \longrightarrow \mathbb{C}$ is holomorphic on a simply connected domain $U$, if

$$\int_C f \, \mathrm{d}z = 0 \qquad (1.455)$$

for all closed $C^1$-Jordan curves $C$ in $U$.

*Example 2:* If $C$ is a closed Jordan curve of class $C^1$ (for example a circle) encircling the origin, positively oriented in the mathematical sense (i.e., counter clockwise), then

$$\int_C z^k \mathrm{d}z = \begin{cases} 0 & \text{for } k = 0, 1, 2, \ldots \\ 2\pi \mathrm{i} & \text{for } k = -1, \\ 0 & \text{for } k = -2, -3, \ldots \end{cases} \qquad (1.456)$$

If $k = 0, 1, 2, \ldots$, the function $f(z) := z^k$ is holomorphic in $\mathbb{C}$. In this case (1.456) follows from (1.455). For $k = -1$, the function $f(z) := z^{-1}$ is homomorphic in the non-simply

[166] A *path* is a $C^1$-curve or one which is pasted together of finitely many such curves (Figure 1.170(b)). This includes for example the circumferences of polygons.

connected domain $\mathbb{C} - \{0\}$, while this function is not holomorphic in all of $\mathbb{C}$. The relation which arises from (1.456),

$$\boxed{\int_C \frac{dz}{z} = 2\pi i,}$$

shows therefore that the assumption that $U$ is simply connected in (1.455) cannot be weakened (Figure 1.170(c)).

In what follows, let $C$ denote a $C^1$-curve in $U$ with starting point $z_0$ and endpoint $z$.

**The fundamental theorem of calculus:** Let a continuous function $f : U \subseteq \mathbb{C} \longrightarrow \mathbb{C}$ on a domain $U$ be given and let $F$ be a primitive of $f$ on $U$, i.e., $F' = f$ on $U$. Then

$$\boxed{\int_C f\,dz = F(z) - F(z_0).}$$

Two primitives of $f$ on $U$ differ by a constant.

*Example 3:* $\int_C e^z dz = e^z - e^{z_0}$.

**Corollary:** If the function $f : U \subseteq \mathbb{C} \longrightarrow \mathbb{C}$ is holomorphic in a simply connected domain $U$, then the function

$$F(z) := \int_{z_0}^{z} f(\zeta)d\zeta$$

is a primitive of $f$ on $U$.[167]

A fundamental (topological) property of integrals of holomorphic functions is the fact that they are unchanged upon passing to $C^1$-homotopic and $C^1$-homologous paths.

**$C^1$-homotopic paths:** Let a domain $U$ in the complex plane $\mathbb{C}$ be given. Two $C^1$-curves $C$ and $C'$ are said to be *$C^1$-homotopic*, if the following two conditions are satisfied:

(i) $C$ and $C'$ have the same start and end points.

(ii) $C$ can be deformed differentiably into $C'$ (Figure 1.170(a)).[168]

**Theorem 1:** If $f : U \subseteq \mathbb{C} \longrightarrow \mathbb{C}$ is holomorphic on the domain $U$, then one has

$$\boxed{\int_C f\,dz = \int_{C'} f\,dz,} \qquad (1.457)$$

if the curves $C$ and $C'$ are $C^1$-homotopic.

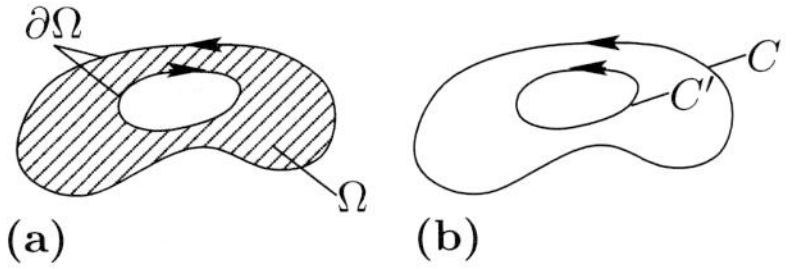

*Figure 1.171. Homotopic paths.*

**$C^1$-homologous paths:** Two $C^1$-curves $C$ and $C'$ in the domain $U$ are said to be *$C^1$-homologous*, if they differ by a *boundary*, i.e.,

---

167 $\int_{z_0}^{z} f d\zeta$ stands for $\int_C f d\zeta$. Because of the independence of the integral from the path of integration, one can choose for $C$ any curve in $U$ which connects $z_0$ with $z$.

168 This means there exists a $C^1$-function $z = z(t,\tau)$ from $[a,b] \times [0,1]$ into $U$ such that for $\tau = 0$ the image is $C$ and for $\tau = 1$ the image is $C'$.

there is a domain $\Omega$, whose closure is contained in $U$, such that

$$\boxed{C = C' + \partial\Omega.}$$

The boundary curve $\partial\Omega$ is oriented in such a way that the domain $\Omega$ is to the left of this curve (Figure 1.171(a)). We write in this case $C \sim C'$.

*Example 4:* The boundary curve $\partial\Omega$ of the domain $\Omega$ in Figure 1.171 consists of the two curves $C$ and $-C'$, in other words, $\partial\Omega = C - C'$, hence $C = C' + \partial\Omega$.

In a similar manner it follows that $C \sim C'$ in Figure 1.172.

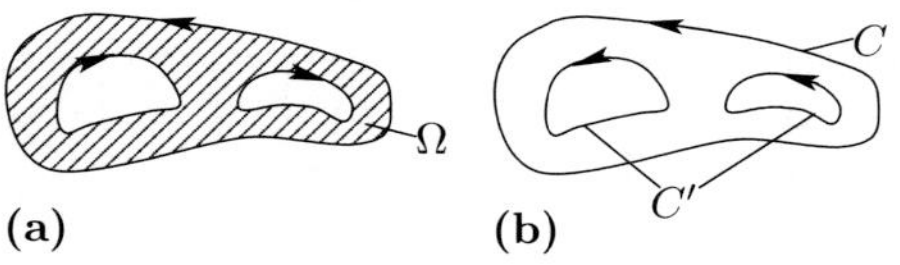


Figure 1.172. Homologous paths.

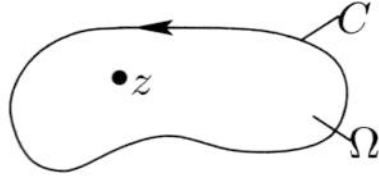


Figure 1.173.

**Theorem 2:** The equation (1.457) attains for $C^1$-homologous paths $C$ and $C'$.

**The integral formula of Cauchy (1831):** Let $U$ be a domain in the complex plane which contains the disc $\Omega := \{z \in \mathbb{C} : |z-a| < r\}$ together with its (mathematically positively oriented) boundary curve $C$. If the function $F : U \longrightarrow \mathbb{C}$ is holomorphic, then for all $z \in \Omega$ we have the relations:

$$\boxed{f(z) = \frac{1}{2\pi \mathrm{i}} \int\limits_C \frac{f(\zeta)}{\zeta - z}\, \mathrm{d}\zeta}$$

and

$$f^{(n)}(z) = \frac{n!}{2\pi \mathrm{i}} \int\limits_C \frac{f(\zeta)}{(\zeta - z)^{n+1}}\, \mathrm{d}\zeta, \qquad n = 1, 2, \ldots .$$

This result continues to hold if $\Omega$ is a domain whose boundary curve $C$ is a (positively oriented in the mathematical sense) closed $C^1$-Jordan curve. Moreover $\Omega$ and $C$ are required to lie in $U$ (cf. Figure 1.173).

> The theory of complex-valued functions of one variable contains the seed of the general theories of homotopy and homology of algebraic topology, which was created (in this context) by Poincaré at the end of the nineteenth century.

Algebraic topology is considered in much more detail in [212].

### 1.14.5 The language of differential forms

*So that is the secret!*
*Faust*

A deeper understanding of the theorems of Cauchy–Riemann and of Cauchy–Morera in the preceding section becomes possible if one uses the language of differential forms. The starting point of our considerations is the one-form

$$\boxed{\omega = f(z)\mathrm{d}z.}$$

We use the decomposition $z = x+\mathrm{i}y$ of the complex number $z$ and $f(z) = u(x,y)+\mathrm{i}v(x,y)$ of the function $f$ into their real and imaginary parts, respectively.[169]

Results of function theory which are deeper, like the Riemann–Roch theorem, require the notion of Riemann surfaces. In this context differential forms are fundamental objects, which are both natural and indispensable (cf. [212]).

**Theorem 1:** $\displaystyle\int_C f(z)\mathrm{d}z = \int_C \omega.$

This result shows us that the definition of the integral $\int_C f\mathrm{d}z$ in 1.14.4 has exactly the same meaning as the integral one gets from the context of differential forms.

*Proof:*

$$\int_C \omega = \int_C (u+\mathrm{i}v)(\mathrm{d}x+\mathrm{i}\mathrm{d}y) = \int_a^b (u+\mathrm{i}v)(x'(t)+\mathrm{i}y'(t))\mathrm{d}t = \int_a^b f(z(t))z'(t)\mathrm{d}t. \qquad \square$$

**Theorem 2:** Let two $C^1$-functions $u, v : U \longrightarrow \mathbb{R}$ (real and imaginary parts of a complex-valued function $f$) be given on an open set $U$. Then the following statements are equivalent:

(i) $\mathrm{d}\omega = 0$ on $U$.

(ii) $f$ is holomorphic on $U$.

This is the theorem of Cauchy–Riemann of 1.14.3 in the language of differential forms.

*Proof:* Because of the relations $\mathrm{d}u = u_x\mathrm{d}x + u_y\mathrm{d}y$ and $\mathrm{d}v = v_x\mathrm{d}x + v_y\mathrm{d}y$ one gets

$$\mathrm{d}\omega = (\mathrm{d}u + \mathrm{i}\mathrm{d}v)(\mathrm{d}x + \mathrm{i}\mathrm{d}y) = \{(u_y + v_x) + \mathrm{i}(v_y - u_x)\}\mathrm{d}y \wedge \mathrm{d}x.$$

Therefore $\mathrm{d}\omega = 0$ is equivalent to the Cauchy–Riemann differential equations

$$u_y + v_x = 0, \qquad v_y - u_x = 0. \qquad \square$$

**Theorem 3:** Let two $C^1$-functions $u$ and $v$ be given on a simply connected domain $U$. Then the following statements are equivalent:

(i) $\mathrm{d}\omega = 0$ on $U$.

(ii) $\displaystyle\int_C \omega$ is independent of the path of integration $C$ in $U$.

---

[169] If we write $f(x,y)$ instead of $f(z)$, then the relation is written as

$$\omega = f(x,y)(\mathrm{d}x + \mathrm{i}\mathrm{d}y).$$

This is the theorem of Cauchy–Morera of 1.14.4 (up to an additional regularity assumption).

**Sketch of proof:** (i)⇒(ii). We consider the situation depicted in Figure 1.174. We have $\partial\Omega = C - C'$. If $\mathrm{d}\omega = 0$ on $U$, then *Stokes theorem* yields

$$0 = \int_{\Omega} \mathrm{d}\omega = \int_{\partial\Omega} \omega = \int_{C} \omega - \int_{C'} \omega.$$

This implies $\int_C \omega = \int_{C'} \omega$, i.e., the integral of $\omega$ in $U$ is independent of the path of integration.

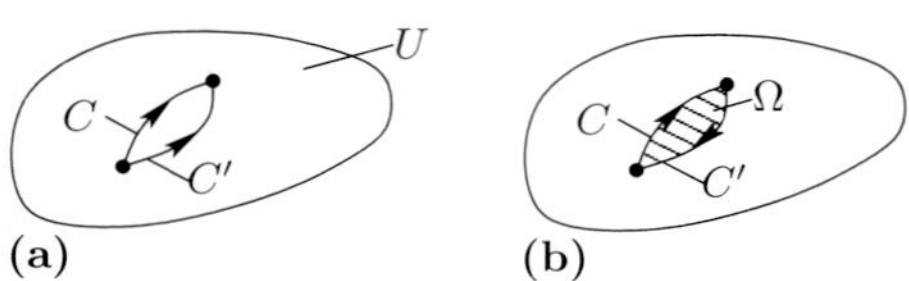


*Figure 1.174. Proof of Theorem 3.*

(ii)⇒(i). If conversely the integral of $\omega$ in $U$ is independent of the path of integration, then one has

$$\int_{\partial\Omega} \omega = 0$$

for all domains $\Omega$ in $U$ which have a closed $C^1$-Jordan curve as boundary $\partial\Omega$. It follows from this that $\mathrm{d}\omega = 0$ on $U$ by de Rham's theorem (cf. [212]). □

**Theorem 4:** If $f : U \subseteq \mathbb{C} \longrightarrow \mathbb{C}$ is holomorphic on a domain $U$, then one has the relation

$$\int_{C} \omega = \int_{C'} \omega$$

for $C^1$-homologous paths $C$ and $C'$ in $U$.

*Proof:* From $\mathrm{d}\omega = 0$ on $U$ and $C = C' + \partial\Omega$ we get

$$\int_{C} \omega = \int_{C'} \omega + \int_{\partial\Omega} \omega = \int_{C'} \omega,$$

since it follows from Stokes theorem that $\int_{\partial\Omega} \omega = \int_{\Omega} \mathrm{d}\omega = 0.$ □

The preceding considerations form the basis for the de Rham cohomology theory, which is at the center of modern differential geometry and has important applications to modern physics in the theory of elementary particles.

**Symplectic geometry of the complex plane $\mathbb{C}$:** The space $\mathbb{R}^2$ carries a natural symplectic structure, given by the volume form

$$\mu = \mathrm{d}x \wedge \mathrm{d}y.$$

We can identify $\mathbb{R}^2$ with $\mathbb{C}$ by means of the map $(x, y) \mapsto x + \mathrm{i}y$. In this way also $\mathbb{C}$ is a symplectic space. One has

$$\boxed{\mu = \frac{\mathrm{i}}{2} \mathrm{d}z \wedge \mathrm{d}\overline{z},}$$

since $\mathrm{d}z \wedge \mathrm{d}\overline{z} = (\mathrm{d}x + \mathrm{i}\mathrm{d}y) \wedge (\mathrm{d}x - \mathrm{i}\mathrm{d}y) = -2\mathrm{i}\mathrm{d}x \wedge \mathrm{d}y$.

**The Riemannian metric on the complex plane $\mathbb{C}$:** The classical Euclidean metric on $\mathbb{R}^2$ is given by

$$\mathbf{g} := \mathrm{d}x \otimes \mathrm{d}x + \mathrm{d}y \otimes \mathrm{d}y.$$

If $u = (u_1, u_2)$ is a point in $\mathbb{R}^2$, then one has $\mathrm{d}x(u) = u_1$ and $\mathrm{d}y(u) = u_2$. From this we get the expression

$$\mathbf{g}(u, v) = u_1 v_1 + u_2 v_2 \quad \text{for all } u, v \in \mathbb{R}^2.$$

This is the standard scalar product in $\mathbb{R}^2$.

By identifying $\mathbb{R}^2$ with $\mathbb{C}$ the complex plane $\mathbb{C}$ also becomes a Riemannian manifold. From $z = x + \mathrm{i}y$ and $\overline{z} = x - \mathrm{i}y$ it follows that

$$\boxed{\mathbf{g} = \frac{1}{2}(\mathrm{d}z \otimes \mathrm{d}\overline{z} + \mathrm{d}\overline{z} \otimes \mathrm{d}z).}$$

**The complex plane $\mathbb{C}$ as a Kähler manifold:** The space $\mathbb{R}^2$ carries an *almost complex structure*, which is given by the linear operator $J : \mathbb{R}^2 \longrightarrow \mathbb{R}^2$ with

$$\boxed{J(x, y) := (-y, x) \qquad \text{for all } (x, y) \in \mathbb{R}^2.}$$

If one identifies $(x, y)$ with $z = x + \mathrm{i}y$, then $J$ corresponds to the mapping $z \mapsto \mathrm{i}z$ (multiplication by the complex number i). The metric $\mathbf{g}$ is compatible with the almost complex structure $J$, i.e., one has

$$\mathbf{g}(Ju, Jv) = \mathbf{g}(u, v) \quad \text{for all } u, v \in \mathbb{R}^2.$$

Moreover, we denote by

$$\boxed{\Phi(u, v) := \mathbf{g}(u, Jv) \quad \text{for all } u, v \in \mathbb{R}^2}$$

the two-form which is called the *fundamental form* of $\mathbf{g}$. One has $\Phi(u, v) = u_2 v_1 - u_1 v_2$ for all $u, v \in \mathbb{R}^2$, i.e.,

$$\Phi = \mathrm{d}y \wedge \mathrm{d}x.$$

It follows from this that $\mathrm{d}\Phi = 0$. In this way $\mathbb{R}^2$ becomes a Kähler manifold.[170]

If we now identify $\mathbb{C}$ with $\mathbb{R}^2$, then also the complex plane $\mathbb{C}$ becomes a Kähler manifold. Manifolds of this kind play a central role in modern string theory as the configuration spaces of strings. This theory has the declared goal of explaining all four elementary forces in a unified theory (theory of everything).

An important theorem concerning Kähler manifolds is the famous theorem of Shing–Tung Yau, for which (together with other important results due to him) he received the Fields medal in 1982 (cf. [212]).

## 1.14.6 Representations of functions

### 1.14.6.1 Power series

An extensive list of important power series is contained in the tables of section 0.7.2. The properties of power series are collected in section 1.10.3.

[170] The general definition of Kähler manifolds is given in [212]. These manifolds were introduced by Erich Kähler (1906–2000) in 1932.

**Definition:** A function $f : U \subseteq \mathbb{C} \longrightarrow \mathbb{C}$ on an open set $U$ is called *analytic* if and only if for every point of $U$ there is a neighborhood in which $f$ can be expanded in a power series.

**Main theorem of Cauchy (1831):** A function $f : U \subseteq \mathbb{C} \longrightarrow \mathbb{C}$ is analytic if and only if it is holomorphic.

**Consequence:** A holomorphic function on $U$ has derivatives of arbitrary order.

**Cauchy's expansion of holomorphic functions:** If a function $f$ is holomorphic in the neighborhood of a point $a \in \mathbb{C}$, then one has the following expansion as a power series:

$$\boxed{f(z) = f(a) + f'(a)(z-a) + \frac{f''(a)}{2!}(z-a)^2 + \ldots .}$$

The domain of convergence is the largest open disc around $a$ in which the function $f$ is holomorphic.

*Example 1:* Consider $f(z) := \dfrac{1}{1-z}$. The largest open disc around the origin in which $f$ is holomorphic is the disc of radius 1. Therefore the geometric series

$$\frac{1}{1-z} = 1 + z + z^2 + \ldots$$

has the convergence radius $r = 1$.

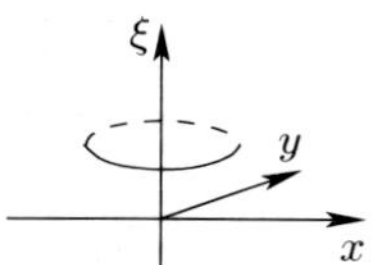


*Figure 1.175.*

**The analytic landscape:** Every complex-valued function $w = f(z)$ is associated with an *analytic landscape* over the complex plane, by choosing a *Cartesian* $(x, y, \zeta)$-coordinate system and taking the value $\zeta := |f(z)|$ as the height of the landscape at the point $z = x + \mathrm{i}y$.

*Example 2:* The analytic landscape of the function $f(z) := z^2$ is the paraboloid $\zeta = x^2 + y^2$ (Figure 1.175).

**The maximum principle:** If a non-constant function $f : U \subseteq \mathbb{C} \longrightarrow \mathbb{C}$ on an open set $U$ is holomorphic, then the function $\zeta := |f(z)|$ has no maximum on $U$.

If the function $\zeta = |f(z)|$ on $U$ has an absolute minimum at a point $a$ in $U$, then $f(a) = 0$.

Intuitively this means that the analytic landscape of $f$ over $U$ has no highest peak. If there is a lowest point, then the height there is 0.

**Sequences of holomorphic functions:** Suppose we are given functions $f_n : U \subseteq \mathbb{C} \longrightarrow \mathbb{C}$ which are holomorphic on an open set $U$. If the sequence

$$\lim_{n\to\infty} f_n(z) = f(z) \quad \text{for all } z \in U$$

converges *uniformly*[171] on $U$, then the limit function $f$ is also holomorphic on $U$. Moreover one has

$$\lim_{n\to\infty} f_n^{(k)}(z) = f^{(k)}(z) \quad \text{for all } z \in U$$

for all derivatives of orders $k = 1, 2, \ldots$.

[171] This means $\lim\limits_{n\to\infty} \sup\limits_{z\in U} |f_n(z) - f(z)| = 0$.

### 1.14.6.2 Laurent series and singularities

Let $r$, $\varrho$ and $R$ be given with $0 \le r < \varrho < R \le \infty$. We consider the annulus $\Omega := \{z \in \mathbb{C} \,|\, r < |z| < R\}$.

**The expansion theorem of Laurent (1843):** Let $f : \Omega \longrightarrow \mathbb{C}$ be holomorphic. Then for all $z \in \Omega$ one has the absolutely convergent series expansion

$$\boxed{\begin{aligned} f(z) &= a_0 + a_1(z-a) + a_2(z-a)^2 + \ldots \\ &\qquad + \frac{a_{-1}}{(z-a)} + \frac{a_{-2}}{(z-a)^2} + \frac{a_{-3}}{(z-a)^3} + \ldots \end{aligned}} \tag{1.458}$$

Here the coefficients are given by the formula

$$a_k = \frac{1}{2\pi \mathrm{i}} \int_C f(\zeta)(\zeta - a)^{-k-1} \mathrm{d}\zeta, \qquad k = 0, \pm 1, \pm 2, \ldots$$

Let $C$ denote a circle of radius $\varrho$ centered at $a$. The so-called *Laurent series* (1.458) can be termwise integrated and differentiated in $\Omega$.

**Isolated singularities:** A point $a$ is said to be an *isolated singularity* of the function $f$, if there is an open neighborhood $U$ of $a$ such that $f$ is holomorphic on $U - \{a\}$. In this case (1.458) holds with $r = 0$ and a sufficiently small number $R$.

(i) $a$ is called a *pole of order* $m$ of $f$, if (1.458) holds with $a_{-m} \neq 0$ and $a_{-k} = 0$ for all $k > m$.

(ii) $a$ is called a *removable singularity* of $f$, if (1.458) holds with $a_{-k} = 0$ for all $k \ge 1$. In this case $f$ is turned into a holomorphic function on all of $U$ by setting $f(a) := a_0$.

(iii) $a$ is said to be an *essential singularity*, if neither case (i) nor case (ii) holds, i.e., the Laurent series contains infinitely many terms with negative exponents.

The coefficient $a_{-1}$ in (1.458) is called the *residue* of $f$ at the point $a$ and is denoted by the symbol

$$\boxed{\mathrm{Res}_a f := a_{-1}.}$$

*Example 1:* The function

$$f(z) := z + \frac{a}{z-1} + \frac{b}{(z-1)^2}$$

has a pole of second order at the point $z = 1$, whose residue at this point is $a_{-1} = a$.

*Example 2:* The function

$$\sin \frac{1}{z} = \frac{1}{z} - \frac{1}{3!z^3} + \frac{1}{5!z^5} - \ldots$$

has an essential singularity at the point $z = 0$, with the residue $a_{-1} = 1$ there.

**Bounded functions:** A function $f : U \subseteq \mathbb{C} \longrightarrow \mathbb{C}$ is said to be *bounded*, if there is a number $S$ such that $|f(z)| \le S$ for all $z \in U$.

**Theorem:** Suppose the function $f$ has a singularity at the point $a$.

(i) If the function $f$ is bounded in a neighborhood of $a$, then the singularity of $f$ at $a$ is removable.

(ii) If one has $|f(z)| \longrightarrow \infty$ for $z \to a$, then the function $f$ has a pole at the point $a$.

**Theorem of Picard (1879):** If a function $f$ has an essential singularity at the point $a$, then $f$ takes on *all* complex numbers as values, with at most finitely many exceptions, in every neighborhood of $a$.

This means that the function is quite pathological near any essential singularity.

*Example 3:* The function $w = e^{1/z}$ has an essential singularity at the point $z = 0$, and attains every complex number as a value in a neighborhood of the origin, with the exception of $w = 0$.

### 1.14.6.3 Entire functions and their product expansions

Entire functions are a generalization of polynomials.

**Definition:** The functions which are holomorphic on the *entire* complex plane, and only these functions, are called *entire.*

*Example 1:* The functions $w = e^z$, $\sin z$, $\cos z$, $\sinh z$ and $\cosh z$, as well as every polynomial, are entire.

**Theorem of Liouville (1847):** A bounded entire function is constant.

Intuitively this means that the height of the analytic landscape of a non-constant entire function grows beyond all bounds.

**Theorem of Picard:** A non-constant entire function attains every complex number as a value with finitely many exceptions.

*Example 2:* The function $w = e^z$ attains every complex number as a value with the exception of $w = 0$.

**Zeros of entire functions:** For an entire function $f : \mathbb{C} \longrightarrow \mathbb{C}$ the following statements hold.

(i) Either $f \equiv 0$ of $f$ has in every disc of the plane at most finitely many zeros.

(ii) The function $f$ is a polynomial if and only if it has finitely many zeros in the entire plane.

**Multiplicity of a zero:** If a function $f$ is holomorphic in a neighborhood at a point $a$ with $f(a) = 0$, we say that the zero $a$ of $f$ has by definition *multiplicity* $m$, if the power series expansion of $f$ in a neighborhood of $a$ has the form

$$f(z) = a_m(z-a)^m + a_{m+1}(z-a)^{m+1} + \ldots$$

with $a_m \neq 0$. This is equivalent to the condition

$$f^{(m)}(a) \neq 0 \quad \text{and} \quad f'(a) = f''(a) = \ldots = f^{(m-1)}(a) = 0.$$

The *fundamental theorem of algebra* says that for a non-constant polynomial $f$, one has

$$f(z) = a \prod_{k=1}^{n} (z - z_k)^{m_k} \qquad \text{for all } z \in \mathbb{C}.$$

Here the points $z_1, \ldots, z_n$ are the distinct zeros of $f$ and $m_1, \ldots, m_n$ are their multiplicities; $a$ denotes some non-vanishing complex number.

The following theorem is a generalization to more general functions:

**The product formula of Weierstrass (1876):** Let $f : \mathbb{C} \longrightarrow \mathbb{C}$ denote a non-constant entire function, with infinitely many zeros $z_1, z_2, \ldots$, and corresponding multiplicities $m_1, m_2 \ldots$. Then for $f$ we have the formula

$$\boxed{f(z) = \mathrm{e}^{g(z)} \prod_{k=1}^{\infty} (z - z_k)^{m_k} \mathrm{e}^{p_k(z)} \qquad \text{for all } z \in \mathbb{C}.} \tag{1.459}$$

Here $p_1, p_2, \ldots$ are polynomials and $g$ denotes an entire function.[172]

*Example 3:* $\sin \pi z = \pi z \prod_{k=1}^{\infty} \left(1 - \frac{z^2}{k^2}\right)$ for all $z \in \mathbb{C}$. This formula was first discovered by Euler (1707–1783) when he was still quite young.

**Corollary:** If one prescribes an at most countable number of zeros and their multiplicities, then there is an entire function which has these zeros and multiplicities.

### 1.14.6.4 Meromorphic functions and partial fraction decompositions

Meromorphic functions are a generalization of rational functions (quotients of polynomials).

**Definition:** A function $f$ is said to be *meromorphic*, if it is holomorphic on $\mathbb{C}$ with the exception of isolated singularities, all of which are poles.

We associate the value $\infty$ to the poles of $f$, and consider the compactification of $\mathbb{C}$ obtained by adjoining the point $\infty$, $\overline{\mathbb{C}} := \mathbb{C} \cup \{\infty\}$ (this space can be identified with the two-dimensional sphere, cf. 1.14.11.4 below, which carries a natural complex structure, but here this just refers to a space consisting of the union of $\mathbb{C}$ and the point denoted $\infty$).

*Example 1:* The functions $w = \tan z$, $\cot z$, $\tanh z$ and $\coth z$ as well as arbitrary rational functions, and all the more all entire functions, are meromorphic.

**Poles of meromorphic functions:** For every meromorphic function $f : \mathbb{C} \longrightarrow \overline{\mathbb{C}}$ the following statements hold:

(i) The function $f$ has in every disc at most finitely many poles.

(ii) $f$ is a rational function if and only if it has only finitely many poles and finitely many zeros in the entire complex plane.

(iii) $f$ is the quotient of two entire functions (generalizing the fact that rational functions are, by definition, quotients of polynomials).

(iv) The set of all meromorphic functions form a field (in the sense of section 2.5.3), which is the quotient field of the ring (in the sense of section 2.5.2) of entire functions.

The theorem on *partial fraction decompositions* of rational functions states that every rational function $f$ can be written as a linear combination of polynomials and expressions of the form

$$\frac{b}{(z-a)^k},$$

where the points $a$ are the poles of $f$ and $b$ is some complex number.

[172] Since $\mathrm{e}^w \neq 0$ for all $w \in \mathbb{C}$, the exponential factors in (1.459) do not contribute zeros to $f$, but are present to insure convergence of the product.

**Theorem of Mittag–Leffler (1877):** Let $f : \mathbb{C} \longrightarrow \overline{\mathbb{C}}$ be a meromorphic function, with (infinitely many) poles $z_1, z_2, \ldots$, ordered such that $|z_1| \leq |z_2| \leq \cdots$. Then there is an expression

$$\boxed{f(z) = \sum_{k=1}^{\infty} g_k\left(\frac{1}{z - z_k}\right) - p_k(z),} \tag{1.460}$$

valid for all $z \in \mathbb{C}$ with the exception of the poles of $f$. Here $g_1, g_2, \ldots$ and $p_1, p_2, \ldots$ are polynomials.

*Example:* One has

$$\frac{1}{\sin \pi z} = \frac{1}{z} + \sum_{k=1}^{\infty}(-1)^k\left(\frac{1}{z-k} + \frac{1}{z+k}\right)$$

for all complex numbers $z$ different from the poles $k \in \mathbb{Z}$ of the function $w = \sin \pi z$.

**Corollary:** If one prescribes the poles and their principal parts in a Laurent expansion (i.e., the terms with negative powers), then there is a meromorphic function with the given poles and principal parts.

#### 1.14.6.5 Dirichlet series

Dirichlet series are important in analytic number theory.

**Definition:** The infinite series

$$\boxed{f(s) := \sum_{n=1}^{\infty} a_n \mathrm{e}^{-\lambda_n s}} \tag{1.461}$$

is said to be a *Dirichlet series*, if all $a_n$ are complex numbers and the real exponents $\lambda_n$ form a strictly increasing sequence with $\lim\limits_{n\to\infty} \lambda_n = +\infty$.

We set

$$\sigma_0 := \overline{\lim_{N\to\infty}} \frac{\ln|A(N)|}{\lambda_N}.$$

Here the quantities $A(N)$ are defined by the relations

$$A(N) := \begin{cases} \sum\limits_{n=1}^{N} a_n, & \text{if } \sum\limits_{n=1}^{\infty} a_n \text{ diverges,} \\ \sum\limits_{n=N}^{\infty} a_n, & \text{if } \sum\limits_{n=1}^{\infty} a_n \text{ converges.} \end{cases}$$

Moreover, we let $\mathscr{H} = \mathscr{H}_{\sigma_0} := \{s \in \mathbb{C} \,|\, \mathrm{Re}\, s > \sigma_0\}$ denote the part of $\mathbb{C}$ to the 'right' of $\sigma_0$.

*Example 1:* If $\lambda_n := \ln n$ and $a_n := 1$, we get the *Riemann $\zeta$-function*

$$\zeta(s) = \sum_{n=1}^{\infty} \frac{1}{n^s}$$

with $\sigma_0 = \lim\limits_{N\to\infty} \dfrac{\ln N}{\ln N} = 1$. The following three statements hold for Dirichlet series, for example, for the Riemann $\zeta$-function.

**Theorem:** (i) The Dirichlet series (1.461) converges in the open half-space $\mathscr{H}$ and diverges in the open half-space $\mathbb{C} - \overline{\mathscr{H}}$.

The convergence is uniform on compact subsets of $\mathscr{H}$.

(ii) The function $f$ is holomorphic on $\mathscr{H}$. The series (1.461) may be differentiated termwise arbitrarily often.

(iii) If $a_n \geq 0$ and $\lambda_n = \ln n$ for all $n$, then $f$ has a singularity in the point $s = \sigma_0$.

**The connection with the theory of prime numbers:** Let $g : \mathbb{N}_+ \longrightarrow \mathbb{C}$ be a function defined on the set of positive natural numbers. $g$ is called *multiplicative*, if and only if $g(mn) = g(m)g(n)$ for all relatively prime natural numbers $m$ and $n$. If the series

$$f(s) = \sum_{n=1}^{\infty} \frac{g(n)}{n^s}$$

converges absolutely, then one has an *Euler product*

$$f(s) = \prod_p \left(1 + \frac{g(1)}{p^s} + \frac{g(2)}{p^{2s}} + \ldots\right),$$

where the product is to be taken over all prime numbers $p$. This product is always absolutely convergent.

*Example 2:* In the special case $g \equiv 1$ we get

$$\zeta(s) = \prod_p \left(1 - \frac{1}{p^s}\right)^{-1} \qquad \text{for all } s \in \mathbb{C} \text{ with } \operatorname{Re} s > 1.$$

A more detailed discussion of the Riemann $\zeta$-function and the famous Riemann hypothesis (a conjecture) is found in 2.7.3.

## 1.14.7 The calculus of residues and the calculation of integrals

*Mathematics is the art of avoiding calculations.*
*Folklore*

The following theorem is of great importance. It shows that, in order to calculate complex integrals, just the behavior of the integrand at its singularities which lie inside the path of integration is sufficient to do the calculation of the whole integral. The calculation of integrals can be very tedious. It must have been a God's send for Cauchy when he discovered the beautiful trick with the residues, which reduces the amount of calculation in many cases to a minimum.

**The residue theorem of Cauchy (1826):** Let the function $f : U \subseteq \mathbb{C} \longrightarrow \mathbb{C}$ be holomorphic on the open set $U$ except for finitely many poles in the points $z_1, \ldots, z_n$. Then one has

$$\boxed{\int_C f \, \mathrm{d}z = 2\pi \mathrm{i} \sum_{k=1}^{n} \operatorname{Res}_{z_k} f.} \qquad (1.462)$$

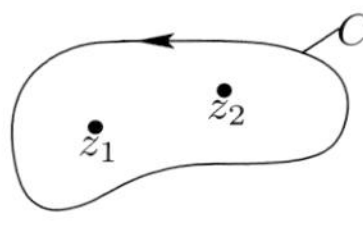


*Figure 1.176.*

Here $C$ is closed $C^1$-Jordan curve in $U$, which has all points $z_1, \ldots, z_n$ in its interior and which is in addition positively oriented (in the mathematical sense, cf. Figure 1.176).

*Example 1:* Let

$$f(z) = \frac{1}{z-1} - \frac{2}{z+1},$$

hence $\text{Res}_{z=1} f = 1$ and $\text{Res}_{z=-1} = -2$. For a circle $C$, which goes around both points $z = \pm 1$, one has

$$\int_C f\,\mathrm{d}z = 2\pi\mathrm{i}(\text{Res}_{z=1} f + \text{Res}_{z=-1} f) = -2\pi\mathrm{i}.$$

**Rules for calculations:** If the function $f$ has a pole of $m^{th}$ order at the point $a$, then

$$\text{Res}_a f = \lim_{z \to a} (z-a) f(z) \qquad \text{for } m = 1 \tag{1.463}$$

and

$$\text{Res}_a f = \lim_{z \to a} F^{(m-1)}(z) \qquad \text{for } m \geq 2$$

with $F(z) := (z-a)^m f(z)/(m-1)!$.

*Example 2:* A rational function $\dfrac{g(z)}{h(z)}$ with $g(a) \neq 0$ has a pole of the $m^{th}$ order at $a$ if and only if the denominator $h$ has a zero of $m^{th}$ order at $a$.

**Standard examples:** One has

$$\boxed{\int_{-\infty}^{\infty} \frac{g(x)}{h(x)}\,\mathrm{d}x = 2\pi\mathrm{i} \sum_{k=1}^{n} \text{Res}_{z_k} \frac{g}{h}.} \tag{1.464}$$

In the calculation of the real integral it is assumed that $g$ and $h$ are polynomials with $\deg h \geq \deg g + 2$. The denominator $h$ should have no zeros on the real axis. We denote by $z_1, \ldots, z_n$ the zeros of $h$ in the upper half-space, that is, in the set of complex numbers in the complex plane with positive imaginary part.

*Example 3:* $\displaystyle\int_{-\infty}^{\infty} \frac{\mathrm{d}x}{1+x^2} = \pi.$

*Proof:* The polynomial $h(z) := 1 + z^2$ has, because of the decomposition $h(z) = (z - \mathrm{i})(z+\mathrm{i})$, a simple zero at the point $z = \mathrm{i}$ in the upper half plane. From (1.463) it follows

$$\text{Res}_\mathrm{i} \frac{1}{1+z^2} = \lim_{z \to \mathrm{i}} \frac{(z-\mathrm{i})}{(z+\mathrm{i})(z-\mathrm{i})} = \frac{1}{2\mathrm{i}}.$$

The relation (1.464) therefore yields

$$\int_{-\infty}^{\infty} \frac{\mathrm{d}x}{1+x^2} = 2\pi\mathrm{i}\,\text{Res}_\mathrm{i} \frac{1}{1+z^2} = \pi. \qquad \square$$

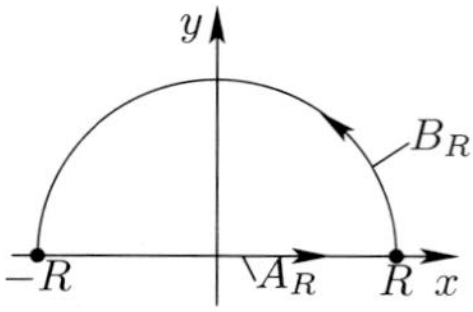

Figure 1.177.

*Sketch of proof of* (1.464)*:* We set $f := \dfrac{g}{h}$. The boundary $A_R + B_R$ of the half-circle in Figure 1.177 is chosen sufficiently large that all zeros of $h$ in the upper half-plane are contained in the interior. From (1.462) one gets

$$\int_{A_R} f\,\mathrm{d}z + \int_{B_R} f\,\mathrm{d}z = \int_{A_R+B_R} f\,\mathrm{d}z = 2\pi\mathrm{i} \sum_{k=1}^{n} \text{Res}_{z_k} f. \tag{1.465}$$

Moreover one has

$$\lim_{R \to \infty} \int_{A_R} f\,\mathrm{d}z = \int_{-\infty}^{\infty} f(x)\mathrm{d}x$$

and

$$\lim_{R\to\infty} \int\limits_{B_R} f\,\mathrm{d}z = 0. \tag{1.466}$$

The claim (1.464) now follows from (1.465) for $R \longrightarrow \infty$.

The limiting relation (1.466) can be obtained from the estimate[173]

$$|f(z)| \le \frac{\text{const}}{|z^2|} = \frac{\text{const}}{R^2} \qquad \text{for all } z \text{ with } |z| = R,$$

which follows from the relation $\deg h \ge \deg g + 2$ for the degrees, and from the triangle inequality for contour integrals:

$$\left|\int\limits_{B_R} f\,\mathrm{d}z\right| \le (\text{length of the half-circle } B_R) \sup_{z\in B_R} |f(z)|$$
$$\le \frac{\pi R\cdot \text{const}}{R^2} \longrightarrow 0 \qquad \text{for } R \longrightarrow \infty.$$

□

### 1.14.8 The mapping degree

Let a bounded domain $\Omega$ in the complex plane be given, whose boundary $\partial\Omega$ is formed from finitely many closed $C^1$-Jordan curves, which are oriented in such a way that $\Omega$ lies to its left (Figure 1.178). We write $f \in \mathscr{C}(\Omega)$ if:

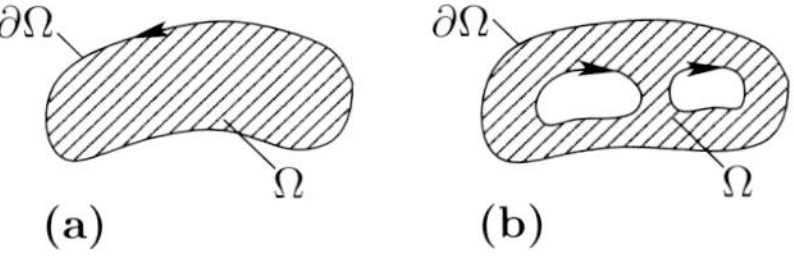


*Figure 1.178. The domain of definition of f for the definition of the mapping degree.*

(i) The function $f$ is holomorphic in an open neighborhood of $\overline{\Omega}$ except for finitely many poles, which all lie in $\Omega$.

(ii) There are no zeros of $f$ on the boundary $\partial\Omega$.

**Definition:** The *mapping degree* of $f$ on $\Omega$ is defined by

$$\boxed{\deg(f,\Omega) := N - P.}$$

Here $N$ (resp. $P$) is the sum of the multiplicities of zeros (resp. poles) of $f$ in $\Omega$.

*Example:* For $f(z) := z^k$ and the disc $\Omega := \{z \in \mathbb{C} : |z| < R\}$ one has

$$\deg(f,\Omega) = k, \qquad k = 0, \pm 1, \pm 2, \ldots$$

**Theorem:** Let $f,\ g \in \mathscr{C}(\Omega)$. Then one has the following:

(i) *Representation formula:*

$$\deg(f,\Omega) = \frac{1}{2\pi \mathrm{i}} \int\limits_{\partial\Omega} \frac{f'(z)}{f(z)}\mathrm{d}z.$$

[173] The constant here is independent of $R$.

(ii) *Existence principle:* If $\deg(f, \Omega) \neq 0$, then the function $f$ has a zero or a pole on $\Omega$.

(iii) *Stability of the mapping degree:* From the relation

$$|g(z)| < \max_{z \in \partial\Omega} |f(z)| \tag{1.467}$$

it follows that $\deg(f, \Omega) = \deg(f + g, \Omega)$.

**The principle of Rouché on zeros (1862):** Assume the two functions $f$ and $g$ are holomorphic on an open neighborhood of $\overline{\Omega}$, and that (1.467) holds. Then if $f$ has a zero on $\Omega$, it follows that $f + g$ also has a zero on $\Omega$.

*Proof:* Since $f$ has a zero and no pole ($f$ is holomorphic), one has $\deg(f, \Omega) \neq 0$. From (iii) above it follows that $\deg(f + g, \Omega) \neq 0$, and (ii) yields the statement. □

The general theory of mapping degrees, which can be found in [212], makes it possible to prove the existence of solutions for a large class of problems in mathematics (systems of equations, ordinary and partial differential equations, integral equations) without having to explicitly construct these solutions.

### 1.14.9 Applications to the fundamental theorem of algebra

*From a modern point of view we would say that the proof of the fundamental theorem of algebra given by Gauss in 1799 is in principle correct, but not complete.*

*Felix Klein* (1849–1925)

*However – so we ask – will it become impossible, because of the ever-expanding volume of material in mathematics, for a single researcher to be acquainted with all of mathematics? As an answer to this I would like to remind you how important it is for the essence of the mathematical sciences that each bit of progress is accompanied by a corresponding extension of the tools as well as simpler methods, which makes the understanding of earlier results easier and makes complicated methods used previously unnecessary, and that a single researcher, by using these new tools and simpler methods, has an easier time orienting himself in the different branches of mathematics, more than this is the case for any other science.*

*David Hilbert,*
*Paris lecture, 1900*

**The fundamental theorem of algebra:** Every polynomial

$$p(z) := z^n + a_{n-1}z^{n-1} + \ldots + a_1 z + a_0$$

with degree deg $\geq n$ with complex coefficients $a_j$ has a (complex) zero.

Gauss used the decomposition $p(z) = u(x, y) + \mathrm{i}v(x, y)$ of $p$ into its real and imaginary part and studied the properties of the plane algebraic curves $u(x, y) = 0$ and $v(x, y) = 0$. A proof of this kind is by nature tedious and requires a developed apparatus for algebraic curves, which is available today but not in Gauss' days.

The intuitively clear idea of Gauss is the following: we consider the disc of radius $R$

$$D_R := \{z \in \mathbb{C} : |z| < R\}.$$

The polynomial $p$ yields a map (also denoted by $p$)

$$p : \overline{D}_R \longrightarrow \mathbb{C},$$

where the mathematically positively oriented boundary curve $\partial D_R$ is mapped to a circle $p(\partial D_R)$ which wraps around the origin $n$ times in the mathematically positive sense (cf. Figure 1.179 for $n = 2$). Hence there must be a point $z_1 \in D_R$ which is mapped to the origin, i.e., for which one has $p(z_1) = 0$. In order to show that the image curve $p(\partial D_R)$ wraps around the origin $n$ times, we consider the polynomial $w = f(z)$ defined by $f(z) := z^n$. From the relation $z = Re^{i\varphi}$ it follows that

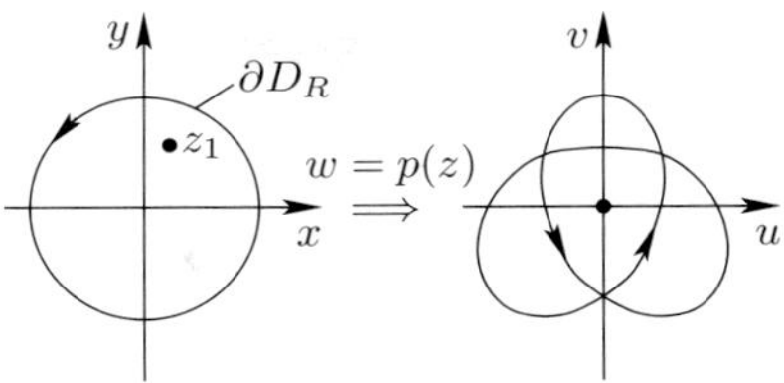


*Figure 1.179.*

$$w = R^n e^{in\varphi}, \qquad 0 \le \varphi \le 2\pi.$$

Hence the circle $\partial D_R$ of radius $R$ is mapped to the circle $f(\partial D_R)$ of radius $R^n$, which wraps around the origin $n$ times. If $R$ is sufficiently large, then this wrapping behavior holds also for $p$, as $p$ and $f$ differ only by terms of lower order.

The following proof is the rigorous formulation of the above idea.

**First proof of the fundamental theorem of algebra (mapping degree):** We write

$$p(z) := f(z) + g(z)$$

with $f(z) := z^n$. For all $z$ with $|z| = R$, we have

$$|f(z)| = R^n \quad \text{and} \quad |g(z)| \le \text{const} \cdot R^{n-1}.$$

If $R$ is sufficiently large, then one has

$$|g(z)| < |f(z)| \quad \text{for all } z \text{ with } |z| = R.$$

The function $f$ obviously has a zero. According to the theorem of Rouché in 1.14.8 it follows that the function $f + g = p$ has a zero also. □

The following proof is shorter yet.

**Second proof of the fundamental theorem of algebra (theorem of Liouville):** Suppose the polynomial $p$ has *no* zero. Then the inverse function $1/p$ is entire, and because of

$$\lim_{|z|\to\infty} \left| \frac{1}{p(z)} \right| = 0,$$

it is also bounded. According to the theorem of Liouville $1/p$ must be constant. This is a contradiction and proves that $p$ has a zero. □

**Corollary:** The power series expansion of $p$ at the point $z_1$ yields, because of $p(z_1) = 0$,

$$p(z_1) = a_1(z - z_1) + a_2(z - z_2)^2 + \ldots,$$

hence $p(z) = (z - z_1)q(z)$. The polynomial $q$ has a zero $z_2$, hence $q(z) = (z - z_2)r(z)$, etc. Putting these facts together obtain a factorization

$$\boxed{p(z) = (z - z_1)(z - z_2) \ldots (z - z_n).}$$

### 1.14.10 Biholomorphic maps and the Riemann mapping theorem

The class of biholomorphic maps has the important property that it transforms holomorphic functions into holomorphic functions. Moreover, biholomorphic maps are angle preserving (conformal).

**Definition:** Let $U$ and $V$ be open sets in the complex plane $\mathbb{C}$. A function $f : U \longrightarrow V$ is *biholomorphic*, if it is bijective and both $f$ and $f^{-1}$ are holomorphic.

**Local theorem on inverse functions:** Let a holomorphic function $f : U \subseteq \mathbb{C} \longrightarrow \mathbb{C}$ be given in a neighborhood of a point $a$ with

$$\boxed{f'(a) \neq 0.}$$

Then $f$ is a biholomorphic map from a neighborhood of the point $a$ to a neighborhood of the point $f(a)$.

For the inverse function $w = f(z)$ one has, just as in the real case, the Leibniz rule

$$\boxed{\frac{\mathrm{d}z(w)}{\mathrm{d}w} = \frac{1}{\frac{\mathrm{d}w(z)}{\mathrm{d}z}}.} \tag{1.468}$$

**Global theorem on inverse functions:** Suppose a function $f : U \subseteq \mathbb{C} \longrightarrow \mathbb{C}$ is holomorphic and injective on a domain $U$. Then the image $f(U)$ is again a domain, and $f$ is a biholomorphic function from $U$ to $f(U)$.

Moreover one has $f'(z) \neq 0$ on $U$, and the derivative of the inverse function of $f$ on $f(U)$ is given by the formula (1.468).

**The principle of holomorphic transformations:** Let a holomorphic function

$$f : U \subseteq \mathbb{C} \longrightarrow \mathbb{C}$$

be given on an open set $U$. Moreover, let $b : U \longrightarrow V$ be a biholomorphic map. Then $f$ can be transformed to the set $V$ in a natural manner.[174] For this transformed function $f_*$ one has:

(i) $f_* : V \subseteq \mathbb{C} \longrightarrow \mathbb{C}$ is holomorphic.

(ii) The integral is invariant, i.e., one has

$$\boxed{\int_C f(z)\mathrm{d}z = \int_{C_*} f_*(w)\mathrm{d}w}$$

for all $C^1$-curves $C$ in $U$ and $C_* := b(C)$.

**Remarks:** This important result makes it possible to introduce complex manifolds. Roughly speaking, one has the following statements

(i) A one-dimensional complex manifold $M$ is constructed in such a way that around every point $P \in M$, it is locally described by a coordinate $z$ which belongs to an open set $U$ of the complex plane $\mathbb{C}$.

[174] Explicitly one has $f_* := f \circ b^{-1}$, i.e., $f_*(w) = f(b^{-1}(w))$.

(ii) The change from the local coordinate $z$ to the local coordinate $w$ is described by a biholomorphic function $w = b(z)$ from the open set $U$ to the open set $V$.

(iii) Only those properties of $M$ are considered important which are *invariant* under such a change of local coordinate.

(iv) The principle of holomorphic transformation shows that the notion of *holomorphic function* and of *integral* are defined in an invariant manner on complex manifolds.

(v) Connected one-dimensional complex manifolds are called *Riemann surfaces.*

The precise definitions of these objects are found in [212].

**The main theorem of Riemann (1851) (Riemann mapping theorem):** Every simply connected domain in the complex plane $\mathbb{C}$ which is not equal to $\mathbb{C}$ itself can be biholomorphically mapped to the interior of the unit circle.[175]

## 1.14.11 Examples of conformal maps

In order to interpret the properties of holomorphic functions $f : U \subseteq \mathbb{C} \longrightarrow \mathbb{C}$ geometrically, we view

$$\boxed{w = f(z)}$$

as a mapping, which associates to each point $z$ of the $z$-plane a point $w$ of the $w$-plane.

**Conformal mappings:** The map defined by $f$ is said to be *angle preserving* or *conformal* in the point $z = a$, if the intersection angle of two $C^1$-curves through the point (including the sense of direction) is preserved under the map (Figure 1.180).

A map is said to be angle preserving or conformal, if it is so in every point of its domain of definition.

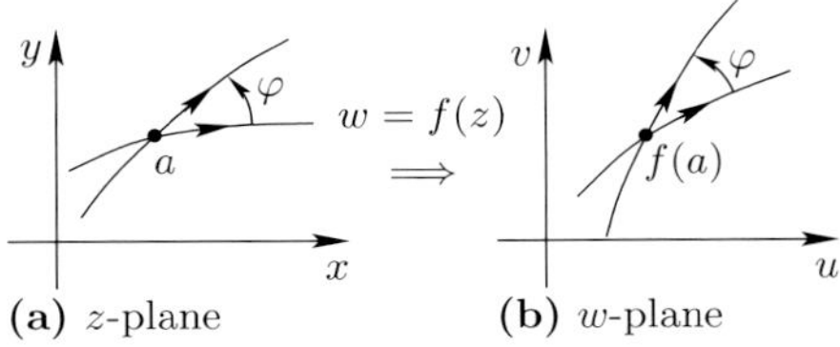


*Figure 1.180. Conformal maps.*

**Theorem:** A holomorphic function $f : U \subseteq \mathbb{C} \longrightarrow \mathbb{C}$ from a neighborhood $U$ of a point $a$ determines a conformal mapping at the point $a$ if and only if $f'(a) \neq 0$.

Every biholomorphic map $f : U \longrightarrow V$ is conformal.

### 1.14.11.1 The group of similarity transformations

Let $a$ and $b$ be fixed complex numbers with $a \neq 0$. Then the association

$$\boxed{w = az + b \quad \text{for all } z \in \mathbb{C}} \tag{1.469}$$

determines a biholomorphic (and hence conformal) mapping $w : \mathbb{C} \longrightarrow \mathbb{C}$ of the complex plane $\mathbb{C}$ to itself.

*Example 1:* For $a = 1$, the map (1.469) is a translation.

[175] Every biholomorphic map is angle preserving (conformal).

The deep uniformization theorem of Poincaré and Koebe from 1907 generalizes the Riemann mapping theorem in the following way: every simply connected Riemann surface can be mapped biholomorphically to exactly one of the following: the interior of the unit circle, the complex plane itself, or the closed complex plane $\overline{\mathbb{C}}$ (Riemann sphere), which is described in section 1.14.11.3. Details on this generalization can be found in [212].

*Example 2:* We set $z = r\mathrm{e}^{\mathrm{i}\varphi}$. If $b = 0$ and $a = |a|\mathrm{e}^{\mathrm{i}\alpha}$, then one has

$$w = |a|r\,\mathrm{e}^{\mathrm{i}(\varphi+\alpha)}.$$

Consequently the mapping $w = az$ is a rotation by an angle of $\alpha$ and simultaneously multiplies the lengths by a factor of $|a|$.

In case $b = 0$ and $a > 0$ one gets the proper similarity transformations (multiplying lengths by the factor of $a$).

The set of all transformations (1.469) forms the group of orientation-preserving similarity transformations of the complex plane $\mathbb{C}$ into itself.

#### 1.14.11.2 Inversion on the unit circle

The mapping

$$\boxed{w = \frac{1}{z} \quad \text{for all } z \in \mathbb{C} \text{ with } z \neq 0}$$

is a biholomorphic and hence conformal mapping of the punctured complex plane $\mathbb{C} - \{0\}$ into itself. If we set $z = r\mathrm{e}^{\mathrm{i}\varphi}$, then we have

$$w = \frac{1}{r}\mathrm{e}^{-\mathrm{i}\varphi}.$$

Because of $|w| = \dfrac{1}{|z|}$ the point $w$ is obtained from $z$ through the reflection on the unit circle and a simultaneous reflection on the real axis (Figure 1.181).

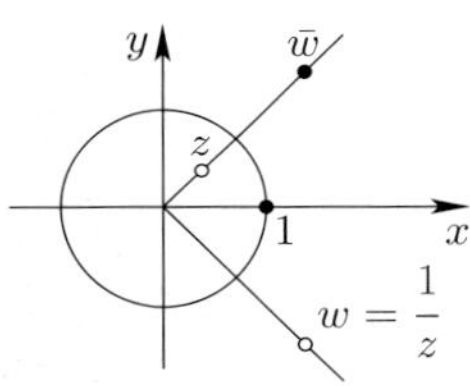


*Figure 1.181.*

#### 1.14.11.3 The closure of the complex plane

We set

$$\boxed{\overline{\mathbb{C}} := \mathbb{C} \cup \{\infty\},}$$

i.e., we add to the complex plane $\mathbb{C}$ a point $\infty$ and call $\overline{\mathbb{C}}$ the completed complex plane. The following construction is typical for the construction of complex manifolds. It is our goal to give local coordinates on $\overline{\mathbb{C}}$.

**Definition of local coordinates:** (i) For any point $a \in \mathbb{C}$ we take the neighborhood $U := \mathbb{C}$ as neighborhood, with coordinate $\zeta := z$ for $z \in \mathbb{C}$.

(ii) As a neighborhood of the point $\infty$ we take the set $V := \overline{\mathbb{C}} - \{0\}$ with the local coordinate

$$\zeta' := \begin{cases} \dfrac{1}{z} & \text{for } z \in \mathbb{C} \text{ with } z \neq 0, \\ 0 & \text{for } z = \infty. \end{cases}$$

Thus we have for every point $z \in \overline{\mathbb{C}} - \{0\}$ two different local coordinates, $\zeta = z$ and $\zeta' = 1/z$, and these two local coordinates are related by the relation $\zeta = 1/\zeta'$. The point 0 has only the local coordinate $\zeta = 0$, while the point $\infty$ has the local coordinate $\zeta' = 0$. Thus we get a rigorous expression of the notation $0 = 1/\infty$.

**Mappings of the completed complex plane:** The properties of a map

$$\boxed{f : U \subseteq \overline{\mathbb{C}} \longrightarrow \overline{\mathbb{C}}}$$

are defined by passing to local coordinates. For example, $f$ is holomorphic if and only if it is so upon passing to local coordinates.

*Example 1:* Let $n = 1, 2, \ldots$ The mapping

$$f(z) = \begin{cases} z^n & \text{for } z \in \mathbb{C}, \\ \infty & \text{for } z = \infty \end{cases}$$

is a holomorphic mapping $f : \overline{\mathbb{C}} \longrightarrow \overline{\mathbb{C}}$.

*Proof:* First of all, $f$ is clearly holomorphic on the open set $U := \mathbb{C}$. Transforming the equation

$$w = f(z), \qquad z \in \overline{\mathbb{C}} - \{0\}$$

to local coordinates $\zeta = \dfrac{1}{z}$ and $\mu = \dfrac{1}{w}$ yields $\dfrac{1}{\mu} = \dfrac{1}{\zeta^n}$, hence, on the other open set $V$,

$$\boxed{\mu = \zeta^n.} \tag{1.470}$$

This is a holomorphic function on $\mathbb{C}$. Hence $f$ is by definition holomorphic on $V = \overline{\mathbb{C}} - \{0\}$. □

According to equation (1.470) we have in local coordinates a zero of order $n$. Since $\zeta = 0$ corresponds to the point $z = \infty$ and $f(\infty) = \infty$, we say that $f$ has an infinity (a pole) of order $n$ at $\infty$.

*Example 2:* The mapping

$$f(z) = \begin{cases} z & \text{for } z \in \mathbb{C}, \\ \infty & \text{for } z = \infty \end{cases}$$

is a biholomorphic mapping $f : \overline{\mathbb{C}} \longrightarrow \overline{\mathbb{C}}$.

*Proof:* According to the first example, $f$ is holomorphic. Moreover, $f : \overline{\mathbb{C}} \longrightarrow \overline{\mathbb{C}}$ is bijective and $f^{-1} = f$. Hence we conclude that $f^{-1} : \overline{\mathbb{C}} \longrightarrow \overline{\mathbb{C}}$ is holomorphic, as was to be shown. □

*Example 3:* If $w = p(z)$ is a polynomial of $n^{th}$ degree and if we set $p(\infty) := \infty$, then $p : \overline{\mathbb{C}} \longrightarrow \overline{\mathbb{C}}$ is a holomorphic function with a pole of order $n$ at the point $z = \infty$.

*Example 4:* Let $n = 1, 2\ldots$. We set

$$f(z) = \begin{cases} \dfrac{1}{z^n} & \text{for all} \quad z \in \mathbb{C} \text{ with } z \neq 0, \\ \infty & \text{for} \quad z = 0, \\ 0 & \text{for} \quad z = \infty. \end{cases}$$

Then $f : \overline{\mathbb{C}} \longrightarrow \overline{\mathbb{C}}$ is a holomorphic mapping with a pole of order $n$ in the point $z = 0$ and a zero of $n^{th}$ order in the point $z = \infty$.

For $n = 1$, $f : \overline{\mathbb{C}} \longrightarrow \overline{\mathbb{C}}$ is biholomorphic.

*Proof:* To study $w = f(z)$ in a neighborhood of $z = 0$, we use the local coordinates $w = 1/\mu$ and $\zeta = z$. The function formed from this,

$$\mu = \zeta^n,$$

is holomorphic in a neighborhood of $\zeta = 0$ and has a zero of order $n$. Consequently, $f$ has a pole of order $n$ at $z = 0$. □

**Neighborhoods:** Let $\varepsilon > 0$ be given.
For every point $p \in \overline{\mathbb{C}}$ we define its $\varepsilon$-neighborhood by

$$U_\varepsilon(p) := \{z \in \mathbb{C} : |z - p| < \varepsilon\} \qquad \text{in case } p \in \mathbb{C},$$

and set $U_\varepsilon(\infty)$ to be the set of all complex numbers $z$ with $|z| > \varepsilon^{-1}$ together with the point $\infty$.

**Open sets:** A set $U$ of the completed number sphere $\overline{\mathbb{C}}$ is called *open*, if it contains, for each of its points, at least one $\varepsilon$-neighborhood of that point.

**Remarks:** (i) With the help of these open sets we can give the completed complex number sphere $\overline{\mathbb{C}}$ the structure of a *topological space*, and then all the notions of topological spaces can be applied to it (cf. [212]). In particular, $\overline{\mathbb{C}}$ is *compact* and *connected*.

(ii) With respect to the local coordinates we have introduced, the space $\overline{\mathbb{C}}$ becomes a one-dimensional *complex manifold*, and all the notions of complex manifolds can be applied to it (cf. [212]).

(iii) By definition, one-dimensional complex manifolds are Riemann surfaces. Hence $\overline{\mathbb{C}}$ is also a *compact Riemann surface*.

In the middle of the nineteenth century, Riemann worked with a quite intuitive notion of Riemann surface (cf. 1.14.11.6). Historically, the struggle for a mathematically rigorous concept of 'Riemann surface' contributed much to the development of topology and the theory of manifolds. A decisive step in this direction was taken by Hermann Weyl with the appearance of his book *Die Idee der Riemannschen Fläche* (The idea of a Riemann surface), which was published in 1913.

### 1.14.11.4 The Riemann sphere

Let $(x, y, \zeta)$ be a given Cartesian coordinate system. The sphere

$$S^2 := \{(x, y, \zeta) \in \mathbb{R}^3 \mid x^2 + y^2 + \zeta^2 = 1\}$$

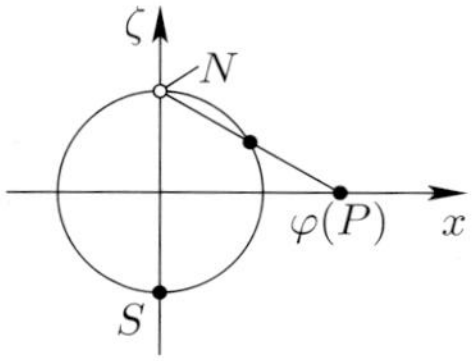

is called the Riemann sphere. Let $N$ denote the north pole, i.e., the point with the coordinates $(0, 0, 1)$. The *stereographic projection*

$$\boxed{\varphi : S^2 - \{N\} \longrightarrow \mathbb{C}}$$

is defined by sending a given point $P$ of $S^2 - \{N\}$ to the intersection point $z = \varphi(P)$ of the line $NP$ connecting $N$ with $P$ with the $(x, y)$-plane (Figure 1.182 shows the intersection of $S^2$ with the $(x, \zeta)$-plane). This map can be extended to a map $\varphi : S^2 \longrightarrow \overline{\mathbb{C}}$ by sending $N$ to the point $\infty$.



*Figure 1.182.*

*Example:* The south pole $S$ of $S^2$ (with coordinates $(0, 0, -1)$) is mapped under $\varphi$ to the origin of the complex plane $\mathbb{C}$, while the equator is mapped to the unit circle.

**Theorem:** The map $\varphi : S^2 \longrightarrow \overline{\mathbb{C}}$ is a homeomorphism, i.e., it maps $S^2 - \{N\}$ conformally to $\mathbb{C}$.

**Corollary:** If we transfer the local coordinates from $\overline{\mathbb{C}}$ to $S^2$, the Riemann sphere $S^2$ becomes a one-dimensional complex manifold, and the map $\varphi : S^2 \longrightarrow \overline{\mathbb{C}}$ is biholomorphic. More precisely, $S^2$ is a compact Riemann surface.

#### 1.14.11.5 The group of Möbius transformations

**Definition:** The set of all biholomorphic maps $f : \overline{\mathbb{C}} \longrightarrow \overline{\mathbb{C}}$ form a group, which is called the *automorphism group* $\mathrm{Aut}(\overline{\mathbb{C}})$ of $\overline{\mathbb{C}}$.[176]

**Conformal geometry on $\overline{\mathbb{C}}$:** The group $\mathrm{Aut}(\overline{\mathbb{C}})$ determines the notion of conformal symmetry on the completed plane $\overline{\mathbb{C}}$. A property belongs by definition to the conformal geometry of $\overline{\mathbb{C}}$, if it is invariant under all transformations of the group $\mathrm{Aut}(\overline{\mathbb{C}})$.

*Example 1:* A generalized circle on $\overline{\mathbb{C}}$ is (the image in $\overline{\mathbb{C}}$ of) a circle in $\mathbb{C}$ or a line in $\mathbb{C}$ together with the point $\infty$.

The elements of $\mathrm{Aut}(\overline{\mathbb{C}})$ map generalized circles into generalized circles.

**Möbius transformations:** If $a, b, c$ and $d$ are complex numbers with $ad - bc \neq 0$, then the transformation

$$\boxed{f(z) := \frac{az+b}{cz+d}} \tag{1.471}$$

is called a *Möbius transformation*, where we agree to the following:

(i) For $c = 0$, we set $f(\infty) := \infty$.

(ii) For $c \neq 0$, we set $f(\infty) := a/c$ and $f(-d/c) := \infty$.

These transformations were first studied by August Ferdinand Möbius (1790–1868).

**Theorem 1:** The group $\mathrm{Aut}(\overline{\mathbb{C}})$ of automorphisms of $\overline{\mathbb{C}}$ consists precisely of the Möbius transformations.

*Example 2:* The set of Möbius transformations which map the *upper half-plane* $\mathscr{H}_+ := \{z \in \mathbb{C} \,|\, \mathrm{Im}\, z > 0\}$ into itself in a conformal (angle-preserving) manner are those transformations of the form (1.471) for which $a, b, c$ and $d$ are real and $ad - bc > 0$.

*Example 3:* The set of Möbius transformations which map the upper half-plane conformally to the unit disc (interior of the unit circle) are those of the form

$$a\frac{z-p}{z-\overline{p}}$$

with complex numbers $a$ and $p$ for which $|a| = 1$ and $\mathrm{Im}\, p > 0$.

*Example 4:* The set of all Möbius transformations which map the interior of the unit disc conformally to itself is the set of mappings given by

$$a\frac{z-p}{\overline{p}z-1}$$

with complex numbers $a$ and $p$ for which $|a| = 1$ and $|p| < 1$.

**Properties of Möbius transformations:** For a Möbius transformation $f$ one has:

[176] The notion of automorphism group depends on the structure which is to be preserved. Here we require preserving the structure as a complex manifold, hence the biholomorphic maps. There is also a group of diffeomorphism, homeomorphism, etc., each of which is distinct.

(i) $f$ can be composed of a translation, a rotation, a proper similarity transformation and an inversion on the unit circle. Conversely, every such composition of mappings is a Möbius transformation.

(ii) $f$ is conformal and maps generalized circles to generalized circles.

(iii) $f$ preserves the *double ratio*

$$\frac{z_4 - z_3}{z_4 - z_2} : \frac{z_1 - z_3}{z_1 - z_2}$$

of four points in $\overline{\mathbb{C}}$.[177]

(iv) Every Möbius transformation other than the identity has at least one and at most two fixed points (points $P$ with $f(P) = P$).

We let $GL(2,\mathbb{C})$ denote the group of all complex invertible $(2 \times 2)$-matrices. Moreover let $D$ denote the subgroup[178] of all such matrices of the form $\lambda I$ with $\lambda \neq 0$ in $GL(2,\mathbb{C})$ (here $I$ denotes the *identity matrix* $I = \begin{pmatrix} 1 & 0 \\ 0 & 1 \end{pmatrix}$ ).

**Theorem 2:** The mapping defined by the prescription

$$\begin{pmatrix} a & b \\ c & d \end{pmatrix} \mapsto \frac{az+b}{cz+d}$$

is a group homomorphism[179] from $GL(2,\mathbb{C})$ to $\mathrm{Aut}(\overline{\mathbb{C}})$ with the kernel $D$. Hence one has a group isomorphism (a bijective group homomorphism)

$$GL(2,\mathbb{C})/D \cong \mathrm{Aut}(\overline{\mathbb{C}}),$$

i.e., $\mathrm{Aut}(\overline{\mathbb{C}})$ is isomorphic to the complex projective group $PGL(2,\mathbb{C})$.

### 1.14.11.6 The Riemann surface of the square root

The ingenious idea of Riemann is to consider *many-valued* complex functions defined on the complex plane $\mathbb{C}$ (as for example $z = \sqrt{w}$), which can be made single-valued by changing the domain of definition to a more complicated object $D$ than $\mathbb{C}$ itself. In simple cases one obtains $D$ by cutting several copies of the complex plane along certain segments and gluing them together along these cuts. This leads to the notion of Riemann surface.

**The map $w = z^2$:** We set $z = r\mathrm{e}^{\mathrm{i}\varphi}$, $-\pi < \varphi \leq \pi$, and have accordingly

$$\boxed{w = r^2 \mathrm{e}^{2\mathrm{i}\varphi}.}$$

The map $w = z^2$ squares the distance $r$ of the point $z$ from the origin and doubles the angle $\varphi$ of $z$.

To study the behavior of $w = z^2$ more precisely, we consider in the $z$-plane a circle $C$ around the origin of radius $r$, where $C$ is given the mathematically positive orientation.

[177] One uses the natural method of calculation for the point $\infty$, i.e., $1/\infty = 0$, $1/0 = \infty$ and $\infty \pm z = \infty$ for $z \in \mathbb{C}$.

[178] Groups and subgroups are defined in section 2.5.1 below.

[179] Group homomorphisms are defined in section 2.5.1.2 below.

If one runs through $C$ once in the $z$-plane, then the image in the $w$-plane is a circle of radius $r^2$ which runs around the origin *twice*.

To study the converse mapping $z = \sqrt{w}$ it is advantageous to take two copies of the $w$-plane and cut them along the *negative real axis* (Figure 1.183).

(i) If we run through $C$ in the $z$-plane from the point $z = r$ to the point $z = \mathrm{i}r$, then the image points run on the *first copy* of the $w$-plane from the point $r^2$ to the point $-r^2$, or, as one says, along the *first sheet* of the Riemann surface.

(ii) We continue along $C$ from the point $\mathrm{i}r$ to the point $-\mathrm{i}r$. The image points now run through the *second sheet* from the point $-r^2$ through the point $r^2$ to the point $-r^2$.

(iii) If we finally pass along $C$ from $-\mathrm{i}r$ to the point $r$, then the image points run along the *first sheet* from $-r^2$ to $r^2$.

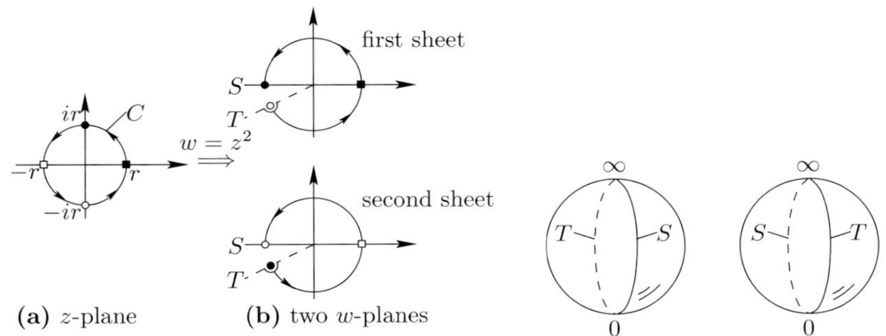


*Figure 1.183. The Riemann surface for $z = \sqrt{w}$.*

*Figure 1.184. The two 'sheets' of the Riemann surface.*

**The inverse map $z = \sqrt{w}$:** The important observation for the following is:

> To every point $w \neq 0$ on one of the two sheets of the $w$-plane there is exactly one point $z$ in the $z$-plane with $w = z^2$.

This makes the function $z = \sqrt{w}$ on the union of the two sheets of the $w$-plane well-defined (single-valued). One has explicitly for a point $w = R\mathrm{e}^{\mathrm{i}\psi}$ with $-\pi < \psi \leq \pi$ the relation

$$\sqrt{w} := \begin{cases} \sqrt{R}\mathrm{e}^{\mathrm{i}\psi/2} & \text{for } w \text{ on the first sheet,} \\ -\sqrt{R}\mathrm{e}^{\mathrm{i}\psi/2} & \text{for } w \text{ on the second sheet.} \end{cases}$$

Here $\sqrt{R} \geq 0$. The value of $\sqrt{w}$ on the first sheet is called the *principal value* of $\sqrt{w}$ and is denoted $_+\sqrt{w}$.

**The intuitive Riemann surface $\mathscr{F}$ of $z = \sqrt{w}$:** If we glue the two sheets as depicted in Figure 1.183(b) (i.e., $T$ with $T$ and $S$ with $S$), then we get the intuitive Riemann surface $\mathscr{F}$ of $z = \sqrt{w}$.

**The topological type of the Riemann surface $\mathscr{F}$:** The situation becomes easier to understand if instead of the two $w$-planes one uses two Riemann spheres, cutting these along a curve from the south pole to the north pole, and then gluing as depicted in

Figure 1.184. The object we get in this manner can be blown up like a balloon to a sphere. Thus $\mathscr{F}$ is homeomorphic to a sphere, which in turn is homeomorphic to the Riemann sphere.

The intuitive Riemann surface $\mathscr{F}$ of the two-valued function $z = \sqrt{w}$ is homeomorphic to the Riemann sphere.

Similarly one can treat the maps $w = z^n$, $n = 3, 4, \ldots$. In this case one requires $n$ copies of the $w$-plane to get the intuitive Riemann surface of the function $z = \sqrt[n]{w}$.

**Remark:** The representation of Riemann surfaces with paper, scissors and glue is quite instructive and intuitive for simple cases. However, for more complicated functions this method meets its limits. A satisfying mathematical construction of the Riemann surface of an arbitrary analytic function can be found in [212].

### 1.14.11.7 The Riemann surface of the logarithm

The equation

$$w = \mathrm{e}^z$$

has a many-valued inverse function, which we denote by $z = \operatorname{Ln} w$. To describe this function, we choose for each integer $k$ a copy $B_k$ of the $w$-plane which we cut along the negative real axis. For $w = R\mathrm{e}^{\mathrm{i}\psi}$ with $-\pi < \psi \leq \pi$ and $w \neq 0$ we set

$$\operatorname{Ln} w := \ln R + \mathrm{i}\psi + 2k\pi\mathrm{i} \quad \text{on } B_k,\ k = 0, \pm 1, \pm 2, \ldots$$

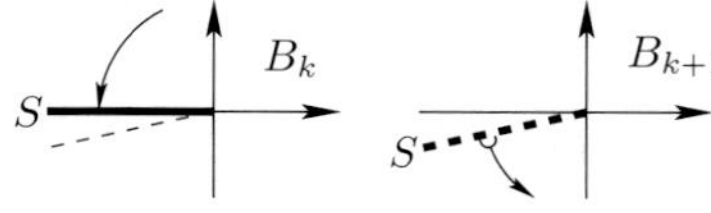


*Figure 1.185. The Riemann surface of the logarithm* Ln.

If we glue the sheet $B_k$ with the sheet $B_{k+1}$ along the cut denoted $S$ in Figure 1.185, and let $k$ run through all integers, then we get the 'infinite round staircase' $\mathscr{F}$, on which the function $z = \operatorname{Ln} w$ is well-defined. We call $\mathscr{F}$ the intuitive Riemann surface of the logarithm.

**Branching point:** We call $w = 0$ a *branching point* of infinite order of the Riemann surface $\mathscr{F}$. In the case of the function $z = \sqrt{w}$, the point $w = 0$ is called a branching point of second order.

**Principal value of the logarithm:** Let $w = R\mathrm{e}^{\mathrm{i}\psi}$ with $-\pi < \psi \leq \pi$ and $w \neq 0$. We set

$$\ln w := \ln R + \mathrm{i}\psi$$

and call $\ln w$ the *principal value* of the logarithm of $w$. This value corresponds to $\operatorname{Ln} w$ on the sheet $B_0$ (the important point here is the assumption $-\pi < \psi \leq \pi$).

*Example 1:* For all $z \in \mathbb{C}$ with $|z| < 1$ one has

$$\ln(1+z) = z - \frac{z^3}{3} + \frac{z^5}{5} - \ldots$$

The principal value of $\sqrt[n]{w}$ for $w = R\mathrm{e}^{\mathrm{i}\psi}$ with $-\pi < \psi \leq \pi$ and $n = 2, 3, \ldots$ is defined by $\sqrt[n]{R}\mathrm{e}^{\mathrm{i}\psi/n}$.

*Example 2:* For all $z \in \mathbb{C}$ with $|z| < 1$ one has

$$\sqrt[n]{1+z} = 1 + \alpha z + \binom{\alpha}{2} z^2 + \binom{\alpha}{3} z^3 + \ldots$$

with $\alpha = 1/n$ in the sense of the principal value of the $n^{th}$ root.

#### 1.14.11.8 The Schwarz–Christoffel mapping formula

$$w = \int_{i}^{z} (\zeta - z_1)^{\gamma_1 - 1}(\zeta - z_2)^{\gamma_2 - 1} \ldots (\zeta - z_n)^{\gamma_n - 1} \mathrm{d}\zeta.$$

This function maps the upper half-space $\{z \in \mathbb{C} \,|\, \operatorname{Im} x > 0\}$ biholomorphically (and hence conformally) to the interior of an $n$-gon with the inside angles $\gamma_j \pi$, $j = 1, \ldots, n$ (Figure 1.186 shows the case $n = 3$). It is assumed that all $z_j$ are real numbers with $z_1 < z_2 < \cdots < z_n$ and that $0 < \gamma_j \pi < 2\pi$ for all $j$ and moreover that the relation $\gamma_1 \pi + \cdots + \gamma_n \pi = (n-2)\pi$ (the sum of all angles in the $n$-gon). The points $z_1, \ldots, z_n$ are mapped to the corners (vertices) of the $n$-gon.

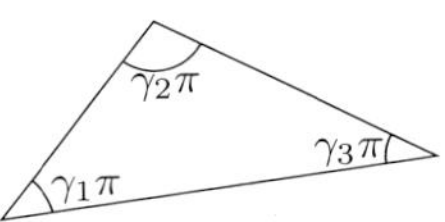


*Figure 1.186.*

### 1.14.12 Applications to harmonic functions

Let $\Omega$ be a domain in $\mathbb{R}^2$. We identify $\mathbb{R}^2$ with the complex plane $\mathbb{C}$, by identifying $z = x + \mathrm{i}y \in \mathbb{C}$ with the point $(x, y) \in \mathbb{R}^2$.

**Definition:** A function $u : \Omega \longrightarrow \mathbb{R}$ is called *harmonic*, if

$$\Delta u = 0 \quad \text{on } \Omega.$$

Here we have set $\Delta u := u_{xx} + u_{yy}$. We use in addition the decomposition

$$f(z) = u(x, y) + \mathrm{i}v(x, y)$$

of a complex-valued function $f$ into its real and imaginary parts.

**Theorem 1:** (i) If the function $f : \Omega \subseteq \mathbb{C} \longrightarrow \mathbb{C}$ is holomorphic on the domain $\Omega$, then $u$ and $v$ are harmonic on $\Omega$.[180]

If $f, g : \Omega \longrightarrow \mathbb{C}$ are holomorphic functions with identical real part $u$ on $\Omega$, then the imaginary parts of $f$ and $g$ differ by at most a constant.

(ii) Conversely, if the function $u : \Omega \longrightarrow \mathbb{R}$ is harmonic on a *simply connected* domain $\Omega$, then the curve integral

$$v(x, y) = \int_{z_0}^{z} -u_y \mathrm{d}x + u_x \mathrm{d}y + \text{const}$$

[180] This arises from the fact that the Cauchy–Riemann differential equations

$$u_x = v_y, \qquad u_y = -v_x$$

hold; it follows that $u_{xx} = v_{yx}$ and $u_{yy} = -v_{xy}$, hence $u_{xx} + u_{yy} = 0$.

is independent of the path of integration for fixed start and end points $z_0 \in \Omega$, and the function $f = u + \mathrm{i}v$ is holomorphic on $\Omega$.

The function $v$ is said to be a *conjugate harmonic function* to $u$.

*Example 1:* Let $\Omega = \mathbb{C}$. For $f(z) = z$ we get from $z = x + \mathrm{i}y$ on $\Omega$ the harmonic functions $u(x, y) = x$ and $v(x, y) = y$.

*Example 2:* Let $\Omega = \mathbb{C} - \{0\}$ and $z = r\mathrm{e}^{\mathrm{i}\varphi}$ with $-\pi < \varphi \leq \pi$. For the principal value of the logarithm one has

$$\ln z = \ln r + \mathrm{i}\varphi.$$

Hence

$$u(x, y) := \ln r, \qquad r = \sqrt{x^2 + y^2},$$

and this is a harmonic function on $\Omega$. The function $v(x, y) := \varphi$ is harmonic on every subdomain $\Omega'$ of $\Omega$ which does not contain the negative real axis $A$. On the other hand, $v$ is discontinuous on $\Omega$ with jumps along $A$.

This example shows that the assumptions on the simple-connectedness of the domain $\Omega$ in Theorem 1 cannot be weakened.

**The Green's function:** Let $\Omega$ be a bounded domain in the complex plane $\mathbb{C}$ with smooth boundary. The Green's function $w = G(z, z_0)$ of $\Omega$ is by definition a function with the following properties:

(i) For every fixed point $z_0 \in \Omega$ one has

$$G(z, z_0) = -\frac{1}{2\pi} \ln |z - z_0| + h(z)$$

with a continuous function $h : \Omega \longrightarrow \mathbb{R}$, which is harmonic on $\Omega$.

(ii) $G(z, z_0) = 0$ for all $z \in \partial\Omega$.

In the language of *distributions*, one has for a fixed point $z_0 \in \Omega$:

$$\begin{aligned} -\Delta G(z, z_0) &= \delta_{z_0} \text{ on } \Omega, \\ G &= 0 \text{ on } \partial\Omega. \end{aligned}$$

The first equation means

$$-\int_{\Omega} G(z, z_0) \Delta\varphi(x, y) \mathrm{d}x\mathrm{d}y = \varphi(z_0) \quad \text{for all } \varphi \in C_0^\infty(\Omega).$$

**Theorem 2:** (i) There is a unique Green's function $G$ of $\Omega$.

(ii) One has the symmetry property

$$G(z, z_0) = G(z_0, z)$$

and the positivity $G(z, z_0) > 0$ for all $z, z_0 \in \Omega$ with $z \neq z_0$.

(iii) If $g : \partial\Omega \longrightarrow \mathbb{R}$ is a given continuous function, the first boundary value problem

$$\boxed{\Delta u = 0 \text{ on } \Omega \text{ and } u = g \text{ on } \partial\Omega} \tag{1.472}$$

has a unique solution $u$, which is continuous on $\overline{\Omega}$ and smooth on $\Omega$. For all $z \in \Omega$ one has the formula:

$$\boxed{u(z) = -\int_{\partial\Omega} g(\zeta) \frac{\partial G(z, \zeta)}{\partial n_\zeta} \mathrm{d}s.} \tag{1.473}$$

Here $\partial/\partial n_\zeta$ denotes the outer normal derivative with respect to $\zeta$, and $s$ is the arc length of the boundary curve $\partial\Omega$, which is oriented in such a way that $\Omega$ lies to its left.

**Main theorem:** Let $\Omega$ be a bounded, simply connected domain in the complex plane with smooth boundary. Suppose we are given a biholomorphic (and hence conformal) mapping $f$ from $\Omega$ onto the interior of the unit disc with $f(z_0) = 0$. Then the formula

$$\boxed{G(z, z_0) = -\frac{1}{2\pi} \ln |f(z)|}$$

defines the Green's function of $\Omega$.

*Example 3:* Let $\Omega := \{z \in \mathbb{C} : |z| < 1\}$. The Möbius transformation

$$f(z) = \frac{z - z_0}{\overline{z}_0 z - 1}$$

maps the unit disc $\Omega$ to itself, with $f(z_0) = 0$. The formula (1.473) for the solution becomes here explicitly for the unit disc $\Omega$:

$$\boxed{u(z) = \frac{1}{2\pi} \int_{-\pi}^{\pi} \frac{g(\varphi)(1 - r^2)}{1 + r^2 - 2r\cos\varphi} \mathrm{d}\varphi.}$$

This is the so-called *Poisson formula.* Here we have set $z = \mathrm{e}^{\mathrm{i}\varphi}$ with $0 \le r < 1$.

**The Dirichlet principle:** A smooth solution $u$ of the variational problem

$$\int_\Omega (u_x^2 + u_y^2)\mathrm{d}x\mathrm{d}y \stackrel{!}{=} \min., \qquad u = g \ \text{ on } \partial\Omega \tag{1.474}$$

is the unique solution of the first boundary value problem (1.472).

This result was originally derived by Gauss and Dirichlet.

**Historical remark:** The previous considerations show that there is very close relationship between harmonic functions and conformal mappings. This connection was used in an important way by Riemann in his construction of geometric complex analysis in 1851. In section 1.14.10 one can find the famous Riemann mapping theorem. Riemann was able to reduce the proof of this theorem to the first boundary value problem for the Laplace equation (1.472). To solve this equation, he used the variational problem (1.474). It seems he viewed the existence of a solution of (1.474) as being evident from physical considerations.

Weierstrass pointed out this gap in Riemann's proof. It wasn't until half a century later in 1900 that Hilbert was able to give a rigorous proof of the solution of (1.474) in a famous paper of his, thus completing Riemann's proof of the mapping theorem. This paper of Hilbert was the starting point of an intensive period of progress of direct methods of the calculus of variations in the context of functional analysis. A more detailed discussion can be found in [212].

### 1.14.13 Applications to hydrodynamics

**The basic equations for plane flows:**

$$\boxed{\begin{aligned} &\varrho(\mathbf{v}\,\mathbf{grad}\,)\mathbf{v} = \mathbf{f} - \mathbf{grad}\ p, \\ &\operatorname{div}\mathbf{v} = 0, \quad \mathbf{curl}\,\mathbf{v} = \mathbf{0} \quad \text{on } \Omega. \end{aligned}} \tag{1.475}$$

These equations describe a plane stationary (time-independent) flow free of sources of an ideal[181] fluid of constant density $\varrho$. We identify the point $(x, y)$ with $z = x + \mathrm{i}y$. The notations in (1.475) are as follows. $\mathbf{v}(z)$ is the velocity vector of the fluid particle at the point $z$, $p(z)$ is the pressure at the point $z$, $\mathbf{f} = -\mathbf{grad}\,W$ is the density of the exterior force with the potential $W$.

**Circulation and source strength:** Let $C$ be a closed, mathematically positively oriented curve. The number

$$\boxed{Z(C) := \int_C \mathbf{v}\,\mathrm{d}\mathbf{x}}$$

is called the *circulation* of $C$. Moreover, one calls

$$Q(C) := \int_C \mathbf{vn}\,\mathrm{d}s$$

the *source strength* of the domain surrounding $C$ (where $\mathbf{n}$ is the outer normal vector and $s$ denotes arc length).

**Integral curves:** The fluid particles move along the integral curves (or flow lines), i.e., the velocity vector $\mathbf{v}$ is tangent to the integral curves. In the following we will identify the velocity vector $\mathbf{v} = a\mathbf{i} + b\mathbf{j}$ with the complex number $a + ib$.

**Connection with holomorphic functions:** Every holomorphic function

$$f(z) = U(z) + \mathrm{i}V(z)$$

on the domain $\Omega$ of the complex plane $\mathbb{C}$ corresponds to a plane flow, i.e., it is a solution of the basic equations (1.475), in the following way.

(i) The *velocity field* $\mathbf{v}$ is obtained as $\mathbf{v} = -\mathbf{grad}\,U$, hence

$$\boxed{\mathbf{v}(z) = -\overline{f'(z)}.}$$

The function $U$ is called the *velocity potential*; $f$ is called a *complex velocity potential.* Moreover, $|\mathbf{v}(z)| = |f'(z)|$.

(ii) The *pressure* $p$ is calculated from the *Bernoulli equation*:

$$\frac{\mathbf{v}^2}{2} + \frac{p}{\varrho} + \frac{W}{\varrho} = \text{const} \ \text{ on } \Omega.$$

The constant is determined by prescribing the pressure at some fixed point.

(iii) The curves $V(x, y) = \text{const}$ are the *integral curves.*

[181] One neglects the inner friction for ideal fluids.

(iv) The curves $U(x,y) = \text{const}$ are called *equipotential curves*. In the points $z$ with $f'(z) \neq 0$ the integral curves are orthogonal to the equipotential curves.

(v) *Circulation* and *source strength* follow from the formula

$$Z(C) + \mathrm{i}Q(C) = -\int_C f'(z)\mathrm{d}z.$$

This integral can be conveniently calculated with help of the residue theorem.

**Pure parallel flow** (Figure 1.187(a)): Let $c > 0$. The function

$$\boxed{f(z) := -cz}$$

with $U = -cx$ and $V = -cy$ corresponds to the parallel flow

$$\mathbf{v} = c,$$

that is $\mathbf{v} = c\mathbf{i}$. The lines $y = \text{const}$ are the integral curves, the lines $x = \text{const}$ are the equipotential curves, orthogonal to the integral curves.

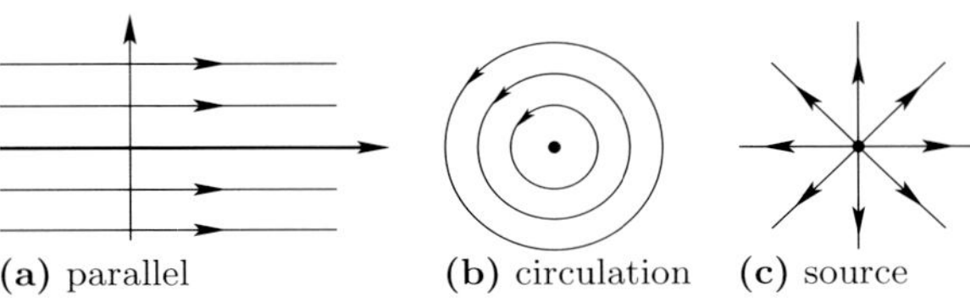


*Figure 1.187. Hydrodynamical flows.*

**Pure circulation flow** (Figure 1.187(b)): Let $\Gamma$ be a real number. We set $z = re^{\mathrm{i}\varphi}$. For the function

$$\boxed{f(z) := -\frac{\Gamma}{2\pi\mathrm{i}}\ln z,}$$

one has $U = -\dfrac{\Gamma}{2\pi}\varphi$, $V = \dfrac{\Gamma}{2\pi}\ln r$ and $\mathbf{v} = -\overline{f'(z)}$. This yields for the velocity field

$$\mathbf{v}(z) = \frac{\mathrm{i}\Gamma z}{2\pi r^2}.$$

Let $C$ be a circle around the origin. For the circulation we get from the residue theorem

$$Z(C) = \frac{\Gamma}{2\pi\mathrm{i}}\int_C \frac{\mathrm{d}z}{z} = \Gamma.$$

The integral curves $V = \text{const}$ are concentric circles around the origin, and the equipotential curves $U = \text{const}$ are rays originating in the origin.

**Pure source flow** (Figure 1.187(c)): Let $q > 0$. The function

$$f(z) = -\frac{q}{2\pi}\ln z$$

corresponds to a source flow with the velocity field

$$\mathbf{v}(z) = \frac{qz}{2\pi r^2},$$

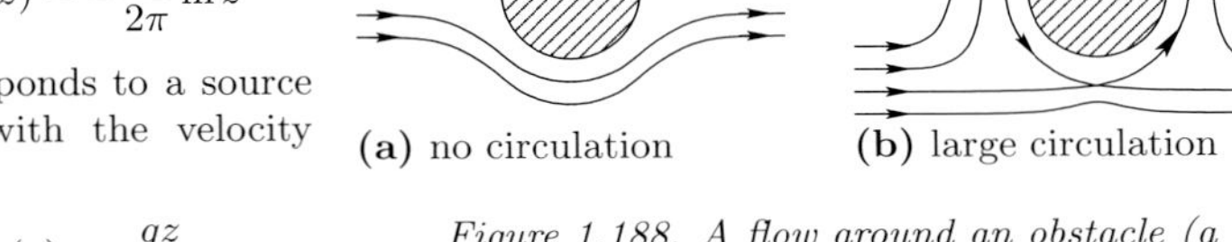


*Figure 1.188. A flow around an obstacle (a disc).*

whose potential is $U = -\dfrac{q}{2\pi} \ln r$ and whose source strength is equal to

$$Q(C) = \frac{q}{2\pi \mathrm{i}} \int_C \frac{\mathrm{d}z}{z} = q.$$

The integral curves are rays originating in the origin, the equipotential curves are concentric circles around the origin.

**Flow with obstacle a disc** (Figure 1.188): Let $c > 0$ and $\Gamma \geq 0$. The function

$$\boxed{f(z) = -c\left(z + \frac{R^2}{z}\right) - \frac{\Gamma}{2\pi \mathrm{i}} \ln z} \qquad (1.476)$$

describes the flow past a disc of radius $R$; this flows is composed of a parallel flow with the speed $c$ and a circulation flow determined by $\Gamma$.

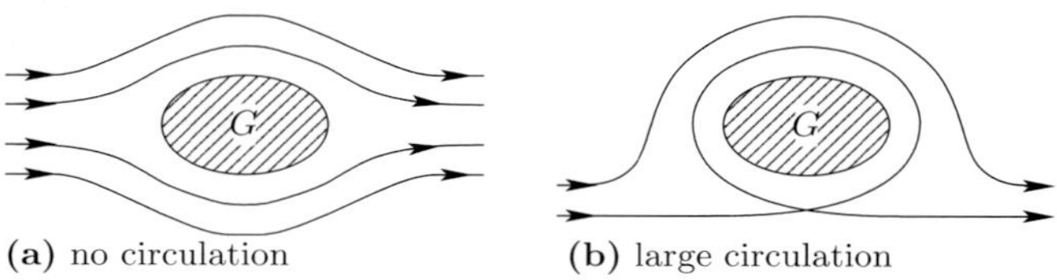


*Figure 1.189. A flow around an obstacle $G$, an arbitrary simply connected domain.*

**The trick with conformal mappings:** Since biholomorphic maps transform holomorphic functions into holomorphic functions, one gets at the same time a map of flows which transforms integral curves into integral curves. Recall that biholomorphic maps are always conformal.

This explains the importance of conformal mappings for physics and technology. The same principle can also be applied in electrostatics and magnetostatics (cf. 1.14.14).

**Flow with obstacle a domain $G$** (Figure 1.189): Suppose we are given a simply connected domain $G$ with a smooth boundary. Let $g$ be a biholomorphic map from $G$ onto a disc. A map of this kind always exists by virtue of the Riemann mapping theorem. We choose the function $f$ as in (1.476) with $R = 1$. Then the composed function

$$w = f(g(z))$$

is a flow around the domain $G$.

## 1.14.14 Applications in electrostatics and magnetostatics

**The basic equations of plane electrostatics:**

$$\boxed{\operatorname{div} \mathbf{E} = 0, \quad \mathbf{curl}\, \mathbf{E} = \mathbf{0} \ \text{ on } \Omega.} \qquad (1.477)$$

These are the Maxwell equations for a stationary electric field $\mathbf{E}$ in the absence of electric charges and flows as well as the absence of a magnetic field.

**The analogy principle:** Every fluid flow from 1.14.13 corresponds to an electrostatic field, if one uses the following dictionary to translate the two notions:

| | | |
|---|---|---|
| velocity field $\mathbf{v}$ | $\Rightarrow$ | electrical field $\mathbf{E}$, |
| velocity potential $U$ | $\Rightarrow$ | electric potential (voltage) $U$, |
| integral curves | $\Rightarrow$ | field curves of $\mathbf{E}$, |
| source strength $Q(C)$ | $\Rightarrow$ | plane charge $q$ on the interior of $C$, |
| circulation $Z(C)$ | $\Rightarrow$ | circulation $Z(C)$. |

At a point of charge $Q$ the force of $Q\mathbf{E}$ acts in the direction of the field curves. The electric field vector is perpendicular to the equipotential curves. Electrical conductors (like metals) correspond to a constant value of the potential $U$.

One of the strengths of mathematics is that the same mathematical theory can be applied to completely different situations occuring in nature.

**Point charges:** The pure source flow $f(z) := -\frac{q}{2\pi}\ln z$ in 1.14.13 corresponds to an electrostatic field $\mathbf{E}$ with the potential

$$U(z) = -\frac{q}{2\pi}\ln r.$$

The field $\mathbf{E}(z) = \frac{qz}{2\pi r^2}$ is generated by a plane charge of strength $q$ at the origin (Figure 1.187(c)).

**Metallic circular cylinders of radius $R$:** The electrostatic field of a cylinder corresponds in each plane perpendicular to the axis of the cylinder the source flow

$$f(z) = -\frac{q}{2\pi}\ln z,$$

with the electrostatic potential $U = -\frac{q}{2\pi}\ln r$ for $r \geq R$. The equipotential curves are concentric circles. In the cylinder we have $U = \text{const} = U(R)$ (Figure 1.190).

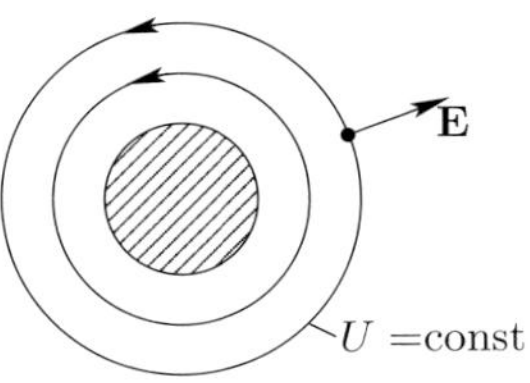

*Figure 1.190.*

**Magnetostatics:** If one replaces the electric field $\mathbf{E}$ in (1.477) by the magnetic field $\mathbf{B}$, then one gets the basic equations of magnetostatics.

### 1.14.15 Analytic continuation and the identity principle

One of the most wonderful properties of holomorphic functions is the fact that one can analytically continue equations and differential equations in a unique manner to larger domains of definition without changing the form of the equations.

**Definition:** Let two holomorphic functions $f : U \subseteq \mathbb{C} \longrightarrow \mathbb{C}$ and $F : V \subseteq \mathbb{C} \longrightarrow \mathbb{C}$ be given on the domains $U$ and $V$ with $U \subset V$. If we have the identity

$$f = F \quad \text{on } U,$$

then $F$ is completely determined by $f$ and is called the *analytic continuation* of $f$.

*Example 1:* We set

$$f(z) := 1 + z + z^2 + \ldots \qquad \text{for all } z \in \mathbb{C} \text{ with } |z| < 1.$$

Moreover let

$$F(z) := \frac{1}{1-z} \qquad \text{for all } z \in \mathbb{C} \text{ with } z \neq 1.$$

Then $F$ is the analytic continuation of $f$ on $\mathbb{C} - \{1\}$.

**Identity principle:** Let two holomorphic functions $f, g : \Omega \longrightarrow \mathbb{C}$ be given on a domain $\Omega$, and suppose that

$$\boxed{f(z_n) = g(z_n) \qquad \text{for all } n = 1, 2, \ldots\ ,}$$

where $(z_n)$ is a sequence with $z_n \to a$ for $n \to \infty$ and $a \in \Omega$. Assume in addition that $z_n \neq a$ for all $n$. Then $f = g$ on $\Omega$.

*Example 2:* Suppose we have the addition theorem

$$\sin(x + y) = \sin x \cos y + \cos x \sin y \tag{1.478}$$

for all $x, y \in ]-\alpha, \alpha[$ for small angles with $\alpha > 0$. Since $w = \sin z$ and $w = \cos z$ are holomorphic functions on $\mathbb{C}$, we know (without any calculation) that the addition theorem holds for all complex numbers $x$ and $y$.

*Example 3:* Suppose we have proven a formula for the derivative

$$\frac{\mathrm{d} \sin x}{\mathrm{d}x} = \cos x$$

for all $x \in ]-\alpha, \alpha[$. Then the validity of this formula for all complex numbers $z$ follows.

**Analytic continuation along a curve:** Suppose we are given the power series

$$f(z) = a_0 + a_1(z - a) + a_2(z - a)^2 + \ldots \tag{1.479}$$

with the domain of convergence $D = \{z \in \mathbb{C} : |z - a| < r\}$. We choose a point $b \in D$ and set $z - a = (z - b) + b - a$. The series (1.479) may be reordered, and we get the new power series

$$g(z) = b_0 + b_1(z - b) + b_2(z - b)^2 + \ldots$$

with the domain of convergence $D' := \{z \in \mathbb{C} : |z - b| < R\}$ (Figure 1.191(a)). Here we have $f = g$ on the intersection $D \cap D'$. If $D'$ contains points which are not in $D$, then we get an analytic continuation $F$ of $f$ to the union $D \cup D'$, by setting

$$F := f \text{ on } D \quad \text{and} \quad F := g \text{ on } D'.$$

One can try to continue this process (Figure 1.191(b)).

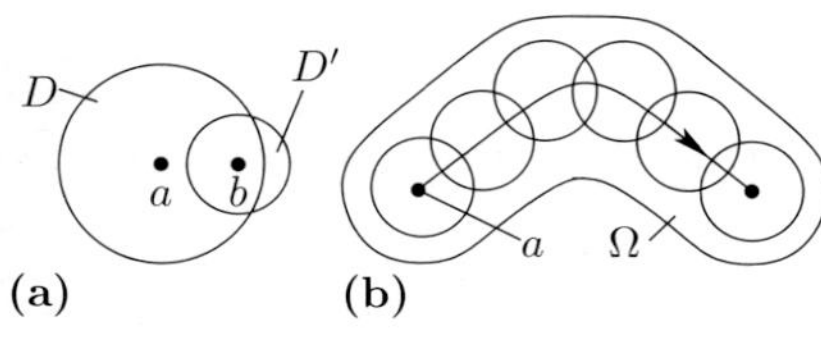


*Figure 1.191. Analytic continuation.*

*Example 4:* The function

$$f(z) := \sum_{n=1}^{\infty} z^{2^n}$$

is holomorphic on the unit disc. However, it *cannot* be analytically continued to a larger domain.[182]

**Monodromy theorem:** Let $\Omega$ be a *simply connected* domain and $f$ a function which is holomorphic in a circular neighborhood of a point $a \in \Omega$, so that $f$ can locally be expanded in a power series around $a$.

If the function $f$ can be analytically continued along a $C^1$-curve in $\Omega$, then this yields a uniquely determined holomorphic function $F$ on the domain $\Omega$ (Figure 1.191(b)).

**Analytic continuation and Riemann surfaces:** If the domain $\Omega$ is not simply connected, then it may happen that the analytic continuation leads to a many-valued function.

*Example 5:* Let $C$ be a circle in the $w$-plane, positively oriented in the mathematical sense. We start the analytic continuation along a curve for the principal value of the function

$$z = {}_{+}\sqrt{w}$$

in a neighborhood of the point $w = 1$. After going around $C$ once, the analytic continuation along the curve $C$ gives rise to a power series expansion at $w = 1$ of $-{}_{+}\sqrt{w}$ (the negative of the principal value). Running along $C$ one more time we return to the power series expansion of ${}_{+}\sqrt{w}$ again.

The situation becomes understandable if we use the intuitive Riemann surface of the function $z = \sqrt{w}$ (cf. 1.14.11.6). We start at the point $w = 1$ on the first sheet and land, after running through $C$ once, on the second sheet. Passing a second time along $C$, we land again on the first sheet.

*Example 6:* If we start the analytic continuation with the power series expansion of the principal value

$$z = \ln w$$

in a neighborhood of the point $w = 1$, then we get after $m$ revolutions around $w = 1$ along $C$ the power series expansion of

$$z = \ln w + 2\pi m i, \qquad m = 0, \pm 1, \pm 2, \ldots$$

Here the number $m = -1$ corresponds to a revolution around $C$ in a negative mathematical sense, i.e., with the reverse orientation, etc. This procedure yields the many-valued function $z = \operatorname{Ln} w$ discussed above.

With the help of the intuitive Riemann surface of $z = \operatorname{Ln} w$ in 1.14.11.7, we can interpret the analytic continuation above as follows. We start at the point $w = 1$ on the zeroth sheet, land on the first sheet after the first revolution along $C$, on the second sheet after the second revolution, etc.

In general one can utilize this procedure to continue a given power series expansion of $f$ to the Riemann surface of the maximal analytic continuation of $f$. This is described in [212].

**Analytic continuation using the Schwarz reflection principle:** Suppose we are given a domain

$$\Omega = \Omega_+ \cup \Omega_- \cup S,$$

which consists of two domains $\Omega_+$ and $\Omega_-$ together with a segment $S$. Here $\Omega_-$ should be the image of $\Omega_+$ under a reflection on the segment $S$ (Figure 1.192). We make the following assumptions:

[182]The symbol $z^{2^n}$ stands for $z^{(2^n)}$ and should be not be confused with $(z^2)^n = z^{2\cdot n}$. In general $a^{b^c}$ means the same thing as $a^{(b^c)}$.

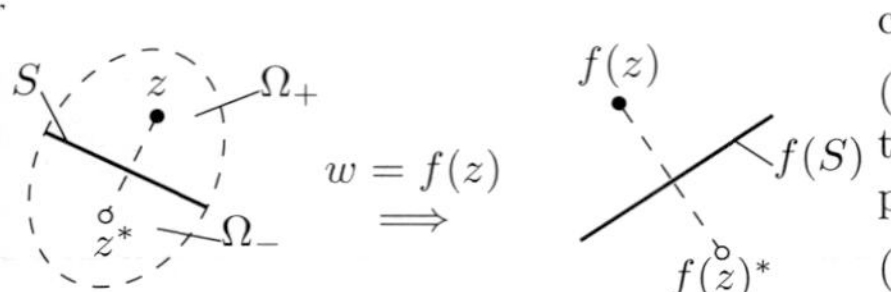


*Figure 1.192.*

(i) The function $f$ is holomorphic in $\Omega_+$ and continuous on the union $\Omega_+ \cup S$.

(ii) The image $f(S)$ of the segment $S$ under the map $w = f(z)$ is a segment in the $w$-plane.

(iii) We set

$$\boxed{f(z^*) := f(z)^* \quad \text{for all } z \in \Omega_+.}$$

Here the star in the notion means the reflection on the segment $S$ in the $z$-plane (resp. on the segment $f(S)$ in the $w$-plane).

This construction leads to an analytic continuation of $f$ to the entire domain $\Omega$.

**The general power function:** Let $\alpha \in \mathbb{C}$. Then one has

$$\boxed{z^\alpha = \mathrm{e}^{\alpha \ln z} \quad \text{for all } z \in \mathbb{R} \text{ with } z > 0.}$$

The function on the right-hand side can be analytically continued. This continuation is then the function $w = z^\alpha$.

(i) $w = z^\alpha$ is well-defined on $\mathbb{C}$, if $\operatorname{Re}\alpha$ and $\operatorname{Im}\alpha$ are integers.

(ii) $w = z^\alpha$ is finitely-valued (many-valued with finitely many values for a give argument), if $\operatorname{Re}\alpha$ and $\operatorname{Im}\alpha$ are rational numbers which are not integers.

(iii) $w = z^\alpha$ is infinitely-valued, if $\operatorname{Re}\alpha$ or $\operatorname{Im}\alpha$ are irrational.

In case (ii) the intuitive Riemann surface of $w = z^\alpha$ is the same as that of the function $w = \sqrt[n]{z}$ for some natural number $n \geq 2$.

In case (iii) the intuitive Riemann surface of the $w = z^\alpha$ is the same as that of the function $w = \operatorname{Ln} z$.

## 1.14.16 Applications to the Euler gamma function

We define

$$\Gamma(n+1) := n!, \qquad n = 0, 1, 2, \ldots$$

This trivially implies the relation

$$\boxed{\Gamma(z+1) = z\Gamma(z)} \tag{1.480}$$

for $z = 1, 2, \ldots$ Euler (1707–1783) asked whether one can define the number $n!$ in a meaningful way for other values of $z$. To do this, he sought a solution $\Gamma$ of the functional equation (1.480) and found the convergent integral

$$\Gamma(x) := \int_0^\infty \mathrm{e}^{-t} t^{x-1} \mathrm{d}t \quad \text{for all } x \in \mathbb{R} \text{ with } x > 0 \tag{1.481}$$

as a solution.

**Analytic continuation:** The real-valued function $\Gamma$ defined by (1.481) can be uniquely extended to a meromorphic function in the complex plane $\mathbb{C}$. The poles of this function are precisely in the points $z = 0, -1, -2, \ldots$; all of these poles have order one.

The Laurent series in a neighborhood of the pole $z = -n$ with $n = 0, 1, 2, \ldots$ is as follows:

$$\Gamma(z) = \frac{(-1)^n}{n!(z+n)} + \text{power series in } (z+n).$$

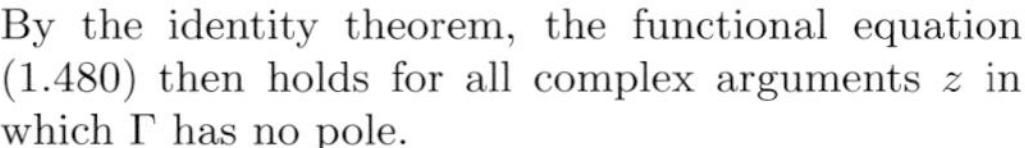

By the identity theorem, the functional equation (1.480) then holds for all complex arguments $z$ in which $\Gamma$ has no pole.

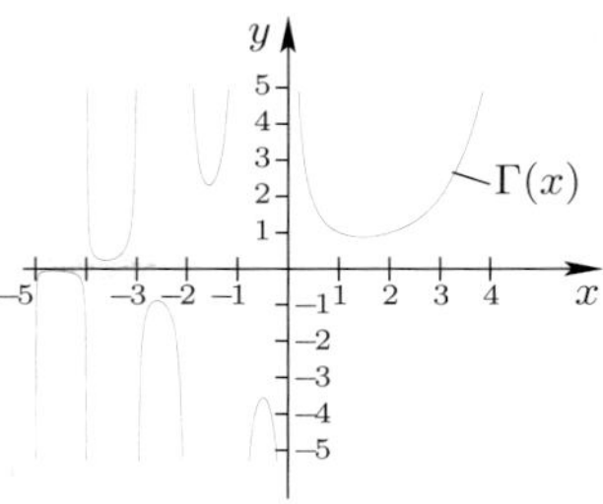


*Figure 1.193. Euler's* $\Gamma$*-function.*

The graph of the gamma function for real values $x$ is pictured in Figure 1.193.

**The Gaussian product formula:** The function $\Gamma$ has no zeros. It follows that the inverse function $1/\Gamma$ is an entire function. One has the product formula

$$\boxed{\frac{1}{\Gamma(z)} = \lim_{n\to\infty} \frac{1}{n^z n!} z(z+1)\ldots(z+n) \quad \text{for all } z \in \mathbb{C}.}$$

**The Gaussian multiplication formula:** For $k = 1, 2, \ldots$ and all $z \in \mathbb{C}$ for which $\Gamma$ has no pole, one has:

$$\Gamma\left(\frac{z}{k}\right)\Gamma\left(\frac{z+1}{k}\right)\ldots\Gamma\left(\frac{z+k-1}{k}\right) = \frac{(2\pi)^{(k-1)/2}}{k^{z-1/2}}\Gamma(z).$$

In particular, for $k = 2$ one gets the *doubling formula of Lagrange*:

$$\Gamma\left(\frac{z}{2}\right)\Gamma\left(\frac{z}{2}+\frac{1}{2}\right) = \frac{\sqrt{\pi}}{2^{z-1}}\Gamma(z).$$

**The complementary theorem of Euler:** For all complex numbers $z$ which are not integers, one has

$$\Gamma(z)\Gamma(1-z) = \frac{\pi}{\sin \pi z}.$$

**The Stirling formula:** For every positive real number $x$ there is a number $\vartheta(x)$ with $0 < \vartheta(x) < 1$, such that the following relation is satisfied:

$$\Gamma(x+1) = \sqrt{2\pi}\, x^{x+1/2} \mathrm{e}^{-x} \mathrm{e}^{\vartheta(x)/12x}.$$

For every complex number $z$ with $\operatorname{Re} z > 0$ one has $|\Gamma(z)| \le |\Gamma(\operatorname{Re} z)|$. In particular, one has

$$\boxed{n! = \sqrt{2\pi n}\left(\frac{n}{\mathrm{e}}\right)^n \mathrm{e}^{\vartheta(n)/12n}, \qquad n = 1, 2, \ldots}$$

**Further properties of the gamma function:** (i) The Euler integral representation (1.481) holds for all complex numbers $z$ with $\operatorname{Re} z > 0$.

(ii) $\Gamma(1) = 1$, $\Gamma(1/2) = \sqrt{\pi}$, $\Gamma(-1/2) = -2\sqrt{\pi}$.

(iii) $\Gamma(z)\Gamma(-z) = -\dfrac{\pi}{z\sin(\pi z)}$ for all complex numbers $z$ which are not integers.

(iv) $\Gamma\left(\frac{1}{2}+z\right)\Gamma\left(\frac{1}{2}-z\right)=\frac{\pi}{\cos(\pi z)}$ for all complex numbers $z$, for which $z+\frac{1}{2}$ is not an integer.

**The uniqueness result of Wielandt (1939):** Let a domain $\Omega$ in the complex plane $\mathbb{C}$ be given, which contains the vertical strip $S := \{z \in \mathbb{C} \mid 1 \le \operatorname{Re} z < 2\}$. Suppose $f$ is a holomorphic function $f : \Omega \longrightarrow \mathbb{C}$ with the following properties:

(i) $f(z+1) = zf(z)$ for all complex numbers $z$ in $\Omega$, for which $z+1$ also belongs to $\Omega$.

(ii) $f$ is bounded on $S$ and $f(1) = 1$.

Then the function $f$ is identical to the gamma function $\Gamma$.

## 1.14.17 Elliptic functions and elliptic integrals

### 1.14.17.1 Basic ideas

**The addition theorem of Fagnano (1781):** The equation

$$\boxed{r^2 = \cos 2\varphi}$$

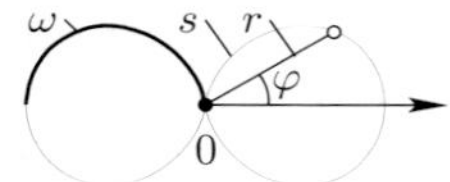


*Figure 1.194.*

describes in polar coordinates the lemniscate of Jakob Bernoulli (1654–1705) with the arc length

$$s(r) = \int_0^r \frac{\mathrm{d}\varrho}{\sqrt{1-\varrho^4}} \tag{1.482}$$

between the origin $O$ and the lemniscate point which has the distance $r$ from $O$ (Figure 1.194). The Italian mathematician Fagnano (1682–1766) found in 1718 the *doubling formula*

$$2s(r) = s(R) \qquad \text{for } R = \frac{2r\sqrt{1-r^4}}{1+r^4}. \tag{1.483}$$

This formula contains a prescription of how to double the lemniscate using a compass and ruler.

In 1753, Euler found many additional formulas for elliptic integrals, which are referred to as addition theorems.

**The discovery of Gauss:** Gauss, when he was only 19 years old, studied the lemniscate in 1796.[183] He asked how one could calculate the distance $r$ of a point on the lemniscate from the arc length $s$. In other words, he was interested in the *inverse function* $r = r(s)$ of the elliptic integral (1.482). Upon differentiation of (1.482) one gets

$$s'(r) = \frac{1}{\sqrt{1-r^4}}, \qquad -1 < r < 1.$$

Hence the function $s : ]-1,1[ \longrightarrow \mathbb{R}$, for which one has $s'(r) > 0$, is strictly increasing and has an inverse function, which Gauss denoted by

$$r = \operatorname{sl} s, \qquad -\omega < s < \omega$$

[183] This research was not published until after his death.

and called the *lemniscate sine function.* Here the number

$$\omega := \int_0^1 \frac{\mathrm{d}\varrho}{\sqrt{1-\varrho^4}}$$

is the length of a half-arc of the lemniscate (Figure 1.194). Moreover Gauss introduced through the relation

$$\mathrm{cl}\, s := \mathrm{sl}(\omega - s) \tag{1.484}$$

the *lemniscate cosine function.* One has

$$\mathrm{sl}^2 s + \mathrm{cl}^2 s + \mathrm{sl}^2 s\, \mathrm{cl}^2 s = 1. \tag{1.485}$$

This relation indicates that Gauss was quite aware of the analogy to the trigonometric functions. This analogy becomes even more evident if one considers the integral

$$s(r) := \int_0^r \frac{\mathrm{d}\varrho}{\sqrt{1-\varrho^2}}.$$

The inverse of this function $s = s(r)$ yields the trigonometric sine function

$$r = \sin s.$$

If we choose the number

$$\omega := \int_0^1 \frac{\mathrm{d}\varrho}{\sqrt{1-\varrho^2}} = \frac{\pi}{2},$$

then we get parallel to (1.484) the trigonometric cosine function

$$\cos s = \sin(\omega - s)$$

with the familiar relation

$$\sin^2 s + \cos^2 s = 1,$$

which is generalized by (1.485). Just as there is an addition theorem for the trigonometric functions

$$\sin(x+y) = \sin x \cos y + \cos x \sin y$$

there are algebraic addition theorems for the lemniscate sine and cosine functions. With these formulas, Gauss introduced the functions $r = \mathrm{sl}\, s$ and $r = \mathrm{cl}\, s$ for all real arguments $s$. The brilliant idea which Gauss had at this point was to *extend* these functions to complex arguments. To do this, he first utilized the substitution $t = \mathrm{i}\varrho$ and derived, formally,

$$\int_0^{\mathrm{i}r} \frac{\mathrm{d}t}{\sqrt{1-t^4}} = \mathrm{i}\int_0^r \frac{\mathrm{d}\varrho}{\sqrt{1-\varrho^4}},$$

hence $s(\mathrm{i}r) = \mathrm{i}s(r)$. This led him to the definition

$$\mathrm{sl}(\mathrm{i}s) := \mathrm{i}(\mathrm{sl}\, s) \qquad \text{for all } s \in \mathbb{R}.$$

With the help of this relation and the addition theorem he was then easily able to define $\mathrm{sl}\, s$ for all complex numbers $s$ and obtained, as a corollary, the two *fundamental periodicity relations*

$$\boxed{\mathrm{sl}(s + 4\omega) = \mathrm{sl}\, s \quad \text{and} \quad \mathrm{sl}(s + 4\omega\mathrm{i}) = \mathrm{sl}\, s \qquad \text{for all } s \in \mathbb{C}.}$$

As opposed with the trigonometric sine the lemniscate sine has not only a real period of $4\omega$, but also a *second, purely imaginary period* $4\omega\mathrm{i}$. This means that $\mathrm{sl}\, s$ is a meromorphic function which is doubly periodic. Functions of this kind are called *elliptic functions.*

Gauss had discovered already in 1796 the existence of elliptic functions.

**General elliptic integrals:** An integral of the form

$$\boxed{\int R(z, \sqrt{p(z)})\, \mathrm{d}z} \qquad (1.486)$$

is called *rational*, if $p$ is a quadratic polynomial with two distinct zeros. Integrals of this kind can always be solved by substitutions involving the trigonometric functions (with real or complex arguments). Trigonometric functions are periodic.

If $p$ is a polynomial of third or fourth degree with pairwise distinct zeros, then we call (1.486) an *elliptic integral.* Integrals of this kind can be solved by substitutions, in which elliptic, that is doubly periodic functions, are used.

> Elliptic integrals and elliptic functions generalize the rational integrals and the trigonometric functions.

The integral (1.486) is called *hyperelliptic* if $p$ is polynomial of fifth or sixth degree with pairwise distinct zeros. Integrals of the form

$$\int R(z, w)\mathrm{d}z,$$

where $w$ is an algebraic function,[184] are called *Abelian integrals.* Such integrals were studied by Niels H. Abel (1802–1829).

**A general theory of elliptic integrals:** Elliptic integrals were studied systematically by Legendre (1752–1833) and Jacobi (1804–1851), whereby Jacobi used rapidly converging theta functions and derived the Jacobian sine and cosine functions $w = \mathrm{sn}\, z$ and $w = \mathrm{cn}\, z$, generalizing the trigonometric functions.

However, a deeper understanding of the theory of elliptic integrals comes about only when one takes the elliptic functions as the starting point of the theory. This was the way that Weierstrass systematically presented the material in a famous series of lectures at Berlin University in 1862. His starting point was the $\wp$-function which he introduced. One can obtain all elliptic functions from this one and its derivative by means of rational operations.

The *basic ideas* of the general theory are the following:

(i) Elliptic integrals can be solved with the help of a *universal substitution*, in which the Weierstrass $\wp$-function is used.

(ii) Elliptic integrals have *local* inverses. Analytic continuation of these local inverse functions yields elliptic functions on the complex plane $\mathbb{C}$.

(iii) The *global* behavior of elliptic integrals are influenced by the many-valuedness of the function $\sqrt{p(z)}$ under the integral. In order for the integral (1.486) to be well-defined as a curve integral, one must utilize paths of integration which lie on the *Riemann surface* of the function $\sqrt{p(z)}$. On this Riemann surface, the function $w = \sqrt{p(z)}$ is single-valued. In this way, the function under the integral also becomes single-valued.

[184] This means that the (many-valued) function $w = w(z)$ satisfies an equation $p(w, z) = 0$, where $p$ is a polynomial of arbitrary degree.

(iv) The Weierstrass $\wp$-function has an *algebraic addition theorem.*

This is the basis of the doubling formula (1.483) found by Fagnano, the general addition theorems of Euler as well as the addition theorems for general elliptic functions, such as the Gaussian lemniscate sine and cosine functions.

The deeper reason for the existence of an addition theorem of the $\wp$-function is the group structure on an elliptic curve (cf. 3.8.1.3).

(v) In order to apply the $\wp$-function to the calculation of arbitrary elliptic integrals, one must solve the inversion problem for the $\wp$-function, in other words, the calculation of a period lattice from certain given invariants of the $\wp$-function. This leads to the theory of modular forms, which we will consider in 1.14.18. Modular forms play an important role in areas as diverse as number theory and string theory in modern particle physics (cf. [212]).

**The famous Jacobi problem of inversion of hyperelliptic integrals:** In 1832, Jacobi formulated the following conjecture. Let $w = p(z)$ be a polynomial of the sixth degree, which has only simple zeros. We consider the two functions $u = u(a,b)$ and $v = v(a,b)$ which are solutions of the following system of equations:

$$\int_{u_0}^{u} \frac{\mathrm{d}z}{\sqrt{p(z)}} + \int_{v_0}^{v} \frac{\mathrm{d}z}{\sqrt{p(z)}} = a,$$

$$\int_{u_1}^{u} \frac{z\mathrm{d}z}{\sqrt{p(z)}} + \int_{v_1}^{v} \frac{z\mathrm{d}z}{\sqrt{p(z)}} = b,$$

where $u_j$ and $v_j$ are fixed given complex numbers, $j = 0, 1$.

Then the two functions $u + v$ and $uv$ are unique and possess four distinct periods.

Riemann and Weierstrass worked intensively on finding a solution to this problem and developed in the process essential parts of the theory of complex functions. Both found a solution to the problem, with different methods, and showed that this is but a special case of more general properties of Abelian integrals.

**Automorphic functions:** In place of the elliptic functions in the case of elliptic integrals, automorphic functions arise in the case of hyperelliptic integrals. Automorphic functions are meromorphic functions on a domain (for example on the upper half-plane or on a disc), which are invariant under the action of a discrete group of automorphisms of that domain. The importance of automorphic functions in the calculation of Abelian integrals is based on the fact that for every compact Riemann surface $\mathscr{R}$ of genus $g \geq 2$, there is an automorphic map $p : \mathscr{B} \longrightarrow \mathscr{R}$ from the open disc $\mathscr{B}$ to $\mathscr{R}$, with the property that $p$ in invariant under the group of *deck transformations* of $\mathscr{B}$ (cf. [212]).

In the case of elliptic integrals one uses elliptic functions $p : \mathbb{C} \longrightarrow \mathscr{R}$, where $\mathscr{R}$ is a Riemann surface of genus $g = 1$, which is hence homeomorphic to a torus. The group of deck transformations is formed by the set of all translations of the complex plane $\mathbb{C}$ which leave the period lattice invariant.

The general theory of Abelian integrals is determined by an extraordinarily harmonious interaction of *analysis, algebra and geometry.* The rich collections of ideas contained in this theory has turned out to be fruitful also in many other areas and has influenced the development of mathematics in the twentieth century in an essential way.

#### 1.14.17.2 Properties of elliptic functions

**Definition:** An *elliptic* function is a doubly periodic meromorphic function $f : \mathbb{C} \longrightarrow \overline{\mathbb{C}}$, i.e., there are two complex numbers $\omega_1,\ \omega_2$, both non-vanishing, such that

$$\boxed{f(z+\omega_j) = f(z) \quad \text{for all } z \in \mathbb{C} \text{ and } j = 1, 2.} \tag{1.487}$$

We set $\tau := \omega_2/\omega_1$ and assume henceforth that the two periods are ordered in such a way that $\operatorname{Im}\tau > 0$. From (1.487) we get

$$f(z + n\omega_1 + m\omega_2) = f(z) \text{ for all } z \in \mathbb{C},$$

where $n$ and $m$ are arbitrary integers.

**The period lattice:** The set

$$\Gamma := \{n\omega_1 + m\omega_2 \mid n \text{ and } m \text{ are integers}\}$$

is called the *period lattice* generated by $\omega_1$ and $\omega_2$; $\Gamma$ is a subgroup of the additive group $\mathbb{C}$. The set

$$\mathscr{F} := \{\lambda\omega_1 + \mu\omega_2 \mid 0 \le \lambda,\ \mu < 1\}$$

is called the *fundamental domain*. This is a parallelogram spanned by $\omega_1$ and $\omega_2$ (Figure 1.195). We write

$$z_1 \equiv z_2 \bmod \Gamma,$$

if and only if $z_1 - z_2 \in \Gamma$, and in this case we say that $z_1$ and $z_2$ are equivalent. Doubly periodic functions take on identical values at equivalent points.

*Example:* In Figure 1.195 four equivalent points are indicated by the open circles.

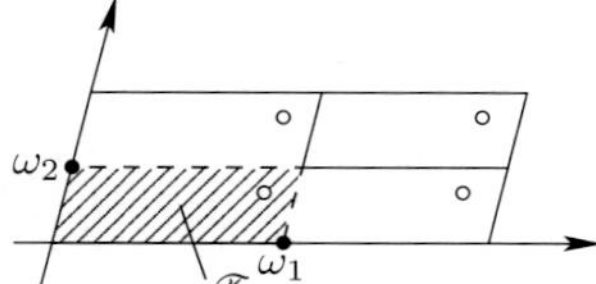

*Figure 1.195. A period lattice.*

**Theorem:** For every point $z_1$ in the complex plane $\mathbb{C}$ there is a unique point $z_2$ in the fundamental domain $\mathscr{F}$ which is equivalent to $z_1$.

Therefore it is sufficient to know the values of an elliptic function in its fundamental domain to know its values everywhere.

**Liouville's theorem (1847):** For a non-constant elliptic function $f : \mathbb{C} \longrightarrow \overline{\mathbb{C}}$ one has:

(i) $f$ has at least one and at most finitely many poles in the fundamental domain $\mathscr{F}$.

(ii) The sum of the residues of all poles of $f$ in $\mathscr{F}$ is zero.

(iii) $f$ takes on every value $w \in \mathbb{C}$ and $w = \infty$ with the same multiplicity at all points of $\mathscr{F}$.[185]

#### 1.14.17.3 The Weierstrass ℘-function

**Definition:** Let $\omega_1$ and $\omega_2$ be given, spanning a period lattice. For all $z \in \mathbb{C} - \Gamma$ we set

$$\boxed{\wp(z) := \frac{1}{z^2} \sum_{g\in\Gamma}{}' \left( \frac{1}{(z-g)^2} - \frac{1}{g^2} \right).}$$

[185] The sum of the multiplicities of all zeros of $f - w$ in $\mathscr{F}$ and the sum of all multiplicities of all poles of $f$ in $\mathscr{F}$ coincide.

The summuation symbol with the apostrophe inidicates that the lattice point $g = 0$ is not a summand.

In addition we set

$$e_1 := \wp\left(\frac{\omega_1}{2}\right), \qquad e_2 := \wp\left(\frac{\omega_1+\omega_2}{2}\right), \qquad e_3 := \wp\left(\frac{\omega_2}{2}\right),$$
$$g_2 := -4(e_1e_2 + e_1e_3 + e_2e_3), \qquad g_3 := 4e_1e_2e_3. \tag{1.488}$$

One has:

$$4(z-e_1)(z-e_2)(z-e_3) = 4z^3 - g_2z - g_3.$$

The (non-normalized) discriminant of this polynomial is

$$\Delta := (g_2)^3 - 27(g_3)^2.$$

If the numbers $e_1$, $e_2$ and $e_3$ are pairwise distinct, then $\Delta \neq 0$.

**Theorem 1:** (i) The $\wp$-function is elliptic with the periods $\omega_1$ and $\omega_2$.

(ii) $\wp$ has exactly one pole in the fundamental domain $\mathscr{F}$. This is at the point $z = 0$ and has multiplicity two.

(iii) The $\wp$-function attains every value $w \in \mathbb{C}$ in $\mathscr{F}$ twice and is an even function, i.e., $\wp(-z) = \wp(z)$ for all $z \in \mathbb{C}$.

(iv) For all $z \in \mathbb{C} - \Gamma$ the function $w = \wp(z)$ obeys the differential equation

$$\boxed{w'^2 = 4(w-e_1)(w-e_2)(w-e_3).} \tag{1.489}$$

(v) One has the addition theorem

$$\boxed{\wp(u+v) = -\wp(u) - \wp(v) + \frac{1}{4}\left(\frac{\wp'(u)-\wp'(v)}{\wp(u)-\wp(v)}\right)^2}$$

for all $u, v \in \mathbb{C} - \Gamma$ with $u \neq v$. Moreover, one has the relation

$$\wp(2u) = -2\wp(u) + \frac{1}{4}\left(\frac{\wp''(u)}{\wp'(u)}\right)^2.$$

**The field of elliptic functions:** The set of all elliptic functions with given periods $\omega_1$ and $\omega_2$ form a *field*[186] $\mathscr{K}$. This field is generated by the $\wp$-function and its derivative. Explicitly $\mathscr{K}$ consists of all functions of the form

$$\boxed{R(\wp, \wp'),}$$

where $R$ is an arbitrary rational function in two variables (a quotient of two polynomials in two variables).

**The Eisenstein (1832–1852) series:** For $n = 3, 4, \ldots$ the series

$$G_n := {\sum_{\omega \in \Gamma}}' \frac{1}{\omega^n}$$

[186] Fields are defined in section 2.5.3

converge. Here again the summation with the apostrophe indicates that the lattice point $\omega = 0$ is not included.

**Theorem 2:** One has $g_2 = 60G_4$ and $g_3 = 140G_6$. In a neighborhood of $z = 0$ one has the Laurent expansion

$$\wp(z) = \frac{1}{z^2} + \sum_{n=1}^{\infty}(2n+1)G_{2n+2}z^{2n}.$$

### 1.14.17.4 The Jacobian theta functions

**Definition:**

$$\boxed{\vartheta_0(z;\tau) = 1 + 2\sum_{n=1}^{\infty}(-1)^n q^{n^2}\cos 2\pi n z, \qquad z \in \mathbb{C}.}$$

Here $q$ is defined as $q := \mathrm{e}^{\mathrm{i}\pi\tau}$ with $\tau \in \mathbb{C}$ and $\operatorname{Im}\tau > 0$.[187]

**Theorem:** For fixed parameter $\tau$ the function $\vartheta_0$ is entire. It has a period of 1 and zeros at exactly the points

$$\frac{\tau}{2} + n + m\tau, \qquad n, m \in \mathbb{Z}.$$

As a function of $z$ and $\tau$ the function $\vartheta_0$ satisfies the complex heat equation

$$\frac{\partial^2\vartheta_0}{\partial z^2} = 4\pi\mathrm{i}\frac{\partial\vartheta_0}{\partial\tau}.$$

**Definition:** From $\vartheta_0$ one gets the other *Jacobian theta functions*:

$$\vartheta_1(z;\tau) := -\mathrm{i}q^{1/4}\mathrm{e}^{\mathrm{i}\pi z}\vartheta_0\left(z + \frac{\tau}{2};\tau\right) = 2\sum_{n=0}^{\infty}(-1)^n q^{(n+\frac{1}{2})^2}\sin(2n+1)z,$$

$$\vartheta_2(z;\tau) := \vartheta_1\left(z + \frac{1}{2};\tau\right) = 2\sum_{n=0}^{\infty} q^{(n+\frac{1}{2})^2}\cos(2n+1)\pi z,$$

$$\vartheta_3(z;\tau) := \vartheta_0\left(z - \frac{1}{2};\tau\right) = 1 + 2\sum_{n=1}^{\infty} q^{n^2}\cos 2\pi n z.$$

### 1.14.17.5 The Jacobian elliptic functions

**Definition:** Let $0 < k,\ k' < 1$ with $k^2 + k'^2 = 1$. We set

$$\operatorname{sn}(z;k) := \frac{1}{\sqrt{k}}\frac{\vartheta_1\left(\frac{z}{2K};\tau\right)}{\vartheta_0\left(\frac{z}{2K};\tau\right)} \qquad \text{(sinus amplitudinis)},$$

$$\operatorname{cn}(z;k) := \sqrt{\frac{k'}{k}}\frac{\vartheta_2\left(\frac{z}{2K};\tau\right)}{\vartheta_0\left(\frac{z}{2K};\tau\right)} \qquad \text{(cosinus amplitudinis)},$$

$$\operatorname{dn}(z;k) := \sqrt{k'}\frac{\vartheta_3\left(\frac{z}{2K};\tau\right)}{\vartheta_0\left(\frac{z}{2K};\tau\right)} \qquad \text{(delta amplitudinis)}.$$

---

[187] $q^{n^2} = q^{(n^2)}$.

The quantities used here are:

$$K(k) = \int_0^{\pi/2} \frac{\mathrm{d}\varphi}{\sqrt{1 - k^2 \sin \varphi}}$$

and $\tau := \mathrm{i}K'(k)/K(k)$ with $K'(k) := K(k')$.[188] In what follows we keep $k$ fixed and write more briefly $\operatorname{sn} z$ and $\operatorname{cn} z$ as well as $K$ and $K'$.

**Theorem:** The three functions $w = \operatorname{sn} z$, $\operatorname{cn} z$ and $\operatorname{dn} z$ are elliptic with the properties listed in Table 1.8. Moreover, one has:

$$\operatorname{sn}^2 z + \operatorname{cn}^2 z = 1 \quad \text{and} \quad (\operatorname{sn} z)' = \operatorname{cn} z \operatorname{dn} z$$

for all $z \in \mathbb{C}$, provided there are no poles there. The function $\operatorname{sn} z$ is odd, while $\operatorname{cn} z$ and $\operatorname{dn} z$ are even.

**Differential equations:** The general non-constant solution of the differential equations

$$\begin{aligned} u'^2 &= (1 - u^2)(1 - k^2 u^2), \\ v'^2 &= (1 - v^2)(k'^2 + k^2 v^2), \\ w'^2 &= -(1 - w^2)(k'^2 - w^2) \end{aligned}$$

is

$$\begin{aligned} u &= \pm \operatorname{sn}(z + \text{const}), \\ v &= \pm \operatorname{cn}(z + \text{const}), \\ w &= \pm \operatorname{dn}(z + \text{const}). \end{aligned}$$

**Addition theorems:**

$$\begin{aligned} \operatorname{sn}(u + v) &= \frac{\operatorname{sn}u \operatorname{cn}v \operatorname{dn}v + \operatorname{sn}v \operatorname{cn}u \operatorname{dn}u}{1 - k^2 \operatorname{sn}^2 u \operatorname{sn}^2 v}, \\ \operatorname{cn}(u + v) &= \frac{\operatorname{cn}u \operatorname{cn}v - \operatorname{sn}u \operatorname{sn}v \operatorname{dn}u \operatorname{dn}v}{1 - k^2 \operatorname{sn}^2 u \operatorname{sn}^2 v}, \\ \operatorname{dn}(u + v) &= \frac{\operatorname{dn}u \operatorname{dn}v - k^2 \operatorname{sn}u \operatorname{sn}v \operatorname{cn}u \operatorname{cn}v}{1 - k^2 \operatorname{sn}^2 u \operatorname{sn}^2 v}. \end{aligned}$$

### 1.14.18 Modular forms and the inversion problem for the $\wp$-function

Different periods $\omega_1$, $\omega_2$ can lead to the same lattice.

*Example:* The two pairs of periods $(1, \mathrm{i})$ and $(1, 1 + \mathrm{i})$ generate the same lattice in $\mathbb{C}$ (Figure 1.196), as the set of lattice points for both coincide.

**Main theorem:** (i) The Weierstrass $\wp$-function depends only on the period lattice and not on the periods generating the lattice.

(ii) If three distinct complex numbers $e_1$, $e_2$ and $e_3$ are given, there is a lattice and a $\wp$-function which belong to these values (in the sense of (1.488)).

[188] The standard symbol $K'$ means here a real number and not the derivative of a function.

*Table 1.8. The elliptic functions of Jacobi.*

| | *Periods* | *Zeros* | *Poles* | *Residues* |
|---|---|---|---|---|
| $\operatorname{sn} z$ | $4K,\ 2K'\mathrm{i}$ | $2mK + 2nK'\mathrm{i}$ | $2mK + (2n+1)K'\mathrm{i}$ | $\frac{1}{k}, -\frac{1}{k}$ |
| $\operatorname{cn} z$ | $4K,\ 2(K + K'\mathrm{i})$ | $(2m+1)K + 2nK'\mathrm{i}$ | $2mK + (2n+1)K'\mathrm{i}$ | $\frac{\mathrm{i}}{k}, -\frac{\mathrm{i}}{k}$ |
| $\operatorname{dn} z$ | $2K,\ 4K'\mathrm{i}$ | $(2m+1)K + (2n+1)K'\mathrm{i}$ | $2mK + (2n+1)K'\mathrm{i}$ | $-\mathrm{i},\ \mathrm{i}$ |
| $n$ and $m$ are integers | | | | |

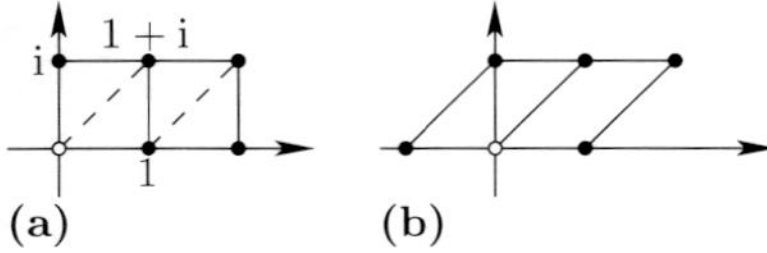


*Figure 1.196. Different periods which define the same lattice.*

**Remark:** The universal method of solving elliptic integrals is based on this result, namely making the substitution $w = \wp'(t),\ z = \wp(t)$ (cf. 1.14.19).

The theorem above follows from the theory of modular forms which is presented in the sequel. In particular one uses the fact that the modular $J$-function of Klein takes on every complex number as one of its values. The theory of modular forms has important applications in number theory, in algebraic geometry (cf. the deep Shimura–Taniyama–Weil conjecture in 3.8.6.2), in numerics for the calculation of $\pi$, as well as in string theory in high-energy physics.

**The modular group:** Let $\mathscr{H}_+ := \{z \in \mathbb{C} \mid \operatorname{Im} z > 0\}$ denote the upper half-plane. A *modular transformation* is by definition a Möbius transformation of the form

$$\tau' = \frac{a\tau + b}{c\tau + d},$$

where $a,\ b,\ c$ and $d$ are integers with $ad - bc = 1$. These transformations map $\mathscr{H}_+$ biholomorphically (and hence also conformally) into itself.

The set of all modular transformations form a subgroup of the automorphic group (of Möbius transformations, see 1.14.11.5), which one calls the *modular group* $\mathscr{M}$.

Let $\tau,\ \tau' \in \mathscr{H}_+$. We write

$$\tau \equiv \tau' \bmod \mathscr{M},$$

if and only if there is a modular transformation which maps $\tau$ to $\tau'$. This is an equivalence relation which is denoted by $\mathscr{M}\backslash\mathscr{H}_+$ (the operation is from the left).

The modular group is generated by the transformations $\tau' = \tau + 1$ and $\tau' = -1/\tau$.

**The fundamental domain of the modular group:** We set

$$\mathscr{F}(\mathscr{M}) := \left\{ z \in \mathscr{H}_+ \,\middle|\, -\frac{1}{2} \le \operatorname{Re} z \le \frac{1}{2},\ \ |z| \ge 1 \right\}$$

(see Figure 1.197). Then we have:

(i) Every point of the upper half-plane is equivalent to a point of $\mathscr{F}(\mathscr{M})$ modulo $\mathscr{M}$ (this means that there is an element of $\mathscr{M}$ transforming the given point to a point in the fundamental domain).

(ii) Two arbitrary points of $\mathscr{F}(\mathscr{M})$ are inequivalent modulo $\mathscr{M}$ if and only if they lie on the boundary.

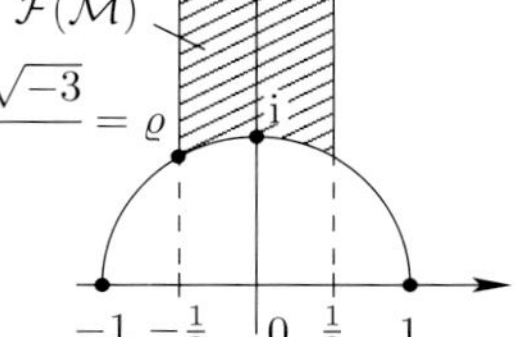


*Figure 1.197. The fundamental domain of the modular group.*

**Equivalent lattices:** Two lattices $\Gamma$ and $\Gamma'$ are said to be *equivalent* if there is a complex number $a \neq 0$ such that $\Gamma' = a\Gamma$.

**Theorem 1:** Two pairs of periods $(\omega_1, \omega_2)$ and $(\omega_1', \omega_2')$ generate equivalent lattices if and only if the quotients $\omega_2/\omega_1$ and $\omega_2'/\omega_1'$ are equivalent modulo $\mathscr{M}$. Recall that we agree that $\operatorname{Im}\tau > 0$ for $\tau := \omega_2/\omega_1$.

The association $(\omega_1, \omega_2) \mapsto \omega_2/\omega_1$ yields a bijective map from the set of equivalence classes of lattices to the fundamental domain $\mathscr{F}(\mathscr{M})$ of the modular group.

**Modular forms:** A *modular form of weight* $k$ is a meromorphic function $f : \mathscr{H}_+ \longrightarrow \overline{\mathbb{C}}$ with the property that

$$f\left(\frac{a\tau + b}{c\tau + d}\right) = (c\tau + d)^k f(\tau) \quad \text{for all } \tau \in \mathscr{H}_+$$

and for all modular transformations. In the case of $k = 0$ we speak of *modular functions.*

In the sequel we shall be using the quantities

$$g_2(\omega_1, \omega_2), \quad g_3(\omega_1, \omega_2), \quad \text{and} \quad \Delta(\omega_1, \omega_2),$$

which were defined in (1.488), where we indicate the dependency of these on the periods $\omega_1$, $\omega_2$.

**Definition:** Let $g_j(\tau) := g_j(1, \tau)$, $j = 1, 2$ and $\Delta(\tau) := \Delta(1, \tau)$. The *modular (Klein) J-function* is defined by the relation

$$\boxed{J(\tau) := \frac{g_2(\omega_1, \omega_2)^3}{\Delta(\omega_1, \omega_2)}.}$$

The function $J$ only depends on the period ratio $\tau = \omega_2/\omega_1$.

**Theorem:** (i) The functions $w = J(\tau)$, $\Delta(\tau)$, $g_2(\tau)$ and $g_3(\tau)$ are holomorphic on the upper half-plane.

(ii) $J$ is a modular function, which maps the fundamental domain $\mathscr{F}(\mathscr{M})$ of the modular group bijectively to the complex plane $\mathbb{C}$.

(iii) The function $w = \Delta(\tau)$ is a modular form of weight 12.

**The Dedekind (1831–1916) eta function:**

$$\boxed{\eta(\tau) := \mathrm{e}^{\pi \mathrm{i}\tau/12} \prod_{n=1}^{\infty} \left(1 - \mathrm{e}^{2\pi \mathrm{i} n\tau}\right), \qquad \tau \in \mathscr{H}_+.}$$

This function, which is important in number theory as well as, for example, in string theory, is holomorphic on the upper half-plane and satisfies the relation

$$\eta\left(\frac{a\tau+b}{c\tau+d}\right) = \varepsilon(c\tau+d)^{1/2}\eta(\tau) \quad \text{for all } \tau \in \mathscr{H}_+$$

and all modular transformations. Here $\varepsilon$ is a 24th root of unity, i.e., $\varepsilon^{24} = 1$. Moreover, one has

$$\Delta(\tau) = (2\pi)^{12}\eta(\tau)^{24}.$$

## 1.14.19 Elliptic integrals

In order to get a better grasp of the general theory, we begin with an important example. With the example we want to elucidate the following basic principle:

> Riemann surfaces are of great practical value in the calculation of integrals of many-valued algebraic functions (for example elliptic integrals).

### 1.14.19.1 The Legendre normal form of integrals of the first kind and the Jacobian sine function

**The basic integral**

Consider the real integral

$$f(z) := \int_0^z \frac{\mathrm{d}x}{\sqrt{(1-x^2)(1-k^2x^2)}}, \qquad -1 < z < 1.$$

Here the numbers $k$ and $k'$ are given and satisfy $0 < k,\ k' < 1$ and $k^2 + k'^2 = 1$. We furthermore set

$$K(k) := \int_0^1 \frac{\mathrm{d}x}{\sqrt{(1-x^2)(1-k^2x^2)}}$$

and $K' := K(k')$. If we use the integral

$$F(k,\varphi) := \int_0^\varphi \frac{\mathrm{d}\psi}{\sqrt{1-k^2\sin^2\psi}},$$

then we have

$$f(\sin\varphi) = F(k,\varphi) \qquad \text{for } -\frac{\pi}{2} < \varphi < \frac{\pi}{2}$$

and $K(k) = F\left(k, \frac{\pi}{2}\right)$. The elliptic integral $F(k,\varphi)$ is tabulated in section 0.5.4. For all $z \in ]-1,1[$ one has

$$f'(z) = \frac{1}{\sqrt{(1-z^2)(1-k^2z^2)}}.$$

Therefore the function $f : ]-1,1[ \longrightarrow ]-K,K[$ is strictly increasing and has a unique inverse function, which Jacobi (1804–1851) denoted by

$$\boxed{z = \operatorname{sn}(t;k) \qquad -K < t < K.} \tag{1.490}$$

Instead of $\operatorname{sn}(t;k)$ and $K(k)$ we write in what follows more briefly $\operatorname{sn} t$ and $K$. Hence we have

$$t = \int_0^{\operatorname{sn} t} \frac{\mathrm{d}x}{\sqrt{(1-x^2)(1-k^2x^2)}}, \qquad -K < t < K.$$

**The amplitude functions:** For every fixed $k \in ]-1,1[$ the function

$$t = F(k,\varphi), \qquad -\frac{\pi}{2} < \varphi < \frac{\pi}{2}$$

is strictly increasing and has the smooth inverse function

$$\varphi = \operatorname{am}(k,t), \qquad -K < t < K,$$

which one calls the *amplitude function.*

**Theorem 1:** For all $t \in ]-K,K[$ one has

$$\operatorname{sn} t = \sin \operatorname{am} t, \quad \operatorname{cn} t = \cos \operatorname{am} t \quad \text{and} \quad \operatorname{dn} t = \sqrt{1-k^2\operatorname{sn}^2 t}.$$

This explains the term sinus amplitudinis and cosinus amplitudinus (Latin) for the Jacobian functions $\operatorname{sn} t$ and $\operatorname{cn} t$.

**The limiting case $k=0$:** If $k=0$, then one has $K = \pi/2$ and

$$\operatorname{am} t = t, \quad \operatorname{sn} t = \sin t, \quad \operatorname{cn} t = \cos t, \quad \operatorname{dn} t = 1 \quad \text{for all } t \in \left]-\frac{\pi}{2}, \frac{\pi}{2}\right[.$$

**Analytic continuation**

**Theorem (Jacobi):** The functions $z = \operatorname{sn} t$, $\operatorname{cn} t$ and $\operatorname{dn} t$, which are defined on the interval $]-K,K[$, can be extended in a unique manner to the complex plane $\mathbb{C}$. The result of this extension is a set of elliptic functions.

The corresponding formulas expressing these functions in terms of the theta functions can be found in 1.14.17.5. In particular one has

$$\operatorname{sn}(t+4K) = \operatorname{sn} t \quad \text{and} \quad \operatorname{sn}(t+2K'\mathrm{i}) = \operatorname{sn} t \quad \text{for all } t \in \mathbb{C},$$

i.e., the function $z = \operatorname{sn} t$ has a real period of $4K$ and a purely imaginary period of $2K'\mathrm{i}$.

**The general contour integral**

The contour integral

$$I(C) = \int_C \frac{\mathrm{d}z}{\sqrt{(1-z^2)(1-k^2z^2)}}$$

only makes sense if one declares how the many-valued square root along the curve $C$ is to be understood (i.e., along which sheet does one take the function?). Since the function

$$w = \sqrt{(1-z^2)(1-k^2z^2)}$$

is single valued on its Riemann surface $\mathscr{R}$, it is natural to consider the integrand of the integral $I(C)$ as living on $\mathscr{R}$. In order to amplify this point we write

$$I(C) = \int_C \frac{\mathrm{d}z}{w}$$

with the algebraic function

$$w^2 = (1 - z^2)(1 - k^2 z^2).$$

**The Riemann surface $\mathscr{R}$:** We choose two copies of the complex $z$-plane, cut these along the intervals

$$\left[-\frac{1}{k}, -1\right] \qquad \text{and} \qquad \left[1, \frac{1}{k}\right],$$

and glue the corresponding edges criss–cross (Figure 1.198). We denote the surface constructed in this manner by $\mathscr{R}$. We define[189]

$$w := \begin{cases} {}_+\sqrt{(1-z^2)(1-k^2z^2)} & \text{for } z \text{ on the first cut sheet,} \\ -{}_+\sqrt{(1-z^2)(1-k^2z^2)} & \text{for } z \text{ on the second cut sheet,} \\ \mathrm{i}_+\sqrt{(1-z^2)(k^2z^2-1)} & \text{for } z \text{ on the cut intervals } \left[-\frac{1}{k}, -1\right] \text{ and } \left[1, \frac{1}{k}\right]. \end{cases}$$

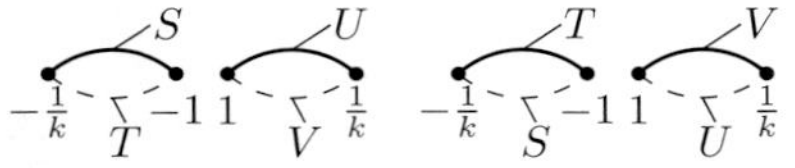

(a) first sheet (b) second sheet

*Figure 1.198. Sheets of a Riemann surface.*

With this definition, the function $w = \sqrt{(1-z^2)(1-k^2z^2)}$ is single-valued on $\mathscr{R}$, i.e., $\mathscr{R}$ is the intuitive Riemann surface of this function.

**The parameterization of paths on the Riemann surface (global uniformization)**

**Theorem 2:** Every continuous compact path on the Riemann surface $\mathscr{R}$ corresponds in a one to one manner to a path

$$C_* : t = t(\tau), \qquad \tau_0 \le \tau \le \tau_1,$$

in the complex $t$-plane, so that $C$ has a parameterization

$$z = \operatorname{sn} t(\tau), \qquad \tau_0 \le \tau \le \tau_1, \tag{1.491}$$

where the corresponding values of $w$ are given by $w(\tau) = (\operatorname{sn})'(t(\tau))$. This clarifies the question as to which sheet of the Riemann surface gives the values for $\operatorname{sn} t(\tau)$.[190]

**The calculation of the integral**

From Theorem 2 it follows that

$$\boxed{I(C) = t(\tau_0) - t(\tau_1).}$$

[189] We denote the principal value of the square root by ${}_+\sqrt{u}$ (cf. 1.14.11.6).

[190] The map

$$z = \operatorname{sn} t, \qquad w = (\operatorname{sn})'(t) \tag{1.493}$$

is closely related to the equation

$$w^2 = (1 - z^2)(1 - k^2 z^2)$$

and the differential equation

$$\left(\frac{\mathrm{d} \operatorname{sn} t}{\mathrm{d}t}\right)^2 = (1 - \operatorname{sn}^2 t)(1 - k^2 \operatorname{sn}^2 t).$$

From a general topological point of view the complex $t$-plane is the universal covering surface of the Riemann surface $\mathscr{R}$. The corresponding covering map $p : \mathbb{C} \longrightarrow \mathscr{R}$ is given by (1.493). Covering spaces of manifolds are considered in [212].

Indeed, the substitution (1.491) yields

$$I(C) = \int_C \frac{\mathrm{d}z}{w} = \int_{\tau_0}^{\tau_1} \frac{z'(t(\tau))t'(\tau)}{w(t(\tau))}\mathrm{d}\tau = \int_{\tau_0}^{\tau_1} t'(\tau)\mathrm{d}\tau = t(\tau_1) - t(\tau_0).$$

One writes for this more briefly $z = \operatorname{sn} t,\ w = z'(t)$ and

$$I(C) = \int_C \frac{\mathrm{d}z}{w} = \int_{C_*} \frac{z'(t)}{w(t)}\mathrm{d}t = \int_{C_*} \mathrm{d}t = t(\tau_1) - t(\tau_0).$$

**Conformal maps**

In order to explain the connection between the paths $C$ on the Riemann surface and the corresponding paths $C_*$ in the complex $t$-plane, we investigate the function

$$\boxed{z = \operatorname{sn} t}$$

from the complex $t$-plane to the complex $z$-plane. We again denote the upper half-plane (of the $z$-plane) by $\mathscr{H}_+ = \{z \in \mathbb{C} \mid \operatorname{Im} z > 0\}$.

**Theorem 3:** (i) The function

$$t = \int_0^z \frac{\mathrm{d}\zeta}{+\sqrt{(1-\zeta^2)(1-k^2\zeta^2)}}, \qquad z \in \mathscr{H}_+, \tag{1.492}$$

maps the upper half-plane $\mathscr{H}_+$ biholomorphically (hence conformally) to the open rectangle $Q$ in the complex $t$-plane, which has corners $\pm K$ and $\pm K + K'\mathrm{i}$.

(ii) Moreover, the closed lower half-plane (including the point $\infty$) is mapped homeomorphically to the closed rectangle $\overline{Q}$, where the mapping sends the points

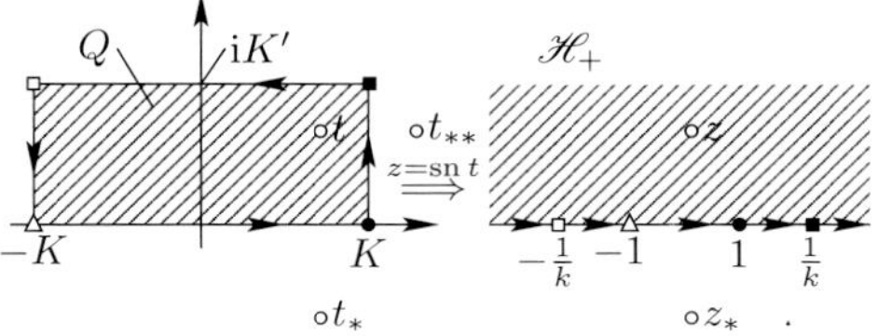


Figure 1.199. The function sinus amplitudinis.

$$-\frac{1}{k}, -1, 1, \frac{1}{k} \quad \text{to the vertices} \quad -K + K'\mathrm{i}, -K, K, K + K'\mathrm{i}$$

and the edges between these points to the corresponding edges of $Q$.

(iii) The inverse map to (1.492) from $\overline{Q}$ to $\overline{\mathscr{H}_+}$ is given by $z = \operatorname{sn} t$ (Figure 1.199).

We consider the period rectangle

$$\mathscr{P} := \{t \in \mathbb{C} : -K \le \operatorname{Re} t < 3K,\ -K' \le \operatorname{Im} t < K'\}.$$

**Theorem 4:** The function $z = \operatorname{sn} t$ maps $\mathscr{P}$ bijectively to the Riemann surface $\mathscr{R}$.

*Example 1:* The path $C$ in the $t$-plane is transformed to a path $\mathscr{C}$ on $\mathscr{R}$, during which a transition from the first to the second sheet is made (Figure 1.200).

*Proof:* We consider the point $t$ in Figure 1.199. Reflection of $t$ on the real axis gives us a point $t_*$. According to the Schwarz reflection principle, the image point of $z = \operatorname{sn} t$, namely $z_* = \operatorname{sn} t_*$, is obtained by reflection on the real axis (cf. 1.14.15).

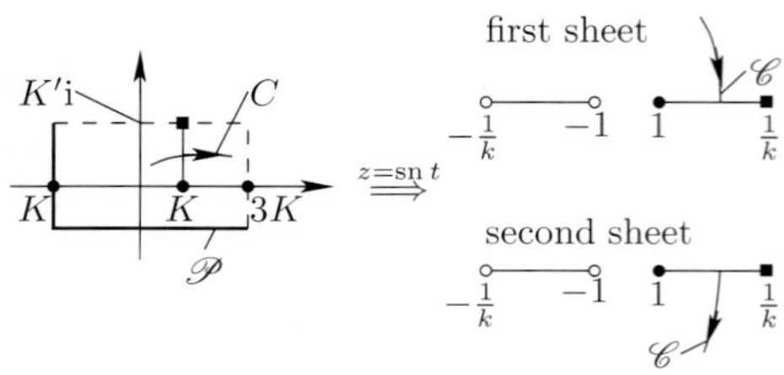


*Figure 1.200. The function sn is a uniformization.*

Furthermore, the reflection of $t$ on the segment passing through the two points $K \pm K'\mathrm{i}$ yields a point $t_{**}$. The image point $z_{**} = \mathrm{sn}\, t_{**}$ is obtained by reflection of $z = \mathrm{sn}\, t$ on the real axis, where we this time view $z_{**}$ as a point on the second sheet of the $z$-plane.

### The deformation trick for simplifying the integral

We have the following two important rules at our disposal:

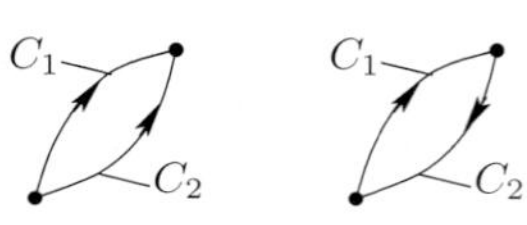

*Figure 1.201.*

(i) If $C$ is a closed path on the Riemann surface $\mathscr{R}$ which can be continuously deformed to a point on $\mathscr{R}$, then one has $I(C) = 0$.

(ii) If $C_1$ and $C_2$ are two paths on $\mathscr{R}$ with the same start and end points, then one has

$$\boxed{I(C_1) = I(C_2),}$$

if the closed curve $C = C_1 - C_2$ has the property (i) (Figure 1.201).

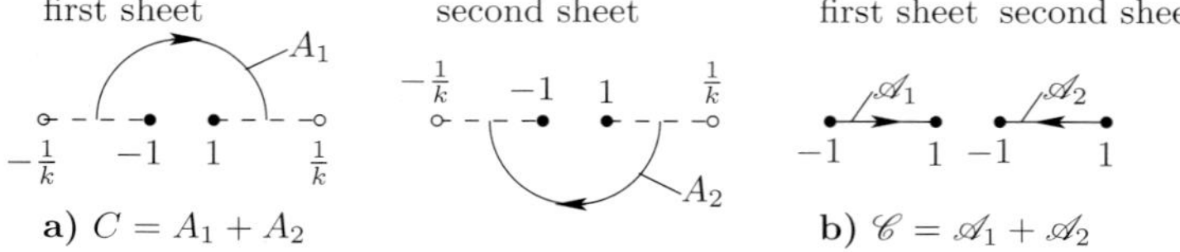


*Figure 1.202. Deforming a path to calculate a curve integral.*

*Example 2:* For the path $C = A_1 + A_2$ in Figure 1.202(a) we get

$$\boxed{I(C) = 4K.}$$

This is a period of the function $z = \mathrm{sn}\, t$.

*Proof:* The second rule (ii) above gives us $I(A_j) = I(\mathscr{A}_j)$ (Figure 1.202(b)). This yields

$$I(C) = I(\mathscr{A}_1) + I(\mathscr{A}_2) = \int_{-1}^{1} \frac{\mathrm{d}z}{+\sqrt{g(z)}} + \int_{1}^{-1} \frac{\mathrm{d}z}{-+\sqrt{g(z)}} = 2\int_{-1}^{1} \frac{\mathrm{d}z}{+\sqrt{g(z)}} = 4K,$$

where we have set $g(z) := (1 - z^2)(1 - k^2 z^2)$. □

### The topological structure of the Riemann surface

*Following Riemann's example one should master proofs through ideas and not through brutal computation.*

*David Hilbert* (1862–1943)

We set $e_1 := -1/k$, $e_2 := -1$, $e_3 := 1$ and $e_4 := 1/k$. Instead of two $z$-planes we consider two Riemann spheres, cut these from $e_1$ to $e_2$ (resp. from $e_3$ to $e_4$), and glue them diagonally. If we blow the resulting surface up like a balloon, we get a torus (Figure 1.203). The fundamental *topological* property of a torus $\mathscr{T}$ is that on such a surface there are two different types of closed curves, which *cannot* be continuously deformed to a point. Such closed curves are, for example, the longitude $L$ and an arbitrary meridian $M$ (Figure 1.203). From Example 2 we infer

$$I(M) = 4K.$$

Moreover one has

$$I(L) = 2K'\mathrm{i}.$$

These are precisely the two periods $4K$ and $2K'\mathrm{i}$ of the function $z = \operatorname{sn} t$. If we denote by $mM$ a curve which is obtained by running $m$ times around $M$, then we have:

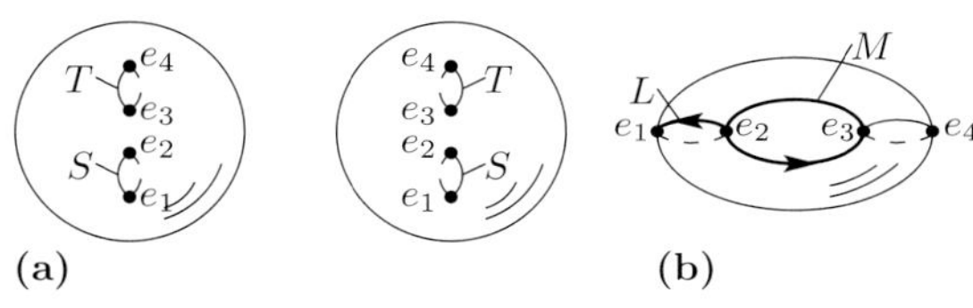


*Figure 1.203. Glueing two spheres to obtain a torus.*

$$\boxed{I(mM) = mI(M) = 4mK, \quad I(mL) = mI(L) = 2mK'\mathrm{i}, \quad m = \pm 1, \pm 2, \ldots}$$

Hence the elliptic integral $I = \int_C \frac{\mathrm{d}z}{w}$ has two additive periods, which are the two periods of the inverse function $z = \operatorname{sn} t$. These considerations are true more generally:

> The Riemann surface of the integrand of an elliptic integral always has the topological structure of a torus, which is the reason that the inverse function of this elliptic integral is an elliptic function (i.e., is a doubly periodic function).

This elegant topological argument allowed Riemann to understand the behavior of arbitrary Abelian integrals *without any computation.* The corresponding Riemann surfaces are homeomorphic to a sphere with $g$ handles, where $g$ is the called the *genus* of the Riemann surface. For example a sphere with one handle as in Figure 1.204 can be turned into a torus by a cut and paste procedure as above. Note that the torus corresponds to this construction in

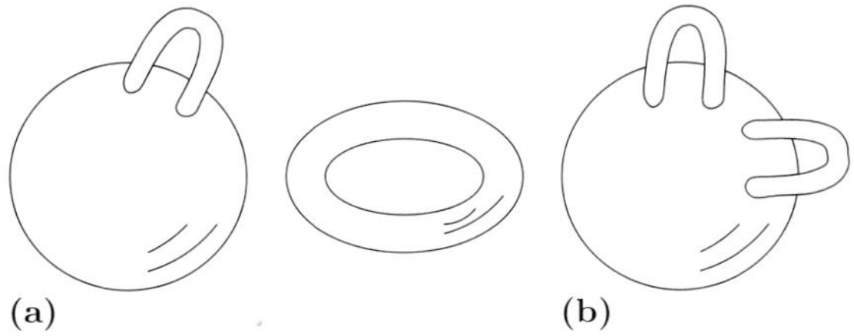


*Figure 1.204. Riemann surfaces.*

the case $g = 1$. In a similar manner, the following general result can be proved:

If the Riemann surface of the integrand has genus $g$, then the Abelian integral $\int_C R(z,w)\mathrm{d}z$ has $2g$ additive periods.

With these topological ideas, Riemann opened completely new perspectives in mathematics. In the twentieth century, topology has triumphantly marched into the center of mathematics and even plays also a central role in theoretical physics.

### 1.14.19.2 The general method of substitutions of Weierstrass

The considerations for the Legendre normal form for integrals of the first kind can be generalized to arbitrary elliptic integrals

$$\int_C R(\zeta, \sqrt{(\zeta-a_1)(\zeta-a_2)(\zeta-a_3)(\zeta-a_4)}\,)\,\mathrm{d}\zeta.$$

Here $a_1,\ a_2,\ a_3,\ a_4$ are four distinct complex numbers, and $u = R(z,w)$ denotes a rational function of the complex arguments $z$ and $w$. The substitution

$$z = \frac{1}{\zeta - a_4}$$

maps the point $a_4$ to the point $z = \infty$, and we get the elliptic integral

$$I(C) = \int_C R(z, \sqrt{4(z-e_1)(z-e_2)(z-e_3)}\,)\mathrm{d}z$$

with a new rational function $R$. here $e_1,\ e_2$ and $e_3$ are three distinct complex numbers. This integral can be written in the form

$$I(C) = \int_C R(z,w)\mathrm{d}z$$

together with the algebraic equation

$$w^2 = 4(z-e_1)(z-e_2)(z-e_3).$$

The fact that a Weierstrass $\wp$-function with the periods $\omega_1$ and $\omega_2$ belongs to this equation is important. The $\wp$-function satisfies the differential equation

$$\wp'(t)^2 = 4(\wp(t)-e_1)(\wp(t)-e_2)(\wp(t)-e_3)$$

on $\mathbb{C}$. We consider the path

$$C_* : t = t(\tau), \qquad \tau_0 \le \tau \le \tau_1$$

in the complex $t$-plane and use the substitution

$$z = \wp(t), \qquad w = \wp'(t).$$

This maps the curve $C_*$ to the curve $C$ and we get the decisive formula

$$I(C) = \int_{C_*} R(\wp(t), \wp'(t))\wp'(t)\,\mathrm{d}t.$$

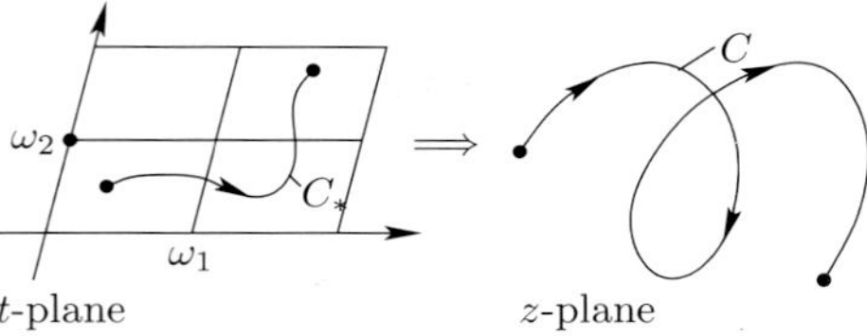


*Figure 1.205. Transformation of an elliptic integral.*

This is an integral over a well-defined elliptic function with the periods $\omega_1$ and $\omega_2$ in the $t$-plane (Figure 1.205).

**The Riemann surface:** As in 1.14.19.1 the curve $C$ is actually a curve on the Riemann surface $\mathscr{R}$ of the integrand, and conversely, to every curve $C$ on $\mathscr{R}$ there corresponds a curve $C_*$ in the $t$-plane. The Riemann surface $\mathscr{R}$ is obtained as shown in Figure 1.203 with $e_4 = \infty$.

This construction for the Riemann sphere corresponds to the choice of two $z$-planes, which are cut along a segment from $e_1$ to $e_2$ and half-line from $e_3$ to $\infty$. The two edges are then glued together cross-wise (Figure 1.206). One can find a treasure of concrete methods for calculating elliptic integrals in the classical monograph [226].

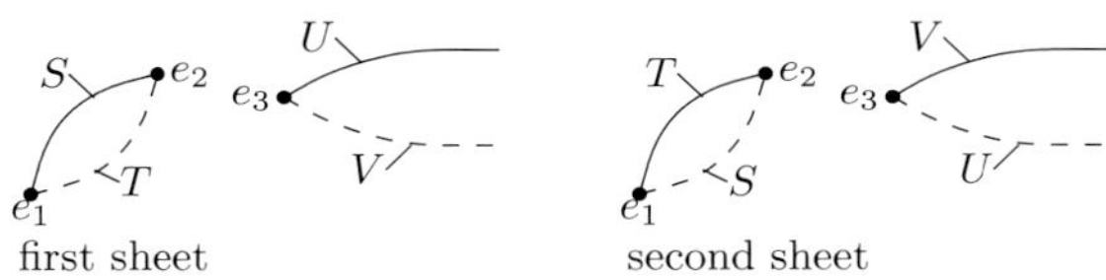


*Figure 1.206. Constructing the Riemann surface $\mathscr{R}$.*

### 1.14.19.3 Applications

**The length of an elliptic arc:** Suppose we are given an ellipse

$$x = a\cos\varphi, \quad y = b\sin\varphi, \qquad 0 \le \varphi < 2\pi$$

with the major and minor axi $a$, $b$ and the numerical eccentricity $\varepsilon = \sqrt{1-(b/a)^2}$. Then the length $L$ of the elliptic arc between the point $(a, 0)$ and the point $(x, y)$ is given by the elliptic integral

$$L = a\int_0^{\varphi} \sqrt{1-\varepsilon^2\sin^2\psi}\,\mathrm{d}\psi$$

(Figure 1.207(a)). The values of this integral are tabulated in section 0.5.4.

The geometric relationship has led to the notion of 'elliptic integral' for a more general class of integrals.

**The spherical pendulum:** The motion $\varphi = \varphi(t)$ of a spherical pendulum of length $l$ swinging under the influence of gravity leads to the differential equation

$$\varphi'' + \omega^2\sin\varphi = 0, \qquad \varphi(0) = \varphi_0,\ \varphi'(0) = 0$$

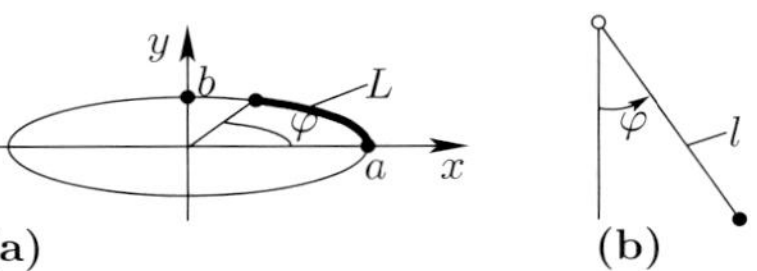


*Figure 1.207. The spherical pendulum.*

with $\omega^2 = g/l$ and the acceleration of gravity g (Figure 1.207(b)). The solution is obtained by solving the equation

$$\boxed{\sin\frac{\varphi}{2} = k\,\mathrm{sn}(\omega t, k)}$$

for $\varphi$, where $k := \sin(\varphi_0/2)$. The period of the pendulum is given by $T = 4\omega K(k)$. More details can be found in section 5.1.2.

### 1.14.20 Singular differential equations

Consider the differential equation

$$\boxed{w'' + p(z)w' + q(z)w = 0.} \qquad (1.494)$$

We assume that there is a neighborhood $U$ of the point $a$ such that $p$ and $q$ have an isolated singularity at the point $z = a$. The general solution in a neighborhood of $z = a$ can be written in the form

$$\boxed{w(z) = C_1 w_1(z) + C_2 w_2(z), \qquad 0 < |z - a| < r,}$$

with complex constants $C_1$ and $C_2$ and a positive real number $r$.

*Case 1:* The functions $w_1$ and $w_2$ have the form

$$\boxed{w_j(z) = (z-a)^{\varrho_j}\mathscr{L}_j(z-a), \qquad j = 1, 2,} \qquad (1.495)$$

with real coefficients $\varrho_1$ and $\varrho_2$. Here $\mathscr{L}_j$ denotes the Laurent series of $w_j$ at the point $z = a$.

*Case 2:* The function $w_1$ is given by (1.495), while we have for $w_2$:

$$\boxed{w_2(z) = (z-a)^{\varrho_1}\mathscr{L}_2(z-a) + w_1(z)\ln(z-a).}$$

**Theorem:** The two following statements are equivalent:

(i) The singularity $z = a$ is a *regular singular point*, i.e., $\mathscr{L}_1$ and $\mathscr{L}_2$ are power series.

(ii) $q$ has a pole of order at most two in $a$, and $p$ has a pole of order at most one in $a$.

**Construction of solutions in the case of regular singular points:** The ansatz

$$w(z) = (z-a)^{\varrho}(a_0 + a_1(z-a) + a_2(z-a)^2 + \ldots) \qquad (1.496)$$

leads in (1.494) to the quadratic equation

$$\boxed{\varrho^2 + \varrho(A-1) + B = 0} \qquad (1.497)$$

with the two solutions $\varrho_1$ and $\varrho_2$. The quantities $A$ and $B$ satisfy

$$A = \lim_{z\to a} p(z)(z-a)^2, \qquad B = \lim_{z\to a} q(z)(z-a).$$

The equation (1.497) is called the *index equation* of the differential equation (1.494).

*Case 1:* The difference $\varrho_1 - \varrho_2$ is a small number. Then one gets the solutions $w_1$ and $w_2$ by using the Ansatz (1.496) with $\varrho = \varrho_1,\ \varrho_2$ and comparing coefficients.

*Case 2:* The difference $\varrho_1 - \varrho_2$ is an integer. Then we get $w_1$ as in the first case, while the second solution $w_2$ is obtained through integration, applying the formula

$$\frac{\mathrm{d}}{\mathrm{d}z}\left(\frac{w_2}{w_1}\right)(z) = \frac{1}{w_1^2(z)}\ \exp\left(-\int_a^z p(t)\mathrm{d}t\right).$$

**Behavior at $\infty$:** By means of the substitution $z = \dfrac{1}{\zeta}$ the original differential equation (1.494) leads in a neighborhood of the point $z = \infty$ to a differential equation in a neighborhood of the point $\zeta = 0$, with can be studied with the previous methods.

In what follows we shall consider some important examples. A more detailed investigation of large classes of special functions on the basis of singular differential equations can be found in [229].

### 1.14.21 Applications to the Gaussian hypergeometric differential equation

Let $\alpha,\ \beta$ and $\gamma$ be complex numbers. The *Gaussian hypergeometric differential equation*

$$\boxed{z(z-1)w'' + ((\alpha+\beta+1)z - \gamma)w' + \alpha\beta w = 0} \qquad (1.498)$$

has precisely three regular singular points, located at the points $z = 0,\ 1,\ \infty$.

We consider the point $z = 0$ and assume that $\gamma$ is not an integer. The index equation is then

$$\varrho(\varrho + \gamma - 1) = 0$$

and has the two solutions $\varrho_1 = 0$ and $\varrho_2 = 1 - \gamma$. Equation (1.496), together with a comparison of coefficients yield for $\varrho_1 = 0$ the *Gaussian hypergeometric function*

$$\boxed{F(z;\alpha,\beta,\gamma) = 1 + \sum_{k=1}^{\infty} \frac{(\alpha)_k(\beta)_k}{(\gamma)_k k!} z^k.}$$

This series converges for all $z \in \mathbb{C}$ with $|z| < 1$. Here we have set $(\alpha)_k := \alpha(\alpha+1)\cdots(\alpha+k-1)$. For $\varrho_2 = 1 - \gamma$ we get in addition to $w_1(z) = F(z;\alpha,\beta,\gamma)$ the second linearly independent solution

$$\boxed{w_2(z) = z^{1-\gamma}F(z;\alpha-\gamma+1,\ \beta-\gamma+1,\ 2-\gamma)}$$

of (1.498) in a neighborhood of the point $z = 0$.

Important special cases of the function $w = F(z;\alpha,\beta,\gamma)$ are given in section 0.7.2.

### 1.14.22 Application to the Bessel differential equation

Let $\alpha$ be a complex number with $\operatorname{Re}\alpha \geq 0$. The *Bessel differential equation*

$$\boxed{z^2w'' + zw' + (z^2 - \alpha^2)w = 0} \qquad (1.499)$$

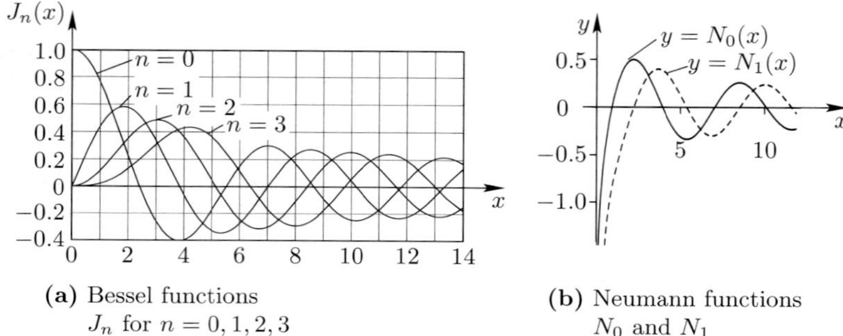


**(a)** Bessel functions $J_n$ for $n = 0, 1, 2, 3$

**(b)** Neumann functions $N_0$ and $N_1$

*Figure 1.208. Solutions of the Bessel differential equation.*

has a regular singular point at $z = 0$ and a non-regular singular point at $z = \infty$. We consider the point $z = 0$. The index equation

$$\varrho^2 - \alpha^2 = 0$$

has the two solutions $\varrho_1 = \alpha$ and $\varrho_2 = -\alpha$. The general solution of (1.499) is

$$w = C_1 w_1 + C_2 w_2.$$

*Case 1:* If $\alpha$ is not an integer, one gets the two linearly independent solutions $w_1 = J_\alpha$ and $w_2 = J_{-\alpha}$, where the *Bessel function* $J_\alpha$ is defined by the formula

$$J_\alpha(z) := \left(\frac{z}{2}\right)^\alpha \sum_{k=0}^{\infty} \frac{(-1)^k}{k!\,\Gamma(k+\alpha+1)} \left(\frac{z}{2}\right)^{2k}.$$

The power series converges for all $z \in \mathbb{C}$ (see Figure 1.208(a)).

*Case 2:* If $\alpha = n$ with $n = 0, 1, 2, \ldots$, then one has $J_{-n}(z) = (-1)^n J_n(z)$. Therefore the functions $J_{-n}$ and $J_n$ are linealy dependent. In this case one chooses $w_1 = J_n$ and $w_2 = N_n$, where $N_n$ is the *Neumann function*, defined by

$$N_n(z) := \frac{1}{\pi}\left(\frac{\partial J_\alpha(z)}{\partial \alpha} - (-1)^n \frac{\partial J_{-\alpha}(z)}{\partial \alpha}\right)\bigg|_{\alpha=n}.$$

More explicitly, this implies the formula:

$$N_n(z) = \frac{2}{\pi}(\mathrm{C} + \ln z - \ln 2)J_n(z) - \frac{1}{\pi}\sum_{k=0}^{n-1} \frac{(n-k-1)!}{k!}\left(\frac{2}{z}\right)^{n-2k}$$
$$-\frac{1}{\pi}\sum_{k=0}^{\infty} \frac{(-1)^k}{k!(n+k)!}\left(\frac{z}{2}\right)^{n+2k}(H_{n+k} + H_k)$$

with the Euler constant $\mathrm{C} = 0.5772\ldots$ and $H_k := \sum_{r=1}^{k} \frac{1}{r}$ as well as $H_0 := 0$ (Figure 1.208(b)). The power series converges for all $z \in \mathbb{C}$.

More formulas for the Hankel function, Bessel function with imaginary argument and the MacDonald function can be found in section 0.7.2.

**Application to an eigenvalue problem:** Let $D := \{(x, y) \in \mathbb{R}^2 \,|\, x^2 + y^2 < 1\}$ be the open unit disc. The eigenvalue problem

$$\boxed{\begin{aligned} -v_{xx} - v_{yy} = \lambda v \quad & \text{on } D, \\ v = 0 \quad & \text{on } \partial D \end{aligned}} \tag{1.500}$$

has the eigensolutions

$$v_{km}(x, y) := \frac{1}{\sqrt{\pi}\,|J_k'(\lambda_{km})|} J_k(\lambda_{km} r)\mathrm{e}^{\mathrm{i}k\varphi}, \qquad \lambda = \lambda_{km}$$

with $k = 0, 1, \ldots$ and $m = 1, 2, \ldots$ as well as $x = r\cos\varphi$, $y = r\sin\varphi$. We denote the zeros of the Bessel function $J_k$ by $0 < \lambda_{k_1} < \lambda_{k_2} < \cdots$. The functions $\{v_{km}\}$ form a complete orthonormal system in the Hilbert space $L_2(D)$ with the scalar product

$$(u, v) := \int_D u(x, y)v(x, y)\mathrm{d}x\mathrm{d}y$$

(cf. 1.13.2.13).

**Application to the vibration of a spanned membrane:** Let $u(x, y, t)$ be the displacement of a membrane at the point $(x, y)$ at time $t$. The vibration problem is

$$\boxed{\begin{aligned} & \frac{1}{c^2}u_{tt} - u_{xx} - u_{yy} = 0, \qquad (x, y) \in D,\ t > 0, \\ & u = 0 \quad \text{on } \partial D, \qquad \text{(boundary value)}, \\ & u(x, y, 0) = u_0(x, y), u_t(x, y, 0) = u_1(x, y), \qquad \text{(initial value)}. \end{aligned}} \tag{1.501}$$

We set $c = 1$.

**Theorem:** If the prescribed continuous functions $u_0, u_1 : \overline{D} \longrightarrow \mathbb{R}$ are smooth, then the boundary–initial value problem (1.501) has the unique solution

$$\boxed{u(x, y) = \sum_{k,m=0}^{\infty} (a_{km}\cos(\lambda_{km}t) + b_{km}\sin(\lambda_{km}t))v_{km}(x, y)}$$

with

$$a_{km} = (u_0, v_{km}), \qquad \lambda_{km}b_{km} = (u_1, v_{km}).$$

This solution is obtained by utilizing the Fourier method, taking (1.500) into account.

## 1.14.23 Functions of several complex variables

The theory of functions of several complex variables is in elementary aspects similar to the corresponding theory of a function of one variable. More complicated questions, however, for example those depending on the existence of domains of holomorphy, are completely different. The modern theory is dominated by the abstract theory of sheaves. This circle of questions will be considered in [212].

**The space $\mathbb{C}^n$ as a metric space:** Let $\mathbb{C}^n$ denote the set of all $n$-tuples $(z_1, \ldots, z_n)$ of complex numbers $z_1, \ldots, z_n$. We set

$$|z| := \sqrt{|z_1|^2 + \ldots + |z_n|^2}.$$

By defining the *distance* of complex numbers by

$$d(z, w) := |z - w| \quad \text{for all } z, w \in \mathbb{C}^n,$$

the set $\mathbb{C}^n$ acquires the structure of a metric space. Hence all the general notions for metric spaces (cf. 1.3.2) can be applied to $\mathbb{C}^n$.

The notion of holomorphy for functions of several variables is reduced to the notion of holomorphy of functions of a single complex variables by means of the following definition.

**Holomorphic functions:** Let $U$ be an open subset of $\mathbb{C}^n$. A function

$$\boxed{w = f(z_1, \ldots, z_n)}$$

is said to he *holomorphic* on $U$, if it is holomorphic with respect to all of its variables $z_i$.[191]

*Example 1:* The function $w = e^{z_1+z_2}$ is holomorphic on $\mathbb{C}^2$. Indeed, if we keep $z_1$ (resp. $z_2$) fixed, then we get a holomorphic function on $\mathbb{C}$ with respect to $z_2$ (resp. $z_1$).

**Power series:** Let $a, b$ and $a_{km}$ denote complex numbers. The series

$$\sum_{k,m=0}^{\infty} a_{km}(z_1 - a)^k (z_2 - b)^m \tag{1.502}$$

converges by definition at the point $z = (z_1, z_2)$, if the series of absolute values

$$\sum_{k,m=0}^{\infty} |a_{km}(z_1 - a)^k (z_2 - b)^m|$$

converges. In (1.502) the order of the terms is irrelevant.

Similarly one defines power series of $n$ variables $z_1, \ldots, z_n$.

**Analyticity:** According to definition the function $f : U \subseteq \mathbb{C}^n \longrightarrow \mathbb{C}$ is *analytic* on the open set $U$ if and only if for every point in $U$ there is a neighborhood in which the function $f$ can be represented by a convergent power series.

**Theorem:** A function $f : U \subseteq \mathbb{C}^n \longrightarrow \mathbb{C}$ is holomorphic on an open set $U$ if and only if it is analytic there.

**The Weierstrass preparation theorem:** This fundamental result allows the application of the algebraic methods of ideal theory to the theory of several complex variables. This theory of ideals is found in section 3.8.7.

**Domains of holomorphy:** A domain $D$ in $\mathbb{C}^n$ is called a *domain of holomorphy*, if there is a (!) holomorphic function $f : D \longrightarrow \mathbb{C}$ which cannot be extended to a holomorphic function on any larger domain.

---

[191] Let $p$ be an arbitrary point of $U$. We consider the function

$$g(z) := f(z, p_2, \ldots, p_n)$$

and require that $g$ is holomorphic in a neighborhood of $z = p_1$. In a similar way one defines the holomorphicity with respect to the other variables $z_j$.

*Example 2:* $\mathbb{C}^n$ is trivially a domain of holomorphy, as there are holomorphic functions on $\mathbb{C}^n$ (for example $f(z) := z_1 + \cdots + z_n$). The question of extension of the function is vacant in this case.

*Example 3:* The function

$$f(z) := \sum_{n=1}^{\infty} z^{n!}$$

has the interior of the unit disc in the complex plane as a domain of holomorphy.

**Theorem:** (i) For $n = 1$ every domain in the complex plane $\mathbb{C}$ is a domain of holomorphy.

(ii) For $n \geq 2$ not every domain in $\mathbb{C}^n$ is a domain of holomorphy.

(iii) Convex domains in $\mathbb{C}^n$ are domains of holomorphy.

*Example 4:* We consider the annulus

$$A := \left\{ z \in \mathbb{C}^2 \,\middle|\, \frac{1}{2} < |z| < 1 \right\}.$$

Then every holomorphic function $f : A \longrightarrow \mathbb{C}$ can be extended to a holomorphic function on $\{z \in \mathbb{C}^2 : |z| < 1\}$. This means that $A$ is not a domain of holomorphy.

**Pseudoconvex domains:** Let $D$ be a domain in $\mathbb{C}^n$ with at least one boundary point. We set

$$d(z, \partial D) := \text{the distance of the point } z \text{ to the boundary } \partial D$$

According to definition the domain $D$ is *pseudoconvex*, if the function

$$g(z) := -\ln d(z, \partial D)$$

is *subharmonic* on $D$, i.e., to every point $z \in D$ there is a number $r_0 > 0$ such that the mean value inequality

$$g(z) \leq \frac{1}{2\pi} \int_0^{2\pi} g(z + r\mathrm{e}^{\mathrm{i}\varphi})\, \mathrm{d}\varphi$$

is satisfied for all radii $r$ with $0 < r \leq r_0$.[192]

We also agree that $\mathbb{C}^n$ is pseudoconvex by definition.

**Main theorem of Oka (1942):** Let $n \geq 2$. A domain in $\mathbb{C}^n$ is a domain of holomorphy if and only if it is pseudoconvex.

*Example 6:* The annulus in Example 4 is not a domain of holomorphy and hence is also not pseudoconvex.

[192] We write $z + r\mathrm{e}^{\mathrm{i}\varphi}$ for $(z_1 + r\mathrm{e}^{\mathrm{i}\varphi}, \ldots, z_n + r\mathrm{e}^{\mathrm{i}\varphi})$.

# 2. Algebra

*Algebra is the study of the four basic arithmetic operations – addition, subtraction, multiplication and division – and the solution of equations which arise in this quest. Such a theory is possible because the objects on which these operations act can all be left indefinite to a great extent.*

*In the early days of algebra, the symbols which were used instead of actual numbers were just viewed as not determined numbers. That is, the quantity which each symbol represented was left indefinite, while the quality of the object it represented in algebraic calculations was fixed.*

*It is typical of modern algebra, which has been developed in the last century, and in particular typical of what is now known as 'abstract algebra', that even the quality of the symbols used can be left indeterminate, leading to a genuine theory of the operations.*

*Erich Kähler* (1953)

A important prerequisite for the development of algebraic thinking was the transition from the calculation with numbers to the use of letters representing indefinite quantities. This revolution in mathematics was carried out by the Frenchman François Viète (Vieta) in the second half of the sixteenth century.

The modern structural theory of algebra has its origin in lectures of Emmy Noether (1882–1935) in Göttingen and Emil Artin (1898–1962) in Hamburg in the twenties of the twentieth century and was presented in a monograph for the first time in the book *Modern Algebra* by Bartel Leendert van der Waerden which appeared in 1930. Since then this book has been published in many editions and is still today a readable standard reference for modern algebra.

However, the basis for this work was laid in the nineteenth century. Important impulses were given by Gauss (cyclotomic fields), Abel (algebraic functions), Galois (group theory and algebraic equations), Riemann (genus and divisors of algebraic functions), Kummer and Dedekind (ideal theory), Kronecker (number fields), Jordan (group theory) and Hilbert (number fields and invariant theory).

## 2.1 Elementary algebra

### 2.1.1 Combinatorics

Combinatorics is the study of the number of ways one can combine a certain number of elements. Here the symbol

$$n! := 1 \cdot 2 \cdot 3 \cdot \ldots \cdot n, \qquad 0! := 1, \qquad n = 1, 2, \ldots$$

is used, which is read '$n$ factorial', as are the binomial coefficients[1]

$$\binom{n}{k} := \frac{n(n-1)\dots(n-k+1)}{1\cdot 2\cdot\ldots\cdot k}, \qquad \binom{n}{0} := 1.$$

*Example 1:* $3! = 1\cdot 2\cdot 3 = 6$, $4! = 1\cdot 2\cdot 3\cdot 4 = 24$, $\binom{4}{2} = \frac{4\cdot 3}{1\cdot 2} = 6$, $\binom{4}{3} = \frac{4\cdot 3\cdot 2}{1\cdot 2\cdot 3} = 4$.

**Binomial coefficients and the binomial formula:** This is described in section 0.1.10.3.

**Factorial and the gamma function:** This is described in section 1.14.6.

**Basic problems in combinatorics:** These are:[2]

(i) permutations,

(ii) permutations with repetitions (also known as the *book problem*),

(iii) combinations without repetitions

(a) without taking order into consideration (the *lottery problem*),

(b) taking the order into consideration (the *modified lottery problem*),

(iv) combinations with repetitions

(a) without taking order into account (the *modified word problem*),

(b) taking order into account (the *word problem*).

**Permutations:** There are exactly

$$\boxed{n!} \tag{2.1}$$

different possibilities to combine $n$ different elements. Such a combination of the elements is referred to as a *permutation.*

*Example 2:* For the numbers 1 and 2 there are $2! = 2\cdot 1$ possible combinations, namely

$$12, \qquad 21.$$

For the three numbers $1, 2, 3$ there are $3! = 1\cdot 2\cdot 3$ possible combinations, namely

$$\begin{matrix} 123, & 213, & 312, \\ 132, & 231, & 321. \end{matrix} \tag{2.2}$$

**The book problem:** Let $n$ books be given, not necessarily different, of which there are $m_1, \dots, m_s$ copies (i.e., $s$ of the books are distinct). Then there are

$$\boxed{\frac{n!}{m_1! m_2! \dots m_s!}}$$

different possibilities to order these books in a row, where the different copies of each book are not distinguished from one another.

*Example 3:* For three books, two of which are the same, there are

$$\frac{3!}{2!\cdot 1!} = \frac{1\cdot 2\cdot 3}{1\cdot 2} = 3$$

[1] This definition is valid for real or complex numbers $n$ and for $k = 0, 1, \dots$

[2] Combinatorics which take the order of the elements into account are also called *variations*.

possible ways to order them in a row. One obtains these different orderings by replacing the number 2 by 1 in (2.2) and eliminating combinations which are identical to already existing ones. This gives the three possibilities

$$113, \quad 311, \quad 131.$$

**The word problem:** Starting from $k$ letters, one can form exactly

$$\boxed{k^n}$$

different words of the length $n$.

If one defines two words to be *equivalent* if they differ only by a permutation of the letters, then the number of classes of equivalent words is equal to

$$\boxed{\binom{n+k-1}{n}}$$

(modified word problem).

*Example 4:* From the two symbols 0 and 1 one can form $2^2 = 4$ words of length four, namely

$$00, \quad 01, \quad 10, \quad 11.$$

The number $A$ of the classes of equivalent words is $\binom{n+k-1}{n}$ with $n = k = 2$, that is $A = \binom{3}{2} = \frac{3 \cdot 2}{1 \cdot 2} = 3$. Representatives of these classes are

$$00, \quad 01, \quad 11.$$

Moreover there are $2^3 = 8$ words of length 3:

$$\begin{array}{cccc} 000, & 001, & 010, & 011, \\ 100, & 101, & 110, & 111. \end{array}$$

The number of classes of equivalent words is $A = \binom{n+k-1}{n}$ with $n = 3$ and $k = 2$, i.e., $A = \binom{4}{3} = \frac{4 \cdot 3 \cdot 2}{1 \cdot 2 \cdot 3} = 4$. As representatives of these classes we can take

$$000, \quad 001, \quad 011, \quad 111.$$

**The lottery problem:** There are

$$\boxed{\binom{n}{k}}$$

possible ways of choosing $k$ numbers from $n$ without taking the order of the chosen numbers into account.

If one does take the order into account, then there are

$$\boxed{\binom{n}{k} k! = n(n-1)\ldots(n-k+1)}$$

possibilities.

*Example 5:* In the German game "6 from 49" (one chooses 6 numbers from a field of 49) one would have to fill out exactly

$$\binom{49}{6} = \frac{49 \cdot 48 \cdot 47 \cdot 46 \cdot 45 \cdot 44}{1 \cdot 2 \cdot 3 \cdot 4 \cdot 5 \cdot 6} = 13\,983\,816$$

lottery tickets, in order to have a sure bet.

*Example 6:* There are $\binom{3}{2} = 3$ possibilities to choose two numbers from the set $1, 2, 3$ without taking the order of the chosen two into account. These are

$$12, \quad 13, \quad 23.$$

If one does take the order into account, there are $\binom{3}{2} \cdot 2! = 6$ possibilities:

$$12, \quad 21, \quad 13, \quad 31, \quad 23, \quad 32.$$

**The sign of a permutation:** Let the $n$ numbers $1, 2, \ldots, n$ be given. The natural order $12 \cdots n$ is by definition an even permutation. A permutation of the $n$ numbers is called *even* (resp. *odd*), if it results from the natural order by an even (resp. odd) number of transpositions of two elements.[3] By definition, the *sign* of a permutation $\sigma$, written $\operatorname{sgn} \sigma$, is $+1$ if the permutation is even and $-1$ if it is odd.

*Example 7:* The permutation 12 of the numbers 1,2 is even and 21 is odd.

For permutations of the three elements 1,2,3 one has:

(i) the even permutations are the permutations 123, 312, 231;

(ii) the odd permutations are the permutations 213, 132, 321.

**The drawer principle of Dirichlet:** Sorting more than $n$ objects into $n$ drawers results in at least one of the drawers containing more than one object.

This simple principle due to Dirichlet (1805–1859) has been applied successfully in number theory.

### 2.1.2 Determinants

**Basic idea:** A two-rowed determinant is calculated by means of the formula

$$\begin{vmatrix} a & b \\ c & d \end{vmatrix} := ad - bc. \tag{2.3}$$

The calculation of three-rowed determinants is done by using the conceptually clear development rule, developing the determinant by the first row

$$\begin{vmatrix} a & b & c \\ d & e & f \\ g & h & k \end{vmatrix} := a \begin{vmatrix} e & f \\ h & k \end{vmatrix} - b \begin{vmatrix} d & f \\ g & k \end{vmatrix} + c \begin{vmatrix} d & e \\ g & h \end{vmatrix}, \tag{2.4}$$

which reduces the calculation to that of two-rowed determinants. This rule is gotten by striking certain rows or columns from the three-by-three determinant: for example

[3] A *transposition* is the permutation which just switches two of the elements and leaves the others unchanged. The definition of even and odd just given in independent of the choice of transposition used to arrive at the given permutation from the natural order.

the determinant next to the $a$ in the formula is gotten by striking the row and column of $a$. In a similar manner, the calculation of higher determinants are reduced to the calculation of smaller ones by a general rule for developing determinants. The general rule is given by the *Laplacian rule for developing determinants*, see (2.6).

*Example 1:* Let

$$\begin{vmatrix} 2 & 3 \\ 1 & 4 \end{vmatrix} := 2 \cdot 4 - 3 \cdot 1 = 8 - 3 = 5$$

and

$$\begin{vmatrix} 1 & 2 & 3 \\ 2 & 2 & 3 \\ 4 & 1 & 4 \end{vmatrix} = 1 \cdot \begin{vmatrix} 2 & 3 \\ 1 & 4 \end{vmatrix} - 2 \cdot \begin{vmatrix} 2 & 3 \\ 4 & 4 \end{vmatrix} + 3 \cdot \begin{vmatrix} 2 & 2 \\ 4 & 1 \end{vmatrix}$$

$$= 1 \cdot 5 - 2 \cdot (-4) + 3 \cdot (-6) = 5 + 8 - 18 = -5.$$

**Definition:** The *determinant*

$$\begin{vmatrix} a_{11} & a_{12} & a_{13} & \ldots & a_{1n} \\ a_{21} & a_{22} & a_{23} & \ldots & a_{2n} \\ \ldots & & & & \\ a_{n1} & a_{n2} & a_{n3} & \ldots & a_{nn} \end{vmatrix} \tag{2.5}$$

is the number

$$\boxed{D := \sum_{\pi} \operatorname{sgn} \pi \; a_{1m_1} a_{2m_2} \ldots a_{nm_n}.}$$

Here the summation is carried out over all permutations $m_1 m_2 \cdots m_n$ of the numbers $1, 2, \ldots, n$, where $\operatorname{sgn} \pi$ denotes the sign of the corresponding permutation.

All $a_{jk}$ are real or complex numbers.

*Example 2:* For $n = 2$ we have the even permutation 12 and the odd permutation 21. Therefore we have

$$D = a_{11} a_{22} - a_{12} a_{21}.$$

This coincides with the formula (2.3).

**Properties of determinants:**

(i) A determinant is unchanged when one permutes a row with a column.

(ii) A determinant changes its sign if two rows or two columns are permuted with each other (i.e., if they are *transposed*).

(iii) A determinant vanishes if there are two equal rows or two equal columns.

(iv) A determinant is left unchanged upon adding a multiple of one row to another row.

(v) A determinant is unchanged when one adds a multiple of a column to another column.

(vi) A determinant is multiplied by a number by multiplying a fixed row or column of the determinant by this number.

*Example 3:*

(a) $\begin{vmatrix} a & b \\ c & d \end{vmatrix} = \begin{vmatrix} a & c \\ b & d \end{vmatrix}$ (by(i));

(b) $\begin{vmatrix} a & b \\ c & d \end{vmatrix} = -\begin{vmatrix} c & d \\ a & b \end{vmatrix}$ (by(ii)); $\begin{vmatrix} a & b \\ a & b \end{vmatrix} = 0$ (by(iii));

(c) $\begin{vmatrix} a & b \\ c & d \end{vmatrix} = \begin{vmatrix} a & b \\ c+\lambda a & d+\lambda b \end{vmatrix}$ (by (iv));

(d) $\begin{vmatrix} \alpha a & \alpha b \\ c & d \end{vmatrix} = \alpha\begin{vmatrix} a & b \\ c & d \end{vmatrix}$ (by(vi)).

**Triangular form:** If in (2.5) all elements underneath (resp. above) the main diagonal (from upper left to lower right) vanish, then one has

$$D = a_{11}a_{22}\cdots a_{nn}.$$

*Example 4:*

$$\begin{vmatrix} a & \alpha & \beta \\ 0 & b & \gamma \\ 0 & 0 & c \end{vmatrix} = abc, \qquad \begin{vmatrix} a & 0 & 0 \\ \alpha & b & 0 \\ \beta & \gamma & c \end{vmatrix} = abc.$$

An important strategy for calculating large determinants is to apply the operations (ii) and (iii) in order to bring the determinant into a triangular form and then apply the above formula. This can always be done.

*Example 5:* For $\lambda = -2$ one has

$$\begin{vmatrix} 2 & 3 \\ 4 & 1 \end{vmatrix} = \begin{vmatrix} 2 & 3 \\ 4+2\lambda & 1+3\lambda \end{vmatrix} = \begin{vmatrix} 2 & 3 \\ 0 & -5 \end{vmatrix} = -10.$$

**Laplacian rule for the development of determinants:** For the determinant $D$ in (2.5) one has:

$$\boxed{D = a_{k1}A_{k1} + a_{k2}A_{k2}\ldots a_{kn}A_{kn}.} \tag{2.6}$$

The notations are as follows. $k$ is some fixed row number[4]. $A_{jk}$ denotes the so-called *adjoint* to the element $a_{jk}$, which means it is the determinant which is obtained by striking the $k^{th}$ row and the $j^{th}$ column from (2.5) and multiplying by $(-1)^{j+k}$.

*Example 6:* The formula (2.4) is a special case of this formula.

**Multiplication of two determinants:** If $A = (a_{jk})$ and $B = (b_{jk})$ are two square matrices (dealt with in section 2.1.3 below) with $n$ rows and columns, then one has

$$\boxed{\det A \det B = \det(AB).}$$

Here $\det A$ denotes the determinant of $A$ (i.e., $\det A = D$ in (2.5)), and $AB$ denotes the product of the two matrices (see section 2.1.3 below). Moreover, one has

$$\boxed{\det A = \det A^{\mathsf{T}},}$$

[4]A similar statement also holds for columns instead of rows.

where $A^\mathsf{T}$ denotes the *transpose* of $A$ (again, this is defined in section 2.1.3).

**Differentiation of a determinant:** If the elements of a determinant depend on a variable $t$, then one obtains the derivative $D'(t)$ of the determinant $D(t)$ by differentiating each row with respect to $t$, leading to $n$ new determinants, and then adding all of these $n$ determinants together.

*Example 7:* For the derivative of

$$D(t) := \begin{vmatrix} a(t) & b(t) \\ c(t) & d(t) \end{vmatrix}$$

one gets

$$D'(t) := \begin{vmatrix} a'(t) & b'(t) \\ c(t) & d(t) \end{vmatrix} + \begin{vmatrix} a(t) & b(t) \\ c'(t) & d'(t) \end{vmatrix}.$$

**Rule of multiplication of functional determinants:** One has

$$\boxed{\frac{\partial(f_1, \ldots, f_n)}{\partial(u_1, \ldots, u_n)} = \frac{\partial(f_1, \ldots, f_n)}{\partial(v_1, \ldots, v_n)} \cdot \frac{\partial(v_1, \ldots, v_n)}{\partial(u_1, \ldots, u_n)}.}$$

Here $\dfrac{\partial(f_1, \ldots, f_n)}{\partial(u_1, \ldots, u_n)}$ denotes the determinant of the first partial derivatives $\partial f_j/\partial u_k$ (cf. 1.5.3).

**The Vandermonde determinant:**

$$\begin{vmatrix} 1 & a & a^2 \\ 1 & b & b^2 \\ 1 & c & c^2 \end{vmatrix} = (b-a)(c-a)(c-b).$$

More generally, the determinant

$$\begin{vmatrix} 1 & a_1 & a_1^2 & a_1^3 & \ldots & a_1^{n-1} \\ 1 & a_2 & a_2^2 & a_2^3 & \ldots & a_2^{n-1} \\ \ldots & & & & & \\ 1 & a_n & a_n^2 & a_n^3 & \ldots & a_n^{n-1} \end{vmatrix}$$

is equal to the difference product

$$\begin{aligned} (a_2 - a_1)(a_3 - a_1)(a_4 - a_1)\ldots(a_n - a_1)\times \\ (a_3 - a_2)(a_4 - a_2)\ldots(a_n - a_2)\times \\ \ldots\ldots\ldots\ldots\ldots\ldots \\ (a_n - a_{n-1}). \end{aligned}$$

## 2.1.3 Matrices

**Definition:** A *matrix* $A$ of type $(m, n)$ is a rectangular scheme of numbers

$$A = \begin{pmatrix} a_{11} & a_{12} & a_{13} & \ldots & a_{1n} \\ a_{21} & a_{22} & a_{23} & \ldots & a_{2n} \\ \ldots & \ldots & \ldots & \ldots & \ldots \\ a_{m1} & a_{m2} & a_{m3} & \ldots & a_{mn} \end{pmatrix}$$

with $m$ rows and $n$ columns. Here the elements $a_{jk}$ can be real or complex numbers.[5] A *square* matrix is a matrix $A$ for which $m = n$.

[5] The matrix $A$ is said to be *real*, if all $a_{jk}$ are real. Another notation for a matrix of type $(m, n)$ is a $(m \times n)$-matrix.

The set of all matrices of type $(m, n)$ is denoted $\mathrm{Mat}(m, n)$.

**Goal:** We would like to define algebraic operations like addition and multiplication for matrices. These operations will no longer possess all the familiar properties which we are accustomed to from the real or complex numbers. For instance, for the matrix multiplication the rule $AB = BA$ does not in general hold, in contrast to the case of commutative multiplication of the real or complex numbers. Algebraically, we are observing here the fact that the set of matrices form a (non-commutative) *ring* instead of a *field*. These algebraic notions are defined in section 2.5 below.

**Addition of two matrices:** If $A$ and $B$ both belong to $\mathrm{Mat}(m, n)$, then we define the sum matrix $A + B$ by forming the matrix whose elements are the sums of the elements of $A$ with those of $B$. This matrix again belongs to $\mathrm{Mat}(m, n)$.

*Example 1:*

$$\begin{pmatrix} a & b & c \\ d & e & z \end{pmatrix} + \begin{pmatrix} \alpha & \beta & \gamma \\ \delta & \varepsilon & \zeta \end{pmatrix} = \begin{pmatrix} a+\alpha & b+\beta & c+\gamma \\ d+\delta & e+\varepsilon & z+\zeta \end{pmatrix},$$

$$(a, b) + (\alpha, \beta) = (a + \alpha, b + \beta), \qquad (1, 2) + (3, 1) = (4, 3).$$

**Multiplication of a matrix by a number (scalar multiplication):** Let $A \in \mathrm{Mat}(m, n)$ and let $\alpha$ be a (real or complex) number. We define the product $\alpha A$, the product of $A$ by the *scalar* $\alpha$, by multiplying every element of the matrix $A$ by the number $\alpha$. This matrix again belongs to $\mathrm{Mat}(m, n)$.

*Example 2:*

$$\alpha \begin{pmatrix} a & b & c \\ d & e & z \end{pmatrix} = \begin{pmatrix} \alpha a & \alpha b & \alpha c \\ \alpha d & \alpha e & \alpha z \end{pmatrix}, \qquad 4 \begin{pmatrix} 1 & 3 & 2 \\ 1 & 2 & 1 \end{pmatrix} = \begin{pmatrix} 4 & 12 & 8 \\ 4 & 8 & 4 \end{pmatrix}.$$

**The zero matrix:** The $(m \times n)$-matrix

$$O := \begin{pmatrix} 0 & 0 & \ldots & 0 \\ \ldots & \ldots & \ldots & \ldots \\ 0 & 0 & \ldots & 0 \end{pmatrix},$$

all of whose elements are 0, is called the *zero matrix*.

**Rules for calculations:** For $A$, $B$, $C \in \mathrm{Mat}(m, n)$ and $\alpha \in \mathbb{C}$, one has:

$$A + B = B + A, \qquad (A + B) + C = A + (B + C), \qquad A + O = A,$$
$$\alpha(A + B) = \alpha A + \alpha B.$$

More precisely, the set $\mathrm{Mat}(m, n)$ forms a linear space over the field of complex numbers (cf. 2.3.2).

**Multiplication of two matrices:** The basic idea of matrix multiplication is contained in the formula

$$(a, b) \begin{pmatrix} \alpha \\ \beta \end{pmatrix} := a\alpha + b\beta. \tag{2.7}$$

*Example 3:* One has $(1, 3) \begin{pmatrix} 2 \\ 4 \end{pmatrix} = 1 \cdot 2 + 3 \cdot 4 = 2 + 12 = 14.$

The natural generalization of formula (2.7) is

$$(a_1, a_2, \ldots, a_n)\begin{pmatrix} \alpha_1 \\ \alpha_2 \\ \vdots \\ \alpha_n \end{pmatrix} := a_1\alpha_1 + a_2\alpha_2 + \ldots + a_n\alpha_n.$$

The product of a matrix $A \in \mathrm{Mat}(m,n)$ with a matrix $B \in \mathrm{Mat}(n,l)$ is a matrix $C = AB \in \mathrm{Mat}(m,l)$, whose elements $c_{jk}$ are defined by the following rule:

$$\boxed{c_{jk} := \text{the } j^{th} \text{ row of } A \text{ times the } k^{th} \text{ column of } B.} \tag{2.8}$$

If we denote the elements of $A$ (resp. of $B$) by $a_{..}$ (resp. by $b_{..}$), then this formula is

$$\boxed{c_{jk} = \sum_{s=1}^{n} a_{js}b_{sk}.}$$

*Example 4:* Let

$$A := \begin{pmatrix} 1 & 2 \\ 3 & 4 \end{pmatrix}, \qquad B := \begin{pmatrix} 2 & 1 \\ 4 & 1 \end{pmatrix}.$$

The product matrix $C := AB$ is written as

$$C := \begin{pmatrix} c_{11} & c_{12} \\ c_{21} & c_{22} \end{pmatrix}.$$

Then one has:

$$c_{11} = \text{the first row of } A \text{ times the first column of } B = (1,2)\begin{pmatrix} 2 \\ 4 \end{pmatrix} = 1\cdot 2 + 2\cdot 4 = 10,$$

$$c_{12} = \text{the first row of } A \text{ times the second column of } B = (1,2)\begin{pmatrix} 1 \\ 1 \end{pmatrix} = 1\cdot 1 + 2\cdot 1 = 3,$$

$$c_{21} = \text{the second row of } A \text{ times the first column of } B = (3,4)\begin{pmatrix} 2 \\ 4 \end{pmatrix} = 3\cdot 2 + 4\cdot 4 = 22,$$

$$c_{22} = \text{the second row of } A \text{ times the second column of } B = (3,4)\begin{pmatrix} 1 \\ 1 \end{pmatrix} = 3\cdot 1 + 4\cdot 1 = 7.$$

Altogether we then have

$$AB = C = \begin{pmatrix} 10 & 3 \\ 22 & 7 \end{pmatrix}.$$

Moreover one gets

$$\begin{pmatrix} 1 & 2 \\ 0 & 1 \end{pmatrix}\begin{pmatrix} 1 & 0 & 2 \\ 0 & 1 & 1 \end{pmatrix} = \begin{pmatrix} 1 & 2 & 4 \\ 0 & 1 & 1 \end{pmatrix}.$$

This is because we have

$$(1,2)\begin{pmatrix} 1 \\ 0 \end{pmatrix} = 1\cdot 1 + 2\cdot 0 = 1, \qquad (0,1)\begin{pmatrix} 1 \\ 0 \end{pmatrix} = 0\cdot 1 + 1\cdot 0 = 0 \quad \text{etc.}$$

| The matrix product $AB$ exists if and only if $A$ and $B$ *fit*, i.e., the number of columns of $A$ is equal to the number of rows of $B$. |
|---|

**Identity matrix:** The quadratic $(3 \times 3)$-matrix

$$E := \begin{pmatrix} 1 & 0 & 0 \\ 0 & 1 & 0 \\ 0 & 0 & 1 \end{pmatrix}$$

is called the $(3 \times 3)$-identity matrix.[6] Similarly, the $(n \times n)$-identity matrix is defined as the matrix with 1's in the main diagonal and zeros everywhere else.

**Calculations with square matrices:** Let $A,\ B,\ C \in \mathrm{Mat}(n,n)$, and let $E$ denote the $(n \times n)$-identity matrix and $O$ the $(n \times n)$-zero matrix. Then one has:

$$\boxed{\begin{array}{lll} A(BC) = (AB)C, & A(B+C) = AB + AC, & \\ AE = EA = A, & AO = OA = O, & A + O = A. \end{array}}$$

More precisely, the set $\mathrm{Mat}(n,n)$ forms a (non-commutative) ring and in addition an algebra of the field of complex numbers (cf. 2.4.1 and 2.5.2).

**Non-commutativity of the product of matrices:** For $A,\ B \in \mathrm{Mat}(n,n)$ with $n \geq 2$, it in general does not hold that $AB = BA$ (this is the non-commutativity just referred to).

*Example 5:* One has

$$\begin{pmatrix} 0 & 1 \\ 0 & 0 \end{pmatrix}\begin{pmatrix} 1 & 0 \\ 0 & 0 \end{pmatrix} = \begin{pmatrix} 0 & 0 \\ 0 & 0 \end{pmatrix}, \quad \text{but} \quad \begin{pmatrix} 1 & 0 \\ 0 & 0 \end{pmatrix}\begin{pmatrix} 0 & 1 \\ 0 & 0 \end{pmatrix} = \begin{pmatrix} 0 & 1 \\ 0 & 0 \end{pmatrix}.$$

**Zero divisors for the matrix product:** If $AB = O$ for two matrices $A$ and $B$, then this does not necessarily imply that $A = O$ or $B = O$. Let $A,\ B \in \mathrm{Mat}(n,n)$. If $AB = O$ for $A \neq O$ and $B \neq O$, then one calls the matrices $A$ and $B$ *zero divisors* in the ring $\mathrm{Mat}(n,n)$.

*Example 6:* For

$$A := \begin{pmatrix} 0 & 1 \\ 0 & 0 \end{pmatrix}$$

one has $A \neq O$ but $AA = O$. Indeed, one has

$$\begin{pmatrix} 0 & 1 \\ 0 & 0 \end{pmatrix}\begin{pmatrix} 0 & 1 \\ 0 & 0 \end{pmatrix} = \begin{pmatrix} 0 & 0 \\ 0 & 0 \end{pmatrix}.$$

**Inverse matrices:** Let $A \in \mathrm{Mat}(n,n)$. An *inverse matrix* of a given matrix $A$ is by definition a matrix $B \in \mathrm{Mat}(n,n)$ for which

$$AB = BA = E,$$

where $E$ is the $(n \times n)$-identity matrix. A matrix $B$ of this kind exists if and only if $\det A \neq 0$, i.e., when the determinant of $A$ is non-vanishing. In this case $B$ is uniquely determined and is denoted by $A^{-1}$. Hence, in case $\det A \neq 0$, we have

$$\boxed{AA^{-1} = A^{-1}A = E.}$$

[6] This name comes from the fact that this matrix is the identity element in the ring of matrices.

*Example 7:* The matrix $A^{-1}$ inverse to the matrix

$$A := \begin{pmatrix} a & b \\ c & d \end{pmatrix}$$

exists if and only if $\det A \neq 0$, i.e., when $ad - bc \neq 0$. In this case the inverse is given by

$$A^{-1} = \frac{1}{ad - bc} \begin{pmatrix} d & -b \\ -c & a \end{pmatrix}.$$

Indeed, a calculation shows that

$$AA^{-1} = A^{-1}A = \begin{pmatrix} 1 & 0 \\ 0 & 1 \end{pmatrix}.$$

**Theorem:** For an arbitrary $(n \times n)$-matrix $A$ with $\det A \neq 0$ one has

$$\boxed{(A^{-1})_{jk} = (\det A)^{-1} A_{kj}.}$$

Here $(A^{-1})_{jk}$ denotes the element of $A^{-1}$ in the $j^{th}$ row and $k^{th}$ column. Moreover $A_{kj}$ is the adjoint to $a_{kj}$ in the determinant of $A$ (cf. 2.1.2).

**The group $GL(n, \mathbb{C})$:** A matrix $A \in \mathrm{Mat}(n, n)$ is said to be *regular* (or *invertible*), if $\det A \neq 0$, so that the inverse matrix $A^{-1}$ exists. The set of all regular $(n \times n)$-matrices is denoted by $GL(n, \mathbb{C})$.

More precisely, the set $GL(n, \mathbb{C})$ forms a group (cf. 2.5.1), which one calls the (complex) general linear group (hence the notation).[7]

**Applications to systems of equations:** See 2.1.4 for this.

**Transposed and adjoint matrices:** Let a real or complex $(m \times n)$-matrix $A = (a_{jk})$ be given. The *transposed matrix* $A^{\mathsf{T}}$ of $A$ results by exchanging the rows and columns of $A$. If one in addition takes the complex conjugate entries, one gets the *adjoint* matrix $A^*$ to $A$.

If one denotes the elements of $A^{\mathsf{T}}$ (resp. $A^*$) by $a^{\mathsf{T}}_{jk}$ (resp. by $a^*_{jk}$), then one has:

$$a^{\mathsf{T}}_{kj} := a_{jk}, \qquad a^*_{kj} := \overline{a_{jk}}, \qquad k = 1, \ldots, n, \quad j = 1, \ldots, m.$$

Hence $A^{\mathsf{T}}$ and $A^*$ are $(n \times m)$-matrices.

*Example 8:*

$$A := \begin{pmatrix} 1 & 2 & 3\mathrm{i} \\ 4 & 5 & 6 \end{pmatrix}, \qquad A^{\mathsf{T}} = \begin{pmatrix} 1 & 4 \\ 2 & 5 \\ 3\mathrm{i} & 6 \end{pmatrix}, \qquad A^* = \begin{pmatrix} 1 & 4 \\ 2 & 5 \\ -3\mathrm{i} & 6 \end{pmatrix}.$$

For all real or complex numbers $\alpha$ and $\beta$ one has:

$$\boxed{\begin{aligned} (A^{\mathsf{T}})^{\mathsf{T}} &= A, & (A^*)^* &= A, \\ (\alpha A + \beta B)^{\mathsf{T}} &= \alpha A^{\mathsf{T}} + \beta B^{\mathsf{T}}, & (\alpha A + \beta B)^* &= \overline{\alpha} A^* + \overline{\beta} B^*, \\ (CD)^{\mathsf{T}} &= D^{\mathsf{T}} C^{\mathsf{T}}, & (CD)^* &= D^* C^*, \\ (Q^{-1})^{\mathsf{T}} &= (Q^{\mathsf{T}})^{-1}, & (Q^{-1})^* &= (Q^*)^{-1}. \end{aligned}}$$

[7] Similarly, the set $GL(n, \mathbb{R})$ of all real regular $(n \times n)$-matrices forms the so-called real general linear group. Both $GL(n, \mathbb{C})$ and $GL(n, \mathbb{R})$ are very important examples of Lie groups and will be investigated in more detail in [212] in connection with applications to elementary particle physics.

Here we are assuming that the matrices $A$ and $B$ have the same number of rows and columns and that the matrix product $CD$ exists (recall that this requires that $C$ and $D$ fit). Moreover it is assumed that the inverse matrix $Q^{-1}$ of $Q$ exists (recall that this requires that $\det Q \neq 0$); in this case the inverse matrices to $Q^\mathsf{T}$ and $Q^*$ also exist.

The matrix $(Q^{-1})^\mathsf{T}$ is called the *contragredient* matrix to $Q$.

**The trace of a matrix:** The *trace* of a $(n \times n)$-matrix $A = (a_{jk})$, denoted $\operatorname{tr} A$, is the sum of the diagonal elements of $A$, i.e.,

$$\boxed{\operatorname{tr} A := a_{11} + a_{22} + \ldots + a_{nn}.}$$

*Example 9:*

$$\operatorname{tr} \begin{pmatrix} a & 2 \\ 3 & b \end{pmatrix} = a + b.$$

For all complex numbers $\alpha$, $\beta$ and all $(n \times n)$-matrices $A$, $B$, one has:

$$\boxed{\begin{array}{ll} \operatorname{tr}(\alpha A + \beta B) = \alpha \operatorname{tr} A + \beta \operatorname{tr} B, & \operatorname{tr}(AB) = \operatorname{tr}(BA), \\ \operatorname{tr} A^\mathsf{T} = \operatorname{tr} A, \qquad \operatorname{tr} A^* = \overline{\operatorname{tr} A}. & \end{array}}$$

*Example 10:* If the $(n \times n)$-matrix $C$ is invertible, then one has the relation $\operatorname{tr}(C^{-1}AC) = \operatorname{tr}(ACC^{-1}) = \operatorname{tr} A$.

### 2.1.4 Systems of linear equations

**Basic ideas:** Linear systems of equations can be solvable or not. In case they are solvable, the solution is unique or there is a whole family of solutions which depend on infinitely many parameters.

*Example 1* (solutions depending on parameters): In order to solve the linear system of equations

$$\begin{aligned} 3x_1 + 3x_2 + 3x_3 &= 6, \\ 2x_1 + 4x_2 + 4x_3 &= 8 \end{aligned} \tag{2.9}$$

in the real numbers, we multiply the first row by $-2/3$. This yields the modified first row

$$-2x_1 - 2x_2 - 2x_3 = -4.$$

This expression is now added to the second row of (2.9). Thus from (2.9) we get the new system of equations

$$\begin{aligned} 3x_1 + 3x_2 + 3x_3 &= 6, \\ 2x_2 + 2x_3 &= 4. \end{aligned} \tag{2.10}$$

From the second equation of (2.10) it follows that $x_2 = 2 - x_3$. If we insert this expression into the first equation in (2.10), we get $x_1 = 2 - x_2 - x_3 = 0$. The general solution of (2.9) in real numbers then takes the form[8]

$$x_1 = 0, \quad x_2 = 2 - p, \quad x_3 = p. \tag{2.11}$$

Here $p$ is a real number.

---

[8]These considerations establish that any solution of (2.9) must have the form (2.11). By inverting the reasoning, it is seen that (2.11) is indeed a solution of (2.9).

If one chooses $p$ to be an arbitrary complex number, then (2.11) is the general solution of (2.9) in complex numbers.

*Example 2* (unique solution): If we apply the same method as used in the first example to the system

$$\begin{aligned} 3x_1 + 3x_2 &= 6, \\ 2x_1 + 4x_2 &= 8 \end{aligned}$$

then we get

$$\begin{aligned} 3x_1 + 3x_2 &= 6, \\ 2x_2 &= 4 \end{aligned}$$

with the unique solution $x_2 = 2,\ x_1 = 0$.

*Example 3* (no solution): Suppose the system

$$\begin{aligned} 3x_1 + 3x_2 + 3x_3 &= 6, \\ 2x_1 + 2x_2 + 2x_3 &= 8 \end{aligned} \tag{2.12}$$

had a solution $x_1,\ x_2,\ x_3$. Applying the method of Example 1 leads to the contradiction

$$\begin{aligned} 3x_1 + 3x_2 + 3x_3 &= 6, \\ 0 &= 4. \end{aligned}$$

Consequently, (2.12) has no solutions at all.

**Solutions of linear systems of equations with Mathematica:** This software package is able to find the general solution[9] of arbitrary linear systems of equations.

#### 2.1.4.1 The general solution of a system of linear equations

A real linear system of equations has the form

$$\begin{aligned} a_{11}x_1 + a_{12}x_2 + \ldots + a_{1n}x_n &= b_1, \\ a_{21}x_1 + a_{22}x_2 + \ldots + a_{2n}x_n &= b_2, \\ \ldots \\ a_{m1}x_1 + a_{m2}x_2 + \ldots + a_{mn}x_n &= b_m. \end{aligned} \tag{2.13}$$

Let the real numbers $a_{jk},\ b_j$ be given. We are looking for the real numbers $x_1, \ldots, x_n$ which solve (2.13). This corresponds to the solutions of the matrix equation

$$\boxed{Ax = b.} \tag{2.14}$$

In more detail, this equation is:

$$\begin{pmatrix} a_{11} & a_{12} & \ldots & a_{1n} \\ a_{21} & a_{22} & \ldots & a_{2n} \\ \vdots & \vdots & \ldots & \vdots \\ a_{m1} & a_{m2} & \ldots & a_{mn} \end{pmatrix} \begin{pmatrix} x_1 \\ x_2 \\ \vdots \\ x_n \end{pmatrix} = \begin{pmatrix} b_1 \\ b_2 \\ \vdots \\ b_n \end{pmatrix}.$$

**Definition:** The system (2.13) is called *homogenous*, if all the coefficients $b_j$ of the right-hand side vanish, otherwise (2.13) is said to be *inhomogenous*. A homogenous system always has at least the trivial solution $x_1 = x_2 = \cdots = x_n = 0$.

---

[9] The *general solution* of a linear system of equations is explained in the next section.

**Superposition principle:** If one knows a particular solution $x_{\text{part}}$ of the inhomogenous system (2.14), then one gets the set of all solutions of (2.14) by setting

$$\boxed{x_1 = x_{\text{part}} + y,}$$

where $y$ is an arbitrary solution of the homogenous system $Ay = 0$. In sum:

| the general solution of the inhomogenous system<br>= the special solution of the inhomogenous system<br>+ the general solution of the homogenous system. |
|---|

This principle is valid for all linear problems in mathematics (for example, for linear differential or integral equations).

### 2.1.4.2 The Gaussian algorithm

The Gaussian algorithm is a universal method to find the general solution of (2.13) or to determine the non-solvability of (2.13). It is just a natural generalization of the method used in (2.9) to (2.12).

**Triangular form:** The idea of the Gaussian algorithm is to bring the initial system of equations (2.13) into the following equivalent form:

$$\begin{aligned}
\alpha_{11}y_1 + \alpha_{12}y_2 + \ldots + \alpha_{1n}y_n &= \beta_1, \\
\alpha_{22}y_2 + \ldots + \alpha_{2n}y_n &= \beta_2, \\
\ldots \qquad\qquad &\;\;\ldots \\
\alpha_{rr}y_r + \ldots + \alpha_{rn}y_n &= \beta_r, \\
0 &= \beta_{r+1}, \\
&\;\;\ldots \\
0 &= \beta_m.
\end{aligned} \tag{2.15}$$

Here $y_k = x_k$ for all $k$ or $y_1, \ldots, y_n$ can be obtained from $x_1, \ldots, x_n$ by a renumbering (permutation of the indices). Moreover one has

$$\alpha_{11} \neq 0, \quad \alpha_{22} \neq 0, \quad \ldots, \quad \alpha_{rr} \neq 0.$$

The system (2.15) is obtained as follows.

(i) Suppose at least one of the $a_{jk}$ is not zero. After a permutation of the rows or columns we may then assume that $a_{11} \neq 0$.

(ii) We multiply the first row of (2.13) with $-a_{k1}/a_{11}$ and adding this row to the $k^{th}$ row for $k = 2, \ldots, m$. This yields a system of equations whose first and second row are of the form as in (2.15) with $\alpha_{11} \neq 0$.

(iii) We apply the same procedure to rows 2 to $m$ of the new system (iteration of the first two steps), etc.

**Calculation of the solutions:** The solution of (2.15) is easily written down. This in turn yields a solution to the intitial system of equations (2.13).

*Case 1:* We have $r < m$, and not all of $\beta_{r+1}, \beta_{r+2}, \ldots, \beta_m$ are zero. Then the equations (2.15), hence also (2.13), have no solutions.

*Case 2:* One has $r = m$. From $\alpha_{rr} \neq 0$ we can solve the $r^{th}$ equation in (2.15) for $y_r$, where we view $y_{r+1}, \ldots, y_n$ as parameters. Then we use the $(r-1)^{st}$ equation to calculate $y_{r-1}$. Similarly we obtain the solutions for $y_{r-2}, \ldots, y_1$.

This shows that the general solution of (2.15) and (2.13) depends on $n-r$ real parameters.

*Case 3:* One has $r < m$ and $\beta_{r+1} = \ldots = \beta_m = 0$. In this case we proceed as in case 2 and get the general solution of (2.15) and (2.13), which depends on $n - r$ real parameters.

The number $r$ is equal to the rank of the matrix $A$ (cf. 2.1.4.5).

#### 2.1.4.3 Cramer's rule:

**Theorem:** Let $n = m$ and $\det A \neq 0$. Then the linear system of equations (2.13) has the unique solution

$$\boxed{x = A^{-1}b.}$$

Explicitly this means:

$$\boxed{x_j = \frac{(\det A)_j}{\det A}, \qquad j = 1, \ldots, n.} \tag{2.16}$$

Here the determinant $(\det A)_j$ is determined as the determinant of the matrix which results from $A$ upon replacing the $j^{th}$ column by $b$. One calls this formula *Cramer's rule.*

*Example:* The linear system of equations

$$\begin{aligned} a_{11}x_1 + a_{12}x_2 &= b_1, \\ a_{21}x_1 + a_{22}x_2 &= b_2 \end{aligned}$$

has in case $a_{11}a_{22} - a_{12}a_{21} \neq 0$ the following unique solution:

$$x_1 = \frac{\begin{vmatrix} b_1 & a_{12} \\ b_2 & a_{22} \end{vmatrix}}{\begin{vmatrix} a_{11} & a_{12} \\ a_{21} & a_{22} \end{vmatrix}} = \frac{b_1a_{22} - a_{12}b_2}{a_{11}a_{22} - a_{12}a_{21}},$$

$$x_2 = \frac{\begin{vmatrix} a_{11} & b_1 \\ a_{21} & b_2 \end{vmatrix}}{\begin{vmatrix} a_{11} & a_{12} \\ a_{21} & a_{22} \end{vmatrix}} = \frac{a_{11}b_2 - b_1a_{21}}{a_{11}a_{22} - a_{12}a_{21}}.$$

#### 2.1.4.4 The Fredholm alternative

**Theorem:** The linear system of equations $Ax = b$ has a solution $x$ if and only if

$$\boxed{b^{\mathsf{T}}y = 0}$$

for all solutions $y$ of the homogenous dual equation $A^{\mathsf{T}}y = 0$.

### 2.1.4.5 The criterion on the rank

**Linearly independent row-matrices:** Let $m$ row-matrices[10] $A_1, \ldots, A_m$ of length $n$ with real entries be given. If the equation

$$\alpha_1 A_1 + \ldots + \alpha_m A_m = 0$$

is satisfied for real numbers $\alpha_j$ only if $\alpha_1 = \alpha_2 \cdots = \alpha_m = 0$, then one says $A_1, \ldots, A_m$ are *linear independent*. Otherwise they are said to be *linearly dependent*.

A similar definition holds for column-matrices (i.e., $(n \times 1)$-matrices).

*Example 1:* (i) *Linear independence.* For $A_1 := (1, 0)$ and $A_2 := (0, 1)$ it follows from

$$\alpha_1 A_1 + \alpha_2 A_2 = 0$$

that $(\alpha_1, \alpha_2) = (0, 0)$, hence that $\alpha_1 = \alpha_2 = 0$. This means that $A_1$ and $A_2$ are linearly independent.

(ii) *Linear dependence.* For $A_1 := (1, 1)$ and $A_2 := (2, 2)$ one has

$$2A_1 - A_2 = (2, 2) - (2, 2) = 0,$$

i.e., $A_1$ and $A_2$ are linearly dependent.

**Definition:** The *rank* of a matrix $A$ is equal to the maximal number of linearly independent columns (i.e., column-matrices which make up columns of $A$).

Each determinant which is obtained from the matrix $A$ by striking a certain number of rows and/or columns is called a *subdeterminant* of (the determinant of) $A$.

**Theorem:** (i) The rank of a matrix is equal to the maximal number of linear independent rows, in other words, one has $\text{rank}(A) = \text{rank}(A^\mathsf{T})$.

(ii) The rank of matrix is equal to the maximal size of a non-vanishing subdeterminant.

**The rank theorem:** A linear system of equations $Ax = b$ has a solution if and only if the rank of the coefficient matrix $A$ is equal to the rank of the extended matrix $(A, b)$ (this is a $(m \times (n+1))$-matrix). In this case the general solution depends on $n - r$ real parameters, where $n$ is the number of indeterminants and $r$ denotes the rank of the matrix $A$.

*Example 2:* We consider the system of equations

$$\begin{aligned} x_1 + x_2 &= 2, \\ 2x_1 + 2x_2 &= 4. \end{aligned} \tag{2.17}$$

Then we have

$$A := \begin{pmatrix} 1 & 1 \\ 2 & 2 \end{pmatrix}, \qquad (A, b) := \begin{pmatrix} 1 & 1 & 2 \\ 2 & 2 & 4 \end{pmatrix}.$$

(i) *Linear dependence of the rows.* The second row of $A$ is equal to twice the first row, i.e.,

$$2(1, 1) - (2, 2) = 0.$$

Hence the first and second row of $A$ are linearly dependent. Consequently $r = \text{rank}(A) = 1$. Similarly one finds $\text{rank}(A, b) = 1$. On the other hand $n - r = 2 - 1 = 1$.

Therefore the system of equations (2.17) has a solution, which depends on a real parameter.

[10]These are $(1 \times n)$-matrices.

This result is also easy to verify directly. Since the second equation in (2.17) is just twice the first equation, one may omit the second equation. The first equation in (2.17) has the general solution

$$x_1 = 2 - p, \qquad x_2 = p,$$

with the real parameter $p$.

(ii) *Determinant criterion.* Because of

$$\begin{vmatrix} 1 & 1 \\ 2 & 2 \end{vmatrix} = 0, \qquad \begin{vmatrix} 1 & 2 \\ 2 & 4 \end{vmatrix} = 0,$$

the subdeterminants of $A$ and of $(A, b)$ of size 2 vanish. However, there are non-vanishing subdeterminants of size 1. Hence we have $\operatorname{rank}(A) = \operatorname{rank}(A, b) = 1$.

**Algorithms for determining the rank:** We consider the matrix

$$\begin{pmatrix} a_{11} & a_{12} & \ldots & a_{1n} \\ \ldots & & & \\ a_{m1} & a_{m2} & \ldots & a_{mn} \end{pmatrix}.$$

If all $a_{jk}$ vanish, then we have $\operatorname{rank}(A) = 0$.

In all other cases we can by permuting rows and/or columns and by adding multiples of rows to other rows achieve a triangular form

$$\begin{pmatrix} \alpha_{11} & \alpha_{12} & & & \ldots & \alpha_{1n} \\ 0 & \alpha_{22} & & & \ldots & \alpha_{2n} \\ & & \ddots & & & \\ 0 & \ldots & 0 & \alpha_{rr} & \ldots & \alpha_{rn} \\ 0 & \ldots & 0 & 0 & \ldots & 0 \\ \ldots & & & & & \\ 0 & \ldots & 0 & 0 & \ldots & 0 \end{pmatrix},$$

where all $\alpha_{jj}$ are non-vanishing. Then we have $\operatorname{rank}(A) = r$.

*Example 3:* Let the matrix

$$A := \begin{pmatrix} 1 & 1 & 1 \\ 2 & 4 & 2 \end{pmatrix}$$

be given. We subtract twice the first row from the second row, and get

$$\begin{pmatrix} 1 & 1 & 1 \\ 0 & 2 & 0 \end{pmatrix},$$

i.e., $\operatorname{rank}(A) = 2$.

**Complex systems of equations:** If the coefficients $a_{jk}$ and $b_j$ of the linear system of equations (2.13) are complex numbers, then we seek complex numbers $x_1, \ldots, x_n$ as solutions. All statements made about linear systems of equations still hold in this case. In the definition of linear independence one must allow $\alpha_1, \ldots, \alpha_k$ to be complex also.

### 2.1.5 Calculations with polynomials

A *polynomial of degree n* with real (resp. complex) coefficients is an expression

$$\boxed{a_0 + a_1 x + a_2 x^2 + \ldots + a_n x^n.} \tag{2.18}$$

where $a_0, \ldots, a_n$ are real (resp. complex) numbers with[11] $a_n \neq 0$.

**Equality:** By definition

$$a_0 + a_1 x + \ldots + a_n x^n = b_0 + b_1 x + \ldots + b_m x^m$$

if and only if $n = m$ and $a_j = b_j$ for all $j$ (i.e., the polynomials have the same degree and the same coefficients).

**Addition and multiplication:** One uses the natural rules (1.1.4) to calculate these operations, collecting terms of like degree.

*Example 1:* $(x^2+1)+(2x^3+4x^2+3x+2) = 2x^3+5x^2+3x+3$,
$(x+1)(x^2-2x+2) = x^3-2x^2+2x+x^2-2x+2 = x^3-x^2+2$.

**Division:** Instead of $7 \div 2 = 3$ with remainder 1 one can also write $7 = 2 \cdot 3 + 1$ or $\frac{7}{2} = 3 + \frac{1}{2}$. One does the same in the case of polynomials.

Let $N(x)$ and $D(x)$ be polynomials, where we assume that the degree of $D(x)$ is at least one. Then there are uniquely determined polynomials $Q(x)$ and $R(x)$ such that

$$\boxed{N(x) = D(x)Q(x) + R(x),} \tag{2.19}$$

where the degree of the *remainder* polynomial $R(x)$ is strictly less than the degree of the *denominator* $D(x)$. Instead of (2.19) we also write

$$\frac{N(x)}{D(x)} = Q(x) + \frac{R(x)}{D(x)}. \tag{2.20}$$

One calls $N(x)$ the *numerator* and $Q(x)$ the *quotient* of the two.

*Example 2* (division without remainder): For $N(x) := x^2 - 1$ and $D(x) := x - 1$ (2.19) is satisfied with $Q(x) = x + 1$ and $R(x) = 0$. Indeed, $x^2 - 1 = (x-1)(x+1)$. This means

$$\frac{x^2-1}{x-1} = x + 1.$$

*Example 3* (division with remainder): One has

$$\frac{3x^4 - 10x^3 + 22x^2 - 24x + 10}{x^2 - 2x + 3} = 3x^2 - 4x + 5 + \frac{-2x-5}{x^2-2x+3}.$$

To obtain the corresponding decomposition

$$3x^4 - 10x^3 + 22x^2 - 24x + 10 = (x^2 - 2x + 3)(3x^2 - 4x + 5) + (-2x - 5)$$

[11]In a rigorous formal development of mathematics (2.18) is an abstract expression, which can be associated to complex numbers, if one substitutes fixed complex numbers in $a_0, \ldots, a_n, x$. One also says in this case that $a_0, \ldots, a_n, x$ are occupied by complex numbers.

with the remainder polynomial $R(x) = -2x - 5$, we use the following scheme:

$$
\begin{array}{rrrrrl}
3x^4 & -10x^3 & +22x^2 & -24x & +10 & \text{(division: } 3x^4 \div x^2 = \boxed{3x^2}\text{)} \\
3x^4 & -\ 6x^3 & +\ 9x^2 & & & \text{(multiplication: } (x^2 - 2x + 3)\boxed{3x^2}\text{)} \\
\hline
 & -\ 4x^3 & +13x^2 & -24x & +10 & \text{(subtraction + division: } -4x^3 \div x^2 = \boxed{-4x}\text{)} \\
 & -\ 4x^3 & +\ 8x^2 & -12x & & \text{(multiplication: } (x^2 - 2x + 3)\boxed{(-4x)}\text{)} \\
\hline
 & & 5x^2 & -12x & +10 & \text{(subtraction + division: } 5x^2 \div x^2 = \boxed{5}\text{)} \\
 & & 5x^2 & -10x & +15 & \text{(multiplication: } (x^2 - 2x + 3)\boxed{5}\text{)} \\
\hline
 & & & -\ 2x & -\ 5 & \text{(subtraction).}
\end{array}
$$

This method is summarized in the statement: divide the terms of highest order, multiply back, subtract, divide the resulting terms of highest order, etc. This method comes to an end when the terms of highest order can no longer be divided.

*Example 4:* The decomposition

$$x^3 - 1 = (x - 1)(x^2 + x + 1)$$

follows from the scheme:

$$
\begin{array}{rrrr}
x^3 & -\ 1 & & \\
x^3 & -\ x^2 & & \\
\hline
 & x^2 & -\ 1 & \\
 & x^2 & -\ x & \\
\hline
 & & x & -\ 1.
\end{array}
$$

**The greatest common divisor (gcd) of two polynomials (the Euclidean algorithm):** Let $N(x)$ and $D(x)$ be polynomials of degrees at least one. Similarly as in (2.19) we form the division chain with remainder

$$
\begin{aligned}
N(x) &= D(x)Q(x) + R_1(x), \\
D(x) &= R_1(x)Q_1(x) + R_2(x), \\
R_1(x) &= R_2(x)Q_2(x) + R_3(x), \qquad \text{etc.}
\end{aligned}
$$

From the relation $\deg(R_{j+1}) < \deg(R_j)$ we get, after finitely many steps, at some point a vanishing remainder polynomial, i.e., we have

$$R_m(x) = R_{m+1}(x)Q_{m+1}(x).$$

Then $R_{m+1}(x)$ is the greatest common divisor of $N(x)$ and $D(x)$.

*Example 5:* For $N(x) := x^3 - 1$ and $D(x) := x^2 - 1$ we get:

$$
\begin{aligned}
x^3 - 1 &= (x^2 - 1)x + x - 1, \\
x^2 - 1 &= (x - 1)(x + 1).
\end{aligned}
$$

Hence $x - 1$ is the greatest common divisor of $x^3 - 1$ and $x^2 - 1$.

### 2.1.6 The fundamental theorem of algebra according to Gauss

**Fundamental theorem:** Every polynomial of $n^{th}$ degree, $p(x) := a_0 + a_1x + \cdots + a_nx^n$ with complex coefficients and $a_n \neq 0$ has the *product representation*

$$\boxed{p(x) = a_n(x - x_1)(x - x_2) \cdot \ldots \cdot (x - x_n).} \tag{2.21}$$

The complex numbers $x_1, \ldots, x_n$ are unique up to order.

This famous theorem of Gauss was proved in his dissertation in 1799. However this proof still had a gap. A very elegant function-theoretic proof has been given in section 1.14.9.

**Zeros:** The equation

$$\boxed{p(x) = 0}$$

has exactly the solutions $x = x_1, \ldots, x_n$. The numbers $x_j$ are called the *zeros* of $p(x)$. If a number $x_j$ occurs in the decomposition (2.21) exactly $m$ times, then one refers to $m$ as the *multiplicity of the zero* $x_j$.

**Theorem:** If the coefficients of $p(x)$ are real, then for every complex zero $x_j$, also the conjugate complex number $\overline{x}_j$ is a zero, and these two zeros have the same multiplicity.

*Example 1:* (i) For $p(x) := x^2 - 1$ one has the decomposition $p(x) = (x - 1)(x + 1)$. Hence $p(x)$ has the simple zeros $x = 1$ and $x = -1$.

(ii) For $p(x) = x^2 + 1$ one has the decomposition $p(x) = (x - \mathrm{i})(x + \mathrm{i})$. Hence $p(x)$ has the simple zeros $x = \mathrm{i}$ and $x = -\mathrm{i}$. Note that these are conjugates.

(iii) The polynomial $p(x) := (x - 1)^3(x + 1)^4(x - 2)$ has a triple zero (i.e., a zero of multiplicity 3) at $x = 1$, a quadruple zero at $x = -1$ and a simple zero at $x = 2$.

**Calculating zeros with the division algorithm:** If we know a zero $x_1$ of the polynomial $p(x)$, then we can divide the factor $(x - x_1)$ into $p(x)$ with vanishing remainder, i.e., one has

$$p(x) = (x - x_1)q(x).$$

The other zeros of $p(x)$ are then equal to the zeros of $q(x)$. In this manner one can reduce the problem of determining the zeros of a polynomial to a problem of lower degree.

*Example 2:* Let $p(x) := x^3 - 4x^2 + 5x - 2$. Obviously $x_1 := 1$ is a zero of $p(x)$. Performing division according to (2.19) yields

$$p(x) = (x - 1)q(x) \qquad \text{with} \quad q(x) := x^2 - 3x + 2.$$

The zeros of the quadratic equation $q(x) = 0$ can be solved with the help of 2.1.6.1. But in the case at hand one sees easily that $x_2 = 1$ is again a zero of $q(x)$, hence performing division once again, we get

$$q(x) = (x - 1)(x - 2).$$

Hence we have $p(x) = (x - 1)^2(x - 2)$, i.e., $p(x)$ has the double zero $x = 1$ and the simple zero $x = 2$.

**Numerical calculation of zeros with Mathematica:** This software package is able to calculate the zeros of a given degree $n$ polynomial to an arbitrary degree of precision.

**Explicit formulas for solutions:** For equations of $n^{th}$ degree with $n = 2, 3, 4$, one knows since the sixteenth century formulas for calculating the zeros, yielding formulas with roots (cf. 2.1.6.1 ff.). For $n \geq 5$ formulas of this kind no longer exist (Theorem of Abel, 1825) The general instrument for investigating the solvability of algebraic equations is Galois theory (cf. 2.6).

#### 2.1.6.1 Quadratic equations:

The equation

$$\boxed{x^2 + 2px + q = 0} \tag{2.22}$$

with complex coefficients $p$ and $q$ has the two solutions

$$\boxed{x_{1,2} = -p \pm \sqrt{D}\,.}$$

Here $D := p^2 - q$ is the so-called *discriminant.* Moreover we denote by $\sqrt{D}$ a fixed root, i.e., a fixed solution of the equation $y^2 = D$. One always has the relations

$$-2p = x_1 + x_2, \quad q = x_1 x_2, \quad 4D = (x_1 - x_2)^2 \qquad \text{(theorem of Vieta).}$$

This follows from the decomposition $(x - x_1)(x - x_2) = x^2 + 2px + q$. The theorem of Vieta can be used as a check for pre-determined zeros.

The behavior of the solutions in (2.22) for real coefficients is listed in Table 2.1.

*Table 2.1: Quadratic equations with real coefficients.*

| | |
|---|---|
| $D > 0$ | two real zeros |
| $D = 0$ | one double real zero |
| $D < 0$ | two conjugate complex zeros. |

*Example:* The equation $x^2 - 2x - 3 = 0$ has the discriminant $D = 4$ and the two solutions $x_{1,2} = 1 \pm 2$, that is $x_1 = 3$ and $x_2 = -1$.

#### 2.1.6.2 Cubic equations

**Normal forms:** The general cubic equation

$$x^3 + ax^2 + bx + c = 0 \tag{2.23}$$

with complex coefficients $a$, $b$ and $c$ can be transformed by means of the substitution $y = x + \frac{a}{3}$ into the normal form

$$\boxed{y^3 + 3py + 2q = 0.} \tag{2.24}$$

Here one has the relations

$$2q = \frac{2a^3}{27} - \frac{ab}{3} + c, \qquad 3p = b - \frac{a^2}{3}.$$

The quantity $D := p^3 + q^2$ is called the *discriminant* of (2.24). Table 2.2 describes the behavior of the solutions of (2.24), hence also of (2.23), in the case that the coefficients are real.

*Table 2.2: Cubic equations with real coefficients.*

| | |
|---|---|
| $D > 0$ | one real and two conjugate complex zeros |
| $D < 0$ | three distinct real zeros |
| $D = 0,\ q \neq 0$ | two real zeros, one of which is double |
| $D = 0,\ q = 0$ | one triple real zero. |

**The formulas of Cardano:** The solutions of (2.24) are:

$$\boxed{y_1 = u_+ + u_-, \quad y_2 = \rho_+ u_+ + \rho_- u_-, \quad y_3 = \rho_- u_+ + \rho_+ u_-.} \tag{2.25}$$

Here one has

$$u_\pm := \sqrt[3]{-q \pm \sqrt{D}}, \qquad \rho_\pm := \frac{1}{2}(-1 \pm \mathrm{i}\sqrt{3}).$$

For a real discriminant $D \geq 0$ the two roots $u_\pm$ are uniquely determined. In general one must determine the two complex third roots $u_\pm$ in such a way that $u_+ u_- = -p$.

*Example 1:* For the cubic equation

$$y^3 + 9y - 26 = 0$$

we get $p = 3,\ q = -13,\ D = p^3 + q^2 = 196$. According to (2.25) we have

$$u_\pm = \sqrt[3]{13 \pm 14}, \qquad u_+ = 3, \quad u_- = -1.$$

From this the zeros can be determined to be $y_1 = u_+ + u_- = 2,\ y_{2,3} = -1 \pm 2\mathrm{i}\sqrt{3}$.

*Example 2:* The equation $x^3 - 6x^2 + 21x - 52 = 0$ is transformed by the substitution $x = y + 2$ into the equation of the first example. The zeros are therefore $x_j = y_j + 2$, i.e., $x_1 = 4,\ x_{2,3} = 1 \pm 2\mathrm{i}\sqrt{3}$.

**The importance of the *casus irreducibilis* in the history of mathematics:** The formulas (2.25) for $y_1$ were proven by the Italian mathematician Cardano in his book *Ars Magna* (Great Art) which was published in 1545.[12] In the monograph *Geometry*, Raffael Bombelli introduced the symbol $\sqrt{-1}$, in order to handle the so-called *'casus irreducibilis'*. This corresponds to the case of real coefficients $p,\ q$ with $D < 0$. Although in this case all three zeros $y_1,\ y_2,\ y_3$ are real, they are built up of the complex quantities $u_+$ and $u_-$. This surprising turn of events was an important factor in the introduction of complex numbers in the sixteenth century.

The detour through complex numbers can be avoided by applying the trigonometric formula listed in Table 2.3.

**Theorem of Vieta:** For the solutions $y_1,\ y_2$ and $y_3$ of (2.24) one has:

$$y_1 + y_2 + y_3 = 0, \quad y_1y_2 + y_1y_3 + y_2y_3 = 3p, \quad y_1y_2y_3 = -2q,$$
$$(y_1 - y_2)^2(y_1 - y_3)^2(y_2 - y_3)^2 = -108D.$$

**The trigonometric formulas for solution:** In the case of real coefficients one can obtain the solution of (2.24) by means of the relations listed in Table 2.3.

*Table 2.3: cubic equation ($p, q$ real, $q \neq 0,\ P := (\operatorname{sgn} q)\sqrt{|p|}$)*

| | $p < 0,\ D \leq 0$ | $p < 0,\ D > 0$ | $p > 0$ |
|---|---|---|---|
| | $\beta := \frac{1}{3}\arccos\frac{q}{P^3}$ | $\beta := \frac{1}{3}\operatorname{arcosh}\frac{q}{P^3}$ | $\beta := \frac{1}{3}\operatorname{arsinh}\frac{q}{P^3}$ |
| $y_1$ | $-2P\cos\beta$ | $-2P\cosh\beta$ | $-2P\sinh\beta$ |
| $y_{2,3}$ | $2P\cos\left(\beta \pm \frac{\pi}{3}\right)$ | $P(\cosh\beta \pm \mathrm{i}\sqrt{3}\sinh\beta)$ | $P(\sinh\beta \pm \mathrm{i}\sqrt{3}\cosh\beta)$ |

[12] This solution was not found by him, but rather by Nicol of Brescia, who because of an injury to his tongue was called *Tartaglia*, "the stammerer". Cardano had vowed to keep the formula secret, but then published it.

#### 2.1.6.3 Biquadratic equations

Solutions of biquadratic equations can be reduced to those of cubic equations. This can already be found in Cardano's *Ars Magna*.

**Normal form:** The general equation of fourth degree

$$x^4 + ax^3 + bx^2 + cx + d = 0$$

with complex coefficients $a,\ b,\ c,\ d$ can be brought into the normal form

$$\boxed{y^4 + py^2 + qy + r = 0} \tag{2.26}$$

by means of the substitution $y = x + \frac{a}{4}$. The behavior of solutions of (2.26) depends on the behavior of solutions of the so-called *cubic resolvent*:

$$z^3 + 2pz^2 + (p^2 - 4r)z - q^2 = 0.$$

If $\alpha,\ \beta$ and $\gamma$ denote the solutions of this cubic equation, then one gets the zeros $y_1, \ldots, y_4$ of (2.26) from the formula:

$$\boxed{2y_1 = u + v + w, \quad 2y_2 = u - v + w, \quad 2y_3 = -u + v + w, \quad 2y_4 = -u - v - w.}$$

Here $u,\ v,\ w$ are solutions of the equations $u^2 = \alpha,\ v^2 = \beta,\ w^2 = \gamma$, where in addition it is required that $uvw = q$.

Table 2.4. describes the behavior of solutions of (2.26) in the case of real coefficients.

*Table 2.4: Solutions of biquadratic equations with real coefficients.*

| cubic resolvent | biquadratic equation |
|---|---|
| $\alpha,\ \beta,\ \gamma > 0$ | four real zeros |
| $\alpha > 0,\ \beta,\ \gamma < 0$ | two pairs of conjugate complex zeros |
| $\alpha$ real, $\beta$ and $\gamma$ conjugate complex | two real and two conjugate complex zeros |

*Example:* Suppose we are given the biquadratic equation

$$y^4 - 25y^2 + 60y - 36 = 0.$$

The corresponding cubic resolvent is $z^3 - 50z^2 + 769z - 3600$, which has the zeros $\alpha = 9,\ \beta = 16,\ \gamma = 25$. From this it follows that $u = 3,\ v = 4$ and $w = 5$. Hence we get the zeros

$$y_1 = \frac{1}{2}(u + v - w) = 1, \qquad y_2 = 2, \quad y_3 = 3, \quad y_4 = -6.$$

#### 2.1.6.4 General properties of algebraic equations

We consider the equation

$$p(x) := a_0 + a_1x + \ldots + a_{n-1}x^{n-1} + x^n = 0 \tag{2.27}$$

with $n = 1, 2, \ldots$ Important properties of the solutions of algebraic equations of arbitrary degree can be read off of the coefficients $a_0, \ldots, a_{n-1}$ above. Assume first that the coefficients are all real. Then one has:

(i) If $x_j$ is a zero of (2.27), then so is also the conjugate complex number $\overline{x}_j$.

(ii) If the degree $n$ is odd, then (2.27) has at least one real zero.

(iii) If $n$ is even and one has $a_0 < 0$, then (2.27) has at least two real zeros with different signs.

(iv) If $n$ is even and (2.27) has no real zeros, then $p(x) > 0$ for all real numbers $x$.

**The rule of signs of Descartes (1596–1650):** (i) The number of positive zeros of (2.27), counted with multiplicities, is equal to the number $A$ of changes in sign in the sequence $1, a_{n-1}, \ldots, a_0$, or less than this by some even number.

(ii) If the equation (2.27) has only real zeros, then $A$ is equal to the number of positive zeros.

*Example 1:* For

$$p(x) := x^4 + 2x^3 - x^2 + 5x - 1$$

the sequence $1, 2, -1, 5, -1$ of coefficients has three changes of sign, hence $p(x)$ has one or three positive zeros.

If we replace $x$ by $-x$, we obtain $q(x) := p(-x) = x^4 - 2x^3 - x^2 - 5x - 1$. In this case there is but a single change of sign in the sequence of coefficients $1, -2, -1, -5, -1$. Hence $q(x)$ has only one positive real zero, i.e., $p(x)$ has at least one negative zero.

If we replace $x$ by $x + 1$, i.e., we consider $r(x) := p(x+1) = x^4 + 6x^3 + 11x^2 + 13x + 6$, then $r(x)$ has according to the rule of signs no positive zero, i.e., $p(x)$ has no zero $> 1$.

**The rule of Newton (1643–1727):** The real number $S$ is an upper bound for the real zeros of (2.27), provided one has the relations

$$p(S) > 0, \quad p'(S) > 0, \quad p''(S) > 0, \quad \ldots, \quad p^{(n-1)}(S) > 0. \tag{2.28}$$

*Example 2:* Let $p(x) := x^4 - 5x^2 + 8x - 8$. Then one has

$$p'(x) = 4x^3 - 10x + 8, \quad p''(x) = 12x^2 - 10, \quad p'''(x) = 24x.$$

From

$$p(2) > 0, \quad p'(2) > 0, \quad p''(2) > 0, \quad p'''(2) > 0$$

one then gets $S = 2$ as an upper bound for the set of real zeros of $p(x)$.

If one applies this procedure to $q(x) := p(-x)$, then one obtains the result that all real zeros of $q(x)$ are less than or equal to three.

Hence all real zeros of $p(x)$ are in the interval $[-3, 2]$.

**Sturm's theorem (1803–1855):** Let $p(a) \neq 0$ and $p(b) \neq 0$ with $a < b$. We apply a slight modification of the Euclidean algorithm (cf. 2.1.5) to the polynomial $p(x)$ and to its derivative $p'(x)$:

$$\begin{aligned} p &= p'q - R_1, \\ p' &= R_1 q_1 - R_2, \\ R_1 &= R_2 q_2 - R_3, \\ &\ldots \\ R_m &= R_{m+1} q_{m+1}. \end{aligned}$$

With $W(a)$ we denote the number of sign changes in the sequence $p(a),\ p'(a),\ R_1(a), \ldots, R_{m+1}(a)$. Then $W(a) - W(b)$ is equal to the number of distinct zeros of the polynomial $p(x)$ in the interval $[a, b]$, where multiple zeros are only counted once for this purpose.

If $R_{m+1}$ is a real number, then $p$ has no multiple roots.

*Example 3:* For the polynomial $p(x) := x^4 - 5x^2 + 8x - 8$ one can choose $a = -3$ and $b = 2$ (cf. Example 2). We get

$$p'(x) = 4x^3 - 10x + 8,$$
$$R_1(x) = 5x^2 - 12x + 16, \quad R_2(x) = -3x + 284, \quad R_3(x) = -1.$$

Since $R_3$ is a number, $p(x)$ has no multiple zeros. The Sturm sequence $p(x)$, $p'(x)$, $R_1(x), \ldots, R_3(x)$ is for $x = -3$: $4, -70, 97, 293, -1$, which has three changes of sign, hence $W(-3) = 3$. Similarly for $x = 2$ one obtains the series $4, 20, 12, 278, -1$, hence $W(2) = 1$.

Because of $W(-3) - W(2) = 2$ the polynomial $p(x)$ has two real zeros in the interval $[-3, 2]$. According to Example 2 all real zeros lie in this interval.

A similar consideration yields $W(0) = 2$. From $W(-3) - W(0) = 1$ and $W(0) - W(2) = 1$ we get that the polynomial $p(x)$ has exactly one zero in each of the intervals $[-3, 0]$ and $[0, 2]$. The other zeros of $p(x)$ are not real, but rather conjugate complex numbers.

**Elementary symmetric functions:** The functions

$$e_1 := x_1 + \ldots + x_n,$$
$$e_2 := \sum_{j<k} x_j x_k = x_1 x_2 + x_2 x_3 + \ldots + x_{n-1} x_n,$$
$$e_3 := \sum_{j<k<m} x_j x_k x_m = x_1 x_2 x_3 + \ldots + x_{n-2} x_{n-1} x_n,$$
$$\ldots\ldots$$
$$e_n := x_1 x_2 \cdot \ldots \cdot x_n,$$

are called the *elementary symmetric functions* of the variables $x_1, \ldots, x_n$.

**Theorem of Vieta (1540–1603):** If $x_1, \ldots, x_n$ are complex zeros of a polynomial $p(x) := a_0 + a_1 x + \cdots + a_{n-1} x^{n-1} + x^n$ with complex coefficients, then one has:

$$\boxed{a_{n-1} = -e_1, \quad a_{n-2} = e_2, \quad \ldots, \quad a_0 = (-1)^n e_n.}$$

This follows from $p(x) = (x - x_1) \cdots (x - x_n)$. Hence the coefficients of the polynomial can be expressed in terms of its zeros.

A polynomial is said to be *symmetric*, if it is invariant under an arbitrary permutation of its variables. For example the polynomials $e_1, \ldots, e_n$ are symmetric.

**Main theorem on symmetric polynomials:** Every symmetric polynomial in the variables $x_1, \ldots, x_n$ with complex coefficients can be expressed as a polynomial (with complex coefficients) in the elementary symmetric functions $e_1, \ldots, e_n$.

**Application to the discriminant:** The symmetric polynomial

$$\Delta := \prod_{j<k} (x_j - x_k)^2$$

is called the *(normalized) discriminant.*

If we denote by $x_1, \ldots, x_n$ the zeros of a polynomial $p$, then $\Delta$ is called the (normalized) discriminant of $p$. This quantity can always be expressed in terms of the coefficients of $p$.

*Example 4:* For $n = 2$ one has $\Delta = (x_1 - x_2)^2$, hence

$$\Delta = (x_1 + x_2)^2 - 4x_1 x_2 = e_1^2 - 4e_2.$$

For $p(x) := a_0 + a_1x + x^2$ one has $p(x) = (x - x_1)(x - x_2) = x^2 - (x_1 + x_2)x + x_1x_2$. Therefore we have $a_1 = -(x_1 + x_2) = -e_1$ and $a_0 = x_1x_2 = e_2$, that is

$$\boxed{\Delta = a_1^2 - 4a_0.}$$

For the (non-normalized) discriminant $D$ used in 2.1.6.1 we have $D = \Delta/4$.

**The resultant of two polynomials:** Let the two polynomials

$$p(x) := a_0 + a_1x + \ldots + a_nx^n, \qquad q(x) := b_0 + b_1x + \ldots + b_mx^m$$

with complex coefficients be given, where $n, m \geq 1$ and $a_n \neq 0$, $b_m \neq 0$. The *resultant* $R(p, q)$ of $p$ and $q$ is defined as the following determinant:

$$R(p,q) := \begin{vmatrix} a_n & a_{n-1} & \ldots & a_0 & & & & \\ & a_n & a_{n-1} & \ldots & a_0 & & & \\ & & \ldots & \ldots & \ldots & & & \\ & & & & a_n & a_{n-1} & \ldots & a_0 \\ b_m & b_{m-1} & \ldots & b_0 & & & & \\ & b_m & b_{m-1} & \ldots & b_0 & & & \\ & & \ldots & \ldots & \ldots & & & \\ & & & & b_m & b_{m-1} & \ldots & b_0 \end{vmatrix}. \qquad (2.29)$$

The vacant spots in this determinant are all zero.

**Main theorem on common zeros:** Two given polynomials $p$ and $q$ have a common complex zero if and only if one of the following two conditions is satisfied:

(i) $p$ and $q$ have a greatest common divisor of degree $n \geq 1$.

(ii) $R(p, q) = 0$.

The greatest common divisor of the two polynomials can be easily determined with the help of the Euclidean algorithm (cf. 2.1.5).

**Main theorem on multiple zeros:** A polynomial $p$ has a multiple zero if and only if one of the following three conditions is satisfied:

(i) The greatest common divisor of $p$ and the derivative $p'$ of $p$ has a degree $n \geq 1$.

(ii) The discriminant $\Delta$ of $p$ vanishes.

(iii) $R(p, p') = 0$.

Up to a non-vanishing constant factor $\Delta$ and $R(p, p')$ coincide.

### 2.1.7 Partial fraction decomposition

The method of partial fraction decompositions makes it possible to give an *additive* decomposition of rational functions in terms of polynomials and terms of the form

$$\boxed{\frac{A}{(x-a)^k}.}$$

**Basic ideas:** In order to decompose the function $f(x) := \dfrac{x}{(x-1)(x-2)^2}$, we start with the Ansatz

$$\boxed{f(x) = \frac{A}{x-1} + \frac{B}{x-2} + \frac{C}{(x-2)^2}.}$$

Multiplying by the denominator $(x-1)(x-2)^2$ then yields the relation

$$\boxed{x = A(x-2)^2 + B(x-1)(x-2) + C(x-1).} \tag{2.30}$$

*First method* (comparison of coefficients): From (2.30) we get

$$x = A(x^2 - 4x + 4) + B(x^2 - 3x + 2) + C(x - 1).$$

Comparison of the coefficients of $x^2$, $x$ and 1 then yields

$$0 = A + B, \quad 1 = -4A - 3B + C, \quad 0 = 4A + 2B - C.$$

This linear system of equations has the solution $A = 1,\ B = -1,\ C = 2$.

*Second method* (inserting special values): We choose the values $x = 2, 1, 0$ in (2.30), yielding the linear system of equations

$$2 = C, \quad 1 = A, \quad 0 = 4A + 2B - C$$

with the solution $A = 1,\ C = 2,\ B = -1$. The second method is generally quicker.

**Definition:** A *rational function in lowest terms* is an expression

$$f(x) := \frac{N(x)}{D(x)},$$

where $N$ and $D$ (numerator and denominator, respectively) are polynomials with complex coefficients satisfying the condition $0 \le \deg(N) < \deg(D)$.

Let the pairwise distinct zeros of the denominator $D$ be $x_1, \ldots, x_r$ with multiplicities $\alpha_1, \ldots, \alpha_r$, so that for $D$ we get the expression

$$D(x) = (x - x_1)^{\alpha_1} \cdot \ldots \cdot (x - x_r)^{\alpha_r}.$$

**Theorem:** Let $f$ be a rational function in lowest terms. For all complex numbers $x \neq x_1, \ldots, x_r$, we have the decomposition

$$\boxed{f(x) = \sum_{j=1}^{r} \left( \sum_{\beta=1}^{\alpha_j} \frac{A_{j\beta}}{(x - x_j)^\beta} \right)}$$

with the uniquely determined complex numbers $A_{j\beta}$.

If $N$ and $D$ have real coefficients, then the zeros of $D$ occur in conjugate complex pairs with equal multiplicities. The corresponding coefficients $A_{j\beta}$ are in this case also complex conjugate.

The coefficients $A_{j\beta}$ can at any rate be calculated using either of the previously described methods.

**General rational functions:** If one has $\deg(N) \ge \deg(D)$, then the Euclidean algorithm in 2.1.5 yields a decomposition

$$\frac{N(x)}{D(x)} = Q(x) + \frac{R(x)}{D(x)}$$

with a polynomial $Q(x)$ and a rational function $R(x)/D(x)$ in lowest terms.

*Example:*

$$\frac{x^2}{x^2+1} = 1 - \frac{1}{x^2+1} = 1 - \frac{1}{2\mathrm{i}} \left( \frac{1}{x - \mathrm{i}} - \frac{1}{x + \mathrm{i}} \right).$$

**Partial fraction decomposition with Mathematica:** This software package is able to determine the partial fraction decompositions of arbitrary rational functions.

## 2.2 Matrices

Elementary operations with matrices are described in section 2.1.3. All deeper statements about matrices are based on their spectrum. Spectral theory of matrices makes deep generalizations to operator equations possible (for example differential and integral equations) in the context of functional analysis. This is described in [212].

### 2.2.1 The spectrum of a matrix

**Notations:** We denote by $\mathbb{C}^n_{\mathrm{S}}$ the set of all column matrices

$$\begin{pmatrix} \alpha_1 \\ \vdots \\ \alpha_n \end{pmatrix}$$

with complex numbers $\alpha_1, \ldots, \alpha_n$. On the other hand, the symbol $\mathbb{C}^n$ will always denote the set of row matrices $(\alpha_1, \ldots, \alpha_n)$ with complex entries. If $\alpha_1, \ldots, \alpha_n$ are real, then we get in a similar manner the spaces $\mathbb{R}^n_{\mathrm{S}}$ and $\mathbb{R}^n$.

**Eigenvalues and eigenvectors:** Let $A$ be a complex $(n \times n)$-matrix. An *eigenvalue* of $A$ is a complex number $\lambda$, for which the equation

$$\boxed{Ax = \lambda x}$$

has a solution $x \in \mathbb{C}^n_{\mathrm{S}}$ with $x \neq 0$. We then call $x$ a *eigenvector* for the eigenvalue $\lambda$.

**Spectrum:** The set of all eigenvalues of $A$ is called the *spectrum* $\sigma(A)$ of $A$. The set of complex numbers which do not belong to $\sigma(A)$ is called by definition the *resolvent set* $\rho(A)$ of $A$.

The largest absolute value $|\lambda|$ of all eigenvalues of $A$ is called the *spectral radius* of $A$, denoted $r(A)$.

*Example 1:* From

$$\begin{pmatrix} a & 0 \\ 0 & b \end{pmatrix}\begin{pmatrix} 1 \\ 0 \end{pmatrix} = a\begin{pmatrix} 1 \\ 0 \end{pmatrix}, \qquad \begin{pmatrix} a & 0 \\ 0 & b \end{pmatrix}\begin{pmatrix} 0 \\ 1 \end{pmatrix} = b\begin{pmatrix} 0 \\ 1 \end{pmatrix}$$

it follows that the matrix $A := \begin{pmatrix} a & 0 \\ 0 & b \end{pmatrix}$ has the eigenvalues $\lambda = a,\ b$ with the corresponding eigenvectors $x = (1,0)^{\mathsf{T}},\ (0,1)^{\mathsf{T}}$. The $(n \times n)$-unit matrix $E$ has $\lambda = 1$ as its sole eigenvalue. Every column vector $x \neq 0$ of length $n$ is eigenvector for $E$.

**Characteristic equation:** The eigenvalues $\lambda$ of $A$ are exactly the solutions of the so-called *characteristic equation*

$$\boxed{\det(A - \lambda E) = 0}$$

(or the zeros of the polynomial $\det(A - \lambda E)$). The multiplicity of the zero $\lambda$ is called the *algebraic multiplicity* of the eigenvalue.

The inverse matrix $(A - \lambda E)^{-1}$ exists if and only if $\lambda$ belongs to the resolvent set $\rho(A)$ of $A$. One calls the matrix $(A - \lambda E)^{-1}$ the *resolvent* of the matrix $A$. The equation

$$\boxed{Ax - \lambda x = y} \tag{2.31}$$

has, for a given right-hand side $y \in \mathbb{C}^n_{\mathrm{S}}$, the following behavior:

(i) *Regular (non-singular) case:* If the complex number $\lambda$ is not an eigenvalue of $A$, i.e., $\lambda \in \rho(A)$, then (2.31) has the unique solution $x = (A - \lambda E)^{-1}y$.

(ii) *Singular case:* If $\lambda$ is a eigenvalue of $y$, i.e., $\lambda \in \sigma(A)$, then (2.31) has no solution at all, or the solution, if it exists, is not unique.

*Example 2:* Let $A := \begin{pmatrix} 0 & 1 \\ 1 & 0 \end{pmatrix}$. Because of $\det(A - \lambda E) = \begin{vmatrix} -\lambda & 1 \\ 1 & -\lambda \end{vmatrix} = \lambda^2 - 1$, the characteristic equation is

$$\lambda^2 - 1 = 0.$$

The zeros $\lambda = \pm 1$ are the eigenvalues of $A$ with the algebraic multiplicity one. The corresponding eigenvectors are $x_+ = (1, 1)^\mathsf{T}$ and $x_- = (1, -1)^\mathsf{T}$.

**Special matrices:** Let $A$ be a complex $(n \times n)$-matrix. We denote by $A^*$ the adjoint matrix (section 2.1.3).

(i) $A$ is called *self-adjoint*, if $A = A^*$.

(ii) $A$ is called *skew-adjoint*, if $A = -A^*$.

(iii) $A$ is called *unitary*, if $AA^* = A^*A = E$.

(iv) $A$ is called *normal*, if $AA^* = A^*A$.

The matrices in (i) to (iii) are all normal.

The matrix $AA^*$ is always self-adjoint. The matrix $A$ is skew-adjoint if and only if $\mathrm{i}A$ is self-adjoint.

If $A$ is real, then one has $A^* = A^\mathsf{T}$. In this case one speaks, in the cases (i), (ii) and (iii), respectively, of *symmetric*, *skew-symmetric* and *orthogonal* matrices, respectively.

*Example 3:* We consider the matrices

$$A := \begin{pmatrix} a_{11} & a_{12} \\ a_{21} & a_{22} \end{pmatrix}, \qquad U := \begin{pmatrix} \cos\varphi & \sin\varphi \\ -\sin\varphi & \cos\varphi \end{pmatrix}.$$

If all $a_{jk}$ are real, then $A$ is *symmetric* if an only if $a_{12} = a_{21}$. If the elements $a_{jk}$ are complex numbers, then $A$ is *self-adjoint*, if and only if $a_{jk} = \overline{a}_{kj}$ for all $j$ and $k$. In particular, this implies that $a_{11}$ and $a_{22}$ must be real.

For every real number $\varphi$ the matrix $U$ is orthogonal (or unitary). The spectrum of $U$ consists of the numbers $\mathrm{e}^{\pm\mathrm{i}\varphi}$.

The transformation $x' = Ux$, i.e.,

$$\begin{aligned} x_1' &= x_1 \cos\varphi + x_2 \sin\varphi, \\ x_2' &= -x_1 \sin\varphi + x_2 \cos\varphi, \end{aligned}$$

corresponds to a rotation of the Cartesian coordinate system by an angle of $\varphi$ in the mathematically positive sense (cf. 3.4.1).

**Spectral theorem:**

(i) The spectrum of a self-adjoint matrix lies on the real line.

(ii) The spectrum of a skew-adjoint matrix lies on the imaginary axis.

(iii) The spectrum of a unitary matrix lies on the unit circle.

**Theorem of Perron:** If all elements of a real quadratic matrix $A$ are positive, then the spectral radius $r(A)$ is an eigenvalue of $A$ of the algebraic multiplicity one, and all other eigenvalues of $A$ are on the inside of a circle of radius $r(A)$ (centered at the origin).

The eigenvector corresponding to the eigenvalue $r(A)$ has entries all of which are positive.

**Calculation of eigenvalues and eigenvectors with Mathematica:** This software system can also calculate the eigenvalues and eigenvectors for matrices of arbitrary size.

## 2.2.2 Normal forms for matrices

**Basic idea:** Let $A$ and $B$ be complex $(n \times n)$-matrices; these are called *similar* if there is a complex, invertible $(n \times n)$-matrix $C$ such that

$$C^{-1}AC = B.$$

The matrix $A$ is said to be *diagonalizable*, if there is a diagonal matrix $B$ which is similar to $A$. In this case the diagonal entries of $B$ are precisely the eigenvalues of $A$ (occuring as often as the corresponding multiplicity).

**Theorem:** A $(n \times n)$-matrix is diagonalizable if and only if it has $n$ linearly independent eigenvalues. If this is the case, then:

$$\boxed{C^{-1}AC = \begin{pmatrix} \lambda_1 & & 0 \\ & \ddots & \\ 0 & & \lambda_n \end{pmatrix},} \tag{2.32}$$

and $\lambda_1, \ldots, \lambda_n$ are the eigenvalues of $A$.

**Application:** The linear transformation

$$\boxed{x^+ = Ax}$$

is mapped through the introduction of new coordinates $y = C^{-1}x$ into

$$y^+ = (C^{-1}AC)y. \tag{2.33}$$

Let $x = (\xi_1, \ldots, \xi_n)^\mathsf{T}$ and $z = (\eta_1, \ldots, \eta_n)^\mathsf{T}$. Because of (2.32) the transformation takes, in the new coordinates, the particularly simple form

$$\boxed{\eta_j^+ = \lambda_j \eta_j, \qquad j = 1, \ldots, n.}$$

Considerations of this kind are often used in geometry, for example to simplify rotations or projective maps.

Every normal matrix is diagonalizable. The goal of the theory of normal forms for square matrices is to obtain particularly simple forms (Jordan normal form) through the use of similarity transformations. In this way one is able to obtain geometric statements.

### 2.2.2.1 Diagonalization of self-adjoint matrices

The theory of normal forms of self-adjoint matrices is dominated by the notion of orthogonality. The following considerations make it possible, through the choice of new coordinates, to give conic sections and surfaces of degree two particularly simple normal forms. This is based on the notion of *major axis*, which are perpendicular to each other (cf. 3.4.2 and 3.4.3).

The generalization of these major axis transformations to functional analysis, performed by Hilbert and von Neumann, is the basis for the mathematical treatment of quantum theory (cf. [212]).

**Orthogonality:** For $x,\ y \in \mathbb{C}^n_{\mathrm{S}}$ we define[13]

$$\boxed{(x|y) := x^* y} \tag{2.34}$$

and $||x|| := \sqrt{(x|x)}$.

We say that $x$ and $y$ are *orthogonal*, if $(x|y) = 0$. Moreover, the vectors $x_1, \ldots, x_n$ form by definition an *orthonormal system*, if

$$\boxed{(x_j|x_k) = \delta_{jk}}$$

for $j,\ k = 1, \ldots, r$.[14] In case $r = n$ we speak of an *orthonormal basis*.

**The Schmidt orthogonalization procedure:** If $x_1, \ldots, x_r \in \mathbb{C}^n_{\mathrm{S}}$ are linearly independent, then one can obtain an orthonormal system $y_1, \ldots, y_r$ by passing to an appropriate linear combination. Explicitly we choose $z_1 := x_1$ and define for $k = 2, \ldots, r$ inductively

$$z_k := x_k - \sum_{j=1}^{k-1} \frac{(x_k|z_j)}{(z_j|z_j)} z_j .$$

Finally we set $y_j := z_j/||z_j||$ for $j = 1, \ldots, r$.

**Main theorem:** For every self-adjoint $(n \times n)$-matrix $A$ one has:

(i) All eigenvalues are real.

(ii) Eigenvectors to different eigenvalues are orthogonal.

(iii) For every eigenvalue of algebraic multiplicity $s$ there are exactly $s$ linearly independent eigenvectors.

(iv) If one applies the Schmidt orthogonalization procedure to (iii), one gets an orthonormal basis of eigenvectors $x_1, \ldots, x_n$ to the eigenvalues $\lambda_1, \ldots, \lambda_n$.

(v) If we set $U := (x_1, \ldots, x_n)$, then we have

$$\boxed{U^{-1}AU = \begin{pmatrix} \lambda_1 & & 0 \\ & \ddots & \\ 0 & & \lambda_n \end{pmatrix}} . \tag{2.35}$$

---

[13] If one has $x = (\xi_1, \ldots, \xi_n)^{\mathsf{T}}$ and $y = (\eta_1, \ldots, \eta_n)^{\mathsf{T}}$, then:

$$(x|y) = \overline{\xi}_1 \eta_1 + \cdots + \overline{\xi}_n \eta_n .$$

[14] The so-called *Kronecker symbol* is defined by the relation:

$$\delta_{jk} = \begin{cases} 1 \text{ for } & j = k \\ 0 \text{ for } & j \neq k \end{cases} .$$

Here $U$ is unitary, i.e, $U^{-1} = U^*$.

(vi) We have the formulas $\det A = \lambda_1 \lambda_2 \cdots \lambda_n$ and $\operatorname{tr} A = \lambda_1 + \cdots + \lambda_n$.

(vii) If $A$ is real, then so are the eigenvectors $x_1, \ldots, x_n$, and the matrix $U$ is orthogonal, i.e., $U$ is real and $U^{-1} = U^\mathsf{T}$.

*Example 1:* The symmetric matrix $A := \begin{pmatrix} 0 & 1 \\ 1 & 0 \end{pmatrix}$ has the eigenvalues $\lambda_\pm = \pm 1$ and the eigenvalues $x_+ = (1,1)^\mathsf{T}$, $x_- = (1,-1)^\mathsf{T}$. Because of $||x_\pm|| = \sqrt{2}$, the corresponding orthonormal basis is given by $x_1 = x_+/\sqrt{2}$, $x_2 = x_-/\sqrt{2}$. The matrix

$$U := (x_1, x_2) = \frac{1}{\sqrt{2}} \begin{pmatrix} 1 & 1 \\ 1 & -1 \end{pmatrix}$$

is orthogonal and we have the relation

$$U^{-1}AU = \begin{pmatrix} 1 & 0 \\ 0 & -1 \end{pmatrix}.$$

**Morse index and the signature:** The number $m$ of negative eigenvalues of $A$ is called the *Morse index*[15]of $A$.

The number of non-vanishing eigenvalues of $A$ is equal to the rank $r$ of the matrix $A$. Hence $A$ has precisely $m$ negative and $r - m$ positive eigenvalues. The pair $(r - m, m)$ is called the *signature* of $A$.

Let a real symmetric $(n \times n)$-matrix $A := (a_{jk})$ be given. The signature of $A$ can be calculated directly from the entries of the matrix $A$. We consider for the the so-called $s$-rowed *major subdeterminants* of $A$,

$$D_s := \det(a_{jk}), \qquad j, k = 1, \ldots, s.$$

After perhaps renumbering rows and columns of $A$ we may assume that

$$D_1 \neq 0, \quad D_2 \neq 0, \quad \ldots, \quad D_\rho \neq 0, \quad D_{\rho+1} = \ldots = D_n.$$

**The signature criterion of Jacobi:** The rank of $A$ is equal to $\rho$ and the Morse index $m$ of $A$ is equal to the number of changes in sign of the sequence $1, D_1, \ldots, D_\rho$. Moreover, for the signature of $A$ we have $\operatorname{sig}(A) = (\rho - m, m)$.

*Example 2:* For $A := \begin{pmatrix} 1 & 2 \\ 2 & 1 \end{pmatrix}$ one has $D_1 = 1$ and $D_2 = \begin{vmatrix} 1 & 2 \\ 2 & 1 \end{vmatrix} = -3$. The sequence $1, D_1, D_2$ has only one change of sign. Thus the Morse index is $\mathrm{Morse}(A) = 1$ and $\operatorname{sig}(A) = (1,1)$. In fact $A$ has the eigenvalues $\lambda = 3,\ -1$.

**Applications to quadratic forms:** We consider the real equation

$$x^\mathsf{T} A x = b \tag{2.36}$$

with the real *symmetric* $(n \times n)$-matrix $A = (a_{jk})$, the real column matrix $x = (x_1, \ldots, x_n)^\mathsf{T}$ and the real number $b$. Explicitly (2.36) means:

$$\sum_{j,k=1}^{n} a_{jk} x_j x_k = b. \tag{2.37}$$

[15]The importance of the Morse index for the theory of catastrophes and for the topological theory of extremal problems of functions on manifolds can be found in [212].

The real coefficients $a_{jk}$ satisfy the symmetry condition $a_{jk} = a_{kj}$ for all $j, k$. For $n = 2$ (resp. $n = 3$), this is the equation of a conic section (resp. the equation of a quadric surface) (cf. 3.4.2 and 3.4.3).

Through the transformation $x = Uy$ the equation $(y^\top U^\top)AUy = y^\top U^{-1}AUy = b$ results from (2.36), taking $U^{-1} = U^\top$ into account. By (2.35) the following formula is obtained:

$$\lambda_1 y_1^2 + \ldots + \lambda_n y_n^2 = b.$$

To further simplify this equation, we set $z_j := \sqrt{\lambda_j} y_j$ for $\lambda_j \geq 0$ and $z_j := -\sqrt{-\lambda_j} y_j$ for $\lambda_j < 0$. After a possible renumbering of the variables we then obtain from (2.37) the final *normal form*

$$-z_1^2 - \ldots - z_m^2 + z_{m+1}^2 + \ldots + z_r^2 = b. \tag{2.38}$$

Here we have $\mathrm{Morse}(A) = m$, $\mathrm{rank}(A) = r$, $\mathrm{sig}\,(A) = (r - m, m)$.

**Uniqueness of the normal form (Sylvester's law of inertia):** If one has obtained a normal form

$$\alpha_1 z_1^2 + \ldots + \alpha_n z_n^n = b \tag{2.39}$$

of (2.37) through a transformation $x = Bz$ with a real invertible $(n \times n)$-matrix $B$, where for the form (2.39) $\alpha_j = \pm 1$ or $\alpha_j = 0$, then (2.39) and (2.38) coincide (after a possible renumbering of the variables).

**Remark:** This theorem is usually formulated more simply as follows: the signature of the quadratic form $x^\top Ax$, i.e., the number of positive and negative eigenvalues of $A$, is independent of the basis chosen with respect to which the matrix $A$ is formed (the matrix $U$ above represents a change of basis).

**Definiteness:** The quadratic form $x^\top Ax$ in (2.36) is called *positive definite*, if

$$x^\top Ax > 0 \qquad \text{for all } x \neq 0.$$

This is equivalent to one of the following conditions being satisfied:

(i) All eigenvalues of $A$ are positive.

(ii) All major subdeterminants of $A$ are positive.

#### 2.2.2.2 Normal matrices

**Main theorem:** Every normal $(n \times n)$-matrix $A$ has a complete orthonormal basis $x_1, \ldots, x_n$ of eigenvectors to the eigenvalues $\lambda_1, \ldots, \lambda_n$. If we set $U := (x_1, \ldots, x_n)$, then $U$ is unitary, and one has

$$\boxed{U^{-1}AU = \begin{pmatrix} \lambda_1 & & 0 \\ & \ddots & \\ 0 & & \lambda_n \end{pmatrix}.} \tag{2.40}$$

Every self-adjoint, skew-adjoint, unitary or real symmetric, skew-symmetric or orthogonal matrix is normal.

**Application to orthogonal matrices (rotations):** If a real $(n \times n)$-matrix $A$ is orthogonal, then one has (2.40), where the eigenvalues $\lambda_j$ are equal to $\pm 1$ or occur pairwise in the form $\mathrm{e}^{\pm \mathrm{i}\varphi}$ with a real number $\varphi$. The matrix $U$ is in general not real.

There is however always a real orthogonal matrix $B$, such that

$$B^{-1}AB = \begin{pmatrix} A_1 & & 0 \\ & \ddots & \\ 0 & & A_s \end{pmatrix}. \tag{2.41}$$

The matrices $A_j$ are either $(1 \times 1)$-matrices consisting of $\pm 1$, or they are of the form

$$A_j = \begin{pmatrix} \cos\varphi & \sin\varphi \\ -\sin\varphi & \cos\varphi \end{pmatrix}, \tag{2.42}$$

where $\varphi$ is real. The numbers $\pm 1$ occuring are the eigenvalues of $A$. In case (2.42) occurs, $e^{\pm i\varphi}$ are a pair of conjugate complex eigenvalues of $A$. The block matrices $A_j$ occur as often as the corresponding algebraic multiplicity of the eigenvalues.

For an arbitrary orthogonal matrix $A$ one has $\det A = \pm 1$.

The most general orthogonal $(2 \times 2)$-matrix $A$ with $\det A = 1$ has the form (2.42), where $\varphi$ is an arbitrary real number.

*Example:* For $n = 3$ and $\det A = 1$ one has the normal form

$$B^{-1}AB = \begin{pmatrix} 1 & 0 & 0 \\ 0 & \cos\varphi & \sin\varphi \\ 0 & -\sin\varphi & \cos\varphi \end{pmatrix}. \tag{2.43}$$

Geometrically this corresponds to the fact that every rotation of the three-dimensional space at a point is a rotation on a fixed axis with an angle $\varphi$. The rotation axis is the eigenvector of $A$ for the eigenvalue $\lambda = 1$ (Euler–d'Alembert's law).

In case $\det A = -1$ one has to replace the number 1 in (2.43) by $-1$. This corresponds to an additional reflection on the plane of rotation, which passes through the center of rotation and is perpendicular to the axis of rotation.

**Application to skew-symmetric matrices:** Let $A = (a_{jk})$ be a real $(n \times n)$-skew-symmetric-matrix[16], which means that $a_{jj} = 0$ for all $j$ and $a_{jk} = -a_{kj}$ for all $j, k$ with $j \neq k$. Then (2.40) holds, where the eigenvalues $\lambda_j$ are zero or occur pairwise as $\pm\alpha i$. The matrix $U$ is in general not real.

There is however always a real invertible matrix $B$ with

$$B^{-1}AB = \begin{pmatrix} A_1 & & 0 \\ & \ddots & \\ 0 & & A_s \end{pmatrix}.$$

The matrices $A_j$ are either $(1 \times 1)$-matrices with entry 0, or are of the form

$$A_j = \begin{pmatrix} 0 & 1 \\ -1 & 0 \end{pmatrix}.$$

[16]The most general skew-symmetric $(2 \times 2)$-matrix has the form

$$\begin{pmatrix} 0 & a \\ -a & 0 \end{pmatrix},$$

where $a$ is an arbitrary real number.

The total length of these $(2 \times 2)$-blocks is equal to the rank of $A$.

**Application to symplectic forms:** We consider the real equation

$$\boxed{x^\mathsf{T} Ay = b} \tag{2.44}$$

with the real *skew-symmetric* $(n \times n)$-matrix, real column vectors $x$, $y$ and a real number $b$. Explicitly, (2.44) is the equation

$$\boxed{\sum_{j,k=1}^{n} a_{jk} x_j y_k = b.}$$

The real coefficients $a_{jk}$ satisfy the condition $a_{jj} = 0$ for all $j$ and $a_{jk} = -a_{kj}$ for all $j \neq k$. Then there is a real invertible matrix $B$, such that the equation is transformed by the change of coordinates $u = Bx$, $v = By$ into the following normal form:

$$\boxed{(v_2 u_1 - u_2 v_1) + (v_4 u_3 - u_4 v_3) + \ldots + (v_{2s} u_{2s-1} - u_{2s} v_{2s-1}) = b.} \tag{2.45}$$

Here $2s$ is the rank of $A$.

We call $x^\mathsf{T} Ay$ a *symplectic form*, if $A$ is also invertible. Then $n$ is even, and one has the normal form (2.45) with $2s = n$.

Symplectic forms are the basis of symplectic geometry (cf. 3.9.8), which is the fundamental notion for classical mechanics or geometric optics, as well as of the theory of Fourier integral operators (cf. [212]). Many physical theories can be formulated parallel to classical Hamiltonian mechanics in terms of Hamiltonian equations (cf. 1.3.1). All of these theories are based on the notion of symplectic form.

**Simultaneous diagonalization:** Let $A_1, \ldots, A_r$ be complex $(n \times n)$-matrices, which pairwise commute with each other, i.e., $A_j A_k = A_k A_j$ for all $j, k$. Then all of these matrices have a common eigenvector.

If these matrices are in addition *normal*, then they in fact possess a common orthonormal basis $x_1, \ldots, x_n$ of eigenvectors. If we form the matrix $U := (x_1, \ldots, x_n)$, then $U$ is unitary and the matrices

$$U^{-1} A_j U$$

have diagonal form for all $j$, where the eigenvalues of $A_j$ are the diagonal entries.

#### 2.2.2.3 The Jordan normal form

The Jordan normal form is the most general normal form for complex matrices. It originated in work of the French mathematician Camille Jordan (1838–1922). The theory of elementary divisor goes back to work of Karl Weierstrass (1815–1897) in 1868.

**Jordan blocks:** The matrices

$$\begin{pmatrix} \lambda & 1 \\ 0 & \lambda \end{pmatrix}, \qquad \text{and} \qquad \begin{pmatrix} \lambda & 1 & 0 \\ 0 & \lambda & 1 \\ 0 & 0 & \lambda \end{pmatrix}$$

are called Jordan blocks of size two resp. three. The number $\lambda$ is the sole eigenvalue of these matrices. In general one calls a matrix of the form

$$J(\lambda) := \begin{pmatrix} \lambda & 1 & & & 0 \\ & \lambda & 1 & & \\ & & \ddots & \ddots & \\ & & & & 1 \\ 0 & & & & \lambda \end{pmatrix}$$

a *Jordan block.*

**Main theorem:** For an arbitrary complex $(n \times n)$-matrix $A$ there is an invertible complex $(n \times n)$-matrix $C$, such that

$$\boxed{C^{-1}AC = \begin{pmatrix} J_1(\lambda_1) & & 0 \\ & \ddots & \\ 0 & & J_s(\lambda_s) \end{pmatrix}}. \tag{2.46}$$

There exists at least one Jordan block for every eigenvalue $\lambda_j$ of $A$.

The matrix on the right in (2.46) is called the *Jordan normal form* of $A$. This matrix is unique up to a permutation of the Jordan blocks.

**Geometric and algebraic multiplicity of an eigenvalue:** By definition the *geometric multiplicity* of an eigenvalue $\lambda$ is the number of linearly independent eigenvectors to the eigenvalue $\lambda$. The geometric multiplicity of $\lambda$ is equal to the number of Jordan blocks in (2.46). The algebraic multiplicity of $\lambda$ on the other hand is equal to the total length of all Jordan blocks to that eigenvalue.

*Example 1:* The matrix

$$A := \begin{pmatrix} \lambda_1 & 1 & 0 & 0 \\ 0 & \lambda_1 & 0 & 0 \\ 0 & 0 & \lambda_1 & 0 \\ 0 & 0 & 0 & \lambda_2 \end{pmatrix}$$

is already in Jordan normal form. The numbers $\lambda_1,\ \lambda_2$ are the eigenvalues of $A$, where it is possible that $\lambda_1 = \lambda_2$.

Suppose that $\lambda_1 \neq \lambda_2$. Then $\lambda_1$ has algebraic multiplicity three and geometric multiplicity two. For $\lambda_2$ the algebraic and the geometric multiplicity are both one.

For many considerations the algebraic multiplicity is more important than the geometric multiplicity.

The lengths of the Jordan blocks can be determined from the coefficients of $A$, as we shall see presently.

**Elementary divisors:** The greatest common divisor $\mathscr{D}_s(\lambda)$ of all $s$-row subdeterminants of the characteristic matrix $A - \lambda E$ of $A$ is called the $s^{th}$ determinant divisor[17] of $A$. We set $\mathscr{D}_0 := 1$. The quotients $\mathscr{J}_s(\lambda) := \mathscr{D}_s(\lambda)/\mathscr{D}_{s-1}(\lambda),\ s = 1, \ldots, n$ are polynomials and are called the combined elementary divisors of $A$. The factors in the product decomposition

$$\boxed{\mathscr{J}_s(\lambda) = (\lambda - \lambda_1)^{r_1} \cdot \ldots \cdot (\lambda - \lambda_k)^{r_k}}$$

are called the *elementary divisors* of $A$. The numbers $\lambda_1, \ldots, \lambda_n$ are always eigenvalues of $A$.

[17] $\mathscr{D}_s$ is a polynomial in $\lambda$, whose largest term has by definition the coefficient 1.

**Corollary to the main theorem:** To every elementary divisor $(\lambda - \lambda_m)^r$ of $A$ there is a Jordan block of size $r$ in the Jordan normal form of (2.46). In this manner one obtains all Jordan blocks.

**Criterion for diagonalizability:** The Jordan normal form of a square matrix $A$ has diagonal form if and only if one of the following three conditions is satisfied.

(i) All elementary divisors of $A$ have the degree 1.

(ii) For all eigenvalues of $A$ the algebraic and geometric multiplicity coincide.

(iii) The number of linearly independent eigenvalues of $A$ is equal to the number of rows of $A$.

**Theorem of trace:** The trace $\operatorname{tr} A$ of a square matrix $A$ is equal to the sum of all eigenvalues of $A$, each counted with the corresponding algebraic multiplicity.

This follows from (2.46) and the relation $\operatorname{tr} A = \operatorname{tr}(C^{-1}AC)$.

**Theorem of determinant:** The determinant $\det A$ of a square matrix $A$ is equal to the product of all eigenvalues, each counted with the corresponding algebraic multiplicity.

**Similarity theorem:** Two complex square matrices $A$ and $B$ are similar if and only if they have the same elementary divisors.

**Methods for calculating the Jordan normal form:** Methods of this kind can be found in [256].

## 2.2.3 Matrix functions

In this section we let $A$, $B$, $C$ denote complex $(n \times n)$-matrices, and $r(A)$ (resp. $\sigma(A)$) denote the spectral radius (resp. spectrum) of $A$ (cf. 2.2.1).

### 2.2.3.1 Power series

**Definition:** Let the power series

$$\boxed{f(z) := a_0 + a_1 z + a_2 z^2 + \ldots}$$

be given, with the radius of convergence $\rho$ around the origin. If $r(A) < \rho$, then we define

$$\boxed{f(A) := a_0 E + a_1 A + a_2 A^2 + \ldots\,.} \tag{2.47}$$

This series converges absolutely for every matrix element of $f(A)$.

In particular, if $\rho = \infty$ (for example, this is satisfied for $f(z) = \mathrm{e}^z$, $\sin z$, $\cos z$ or $f(z) =$ a polynomial in $z$), then (2.47) is valid for all square matrices $A$.

**Theorem:** (i) $C^{-1}f(A)C = f(C^{-1}AC)$, in case $C^{-1}$ exists.

(ii) $f(A)^\mathsf{T} = f(A^\mathsf{T})$, $Af(A) = f(A)A$.

(iii) $f(A)^* = f^*(A^*)$, where $f^*(z) := \overline{f(\overline{z})}$.

(iv) If $\lambda_1, \ldots, \lambda_n$ are the eigenvalues of $A$, then $f(\lambda_1), \ldots, f(\lambda_n)$ are the eigenvalues of $f(A)$, counted with the corresponding algebraic multiplicities.

**Diagonalizable matrices:** From the relation

$$C^{-1}AC = \begin{pmatrix} \lambda_1 & & 0 \\ & \ddots & \\ 0 & & \lambda_n \end{pmatrix} \tag{2.48}$$

it follows that

$$f(A) = C \begin{pmatrix} f(\lambda_1) & & 0 \\ & \ddots & \\ 0 & & f(\lambda_n) \end{pmatrix} C^{-1}. \qquad (2.49)$$

**Jordan normal form:** From the relation

$$C^{-1}AC = \begin{pmatrix} J_1(\lambda_1) & & 0 \\ & \ddots & \\ 0 & & J_s(\lambda_s) \end{pmatrix}$$

it follows that

$$f(A) = C \begin{pmatrix} J_1(\mu_1) & & 0 \\ & \ddots & \\ 0 & & J_s(\mu_s) \end{pmatrix} C^{-1}$$

with $\mu_j := f(\lambda_j)$.

**The exponential function:** For every square matrix $A$ and every complex number $t$ one has[18]:

$$\mathrm{e}^{tA} = E + tA + \frac{t^2}{2!}A^2 + \frac{t^3}{3!}A^3 + \ldots .$$

The exponential function has the following properties:

(i) $\mathrm{e}^A\mathrm{e}^B = \mathrm{e}^{A+B}$, in case $A$ and $B$ commute, i.e., $AB = BA$ (addition theorem),

(ii) $\det \mathrm{e}^A = \mathrm{e}^{\operatorname{tr} A}$ (determinant formula),

(iii) $(\mathrm{e}^A)^{-1} = \mathrm{e}^{-A}$, $(\mathrm{e}^A)^\mathsf{T} = \mathrm{e}^{A^\mathsf{T}}$, $(\mathrm{e}^A)^* = \mathrm{e}^{A^*}$.

*Example:* For $A = \begin{pmatrix} 0 & 1 \\ 0 & 0 \end{pmatrix}$ one has $A^2 = A^3 = \cdots = 0$, hence $\mathrm{e}^{tA} = E + tA$.

**The logarithm:** If $r(B) < 1$, then the series

$$\ln(E + B) = B - \frac{1}{2}B^2 + \frac{1}{3}B^3 - \ldots = \sum_{k=1}^{\infty} \frac{(-1)^{k+1}}{k} B^k.$$

exists.

The equation

$$\mathrm{e}^C = E + B$$

has in this case the unique solution $C = \ln(E + B)$.

### 2.2.3.2 Functions of normal matrices

If $A$ is normal, then there exists a complete orthonormal basis $x_1, \ldots, x_n$ of eigenvectors to the eigenvalues $\lambda_1, \ldots, \lambda_n$. If we set $C := (x_1, \ldots, x_n)$, then $C$ is unitary and (2.48) holds.

[18]Important applications of expressions like $\mathrm{e}^{tA}$ to ordinary differential equations (resp. Lie groups and Lie algebras) may be found in [212].

**Definition:** For an arbitrary function $f : \sigma(A) \longrightarrow \mathbb{C}$ we define[19] $f(A)$ by the relation (2.49).

This definition is independent of the choice of complete orthonormal basis of eigenvectors of $A$.

**Theorem:** (i) The statements of the theorem in section 2.2.3.1 continues to hold.

(ii) If $A$ is self-adjoint and $f$ is real for real arguments, then $f(A)$ is also self-adjoint.

**The square root:** If a self-adjoint matrix $A$ has only non-negative eigenvalues, then the square root $\sqrt{A}$ exists. This matrix is again self-adjoint.

**Polar decomposition:** Every square complex $(n \times n)$-matrix $A$ can by written as a product

$$\boxed{A = US}$$

where $U$ is unitary and $S$ is self-adjoint, with eigenvalues all of which are non-negative.

If $A$ is in addition real, then $U$ and $S$ are also real.

If $A$ is invertible, then one can choose $U$ and $S$ as $S = \sqrt{AA^*}$ and $U = AS^{-1}$.

## 2.3 Linear algebra

### 2.3.1 Basic ideas

**The idea of linearity:** Differentiation and integration of functions are linear operations, meaning that one has relations

$$(\alpha f + \beta g)'(x) = \alpha f'(x) + \beta g'(x),$$
$$\int_G (\alpha f + \beta g)\mathrm{d}x = \alpha \int_G f\ \mathrm{d}x + \beta \int_G g\ \mathrm{d}x$$

where $\alpha$ and $\beta$ are numbers. In general a linear operator $L$ satisfies the condition

$$L(\alpha f + \beta g) = \alpha L f + \beta L g$$

where $f, g$ can be functions, vectors, matrices or whatever.

The *idea of linearity* is important in many mathematical and physical problems. *Linear algebra* collects in a comprehensive and unified manner the experiences of generations of mathematicians and physicists with linear structures.

**The principle of superposition:** By definition, in a physical system a principle of superposition holds if for any two states of the system the linear combination of these is again a state in the system. For example, the two functions $x = f(t),\ g(t)$ are both solutions of the differential equation

$$x'' + \omega^2 x = 0$$

of the harmonic oscillator, and the linear combination $\alpha f + \beta g$ is again a solution.

**The principle of linearization:** An important and often recurring principle in mathematics consists in the method of linearizing a problem. A typical example of this is the notion of derivative of a function. Closely connected with this notion is that of tangents

[19] If, in addition, we have the situation of 2.2.3.1, then both definitions of $f(A)$ given coincide.

of a curve, tangent plane on a surface or more generally tangent spaces on manifolds. The basic idea of the principle of linearization is contained in the Taylor expansion

$$f(x) = f(x_0) + f'(x_0)(x - x_0) + \ldots$$

where the right hand side is a linear approximation of the function $f$ in a neighborhood of the point $x_0$. Adding additional terms

$$f(x) = f(x_0) + f'(x_0)(x - x_0) + \frac{f''(x_0)}{2}(x - x_0)^2 + \ldots$$

means adding quadratic and higher terms. This kind of multi-linear structures are considered in the theory of multi-linear algebra (cf. 2.4).

The branch of mathematics known as topology is concerned with the qualitative behavior of systems. An important method used in that science is to associate to topological spaces certain linear spaces (for example, the de Rham cohomology groups) or vector bundles, and to study the properties of these linear spaces with the tools of linear algebra (cf. [212]).

**Infinite-dimensional function spaces and functional analysis:** In classical geometry one is generally concerned with finite-dimensional spaces (cf. Chapter 3). More modern methods of investigations of differential and integral equations in the context of functional analysis are based on infinite-dimensional spaces, whose elements are functions (for example, metric spaces, Banach spaces, Hilbert spaces, locally convex spaces; cf. [212]).

**Quantum systems and Hilbert spaces:** If one gives a linear space the additional structure of a scalar product, then one gets a class of spaces known as Hilbert spaces, which are fundamental for the mathematical description of quantum systems. The states of the quantum system correspond to elements of a Hilbert space $\mathscr{H}$ and physical quantities are given by appropriately defined linear operators on $\mathscr{H}$, which are called observables (cf. [212]).

The beginnings of linear algebra go back to the book *Der baryzentrische Kalkül* (The Barycentrical Calculus) by August Ferdinand Möbius (1790–1868),which appeared in 1827, and *Die lineare Ausdehnungslehre* (The theory of linear extension) by Hermann Grassmann (1809–1877), which appeared in 1844.

## 2.3.2 Linear spaces

In what follows, the symbol $\mathbb{K}$ stands for $\mathbb{R}$ (the set of real numbers) or $\mathbb{C}$ (the set of complex numbers). In a linear space $X$ over $\mathbb{K}$ *linear combinations* of the form

$$\boxed{\alpha u + \beta v}$$

are declared for $u, v \in X$ and $\alpha, \beta \in \mathbb{K}$.

**Definition:** A set $X$ is called a *linear space* (or *vector space*) over $\mathbb{K}$, if for every ordered pair $(u, v)$ with $u \in X$ and $v \in X$ there is a uniquely determined element in $X$, denoted $u + v$, and for every pair $(\alpha, u)$ with $\alpha \in \mathbb{K}$ and $u \in X$ a uniquely determined element in $X$, denoted $\alpha u$, such that for all $u, v, w \in X$ and all $\alpha, \beta \in \mathbb{K}$ one has the following properties[20]:

[20]Instead of $u + (-v)$ we write forthwith more briefly $u - v$. One can show that, moreover, the properties above imply the relations '$0u = o$ for all $u \in X$' as well as '$\alpha o = o$ for all $\alpha \in \mathbb{K}$'. For this reason we shall in what follows always denote both the element '0' in $\mathbb{K}$ and '$o$' in $X$ by the symbol '0'. Because of the rules laid down here this results in no contradiction. In a similar manner linear spaces over arbitrary fields $\mathbb{K}$ (cf. 2.5.3) can be defined.

(i) $u + v = v + u$ (commutativity).

(ii) $(u + v) + w = u + (v + w)$ (associativity).

(iii) There is a uniquely determined element in $X$, which we denote by $o$, such that

$$z + o = z \qquad \text{for all } z \in X \text{ (neutral element).}$$

(iv) For every $z \in X$ there is a uniquely determined element in $X$, which we denote by $(-z)$, such that

$$z + (-z) = o \qquad \text{for all } z \in X \text{ (inverse element).}$$

(v) $\alpha(u + v) = \alpha u + \alpha v$ and $(\alpha + \beta)u = \alpha u + \beta u$ (distributivity).

(vi) $(\alpha\beta)u = \alpha(\beta u)$ (associativity) and $1u = u$.

Linear spaces over $\mathbb{R}$ (resp. $\mathbb{C}$) are called *real* (resp. *complex*) (vector) spaces.

**Linear independence:** Elements $u_1, \ldots, u_n$ of a linear space over $\mathbb{K}$ are said to be *linearly independent* if and only if a relation

$$\alpha_1 u_1 + \ldots + \alpha_n u_n = 0, \qquad \alpha_1, \ldots, \alpha_n \in \mathbb{K}$$

always implies the relation $\alpha_1 = \cdots = \alpha_n$. Otherwise $u_1, \ldots, u_n$ are said to be *linearly dependent.*

**Dimension:** The maximal number of linearly independent elements of a linear space $X$ is called its *dimension* and denoted $\dim X$. The symbol $\dim X = \infty$ means that there are arbitrarily many linearly independent elements in $X$. If $X$ consists only of the neutral element (check that this is consistent with the definition!), then $\dim X = 0$.

**Basis:** Let $\dim X < \infty$. A system $b_1, \ldots, b_n$ of elements of a linear space $X$ over $\mathbb{K}$ is said to be a *basis* of $X$, if every element $u \in X$ can be written uniquely in the form

$$u = \alpha_1 b_1 + \ldots + \alpha_n b_n, \qquad \alpha_1, \ldots, \alpha_n \in \mathbb{K}.$$

In this case we call the $\alpha_1, \ldots, \alpha_n$ the *coordinates* of $u$ with respect to the basis.

**Theorem on basis:** Let $n$ be a positive natural number. We have the equality $\dim X = n$ if every system of $n$ linear independent elements forms a basis of $X$.

**Steinitz' theorem on exchange of basis:** If $b_1, \ldots, b_n$ form a basis of a linear space $X$ and if $u_1, \ldots, u_r$ are linearly independent elements in $X$, then

$$u_1, \ldots, u_r, b_{r+1}, \ldots, b_n$$

forms a new basis of $X$ (after a renumbering if necessary).

*Example 1* (the linear space $\mathbb{R}^n$): We denote by $\mathbb{R}^n$ the set of all $n$-tuples $(\xi_1, \ldots, \xi_n)$ of real numbers $\xi_j$. Upon setting

$$\boxed{\alpha(\xi_1, \ldots, \xi_n) + \beta(\eta_1, \ldots, \eta_n) = (\alpha\xi_1 + \beta\eta_1, \ldots, \alpha\xi_n + \beta\eta_n)}$$

for all $\alpha, \beta \in \mathbb{R}$, $\mathbb{R}^n$ becomes an $n$-dimensional real linear space. As basis one may take

$$b_1 := (1, 0, \ldots, 0), \quad b_2 := (0, 1, 0, \ldots, 0), \quad \ldots, \quad b_n := (0, 0 \ldots, 0, 1)$$

since

$$(\xi_1, \ldots, \xi_n) = \xi_1 b_1 + \ldots + \xi_n b_n.$$

If one admits complex numbers $\xi_j$, $\eta_j$, $\alpha$ and $\beta$, then one gets the $n$-dimensional complex linear space $\mathbb{C}^n$.

*Example 2:* Let $n$ be a positive natural number. The set of all polynomials

$$a_0 + a_1 x + \ldots + a_{n-1} x^{n-1}$$

with real (resp. complex) coefficients $a_0, \ldots, a_{n-1}$ forms a real (resp. complex) $n$-dimensional linear space. A basis is given for example by the elements $1, x, x^2, \ldots, x^{n-1}$.

*Example 3:* The set of all functions $f : \mathbb{R} \longrightarrow \mathbb{R}$ forms with respect to the usual linear combination $\alpha f + \beta g$ a real linear space. This space is infinite-dimensional, since the power functions $b_j(x) := x^j$, $j = 0, 1, \ldots, n$ are linearly independent for every $n$, i.e., from the relation

$$\alpha_0 b_0(x) + \ldots + \alpha_n b_n(x) = 0, \qquad \alpha_0, \ldots, \alpha_n \in \mathbb{R}$$

it follows that $\alpha_0 = \cdots = \alpha_n = 0$.

*Example 4:* All continuous functions $f : [a, b] \longrightarrow \mathbb{R}$ on the compact interval $[a, b]$ form a real (infinite-dimensional) linear space $C[a, b]$.

This statement is based on the fact that for any two continuous functions $f$ and $g$, the linear combination $\alpha f + \beta g$ is again continuous.

**Linear combinations of sets:** Let $\alpha, \beta \in \mathbb{K}$. If $U$ and $V$ are non-empty sets of a linear space $X$ over $\mathbb{K}$, then we set

$$\alpha U + \beta V := \{\alpha u + \beta v : \ u \in U \text{ and } v \in V\}.$$

**Subspace:** A subset $Y$ of a linear space $X$ over $\mathbb{K}$ is called a *subspace*, if for all $u, v \in Y$ and all $\alpha, \beta \in \mathbb{K}$,

$$\boxed{\alpha u + \beta v \in Y\,.}$$

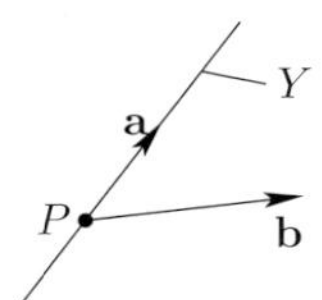


*Figure 2.1.*

*Example 5:* We draw the two vectors $\mathbf{a}$ and $\mathbf{b}$ at the point $P$ (Figure 2.1). The set $\{\alpha\mathbf{a} + \beta\mathbf{b} \,|\, \alpha,\ \beta\ \in \mathbb{R}\}$ of all real linear combinations forms a linear space $X$, which corresponds to the plane through the point $P$ spanned by the elements $\mathbf{a}$ and $\mathbf{b}$. The subspace $Y := \{\alpha\mathbf{a} \,|\, \alpha \in \mathbb{R}\}$ is the line through $P$ in the direction of the vector $\mathbf{a}$.

One has $\dim X = 2$ and $\dim Y = 1$.

**Dimension theorem:** If $Y$ and $Z$ are subspaces of a linear space $X$, then one has:

$$\boxed{\dim (Y + Z) + \dim (Y \cap Z) = \dim Y + \dim Z.}$$

Here the sum $Y + Z$ and the intersection $Y \cap Z$ are again subspaces of $X$.

**Codimension:** Let $Y$ be a subspace of $X$. In addition to the dimension $\dim Y$, often the codimension $\operatorname{codim} Y$ of $Y$ is important. By definition one has[21] $\operatorname{codim} Y := \dim X/Y$. In case $\dim X < \infty$, one has

$$\boxed{\operatorname{codim} Y = \dim X - \dim Y.}$$

[21] The quotient space $X/Y$ is introduced in section 2.3.4.2.

### 2.3.3 Linear operators

**Definition:** If $X$ and $Y$ are linear spaces over $\mathbb{K}$, then a *linear operator* $A : X \longrightarrow Y$ is a map with the property

$$\boxed{A(\alpha u + \beta v) = \alpha Au + \beta Av}$$

for all $u, v \in X$ and all $\alpha, \beta \in \mathbb{K}$.

**Isomorphism:** Two linear spaces $X$ and $Y$ are said to be *isomorphic* if there is a bijective linear mapping $A : X \longrightarrow Y$. Maps of this kind are called *isomorphisms.*[22]

Calculations can be carried out the same way in isomorphic linear spaces. Therefore, from an abstract point of view, there is no difference between isomorphic linear spaces.

**Theorem:** Two finite-dimensional linear spaces over $\mathbb{K}$ are isomorphic, if and only if they have the same dimension.

Thus the dimension of a linear space is the *only characteristic* of finite-dimensional linear spaces.

If $b_1, \ldots, b_n$ is a basis in an $n$-dimensional linear space $X$ over $\mathbb{K}$, then the map defined by

$$A(\alpha_1 b_1 + \ldots + \alpha_n b_n) := (\alpha_1, \ldots, \alpha_n)$$

is a linear isomorphism of $X$ to $\mathbb{K}^n$.

*Example 1:* Let $[a, b]$ be a compact interval. If we set

$$Au := \int_a^b u(x)\mathrm{d}x,$$

then $A : C[a, b] \longrightarrow \mathbb{R}$ is a linear operator.

*Example 2:* We defined a derivation operator by

$$(Au)(x) := u'(x) \qquad \text{for all } x \in \mathbb{R}.$$

Then $A : X \longrightarrow Y$ is a linear operator, if $X$ is the space of all differentiable functions $u : \mathbb{R} \longrightarrow \mathbb{R}$ and $Y$ is the space of all functions $v : \mathbb{R} \longrightarrow \mathbb{R}$.

*Example 3* (matrices): Let $\mathbb{R}^n_{\mathrm{S}}$ denote the real linear $n$-dimensional space of all real column matrices

$$u = \begin{pmatrix} u_1 \\ \vdots \\ u_n \end{pmatrix}.$$

The linear combination $\alpha u + \beta v$ corresponds to the usual matrix operation. The set of all linear operators $A : \mathbb{R}^n_{\mathrm{S}} \longrightarrow \mathbb{R}^m_{\mathrm{S}}$ consists precisely of the set of real $(m \times n)$-matrices $A = (a_{jk})$. The equation $Au = v$ corresponds to the matrix equation

$$\begin{pmatrix} a_{11} & a_{12} & \ldots & a_{1n} \\ a_{21} & a_{22} & \ldots & a_{2n} \\ \ldots & & & \\ a_{m1} & a_{m2} & \ldots & a_{mn} \end{pmatrix} \begin{pmatrix} u_1 \\ u_2 \\ \vdots \\ u_n \end{pmatrix} = \begin{pmatrix} v_1 \\ v_2 \\ \vdots \\ v_m \end{pmatrix}.$$

[22] The Greek word 'isomorph' means 'of the same form'.

#### 2.3.3.1 Calculations with linear operators

We consider linear operators $A, B : X \longrightarrow Y$ and $C : Y \longrightarrow Z$, where $X, Y$ and $Z$ are linear spaces over $\mathbb{K}$.

**Linear combinations:** Let $\alpha,\ \beta \in \mathbb{K}$ be given. We define the linear operator $\alpha A+\beta B : X \longrightarrow Y$ by means of the formula

$$\boxed{(\alpha A + \beta B)u := \alpha Au + \beta Bu \qquad \text{for all } u \in X.}$$

**Product:** The product $AC : X \longrightarrow Z$ is a linear operator, which is defined by composition, i.e., by setting

$$\boxed{(AC)u = A(Cu) \qquad \text{for all } u \in X.}$$

**Unity operator:** The operator defined by $Iu := u$ for all $u \in X$ is a linear operator $I : X \longrightarrow X$ called the *unity operator* and is also denoted by $\mathrm{id}_X$. For all linear operators $A : X \longrightarrow X$ one has the relation

$$\boxed{AI = IA = A.}$$

#### 2.3.3.2 Linear operator equations

Let a linear operator $A : X \longrightarrow Y$ be given. We consider the equation

$$\boxed{Au = v.} \tag{2.50}$$

**Kernel and image:** We define the *kernel* $\mathrm{Ker}(A)$ and the image $\mathrm{Im}(A)$ by the equations

$$\mathrm{Ker}(A) := \{u \in X \,|\, Au = 0\} \quad \text{and} \quad \mathrm{Im}(A) := \{Au \,|\, u \in X\}.$$

**Definition:** (i) $A$ is said to be *surjective*, if $\mathrm{Im}(A) = Y$, i.e., the equation (2.50) has a solution $u \in X$ for every $v \in Y$.

(ii) $A$ is said to be *injective*, if (2.50) has at most one solution $u \in X$ for every $v \in Y$.

(iii) $A$ is called *bijective*, if $A$ is both surjective and injective, i.e., the equation (2.50) has for every $v \in Y$ a unique solution $u \in X$.

In the last case the relation $A^{-1}v := u$ defines the *inverse linear operator* $A^{-1} : Y \longrightarrow X$.

**The superposition principle:** If $u_0$ is a special solution of (2.50), then the set $u_0 + \mathrm{Ker}(A)$ is the set of all solutions of (2.50).

In particular, for $v = 0$ the subspace $\mathrm{Ker}(A)$ is the solution space of (2.50). Hence we have: if $u$ and $w$ are solutions of the homogenous equation (2.50) with $v = 0$, then this is also the case for every linear combination $\alpha u + \beta w$ with $\alpha,\ \beta \in \mathbb{K}$.

**Criterion for surjectivity:** A linear operator $A$ is surjective if and only if there is a linear operator $B : Y \longrightarrow X$ with the property that

$$\boxed{AB = I_Y.}$$

**Criterion for injectivity:** A linear operator $A : X \longrightarrow Y$ is injective if and only if one of the two following conditions is satisfied:

(a) From $Au = 0$ it follows that $u = 0$, i.e., $\dim \operatorname{Ker}(A) = 0$.

(b) There is a linear operator $B : Y \longrightarrow X$ with

$$\boxed{BA = I_X.}$$

**Rank and index:** We set:

$$\boxed{\operatorname{rank}(A) := \dim \operatorname{Im}(A) \quad \text{and} \quad \operatorname{ind}(A) := \dim \operatorname{Ker}(A) - \operatorname{codim} \operatorname{Im}(A).}$$

The index $\operatorname{ind}(A)$ is only defined if one of $\dim \operatorname{Ker}(A)$ and $\operatorname{codim} \operatorname{Im}(A)$ is finite.

*Example 1:* For a linear operator $A : X \longrightarrow Y$ between two finite-dimensional linear spaces $X$ and $Y$ one has:

$$\boxed{\operatorname{ind}(A) = \dim X - \dim Y, \qquad \dim \operatorname{Ker}(A) = \dim X - \operatorname{rank}(A).}$$

(a) The second statements includes the fact that the dimension of the linear space[23] of solutions $u_0 + \operatorname{Ker}(A)$ of (2.50) is equal to $\dim X - \operatorname{rank}(A)$.

(b) If one has $\operatorname{ind}(A) = 0$, then from $\dim \operatorname{Ker}(A) = 0$ it follows immediately that $\operatorname{Im}(A) = Y$. Consequently $A$ is bijective, i.e., the equation $Au = v$ has for every $v \in Y$ a unique solution $u = A^{-1}v$.

**Importance of the index:** The index plays a fundamental role. This is seen clearly when one studies the behavior of solutions of differential and integral equations in the case of infinite-dimensional linear spaces. One of the deepest results of mathematics in the twentieth century is the *Atiyah–Singer index theorem.* This theorem states that one can calculate the index of an important class of linear differential and integral operators on compact manifolds solely in terms of topological (qualitative) data on that manifold and the so-called *symbol* of the operator. This has the consequence that the index of an operator remains the same under relatively strong perturbations of the operator and the manifold (cf. [212]).

### 2.3.3.3 Exact sequences

Modern linear algebra and algebraic topology are often formulated in the language of exact sequences. A sequence

$$X \xrightarrow{A} Y \xrightarrow{B} Z$$

of linear operators $A$ and $B$ is said to be *exact*, if $\operatorname{Im}(A) = \operatorname{Ker}(B)$.

More generally, a sequence

$$\ldots \longrightarrow X_k \xrightarrow{A_k} X_{k+1} \xrightarrow{A_{k+1}} X_{k+2} \longrightarrow \ldots$$

is called exact, if for all $k$ we have $\operatorname{Im}(A_k) = \operatorname{Ker}(A_{k+1})$.

**Theorem:** For a linear operator $A : X \longrightarrow Y$ we have:[24]

(i) $A$ is surjective if and only if the sequence $X \xrightarrow{A} Y \longrightarrow 0$ is exact.

(ii) $A$ is injective if and only if $0 \longrightarrow X \xrightarrow{A} Y$ is exact.

(iii) $A$ is bijective if and only if $0 \longrightarrow X \xrightarrow{A} Y \longrightarrow 0$ is exact.

[23] This is an example of what is called an *affine* subspace of the linear space $X$, see 2.3.4.2.

[24] We denote by 0 the trivial linear space $\{0\}$ consisting only of the neutral element. Moreover, $0 \longrightarrow X$ and $Y \longrightarrow 0$ stand for zero operators.

### 2.3.3.4 The connection with the calculus of matrices

**The matrix $\mathscr{A}$ associated to a linear operator $A$:** Let $A : X \longrightarrow Y$ be a linear operator, where we are assuming that $X$ and $Y$ are finite-dimensional linear spaces over $\mathbb{K}$.

We choose a fixed basis $b_1, \ldots, b_n$ in $X$ and a fixed basis $c_1, \ldots, c_m$ in $Y$. For $u \in X$ and $v \in Y$ one has the uniquely determined decompositions:

$$u = u_1 b_1 + \ldots + u_n b_n, \qquad v = v_1 c_1 + \ldots + v_m c_m$$

and

$$\boxed{Ab_k = \sum_{j=1}^{m} a_{jk} c_j, \qquad k = 1, \ldots, n.}$$

Here $b_j$, $c_k$ and $a_{jk}$ are elements of $\mathbb{K}$. We call the $(m \times n)$-matrix

$$\mathscr{A} := (a_{jk}), \qquad j = 1, \ldots, m, \quad k = 1, \ldots, n$$

the matrix associated to the linear operator $A$ (with respect to the chosen basis). Moreover we introduce the coordinate column matrices associated to $u$ and $v$ by

$$\mathscr{U} := (u_1, \ldots, u_n)^{\mathsf{T}} \qquad \text{and} \qquad \mathscr{V} := (v_1, \ldots, v_m)^{\mathsf{T}}.$$

Then the operator equation

$$\boxed{Au = v}$$

corresponds to the matrix equations

$$\boxed{\mathscr{A}\mathscr{U} = \mathscr{V}.}$$

**Theorem:** The sum (resp. product) of linear operators correspond to the sum (resp. product) of the corresponding matrices.

The rank of a linear operator is the same as the rank of the corresponding matrix (this latter rank is independent of the chosen basis).

**Change of base:** By means of the transformation formulas

$$b_k = \sum_{r=1}^{n} t_{rk} b'_r, \qquad c_j = \sum_{i=1}^{m} s_{ij} c'_i, \qquad k = 1, \ldots, n, \quad j = 1, \ldots, m$$

we pass to a new basis $b'_1, \ldots, b'_n$ (resp. $c'_1, \ldots, c'_m$) in $X$ (resp. in $Y$). Here we require that the $(n \times n)$-matrix $\mathscr{T} = (t_{rk})$ and the $(m \times m)$-matrix $\mathscr{S} = (s_{ij})$ are invertible. The new coordinates $u'_k$ (resp. $v'_j$) of $u$ (resp. $v$) are determined by the decomposition

$$u = u'_1 b'_1 + \ldots + u'_n b'_n, \qquad v = v'_1 c'_1 + \ldots + v'_m c'_m.$$

From this we obtain the transformation formulas for the coordinates of $u$ and $v$, respectively:

$$\boxed{\mathscr{U} = \mathscr{T}\mathscr{U}', \quad \mathscr{V} = \mathscr{S}\mathscr{V}'.}$$

The operator equation $Au = v$ corresponds to the matrix equation $\mathscr{A}'\mathscr{U}' = \mathscr{V}'$ with

$$\boxed{\mathscr{A}' = \mathscr{S}^{-1}\mathscr{A}\mathscr{T}.}$$

In the special case in which $X = Y$ and $b_j = c_j$ for all $j$, one has $\mathscr{S} = \mathscr{T}$. Then we get a similarity transformation $\mathscr{A}' = \mathscr{T}^{-1}\mathscr{A}\mathscr{T}$ of the matrix $\mathscr{A}$.

**Trace and determinant:** Let $\dim X < \infty$. We define the *trace* $\operatorname{tr} A$ and the *determinant* $\det A$ of the linear operator $A : X \longrightarrow X$ by the relation

$$\boxed{\operatorname{tr} A := \operatorname{tr}(a_{jk}), \qquad \det A := \det(a_{jk}).}$$

These definitions are all independent of the choice of the basis chosen for forming the matrix.

**Theorem:** For two linear operators $A,\ B : X \longrightarrow X$ we have:

(i) $\det(AB) = (\det A)(\det B)$.

(ii) $A$ is bijective if and only if $\det A \neq 0$.

(iii) $\operatorname{tr}(\alpha A + \beta B) = \alpha \operatorname{tr} A + \beta \operatorname{tr} B$ for all $\alpha,\ \beta \in \mathbb{K}$.

(iv) $\operatorname{tr}(AB) = \operatorname{tr}(BA)$.

(v) $\operatorname{tr} I_X = \dim X$.

## 2.3.4 Calculating with linear spaces

From given linear spaces we can construct new ones. The following constructions are models for all algebraic structures (for example, groups, rings and fields). Tensor products of linear spaces will be handled in 2.4.3.1.

### 2.3.4.1 Cartesian products

If $X$ and $Y$ are linear spaces over $\mathbb{K}$, then the product set $X \times Y := \{(u,v) \,|\, u \in X,\ v \in Y\}$ is made into a linear space over $\mathbb{K}$ by setting

$$\alpha(u,v) + \beta(w,z) := (\alpha u + \beta w, \alpha v + \beta z), \qquad \alpha, \beta \in \mathbb{K},$$

which is called the *Cartesian product* of $X$ and $Y$.

If $X$ and $Y$ are finite-dimensional, then one has the dimension formula for the Cartesian product,

$$\boxed{\dim(X \times Y) = \dim X + \dim Y.}$$

If $b_1, \ldots, b_n$ (resp. $c_1, \ldots, c_m$) is a basis in $X$ (resp. in $Y$), then the set of all possible pairs $(b_j, c_k)$ form a basis of $X \times Y$.

*Example:* For $X = Y = \mathbb{R}$ one has $X \times Y = \mathbb{R}^2$.

### 2.3.4.2 Quotient spaces

**Linear manifolds (affine linear spaces):** Let $Y$ be a subspace of a linear space $X$ over $\mathbb{K}$. Every set

$$u + Y := \{u + v \,|\, v \in Y\}$$

with fixed $u \in X$ is called a *linear manifold (affine subspace)* (parallel to $Y$). One has

$$u + Y = w + Y$$

if and only if $u - w \in Y$. We set $\dim(u + Y) := \dim Y$.

**Quotient space (factor space):** We denote by $X/Y$ the set of all linear manifolds in $X$ parallel to $Y$. By defining the linear combinations

$$\alpha U + \beta V$$

for $U,\ V \in X/Y$ and $\alpha,\ \beta \in \mathbb{K}$, we make $X/Y$ to a linear space, which is called the *quotient space (factor space)* of $X$ modulo $Y$. Explicitly we have

$$\alpha(u+Y) + \beta(v+Y) = (\alpha u + \beta v) + Y.$$

For $\dim X < \infty$ we have

$$\boxed{\dim X/Y = \dim X - \dim Y.}$$

*Example:* Let $X := \mathbb{R}^2$. If $Y$ is a line through the origin, then $X/Y$ consists of all lines parallel to $Y$ (Figure 2.2).

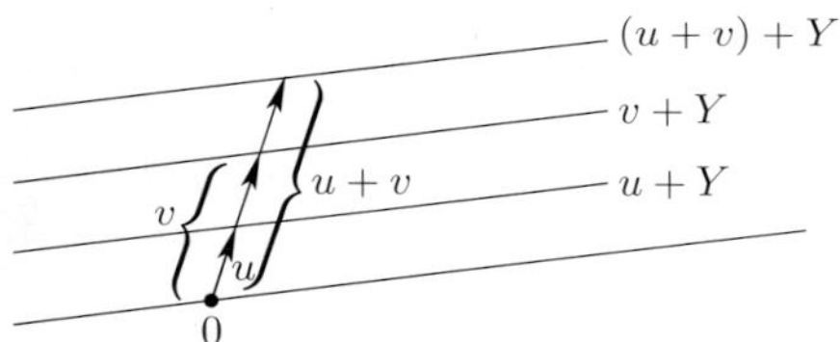


*Figure 2.2. A quotient space.*

**Alternative definition:** Let $u,\ w \in X$. We write

$$u \sim w \qquad \text{if and only if} \quad u - w \in Y.$$

This is an equivalence relation (cf. 4.3.5.1) on $X$. The corresponding equivalence classes $[u] := u + Y$ form the set $X/Y$. By setting

$$\alpha[u] + \beta[z] := [\alpha u + \beta z],$$

the set $X/Y$ acquires the structure of a linear space. This definition does not depend on the choice of representatives for $u$ and $z$.

The map $u \mapsto [u]$ is called the *canonical map* from $X$ onto $X/Y$.

**Structure theorem:** If $Y$ and $Z$ are two subspaces of $X$, then one has the isomorphism

$$\boxed{(Y+Z)/Z \cong Y/(Y \cap Z).}$$

In case $Y \subseteq Z \subseteq X$ one has in addition

$$\boxed{X/Y \cong (X/Z)/(Z/Y).}$$

### 2.3.4.3 Direct sums

**Definition:** Let $Y$ and $Z$ be two subspaces of a linear space $X$. We write

$$\boxed{X = Y \oplus Z,}$$

if and only if every $u \in X$ can be written uniquely in the form

$$u = y + z, \qquad y \in Y,\ z \in Z.$$

The space $X$ is referred to as the *direct sum* of $Y$ and $Z$, and one calls $Z$ the *algebraic complement* to $Y$ in $X$. One has

$$\boxed{\dim(Y \oplus Z) = \dim Y + \dim Z.}$$

**Theorem:** (i) From $X = Y \oplus Z$ it follows that $Z \cong X/Y$.

(ii) One has $\dim Z = \operatorname{codim} Y$.

Intuitively speaking the codimension of $Y$, $\operatorname{codim} Y$, is equal to the number of dimensions which are missing in $Y$, compared with the whole space $X$.

*Example 1:* For $X = \mathbb{R}^3$ the origin 0, a line through the origin and a plane through the origin have, respectively, the dimensions 0, 1 and 2 and the codimensions 3, 2 and 1.

*Example 2:* In Figure 2.3 the decomposition $\mathbb{R}^2 = Y \oplus Z$ is depicted.

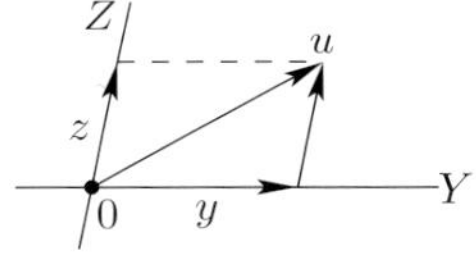

*Figure 2.3.*

**Existence theorem:** For every subspace $Y$ of a linear space $X$ there is an algebraic complement $Z$, i.e., a subspace for which $X = Y \oplus Z$.

**Linear hull:** If $M$ is a set in a linear space $X$ over $\mathbb{K}$, then we call the set

$$\operatorname{span} M := \{\alpha u_1 + \ldots + \alpha_n u_n \mid u_j \in M,\ \alpha_j \in \mathbb{K},\ j = 1, \ldots, n \text{ and } n \geq 1\}$$

the *linear hull* (or *span*) of $M$.

The space $\operatorname{span} M$ is the smallest subspace of $X$ which contains $M$.

**Construction of complementary spaces:** Let $Y$ be an $m$-dimensional subspace of the $n$-dimensional linear space $X$ with $0 < m < n < \infty$. We choose a basis $u_1, \ldots, u_n$ of $X$. Then one has

$$X = Y \oplus \operatorname{span}\{u_{m+1}, \ldots, u_n\}.$$

**The direct sum of arbitrarily many subspaces:** Let $\{X_\alpha\}_{\alpha \in A}$ be a family of subspaces of a linear space $X$. We write

$$\boxed{X = \bigoplus_{\alpha \in A} X_\alpha,}$$

if and only if every $x \in X$ can be written uniquely in the form

$$x = \sum_{\alpha \in A} x_\alpha, \qquad x_\alpha \in X_\alpha,$$

where only finitely many summands are allowed to occur. If the index set $A$ is finite, then one has the formula

$$\boxed{\dim X = \sum_{\alpha \in A} \dim X_\alpha}$$

for the dimensions.

**The outer direct sum of linear spaces:** Let $\{X_\alpha\}_{\alpha \in A}$ be a family of linear spaces $X_\alpha$ over $\mathbb{K}$. Then the *Cartesian product* $\prod_{\alpha \in A} X_\alpha$ consists of the set of all tuples $(x_\alpha)$, which are added together and multiplied by with scalars from $\mathbb{K}$ component-wise. The *outer direct sum*

$$\boxed{\bigoplus_{\alpha \in A} X_\alpha}$$

of the linear spaces $X_\alpha$ is by definition the subset of $\prod_{\alpha \in A} X_\alpha$, which consists of all tuples which differ from zero *for only finitely many values of* $\alpha$.

If one identifies $X_\beta$ with all tuples $(x_\alpha)$ for which $x_\alpha = 0$ for $\alpha \neq \beta$, then $\bigoplus_{\alpha \in A} X_\alpha$ corresponds to the direct sum of the subspaces $X_\alpha$.

**Grading:** One says that the space $\bigoplus_{\alpha \in A} X_\alpha$ is *graded* by the subspaces $X_\alpha$.

#### 2.3.4.4 Application to linear operators

**The rank theorem:** Let a linear operator $A : X \longrightarrow Y$ be given. We choose some decomposition $X = \operatorname{Ker}(A) \oplus Z$; then the restriction of $A$

$$A : Z \longrightarrow \operatorname{Im}(A)$$

is bijective. From this it follows that

$$\boxed{\operatorname{codim} \operatorname{Ker}(A) = \dim \operatorname{Im}(A) = \operatorname{rank}(A).}$$

**Invariant subspaces:** Let a linear operator $A : X \longrightarrow X$ be given, where $X$ is a linear space over $\mathbb{K}$. The subspace $Y$ in $X$ is said to be *invariant with respect to* $A$, if $u \in Y$ implies $Au \in Y$.

In addition $Y$ is called *irreducible* (with respect to $A$), if $Y$ has no genuine invariant subspaces other that the trivial one 0.

**The fundamental decomposition theorem:** If $\dim X < \infty$, then there is, for each $A$, a decomposition

$$\boxed{X = X_1 \oplus X_2 \oplus \ldots \oplus X_k}$$

of $X$ into (non-trivial) irreducible, invariant subspaces $X_1, \ldots, X_k$, i.e., there are operators $A : X_j \longrightarrow X_j$ for all $j$.

If $X$ is a complex linear space, then one can choose a basis in each subspace such that $A$, restricted to $X_j$, is given by a Jordan block matrix. The matrix of $A$ on $X$ with respect to such a basis is then in Jordan normal form.

The sizes of the Jordan blocks are then the dimensions of the subspaces $X_j$. These dimensions can be calculated by the method of elementary divisors as described in section 2.2.2.3, by applying this method to a matrix of $A$ with respect to some basis.

### 2.3.5 Duality

The notion of duality is important in many areas of mathematics (for example in projective geometry and in functional analysis).[25]

[25] A more detailed investigation of the theory of duality for linear spaces and its applications can be found in [212].

**Linear forms:** A *linear form* (or *linear functional*) on a linear space $X$ over $\mathbb{K}$ is a linear map $u^* : X \longrightarrow \mathbb{K}$.

*Example 1:* Setting

$$u^*(u) := \int_a^b u(x)\mathrm{d}x$$

defines a linear form on the space $C[a,b]$ of continuous functions $u : [a,b] \longrightarrow \mathbb{R}$.

**The dual space:** We denote by $X^*$ the set of all linear forms on $X$. Defining a linear combination $\alpha u^* + \beta v^*$ by

$$(\alpha u^* + \beta v^*)(u) := \alpha u^*(u) + \beta v^*(u)$$

for all $u \in X$, the space $X^*$ acquires the structure of a linear space over $\mathbb{K}$, which is called the *dual space* to $X$.

We set $X^{**} := (X^*)^*$.

*Example 2:* Let $X$ be a $n$-dimensional linear space over $\mathbb{K}$. Then one has an isomorphism

$$\boxed{X^* \cong X,}$$

which is however not canonical, but rather depends on the choice of a basis $b_1, \ldots, b_n$ of $X$. In order to show this, we set

$$b_j^*(u_1 b_1 + \ldots + u_n b_n) := u_j, \qquad j = 1, \ldots, n.$$

Then the linear functionals $b_1^*, \ldots, b_n^*$ form a basis of the dual space $X^*$, which we call the *dual basis*.

Each linear form $u^*$ on $X$ can be written in the form

$$u^* = \alpha_1 b_1^* + \ldots + \alpha_n b_n^*, \qquad \alpha_1, \ldots, \alpha_n \in \mathbb{K}.$$

If we set $A(u^*) := \alpha_1 b_1 + \cdots + \alpha_n b_n$, then $A : X^* \longrightarrow X$ is a bijective linear map, which yields the isomorphism $X^* \cong X$.

On the other hand, the isomorphism

$$\boxed{X^{**} \cong X}$$

*is* canonical, i.e., does not depend on the choice of basis, by setting

$$u^{**}(u^*) := u^*(u) \qquad \text{for all } u^* \in X^*.$$

Then to every $u \in X$ is associated a $u^{**}$, and this map of $X$ to $X^{**}$ is linear and bijective.

For infinite-dimensional linear spaces, the relation between $X$ and $X^*$ as well as $X^{**}$ is no longer so clear as in the finite-dimensional case (cf. for example the theory of reflexive Banach spaces in [212]).

**The dual operator:** Let $A : X \longrightarrow Y$ be a linear operator, where $X$ and $Y$ are linear spaces over $\mathbb{K}$. We define the *dual operator* $A^*$ by the formula

$$(A^*v^*)(u) := v^*(Au) \qquad \text{for all } u \in X.$$

This is a linear operator $A^* : Y^* \longrightarrow X^*$.

**Product rule:** If $A : X \longrightarrow Y$ and $B : Y \longrightarrow Z$ are linear operators, then one has

$$\boxed{(AB)^* = B^*A^*.}$$

## 2.4 Multilinear algebra

Let $X$, $Y$ and $Z$ be linear spaces over $\mathbb{K}$. Multilinear algebra investigates products

$$\boxed{uv}$$

with values in $Z$, i.e., for which $u \in X$, $v \in Y$ and $uv \in Z$. The typical property of such a product is the relation

$$\boxed{(\alpha u + \beta w)v = \alpha(uv) + \beta(wv), \qquad u(\alpha v + \beta z) = \alpha(uv) + \beta(uz)}$$

for all $u, w \in X$, $v, z \in Y$ and $\alpha, \beta \in \mathbb{K}$. Important examples of products of this kind are the *tensor product* $u \otimes v$, the *exterior product* $u \wedge v$ and the *inner product* $u \vee v$ (Clifford multiplication).

All products of this sort can be expressed in terms of the tensor product, i.e., there is always a uniquely determined, linear operator $L : X \otimes Y \longrightarrow Z$ with the property that

$$\boxed{uv = L(u \otimes v)}$$

(cf. 2.4.3; this is the *universal property* of the tensor product). For example, the exterior product can be expressed as $u \wedge v := u \otimes v - v \otimes u$. This yields an anti-symmetric relation

$$\boxed{u \wedge v = -(v \wedge u).}$$

The exterior product is closely related to the theory of determinants. In quantum theory, the tensor product $a \otimes b$ describes composite states (for example $a \otimes b \otimes c$ correspond to the proton being the composite of three quarks, cf. [212]). The inner product is used to describe particles with half-integer spin (fermions).

### 2.4.1 Algebras

**Definition:** An *algebra* $\mathscr{A}$ over $\mathbb{K}$ is a linear space, in which a distributive and associative multiplication (product) is defined.

More explicitly, this means there to every ordered pair $(a, b)$ of elements $a$ and $b$ in $\mathscr{A}$ there is a uniquely determined third element of $\mathscr{A}$ associated to them, which is denoted $ab$, such that for all $a, b, c \in \mathscr{A}$ and $\alpha, \beta \in \mathbb{K}$ one has:

(i) $(\alpha a + \beta b)c = \alpha(ac) + \beta(bc)$ and $c(\alpha a + \beta b) = \alpha(ca) + \beta(cb)$, and

(ii) $a(bc) = (ab)c$.

**Homomorphisms:** A *homomorphism* $\varphi : \mathscr{A} \longrightarrow \mathscr{B}$ from an algebra $\mathscr{A}$ to an algebra $\mathscr{B}$ is a linear map which respects the product, i.e., one has

$$\boxed{\varphi(ab) = \varphi(a)\varphi(b) \qquad \text{for all } a, b \in \mathscr{A}.}$$

**Isomorphism:** An algebra $\mathscr{A}$ is said to be *isomorphic* to another algebra $\mathscr{B}$, if there is a bijective homomorphism $\varphi : \mathscr{A} \longrightarrow \mathscr{B}$. Bijective homomorphisms are called *isomorphisms*.

Isomorphic algebras can, for all practical purposed, be viewed as identical objects.

### 2.4.2 Calculations with multilinear forms

Let $X, Y, Z$ and $X_1, \ldots, X_n, Y_1, \ldots, Y_m$ denote linear spaces over $\mathbb{K}$.

**Bilinear forms:** A *bilinear form* is a map $B : X \times Y \longrightarrow Z$, which is linear in each argument separately, i.e., for which

$$\boxed{\begin{aligned} B(\alpha u + \beta w, v) &= \alpha B(u, v) + \beta B(w, v), \\ B(u, \alpha v + \beta z) &= \alpha B(u, v) + \beta B(u, z) \end{aligned}}$$

for all $u,\ w \in X,\ \ v,\ z \in Y$ and $\alpha,\ \beta \in \mathbb{K}$.

**Products:** If we set $uv := B(u, v)$, then we get a product between elements of the linear spaces $X$ and $Y$ with values in the linear space $Z$.

**Special properties:** Let $B : X \times X \longrightarrow Z$ be a bilinear form, in the special case in which $X = Y$.

(i) $B$ is called *symmetric*, if $B(u, v) = B(v, u)$ for all $u, v \in X$.

(ii) $B$ is called *anti-symmetric*, if $B(u, v) = -B(v, u)$ for all $u, v \in X$.

(iii) $B$ is said to be *non-degenerate*, if $B(u, v) = 0$ (or $B(v, u) = 0$) for all $v \in X$ implies that $u = 0$.

**Multilinear forms:** A *n-multilinear form* is a map

$$M : X_1 \times \ldots \times X_n \longrightarrow \mathbb{K},$$

which is linear in each argument. The number $n$ is called the *degree* of $M$.

If we set $u_1 u_2 \cdots u_n := M(u_1, \ldots, u_n)$, there we get a product on the $n$-fold Cartesian product.

**Symmetry properties:** A given $n$-linear form $M : X \times \cdots \times X \longrightarrow Z$ is said to be *symmetric*, if $M(u_1, \ldots, u_n)$ is invariant under an arbitrary permutation of its $n$ arguments.

$M$ is said to be *anti-symmetric*, if $M(u_1, \ldots, u_n)$ is invariant under an even permutation of the $n$ arguments, and under an odd permutation, $M$ gets multiplied by $(-1)$.

**Determinants:** Let $A : X \longrightarrow X$ be a linear operator on a $n$-dimensional linear space $X$ over $\mathbb{K}$. Then one has

$$\boxed{M(Au_1, \ldots, Au_n) = (\det A) M(u_1, \ldots, u_n)}$$

for all anti-symmetric $n$-linear forms $M : X \times \cdots \times X \longrightarrow \mathbb{K}$ and all $u_1, \ldots, u_n \in X$.

**The tensor product of multilinear forms:** Let $M : X_1 \times \cdots \times X_m \longrightarrow \mathbb{K}$ be a $m$-linear form, and $N : Y_1 \times \cdots \times Y_n \longrightarrow \mathbb{K}$ be a $n$-linear form. Through the formula

$$\boxed{(M \otimes N)(u_1, \ldots, u_m, v_1, \ldots, v_n) := M(u_1, \ldots, u_m) N(v_1, \ldots, v_n)}$$

for all $u_j \in X_j$ and $v_k \in Y_k$ one obtains a $(m+n)$-linear form

$$M \otimes N : X_1 \times \ldots \times X_m \times Y_1 \times \ldots \times Y_n \longrightarrow \mathbb{K},$$

which is called the *tensor product* of $M$ and $N$.

**Properties:** For arbitrary multilinear forms $M, N, K$ and arbitrary numbers $\alpha, \beta \in \mathbb{K}$, one has:

(i) $(\alpha M+\beta N)\otimes K = \alpha(M\otimes K)+\beta(N\otimes K)$ and $K\otimes(\alpha M+\beta N) = \alpha(K\otimes M)+\beta(K\otimes N)$, and

(ii) $(M \otimes N) \otimes K = M \otimes (N \otimes K)$.

In (i) $M$ and $N$ must of course have the same degree for the formula to make sense.

### 2.4.2.1 Anti-symmetric multilinear forms

**The exterior product:** Let $X$ be a linear space over $\mathbb{K}$. By $\mathscr{A}^q(X)$ we denote the set of all anti-symmetric $q$-linear forms

$$M : X \times \ldots \times X \longrightarrow \mathbb{K}.$$

Moreover we set $\mathscr{A}^0(X) := \mathbb{K}$. The space $\mathscr{A}^q(X)$ is naturally a linear space over $\mathbb{K}$. For $M \in \mathscr{A}^q(X)$ and $N \in \mathscr{A}^p(X)$ with $q, p \geq 1$ we define

$$\boxed{(M \wedge N)(u_1, \ldots, u_{q+p}) := \sum_\pi (\operatorname{sgn} \pi) M(u_{\pi(1)}, \ldots, u_{\pi(q)}) N(u_{\pi(q+1)}, \ldots, u_{\pi(q+p)})}$$

for all $u_1, \ldots, u_n \in X$. The summation is extended over all possible permutations $\pi$ of the indices with $\pi(1) < \pi(2) < \cdots < \pi(q)$ and $\pi(q+1) < \cdots < \pi(q+p)$. We denote by $\operatorname{sgn} \pi$ the sign of a permutation $\pi$.

The anti-symmetric $(p+q)$-linear form $M \wedge N$ which is obtained in this manner is called the *exterior product* of $M$ and $N$. One has $M \wedge N \in \mathscr{A}^{p+q}(X)$.

For $\alpha,\ \beta \in \mathbb{K}$ and $M \in \mathscr{A}^q(X)$ with $q \geq 1$ we define

$$\boxed{\alpha \wedge M = M \wedge \alpha = \alpha M, \qquad \alpha \wedge \beta = \alpha\beta.}$$

*Example 1:* In case $q = p = 1$ one has

$$(M \wedge N)(u, v) = M(u)N(v) - M(v)N(u) \qquad \text{for all } u, v \in X.$$

If $q = 1$ and $p = 2$, then we get

$$(M \wedge N)(u, v, w) = M(u)N(v, w) - M(v)N(u, w) + M(w)N(u, v).$$

*Example 2:* Let $a, b, c \in X^*$. Then we get

$$(a \wedge (b \wedge c))(u, v, w) = a(u)\begin{vmatrix} b(v) & b(w) \\ c(v) & c(w) \end{vmatrix} - a(v)\begin{vmatrix} b(u) & b(w) \\ c(u) & c(w) \end{vmatrix} + a(w)\begin{vmatrix} b(u) & b(v) \\ c(u) & c(v) \end{vmatrix}.$$

The development theorem of determinants then yields

$$(a \wedge (b \wedge c))(u, v, w) = \begin{vmatrix} a(u) & a(v) & a(w) \\ b(u) & b(v) & b(w) \\ c(u) & c(v) & c(w) \end{vmatrix} \qquad \text{for all } u, v, w \in X.$$

Similarly one gets the same expression for $((a \wedge b) \wedge c)(u, v, w)$. This gives rise to an *associative law*

$$a \wedge (b \wedge c) = (a \wedge b) \wedge c.$$

We also abbreviate this by writing $a \wedge b \wedge c$.

If we use the tensor product, then we have the relation

$$\boxed{a \wedge b = a \otimes b - b \otimes a}$$

and

$$\boxed{a \wedge b \wedge c = a \otimes b \otimes c - a \otimes c \otimes b + b \otimes c \otimes a - b \otimes a \otimes c + c \otimes a \otimes b - c \otimes b \otimes a.}$$

This corresponds to the sum over all permutations of $a \otimes b \otimes c$, each permutation added with its sign.

**Properties:** For all anti-symmetric multilinear forms $M, N$ and $K$ of degree $\geq 0$ and all numbers $\alpha,\ \beta \in \mathbb{K}$ we have:

(i) $(\alpha M + \beta N) \wedge K = \alpha(M \wedge K) + \beta(N \wedge K)$,
$K \wedge (\alpha M + \beta N) = \alpha(K \wedge M) + \beta(K \wedge N)$;

(ii) $(M \wedge N) \wedge K = M \wedge (N \wedge K)$;

(iii) $\boxed{M \wedge N = (-1)^{qp} N \wedge M.}$

In (i) it is assumed that $M$ and $N$ have the same degree. In (iii) we let $q = \deg M$ and $p = \deg N$. This graded multiplication rule says that the commutativity or anticommutativity of the product $M \wedge N$ depends on the degrees of the factors.

**The algebra $\mathscr{A}(X)$:** The exterior direct sum

$$\mathscr{A}(X) := \bigoplus_{p=0}^{\infty} \mathscr{A}^p(X)$$

is with respect to the $\wedge$-multiplication an algebra over $\mathbb{K}$, which one calls the algebra of anti-symmetric multilinear forms over $X$. This algebra is graded by the linear spaces $\mathscr{A}^p(X)$. The elements of $\mathscr{A}(X)$ are sums

$$M_0 + M_1 + M_2 + \ldots$$

with $M_q \in \mathscr{A}^q(X)$, where at most finitely many of the $M_q$ are non-vanishing. The sum and $\wedge$-product are formed in the usual manner, taking care of the order of the factors.

*Example 3:*

$$(M_0 + M_1) \wedge (N_0 + N_1) = M_0 \wedge N_0 + (M_0 \wedge N_1 + M_1 \wedge N_0) + M_1 \wedge N_1.$$

Because $M_0,\ N_0 \in \mathbb{K}$, this expression is equal to $M_0N_0 + M_0N_1 + N_0M_1 + M_1 \wedge N_1$.

**Finite-dimensional spaces:** Let $b_1, \ldots, b_n$ be a basis of $X$. The basis which is dual to this in $X^*$ is denoted $b^1, \ldots, b^n$, i.e., we have

$$b^j(\alpha_1 b_1 + \ldots + \alpha_n b_n) = \alpha_j, \qquad j = 1, \ldots, n,$$

for all $\alpha_1, \ldots, \alpha_n \in \mathbb{K}$. Then one has

$$\boxed{b^j \wedge b^k = -b^k \wedge b^j}$$

for all $j, k = 1, \ldots, n$. In particular, $b^k \wedge b^k = 0$.

*Example 4:* For $n = 2$ and $q = 2$, all elements of $\mathscr{A}^2(X)$ are of the form

$$M = \alpha(b^1 \wedge b^2) + \beta(b^2 \wedge b^1) = (\alpha - \beta) b^1 \wedge b^2$$

with arbitrary numbers $\alpha,\ \beta \in \mathbb{K}$. This means that $\dim \mathscr{A}^2(X) = 1$. One can also write $M$ uniquely in the form

$$M = \frac{1}{2!}(\alpha_{12} b^1 \wedge b^2 + \alpha_{21} b^2 \wedge b^1) = \frac{1}{2!} \sum_{j,k=1}^{2} \alpha_{jk} b^j \wedge b^k,$$

where $\alpha_{jk}$ is anti-symmetric with respect to the indices, i.e., one has $\alpha_{jk} \in \mathbb{K}$ and

$$\alpha_{jk} = -\alpha_{kj} \qquad \text{for all } j, k.$$

*Example 5:* In case $\dim X = n$, the form

$$b^{j_1} \wedge \ldots \wedge b^{j_q} \tag{2.51}$$

is anti-symmetric with respect to all indices. All products in (2.51) with $j_1 < j_2 < \cdots < j_q$ and $j_k = 1, \ldots, n$ for all $k$ form a basis of $\mathscr{A}^q(X)$, and one has

$$\boxed{\dim \mathscr{A}^q(X) = \binom{n}{q}, \qquad \dim \mathscr{A}(X) = 2^n.}$$

Each $M \in \mathscr{A}^q(X)$ can be written uniquely in the form

$$\boxed{M = \frac{1}{q!} \alpha_{j_1 \ldots j_q} b^{j_1} \wedge \ldots \wedge b^{j_q},} \tag{2.52}$$

where $\alpha_{\ldots}$ are elements of $\mathbb{K}$ and are anti-symmetric with respect to all indices. Here the *Einstein summation convention* is being used, whereby like indices which occur both as superscripts and as subscripts are summed over.

Under a change of basis the coefficients $\alpha_{\ldots}$ of $M$ transform like a $q$-covariant, anti-symmetric tensor (cf. [212]).

**Applications to differential forms in $\mathbb{R}^n$:** Let $X = \mathbb{R}^n$. We choose the natural (canonical) basis

$$b_1 := (1, 0, \ldots, 0), \ \ldots, \ b_n := (0, 0, \ldots, 1)$$

in $\mathbb{R}^n$. The dual basis is denoted

$$\boxed{\mathrm{d}x^1, \ldots, \mathrm{d}x^n,}$$

i.e., for all $\alpha_1, \ldots, \alpha_n \in \mathbb{R}$ one has:

$$\boxed{\mathrm{d}x^j(\alpha_1 b_1 + \ldots + \alpha_n b_n) = \alpha_j, \qquad j = 1, \ldots, n.}$$

In all the formulas of Example 5 one just replaces the symbols $b^j$ by $\mathrm{d}x^j$. Then one refers to $M$ in (2.52) as a *differential form* of degree $q$ (with constant coefficients).

Differential forms of this kind play a fundamental role in modern analysis and geometry (cf. [212]).

*Example 6:* In $\mathbb{R}^2$ the canonical basis consists of $b_1 := (1,0)$ and $b_2 := (0,1)$. The dual basis $\mathrm{d}x^1,\ \mathrm{d}x^2$ is determined by the rule

$$\mathrm{d}x^1(\alpha b_1 + \beta b_2) := \alpha, \quad \mathrm{d}x^2(\alpha b_1 + \beta b_2) := \beta, \qquad \alpha, \beta \in \mathbb{R}.$$

For all $u, v \in \mathbb{R}^2$, one has

$$(\mathrm{d}x^j \wedge \mathrm{d}x^k)(u,v) = \mathrm{d}x^j(u)\mathrm{d}x^k(v) - \mathrm{d}x^j(v)\mathrm{d}x^k(u)$$

and $(\mathrm{d}x^j \otimes \mathrm{d}x^k)(u,v) = \mathrm{d}x^j(u)\mathrm{d}x^k(v)$. This yields

$$\mathrm{d}x^j \wedge \mathrm{d}x^k = \mathrm{d}x^j \otimes \mathrm{d}x^k - \mathrm{d}x^k \otimes \mathrm{d}x^j .$$

From this it follows that

$$\mathrm{d}x^1 \wedge \mathrm{d}x^2 = -\mathrm{d}x^2 \wedge \mathrm{d}x^1, \qquad \mathrm{d}x^1 \wedge \mathrm{d}x^1 = \mathrm{d}x^2 \wedge \mathrm{d}x^2 = 0.$$

All $\wedge$-products with more than two factors vanish.

The two-dimensional space $\mathscr{A}^1(\mathbb{R}^2)$ consists of the linear combinations

$$\beta \mathrm{d}x^1 + \gamma \mathrm{d}x^2, \qquad \beta, \gamma \in \mathbb{R},$$

while the one-dimensional space $\mathscr{A}^2(\mathbb{R}^2)$ consist of the expressions $\delta(\mathrm{d}x^1 \wedge \mathrm{d}x^2)$ with $\delta \in \mathbb{R}$. The algebra $\mathscr{A}(\mathbb{R}^2)$ consists of all expressions of the form

$$\alpha + \beta \mathrm{d}x^1 + \gamma \mathrm{d}x^2 + \delta(\mathrm{d}x^1 \wedge \mathrm{d}x^2), \qquad \alpha, \beta, \gamma, \delta \in \mathbb{R}.$$

#### 2.4.2.2 Covariant and contravariant tensors

Tensors play a fundamental role in differential geometry and in mathematical physics (cf. [212]).

**Tensors:** Let $X$ be a finite-dimensional linear space over $\mathbb{K}$. The set $\mathscr{T}_q^p(X)$ consists by definition of the set of all multilinear forms

$$M : X \times \ldots \times X \times X^* \times \ldots \times X^* \longrightarrow \mathbb{K},$$

where the space $X$ occurs in the Cartesian product $q$ times and the dual space $X^*$ occurs $p$ times. The elements of $\mathscr{T}_q^p(X)$ are referred to as *$q$-covariant, $p$-contravariant tensors* on $X$. Moreover one sets $\mathscr{T}_0^0 := \mathbb{K}$.

**Tensor products:** For all $M \in \mathscr{T}_q^p(X)$ and $N \in \mathscr{T}_s^r(X)$ with $p+q \geq 1$ and $r+s \geq 1$ we define the natural product ("natural" meaning independent of any choices)

$$\begin{aligned}&(M \otimes N)(u_1, \ldots, u_{q+s}, v_1, \ldots, v_{p+r})\\ &\qquad := M(u_1, \ldots, u_q, v_1, \ldots, v_p)N(u_{q+1}, \ldots, u_{q+s}, v_{p+1}, \ldots, v_{p+r})\end{aligned}$$

for all $u_j \in X$ and $v_k \in X^*$. This is the usual tensor product, where however the arguments are arranged in such a way that first the elements of $X$, then the elements of $X^*$ occur in the product. Moreover, let

$$\alpha \otimes M = M \otimes \alpha = \alpha M \qquad \text{for all } \alpha \in \mathbb{K},\ M \in \mathscr{T}_q^p(X)$$

with $p, q \geq 0$. In general one has:

$$\text{From } M \in \mathscr{T}_q^p(X) \quad \text{and} \quad N \in \mathscr{T}_s^r(X) \quad \text{it follows that} \quad M \otimes N \in \mathscr{T}_{q+s}^{p+r}(X).$$

**Einstein summation convention:** In what follows we shall always be assuming summation over like sub- and superscripts (upper and lower indices). This summation will be from 1 to $n$.

**Basis representation:** Let $b_1, \ldots, b_n$ be a basis over $\mathbb{K}$ of the linear space $X$. Every tensor $M \in \mathscr{T}_q^p(X)$ with $p + q \geq 1$ can be written *uniquely* in the form

$$M = t_{i_1 \ldots i_q}^{j_1 \ldots j_p} b^{i_1} \otimes \ldots \otimes b^{i_q} \otimes b_{j_1} \otimes \ldots \otimes b_{j_p} \tag{2.53}$$

with coefficients $t_{\cdots}^{\cdots}$ in $\mathbb{K}$. Here $b^1, \ldots, b^n$ denotes the basis which is dual to $b_1, \ldots, b_n$. In (2.53) we identify $b_j$ with the linear form $b_j(u^*) := u^*(b_j)$ for all $u^* \in X^*$.

**Change of basis:** For a basis transformation

$$b_{j'} = A_{j'}^j b_j$$

one has the transformation formula

$$b^{k'} = A_k^{k'} b^k$$

for the dual basis. Here $A_k^{k'}$ is uniquely determined as a solution of the system of equations

$$A_k^{k'} A_{k'}^j = \delta_k^j, \qquad j, k = 1, \ldots, n.$$

**Theorem:** The coordinates $t_{i_1 \ldots i_q}^{j_1 \ldots j_p}$ of a tensor transform under a change of basis the same way as

$$b^{j_1} \otimes \ldots \otimes b^{j_p} \otimes b_{i_1} \otimes \ldots \otimes b_{i_q}.$$

*Example 1:* Let $M = t_r^s b^r \otimes b_s$. Then $M = t_{r'}^{s'} b_{s'} \otimes b^{r'}$ with

$$t_{r'}^{s'} = A_{r'}^r a_s^{s'} t_r^s.$$

**Contraction:** From $M$ in (2.53) one gets a new tensor, by setting one of the upper indices equal to one of the lower ones in the coordinate $t_{\cdots}^{\cdots}$ (for example $j$), and then eliminating the basis vectors $b_j$ and $b^j$ from the linear combination. This operation is independent of the chosen basis and is referred to as the *contraction* of $M$ on the index $j$.

*Example 2:* From $M = t_{jk}^i b^j \otimes b^k \otimes b_i$ one gets $N = t_{ik}^i b^k$.

**The algebra of covariant and contravariant tensors:** The (outer) direct sum

$$\mathscr{T}(X) := \bigoplus_{p,q=0}^{\infty} \mathscr{T}_q^p(X)$$

becomes an algebra with the operations $+$ and $\otimes$.

*Example 3:* From $M = t_k^j b^k \otimes b_j$ and $N = s_k^j b^k \otimes b_j$ it follows that

$$M + N = (t_k^j + s_k^j)(b^k \otimes b_j)$$

and

$$M \otimes N = t_k^j s_q^p (b^k \otimes b^q \otimes b_j \otimes b_p).$$

## 2.4.3 Universal products

### 2.4.3.1 The tensor product of linear spaces

Let $X$ and $Y$ be linear spaces over $\mathbb{K}$. To every $u \in X$ we associate via

$$\boxed{u(u^*) := u^*(u) \qquad \text{for all } u^* \in X^*} \tag{2.54}$$

a linear form on $X^*$. Similarly we define linear forms on $Y^*$. By

$$\boxed{u \otimes v}$$

we denote the $\otimes$-product of these linear forms, i.e., we have the relation

$$(u \otimes v)(u^*, v^*) = u(u^*)v(v^*) \qquad \text{for all } u^* \in X^*,\ v^* \in Y^*.$$

**The tensor product $X \otimes Y$:** The set of all finite sums

$$u_1 \otimes v_1 + \ldots + u_k \otimes v_k \tag{2.55}$$

with $u_j \in X$, $v_j \in Y$ for all $j$ and $k = 1, \ldots$ forms a linear space over $\mathbb{K}$, which is called the *tensor product* of $X$ and $Y$.

Two sums of the form (2.55) are by definition identical, if they have the same values as bilinear forms. This can occur for quite different linear combinations.

**Basis theorem:** If $u_1, \ldots, u_k$ are linearly independent in $X$ and $v_1, \ldots, v_m$ are linearly independent in $Y$, then the set of all products

$$u_\alpha \otimes v_\beta, \qquad \alpha = 1, \ldots, k, \quad \beta = 1, \ldots, m$$

are linearly independent in $X \otimes Y$. Moreover, these products form a basis of $X \otimes Y$, if $u_1, \ldots, u_k$ is a basis of $X$ and $v_1, \ldots, v_m$ is a basis of $Y$.

**The tensor product $X_1 \otimes X_2 \otimes \cdots \otimes X_r$:** If $X_1, \ldots, X_r$ are linear spaces over $\mathbb{K}$, then $X_1 \otimes \cdots \otimes X_r$ denotes the set of all finite sums of terms of the form

$$u_1 \otimes u_2 \otimes \ldots \otimes u_r$$

with $u_j \in X_j$ for all $j$. This space is called the *tensor product* of the $X_j$, $j = 1, \ldots, r$.

The basis theorem above for two spaces holds analogously for $r$ factors, and one has

$$\dim (X_1 \otimes X_2 \otimes \ldots \otimes X_r) = \dim X_1 \cdot \dim X_2 \cdot \ldots \cdot \dim X_r.$$

The product on the right is by definition zero if one of the factors (i.e., one of the dimensions $\dim X_j$) vanishes.

**The universal property of the tensor product:** Let $X_1, \ldots, X_n$ and $Z$ be linear spaces over $\mathbb{K}$. If $M : X_1 \times \cdots \times X_n \longrightarrow Z$ is an $n$-linear mapping, then there is a linear map $L : X_1 \otimes \cdots \otimes X_n \longrightarrow Z$ with

$$\boxed{M(u_1, \ldots, u_n) = L(u_1 \otimes \ldots \otimes u_n)}$$

for all $u_j \in X_j$, $j = 1, \ldots, n$. In this manner all products $u_1 u_2 \cdots u_n := M(u_1, \ldots, u_n)$ can be reduced to the tensor product.[26]

[26] The tensor product of linear spaces can equivalently be described in terms of factor spaces (cf. [212]).

**Complexification of a linear space:** If $X$ is a real linear space, then one calls the tensor product $X \otimes \mathbb{C}$, a complex linear space, the *complexification* of the real linear space $X$. Because $u \otimes (\alpha + \beta \mathrm{i}) = (\alpha u) \otimes 1 + (\beta u) \otimes \mathrm{i}$, $X \otimes \mathbb{C}$ consists of all expressions of the form

$$u \otimes 1 + v \otimes \mathrm{i}$$

with $u, v \in X$. One has $u \otimes 1 + v \otimes \mathrm{i} = u' \otimes 1 + v' \otimes \mathrm{i}$ if and only if $u = u'$ and $v = v'$. The formula $\varphi(u) := u \otimes 1$ gives rise to a injective linear map $\varphi : X \longrightarrow X \otimes \mathbb{C}$. Hence $u$ can be identified with $u \otimes 1$. Moreover, one has $\dim (X \otimes \mathbb{C}) = \dim X$ (note that on the left hand side we are talking about dimensions of complex vector space, on the right hand side of real ones).

### 2.4.3.2 The tensor algebra of a linear space

Let $X$ be a linear space over $\mathbb{K}$. We denote by $\otimes^p X$ the $p$-fold tensor product of $X$ with itself, i.e., $X \otimes \cdots \otimes X$. Moreover we set $\otimes^0 X := \mathbb{K}$. The outer direct sum

$$\boxed{\otimes(X) := \bigoplus_{p=0}^{\infty} (\otimes^p X)}$$

consists of all finite sums $M_0 + M_1 + \cdots$ of $p$-linear forms $M : X \times \cdots \times X \longrightarrow \mathbb{K}$, for which a $\otimes$-product is defined in 2.4.2. In this way $\otimes(X)$ becomes an algebra over $\mathbb{K}$, which is called the *tensor algebra* (over $\mathbb{K}$) of the linear space $X$.

*Example:* For $u, v, w \in X$ one has

$$(2 + u) \otimes (3 + v \otimes w) = 6 + 3u + 2v \otimes w + u \otimes v \otimes w.$$

### 2.4.3.3 The exterior product of a linear space (Grassmann algebra)

**The exterior product:** Let $X$ be a linear space over $\mathbb{K}$. For $u, v \in X$ we let

$$\boxed{u \wedge v}$$

denote the $\wedge$-product in the sense of linear forms as in (2.54), i.e., such that

$$(u \wedge v)(u^*, v^*) = u(u^*)v(v^*) - u(v^*)v(u^*)$$

for all $u^*, v^* \in X^*$. This means $u \wedge v = u \otimes v - v \otimes u$.

**The exterior product $X \wedge X$:** The set of all finite sums

$$u_1 \wedge v_1 + \ldots + u_k \wedge v_k$$

with $u_j, v_j \in X$ for all $j$ and $k = 1, \ldots$ forms a linear space over $\mathbb{K}$, which is called the *exterior product* of $X$ with itself.

$X \wedge X$ is the subspace of $X \otimes X$ consisting of exactly the anti-symmetric bilinear forms $M : X \times X \longrightarrow \mathbb{K}$ of $X \otimes X$.

**Basis theorem:** If $u_1, \ldots, u_k$ are linearly independent in $X$, then all the products

$$u_i \wedge u_j \tag{2.56}$$

with $i < j$ and $i, j = 1, \ldots, k$ are linearly independent. If $u_1, \ldots, u_k$ in fact form a basis of $X$, then the products in (2.56) form a basis of $X \wedge X$.

**The exterior product $\wedge^p X$:** Let $p = 2, 3, \ldots$ The set of all finite sums of products of the form

$$u_{j_1} \wedge u_{j_2} \wedge \ldots \wedge u_{j_p} \tag{2.57}$$

form a linear space over $\mathbb{K}$, which is called the *exterior product* $\wedge^p X$.

Moreover we set $\wedge^0 X := \mathbb{K}$ and $\wedge^1 X := X$.

**Basis theorem:** If $u_1, \ldots, u_k$ are linearly independent in $X$, then the set of all products of the form (2.57) with $j_1 < j_2 < \cdots < j_p$ and $j_1, \ldots, j_p = 1, \ldots, k$ are linearly independent. If $u_1, \ldots, u_k$ forms a basis of $X$, then the products of the form (2.57) form a basis of $\wedge^p X$.

**Criterion for dependency:** The elements $u_1, \ldots, u_n$ in $X$ are linearly independent if $u_1 \wedge \cdots \wedge u_n = 0$.

**The universality of the exterior product:** If $X$ and $Z$ are linear spaces over $\mathbb{K}$, and $M : X \times \cdots \times X \longrightarrow Z$ is an anti-symmetric $p$-linear form, then there is a linear mapping $L : \wedge^p X \longrightarrow Z$ with

$$\boxed{M(u_1, \ldots, u_p) = L(u_1 \wedge \ldots \wedge u_p) \qquad \text{for all } u_j \in X.}$$

**The exterior algebra:** The *outer direct sum*

$$\boxed{\wedge(X) := \bigoplus_{p=0}^{\infty} (\wedge^p X)}$$

consists of all finite sums $M_0 + M_1 + \cdots$ of anti-symmetric $p$-linear forms $M_p : X \times \cdots \times X \longrightarrow \mathbb{K}$, on which a $\wedge$-product is defined in 2.4.2. In this manner $\wedge(X)$ becomes an algebra over $\mathbb{K}$, which is called the *exterior algebra* of the space $X$.

If $X$ is finite-dimensional with $\dim X = n$, then

$$\dim \wedge^p X = \binom{n}{p} \quad \text{and} \quad \dim \wedge (X) = 2^n.$$

*Example:* For $u, v, w \in X$ one has

$$(2 + u) \wedge (3 + v \wedge w) = 6 + 3u + 2v \wedge w + u \wedge v \wedge w.$$

#### 2.4.3.4 The inner algebra of a linear space (Clifford algebra)

Let $X$ be an $n$-dimensional linear space over $\mathbb{K}$, and $B : X \times X \longrightarrow \mathbb{K}$ a bilinear form on $X$. Our goal is to introduce the so-called *inner multiplication* $u \vee w$ on $X$ with the property

$$\boxed{u \vee w + w \vee u = 2B(u, w)} \tag{2.58}$$

for all $u, w \in X$. In addition, it should hold that

$$\alpha \vee u = u \vee \alpha = \alpha u \tag{2.59}$$

for all $\alpha \in \mathbb{K}$ and $u \in X$.

Clifford algebras play a central role in modern physics, in order to describe the spin of elementary particles (cf. 3.9.6).

**Existence theorem:** There is an algebra $\mathscr{C}(X)$ over $\mathbb{K}$, whose multiplication is denoted by $\vee$, such that the following conditions are satisfied:

(i) $\mathscr{C}(X)$ contains $\mathbb{K}$ and $X$ and (2.58) and (2.59) hold.

(ii) If $b_1, \ldots, b_n$ is a basis, then the ordered products

$$1, b_1, \ldots, b_n, \qquad b_{i_1} \vee b_{i_2} \vee \ldots \vee b_{i_r}, \qquad r = 2, \ldots, n \tag{2.60}$$

form a basis of $\mathscr{C}(X)$, if $i_1 < i_2 < \cdots < i_r$ and $i_k = 1, \ldots, n$ for all $k$.

From (ii) we see that every element of $\mathscr{C}(X)$ can be written as a linear combination of elements as in (2.60) with uniquely determined coefficients in $\mathbb{K}$. The number of these elements is $2^n$. Hence we have

$$\boxed{\dim \mathscr{C}(X) = 2^n.}$$

**Uniqueness theorem:** The algebra $\mathscr{C}(X)$ is uniquely determined up to isomorphism by the conditions (i) and (ii). We call $\mathscr{C}(X)$ the *Clifford algebra* of $X$ with respect to the bilinear form $B(.,.)$.

**Universal property of the Clifford algebra:** Let $\mathscr{A}$ be an algebra over $\mathbb{K}$, whose multiplication we denote by $\vee$, such that $\mathbb{K}$ and $X$ are contained in $\mathscr{A}$ and the multiplication rules (2.58) and (2.59) hold.

Then there is an algebra homomorphism from $\mathscr{C}(X)$ to $\mathscr{A}$.

*Example 1* (quaternions): Let $b_1$, $b_2$ be a basis of $\mathbb{R}^2$. Then the Clifford algebra $\mathscr{C}(\mathbb{R}^2)$ consists of all expressions of the form

$$\alpha + \beta b_1 + \gamma b_2 + \delta b_1 \vee b_2$$

with $\alpha, \beta, \gamma, \delta \in \mathbb{R}$. The multiplication table is given by the rules

$$b_j \vee b_k + b_k \vee b_j = 2B(b_j, b_k), \qquad j, k = 1, 2.$$

The sum is formed in the natural way.

In the special case that $B(b_j, b_k) = -\delta_{jk}$, one has

$$b_j \vee b_k + b_k \vee b_j = -2\delta_{jk}, \qquad j, k = 1, 2.$$

In this case the algebra $\mathscr{C}(\mathbb{R}^2)$ is isomorphic to the space of *quaternions* $\mathbb{H}$. Classically, the quaternions (R. Hamilton's claim to fame) are given by the elements

$$\alpha + \beta \mathbf{i} + \gamma \mathbf{j} + \delta \mathbf{k}$$

with $\alpha, \beta, \gamma, \delta \in \mathbb{R}$, together with the multiplication table

$$\mathbf{i}^2 = \mathbf{j}^2 = \mathbf{k}^2 = -1$$
$$\mathbf{ij} = -\mathbf{ji} = \mathbf{k}, \quad \mathbf{jk} = -\mathbf{kj} = \mathbf{i}, \quad \mathbf{ki} = -\mathbf{ik} = \mathbf{j}.$$

The isomorphism of $\mathscr{C}(\mathbb{R}^2)$ to $\mathbb{H}$ is obtained by mapping $b_1 \mapsto \mathbf{i}$, $b_2 \mapsto \mathbf{j}$, $b_1 \vee b_2 \mapsto \mathbf{k}$.

*Example 2* (Grassmann algebra): Let $b_1, \ldots, b_n$ be basis of a linear space $X$ over $\mathbb{K}$. If we choose $B \equiv 0$ (i.e., $B(b_j, b_k) = 0$ for all $j, k$), then in the Clifford algebra $\mathscr{C}(X)$ we have the multiplication rule

$$b_j \vee b_k + b_k \vee b_j = 0 \qquad \text{for all } j, k = 1, \ldots, n.$$

$\mathscr{C}(X)$ is isomorphic to the Grassmann algebra $\wedge(X)$, simply by replacing the $\vee$ product symbol by $\wedge$.

*Example 3* (the *spinor algebra* of Dirac): Let $b_1, \ldots, b_4$ be a basis of a complex linear space $X$. We choose the Minkowski metric

$$g_{jk} = \begin{cases} 1 & \text{for } j = k = 1,2,3, \\ -1 & \text{for } j = k = 4, \\ 0 & \text{for } j \neq k \end{cases}$$

and set $B(b_j, b_k) := g_{jk}$. In the corresponding Clifford algebra $\mathscr{C}(X)$ one then has the multiplication rule

$$b_j \vee b_k + b_k \vee b_j = 2g_{jk}, \qquad j, k = 1, 2, 3, 4.$$

$\mathscr{C}(X)$ is isomorphic to the algebra $M(4,4)$ of complex $(4\times 4)$-matrices. This isomorphism is given by the map $b_j \mapsto \gamma_j$, where the $\vee$-product is replaced by the matrix product. In particular one has

$$\gamma_j \gamma_k + \gamma_k \gamma_j = 2g_{jk}, \qquad j, k = 1, 2, 3, 4,$$

with the *Pauli matrices*

$$\sigma_1 = \begin{pmatrix} 0 & 1 \\ 1 & 0 \end{pmatrix}, \quad \sigma_2 = \begin{pmatrix} 0 & -\mathrm{i} \\ \mathrm{i} & 0 \end{pmatrix}, \quad \sigma_3 = \begin{pmatrix} 1 & 0 \\ 0 & -1 \end{pmatrix}, \quad \sigma_4 = I = \begin{pmatrix} 1 & 0 \\ 0 & 1 \end{pmatrix},$$

and the *Dirac matrices*

$$\gamma_j = \mathrm{i}\begin{pmatrix} 0 & -\sigma_j \\ \sigma_j & 0 \end{pmatrix}, \quad j = 1, 2, 3, \qquad \gamma_4 = \mathrm{i}\begin{pmatrix} 0 & \sigma_4 \\ \sigma_4 & 0 \end{pmatrix}.$$

These matrices play a fundamental role in the formulation of the *Dirac equation* for the relativistic electron. From the Dirac equation, the spin of the electron *follows* as a property of the solution (cf. 3.9.6).

### 2.4.4 Lie algebras

**Definition:** A *Lie algebra* over $\mathbb{K}$ is a linear space $\mathscr{L}$ over $\mathbb{K}$, in which for every ordered pair $(A, B)$ of elements in $\mathscr{L}$ there is an element in $\mathscr{L}$, denoted $[A, B]$ and referred to as the *bracket* of $A$ and $B$, such that for all $A, B, C \in \mathscr{L}$ and $\alpha, \beta \in \mathbb{K}$, one has:

(i) $[\alpha A + \beta B, C] = \alpha[A, C] + \beta[B, C]$ (linearity),

(ii) $[A, B] = -[B, A]$ (anti-commutativity), and

(iii) $[A, [B, C]] + [B, [C, A]] + [C, [A, B]] = 0$ (Jacobi identity).

The Jacobi identity (iii) is a replacement for the missing associativity of the Lie product $[A, B]$.

*Example 1:* If $\mathrm{gl}(X)$ denotes the set of all linear operators $A : X \longrightarrow X$ on a linear space $X$ over $\mathbb{K}$, we can make $\mathrm{gl}(X)$ to a Lie algebra over $\mathbb{K}$ by defining the product as

$$[A, B] := AB - BA. \tag{L}$$

*Example 2:* The set $\mathrm{gl}(n, \mathbb{R})$ of all real $(n \times n)$-matrices is a real Lie algebra over $\mathbb{R}$ with respect to (L).

**The Virasoro algebra:** Let $C^\infty(S^1)$ denote the linear space[27] of all functions $f : S^1 \longrightarrow \mathbb{C}$ on the unit circle $S^1 := \{z \in \mathbb{C} : |z| = 1\}$, which are holomorphic in a neighborhood of the origin. We set

$$L_n(f) := -z^{n+1}\frac{\mathrm{d}f}{\mathrm{d}z}, \qquad n = 0, \pm 1, \pm 2, \ldots$$

If $W$ denotes the complex linear hull of all $L_n$, then $W$ is with respect to the bracket

$$[L_n, L_m] = (n-m)L_{n+m}, \qquad n, m = 0, \pm 1, \pm 2, \ldots$$

an infinite-dimensional complex Lie algebra. One has $[L_n, L_m] = L_n L_m - L_m L_n$.

We choose an one-dimensional complex linear space $Y := \mathrm{span}\{Q\}$. Then the outer direct sum $\mathrm{Vir} := W \oplus Y$ with respect to the product

$$\boxed{\begin{aligned} &[L_n, L_m] = (n-m)L_{n+m} + \delta_{n,-m}\frac{n^3-n}{12}Q, \qquad n, m = 0, \pm 1, \ldots, \\ &[L_n, Q] = 0 \end{aligned}} \tag{Vir}$$

becomes an infinite-dimensional complex Lie algebra, which is called the *Virasoro algebra*; this is a *central extension*[28] of $W$.

The Virasoro algebra plays an exceptionally important role in modern string theory and conformal theory.

**The Heisenberg algebra:** Let $X$ be a complex linear space which is the linear hull of linearly independent elements $b, a_0, a_{\pm 1}, a_{\pm 2}, \ldots$ Then $X$ becomes an infinite-dimensional complex Lie algebra by means of the bracket

$$[a_n, a_m] = m\delta_{n,-m}a_0, \quad [b, a_n] = 0, \qquad n, m = 0, \pm 1, \pm 2, \ldots,$$

which is called the *Heisenberg algebra*.

A number of important Lie algebras and applications of these to geometry and modern high-energy physics is contained in [212].

## 2.4.5 Superalgebras

A *superalgebra* is an algebra $\mathscr{A}$ with a decomposition $\mathscr{A} = \mathscr{A}_0 \oplus \mathscr{A}_1$, such that the product in $\mathscr{A}$ respects the grading, i.e., one has

(a) From $u, \ v \in \mathscr{A}_0$ it follows that $uv \in \mathscr{A}_0$.

(b) From $u, \ v \in \mathscr{A}_1$ it follows that $uv \in \mathscr{A}_0$.

(c) From $u \in \mathscr{A}_0, \ v \in \mathscr{A}_1$ or $u \in \mathscr{A}_1, \ v \in \mathscr{A}_0$ it follows that $uv \in \mathscr{A}_1$.

A superalgebra is said to be *supercommutative*, if

$$\boxed{uv = (-1)^{jk}vu \qquad \text{for all } u \in \mathscr{A}_j, \ v \in \mathscr{A}_k, \ j, k = 0, 1.}$$

Supercommutative algebras play an important role in the modern supersymmetric theory of elementary particles. The commutative elements in $\mathscr{A}_0$ correspond to *bosons*

[27] The linear structure is the natural one: the sum of two such functions is the function whose value is the sum, i. e., $(f+g)(z) = f(z) + g(z)$, and similarly with scalar multiplication. The reader may easily verify all the necessary properties for this to make $C^\infty(S^1)$ a linear space.

[28] This abstract notion is a standard one in the theory of algebras; in the case at hand it is defined precisely by the expressions in (Vir) involving $Q$.

(particles with integer spin, for example photons) and the anti-commutative elements in $\mathscr{A}_1$ correspond to the *fermions* (particles with half-integer spin, for example electrons).

*Example:* The Grassmann algebra is made into a supercommutative superalgebra by the gradation

$$\wedge(X) = \bigoplus_{p=0}^{\infty} \wedge^{2p}(X) \oplus \left( \bigoplus_{p=0}^{\infty} \wedge^{2p+1} X \right).$$

## 2.5 Algebraic structures

Real numbers can be added and multiplied. Operations of this sort can however be declared for many other mathematical objects. This leads to the notions of group, ring and field, which arose in the context of the solutions of algebraic equations and the solution of number theoretic as well as geometric problems in the nineteenth century.

### 2.5.1 Groups

Groups are sets on which a product $gh$ (of elements $g, h$ of the set) has been declared. One uses groups to describe the geometric phenomenon of symmetry mathematically.

**Definition:** A *group* $G$ is a set, in which there is a product $gh$ (also referred to as the *group operation*) assigned to any ordered pair $(g, h)$ of elements in $G$, such that the following hold:

(i) $g(hk) = (gh)k$ for all $g, h, k \in G$ (associative law).

(ii) There is exactly one element $e$ such that for all $g \in G$, the relation $eg = ge = g$ is satisfied (neutral element).

(iii) For every $g \in G$ there is exactly one element $h \in G$ with $gh = hg = e$. Instead of $h$ one writes for this element $g^{-1}$ (inverse element).

A group $G$ is said to be *commutative* (or Abelian), if the commutative law $gh = hg$ is satisfied for all $g, h \in G$.

*Example 1* (groups of numbers): The set of all non-vanishing real numbers forms, with respect to multiplication as the product, a commutative group, which is called the *multiplicative group* of real numbers.

*Example 2* (matrix groups): The set $GL(n, \mathbb{R})$ of all real $(n \times n)$-matrices $A$ with non-vanishing determinant form, with the usual matrix multiplication $AB$ as the group operation, a group; this group is non-commutative for $n \geq 2$. The neutral element in this group is the identity matrix $E$. This group is referred to as the *general linear group* (over $\mathbb{R}$), hence the notation.

**Symmetry** (rotation group): The set $\mathscr{D}$ of all rotations around a fixed point $O$ of three-dimensional space form a non-commutative group, which is referred to as the three-dimensional *rotation group*. The group operation is given by the *composition* of two roations (see below). The neutral element is the transformation which acts trivially on all points, while the inverse of a rotation is the inverse rotation.

The intuitive symmetry group of a ball $B$ with center at $O$ can be described group-theoretically as the set of elements of $\mathscr{D}$ which map $B$ into itself.

**Transformation groups:** If $X$ is a non-empty set, the set of all bijective maps $g : X \longrightarrow X$ forms a group $G(X)$. The group operation corresponds to the *composition* of maps, i.e., for $g, h \in G(X)$, and all $x \in X$,

$$\boxed{(gh)(x) := g(h(x)).}$$

The neutral element corresponds to the identity map $\mathrm{id} : X \longrightarrow X$ with $\mathrm{id}(x) = x$ for all $x \in X$. Moreover, the inverse element $g^{-1}$ of $g$ is the inverse map (see 4.3.3), which exists because $g$ is bijective.

**Permutation groups:** Let $X = \{1, \ldots, n\}$. The set of all bijective maps $\pi : X \longrightarrow X$ is called the *symmetric group* and denoted $\mathscr{S}_n$. One also refers to $\mathscr{S}_n$ as the *permutation group* on $n$ letters. Every element $\pi \in \mathscr{S}_n$ can be represented by a symbol

$$\pi = \begin{pmatrix} 1 & 2 & \ldots & n \\ i_1 & i_2 & \ldots & i_n \end{pmatrix},$$

which tells us that the element $k$ is mapped under $\pi$ to $i_k$, i.e., $\pi(k) = i_k$ for all $k$. The product of two permutations $\pi_2\pi_1$ corresponds to the composition of the two permutations, i.e., first $\pi_1$ is applied to $X$, then $\pi_2$. The neutral element $e$ and the inverse element $\pi^{-1}$ to $\pi$ are given by

$$e = \begin{pmatrix} 1 & 2 & \ldots & n \\ 1 & 2 & \ldots & n \end{pmatrix} \quad \text{and} \quad \pi^{-1} = \begin{pmatrix} i_1 & i_2 & \ldots & i_n \\ 1 & 2 & \ldots & n \end{pmatrix}.$$

The number of elements[29] of $\mathscr{S}_n$ is $n!$ In the special case in which $n = 3$, one gets for example for the two elements

$$\pi_2 = \begin{pmatrix} 1 & 2 & 3 \\ 1 & 3 & 2 \end{pmatrix}, \qquad \pi_1 = \begin{pmatrix} 1 & 2 & 3 \\ 3 & 2 & 1 \end{pmatrix}$$

the product

$$\pi_2\pi_1 = \begin{pmatrix} 1 & 2 & 3 \\ 2 & 3 & 1 \end{pmatrix},$$

since $\pi_1$ maps 1 to 3 and $\pi_2$ maps 3 to 2, i.e., $(\pi_2\pi_1)(1) = \pi_2(\pi_1(1)) = \pi_2(3) = 2$. For $n \geq 2$ $\mathscr{S}_n$ is not commutative.

**Transpositions:** A *transposition* $(km)$ with $k \neq m$ is the permutation which maps $k$ to $m$ and $m$ to $k$, fixing all other elements of $X$. Every permutation $\pi$ can be written (in more than one way!) as a product of $r$ transpositions, where $r$ is either even or odd (i.e., independent of the particular choice of product decomposition). Therefore we can define the *sign* of $\pi$ by the rule

$$\operatorname{sgn} \pi := (-1)^r.$$

For $\pi_1, \pi_2 \in \mathscr{S}_n$ one has

$$\boxed{\operatorname{sgn}(\pi_1\pi_2) = \operatorname{sgn} \pi_1 \operatorname{sgn} \pi_2.} \tag{2.61}$$

In the sense of section 2.5.1.2 this means that the map $\pi \mapsto \operatorname{sgn} \pi$ is a homomorphism of the permutation group $\mathscr{S}_n$ to the multiplicative group of real numbers.

A permutation $\pi$ is *even* (resp. *odd*), $\operatorname{sgn} \pi = 1$ (resp. $\operatorname{sgn} \pi = -1$). Every transposition is odd.

[29]For a finite group $G$, the number of elements in the group is called the *order* of the group and denoted $\operatorname{ord} G$.

**Cycles:** Let $(abc)$ denote the permutation which maps $a$ to $b$, $b$ to $c$ and $c$ to $a$, fixing the rest of $X$. One says that this element *cyclically permutes* the set $\{a, b, c\}$. Similarly one can define a *cycle* $(z_1 z_2 \cdots z_k)$. Every permutation $\pi$ can be written uniquely (up to a reordering) as a product of cycles. For example one gets for the $3! = 6$ elements of $\mathscr{S}_3$ as the following cycles:

$$(1), \quad (12), \quad (13), \quad (23), \quad (123), \quad (132).$$

#### 2.5.1.1 Subgroups

**Definition:** A subset $H \subset G$ of a group $G$ is called a *subgroup*, if $H$ is a group with respect to the multiplication induced on $H$ by the multiplication in $G$. This is equivalent to the condition that for all $g, h \in H$, one has $gh^{-1} \in H$ (closure of the operation).

**Normal subgroups:** A *normal subgroup* $H$ of $G$ is a subgroup $H$ of $G$ which the additional property that

$$\boxed{ghg^{-1} \in H \qquad \text{for all} \;\; g \in H, \;\; h \in H.}$$

The group $G$ itself, as well as the *trivial subgroup* $\{e\}$ are always normal subgroups; these are referred to as trivial normal subgroups.

Every subgroup of a commutative group is normal (since $ghg^{-1} = gg^{-1}h = h \in H$).

**Simple groups:** A group $G$ is said to be *simple*, if it has only the trivial normal subgroups.

*Example 1:* All positive real numbers form a subgroup (in fact a normal subgroup) of the multiplicative group of all non-vanishing real numbers.

**The order theorem of Lagrange:** The order (defined in the last footnote) of every subgroup of a finite group is a divisor of the order of the group.

*Example 2* (permutations): The set of even permutations of $\mathscr{S}_n$ form a normal subgroup $\mathscr{A}_n$, which is called the *alternating group on $n$ letters*. For $n \geq 2$ one has

$$\operatorname{ord} \mathscr{A}_n = \frac{1}{2} \operatorname{ord} \mathscr{S}_n = \frac{n!}{2}.$$

(i) The group $\mathscr{S}_2$ consists of the elements $(1)$, $(12)$ and has only the trivial normal subgroups $\mathscr{A}_2 = (1)$ and itself.

(ii) The six subgroups of $\mathscr{S}_3$ are:

$$\begin{array}{ll} \mathscr{S}_3: & (1),\ (12),\ (13),\ (23),\ (123),\ (132),\ \mathscr{E}: (1), \\ \mathscr{A}_3: & (1),\ (123),\ (132), \\ \mathscr{S}_2: & (1),\ (12), \quad \mathscr{S}_2': (1),\ (13), \quad \mathscr{S}_2'': (1),\ (23). \end{array}$$

Here $\mathscr{A}_3$ is the only non-trivial normal subgroup of $\mathscr{S}_3$.

(iii) $\mathscr{S}_4$ has $\mathscr{A}_4$ and the commutative *Klein four-group*:

$$\mathscr{K}_4: \; (1),\ (12)(34),\ (13)(24),\ (14)(23)$$

as normal subgroups.

(iv) For $n \geq 5$, $\mathscr{A}_n$ is the only non-trivial normal subgroup of $\mathscr{S}_n$, and the group $\mathscr{A}_n$ is simple.

**Additive groups:** An *additive* group is a set $G$ with an operation associating to every ordered pair $(g, h)$ of elements in $G$ a *sum* $g + h \in G$, such that

(i) $g + (h + k) = (g + h) + k$ for all $g, h, k \in G$ (associative law).

(ii) There is precisely one element, denoted 0, such that for all $g \in G$ we have $0 + g = g + 0 = g$ (neutral element).

(iii) For every $g \in G$ there is exactly one element $h \in G$ such that $h + g = g + h = 0$. Instead of $h$ we write for this element $-g$ (inverse element).

(iv) $g + h = h + g$ for all $g, h \in G$ (commutative law).

Hence an additive group is nothing but a commutative group in which the group operation is written + and the neutral element is written 0.

*Example 3:* The set $\mathbb{R}$ of all real numbers is an additive group with respect to the usual addition. The set $\mathbb{Z}$ of integers is an additive subgroup of $\mathbb{R}$.

*Example 4:* Every linear space is an additive group.

#### 2.5.1.2 Group homomorphisms

**Definition:** A *homomorphism*[30] between two group $G$ and $H$ is a map $\varphi : G \longrightarrow H$ which respects the group operations in both groups, i.e., for which

$$\boxed{\varphi(gh) = \varphi(g)\varphi(h) \qquad \text{for all } g, h \in G.}$$

The bijective homomorphism are referred to as group *isomorphisms*.

Two groups $G$ and $H$ are said to be *isomorphic*, if there is an isomorphism $\varphi : G \longrightarrow H$. Isomorphic groups have the same structure and may be identified.[31]

Surjective (resp. injective) homomorphisms are also referred to as *epimorphism* (resp. monomorphisms). An *automorphism* of a group $G$ is an isomorphism of $G$ to itself.

*Example 1:* The group $G := \{1, -1\}$ is isomorphic to the group $H := \{E, -E\}$ with

$$E := \begin{pmatrix} 1 & 0 \\ 0 & 1 \end{pmatrix}.$$

The isomorphism $\varphi : G \longrightarrow H$ is given by $\varphi(\pm 1) := \pm E$.

**Group symmetries:** With respect to the composition, all automorphisms of a group $G$ form a new group, which is called the *automorphism group* of $G$ and denoted $\mathrm{Aut}(G)$. This group describes the symmetries of the group $G$.

**Inner automorphisms:** Let $g$ be a fixed element of a group $G$. We set

$$\varphi_g(h) := ghg^{-1} \qquad \text{for all } h \in G.$$

Then $\varphi_g : G \longrightarrow G$ is an automorphism of $G$. The set of all such automorphisms is referred to as the set of *inner automorphisms* of $G$.

The inner automorphisms of $G$ form a subgroup of $\mathrm{Aut}(G)$. A subgroup $H$ of $G$ is normal if and only if it is mapped by all inner automorphisms into itself. This is just a restatement of the definition.

---

[30] In the language of category theory (cf. [212]) this is a *morphism in the category of groups*.

[31] In the language of category theory, isomorphic groups are *equivalent* objects in the category of groups.

**Factor groups:** Let $N$ be a normal subgroup of a group $G$. For $g, h \in G$ we write

$$g \sim h$$

if and only if $gh^{-1} \in N$. This is an equivalence relation on the group $G$ (cf. 4.3.5.1). The equivalence classes are denoted by $[g]$, and defining an operation by

$$\boxed{[g][f] := [gf],}$$

this set of equivalence classes acquires the structure of a group[32], which is called the *factor group* of $G$ modulo $N$, and denoted $G/N$. In the case of an additive group we write $g \sim h$ if and only if $g - h \in N$. In this case the group operation on $G/N$ is given by $[g] + [h] := [g + h]$.

*Example 2:* Let $G$ be the multiplicative group of real numbers, and let $N := \{x \in \mathbb{R} \,|\, x > 0\}$. Then one has $g \sim h$ if and only if $g$ and $h$ have the same sign. Hence $G/N$ has one element for each sign, $[1]$ and $[-1]$; the multiplication table is $[1][-1] = [-1]$, etc. This means that $G/N$ is isomorphic to the group $\{1, -1\}$.

If $N$ is a normal subgroup of a group $G$, then one has

$$\boxed{\operatorname{ord}(G/N) = \frac{\operatorname{ord} G}{\operatorname{ord} N}\,.}$$

The importance of factors groups lies in the fact that they describe (up to isomorphism) all epimorphic images[33], as seen in the following result.

**Structure theorem for group homomorphisms:** (i) If $\varphi : G \longrightarrow H$ is an epimorphism of groups, then the *kernel*, defined as $\ker\varphi := \varphi^{-1}(e)$, is a normal subgroup of $G$ and one has the isomorphism

$$\boxed{H \cong G/\ker\varphi.}$$

(ii) If, conversely, $N$ is a normal subgroup of $G$, then the map defined by

$$\varphi(g) := [g]$$

is an epimorphism $\varphi : G \longrightarrow G/N$ with $\ker\varphi = N$.

In particular, a group $G$ is simple if and only if every epimorphic image of $G$ is isomorphic to either $G$ or $\{e\}$, in other words, when $G$ has only trivial epimorphic images.

*Example 3:* Let $\mathbb{Z}$ be the additive group of integers. A group $H$ is an epimorphic image of $\mathbb{Z}$ if and only if the group is *cyclic* (cf. 2.5.1.3).

**First isomorphism law for groups:** If $N$ is a normal subgroup of a group $G$ and $H$ is a subgroup of $G$, then $N \cap H$ is normal subgroup of $G$, and one has the isomorphism

$$\boxed{HN/N \cong H/(H \cap N).}$$

Here we have set $HN := \{hg \,|\, h \in H,\ g \in N\}$.

**Second isomorphism law for groups:** Let $N$ and $H$ be normal subgroups of $G$ with $N \subseteq H \subseteq G$. Then $H/N$ is normal subgroup of $G/N$ and one has an isomorphism

$$\boxed{G/H \cong (G/N)/(H/N).}$$

[32] The definition $[g][h]$ does not depend on the choice of representatives of the equivalence classes $[g]$ and $[f]$, as is easily seen.

[33] An *epimorphic image* is the image of $G$ under some epimorphism.

### 2.5.1.3 Cyclic groups

A group $G$ is said to be *cyclic*, if every element $g \in G$ can be written in the form[34]

$$g = a^n, \qquad n = 0, \pm 1, \ldots .$$

The element $a$ is called the *generating element* or *generator* of $G$.

(i) If $a^n \neq e$ for all natural numbers $n \geq 1$, then $G$ contains infinitely many elements.

(ii) If $a^n = e$ for some natural number $n \geq 1$, then $G$ consists of finitely many elements.

To every natural number $m \geq 1$ there is a cyclic group of order $m$.

Two cyclic groups are isomorphic if and only if they have the same number (finite or infinite) of elements. The isomorphism is given by $a^n \mapsto b^n$, where $a$ and $b$ denote the corresponding generators.

**Theorem:** (a) Every cyclic group is commutative.

(b) Every finite group of prime order[35] is cyclic.

(c) Two finite groups of the same prime order are cyclic and isomorphic to each other.

*Example 1:* A cyclic group of order 2 consists of the elements $e$ and $a$ with

$$a^2 = e.$$

Then one has $a^{-1} = a$. A cyclic group of order 3 consists of elements $e, a, a^2$, where $a^3 = e$. From this it follows that $a^{-1} = a^2$ and $a^{-2} = a$.

*Example 2:* The additive group $\mathbb{Z}$ of integers is an infinite additive cyclic group, which is generated by the element 1. Any infinite cyclic group is isomorphic to $\mathbb{Z}$.

*Example 3:* An additive cyclic group of order $m \geq 2$ can be displayed by the symbols $0, a, 2a, \ldots, (m-1)a$; in this group we generally do calculations in the natural manner, taking

$$ma = 0$$

into account.[36]

**Main theorem on additive groups:** Let $G$ be an additive group with finitely many generators $a_1, \ldots, a_s \in G$, i.e., every element $g \in G$ can be written as a linear combination $m_1 a_1 + \cdots + m_s a_s$ with integral coefficients $m_j$. Then $G$ is a direct sum[37]

$$G = G_1 \oplus G_2 \oplus \ldots \oplus G_r \oplus G_{r+1} \oplus \ldots \oplus G_s$$

of (finitely many) additive cyclic groups $G_j$. Moreover, we have

(i) $G_1, \ldots, G_r$ are isomorphic to $\mathbb{Z}$, and

(ii) $G_{r+1}, \ldots, G_{r+s}$ are cyclic of the finite orders $\tau_{r+1}, \ldots, \tau_{r+s}$, where $\tau_j$ divides $\tau_{j+1}$ for all $j$. One refers to $r$ as the *rank* of the cyclic group $G$, and $\tau_{r+1}, \ldots, \tau_{r+s}$ are called the *torsion coefficients* of $G$.

[34] We set $a^0 := e$, $a^{-2} := (a^{-1})^2$, etc.

[35] This means the order of the group is a prime number.

[36] This group is isomorphic to the Gaussian residue class group $\mathbb{Z}/m\mathbb{Z}$ (cf. 2.5.2).

[37] This means that every element $g \in G$ has a unique decomposition $g = g_1 + \cdots + g_s$ with $g_j \in G_j$ for all $j$.

Two additive groups are isomorphic if and only if they have the same rank and the same torsion coefficients.

In classical combinatorial topology, groups $G$ of this kind occur as Betti groups (homology groups). In this case the rank $r$ is also referred to as the *Betti number* of $G$.

#### 2.5.1.4 Solvable groups

**Solvable groups:** A group $G$ is said to be *solvable* if and only if there is a sequence

$$G_0 \subseteq G_1 \subseteq \ldots \subseteq G_{n-1} \subseteq G_n := G$$

of subgroups $G_j$ of $G$ with $G_0 = \{e\}$, and for which, for all $j$,

$G_j$ is a normal subgroup of $G_{j+1}$ and $G_{j+1}/G_j$ is commutative.

*Example 1:* Every commutative group if solvable.

*Example 2* (permutation groups):

(i) The commutative group $\mathscr{S}_2$ is solvable by Example 1.

(ii) The group $\mathscr{S}_3$ is solvable. Take $\{e\} \subseteq \mathscr{A}_3 \subseteq \mathscr{S}_3$ as the sequence of subgroups.

(iii) The group $\mathscr{S}_4$ is solvable. Take $\{e\} \subseteq \mathscr{K}_4 \subseteq \mathscr{A}_4 \subseteq \mathscr{S}_4$ as the sequence.

Note that $\mathrm{ord}(\mathscr{A}_3) = 3$, $\mathrm{ord}(\mathscr{S}_j/\mathscr{A}_j) = 2$ and $\mathrm{ord}(\mathscr{A}_4/\mathscr{K}_4) = 3$. These are prime numbers. Therefore these groups are all cyclic and hence commutative by the theorem in the previous section.

(iv) The group $\mathscr{S}_n$ is not solvable for $n \geq 5$. This is caused by the simplicity of $\mathscr{A}_5$.

By Galois theory, these statements are responsible for the fact that algebraic equations of order $\geq 5$ are not solvable by radicals (cf. 2.6.5).

### 2.5.2 Rings

In a ring both the sum $a + b$ and the product $ab$ of two elements are defined. In rings one has a theory of divisibility (cf. 2.7.11).

**Definition:** A set $R$ is called a *ring* if $R$ is an additive group and for every ordered pair $(a, b)$ with $a, b \in R$ there is an element $ab \in R$, such that for all $a, b, c \in R$ one has:

(i) $a(bc) = (ab)c$[38] (associative law).

(ii) $a(b + c) = ab + ac$ and $(b + c)a = ba + ca$ (distributive law).

The ring $R$ is *commutative* if, moreover, for all $a, b \in R$, $ab = ba$.

If the ring $R$ has an element $e$ such that $ae = ea = a$ for all $a \in R$, then the element $e$ is uniquely determined by this property and is called a *unit* and $R$ is called a *ring with unit.*

A *zero divisor* $a$ in a ring $R$ is an element $a \neq 0$ such that for some element $b \neq 0$, $ab = 0$.

**Integral domains:** The commutative rings with unit and with no zero divisors are called *integral domains.*

[38] In the expression $a(bc)$ the brackets indicate that one first performs the multiplication $bc$, then multiplies the result from the left with $a$.

*Example 1:* The set $\mathbb{Z}$ of all integers is an integral domain (thus the name).

*Example 2:* The set of all real $(n \times n)$-matrices form a ring with the unit matrix $E$ as unit. For $n \geq 2$ this ring is not commutative and has zero divisors. For example, for $n = 2$, the product

$$\begin{pmatrix} 0 & 1 \\ 0 & 0 \end{pmatrix}\begin{pmatrix} 1 & 0 \\ 0 & 0 \end{pmatrix} = \begin{pmatrix} 0 & 0 \\ 0 & 0 \end{pmatrix}$$

yields the zero element although both factors are non-trivial.

**Subrings:** A subset $U$ of a ring is called a *subring* of $R$ if $U$ (with the operations induced on it by those of $R$) is itself a ring. This is the same thing as requiring $a - b \in U$ and $ab \in U$ for all $a, b \in U$.

**Ideals:** A subset $J$ of a ring $R$ is called an *ideal*, if $J$ is a subring with the following additional property:

From $r \in R$ and $a \in J$ it follows that $ra \in J$ and $ar \in J$.

*Example 3:* All ideals of $\mathbb{Z}$ are obtained in the form $m\mathbb{Z} := \{mz \mid z \in \mathbb{Z}\}$ for an arbitrary natural number $m$.

**The ring of polynomials $P[x]$:** Let $P$ be a ring. We denote by $P[x]$ the set of all expressions of the form

$$a_0 + a_1x + a_2x^2 + \ldots + a_kx^k$$

with $k = 0, 1, \ldots$ and $a_k \in P$ for all $k$. With respect to the addition and multiplication of polynomials, $P[x]$ is a ring, referred to as the *polynomial ring* (in one variable $x$).

If $P$ is an integral domain, then the polynomial ring $P[x]$ is also.

**Ring homomorphisms:** A *homomorphism* $\varphi : R \longrightarrow S$ between two rings $R$ and $S$ is a map which respects both operations in the rings, i.e., for which

$$\varphi(ab) = \varphi(a)\varphi(b) \quad \text{and} \quad \varphi(a+b) = \varphi(a) + \varphi(b) \qquad \text{for all } a, b \in R.$$

The bijective homomorphisms are referred to, just as in the case of groups, as *isomorphisms*.

Surjective (resp. injective) homomorphisms are called *epimorphisms* (resp. *monomorphisms*). A *ring automorphism* is an isomorphism of a ring to itself.

**Factor rings:** Let $J$ be an ideal of the ring $R$. For $a, b \in R$ we write

$$a \sim c$$

if $a - c \in J$. This is an equivalence relation on the ring $R$ (cf. 4.3.5.1). The corresponding equivalence classes are denoted $[a]$; the set of all these equivalence classes forms a ring with the operations

$$[a][b] := [ab] \quad \text{and} \quad [a] + [b] := [a+b];$$

this ring is called the *factor ring* and is denoted $R/J$.[39] The elements $[a]$ of the factor ring are also called *residue classes*, and the factor ring itself is in some cases referred to as the *residue class ring*.

[39] As always, it must be verified that the definitions of $[a][b]$ and $[a] + [b]$ do not depend on the choice of representatives (here $a$ and $b$).

**The Gaussian residue class ring $\mathbb{Z}/m\mathbb{Z}$:** Let $\mathbb{Z}$ be the additive ring of integers, and choose some $m \in \mathbb{Z}$, $m > 0$. Consider the ideal $m\mathbb{Z}$. Then we have

$$z \sim w$$

if and only if $z - w \in m\mathbb{Z}$, i.e., the difference $z - w$ is divisible by $m$. Following Gauss, we also use the following notation to denote this:[40]

$$\boxed{z \equiv w \bmod m.}$$

For the residue classes one has $[z] = [w]$ if and only if the difference $z - w$ is divisible by $m$. The ring of residue classes $\mathbb{Z}/m\mathbb{Z}$ consists of precisely the $m$ classes

$$[0], [1], \ldots, [m-1],$$

for which the rules

$$[a] + [b] = [a+b], \quad \text{and} \quad [a][b] = [ab]$$

hold by definition.

*Example 4:* For $m = 2$, the ring $\mathbb{Z}/2\mathbb{Z}$ consists of the two residue classes $[0]$ and $[1]$. One has $[z] = [w]$ if and only if $z - w$ is divisible by 2. Hence the residue class $[0]$ (resp. $[1]$) corresponds to the set of even (resp. odd) integers. The set of all operations (complete addition and multiplicition tables) is:

$$\begin{aligned}
&[1] + [1] = [2] = [0], \quad [0] + [0] = [0], \quad [0] + [1] = [1] + [0] = [1], \\
&[1][1] = [1], \quad [0][1] = [1][0] = [0][0] = [0].
\end{aligned}$$

For $m = 3$ the ring $\mathbb{Z}/3\mathbb{Z}$ consists of three residue classes $[0]$, $[1]$, $[2]$. One has $[z] = [w]$ if and only if the difference $z - w$ is divisible by 3. For example one has

$$[2][2] = [4] = [1].$$

For $m = 4$ the ring $\mathbb{Z}/4\mathbb{Z}$ consists of four residue classes $[0]$, $[1]$, $[2]$, $[3]$. From the decomposition $4 = 2 \cdot 2$ it follows that $[2][2] = [4]$ and hence

$$[2][2] = [0].$$

Thus, $\mathbb{Z}/4\mathbb{Z}$ has zero divisors.

The residue class ring $\mathbb{Z}/m\mathbb{Z}$ for $m \geq 2$ is free of zero divisors if and only if $m$ is a prime number. In this case $\mathbb{Z}/m\mathbb{Z}$ is in fact a field (see below).

The following theorem shows that all epimorphisms of a ring can be constructed by knowing all of the ideals of the ring.

**Structure theorem for ring homomorphisms:** (i) If $\varphi : R \longrightarrow S$ is an epimorphism of the ring $R$ onto the ring $S$, then the *kernel* $\ker \varphi := \varphi^{-1}(0)$ is an ideal of $R$, and $S$ is isomorphic to the factor ring $R/\ker \varphi$.

(ii) If, conversely, $J$ is an ideal of $R$, then

$$\varphi(a) := [a]$$

defines an epimorphism $\varphi : R \longrightarrow R/J$ with $\ker \varphi = J$.

[40] This reads: $z$ is congruent to $w$ modulo $m$.

### 2.5.3 Fields

Fields are rings in which multiplicative inverses exists, and the usual rules for calculations, familiar from the real numbers, hold in them. In a sense fields are the most perfect algebraic structures. A central theme in field theory is the investigation of field extensions. Galois theory reduces this investigation to the theory of the symmetry groups of these extensions.

**Definition:** A set $K$ is called a *skew field*, if it satisfies:

(i) $K$ is an additive group with zero element 0.

(ii) $K - \{0\}$ is a multiplicative group with unit element $e$.

(iii) $K$ is a ring.

A *field* is a skew field for which the multiplication is commutative.

**Equations:** Let $K$ be a skew field. Let elements $a, b, c, d \in K$ with $a \neq 0$ be given. Then the equations

$$\boxed{ax = b, \quad ya = b, \quad c + z = d, \quad z + c = d}$$

have unique solutions in $K$, which are given by $x = a^{-1}b,\ y = ba^{-1}$ and $z = d - c$.

**Subfields:** A subset $U$ of a skew field $K$ is called a *subfield* of $K$, if $U$ is a skew field itself with respect to the induced operations (closure of addition and multiplication).

A subfield of $K$ is referred to as non-trivial, if it is not equal to $\{e\}$ or $K$ itself.

**Characteristic:** By definition a skew field $K$ has the *characteristic zero*, if

$$ne \neq 0 \qquad \text{for all } n = 1, 2, \ldots$$

A skew field has the *characteristic* $m > 0$, if

$$me = 0 \quad \text{and} \quad ne \neq 0 \qquad \text{for } n = 1, \ldots, m - 1.$$

In this case $m$ is necessarily a prime number.

*Example 1:* The set $\mathbb{R}$ of real numbers and the set $\mathbb{Q}$ of rational numbers are fields of characteristic zero. $\mathbb{Q}$ is a subfield of $\mathbb{R}$.

*Example 2:* Let $p \geq 2$ be prime. Then the Gaussian residue class ring $\mathbb{Z}/p\mathbb{Z}$ is a field of characteristic $p$ (cf. 2.5.2).

**Homomorphisms:** A *homomorphism* of skew fields $\varphi : K \longrightarrow M$ is a homomorphism of the corresponding rings. For all $a, b, c \in K$ one then has

$$\boxed{\varphi(a + b) = \varphi(a) + \varphi(b), \quad \varphi(ab) = \varphi(a)\varphi(b), \quad \varphi(c^{-1}) = \varphi(c)^{-1}}$$

if $c \neq 0$. Moreover one has $\varphi(e) = e$ and $\varphi(0) = 0$.

The bijective homomorphisms are referred to, just as in the case of groups, as *isomorphisms.*

Surjective (resp. injective) homomorphisms are called *epimorphisms* (resp. monomorphisms). A *field automorphism* is an isomorphism of a field to itself.

**Prime fields:** A skew field which contains no non-trivial subfields is called a *prime field.*

(i) In every skew field there is precisely one subfield which is a prime field.

(ii) This prime field is isomorphic to either $\mathbb{Q}$ or to $\mathbb{Z}/p\mathbb{Z}$ for some prime number $p$.

(ii) The characteristic of a skew field is equal to 0 (resp. $p$), if the prime field just mentioned is equal to $\mathbb{Q}$ (resp. $\mathbb{Z}/p\mathbb{Z}$).

**Galois fields:** The finite skew fields are also referred to as *Galois fields*. Every Galois field is in fact a field.

For every prime number $p$ and $n = 1, 2, \ldots$ there is a field with $p^n$ elements. In this manner one gets all of the Galois fields.

Two Galois fields are isomorphic if and only if they have the same number of elements.

**Complex numbers:** We will show, following Hamilton (1805–1865), that one can construct algebraic objects which give a rigorous foundation for the theory of complex numbers. We denote by $\mathbb{C}$ the set of all ordered pairs $(a, b)$ with $a, b \in \mathbb{R}$. Defining the operations

$$(a,b) + (c,d) := (a+c, b+d)$$

and

$$(a,b)(c,d) := (ac - bd, ad + bc)$$

we get the structure of a field on $\mathbb{C}$. If we set

$$\boxed{\mathrm{i} := (0,1),}$$

then we have $\mathrm{i}^2 = (-1, 0)$. For every element $(a, b) \in \mathbb{C}$ one has the unique decomposition[41]

$$(a,b) = (a,0) + (b,0)\mathrm{i}.$$

The map $\varphi(a) := (a, 0)$ is a monomorphism of $\mathbb{R}$ into $\mathbb{C}$. Thus we may view any element $a \in \mathbb{R}$ as the element $(a, 0) \in \mathbb{C}$. In this sense we can write any element $(a, b)$ uniquely in the form

$$\boxed{a + b\mathrm{i}}$$

with $a, b \in \mathbb{R}$. In particular one has

$$\boxed{\mathrm{i}^2 = -1.}$$

Thus we have arrived at the usual notation, showing that $\mathbb{C}$ contains $\mathbb{R}$ as a subfield.[42]

**Quaternions:** The set $\mathbb{H}$ of quaternions $\alpha+\beta\mathbf{i}+\gamma\mathbf{j}+\delta\mathbf{k}$ with $\alpha, \beta, \gamma, \delta \in \mathbb{R}$ is a skew field, which contains the field $\mathbb{C}$ of complex numbers as a subfield. For $\alpha^2 + \beta^2 + \gamma^2 + \delta^2 \neq 0$ one obtains the inverse element through

$$(\alpha + \beta\mathbf{i} + \gamma\mathbf{j} + \delta\mathbf{k})^{-1} = \frac{\alpha - \beta\mathbf{i} - \gamma\mathbf{j} - \delta\mathbf{k}}{\alpha^2 + \beta^2 + \gamma^2 + \delta^2}$$

(cf. Example 1 in 2.4.3.4).

**Quotient fields:** Let $P$ be an integral domain, which does not consist of only the zero element (cf. 2.5.2). Then there is a field $Q(P)$ with the following properties:

(i) $P$ is contained in $Q(P)$.

---

[41] This means that $(0, b) = (b, 0)\mathrm{i}$.

[42] Note also that this shows that $\mathbb{C}$ is the complexification of the one-dimensional $\mathbb{R}$-vector space $\mathbb{R}$, i.e., $\mathbb{C} = \mathbb{R} \otimes \mathbb{C}$ (cf. 2.4.3.1).

(ii) If $P$ is contained in a field $K$, then the smallest subfield of $K$ which contains $P$ is the field $Q(P)$. The field $Q(P)$ is called the *quotient field* or the *field of fractions* of $P$.

Explicitly one obtains $Q(P)$ through the following construction. We consider the set of all ordered pairs $(a, b)$ with $a, b \in P$ and $b \neq 0$. We write

$$(a,b) \sim (c,d) \quad \text{if and only if} \quad ad = bc.$$

This is an equivalence relation (cf. 4.3.5.1). The set $Q(P)$ of corresponding equivalence classes $[(a, b)]$ is a field with respect to the operations defined by[43]

$$[(a,b)][(c,d)] := [(ac, bd)],$$
$$[(a,b)] + [(c,d)] = [(ad + bc, bd)].$$

Instead of $[(a, b)]$ one also writes $\frac{a}{b}$. Then $Q(P)$ consists of the set of all symbols

$$\boxed{\frac{a}{b}}$$

with $a, b \in P$ and $b \neq 0$, where the calculations with these symbols are done in the usual way one calculates with fractions, i.e.,

$$\frac{a}{b} = \frac{c}{d} \quad \text{if and only if} \quad ad = bc$$

and

$$\frac{a}{b}\frac{c}{d} = \frac{ac}{bd}, \qquad \frac{a}{b} + \frac{c}{d} = \frac{ad+bc}{bd}.$$

If we choose an $r$ in $P$ with $r \neq 0$ and set $\varphi(a) := \frac{ar}{r}$, then we obtain a monomorphism $\varphi : P \longrightarrow Q(P)$; this is independent of the choice of $r$. In this sense we can identify the elements $a$ of $P$ with the elements $\frac{ar}{r}$ of $Q(P)$. Then $\frac{r}{r}$ is the unit element in $Q(P)$ and we have

$$\boxed{\frac{a}{b} = ab^{-1}.}$$

As in the case of the complex numbers above, the seemingly complicated method of using the residue classes $[(a, b)]$ is only done to guarantee that the formal calculations with fractions $\frac{a}{b}$ does not lead to any contradictions.

*Example 3:* The quotient field of the integral domain $\mathbb{Z}$ is the field $\mathbb{Q}$ of rational numbers.

*Example 4:* Let $P[x]$ be the polynomial ring over an integral domain $P \neq \{0\}$. Then the corresponding quotient field $Q(P[x])$, which is usually denoted $P(x)$, is the field of *rational functions* with coefficients in $P$, i.e., the elements of $P(x)$ are quotients

$$\frac{p(x)}{q(x)}$$

of polynomials $p(x)$ and $q(x)$ with coefficients in $P$ (cf. 2.5.2), where one requires that $q \neq 0$, i.e., $q(x)$ is not the zero polynomial.

[43] Again, these operations are independent of the choices used in their definitions.

## 2.6 Galois theory and algebraic equations

*The Paris circles with their intensive mathematical enterprise, produced around 1830 with Évariste Galois a genius of the highest quality, who, like a comet, vanished just as fast as he had appeared.*[44]

*Dirk J. Struik*

### 2.6.1 The three famous ancient problems

In classical Greek mathematical culture there were three famous problems, whose non-solvability was not shown until the nineteenth century:

(i) the squaring of the circle,

(ii) the Delian problem of doubling the cube, and

(iii) the trisection of an arbitrary angle.

In all these cases only constructions utilizing a ruler and a compass are allowed. In problem (i), one is to construct a square whose area is the same as that of some given circle. In problem (ii), one is to construct from a given cube the length of the sides of a new cube which has twice the volume.[45]

Besides these problems the general problem of constructing regular polygons with a ruler and a compass was important in ancient times.

All of these problems can be reduced to problems concerning the solvability of algebraic equations (cf. 2.6.6). The investigation of solutions of algebraic equations is done with the help of a general theory which was created by the French mathematician Everisté Galois (1811–1832). Galois theory paved the way for modern algebraic thinking.

Galois theory contains as special cases results of the young Gauss (1777–1855) on cyclotomic fields and the construction of regular polygons, as well as the theorem of the Norwegian mathematician Niels Henrik Abel (1802–1829) on the non-solvability of the general equation of fifth and higher degree through radicals (cf. 2.6.5 and 2.6.6).

### 2.6.2 The main theorem of Galois theory

**Field extensions:** A *field extension* $K \subseteq E$ (which is also denoted by $(E|K)$) is a field $E$, which contains a given field $K$ as a subfield. Every subfield $Z$ of $E$ with

$$\boxed{K \subseteq Z \subseteq E}$$

is called an *intermediate* field of the extension $K \subseteq E$. A *chain of field extensions*

$$K_0 \subseteq K_1 \subseteq \ldots \subseteq E$$

is a set of subfields $K_j$ of the field $E$ with the indicated inclusion relations.

[44]The tragical life of Galois, who died in a dual at the age of 21 and wrote down the most important results of his theory on the eve of that dual in a letter to a friend, is described in the book of the student of Einstein, Leopold Infeld, *Wen die Götter lieben* (Whom the gods love), which appeared in 1954 in the Schönbrunn-Verlag, Vienna.

[45]On the Greek island Delos in the Aegean Sea one of the most famous shrines of the ancient world was located, dedicated to Artemis and Apollo. The Delian problem supposedly originated as the problem of constructing a shrine with twice the volume. Giovanni Casanova (1725–1798), the protagonist in Mozart's *Don Giovanni*, worked on this problem.

**The basic idea of Galois theory:** Galois theory considers an important class of field extensions (Galois extensions) and determines *all intermediate fields* with the help of all subgroups of the symmetry group (Galois group) of this field extension.

In this way a field-theoretic problem (quite difficult in general) is reduced to a much easier group-theoretic problem.

It is a general strategy in modern mathematics to reduce the investigation of complicated structures to those of simpler structures. For example one can reduce topological problems to algebraic problems (which are easier), and the investigation of continuous Lie Groups to the study of Lie algebras, objects of linear algebra (cf. [212]).

**The degree of a field extension:** If $K \subseteq E$ is a field extension, then one may view $E$ as a linear space over $K$. The dimension of this space[46] is called the *degree* of the field extension and is denoted $[E : K]$. If this degree is finite, we speak of a *finite* field extension.

*Example 1:* The extension $\mathbb{Q} \subseteq \mathbb{R}$ of the field of rational numbers to the field of real numbers is infinite.

*Example 2:* The extension $\mathbb{R} \subseteq \mathbb{C}$ from the field $\mathbb{R}$ of real numbers to the field $\mathbb{C}$ of complex numbers is finite and has degree two, since it has been shown in section 2.5.3 that $\mathbb{C}$ may be viewed as a two-dimensional vector space over $\mathbb{R}$, with basis consisting of 1 and i.

The extension $\mathbb{Q} \subseteq \mathbb{R}$ is moreover a transcendental extension in the sense of section 2.6.3, while the extension $\mathbb{R} \subseteq \mathbb{C}$ is a simple algebraic extension.

**The theorem on degrees:** If $Z$ is an intermediate field of a finite field extension $K \subseteq E$, then $K \subseteq Z$ is also a finite field extension, and one has the relation

$$[E : K] = [Z : K][E : Z].$$

This relation is easy to remember if one thinks of the symbols as defining fractions.

**The Galois group of a field extension:** Let $K \subseteq E$ be a field extension. Then the *Galois group* of this extension, denoted $G_K^E$ or $\mathrm{Gal}(E|K)$, is by definition the group of all automorphisms of the field $E$ which act trivially on all elements of $K$.

A *finite* field extension $K \subseteq E$ is called *Galois*, if

$$\boxed{\operatorname{ord} G_K^E = [E : K],}$$

i.e., if there are as many symmetries of $E$ over $K$ as the degree of the field extension.

**Main theorem of Galois theory:** Let $K \subseteq E$ be a finite Galois field extension. There there is a one-to-one correspondence between the intermediate fields $Z$ of this extension and the subgroups $H \subset G_K^E$; this correspondence is such that the subgroup $H$ corresponding to the intermediate field $Z$ consists of all automorphisms which fix all elements of $Z$.

In this way one obtains a *bijective map* from the set of all intermediate fields and the set of all subgroups of the Galois group. Note that an intermediate field $Z$ is Galois over $K$ if and only if the subgroup $H$ corresponding to it, which is isomorphic to $G_K^Z$, is a *normal* subgroup of $G_K^E$.

**Corollary:** A finite, Galois field extension $K \subseteq E$ has no intermediate *Galois* field extensions if and only if the Galois group of the extension is a *simple* group (see section 2.5.1.1 for this notion).

[46]This is defined by using the general notions of linear algebra, replacing the field $\mathbb{K}$ used in section 2.3.2 with the field $K$.

*Example 3:* We consider the classical field extension

$$\boxed{\mathbb{R} \subseteq \mathbb{C}}$$

of the field $\mathbb{R}$ of real numbers to the field $\mathbb{C}$ of complex numbers.

First we determine the Galois group $\mathrm{Gal}(\mathbb{C}|\mathbb{R})$ of this extension. Let $\varphi : \mathbb{C} \longrightarrow \mathbb{C}$ be an automorphism which fixes all real numbers. From $\mathrm{i}^2 = -1$ it follows that $\varphi(\mathrm{i})^2 = -1$. Hence either $\varphi(\mathrm{i}) = \mathrm{i}$ or $\varphi(\mathrm{i}) = -\mathrm{i}$. These two possibilities correspond to the identity automorphism $\mathrm{id}(a + b\mathrm{i}) := a + b\mathrm{i}$ on $\mathbb{C}$ and the automorphism

$$\boxed{\varphi(a + b\mathrm{i}) := a - b\mathrm{i}}$$

of $\mathbb{C}$ (the transition to the complex conjugate variable), respectively.

Hence $G_{\mathbb{R}}^{\mathbb{C}} = \{\mathrm{id}, \varphi\}$ with $\varphi^2 = \mathrm{id}$. According to Example 2

$$[\mathbb{C} : \mathbb{R}] = \mathrm{ord}\, G_{\mathbb{R}}^{\mathbb{C}} = 2.$$

Hence the field extension $\mathbb{C}|\mathbb{R}$ is Galois.

The Galois group $G_{\mathbb{R}}^{\mathbb{C}}$ is cyclic of order two. From the simplicity of the Galois group the simplicity of the field extension $\mathbb{C}|\mathbb{R}$ follows. Note that in this case, as the cyclic group of order two has *no* subgroups at all, there are also *no* intermediate field extensions between $\mathbb{R}$ and $\mathbb{C}$.

*Example 4:* The equation

$$\boxed{x^2 - 2 = 0}$$

has in $\mathbb{Q}$ no solutions. In order to construct a field in which this equation does have a solution, we set $\vartheta := \sqrt{2}$. We denote by $\mathbb{Q}(\vartheta)$ the smallest subfield of $\mathbb{C}$ which contains both $\mathbb{Q}$ and $\vartheta$. One then has $\vartheta, -\vartheta \in \mathbb{Q}(\vartheta)$ and

$$x^2 - 2 = (x - \vartheta)(x + \vartheta),$$

i.e., $\mathbb{Q}$ is the splitting field for $x^2 - 2$ (cf. 2.6.3 for this notion). This field consists of all expressions

$$\frac{p(\vartheta)}{q(\vartheta)}$$

where $p$ and $q$ are polynomials over $\mathbb{Q}$ with $q \neq 0$. From $\vartheta^2 = 2$ and $(c + d\vartheta)(c - d\vartheta) = c^2 - 2d^2$ one can reduce all of these expressions to ones of the form[47]

$$\boxed{a + b\vartheta, \qquad a, b \in \mathbb{Q}.}$$

As in Example 3, we see that the field extension $\mathbb{Q} \subseteq \mathbb{Q}(\vartheta)$ is Galois of degree 2 and hence simple. The corresponding Galois group

$$G_{\mathbb{Q}}^{\mathbb{Q}(\vartheta)} = \{\mathrm{id}, \varphi\} \qquad \text{with} \quad \varphi^2 = \mathrm{id}$$

is generated by the permutations of the zeros $\vartheta, -\vartheta$ of $x^2 - 2$. The identity permutation corresponds to the identity automorphism $\mathrm{id}(a + b\vartheta) := a + b\vartheta$ of $\mathbb{Q}(\vartheta)$. On the other

---

[47] For example we have $\dfrac{1}{1 + 2\vartheta} = \dfrac{1 - 2\vartheta}{(1 + 2\vartheta)(1 - 2\vartheta)} = -\dfrac{1}{7} + \dfrac{2\vartheta}{7}$.

hand, the transposition of $\vartheta$ and $-\vartheta$ corresponds to the automorphism $\varphi(a+b\vartheta) := a-b\vartheta$ of $\mathbb{Q}(\vartheta)$.

*Example 5* (cyclotomic equation): If $p$ is a prime number, then the equation

$$\boxed{x^p - 1 = 0, \qquad x \in \mathbb{Q}}$$

has the solution $x = 1$. To construct the smallest field $E$ which contains all solutions of this equation, we set $\vartheta := \mathrm{e}^{2\pi\mathrm{i}/p}$ and denote by $\mathbb{Q}(\vartheta)$ the smallest subfield of $\mathbb{C}$, which contains both $\mathbb{Q}$ and $\vartheta$. Because of the relation

$$x^p - 1 = (x-1)(x-\vartheta)\ldots(x-\vartheta^{p-1}),$$

the field $E := \mathbb{Q}(\vartheta)$ is the splitting field of the polynomial $x^p - 1$ (cf. 2.6.3).

The field $\mathbb{Q}(\vartheta)$ consists of all expressions of the form

$$\boxed{a_0 + a_1\vartheta + a_2\vartheta^2 + \ldots + a_{p-1}\vartheta^{p-1}, \qquad a_j \in \mathbb{Q},}$$

which are added and multiplied in the usual manner, taking $\vartheta^p = 1$ into account. Since the elements $1, \vartheta, \ldots, \vartheta^{p-1}$ are linearly independent over $\mathbb{Q}$, one has $[\mathbb{Q}(\vartheta) : \mathbb{Q}] = p$. Moreover the extension $\mathbb{Q} \subseteq \mathbb{Q}(\vartheta)$ is Galois. The corresponding Galois group $G_{\mathbb{Q}}^{\mathbb{Q}(\vartheta)} = \{\varphi_0, \ldots, \varphi_{p-1}\}$ consists of all automorphisms $\varphi_k : \mathbb{Q}(\vartheta) \longrightarrow \mathbb{Q}(\vartheta)$, which are generated by

$$\boxed{\varphi_k(\vartheta) := \vartheta^k.}$$

One has $\mathrm{id} = \varphi_0$ and $\varphi_1^k = \varphi_k$, $k = 1, \ldots, p-1$ as well as $\varphi_1^p = \mathrm{id}$.

Hence the Galois group $G_{\mathbb{Q}}^{\mathbb{Q}(\vartheta)}$ is cyclic of prime order $p$ and simple, which implies the simplicity of the extension $\mathbb{Q} \subseteq \mathbb{Q}(\vartheta)$.

### 2.6.3 The generalized fundamental theorem of algebra

**Algebraic and transcendental elements:** Let $E|K$ be a field extension. We denote by $K[x]$ the ring of all polynomials

$$\boxed{p(x) := a_0 + a_1 x + \ldots + a_n x^n}$$

with $a_k \in K$ for all $k$. These expressions are called *polynomials over* $K$ (or polynomials *defined over* $K$). Moreover we denote by $K(x)$ the field of all rational functions

$$\boxed{\frac{p(x)}{q(x)},}$$

where $p$ and $q$ are polynomials over $K$ and $q \neq 0$.

An element $\vartheta$ of $E$ is said to be *algebraic over* $K$, if $\vartheta$ is the zero of a polynomial over $K$. Otherwise $\vartheta$ is *transcendental.*

The field extension $E|K$ is said to be *algebraic over* $K$, if every element of $E$ is algebraic over $K$. Otherwise $E$ is called *transcendental over* $K$.

A polynomial over $K$ is *irreducible*, if it is not the product of two polynomials over $K$, each of which has a degree $\geq 1$. It is an important general problem in algebra to find an extension $E$ of the field $K$ in which a given polynomial $p(x)$ splits into linear factors:

$$p(x) = a_n(x - x_1)(x - x_2) \cdot \ldots \cdot (x - x_n)$$

with $x_j \in E$ for all $j$.

**Algebraic closure:** An extension field $E$ of $K$ is said to be *algebraically closed*, if every polynomial over $K$ splits into linear factors over $E$ (meaning all of its zeros are contained in $E$).

An *algebraic closure* of $K$ is an algebraically closed extension $E$ of $K$, which contains no non-trivial (not equal to $E$ itself) subfield with this property.

**The generalized fundamental theorem of algebra due to Steinitz (1910):** Every field $K$ has a unique (up to an isomorphism which fixes all elements of $K$.) algebraic closure; it is denoted $\overline{K}$.

**Splitting field of a polynomial:** The smallest subfield of $\overline{K}$ in which a given polynomial $p(x)$ splits into linear factors is called the *splitting field* of the polynomial $p(x)$.

*Example:* The field $\mathbb{C}$ of complex numbers is the algebraic closure of the field $\mathbb{R}$ of real numbers. Because of the relation

$$\boxed{x^2 + 1 = (x - \mathrm{i})(x + \mathrm{i}),}$$

$\mathbb{C}$ is at the same time the splitting field of the polynomial $x^2 + 1$, which is defined over $\mathbb{R}$ and is irreducible over $\mathbb{R}$.

### 2.6.4 Classification of field extensions

The notions of finite, simple, algebraic and transcendental field extensions were defined in section 2.6.2 and 2.6.3.

**Definition:** Let $E|K$ be a field extension.

(a) An irreducible polynomial over $K$ is called *separable*, if it contains no multiple zeros (recall these zeros are elements of the algebraic closure $\overline{K}$ of $K$).

(b) $E|K$ is said to be a *separable* extension, if the extension is algebraic and every element of $E$ is the zero of an irreducible and separable polynomial in $K$.

(c) The extension $E|K$ is said to be *normal*, if the extension is algebraic and every irreducible polynomial over $K$ has either no zero in $E$ or splits in $E$ completely into linear factors.

**Theorem:** (i) Every finite field extension is algebraic.

(ii) Every finite separable extension is simple.

(iii) Every algebraic extension of a field of characteristic 0 or of a finite field is separable.

**Characterization of Galois field extensions:** For a field extension $E|K$ one has:

(i) $E|K$ is Galois, if the extension is finite, separable and normal.

(ii) $E|K$ is Galois if and only if $E$ is the splitting field of an irreducible polynomial defined over $K$.

(iii) If a field $K$ has characteristic 0 or is finite, then $E|K$ is Galois if and only if $E$ is the splitting field of a polynomial defined over $K$.

The following results give a complete description of all simple field extensions.

**Simple transcendental field extensions:** Every simple transcendental extension of a field $K$ is isomorphic to the field $K(x)$ of rational functions over $K$.

**Simple algebraic field extensions:** Let $p(x)$ be an irreducible polynomial defined over a field $K$. We consider a symbol $\vartheta$ and all expressions of the form

$$\boxed{a_0 + a_1\vartheta + \ldots + a_m\vartheta^m, \qquad a_j \in K,} \tag{2.62}$$

which we add and multiply in the usual manner, taking $p(\vartheta) = 0$ into account (we take $\vartheta$ to be a zero of $p(x)$). The set of all of these expressions with the operations just introduced yields a field $K(\vartheta)$, which is a simple algebraic field extension of $K$ and has the important property that $\vartheta$ is a zero of $p(x)$. Moreover one has

$$\boxed{[E : K] = \deg p(x).}$$

If $(p(x))$ denotes the set of all polynomials over $K$, which can be written in the form $q(x)p(x)$ with a polynomial $q(x)$ (note that this is the ideal in the ring $K[x]$ generated by the element $p(x)$), then the factor ring $K[x]/(p(x))$ is a field, which is isomorphic to $K(\vartheta)$.

If one takes all irreducible polynomials $p(x)$ defined over $K$, then one gets (up to isomorphism) all simple algebraic extensions of $K$.

*Example:* Let $K = \mathbb{R}$ be the field of real numbers. We choose $p(x) := x^2 + 1$. If we use all expressions of the form (2.62) with $p(\vartheta) = 0$, then we have $\vartheta^2 = -1$. This means that $\vartheta$ corresponds to the imaginary unit i.

The extension field $K(\vartheta)$ is the field $\mathbb{C}$ of complex numbers, hence $\mathbb{C} = \mathbb{R}[x]/(x^2 + 1)$.

### 2.6.5 The main theorem on equations which can be solved by radicals

Let $K$ be a given field. We consider the equation

$$\boxed{p(x) := a_0 + a_1x + \ldots + a_{n-1}x^{n-1} + x^n = 0} \tag{2.63}$$

with $a_j \in K$ for all $j$. We assume that the polynomial $p(x)$ is irreducible over $K$ and also that it is separable[48]. We denote by $E$ the splitting field of the $p(x)$, i.e., such that

$$p(x) = (x - x_1)(x - x_2) \cdot \ldots \cdot (x - x_n)$$

with $x_j \in E$ for all $j$.

**Goal:** One would like to express the zeros $x_1, \ldots, x_n$ of the equation (2.63) in as simple a manner as possible through the elements of $K$ and a certain number of additional quantities $\vartheta_1, \ldots, \vartheta_k$. The classical formulas for solving equations of the second, third and fourth degrees achieve this with expressions $\vartheta_j$ which are roots of the coefficients $a_0, \ldots, a_n$. After these formulas were found in the sixteenth century, it was natural to try to find similar expressions for equations of higher degrees. Building on work of Lagrange (1736–1813) and Cauchy (1789–1857), the 22-year old Norwegian mathematician Abel

[48]If the field $K$ has characteristic 0 (for example $K = \mathbb{R}$ oder $K = \mathbb{Q}$) or if $K$ is finite, then every irreducible polynomial over $K$ is automatically separable.

proved in 1824 for the first time that there simply *is no such formula* for general[49] equations of the fifth degree. The same result was obtained in 1830 independently by Galois.

**Definition:** The equation (2.63) is said to be *solvable by radicals*, if the splitting field $E$ can be obtained from $K$ by successively adding quantities $\vartheta_1, \ldots, \vartheta_k$, each of which satisfies an equation of the form

$$\boxed{\vartheta_j^{n_j} = c_j,} \tag{2.64}$$

where $n_j \geq 2$ and $c_j$ is in the extension constructed previously. If the characteristic $p$ of the field $K$ is non-zero, then we assume in addition that $p$ is not a divisor of $n_j$.

**Remark:** Instead of (2.64) one also writes

$$\vartheta_j = \sqrt[n_j]{c_j}.$$

This legitimates the use of the word radical (root). The above definition then corresponds to a chain of field extensions

$$\boxed{K =: K_1 \subseteq K_2 \subseteq \ldots \subseteq K_{k+1} =: E}$$

with $K_{j+1} = K_j(\vartheta_j)$ and $c_j \in K_j$ for $j = 1, \ldots, k$. Since all zeros $x_1, \ldots, x_n$ of (2.63) are in $E$, we get

$$\boxed{x_j = P_j(\vartheta_1, \ldots, \vartheta_k), \qquad j = 1, \ldots, n,}$$

where $P_j$ is a polynomial in $\vartheta_1, \ldots, \vartheta_k$, whose coefficients lie in $K$.

**Main theorem:** The algebraic equation (2.63) can be solved by radicals if and only if the Galois group $G_K^E$ of the extension $E|K$ is solvable.

**Definition:** A *general* equation of $n$th degree is an equation (2.63) over the field $K := \mathbb{Z}(a_0, \ldots, a_{n-1})$, which consists of all rational functions

$$\frac{p(a_0, \ldots, a_{n-1})}{q(a_0, \ldots, a_{n-1})},$$

where $p$ and $q$ are polynomials in the variables $a_0, \ldots, a_{n-1}$ with integral coefficients and $q$ is not the zero polynomial.

**Theorem of Abel–Galois:** The general equation of $n$th degree is not solvable by radicals for $n \geq 5$.

*Sketch of proof:* We view $x_1, \ldots, x_n$ as variables and consider the field $E := \mathbb{Z}(x_1, \ldots, x_n)$ of all rational functions

$$\frac{P(x_1, \ldots, x_n)}{Q(x_1, \ldots, x_n)} \tag{2.65}$$

with integral coefficients. Through multiplication and comparison of coefficients in

$$\begin{aligned} p(x) &= (x - x_1)(x - x_2) \cdot \ldots \cdot (x - x_n) \\ &= a_0 + a_1 x + \ldots + a_{n-1}x^{n-1} + x^n, \end{aligned}$$

[49]This means that the coefficients of the equation are general, that is, not special; for special coefficients there may very well be solutions given by simple formulas.

we get the quantities $a_0, \ldots, a_{n-1}$ (up to sign) as the elementary symmetric functions of $x_1, \ldots, x_n$ (cf. 2.1.6.4). For example one gets

$$-a_{n-1} = x_1 + \ldots + x_n.$$

All expressions of the form (2.65) with the property that $P$ and $Q$ are symmetric with respect to $x_1, \ldots, x_n$, form a subfield of $E$, which is isomorphic to $K$ and hence may be identified with $K$.

The polynomial $p(x)$ is irreducible and separable in $K$, and $E$ is its splitting field. The field extension $E|K$ is Galois. By a permutation of $x_1, \ldots, x_n$ one gets an automorphism of $E$ which fixes all elements of $K$. Two different such permutations correspond to different automorphisms of $E$, hence the Galois group of the field extension $E|K$ is isomorphic to the symmetric group on $n$ letters, i.e.,

$$\boxed{\mathrm{Gal}(E|K) \cong \mathscr{S}_n.}$$

For $n = 2, 3, 4$ the group $\mathscr{S}_n$ is solvable, while for $n \geq 5$ it is not solvable (cf. 2.5.1.4). The theorem of Abel–Galois is therefore a consequence of the main theorem above.

### 2.6.6 Constructions with a ruler and a compass

*Every beginner in geometry knows that different regular polygons, namely the triangle, the pentagon, the 15-gon and those which are obtained from these by doubling the number of sides of these, can be constructed geometrically. This was already known to Euclid, and it seems that one had convinced oneself since then that these cases represent the limit of the possible: at least I am not aware of any successful attempt in extending these results.*

*All the more I am convinced that the discovery, that besides the above mentioned polygons a series of further ones, among them the 17-gon, are capable of a geometric construction, is worthy of mention. This discovery is actually but a corollary of a not yet completed wider-ranging theory, which shall be presented to the public as soon as it has been completed.*

*C. F. Gauss, from Braunschweig*
*Student of mathematics in Göttingen*
*(Intelligenzblatt der allgemeinen Literaturzeitung, June 1, 1796*

We consider in the plane a Cartesian system of coordinates and finitely many points

$$P_1 = (x_1, y_1), \quad \ldots, \quad P_n = (x_n, y_n).$$

We can always choose the coordinate system in such a way that $x_1 = 1$ and $y_1 = 0$. We denote by $K$ the smallest subfield of the field of real numbers which contains all the numbers $x_j$ and $y_j$. Moreover, let $\mathbb{Q}(y)$ be the smallest subfield of $\mathbb{R}$ which contains all rational numbers as well as the number $y$.

**Main theorem:** A point $(x, y)$ can be constructed from the points $P_1, \ldots, P_n$ with a ruler and compass if and only if $x$ and $y$ belong to a Galois field extension $E$ of $K$ with

$$[E : K] = 2^m, \tag{2.66}$$

where $m$ is some natural number.

A segment $\vartheta$ can be constructed from a segment of the length $y$ and from the unit segment with ruler and compass, if and only if $\vartheta$ belongs to a Galois field extension $E$ with $K := \mathbb{Q}(y)$ for which (2.66) holds.

**The non-solvability of the problem of squaring the circle:** We want to construct a square with ruler and compass; we require it to have the same surface area as the unit circle. Letting $\vartheta$ denote the length of the sides of the square, one has

$$\boxed{\vartheta^2 = \pi}$$

(cf. Figure 2.4(a)). In 1882, the thirty-year old Ferdinand Lindemann (Hilbert's teacher) proved the transcendence of the number $\pi$ over the field $\mathbb{Q}$ of rational numbers. Consequently, $\pi$ (and hence also $\vartheta$) cannot be an element of any algebraic field extension of $\mathbb{Q}$.

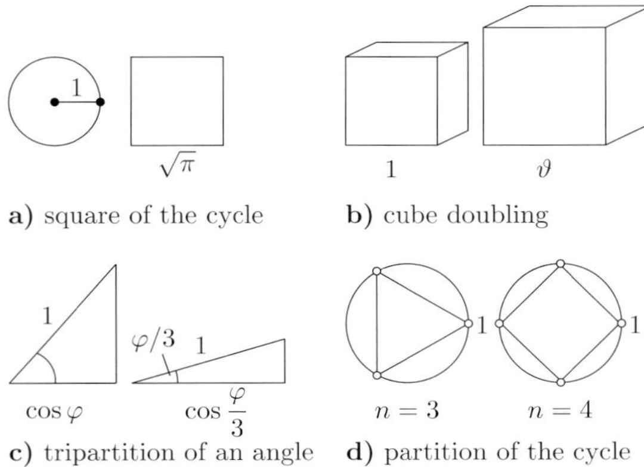


*Figure 2.4. The Delian problems.*

**The non-solvability of the problem of doubling the cube:** The length $\vartheta$ of a cube of volume 2 is a solution of the equation

$$\boxed{x^3 - 2 = 0.}$$

The Delian problem requires the construction of this number $\vartheta$ from the length of the unit segment using only a ruler and a compass (Figure 2.4(b)). The Delian problem is, according to the main theorem, solvable when $\vartheta$ belongs to a Galois extension field $E$ of $\mathbb{Q}$ for which $[E : \mathbb{Q}] = 2^m$. Since, however, the polynomial $x^3 - 2$ is irreducible over $\mathbb{Q}$, one has the chain of field extensions

$$\mathbb{Q} \subset \mathbb{Q}(\vartheta) \subseteq E$$

with $[\mathbb{Q}(\vartheta) : \mathbb{Q}] = 3$. From the theorem on the degrees of extensions, we get from this

$$[E : \mathbb{Q}] = [\mathbb{Q}(\vartheta) : \mathbb{Q}][E : \mathbb{Q}(\vartheta)] = 3[E : \mathbb{Q}(\vartheta)]$$

(cf. 2.6.2). Hence $[E : \mathbb{Q}]$ cannot possibly be of the form $2^m$.

**The non-solvability of the general trisection with a ruler and a compass:** According to Figure 2.4(c), this problem can be reduced to the problem of constructing a segment with the length $\vartheta = \cos\frac{\varphi}{3}$ from a segment of length $\cos\varphi$ and the unit segment. This means that $\vartheta$ is a solution to the equation

$$\boxed{4x^3 - 3x - \cos\varphi = 0.} \tag{2.67}$$

For $\varphi = 60°$, one has $\cos\varphi = \frac{1}{2}$, and the polynomial on the left in (2.67) is irreducible over $\mathbb{Q}$. The same argument with the degrees of the field extensions just given shows that $\vartheta$ cannot belong to a field extension $E$ of $\mathbb{Q}$ of degree $2^m$. Consequently, the trisection of an angle of 60° cannot be constructed with a ruler and a compass.

**The construction of regular polygons with a ruler and a compass:** The set of complex solutions of the so-called *cyclotomic equation*

$$\boxed{x^n - 1 = 0}$$

contains the number 1 and divides the unit circle into $n$ equal parts (cf. Figure 2.4(d)). The main theorem above together with the properties of the cyclotomic equation yields the following result.

**Theorem of Gauss:** A regular $n$-gon can be constructed with ruler and compass if and only if

$$\boxed{n = 2^m p_1 p_2 \ldots p_r.} \tag{2.68}$$

Here $m$ is a natural number, and the $p_j$'s are pairwise distinct prime numbers of the form[50]

$$2^{2^k} + 1, \qquad k = 0, 1, \ldots . \tag{2.69}$$

It is presently known that, for $k = 0, 1, 2, 3, 4$, the above number is prime.[51] Consequently one can construct regular $n$-gons for $n$ in the list of prime numbers

$$\boxed{2, \quad 3, \quad 5, \quad 17, \quad 257, \quad 65\,537.}$$

For $n \leq 20$ one gets in this way the constructibility of all regular $n$-gons with

$$n = 3, 4, 5, 6, 8, 10, 12, 15, 16, 17, 20$$

using only a ruler and a compass.

[50] The number $2^{2^k}$ is to be interpreted as $2^{(2^k)}$.

[51] Fermat (1601–1665) already conjectured that the numbers (2.69) are all prime numbers. However, Euler discovered that for $k = 5$ there is a decomposition of the number in (2.69) into the product of two primes: $641 \cdot 6700417$.

## 2.7 Number theory

*Your Disqusitiones artihmeticae has placed you among the top mathematicians, and I see that the last section*[52] *contains the most beautiful analytic discovery which has been made in a long time.*

*The nearly seventy-year old Lagrange*
*in a letter to the youthful Gauss in 1804*

*It is well known that Fermat claimed that the Diophantine equation*

$$x^n + y^n = z^n$$

*– with trivial exceptions – has no solutions in integers $x, y, z$. The problem of showing this non-solvability result gives an excellent example of how a special and seemingly meaningless problem can give incredible impetus to scientific research. In fact, roused by the challenge of this Fermat conjecture, Kummer was led to his introduction of ideal numbers and to the discovery of the theorem of the unique decomposition of numbers of a cyclotomic field into ideal prime factors – a theorem which, in the form of the generalization of the result due to Dedekind and Kronecker to general algebraic systems, is at the heart of modern number theory and has importance far beyond the boundaries of number theory in the areas of algebra and function theory.*[53]

*David Hilbert (*Paris 1900*)*

Number theory is often called the queen of mathematics. Number-theoretic problems can often be formulated very easily, but solved only with a considerable amount of effort. For the proof of the Fermat conjecture mentioned above, mathematicians required the work of 350 years. It wasn't until the complete restructuring of mathematics in the twentieth century and the development of incredibly abstract tools that Andrew Wiles (Princeton, USA) was finally able to give a complete proof in 1994.

The most famous open problem in mathematics – the Riemann hypothesis – is closely connected with the distribution of prime numbers (cf. 2.7.3). The greatest mathematicians of all times have time and again put all their energy into solving number-theoretic questions and developed important mathematical tools in the process, which have then led to progress in other areas of mathematics.

The basic classics in number theory are Diophant's *Arithmetica* of ancient times, Gauss' *Disquisitiones arithmeticae*, which appeared in 1801 and founded modern number theory, and Hilbert's *Zahlbericht* of 1897, which discussed algebraic number fields. Number theory in the twentieth century has been decisively influenced by the problems which Hilbert posed at the mathematical world congress in 1900.

[52]This section is concerned with cyclotomic fields and the construction of regular polygons with a ruler and a compass (cf. 2.6.6).

[53]Fermat (1601–1665), Gauss (1777–1855), Kummer (1810–1893), Kronecker (1823–1891), Dedekind (1831–1916) and Hilbert (1862–1943).

## 2.7.1 Basic ideas

### 2.7.1.1 Different forms of mathematical thought

One makes the distinction in mathematics between:
(i) continuous thinking (for example real numbers and limits), and
(ii) discrete thinking (for example natural numbers and number theory).

Experience shows that continuous problems are often easier to treat than discrete ones. The great successes of the continuous way of thinking are based on the notion of limits and the theories connected with this notion (calculus, differential equations, integral equations and the calculus of variations) with diverse applications in physics and other natural sciences.

In contrast, number theory is the prototype for the creation of effective mathematical methods for treating discrete problems, arising in today's world in computer science, optimization of discrete systems and lattice models in theoretical physics for studying elementary particles and strings.

The epochal discovery by Max Planck in 1900 that the energy of the harmonic oscillator is not continuous but rather discrete (quantized), led to the important mathematical problem of generating discrete structures from continuous ones by an appropriate, non-trivial quantization process.

### 2.7.1.2 The modern strategy of number theory

A the end of the nineteenth century, Hilbert suggested the program of extending the, at the time already highly developed, methods of complex analysis (algebraic functions, Riemann surfaces) to number-theoretic questions. The challenge here was to formulate notions from continuous mathematics in such a way that they could also be applied to discrete systems. Number theory in the twentieth century has been molded by this program; this has led to very abstract, but at the same time very powerful methods. Important impulses in this direction are due to André Weil (1902–1998) and Alexander Grothendieck (born 1928), who revolutionized algebraic geometry and number theory with his theory of schemes.

Climaxes of number theory of the twentieth century are:
(a) the proof of the general reciprocity law for algebraic number fields by Emil Artin in 1928,
(b) the proof of André Weil's analog of the Riemann hypothesis for the $\zeta$-function of algebraic varieties over a finite field by Pierre Deligne in 1973 (Fields medal 1978).[54]
(c) the proof of the Mordell conjecture for Diophantine equations by Gerd Faltings in 1983 (Fields medal 1986) and
(d) the proof of the Fermat's last theorem (Fermat conjecture) and a more general part of the Shimura–Taniyama conjecture in 1994 by Andrew Wiles.

### 2.7.1.3 Applications of number theory

Nowadays supercomputers are applied to test number-theoretic conjectures. Ever since 1978 number-theoretic methods are applied to give subtle coding of data and digital

[54]The Fields medals have been awarded every four years at the International Congress of Mathematicians for pioneering new mathematical results. This award can be compared to the Nobel prizes. In contrast with the latter, however, the mathematicians awarded the Fields medal are not allowed to be older than 40. For outstanding life achievements, mathematicians can be awarded the Wolf prize.

information (cf. 2.7.8.1). Classical statements about the approximation of rational numbers by sequences of irrational numbers play an important role today in the investigation of chaotic and non-chaotic states in the theory of dynamical systems (for example in celestial mechanics). A bridge between number theory and theoretical physics has been spanned by superstring theory in the last few years, which has led to a fruitful transfer of ideas from pure mathematics to physics and back.[55]

#### 2.7.1.4 Compression of information in mathematics and physics

In order to understand the interaction between number theory and physics from a philosophical point of view, we mention that mathematicians have learned over the centuries, in a long process of trial and error, to code information in very compressed form using the structure of of discrete systems, making it possible to make important and deep statements about the systems. A typical example for this is the Riemann $\zeta$-function, which codes the structure of the set of prime numbers, and the law of distribution of prime numbers. Other important examples of this are given by the Dirichlet $L$-series (cf. 2.7.3) and modular forms, which grew out of the theory of elliptic functions (cf. 1.14.18). The most important information about the fine structure of a real number is coded in its continued fraction representation (cf. 2.7.5).

On the other hand, physicists have come to the notion of *partition number* from a completely different point of view, which codes the behavior of systems a large number of particles (statistical systems). If one knows the partition function, one can derive from it all relevant physical quantities of the system.[56] The Riemann $\zeta$-function may be viewed as a special kind of partition function.

The fruitful exchange of ideas between mathematics and physics is based, for example, on the fact that mathematical problems can be looked at with physical intuition, when they are translated into a physical language. A pioneer in applying this idea is Edward Witten (Institute for advanced studies, Princeton), who was awarded the Fields medal in 1990, although he is a physicist and not a mathemtician. It is interesting that many great number theorists like Fermat, Euler, Lagrange, Gauss and Minkowski also made important contributions to physics. When *Disquisitiones arithmeticae* by Gauss appeared in 1801, it represented a profound change in mathematics from a universal science, as Gauss considered it, to a specialized science. In particular, number theory took its own course for a long time. Presently one observes again a fortunate convergence of the methods of mathematics and physics.

### 2.7.2 The Euclidean algorithm

**Divisors:** If $a, b$ and $c$ are integers with the property that

$$\boxed{c = ab,}$$

then one refers to $a$ and $b$ as *divisors* of $c$. For example it follows from $12 = 3 \cdot 4$ that the numbers 3 and 4 are divisors of 12.

An integer is said to be *even*, if it is divisible by 2, otherwise it is called *odd.*

*Example 1:* Even numbers are $2, 4, 6, 8, \ldots$. Odd numbers are $1, 3, 5, 7, \ldots$.

[55] Numerous applications of number theory in the natural sciences and in computer science can be found in [286]. The relationship between number theory and modern physics is described in [276] and [288].

[56] In quantum field theory, the partition function is given by the Feynman path integral.

**An elementary criterion for divisibility:** The following statements are valid for the decimal representation of natural numbers $n$.
(i) $n$ is divisible by 3 if and only if the cross sum (i.e., the sum of the digits) is divisible by 3.
(ii) $n$ is divisible by 4 if and only if the last two digits of $n$ form a number divisible by 4.
(iii) $n$ is divisible by 5 if an only if the last digit is 5 or 0.
(iv) $n$ is divisible by 6 if and only if $n$ is even and the cross sum is divisible by 3.
(v) $n$ is divisible by 9 if and only if the cross sum is divisible by 9.
(vi) $n$ is divisible by 10 if and only if the last digit is 0.

*Example 2:* The cross sum of 4,656 is $4 + 6 + 5 + 6 = 21$. The cross sum of 21 is 3; therefore 21 and hence also 4,656 is divisible by 3, but not by 9.

The cross sum of $n = 1,234,656$ is $1 + 2 + 3 + 4 + 6 + 5 + 6 = 27$. The next cross sum is 9, hence 27 and $n$ are divisible by 9.

The last two digits of the number $m = 1,234,567,897,216$ are 16, which is divisible by 4. Hence $m$ is divisible by 4. The number $1,456,789,325$ is divisible by 5, but not by 10.

**Prime numbers:** A natural number $p$ is said to be a *prime* number, when $p \geq 2$ and the only divisors of $p$ are 1 and $p$ itself. The first prime numbers are 2,3,5,7,11.

**The sieve of Eratosthenes (around 300 BC):** Let a natural number $n > 11$ be given. To determine all prime numbers

$$\boxed{p \leq n}$$

one does the following:
(i) One writes down all natural numbers $\leq n$ (for convenience one can skip those divisible by 2, 3 or 5 to start with).
(ii) One considers all numbers $\leq \sqrt{n}$ in this list and erases all those which are multiples of these.

All numbers remaining are then prime numbers.

*Example 3:* Let $n = 100$. All prime numbers $\leq 10 = \sqrt{100}$ are $2, 3, 5$ and 7. We only have to erase the numbers which are divisible by 7 (in the following list they are underlined) and get

$$\begin{array}{l} 2 \quad 3 \quad 5 \quad 7; \quad 11 \quad 13 \quad 17 \quad 19 \quad 23 \quad 29 \quad 31 \quad 37 \quad 41 \\ 43 \quad 47 \quad \underline{49} \quad 53 \quad 59 \quad 61 \quad 67 \quad 71 \quad 73 \quad \underline{77} \quad 79 \quad 83 \quad 89 \quad \underline{91} \quad 97. \end{array}$$

All prime numbers $\leq 100$ are obtained as those numbers which are not underlined.

A table of all prime numbers $\leq 4000$ is given in section 0.6.

**Theorem of Euclid (around 300 BC):** There are infinitely many prime numbers.

This theorem is proved in Euclid's *Elements*. The following theorem can be found there also, but only implicitly.

**The fundamental theorem of arithmetic:** Every natural number $n \geq 2$ is the product of prime numbers. This decomposition is unique, if the prime numbers are ordered according to their size.

*Example 4:* One has

$$24 = 2 \cdot 2 \cdot 2 \cdot 3 \quad \text{and} \quad 28 = 2 \cdot 2 \cdot 7.$$

**Least common multiple:** If $m$ and $n$ are two positive natural numbers, then one obtains the *least common multiple* of $n$ and $m$, denoted

$$\boxed{\operatorname{lcm}(m, n)\ ,}$$

by multiplying all *distinct* prime numbers in the prime number decompositions of $m$ and $n$.

*Example 5:* From Example 4 we have

$$\operatorname{lcm}(24, 28) = 2 \cdot 2 \cdot 2 \cdot 3 \cdot 7 = 168.$$

**Greatest common divisor:** If $m$ and $n$ are two positive natural numbers, then one denotes by

$$\boxed{\gcd(m, n)}$$

the *greatest common divisor* of $m$ and $n$. This gcd is obtained by taking the product of all prime numbers which occur in both the prime number decompositions of $m$ and $n$.

*Example 6:* From Example 4 we have $\gcd(24, 28) = 2 \cdot 2 = 4$.

**The Euclidean algorithm for the calculation of the greatest common divisor:** If $n$ and $m$ are two given integers, both non-vanishing, then we set $r_0 := |m|$ and use the following scheme of division with remainder:

$$\boxed{\begin{array}{rll} n &= \alpha_0 r_0 + r_1, & 0 \le r_1 < r_0, \\ r_0 &= \alpha_1 r_1 + r_2, & 0 \le r_2 < r_1, \\ r_1 &= \alpha_2 r_2 + r_3, & 0 \le r_3 < r_2 \quad \text{etc.} \end{array}}$$

Here $\alpha_0, \alpha_1, \ldots$ as well as the remainders $r_1, r_2, \ldots$ are uniquely determined integers. After finitely many steps one obtains at some step $r_k = 0$. Thus

$$\boxed{\gcd(m, n) = r_{k-1}.}$$

This process is referred to as the Euclidean algorithm applied to $m$ and $n$. Briefly:

> *The greatest common divisor is the last non-vanishing remainder in the Euclidean algorithm.*

*Example 7:* For $m = 14$ and $n = 24$ one gets:

$$\boxed{\begin{array}{rll} 24 &= 1 \cdot \mathbf{14} + \mathbf{10} & \text{(remainder } 10), \\ \mathbf{14} &= 1 \cdot \mathbf{10} + 4 & \text{(remainder } 4), \\ \mathbf{10} &= 2 \cdot \mathbf{4} + \boxed{2} & \text{(remainder } 2), \\ \mathbf{4} &= 2 \cdot \mathbf{2} & \text{(remainder } 0). \end{array}}$$

Hence $\gcd(14, 24) = 2$.

**Relatively prime numbers:** Two positive natural numbers $m$ and $n$ are said to *relatively prime*, if

$$\boxed{\gcd(m, n) = 1.}$$

*Example 8:* The number 5 is relatively prime to 6,7,8,9, but not to 10.

**The Euler $\varphi$-function:** Let $n$ be a positive natural number. We denote by $\varphi(n)$ the number of all positive natural numbers $m \leq n$ which are relatively prime to $n$. For $n = 1, 2, \ldots$ one has:

$$\boxed{\varphi(n) = n \prod_{p|n} \left(1 - \frac{1}{p}\right).}$$

The product in this expression is carried out over all prime numbers $p$ which divide $n$.

*Example 9:* One has $\varphi(1) = 1$. For $n \geq 2$ one has $\varphi(n) = n - 1$ if and only if $n$ is a prime number.

From $\gcd(1,4) = \gcd(3,4) = 1$ and $\gcd(2,4) = 2$, $\gcd(4,4) = 4$, it follows that

$$\varphi(4) = 2.$$

**The Möbius function:** Let $n$ be a positive natural number. We set

$$\mu(n) := \begin{cases} 1 & \text{for } \quad n = 1 \\ (-1)^r & \text{if the decomposition of } n \text{ into prime numbers} \\ & \text{contains exactly } (r) \text{ distinct primes} \\ 0 & \text{otherwise.} \end{cases}$$

*Example 10:* From $10 = 2 \cdot 5$ it follows that $\mu(10) = 1$. Because of $8 = 2 \cdot 2 \cdot 2$ one has $\mu(8) = 0$.

**Calculations with number-theoretic functions from their sums:** Let a function $f$ be given which associates to every natural number $n$ an integer $f(n)$. Then one has

$$f(n) = \sum_{d|n} \mu\left(\frac{n}{d}\right) s(d) \qquad \text{for } n = 1, 2, \ldots$$

with

$$s(d) := \sum_{c|d} f(c).$$

Here the sum is extended over all divisors $d \geq 1$ of $n$ and all divisors $c \geq 1$ of $d$ (this is the *inversion formula* of Möbius (1790–1868)).

This formula tells us that the value $f(n)$ can be constructed from the sums $s(d)$.

## 2.7.3 The distribution of prime numbers

> *I feel that I am expressing my gratitude for the honor that the Berlin Academy has endowed upon me by appointing me as one of its correspondents, by making immediate use of my privileges thereof and presenting the results of an investigation on the frequency of prime numbers; a subject which, by the interest which Gauss and Dirichlet have given it over a long period of time, seems to be worthy of anewed mention.*[57]
>
> *Bernhard Riemann* (1859)

[57] This is the beginning of one of the most famous works in all of mathematics. On page 8 of this paper Riemann develops his new ideas and presents the celebrated "Riemann hypothesis".

The collected works of Riemann, which, together with extensive up-to-date commentaries, are contained in [182], are all in all just one volume. However, every single one of these papers is a jewel of mathematics. Riemann has profoundly influenced the mathematics of the twentieth century with his treasure of ideas.

One of the main problems of number theory is to discover laws about the distribution of prime numbers.

**The gap theorem:** The numbers $n! + 2, n! + 3, \ldots, n! + n$ are, for $n = 2, 3, 4, \ldots$, not prime numbers, and for growing $n$ the sets of non-primes (gaps in the set of primes) get longer and longer.

**Theorem of Dirichlet (1837) on arithmetic progressions:** The sequence[58]

$$\boxed{a, \quad a+d, \quad a+2d, \quad a+3d, \quad \ldots} \tag{2.70}$$

contains infinitely many prime numbers, provided that $a$ and $d$ are two relatively prime natural numbers.

*Example 1:* One may choose $a = 3$ and $d = 5$. Then one gets the sequence

$$3, \quad 8, \quad 13, \quad 18, \quad 23, \quad \ldots$$

which contains infinitely many prime numbers.

**Corollary:** For an arbitrary real number $x \geq 2$ we define

$$P_{a,d}(x) := \text{set of all prime numbers } p \text{ in (2.70) with } p \leq x.$$

Then one has:

$$\boxed{\sum_{p \in P_{a,d}(x)} \frac{\ln p}{p} = \frac{1}{\varphi(d)} \ln x + O(1), \qquad x \to +\infty,} \tag{P}$$

where $\varphi$ denotes the Euler function (cf. 2.7.2). The remainder term $O(1)$ does not depend on $a$. In Example 1 we have $\varphi(d) = 4$.

The formula (P) makes the statement precise that all sequences (2.70) with a constant difference $d$ contain "asymptotically" the same number of prime numbers, independent of the value of $a$.

In particular for $a = 2$ and $d = 1$ (2.70) contains all prime numbers. In this case one has $\varphi(d) = 1$.

**Analytic number theory:** In his proof of this theorem, Dirichlet introduced totally new methods into number theory (Fourier series, Dirichlet series and $L$-series), which have also turned out to be fundamental in the theory of algebraic numbers. In so doing he founded a new branch of mathematics – analytic number theory.[59]

#### 2.7.3.1 The prime number theorem

**Prime number distribution functions:** For an arbitrary real number $x \geq 2$ we

[58] The sequence in (2.70) is an *arithmetic progression*, i.e., the difference of two successive terms is constant.

If $a$ and $d$ are not relatively prime, then in (2.70) no prime numbers appear at all, unless $a$ itself is a prime.

[59] After Gauss' death in 1855, Dirichlet (1805–1859) became his successor in Göttingen. In 1859, Riemann was appointed to this famous chair in Göttingen. From 1886 until his death in 1925, Felix Klein was also in Göttingen; in 1895 Klein brought David Hilbert to Göttingen, who worked there until his retirement in 1930.

In the twenties of the twentieth century, Göttingen was the leading center of mathematics and physics in the world. In 1933 many of the leading scientists emigrated and Göttingen lost its supreme position among the intellectual centers of the world.

define

$$\boxed{\pi(x) := \text{ number of the prime numbers } \le x.}$$

**Theorem of Legendre (1798):** $\lim_{x\to\infty} \frac{\pi(x)}{x} = 0.$

Hence there are considerably fewer prime numbers than natural numbers.

**The fundamental prime number theorem:** For large numbers $x$ one has the following asymptotic equality:[60]

$$\boxed{\pi(x) \sim \frac{x}{\ln x} \sim \operatorname{li} x, \qquad x \to +\infty.}$$

This is the most famous asymptotic formula in mathematics. Table 2.5 compares $\pi(x)$ with $\operatorname{li} x$.

*Table 2.5.*

| $x$ | $\pi(x)$ | $\operatorname{li} x$ |
|---|---|---|
| $10^3$ | 168 | 178 |
| $10^6$ | 78 498 | 78 628 |
| $10^9$ | 50 847 534 | 50 849 235 |

Euler (1707–1783) still believed that prime numbers are distributed totally irregularly. The asymptotic distribution for $\pi(x)$ was found independently by the 33-year old Legendre in 1785 and the 14-year old Gauss in 1792 through an intensive study of tables of logarithms.

A rigorous proof of the prime number theorem was given independently by the thirty-year old Jacques Hadamard (1865–1963) and the thirty-year old Charles de la Valleé–Poussin in 1896. If $p_n$ denotes the $n$th prime number, then one has

$$\boxed{p_n \sim n \cdot \ln n, \qquad n \to +\infty.}$$

**Error estimate:** There are positive constants $A$ and $B$ so that for all $x \ge 2$ we have the relation

$$\pi(x) = \frac{x}{\ln x}(1 + r(x))$$

with

$$\frac{A}{\ln x} \le r(x) \le \frac{B}{\ln x}.$$

**Statement of Riemann:** We have the considerably more precise estimate

$$\boxed{|\pi(x) - \operatorname{li} x| \le \text{const} \cdot \sqrt{x} \ln x \qquad \text{for all } x \ge 2,}$$

provided the Riemann hypothesis (2.72) turns out to be true.[61]

[60] Explicitly this corresponds to the limiting relation

$$\lim_{x\to+\infty} \frac{\pi(x)}{\left(\frac{x}{\ln x}\right)} = \lim_{x\to+\infty} \frac{\pi(x)}{\operatorname{li} x} = 1.$$

The definition of logarithmic integral is:

$$\operatorname{li} x := \mathrm{PV} \int_0^x \frac{\mathrm{d}t}{\ln t} = \lim_{\varepsilon\to+0} \left( \int_0^{1-\varepsilon} \frac{\mathrm{d}t}{\ln t} + \int_{1+\varepsilon}^{x} \frac{\mathrm{d}t}{\ln t} \right).$$

[61] In 1914 Littlewood proved that the difference $\pi(x) - \operatorname{li} x$ changes sign infinitely often for growing $x$. If, however, $x_0$ denotes the value of $x$ for the first such change of sign, then according to Skewes (1955) we have $10^{700} < x_0 < 10^{10^{10^{34}}}$.

**The Riemann $\zeta$-function:** Riemann considered the function

$$\boxed{\zeta(s) := \sum_{n=1}^{\infty} \frac{1}{n^s}, \qquad s \in \mathbb{C},\ \mathrm{Re}\, s > 1.}$$

The surprising relation of this function with the theory of prime numbers arises from the following result.

**Euler's theorem (1737):** For all real numbers $s > 1$ one has

$$\boxed{\prod_p \left(1 - \frac{1}{p^s}\right)^{-1} = \zeta(s),} \qquad (2.71)$$

where the product is extended over all prime numbers $p$. This means:

> The Riemann $\zeta$-function encodes the structure of the set of all prime numbers.

**Riemann's theorem (1859):**

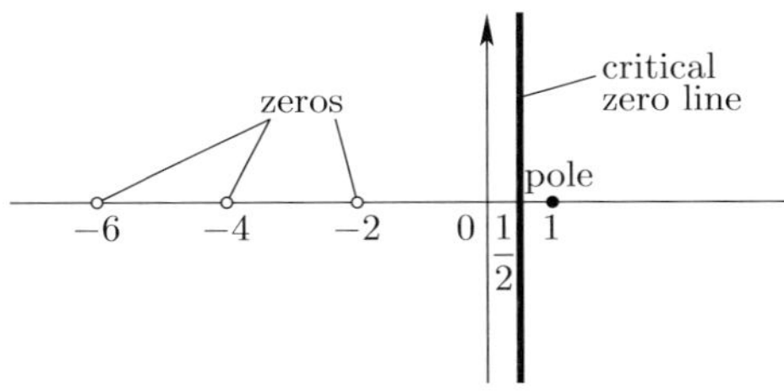


*Figure 2.5. Riemann's $\zeta$-function.*

(i) The $\zeta$-function can be extended to an analytic function on $\mathbb{C} - \{1\}$, and it has a pole of the first order at $s = 1$ and residue there equal to unity, i.e., for all complex numbers $s \neq 1$, one has

$$\zeta(s) = \frac{1}{s-1} + \text{power series centered at the point } s.$$

For example, $\zeta(0) = -\frac{1}{2}$.

(ii) For all complex numbers $s \neq 1$ one has a *functional equation*

$$\boxed{\pi^{-s/2}\Gamma\left(\frac{s}{2}\right)\zeta(s) = \pi^{(s-1)/2}\Gamma\left(\frac{1}{2} - \frac{s}{2}\right)\zeta(1-s).}$$

(iii) The $\zeta$-function has the so-called *trivial zeros* at $s = -2k$ for $k = 1, 2, 3, \ldots$ (Figure 2.5).

### 2.7.3.2 The celebrated Riemann hypothesis

In 1859 Riemann formulated the following conjecture:

> All non-trivial zeros of the $\zeta$-function lie on the *critical line* $\mathrm{Re}\, s = \frac{1}{2}$ in the complex plane. (2.72)

**Hardy's theorem (1914):** On the critical line $\mathrm{Re}\, s = \frac{1}{2}$ there are infinitely many zeros of the Riemann $\zeta$-function.[62]

One today knows of a series of statements of different kinds, all of which are equivalent to the Riemann hypothesis. Extensive, imaginative projects and experiments with supercomputers have given no clue as of yet that the Riemann hypothesis might not hold. The calculations yield billions of zeros on the critical line. Precise asymptotic estimates show that at least one-third of all of the zeros of the $\zeta$-function must lie on that critical line.

#### 2.7.3.3 The Riemann $\zeta$-function and statistical physics

A basic realization of statistical physics is that one can obtain all physical properties of a statistical system from the energy states $E_1, E_2, \ldots$ for a fixed number of particles from the partition function

$$\mathscr{Z} = \sum_{n=1}^{\infty} \mathrm{e}^{-E_n/\mathrm{k}T}. \tag{2.73}$$

Here $T$ denotes the absolute temperature of the system, and k is the Boltzmann constant. If we set

$$E_n := \mathrm{k}Ts \cdot \ln n, \qquad n = 1, 2, \ldots,$$

then we have

$$\boxed{\mathscr{Z} = \zeta(s).}$$

This means that the Riemann $\zeta$-function is a *particular partition function.* This explains the fact that functions of type similar to the $\zeta$-function are important in treating models in statistical physics precisely (and not just approximately on supercomputers).

#### 2.7.3.4 The Riemann $\zeta$-function and renormalization in physics

Unfortunately divergent expressions occur often in statistical physics and in quantum field theory. Physicists have developed ingenious methods to overcome these difficulties and make some sense of these initially meaningless (divergent) expressions. This is the broad area of renormalization in physics, which in quantum electrodynamics and the standard model, in spite of the mathematical dubiosity of the arguments, agree phenomenally well with experimental evidence.[63]

*Example:* The trace of an $(n \times n)$-unity matrix $I_n$ has the value

$$\mathrm{tr}\, I_n = 1 + 1 + \ldots + 1 = n.$$

For the infinite unity matrix $I$ we get for the trace

$$\mathrm{tr}\, I = \lim_{n \to \infty} n = \infty.$$

[62] In the posthumous works of Riemann at the university library at Göttingen it has been discovered that Riemann actually had proved this theorem, although he never published it.

[63] Euler often worked with divergent series. It was his keen mathematical intuition which led him to correct results in doing this. Renormalization is in a sense an extension of Euler's method. A famous open problem in mathematical physics is to present rigorous foundations of quantum field theory in which renormalization is better understood and justified.

In order to associate a sensible finite value to $\operatorname{tr} I$, we consider the equation

$$\zeta(s) = \sum_{n=1}^{\infty} \frac{1}{n^s}. \tag{2.74}$$

For $\operatorname{Re} s > 1$ this is a correct formula in the sense of convergent series. The left-hand side can, by Riemann's theorem, be extended to an analytic function which associates to every complex number $s \neq 1$ a definite value. Using this fact, we can define the right-hand side of (2.74) as this definite value. In the special case in which $s = 0$, the right-hand side is formally $1 + 1 + 1 + \cdots$. Hence we can defined the renormalized value of $\operatorname{tr} I$ by

$$\boxed{(\operatorname{tr} I)_{\text{ren}} := \zeta(0) = -\frac{1}{2}.}$$

The "infinite sum" $1 + 1 + 1 + \cdots$ of positive numbers surprisingly gives as a result a negative number!

#### 2.7.3.5 Dirichlet's localization principle for prime numbers modulo $m$

Let $m$ be a positive natural number. To prove his fundamental theorem on the distribution of prime numbers, Dirichlet put Euler's theorem (2.71) in the following form:

$$\boxed{\prod_p \left(1 - \frac{\chi_m(p)}{p^s}\right)^{-1} = L(s, \chi_m) \qquad \text{for all } s > 1.}$$

The product is to be taken over all prime numbers $p$. Here we have set

$$L(s, \chi_m) := \sum_{n=1}^{\infty} \frac{\chi_m(n)}{n^s}.$$

This series is called a *Dirichlet L-series.* The symbol $\chi_m$ denotes a *Dirichlet character*, which is a character modulo $m$, meaning:

(i) The map $\chi' : \mathbb{Z}/m\mathbb{Z} \longrightarrow \mathbb{C} - \{0\}$ is a group homomorphism.[64]

(ii) For all $g \in \mathbb{Z}$ we set

$$\chi_m(g) := \begin{cases} \chi'([g]) & \text{if } \gcd(g, m) = 1, \\ 0 & \text{otherwise.} \end{cases}$$

*Example:* If $m = 1$, we get $\chi_1(g) = 1$ for all $g \in \mathbb{Z}$. Then

$$L(s, \chi_1) = \zeta(s) \qquad \text{for all } s > 1.$$

Thus the Dirichlet $L$-series generalize the Riemann $\zeta$-function. Roughly speaking, one has

> The Dirichlet function $L(\cdot, \chi_m)$ encodes the structure of the set of prime numbers modulo $m$.

The theory of $L$-functions can be generalized to other algebraic objects, in particular to algebraic number fields.

---

[64] If $[g] = g + m\mathbb{Z}$ denotes the residue class of $g$ modulo $m$ (cf. 2.5.2), then one has $\chi'([g]) \neq 0$ and $\chi'([g][h]) = \chi'([g])\chi'([h])$ for all $g, h \in \mathbb{Z}$.

#### 2.7.3.6 The conjecture on prime twins

Two different prime numbers which differ just by two are called *twins*. Examples are given by the pairs 3, 5 as well as 5, 7 and 11, 13. It is generally conjectured that there are infinitely many of these pairs.

### 2.7.4 Additive decompositions

Additive number theory begins with the decomposition of numbers into sums.

#### 2.7.4.1 The Goldbach conjecture

In 1742, Goldbach formulated the following two conjectures in a letter to Euler:
(G1) Every even number $n > 2$ is the sum of two prime numbers.
(G2) Every odd number $n > 5$ is the sum of three prime numbers.

*Example:* One has the following decompositions:

$$4 = 2 + 2, \quad 6 = 3 + 3, \quad 8 = 5 + 3, \quad 10 = 7 + 3, \quad \ldots$$

and

$$7 = 3 + 2 + 2, \quad 9 = 5 + 2 + 2, \quad 11 = 7 + 2 + 2, \quad 13 = 7 + 3 + 3, \quad \ldots .$$

If one uses the decomposition $n = 3 + m$, then (G2) follows immediately from (G1).

Computer experiments have shown that the two statements (G1) and (G2) are correct for all $n \leq 10^8$.

The proof of the conjecture (G1) is completely open to the present day. On the other hand, Vinogradov showed in 1937 that (G2) holds for all $n$ with

$$\boxed{n \geq 3^{3^{15}}.}$$

This lower bound has more that 6 million decimal places.

#### 2.7.4.2 The Waring problem

It was proven by Lagrange in 1770 that every natural number is the sum of four squares.

*Example 1:* One has

$$2 = 1^2 + 1^2 + 0^2 + 0^2 \quad \text{and} \quad 7 = 2^2 + 1^2 + 1^2 + 1^2.$$

Also in 1770, Waring formulated the conjecture that for every natural number $k \geq 2$ there is a natural number $g(k) \geq 1$ such that every natural number $n$ can be written in the form

$$\boxed{n = m_1^k + \ldots + m_{g(k)}^k}$$

with integers $m_1, m_2, \ldots$.

This conjecture was proved by Hilbert in 1909. The minimal amounts are $g(2) = 4$ (four squares), $g(3) = 9$ (nine cubes), $g(4) = 19$ (19 fourth powers). In general one has the estimate

$$g(k) \geq 2^k + \left[\left(\frac{3}{2}\right)^k\right] - 2 \qquad \text{for all } k \geq 2.$$

Here $[m]$ denotes the largest integer $\leq m$ (Gauss bracket).

**The special case of two squares:** A natural number $n \geq 2$ can be written as the sum of two squares of integers if and only if, in the primary decomposition of $n$, all appearing primes of the form

$$4m + 3, \qquad m = 0, 1, 2, \ldots$$

occur only with odd powers.

**Fermat's theorem (1659):** A prime number is a sum of two squares of natural numbers if and only if it is of the form

$$\boxed{4m + 1, \qquad m = 1, 2, \ldots\,.}$$

This decomposition is unique up to the order of the squares.

*Example 2:* The prime number $13 = 4 \cdot 3 + 1$ has the unique decomposition

$$13 = 2^2 + 3^2.$$

We denote by $N(n)$ the number of different possibilities to represent a positive natural number $n$ as the sum of four squares.

**Jacobi's theorem (1829):**

$$\boxed{N(n) = 8 \cdot \left\{ \begin{array}{l} \text{sum of all positive divisors of } n, \\ \text{which are not divisible by 4.} \end{array} \right\}}$$

*Example 3:* One has $N(1) = 8 \cdot 1$. In fact, one has:

$$\begin{array}{ll} 1 = (\pm 1)^2 + 0^2 + 0^2 + 0^2, & 1 = 0^2 + (\pm 1)^2 + 0^2 + 0^2 + 0^2, \\ 1 = 0^2 + 0^2 + (\pm 1)^2 + 0^2, & 1 = 0^2 + 0^2 + 0^2 + (\pm 1)^2. \end{array}$$

### 2.7.4.3 Partitions

Let $n$ be a positive natural number. We define

$$\boxed{p(n) := \left\{ \begin{array}{l} \text{the number of decompositions of } n \text{ into} \\ \text{a sum of positive natural numbers.} \end{array} \right\}}$$

*Example 1:* One has $p(3) = 3$, since

$$3 = 1 + 1 + 1, \qquad 3 = 2 + 1, \qquad 3 = 1 + 2.$$

**Coding information:** We define the partition function

$$P(q) := \sum_{n=0}^{\infty} p(n) q^n$$

with $p(0) := 1$. This series converges for all complex numbers $q$ with $|q| < 1$. Thus all information on partitions are coded by the function $P$. The problem now is to obtain information on partitions by elegant manipulations with $P$. This problem was solved by Euler.

**Euler's theorem:** For all $q \in \mathbb{C}$ with $|q| < 1$ one has the convergent product representation

$$P(q) = \frac{1}{\prod\limits_{n=1}^{\infty}(1-q^n)}$$

together with

$$\prod_{n=1}^{\infty}(1-q^n) = \sum_{n=-\infty}^{\infty}(-1)^n q^{\frac{3n^2+n}{2}} = 1 - q - q^2 + q^5 + q^7 + \ldots .$$

These surprisingly simple formulas were found by Euler numerically. After finding the formulas, he had to work a long time to get a proof.

**The Euler recursion formula:** We set $p(n) := 0$ for $n < 0$. For all $n = 1, 2, \ldots$ one has

$$p(n) = \sum_{k=1}^{n}(-1)^{k+1}\{p(n-\omega(k)) + p(n-\omega(-k))\}$$

with $\omega(k) := \frac{1}{2}(3k^2 - k)$. The first few terms of this series are explicitly

$$p(n) = p(n-1) + p(n-2) - p(n-5) - p(n-7) + \ldots .$$

*Example 2:*

$$\begin{aligned} p(2) &= p(1) + p(0) = 2, \\ p(3) &= p(2) + p(0) = 3, \\ p(4) &= p(3) + p(2) = 5, \\ &\ldots \\ p(200) &= 3\,972\,999\,029\,388. \end{aligned}$$

**The asymptotic formulas of Hardy and Ramanunjan (1918)**[65]**:** Let $K := \pi\sqrt{(2/3)}$. One has for $n \to \infty$ the asymptotic equalities

$$p(n) \approx \frac{e^{K\sqrt{n}}}{4n\sqrt{3}},$$

$$\ln p(n) \approx \pi\sqrt{\frac{2n}{3}}.$$

**The product formula of Jacobi (1829):** For all complex numbers $q$ and $z \neq 0$ with $|q| < 1$ one has the equality

$$\prod_{n=1}^{\infty}(1-q^{2n})(1+q^{2n-1}z)(1+q^{2n-1}z^{-1}) = \sum_{n=-\infty}^{\infty} q^{n^2} z^n.$$

---

[65] Rademacher discovered in 1937 that $p(n)$ can be expanded in a convergent series in $n$. His proof used the Dedekind $\eta$-function

$$\eta(\tau) := e^{\pi\tau i/12}\prod_{n=1}^{\infty}(1-q^n) \qquad \text{with} \quad q := e^{2\pi\tau i},$$

which is holomorphic in the open upper half plane (cf. [216]). This is a typical example for the fruitfulness of the deep theory of modular forms (cf. 1.14.18).

### 2.7.5 The approximation of irrational numbers by rational numbers and continued fractions

We consider the question of how one can approximate irrational numbers by rational ones. In these considerations, continued fractions play a central role. Continued fractions first occurred in the seventeenth century. For example, Christian Huygens (1629–1695) ran across continued fractions in his construction of a cogwheel model of the solar system and to this end tried to approximate the relations between the periods of the planets with as few teeth as possible. The theory of continued fractions goes back to Euler (1707–1783). In contrast to decimal decompositions, continued fractions give information on the *fine structure* of real numbers. For example, the fundamental problem of the best approximation of irrational numbers by rational numbers is solved with the help of continued fractions (cf. 2.7.5.3). Often continued fractions are significantly more effective than power series. They are also applied in many ways in computer algorithms.

**Basic idea:** From the identity

$$\sqrt{2} = 1 + \frac{1}{1+\sqrt{2}}$$

it follows by repeated insertion of factors

$$\sqrt{2} = 1 + \frac{1}{1+\left(1+\dfrac{1}{1+\sqrt{2}}\right)}$$

and

$$\sqrt{2} = 1 + \cfrac{1}{2+\cfrac{1}{2+\cfrac{1}{1+\sqrt{2}}}} \quad \text{etc.} \tag{2.75}$$

#### 2.7.5.1 Finite continued fractions

**Definition:** A *finite continued fraction* is an expression of the form

$$a_0 + \cfrac{1}{a_1 + \cfrac{1}{a_2 + \cfrac{1}{a_3 + \cdots \cfrac{\ddots}{\cfrac{1}{a_{n-1} + \cfrac{1}{a_n}}}}}}\ .$$

For this we also use the symbol

$$[a_0, a_1, \ldots, a_n].$$

Here $a_0, a_1, \ldots, a_n$ are real or complex numbers, which with the exception of $a_0$ are all assumed to be non-vanishing.

*Example 1:*

$$1+\cfrac{1}{1+\cfrac{1}{1+\cfrac{1}{2}}}=1+\cfrac{1}{1+\cfrac{2}{3}}=1+\frac{3}{5}=\frac{8}{5}.$$

*Example 2:*

$$[a_0,a_1]=a_0+\frac{1}{a_1},$$

$$[a_0,a_1,a_2]=a_0+\cfrac{1}{a_1+\cfrac{1}{a_2}}=a_0+\frac{a_2}{a_1a_2+1}.$$

**Recursion formula:** One has

$$\boxed{[a_0,a_1,\ldots,a_n]=a_0+\frac{1}{[a_1,\ldots,a_n]}.}$$

**Effective algorithm:** If one uses the iteration process

$$\begin{aligned} p_k &= a_kp_{k-1}+p_{k-2}, \\ q_k &= a_kq_{k-1}+q_{k-2}, \end{aligned} \qquad k=0,1,\ldots,n, \tag{2.76}$$

with the initial values $p_{-2}:=0,\ p_{-1}:=1$ and $q_{-2}:=1,\ q_{-1}:=0$, then one has

$$\boxed{[a_0,a_1,\ldots,a_n]=\frac{p_n}{q_n}, \qquad n=0,1,\ldots.}$$

*Example 3:* In order to calculate

$$[2,1,2,1]=\frac{11}{4}$$

conveniently with the help of (2.76), we use the scheme presented in Table 2.6. Every number of the third line arises as the product of the number $a_n$ above it with the previous number of the third line, plus the number of the third line preceeding that previous number. Similarly the fourth line is calculated.

*Table 2.6. Calculation of a continued fraction.*

| $n$ | $-2$ | $-1$ | 0 | 1 | 2 | 3 | length of continued fraction |
|---|---|---|---|---|---|---|---|
| $a_n$ | | | 2 | 1 | 2 | 1 | components of continued fraction |
| $p_n$ | 0 | 1 | 2 | 3 | 8 | 11 | iteration values |
| $q_n$ | 1 | 0 | 1 | 1 | 3 | 4 | iteration values |
| $\frac{p_n}{q_n}$ | | | $\frac{2}{1}$ | $\frac{3}{1}$ | $\frac{8}{3}$ | $\frac{11}{4}$ | canonical approximation fractions |

### 2.7.5.2 Infinite continued fractions

**Definition:** An *infinite continued fraction*, denoted

$$\boxed{[a_0,a_1,\ldots]} \tag{2.77}$$

is a sequence $\left(\frac{p_n}{q_n}\right)$ of finite continued fractions

$$\frac{p_n}{q_n} = [a_0, a_1, \ldots, a_n].$$

The infinite continued fraction (2.77) is said to be *convergent*, if the finite limit

$$\boxed{\lim_{n\to\infty} \frac{p_n}{q_n} = \alpha}$$

exists. Then we associate the number $\alpha$ to the infinite continued fraction (2.77).

**Criterion for convergence:** The infinite continued fraction (2.77) is convergent if and only if the infinite series

$$\sum_{n=1}^{\infty} a_n$$

diverges. Then one has in addition the interlocking of intervals

$$\boxed{\frac{p_{2m}}{q_{2m}} < \alpha < \frac{p_{2m+1}}{q_{2m+1}}, \qquad m = 0, 1, 2, \ldots} \tag{2.78}$$

and the (generally) much sharper error estimate (2.83).

*Example 1:* A continued fraction $[a_0, a_1, \ldots]$ is said to be *regular*, if all $a_j$ are integers with $a_j > 0$ for all $j \geq 1$. Every continued fraction of this type converges.

*Example 2:* The continued fraction $[1, \overline{2}] = [1, 2, 2, 2, \ldots]$ converges.[66] According to (2.75) one has:

$$\boxed{\sqrt{2} = [1, \overline{2}].}$$

**Unique representation of of real numbers by continued fractions:** Every real number $\alpha$ can be written uniquely as a continued fraction. For this expression one has:

(i) $\alpha$ is rational if the corresponding continued fraction is finite.[67]

(ii) $\alpha$ is irrational, if the corresponding continued fraction is infinite.

This theorem gives us a general tool to determine the irrationality of real numbers. One calculates their continued fraction and checks whether the latter is infinite.

*Example 3:* The number $\sqrt{2}$ is irrational, since its continued fraction is the infinite one of Example 2.

**Constructive iteration:** Let $\rho_0 := \alpha$. The determination of the continued fraction of

[66] For simplicity we indicate periods by an over-line. For example

$$[a, b, \overline{c, d}] = [a, b, c, d, c, d, c, d, \ldots].$$

[67] To avoid trivial ambiguities in the representation $\alpha = [a_0, a_1, \ldots, a_n]$ for a rational number $\alpha$, we assume in addition that $a_n \neq 1$ for all $n \geq 1$. The given algorithms automatically take this convention into account.

$\alpha$ can be done with the help of the following (modified) Euclidean algorithm:[68]

$$\boxed{\begin{aligned} \rho_0 &= [\rho_0] + \frac{1}{\rho_1}, \\ \rho_1 &= [\rho_1] + \frac{1}{\rho_2}, \\ \rho_2 &= [\rho_2] + \frac{1}{\rho_3} \quad \text{etc.} \end{aligned}} \tag{2.79}$$

The iteration ends when at some stage there is no remainder. If we set $a_0 := [\rho_0]$, $a_1 := [\rho_1], \ldots$, then we have

$$\boxed{\alpha = [a_0, a_1, \ldots].}$$

*Example 4:* For $\alpha = \frac{10}{7}$ we get

$$\begin{aligned} \frac{10}{7} &= \boxed{1} + \frac{\mathbf{3}}{\mathbf{7}}, \\ \frac{\mathbf{7}}{\mathbf{3}} &= \boxed{2} + \frac{\mathbf{1}}{\mathbf{3}}, \\ \frac{\mathbf{3}}{\mathbf{1}} &= \boxed{3}. \end{aligned}$$

This yields $\frac{10}{7} = [1, 2, 3]$.

*Example 5:* Euler determined for the number e the neat continued fraction expression

$$\boxed{\mathrm{e} = [2, \overline{1, 2n, 1}]_{n=1}^{\infty}.} \tag{2.80}$$

This means $\mathrm{e} = [2, 1, 2, 1, 1, 4, 1, 1, 6, 1, \ldots]$. Since this continued fraction is infinite, Euler was able to prove the irrationality of e in 1737. It wasn't until 150 years later that Hermite was able to prove the transcendence of e.

In fact, Euler derived first the formula

$$\frac{\mathrm{e}^{k/2} + 1}{\mathrm{e}^{k/2} - 1} = [k, 3k, 5k, \ldots], \qquad k = 1, 2, \ldots$$

and then guessed the formula (2.80) before he was able to give it a rigorous proof. More continued fractions can be found in Table 2.7.

**The golden ratio:** If one divides the unit segment $[0, 1]$ by the point $x$, where

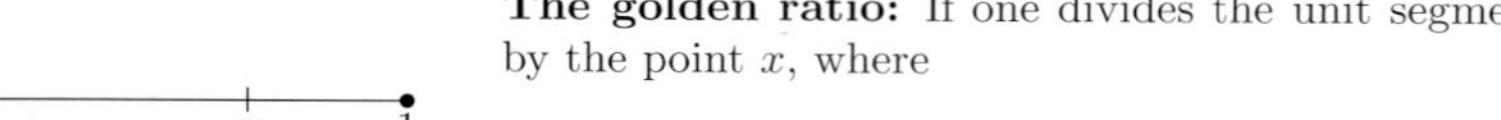


*Figure 2.6. The golden ratio.*

$$\boxed{\frac{1}{x} = \frac{x}{1 - x},} \tag{2.81}$$

then this division has be known since ancient times as the golden ratio and is considered to be particularly esthetic for

[68]Here again $[x]$ denotes the largest integer $\leq x$ (Gauss bracket).

*Table 2.7. Continued fractions of some special real numbers.*

| ***Real number*** | ***Continued fraction*** |
|---|---|
| $\frac{1}{2}(\sqrt{5}-1)$ (the golden ratio) | $[0, \overline{1}]$ |
| $\frac{1}{2}(\sqrt{5}+1)$ | $[\overline{1}]$ |
| $\sqrt{2}$ | $[1, \overline{2}]$ |
| $\sqrt{3}$ | $[1, \overline{1, 2}]$ |
| $\sqrt{4}$ | $[2]$ |
| $\sqrt{5}$ | $[2, \overline{4}]$ |
| $\sqrt{6}$ | $[2, \overline{2, 4}]$ |
| $\sqrt{7}$ | $[2, \overline{1, 1, 1, 4}]$ |
| e | $[2, \overline{1, 2n, 1}]_{n=1}^{\infty}$ |
| $\pi$ | $[3, 7, 15, 1, 292, 1, 1, 1, 2, 1, 3, 1, 14, \ldots]$ (no visible pattern) |

sculptures, paintings and buildings (Figure 2.6). From (2.81) it follows that $x^2+x-1=0$, that is

$$x = \frac{1}{2}(\sqrt{5}-1) = 0,618\ldots$$

with the continued fraction decomposition

$$x = [0, \overline{1}] = [0, 1, 1, 1, \ldots].$$

It is interesting to note that this number has the *simplest possible continued fraction.*

#### 2.7.5.3 Best rational approximation

**Main theorem:** Let $\alpha$ be an irrational real number and $n \geq 2$. Then one has[69]

$$\boxed{\left|\alpha - \frac{p_n}{q_n}\right| < \left|\alpha - \frac{p}{q}\right|,} \tag{2.82}$$

if $p$ and $q$ are integers which satisfy the relation

$$0 < q \leq q_n \quad \text{and} \quad \frac{p}{q} \neq \frac{p_n}{q_n}.$$

[69] One has even stronger statements such as $|\alpha q_n - p_n| < |\alpha q - p|$.

**Corollary:** Let $n = 0, 1, 2, \ldots$ For the error of the approximation of a real number by rational numbers, one has the estimate

$$\boxed{\frac{1}{q_n(q_n + q_{n+1})} < \left|\alpha - \frac{p_n}{q_n}\right| \le \frac{1}{q_n q_{n+1}}.} \tag{2.83}$$

*Example 1* (golden ratio): The golden ratio number $\alpha_{\mathrm{g}} = \frac{1}{2}(\sqrt{5} - 1)$ has the representation $\alpha_{\mathrm{g}} = [0, \overline{1}]$. For $n = 0, 1, \ldots$ the canonical approximation fractions are

$$\frac{p_n}{q_n} = \frac{0}{1}, \frac{1}{1}, \frac{1}{2}, \frac{2}{3}, \frac{3}{5}, \frac{5}{8}, \frac{8}{13}, \frac{13}{21}, \ldots.$$

Hence $\frac{8}{13}$ is the best rational approximation to $\alpha_{\mathrm{g}}$ with a denominator which is $\le 13$. From (2.83) it follows that one has the error estimate

$$\left|\alpha_{\mathrm{g}} - \frac{8}{13}\right| \le \frac{1}{q_6 q_7} = \frac{1}{13 \cdot 21} < \frac{4}{1\,000}.$$

*Example 2:* For $\sqrt{2} = [1, 2, 2, 2, \ldots]$ we get the canonical approximation fractions

$$1, \quad \frac{3}{2}, \quad \frac{7}{5}, \quad \frac{17}{12}, \quad \frac{41}{29}, \quad \ldots.$$

Hence $\frac{17}{12}$ is the best rational approximation of $\sqrt{2}$ with a denominator $\le 12$. From (2.83) we get the error estimate

$$\left|\sqrt{2} - \frac{17}{12}\right| \le \frac{1}{12 \cdot 29} < \frac{3}{1\,000}.$$

*Example 3:* For $\mathrm{e} = [2, 1, 2, 1, 1, 4, 1, \ldots]$ the canonical approximations are

$$\frac{2}{1}, \quad \frac{3}{1}, \quad \frac{8}{3}, \quad \frac{11}{4}, \quad \frac{19}{7}, \quad \frac{87}{32}, \quad \frac{106}{39}, \quad \ldots.$$

Hence $\frac{87}{32}$ is the best rational approximation for e with a denominator $\le 32$. Moreover one has the estimate of the error

$$\left|\mathrm{e} - \frac{87}{32}\right| \le \frac{1}{32 \cdot 39} < \frac{1}{1\,000}.$$

The determination of the optimal rational approximation has played an exceptional role in the history of mathematics in connection with the approximation of the circumference of a circle (approximation of $\pi$, cf. 2.7.7).

**Diophantine approximation theorem of Dirichlet (1842):** A real number $\alpha$ is irrational if and only if the inequality

$$|q\alpha - p| < \frac{1}{q}$$

has infinitely many solutions for relatively prime integers $p$ and $q > 0$.

**The optimal approximation theorem of Hurwitz (1891):** For every irrational number $\alpha$ the inequality

$$\left|\alpha - \frac{p}{q}\right| < \frac{1}{\sqrt{5}q^2} \tag{2.84}$$

has an infinite number of rational solutions $\frac{p}{q}$.

The constant $\sqrt{5}$ is optimal[70]. The number which can be approximated the *worst* is the golden ratio[71] $\alpha_{\mathrm{g}} = \frac{1}{2}(\sqrt{5} - 1)$. In this sense $\alpha_{\mathrm{g}}$ is the "most irrational" of all real numbers.

**The role of the golden ratio in chaos theory:** If one has two coupled oscillating systems with the angular frequencies $\omega_1$ and $\omega_2$, then the *resonance case*

$$\boxed{\frac{\omega_1}{\omega_2} = \text{rational number}}$$

is particularly dangerous. On a computer there are, practically speaking, only rational numbers. However, experience has shown that the irrationality of the quotients $\omega_1/\omega_2$ can be simulated on a computer by using the canonical approximation fractions of the golden ratio from Example 1 for $\omega_1/\omega_2$ (cf. KAM theory in [212]).

### 2.7.6 Transcendental numbers

**The classification of real numbers** (cf. Figure 2.7): A real number is said to be *rational*, if it is the solution of an equation of the form

$$c_1 x + c_0 = 0$$

with integral coefficients $c_0$ and $c_1 \neq 0$. It was already known to the Pythagoreans (around 500 BC) that the number $\sqrt{2}$ is not rational. A real or complex number is said to be *algebraic*, if it is a solution of an equation of the form

$$c_n x^n + c_{n-1} x^{n-1} + \ldots + c_1 x + c_0 = 0, \qquad n \geq 1 \tag{2.85}$$

with integral coefficients $c_j$ for all $j$ and $c_n \neq 0$. The lowest degree of a polynomial which and algebraic number $\alpha$ satisfies is called the *degree* of the number. Algebraic numbers of degree 2 are also called *quadratic*.

The deeper investigation of algebraic numbers is the topic of algebraic number theory. The basic constituents of this theory are ideal theory, Galois theory and the theory of $p$-adic numbers.

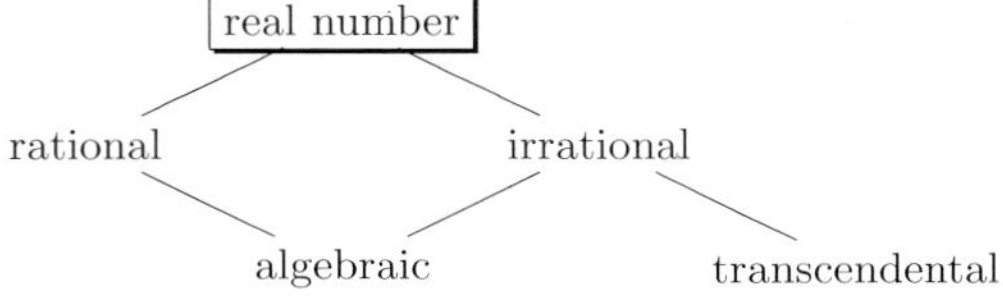


*Figure 2.7. The different types of real numbers.*

[70] If one replaces $\sqrt{5}$ in (2.84) by a larger number, then one can always find an irrational number $\alpha$ so that the new inequality (2.84) has only finitely many rational solutions $p/q$.

[71] One has $\lim_{n\to\infty} \left|\alpha_{\mathrm{g}} - \frac{p_n}{q_n}\right| q_n^2 = \frac{1}{\sqrt{5}}$.

**Criteria for irrationality:** A real number $\alpha$ is irrational if one of the following conditions is satisfied:

(i) The continued fraction of $\alpha$ is infinite.

(ii) In the expansion of $\alpha$ as a decimal there are no periods.

(iii) (Theorem of Gauss) The number $\alpha$ is the solution of an algebraic equation

$$x^n + c_{n-1}x^{n-1} + \ldots + c_1 x + c_0 = 0, \qquad n \geq 2$$

with integral coefficients, and this equation has no integral solutions.

A further criterion is the Diophantine approximation theorem of Dirichlet in section 2.7.5.3.

*Example 1:* The number $\sqrt{2}$ is a solution of the algebraic equation

$$x^2 - 2 = 0$$

and hence a (quadratic) algebraic number. Since this equation obviously has no integral solutions, $\sqrt{2}$ is irrational by the theorem of Gauss.

Real or complex numbers which are not algebraic are called *transcendental.* The existence of transcendental numbers was first proved by Liouville in 1844 with the aid of his approximation theorem (see Example 3 below).

**Theorem of Euler–Lagrange:** A real number is a quadratic algebraic number if and only if it has a periodic continued fraction.

*Example 2:* The continued fractions of $\sqrt{2}$, $\sqrt{3}$ and $\sqrt{5}$ are periodic (see Table 2.7).

**The existence of transcendental numbers according to Cantor (1874):** A first sensational success of the set theory developed by Cantor was that he was, in contrast to Liouville, able to give a completely elementary proof of the existence of transcendental numbers. He showed:

(i) The set of algebraic numbers is countable.

(ii) The set of real numbers is not countable.

Consequently there must be transcendental numbers. If one chooses an arbitrary compact interval of the real numbers, then the probability is one that a transcendental number is contained in that interval. In this sense almost all real numbers are transcendental.

**Approximation order:** An irrational number $\alpha$ has by definition a real number $\kappa > 0$ as its *approximation order*, if the inequality

$$\left|\alpha - \frac{p}{q}\right| < \frac{1}{q^\kappa}$$

is satisfied for infinitely many rational numbers $p/q$ with $q > 0$. In particular it follows from this that there exists an infinite sequence of rational numbers $(P_n/Q_n)$ with

$$\left|\alpha - \frac{P_n}{Q_n}\right| < \frac{1}{Q_n^\kappa} \quad \text{and} \quad Q_n \geq n \qquad \text{for all } n = 1, 2, \ldots .$$

**Theorem:** Every irrational number has at least the approximation order 2.

**Liouville's approximation theorem (1844):** An algebraic number of degree $n \geq 1$ can have at most the approximation order $n$.

*Example 3:* The number

$$\frac{1}{10^{1!}} + \frac{1}{10^{2!}} + \frac{1}{10^{3!}} + \ldots$$

has arbitrarily high approximation orders. Hence this number must be transcendental.

In this way Liouville was able to prove the existence of transcendental numbers. A much stronger result is the following.

**Approximation theorem of Roth (1955)**[72]**:** The maximal approximation order of an irrational number is two.

Roughly speaking this means the following:

> *Algebraic irrational numbers can only poorly be approximated by rational numbers, whereas transcendental numbers can be efficiently approximated by rational numbers.*

**Theorem of Hermite (1873):** The number e is transcendental..

**Theorem of Lindemann (1882):** The number $\pi$ is transcendental.

Both of these famous theorems are special cases of the following result (cf. Example 4).

**Theorem of Lindemann–Weierstrass (1882):** If $\alpha_1, \ldots, \alpha_n$ are pairwise distinct complex algebraic numbers, then from

$$\beta_1 \mathrm{e}^{\alpha_1} + \ldots + \beta_n \mathrm{e}^{\alpha_n} = 0,$$

it follows that at least one of the complex coefficients $\beta_j$ is transcendental or the case is trivial, i.e., $\beta_k = 0$ for all $k$.

**Corollary:** The complex number

$$\boxed{\mathrm{e}^z \quad \text{is transcendental,}}$$

if the complex number $z \neq 0$ is algebraic.

*Proof:* We set $\alpha_1 := 0$ and $\alpha_2 := z$. According to the theorem of Lindemann–Weierstrass, from the non-trivial relation

$$(\mathrm{e}^z)\mathrm{e}^0 + (-1)\mathrm{e}^z = 0$$

it follows that the coefficient $\mathrm{e}^z$ is transcendental.

*Example 4:* (i) It we choose the algebraic number $z = 1$, then we get the statement that e is transcendental.

(ii) For $z = 2\pi\mathrm{i}$ the number $\mathrm{e}^z = 1$ is not transcendental, hence $2\pi\mathrm{i}$ cannot be algebraic, hence also $\pi$ cannot be algebraic and is consequently transcendental.

**Theorem of Gelfond–Schneider (1934):** At least one of the complex numbers

$$\boxed{\alpha, \ \beta, \ \alpha^\beta}$$

is transcendental, if one excludes the trivial cases ($\alpha = 0$ or $\ln \alpha = 0$ or $\beta =$ rational number).

This famous theorem provided a solution of the seventh Hilbert problem (posed in 1900).

*Example 5:* The number

$$2^{\sqrt{5}}$$

[72]Klaus Roth received the Fields medal in 1958 for the proof of this fundamental result.

is transcendental, since in the sequence 2, $\sqrt{5}$, $2^{\sqrt{5}}$ only the last number can be transcendental. Statements of this kind were already suspected to hold by Euler.

*Example 6:* $e^\pi$ is transcendental.

*Proof:* One has $e^\pi = i^{-2i}$. Of the three numbers i, $-2i$ and $i^{-2i}$, only the last can be transcendental.

### 2.7.7 Applications to the number $\pi$

The calculation of the circumference of a circle, i.e., the calculation of the number $\pi$, has it origins in ancient times and has always engulfed the fantasy of mathematicians.

In the Old Testament one can find the number 3 as an approximation to $\pi$. In the first Book of Kings, verse 7.23, one can read: "And he made a sea, poring from one edge to the other ten ellens across, and five ellens high, and a string of 30 ellens' length measured the extension of the shore."

In papyrus roles of the ancient Egyptians (dating from about 1650 BC) one can read: "Take $\frac{1}{9}$ from a diameter and construct a square of what remains, and it will have the same area as the circle." This gives are more precise approximation:

$$\pi = \left(\frac{16}{9}\right)^2 = 3.16\ldots.$$

**Progress due to Archimedes (287–212 BC):** He approximated the circle by polygons and found with the help of a 96-gon the famous estimate:[73]

$$\boxed{\frac{223}{71} < \pi < \frac{22}{7}.}$$

Here one has $\frac{22}{7} = 3.14\ldots$

**Best rational approximation of $\pi$:** We will show below that $\pi$ has the continued fraction

$$\pi = [3, 7, 15, 1, 292, 1, 1, 1, \ldots], \tag{2.86}$$

which has no regular behavior whatsoever. The canonical rational approximations coming from this continued fraction are

$$3, \quad \frac{22}{7}, \quad \frac{333}{106}, \quad \frac{355}{113}, \quad \frac{103\,993}{33\,102}, \quad \ldots.$$

As second fraction occuring we have already the approximation $\frac{22}{7}$ of Archimedes. From the main theorem in section 2.7.5.3 it follows that $\frac{22}{7}$ is the best rational approximation of $\pi$ if we bound the denominator of the fraction to be $\leq 7$. Similarly

$$\boxed{\frac{355}{113}} \tag{2.87}$$

[73]The symbol $\pi$ was introduced by Euler in 1737. Possibly Euler was thinking of the Greek word περιφέρεια for periphery.

is the best rational approximation if we admit denominators $\leq 113$. The surprisingly good estimate given by this fraction follows from (2.83) and the estimate

$$\left|\pi - \frac{355}{113}\right| \leq \frac{1}{q_3 q_4} = \frac{1}{113 \cdot 33\,102} < 10^{-6}.$$

Astonishingly one can find the fraction (2.87) as an approximation for $\pi$ already in the works of the Chinese mathematician Zu Chong–Zhi (430–501).

In 1766 the following estimate was discovered in Japan:

$$\frac{5\,419\,351}{1\,725\,033} < \pi < \frac{428\,224\,593\,349\,304}{136\,308\,121\,570\,117}.$$

These are in fact canonical approximations stemming from the continued fraction expression for $\pi$.

We now want to show how the continued fraction (2.86) of $\pi$ is derived. One needs for this the decimal estimate

$$3.1415926535\mathbf{8} < \pi < 3.1415926535\mathbf{9}, \tag{2.88}$$

which one can obtain with the numerical procedure described below. The continued fraction algorithm (2.79) yields in this case

$$\begin{aligned} 3.14159265358 &= [3, 7, 15, 1, 292, 1, 1, 1, 1, \ldots], \\ 3.14159265359 &= [3, 7, 15, 1, 292, 1, 1, 1, 2, \ldots]. \end{aligned}$$

This yields (2.86).

**The product formula of Vieta:** An analytical formula for $\pi$ was first given in 1579 by Francis Vieta. This formula is:

$$\boxed{\frac{2}{\pi} = \prod_{n=1}^{\infty} a_n}$$

with

$$a_{n+1} = \sqrt{\frac{1}{2}\left(1 + \sqrt{a_n}\right)} \quad \text{and} \quad a_0 := 0.$$

If we set $b_n := \prod_{k=1}^{n} a_k$, then we have the estimate

$$0 < b_n - \frac{2}{\pi} < \frac{1}{4^n}, \qquad n = 1, 2, \ldots .$$

From $b_{22}$ one gets the estimate (2.88) for $\pi$.

**The product formula of Wallis:** In 1655 John Wallis published in his monograph *Arithmetica infinitorum* the following infinite product formula:

$$\boxed{\frac{\pi}{2} = \prod_{n=1}^{\infty} \frac{(2n)^2}{(2n-1)(2n+1)}.} \tag{2.89}$$

**The infinite summation formula of Newton:** In 1665 the then 22-year-old Isaac Newton used the series

$$\arcsin x = \sum_{n=1}^{\infty} \frac{1 \cdot 3 \cdot \ldots \cdot (2n-1)}{2 \cdot 4 \cdot \ldots \cdot (2n)} \cdot \frac{z^{2n+1}}{2n+1}$$

and obtained, in the particular case where $x = 1/2$, the formula

$$\boxed{\frac{\pi}{6} = \frac{1}{2} + \frac{1}{2 \cdot 3 \cdot 8} + \frac{1 \cdot 3}{2 \cdot 4 \cdot 5 \cdot 32} + \frac{1 \cdot 3 \cdot 5}{2 \cdot 4 \cdot 6 \cdot 7 \cdot 128} + \ldots}$$

from which he was able to derive the first 14 decimal places of $\pi$.

**The infinite summation formula of Leibniz:** The then 28 year-old Leibniz made the discovery in 1674, through geometric considerations, of the summation formula

$$\boxed{\frac{\pi}{4} = 1 - \frac{1}{3} + \frac{1}{5} - \frac{1}{7} + \ldots,} \tag{2.90}$$

which is distinguished by its great simplicity. The error is determined by the first discarded term. Thus this series converges only very slowly and is not of any use for the calculation of $\pi$.[74]

**The product formula of Euler:** Roughly 80 years after Wallis, Euler found his famous product formula:

$$\boxed{\sin \pi z = \pi z \prod_{k=1}^{\infty} \left(1 - \frac{z^2}{k^2}\right) \qquad \text{for all } z \in \mathbb{C}.} \tag{2.91}$$

The product formula of Wallis (2.89) is a special case of this formula, when $z = 1/2$.

**Exact calculation of $\pi$:** Ludolf van Ceulen (1540–1610) was the first to calculate the number $\pi$ up to 35 decimal places. For this reason $\pi$ is sometimes referred to as the Ludolf number.

Starting in the eighteenth century the formula of John Machin

$$\boxed{\frac{\pi}{4} = 4 \arctan\left(\frac{1}{5}\right) - \arctan\left(\frac{1}{239}\right)}$$

was used for the calculation of $\pi$, together with the power series for $\arctan x$ (cf. (2.93)).

**Ramanunjan's formula:** The Indian mathematician Srinivasa Ramanunjan[75] discov-

[74] In fact this series was discovered three years before Leibniz discovered it, by the British mathematician James Gregory.

[75] Ramanunjan (1887–1920) is, in his genius, one of the most amazing personalities in the history of mathematics. He discovered mathematical formulas of incredible complexity. Through a letter which he directed to the great English mathematician Godefroy Harold Hardy (1877–1947), the latter became aware of the mathematical talent of Ramanunjan and invited him to England in 1914. Ramanunjan's Notebook is, with its treasury of mathematical formulas which appear to be coming from another world, a unique document in the mathematical literature. For those who are interested in the fascination of deep mathematics this notebook, in its present edition containing proofs and commentaries, is highly recommended: [263].

ered in 1914 the formula

$$\frac{1}{\pi} = \frac{\sqrt{8}}{9\,801} \sum_{n=0}^{\infty} \frac{(4n)!\,[1\,103 + 26\,390n]}{(n!)^4\, 396^{4n}}.$$

This formula is closely connected with the deep theory of modular forms (cf. 1.14.18).

The brothers David and Gregory Chudnovsky of Columbia University (New York) calculated, with a modified and very complicated formula similar to the above formula of Ramanunjan, the number $\pi$ up to more than two billion decimal places (2,260,321,336, to be precise).

**The iteration scheme of the brothers Borwein:** The following spectacular iteration scheme was derived by Peter and Jonathan Borwein (University of Waterloo, Canada):

$$\begin{aligned} y_{n+1} &= \frac{1-(1-y_n^4)^{1/4}}{1+(1-y_n^4)^{1/4}}, \qquad n = 0, 1, \ldots, \\ \alpha_{n+1} &= (1+y_{n+1})^4\alpha_n - 2^{2n+3}y_{n+1}(1+y_{n+1}+y_{n+1}^2) \end{aligned}$$

with the initial values $y_0 := \sqrt{2}-1$, $\alpha_0 := 6 - 4\sqrt{2}$. One has

$$\lim_{n\to\infty} \frac{1}{\alpha_n} = \pi.$$

This sequence has such an amazing speed of convergence that already 15 iterations suffice to get the value of $\pi$ up to two billion decimal places.

The scheme is the expression of a *revolution* in modern mathematics. Powerful computers raise the need of completely new methods and algorithms, which one describes with the phrase "scientific calculation".

**General continued fractions:** A *general continued fraction* is an expression of the form

$$a_0 + \cfrac{b_1}{a_1 + \cfrac{b_2}{a_2 + \cfrac{b_3}{a_3 + \cdots \ddots \cfrac{b_{n-1}}{a_{n-1} + \cfrac{b_n}{a_n}}}}}.$$

For this expression one also uses the symbol

$$a_0 + \frac{b_1|}{|a_1} + \frac{b_2|}{|a_2} + \ldots + \frac{b_n|}{|a_n}.$$

In the special case in which $b_j = 1$ for all $j$ one gets the symbol $[a_0, \ldots, a_n]$ of section 2.7.5.1.

**Legendre's proof of the irrationality of $\pi$ (1806):** In 1766 Johann Lambert (1728–1777) found the (general) continued fraction:[76]

$$\tan z = \frac{z^2 \rfloor}{\lceil 1} - \frac{z^2 \rfloor}{\lceil 3} - \frac{z^2 \rfloor}{\lceil 5} - \ldots .$$

From this expression he concluded the irrationality of $\tan z$, if $z \neq 0$ is rational. In particular, from the rationality of the expression

$$\tan \frac{\pi}{4} = 1$$

he concluded that $\pi$ is irrational. The proof given by Lambert contains however a gap, which was not closed until 1806 by Legendre.

Already 2000 years before, Aristotle had conjectured the irrationality of $\pi$, in the form of stating that the radius and the circumference of a circle are not commensurable.

**The theorem of Lindemann on the impossibility of the quadrature of the circle (1882):** Ferdinand Lindemann showed in 1882 that the number $\pi$ is transcendental. This has as a consequence the negative solution of the ancient problem of the quadrature of the circle (cf. 2.6.6).

**The general continued fraction of $\pi$:** For all real numbers $x$ one has the convergent continued fraction

$$\arctan x = \frac{x \rfloor}{\lceil 1} + \frac{1^2 \cdot x^2 \rfloor}{\lceil 3} + \frac{2^2 \cdot x^2 \rfloor}{\lceil 5} + \frac{3^2 \cdot x^3 \rfloor}{\lceil 7} + \ldots . \qquad (2.92)$$

In contrast, the power series

$$\arctan x = x - \frac{x^3}{3} + \frac{x^5}{5} - \ldots \qquad (2.93)$$

only converges for $-1 \leq x \leq 1$. If we use

$$\frac{\pi}{4} = \arctan 1,$$

then we get from (2.93) for $x = 1$ the Leibniz series (2.90), which converges so slowly. In order to calculate $\pi$ up to seven decimals, one requires roughly $10^6$ terms of the Leibniz series. The same precision of $\pi$ is obtained by setting $x = 1$ in (2.92) and calculating nine terms. In general one gets the very regular expression

$$\frac{\pi}{4} = \frac{1 \rfloor}{\lceil 1} + \frac{1^2 \rfloor}{\lceil 3} + \frac{2^2 \rfloor}{\lceil 5} + \frac{3^3 \rfloor}{\lceil 7} + \ldots . \qquad (2.94)$$

## 2.7.8 Gaussian congruences

**Definition:** if $a$, $b$ and $m$ are integers, then one writes, following Gauss, the symbol

$$\boxed{a \equiv b \bmod m}$$

if and only if the difference $a - b$ is divisible by $m$. In this case, $a$ and $b$ are said to be *congruent modulo m*.

*Example 1:* $5 \equiv 2 \bmod 3$, since $5 - 2$ is divisible by 3.

[76]This continued fraction converges for all complex numbers $z$ which are not poles of $\tan z$.

**Manipulations:** (i) $a \equiv a \bmod m$.

(ii) From $a \equiv b \bmod m$ it follows that $b \equiv a \bmod m$.

(iii) From $a \equiv b \bmod m$ and $b \equiv c \bmod m$ it follows that $a \equiv c \bmod m$.

(iv) From $a \equiv b \bmod m$ together with $c \equiv d \bmod m$ one obtains

$$a + c \equiv b + d \bmod m \quad \text{and} \quad ac \equiv bd \bmod m.$$

**Theorem:** If $s$ and $m$ are relatively prime positive natural numbers, then the equation

$$ts \equiv 1 \bmod m$$

has a solution $t$ in positive integers.

**Theorem of Fermat (1640) and Euler (1760):** For positive natural numbers $a$ and $m$, which are assumed to be relatively prime, one has

$$\boxed{a^{\varphi(m)} \equiv 1 \bmod m.}$$

Here $\varphi(m)$ denotes the number of integers $1, \ldots, m$ which are relatively prime to $m$ (the Euler $\varphi$-function, see section 2.7.2).

*Example 2:* For a prime number $p$, which does not divide $a$, one has

$$a^{p-1} \equiv 1 \bmod p.$$

In this form the theorem was originally formulated by Fermat.

### 2.7.8.1 Application of the theorem of Fermat–Euler in coding theory

More that 200 years the theorem of Fermat–Euler was considered to be only a result in pure mathematics. In 1977 however, Rivest, Shamir and Adlemann published the following bafflingly simple and yet extraordinarily secure code, which is based on the theorem above and is often used today.

**Preparations by the operator:** (i) Here two prime numbers $p$ and $q$, roughly of the size $10^{100}$, are chosen and kept secret.

(ii) One forms the product

$$\boxed{m = pq}$$

and calculates $\varphi(m) = (p-1)(q-1)$.

(iii) One chooses an additional natural number $s$ with $1 < s < \varphi(m)$.

(iv) The person sending the message is given (publicly) the two numbers

$$\boxed{m \quad \text{and} \quad s.}$$

**Encoding the message:** The message is simply encoded in a single natural number

$$\boxed{n}$$

as follows. One associates to every letter (for example) a two-digit number $10, 11, 12 \ldots$ and replaces in the message all occurrences of the letter by that two-digit number. Then forming the concatenation of all of these, one gets a big number $n$. The number $n^s$ is then divided by $m$, and the remainder $r$ is sent to the operator, i.e., the number $r$ satisfying

$$\boxed{n^s \equiv r \bmod m}$$

is the only information sent.

**Decoding the message by the operator:** Here one must construct from the remainder $r$ the original number $n$.

Since $\varphi(m)$ and $s$ are relatively prime, there is a natural number $t \geq 1$ with

$$ts \equiv 1 \bmod \varphi(m). \tag{2.95}$$

**Theorem:** $r^t \equiv n \bmod m$.

The operator now just divides $r^t$ by $m$. The remainder here is the sought for number $n$. Note that from the size of $m$ one always has $n < m$.

*Proof of the theorem:* By the Fermat–Euler theorem one has

$$n^{\varphi(m)} \equiv 1 \bmod m.$$

From (2.95) there exists an integer $k$ with $ts = 1 + k\varphi(m)$. It follows that

$$r^t \equiv n^{st} \equiv n^{1+k\varphi(m)} \equiv n \bmod m.$$

**The security of this method:** If an intruder wants to decode the message, he needs the number $t$, that is $\varphi(m) = (p-1)(q-1)$. To get this number he must determine the prime number decomposition of the number $m$, which is known to him. The trick of this method is simply that because of the size chosen for $p$ and $q$, no computer is as yet able to determine the factors $p$ and $q$ in a reasonable amount of time.

Since computers are becoming more and more powerful all the time, the security of the method is only guaranteed if one chooses new, larger numbers $p$ and $q$ from time to time.

The method could become unsafe if some algorithm were found for quickly factoring large numbers into their prime decompositions. For this reason all mathematicians who consider problems in this area are closely monitored by the National Security Council.

#### 2.7.8.2 The quadratic reciprocity law

We study the solvability of the two equations

$$\boxed{x^2 \equiv q \bmod p} \tag{2.96}$$

and

$$\boxed{x^2 \equiv p \bmod q.} \tag{2.97}$$

**Legendre symbol:** We set

$$\left(\frac{q}{p}\right) := \begin{cases} 1, & \text{if (2.96) has a solution,} \\ -1, & \text{if (2.96) has no solution.} \end{cases}$$

**The quadratic reciprocity law of Gauss (1796):** If $p$ and $q$ are prime numbers greater than two, then one has

$$\boxed{\left(\frac{q}{p}\right) = (-1)^{(p-1)(q-1)/4}\left(\frac{p}{q}\right).}$$

Moreover, for the Legendre symbol we have

$$\left(\frac{-1}{p}\right) = (-1)^{(p-1)/2}, \qquad \left(\frac{2}{p}\right) = (-1)^{(p^2-1)/8}.$$

**Historical remark:** This theorem was discovered empirically independently by Euler (1722), Legendre (1785) and Gauss. The first complete proof was given by Gauss. This law, together with its many generalizations, expresses the deepest elementary behavior known in number theory.

*Example:* Let $p = 4n+1$ be prime, where $n$ denotes some positive natural number. Then the equation

$$x^2 \equiv -1 \bmod p$$

has a solution, since $\left(\frac{-1}{p}\right) = 1$.

**Theorem of Gauss (1808):** Let $\varepsilon := \mathrm{e}^{2\pi \mathrm{i}/p}$, where $p$ denotes a prime. Then one has

$$\sum_{k=1}^{p-1}\left(\frac{k}{p}\right)\varepsilon^k = \begin{cases} \sqrt{p} & \text{for} \quad p \equiv 1 \bmod 4, \\ \mathrm{i}\sqrt{p} & \text{for} \quad p \equiv 3 \bmod 4. \end{cases}$$

Sums of this type are called *Gaussian sums.* It took even Gauss a long time to prove this result.

### 2.7.9 Minkowski's geometry of numbers

**Lattices:** Let $b_1, \ldots, b_n$ be linearly independent column vectors of $\mathbb{R}^n$ with $n \geq 2$. The set

$$L := \left\{ \sum_{k=1}^{n} \alpha_k b_k \;\middle|\; \alpha_1, \ldots, \alpha_n \in \mathbb{Z} \right\}$$

is called a *lattice* in $\mathbb{R}^n$. The number $\mathrm{Vol}\,(L) := |\det(b_1, \ldots, b_n)|$ is equal to the volume of the $n$-dimensional cube spanned by $b_1, \ldots, b_n$ and is called the *lattice volume.*

**Lattice point theorem of Minkowski (1891):** Let $L$ be a lattice, and let $C$ be convex set which is centrally symmetric with respect to the origin, i.e., from $x \in C$ it follows that $-x \in C$. If for the volume $\mathrm{Vol}\,(C)$ of $C$ the inequality

$$\mathrm{Vol}\,(C) \geq 2^n \mathrm{Vol}\,(L)$$

is satisfied, then $C$ contains a lattice point $x \neq 0$.

*Example:* If $C$ is a ball of $\mathbb{R}^3$ centered at the origin with $\mathrm{Vol}\,(C) \geq 8$, then $C$ contains some lattice point $x \neq 0$.

## 2.7.10 The fundamental local–global principle in number theory

### 2.7.10.1 Valuations

**Definition:** Let $K$ be a given field. A (real) *valuation* on the field $K$ associates to every element $x \in K$ a real number $\nu(x) \geq 0$ with the following properties:

(i) $\nu(x) = 0$ if and only if $x = 0$.

(ii) $\nu(xy) = \nu(x)\nu(y)$ and $\nu(x+y) \leq \nu(x) + \nu(y)$ for all $x, y \in K$.

*Example 1:* The trivial valuation is $\nu(0) = 0$ and $\nu(x) = 1$ for all $x \neq 0$.

*Example 2:* Let $\mathbb{Q}$ be the field of rational numbers. The prescription

$$\boxed{\nu_\infty(x) := |x|}$$

yields a valuation on $\mathbb{Q}$.

If $p$ is a prime, then every number $x \neq 0$ in $\mathbb{Q}$ can be written in the form

$$x = \frac{a}{b}p^m,$$

where $m$ is an integer, and the two numbers $a$ and $b$ are not divisible by $p$. If we set

$$\boxed{\nu_p(x) := p^{-m},}$$

then we get the so-called *p-adic valuation* on $\mathbb{Q}$.

### 2.7.10.2 $p$-adic numbers

Every metric space $X$ can be extended to a complete metric space. This extension is unique up to an isometry (cf. [212]).

The field $\mathbb{Q}$ of rational numbers is made into a metric space by setting

$$\boxed{d(x,y) := \nu_\infty(x-y).}$$

The completion of this metric space yields the field $\mathbb{R}$ of real numbers.

If we instead use the $p$-adic valuation in the above procedure, then the completion of $\mathbb{Q}$ with respect to the metric

$$d(x,y) := \nu_p(x-y)$$

yields the field $\mathbb{Q}_p$ of *p-adic numbers.*

*Example:* An infinite series $\sum_{n=0}^{\infty} a_n$ is convergent in $\mathbb{Q}_p$ if and only if $(a_n)$ converges to zero.

A result of that simplicity is not available in $\mathbb{R}$.

**Theorem of Ostrowski (1918):** If one has a non-trivial valuation $\nu$ on the field $\mathbb{Q}$ of rationals, then the completion of $\mathbb{Q}$ with respect to the metric $d(x,y) := \nu(x-y)$ is either the field $\mathbb{R}$ or one of the $p$-adic fields $\mathbb{Q}_p$.

**Remark:** If one takes the point of view that rational numbers are uniquely determined by natural numbers, then the theorem of Ostrowski shows that the classical abstraction from $\mathbb{Q}$ to $\mathbb{R}$ is not necessary; rather one has all the $p$-adic fields $\mathbb{Q}_p$ as possible completions of the same field (metric space).

The $p$-adic numbers were introduced in 1904 by Kurt Hensel in number theory and have since proved themselves to be fundamental. There are mathematicians and physicists who think that theoretical physics is so incomplete today because it has up to now been restricted to the real numbers for historical reasons and hence does not take the other fields $\mathbb{Q}_p$ into account.

#### 2.7.10.3 The theorem of Minkowski–Hasse

We consider the equation

$$\boxed{a_1 x_1^2 + \ldots + a_n x_n^2 = 0.} \tag{2.98}$$

Here $a_1, \ldots, a_n$ are rational numbers, all non-vanishing.

**Theorem:** If the equation (2.98) has a non-trivial solution

$$x_1, \ldots, x_n \in K \tag{2.99}$$

for $K = \mathbb{R}$ and $K = \mathbb{Q}_p$ ($p$ an arbitrary prime number), then it also has a non-trivial solution

$$x_1, \ldots, x_n \in \mathbb{Q}. \tag{2.100}$$

**Remark:** One calls solutions of (2.99) local and the solutions of (2.100) global. The theorem thus tells us that from the local solvability of the equation (2.98), the global solvability follows. This is a special case of the general *local–global principle* in number theory, which documents the fundamental nature of the $p$-adic number fields.

### 2.7.11 Ideals and the theory of divisors

Every integer can be written as a product of prime numbers. This fact is not true for arbitrary rings. In this more abstract setting, one must use the theory of divisors (cf. Figure 2.8).

The starting point for the theory of ideals is the work of Kummer from 1843, which contains an incorrect proof of Fermat's last theorem. Dirichlet recognized this mistake, which was precisely that the prime decomposition of numbers does not hold in arbitrary rings. Following this, Kummer studied the problem of decomposition. By introducing ideal numbers he was able to prove generalized decomposition theorems, which made a correct proof of Fermat's last theorem possible in some special cases. Dedekind introduced the general notion of ideal in 1871 and so founded ideal theory, which today plays an important role in modern mathematical physics in the context of operator algebras (cf. [212]).

#### 2.7.11.1 Basic notions

Let $R$ be an integral domain with unit, i.e., $R$ is a commutative ring without divisors of zero.

**Units:** An element $\varepsilon$ of $R$ is called a *unit*, if $\varepsilon^{-1}$ also belongs to $R$.

*Example 1:* In the ring $\mathbb{Z}$ of integers, the only units are $\pm 1$.

**Prime elements:** A ring element $p$ is called *prime* or *irreducible*, if $p \neq 0$, $p$ is not a unit, and from a decomposition

$$p = ab$$

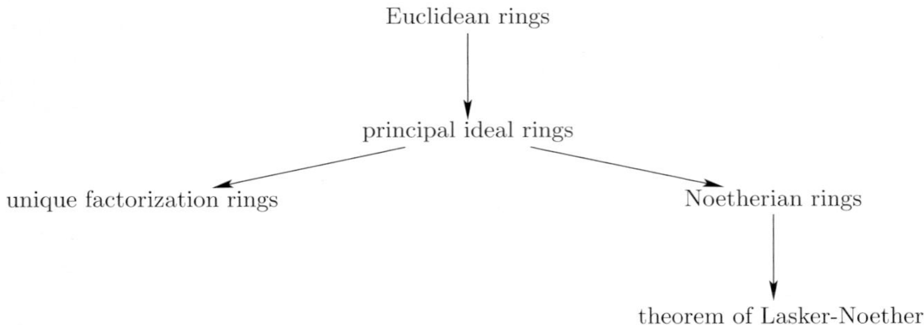


*Figure 2.8. Relations between rings of different types*

it follows that $a$ or $b$ is a unit.

**Unique decomposition into prime elements:** A ring $R$ is called *factorial* (or a *unique factorization ring*), if and only if every non-vanishing ring element can be uniquely (up to units and up to the order of factors) written as a product of prime elements.

*Example 2:* The ring $\mathbb{Z}$ of integers has this property.

### 2.7.11.2 Principal ideal domains and Euclidean rings

Let $R$ be a commutative ring with unit.

**Ideals:** A non-empty set $\mathscr{A}$ of $R$ is said to be an *ideal*, if the following two properties are satisfied:

(i) From $a, b \in \mathscr{A}$ it follows that $a - b \in \mathscr{A}$.

(ii) From $a \in \mathscr{A}$ and $r \in R$ it follows that $ra \in \mathscr{A}$.

Note that this property just states that $\mathscr{A}$ is closed with respect to addition and scalar multiplication by elements in $R$. In this respect, an ideal is the ring analog of a normal subgroup in the theory of groups. In fact, an analog of the homomorphism theorem in group theory characterizing the normal subgroups as the set of kernels of group homomorphisms holds for ideals: a subset $\mathscr{A}$ of $R$ is an ideal if and only if it is the kernel of a *ring homomorphism*.

We denote by $(a)$ the smallest ideal which contains an element $a$. Explicitly one has $(a) = \{ra \,|\, r \in R\}$. Ideals of this kind are called *principal*.

**Principal ideal rings:** An integral domain with unit is called a *principal ideal ring*, if and only if every ideal of the ring is a principal ideal.

**Theorem 1:** Every principal ideal ring has the property of unique decomposition into prime elements.

**Euclidean rings:** An integral domain with unit $R$ is called a *Euclidean ring*, if to every ring element $r \neq 0$ there is an integer $h(r) \geq 0$ such that

(i) $h(rs) \geq h(r)$ for all $r \neq 0$ and $s \neq 0$.

(ii) Two every two ring elements $a$ and $b$ with $b \neq 0$ there is a representation

$$a = qb + r,$$

for which either $r = 0$ or $h(r) < h(b)$. The function $h : R \longrightarrow \mathbb{Z}$ is called a *height*.

**Theorem 2:** Every Euclidean ring is a principal ideal ring.

*Example 1:* The ring $\mathbb{Z}$ of integers is a Euclidean ring with $h(r) := |r|$.

*Example 2:* If $K$ is a field, then the ring $K[x]$ of all polynomials in the indeterminant $x$ with coefficients in $K$ is a Euclidean ring. The height of a polynomial is its degree.

The units in the ring $K[x]$ are the non-vanishing elements of $K$. The prime elements in $K[x]$ are called *irreducible polynomials.*

The polynomial ring $\mathbb{Z}[x]$ is not a principal ideal ring.

**Prime ideals and primary ideals:** Let $\mathscr{A}$ be an ideal of the ring $R$.

(i) $\mathscr{A}$ is called a *prime ideal,* if the factor ring $R/\mathscr{A}$ has no divisors of zero.

(ii) $\mathscr{A}$ is said to be a *primary ideal,* if the zero divisors of $R/\mathscr{A}$ are idempotent (i.e., some power of the element vanishes).

**Theorem 3:** To every primary ideal $\mathscr{A}$ there is prime ideal $\mathscr{A}'$ which consists of all elements of $R$ which are powers of elements of $\mathscr{A}$.

*Example 3:* In the ring $\mathbb{Z}$ of integers, the ideal $(p)$ is a prime ideal if and only if $p$ is a prime number (this fact explains the terminology).

Moreover, $(a)$ is primary if $a$ is a power of a prime number.

#### 2.7.11.3 The theorem of Lasker–Noether

We denote by $(a_1, \ldots, a_s)$ he smallest ideal which contains the elements $a_1, \ldots, a_s$ (one also speaks of the ideal *generated* by the elements).

**Noetherian rings:** A ring is said to be *Noetherian,* if it is commutative and every ideal is generated by finitely many elements.

**The Hilbert basis theorem (1983):** If $R$ is a Noetherian ring with unit, then this is also true for every ring of polynomials $R[x_1, \ldots, x_n]$ in $n$ variables with coefficients in $R$.

The following theorem is the main result in the theory of divisors.

**Theorem of Emanuel Lasker (1905) and Emmy Noether (1926):** Let $R$ be a Noetherian ring. Then every ideal of $R$ can be written in a non-redundant way as the intersection of primary ideals, whose associated prime ideals are distinct.

Any two such representations have the same number of primary ideals and (up to order) the same associated prime ideals.

**Products of ideals:** Let $\mathscr{A}$ and $\mathscr{B}$ be ideals of a ring $R$; one denotes by

$$\mathscr{A}\mathscr{B}$$

the smallest ideal which contains all products $ab$ with $a \in \mathscr{A}$ and $b \in \mathscr{B}$. Moreover, the intersection $\mathscr{A} \cap \mathscr{B}$ and the ideal-theoretic sum $\mathscr{A} + \mathscr{B} := \{a + b \,|\, a \in \mathscr{A},\ b \in \mathscr{B}\}$ are ideals.

### 2.7.12 Applications to quadratic number fields

**The field $\mathbb{Q}(\sqrt{d})$:** Let $d$ be an integer with $d \neq 0$ and $d \neq 1$. Moreover suppose that $d$ is *square-free,* i.e., not divisible by a square. The *quadratic number field* $\mathbb{Q}(\sqrt{d})$ consists by definition of all numbers of the form

$$\boxed{a + b\sqrt{d} \qquad \text{with} \quad a, b \in \mathbb{Q},}$$

where $\mathbb{Q}$ is the field of rational numbers. The *conjugate* number $z'$ to $z := a + b\sqrt{d}$ is the number

$$z' := a - b\sqrt{d}.$$

Moreover, we define the *norm* $\mathrm{N}(z)$ and the *trace* $\mathrm{tr}\,(z)$ of the number $z$ by the formulas

$$\mathrm{N}(z) := zz' \quad \text{and} \quad \mathrm{tr}\,(z) := z + z'.$$

**Theorem 1:** Every extension field $K$ of the field $\mathbb{Q}$ of degree 2 is isomorphic to a quadratic number field $\mathbb{Q}(\sqrt{d})$ for some $d$.

The Galois group of $K$ over $\mathbb{Q}$ consists of the automorphisms $\varphi_\pm : K \longrightarrow K$ defined by

$$\varphi_+(z) := z \quad \text{and} \quad \varphi_-(z) := z'.$$

**Integers:**[77] A number $z \in \mathbb{Q}(\sqrt{d})$ is said to be *integer*, if it satisfies an equation of the form

$$z^n + a_{n-1}z^{n-1} + \ldots + a_1 z + a_0 = 0$$

with (rational) integral coefficients $a_0, \ldots, a_{n-1}$ and a positive natural number $n$. The set of integers in $\mathbb{Q}(\sqrt{d})$ is denoted by $\mathscr{O}$ (or $\mathscr{O}_d$ to make the dependency on $d$ explicit). It is a ring and is called the *ring of integers* of $K$.

**Theorem 2:** For $d \equiv 2 \bmod 4$ and $d \equiv 3 \bmod 4$ one has

$$\mathscr{O}_d := \{a + b\sqrt{d} \mid a, b \in \mathbb{Z}\}, \qquad D := 4d,$$

and for $d \equiv 1 \bmod 4$ one has

$$\mathscr{O}_d := \left\{ a + b\frac{1+\sqrt{d}}{2} \,\middle|\, a, b \in \mathbb{Z} \right\}, \qquad D := d.$$

The number $D$ occuring here is called the *discriminant* of the field $\mathbb{Q}(\sqrt{d})$.[78]

**Corollary:** The units in the ring $\mathscr{O}$ are:

$$\begin{array}{lll} \pm 1, \pm \mathrm{i} & \text{for} & d = -1, \\ 1, \eta, \ldots, \eta^5 & \text{for} & d = -3 \quad \text{with} \quad \eta := \mathrm{e}^{\mathrm{i}\pi/3}, \\ 1, -1 & \text{for} & d < 0, \ d \neq -1, -3, \\ \pm \varepsilon^k, \ k \in \mathbb{Z} & \text{for} & d > 0. \end{array}$$

Here one has $\varepsilon := x + y\sqrt{d}$, where $(x, y)$ is the smallest solution of the Fermat equation $x^2 - dy^2 = 1$ with $x, y \in \mathbb{N}$ (cf. 3.8.6.1). This unit is called the *fundamental unit* of $\mathbb{Q}(\sqrt{d})$.

**The fundamental decomposition theorem of Dedekind (1871):** $\mathscr{O}_d$ is a ring, in which every ideal $\mathscr{A} \neq 0$ can be written uniquely (up to order) as a product of prime ideals.

*Example 1:* Let $d = -5$. In the field $\mathbb{Q}(\sqrt{-5})$ one has the two decompositions

$$9 = 3 \cdot 3$$

[77]To discriminate between the usual integers and the integers of a number field, one refers to the former as *rational integers*.

[78]The notions norm, trace and discriminant as well as ring of integers are quite general and can be defined for arbitrary number fields.

and

$$9 = (2 + \sqrt{-5})(2 - \sqrt{-5})$$

of the number 9 into prime elements, i.e., the decomposition into prime elements in the ring $\mathcal{O}_{-5}$ is not unique. However, if one considers instead of the number 9 the principal ideal (9), one gets the unique decomposition

$$\boxed{(9) = \mathscr{P}\mathscr{Q}}$$

with the two prime ideals $\mathscr{P} := (3, 2 + \sqrt{-5})$ and $\mathscr{Q} := (3, 2 - \sqrt{-5})$.

*Example 2:* (i) Let $d < 0$. Then $\mathcal{O}_d$ is a unique factorization ring only for the values

$$d = -1, -2, -3, -7, -11, -19, -43, -67, -163.$$

For exactly the values $d = -1, -2, -3, -7$ and $-11$, $\mathcal{O}_d$ is a Euclidean ring.

(ii) Let $d > 0$. Then $\mathcal{O}_d$ is a Euclidean ring if and only if

$$d = 2, 3, 5, 6, 7, 11, 13, 17, 19, 21, 29, 33, 37, 41, 57, 73.$$

These rings $\mathcal{O}_d$ are also unique factorization rings. A complete description of the rings $\mathcal{O}_d$ which are factorial is even today an unsolved problem.

**Fractional ideals:** A subset $\mathscr{A}$ of $\mathbb{Q}(\sqrt{d})$ is said to be a *fractional ideal,* if one has:

(i) The elements of $\mathscr{A}$ are exactly the elements of $\mathbb{Q}(\sqrt{d})$ which can be written in the form

$$a_1 z_1 + \ldots + a_n z_n$$

with fixed numbers $z_1, \ldots, z_n$ in $\mathbb{Q}(\sqrt{d})$ and arbitrary coefficients $a_1, \ldots, a_n \in \mathbb{Z}$.

(ii) From $z \in \mathscr{A}$ and $r \in \mathcal{O}_d$ it follows that $rz \in \mathscr{A}$.

*Example 3:* $\mathcal{O}_d$ is a fractional ideal.

Two fractional ideals $\mathscr{A}$ and $\mathscr{B}$ are said to be *equivalent,* if $\mathscr{A} = k\mathscr{B}$ with a fixed number $k \neq 0$ in $\mathbb{Q}(\sqrt{d})$. If $\mathscr{A}$ and $\mathscr{B}$ are fractional ideals, then one denotes by

$$\mathscr{A}\mathscr{B}$$

the smallest fractional ideal which contains all products $ab$ with $a \in \mathscr{A}$ and $b \in \mathscr{B}$.

**The fundamental class number of $\mathbb{Q}(\sqrt{d})$:** The set of equivalence classes of fractional ideals of $\mathbb{Q}(\sqrt{d})$ forms a group with respect to the multiplication $\mathscr{A}\mathscr{B}$, which one calls the *class group* of $\mathbb{Q}(\sqrt{d})$; the order of this group is called the *class number* of the field. The class number is usually denoted by $h$, or in this case $h(d)$ to indicate the dependency on the field $\mathbb{Q}(\sqrt{d})$.

These notions can all be found in Gauss' *Disquisitiones arithmeticae* published in 1801. The larger the class number, the more complex the structure of the ring $\mathcal{O}_d$ and of the field $\mathbb{Q}(\sqrt{d})$.

**Theorem 3:** $\mathcal{O}_d$ is a principal ideal ring if and only if the class number of $\mathbb{Q}(\sqrt{d})$ is equal to unity, i.e., up to equivalence there is but a single fractional ideal in $\mathbb{Q}(\sqrt{d})$.

*Example 4:* All Euclidean rings in Example 2 are principal ideal rings and hence have class number one.

## 2.7.13 The analytic class number formula

> *In 1855, after Gauss' death, the University of Göttingen tried to keep the reputation it had gained through having the first of all living mathematicians for half a century by appointing Dirichlet as Gauss' successor.*
>
> *Eduard Kummer* (1810–1893)

Dirichlet (1805–1859) was the first to systematically apply analytic methods in the theory of numbers and so created what is now known as analytic number theory. Among other things he got formulas for the class numbers using his $L$-series.

**Class number formula:** For the class number $h(d)$ of $\mathbb{Q}(\sqrt{d})$ one has

$$h(d) = \begin{cases} 1 & \text{for} \quad d = -1, -3, \\ \dfrac{1}{\pi}\sqrt{|D|}L(1,\chi) & \text{for} \quad d < 0,\ d \neq -1, -3, \\ \dfrac{\sqrt{D}}{2\ln\varepsilon}L(1,\chi) & \text{for} \quad d > 0. \end{cases}$$

Here $D$ is the discriminant of the field $\mathbb{Q}(\sqrt{d})$ and $\varepsilon$ is the fundamental unit. Moreover

$$L(s,\chi) := \sum_{n=1}^{\infty} \frac{\chi(n)}{n^s}$$

with

$$\chi(n) := \begin{cases} \displaystyle\prod_{p|d}\left(\frac{n}{p}\right) & \text{for} \quad d \equiv 1 \bmod 4, \\ \displaystyle(-1)^{(n-1)/2}\prod_{p|d}\left(\frac{n}{p}\right) & \text{for} \quad d \equiv 3 \bmod 4, \\ \displaystyle(-1)^{\rho}\prod_{p|\delta}\left(\frac{n}{p}\right) & \text{for} \quad d = 2\delta \quad \text{and} \quad \delta \text{ odd}, \end{cases}$$

as well as $\rho := \dfrac{n^2-1}{8} + \dfrac{(n-1)(\delta-1)}{4}$. The products extend over all prime numbers $p$ which are divisors of $d$ (resp. of $\delta$ in the third case). Moreover, $\left(\dfrac{n}{p}\right)$ is the Legendre symbol (cf. 2.7.8.2). One calls $\chi$ a *character* of $\mathbb{Q}(\sqrt{d})$.

## 2.7.14 Hilbert's class field theory for general number fields

> *The theory of number fields is like an architectural masterpiece of wonderful beauty and harmony.*
>
> *David Hilbert,*
> *Zahlbericht (*1895)

The ultimate goal of class field theory is to give a complete classification of all fields. Already in the apparently simple case of algebraic number fields this turns out to be quite a challenge.

**Abelian field extensions of number fields:** A field extension from $K$ to $L$ is called *Abelian*, if the Galois group of the extension (cf. 2.6.2) is Abelian, i.e., commutative.

An *algebraic number field* is a finite field extension of the rational numbers. If $K$ is an algebraic number field, then one would like to determine all the Abelian field extensions $L$ of $K$.

To this end, on considers a particular finite field extension $H(K)$ of $K$, which is called the *Hilbert class field* of $K$. This field $H(K)$ contains important information on the Abelian extensions of $K$.

*Example:* The Hilbert class field of $K = \mathbb{Q}(\sqrt{-5})$ is $H(K) = \mathbb{Q}(\mathrm{i}, \sqrt{5})$, i.e., $H(K)$ is the smallest field extension of $\mathbb{Q}$ which contains i and $\sqrt{5}$.

A presentation of modern class field theory on the basis of homological algebra (cohomology of groups) including the deep reciprocity laws can be found in [287] and [282]. In this theory broad generalizations of the local–global principle are applied, which connect ideal theory with the theory of valuations and which generalize the theory of $p$-adic numbers (cf. 2.7.10.3).

The starting point of Hilbert's theory is the following classical result of Kronecker and Weber (1887).

**Theorem:** Every finite Abelian extension $L$ of the field $\mathbb{Q}$ of rational numbers is contained in the cyclotomic field $\mathbb{Q}(\zeta_n)$.

**Remarks:** 1. The notation is $\zeta_n := \mathrm{e}^{2\pi \mathrm{i}/n}$, an $n^{th}$ root of unity, and $\mathbb{Q}(\zeta_n)$ denotes the smallest subfield of $\mathbb{C}$ which contains $\zeta_n$. The Galois group of the extension $\mathbb{Q}(\zeta_n)|\mathbb{Q}$ is equal to $(\mathbb{Z}/n\mathbb{Z})^\times$ (the group of units in the residue class ring $\mathbb{Z}/n\mathbb{Z}$ of $\mathbb{Z}$ modulo $n$). It follows from Galois theory that a bijective mapping

$$\boxed{U \mapsto L}$$

exists between the set $U$ of all subgroups of $(\mathbb{Z}/n\mathbb{Z})^\times$ and the set of Abelian field extensions $L$ of $\mathbb{Q}$ which are contained in $\mathbb{Q}(\zeta_n)$.

2. The extension of this classical theory to *non-Abelian* extensions, laid down in the so-called "Langlands program", reaches into the furthest frontiers of modern mathematics, creating links between number theory, commutative algebra, algebraic geometry, representation theory of Lie groups and many other areas. In particular, *Shimura varieties* occur, which in turn are related to the *Shimura–Taniyama–Weil conjecture*, proved by Andrew Wiles for semi-stable curves; the proof of Fermat's last theorem is a corollary of this result.

# 3. Geometry

*He who understands geometry, understands anything in the universe.*
*Galileo Galilei* (1564–1642)

*Geometry is the invariant theory of groups of transformations.*
*Felix Klein,*
*Erlanger Program* 1872

## 3.1 The basic idea of geometry epitomized by Klein's Erlanger Program

The geometry known in ancient times was Euclidean geometry, and it dominated mathematics for over 2000 years. The famous question as to the existence of non-Euclidean geometries led in the nineteenth century to the description of a series of different geometries. This being established, it was natural to consider the classification of possible geometries. Felix Klein at the age of 23 solved this problem and showed in 1872 with his Erlanger Program that geometries can be conveniently classified by means of group theory. A geometry requires a group $G$ of transformations. Every property or quantity remaining invariant under the action of this group $G$ is a property of the associated geometry, which is therefore also referred to as a $G$-geometry. We will make continual use of this classification principle in this chapter. We explain the basic idea with the example of Euclidean geometry and the so-called similarity geometry.

**Euclidean geometry (geometry of motion):** We consider a plane $E$. We denote by $\mathrm{Aut}\,(E)$ the set of all maps of $E$ into itself which are composed of the following types of transformations (a special case of *automorphisms*, hence the notation):

(i) translations,

(ii) rotations around a point, and

(iii) reflections on a fixed line (Figure 3.1).

Compositions of these transformations, i.e., the elements of $\mathrm{Aut}\,(E)$, are called *motions*[1] of $E$. We denote by

$$\boxed{hg}$$

the transformation which is the composition of $g$ and $h$, i.e., first applying $g$, then $h$. With this multiplication $hg$ the set

$$\boxed{\mathrm{Aut}\,(E)}$$

[1]Specializing to those transformations which are compositions of only the translations and the rotations, we obtain the set of *proper motions*.

inherits the structure of a *group* (a special case of *automorphism group*). The neutral element $e$ in the group is the identical motion (no motion at all).

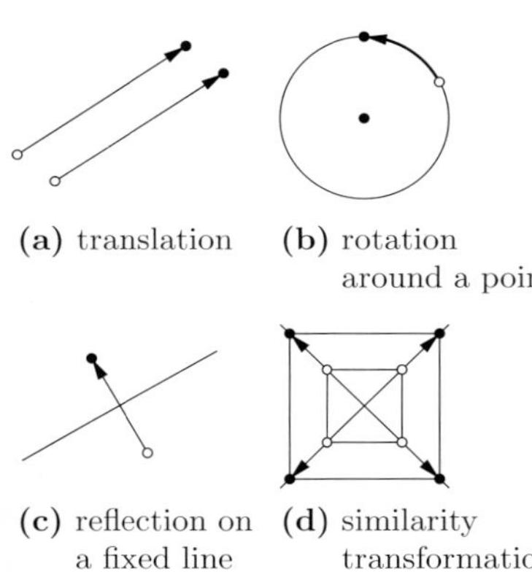

**(a)** translation **(b)** rotation around a point

**(c)** reflection on a fixed line **(d)** similarity transformation

*Figure 3.1. Plane motions.*

By definition the set of properties and quantities which are invariant under this group belong to the *Euclidean geometry* of the plane. Examples of this are 'the length of a segment' or 'the radius of a circle'.

**Congruence:** Two subsets of the plane (for example, two triangles) are said to be *congruent*, if they can be mapped into one another by a transformation in Aut $(E)$. The well-known theorems on the congruences of triangles in the plane are results about Euclidean geometry (see section 3.2.1.5).

**Similarity geometry:** A *special similarity transformation* is a transformation of the plane $E$ which maps the set of all lines through a chosen point $P$ into itself and which at the same time multiplies all distances from $P$ to somewhere else with a fixed (positive) number. We denote by $\mathrm{Sim}(E)$ the set of all transformations of $E$ which are composed of the motions above and the similarity transformations. Then

$$\boxed{\mathrm{Sim(E)}}$$

forms with respect to the same product $hg$ as above a group, which is called the *group of similarity transformations*.

The notion 'length of a segment' is *not* a notion of this geometry. However, the notion 'the ratio of two segments' *is*.

**Similarity:** Two subsets of $E$ (for example, two triangles) are said to be *similar*, if they are related by a similarity transformation, i.e., if there is a similarity transformation which maps one to the other. The well-known theorems on similarity of triangles in the plane are theorems of this geometry.

Every technical drawing is similar to the object it is rendering.

## 3.2 Elementary geometry

Unless we explicitly state the contrary, all angles will be measured in radians (cf. 0.1.2).

### 3.2.1 Plane trigonometry

**Notations:** A plane triangle consists of three points which do not lie on a line called *vertices*, as well as the three segments joining these vertices in pairs, called *edges* or *sides*. We denote the sides[2] by $a, b, c$ and the angles opposite those sides by $\alpha, \beta, \gamma$,

[2] We will also denote by the same symbols the lengths of the corresonding sides, provided no confusion is caused by this.

respectively (Figure 3.2(a)). Moreover, we set:

$s = \frac{1}{2}(a+b+c)$ (half the circumference),

$F$ surface area, $h_a$ height of the triangle over the side $a$,

$R$ radius of the circumscribed circle, $r$ radius of the inscribed circle.

The circumscribed circle is the smallest circle which contains the entire triangle and which passes through the three vertices. The inscribed circle is the largest circle which is contained in the triangle.

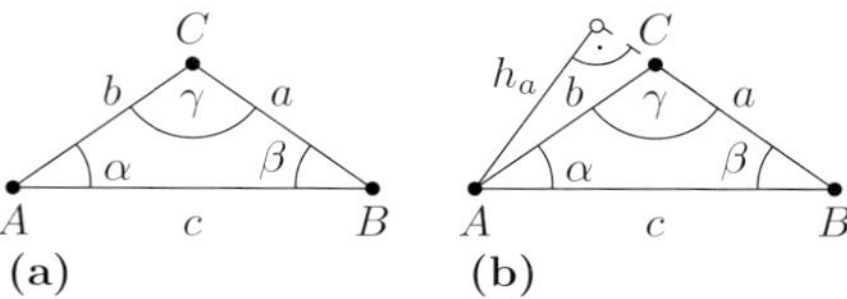


*Figure 3.2. Quantities of a plane triangle.*

#### 3.2.1.1 Four fundamental laws for triangles

*Law on the sum of the angles:*
$$\alpha + \beta + \gamma = \pi. \tag{3.1}$$

*Cosine law:*
$$c^2 = a^2 + b^2 - 2ab\cos\gamma. \tag{3.2}$$

*Sine law:*
$$\frac{a}{b} = \frac{\sin\alpha}{\sin\beta}. \tag{3.3}$$

*Tangent law:*
$$\frac{a-b}{a+b} = \frac{\tan\frac{\alpha-\beta}{2}}{\tan\frac{\alpha+\beta}{2}} = \frac{\tan\frac{\alpha-\beta}{2}}{\cot\frac{\gamma}{2}}. \tag{3.4}$$

**Triangle inequality:** $c < a + b$.

**Circumference:** $C = a + b + c = 2s$.

**Height:** For the height of the triangle over the side $a$ one has (Figure 3.2(b)):

$$h_a = b\sin\gamma = c\sin\beta.$$

**Surface area:** From the height formula one gets

$$A = \frac{1}{2}h_a\,a = \frac{1}{2}ab\cdot\sin\gamma.$$

In words: the surface area of a triangle is equal to half the product of the length of the side and height over that side.

Moreover, one can also use the *Heronian formula*[3]:

$$A = \sqrt{s(s-a)(s-b)(s-c)} = rs.$$

In words: The surface area of a triangle is equal to half the product of the radius of the inscribed circle and the perimeter of the triangle.

[3]This formula is named after Heron of Alexandria (first century AD), one of the most important mathematicians of ancient times who wrote numerous books on applied mathematics and engineering.

**More formulas on triangles:**

*Laws of half-angle:*
$$\sin\frac{\gamma}{2} = \sqrt{\frac{(s-a)(s-b)}{ab}}\,,$$
$$\cos\frac{\gamma}{2} = \sqrt{\frac{s(s-c)}{ab}}, \qquad \tan\frac{\gamma}{2} = \frac{\sin\frac{\gamma}{2}}{\cos\frac{\gamma}{2}}\,.$$

*Mollweidian formulas:*
$$\frac{a+b}{c} = \frac{\cos\frac{\alpha-\beta}{2}}{\cos\frac{\alpha+\beta}{2}} = \frac{\cos\frac{\alpha-\beta}{2}}{\sin\frac{\gamma}{2}}\,,$$
$$\frac{a-b}{c} = \frac{\sin\frac{\alpha-\beta}{2}}{\sin\frac{\alpha+\beta}{2}} = \frac{\sin\frac{\alpha-\beta}{2}}{\cos\frac{\gamma}{2}}\,.$$

*tangent formula:*
$$\tan\gamma = \frac{c\,\sin\alpha}{b-c\,\cos\alpha} = \frac{c\,\sin\beta}{a-c\,\cos\beta}\,.$$

*projection law:*
$$c = a\,\cos\beta + b\,\cos\alpha\,. \tag{3.5}$$

**Cyclic permutation:** More formulas can be obtained from (3.1) to (3.5) by cyclically permuting the sides and the angles:

$$a \longrightarrow b \longrightarrow c \longrightarrow a \qquad \text{and} \qquad \alpha \longrightarrow \beta \longrightarrow \gamma \longrightarrow \alpha.$$

**Special triangles:** A triangle is called a *right triangle*, if one of the angles is $\pi/2$ (that is, $90°$) (Figure 3.3).

**(a)** $\gamma = 90°$ **(b)** $h^2 = pq$

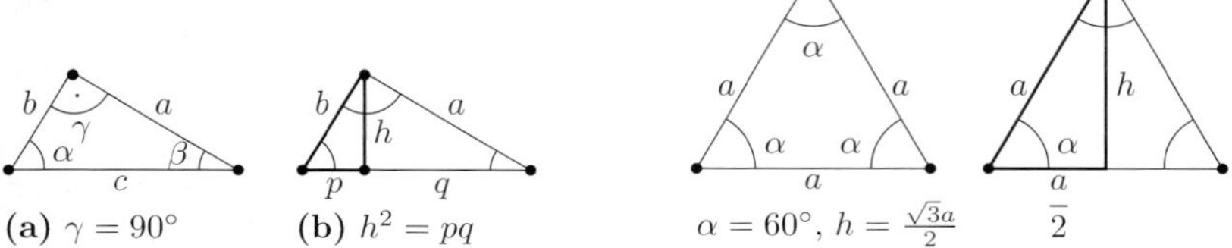

*Figure 3.3. A right triangle.*

$\alpha = 60°$, $h = \frac{\sqrt{3}a}{2}$

*Figure 3.4. An equilateral triangle.*

**(a)** **(b)** $\alpha = \beta = 45°$, $\gamma = 90°$, $h = \frac{a\sqrt{2}}{2}$

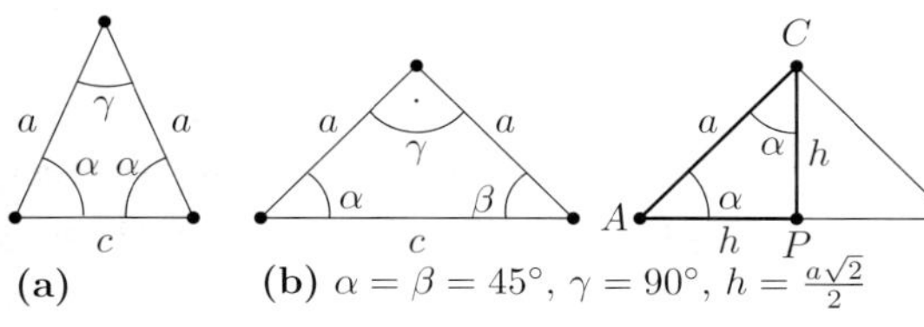

*Figure 3.5. A symmetric triangle.*

A triangle is called *symmetric* (resp. *equilateral*) if two (resp. all three) of its sides are equal (Figure 3.5 and Figure 3.4, respectively).

**Acute and obtuse angles:** An angle $\gamma$ is said to be *acute* (resp. *obtuse*), if $\gamma$ is between 0 and 90° (resp. between 90° and 180°).

**Calculations for triangles on a calculator:** In order to calculate the formulas which occur below, one requires the values $\sin\alpha$, $\cos\alpha$, etc. These are provided by most calculators.

#### 3.2.1.2 The right triangle

In a right triangle the side opposite the right angle is called the *hypotenuse*. The other sides are referred to as *catheti* (cathetus is the singular) or just *legs*. Both of these word are taken from Greek. In what follows we use the notations as shown in Figure 3.3.

**Surface area:**

$$\boxed{A = \frac{1}{2}ab = \frac{a^2}{2}\tan\beta = \frac{c^2}{4}\sin 2\beta.}$$

**The theorem of Pythagoras:**

$$\boxed{c^2 = a^2 + b^2.} \tag{3.6}$$

In words: the square of the length of the hypotenuse is equal to the sum of the squares of the lengths of the other two legs.

Because of $\gamma = \pi/2$ and $\cos\gamma = 0$, the result (3.6) is actually a special case of the cosine law (3.2).

**Euclid's law of height:**

$$\boxed{h^2 = pq.}$$

In words: the square of the height is equal to the product of the lengths of the segments of the hypotenuse which arise from projection of the legs on the hypotenuse.

**Euclid's law on catheti:**

$$a^2 = qc, \qquad b^2 = pc.$$

In words: the square of the length of one of the legs is equal to the product of the lengths of the segments of the hypotenuse which arise when the leg is projected onto the hypotenuse.

**Angle relations:**

$$\boxed{\begin{array}{llll} \sin\alpha = \dfrac{a}{c}, & \cos\alpha = \dfrac{b}{c}, & \tan\alpha = \dfrac{a}{b}, & \cot\alpha = \dfrac{b}{a}, \\ \sin\beta = \cos\alpha, & \cos\beta = \sin\alpha, & \alpha + \beta = \dfrac{\pi}{2}\,. & \end{array}} \tag{3.7}$$

Because of the relation $\sin\beta = \cos\alpha$ the sine theorem (3.3) is transformed into the result $\tan -\alpha = \dfrac{a}{b}$. One refers to $a$ (resp. $b$) as the opposite (resp. adjacent) leg to the angle $\alpha$.

**Calculations on right triangles:** All of the problems which are posed about right angles can be solved with the help of (3.7) (cf. Table 3.1).

*Example 1* (Figure 3.5(b)): In a right triangle *with equal sides* one has the relation for the height over side $c$:

$$\boxed{h_c = \frac{a\sqrt{2}}{2}\,.}$$

*Table 3.1. Formulas for right triangles.*

| ***Given quantities*** | ***Formulas for the remaining quantities for a right angle*** | | 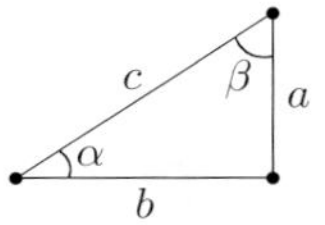 |
|---|---|---|---|
| $a, b$ | $\alpha = \arctan \frac{a}{b}$, | $c = \frac{a}{\sin \alpha}$, | $\beta = \frac{\pi}{2} - \alpha$ |
| $a, c$ | $\alpha = \arcsin \frac{a}{c}$, | $b = c \cos \alpha$, | $\beta = \frac{\pi}{2} - \alpha$ |
| $b, c$ | $\alpha = \arccos \frac{b}{c}$, | $a = b \tan \alpha$, | $\beta = \frac{\pi}{2} - \alpha$ |
| $a, \alpha$ | $b = a \cot \alpha$, | $c = \frac{a}{\sin \alpha}$, | $\beta = \frac{\pi}{2} - \alpha$ |
| $a, \beta$ | $\alpha = \frac{\pi}{2} - \beta$, | $b = a \cot \alpha$, | $c = \frac{a}{\sin \alpha}$ |
| $b, \alpha$ | $a = b \tan \alpha$, | $c = \frac{a}{\sin \alpha}$, | $\beta = \frac{\pi}{2} - \alpha$ |
| $b, \beta$ | $\alpha = \frac{\pi}{2} - \beta$, | $a = b \tan \alpha$, | $c = \frac{a}{\sin \alpha}$ |

*Proof:* The triangle $APC$ in Figure 3.5(b) is a right triangle. Since the sum of angles in a triangle is always 180°, one has $\alpha = \beta = 45°$. Because of $\frac{\gamma}{2} = 45°$, the triangle $APC$ has equal sides. It then follows from the Pythagorean theorem that $a^2 = h^2 + h^2$. This implies $h^2 = a^2/2$, hence $h = a/\sqrt{2} = a\sqrt{2}/2$. □

Moreover we get

$$\sin 45° = \frac{h}{a} = \frac{\sqrt{2}}{2}, \qquad \cos 45° = \sin 45°.$$

*Example 2* (Figure 3.4(b)): In an *equilateral triangle* we have for the height over the side $c$:

$$\boxed{h_c = \frac{a\sqrt{3}}{2}.}$$

*Proof:* The Pythagorean theorem yields $a^2 = h^2 + \left(\frac{a}{2}\right)^2$. From this it follows that $4a^2 = 4h^2 + a^2$, hence $4h^2 = 3a^2$. This implies $2h = \sqrt{3}a$.

Moreover, we get

$$\sin 60° = \frac{h}{a} = \frac{\sqrt{3}}{2}, \qquad \cos 30° = \sin 60°.$$

### 3.2.1.3 Four basic problems on triangles

From the equation $\sin \alpha = d$ one cannot determine the angle $\alpha$ uniquely, since $\alpha$ could be acute or obtuse and $\sin(\pi - \alpha) = \sin \alpha$. The following methods for the first and third of the problems now presented however do yield unique angles.

**First problem:** Let the side $c$ and the adjacent angles $\alpha$ and $\beta$ be given. The problem is to find the other sides and angle of the triangle (Figure 3.2).

(i) The angle $\gamma = \pi - \alpha - \beta$ is determined by an application of the law on the sum of angles.

(ii) Both sides $a$ and $b$ can be determined from the sine law:

$$a = c\,\frac{\sin\alpha}{\sin\gamma}\,, \qquad b = c\,\frac{\sin\beta}{\sin\gamma}\,.$$

(iii) For the surface area we have $A = \frac{1}{2}ab\sin\gamma$.

**Second problem:** Suppose now that the sides $a$ and $b$, as well as the angle $\gamma$ between them are given.

(i) One calculates $\frac{\alpha-\beta}{2}$ uniquely from the tangent law:

$$\tan\frac{\alpha-\beta}{2} = \frac{a-b}{a+b}\cot\frac{\gamma}{2}, \qquad -\frac{\pi}{4} < \frac{\alpha-\beta}{2} < \frac{\pi}{4}\,.$$

(ii) From the law on the sum of angles one has:

$$\alpha = \frac{\alpha-\beta}{2} + \frac{\pi-\gamma}{2}, \qquad \beta = \frac{\pi-\gamma}{2} - \frac{\alpha-\beta}{2}\,.$$

(iii) The side $c$ follows now from the sine law:

$$c = \frac{\sin\gamma}{\sin\alpha}\,a\,.$$

(iv) For the surface area we have $A = \frac{1}{2}ab\sin\gamma$.

**Third problem:** Now suppose that all three sides $a, b$ and $c$ are given.

(i) One calculates half the perimeter of the triangle $s = \frac{1}{2}(a+b+c)$ and the radius of the inscribed circle

$$r = \sqrt{\frac{(s-a)(s-b)(s-c)}{s}}\,.$$

(ii) The angles $\alpha$ and $\beta$ are now uniquely determined by the equations:

$$\tan\frac{\alpha}{2} = \frac{r}{s-a}\,, \qquad \tan\frac{\beta}{2} = \frac{r}{s-b}\,, \qquad 0 < \frac{\alpha}{2}, \frac{\beta}{2} < \frac{\pi}{2}\,.$$

(iii) The angle $\gamma = \pi - \alpha - \beta$ is again determined by the law on the sum of the angles.

(iv) For the surface area we get the easy formula $A = rs$.

**Fourth problem:** Finally, suppose that the two sides $a$ and $b$ as well as the one opposite angle $\alpha$ are given.

(i) We first determine the angle $\beta$.

*Case 1:* $a > b$. Then $\beta < 90°$, and $\beta$ is determined from the sine law uniquely from the equation

$$\sin\beta = \frac{b}{a}\sin\alpha\,. \tag{3.8}$$

*Case 2:* $a = b$. Here one just has $\alpha = \beta$.

*Case 3:* $a < b$. If $b \sin\alpha < a$, then the equation (3.8) yields two angles $\beta$ as solutions, one acute and one obtuse. In case $b \sin\alpha = b$ we have $\beta = 90°$. For $b \sin\alpha > a$ there is no triangle with the given conditions.

(ii) The angle $\gamma = \pi - \alpha - \beta$ is again determined by the law on the sum of the angles.

(iii) The side $c$ is determined from the sine law:

$$c = \frac{\sin\gamma}{\sin\alpha}\, a\,.$$

(iv) For the surface area we have $A = \frac{1}{2} ab \sin\gamma$.

#### 3.2.1.4 Special lines in a triangle

**Medians and the center:** A *median* is by definition a line through the midpoint of the one of the sides and the opposite vertex.

All three medians of a triangle meet in the *center*. One knows that in addition the center cuts each median in a ratio of 2 : 1 (measured from the vertex, see Figure 3.6(a)).

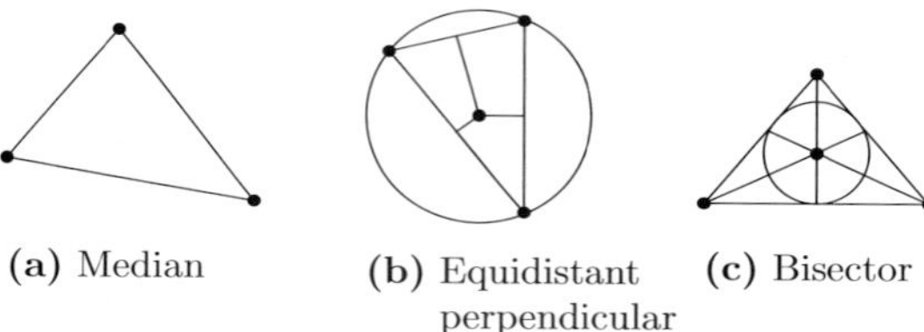

**(a)** Median **(b)** Equidistant perpendicular **(c)** Bisector

*Figure 3.6. Geometric properties of circles and triangles.*

*The length of the median meeting the side c in its midpoint:*

$$s_c = \frac{1}{2}\sqrt{a^2 + b^2 + 2ab\cos\gamma} = \frac{1}{2}\sqrt{2(a^2 + b^2) - c^2}\,.$$

**Equidistant perpendicular and circumscribing circle:** An *equidistant perpendicular* is by definition a segment which is perpendicular to one of the sides and passes through the midpoint of that side. The three equidistant perpendiculars meet at the center of the circumscribing circle.

*Radius of the circumscribing circle:* $R = \dfrac{a}{2\sin\alpha}$.

**Bisectors and the inscribed circle:** A *bisector* passes through one of the vertices and the opposite side, dividing the angle into two equal angles (bisects the angle).

All three bisectors meet at the center of the inscribed circle.

*Radius of the inscribed circle:* $r = (s-a)\tan\dfrac{\alpha}{2} = \dfrac{A}{s} = \sqrt{\dfrac{(s-a)(s-b)(s-c)}{s}}$,

$$r = s\,\tan\frac{\alpha}{2}\,\tan\frac{\beta}{2}\,\tan\frac{\gamma}{2} = 4R\,\sin\frac{\alpha}{2}\,\sin\frac{\beta}{2}\,\sin\frac{\gamma}{2}\,.$$

*Length of the bisector to the angle* $\gamma$:

$$w_\gamma = \frac{2ab}{a+b}\cos\frac{\gamma}{2} = \frac{\sqrt{ab\left((a+b)^2 - c^2\right)}}{a+b}\,.$$

**Theorem of Thales**[4]**:** If three points lie on a circle (with center $M$), then the central angle $2\gamma$ displayed in Figure 3.7 is equal to twice the periphery angle $\gamma$.

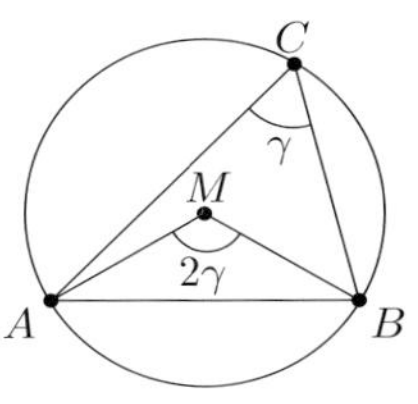


*Figure 3.7.*

#### 3.2.1.5 Theorems on congruent triangles

Two triangles are congruent (that is, they are related by one of the transformations described in section 3.1) if and only if one of the following four cases holds (Figure 3.8(a)):

(i) Two sides and the angle between them are equal.
(ii) One side and the two adjacent angles are equal.
(iii) Three sides are equal.
(iv) Two sides and the larger of the opposite angles are equal.

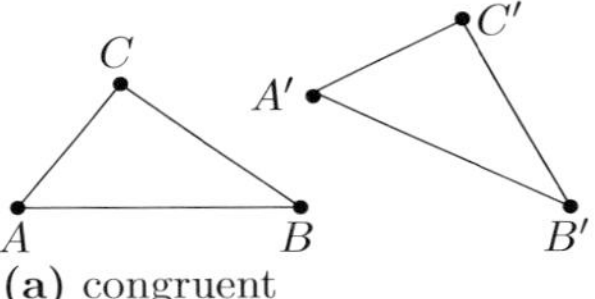


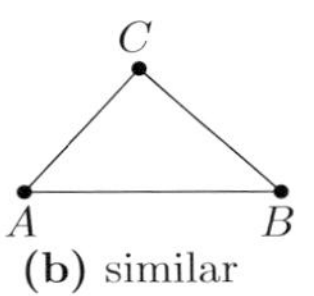


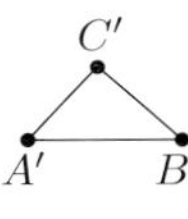


*Figure 3.8. Congruent and similar triangles.*

#### 3.2.1.6 Theorems on similar triangles

Two triangles are *similar* (i.e., they can be transformed into one another by one of the similarity transformations of section 3.1), if one of the following four cases holds (Figure 3.8(b)):

(i) Two angles are equal.
(ii) Two ratios of side lengths are equal.
(iii) One ratio of side lengths and the enclosed angle are equal.
(iv) One ratio of side lengths and the angle opposite the longer of these two sides are equal.

**Theorem of Thales** (the ray theorem): Let two lines be given, which intersect one another in a point $C$. If two parallel lines meet these two lines, the corresponding triangles $ABC$ and $A'B'C$ are similar (Figure 3.9).

For this reason the angles of both triangles and the ratios of the corresponding sides equal. For example, one has:

$$\frac{CA}{CA'} = \frac{CB}{CB'} .$$

*Figure 3.9.*

### 3.2.2 Applications to geodesy

Geodesy is the science of making measurement on the surface of the earth. One uses triangles for this (triangulation). Strictly speaking the triangles

[4]Thales of Milet (624–548 BC) is regarded as the founder of Greek mathematics.

here are triangles on the surface of a sphere (spherical triangles). If, however, these triangles are small enough (compared with the sphere), then one may treat them as plane triangles and apply the formulas of *plane trigonometry*. This is the case for most applications in geodesy. In sea and air travel, however, the triangles used are so large that one must use the formulas for *spherical trigonometry* (cf. 3.2.4).

**Height of a tower:** One is trying to determine the height $h$ of a tower (Figure 3.10).

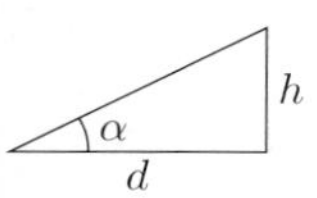


Figure 3.10.

*Measured quantities:* We measure the distance $d$ from the tower and the angle $\alpha$ of inclination.

*Calculation:* $h = d \tan \alpha$.

**Distance to a tower:** Here one is trying to determine the distance $d$ to a tower whose height is known.

*Measured quantities:* We measure the angle of inclination $\alpha$ and know the height $h$ of the tower.

*Calculation:* $d = h \cot \alpha$.

**The basic formulas of geodesy:** Let two point $A$ and $B$ be given with Cartesian coordinates $(x_A, y_A)$ and $(x_B, y_B)$, where we make the assumption that $x_A < x_B$ (Figure 3.11). Then we get for the distance $d = AB$ and the angle $\alpha$ the formulas:

$$d = \sqrt{(x_A - x_B)^2 + (y_A - y_B)^2}\,, \qquad \alpha = \arctan \frac{y_B - y_A}{x_B - x_A}\,.$$

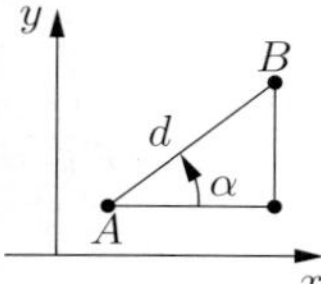


(a) $y_B > y_A$, $\alpha > 0$

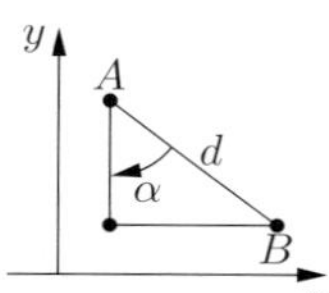


(b) $y_B < y_A$, $\alpha < 0$

Figure 3.11. The basic idea of geodesy.

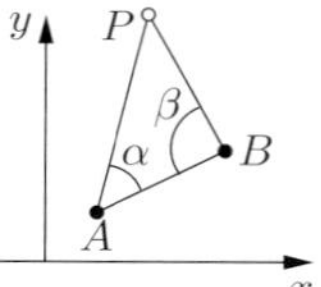


(a) forward cutting

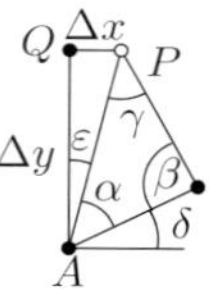


(b) $b = \overline{AP}$, $c = \overline{AB}$

Figure 3.12. Formulating problems in geodesy.

#### 3.2.2.1 The first basic problem (forward cutting)

**Problem:** Let two points $A$ and $B$ be given, with Cartesian coordinates $(x_A, y_A)$ and $(x_B, y_B)$. The Cartesian coordinates $(x, y)$ of a third point $P$ (as in Figure 3.12(a)) are sought for.

*Measured quantities:* We measure the angles $\alpha$ and $\beta$ as in Figure 3.12.

*Calculation:* We determine $b$ and $\delta$ by means of the formulas

$$c = \sqrt{(x_B - x_A)^2 + (y_B - y_A)^2}\,,$$

$$b = c\,\frac{\sin \beta}{\sin(\alpha + \beta)}, \qquad \delta = \arctan \frac{y_B - y_A}{x_B - x_A}$$

and get

$$\boxed{x = x_A + b\cos(\alpha+\delta), \qquad y = y_A + b\sin(\alpha+\delta)\,.}$$

*Proof:* We use the right triangle $APQ$ in Figure 3.12. Then we have

$$x = x_A + \triangle x = x_A + b\sin\varepsilon, \qquad y = y_A + \triangle y = y_A + b\cos\varepsilon\,.$$

From $\varepsilon = \dfrac{\pi}{2} - \alpha - \delta$ one has $\sin\varepsilon = \cos(\alpha+\delta)$ and $\cos\varepsilon = \sin(\alpha+\delta)$. From the sine law it follows that

$$b = c\,\frac{\sin\beta}{\sin\gamma}\,.$$

Finally one gets for $\gamma = \pi - \alpha - \beta$ (sum of angles in a triangle) the relation $\sin\gamma = \sin(\alpha+\beta)$.

□

#### 3.2.2.2 The second basic problem (backwards cutting)

**Problem:** We are now given three points $A, B$ and $C$ with Cartesian coordinates $(x_A, y_A)$, $(x_B, y_B)$ and $(x_C, y_C)$. We are looking for the Cartesian coordinates $(x, y)$ of a point $P$ as in Figure 3.13.

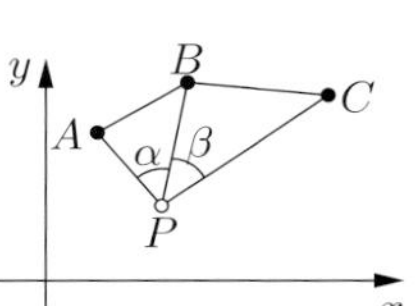

*Figure 3.13.*

*Measured quantities:* We measure the angles $\alpha$ and $\beta$.

The problem can only be solved if the four points do not lie on a circle.

*Calculation:* From the auxiliary quantities

$$x_1 = x_A + (y_C - y_A)\cot\alpha, \qquad y_1 = y_A + (x_C - x_A)\cot\alpha,$$
$$x_2 = x_B + (y_B - y_C)\cot\beta, \qquad y_2 = y_B + (x_B - x_C)\cot\beta$$

we calculate $\mu$ and $\eta$ as

$$\mu = \frac{y_2 - y_1}{x_2 - x_1}, \qquad \eta = \frac{1}{\mu}$$

and get

$$\boxed{\begin{aligned} y &= y_1 + \frac{x_C - x_1 + (y_C - y_1)\mu}{\mu + \eta}, \\ x &= \begin{cases} x_C - (y - y_C)\mu & \text{for } \mu < \eta \\ x_1 + (y - y_1)\eta & \text{for } \eta < \mu\,. \end{cases} \end{aligned}}$$

#### 3.2.2.3 The third basic problem (calculation of a distance which cannot be directly measured)

**Problem:** We are looking for the distance $\overline{PQ}$ between the two points $P$ and $Q$ as in Figure 3.14, which are for example separated by a lake. Therefore the distance cannot be directly measured.

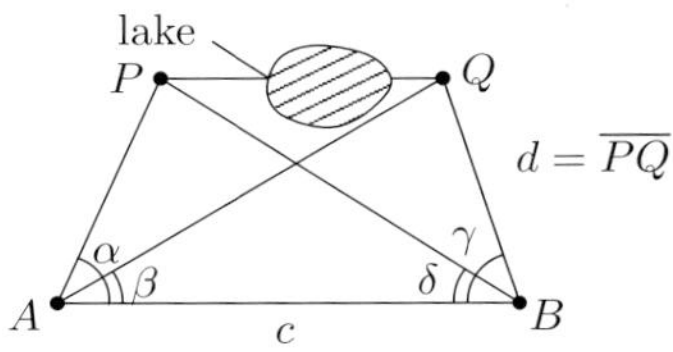

*Figure 3.14.*

*Measured quantities:* One measures the distance $c = \overline{AB}$ between two other points $A$ and $B$, as well the four angles $\alpha, \beta, \gamma$ and $\delta$ (as in Figure 3.14).

*Calculation:* From the auxiliary quantities

$$\varrho = \frac{1}{\cot\alpha + \cot\delta}, \qquad \sigma = \frac{1}{\cot\beta + \cot\gamma}$$

and $x = \sigma\cot\beta - \varrho\cot\alpha,\ y = \sigma - \varrho$ we get

$$\boxed{d = \sqrt{x^2 + y^2}\,.}$$

## 3.2.3 Spherical geometry

Spherical geometry is concerned with the geometry on a sphere (the surface of a ball). In the case of the surface of the earth on can apply the methods and formulas of plane trigonometry in good approximation provided the triangles (i.e., distances) are small enough. However, for calculations involving larger distances (for example trans-Atlantic flights or long ship journeys), the curvature of the earth plays an important role; in other words, in these cases one must use formulas of spherical trigonometry instead of plane trigonometry.

In what follows we will view the earth as a round ball, i.e., we ignore the flattening near the poles. The word geometry comes from Greek and means measuring the earth.

### 3.2.3.1 Measuring distances and great circles

We consider a ball of radius $R$ and denote the surface of this ball by $\mathscr{S}_R$ (sphere of radius $R$). We agree to call the circles on $\mathscr{S}_R$ which are centered at the center of the ball *great circles.*

> Instead of the lines of plane geometry, in spherical geometry one has the great circles.

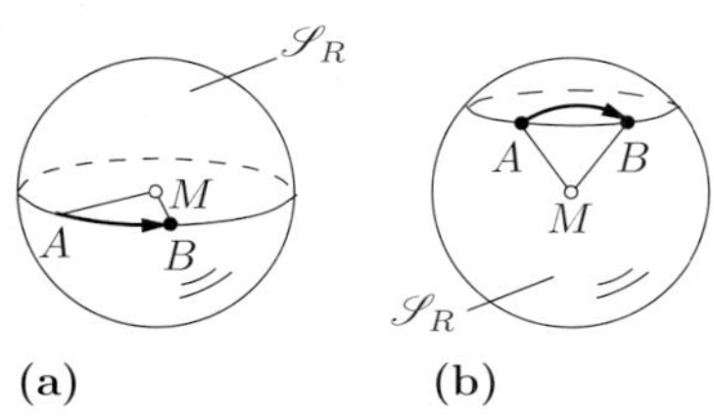


*Figure 3.15. Distances on a sphere.*

**Definition:** If $A$ and $B$ are points on $\mathscr{S}_R$, one gets the *great circle* which passes through $A$ and $B$ by intersecting $\mathscr{S}_R$ with the plane spanned by $A$, $B$ and the center $M$ of the ball.

*Example 1:* The equator and the circles of longitude on earth are great circles. Parallels of latitude are *not* great circles.

**Measuring distances on a sphere:** The shortest segment on a sphere joining two points $A$ and $B$ on $\mathscr{S}_R$ is obtained by considering the great circle between $A$ and $B$ (as above) and taking on this the shorter of the two segments into which this great circle is divided by $A$ and $B$ (Figure 3.15).

> The distance between two points $A$ and $B$ on a sphere is by definition the shortest distance between the two points on the sphere.

*Example 2:* If a ship (or a plane) wishes to take the shortest route between two points $A$ and $B$, then it must travel on the great circle joining these two points.

(i) If $A$ and $B$ both lie on the equator, then the ship just needs to travel along the equator (provided this is possible), see Figure 3.15(a).

(ii) If $A$ and $B$ are on the same parallel of latitude (which are not great circles), then the route which is shortest would be along a great circle joining them and not along the parallel of latitude, see Figure 3.15(b).

**Uniqueness of the shortest path:** If $A$ and $B$ are not on exactly opposite point of the sphere (the line between them passes through the center of the ball), then there is a uniquely determined shortest path.

If however $A$ and $B$ are opposite, then there are infinitely many shortest paths, all of the same length.

*Example 3:* The shortest paths from the north to the south pole consist of all great circles of longitude.

**Geodesics:** All segments of the great circles are called geodesics.

### 3.2.3.2 Measuring angles

**Definition:** If two great circles intersect in a point $A$, then the *angle* between them is by definition the angle between the tangents to the great circles at the point $A$ (Figure 3.16).

**Spherical diangle:** If one joins two pints $A$ and $B$ on a sphere $\mathscr{S}_R$ by means of two great circles, then one gets what is called a *spherical diangle* with the surface area

$$\boxed{S = 2R^2\alpha}$$

where $\alpha$ is the angle between the two great circles (Figure 3.17).

Figure 3.16.

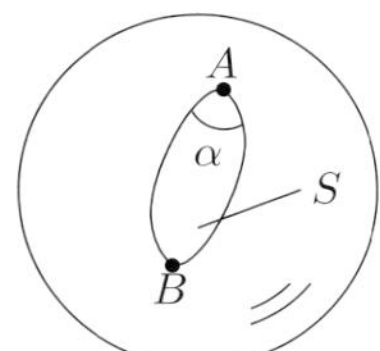


Figure 3.17.

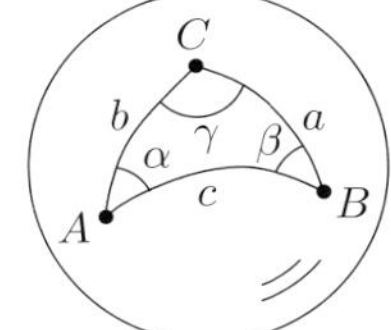


Figure 3.18.

### 3.2.3.3 Spherical triangles

**Definition:** A *spherical triangle* is formed by three points $A$, $B$ and $C$ on the sphere $\mathscr{S}_R$ and the shortest paths joining these points.[5] The angles are denoted by $\alpha$, $\beta$ and $\gamma$. The length of the sides are denoted $a$, $b$ and $c$ (Figure 3.18).

**Cyclic permutations:** All the formulas which follow remain correct when the following cyclic permutation is performed:

$$a \longrightarrow b \longrightarrow c \longrightarrow a \qquad \text{and} \qquad \alpha \longrightarrow \beta \longrightarrow \gamma \longrightarrow \alpha.$$

[5] We assume in addition that no two of the points are dimetral (i.e., opposite) and that the three points $A$, $B$ and $C$ do not lie on a single great circle.

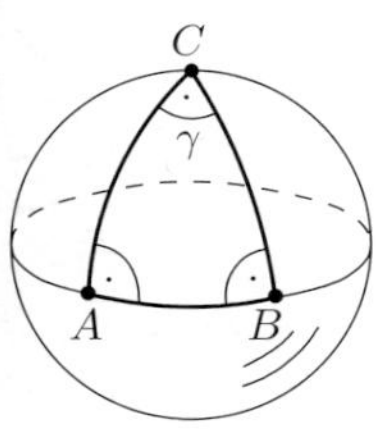


*Figure 3.19.*

**Surface area $S$ of a spherical triangle:**

$$S = (\alpha + \beta + \gamma - \pi)R^2.$$

Since the surface of the sphere itself is equal to $4\pi R^2$, we get from $0 < S < 4\pi R^2$ for the sum of the angles the inequality

$$\pi < \alpha + \beta + \gamma < 3\pi\,.$$

If one is not sure whether one lives on a sphere or in a plane, one can answer this question by measuring the sum of angles in triangles. For plane triangles one always has $\alpha + \beta + \gamma = \pi$.

The difference $\alpha + \beta + \gamma - \pi$ is called the *spherical excess.*

*Example 1:* The triangle in Figure 3.19 is formed by the north pole $C$ and two points $A$ and $B$ on the equator. Here one has $\alpha = \beta = \pi/2$. For the sum of the angles we get $\alpha + \beta + \gamma = \pi + \gamma$. The surface area is given by $S = R^2\gamma$.

**Triangle inequality:**

$$|a - b| < c < a + b\,.$$

**Ratios of the sides:** The longest side is always opposite the largest angle. Explicitly one has:

$$\alpha < \beta \iff a < b, \qquad \alpha > \beta \iff a > b, \qquad \alpha = \beta \iff a = b\,.$$

**Convention:**[6] We set

$$a_* := \frac{a}{R}, \qquad b_* := \frac{b}{R}, \qquad c_* := \frac{c}{R}\,.$$

**Law of sines:**[7]

$$\frac{\sin\alpha}{\sin\beta} = \frac{\sin a_*}{\sin b_*}\,. \tag{3.9}$$

**Law of cosines of sides and angles:**

$$\cos c_* = \cos a_* \cos b_* + \sin a_* \sin b_* \cos\gamma\,, \tag{3.10}$$

$$\cos\gamma = \sin\alpha \sin\beta \cos c_* - \cos\alpha \cos\beta. \tag{3.11}$$

**Half-angle law:** We set $s_* := \frac{1}{2}(a_* + b_* + c_*)$. Then one has

$$\tan\frac{\gamma}{2} = \sqrt{\frac{\sin(s_* - a_*)\sin(s_* - b_*)}{\sin s_* \sin(s_* - c_*)}}\,, \qquad 0 < \gamma < \pi\,, \tag{3.12}$$

[6] Often one takes $R = 1$. Then one has $a_* = a$, etc. We keep the radius $R$ in the formulas in order to be able to pass to the limit $R \to \infty$ (Euclidean geometry) and replace $R \mapsto \mathrm{i}R$ (the transition to non-Euclidean hyperbolic geometry) (cf. 3.2.8).

[7] If $\alpha$ is a right angle, then one has $\sin\alpha = 1$ and $\cos\alpha = 0$.

$$\sin\frac{\gamma}{2} = \sqrt{\frac{\sin(s_* - a_*)\,\sin(s_* - b_*)}{\sin a_*\,\sin b_*}}\,, \qquad \cos\frac{\gamma}{2} = \sqrt{\frac{\sin s_*\,\sin(s_* - c_*)}{\sin a_*\,\sin b_*}}\,.$$

**Formula for the surface area $A$ of a spherical triangle:**

$$\tan\frac{A}{4} = \sqrt{\tan\frac{s_*}{2}\,\tan\frac{s_* - a_*}{2}\,\tan\frac{s_* - b_*}{2}\,\tan\frac{s_* - c_*}{2}}$$

(generalized Heronian formula).

**Half-side law:** Let $\sigma := \frac{1}{2}(\alpha + \beta + \gamma)$. Then one has:

$$\boxed{\tan\frac{c_*}{2} = \sqrt{\frac{-\cos\sigma\,\cos(\sigma - \gamma)}{\cos(\sigma - \alpha)\,\cos(\sigma - \beta)}}\,, \qquad 0 < c_* < \pi\,,} \tag{3.13}$$

$$\sin\frac{c_*}{2} = \sqrt{-\frac{\cos\sigma\,\cos(\sigma - \gamma)}{\sin\alpha\,\sin\beta}}\,, \qquad \cos\frac{c_*}{2} = \sqrt{\frac{\cos(\sigma - \alpha)\,\cos(\sigma - \beta)}{\sin\alpha\,\sin\beta}}\,.$$

**Neperian formulas:**

$$\begin{aligned}
\tan\frac{c_*}{2}\,\cos\frac{\alpha - \beta}{2} &= \tan\frac{a_* + b_*}{2}\,\cos\frac{\alpha + \beta}{2}\,,\\
\tan\frac{c_*}{2}\,\sin\frac{\alpha - \beta}{2} &= \tan\frac{a_* - b_*}{2}\,\sin\frac{\alpha + \beta}{2}\,,\\
\cot\frac{\gamma}{2}\,\cos\frac{a_* - b_*}{2} &= \tan\frac{\alpha + \beta}{2}\,\cos\frac{a_* + b_*}{2}\,,\\
\cot\frac{\gamma}{2}\,\sin\frac{a_* - b_*}{2} &= \tan\frac{\alpha - \beta}{2}\,\sin\frac{a_* + b_*}{2}\,.
\end{aligned}$$

**Mollweidian formulas:**

$$\begin{aligned}
\sin\frac{\gamma}{2}\,\sin\frac{a_* + b_*}{2} &= \sin\frac{c_*}{2}\,\cos\frac{\alpha - \beta}{2}\,,\\
\sin\frac{\gamma}{2}\,\cos\frac{a_* + b_*}{2} &= \cos\frac{c_*}{2}\,\cos\frac{\alpha + \beta}{2}\,,\\
\cos\frac{\gamma}{2}\,\sin\frac{a_* - b_*}{2} &= \sin\frac{c_*}{2}\,\sin\frac{\alpha - \beta}{2}\,,\\
\cos\frac{\gamma}{2}\,\cos\frac{a_* - b_*}{2} &= \cos\frac{c_*}{2}\,\sin\frac{\alpha + \beta}{2}\,.
\end{aligned}$$

**Radii $r$ and $\varrho$ of an inscribed circle and a circumscribing circle of a spherical triangle:**

$$\tan r = \sqrt{\frac{\sin(s_* - a_*)\,\sin(s_* - b_*)\,\sin(s_* - c_*)}{\sin s_*}} = \tan\frac{\alpha}{2}\,\sin(s_* - a_*)\,,$$

$$\cot\varrho = \sqrt{-\frac{\cos(\sigma - \alpha)\,\cos(\sigma - \beta)\,\cos(\sigma - \gamma)}{\cos\sigma}} = \cot\frac{a_*}{2}\,\cos(\sigma - \alpha)\,.$$

**Passage of limit to plane trigonometry:** If one carries out the limit $R \to \infty$ in the formulas above (meaning that the radius of the sphere grows beyond all bounds), then the curvature of the surface of the sphere becomes smaller and smaller. In the limit one gets the familiar formulas of plane trigonometry.

*Example 2:* From the cosine law (3.10) it follows from $\cos x = 1 - \frac{x^2}{2} + o(x^2),\ x \to 0$, and $\sin x = x + o(x),\ x \to 0$ that

$$1 - \frac{c^2}{2R^2} + \ldots = \left(1 - \frac{a^2}{2R^2} + \ldots\right)\left(1 - \frac{b^2}{2R^2} + \ldots\right) + \left(\frac{a}{R} + \ldots\right)\left(\frac{b}{R} + \ldots\right)\cos\gamma\,.$$

After multiplying the formula by $R^2$ we get for $R \to +\infty$ the expression

$$c^2 = a^2 + b^2 - 2ab\cos\gamma\,.$$

This is the cosine law of plane trigonometry.

### 3.2.3.4 The calculation of spherical triangles

Note in what follows that $a_* := a/R$, etc. We consider here only triangles in which all angles and sides lie between zero and $\pi$.

**First basic problem:** Two sides $a$ and $b$ are given, together with the enclosed angle $\gamma$. One calculates the other side $c$ and the other angles $\alpha$ and $\beta$ with the help of the cosine law for the sides:

$$\cos c_* = \cos a_* \cos b_* + \sin a_* \sin b_* \cos\gamma\,,$$

$$\cos\alpha = \frac{\cos a_* - \cos b_* \cos c_*}{\sin b_* \sin c_*}\,,$$

$$\cos\beta = \frac{\cos b_* - \cos c_* \cos a_*}{\sin c_* \sin a_*}\,.$$

**Second basic problem:** Here we are given all three sides $a,\ b$ and $c$, all of which are to lie between 0 and $\pi$. The angles $\alpha,\ \beta$ and $\gamma$ are calculated by means of the half-side laws:

$$\tan\frac{\alpha}{2} = \sqrt{\frac{\sin(s_* - b_*)\sin(s_* - c_*)}{\sin s_* \sin(s_* - a_*)}} \qquad \text{and so on.}$$

The formulas for $\tan\frac{\beta}{2}$ and $\tan\frac{\gamma}{2}$ are obtained from $\tan\frac{\alpha}{2}$ by cyclic permutation.

**Third basic problem:** Here the three angles $\alpha,\ \beta$ and $\gamma$ are given. The sides $a,\ b$ and $c$ can be obtained from the half-side laws:

$$\tan\frac{a_*}{2} = \sqrt{\frac{-\cos\sigma\cos(\sigma - \alpha)}{\cos(\sigma - \beta)\cos(\sigma - \gamma)}} \qquad \text{and so on.} \tag{3.14}$$

The formulas for $\tan\frac{b_*}{2}$ and for $\tan\frac{c_*}{2}$ are obtained from $\tan\frac{a_*}{2}$ by cyclic permutation.

**Fourth basic problem:** The side $c$ is given, together with the two incident angles $\alpha$ and $\beta$. The missing angle $\gamma$ is obtained from the cosine law:

$$\cos\gamma = \sin\alpha\sin\beta\cos c_* - \cos\alpha\cos\beta.$$

The other sides are then calculated by an application of (3.14).

### 3.2.4 Applications to sea and air travel

In order to exemplify the principle, we calculate with rounded values.

**A sea journey from San Diego to Honolulu:** How long is the shortest route between these two cities? With which angle $\beta$ do we have to start at San Diego?

*Answer:* We consider Figure 3.20:

$$c = \text{distance} = 4\,100\,\text{km}, \qquad \beta = 97^\circ\,.$$

*Solution:* The two cities have the following geographical coordinates:

| | |
|---|---|
| $A$ (Honolulu) : | 22° northern latitude, 157° western longitude, |
| $B$ (SanDiego) : | 33° northern latitude, 117° western longitude. |

We use angular measurements. With the notations as in Figure 3.20, one has:

$$\gamma = 157^\circ - 117^\circ = 40^\circ, \qquad a_* = 90^\circ - 33^\circ = 57^\circ, \qquad b_* = 90^\circ - 22^\circ = 68^\circ\,.$$

The first basic problem in 3.2.3.4 yields:

$$\cos c_* = \cos a_* \, \cos b_* + \sin a_* \, \sin b_* \, \cos\gamma, \tag{3.15}$$

$$\cos\beta = \frac{\cos b_* - \cos a_* \, \cos c_*}{\sin a_* \, \sin c_*}. \tag{3.16}$$

This yields $c_* = 37^\circ$ and $\beta = 97^\circ$. The radius of the earth is $R = 6370$ km. Hence the triangle side $c$ is given by

$$c = R\frac{2\pi c_*^\circ}{360^\circ} = 4100.$$

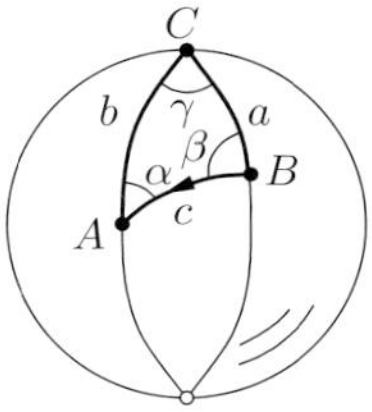

*Figure 3.20.*

**Transatlantic flight from Copenhagen to Chicago:** How far is the shortest route (by air) between these two cities? With which angle $\beta$ does on have to start from Copenhagen?

*Answer:* We consider Figure 3.20 again.

$$c = \text{distance} = 6\,000\,\text{km}, \qquad \beta = 82^\circ\,.$$

*Solution:* The two cities have the geographic coordinates:

| | |
|---|---|
| $A$ (Chicago) : | 42° northern latitude, 88° western longitude, |
| $B$ (Copenhagen) : | 56° northern latitude, 12° eastern longitude. |

We again use angular measurements. In the notations of Figure 3.20, we have:

$$\gamma = 12^\circ + 88^\circ = 100^\circ, \qquad a_* = 90^\circ - 56^\circ = 34^\circ, \qquad b_* = 90^\circ - 42^\circ = 38^\circ.$$

From (3.15) it follows that $c_* = 54^\circ$, hence $c = R\dfrac{2\pi c_*^\circ}{360^\circ} = 6000$. The angle is obtained from (3.16); it is $\beta$.

### 3.2.5 The Hilbert axioms of geometry

*Hence all human knowledge begins with perception, passes from that to notions and ends in ideas.*

*Immanuel Kant* (1724–1804)
*Kritik der reinen Vernunft, Elementarlehre*[9]

*Geometry needs – just as does arithmetic with numbers – to be put on rigorous foundations consisting of only a few simple principles. These principles are called the axioms of geometry. The derivation of these axioms and the study of there interconnections is a task which has led to many exceptional treatises in mathematics ever since Euclid. The task just mentioned amounts in the logical analysis of our spatial perception.*

*David Hilbert* (1862–1943)
*Principles of Geometry*

The first systematic presentation of geometry was given in the famous *Elements* of Euclid (365–300 BC), which have been taught unchanged for over 2000 years. An axiomatic presentation which is completely rigorous from a modern point of view was given by David Hilbert in his *Principles of Geometry* which appeared in 1899. This book has lost none of its intellectual freshness since then and in 1987 was published in a 13th edition by Teubner-Verlag. The following rather formal and seemingly dry axioms are the result of a long and tedious epistemological path, which was cluttered with errors and misconceptions all along the way. They are closely related to the Euclidean parallel axiom, which will be discussed in section 3.2.6. For clarity of presentation we restrict ourselves here to the axioms of plane geometry. To make the presentation more understandable we include figures illustrating the axioms. We would like to explicitly bring to the reader's attention that visualization methods like this have been used for 2000 years by mathematicians but helped to conceal the true nature of geometry (cf. 3.2.6 to 3.2.8).

**Basic notions of plane geometry:** For emphasis we state the most important notions of plane geometry at the start.

| Point, line, incident,[10] between, congruent. |
|---|

In laying the foundations of geometry, these notions *are not described.* This is a radical point of view, as Hilbert was the first to point out, and is the basis for every modern axiomatic treatment in mathematics. The missing contextual interpretation of the modern mathematical form of geometry is an apparent philosophical weakness; in fact it is however one of the great *strengths* of this approach and is typical of mathematical thinking. By restraining from trying to give these notions a concrete meaning one is

[9] Translated this means "A critique of pure reason, elementary theory".

[10] Instead of the statement 'the point $P$ is incident to the line $l$', one also says '$P$ lies on $l$' or '$l$ passes through $P$'. If $P$ lies on two lines $l$ and $m$, then one says that the lines $l$ and $m$ intersect in $P$.

in a position to all at once deal with a myriad of different situations with one logical construction (cf. 3.2.6 to 3.2.8).

**Incidence axiom** (Figure 3.21(a)): (i) *To two different points $A$ and $B$ there is exactly one line $l$ which passes through both $A$ and $B$.*
(ii) *There are at least two points lying on each line.*
(iii) *There exist three points, not all of which lie on the same line.*
**Order axiom** (Figure 3.21(b)): (i) *If a point $B$ lies between the points $A$ and $C$, then $A$, $B$ and $C$ are three different points, which lie on a line, and the point $B$ also lies between $C$ and $A$.*
(ii) *For every two distinct points $A$ and $C$ there is a point $B$ which lies between $C$ and $A$.*
(iii) *If three distinct points lie on a line, then there is exactly one of these points which lies between the others.*

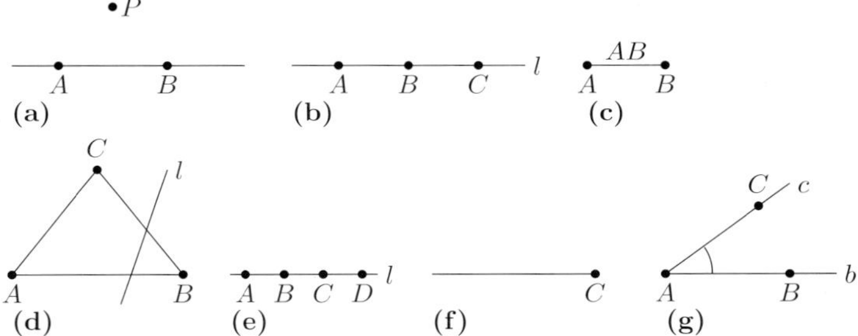


*Figure 3.21. The Hilbert axioms of geometry.*

**Definition of a segment** (Figure 3.21(c)): Let $A$ and $B$ be two distinct points, which lie on a line $l$. The *segment* $AB$ is the set of all points of $l$ which lie between $A$ and $B$. The points $A$ and $B$ themselves are counted for this purpose.

**The axiom of Pasch** (Figure 3.21(d)): *Let $A$, $B$ and $C$ be three distinct points which do not all line on a single line. Furthermore, let $l$ be a line on which none of the points $A, B$ and $C$ lie. If the line $l$ intersects the segment $AB$, then $l$ also intersects the segment $BC$ or the segment $AC$.*

**Definition of a ray** (Figure 3.21(e),(f)): Let $A, B, C$ and $D$ be four distinct points, all of which lie on a line $l$, where $C$ is between $A$ and $D$, but not between $A$ and $B$. Then we say that the points $A$ and $B$ *lie on the same side* of $C$, while $A$ and $D$ *lie on different sides* of $C$.

The set of all the points which lie on one side of $C$ is called a *ray*.

**Definition of angle** (Figure 3.21(g)): An angle $\measuredangle(b, c)$ is a set $\{b, c\}$ of two rays $b$ and $c$ which belong to different lines and initiate from a common point $A$. Instead of $\measuredangle(b, c)$ also the notion $\measuredangle(c, b)$ is used.[11]

[11] According to this convention, the rays $b$ and $c$ are treated equally. Intuitively one will choose the angle formed by $b$ and $c$ which is less than $180°$.

If $B$ (resp. $C$) is a point on a ray $b$ (resp. $c$), where $B$ and $C$ are distinct from the point $A$, then we write $\measuredangle BAC$ or $\measuredangle CAB$ instead of $\measuredangle(b, c)$.

With the help of the axioms presented thusfar, we can prove the following result.

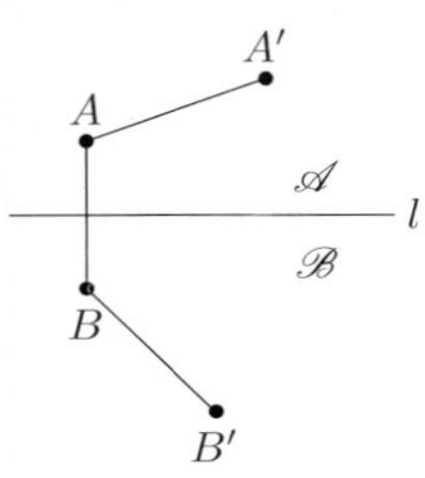


Figure 3.22.

**The theorem on the decomposition of the plane by a line** (Figure 3.22): If $l$ is a line, then all points either lie on the line $l$ or in one of the two sets $\mathscr{A}$ and $\mathscr{B}$ which have the following properties:

(i) If the point $A$ lies in $\mathscr{A}$ and the point $B$ in $\mathscr{B}$, then the segment $AB$ intersects the line $l$.

(ii) If two points $A$ and $A'$ (resp. $B$ and $B'$) lie in $\mathscr{A}$ (resp. $\mathscr{B}$), then the segment $AA'$ (resp. $BB'$) does not intersect the line $l$.

**Definition:** The points of $\mathscr{A}$ (resp. $\mathscr{B}$) lie on one (resp. the other) side of the line $l$.

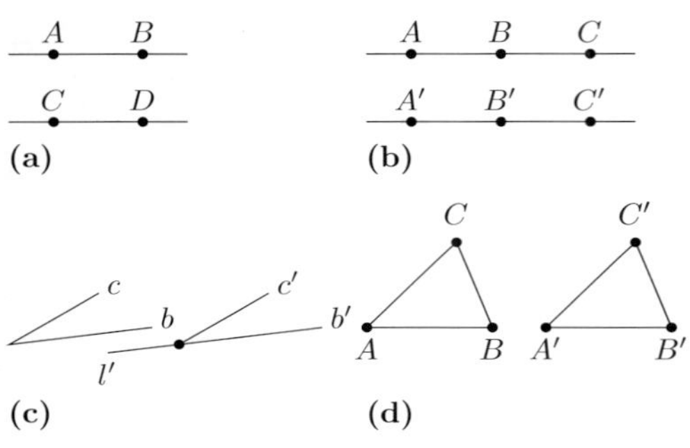


Figure 3.23. Congruence axioms for segments and angles.

A *congruence* of segments and angles are notions which are also not given a more precise meaning. Intuitively speaking, congruent objects are ones which can be transformed into one another by motions. The symbol

$$\boxed{AB \simeq CD}$$

means that the segment $AB$ is congruent to the segment $CD$, and

$$\boxed{\measuredangle ABC \simeq \measuredangle EFG}$$

means that the angle $\measuredangle ABC$ is congruent to the angle $\measuredangle EFG$.

**Congruence axiom for segments** (Figure 3.23(a),(b)): (i) *Assume that two points $A$ and $B$ lie on a line $l$, and the point $C$ lies on a line $m$. Then there is a point $D$ on $m$ such that*

$$AB \simeq CD.$$

(ii) *If two segments are congruent to a third segment, then they are also congruent to one another.* (iii) *Let $AB$ and $BC$ be two segments on a line $l$, which have no common point other than $B$. Moreover let $A'B'$ and $B'C'$ be two segments on a line $l'$, which have no common points other than $B'$. Then the relations*

$$AB \simeq A'B' \quad \text{and} \quad BC \simeq B'C'$$

*always imply*

$$AC \simeq A'C'.$$

**Definition of a triangle:** A *triangle $ABC$* consists of three distinct points $A$, $B$ and

$C$, which do not all lie on a line.

**Congruence axioms for angles** (Figure 3.23(c),(d)): (i) *Every angle is congruent to itself, i.e.,* $\measuredangle(b,c) \simeq \measuredangle(b,c)$. (ii) *Let* $\measuredangle(b,c)$ *be an angle and let* $b'$ *be a ray on the line* $l'$. *Then there is a ray* $c'$ *with*

$$\measuredangle(b,c) \simeq \measuredangle(b',c'),$$

*and all inner points of* $\measuredangle(b',c')$ *lie on one side of the line* $l'$. (iii) *Let two triangles* $ABC$ *and* $A'B'C'$ *be given. Then from*

$$AB \simeq A'B', \qquad AC \simeq A'C' \qquad \text{and} \qquad \measuredangle BAC \simeq B'A'C'$$

*it always follows that*

$$\measuredangle ABC \simeq \measuredangle A'B'C'.$$

**The axiom of Archimedes** (Figure 3.24): *If* $AB$ *and* $CD$ *are two given segments, then on the line through* $A$ *and* $B$ *there are points*

$$A_1, A_2, \ldots, A_n$$

*such that the segments* $AA_1, A_1A_2, \ldots, A_{n-1}A_n$ *are all congruent to the segment* $CD$ *and* $B$ *lies between* $A$ *and* $A_n$.[12]

**Completeness axiom:** *It is not possible to extend the system by adding points or lines such that the axioms all continue to hold.*

**Theorem of Hilbert (1899):** If the theory of real numbers is free of contradictions, then geometry as defined by the axioms is free of contradictions.

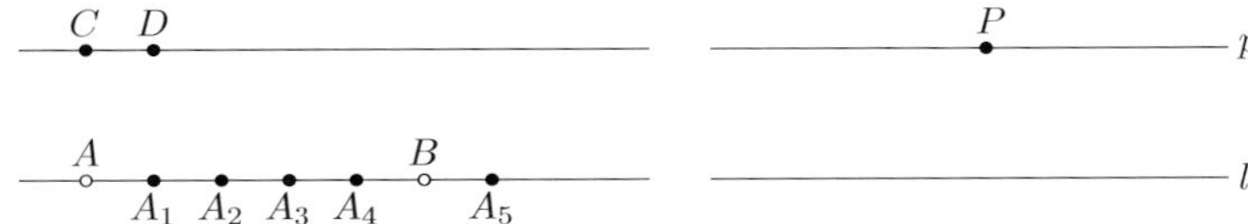


*Figure 3.24. The axiom of Archimedes.*

*Figure 3.25. The parallel axiom of Euclid.*

### 3.2.6 The parallel axiom of Euclid

**Definition of parallel lines:** Two lines $l$ and $m$ are said to be *parallel*, if they do not intersect in a point.

**Euclidean parallel axiom** (Figure 3.25): *If a point* $P$ *does not lie on a line* $l$, *then there is exactly one parallel line* $p$ *to* $l$ *containing the point* $P$.

[12]Intuitively this means that by drawing the segment $CD$ $n$ times one gets a segment which contains $AB$.
There are geometries in which all of the axioms except the axiom of Archimedes hold. Geometries of this kind are called *non-Archimedean*.

**Historical comment:** The *parallel problem* is:

| Can the parallel axiom be proved from the remaining axioms of Euclid? |
|---|

This was a famous unsolved problem in mathematics for over 2000 years. Karl Friedrich Gauss (1777–1855) was the first to realize that the parallel axiom cannot be proven from the other axioms. However, in order to avoid possible irrational quibbles he never published this result. The Russian mathematician Nikolai Ivanovich Lobachevski (1793–1856) published in 1830 a book on a new kind of geometry in which the Euclidean parallel axiom simply didn't hold. This is the Lobachevski geometry (or hyperbolic non-Euclidean geometry). The Hungarian mathematician Janos Bólyai (1802–1860) derived independently similar results.

**The Euclidean geometry of the plane:** The Hilbert axioms in section 3.2.5, including the parallel axiom, hold for the usual geometry in the plane, as depicted in Figures 3.21 to 3.25.

**Visualization and intuition can be misleading:** Figure 3.25 suggest that the parallel axiom is obviously correct. This view is however false! The mistake is based on the fact that we intuitively think of a line as being something 'straight'. But none of the axioms of geometry state that this should be so. The two geometries which follow in sections 3.2.7 and 3.2.8 illustrate this point.

### 3.2.7 The non-Euclidean elliptic geometry

We consider a sphere $\mathscr{S}$ of radius $R = 1$. As a 'plane of reference' $E_{\text{ellip}}$ we choose the northern hemisphere including the equator.

(i) 'Points' are either classical points which do not lie on the equator, or pairs of points $\{A, B\}$ which lie on opposite sides of the equator.

(ii) 'Lines' are the great circles on the sphere.

(iii) 'Angles' are the usual angles of great circles (Figure 3.26).

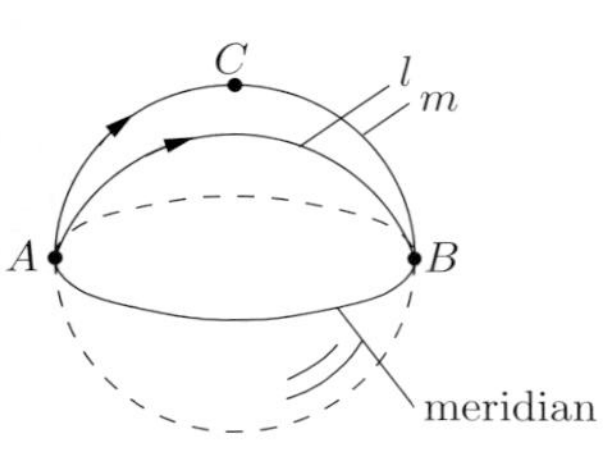


*Figure 3.26.*

**Theorem:** In this geometry, the parallel axiom does not hold.

*Example:* Let a line $l$ be given as well as a point $P$ which does not lie on the line (north pole in the figure). Every line through $P$ is a circle of longitude. All of these lines intersect $l$ in one point. In Figure 3.26 for example the line $l$ intersects the line $m$ in the point $\{A, B\}$.

**Congruence:** 'Motions' in this geometry are the rotations on the axis passing through the north and south pole. Congruent segments and angles are by definition such which can be transformed into one another by such a motion.

This geometry satisfies all the Hilbert axioms of geometry except for the Euclidean parallel axiom. For this reason this geometry is referred to as non-Euclidean. It is amazing that for 2000 years no mathematician came across the idea of using this simple model for proving that the parallel axiom does not follow from the other axioms. Obviously there was a barrier in the thinking of the time. One was too rigid in imagining points as usual points, lines as usual (straight) lines, etc. In fact this kind of intuitive visualization plays no role in the proof of geometric theorems with the help of the axioms and the usual rules of logic.

### 3.2.8 The non-Euclidean hyperbolic geometry

**The Poincaré model:** We choose a Cartesian $(x, y)$-coordinate system and consider the open *upper half-plane*

$$\mathcal{H} = \mathcal{H}_{\text{hyp}} := \{(x, y) \in \mathbb{R}^2 \mid y > 0\},$$

which we refer to as the *hyperbolic plane*. We use the following conventions:

(i) 'Points' are classical points in the upper half-plane.

(ii) 'Lines' are half-circles in the upper half-plane whose centers lie on the $x$-axis (Figure 3.27).

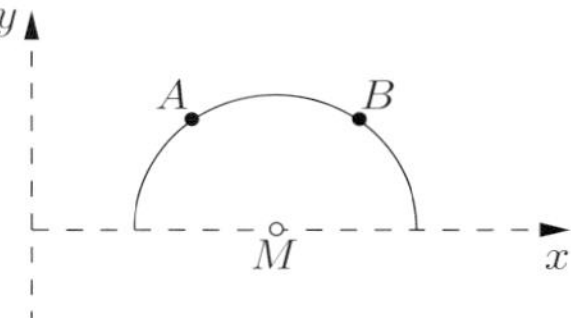


*Figure 3.27.*

**Theorem** (Figure 3.28): (i) There is exactly one line $l$ through two arbitrary distinct points $A$ and $B$ of $\mathcal{H}$.

(ii) If $l$ is a line, then through every other point $P$ outside of $l$ there are infinitely many lines $p$ which do not intersect $l$, i.e., there are infinitely many parallels to $l$ through the point $P$.

| The Euclidean parallel axiom does not hold in hyperbolic geometry. |
|---|

**Angles:** The 'angle' between to lines is equal to the angle between the corresponding circular arcs (Figure 3.28(b)).

**Distance:** The length of a curve $y = y(x)$, $a \le x \le b$ in the hyperbolic plane $\mathcal{H}$ is given by the integral

$$L = \int_a^b \frac{\sqrt{1 + y'(x)^2}}{y(x)}\, \mathrm{d}x.$$

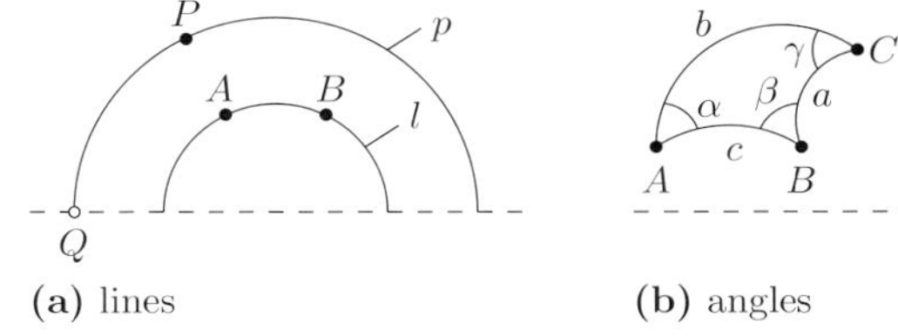


**(a)** lines **(b)** angles

*Figure 3.28. Geometry in the Poincaré plane.*

With respect to this distance, the lines are the shortest paths (geodesics).

*Example 1:* The distance between the points $P$ and $Q$ in Figure 3.28(a) is infinite. Therefore one calls the $x$-axis in Figure 3.28(a) the *line at infinity* of the hyperbolic plane.

**Hyperbolic trigonometry:** This is the science of calculation of triangles in the hyperbolic plane. All formulas of hyperbolic geometry can be elegantly derived from the formulas for spherical trigonometry using the following *translation principle*:

| One replaces in all formulas of spherical trigonometry the radius $R$ by $\mathrm{i}R$ (where i is the imaginary unit with $\mathrm{i}^2 = -1$) and sets $R = 1$. |
|---|

*Example 2:* From the cosine law for sides of spherical trigonometry

$$\cos\frac{c}{R} = \cos\frac{a}{R}\,\cos\frac{b}{R} + \sin\frac{a}{R}\,\sin\frac{b}{R}\,\cos\gamma$$

one gets upon replacement of $R$ by $\mathrm{i}R$ the relation[13]

$$\cosh\frac{c}{R} = \cosh\frac{a}{R}\,\cosh\frac{b}{R} - \sinh\frac{a}{R}\,\sinh\frac{b}{R}\,\cos\gamma.$$

For $R = 1$ we get the cosine law for sides in hyperbolic geometry

$$\cosh c = \cosh a\,\cosh b - \sinh a\,\sinh b\,\cos\gamma.$$

If $\gamma$ is a right angle, then $\cos\gamma = 0$, and we get the *theorem of Pythagoras of hyperbolic geometry*

$$\boxed{\cosh c = \cosh a\,\cosh b.}$$

More important formulas can be found in Table 3.2. The formulas for elliptic geometry correspond to those for spherical trigonometry on a sphere of radius $R = 1$.

*Table 3.2. Formulas in various geometries.*

| | ***Euclidean geometry*** | ***Elliptic geometry*** | ***Hyperbolic geometry*** |
|---|---|---|---|
| sum of the angles of the triangle ($A$ the area) | $\alpha+\beta+\gamma=\pi$ | $\alpha+\beta+\gamma=\pi+A$ | $\alpha+\beta+\gamma=\pi-A$ |
| area of a circle of radius $r$ | $\pi r^2$ | $2\pi(1-\cos r)$ | $2\pi(\cosh r-1)$ |
| circumference of a circle of radius $r$ | $2\pi r$ | $2\pi\,\sin r$ | $2\pi\,\sinh r$ |
| theorem of Pythagoras | $c^2=a^2+b^2$ | $\cos c=\cos a\,\cos b$ | $\cosh c=\cosh a\,\cosh b$ |
| cosine law | $c^2=a^2+b^2$ $-2ab\,\cos\gamma$ | $\cos c=\cos a\,\cos b$ $+\sin a\,\sin b\,\cos\gamma$ | $\cosh c=\cosh a\,\cosh b$ $-\sinh a\,\sinh b\,\cos\gamma$ |
| sine law | $\dfrac{\sin\alpha}{\sin\beta}=\dfrac{a}{b}$ | $\dfrac{\sin\alpha}{\sin\beta}=\dfrac{\sin a}{\sin b}$ | $\dfrac{\sin\alpha}{\sin\beta}=\dfrac{\sinh a}{\sinh b}$ |
| Gaussian curvature | $K=0$ | $K=1$ | $K=-1$ |

Further formulas can be obtain upon perfoming cyclic permutations

$$a\longrightarrow b\longrightarrow c\longrightarrow a \qquad \text{and} \qquad \alpha\longrightarrow\beta\longrightarrow\gamma\longrightarrow\alpha.$$

[13]Note that $\cos\mathrm{i}x = \cosh x$ and $\sin\mathrm{i}x = \mathrm{i}\sinh x$.

**Motions:** We set $z = x + \mathrm{i}y$ and $z' = x' + \mathrm{i}y'$. The 'motions' of hyperbolic geometry are the special *Möbius transformations*

$$\boxed{z' = \frac{\alpha z + \beta}{\gamma z + \delta}}$$

with real numbers $\alpha, \beta, \gamma$ and $\delta$, where in addition $\alpha\delta - \beta\gamma > 0$ is assumed to hold. The set of all such transformations form a group, which is called the group of motions of the hyperbolic plane.

(i) Lines of $\mathcal{H}$ are mapped under hyperbolic motion into other lines.

(ii) Hyperbolic motions are angle-preserving and distance-preserving.

According to Klein's Erlanger Program, properties of hyperbolic geometry are those which are preserved under the group of hyperbolic motions.

**Theorem:** The hyperbolic geometry just defined satisfies all the Hilbert axioms except for the Euclidean parallel axiom.

**Riemannian geometry:** Hyperbolic geometry is a Riemannian geometry with the metric

$$\boxed{\mathrm{d}s^2 = \frac{\mathrm{d}x^2 + \mathrm{d}y^2}{y^2}, \qquad y > 0}$$

and the (negative) constant Gaussian curvature $K = -1$ (cf. [212]).

**Physical interpretation:** A simple interpretation of the Poincaré model in the context of geometric optics can be found in section 5.1.2.

## 3.3 Applications of vector algebra in analytic geometry

> *The discovery of the method of Cartesian coordinates by Descartes (1596–1650) and Fermat (1601–1665), which was referred to at the end of the eighteenth century as 'analytic geometry', increased the importance of algebra in geometric considerations.*
>
> *Jean Dieudonné*

Vector algebra makes it possible to describe geometric objects by means of equations, which are independent of the chosen coordinate system.

Let $O$ be fixed point. We denote by $\mathbf{r} = \overrightarrow{OP}$ the radius vector of the point $P$. If we choose three pairwise perpendicular unit vectors $\mathbf{i}$, $\mathbf{j}$ and $\mathbf{k}$ which form a right-handed system, then we have

$$\boxed{\mathbf{r} = x\mathbf{i} + y\mathbf{j} + z\mathbf{k}.}$$

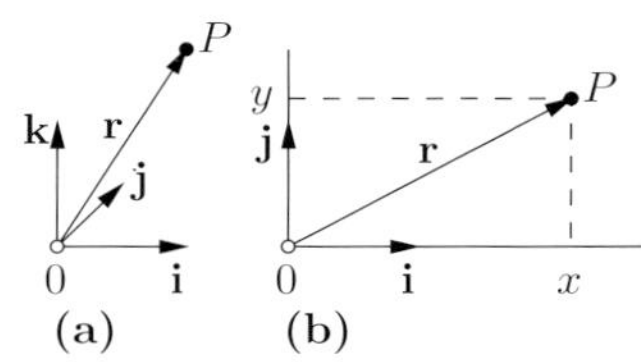


*Figure 3.29. Cartesian coordinates.*

The real numbers $x, y, z$ are called the *Cartesian coordinates* of the point $P$ (Figure 3.29 and Figure 1.85). Moreover we set $\mathbf{a} = a_1\mathbf{i} + a_2\mathbf{j} + a_3\mathbf{k}$ etc.

All the formulas which follow contain the vectorial formulation and the representation in Cartesian coordinates.

### 3.3.1 Lines in the plane

**The equation of a line through a point $P_0(x_0, y_0)$ in the direction of the vector v** (Figure 3.30(a)):

$$\boxed{\mathbf{r} = \mathbf{r}_0 + t\mathbf{v}, \qquad -\infty < t < \infty,}$$

$$x = x_0 + tv_1, \qquad y = y_0 + tv_2.$$

If we view the real parameter $t$ as the time, then this is the equation of a motion of a point with the velocity vector $\mathbf{v} = v_1\mathbf{i} + v_2\mathbf{j}$ and $\mathbf{r}_j = x_j\mathbf{i} + y_j\mathbf{j}$.

**The equation of a line through the two points $P_j(x_j, y_j)$, $j = 0, 1$** (Figure 3.30(b)):

$$\boxed{\mathbf{r} = \mathbf{r}_0 + t(\mathbf{r}_1 - \mathbf{r}_0), \qquad -\infty < t < \infty,}$$

$$x = x_0 + t(x_1 - x_0), \qquad y = y_0 + t(y_1 - y_0).$$

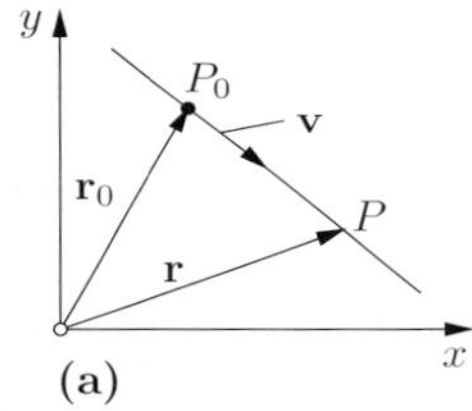


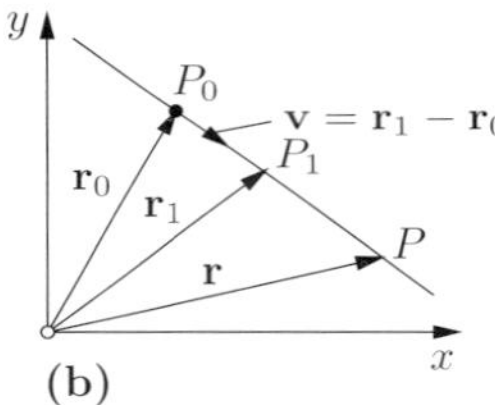


*Figure 3.30. A line through two points.*

**The equation of the line $l$ through the point $P_0(x_0, y_0)$ perpendicular to the unit vector n** (Figure 3.31(a)):

$$\boxed{\mathbf{n}(\mathbf{r} - \mathbf{r}_0) = 0,}$$

$$n_1(x - x_0) + n_2(y - y_0) = 0.$$

Here one has $\sqrt{n_1^2 + n_2^2} = 1$.

**Distance of a point $P_*$ from the line $l$:**

$$\boxed{d = \mathbf{n}(\mathbf{r}_* - \mathbf{r}_0),}$$

$$d = n_1(x_* - x_0) + n_2(y_* - y_0).$$

Here we have set $\mathbf{r}_* = \overrightarrow{OP_*}$. One has $d > 0$ (resp. $d < 0$) if the point $P_*$ is on the positive (resp. negative) side of $l$ with respect to $\mathbf{n}$ (Figure 3.31(b)). Moreover one has $\mathbf{n} = n_1\mathbf{i} + n_2\mathbf{j}$.

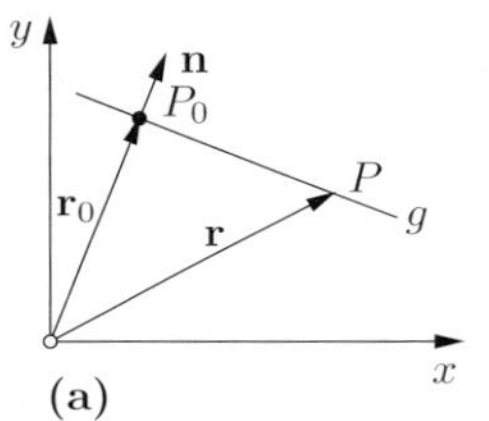


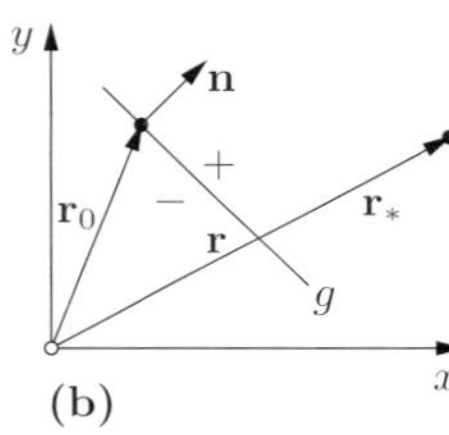


*Figure 3.31. A line perpendicular to a vector.*

**Distance $d$ between two points $P_1$ and $P_0$** (Figure 3.32(a)):

$$\boxed{d = |\mathbf{r}_1 - \mathbf{r}_0|,}$$

$$d = \sqrt{(x_1 - x_0)^2 + (y_1 - y_0)^2}\,.$$

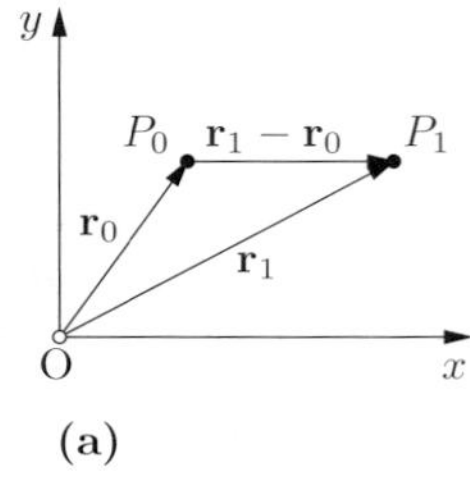


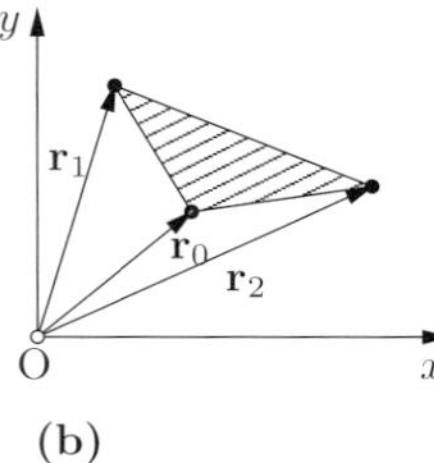


*Figure 3.32. Distance and area.*

**Surface area $A$ of a triangle with vertices $P_j(x_j, y_j)$, $j = 0, 1, 2$** (Figure 3.32(b)):

$$\boxed{A = \frac{1}{2}\mathbf{k}\big((\mathbf{r}_1 - \mathbf{r}_0) \times (\mathbf{r}_2 - \mathbf{r}_0)\big).}$$

Explicitly one has:

$$A = \frac{1}{2} \begin{vmatrix} x_1 - x_0 & y_1 - y_0 \\ x_2 - x_0 & y_2 - y_0 \end{vmatrix}.$$

### 3.3.2 Lines and planes in space

**Equation of a line through the point $P_0(x_0, y_0, z_0)$ in the direction of the vector v** (Figure 3.33(a)):

$$\boxed{\mathbf{r} = \mathbf{r}_0 + t\mathbf{v}, \qquad -\infty < t < \infty,}$$

$$x = x_0 + tv_1, \qquad y = y_0 + tv_2, \qquad z = z_0 + tv_3.$$

If we view the real parameter $t$ as the time, then this is the equation for the motion of a point with velocity vector $\mathbf{v} = v_1\mathbf{i} + v_2\mathbf{j} + v_3\mathbf{k}$ and $\mathbf{r} = x\mathbf{i} + y\mathbf{j} + z\mathbf{k}$.

**Equation of a line through two points $P_j(x_j, y_j, z_j)$, $j = 0, 1$** (Figure 3.33(a)):

$$\boxed{\mathbf{r} = \mathbf{r}_0 + t(\mathbf{r}_1 - \mathbf{r}_0), \qquad -\infty < t < \infty,}$$

$$x = x_0 + t(x_1 - x_0), \qquad y = y_0 + t(y_1 - y_0), \qquad z = z_0 + t(z_1 - z_0).$$

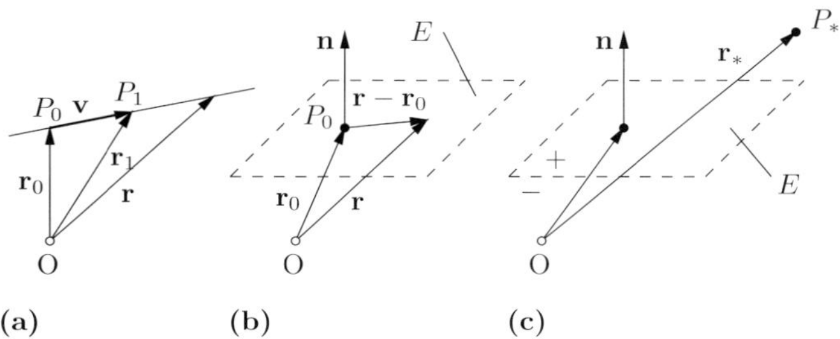


*Figure 3.33. Equations of objects in space.*

**Equation for a plane through three given points $P(x_j, y_j, z_j)$, $j = 0, 1, 2$:**

$$\boxed{\mathbf{r} = \mathbf{r}_0 + t(\mathbf{r}_1 - \mathbf{r}_0) + s(\mathbf{r}_2 - \mathbf{r}_0), \qquad -\infty < t, s < \infty,}$$

$$x = x_0 + t(x_1 - x_0) + s(x_2 - x_0),$$

$$y = y_0 + t(y_1 - y_0) + s(y_2 - y_0),$$

$$z = z_0 + t(z_1 - z_0) + s(z_2 - z_0).$$

**Equation of plane $E$ through the point $P(x_0, y_0, z_0)$ perpendicular to the unit vector n** (Figure 3.33(b)):

$$\boxed{\mathbf{n}(\mathbf{r} - \mathbf{r}_0) = 0,}$$

$$n_1(x - x_0) + n_2(y - y_0) + n_3(z - z_0) = 0.$$

Here one has $\sqrt{n_1^2+n_2^2+n_3^2}=1$. One calls $\mathbf{n}$ the *unit normal vector* of the plane $E$. If three points $P_1$, $P_2$ and $P_3$ are given on $E$, then one obtains $\mathbf{n}$ by means of the formula

$$\mathbf{n}=\frac{(\mathbf{r}_1-\mathbf{r}_0)\times(\mathbf{r}_2-\mathbf{r}_0)}{|(\mathbf{r}_1-\mathbf{r}_0)\times(\mathbf{r}_2-\mathbf{r}_0)|}\,.$$

**Distance of a point $P_*$ from a plane $E$:**

$$\boxed{d=\mathbf{n}(\mathbf{r}_*-\mathbf{r}_0),}$$

$$d=n_1(x_*-x_0)+n_2(y_*-y_0)+n_3(z_*-z_0).$$

Here one has $d>0$ (resp. $d<0$) if the point $P_*$ is on the positive (resp. negative) side of the plane $E$ with respect to the unit normal $\mathbf{n}$ (Figure 3.33(c)).

**Distance between two points $P_0$ and $P_1$:**

$$\boxed{d=|\mathbf{r}_1-\mathbf{r}_0|,}$$

$$d=\sqrt{(x_1-x_0)^2+(y_1-y_0)^2+(z_1-z_0)^2}\,.$$

**Angle $\varphi$ between two vectors a and b:**

$$\boxed{\cos\varphi=\frac{\mathbf{ab}}{|\mathbf{a}|\,|\mathbf{b}|}\,,}$$

$$\cos\varphi=\frac{a_1b_1+a_2b_2+a_3b_3}{\sqrt{a_1^2+a_2^2+a_3^2}\,\sqrt{b_1^2+b_2^2+b_3^2}}\,.$$

### 3.3.3 Volumes

**Volume of the parallelopid spanned by the vectors a, b and c** (Figure 3.34(a)):

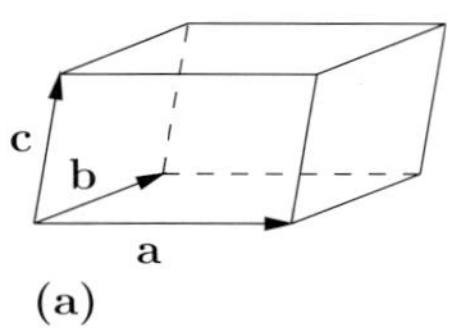


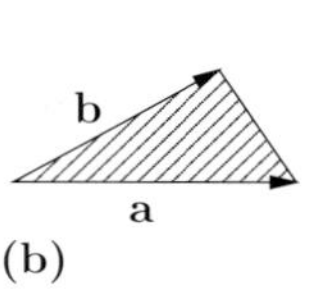


*Figure 3.34. Two and three-dimensional volumes.*

$$\boxed{V=(\mathbf{a}\times\mathbf{b})\mathbf{c},}$$

$$V=\begin{vmatrix} a_1 & a_2 & a_3\\ b_1 & b_2 & b_3\\ c_1 & c_2 & c_3\end{vmatrix}.$$

Here one has $V>0$ (resp. $V<0$) if $\mathbf{a}$, $\mathbf{b}$, $\mathbf{c}$ form a right-handed (resp. left-handed) system.

**Volume of the parallelopid spanned by the points $P_j(x_j,y_j,z_j)$, $j=0,1,2,3$:** Set $\mathbf{a}:=\mathbf{r}_1-\mathbf{r}_0$, $\mathbf{b}:=\mathbf{r}_2-\mathbf{r}_0$, $\mathbf{c}:=\mathbf{r}_3-\mathbf{r}_0$.

**Surface area of the triangle spanned by the vectors a and b** (Figure 3.34(b)):

$$\boxed{A=\frac{1}{2}\,|\mathbf{a}\times\mathbf{b}|,}$$

$$A=\sqrt{\begin{vmatrix} a_2 & a_3\\ b_2 & b_3\end{vmatrix}^2+\begin{vmatrix} a_3 & a_1\\ b_3 & b_1\end{vmatrix}^2+\begin{vmatrix} a_1 & a_2\\ b_1 & b_2\end{vmatrix}^2}\,.$$

# 3.4 Euclidean geometry (geometry of motion)

## 3.4.1 The group of Euclidean motions

We denote by $x_1, x_2, x_3$ and $x'_1, x'_2, x'_3$ two Cartesian coordinate systems. A *Euclidean motion* is a transformation

$$\boxed{x' = Dx + a}$$

with a constant column vector $a = (a_1, a_2, a_3)^\mathsf{T}$ and real orthogonal $(3 \times 3)$-matrix $D$, i.e., such that $DD^\mathsf{T} = D^\mathsf{T}D = E$ (where $E$ denotes the identity matrix). Explicitly these transformations are given by the equations

$$x_j{}' = d_{j1}x_1 + d_{j2}x_2 + d_{j3}x_3 + a_j, \qquad j = 1, 2, 3.$$

**Classification:** (i) *Translation*: $D = E$.

(ii) *Rotation*: $\det D = 1,\ a = 0$.

(iii) *Rotational reflection*: $\det D = -1,\ a = 0$.

(iv) *Proper motion*: $\det D = 1$.

**Definition:** The set of all motions forms with respect to the composition a group, which is called the *group of Euclidean motions.*

All rotations form a subgroup, which is referred to as the *group of rotations.*[14]

All translations (resp. all proper motions) for a subgroup of the group of Euclidean motions, which is referred to as the *group of translations* (resp. the *group of proper Euclidean motions*).

*Example 1:* A rotation around the $\zeta$-axis with an angle of rotation $\varphi$ (taken in positive mathematical sense) in a Cartesian $(\xi, \eta, \zeta)$-system is:

$$\boxed{\begin{aligned} \xi' &= \xi \cos\varphi + \eta \sin\varphi, \qquad \zeta' = \zeta, \\ \eta' &= -\xi \sin\varphi + \xi \cos\varphi. \end{aligned}}$$

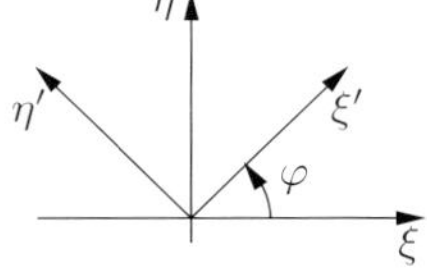


*Figure 3.35.*

Figure 3.35 shows the rotation in the $(\xi, \eta)$-plane.

*Example 2:* A reflection on the $(\xi, \eta)$-plane is given by the relations:

$$\xi' = \xi, \qquad \eta' = \eta, \qquad \zeta' = -\zeta.$$

**Structure theorem:** (i) Every rotation can be transformed in an appropriate Cartesian coordinate system into the rotation around the $\zeta$-axis.

(ii) Every rotational reflection can be represented in an appropriately chosen $(\xi, \eta, \zeta)$-coordinate system as the composition of a rotation around the $\zeta$-axis and a reflection on the $(\xi, \eta)$-plane.

**Euclidean geometry:** According to Felix Klein's Erlanger Program the properties of Euclidean geometry are exactly those properties which are invariant under the group of Euclidean motions (for example, the length of a segment).

[14] This is a three-dimensional Lie group (see [212] for a definition of these).

### 3.4.2 Conic sections

The elementary theory of conic sections was presented in section 0.1.7.

**Quadratic forms:** We consider an equation

$$\boxed{x^{\mathsf{T}} A x = b.}$$

Explicitly this equation is

$$a_{11}x_1^2 + 2a_{12}x_1x_2 + a_{22}x_2^2 = b \tag{3.17}$$

with the real symmetric matrix $A = \begin{pmatrix} a_{11} & a_{12} \\ a_{21} & a_{22} \end{pmatrix}$. Assume $A \neq 0$. Then one has $\det A = a_{11}a_{22} - a_{12}a_{21}$ and $\operatorname{tr} A = a_{11} + a_{22}$.

**Theorem:** By applying a rotation in the Cartesian $(x_1, x_2, x_3)$-coordinate system one can always put the equation (3.17) into the normal form

$$\boxed{\lambda x^2 + \mu y^2 = b.} \tag{3.18}$$

Here $\lambda$ and $\mu$ are eigenvalues of $A$, i.e., one has

$$\begin{vmatrix} a_{11} - \zeta & a_{12} \\ a_{21} & a_{22} - \zeta \end{vmatrix} = 0$$

with $\zeta = \lambda, \mu$. One has $\det A = \lambda\mu$ and $\operatorname{tr} A = \lambda + \mu$.

*Proof:* We determine two eigenvectors $u$ and $v$ of the matrix $A$, i.e., such that

$$Au = \lambda u \qquad \text{and} \qquad Av = \mu v.$$

Here we can choose $u$ and $v$ in such a way that $u^{\mathsf{T}}v = 0$ and $u^{\mathsf{T}}u = v^{\mathsf{T}}v = 1$ hold. We set $D := (u, v)$. Then

$$x = Dx'$$

is a rotation. From (3.17) it follows that

$$b = x^{\mathsf{T}}Ax = x'^{\mathsf{T}}(D^{\mathsf{T}}AD)x' = x'^{\mathsf{T}} \begin{pmatrix} \lambda & 0 \\ 0 & \mu \end{pmatrix} x' = \lambda\, x_1'^{\,2} + \mu\, x_2'^{\,2}.$$

□

**General conic sections:** We now study the equation

$$\boxed{x^{\mathsf{T}}Ax + x^{\mathsf{T}}a + a_{33} = 0}$$

with $a = (a_{13}, a_{23})^{\mathsf{T}}$, i.e.,

$$a_{11}x_1^2 + 2a_{12}x_1x_2 + a_{22}x_2^2 + a_{13}x_1 + a_{23}x_2 + a_{33} = 0 \tag{3.19}$$

with the real symmetric matrices

$$A = \begin{pmatrix} a_{11} & a_{12} \\ a_{21} & a_{22} \end{pmatrix}, \qquad \mathscr{A} = \begin{pmatrix} a_{11} & a_{12} & a_{13} \\ a_{21} & a_{22} & a_{23} \\ a_{31} & a_{32} & a_{33} \end{pmatrix}.$$

**First main case:** Equation for the center. If $\det A \neq 0$, then the system of linear equations

$$a_{11}\alpha_1 + a_{12}\alpha_2 + a_{13} = 0,$$
$$a_{21}\alpha_1 + a_{22}\alpha_2 + a_{23} = 0$$

has a unique solution $(\alpha_1, \alpha_2)$. By translating $X_j := x_j - \alpha_j$, $j = 1, 2$, the equation (3.19) transforms into

$$a_{11}X_1^2 + 2a_{12}X_1X_2 + a_{22}X_2^2 = -\frac{\det \mathscr{A}}{\det A}.$$

Similarly as in (3.17) we get from this a rotation

$$\boxed{\lambda x^2 + \mu y^2 = -\frac{\det \mathscr{A}}{\det A}\,.}$$

Because of $\det A = \lambda\mu$ and $\operatorname{tr} A = \lambda + \mu$ one obtains the normal forms listed in Table 3.3.

**Second main case:** No equation for the center. We have in this case $\det A = 0$, hence $\lambda \neq 0$ and $\mu = 0$. By applying a rotation to (3.19) we get

$$\boxed{\lambda x^2 + 2qx + py + c = 0.}$$

Quadratic completion applied to this yields

$$\lambda\left(x + \frac{q}{\lambda}\right)^2 + py + c - \frac{q^2}{\lambda} = 0.$$

1. If $p \neq 0$, the the curve is a parabola.
2. If $p = 0$, then by applying a translation we get $x^2 = 0$ or $x^2 = \pm a^2$ (cf. Table 3.4).

### 3.4.3 Quadratic surfaces

**Quadratic forms:** We consider the equation

$$\boxed{x^\top A x = b\,.}$$

Explicitly, this equation is

$$a_{11}x_1^2 + 2a_{12}x_1x_2 + 2a_{13}x_1x_3 + 2a_{23}x_2x_3 + a_{22}x_2^2 + a_{33}x_3^2 = b \qquad (3.20)$$

with a real symmetric matrix $A = (a_{jk})$. Suppose that $A \neq 0$. Then for the trace we have $\operatorname{tr} A = a_{11} + a_{22} + a_{33}$.

**Theorem:** By applying a rotation to the Cartesian $(x_1, x_2, x_3)$-coordinate system we can put the above equation in the following normal form:

$$\boxed{\lambda x^2 + \mu y^2 + \zeta z^2 = b\,.}$$

Here $\lambda$, $\mu$ and $\zeta$ are the eigenvalues of $A$, i.e., one has $\det(A - \nu E) = 0$ with $\nu = \lambda, \mu$ or $\zeta$. One has $\det A = \lambda\mu\zeta$ and $\operatorname{tr} A = \lambda + \mu + \zeta$.

*Proof:* We determine the three eigenvectors $u, v$ and $w$ of the matrix $A$, i.e., vectors such that

$$Au = \lambda u, \qquad Av = \mu v, \qquad Aw = \zeta w\,.$$

*Table 3.3 Centered conic sections.*

| **det $A$** | **det $\mathscr{A}$** | ***Normal form*** **($a>0,\ b>0,\ c>0$)** | ***Name*** | ***Diagram*** |
|---|---|---|---|---|
| $>0$ | $<0$ | $\frac{x^2}{a^2}+\frac{y^2}{b^2}=1$ | ellipse | 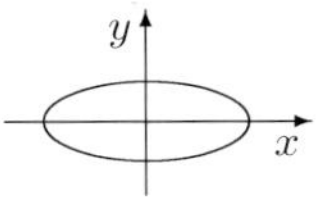 |
| | $>0$ | $\frac{x^2}{a^2}+\frac{y^2}{b^2}=-1$ | imaginary ellipse | |
| | $=0$ | $\frac{x^2}{a^2}+\frac{y^2}{b^2}=0$ | double point | 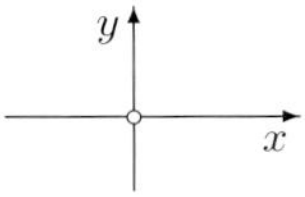 |
| $<0$ | $<0$ | $\frac{x^2}{a^2}-\frac{y^2}{b^2}=1$ | hyperbola | 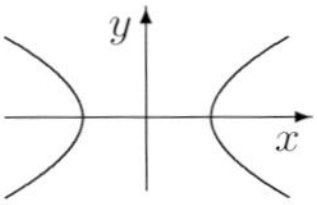 |
| | $>0$ | $\frac{y^2}{b^2}-\frac{x^2}{a^2}=1$ | hyperbola | 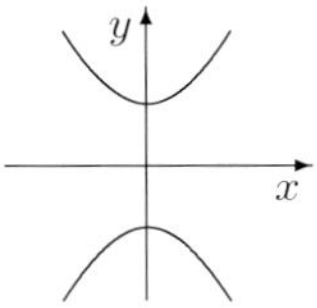 |
| | $=0$ | $\frac{x^2}{a^2}-\frac{y^2}{b^2}=0$ | double line | 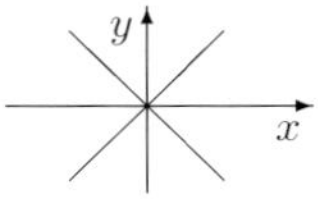 |

*Table 3.4 Non-centered curves ($\det A=0$).*

| ***Normal form*** **($a>0$)** | ***Name*** | ***Diagram*** |
|---|---|---|
| $y=ax^2$ | parabola | 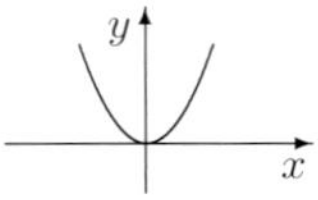 |
| $y^2=0$ | double line | 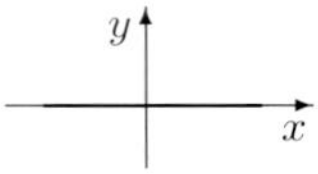 |
| $y^2=a^2$ | two lines $y=\pm a$ | 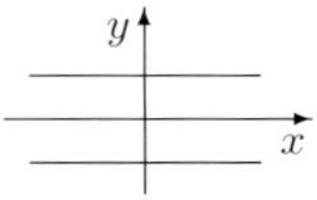 |
| $y^2=-a^2$ | two imaginary lines | |

Table 3.5. Centered surfaces.

| ***Normal form*** **($a > 0,\ b > 0,\ c > 0$)** | ***Name*** | ***Diagram*** |
|---|---|---|
| $\frac{x^2}{a^2} + \frac{y^2}{b^2} + \frac{z^2}{c^2} = 1$ | ellipsoid | 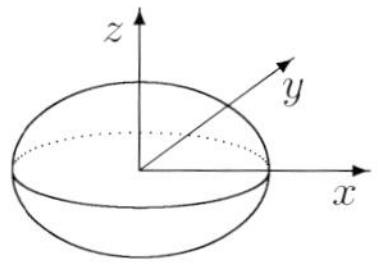  |
| $\frac{x^2}{a^2} + \frac{y^2}{b^2} + \frac{z^2}{c^2} = -1$ | imaginary ellipsoid | |
| $\frac{x^2}{a^2} + \frac{y^2}{b^2} + \frac{z^2}{c^2} = 0$ | origin | 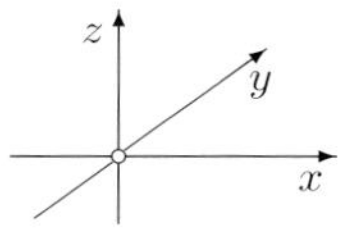  |
| $\frac{x^2}{a^2} + \frac{y^2}{b^2} - \frac{z^2}{c^2} = 1$ | single-sheeted hyperboloid | 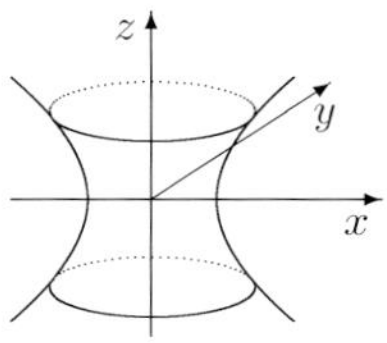  |
| $\frac{x^2}{a^2} + \frac{y^2}{b^2} - \frac{z^2}{c^2} = 0$ | double cone | 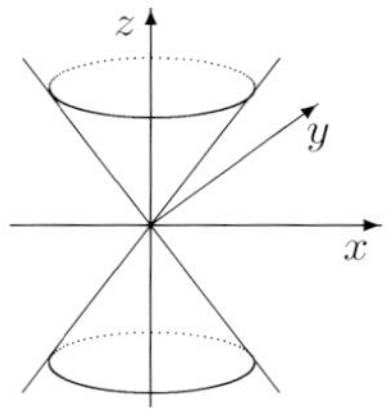  |
| $\frac{z^2}{c^2} - \frac{x^2}{a^2} - \frac{y^2}{b^2} = 1$ | two-sheeted hyperboloid | 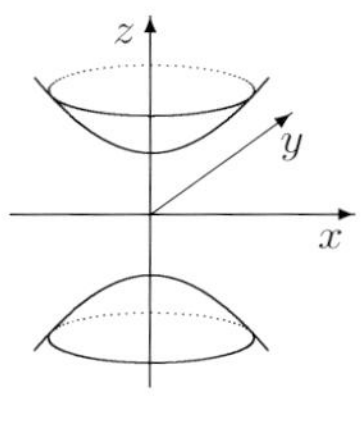  |

*Table 3.6. Non-centered surfaces* ($\det A = 0$).

| ***Normal form*** **(*a* > 0, *b* > 0, *c* > 0)** | ***Name*** | ***Diagram*** |
|---|---|---|
| $\frac{x^2}{a^2} + \frac{y^2}{b^2} = 2cz$ | elliptic paraboloid | 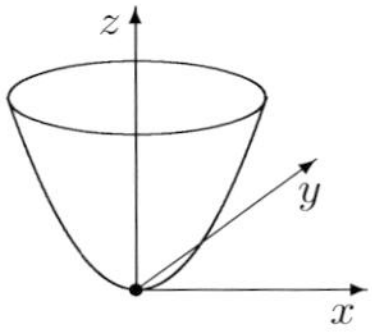 |
| $\frac{x^2}{a^2} - \frac{y^2}{b^2} = 2cz$ | hyperbolic paraboloid (saddle) | 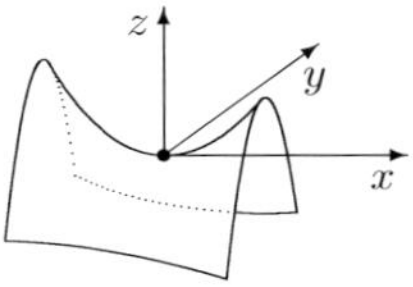 |
| $\frac{x^2}{a^2} + \frac{y^2}{b^2} = 1$ | elliptic cylinder | 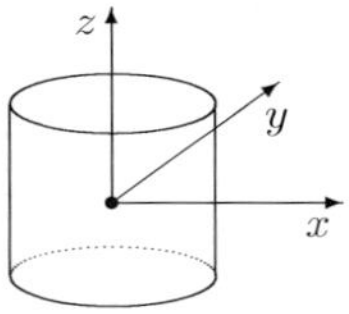 |
| $\frac{x^2}{a^2} - \frac{y^2}{b^2} = 1$ | hyperbolic cylinder | 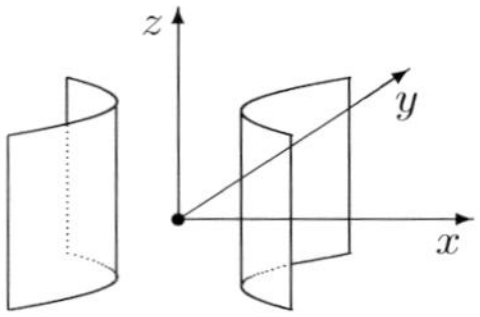 |
| $\frac{x^2}{a^2} - \frac{y^2}{b^2} = 0$ | two intersecting planes | 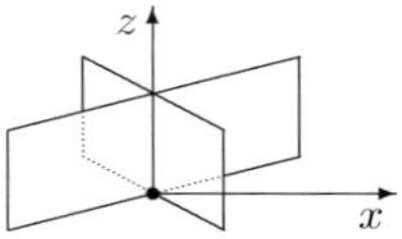 |
| $x = 2cy^2$ | parabolic cylinder | 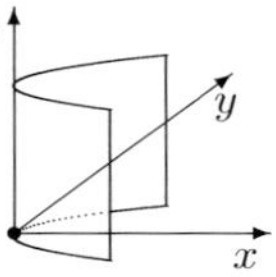 |

*Table 3.6. (Continued)*

| ***Normal form* ($a > 0,\ b > 0,\ c > 0$)** | ***Name*** | ***Diagram*** |
|---|---|---|
| $x^2 = a^2$ | two parrallel planes ($x = \pm a$) | |
| $x^2 = 0$ | double plane | |
| $\frac{x^2}{a^2} + \frac{y^2}{b^2} = -1$ | imaginary elliptic cylinder | |
| $\frac{x^2}{a^2} + \frac{y^2}{b^2} = 0$ | degenerate elliptic cylinder (the $z$-axis) | $z$, $y$, $x$ |

Here one can choose $u, v$ and $w$ such that $u^\top v = u^\top w = v^\top w = 0$ and $u^\top u = v^\top v = w^\top w = 1$. Setting $D := (u, v, w)$, the transformation

$$x = Dx'$$

is a rotation. It follows from (3.20) that

$$b = x^\top Ax = x'^\top(D^\top AD)x' = x'^\top \begin{pmatrix} \lambda & 0 & 0 \\ 0 & \mu & 0 \\ 0 & 0 & \zeta \end{pmatrix} x' = \lambda\, {x'_1}^2 + \mu\, {x'_2}^2 + \zeta\, {x'_3}^2 .$$

□

**The general quadratic surface:** We now study the equation

$$\boxed{x^\top Ax + x^\top a + a_{44} = 0}$$

with $a = (a_{14}, a_{24}, a_{34})^\top$. Explicitly this equation is

$$a_{11}x_1^2 + 2a_{12}x_1x_2 + 2a_{13}x_1x_3 + 2a_{23}x_2x_3 + a_{22}x_2^2 + a_{33}x_3^2 + a_{14}x_1 + a_{24}x_2 + a_{34}x_3 + a_{44} = 0 \qquad (3.21)$$

with the real symmetric matrices

$$A = \begin{pmatrix} a_{11} & a_{12} & a_{13} \\ a_{21} & a_{22} & a_{23} \\ a_{31} & a_{32} & a_{33} \end{pmatrix}, \qquad \mathscr{A} = \begin{pmatrix} a_{11} & a_{12} & a_{13} & a_{14} \\ a_{21} & a_{22} & a_{23} & a_{24} \\ a_{31} & a_{32} & a_{33} & a_{34} \\ a_{41} & a_{42} & a_{43} & a_{44} \end{pmatrix}.$$

**First main case:** Equation for the center. If $\det A \neq 0$, then the system of linear equations

$$\begin{aligned} a_{11}\alpha_1 + a_{12}\alpha_2 + a_{13}\alpha_3 + a_{14} &= 0, \\ a_{21}\alpha_1 + a_{22}\alpha_2 + a_{23}\alpha_3 + a_{24} &= 0, \\ a_{31}\alpha_1 + a_{32}\alpha_2 + a_{33}\alpha_3 + a_{34} &= 0 \end{aligned}$$

has a unique solution $(\alpha_1, \alpha_2, \alpha_3)$. Through the translation $X_j := x_j - \alpha_j$, $j = 1, 2, 3$, (3.21) is transformed into the equation

$$a_{11}X_1^2 + 2a_{12}X_1X_2 + 2a_{13}X_1X_3 + 2a_{23}X_2X_3 + a_{22}X_2^2 + a_{33}X_3^2 = -\frac{\det \mathscr{A}}{\det A}.$$

Similarly as in (3.20) we get from this a rotation:

$$\boxed{\lambda x^2 + \mu y^2 + \zeta z^2 = -\frac{\det \mathscr{A}}{\det A}.}$$

**Second main case:** No equation for the center. This occurs in the case when $\det A = 0$. By applying a rotation to (3.21) we get in this case

$$\boxed{\lambda x^2 + \mu y^2 + px + ry + sz + c = 0.}$$

Forming the quadratic completion and translating yields then the different normal forms listed in Table 3.6.

## 3.5 Projective geometry

### 3.5.1 Basic ideas

In the geometry of the Euclidean plane there is no such thing as duality between points and lines. Instead, one has:

(i) There is always exactly one line passing through two distinct points.

(ii) However, it is not true that two distinct lines meet in a point.

In order to eliminate this asymmetry, one defines:

$$\boxed{\text{a point at infinity} = \text{a non-oriented direction.}}$$

Two parallel lines always have the same direction, i.e., the have the same point at infinity. Using this convention, one has: any two distinct lines meet in a point (which may be a point on the 'line at infinity', the set of all directions).

**Homogenous coordinates:** To make this idea amenable to calculations, we replace for example in the equation of a line

$$y = 2x + 1$$

the quantities $x$ (resp. $y$) by $x/u$ (resp. by $y/u$) an multiply through by $u$. This yields

$$y = 2x + u.$$

To every point $(x, y)$ we associate the set of *all* homogenous coordinates $(xu, yu)$, where $u$ is an arbitrary non-vanishing real number. The two parallel lines

$$y = 2x + 1, \qquad y = 2x + 3$$

correspond to the equations

$$y = 2x + u, \qquad y = 2x + 3u$$

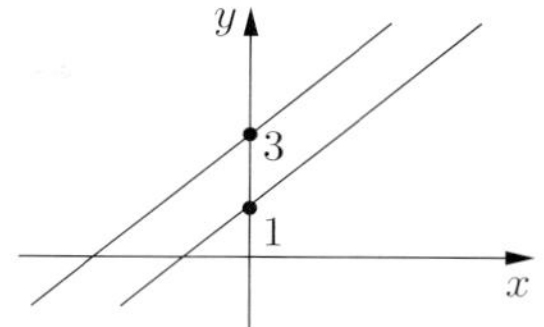

*Figure 3.36.*

with the solution $x = 1,\ y = 2,\ u = 0$, which corresponds to the point at infinity (Figure 3.36). The equation

$$\boxed{u = 0}$$

is the equation of the *line at infinity.*

**Projective points of the plane:** A *projective point* is a set

$$\big[(x, y, u)\big] := \big\{(\lambda x, \lambda y, \lambda u) \mid \lambda \in \mathbb{R},\ \lambda \neq 0\big\}$$

with $x^2 + y^2 + u^2 \neq 0$ (note that this is just requiring that at least one of $x, y, u$ is non-zero). The set of all these projective points is denoted by $\mathbb{R}P^2$ and is called the *real projective plane.*

Every tuple $(\lambda x, \lambda y, \lambda u)$ with $\lambda \neq 0$ is called a set of homogenous coordinates for the point $[(x, y, u)]$. The point $[(x, y, u)]$ with $u = 0$ is the *point at infinity.*

**Projective lines:** The set of all projective points $[(x, y, u)]$ which satisfy the equation

$$ax + by + cu = 0$$

is called a *projective line*. Here $a, b$ and $c$ are real coefficients with $a^2 + b^2 + c^2 \neq 0$.

Two projective lines

$$a_1x + b_1y + c_1u = 0,$$
$$a_2x + b_2y + c_2u = 0$$

are distinct from one another, when $\operatorname{rank}\begin{pmatrix} a_1 & b_1 & c_1 \\ a_2 & b_2 & c_2 \end{pmatrix} = 2.$

**Duality principle:** (i) Two distinct projective points determine a unique line containing them both.

(ii) Two distinct projective lines determine a unique point, namely the point of intersection.

**Realizing the projective plane with the unit circle:** We denote by $\mathbb{R}P^2_*$ the set of all points of the open unit disc. Here we identify two points $A, B$ on the boundary (the unit circle) with each other if they are opposite points on the unit circle (Figure 3.37).

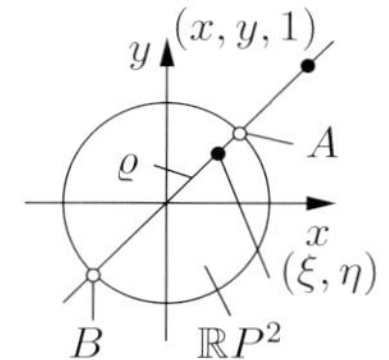

*Figure 3.37.*

**Theorem:** There is a bijective mapping from $\mathbb{R}P^2$ to $\mathbb{R}P^2_*$.

*Proof:* We associate to the projective point $[(x, y, 1)]$, which corresponds to the point with Cartesian coordinates $(x, y)$, the point $(\xi, \eta)$ which is the point on the line from $(x, y)$ and $(0, 0)$ whose distance from the origin $(0, 0)$ is equal to

$$\varrho = \frac{2}{\pi} \arctan r$$

with $r := \sqrt{x^2 + y^2}$. In contrast, the point at infinity $[(1, m, 0)]$ corresponds to the pair $\{A, B\}$ of opposite points which are obtained as the intersection of the boundary of the unit circle with the line $y = mx$ (Figure 3.37). □

**The theorem of Desargues (1593–1662):** If the lines joining the corresponding vertices of two triangles pass through a point, then the intersections of the corresponding sides lie on a line (Figure 3.38).

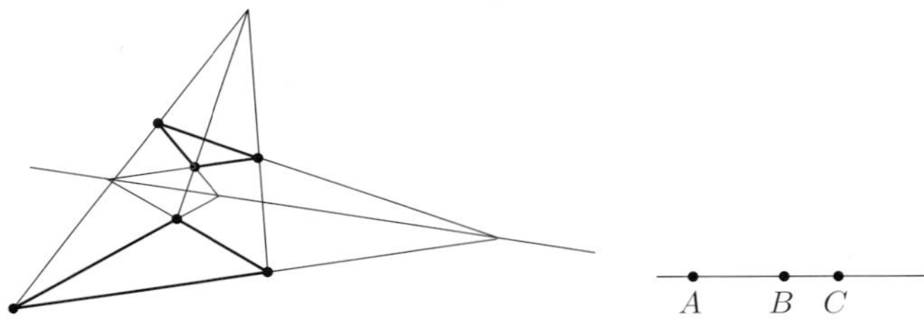


*Figure 3.38. Theorem of Desargues.*

*Figure 3.39. The double ratio.*

**Double ratio:** If four points $A, B, C$ and $D$ lie on a line, then one calls the real number

$$\boxed{\frac{AC}{AD} : \frac{BC}{BD}}$$

the *double ratio* of these four points (Figure 3.39). Here $AC$ is the length of the segment from $A$ to $C$, etc.

The double ratio is the most important invariant in projective geometry.

### 3.5.2 Projective maps

**Projection of two lines onto each other:** Figure 3.40 shows a parallel projection and a central projection. Under these projections, the lengths of segments may be changed. However, one does have the following fundamental result:

| The double ratio of four points is preserved. |
|---|

**Projection of two planes onto one another:** Figure 3.41 shows a central projection of the plane $\mathscr{E}$ onto the plane $\mathscr{E}'$. Here once again the double ratios of four points on a line are preserved.

**Collineation:** If we introduce homogenous coordinates into $\mathscr{E}$ and $\mathscr{E}'$, $(x, y, u)$ and $(x', y', u')$, respectively, then a collineation is defined by a mapping of the form

$$\begin{pmatrix} x' \\ y' \\ u' \end{pmatrix} = \begin{pmatrix} a_{11} & a_{12} & a_{13} \\ a_{21} & a_{22} & a_{23} \\ a_{31} & a_{32} & a_{33} \end{pmatrix} \begin{pmatrix} x \\ y \\ u \end{pmatrix}$$

with a real $(3 \times 3)$-matrix $(a_{jk})$, whose determinant does not vanish.

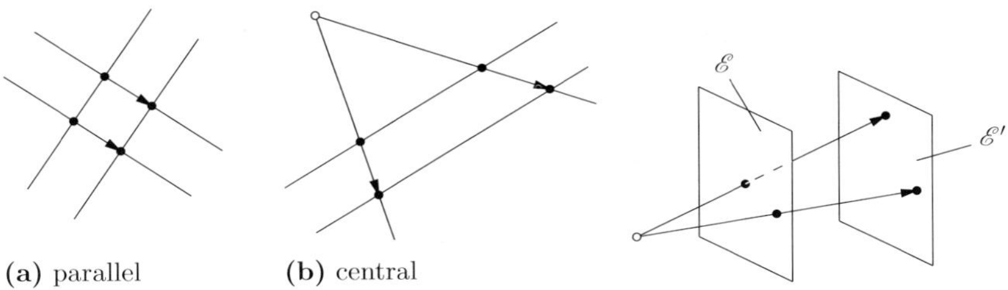


*Figure 3.40.* *Projections of two lines.*

*Figure 3.41.* *Central projection.*

**Structure theorem:** (i) Every map from $\mathscr{E}$ to $\mathscr{E}'$ which is combined from finitely many parallel projections and central projections is a collineation. In this manner one obtains all collineations from $\mathscr{E}$ to $\mathscr{E}'$.

(ii) The double ratio of four points on a line are preserved under collineations.

*Example:* During photographic work on landscapes, the double ratio of four points lying on a line is equal to the double ratio of the points on the photography. Hence, if one know the coordinates of three of the points in the landscape, the fourth can be calculated by measuring the double ratio of the four points on the photography.

**Projective properties:** According to Felix Klein's Erlanger Program the projective properties are precisely those which are preserved under the group of collineations.

### 3.5.3 The $n$-dimensional real projective space

Let $n = 1, 2, \ldots$

**Projective points:** A *projective point* in $n$ dimensions is a set

$$\big[x_1, \ldots, x_{n+1}\big] := \big\{(\lambda x_1, \ldots, \lambda x_{n+1}) \mid \lambda \in \mathbb{R},\ \lambda \neq 0\big\}.$$

Here the $x_j$ are real numbers such that not all $x_i$ vanish simultaneously. We call $(\lambda x_1, \ldots, \lambda x_{n+1})$ with $\lambda \neq 0$ *homogenous coordinates* of the projective point $[x_1, \ldots, x_{n+1}]$.

The set of all of these points is denoted $\mathbb{R}P^n$ and is called the *$n$-dimensional real projective space*. The points for which $x_{n+1} = 0$ are called *points at infinity*.

**Projective subspaces:** Let $m$ linearly independent vectors $p_1, \ldots, p_m \in \mathbb{R}^{n+1}$ be given. The set of all projective points $[x]$, whose homogenous coordinates $x$ can be written in the form

$$\boxed{x = t_1 p_1 + \ldots + t_m p_m}$$

with arbitrary real parameters $t_1, \ldots, t_m$, is by definition an $m$-dimensional *projective subspace* of $\mathbb{R}P^n$, which is generated by the points $[p_1], \ldots, [p_m]$.

If $m = 1$ (resp. $m = n - 1$), then we speak of a projective *line* (resp. projective *hyperplane*).

**Collineations:** These are mappings

$$\boxed{x' = Ax}$$

between sets of projective coordinates $x$ and $x'$. Here $A$ is a real square matrix with $n+1$ rows and $\det A \neq 0$.

Every such collineation corresponds to a map

$$\varphi_A : \mathbb{R}P^n \longrightarrow \mathbb{R}P^n .$$

One has $\varphi_A = \varphi_B$ if and only if the matrices $A$ and $B$ are multiples of one another by a non-zero real scalar.

**Projective group:** The set of all of these mappings $\varphi$ forms the projective group $PGL(n+1, \mathbb{R})$ with respect to the composition of mappings. This group is itself a *factor group* of the form

$$\boxed{PGL(n+1, \mathbb{R}) = GL(n+1, \mathbb{R})/D.}$$

Here $GL(n+1, \mathbb{R})$ denotes the group of all real invertible $(n+1)$-matrices, while $D$ stands for the subgroup of all diagonal matrices $\lambda I$ with $\lambda \neq 0$.

**Theorem:** Collineations map $m$-dimensional projective subspaces to $m$-dimensional projective subspaces and preserve incidence relations.

**The topological structure of $\mathbb{R}P^n$:** The space $\mathbb{R}P^n$ is an $n$-dimensional, connected, compact, smooth real manifold.[15] If we denote by

$$S^n := \{x \in \mathbb{R}^{n+1} \mid |x| = 1\}$$

the $n$-dimensional unit sphere and by $\mathbb{Z}_2$ the cyclic group of order 2, then $S^n/\mathbb{Z}_2$ is diffeomorphic to $\mathbb{R}P^n$. We indicate this by writing

$$\boxed{\mathbb{R}P^n \simeq S^n/\mathbb{Z}_2 .}$$

Here the quotient $S^n/\mathbb{Z}_2$ is obtained from $S^n$ by identifying antipodal points of $S^n$ with each other. Another representation of $\mathbb{R}P^n$ as a quotient is gotten by setting

$$S^n_+ := \{x \in S^n \mid x_{n+1} \geq 0\}$$

for the closed northern hemisphere; then we have also

$$\boxed{\mathbb{R}P^n \simeq S^n_+/\mathbb{Z}_2 .}$$

The quotient in this case is obtained by identifying antipodal points on the equator of $S^n_+$ with each other (and making no further identifications).

*Example 1:* For $n = 1$, $S^1_+$ is the upper half of the unit circle $\{x^2 + y^2 = 1 \mid y \geq 0\}$, and the quotient $S^1_+/\mathbb{Z}_2$ is obtained by identifying the two boundary points $(-1, 0)$ and $(0, 1)$. In this way we get a deformed circle, which can be stretched into the usual unit circle $S^1$. The motivates the homeomorphism

$$\boxed{\mathbb{R}P^1 \simeq S^1}$$

which is in fact a diffeomorphism.

[15] This basic notions will be introduced in [212] and studied in detail there.

*Example 2:* In case $n = 2$ we have the homeomorphisms

$$\boxed{\mathbb{R}P^2 \simeq S_+^2/\mathbb{Z}_2 \simeq \mathbb{R}P_*^2\,,}$$

where $\mathbb{R}P_*^2$ is obtained from the (closed) unit disc $\{(x,y) \in \mathbb{R}^2 \mid x^2+y^2 \leq 1\}$ by identifying the points on the boundary (i.e., on the unit circle) which are opposite with respect to the origin (see Figure 3.37).

**Theorem:** $\mathbb{R}P^2$ is a non-orientable surface.

For a usual surface one has two different 'sides', an inside and an outside. For example every sphere has this property. In the case of non-orientable surfaces, there is no such thing as different sides.

### 3.5.4 The $n$-dimensional complex projective space

If, in the above definitions, we allow the variables $x_1, \ldots, x_{n+1}$ as well as $\lambda$ to be complex, then we get what is called the *complex projective space* $\mathbb{C}P^n$ in the same way as $\mathbb{R}P^n$ was obtained in section 3.5.3.

Also, collineations of $\mathbb{C}P^n$ are defined just as for $\mathbb{R}P^n$, by allowing the matrices $A$ to be complex. The *complex projective group* $PGL(n+1,\mathbb{C})$ is defined as the quotient group

$$\boxed{PGL(n+1,\mathbb{C}) = GL(n+1,\mathbb{C})/D.}$$

Here $GL(n+1,\mathbb{C})$ is the group of complex invertible $(n+1)$-matrices, and $D$ denotes the subgroup of all diagonal matrices $\lambda I$ with $\lambda \neq 0$.

**Projective properties:** A property belongs to complex $n$-dimensional projective geometry, if it is invariant under all collineations of $\mathbb{C}P^n$, i.e., if the property is invariant under the action of $PGL(n+1,\mathbb{C})$.

**The topological structure of $\mathbb{C}P^n$:** The space $\mathbb{C}P^n$ is a $n$-dimensional connected, compact, smooth complex manifold. If we denote by

$$S_{\mathbb{C}}^n := \left\{ x \in \mathbb{C}^{n+1} \;\middle|\; \sum_{j=1}^{n} |x_j|^2 = 1 \right\}$$

the $n$-dimensional complex unit sphere, then we have a diffeomorphism

$$\boxed{\mathbb{C}P^n \simeq S_{\mathbb{C}}^n / S_{\mathbb{C}}^1\,.}$$

In section 3.8.4 we will see the fruitfulness of the methods of complex projective geometry for the investigation of plane algebraic curves. Without such methods, the theory of algebraic curves would be a basket of isolated results, whose formulation would be continuously cluttered by exceptional cases.

> The ideal-theoretic and topological properties of $\mathbb{C}P^n$ are the basis for the success of projective methods in algebraic geometry.

### 3.5.5 The classification of plane geometries

We consider a plane $P$. The classification of the plane geometries is affected under utilization of the Erlanger Program of Felix Klein (1872), by considering the possible groups of transformations in the plane $P$ and their properties.

### 3.5.5.1 Euclidean geometry

**The Euclidean group of motions:** The group of this geometry is the Euclidean group of motions of $P$, which is composed of translations, rotations and reflections. Analytically this group is given by transformations $x \mapsto x'$ of the plane, which are of the form

$$\boxed{x' = Ax + a} \tag{T}$$

with $x = (x_1, x_2)^{\mathsf{T}}$ and $a = (a_1, a_2)^{\mathsf{T}}$. Here $A$ is an orthogonal matrix, that is, $A^{\mathsf{T}}A = AA^{\mathsf{T}} = E$ (where $E$ is the unit matrix). For $A = E$, the formula (T) describes a translation.[16]

**Congruence:** Two figures in the plane $P$ are called *congruent*, if and only if they can be mapped into each other through a Euclidean motion.

### 3.5.5.2 Similarity geometry

**The group of similarity transformations:** The group of this geometry is the group of similarities, which are the transformations as in (T) for which $A$ is a product of an orthogonal matrix and a diagonal matrix with positive elements.

**Geometric characterization:** A *proper* similarity transformation is obtained by performing a central projection on two parallel planes $P$ and $Q$ and then identifying the corresponding points of $P$ and $Q$. This corresponds to (T) with a diagonal matrix

$$A = \begin{pmatrix} \lambda & 0 \\ 0 & \lambda \end{pmatrix}.$$

The positive number $\lambda$ serves as a multiplication factor of the similarity transformation. Arbitrary similarity transformations consist of compositions of proper similarity transformations and Euclidean motions.

**Similar figures:** Two figures in the plane $P$ are said to be *similar*, if they can be transformed into one another by means of a similarity transformation.

### 3.5.5.3 Affine geometry

**The affine group:** This is the transformation group of the plane $P$ consisting of those mappings (T) for which the matrix $A$ is invertible. Such transformations are called *affine* transformations or *affinities*.

**Geometric characterization:** Geometrically one gets an affinity by taking finitely many planes $P, P_1, \ldots, P_n, Q$ in space and performing a parallel projection on each, then identifying the points of $P$ and $Q$ which are mapped to each other.

**Affine equivalence:** Two figures in the plane $P$ are said to be *affinely equivalent*, if they can be mapped into one another by a affinity.

*Example:* Under an affinity, circles are mapped to ellipses, while ellipses are mapped to ellipses. Thus, the notion of ellipse is a notion of affine geometry.

---

[16] Explicitly one has

$$x_1' = a_{11}x_1 + a_{12}x_2 + a_1,$$
$$x_2' = a_{21}x_1 + a_{22}x_2 + a_2,$$

with real coefficients $a_{jk}$, $a_j$ and real variables $x_j$ and $x_j'$.

#### 3.5.5.4 Projective geometry

**The group of collineations (projectivities):** The group of the projective plane consists of the *collineations*[17]

$$\boxed{y' = By\,.}$$

with the homogenous coordinates $y = (y_1, y_2, y_3)^\top$ and $y' = (y_1', y_2', y_3')^\top$. Here $y \neq 0$ and $y' \neq 0$. Moreover, $B$ is an invertible real $(3 \times 3)$-matrix. Collineations are also called *projectivities.*

*Points at infinity:* By adding the points at infinity $y$ with $y_3 = 0$, the plane $P$ is extended to the projective plane $P_\infty$, and this corresponds to the two-dimensional projective space $\mathbb{R}P^2$. Two tuples $y$ and $y'$ are homogenous coordinates of one and the same point in the projective plane $P_\infty$, if $y = \alpha y'$ for some real number $\alpha \neq 0$.

**Geometric characterization:** Geometrically one gets a projectivity when one performs a *central projection* on finitely many planes $P, P_1, \ldots, P_n, Q$ and then identifies the points of $P$ and $Q$ which are mapped under this composition.

**Projective equivalence:** Two figures in the plane $P$ are said to be *projectively equivalent,* if they can be mapped into one another under a projectivity.

*Example:* Under a projectivity, circles, ellipses, parabolas and hyperbolas can all be mapped into one another. For this reason, the notion of (non-degenerate) *conic section* is a notion of projective geometry.

Moreover, the following notions also belong to projective geometry: 'line', 'point', 'a point lying on a line', 'two lines intersect in a point' and 'the double ratio of four points on a line'.

#### 3.5.5.5 Historical remarks:

Euclidean geometry was presented in great detail by Euclid of Alexandra (ca. 365-300 BC) in his famous treatise *The elements.* These books dominated all school courses for over 2000 years.

Affinities are due to Euler (1707–1783).

**Descriptive geometry and orthogonal projections:** Descriptive geometry which uses orthogonal projections on (one or two) planes was created by Monge (1746–1818) between 1766 and 1770 in connection with work in the construction of fortifications. In 1798 his fundamental work *Géométrie descriptive* appeared.

**Projective geometry and central projections:** During the Renaissance there was a revolution in European painting caused by the introduction of *perspective*, which is based on central projections. Several great artists were involved in this: the painter Leon Batista Alberti (1404–1472) (architect of Saint Peters Cathedral in Rome), Leonardo da Vinci (1452–1519) and Albrecht Dürer (1471–1528). He published the book *Unterweisung der Messung mit Zirkel und Richtscheit* (Instructions in making measurements

[17] Explicitly one has

$$\begin{aligned} y_1' &= b_{11}y_1 + b_{12}y_2 + b_{13}y_3,\\ y_2' &= b_{21}y_1 + b_{22}y_2 + b_{23}y_3,\\ y_3' &= b_{31}y_1 + b_{32}y_2 + b_{33}y_3. \end{aligned}$$

The coefficients $b_{jk}$ and all variables $y_j$ and $y_j'$ are real.

The transition to affine coordinates and affinities is obtained when one sets $y_3 = 1$, $b_{31} = b_{32} = 0$ and $b_{33} = 1$.

with compass and ruler). The use of perspective prevailed until about 1900.

Modern photography is also based on central projections. The first camera was described by Leonardo da Vinci in 1500. Niecéphore Niepce generated the first actual photographs in 1822 with the 'camera obscura'.

*Synthetic projective geometry:* The development of projective geometry as a mathematical discipline began in 1822 with the appearance of the book *Traité des proprietés pojectives des figures* (Textbook on the projective properties of figures) by the French mathematician Poncelet (1788–1867). Poncelet used in this work only the method of drawings, which is often referred to as *synthetic geometry*, as opposed to the *analytic geometry* due to Descartes (1596–1650). Poncelet stands here in the tradition of the French mathematicians Desargues (1591–1661), Pascal (1623–1662) and Monge (1746–1818).

*The barycentric coordinates of Möbius* Our ability to do calculations with projectivities is due to the work of Möbius, presented in his book *Der baryzentrische Calcul, ein neues Hilfsmittel zur analytischen Behandlung der Geometrie* (The barycentric calculus, a new tool for analytic treatment of geometry), which appeared in 1827.[18] Möbius introduced in this work *barycentric coordinates* $(m_1, m_2, m_3)$. If $p_1, p_2, p_3$ are three points of the plane $P$ which do not lie on a line, then one can describe every point $p$ of $P$ uniquely by real coordinates $m_1, m_2, m_3$ with $m_1 + m_2 + m_3 = 1$. If $\mathbf{x}_1, \mathbf{x}_2, \mathbf{x}_3$ and $\mathbf{x}$ are the corresponding vectors of the points $p_1, p_2, p_3$ and $p$, then one has:

$$\boxed{\mathbf{x} = m_1\mathbf{x}_1 + m_2\mathbf{x}_2 + m_3\mathbf{x}_3 \,.}$$

*Example:* We can take for the points $p_1 = (1, 0, 0)$, $p_2 = (0, 1, 0)$ and $p_3 = (0, 0, 1)$.

The barycentric coordinates can be given a simple physical interpretation. If one puts the masses $m_1, m_2, m_3$ at the three points $p_1, p_2$ and $p_3$, then $p$ is the *center of mass* of these mass points and lies in the interior of the triangle spanned by the three points. If one allows also negative or vanishing masses, then one gets the rest of the points of the plane.

If one drops the condition on the $m_i$ that $m_1 + m_2 + m_3 = 1$, then one also gets the points at infinity.

Further important contributors to the theory of projective geometry in the nineteenth century were Steiner (1796–1863), von Staudt (1798–1867), Plücker (1801–1868), Cayley (1821–1895) and Klein (1849–1925). The notion of projective space had gained general acceptance as a fundamental geometric notion by 1870. Between the appearance of Euclid's *Elements* and this point in time lay over 2000 years.[19]

---

[18] Augustus Ferdinand Möbius (1790–1868) worked from 1816 until his death at the university of Leipzig and was the head of the Leipzig observatorium.

[19] The history of geometry, which exploded in the nineteenth century, is masterly described in Felix Klein's *Vorlesungen über die Entwicklung der Mathematik im 19. Jahrhundert* (Lectures on the developement of mathematics in the nineteenth century) (cf. [492])

## 3.6 Differential geometry

*There is no science which is not developed from knowledge of phenomena, but to obtain advantage from this knowledge, it is necessary to be a mathematician.*

*Daniel Bernoulli* (1700–1782)

Differential geometry studies the properties of curves and surfaces with the methods of calculus. The most important differential geometric property is *curvature*. During the nineteenth and twentieth centuries mathematicians have attempted to generalize this intuitive notion to higher dimensions and to more and more abstract situations (theory of principal bundles).

Physicists, on the other hand, have tried ever since Newton to understand the forces acting on our world and in the universe. Surprisingly, the four basic forces in cosmology and elementary particle physics (gravity and the weak, strong and electromagnetic forces) are all based on the basic relation

$$\boxed{\text{force} = \text{curvature},}$$

which is the *deepest known connection* between mathematics and physics. This will be discussed in detail in [212].

In this section we consider the classical theory of differential geometry of curves and surfaces in three-dimensional space. The theory of curves was created in the eighteenth century by Clairaut, Monge and Euler and was further developed in the nineteenth century by Cauchy, Frenet and Serret. From 1821 to 1825 Gauss carried out extremely strenuous geodesic measurements in the Kingdom of Hannover. This was motivation for him, who always connected theory and application in an exemplary manner, to intensive studies of curved surfaces. In 1827 his epochal treatise *Disquisitiones generales circa superficies curvas*, in which he created the theory of differential geometry of surfaces, appeared, and in center stage of that theory stood his *theorema egregium*. This deep mathematical theorem says that the curvature of a surface can be determined by measurements taken only on that surface, without using the ambient space. With this result Gauss laid the foundations for the general theory of differential geometry, whose further development by Riemann and Élie Cartan led to the crowning achievement: Einstein's general theory of relativity and gravity (cosmology) as well as the standard model of elementary particle physics.[20] The standard model is based on a gauge theory, which from a mathematical point of view corresponds to the curvature of appropriately defined principal bundles.

**Local behavior:** To study the behavior of curves and surfaces in the neighborhood of a point one uses the *Taylor expansion*.[21] This leads to the notions of 'tangent, curvature and torsion' of a curve as well as the notions of 'tangential plane and curvature' of surfaces.

**Global behavior:** Besides the local behavior 'in the small' just mentioned, one is interested in the behavior in the large. A typical result of this type is the Gauss–Bonnet formula in the theory of surfaces, which is the starting point of modern differential geometry in the context of the theory of characteristic classes (cf. [212]).

[20] In connection with the theory of relativity, Henri Poincaré's name should be mentioned. In fact, for his work in this area, he was nominated several times for the Nobel Prize in physics. – *The translator*

[21] We tacitly assume that all functions are sufficiently smooth.

### 3.6.1 Plane curves

**Parameter representation:** A plane curve is given by an equation of the form

$$x = x(t), \quad y = y(t), \qquad a \le t \le b, \tag{3.22}$$

in Cartesian coordinates $(x, y)$ (Figure 3.42).

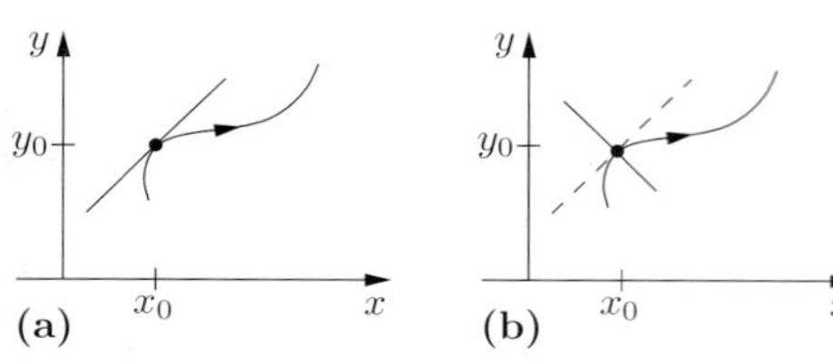


*Figure 3.42. Plane curves.*

If we interpret the real parameter $t$ as the time, then (3.22) describes the motion of a point which at time $t$ has the coordinates $(x(t), y(t))$.

*Example 1:* In the special case $t = x$, we get the curve equation $y = y(x)$ (the equation of a graph of a function in $\mathbb{R}^2$).

**Arc length $s$ of a curve:**

$$s = \int_a^b \sqrt{x'(t)^2 + y'(t)^2}\,\mathrm{d}t.$$

**Equation of a tangent at the point $(x_0, y_0)$** (Figure 3.42(a)):

$$x = x_0 + (t - t_0)x_0', \quad y = y_0 + (t - t_0)y_0', \qquad -\infty < t < \infty.$$

Here we have set $x_0 := x(t_0)$, $x_0' := x'(t_0)$, $y_0 := y(t_0)$, $y_0' := y'(t_0)$.

**Equation of the normal vector to the curve at the point $(x_0, y_0)$** (Figure 3.42(b)):

$$x = x_0 - (t - t_0)y_0', \quad y = y_0 + (t - t_0)x_0', \qquad -\infty < t < \infty. \tag{3.23}$$

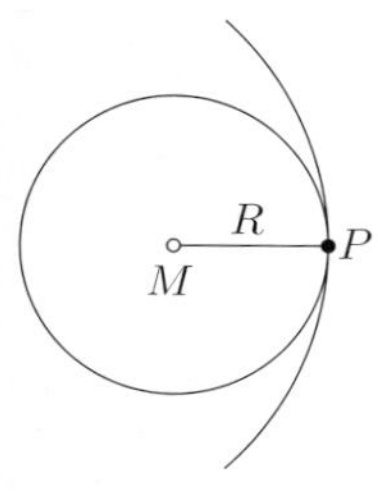


*Figure 3.43.*

**Curvature radius $R$:** If the curve (3.22) is of type $C^2$ in a neighborhood of a point $P(x_0, y_0)$, then there is a uniquely determined circle of radius $R$ centered at a point $M(\xi, \eta)$, and $R$ is called the *curvature radius of the curve*, which coincides with the curve at the point $P$ up to order 2 (one also says the circle *touches* the curve to order 2, or that the curve and the circle have a *point of contact* of order 2, see Figure 3.43).

**Curvature $K$:** This quantity is introduced as the number $K$ whose absolute value is inverse to the curvature radius $R$:

$$|K| := \frac{1}{R}.$$

The sign of $K$ at a point $P$ is by definition positive (resp. negative), if the curve lies above (resp. below) the tangent at the point $P$ (Figure 3.44). For the curve (3.22) one has

$$K = \frac{x_0' y_0'' - y_0' x_0''}{\left({x_0'}^2 + {y_0'}^2\right)^{3/2}}$$

with the center of curvature $M(\xi, \eta)$:

$$\xi = x_0 - \frac{y_0'({x_0'}^2 + {y_0'}^2)}{x_0' y_0'' - y_0' x_0''}, \qquad \eta = y_0 + \frac{x_0'({x_0'}^2 + {y_0'}^2)}{x_0' y_0'' - y_0' x_0''}.$$

**Inflection point:** By definition a curve has an *inflection point* at a point $P$, if $K(P) = 0$ and $K$ changes its sign at the point $P$ (Figure 3.44(c)).

**Extremal points:** Points on a curve where the curvature $K$ has a minimum or a maximum are called *extremal points* of the curve.

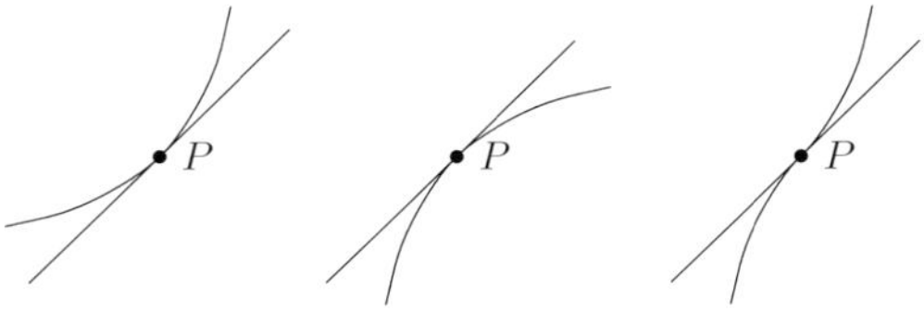

**(a)** $K > 0$ **(b)** $K < 0$ **(c)** $K = 0$

*Figure 3.44. The curvature $K$ of a curve.*

**Angle $\varphi$ between two curves $x = x(t)$, $y = y(t)$ and $X = X(\tau)$, $Y = Y(\tau)$ in a point of intersection:**

$$\cos\varphi = \frac{x_0' X_0' + y_0' Y_0'}{\sqrt{{x_0'}^2 + {y_0'}^2}\,\sqrt{{X_0'}^2 + {Y_0'}^2}}, \qquad 0 \le \varphi < \pi .$$

Here $\varphi$ is the angle between the tangents in the intersection point measured in the mathematically positive sense, see Figure 3.45.

The values $x_0$, $X_0$, etc., are those of the point of intersection.

**Applications:**

*Example 2:* The equation of circle of radius $R$ centered at the origin $(0, 0)$ is:

$$x = R\cos t, \quad y = R\sin t, \qquad 0 \le t < 2\pi$$

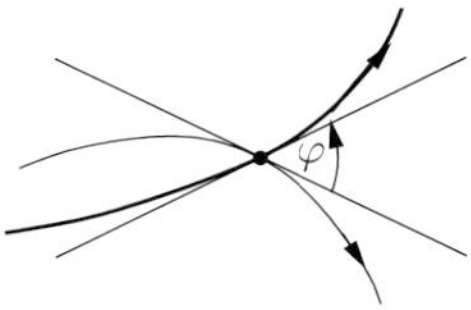

*Figure 3.45.*

(Figure 3.46). The point $(0, 0)$ is the center of curvature, and $R$ is the curvature radius. This yields for the curvature

$$K = \frac{1}{R}.$$

From the derivatives $x'(t) = -R\sin t$ and $y'(t) = R\cos t$ we get the parametrized equation of the tangent:

$$x = x_0 - (t - t_0)y_0, \quad y = y_0 + (t - t_0)x_0, \qquad -\infty < t < \infty.$$

For the arc length $s$ we get from $\cos^2 t + \sin^2 t = 1$ the relation

$$s = \int_0^{\varphi} \sqrt{x'(t)^2 + y'(t)^2}\,\mathrm{d}t = R\varphi .$$

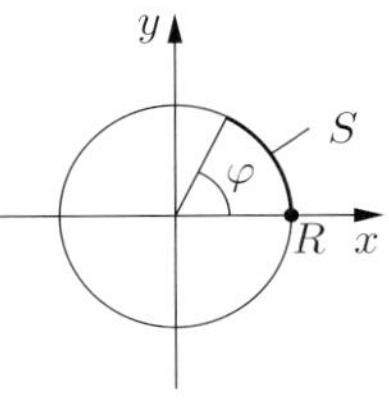


*Figure 3.46.*

If we write the circle equation in the implicit form

$$x^2 + y^2 = R^2$$

then we get from Table 3.8 with $F(x, y) := x^2 + y^2 - R^2$ the equation of the tangent at the point $(x_0, y_0)$:

$$\boxed{x_0(x - x_0) + y_0(y - y_0) = 0,}$$

i.e., $x_0 x + y_0 y = R^2$. In polar coordinates the equation for the circle is

$$r = R.$$

*Table 3.7. Formulae for explicitly given plane curves.*

| | | |
|---|---|---|
| ***Equation of the curve*** | $y = y(x)$ (explicit form) | $r = r(\varphi)$ (polar coordinates) |
| ***Arc length*** | $s = \int_a^b \sqrt{1 + y'(x)^2}\,\mathrm{d}x$ | $\int_\alpha^\beta \sqrt{r^2 + r'^2}\,\mathrm{d}\varphi$ |
| ***Tangent at a point $P(x_0, y_0)$*** | $y = x_0 + y_0'(x - x_0)$ | |
| ***Normal vector at a point $P(x_0, y_0)$*** | $-y_0'(y - y_0) = x - x_0$ | |
| ***Curvature $K$ at a point $P(x_0, y_0)$*** | $\dfrac{y_0''}{\left(1 + y_0'^2\right)^{3/2}}$ | $\dfrac{r^2 + 2r'^2 - rr''}{\left(r^2 + r'^2\right)^{3/2}}$ |
| ***Center of curvature*** | $\xi = x_0 - \dfrac{y_0'(1 + y_0'^2)}{y_0''}$<br>$\eta = y_0 + \dfrac{1 + y_0'^2}{y_0''}$ | $\xi = x_0 - \dfrac{(r^2 + r'^2)(x_0 + r'\sin\varphi)}{r^2 + 2r'^2 - rr''}$<br>$\eta = y_0 - \dfrac{(r^2 + r'^2)(y_0 - r'\cos\varphi)}{r^2 + 2r'^2 - rr''}$ |

*Table 3.8. Formulae for implicitly given curves.*

| | |
|---|---|
| ***Implicit equation*** | $F(x, y) = 0$ |
| ***Tangent at a point $P(x_0, y_0)$*** | $F_x(P)(x - x_0) + F_y(P)(y - y_0) = 0$ |
| ***Normal vector at a point $P(x_0, y_0)$*** | $F_x(P)(y - y_0) - F_y(P)(x - x_0) = 0$ |
| ***Curvature $K$ at a point $P$ ($F_x := F_x(P)$ etc.)*** | $\dfrac{-F_y^2 F_{xx} + 2F_x F_y F_{xy} - F_x^2 F_{yy}}{\left(F_x^2 + F_y^2\right)^{3/2}}$ |
| ***Center of curvature*** | $\xi = x_0 - \dfrac{F_x(F_x^2 + F_y^2)}{F_y^2 F_{xx} - 2F_x F_y F_{xy} + F_x^2 F_{yy}}$<br>$\eta = y_0 - \dfrac{F_y(F_x^2 + F_y^2)}{F_y^2 F_{xx} - 2F_x F_y F_{xy} + F_x^2 F_{yy}}$ |

*Example 3:* The parabola

$$\boxed{y = \frac{1}{2}ax^2}$$

with $a > 0$ has according to Table 3.7 at the point $x = 0$ the curvature $K = a$.

*Example 4:* Let $y = x^3$. From Table 3.7 we get

$$K = \frac{y''(x)}{\left(1 + y'(x)^2\right)^{3/2}} = \frac{6x}{\left(1 + 9x^4\right)^{3/2}}\,.$$

At the point $x = 0$ the sign of $K$ changes. Hence this is an inflection point.

**Singular points:** We consider the curve $x = x(t)$, $y = y(t)$. By definition a point $(x(t_0), y(t_0))$ is called a *singular point* of the curve, if

$$\boxed{x'(t_0) = y'(t_0) = 0.}$$

At such a point the tangent is not well-defined. The study of the behavior of the curve in a neighborhood of a singular point is affected by applying the Taylor series

$$x(t) = x(t_0) + \frac{(t-t_0)^2}{2}x''(t_0) + \frac{(t-t_0)^3}{6}x'''(t_0) + \ldots\,,$$
$$y(t) = y(t_0) + \frac{(t-t_0)^2}{2}y''(t_0) + \frac{(t-t_0)^3}{6}y'''(t_0) + \ldots\,.$$

*Example 5:* For

$$x = x_0 + (t-t_0)^2 + \ldots\,, \qquad y = y_0 + (t-t_0)^3 + \ldots$$

one has a point of return (Figure 3.47(a)). In case of

$$x = x_0 + (t-t_0)^2 + \ldots\,, \qquad y = y_0 + (t-t_0)^2 + \ldots$$

the curve ends at the point $(x_0, y_0)$ (Figure 3.47(b)).

**Singular points for curves given by implicit equations:** If the curve is given by an equation $F(x, y) = 0$ with $F(x_0, y_0) = 0$, then by definition $(x_0, y_0)$ is a singular point if and only if

$$\boxed{F_x(x_0, y_0) = F_y(x_0, y_0) = 0.}$$

$(t > t_0)$ $(x_0, y_0)$ $(x_0, y_0)$ **(a)** point of return

$t \gtrless t_0$ $(x_0, y_0)$ **(b)** end point

*Figure 3.47. Singular points.*

The behavior of the curve in a neighborhood of $(x_0, y_0)$ is again studied by means of the Taylor expansion

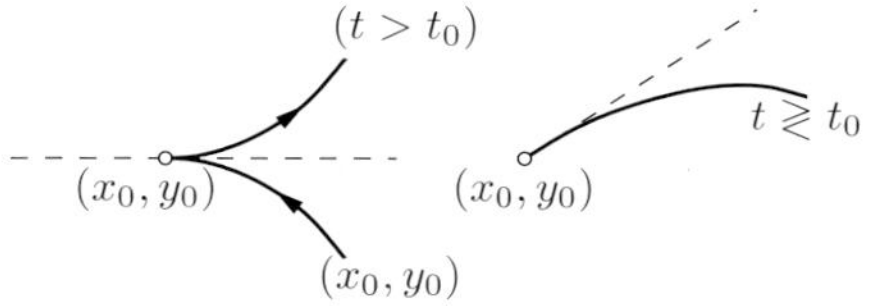

$$F(x, y) = a(x-x_0)^2 + 2b(x-x_0)(y-y_0) + c(y-y_0)^2 + \ldots\,. \tag{3.24}$$

Here we have set $a := \frac{1}{2}F_{xx}(x_0, y_0)$, $b := \frac{1}{2}F_{xy}(x_0, y_0)$ and $c := \frac{1}{2}F_{yy}(x_0, y_0)$. Moreover let $D := ac - b^2$.

*Case 1:* $D > 0$. Then $(x_0, y_0)$ is an isolated point (Figure 3.48(a)).

*Case 2:* $D < 0$. Here the two branches of the curve intersect at the point $(x_0, y_0)$ (Figure 3.48(b)).

If $D = 0$, then we must consider also terms of higher order in the expansion (3.24). For example the following situations can occur: point of contact, point of return, end point, triple point or more generally a point of order $n$ (cf. Figure 3.47 and Figure 3.48).

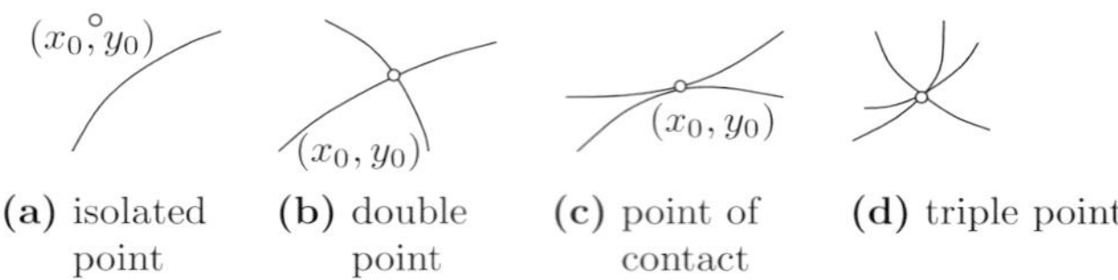


*Figure 3.48. Singular points on curves.*

**Catastrophe theory:** A discussion of singularities in the context of catastrophe theory is to be found in [212].

**Asymptotes:** If a curve approaches a line as the distance from the origin grows beyond bounds, then one calls the line an *asymptote* of the curve.

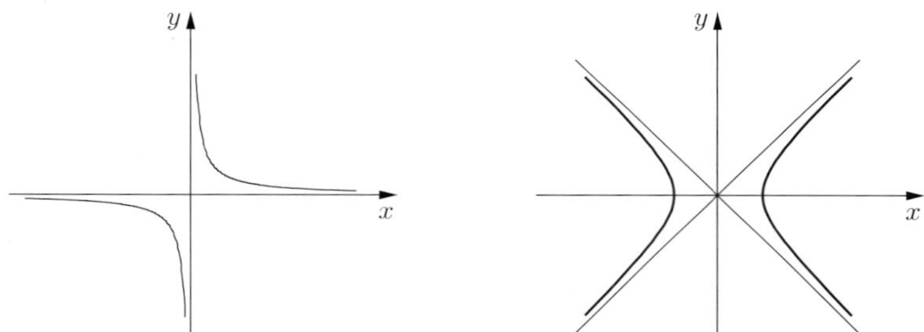


*Figure 3.49. Asymptotes: the hyperbola* $xy = 1$.

*Figure 3.50. Asymptotes: the hyperbola* $\frac{x^2}{a^2} - \frac{y^2}{b^2} = 1$.

*Example 6:* The $x$-axis and the $y$-axis are asymptotes of the hyperbola given by the equation $xy = 1$ (Figure 3.49).

Let $a \in \mathbb{R}$. We consider the curve $x = x(t),\ y = y(t)$ and the limit as $t$ goes to $t_0 + 0$.

(i) For $y(t) \to \pm\infty,\ x(t) \to a$, the line $x = a$ is a vertical asymptote.

(ii) For $x(t) \to \pm\infty,\ y(t) \to a$, the line $y = a$ is a horizontal asymptote.

(iii) Suppose that $x(t) \to +\infty$ and $y(t) \to +\infty$. If the two limits

$$m = \lim_{t \to t_0 + 0} \frac{y(t)}{x(t)} \qquad \text{and} \qquad c = \lim_{t \to t_0 + 0} \big(y(t) - mx(t)\big)$$

exist, then the line $y = mx + c$ is an asymptote.

Similarly one can treat the cases $t \to t_0 - 0$ and $t \to \pm\infty$.

*Example 7:* The hyperbola

$$\frac{x^2}{a^2} - \frac{y^2}{b^2} = 1$$

has the parameterization $x = a\cosh t,\ y = b\sinh t$. The two lines

$$y = \pm\frac{b}{a}x$$

are asymptotes (Figure 3.50).

*Proof:* For example one has

$$\lim_{t\to+\infty}\frac{b\sinh t}{a\cosh t} = \frac{b}{a}, \qquad \lim_{t\to+\infty}(b\sinh t - b\cosh t) = 0.$$

### 3.6.2 Space curves

**Parameterizations:** Let $x, y$ and $z$ be Cartesian coordinates with the basis vectors $\mathbf{i}, \mathbf{j}, \mathbf{k}$ and the radius vector $\mathbf{r} = \overrightarrow{OP}$ of a point $P$. A *space curve* is given by an equation

$$\boxed{\mathbf{r} = \mathbf{r}(t), \qquad a \le t \le b,}$$

i.e., $x = x(t),\ y = y(t),\ z = z(t)$ and $a \le t \le b$.

**Equation of a tangent at a point $\mathbf{r}_0 := \mathbf{r}(t_0)$:**

$$\boxed{\mathbf{r} = \mathbf{r}_0 + (t - t_0)\mathbf{r}'(t_0), \qquad t \in \mathbb{R}.}$$

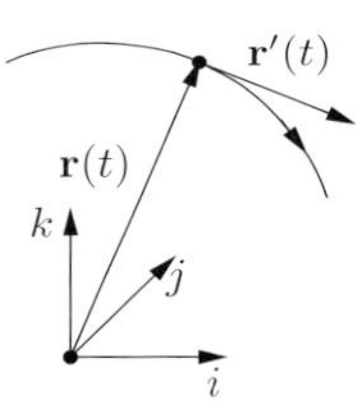


*Figure 3.51.*

**Physical interpretation:** If $t$ denotes the time, then the space curve $\mathbf{r} = \mathbf{r}(t)$ describes the motion of a mass point with the velocity vector $\mathbf{r}'(t)$ and the acceleration vector $\mathbf{r}''(t)$ at the time $t$ (Figure 3.51).

#### 3.6.2.1 Curvature and torsion

**Arc length $s$ of the curve:**

$$\boxed{s := \int_a^b |\mathbf{r}'(t)|\mathrm{d}t = \int_a^b \sqrt{x'(t)^2 + y'(t)^2 + z'(t)^2}\,\mathrm{d}t\,.}$$

If one replace $b$ by $t_0$, then one gets the arc length between the starting point and the point of the curve at time $t_0$.

In what follows we will consider the space curve $\mathbf{r}(s)$ as a function of the arc length $s$ and denote by $\mathbf{r}'(s)$ the derivative with respect to $s$.

**Taylor expansion:**

$$\mathbf{r}(s) = \mathbf{r}(s_0) + (s - s_0)\mathbf{r}'(s_0) + \frac{(s-s_0)^2}{2}\mathbf{r}''(s_0) + \frac{(s-s_0)^3}{6}\mathbf{r}'''(s_0) + \dots\,. \qquad (3.25)$$

The following definitions are based on this formula.

**Tangent unit vector:**

$$\boxed{\mathbf{t} := \mathbf{r}'(s_0).}$$

**Curvature:**

$$k := |\mathbf{r}''(s_0)|.$$

The number $R := 1/k$ is called the curvature radius.

**Normal unit vector:**

$$\mathbf{n} := \frac{1}{k}\mathbf{r}''(s_0).$$

**Binormal vector:**

$$\mathbf{b} := \mathbf{t} \times \mathbf{n}.$$

**Torsion:**

$$w := R\mathbf{b}\mathbf{r}'''(s_0).$$

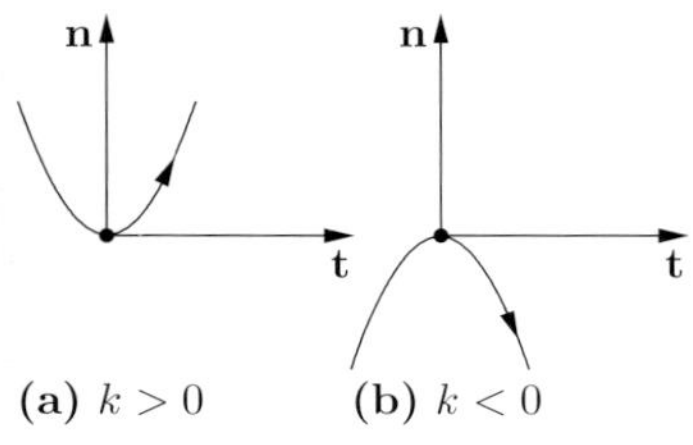


(a) $k > 0$ (b) $k < 0$

*Figure 3.52. Minimum and maximum.*

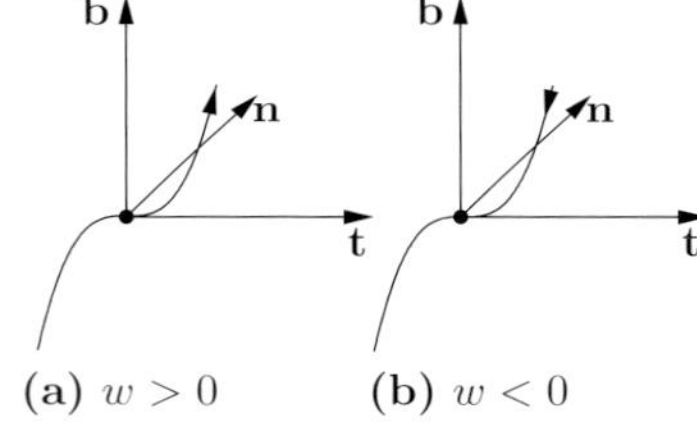


(a) $w > 0$ (b) $w < 0$

*Figure 3.53. Inflection points.*

**Geometric interpretation:** The three vectors $\mathbf{t}, \mathbf{n}, \mathbf{b}$ form the so-called *accompanying three-frame* at the point $P_0$ of the curve.[22] This is a right-handed system of pairwise orthogonal unit vectors (Figure 3.53).

(i) The *contact plane* of the curve at the point $P_0$ is the plane spanned by $\mathbf{t}$ and $\mathbf{n}$.

(ii) The *normal plane* to the curve at the point $P_0$ is the plane spanned by $\mathbf{n}$ and $\mathbf{b}$.

(iii) The *rectifying plane* to the curve at the point $P_0$ is the plane spanned by the vectors $\mathbf{t}$ and $\mathbf{b}$.

According to (3.25) the curve at the point $P_0$ lies to second order (has a contact of order 2) in the contact plane (Figure 3.52).

If $w = 0$ for the point $P_0$, then the curve is according to (3.25) a plane curve to third order.

If $w > 0$ (resp. $w < 0$) at $P_0$, then the curve moves in a neighborhood of $P_0$ in the direction of $\mathbf{b}$ (resp. $-\mathbf{b}$) (Figure 3.53).

**General parameterization:** If the curve is given in the form $\mathbf{r} = \mathbf{r}(t)$ with a real parameter $t$, then one has

$$k^2 = \frac{1}{R^2} = \frac{\mathbf{r}'^2\mathbf{r}''^2 - (\mathbf{r}'\mathbf{r}'')^2}{\left(\mathbf{r}'^2\right)^3} = \frac{(x'^2 + y'^2 + z'^2)(x''^2 + y''^2 + z''^2) - (x'x'' + y'y'' + z'z'')^2}{\left(x'^2 + y'^2 + z'^2\right)^3}$$

[22] In German, *bein* means leg; in French, *repère* means *marking*; both terms have been adapted and are used: one also calls an $n$-frame an $n$-bein or $n$-repère.

$$w = R^2 \frac{(\mathbf{r}' \times \mathbf{r}'')\mathbf{r}'''}{\left(\mathbf{r}'^2\right)^3} = R^2 \frac{\begin{vmatrix} x' & y' & z' \\ x'' & y'' & z'' \\ x''' & y''' & z''' \end{vmatrix}}{\left(x'^2 + y'^2 + z'^2\right)^3} . \tag{3.26}$$

*Example:* We consider the spiral

$$\boxed{x = a \cos t, \quad y = a \sin t, \quad z = bt, \qquad t \in \mathbb{R}}$$

with $a > 0$ and $b > 0$ (right-handed, see Figure 3.54) resp. $b < 0$ (left-handed). One has:

$$\boxed{k = \frac{1}{R} = \frac{a}{a^2 + b^2}, \qquad w = \frac{b}{a^2 + b^2} .}$$

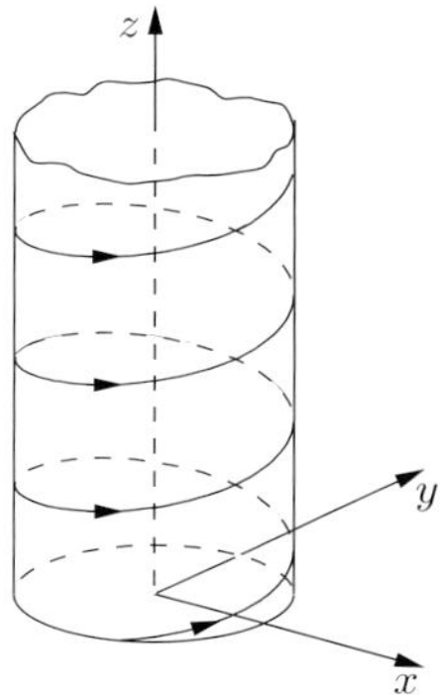

*Figure 3.54.*

*Proof:* We replace the parameter $t$ by the arc length

$$s = \int_0^t \sqrt{\dot{x}^2 + \dot{y}^2 + \dot{z}^2}\, \mathrm{d}t = t\sqrt{a^2 + b^2} .$$

Then one has

$$x = a \cos \frac{s}{\sqrt{a^2 + b^2}} , \qquad y = a \sin \frac{s}{\sqrt{a^2 + b^2}} , \qquad z = \frac{bs}{\sqrt{a^2 + b^2}} ,$$

hence

$$k = \frac{1}{R} = \sqrt{\left(\frac{\mathrm{d}^2 x}{\mathrm{d}s^2}\right)^2 + \left(\frac{\mathrm{d}^2 y}{\mathrm{d}s^2}\right)^2 + \left(\frac{\mathrm{d}^2 z}{\mathrm{d}s^2}\right)^2} = \frac{a}{a^2 + b^2} .$$

The curvature $k$ is thus constant. For the torsion one has by (3.26) the value

$$w = \left(\frac{a^2 + b^2}{a}\right)^2 \frac{\begin{vmatrix} -a \sin t & a \cos t & b \\ -a \cos t & -a \sin t & 0 \\ a \sin t & -a \cos t & 0 \end{vmatrix}}{\left[(-a \sin t)^2 + (a \cos t)^2 + b^2\right]^3} = \frac{b}{a^2 + b^2} .$$

This means that the torsion is also constant. □

#### 3.6.2.2 The main theorem in the theory of curves

**The formulas of Frenet:** For the derivatives of the vectors $\mathbf{t}, \mathbf{n}, \mathbf{b}$ with respect to the arc length one has:

$$\boxed{\mathbf{t}' = k\mathbf{n}, \qquad \mathbf{n}' = -k\mathbf{t} + w\mathbf{b}, \qquad \mathbf{b}' = -w\mathbf{n}.} \tag{3.27}$$

**Main theorem:** If two continuous functions

$$k = k(s) \qquad \text{and} \qquad w = w(s)$$

are given in the interval $a \leq s \leq b$ and $k(s) > 0$ for all $s$, then there is, up to transformations of the entire space, exactly one curve segment $\mathbf{r} = \mathbf{r}(s)$, $a \leq s \leq b$, which has arc length equal to $s$ and has the curvature $k$ and torsion $w$.

**Construction of the curve:** (i) The equation (3.27) consists of a system of nine ordinary differential equations for the three components each of $\mathbf{t}, \mathbf{n}$ and $\mathbf{b}$. By prescribing the values $\mathbf{t}(0), \mathbf{n}(0)$ and $\mathbf{b}(0)$ (which amounts to the prescription of the accompanying three-frame at the point $s = 0$), the solution of these differential equations (3.27) is uniquely determined.

(ii) If one in addition prescribes the vector $\mathbf{r}(0)$, then one gets

$$\mathbf{r}(s) = \mathbf{r}(0) + \int_0^s \mathbf{t}(s)\,\mathrm{d}s\,.$$

### 3.6.3 The Gaussian local theory of surfaces

*From time to time in the past, certain brilliant, unusually gifted personalities have arisen from their environment, who by virtue of the creative power of their thoughts and the energy of their actions have had such an overall positive influence on the intellectual development of mankind, that they at the same time stand tall as markers between the centuries...Such epoch-making mental giants in the history of mathematics and the natural sciences are Archimedes of Syracuse in ancient times, Newton toward the end of the dark ages and Gauss in our present day, whose shining, glorious career has come to an end after the cold hand of death touched his at one time deeply-thinking head on February 23 of this year.*

*Sartorius von Waltershausen,* 1855
*in honor of Gauss*

**Parameterization of a surface:** Let $x, y$ and $z$ be Cartesian coordinates with the basis vectors $\mathbf{i}, \mathbf{j}, \mathbf{k}$ and the radius vector $\mathbf{r} = \overrightarrow{OP}$ of the point $P$. A *surface* is given by an equation

$$\boxed{\mathbf{r} = \mathbf{r}(u, v)}$$

with real parameters $u$ and $v$, i.e.,

$$x = x(u, v), \qquad y = y(u, v), \qquad z = z(u, v).$$

**The accompanying three-frame:** We set

$$\boxed{\mathbf{e}_1 := \mathbf{r}_u(u_0, v_0), \qquad \mathbf{e}_2 := \mathbf{r}_v(u_0, v_0), \qquad \mathbf{N} = \frac{\mathbf{e}_1 \times \mathbf{e}_2}{|\mathbf{e}_1 \times \mathbf{e}_2|}\,.}$$

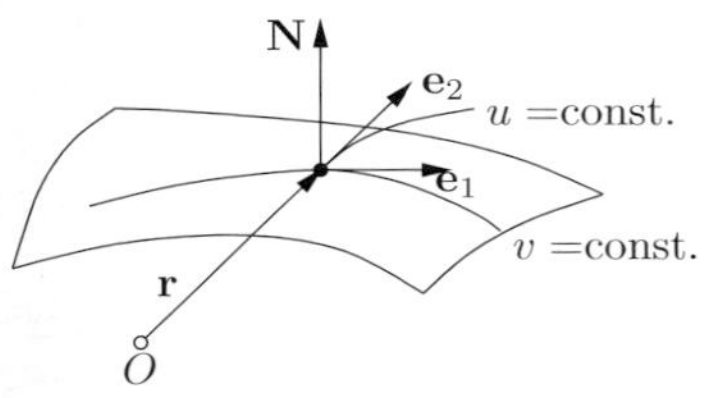


Figure 3.55.

Then $\mathbf{e}_1$ (resp. $\mathbf{e}_2$) is the tangent vector to the coordinate line $v = \text{const}$ (resp. $u = \text{const}$) through the point $P_0(u_0, v_0)$ of the surface. Furthermore, $\mathbf{N}$ is the normal unit vector to the surface at $P_0$ (Figure 3.55).

Explicitly one has

$$\begin{aligned}\mathbf{e}_1 &= x_u(u_0, v_0)\mathbf{i} + y_u(u_0, v_0)\mathbf{j} + z_u(u_0, v_0)\mathbf{k},\\ \mathbf{e}_2 &= x_v(u_0, v_0)\mathbf{i} + y_v(u_0, v_0)\mathbf{j} + z_v(u_0, v_0)\mathbf{k}.\end{aligned}$$

**Equation of the tangent plane at the point $P_0$:**

$$\boxed{\mathbf{r} = \mathbf{r}_0 + t_1\mathbf{e}_1 + t_2\mathbf{e}_2, \qquad t_1, t_2 \in \mathbb{R}\,.}$$

**Implicit equation of the surface:** If the surface is given by an equation $F(x,y,z) = 0$, then one gets the unit normal vector at the point $P_0(x_0, y_0, z_0)$ by the formula

$$\mathbf{N} = \frac{\mathbf{grad}\, F(P_0)}{|\mathbf{grad}\, F(P_0)|}\,.$$

The equation of the tangent plane at the point $P_0$ is:

$$\boxed{\mathbf{grad}\, F(P_0)(\mathbf{r} - \mathbf{r}_0) = 0.}$$

Explicitly this means

$$F_x(P_0)(x - x_0) + F_y(P_0)(y - y_0) + F_z(P_0)(z - z_0) = 0.$$

**Explicit equations for surfaces:** The equation $z = z(x,y)$ can be written in the form $F(x,y,z) = 0$ as follows: $F(x,y,z) := z - z(x,y)$.

*Example 1:* The equation of a sphere of radius $R$ is:

$$x^2 + y^2 + z^2 = R^2.$$

If we set $F(x,y,z) := x^2 + y^2 + z^2 - R^2$, then we get the equation for the tangent plane at the point $P_0$:

$$x_0(x - x_0) + y_0(y - y_0) + z_0(z - z_0) = 0$$

with the unit normal vector $\mathbf{N} = \mathbf{r}_0/|\mathbf{r}_0|$.

**Singular points on surfaces:** For a surface $\mathbf{r} = \mathbf{r}(u,v)$ the point $P_0(u_0, v_0)$ is said to be *singular*, if and only if, $\mathbf{e}_1$ and $\mathbf{e}_2$ do not span a plane.

In the case of an implicit equation $F(x,y,z) = 0$, a point $P_0$ is by definition singular if there is no unit normal vector, i.e., $\mathbf{grad}\, F(P_0) = 0$. Explicitly this means

$$F_x(P_0) = F_y(P_0) = F_z(P_0) = 0.$$

*Example 2:* The cone $x^2 + y^2 - z^2 = 0$ has the singular point $x = y = z = 0$, which is just the vertex of the cone.

**Change of parameters and tensor calculus:** We set $u^1 = u$, $u^2 = v$. If on the surface we are given two functions $a_\alpha(u^1, u^2)$, $\alpha = 1, 2$, which transform under the change of coordinates given by passing from the $u^\alpha$-system to a $u'^\alpha$-system on the surface as follows:

$$a'_\alpha(u'^1, u'^2) = \frac{\partial u^\gamma}{\partial u'^\alpha}\, a_\gamma(u^1, u^2) \tag{3.28}$$

then we call $a_\alpha(u^1, u^2)$ a *simple covariant tensor field* on the surface. In (3.28) the *Einstein summation convention* is used (this will be used for the rest of section 3.6.3), which says that sums are formed over same indices which occur both as superscripts and subscripts (here of course the summation is from 1 to 2; the convention makes sense in a more general situation).

The $2k+2l$ functions $a^{\beta_1\cdots\beta_l}_{\alpha_1\cdots\alpha_k}(u^1,u^2)$ form by definition a *k-covariant and l-contravariant tensor field* on the surface, if they transform under the coordinate change from $u^\alpha$ to $u'^\alpha$ as follows:

$$a'^{\beta_1\cdots\beta_l}_{\alpha_1\cdots\alpha_k}(u'^1,u'^2) = \frac{\partial u^{\gamma_1}}{\partial u'^{\alpha_1}}\frac{\partial u^{\gamma_2}}{\partial u'^{\alpha_2}}\cdots\frac{\partial u^{\gamma_k}}{\partial u'^{\alpha_k}}\frac{\partial u'^{\beta_1}}{\partial u^{\delta_1}}\cdots\frac{\partial u'^{\beta_l}}{\partial u^{\delta_l}}a^{\delta_1\cdots\delta_l}_{\gamma_1\cdots\gamma_k}(u^1,u^2)\,.$$

**Advantage of tensor calculus:** If one applies tensor calculus in the theory of surfaces, then one recognizes immediately when some given (analytic-algebraic) expression has a geometrical meaning, i.e., is independent of the chosen parameterization. This goal can be achieved by utilizing tensors and constructing scalars (cf. [212]).

#### 3.6.3.1 The first Gaussian fundamental form and metrical properties of surfaces

**First fundamental form:** According to Gauss, this form can be written on a surface given by an equation $\mathbf{r}=\mathbf{r}(u,v)$ as

$$\boxed{\mathrm{d}s^2 = E\mathrm{d}u^2 + F\mathrm{d}u\mathrm{d}v + G\mathrm{d}v^2}$$

with

$$E=\mathbf{r}_u^2 = x_u^2+y_u^2+z_u^2, \qquad G=\mathbf{r}_v^2=x_v^2+y_v^2+z_v^2,$$
$$F=\mathbf{r}_u\mathbf{r}_v = x_ux_v+y_uy_v+z_uz_v.$$

If the surface is given in the form $z=z(x,y)$, then one has $E=1+z_x^2$, $G=1+z_y^2$, $F=z_xz_y$.

| The first fundamental form encodes all metrical properties of the surface. |
|---|

**Arc length:** The arc length of a curve $\mathbf{r}=\mathbf{r}(u(t),v(t))$ on the surface between the points with the parameter values $t_0$ and $t$ is equal to

$$s=\int_{t_0}^{t}\mathrm{d}s = \int_{t_0}^{t}\sqrt{E\left(\frac{\mathrm{d}u}{\mathrm{d}t}\right)^2+2F\frac{\mathrm{d}u}{\mathrm{d}t}\frac{\mathrm{d}v}{\mathrm{d}t}+G\left(\frac{\mathrm{d}v}{\mathrm{d}t}\right)^2}\,\mathrm{d}t\,.$$

**Surface area:** The piece of the surface described by allowing the parameters $u,v$ to vary in a domain $D$ of the $u,v$-plane has the surface area

$$\iint_D \sqrt{EG-F^2}\,\mathrm{d}u\mathrm{d}v.$$

**Angle between two curves on the surface:**
If $\mathbf{r}=\mathbf{r}(u_1(t),v_1(t))$ and $\mathbf{r}=\mathbf{r}(u_2(t),v_2(t))$ are two curves on the surface $\mathbf{r}=\mathbf{r}(u,v)$ which intersect in a point $P$, then the angle of intersection $\alpha$ (the angle taken in the positive tangent direction at the point $P$) is determined as

$$\cos\alpha = \frac{E\dot u_1\dot u_2+F(\dot u_1\dot v_2+\dot v_1\dot u_2)+G\dot v_1\dot v_2}{\sqrt{E\dot u_1^2+2F\dot u_1\dot v_1+G\dot v_1^2}\,\sqrt{E\dot u_2^2+2F\dot u_2\dot v_2+G\dot v_2^2}}\,.$$

Here $\dot{u}_1$ and $\dot{u}_2$, respectively, are the first derivatives of $u_1(t)$ and $u_2(t)$, respectively, in the parameter values corresponding to $P$, etc.

**Maps between two surfaces:** Suppose we are given two surfaces

$$\mathscr{F}_1 : \mathbf{r} = \mathbf{r}_1(u,v) \qquad \text{and} \qquad \mathscr{F}_2 : \mathbf{r} = \mathbf{r}_2(u,v)$$

both of which are given with respect to the same parameters $u$ and $v$ (perhaps after reparameterizations). If one associates to the point $P_1$ of $\mathscr{F}_1$ with the radius vector $\mathbf{r}_1(u,v)$ the point $P_2$ of $\mathscr{F}_2$ with the radius vector $\mathbf{r}_2(u,v)$, then one gets a bijective map $\varphi : \mathscr{F}_1 \longrightarrow \mathscr{F}_2$ between these two surfaces.

(i) $\varphi$ is called *distance-preserving*, if the length of an arbitrary curve segment is preserved under $\varphi$.

(ii) $\varphi$ is said to be *angle-preserving* (or *conformal*), if the angle between two arbitrary curves which intersect is preserved under $\varphi$.

(iii) $\varphi$ is called *area-preserving*, if the area of an arbitrary piece of surface is preserved under $\varphi$.

In Table 3.9 the quantities $E_j, F_j, G_j$ are the coefficients of the first fundamental form of $\mathscr{F}_j$, taken with respect to the parameters $u$ and $v$. The conditions listed in Table 3.9 must be satisfied at every point of the surface.

**Theorem:** (i) Every length-preserving map is conformal and area-preserving.

(ii) Every area-preserving map and every conformal map is length-preserving.

(iii) A length-preserving map preserves the Gaussian curvature $K$ at every point (cf. 3.6.3.3).

*Table 3.9. The algebraic conditions for geometric properties of surfaces.*

| ***Property*** | ***Necessary and sufficient conditions on*** $\mathbf{E_j}$, $\mathbf{F_j}$, $\mathbf{G_j}$ | | | |
|---|---|---|---|---|
| distance-preserving | $E_1 = E_2,$ | $F_1 = F_2,$ | $G_1 = G_2$ | |
| angle-preserving (conformal) | $E_1 = \lambda E_2,$ | $F_1 = \lambda F_2,$ | $G_1 = \lambda G_2,$ | $\lambda(u,v) > 0$ |
| volume-preserving | $E_1 G_1 - F_1^2 = E_2 G_2 - F_2^2$ | | | |

**The metric tensor** $g_{\alpha\beta}$**:** If one sets $u^1 = u$, $u^2 = v$, $g_{11} = E$, $g_{12} = g_{21} = F$ and $g_{22} = G$, then one can write, under utilization of the Einstein summation convention, the metric as

$$\boxed{\mathrm{d}s^2 = g_{\alpha\beta}\mathrm{d}u^\alpha \mathrm{d}u^\beta\,.}$$

By passing to another set of coordinates $u'^\alpha$ on the surface, one has $\mathrm{d}s^2 = g'_{\alpha\beta}\mathrm{d}u'^\alpha \mathrm{d}u'^\beta$ with

$$g'_{\alpha\beta} = \frac{\partial u^\gamma}{\partial u'^\alpha}\frac{\partial u^\delta}{\partial u'^\beta}\, g_{\gamma\delta}\,.$$

Hence the $g_{\alpha\beta}$ are manifestly the coordinates of a two-covariant tensor field (*metric tensor*).

Moreover, one sets

$$g = \det g_{\alpha\beta} = EG - F^2$$

and

$$g^{11} = \frac{G}{g}, \qquad g^{12} = g^{21} = \frac{-F}{g}, \qquad g^{22} = \frac{E}{g}.$$

One has $g^{\alpha\beta}g_{\beta\gamma} = \delta^\alpha_\gamma$. By transition from the $u^\alpha$-coordinate system to coordinates $u'^\alpha$, $g^{\alpha\beta}$ and $g$ transform as

$$g'^{\alpha\beta} = \frac{\partial u'^\alpha}{\partial u^\gamma}\frac{\partial u'^\beta}{\partial u^\delta}\, g^{\gamma\delta}$$

(a two-contravariant tensor field) and

$$g' = \left(\frac{\partial(u^1, u^2)}{\partial(u'^1, u'^2)}\right)^2 g$$

with the functional determinant

$$\frac{\partial(u^1, u^2)}{\partial(u'^1, u'^2)} = \frac{\partial u^1}{\partial u'^1}\frac{\partial u^2}{\partial u'^2} - \frac{\partial u^2}{\partial u'^1}\frac{\partial u^1}{\partial u'^2}\,.$$

In order to give a curved coordinate system an *orientation* $\eta = \pm 1$, one fixes a coordinate system $u_0^\alpha$ and declares it as positive ($\eta = +1$), and declares that $\eta$ is given by the sign of the functional determinant $\eta = \operatorname{sgn}\dfrac{\partial(u^1, u^2)}{\partial(u_0^1, u_0^2)}$. If we set $\varepsilon^{11} = \varepsilon^{22} = 0$ and $\varepsilon^{12} = -\varepsilon^{21} = 1$, then the following *Levi–Civita tensors*

$$E^{\alpha\beta} := \frac{\eta}{\sqrt{g}}\,\varepsilon^{\alpha\beta} \qquad \text{resp.} \qquad E_{\alpha\beta} := \eta\sqrt{g}\,\varepsilon^{\alpha\beta}$$

are just the same as $g^{\alpha\beta}$ and $g_{\alpha\beta}$, respectively.

#### 3.6.3.2 The second Gaussian fundamental form and the curvature properties of the surface

**Second fundamental form:** According to Gauss, for a surface $\mathbf{r} = \mathbf{r}(u, v)$, this form is given by the relation

$$\boxed{-\mathrm{d}\mathbf{N}\mathrm{d}\mathbf{r} = L\mathrm{d}u^2 + 2M\mathrm{d}u\mathrm{d}v + N\mathrm{d}v^2}$$

with

$$L = \mathbf{r}_{uu}\mathbf{N} = \frac{l}{\sqrt{EG - F^2}}\,, \qquad N = \mathbf{r}_{vv}\mathbf{N} = \frac{n}{\sqrt{EG - F^2}}\,, \qquad M = \frac{m}{\sqrt{EG - F^2}}$$

and

$$l := \begin{vmatrix} x_{uu} & y_{uu} & z_{uu} \\ x_u & y_u & z_u \\ x_v & y_v & z_v \end{vmatrix}, \qquad n := \begin{vmatrix} x_{vv} & y_{vv} & z_{vv} \\ x_u & y_u & z_u \\ x_v & y_v & z_v \end{vmatrix}, \qquad m := \begin{vmatrix} x_{uv} & y_{uv} & z_{uv} \\ x_u & y_u & z_u \\ x_v & y_v & z_v \end{vmatrix}.$$

If one sets $u^1 = u,\ u^2 = v$ and $b_{11} = L,\ b_{12} = b_{21} = M,\ b_{22} = N$, then one can write

$$-\mathrm{d}\mathbf{N}\mathrm{d}\mathbf{r} = b_{\alpha\beta}\mathrm{d}u^\alpha\mathrm{d}u^\beta.$$

Under a coordinate transformation from the $u^\alpha$-coordinates to new coordinates $u'^\alpha$, one has $-\mathrm{d}\mathbf{N}\mathrm{d}\mathbf{r} = b'_{\alpha\beta}\mathrm{d}u'^\alpha\mathrm{d}u'^\beta$ with

$$b'_{\alpha\beta} = \varepsilon\,\frac{\partial u^\gamma}{\partial u'^\alpha}\frac{\partial u^\delta}{\partial u'^\beta}\, b_{\gamma\delta}\,,$$

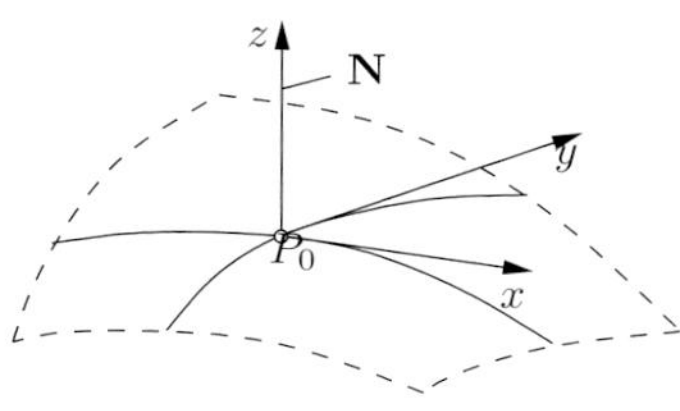


Figure 3.56.

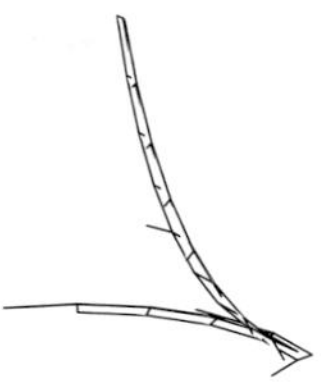

Figure 3.57.

where $\varepsilon$ is the sign of the functional determinant: $\varepsilon = \operatorname{sgn} \dfrac{\partial(u'^1, u'^2)}{\partial(u^1, u^2)}$. Hence the $b_{\alpha\beta}$ are the coordinates of a two-covariant pseudo-tensor. One further sets $b := \det b_{\alpha\beta} + LN - M^2$. The quantity $b$ transforms according to the same law as $g$.

| The second fundamental form encodes the curvature properties of the surface. |
|---|

**The canonical Cartesian coordinate system at a point $P_0$ on the surface:** Near a given point $P_0$ one can always choose a Cartesian coordinate $x, y, z$-system whose origin is at $P_0$ and whose $x, y$-plane is the tangent plane of the surface at the point $P_0$ (Figure 3.56). In this $x, y, z$-system the surface can be, near to $P_0$, described by $z = z(x, y)$ with $z(0,0) = z_x(0,0) = z_y(0,0) = 0$. This corresponding three-frame at the point $P_0$ consists of the three unit vectors $\mathbf{i}, \mathbf{j}$ and $\mathbf{N} = \mathbf{i} \times \mathbf{j}$. The Taylor expansion in a neighborhood of $P_0$ is:

$$z = \frac{1}{2}\, z_{xx}(0,0)x^2 + z_{xy}(0,0)xy + \frac{1}{2}\, z_{yy}(0,0)y^2 + \ldots .$$

By making an additional rotation of the Cartesian coordinate system around the $z$-axis one can furthermore achieve the situation in which

$$\boxed{z = \frac{1}{2}(k_1 x^2 + k_2 y^2) + \ldots .}$$

This $x, y$-system is called the *canonical Cartesian coordinate system* of the surface at the point $P_0$. One calls the $x$-axis and the $y$-axis the *principal curvature directions* and $k_1, k_2$ the *principal curvatures* of the surface at the point $P_0$.

Moreover $R_1 := 1/k_1$ and $R_2 := 1/k_2$ are called the *principal curvature radii*.

**The Gaussian curvature $K$ at a point $P_0$:** We define

$$\boxed{K := k_1 k_2 .}$$

This is the fundamental curvature quantity of a surface.

*Example:* For a sphere of radius $R$ one has $R_1 = R_2 = R$ and $K = 1/R^2$.

Surfaces with $K =$ const are called surfaces of *constant Gaussian curvature*. Examples of this are:

(a) a sphere, for which $K > 0$, and

(b) a pseudo-sphere, for which $K < 0$; such a surface can be formed by rotation of a tractrix around the $z$-axis as in Figure 3.57.

**The mean curvature $H$ at a point $P_0$:** We set

$$\boxed{H := \frac{1}{2}(k_1 + k_2)}$$

and refer to $H$ as the *mean curvature* of the surface. Surfaces with $H \equiv 0$ are called *minimal surfaces* (cf. [212]).

If we change coordinates by the transformation $x \mapsto y,\ y \mapsto x,\ z \mapsto -z$, then the principal curvatures are transformed as $k_1 \mapsto -k_2$ and $k_2 \mapsto -k_1$. From this it follows that $K \mapsto K$ and $H \mapsto -H$. This means that $K$ is a genuinely geometrical quantity, while this is not true for $H$ but only for $|H|$.

Table 3.10 gives the geometric interpretation of the different signs of $K$.

*Table 3.10. Possible values for surface curvatures.*

| ***Type of point $P_0$*** | ***Analytic definition*** | ***Surface behavior near $P_0$ to second order*** |
|---|---|---|
| elliptic point | $K = k_1k_2 > 0$ (i.e. $LN - M^2 > 0$) | ellipsoid |
| umbilical point | $K = k_1k_2 > 0, \quad k_1 = k_2$ | sphere |
| hyperbolic point | $K = k_1k_2 < 0$ (i.e. $LN - M^2 < 0$) | single-sheeted hyperboloid |
| parabolic point | $K = k_1k_2 = 0$ (i.e. $LN - M^2 = 0$) | |
| | (a) $k_1^2 + k_2^2 \neq 0$ | cylinder |
| | (b) $k_1 = k_2 = 0$ | plane |

**Theorem:** Suppose a surface is given in the parameterization $\mathbf{r} = \mathbf{r}(u, v)$.

(i) One has

$$\boxed{K = \frac{LN - M^2}{EG - F^2}, \qquad H = \frac{LG - 2FM + EN}{2(EG - F^2)}\,.}$$

(ii) The principal curvatures $k_1$ and $k_2$ are solutions of the quadratic equation

$$k^2 - 2Hk + K = 0.$$

(iii) If $\mathbf{e}_1, \mathbf{e}_2$ and $\mathbf{N}$ denote the three-frame on the surface, the principal curvature directions are of the form $\lambda_1\mathbf{e}_1 + \mu\mathbf{e}_2$, where $\lambda$ and $\mu$ are solutions of the equation

$$\lambda^2(FN - GM) + \lambda\mu(EN - GL) + \mu^2(EM - FL) = 0. \tag{3.29}$$

*Sketch of proof:* In the canonical Cartesian coordinate system one has for the first and second fundamental forms the very simple expressions

$$\mathrm{d}s^2 = \mathrm{d}x^2 + \mathrm{d}y^2, \qquad -\mathrm{d}\mathbf{N}\mathrm{d}\mathbf{r} = k_1\mathrm{d}x^2 + k_2\mathrm{d}y^2.$$

From this one gets

$$K = \frac{b}{g}, \qquad H = \frac{1}{2}g^{\alpha\beta}b_{\alpha\beta}.$$

It then follows from tensor calculus hat these expressions are valid in an arbitrary $u^\alpha$-coordinate system, since $K$ is a scalar and $H$ is a pseudo-scalar. Equation (3.29) corresponds to the result $E^{\alpha\beta}g_{\alpha\sigma}b_{\beta\mu}\lambda^\sigma\lambda^\mu = 0$.

**Developable ruled surfaces:** A surface is called a *ruled surface*, if it can be generated by the motion of lines in space (for example a cone, cylinder, hyperboloid or hyperbolic paraboloid). If the ruled surface can in fact be 'peeled off' to a plane, then one calls the surface *developable* (for example a cone or a cylinder). In all points of a developable surface one has $K = 0$, hence $LN - M^2 = 0$.

**Surface sections:** Let $\mathbf{e}_1, \mathbf{e}_2, \mathbf{N}$ be the accompanying three-frame of the surface $\mathbf{r} = \mathbf{r}(u, v)$ at the point $P_0$. We cut the surface with a plane $E$, which passes through the point $P_0$ and the line which is generated by $\lambda\mathbf{e}_1 + \mu\mathbf{e}_2$. Moreover $E$ and the normal vector $\mathbf{N}$ form an angle we denote by $\gamma$. Then the curvature $k$ of the curve of intersection at the point $P_0$ is given by

$$\boxed{k = \frac{k_N}{\cos\gamma}}$$

with

$$k_N = \frac{L\lambda^2 + 2M\lambda\mu + N\mu^2}{E\lambda^2 + 2F\lambda\mu + G\mu^2}\,.$$

**Curvature of a curve on a surface:** A curve on a surface which passes through a point $P_0$ of the surface has the same curvature $k$ as the curve of intersection of the surface with the contact plane at the point $P_0$. If the contact plane forms the angle $\gamma$ with $\mathbf{N}$ and the angle $\alpha$ with the principal curvature direction corresponding to $k_1$, then one has

$$\boxed{k = \frac{k_1\cos^2\alpha + k_2\sin^2\alpha}{\cos\gamma}}$$

(Theorem of Euler–Meusnier).

#### 3.6.3.3 The main theorem of the theory of surfaces and Gauss' *theorema egregium*

**The equations of Gauss and Weingarten for derivatives:** A change in the accompanying three-frame is described by the so-called *derivative equations*:

$$\boxed{\begin{aligned} \frac{\partial\mathbf{e}_\alpha}{\partial u^\beta} &= \Gamma^\sigma_{\alpha\beta}\mathbf{e}_\sigma + b_{\alpha\beta}\mathbf{N} \quad &\text{(Gauss)},\\ \frac{\partial\mathbf{N}}{\partial u^\alpha} &= -g^{\sigma\gamma}b_{\gamma\alpha}\mathbf{e}_\sigma \quad &\text{(Weingarten)}.\end{aligned}}$$

All indices here run from 1 to 2. Indices which appear both in subscripts and in superscripts are summed over. The *Christoffel symbols* are given by the formula

$$\Gamma^\sigma_{\alpha\beta} := \frac{1}{2}g^{\sigma\delta}\left(\frac{\partial g_{\alpha\delta}}{\partial u^\beta} + \frac{\partial g_{\beta\delta}}{\partial u^\alpha} - \frac{\partial g_{\alpha\beta}}{\partial u^\delta}\right).$$

These symbols do not represent tensors (they transform incorrectly under coordinate transformations). The derivative equations form a system of 18 partial differential equations of the first order for the three components of each of $\mathbf{e}_1, \mathbf{e}_2$ and $\mathbf{N}$; these equations can be solved with the aid of theorem of Frobenius (cf. 1.13.5.3).

**Integrability condition:** From $\dfrac{\partial^2 \mathbf{e}_\alpha}{\partial u^\beta \partial u^\gamma} = \dfrac{\partial^2 \mathbf{e}_\alpha}{\partial u^\gamma \partial u^\beta} = \dfrac{\partial^2 \mathbf{N}}{\partial u^\alpha \partial u^\beta} = \dfrac{\partial^2 \mathbf{N}}{\partial u^\beta \partial u^\alpha}$ one gets the so-called *integrability conditions*

$$\begin{aligned} &\frac{\partial b_{11}}{\partial u^2} - \frac{\partial b_{12}}{\partial u^1} - \Gamma^1_{12} b_{11} + (\Gamma^1_{11} - \Gamma^2_{12}) b_{12} + \Gamma^2_{11} b_{22} = 0, \\ &\frac{\partial b_{12}}{\partial u^2} - \frac{\partial b_{22}}{\partial u^1} - \Gamma^1_{22} b_{11} + (\Gamma^1_{12} - \Gamma^2_{22}) b_{12} + \Gamma^2_{12} b_{22} = 0 \end{aligned} \tag{3.30}$$

(*equation of Mainardi–Codazzi*) and

$$\boxed{K = \frac{R_{1212}}{g}} \tag{3.31}$$

(Gauss' *theorema egregium*).

Here $R_{\alpha\beta\gamma\delta} = R^{\cdot\cdot\cdot\nu}_{\alpha\beta,\gamma\cdot} g_{\nu\delta}$ is the *Riemannian curvature tensor* with

$$R^{\cdot\cdot\cdot\nu}_{\alpha\beta,\gamma\cdot} = \frac{\partial \Gamma^\nu_{\alpha\gamma}}{\partial x^\beta} + \Gamma^\nu_{\beta\sigma}\Gamma^\sigma_{\alpha\gamma} - \frac{\partial \Gamma^\nu_{\beta\gamma}}{\partial x^\alpha} - \Gamma^\nu_{\alpha\sigma}\Gamma^\sigma_{\beta\gamma}\,.$$

**Main theorem:** If one is given functions

$$g_{11}(u^1, u^2) \equiv E(u, v), \quad g_{12}(u^1, u^2) = g_{21}(u^1, u^2) \equiv F(u, v), \quad g_{22}(u^1, u^2) \equiv G(u, v)$$

(which are assumed to be twice continuously differentiable) and

$$b_{11}(u^1, u^2) \equiv L(u, v), \quad b_{12}(u^1, u^2) = b_{21}(u^1, u^2) \equiv M(u, v), \quad b_{22}(u^1, u^2) \equiv N(u, v)$$

(which are assumed to be continuously differentiable), which moreover satisfy the integrability conditions (3.30) and (3.31) and which for arbitrary real numbers $\lambda, \mu$ with $\lambda^2 + \mu^2 \neq 0$ satisfy $E\lambda^2 + 2F\lambda\mu + G\mu^2 > 0$, there is a surface $\mathbf{r} = \mathbf{r}(u, v)$ which is three times continuously differentiable and whose coefficients in the first and second fundamental form are the given functions. This surface is unique up to translations and rotations in space.

The construction of the surface is as follows. 1. From the derivative formulas of Gauss and Weingarten one gets uniquely the accompanying three-frame $\mathbf{e}_1, \mathbf{e}_2, \mathbf{N}$, if the values of these are given in a fixed point $P_0(u_0, v_0)$. 2. From $\partial \mathbf{r}/\partial u^\alpha = \mathbf{e}_\alpha$ one can calculate $\mathbf{r}(u, v)$; $\mathbf{r}(u, v)$ is uniquely determined if one requires that the surface passes through the point $P_0$.

**The fundamental theorema egregium:** The Gaussian curvature $K$ of a surface was introduced in section 3.6.3.2 with the aid of the ambient space. According to (3.29) however, $K$ depends only on the metric tensor $g_{\alpha\beta}$ and its derivatives, hence only on the first fundamental form.

> The Gaussian curvature $K$ can be determined by measurements on the surface alone.

Hence the curvature $K$ is an *intrinsic property* of the surface, i.e., it does not depend on the ambient space. This is the starting point for the theory of curvature of manifolds. In the general theory of relativity for example the curvature of four-dimensional space-time is responsible for the gravitational force.

It was a long struggle for Gauss to complete the proof of the theorema egregium (which means the 'exquisite theorem'). It is the culmination of his theory of surfaces.

*Example:* For a distance-preserving map the first fundamental form and hence also the Gaussian curvature is preserved. For the sphere (resp. the plane) one has $K = 1/R^2$ (resp. $K = 0$). Hence the sphere cannot be mapped in a distance-preserving manner onto the plane. Consequently there is no distance-preserving map of the surface of the earth.

However, one can make angle-preserving maps. This is of great importance for navigation.

#### 3.6.3.4 Geodesic lines

**Geodesic lines:** A curve on a surface is called a *geodesic line*, if at every point of the curve, the principal normal to the curve and the principal normal to the surface are either parallel or anti-parallel. The shortest segment between two points on a surface is always part of a geodesic line. In the plane these geodesic lines are just the lines in the usual sense. On the sphere, the geodesic lines are the great circles (for example longitudes and the equator. A geodesic line $\mathbf{r} = \mathbf{r}(u^1(s), u^2(s))$ (where $s$ denotes arc length) satisfies the differential equation

$$\boxed{\frac{\mathrm{d}^2 u^\alpha}{\mathrm{d}s^2} + \Gamma^\alpha_{\beta\gamma}\frac{\mathrm{d}u^\beta}{\mathrm{d}s}\frac{\mathrm{d}u^\gamma}{\mathrm{d}s} = 0, \qquad \alpha = 1, 2}$$

and conversely, solutions of these equations are always geodesic lines.

If the surface is given in the form $z = z(x, y)$, then the differential equation for geodesic lines $z = z(x, y(x))$ are:

$$(1 + z_x^2 + z_y^2)y'' = z_x z_{yy}(y')^3 + (2z_x z_{xy} - z_y z_{yy})(y')^2 + (z_x z_{xx} - 2z_y z_{xy})y' - z_y z_{xx}.$$

**Geodesic curvature:** For a curve on a surface given by $\mathbf{r} = \mathbf{r}(u(s), v(s))$ ($s$ arc length) there is always a decomposition of the form

$$\boxed{\frac{\mathrm{d}^2\mathbf{r}}{\mathrm{d}s^2} = k_N\mathbf{N} + k_g\left(\mathbf{N} \times \frac{\mathrm{d}\mathbf{r}}{\mathrm{d}s}\right)}$$

with

$$k_N = \mathbf{N}\,\frac{\mathrm{d}^2\mathbf{r}}{\mathrm{d}s^2}, \qquad k_g = \left(\mathbf{N} \times \frac{\mathrm{d}\mathbf{r}}{\mathrm{d}s}\right)\frac{\mathrm{d}^2\mathbf{r}}{\mathrm{d}s^2}\,.$$

The number $k_g$ is called the *geodesic curvature*. A curve on a surface is a geodesic line if and only if one has $k_g \equiv 0$.

### 3.6.4 Gauss' global theory of surfaces

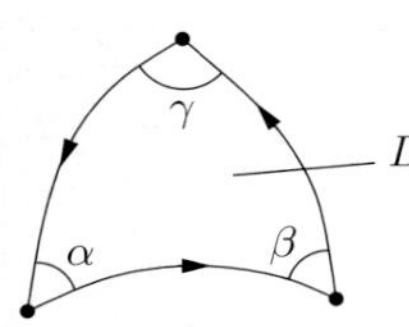


Figure 3.58. A triangle on a surface.

**The theorem of Gauss (1827) on the sum of angles in a triangle:** Let $D$ be a geodesic triangle on a surface with angles $\alpha, \beta$ and $\gamma$, i.e., the sides of this triangle are geodesic lines. Then one has

$$\boxed{\int_D K \, \mathrm{d}F = \alpha + \beta + \gamma - \pi} \tag{3.32}$$

where $\mathrm{d}F$ denotes the surface measure (cf. Figure 3.58).

*Example:* For the unit sphere one has $K = 1$, i.e., $\int_D K\mathrm{d}F$ is equal to the surface area of the triangle.

**The total curvature:** For an arbitrary sphere one has

$$\boxed{\int_F K \, \mathrm{d}F = 4\pi.} \tag{3.33}$$

This relation remains valid for every closed smooth surface $F$ which is diffeomorph to a sphere. In this fundamental result differential geometry and topology meet. The connection with the Euler characteristic and the theory of characteristic classes is explained in [212].

If the surface $F$ is a torus or diffeomorphic to a torus, then one has to replace the right-hand side of (3.33) by '$= 0$'.

**The theorem of Bonnet (1848):** If the sides of the triangle in Figure 3.58 are arbitrary curves with arc length $s$, then one must replace (3.32) by the relation

$$\boxed{\int_D K \, \mathrm{d}F + \int_{\partial D} k_g \, \mathrm{d}s = \alpha + \beta + \gamma - \pi}$$

where the boundary curve $\partial D$ is transversed in a mathematically positive sense.

## 3.7 Examples of plane curves

### 3.7.1 Envelopes and caustics

We consider a family of curves depending on a real parameter $c$:

$$\boxed{F(x, y, c) = 0.}$$

The *envelope* of this family is obtained by eliminating $c$ from the equations

$$F_c(x, y, c) = 0, \qquad F(x, y, c) = 0.$$

*Example:* For the family of curves $(x - c)^2 + y^2 - 1 = 0$ we get $x - c = 0$ and hence

$$y^2 = 1.$$

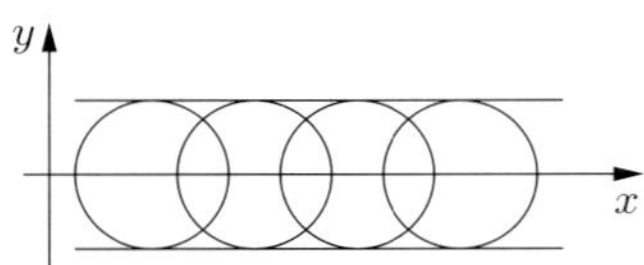


*Figure 3.59. An envelope.*

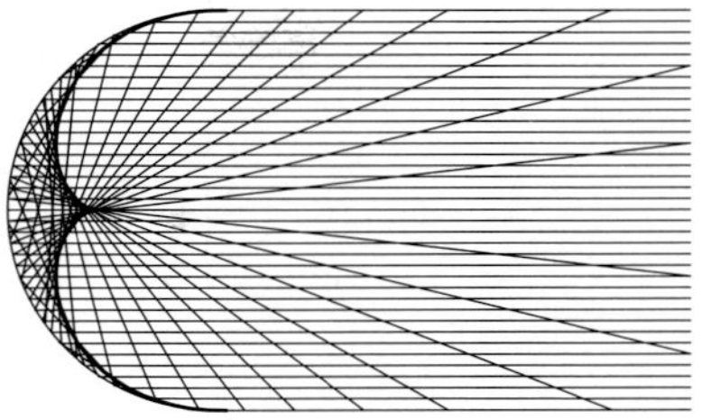

*Figure 3.60. Caustic of the circle reflecting parallel light rays.*

The solution set consists of the two lines $y = \pm 1$ (Figure 3.59).

**Caustics:** Figure 3.60 shows a circular mirror, which reflects parallel light rays falling into it. The envelope of the reflected light rays is called a *caustic*. The appearance of caustics typically causes great difficulties in geometrical optics and, more generally, in the calculus of variations.

Caustics were already known in ancient Greece.

### 3.7.2 Evolutes

**Definition:** The geometric locus of all centers of curvature of a given curve $C$ is called the *evolute* $E$ of $C$.

If the curve $C$ is given in the form $y = f(x)$, then one obtains the equation for the evolute in the parameter form:

$$\boxed{x = t - \frac{f'(t)\big(1 + f'(t)^2\big)}{f''(t)}\,, \qquad y = f(t) + \frac{1 + f'(t)^2}{f''(t)}}$$

(cf. Table 3.7).

**Theorem:** The normal to the curve $C$ in a point is equal to the tangent to the evolute $E$ at the corresponding center of curvature (cf. the segment $PQ$ in Figure 3.64).

*Example 1:* For the parabola $C:\ y = \frac{1}{2}x^2$ we get

$$x = -t^3, \qquad y = 1 + \frac{3}{2}\,t^2\,.$$

By eliminating $t$ one gets the semicubical parabola

$$y = 1 + \frac{3}{2}\,x^{2/3}$$

as the evolute of the parabola (Figure 3.61).

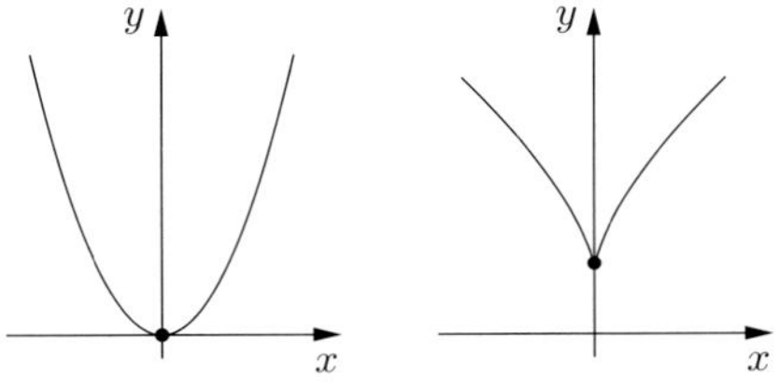

**(a)** parabola **(b)** evolute

*Figure 3.61. The evolute of a parabola.*

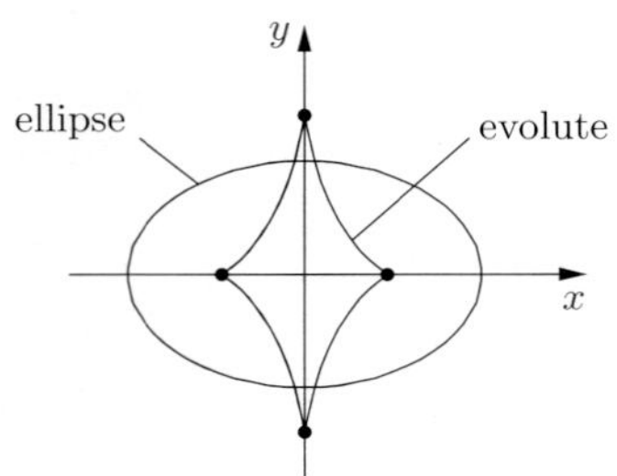


*Figure 3.62. The evolute of an ellipse.*

*Example 2:* The evolute of the ellipse

$$\frac{x^2}{a^2} + \frac{y^2}{b^2} = 1$$

is the astroid

$$\left(\frac{ax}{c^2}\right)^2 + \left(\frac{by}{c^2}\right)^2 = 1$$

with $c^2 = a^2 - b^2$ (cf. Figure 3.62).

### 3.7.3 Involutes

**Definition:** If a curve $E$ is given, then one gets the *involute* $C$ of $E$ by wrapping (or unwrapping) a chord of constant length (Figure 3.63).

**Theorem:** If $E$ is the evolute of $C$, then $C$ is the involute of $E$.

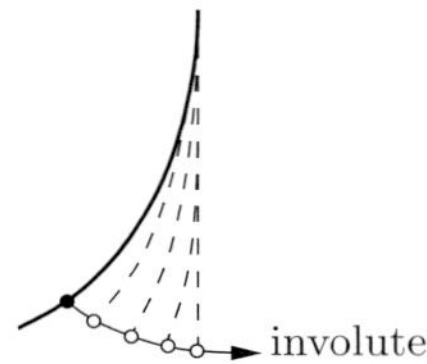


*Figure 3.63.*

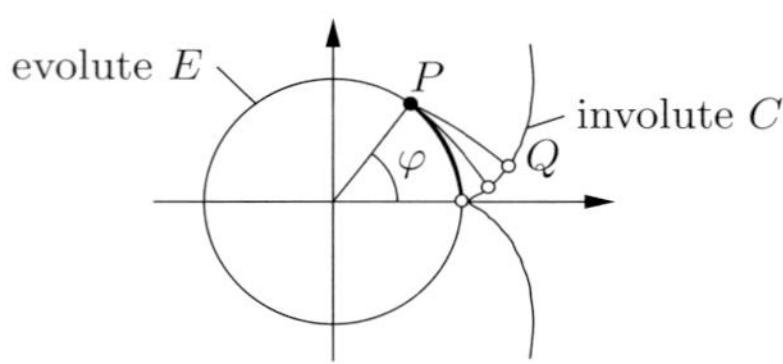


*Figure 3.64.*

*Example:* The involute of the circle

$$E: \ x^2 + y^2 = R^2 \tag{3.34}$$

is

$$C: \ x = R(\cos\varphi + \varphi\,\sin\varphi), \qquad y = R(\sin\varphi - \varphi\,\cos\varphi) \tag{3.35}$$

(Figure 3.64). More precisely, one has the following situation: the circle (3.34) is the evolute of the curve (3.35). If we consider in Figure 3.64 the tangent $PQ$ to $E$, then one gets by unwrapping this segment an arc of the curve $C$.

### 3.7.4 Huygens' tractrix and the catenary curve

**The tractrix** (Figure 3.65(a)): This curve is given by the equation

$$\boxed{x = \pm\left(a \operatorname{arcosh}\frac{a}{y} - \sqrt{a^2 - y^2}\right).}$$

*Geometric characterization:* If we consider the tangent to the curve through a point $K$, which intersects the $x$-axis in a point $H$, then the length of the segment $KH$ is a constant.

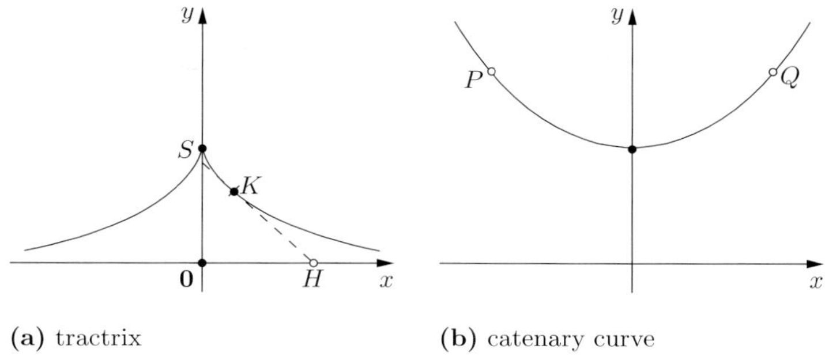


(a) tractrix (b) catenary curve

*Figure 3.65. Two curves involving the hyperbolic trigonometric functions.*

The name tractrix comes from the fact that the curve arises when an ox at the point $H$ pulls a cart at the point $K$ in the direction of the $x$-axis. This is a situation which earlier often actually occurred in mining.

The $x$-axis is the asymptote of the tractrix. If $s(y) := SK$ denotes the length of the arc from the cusp of the curve to the cart at $K$, and $x(y) = OH$ denotes the distance of the ox from the origin, then one has

$$\lim_{y \to 0} |s(y) - x(y)| = a(1 - \ln 2).$$

This value is approximately $0.3069 \cdot a$.

By rotating the tractrix around the $x$-axis one gets a surface of constant negative curvature, which is known as the pseudosphere (Figure 3.57).

**The catenary curve** (Figure 3.65(b)): This curve has the equation

$$\boxed{y = a \cosh \frac{x}{a}\,.}$$

*Physical characterization:* This curve corresponds in form to a clothesline or chain, which is what gives rise to the name: catenary means chain line in Latin; the line is hung at the points $P$ and $Q$ and sags under the force of gravity.

The length of the arc between $P$ and $Q$ is equal to $2a \cdot \sinh(x/a)$.

*Geometric characterization:* The catenary curve is the evolute of the tractrix.

By rotating the catenary curve around the $x$-axis, one obtains a surface whose mean curvature vanishes (a minimal surface), which is known as the *catenoid*.

### 3.7.5 The lemniscate of Jakob Bernoulli and Cassini's oval

**The equation of the lemniscate in Cartesian coordinates** (Figure 3.66(a)):

$$\boxed{\left(x^2 + y^2\right)^2 - 2a^2\left(x^2 - y^2\right) = 0.}$$

The constant $a$ appearing in this equation is assumed to be positive. This curve was first described by Jakob Bernoulli in in his *Acta eruditorum*, which appeared in 1694.

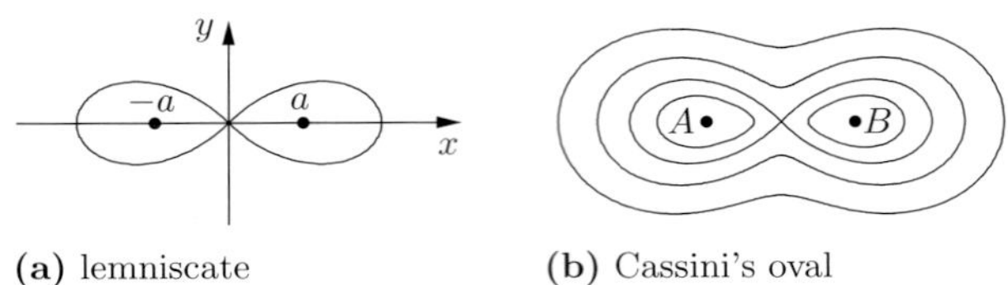


**(a)** lemniscate **(b)** Cassini's oval

*Figure 3.66. The lemniscate, which is a special case of Cassini's oval.*

*Geometric characterization:* The lemniscate is the geometric locus of all points for which the product of the distances from two points $A = (a, 0)$ and $B = (-a, 0)$ is equal to $a^2$.

The two lines $y = \pm x$ are tangent to the lemniscate at the origin.

*Total area:* $2a^2$.

*Length of the curve:* For $a = 1$, the lemniscate has the total length

$$L = 2\int_0^1 \frac{\mathrm{d}x}{\sqrt{1-x^4}} .$$

This is an elliptic integral. The 22-year-old Gauss discovered in 1799 the formula

$$\boxed{\frac{\pi}{L} = M(1, \sqrt{2}).} \tag{3.36}$$

Here the quantity $M(a_0, b_0)$ denotes the *arithmetic-geometric mean* of the two positive numbers $a_0$ and $b_0$, that is to say that

$$M(a_0, b_0) = \lim_{n\to\infty} a_n = \lim_{n\to\infty} b_n$$

with

$$a_{n+1} := \frac{a_n + b_n}{2}, \qquad b_{n+1} := \sqrt{a_n b_n} .$$

The Gaussian formula (3.36) is a special case of the more general formula

$$\boxed{K(k) := \int_0^{\pi/2} \frac{\mathrm{d}\varphi}{\sqrt{1-k^2\sin^2\varphi}} = \frac{\pi}{2M(1, \sqrt{1-k^2}\,)}} \tag{3.37}$$

for the complete elliptic integral $K(k)$ of the first kind with $0 < k < 1$.

**The equation of Cassini's oval in Cartesian coordinates** (Figure 3.66(b)):

$$\boxed{\left(x^2+y^2\right)^2 - 2a^2\left(x^2-y^2\right) + a^4 - c^4 = 0.}$$

Here the constants $a$ and $c$ are positive. This curve was described by the astronomer Jean–Dominique Cassini (1625–1712) in his *Eléments d'astronomie*, which appeared in 1749.

*Geometric characterization:* Cassini's oval is the geometric locus of all points for which the product of the distances to two fixed points $A = (a, 0)$ and $B = (-a, 0)$ is a constant, equal to $c^2$.

The lemniscate is for $a = c$ a special case of Cassini's oval, which was later realized in 1782 by Pietro Ferroni.

### 3.7.6 Lissajou figures

**Parameterization of Lissajou figures in Cartesian coordinates** (Figure 3.67):

$$x = a \sin \omega t, \qquad y = b \sin(\omega' t + \alpha).$$

If we interpret $t$ as the time, then these are oscillations, which were studied in 1815 by the American astronomer Nathaniel Bowditch[23] (1773–1838) and in 1850 by Lissajou. By varying the angular frequencies $\omega$ and $\omega'$ as well as a changing the phase by $\alpha$, one can generate with a laser a great variety of curves, something which is occasionally shown where laser performances are set up.

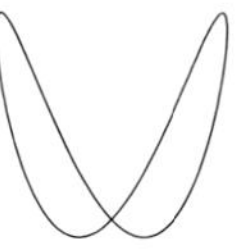

(a) $\dfrac{\omega}{\omega'} = \dfrac{1}{2}$

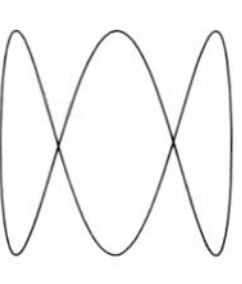

(b) $\dfrac{\omega}{\omega'} = \dfrac{1}{3}$

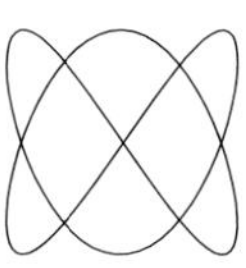

(c) $\dfrac{\omega}{\omega'} = \dfrac{2}{3}$

*Figure 3.67. Lissajou curves.*

### 3.7.7 Spirals

**Archimedean spirals** (Figure 3.68(a)): The equation of these curves in polar coordinates is:

$$r = a\varphi, \qquad \varphi > 0.$$

The constant $a$ which occurs here is assumed to be positive.

**Logarithmic spirals** (Figure 3.68(b)): These are given by the equation

$$r = a\,\mathrm{e}^{b\varphi}, \qquad -\infty < \varphi < \infty.$$

The constants $a$ and $b$ occuring are again assumed to be positive.

*Geometrical property:* Each ray issuing from the origin $\mathbf{0}$ cuts a logarithmic spiral in a constant angle $\alpha$, where $\cot \alpha = b$.

*Length of the arc determined by the condition $\beta \le \varphi \le \gamma$:* $L = (\gamma - \beta)\dfrac{\sqrt{1+b^2}}{b}$.

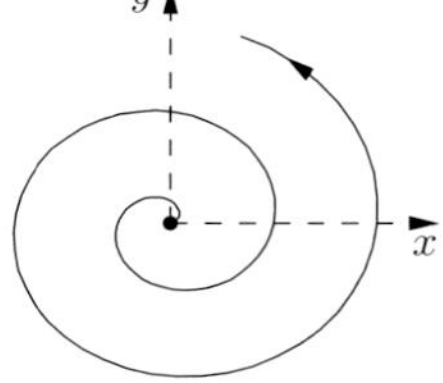


(a) Archimedean spiral

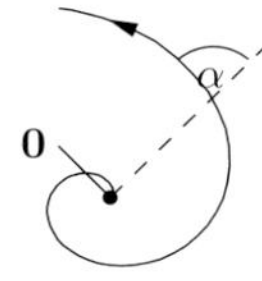


(b) logarithmic spiral

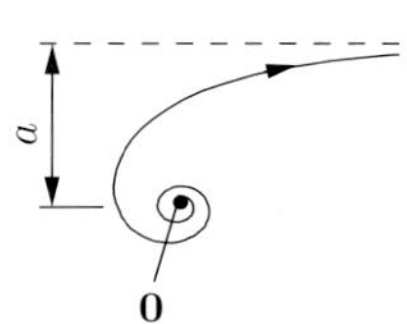


(c) hyperbolic spiral

*Figure 3.68. Various spiral curves.*

*Curvature radius:* $r\sqrt{1+b^2}$.

For $b = 0$ the curve is a circle of radius $a$.

[23]He also translated Laplace's *Méchanique Céleste* into English.

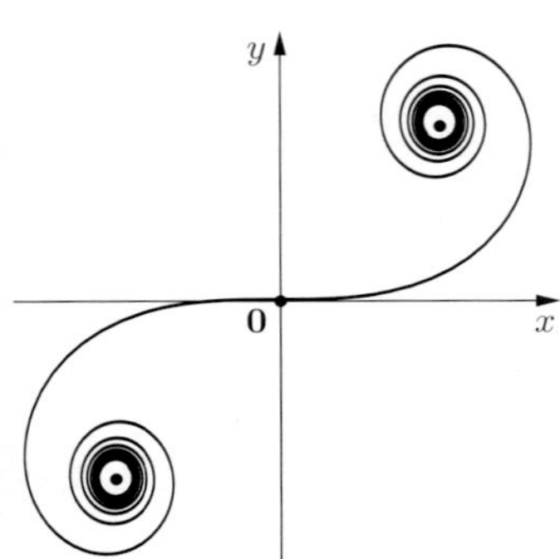


*Figure 3.69. Spider curve.*

**Hyperbolic spirals** (Figure 3.68(c)): These curves have the equation

$$r = \frac{a}{\varphi}, \qquad \varphi > 0.$$

The constant $a$ is again positive.

**Spider curves (clothoids)** (Figure 3.69):

$$R = \frac{a^2}{s}. \tag{3.38}$$

In this equation, $R$ is the curvature radius and $s$ represents the arc length between the point of the curve and the origin $\mathbf{0}$. The constant $a$ is positive.

If a curve is described by purely geometrical quantities as in (3.38), then one speaks of a *natural* curve equation.

*Parameterization in Cartesian coordinates:*

$$x = a\sqrt{\pi}\int_0^t \cos\frac{\pi u^2}{2}\,du, \qquad y = a\sqrt{\pi}\int_0^t \sin\frac{\pi u^2}{2}\,du, \qquad -\infty < t < \infty.$$

This curve is situated symmetrically with respect to the origin $\mathbf{0} = (0,0)$.

*Arc length:* $s = at\sqrt{\pi}$.

*Asymptotic points:* $A = \left(\frac{a\sqrt{\pi}}{2}, \frac{a\sqrt{\pi}}{2}\right), \quad B = \left(-\frac{a\sqrt{\pi}}{2}, -\frac{a\sqrt{\pi}}{2}\right).$

## 3.7.8 Ray curves (chonchoids)

**Definition:** If a curve $C$

$$r = f(\varphi)$$

is given in polar coordinates, then the *chonchoid* of $C$ is the curve with the equation

$$r = f(\varphi) + b$$

(or $r = f(\varphi) - b$). Here again $b$ is a positive constant.

### 3.7.8.1 The chonchoid of Nikomedes

*Equation in polar coordinates* (Figure 3.70):

$$r = \frac{a}{\cos\varphi} \pm b, \qquad -\frac{\pi}{2} < \varphi < \frac{\pi}{2}. \tag{3.39}$$

The two constants $a$ and $b$ are both assumed to be constant. For '+' (resp. '−') one gets the positive (resp. negative) branch of the curve. If one sets $b = 0$ in (3.39), then

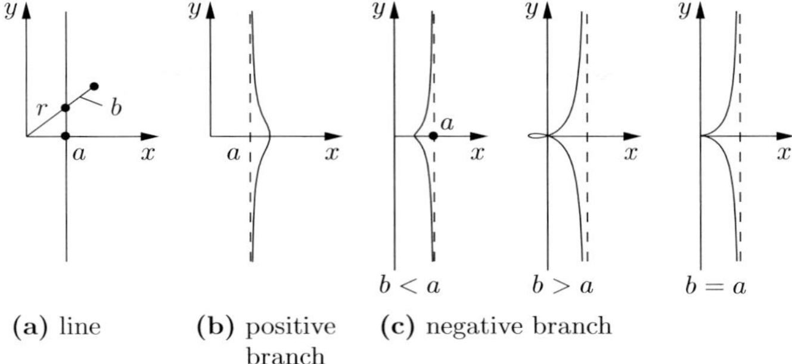


*Figure 3.70. Chonchoid of Nikomedes.*

one gets the equation of the line $x = a$ (Figure 3.70(a)). The chonchoid of Nikomedes is thus the chonchoid of a line.

This curve was invented around 180 BC, in an attempt to solve the Delphian problem of the doubling of the cube and the trisection problem graphically.

### 3.7.8.2 Pascal's snail and the cardioid

*Equation in polar coordinates* (Figure 3.71):

$$\boxed{r = a\,\cos\varphi + b, \qquad -\pi < \varphi \leq \pi.} \tag{3.40}$$

Here the constants $a$ and $b$ are positive. The equation of the circle $(x-a)^2 + y^2 = a^2$ is

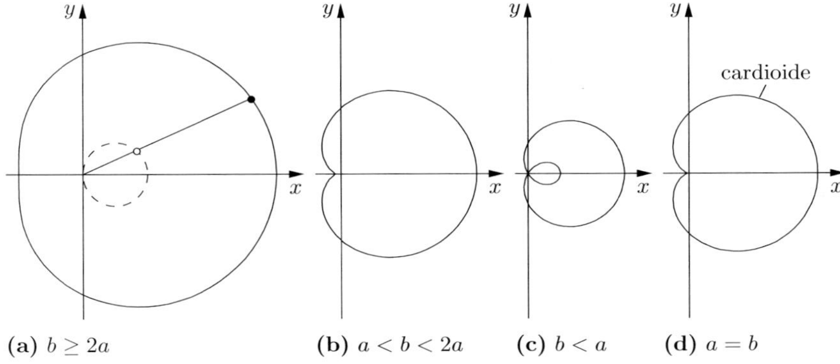


*Figure 3.71. Pascal's snail.*

in polar coordinates $r = a\cos\varphi$, where $-\dfrac{\pi}{2} < \varphi \leq \dfrac{\pi}{2}$. This is why one refers to (3.40) as the chonchoid of the circle.

*Enclosed surface area:*[24] $A = \dfrac{\pi a^2}{2} + \pi b^2$.

[24] In case $b < a$, the area enclosed by the inner cycle is counted twice as shown in Figure 3.71(c).

*Equation in Cartesian coordinates:* $(x^2+y^2-ax)^2 = b^2(x^2+y^2)$.

**The special case of the the cardioid** occurs for $a = b$. In this case we have

*Enclosed surface area:* $A = \dfrac{3\pi a^2}{2}$.

*Inflection point:* $x = 2a,\ y = 0$.

## 3.7.9 Wheel curves

Many curves can be generated by rotating a wheel along a given curve with a constant angular velocity and taking the trajectory of a fixed point $P$ on a spike of the wheel (for example on the periphery). This is illustrated in Figures 3.72 to 3.74. Curves of this kind, with which many mathematicians have dealt since the renaissance, are used often in all areas of technology.

### 3.7.9.1 Rolling a wheel along a line (cycloids)

*Parameterization in Cartesian coordinates* (Figure 3.72):

$$x = a(\varphi - \mu \sin\varphi), \qquad y = a(\varphi - \mu \cos\varphi), \qquad -\infty < \varphi < \infty.$$

Here $a$ is the radius, and $\varphi$ is the relative angle of the rolling wheel.

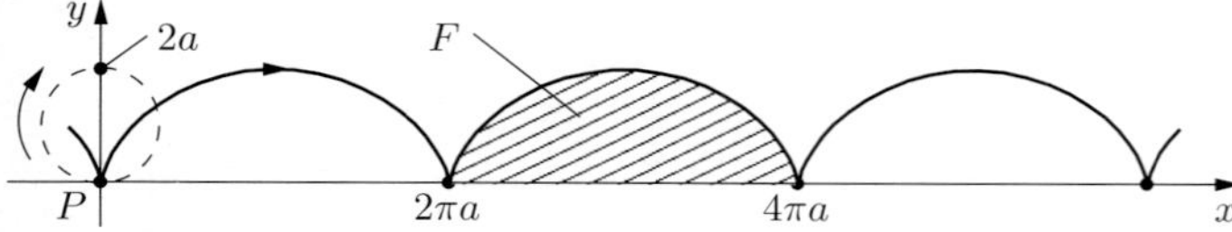


**(a)** cycloid

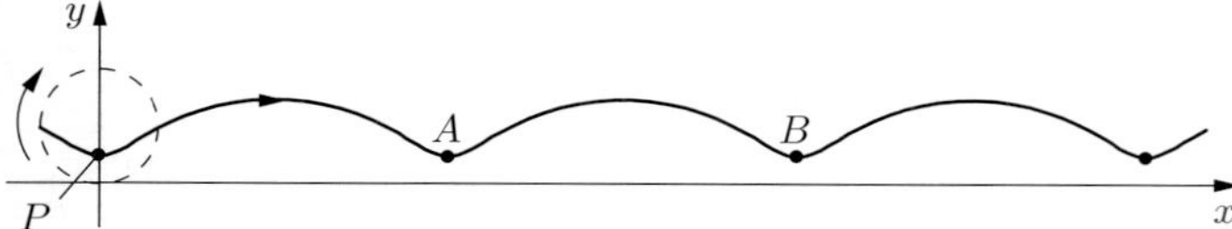


**(b)** shortened cycloid

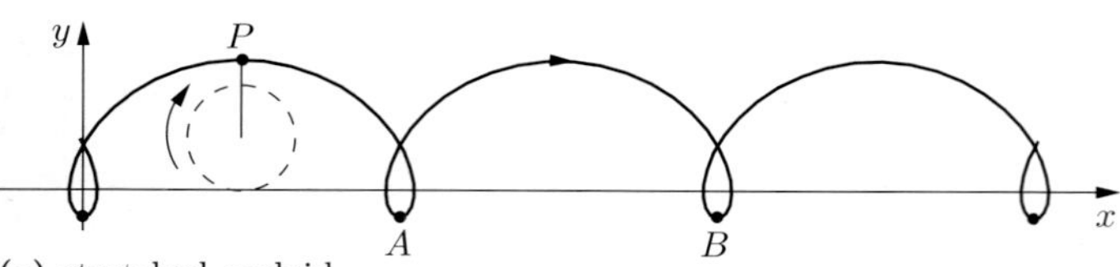


**(c)** stretched cycloid

*Figure 3.72. Wheel curves or cycloids for various parameters.*

*Classification:* (i) $\mu = 1$ (cycloid),

(ii) $0 < \mu < 1$ (shortened cycloid),

(iii) $\mu > 1$ (stretched cycloid).

In case (i), the point $P$ is on the periphery of the wheel, while in case (ii) (resp. case (iii)) it lies inside (resp. outside) of the periphery of the wheel. Curves of the latter two cases are also called *trochoids*.

*Surface area between a cycloid arc and the x-axis:* $A = 3\pi a^2$.

*Length of a cycloid arc between the points* $x = 0$ *and* $x = a$: $L = 8a$.

*Curvature radius of the cycloid:* $4a \sin \dfrac{\varphi}{2}$.

*Curvature radius of the trochoid:* $\dfrac{a(1+\mu^2 - 2\mu\cos\varphi)^{3/2}}{\mu(\cos\varphi - \mu)}$.

*Length of a trochoid arc from A to B:* $L = a \displaystyle\int_0^{2\pi} \sqrt{1+\mu^2 - 2\mu\cos\varphi}\,\mathrm{d}\varphi$.

The cycloid is the solution of the famous problem of the brachystochrone due to Johann Bernoulli (cf. 5.1.2).

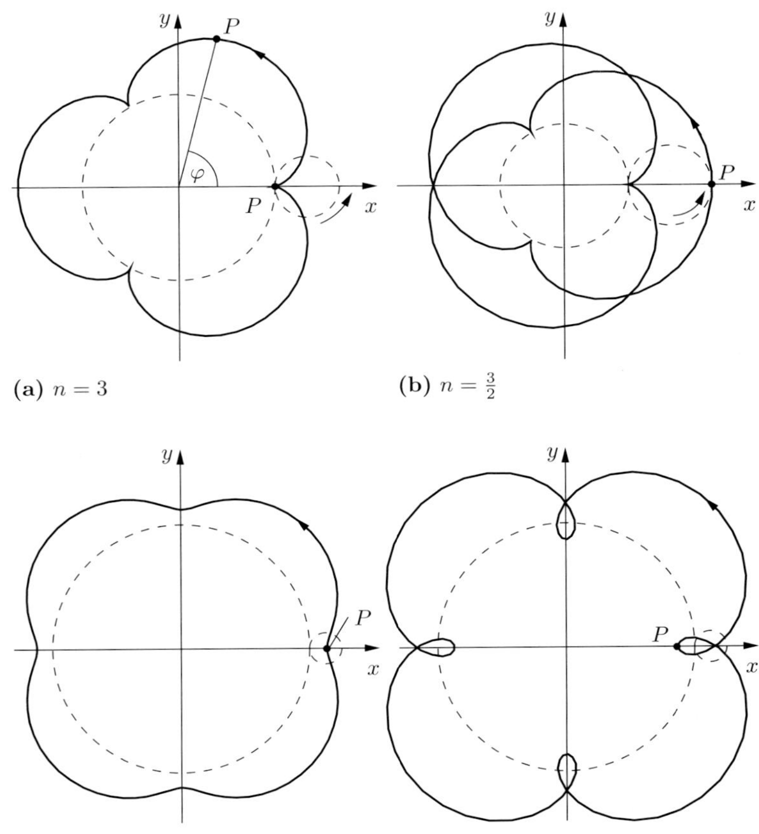


*Figure 3.73. Epicycoids (generated by rotating a wheel on a circle).*

#### 3.7.9.2 Rotating the wheel on the periphery of a circle (epicycloids)

*Parameterization in Cartesian coordinates* (Figure 3.73):

$$x = (A+a)\cos\varphi - \mu a \cos\frac{A+a}{a}\varphi,$$
$$y = (A+a)\sin\varphi - \mu a \sin\frac{A+a}{a}\varphi.$$

Here a wheel of radius $a$ is rotating on a circle of radius $A$. The point $P$ has polar coordinates $\varphi$ and $r$.

*Classification:* (i) $\mu = 1$ (epicycloid),

(ii) $0 < \mu < 1$ (shortened epicycloid),

(iii) $\mu > 1$ (extended epicycloid).

*Curvature radius of an epicycloid:* $\dfrac{4a(A+a)}{A+2a}\sin\dfrac{A\varphi}{2a}$.

*Interesting phenomenon:* Set $n = \dfrac{A}{a}$.

(i) If $n$ is a natural number, then the curve closes after one revolution around the circle.

(ii) If $n$ is a rational number, then the curve closes after finitely many revolutions around the circle.

(iii) If on the other hand $n$ is irrational, then the curve does not close at all.

*Length of an arc from one cusp to the next for the epicycloid:* $8(A+a)/n$.

#### 3.7.9.3 Rotating the wheel on the inside of a circle (hypocycloids)

*Parameterization in Cartesian coordinates* (Figure 3.74):

$$x = (A-a)\cos\varphi - \mu a \cos\frac{A-a}{a}\varphi,$$
$$y = (A-a)\sin\varphi - \mu a \sin\frac{A-a}{a}\varphi.$$

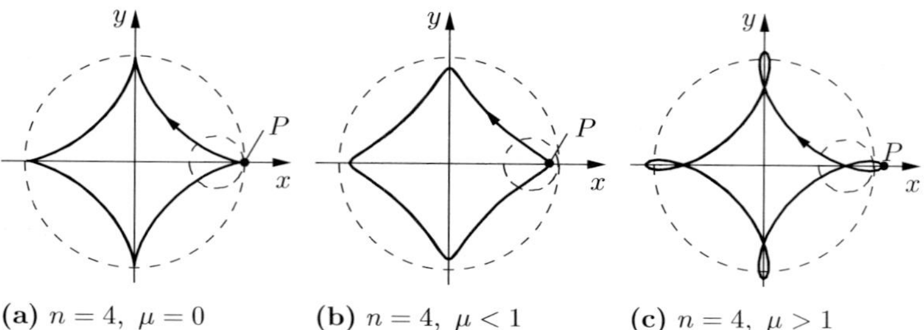


**(a)** $n = 4,\ \mu = 0$ **(b)** $n = 4,\ \mu < 1$ **(c)** $n = 4,\ \mu > 1$

*Figure 3.74. Hypocycloids (generating by rotating a wheel inside a circle).*

Here a wheel of radius $a$ rotates on the inner periphery of a circle of radius $A$. The point $P$ has polar coordinates $\varphi$ and $r$.

*Classification:* (i) $\mu = 1$ (hypocycloid),

(ii) $0 < \mu < 1$ (shortened hypocycloid),

(iii) $\mu > 1$ (extended hypocycloid).

*Curvature radius of a hypocycloid:* $\dfrac{4a(A-a)}{(A-2a)} \sin \dfrac{A\varphi}{2a}$.

*Example 1:* For $A = 2a$ and $\mu > 0$ one gets an ellipse, which in the special case $\mu = 1$ degenerates to a line segment.

*Example 2:* In the special case $n = \dfrac{A}{a} = 4$ and $\mu = 1$ we get the parameterization

$$x = A\cos^3\varphi, \qquad y = A\sin^3\varphi, \qquad 0 \le \varphi < 2\pi.$$

This is an *astroid*, whose equation in Cartesian coordinates is given by

$$\boxed{x^{2/3} + y^{2/3} = A^{2/3}}$$

or $(x^2 + y^2 - A^2)^3 + 27x^2y^2A^2 = 0$ (Figure 3.74(a)). The astroid is thus an algebraic curve of order 6 (see section 3.8.2 below).

#### 3.7.9.4 The epicycles of Hipparchos

*Parameterization in Cartesian coordinates* (Figure 3.75):

$$\boxed{x = A\cos\omega t + a\cos\omega' t, \qquad y = A\sin\omega t + a\sin\omega' t.} \tag{3.41}$$

Interpreting $t$ as the time, (3.41) describes the motion of a particle at the point $P$ on a circle $K_a$ of radius $a$. During this motion, the center of $K_a$ moves on a circle $K_A$ of radius $A$ with the angular velocity $\omega$, while $K_a$ rotates with the angular velocity $\omega'$.

These epicycles were used by the great astronomers Hipparchos (180–125 BC) and Ptolemy (around 150 BC) in ancient times to describe the complicated motion of the planets in the heavens.

The theory of epicycles warns us: with a sufficiently flexible model, one can approximately describe reality, even though the model is definitely wrong.

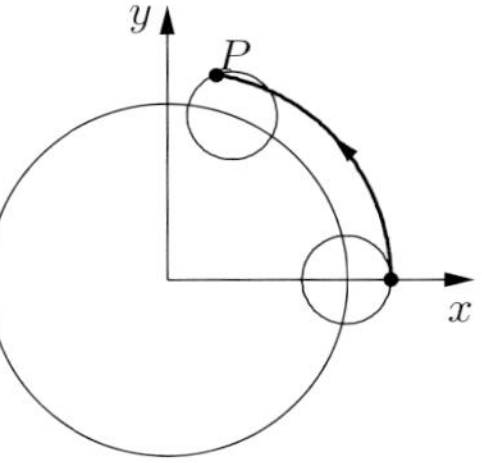


*Figure 3.75.*

## 3.8 Algebraic geometry

*What the geometer loves about his science is that he sees what he thinks.*
*Felix Klein* (1849–1925)

### 3.8.1 Basic ideas

#### 3.8.1.1 The basic problem

Let $m$ polynomials $p_j = p_j(z)$ with complex coefficients in $n$ (complex) variables $z_1, \ldots, z_n$ be given. Set $z = (z_1, \ldots, z_n)$. Algebraic geometry is concerned with finding solutions of systems of equations of the form

$$\boxed{p_j(z) = 0, \qquad z \in \mathbb{C}^n, \quad j = 1, \ldots, m\,.} \tag{3.42}$$

This a central problem in all of mathematics, the solution of which is of importance to many problems.[25]

#### 3.8.1.2 Singularities and their relevance in physics

A typical difficulty in algebraic geometry is the appearance of singularities.

**Definition:** Let $m < n$. A solution $z$ of (3.42) is said to be a *regular point*, if

$$\boxed{\text{Rank}\left(\frac{\partial p_j(z)}{\partial z_k}\right) = m}$$

(i.e., the Jacobian matrix of $f$ has maximal rank). Otherwise $z$ is a *singular point* or *singularity* of the set (3.42).

If all solutions of the system (3.42) are regular, then the set of solutions form a smooth *manifold.* Manifolds are studied in differential topology and differential geometry (cf. [212]).

*Example 1* (manifold): The equation for a line

$$\boxed{y = mx + b}$$

and the equation for a circle

$$\boxed{x^2 + y^2 = r^2}$$

with $r > 0$ describe curves without singularities, i.e., these curves form one-dimensional manifolds. A characterizing property of such a manifold is that there is a unique tangent at every point on the curve (cf. Figure 3.76(a)).

*Example 2* (a double point as a singularity): The equation

$$\boxed{x^2 - y^2 = 0}$$

decomposes as $x^2 - y^2 = (x - y)(x + y) = 0$. This implies that the curve which the equation describes splits off into two separate lines, with the equations $x - y = 0$ and $x + y = 0$. These two lines intersect at the point $(0, 0)$, and this point is called an *ordinary double point.* Clearly, at this points the curve has two tangents (the two lines), and thus it is not a manifold (Figure 3.76(b)).

Double points can also occur where a curve intersects itself, as in the case of the Cartesian leaf (cf. Figure 3.82 in section 3.8.2.3).

---

[25] An additional important generalization of this problem is to replace the field $\mathbb{C}$ of complex numbers as the domain of coefficients and solutions by an arbitrary field $K$.

In the case of the field $K = \mathbb{Q}$ of rational numbers for example, this leads to Diophantine geometry, which has been studied for almost 2000 years by the most astute minds in mathematics. See section 3.8.6 for more information on this.

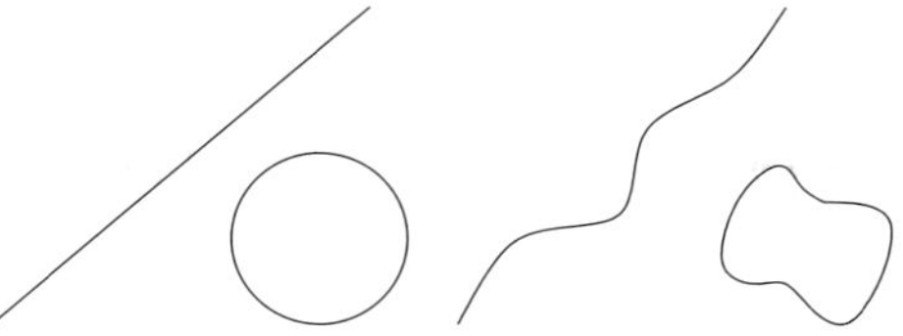

**(a)** manifolds (no singular points)

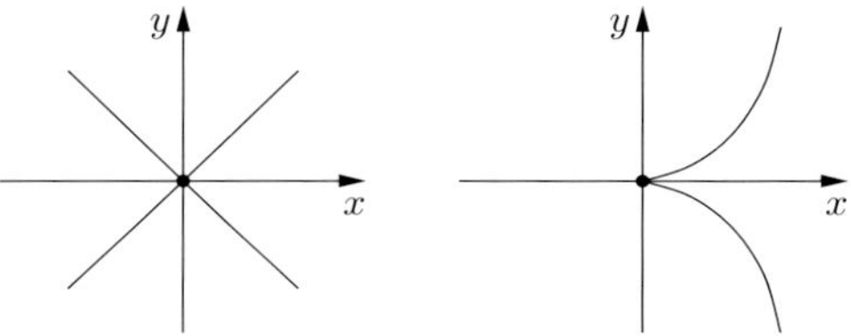


**(b)** double point (singular point)

**(c)** semi-cubical parabola with cusp (singular point)

*Figure 3.76. Smooth manifolds and singular points.*

*Example 3* (a cusp as a singularity): The *semi-cubical parabola*

$$\boxed{y^2 - x^3 = 0\,,} \tag{3.43}$$

has at the point $(0, 0)$ a so-called *cusp*. At the cusp there is no tangent, so again, this is not a manifold.

Double points and cusps are the simplest singularities which can occur.

**The relevance of singularities in physics:** A fundamental phenomenon in nature is that systems can drastically change their qualitative behavior under critical external influences. In this case one speaks of a *bifurcation* (branching). These can belong to, for example, ecological catastrophes or economic crises.

*Example 4* (bifurcation of equilibrium positions): If a system is in equilibrium, it can, under the influence of external forces, pass over to a new equilibrium position. If, for example, forces act on a rod over its length, then at some critical level a bulge in the rod can occur.

*Example 5* (Hopf bifurcation): A dynamical system which is in equilibrium can, under the influence of external forces, start to vibrate.

> Bifurcations which occur in nature can be mathematically modeled with the aid of singularities.

This is the content of *bifurcation theory*, which we shall present in [212]. Also the so-called *catastrophe theory*, due to the French mathematician René Thom, belongs to this area.

We now consider the easiest special cases of (3.42).

### Linear equations

If all polynomials $p_j$ are linear (have degree one), then (3.42) is a system of linear equations, for which there is a complete theory of solutions (cf. 2.3).

From a geometric point of view, the study of linear equations corresponds to the investigation of the intersection behavior of lines, planes and hyperplanes.

**Functional analysis:** Even the simplest case of linear equations, which from a modern point of is view trivial, led to the development of the theory of linear algebra. This in turn forms the basis for the theory known as functional analysis. Both the modern theory of partial differential equations and quantum theory are formulated in the language of functional analysis.

**Topology:** A basic strategy in modern mathematics is to study complicated structures by associating to them simple structures belonging to linear algebra. This is the method which is applied for example in algebraic topology, in which (topological) spaces are represented by their de Rham cohomology groups, which forms the basis of modern differential topology. A not so sophisticated example is the use of tangent spaces to study manifolds (see [212]).

### Quadratic equations

If the polynomials $p_j$ are quadratic, i.e., have degree two, then the basic equation (3.42) describes conic sections and the intersection behavior of these. If we consider a single equation

$$\boxed{p(x,y) = 0}$$

of second degree, supposing that the polynomial $p$ is irreducible, we get a smooth conic section. Singularities can occur only in the case that the polynomial $p$ splits (i.e., factors into a product), in which case we have the situation described in Example 2.

**Number theory:** Quadratic equations correspond to quadratic forms, which are intensively studied in number theory. The basis for this is the theory of quadratic forms due to Gauss, which was developed in his treatise "Disquisitiones arithmeticae", which appeared in 1801. This in turn was the foundation for the theory of quadratic number fields as well as the modern theories of algebraic and analytic number theory.

**Spectral theory:** A quadratic equation in $n$ variables

$$\sum_{j,k=1}^{n} a_{jk} x_j x_k = \text{const}$$

can be elegantly written in matrix form as

$$\boxed{x^{\mathsf{T}} A x = \text{const.}}$$

The investigation of this kind of equations involves finding normal forms for the matrix $A$. This theory was developed in the second half of the nineteenth century. The basis was formed by investigations of Euler (1765) and Lagrange (1773) on the axis of inertia of rotating rigid bodies, which led to special transformations of the principle axis. The general form of this transformation was laid down by Cauchy in 1829. In 1904, Hilbert generalized this in connection with his theory of integral equations to infinite-dimensional matrices. John von Neumann recognized in 1928 that the considerations used by Hilbert actually also apply to the spectral theory of unbounded, self-adjoint operators in Hilbert spaces, which in turn is the basis for quantum theory. The spectrum of the Hamiltonian operator generalizes the eigenvalues of symmetric matrices and describes exactly the possible energy levels of a quantum system.

**Quadratic forms and geometries of modern physics:** Groups of transformations which preserve a quadratic form, and in particular lead to normal forms, are the basis of important geometries, which are considered in detail in section 3.9.

**Special functions:** If one seeks the parameter representation of the circle

$$\boxed{x^2 + y^2 = 1\,,}$$

then one is lead to the trigonometric functions. The global parameter representation (uniformization) of the circle is given by

$$\boxed{x = \cos t\,, \quad y = \sin t\,, \qquad t \in \mathbb{R}\,.}$$

The fact that one requires periodic functions for this description has a deeper topological reason, which is that the circle is an irreducible algebraic curve of second degree and genus 0.

From the parameterization of the circle one gets immediately the global parameterization (uniformization) of the hyperbola

$$\boxed{x^2 - y^2 = 1}$$

by replacing $y$ by $\mathrm{i}y$ and $t$ by $\mathrm{i}s$:

$$\boxed{x = \cos \mathrm{i}s\,, \quad y = -\mathrm{i}\sin \mathrm{i}s\,, \qquad s \in \mathbb{R}\,.}$$

This is identical with the parameterization of the hyperbola

$$\boxed{x = \cosh\, s\,, \quad y = \sinh\, s\,, \qquad s \in \mathbb{R}}$$

in terms of hyperbolic functions.

Looking only at the real values, the periodicity of the functions is not obvious. Euler discovered that the inverse functions of certain elliptic integrals are periodic. During his investigations of the lemniscate, Gauss, as a twenty-year old, made in 1796 the discovery, which was to have great consequences, that the inverse functions of certain elliptic integrals have, in addition to the real periods derived by Euler, two purely imaginary periods. This fact later turned out to be, in the theory of doubly periodic (elliptic) functions developed by Weierstrass, the key to a complete understanding of elliptic integrals. The deeper topological reason for the double-periodicity of the inverse functions of elliptic integrals was explained in section 1.14.19.

**Uniformization and the resolution of singularities:** An interesting problem is to find a global parameterization of general curves and surfaces. Here the notion of uniformization is what is also referred to as resolution of singularities. This area of problems, at least for quite general objects which one would like to uniformize, is among the most difficult in all of mathematics.

(i) The uniformization of all algebraic curves of degree three (cubic curves) leads to the theory of elliptic functions and elliptic integrals (cf. 3.8.1.3).

(ii) The uniformization of arbitrary algebraic curves is the content of the famous uniformization theorem due to Koebe and Poincaré, proved in 1907. This uniformization allows the calculation of Abelian integrals in terms of automorphic functions.

(iii) In 1964, Hironaka succeeded in proving the general resolution of singularities of projective algebraic varieties of arbitrary dimension.

**Cubic equations**

Algebraic curves of degree three can (even if the equation is irreducible) have singularities. These are points at which there is not a uniquely defined tangent. The simplest example of this is the semi-cubical parabola (3.43).

### 3.8.1.3 Elliptic curves and elliptic integrals

**Elliptic curves:** The set of complex solutions $(z, w)$ of the equation

$$\boxed{w^2 = 4z^3 - g_2 z - g_3} \tag{3.44}$$

are, according to Weierstrass (1815–1897), given by the parameter representation

$$\boxed{z = \wp(t)\,, \quad w = \wp'(t)\,, \qquad t \in \mathbb{C}\,.} \tag{3.45}$$

Here $\wp$ denotes the elliptic function defined by Weierstrass with the two complex periods $2\omega_1$, $2\omega_2$ and the constants

$$\begin{aligned} &e_1 := \wp(\omega_1)\,, \qquad e_2 := \wp(\omega_1 + \omega_2)\,, \qquad e_3 := \wp(\omega_2)\,, \\ &g_2 := -4(e_1e_2 + e_1e_3 + e_2e_3)\,, \qquad g_3 := 4e_1e_2e_3\,. \end{aligned}$$

**Elliptic integrals:** The integral

$$J = \int R\left(z, \sqrt{4z^3 - g_2 z - g_3}\right) \mathrm{d}z$$

is to be understood in the sense

$$J = \int R(z, w)\, \mathrm{d}z \tag{3.46}$$

where $(z, w)$ is a solution of equation (3.44). Through the substitution $z = \wp(t)$, $w = \wp'(t)$ we get

$$J = \int R(\wp(t), \wp'(t))\wp'(t)\, \mathrm{d}t\,.$$

Because of the connection with the theory of elliptic functions, one denotes (3.44) as an *elliptic curve.* In the theory of functions of a complex variable (function theory) on the other hand, one speaks in this case of the *Riemann surface*[26] of the "many-valued function" $w = w(z)$ given by (3.44). Thus, one has

> The investigation of elliptic curves leads to the theory of elliptic functions and elliptic integrals.

[26]By definition, a Riemann surface is a connected, smooth, one-dimensional complex manifold. The simplest Riemann surface is just the complex plane $\mathbb{C}$, which is one-dimensional over the complex numbers, but viewed as a real space of course, being isomorphic to $\mathbb{R}^2$, is two-dimensional. This explains that fact that the notion "curve" and "surface" are both applied to the same object, (cf. [212]).

**The topological structure of an elliptic curve:** The set of all pairs of complex numbers $(z, w)$ which satisfy (3.44) form by definition an elliptic curve $C$. Since the $\wp$-functions has complex periods $\omega_1$ and $\omega_2$, we can, when seeking a parameter representation in (3.45), restrict ourselves to the $t$-values in a parallelogram $T$, where the opposite points on the boundary of $T$ are identified (Figure 3.77(a)). With the aid of the formula

$$\boxed{z = \wp(t)\,, \quad w = \wp'(t)\,, \quad t \in T\,,}$$

we get a bijective mapping between the elliptic curve $C$ and $T$.

If we glue the opposite points of $T$ together, we get a torus $\mathscr{T}$ (Figure 3.77(b)). It follows that the elliptic curve $C$ is bijectively related to the torus $\mathscr{T}$. If we endow $C$ with the topology obtained from $\mathscr{T}$, then $C$ is homeomorphic to a torus and thus has the genus

$$\boxed{p = 1}$$

(cf. [212]).

**The group structure on an elliptic curve:** There is a natural group action of the group $\mathscr{G}$, which is generated by the addition rule

$$\boxed{t_1 + t_2 = t_3 \bmod T}$$

**(a)** parallelogram

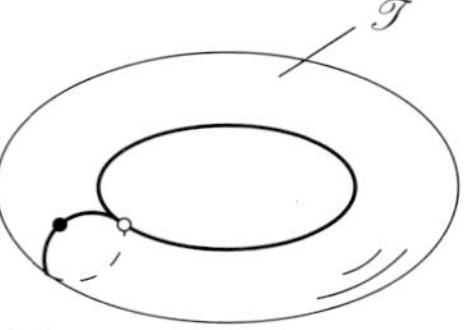

**(b)** the torus

**(c)** equivalent points

*Figure 3.77. The definition of elliptic curves.*

on the space $T$. This is defined by first defining the sum $t_1 + t_2$ as the usual sum of the two complex numbers $t_1$ and $t_2$. If this sum is in $T$, then we set $t_1 + t_2 := t_3$. If the sum lies outside of $T$, this means there are two (uniquely determined) complex numbers $m_1$ and $m_2$ such that the point $t_3^* := t_1 + t_2 - 2m_1\omega_1 - 2m_2\omega_2$ does lie in $T$, and we set accordingly $t_1 + t_2 := t_3^*$ (Figure 3.77(c)).

The group $\mathscr{G}$ is easily made into a group $\mathscr{G}'$ for the elliptic curve $C$, simply by defining the sum by the formula

$$\boxed{(z_1, w_1) + (z_2, w_2) = (z_3, w_3).} \tag{3.47}$$

Here we have set

$$z_j = \wp(t_j), \quad w_j = \wp'(t_j) \quad \text{and} \quad t_3 = t_1 + t_2\,.$$

It then follows from the *addition theorem*:

$$\wp(u+v) = -\wp(u) - \wp(v) + \frac{1}{4}\left(\frac{\wp'(u) - \wp'(v)}{\wp(u) - \wp(v)}\right)^2$$

that

$$\boxed{z_3 = -z_1 - z_2 + \frac{1}{4}\left(\frac{w_1 - w_2}{z_1 - z_2}\right)^2.} \tag{3.48}$$

The group operation (3.47) was discovered in 1834 by Jacobi. To prove it, he used the famous addition theorem for elliptic integrals discovered by Euler in 1753, on which many addition theorems for elliptic functions are based.

**The intuitive interpretation of the group structure on an elliptic curve:** We consider an elliptic curve $C$, i.e., a cubic curve without singular points with real coefficients in the real plane $\mathbb{R}^2$.

(i) We fix a point $P_0$ on that curve.

(ii) The sum of two points $P_1$ and $P_2$ is the point $P_3$ obtained by the geometric construction pictured in Figure 3.78

This means we first determine the intersection $S$ of the line $P_1P_2$ and the curve $C$. Then $P_3$ is the intersection of the line $P_0S$ with the curve $C$. The 'sum' of $P_1$ and $P_2$ is then defined by

$$\boxed{P_1 + P_2 := P_3 .} \tag{3.49}$$

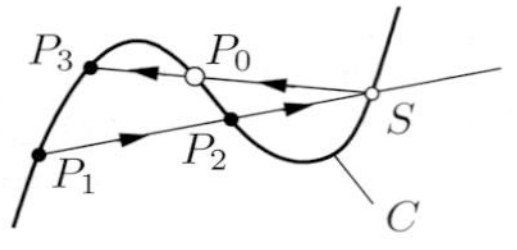


*Figure 3.78. The group stucture on an elliptic curve.*

**Theorem:** (a) The curve $C$ becomes a group with the addition as just defined, the neutral element of the addition being $P_0 = 0$.

(b) If one chooses $P_0$ as one of the inflection points of $C$, then three points $P, Q, R$ on the curve lie on a line if and only if

$$\boxed{P + Q + R = 0.} \tag{3.50}$$

The statement (b) was discovered by Poincaré in 1901 and plays an important role in the investigation of Diophantine geometry (cf. 3.8.6).

**The principle of analogy as a cause of developments in mathematics:** An elliptic curve $C$ has the typical property that every line through two distinct points of the curve meets the curve in precisely one more point.[27] This is the geometric basis for the construction of the sum $P_1 + P_2$ of two points on the curve in (3.47). The notation 'sum' might at first sight seem unnatural, since one usually associates linear objects with this notion, like lines or planes. But here we are considering a curved space. One of the strengths of mathematics is the possibility of introducing compositions into new objects through analogy with compositions on known objects. In this case the known addition is that of numbers, which allows us by analogy to introduce an addition on a curve.

In this way the known results of mathematics can be carried over to more and more complicated objects resp. situations, leading to new findings and discoveries. It turns out that the number of basic structures is relatively small. Thus, just a few basic structures are sufficient (for example, groups, rings, fields, topological spaces, manifolds). The next step in abstraction is the

$$\boxed{\text{combination of the basic stuctures.}}$$

For example, combing the basic notions of *group* and of *manifold*, we get the structure known as a *Lie group*, which is of basic importance in physics (cf. [212]).

However, the history of mathematics is not clear and smooth like this, but rather takes its winding paths to get to the objective. The main impetus for new developments in mathematics is the solution of complicated problems. The mathematician is forced

[27] This statement is only correct if one also admits a point at infinity, i.e., works with the methods of projective geometry, see 3.8.4.

to look for new and powerful ideas. The first proofs of deep results are often very complicated and difficult to follow, leading to the desire to simplify the proofs. To do so, often new theories are developed, the application of which allows the complicated problems to brought to higher level of abstraction, in which they turn out to be much more tractable and simple. Starting from this new level of abstraction in turn, more and more complicated problems can be dealt with.

Comparing this development in mathematics with mountain climbing (which is interestingly a popular sport among mathematicians), a group of mountain climbers is proceeding from one plateau to the next. While some especially daring individuals rush ahead without ropes to secure them, the greater part of the company investigates each plateau, clearing it of boulders and so allowing those to follow an easier climb.

### 3.8.1.4 Algebraic curves of higher degree and Abelian integrals

Up till now we have only considered elliptic curves, which are closely connected with elliptic integrals and elliptic functions. The investigation of more complicated integrals, which at first sight appear to be untractable, is elegantly dealt with by taking advantage of investigations of complicated curves, first studied by Riemann (1857). These are integrals of the form

$$\boxed{\int R(z, w)\,\mathrm{d}z\,,}$$

where the point $(z, w)$ is a point on the curve

$$\boxed{p(z, w) = 0.} \tag{3.51}$$

Here $p$ is a polynomial in $z$ and $w$. The "many-valued function" $w = w(z)$, which satisfies (3.51), is referred to as an *algebraic function*. Integrals of this kind were first studied in generality by the young Norwegian mathematician Niels Henrik Abel (1802–1829).

**Historical remarks:** The investigation of Abelian integrals played a fundamental role in the mathematics of the nineteenth century and lead to the development of function theory by Riemann and Weierstrass. Riemann, who lived from 1826 to 1866, recognized in an ingenious manner that the treatment of Abelian integrals becomes quite simple and clear by studying the qualitative behavior, i.e., the topology, of the Riemann surface belonging to the equation (3.51). The decisive role here is played by the *genus* of the Riemann surface, because the genus is the only topological invariant of a compact Riemann surface.

The equation (3.45) induces a parameter representation of the elliptic curve (3.44). Felix Klein (1849–1925) and Henri Poincaré (1854–1912) both tried, as young men, to solve the difficult problem of finding an appropriate parametrization for an arbitrary algebraic function (3.51). This lead to the development of the theory of *automorphic functions*, which generalize elliptic functions.

The final solution of the problem of parametrizing (3.51) was obtained independently by Paul Koebe (1882–1945) and Henri Poincaré in 1907 with the proof of the famous *uniformization theorem* (cf. [212]).

**The language of schemes:** In modern algebraic geometry one considers the system of equations (3.42) over arbitrary fields (instead of over the complex numbers $\mathbb{C}$). One of the modern tools in this study are objects called schemes. This notion, which is among the most important in all of mathematics (for example, in its most general form, it contains the notion of manifold), connects topology, differential topology, algebraic geometry and number theory. The basic objects of all these central mathematical disciplines are schemes (cf. 3.8.9.4).

**The Fermat conjecture (Fermat's last theorem) and the Shimura–Taniyama–Weil conjecture:** In 1994, Andrew Wiles at Princeton succeeded in proving Fermat's last theorem, which had been one of the central open problems for over 300 years. One part of the proof was to reduce it to the proof of a partial verification of a much deeper geometric conjecture about elliptic curves (the Shimura–Taniyama–Weil conjecture).

**String theory:** The contemporary attempts to unify all the basic forces of nature (gravitation, the weak and strong forces and electromagnetism) in the context of *string theory* has lead to extremely fruitful interactions of ideas between physics and mathematics. The methods of algebraic geometry play a predominant role in these modern developments.

Algebraic geometry is a mathematical discipline which has strong interactions to many areas of mathematics and also to the natural sciences, and no doubt belongs to the fundamental and basic pillars of mathematics. In what follows we shall attempt to build the bridge from the intuitive and geometric origins of the theory over to the modern abstract aspects of the theory, which were not developed for their own sake, but to prove difficult results.

## 3.8.2 Examples of plane curves

The most important property of a plane algebraic curve is its genus $p$ (cf. 3.8.5).

In what follows, $a, b$ and $c$ are positive (real) constants.

### 3.8.2.1 Curves of the first and second degrees

Algebraic curves of degree one (linear curves) are lines; they have the genus $p = 0$.

Irreducible algebraic curves of the second degree (quadratic curves) are non-degenerate conic sections (circles, ellipses, parabolas and hyperbolas); they again have the genus $p = 0$.

Quadratic curves which are reducible are necessarily pairs of lines.

All the curves introduced in this and the next section are irreducible. An irreducible curve of degree three (cubic curve) has the genus $p = 1$, provided there are no singularities[28], otherwise the genus is $p = 0$.

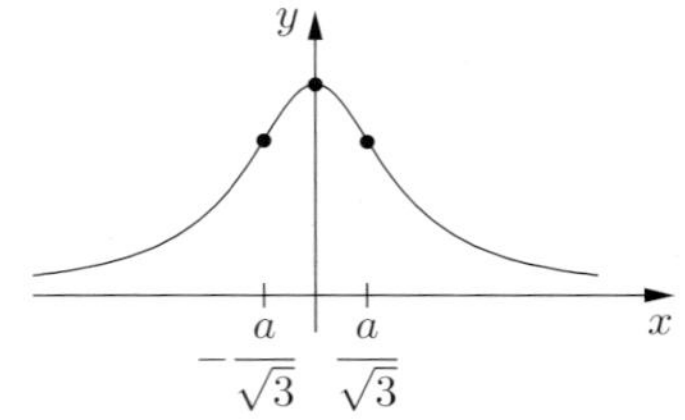


*Figure 3.79. Versiera of Maria Agnesi.*

An irreducible curve of the fourth degree (quartic curve) has the genus $p = 3, 2, 1$ or $p = 0$. The first case corresponds to the curve being *smooth*, i.e., the absence of singularities, the other cases occur depending on the number and kind of singularities. For example, a famous curve, the *Klein quartic*, has three ordinary double points and genus $p = 0$.

[28]This statement is made with respect to the projective representation of the curve in the complex domain (cf. 3.8.4). The graphical representation of the curves in a real plane thus gives an incomplete picture of the singularities of the complex curve.

#### 3.8.2.2 Cubic curves

**The versiera of Maria Agnesi** (Figure 3.79):

$$\boxed{(x^2+a^2)y-a^3=0\,.} \tag{3.52}$$

*Asymptote:* $y=0$.

*Surface area between the curve and its asymptote:* $\pi a^2$.

*Curvature radius at the vertex* $(0,a)$*:* $R=a/2$.

*Two inflection points:* $(\pm a/\sqrt{3}, 3a/4)$.

*Genus:* $p=0$.

**The equation of the versiera in the complex projective plane $\mathbb{C}P^2$:** The equation of the curve (3.52), written in projective coordinates of the projective plane, is

$$\boxed{x^2y+a^2yu^2-a^3u^3=0\,.} \tag{3.53}$$

This equation is obtained by replacing the variables $x$ and $y$ in (3.52) by $x/u$ and $y/u$ and then multiplying through by $u^3$ (this is referred to as *homogenizing* the polynomial or the equation).

Here, $x, y$ and $u$ are complex variables, for which the case $x=y=u=0$ (all simultaneously vanishing) has been excluded. Two solutions $(x_j, y_j, u_j)$ of (3.53) correspond to the same point on the curve if there is a complex number $\lambda\neq 0$, such that

$$(x_1,y_1,u_1)=\lambda(x_2,y_2,u_2):=(\lambda x_2,\lambda y_2,\lambda u_2)\,.$$

If one sets $u=1$, one gets the so-called *affine* form (3.52) of (3.53). This is the inverse of the homogenization procedure, also called dehomogenization.

*The infinitely far point:* The curve (3.53) intersects the line at infinity, defined by the equation $u=0$, at the points

$$(1,0,0) \qquad \text{and} \qquad (0,1,0)\,,$$

which correspond in Figure 3.79 to the directions of the $x$-axis and the $y$-axis.

*Singularities:* The only singularity of the curve is the infinitely far double point (this is just another way of saying "the double point at the line of infinity") $(1,0,0)$. This corresponds to the asymptote $y=0$ of the curve in Figure 3.79.

*Genus:* $p=0$.

**The cissoid of Diocles** (Figure 3.80):

$$\boxed{x^3+(x-a)y^2=0\,.}$$

*Rational parametrization:* $x=\dfrac{at^2}{1+t^2}$, $y=\dfrac{at^3}{1+t^2}$, $-\infty<t<\infty$. In polar coordinates one has $t=\tan\varphi$.

*Representation in polar coordinates:* $r=\dfrac{a\sin^2\varphi}{\cos\varphi}$.

*Asymptote:* $x=a$.

*Surface area between the curve and the asymptote:* $3\pi a^2/4$.

*Geometric characterization:* Let $K$ be a circle of radius $a/2$, centered at the point $(a/2, 0)$, and let $g$ be the line $x = a$. A ray emanating at the origin $O$ intersects $K$ at a point $A$ and the line $g$ at a point $B$. The point $P$ on the cissoid has the property that

$$\boxed{OP = AB.}$$

The word *cissoid* is of Greek origin and means the outline of a leaf of ivy ($\kappa\iota\sigma\sigma o\varsigma$) (cissoz meaning ivy).

The cissoid has a cusp at the origin $(0, 0)$ as its sole singularity.

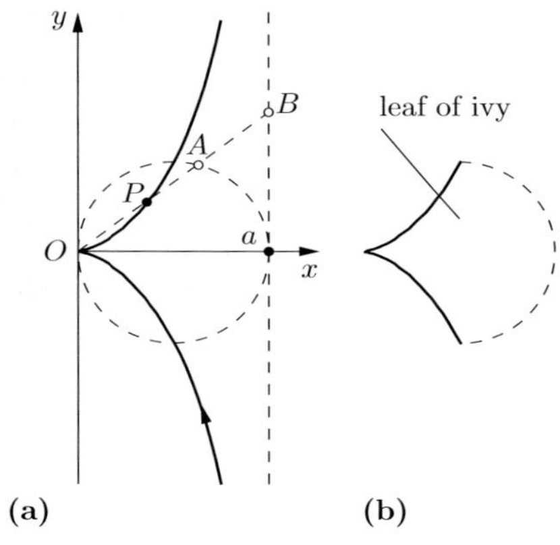


Figure 3.80. Cissoid.

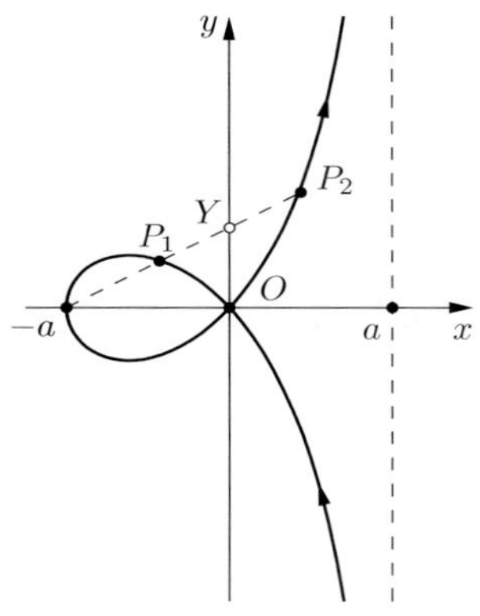


Figure 3.81. Strophoid.

**Strophoid** (Figure 3.81):

$$\boxed{(x + a)x^2 + (x - a)y^2 = 0\,.}$$

*Rational parametrization:* $x = \dfrac{a(t^2 - 1)}{1 + t^2}$, $y = \dfrac{at(t^2 - 1)}{1 + t^2}$, $-\infty < t < \infty$. In polar coordinates one has $t = \tan\varphi$.

*Representation in polar coordinates:* $r = -\dfrac{a\cos 2\varphi}{\cos\varphi}$.

*Asymptote:* $x = a$.

*Tangent at the origin O:* $y = \pm x$.

*Surface area of the loop:* $\left(2 - \dfrac{\pi}{2}\right) a^2$.

*Surface area between the curve and the asymptote:* $\left(2 + \dfrac{\pi}{2}\right) a^2$.

*Geometric characterization:* We fix a ray originating at a point $(-a, 0)$, which intersects the $y$-axis at a point $Y$. The points $P_1$ and $P_2$ on the strophoid satisfy the condition

$$\boxed{P_jY = OY\,, \qquad \text{j=1,2}\,.}$$

It is this property which gives rise to the Greek name strophoid.

The strophoid has a double point at the origin $(0,0)$ as its sole singularity.

**Cartesian leaf** (Figure 3.82):

$$x^3 + y^3 - 3axy = 0\,.$$

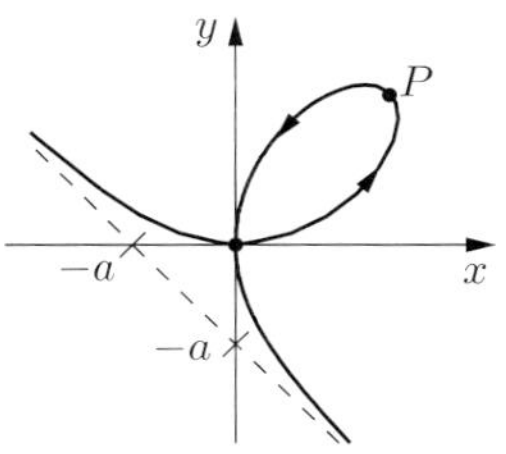


Figure 3.82. Cartesian leaf.

*Rational parametrization:* $x = \dfrac{3at}{1+t^3}$, $y = \dfrac{3at^2}{1+t^3}$, $-\infty < t < -1$ and $-1 < t < \infty$. In polar coordinates one has $t = \tan\varphi$.

*Asymptote:* $x + y + a = 0$.

*Tangent at the origin O:* $y = 0$ and $x = 0$.

*Surface area of the loop:* $3a^2/2$.

*Surface area between the curve and the asymptote:* $3a^2/2$.

*Apex P:* $(3a/2, 3a/2)$.

*Curvature radius at the origin for both branches:* $R = 3a/2$.

*Genus:* $p = 0$. The Cartesian leaf has a double point at the origin as its only singularity.

### 3.8.2.3 Curves of the fourth order (quartic curves)

The following curves are discussed in detail in section 3.7:

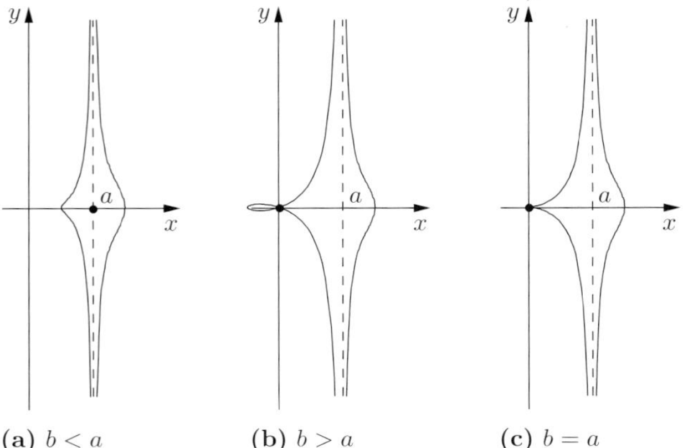


Figure 3.83. Conchoid of Nicomedes.

**Conchoid of Nicomedes** (Figure 3.83):

$$(x-a)^2(x^2+y^2) - b^2x^2 = 0\,.$$

*Genus:* $p = 2$.

The name conchoid is again of Greek origin and means shell curve ($\kappa\acute{o}\nu\chi\eta$ =shell). The conchoid has, depending on the values of the parameters $a$ and $b$, a cusp or a double point as its only singular point.

**Pascal's snail** (Figure 3.71):

$$(x^2 + y^2 - ax)^2 - b^2\,(x^2 + y^2) = 0\,.$$

*Genus:* $p = 2$ or $p = 3$, depending on the values of the parameters $a$ and $b$.

**Cardioid** (Figure 3.84(a)): This is a Pascal's snail with $a = b$.

*Genus:* $p = 2$.

The cardioid has a cusp at the origin $(0,0)$ as its only singularity.

**Cassini's oval** (Figure 3.66(b)):

$$\boxed{(x^2 + y^2)^2 - 2c^2(x^2 - y^2) + c^4 - a^4 = 0\,.}$$

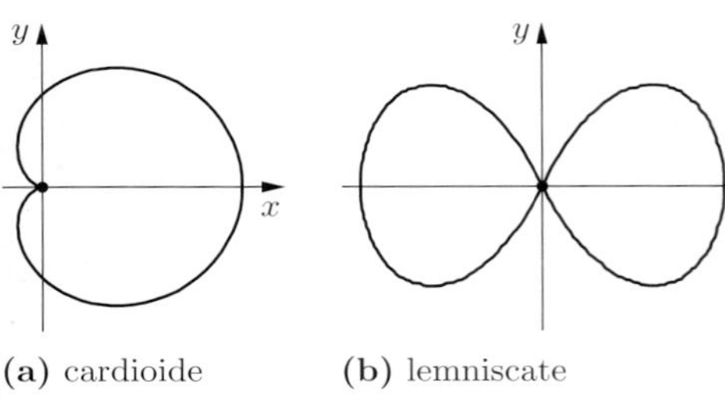


**(a)** cardioide **(b)** lemniscate

*Figure 3.84. Two famous curves.*

*Genus:* $p = 3$ or $p = 2$, depending on the value of the parameters $a$ and $c$.

**Lemniscate of Jakob Bernoulli** (Figure 3.84(b)): This is a Cassini's oval for the parameters $a = c$.

*Genus:* $p = 2$.

The lemniscate has a double point at the origin $(0,0)$.

**Historical remarks:** The cissoid of Diocles (the ivy curve) and the conchoid of Nicomedes (the shell curve) are the oldest known algebraic curves with singularities, discovered around 180 BC. They were invented (discovered) to solve the two famous Delian problems of the doubling of the cube and the trisectomy of an angle by graphical methods (cf. 2.6.1). In ancient times, the conchoid of Nicomedes was also used for the construction of profiles of pillars (Figure 3.83(c)).

Properties of Pascal's snail were studied by the father of the famous mathematician Blaise Pascal (1623–1662).

The Cartesian leaf (folium Cartesii) was introduced by Descartes (1596–1650), who made important contributions towards algebraic geometry by representing geometric figures in 'Cartesian coordinates'. Cassini's curves originate with the work of the Italian astronomer Cassini (1650–1700).

Newton (1643–1727) carried out various investigations on the behavior of algebraic curves. The lemniscate of Jakob Bernoulli (1655–1705) played an important role in the development of the theory of elliptic integrals (cf. 3.7.5).

In 1748 the Italian mathematician Maria Gaetana Agnesi (1718–1799) wrote a book with the title *Instituzioni analittiche*, which collected the knowledge of the day in algebra and analysis and because of the clear presentation was translated into several other languages. It was in this book that the curve now known as the versiera of Agnesi was treated. The Italian word *versiera* means *nightmare*.

### 3.8.3 Applications to the calculation of integrals

We consider the integral

$$\boxed{J = \int R(x, w)\,\mathrm{d}x\,.} \tag{3.54}$$

Here $R$ denotes a rational function of the variables $x$ and $w$. Moreover, $w$ is an algebraic function of $x$, i.e., the pairs $(x, w)$ are points on a plane algebraic curve

$$\boxed{C: \quad p\,(x, w) = 0\,.}$$

If the parameter representation

$$x = x(t), \qquad w = w(t) \tag{3.55}$$

of the curve $C$ has been given, then we get for the integral (3.54) via substitution the expression

$$J = \int R(x(t), w(t))x'(t)\,\mathrm{d}t\,. \tag{3.56}$$

Integrals like this containing rational functions can be calculated with the aid of a partial fraction decomposition. Ever since Newton's and Leibniz' time around 1700, the search has been on for new rational substitutions to transform integrals of the form (3.54) into integrals like (3.56) with a rational integrand, which can be calculated as just mentioned. After some time the following question crystallized itself as the critical issue: for which integrals do rational substitutions exist?

The rule of thumb is:[29]

> The curve $C$ has a rational representation as in (3.55) if and only if the genus of the curve is zero.

*Example 1:* The integral

$$J = \int R\left(x, \sqrt{1 - x^2}\right)\,\mathrm{d}x \tag{3.57}$$

can be written in the form $\int R(x, w)\mathrm{d}x$ with

$$C: \quad x^2 + w^2 = 1\,.$$

This is a circle, which has genus zero. A rational parameterization of the circle is given by

$$x = \frac{t^2 - 1}{t^2 + 1}\,, \qquad w = \frac{2t}{t^2 + 1}\,.$$

If we set

$$\boxed{t = \tan\frac{\varphi}{2}\,,} \tag{3.58}$$

then we have

$$x = \cos\varphi\,, \qquad y = \sin\varphi\,. \tag{3.59}$$

[29] The precise answer is given by the theorem of Poincaré (1901), described in section 3.8.5. For this one requires the projective complex form of the curve $C$, which will be introduced in the next session.

This explains the universal success of the substitution (3.58) for the calculation of integrals of the form

$$\int R(\cos\varphi, \sin\varphi)\,\mathrm{d}\varphi\,.$$

*Example 2:* Let $-\infty < e_1 < e_2 < e_3 < \infty$. The integral

$$\int R\big(x, \sqrt{(x-e_1)(x-e_2)(x-e_3)}\,\big)\,\mathrm{d}x \qquad (3.60)$$

can be written in the form $\int R(x,w)\mathrm{d}x$ with

$$w^2 = (x-e_1)(x-e_2)(x-e_3)\,.$$

This curve is of the third order and has no singularities; it therefore has genus one. This is the deeper reason why elliptic integrals can*not* be solved by rational substitutions. This fact was gradually realized in the course of the eighteenth century and lead in the nineteenth century to the development of the theory of elliptic functions.

The integral (3.57) can be solved with the substitution (3.58) with the help of *(simply) periodic* trigonometric functions. To calculate the integral (3.60), one requires the substitution

$$x = \wp(t), \qquad w = \wp'(t)\,,$$

with the *doubly periodic* (elliptic) Weierstrass $\wp$-function (cf. 1.14.17.3).

### 3.8.4 The projective complex form of a plane algebraic curve

**Basic idea:** The only way to clearly organize the theory of plane algebraic curves is to pass over to homogenous complex coordinates, meaning that the curve is considered in the complex plane $\mathbb{C}P^2$ (cf. 3.5.4).

*Example 1:* In the equation of the circle

$$\boxed{x^2 + y^2 = 1}$$

we replace $x$ and $y$ by $x/u$ and $y/u$. After multiplying through by $u^2$ we get the complex projective form

$$\boxed{x^2 + y^2 = u^2}$$

of the circle. Here $x, y, u$ are complex numbers, for which the triple $(0,0,0)$ is excluded, i.e., not all of the variables are allowed to vanish at the same time. Two such tuples $(x,y,u)$ and $(x_*, y_*, u_*)$ define the same point (in the projective plane), if they differ by a non-vanishing constant factor $\lambda$, i.e., if we have $(x,y,u) = \lambda(x_*, y_*, u_*)$.

**Definition:** Every plane algebraic curve of degree $n$ can be written in the form

$$\boxed{p\,(x,y,u) = 0} \qquad (3.61)$$

where $p$ is a *homogenous polynomial* of degree $n$ with complex coefficients in the complex variables $x, y$ and $u$. Such curves are called *algebraic curves* in $\mathbb{C}P^2$. The number $n$ is called the *degree* of the curve.

**Irreducible curves:** The curve defined by the equation (3.61) is said to be *irreducible*, if the polynomial $p$ is irreducible over $\mathbb{C}$. This means intuitively that the curve consists of "a single piece", i.e., it does not split into more than one component.

*Example 2:* The equation $x = 0$ is irreducible and describes a line. The equation

$$\boxed{xy = 0}$$

is of second degree and is *reducible* (i.e., not irreducible). This curve is a degenerate conic section, which splits into (consists of the union of) the two lines $x = 0$ and $y = 0$.

The irreducible quadric curves are precisely the ellipses, parabolas and hyperbolas (where, however, there is no difference among these from the point of view of the complex projective plane). These are the non-degenerate conic sections.

### 3.8.4.1 The theorem of Bézout on the intersections of curves

The following theorem is one of the most important results in the theory of algebraic curves.

**Generic intersections:** If two irreducible curves intersect at a point $P$, then this point is said to be *regular* if both curves have a unique tangent at this point and the two tangents do not coincide (Figure 3.85). This situation "almost always" occurs. This adverb is the content of the mathematical term *generic*, which means "the exceptions occur very seldom", in more mathematical language, "the exceptions occur on a set of lower dimension".

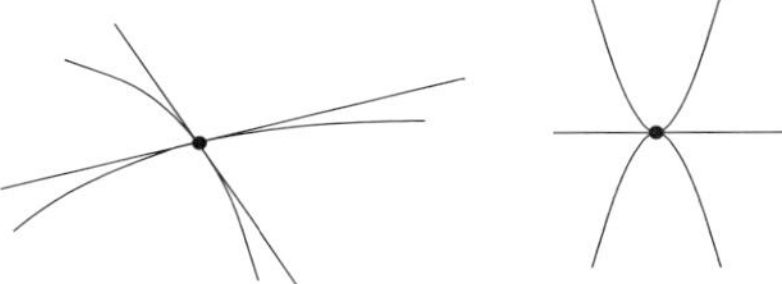

(a) regular intersection point (b) irregular intersection point

*Figure 3.85. Intersection points of plane curves.*

**Theorem of Bézout (1779):** Let two distinct irreducible algebraic curves $C$ resp. $D$ be given in $\mathbb{C}P^2$, and let $m$ resp. $n$ be their degrees.[30] Then there are at most $mn$ intersection points of the two curves. Moreover, if all intersection points are regular, then there are exactly $mn$ points of intersection.[31]

*Example 1:* Two conic sections (without common components) can intersect at most at $mn = 2 \cdot 2 = 4$ points.

*Example 2:* The unit circle $x^2 + y^2 = 1$ and the line $x = 2$ do not intersect in the real plane. Passing to homogenous coordinates we get

$$x^2 + y^2 = u^2, \qquad x = 2u$$

with the two points of intersection $(2, \pm i\sqrt{3}, 1)$. Note that these intersection points are finite, but imaginary.

---

[30]The same statement holds if $C$ and $D$ are reducible, but we assume they have no component in common.

[31]More generally, one can introduce the notion of multiplicity of a point of intersection (see below). Then the statement is that, counted with multiplicities, there are always exactly $mn$ points of intersection.

**Corollary:** Every point of intersection can be assigned a *multiplicity*, such that:[32]

| The sum of all multiplicities of the points of intersections of the two curves is precisely $mn$. |
|---|

Regular points of intersection have the multiplicity one.

### 3.8.4.2 Rational transformations of curves

In mathematics there is for every class of objects a corresponding class of transformations. In the theory of algebraic curves in the projective plane $\mathbb{C}P^2$ the first thing one thinks of are projective maps of $\mathbb{C}P^2$ to itself. A classification of the curves with respect to this relation turns out, however, to be too difficult and complex to be of any use. Instead, what has turned out to be useful is the classification with respect to birational transformations.

**Rational maps of curves:** Let two algebraic curves $C$ and $C'$ in $\mathbb{C}P^2$ be given. The map defined by

$$\boxed{x' = X(x,y,u)\,,\quad y' = Y(x,y,u)\,,\quad u' = U(x,y,u)} \tag{3.62}$$

from $C$ to $C'$ is said to be *rational*, if $X, Y$ and $U$ are homogenous polynomials in the variables $x, y$ and $u$, which all have the same degree.

The map (3.62) is said to be *birational* if the map is rational and bijective and the inverse mapping is also rational.

**Rational curves:** A curve is said to be *rational* if it is the rational image of a line. Explicitly this corresponds to a representation

$$x = X(\lambda,\mu)\,,\qquad y = Y(\lambda,\mu)\,,\qquad u = U(\lambda,\mu) \tag{3.63}$$

of the curve, where $X, Y$ and $U$ are homogenous polynomials in the complex variables $\lambda$ and $\mu$, which all have the same degree.

*Example 1:* Lines and conic sections are rational curves.

**Birational equivalence of curves:** Two algebraic plane curves are said to be *birationally equivalent* or *birational* to one another if they can be transformed into one another by means of a birational mapping.

| The algebraic geometry of algebraic plane curves is the invariant theory of these curves under birational transformations. |
|---|

*Example 2:* The degree of a curve is not a birational invariant, but the genus of the curve (cf. 3.8.5) is.

---

[32] Suppose the two curves are given by the equations

$$p(x,y,u) = 0 \quad \text{and} \quad q(x,y,u) = 0.$$

With the help of the coordinate transformation (collineation) we may assume that the origin $(0,0,1)$ does not lie on the line joining two points of intersection.

We set $\mathscr{R} := \mathbb{C}[x,u]$ (this is the polynomial ring over $\mathbb{C}$ with variables $x$ and $u$). Then we have

$$p, q \in \mathscr{R}[y]\,.$$

The resultant $R(p,q)$ of $p$ and $q$ vanishes at the point of intersection $P$ of the two curves. The *multiplicity* of the corresponding zero $y$ of the resultant over $\mathscr{R}[y]$ is called the multiplicity of the point of intersection.

**Cremona group:** The set of birational maps of $\mathbb{C}P^2$ into itself form a group, which was first studied in detail by the Italian geometer Luigi Cremona between 1863 and 1865. That is why the group carries his name today.

#### 3.8.4.3 Singularities

**Tangents:** The tangents to the curve $p(x, y, u) = 0$ at the point $P := (x_0, y_0, u_0)$ is given by the equation

$$\boxed{p_x(P)(x - x_0) + p_y(P)(y - y_0) + p_u(P)(u - u_0) = 0.} \tag{3.64}$$

*Example 1:* The equation of the tangents to the unit circle $x^2 + y^2 - u^2 = 0$ is:

$$2x(x - x_0) + 2y(y - y_0) - 2u(u - u_0) = 0\,.$$

This is equivalent to the so-called *polar equation*

$$xx_0 + yy_0 - uu_0 = 0\,.$$

**Regular point:** A point $P$ on a curve is called a *regular* point, if and only if there is a uniquely determined tangent at $P$, i.e., one has

$$(p_x(P), p_y(P), p_u(P)) \neq (0, 0, 0)\,.$$

**Singular point:** A point which is not regular is said to be *singular*. A singular point $P$ has by definition the *multiplicity* $s$, if all partial derivatives of the polynomial $p$ up to order $s - 1$ vanish at the point $P$, while at least one partial derivative of order $s$ is non-vanishing at $P$.

**Double points and cusps:** Singular points of multiplicity $s = 2$ (resp. $s = 3$) are called *double points* (resp. *cusps*).

*Example 2* (double point): For $p\,(x, y) := x^2 - y^2$ and $P := (0, 0, 1)$ one has

$$p_x(P) = p_y(P) = p_u(P) = 0 \qquad \text{and} \qquad p_{xx}(P) = 2\,.$$

Thus the intersection point $P$ of the two lines $y = \pm x$ into which $x^2 - y^2 = 0$ splits is a double point (cf. Figure 3.76 in section 3.8.1.2).

*Example 3* (cusp): For $p\,(x, y, u) := x^3 - y^2 u$ and $P := (0, 0, 1)$, we have

$$p_{xxx}(P) = 6\,,$$

while all partial derivatives up to the second order vanish at $P$. Thus $P$ is a cusp. This point corresponds to the origin $(0, 0)$ of the semi-cubical parabola $x^3 - y^2 = 0$ (cf. Figure 3.76 in section 3.8.1.2).

**Theorem:** Regular points and singular points including their multiplicities are invariants with respect to the birational equivalence of curves.

**Application to the versiera of Agnesi:** The equation of this curve is

$$\boxed{C: \quad x^2 y + yu^2 - u^3 = 0}$$

(cf. Figure 3.79 in section 3.8.2.2). We set $p = x^2 y + yu^2 - u^3$.

(i) The points of intersection of this curve with the line at infinity $u = 0$ are $(1, 0, 0)$ and $(0, 1, 0)$.

(ii) The singular points on this curve are determined from the common solutions of the set of equations

$$p_x = 2xy = 0\,, \qquad p_y = x^2 + u^2\,, \qquad p_u = 2uy - 3u^2 = 0\,, \qquad p = 0\,.$$

This yields the solution $(0, 1, 0)$ and the geometrically irrelevant point $(0, 0, 0)$ (recall that projective space does not contain the point $(0, 0, 0)$). Because $p_{xx}(0, 1, 0) = 2$, the point at infinity $(0, 1, 0)$ is a double point of $C$, which corresponds to the asymptote of $C$ in Figure 3.79.

#### 3.8.4.4 Duality

**The dual curve:** Let an algebraic curve $C: \ p(x, y, u) = 0$ be given. The mapping

$$C_*: \quad x_* = p_x(x, y, u)\,, \qquad y_* = p_y(x, y, u)\,, \qquad u_* = p_u(x, y, u)\,,$$

which is considered for all regular points $(x, y, u)$ of $C$, describes a curve in $\mathbb{C}P^2$, which is called the *dual curve*[33] of the curve $C$. The degree of the dual curve is called the *class* of the curve $C$.

From a geometric point of view, the point coordinates of the dual curve are the coordinates of the tangent lines of the original curve.

**Theorem:** Dualizing a curve twice leads again to the curve, i.e., we have $(C_*)_* = C$.

*Example:* Let $p(x, y, u) := x^2 + y^2 - u^2$. The corresponding algebraic curve $C$ : $p(x, y, u) = 0$ is the unit circle. For this curve we get the parameterization

$$x_* = 2x\,, \qquad y_* = 2y\,, \qquad u_* = -2u$$

for the dual curve $C_*$. From $x_*^2 + y_*^2 - u_*^2 = 0$, we see that $C = C_*$, i.e., the unit circle is dual to itself.

The previous example should not give the false impression that determining the dual of a given curve is easy; on the contrary, it is a difficult algebraic problem.

### 3.8.5 The genus of a curve

In this section we consider irreducible algebraic plane curves $C$ in $\mathbb{C}P^2$, i.e., we pass to projective coordinates. The most important and fundamental characteristic of an algebraic plane curve is it genus. The definition of the genus is based on an appropriate parameterization of the curve, which is given by the uniformization theorem.

**The uniformization theorem for algebraic plane curves:** Every curve

$$C: \quad p\,(x, y, u) = 0$$

has a global parameterization

$$\boxed{x = x(t)\,, \quad y = y(t)\,, \quad u = u(t)\,, \qquad t \in \mathscr{T}} \tag{3.65}$$

with the following properties.

[33] A more usual notation for the dual curve is $C^*$ or $C^\vee$.

(i) The parameter space $\mathscr{T}$ is a compact, connected, one-dimensional complex manifold, in other words, a Riemann surface.

(ii) The mapping defined by (3.65) $\pi : \mathscr{T} \longrightarrow C$ is holomorphic and surjective.

We denote by $S$ the necessarily finite set of singular points of the curve $C$. The inverse image $\mathscr{S} := \pi^{-1}(S)$ is called the *critical set* of parameter values.

(iii) The map

$$\pi : \quad \mathscr{T} \backslash \mathscr{S} \to C \backslash S$$

is biholomorphic, and the critical set of parameter values is compact and has no interior points, i.e., $\mathscr{S}$ is "thin".

**Remark:** We interpret the curve parameter $t$ as the time. Then (3.65) describes the curve $C$ as the (generalized) trajectory of a point. It is important that the parameter space $\mathscr{T}$ has no singularities. For this reason one also refers to the parameterization (3.65) as a *resolution of singularities* of the curve $C$.

*Example 1:* We first consider the *strophoid* $(x+u)x^2 + (x-u)y^2 = 0$ in $\mathbb{R}^2$. There we have the parametrization

$$x = \frac{t^2-1}{t^2+1}, \quad y = \frac{t(t^2-1)}{t^2+1}, \quad u = 1, \qquad -\infty < t < \infty . \tag{3.66}$$

The parameter space here is the real axis $\mathbb{R}$, which is free of singularities. Ift the time $t$ runs through all real values, the curve in Figure 3.86 is transcribed exactly once. However, the parametrization (3.66) is not appropriate for our needs. This is because we need to understand all points $(x, y, u)$ of the complex curve including the points at infinity. Furthermore, the parameter space $\mathscr{T}$ is required to be compact, which $\mathbb{R}$ is not.

This simple example already demonstrates how non-trivial the statement of the uniformization theorem is. On the contrary, it is an extremely deep mathematical result.

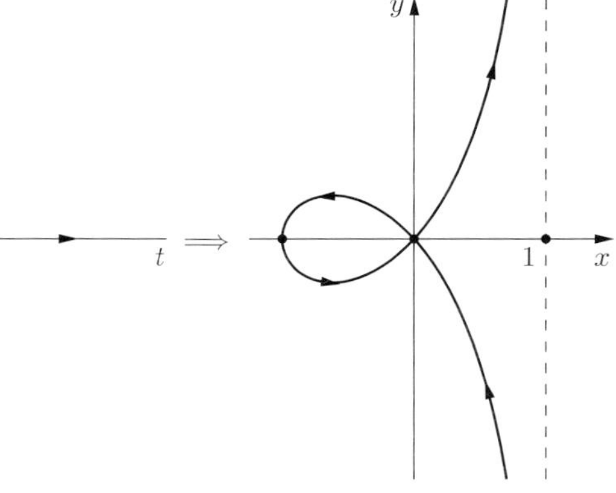


*Figure 3.86. Uniformization of the cuspidal cubic.*

**Definition of the genus:** The *genus* of a curve $C$ is the genus $p$ of the parameter space $\mathscr{T}$ of the uniformization theorem.[34]

(i) If $\mathscr{T}$ is homeomorphic to the Riemannian sphere, then $p = 0$ (Figure 3.87(a)).

(ii) If $\mathscr{T}$ is homeomorphic to a torus, then $p = 1$ (Figure 3.87(b)).

(iii) If neither of the two preceeding cases holds, then $\mathscr{T}$ is homeomorphic to a surface which is obtained from the Riemannian sphere by adding $p$ handles. One calls $p$ the *genus* of the

curve (Figure 3.87(c)).

[34] A general result of topology is the following. Every oriented, connected, compact, two-dimensional real manifold is homeomorphic to a sphere with $p$ handles; the number $p$ is defined to be the *genus* of the manifold. Two such manifolds are homeomorphic if and only if they have the same genus $p$.

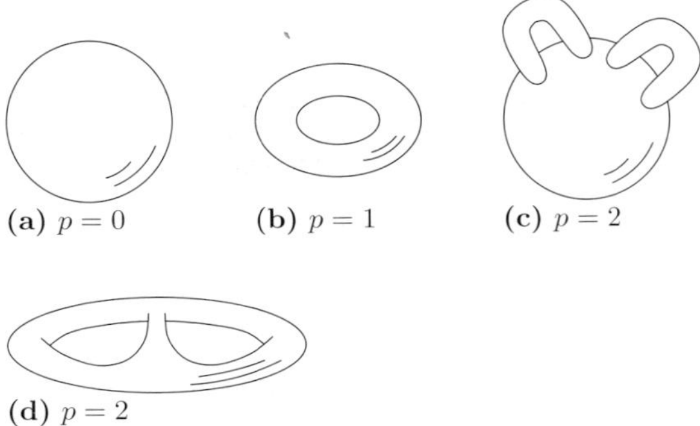


*Figure 3.87. Curves of various genera.*

**Remark:** The genus is well-defined, since different parametrizations of the kind occuring in the uniformization theorem lead to homeomorphic parameter spaces $\mathscr{T}$ with the same genus.

*Example 2:* The two surfaces in Figure 3.87(c) and (d) are homeomorphic, i.e., they can be deformed by an elastic motion into one another, thus they have the same genus $p = 2$.

The genus of Riemann surfaces was introduced, at least in spirit, by Riemann in 1857 during his fundamental investigations of Abelian integrals. The name *genus* was introduced seven years later by Clebsch.

**The fundamental invariance of the genus:** The genus of a curve is invariant under birational transformations.

There is the following rule of thumb:

> The greater the genus of a surface is, the more complicated is its structure.

**Examples of the determination of the genus:** (i) *The theorem of Poincaré:* A curve is rational if and only if it has the genus $p = 0$.

Among these curves are the lines, the conic sections (quadratic curves) and cubic curves with singularities.

(ii) Smooth cubic curves (curves of third degree) have the genus $p = 1$. By definition, these are the *elliptic curves.*

(iii) A non-singular curve $C$ of degree $n$ has the genus

$$\boxed{p = \frac{(n-1)(n-2)}{2}, \qquad \text{n=1,2, \ldots}}$$

The Euler characteristic $\chi$ of $C$ is given by the relation

$$\boxed{\chi = 2 - 2p = 2 - (n-1)(n-2)\,.}$$

(iv) *The Clebsch formula:* If an irreducible curve of degree $n$ has only double points or cusps as singularities, then for the genus we have the relation

$$\boxed{p = \frac{(n-1)(n-2)}{2} - c - d\,,} \tag{3.67}$$

where $d$ is the number of double points and $c$ is the number of cusps.

(v) *The theorem of Harnack:* An (irreducible) algebraic curve with real coefficients of genus $p$ has in the real domain at most $p+1$ components.

*Example 3:* Let $e_1, e_2$ and $e_3$ be real numbers with $e_1 < e_2 < e_3$. Then the equation

$$\boxed{y^2 = (x-e_1)(x-e_2)(x-e_3)}$$

defines a curve of genus $p = 1$, which consists of $p+1 = 2$ components (Figure 3.88).

*Example 4:* If an irreducible cubic curve $C$ has only double points or cusps, then from (3.67) we get the inequality

$$p = 1 - c - d \geq 0\,,$$

i.e., for $C$ only the following three cases are possible:

(i) $C$ is regular (no singular points) and $p = 1$.

(ii) $C$ has exactly one double point ($d = 1$ and $c = 0$), and this curve is rational ($p = 0$).

(iii) $C$ has exactly one cusp and no double points ($d = 0$ and $c = 1$), and $C$ is rational ($p = 0$).

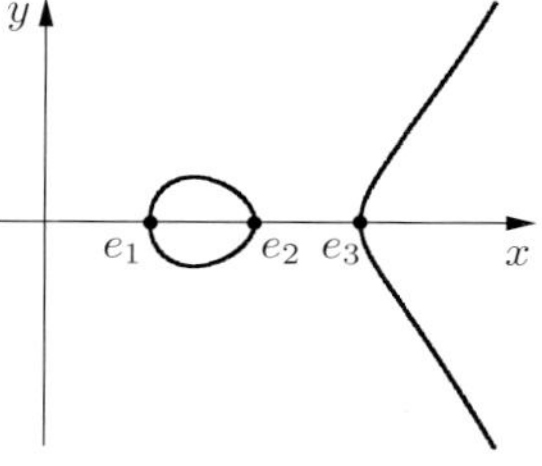


*Figure 3.88. Real points of a cubic curve.*

**Remark:** Using the notion of multiplicity of a singularity, a more general formula than (3.67) can be derived, from which it follows that a cubic curve can have no other singularities than double points or cusps, i.e., the assumption above that $C$ has only double points or cusps is always satisfied. Thus the three cases above are the only possibilities for cubic curves.

*Example 5:* The versiera of Agnesi is a cubic curve which has a double point, which is located at the line at infinity, corresponding to the real asymptote as depicted in Figure 3.79. Thus $d = 1$, $c = 0$ and $p = 0$.

## 3.8.6 Diophantine Geometry

*There are no unfinished symphonies in mathematics or the natural sciences. For hundreds of years certain problems can be studied by generation after generation without losing their dynamics. Looking back onto problems of this kind, the long, continuously developing set of ideas for studying the problem form a fascinating example of the continuity of human thinking.*

*Hans Wussing,* 1974

*Diophantus is one of the greatest riddles in the history of science. We do not know exactly when he lived, nor do we know who his contemporaries or predecessors were, who worked on similar problems as he did.*

*The time at which he lived in Alexandria cannot be determined more exactly than to say that it could have been anytime during half a millennium. In his book on polygonal numbers, Diophantus mentions several times the mathematician Hypsicles of Alexandria, from whom we know that he lived during the second century BC. On the other hand, the commentaries of Theon of Alexandria on the "Almagest" of the astronomer Ptolemy contain excerpts of work of Diophantus. Theon lived during the fourth century AD.*

*Isabella Baschmakowa,* 1974

### 3.8.6.1 Elementary Diophantine equations

The basic idea of Diophantine equations is to find integral (resp. rational) solutions of equations given by polynomials with integral (resp. rational). In this section we first consider the case of integral solutions.

**Linear Diophantine equations and the Euclidean algorithm:** Let integers $a, b$ and $c$ be given, not all of which vanish. We are looking for integral numbers $x$ and $y$ such that

$$\boxed{ax + by = c\,.}$$

This is a linear Diophantine equation.

(i) This equation has a solution if and only if the greatest common divisor $d$ of $a$ and $b$ also divides $c$.

(ii) The general solution of the equation is obtained by setting

$$x = \frac{cx_0 - bg}{d}\,, \qquad y = \frac{cy_0 + ag}{d}$$

with an arbitrary integer $g$ and $x_0 := \alpha_n \operatorname{sgn} a$ and $y_0 := \beta_n \operatorname{sgn} b$. For the calculation of $\alpha_n$ and $\beta_n$ one uses the Euclidean algorithm

$$r_0 = q_0 r_1 + r_2\,, \quad \ldots\,, \quad r_{n-1} = q_{n-1} r_n + r_{n+1}\,, \quad r_n = q_n r_{n+1}$$

with $r_0 := |a|,\ r_1 := |b|$; one then sets

$$\alpha_0 := 0, \quad \beta_0 := 1, \quad \alpha_k := \beta_{k-1}\,, \quad \beta_k := \alpha_{k-1} - q_{n-k}\beta_{k-1}\,, \quad k = 1, \ldots, n\,.$$

*Example 1:* The Diophantine equation

$$\boxed{9973x - 2137y = 1}$$

has the general solution $x = 3 + 2137g,\ y = 14 + 9973g$ with an arbitrary integer $g$.

*Proof:* The Euclidean algorithm in this case yields the relations

$$\begin{aligned} 9973 &= 4 \cdot 2137 + 1425\,, \\ 2137 &= 1 \cdot 1425 + 712\,, \\ 1425 &= 2 \cdot 712 + 1\,, \\ 712 &= 712 \cdot 1\,. \end{aligned}$$

Therefore $n = 3,\ q_0 = 4,\ q_1 = 1$ and $q_2 = 2$. From this it follows that

$$\begin{array}{ll} \alpha_1 = 1, & \beta_1 = -q_2 = -2, \\ \alpha_2 = \beta_1 = -2, & \beta_2 = \alpha_1 - q_1\beta_1 = 3, \\ \alpha_3 = \beta_2 = 3, & \beta_3 = \alpha_2 - q_0\beta_2 = -14\,, \end{array}$$

hence $x_0 = \alpha_3 = 3,\ y_0 = -\beta_3 = 14$.

This method of solution was already used by Indian astronomers in the sixth century AD.

**Pythagorean numbers:** Because

$$3^2 + 4^2 = 5^2$$

every triangle with sides of lengths $x = 3,\ y = 4$ and $z = 5$ is a right triangle. This fact can be used to construct right angles, a method used by the ancient Egyptians.

It is interesting to note that three strings with lengths in ratios $3 : 4 : 5$ corresponds to the accord known as the quartic-sextic accord, given by the base tone, the quartic and the sextic tones above the given base.

**Theorem:** The quadratic Diophantine equation

$$\boxed{x^2 + y^2 = z^2}$$

has the general solution

$$x = 2ab\,, \quad y = a^2 - b^2\,, \quad z = a^2 + b^2$$

with arbitrary natural numbers $a$ and $b$, with $0 < b \le a$, provided one restricts to pairwise prime natural numbers $x, y, z$ as solutions.

This result was already known by Babylonian mathematicians 3,500 years ago.

*Example 2:* For $a = 11,\ b = 10$ we get the Pythagorean numbers $x = 220,\ y = 21$ and $z = 221$.

**The Fermat or Pell equation and continued fractions:** Let $d > 0$ be a natural number which divides no square of a prime number. We seek natural numbers $x$ and $y$ which satisfy

$$\boxed{x^2 - dy^2 = 1.}$$

(i) All solutions $(x_n, y_n)$ are obtained by the formula

$$x_n + y_n\sqrt{d} = (x_1 + y_1\sqrt{d})^n\,, \qquad n = 2, 3, \ldots$$

(ii) The smallest solution $(x_1, y_1)$ is obtained as follows.

The number $\sqrt{d}$ has a representation as a continued fraction of period $k$.

If $p_j/q_j$ for $j = 0, 1, \ldots$ are the corresponding finite parts of the continued fraction, then

$$x_1 = p_{k-1}, \quad y_1 = q_{k-1} \quad \text{for even } k$$

and

$$x_1 = p_{2k-1}, \quad y_1 = q_{2k-1} \quad \text{for odd } k.$$

This result is due to Lagrange (1736–1813), who became the successor of Euler at the Berlin Academy but returned to Paris in 1787. His corps lies in the Pantheon in Paris along with those of other great French personalities.

*Example 3:* For $d = 8$, we have $x_1 = 3$, $y_1 = 1$ and for $d = 13$ we have

$$x_1 = 649, \qquad y_1 = 180.$$

In general, the solutions $x_1, y_1$ are very irregular, yielding small and very large values. For $d = 60$, for example, one gets $x_1 = 31$, $y_1 = 4$, while for the next solution $d = 61$ one has

$$x_1 = 1,766,319,049, \quad y_1 = 226,153,980$$

### 3.8.6.2 Rational points on curves and the role of the genus

**The basic problem:** We consider the equation

$$p(x, y) = 0, \tag{3.68}$$

where $p$ is a polynomial in $x$ and $y$. The decisive assumption which is made here is that the coefficients of the polynomial are *rational* numbers.

| The goal is to find all rational numbers $x$ and $y$ which satisfy the equation (3.68). |
|---|

**Geometric interpretation:** If we view (3.68) as a (Diophantine) equation of a curve in $\mathbb{R}^2$, then we seek all rational points which lie on this curve. A point in $\mathbb{R}^2$ is said to be *rational*, if both coordinates $x$ and $y$ are rational numbers.

The set of rational points are dense in the plane, but there are nonetheless many more irrational points. More precisely, the set of rational points is, according to Cantor (1845–1918), of the same cardinality as the set of integral points in the plane (this means that there is a bijective mapping between the two sets), while the set of irrational points has the same cardinality as the whole plane (cf. 4.3.4). Thus there is no intuitive way of seeing how many rational points are on a complicated curve. It is to be expected that the genus of the curve is an important aspect of the answer, as the higher the genus, the more complicated the curve.

*Example 1:* On the line

$$\boxed{y = x}$$

there are infinitely many rational points and infinitely many irrational points, depending only on whether $x$ is rational or irrational.

*Example 2:* On the circle

$$\boxed{x^2 + y^2 = 1}$$

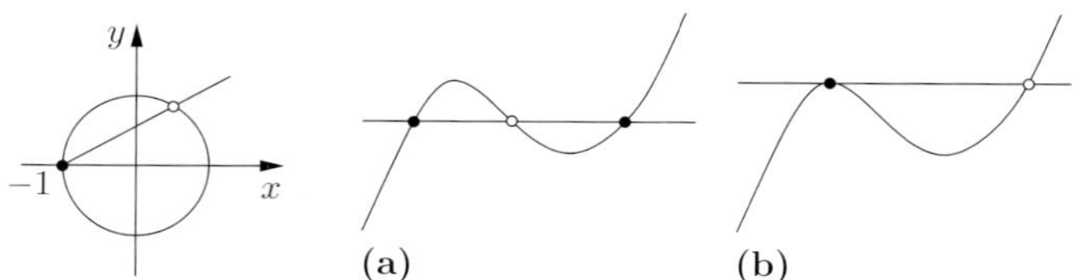


*Figure 3.89.* *Figure 3.90. Rational points on cubic curves.*

there are infinitely many rational points.

*Proof* (following Diophantus): The line $y = m(x+1)$ intersects the circle for every rational slope $m$ in a rational point (Figure 3.89).

**Theorem of Diophantus:** (i) On every Diophantine line there are infinitely many rational points.

(ii) On every Diophantine conic section there are either none or infinitely many rational points.

We now consider a smooth, cubic Diophantine curve $C$. This means that the genus of the curve is $p = 1$.

(iii) *Secant method of Diophantus:* If two rational points lie on $C$, then the line joining these points intersects the curve $C$ in a third rational point of the curve (Figure 3.90(a)).

(iv) *Tangent method of Diophantus:* If a rational point $P$ lies on $C$, then the tangent to $C$ through $P$ intersects the curve $C$ in a further rational point (Figure 3.90(b)).[35]

**Remark:** These results can be found in the various examples given in Diophantus' *Arithmetica*, the first great treatise on number theory in the history of mathematics. In this monograph, Diophantus used positive and negative rational numbers, as well as the symbols

$$\zeta, \quad \Delta^{\tilde{v}}, \quad K^{\tilde{v}}, \quad \Delta^{\tilde{v}}\Delta, \quad \Delta K^{\tilde{v}}, \quad K^{\tilde{v}}K$$

which correspond to the symbols $x, x^2, x^3, x^4, x^5$ and $x^6$ used today.[36]

Decisive advances in Diophantine geometry were not made until Poincaré published in 1901 a fundamental paper on this topic. This paper completely solved the problem discussed above in the case of genus $p = 0$. He was also the first to realize the importance of the secant and tangent method of Diophantus for the elliptic case $p = 1$. He discovered that both these results were an expression of the structure of a group on an elliptic curve (cf. (3.47)). The methods of Diophantus are in fact quite ingenious, which is seen in the following theorems of Poincaré and Mordell; these theorems show that Diophantus' methods are universal for the cases of genus $p = 0$ or $p = 1$.

[35] In (iii) and (iv), one must also consider the points of the curve $C$ at infinity.

[36] The symbol $\Delta^{\tilde{v}}$ (for what we denote today as $x^2$) is derived from the Greek word $\Delta\acute{\upsilon}\nu\alpha\mu\iota\zeta$ (dynamis), which means power. The symbol $K^{\tilde{v}}$ representing the third power stands for $K\acute{\upsilon}\beta o\zeta$ (kubos), which means cube.

For numbers, which for Diophantus always meant rational numbers, he uses the word ἀριθμός (arithmos). From this the notation arithmetic (the study of the rules of calculations with numbers and letters) is derived.

Negative numbers were denoted by Diophantus λεῖψις (leipsis), which means deficiency. Moreover, Diophantus also introduced symbols for our present-day negative powers $x^{-1}, \ldots, x^{-6}$.

**The Diophantine birational transformation due to Poincaré:** A transformation

$$x = f(\xi, \eta), \quad y = g(\xi, \eta)$$

is said to be *Diophantine rational* (in modern language *rational over* $\mathbb{Q}$ or *defined over* $\mathbb{Q}$), if $f$ and $g$ are rational functions with rational coefficients.

If a Diophantine rational transformation is invertible such that the inverse transformation is also Diophantine rational, then we refer to the transformation as *Diophantine birational* transformation.

**Diophantine equivalence:** Two Diophantine curves are said to *(Diophantine) equivalent* , if there is a Diophantine birational transformation between the two of them.

**Theorem of Poincaré (1901) for the genus p=0:** (i) Every Diophantine curve of genus $p = 0$ and odd degree is equivalent to a Diophantine line. Thus, it has infinitely many rational points.

(ii) Every Diophantine curve of genus $p = 0$ of even degree is equivalent to a Diophantine conic section and thus has either none or infinitely many rational points.

Poincaré further conjectured the following result which was later proved by Mordell.

**The theorem of Mordell (1922) for genus p=1:** On a Diophantine elliptic curve ($p = 1$) there are either no rational points on the curve or there are finitely many rational points $P_1, \ldots, P_n$, such that every rational point on the curve can be obtained from these by applying the secant and tangent method.

In the language of group theory this means that the subgroup of rational points of the additive group of the elliptic curve is generated by these finitely many point $P_1, \ldots, P_n$, i.e. every rational point can be written in the form

$$\boxed{P = m_1 P_1 + \ldots + m_n P_n}$$

with integers $m_1, \ldots, m_n$. The addition in this representation is to be understood in the sense of (3.47).

The following result was conjectured by Mordell in 1922 and remained for a long time a conjecture.

**The fundamental theorem of Diophantine geometry proved by Faltings in 1983 for genus p≥2:**

> On every Diophantine curve of genus $p \geq 2$ there are at most finitely many rational points.

For this fundamental result, Gerd Faltings (born 1954) was awarded at the International Congress of Mathematicians in 1986 in Berkeley the *Fields Medal*, which is for mathematics what the Nobel Prize is to the natural sciences. Faltings' proof is based on very abstract mathematical models, which have only been developed since the middle of the twentieth century.

**The Shimura–Taniyama–Weil conjecture (1955):** Let $y^2 = ax^3 + bx^2 + cx + d$ be a Diophantine elliptic curve. For every prime number $p$ let $n_p$ denote the number of solutions $(x, y)$ of the equation

$$\boxed{y^2 \equiv ax^3 + bx^2 + cx + d \mod p\,.}$$

*Conjecture:* There is a modular form[37] $f$ with the Fourier expansion

$$f(z) = \sum_{n=0}^{\infty} b_n \, e^{2\pi i n \, z}$$

for all $z$ in the upper half-plane, such that for sufficiently large prime numbers $p$, we have the surprisingly simple relation

$$\boxed{b_p = p + 1 - n_p}$$

This conjecture follows the general philosophy that modular forms are a fundamental instrument for encoding countably-infinite systems, see section 1.14.18.

#### 3.8.6.3 Fermat's last theorem

*The age of modern mathematics began with the four great French mathematicians Girard Desargues (1591–1661), René Descartes (1596–1650), Pierre de Fermat (1601–1665) and Blaise Pascal (1623–1662).*

*Four people with less in common that this quartet is difficult to imagine; Desargues – the most original of the four and an architect – is described as a peculiar person, who wrote his most important work in a kind of secret language and printed it with microscopically small letters.*

*Descartes – the most famous – was first an enlisted soldier and could, if need be, defend himself with a dagger against thieves among the sailors of the Rhein; in the manner of a soldier he also prepared his general attack (Discourse sur la méthode) on the foundations of the sciences.*

*Pascal – the most ingenious – turned his back to mathematics and became in later life a religious visionary; throughout his life he suffered under constipation.*

*Finally Fermat – the most important – was employed in the king's court as advisor to the Parliament of Toulouse, a position which might best be compared with a senior civil servant today. He consequently had sufficient leisure to spend time thinking about mathematics . . .*

*Winfried Scharlau and Hans Opolka*
*in: "Von Fermat bis Minkowski", 1990*[38]

Fermat belongs to the founding fathers of analytic geometry and probability theory; his method for calculating the minima and maxima is a precursor of the differential calculus of Newton and Leibniz. In Fermat's copy of the monograph *Arithmetica* by Diophantus, the following result is written in the margin:

*Cubum autem in duos cubos, aut quadrato-quadratum in duos quadrato-quadratos, et generaliter nullam in infinitum ultra quadratum potestatem in duas ejusdem nominis fas est dividere; cujus rei demonstrationem mirabilem sane detexi. Hanc marginis exiguitas non caperet.*

[37] Modular forms are holomorphic functions $f$ defined on the upper half-plane and which satisfy $f(z+1) = f(z)$ and $f(z^{-1}) = z^{-k} f(z)$ for all complex numbers $z$ in the upper half-plane and some fixed natural number $k$.

[38] We recommend [285] as a very vivid introduction to number theory with many historical comments.

In modern terminology, Fermat is claiming here that the equation

$$\boxed{x^n + y^n = z^n}$$

for $n = 3, 4, \ldots$ has no solution in integers $x, y$ and $z$. Furthermore he writes that he has found a wonderful proof for this, but the margin is too small to write it down here. This margin remark of his seems to have been written sometime between 1631 and 1637. Since this time, this result has become known as *Fermat's last theorem*, although more correctly it should be called Fermat's conjecture.

Euler proved in 1760 this conjecture for the case $n = 3$. Between 1825 and 1830 Dirichlet and Legendre were able to complete the proof for the case $n = 4$. In 1843, Kummer sent a paper to Dirichlet, in which he claimed to have found a proof for all $n \geq 3$. However, Dirichlet found a deep gap in the proof, which amounted to Kummer taking for granted the validity of the uniqueness of the prime decomposition of integers in all number fields, which is incorrect. In order to correct this mistake, Kummer worked intensively on the laws of divisibility in number theories, laying the foundations for the later theory of divisors developed by Dirichlet. With the aid of his theory of divisibility, Kummer was able to show that Fermat's conjecture is correct for all so-called regular primes, for example for all primes $n < 100$ except for $n = 37$, 59 and 67. In 1977, Wagstaff proved with the help of computer calculations that Fermat's last theorem is correct for all primes $2 < p < 125,000$. Fermat's last theorem is equivalent to the following statement of Diophantine geometry. On the curve

$$\boxed{x^n + y^n = 1}$$

there are, for $n \geq 3$, no rational points.

**Theorem of Wiles (1994):** Fermat's last theorem is a theorem.

Andrew Wiles proved this statement by proving the Shimura–Taniyama–Weil conjecture for all so-called semistable curves. This displays the geometric essence of Fermat's last theorem. The fact that the Shimura–Taniyama–Weil conjecture implies Fermat's last theorem is itself quite deep, and was originally suggested by Gerhard Frey The idea is, supposing Fermat's last theorem were false, a solution would produce, by a complicated construction, an elliptic curve which would contradict the Shimura–Taniyama–Weil conjecture.

### 3.8.7 Analytic sets and the Weierstrass preparation theorem

Analytical sets are locally zero sets of finitely many holomorphic functions.

**Definition:** A subset $X$ of $n$-dimensional complex space $\mathbb{C}^n$ is said to be analytic, if for every point $z_0 \in X$, there is an open set $U$ of $z_0$ and finitely many holomorphic functions $f_1, \ldots, f_k : U \subseteq \mathbb{C}^n \longrightarrow \mathbb{C}$, such that $X \cap U$ is the set of all solutions of the set of equations

$$\boxed{f_1(z) = 0, \ldots, f_k(z) = 0\,, \qquad z \in U\,.}$$

**Analytic varieties:** Irreducible analytic sets are called *analytic varieties*.[39]

Since a smooth complex manifold $\mathscr{M}$ is locally isomorphic to an open set in $\mathbb{C}^n$, the notion of analytic set is carried over naturally to $\mathscr{M}$.

[39] An analytic set $X$ is said to be *irreducible*, if there is no disjoint decomposition $X = Y \cup Z$ with non-empty analytic sets $Y$ and $Z$.

**The factorization problem:** *Example:* The equation

$$\sin z = 0\,, \qquad z \in \mathbb{C} \tag{3.69}$$

has, because $\sin z = z\left(1 - \frac{z^2}{3!} + \cdots\right)$ the factorization

$$zg(z) = 0\,, \qquad z \in \mathbb{C}\,,$$

with $g(0) \neq 0$. Therefore the original problem (3.69) is, in a sufficiently small neighborhood of the origin, equivalent to the much simpler equation $z = 0$.

More generally we seek a factorization for the equation

$$\boxed{f(z,t) = 0\,,} \qquad z \in \mathbb{C}^{n-1}, \quad t \in \mathbb{C}, \quad n \geq 1$$

of the form

$$\boxed{f(z,t) = p(z,t)g(z,t)\,, \quad g(0,0) \neq 0} \tag{3.70}$$

with a holomorphic function $g : V \subseteq \mathbb{C}^n \longrightarrow \mathbb{C}$ from a neighborhood $V$ of the origin, and a $t$-polynomial

$$p(z,t) := t^k + a_{k-1}(z)t^{k-1} + \ldots + a_1(z)t + a_0(z)\,, \qquad k \geq 1\,.$$

The coefficients $a_1, \ldots, a_{k-1}$ as assumed to be holomorphic in a neighborhood of the origin.

**The Weierstrass preparation theorem:** Let $f : U \subseteq \mathbb{C}^n \longrightarrow \mathbb{C}$ be holomorphic in a neighborhood of the origin, such that the function $w = f(0,t)$ is a power series beginning with the term $t^k$. Then there is a uniquely determined factorization of the form (3.70).

### 3.8.8 The resolution of singularities

Let an analytic set $X$ in $\mathbb{C}^n$ be given, with an isolated singularity at the origin $z = 0$. We would like to find a local parametrization

$$\boxed{z = \pi(t)\,, \qquad t \in \mathscr{T}}$$

of the set $X$, such that the parameter space $\mathscr{T}$ has no singularities. We denote by $\mathscr{S} := \pi^{-1}(0)$ the critical set of parameters.

**The local uniformization theorem of Hironaka (1964):**[40]

There exists a parametrization $\pi : \mathscr{T} \longrightarrow X$ with the following properties.

(i) The parameter space $\mathscr{T}$ is a smooth complex manifold.

(ii) $\pi$ is a proper[41], holomorphic mapping from $\mathscr{T}$ to a neighborhood of the origin in $X$.

(iii) The mapping $\pi : \mathscr{T} \backslash \mathscr{S} \longrightarrow X \backslash \{0\}$ is biholomorphic.

(iv) The critical set $\mathscr{S}$ of parameters is an analytic set of $\mathscr{T}$ of codimension 1, i.e., $\dim \mathscr{S} = \dim \mathscr{T} - 1$.

[40] The Japanese mathematician Heisuke Hironaka (born 1931) received the Fields Medal in 1970 for his fundamental results on local and global uniformization. The global uniformization result is in section 3.8.9.2.

[41] A *proper* map has the important property that the inverse images of compact sets are again compact.

One refers to $\pi : \mathscr{T} \longrightarrow X$ as a *resolution of singularities* at the origin (the isolated singularity) of $X$.

**Blowing up at singularities:** In the course of it's resolution, a singularity gets *blown up* finitely many times. We now describe this process, which is quite fundamental for algebraic geometry, with a couple of examples.

*Example* (Figure 3.91): We consider the curve

$$\boxed{X : x^2 - y^2 = 0\,, \qquad (x,y) \in \mathbb{C}^2\,,} \tag{3.71}$$

which has a double point at $P = (0,0)$.

*Step 1:* We blow up the double point $P$ to a line $g_P$ (Figure 3.91(b)), i.e., replace the point by a projective line. This is effected by utilizing the parametrization

$$(x,y) = f(u,v), \qquad (u,v) \in \mathbb{C} \tag{3.72}$$

with $f(u,v) := (v, uv)$. From (3.71) we get the equation

$$\boxed{v^2(1-u^2) = 0\,, \qquad (u,v) \in \mathbb{C}^2\,.}$$

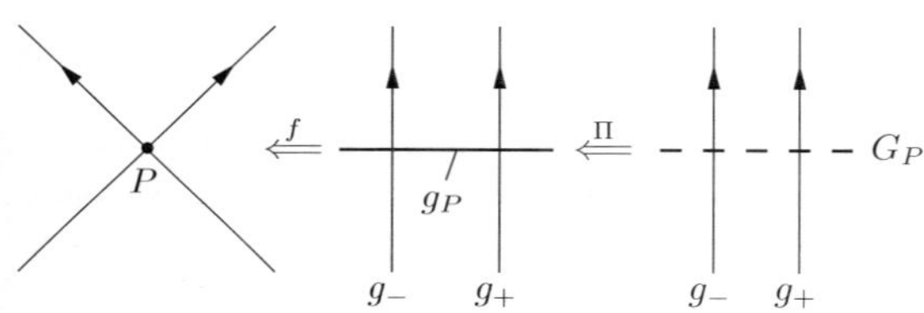


**(a)** curve with double point **(b)** blow up of the double point **(c)** embedding

*Figure 3.91. Resolving a double point.*

This leads to three lines, $g_\pm$ : $u = \pm 1$ and $g_P : v = 0$.

The lines $g_+$ and $g_-$ (resp. the line $g_P$) are transformed by (3.72) into the two lines which are components of $X$ (resp. the double point of $X$).

*Step 2:* We now embed the line $g_P$ in a three-dimensional space and obtain a line in $\mathbb{C}^3$ which we denote by $G_P$ (Figure 3.91(c)). This can be done, for example, by setting

$$G_P := ((u,v,w) \in \mathbb{C}^3 : v = 0, w = 1)\,, \quad g_\pm := \{(u,v,w) \in \mathbb{C}^3 : u = \pm 1, w = 0\}\,.$$

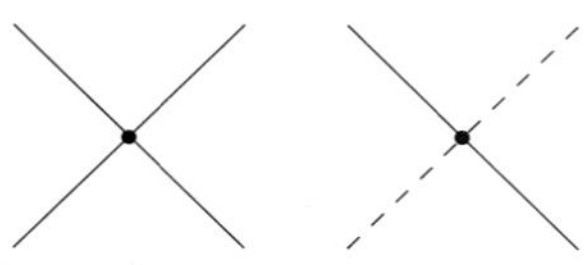

*Figure 3.92.*

*Step 3 – Construction of the resolution:* We choose as parameter space $\mathscr{T}$ the union of *three non-intersecting lines* $g_+$, $g_-$ and $G_P$. The resolution mapping $\pi : \mathscr{T} \longrightarrow X$ is given simply by setting

$$\boxed{\pi : \mathscr{T} \xrightarrow{\Pi} \mathbb{C}^2 \xrightarrow{f} X}$$

with the projection $\Pi(u,v,w) := (u,v)$.

In this example one could easily define the resolution explicitly by embedding the local construction into three-space (Figure 3.92). The advantage of the general construction of blowing up is that it can be extended to a universal kind of construction.

### 3.8.9 The algebraization of modern algebraic geometry

*Algebraic geometry has developed in waves, each with its own language and point of view. The late nineteenth century saw the function-theoretic approach of Riemann, the more geometric approach of Brill and Noether, and the purely algebraic approach of Kronecker, Dedekind, and Weber. The Italian school followed with Castelnuovo, Enriques, and Severi, culminating in the classification of algebraic surfaces. Then came the twentieth-century 'American' school of Chow, Weil, and Zariski, which gave firm algebraic foundation to the Italian intuition,. Most recently, Serre and Grothendieck initiated the French school, which has rewritten the foundation of algebraic geometry in terms of schemes and cohomology, and which has an impressive record of solving old problems with new techniques.*

*Robin Hartshorne* (1977),
*University of California at Berkeley*

Modern algebraic geometry has a strong algebraic flavor to it. This very fruitful tendency turned up in the development of the theory in the second half of the nineteenth century. Kummer (1810–1893) developed in connection with divisibility questions in number fields and his attempts to prove, with these results, Fermat's last theorem, the theory of what he called 'ideal numbers', which gave rise to Dedekind's development of ideal theory. Next, Dedekind (1831–1916) and Weber (1843–1913) discovered a purely field-theoretic formulation of the deep theorem of Riemann–Roch and exposed thus the algebraic basis of this geometric theorem.[42] Hilbert (1862–1943) proposed towards the end of the nineteenth century the general program of formulating the well-developed methods of continuous mathematics as used in analysis in such a way that they could be applied to discrete questions, in particular to number-theoretic problems and to problems in algebraic geometry, where singularities regularly occur. There was intensive work in the twentieth century in realizing this program. One of the central notions developed in this respect is that of *scheme*, which was devised by the French mathematician Alexandre Grothendieck in 1960; this notion has proved itself to be exceptionally effective in solving difficult problems. Schemes, in turn, are based on the notion of *sheaf*, which was invented in 1945 by the French mathematician Jean Leray. Modern books on algebraic geometry are written in the language of schemes (an important exception is the book *Foundations of Algebraic Geometry* by Phillip Griffiths and John Harris, which is written in the language of complex analysis in several variables and complex differential geometry, the other basic possible approach to algebraic geometry). In what follows we try to make some of these ideas accessible through a series of concrete examples.

#### 3.8.9.1 The connection with field theory

Let a plane algebraic curve

$$\boxed{C : p(x, y) = 0\,,}$$

be given, where $p$ is an irreducible polynomial in the polynomial ring $\mathbb{C}[x, y]$. We want to construct the 'quotient field mod $p$'. By definition, '$a \equiv b \mod p$' precisely when the

[42] The theorem of Riemann–Roch and its modern generalization due to Hirzebruch (accordingly known as the theorem of Riemann–Roch–Hirzebruch) are special cases of the Atiyah–Singer index theorem, one of the deepest results of mathematics in the twentieth century. This area of problems will be addressed in [212].

difference $a - b$ is divisible by $p$.

**Definition:** We consider quotients

$$\frac{f}{g}$$

with $f, g \in \mathbb{C}[x, y]$, where we assume that $g \not\equiv 0 \bmod p$, and write $\frac{f_1}{g_1} \sim \frac{f_2}{g_2}$ if and only if $f_1 g_2 \equiv f_2 g_1 \bmod p$.

**Theorem:** The set of equivalence classes with respect to this relation form a field $K(C)$, which is refereed to as the *field of rational functions on the curve* $C$.

**Rational curves:** The curve $C$ is by definition rational if it admits a parametrization

$$x = X(t)\,, \qquad y = Y(t)\,, \qquad t \in \mathbb{C}$$

with complex rational functions $X$ and $Y$.

**Main theorem:** The curve $C$ is rational if and only if the field $K(C)$ is isomorphic to a subfield of the field $\mathbb{C}(x)$ of rational functions in a variable $x$ with complex coefficients.

*Example:* For the line defined by $C:\ y = 0$, we have $K(C) = \mathbb{C}(x)$.

### 3.8.9.2 The connection with ideal theory and the theorem of Hironaka

Let $\mathscr{R} := \mathbb{C}[x_1, \ldots, x_{n+1}]$ denote the ring of polynomials in the variables $x_1, \ldots, x_{n+1}$ with complex coefficients. An ideal $\mathscr{I}$ of $\mathscr{R}$ is said to be *homogenous*, if it has a basis consisting of homogenous elements.

**Definition:** A point $P \in \mathbb{C}P^n$ is said to *annihilate* a polynomial $p \in \mathscr{R}$, if

$$p(x_1, \ldots, x_{n+1}) = 0$$

for every set of homogenous coordinates $(x_1, \ldots, x_{n+1})$ representing $P$.

**Algebraic sets:** A set $X$ of projective space $\mathbb{C}P^n$ is called *algebraic* if and only if it is the zero set of a finite set of homogenous polynomials in $\mathscr{R}$.[43]

*Example:* Every algebraic curve in $\mathbb{C}P^2$ is an algebraic set.

**Theorem:** The mapping

$$\boxed{X \mapsto \mathscr{I}_X, \quad \mathscr{I} \mapsto Z(\mathscr{I}),}$$

where $\mathscr{I}_X$ denotes the set of polynomials vanishing precisely at $X$, or, put differently, the set of polynomials annihilated by all points of $X$, which in fact forms an ideal in the polynomial ring, and $Z(\mathscr{I})$ denotes the *zero set* of $\mathscr{I}$, i.e., the set of points $P$ which annihilate all $f \in \mathscr{I}$), defines a bijective correspondence between the set of algebraic subsets $X$ in $\mathbb{C}P^n$ and the set of homogenous ideals $\mathscr{I}_X$ in $\mathscr{R}$.

> The investigation of algebraic sets in projective space can be reduced to the study of homogenous ideals in polynomial rings.

**Remark:** This mapping "reverses inclusions", in the following sense. The union and intersection of algebraic sets is well-defined as the usual union and intersection of sets.

[43] In other words, there are polynomials $p_1, \ldots, p_k \in \mathscr{R}$, such that $P \in X$ if and only if $P$ annihilates all $p_1, \ldots, p_k$.

The union and intersection of ideals is defined in the same manner. Then under the bijective mapping above, the intersection of two algebraic sets maps to the union of the ideals:

$$X \cap Y \mapsto \mathscr{I}_X \cup \mathscr{I}_Y,$$

or, put differently, $\mathscr{I}_{X\cap Y} = \mathscr{I}_X \cup \mathscr{I}_Y$. This is because a point $P \in X \cap Y$ if it is in $X$ and hence annihilates all $f \in \mathscr{I}_X$ *and* it is in $Y$ and hence annihilates all $f \in \mathscr{I}_Y$. Similarly, one has $\mathscr{I}_{X\cup Y} = \mathscr{I}_X \cap \mathscr{I}_Y$, since a point is in the union if it is either in $X$ and hence annihilates $\mathscr{I}_X$ *or* it is in $Y$ and hence annihilates $\mathscr{I}_Y$. Thus, the set of polynomials annihilated by all points in the union are those polynomials which are in both $\mathscr{I}_X$ and $\mathscr{I}_Y$.

Similarly, we have $Z(\mathscr{I} \cap \mathscr{J}) = Z(\mathscr{I}) \cup Z(\mathscr{J})$, as a point $P$ is in $Z(\mathscr{I}) \cup Z(\mathscr{J})$ if it is in $Z(\mathscr{I})$, hence annihilates $\mathscr{I}$, *or* it is in $Z(\mathscr{J})$, hence it annihilates $\mathscr{J}$. Hence it either in $Z(\mathscr{I})$ or in $Z(\mathscr{J})$. In the same way we have $Z(\mathscr{I} \cup \mathscr{J}) = Z(\mathscr{I}) \cap Z(\mathscr{J})$.

The two following theorems of Hilbert are fundamental and revolutionized invariant theory toward the end of the nineteenth century.

**Hilbert's theorem on the finite generation of ideals (1893):** Every ideal $\mathscr{I}$ in the polynomial ring $\mathscr{R}$ is finitely generated, i.e., has a finite basis.

**Hilbert's Nullstellensatz (1893):** Let $\mathscr{I}$ be an ideal in $\mathscr{R}$. If a polynomial $p \in \mathscr{R}$ vanishes at every point of the zero set of all polynomials in $\mathscr{I}$, then some power of $p$ belongs to $\mathscr{I}$. In other words, if $p \in \mathscr{I}_{Z(\mathscr{I})}$, then $p^k \in \mathscr{I}$ for some $k$.

**A more general theorem by Bézout:** Let $p_1, \ldots, p_n$ be homogenous polynomials. Then the number of points $x = (x_1, \ldots, x_{n+1})$ which satisfy

$$\boxed{p_1(x_1, \ldots, x_{n+1}) = 0, \ldots, p_{n+1}(x_1, \ldots, x_{n+1}) = 0}$$

is either infinite or at most equal to the product of the degrees of the $p_j$.

**The Zariski-topology on projective spaces:** A subset of projective space $\mathbb{C}P^n$ is said to be *closed* if it is algebraic. A subset of projective space $\mathbb{C}P^n$ is called *open* it its complement in projective space is closed.

**Theorem:** The set of open sets as just defined give rise to a topology, which is called the *Zariski topology* on $\mathbb{C}P^n$.[44]

**Projective varieties:** An irreducible algebraic set $X$ in $\mathbb{C}P^n$ is called a *projective variety*.[45]

**Rational maps:** If $X$ and $Y$ are two projective varieties in $\mathbb{C}P^n$, then a *rational map* $\varphi : X \longrightarrow Y$ is a map of the form

$$x'_j = P_j(x_1, \ldots, x_{n+1}), \qquad j = 1, \ldots, n+1.$$

Here all $P_j$ are homogenous polynomials of the same degree.

A map $\varphi : X \longrightarrow Y$ is said to be *birational*, if $\varphi$ is bijective and both $\varphi$ and $\varphi^{-1}$ are rational.

---

[44]The notion of topology is defined and discussed in [212]. It amounts to an axiomization of the situation of open sets in $\mathbb{R}^n$, which can, however, be applied to many mathematical objects. As soon as a topology exists on a space, the entire apparatus of topology theory may be applied to this space.

[45]Irreducibility here means that there is no disjoint decomposition $X = Y \cup W$ into non-empty algebraic sets $Y$ and $W$.

**The category of projective varieties:** The objects of this category are projective varieties, and the morphisms are the rational maps between these varieties. Isomorphisms in the category are birational maps.

The following theorem is a very deep result of modern mathematics.

**The global uniformization theorem of Hironaka (1964):** Let $X$ be a projective variety with the singular set $S \subset X$. Then there is a surjective mapping

$$\pi : \mathscr{T} \longrightarrow X$$

with the following properties:

(i) $\mathscr{T}$ is a non-singular projective variety.

(ii) $\pi$ is a morphism.

(iii) Let $\mathscr{S} := \pi^{-1}(S)$. Then $\pi : \mathscr{T} \backslash \mathscr{S} \longrightarrow X \backslash S$ is an isomorphism.

The space $\mathscr{T}$ which exists by virtue of this theorem is called the *resolution of singularities of* $X$.

### 3.8.9.3 Local rings

Local rings make the use of notions like locally near a point possible through the use of algebraic tools.

Let $\mathscr{I}$ be an ideal in a ring $\mathscr{R}$. Then $\mathscr{I}$ is by definition *trivial* if $\mathscr{I} = \{0\}$ or $\mathscr{I} = \mathscr{R}$.

**Maximal ideals:** The ideal $\mathscr{I}$ is *maximal* if it cannot be extended to a larger ideal.

**Basic idea:** Let $C(X)_{\mathbb{C}}$ denote the ring of complex continuous functions

$$\boxed{f : X \longrightarrow \mathbb{C}}$$

on a non-empty compact topological space $X$ (for example, $X$ could be a compact subset of $\mathbb{R}^n$). We associate to every point $P \in X$ the set

$$\mathscr{I}_P := \{f \in C(X) : f(P) = 0.\}$$

A $*$-ideal of $C(X)_{\mathbb{C}}$ is an ideal which contains for every function $f$ also the complex conjugate function $\overline{f}$.

**Theorem:** Every set $\mathscr{I}_P$ as just defined is a maximal $*$-ideal in $C(X)_{\mathbb{C}}$. The mapping

$$\boxed{P \longmapsto \mathscr{I}_P}$$

yields a bijective map from the topological space $X$ and the set of maximal $*$-ideals in $C(X)_{\mathbb{C}}$.

With this result one can associate to the points of a geometric object corresponding algebraic objects, namely maximal $*$-ideals.[46]

**Local rings:** A *local ring* is a Noetherian ring $\mathscr{R}$ with a non-trivial ideal $\mathscr{I}$ which contains all non-trivial ideals of $\mathscr{R}$.

**The local ring at a point $P$ on an algebraic curve:** Suppose we are given an algebraic curve

$$\boxed{C : p(x, y) = 0}$$

[46]This is the starting point for modern non-commutative geometry.

where $p$ denotes an irreducible polynomial in $\mathbb{C}[x, y]$.

(i) A function $f : C \longrightarrow \mathbb{C}$ on the curve is said to be *regular*, if it is the restriction of a polynomial in $\mathbb{C}[x, y]$ to the curve $C$.

(ii) We choose a fixed point $P$ on the curve. If $f$ and $g$ are two regular functions on $C$, then we write

$$f \sim g \bmod P\,,$$

if and only if $f$ and $g$ coincide in some neighborhood of the point $P$. This is an equivalence relation. The set of equivalence classes $[f]$ carry the natural structure of a ring, which we denote by $\mathscr{K}_P$ and call the *ring of germs of regular functions at* $P$.

**Theorem:** $\mathscr{K}_P$ is a local ring.

This ring is an algebraic replacement or representative of the point $P$.

**The localization of a ring:** If $\mathscr{P}$ is a non-trivial prime ideal in a Noetherian ring $\mathscr{R}$, then the *quotient ring* $\mathscr{R}_{\mathscr{P}}$ is a local ring. This ring consists of all fractioins

$$\frac{r}{s}$$

of elements $r$ and $s$ in $\mathscr{R}$; the usual rules for calculations with fractions are applied to such expressions. We require in addition that the demoninator $s$ is not in the prime ideal $\mathscr{P}$.

#### 3.8.9.4 Schemes

**Ringed spaces:** A *ringed space* consists of a topological space $X$ (for example, a metric space) and a sheaf $\mathscr{G}$, which associates to each open set $U$ of $X$ a ring $R(U)$.[47]

**Standard example 1:** Every topologicial space $X$ forms with the sheaf $\mathscr{G}$ of continuous functions a ringed space. This sheaf associates to each open set $U \subseteq X$ the ring $R(U)$ consistsing of all continuous functions

$$\boxed{f : U \longrightarrow \mathbb{R}.}$$

To obtain the localization at a point $x \in X$, we write

$$f \sim g \bmod x\,,$$

if and only if the continuous real-valued functions $f$ and $g$ coincide in a neighborhood of the point $x$. The set of corresponding equivalence classes $[f]$ forms a ring, the so-called

$$\boxed{\textit{stalk } \mathscr{G}_x \text{ of the sheaf } \mathscr{G} \text{ at the point } x.}$$

**The fundamental notion of a scheme:** A ringed space

$$\boxed{(X, \mathscr{G})}$$

is said to be a *scheme* if it looks locally like a given ringed space $(X_0, \mathscr{G}_0)$.

[47] Topological spaces and sheaves are introduced in [212]. These notions present a kind of axiomatization of the Standard example 1 which follows.

Explicitly this means that for every open set $U \subseteq X$ the restriction $(U, \mathscr{G})$ is isomorphic to $(X_0, \mathscr{G}_0)$.

We denote by $C^\infty(U)$ the ring of all $C^\infty$-functions $f : U \longrightarrow \mathbb{R}$.

**Standard example 2:** Every $n$-dimensional $C^\infty$-manifold $X$ defines a scheme.[48]

(i) The manifold $X$ is made into a ringed space by taking $\mathscr{G}$ to be the sheaf of all smooth functions on $X$. This sheaf associates to every open set $U$ in $X$ the ring $C^\infty(U)$.

(ii) As the "comparison ring" $(X_0, \mathscr{G}_0)$ we choose $X_0 := \mathbb{R}^n$ with the sheaf $\mathscr{G}_0$ of smooth functions on $\mathbb{R}^n$. We have:

$$\boxed{(U, C^\infty(U)) \text{ is isomorphic to } (\mathbb{R}^n, C^\infty(\mathbb{R}^n)).}$$

### 3.8.9.5 Affine schemes

**Definition:** A scheme is said to be *affine* if the given "comparison ring" $(X_0, \mathscr{G}_0)$ is the spectrum $(\operatorname{Spec} R, \mathscr{G}_0)$ of a ring $R$.

We now explain the notion of spectrum of a ring $R$. For this, suppose that $R$ is a commutative ring with unit element.

**The underlying space Spec $R$ of the spectrum:** We denote by $\operatorname{Spec} R$ the set of all prime ideals of $R$.

*Example:* The space $\operatorname{Spec} \mathbb{Z}$ of the ring $\mathbb{Z}$ of integers consists of the set of all principal ideals

$$\boxed{(p), \quad p \text{ a prime number.}}$$

Moreover, the zero ideal $\{0\}$ belongs to $\operatorname{Spec} \mathbb{Z}$.

**The topology on Spec $R$:** If $\mathscr{I}$ is an ideal of $R$, then we denote by $V(\mathscr{I})$ the set of all prime ideals of $R$ which contain $\mathscr{I}$.

The sets of the form $V(\mathscr{I})$ are defined to be the *closed* sets. The open sets are then by definition the complements of these. This defines a topology on $\operatorname{Spec} R$.

**The sheaf $\mathscr{G}_0$ on** $\operatorname{Spec} R$**:** If $\mathscr{P} \in \operatorname{Spec} R$, then we denote by $R_{\mathscr{P}}$ the localization of the ring $R$ with respect to the prime ideal $\mathscr{P}$ (cf. 3.8.9.3).

Let an open set $U$ in $\operatorname{Spec} R$ be given. We assoiciate to $U$ the ring $\mathscr{R}(U)$ of all functions $f$ on $U$ with

$$\boxed{f(\mathscr{P}) \in R \quad \text{for all} \quad \mathscr{P} \in U.}$$

These functions $f$ are assumed to be locally quotients of elements of the original ring $R$, i.e., for every $\mathscr{P} \in U$ there is a neighborhood $V$ of $\mathscr{P}$ and ring elements $r, s \in R$, such that for every $\mathscr{Q} \in V$ we have

$$\boxed{f(\mathscr{Q}) = \frac{r}{s} \text{ with } s \notin \mathscr{Q}.}$$

**Localized schemes:** Because of the great importance of local rings for algebraic geometry, one often assumes in the definition of a scheme in addition to the above that the stalk $\mathscr{G}_x$ of the sheaf $\mathscr{G}$ on the ringed space $(X, \mathscr{G})$ is a local ring for every point $x \in X$.

[48] The notion of *manifold* is introduced in [212].

## 3.9 Geometries of modern physics

*The vision of space and time which I will now present to you are based on experimentally verified facts of physics. This is its strength. It gives a radical departure from our previous notions.*

*Starting with this lecture, space and time as individual entities shall cease to exist, and there will only remain a union of both as the sole notion of reality.*

*Hermann Minkowski,*
*Conference of the Society of German Scientists and Doctors in Cologne in 1908*

Modern physics is formulated in the language of geometry. Physical phenomena correspond to geometric objects. The description of physical observations in different frames of reference is given by the coordinates of geometric objects in different coordinate systems.

### 3.9.1 Basic ideas

**Pseudo-unitary geometry and the theory of relativity:** The geometrization of physics has its birth in work of Minkowski, who in 1908 interpreted the special theory of relativity, which Einstein had laid down three years previously, as a pseudo-unitary geometry of four-dimensional space-time and showed that the Lorentz transformations used by Einstein to connect different inertial frames form the *symmetry group* of Minkowski space.

Einstein's theory of gravity (the general theory of relativity) of 1915 geometrizes the force of gravity, which corresponds to the curvature of the four-dimensional pseudo-Riemannian space-time manifold. In the case of vanishing curvature, one gets back the Minkowski space of the special theory of relativity.

**Unitary geometry and quantum theory:** Modern quantum mechanics was created in 1925 by Heisenberg (1901–1976) as a matrix-mechanical system and in 1926 by Schrödinger (1887–1961) as a wave-mechanical system. In 1928, Dirac (1902–1984) invented a mathematical formalism which showed that these two different systems of mechanics were but different views of one and the same theory, namely the theory of abstract Hilbert spaces. At about the same time, John von Neumann (1903–1957) recognized that quantum theory can be formulated as a rigorous mathematical theory of self-adjoint operators in a Hilbert space. The spectrum of the energy operator (the Hamiltonian) is identical with the possible energy values of the quantum system. The Heisenberg uncertainty principle states that one can not at the same time exactly measure both the position and the velocity of a quantum particle. This fundamental fact of quantum theory has a geometric origin. It follows from an infinite-dimensional analog of the fact that for the scalar product

$$\mathbf{ab} = |\mathbf{a}|\,|\mathbf{b}|\,\cos\gamma$$

in three-space one always has (from $|\cos\gamma| \le 1$) the Schwarz inequality

$$\boxed{|\mathbf{ab}| \le |\mathbf{a}|\,|\mathbf{b}|}$$

(cf. [212]). Behind quantum theory, the unitary geometry of Hilbert spaces is lurking.

**Spin geometry, Clifford algebras and the spin of the electron:** To give an interpretation to the splitting of the spectral lines of atoms in magnetic fields, which was experimentally observed at the beginning of the twentieth century, Uhlenbeck and Goudsmit postulated in 1924 the existence of a self-rotation impulse of the electron, which they dubbed the *spin*. Four years later, Dirac formulated his famous fundamental equation for the relativistic electron with the help of the Clifford algebra which derives from the Minkowski metric (see section 2.4.3.4). The spin of the electron arises quite naturally from this equation, so that the spin can be viewed as a relativistic effect. In close connection with the formalism of the spin, there is a spin geometry which can be elegantly described with the help of inner algebra (Clifford algebras) and vector spaces.

The most simple spin geometry arises from the Clifford algebra of a Hilbert space. This geometry is the theory of invariants of the spin group $\mathrm{Spin}(n)$. In particular, one has

$$\boxed{\mathrm{Spin}(3) = SU(2),}$$

and this is the group describing the spin of the electron.

Clifford algebras also play a central role in the formulation of the contemporary standard model of elementary particles, which unifies electromagnetism with the weak and strong forces using gauge theory.

**Symplectic and classical mechanics:** Symplectic geometry is based on a skew-symmetric bilinear form. This geometry is the basis of classical geometric optics, classical mechanics (for example celestial mechanics) and the classical statistical physics originating with Gibbs.

**Kähler geometry and string theory:**

> The importance of Kähler geometry is the fact that it is a synthesis of symplectic and unitary geometry.

This geometry, which has its origin in the work of Erich Kähler (born in 1906) in 1932, is decisive in the formulation of string theory, whose goal it is to unify *all* the fundamental forces of nature, including the general theory of relativity (i.e., the theory contains a graviton). The theorem of Yau and the so-called Calabi–Yau spaces are fundamental in this theory as the configuration spaces of strings (cf. [212]).

**Conformal geometry:** The theory of complex functions of one variable is a specialization of a more basic structure, conformal geometry, since biholomorphic mappings are conformal (angle-preserving). The group of all biholomorphic mappings of the Riemannian sphere into itself is the automorphic group, consisting of all Möbius transformations (cf. 1.14.11.5).

The proper Lorentz group $SO^+(3,1)$ can be described as the group $SL(2,\mathbb{C})$ of complex $(2 \times 2)$-matrices with determinant equal to one (this is one of the so-called *exceptional isomorphisms* between Lie groups of low dimension). This group is isomorphic to the subgroup of all Möbius transformations which map the upper half-plane conformally onto itself. Meromorphic functions which are invariant under discrete subgroups of the automorphic group are called *automorphic functions*. This is an exceptionally important class of functions, which contains for example the elliptic modular function $J$ (see section 1.14.18) and occurs also during the calculation of Abelian integrals.

The mathematical richness of string theory is based on the fact that this theory is invariant under conformal transformations (conformal quantum field theory) and that the group of conformal transformation on two-dimensional Riemannian manifolds is very large, compared with higher dimensions.

**Infinitesimal symmetries:** If one linearizes the symmetry group $G$ of a geometry in a neighborhood of the identity element, one gets, in the language of physicists, the so-called infinitesimal symmetries. In the language of mathematics, symmetries of this kind are just elements of the Lie algebra $\mathscr{L}(G)$ of the Lie group $G$.

The most important result of the theory of Lie groups, founded by Sophus Lie (1842–1899) is the following statement:

> The Lie algebra $\mathscr{L}(G)$ contains all information on the structure of the Lie group $G$ near the identity element.

However, several different Lie groups can belong to the same Lie algebra, i.e., have an identical structure near the identity element. There is a privileged group among these, the universal covering group of the Lie algebra, which has the special property of being simply connected (cf. 1.3.2.4).

*Examples:* (i) The universal covering group of the group $SO(3)$ of all rotations in three-space is the group $SU(2)$ which is responsible for the spin of the electron.

(ii) The universal covering group of $SO(n)$ for $n \geq 3$ is the spin group $\mathrm{Spin}(n)$; for $n = 3$ we have $\mathrm{Spin}(3) = SU(2)$.

(iii) The universal covering group of $SO^+(3,1)$ of proper Lorentz transformations is the group $SL(2,\mathbb{C})$ (cf. [212]).

**Manifolds:** In this section we consider the geometries which are relevant for modern physics, formulated in linear spaces (linear manifolds). These geometries are all related to bilinear forms, which can all be viewed as generalized scalar products.

However, it is decisive in physics to have not only this theory in the linear spaces, but also on manifolds.

> A manifold is a global geometric object which locally looks like a linear space. Using this fact, all geometries on linear spaces can be extended to manifolds.[49]

It might be helpful to think of the curved face of the earth when trying to visualize the notion of manifold, which is locally mapped through the maps of a geographical atlas.

**The mathematical efficacy of nature:** One observation is imposed upon us while studying the geometries which occur in modern physics. From a mathematical point of view there is an incredible spectrum of symmetry groups (Lie groups and Lie algebras), to which there correspond geometries by the Erlanger Program of Felix Klein (cf. 3.1).

| | |
|---|---|
| $SL(2,\mathbb{C})$ | (relatvity), |
| $U(1)$ | (electromagnatism), |
| $SU(2)$ | (electron spin), |
| $SU(3)$ | (quark structure), |
| $U(1)\times SU(2)\times SU(3)$ | (the standard model). |

Even today there is not a complete classification of all possible groups which can occur. At our present state of understanding, however, just a few of these are sufficient to describe all physical phenomena of nature. These are the groups indicated in the box above. The physicist and Nobel price winner Eugene

[49]In particular, the fundamental principle *force* = *curvature* is discussed in detail in [212].

Wigner (1902–1995), who worked for a long time at Princeton, spoke in this connection of the "unreasonable efficiency of mathematics".

**Convention:** In what follows we consider *finite-dimensional* linear spaces $X$ over a field $\mathbb{K}$ with

$$\boxed{\dim X = n\,.}$$

The elements of $X$ will be referred to as vectors. We will assume that $\mathbb{K} = \mathbb{R}$ (the reals) or $\mathbb{K} = \mathbb{C}$ (complex numbers).

**Leibniz' vision:** Following a suggestion of Leibniz (1646–1716) one works in geometry directly with the geometrical objects and avoids whenever possible the use of coordinates. Only in this manner is it possible to generalize the results of the following sections to infinite-dimensional spaces, which describe in physics systems with an infinite number of degrees of freedom. This is described in relation with functional analysis in [212].

### 3.9.2 Unitary geometry, Hilbert spaces and elementary particles

Unitary geometry is based on the notion of a positive definite scalar product, which generalizes the classical scalar product **uv** for vectors **u** and **v** (cf. 1.8.3). All important notions of unitary geometry can be, as we shall see, directly interpreted in the quark model of elementary particles.

**Definition:** A *unitary space* is a linear space over $\mathbb{K}$, on which we are given a *scalar product*. This means that we associate to every pair $u, v \in X$ of vectors a number $(u, v) \in \mathbb{K}$, such that for all $u, v, w \in X$ and all scalars $\alpha, \beta \in \mathbb{K}$ we have:

(i) $(u, v) \geq 0\,;\ (u, u) = 0$ if and only if $u = 0\,;$

(ii) $(w, \alpha u + \beta v) = \alpha(w, u) + \beta(w, v)\,;$

(iii) $\overline{(u, v)} = (v, u)\,.$

From (ii) and (iii) we get

$$(\alpha u + \beta v, w) = \overline{\alpha}(u, w) + \overline{\beta}(v, w) \quad \text{for all} \quad u, v, w \in X\,, \qquad \alpha, \beta \in \mathbb{K}\,.$$

Here, $\overline{\alpha}$ denotes the complex conjugate of a complex number $\alpha$, i.e., if $\mathbb{K} = \mathbb{R}$, this is the identity and the bars can be omitted.

**Hilbert space:** Every finite-dimensional unitary space is at the same time a Hilbert space in the sense of the general definition which can be found in [212].

**The adjoint operator:** If $A : X \longrightarrow X$ is a linear operator, then there is a naturally associated linear operator $A^* : X \longrightarrow X$ which satisfies the relation

$$\boxed{(u, Av) = (A^*u, v)}$$

for all $u, v \in X$. $A^*$ is called the *adjoint operator.*

**The unitary group $U(n, X)$:** An operator $U : X \longrightarrow X$ is said to be *unitary,* if it preserves the scalar product, i.e., if $U$ is linear and the relation

$$\boxed{(Uv, Uw) = (v, w)}$$

is satisfied for all $v, w \in X$. The set of all unitary operators on $X$ forms a group, which is called the *unitary group* and is denoted $U(n, X)$. All operators in the group which satisfy in addition

$$\det U = 1$$

form a subgroup of $U(n, X)$, which is denoted by $SU(n, X)$ and is called the *special unitary group*. A linear operator $U : X \longrightarrow X$ belongs to $U(n, X)$ if and only if

$$\boxed{UU^* = U^*U = I.}$$

In the case of a real space (i.e., $\mathbb{K} = \mathbb{R}$), we write $O(n, X)$ (resp. $SO(n, X)$) instead of $U(n, X)$ (resp. $SU(n, X)$) and call the group in this case the *orthogonal group* (resp. the *special orthogonal group*).[50]

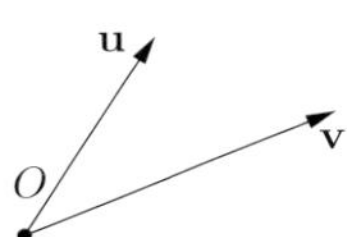


Figure 3.93.

*Example 1:* The position vectors $\mathbf{u}, \mathbf{v}$ with respect to some origin $O$ of the three-space of our world form a real, three-dimensional Hilbert space $H$ with the usual scalar product

$$\boxed{(\mathbf{u}, \mathbf{v}) = \mathbf{uv}}$$

(see Figure 3.93). The group $SO(3, H)$ consists of all rotations around the point $O$. If one adds the reflection $\mathbf{u} \mapsto -\mathbf{u}$ at the point $O$ (that is, one considers the group of all transformation which are compositions of elements of $SO(3, H)$ and the reflection), then one gets the group $O(3, H)$.

**Unitary geometry:** By definition, a property belongs to the unitary geometry of a Hilbert space $X$ if and only if it is invariant under all operators of the group $U(n, X)$. All of the following properties belong to unitary geometry in this sense. An exception is the volume, which is only invariant under transformations which preserve the orientation.

Unitary geometry generalizes the intuitive Example 1 to arbitrary dimensions.

**Orthogonality:** Two vectors $u, v \in X$ are said to be *orthogonal*, if they satisfiy the relation

$$\boxed{(u, v) = 0.}$$

If $L$ is a linear subspace of $X$, then we denote by $L^\perp$ the *orthogonal complement* of $L$. By definition, this means

$$L^\perp := \{w \in X \mid (v, w) = 0 \quad \text{for all} \quad v \in L\}.$$

For an arbitrary vector $u \in X$ there is a unique decomposition

$$\boxed{u = v + w\,, \qquad v \in L\,,\, w \in L^\perp\,.}$$

In particular, $X = L \oplus L^\perp$, and one has the formula for the dimensions,

$$\dim L + \dim L^\perp = \dim X\,.$$

[50]The groups $U(n, X)$, $SU(n, X)$, $O(n, X)$ and $SO(n, X)$ are real compact Lie groups of dimensions

$$\dim U(n, X) = n^2, \qquad \dim SU(n, X) = n^2 - 1\,,$$

$$\dim O(n, X) = \dim SO(n, X) = \frac{n(n-1)}{2}\,.$$

These dimensions indicate on how many parameters the groups depend (cf. [212]).

**Length and distance:** To every vector $u \in X$ we associate a *length* $||u||$ by setting

$$\boxed{||u|| := \sqrt{(u,u)}\,.}$$

One has $||u|| > 0$ if and only if $u \neq 0$. Moreover, $||u|| = 0$ for $u = 0$. Then the number

$$\boxed{d(u,v) := ||u - v||}$$

is called the *distance* between the vectors $u$ and $v$. With this notion of distance, every Hilbert space becomes a metric space.

**Angles:** We further associate to two given vectors $u, v \in X$ in a real Hilbert space $X$ a uniquely defined *angle* $\alpha$ between the vectors by the relation

$$\cos\alpha = \frac{(u,v)}{||u||\,||v||}, \qquad 0 \leq \alpha \leq \pi\,.$$

Here $u \neq 0$ and $v \neq 0$ are assumed. In an arbitrary real or complex Hilbert space $X$ one has the *Schwarz inequality*

$$\boxed{|(u,v)| \leq ||u||\,||v|| \quad \text{for all} \quad u, v \in X\,.}$$

**Orthonormal basis:** $n$ given vectors $e_1, \ldots, e_n$ in $X$ form an *orthonormal basis*, if

$$\boxed{(e_j, e_k) = \delta_{jk}\,, \qquad j, k = 1, \ldots, n\,.}$$

In this case one has for every vector $u \in X$ the Fourier representation

$$u = \sum_{j=1}^{n} u_j e_j \tag{3.73}$$

with the Fourier coefficients

$$u_j := (e_j, u)\,.$$

The tuple $(u_1, \ldots, u_n)$ forms by definition the Cartesian coordinates of the vector $u$ with respect to the basis $e_1, \ldots, e_n$.

**Basis theorem:** Every Hilbert space has an orthonormal basis.

**Construction of unitary operators:** If $e_1, \ldots, e_n$ and $e'_1, \ldots, e'_n$ are two arbitrary basis in a Hilbert space $X$, then the map defined by

$$Ue_j := e'_j\,, \qquad j = 1, \ldots, n \tag{3.74}$$

is a unitary operator $U : X \longrightarrow X$. In this manner one obtains all unitary operators on $X$.

**Theorem 1:** A linear operator $U : X \longrightarrow X$ is unitary if and only if every orthonormal basis is again mapped to an orthonormal basis.

**Orientation:** An *orientation* of a Hilbert space $X$ is provided by choosing a fixed orthonormal basis $e_1, \ldots, e_n$ and defining the *volume form*

$$\boxed{\mu := \mathrm{d}x^1 \wedge \mathrm{d}x^2 \wedge \ldots \wedge \mathrm{d}x^n\,.}$$

Here $\mathrm{d}x^j : X \longrightarrow \mathbb{R}$ is a linear map with $\mathrm{d}x^j(e_k) = \delta_{jk}$ for all $j, k = 1, \ldots, n$. If $b_1, \ldots, b_n$ is an arbitrary basis in $X$, then we call the number

$$\boxed{\alpha := \operatorname{sgn} \mu(b_1, \ldots, b_n)}$$

the orientation of the basis. One always has $\alpha = 1$ (positive orientation) or $\alpha = -1$ (negative orientation).

**Theorem 2:** (i) For an arbitrary orthonormal basis $e'_1, \ldots, e'_n$ one has

$$\mu = \alpha \mathrm{d}x'^1 \wedge \mathrm{d}x'^2 \wedge \ldots \wedge \mathrm{d}x'^n ,$$

i.e., the definition of the volume form $\mu$ depends only on the orientation of the Hilbert space.

(ii) A unitary transformation $U$ preserves the orientation if and only if $\det U = 1$, i.e., if $U \in SO(n, X)$.

**Volume:** Let $\mathscr{G}$ denote a bounded domain in a real oriented Hilbert space $X$. The volume of $\mathscr{G}$ is defined by

$$\boxed{\operatorname{Vol}(\mathscr{G}) := \int_{\mathscr{G}} \mu.}$$

To interpret this formula, we use the decomposition (3.73). Let $G$ denote the set of all Cartesian coordinates $(x_1, \ldots, x_n)$ of the points in $\mathscr{G}$. We then get the classical formula

$$\int_{\mathscr{G}} \mu = \int_{\mathscr{G}} G \, \mathrm{d}x_1 \mathrm{d}x_2 \ldots \mathrm{d}x_n .$$

This volume depends only on the choice of an orientation of the Hilbert space $X$. Reversing the orientation leads to a sign change of the volume.

If $U : X \longrightarrow X$ is an operator in $O(n, X)$, then we have

$$\boxed{\operatorname{Vol}(U\mathscr{G}) = (\operatorname{sgn} U) \operatorname{Vol}(\mathscr{G})}$$

with $\operatorname{sgn} U = \pm 1$ and $\operatorname{sgn} U = 1$ if $U \in SO(n, X)$.

**Infinitesimal unitary operators and the Lie algebra $u(n, X)$:** An operator $A : X \longrightarrow X$ is said to be *infinitesimally unitary*, if it is linear and satisfies the relation

$$\boxed{(u, Av) = -(Au, v) \qquad \text{for all} \quad u, v \in X.}$$

This is equivalent to $A^* = -A$ ($A$ is skew-hermitian). The set of all such operators is denoted $u(n, X)$, with the set of all such operators satisfying in addition $\operatorname{tr} A = 0$ (the infinitesimal version of $\det A = 0$) forming by definition the set $su(n, X)$.

If $X$ is a real Hilbert space, then we denote $u(n, X)$ (resp. $su(n, X)$) by $o(n, X)$ (resp. $so(n, X)$).

**Theorem 3:** Let $X$ be a complex Hilbert space with $\dim X = n$.

(i) With respect to the Lie bracket

$$[A, B] := AB - BA \tag{3.75}$$

and the usual linear combinations $\alpha A + \beta B$, $\alpha, \beta \in \mathbb{R}$ of linear operators, $u(n, X)$ becomes a real vector space and a Lie algebra with $\dim u(n, X) = n^2$.

Moreover, $su(n, X)$ is a Lie subalgebra of $u(n, X)$ with $\dim su(n, X) = n^2 - 1$.

(ii) From $A \in u(n, X)$ it follows that $\exp(A) \in U(n, X)$ ($\exp(A)$ is a power series of operators). Conversely, there is a number $\varepsilon > 0$ such that for each $U \in U(n, X)$ with $||I - U|| < \varepsilon$ there is a unique operator $A \in u(n, X)$ such that

$$\boxed{U = e^A,} \tag{3.76}$$

i.e., $A = \ln U$ (this is again a power series in operators).

(iii) From $A \in su(n, X)$ it follows that $\exp(A) \in SU(n, X)$. Conversely, there is a number $\varepsilon > 0$ such that for each $U \in SU(n, X)$ with $||I - U|| < \varepsilon$ there is a unique operator $A \in su(n, X)$ which satisfies the equation (3.76).

The statements (ii) (resp. (iii)) mean that $u(n, X)$ (resp. $su(n, X)$) represents the Lie algebra of the Lie group $U(n, X)$ (resp. $SU(n, X)$) (cf. [212]).

**Theorem 4:** Let $X$ be a real Hilbert space with $\dim X = n$.

(i) With respect to the Lie bracket (3.75), $o(n, X)$ is a real vector space and a Lie algebra with $\dim o(n, X) = n(n-1)/2$.

Moreover, $so(n, X)$ is a Lie subalgebra of $o(n, X)$ with $\dim so(n, X) = n(n-1)/2$.

(ii) From $A \in o(n, X)$ it follows that $\exp(A) \in O(n, X)$. Conversely, there is a number $\varepsilon > 0$ such that for each $U \in O(n, X)$ with $||I - U|| < \varepsilon$ there is a unique operator $A \in o(n, X)$ such that $U = \exp(A)$, i.e., $A = \ln U$.

(iii) From $A \in so(n, X)$ it follows that $\exp(A) \in SO(n, X)$. Conversely, there is a number $\varepsilon > 0$ such that for each $U \in SO(n, X)$ with $||I - U|| < \varepsilon$ there is a unique operator $A \in so(n, X)$ which satisfies the equation (3.76).

According to (ii) (resp. (iii)) $o(n, X)$ (resp. $so(n, X)$) is the Lie algebra of the Lie group $O(n, X)$ (resp. $SO(n, X)$).

**Construction of a Hilbert space:** Let $e_1, \ldots, e_n$ be a basis of a linear space $X$. We define

$$(e_j, e_k) := \delta_{jk} \text{ for all } j, k\,. \tag{3.77}$$

In this way, $X$ becomes a Hilbert space with the orthonormal basis $e_1, \ldots, e_n$. More explicitly, we have

$$(u, v) = \overline{x}_1 y_1 + \overline{x}_2 y_2 + \ldots + \overline{x}_n y_n\,,$$

where $(x_j)$ and $(y_j)$, respectively, denote the Cartesian coordinates of $u$ and $v$, respectively.

This gives every linear space the structure of a Hilbert space. The scalar product depends, however, on the choice of basis.

**Application to the quark model of particle physics:**
We consider a three-dimensional complex Hilbert space $X$ with an orthonormal basis

$e_1, e_2, e_3$. We interpret[51]

$e_1$ as the $u$-quark (up), $e_2$ as the $d$-quark (down), and $e_3$ as the $s$-quark (strange).

The physical states correspond to the vectors in $X$ of unit length. If $u \in X$ has $||u|| = 1$, then the Fourier representation

$$\boxed{u = (e_1, u)e_1 + (e_2, u)e_2 + (e_3, u)e_3}$$

allows the following physical interpretation

$$\boxed{|(e_j, u)|^2 \text{ is the probability that a quark } e_j \text{ exists in the state } u.}$$

This uses the relation

$$|(e_1, u)|^2 + |(e_2, u)|^2 + |(e_3, u)|^2 = (u, u) = 1.$$

Two unit vectors $u$ and $v$ represent by definition the same physical state if $u = \lambda v$ with a complex number $\lambda$ of absolute value $|\lambda| = 1$.

The *physical quantities* (observables) are represented by *self-adjoint operators* $A : X \longrightarrow X$, i.e., such that $A = A^*$. The number

$$\boxed{\overline{A} := (u, Au)}$$

which is always real, is the expectation value of the physical observable $A$ being in the state $u$ upon measurement. The corresponding variance $\Delta A \geq 0$ follows from

$$(\Delta A)^2 := \overline{(A - \overline{A})^2} = \left(u, (A - \overline{A})^2\, u\right).$$

**Hypercharge and the isospin of a quark:** The decisive role in the mathematical model of the quarks is played by the group $SU(3, X)$ and its Lie algebra $su(3, X)$. By definition, the *Cartan algebra* $\mathscr{C} = \mathscr{C}(\mathscr{L})$ of a Lie algebra $\mathscr{L}$ is the largest commutative subalgebra of $\mathscr{L}$.

For $su(3, X)$, one has $\dim \mathscr{C} = 2$. A basis of $\mathscr{C}$ is obtained by taking $\mathrm{i}\mathscr{T}_3$ and $\mathrm{i}\mathscr{Y}$, where $\mathscr{T}_3$, $\mathscr{Y} : X \longrightarrow X$ are self-adjoint linear operators. One has explicitly:

$$\boxed{\begin{array}{lll} \mathscr{T}_3 e_1 = \frac{1}{2} e_1\,, & \mathscr{T}_3 e_2 = -\frac{1}{2} e_2\,, & \mathscr{T}_3 e_3 = 0\,, \\ \mathscr{Y} e_1 = \frac{1}{3} e_1\,, & \mathscr{Y} e_2 = \frac{1}{3} e_2\,, & \mathscr{Y} e_3 = -\frac{2}{3} e_3\,. \end{array}} \tag{3.79}$$

One refers to $\mathscr{T}_3$ (resp. $\mathscr{Y}$) as the operator of the third component of the isospin (resp. hypercharge). The eigenvalues $T_3$ (resp. $Y$) of $\mathscr{T}_3$ (resp. $\mathscr{Y}$) are referred to as the third component of the isospin (resp. hypercharge) of the corresponding quark particle.

---

[51] In nature there are six quarks. The last of these, the top quark, was not experimentally verified until 1994. A proton consists of two $u$-quarks and one $d$-quark. This corresponds to the state

$$\boxed{p = \frac{1}{\sqrt{2}}(e_1 \otimes e_1 \otimes e_2 - e_2 \otimes e_1 \otimes e_1)\,.} \tag{3.78}$$

A more detailed discussion of the physics and mathematics of the quark model will be given in [212].

*Example 2:* According to (3.79), we have $T_3 = 1/2$ and $Y = 1/3$ for the $u$-quark $e_1$ (Figure 3.94).

**The charge operator for quarks:** According to Gell–Mann and Nishijima (1953), the charge operator $\mathscr{Q}$ for elementary particles is given by the famous formula

$$\boxed{\mathscr{Q} := \left(\mathscr{T}_3 + \frac{1}{2}(\mathscr{Y} + \mathscr{S})\right)|e|\,.}$$

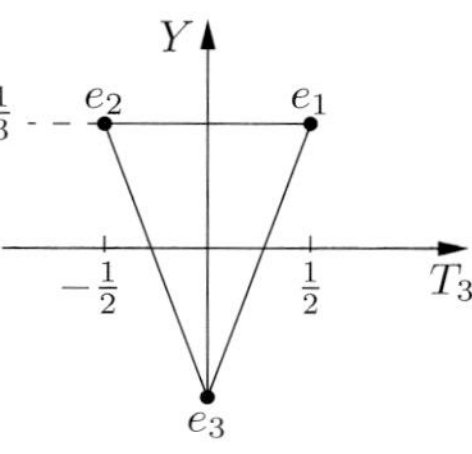


*Figure 3.94. Quarks.*

Here, $\mathscr{S}$ is the *strangeness* operator[52], and $e$ denotes the charge of the electron. In the Hilbert space $X$ one has for the three quarks $e_1$, $e_2$ and $e_3$ the relation $\mathscr{S} = 0$. Thus, we have

$$\boxed{\mathscr{Q}e_1 = \frac{2}{3}|e|e_1\,, \quad \mathscr{Q}e_2 = \frac{1}{3}|e|e_2\,, \quad \mathscr{Q}e_3 = -\frac{1}{3}|e|e_3\,.}$$

*Example 3* (charge of a proton): The 27 tensor products $e_i \otimes e_j \otimes e_k$, $i, j, k = 1, 2, 3$, form a basis of the space $Z := X \otimes X \otimes X$. Let a linear operator $A : X \to X$ act on $Z$ by means of the formula

$$A(e_i \otimes e_j \otimes e_k) = (Ae_i) \otimes e_j \otimes e_k + e_i \otimes (Ae_j) \otimes e_k + e_i \otimes e_j \otimes (Ae_k).$$

Thus, we get the eigenvalue formula for the proton state from (3.78)

$$\mathscr{Q}p = \frac{1}{\sqrt{2}}(\mathscr{Q}e_1 \otimes e_1 \otimes e_2 + \ldots) + \ldots$$

which yields

$$\boxed{\mathscr{Q}p = |e|p\,,}$$

in other words, the proton has the charge $|e|$.

**Relation to the calculus of matrices:** Let $X$ be an $n$-dimensional Hilbert space over $\mathbb{K}$. We choose an orthonormal basis $e_1, \ldots, e_n$ and associate to every linear operator $A : X \to X$ a matrix $\mathscr{A} = (a_{ij}$ by setting

$$\boxed{a_{jk} := (e_j, Ae_k)\,.}$$

Then we have

$$A(\alpha_1 e_1 + \ldots + \alpha_n e_n) = \sum_{j=1}^{n} (\mathscr{A}\,\alpha)_j\, e_j = \sum_{j,k=1}^{n} a_{jk}\alpha_k e_j\,.$$

We denote by $L(X, X)$ the ring composed of all linear operators $A : X \to X$. Moreover, let $L(\mathbb{K}^n, \mathbb{K}^n)$ denote the ring consisting of all $(n \times n)$-matrices with coefficients in $\mathbb{K}$.

**Theorem 5:** By means of the association $A \mapsto \mathscr{A}$ a bijective linear map

$$\boxed{\varphi : \quad L(X, X) \mapsto L(\mathbb{K}^n, \mathbb{K}^n)} \tag{3.80}$$

[52]The eigenvalues of $\mathscr{S}$ are in correspondence with a quantum number $s$; if $s \neq 0$, the particles are referred to as *strange*. The three quarks $e_1$, $e_2$ and $e_3$ are not strange.

is defined, which respects the multiplicative structure as well as the $*$-structure, meaning that for all $A, B \in L(X,X)$ and $\alpha, \beta \in \mathbb{K}$, one has:

(i) $\varphi(\alpha A + \beta B) = \alpha\varphi(A) + \beta\varphi(B)$;

(ii) $\varphi(AB) = \varphi(A)\varphi(B)$;

(iii) $\varphi(A^*) = \varphi(A)^*$.

Note that for real matrices $A^* = A^\mathsf{T}$.

*Example 4:* Let $X$ be a complex Hilbert space. $\varphi$ induces a group isomorphism

$$\boxed{U(n,X) \simeq U(n) \quad \text{and} \quad SU(n,X) \simeq SU(n)\,.}$$

Here $U(n)$ denotes the group of all *unitary* complex $(n \times n)$-matrices, which are defined by the relation $U^*U = UU^* = E$. The symbol $SU(n)$ denotes the subgroup consisting of all matrices $U$ of $U(n)$ for which $\det U = 1$.

*Example 5:* Let $X$ be a real Hilbert space. Then $\varphi$ induces a group isomorphism

$$\boxed{O(n,X) \simeq O(n) \quad \text{and} \quad SO(n,X) \simeq SO(n)\,.}$$

Here $O(n)$ denotes the group of all real orthogonal $(n \times n)$-matrices which are characterized by the validity of the relation $U^\mathsf{T}U = UU^\mathsf{T} = E$. Moreover, $SO(n)$ denotes the subgroup of matrices $U \in O(n)$ for which $\det U = 1$.

### 3.9.3 Pseudo-unitary geometry

Pseudo-unitary geometry is not a real geometry of our experience of the everyday world, however, it is the geometry of Einstein's special theory of relativity.

*Example 1:* The space $\mathbb{R}^2$ can be made a pseudo-unitary space of Morse index $m = 1$ and signature $(1,1)$ in the sense of the definition below by setting

$$B(u,v) := u_1v_1 - u_2v_2 \quad \text{for all} \quad u, v \in \mathbb{R}^2.$$

The corresponding pseudo-orthogonal group $O(1,1)$ consists of the set of all transformations

$$\begin{pmatrix} u_1' \\ u_2' \end{pmatrix} = (\pm 1) \begin{pmatrix} \cosh\alpha & -\sinh\alpha \\ -\sinh\alpha & \cosh\alpha \end{pmatrix} \begin{pmatrix} u_1 \\ u_2 \end{pmatrix}, \qquad \alpha \in \mathbb{R}\,.$$

An important fact about this geometry is that there exist vectors $u \neq 0$ with $B(u,u) = 0$; for example this is the case for $u = (1,1)^\mathsf{T}$. Vectors with this property are referred to as *isotropic*.

**Definition:** A *pseudo-unitary space* is defined to be a finite-dimensional space $X$ over $\mathbb{K}$, endowed with a map $B : X \times X \longrightarrow \mathbb{K}$, such that for all $u, v, w \in X$ and all $\alpha, \beta \in \mathbb{K}$, the following properties hold:

(i) $B(w, \alpha u + \beta v) = \alpha B(w,u) + \beta B(w,v)$.

(ii) $\overline{B(u,v)} = B(v,u)$.

(iii) If $B(u,v) = 0$ for all $v \in X$, then $u = 0$.

Condition (iii) means that the form $B$ is *non-degenerate.* From (i) and (ii) it follows that

$$B(\alpha u + \beta v, w) = \overline{\alpha}B(u,w) + \overline{\beta}B(v,w)\,.$$

Here $\overline{\alpha}$ denotes the complex conjugate of $\alpha$. In a real space $X$ (i.e., $\mathbb{K} = \mathbb{R}$), the bar can be omitted throughout.

**Morse index and signature:** If $e_1, \ldots, e_n$ denotes a basis of $X$, then we construct a matrix $\mathscr{B} = (b_{jk})$ by setting

$$b_{jk} := B(e_j, e_k), \qquad j, k = 1, \ldots, n.$$

Then $\mathscr{B}$ is self-adjoint, i.e., $\mathscr{B}^* = \mathscr{B}$. All eigenvalues of $\mathscr{B}$ are real and non-vanishing. The number $m$ of negative eigenvalues of $\mathscr{B}$ is referred to as the *Morse index* of $B$ (note that this number is independent of the basis by the Sylvester inertia theorem), and $(n - m, m)$ is called the *signature* of $B$.

**Pseudo-orthonormal basis:** A basis $e_1, \ldots, e_n$ of the space $X$ is said to be *pseudo-orthonormal*, if:

$$B(e_j, e_k) = \begin{cases} 0 & \text{for} \quad j \neq k, \\ 1 & \text{for} \quad j = k,\ j = 1, \ldots, n - m, \\ -1 & \text{for} \quad j = k,\ j = n - m + 1, \ldots, n. \end{cases}$$

**Basis theorem:**

(i) There always exists a pseudo-orthonormal basis $e_1, \ldots, e_n$ of $X$.

(ii) Every vector $u \in X$ can be written uniquely in the form

$$u = x_1 e_1 + \ldots + x_n e_n.$$

The numbers $x_1, \ldots, x_n \in \mathbb{K}$ are called *pseudo-Cartesian* coordinates of $u$. These depend of course on the choice of pseudo-orthonormal basis.

(iii) If we define a linear map $\mathrm{d}x^j : X \longrightarrow \mathbb{K}$ by

$$\mathrm{d}x^j(\alpha_1 e_1 + \ldots + \alpha_n e_n) = \alpha_j,$$

then we have

$$\boxed{B = \overline{\mathrm{d}x^1} \otimes \mathrm{d}x^1 + \overline{\mathrm{d}x^2} \otimes \mathrm{d}x^2 + \ldots + \overline{\mathrm{d}x^n} \otimes \mathrm{d}x^n.}$$

**The pseudo-unitary group $U(n - m, m; X)$:** An operator $U : X \longrightarrow X$ is said to be *pseudo-unitary* , if $U$ preserves the hermitian form $B$, i.e., $U$ is linear and

$$\boxed{B(Uv, Uw) = (v, w) \quad \text{for all} \quad v, w \in X.}$$

The set of pseudo-unitary operators on $X$ form a group, which is referred to as the *pseudo-unitary group* $U(n - m, m; X)$. All operators in $U(n - m, m; X)$ with

$$\det U = 1$$

form a subgroup, which is denoted $SU(n - m, m; X)$ and called the *special pseudo-unitary group*. In case the space $X$ is real, we write $O(n - m, m; X)$ (resp. $SO(n - m, m; X)$) instead of $U(n - m, m; X)$ (resp. $SU(n - m, m; X)$) and speak of the *pseudo-orthogonal group* (resp. of the *special pseudo-orthognal group*).[53]

[53] The groups $U(n - m, m; X)$, $SU(n - m, m; X)$, $O(n - m, m; X)$ and $SO(n - m, m; X)$ are examples of Lie groups of dimensions

$$\dim U(n - m, m; X) = n^2, \qquad \dim SU(n - m, m; X) = n^2 - 1,$$

$$\dim O(n - m, m; X) = \dim SO(n - m, m; X) = \frac{n(n-1)}{2}.$$

These dimensions indicate the number of real parameters on which the groups depend (cf. [212]).

**Remark:** In the literature, the prefix "pseudo" is usually omitted, and one speaks of operators which are *unitary with respect to* $B$, regardless of the signature of $B$. A more precise notation, which is also used in the literature, is $U(B, X)$, in words: the unitary group of linear transformations of $X$ preserving $B$. The difference is important if one is considering changes of basis and thinking of the groups as explicit matrix groups. The notation $U(n-m, m; X)$ indicates that the form $B$ is in diagonal form with respect to the given basis. Similarly, the more precise notation for the special pseudo-orthogonal group is $SU(B, X)$, and this subgroup is usually just referred to as the *special unitary group of* $B$.

**Theorem:** Let $X$ be a complex linear space. A linear operator $U : X \longrightarrow X$ belongs to $U(n-m, m; X)$, if $U$ transforms an arbitrary pseudo-unitary basis of $X$ into another such.

If $X$ is real, corresponding statements with $U$ replace by $O$ hold.

**Pseudo-unitary geometry:** A property belongs to the pseudo-unitarian geometry, if it is invariant under all operations in the group $U(n-m, m; X)$.

*Example 2:* The notions of *orthogonal vectors, pseudo-orthonormal basis* and *isotropic vectors* are all notions of the pseudo-unitarian geometry, i.e., the defining properties (orthonormality, isotropic, etc.) are preserved by all $U \in U(n-m, m; X)$.

These notions are defined as follows. Two vectors $u$ and $v$ are said to be *orthogonal*, if

$$B(u, v) = 0.$$

A vector $u$ is called *isotropic*, if it is orthogonal to itself, i.e., $B(u, u) = 0$.

**Infinitesimal pseudo-unitary operators and the Lie algebra $u(n-m, m; X)$:** An operator $A : X \longrightarrow X$ is referred to as *infinitesimally pseudo-unitary*, if it is linear and the relation

$$\boxed{B(u, Av) = -B(Au, v)}$$

for all $u, v \in X$. The set of the operators form by definition the space $u(n-m, m; X)$. The set of operators $A \in u(n-m, m; X)$ for which $\operatorname{tr} A = 0$ form by definition the set $su(n-m, m; X)$.

If $X$ is real, then we denote $u(n-m, m; X)$ (resp. $su(n-m, m; X)$ by $o(n-m, m; X)$) (resp. $so(n-m, m; X)$).

Theorems 3 and 4 in section 3.9.2 continue to hold when the notion 'Hilbert space' is replaced by 'pseudo-unitary space'. One needs only replace $U(n, X)$ by $U(n-m, m; X)$. Similarly for all the other groups and algebras occuring here one simply replaces '$n$' by '$n-m, m$'.

**The relation to the calculus of matrices:** We choose a fixed orthonormal basis $e_1, \ldots, e_n$ and endow the linear space $X$ with a scalar product by setting $(e_j, e_k) := \delta_{jk}$. To every linear operator $A : X \longrightarrow X$ we associate a matrix $\mathscr{A} := (a_{jk})$ by setting

$$a_{jk} := (e_j, Ae_k).$$

If $X$ is a complex linear space, then the map defined by $A \mapsto \mathscr{A}$ induces group isomorphisms

$$\boxed{U(n-m, m; X) \simeq U(n-m, m) \quad \text{and} \quad SU(n-m, m; X) \simeq SU(n-m, m)\,.}$$

Similarly, one gets for a real space $X$ group isomorphisms

$$\boxed{O(n-m, m; X) \simeq O(n-m, m) \quad \text{and} \quad SO(n-m, m; X) \simeq SO(n-m, m)\,.}$$

**Definition:** The group $U(n-m,m)$ consists of precisely the set of $(n \times n)$-matrices $\mathscr{A}$ with the property

$$\mathscr{A}^* \mathscr{D}_{n-m,m} \mathscr{A} = \mathscr{D}_{n-m,m} \, .$$

Here $\mathscr{D}_{n-m,m}$ is a diagonal matrix of the form

$$\mathscr{D}_{n-m} := \begin{pmatrix} I_{n-m} & 0 \\ 0 & -I_m \end{pmatrix} ,$$

where $I_r$ denotes the unit matrix of size $r$ (i.e., a $(r \times r)$-matrix). The subset of matrices $\mathscr{A}$ for which $\det \mathscr{A} = 1$ forms by definition the linear (matrix) group $SU(n-m,m)$. Similarly, $O(n-m,m)$ (resp. $SO(n-m,m)$) consists of the set of real matrices in $U(n-m,m)$ (resp. $SU(n-m,m)$).

### 3.9.4 Minkowski geometry

Minkowski geometry is the geometry of a four-dimensional real vector space with an indefinite scalar product

$$\boxed{uv := B(u,v) \qquad \text{for all} \qquad u, v \in M_4 \, .} \tag{3.81}$$

In what follows we only use geometric notions which are independent of a chosen coordinate system. In 3.9.5 we will show how this leads to an elegant formulation of the special theory of relativity, which is fitting for the most important principle in physics – *Einstein's principle of relativity.*

**Definition:** Minkowski space $M_4$ is the real pseudo-unitary (i.e., pseudo-orthogonal) space for which the signature of the scalar form is (3,1).

**Pseudo-orthonormal basis:** A given basis $e_1$, $e_2$, $e_3$, $e_4$ of $M_4$ is said to be *pseudo-orthonormal*, if[54]

$$\boxed{\begin{array}{c} e_1^2 = e_2^2 = e_3^2 = 1 \, , \quad e_4^2 = -1 \, , \\ e_j e_k = 0 \qquad \text{for} \quad j \neq k . \end{array}} \tag{3.82}$$

From the decomposition

$$u = x_1 e_1 + x_2 e_2 + x_3 e_3 + x_4 e_4 \tag{3.83}$$

we get the component representation of the scalar product $uu^*$, which is called the *Lorentzian scalar product*:

$$\boxed{uu^* = x_1 x_1^* + x_2 x_2^* + x_3 x_3^* - x_4 x_4^* \qquad \text{for all} \qquad u,\, u^* \in M_4 \, .} \tag{3.84}$$

**Symmetry groups:** The group $O(3,1;M_4)$ belonging to $M_4$ is called the *Lorentz group.* By definition, a linear operator $A : M_4 \longrightarrow M_4$ belongs to the Lorentz group if and only if it preserves the Lorentzian scalar product, i.e.,

$$\boxed{(Au)(Av) = uv \qquad \text{for all} \qquad u,\, v \in M_4 \, .}$$

[54]In the literature, the scalar product which corresponds to $-uv$ is used. The convention which we have chosen here has the advantage of yielding the standard Euclidean scalar product as a special case.
The other convention, on the other hand, has the advantage of a direct physical interpretation of the pseudo-Riemannian arc length of the four-dimensional world-line of a moving particle as being proportional to the eigentime $\tau$ of the particle.

The group $SO(3,1;M_4)$ consists by definition of all transformations $A \in O(3,1;M_4)$ for which $\det A = 1$.

The *proper Lorentz group* $SO^+(3,1;M_4)$ consists by definition of all $A \in SO(3,1;M_4)$ for which

$$\boxed{\operatorname{sgn}(Au)e_4 = \operatorname{sgn} ue_4 \quad \text{for all} \quad u \in M_4\,.}$$

This definition is independent of the choice of pseudo-orthonormal basis $e_1, e_2, e_3, e_4$.

As we will see, the elements of $SO(3,1;M_4)$ preserve the orientation, while the elements of $SO^+(3,1;M_4)$ in addition preserve the direction of time.

**The Poincaré group** $P(M_4)$**:** This group, which is the most important group in quantum field theory, consists by definition of the set of all transformations

$$\boxed{u' = Au + a\,, \qquad u \in M_4}$$

of $M_4$ to itself, for which $A \in O(3,1;M_4)$ and $a \in M_4$.

**Classification of vectors:** Let $u \in M_4$ be a given vector.

(i) $u$ is said to be *space-like* , if $u^2 > 0$.

(ii) $u$ is said to be *time-like* , if $u^2 < 0$.

(i) $u$ is said to be *light-like* , if $u^2 = 0$.

**Arc length:** If $u = u(\sigma)$, $\sigma_1 \le \sigma \le \sigma_2$ a curve on $M_4$, the we define the *arc length* $s$ with respect to the curve parameter $\sigma$ by the relation

$$\frac{\mathrm{d}s}{\mathrm{d}\sigma} = \begin{cases} \sqrt{u'(\sigma)^2} & \text{for } u'(\sigma)^2 \ge 0\,, \\ \mathrm{i}\sqrt{-u'(\sigma)^2} & \text{for } u'(\sigma)^2 < 0\,. \end{cases}$$

For this, one writes more briefly:

$$\boxed{\mathrm{d}s^2 = u'(\sigma)^2 \mathrm{d}\sigma^2\,.} \tag{3.85}$$

*Case 1:* If the curve $u = u(\sigma)$ is space-like, i.e., $u'(\sigma)$ is space-like for all $\sigma$, then one can introduce the arc length

$$s = \int_{\sigma_1}^{\sigma} \sqrt{u'(\sigma)^2}\, \mathrm{d}\sigma$$

as a new parameter on the curve.

*Case 2:* If the curve $u = u(\sigma)$ is time-like, i.e., $u'(\sigma)$ is time-like for all $\sigma$, then we can use the so-called *eigentime*

$$\boxed{\tau := \frac{1}{\mathrm{c}} \int_{\sigma_1}^{\sigma} \sqrt{-u'(\sigma)^2}\, \mathrm{d}\sigma} \tag{3.86}$$

as a new parameter on the curve. Here c denotes the speed of light in a vacuum.[55]

[55] In the special theory of relativity, time-like curves represent motions of particles moving at more than the speed of light. The eigentime $\tau$ is then time as rendered by a clock which is in the same frame as the particle.

**Orientation:** We orient Minkowski space by distinguishing a fixed pseudo-orthonormal basis $e_1, e_2, e_3, e_4$ and defining the *volume form*

$$\boxed{\mu := \mathrm{d}x^1 \wedge \mathrm{d}x^2 \wedge \mathrm{d}x^3 \wedge \mathrm{d}x^4 .} \tag{3.87}$$

Here, $\mathrm{d}x^j : X \longrightarrow \mathbb{R}$ is a linear mapping with $\mathrm{d}x^j(e_k) = \delta_{jk}$ for all $j, k = 1, 2, 3, 4$. If $b_1, b_2, b_3, b_4$ is an arbitrary basis of $M_4$, we define the number

$$\boxed{\alpha := \operatorname{sgn} \mu(b_1, b_2, b_3, b_4)}$$

to be the *orientation* of the basis. This number is either $\alpha = 1$, in which case one speaks of a *positive* orientation, or $\alpha = -1$, in which case one speaks of a *negative* orientation.

**Theorem:** (i) For an arbitrary pseudo-orthonormal basis $e'_1, \ldots, e'_4$ we have

$$\mu = \alpha' \mathrm{d}x'^1 \wedge \mathrm{d}x'^2 \wedge \mathrm{d}x'^3 \wedge \mathrm{d}x'^4 .$$

The constant $\alpha'$ is determined to be $\alpha' = 1$ (resp. $\alpha' = -1$) if the basis is positively (resp. negatively) oriented.

(ii) A Lorentz transformation $A$ is orientation-preserving if and only if $\det A = 1$, in other words, $A \in SO(3, 1; M_4)$.

**Multilinear algebra on $M_4$:** Minkowski space $M_4$ is a linear space. For this reason, all notions of multilinear algebra can be applied to it. Among these are

(i) tensor algebra,

(ii) exterior algebra (the Grassmann algebra),

(iii) inner algebra (the Clifford algebra),

(iv) the differential calculus of Cartan, and

(v) the Hodge operator $*$ (duality operator).

In what follows we define a series of operators with the help of a pseudo-orthonormal basis. This formulas are special cases of general formulas of tensor calculus, which can be found in [212]. In particular, this calculus shows that all operators $d, \delta, \mathrm{Div}$ have an invariant meaning and do not depend on the choice of the pseudo-orthonormal basis used to define them. Also, the $*$-operator has an invariant meaning, provided one uses only positively oriented pseudo-orthonormal basis in its definition. A change of orientation of the basis has the effect of changing $*$ to $(-1)*$.

**The tensor algebra of $M_4$:** For any two vectors $u, v \in M_4$, the tensor product

$$\boxed{u \otimes v}$$

is defined (cf. 2.4.3.1).

**The exterior algebra of $M_4$:** For two vectors $u, v \in M_4$, the exterior product $u \wedge v = u \otimes v - v \otimes u$ is defined. One has

$$\boxed{u \wedge v = -v \wedge u .}$$

This product corresponds to the usual vector product of three-dimensional space. However, $u \wedge v$ no longer belongs to $M_4$, but rather to the vector space of anti-symmetric bilinear forms on the dual space $M_4^*$.

**Duality:** The Hodge $*$-operator acts on the space of exterior products. If $e_1, e_2, e_3, e_4$ is a pseudo-orthonormal basis of $M_4$, then one has:

$$*(e_1 \wedge e_2) = e_4 \wedge e_3\,, \quad *(e_2 \wedge e_3) = e_4 \wedge e_1\,, \quad *(e_3 \wedge e_1) = e_4 \wedge e_2\,,$$
$$*\,(e_1 \wedge e_4) = e_2 \wedge e_3\,, \quad *(e_2 \wedge e_4) = e_3 \wedge e_1\,, \quad *(e_3 \wedge e_4) = e_1 \wedge e_2\,.$$

The general action of the $*$-operator on $u \wedge v$ can be determined from these formulas, using its linearity. One has

$$\boxed{**\,(u \wedge v) = v \wedge u \qquad \text{for all} \quad u, v \in M_4\,.}$$

*Example 1:* $*(e_1 \wedge (ae_1 + be_2)) = *(be_1 \wedge e_2) = b * (e_1 \wedge e_2) = b(e_4 \wedge e_3)\,.$

**The inner algebra (Clifford algebra) of $M_4$:** For all $u, v \in M_4$, one has

$$\boxed{u \vee v + v \vee u = 2uv\,.}$$

**Differential forms on $M_4$:** Let $e_1, e_2, e_3, e_4$ be a pseudo-orthonormal basis with the corresponding *dual basis* $\mathrm{d}x^1, \ldots, \mathrm{d}x^4$, which is given by the relations

$$\boxed{\mathrm{d}x^j(e_k) = \delta_{jk}\,, \qquad j,k = 1,2,3,4\,.}$$

From this we get $p$-forms for $p = 1,2,3,4$.

*Example 2:* One-forms have the form

$$\omega = a_1\mathrm{d}x^1 + a_2\mathrm{d}x^2 + a_3\mathrm{d}x^3 + a_4\mathrm{d}x^4$$

with real-valued functions $a_j : M_4 \longrightarrow \mathbb{R}$. Linear combinations of the products $\mathrm{d}x^j \wedge \mathrm{d}x^k$ yield two-forms, etc. The volume form

$$\mu = \mathrm{d}x^1 \wedge \mathrm{d}x^2 \wedge \mathrm{d}x^3 \wedge \mathrm{d}x^4$$

is an example of a four-form.

**The exterior derivative d:** For a $p$-form $\omega$, the exterior derivative

$$\boxed{\mathrm{d}\omega}$$

is defined in an invariant manner (cf. 1.5.10.5).

*Example 3:* For a function $a : M_4 \longrightarrow \mathbb{R}$ one has

$$\mathrm{d}a = \partial_1 a\mathrm{d}x^1 + \ldots + \partial_4 a\mathrm{d}x^4$$

with $\partial_j := \partial/\partial x^j$. From this we get

$$\mathrm{d}\,(a\mathrm{d}x^1) = \mathrm{d}a \wedge \mathrm{d}x^1 = \partial_2 a\mathrm{d}x^2 \wedge \mathrm{d}x^1 + \partial_3 a\mathrm{d}x^3 \wedge \mathrm{d}x^1 + \partial_4 a\mathrm{d}x^4 \wedge \mathrm{d}x^1\,.$$

For the volume form $\mu$ we get the relation $\mathrm{d}\mu = 0$.

**The Hodge $\delta$-operator:** For an arbitrary $p$-form we set

$$\boxed{\delta\omega := -(-1)^{(4-p)p} * \mathrm{d} * \omega\,.} \qquad (3.88)$$

The linear $*$-operator is defined by the following relations:

(i) dualization of zero-forms: $*1 = \mathrm{d}x^1 \wedge \mathrm{d}x^2 \wedge \mathrm{d}x^3 \wedge \mathrm{d}x^4$.

(ii) dualization of one-forms:[56] $*\mathrm{d}x^1 = \mathrm{d}x^2 \wedge \mathrm{d}x^3 \wedge \mathrm{d}x^4$.

(iii) dualization of two-forms: $*\,(\mathrm{d}x^1 \wedge \mathrm{d}x^2) = \mathrm{d}x^3 \wedge \mathrm{d}x^4\,, *(\mathrm{d}x^2 \wedge \mathrm{d}x^3) = \mathrm{d}x^1 \wedge \mathrm{d}x^4\,,$ $*\,(\mathrm{d}x^3 \wedge \mathrm{d}x^1) = \mathrm{d}x^2 \wedge \mathrm{d}x^4$.

The missing expressions required to completely define the operator are obtained from the *dualization formula*

$$\boxed{* * \omega = -(-1)^{(4-p)p}\omega}$$

which holds for arbitrary $p$-forms $\omega$.

*Example 4:* $*(\mathrm{d}x^3 \wedge \mathrm{d}x^4) = * * (\mathrm{d}x^1 \wedge \mathrm{d}x^2) = \mathrm{d}x^2 \wedge \mathrm{d}x^1\,,$

$*\,(\mathrm{d}x^2 \wedge \mathrm{d}x^3 \wedge \mathrm{d}x^4) = * * \mathrm{d}x^1 = \mathrm{d}x^1\,, *\mu = *(\mathrm{d}x^1 \wedge \mathrm{d}x^2 \wedge \mathrm{d}x^3 \wedge \mathrm{d}x^4) = * * 1 = -1\,.$

The application of the $*$-operator to arbitrary forms is obtained by combining the above actions with the linearity of the operator.

*Example 5:* $*(a\mathrm{d}x^1 + b\mathrm{d}x^2) = a * \mathrm{d}x^1 + b * \mathrm{d}x^2 = a\mathrm{d}x^2 \wedge \mathrm{d}x^3 \wedge \mathrm{d}x^4 + b\mathrm{d}x^3 \wedge \mathrm{d}x^4 \wedge \mathrm{d}x^1$.

**The divergence operator:** We set for $F = T^{jk} e_j \otimes e_k$

$$\boxed{\operatorname{Div} F := \partial_j T^{jk} e_k\,.}$$

Here, one sums over the indices which appear both as upper and lower indices (Einstein convention), which run from 1 to 4.

**The operator** $\mathrm{Alt}\,(\mathbf{B})$**:** For a vector $\mathbf{B} = B^1 e_1 + B^2 e_2 + B^3 e_3$ we define

$$\boxed{\mathrm{Alt}(\mathbf{B}) := B^1(e_2 \wedge e_3) + B^2(e_3 \wedge e_1) + B^3(e_1 \wedge e_2)\,.}$$

### 3.9.5 Applications to the special theory of relativity

**The relativity principle of Einstein (1905):**

> In any two inertial systems, with identical initial and boundary conditions, all physical processes transpire in identical manners. (3.89)

An *inertial system* $\Sigma$ is a Cartesian $(x, y, z)$-coordinate system with an orthonormal basis $\mathbf{i}, \mathbf{j}, \mathbf{k}$ such that, with respect to time $t$, a body free from external forces is at rest or moves along a line with constant velocity.

*Example 1:* In every inertial system the speed of light in a vacuum is identical, the constant c.

The Einstein principle of relativity replaces the classical principle of relativity due to Galilei, according to which (3.89) holds for physical processes of mechanics.

*Example 2:* Let $\Sigma$ and $\Sigma'$ be two inertial systems which have parallel axi, in which an observer in $\Sigma$ observes a motion of the origin of $\Sigma'$ with the equation

$$x = vt$$

[56] On gets $*\mathrm{d}x^j$ from this by cyclic permutation.

(Figure 3.95). According to Galilei one then has the transformation formula

$$x' = x - vt\,, \quad y' = y\,, \quad z = z'\,, \quad t = t' \tag{3.90}$$

between the coordinates $x, y, z, t$ in $\Sigma$ and the coordinates $x', y', z', t'$ in $\Sigma'$ (a so-called *Galilei transformation*). The light ray

$$x = \mathrm{c}t$$

in $\Sigma$ has the equation

$$x' = (\mathrm{c} - v)t'$$

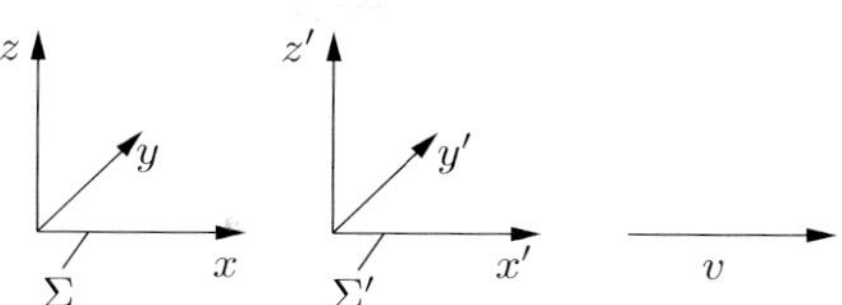


*Figure 3.95. Frames of reference.*

in $\Sigma'$.

This means that if the speed of light in $\Sigma$ is c, then its speed in $\Sigma'$ is $\mathrm{c} - v$, which implies that the statement of Example 1 does not hold. This is why Einstein replaced the Galilei transformation by the *special Lorentz transformation*

$$x' = \frac{x - vt}{\sqrt{1 - v^2/\mathrm{c}^2}}\,, \qquad y' = y\,, \qquad z' = z\,, \qquad t' = \frac{t - vx/\mathrm{c}^2}{\sqrt{1 - v^2/\mathrm{c}^2}}\,. \tag{3.91}$$

From $x = ct$ it follows that $x' = ct'$, i.e., the statement of Example 1 is now actually verified.

If $v$ is very small compared with the speed of light c, then the Lorentz transformation (3.91) approximates the Galilei transformation (3.90). More precisely, the limit of (3.91) as $\mathrm{c} \to \infty$ is (3.90). More generally, one has:

The limit of the theory of relativity as $\mathrm{c} \to \infty$ is classical physics.

*Example 3:* If the inertial system $\Sigma'$ is not parallel to $\Sigma$, then by applying a rotation $D$ one can bring $\Sigma'$ into a position so that its axi are parallel to those of $\Sigma$. Following this, one applies the Lorentz transformation (3.91), then the inverse rotation $D^{-1}$ to reverse the original rotation $D$.

The special Lorentz transformation (3.91) can be described particularly elegantly by using the hyperbolic functions in the form

$$x' = x\cosh\alpha - ct\sinh\alpha\,, \qquad y = y'\,, \qquad z = z'\,, \qquad \mathrm{c}t' = \mathrm{c}t\cosh\alpha - x\sinh\alpha.$$

Here, $\alpha$ is defined by

$$\tanh\alpha = \frac{v}{\mathrm{c}}, \text{ i.e., } \alpha := \operatorname{arctanh}\frac{v}{\mathrm{c}}$$

and $|v| < \mathrm{c}$.

From our point of view today, the Einstein principle of relativity (3.89) is more natural than the principle of relativity of Galilei, which refers to a particular physical discipline – mechanics. In fact, the Einstein principle implies a complete revision of the classical ideas of time and space. There is no longer such a thing as an absolute world-time, but rather, every inertial system has its own notion of time. Moreover, only velocities

$v$ with $|v| < \mathrm{c}$ are possible. Einstein postulated in this connection the much stronger statement:

> An arbitrary physical action can, in a given inertial system, propagate with at most the velocity c.

**Lorentz transformations:** We consider the transformations

$$\begin{pmatrix} x' \\ y' \\ z' \\ \mathrm{c}t' \end{pmatrix} = \mathscr{A} \begin{pmatrix} x \\ y \\ z \\ \mathrm{c}t \end{pmatrix} + \begin{pmatrix} x_0 \\ y_0 \\ z_0 \\ \mathrm{c}t_0 \end{pmatrix} \tag{3.92}$$

and set

$$\mathscr{L}_v := \begin{pmatrix} \beta & 0 & 0 & -\beta v/\mathrm{c} \\ 0 & 1 & 0 & 0 \\ 0 & 0 & 1 & 0 \\ -\beta v/\mathrm{c} & 0 & 0 & \beta \end{pmatrix}, \quad \mathscr{D} := \begin{pmatrix} d_{11} & d_{12} & d_{13} & 0 \\ d_{21} & d_{22} & d_{23} & 0 \\ d_{31} & d_{32} & d_{33} & 0 \\ 0 & 0 & 0 & 1 \end{pmatrix}, \quad \mathscr{S} := \begin{pmatrix} s_1 & 0 & 0 & 1 \\ 0 & s_2 & 0 & 0 \\ 0 & 0 & s_3 & 0 \\ 0 & 0 & 0 & s_4 \end{pmatrix}.$$

Here one has $\beta := 1/\sqrt{1 - v^2/\mathrm{c}^2}$ and $s_j = \pm 1$ for all $j$. Moreover, $\mathscr{D}$ is a rotation, i.e, one has $\mathscr{D}\mathscr{D}^\mathsf{T} = \mathscr{D}^\mathsf{T}\mathscr{D} = E$ and $\det \mathscr{D} = 1$. The matrix $\mathscr{S}$ describes for $s_4 = 1$ the identical transformation or a reflection in space. In case $s_4 = -1$, one gets a reflection in time.

**Definition:** The *Lorentz group* $O(3,1)$ consists of all matrices $\mathscr{A}$ of the form

$$\boxed{\mathscr{A} = \mathscr{S}\mathscr{D}_1\mathscr{L}_v\mathscr{D}_2\,.}$$

From this it follows that $\det \mathscr{A} = \det \mathscr{S} = \pm 1$. Moreover, the group $SO(3,1)$ (resp. $SO^+(3,1)$) consists of all matrices $\mathscr{A} \in O(3,1)$ for which $\det \mathscr{S} = 1$ (resp. $\mathscr{S} = E$). If $\mathscr{A}$ is a Lorentz transformation, i.e., $\mathscr{A} \in O(3,1)$, then (3.92) is called a *Poincaré transformation.* The set of all such Poincaré transformations forms by definition the *Poincaré group.* Moreover, the Lorentz transformations for which $\mathscr{A} \in SO^+(3,1)$ are called *proper*, i.e., reflections in space and time are not allowed. For the theory of elementary particles one requires the full Poincaré group.[57]

**Geometric interpretation:** We consider Minkowski space. A pseudo-orthonormal basis $e_1, e_2, e_3, e_4$ with the decomposition[58]

$$u = x_1e_1 + x_2e_2 + x_3e_3 + x_4e_4 \tag{3.93}$$

corresponds to an inertial system with Cartesian coordinates

$$x = x_1\,, \quad y = x_2\,, \quad z = x_3$$

and $x_4 = \mathrm{c}t$. One then also has $e_1 = \mathbf{i}$, $e_2 = \mathbf{j}$, $e_3 = \mathbf{k}$. From this it follows that

$$u = \mathbf{x} + \mathrm{c}te_4 \quad \text{with} \quad \mathbf{x} = x\mathbf{i} + y\mathbf{j} + z\mathbf{k}\,.$$

[57] The Poincaré group $\mathscr{P}$ is a ten-dimensional real Lie group. In the Lie algebra of $\mathscr{P}$ one can find ten basis operators, which correspond in quantum field theory to the conservation of energy, momentum and angular momentum.

[58] Instead of $x_j$ one also writes here $x^j$.

**Theorem:** The group $\mathscr{P}$ of Poincaré transformations (3.92) is isomorphic to the group $P(M_4)$ of the Poincaré transformations

$$\boxed{u' = Au + a}$$

on $M_4$. This isomorphism is obtained by setting

$$u' = x_1' e_1 + x_2' e_2 + x_3' e_3 + x_4' e_4 \,, \quad a = x_0 e_1 + y_0 e_2 + z_0 e_3 + \mathrm{c} t_0 e_4$$

and using formula (3.92) to calculate $x_j'$.

**The eigentime:** We consider the motions of a mass particle

$$\mathbf{x} = \mathbf{x}(t)$$

with the velocity less than c in an inertial system. Then the curve

$$u = u(t) = \mathbf{x}(t) + \mathrm{c} t e_4$$

belongs to $M_4$. One has $u'(t) = \mathbf{x}'(t) + \mathrm{c} e_4$ and $u'(t)^2 = \mathbf{x}'(t)^2 - \mathrm{c}^2$. From (3.86) the eigentime

$$\boxed{\tau = \int_{t_0}^{t} \sqrt{1 - \mathbf{x}'(\sigma)^2/\mathrm{c}^2}\, \mathrm{d}\sigma} \tag{3.94}$$

follows. This is the time that a clock, which is fixed with respect to the moving particle, displays.

**The Einstein twin paradox:** We assume that two twins $Z$ and $Z'$ at the origin $\mathbf{x} = 0$ of an inertial system are born at time $t_0 = 0$.

While $Z'$ remains at the origin, $Z$ travels with a space ship and returns at time $t$. Then $Z$ has experienced the eigentime $\tau$, while $Z'$ has experienced the eigentime $\tau' = t$ because of $\mathbf{x} = 0$. From (3.94) we get

$$\boxed{\tau < t \,,}$$

i.e., $Z$ is younger that $Z'$ upon his return. The difference in age is more pronounced the greater the velocity $|\mathbf{x}'(t)|$ of travel had been.

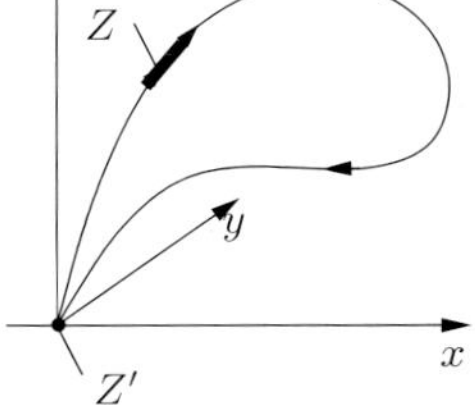


*Figure 3.96. The twin paradox.*

**The Maxwell equations of electrodynamics:** In an arbitrary inertial system we consider the following quantities:

$\mathbf{E}$, the vector of the electric field strength,

$\mathbf{B}$, the vector of the magnetic field strength,

$\varrho$, the electric charge density,

$\mathbf{J}$, the electric current density vector.

These quantities are associated to the following geometric objects on $M_4$:

$$F = \mathbf{E} \wedge e_4 - \mathrm{Alt}(\mathbf{B}) \,, \quad J = \mathbf{J} + \varrho e_4 \,. \tag{3.95}$$

It follows that

$$*F = -\mathbf{B} \wedge e_4 - \mathrm{Alt}(\mathbf{E}) \,. \tag{3.96}$$

One refers to $F$ as the *electromagnetic field tensor.* Then the Maxwell equations of electrodynamics are transformed into the surprisingly simple and elegant form

$$\boxed{\operatorname{Div} F = J\,, \qquad \operatorname{Div} *F = 0\,.} \tag{3.97}$$

The first equation $\operatorname{Div} F = J$ expresses the fact that electric charges and currents are the source of the electromagnetic field tensor $F$. The second equation $\operatorname{Div} *F = 0$ reflects a duality between electric and magnetic fields and at the same time an asymmetry which arises from the fact that there is no inhomogenous term. The cause is that in classical electrodynamics there is no such thing as isolated magnetic charges (monopoles). This asymmetry bothered Dirac. In order to get rid of it, he assumed the existence of monopoles. In modern gauge field theories the existence of such magnetic monopoles follows purely mathematically. Currently researchers are searching in space for the existence of these creatures.

**Discussion:** The formulation (3.97) is valid in Minkowski space $M_4$. If $F$ and $J$ are described with respect to a basis $b_1, \ldots, b_4$ of $M_4$, then one gets the form of the Maxwell equations in a system of reference with respect to the basis $b_1, \ldots, b_4$. It is possible that electric fields transform into magnetic fields and conversely.[59]

For every pseudo-orthonormal basis, $F$ and $J$ have the same structure. For this reason, the Maxwell equations have the same form in an arbitrary inertial system. This is explicitly:[60]

$$\boxed{\begin{array}{ll} \operatorname{div}\mathbf{E} = \varrho\,, & \operatorname{div}\mathbf{B} = 0\,, \\ \mathbf{rot}\,\mathbf{B} = \mathbf{E}_t + \mathbf{J}\,, & \mathbf{rot}\,\mathbf{E} = -\mathbf{B}_t\,. \end{array}} \tag{3.98}$$

The transition from an inertial system $\Sigma$ to another inertial system $\Sigma'$ with the help of the special Lorentz transformation (3.91), the electromagnetic field and the charges and currents have to be transformed as follows:

$$\begin{array}{lll} \varrho' = \beta(\varrho - \mathbf{vJ})\,, & \mathbf{J}' = \beta(\mathbf{J} - \mathbf{v}\varrho)\,, & Q' = Q\,, \\ \mathbf{E}' = \beta(\mathbf{E}_* + \mathbf{v}\times\mathbf{B}_*)\,, & \mathbf{B}' = \beta(\mathbf{B}_* + \mathbf{E}_*\times\mathbf{v})\,. & \end{array}$$

Here $\mathbf{v} := v\mathbf{i}$, $\mathbf{E} = E_1\mathbf{i} + E_2\mathbf{j} + E_3\mathbf{k}$ and $\mathbf{E}_* = \beta^{-1}(E_1\mathbf{i} + E_2\mathbf{j} + E_3\mathbf{k})$ with $\beta^{-1} := \sqrt{1 - v^2}$. These equations are written in dimensionless quantities.

**Duality symmetry of the Maxwell equations:** Because of $**F = -F$ the Maxwell equations (3.97) possesses a duality, which arises from (3.95) and (3.96), in case $\varrho = 0$

[59] An electric charge which is at rest generates only an electric field. However, an observer in motion sees a charge in motion, which corresponds to an electric field which generates a magnetic field.

Around 1900 the electrodynamics of mediums in motion was an important open problem in physics. It was not clear how to describe the transformation of the electromagnetic field under a change of systems of reference. Einstein recognized in 1905 that the Maxwell equations can be left *unchanged*, replacing only the Galilei transformation by a Lorentz transformation.

[60] This formulation is so appealing, as it contains no physical constants. This is achieved by passing to the international MKSA-system, leading to dimensionless quantities, by making the following substitutions in Table 1.5:

$$\mathbf{x} \Rightarrow r_e\mathbf{x}\,, \quad t \Rightarrow r_e t/c\,, \quad m_0 \Rightarrow m_e m_0\,, \quad Q \Rightarrow eQ\,, \quad \mathbf{v} \Rightarrow c\mathbf{v}\,,$$

$$\mathbf{E} \Rightarrow \frac{e}{4\pi\varepsilon_0 r_e^2}\mathbf{E}\,, \quad \mathbf{B} \Rightarrow \frac{1}{4\pi\varepsilon_0 r_e^2 c}\mathbf{B}\,, \quad \mathbf{J} \Rightarrow \frac{ec}{4\pi r_e^3}\mathbf{J}\,, \quad \varrho \Rightarrow \frac{e}{4\pi\varepsilon_0 r_e^3 c}\varrho$$

(where $e$ is the charge of the electron, $m_e$ is its mass, $r_e$ its radius and $\varepsilon_0$ is the dielectric constant in a vacuum).

and $\mathbf{j} = 0$, and from the substitution

$$\mathbf{E} \Rightarrow -\mathbf{B}\,, \qquad \mathbf{B} \Rightarrow \mathbf{E}.$$

**The equation of motion of a charged particle in electromagnetic field:** The motion $u = u(\tau)$ of a particle with rest mass $m$ and the charge $Q$ is given by the equation[61]

$$\boxed{m_0 u''(\tau) = QF(u(\tau))u'(\tau).} \tag{3.99}$$

This corresponds to the differential equation

$$\boxed{\mathbf{p}'(t) = Q\mathbf{E}\,(\mathbf{x}(t), t) + Q\mathbf{x}'(t) \times \mathbf{B}(\mathbf{x}(t), t)}$$

for the trajectory $\mathbf{x} = \mathbf{x}(t)$ in an arbitrary inertial system. Here $\mathbf{p} := m\mathbf{x}'$ is the momentum vector, and the quantity

$$m = \frac{m_0}{\sqrt{1 - \mathbf{x}'(t)^2/c^2}}$$

is called the *relativistic mass* of the particle. It grows as the speed $|\mathbf{x}'(t)|$ of the particle approaches the speed of light c.

The fourth component of the equation of motion (3.99) is in an inertial system:

$$\boxed{E'(t) = Q\mathbf{E}\,(\mathbf{x}(t), t)\mathbf{x}'(t)\,.}$$

This means that the change in time of the energy

$$\boxed{E = mc^2}$$

is equal to the power with which the electric force $\mathbf{F} = Q\mathbf{E}$ applies to the moving particle. If the particle is at rest, then $m = m_0$, and we get as a special case the famous mass–energy relation of Einstein

$$\boxed{E_0 = m_0 c^2}$$

for the energy of the relativistic particle which is at rest. This energy is freed in the fusion process which drives the energy production of the sun.

**Remark:** The Maxwell equations can, in addition to using the language of four-dimensional vector analysis used in (3.97), also be formulated in the language of general tensor analysis, in the language of differential forms and in the language of principle fiber bundles. This is will be discussed in detail in [212]. There one can also find the

[61] If $b_1, b_2, b_3, b_4$ is a basis of $M_4$, then one has

$$F = F^{jk} b_j \wedge b_k\,, \quad u = x^j b_j\,, \quad Fu = (F^{jk} x_k)\, b_j$$

with $x_k = g_{kr}x^r$ and $g_{kr} = b_k b_r$. Here, one sums over indices which appear as both sub- and superscripts (Einstein convention), with indices running from 1 to 4. This definition of $Fu$ is independent of the choice of the basis.

relativistic addition theorem for velocities as well as the relativistic contraction of lengths and dilation of time.

> The Maxwell equations of electrodynamics have given important impulses to the development of both physics and mathematics.

In particular one can formulate the Maxwell equations in the context of $U(1)$ gauge theory. If one replaces the Abelian group $U(1)$ by other Lie groups (for example $SU(2)$ or $SU(3)$), then one gets non-Abelian gauge theories, which are used in the *standard model* of the theory of elementary particle (cf. [212]).

## 3.9.6 Spin geometry and fermions

The relativistic electron is closely related to the Clifford algebra of Minkowski space $M_4$. The spin of the electron is the symmetry given by the universal covering group $Spin(3)$ of the rotation group $SO(3)$ of three-dimensional space; $Spin(3)$ is isomorphic to the special unitary group $SU(2)$. One is naturally lead to Clifford algebras when one tries to generalize the structure of the complex numbers (a real two-dimensional space) to higher dimensions. William Hamilton (1805–1865) first discovered in 1843 the four-dimensional space of quanternions, after vain attempts to find a three-dimensional analogue of the complex numbers.[62] The Clifford algebra of Euclidean space was introduced by William Clifford (1845–1879) a year before his death. These algebras have dimensions $2^n$ with $n = 1, 2, 3, \ldots$

**Notations:** The symbol $SL(2, \mathbb{C})$ stands for the group of all complex $(2 \times 2)$-matrices $A$ with $\det A = 1$. The unitary matrices in $SU(2, \mathbb{C})$ form by definition the group $SU(2)$. Moreover, $U(1)$ denotes the (multiplicative) group of all complex numbers $z$ with $|z| = 1$.

The group $SO(n)$ is defined as the group of all real orthogonal matrices $A$ of size $(n \times n)$ with $\det A = 1$. In particular, $SO(2)$ consists of the real matrices in $SU(2)$.

### 3.9.6.1 The Clifford algebra of Minkowski space

Let $\mathscr{C}(M_4)$ be the Clifford algebra (see 2.4.3.4) of Minkowski space $M_4$. Then $M_4 \subset \mathscr{C}(M_4)$. For all $u, v \in M_4$ we have

$$\boxed{u \vee v + v \vee u = 2uv.}$$

If one chooses a pseudo-orthonormal basis $e_1, \ldots, e_4$, then one has

$$e_j \vee e_k + e_k \vee e_j = 2e_j e_k\,, \tag{3.100}$$

i.e., $e_1 \vee e_1 = e_2 \vee e_2 = e_3 \vee e_3 = 1$ and $e_4 \vee e_4 = -1$ as well as

$$e_j \vee e_k = -e_k \vee e_j\,, \quad j = 1, 2, 3, 4\,.$$

[62] According to Hamilton himself, he was lead to the discovery of quaternions by a question asked by his two sons at breakfast: can you multiply triples $(a, b, c)$ of numbers? He was unable to do this in spite of long deliberations, but he did discover that there is a way to multiply real four-tuples $(a, b, c, d)$, i.e., *quaternions* (cf. 3.9.6.3).

The intutive reason for the missing multiplication of triples of numbers is because of the fact that the classical vector product is not associative. The notion of *vector* was also introduced by Hamilton, in 1845.

Over and again in the history of mathematics it occurs that simple inquisitiveness lead to mathematical results which later turn out to be very fruitful for the description of nature.

One has $\dim \mathscr{C}(M_4) = 16$. A basis of $\mathscr{C}(M_4)$ can be taken to be the following 16 elements:

(i) $1, e_1, e_2, e_3, e_4$;

(ii) $e_1 \vee e_2\,,\quad e_1 \vee e_3\,,\quad e_1 \vee e_4\,,\quad e_2 \vee e_3\,,\quad e_2 \vee e_4\,,\quad e_3 \vee e_4$;

(iii) $e_1 \vee e_2 \vee e_3\,,\quad e_1 \vee e_2 \vee e_4\,,\quad e_1 \vee e_3 \vee e_4\,,\quad e_2 \vee e_3 \vee e_4$;

(iv) $e_1 \vee e_2 \vee e_3 \vee e_4\,.$

All other products can be expressed in terms of these with the aid of (3.100).

**The Pauli matrices and the Lie algebra $su(2)$:** By definition the algebra $su(2)$ consists of all complex $(2 \times 2)$-matrices $A$ with $A^* = -A$ and $\operatorname{tr} A = 0$. The matrices

$$\sigma_1 := \begin{pmatrix} 0 & 1 \\ 1 & 0 \end{pmatrix}, \quad \sigma_2 := \begin{pmatrix} 0 & -\mathrm{i} \\ \mathrm{i} & 0 \end{pmatrix}, \quad \sigma_3 := \begin{pmatrix} 1 & 0 \\ 0 & -1 \end{pmatrix}, \quad \sigma_4 := \begin{pmatrix} 1 & 0 \\ 0 & 1 \end{pmatrix}$$

are called *Pauli matrices*; $\mathrm{i}\sigma_1, \mathrm{i}\sigma_2$ and $\mathrm{i}\sigma_3$ form a basis of the real linear space $su(2)$, which is made into a Lie algebra by setting[63]

$$[A, B] := AB - BA.$$

**The Pauli–Dirac matrices:** These complex $(4 \times 4)$-matrices are defined by setting

$$\gamma_\alpha := \mathrm{i}\begin{pmatrix} 0 & -\sigma_\alpha \\ \sigma_\alpha & 0 \end{pmatrix}, \quad \gamma_4 := \mathrm{i}\begin{pmatrix} 0 & \sigma_4 \\ \sigma_4 & 0 \end{pmatrix}, \qquad \alpha = 1, 2, 3.$$

In addition we set $\gamma^\alpha := \gamma_\alpha$ for $\alpha = 1, 2, 3$ and $\gamma^4 := -\gamma_4$. Let $\mathbf{Mat}(4,4)$ denote the set of all complex $(4 \times 4)$-matrices. One has

$$\boxed{\gamma_j\gamma_k + \gamma_k\gamma_j = 2\gamma_j\gamma_k\,,} \qquad j = 1, 2, 3, 4\,. \tag{3.101}$$

One sees the analogy of (3.100) with (3.101) immediately.

**Theorem:** With respect to the matrix multiplication, $\mathbf{Mat}(4,4)$ is a Clifford algebra. By means of the association

$$\boxed{e_j \longmapsto \gamma_j}$$

one gets an isomorphism between the Clifford algebra $\mathscr{C}(M_4)$ of Minkowski space and the Clifford algebra $\mathbf{Mat}(4,4)$.

*Example:* Under this isomorphism, $e_j \vee e_k$ maps to $\gamma_j\gamma_k$, etc.

### 3.9.6.2 The Dirac equation and relativistic electrons

In this section we use the Einstein summation convention, whereby like indices which occur both as superscripts and as subscripts are summed over. Latin indices run from 1 to 4, Greek indices from 1 to 3.

**Dirac equation:** The equation which Dirac derived in 1928 for the free electron is

$$\partial \vee \Psi + \frac{m_0 c}{\hbar}\Psi = 0\,. \tag{3.102}$$

Here the complex four-column matrix $\Psi = (\varphi_1, \varphi_2, \mathrm{i}\chi_1, \mathrm{i}\chi_2)^\mathsf{T}$ is called the *wave function* of the electron. We abreviate this by writing $\Psi = (\varphi, \mathrm{i}\chi)^\mathsf{T}$.

[63] Letting $SU(n)$ denote the Lie group of complex unitary matrices $B$ of size $(n \times n)$ with $\det B = 1$, then $su(2)$ is the Lie algebra of $SU(2)$ (cf. [212]).

The Dirac equation (3.102) is written with respect to an inertial system with coordinates $x = (x^1, x^2, x^3, x^4)$ with $x^4 = ct$ (cf. 3.9.5). The function $\Psi = \Psi(x)$ depends on the space-time point $x$. Moreover, we set

$$\partial \vee \not{\psi} := \gamma^k \partial_k$$

with $\partial_k := \partial/\partial x^k$. Furthermore, c is the speed of light in a vacuum, $m_0$ is the rest mass of the electron, h is the Planck action quantum and $\hbar = \mathrm{h}/2\pi$.

The Dirac equation (3.102) corresponds to the system of equations[64]

$$\boxed{\begin{aligned} \sigma_\alpha \partial_\alpha \varphi - \sigma_4 \partial_4 \varphi + \frac{m_0 c}{\hbar} \chi = 0\,, \\ \sigma_\alpha \partial_\alpha \chi + \sigma_4 \partial_4 \chi + \frac{m_0 c}{\hbar} \varphi = 0\,. \end{aligned}} \qquad (3.103)$$

**Behavior under transformations:** Upon the passage from one inertial system to another by means of a proper Lorentz transformation, one has:

$$\boxed{\begin{aligned} x' = L(A)x\,, \\ \varphi' = A^*\varphi\,, \quad \chi' = A^{-1}\chi\,, \quad A \in SL(2, \mathbb{C})\,. \end{aligned}}$$

This transforms the Dirac equation to the corresponding equation for $\Psi'$. For every matrix $A \in SL(2, \mathbb{C})$ there is a Lorentz transformation $x' = L(A)x$ corresponding to it, where $x'$ is derived from the equation

$$\boxed{\sigma_j x^{j'} = A^{-1}(\sigma_j x^j)(A^*)^{-1}.}$$

One calls $\varphi$ and $\chi$ *spinors* and $\Psi$ a *bispinor.*

**Theorem:** (i) The association $A \mapsto L(A)$ gives rise to a homomorphism from the group $SL(2, \mathbb{C})$ to the proper Lorentz group $SO^+(3, 1)$ with kernel $N = \{E, -E\}$.

(ii) There is an isomorphism

$$\boxed{SL(2, \mathbb{C})/N \cong SO^+(3, 1)\,.}$$

(iii) If the matrix $A$ in $SL(2, \mathbb{C})$ is in addition unitary, then the Lorentz transformation $x' = L(A)x$ corresponds to a spatial rotation. In this way the association $A \mapsto L(A)$ yields a homomorphism from the group $SU(2)$ to the rotation group $SO(3)$ of three-space.

(iv) There is an isomorphism[65]

$$\boxed{SU(2)/N \cong SO(3)\,.}$$

[64] In the liturature, many different formulations of the Dirac equation can be found. For example, different definitions are used for the Dirac matrices $\gamma_j$, and the scalar product of Minkowski space $M_4$ is often $-e_j e_k$ instead of $e_j e_k$. However, all of these formulations can be easily converted into one another and yield identical physics (as they should!).

[65] The group $SL(2, \mathbb{C})$ (resp. $SU(2)$) is the universal covering group of $SO^+(3, 1)$ (resp. $SO(3)$), see [212] for more details.

**Spatial reflections $P$:**

$$x^{\alpha'} = -x^{\alpha}\,, \quad \alpha = 1,2,3, \quad x^{4'} = x^4\,,$$
$$\varphi' = \chi\,, \quad \chi' = -\varphi\,.$$

**Reflections in time $T$:**

$$x^{\alpha'} = x^{\alpha}\,, \quad \alpha = 1,2,3, \quad x^{4'} = -x^4\,,$$
$$\varphi' = \chi\,, \quad \chi' = \varphi\,.$$

**Charge conjugation $C$:**

$$x^{j'} = x^j\,, \quad j = 1,2,3,4\,,$$
$$\varphi' = -\sigma_2\overline{\chi}\,, \; \chi' = \sigma_2\overline{\varphi}\,.$$

The Dirac equation is invariant under each individual of these three transformations. For general processes involving elementary particles there is an invariance under the *composed* transformation $PCT$. The following intuitive principle is behind this invariance:

> If some process $\mathscr{P}$ of elementary particles in possible in nature, then there is a process $\mathscr{P}'$ which is also possible, which is obtained from $\mathscr{P}$ by a spatial reflection, a reflection in time and the transition from the particle to its anti-particle, all at the same time.

**The electrospin:** The operator

$$\boxed{D\Psi = (\mathbf{x} \times \mathbf{p})\Psi + \frac{\hbar}{2}\mathbf{s}\Psi}$$

is called the operator of *total angular momentum*. The expressions involved are defined as

$$\mathbf{p} := \frac{\hbar}{\mathrm{i}} e_\alpha \partial_\alpha \qquad \text{and} \qquad \mathbf{s} := \begin{pmatrix} \sigma_\alpha & 0 \\ 0 & \sigma_\alpha \end{pmatrix} e_\alpha\,,$$

where $\alpha$ is summed from 1 to 3. If we write the Dirac equation (3.102) in the form $\mathrm{i}\hbar\partial_4\Psi = \mathrm{H}\Psi$, then we get

$$\boxed{HD - DH = 0\,.} \tag{3.104}$$

Here $D$ represents an conserved quantity, a property which does not hold for the expression $\mathbf{x} \times \mathbf{P}$ without the spin operator $\mathbf{s}$. For the $e_3$-component of $\mathbf{s}$ one has

$$\boxed{s_3\Psi_\pm = \pm\frac{\hbar}{2}\psi_\pm}$$

with $\psi_+ := 2^{-1/2}(1,0,1,0)^\mathsf{T}$ and $\Psi_- = 2^{-1/2}(0,1,0,1)^\mathsf{T}$. The functions $\Psi_\pm$ represent electron states with a spin (self-angular momentum) in the direction of the $e_3$-axis of the size $\pm\hbar/2$.

The natural requirement (3.104) of the conservation of the total momentum thus forces the existence of the electrospin.

**Chirality:** We define the *chirality operator* by the expression

$$\boxed{P := \frac{1}{2}(1 + \gamma_5)}$$

with $\gamma_5 := -\mathrm{i}\gamma_1\gamma_2\gamma_3\gamma_4$. One has $\gamma_5^2 = I$ and

$$P^2 = P.$$

The states $\Psi$ for which $P\Psi = \Psi$ are assigned a *chirality* of $= 1$ and the states for which $(I - P)\Psi = \Psi$ are assigned the chirality $-1$. More explicitly, we have

$$\gamma_5 = \begin{pmatrix} \sigma_4 & 0 \\ 0 & -\sigma_4 \end{pmatrix}.$$

It is precisely the set of eigenvectors $\Psi$ of $\gamma_5$ which have a definite chirality. From $\gamma_5\Psi = \alpha\Psi$ the chirality ($\alpha = \pm 1$) is determined.

*Example:* $\Psi := (\varphi, 0)$ has the chirality $= +1$, while $\Psi := (0, \mathrm{i}\chi)$ has the chirality $-1$. In the case of spatial reflections $\varphi \mapsto \varphi'$, $\chi \mapsto \chi'$, one has

$$\varphi' = \chi\,, \qquad \chi' = -\varphi\,,$$

i.e., the chirality changes sign.

In 1956, Lee and Yang introduced the hypothesis that the neutrino only naturally occurs with chirality $-1$. This means that during a process involving neutrinos, no spatial reflection can occur. This effect is referred to as the *loss of parity* of the weak interaction (force). This spatial asymmetry was verified experimentally by Mrs. Wu in 1957 during the observation of $\beta$-decay of cobalt. That same year Lee and Yang won the Nobel award for physics for the theory of violation of parity.

**Fermions, bosons and the standard model of elementary particles:** All particles with a half-integral spin

$$k\hbar/2\,, \qquad \mathrm{k} = 1, 3, 5, \ldots\,,$$

are referred to as *fermions*; the particles for which the spin is integral

$$m\hbar\,, \qquad \mathrm{m} = 0, 1, \ldots\,,$$

are called *bosons*. The standard theory in its present form uses as elementary particles

| 6 quarks and 6 leptons (such as electrons and neutrinos) |
|---|

and their anti-particles.[66] All these particles are fermions and are described by equations of the same type as the Dirac equation. According to the principle of local gauge invariance, these Dirac equations must be coupled to additional fields which correspond to the gauge bosons, which describe (or mediate) the interactions between the 12 basic fermions (cf. [212]). For example, the electromagnetic interaction is mediated by the photon (see Table 3.11).

The existence of anti-particles is only required after the second quantization of the Dirac equation in the context of quantum field theory.

[66] The sixth quark (the top quark) was long sought for and finally verified to exist in experiments at the Fermilab (near Chicago) in 1994.

Table 2.10. Gluons of the four elementary forces.

| *Interaction occuring in nature* | *Gauge boson* | *Spin* |
|---|---|---|
| electromagnetic | photon | $\hbar$ |
| strong (nuclear forces) | 8 gluons | $\hbar$ |
| weak (radioactive decay) | $W^{\pm}, Z$ | $\hbar$ |
| gravity | graviton | $2\hbar$ |

**Supersymmetry (SUSY):** It is generally agreed that a realistic model of the universe incorperates the notion that the physics was supersymmetric shortly following the big bang, i.e., to each boson there was a partner particle which was fermionic. For example, the fermion corresponding to the graviton is the *gravitino* with spin $3\hbar/2$.

It is still hoped that the next generation of accelerators will be able to verify the existence of supersymmetric particles, which have not yet been observed; the higher the energy of an accelerator, the better the conditions following the big bang can be simulated.

### 3.9.6.3 The Clifford algebra of a Hilbert space and the spin groups

Let $X$ be a real Hilbert space with $\dim X = n$. Then the Clifford algebra (cf. section 2.4.3.4) $\mathscr{C}(X)$ of $X$ is a subset of $X$, $\mathscr{C}(X) \subseteq X$, and

$$\boxed{u \vee v + v \vee u = -2(u,v)_X \qquad \text{for all} \quad u, v \in X\,,}$$

where $(u,v)_X$ denotes the scalar product on $X$.

We choose an orthonormal basis $e_1, \ldots, e_n$ of $X$. Then we have

$$e_j \vee e_k + e_k \vee e_j = -2\delta_{jk}\,, \qquad j,k = 1,\ldots,n\,,$$

i.e., $e_j \vee e_k = -e_k \vee e_j$ for $j \neq k$ and

$$\boxed{e_j \vee e_j = -1\,, \quad j = 1,\ldots,n\,.}$$

This relation generalizes the equation $\mathrm{i}^2 = -1$ for the imaginary unit. We set $e_0 := 1$ and

$$e_\alpha := e_{\alpha_1} \vee e_{\alpha_2} \vee \ldots \vee e_{\alpha_k}$$

for $1 \leq \alpha_1 < \cdots < \alpha_k \leq n$, $\alpha = (\alpha_1, \ldots, \alpha_n)$ and $|\alpha| = \alpha_1 + \cdots + \alpha_n$.

**Basis theorem:** One has $\dim \mathscr{C}(X) = 2^n$. A basis of $\mathscr{C}(X)$ is formed by $e_0$ and the set of the $e_\alpha$.

*Example 1:* Let $\dim X = 1$. Then $e_1 \vee e_1 = -1$. Thus in this case one can do calculations with elements $a + be_1$ in the Clifford algebra $\mathscr{C}(X)$ just as with complex numbers, i.e., there is an isomorphism

$$\mathscr{C}(X) \cong \mathscr{C}(\mathbb{R}) \cong \mathbb{C}$$

between the Clifford algebra $\mathscr{C}(\mathbb{R})$ and the field $\mathbb{C}$ of complex numbers.

**The Hilbert space $\mathscr{C}(X)$:** We endow $\mathscr{C}(X)$ with the uniquely determined scalar product $(.,.)_{\mathscr{C}(X)}$, with respect to which the basis $e_0, e_\alpha, \ldots$ is orthonormalized, i.e., one has $(e_\beta, e_\gamma)_{\mathscr{C}(X)} = 0$ for $\beta \neq \gamma$ and $(e_\beta, e_\gamma)_{\mathscr{C}(X)} = 1$ for $\beta = \gamma$.

*Example 2:* For $\mathscr{C}(\mathbb{R})$ one has

$$(a + be_1, c + e_1 d)_{\mathscr{C}(X)} = ac + bd\,.$$

**The Clifford norm:** For every $u \in \mathscr{C}(X)$ we define

$$\boxed{|u| := \sup ||u \vee w||_{\mathscr{C}(X)}\,.}$$

Here, one forms the supremum over all $w \in \mathscr{C}(X)$ for which $||w||_{\mathscr{C}(X)} \le 1$.

*Example 3:* For $\mathscr{C}(\mathbb{R})$ we get

$$|a + be_1| = \sqrt{a^2 + b^2}\,.$$

This formula corresponds to the absolute value $|a + bi|$ of the complex number $a + bi$ which corresponds to $a + be_1$ as in Example 1 above.

**The conjugation operator:** By making the assignment

$$\varphi(e_0) := e_0\,, \quad \varphi(e_j) := -e_j\,, \quad j = 1, \ldots, n\,,$$

one gets an automorphism $\varphi : \mathscr{C}(X) \longrightarrow \mathscr{C}(X)$ of the Clifford algebra $\mathscr{C}(X)$, i.e., the map $\varphi$ respects linear combinations and the $\vee$-product. Instead of $\varphi(u)$ we also write $\overline{u}$.

*Example 4:* For $j, k = 1, \ldots, n$ one has

$$\overline{e}_j = -e_j\,, \quad \overline{e_j \vee e_k} = \overline{e}_j \vee \overline{e}_k = e_j \vee e_k\,.$$

*Example 5:* For $\mathscr{C}(\mathbb{R})$ one gets in particular

$$\overline{(a + e_1 b)} = a - be_1\,.$$

This corresponds to the passage to the complex conjugate number $\overline{a + bi} = a - bi$.

**The $*$-operator:** We set

$$e_0^* := e_0\,, \quad e_j^* := e_j\,, \quad j = 1, \ldots, n\,.$$

The action on the other elements of the basis are determined by taking the products in the opposite order:

$$(e_{\alpha_1} \vee e_{\alpha_2} \vee \ldots \vee e_{\alpha_k})^* := e_{\alpha_k} \vee \ldots \vee e_{\alpha_2} \vee e_{\alpha_1}\,.$$

Finally, the $*$-operator is uniquely determined by the further requirement that it is linear on $\mathscr{C}(X)$.

*Example 6:* $(a + be_1 + ce_1 \vee e_2)^* = a + be_1 + c(e_2 \wedge e_1)$.

**The algebra of Hamiltonian quaternions (1843):** We consider all formal sums

$$\boxed{a + \mathbf{v}}$$

consisting of real numbers $a$ and classical (three-) vectors $\mathbf{v}$. We define on this set an addition by the rule

$$(a + \mathbf{v}) + (b + \mathbf{w}) := (a + b) + (\mathbf{v} + \mathbf{w})$$

and a multiplication by the rule

$$\boxed{(a+\mathbf{v}) \vee (b+\mathbf{w}) := ab + a\mathbf{w} + b\mathbf{v} + (\mathbf{v} \times \mathbf{w}) - \mathbf{vw}\,.}$$

Here, $\mathbf{v} \times \mathbf{w}$ (resp. $\mathbf{vw}$) denotes the classical vector product (resp. the classical scalar product).

These operations satisfy the associative and distributive laws. We set

$$\overline{a+\mathbf{v}} := a - \mathbf{v}\,.$$

In particular one gets

$$(a+\mathbf{v}) \vee \overline{(a+\mathbf{v})} = a^2 + \mathbf{v}^2\,.$$

Finally, we set

$$|a+\mathbf{v}| = \sqrt{a^2+\mathbf{v}^2}\,.$$

Then the set of all of these elements define a skew-field with

$$(a+\mathbf{v})^{-1} = \frac{a-\mathbf{v}}{|a+\mathbf{v}|}\,.$$

*Example 7:* Let $\mathbf{i}, \mathbf{j}, \mathbf{k}$ be an orthonormal basis. Then we have

$$\boxed{\mathbf{i} \vee \mathbf{j} = -\mathbf{j} \vee \mathbf{i} = \mathbf{k}\,, \quad \mathbf{i} \vee \mathbf{i} = -1\,.} \tag{3.105}$$

The other products can be obtained from these by cyclically permuting: $\mathbf{i} \to \mathbf{j} \to \mathbf{k} \to \mathbf{i}$. If one writes

$$a + \mathbf{v} = a + \alpha\mathbf{i} + \beta\mathbf{j} + \gamma\mathbf{k}\,,$$

then the multiplication be also be easily obtained with the aid of (3.105). For example one has

$$(1+\mathbf{i}) \vee (\mathbf{j}+\mathbf{k}) = \mathbf{j} + \mathbf{k} + \mathbf{i} \vee \mathbf{j} + \mathbf{i} \vee \mathbf{k} = \mathbf{j} + \mathbf{k} + \mathbf{k} - \mathbf{j} = 2\mathbf{k}\,.$$

**Theorem:** If $X$ is a real Hilbert space with $\dim X = 2$ and orthonormal basis $\mathbf{i}$ and $\mathbf{j}$, then the Clifford algebra $\mathscr{C}(X)$ is isomorphic to the algebra of (Hamiltonian) quaternions. This isomorphism is given explicitly by mapping

$$a + \alpha\mathbf{i} + \beta\mathbf{j} + \gamma(\mathbf{i} \vee \mathbf{j}) \mapsto a + \alpha\mathbf{i} + \beta\mathbf{j} + \gamma\mathbf{k}.$$

This preserves absolute values (norms) and maps each element to its conjugate element.

**The rotation formula of Hamilton:** Let

$$Q := \cos\frac{\varphi}{2} + \mathbf{e}\sin\frac{\varphi}{2}$$

with $\mathbf{e}^2 = 1$. Then the elegant formula

$$\boxed{\mathbf{x}' = Q \vee \mathbf{x} \vee \overline{Q}} \tag{3.106}$$

corresponds to a rotation of the vector $\mathbf{x}$ by an angle $\varphi$ in the mathematical positive sense around the axis $\mathbf{e}$ (Figure 3.97).

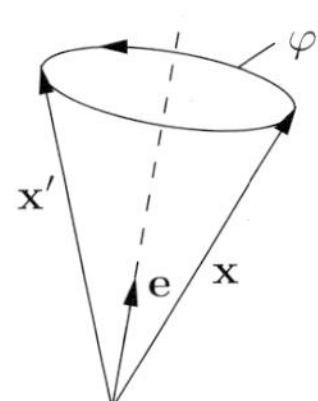

*Figure 3.97.*

It is our goal to generalize the formula (3.106) to higher dimensions with the help of Clifford algebras.

**The group** $Spin(n,X)$, $n \geq 2$**:** The set of all products

$$a = u_1 \vee u_2 \vee \ldots \vee u_{2k}\,,$$

which contain an even number of factors and in addition satisfy the two conditions

$$a \vee a^* = 1 \quad \text{and} \quad |a| = 1$$

form with respect to the product "$\vee$" a group, which is denoted by $Spin(n,X)$.

**Theorem of Brauer and Weyl (1935):** Every element $a \in Spin(n,X)$ generates via the prescription

$$D(a)u := a \vee u \vee a^* \qquad \text{for all} \quad u \in X$$

a unitary transformation (rotation) $D(a) : X \longrightarrow X$. The mapping

$$a \mapsto D(a)$$

gives rise to a homomorphism of the group $Spin(n,X)$ to the group $SO(n,X)$ with kernel $N = \{I, -I\}$. This in turn gives rise to an isomorphism[67]

$$\boxed{\text{Spin}\,(n,X)/N \cong SO(n,X)\,.}$$

For $X = \mathbb{R}^n$ we write $Spin(n)$ instead of $Spin(n,X)$.

*Example 8:* One has the following isomorphisms[68]

(i) $SO(2) \cong U(1)$, $\mathbb{R}/\mathbb{Z} \cong SO(2)$; the additive group $\mathbb{R}$ of real numbers is the universal covering group of $SO(2)$.

(ii) $Spin(3) \cong SU(2)$.

(iii) $Spin(4) \cong SU(2) \times SU(2)$.

(iv) $Spin(n,X) \cong Spin(n)$ for every real Hilbert space $X$ with $\dim X = n$.

**The elliptic Dirac operator:** This operator is defined by

$$D\psi := \sum_{j=1}^{n} e_j \partial_j \psi.$$

It acts on functions $\psi : \mathbb{R}^n \longrightarrow \mathscr{C}(X)$ with values in the Clifford algebra $\mathscr{C}(X)$. Moreover, we define the *Laplace operator*[69]

$$\boxed{\Delta := \sum_{j=1}^{n} \partial_j^2\,.}$$

**Theorem:** One has

$$\boxed{D \vee D = -\Delta\,.}$$

**Remark:** This relation shows that the Laplace operator can be constructed in terms of the Dirac operator. This is one of the reasons that the Dirac operator plays an extremely

[67] $Spin(n,X)$ is for $n \geq 2$ a compact real Lie group of dimension $n(n-1)/2$. Moreover, for $n \geq 3$, $Spin(n,X)$ is the universal covering group of $SO(n,X)$.

[68] (i) – (ii) are example of so-called exceptionial isomorphisms which exist for certain low-dimensional Lie groups.

[69] The operators $D$ and $\Delta$ are often defined in the literature with the opposite sign.

fundamental role in modern analysis. The problem of calculating the index of the Dirac operator on manifolds was one of the questions which led to the discovery of the Atiyah–Singer index theorem in 1960. This result is among the deepest of the twentieth century (cf. [212]).[70]

### 3.9.7 Almost complex structures

**Definition:** A real linear space $X$ of even dimension $2n$ is said to be *almost complex*, if there is a bijective linear operator $J : X \longrightarrow X$ such that

$$\boxed{J^2 = -I\,.}$$

**Theorem:** Every almost complex (real) linear space $X$ with $\dim X = 2n$ can be made into a complex linear space of dimension $n$ by setting

$$(\alpha + \beta \mathrm{i})u := \alpha u + \beta J u$$

for all $\alpha, \beta \in \mathbb{R}$ and $u \in X$.

### 3.9.8 Symplectic geometry

**Definition:** A *symplectic linear space* $X$ is a real linear space $X$ of even dimension $2n$ on which is given a bilinear mapping $\omega : X \times X \longrightarrow \mathbb{R}$, fulfilling the following two conditions:

(i) $\omega$ is *skew-symmetric*, i.e., one has $\omega(u, v) = -\omega(v, u)$ for all $u, v \in X$.

(ii) $\omega$ is *non-degenerate*, i.e., from $\omega(u, v) = 0$ for all $u \in X$ it follows that $v = 0$.

**Basis theorem:** An arbitrary linear symplectic space has a basis $e_1, \ldots, e_n, f_1, \ldots, f_n$, such that

$$\omega(e_j, e_k) = \omega(f_j, f_k) = 0\,,$$
$$\omega(e_j, f_k) = \delta_{jk}\,, \qquad j, k = 1, \ldots, n\,.$$

A basis with these properties is referred to as a *symplectic basis.*

**Normal form for $\omega$:** If we set

$$\mathrm{d}q^i(e_j) = \mathrm{d}p^i(f_j) = \delta_{ij}\,, \quad \mathrm{d}q^i(f_j) = \mathrm{d}p^i(e_j) = 0\,, \qquad i, j = 1, \ldots, n\,,$$

then we have (Darboux's theorem)

$$\boxed{\omega = \sum_{j=1}^{n} \mathrm{d}q^j \wedge \mathrm{d}p^j\,.}$$

In particular this implies

$$\boxed{\mathrm{d}\omega = 0\,.}$$

Now setting $u = q_1 e_1 + \cdots + q_n e_n + p_1 f_1 + \cdots + p_n f_n$, we get

$$\omega(u, u') = \sum_{j=1}^{n} q_j p_j' - p_j q_j' \qquad \text{for all} \quad u, u' \in X\,.$$

[70] This circle of problems together with important applications can be found in [296].

In the langauge of matrices, this means

$$\omega(u,u') = (q_1,\ldots,q_n,p_1,\ldots,p_n)\begin{pmatrix} 0 & I_n \\ -I_n & 0 \end{pmatrix}\begin{pmatrix} q_1' \\ \vdots \\ p_n' \end{pmatrix}$$

where $I_n$ denotes the $n$-dimensional unity matrix.

**Symplectic mappings:** A *symplectic mapping* $A : X \longrightarrow X$ is a linear map such that

$$\boxed{\omega\,(Au, Av) = \omega(u,v) \qquad \text{for all} \quad u,v \in X\,.}$$

A linear mapping $A : X \longrightarrow X$ is symplectic if and only if it transforms every symplectic basis into another symplectic basis.

**The symplectic group $Sp(2n,X)$:** The set of all symplectic mappings $A : X \longrightarrow X$ form a group which is referred to as the *symplectic group of* $X$ and which is denoted $Sp(2n,X)$.

The invariants of this group are, according to the general philosophy of the Erlanger Programm (cf. 3.1) the properties of the symplectic geometry of $X$.

**Volume:** We define the volume form

$$\mu := \omega \wedge \omega \wedge \ldots \wedge \omega \qquad (n \text{ factors})\,.$$

The volume of a subset $\mathscr{G}$ of $X$ is then defined, in analogy with 3.9.2, by

$$\mathrm{Vol}(\mathscr{G}) := \int_{\mathscr{G}} \mu.$$

For $A \in Sp(2n,X)$ one has $\mathrm{Vol}\,(A(\mathscr{G})) = \mathrm{Vol}\,(\mathscr{G})$.

With respect to a symplectic basis as above, the form $\mu$ can be written

$$\mu = \alpha \mathrm{d}q^1 \wedge \mathrm{d}q^2 \wedge \ldots \wedge \mathrm{d}q^n \wedge \mathrm{d}p^1 \wedge \ldots \wedge \mathrm{d}p^n$$

with an appropriately chosen real factor $\alpha$. Thus the symplectic volume coincides with the classical volume up to a constant factor.

**Orthogonality:** Two vectors $u,v \in X$ are reffered to as being *orthogonal* to one another if $\omega(u,v) = 0$.

**Lagrangian subspaces:** A linear subspace $L$ of $X$ is said to be *isotropic* if

$$\omega(u,v) = 0 \qquad \text{for all} \quad u,v \in L.$$

A subspace with this property is called a *Lagrangian subspace*, if and only if it cannot be extended to a larger isotropic subspace (in other words, if it is isotropic and maximal with that property). The dimension of a Lagrangian subspace is always $n$.

**The almost complex structure of a symplectic space:** Let $e_1,\ldots,e_n,f_1,\ldots,f_n$ be a symplectic basis of $X$. By means of the formula

$$(u,u') := \sum_{j=1}^{n} q_j q_j' + p_j p_j'$$

with $u = q_1e_1 + \cdots + q_ne_n + p_1f_1 + \cdots + p_nf_n$, a scalar product is defined on $X$. Then there is a uniquely defined linear operator $J : X \longrightarrow X$ with the property

$$\boxed{\omega(u,v) = (Ju,v) \qquad \text{for all} \quad u,v \in X\,.}$$

Because of $J^2 = -I$, this gives rise to an almost complex structure on $X$ (cf. 3.9.7).

**Theorem:** A linear mapping $A : X \longrightarrow X$ is symplectic if and only if it satisfies the relation

$$\boxed{A^* J A = J\,.}$$

**Symplectic matrices:** If we set $b_j := e_j$ and $b_{n+j} := f_j$ for $j = 1, \ldots, n$, and if we associate with every linear operator $A : X \longrightarrow X$ the matrix $\mathscr{A} := (a_{jk})$ with

$$a_{jk} := (b_j, Ab_k)\,,$$

then the operator $J$, viewed as a linear map, corresponds to the matrix

$$\mathscr{S} := \begin{pmatrix} 0 & I_n \\ -I_n & 0 \end{pmatrix}.$$

Moreover, $A$ is symplectic if and only if the corresponding matrix is symplectic, i.e., if

$$\boxed{\mathscr{A}^{\mathsf{T}} \mathscr{S} \mathscr{A} = \mathscr{S}\,.}$$

The set of all of these matrices forms the symplectic matrix group $Sp(2n)$. The association $A \mapsto \mathscr{A}$ gives an isomorphism between $Sp(2n, X)$ and $Sp(2n)$.

# 4. Foundations of Mathematics

*We need to know,*
*we shall know.*[1]
*David Hilbert* (1862–1943)

## 4.1 The language of mathematics

As opposed to our everyday form of communication, mathematics uses a very precise language; we first will explain some of the basic notions concerning this.

### 4.1.1 True and false statements

A *statement* is a construction of language which makes sense and can either be true or false.

*Example 1:* The statement '2 divides 4 or 3 divides 5' is true, since the first part of the statement is true. On the other hand, the statement '2 divides 5 or 3 divides 7' is false, since both parts of the statement are so.

**The alternative:** If $\mathscr{A}$ and $\mathscr{B}$ denote statements, then there is a *composed* statement

$$\boxed{\mathscr{A} \text{ or } \mathscr{B}}$$

which is true if either one of the two components $\mathscr{A}$ or $\mathscr{B}$ is true. If both statements $\mathscr{A}$ and $\mathscr{B}$ are false, then the composed statement '$\mathscr{A}$ or $\mathscr{B}$' is also false. This statement is called an *alternative* or an *or statement.*

**The strict alternative:** The statement

$$\boxed{\text{either } \mathscr{A} \text{ or } \mathscr{B}}$$

is true if one of the two statements is true *and* (at the same time) the other is false. In all other cases, the statement 'either $\mathscr{A}$ or $\mathscr{B}$' is false.

*Example 2:* Let $m$ be an integer. The statement 'either $m$ is even or $m$ is odd' is true. In contrast, the statement 'either $m$ is even or $m$ is divisible by 3' is false.

**The conjunction:** The statement

$$\boxed{\mathscr{A} \text{ and } \mathscr{B}}$$

[1]These words of Hilbert are written on his gravestone in Göttingen. Although Hilbert was quite aware of the limits of human understanding, these words express an optimistic epistemological point of view.

is defined to be true if both parts of the statement are true. Otherwise it is false. This statement is called a *conjunction* or an *and statement.*

*Example 3:* The statement '2 divides 4 and 3 divides 5' is false.

**Negation:** The statement

$$\boxed{\text{not } \mathscr{A}}$$

is true (resp. false) if $\mathscr{A}$ is false (resp. true).

**Statements on existence:** Let $D$ be a fixed set of objects. For example, $D$ could denote the set of real numbers.

The somewhat long-winded statement

$$\boxed{\text{there is an object } x \text{ in } D \text{ with the property } E}$$

is abbreviated by writing

$$\boxed{\exists_{x\in D} : E \text{ (or also } \exists_{x\in D} \,|\, E).}$$

This statement is true if there is an object $x$ in $D$ with the property $E$, whereas the statement is false if there is no such object $x$ in $D$ with the mentioned property. A statement of this kind is referred to as an *existence statement.*

*Example 4:* The statement 'there is a real number $x$ with the property that $x^2 + 1 = 0$' is false.

**Generalizers:** As above, instead of the statement

$$\boxed{\text{all objects } x \text{ in } D \text{ have the property } E}$$

we also write more briefly

$$\boxed{\forall_{x\in D} : E \text{ (or again also } \forall_{x\in D} \,|\, E).}$$

The statement is true if all elements $x$ in $D$ have the property $E$, and the statement is false if there is at least one object $x$ in $D$ which does not have the property. This kind of a statement is referred to as a *generalizer* or, more often, as a *for all* statement.

*Example 5:* The statement 'all integers are prime numbers' is false, since (for example) 4 is not prime.

### 4.1.2 Implications

Instead of the statement

$$\boxed{\mathscr{A} \text{ implies } \mathscr{B} \text{ ( or } \mathscr{B} \text{ follows from } \mathscr{A})}$$

we use the symbolic notation

$$\boxed{\mathscr{A} \Rightarrow \mathscr{B}.} \tag{4.1}$$

A composed statement of this kind is referred to as a *conclusion* (or *implication*). Instead of (4.1) the following terminology has become commonplace:

(i) $\mathscr{A}$ is *sufficient* for $\mathscr{B}$;

(ii) $\mathscr{B}$ is *necessary* for $\mathscr{A}$.

The implication '$\mathscr{A} \Rightarrow \mathscr{B}$' is false if the assumption $\mathscr{A}$ is true and the implied statement $\mathscr{B}$ is false. Otherwise the implication is true. This corresponds to the at first sight somewhat surprising convention in mathematics that any conclusion made from a false premise (assumption) is to be considered true. In other words, with a false assumption one can prove *anything*. The following example shows that this convention is quite natural and corresponds to the usual formulation of mathematical statements.

*Example 1:* Let $m$ be an integer. The statement $\mathscr{A}$ (resp. $\mathscr{B}$) is $m$ is divisible by 6 (resp. $m$ is divisible by 2). The implication '$\mathscr{A} \Rightarrow \mathscr{B}$' can then be formulated as follows.

From the divisibility of $m$ by 6 the divisibility of $m$ by 2 follows. (4.2)

This is also expressed by saying

(a) The divisibility of $m$ by 6 is sufficient for the divisibility of $m$ by 2;

(b) The divisibility of $m$ by 2 is necessary for the divisibility of $m$ by 6.

Intuitively you would think that the statement (4.2) is always true, i.e., that is expresses a *mathematical theorem.* To actually prove that this is the case, we have to consider two cases.

*Case 1:* $m$ is divisible by 6.

This means that there is an integer $k$ such that $m = 6k$. This implies that $m = 2(3k)$, which expresses the fact that $m$ is divisible by 2; the statement (4.2) is correct in this case.

*Case 2:* $m$ is not divisible by 6.

This means that the premise is false (from which it follows that any conclusion is correct), so that the conclusion (4.2) is also correct in this case.

Taking both cases together, we see that the statement (4.2) is always true. The argumentation we produced above is what one refers to as a *proof* of (4.2).

**Logical equivalences:** It is false that '$\mathscr{A} \Rightarrow \mathscr{B}$' implies '$\mathscr{B} \Rightarrow \mathscr{A}$'. For example, the converse implication of statement (4.2) is false. This is also expressed by saying that the divisibility of $m$ by 2 is necessary, but not sufficient, for $m$ to be divisible by 6.

A statement of the form

$$\mathscr{A} \Longleftrightarrow \mathscr{B} \tag{4.3}$$

means that the two implications '$\mathscr{A} \Rightarrow \mathscr{B}$' and '$\mathscr{B} \Rightarrow \mathscr{A}$' are true. Instead of the so-called *logical equivalence* (4.3), the same state of affairs is described by any of the following:

(i) $\mathscr{A}$ holds if and only if $\mathscr{B}$ holds;

(ii) $\mathscr{A}$ is sufficient and necessary for $\mathscr{B}$;

(iii) $\mathscr{B}$ is sufficient and necessary for $\mathscr{A}$.

*Example 2:* Let $m$ be an integer. The statement $\mathscr{A}$ (resp. $\mathscr{B}$) is '$m$ is divisible by 6' (resp. '$m$ is divisible by 2 and 3'). Then the equivalence '$\mathscr{A} \Longleftrightarrow \mathscr{B}$' means.

The integer $m$ is divisible by 6 if and only if $m$ is divisible by 2 and 3.

One can also say: the divisibility of $m$ by 2 and 3 is necessary and sufficient for the divisibility of $m$ by 6.

Mathematical statements in the form of logical equivalences always contain a conclusion and are therefore of particular interest for mathematics.

**Counter-position of an implication:** A given implication '$\mathscr{A} \Rightarrow \mathscr{B}$' always implies and is implied by the new (equivalent) implication

$$\boxed{\text{not } \mathscr{B} \Rightarrow \text{not } \mathscr{A},}$$

where the statement 'not $\mathscr{A}$' means that $\mathscr{A}$ is assumed *not* to hold, and similarly with 'not $\mathscr{B}$'. Note that each statement ($\mathscr{A}$ and $\mathscr{B}$) are negated, while the *direction* of the implication is inverted. This is called the *counter-position* or *negation* of the implication '$\mathscr{A} \Rightarrow \mathscr{B}$'.

*Example 3:* From the statement (4.2) we get the new statement: If $m$ is not divisible by 2, then $m$ is also not divisible by 6.

**Counter-position of a logical equivalence:** From the logical equivalence '$\mathscr{A} \iff \mathscr{B}$' we get the new, equivalent logical equivalence

$$\boxed{\text{not } \mathscr{A} \iff \text{not } \mathscr{B}.}$$

*Example 4:* Let $(a_n)$ be an increasing (or decreasing) sequence of real numbers. Consider the theorem:[2]

$$(a_n) \text{ is convergent } \iff (a_n) \text{ is bounded.}$$

By applying the counter-position to this equivalence we get the new theorem:

$$(a_n) \text{ is not convergent } \iff (a_n) \text{ is not bounded.}$$

In other words, (i) an increasing (or decreasing) sequence is convergent if and only if it is bounded, and (ii) an increasing (or decreasing) sequence is not convergent if and only if it is not bounded.

### 4.1.3 Tautological and logical laws

*Tautologies* are composed statements, which are *always true*, regardless of the truth of the partial statements composing it. Our entire logical thinking, which is the basis for mathematics, is itself based on the application of tautologies. These tautologies can also be viewed as logical laws. The most important tautologies are the following.

(i) The distributive law for alternatives and conjunctions:

$$\begin{aligned} \mathscr{A} \text{ and } (\mathscr{B} \text{ or } \mathscr{C}) &\iff (\mathscr{A} \text{ and } \mathscr{B}) \text{ or } (\mathscr{A} \text{ and } \mathscr{C}), \\ \mathscr{A} \text{ or } (\mathscr{B} \text{ and } \mathscr{C}) &\iff (\mathscr{A} \text{ or } \mathscr{B}) \text{ and } (\mathscr{A} \text{ or } \mathscr{C}). \end{aligned}$$

(ii) The negation of a negation:

$$\text{not}\,(\text{not}\,\mathscr{A}) \iff \mathscr{A}.$$

(iii) The counter-position of an implication:

$$(\mathscr{A} \Rightarrow \mathscr{B}) \iff (\text{not}\,\mathscr{B} \Rightarrow \text{not}\,\mathscr{A}).$$

(iv) The counter-position of a logical equivalence:

$$(\mathscr{A} \iff \mathscr{B}) \iff (\text{not}\,\mathscr{A} \iff \text{not}\,\mathscr{B}).$$

[2]Here we mean by convergent sequence one which has a finite limit.

(v) The negation of an alternative (*rule of de Morgan*):[3]

$$\boxed{\text{not}\,(\mathscr{A}\ \textit{or}\ \mathscr{B})\iff(\text{not}\,\mathscr{A}\ \textit{and}\ \text{not}\,\mathscr{B})\,.}$$

(vi) The negation of a conjunction (*rule of de Morgan*):[3]

$$\boxed{\text{not}\,(\mathscr{A}\ \textit{and}\ \mathscr{B})\iff(\text{not}\,\mathscr{A}\ \textit{or}\ \text{not}\,\mathscr{B})\,.}$$

(vii) The negation of a statement of existence:

$$\boxed{(\text{not}\,\exists_x \mid E)\iff(\forall_x \mid \text{not}\,E)\,.}$$

(viii) The negation of a generalized statement:

$$\boxed{(\text{not}\,\forall_x \mid E)\iff(\exists_x \mid \text{not}\,E)\,.}$$

(ix) The negation of an implication:

$$\boxed{\text{not}\,(\mathscr{A}\Rightarrow\mathscr{B})\iff\{\mathscr{A}\ \text{and not}\ \mathscr{B}\}\,.}$$

(x) The negation of a logical equivalence:

$$\text{not}\,(\mathscr{A}\iff\mathscr{B})\iff\{(\mathscr{A}\ \text{and not}\ \mathscr{B})\ \text{or}\ (\mathscr{B}\ \text{and not}\ \mathscr{A})\}\,.$$

(xi) The fundamental rule of separation (*modus ponens*) of Theophrast of Eresos (372–287 BC):

$$\boxed{\{(\mathscr{A}\Rightarrow\mathscr{B})\ \text{and}\ \mathscr{A}\}\Rightarrow\mathscr{B}\,.}$$

The tautologies in (i) are responsible for the distributive laws (with respect to unions and intersections) in set theory (see section 4.3.2).

The law of negation of a negation means that the double negation of a statement is logically equivalent to the statement one started with.

We already used the tautologies (iii) and (iv) in the Examples 3 and 4 in section 4.1.2. The tautologies (iv) to (x) are utilized quite often in mathematical induction proofs (see 4.2.1).

The rule of separation (xi) implies the following logical law, which is often used in all of mathematics and it therefore referred to sometimes as a *Fundamental Theorem of Logic*:

> If a statement $\mathscr{B}$ follows from the assumption $\mathscr{A}$ and if the assumption $\mathscr{A}$ is satisfied, then the statement $\mathscr{B}$ holds.

The rule of de Morgan (v) implies the following logical law:

> The negation of an alternative is logically equivalent to the conjunction of the negated alternative statements.

*Example:* Let $m$ be an integer. Then:

(i) If the number $m$ does not satisfy: it is even *or* divisible by 3, then $m$ is not even *and* $m$ is not divisible by 3.

[3] Note the exchange of 'or' and 'and' on the two sides!

(ii) If the number $m$ does not satisfy: it is even *and* it is divisible by 3, then $m$ is not even *or* $m$ is not divisible by 3.

The tautology (vii) is, expressed in words, the statement: If there is no object $x$ with the property $E$, then all objects do not have property $E$, and the converse statement is true also.

The tautology (viii), expressed in words, states: If it is not true that all objects $x$ have the property $E$, then there exists an object $x$, which does not have property $E$, and the converse statement also holds.

## 4.2 Methods of proof

### 4.2.1 Indirect proofs

Many proofs in mathematics are carried out as follows. One assumes that the statement (which one is trying to prove) is false, then leads this assumption to a contradiction.

The following proof is due to Aristotle.

*Example:* The number $\sqrt{2}$ is not rational.

*Proof* (indirect): We assume that the statement is false. Then $\sqrt{2}$ is a rational number and can be written in the form

$$\sqrt{2} = \frac{m}{n} \tag{4.4}$$

with integers $m$ and $n \neq 0$. We may further assume that $m$ and $n$ are relatively prime (otherwise reduce the fraction).

We utilize the following elementary facts which hold for arbitrary integers $p$.

(i) If $p$ is even, then $p^2$ is divisible by 4.

(ii) If $p$ is odd, then also $p^2$ is odd.[4]

If we now square (4.4), then we get

$$2n^2 = m^2 . \tag{4.5}$$

The square $m^2$ is thus even. Consequently, $m$ must also be even according to the counter-position of (ii). But then $m^2$ must be divisible by 4 according to (i), which means by (4.5) that $n^2$ is even. Thus, both $m$ and $n$ are even. But this is a contradiction to the fact that $m$ and $n$ are relatively prime.

This contradiction shows that the assumption made, namely that $\sqrt{2}$ is rational, is false. Thus, $\sqrt{2}$ is not rational. □

### 4.2.2 Induction proofs

The principle of induction, presented in section 1.2.2.2, is often used in the following way. Let a statement $\mathscr{A}(n)$ be given, which depends on an integer $n$ with $n \geq n_0$ for some fixed integer $n_0$. Furthermore, assume that the following hold:

(i) The statement $\mathscr{A}(n)$ is true for $n = n_0$;

(ii) The validity of $\mathscr{A}(n)$ implies the validity of $\mathscr{A}(n+1)$.

Then the statement $\mathscr{A}(n)$ is true for all integers $n$ with $n \geq n_0$.

[4] This follows from $(2k)^2 = 4k^2$ and $(2k+1)^2 = 4k^2 + 4k + 1$.

*Example:* The following equation

$$1 + 2 + \ldots + n = \frac{n(n+1)}{2} \tag{4.6}$$

is valid for all positive natural numbers $n$, i.e., for $n = 1, 2, \ldots$

*Proof:* The statement $\mathscr{A}(n)$ is: (4.6) holds for the positive natural number $n$.

Step 1: $\mathscr{A}(n)$ is clearly true for $n = 1$.

Step 2: Let $n$ be any fixed chosen positive natural number. Assume that $\mathscr{A}(n)$ holds; we must conclude that this assumption implies that $\mathscr{A}(n+1)$ holds also.

If we add $n + 1$ to both sides of (4.6), then we obtain

$$1 + 2 + \ldots + n + (n+1) = \frac{n(n+1)}{2} + n + 1\,.$$

Moreover,

$$\frac{(n+1)(n+2)}{2} = \frac{n^2 + n + 2n + 2}{2} = \frac{n(n+1)}{2} + n + 1\,.$$

This implies that

$$1 + 2 + \ldots + n + (n+1) = \frac{(n+1)(n+2)}{2}$$

which is nothing but $\mathscr{A}(n+1)$.

Step 3: We conclude that the statement $\mathscr{A}(n)$ is true for all natural numbers $n \geq 1$. □

### 4.2.3 Uniqueness proofs

A uniqueness statement expresses the fact that there are only finitely many objects with a given property.

*Example:* There is at most one positive real number $x$ with

$$x^2 + 1 = 0\,. \tag{4.7}$$

*Proof:* Suppose that there are two solutions $a$ and $b$. But $a^2 + 1 = 0$ and $b^2 + 1 = 0$ imply $a^2 - b^2 = 0$. Thus,

$$(a - b)(a + b) = 0\,.$$

Because of $a > 0$ and $b > 0$, we get $a + b > 0$. Dividing the left-hand side by $a + b$, this then implies that $a - b = 0$. Thus $a = b$. □

### 4.2.4 Proofs of existence

It is important to make a clear distinction between the uniqueness and the existence of a solution. The equation (4.7) has at most one positive real solution. This means that it either has one or it has no solution at all. In fact, equation (4.7) has no real solution. Indeed, were $x$ a real solution of (4.7), then from $x^2 \geq 0$ we get immediately the relation $x^2 + 1 > 0$, which contradicts $x^2 + 1 = 0$. Generally speaking, proofs of existence are much more difficult than proofs of uniqueness. There are two types of these proofs:

(i) abstract proofs of existence, and

(ii) constructive proofs of existence.

*Example 1:* The equation

$$x^2 = 2 \tag{4.8}$$

has a solution $x$ in the real numbers.

**Abstract proof of existence:** We set

$$A := \{a \in \mathbb{R} : a < 0 \quad \text{or} \quad \{a \geq 0 \quad \text{and} \quad a^2 \leq 2\}\},$$
$$B := \{a \in \mathbb{R} : a \geq 0 \quad \text{and} \quad a^2 > 2\}.$$

In words: the set $A$ consists of those real numbers $a$ which satisfy one of the two conditions '$a < 0$' or '$a \geq 0$ and $a^2 \leq 2$'.

Similarly, the set $B$ consists of all real numbers $a$ with $a \geq 0$ and $a^2 > 2$.

Clearly every real number belongs to one of the two sets $A$ or $B$. Because of $0 \in A$ and $2 \in B$, both sets are not empty. Therefore, according to the completeness axiom in 1.2.2.1, there is a real number $\alpha$ with the property:

$$a \leq \alpha \leq b \quad \text{for all } a \in A \text{ and all } b \in B. \tag{4.9}$$

We show that $\alpha^2 = 2$. Otherwise, we would have from $(\pm\alpha)^2 = \alpha^2$ the following two cases.

*Case 1:* $\alpha^2 < 2$ and $\alpha > 0$.

*Case 2:* $\alpha^2 > 2$ and $\alpha > 0$.

In the first case we choose a number $\varepsilon > 0$ sufficiently small, so that

$$(\alpha + \varepsilon)^2 = \alpha^2 + 2\varepsilon\alpha + \varepsilon^2 < 2.$$

For example, we could take $\varepsilon = \min\left(\dfrac{2-\alpha^2}{2\alpha+1}, 1\right)$. Then we have $\alpha + \varepsilon \in A$. By (4.9) it would follow that $\alpha + \varepsilon \leq \alpha$, which is impossible. Therefore case 1 is impossible. Similarly one can show that case 2 is impossible. Thus, the assumption that $\alpha^2 \neq 2$ is false, hence the proof is complete. □

**Uniqueness theorem:** As in 4.2.3 one can show that the equation (4.8) has at most one positive solution.

**The conclusion on existence and uniqueness:** It follows from the above that the equation $x^2 = 2$ has a unique positive solution $x$, which is usually denoted by $\sqrt{2}$.

**Constructive proof:** We show that the iteration

$$a_{n+1} = \frac{a_n}{2} + \frac{1}{a_n}, \qquad n = 1, 2, \ldots \tag{4.10}$$

with $a_1 := 2$ converges to a number $\sqrt{2}$, i.e., $\lim\limits_{n\to\infty} a_n = \sqrt{2}$.

*Step 1:* We show that $a_n > 0$ for $n = 1, 2, \ldots$

This is true for $n = 1$. Moreover, from $a_n > 0$ for some fixed $n$, together with (4.10), we see that also $a_{n+1} > 0$. By the principle of induction we get $a_n > 0$ for $n = 1, 2, \ldots$

*Step 2:* We show that $a_n^2 \geq 2$ for $n = 1, 2, \ldots$

This statement is true for $n = 1$. If $a_n^2 \geq 2$ for a fixed $n$, then it follows from the Bernoulli inequality:[5]

$$a_{n+1}^2 = a_n^2\left(1 + \frac{2 - a_n^2}{2a_n^2}\right)^2 \geq a_n^2\left(1 + \frac{2-a_n^2}{a_n^2}\right) = 2.$$

[5] For all real numbers $r$ with $r \geq -1$ and all natural numbers $n$ one has

$$(1+r)^n \geq 1 + nr.$$

Thus, $a_{n+1}^2 \geq 2$. By the principle of induction we then have the relation $a_n^2 \geq 2$ for $n = 1, 2, \ldots$

*Step 3:* From (4.10) we get

$$a_n - a_{n+1} = a_n - \frac{1}{2}\left(a_n + \frac{2}{a_n}\right) = \frac{1}{2a_n}\left(a_n^2 - 2\right) \geq 0\,, \qquad n = 1, 2, \ldots$$

Thus the sequence $(a_n)$ is *decreasing* and bounded below. The convergence criterion for decreasing sequences in section 1.2.4.1 then shows the existence of the limit

$$\lim_{n \to \infty} a_n = x\,.$$

If we pass to the limit in the equation (4.10), we obtain

$$x = \lim_{n \to \infty} a_{n+1} = \frac{x}{2} + \frac{1}{x}\,.$$

This means that $2x^2 = x^2 + 2$, hence

$$x^2 = 2.$$

From $a_n \geq 0$ for all $n$, we conclude that $x \geq 0$. Thus $x = \sqrt{2}$. □

### 4.2.5 The necessity of proofs in the age of computers

One might think that theoretical considerations in mathematics are superfluous by the incredible computing power available today through modern computers. On the contrary, given a mathematical problem, the following steps can be used for solving the problem.

(i) Prove the existence of the solution (abstract proof of existence);

(ii) Prove the uniqueness of the solution;

(iii) Study the stability of the solution (which is now known to exist) with respect to small perturbations of parameters of the problem;

(iv) Develop an algorithm for calculating (an approximation of) the solution on a computer;

(v) Prove the convergence of the algorithm, i.e., show that the algorithm converges to the unique solution under appropriate (reasonable) assumptions;

(vi) Prove an estimate for the approximation delivered by the algorithm;

(vii) Study the speed of convergence of the algorithm;

(viii) Prove the numerical stability of the algorithm.

In (v) and (vi) it is important that the existence of a unique solution is already known, as otherwise a computer calculation could yield an apparent solution, which, however, does not exist in reality (so-called ghost solutions).

A problem is said to be *well-posed*, if (i), (ii) and (iii) are assured. Estimates of a given approximation are of two kinds:

(a) *a priori* estimates, and

(b) *a posteriori* estimates.

These notions are similar to the philosophy of Immanuel Kant (1724–1804). An *a priori* estimate yields information on the error of approximations *before* the calculation is done (on a computer). On the other hand, an *a posteriori* estimate uses the information one has obtained *from* an already done calculation (on a computer). As a rule of thumb, one has:

| *A posteriori* estimates are more precise that *a priori* estimates. |
|---|

Algorithms must be numerically stable, i.e., they must be robust with respect to rounding errors which occur in the process of computer calculations. As it turns out, iteration procedures are particularly well-behaved for numerical calculations.

*Example:* We consider the sequence $(a_n)$ from (4.10) for the iterative calculation of $\sqrt{2}$. The total error $a_n - \sqrt{2}$ is denoted by $\Delta_n$. For $n = 1, 2, \ldots$, the following estimates hold.

(i) Speed of convergence:

$$\boxed{\Delta_{n+1} \leq \Delta_n^2 .}$$

This is what is known as a *quadratic convergence*, i.e., the process converges very quickly.[6]

(ii) *A priori* estimate:

$$\boxed{\Delta_{n+2} \leq 10^{-2^n} .}$$

(iii) *A posteriori* estimate:

$$\boxed{\frac{2}{a_n} \leq \sqrt{2} \leq a_n .}$$

A glance at Table 4.1 shows that

$$\sqrt{2} = 1.414213562 \pm 10^{-9} .$$

*Table 4.1: Successive approximations to* $\sqrt{2}$

| $n$ | $a_n$ | $\dfrac{2}{a_n}$ |
|---|---|---|
| 1 | 2 | 1 |
| 2 | 1.5 | 1.33 |
| 3 | 1.4118 | 1.4116 |
| 4 | 1.414215 | 1.414211 |
| 5 | 1.414213562 | 1.414213562 |

## 4.2.6 Incorrect proofs

The two most predominant mistakes in proofs are making a step in the proof where one 'divides by zero', and proving statements 'in the wrong direction'.

### 4.2.6.1 Division by zero

During manipulations with equations, you must always watch out for situations where you could, depending on the values of certain variables, divide by zero.

*False claim:* The equation

$$(x-2)(x+1) + 2 = 0 \tag{4.11}$$

[6]The method of iteration of (4.10) corresponds to the Newton iteration method for the equation $f(x) := x^2 - 2 = 0$ (see also section 7.4.1).

has exactly one real solution $x = 1$.

*'Proof' of this claim:* It is easily verified that $x = 1$ solves the equation (4.11). In order to show that this is the only solution, we assume that $x_0$ is a further solution of (4.11). Then we have

$$x_0^2 - 2x_0 + x_0 - 2 + 2 = 0.$$

From this we get $x_0^2 - x_0 = 0$, hence

$$x_0(x_0 - 1) = 0. \tag{4.12}$$

Division by $x_0$ yields $x_0 - 1 = 0$, which means that $x_0 = 1$.

Clearly the claim is false, as $x = 1$ and $x = 0$ are both solutions of (4.11). The mistake in the proof is in (4.12) we can only divide by $x_0$ under the assumption that $x_0 \neq 0$.

*Correct claim:* The equation (4.11) has exactly the two solutions $x = 1$ and $x = 0$.

*Proof:* If $x$ is a solution of (4.11), then (4.12) follows; this implies that either $x = 0$ or $x - 1 = 0$.

Checking both values shows that both indeed satisfy (4.11). □

#### 4.2.6.2 Proof in the wrong direction

A mistake which is often made consists of trying to prove the statement

$$\boxed{\mathscr{A} \Rightarrow \mathscr{B}}$$

by proving the *converse* implication $\mathscr{B} \Rightarrow \mathscr{A}$. The so-called '0 = 0' proofs belong to this type of mistake. We show this in the following example.

*False claim:* Every real number $x$ is a solution of the equation

$$x^2 - 4x + 3x + 1 = (x - 1)^2 + 3x\,. \tag{4.13}$$

*'Proof' of this claim:* Let $x \in \mathbb{R}$. Then from (4.13) we have

$$x^2 - x + 1 = x^2 - 2x + 1 + 3x = x^2 + x + 1\,. \tag{4.14}$$

Adding $-x^2 - 1$ to both sides yields

$$-x = x\,. \tag{4.15}$$

By squaring this result we get

$$x^2 = x^2\,. \tag{4.16}$$

This implies

$$0 = 0\,. \tag{4.17}$$

This correct chain of implications shows:

$$(4.13) \Rightarrow (4.14) \Rightarrow (4.15) \Rightarrow (4.16) \Rightarrow (4.17)\,.$$

However, this is not a proof of the claim above. Rather, we would have to prove the chain of implications

$$(4.17) \Rightarrow (4.16) \Rightarrow (4.15) \Rightarrow (4.14) \Rightarrow (4.13)\,.$$

But this is impossible. as the implication (4.15) $\Rightarrow$ (4.16) cannot be inverted. The implication (4.16) $\Rightarrow$ (4.15) is only true for $x = 0$.

*Correct claim:* The equation (4.13) has the unique solution $x = 0$.

*Proof:Step 1:* Suppose that a real number $x$ is a solution of (4.13). The chain of implications above yields

$$(4.13) \Rightarrow (4.14) \Rightarrow (4.15) \Rightarrow x = 0\,.$$

Hence (4.13) can have at most the solution $x = 0$.

*Step 2:* We show that

$$x = 0 \Rightarrow (4.15) \Rightarrow (4.14) \Rightarrow (4.13)\,.$$

This chain of implications is correct, and hence $x = 0$ is a solution. □

In this simple example we could have left out the second step and directly have checked that $x = 0$ is a solution of (4.13). In more complicated cases one often tries to make only logically equivalent manipulations, so that both directions are shown at once. In the example above, this would be

$$\boxed{(4.13) \iff (4.14) \iff (4.15) \iff x = 0\,.}$$

## 4.3 Naive set theory

In this section we describe the naive handling of sets. An axiomatic foundation of this can be found in section 4.4.3.

### 4.3.1 Basic ideas

**Classes and elements:** A *class* is a collection of objects. The symbol

$$\boxed{a \in A}$$

means that the object $a$ belongs to the class $A$. The symbol

$$\boxed{a \notin A}$$

means that the object $a$ is not an element of the class $A$. Either $a \in A$ is true or $a \notin A$ is true.

In mathematics, one often considers classes of classes, i.e., classes are iterated. For example, a plane is a class whose elements are points. On the other hand, the set of all planes through the origin in three-dimensional space is a class called a Grassmann manifold. There are two types of classes (Figure 4.1):

(i) The class is called a *set*, if it itself is an element in a new class.

(i) The class is called a *non-set*, if it cannot be an element in a new class.

Intuitively you can think of non-sets as classes which are so huge that no further classes could hold them as an element. For example the class of all sets is a non-set.

Set theory was created by Georg Cantor (1845–1918) during the last quarter of the nineteenth century. The boldest idea which Cantor had was structuring infinity by introducing transfinite cardinalities (see section 4.3.4) and developing an arithmetic of transfinite ordinal- and cardinal numbers (see 4.4.4 for this). Cantor made the following definition:

> *"A set is a collection of well-defined objects of our imagination or our intuition (which then are the elements of the set)."*

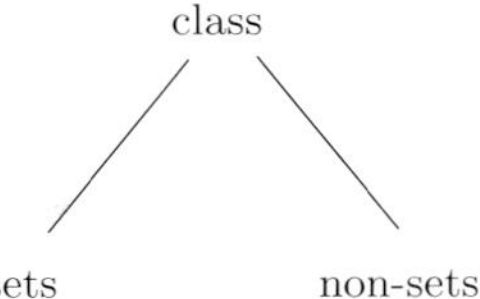


*Figure 4.1. Sets and non-sets.*

In 1901, the English philosopher and mathematician Bertrand Russel discovered that the notion of 'the set of all sets' is contradictory (Russel's antinomy). This created a crisis in the foundation of mathematics, which however could be overcome by

(i) making the distinction between sets and non-sets, and

(ii) giving set theory an axiomatic foundation.

The Russel contradiction is solved by defining that the collection of all sets is not a set, but rather a non-set. This will be discussed in more detail in section 4.4.3.

**Subsets and equality of sets:** If $A$ and $B$ are sets, then the symbol

$$\boxed{A \subseteq B}$$

means that every element of $A$ is at the same time an element of $B$. One also says that $A$ is a *subset* of $B$ (Figure 4.2). Two sets $A$ and $B$ are said to be *equal*, denoted by the symbol

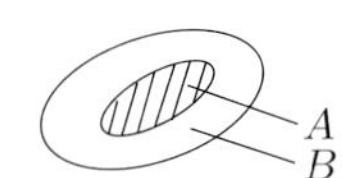


*Figure 4.2. Subsets.*

$$\boxed{A = B,}$$

if both conditions $A \subseteq B$ and $B \subseteq A$ are satisfied. Moreover, we write

$$\boxed{A \subset B}$$

if $A \subseteq B$ *and* $A \neq B$. For sets $A, B$ and $C$, the following rules are valid.

(i) $A \subseteq A$ (reflexiveness);

(ii) The conditions $A \subseteq B$ and $B \subseteq C$ imply the relation $A \subseteq C$ (transitivity);

(iii) The conditions $A \subseteq B$ and $B \subseteq A$ imply the relation $A = B$ (anti-symmetry).

The property (iii) is used to prove (and in fact is often the only way of showing) the equality of sets (see the example in 4.3.2).

**Definition of sets:** The most important method of defining sets is by means of a formula

$$\boxed{A := \{x \in B \,|\, \text{for } x \text{ the statement } \mathscr{A}(x) \text{ holds}\}.}$$

This means that the set $A$ consists of those elements $x$ of the set $B$, for which the statement $\mathscr{A}(x)$ is true.

*Example:* Let $\mathbb{Z}$ denote the set of integers. Then the formula

$$A := \{x \in \mathbb{Z} \,|\, x \text{ is divisible by } 2\}$$

defines the set of even numbers. If $a$ and $b$ are objects, then $\{a\}$ (resp. $\{a, b\}$) denotes the set consisting only of the element $a$ (resp. of the elements $a$ and $b$).

**The empty set:** The symbol $\emptyset$ denotes the *empty set*, which is the set consisting of no elements at all (note that this set is uniquely defined).

### 4.3.2 Calculations with sets

In this section we let $A, B, C, X$ denote sets.

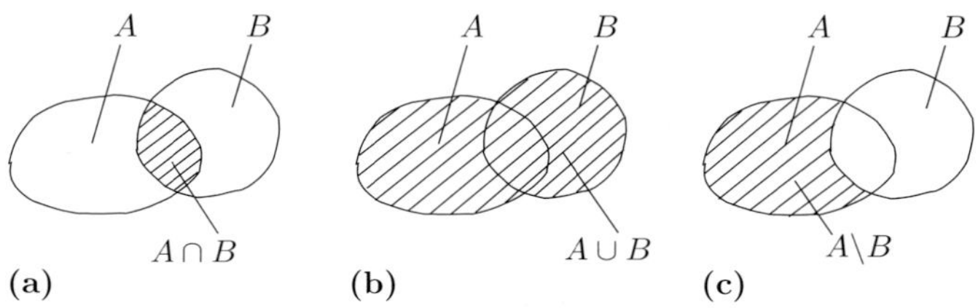


*Figure 4.3. The intersection, union and difference of sets.*

**The intersection of two sets:** The *intersection*

$$\boxed{A \cap B}$$

of the two sets $A$ and $B$ consists by definition of all elements which belong to both $A$ and $B$ (Figure 4.3(a)). Using the formula notation above, this is $A \cap B := \{x \,|\, x \in A \text{ and } x \in B\}$.

Two sets $A$ and $B$ are said to be *intersection-free* or *disjoint*, if $A \cap B = \emptyset$.

**The union of two sets:** The *union* of two sets $A$ and $B$, denoted

$$\boxed{A \cup B}$$

consists by definition of the set of elements which belong to *either* $A$ *or* $B$ (Figure 4.3(b)). One has a relation of inclusions among the intersection and union:

$$\boxed{(A \cap B) \subseteq A \subseteq (A \cup B)\,.}$$

Moreover, $A \subseteq B$ is equivalent to $A \cap B = A$ as well as to $A \cup B = B$.

The intersection $A \cap B$ (resp. the union $A \cup B$) behave in a similar manner to the product (resp. sum) of two numbers; the empty set $\emptyset$ plays the role of the number 0. Explicitly, we have the following rules.

(i) The law of commutativity:

$$A \cap B = B \cap A\,, \quad A \cup B = B \cup A\,.$$

(ii) The law of associativity:

$$A \cap (B \cap C) = (A \cap B) \cap C\,, \quad A \cup (B \cup C) = (A \cup B) \cup C\,.$$

(iii) The law of distributivity:

$$\boxed{A \cap (B \cup C) = (A \cap B) \cup (A \cap C)\,, \quad A \cup (B \cap C) = (A \cup B) \cap (A \cup C)\,.}$$

(iv) Neutral element:

$$A \cap \emptyset = \emptyset\,, \quad A \cup \emptyset = A\,.$$

*Example 1:* We wish to show $A \cap B = B \cap A$.

*Step 1:* We show that $(A \cap B) \subseteq (B \cap A)$. In fact, we have

$$a \in (A \cap B) \implies (a \in A \text{ and } a \in B) \implies (a \in B \text{ and } a \in A) \implies a \in (B \cap A)\,.$$

*Step 2:* We show that $(B \cap A) \subseteq (A \cap B)$. This follows in the same way as we showed Step 1 (exchanging $A$ and $B$).

From these two steps we get $A \cap B = B \cap A$. □

**The difference of two sets:** The *difference* set

$$\boxed{A \backslash B}$$

consists by definition of those elements of $A$ which are not elements of $B$ (Figure 4.3(c)). In the formula notation of above, this is $A \backslash B := \{x \in A \,|\, x \notin B\}$. This notation is not universal; in some situations, it could be misunderstood as a coset notation. In order to avoid misunderstandings, the notation $A - B$ is also often used.

In addition to the obvious inclusion relation

$$A \backslash B \subseteq A$$

and the relation $A \backslash A = \emptyset$, the following rules hold.

(i) Distributive law:

$$(A \cap B) \backslash C = (A \backslash C) \cap (B \backslash C)\,, \quad (A \cup B) \backslash C = (A \backslash C) \cup (B \backslash C)\,.$$

(ii) Generalized distributive law:

$$A \backslash (B \cap C) = (A \backslash B) \cap (A \backslash C) \quad A \backslash (B \cup C) = (A \backslash B) \cap (A \backslash C)\,.$$

(iii) Generalized associative law:

$$(A \backslash B) \backslash C = A \backslash (B \cup C)\,, \quad A \backslash (B \backslash C) = (A \backslash B) \cup (A \cap C)\,.$$

**The complement of a set:** Let $A$ and $B$ be subsets of a set $X$. The *complement*

$$\boxed{C_X A}$$

of $A$ in $X$ is by definition the set of elements of $X$ which do not belong to $A$, i.e., $C_X A := X \backslash A$ (Figure 4.4(b)). One has a disjoint decomposition

$$X = A \cup C_X A, \quad A \cap C_X A = \emptyset.$$

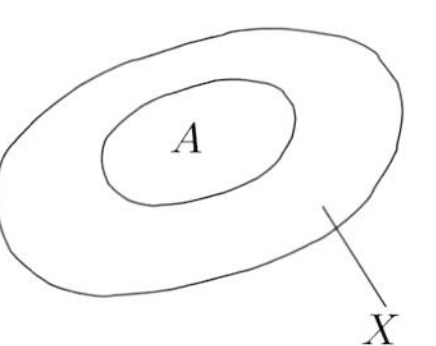


**(a)** $A \subseteq X$

**(b)** $X = A \cup C_X A$

*Figure 4.4. The complement of a set.*

Moreover, one has the so-called *rule of de Morgan*:

$$\boxed{C_X(A\cap B) = C_XA\cup C_XB\,,\quad C_X(A\cup B) = C_XA\cap C_XB\,.}$$

In addition, $C_XX = \emptyset$, $C_X\emptyset = X$ and $C_X(C_XA) = A$. The inclusion $A\subseteq B$ is equivalent to $C_XB\subseteq C_XA$.

**Ordered pairs:** Intuitively, an *ordered pair* $(a,b)$ is a collection of the two objects $a$ and $b$, with the addition of an ordering ('$a$ is first'). The precise mathematical definition follows:[7]

$$(a,b) := \{a,\ \{a,b\}\}.$$

It follows that $(a,b) = (c,d)$ if and only if $a = c$ and $b = d$.

For $n = 1,2,3,4,\ldots$ one defines successively $n$-tuples by the conditions

$$(a_1,\ldots,a_n) := \begin{cases} a_1 & \text{for} \quad n = 1\,, \\ (a_1,a_2) & \text{for} \quad n = 2\,, \\ ((a_1,\ldots,a_{n-1}),a_n) & \text{for} \quad n > 2\,. \end{cases}$$

**The Cartesian product of two sets:** If $A$ and $B$ are sets, then the *Cartesian product* of the two sets, denoted

$$\boxed{A\times B}$$

consists by definition of all ordered pairs $(a,b)$ with $a\in A$ and $b\in B$. For arbitrary sets $A,B,C,D$, one has the following distributive laws:

$$\begin{aligned} A\times(B\cup C) &= (A\times B)\cup(A\times C), & A\times(B\cap C) &= (A\times B)\cap(A\times C)\,, \\ (B\cup C)\times A &= (B\times A)\cup(C\times A), & (B\cap C)\times A &= (B\times A)\cap(C\times A)\,, \\ A\times(B\backslash C) &= (A\times B)\backslash(A\times C), & (B\backslash C)\times A &= (B\times A)\backslash(C\times A)\,. \end{aligned}$$

One has $A\times B = \emptyset$ if and only if $A = \emptyset$ or $B = \emptyset$.

Similarly, for $n = 1,2,\ldots$ one defines the Cartesian product

$$\boxed{A_1\times\cdots\times A_n}$$

as the set of all $n$-tuples $(a_1,\ldots,a_n)$ with $a_j\in A_j$ for all $j = 1,\ldots,n$. This is also denoted by using the product symbol $\prod\limits_{j=1}^{n} A_j$.

**Disjoint union:** A *disjoint union*, denoted

$$\boxed{A\cup_{\mathrm{d}} B}$$

is the union $(A\times\{1\})\cup(B\times\{2\})$. The notation $A\dot{\cup}B$ is also often used for the disjoint union.

*Example 2:* For $A := \{a,b\}$, one has $A\cup A = A$, but

$$A\cup_{\mathrm{d}} A = \{(a,1),(b,1),(a,2),(b,2)\}\,.$$

[7] This means that $(a,b)$ is a set, which consists of the singleton (set with one element) $\{a\}$ and the set $\{a,b\}$.

### 4.3.3 Maps

**Intuitive definition:** A *map* (or *mapping*) between two sets $X$ and $Y$, denoted

$$\boxed{f : X \longrightarrow Y}$$

is an association of a uniquely defined element $y = f(x)$ to every $x \in X$; $f(x)$ is called the *image* point of $x$. Maps are also often referred to as *functions*, although in general the latter term is more often used for maps whose image is the set of real or complex numbers.

If $A \subseteq X$, then we define the *image* of $A$ (under $f$) by

$$f(A) := \{f(a) \mid a \in A\}.$$

The set $X$ is called the *domain* of $f$, and the set $f(X)$ is called the *range* of $f$.

The domain of $f$ is also denote by $D(f)$ or $\operatorname{Dom} f$; the range is also denoted $R(f)$ or $\operatorname{Im} f$, the later indicating the image of $f$ (more literally the 'image of the domain of $f$', i.e., the image set of the entire set on which $f$ is defined).

The set

$$G(f) := \{(x, f(x)) \mid x \in X\}$$

is called the *graph* of $f$.

*Example 1:* Through the prescription $f(x) := x^2$ for all real numbers $x$ we obtain a function $f : \mathbb{R} \longrightarrow \mathbb{R}$ with $D(f) = \mathbb{R}$ and $R(f) = [0, \infty[$. The graph $G(f)$ is given by the parabola depicted in Figure 4.5(a).

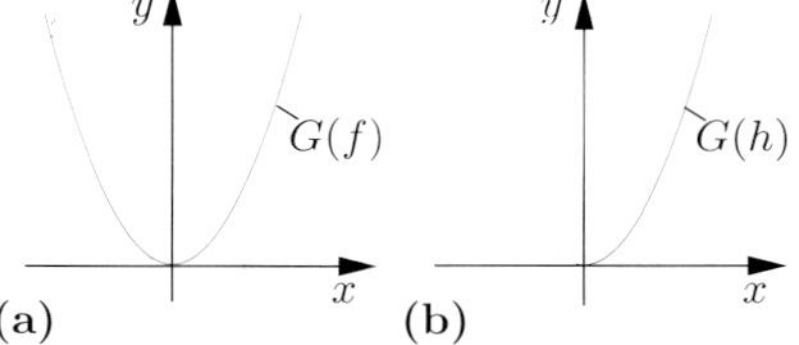


*Figure 4.5. The graph of functions.*

**Classification of functions:** Let the map $f : X \longrightarrow Y$ be given. We consider the equation

$$\boxed{f(x) = y\,.} \tag{4.18}$$

(i) $f$ is said to be *surjective*, if and only if the equation (4.18) has a solution $x \in X$ for every $y \in Y$, i.e., $f(X) = Y$.

(ii) $f$ is said to be *injective*, if and only if the equation (4.18) has for every $y \in Y$ *at most* one solution, i.e., $f(x_1) = f(x_2)$ implies $x_1 = x_2$.

(iii) $f$ is said to be *bijective*, if and only if it is both injective and surjective, i.e., for every $y \in Y$ is there is exactly one solution $x \in X$ to (4.18).[8]

**Inverse maps:** If a map $f : X \longrightarrow Y$ is injective, then we refer to the unique solution $x \in X$ of (4.18) (for a given $y \in Y$) by $f^{-1}(y)$ and call $f^{-1} : f(X) \longrightarrow X$ the *inverse map* to $f$.

*Example 2:* Let $X := \{a, b\}$, $Y := \{c, d, e\}$. We set

$$f(a) := c\,, \quad f(b) := d\,.$$

Then $f : X \longrightarrow Y$ is injective, but not surjective. The inverse map $f^{-1} : f(X) \longrightarrow X$ is given by

$$f^{-1}(c) = a\,, \quad f^{-1}(d) = b.$$

[8]In older literature one uses for 'injective', 'surjective' and 'bijective' respectively the terms 'one-to-one', 'onto' and 'one-to-one onto', respectively.

*Example 3:* If we set $f(x) := x^2$ for all real numbers $x$, then the map $f : \mathbb{R} \longrightarrow [0, \infty[$ is surjective, but not injective, since for example the equation $f(x) = 4$ has two solutions, $x = 2$ and $x = -2$, so (4.18) is not uniquely solvable (see Figure 4.5(a)).

If, on the other hand, we define $h(x) := x^2$ for all *non-negative* real numbers $x$, then the map $h : [0, \infty[ \longrightarrow [0, \infty[$ is bijective. The corresponding inverse map is given by $h^{-1}(y) = \sqrt{y}$ (see Figure 4.5(b)).

*Example 4:* If we set $\mathrm{pr}_1(a, b) := a$, then we get the so-called *projection* mapping $\mathrm{pr}_1 : A \times B \longrightarrow A$, which is surjective.

**Identity map; composition:** The map $\mathrm{id}_X : X \longrightarrow X$ defined by

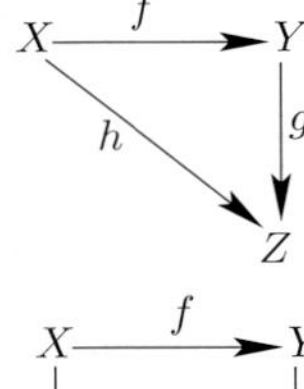


$$\mathrm{id}_X(x) := x \quad \text{for all} \quad x \in X$$

is called the *identity mapping* of $X$. It is also denoted by $I$ or $I_X$. If

$$f : X \to Y \quad \text{and} \quad g : Y \to Z$$

are two maps, then the *composed map* $g \circ f : X \longrightarrow Z$ is defined by the relation

$$\boxed{(g \circ f)(x) := g(f(x)).}$$

An associative law holds:

$$h \circ (g \circ f) = (h \circ g) \circ f\,.$$

**Commutative diagrams:** Many relations between mappings can be easily visualized with the help of commutative diagrams. A diagram is said to be *commutative*, if $h = g \circ f$, i.e, it is irrelevant whether one follows the diagram from $X$ via $Y$ to $Z$ or directly to $Z$. Similarly, a diagram is said to be commutative, if $g \circ f = s \circ r$.

*Example 5:* If the map $f : X \longrightarrow Y$ is bijective, then

$$f^{-1} \circ f = \mathrm{id}_X \quad \text{and} \quad f \circ f^{-1} = \mathrm{id}_Y\,.$$

**Theorem:** Assume a map $f : X \longrightarrow Y$ is given. Then:

(i) $f$ is surjective if and only if there is a map $g : Y \longrightarrow X$ such that

$$f \circ g = \mathrm{id}_Y,$$

i.e., the diagram in Figure 4.6(a) is commutative.

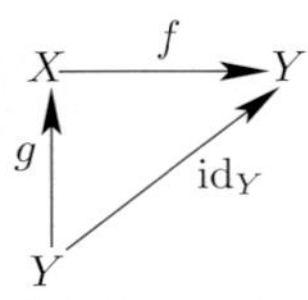


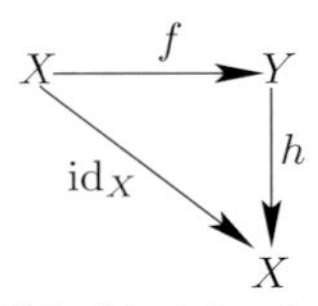


**(a)** $f$ is surjective **(b)** $f$ is injective

*Figure 4.6. Definition of properties of maps through diagrams.*

(ii) $f$ is injective if and only if there is a map $h : Y \longrightarrow X$ such that

$$h \circ f = \mathrm{id}_X,$$

i.e., the diagram in Figure 4.6(b) is commutative.

(iii) $f$ is bijective if and only if there are maps $g : Y \longrightarrow X$ and $h : Y \longrightarrow X$ with

$$f \circ g = \mathrm{id}_Y \quad \text{and} \quad h \circ f = \mathrm{id}_X,$$

i.e., both diagrams in Figure 4.6 are commutative. In this case one has also $h = g = f^{-1}$.

**Inverse images:** Let $f : X \longrightarrow Y$ be a map, and let $B$ be a subset of $Y$. We set

$$f^{-1}(B) := \{x : f(x) \in B\},$$

i.e., $f^{-1}(B)$ consists of the set of all points of $X$ whose images lie in the set $B$. We call $f^{-1}(B)$ the *inverse image* of the set $B$.

For arbitrary subsets $B$ and $C$ of $Y$, we have:

$$f^{-1}(B \cup C) = f^{-1}(B) \cup f^{-1}(C), \quad f^{-1}(B \cap C) = f^{-1}(B) \cap f^{-1}(C).$$

From $B \subseteq C$ it follows that $f^{-1}(B) \subseteq f^{-1}(C)$ and $f^{-1}(C \backslash B) = f^{-1}(C) \backslash f^{-1}(B)$.

**Power sets:** If $A$ is a set, then $2^A$ denotes the *power set* of $A$, i.e, the set of all subsets of $A$.

**Correspondences:** A *correspondence* $c$ between two sets $A$ and $B$ is a map

$$\boxed{c : A \longrightarrow 2^B}$$

from the set $A$ into the power set $2^B$, i.e., every point $a \in A$ is mapped to a uniquely determined *subset* of $B$, which we denote by $c(a)$.

The *image set* of a correspondence $c$ is the union of all subsets $c(a)$ in the image of $c$.

**The precise set-theoretic definition of a map:** A map $f$ from $X$ to $Y$ is a subset of the Cartesian product $X \times Y$ with the two following properties:

(i) For every $x \in X$ there is a $y \in Y$ with $(x, y) \in f$.

(ii) The two relations $(x, y_1) \in f$ and $(x, y_2) \in f$ imply $y_1 = y_2$.

(i) and (ii) together state that for every $x \in X$ there is (condition (i)) a uniquely determined $y \in Y$ (condition (ii)) defined by the subset. The unique element $y$ is also denoted by $f(x)$, and one also writes $f : X \longrightarrow Y$ or $G(f)$ for the map $f$.

## 4.3.4 Cardinality of sets

**Definition:** Two sets $A$ and $B$ are said to have the same *cardinality*, or a said to be *isomorphic* (as sets), denoted

$$\boxed{A \cong B}$$

if and only if there is a bijective map $\varphi : A \longrightarrow B$.

For arbitrary sets $A, B, C$, the following laws are valid:

(i) $A \cong A$ (reflex law);

(ii) $A \cong B$ implies $B \cong A$ (symmetry);

(iii) $A \cong B$ and $B \cong C$ imply $A \cong C$ (transitive law).

**Finite sets:** A set $A$ is said to be *finite*, if it is either empty or there is a natural number $n$, so that $A$ has the cardinality of $B := \{k \in \mathbb{N} \,|\, 1 \leq k \leq n\}$. In this case $n$ is called the number of elements in $A$ (in the first case, $A$ has zero elements).

If $A$ consists of $n$ elements, then the power set $2^A$ has exactly $2^n$ elements (explaining the notation).

**Infinite sets:** A set is said to be *infinite* (or *unbounded*), if it is not finite.

*Example 1:* The set $\mathbb{N}$ of natural numbers is infinite.

**Theorem:** A set is infinite if and only if it has the same cardinality as a strict subset.

A set is infinite if and only if it contains an infinite subset.

**Countable sets:** A set $A$ is said to be *countable*, if it has the same cardinality as the set $\mathbb{N}$ of natural numbers.

A set is said to be *at most countable* if it is finite or it is countable.

A set is said to be *uncountable*, if it is not at most countable.

Clearly, both countable and uncountable sets are infinite.

*Example 2:* (a) The following sets are countable: the set of integers, the set of rational numbers, the set of algebraic real numbers.

(b) The following sets are uncountable: the set of real numbers, the set of irrational real numbers, the set of transcendental real numbers, the set of complex numbers.

(c) The union of $n$ countable sets $M_1, \ldots, M_n$ is again countable.

(d) The union $M_1, M_2, \ldots$ of countably many countable sets is again countable.

*Example 3:* The set $\mathbb{N} \times \mathbb{N}$ is countable.

*Proof:* We order the elements $(m, n)$ of $\mathbb{N} \times \mathbb{N}$ in the form of a matrix:

$$\begin{array}{ccccccc}
(0,0) & \longrightarrow & (0,1) & & (0,2) & \ldots \\
 & & \downarrow & & \downarrow & \\
(1,0) & \longleftarrow & (1,1) & & (1,2) & \ldots \\
 & & & & \downarrow & \\
(2,0) & \longleftarrow & (2,1) & \longleftarrow & (2,2) & \ldots \\
\ldots & & \ldots & & \ldots & 
\end{array}$$

If one follows the matrix elements as indicated by the arrows, then every matrix element is associated to a unique natural number and conversely, every natural number is represented by precisely one matrix element. □

**Remark:** Although seemingly trivial, this proof is important and is the usual way of showing the countability of any set.

We use the notation

$$A \lesssim B$$

to indicate that $B$ contains a subset which has the same cardinality as $A$. We say that $B$ has a larger cardinality than $A$ if $A \lesssim B$ and $A$ does not have the same cardinality as $B$.

**Theorem of Schröder–Bernstein:** For arbitrary sets $A, B$, the following laws concerning the cardinalities hold.

(i) $A \lesssim A$ (reflexive law);

(ii) the two relations $A \lesssim B$ and $B \lesssim A$ imply $A \cong B$ (anti-symmetry);

(iii) the two relations $A \lesssim B$ and $B \lesssim C$ imply $A \lesssim C$ (transitive law).

**Cantor's theorem:** The power set of a set always has a higher cardinality than the set itself.

### 4.3.5 Relations

Intuitively speaking, a relation on a set $X$ is an association which either holds or does not hold between any two elements of $X$. Formally, one defines a relation on $X$ to be a subset $R$ of the Cartesian product $X \times X$.

*Example:* We set $R$ to consist of the ordered pairs $(x, y)$ of real numbers $x$ and $y$ with $x \leq y$. This subset $R$ of $\mathbb{R} \times \mathbb{R}$ then corresponds to the ordering relation.

#### 4.3.5.1 Equivalence relations

**Definition:** An *equivalence relation* on a set $X$ is a subset of $X \times X$ with the following three properties.

(i) $(x, x) \in R$ for all $x \in X$ (reflexiveness of the relation);

(ii) from $(x, y) \in R$ it follows that $(y, x) \in R$ (symmetry of the relation);

(iii) from $(x, y) \in R$ and $(y, z) \in R$ it follows that $(x, z) \in R$ (transitivity of the relation).

In other words, an equivalence relation is a relation which is reflexive, symmetric and transitive. Instead of $(x, y) \in R$ one often writes $x \sim y$. In this notation the conditions above are

(a) $x \sim x$ for all $x \in X$ (reflexiveness of the relation);

(b) from $x \sim y$ it follows that $y \sim x$ (symmetry of the relation);

(c) from $x \sim y$ and $y \sim z$ it follows that $x \sim z$ (transitivity of the relation).

Let $x \in X$. Then we denote by $[x]$ the *equivalence class* of $x$, which is by definition the set of all elements equivalent to $x$, i.e.,

$$[x] := \{y \in X \mid x \sim y\}.$$

The elements of $[x]$ are called *representatives* of the equivalence class.

**Theorem:** Every set $X$ decomposes into pairwise disjoint equivalence classes (under an arbitrary equivalence relation on $X$).

**Factor set:** We denote by $X/\sim$ the set of all equivalence classes. This set is also often called the *factor set* (or *factor space*).

If there are operations which are defined on the set $X$, then one can carry these over to the factor space, provided the operations are *compatible* with the equivalence relation; this means that the operation preserves the relations provided by the equivalence relation (see the example below). This is a general principle in mathematics for constructing new structures (factor structures[9]).

*Example:* If $x$ and $y$ are two integers, then we write $x \sim y$ if and only if $x - y$ is divisible by 2. With respect to this equivalence relation, we have

$$[0] = \{0, \pm 2, \pm 4, \ldots\}, \quad [1] = \{\pm 1, \pm 3, \ldots\},$$

i.e., the equivalence class [0] (resp. [1]) consist of all the even (resp. odd) integers.

Moreover, from $x \sim y$ and $u \sim v$ it follows that $x + u \sim y + v$. Therefore, the definition

$$[x] + [y] := [x + y]$$

is independent of the choice of representatives. Thus,

$$[0] + [1] = [1] + [0] = [1], \quad [1] + [1] = [0], \quad [0] + [0] = [0].$$

Every cognitive process is based on the fact that different things are identified with one another and statements are made about the corresponding identifications made (for example the classification of animals by overriding notions of mammals, fish, birds and so forth). Equivalence relations are the precise mathematical formulation of the process.

[9]We considered factor groups in section 2.5.1.2 above.

#### 4.3.5.2 Ordering relations

**Definition:** An *order relation* on a set $X$ is a subset $R$ of the Cartesian product $X \times X$ with the following three properties.

(i) One has $(x, x) \in R$ for all $x \in X$ (*reflexive law*).

(ii) From the relation $(x, y) \in R$ and $(y, x) \in R$ it follows that $x = y$ (*anti-symmetry*).

(iii) From $(x, y) \in R$ and $(y, z) \in R$ it follows that $(x, y) \in R$ (*transitive law*).

Instead of $(x, y)$ one also writes $x \leq y$. Using this notion, we have

(a) For all $x \in X$, $x \leq x$ (reflexive law).

(b) From $x \leq y$ and $y \leq x$ it follows that $x = y$ (anti-symmetry).

(c) From $x \leq y$ and $y \leq z$ it follows that $x \leq z$ (transitive law).

The symbol $x < y$ means precisely that $x \leq y$ *and* $x \neq y$.

An ordered set $X$ is said to be *totally ordered*, if for all elements $x$ and $y$ of $X$ either the relation $x \leq y$ or $y \leq x$ is valid.

Let $x \in X$ be given. If $x \leq z$ always implies $x = z$, then $x$ is said to be a *maximal element* of $X$. Suppose $M \subseteq X$. If $y \leq s$ for all $y \in M$ and some fixed $s \in X$, the $s$ is called an *upper bound* of the subset $M$.

Finally, $u$ is called a *minimal element* of $M$, if $u \in M$ and $u \leq z$ for all $z \in M$.

**Zorn's Lemma:** An ordered set $X$ has a maximal element provide every totally ordered subset of $X$ has an upper bound.

**Complete order:** An ordered set is said to be *completely ordered*, if every non-empty subset contains a smallest element.

*Example:* The set of natural numbers is completely ordered with respect to the usual order relation.

On the other hand, the set of real numbers $\mathbb{R}$ is, with respect to the usual order relation, not completely ordered, since the open set $]0, 1]$ does not contain a smallest element.

**Zermelo's theorem on completely ordered sets:** Every set can be completely ordered.

**The principle of transfinite induction:** Let $M \subseteq X$ be a subset of an ordered set $X$. Then $M = X$ if for all $a \in X$, we have

$$\text{If the set } \{x \in X \mid x < a\} \text{ belongs to } M, \text{ then } a \in M \text{ as well.}$$

**Order-preserving maps:** A map $\varphi : X \longrightarrow Y$ between two ordered sets $X$ and $Y$ is said to be *order-preserving*, if from $x \leq y$ it follows that $\varphi(x) \leq \varphi(y)$.

Two ordered sets $X$ and $Y$ are said to be ordered in the same way if there is a bijective map $\varphi : X \longrightarrow Y$ such that both $\varphi$ and $\varphi^{-1}$ are order-preserving.

#### 4.3.5.3 $n$-fold relations

An *$n$-fold relation* $R$ on a set $X$ is a subset of the $n$-fold product $X \times \cdots \times X$.

This kind of relation is often used to describe operations.

*Example:* Let the set $R$ consist of all 3-tuples $(a, b, c)$ of real numbers $a, b, c$ with $ab = c$. Then the 3-fold relation $R \subseteq \mathbb{R} \times \mathbb{R} \times \mathbb{R}$ is the multiplication of real numbers.

### 4.3.6 Systems of sets

A *system of sets* $\mathscr{M}$ is a set $\mathscr{M}$ whose elements are sets $X$. The union

consists by definition of those elements which belong to one of the sets $X$ of $\mathscr{M}$.

If $\mathscr{M}$ contains at least one set, then the intersection

$$\boxed{\cap\mathscr{M}}$$

consists by definition of the elements which are contained in all elements (sets) $X$ of $\mathscr{M}$.

By definition a *family of sets* $(X_\alpha)_{\alpha\in A}$ is a function which is defined on the index set $A$ and which associates to each $\alpha \in A$ a set $X_\alpha$.

An $A$-tuple $(x_\alpha)$ (also denoted $(x_\alpha)_{\alpha\in A}$) is a function on $A$ which associates to each $\alpha \in A$ an element $x_\alpha \in X_\alpha$. The *Cartesian product*

$$\boxed{\prod_{\alpha\in A} X_\alpha}$$

consists of all $A$-tuples $(x_\alpha)$. One also calls $(x_\alpha)$ a *selection function.* A *union*

$$\boxed{\bigcup_{\alpha\in A} X_\alpha}$$

consists of all elements which are contained in at least one of the sets $X_\alpha$.

Suppose that $A$ is non-empty. The *intersection*

$$\boxed{\bigcap_{\alpha\in A} X_\alpha}$$

consists of all elements which belong to all the sets $X_\alpha$.

## 4.4 Mathematical logic

*Truth occurs when thought and reality coincide.*

*Aristotle* (384–322 BC)

*Theoretical logic, also called mathematical or symbolic logic, is the extension of formal methods of mathematics to the area of logic. It applies to logic a language and syntax similar to that long used to express mathematical relations.*[10]

*David Hilbert and Wilhelm Ackermann* (1928)

Logic is the science of thinking. Mathematical logic is the most precise form of logic. It uses a strict and formalized calculus to express statements and relations, and in the age of computers forms the foundation of computer science.

[10] Taken from the preface of the first edition of *Foundations of Theoretical Logic.*

### 4.4.1 Propositional calculus

**Basic symbols:** We use symbols $q_1, q_2, \ldots$ (*statements*), the symbol ( , ) as well as symbols

$$\neg, \quad \wedge, \quad \vee, \quad \longrightarrow, \quad \longleftrightarrow . \tag{4.19}$$

Instead of the statement variables $q_j$ we also use $q, p, r.\ldots$

**Expressions:** (i) Every chain of symbols which contains a single statement is an *expression.*

(ii) If $A$ and $B$ are expressions, then so are the chains of symbols

$$\neg A\,, \quad (A \wedge B)\,, \quad (A \vee B)\,, \quad (A \longrightarrow B)\,, \quad (A \longleftrightarrow B).$$

These expressions are called in order *negation*, *conjunction* (both one and the other), *alternative* (either one or the other), *implication* and *equivalence.*

(iii) A chain of symbols is an expression only if it is formed as in (i) and (ii) above.

*Example 1:* The following are examples of expressions:

$$(p \longrightarrow q)\,, \quad ((p \longrightarrow q) \longleftrightarrow r)\,, \quad ((p \wedge q) \longrightarrow r)\,, \quad ((p \wedge q) \vee r)\,. \tag{4.20}$$

**Dropping parenthesis:** For real numbers $a, b, c$, the product symbol has priority over the addition symbol. Thus there is no ambiguity in writing $((ab) + c)$ more briefly as $ab + c$.

Similarly we agree that the symbols displayed in (4.19) have priority starting from the left, allowing us to discard of unnecessary parenthesis.

*Example 2:* Instead of (4.20) we may write

$$p \longrightarrow q\,, \quad p \longrightarrow q \longleftrightarrow r\,, \quad p \wedge q \longrightarrow r\,, \quad p \wedge q \vee r\,.$$

**True and false statements:** We set

$$\text{non(T)}{:=}\text{F}\,, \quad \text{non(F)}{:=}\text{T}\,. \tag{4.21}$$

The symbols 'T' and 'F' used here represent 'true' and 'false'. The background of (4.21) is that the negation of a true (resp. false) statement is false (resp. true).

Moreover, the functions et, vel, seq, eq are defined by the values given in Table 4.2. The expressions $\text{et}(X,Y)$, $\text{vel}(X,Y)$, $\text{seq}(X,Y)$ and $\text{eq}(X,Y)$ have in order the following

*Table 4.2. Functions of logic.*

| $X$ | $Y$ | $\text{et}(X,Y)$ | $\text{vel}(X,Y)$ | $\text{seq}(X,Y)$ | $\text{eq}(X,Y)$ |
|---|---|---|---|---|---|
| T | T | T | T | T | T |
| T | F | F | T | F | F |
| F | T | F | T | T | F |
| F | F | F | F | T | T |

meanings: '$X$ and $Y$', '$X$ or $Y$', '$X$ implies $Y$' and '$X$ is equivalent to $Y$'. For example $\text{et}(\text{T}, \text{F}) = \text{F}$, meaning that if $X$ is true and $Y$ is false, then the combined statement $X$ and $Y$ is also false.

**Truth values of expressions:** A map

$$b : \{q_1, q_2, \ldots\} \longrightarrow \{\mathrm{T,F}\},$$

which associates to each statement variable $q_j$ either a value T (true) or F (false), is called a *truth value* function. If $b$ is given, then every expression $A, B, \ldots$ can be assigned a value T or F according to the following rules:

(i) value $(q_j) := b(q_j)$.

(ii) value $(\neg A) :=$ non(value$(A)$).

(iii) value $(A \wedge B) :=$ et(value$(A)$,value$(B)$).

(iv) value $(A \vee B) :=$ vel(value$(A)$,value$(B)$).

(v) value $(A \longrightarrow B) :=$ seq(value$(A)$,value$(B)$).

(vi) value $(A \longleftrightarrow B) :=$ eq(value$(A)$,value$(B)$).

*Example 1:* For $b(q) :=$T and $b(p) :=$ F, we get value$(\neg p) =$ non(F) $=$ T and

$$\text{value}(q \longrightarrow \neg p) = \text{seq(T,T)=T}.$$

**Tautologies:** An expression is called a *tautology*, if for every truth value function this expression carries the value T.

*Example 2:* The expression

$$q \vee \neg q$$

is a tautology.

*Proof:* For $b(q) =$ T we get

$$\text{value}(q \vee \neg q) = \text{vel(T,F)=T},$$

and $b(q) =$ F yields

$$\text{value}(q \vee \neg q) = \text{vel(F,T)=T}.$$

□

**Equivalent expressions:** Two expressions $A$ and $B$ are said to be (logically) *equivalent* if they yield the same value for every truth value function. This is denoted symbolically by

$$\boxed{A \cong B.}$$

**Theorem:** We have $A \cong B$ if and only if $A \longleftrightarrow B$ is a tautology.

*Example 3:*

$$\boxed{p \longrightarrow q \cong \neg p \vee q.}$$

Thus,

$$p \longrightarrow q \longleftrightarrow \neg p \vee q$$

is a tautology.

Important tautologies of classical logic can be found in section 4.1.3.

#### 4.4.1.1 The fundamental axioms

The goal of the logic of statements, or as it is also referred to, of expression logic, is to describe all tautologies. This is done in a purely formal manner by making axioms and rules for making deductions from these. We now present these axioms.

(A1) $p \longrightarrow \neg\neg p$

(A2) $\neg\neg p \longrightarrow p$

(A3) $p \longrightarrow (q \longrightarrow p)$

(A4) $((p \longrightarrow q) \longrightarrow p) \longrightarrow p$

(A5) $(p \longrightarrow q) \longrightarrow ((q \longrightarrow r) \longrightarrow (p \longrightarrow r))$

(A6) $p \wedge q \longrightarrow p$

(A7) $p \wedge q \longrightarrow q$

(A8) $(p \longrightarrow q) \longrightarrow ((q \longrightarrow r) \longrightarrow (p \longrightarrow q \wedge r))$

(A9) $p \longrightarrow p \vee q$

(A10) $q \longrightarrow p \vee q$

(A11) $(p \longrightarrow r) \longrightarrow ((q \longrightarrow r) \longrightarrow (p \vee q \longrightarrow r))$

(A12) $(p \longleftrightarrow q) \longrightarrow (p \longrightarrow q)$

(A13) $(p \longleftrightarrow q) \longrightarrow (q \longrightarrow p)$

(A14) $(p \longrightarrow q) \longrightarrow ((q \longrightarrow p) \longrightarrow (p \longleftrightarrow q))$

(A15) $(p \longrightarrow q) \longrightarrow (\neg q \longrightarrow \neg p)$.

#### 4.4.1.2 The rules of deduction

We denote by $A, B, C$ arbitrary expressions. The rules for deduction are

(R1) Every axiom can be deduced.

(R2) (*modus ponens* – reduction rule) If $(A \longrightarrow B)$ and if $A$ is deducible from the axioms, then $B$ is also deducible from the axioms.

(R3) (replacement rule) If $A$ can be deduced from the axioms and if $B$ arises by replacing a statement variable $q_j$ in $A$ throughout by a fixed expression $C$, then $B$ can also be deduced from the axioms.

(R4) An expression can be deduced from the axioms if and only if it can be so deduced by virtue of rules (R1), (R2) and (R3).

#### 4.4.1.3 The main theorem of expression logic

(i) *Completeness of the axioms:* An expression is a tautology if and only if it can be deduced from the axioms.

In particular, all axioms are tautologies.

(ii) *Classical freedom from contradictions:* It is impossible to deduce from the axioms an expression and at the same time its negation.[11]

(iii) *Independence of the axioms:* None of the axioms can be deduced from the remaining ones.

(iv) *Decidability:* There is an algorithm which terminates after a finite number of steps to decide whether an expression is a tautology or not.

### 4.4.2 Predicate logic

Intuitively speaking, predicate calculus investigates properties of entities and their relationships. For this one applies the expressions 'for all entities it is true ...' and 'there is an entity for which ...'.

**Entity domain:** Working formally, we hypothesize the existence of a set $M$, referred to as the *entity domain* or *domain of entities*. This set is also referred to as an *alphabet* or a *base set*.

**Relations:** Properties and relations between entities are described by $n$-fold relations on the set $M$. Such a relation $R$ is a subset of the $n$-fold Cartesian product $M \times \cdots \times M$. The symbol

$$\boxed{(a_1, \ldots, a_n) \in R}$$

indicates by definition that there is a relation $R$ between the entities $a_1, \ldots, a_n$. If $n = 1$, one also says simply that $a_1$ has the property $R$.

*Example 1:* The entity domain is assumed to be the set $\mathbb{R}$ of real numbers. Let $R$ be the set of natural numbers. Then the statement

$$a \in R$$

expresses the fact that the real number $a$ is in fact a natural number.

If $x$ denotes a so-called entity variable, then the expression

$$\boxed{\forall_x \, Rx}$$

is considered to be true if all entities in $M$ belong to $R$ (i.e., have the property $R$). If this statement is however false, then there is an entity in $M$ which does not belong to $R$. The expression

$$\boxed{\exists_x \, Rx}$$

on the other hand is true if there is an entity in $M$ which also lies in $R$, and false if none of the entities of $M$ lie in $R$. Similarly one can define 2-fold relations like

$$\forall_x \forall_y \, Rxy \quad \text{etc.}$$

---

[11] A system of axioms is said to have *semantics free from contradictions* if only tautologies can be deduced from them. Furthermore, a system of axioms is said to have *syntax free from contradictions*, if not all expressions can be deduced from it.

The system of axioms (A1) to (A15) has both semantics and syntax free from contradictions.

*Example 2:* As the domain of entities we choose the set $\mathbb{R}$ of real numbers, and we set $R := \{(a,a), \mid a \in \mathbb{R}\}$. Then $R$ is a subset of $\mathbb{R} \times \mathbb{R}$. The symbol $Rab$ means $(a,b) \in R$, and the formalization

$$\forall_a \forall_b (Rab \longrightarrow Rba)$$

means: for all real numbers $a$ and $b$, the relation $a = b$ implies $b = a$.

**Basic notations:** In the predicate calculus of the first level, the following notations are used:

(a) entity variables $x_1, x_2, \ldots$;

(b) relation variables $R_1^{(k)}, R_2^{(k)}, \ldots$ with $k = 1, 2, \ldots$;

(c) logical functors $\neg, \wedge, \vee, \longrightarrow, \longleftrightarrow$;

(d) the generalizer $\forall$ and the existence declarator $\exists$;

(e) parenthesis ( , ).

The relation variables $R_n^{(k)}$ applies by definition to $n$ entity variables, for example $R_2^{(k)} x_1 x_2$.

**Expressions:** (i) Every chain of symbols $R_n^{(k)} x_{i_1} x_{i_2} \cdots x_{i_n}$ is an expression $(k, n = 1, 2, \ldots)$.

(ii) If $A$ and $B$ are expressions, then so are also

$$\neg A, \quad (A \wedge B), \quad (A \vee B), \quad (A \longrightarrow B), \quad (A \longleftrightarrow B).$$

(iii) If $A(x_j)$ is an expression in which the entity variable $x_j$ occurs *completely freely*, meaning that although $x_j$ occurs, neither $\forall_{x_j}$ nor $\exists_{x_j}$ occur, then so also are $\forall_{x_j} A(x_j)$ and $\exists_{x_j} A(x_j)$ expressions.

(iv) A chain of symbols is an expression only if it is so because of (i) to (iii) above.

**Completeness theorem of Gödel (1930):** In first order predicate logic there are explicitly exhibitable axioms and rules of deduction, so that an expression is a tautology if and only if it can be deduced from these axioms.

The situation is similar to that of propositional logic[12] in section 4.4.1.3.

**Theorem of Church (1936):** As opposed to propositional logic there is no algorithm in first order predicate logic which can decide after a finite number of steps whether a statement is a tautology or not.

### 4.4.3 The axioms of set theory

In order to create a rigorous axiomatic set theory, one uses, according to Zermelo (1908) and Fraenkel (1925), the basic notions 'set' and 'element of a set', and requires the following axioms to be satisfied.

(i) *Existence axiom:* There is a set.

(ii) *Identity axiom:* Two sets are equal if and only if they have the same elements.

(iii) *Condition axiom:* Given any set $M$ and any statement $\mathscr{A}(x)$, there is a set $A$ whose elements are precisely those elements of $M$ for which $\mathscr{A}(x)$ is true.[13]

*Example 1:* We choose a set $M$, and let $\mathscr{A}(x)$ be the condition (statement) $x \neq x$. Then there is a set $A$ which consists of those elements $x \in M$ for which $x \neq x$. This set is

[12] More details can be found in [349].

[13] We assume hereby that $x$ occurs in $\mathscr{A}(x)$ at least once without the quantifiers $\forall$ and $\exists$, i.e., that $x$ is *free*.

denoted by $\emptyset$ and is called the *empty set.* According to the identity axiom, this set is unique.

*Example 2:* There is no set whose elements are all sets (no 'set of sets').

*Proof:* Suppose there did exists such a set $X$ of all sets. According to the condition axiom, the set

$$A := \{x \in X \mid x \notin x\}$$

is again a set.

*Case 1:* If $A \notin A$, then by construction $A \in A$.

*Case 2:* We have $A \in A$. But then by construction $A \notin A$.

In both cases we have produced a contradiction, thus the assumption that such an $X$ exists is false. □

(iv) *Binary set axiom:* If $M$ and $N$ are sets, then there is a set which contains precisely $M$ and $N$ as its elements.

(v) *Union axiom:* To every system of sets $\mathscr{M}$ there exists a set which consists precisely of those elements which belong to a set of $\mathscr{M}$.

(vi) *Power set axiom:* Given a set $M$ there is a system of sets $\mathscr{M}$ which contains precisely all subsets of $M$.

We define the *successor* $X^+$ of a set $X$ by the condition[14]

$$X^+ := X \cup \{X\}.$$

(vii) *Infinity axiom:* There is a system of sets $\mathscr{M}$ which contains the empty set and with each set also its successor.

(viii) *Axiom of choice:* The Cartesian product of a non-empty family of non-empty sets is not empty.[15]

(ix) *Replacement axiom:* Let $\mathscr{A}(a, b)$ be a binary statement, so that for every element $a$ of a set $A$ we can form the set $M(a) := \{b \mid \mathscr{A}(a, b)\}$. Then there is a unique function $F$ with the domain of definition $A$ such that $F(a) = M(a)$ for all $a \in A$.

The construction of set theory from these axioms is described in [347]. The extremely careful formulation of these axioms is necessary to prevent pathologies like the Russel paradox of the set of all sets.

### 4.4.4 Cantor's structure at infinity

In his work on set theory, Cantor introduced transfinite ordinal numbers and cardinal numbers. Ordinal numbers correspond to our intuitive feeling of 'counting on and on', while cardinal numbers describe the 'number of elements', also called the cardinality, of sets.

#### 4.4.4.1 Ordinal numbers

**The set $\omega$:** We set

$$0 := \emptyset, \quad 1 := 0^+, \quad 2 := 1^+, \quad 3 := 2^+, \quad \ldots,$$

[14] Here $\{X\}$ denotes the set whose sole element is the set $X$.

[15] Functions, families of sets and Cartesian products are defined as in section 4.3.

where $x^+ = x \cup \{x\}$ is the successor to $x$. Then we have[16]

$$1 = \{0\}, \quad 2 = \{0,1\}, \quad 3 = \{0,1,2\}, \quad \ldots$$

A set $M$ is said to be a *successor set* if it contains the empty set and, for each set it contains, also its successor. There is a unique successor set $\omega$ which is a subset of every other successor set (a 'smallest' successor set).

**Definition:** The elements of $\omega$ are called *natural numbers.*

**The recursion formula of Dedekind:** Let a function $\varphi : X \longrightarrow X$ on a set $X$ be given, and let $m$ be a fixed element of $X$. Then there is precisely one function

$$\boxed{R : \omega \longrightarrow X}$$

with $R(0) = m$ and $R(n^+) = \varphi(R(n))$. One refers to $R$ as a *recursive function.*

*Example 1* (addition of natural numbers): We chose $X := \omega$ and let $\varphi$ be defined by $\varphi(x) := x^+$ for all $x \in \omega$. Then for every natural number $m$ there is a function $R : \omega \longrightarrow \omega$ with $R(0) = m$ and $R(n^+) = R(n)^+$ for all $n \in \omega$. We set $m + n := R(n)$. This means that

$$\boxed{m + 0 = M \quad \text{and} \quad m + n^+ = (m+n)^+}$$

for all natural numbers $n$ and $m$. In particular, $m + 1 = m^+$, since $m + 1 = m + 0^+ = (m+0)^+ = m^+$.

In this manner it is possible, with the aid of the axioms of set theory, to introduce the set of natural numbers as the set $\omega$ and introduce an addition on this set.[17] Similarly, we can define a multiplication on the set of natural numbers. By using the construction of appropriate equivalence relations, we can form from this set the set of integers, rational numbers, and finally real and complex numbers.

*Example 2* (integers): The set of integers can be obtained with the following construction. We consider all pairs $(m, n)$ of natural numbers $m, n \in \omega$ and write

$$(m,n) \sim (a,b) \quad \text{if and only if} \quad m + b = a + n.$$

The corresponding equivalence classes $[(m,n)]$ are called *integers.*[18] For example, we have $(1,3) \sim (2,4)$.

**Definition:** An *ordinal number* or *ordinal* is a well-ordered set $X$ with the property that for all $a \in X$, the set

$$\{x \in X : x < a\}$$

coincides with $a$.

*Example 3:* The sets defined above, $0, 1, 2 \ldots$ and $\omega$ are ordinals. Moreover,

$$\omega + 1 := \omega^+, \quad \omega + 2 := (\omega + 1)^+, \quad \ldots$$

are all ordinals. They correspond to counting, starting at $\omega$.

If $\alpha$ and $\beta$ are ordinals, we write

$$\boxed{\alpha < \beta,}$$

---

[16] Note that $1 = \emptyset \cup \{\emptyset\} = \{\emptyset\} = \{0\}$, $2 = 1 \cup \{1\} = \{0\} \cup \{1\} + \{0,1\}$ and so forth.

[17] In order to emphasize the ordinal character of this construction, we, following the tradition of set theory, denote the natural numbers here by the symbol $\omega$ instead of $\mathbb{N}$.

[18] Intuitively, the equivalence class $[(m,n)]$ is the number $m - n$.

if $\alpha$ is a subset of $\beta$.

**Theorem:** (i) For any two ordinals $\alpha$ and $\beta$, precisely one of the three conditions $\alpha < \beta$, $\alpha = \beta$, $\alpha > \beta$, is true.

(ii) Every set of ordinals is well-ordered.

(iii) Every well-ordered set $X$ is ordered in the same way as exactly one ordinal, which we denote by $\operatorname{ord} X$ and call the *ordinal (number) of $X$*.

**Ordinal sums:** If $A$ and $B$ are two disjoint, well-ordered sets, then we define the *ordinal sum*

$$\text{`}A \cup B\text{'}$$

as the union $A \cup B$ with the following natural order:

$$a \leq b \quad \text{for} \quad a \in A \quad \text{and} \quad b \in B\,.$$

This means in particular that $a \leq b$ corresponds to the order in $A$ (resp. $B$) if both $a$ and $b$ belong to $A$ (resp. to $B$).

**Ordinal product:** If $A$ and $B$ are two well-ordered sets, then the *ordinal product* of these, denoted

$$\text{`}A \times B\text{'}$$

is defined as the product set $A \times B$ with the lexicographical order, i.e., $(a, b) < (c, d)$ if and only if either $a < c$ or $a = c$ and $b < d$.

**Arithmetic of ordinals:** Let $\alpha$ and $\beta$ be two ordinals. Then there is precisely one ordinal $\gamma$ which is ordered as the ordinal sum '$\alpha \cup \beta$'. We define the *sum of ordinal numbers* by

$$\alpha + \beta := \gamma\,.$$

*Example 4:* We have $\alpha + 1 = \alpha^+$ for all ordinals $\alpha$.

Moreover, there is precisely one ordinal $\delta$, which is ordered as the ordinal product '$\alpha \times \beta$'. We define the *product of ordinal numbers* by

$$\alpha\beta := \delta\,.$$

*Example 5:* We consider the lexicographical order on the set $\omega \times \omega$:

$$\begin{array}{lllll} (0,0) & (0,1) & (0,2) & (0,3) & \ldots \\ (1,0) & (1,1) & (1,2) & (1,3) & \ldots \\ \ldots & & & & \end{array}$$

(i) The first row is ordered as $\omega$, i.e., the ordinal number of the first row is $\omega$.

(ii) The first row together with $(1, 0)$ is ordered as $\omega + 1$, i.e., the ordinal of this set is $\omega + 1$.

(iii) The ordinal of the first row together with the second row is $2\omega$.

(iv) If we endow the first and second column with the lexicographical order $(0, 0)$, $(0, 1)$, $(1, 0)$, $(1, 1), \ldots$, then we get a set $M$ whose ordinal is $\omega 2$. On the other hand, $M$ is ordered as the set of natural numbers. Hence $\omega 2 = \omega$, i.e., $2\omega \neq \omega 2$.

(v) The ordinal number of the entire matrix is $\omega\omega$, i.e., $\operatorname{ord}(\omega \times \omega) = \omega\omega$.

**Paradox of Burali–Forti:** The totality of all ordinals is not a set.

#### 4.4.4.2 Cardinal numbers

Let an arbitrary set $A$ be given. All ordinals which can be bijectively mapped to $A$ form a well-ordered set. The smallest ordinal of this set is called the *cardinality* of $A$, denoted $\operatorname{card} A$.

The relation $\operatorname{card} A \leq \operatorname{card} B$ corresponds to the relation between ordinal numbers.

**Theorem:** (i) We have $\operatorname{card} A = \operatorname{card} B$ if and only if $A$ and $B$ are in bijective relation.

(ii) We have $\operatorname{card} A < \operatorname{card} B$ if and only if $A$ is bijective to a subset of $B$.

*Example:* For a finite set $A$, the cardinality $\operatorname{card} A$ is just the number of elements of the set.

**Arithmetic of cardinal numbers:** Let $A$ and $B$ be disjoint sets. We define the *sum of cardinal numbers* by the relation

$$\operatorname{card} A + \operatorname{card} B := \operatorname{card}(A \cup B)\,.$$

For two arbitrary sets $A$ and $B$ we define the *product of cardinal numbers* by

$$(\operatorname{card} A)(\operatorname{card} B) := \operatorname{card}(A \times B)\,.$$

This definition is independent of the choice of representatives for $A$ and $B$.

**Cantor's paradox:** The totality of cardinal numbers is not a set.

#### 4.4.4.3 The continuum hypothesis

It is customary to denote the cardinal number $\operatorname{card} \omega$ of the natural numbers $\omega$ by[19] $\aleph_0$. There is also a smallest cardinal number $\aleph_1$, which is genuinely larger than $\aleph_0$. Either of the two following situations are conceivable:

(i) $\aleph_0 < \aleph_1 = \operatorname{card} \mathbb{R}$.

(ii) $\aleph_0 < \aleph_1 < \operatorname{card} \mathbb{R}$.

One denotes the cardinal number $\operatorname{card} \mathbb{R}$ of the set of real numbers as the *cardinality of the continuum.* Cantor tried in vain to prove the so-called *continuum hypothesis* (i). Intuitively (i) tells us that there is no further cardinality between that of the natural numbers and that of the continuum.

In 1940 it was proved by Gödel that the continuum hypothesis (i) is compatible with the rest of the axioms of set theory; in 1963 Cohen proved that (ii) is also.

**Theorem of Gödel–Cohen:** The axiom of choice and the continuum hypothesis are independent of the remaining axioms of set theory.

More precisely this means that if the axioms of set theory presented in 4.4.3 are free of contradictions, then both (i) and (ii) can hold without contradiction. The same result holds if one replaces the axiom of choice by its negation!

This surprising result shows that there is not a unique set theory, but rather several possible set theories, and that contrary to what you would be inclined to think, the very natural axioms of 4.4.3 do not completely determine the structure of infinity.

One of the profound findings of physics and mathematics of the twentieth century is that our usual 'common sense' can be so totally inadequate when dealing with ideas which are far from our daily doings. This is true of quantum theory (atomic dimensions), relativity (high velocities and forces) as well as set theory (the notion of infinity).

[19]The symbol $\aleph$ is *aleph*, the first letter of the Hebrew alphabet.

## 4.5 The history of the axiomatic method and its relation to philosophical epistemology

*Before you start to axiomize things, be sure that you first have something of mathematical substance.*

*Hermann Weyl*[20] (1885–1955)

In the history of mathematics, two fundamental trends can be observed:

(i) the interaction with the other natural sciences, and

(ii) the interaction with philosophical epistemology.

In this section we would like to briefly consider (ii). A detailed discussion of (i) is given in [212] including the fascinating (more recent) interaction between geometry and modern physics (elementary particle theory and cosmology). At the end of this volume there is a collection of names and events which have made an impact in the history of mathematics.

**Axioms:** The axiomatic representation of a mathematical discipline corresponds to a scheme as follows:

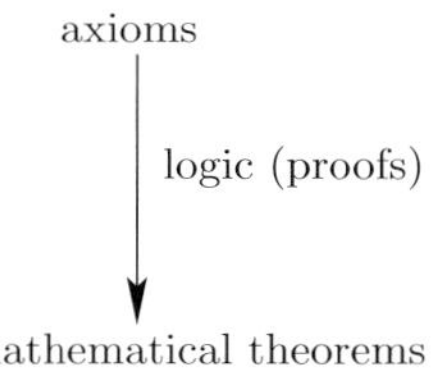


At the top we have the so-called *postulates* or *axioms*. These are assumptions which are not to be proven but rather form the basis of the deductive reasoning process and are taken for granted. However, these axioms are not arbitrarily stated, but are generally formulating the result of a hard and long mathematical process of gaining insight into the heart of some topic. With the help of these axioms, one makes logical deductions, which are called proofs, resulting in mathematical results (lemmas, propositions, theorems) Particular importance is given certain results which are then called theorems.[21]

Definitions give a name to a notion which is repeatedly used. Often the key to making proofs workable is to make the right definitions.

**Euclid's *Elements*:** The famous book *The Elements* by Euclid is a shining example for the axiomatic method and has served as a model for the next 2000 years. This book was written about 325 BC in Alexandria. This book begins with the following definition: 1. A point is something which has no parts. 2. A line is extended length. 3....

The *parallel axiom* is particularly famous. Expressed in modern language it states the following.

(P) If a point $P$ is not on a line $l$, then there in the plane which is spanned by $P$ and $l$ precisely one line which passes through $P$ but does not intersect $l$.

It wasn't until the nineteenth century that Bolyai, Gauss and Lobachevsky proved that the parallel axiom is independent of the other Euclidean axioms. This implies that there are geometries for which (P) holds and there are also geometries for which it does not hold (see section 3.2.7).

[20] Hermann Weyl became the successor to David Hilbert in Göttingen in 1930. In 1933 he emigrated to the United States and worked together with Albert Einstein at the famous Institute for Advanced Studies in Princeton, New Jersey. Richard Courant, who also emigrated in 1933, founded in New York the famous institution now called the Courant Institute.

[21] To structure a proof it is often helpful to formulate intermediate results which are often referred to as lemmas.

**Hilbert's foundations of Geometry:** The modern axiomatic method was created by Hilbert in his *Foundations of Geometry*, which was written in 1899. As compared with Euclid, Hilbert's presentation represents a much more radical development. He does not attempt to define what a point is, but rather bases the theory on the notions 'point', 'line' and 'plane' together with relations 'passes through', 'congruent' and 'lies between', without trying to give these some fixed content. He then formulates axioms expressed in these notions, thus creating the notion of a geometry. For example, his first axiom is simply: Through any two points, a line passes. This makes it possible to consider completely *different models*. For example, in the Poincaré model of hyperbolic geometry, a 'line' is a circle whose center is located on the $x$-axis (see section 3.2.8).

**Hilbert's proof of geometry's relative freedom from contradictions:** Under the assumption that certain parts of algebra and analysis are free of contradictions, Hilbert was able to prove the non-contradictory nature of geometry, by utilizing Cartesian coordinates and then translating geometrical statements into algebraic and analytic statements. For example, the theorem in Euclidean geometry that 'two non-parallel lines intersect in exactly one point' corresponds to the fact that the system of equations

$$Ax + By + C = 0\,,$$
$$Dx + Ey + F = 0$$

for given real numbers $A, B, C, D, E, F$ with $AE - BD \neq 0$ has in the real numbers precisely one solution $(x, y)$.

**Hilbert's program of an absolute proof for the freeness from contradictions of mathematics:** Around 1920, Hilbert developed a program whose object is was to prove that all of mathematics is free of contradictions. The main theorem of statement logic in 4.4.1.3 serves as model for the proof. The basic idea is that one can obtain all 'theorems of a theory' from a fixed number of axioms by applying, in a purely formal manner, a fixed number of rules for deductions.

**The non-completeness theorem of Gödel:** In 1931, Gödel's foundational work *On Formally Decidable Theorems in the Principia Mathematica and Related Theorems.* In this paper he shows that in any theory based on axioms which is sufficiently rich to encompass number systems there are always theorems which cannot be deduced purely from the axioms, although they are definitely true and can be proved.

Moreover, Gödel shows that the only way to prove that such a system of axioms is free of contradictions is by passing to a larger system. With this Hilbert's program for a proof of the non-contradictory nature of mathematics turned out to be unattainable. According to Gödel's insight, mathematics is more than just a formal system of axioms and rules for making deductions.

**Mathematical logic:** Gödel's work represents a highlight of mathematical logic. Formal logic – an important part of philosophy – originated with Aristotle (384–322 BC).

The first basic principle of thought is, according to Aristotle, the *Theorem of contradiction.* He writes: "It is unacceptable the something should be able to exist and at the same time not exist".

The second basic principle of Aristotle is the *Theorem of the excluded third:* "Either a statement is true or it is false". All indirect proofs in mathematics use this principle in the following form: if the negation of a statement is false, then the statement itself must be true.[22]

[22]Not all mathematicians accept this principle. In the 1920's Brouwer founded the so-called intuitionist mathematics, which refuses to accept indirect proofs and accepts only constructive ones.

# 6. Stochastic Calculus – Mathematics of Chance

*I believe that the astute reader will notice in what follows that this topic is not only a question of games of chance, but forms the basis of a very interesting and rewarding theory.*

*Christian Huygens* (1654)
*De Rationciniis in Aleae Ludo*[1]

*The true logic of this world is to be found in the theory of probability.*

*James Clerk Maxwell* (1831–1897)

Stochastics is concerned with the *mathematical laws of chance*, or as one prefers to say, of *randomness.* While the theory of probability is concerned with theoretical foundations, the theory of statistics is concerned with the ways in which large amounts of raw data can be gleaned for laws or rules of behavior of the object under investigation. Therefore, mathematical statistics is a mathematical instrument which all sciences dealing with empirical data (medicine, natural sciences, social sciences and economy) cannot do without.

> A useful compilation of the most important procedures of *mathematical statistics* which require a minimum amount of knowledge on the part of the reader to apply can be found in section 0.4.

It is typical for probability theory and mathematical statistics to create and study models which apply to situations of differing concreteness. As in other sciences, it is therefore important to *choose the model* to by studied *very carefully.* The application of different models to the same situation will in general lead to different results and hence to different conclusions.

In the nineteenth century James Clerk Maxwell and Ludwig Boltzmann (1844–1906) founded statistical physics. They applied methods of probability theory to describe systems with *large numbers of particles* (like gases). In this theory the physicists of the nineteenth century assumed that the particles obey the laws of classical mechanics and thus move along well-defined trajectories. These trajectories are determined for

[1]The translation of the title of this book is *On Calculations of Games of Chance.* This was the first book ever about about probability theory. The mathematical investigation of games of chance (for example games with dice) began in Italy during the fifteenth century.

The mathematical discipline of probability theory was founded by Jakob Bernoulli with his famous treatise *Ars Conjectandi*, in which he gave a mathematical proof of the "law of large numbers". This book appeared in 1713, which was eight years after Jakob Bernoulli's death.

The classical standard reference for the theory of probability was *Théorie analytique des probabilités* by the French mathematician and physicist Pierre Simon Laplace, which appeared in 1812. The modern axiomatic theory was founded by the Russian mathematician Andrei Nikolajevic Kolmogorov in his book *Foundations of Probability Theory*, which appeared in 1933.

all times by the initial conditions (location and velocity). In fact it is impossible to determine these initial conditions for a collection of $10^{23}$ gas molecules (a mole). In order to compensate for this impossibility, methods of statistics are applied.

The situation was radically changed with the introduction of quantum mechanics by Heisenberg and Schrödinger around 1925. This theory is statistical from the very onset. According to the Heisenberg uncertainty principle, it is impossible to know (measure) the location and simultaneously know (measure) the velocity of a particle. Most physicists today are convinced that the fundamental processes of elementary particles are stochastic *by nature*, as opposed to being unrecognizable because of the inability to determine hidden parameters. Thus the theory of stochastics is of fundamental importance for modern physics.

**Basic notions:** In probability theory there are the following basic notions:

(i) random event (for example the sex of a baby at birth (conception));

(ii) random variable (for example the height of people);

(iii) random function (for example, the temperature in New York City during a given year).

The notion (iii) is also referred to as a *stochastic process*. In addition there is a notion of 'independent' which applies to (i) – (iii).

**Standard notations:**

| $P(A)$ denotes the probability that the event $A$ will occur. |
|---|

It is a convention that probabilities lie between 0 and 1, meaning the following.

(a) If $P(A) = 0$, one says the event $A$ is *almost impossible*.

(b) If $P(A) = 1$, one says that the event $A$ is *almost certain*.

*Example 1:* The probability that a child is female (resp. male) is $p = 0.485$ (resp. $p = 0.515$). This means that of 1000 born children, 485 are female and 515 are male.

The investigation of the relationship between probability and frequency is one of the aims of mathematical statistics (see section 6.3).

*Example 2:* Let a needle fall upright onto a tabletop; then it is almost impossible that the needle will hit a given point $Q$ with its tip, while it is almost certain that it will not.

Let $X$ be a random variable (defined in 6.2.2 below).

| $P(a \leq X \leq b)$ denotes the probability that a measurement of $X$ yields a value $x$ which satisfies $a \leq x \leq b$. |
|---|

**Mathematization of phenomena:** *The theory of probability (stochastics) is a typical example of how a phenomenon of daily experience ('chance') can be put into a mathematical framework and how this can lead us to a deeper understanding of our views of reality.*

## 6.1 Elementary stochastics

We now discuss several basic patterns of probability theory which were of fundamental importance in the development of the theory.

### 6.1.1 The classical probability model

**Basic model:** We consider a random experiment and denote all possible events of the experiment by

$$\boxed{e_1\,,\, e_2\,, \ldots,\, e_n\,.}$$

We call $e_1, \ldots, e_n$ the *elementary events* of this experiment.

Moreover, we use the following notations.

(i) *Totality of events* $E$, meaning the set of all $e_j$. This is also referred to as a *field of events.*

(ii) *Event* $A$, meaning a subset of $E$.

We associate to a given event $A$ a *probability* $P(A)$ by the following rule:[2]

$$\boxed{P(A) := \frac{\text{number of elements of } A}{n}.} \tag{6.1}$$

In the classical literature one refers to the elementary events in $A$ as 'positive outcomes', while the set of all elementary events is the set of 'possible outcomes'. Then we have

$$P(A) = \frac{\text{number of positive outcomes}}{\text{number of possible outcomes}}\,. \tag{6.2}$$

This formulation of the notion of probability was introduced at the end of the seventeenth century by Jakob Bernoulli. We consider a few examples.

**Throwing a die:** The possible outcome of this random experiment consists of the elementary events

$$e_1\,,\, e_2\,, \ldots\,,\, e_6\,,$$

where $e_j$ corresponds to the outcome that $j$ eyes are on the top face of the die.

(i) The event $A := \{e_1\}$ is that in which the top face has one eye. According to (6.1) we have $P(A) = \frac{1}{6}$.

(ii) The event $B := \{e_2, e_4, e_6\}$ is the appearance of an even number on the top face of the die. According to (6.1) we have $P(B) = \frac{3}{6} = \frac{1}{2}$, which proves our intuition that there is a 50-50 chance of throwing an even (or odd) number.

**Throwing two dice:** The possible outcomes in this case consist of elementary events

$$e_{ij}\,, \qquad i, j = 1, \ldots, 6\,.$$

Here $e_{23}$ for example means that the first die has a 2, the second a 3. There are 36 elementary events.

(i) For $A := \{e_{ij}\}$ (for any fixed $i, j$), we get $P(A) = \frac{1}{36}$ from (6.1).

(ii) The event $B := \{e_{11}, e_{22}, e_{33}, e_{44}, e_{55}, e_{66}\}$ consists in both die showing the same number. From (6.1) again, we get $P(B) = \frac{6}{36} = \frac{1}{6}$.

**The lottery problem:** We consider the lottery game '6 from 45', in which six numbers out of 45 are checked, and six numbers are drawn to determine the winner. What is the probability of checking $n$ correct numbers? The results are gathered in Table 6.1.

[2] The notation $P(A)$ goes back to the French word *probabilité* for probability.

*Table 6.1. The lottery game '6 from 45'.*

| *Number of correctly checked numbers* | *Probability* | | *Number of winners among 10 million participants* |
|---|---|---|---|
| 6 | $a := \frac{1}{\binom{45}{6}}$ | $= 10^{-7}$ | 1 |
| 5 | $\binom{6}{5} 39a$ | $= 2 \cdot 10^{-5}$ | 200 |
| 4 | $\binom{6}{4}\binom{39}{2} a$ | $= 10^{-3}$ | 10 000 |
| 3 | $\binom{6}{3}\binom{39}{3} a$ | $= 2 \cdot 10^{-2}$ | 200 000 |

The elementary events in this example have the form

$$\boxed{e_{i_1 i_2 \ldots i_6}}$$

with $i_j = 1, \ldots, 45$ and $i_1 < i_2 < \cdots < i_6$. There are $\binom{45}{6}$ elementary events of this kind (see Example 5 in 2.1.1). If, for example, the winning numbers are $1, 2, 3, 4, 5, 6$, then $A := \{e_{123456}\}$ represents the event of having six correct numbers. According to (6.1) we have

$$P(A) = \frac{1}{\binom{45}{6}}.$$

To determine all the elementary events which correspond to five correctly checked numbers, we have to choose five among $1, 2, 3, 4, 5, 6$. From the remaining 39 wrong numbers $7, 8, \ldots, 45$ we have to choose one. This means that there are $\binom{6}{5} \cdot 39$ positive elementary events of this type.

In a similar manner one can check all the values listed in Table 6.1.

If we multiply the probabilities by the number of participants, we get roughly the number of winners in each class, listed in the last column.

**The birthday problem:** At a party there are $n$ guests. How large is the probability $p$ that two of the guests have their birthdays on the same day? According to Table 6.2 one can wager a relatively safe bet even if the number of guests is only about 30. One gets

$$\boxed{p = \frac{365^n - 365 \cdot 364 \ldots (365 - n + 1)}{365^n}.} \tag{6.3}$$

The elementary events are given by

$$\boxed{e_{i_1 \ldots i_n}, \qquad i_j = 1, \ldots, 365.}$$

For example, $e_{12,14,\ldots}$ represents the event that the first guest has birthday on the $12^{th}$ day of the year, the second on the $14^{th}$ and so forth. There are $365^n$ elementary events. Moreover, there are $365 \cdot 364 \cdots (365 - n + 1)$ elementary events corresponding to the event that there are no coincidences among the $n$ birthdays. Thus the numerator in (6.3) represents the number of positive outcomes.

*Table 6.2. The birthday problem.*

| *Number of guests* | 20 | 23 | 30 | 40 |
|---|---|---|---|---|
| *Probability that at least two guests have coincident birthdays* | 0.4 | 0.5 | 0.7 | 0.9 |

### 6.1.2 The law of large numbers due to Jakob Bernoulli

A fundamental experience is that upon repeating a random experiment many times the relative frequencies obtained get nearer and nearer to the probabilities. This is the basis for a great number of applications of probability theory. Mathematically this fact can be proved as the 'law of large numbers', first put forward and proved by Jakob Bernoulli. We explain this using the example of throwing a coin.

**Throwing a coin:** This is probably the easiest of all experiments: throw a coin; the elementary events are

$$\boxed{e_1\,,\ e_2\,,} \tag{6.4}$$

where $e_1$ denotes 'heads' and $e_2$ denotes 'tails'. The event $A = \{e_1\}$ corresponds to throwing heads; its probability is, according to (6.1), given by

$$\boxed{P(A) = \frac{1}{2}\,.}$$

**Relative frequency:** Our experience tells us that throwing a coin $n$ times for a large number of trials will yield a frequency near to $1/2$ for heads and the same for tails. We now give a mathematical discussion of this.

**Throwing a coin $n$ times:** The elementary events are

$$\boxed{e_{i_1 i_2 \ldots i_n}\,, \qquad i_1\,, \ldots, i_n = 1, 2\,.}$$

Here $e_{i_1 i_2 \ldots i_n}$ means that the first throw yields the elementary event $e_{i_1}$ of (6.4), the second the event $e_{i_2}$ of (6.4) and so on. Thus $i_j = 1$ or $i_j = 2$ for all $j$. We associate to this elementary event a *relative frequency* as follows (here 'heads'):

$$\boxed{H(e_{i_1 i_2 \ldots i_n}) := \frac{\text{number of occurrences of heads}}{\text{number of trials}\, n}\,.}$$

The number in the numerator is thus equal to the number of occurrences of '1' as index in the elementary event $e_{i_1 i_2 \ldots i_n}$.

**The law of large numbers of Jakob Bernoulli:**[3] Let an arbitrary real number $\varepsilon > 0$ be given. We denote by $A_n$ the set of all elementary events $e_{i_1 i_2 \ldots}$ for which

$$\left| H(e_{\ldots}) - \frac{1}{2} \right| < \varepsilon.$$

[3]This famous law was published eight years after Jakob Bernoulli's death in 1705.

This means that among the indices, the difference between the percentage of '1's (and also '2's) and 1/2 (50%) is less than the given number. Bernoulli calculated the probability $P(A_n)$, using (6.1), and showed that

$$\lim_{n\to\infty} P(A_n) = 1\,.$$

This theorem is also described in the single formula

$$\boxed{\lim_{n\to\infty} P\left(\left|H_n - \frac{1}{2}\right| < \varepsilon\right) = 1\,.}$$

### 6.1.3 The limit theorem of de Moivre

One of the most import insights of probability theory is that one obtains results which are easy to understand upon taking the limit $n \to \infty$, where $n$ denotes the number of trials performed. We explain this again using the example of throwing a coin. Let $A_{n,k}$ denote the set of all elementary events $e_{i_1 i_2 \ldots i_n}$ for which the index '1' occurs exactly $k$ times. This corresponds to the set of all trials for which heads turns up $k$ times in $n$ throws. This implies that for the relative frequency $H_n$ of heads occuring in each elementary event in $A_{n,k}$ we have the relation

$$H_n = \frac{k}{n}\,.$$

Also, $P(A_{n,k}) = \binom{n}{k}\frac{1}{2^n}$. This is indicated again in a single formula:

$$\boxed{P\left(H_n = \frac{k}{n}\right) = \binom{n}{k}\frac{1}{2^n}\,.}$$

This is the probability that upon throwing the coin $n$ times, the relative frequency $H_n$ of throwing heads is $k/n$.

**Theorem of de Moivre (1730):** For a large number $n$ of coin tosses one has the asymptotic equality[4]

$$\boxed{P\left(H_n = \frac{k}{n}\right) \sim \frac{1}{\sigma\sqrt{2\pi}}\,\mathrm{e}^{-(k-\mu)^2/2\sigma^2}\,, \qquad k = 1, 2, \ldots, n\,,} \tag{6.5}$$

with the parameters $\mu = n/2$ and $\sigma = \sqrt{n/4}$. In (6.5) the function on the right-hand side is the so-called Gaussian normal distribution (see Figure 6.2). As expected, the probability $P$ in (6.5) is largest for $k = n/2$. ;

### 6.1.4 The Gaussian normal distribution

**The basic model of a measurement process:** Let a continuous (or more generally almost everywhere continuous) function $\varphi : \mathbb{R} \longrightarrow \mathbb{R}$ be given, with

$$\boxed{\int_{-\infty}^{\infty} \varphi\,\mathrm{d}x = 1\,.}$$

[4]The quotient of the two expressions of (6.5) approaches 1 for fixed $k$ as $n \longrightarrow \infty$.

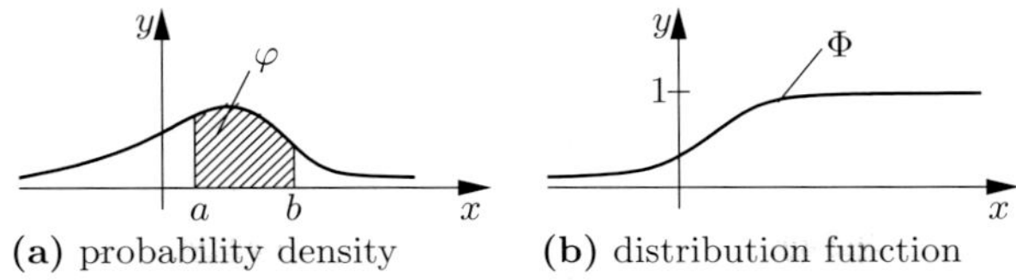


*Figure 6.1. A probability density and corresponding distribution function.*

This situation can be interpreted in a probabilistic manner as follows.

(i) Suppose we are given a random measurement variable $X$ determined by the measurement of some real quantity. An example could be the height of people.

(ii) We set

$$\boxed{P(a \le X \le b) := \int_a^b \varphi(x)\,\mathrm{d}x\,,}$$

in which this expression is the probability that the measured quantity $X$ lies in the interval $[a, b]$. Intuitively, $P(a \le X \le b)$ corresponds to the surface area under the curve $\varphi$ in the interval $[a, b]$ (see Figure 6.1(a)). We call $\varphi$ a *probability density.* The function

$$\Phi(x) := \int_{-\infty}^{x} \varphi(\xi)\,\mathrm{d}\xi\,, \qquad x \in \mathbb{R}\,,$$

is called the *distribution function* determined by $\varphi$ (see Figure 6.1(b)).

(iii) The quantity

$$\boxed{\overline{X} := \int_{-\infty}^{\infty} x\varphi(x)\,\mathrm{d}x}$$

is called the *mean* (or *expectation*) of $X$. Moreover,

$$\boxed{(\Delta X)^2 := \int_{-\infty}^{\infty} \left(x - \overline{X}\right)^2 \varphi(x)\,\mathrm{d}x}$$

is referred to as the *variance* or square deviation, and the non-negative quantity $\Delta X$ is called the *standard deviation* of $X$.

If we interpret $\varphi$ as a mass density, then $\overline{X}$ is the center of mass.

**The Chebychev inequality:** For all $\beta > 0$ we have:

$$\boxed{P(|X - \overline{X}| > \beta\Delta X) \le \frac{1}{\beta^2}\,.}$$

In particular for $\Delta X = 0$ we have $P(X = \overline{X}) = 1$.

**Confidence interval:** Let $0 < \alpha < 1$. The measured value of the random variable $X$ lies with a probability $> 1 - \alpha$ in the interval

$$\left[\overline{X} - \frac{\Delta X}{\sqrt{\alpha}}, \overline{X} + \frac{\Delta X}{\sqrt{\alpha}}\right].$$

*Example 1:* Let $\alpha = 1/16$. The measured value of $X$ lies with a probability $> \frac{15}{16}$ in the interval $[\overline{X} - 4\Delta X, \overline{X} + 4\Delta X]$.

This makes in meaning of the mean and the standard deviation more precise:

> The smaller the standard deviation $\Delta X$ is, the more the measured values of $X$ are concentrated around the mean $X$.

**The Gaussian normal distribution $N(\mu, \sigma)$:** This distribution is given by the probability density

$$\varphi(x) := \frac{1}{\sigma\sqrt{2\pi}} \mathrm{e}^{-(x-\mu)^2/2\sigma^2}, \qquad x \in \mathbb{R}$$

with real parameters $\mu$ and $\sigma > 0$ (Figure 6.2). One has

$$\overline{X} = \mu, \qquad \Delta X = \sigma.$$

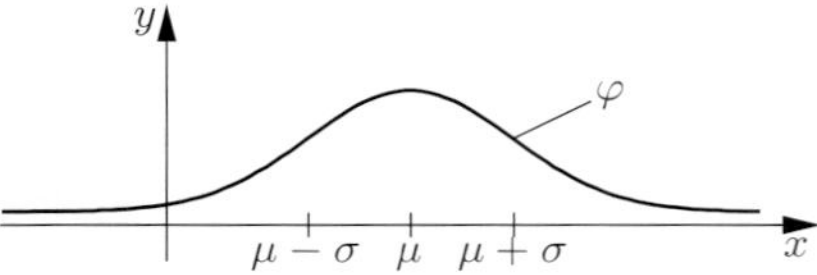


*Figure 6.2. The Gaussian normal distribution.*

This normal distribution is the most important distribution of probability theory. The reason for this is the *central limit theorem*, according to which any random variable is nearly normally distributed, provided it is composed of many independent random quantities (see section 6.2.4).

*Table 6.3. Several continuous probability distributions.*

| ***Name of distribution*** | ***Probability density*** $\varphi$ | ***Mean*** $\overline{X}$ | ***St. dev.*** $\Delta X$ |
|---|---|---|---|
| normal distribution $N(\mu, \sigma)$ | $\frac{1}{\sigma\sqrt{2\pi}} \mathrm{e}^{-(x-\mu)^2/2\sigma^2}$ | $\mu$ | $\sigma$ |
| exponential distribution (Figure 6.3) | $\frac{1}{\mu}\mathrm{e}^{-x/\mu}$ for $x \geq 0$ $(\mu > 0)$<br>$0$ for $x < 0$ | $\mu$ | $\mu$ |
| equidistant distribution (Figure 6.4) | $\frac{1}{b-a}$ for $a \leq x \leq b$<br>$0$ otherwise | $\frac{b+a}{2}$ | $\frac{b-a}{\sqrt{12}}$ |

**Exponential distribution:** The corresponding density is listed in Table 6.3. This distribution is used, for example, to describe the life span of a product (for example a

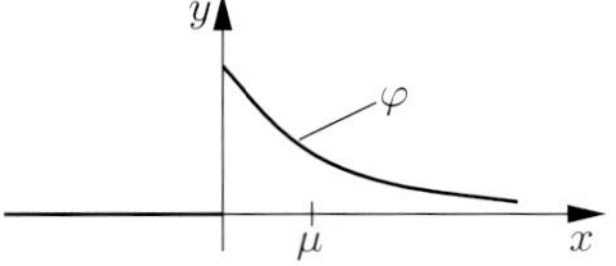

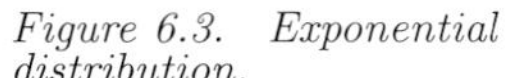

*Figure 6.3. Exponential distribution.*

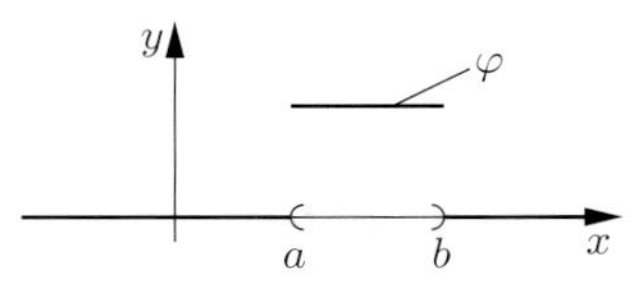


*Figure 6.4. Equidistant distribution.*

light bulb). Then

$$\int_a^b \frac{1}{\mu} \mathrm{e}^{-x/\mu} \, \mathrm{d}x$$

is the probability that the life span of the product is contained in the interval $[a, b]$. The mean life span is equal to $\mu$.

**Mean value of functions of random variables:** Let $Z = F(X)$ be a function of a random variable $X$. Every measurement of $X$ yields a value for $Z$. The mean $\overline{Z}$ and the standard deviation $(\Delta Z)^2$ of $Z$ is given by

$$\overline{Z} = \int_{-\infty}^{\infty} F(x)\varphi(x) \, \mathrm{d}x \,, \qquad (\Delta Z)^2 = \int_{-\infty}^{\infty} (F(x) - \overline{Z})^2 \varphi(x) \, \mathrm{d}x \,.$$

*Example 2:* $(\Delta X)^2 = \overline{(X - \overline{X}\,)^2} = \int_{-\infty}^{\infty} (x - \overline{X})^2 \varphi(x) \, \mathrm{d}x \,.$

**The addition formula for means:**

$$\overline{F(X) + G(X)} = \overline{F(X)} + \overline{G(X)} \,.$$

### 6.1.5 The correlation coefficient

The most important quantities for arbitrary measurements are mean, standard deviation and correlation coefficient $r$ with $-1 \leq r \leq 1$. For this last quantity we have the following fact.

> The larger the absolute value $|r|$ of the correlation coefficient, the more two measured values are correlated, i.e., depend on one another.

**The basic model for the measurement of two random variables:** Let a function $\varphi : \mathbb{R}^2 \longrightarrow \mathbb{R}$ be given, which is almost everywhere continuous and non-negative, and fulfills:

$$\int_{\mathbb{R}^2} \varphi(x, y) \, \mathrm{d}x\mathrm{d}y = 1 \,.$$

This situation can be given a probabilistic interpretation as follows.

(i) Let two random variables $X$ and $Y$ be given, which have real values given by two measurements. One refers to the pair $(X, Y)$ as a *random vector.*

(ii) Probability: We set

$$\boxed{P((X,Y) \in G) := \int_G \varphi(x,y)\,\mathrm{d}x\mathrm{d}y\,.}$$

This is the probability that upon taking measurements of $X$ and $Y$ the point $(X, Y)$ lies in the set $G$. We call $\varphi$ the probability density of the random variable $(X, Y)$.

(iii) The probability densities $\varphi_X$ and $\varphi_Y$ for $X$ and $Y$ are:

$$\boxed{\varphi_X(x) := \int_{-\infty}^{\infty} \varphi(x,y)\,\mathrm{d}y\,, \qquad \varphi_Y(y) := \int_{-\infty}^{\infty} \varphi(x,y)\,\mathrm{d}x\,.}$$

(iv) The mean $\overline{X}$ and standard deviation $(\Delta X)^2$ of $X$ are given by:

$$\overline{X} = \int_{-\infty}^{\infty} x\varphi_X(x)\,\mathrm{d}x\,, \qquad (\Delta X)^2 = \int_{-\infty}^{\infty} (x-\overline{X})^2\,\varphi_X(x)\,\mathrm{d}x\,.$$

Similarly one calculates $\overline{Y}$ and $(\Delta Y)^2$.

(v) The mean value for a function $Z = F(X, Y)$ is given by:

$$\boxed{\overline{Z} := \int_{\mathbb{R}^2} F(x,y)\varphi(x,y)\,\mathrm{d}x\mathrm{d}y\,.}$$

(vi) The standard deviation $(\Delta Z)^2$ of $Z = F(X, Y)$ is given by:

$$\boxed{(\Delta Z)^2 := \overline{\left(Z - \overline{Z}\,\right)^2} = \int_{\mathbb{R}^2} \left(F(x,y) - \overline{Z}\,\right)^2 \varphi(x,y)\,\mathrm{d}x\mathrm{d}y\,.}$$

(vii) The addition formula for the means is:

$$\boxed{\overline{F(X,Y) + G(X,Y)} = \overline{F(X,Y)} + \overline{G(X,Y)}\,.}$$

**Covariance:** The number

$$\mathrm{Cov}(X,Y) := \overline{(X - \overline{X}\,)(Y - \overline{Y}\,)}$$

is called the *covariance* of $X$ and $Y$. Explictly we have

$$\mathrm{Cov}(X,Y) = \int_{\mathbb{R}^2} (x - \overline{X}\,)(y - \overline{Y}\,)\varphi(x,y)\,\mathrm{d}x\mathrm{d}y\,.$$

**The correlation coefficient:** A fundamental question coming up at this point is: *Are $X$ and $Y$ strongly or weakly correlated?* The answer to this question is given by the *correlation coefficient* $r$, which we define by the relation

$$\boxed{r = \frac{\operatorname{Cov}(X,Y)}{\Delta X \Delta Y}.}$$

This coefficient always satisfies $-1 \le r \le 1$, i.e., $r^2 \le 1$.

**Definition:** The larger $r^2$ is, the *stronger* $X$ and $Y$ are *correlated.*

**Motivation:** We consider the minimization problem

$$\boxed{\overline{(Y - a - bX)^2} \overset{!}{=} \min\,, \qquad a, b \in \mathbb{R}\,.} \tag{6.6}$$

This means that we seek a linear function $a + bX$ which approximates $Y$ closely. The minimization problem corresponds to the method of least squares of Gauss.

(a) The solution of (6.6) is the so-called *regression line*

$$\overline{Y} + r\frac{\Delta Y}{\Delta X}(X - \overline{X})\,.$$

(b) For this solution we have

$$\overline{(Y - a - bX)^2} = (\Delta Y)^2\,(1 - r^2)\,.$$

The best (resp. worst) approximation is obtained for $r^2 = 1$ (resp. $r = 0$).

*Example:* In practical cases one is given measurements $x_1, \ldots, x_n$ and $y_1, \ldots, y_n$ of $X$ and $Y$. These measurements $(x_j, y_j)$ are plotted in the $(x, y)$-plane. The regression line

$$\boxed{y = \overline{Y} + r\frac{\Delta Y}{\Delta X}(x - \overline{X})}$$

is the line which is nearest to all these points (see Figure 6.5).

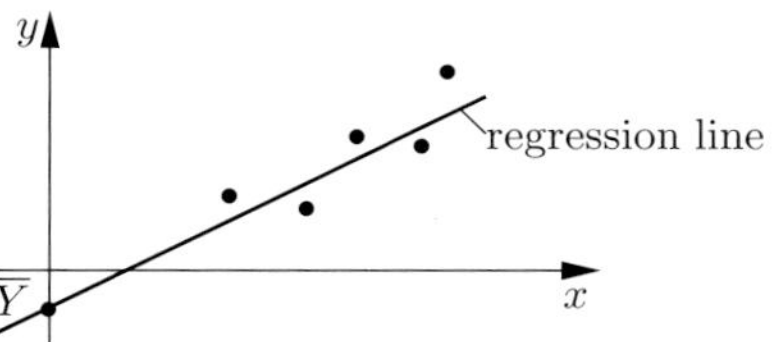


*Figure 6.5. A regression line.*

The true quantities $\Delta X$, $\Delta Y$ and $r$ are not known. But we can estimate them from our measurements as follows.

$$\boxed{\begin{gathered}\overline{X} = \frac{1}{n}\sum_{j=1}^{n} x_j\,, \qquad (\Delta X)^2 = \frac{1}{n-1}\sum_{j=1}^{n}(x_j - \overline{X}\,)^2\,,\\ r = \frac{1}{(n-1)\Delta X \Delta Y}\sum_{j=1}^{n}(x_j - \overline{X}\,)(y_j - \overline{Y}\,)\,.\end{gathered}}$$

**Independence of random variables:** By definition random variables $X$ and $Y$ are said to be *independent* , if the (joint) probability density $\varphi$ has a product decomposition of the form

$$\boxed{\varphi(x,y) = a(x)b(y)\,, \qquad (x,y) \in \mathbb{R}^2.}$$

Then we have:

(i) $\varphi_X(x) = a(x)$ and $\varphi_Y(y) = b$.

(ii) The product formula for the probabilities is:

$$P(a \le X \le b,\, c \le Y \le d) = P(a \le X \le b)P(c \le Y \le d)\,.$$

(iii) The product formula for the means is given by:

$$\overline{F(X)G(Y)} = \overline{F(X)} \cdot \overline{G(X)}\,.$$

(iv) The correlation coefficient $r$ vanishes.[5]

(v) The addition formula for the standard deviation is:

$$(\Delta(X+Y))^2 := (\Delta X)^2 + (\Delta Y)^2\,.$$

**The Gaussian normal distribution:**

$$\varphi(x,y) := \frac{1}{\sigma_x\sqrt{2\pi}}\mathrm{e}^{-(x-\mu_x)/2\sigma_x^2} \cdot \frac{1}{\sigma_y\sqrt{2\pi}}\mathrm{e}^{-(y-\mu_y)^2/2\sigma_y^2}\,.$$

This distribution, which is a product of one-dimensional normal distributions, is a probability density which corresponds to two independent random variables $X$ and $Y$. One has

$$\overline{X} = \mu_x\,, \qquad \Delta X = \sigma_x\,, \qquad \overline{Y} = \mu_y\,, \qquad \Delta Y = \sigma_y$$

and

$$(\Delta(X+Y))^2 = \sigma_x^2 + \sigma_y^2\,.$$

### 6.1.6 Applications to classical statistical physics

The whole of classical statistical physics can be described quite elegantly and briefly with the help of the results of the last paragraph. We consider a system which consists of $N$ particles of mass $m$. The starting point for the description is the following expression for the energy $E$ of the system.

$$E = H(q,p)\,.$$

The function $H$ is the *Hamilton function* of the system. Every particle is assumed to have $f$ degrees of freedom (for example three translations and/or additional degrees of freedom coming from rotations or from vibrations). We set

$$q = (q_1, \ldots, q_{fN})\,, \qquad p = (p_1, \ldots, p_{fN})\,.$$

Here the $q_j$ are the coordinates of the particles, and the $p_j$ are (generalized) momentum variables, which are related to the velocity of the particles.

---

[5] This follows from $\overline{X - \overline{X}} = 0$ and $\overline{(X-\overline{X})(Y-\overline{Y})} = \overline{(X-\overline{X})} \cdot \overline{(Y-\overline{Y})} = 0$.

**Classical mechanics:** The equation for the motion $q = q(t)$ and $p = p(t)$ of the particles at time $t$ is given by

$$\boxed{q_j'(t) = H_{p_j}(q(t), p(t))\,, \qquad p_j'(t) = -H_{q_j}(q(t), p(t))\,, \qquad j = 1, \ldots, fN\,.}$$

The variables $(q, p)$ are supposed to move in a domain $\Pi$ of $\mathbb{R}^{fN}$, which we call the phase space of the system (for these basic notions on Hamiltonian mechanics, see section 5.1.3).

**Classical statistical mechanics:** We start with a probability density

$$\boxed{\varphi(q, p) := C\mathrm{e}^{-H(q,p)/\mathrm{k}T}\,,}$$

where the constant $C$ is to be determined in such a way that $\int_\Pi \varphi \mathrm{d}q\mathrm{d}p = 1$. Here $T$ denotes the absolute temperature of the system, and k is a constant of nature called the *Boltzmann constant.* The constant is responsible for the dimensionlessness of $H/\mathrm{k}T$. With the aid of $\varphi$ we can now introduce the fundamental quantities.

(i) The system is in a subdomain $G$ of the phase space with the probability $P(G)$:

$$\boxed{P(G) = \int_G \varphi(q, p)\mathrm{d}q\mathrm{d}p\,.} \tag{6.7}$$

(ii) The *mean* and *standard deviation* of a function $F = F(q, p)$ are given by:

$$\overline{F} = \int_\Pi F(q, p)\varphi(q, p)\mathrm{d}q\mathrm{d}p\,, \qquad (\Delta F)^2 = \int_\Pi (F(q, p) - \overline{F})^2 \varphi(q, p)\mathrm{d}q\mathrm{d}p\,.$$

(iii) The *correlation coefficient* $r$ of two given functions $A = A(q, p)$ and $B = B(q, p)$ is:

$$\boxed{r = \frac{1}{\Delta A \Delta B}\overline{(A - \overline{A})(B - \overline{B})}\,.}$$

(iv) The *entropy* of the system at the absolute temperature $T$ is:[6]

$$\boxed{S = -\mathrm{k}\,\overline{\ln \varphi}\,.}$$

**Applications to the Maxwell velocity distribution:** We consider an ideal gas consisting of $N$ particles of mass $m$, which moves in a bounded domain $\Omega$ of $\mathbb{R}^3$. Let $V$ denote the volume of the domain $\Omega$. The $j^{th}$ particle is described by a coordinate vector $\mathbf{x}_j = \mathbf{x}_j(t)$ and the momentum vector

$$\mathbf{p}_j(t) = m\mathbf{x}_j'(t),$$

where $\mathbf{x}_j'(t)$ is the velocity of the particle at time $t$. If $v$ denotes an arbitrary component of the velocity vector of the $j^{th}$ particle in a Cartesian coordinate system, then we have:

$$\boxed{P(a \le mv \le b) = \int_a^b \frac{1}{\sigma\sqrt{2\pi}}\mathrm{e}^{-x^2/2\sigma^2}\mathrm{d}x\,.} \tag{6.8}$$

[6]Note that $\varphi$ depends on $T$.

This is the probability that $mv$ lies in the interval $[a, b]$. The corresponding probability density is a Gaussian normal distribution with the mean $\overline{mv} = 0$ and the standard deviation

$$\Delta(mv) = \sigma = \sqrt{m\mathrm{k}T}\,.$$

This law was laid down by Maxwell in 1860. With it he laid the foundations on which Boltzmann built statistical mechanics.

*Reasoning:* In a Cartesian coordinate system we set $\mathbf{p}_1 = p_1\mathbf{i}+p_2\mathbf{j}+p_3\mathbf{k}$, $\mathbf{p}_2 = p_4\mathbf{i}+p_5\mathbf{j}+p_6\mathbf{k}, \ldots$ and $\mathbf{x}_1 = q_1\mathbf{i}+q_2\mathbf{j}+q_3\mathbf{k}, \ldots$ The total energy $E$ of the ideal gas is, because there is no interaction between the particles (ideal gas), the sum of all the kinetic energies of the individual particles:

$$E = \frac{\mathbf{p}_1^2}{2m} + \ldots + \frac{\mathbf{p}_N^2}{2m} = \sum_{j=1}^{3N} \frac{p_j^2}{2m}\,.$$

Consider for example $p_1 = mv$. According to (6.7) we have[7]

$$P(a \le p_1 \le b) = C \int_a^b \exp\left(-\frac{p_1^2}{2m\mathrm{k}T}\right) \mathrm{d}p_1 \cdot J\,.$$

where the number $J$ is obtained from integration over $p_2 \cdots p_{3N}$ from $-\infty$ to $\infty$ and from the integration over the coordinates $q_j$. The value of $CJ$ is then deduced from the normalization condition $P(-\infty < p < \infty) = 1$. This yields (6.8).

**The principle of fluctuations:** A decisive question is: Why is it necessary to have very sensitive measurement capabilities to see the statistical nature of a gas? The answer is to be found in the fundamental formula

$$\boxed{\frac{\Delta E}{\overline{E}} = \frac{1}{\sqrt{N}}\frac{\Delta\varepsilon}{\overline{\varepsilon}}} \tag{6.9}$$

which is valid for an ideal gas. The notations are: $N$ is the number of particles, $E$ is the total energy, $\varepsilon$ is the energy of a single particle. Since $\Delta\varepsilon/\overline{\varepsilon}$ is roughly unity and $N$ has the size of $10^{23}$, the relative variations in the energy $\Delta E/\overline{E}$ of a gas are extremely small and do not play any observable role in everyday life.

*Reasoning:* Since the particles of an ideal gas do not interact with each other, the energies of the individual particles are *independent* random variables. Therefore we may apply the addition formula for the mean and the standard deviation. This yields

$$\overline{E} = N\varepsilon\,, \qquad (\Delta E)^2 = N(\Delta\,\varepsilon)^2\,.$$

(6.9) follows from this.

**Systems with variable numbers of particles and the chemical potential:** During chemical reactions, the number of particles can change. The corresponding statistical physics then works with the parameter $T$ (the absolute temperature) and the parameter $\mu$ (chemical potential). This is discussed in [212]. There we consider a general scheme which can also be applied to modern quantum statistics (the statistics of atoms, molecules, photons and elementary particles).

[7] For clarity we use the notation $\exp(x) := \mathrm{e}^x$ here.

## 6.2 Kolmogorov's axiomatic foundation of probability theory

**The general probability model of Kolmogorov:** Suppose we are given a non-empty set $E$ which we view as a given field of events. The elements $e$ of $E$ are called the *elementary events.* Let $P$ be a measure on $E$ satisfying

$$\boxed{P(E) = 1\,.}$$

The general *events* in this picture are the subsets $A \subseteq E$ for which the measure $P(A)$ is defined.

**Connection with measure theory:** With this Ansatz, probability theory becomes just a special case of the modern mathematic branch of measure theory, which is developed in [212]. A measure on a set $E$ with the property $P(E) = 1$ is called a *probability measure.* The events correspond to measurable sets. In what follows we formulate explicitly the definition of a probability measure.

**Explicit formulation of the Kolmogorov axioms:** On the set $E$ we assume we are given a system $\mathscr{S}$ of subsets $A \subset E$, which satisfy the following conditions.

(i) The empty set $\emptyset$ and the set $E$ are elements of $\mathscr{S}$.

(ii) If $A$ and $B$ belong to $\mathscr{S}$, then this is also true for the union $A \cup B$, the intersection $A \cap B$, the difference set $A - B$ and the complement $C_E A := E - A$.

(iii) If $A_1, A_2, \ldots$ belong to $\mathscr{S}$, then the (infinite) union $\bigcup_{n=1}^{\infty} A_n$ and the (infinite) intersection $\bigcap_{n=1}^{\infty} A_n$ also belong to $\mathscr{S}$.

The *events* are the elements of $\mathscr{S}$. To every event $A$ we associate a real number $P(A)$ which satisfies:

(a) $0 \leq P(A) \leq 1$.

(b) $P(E) = 1$ and $P(\emptyset) = 0$.

(c) For any two events $A$ and $B$ with $A \cap B = \emptyset$ we have

$$\boxed{P(A \cup B) = P(A) + P(B)\,.}$$

(d) If $A_1, A_2, \ldots$ are countably many events with $A_j \cap A_k = \emptyset$ for all indices $j \neq k$, then we have

$$\boxed{P\left(\bigcup_{n=1}^{\infty} A_n\right) = \sum_{n=1}^{\infty} P(A_n)\,.} \tag{6.10}$$

**Interpretation:** The elementary events correspond to possible outcomes of a random experiment, and $P(A)$ is the *probability* for the occurrence of the outcome.

**Definition:** The triple $(E, \mathscr{S}, P)$ as above is called a *probability space.*

**Philosophical context:** This general approach to probability theory, proposed by Kolmogorov in 1933, assumes that each outcome has a well-defined probability of occurrence independent of any measurements which are made of the experiment.

Attempts to create a theory of probability based on measurements and the ensuing relative frequencies have not had any success.

Kolmogorov's approach is also compatible with the philosophy of Immanuel Kant (1724–1804), by assuming that probabilities exist *a priori*. Relative frequencies, on the other hand, are products of measurements and thus *a posteriori*.

**Three facts of our experience:** In our daily life we use the following three basic facts.

(i) Events with a very small probability occur very seldom.

(ii) Probabilities can be estimated by relative frequencies.

(iii) Relative frequencies stabilize after a certain number of measurements have been made and recorded.

The law of large numbers shows mathematically that (ii) and (iii) can in fact be derived from (i).

*Example 1:* The probability for having six winning numbers in '6 from 45' is $10^{-7}$. Everybody knows that the chances of winning are negligible.

*Example 2:* Life insurance companies require the probabilities for the deaths of their customers depending on their ages. Probabilities of this kind cannot be derived as in 6.1.1 with combinatorial methods, but rather require a detailed analysis of large amounts of data. In order to determine the probability $p$ that a person will reach the age of 70, choose $n$ people. If $k$ of these are 70 or older, then we can say approximately

$$\boxed{p = \frac{k}{n}\,.}$$

*Example 3:* To determine the probability that a newborn child is a girl or a boy, one also must use data analysis. Already Laplace (1749–1827) investigated the data available from cities like London, Berlin, St. Petersburg and more data from France. He found a relative frequency of girls being born of about

$$\boxed{p = 0.49\,.}$$

One the other hand, in the city of Paris the value was about $p = 0.5$. Having faith in the universality of laws of chance, Laplace sought to explain this difference. He discovered that in Paris children who had been found by others also contributed to the statistics; at the time people in Paris would abandon mostly female babies. Once he accounted for these children, also Paris turned out to have a value near $p = 0.49$.

**Finitely many possible outcomes:** If a random experiment has a finite number $n$ of possible outcomes, then we choose a set $E$ with elements

$$\boxed{e_1, \ldots, e_n}$$

and associate to each elementary event a number $P(e_j)$ with $0 \le P(e_j) \le 1$, and such that

$$P(e_1) + P(e_2) + \ldots + P(e_n) = 1\,.$$

All subsets $A$ of $E$ are called events. To each event $A = \{e_{i_1}, \ldots, e_{i_k}\}$ we associate a probability

$$\boxed{P(A) := P(e_{i_1}) + \ldots + P(e_{i_k}).}$$

*Example 4* (throwing a die): This experiment has $n = 6$. If

$$P(e_j) = \frac{1}{6}, \qquad j = 1, \ldots, 6,$$

then we have a fair die, otherwise this is probably the die of a cheater.

**The needle experiment, infinitely many possible outcomes and the Monte Carlo method:** We let a needle fall perpendicularly onto a unit square $E : \{(x, y) \mid 0 \leq x, y \leq 1\}$. The probability that a subset $A$ of $E$ is hit by the tip of the falling needle is

$$\boxed{P(A) := \text{surface area of } A}$$

(see Figure 6.6(a)). The set $E$ is again referred to as a field of events. An elementary event $e$ is any point of $E$. In this case, we have the two following surprising facts which can be observed.

(i) Not every subset $A$ of $E$ is an event.

(ii) One has $P(\{e\}) = 0$.

In fact it is not possible to associate to every subset $A$ of $E$ a surface area in such a way that we obtain a measure on $E$ fulfilling (6.10). The natural candidate for such a measure would be the Lebesgue measure on $\mathbb{R}^2$. For sufficiently reasonable sets $A$ the probability $P(A)$ is precisely the surface area of $A$. However, there are also 'wild' subsets $A$ of $E$, which are not Lebesgue measurable and thus not events. These sets cannot in any sensible way be assigned a probability of being hit by the tip of the needle.

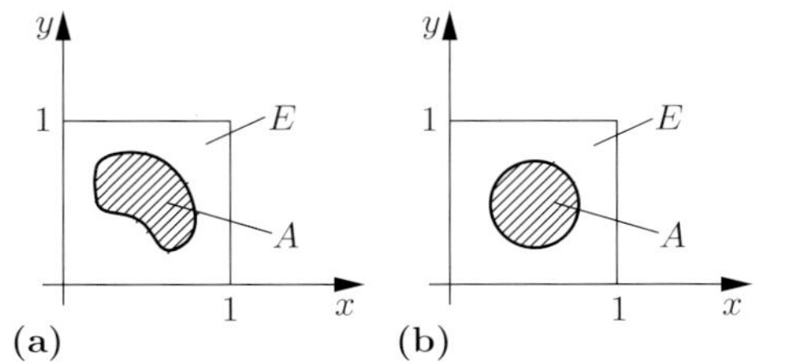


*Figure 6.6. Probability as surface area.*

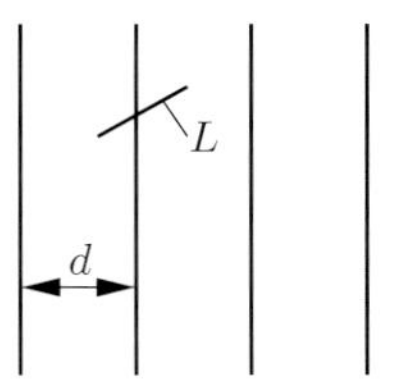


*Figure 6.7. Buffon's needle problem.*

If we consider the set $A$ consisting of the unit square $E$ without a single point $e$, then we have

$$P(A) = 1 - P(\{e\}) = 1.$$

This is paraphrased by saying that it is almost certain that the tip of the needle will hit $A$.

This motivates the following definitions.

**Almost impossible events:** An event $A$ is said to be *almost impossible* if $P(A) = 0$.

**Almost certain events:** An event $A$ is said to be *almost certain* if $P(A) = 1$.

*Example 5:* We choose a circle $A$ of radius $r$. Then we have $P(A) = \pi r^2$. Thus we can determine an experimental value of $\pi$ with the aid of this experiment.

The falling needle can be simulated with the help of a random number generator on a computer. This is the basic idea behind Monte Carlo simulations, which are used to calculate high-dimensional integrals in atomic physics, elementary particle theory and quantum chemistry.

*Example 6* (Buffon's needle problem): In 1777 the following problem was posed by the French natural scientist Buffon. In the plane we draw parallel lines separated at a distance of $d$ from each other (Figure 6.7). Now throw a needle of length $L$ with $L < d$ at the plane. What is the probability that the needle will hit one of the lines? The answer is

$$\boxed{p = \frac{2L}{d\pi}\,.}$$

In 1850 the astronomer Wolf in Zürich (Switzerland) threw a needle 5000 times and thus determined (an approximation of) the probability $p$. From this he obtained the approximation $\pi \sim 3.16$, which approximates the true value (3.14 to two decimal places) relatively well.

### 6.2.1 Calculations with events and probabilities

Events are sets. Every set-theoretic construction corresponds to another such in a probability theory; these are collected in Table 6.4. The calculation with events is done according to the rules of set theory (explained in section 4.3.2).

**Monotonous property of probabilities:** If $A_1, A_2, \ldots$ are events, then one has the inequality

$$P\left(\bigcup_{n=1}^{N} A_n\right) \leq \sum_{n=1}^{N} P(A_n)$$

for $N = 1, 2, \ldots$ and $N = \infty$. According to (6.10) we have equality if $A_j$ and $A_k$ for all $k \neq j$ have no elements in common, i.e., the events are incompatible with one another.

**Limit properties:**

(i) $A_1 \subseteq A_2 \subseteq \ldots$ implies $\lim\limits_{n\to\infty} P(A_n) = P\left(\bigcup\limits_{n=1}^{\infty} A_n\right)$.

(ii) $A_1 \supseteq A_2 \supseteq \ldots$ implies $\lim\limits_{n\to\infty} P(A_n) = P\left(\bigcap\limits_{n=1}^{\infty} A_n\right)$.

#### 6.2.1.1 Conditional probabilities

We choose a fixed field of events $E$ and consider events $A, B, \ldots$ which belong to $E$.

**Definition:** Let $P(B) \neq 0$. The number

$$\boxed{P(A|B) := \frac{P(A \cap B)}{P(B)}} \tag{6.11}$$

is called the *conditional probability* of $A$ occuring under the condition that $B$ has already occurred.

*Table 6.4. The algebra of events.*

| ***Event*** | ***Interpretation*** | ***Probability*** |
|---|---|---|
| $E$ | collective event | $P(E) = 1$ |
| $\emptyset$ | impossible event | $P(\emptyset) = 0$ |
| $A$ | arbitrary event | $0 \le P(A) \le 1$ |
| $A \cup B$ | the event $A$ or $B$ occur | $P(A \cup B) = P(A) + P(B) - P(A \cap B)$ |
| $A \cap B$ | both $A$ and $B$ occur | $P(A \cap B) = P(A) + P(B) - P(A \cup B)$ |
| $A \cap B = \emptyset$ | the events $A$ and $B$ can not occur simultaneously | $P(A \cup B) = P(A) + P(B)$ |
| $A - B$ | $A$ occurs, $B$ does not | $P(A - B) = P(A) - P(B)$ |
| $C_E A$ | the event $A$ does not occur ($C_E A := E - A$) | $P(C_E A) = 1 - P(A)$ |
| $A \subseteq B$ | if $A$ occurs, then so does $B$ | $P(A) \le P(B)$ |
| | the events $A$ and $B$ are independent | $P(A \cap B) = P(A)P(B)$ |

**Motivation:** We choose the set $B$ as a new field of events and consider the subsets $A \cap B$ of $B$, where $A$ is an event with respect to $E$ (i.e., $A \subseteq E$, Figure 6.8).

We construct a probability measure $P_B$ on $B$ with $P_B(B) := 1$ and $P_B(A \cap B) := P(A \cap B)/P(B)$. Then we have $P(A|B) = P_B(A)$.

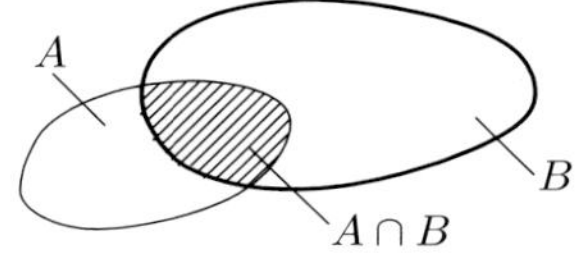


*Figure 6.8.*

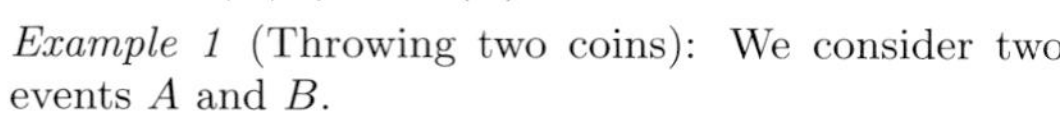

*Example 1* (Throwing two coins): We consider two events $A$ and $B$.

$A$: Both coins are heads.

$B$: The first coin shows heads.

Then we have

$$\boxed{P(A) = \frac{1}{4}, \qquad P(A|B) = \frac{1}{2}.}$$

(i) Intuitive determination of the probabilities: the outcomes of the experiment (elementary events) are given by

$$HH\,,\ HT\,,\ TH\,,\ TT\,.$$

This means that, for example, $HT$ denotes the event in which the first count is heads, the second tails, and so forth. We have

$$A = \{HH\}\,, \qquad B = \{HH\,,\ HT\}\,.$$

From this it follows that $P(A) = 1/4$. If one knows that $B$ has already occurred, then only $HH$ and $HT$ have a non-vanishing conditional probability. This means $P(A|B) = 1/2$.

(ii) Using the definition (6.11): From $A \cap B = \{HH\}$ and $P(A \cap B) = 1/4$ as well as $P(B) = 1/2$, we get

$$P(A|B) = \frac{P(A \cap B)}{P(B)} = \frac{1}{2}\,.$$

It is important to make a clear distinction between normal probabilities and conditional probabilities.

**The law of total probability:** If we have the conditions

$$E = \bigcup_{j=1}^{n} B_j \quad \text{with} \quad B_j \cap B_k = \emptyset \quad \text{for all} \quad j \neq k \tag{6.12}$$

then we have for every event $A$ the relation

$$\boxed{P(A) = \sum_{j=1}^{n} P(B_j)P(A|B_j)\,.}$$

*Example 2:* We draw a marble from one of two identical urns.

(i) The first urn contains one white and four black marbles.

(ii) The second urn contains one white and two black marbles.

We consider the following events:

$A$: The marble we have drawn is black.

$B$: The marble we have drawn comes from the $j^{th}$ urn.

For the probability $P(A)$ of drawing a black marble we have

$$P(A) = P(B_1)P(A|B_1) + P(B_2)P(A|B_2) = \frac{1}{2}\cdot\frac{4}{5} + \frac{1}{2}\cdot\frac{2}{3} = \frac{11}{15}\,.$$

**The theorem of Bayes (1763):** Let $P(A) \neq 0$. Then under the assumption (6.12) we have

$$\boxed{P(B_j|A) = \frac{P(B_j)P(A|B_j)}{P(A)}\,.}$$

*Example 3:* In Example 2 suppose we have already drawn a black marble. What is the probability that it comes from the first urn?

From $P(B_1) = 1/2$, $P(A|B_1) = 4/5$ and $P(A) = 11/15$, we have

$$P(B_1|A) = \frac{P(B_1)P(A|B_1)}{P(A)} = \frac{6}{11}\,.$$

#### 6.2.1.2 Independent events

One of the most important problems for probability theory is to give a precise mathematical definition to the intuitive notion of events being independent.

**Definition:** Two events $A$ and $B$ in a field of events $E$ are said to be *independent*, if

$$\boxed{P(A \cap B) = P(A)P(B).}$$

Similarly, $n$ events $A_1, \ldots, A_n$ of $E$ are independent if the product property

$$\boxed{P(A_{j_1} \cap A_{j_2} \cap \cdots \cap A_{j_m}) = P(A_{j_1})P(A_{j_2})\cdots P(A_{j_m})}$$

holds for all possible $m$-tuples of indices $j_1 < j_2 < \cdots < j_m$ and all $m = 2, \ldots, n$.

**Theorem:** Let $P(B) \neq 0$. Then the events $A$ and $B$ are independent if and only if their conditional probabilities satisfy

$$\boxed{P(A|B) = P(A).}$$

**Motivation:** In our daily life one often works with relative frequencies instead of probabilities. We expect that from $n$ cases of an event $A$ (resp. $B$) the occurrence will have a relative frequency of $nP(A)$ (resp. $nP(B)$).

If $A$ and $B$ are independent, then our intuition tells us that the event '$A$ and $B$ both occur' has the relative frequency $(nP(A)) \cdot P(B)$.

*Example:* We roll a pair of dice and consider the following events.

$A$: The first of the dice shows a '1'.

$B$: The second of the dice shows a '3' or a '6'.

There are 36 elementary events

$$(i, j)\,, \qquad i, j = 1, \ldots, 6\,.$$

Here $(i, j)$ denotes the event that the first die has a $i$ and the other has a $j$. The events $A, B$ and $A \cap B$ are associated to the following elementary events

$$\begin{array}{ll} A: & (1,1)\,,\ (1,2)\,,\ (1,3)\,,\ (1,4)\,,\ (1,5)\,,\ (1,6)\,. \\ B: & (1,3)\,,\ (2,3)\,,\ (3,3)\,,\ (4,3)\,,\ (5,3)\,,\ (6,3)\,, \\ & (1,6)\,,\ (2,6)\,,\ (3,6)\,,\ (4,6)\,,\ (5,6)\,,\ (6,6)\,. \\ A \cap B: & (1,3)\,,\ (1,6)\,. \end{array}$$

Therefore $P(A) = 6/36 = 1/6$, $P(B) = 12/36 = 1/3$ and $P(A \cap B) = 2/36 = 1/18$. In fact we have $P(A \cap B) = P(A)P(B)$, i.e., these events are independent.

### 6.2.2 Random variables

In this section we introduce the notion of random variable, which is used to model measured quantities which have random character, like for example the height of people.

#### 6.2.2.1 Basic ideas

Let $E = \{e_1, \ldots, e_n\}$ be a finite field of events with probabilities $p_1, \ldots, p_n$ for the outcomes $e_1, \ldots, e_n$. A *random function* on $E$ is a function

$$\boxed{X : E \longrightarrow \mathbb{R}}$$

which maps each elementary event $e_j$ to a real number $X(e_j) := x_j$. Making a measurement of $X$ will yield the value $x_j$ with probability $p_j$. The important quantities of this function are the *mean* $\overline{X}$ and the *square of the standard deviation* $(\Delta X)^2$, given as follows:

$$\boxed{\overline{X} := \sum_{j=1}^{n} x_j p_j\,, \qquad (\Delta X)^2 := \overline{\left(X - \overline{X}\right)^2} = \sum_{j=1}^{n} \left(x_j - \overline{X}\right)^2 p_j\,.}$$

The square of the standard deviation is more commonly referred to as the *variance*. The quantity $\Delta X = \sqrt{(\Delta X)^2}$ is called the *standard deviation*, as it is the deviation of the measured value from the expected one.

From the Chebychev inequality we get as a special case the following statement, which explains the importance of the variation and the standard deviation: upon taking a measurement of $X$, the probability of it lying in the interval

$$\boxed{[\overline{X} - 4\Delta X\, ,\, \overline{X} + 4\Delta X]} \tag{6.13}$$

is greater than 0.93 (see 6.2.2.4).

*Example:* An (imaginary) cassino allows the player to throw a die and pays out the amount listed in Table 6.5. Negative (resp. positive) amounts are the wins (losses) for

*Table 6.5. Win/loss in a casino game.*

| ***Number thrown by player*** | 1 | 2 | 3 | 4 | 5 | 6 |
|---|---|---|---|---|---|---|
| ***Amount paid out ($)*** | 1 | 2 | 3 | −4 | −5 | −6 |
| $x_j$ | $x_1$ | $x_2$ | $x_3$ | $x_4$ | $x_5$ | $x_6$ |

the cassino. Daily the game is played 10,000 times.

*What are the average daily earnings for the cassino?*

*Answer:* We construct a field of events

$$E = \{e_1, \ldots, e_n\}\,.$$

Here $e_i$ is the event that an '$i$' is thrown. Moreover, we set

$$X(e_j) := \text{earnings for the cassino when a } j \text{ is thrown.}$$

As mean we obtain

$$\overline{X} = \sum_{j=1}^{6} x_j p_j = (x_1 + \ldots + x_6)\frac{1}{6} = -1.5\,.$$

Thus the cassino makes an average daily earnings of $\$\,1.5 \cdot 10{,}000 = \$\,15{,}000$. However, since the standard deviation $\Delta X = 3.6$ is quite large, the profit for the cassino can change drastically from day to day. Generally this would motivate the cassino owner to offer a more advantageous game.

The importance of the fundamental notion of mean (expectation) $\overline{X}$ was realized during games of chance played in the seventeenth century. An important role in this was played by a famous correspondence between two of the best mathematicians of the day, Pascal (1623–1662) and Fermat (1601–1665).

#### 6.2.2.2 The distribution function

**Definition:** Let $(E, \mathscr{S}, P)$ be a probability space with probability measure $P$. A *random variable* on $E$ is a function $X : E \longrightarrow \mathbb{R}$ such that for each $x \in \mathbb{R}$, the set

$$A_x := \{e \in E : X(e) < x\}$$

is an event.[8] Thus, the distribution function

$$\boxed{\Phi(x) := P(X < x)}$$

is well-defined. Here $P(X < x)$ stands for $P(A_x)$.

**Strategy:** We reduce the investigation of random variables completely to the investigation of distributions.

**Intuitive interpretation of the distribution function:** Suppose that the real axis has been endowed with a mass distribution, such that the total mass of the axis is 1. The value $\Phi(x)$ of the distribution function tells us how much of the mass is contained in the interval $J := ]-\infty, x[$. This mass is also the probability that the measured value of $X$ lies in $J$.

> The larger $\Phi(x)$ is, the larger the probability that the measured value of $X$ lies in the interval $]-\infty, x[$.

*Example 1:* If the point $x_1$ has the mass $p = 1$, then the corresponding distribution function looks like Figure 6.9.

*Example 2:* If we have masses $p_1$ and $p_2$ at the points $x_1$ and $x_2$, and if $p_1 + p_2 = 1$, then the distribution function looks like that in Figure 6.10.

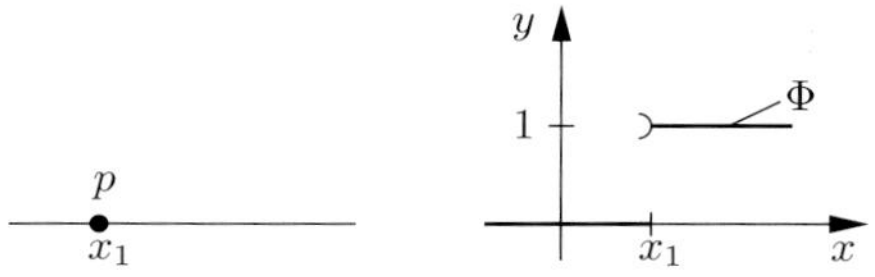

*Figure 6.9. A point p with unit mass.*

More explicitly we have

$$\Phi(x) = \begin{cases} 0 & \text{for } x \leq x_1 \\ p_1 & \text{for } x_1 < x \leq x_2 \\ p_1 + p_2 = 1 & \text{for } x_2 < x\,. \end{cases}$$

*Example 3:* If the distribution function $\Phi : \mathbb{R} \longrightarrow \mathbb{R}$ is continuously differentiable, then the derivative

$$\boxed{\varphi(x) := \Phi'(x)}$$

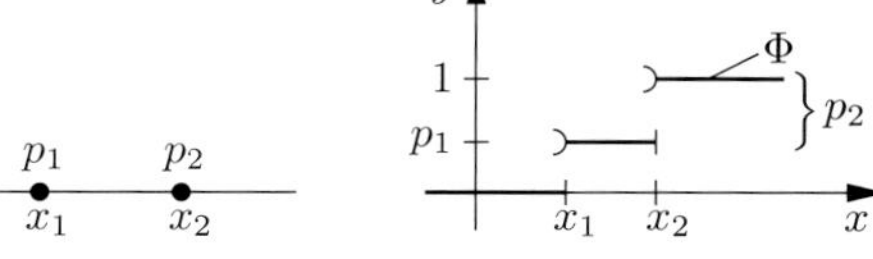


*Figure 6.10. A two point distribution.*

is a continuous mass density $\varphi : \mathbb{R} \longrightarrow \mathbb{R}$, and one has

$$\boxed{\Phi(x) = \int_{-\infty}^{x} \varphi(\xi)\, \mathrm{d}\xi\,, \qquad x \in \mathbb{R}\,.}$$

The mass which is contained in the interval $[a, b]$ is equal to the surface area of the hatched area in Figure 6.11.

The function $\varphi$ is called the *mass density* (or *probability density*). A standard example is the Gaussian normal distribution:[9]

$$\boxed{\varphi(x) := \frac{1}{\sigma\sqrt{2\pi}} \mathrm{e}^{-(x-\mu)^2/2\sigma^2}\,.}$$

[8] $X$ is a random variable if and only if the inverse image $X^{-1}(M)$ for every set $M$ in the Borel algebra $\mathscr{B}(\mathbb{R})$ represents an event.

[9] The relation between the distribution and the density was explained in section 6.1.4.

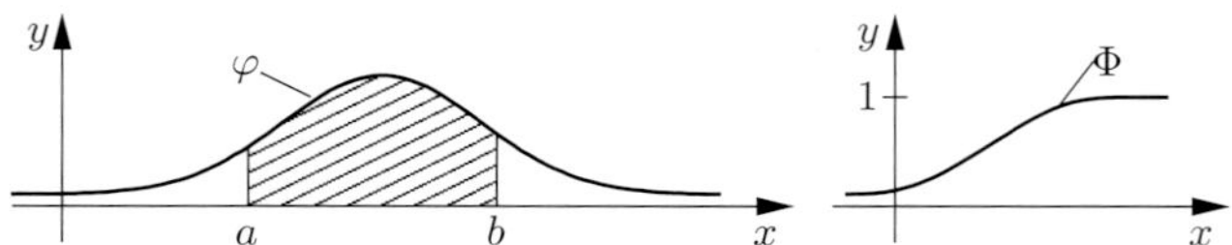


*Figure 6.11. The probability density and distribution viewed as mass functions.*

**Discrete and continuous random variables:** A random variable is said to be *discrete*, if the corresponding distribution only takes on finitely many values.

On the other hand, $X$ is said to be a *continuous* random variable, if the distribution function is continuously differentiable as in Example 3.

We set $\Phi(x \pm 0) := \lim\limits_{t \to x \pm 0} \Phi(t)$.

**Theorem 1:** A distribution function $\Phi : \mathbb{R} \longrightarrow \mathbb{R}$ has the following properties.

(i) $\Phi$ is an increasing function and is continuous from the left, i.e., one has $\Phi(x-0) = \Phi(x)$ for all $x \in \mathbb{R}$.

(ii) $\lim\limits_{x \to -\infty} \Phi(x) = 0$ and $\lim\limits_{x \to +\infty} \Phi(x) = 1$.

**Theorem 2:** For all real numbers $a$, $b$ with $a < b$ we have:

(i) $P(a \leq X < b) = \Phi(b) - \Phi(a)$ .

(ii) $P(a \leq X \leq b) = \Phi(b+0) - \Phi(a)$ .

(iii) $P(X = a) = \Phi(a+0) - \Phi(a-0)$ .

**The Stieltjes integral:** For the calculation with random variables, the *Stieltjes integral*

$$S := \int_{-\infty}^{\infty} f(x)\,\mathrm{d}\Phi(x)$$

is an instrument of fundamental importance (see 6.2.2.3). This integral is the integral over a measure formed by the mass density corresponding to $\Phi$ on the real line. Intuitively, we have approximately

$$S = \sum_j f(x_j)\Delta m_j \,.$$

This means that we subdivide the real line into intervals $[x_j, x_{j+1}[$ with the mass $\Delta m_j$, form the product $f(x_j)\Delta m_j$ and sum over all these factors along the real line (Figure 6.12). Finally, we pass to a limit by allowing the intervals to shrink in size. Thus, we have

$$\boxed{\int_{-\infty}^{\infty} \mathrm{d}\Phi = \text{total mass on } \mathbb{R} = 1 \,.}$$

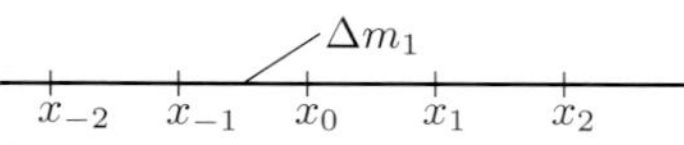


*Figure 6.12.*

The rigorous definition of the Stieltjes integral can be found in [212]. For most practical purposes, the following results are sufficient.

**The calculation of the Stieltjes integral:** Assume we are given a continuous function $f : \mathbb{R} \longrightarrow \mathbb{R}$.

(i) If the distribution function $\Phi : \mathbb{R} \longrightarrow \mathbb{R}$ is differentiable, then we have

$$\boxed{\int_{-\infty}^{\infty} f(x)\,\mathrm{d}\Phi = \int_{-\infty}^{\infty} f(x)\Phi'(x)\,\mathrm{d}(x)\,,}$$

if the classical integral on the right-hand side converges.

(ii) If $\Phi$ only takes on finitely many values, then we have

$$\boxed{\int_{-\infty}^{\infty} f(x)\,\mathrm{d}\Phi = \sum_{j=1}^{n} f(x_j)(\Phi(x_j+0) - \Phi(x_j-0))\,,}$$

where the sum is over all points of discontinuity $x_j$ of $\Phi$.

(iii) If $\Phi$ only takes on countably many values, and for the points of discontinuity $x_j$ we have the relation $\lim\limits_{n\to\infty} x_j = +\infty$, then

$$\boxed{\int_{-\infty}^{\infty} f(x)\,\mathrm{d}\Phi = \sum_{j=1}^{\infty} f(x_j)(\Phi(x_j+0) - \Phi(x_j-0))\,,}$$

if the infinite series on the right-hand side converges.

(iv) If the distribution function $\Phi$ is differentiable except for finitely many points of discontinuity $x_1, \ldots, x_n$, then

$$\boxed{\int_{-\infty}^{\infty} f(x)\,\mathrm{d}\Phi = \int_{-\infty}^{\infty} f(x)\Phi'(x)\,\mathrm{d}x + \sum_{j=1}^{n} f(x_j)(\Phi(x_j+0) - \Phi(x_j-0))\,,}$$

provided the integral on the right-hand side converges.

#### 6.2.2.3 The expectation value (mean)

The expectation value is the most important quantity associated with a random variable. All other quantities of importance (for example, standard deviation, higher moments, correlation coefficients, covariance) can be obtained by the construction of appropriate expectations from this one.

**Definition:** The *expectation* (*mean*) of a random variable $X : E \longrightarrow \mathbb{R}$ is given by

$$\boxed{\overline{X} = \int_E X(e)\,\mathrm{d}P} \qquad (6.14)$$

if this integral exists.

This integral is to be understood in the sense of an abstract measure-theoretic integral (see [212]). However, it can be reduced to the Stieltjes integral with respect to the distribution function $\Phi$ of $X$. For this we have

$$\boxed{\overline{X} = \int_{-\infty}^{\infty} x \, \mathrm{d}\Phi \, .}$$

**Intuitive meaning:** The expectation $\overline{X}$ is the center of mass of the mass distribution associated with $\Phi$.

**Calculations:** We have the following rules concerning calculations with the expectation.

(i) *Additivity:* If $X$ and $Y$ are random variables on $E$, then we have

$$\boxed{\overline{X+Y} = \overline{X} + \overline{Y} \, .}$$

(ii) *Functions of random variables:* Let $X : E \longrightarrow \mathbb{R}$ be a random variable with distribution function $\Phi$. If $F : \mathbb{R} \longrightarrow \mathbb{R}$ is a continuous function, then the composed function $Z := F(X)$ is also a random variable on $E$ with the expectation

$$\boxed{\overline{Z} = \int_E F(X(e)) \, \mathrm{d}P = \int_{-\infty}^{\infty} F(x) \, \mathrm{d}\Phi \, ,}$$

provided the integral on the right-hand side converges.

#### 6.2.2.4 The variance and Chebychev's inequality

**Definition:** If $X : E \longrightarrow \mathbb{R}$ is a random variable, then we define the *variance* of $X$ by

$$\boxed{(\Delta X)^2 := \overline{(X - \overline{X})^2} \, .}$$

If $\Phi$ denotes the distribution function of $X$, then we have

$$(\Delta X)^2 = \int_E (X(e) - \overline{X})^2 \, \mathrm{d}P = \int_{-\infty}^{\infty} (x - \overline{X})^2 \, \mathrm{d}\Phi \, ,$$

if the integral on the right-hand side converges.

The *standard deviation* $\Delta X$ of $X$ is defined by

$$\Delta X := \sqrt{(\Delta X)^2}.$$

*Example 1* (continuous random variable): If $\Phi$ has a continuous derivative $\varphi = \Phi'$ on $\mathbb{R}$, then

$$\boxed{\overline{X} = \int_{-\infty}^{\infty} x\varphi(x) \, \mathrm{d}x \, , \qquad (\Delta X)^2 = \int_{-\infty}^{\infty} (x - \overline{X})^2 \varphi(x) \mathrm{d}x \, .}$$

*Example 2* (discrete random variable): If $X$ only attains finitely many values $x_1, \ldots, x_n$, and if we set $p_j := P(X = x_j)$, then

$$\overline{X} = \sum_{j=1}^{n} x_j p_j\,, \qquad (\Delta X)^2 = \sum_{j=1}^{n} (x_j - \overline{X})^2 p_j\,.$$

**Chebychev's inequality:** If $X : E \longrightarrow \mathbb{R}$ is a random variable with $\Delta X < \infty$, then for every real number $\beta > 0$ we have the fundamental inequality

$$P\left(|X - \overline{X}| > \beta \Delta X\right) \leq \frac{1}{\beta^2}\,.$$

In particular, for $\Delta X = 0$ we have $P(X = \overline{X}) = 1$, i.e., the value of $X$ is almost surely its expected value.

**Applications to confidence intervals:** If we choose are real number $\alpha$ with $0 < \alpha < 1$, then the observed values of $X$ lie with probability $> 1 - \alpha$ in the interval

$$\left[\overline{X} - \frac{\Delta X}{\sqrt{\alpha}}\,, \overline{X} + \frac{\Delta X}{\sqrt{\alpha}}\right].$$

*Example 3* ($4\Delta X$-rule): Let $\alpha = 1/16$. With a probability greater than 0.93, all observed values of $X$ lie in the interval

$$\left[\overline{X} - 4\Delta X\,, \overline{X} + 4\Delta X\right].$$

**Moments of a random variable:** The expectation

$$\alpha_k := \overline{X^k}\,, \qquad k = 0, 1, 2, \ldots$$

of the $k^{th}$ power of $X$ is called the $k^{th}$ *moment of* $X$. If $\Phi$ denotes the distribution of $X$, then we have

$$\mu_k = \int_E X^k \,\mathrm{d}P = \int_{-\infty}^{\infty} x^k \,\mathrm{d}\Phi\,.$$

The famous 'moment problem' is: do the values of all the moments of $X$ determine the distribution function uniquely? Under appropriate assumptions, there is an affirmative answer (see [212]).

## 6.2.3 Random vectors

In order to deal with series of measured (observed) values of a random variable in the context of mathematical statistics, it is necessary to consider vectors $(X_1, \ldots, X_n)$ whose components $X_j$ are random variables. Intuitively, $X_j$ is then the measured value of $X$ in the $j^{th}$ trial.

### 6.2.3.1 The joint distribution

**Definition:** Let $(E, \mathscr{S}, P)$ be a probability space. A *random vector* $(X, Y)$ on $E$ is a pair of functions $X, Y : E \longrightarrow \mathbb{R}$, such that for each pair $(x, y)$ of real numbers, the set

$$A_{x,y} := \{e \in E : X(e) < x,\, Y(e) < y\}$$

is an event. If this is the case, the distribution function

$$\boxed{\Phi(x,y) := P(X < x,\, Y < y)}$$

is well-defined. Here $P(X < x, Y < y)$ stands for $P(A_{x,y})$. We call this the *joint distribution function* of $X$ and $Y$.

**Strategy:** As above we reduce the investigation of random vectors to the investigation of distribution functions.

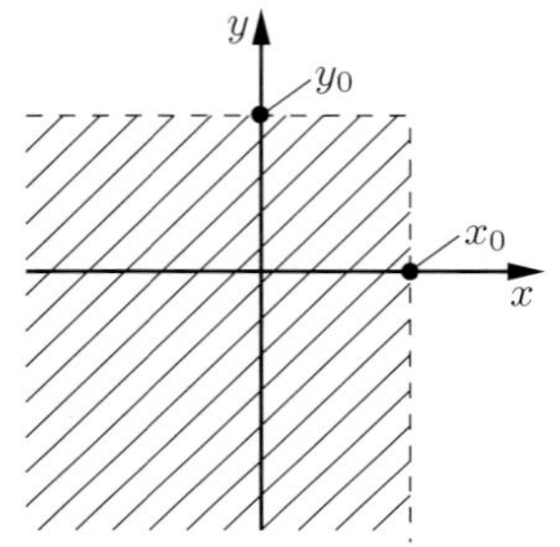


*Figure 6.13.*

**Intuitive interpretation of the distribution function:** Suppose that the plane is endowed with a mass density with a total mass of unity. The value $\Phi(x_0, y_0)$ of the distribution function describes the amount of mass which is contained in the set

$$\{(x,y) \in \mathbb{R}^2 : x < x_0\,,\ y < y_0\}$$

(Figure 6.13). This mass is equal to the probability that the measured values of $X$ and $Y$ are in the corresponding open intervals $]-\infty, x_0[$ and $]-\infty, y_0[$.

**Theorem:** The components $X$ and $Y$ of a random vector are random variables with distribution functions

$$\Phi_X(x) = \lim_{y\to+\infty} \Phi(x,y)\,, \qquad \Phi_Y(y) = \lim_{x\to+\infty} \Phi(x,y)\,.$$

**Probability density:** If there is a non-negative continuous function $\varphi : \mathbb{R} \longrightarrow \mathbb{R}$ with $\int_{\mathbb{R}^2} \varphi(x,y)\mathrm{d}x\mathrm{d}y = 1$ and

$$\Phi(x,y) = \int_{-\infty}^{x}\int_{-\infty}^{y} \varphi(\xi,\eta)\,\mathrm{d}\xi\mathrm{d}\eta\,, \qquad x, y \in \mathbb{R}\,,$$

then we call $\varphi$ a (joint) *probability density* of the random vector $(X, Y)$. In this case the variables $X$ and $Y$ have probability densities with

$$\varphi_X(x) := \int_{-\infty}^{\infty} \varphi(x,y)\,\mathrm{d}y\,, \qquad \varphi_Y(y) := \int_{-\infty}^{\infty} \varphi(x,y)\,\mathrm{d}x\,.$$

**Random vectors $(X_1, \ldots, X_n)$:** All of these considerations can be easily extended to the case of vectors of $n$ components.

### 6.2.3.2 Independent random variables

**Definition:** Two random variables $X, Y : E \longrightarrow \mathbb{R}$ are said to be *independent*, if $(X, Y)$ is a random vector which has the following product property:

$$\boxed{\Phi(x,y) = \Phi_X(x)\Phi_Y(y) \quad \text{for all} \quad x, y \in \mathbb{R}.} \tag{6.15}$$

Here $\Phi$, $\Phi_X$ and $\Phi_Y$, respectively, denote the distribution functions of $(X, Y)$, $X$ and $Y$, respectively.

**Rules for calculations:** For independent random variables $X$ and $Y$ we have:

(i) $\overline{XY} = \overline{Y}\,\overline{X}$.

(ii) $(\Delta(X+Y))^2 = (\Delta X)^2 + (\Delta Y)^2$.

(iii) The correlation coefficient (see next paragraph) $r$ of $X$ and $Y$ vanishes.

(iv) If $J$ and $K$ are real intervals, then we have

$$P(X \in J, Y \in K) = P(X \in J)P(Y \in K).$$

**Theorem:** If a random variable $(X, Y)$ has a continuous distribution density $\varphi$, then $X$ and $Y$ are independent if and only if there is a product representation

$$\boxed{\varphi(x, y) = \varphi_X(x)\varphi_Y(y),}$$

valid for all $x, y \in \mathbb{R}$.

**Dependence of random quantities:** Practical applications will often lead one to suspect that two given random variables $X$ and $Y$ are in some way dependent on one another. There are two methods of mathematically verifying whether this is the case.

(i) Use the correlation coefficient (see 6.2.3.3).

(ii) Use the regression line (see 6.2.3.4).

#### 6.2.3.3 Dependent random variables and the correlation coefficient

**Definition:** For a random vector $(X, Y)$ we define the *covariance*

$$\operatorname{Cov}(X, Y) := \overline{(X - \overline{X})(Y - \overline{Y})}$$

and the *correlation coefficient*

$$\boxed{r := \frac{\operatorname{Cov}(X, Y)}{\Delta X \Delta Y}\,.}$$

We always have the relation $-1 \leq r \leq 1$.

**Definition:** The larger $r^2$ is, the *stronger* $X$ and $Y$ are *correlated.*

**Motivation:** The minimization problem

$$\boxed{\overline{(Y - a - bX)^2} \overset{!}{=} \text{min.}, \quad a, b \in \mathbb{R}}$$

has the so-called *regression line*:

$$\overline{Y} + r\frac{\Delta Y}{\Delta X}(X - \overline{X})$$

with a minimal value at $(\Delta Y)^2(1 - r^2)$ as solution (see the discussion in section 6.1.5).

For the covariance we have

$$\boxed{\operatorname{Cov}(X, Y) := \int_E (X(e) - \overline{X})(Y(e) - \overline{Y})\,\mathrm{d}P = \int_{\mathbb{R}^2} (x - \overline{X})(y - \overline{Y})\,\mathrm{d}\Phi\,,}$$

where $\Phi$ denotes the joint distribution of $X$ and $Y$.

*Example 1* (discrete random variable): If $\Phi$ only attains finitely many values $(x_j, y_k)$ with probabilities $p_{jk} := P(X = x_j,\, Y = y_k)$, then we have

$$\operatorname{Cov}(X,Y) = \sum_{j=1}^{n}\sum_{k=1}^{m}(x_j - \overline{X})(y_k - \overline{Y})p_{jk}$$

with

$$\overline{X} = \sum_{j=1}^{n} x_j p_j\,, \quad (\Delta X)^2 = \sum_{j=1}^{n}(x_j - \overline{X})^2 p_j\,, \quad p_j := \sum_{k=1}^{m} p_{jk}\,,$$

and

$$\overline{Y} = \sum_{k=1}^{m} y_k q_k\,, \quad (\Delta Y)^2 = \sum_{k=1}^{m}(x_k - \overline{Y})^2\, q_k\,, \quad q_k := \sum_{j=1}^{n} p_{jk}\,.$$

*Example 2:* If $(X,Y)$ has a continuous probability density $\varphi$, then $\operatorname{Cov}(X,Y)$ and $r$ are calculated as in section 6.1.5.

**The covariance matrix:** If we are given a random vector $(X_1, \ldots, X_n)$, then the $(n \times n)$-*covariance matrix* is defined as $C = (c_{jk})$, where $c_{jk}$ is given by

$$\boxed{c_{jk} := \operatorname{Cov}(X_j\,, X_k)\,, \qquad j,k = 1,\ldots,n\,.}$$

This matrix is symmetric, and all eigenvalues are non-negative.

**Interpretation:** (i) $c_{jj} = (\Delta X_j)^2$, $j = 1, \ldots, n$.

(ii) For $j \neq k$ the number

$$r_{jk}^2 := \frac{c_{jk}^2}{c_{jj}c_{kk}}$$

is the square of the correlation coefficient of $X_j$ and $X_k$.

(iii) If $(X_1, \ldots, X_n)$ are independent, then we have $c_{jk} = 0$ for all $j \neq k$, i.e., the covariance matrix is a diagonal matrix, the entries of which are the standard deviations of the individual variables.

**The general Gauss distribution:** Let $A$ be a real, symmetric, positive definite $(n\times n)$-matrix. Then the probability density

$$\boxed{\varphi(x) := K\mathrm{e}^{-Q(x,x)}\,, \qquad x \in \mathbb{R}^n}$$

with $Q(x,x) := \frac{1}{2}x^{\mathsf{T}}Ax$ and $K^2 := \frac{\det A}{(2\pi)^n}$ defines by definition the *general Gauss distribution* of the random vector $(X_1, \ldots, X_n)$ with the covariance matrix

$$\boxed{(\operatorname{Cov}(X_j, X_k)) = A^{-1}}$$

and the expectations $\overline{X}_j = 0$ for all $j$.

If $A = \operatorname{diag}(\lambda_1, \ldots, \lambda_n)$ is a diagonal matrix with eigenvalues $\lambda_j$, then the random variables $X_1, \ldots, X_n$ are independent. Moreover, we have

$$\operatorname{Cov}(X_j, X_k) = \begin{cases} (\Delta X_j)^2 = \lambda_j^{-1} & \text{for } j = k\,, \\ 0 & \text{for } j \neq k\,. \end{cases}$$

#### 6.2.3.4 The dependency curve between two random variables

**Conditional distributions:** Let $(X, Y)$ be a random vector. We fix a real number $x$ and set

$$\Phi_x(y) := \lim_{h \to +0} \frac{P(x \leq X < x+h, Y < y)}{P(x \leq X < x+h)} \quad \text{for all} \quad y \in \mathbb{R}.$$

If this limit exists, $\Phi_x$ is called the *conditional distribution* of the random variable $Y$ under the assumption that $X$ takes on the value $x$.

**Dependency curve (regression curve):** The curve defined by

$$\overline{y}(x) := \int_{-\infty}^{\infty} y \, \mathrm{d}\Phi_x(y)$$

is called the *dependency curve* (or *regression curve*) of the random variable $Y$ with respect to the random variable $X$.

**Interpretation:** The number $\overline{y}(x)$ is the expectation of $Y$ under the assumption that $X$ takes on the value $x$ (Figure 6.14). If at $x = x_0$ there are observed values $y_1, \ldots, y_n$ for $Y$, then one can take

$$\frac{y_1 + \ldots + y_n}{n}$$

as an approximation for $\overline{y}(x_0)$ (Figure 6.14).

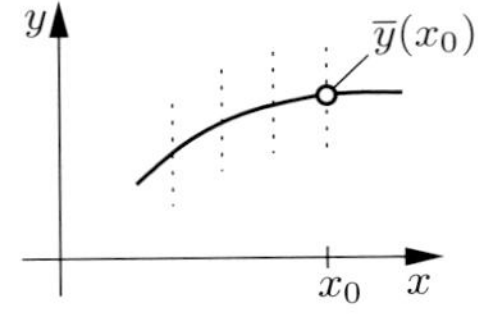


*Figure 6.14.*

**Probability density:** If $(X, Y)$ has a continuous probability density $\varphi$, then we have

$$\Phi_x(y) = \frac{\int_{-\infty}^{y} \varphi(x, \eta) \mathrm{d}\eta}{\int_{-\infty}^{\infty} \varphi(x, y) \mathrm{d}y}$$

and

$$\overline{y}(x) = \frac{\int_{-\infty}^{\infty} y\varphi(x, y) \, \mathrm{d}y}{\int_{-\infty}^{\infty} \varphi(x, y) \, \mathrm{d}y}.$$

### 6.2.4 Limit theorems

Limit theorems generalize the classical law of large numbers due to Jakob Bernoulli in 1713 and are among the most important results of all in probability theory.

#### 6.2.4.1 The weak law of large numbers

**Theorem of Chebychev (1867):** Let $X_1, X_2, \ldots$ be independent random variables on a probability space. We set

$$Z_n := \frac{1}{n}\sum_{j=1}^{n}(X_j - \overline{X}_j)\,.$$

If the standard deviation are uniformly bounded (i.e., $\sup\limits_n \Delta X_n < \infty$), then we have the relation

$$\boxed{\lim_{n\to\infty} P(|Z_n| < \varepsilon) = 1} \tag{6.16}$$

for arbitrary small numbers $\varepsilon > 0$.

This theorem is a generalization of the law of large numbers as just mentioned (see 6.2.5.7).

#### 6.2.4.2 The strong law of large numbers

**Theorem of Kolmogorov (1930):** Let $X_1, X_2, \ldots$ be independent random variables on a probability space, whose standard deviations satisfy

$$\sum_{n=1}^{\infty}\frac{(\Delta X_n)^2}{n^2} < \infty$$

(for example $\sup\limits_n \Delta X_n < \infty$). Then the following limit relation

$$\boxed{\lim_{n\to\infty} Z_n = 0} \tag{6.17}$$

holds almost surely.[10] Moreover, (6.16) is a consequence of (6.17).

A weaker statement had already been proved in 1905 by Borel and in 1917 by Cantelli.

**The importance of the expectation:** If the assumptions of the Theorem of Kolmogorov are satisfied and if $\overline{X}_j = \mu$ for all $j$, then almost surely we have

$$\boxed{\lim_{n\to\infty}\frac{1}{n}\sum_{j=1}^{n} X_j = \mu\,.}$$

#### 6.2.4.3 The central limit theorem

Let $X_1, X_2, \ldots$ be independent random variables on a probability space. We set

$$Y_n := \frac{1}{\Delta_n\sqrt{n}}\sum_{j=1}^{n}(X_j - \overline{X}_j)$$

[10] If $A$ denotes the set of all elementary events $e$ with

$$\lim_{n\to\infty} Z_n(e) = 0,$$

then $P(A) = 1$.

with the average standard deviation $\Delta_n := \left(\frac{1}{n}\sum_{j=1}^{n}(\Delta X_j)^2\right)^{1/2}$.

**Central Limit Theorem:**[11] The following two conditions are equivalent:

(i) The distribution function $\Phi_n$ of $Y_n$ converges for $n \to \infty$ to the normal distribution $N(0,1)$, i.e., we have

$$\boxed{\lim_{n\to\infty} \Phi_n(x) = \frac{1}{\sqrt{2\pi}} \int_{-\infty}^{x} \mathrm{e}^{-t^2/2} \mathrm{d}t \quad \text{for all } x \in \mathbb{R}.}$$

(ii) The distribution function $\Phi_n$ of $X_n$ satisfies the Lindeberg condition:

$$\lim_{n\to\infty} \frac{1}{n\Delta_n^2} \sum_{j=1}^{n} \int\limits_{|x-\overline{X}_j|>\tau n\Delta_n} (x-\overline{X}_j)^2 \, \mathrm{d}\Phi_j(x) = 0 \tag{L}$$

for all $\tau > 0$.

**Remark:** The Lindeberg condition (L) is satisfied if all $X_j$ have the same distribution $\Phi$ with mean $\mu$ and standard deviation $\sigma$. In that case (L) is equivalent with

$$\lim_{n\to\infty} \frac{1}{\sigma^2} \int\limits_{|x-\mu|>n\tau\sigma} (x-\mu)^2 \, \mathrm{d}\Phi(x) = 0 \, .$$

Condition (L) is also satisfied if all distribution functions $\Phi_k$ of the $X_k$ have a similar structure at infinity and regarding expectations and standard deviations.

**The importance of the central limit theorem:** The central limit theorem is the most important result in probability theory. It explains why the Gaussian normal distribution occurs so often. The central limit theorem also makes the following intuitive principle mathematically rigorous:

| If a random variable $X$ is a superposition of many other random variables, which are treated on an equal basis, then $X$ is normally distributed. |
|---|

## 6.2.5 Applications to the Bernoulli model for successive independent trials

Jakob Bernoulli developed the following model, which is applicable in many situations. It is among the most important models used in probability theory, and in particular is of use in deriving a relationship between theoretical probabilities and relative frequencies.

### 6.2.5.1 The basic idea

**Intuitive situation:**

[11]This fundamental result has a long history. Among others, the following have contributed: Chebychev (1887), Markov (1898), Liapunov (1900), Lindeberg (1922) and Feller (1934).

(i) We first carry out one trial (the reference or ground trial), which has two possible outcomes,

$$\boxed{e_1\,,\ e_2.}$$

Suppose the probability of outcome $e_j$ is $p_j$. Moreover, we set $p := p_1$, so that $p_2 = 1-p$. We call $p$ the *probability of the ground trial.*

(ii) Now we carry out $n$ trials of the experiment.

(iii) All of these trials are independent, meaning that their outcomes do not influence one another.

*Example:* The reference trial is a throw of a coin, where $e_1$ denotes an occurrence of 'heads' and $e_2$ an occurrence of 'tails'. If $p = 1/2$, we think of the coin as being fair; if $p \neq 1/2$, then this is clearly not a fair coin. In 6.2.5.5 we will show how to expose a cheater by evaluating a series of trials.

#### 6.2.5.2 The probability model

**The probability space:** The total event $E$ consists of the elementary events

$$\boxed{e_{i_1 i_2 \ldots i_n}\,, \quad i_j = 1,2 \quad \text{and} \quad j = 1,\ldots,n}$$

with probabilities

$$\boxed{P(e_{i_1 i_2 \ldots i_n}) := p_{i_1} p_{i_2} \ldots p_{i_n}\,.} \tag{6.18}$$

**Interpretation:** $e_{121\ldots}$ means that the series of trials has outcomes $e_1, e_2, e_1, \ldots$ in that order. For example $P(e_{121}) = p(1-p)p = p^2(1-p)$.

**Independence of the trials:** We define an event

$$A_i^{(k)}\ :\ \text{the event } e_i \text{ occurs in the } k^{th} \text{ trial.}$$

Then the events

$$A_{i_1}^{(1)}\,,\ A_{i_2}^{(2)}\,,\ \ldots\ ,\ A_{i_n}^{(n)}$$

are independent for all possible indices $i_1, \ldots, i_n$.

*Proof:* We consider the special case $n = 2$. The event $A_i^{(1)} = \{e_{i1}, e_{i2}\}$ consists of the elementary events $e_{i1}$ and $e_{i2}$. Hence

$$P(A_i^{(1)}) = P(e_{i1}) + P(e_{i2}) = p_i p_1 + p_i p_2 = p_i\,.$$

Because of $A_j^{(2)} = \{e_{1j}, e_{2j}\}$ we have $A_i^{(1)} \cap A_j^{(2)} = \{e_{ij}\}$. From (6.18) we obtain the required product property

$$P(A_i^{(1)} \cap A_j^{(2)}) = P(A_i^{(1)}) P(A_j^{(2)})\,,$$

since on the left we have $P(e_{ij}) = p_i p_j$ and on the right we also have $p_i p_j$. □

**The relative frequency as a random variable on $E$:** We define a function $H_n : E \longrightarrow \mathbb{R}$ by

$$\boxed{H_n(e_{i_1} \ldots e_{i_n}) = \frac{1}{n} \cdot \Big(\text{number of indices of } e_{\ldots} \text{ which are equal to } 1\Big)\,.}$$

Then $H_n$ is the relative frequency for the occurrence of the event $e_1$ in the series of trials (for example the relative number of 'heads' occuring in a trial of throwing a coin).

Our objective is to investigate the random variables $H_n$.

**Theorem 1:** (i) $P\left(H_n = \frac{k}{n}\right) = \binom{n}{k} p^k (1-p)^{n-k}$ for $k = 0, \ldots, n$.

(ii) $\overline{H}_n = p$ (expectation).

(iii) $\Delta H_n = \frac{\sqrt{p(1-p)}}{\sqrt{n}}$ (standard deviation).

(iv) $P(|H_n - p| \le \varepsilon) \ge 1 - \frac{p(1-p)}{n\varepsilon^2}$ (inequality of Chebychev).

In (iv) the number $\varepsilon > 0$ must be sufficiently small. You will see that the relative frequency varies less and less from the expected value as the number of trials $n$ grows. This expectation is the probability $p$ that $e_1$ occurs in the reference trial. As mentioned above, in the case of a fair coin we have $p = 1/2$.

This probability model was considered first by Jakob Bernoulli (1654–1705). The expression in (i) of this result is not convenient for calculations. To improve this matter of affairs, de Moivre (1667–1754), Laplace (1749–1827) and Poisson (1781–1840) looked to find appropriate approximations (see section 6.2.5.3). The inequality of Chebychev (1821–1894) holds for arbitrary random variables.

**The absolute frequency:** The function $A_n := nH_n$ describes the number of occurrences of the event $e_1$ during the series of trials (for example the total number of 'heads' thrown in a series of trials).

**Theorem 2:** (i) $P(A_n = k) = P\left(H_n = \frac{k}{n}\right) = \binom{n}{k} p^k (1-p)^{n-k}$ for $k = 0, \ldots, n$.

(ii) $\overline{A}_n = n\overline{H}_n = np$ (expectation).

(iii) $\Delta A_n = \sqrt{np(1-p)}$ (standard deviation).

**The indicator function:** We define a random variable $X_j : E \longrightarrow \mathbb{R}$ by the following rule:

$$X_j(e_{i_1 \cdots i_n}) := \begin{cases} 1, & \text{if } i_j = 1, \\ 0, & \text{otherwise.} \end{cases}$$

Thus $X_j$ is 1 if and only if the event $e_1$ occurs in the $j^{th}$ trial.

**Theorem 3:** (i) $P(X_j = 1) = p$.

(ii) $\overline{X}_j = p$ and $\Delta X_j = \sqrt{p(1-p)}$.

(iii) $X_1, \ldots, X_n$ are independent.

(iv) $H_n = \frac{1}{n}(X_1 + \ldots + X_n)$.

(v) $A_n = X_1 + \ldots + X_n$.

The frequency $A_n$ is thus a superposition of equally treated independent random variables. Therefore we expect, taking the central limit theorem into consideration, that $A_n$ is nearly normally distributed for large $n$. This statement is the content of the theorem of de Moivre–Laplace.

#### 6.2.5.3 Approximation theorems

**The law of large numbers (Bernoulli):** For every $\varepsilon > 0$ we have

$$\boxed{\lim_{n\to\infty} P(|H_n - p| < \varepsilon) = 1\,.} \tag{6.19}$$

Jakob Bernoulli found this law through voluminous computations. In fact (6.10) follows from the Chebychev inequality (Theorem 1 in 6.2.5.2).

If one uses Theorem 3, (iv) of 6.2.5.2, then (6.19) is a special case of the weak law of large numbers due to Chebychev (see 6.2.4.1).

**The local limit theorem of de Moivre and Laplace:** For $n \to \infty$, the absolute frequency has the asymptotic form

$$\boxed{P(A_n = k) \sim \frac{1}{\sigma\sqrt{2\pi}}\mathrm{e}^{-(k-\mu)^2/2\sigma^2}} \tag{6.20}$$

with $\mu = \overline{A}_n = np$ and $\sigma = \Delta A_n = \sqrt{np(1-p)}$.

This means that for each $k = 0, 1, \ldots$ the quotient of the expression on the left and on the right in (6.20) approaches unity for $n \to \infty$.[12]

We now investigate the relative frequency

$$\mathscr{H}_n := \frac{H_n - \overline{H}_n}{\Delta H_n}\,.$$

Then we have $\overline{\mathscr{H}}_n = 0$ and $\Delta\mathscr{H}_n = 1$. We denote the distribution function for $\mathscr{H}_n$ by $\Phi_n$. Let $\Phi$ be the distribution function for the Gaussian normal distribution $N(0,1)$ with mean $\mu = 0$ and standard deviation $\sigma = 1$. The normalized absolute frequency

$$\mathscr{A}_n := \frac{A_n - \overline{A}_n}{\Delta A_n}$$

is equal to the normalized relative frequency $\mathscr{H}_n$ and therefore also has $\Phi_n$ as its distribution function.

**The global limit theorem of de Moivre and Laplace:**[13] For all $x \in \mathbb{R}$ we have

$$\boxed{\lim_{n\to\infty} \Phi_n(x) = \Phi(x)\,.}$$

For all intervals $[a, b]$ we obtain from this the equality

$$\lim_{n\to\infty} P(a \le \mathscr{H}_n \le b) = \frac{1}{\sqrt{2\pi}}\int_a^b \mathrm{e}^{-z^2/2}\,\mathrm{d}z\,. \tag{6.22}$$

[12] For the proof, Abraham de Moivre (1667–1754), who lived in London, used the approximation

$$n! = C\sqrt{n}\left(\frac{n}{\mathrm{e}}\right)^n, \qquad n \to \infty \tag{6.21}$$

for large $n$; the value of $C$ is approximately $C \approx 2.5074$. When de Moivre asked Stirling (1692–1770) for help, the latter found the precise value of $C = \sqrt{2\pi}$. The corresponding formula (6.21) is known as the Stirling formula.

[13] De Moivre found this formula for $p = 1/2$ and symmetrical boundaries $b = -a$. The general formula was proved by Laplace in his fundamental book *Théorie analytique des probabilités*, which appeared in 1812.

In fact, there is a very precise estimate:

$$\sup_{x\in\mathbb{R}} |\Phi_n(x) - \Phi(x)| \leq \frac{p^2 + (1-p)^2}{\sqrt{np(1-p)}}\,, \qquad n = 1, 2, \ldots \tag{6.23}$$

**Remark:** For large $n$ the relative frequency is almost normally distributed with expectation $\overline{H}_n = p$ and standard deviation $\Delta H_n = \sqrt{p(p-1)/n}$. For every interval $[a, b]$ and large $n$ we therefore have the fundamental relation

$$P(p + a\Delta H_n \leq H_n \leq p + b\Delta H_n) = \Phi_0(b) - \Phi_0(a) = \frac{1}{\sqrt{2\pi}} \int_a^b \mathrm{e}^{-z^2/2}\,\mathrm{d}z\,. \tag{6.24}$$

On the left-hand side we have the probability that the measured value of the relative frequency $H_n$ lies in the interval $[p+a\Delta H_n, p+b\Delta H_n]$. This statement gives us a precise expression of the law of large numbers of Bernoulli.

The values of $\Phi_0$ can be found in Table 0.34.

For negative $z$ we have $\Phi_0(z) = -\Phi_0(-z)$.

Formula (6.24) is equivalent to the statement

$$P(x \leq H_n \leq y) = \Phi_0\left(\frac{y-p}{\Delta H_n}\right) - \Phi_0\left(\frac{x-p}{\Delta H_n}\right).$$

The absolute frequency $A_n$ thus satisfies, because of $A_n = nH_n$, the relation

$$P(u \leq A_n \leq v) = \Phi_0\left(\frac{v-np}{\sqrt{np(1-p)}}\right) - \Phi_0\left(\frac{u-np}{\sqrt{np(1-p)}}\right).$$

Here we have $-\infty < x < y < \infty$ and $-\infty < u < v < \infty$.

**Small probabilities $p$ of the reference trial:** If the probability $p$ is very small, then formula (6.23) shows us that the approximation through a normal distribution does not take effect until very large $n$. Poisson (1781–1840) discovered that for small $p$ there is a better approximation.

**Definition of the Poisson distribution:** In the points $x = 0, 1, 2, \ldots$ on the real axis we attach masses $m_0, m_1, \ldots$, where we take

$$m_r := \frac{\lambda^r}{r!}\mathrm{e}^{-\lambda}\,, \qquad r = 0, 1, \ldots$$

The number $\lambda > 0$ is a parameter. The corresponding mass distribution function is

$$\Phi(x) := \text{mass on} \quad ]-\infty, x[$$

and is called the *Poisson distribution function* (Figure 6.15).

**Theorem:** If a random variable $x$ is Poisson distributed (i.e., its distribution function is the Poisson distribution), then we have

$$\overline{X} = \lambda \quad \text{(mean)} \quad \text{and} \quad \Delta X = \sqrt{\lambda} \quad \text{(standard deviation)}\,.$$

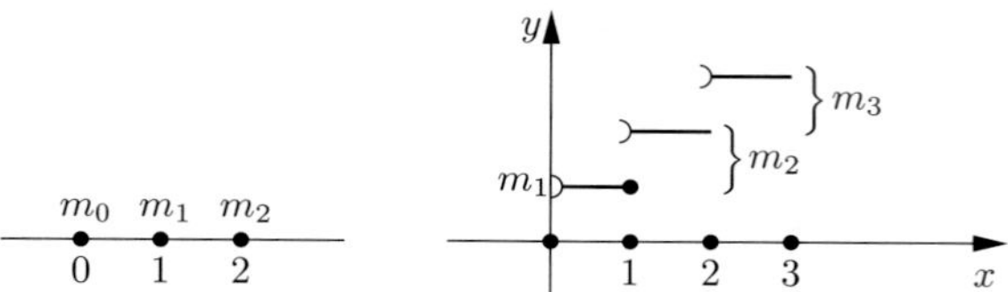


*Figure 6.15. The Poisson distribution function.*

**The approximation theorem of Poisson (1837):** If the probability $p$ of the reference trial is very small, then for the absolute frequency $A_n$ we have approximately:

(i) $P(A_n = r) = \dfrac{\lambda^r}{r!}\mathrm{e}^{-\lambda}$ with $\lambda = np$ and $r = 0, 1, \ldots, n$.

(ii) The distribution function $\Phi_n$ of $A_n$ is nearly Poisson distributed with the parameter $\lambda = np$. More precisely, we have the estimate

$$\sup_{x\in\mathbb{R}} |\Phi_n(x) - \Phi(x)| \le 3\sqrt{\frac{\lambda}{n}}\,.$$

| The values of $\dfrac{\lambda^r}{r!}\mathrm{e}^{-\lambda}$ can be found in 0.4.6.9. |
|---|

### 6.2.5.4 Applications to quality control

Suppose a factory produces a product $\mathscr{P}$ (for example light bulbs). The probability that $\mathscr{P}$ is defective is very small; let it be $p$ (for example $p = 0.001$). Suppose further that in a transport container there are $n$ pieces of the product.

(i) According to the model developed in 6.2.5.2, the probability that the container contains exactly $r$ defective parts is given by the following formula:

$$P(A_n = r) = \binom{n}{r} p^r (1-p)^{n-r}\,.$$

(ii) The probability that the number of defective parts in the container is between $k$ and $m$ is obtained from the relation

$$P(k \le A_n \le m) = \sum_{r=k}^{m} P(A_n = r)\,.$$

**Approximation:** We now wish to give a practical formula for these probabilities. For this we note that $p$ is small and apply the Poisson approximation

$$P(A_n = r) = \frac{\lambda^r}{r!}\mathrm{e}^{-\lambda}$$

with $\lambda = np$. These values are tabulated in section 0.4.6.9 of the book.

*Example 1:* Suppose we have 1000 light bulbs in our container, and that the probability of a light bulb being defective is $p = 0.001$. From 0.4.6.9 with $\lambda = np = 1$ we obtain

$$P(A_{1000} = 0) = 0.37\,,$$
$$P(A_{1000} = 1) = 0.37\,, \quad P(A_{1000} = 2) = 0.18\,.$$

It follows that

$$P(A_{1000} \leq 2) = 0.37 + 0.37 + 0.18 = 0.92\,.$$

The probability that there are no defective bulbs in the container is thus 0.37 (37%). There are at most two defective bulbs in the container with probability 0.92.

If $n$ is sufficiently large, then we can assume that $A_n$ is normally distributed. From (6.24) and the equations following it, we get

$$P(k \leq A_n \leq m) = \Phi_0\left(\frac{m - np}{\sqrt{np(1-p)}}\right) - \Phi_0\left(\frac{k - np}{\sqrt{np(1-p)}}\right).$$

The values of $\Phi_0$ can be found in Table 0.34.

*Example 2:* Suppose now that the probability that a light bulb is defective is 0.005. The probability that in a shipment of 10,000 bulbs there are at most 100 which are defective is given by the relation[14]

$$P(A_{10\,000} \leq 100) = \Phi_0(7) - \Phi_0(-7) = 2\Phi_0(7) = 1\,.$$

Thus we can be almost certain that there are at most 100 defective bulbs.

#### 6.2.5.5 Applications to testing hypothesis

It is our goal to expose a cheater using a coin which is not fair, just by making observations of a trial of experiments. For this we use a mathematical argument which *is typical for mathematical statistics.* One of the prime properties of mathematical statistics is that, in the situation at hand, it accepts that 'exposing a cheater' can only be done with a certain residual probability, say $\alpha$, that our conclusion is incorrect. This means, for example for $\alpha = 0.05$, that if we attempt to expose a cheater 100 times, then on average we will come to an incorrect conclusion about 5 times, in this case accusing a fair person of being a swindler.

**Trial for exposing an unfair coin (thrower):** We flip the coin $n$ times, and observe that 'heads' appears $k$ times. We call $h_n = k/n$ the realization of the random variable $H_n$ (the relative frequency). We denote by $p$ the probability for the appearance of 'heads'. We make the hypothesis:

(H) *The coin is fair, i.e.,* $p = 1/2$.

**Basic principle for mathematical statistics:** We will discard our hypothesis (H) with an error probability $\alpha$ if:

$$h_n \text{ does not lie in the confidence interval } \left[\frac{1}{2} - z_\alpha \Delta H_n\,, \frac{1}{2} + z_\alpha \Delta H_n\right]. \tag{6.25}$$

Here $\Delta H_n := 1/2\sqrt{n}$. The number $z_\alpha$ is determined from Table 0.34 from the equation $2\Phi_0(z_\alpha) = 1 - \alpha$. For $\alpha = 0.01$ (resp. 0.05 and 0.1) we have $z_\alpha = 1.6$ (resp. 2.0 and 2.6).

**Reasoning:** According to (6.24) the probability that the measured value $h_n$ of the theoretical quantity $H_n$ lies in the confidence interval (6.25) is for large $n$

$$\Phi_0(z_\alpha) - \Phi_0(-z_\alpha) = 1 - \alpha\,.$$

[14]The value $\Phi_0(7)$ can not be found in Table 0.34. The value is near to 0.5.

If the measured value does not lie in this confidence interval, then we discard our hypothesis, with a probability of error $\alpha$.

*Example:* For $n = 10,000$ flips of a coin we have $\Delta H_n = 1/200 = 0.005$. The confidence interval for an error probability of $\alpha = 0.05$ is

$$[0.49,\, 0.51]. \tag{6.26}$$

If in 10,000 flips we observe 5,200 occurrences of 'heads', then $h_n = 0.52$. This value lies outside of the confidence interval (6.26), and our conclusion, with an error probability of 0.05, is that the coin is not fair.

On the other hand, if we observe 5,050 occurrences of 'heads' in our 10,000 flips, then we have $h_n = 0.505$, and $h_n$ does lie in the confidence interval (6.26), so we are convinced that the coin in this case is indeed fair. This conclusion will be correct with a probability of 95%, i.e., the error probability is $\alpha = 0.05$.

#### 6.2.5.6 Application to the confidence interval for the probability $p$

We consider a coin. Let $p$ denote the probability for the occurrence of 'heads' in a reference trial. We flip the coin $n$ times and measure the relative frequency $h_n$ of the occurrence of 'heads'. We now consider $p$ as unknown, and wish to determine its value with a certain degree of certainty.

**Basic principle of mathematical statistics:** With an error probability of $\alpha$, we assume that the unknown probability $p$ lies in the interval

$$\boxed{[p_-\,, p_+]\,.}$$

Here we have[15]

$$\left(1+\frac{z_\alpha^2}{n}\right)p_\pm = h_n + \frac{z_\alpha^2}{2n} \pm \sqrt{\frac{h_n z_\alpha^2}{n} + \frac{z_\alpha^2}{4n^2}}\,.$$

This statement is true in general for the estimation of probabilities $p$ occurring in Bernoulli's model 6.2.5.2.

**Reasoning:** According to (6.24), for large $n$ we have the inequality

$$\left|\frac{h_n - p}{\Delta H_n}\right| < z_\alpha \tag{6.27}$$

with probability $\Phi_0(z_\alpha) - \Phi_0(-z_\alpha) = 1 - \alpha$. Since $(\Delta H_n)^2 = p(1-p)/n$, (6.27) is equivalent to

$$(h_n - p)^2 < z_\alpha^2 \frac{p\,(1-p)}{n}\,,$$

in other words,

$$p^2\left(1+\frac{z_\alpha^2}{n}\right) - \left(2h_n + \frac{z_\alpha^2}{n}\right)p + h_n^2 < 0\,. \tag{6.28}$$

This inequality holds if and only if $p$ lies between the zeros $p_-$ and $p_+$ of the corresponding quadratic equation.

*Example:* Flipping a coin 10,000 times, suppose that 'heads' turns up 5,010 times. Then $h_n = 0.501$, and the unknown probability $p$ for the occurrence of 'heads' lies in the interval $[0.36, 0.64]$ with a residual probability of error of $\alpha = 0.05$.

[15] The meaning of $\alpha$ and $z_\alpha$ was explained in 6.2.5.5.

This estimate is, however, still quite crude. Making 1,000,000 flips and obtaining a relative frequency 0.501 for 'heads', we can assert that $p$ lies, with the same residual error probability as before, in the interval $[0.500, 0.503]$.

#### 6.2.5.7 The strong law of large numbers

**Infinite trials of experiments:** Up until now we have considered Bernoulli's model for $n$ trials. In order to formulate the strong law of large numbers we need to pass over to an infinite number of trials.

Thus, we consider a total event $E$ consisting of the elementary events

$$\boxed{e_{i_1 i_2 \cdots}\,,}$$

where each index $i_j$ can take the value 1 or 2. The symbol $e_{12\ldots}$ means that in the first trial $e_1$ occurs, in the second $e_2$ and so forth. We denote by $A_{i_1\cdots i_n}$ the set of all elementary events of the form $e_{i_1\cdots i_n\cdots}$, and set

$$\boxed{P(A_{i_1 i_2\cdots i_n}) = p_{i_1 i_2\cdots i_n}} \tag{6.29}$$

with $p_1 := p$ and $p_2 := 1 - p$ (see (6.18)). We denote by $\mathscr{S}$ the smallest $\sigma$-algebra of $E$ which contains all sets $A_{i_1\cdots i_n}$ for all $n$.

**Theorem:** There is a uniquely determined probability measure $P$ on the subsets of $\mathscr{S}$ which has property (6.29). $(E, \mathscr{S}, P)$ is thus a probability space.

**Relative frequency:** We define a random variable by $H_n : E \longrightarrow \mathbb{R}$ by

$$H_n(e_{i_1\cdots i_n\cdots}) = \frac{1}{n}\cdot(\text{number of indices with } i_j = 1 \text{ and } 1 \le j \le n)\,.$$

**The strong law of large numbers of Borel (1909) and Cantelli (1917):** The limit relation

$$\boxed{\lim_{n\to\infty} H_n = p}$$

holds almost surely[16] on $E$.

## 6.3 Mathematical statistics

*Don't trust any statistic which you haven't tampered with yourself.*

*Old proverb*

Mathematical statistics investigates the properties of random phenomena of our daily world on the basis of series of measurements of random variables. This requires a very responsible use of statistical procedures. Different methods and models will lead to completely different conclusions. Therefore, you should always act by the following *golden rule of mathematical statistics*:

> Every statement of mathematical statistics is based on certain assumptions. Without a complete enunciation of all assumptions on which a conclusion is based, the statement is worthless.

[16] If $A$ denotes the set of all $e \in E$ with $\lim_{n\to\infty} H_n(e) = p$, then $P(A) = 1$.

### 6.3.1 Basic ideas

**Confidence intervals:** Let $\Phi$ be a distribution function of a random variable $X$. An $\alpha$-confidence interval $[x_\alpha^-, x_\alpha^+]$ is given by definition by the requirement

$$\boxed{P(x_\alpha^- \le X \le x_\alpha^+) = 1 - \alpha\,.}$$

*Interpretation:* The value of a measurement of $X$ lies with the probability $1-\alpha$ in the confidence interval $[x_\alpha^-, x_\alpha^+]$.

*Example 1:* If $X$ has a continuous probability density $\varphi$, then the hatched area in Figure 6.16 over the confidence interval $[x_\alpha^-, x_\alpha^+]$ is equal to $1-\alpha$, i.e., we have

$$\int_{x_\alpha^-}^{x_\alpha^+} \varphi(x)\,\mathrm{d}x = 1 - \alpha\,.$$

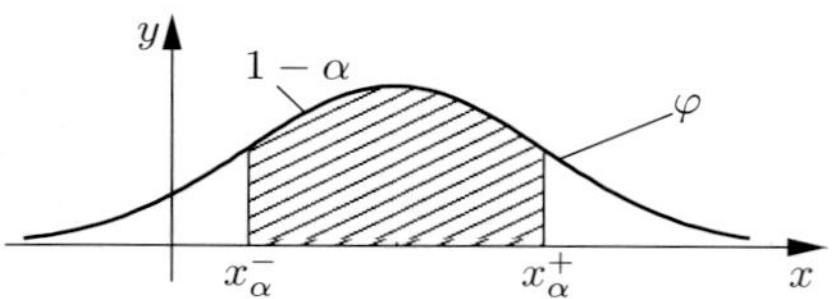


*Figure 6.16. The normal distribution.*

*Example 2:* For a normal distribution $N(\mu, \sigma)$ with mean $\mu$ and standard deviation $\sigma$, the confidence interval $[x_\alpha^-, x_\alpha^+]$ is given by

$$x_\alpha^\pm = \mu \pm \sigma z_\alpha.$$

The value $z_\alpha$ is obtained from the equation $\Phi_0(z_\alpha) = \dfrac{1-\alpha}{2}$ with the help of Table 0.34. In particular we have $z_\alpha = 1.6,\ 2.0$ and $2.6$ for the values $\alpha = 0.01,\ 0.05$ and $0.1$.

**Variational series:** Let $X$ be a given random variable. In practical cases $X$ will be measured $n$ times in a series of measurements or experiments, and one obtains $n$ real numbers

$$\boxed{x_1, x_2, \ldots, x_n}$$

as measured values of $X$. Such a series is referred to as a *variational series.* Our basic assumption at this point is that the measurements are independent of one another, i.e., the individual measurements have no influence on measurements following them.

**Mathematical random samples:** The measurement will vary from trial to trial. In order to describe this fact mathematically, we consider the random vector

$$\boxed{(X_1, \ldots, X_n)}$$

of independent variables, where all $X_j$ have the same distribution function as $X$ does.

**The basic strategy of mathematical statistics:**

(i) We formulate the hypothesis:

$$\boxed{\text{The distribution function } \Phi \text{ of } X \text{ has the property } \mathscr{E}.} \qquad \text{(H)}$$

(ii) We construct a so-called *sample function*

$$Z = Z(X_1, \ldots, X_n)$$

and determine the distribution function $\Phi_Z$ under the assumption (H).

(iii) After a series of measurements has been made with measured values $x_1, \ldots, x_n$, we calculate the real number $z := Z(x_1, \ldots, x_n)$. We call $z$ a realization of the sample function $Z$.

(iv) The hypothesis (H) will be discarded with a probability of error of $\alpha$, if $z$ does not lie in the $\alpha$-confidence interval of $Z$.

(v) If $z$ does lie in the $\alpha$-confidence interval of $Z$, then we say that the observations do not contradict the hypothesis, again with a probability of error of $\alpha$.

*Example 3:* The hypothesis might, for example, be: '$\Phi$ is a normal distribution'.

**Estimate of parameters:** If the distribution function $\Phi$ depends on parameters, then one would like to know intervals in which these parameters lie. A typical example of this can be found in 6.2.5.6.

**Comparison of two variational series:** If we are working with two random variables $X$ and $Y$, then the hypothesis (H) will be an assumption on the distribution functions of $X$ and of $Y$. The sample function will then have the form

$$Z = Z(X_1, \ldots, X_n, Y_1, \ldots, Y_n).$$

The measured values $x_1, \ldots, x_n, y_1, \ldots, y_n$ will then determine a realization of the sample function $z := Z(x_1, \ldots, x_n, y_1, \ldots, y_n)$. This reduces steps (iv) and (v) in our strategy to checking it for a single function, namely $Z$, and drawing our conclusions from this.

## 6.3.2 Important estimators

Let $(X_1, \ldots, X_n)$ be a mathematical sample for a random variable $X$.

**Estimation of the expectation:** The sample function

$$\boxed{M := \frac{1}{n}\sum_{j=1}^{n} X_j}$$

is called the *estimator for the expectation* $\overline{X}$ of $X$. The dependency on $n$ will be indicated, if necessary, by the notation $M_n$.

(i) The estimator $M$ respects expectations, i.e., we have

$$\boxed{\overline{M} = \overline{X}\,.}$$

(ii) If $X$ is normally distributed with distribution $N(\mu, \sigma)$, then $M$ is normally distributed with distribution $N(\mu, \sigma/\sqrt{n})$.

(iii) Assume that $\Delta X < \infty$. If $\Phi_n$ denotes the distribution function of $M_n$, then the limit function

$$\Phi(x) := \lim_{n\to\infty} \Phi_n(x)$$

is normally distributed of type $N(\overline{X}, \Delta X/\sqrt{n})$.

**Estimation of the variance:** The sample function

$$\boxed{S^2 = \frac{1}{n-1}\sum_{j=1}^{n}(X_j - \overline{X})^2}$$

is called the *estimator of the variance* of $X$.

(i) This estimator respects expectations, i.e.,

$$\boxed{\overline{S^2} = (\Delta X)^2 .}$$

(ii) If $X$ is normally distributed of type $N(\mu, \sigma)$, then the distribution function of

$$T := \frac{M - \mu}{S}\sqrt{n}$$

is a $t$-distribution with $n-1$ degrees of freedom. Moreover, the distribution function of

$$\chi^2 := \frac{(n-1)S^2}{\sigma^2}$$

is a $\chi^2$-distribution with $n-1$ degrees of freedom (see Table 6.6).

*Table 6.6. Probability densitiy of probability functions.*

| ***Name of the distribution*** | ***Probability density*** |
|---|---|
| $t$-distribution with $n$ degrees of freedom | $\dfrac{\Gamma(\frac{n+1}{2})}{\sqrt{\pi n}\,\Gamma\left(\frac{n}{2}\right)}\left(1+\dfrac{x^2}{n}\right)^{-\frac{n+1}{2}}$ |
| $\chi^2$-distribution with $n$ degrees of freedom | $\dfrac{x^{(n/2)-1}\mathrm{e}^{-x/2}}{2^{n/2}\Gamma\left(\frac{n}{2}\right)}$ |

## 6.3.3 Investigating normally distributed measurements

The assumption is often made that a given random variable $X$ is normally distributed. The theoretical legitimization of this results from the central limit theorem (see 6.2.4.3). Examples for the following procedures can be found in 0.4.

### 6.3.3.1 The confidence interval for the expectation

**Assumption:** $X$ is normally distributed of type $N(\mu, \sigma)$.

**Variational series:** From the measured values $x_1, \ldots, x_n$ of $X$ we calculate the empirical expectation

$$\overline{x} = \frac{1}{n}\sum_{j=1}^{n} x_j$$

and the empirical standard deviation

$$\Delta x = \sqrt{\frac{1}{n-1}\sum_{j=1}^{n}(x_j - \overline{x})^2}\ .$$

**The statistical statement:** With a possibility of making an error of $\alpha$, the following inequality for the expectation $\mu$ is valid:

$$\boxed{|\overline{x}-\mu| \le \frac{\Delta x}{\sqrt{n}} t_{\alpha,n-1}\,.} \tag{6.30}$$

The value $t_{\alpha,n-1}$ can be found in section 0.4.6.3.

**Reasoning:** The random variable $\sqrt{n}(M-\mu)/S$ is $t$-distributed with $n-1$ degrees of freedom. We have $P(|T| \le t_{\alpha,n-1}) = 1-\alpha$. Hence the inequality

$$\frac{|\overline{x}-\mu|}{\Delta x}\sqrt{n} \le t_{\alpha,n-1}$$

holds with a probability of $1-\alpha$. This yields (6.30).

#### 6.3.3.2 The confidence interval for the standard deviation

**Assumption:** $X$ is normally distributed of type $N(\mu,\sigma)$.

**The statistical statement:** With an error probability of $\alpha$, the following inequality for the standard deviation holds:

$$\boxed{\frac{(n-1)(\Delta x)^2}{b} \le \sigma^2 \le \frac{(n-1)(\Delta x)^2}{a}\,.} \tag{6.31}$$

The values $a := \chi^2_{1-\alpha/2}$ and $b := \chi^2_{\alpha/2}$ can be found in section 0.4.6.4 for the value $m = n-1$ degrees of freedom.

**Reasoning:** The quantity $A := (n-1)S^2/\sigma^2$ is $\chi^2$-distributed with $n-1$ degrees of freedom. According to Figure 0.50 we have $P(a \le A \le b) = P(A \ge b) - P(A \ge a) = 1 - \frac{\alpha}{2} - \frac{\alpha}{2} = 1-\alpha$. Therefore, the inequality

$$a \le \frac{(n-1)(\Delta x)^2}{\sigma^2} \le b$$

holds with a probability of $1-\alpha$, and (6.31) is a consequence of this.

#### 6.3.3.3 The fundamental significance test ($t$-test)

It is the objective of this test to determine from two variational series of random variables $X$ and $Y$, whether $X$ and $Y$ have the same expectation, i.e., whether there is a significant difference between $X$ and $Y$.

**Assumption:** $X$ and $Y$ are both normally distributed with the same standard deviation.[17]

**Hypothesis:** $X$ and $Y$ have the same expectation.

**Variational series:** From the observed values

$$\boxed{x_1,\ldots,x_{n_1} \quad \text{and} \quad y_1,\ldots,y_{n_2}} \tag{6.32}$$

[17] This assumption can be verified with the aid of the $F$-test, see 6.3.3.4.

of $X$ and $Y$ we calculate the empirical expectations (means) $\overline{x}$ and $\overline{y}$ as well as the empirical standard deviations $\Delta x$ and $\Delta y$ (see 6.3.3.1). Moreover, we calculate the number

$$t := \frac{\overline{x} - \overline{y}}{\sqrt{(n_1 - 1)(\Delta x)^2 + (n_2 - 1)(\Delta y)^2}} \sqrt{\frac{n_1 n_2 (n_1 + n_2 - 2)}{n_1 + n_2}} . \tag{6.33}$$

**The statistical statement:** With an error probability $\alpha$ the hypothesis is false, i.e., there is a significant difference between $X$ and $Y$, provided

$$\boxed{|t| > t_{\alpha,m} .}$$

The value $t_{\alpha,m}$ can be found in 0.4.6.3 with $m = n_1 + n_2 - 1$ degrees of freedom.

**Reasoning:** If we replace in (6.33) the quantities $\overline{x}, \overline{y}, (\Delta x)^2, (\Delta y)^2$ by $\overline{X}, \overline{Y}, S_X^2, S_Y^2$, then we obtain a random variable $T$, whose distribution is a $t$-distribution with $m$ degrees of freedom. Since $P(|T| > t_\alpha) = \alpha$, the hypothesis will be discarded if $|t| > t_\alpha$.

#### 6.3.3.4 The *F*-test

This test is to determine whether two normally distributed random variables have different standard deviations.

**Assumption:** Both random variables $X$ and $Y$ are normally distributed.

**Hypothesis:** $X$ and $Y$ have the same standard deviation.

**Variational series:** From the observed values (6.32) we calculate the empirical standard deviations $\Delta x$ and $\Delta y$. We suppose that $\Delta x \geq \Delta y$.

**The statistical statement:** The hypothesis will be discarded with a probability of error equal to $\alpha$, provided

$$\left(\frac{\Delta x}{\Delta y}\right)^2 > F_{\frac{\alpha}{2}} . \tag{6.34}$$

The value $F_{\frac{\alpha}{2}}$ can be taken form 0.4.6.5 with $m_1 = n_1 - 1$ and $m_2 = n_2 - 1$.

If, on the other hand, (6.34) is true but with '$\leq F_\alpha$' on the right-hand side, then the observations do not contradict the hypothesis, with an error probability $\alpha$.

**Reasoning:** If the hypothesis is true, then the random variable $F := S_X^2 / S_Y^2$ is $F$-distributed with the degrees of freedom $(m_1, m_2)$. Since $P(F \geq F_\alpha) = \alpha$, we will discard the hypothesis if (6.34) holds, with an error probability $\alpha$.

#### 6.3.3.5 The correlation test

The correlation test can be used to check whether two random variables $X$ and $Y$ are correlated.

**Assumption:** $X$ and $Y$ are normally distributed.

**Hypothesis:** For the correlation coefficient we have $r = 0$, i.e., there is no dependency between $X$ and $Y$.

**Variational series:** From the measured values

$$x_1, \ldots, x_n \quad \text{and} \quad y_1, \ldots, y_n \tag{6.35}$$

we calculate the empirical correlation coefficient

$$\rho = \frac{m_{XY}}{\Delta x \Delta y}$$

with the empirical covariance $m_{XY} := \frac{1}{n-1}\sum\limits_{j=1}^{n}(x_j - \overline{x})(y_j - \overline{y})$.

**The statistical statement:** With an error probability $\alpha$, the independency hypothesis will be discarded when

$$\boxed{\frac{\rho\sqrt{n-2}}{\sqrt{1-\rho^2}} > t_{\alpha,m}\,.} \tag{6.36}$$

The value $t_{\alpha,m}$ can be found in 0.4.6.3 with $m = n - 2$.

**Reasoning:** We set

$$R := \frac{\sum\limits_{j=1}^{n}(X_j - \overline{X})(Y_j - \overline{Y})}{\left(\sum\limits_{j=1}^{n}(X_j - \overline{X})^2 \sum\limits_{j=1}^{n}(Y_j - \overline{Y})^2\right)^{1/2}}\,.$$

The random variable $\frac{R\sqrt{n-2}}{\sqrt{1-R^2}}$ is $t$-distributed with $n-2$ degrees of freedom. The observations (6.35) yield a realization $\rho$ of $R$. Since $P(|R| \geq t_\alpha) = \alpha$, we discard the hypothesis if (6.36) holds with a residual probability $\alpha$ of making an error.

**Test for normal distribution:** To determine whether a random variable is normally distributed one can apply the $\chi^2$-suitability test (see 6.3.4.4).

## 6.3.4 The empirical distribution function

The empirical distribution function is an approximation of the actual distribution function of a random variable. This statement is made precise in the main theorem of mathematical statistics.

### 6.3.4.1 The Main Theorem of mathematical statistics and the Kolmogorov–Smirnov test for distribution functions

**Definition:** Let observed values $x_1, \ldots, x_n$ of a random variable $X$ be given. We set

$$\boxed{F_n(x) := \frac{1}{n}\cdot(\text{number of observations} < x)}$$

and call the step function $F_n$ the *empirical distribution function.*

*Example:* Suppose our observations have the values $x_1 = x_2 = 3.1, x_3 = 5.2$ and $x_4 = 6.4$. The empirical distribution function (see Figure 6.17):

$$F_4(x) = \begin{cases} 0 & \text{for} \quad x \leq 3.1, \\ \frac{1}{2} & \text{for} \quad 3.1 < x \leq 5.2, \\ \frac{3}{4} & \text{for} \quad 5.2 < x \leq 6.4, \\ 1 & \text{for} \quad 6.4 < x\,. \end{cases}$$

The difference between the empirical distribution function $F_n$ and the actual distribution function $\Phi$ of the random variable $X$ is measured by the quantity

$$d_n := \sup_{x \in \mathbb{R}} |F_n(x) - \Phi(x)|.$$

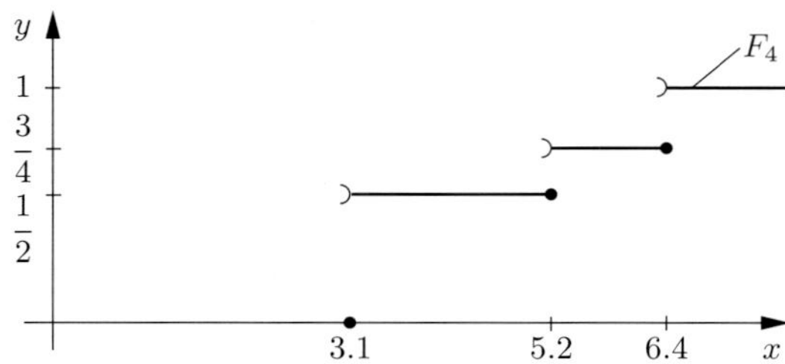


*Figure 6.17. An empirical distribution function.*

**The Main Theorem of mathematical statistics (Glivenko, 1933):** We have almost surely

$$\boxed{\lim_{n \to \infty} d_n = 0\,.}$$

**The theorem of Kolmogorov–Smirnov:** For all real numbers $\lambda$ we have

$$\lim_{n \to \infty} P(\sqrt{n} d_n < \lambda) = Q(\lambda)$$

with $Q(\lambda) := \sum\limits_{k=-\infty}^{\infty} (-1)^k \mathrm{e}^{-2k^2\lambda^2}$.

**The Kolmogorov–Smirnov test:** We choose a distribution function $\Phi : \mathbb{R} \longrightarrow \mathbb{R}$. With an error probability of $\alpha$ we discard $\Phi$ as the distribution of $X$ provided

$$\boxed{\sqrt{n} d_n > \lambda_\alpha.}$$

Here $\lambda_\alpha$ is a solution of the equation $Q(\lambda_\alpha) = 1 - \alpha$, using 0.4.6.8.

In case $\sqrt{n} d_n \le \lambda_\alpha$, the observations do not contradict the hypothesis that $\Phi$ is the distribution of $X$, with an error probability of $\alpha$.

This test can only be used for a sufficiently large number $n$ of measurements. Moreover, one cannot apply it to distribution functions $\Phi$ depending on parameters, when these parameters themselves have to be estimated from the measurements. For this one uses instead the $\chi^2$-test (see 6.3.4.4).

**Application to the equidistribution:** A random variables $X$ is assumed given, with values $a_1 < a_2 < \cdots < a_k$, where we assume an equidistribution, i.e., each value $a_j$ has probability $\frac{1}{k}$. In this case the Kolmogorov–Smirnov test works as follows:

(i) We determine the observations $x_1, \ldots, x_n$.

(ii) Let $m_r$ be the number of measurements in the interval $[a_r, a_{r+1}[$.

(iii) Given the error probability $\alpha$, we determine $\lambda_\alpha$ from Table 0.4.6.8 such that $Q(\lambda_\alpha) = 1 - \alpha$.

(iv) We calculate the test quantity

$$d_n := \max_{r=1,\ldots,k} \left| \frac{m_r}{n} - \frac{1}{k} \right| .$$

**The statistical statement:** If $\sqrt{n}d_n > \lambda_\alpha$, then with error probability $\alpha$ the hypothesis that the distribution is an equidistribution is false.

If $\sqrt{n}d_n \le \lambda_\alpha$ we accept our hypothesis of equidistribution, with the probability of making an error of $\alpha$.

**Testing an apparatus for drawing lottery numbers (6 from 45):** We lay 6 balls with numbers $r = 1, \ldots, 6$ into the apparatus and make a series of 600 drawings. The observed frequencies $m_r$ of drawing the number $r$ are displayed in Table 6.7.

*Table 6.7. Observed frequencies.*

| $\boldsymbol{r}$ | 1 | 2 | 3 | 4 | 5 | 6 |
|---|---|---|---|---|---|---|
| $\boldsymbol{m_r}$ | 99 | 102 | 101 | 103 | 98 | 97 |

Let $\alpha = 0.05$. According to 0.4.6.8 it follows from $Q(\lambda_\alpha) = 0.95$ that $\lambda_\alpha = 1.36$. From Table 6.7 we obtain $d_n = \dfrac{3}{600} = 0.005$. Since $\sqrt{600}d_n = 0.12 < \lambda_\alpha$ we have (with an error probability $\alpha = 0.05$) no reason to doubt that the apparatus is a fair one.

### 6.3.4.2 The histogram

Histograms correspond to empirical probability densities.

**Definition:** Let the observations

$$x_1, \ldots, x_n$$

be given. We choose numbers $a_1 < a_2 < \cdots < a_k$ with the corresponding intervals $\Delta_r := [a_r, a_{r+1}[$, so that every measurement lies in at least one of these intervals. The quantity

$$\boxed{m_r := \text{number of observations in the interval}\, \Delta_r}$$

is called the *frequency* of the $r^{th}$ class. The empirical distribution function is given by

$$\varphi_n(x) := \frac{m_r}{n} \qquad \text{for all} \quad x \in \Delta_r.$$

A graphical representation of this function is called a *histogram.*

*Table 6.8*

| $x_1$ | $x_2$ | $x_3$ | $x_4$ | $x_5$ | $x_6$ | $x_7$ | $x_8$ | $x_9$ | $x_{10}$ |
|---|---|---|---|---|---|---|---|---|---|
| 1 | 1.2 | 2.1 | 2.2 | 2.3 | 2.3 | 2.8 | 2.9 | 3.0 | 4.9 |

*Table 6.9*

| $r$ | $\Delta_r$ | $m_r$ | $\frac{m_r}{n}$ $(n = 10)$ |
|---|---|---|---|
| 1 | $1 \leq x < 2$ | 2 | $\frac{2}{10}$ |
| 2 | $2 \leq x < 3$ | 6 | $\frac{6}{10}$ |
| 3 | $3 \leq x < 5$ | 2 | $\frac{2}{10}$ |

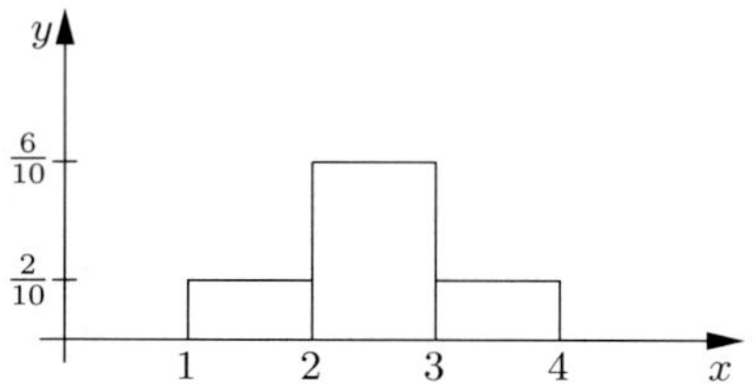

*Figure 6.18.*

*Example:* If the measured values $x_1, \ldots, x_{10}$ are as in Table 6.8, then Table 6.9 gives a possible division of these values into classes, and the corresponding histogram is shown in Figure 6.18.

### 6.3.4.3 The $\chi^2$-suitability test for distribution functions

This test is used to determine whether a function $\Phi$ is the distribution function of a given random variable $X$.

**Hypothesis:** $X$ has the distribution function $\Phi$.

**Variational series:** (i) We observe values $x_1, \ldots, x_n$ of $X$ and divide these observations into classes $\Delta_r := [a_r, a_{r+1}[$ with $r = 1, \ldots, k$.

(ii) We determine the number $m_r$ of measurements in the interval $\Delta_r$.

(iii) We set $p_r := \Phi(a_{r+1}) - \Phi(a_r)$ and calculate the test quantity

$$c^2 = \sum_{r=1}^{k} \frac{(m_r - np_r)^2}{np_r}.$$

(iv) We choose an error probability $\alpha$ and determine the value $\chi^2_\alpha$ from 0.4.6.4 with $m = k - 1$ degrees of freedom.

**The statistical statement:** If $c^2 > \chi^2_\alpha$, then $\Phi$ is with an error probability of $\alpha$ *not* the distribution function of $X$.

If $c^2 \leq \chi^2_\alpha$, then, with a probability $\alpha$ of making an error, we may assume that our hypothesis is correct and $\Phi$ is the distribution function of $X$.

**Reasoning:** For $n \to \infty$ the test quantity $c^2$ is $\chi^2$-distributed with $m = k - 1$ degrees of freedom.

**$\chi^2$-test for distributions depending on parameters:** If the distribution function $\Phi$ depends on parameters $\beta_1, \ldots, \beta_s$, then these parameters have to be estimated from the same observations.

For $k$ classes of measurements we need to set $m = k - 1 - s$ in using 0.4.6.4 to determine $\chi^2_\alpha$.

*Example:* Mean and standard deviation of a normal distribution can be estimated by empirical means $\overline{x}$ and empirical standard deviations $\Delta x$ (see 6.3.4.4).

In the general case, one can use the maximal likelihood method to estimate the values of the parameters (see section 6.3.5).

#### 6.3.4.4 The $\chi^2$-suitability test for normal distributions

The test presented in this paragraph is used to test whether a random variable, which one expects from observations is normally distributed, actually is so.

(i) We make measurements $x_1, \ldots, x_n$ of $X$ and determine the empirical expectation $\overline{x}$ and the empirical standard deviation $\Delta x$:

$$\overline{x} = \frac{1}{n}\sum_{j=1}^{n} x_j\,, \quad (\Delta x)^2 = \frac{1}{n-1}\sum_{j=1}^{n}(x_j - \overline{x})^2\,.$$

(ii) We choose values $a_1 < a_2 < \cdots < a_k$ and determine the number $m_r$ of observations which lie in the interval $[a_r, a_{r+1}[$.

(iii) With the aid of Table 0.34 we calculate

$$p_r := \Phi\left(\frac{a_{r+1} - \overline{x}}{\Delta x}\right) - \Phi\left(\frac{a_r - \overline{x}}{\Delta x}\right)\,.$$

(iv) We form the test quantities

$$c^2 = \sum_{r=1}^{n} \frac{(m_r - np_r)^2}{np_r}\,.$$

(v) Given an error probability $\alpha$ we determine with the help of 0.4.6.4 for $m = k - 3$ degrees of freedom the value $\chi^2_\alpha$.

**The statistical statement:** If

$$\boxed{c^2 \leq \chi^2_\alpha\,,}$$

we accept the hypothesis that $X$ is normally distributed, with an error probability of $\alpha$.

If $c^2 > \chi^2_\alpha$ we discard the hypothesis that $X$ is normally distributed as incorrect, again with a probability of having made an error of $\alpha$.

Here it is important to assure $np_r \geq 5$. This can be achieved by an appropriate choice of the $a_r$.

**Errors of a measurement instrument:** We consider an instrument which makes some kind of measurement (for example length). We want to verify that errors in these measurements are normally distributed. For this we make 100 measurements of some normed object (i.e., for which we know whether the observed measurement is correct or not) and enter any errors in these measurements in Table 6.10. Suppose for example that $\overline{x} = 1$ and $\Delta x = 10$. In this case we obtain

$$p_8 = \Phi\left(\frac{15 - \overline{x}}{\Delta x}\right) - \Phi\left(\frac{10 - \overline{x}}{\Delta x}\right) = \Phi_0(1.4) - \Phi_0(0.9) = 0.42 - 0.32 = 0.10$$

by Table 0.34. If we choose $\alpha = 0.05$, then 0.4.6.4 yields for $m = 10 - 3 = 7$ degrees of freedom the value $\chi^2_\alpha = 14.1$. Since $c^2 < 14.1$, we can assume, with an error probability of $\alpha$, that errors in measurements of our instrument are indeed normally distributed.

Actually, the condition $np_r \geq 5$ is not satisfied in Table 6.10. In order to achieve this we would group the classes $r = 1, 2, 3$ and $r = 9, 10$ together.

*Table 6.10. Errors of a measurement instrument.*

| $r$ | *Interval of measurement* | *Frequency $m_r$ of the measurement* | $p_r$ | $np_r$ | $\frac{(m_r - np_r)^2}{np_r}$ |
|---|---|---|---|---|---|
| 1 | $x < -20$ | 1 | 0.01 | 1 | 0 |
| 2 | $-20 \leq x < -15$ | 4 | 0.03 | 3 | 0.3 |
| 3 | $-15 \leq x < -10$ | 9 | 0.09 | 9 | 0 |
| 4 | $-10 \leq x < -5$ | 10 | 0.14 | 14 | 1.1 |
| 5 | $-5 \leq x < 0$ | 24 | 0.19 | 19 | 1.3 |
| 6 | $0 \leq x < 5$ | 26 | 0.20 | 20 | 1.8 |
| 7 | $5 \leq x < 10$ | 16 | 0.16 | 16 | 0 |
| 8 | $10 \leq x < 15$ | 6 | 0.10 | 10 | 1.6 |
| 9 | $15 \leq x < 20$ | 2 | 0.06 | 6 | 2.7 |
| 10 | $20 \leq x$ | 2 | 0.02 | 2 | 0 |
| ***Sum*** | | 100 | 1.0 | 100 | $c^2 = 8.8$ |

### 6.3.4.5 Comparing two distribution functions with the Wilcoxon test

The Wilcoxon test allows us to check whether two given variational series belong to truly different statistical quantities. The decisive advantage of this test is that it makes no assumptions on the distribution functions and it is very easy to apply.

**Hypothesis:** The distribution functions of two random variables $X$ and $Y$ coincide.

**Variational series:** Suppose the observed values for $X$ and $Y$ are

$$x_1, \ldots, x_{n_1} \quad \text{and} \quad y_1, \ldots, y_{n_2}\,.$$

By definition a pair of values $(x_j, y_k)$ are *in inversion* if $y_k < x_j$. We measure the quantities

$$u := \text{number of inversions.}$$

From 0.4.6.7 we determine the value $u_\alpha$ for a given $\alpha$[18].

**The statistical statement:** If

$$\boxed{\left| u - \frac{n_1 n_2}{2} \right| > u_\alpha\,,}$$

then the hypothesis is discarded, with an error probability of $\alpha$, i.e., we conclude the random variables are significantly different.

**Testing two medicines:** Let $A$ and $B$ be two medicines for treating a disease. Table 6.11 displays the number of days which pass before recuperation. Here $y_j$ is inverse to

[18]For large $n_1$ and $n_2$ we have

$$u_\alpha = z_\alpha \sqrt{\frac{n_1 n_2 (n_1 + n_2 + 1)}{12}}\,.$$

Here $z$ is determined from Table 0.34 from the equation

$$\Phi_0(z_\alpha) = \frac{1}{2}(1 - \alpha)\,.$$

Table 6.11.

| | | | | | |
|---|---|---|---|---|---|
| ***Medicine A*** | $x_1$ | $x_2$ | $x_3$ | $x_4$ | $x_5$ |
| | 3 | 3 | 3 | 4 | 4 |
| ***Medicine B*** | $y_1$ | $y_2$ | $y_3$ | $y_4$ | – |
| | 2 | 1 | 2 | 1 | – |

all $x_k$ The number of inversion is therefore equal to $u = 4 \cdot 5 = 20$. From 0.4.6.7 with $\alpha = 0.05$ and $n_1 = 5,\ n_2 = 4$, it follows that $u_\alpha = 9$. From

$$\left| u - \frac{n_1 n_2}{2} \right| = |20 - 10| = 10 > u_\alpha$$

we can assert, with an $\alpha$-chance of being mistaken, that these two medicines are quite different in their efficacy against this disease. In other words, the much quicker effect seen from medicine $B$ is not just a product of coincidence.

### 6.3.5 The maximal likelihood method for estimation of parameters

The fundamental maximal likelihood method of mathematical statistics makes it possible to obtain estimates for unknown parameters which are, in a sense, optimal.

**Continuous random variables:** Let $\varphi = \varphi(x, \boldsymbol{\beta})$ be a probability density of a random variable $X$, which depends on parameters $(\beta_1, \ldots, \beta_k)$, which we will denote by $\boldsymbol{\beta}$ for brevity. Then one obtains a *maximal likelihood function* for $\boldsymbol{\beta} = \boldsymbol{\beta}(x_1, \ldots, x_n)$ by solving the system of equations[19]

$$\boxed{\sum_{j=1}^{n} \frac{1}{\varphi(x_j, \boldsymbol{\beta})} \frac{\partial \varphi(x_j, \boldsymbol{\beta})}{\partial \beta_r} = 0\,, \qquad r = 1, \ldots, k} \tag{6.37}$$

for $\beta_1, \ldots, \beta_k$. The quantities $x_1, \ldots, x_n$ are measured values.

*Example 1* (normal distribution): Let

$$\varphi(x, \mu, \sigma) = \frac{1}{\sigma\sqrt{2\pi}} \mathrm{e}^{-(x-\mu)^2/2\sigma^2}$$

be the probability density for a normal distribution with mean $\mu$ and standard deviation $\sigma$. Because of

$$\frac{\partial \varphi}{\partial \mu} = \frac{\mu - x}{\sigma^2}\varphi\,, \qquad \frac{\partial \varphi}{\partial \sigma} = \left( -\frac{1}{\sigma} + \frac{(x-\mu)^2}{\sigma^3} \right) \varphi$$

the equations (6.37) with $\beta_1 = \mu$ and $\beta_2 = \sigma$ correspond to the system

$$\sum_{j=1}^{n} (\mu - x_j) = 0\,, \qquad -\frac{n}{\sigma} + \sum_{j=1}^{n} \frac{(x_j - \sigma)^2}{\sigma^3} = 0\,.$$

[19]The name of this procedure comes from the fact that the condition

$$L \stackrel{!}{=} \max$$

for the so-called *likelihood function* $L = \varphi(x_1, \boldsymbol{\beta})\varphi(x_2, \boldsymbol{\beta}) \cdots \varphi(x_n, \boldsymbol{\beta})$ leads to the equation

$$\frac{\partial L}{\partial \beta_r} = 0, \quad r = 1, \ldots, k,$$

which after dividing by $L$ is identical to (6.37).

From this the maximal likelihood estimators

$$\mu = \frac{1}{n}\sum_{j=1}^{n} x_j\,, \qquad \sigma^2 = \frac{1}{n}\sum_{j=1}^{n}(x_j-\mu)^2$$

result.

**Discrete random variables:** Let a random variable $X$ have finitely many values $a_1, \ldots, a_k$ with probabilities $p_1(\boldsymbol{\beta}), \ldots, p_k(\boldsymbol{\beta})$. We denote by $h_j$ the relative frequency of the appearance of $a_j$ in a variational series $x_1, \ldots, x_n$. Then the maximal likelihood estimators $\beta_r = \beta_r(x_1, \ldots, x_n)$, $r = 1, \ldots, k$ are obtained from the equation

$$\sum_{j=1}^{k} \frac{h_j}{p_j(\boldsymbol{\beta})}\frac{\partial p_j(\boldsymbol{\beta})}{\partial \beta_r} = 0\,, \qquad r = 1, \ldots, k \tag{6.38}$$

with $h_1 + \cdots + h_k = 1$.

*Example 2:* Let the event $A$ occur with a probability of $p$. We set

$$X := \begin{cases} 1, & \text{if } A \text{ occurs,} \\ 0, & \text{if } A \text{ does not occur.} \end{cases}$$

Then $X$ is a random variable with $P(X = 1) = p$ and $P(X = 0) = 1 - p$. We use a parameter $\beta = p$. From (6.38) we obtain

$$\frac{h_1}{p} - \frac{h_2}{1-p} = 0\,, \qquad h_1 + h_2 = 1\,.$$

This yields the maximal likelihood estimator

$$p = h_1\,.$$

This means that the probability $p$ is estimated by the relative frequency $h_1$ of the occurrence of $A$.

*Example 3:* Let $X$ be a random variable taking on the values $j = 0, 1, \ldots$ with probabilities

$$p_j := \frac{\beta^j}{j!}\mathrm{e}^{-\beta}\,, \qquad j = 0, 1, \ldots\,,$$

which means that $X$ is Poisson distributed. If $h_j$ denotes the relative frequency for the occurrence of $j$ among the measured values $x_1, \ldots, x_n$ of $X$, then the maximal likelihood function for $\beta$ is:

$$\beta = \frac{1}{n}\sum_{j=1}^{n} x_j\,. \tag{6.39}$$

*Reasoning:* Equation (6.38) is here:

$$\sum_{j=0}^{n}\left(\frac{j}{\beta} - 1\right)h_j = 0\,.$$

This implies $\beta = \sum_{j=0}^{\infty} h_j j$, which corresponds to (6.39), as $nh_j$ is the number of measured quantities $x_r$ which are equal to $j$.

## 6.3.6 Multivariate analysis

If we are given voluminous data, a central question is:

| Are the given measurements purely random or do they depend on a finite number of variables of some kind? |
|---|

To answer this question the methods of multivariate analysis are applied, which originate in the work of the American statistician Ronald Fischer (1890–1962). The most important influencing quantities are also called *factors.*

In what follows we shall merely describe some of the basic ideas. The statistical software SPSS or the various software packages of the company SAS can be used to apply these methods to concrete situations.

### 6.3.6.1 Variance analysis

**Known factors and clusters:** The methods of variance analysis are used to investigate the influence of *known* factors on a set of data. For this, the measurements are split into *clusters.*

The basic idea is that the standard deviation (resp. variance) of the factors is much larger than the standard deviation (resp. variance) of a random perturbation of the data. From the theoretical point of view, this approach is related to the $F$-test (cf. 6.3.3.4).

*Example:* We wish to investigate how fertilizers affect the yearly yield of a crop of grain. For this we choose $n$ different fertilizers

$$\boxed{F_1, \ldots, F_n}$$

and use differing amounts of each on the fields.

We allow several of the fertilizers to be applied to a single field.

(i) The *factors* are the different kinds of fertilizers $F_1, \ldots, F_n$.

(ii) The *measured quantity* is the yearly yield of all fields.

(iii) All fields to which the same fertilizers have been applied in identical concentrations are the *clusters* of the analysis.

The method of variance analysis makes it possible to first of all ascertain whether the measured quantity (yearly yield) is at all affected by the fertilizers.

If this is the case, then one can use the method of multiple means to obtain the clusters which have been affected most strongly by the corresponding fertilizers.

### 6.3.6.2 Factor analysis

**Factors:** As opposed to variance analysis, factor analysis uses factors which are *a priori* not known. The goal is to find as small a group as possible of background variables (factors) which determine the data (and the large number of variables on which it depends) as far as possible. For this several variables which are strongly correlated are collected, or grouped, into one factor.

*Example:* We would like to know which factors determine damage to forests. To start the investigation, $k$ properties

$$\boxed{M_1, \ldots, M_k}$$

are chosen and measured in different forests (examples could be number of leaves on a certain type of tree, the thickness of bark, etc.).

With the aid of factor analysis, we can, from our measurements, determine whether $n$ factors

$$\boxed{F_1, \ldots, F_n}$$

have an essential influence on the data. Working with computer programs, also different possible values for $n$ can be tried out.

It is important to note that statistics of this kind do not make any statement about the kind of factors which will be important. Rather, this is the job of the investigator, and requires a certain amount of experience and skill. However, once these have been determined, statistics does provide an additional aid in the investigation: the so-called *factor weights* describe the strength of the influence of the different factors $F_1, \ldots, F_n$ on the properties $M_1, \ldots, M_k$.

### 6.3.6.3 Cluster analysis

**Division into clusters:** In order to divide a given set of data into clusters, one uses cluster analysis. The clusters are supposed to collect data of like properties.

*Example:* A bank wishes to divide the set of its customers into clusters:

$$\boxed{G_1, G_2, G_3, G_4 .}$$

These clusters correspond to the following credit ratings of the customers: *very trustworthy, trustworthy, partially trustworthy* and *not trustworthy.*

In order to achieve a sensible division, we require much information about the customers, which may perhaps have to be obtained from other sources, like other banks, information to persons (age, earnings, etc.). Cluster analysis is the right tool to carry out such a division.

### 6.3.6.4 Discriminant analysis

Following a division into clusters by an application of cluster analysis, one can apply discriminant analysis. This statistical method makes it possible to determine the rew traits which characterize the clusters in an optimal way. New data can then be immediately divided among the existing clusters. An important assumption for discriminant analysis is that a clustering already exists.

*Example 1:* If, as in 6.3.6.3, we have a division of bank customers into clusters according to their credit ranking, discriminant analysis makes it then possible to determine the (measurable) factors which characterize the various clusters. Thus, for a new customer, it is easy to put her or him into one of the clusters, i.e., ascertain her or his credit rating.

*Example 2:* Discriminant analysis is often applied in medicine. Those patients which suffer from certain diseases are divided into clusters according to their reaction to certain medicines. The properties of the corresponding measured data (degree of fever, composition of blood, etc.) are then characteristics of these clusters. With the aid of discriminant analysis, the physicians can make some prediction on the chance of a new patient for recovering from the illness.

#### 6.3.6.5 Multiple regression

Let random variables

$$X_1, \ldots, X_n \quad \text{and} \quad Y_1, \ldots, Y_m$$

be given. We seek a functional correspondence of the form

$$Y_j = F_j(X_1, \ldots, X_n) + \varepsilon_j\,, \qquad j = 1, \ldots, m\,.$$

To estimate the (unknown) functions $F_1, \ldots, F_m$ in the context of a given class of functions, we use the measurements $X_1, \ldots, X_n$ and $Y_1, \ldots, Y_m$. The random quantities $\varepsilon_j$ describe small perturbations.

**Linear regression:** If all functions $F_j$ are supposed to be linear, then we obtain a (linear) system of equations

$$Y_j = \sum_{k=1}^{m} a_{jk} X_k + b_j + \varepsilon_j\,, \qquad j = 1, \ldots, m\,.$$

The coefficients $a_{jk}$ and $b_j$, which are assumed to be real, need to be estimated in terms of the measured data. What we do here is a generalization of the degrees of regression and the correlation coefficients discussed in section 6.2.3.3.

Computer programs are very flexible and make it possible to determine the most important factors by deleting some of the coefficients $a_{jk}$ and $b_j$ which are sufficiently small.

**Recommendation for practical applications:** In order to determine which of the various statistical methods is appropriate for a given problem, a certain amount of experience is required. One should take note of the fact that each of the methods depends on certain assumptions, which often are only approximately satisfied.

If you are unsure how to proceed, the best thing to do is to consult an expert, i.e., someone with experience with the application of statistics to practical problems. Such experts can be contacted at university computer centers or other scientific institutions.

## 6.4 Stochastic processes

*In this thesis we shall show that, according to the theory of molecular thermal kinematics, microscopic particles suspended in liquids must, as a result of thermal motion, necessarily carry out motions of such magnitude that they are simply visible in a microscope. It is possible that the motions described here coincide with the so-called 'Brownian motion';*[20] *however, as the information I have about the latter is so imprecise, I do not wage to make a more definite statement.*

*Albert Einstein, 1905*

Stochastic processes are time-dependent processes in nature, engineering and economy, which depend to some extent on random factors. Already the ancient Egyptians attempted to determine the laws for the flooding properties of the Nile river to protect their civilization from damage due to flooding. More examples for stochastic processes

[20] This motion was discovered in 1827 by the English botanic scientist Robert Brown (1773–1858) during observations through a microscope.

are the temperature measurements at some fixed place taken over time, the development of populations, forces acting on a car driving along a rocky road, price development of stocks or commodities and the change in the gross national product of a nation in time.

In classical thermodynamics stochastic processes are used to account for the imprecision of dealing with the millions of gas particles in a gaseous volume. The situation has been totally changed since the advent of quantum mechanics. This is because quantum processes are stochastic by very nature.

**Historical remarks:** The investigation of stochastic processes in physics goes back to the fundamental paper of Einstein in 1905 cited from above; in 1922 Norbert Wiener (1894–1964) began the systematic mathematical investigation of Brownian motion. He realized:

> The precise mathematical description of Brownian motion utilizes a probability measure $\mu$ on the space of trajectories.

The typical jittery motions involved in the Brownian motion observed under the microscope is mathematically described by the fact that the trajectories, even though they are continuous, are, with probability one, not differentiable.[21]

For the calculation of expectations, one requires the notion of integral with respect to a measure $\mu$ which Wiener introduced and today carries his name.

In 1933, Andrei Kolmogorov (1903–1987) created modern (set-theoretic) axiomatic probability theory and at the same time, following the lead of Wiener, laid the foundations for a general theory of stochastic processes. The elementary events in this theory are the possible trajectories (realization of the process). The probability that a certain set of trajectories will occur is a measure on the space of all trajectories. Thus the theory of stochastic processes leads naturally to measure theory, including integrals over subsets of spaces of functions.

During the second world war, Norbert Wiener, then working at MIT, developed prediction theory, in order to shoot down enemy planes over England with a maximum of precision.

In his famous 1941 dissertation at Princeton University, the ingenious physicist Richard Feynman (1918–1988) presented a completely new approach to quantum mechanics by introducing the *Feynman integral*, which sums over all possible trajectories of quantum particles and takes an average of these. From a mathematical point of view this is a formal Wiener integral in imaginary time. Many calculations of quantum field theory (used in the theory of elementary particles) are based on the very successful application of Feynman's integral, although there still is not a rigorous mathematical justification for this integral even today.

The physicist Edward Witten, who works at the Institute for Advanced Studies in Princeton, has masterly applied the Feynman integral to gain totally new insights in deep mathematical topics. He was awarded the Fields Medal for this work at the International Congress of Mathematicians in Kyoto in 1990 (see also the historical outline of

[21] Ampère attempted to prove in 1806 that every continuous function is in fact differentiable. It wasn't until 50 years later that Weierstrass constructed a function which is continuous everywhere, but differentiable nowhere. The precise formulation of this result can be found in [212]. The investigations of Wiener proved that this is by far not a mathematical sophistry, but that functions of this kind occur in nature.

Modern chaos theory has shown that a series of mathematical phenomena, all of which were theoretically constructed during the second half of the nineteenth century using set theory and at the time were considered to be pathologies, play an important role in nature. This is for example also the case with so-called *fractals* with non-integral dimension.

mathematics at the end of this volume).

**Goal:** The goal of the theory of stochastic processes is to derive theoretical laws for the interpretation and future calculation of these processes using only the empirical data at hand.

## 6.4.1 Time series

**Basic ideas:** We consider the points in time

$$t = 0, \Delta t, 2\Delta t, \ldots$$

with $\Delta t > 0$. To every time $t = n\Delta t$ is associated a random variable $X_n$. The measured values of this random variable $X_n$ at time $t = n\Delta t$ will be denoted by $x_n$.

*Example 1:* An analysis of the price for wheat during the years 1500 and 1870 showed that there was a period of 13.3 years in these prices. An excerpt of this data is shown in Figure 6.19.

> An important problem in the analysis of time series is to discover periodic behavior in time or to negate the existence of such.

For this one uses the elementary method of autocorrelation coefficients or the more modern method of spectral analysis (Fourier analysis of the autocovariance function).

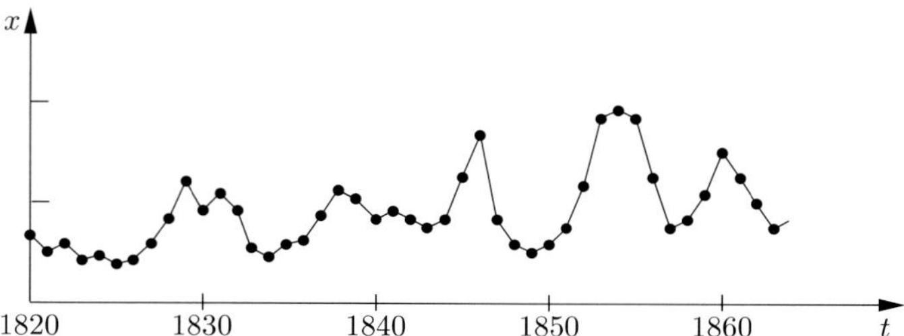


*Figure 6.19. Periodic behavior of wheat prices from 1820–1860.*

**Evaluation of time series on computers:** We recommend using the software package SPSS Statistic.[22]

### 6.4.1.1 Empirical autocorrelation coefficients

Let a variational series $x_0, x_1, \ldots, x_N$ be given. We fix a number $k = 1, 2, \ldots, N/2$ and consider the two relatively shifted sub-series

$$\boxed{x_0, x_1, \ldots, x_{N-k}}$$

and

$$\boxed{x_k, x_{k+1}, \ldots x_N\,.}$$

[22]Since the publication of the original in 1996 there has of course been a great deal of movement in this market. – *The translator*

Recall the notion of empirical correlation coefficient $r_k$ of these two variational series; here it is called the $k^{th}$-*autocorrelation coefficient* of the variational series. Explicitly we have

$$r_k := \frac{\sum_{j=0}^{N-k}(x_j-\overline{x})(y_i-\overline{y})}{\left(\sum_{j=0}^{N-k}(x_j-\overline{x})^2\sum_{j=0}^{N-k}(y_j-\overline{y})^2\right)^{1/2}}.$$

Here we have set $y_j := x_{j+k}$ and

$$\overline{x} = \frac{1}{N-k+1}\sum_{j=0}^{N-k} x_j\,, \qquad \overline{y} = \frac{1}{N-k+1}\sum_{j=0}^{N-k} y_j\,.$$

**Interpretation of $r_k$:** First of all, we have $-1 \le r_k \le 1$. The larger $|r_k|$ is, the larger the correlation of the values which are obtained by shifting the time by $k\Delta t$.

(i) If the values of $|r_k|$ for $k = m, 2m, 3m, \ldots$ are particularly large compared with the other values, this is a strong indication of the existence of a period $T = 2m\Delta t$ in the time series (see Example 2 below).

(ii) If all values of $|r_k|$ are small, then there is practically no correlation between the values measured at different times. This is a strong indication of an aperiodic behavior of the time series.

*Example 2:* For a purely periodic series

$$x_j = a\cos\left(\frac{2\pi j\Delta t}{T}\right), \qquad j = 0, 1, \ldots, N$$

of period $T = 2m\Delta t$ and amplitude $a \neq 0$ we have for a large number $N$ of measured values approximately

$$r_k = \cos\left(\frac{2\pi k\Delta t}{T}\right), \qquad k = 1, 2, \ldots$$

In particular, we get (see Figure 6.20(a)):

$$r_0 = 1\,, \quad r_m = -1\,, \quad r_{2m} = 1\,, \quad r_{3m} = -1\,, \quad r_{4m} = 1\,, \quad \ldots$$

*Example 3:* Suppose that $x_j = \text{const}$ for all $j$. Then we have $r_k = 0$ for all $k$.

*Example 4:* In Figure 6.20(b) we have drawn a time series which is only correlated for short times. For large time shifts $k\Delta t$ the value of $|r_k|$ gets smaller and smaller, hence there is no longer any correlation at all.

**Approximate values:** For a large number $N$ of measurements we have approximately

$$\boxed{r_k = \frac{c_k}{c_0}} \tag{6.40}$$

with

$$c_k := \frac{1}{N}\sum_{j=0}^{N-k}(x_j-\overline{x})(x_{j+k}-\overline{x})\,, \qquad \overline{x} := \frac{1}{N}\sum_{j=0}^{N} x_j\,.$$

The approximation (6.40) is often applied to practical cases.

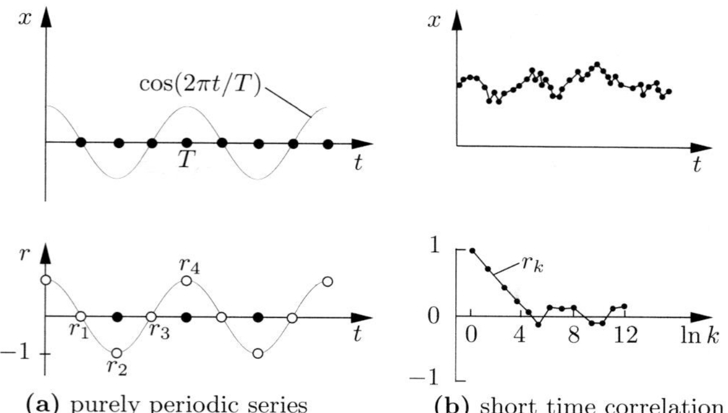


(a) purely periodic series (b) short time correlation

*Figure 6.20. Time series and the autocorrelation coefficients.*

#### 6.4.1.2 The spectral analysis of discrete time series

This method is based on the use of Fourier series, whose coefficients are the autocovariance coefficients.

**Stationary time series:** Let $E$ be a probability space and

$$X_n : E \longrightarrow \mathbb{R}\,, \qquad n = 0, \pm 1, \pm 2, \ldots$$

a series of random variables $X_n$ with $0 < \Delta X_0 < \infty$. We interpret $X_n$ as a random variable at time $t = n\Delta t$ and say the series is *stationary* (in a broad sense) if

$$\boxed{\overline{X}_n = \overline{X}_0}$$

and

$$\boxed{\mathrm{Cov}(X_n, X_{n+k}) = \mathrm{Cov}(X_0, X_k)}$$

for all $n, k = 0, \pm 1, \pm 2, \ldots$ Here $\mathrm{Cov}\,(X,Y) = \overline{(X - \overline{X})(Y - \overline{Y})}$, so that in particular $\mathrm{Cov}\,(X,X) = (\Delta X)^2$.

**Definition:** We define the *autocorrelation coefficients*

$$r_k := \frac{\mathrm{Cov}(X_n, X_{n+k})}{\Delta X_n \Delta X_{n+k}}\,, \qquad k = 1, 2, \ldots$$

**Interpretation:** The number $r_k$ is the correlation coefficient between the random variable $X_n$ at time $n\Delta t$ and the random variable $X_{n+k}$ at time $(n+k)\Delta t$. That the time series is stationary is expressed in the fact that the expectations $\overline{X}_n$ and the standard deviation $\Delta X_n$ do not depend on the time $t = n\Delta t$. Moreover, the correlation coefficient $r_k$ does not depend on the chosen time $t = n\Delta t$, but rather depends only on the difference $k\Delta t$.

**Theorem:** If we set $R(k) := \mathrm{Cov}\,(X_0, X_k)$, then we have

$$\boxed{r_k = \frac{R(k)}{R(0)}\,, \qquad k = 0, 1, \ldots}$$

In addition, $R(-k) = R(k)$ for all $k$.

**Spectral theorem:** Let $\sum_{k=0}^{\infty} |R(k)| < \infty$. Then the function

$$f(\lambda) := \frac{1}{2\pi} \sum_{k=-\infty}^{\infty} R(k)\mathrm{e}^{-\mathrm{i}k\lambda}$$

is continuous, and we have

$$\boxed{R(k) = \int_{-\pi}^{\pi} f(\lambda)\, \mathrm{e}^{\mathrm{i}k\lambda}\, \mathrm{d}\lambda, \quad k = 0, \pm 1, \pm 2 \ldots} \tag{6.41}$$

In particular, we get for the variance the expression

$$(\Delta X_n)^2 = R(0) = \int_{-\pi}^{\pi} f(\lambda)\, \mathrm{d}\lambda \quad \text{for all} \quad n\,. \tag{6.42}$$

The $2\pi$-periodic function $f$ is called the *spectral density* of the stationary time series. According to (6.41) and (6.42), it contains all information on the standard deviation $\Delta X$ and the autocorrelation coefficients $r_k = R(k)/R(0)$.

*Example 1:* For a fixed natural number $n \geq 1$, let

$$R(\pm n) \neq 0 \quad \text{and} \quad R(k) = 0 \quad \text{for all} \quad k \neq n\,.$$

Then we have

$$f(\lambda) = \frac{R(n)}{2\pi}(\mathrm{e}^{-\mathrm{i}n\lambda} + \mathrm{e}^{\mathrm{i}n\lambda}) = \frac{1}{\pi}\cos n\lambda\,, \qquad \lambda \in \mathbb{R}\,.$$

*Example 2:* Let

$$R(0) \neq 0 \quad \text{and} \quad R(k) = 0 \quad \text{for all} \quad k \neq 0\,.$$

Then

$$f(\lambda) = \frac{R(0)}{2\pi}\,, \qquad \lambda \in \mathbb{R}\,.$$

Uncorrelated time series of this kind are referred to as *white noise.*

#### 6.4.1.3 Statistics of discrete time series

Let a time series $(X_n)$ be given, which is stationary (in the broad sense). We now wish to estimate important quantities from measured values

$$x_0, x_1, \ldots, x_{N-1}$$

provided that $N$ is sufficiently large.

(i) *Estimate of the expectation* $\overline{X}_n$ *for all* $n$*:*

$$\mu = \frac{1}{N}\sum_{j=0}^{N-1} x_j\,.$$

(ii) *Estimate of $R(k)$:*

$$R_N(k) = \frac{1}{N-k} \sum_{j=0}^{N-k-1} (x_j - \mu)(x_{j+k} - \mu)\,.$$

In particular, $R_N(0)$ is the estimate for the standard deviation $\Delta X_n$ for all $n$. Moreover,

$$\frac{R_N(k)}{R_N(0)}\,, \qquad k = 1, 2, \ldots$$

is an estimate for the autocorrelation coefficient $r_k$.

(iii) *Estimate of the spectral density:*

$$f_N(\lambda) = \frac{1}{2\pi N} \left| \sum_{n=0}^{N-1} x_j \, \mathrm{e}^{-\mathrm{i}n\lambda} \right|^2 .$$

### 6.4.1.4 Herglotz' spectral theorem

**Theorem:** If $(X_n)$ is a stationary (in the broad sense) time series, then there is a non-decreasing, bounded function $F : [-\pi, \pi] \longrightarrow \mathbb{R}$ such that

$$\boxed{R(k) = \int_{-\pi}^{\pi} \mathrm{e}^{\mathrm{i}\lambda k} \, \mathrm{d}F(\lambda)\,, \qquad k = 0\,, \pm 1\,, \pm 2\,, \ldots} \tag{6.43}$$

giving a representation of $R(k)$ as a Lebesgue–Stieltjes integral. If $\sum_{k=0}^{\infty} |R(k)| < \infty$, then the derivative $F'(\lambda) =: f(\lambda)$ exists almost everywhere and we can replace the symbol $\mathrm{d}F(\lambda)$ in (6.43) by $f(\lambda)\mathrm{d}\lambda$.

### 6.4.1.5 Spectral analysis of continuous time series and white noise

Let $E$ be a probability space and

$$X_t : E \longrightarrow \mathbb{R}\,, \qquad t \in \mathbb{R}$$

a continuous series of random variables $X_t$ with $0 < \Delta X_0 < \infty$. We interpret $X_t$ as a random variable at time $t$. This series will be called *stationary* (in the broad sense) if

$$\boxed{\overline{X}_t = \overline{X}_0}$$

and

$$\boxed{\mathrm{Cov}(X_t, X_{t+s}) = \mathrm{Cov}(X_0, X_s)}$$

for all times $t, s \in \mathbb{R}$. The expression

$$R(s) := \mathrm{Cov}(X_t, X_{t+s})$$

defines a function $R : \mathbb{R} \longrightarrow \mathbb{R}$ which is called the *autocovariance function.* The number

$$r(s) = \frac{R(s)}{R(0)}, \quad s \in \mathbb{R},$$

is the correlation coefficient between the random variables $X_t$ and $X_{t+s}$. We have $R(-s) = R(s)$ for all $s$.

**Spectral theorem:** Let $\int_{-\infty}^{\infty} |R(s)| \mathrm{d}s < \infty$. Then the function

$$f(\lambda) := \frac{1}{2\pi} \int_{-\infty}^{\infty} R(s)\mathrm{e}^{-\mathrm{i}s\lambda}\, \mathrm{d}s$$

is continuous and we have

$$\boxed{R(s) = \int_{-\pi}^{\pi} f(\lambda)\mathrm{e}^{\mathrm{i}s\lambda}\, \mathrm{d}\lambda, \qquad s \in \mathbb{R}.}$$

Thus the spectral density function $f$ is the Fourier transform of the autocovariance function $R$.

**Short time correlation, white noise and distributions:** For some fixed, small $\varepsilon > 0$, let

$$R_\varepsilon(s) := \begin{cases} \dfrac{1}{2\varepsilon}, & \text{if} \quad -\varepsilon \le s \le \varepsilon, \\ 0, & \text{otherwise}. \end{cases}$$

Then the correlation function satisfies $r_\varepsilon(s) = 1$ on the small time interval $[-\varepsilon, \varepsilon]$ and $r(s) = 0$ outside of this interval. This is a typical short time correlation. For the spectral density $f_\varepsilon$ we obtain the expression

$$f_\varepsilon(\lambda) = \frac{1}{4\pi\varepsilon} \int_{-\varepsilon}^{\varepsilon} \mathrm{e}^{-\mathrm{i}\lambda s}\, \mathrm{d}s, \qquad \lambda \in \mathbb{R}.$$

In the limiting case $\varepsilon \longrightarrow +0$, we get

$$\boxed{\lim_{\varepsilon \to +0} f_\varepsilon(\lambda) = \frac{1}{2\pi}.}$$

This limiting case is called *white noise.* In the language of physicists and engineers, we have for the autocovariance function the relation

$$R(s) = \lim_{\varepsilon \to +0} R_\varepsilon(s) = \delta(s), \tag{6.44}$$

where $\delta$ denotes the Dirac delta distribution (see [212]). In fact, $\delta$ is not a classical function but a distribution. The formal relation (6.44) is more rigorously to be interpreted as the relation

$$R = \lim_{\varepsilon \to +0} R_\varepsilon = \delta_0 \tag{6.45}$$

in the sense of distributions. This means

$$\lim_{\varepsilon \to +0} \int_{-\infty}^{\infty} R_\varepsilon(s)\varphi(s)\,\mathrm{d}s = \delta_0(\varphi) = \varphi(0)$$

for all test functions $\varphi \in C_0^\infty(\mathbb{R})$.

White noise is a model of stochastic processes occuring in nature and engineering, for which there is only a correlation among very small time shifts.

### 6.4.2 Markov chains and stochastic matrices

The Bernoulli experimental scheme is based on the assumption that the individual trials are independent (see 6.2.5). The Markov chains to be introduced in this section, which were first investigated by Andrei Markov (1856–1922) in 1906, are the simplest possible generalization of Bernoulli's scheme to events which are dependent.

**The basic model:** (i) We consider a system of discrete points in time

$$t = 0,\ \Delta t,\ 2\Delta t,\ 3\Delta t, \ldots$$

(ii) To each of these times we consider a system of states

$$Z_1, Z_2, \ldots, Z_k.$$

(iii) We define

$p_{ij} :=$ transition probability of the state $Z_i$ at time $t = n\Delta t$ changing into the state $Z_j$ at time $t = (n+1)\Delta t$.

This transition probability is equal to the conditional probability that the system is in the state $Z_j$ at time $t = (n+1)\Delta t$, provided it had been in the state $Z_i$ at time $t = n\Delta t$. This system is what is referred to as a *Markov chain.*

It is typical for Markov chains that the transition probabilities do not depend on the point in time at which they are considered (homogeneity in time).

(iv) We collect these transition probabilities into a so-called square *transition matrix*

$$P := \begin{pmatrix} p_{11}\, p_{12} & \cdots & p_{1k} \\ \cdots & & \\ p_{k1}\, p_{k2} & \cdots & p_{kk} \end{pmatrix}.$$

We assume that $P$ is a *stochastic matrix,* meaning that the elements of $P$ satisfy the inequality

$$0 \le p_{ij} \le 1$$

and all sum of rows are equal to unity, i.e., one has

$$\sum_{j=1}^{k} p_{ij} = 1 \qquad \text{for} \quad i = 1, \ldots, k\,.$$

**Definition:** Let $p_{ij}^{(k)}$ denote the transition probability for the state $Z_i$ at time $t = n\Delta t$ moving to the state $Z_j$ at time $t = (n+k)\Delta t$. The elements $(p_{ij}^{(k)})$ are collected in a matrix which we denote by $P^{(k)}$, and which could naturally be called the $k^{th}$ transition matrix.

**The Chapman–Kolmogorov equality:** For all $k, m = 1, 2, \ldots$ we have

$$\boxed{P^{(k+m)} = P^{(k)}P^{(m)}\,.} \tag{6.46}$$

**Corollary:** $P^{(k)} = P^k$.

In the special case $m = 1$ (resp. $k = 1$), (6.46) is referred to as the *forward equation* (resp. *backward equation*).

#### 6.4.2.1 Ergodic behavior

**Definition:** A Markov chain is said to be *ergodic*, if the limits

$$p_j := \lim_{n\to\infty} p_{ij}^{(n)}\,, \qquad i = 1, \ldots, N$$

exist and are independent of $i$. Moreover, we assume that $p_j > 0$ for all $j$ such that $\sum_{j=0}^{N} p_j = 1$.

**Interpretation:** An ergodic Markov chain completely forgets the situation at the initial time $t = 0$. The number $p_j$ is the probability that for large times the state $Z_j$ is realized.

**The Ergodic Theorem of Markov (1906):** A Markov chain is ergodic if and only if there is a natural number $n \geq 1$ such that all entries of $P^n$ are positive.

**A model for the spreading of rumors:** We denote by

$$Z_1, Z_2$$

two possible variations of a message. For example $Z_1$ (resp. $Z_2$) could mean that Mr $X$ will turn in his notice (resp. will not turn in his notice). We assume the following transition matrix

$$P = \begin{pmatrix} 1-p & p \\ q & 1-q \end{pmatrix}.$$

This means

(i) If a person hears the message $Z_1$, then with a probability of $p$ he will spread the message $Z_2$, while the probability that he spreads it correctly is $1-p$ ($p$ is the probability that the message will be incorrectly reproduced, i.e., becomes a rumor).

(ii) Similarly the message $Z_2$ will be incorrectly reproduced with a probability of $q$.

The assumptions $0 < p < 1$ and $0 < q < 1$ are realistic. Then we have

$$\lim_{n\to\infty} P^n = \frac{1}{p+q}\begin{pmatrix} q & p \\ q & p \end{pmatrix}.$$

This means that $p_1 = \dfrac{q}{p+q}$ and $p_2 = \dfrac{p}{p+q}$. In particular, if $p = q$ we get

$$p_1 = p_2 = \frac{1}{2}\,.$$

After the rumor has been spread for some time the public opinion is that Mr $X$ will turn in his notice with a probability of $p_1 = 1/2$, which has no relation to the actual fact and original message.

#### 6.4.2.2 Recurrence

Suppose that at time $t = 0$ we have a state of our system $Z_i$. We denote by $w_n$ the probability that the system will first return to this state precisely at the time $t = n\Delta t$. Then

$$w = \sum_{j=0}^{\infty} w_n$$

is equal to the probability that the system will return back to the original state after some finite time.

**Definition:** The state $Z_i$ at the initial time $t = 0$ is said to be *recurrent*, if the probability $w$ just introduced is equal to unity. Otherwise we call $Z_i$ *transient.*

**Theorem:** (i) The initial state $Z_i$ is recurrent if and only if

$$\sum_{n=1}^{\infty} p_{ii}^{(n)} = \infty\,.$$

(ii) If the system is in a recurrent state $Z_i$ at the initial time $t = 0$, then with probability one it returns to this state infinitely often.

(iii) If the initial state $Z_i$ is transient, then there is a finite time after which the state $Z_i$ is never again reached.

### 6.4.3 Poisson processes

Poisson processes are used to describe events which occur very rarely in a given short time period.

**Basic model:** We consider an event $\mathscr{E}$ (for example the arrival of a telephone call at the switchboard) and define

$P_n(t,s) :=$ probability that during the time interval $[t,s]$ exactly $n$ events $\mathscr{E}$ (telephone calls in our example) occur.

We make the following assumptions:

(i) The process is homogenous in time, i.e., $P_n(t,s)$ depends only on the difference $s-t$ but not on the starting point $t$.

(ii) $P_1(t,s) = \mu(t-s) + o(t-s)$ for $s-t \to 0$.

(iii) $P_n(t,s) = o(t-s)$ for $s-t \to 0$ and all $n = 2, 3, \ldots$

**Consequence:** Then we have

$$P_n(t-s) = \mathrm{e}^{-\mu(t-s)}\frac{\mu^n(s-t)^n}{n!}, \qquad n = 0, 1, 2, \ldots$$

This is a Poisson distribution for a given fixed interval $[t,s]$.

If $X_{s-t}$ denotes the number of events $\mathscr{E}$ occuring in the time interval $[t, s]$, then we get for the expectation (mean) and the standard deviation

$$\boxed{\overline{X}_{s-t} = \mu(s-t)\,, \quad \Delta X_{s-t} = \sqrt{\mu(s-t)}\,.}$$

### 6.4.4 Brownian motion and diffusion

#### 6.4.4.1 The classical model of random motion

**The transition probabilities for Brownian motion:** We consider a particle moving on the real axis. We denote by $X_t$ the location of the particle at time $t$, and define

$P(y, s; J, t) :=$ conditional probability that the particle is in the interval $J$ at time $t$, provided it had been at the point $y$ at time $s$.

The quantity $P(y, s; J, t)$ is also referred to as a transition probability. For Brownian motion we have

$$\boxed{P(y, s; J, t) = \int_J p(y, s; x, t)\, \mathrm{d}x} \tag{6.47}$$

with

$$\boxed{p(y, s; x, t) := \frac{1}{\sqrt{2\pi(t-s)}} \mathrm{e}^{-(x-y)^2/2(t-s)}\,, \qquad s < t\,.}$$

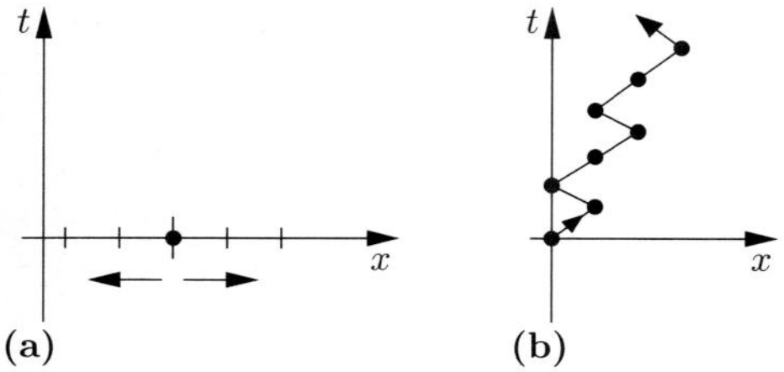

**Motivation:** We partition the real axis by means of the lattice points

$$x = 0, \pm\Delta x, \pm 2\Delta x, \ldots$$

and consider the discrete points in time

$$t = 0, \Delta t, 2\Delta t, \ldots$$

The particle is assumed to be at $x = 0$ at the initial time $t = 0$. Let us assume that if at time $t = n\Delta t$, the particle is at the point

$$x = m\Delta t\,,$$

Figure 6.21. Brownian motion.

then at the next time step $t = (n+1)\Delta t$ it is at the right neighboring point

$$x = (m+1)\Delta x$$

with probability $p = 1/2$, and with the same probability at the left neighboring point

$$x = (m-1)\Delta x$$

(see Figure 6.21). This situation can be obtained with a Bernoulli experiment with $N$ trials (cf. 6.2.5). The reference trials are $e_+$ (motion to the right) and $e_-$ (motion to the left). If the event

$$e_{i_1} e_{i_2} \ldots e_{i_N}$$

with $e_{i_j} = e_\pm$ is such that exactly $k$ symbols are $e_+$ and $N-k$ symbols are $e_-$, then the particle is at time $t = N\Delta t$ at the point

$$x = k\Delta x - (N-k)\Delta x\,.$$

The probability for this is given by

$$\binom{N}{k}\frac{1}{2^k}\frac{1}{2^{N-k}}\,.$$

If one passes to the limit $N \to \infty$ with $\Delta x \to 0$, an finally $\Delta t \to 0$, then we obtain (6.47) by virtue of the limit theorem of de Moivre–Laplace. We refrain from further details on this.

#### 6.4.4.2 The diffusion equation

We now consider, instead of a single particle, a fluid on the $x$-axis with the particle density $\rho(x,t)$ at the point $x$ and time $t$. Then we get from (6.47) the relation

$$\boxed{\rho(x,t) = \int_{-\infty}^{\infty} \rho(y,s)p(y,s;x,t)\,\mathrm{d}y\,.}$$

From this the diffusion equation

$$\boxed{\rho_t = \frac{1}{2}\rho_{xx}\,, \qquad x \in \mathbb{R},\ t > s}$$

follows. Diffusion processes can thus be illuminated with the example of random motion.

#### 6.4.4.3 The Wiener measure and the Wiener process

We set $\mathbb{R}_+ := \{t \in \mathbb{R} \mid t \geq 0\}$ and denote by $C(x_0)$ the space of all continuous functions

$$w : \mathbb{R}_+ \to \mathbb{R} \quad \text{with} \quad w(0) = x_0\,.$$

We interpret $x = w(t)$ as the trajectory of a particle which is at the point $x_0$ at the initial time $t = 0$. Our goal is to construct a $\sigma$-algebra $\mathscr{S}$ of subsets $M$ of $C(x_0)$ and a measure $\mu_{x_0}$ on $\mathscr{S}$ such that

| $\mu_{x_0}(M) :=$ probability that a trajectory $x = w(t)$ of the Brownian motion belongs to the subset $M$ of $C(x_0)$. |
|---|

With this intent we first consider so-called *cylindrical sets*[23]

$$\mathscr{Z} := \{w \in C(x_0) \mid w(t_k) \in J_k\,,\ k = 1,\ldots,n\}\,.$$

[23]Note that this set $\mathscr{Z}$ depends on the $J_k$, so a more precise notation would be $\mathscr{Z}(J_1,\ldots,J_n)$. However for simplicity we refrain from using this more precise notion and trust this will cause no confusion.

The times $t_k$ are assumed to satisfy the condition $0 < t_1 < t_2 < \cdots < t_n$, and $J_k$ is an arbitrary real interval. Thus $\mathscr{Z}$ consists of all continuous trajectories $x = w(t)$ which are in the interval $J_k$ at time $t_k$, $k = 1, \ldots, n$. In case $0 < t_1 < t_2$ we define

$$\mu_{x_0}(\mathscr{Z}) := \int_{J_1}\int_{J_2} p(x_0, t_0; x_1, t_1) p(x_1, t_1; x_2, t_2)\,\mathrm{d}x_1\,\mathrm{d}x_2$$

with $t_0 := 0$. In the general case $0 < t_1 < \cdots < t_n$ we set

$$\boxed{\mu_{x_0}(\mathscr{Z}) := \int_{J_1} \ldots \int_{J_n} p(x_0, t_0; x_1, t_1) \cdots p(x_{n-1}, t_{n-1}; x_n, t_n)\,\mathrm{d}x_1 \cdots \mathrm{d}x_n\,.}$$

Finally, we denote by $\mathscr{S}$ the smallest $\sigma$-algebra of subsets of $C(x_0)$ containing all the cylindrical sets (see [212]).

**Theorem:** The measure $\mu_{x_0}$ can be uniquely extended to a measure on all of $\mathscr{S}$. It is called the *Wiener measure.*

*Example:* We denote by $\mathscr{D}$ the set of all differentiable trajectories in $C(x_0)$. Then

$$\mu_{x_0}(\mathscr{D}) = 0\,.$$

Thus a trajectory of Brownian motion is almost surely not differentiable. This explains the jittery motion of Brownian motion.

**The Wiener process:** We let $X_t$ denote the position $w(t)$ of a trajectory $w \in C(x_0)$ at time $t$. Then we have:

(i) $X_t$ is normally distributed with expectation (mean) $\overline{X}_t = x_0$ and standard deviation $\Delta X_t = \sqrt{t}$.

(ii) The probability that $X_t = x_0$ at time $t = 0$ is one.

(iii) Let $h > 0$ and $0 < s < t$. Then the differences

$$X_{t+h} - X_t \qquad \text{and} \qquad X_{s+h} - X_s$$

are independent. Moreover, these quantities are normally distributed with mean 0 and standard deviation $\sqrt{h}$.

(iv) The quantities $X_{t_1}, X_{t_2}, \ldots, X_{t_n}$ have a common distribution function $\Phi_{t_1 \cdots t_n}$ with density

$$\prod_{j=1}^{n} \varphi_{t_j}(x_j)\,.$$

Here $\varphi_t(x) := \dfrac{1}{\sqrt{2\pi t}}\mathrm{e}^{-(x-x_0)^2/2t}$ is a normal distribution with mean $\mu = x_0$ and standard deviation $\sigma = \sqrt{t}$. One then defines a *Wiener process* as the family

$$\boxed{X_t : C(x_0) \to \mathbb{R}, \quad t \geq 0\,,} \tag{6.48}$$

of random variables $X_t$ on the probability space $(C(x_0), \mathscr{S}, \mu_{x_0})$ (a stochastic process). This process was investigated in 1922 by Norbert Wiener.

#### 6.4.4.4 The formula of Feynman and Kac

Diffusion results from the Brownian motion of particles. The basic idea of the famous Feynman–Kac formula is to obtain the particle density $\varrho$ as an average of stochastic trajectories of these particles. The construction of average utilizes the properties of the Wiener measure $\mu_x$.

We consider the initial value problem

$$\begin{aligned} &\varrho_t = a(\varrho_{xx} - U(x))\,, && x \in \mathbb{R},\ t > 0\,, \\ &\varrho(x,0) = \varrho_0(x)\,, && x \in \mathbb{R} \end{aligned} \tag{6.49}$$

for a diffusion process under an addition external force, which is given by the function $U$. The unknown is the particle density $\varrho = \varrho(x,t)$ at the point $x$ at time $t$. The givens are the initial density $\varrho_0 \in C_0^\infty(\mathbb{R})$ and the function $U \in C_0^\infty(\mathbb{R})$. We choose a unit of time such that $a = 1/2$.

**Theorem:** The unique solution of (6.49) is given by the *Feynman–Kac formula*

$$\varrho(x,t) = \int\limits_{C(x)} \varrho_0(w(t))\mathrm{e}^{-\int\limits_0^t U(w(s))\,\mathrm{d}s}\,\mathrm{d}\mu_x(w)\,, \qquad x \in \mathbb{R},\ t > 0\,. \tag{6.50}$$

Here the integration is over all trajectories $w \in C(x)$. The integral over $C(x)$ with respect to the Wiener measure $\mu_x$ occuring here is to be understood as a measure integral, as explained in [212].

#### 6.4.4.5 The Feynman integral

*Dick Feynman was an exceptional scientist. He invested five years of hard work to formulate his own view of quantum mechanics. The calculation I carried out for Hans Bethe with the help of the orthodox Schrödinger equation took several months and hundreds of pages of paper. For the same result Dick needed, using his methods, only half an hour on the board.*

*Freeman Dyson*, 1979

The initial value problem for the Schrödinger equation

$$\begin{aligned} &-\mathrm{i}\psi_t = a(-\psi_{xx} + U(x))\,, && x \in \mathbb{R},\ t > 0\,, \\ &\psi(x,0) = \psi_0(x)\,, && x \in \mathbb{R}\,, \end{aligned} \tag{6.51}$$

is formally related to the initial value problem (6.49) for the diffusion equation via the formula

$$\psi(x,t) = \varrho(x,\ \mathrm{i}t). \tag{6.52}$$

This allows us to present the following interpretation:

The motion of a quantum particle is Brownian motion in imaginary time.

This is the mathematical background for the Feynman approach to quantum mechanics, which was discovered by Feynman in a completely different way using ingenious physical intuition. Feynman discovered that the quantum motion of a particle can be obtained as a weighted average over its classical trajectories, where the weights correspond to probabilities. This average uses the so-called *Feynman integral*, which at the same time gives the deepest relation between classical and quantum mechanics.

Formally, the Feynman integral is obtained from the Feynman–Kac formula (6.50) by replacing the density $\varrho$ by the Schrödinger (wave) function $\psi$.

The Feynman integral, however, has only in a few exceptional cases been rigorously defined and mathematically justified. Still, it is of immense importance as a tool for calculations of processes of elementary particles in the context of quantum field theory. The secret for the phenomenal success of the Feynman integral lies in its ability to describe *microscopic* effects of the propagation of actions in quantum processes.

An introduction to the Feynman integral together with the physical background can be found in [215].

### 6.4.5 The main theorem of Kolmogorov for general stochastic processes

**The general definition of stochastic processes:** Let $E$ be a probability space with probability measure $P$ and $\sigma$-algebra $\mathscr{S}$. A *stochastic process* is a family

$$\boxed{X_t : E \to \mathbb{R}, \qquad t \in T} \tag{6.53}$$

of random variables $X_t$, where $t$ varies in a non-empty set.

If the index set $T$ is at most countable, then we speak of *discrete* processes, otherwise of *continuous* processes.

*Example 1:* If $T$ is the set of real numbers, then we may view $X_t$ as a random variable at time $t$. For example, $X_t$ could be the temperature at time $t$ in some fixed location.[24]

**Realization of a stochastic process:** Let $e$ be an element of $E$, i.e., an elementary event. We define

$$\boxed{x_e(t) := X_t(e)\,, \qquad t \in T\,.}$$

Then $x_e : T \longrightarrow \mathbb{R}$ is a real function of the time $t$, which we interpret as a *curve of measurements* (Figure 6.22). Let $M$ be a subset of $E$ which is an event. Then we have:

$P(M) =$ probability that one of the curves of measurements $x = x_e(t)$ with an index $e \in M$ has a realization.

The probability space $E$ can therefore be viewed as the index set for all possible curves of measurements, and the probability measure $P$ can be viewed as a measure on the space of all curves of measurements.

[24] If we wish to describe the temperature distribution at every point of the earth, then the random variable will be $X_{(Q,\tau)}$, depending on the location $Q$ as well as the time $\tau$, i.e., the index set $T$ consists of all pairs $(Q, \tau)$.

However, for simplicity and the advantage of concrete intuition, we will stick with $t$ as denoting time.

**Important quantities for applications:** The most important quantities of stochastic processes are the following.

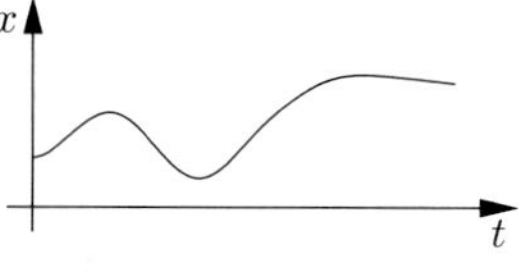


*Figure 6.22.*

(i) The expectation $\overline{X}_t$ and the standard deviation $\Delta X_t$ of the random variable $X_t$ at time $t$.

(ii) The correlation coefficient

$$r(t,s) := \frac{\mathrm{Cov}(X_t, X_s)}{\Delta X_t \Delta X_s}$$

between the random variables $X_t$ and $X_s$ at the times $t$ and $s$.

(iii) The distribution function $\Phi_t$ of $X_t$.

(iv) The common distribution function $\Phi_{t,s}$ of $X_t$ and $X_s$ for times $t$ and $s$ with $t < s$.

**The family of common distribution functions:** Let $\Phi_{t_1 \cdots t_n}$ denote the common distribution function of the random variables $X_{t_1}, \ldots, X_{t_n}$, where the times are to satisfy the inequality

$$t_1 < t_2 < \ldots < t_n. \tag{6.54}$$

More explicitly, we have

$$\boxed{\Phi_{t_1 \ldots t_n}(x_1, \ldots, x_n) := P(X_{t_1} < x_1, \ldots, X_{t_n} < x_n)\,.}$$

In case $t_1 < \cdots < t_n < t_{n+1} < \ldots < t_m$ we have the natural condition

$$\Phi_{t_1 \ldots t_n}(x_1, \ldots, x_n) = \Phi_{t_1 \ldots t_m}(x_1, \ldots, x_n, +\infty, \ldots, +\infty) \tag{6.55}$$

for all $n$ and $m$ with $1 \leq n < m$ and all arguments $x_1, \ldots, x_n \in \mathbb{R}$.

**Gaussian processes:** If the common distribution functions of a stochastic process are Gaussian distributions, then by definition we have a *Gaussian process*.

*Example 2:* The Wiener process (6.48) is a Gaussian process.

**Main theorem of Kolmogorov (1933):** Let an index set $T$ be given as a non-empty subset of the real numbers. Suppose that for $n = 1, 2, \ldots$ and every set of times (6.54), distribution functions $\Phi_{t_1 \cdots t_n}$ are given, which together satisfy the compatibility conditions (6.55).

Then there is a stochastic process of the form (6.53) which has the given functions $\Phi_{t_1 \cdots t_n}$ as its common distribution functions.

# 7. Numerical Mathematics and Scientific Computing

*Already the very early command which Gauss (1777–1855) had of the world of numbers was amazing. He was completely at home in this world and totally in command of the various tools he developed for its investigation. For every number less than a thousand he was able to, according to his friend Sattorius von Waltershausen, "immediately or after very brief reflection name all the particularities of that number". He used this knowledge for elegant calculations. Through new tools he continually developed and through tricks he was able to enliven the, sometimes month-long, calculations he needed. His incredible memory of numbers aided him in this endeavor. He knew the first few decimals of all logarithms and, according to Sattorius, "was able to calculate approximations of these in his head".*
*He carried out one of the most incredibly long and difficult calculations in the second half of 1812, in order to determine the mass of the planets from the perturbations which these caused in the orbits of other planets. According to estimates made later on these calculations, he must have been doing on the order of between 2600 and 4400 digits every day.*

*Erich Worbs*
*Biography of Gauss*

**Numerical mathematics with Mathematica:** With this software package you are able to perform many of the numerical standard procedures on your home PC.

**The basic experience of numerical mathematics:** Many mathematical methods which are by nature constructive, while giving deep insights and making for elegant proofs, are *not* well adapted to numerical calculations on computers. In order to develop effective procedures which can be implemented on computers, a great deal of experience as well as specific knowledge is required.

One rule of thumb is the following:

> For every imaginable numerical procedure, no matter how elegant it appears, there are *counterexamples* for which the method does not work at all.

Therefore it is important not to blindly believe what software systems produce; instead it is important to understand the structure and limitations of numerical methods. That is the object of this chapter.

> The most important property of a numerical procedure is its *numerical stability.*

Already back in 1947, John von Neumann pointed out this fact. Numerical stability is the property of being stable under numerical perturbations (errors in data, rounding errors in computations).

**Complexity:** If the computation time $Z(p)$ required of a numerical procedure depends on a parameter $p$, then the procedure is said to be *complex*, if

$$Z(p) = O(\mathrm{e}^p), \qquad p \to +\infty.$$

This corresponds to exponential growth of the computation time for growing parameter values $p$. Complex algorithms of this kind are useless for large values of $p$, as the computation time can then be measured in millions of years. Modern complexity theory investigates the fundamental question of the complexity of algorithms for solving problems in a given class. The object is to construct optimal algorithms from the point of view of minimal computation time, or to prove that such do not exist.

Complexity theory is a new branch of modern mathematics and computer science, which still has many open questions. For their solution one applies for example deep results of algebraic topology (see [212]). As for modern physics, this means complexity theory lies on the boundary between pure and applied mathematics, once again underlying the unity of mathematics.

# 7.1 Numerical computation and error analysis

## 7.1.1 The notion of algorithm

An important goal of numerics is to develop and apply constructive methods for the most efficient processing and numerical solution of problems from all of the natural sciences and engineering. For this, precisely formulated rules for computations are developed, in the form of *algorithms*, which can be implemented on computers and in this way applied to practical problems. An algorithm is therefore a well-defined series of elementary calculations and decisions, starting from a certain set of knowns, the *input* of the problem, and leading to a result, the *output*. Such algorithms for the numerical treatment of problems must satisfy the following requirements.

(a) Every step is uniquely determined, and must take into account all possible variations and exceptions.

(b) The result must be delivered after finitely many steps by means of elementary calculations which all can be implemented on a computer with maximal, or at least sufficient, numerical accuracy.

(c) The algorithm is generally only applicable to a certain set of problems. Different problems of this set require only changes in the input to be correctly dealt with.

(d) Given the input, the problem will be dealt with resulting in a maximal precision of the result and a minimum of computational expense.

*Table 7.1. Typical representations for real numbers.*

| ***Computer*** | $\beta$ | $t$ | $L$ | $U$ | $x_{\min}$ | $x_{\max}$ | $\delta$ |
|---|---|---|---|---|---|---|---|
| CRAY-1 | 2 | 48 | −8 192 | 8 191 | $4.60 \cdot 10^{-2\,467}$ | $5.50 \cdot 10^{2\,465}$ | $7.11 \cdot 10^{-15}$ |
| (DEC | 2 | 24 | −127 | 127 | $2.94 \cdot 10^{-39}$ | $1.70 \cdot 10^{38}$ | $5.96 \cdot 10^{-8}$ |
| VAX) | 2 | 53 | −1 023 | 1 023 | $5.56 \cdot 10^{-309}$ | $8.99 \cdot 10^{307}$ | $1.11 \cdot 10^{-16}$ |
| IBM 3033 | 16 | 6 | −64 | 63 | $5.40 \cdot 10^{-79}$ | $7.24 \cdot 10^{75}$ | $9.54 \cdot 10^{-7}$ |
| | 16 | 14 | −64 | 63 | $5.40 \cdot 10^{-79}$ | $7.24 \cdot 10^{75}$ | $2.22 \cdot 10^{-16}$ |
| HP 28S | 10 | 13 | −499 | 500 | $1.00 \cdot 10^{-499}$ | $1.00 \cdot 10^{500}$ | $5.00 \cdot 10^{-12}$ |

The implementation of an algorithm on a computer, however, leads to a number of fundamental questions, which are connected with the limited precision a computer can provide, resulting in difficulties with the verification of certain postulates. Therefore the study of sources of errors, error propagation in the course of computations and the influence of this on the result are central problems in numerics. In this respect, a simple algorithm, or one with a minimal number of computations, is not necessarily the optimal solution. Moreover, mathematically elegant solutions like the closed-form formulas of integral tables, etc. are often quite useless.

### 7.1.2 Representing numbers on computers

Most computers use different representations for integers and for real numbers. We consider in what follows only the most important representations for real numbers which are used for computer procedures. In order to obtain as precise a representation as possible, the so called *floating point representation* of a real number $x \in \mathbb{R}$ is used as an approximation

$$\overline{x} = fl(x) = \sigma \cdot (.a_1 a_2 \ldots a_t)_\beta \cdot \beta^e = \sigma \cdot \beta^e \sum_{\nu=1}^{t} a_\nu \beta^{-\nu},$$

where $\beta$ is the base of the number system, $\sigma \in \{+1, -1\}$ is the *sign* of $\overline{x}$, the coefficients $a_j \in \{0, 1, \ldots, \beta - 1\}$ are the digits of the *mantissa*, $t$ is the *length of the mantissa* and $e \in \mathbb{Z}$ denotes the *exponent*. It is assumed that $a_1 \neq 0$, and one speaks of a $t$-digit representation of the number $x$ to the base $\beta$. For $x = 0$, the normalized representation $\sigma = \pm 1$, $a_i = 0$, $i = 1, 2, \ldots, t$ and $e = 0$ is used (cf. 1.1.1.3 and 1.1.1.4).

Computers normally use powers of two to represent numbers, i.e., use the base $\beta = 2$ (dual system), $\beta = 8$ (octal system) or $\beta = 16$ (hexadecimal system); less often $\beta = 10$ (decimal system) is used. The length of the mantissa $t$ is generally some fixed number which depends on the computer, and the set of exponents is bounded by $L \leq e \leq U$ with fixed numbers $L$ and $U$, so that the real numbers which can be represented in this way is bounded according to $x_{\min} \leq |x| \leq x_{\max}$. Some typical combinations are listed in Table 7.1.[1]

If a real number $x \neq 0$ has an infinite representation to the base $\beta$

$$x = \sigma \cdot (.a_1 a_2 \ldots a_t a_{t+1} \ldots)_\beta \cdot \beta^e, \qquad L \leq e \leq U, \qquad a_1 \neq 0,$$

[1] In this chapter, exponential notation like 7.11E − 15 will be used to represent $7.11 \cdot 10^{-15}$.

then some computers (for example CRAY-1 and IBM 3033) drop the digit $a_{t+1}$. For $x$ the chopped-off representation

$$\overline{x} = fl(x) = \sigma \cdot (.a_1 a_2 \dots a_t)_\beta \cdot \beta^e$$

is used as an approximation. This dropping of the digit $a_{t+1}$ is applied after each arithmetic operation. Another procedure is *rounding* the number, which is the representation given by

$$\overline{x} = fl(x) = \begin{cases} \sigma \cdot (.a_1 a_2 \dots a_t)_\beta \cdot \beta^e, & a_{t+1} < \beta/2, \\ \sigma \cdot \left[(.a_1 a_2 \dots a_t)_\beta + (.00\dots01)_\beta\right] \cdot \beta^e, & a_{t+1} \geq \beta/2. \end{cases}$$

In the second case of this definition, the rounding is performed by increasing the digit $a_t$ for $(.00\dots01)_\beta = \beta^{-t}$ by one. This corresponds to the usual rounding of numbers done in the decimal system.

For most real numbers we consequently have $x \neq fl(x)$, hence for $x \neq 0$ the *relative error*, defined as

$$\varepsilon := \frac{fl(x) - x}{x},$$

plays an important role. In the case of the truncated representation we have $|\varepsilon| \leq \beta^{-t+1}$, and for rounding we have $|\varepsilon| \leq \frac{1}{2}\beta^{-t+1}$. In both cases the relation can be written as

$$fl(x) = x(1 + \varepsilon),$$

so that $fl(x)$ can be viewed as a small perturbation of $x$. This definition, which goes back to Wilkinson, is the key to an in-depth study of the errors and their propagation in algorithms. The maximal absolute value of the relative error is called the *relative computational precision*, and this value is equal to the smallest positive floating decimal number $\delta$, for which

$$fl(1 + \delta) > 1.$$

This characteristic value for each computer is dependent on the applied rounding or truncation used (see Table 7.1).

### 7.1.3 Sources of error, finding errors, condition and stability

We now give a rough overview of the possible sources of errors which can occur during the solution of a problem. In what follows, exact values will be denoted by plain symbols such as $a, b, \dots, x, y, z$, while computed values will be denoted by corresponding barred variables such as $\overline{a}, \overline{b}, \dots, \overline{x}, \overline{y}, \overline{z}$. Instead of considering the relative error of $\overline{x}$, in the decimal system one often uses the concept of *significant digits*: $\overline{x}$ has $m$ significant digits with respect to $x \neq 0$, if we have

$$|(x - \overline{x})/x| < 0.5 \cdot 10^{-m}.$$

A first source of errors of an algorithm is that of *input errors*. These occur because of the $t$-digit floating decimal point representation for real numbers $x$, where a relative error of magnitude $\delta$ can ensue through the rounding or truncation, and on the other hand the data themselves can have errors, either through incorrect measurements or through previous computations with raw data.

Further errors can then occur from arithmetical operations with floating point numbers $\overline{x}$ and $\overline{y}$. If $\circ$ denotes one of the operations $+, -, \times, \div$, then in general $\overline{x} \circ \overline{y}$ does not necessarily have a $t$-digit floating point representation; rather, one has

$$fl(\overline{x} \circ \overline{y}) = (\overline{x} \circ \overline{y})(1 + \varepsilon) \qquad \text{with} \quad |\varepsilon| \leq \delta$$

under the assumption that the value $\overline{x} \circ \overline{y}$ is calculated sufficiently precisely in the first place and then is rounded or truncated. These errors, which are generally referred to as *rounding errors*, propagate and multiply during the performance of an algorithm and consequently have an impact on the number of significant digits of the end result of the computation. If one analyzes the magnitude of these errors at each step of the algorithm all the way through to the final result, then one is performing what is called a *forward error analysis*. The resulting estimates of the errors of the result are usually quite pessimistic, but they can give good qualitative insights into the critical steps of the algorithm, which impact the final result the most. The *interval arithmetic* is a method which implements a forward analysis automatically on the computer by working with intervals $[x_a, x_b]$, which guarantee that the actual, correct value $x$ lies in the corresponding interval (cf. [422] and [430]).

Another technique of studying the propagation of errors is based on the *backward analysis*. This principle consists in taking the result of a computation and, for each step, investigating the most precise data which could have produced the result of that step, and estimating the errors which that theoretical computation would have had. In this way one obtains qualitative results on the set of precise data which could have led to the same result without rounding errors; investigation of these quantities and their perturbations yields information on errors which arise invariably or which are inherent in the problem being considered.

Thus the basic problem is to study how errors of the input data result in errors of the output. One should differentiate here between the mathematically caused errors and those due to the particular algorithm used to solve the problem. The relation between changes in the exact results and the errors of the input data is called the *condition* of the problem. This measure can be a single, generally comparable number for all components of a solution in the sense of a norm, or it can be a whole set of individual numbers, each of which is a condition number for a particular part of the algorithm. A quite different notion is that of stability; an algorithm is said to be *stable*, if (small) errors of the input data result in correspondingly small errors of the final result. This *numerical stability* of an algorithm is in the end the single criterion which decides whether an algorithm can be effectively put to use; a given mathematical problem may very well be well-conditioned, but the corresponding algorithm chosen for solving it may be unstable.

Besides the sources of errors already discussed, there are also possible errors which occur during an algorithm from the fact that floating decimal representations limit the actual size of numbers to which it applies; either an *overflow* or an *underflow* can be the result, and these can in certain circumstances render the result unusable. In order to avoid these errors, the algorithm may have to be adapted.

Finally there are *procedural errors* which can occur when precise values are only approximately calculated. This situation can occur for example when an iteration is broken off after finitely many steps or a limiting procedure cannot be carried out or a differential is approximated by the corresponding difference quotient. The analysis of such procedural errors is part of the description of an algorithm.

*Example:* As an illustration of these notions and the difficulties that they represent, consider the following numerical examples which are carried out in a decimal system ($\beta = 10$) with a mantissa length $t = 5$ and usual rounding procedures.

1. Let $\overline{a} = 0.31416 \cdot 10^1$, $\overline{b} = -0.31523 \cdot 10^1$, $\overline{c} = 0.67521 \cdot 10^{-5}$ be decimal floating point numbers. Then we obtain

$$\begin{aligned} fl(\overline{a}+\overline{b}) &= -0.10700 \cdot 10^{-1}, \\ fl\big(fl(\overline{a}+\overline{b})+\overline{c}\big) &= fl(-0.10700 \cdot 10^{-1} + 0.67521 \cdot 10^{-5}) = -0.10693 \cdot 10^{-1}, \\ fl\big(\overline{a} + fl(\overline{c}+\overline{b})\big) &= fl(0.31416 \cdot 10^1 - 0.31523 \cdot 10^1) = -0.10700 \cdot 10^{-1}. \end{aligned}$$

The associative law is in general not valid for floating arithmetic. Moreover, we have

$$\begin{aligned} fl\big(fl(\overline{a}^2) - fl(\overline{b}^2)\big) &= fl(0.98697 \cdot 10^1 - 0.99370 \cdot 10^1) = -0.67300 \cdot 10^{-1}, \\ fl\big(fl(\overline{a}+\overline{b}) \cdot fl(\overline{a}-\overline{b})\big) &= fl\big((-0.10700 \cdot 10^{-1}) \cdot 0.62939 \cdot 10^1\big) = -0.67345 \cdot 10^{-1}. \end{aligned}$$

The first result is only correct to three digits. What happens is that certain digits cancel out upon the subtraction of numbers which are roughly equal.

2. The problem of determining the two solution of the quadratic equation

$$x^2 + 2px - q = 0, \qquad p, q > 0,$$

is well-conditioned. If we calculate the smaller of the two solutions $y$ for $p = 157$ and $q = 2$ according to the formula

$$y = -p + \sqrt{p^2 + q},$$

then we find $fl\big(\sqrt{p^2+q}\,\big) = fl\big(\sqrt{0.24651 \cdot 10^5}\,\big) = 0.15701 \cdot 10^3$ and therefore $\overline{y} = 0.10000 \cdot 10^{-1}$. Using the equivalent formula

$$y = \frac{q}{p + \sqrt{p^2+q}},$$

however, we get $\overline{y} = fl\big(2/(0.31401 \cdot 10^3)\big) = 0.63692 \cdot 10^{-2}$, a result with a relative error of approximately $1.5 \cdot 10^{-5}$, which has four significant digits. The second calculation is numerically stable, while the first is instable due to the cancellation of leading digits.

3. Calculating the integral

$$I_n = \int_0^1 e^{-x} x^n \, dx$$

for $n = 0, 1, 2, \ldots, 8$ can be done with the help of the recursion formula

$$I_n = n \cdot I_{n-1} - \frac{1}{e}, \qquad n = 1, 2, \ldots$$

which results from partial integration, where

$$I_0 = \frac{e-1}{e}.$$

*Table 7.2. Recursive computation of integrals.*

| $n$ | $I_n = n \cdot I_{n-1} - \frac{1}{e}$ <br> $\overline{I}_n$ | $I_{n-1} = \frac{I_n + 1/e}{n}$ <br> $\overline{I}_n$ |
|---|---|---|
| 0 | $0.63212 \cdot 10^0$ | $0.63212 \cdot 10^0$ |
| 1 | $0.26424 \cdot 10^0$ | $0.26424 \cdot 10^0$ |
| 2 | $0.16060 \cdot 10^0$ | $0.16060 \cdot 10^0$ |
| 3 | $0.11392 \cdot 10^0$ | $0.11393 \cdot 10^0$ |
| 4 | $0.87800 \cdot 10^{-1}$ | $0.87836 \cdot 10^{-1}$ |
| 5 | $0.71120 \cdot 10^{-1}$ | $0.71302 \cdot 10^{-1}$ |
| 6 | $0.58840 \cdot 10^{-1}$ | $0.59934 \cdot 10^{-1}$ |
| 7 | $0.44000 \cdot 10^{-1}$ | $0.51656 \cdot 10^{-1}$ |
| 8 | $-0.15880 \cdot 10^{-1}$ | $0.45368 \cdot 10^{-1}$ |

The computation yields after a few recursions incorrect values and for $n = 8$ even a negative value (see Table 7.2). If one considers just the propagation of the input error $\varepsilon_0 := I_0 - \overline{I}_0 \cong 5.59 \cdot 10^{-7}$, then one observes that the resulting error of the final result is $\varepsilon_n := I_n - \overline{I}_n \cong n! \cdot \varepsilon_0$, which is rapidly increasing. The algorithm is very instable. The actual error is in fact, due to superposition of the error of $fl(1/\mathrm{e})$, even larger. On the other hand, if the recursion formula in the form

$$I_{n-1} = \frac{I_n + 1/\mathrm{e}}{n}, \quad n = N, N-1, \ldots, 1$$

is applied, then the error is reduced in the calculation of $I_{n-1}$ over that of $I_n$ by a factor of $n$. In this way, the algorithm is fitted to yield a stable method of computation of the well-conditioned problem. If one notes that $I_n \le 1/(n+1)$ and starts the new recursion at $\overline{I}_{15} = 1/32$, then the initial error satisfies $\varepsilon_{15} \le 1/32$, so that $|\varepsilon_8| \le 10^{-9}$, i.e., the desired results are as precise as the input errors themselves (see Table 7.2).

## 7.2 Linear algebra

### 7.2.1 Linear systems of equations – direct methods

Suppose we want to solve an inhomogeneous linear system of equations

$$\boxed{\mathbf{A}\mathbf{x} + \mathbf{b} = \mathbf{0},} \qquad \text{i.e.,} \qquad \sum_{k=1}^{n} a_{ik}x_k + b_i = 0, \qquad i = 1, 2, \ldots, n$$

with $n$ equations and $n$ unknowns $x_1, \ldots, x_n$. We make the assumption that the $(n \times n)$-matrix $\mathbf{A}$ is regular, i.e., $\det \mathbf{A} \ne 0$, so that the existence and uniqueness of a solution vector $\mathbf{x}$ is assured for every constant $\mathbf{b}$ (cf. 2.1.4.3). We now consider direct elimination methods for the numerical calculation of $\mathbf{x}$.

#### 7.2.1.1 The Gauss algorithm

The Gauss elimination method discussed in section 2.1.4.2 can be put in a form which is convenient for the implementation on a computer. The equation we wish to solve is written in the following self-evident scheme, which for $n = 4$ looks as follows:

| $x_1$ | $x_2$ | $x_3$ | $x_4$ | 1 |
|---|---|---|---|---|
| $a_{11}$ | $a_{12}$ | $a_{13}$ | $a_{14}$ | $b_1$ |
| $a_{21}$ | $a_{22}$ | $a_{23}$ | $a_{24}$ | $b_2$ |
| $a_{31}$ | $a_{32}$ | $a_{33}$ | $a_{34}$ | $b_3$ |
| $a_{41}$ | $a_{42}$ | $a_{43}$ | $a_{44}$ | $b_4$ |

The first part of the *Gauss algorithm* consists of successively making transformations of the system of equations until it has triangle form. The new elements of the matrix which this procedure delivers will be denoted by the same symbols.

In the first step, we first check whether $a_{11} \ne 0$. If $a_{11} = 0$, the there is some $a_{p1} \ne 0$ with $p > 1$, and the two rows with indices 1 and $p$ are switched. The first equation (after this transposition) is now already in final form, and $a_{11}$ is a *pivotal element.* With the help of the quotients

$$l_{i1} := a_{i1}/a_{11}, \qquad i = 2, 3, \ldots, n$$

we then subtract the $l_{i1}$-multiple of the $1^{st}$ row from the $i^{th}$ row. This results in the new scheme

| $x_1$ | $x_2$ | $x_3$ | $x_4$ | 1 |
|---|---|---|---|---|
| $a_{11}$ | $a_{12}$ | $a_{13}$ | $a_{14}$ | $b_1$ |
| 0 | $a_{22}$ | $a_{23}$ | $a_{24}$ | $b_2$ |
| 0 | $a_{32}$ | $a_{33}$ | $a_{34}$ | $b_3$ |
| 0 | $a_{42}$ | $a_{43}$ | $a_{44}$ | $b_4$ |

with
$$a_{ik} := a_{ik} - l_{i1} \cdot a_{1k}, \qquad i,k = 2,3,\ldots,n,$$
$$b_i := b_i - l_{i1} \cdot b_1, \qquad i = 2,3,\ldots,n.$$

The process is now repeated with the last $(n-1)$ rows, which now represent a linear system of equations for the $(n-1)$ unknowns $x_2,\ldots,x_n$. From the assumption that $\mathbf{A}$ is regular, there exists in the $k^{th}$ step among the $a_{kk}, a_{k+1,k}, \ldots, a_{nk}$ at least one element which is non-vanishing, so that again transposing two rows leads to a pivot element $a_{kk} \neq 0$.

After $(n-1)$ steps, the scheme consists of $n$ equations in final form. We now change the notations, denoting the matrix elements in final form by $r_{ik}$, and the values of the constant column by $c_i$. Instead of the 0's in the lower left-hand corner of the matrix, we now write the quotients $l_{ik}$ with $i > k$. The resulting scheme thus has the form

| $x_1$ | $x_2$ | $x_3$ | $x_4$ | 1 |
|---|---|---|---|---|
| $r_{11}$ | $r_{12}$ | $r_{13}$ | $r_{14}$ | $c_1$ |
| $l_{21}$ | $r_{22}$ | $r_{23}$ | $r_{24}$ | $c_2$ |
| $l_{31}$ | $l_{32}$ | $r_{33}$ | $r_{34}$ | $c_3$ |
| $l_{41}$ | $l_{42}$ | $l_{43}$ | $r_{44}$ | $c_4$ |

From this system of equations we can easily calculate the values of the unknowns, the second part of the Gauss algorithm, by using the formulas (in reverse order to the unknowns):

$$x_n = -c_n/r_{nn}, \qquad x_k = -\left(c_k + \sum_{j=k+1}^{n} r_{kj}x_j\right)/r_{kk}, \qquad k = n-1, n-2, \ldots, 1.$$

The quantities occuring in the final equations are related to the original system of equations $\mathbf{Ax}+\mathbf{b} = \mathbf{0}$ in the following way (cf. [432]). With the *right triangle matrix* $\mathbf{R}$ and the *left triangle matrix* $\mathbf{L}$

$$\mathbf{R} := \begin{pmatrix} r_{11} & r_{12} & r_{13} & \ldots & r_{1n} \\ 0 & r_{22} & r_{23} & \ldots & r_{2n} \\ 0 & 0 & r_{33} & \ldots & r_{3n} \\ \vdots & \vdots & \vdots & \ldots & \vdots \\ 0 & 0 & 0 & \ldots & r_{nn} \end{pmatrix}, \qquad \mathbf{L} := \begin{pmatrix} 1 & 0 & 0 & \ldots & 0 \\ l_{21} & 1 & 0 & \ldots & 0 \\ l_{31} & l_{32} & 1 & \ldots & 0 \\ \vdots & \vdots & \vdots & \ldots & \vdots \\ l_{n1} & l_{n2} & l_{n3} & \ldots & 1 \end{pmatrix}$$

and a further *permutation matrix* $\mathbf{P}$, which described the transpositions among the rows carried out during the algorithm, we have

$$\mathbf{P} \cdot \mathbf{A} = \mathbf{L} \cdot \mathbf{R} \qquad \text{(LR-decomposition)}$$

The Gauss algorithm yields for a given matrix $\mathbf{A}$, for which the rows have been appropriately permuted, two matrices: a regular, lower triangular matrix $\mathbf{L}$ with 1's in the diagonals, and a regular, upper triangular matrix $\mathbf{R}$. By virtue of the LR-decomposition, the Gauss algorithm can be described for $\mathbf{P}(\mathbf{A}\mathbf{x}+\mathbf{b}) = \mathbf{P}\mathbf{A}\mathbf{x}+\mathbf{P}\mathbf{b} = \mathbf{L}\mathbf{R}\mathbf{x}+\mathbf{P}\mathbf{b} = -\mathbf{L}\mathbf{c}+\mathbf{P}\mathbf{b} = \mathbf{0}$ with $\mathbf{R}\mathbf{x} = -\mathbf{c}$ as follows:

| | |
|---|---|
| 1. $\mathbf{PA} = \mathbf{LR}$ | (LR-decomposition), |
| 2. $\mathbf{Lc} - \mathbf{Pb} = \mathbf{0}$ | (forward substitution $\rightarrow \mathbf{c}$), |
| 3. $\mathbf{Rx} + \mathbf{c} = \mathbf{0}$ | (reverse substitution $\rightarrow \mathbf{x}$). |

This scheme is particularly useful, if one needs to solve systems of equations for different $\mathbf{b}$ but the same $\mathbf{A}$, as the decomposition of $\mathbf{A}$ only needs to be carried out once.

The computational effort of essential arithmetic operations, i.e., multiplications and divisions, is for the LR-decomposition $Z_{\text{LR}} = (n^3 - n)/3$ and for the processes of the forward and reverse substitution $Z_{\text{VR}} = n^2$. The Gauss algorithm altogether thus requires

$$Z_{\text{Gauss}} = \frac{1}{3}(n^3 + 3n^2 - n)$$

essential operations.

In order to make the Gauss algorithm as stable as possible, a *pivotal strategy* is required, which determines a pivot choice at each step of the algorithm. The *diagonal strategy*, which works on the premise that no row permutations are required, can only lead to success if the matrix $\mathbf{A}$ is *diagonally dominant*, i.e.,

$$|a_{kk}| > \sum_{\substack{j=1 \\ j \neq k}}^{n} |a_{kj}|, \qquad k = 1, 2, \ldots, n.$$

Generally, the *maximal queue strategy* is applied, which determines the element in the $k^{th}$ row which has a maximal absolute value as the pivot element. In the $k^{th}$ step of the elimination, an index $p$ is determined, for which

$$\max_{j \geq k} |a_{jk}| = |a_{pk}|.$$

In case $p \neq k$, the $k^{th}$ and the $p^{th}$ rows are switched. However, this strategy makes the assumption that the matrix is *row scaled*, i.e., the sum of the absolute values of the elements of each row are not equal. Since a scaling of the rows to achieve this is not advisable, as it too can introduce computational errors, one uses instead the so-called *relative queue strategy*, which chooses the element with a maximal absolute value as if the system were row scaled.

#### 7.2.1.2 Gauss–Jordan procedure

A variant of the Gauss algorithm is to eliminate the unknown $x_k$ in the $k^{th}$ step of the algorithm not just in the $k^{th}$ equation, but also in all preceeding equations. After a choice of a pivot element and the necessary permutation of rows, the quotients

$$l_{ik} := a_{ik}/a_{kk}, \qquad i = 1, 2, \ldots, k-1, k+1, \ldots, n$$

are formed and the $l_{ik}$-multiple of the $k^{th}$ row is subtracted from the $i^{th}$ row, so that in the $k^{th}$ row and column, only the pivot element $a_{kk}$ remains. The procedure for calculating the new entries after the $k^{th}$ step of the *Gauss–Jordan procedure* is

$$\begin{aligned} a_{ij} &:= a_{ij} - l_{ik} \cdot a_{kj}, \qquad & i &= 1, \ldots, k-1, k+1, \ldots, n; \quad j = k+1, \ldots, n, \\ b_i &:= b_i - l_{jk} \cdot b_k, \qquad & i &= 1, \ldots, k-1, k+1, \ldots, n. \end{aligned}$$

Since after $n$ steps the result is a scheme which is a diagonal matrix with non-vanishing diagonal entries, the solution itself is then obtained by the simple equations

$$x_k = -b_k/a_{kk}, \qquad k = 1, 2, \ldots, n.$$

Although this algorithm requires, compared with the Gauss-algorithm, more calculations, namely

$$Z_{\mathrm{GJ}} = \frac{1}{2}(n^3 + 2n^2 - n),$$

the procedure itself is simpler and particularly well-adapted to vector calculators.

#### 7.2.1.3 Calculation of determinants

From the LR-decomposition of the Gauss algorithm we get from $\det \mathbf{L} = 1$, $\det \mathbf{R} = \prod_{k=1}^{n} r_{kk}$ and $\det \mathbf{P} = (-1)^V$, where $r_{kk}$ denotes the $k^{th}$ pivot element and $V$ is the total number of row transpositions which need to be carried out, that for the determinant of a $(n \times n)$-matrix $\mathbf{A}$, we have

$$\det \mathbf{A} = (-1)^V \cdot \prod_{k=1}^{n} r_{kk}.$$

The idea of calculating the determinant $\det \mathbf{A}$ as the product of pivot elements of the LR-decomposition is efficient and stable. The evaluation of the defining equation for the determinant presented in section 2.1.2 is too costly and instable because of cancellations.

#### 7.2.1.4 Calculation of inverse matrices

If the inverse matrix $\mathbf{A}^{-1}$ of a regular, quadratic matrix $\mathbf{A}$ needs to be calculated (certainly not for the calculation of solutions of linear systems of equations!), then it is usually determined via the matrix equation $\mathbf{A}\mathbf{A}^{-1} = \mathbf{E}$ by solving a series of systems of linear equations, by finding the solutions of $\mathbf{A}\mathbf{x}_k - \mathbf{e}_k = \mathbf{0}$ for $k = 1, 2, \ldots, n$ with unit vectors $\mathbf{e}_k$; the $k^{th}$ column of $\mathbf{A}^{-1}$ is then $\mathbf{x}_k$. However, this procedure is not optimal, neither from the point of view of memory nor from the number of operations.

It is more convenient to invert $\mathbf{A}$ using a *swaping procedure.* For this we consider the $n$ linear forms in $n$ variables $x_k$:

$$y_i = \sum_{k=1}^{n} a_{ik} x_k, \qquad i = 1, 2, \ldots, n,$$

which belong to $\mathbf{A}$. Assuming $a_{pq} \neq 0$, we solve the $p^{th}$ linear form $y_p$ with respect to $x_q$ and insert the resulting linear form instead of $x_q$ in the other expressions:

$$x_q = \frac{1}{a_{pq}} y_p - \sum_{\substack{k=1\\k\neq q}}^{n} \frac{a_{pk}}{a_{pq}} x_k,$$

$$y_i = \frac{a_{iq}}{a_{pq}} y_p + \sum_{\substack{k=1\\k\neq q}}^{n} \left( a_{ik} - \frac{a_{iq}a_{pk}}{a_{pq}} \right) x_k, \qquad i \neq p.$$

If we now exchange (swap) the linear form $x_q$ and the variable $y_p$, then we get a scheme similar to that above which takes for $n = 4, p = 3$ and $q = 2$ the following form.

$$
\begin{array}{r|cccc|}
 & x_1 & \underline{x_2} & x_3 & x_4 \\
\hline
y_1 = & a_{11} & \underline{a_{12}} & a_{13} & a_{14} \\
y_2 = & a_{21} & \underline{a_{22}} & a_{23} & a_{24} \\
\underline{y_3} = & \underline{a_{31}} & \underline{\underline{a_{32}}} & \underline{a_{33}} & \underline{a_{34}} \\
y_4 = & a_{41} & \underline{a_{42}} & a_{43} & a_{44} \\
\hline
\end{array}
\quad\Longrightarrow\quad
\begin{array}{r|cccc|}
 & x_1 & y_3 & x_3 & x_4 \\
\hline
y_1 = & a'_{11} & a'_{12} & a'_{13} & a'_{14} \\
y_2 = & a'_{21} & a'_{22} & a'_{23} & a'_{24} \\
x_2 = & a'_{31} & a'_{32} & a'_{33} & a'_{34} \\
y_4 = & a'_{41} & a'_{42} & a'_{43} & a'_{44} \\
\hline
\end{array}
$$

The elements of the new scheme are defined by the relations

$$
\boxed{
\begin{aligned}
a'_{pq} &= \frac{1}{a_{pq}}, \\
a'_{pk} &= -\frac{a_{pk}}{a_{pq}}, \qquad k \neq q, \\
a'_{iq} &= \frac{a_{iq}}{a_{pq}}, \qquad i \neq p, \\
a'_{ik} &= a_{ik} - \frac{a_{iq}a_{pk}}{a_{pq}} = a_{ik} + a_{iq}a'_{pk}, \qquad i \neq p,\ k \neq q.
\end{aligned}
}
$$

One calls $a_{pq}$ the pivot element of the exchange step. The computation cost of an exchange step is $n^2$ essential operations. If in this way $n$ successive exchange steps are carried out, swaping $x$ variables with $y$ variables which are on the left, then the inverse linear forms result, which then yield the inverse matrix, after an appropriate permutation of the rows and columns. Pivot strategies are required for this procedure to be stable. This said, the procedure presents the following advantages: every exchange step can be carried out in the allocated space for the matrix $\mathbf{A}$, the operations can easily be vectorized, and the computation cost for calculating the inverse matrix of a $(n \times n)$-matrix $\mathbf{A}$ is $n^3$ essential operations.

*Example:* We invert the matrix

$$
\mathbf{A} = \begin{pmatrix} 2 & 12 & -4 \\ 1 & 2 & -3 \\ -2 & -6 & 5 \end{pmatrix}
$$

with the swap procedure and the maximum queue strategy:

$$
\begin{array}{r|ccc|}
 & \underline{x_1} & x_2 & x_3 \\
\hline
\underline{y_1} = & \underline{\underline{2}} & \underline{12} & \underline{-4} \\
y_2 = & \underline{1} & 2 & -3 \\
y_3 = & \underline{-2} & -6 & 5 \\
\hline
\end{array}
\qquad
\begin{array}{r|ccc|}
 & y_1 & x_2 & x_3 \\
\hline
x_1 = & 0.5 & -6 & 2 \\
y_2 = & 0.5 & -4 & -1 \\
y_3 = & -1 & 6 & 1 \\
\hline
\end{array}
\qquad
\begin{array}{r|ccc|}
 & y_1 & \underline{x_2} & x_3 \\
\hline
x_1 = & 0.5 & \underline{-6} & 2 \\
\underline{y_3} = & \underline{-1} & \underline{\underline{6}} & \underline{1} \\
y_2 = & 0.5 & \underline{-4} & -1 \\
\hline
\end{array}
$$

$$
\begin{array}{r|ccc|}
 & y_1 & y_3 & \underline{x_3} \\
\hline
x_1 = & -\frac{1}{2} & -1 & \underline{3} \\
x_2 = & \frac{1}{6} & \frac{1}{6} & \underline{-\frac{1}{6}} \\
\underline{y_2} = & \underline{-\frac{1}{6}} & \underline{-\frac{2}{3}} & \underline{\underline{-\frac{1}{3}}} \\
\hline
\end{array}
\qquad
\begin{array}{r|ccc|}
 & y_1 & y_3 & y_2 \\
\hline
x_1 = & -2 & -7 & -9 \\
x_2 = & \frac{1}{4} & \frac{1}{2} & \frac{1}{2} \\
x_3 = & -\frac{1}{2} & -2 & -3 \\
\hline
\end{array}
\qquad
\mathbf{A}^{-1} = \begin{pmatrix} -2 & -9 & -7 \\ \frac{1}{4} & \frac{1}{2} & \frac{1}{2} \\ -\frac{1}{2} & -3 & -2 \end{pmatrix}.
$$

#### 7.2.1.5 The Cholesky algorithm

For solving a inhomogeneous linear system of equations $\mathbf{Ax} + \mathbf{b} = \mathbf{0}$ with a symmetric and positive definite matrix $\mathbf{A}$, there is a more efficient and elegant method which takes advantage of the properties of $\mathbf{A}$. Because a positive definite quadratic form $Q(\mathbf{x}) := \mathbf{x}^\mathsf{T}\mathbf{Ax}$ can be written as the sum of squares of linearly independent linear forms, there is a so-called *Cholesky decomposition*

$$\mathbf{A} = \mathbf{LL}^\mathsf{T},$$

where $\mathbf{L}$ is a regular, lower triangular matrix with positive diagonal elements $l_{kk}$. From the matrix equation

$$\begin{pmatrix} a_{11} & a_{12} & \dots & a_{1n} \\ a_{21} & a_{22} & \dots & a_{2n} \\ \vdots & \vdots & \dots & \vdots \\ a_{n1} & a_{n2} & \dots & a_{nn} \end{pmatrix} = \begin{pmatrix} l_{11} & 0 & \dots & 0 \\ l_{21} & l_{22} & \dots & 0 \\ \vdots & \vdots & \dots & \vdots \\ l_{n1} & l_{n2} & \dots & l_{nn} \end{pmatrix} \begin{pmatrix} l_{11} & l_{21} & \dots & l_{n1} \\ 0 & l_{22} & \dots & l_{n2} \\ \vdots & \vdots & \dots & \vdots \\ 0 & 0 & \dots & l_{nn} \end{pmatrix},$$

the elements of $\mathbf{L}$ can be successively calculated from the relations

$$\begin{aligned} a_{ii} &= l_{i1}^2 + l_{i2}^2 + \ldots + l_{ii}^2\,, \\ a_{ik} &= l_{i1}l_{k1} + l_{i2}l_{k2} + \ldots + l_{i,k-1}l_{k,k-1} + l_{ik}l_{kk}\,, \qquad i > k, \end{aligned}$$

according to the formulas

$$l_{11} = \sqrt{a_{11}}\,,$$

$$\left.\begin{aligned} l_{ik} &= \left(a_{ik} - \sum_{j=1}^{k-1} l_{ij}l_{kj}\right)\Big/ l_{kk}, \qquad k = 1, 2, \ldots, i-1, \\ l_{ii} &= \left(a_{ii} - \sum_{j=1}^{i-1} l_{ij}^2\right)^{\frac{1}{2}}, \end{aligned}\right\}, \quad i = 2, 3, \ldots, n.$$

For carrying out the Cholesky decomposition of a symmetric and positive definite matrix $\mathbf{A}$, only the elements on and just below the diagonal are needed, and the matrix $\mathbf{L}$ can be successively computed and stored in the storage area of the matrix $\mathbf{A}$. With this utilization of memory, the procedure is not only efficient from the memory usage point of view, but also numerically very efficient, as the computational cost is

$$Z_{\mathbf{LL}^\mathsf{T}} = \frac{1}{6}\,(n^3 + 3n^2 - 4n)$$

essential operations, plus the much lower number of $n$ square roots which need to be computed. Compared with the LR-decomposition for a general matrix, this means roughly half the cost (which perhaps isn't surprising considering that a symmetric matrix contains only roughly half the number of different elements of a general matrix). In addition, the Cholesky decomposition is stable, because the matrix elements of the matrix $\mathbf{L}$ are bounded in absolute value and cannot become arbitrarily large.

With the help of the Cholesky decomposition of $\mathbf{A}$, the *Cholesky algorithm* for solving $\mathbf{Ax} + \mathbf{b} = \mathbf{0}$ is given by the following three steps.

| | |
|---|---|
| 1. $\mathbf{A} = \mathbf{LL}^\mathsf{T}$ | (Cholesky-decomposition), |
| 2. $\mathbf{Lc} - \mathbf{b} = \mathbf{0}$ | (forward substitution $\to \mathbf{c}$), |
| 3. $\mathbf{L}^\mathsf{T}\mathbf{x} + \mathbf{c} = \mathbf{0}$ | (reverse substitution $\to \mathbf{x}$). |

The computation cost for the substitution steps is altogether $Z_{\text{subCh}} = n^2 + n$ essential operations.

#### 7.2.1.6 Tridiagonal systems of equations

We now treat the important special case of *tridiagonal* systems of equations, but only for the special case that the *diagonal strategy* for the Gauss algorithm applies. In this case the matrices **L** and **R** of the LR-decomposition are both *bidiagonal*, so that we have

$$\mathbf{A} = \begin{pmatrix} a_1 & b_1 & & \\ c_1 & a_2 & b_2 & \\ & c_2 & a_3 & b_3 \\ & & c_3 & a_4 \end{pmatrix} = \begin{pmatrix} 1 & & & \\ l_1 & 1 & & \\ & l_2 & 1 & \\ & & l_3 & 1 \end{pmatrix} \cdot \begin{pmatrix} m_1 & b_1 & & \\ & m_2 & b_2 & \\ & & m_3 & b_3 \\ & & & m_4 \end{pmatrix} = \mathbf{LR}.$$

By comparison of coefficients we obtain from this the algorithm for calculating the LR-decomposition for a tridiagonal $(n \times n)$-matrix $\mathbf{A}$:

$$\begin{aligned} m_1 &= a_1, \\ l_i &= c_i/m_i, \qquad m_{i+1} = a_{i+1} - l_i b_i, \qquad i = 1, 2, \ldots, n-1. \end{aligned}$$

For the system of equations $\mathbf{Ax} - \mathbf{d} = \mathbf{0}$, the forward substitution $\mathbf{Ly} - \mathbf{d} = \mathbf{0}$ is used for calculating the auxiliary vector $\mathbf{y}$:

$$y_1 = d_1, \qquad y_i = d_i - l_{i-1} y_{i-1}, \qquad i = 2, 3, \ldots, n.$$

The solution vector $\mathbf{x}$ is then obtained from $\mathbf{Rx} + \mathbf{y} = \mathbf{0}$ by reverse substitution

$$x_n = -\frac{y_n}{m_n}, \qquad x_i = -\frac{y_i + b_i x_{i+1}}{m_i}, \qquad i = n-1, n-2, \ldots, 1.$$

This extremely simple algorithm requires only $Z_{\text{trid}} = 5n - 4$ essential operations, implying a linear growth in the number of operations with $n$. The algorithm with pivot choice is described in [432].

#### 7.2.1.7 Condition of a linear system of equations

Because of the invariable imprecision of the input, the algorithms presented thus far only really solve nearby systems of equations. But even if one has absolutely precise inputs, the results are not exact, due to rounding errors. If together with a calculated solution $\overline{\mathbf{x}}$, one calculates also the *defect* or the *residue* $\mathbf{r} := \mathbf{A}\overline{\mathbf{x}} + \mathbf{b}$, then $\overline{\mathbf{x}}$ may be viewed as an exact solution of the perturbed set of equations $\mathbf{A}\overline{\mathbf{x}} + (\mathbf{b} - \mathbf{r}) = \mathbf{0}$. Because $\mathbf{Ax} + \mathbf{b} = \mathbf{0}$, the error vector $\mathbf{z} := \mathbf{x} - \overline{\mathbf{x}}$ satisfies the system $\mathbf{Az} + \mathbf{r} = \mathbf{0}$.

If $||\mathbf{x}||$ is a vector norm and $||\mathbf{A}||$ is matrix norm compatible with it, then we have the inequalities

$$||\mathbf{b}|| \leq ||\mathbf{A}|| \cdot ||\mathbf{x}||, \qquad ||\mathbf{z}|| \leq ||\mathbf{A}^{-1}|| \cdot ||\mathbf{r}||.$$

For the relative error we thus obtain

$$\frac{||\mathbf{z}||}{||\mathbf{x}||} = \frac{||\mathbf{x} - \overline{\mathbf{x}}||}{||\mathbf{x}||} \leq ||\mathbf{A}|| \cdot ||\mathbf{A}^{-1}|| \cdot \frac{||\mathbf{r}||}{||\mathbf{b}||} := \kappa(\mathbf{A}) \cdot \frac{||\mathbf{r}||}{||\mathbf{b}||}.$$

The quantity $\kappa(\mathbf{A}) := ||\mathbf{A}|| \cdot ||\mathbf{A}^{-1}||$ is called the *condition number* of the matrix $\mathbf{A}$. In the situation we are considering, $\kappa(\mathbf{A})$ describes how a small change of the constant vector **b** of small norm affects the solution. A relatively small defect thus in general implies little about the precision of the calculated solution $\overline{\mathbf{x}}$.

The size of changes in the solution **x** of $\mathbf{Ax}+\mathbf{b}=\mathbf{0}$ under perturbations $\Delta\mathbf{A}$ and $\Delta\mathbf{b}$ of the inputs to the system $(\mathbf{A}+\Delta\mathbf{A})(\mathbf{x}+\Delta\mathbf{x})+(\mathbf{b}+\Delta\mathbf{b})=\mathbf{0}$ can be estimated by

$$\frac{||\Delta\mathbf{x}||}{||\mathbf{x}||} \leq \frac{\kappa(\mathbf{A})}{1-\kappa(\mathbf{A})\frac{||\Delta\mathbf{A}||}{||\mathbf{A}||}}\left\{\frac{||\Delta\mathbf{A}||}{||\mathbf{A}||}+\frac{||\Delta\mathbf{b}||}{||\mathbf{b}||}\right\},$$

provided $\kappa(\mathbf{A})\cdot||\Delta\mathbf{A}||/||\mathbf{A}||<1$. From this we obtain the following important rule: If $\mathbf{Ax}+\mathbf{b}=\mathbf{0}$ is solved with $d$-digit precision with a condition number $\kappa(\mathbf{A})\sim 10^{\alpha}$, then as a consequence of input errors of **x**, only $d-\alpha-1$ decimal places are significant in the component of largest absolute value. In the other components the relative error can be even larger.

### 7.2.2 Iterative solutions of linear systems of equations

Iterative methods are particularly well adapted to the solution of large systems of equations

$$\boxed{\mathbf{Ax}+\mathbf{b}=\mathbf{0},}$$

in which the matrix **A** is sparse (meaning that most entries vanish), since iteration can take advantage of this property. Iterative procedures are based on either bringing the given system $\mathbf{Ax}+\mathbf{b}=\mathbf{0}$ into the *fix point form*

$$\boxed{\mathbf{x}=\mathbf{Tx}+\mathbf{c},}$$

or by minimizing an appropriate functional. The fixed point relation $\mathbf{x}^{(k+1)}=\mathbf{Tx}^{(k)}+\mathbf{c}$ for a given start vector $\mathbf{x}^{(0)}$ generates a sequence which converges to a solution **x** if and only if the *spectral radius* of the iteration matrix **T** satisfies $\varrho(\mathbf{T})<1$.

#### 7.2.2.1 Classical iteration methods

We now decompose the matrix **A** into the sum $\mathbf{A}=-\mathbf{L}+\mathbf{D}-\mathbf{U}$, where **L** is a strictly lower diagonal matrix, **U** is a strictly upper diagonal matrix and **D** is a diagonal matrix with diagonal non-vanishing elements $a_{kk}$ of **A**:

$$\mathbf{L}:=\begin{pmatrix} 0 & 0 & 0 & \dots & 0\\ -a_{21} & 0 & 0 & \dots & 0\\ -a_{31} & -a_{32} & 0 & \dots & 0\\ \vdots & \vdots & \vdots & \dots & \vdots\\ -a_{n1} & -a_{n2} & -a_{n3} & \dots & 0\end{pmatrix},\ \mathbf{U}:=\begin{pmatrix} 0 & -a_{12} & -a_{13} & \dots & -a_{1n}\\ 0 & 0 & -a_{23} & \dots & -a_{2n}\\ 0 & 0 & 0 & \dots & -a_{3n}\\ \vdots & \vdots & \vdots & \dots & \vdots\\ 0 & 0 & 0 & \dots & 0\end{pmatrix},$$

$$\mathbf{D}:=\operatorname{diag}(a_{11},a_{22},\dots,a_{nn}).$$

The iteration algorithm of the *combined step procedure*, also referred to as the *Jacobi algorithm*, is then

$$\mathbf{Dx}^{(k+1)}=(\mathbf{L}+\mathbf{U})\mathbf{x}^{(k)}-\mathbf{b},\qquad k=0,1,2\dots,$$

and the iteration matrix is defined by the relation

$$\boxed{\mathbf{T}_{\mathrm{J}} := \mathbf{D}^{-1}(\mathbf{L} + \mathbf{U}).}$$

A sufficient condition for the convergence of the Jacobi algorithm is that the matrix $\mathbf{A}$ is diagonally dominant.

The *single step procedure*, also referred to as the *Gauss–Seidel algorithm*, is defined by

$$(\mathbf{D} - \mathbf{L})\mathbf{x}^{(k+1)} = \mathbf{U}\mathbf{x}^{(k)} - \mathbf{b}, \qquad k = 0, 1, 2, \ldots,$$

so that the elements of the fix point iteration are given by

$$\boxed{\mathbf{T}_{\mathrm{E}} := (\mathbf{D} - \mathbf{L})^{-1}\mathbf{U}, \qquad \mathbf{c}_{\mathrm{E}} := -(\mathbf{D} - \mathbf{L})^{-1}\mathbf{b}.}$$

The iteration can be written in components as

$$x_i^{(k+1)} = -\left(b_i + \sum_{j=1}^{i-1} a_{ij}x_j^{(k+1)} + \sum_{j=i+1}^{n} a_{ij}x_j^{(k)}\right)\Big/ a_{ii}, \qquad i = 1, 2, \ldots, n.$$

Sufficient conditions for the convergence of the single step procedure are for example the diagonal dominance or the symmetry and positive definiteness of the matrix $\mathbf{A}$.

It often leads to much better performance (speed of convergence) of the algorithm if one multiplies the individual components with a *relaxation factor* $\omega \neq 1$. If $\omega > 1$, then this is called *over-relaxation*, while the other case $\omega \leq 1$ is called *under-relaxation*. The combined step procedure gives rise to the *JOR-procedure*

$$\begin{aligned}\mathbf{x}^{(k+1)} &= \mathbf{x}^{(k)} + \omega\big[\mathbf{D}^{-1}(\mathbf{L} + \mathbf{U})\mathbf{x}^{(k)} - \mathbf{D}^{-1}\mathbf{b} - \mathbf{x}^{(k)}\big] \\ &= \big[(1 - \omega)\mathbf{E} + \omega\mathbf{D}^{-1}(\mathbf{L} + \mathbf{U})\big]\mathbf{x}^{(k)} - \omega\mathbf{D}^{-1}\mathbf{b}.\end{aligned}$$

The elements of the fix point iteration are therefore

$$\boxed{\mathbf{T}_{\mathrm{JOR}}(\omega) := (1 - \omega)\mathbf{E} + \omega\mathbf{D}^{-1}(\mathbf{L} + \mathbf{U}), \qquad \mathbf{c}_{\mathrm{JOR}} := -\omega\mathbf{D}^{-1}\mathbf{b}.}$$

Similarly one has the *method of successive over-relaxation*, represented by the *SOR-procedure* with

$$\begin{aligned}\mathbf{x}^{(k+1)} &= (\mathbf{D} - \omega\mathbf{L})^{-1}\big[(1 - \omega)\mathbf{D} + \omega\mathbf{U}\big]\mathbf{x}^{(k)} - \omega(\mathbf{D} - \omega\mathbf{L})^{-1}\mathbf{b}, \\ \mathbf{T}_{\mathrm{SOR}}(\omega) &:= (\mathbf{D} - \omega\mathbf{L})^{-1}\big[(1 - \omega)\mathbf{D} + \omega\mathbf{U}\big], \qquad \mathbf{c}_{\mathrm{SOR}} := -\omega(\mathbf{D} - \omega\mathbf{L})^{-1}\mathbf{b}.\end{aligned}$$

The optimal relaxation parameter $\omega_{\mathrm{opt}}$ should be chosen such that the spectral radius $\varrho(\mathbf{T}_{\mathrm{JOR}}(\omega))$ resp. $\varrho(\mathbf{T}_{\mathrm{SOR}}(\omega))$ is minimized. Because of the given properties of the system matrix $\mathbf{A}$, there exist different theoretical possibilities for choosing $\omega_{\mathrm{opt}}$.

#### 7.2.2.2 The method of conjugate gradients

For the iterative solution of very large, sparse systems of equations $\mathbf{A}\mathbf{x} + \mathbf{b} = \mathbf{0}$ with symmetric and positive definite matrix $\mathbf{A}$, which is the type of matrix which occurs for the discretization of elliptic boundary value problems, the *method of conjugate gradients* is especially appropriate. This method is based on the fact that the solution $\mathbf{x}$ is a minimum of the quadratic function

$$\boxed{\mathbf{F}(\mathbf{v}) := \frac{1}{2}\,\mathbf{v}^{\mathsf{T}}\mathbf{A}\mathbf{v} + \mathbf{b}^{\mathsf{T}}\mathbf{v}.}$$

The minimum of $\mathbf{F}(\mathbf{v})$ is successively determined in such a way that for a starting vector $\mathbf{x}^{(0)}$ the negative gradient direction is found and in the $k^{th}$ step a conjugate descent direction $\mathbf{p}^{(k)}$ is found, so that $\mathbf{F}(\mathbf{v})$ is minimized in that direction. The CG-algorithm is:

*Start:* choice of $\mathbf{x}^{(0)}$, $\quad \mathbf{r}^{(0)} = \mathbf{A}\mathbf{x}^{(0)} + \mathbf{b}$, $\quad \mathbf{p}^{(1)} = -\mathbf{r}^{(0)}$,

*Iteration* $(k = 1, 2, \ldots)$ :

$$\text{if} \quad k > 1 : \begin{cases} e_{k-1} = \mathbf{r}^{(k-1)^{\mathsf{T}}}\mathbf{r}^{(k-1)} / \mathbf{r}^{(k-2)^{\mathsf{T}}}\mathbf{r}^{(k-2)}, \\ \mathbf{p}^{(k)} = -\mathbf{r}^{(k-1)} + e_{k-1}\mathbf{p}^{(k-1)}, \end{cases}$$

$$\mathbf{z} = \mathbf{A}\mathbf{p}^{(k)}, \qquad q_k = \mathbf{r}^{(k-1)^{\mathsf{T}}}\mathbf{r}^{(k-1)} / \mathbf{p}^{(k)^{\mathsf{T}}}\mathbf{z},$$

$$\mathbf{x}^{(k)} = \mathbf{x}^{(k-1)} + q_k\mathbf{p}^{(k)}, \qquad \mathbf{r}^{(k)} = \mathbf{x}^{(k-1)} + q_k\mathbf{z},$$

Test of convergence.

$\mathscr{S}$ In the CG-procedure the residue vectors $\mathbf{r}^{(k)}$ are pairwise orthogonal, and the directions of descent $\mathbf{p}^{(k)}$ are pairwise conjugate, i.e., we have $\left(\mathbf{p}^{(k)}\right)^{\mathsf{T}}\mathbf{A}\mathbf{p}^{(j)} = 0$ for $j = 1, 2, \ldots, k-1$. Therefore this method yields a solution $\mathbf{x}$ of the system of equations $\mathbf{A}\mathbf{x} + \mathbf{b} = \mathbf{0}$ with $n$ unknowns after at most $n$ iterations. The $k^{th}$ iterative solution $\mathbf{x}^{(k)}$ of the CG-algorithm is, in fact, the global minimum of the function $\mathbf{F}(\mathbf{v})$ with respect to the subspace $S_k := \operatorname{span}\left(\{\mathbf{p}^{(1)}, \mathbf{p}^{(2)}, \ldots, \mathbf{p}^{(k)}\}\right) = \operatorname{span}\left(\{\mathbf{p}^{(0)}, \mathbf{p}^{(1)}, \ldots, \mathbf{p}^{(k-1)}\}\right)$, so that

$$\mathbf{F}\left(\mathbf{x}^{(k)}\right) = \min_{c_i} \mathbf{F}\left(\mathbf{x}^{(0)} + \sum_{i=1}^{k} c_i\mathbf{p}^{(i)}\right).$$

From this it follows that the error $\mathbf{e}^{(k)} := \mathbf{x}^{(k)} - \mathbf{x}$ is similarly minimized in the energy norm $||\mathbf{z}||_{\mathbf{A}}^2 := \mathbf{z}^{\mathsf{T}}\mathbf{A}\mathbf{z}$ with respect to $S_k$, so that finally we obtain

$$\frac{\|\mathbf{e}^{(k)}\|_{\mathbf{A}}}{\|\mathbf{e}^{(0)}\|_{\mathbf{A}}} \le 2\left(\frac{\sqrt{\kappa(\mathbf{A})} - 1}{\sqrt{\kappa(\mathbf{A})} + 1}\right)^k.$$

The number of CG-steps $k$ which insure that $\|\mathbf{e}^{(k)}\|_{\mathbf{A}} / \|\mathbf{e}^{(0)}\|_{\mathbf{A}} \le \varepsilon$ can be estimated from this as

$$k \le \frac{1}{2}\sqrt{\kappa(\mathbf{A})} \cdot \ln\left(\frac{2}{\varepsilon}\right) + 1.$$

This bound is determined by the tolerance $\varepsilon$ and the square root of the condition number $\kappa(\mathbf{A})$. The convergence of the CG-method can be noticeably improved by a *preconditioning.* This is done by transforming the system of equations to be solved ahead of time, yielding a matrix $\tilde{\mathbf{A}}$ which has a better condition. This uses a regular $(n \times n)$-matrix $\mathbf{C}$ which transforms $\mathbf{Ax} + \mathbf{b} = \mathbf{0}$ into $\tilde{\mathbf{A}}\tilde{\mathbf{x}} + \tilde{\mathbf{b}} = \mathbf{0}$ via

$$\widetilde{\mathbf{A}} := \mathbf{C}^{-1}\mathbf{A}\mathbf{C}^{-\mathsf{T}}, \qquad \widetilde{\mathbf{b}} := \mathbf{C}^{-1}\mathbf{b}, \qquad \widetilde{\mathbf{x}} := \mathbf{C}^{\mathsf{T}}\mathbf{x}.$$

Here $\tilde{\mathbf{A}}$ is symmetric positive definite and similar to

$$\mathbf{K} := \mathbf{C}^{-\mathsf{T}}\widetilde{\mathbf{A}}\mathbf{C}^{\mathsf{T}} = \left(\mathbf{C}\mathbf{C}^{\mathsf{T}}\right)^{-1}\mathbf{A} =: \mathbf{M}^{-1}\mathbf{A}.$$

The *preconditioning matrix* $\mathbf{M} = \mathbf{C}\mathbf{C}^{\mathsf{T}}$ is thus necessarily an approximation to $\mathbf{A}$ in order for the condition number with respect to the spectral norm to satisfy $\kappa_2(\widetilde{\mathbf{A}}) = \kappa_2(\mathbf{K}) = \kappa_2(\mathbf{M}^{-1}\mathbf{A}) \ll \kappa_2(\mathbf{A})$. There are numerous strategies for choosing the preconditioning matrix $\mathbf{M}$ from the specifics of the problem at hand, cf. [425].

### 7.2.3 Eigenvalue problems

For the calculation of eigenvalues $\lambda_j$ and the corresponding eigenvectors $\mathbf{x}_j$ of a matrix $\mathbf{A}$ such that

$$\boxed{\mathbf{A}\mathbf{x}_j = \lambda_j \mathbf{x}_j,}$$

there are many different methods and procedures, which either use specific properties of the problem or specific properties of the matrix $\mathbf{A}$. In what follows we consider the eigenvalue problem under the assumption that the matrix $\mathbf{A}$ has few vanishing entries and the problem is to find all pairs $(\lambda_j, \mathbf{x}_j)$ of eigenvalues and the corresponding eigenvectors.

#### 7.2.3.1 The characteristic polynomial

The theoretical approach of calculating the eigenvalues $\lambda_j$ as the zeros of the characteristic polynomial $P_{\mathbf{A}}(\lambda) = \det(\mathbf{A} - \lambda\mathbf{E})$ and then determining the eigenvectors from the corresponding homogeneous linear system of equations $(\mathbf{A} - \lambda_j\mathbf{E})\mathbf{x}_j = \mathbf{0}$ (cf. 2.2.1) cannot be put into practice for numerical calculations. The largest problem here is that the rounding errors in the calculation of the coefficients of the characteristic polynomial can have drastic effects on the computed values for the eigenvalues (cf. [432]). For this reason this step is in general quite instable. The treatment of eigenvalue problems has to be done with other methods.

#### 7.2.3.2 Jacobi procedure

The eigenvalues $\lambda_j$ of a real symmetric $(n \times n)$-matrix are real, and the $n$ eigenvectors form an orthonormal system of vectors (cf. 2.2.2.1). Therefore there is an *orthogonal* $(n \times n)$-matrix $\mathbf{X}$, whose columns are the eigenvectors $\mathbf{x}_j$, which can be used to bring $\mathbf{A}$ into diagonal form

$$\boxed{\mathbf{X}^{-1}\mathbf{A}\mathbf{X} = \mathbf{X}^{\mathsf{T}}\mathbf{A}\mathbf{X} = \mathbf{D} = \operatorname{diag}(\lambda_1, \lambda_2, \ldots, \lambda_n).}$$

The Jacobi procedure realizes this transformation iteratively, by carrying out an appropriate sequence of orthogonal similarity transformations with *elementary Jacobi rotation matrices*

$$\mathbf{U}(p,q,\varphi) := \begin{pmatrix} 1 & & & & & & & & \\ & \ddots & & & & & & & \\ & & 1 & & & & & & \\ & & & \cos\varphi & & & \sin\varphi & & \\ & & & & 1 & & & & \\ & & & & & \ddots & & & \\ & & & & & & 1 & & \\ & & & -\sin\varphi & & & \cos\varphi & & \\ & & & & & & & 1 & \\ & & & & & & & & \ddots \\ & & & & & & & & & 1 \end{pmatrix} \begin{matrix} \\ \\ \\ \longleftarrow p \\ \\ \\ \\ \longleftarrow q \\ \\ \\ \\ \end{matrix} \qquad \begin{aligned} u_{ii} &= 1,\ i \neq p,q, \\ u_{pp} &= u_{qq} = \cos\varphi, \\ u_{pq} &= \sin\varphi, \\ u_{qp} &= -\sin\varphi, \\ u_{ij} &= 0 \qquad \text{otherwise.} \end{aligned}$$

The pair of indices $(p,q)$ with $1 \leq p < q \leq n$ is called the rotational pair of indices and $\mathbf{U}(p,q,\varphi)$ is called a $(p,q)$ rotation matrix. In the transformed matrix $\mathbf{A}'' := \mathbf{U}^{-1}\mathbf{A}\mathbf{U} = \mathbf{U}^\mathsf{T}\mathbf{A}\mathbf{U}$, only the elements of the $p^{th}$ and $q^{th}$ rows and columns are changed. For the elements of $\mathbf{A}' := \mathbf{U}^\mathsf{T}\mathbf{A}$, we have

$$\left.\begin{aligned} a'_{pj} &= a_{pj}\cos\varphi - a_{qj}\sin\varphi \\ a'_{qj} &= a_{pj}\sin\varphi + a_{qj}\cos\varphi \\ a'_{ij} &= a_{ij} \qquad \text{for } i \neq p,q \end{aligned}\right\} \qquad j = 1,2,\ldots,n.$$

From this we obtain the matrix elements of $\mathbf{A}'' = \mathbf{A}'\mathbf{U}$ by

$$\left.\begin{aligned} a''_{ip} &= a'_{ip}\cos\varphi - a'_{iq}\sin\varphi \\ a''_{iq} &= a'_{ip}\sin\varphi + a'_{iq}\cos\varphi \\ a''_{ij} &= a'_{ij} \qquad \text{for } j \neq p,q \end{aligned}\right\} \qquad i = 1,2,\ldots,n.$$

The matrix elements on the intersection points of the $p^{th}$ and $q^{th}$ rows and columns are given by

$$\begin{aligned} a''_{pp} &= a_{pp}\cos^2\varphi - 2a_{pq}\cos\varphi\sin\varphi + a_{qq}\sin^2\varphi, \\ a''_{qq} &= a_{pp}\sin^2\varphi + 2a_{pq}\cos\varphi\sin\varphi + a_{qq}\cos^2\varphi, \\ a''_{pq} &= a''_{qp} = (a_{pp} - a_{qq})\cos\varphi\sin\varphi + a_{pq}(\cos^2\varphi - \sin^2\varphi). \end{aligned}$$

The angle $\varphi$ of a $(p,q)$ rotation matrix $\mathbf{U}(p,q,\varphi) = \mathbf{U}$ can be chosen in such a way that in the transformed matrix $\mathbf{A}''$ we have $a''_{pq} = a''_{qp} = 0$, i.e., such that

$$\cot(2\varphi) = \frac{a_{qq} - a_{pp}}{2a_{pq}} \qquad \text{with} \qquad -\frac{\pi}{4} < \varphi \leq \frac{\pi}{4}.$$

In the *classical Jacobi procedure* we start with $\mathbf{A}^{(0)} := \mathbf{A}$ and form the orthogonal and similar matrices $\mathbf{A}^{(k)} = \mathbf{U}_k^\mathsf{T}\mathbf{A}^{(k-1)}\mathbf{U}_k$, $k = 1,2,\ldots$, so that in the $k^{th}$ step, a non-diagonal element $a_{pq}^{(k-1)} = a_{qp}^{(k-1)}$ of $\mathbf{A}^{(k-1)}$ can be made by $\mathbf{U}_k = \mathbf{U}(p,q,\varphi)$ to vanish. The vanishing element generated in this way in $\mathbf{A}^{(k)}$ is, in general, destroyed by later rotations.

The *special Jacobi procedure* chooses the following order for the rotation pairs:

$$(1,2),(1,3),\ldots,(1,n),(2,3),(2,4),\ldots,(2,n),(3,4),\ldots,(n-1,n),$$

so that the non-diagonal elements of the upper half of the matrix are, in each step, changed to zero in a whole row.

For both of these procedures one can show that the sum of squares of the non-diagonal elements of $\mathbf{A}^{(k)}$

$$S(\mathbf{A}^{(k)}) := \sum_{i=1}^{n} \sum_{\substack{j=1 \\ j\neq i}}^{n} \left\{a_{ij}^{(k)}\right\}^2, \qquad k = 0, 1, 2, \ldots$$

forms a zero sequence and that therefore the sequence of matrices $\mathbf{A}^{(k)}$ converges to a diagonal matrix $\mathbf{D}$. If $S(\mathbf{A}^{(k)}) \le \varepsilon^2$, then the diagonal elements $a_{ii}^{(k)}$ represent the eigenvalues $\lambda_i$ with an absolute precision of $\varepsilon$, and the columns of the product matrix $\mathbf{V} := \mathbf{U}_1 \cdot \mathbf{U}_2 \cdots \mathbf{U}_k$ of the rotation matrices are orthonormalized approximations to the corresponding eigenvalues.

#### 7.2.3.3 Transformation to the Hessenberg form

Here we start with a non-symmetric $(n \times n)$-matrix $\mathbf{A}$ and transform this matrix with a sequence of similarity transformations into the form of a *Hessenberg matrix*, which is adapted to calculations:

$$\mathbf{H} = \begin{pmatrix} h_{11} & h_{12} & h_{13} & \ldots & h_{1,n-1} & h_{1n} \\ h_{21} & h_{22} & h_{23} & \ldots & h_{2,n-1} & h_{2n} \\ 0 & h_{32} & h_{33} & \ldots & h_{3,n-1} & h_{3n} \\ 0 & 0 & h_{43} & \ldots & h_{4,n-1} & h_{4n} \\ \vdots & \vdots & \vdots & \ldots & \vdots & \vdots \\ 0 & 0 & 0 & \ldots & h_{n,n-1} & h_{nn} \end{pmatrix}$$

The elementary Jacobi rotation matrices are now applied in the *method of Givens* to produce a situation in which the matrix element underneath the first subdiagonal which is to be eliminated does not lie on the intersection point of the $p^{th}$ and $q^{th}$ rows and columns. Moreover, the elements which have been brought to vanish are no longer changed in later steps, so that the transformation can be achieved in $N^* = \dfrac{(n-1)(n-2)}{2}$ steps. The successive elimination of the matrix elements is done in the order

$$a_{31}, a_{41}, \ldots, a_{n1}, a_{42}, a_{52}, \ldots, a_{n2}, a_{53}, \ldots, a_{n,n-2}$$

by rotation matrices $\mathbf{U}(p, q, \varphi)$ with the corresponding rotation pairs of indices

$$(2,3), (2,4), \ldots, (2,n), (3,4), (3,5), \ldots, (3,n), (4,5), \ldots, (n-1,n).$$

The angle $\varphi \in [-\pi/2, \pi/2]$ is determined so that for the elimination of $a_{ij} \neq 0$ with $i \ge j+2$ through the $(j+1, i)$ rotation, we have

$$\boxed{a'_{ij} = a_{j+1,j} \sin\varphi + a_{ij}\cos\varphi = 0.}$$

With the product matrix $\mathbf{Q} := \mathbf{U}_1 \cdot \mathbf{U}_2 \cdot \ldots \cdot \mathbf{U}_{N^*}$ we have $\mathbf{H} = \mathbf{Q}^{\mathsf{T}}\mathbf{A}\mathbf{Q}$, and the eigenvalues of $\mathbf{A}$ are calculated from the eigenvectors $\mathbf{y}_j$ of the Hessenberg matrix $\mathbf{H}$ as $\mathbf{x}_j = \mathbf{Q}\mathbf{y}_j$.

If the transformation is applied to a *symmetric* $(n \times n)$-matrix $\mathbf{A}$, then because of the conservation of the symmetry property under orthogonal similarity transformations, a symmetric tridiagonal matrix $\mathbf{J} := \mathbf{Q}^{\mathsf{T}}\mathbf{A}\mathbf{Q}$ of the form

$$\mathbf{J} = \begin{pmatrix} \alpha_1 & \beta_1 & & & & \\ \beta_1 & \alpha_2 & \beta_2 & & & \\ & \beta_2 & \alpha_3 & \beta_3 & & \\ & & \ddots & \ddots & \ddots & \\ & & & \beta_{n-2} & \alpha_{n-1} & \beta_{n-1} \\ & & & & \beta_{n-1} & \alpha_n \end{pmatrix}$$

results. The orthogonal similarity transformation of **A** into the Hessenberg matrix or a tridiagonal matrix can be done efficiently with the aid of the *fast Givens transformation* or utilizing *Householder matrices* (see 7.2.4.2).

#### 7.2.3.4 The QR-algorithm

For every real $(n \times n)$-matrix **A** there exists by virtue of Schur's Theorem an orthogonal matrix **U** such that **A** is similar to a *quasi triangular matrix* $\mathbf{R} = \mathbf{U}^{\mathsf{T}}\mathbf{A}\mathbf{U}$ of the form

$$\mathbf{J} = \begin{pmatrix} \mathbf{R}_{11} & \mathbf{R}_{12} & \mathbf{R}_{13} & \dots & \mathbf{R}_{1m} \\ \mathbf{0} & \mathbf{R}_{22} & \mathbf{R}_{23} & \dots & \mathbf{R}_{2m} \\ \mathbf{0} & \mathbf{0} & \mathbf{R}_{33} & \dots & \mathbf{R}_{3m} \\ \vdots & \vdots & \vdots & \dots & \vdots \\ \mathbf{0} & \mathbf{0} & \mathbf{0} & \dots & \mathbf{R}_{mm} \end{pmatrix}$$

The matrices $\mathbf{R}_{ii}$, $i = 1, 2, \ldots, m$ are square of size one or two. The real eigenvalues of **A** are thus equal to the elements of the $(1\times1)$-matrices $\mathbf{R}_{ii}$, while the complex conjugate pairs of eigenvalues are equal to those of the $(2\times2)$-matrices $\mathbf{R}_{ii}$. The *QR-algorithm* is a procedure for constructing a sequence of orthogonal similar matrices which converge to a given quasi triangular matrix **R**. It is based on the fact that every real $(n \times n)$-matrix **A** can be written as a product of an orthogonal matrix **Q** and a upper triangular matrix **R** of the form

$$\boxed{\mathbf{A} = \mathbf{Q} \cdot \mathbf{R} \quad \textit{(QR-decomposition).}}$$

The elementary Jacobi rotation matrices are the constructive elements of the QR-decomposition. If one forms the new matrix

$$\mathbf{A}' := \mathbf{R} \cdot \mathbf{Q},$$

from the constituents of the QR-decomposition of **A**, then $\mathbf{A}'$ is orthogonally similar to **A**, and the transition from **A** to $\mathbf{A}'$ is called a QR-step. It is used to construct the sequence of similar matrices, at least in principle. In order to reduce the computation cost, one works with a Hessenberg matrix **H** or with a tridiagonal matrix **J**, since the QR-transformed matrix $\mathbf{H}'$ of a Hessenberg matrix **H** is again a Hessenberg matrix, and after a QR-step, the transformed matrix $\mathbf{J}'$ of a symmetric tridiagonal matrix **J** is again a tridiagonal matrix. For a Hessenberg matrix $\mathbf{H}_1$ the algorithm of the *QR-algorithm of Fansis* is

$$\mathbf{H}_k = \mathbf{Q}_k\mathbf{R}_k, \qquad \mathbf{H}_{k+1} = \mathbf{R}_k\mathbf{Q}_k, \qquad k = 1, 2, 3, \ldots$$

Under certain assumptions the sequence of orthogonal, similar, Hessenberg matrices $\mathbf{H}_k$ converges to a quasi triangular matrix. In order to increase the convergence ability of the algorithm, one applies spectral shifts. The algorithm is then modified to the *QR-algorithm with explicit spectral shift*

$$\mathbf{H}_k - \sigma_k \mathbf{E} = \mathbf{Q}_k \mathbf{R}_k, \qquad \mathbf{H}_{k+1} = \mathbf{R}_k \mathbf{Q}_k + \sigma_k \mathbf{E}, \qquad k = 1, 2, 3, \ldots$$

The spectral shift $\sigma_k$ in the $k^{th}$ QR-step is chosen appropriately, for example as the last diagonal element of $\mathbf{H}_k$. This insures that the last or next to last subdiagonal element of $\mathbf{H}_k$ converges quickly to zero. Thus the matrix $\mathbf{H}_k$ decomposes, and the QR-algorithm can be continued, applied to submatrices of smaller size. In this way the eigenvalues can be successively calculated. If the given matrix $\mathbf{H}_1$ has pairs of complex conjugate eigenvalues, then to avoid the necessity of working with complex numbers one also uses the technique of *QR-double-step*. This applies to successive QR-steps with complex conjugate spectral shifts in such a way that the resulting Hessenberg matrix $\mathbf{H}_{k+2}$ is calculated directly from $\mathbf{H}_k$, where the spectral shifts are implicit.

The QR-algorithm with implicit spectral shifts, applied to Hessenberg matrices or tridiagonal matrices, is a very efficient method for calculating eigenvalues, as for each eigenvalue or conjugate pair of such on average only a few QR-steps are necessary. It is the standard procedure for treating eigenvalue problems for fully filled matrices.

#### 7.2.3.5 The broken inverse vector iteration of Wielandt

This method is used to efficiently calculate approximations of eigenvalues of a matrix like to ones resulting from the QR-algorithm. Let $\overline{\lambda}_k$ be an approximation of the eigenvalue $\lambda_k$ of a Hessenberg matrix $\mathbf{H}$ which satisfies $0 < |\lambda_k - \overline{\lambda}_k| = \varepsilon \ll d := \min_{i \neq k} |\overline{\lambda}_i - \overline{\lambda}_k|$. Then the iteration

$$\boxed{(\mathbf{H} - \overline{\lambda}_k \mathbf{E})\, \mathbf{z}^{(m)} = \mathbf{z}^{(m-1)}, \qquad m = 1, 2, 3, \ldots}$$

generates for a given starting vector $\mathbf{z}^{(0)}$, which has a non-vanishing component of the eigenvector $\mathbf{y}_k$ of $\mathbf{H}$ for the eigenvalue $\lambda_k$, a sequence of vectors $\mathbf{z}^{(m)}$, which converge very quickly to the direction of the eigenvector $\mathbf{y}_k$. For a good starting approximation $\overline{\lambda}_k$ and for a sufficiently well-spaced eigenvalues, a few iteration steps of the *broken inverse vector iteration* are sufficient for a satisfying result. The solution of the linear system of equations $(\mathbf{H} - \overline{\lambda}_k \mathbf{E})\, \mathbf{z}^{(m)} = \mathbf{z}^{(m-1)}$ with respect to $\mathbf{z}^{(m)}$ are obtained with the Gauss algorithm when a column pivot strategy is applied, taking account of the Hessenberg form of the matrix. It is more economical to normalize the iteration vectors.

This procedure can also be applied to determine the eigenvalues of known eigenvalue approximations of symmetric tridiagonal matrices $\mathbf{J}$, where the more specific structure of the set of equations to be solved should be taken into consideration.

### 7.2.4 Fitting and the method of least squares

The basic problem which fitting calculations should solve is to estimate unknown parameters in empirical formulas, which are dictated by known scientific laws or model assumptions. In the simplest case we have a function $f(x; \alpha_1, \alpha_2, \ldots, \alpha_n)$ and its values for $N$ (different) points $x_1, x_2, \ldots, x_N$ coming from measurements $y_1, y_2, \ldots, y_N$ for the values of $f(x_i; \alpha_1, \alpha_2, \ldots, \alpha_n)$. We want to determine the set of parameters $\alpha_1, \alpha_2, \ldots, \alpha_n$ so that the standard deviation or *residues*

$$r_i := f(x_i; \alpha_1, \alpha_2, \ldots, \alpha_n) - y_i, \qquad i = 1, 2, \ldots, N$$

are minimal in a sense to be made precise below. The number $N$ of measurements is larger than the number $n$ of parameters in order to offset the invariable errors in the input data. Under the assumption that the measurements are normally distributed of equal variance, the *Gauss fitting principle* or the *method of least squares* is appropriate for the problem for probability theoretical reasons. The requirement (the statement of being minimal) is

$$\boxed{\sum_{i=1}^{N} r_i^2 \equiv \sum_{i=1}^{N} \left[f(x_i;\alpha_1,\alpha_2,\ldots,\alpha_n) - y_i\right]^2 \stackrel{!}{=} \min.}$$

If the observations have relatively different precisions, i.e., if the variances of the normally distributed errors are different, then this can be accounted for by providing the residues with weights.

In what follows we consider only the case of a function $f(x_i;\alpha_1,\alpha_2,\ldots,\alpha_n)$ which depends linearly on the parameters $\alpha_k$, so that

$$f(x;\alpha_1,\alpha_2,\ldots,\alpha_n) = \sum_{k=1}^{n} \alpha_k \varphi_k(x)$$

for given functions $\varphi_k(x)$, $k = 1,2,\ldots,n$ which do not depend on the parameters $\alpha_k$. The equations we have to solve, then, are the *linear error equations*

$$\alpha_1\varphi_1(x_i) + \alpha_2\varphi_2(x_i) + \ldots + \alpha_n\varphi_n(x_i) - y_i = r_i, \qquad i = 1,2,\ldots,N.$$

With the quantities $c_{ik} := \varphi_k(x_i)$, $i = 1,2,\ldots,N$; $k = 1,2,\ldots,n$, the matrix $\mathbf{C} = (c_{ik}) \in \mathbb{R}^{N\times n}$, the vector $\mathbf{y} \in \mathbb{R}^N$ of the measurements, the parameter vector $\boldsymbol{\alpha} \in \mathbb{R}^N$ and the residue vector $\mathbf{r} \in \mathbb{R}^N$, the system of equations can be compactly written as

$$\mathbf{C}\alpha - \mathbf{y} = \mathbf{r}.$$

#### 7.2.4.1 The method of normal equations

The solution of error equations like the one we have just met, coming from the Gaussian fitting calculation for postulating

$$\mathbf{r}^\mathsf{T}\mathbf{r} = (\mathbf{C}\alpha - \mathbf{y})^\mathsf{T}(\mathbf{C}\alpha - \mathbf{y}) = \alpha^\mathsf{T}\mathbf{C}^\mathsf{T}\mathbf{C}\alpha - 2(\mathbf{C}^\mathsf{T}\mathbf{y})^\mathsf{T}\alpha + \mathbf{y}^\mathsf{T}\mathbf{y} \stackrel{!}{=} \min.$$

leads with $\mathbf{A} := \mathbf{C}^\mathsf{T}\mathbf{C} \in \mathbb{R}^{n\times n}$ and $\mathbf{b} := \mathbf{C}^\mathsf{T}\mathbf{y} \in \mathbb{R}^n$ to the necessary (and at the same time sufficient) condition represented by the linear system of *normal equations*

$$\boxed{\mathbf{A}\alpha + \mathbf{b} = \mathbf{0}}$$

for the parameters to be determined. If the matrix $\mathbf{C}$ of the error equation has maximal rank $n$, then the matrix $\mathbf{A}$ is symmetric and positive definite. Consequently, $\mathbf{A}$ is regular, the parameter vector $\boldsymbol{\alpha}$ is uniquely determined and the system of equations can by solved with the aid of the Cholesky procedure (cf. 7.2.1.5). The residues are obtained from the calculation of the parameter vector $\boldsymbol{\alpha}$ by substitution into the error equation.

The matrix elements $a_{ik}$ and the constants $b_i$ of the normal equations can be calculated from the column vectors $\mathbf{c}_j$ of the matrix $\mathbf{C}$ as scalar products:

$$a_{jk} = \mathbf{c}_j^\top \mathbf{c}_k = \sum_{i=1}^{N} c_{ij} c_{ik} = \sum_{i=1}^{N} \varphi_j(x_i)\varphi_k(x_i), \qquad j,k = 1,2,\ldots,n,$$

$$b_j = \mathbf{c}_j^\top \mathbf{y} = \sum_{i=1}^{N} c_{ij} y_i = \sum_{i=1}^{N} \varphi_j(x_i) y_i, \qquad j = 1,2,\ldots,n.$$

Using Gauss' summation notation $[f(x)g(x)] := \sum_{i=1}^{N} f(x_i)g(x_i)$, the elements of the normal equation are given by $a_{jk} = [\varphi_j(x)\varphi_k(x)]$, $b_j = [\varphi_j(x)y]$. They allow an explicit exhibition of the solutions for special cases which often occur.

**7.2.4.1.1 Fitting of direct observations.** If an unknown quantity $y$ is observed and if we have $N$ measurements $y_i$, then the $N$ error equations $y - y_i = r_i$, $i = 1,2,\ldots,N$ and the resulting single normal equation for the parameter values $y$ which we want to determine are given by

$$[1]y - [y] = 0, \qquad \text{i.e.} \qquad Ny = \sum_{i=1}^{N} y_i, \qquad \text{that is} \qquad y = \frac{1}{N}\sum_{i=1}^{N} y_i.$$

The fitting, i.e., most probable value $y$ according to the method of least squares is then equal to the *arithmetic mean of the measurements*. The resulting residues $r_i$ for the mean are the most probable errors, the quantity $m := \sqrt{[rr]/(N-1)}$ is called the *mean error of the observation*, and $m_y := \sqrt{[rr]/(N(N-1))}$ is called the *mean error of the mean*.

**7.2.4.1.2 The regression line $y = ax + b$.** If measurements $M_i(x_i, y_i)$, $i = 1,2,\ldots,N$ lie nearly on a line (assuming exact abscissas $x_i$ and measured ordinates $y_i$), then the parameters $a$ and $b$ of the *regression line* $y = ax + b$ are to be determined from the error equations

$$\boxed{ax_i + b - y_i = r_i, \qquad i = 1,2,\ldots,N}$$

according to the method of least squares. The corresponding normal equations are

$$\begin{aligned}[x^2]\cdot a + [x]\cdot b - [xy] &= 0,\\ [x]\cdot a + [1]\cdot b - [y] &= 0.\end{aligned}$$

The solution of this equation is often derived from the Cramer rule (cf. 2.1.4.3) in the form

$$\boxed{a = \frac{N\cdot[xy] - [x]\cdot[y]}{N\cdot[x^2] - ([x])^2}, \qquad b = \frac{[x^2]\cdot[y] - [x]\cdot[xy]}{N\cdot[x^2] - ([x])^2}.}$$

These formulas are numerically instable, since in the case of positive $x_i$ and $y_i$ cancellation of numerators and denominators is possible. A stable way to do the calculation is to use the means $\overline{x} := [x]/N$, $\overline{y} := [y]/N$ in the form

$$a = \sum_{i=1}^{N}(x_i - \overline{x})(y_i - \overline{y}) \Big/ \sum_{i=1}^{N}(x_i - \overline{x})^2, \qquad b = \overline{y} - a\cdot\overline{x}.$$

**7.2.4.1.3 Fitted parabola $y = a + bx + cx^2$.** For measurements $M_i(x_i, y_i)$, $i = 1, 2, \ldots, N$ which lie nearly on a parabola, it follows from the corresponding $N$ error equations

$$a + bx_i + cx_i^2 - y_i = r_i, \qquad i = 1, 2, \ldots, N$$

that the following three equations are the normal equations.

$$\begin{aligned} N \cdot a + \ [x] \cdot b + [x^2] \cdot c - \ \ [y] &= 0, \\ [x] \cdot a + [x^2] \cdot b + [x^3] \cdot c - \ [xy] &= 0, \\ [x^2] \cdot a + [x^3] \cdot b + [x^4] \cdot c - [x^2 y] &= 0. \end{aligned}$$

Their solution should be dispensed of with the Cholesky procedure.

**7.2.4.1.4 Fitted polynomials.** In certain situations it is appropriate to use fitted polynomials of higher degree. Using the Ansatz function $\varphi_k(x) = x^{k-1}$, $k = 1, 2, \ldots, n$ for a fitted polynomial of degree $(n-1)$, one obtains the normal equations for the parameters $\alpha_1, \ldots, \alpha_n$

$$\sum_{k=1}^{n} [x^{i+k-2}]\alpha_k - [x^{i-1}y] = 0, \qquad i = 1, 2 \ldots, n.$$

The condition number $\kappa(\mathbf{A})$ of the normal equation matrix $\mathbf{A}$ is often very large, so that the calculation of the solution $\boldsymbol{\alpha}$ has the usual problems as discussed in section 7.2.1.7. An improvement of the situation can be obtained by using, instead of the simple power function, a more appropriate Ansatz function $\varphi_k(x)$ in the form of a Legendre polynomial $P_k(x)$ (cf. 1.13.2.13) or one of the Chebychev polynomials $T_k(x)$ (cf. 7.5.1.3), which fit the approximation interval by a simple substitution of variables.

### 7.2.4.2 The method of orthogonal transformations

In order to treat the error equations, for which the normal equations are often poorly conditioned, in a numerically more stable manner, we can first apply to them an orthogonal transformation with the objective of bringing the system of error equations into a simpler form. An orthogonal transformation is in the context of the method of least squares admissible, since the Euclidean length of the residue vector is not changed by this transformation. Let $\mathbf{Q} \in \mathbb{R}^{N \times N}$ be an orthogonal matrix. Then $\mathbf{C}\boldsymbol{\alpha} - \mathbf{y}$ is transformed into

$$\mathbf{QC}\boldsymbol{\alpha} - \mathbf{Qy} = \mathbf{Qr} =: \widehat{\mathbf{r}}$$

For every matrix $\mathbf{C} \in \mathbb{R}^{N \times n}$ of maximal rank $n < N$ there is an orthogonal matrix $\mathbf{Q} \in \mathbb{R}^{N \times N}$ such that

$$\mathbf{QC} = \widehat{\mathbf{R}} := \left( \frac{\mathbf{R}}{\mathbf{0}} \right),$$

where $\mathbf{R} \in \mathbb{R}^{(N-n) \times n}$ is a regular upper triangular matrix and $\mathbf{0} \in \mathbb{R}^{n \times n}$ is the zero matrix. The orthogonal matrix $\mathbf{Q}$ can be constructed explicitly as a product of $n$ *Householder matrices* of the form

$$\mathbf{U} := \mathbf{E} - 2\mathbf{w}\mathbf{w}^{\mathsf{T}} \in \mathbb{R}^{N \times N} \qquad \text{with} \quad \mathbf{w} \in \mathbb{R}^N, \ \mathbf{w}^{\mathsf{T}}\mathbf{w} = 1.$$

The Householder matrix $\mathbf{U}$ is symmetric and $\mathbf{U}^{\mathsf{T}}\mathbf{U} = \mathbf{E}$, and hence $\mathbf{U}$ is orthogonal. It corresponds to a reflection on a $(N-1)$-dimensional complementary subspace of $\mathbb{R}^N$ which is orthogonal to $\mathbf{w}$. Because of the reflection property, by appropriately choosing the vector $\mathbf{w}$, we can, for every vector $\mathbf{c} \in \mathbb{R}^N, \mathbf{c} \neq \mathbf{0}$, display a Householder matrix $\mathbf{U}$ so that $\mathbf{c}$ maps to an arbitrarily given vector $\mathbf{c}'$ of the same norm as $\mathbf{c}$, by setting $\mathbf{c}' = \mathbf{U}\mathbf{c}$. The vector $\mathbf{w}$ has to be given the direction which bisects the angle between $\mathbf{c}$ and $-\mathbf{c}'$.

In the first transformation step with $\mathbf{U}_1 = \mathbf{E} - 2\mathbf{w}_1\mathbf{w}_1^\mathsf{T}$, we can arrive at the situation in which the first column of the transformed matrix $\mathbf{C}' = \mathbf{U}_1\mathbf{C}$ we have a multiple of the first unit vector $\mathbf{e}_1 \in \mathbb{R}^N$. This means that $\mathbf{U}_1\mathbf{c}_1 = -\gamma\mathbf{e}_1$ with $\gamma = \pm||\mathbf{c}_1||$, where $\mathbf{c}_1$ is the first column of $\mathbf{C}$. Consequently the direction of $\mathbf{w}_1$ is uniquely determined as $\mathbf{h} := \mathbf{c}_1 + \gamma\mathbf{e}_1$. In order to avoid a possible cancellation in the calculation of the first component of $\mathbf{h}$, we choose the sign of $\gamma$ to be the same as the first component $c_{11}$ of $\mathbf{c}$. With the vector $\mathbf{w}_1$ obtained by normalizing $\mathbf{h}$, the first transformation step $\mathbf{C}' = \mathbf{U}_1\mathbf{C} = (\mathbf{E} - 2\mathbf{w}_1\mathbf{w}_1^\mathsf{T})\mathbf{C} = \mathbf{C} - 2\mathbf{w}_1(\mathbf{w}_1^\mathsf{T}\mathbf{C})$ can be carried out with a minimum of computation cost by using the auxiliary quantities

$$p_j = 2\mathbf{w}_1^\mathsf{T}\mathbf{c}_j, \qquad j = 2, 3, \ldots, n$$

according to

$$\mathbf{c}_1' = -\gamma\mathbf{e}_1, \qquad \mathbf{c}_j' = \mathbf{c}_j - p_j\mathbf{w}_1, \qquad j = 2, 3, \ldots, n.$$

In the following $k^{th}$ transformation step we transform the $k^{th}$ column of the matrix $\mathbf{C}^{(k-1)} := \mathbf{U}_{k-1}\ldots\mathbf{U}_1\mathbf{C}$ in the desired form by using a Householder matrix $\mathbf{U}_k = \mathbf{E} - 2\mathbf{w}_k\mathbf{w}_k^\mathsf{T}$, $k = 2, 3, \ldots, n$ with a vector $\mathbf{w}_k \in \mathbb{R}^N$, whose first $(k-1)$ components vanish. This means that the partial vector of the $k^{th}$ column, which consists of the $(N-k+1)$ last components, is mapped to a multiple of the unit vector $\mathbf{e}_1 \in \mathbb{R}^{N-k+1}$. At the same time the first $(k-1)$ columns of $\mathbf{C}^{(k-1)}$ are unchanged. After $n$ such orthogonal transformations we have

$$\mathbf{Q}\mathbf{C} = \widehat{\mathbf{R}} \qquad \text{with} \qquad \mathbf{Q} := \mathbf{U}_n \cdot \mathbf{U}_{n-1} \cdot \ldots \cdot \mathbf{U}_2 \cdot \mathbf{U}_1.$$

If one forms the transformed measured vector

$$\widehat{\mathbf{y}} := \mathbf{Q}\mathbf{y} = \mathbf{U}_n \cdot \mathbf{U}_{n-1} \cdot \ldots \cdot \mathbf{U}_2 \cdot \mathbf{U}_1\mathbf{y}.$$

through successive multiplication with the Householder matrix $\mathbf{U}_k$, then the equivalent, transformed system of error equations becomes

$$\begin{aligned}
r_{11}\alpha_1 + r_{12}\alpha_2 + \ldots + r_{1n}\alpha_n - \widehat{y}_1 &= \widehat{r}_1 \\
r_{22}\alpha_2 + \ldots + r_{2n}\alpha_n - \widehat{y}_2 &= \widehat{r}_2 \\
\ldots\ldots & \\
r_{nn}\alpha_n - \widehat{y}_n &= \widehat{r}_n \\
-\widehat{y}_{n+1} &= \widehat{r}_{n+1} \\
&\vdots \\
-\widehat{y}_N &= \widehat{r}_N.
\end{aligned}$$

Since the last $(N-n)$ residues $\widehat{r}_i$ are determined by the corresponding values $\widehat{y}_i$, the sum of the squares of the residues is minimal if and only if $\widehat{r}_1 = \widehat{r}_2 = \cdots = \widehat{r}_n = 0$. The parameters $\alpha_1, \ldots, \alpha_n$ we wish to determine are obtained from

$$\mathbf{R}\boldsymbol{\alpha} = \widehat{\mathbf{y}}_1, \qquad \widehat{\mathbf{y}}_1 \in \mathbb{R}^n,$$

through the process of reverse substitution with the vector $\widehat{\mathbf{y}}_1$, which consists of the first $n$ components of the $\widehat{\mathbf{y}} \in \mathbb{R}^N$. If the residue vector $\mathbf{r}$ is to be calculated with respect to the given error equations, this can be done most conveniently, by virtue of the relation $\mathbf{Q}\mathbf{r} = \widehat{\mathbf{r}}$ and the symmetry of the Householder matrix $\mathbf{U}_k$, by the formula

$$\mathbf{r} = \mathbf{Q}^\mathsf{T}\widehat{\mathbf{r}} = \mathbf{U}_1 \cdot \mathbf{U}_2 \cdot \ldots \cdot \mathbf{U}_{n-1} \cdot \mathbf{U}_n\widehat{\mathbf{r}}.$$

In $\widehat{\mathbf{r}}$ the first $n$ components vanish, and the last $(N-n)$ components are given by $\widehat{r}_i = -\widehat{y}_i$. The successive multiplication of $\widehat{\mathbf{r}}$ with the matrices $\mathbf{U}_k$ is done with the aid of the efficient calculational technique described above.

The method of orthogonal transformations yields the parameter values with relatively small errors compared with the classical method using the normal equations. This is because the transition from the error equations to the normal equation squares the condition number of $\mathbf{C}$.

Besides the Householder transformations of the error equations with elementary rotation matrices there are orthogonal transformations with elementary rotation matrices, where in each step a single matrix element is eliminated. This variant is costlier, but it makes it possible to take special, irregular ways in which the matrix $\mathbf{C}$ might be occupied into account. In addition, it can be more advantageous from the memory use point of view and takes at each step the particular error equation of the system into account.

#### 7.2.4.3 The method of singular value decomposition

If the rank of the matrix $\mathbf{C}$ of the error equations is not maximal, but rather $\operatorname{rank}\mathbf{C} = \varrho < n$, or if the column vectors of $\mathbf{C}$ are up to the precision of the calculations being made almost linearly dependent, then the methods described up to now are not applicable. In these cases the solution of $\mathbf{C}\boldsymbol{\alpha} - \mathbf{y} = \mathbf{r}$ is not unique with the method of least squares or at least it is ill-defined. Processing problems of this type is based on the facts we now describe.

For every matrix $\mathbf{C} \in \mathbb{R}^{N\times n}$ with $\operatorname{rank}\mathbf{C} = \varrho \le n < N$, there exist orthogonal matrices $\mathbf{U} \in \mathbb{R}^{N\times N}$ and $\mathbf{V} \in \mathbb{R}^{N\times N}$ such that we have the *singular value decomposition*

$$\boxed{\mathbf{C} = \mathbf{U}\widehat{\mathbf{S}}\mathbf{V}^{\mathsf{T}}} \qquad \text{with} \qquad \widehat{\mathbf{S}} = \begin{pmatrix} \mathbf{S} \\ \hline \mathbf{0} \end{pmatrix}, \qquad \widehat{\mathbf{S}} \in \mathbb{R}^{N\times n}, \qquad \mathbf{S} \in \mathbb{R}^{n\times n},$$

in which $\mathbf{S}$ denotes a diagonal matrix with non-negative diagonal elements $s_i$, which can be ordered in such a way that $s_1 \ge s_2 \ge \ldots \ge s_\varrho > s_{\varrho+1} = \ldots = s_n = 0$, and $\mathbf{0} \in \mathbb{R}^{(N-n)\times n}$ is a zero matrix (cf. [432]). Here $s_i$ are the *singular values* of the matrix $\mathbf{C}$. The column vectors $\mathbf{u}_i \in \mathbb{R}^N$ of $\mathbf{U}$ are called the *left-singular*, the column vectors $\mathbf{v}_i \in \mathbb{R}^n$ are called the *right-singular* vectors of $\mathbf{C}$. Because of the singular value decomposition, written as $\mathbf{C}\mathbf{V} = \mathbf{U}\widehat{\mathbf{S}}$ or $\mathbf{C}^{\mathsf{T}}\mathbf{U} = \mathbf{V}\widehat{\mathbf{S}}^{\mathsf{T}}$, one obtains the relations

$$\begin{aligned} \mathbf{C}\mathbf{v}_i &= s_i\mathbf{u}_i, \quad & \mathbf{C}^{\mathsf{T}}\mathbf{u}_i &= s_i\mathbf{v}_i, \quad & i &= 1, 2, \ldots, n, \\ & & \mathbf{C}^{\mathsf{T}}\mathbf{u}_i &= \mathbf{0}, & i &= n+1, n+2, \ldots, N. \end{aligned}$$

The singular value decomposition is related to systems of principal axis of a symmetric, positive definite matrix as follows

$$\mathbf{A} := \mathbf{C}^{\mathsf{T}}\mathbf{C} = \mathbf{V}\widehat{\mathbf{S}}^{\mathsf{T}}\mathbf{U}^{\mathsf{T}}\mathbf{U}\widehat{\mathbf{S}}\mathbf{V}^{\mathsf{T}} = \mathbf{V}\widehat{\mathbf{S}}^{\mathsf{T}}\widehat{\mathbf{S}}\mathbf{V}^{\mathsf{T}} = \mathbf{V}\mathbf{S}^2\mathbf{V}^{\mathsf{T}},$$

$$\mathbf{B} := \mathbf{C}\mathbf{C}^{\mathsf{T}} = \mathbf{U}\widehat{\mathbf{S}}\mathbf{V}^{\mathsf{T}}\mathbf{V}\widehat{\mathbf{S}}^{\mathsf{T}}\mathbf{U}^{\mathsf{T}} = \mathbf{U}\widehat{\mathbf{S}}\widehat{\mathbf{S}}^{\mathsf{T}}\mathbf{U}^{\mathsf{T}} = \mathbf{U}\begin{pmatrix} \mathbf{S}^2 & \vdots & \mathbf{0} \\ \cdots & \cdots & \cdots \\ \mathbf{0} & \vdots & \mathbf{0} \end{pmatrix}\mathbf{U}^{\mathsf{T}}.$$

The squares of the positive singular values are equal to the positive eigenvalues of both $\mathbf{A}$ and $\mathbf{B}$, while the right-singular vectors $\mathbf{v}_i$ are the eigenvectors of $\mathbf{A}$ and the left-singular vectors $\mathbf{u}_i$ are the eigenvectors of $\mathbf{B}$.

With the singular value decomposition we can transform $\mathbf{C}\boldsymbol{\alpha} - \mathbf{y} = \mathbf{r}$ with orthogonal transformations into an equivalent system of error equations :

$$\mathbf{U}^{\mathsf{T}}\mathbf{C}\mathbf{V}\mathbf{V}^{\mathsf{T}}\boldsymbol{\alpha} - \mathbf{U}^{\mathsf{T}}\mathbf{y} = \mathbf{U}^{\mathsf{T}}\mathbf{r} =: \widehat{\mathbf{r}}.$$

With $\boldsymbol{\beta} := \mathbf{V}^{\mathsf{T}}\boldsymbol{\alpha}$, $\widehat{\mathbf{y}} := \mathbf{U}^{\mathsf{T}}\mathbf{y}$ and $\mathbf{U}^{\mathsf{T}}\mathbf{C}\mathbf{V} = \widehat{\mathbf{S}}$, the transformed system of error equations is

$$\begin{aligned} s_i\beta_i - \widehat{y}_i &= \widehat{r}_i, \qquad i = 1, 2, \ldots, \varrho, \\ -\widehat{y}_i &= \widehat{r}_i, \qquad i = \varrho + 1, \varrho + 2, \ldots, N. \end{aligned}$$

Since the last $(N - \varrho)$ residues $\widehat{r}_i$ are determined by the corresponding $\widehat{y}_i$, the sum of the squares of the residues is minimal if $\widehat{r}_1 = \widehat{r}_2 = \cdots = \widehat{r}_\varrho = 0$. Thus the first $\varrho$ auxiliary unknowns $\beta_i$ can be determined by

$$\beta_i = \widehat{y}_i / s_i, \qquad i = 1, 2, \ldots, \varrho,$$

while the rest of the $\beta_{\varrho-1}, \ldots, \beta_n$ in the case $\varrho = \operatorname{rank} \mathbf{C} < n$ are arbitrary. If one further takes account of $\widehat{y}_i = \mathbf{u}_i^{\mathsf{T}}\mathbf{y}$, $i = 1, 2 \ldots, N$, then the solution vector $\boldsymbol{\alpha}$ is given by

$$\boldsymbol{\alpha} = \sum_{i=1}^{\varrho} \frac{\mathbf{u}_i^{\mathsf{T}}\mathbf{y}}{s_i}\mathbf{v}_i + \sum_{i=\varrho+1}^{n} \beta_i \mathbf{v}_i$$

with $(n - \varrho)$ free parameters $\beta_i$, $i = \varrho + 1, \ldots, n$. If $\mathbf{C}$ does not have maximal rank $n$, then the general solution $\boldsymbol{\alpha}$ is the sum of particular solutions from the linear hull which is generated by $\varrho$ right-singular vectors $\mathbf{v}_i$ corresponding to the positive singular values $s_i$ and an arbitrary vector from the kernel of the linear mapping defined by $\mathbf{C}$.

In the set of solutions of a not uniquely solvable system of error equations, one is often interested in the particular solution whose Euclidean norm is minimal. Because of the orthonormality of the right singular vectors $\mathbf{v}_i$, this leads to the equation

$$\boldsymbol{\alpha}^* = \sum_{i=1}^{\varrho} \frac{\mathbf{u}_i^{\mathsf{T}}\mathbf{y}}{s_i}\mathbf{v}_i \qquad \text{with} \qquad \|\boldsymbol{\alpha}^*\| \leq \min_{\mathbf{C}\boldsymbol{\alpha}-\mathbf{y}=\mathbf{r}} \|\boldsymbol{\alpha}\|.$$

In certain applications with very poorly conditioned error equations, which are characterized by a large ratio between the largest and the smallest singular values, it can make sense to drop certain components of $\boldsymbol{\alpha}^*$ if this procedure will make the values of the squares of the residues acceptable.

The actual calculation of the singular value decomposition of a matrix $\mathbf{C}$ is carried out in two steps. First the matrix $\mathbf{C}$ is transformed to a bidiagonal matrix $\mathbf{B}$ which has the same singular values as $\mathbf{C}$ does, by using orthogonal matrices $\mathbf{Q} \in \mathbb{R}^{N \times N}$ and $\mathbf{W} \in \mathbb{R}^{n \times n}$, and calculating $\mathbf{B}$ via $\mathbf{B} = \mathbf{Q}^{\mathsf{T}}\mathbf{C}\mathbf{W}$. In the second step, the singular values of $\mathbf{B}$ are calculated using a special variant of the QR-algorithm.

## 7.3 Interpolation, numerical differentiation and quadrature

### 7.3.1 Interpolation polynomials

Let $(n + 1)$ pairwise different points (interpolation points) $x_0, x_1, \ldots, x_n$ in an interval $[a, b] \subset \mathbb{R}$ be given, together with the corresponding values (interpolation values) $y_0, y_1, \ldots, y_n$ which for example might be the values of a real-valued function $f(x)$ at the interpolation points. The *interpolation problem* is then to find a polynomial $I_n(x)$ of degree at most $n$, such that $I_n$ satisfies the $(n + 1)$ interpolation conditions

$$I_n(x_i) = y_i, \qquad i = 0, 1, 2, \ldots, n.$$

Under these assumptions there is a unique such polynomial. The representation of this uniquely determined polynomial $I_n$ can take various forms.

#### 7.3.1.1 The Lagrange interpolation formula

We introduce the $(n+1)$ special *Lagrange polynomials*

$$L_i(x) := \prod_{\substack{j=0\\j\neq i}}^{n} \frac{x-x_j}{x_i-x_j}$$
$$= \frac{(x-x_0)\cdots(x-x_{i-1})(x-x_{i+1})\cdots(x-x_n)}{(x_i-x_0)\cdots(x_i-x_{i-1})(x_i-x_{i+1})\cdots(x_i-x_n)}, \qquad i=0,1,2,\ldots,n.$$

which correspond to the $(n+1)$ interpolation points. This polynomial has, being a product of $n$ linear factors, degree $n$, and the properties $L_i(x_i)=1$ and $L_i(x_k)=0$ for $k\neq i$. The polynomial $I_n$ can be written in terms of these polynomials as

$$\boxed{I_n(x) = \sum_{i=0}^{n} y_i L_i(x).}$$

In order to calculate the value $I_n(x)$ at a point which is not one of the interpolation points, one uses the *Lagrange interpolation formula*, which is

$$I_n(x) = \sum_{i=0}^{n} y_i \prod_{\substack{j=0\\j\neq i}}^{n} \frac{x-x_j}{x_i-x_j} = \sum_{i=0}^{n} y_i \frac{1}{x-x_i}\cdot\left\{\prod_{\substack{j=0\\j\neq i}}^{n}\frac{1}{x_i-x_j}\right\}\cdot\prod_{k=0}^{n}(x-x_k).$$

Using the *partial coefficients*

$$\lambda_i := 1\Big/\prod_{\substack{j=0\\j\neq i}}^{n}(x_i-x_j), \qquad i=0,1,2,\ldots,n$$

which only depend on the interpolation points, and the *interpolation weights*

$$\mu_i := \lambda_i/(x-x_i), \qquad i=0,1,2,\ldots,n$$

which depend on the point $x$ whose value we wish to determine, one obtains the representation

$$I_n(x) = \left\{\sum_{i=0}^{n}\mu_i y_i\right\}\cdot\prod_{k=0}^{n}(x-x_k).$$

This product of $(n+1)$ linear factors is equal to the reciprocal value of the sum of the $\mu_i$, and hence one obtains the following *barycentric formula* for the calculation of $I_n(x)$

$$I_n(x) = \left\{\sum_{i=0}^{n}\mu_i y_i\right\}\Big/\left\{\sum_{i=0}^{n}\mu_i\right\}.$$

This formula is useful for numerical computations. In the particular case of increasingly ordered and *equidistant* interpolation points, with step size $h>0$,

$$x_0, \quad x_1 = x_0+h, \quad \ldots, \quad x_j = x_0+jh, \quad \ldots, \quad x_n = x_0+nh,$$

the partial coefficients are given by

$$\lambda_i = \frac{(-1)^{n-i}}{h^n n!}\binom{n}{i}, \qquad i=0,1,2,\ldots,n.$$

Since the common factor $(-1)^n/(h^n n!)$ in the barycentric formula can be dropped, we can just as well use the equivalent partial coefficients, the *signed alternating binomial coefficients*:

$$\lambda_0^* = 1, \quad \lambda_1^* = -\binom{n}{1}, \quad \ldots, \quad \lambda_i^* = (-1)^i\binom{n}{i}, \quad \ldots, \quad \lambda_n^* = (-1)^n.$$

#### 7.3.1.2 The Newton interpolation formula

We now consider the $(n+1)$ *Newton polynomials*

$$N_0(x) := 1, \qquad N_i(x) := \prod_{j=0}^{i-1}(x - x_j), \qquad i = 1, 2, \ldots, n,$$

where $N_i(x)$ has degree $i$, being a product of $i$ linear factors. Then the *Newton interpolation formula* is

$$\boxed{I_n(x) = \sum_{i=0}^{n} c_i N_i(x).}$$

The coefficients $c_i$ are determined as the $i^{th}$ *divided difference*, also called the $i^{th}$ *slope*, which is defined by

$$c_i := [x_0 x_1 \ldots x_i] = [x_i x_{i-1} \ldots x_0], \qquad i = 0, 1, \ldots, n.$$

where $[x_i] := y_i, i = 0, 1, \ldots, n$ are the initial values for the recursively defined slopes. Let $j_0, j_1, \ldots, j_i$ be successive index values in $\{0, 1, \ldots, n\}$. Then we have

$$[x_{j_0} x_{j_1} \ldots x_{j_i}] := \frac{[x_{j_1} x_{j_2} \ldots x_{j_i}] - [x_{j_0} x_{j_1} \ldots x_{j_{i-1}}]}{x_{j_i} - x_{j_0}}.$$

For the determination of the needed slopes in the Newton formula, the following *scheme of divided differences*

| | | | | | |
|---|---|---|---|---|---|
| $x_0$ | $[x_0]$ | | | | |
| | | $[x_0x_1]$ | | | |
| $x_1$ | $[x_1]$ | | $[x_0x_1x_2]$ | | |
| | | $[x_1x_2]$ | | $[x_0x_1x_2x_3]$ | |
| $x_2$ | $[x_2]$ | | $[x_1x_2x_3]$ | | $[x_0x_1x_2x_3x_4]$ |
| | | $[x_2x_3]$ | | $[x_1x_2x_3x_4]$ | |
| $x_3$ | $[x_3]$ | | $[x_2x_3x_4]$ | | |
| | | $[x_3x_4]$ | | | |
| $x_4$ | $[x_4]$ | | | | |

turns out to be convenient. Indeed, the coefficients $c_i$ of the Newton formula are just the values of the downward-sloping diagonal. These are the only values of interest, and they can be calculated by a computer program which starts with the points of interpolation $x_i$ and the corresponding values $y_i = [x_i]$, then calculates the values columnwise. The most efficient calculation of an interpolated value at the point $x$ follows from this representation, for example for $n = 4$,

$$I_4(x) = c_0 + (x - x_0)\big[c_1 + (x - x_1)\big\{c_2 + (x - x_2)(c_3 + (x - x_3)c_4)\big\}\big]$$

by successive evaluation of the expressions in brackets from the innermost outward; the computational cost is a mere $n$ multiplicative operations for a general polynomial $I_n(x)$.

For *equidistant interpolation points* $x_j = x_0 + jh,\ j = 0, 1, \ldots, n$, the divided differences simplify, which in turn simplifies the whole interpolation polynomials, as

$$[x_i x_{i+1}] = \frac{y_{i+1} - y_i}{x_{i+1} - x_i} = \frac{1}{h}(y_{i+1} - y_i) =: \frac{1}{h}\Delta^1 y_i, \qquad \textit{(first difference)}$$

$$[x_i x_{i+1} x_{i+2}] = \frac{[x_{i+1}x_{i+2}] - [x_i x_{i+1}]}{x_{i+2} - x_i} = \frac{1}{2h^2}(\Delta^1 y_{i+1} - \Delta^1 y_i) =: \frac{1}{2h^2}\Delta^2 y_i, \qquad \textit{(second difference)}$$

and similarly, for the $k^{th}$ *differences*:

$$[x_i x_{i+1} \ldots x_{i+k}] =: \frac{1}{k!\,h^k}\Delta^k y_i. \qquad (k^{th} \textit{ difference})$$

The *forward differences* are recursively defined by

$$\Delta^k y_i := \Delta^{k-1} y_{i+1} - \Delta^{k-1} y_i, \qquad k = 1, 2, \ldots, n, \quad i = 0, 1, \ldots, n-k,$$

where the initial values are $\Delta^0 y_i := y_i,\ i = 0, 1, \ldots, n$. They again can be calculated with the help of a neat scheme:

| | | | | | |
|---|---|---|---|---|---|
| $x_0$ | $y_0$ | | | | |
| | | $\Delta^1 y_0$ | | | |
| $x_1$ | $y_1$ | | $\Delta^2 y_0$ | | |
| | | $\Delta^1 y_1$ | | $\Delta^3 y_0$ | |
| $x_2$ | $y_2$ | | $\Delta^2 y_1$ | | $\Delta^4 y_0$ |
| | | $\Delta^1 y_2$ | | $\Delta^3 y_1$ | |
| $x_3$ | $y_3$ | | $\Delta^2 y_2$ | | |
| | | $\Delta^1 y_3$ | | | |
| $x_4$ | $y_4$ | | | | |

The Newton interpolation formula thus takes the form

$$I_n(x) = y_0 + \frac{x - x_0}{h}\Delta^1 y_0 + \frac{(x-x_0)(x-x_1)}{2h^2}\Delta^2 y_0 + \frac{(x-x_0)(x-x_1)(x-x_2)}{3!\,h^3}\Delta^3 y_0 + \ldots + \frac{(x-x_0)(x-x_1)\ldots(x-x_{n-1})}{n!\,h^n}\Delta^n y_0,$$

where the coefficients are in the downward-sloping diagonal of the scheme. If we further define $x = x_0 + th,\ t \in \mathbb{R}$, then we get from this the *Newton–Gregory I-interpolation formula*

$$I_n(x) = y_0 + \binom{t}{1}\Delta^1 y_0 + \binom{t}{2}\Delta^2 y_0 + \binom{t}{3}\Delta^3 y_0 + \ldots + \binom{t}{n}\Delta^n y_0.$$

The interpolation polynomial can just as well be developed from the interpolation point the farthest to the right. In this case one has $x_{n-j} = x_n - jh,\ j = 0, 1, \ldots, n$, and one uses this time the *backward differences*:

$$\nabla^k y_{n-j} := \nabla^{k-1} y_{n-j} - \nabla^{k-1} y_{n-j-1}, \qquad k = 1, 2, \ldots, n, \quad j = 0, 1, \ldots, n-k,$$

where the initial values are now $\nabla^0 y_{n-j} := y_{n-j},\ j = 0, 1, \ldots, n$. These again form a scheme:

| | | | | | |
|---|---|---|---|---|---|
| $x_0$ | $y_0$ | | | | |
| | | $\nabla^1 y_1$ | | | |
| $x_1$ | $y_1$ | | $\nabla^2 y_2$ | | |
| | | $\nabla^1 y_2$ | | $\nabla^3 y_3$ | |
| $x_2$ | $y_2$ | | $\nabla^2 y_3$ | | $\nabla^4 y_4$ |
| | | $\nabla^1 y_3$ | | $\nabla^3 y_4$ | |
| $x_3$ | $y_3$ | | $\nabla^2 y_4$ | | |
| | | $\nabla^1 y_4$ | | | |
| $x_4$ | $y_4$ | | | | |

With the backward differences, it is now the lower, upward sloping diagonal in which the desired differences lie, yielding

$$I_n(x) = y_n + \frac{x - x_n}{h}\nabla^1 y_n + \frac{(x-x_n)(x-x_{n-1})}{2h^2}\nabla^2 y_n + \ldots + \frac{(x-x_n)(x-x_{n-1})\ldots(x-x_1)}{n!\,h^n}\nabla^n y_n.$$

From this we get for $x := x_n + sh,\ s \in \mathbb{R}$, the *interpolation formula of Newton–Gregory II*

$$I_n(x) = y_n + \binom{s}{1}\nabla^1 y_n + \binom{s}{2}\nabla^2 y_n + \binom{s}{3}\nabla^3 y_n + \ldots + \binom{s}{n}\nabla^n y_n.$$

#### 7.3.1.3 The Gaussian interpolation formula

In certain situations it makes sense to take as the initial values for developing an interpolation polynomial, not the first or last interpolation point, but rather a point in the middle of the interval. In this case, the interpolation points, assuming they are equidistant, will have the form $x_j = x_0 + jh,\ j = 0, \pm 1, \pm 2, \ldots, \pm m$. The number $n = 2m+1$ of interpolation points is assumed to be uneven. In this case there are *central differences*, which using the initial values $\delta^0 y_j := y_j,\ j = 0, \pm 1, \ldots, \pm m$, are recursively defined by

$$\begin{aligned} \delta^k y_{j+\frac{1}{2}} &:= \delta^{k-1} y_{j+1} - \delta^{k-1} y_j \qquad &&\text{for} \quad k = 1, 3\ldots\,, \\ \delta^k y_j &:= \delta^{k-1} y_{j+\frac{1}{2}} - \delta^{k-1} y_{j-\frac{1}{2}} \qquad &&\text{for} \quad k = 2, 4, \ldots\,. \end{aligned}$$

The scheme for the central differences is

| | | | | | |
|---|---|---|---|---|---|
| $x_{-2}$ | $y_{-2}$ | | | | |
| | | $\delta^1 y_{-1.5}$ | | | |
| $x_{-1}$ | $y_{-1}$ | | $\delta^2 y_{-1}$ | | |
| | | $\delta^1 y_{-0.5}$ | | $\delta^3 y_{-0.5}$ | |
| $x_0$ | $y_0$ | | $\delta^2 y_0$ | | $\delta^4 y_0$ |
| | | $\delta^1 y_{0.5}$ | | $\delta^3 y_{0.5}$ | |
| $x_1$ | $y_1$ | | $\delta^2 y_1$ | | |
| | | $\delta^1 y_{1.5}$ | | | |
| $x_2$ | $y_2$ | | | | |

From the intuitive Ansatz for the polynomial $I_n(x)$

$$I_n^{(\mathrm{I})}(x) = c_0 + c_1(x - x_0) + c_2(x-x_0)(x-x_1) + c_3(x-x_0)(x-x_1)(x-x_{-1}) + \ldots\,,$$

$$I_n^{(\mathrm{II})}(x) = \gamma_0 + \gamma_1(x - x_0) + \gamma_2(x-x_0)(x-x_{-1}) + \gamma_3(x-x_0)(x-x_{-1})(x-x_1) + \ldots\,.$$

we get with $x := x_0 + th,\ t \in \mathbb{R}$, the two *Gaussian interpolation formulas*

$$I_n^{(\mathrm{I})}(x) = y_0 + \binom{t}{1}\delta^1 y_{0.5} + \binom{t}{2}\delta^2 y_0 + \binom{t+1}{3}\delta^3 y_{0.5} + \ldots + \binom{t+m-1}{n}\delta^n y_0,$$

$$I_n^{(\mathrm{II})}(x) = y_0 + \binom{t}{1}\delta^1 y_{-0.5} + \binom{t+1}{2}\delta^2 y_0 + \binom{t+1}{3}\delta^3 y_{-0.5} + \ldots + \binom{t+m}{n}\delta^n y_0.$$

If we now form the arithmetical mean of these two formulas, using the mean values

$$\overline{\delta}^k y_0 := \frac{1}{2}(\delta^k y_{0.5} + \delta^k y_{-0.5}), \qquad k = 1, 3, 5, \ldots,$$

then we obtain the *Stirling interpolation formula*

$$I_n(x) = y_0 + \binom{t}{1}\overline{\delta}^1 y_0 + \frac{t^2}{2}\delta^2 y_0 + \binom{t+1}{3}\overline{\delta}^3 y_0 + \frac{t^2(t^2-1)}{4!}\delta^4 y_0$$
$$+ \ldots + \frac{t^2(t^2-1)\ldots(t^2-(m-1)^2)}{n!}\delta^n y_0,$$

### 7.3.1.4 Interpolation errors

If the function $f(x)$ which we are attempting to approximate with the help of an interpolation polynomial $I_n(x)$ (in the interval $[a, b]$ with $x_i \in [a, b]$ for all $i = 0, \ldots, n$) is $(n + 1)$-times continuously differentiable, then the interpolation error is given by

$$f(x) - I_n(x) = \frac{f^{(n+1)}(\xi)}{(n+1)!}(x - x_0)(x - x_1)\ldots(x - x_n),$$

where $\xi \in ]a, b[$ is a number which is dependent on $x$. We get, using the supremum norm of the $m^{th}$ derivatives in the interpolation interval $[a, b]$

$$M_m := \max_{\xi \in [a,b]} |f^{(m)}(\xi)|, \qquad m = 2, 3, 4, \ldots,$$

for the general interpolation error in the case of equidistant interpolation points with step $h$ for linear, quadratic and cubic approximation polynomials, respectively, the estimates

$$|f(x) - I_1(x)| \le \frac{1}{8}h^2 M_2, \qquad x \in [x_0, x_1],$$

$$|f(x) - I_2(x)| \le \frac{\sqrt{3}}{27}h^3 M_3, \qquad x \in [x_0, x_2],$$

$$|f(x) - I_3(x)| \le \begin{cases} \dfrac{3}{128}h^4 M_4, & x \in [x_1, x_2], \\ \dfrac{1}{24}h^4 M_4, & x \in [x_0, x_1] \cup [x_2, x_3]. \end{cases}$$

### 7.3.1.5 The algorithm of Aitken–Neville and extrapolation

If, for given interpolation points and values, exactly one value of the interpolation polynomial is what we want to calculate, then the *algorithm of Aitken–Neville* is appropriate. Let $S = \{i_0, \ldots, i_k\} \subseteq \{0, 1, \ldots, n\}$ is a subset of $(k + 1)$ pairwise different index values, and let $I^*_{i_0 i_1 \cdots i_k}(x)$ denote the interpolation polynomial for the interpolation points and values $(x_i, y_i)$ with $i \in S$. With an initial polynomial of degree zero, namely $I_k^*(x) := y_k,\ k = 0, 1, \ldots, n$, we have the recursion formula

$$I^*_{i_0 i_1 \ldots i_k}(x) = \frac{(x - x_{i_0}) I^*_{i_1 i_2 \ldots i_k}(x) - (x - x_{i_k}) I^*_{i_0 i_1 \ldots i_{k-1}}(x)}{x_{i_k} - x_{i_0}}, \qquad k = 1, 2, \ldots, n,$$

with which we can successively form interpolation polynomials of higher degrees. It is useful for calculating the interpolated value for a given $x$ as the value $I^*_{01\cdots n}(x) = I_n(x)$. The *Neville algorithm* is a method for successively calculating these values, by utilizing the scheme

| | | | | | |
|---|---|---|---|---|---|
| $x_0$ | $y_0 = I_0^*$ | | | | |
| $x_1$ | $y_1 = I_1^*$ | $I_{01}^*$ | | | |
| $x_2$ | $y_2 = I_2^*$ | $I_{12}^*$ | $I_{012}^*$ | | |
| $x_3$ | $y_3 = I_3^*$ | $I_{23}^*$ | $I_{123}^*$ | $I_{0123}^*$ | |
| $x_4$ | $y_4 = I_4^*$ | $I_{34}^*$ | $I_{234}^*$ | $I_{1234}^*$ | $I_{01234}^* = I_4(x)$ |

Each value of the scheme is a linear combination from the numbers which are to the left and above, for example

$$I_{12}^* = \frac{(x - x_1) I_2^* - (x - x_2) I_1^*}{x_2 - x_1} = I_2^* + \frac{x - x_2}{x_2 - x_1}(I_2^* - I_1^*),$$

$$I_{234}^* = \frac{(x - x_2) I_{34}^* - (x - x_4) I_{23}^*}{x_4 - x_2} = I_{34}^* + \frac{x - x_4}{x_4 - x_2}(I_{34}^* - I_{23}^*),$$

The second representation is more efficient and better for implementation on a computer.

The Neville algorithm is mainly used for *extrapolation.* Often we can only approximate a quantity $A$ with the help of an auxiliary quantity $B(t)$, which depends on a parameter $t$, in the sense that we have an expansion

$$B(t) = A + c_1 t + c_2 t^2 + c_3 t^3 + \ldots + c_n t^n + \ldots\ ,$$

with coefficients $c_1, c_2, \ldots, c_n$ which are independent of $t$. If for some reason $B(t)$ can not be calculated for a sufficiently small value of $t$, which is necessary for $B(t)$ to approximate $A$ closely, then for a series of parameter values $t_0 > t_1 > \cdots > t_n > 0$ we calculate the values $B(t_k),\ k = 0, 1, \ldots, n$ successively and then evaluate the corresponding interpolation polynomials $I_k(t)$ at the point $t = 0$ (which is not one of the interpolation points); this process is called *extrapolation.* For this the Neville scheme is set up row-wise and the decrease of the absolute value of $t$ is stopped as soon as the last extrapolated values change sufficiently little (i.e., near convergence).

*Example:* The number $\pi$ can be approximated from the circumferences $U_n$ of an $n$-gon which has been inscribed in the unit circle. For $U_n,\ n \geq 2$, we have

$$U_n = n \cdot \sin\left(\frac{\pi}{n}\right) = \pi - \frac{\pi^3}{3!}\left(\frac{1}{n}\right)^2 + \frac{\pi^5}{5!}\left(\frac{1}{n}\right)^4 - \frac{\pi^7}{7!}\left(\frac{1}{n}\right)^6 + - \ldots\ .$$

If we set $t := (1/n)^2$, then $U_n = B(t)$ is the approximating function for $A = \pi$. With the circumferences $U_2, U_3, U_4, U_6$ and $U_8$, all of which can be calculated without recourse to trigonometric functions, we get via extrapolation with the Neville scheme the astoundingly accurate approximations for $\pi$:

| | | | | | |
|---|---|---|---|---|---|
| 1/4 | 2.000 000 000 | | | | |
| 1/9 | 2.598 076 211 | 3.076 537 180 | | | |
| 1/16 | 2.828 427 125 | 3.124 592 585 | 3.140 611 053 | | |
| 1/36 | 3.000 000 000 | 3.137 258 300 | 3.141 480 205 | 3.141 588 849 | |
| 1/64 | 3.061 467 459 | 3.140 497 049 | 3.141 576 632 | 3.141 592 411 | 3.141 592 648 |

The parameter values $t_k$ often form a geometric sequence with quotients $q = 1/4$, so that $t_k = t_0 \cdot q^k$, $k = 1, 2, \ldots, n$. In this special case the computation of the Neville scheme is simplified. If we set $p_i^{(k)} := I^*_{i-k,i-k+1,\ldots,i}$, which is the value of the Neville scheme in the $k^{th}$ column, then for $t = 0$ we get

$$p_i^{(k)} = p_i^{(k-1)} + \frac{t_i}{t_{i-k} - t_i}\left(p_i^{(k-1)} - p_{i-1}^{(k-1)}\right)$$

$$= p_i^{(k-1)} + \frac{1}{4^k - 1}\left(p_i^{(k-1)} - p_{i-1}^{(k-1)}\right), \qquad i = k, k+1, \ldots, n, \quad k = 1, 2, \ldots, n.$$

For the $k^{th}$ column the differences should be multiplied with a factor $1/(4^k - 1)$, which for $k \to \infty$ rapidly approaches zero. This special Neville scheme is called a *Romberg scheme.*

#### 7.3.1.6 Spline interpolation

Interpolation polynomials for which we have a large number of interpolation points which are equidistant or nearly so have a strong tendency to oscillate rapid near the ends of the interval. Because of this, they will have a large difference to the function they are supposed to be approximating. In other words, the procedure, which is valid for a given interval, of approximating a function by a polynomial of low degree, leads, in the case of many interpolation points, to a polynomial which no longer is a good approximation, being in general not even continuously differentiable at the end points of the interval.

This state of affairs can be improved by using *spline interpolation*, which always yields a smooth interpolation function. The idea is simply to interpolate between interpolation points by a polynomial of low degree, then glue the pieces together to form a continuous function. We consider here the particular case of *cubic splines*. More precisely, the *natural cubic spline interpolator* $s(x)$ for the interpolation points $x_0 < x_1 < \cdots < x_{n-1} < x_n$ is uniquely determined by the following conditions.

(a) $s(x_j) = y_j$, $j = 0, 1, 2, \ldots, n$,

(b) $s(x)$ is for $x \in [x_i, x_{i+1}]$, $i = 0, 1, 2, \ldots, n-1$ a polynomial of at most degree 3,

(c) $s(x) \in C^2([x_0, x_n])$,

(d) $s''(x_0) = s''(x_n) = 0$.

These conditions uniquely determine a function $s$. The corresponding function is piecewise, i.e., between the interpolation points, a cubic polynomial, and at the interpolation points $s$ is twice continuously differentiable and at the end points of the interval has vanishing second derivatives. To numerically calculate the spline function $s(x)$, we let

$$h_i := x_{i+1} - x_i > 0, \qquad i = 0, 1, 2, \ldots, n-1$$

denote the lengths of the partial intervals $[x_i, x_{i+1}]$, in which for $s(x)$ we require

$$s_i(x) = a_i(x - x_i)^3 + b_i(x - x_i)^2 + c_i(x - x_i) + d_i, \qquad x \in [x_i, x_{i+1}].$$

In addition to the given interpolation values $y_i$ we require also the second derivatives $y_i''$ to determine the partial polynomials $s_i(x)$. For the four coefficients $a_i, b_i, c_i, d_i$ of $s_i(x)$, we thus require

$$a_i = \frac{1}{6h_i}(y_{i+1}'' - y_i''), \qquad b_i = \frac{1}{2}y_i'',$$
$$c_i = \frac{1}{h_i}(y_{i+1} - y_i) - \frac{h_i}{6}(y_{i+1}'' + 2y_i''), \qquad d_i = y_i.$$

These conditions take account of the interpolation conditions as well as the continuity of the second derivatives at the inner interpolation points. The condition of the continuity of the first derivatives at the $(n-1)$ inner interpolation points yields the $(n-1)$ linear equations

$$h_{i-1}y_{i-1}'' + 2(h_{i-1} + h_i)y_i'' + h_i y_{i+1}'' - \frac{6}{h_i}(y_{i+1} - y_i) + \frac{6}{h_{i-1}}(y_i - y_{i-1}) = 0,$$
$$i = 1, 2, \ldots, n-1.$$

If we note also that $y_0'' = y_n'' = 0$ then this system of linear equations are for the $(n-1)$ unknowns $y_1'', \ldots, y_{n-1}''$. The corresponding coefficient matrix is symmetric, tridiagonal and diagonally dominant. The system of equations therefore has a unique solution, which can be calculated with the computational cost of $n$ essential operations (cf. 7.2.1.6). Even for larger values of $n$ we have good numerical properties of the tridiagonal system of equations, as the condition number of the matrix is small provided the partial intervals do not have large differences in size.

The two so-called *natural end conditions* $s''(x_0) = s''(x_n) = 0$ are in most cases, however, not appropriate to the problem at hand. In general they are replaced by two other conditions, so that $s(x)$ is nonetheless uniquely determined. Examples for conditions which can be placed, depending on the problem at hand, are:

$\alpha$) *Prescribe the first derivatives:*

$$s_0'(x_0) = y_0', \qquad s_{n-1}'(x_n) = y_n'.$$

The system of equations is then extended to include the two equations for the further unknowns $y_0''$ and $y_n''$:

$$2h_0y_0'' + h_0y_1'' - \frac{6}{h_0}(y_1 - y_0) + 6y_0' = 0,$$
$$h_{n-1}y_{n-1}'' + 2h_{n-1}y_n'' + \frac{6}{h_{n-1}}(y_n - y_{n-1}) - 6y_n' = 0.$$

The resulting system of equations is still symmetric, tridiagonal and diagonally dominant.

$\beta$) *Smoothing the boundary:*

$$y_0'' = \alpha y_1'', \qquad y_n'' = \beta y_{n-1}'', \qquad \alpha, \beta \in \mathbb{R}.$$

The first and the last equation of the system are adapted to insure that the coefficients of $y_1''$ and $y_{n-1}''$ are additive.

$\gamma$) *Not-a-knot condition:* Here we require that the cubic polynomials $s_0(x)$ and $s_1(x)$ (resp. $s_{n-2}(x)$ and $s_{n-1}(x)$) coincide. This can be achieved by requiring

$$s_0^{(3)}(x_1) = s_1^{(3)}(x_1), \qquad s_{n-2}^{(3)}(x_{n-1}) = s_{n-1}^{(3)}(x_{n-1}).$$

This leads to the following additional equations

$$h_1 y_0'' - (h_0 + h_1) y_1'' + h_0 y_2'' = 0,$$
$$h_{n-1} y_{n-2}'' - (h_{n-2} + h_{n-1}) y_{n-1}'' + h_{n-2} y_n'' = 0.$$

The resulting linear system of equations for the $(n+1)$ unknowns $y_0'', y_1'', \ldots, y_n''$ is no longer symmetric, tridiagonal or diagonally dominant. Still, it can be solved using the Gaussian algorithm with a diagonal strategy, plus special treatment for the first and the last equation.

$\delta$) *Periodicity condition:*

$$s'(x_0) = s'(x_n), \qquad s''(x_0) = s''(x_n),$$

where $T := x_n - x_0$ is the period of the function we are approximating, so that $y_n = y_0$. For the $n$ unknowns $y_0'', y_1'', \ldots, y_{n-1}''$ we have as first and last equation

$$2(h_{n-1} + h_0) y_0'' + h_0 y_1'' + h_{n-1} y_{n-1}'' - \frac{6}{h_0}(y_1 - y_0) + \frac{6}{h_{n-1}}(y_0 - y_{n-1}) = 0,$$
$$h_{n-1} y_0'' + h_{n-2} y_{n-2}'' + 2(h_{n-2} + h_{n-1}) y_{n-1}'' + \frac{6}{h_{n-1}}(y_0 - y_{n-1}) \frac{6}{h_{n-2}}(y_{n-1} - y_{n-2}) = 0,$$

while the other equations remain unchanged. The matrix of the system of equations is symmetric and diagonally dominant, but in general no longer tridiagonal. The special structure of the equations makes applications of appropriate methods possible, leading to a solution.

### 7.3.2 Numerical differentiation

Interpolation polynomials can be used to calculate the derivatives of functions which are given, for example, by a table of values. The formulas obtained for *numerical differentiation* can likewise be used to approximate derivatives of complicated functions, but they are in particular indispensable for approximating the derivatives of solutions of partial differential equations.

For *equidistant interpolation points* $x_i = x_0 - ih$ with corresponding interpolation values $y_i = f(x_i)$, $i = 0, 1, \ldots, n$, one obtains by a $n$-times differentiation process the *Lagrange interpolation formula* (cf. 7.3.1.1)

$$f^{(n)}(x) \approx \frac{1}{h^n}\left[(-1)^n y_0 + (-1)^{n-1}\binom{n}{1} y_1 + (-1)^{n-2}\binom{n}{2} y_2 + \ldots - \binom{n}{n-1} y_{n-1} + y_n\right].$$

For a point $\xi \in (x_0, x_n)$, the expression on the right yields the exact value of the $n^{th}$ derivative of $f(x)$. The corresponding $n^{th}$ *difference quotient* is for $n = 1, 2, 3$:

$$f'(x) \approx \frac{1}{h}(y_1 - y_0),$$
$$f''(x) \approx \frac{1}{h^2}(y_2 - 2y_1 + y_0),$$
$$f^{(3)}(x) \approx \frac{1}{h^3}(y_3 - 3y_2 + 3y_1 - y_0).$$

More generally one can approximate the $p^{th}$ derivative at a particular point $x$ by using the $p^{th}$ derivatives of a interpolation polynomial $I_n(x)$ of higher degree. For $n = 2$ we get in this way for the first derivatives the approximations

$$\begin{aligned} f'(x_0) &\approx \frac{1}{2h}(-3y_0 + 4y_1 - y_2), \\ f'(x_1) &\approx \frac{1}{2h}(y_2 - y_0) \qquad \text{(central difference quotient)}. \end{aligned}$$

Similarly, for $n = 3$ with $x_M := \frac{1}{2}(x_0 + x_3)$, we obtain

$$\begin{aligned} f'(x_0) &\approx \frac{1}{6h}(-11y_0 + 18y_1 - 9y_2 + 2y_3), \\ f'(x_1) &\approx \frac{1}{6h}(-2y_0 - 3y_1 + 6y_2 - y_3), \\ f'(x_M) &\approx \frac{1}{24h}(y_0 - 27y_1 + 27y_2 - y_3), \\ f''(x_0) &\approx \frac{1}{h^2}(2y_0 - 5y_1 + 4y_2 - y_3), \\ f''(x_M) &\approx \frac{1}{2h^2}(y_0 - y_1 - y_2 + y_3). \end{aligned}$$

A few differentiation formulas for five interpolation points are

$$\begin{aligned} f'(x_0) &\approx \frac{1}{12h}(-25y_0 + 48y_1 - 36y_2 + 16y_3 - 3y_4), \\ f'(x_2) &\approx \frac{1}{12h}(y_0 - 8y_1 + 8y_3 - y_4), \\ f''(x_0) &\approx \frac{1}{12h^2}(35y_0 - 104y_1 + 114y_2 - 56y_3 + 11y_4), \\ f''(x_2) &\approx \frac{1}{12h^2}(-y_0 + 16y_1 - 30y_2 + 16y_3 - y_4). \end{aligned}$$

### 7.3.3 Numerical integration

An approximate calculation of a definite integral $I = \int_a^b f(x)\mathrm{d}x$ from known individual or approximate function values of the integrand is known as numerical integration or *quadrature*. The most appropriate method for obtaining an approximation to $I$ depends in an essential way on the properties of the integrand in the interval in which it is to be approximated: is the integrand smooth, or are there singularities of the function $f(x)$ or of one of its derivatives? If we are given values of the function in tabular form, or can we calculate $f(x)$ for arbitrary arguments $x$? What is the desired precision, and are there other, similar integrals which also have to approximated?

#### 7.3.3.1 Interpolative quadrature formulas

One class of quadrature formulas for continuous and sufficiently differentiable integrands arises by first approximating the function $f(x)$ in the desired interval $[a, b]$ by an interpolation polynomial $I_n(x)$ at $(n+1)$ interpolation points $a \leq x_0 < x_1 < \cdots < x_n \leq b$, and then approximating the value of $I$ by approximating the integral of $I_n(x)$. Because of the Lagrange interpolation formula in section 7.3.1.1, this yields

$$I = \int_a^b \sum_{k=0}^n f(x_k) L_k(x)\,\mathrm{d}x + \int_a^b \frac{f^{(n+1)}(\xi)}{(n+1)!} \prod_{i=0}^n (x - x_i)\,\mathrm{d}x.$$

From the first component of the formula, we get the *quadrature formula*

$$\boxed{Q_n = \sum_{k=0}^n f(x_k) \int_a^b L_k(x)\,\mathrm{d}x =: (b-a) \sum_{k=0}^n w_k f(x_k),}$$

depending only on the chosen interpolation points or *knots* $x_0, x_1, \ldots, x_n$, and with corresponding *integral weights* $w_k$ which depend only on the size of the interval $(b-a)$, defined by

$$w_k = \frac{1}{b-a} \int_a^b L_k(x)\,\mathrm{d}x, \qquad k = 0, 1, 2 \ldots, n.$$

The quadrature error $Q_n$ is given by

$$E_n[f] := I - Q_n = \int_a^b \frac{f^{(n+1)}(\xi)}{(n+1)!} \prod_{i=0}^n (x - x_i)\,\mathrm{d}x.$$

This error can be explicitely calculated for equidistant interpolation points. All quadrature formulas for interpolation have by construction the property that $Q_n$ yields the precise value for $I$ in case $f(x)$ is itself a polynomial of degree at most $n$. In certain cases it can even be exact when $f$ is a polynomial of higher degree. This motivates the definition of a *precision degree* $m \in \mathbb{N}$ of an (arbitrary) quadrature interpolation formula $Q_n := (b-a) \sum_{k=0}^n w_k f(x_k)$ as the greatest integer $m$ for which $Q_n$ precisely integrates all polynomials up to and including degree $m$. For given $(n+1)$ interpolation points for the integration $a \leq x_0 < x_1 < \ldots < x_n \leq b$ there is a uniquely determined, interpolation by quadratures formula $Q_n$, whose precision degree is at least $n$.

For *equidistant* knots with $x_0 = a$, $x_n = b$, $x_k = x_0 + kh$, $k = 0, 1, 2, \ldots, n$, $h := (b-a)/n$ we obtain closed formulas known as the *Newton–Cotes quadrature formulas*. If $f_k := f(x_k)$, $k = 0, 1, 2, \ldots, n$ are the interpolation points for the integrand, then some of these formulas, together with the corresponding quadrature errors and with precision degree $m$ are given in the following box.

$$Q_1 = \frac{h}{2}[f_0 + f_1] \quad \text{(trapezoidal rule)}, \quad E_1[f] = -\frac{h^3}{12}f''(\xi), \quad m = 1,$$

$$Q_2 = \frac{h}{3}[f_0 + 4f_1 + f_2] \quad \text{(Simpson rule)}, \quad E_2[f] = -\frac{h^5}{90}f^{(4)}(\xi), \quad m = 3,$$

$$Q_3 = \frac{3h}{8}[f_0 + 3f_1 + 3f_2 + f_3] \quad \text{(Newton 3/8-rule)},$$
$$E_3[f] = -\frac{3h^5}{80}f^{(4)}(\xi), \quad m = 3,$$

$$Q_4 = \frac{2h}{45}[7f_0 + 32f_1 + 12f_2 + 32f_3 + 7f_4], \quad E_4[f] = -\frac{8h^7}{945}f^{(6)}(\xi), \quad m = 5,$$

$$Q_5 = \frac{5h}{288}[19f_0 + 75f_1 + 50f_2 + 50f_3 + 75f_4 + 19f_5],$$
$$E_5[f] = -\frac{275h^7}{12\,096}f^{(6)}(\xi), \quad m = 5.$$

The quadrature formulas for $n = 2l$ and $n = 2l+1$, $l \in \mathbb{N}$ have the same precision degree, namely $m = 2l + 1$. Therefore it is advantageous to use the Newton–Cotes formulas for even $n$, since the precision obtained for the following, odd $n$, is very marginal. Since, as we already mentioned, the interpolation polynomial $I_n(x)$ has the tendency for growing $n$ to oscillate strongly near the end points of the interval, using the Newton–Cotes formula for $n > 6$ is not to be recommended. In particular for $n = 8$ and $n \geq 10$ the value for some of the integration weights become negative.

A better approximation of $I$ can be obtained by subdividing the integration interval $[a, b]$ into $N$ equally large subintervals, to each of which we apply the Newton–Cotes formula. From the simple trapezoidal rule above we get the *summed trapezoidal rule*,

$$S_1 := T(h) := h\left[\frac{1}{2}f_0 + \sum_{k=1}^{N-1} f_k + \frac{1}{2}f_N\right], \qquad h := (b-a)/N.$$

The *summed Simpson rule* is

$$S_2 := \frac{h}{3}\left[f_0 + 4f_1 + f_{2N} + 2\sum_{k=1}^{N-1}\{f_{2k} + 2f_{2k+1}\}\right], \qquad h := (b-a)/2N,$$

$$f_j := f(x_0 + jh), \qquad j = 0, 1, 2, \ldots, 2N,$$

whose quadrature error for an at least four times continuously differentiable integrand $f(x)$ is given by

$$E_{S_2}[f] = -\frac{b-a}{180}\, h^4 f^{(4)}(\xi), \qquad a < \xi < b.$$

The *mean rule* or *tangent trapezoidal rule*

$$Q_0^0 := (b-a)f(x_1), \qquad x_1 = (a+b)/2,$$

is an *open Newton–Cotes formula* with an interpolation point $x_1$ at the midpoint of the interval $[a, b]$. It has the precision degree $m = 1$ and a quadrature error of

$$E_0^0[f] = \frac{1}{24}(b-a)^3 f''(\xi), \qquad a < \xi < b.$$

The *summed mean rule* or *mean sum rule*

$$S_0^0 := M(h) := h \sum_{k=0}^{N-1} f(x_{k+0.5}), \qquad x_{k+0.5} := a + \left(k + \frac{1}{2}\right)h, \qquad h := (b-a)/N,$$

corresponds to a Riemannian sub-sum for a particular decomposition $Z$ of the interval $[a,b]$ (see section 1.6.2).

There is a relation between the trapezoidal rule $T(h)$ and the mean sum rule $M(h)$, which is

$$T\left(\frac{h}{2}\right) = \frac{1}{2}\big[T(h) + M(h)\big],$$

which makes it possible to improve an approximation $T(h)$ through the use of the mean rule sum to an approximation $T\left(\frac{h}{2}\right)$ for half the step width. Each such halving of the step length requires twice the number of functional values.

The trapezoidal rule with successive halving of the step length is particularly appropriate for the calculation of integrals of periodic and analytic integrands over a period interval, since the trapezoidal sums converge very rapidly. The trapezoidal rule is also convenient for the calculation of indefinite integrals over $\mathbb{R}$ for sufficiently fast decreasing (at infinity) functions $f(x)$.

### 7.3.3.2 The Romberg procedure

For an integrand $f(x)$ which is sufficiently often continuously differentiable, we have the *Euler–Maclaurin summation formula* with remainder term

$$T(h) = \int\limits_a^b f(x)\,\mathrm{d}x + \sum_{k=1}^{N} \frac{B_{2k}}{(2k)!}\left[f^{(2k-1)}(b) - f^{(2k-1)}(a)\right]h^{2k} + R_{N+1}(h),$$

where $B_{2k},\ k = 1, 2, \ldots$ are the *Bernoulli numbers* with values

$$B_2 = \frac{1}{6}, \qquad B_4 = -\frac{1}{30}, \qquad B_6 = \frac{1}{42}, \qquad B_8 = -\frac{1}{30}, \qquad B_{10} = \frac{5}{66}, \qquad \ldots$$

Moreover, for the remainder we have $R_{N+1}(h) = O(h^{2N+2})$. The calculated trapezoidal sums $T(h)$ approximate the integral $I$ with an error having an asymptotically valid development in the step length $h$, which only has even powers of $h$. If we successively halve the step length, then the assumptions are satisfied for applying the extrapolation from $t = h^2$ to $t = 0$ using the Romberg scheme (cf. 7.3.1.5). The required trapezoidal sums $T(h_i)$ for the sequence $h_0 = b - a,\ h_i = h_{i-1}/2,\ i = 1, 2, 3, \ldots$ can successively be determined by using the mean sum rule above. In the Romberg scheme, therefore, the values of the important upper diagonal converge to the value of the integral. The halving of the step width can be broken off when two extrapolated values of this upper diagonal are sufficiently near to each other. From this it follows that the Romberg procedure is an efficient, numerically stable integration method, provided the integrand is sufficient smooth.

*Example:* $I = \int\limits_1^2 \frac{\mathrm{e}^x}{x}\,\mathrm{d}x \doteq 3.059\,116\,540.$

| $h$ | $T(h)$ | | | | |
|---|---|---|---|---|---|
| 1 | 3.206 404 939 | | | | |
| 1/2 | 3.097 098 826 | 3.060 663 455 | | | |
| 1/4 | 3.068 704 101 | 3.059 239 193 | 3.059 144 242 | | |
| 1/8 | 3.061 519 689 | 3.059 124 886 | 3.059 117 265 | 3.059 116 837 | |
| 1/16 | 3.059 717 728 | 3.059 117 074 | 3.059 116 553 | 3.059 116 542 | 3.059 116 541 |

#### 7.3.3.3 Gaussian quadrature

Instead of prescribing the interpolation points, we could just as well choose these together with the integration weights in such a way that the resulting quadrature formula has a maximal precision degree. We consider this point of view in this section in the context of a general approximation of an integral

$$I = \int_a^b f(x) \cdot q(x)\,\mathrm{d}x$$

with a given continuous weight function $q(x)$ which is assumed to be positive in the interval $(a, b)$. For every number $n > 0$ there are $n$ interpolation points $x_k \in [a, b]$, $k = 1, 2, \ldots, n$ and weights $w_k$, $k = 1, 2, \ldots, n$ so that

$$\int_a^b f(x) \cdot q(x)\,\mathrm{d}x = \sum_{k=1}^{n} w_k f(x_k) + \frac{f^{(2n)}(\xi)}{(2n)!} \int_a^b \left\{ \prod_{k=1}^{n} (x - x_k) \right\}^2 q(x)\,\mathrm{d}x.$$

for some $a < \xi < b$. The quadrature formula defined by this sum has a maximal precision degree of $m = 2n - 1$, provided the knots $x_k$ are chosen to be the zeros of a polynomial $\varphi_n(x)$ of degree $n$ which belongs to a whole family $\varphi_0(x), \varphi_1(x), \ldots, \varphi_n(x)$ of orthogonal polynomials with the properties

$$\deg \varphi_l(x) = l; \qquad \int_a^b \varphi_k(x)\varphi_l(x)q(x)\,\mathrm{d}x = 0 \qquad \text{for} \quad k \neq l.$$

The zeros of the polynomial $\varphi_k(x)$ are always real, pairwise disjoint and $x_k \in (a, b)$.

The integration weights $w_k$ are determined as the integrals of the weighted Lagrange polynomials by virtue of the corresponding interpolation quadrature formula, i.e., by

$$w_k = \int_a^b \left\{ \prod_{\substack{j=1 \\ j \neq k}}^{n} \left( \frac{x - x_j}{x_k - x_j} \right) \right\} q(x)\,\mathrm{d}x = \int_a^b \left\{ \prod_{\substack{j=1 \\ j \neq k}}^{n} \left( \frac{x - x_j}{x_k - x_j} \right) \right\}^2 q(x)\,\mathrm{d}x > 0,$$

where $k = 1, 2, \ldots, n$. From the second, equivalent representation it follows that the integration weights $w_k$ for all Gaussian quadrature formulas are positive.

Since the orthogonal polynomials $\varphi_k(x)$, $k = 0, 1, 2, \ldots, n$ of the above mentioned family satisfy a three-term recursion formula, the zeros of $\varphi_n(x)$ can easily be calculated as the eigenvalues of a symmetric, tridiagonal matrix. The corresponding integration weights are essentially just the square of the first component of the corresponding normalized eigenvectors of the matrix.

*Table 7.3. Gauss–Legendre quadrature formulas.*

| $n$ | $k$ | $x_k = -x_{n-k+1}$ | $w_k$ | $E_n[f]$ |
|---|---|---|---|---|
| 2 | 1 | 0.5773502692 | 1.0000000000 | $7.4 \cdot 10^{-3} f^{(4)}(\xi)$ |
| 3 | 1<br>2 | 0.7745966692<br>0 | 0.5555555556<br>0.8888888889 | $6.3 \cdot 10^{-5} f^{(6)}(\xi)$ |
| 4 | 1<br>2 | 0.8611363116<br>0.3399810436 | 0.3478548451<br>0.6521451549 | $2.9 \cdot 10^{-7} f^{(8)}(\xi)$ |
| 5 | 1<br>2<br>3 | 0.9061798459<br>0.5384693101<br>0 | 0.2369268851<br>0.4786286705<br>0.5688888889 | $8.1 \cdot 10^{-10} f^{(10)}(\xi)$ |
| 6 | 1<br>2<br>3 | 0.9324695142<br>0.6612093865<br>0.2386191861 | 0.1713244924<br>0.3607615730<br>0.4679139346 | $1.5 \cdot 10^{-12} f^{(12)}(\xi)$ |

The general Gaussian quadrature formulas are, because of their high precision degree, very important for the approximate calculation of definite integrals in the case of (weighted) integrands which we can calculate at an arbitrary point. For applications, the following special cases are particularly important, where for the consideration of the first two cases we assume without restriction of generality that the integration interval is $[-1,+1]$. Indeed, every finite interval $[a,b]$ can, with the aid of a mapping

$$x = 2\,\frac{t-a}{b-a} - 1,$$

be mapped to $[-1,+1]$.

**Gauss–Legendre quadrature formulas:** For the weight function $q(x) = 1$ in $[-1,+1]$, the function $\varphi_n(x) = P_n(x)$ of the previous section are the *Legendre polynomials* (cf. 1.13.2.13). The zeros of the Legendre polynomials $P_n(x)$, $n = 1,2,\dots,$ are symmetric with respect to the origin, and the integration weights $w_k$ for symmetrically situated interpolation points are equal. Table 7.3 contains for a few values of $n$ the most important information. The quadrature error is

$$E_n[f] = \frac{2^{2n+1}(n!)^4}{\big[(2n)!\big]^3(2n+1)}\,f^{(2n)}(\xi), \qquad \xi \in (-1,+1).$$

**Gauss–Chebychev quadrature formulas:** For the weight function $q(x) = 1/\sqrt{1-x^2}$ in $[-1,+1]$, the polynomials $\varphi_n$ are the *Chebychev polynomials* $\varphi_n(x) = T_n(x)$, which are discussed below in section 7.5.1.3. The interpolation points $x_k$ and the weights $w_k$ are

$$x_k = \cos\big((2k-1)\pi/2\pi\big), \qquad w_k = \pi/n, \qquad k = 1,2,\dots,n.$$

For the quadrature error we have the estimate

$$E_n[f] = \frac{2\pi}{2^{2n}(2n)!}\,f^{(2n)}(\xi), \qquad \xi \in ]-1,1[.$$

The Gauss–Chebychev quadrature formula is, in a special case, closely related to the mean sum $M(h)$. Indeed, from

$$\int_{-1}^{1} \frac{f(x)}{\sqrt{1-x^2}}\,\mathrm{d}x = \frac{\pi}{n}\sum_{k=1}^{n} f(x_k) + E_n[f]$$

it follows, setting $x = \cos\theta$,

$$\int_{0}^{\pi} f(\cos\theta)\,\mathrm{d}\theta = \frac{\pi}{n}\sum_{k=1}^{n} f(\cos\theta_k) + E_n[f] = M\left(\frac{\pi}{n}\right) + E_n[f],$$

where $\theta_k = (2k-1)\pi/(2n)$, $k = 1, 2, \ldots, n$ are equidistant interpolation points for the $2\pi$-periodic, even function $f(\cos\theta)$. The mean sum yields for increasing values of $n$ approximation with very small quadrature errors.

**Gauss–Laguerre quadrature formulas:** For the weight function $q(x) = \mathrm{e}^{-x}$ in $[0, \infty]$ the polynomials above are $\varphi_n(x) = L_n(x) := \frac{1}{n!}\mathrm{e}^x \cdot \frac{\mathrm{d}^n}{\mathrm{d}x^n}\{x^n \mathrm{e}^{-x}\}$, $n = 0, 1, 2, \ldots$, which are the *Laguerre polynomials*. The first few of these are

$$L_0(x) = 1, \quad L_1(x) = 1 - x, \quad L_2(x) = 1 - 2x + \frac{1}{2}x^2, \quad L_3(x) = 1 - 3x + \frac{3}{2}x^2 - \frac{1}{6}x^3.$$

They satisfy the recursion relation

$$L_{n+1}(x) = \frac{2n+1-x}{n+1}L_n(x) - \frac{n}{n+1}L_{n-1}(x), \qquad n = 1, 2, 3, \ldots$$

Table 7.4 contains a few interpolation points and the corresponding weights, with the quadrature errors

$$E_n[f] = \frac{(n!)^2}{(2n)!} f^{(2n)}(\xi), \qquad 0 < \xi < \infty.$$

The coefficient of the quadrature error shrinks slowly as $n$ grows in magnitude.

#### 7.3.3.4 Substitution and transformation

An appropriate substitution of variables to calculate the integral $I$ can be used to bring the integral into a form which allows an efficient application of one of the quadrature formulas above. This is of particular interest in the case of integrands which are singular or are indefinite (defined on unbounded intervals) with slowing decreasing integrands. If we use the substitution

$$x = \varphi(t), \qquad \varphi'(t) > 0,$$

with an increasing function $\varphi(t)$ whose inverse maps the given interval of integration $[a, b]$ bijectively to $[\alpha, \beta]$ with $\varphi(\alpha) = a$, $\varphi(\beta) = b$, we get

$$I = \int_{a}^{b} f(x)\,\mathrm{d}x = \int_{\alpha}^{\beta} F(t)\,\mathrm{d}t \qquad \text{with} \quad F(t) := f\big(\varphi(t)\big)\varphi'(t).$$

An *algebraic boundary singularity* like for example

$$I = \int_{0}^{1} x^{pq} f(x)\,\mathrm{d}x, \qquad q = 2, 3, \ldots, \qquad p > -q, \qquad p \in \mathbb{Z},$$

*Table 7.4. Gauss–Laguerre quadrature formulas.*

| $n$ | $k$ | $x_k$ | $w_k$ | $E_n[f]$ |
|---|---|---|---|---|
| 4 | 1 | 0.32254769 | 0.60315410 | $1.43 \cdot 10^{-2} f^{(8)}(\xi)$ |
| | 2 | 1.74576110 | 0.35741869 | |
| | 3 | 4.53662030 | 0.03888791 | |
| | 4 | 9.35907091 | $0.53929471 \cdot 10^{-3}$ | |
| 5 | 1 | 0.26356032 | 0.52175561 | $3.97 \cdot 10^{-3} f^{(10)}(\xi)$ |
| | 2 | 1.41340306 | 0.39866681 | |
| | 3 | 3.59642577 | 0.07594245 | |
| | 4 | 7.08581001 | $0.36117587 \cdot 10^{-2}$ | |
| | 5 | 12.64080084 | $0.23369972 \cdot 10^{-4}$ | |
| 6 | 1 | 0.22284660 | 0.45896467 | $1.08 \cdot 10^{-3} f^{(12)}(\xi)$ |
| | 2 | 1.18893210 | 0.41700083 | |
| | 3 | 2.99273633 | 0.11337338 | |
| | 4 | 5.77514357 | 0.01039920 | |
| | 5 | 9.83746742 | $0.26101720 \cdot 10^{-3}$ | |
| | 6 | 15.98287398 | $0.89854791 \cdot 10^{-6}$ | |

with an analytic (in the interval $[0,1]$) function $f(x)$ can be transformed to the integral

$$I = q \int_0^1 t^{p+q-1} f(t^q)\, \mathrm{d}t$$

by the substitution of variables

$$x = \varphi(t) = t^q, \qquad \varphi'(t) = q\, t^{q-1} > 0 \qquad \text{for} \quad t \in [0,1];$$

note that the integrand of the integral has no singularity because of $p + q - 1 \geq 0$, so that this integral can be efficiently approximated using the Romberg procedure or the Gauss quadrature formulas.

The transformation of intervals of integration which have one or both of its end points infinite can be done as follows. The integral on the interval $[0, \infty)$ is mapped by means of the substitution $x = \varphi(t) := t/(t+1)$ to an integral on the interval $[0,1)$. The substitution $x = \varphi(t) := (\mathrm{e}^t - 1)/(\mathrm{e}^t + 1)$ transforms the interval $]-\infty, \infty[$ to the interval $]-1, 1[$. The resulting integrand is in general not continuous, as it in general will now have singularities at the end points of the interval.

To treat singularities of the integrand at the two ends of the interval $]-1,1[$, whose nature is not known, application of the tanh-*transformation* is appropriate. This is the substitution

$$x = \varphi(t) := \tanh t, \qquad \varphi'(t) = 1/\cosh^2 t$$

which maps the interval $]-1,1[$ to the infinite interval $]-\infty, \infty[$, but it has the advantage that the integrand of the transformed integral

$$I = \int_{-1}^{1} f(x)\, \mathrm{d}x = \int_{-\infty}^{\infty} F(t)\, \mathrm{d}t \qquad \text{with} \quad F(t) := \frac{f(\tanh t)}{\cosh^2 t}$$

often decreases exponentially.

For indefinite integrals with slowly decreasing integrands, the sinh-*transformation* can often be convenient. This is given by

$$x = \varphi(t) := \sinh t, \qquad \varphi'(r) = \cosh t,$$

$$I = \int_{-\infty}^{\infty} f(x)\,\mathrm{d}x = \int_{-\infty}^{\infty} F(t)\,\mathrm{d}t \qquad \text{with} \quad F(t) := f(\sinh t)\cdot\cosh t.$$

A finite number of such sinh-transformations leads to an integrand which decreases exponentially to both sides of the integration interval, so that for a numerical integration the trapezoidal rule is efficient.

## 7.4 Non-linear problems

### 7.4.1 Non-linear equations

Suppose we are given a continuous, non-linear function $f : \mathbb{R} \longrightarrow \mathbb{R}$, and wish to determine the *zeros* of the function as solutions of the equation

$$\boxed{f(x) = 0.}$$

The determination of these zeros can be done iteratively, starting from some known approximations.

Assuming that there are two values $x_1 < x_2$ for which $f(x_1)$ and $f(x_2)$ have opposite signs, there is, according to the last theorem in section 1.3 ('In-between' theorem), at least one zero in the interval $[x_1, x_2]$. This zero can be closed in by shrinking the interval with the *bisection method*, by calculating the value $f(x_3)$ at $x_3 = (x_1 + x_2)/2$, and determining this way in which half of the interval the zero lies. The length of the intervals determined in this way shrink as a geometric series with $q = 0.5$, so that the number of bisections required until one has reached a given precision (length of the interval) only depends on the length of the original interval.

In the method called *regula falsi* approaches the problem by finding a value $x_3$ by a linear approximation, setting $x_3 = (x_1y_2 - x_2y_1)/(y_2 - y_1)$, where $y_i = f(x_i)$ denote the values of the function $f$. The sign of $y_3 = f(x_3)$ determines the subinterval $[x_1, x_3]$ or $[x_3, x_2]$ in which the zero lies. If the function $f(x)$ is either concave or convex in the interval $[x_1, x_2]$, then the sequence of test values converges monotonously to the zero $s$.

The *secant method*, on the other hand, drops the idea of shrinking the interval. Instead, starting from two given approximations $x^{(0)}$ and $x^{(1)}$ for the zero $s$, one forms the iterative sequence of approximations

$$x^{(k+1)} = x^{(k)} - f(x^{(k)})\cdot\frac{x^{(k)} - x^{(k-1)}}{f(x^{(k)}) - f(x^{(k-1)})}, \qquad k = 1, 2, \ldots$$

which is well-defined provide that $f(x^{(k)}) \neq f(x^{(k-1)})$. The point $x^{(k+1)}$ is geometrically the intersection of the secant which approximates $f(x)$ with the $x$-axis.

The *Newton procedure* assumes that $f$ is continuously differentiable and that one can easily calculate the derivative $f'(x)$. Starting with an initial approximation $x^{(0)}$, the iteration formula is

$$\boxed{x^{(k+1)} = x^{(k)} - \frac{f(x^{(k)})}{f'(x^{(k)})}, \qquad f'(x^{(k)}) \neq 0, \qquad k = 1, 2, \ldots .}$$

In this case, $x^{(k+1)}$ is geometrically the intersection of the *tangent* at $f$ with the $x$-axis. Let $f(s) = 0$ with $f'(s) \neq 0$. Then the sequence of the $x^{(k)}$ converges to $s$ for all initial approximations $x^{(0)}$ in a neighborhood of $x$, provided $|f''(x^{(0)})f(x^{(0)})/f'(x^{(0)})^2| < 1$.

As a measure of the quality of the convergence of a procedure of this type, the main parameter is the *order of convergence.* One says that there is at least a *linear* convergence, if for all but finitely many $k \in \mathbb{N}$ we have the estimate

$$|x^{(k+1)} - s| \leq C \cdot |x^{(k)} - s| \qquad \text{with} \quad 0 < C < 1.$$

An iteration procedure has (at least) the order of convergence $p > 1$, if for all but finitely many $k \in \mathbb{N}$ we have the estimate

$$|x^{(k+1)} - s| \leq K \cdot |x^{(k)} - s|^p \qquad \text{with} \quad 0 < K < \infty.$$

The convergence of the bisector method and of the regula falsi is linear. The secant method has a *super-linear* convergence with an order of convergence equal to $p = 1.618$. For the Newton method we have $p = 2$, so that this method has *quadratic convergence.* This means that at each step, the number of correct decimal places of the approximation doubles. Since the secant method does not require the calculation of derivative values, it can often be more efficient, as a double step of this method has $p = 2.618$.

## 7.4.2 Non-linear systems of equations

Let $f_i(x_1, x_2, \ldots, x_n)$, $i = 1, 2, \ldots, n$ be continuous functions of the independent variables $\mathbf{x} := (x_1, x_2, \ldots, x_n)^\mathsf{T}$ in a common domain of definition $D \subseteq \mathbb{R}^n$. We wish to determine the solutions $\mathbf{x} \in D$ of the non-linear system of equations $f_i(\mathbf{x}) = 0,\ i = 1, 2, \ldots, n$, i.e., of

$$\mathbf{f}(\mathbf{x}) = \mathbf{0} \qquad \text{with} \qquad \mathbf{f}(\mathbf{x}) := \big(f_1(\mathbf{x}), f_2(\mathbf{x}), \ldots, f_n(\mathbf{x})\big)^\mathsf{T}.$$

The problem of determining a solution vector $\mathbf{x} \in D$ will be considered for the illustration of two basic procedures in this section.

### 7.4.2.1 Fixed-point iteration

In some applications the system of equations for which we are looking for the solutions are given in the *fixed point form*

$$\boxed{\mathbf{x} = \mathbf{F}(\mathbf{x}) \qquad \text{for} \quad \mathbf{F} : \mathbb{R}^n \to \mathbb{R}^n,}$$

or at least the equations $\mathbf{f}(\mathbf{x}) = 0$ can be brought into this form. The solution vector $\mathbf{x}$ is then a *fixed point* of the map $\mathbf{F}$ in the domain of definition $D \subseteq \mathbb{R}^n$. The idea now is to approximate this solution by starting with a vector $\mathbf{x}^{(0)} \in D$ and performing the fixed point iteration

$$\boxed{\mathbf{x}^{(k+1)} = \mathbf{F}(\mathbf{x}^{(k)}), \qquad k = 0, 1, 2, \ldots .}$$

The Banach fixed point theorem[2], applied here to the $n$-dimensional Euclidean space $\mathbb{R}^n$, yields a necessary convergence statement for the convergence of the approximative sequence $(\mathbf{x}^{(k)})$ to the solution (fixed point) $\mathbf{x}$.

**Theorem:** Let $A \subset D \subseteq \mathbb{R}^n$ be a closed subset of the domain of definition $D$ of a map $\mathbf{F} : A \longrightarrow A$. If $\mathbf{F}$ is *contracting* as a map, i.e., if there is a constant $L < 1$ such that the inequality

$$||\mathbf{F}(\mathbf{x}) - \mathbf{F}(\mathbf{y})|| \le L||\mathbf{x} - \mathbf{y}|| \qquad \text{for all} \quad \mathbf{x}, \mathbf{y} \in A$$

is valid, then we have:

(a) The fixed-point equation $\mathbf{x} = \mathbf{F}(\mathbf{x})$ has a *unique solution* $\mathbf{x} \in A$.

(b) For every initial vector $\mathbf{x}^{(0)} \in A$ the sequence $(\mathbf{x}^{(k)})$ converges to $\mathbf{x}$.

(c) For the error of the approximation, we have (where the norm used here is given by $||\mathbf{x}|| = \left(\sum\limits_{k=1}^{n} x_k^2\right)^{1/2}$ ):

$$||\mathbf{x}^{(k)} - \mathbf{x}|| \le \frac{L^k}{1-L} ||\mathbf{x}^{(1)} - \mathbf{x}^{(0)}||, \qquad k = 1, 2, \ldots,$$

$$||\mathbf{x}^{(k)} - \mathbf{x}|| \le \frac{L}{1-L} ||\mathbf{x}^{(k)} - \mathbf{x}^{(k-1)}||, \qquad k = 1, 2, \ldots.$$

**Fréchet derivative:** Let $\mathbf{F}'(\mathbf{x})$ denote the Fréchet derivative of $\mathbf{F}$ at the point $\mathbf{x}$, i.e.,

$$\mathbf{F}'(\mathbf{x}) = \left(\frac{\partial F_j(\mathbf{x})}{\partial x_k}\right), \qquad j, k = 1, \ldots, n.$$

This matrix of first partial derivatives of the components of $\mathbf{F}$ at the point $\mathbf{x}$ is called the *Jacobian matrix* of $\mathbf{F}$ at the point $\mathbf{x}$.

**Speed of convergence:** (i) The sequence $(\mathbf{x}^{(k)})$ defined by the fixed-point iteration $\mathbf{x}^{(k+1)} = \mathbf{F}(\mathbf{x}^{(k)})$ converges linearly to the fixed point $\mathbf{x}$ under $\mathbf{F}$, provided

$$\mathbf{F}'(\mathbf{x}) \neq 0.$$

(ii) If the functions $f_i$ are at least twice continuously differentiable in $A$ and if $\mathbf{F}'(\mathbf{x}) = 0$, then the order of convergence of the fixed-point iteration is at least quadratic.

#### 7.4.2.2 The method of Newton–Kantorovich

We linearize the equation whose solutions we wish to determine,

$$\boxed{\mathbf{f}(\mathbf{x}) = \mathbf{0},}$$

under the assumption that the $f_i$ are at least continuously differentiable in $D$. For an initial approximation $\mathbf{x}^{(0)}$ of the solution we have the Taylor series with remainder, in the form

$$\mathbf{f}(\mathbf{x}) = \mathbf{f}(\mathbf{x}^{(0)}) + \mathbf{f}'(\mathbf{x})(\mathbf{x} - \mathbf{x}^{(0)}) + \mathbf{R}(\mathbf{x}).$$

If one omits the remainder term $\mathbf{R}(\mathbf{x})$, then as a linear approximation of the non-linear system of equations we obtain

$$\boxed{\mathbf{f}'(\mathbf{x}^{(0)})(\mathbf{x} - \mathbf{x}^{(0)}) + \mathbf{f}(\mathbf{x}^{(0)}) = \mathbf{0}}$$

[2] The general formulation of this important result can be found in [212].

for the correction vector $\mathbf{z} := \mathbf{x} - \mathbf{x}^{(0)}$. The linear system of equations has a unique solution if and only if

$$\det \mathbf{f}'(\mathbf{x}^{(0)}) \neq 0$$

The system to which we have reduced the original system of equations does not in general lead to the correction vector which leads to a solution. Therefore, here also one iteratively improves the approximation $\mathbf{x}^{(0)}$ by applying for $k = 0, 1, \ldots$ the following steps:

1. Calculate $\mathbf{f}(\mathbf{x}^{(k)})$, and test the validity of $\|\mathbf{f}(\mathbf{x}^{(k)})\| \leq \varepsilon_1$.

2. Calculate $\mathbf{f}'(\mathbf{x}^{(k)})$.

3. Solve the system of equations $\mathbf{f}'(\mathbf{x}^{(k)})\mathbf{z}^{(k)} + \mathbf{f}(\mathbf{x}^{(0)}) = 0$ for $\mathbf{z}^{(k)}$, using the Gauss algorithm. This yields, with $\mathbf{x}^{(k+1)} = \mathbf{x}^{(k)} + \mathbf{z}^{(k)}$, a new approximation. The iteration is continued until the condition $\|\mathbf{z}^{(k)}\| \leq \varepsilon_2$ is satisfied.

The *method of Newton–Kantorovich* can be written in the form

$$\boxed{\mathbf{x}^{(k+1)} = \mathbf{F}(\mathbf{x}^{(k)}), \qquad k = 0, 1, \ldots} \tag{N}$$

with

$$\mathbf{F}(\mathbf{x}) := \mathbf{x} - \mathbf{f}'(\mathbf{x})^{-1}\mathbf{f}(\mathbf{x})$$

as a fixed-point iteration. This is a direct generalization of the classical Newton method. What is important here is that the function $\mathbf{F}$ has the property $\mathbf{F}'(\mathbf{x}) = \mathbf{0}$ for a solution $\mathbf{x}$ of the equation $\mathbf{f}(\mathbf{x}) = \mathbf{0}$. This implies a rapid speed of convergence of the method. The following behavior is typical.

> The method of Newton–Kantorovich converges very rapidly for initial approximations which are already sufficiently near the actual solution. The order of convergence in this case is at least quadratic.
> However, this method can be totally useless for initial approximations which are too far from the actual solution.

Note that for complicated problems one in general has no idea as to how close the initial approximations are to the actual solution. Even with poor initial values, the simple iteration

$$\boxed{\mathbf{x}^{(k+1)} = \mathbf{x}^{(k)} - \mathbf{f}(\mathbf{x}^{(k)}), \qquad k = 0, 1, \ldots} \tag{I}$$

at least has a chance of converging, while the Newton–Kantorovich method (N) is completely useless. If both (N) and (I) converge, then in general (I) converges much slower than (N).

**Discrete dynamical systems:** It is important to be aware that also the iteration procedure (I) is not always applicable.

> If we view (I) as a dynamical system, then (I) can only calculate *stable equilibrium states* $\mathbf{x}$ of the system.

Stable equilibrium states are solutions $\mathbf{x}$ of the equation

$$\boxed{\mathbf{f}(\mathbf{x}) = 0}$$

where all eigenvalues of the matrix $\mathbf{E} - \mathbf{f}'(\mathbf{x})$ lie in the interior of the unit circle.

**Simplified Newton–Kantorovich procedure:** This procedure does away with the computationally costly calculation of the Jacobian matrix

$$\mathbf{f}'(x^{(k)})$$

and instead calculates the correction vectors $\mathbf{z}^{(k)}$ from the equation

$$\boxed{\mathbf{f}'(\mathbf{x}^{(0)})\mathbf{z}^{(k)} + \mathbf{f}(\mathbf{x}^{(0)}) = \mathbf{0}, \qquad k = 0, 1, \ldots}$$

with the constant matrix $\mathbf{f}'(\mathbf{x}^{(0)})$ for a good initial approximation $\mathbf{x}^{(0)}$. For this, only a LR-decomposition of $\mathbf{f}'(\mathbf{x}^{(0)})$ is required, and the calculation of $\mathbf{x}^{(0)}$ is carried out using only forward and reverse substitution. The sequence of iterations $(\mathbf{x}^{(0)})$ converges linearly to $\mathbf{x}$.

For large non-linear systems of equations, there are various modifications of Newton's method which can lead to success. For example, if

$$\frac{\partial f_i(x_1, x_2, \ldots, x_n)}{\partial x_i} \neq 0, \qquad i = 1, 2, \ldots, n$$

then the iteration vector $\mathbf{x}^{(k+1)}$ can be calculated component-wise by using the single step procedure (7.2.2.1) for linear systems of equations to successively solve

$$f_i(x_1^{(k+1)}, \ldots, x_{i-1}^{(k+1)}, x_i^{(k+1)}, x_{i+1}^{(k)}, \ldots, x_n^{(k)}) = 0, \qquad i = 1, 2, \ldots, n$$

for the single unknown $x_i^{(k+1)}$. This is the *non-linear single step procedure.* If the unknown $x_i^{(k+1)}$ is determined with Newton's method, where a single iteration step is carried out and the correction is multiplied with the relaxation parameter $\omega \in ]0, 2[$, the result is the *SOR–Newton procedure*:

$$x_i^{(k+1)} = x_i^{(k)} - \omega \cdot \frac{f_i(x_1^{(k+1)}, \ldots, x_{i-1}^{(k+1)}, x_i^{(k)}, \ldots, x_n^{(k)})}{\dfrac{\partial f_i(x_1^{(k+1)}, \ldots, x_{i-1}^{(k+1)}, x_i^{(k)}, \ldots, x_n^{(k)})}{\partial x_i}}, \qquad i = 1, 2, \ldots, n.$$

## 7.4.3 Determination of zeros of polynomials

### 7.4.3.1 Newton's method and the Horner scheme

A polynomial of $n^{th}$ degree

$$P_n(x) = a_0 x^n + a_1 x^{n-1} + a_2 x^{n-2} + \ldots + a_{n-1} x + a_n, \qquad a_0 \neq 0,$$

with real or complex coefficients $a_j$ has $n$ zeros, provided the zeros are counted with their multiplicities (see 2.1.6). Newton's method is an appropriate one for determining the simple zeros. The calculation of the functional values and the values of the first derivatives can be done with the help of the *Horner scheme.* This is based on the process of division with remainder of a polynomial by a linear factor $(x - p)$ for a given value $p$. More precisely, we make the Ansatz

$$\begin{aligned} P_n(x) &= (x - p)P_{n-1}(x) + R \\ &= (x - p)(b_0 x^{n-1} + b_1 x^{n-2} + b_2 x^{n-3} + \ldots + b_{n-2} x + b_{n-1}) + b_n, \end{aligned}$$

yielding the following algorithm for the recursive calculation of the coefficients $b_j$ of the quotient polynomial $P_{n-1}(x)$ and the remainder $R = b_n$:

$$\boxed{b_0 = a_0, \qquad b_j = a_j + pb_{j-1}, \qquad j = 1, 2, \ldots, n.}$$

Then we have

$$P_n(p) = R = b_n$$

The value of the derivative can be obtained from the relations $P_n'(x) = P_{n-1}(x) + (x - p)P_{n-1}'(x)$ for $x = p$, giving us $P_n'(p) = P_{n-1}(p)$. The value of $P_{n-1}(p)$ is calculated in the same way, using the algorithm for division with remainder, by setting

$$\begin{aligned} P_{n-1}(x) &= (x-p)P_{n-2}(x) + R_1 \\ &= (x-p)(c_0x^{n-2} + c_1x^{n-3} + c_2x^{n-4} + \ldots + c_{n-3}x + c_{n-2}) + c_{n-1} \end{aligned}$$

where the coefficients $c_j$ are given recursively by

$$\boxed{c_0 = b_0, \qquad c_j = b_j + pc_{j-1}, \qquad j = 1, 2, \ldots, n-1.}$$

Thus we have $P_n'(p) = P_{n-1}(p) = R_1 = c_{n-1}$. The numerical values occuring here are then collected in the Horner scheme, which for $n = 5$ looks as follows.

| | | | | | | | |
|---|---|---|---|---|---|---|---|
| $P_5(x)$: | | $a_0$ | $a_1$ | $a_2$ | $a_3$ | $a_4$ | $a_5$ |
| | $p)$ | | $pb_0$ | $pb_1$ | $pb_2$ | $pb_3$ | $pb_4$ |
| $P_4(x)$: | | $b_0$ | $b_1$ | $b_2$ | $b_3$ | $b_4$ | $b_5 = P_5(p)$ |
| | $p)$ | | $pc_0$ | $pc_1$ | $pc_2$ | $pc_3$ | |
| $P_3(x)$: | | $c_0$ | $c_1$ | $c_2$ | $c_3$ | $c_4 = P_5'(p)$ . | |

The displayed scheme can be extended to a complete Horner scheme, if for $P_n(x)$ we carry out a total of $n$ divisions with remainder, which then yields the values for all derivatives of $P_n(x)$ for $x = p$.

If we know a zero $x_1$ of $P_n(x)$, then the linear factor $(x - x_1)$ divides $P_n(x)$ exactly. The remaining zeros of $P_n(x)$ are the zeros of the quotient polynomial $P_{n-1}(x)$. In this way the degree of the polynomial whose zeros we must calculate is decreased step by step. If, on the other hand, we attempt to do this with only approximations of a zero, this can have devastating results on the ensuing numerical procedure. There it is in general better to do this splitting off of factors only implicitly, as follows. If $x_1, \ldots, x_n$ are the zeros of $P_n(x)$, then we have the relations

$$P_n(x) = a_0 \prod_{j=1}^{n} (x - x_j), \qquad \frac{P_n'(x)}{P_n(x)} = \sum_{j=1}^{n} \frac{1}{x - x_j}.$$

If $m$ of the zeros $x_1, \ldots, x_m$ are (approximately) known, then we modify the Newton iteration by setting

$$x^{(k+1)} = x^{(k)} - \frac{1}{\dfrac{P_n'(x^{(k)})}{P_n(x^{(k)})} - \displaystyle\sum_{i=1}^{m} \frac{1}{x^{(k)} - x_i}}, \qquad k = 0, 1, 2, \ldots,$$

so that we continue our calculations with the given, unchanged coefficients of $P_n(x)$.

### 7.4.3.2 The Graeffe procedure

The *Graeffe procedure* makes it possible to calculate the zeros of a polynomial simultaneously without using any initial approximations. It is base on a theorem of Vieta, according to which the following relation holds between the $n$ zeros $x_1, \ldots, x_n$ of the polynomial

$$\sum_{i=1}^{n} x_i = -\frac{a_1}{a_0}, \quad \sum_{\substack{i,j=1\\ i<j}}^{n} x_i x_j = \frac{a_2}{a_0}, \quad \sum_{\substack{i,j,k=1\\ i<j<k}}^{n} x_i x_j x_k = -\frac{a_3}{a_0}, \quad \ldots, \quad \prod_{j=1}^{n} x_j = (-1)^n \frac{a_n}{a_0}.$$

In order to present this procedure we make the simplifying assumption that the given polynomial

$$f_0(x) := a_0^{(0)} x^n + a_1^{(0)} x^{n-1} + a_2^{(0)} x^{n-2} + \ldots + a_{n-1}^{(0)} x + a_n^{(0)}, \qquad a_0^{(0)} a_n^{(0)} \neq 0$$

with real coefficients $a_j^{(0)}$ has only simple real zeros with the property that $|x_1| > |x_2| > \cdots > |x_n|$. Starting with the polynomial $f_0(x)$, we form the sequence of polynomials $f_k(x)$, $k = 1, 2, \ldots$, such that $f_k(x)$ has the zeros $x_j^{2^k}$. This has the effect of separating the zeros of $f_k(x)$ more and more as $k$ grows. For the coefficients $a_j^{(k)}$ of $f_k(x)$ we then obtain from the above mentioned theorem of Vieta the relations

$$\left|\frac{a_1^{(k)}}{a_0^{(k)}}\right| \approx |x_1|^{2^k}, \qquad \left|\frac{a_2^{(k)}}{a_0^{(k)}}\right| \approx |x_1 x_2|^{2^k}, \qquad \left|\frac{a_3^{(k)}}{a_0^{(k)}}\right| \approx |x_1 x_2 x_3|^{2^k}, \qquad \ldots$$

Consequently we have for the absolute values of the zeros the estimates

$$x_j \approx \left|\frac{a_j^{(k)}}{a_{j-1}^{(k)}}\right|^{2^{-k}}, \qquad j = 1, 2, \ldots, n.$$

The sign of the root is obtained by making substitutions into the Horner scheme.

The polynomial $f_{k+1}(x)$ is defined by $f_{k+1}(x) := f_k(\mathrm{i}x) \cdot f_k(-\mathrm{i}x)$, $\mathrm{i}^2 = -1$. Comparison of coefficients then reveals

$$\begin{aligned} a_0^{(k+1)} &= \left(a_0^{(k)}\right)^2, \\ a_j^{(k+1)} &= \left(a_j^{(k)}\right)^2 + 2\sum_{l=1}^{j^*} (-1)^l a_{j+l}^{(k)} a_{j-l}^{(k)}, \qquad j = 1, 2, \ldots, n \end{aligned}$$

with $j^* := \min\{j, n-j\}$.

Implementing the Graeffe procedure successfully requires modifications of this simplified model, in particular to avoid unnecessary steps. Other extensions of the method allow the application of the procedure also to polynomials with multiple roots or complex roots with identical absolute values. The rough approximations obtained from this method are good initial approximations for Newton's method.

### 7.4.3.3 Eigenvalue methods

The calculation of zeros of a normalized polynomial $P_n(x) = x^n + a_1 x^{n-1} + a_2 x^{n-2} + \ldots + a_{n-1}x + a_n$, $a_j \in \mathbb{R}$ can also be dealt with using methods of a corresponding eigenvalue problem. Indeed, $P_n(x)$ is the characteristic polynomial of the *Frobenius matrix* given by

$$\mathbf{A} := \begin{pmatrix} 0 & 0 & 0 & \dots & 0 & -a_n \\ 1 & 0 & 0 & \dots & 0 & -a_{n-1} \\ 0 & 1 & 0 & \dots & 0 & -a_{n-2} \\ \vdots & \vdots & \vdots & & \vdots & \vdots \\ 0 & 0 & 0 & \dots & 0 & -a_2 \\ 0 & 0 & 0 & \dots & 1 & -a_1 \end{pmatrix} \in \mathbb{R}^{n\times n},$$

i.e., we have $P_n(x) = (-1)^n \cdot \det(\mathbf{A} - x\mathbf{E})$. Thus the zeros of $P_n(x)$ can be viewed as the eigenvalues of the *Hessenberg* matrix $\mathbf{A}$, which in turn can be found using the QR-algorithm already discussed in section 7.2.3.4.

#### 7.4.3.4 The method of Bernoulli

This method, originally described by Daniel Bernoulli, tries to determine the zero $x_1$ of the normalized polynomial $P_n(x) = x^n + a_1x^{n-1} + \ldots + a_{n-1}x + a_n$ whose absolute value is maximal. The zero with the smallest absolute value, $x_n$, is then obtained by applying the substitution $z = 1/x$, since

$$P_n(x) = a_nx^n\left(z^n + \frac{a_{n-1}}{a_n}z^{n-1} + \ldots + \frac{a_1}{a_n}z + \frac{1}{a_n}\right) =: a_nx^n \cdot Q_n(z)$$

and the zero $z_1$ of $Q_n(z)$ of largest absolute value is just the reciprocal of $x_n$.

Bernoulli was motivated by consideration of a linear homogenous differential equation of degree $n$ and the behavior of a general solution of this equation. The method can also be based on *vector iteration* of the transposed Frobenius matrix $\mathbf{A}^\top$ (cf. 7.4.3.3), which has the same eigenvalues as $\mathbf{A}$. If $\mathbf{A}$ has precisely one eigenvalue $x_1$ of greatest absolute value, and if $\mathbf{z}^{(0)} \in \mathbb{R}^n$ with $\mathbf{z}^{(0)} \neq 0$ is an almost arbitrary initial approximation, then the sequence of vectors

$$\mathbf{z}^{(k)} := \mathbf{A}^\top\mathbf{z}^{(k-1)}, \qquad k = 1, 2, \ldots$$

converges in the direction of the eigenvector of $\mathbf{A}^\top$ which corresponds to $x_1$. For sufficiently large $k$, we therefore have $\mathbf{z}^{(k+1)} \approx x_1\mathbf{z}^{(k)}$, so that the zero of $P_n(x)$ of largest absolute value is approximately the quotient of the corresponding components of successive iteratively determined vectors. For the starting vector

$$\mathbf{z}^{(0)} := (\zeta_1, \zeta_2, \ldots, \zeta_{n-1}, \zeta_n)^\top, \qquad \zeta_n \neq 0,$$

we then have for $k = 1, 2, 3, \ldots,$

$$\mathbf{z}^{(k)} := (\zeta_{k+1}, \zeta_{k+2}, \ldots, \zeta_{k+n-1}, \zeta_{k+n})^\top \qquad \text{with} \quad \zeta_{k+n} := -\sum_{l=0}^{n-1} a_{n-l}\zeta_{k+l}.$$

Consequently the quotients $q_k := \zeta_{k+n}/\zeta_{k+n-1}$ converge as $k \to \infty$ to the dominant zero $x_1$ of $P_n(x)$. The linear convergence of $q_k$ to $x_1$ can be accelerated by applying once or several times the Aitken $\Delta^2$-process.

In addition to this simple method of Bernoulli, there are various variants which can either determine two real zeros of equal absolute value or two complex conjugate zeros. The approximations that these procedures deliver are good starting points for a Newton or Bairstow approximation.

### 7.4.3.5 The method of Bairstow

The zeros of a polynomial $P_n(x)$ with real coefficients $a_j$ are either real or pairwise conjugate complex. To avoid having to do computations with complex numbers in the Newton method, one tries to determine iteratively a quadratic factor of the polynomial with real coefficients, from which one obtains either a pair of real or a complex conjugate pair of zeros. If $z_1 = u + iv \in \mathbb{C}$ is a complex zero of $P_n(x)$, then so is $z_2 = u - iv$. Consequently,

$$(x - z_1)(x - z_2) = x^2 - 2ux + (u^2 + v^2)$$

is a quadratic factor of $P_n(x)$.

The division algorithm can be generalized for a quadratic factor $(x^2 - px - q)$, $p, q \in \mathbb{R}$. From

$$\begin{aligned} &a_0x^n + a_1x^{n-1} + a_2x^{n-2} + \ldots + a_{n-2}x^2 + a_{n-1}x + a_n \\ &\qquad = (x^2 - px - q)(b_0x^{n-2} + b_1x^{n-3} + \ldots + b_{n-3}x + b_{n-2}) + b_{n-1}(x - p) + b_n \end{aligned}$$

with a linear remainder term $R_1(x) := b_{n-1}(x - p) + b_n$, we obtain upon comparing coefficients

$$\boxed{\begin{aligned} &b_0 = a_0, \quad b_1 = a_1 + pb_0, \\ &b_j = a_j + pb_{j-1} + qb_{j-2}, \qquad j = 2, 3, \ldots, n. \end{aligned}}$$

A given quadratic factor $(x^2 - px - q)$ is a divisor of $P_n(x)$ if

$$b_{n-1}(p, q) = 0, \quad b_n(p, q) = 0.$$

These two non-linear conditions for $p$ and $q$ can be solved with the method of Newton for systems of equations. The required partial derivatives of $b_{n-1}(p, q)$ and $b_n(p, q)$ with respect to $p$ and $q$ can be deduced from an analogous recursion formula from the $b_j$ using

$$\boxed{\begin{aligned} &c_0 = b_0, \quad c_1 = b_1 + pc_0, \\ &c_j = b_j + pc_{j-1} + qc_{j-2}, \qquad j = 2, 3, \ldots, n - 1 \end{aligned}}$$

as

$$\frac{\partial b_{n-1}}{\partial p} = c_{n-2}, \qquad \frac{\partial b_{n-1}}{\partial q} = c_{n-3}, \qquad \frac{\partial b_n}{\partial p} = c_{n-1}, \qquad \frac{\partial b_n}{\partial q} = c_{n-2}.$$

If the determinants of the Jacobian are non-vanishing, i.e., if $c_{n-2}^2 - c_{n-3}c_{n-1} \neq 0$ for the approximation pair $(p^{(k)}, q^{(k)})$, then we obtain iteratively

$$p^{(k+1)} = p^{(k)} + \frac{b_nc_{n-3} - b_{n-1}c_{n-2}}{c_{n-2}^2 - c_{n-3}c_{n-1}}, \qquad q^{(k+1)} = q^{(k)} + \frac{b_{n-1}c_{n-1} - b_nc_{n-2}}{c_{n-2}^2 - c_{n-3}c_{n-1}}$$

This *method of Bairstow* has order of convergence $p = 2$. After the determination of a quadratic factor there are, in case the roots are complex conjugate, complex numbers which appear. But instead of calculating them, both of them are split off by dividing $P_n(x)$ by the quadratic factor $(x^2 - px - q)$.

The numbers which need to be calculated in this method can be ordered into a *double-rowed Horner scheme.* In case of $n = 6$, for example, we have

| | | | | | | | | |
|---|---|---|---|---|---|---|---|---|
| $P_6(x)$: | | $a_0$ | $a_1$ | $a_2$ | $a_3$ | $a_4$ | $a_5$ | $a_6$ |
| | $q)$ | | | $qb_0$ | $qb_1$ | $qb_2$ | $qb_3$ | $qb_4$ |
| | $p)$ | | $pb_0$ | $pb_1$ | $pb_2$ | $pb_3$ | $pb_4$ | $pb_5$ |
| $P_4(x)$: | | $b_0$ | $b_1$ | $b_2$ | $b_3$ | $b_4$ | $\boxed{b_5}$ | $\boxed{b_6}$ |
| | $q)$ | | | $qc_0$ | $qc_1$ | $qc_2$ | $qc_3$ | |
| | $p)$ | | $pc_0$ | $pc_1$ | $pc_2$ | $pc_3$ | $pc_4$ | |
| | | $c_0$ | $c_1$ | $c_2$ | $\boxed{c_3}$ | $\boxed{c_4}$ | $\boxed{c_5}$ | |

## 7.5 Approximation

We now consider the problem of finding, for a given function $f$ of a normed space $V$ of real-valued functions on a finite interval $[a, b] \subset \mathbb{R}$, an element $h_0$ of a finite-dimensional subspace $U \subsetneqq V$ with the property

$$\boxed{\|f - h_0\| = \inf_{h \in U} \|f - h\|.}$$

In what follows we consider only the two most important cases for applications, which are when the norm is either the $L_2$-norm or the supremum norm. For both of these norms the existence and uniqueness of a best approximation $h_0 \in U$ can be shown. Using the properties which characterize $h_0$, we discuss its calculation.

### 7.5.1 Approximation in quadratic means

Let $V$ be a real Hilbert space with scalar product $(f, g)$, $f, g, \in V$, and the norm $\|f\| := (f, f)^{1/2}$, and let furthermore $U := \operatorname{span}(\varphi_1, \varphi_2, \ldots, \varphi_n) \subseteq V$ be an $n$-dimensional subspace with basis $\{\varphi_1, \varphi_2, \ldots, \varphi_n\}$. The approximation problem in this context is to find the best approximation $h_0$ of $f$ which is characterized by the relation

$$(f - h_0, u) = 0 \qquad \text{for all} \quad u \in U.$$

The orthogonality condition for $f - h_0$ must be satisfied for all basis elements $\varphi_j$ of $U$. From the given representation for $h_0 \in U$,

$$h_0 = \sum_{k=1}^{n} c_k \varphi_k$$

we obtain for the coefficients $c_k$ the conditions

$$\sum_{k=1}^{n} (\varphi_j, \varphi_k) c_k = (f, \varphi_j), \qquad j = 1, 2, \ldots, n.$$

The matrix $\mathbf{A} \in \mathbb{R}^{n \times n}$ of the system of linear equations with entries $a_{jk} := (\varphi_j, \varphi_k)$, $j, k = 1, 2, \ldots, n$ is called the *Gram matrix*. It is symmetric and positive definite. The coefficients $c_k$ can therefore by uniquely determined for any $f \in V$, and for the resulting best approximation $h_0$ we have

$$\|f - h_0\|^2 = \|f\|^2 - \sum_{k=1}^{n} c_k (f, \varphi_k) = \min_{h \in U} \|f - h\|^2.$$

The Gram matrix **A** for an arbitrary basis $\{\varphi_1, \ldots, \varphi_n\}$ can have a very large condition number $\kappa(\mathbf{A})$, so that the numerical solution of the system of equations can be difficult. This situation is illustrated with great clarity by the approximation problem of finding, for given $f \in V = C_{L_2}([0,1])$, the real vector space of functions which are continuous on $[0,1]$, with scalar product

$$(f,g) := \int_0^1 f(x)g(x)\,\mathrm{d}x,$$

the best approximating polynomial $h_0$ of degree $n$, where the basis of the $(n+1)$-dimensional subspace $U$ is given by $\{x^1, x^2, \ldots, x^n\}$. The elements of the Gram matrix are in this case

$$a_{jk} = (x^{j-1}, x^{k-1}) = \int_0^1 x^{j+k-2}\,\mathrm{d}x = \frac{1}{j+k-1} \qquad j,k = 1,2,\ldots,n+1,$$

so that **A** is equal to the *Hilbert matrix* $\mathbf{H}_{n+1}$, whose condition number grows exponentially with growing $n$.

The problems with the numerical side of the computations can be avoided completely, if, as a basis of $U$, we take a system of *orthogonal elements*, so that

$$(\varphi_j, \varphi_k) = 0 \qquad \text{for all} \quad j \neq k \quad j,k = 1,2,\ldots,n.$$

If, in addition $(\varphi_j, \varphi_j) = \|\varphi_j\|^2 = 1,\ j = 1,2,\ldots,n$, then one speaks of an *orthonormal basis* in $U$. In the case of an orthonormal basis $\{\varphi_1, \varphi_2, \ldots, \varphi_n\}$, the Gram matrix **A** becomes a diagonal matrix, so that the coefficients $c_k$ of the approximation $h_0$ can be obtained directly from the simplified system of equations by

$$c_k = (f, \varphi_k)/(\varphi_k, \varphi_k), \qquad k = 1,2,\ldots,n.$$

If follows from this that increasing the dimension of the subspace $U$ by extending the orthogonal basis does not change the coefficients which have already been calculated, and the mean square error decreases in the weak sense as $n$ grows, as follows from the representation which is valid here:

$$\|f - h_0\|^2 = \|f\|^2 - \sum_{k=1}^{n} \frac{(f,\varphi_k)^2}{(\varphi_k,\varphi_k)}.$$

The cases which follow are all covered by the general theory.

#### 7.5.1.1 Fourier polynomials

In the Hilbert space $V = L_2([-\pi,\pi])$ of measurable functions $f$ on $[-\pi,\pi]$, endowed with the scalar product

$$(f,g) := \int_{-\pi}^{\pi} f(x)g(x)\,\mathrm{d}x$$

$\{1, \sin x, \cos x, \sin 2x, \cos 2x, \ldots, \sin nx, \cos nx\}$ forms an orthogonal basis of the $(2n+1)$-dimensional subspace $U$ they span. Because of the relation

$$(1,1) = 2\pi, \quad (\sin kx,\, \sin kx) = (\cos kx,\, \cos kx) = \pi, \qquad k = 1,2\ldots,n$$

the *Fourier coefficients* of the best approximating

$$h_0(x) = \frac{1}{2}a_0 + \sum_{k=1}^{n} \{a_k \cos kx + b_k \sin kx\}$$

are given by

$$a_k = \frac{1}{\pi} \int_{-\pi}^{\pi} f(x) \cos kx \, \mathrm{d}x, \qquad b_k = \frac{1}{\pi} \int_{-\pi}^{\pi} f(x) \sin kx \, \mathrm{d}x.$$

### 7.5.1.2 Polynomial approximation

Consider now the pre-Hilbert space $V = C_{L_2}([-1,+1])$ of continuous real-valued functions on $[-1,+1]$ with scalar product

$$(f,g) := \int_{-1}^{1} f(x)g(x) \, \mathrm{d}x.$$

We wish to determine the polynomial $h_0$ in the $(n+1)$-dimensional subspace of the polynomials of degree $n$ which best approximates a given $f \in V$. An orthogonal basis in $U$ is provided by the *Legendre polynomials* $P_m(x)$, $m = 0, 1, 2, \ldots$. They are defined by

$$P_m(x) := \frac{1}{2^m m!} \cdot \frac{\mathrm{d}^m}{\mathrm{d}x^m}[(x^2-1)^m], \qquad m = 0, 1, 2, \ldots$$

and have the orthogonality property

$$(P_m, P_l) = \int_{-1}^{1} P_m(x) P_l(x) \, \mathrm{d}x = \begin{cases} 0 & \text{for all} \quad m \neq l, \ m, l \in \mathbb{N}, \\ \dfrac{2}{2m+1} & \text{for} \quad m = l \in \mathbb{N}. \end{cases}$$

The coefficients of the best approximating

$$h_0(x) = \sum_{k=0}^{n} c_k P_k(x),$$

written as linear combination of the Legendre polynomials, are therefore given by

$$c_k = \frac{2k+1}{2} \int_{-1}^{1} f(x) P_k(x) \, \mathrm{d}x, \qquad k = 0, 1, 2, \ldots, n.$$

For an approximate calculation of the integrals occuring here one can apply the Gauss–Legendre quadrature formulas given in section 7.3.3.

The numerical calculation of the values $h_0(x)$ for a given point $x$ with the expansion above in terms of the Legendre polynomials can, because of the recursion formula

$$P_k(x) = \frac{2k-1}{k} x P_{k-1}(x) - \frac{k-1}{k} P_{k-2}(x), \qquad k = 2, 3, \ldots,$$

be carried out by successively eliminating the Legendre polynomial of highest degree with the following algorithm:

$$d_n = c_n, \quad d_{n-1} = c_{n-1} + \frac{2n-1}{n} x d_n,$$
$$d_k = c_k + \frac{2k+1}{k+1} x d_{k+1} - \frac{k+1}{k+2} d_{k+2}, \qquad k = n-2, n-3, \ldots, 0,$$
$$h_0(x) = d_0.$$

#### 7.5.1.3 Weighted polynomial approximation

On the Hilbert space $V = C_{q,L_2}([-1,1])$, the scalar product

$$(f,g) := \int_{-1}^{1} f(x)g(x)q(x)\,\mathrm{d}x$$

is defined with a non-negative weight function $q(x)$. The problem is to determine a polynomial $h_0$ of degree at most $n$ which best approximates a given $f \in V$. For a given weight function $q$, we can explicitly list the corresponding orthogonal basis. For the particularly important case

$$q(x) := 1/\sqrt{1-x^2},$$

these are the *Chebychev polynomials* $T_n(x)$. We have the following trigonometric identity

$$\cos(n+1)\varphi + \cos(n-1)\varphi = 2\cos\varphi\cos n\varphi, \qquad n \geq 1,$$

from which $\cos n\varphi$ can be written as a polynomial of $n^{th}$ degree in $\cos\varphi$, and by definition, the $n^{th}$ Chebychev polynomial $T_n(x)$, $n \in \mathbb{N}$, is given by

$$\cos n\varphi =: T_n(\cos\varphi) = T_n(x) = \cos(n \cdot \arccos x), \quad x = \cos\varphi, \qquad x \in [-1,+1].$$

The first few $T$-polynomials are thus

$$T_0(x) = 1, \quad T_1(x) = x, \quad T_2(x) = 2x^2 - 1, \quad T_3(x) = 4x^3 - 3x, \quad T_4(x) = 8x^4 - 8x^2 + 1.$$

They satisfy a three-step recursion relation

$$T_{n+1}(x) = 2x\,T_n(x) - T_{n-1}(x), \quad n \geq 1; \qquad T_0(x) = 1, \quad T_1(x) = x.$$

The $n^{th}$ $T$-polynomial $T_n(x)$ has in $[-1,1]$ $n$ simple zeros, the so-called *Chebychev abscissas*

$$x_k = \cos\left(\frac{2k-1}{n} \cdot \frac{\pi}{2}\right), \qquad k = 1, 2, \ldots, n,$$

which are crowded at the ends of the interval. From the definition we have the property

$$|T_n(x)| \leq 1 \qquad \text{for} \quad x \in [-1,+1], \quad n \in \mathbb{N},$$

and the extrema $\pm 1$ are realized by $T_n(x)$ at the $n+1$ extremal points $x_j^{(e)}$, as follows.

$$T_n\left(x_j^{(e)}\right) = (-1)^j, \qquad x_j^{(e)} = \cos\left(\frac{j\pi}{n}\right), \qquad j = 0, 1, 2, \ldots, n.$$

The $T$-polynomials have the orthogonality property

$$\int_{-1}^{1} T_k(x) T_j(x) \frac{\mathrm{d}x}{\sqrt{1-x^2}} = \left\{ \begin{array}{ll} 0, & \text{if } k \neq j \\ \frac{\pi}{2}, & \text{if } k = j > 0 \\ \pi, & \text{if } k = j = 0 \end{array} \right\} \quad k, j \in \mathbb{N}.$$

With respect to the orthogonal basis $\{T_0, T_1, T_2, \ldots, T_n\}$ of the subspace $U$ of polynomials of degree $n$, the coefficients $c_k$ of the best approximation

$$h_0(x) = \frac{1}{2} c_0 T_0(x) + \sum_{k=1}^{n} c_k T_k(x)$$

are, for $f \in V$, given by

$$c_k = \frac{2}{\pi} \int_{-1}^{1} f(x) T_k(x) \frac{\mathrm{d}x}{\sqrt{1-x^2}}, \qquad k = 0, 1, 2, \ldots, n.$$

Utilizing the substitution $x = \cos\varphi$, one obtains from this the simpler representation

$$c_k = \frac{2}{\pi} \int_{0}^{\pi} f(\cos\varphi) \cos(k\varphi)\,\mathrm{d}\varphi = \frac{1}{\pi} \int_{-\pi}^{\pi} f(\cos\varphi) \cos(k\varphi)\,\mathrm{d}\varphi, \qquad k = 0, 1, \ldots, n.$$

Consequently, the coefficients $c_k$ of the best approximating weighted polynomial are the Fourier coefficients $a_k$ of the even, $2\pi$-periodic function $F(\varphi) := f(\cos\varphi)$. For an approximation of the integrals, the most appropriate and most efficient method is the trapezoidal formula presented in section 7.3.3, as it generally yields rapid convergence for increasing numbers of integration intervals.

The value of $h_0(x)$ at a point $x$ is, because of the expansion in terms of the $T$-polynomials, numerically reliable and efficient when one applies the *algorithm of Clenshaw*. With the aid of the recursion formula, the $T$-polynomials of highest degree can be eliminated, and one obtains the following set of formulas:

$$\begin{array}{ll} d_n = c_n, \quad y = 2x, \quad d_{n-1} = c_{n-1} + y d_n, & \\ d_k = c_k + y d_{k+1} - d_{k+2}, & k = n-2, n-3, \ldots, 0, \\ h_0(x) = (d_0 - d_2)/2. & \end{array}$$

## 7.5.2 Uniform approximation

The problem of approximating a continuous function $f$ by a function $h_0$ in a subspace $U$ will now be considered with the additional requirement that the maximal difference between $h_0$ and $f$ should be minimal. The space of all functions which are continuous and real-values on $[a, b]$, endowed with the *maximum norm* or the *Chebychev norm*

$$\|f\|_\infty := \max_{x \in [a,b]} |f(x)|,$$

becomes a *Banach space* $V = C([a, b])$. Because the Chebychev norm gives a bound for the maximal difference between two functions on the entire interval, the usual notion in this case is *uniform approximation.*

A subspace $U = \text{span}(\varphi_1, \varphi_2, \ldots, \varphi_n)$ with basis $\{\varphi_1, \varphi_2, \ldots, \varphi_n\}$ is called a *Haar space,* if every element $u \in U, u \neq 0$ in $[a, b]$ has at most $n - 1$ different zeros. In a Haar space $U$ there is a unique best approximating function $h_0 \in U$ for a given continuous function. The *alternate theorem* characterizes the best approximation by the following property: an *alternate* of $f \in C([a, b])$ and $h \in U$ is an ordered set of $(n + 1)$ points $a \leq x_1 < x_2 < \ldots < x_n < x_{n+1} \leq b$ for which the difference $d := f - h$ takes on values with alternating sign, i.e.,

$$\text{sgn}\, d(x_k) = -\text{sgn}\, d(x_{k+1}), \qquad k = 1, 2, \ldots, n.$$

The function $h_0 \in U$ is a best approximation of $f \in C([a, b])$ if there is an alternate of $f$ and $h_0$ such that

$$\boxed{|f(x_k) - h_0(x_k)| = \|f - h_0\|_\infty, \qquad k = 1, 2, \ldots, n + 1.}$$

The alternate theorem forms the basis for the *exchange procedure of Remez* for the iterative construction of a best approximation $h_0 \in U$ in a Haar space $U$ of a given function $f \in C([a, b])$. The space of polynomials of degree $n$, which is important in applications, $U := \text{span}(1, x, x^2, \ldots, x^n)$ with $\dim U = n + 1$ satisfies the condition of a Haar space. The essential steps of the *simple Remez algorithm* are in this case:

1. Prescribe $(n + 2)$ points

$$a \leq x_1^{(0)} < x_2^{(0)} < \ldots < x_{n+1}^{(0)} < x_{n+2}^{(0)} \leq b$$

as a starting approximation of the alternate which is to be determined.

2. Determination of the polynomial $p^{(0)} \in U$ with the property that $[x_k^{(0)}]_{k=1}^{n+2}$ is an alternate of $f$ and $p^{(0)}$, with the additional condition that the absolute value of the defect is identical at all $(n + 2)$ points. For this, we set

$$p^{(0)} := a_0 + a_1 x + a_2 x^2 + \ldots + a_n x^n$$

and reduce the requirement to solution of a system of linear equations

$$a_0 + a_1 x_k^{(0)} + a_2 \left(x_k^{(0)}\right)^2 + \ldots + a_n \left(x_k^{(0)}\right)^2 - (-1)^k r^{(0)} = f(x_k^{(0)}), \qquad k = 1, 2, \ldots, n + 2$$

for the $(n + 2)$ unknowns $a_0, a_1, \ldots, a_n, r^{(0)}$. This system of equations has a unique solution.

3. With the resulting polynomial $p^{(0)}$, find the point $\overline{x} \in [a, b]$ for which

$$\|f - p^{(0)}\|_\infty = |f(\bar{x}) - p^{(0)}(\bar{x})|.$$

If $\overline{x}$ coincides with one of the points $x_k^{(0)}$, $k = 1, 2, \ldots, n + 2$, then according to the alternate theorem we have found in $p^{(0)}$ the best approximation $h_0$.

4. Otherwise we exchange $\overline{x}$ with one of the $x_k^{(0)}$, so that the resulting $(n + 2)$ points

$$a \leq x_1^{(1)} < x_2^{(1)} < \ldots < x_{n+1}^{(1)} < x_{n+2}^{(1)} \leq b$$

are a new alternate for $f$ and $p^{(0)}$. This exchange step forces the absolute value $|r^{(1)}|$ of the defect of the polynomial $p^{(1)}$ which is determined in the analogous second step to strictly decrease. The iteration is continued until the best approximation $h_0$ is sufficiently accurately represented by $p^{(k)}$, i.e., when $\|f - p^{(k)}\|_\infty \approx |r^{(k)}|$ is satisfied.

### 7.5.3 Approximate uniform approximation

For many purposes it is sufficient to have a good approximation of the best approximator $h_0$, which can be obtained in different ways.

If $f$ is twice continuously differentiable on the interval $[-1, 1]$, then the expansion of $f$ with respect to the Chebychev polynomials

$$f(x) = \frac{1}{2}\, c_0 T_0(x) + \sum_{k=1}^{\infty} c_k T_k(x)$$

converges uniformly for all $x \in [-1, +1]$, as in this case the series $\sum\limits_{k=1}^{\infty} |c_k|$ converges. Because $|T_k(x)| \leq 1$ for $x \in [-1, +1]$, every partial sum

$$\tilde{f}_n(x) := \frac{1}{2}\, c_0 T_0(x) + \sum_{k=1}^{n} c_k T_k(x),$$

i.e., every best approximator in the sense of the weighted quadratic mean (see 7.5.1.3), is a good approximation of the best approximator. The calculation of these approximate solutions requires that the coefficients $c_k$ can be easily calculated.

For a function $f$ which is at least $(n+1)$-times continuously differentiable in the interval $[-1, 1]$, there is often a very good approximation of the best uniform approximator, namely the interpolation polynomial $I_n(x)$ for the $(n+1)$ Chebychev abscissas $x_k = \cos\big((2k-1)\pi/(2n+2)\big)$, $k = 1, 2, \ldots, n+1$ of $T_{n+1}(x)$. Indeed, for the interpolation error as discussed in section 7.3.1.4, we have because of $\prod\limits_{k=1}^{n+1}(x - x_k) = T_{n+1}(x)/2^n$ the formula

$$f(x) - I_n(x) = \frac{f^{(n+1)}(\xi)}{2^n \cdot (n+1)!} \cdot T_{n+1}(x), \qquad x \in [-1, +1],$$

with the point $\xi \in ]-1, 1[$ depending on $x$.

If we now write $I_n(x)$ as a linear combination of the $T$-polynomials,

$$I_n(x) = \frac{1}{2}\, c_0 T_0(x) + \sum_{j=1}^{n} c_j T_j(x),$$

then the coefficients $c_j$ are, because of the discrete orthogonality properties of the $T$-polynomials given by

$$c_j = \frac{2}{n+1} \sum_{k=1}^{n+1} f(x_k) T_j(x_k) = \frac{2}{n+1} \sum_{k=1}^{n+1} f\left(\cos\left(\frac{2k-1}{n+1}\frac{\pi}{2}\right)\right) \cos\left(j\,\frac{2k-1}{n+1}\frac{\pi}{2}\right)$$

for $j = 0, 1, 2, \ldots, n$. The term *discrete Chebychev approximation* is used by denote the problem, given $f \in C([a, b])$ and $N$ interpolation points $x_i$ with $a \leq x_1 < x_2 < \ldots < x_{N-1} < x_N \leq b$, of finding the function $h_0 \in U$, $\dim U = n < N$, for which in the *discrete maximum norm* $\|f\|_\infty^{\mathrm{d}} := \max_i |f(x_i)|$ we have

$$\|f - h_0\|_\infty^{\mathrm{d}} = \min_{h \in U} \|f - h\|_\infty^{\mathrm{d}}.$$

If $U$ is a Haar space, then the numerical computation of $h_0$ can be effected with either a discrete version of the Remez algorithm or by applying the methods of linear optimization.

## 7.6 Ordinary differential equations

Since it is in general not possible to explicitly exhibit the general solution of a differential equation or a system of differential equations of order $r$ (see section 1.12), the methods of numerical approximation of solutions is necessary for applications. To treat the problem with the appropriate methods, one must differentiate between initial value problems and boundary value problems, see section 1.12.9 for details on this. In what follows, we shall assume existence and uniqueness of the desired solutions.

### 7.6.1 Initial value problems

Every explicit differential equation of order $r$ or system of such can by an appropriate change of variables be transformed into a system of $r$ differential equations of first order. The *initial value problem* is to determine $r$ functions $y_1(x), y_2(x), \ldots, y_r(x)$ as the solution of the equation

$$y_i'(x) = f_i(x, y_1, y_2, \ldots, y_r), \qquad i = 1, 2, \ldots, r$$

which satisfy at a given point $x_0$ and given values $y_{i0}$, $i = 1, 2, \ldots, r$ the *initial conditions*

$$y_i(x_0) = y_{i0}, \qquad i = 1, 2, \ldots, r.$$

Using the vectors $\mathbf{y}(x) := \big(y_1(x), y_2(x), \ldots, y_r(x)\big)^{\mathsf{T}}$, $\mathbf{y}_0 := (y_{10}, y_{20}, \ldots, y_{r0})^{\mathsf{T}}$ and $\mathbf{f}(x, \mathbf{y}) = \big(f_1(x, \mathbf{y}), f_2(x, \mathbf{y}), \ldots, f_r(x, \mathbf{y})\big)^{\mathsf{T}}$, the *Cauchy problem* can be written

$$\boxed{\mathbf{y}'(x) = \mathbf{f}\big(x, \mathbf{y}(x)\big), \quad \mathbf{y}(x_0) = \mathbf{y}_0.}$$

In order to simplify the notation we consider in what follows the initial value problem for a scalar differential equation of first order

$$y'(x) = f\big(x, y(x)\big), \quad y(x_0) = y_0.$$

The methods presented for this problem are easily carried over to systems of equations.

#### 7.6.1.1 Single step methods

The simplest *Euler method* is to approximate the solution curve $y(x)$ through the initial point $(x_0, y_0)$ by means of using the tangent, whose slope $y'(x_0) = f(x_0, y_0)$ is given through the differential equation itself. At the point $x_1 := x_0 + h$, where $h$ denotes the *step width*, one obtains the approximate value

$$y_1 = y_0 + h\, f(x_0, y_0)$$

for the precise value $y(x_1)$ of the solution vector. If the procedure is continued at the point $(x_1, y_1)$ with the tangent defined by the directional field of the differential equation at that point, then at the *equidistant* interpolation points $x_k := x_0 + kh$, $k = 1, 2, \ldots$ one is lead to successively better approximations $y_k$ by stipulating

$$\boxed{y_{k+1} = y_k + h\, f(x_k, y_x), \qquad k = 0, 1, 2 \ldots.}$$

Because of the geometric interpretable construction of these approximations, the Euler method is also referred to as the *polygonal edge method*. Clearly it is quite rough and will yield useful approximations only for very small values of the step width $h$. But it has the virtue of simplicity, and is the simplest of all single step procedures, using only the known approximation $y_k$ at the point $x_k$ for calculating the next approximation $y_{k+1}$.

A general, explicit single step algorithm is

$$y_{k+1} = y_k + h\,\Phi(x_k, y_k, h), \qquad k = 0, 1, 2, \ldots,$$

where $\Phi(x_k, y_k, h)$ is the prescription for calculating the approximation $y_{k+1}$ at the point $x_{k+1} = x_k + h$ from the pair $(x_k, y_k)$ and the step width $h$. The function $\Phi(x, y, h)$ must be related to the differential equation whose solution we seek for this procedure to work. This motivates the following terminology: a general single step procedure is called *consistent*, if

$$\lim_{h\to 0} \Phi(x, y, h) = f(x, y).$$

Euler's method is consistent.

We define the *local discretization error* of a single step procedure at the point $x_{k+1}$ by

$$d_{k+1} := y(x_{k+1}) - y(x_k) - h\,\Phi\big(x_k, y(x_k), h\big).$$

It describes the error of the algorithm when the precise solution $y(x)$ is inserted. From the Taylor series with remained term

$$y(x_{k+1}) = y(x_k) + h\,y'(x_k) + \frac{1}{2}h^2 y''(\xi), \qquad x_k < \xi < x_{k+1},$$

we get for the Euler method the local discretization error

$$d_{k+1} = \frac{1}{2}h^2 y''(\xi), \qquad x_k < \xi < x_{k+1}.$$

On the other hand, the *global error* $g_k$ at the point $x_k$ is given by the formula

$$g_k := y(x_k) - y_k.$$

It describes the error of the entire single step procedure which sums up all the discretization errors. In case the function $\Phi(x, y, h)$ satisfies the Lipschitz condition in a domain $B$ with respect to a variable $y$, i.e., if we have

$$|\Phi(x, y, h) - \Phi(x, y^*, h)| \le L|y - y^*| \qquad \text{for} \quad x, y, y^*, h \in B, \quad 0 < L < \infty,$$

the global error $g_n$ at the point $x_n = x_0 + nh$, $n \in \mathbb{N}$ can be estimated by the local discretization error. If $\max\limits_{1\le k\le n} |d_k| \le D$, then we have the estimate

$$|g_n| \le \frac{D}{hL}\{\mathrm{e}^{nhL} - 1\} \le \frac{D}{hL}\,\mathrm{e}^{nhL} = \frac{D}{hL}\,\mathrm{e}^{(x_n - x_0)L}.$$

In addition to the Lipschitz constant $L$, the maximal absolute value $D$ of the local discretization error $d_k$ in the interval $[x_0, x_n]$ plays an important role in this formula.

A single step procedure has by definition *an error of order* $p$, if the local discretization error can be estimated as

$$\max_{1\le k\le n} |d_k| \le D = \text{const} \cdot h^{p+1} = O(h^{p+1})$$

so that for the global error we have

$$|g_n| \le \frac{\text{const}}{L}\,\mathrm{e}^{nhL} h^p = O(h^p).$$

The Euler method has the error order $p = 1$, and the global error of the procedure decreases linearly at a fixed point $x := x_0 + nh$ as $h$ approaches zero.

The rounding errors and their propagation in the procedure are, compared with the procedural errors as just discussed, for single step procedures of higher order, of secondary importance.

The *explicit Runge–Kutta procedure* is an important and in many situations applicable method with a higher error order. This procedure start with the integral equation

$$y(x_{k+1}) = y(x_k) + \int_{x_k}^{x_{k+1}} f(x, y(x))\, \mathrm{d}x,$$

which is equivalent to the differential equation. The integral can be approximated by a quadrature formula with $s$ interpolation points $\xi_1, \ldots, \xi_s \in [x_k, x_{k+1}]$ according to the formula

$$\int_{x_k}^{x_{k+1}} f(x, y(x))\, \mathrm{d}x \approx h \sum_{i=1}^{s} c_i f(\xi_i, y_i^*) =: h \sum_{i=1}^{s} c_i k_i.$$

The interpolation points $\xi_i$ are given by

$$\xi_1 = x_k, \quad \xi_i = x_k + a_i h, \qquad i = 2, 3, \ldots, s,$$

and for the unknown functional value $y_i^*$, we assume a relation

$$y_1^* := y_k, \quad y_i^* := y_k + h \sum_{j=1}^{i-1} b_{ij} f(\xi_j, y_j^*), \qquad i = 2, 3, \ldots s.$$

The parameters $c_i, a_i, b_{ij}$ which occur in this formula are determined under further simplifying assumptions in such a way that the Runge–Kutta algorithm with $s$ interpolation points

$$y_{k+1} = y_k + h \sum_{i=1}^{s} c_i f(\xi_i, y_i^*) = y_k + h \sum_{i=1}^{s} c_i k_i, \qquad k = 0, 1, 2, \ldots$$

has as high a order of error $p$ as possible. The parameters are not uniquely determined by this condition, so that it is possible to take account of further factors. Explicit Runge–Kutta algorithms can be displayed as a scheme for the coefficients of the form

$$\begin{array}{c|ccccc}
a_1 & & & & & \\
a_2 & b_{21} & & & & \\
a_3 & b_{31} & b_{32} & & & \\
\vdots & \vdots & \vdots & & & \\
a_s & b_{s1} & b_{s2} & \cdots & b_{s,s-1} & \\
\hline
 & c_1 & c_2 & \cdots & c_{s-1} & c_s
\end{array}$$

Examples of Runge–Kutta methods for low order of error are the following.

$$\begin{array}{c|cc} 0 & & \\ \frac{1}{2} & \frac{1}{2} & \\ \hline & 0 & 1 \end{array}$$

improved polygonal edge method with $p = 2$

$$\begin{array}{c|cc} 0 & & \\ 1 & 1 & \\ \hline & \frac{1}{2} & \frac{1}{2} \end{array}$$

method of Heun ($p = 2$)

$$\begin{array}{c|ccc} 0 & & & \\ \frac{1}{3} & \frac{1}{3} & & \\ \frac{2}{3} & 0 & \frac{2}{3} & \\ \hline & \frac{1}{4} & 0 & \frac{2}{3} \end{array}$$

method of Heun ($p = 3$)

$$\begin{array}{c|ccc} 0 & & & \\ \frac{1}{2} & \frac{1}{2} & & \\ 1 & -1 & 2 & \\ \hline & \frac{1}{6} & \frac{4}{6} & \frac{1}{6} \end{array}$$

method of Kutta with the Simpson rule ($p = 3$)

$$\begin{array}{c|cccc} 0 & & & & \\ \frac{1}{2} & \frac{1}{2} & & & \\ \frac{1}{2} & 0 & \frac{1}{2} & & \\ 1 & 0 & 0 & 1 & \\ \hline & \frac{1}{6} & \frac{2}{6} & \frac{2}{6} & \frac{1}{6} \end{array}$$

classical Runge–Kutta method ($p = 4$)

$$\begin{array}{c|cccc} 0 & & & & \\ \frac{1}{3} & \frac{1}{3} & & & \\ \frac{2}{3} & -\frac{1}{3} & 1 & & \\ 1 & 1 & -1 & 1 & \\ \hline & \frac{1}{8} & \frac{3}{8} & \frac{3}{8} & \frac{1}{8} \end{array}$$

Runge–Kutta method, 3/8 rule ($p = 4$)

In order to estimate the size of the local discretization error, often the simple *principle of Runge* is used; this is also applied to automatically direct the change of the step size. Let $Y_k(x_k) = y_k$ be a solution of $y' = f(x, y)$. We wish to determine the error $y_{k+2}$ compared with $Y_k(x_k + 2h)$ after two integration steps with step width $h$. For this, the value $\tilde{y}_{k+1}$ obtained at $x = x_k + 2h$ with step width $2h$ is used. If the method we are applying has an order of error $p$, then

$$\begin{aligned} Y_k(x_k + h) - y_{k+1} &= d_{k+1} = C_k h^{p+1} + O(h^{p+2}), \\ Y_k(x_k + 2h) - y_{k+2} &= 2C_k h^{p+1} + O(h^{p+2}), \\ Y_k(x_k + 2h) - \tilde{y}_{k+1} &= 2^{p+1} C_k h^{p+1} + O(h^{p+2}). \end{aligned}$$

From this it follows that

$$\begin{aligned} &y_{k+2} - \tilde{y}_{k+1} = 2C_k(2^p - 1)h^{p+1} + O(h^{p+2}), \\ &Y_k(x_k + 2h) - y_{k+2} \approx 2C_k h^{p+1} \approx \frac{y_{k+2} - \tilde{y}_{k+1}}{2^p - 1}. \end{aligned}$$

The estimate is calculated after the double step and requires for the Runge–Kutta method with $s$ interpolation points for the quadrature $(s - 1)$ additional functional evaluations for the calculation of the value $\tilde{y}_{k+1}$.

A different principle also used to estimate the discretization error is based on using a Runge–Kutta procedure with a higher error order. To keep the computational cost as small as possible, the applied method must be embedded in the procedure with the higher error order in such a way that both require the same functional evaluations. The improved polygonal edge method is embedded in the method of Kutta, and as an estimate for the local discretization error, one obtains $d_{k+1}^{(\mathrm{VP})} \approx \frac{h}{6}\{k_1 - 2k_2 + k_3\}$.

This principle was greatly improved by *Fehlberg*, by applying two embedded methods with different error orders in such a way that the two values $y_{k+1}$ obtained is the estimate of the discretization error; this is the *Runge–Kutta–Fehlberg method.* Since a Runge–Kutta method of the fifth order requires six functional evaluations, in [436] a method of fourth order with an exceptionally small discretization error is described.

A further generalization of the procedures described above are the so-called *implicit Runge–Kutta procedures*, in which the interpolation points are more generally given by

$$\xi_i = x_k + a_i h, \qquad i = 1, 2, \ldots, s.$$

The unknown functional values $y_i^*$ are defined by

$$y_i^* = y_k + h \sum_{j=1}^{s} b_{ij} f(\xi_j, y_j^*), \qquad i = 1, 2, \ldots, s,$$

so that in every integration step the system of equations

$$k_i = f\left( x_k + a_i h \, , \, y_k + h \sum_{j=1}^{s} b_{ij} k_j \right), \qquad i = 1, 2, \ldots, s$$

which is generally not linear, has to be solved for the $s$ unknowns $k_i$. One obtains for these

$$y_{k+1} = y_k + h \sum_{i=1}^{s} c_i k_i, \qquad k = 0, 1, 2, \ldots.$$

Among the Runge–Kutta methods with $s$ interpolation points there are also some which have certain stability properties which are important for the solution of *stiff systems of differential equations.* In addition, implicit Runge–Kutta procedures with $s$ interpolation points for the quadrature have for an appropriate choice of the parameters the maximal possible error order, namely $p = 2s$.

Examples of implicit Runge–Kutta algorithms are the *trapezoidal rule*

$$\boxed{y_{k+1} = y_k + \frac{h}{2} \left[ f(x_k, y_k) + f(x_{x+1}, y_{k+1}) \right]}$$

with the error order $p = 2$, the single step procedure

$$\boxed{\begin{aligned} k_1 &= f\left( x_k + \frac{1}{2} h \, , \, y_k + \frac{1}{2} h k_1 \right), \\ y_{k+1} &= y_k + h k_1 \end{aligned}}$$

with the error order $p = 2$ and the two-step procedure

$$\begin{array}{c|cc} \dfrac{3-\sqrt{3}}{6} & \dfrac{1}{4} & \dfrac{3-2\sqrt{3}}{12} \\[2ex] \dfrac{3+\sqrt{3}}{6} & \dfrac{3+2\sqrt{3}}{12} & \dfrac{1}{4} \\[1ex] \hline & \dfrac{1}{2} & \dfrac{1}{2} \end{array}$$

with the maximal error order $p = 4$.

The stability properties of single step procedures are first and foremost analyzed by the linear *test initial value problem*

$$y'(x) = \lambda y(x), \quad y(0) = 1, \quad \lambda \in \mathbb{C},$$

in order to compare numerically computed solutions with, in particular, exponentially or oscillating decreasing solutions $y(x) = e^{\lambda x}$ in case $\operatorname{Re}(\lambda) < 0$. If one applies a Runge–Kutta method to the test initial value problem, then the result is a prescription

$$y_{k+1} = F(h\lambda) \cdot y_k, \qquad k = 0, 1, 2, \ldots,$$

where $F(h\lambda)$ is, for explicit methods, a polynomial in $h\lambda$, and for implicit methods, a rational function in $h\lambda$. In both cases, $F(h\lambda)$ is an approximation of $e^{h\lambda}$ for small arguments. This qualitative behavior of the numerically computed approximation $y_k$ only coincides with $y(x_k)$ for $\operatorname{Re}(\lambda) < 0$ if $|F(h\lambda)| < 1$. For this reason one defines the *domain of absolute stability* of a single step procedure as the set

$$B := \{\mu \in \mathbb{C} : |F(\mu)| < 1\}.$$

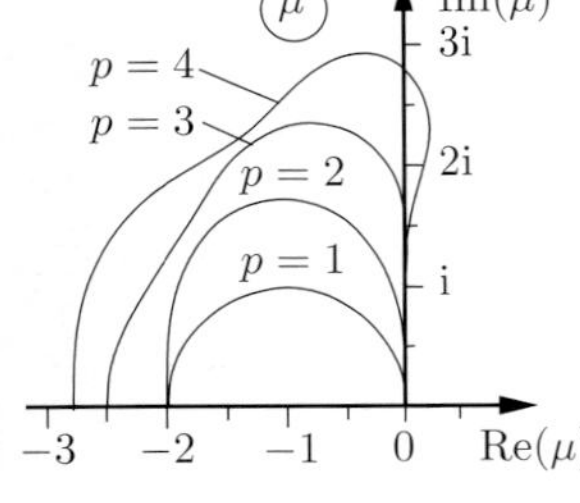


*Figure 7.1.*

For the explicit Runge–Kutta procedure with $s$ interpolation points and error order $p = 4$, we have

$$F(\mu) = 1 + \mu + \frac{1}{2}\mu^2 + \frac{1}{6}\mu^3 + \frac{1}{24}\mu^4, \qquad \mu = h\lambda,$$

which is the same as the first terms for the Taylor series of $e^{\lambda}$. The boundaries of the domains of absolute stability for explicit Runge–Kutta procedures with $x = p = 1, 2, 3, 4$ are shown in Figure 7.1 for symmetry reasons only in the upper half-plane. The domains of stability become larger as the order of error of the procedure grows.

The step width $h$ should be chosen in such a way that for $\operatorname{Re}(\lambda) < 0$ the stability condition $h\lambda = \mu \in B$ is satisfied. Otherwise the explicit Runge–Kutta procedure can yield useless results. The stability condition must be taken into account in the numerical integration of (linear) systems of differential equations, where the step width $h$ must be fixed in such a way as to insure that the constants $\lambda_j$, $j = 1, 2, \ldots, r$ satisfy $h\lambda_j \in B$. If the absolute values of the negative real parts of the $\lambda_j$ vary too much, then one speaks of a *stiff system of differential equations*. The condition on the absolute stability restricts in this case the step width $h$ quite strongly, even if the quantities $\lambda_j$ are very small in absolute value.

For the implicit trapezoidal rule and the single step Runge–Kutta method, we have

$$F(h\lambda) = \frac{2 + h\lambda}{2 - h\lambda} \quad \text{with} \quad |F(\mu)| = \left|\frac{2 + \mu}{2 - \mu}\right| < 1 \quad \text{for all} \quad \mu \quad \text{with} \quad \operatorname{Re}(\mu) < 0.$$

The domain of absolute stability is the entire left half of the complex plane. These two implicit methods are called *absolutely stable*, because the step width $h$ is not required to satisfy a stability condition. Also the Runge–Kutta method with two interpolation points is absolutely stable.

There are further, more refined, stability notions, in particular for non-linear differential equations, see [437] and [438].

#### 7.6.1.2 Multiple step procedures

If we apply the information of the approximation we have obtained from an algorithm with equidistant interpolation points $x_{k-1}, x_{k-2}, \ldots, x_{k-m+1}$ for the determination of the approximation $y_{k+1}$ at the point $x_{k+1}$, then we are working in the direction of *multiple step procedures*, which often yield a more efficient numerical solution of a differential equation. A general linear multiple step procedure is for $s := k - m + 1$

$$\sum_{j=0}^{m} a_j y_{s+j} = h \sum_{j=0}^{m} b_j f(x_{s+j}, y_{s+j}), \qquad m \geq 2$$

for $k \geq m - 1$, i.e., for $s \geq 0$, where the coefficients $a_j$ and $b_j$ are fixed. In the sense of a normalization let $a_m = 1$; then we are dealing with a genuine $m$-step procedure if $a_0^2 + b_0^2 \neq 0$. If $b_m = 0$, then the multiple step procedure is said to be *explicit*, otherwise *implicit*. In order to apply a $m$-step procedure, we require, in addition to the initial value $y_0$, $(m - 1)$ additional starting values $y_1, \ldots, y_{m-1}$, which can be determined for example with the aid of a single step procedure.

A linear multiple step procedure is said to have the *order* $p$, if for the *local discretization error* $d_{k+1}$ we have at an arbitrary point $\overline{x}$ the representation

$$\begin{aligned} d_{k+1} &:= \sum_{j=0}^{m} \{a_j y(x_{s+j}) - h b_i f(x_{s+j}, y(x_{s+j}))\} \\ &= c_0 y(\bar{x}) + c_1 h\, y'(\bar{x}) + c_2 h^2 y''(\bar{x}) + \ldots + c_p h^p y^{(p)}(\bar{x}) + c_{p+1} h^{p+1} y^{(p+1)}(\bar{x}) + \ldots \end{aligned}$$

with $c_0 = c_1 = c_2 = \cdots = c_p = 0$ and $c_{p+1} \neq 0$. Both $p$ and the coefficient $c_{p+1}$ are independent of the chosen point $\overline{x}$. A linear multiple step procedure is said to be *consistent*, if its order of error $p$ is at least unity. There are two characteristic polynomials which are associated with a $m$-step procedure,

$$\varrho(z) := \sum_{j=0}^{m} a_j z^j \quad \text{and} \quad \sigma(z) := \sum_{j=0}^{m} b_j z^j.$$

With these, the consistency condition can be written as

$$c_0 = \varrho(1) = 0, \qquad c_1 = \varrho'(1) - \sigma(1) = 0.$$

The consistence of a multiple step procedure alone is not sufficient to insure the convergence of approximations for $h \to 0$; rather, the procedure must in addition be what is called *null stable*. This means that the characteristic polynomial $\varrho(z)$ has to satisfy the *root condition*, which states that the zeros of this polynomial have absolute value at most unity and any multiple roots which it has lie in the interior of the unit circle (i.e., not on the unit circle itself). For a consistent and null consistent multiple step procedure of error order $p$, we have, under the assumption on the initial values

$$\max_{0 \leq i \leq m-1} |y(x_i) - y_i| \leq K \cdot h^p, \qquad 0 \leq K < \infty,$$

as estimate on the global error $g_n := y(x_n) - y_n$

$$|g_n| = O(h^p), \qquad n \geq m.$$

Under the mentioned assumptions the approximations $y_n$ converge to order $p$ to the solution $y(x_n)$.

The most often occuring multiple step procedure are the *methods of Adams*, which arise from the interpolation quadrature formula for the integral equation which is equivalent to the given differential equation. Examples of these procedures are the *Adams–Bashforth method*, with $f_l := f(x_l, y_l)$:

$$y_{k+1} = y_k + \frac{h}{12}\big[23f_k - 16f_{k-1} + 5f_{k-2}\big],$$

$$y_{k+1} = y_k + \frac{h}{24}\big[55f_k - 59f_{k-1} + 37f_{k-2} - 9f_{k-3}\big],$$

$$y_{k+1} = y_k + \frac{h}{720}\big[1\,901f_k - 2\,774f_{k-1} + 2\,616f_{k-2} - 1\,274f_{k-3} + 251f_{k-4}\big].$$

Every $m$-step procedure in this family has the error order $p = m$. For each integration step, only a single functional evaluation is required.

*Implicit Adams–Moulton methods* are given by the formula

$$y_{k+1} = y_k + \frac{h}{24}\big[9f(x_{k+1}, y_{k+1}) + 19f_k - 5f_{k-1} + f_{k-2}\big],$$

$$y_{k+1} = y_k + \frac{h}{720}\big[251f(x_{k+1}, y_{k+1}) + 646f_k - 264f_{k-1} + 106f_{k-2} - 19f_{k-3}\big],$$

$$y_{k+1} = y_k + \frac{h}{1\,440}\big[475f(x_{k+1}, y_{k+1}) + 1\,427f_k - 798f_{k-1} + 482f_{k-2} - 173f_{k-3} + 27f_{k-4}\big].$$

An Adams–Moulton $m$-step procedure has an order of error $p = m + 1$ and requires at each integration step solving the implicit equation with respect to $y_{k+1}$, for example by application of an appropriate fixed-point iteration (see 7.4.2.1). A good initial value for this iteration can be obtained from the Adams–Bashforth method.

A *prediction–correction procedure* is a combination of an explicit and an implicit multiple step procedure, and works as follows. Using the explicit procedure yields a *prediction formula*, which is substituted into the *correction formula* and improved in the sense of a step of a fixed-point iteration. If one combines multiple step procedures of different orders in this way, then the local discretization error of the prediction–correction method is equal to the smallest absolute value of that of the implicit procedure used.

The ABM43 procedure is a combination of the four-step method of Adams–Bashforth with the three-step method of Adams–Moulton, each of which has order $p = 4$. In this case the algorithm is defined by

$$y_{k+1}^{(\mathrm{P})} = y_k + \frac{h}{24}\big[55f_k - 59f_{k-1} + 37f_{k-2} - 9f_{k-3}\big],$$

$$y_{k+1} = y_k + \frac{h}{24}\big[9f(x_{k+1}, y_{k+1}^{(\mathrm{P})}) + 19f_k - 5f_{k-1} + f_{k-2}\big].$$

Of course it requires initial starting values and for each integration step two functional evaluations. The advantage of prediction–correction procedures like these is that the order can be slightly increased while the number of required functional evaluations stays two.

Another important class of implicit multiple step procedures are the *backward differentiation methods*, BDF for short, because these have special stability properties. The first derivative is approximated in the differential equation at a point $x_{k+1}$ by a differentiation formula, which uses the interpolation values at this interpolation point and the preceding (equidistant) interpolation points (see 7.3.2). The simplest example of this class of procedures is the *backward Euler procedure*

$$\boxed{y_{k+1} - y_k = hf(x_{k+1}, y_{k+1}).}$$

The $m$-step BDF procedures for $m = 2, 3, 4$ are

$$\begin{aligned}
&\frac{3}{2}y_{k+1} - 2y_k + \frac{1}{2}y_{k-1} &&= hf(x_{k+1}, y_{k+1}),\\
&\frac{11}{6}y_{k+1} - 3y_k + \frac{3}{2}y_{k-1} - \frac{1}{3}y_{k-2} &&= hf(x_{k+1}, y_{k+1}),\\
&\frac{25}{12}y_{k+1} - 4y_k + 3y_{k-1} - \frac{4}{3}y_{k-2} + \frac{1}{4}y_{k-3} &&= hf(x_{k+1}, y_{k+1}).
\end{aligned}$$

The basic stability properties of the multiple step procedures are studied with the help of the linear *test initial value problem* $y'(x) = \lambda y(x),\ y(0) = 1$. Substituting the right-hand side in a general $m$-step procedure yields a difference equation of order $m$,

$$\sum_{j=0}^{m} (a_j - h\lambda b_j) y_{s+j} = 0.$$

Setting $y_k = z^k,\ z \neq 0$ leads to the corresponding *characteristic equation*

$$\varphi(z) := \sum_{j=0}^{m} (a_j - h\lambda b_j) z^j = \varrho(z) - h\lambda\sigma(z) = 0.$$

The general solution of the difference equation of order $m$ has, in the only case of interest here, when $\mathrm{Re}\,(\lambda) < 0$, the same asymptotic behavior as the exact solution $y(x)$ if and only if the zeros $z_i$ of the characteristic equation all have absolute value less then unity. The *domain of absolute stability* of a multiple step procedure is the set of complex values $\mu = h\lambda$ for which $\varphi(z)$ only has solutions $z_i \in \mathbb{C}$ which lie in the interior of the unit circle.

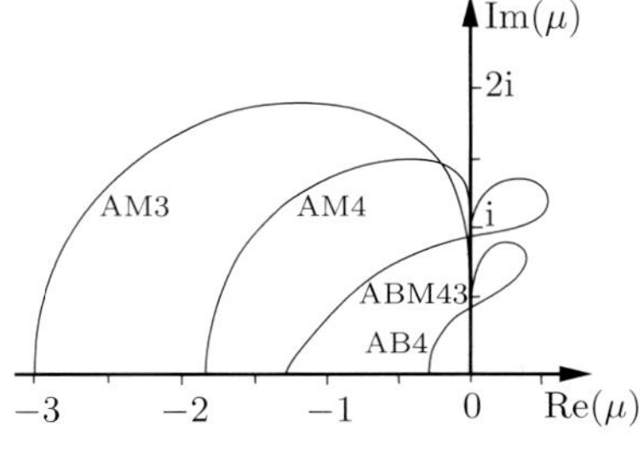


*Figure 7.2.*

The characteristic equation of a prediction–correction method is a combination of the characteristic polynomials of the corresponding multiple step procedures. For example, for the ABM43 method, we have

$$\varphi_{\mathrm{ABM43}}(z) = z[\varrho_{\mathrm{AM}}(z) - \mu\sigma_{\mathrm{AM}}(z)] + b_3^{(\mathrm{AM})}\mu[\varrho_{\mathrm{AB}}(z) - \mu\sigma_{\mathrm{AB}}(z)] = 0,$$

where $b_3^{(AM)}$ is the coefficient of the Adams–Moulton method.

In Figure 7.2 the boundary curves of the domains of absolute stability are depicted (for reasons of symmetry, only the part in the upper half plane are shown); the labels show which method is meant.

The domains of absolute stability of implicit BDF methods have interesting properties. For example, the backward Euler procedure is absolutely stable, as the domain of absolute stability is the left half of the complex plane. The same is true for the two-step BDF method, because in this case the left half of the complex plane is completely contained in the domain of absolute stability. The other BDF methods are no longer absolutely stable, because the boundary curve is partially in the left half of the complex plane. There does exist, however, a maximal angle domain with half-opening angle $\alpha > 0$ for which the vertex is the origin, and these domains lie completely in the domain of absolute stability. For this reason one speaks of $A(\alpha)$*-stable* procedure. The three-step BDF method is $A(88°)$-stable, while the four-step BDF method is $A(72°)$-stable.

## 7.6.2 Boundary value problems

### 7.6.2.1 Analytic methods

As an aid in finding the solution of the linear boundary value problem

$$L[y] := \sum_{i=0}^{r} f_i(x)y^{(i)}(x) = g(x),$$

$$U_i[y] := \sum_{j=0}^{r-1} \{\alpha_{ij}y^{(j)}(a) + \beta_{ij}y^{(j)}(b)\} = \gamma_i, \qquad i = 1, 2, \ldots, r$$

in a given interval $[a, b]$ with continuous functions $f_i(x), g(x)$ and $f_r(x) \neq 0$, one can use the fact that every solution of the differential equation can be written as a linear combination

$$y(x) = y_0(x) + \sum_{k=1}^{r} c_k y_k(x),$$

in which $y_0(x)$ is a special solution of the inhomogeneous differential equation $L[y] = g$ and the functions $y_k(x)$, $k = 1, 2, \ldots, r$ form a *fundamental system* of the homogeneous equation $L[y] = 0$ (see section 1.12.6). These $(r + 1)$ functions can be approximated by a numerical integration of the $(r + 1)$ initial value problems with initial conditions

$$\begin{aligned} &y_0(a) = y_0'(a) = \ldots = y_0^{(r-1)}(a) = 0, \\ &y_k^{(j)}(a) = \delta_{k,j+1}, \qquad k = 1, 2, \ldots, r; \quad j = 0, 1, \ldots, r-1. \end{aligned}$$

Since the *Wronski determinant* $W(a)$ is equal to unity, the constructed functions $y_1(x)$, $y_2(x), \ldots, y_r(x)$ are linear independent. With these $(r+1)$ functions the coefficients $c_k$ in the linear expansion above can by determined from the system of linear inhomogeneous equations

$$\sum_{k=1}^{r} c_k U_i[y_k] = \gamma_i - U_i[y_0], \qquad i = 1, 2, \ldots, r.$$

An approximation of the linear boundary value problem is often determined with the *Ansatz method*, in which the solution $y(x)$ we are to determine is hypothesized to have the form

$$Y(x) := w_0(x) + \sum_{k=1}^{n} c_k w_k(x),$$

where $w_0(x)$ is a function which satisfies the inhomogeneous boundary conditions $U_i[w_0]= \gamma_i$, $i = 1, 2, \ldots, r$, while the linear independent functions $w_k(x)$, $k = 1, 2, \ldots, n$ satisfy the homogeneous boundary condition $U_i[w_k] = 0$. Thus $Y(x)$ satisfies for arbitrary $c_k$ the boundary value condition. If one substitutes this expansion into the differential equation, then one obtains an *error function*

$$\varepsilon(x; c_1, c_2, \ldots, c_n) := L[Y] - g(x) = \sum_{k=1}^{n} c_k L[w_k] + L[w_0] - g(x).$$

The unknown coefficients $c_k$ of the approximation $Y(x)$ can be determined as the solution of a linear system of equations, which is obtained from one of the following conditions on the error function.

1. *Location method:* According to the choice of $n$ appropriately located points $a \le x_1 < x_2 < \ldots < x_n \le b$, we require that

$$\varepsilon(x_i; c_1, c_2, \ldots, c_n) = 0, \qquad i = 1, 2, \ldots, n.$$

2. *Partial interval method:* The interval $[a, b]$ is subdivided into subintervals with $a = x_0 < x_1 < x_2 < \ldots < x_{n-1} < x_n = b$, and we require that the mean of the error function vanishes in each of the subintervals, i.e., that

$$\int_{x_{i-1}}^{x_i} \varepsilon(x; c_1, c_2, \ldots, c_n)\, \mathrm{d}x = 0, \qquad i = 1, 2, \ldots, n.$$

3. *Square mean error:* In the continuous case, this is the requirement

$$\int_a^b \varepsilon^2(x; c_1, c_2, \ldots, c_n)\, \mathrm{d}x \stackrel{!}{=} \min.,$$

while in the discrete case with $N$ interpolation points $x_i \in [a, b]$, $N > n$, the minimization

$$\sum_{i=1}^{N} \varepsilon^2(x_i; c_1, c_2, \ldots, c_n) \stackrel{!}{=} \min.,$$

leads to the corresponding normalized system of equations.

4. *Method of Galerkin:* The error function should be orthogonal to an $n$-dimensional subspace given by $U := \operatorname{span}(\nu_1, \nu_2, \ldots, \nu_n)$, i.e.,

$$\int_a^b \varepsilon(x; c_1, c_2, \ldots, c_n)\, \nu_i(x)\, \mathrm{d}x = 0, \qquad i = 1, 2, \ldots, n.$$

Generally speaking, one has $\nu_i(x) = w_i(x)$, $i = 1, 2, \ldots, n$. The *Finite element method*, in which the functions $w_i(x)$ are chosen very specially, with small support, is the modern form of the method of Galerkin.

#### 7.6.2.2 Reduction to initial value problems

A favorite method of solving a non-linear boundary value problem of second order

$$y''(x) = f\big(x, y(x), y'(x)\big)$$

with *separate* linear boundary conditions

$$\alpha_0 y(a) + \alpha_1 y'(a) = \gamma_1, \qquad \beta_0 y(b) + \beta_1 y'(b) = \gamma_2,$$

is to reduce the problem to an *initial value problem.* For this we consider the initial value conditions

$$y(a) = \alpha_1 s + c_1\gamma_1, \qquad y'(a) = -(\alpha_0 s + c_0\gamma_1),$$

which depend on a parameter $s$ and in which $c_0$ and $c_1$ are constants, satisfying the condition $\alpha_0 c_1 - \alpha_1 c_0 = 1$. The solution of the initial value problem (calculated numerically) is denoted by $Y(x; s)$. This function satisfies, for all $s$ for which $Y(x; s)$ exists, the initial value condition at the point $a$. To find a solution of the boundary value problem, we require that in addition the second boundary condition must be satisfied. Thus, $Y(x; s)$ must satisfy the equation

$$h(s) := \beta_0 Y(b; s) + \beta_1 Y'(b; s) - \gamma_2 = 0.$$

This equation, which is in general non-linear in $s$, can be solved with the regula falsi, with the secant method or with the Newton method. In the last case, the required derivative $h'(s)$ can be approximated as a difference quotient, by also determining $h(s+\Delta s)$ through integration.

The *simple reduction procedure* just discussed can be generalized to systems of differential equations of higher order with correspondingly more parameters in the initial conditions. In addition to the difficulties in finding appropriate starting values for the iteration, in some applications the strong sensitivity of $Y(b; s)$ with respect to small changes in the value of $s$ can create problems. In improve the condition of the problem, in a *multiple step reduction* the interval $[a, b]$ is subdivided into several subintervals, then for each of these subintervals, a set of initial conditions, which are to be determined, are used as parameters and the differential equation is solved in each of the subintervals. The parameters introduced in this procedure can be determined from a non-linear system of equations in such a way that the partial solutions in the subintervals can be put together to yield a solution in the original interval.

#### 7.6.2.3 Difference methods

The principle approach of the difference method will be illustrated on the example of a non-linear boundary value problem of second order

$$\begin{aligned} y''(x) &= f\big(x, y(x), y'(x)\big) \\ y(a) &= \gamma_1, \quad y(b) = \gamma_2. \end{aligned}$$

The given interval $[a, b]$ is subdivided into $(n+1)$ subintervals of length $h := (b-a)/(n+1)$ with equidistant interpolation points $x_i = a + ih$, $i = 0, 1, 2, \ldots, n+1$. The unknown is an approximation $y_i$ for the exact solution $y(x_i)$ of the equation at the $n$ interpolation points. For this, the first and second derivative are approximated by the central (resp. the second) difference quotient

$$y'(x_i) \approx \frac{y_{i+1} - y_{i-1}}{2h}, \qquad \left(\text{resp. } y''(x_i) \approx \frac{y_{i+1} - 2y_i + y_{i-1}}{h^2}\right)$$

at every interior interpolation point, so that both have the same discretization error $O(h^2)$. With this approximation, one obtains a system of non-linear equations

$$\frac{y_{i+1} - 2y_i + y_{i-1}}{h^2} = f\Big(x_i,\, y_i,\, \frac{y_{i+1} - y_{i-1}}{2h}\Big), \qquad i = 1, 2, \ldots, n$$

for the $n$ unknowns $y_1, y_2, \ldots, y_n$, where it needs to be taken account of that $y_0 = \gamma_1$ and $y_{n+1} = \gamma_2$ are given as boundary conditions.

Under appropriate assumptions on the solution $y(x)$ and on the function $f(x, y, y')$, it can be shown that for the resulting approximation we have an error estimate of the form

$$\max_{1 \le i \le n} |y(x_i) - y_i| = O(h^2).$$

The non-linear system of equations is usually solved with the Newton–Kantorovich method or with one of its simplifying variants as discussed in section 7.4.2.2. The special structure

$$\begin{array}{llll}
2y_1 \quad -y_2 & +h^2 f\left(x_1, y_1, \dfrac{y_2 - y_1}{2h}\right) & -\gamma_1 & = 0 \\
-y_1 + 2y_2 \quad -y_3 & +h^2 f\left(x_2, y_2, \dfrac{y_3 - y_1}{2h}\right) & & = 0 \\
\qquad -y_2 + 2y_3 - y_4 & +h^2 f\left(x_3, y_3, \dfrac{y_4 - y_2}{2h}\right) & & = 0 \\
 & \ldots & & \\
\qquad -y_{n-1} + 2y_n & +h^2 f\left(x_n, y_n, \dfrac{y_2 - y_{n-1}}{2h}\right) & -\gamma_2 & = 0.
\end{array}$$

with maximally three successive indexed unknowns in each equation implies that the functional matrix of the system is actually *tridiagonal*. Therefore, the computation cost for the calculation of a correction vector in the Newton–Kantorovich method is only proportional to $n$. If the differential equation is even linear, then the difference method leads directly to a linear system of equations with a tridiagonal matrix for the unknown functional values.

## 7.7 Partial differential equations and scientific computation

*The effective numerical treatment of partial differential equations is not a handicraft, but an art.*

*Folklore*

### 7.7.1 Basic ideas

The breath-taking speed of development of computer technology in the second half of the twentieth century has opened a new chapter in the history of mathematics. Whole new questions centered around the notions of stability, flexible discretization procedures, quick algorithms and adaptivity have been brought into the stage light.

Problems that in the age preceeding computer availability could only be treated for small values of the dimension $n$ can now be calculated also for large $n$, making the behavior as $n \to \infty$ a tractable and fascinating domain. For example, it turns out that the standard polynomial approximation, popular since Newton, is instable as $n \to \infty$ (where $n$ denotes the degree of approximating polynomial) and therefore is a useless procedure for large $n$. The notion of stability is in particular for discretization methods a fundamental one. Not all intuitive discretization methods for dealing with ordinary or partial differential equations are stable (see 7.7.5.4 below). Especially the attempt to defined approximations of higher order leads quickly to instable procedures, as discussed in 7.7.5.6.2 below. For mixed finite element methods, which are discussed in section 7.7.3.2.3, one has the unfortunate situation that precisely very good approximations turn out to instable. In many cases this instability leads to runaway error propagation in the algorithm, which is therefore easy to spot. But in the situation of mixed finite element methods it occurs that a different kind of instability plagues the algorithm, which is then not immediately visible. For this reason, it is always important to accompany the algorithm with a careful numerical analysis.

As computers aquire more and more computational power, more and more complicated problems are attempted to be solved with their help. This complexity can, for example, consists of certain details of the solutions, like boundary layers in the case of singularly perturbed problems, singular behavior of solutions or their derivatives at certain points, microscopic details of solutions in concurrence with turbulence, discontinuity in the case of hyperbolic differential equations and large jumps of coefficients, for example in the case of semiconductors. The treatment of these phenomena requires procedures which are adapted to the specialties of the situation. In reality one is miles away from any kind of an open 'black-box' procedure which would apply to all problems.

The increase in computer capabilities has both the aspect of faster processors and of more and cheaper storage. Paradoxically, precisely this situation leads to an increased need for faster algorithms. If, for example, an algorithm has computational cost of $n^3$ operations for a problem of $n$ dimensions, then the increase of storage by a factor of ten leads to an increase in computational cost of a thousand, which can not be compensated by the increase in computational power of the processors. The fast algorithms which will be introduced in this section are the multi-lattice procedure for the fast solution of system of equations and the fast Fourier and wavelet transformation.

Instead of accelerating the algorithms, one can also try to decrease the dimension of the problem without influencing the quality of solutions negatively. For the solution of partial differential equations this means that instead of a uniform lattice for discretizing a problem, one uses a lattice which is finer "where the action is", leaving it coarse elsewhere. Driving this non-uniform sizing of a lattice is the data produced by previous steps in the algorithms, resulting in an interesting interaction between the numerical analysis and the algorithmic process. We present a brief introduction to this circle of ideas in section 7.7.6.

## 7.7.2 An overview of discretization procedures

### 7.7.2.1 Difference equations

The method of difference equations is based on replacing the derivatives which occur in differential equations by difference quotients. For this one requires a *lattice*, generally chosen to be regular. For the interval $[a, b]$ and the step width $h := (b-a)/N$, $N = 1, 2, \ldots$, the equidistant lattice is

$$\boxed{G_h := \{x_k = a + kh : 0 \le k \le N\}.}$$

For partial differential equations of $d$ independent variables, one requires a $d$-dimensional lattice on the domain of definition $D \subset \mathbb{R}^d$, which in the case of equidistant lattice points takes the form

$$G_h = \{x \in D : x = x_k = kh : k = (k_1, \ldots, k_d),\ k_i \text{ integer}\}$$

(see Figure 7.3).

For a difference method, as we shall refer to this method henceforth, the *lattice points* are the main object of interest, rather than the associated rectangles or edges (for $d = 2$, see Figure 7.3). One uses the functional values $u(x_k)$ at the lattice points $x_k \in G_h$ for approximating the derivatives. Since most difference approximations are one-dimensional, it is generally sufficient, at least as an introduction, to discuss functions of a single variable.

The first derivative of a (smooth) function $u$ can be approximated in different ways. The *forward difference*

$$\boxed{\partial_h^+ u(x_k) := \frac{1}{h}\big(u(x_{k+1}) - u(x_k)\big)} \tag{7.1a}$$

as well as the *backward difference*

$$\boxed{\partial_h^- u(x_k) := \frac{1}{h}\big(u(x_k) - u(x_{k-1})\big)} \tag{7.1b}$$

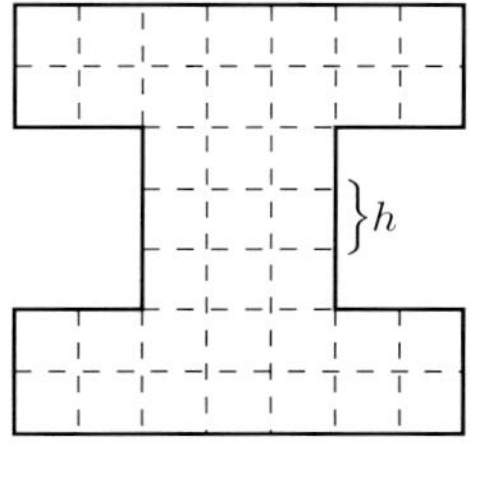


*Figure 7.3. Grid with step h.*

are examples of so-called one-sided differences. They are only of first order, i.e., satisfy

$$u'(x_k) - \partial_h^{\pm} u(x_k) = O(h), \qquad h \to 0. \tag{7.1c}$$

The *central difference* or *symmetric difference*

$$\boxed{\partial_h^0 u(x_k) := \frac{1}{2h}\big(u(x_{k+1}) - u(x_{k-1})\big)} \tag{7.2a}$$

is of second order, i.e.,

$$u'(x_k) - \partial_h^0 u(x_k) = O(h^2), \qquad h \to 0. \tag{7.2b}$$

The second derivative can be approximated by

$$\boxed{\partial_h^2 u(x_k) := \frac{1}{h^2}\big(u(x_{k+1}) - 2u(x_k) + u(x_{k-1})\big).} \tag{7.3a}$$

This second difference is also of second order:

$$u''(x_k) - \partial_h^2 u(x_k) = O(h^2), \qquad h \to 0. \tag{7.3b}$$

One should note that not all differences at the boundary points $x_0 = a$ or $x_N = b$ are well-defined, since the necessary neighboring lattice points are missing.

In the two-dimensional case one needs to use the subset $G_h$ of the infinite lattice $\{(x, y) = (kh, lh) : k, l \text{ integer}\}$. The differences (7.1a,b), (7.2a), (7.3a) can be applied in the $x$ as well as in the $y$ direction; correspondingly we use the notations $\partial_{h,x}^+$, $\partial_{h,y}^+$, etc. In Figure 7.4 the second difference $\partial_{h,x}^2 u$, which uses the values at the lattice points $A, B, C$, and the second difference $\partial_{h,y}^2 u$, which uses the lattice points $D, E, F$, are used. The sum of these two differences yield an approximation for the Laplace operator $\Delta u = u_{xx} + u_{yy}$:

$$\boxed{\begin{aligned} \Delta_h u(kh, lh) &= \big(\partial_{h,x}^2 + \partial_{h,y}^2\big) u(kh, lh) \\ &= \frac{1}{h^2}(u_{k-1,l} + u_{k+1,l} + u_{k,l-1} + u_{k,l+1} - 4u_{kl}), \end{aligned}} \tag{7.4}$$

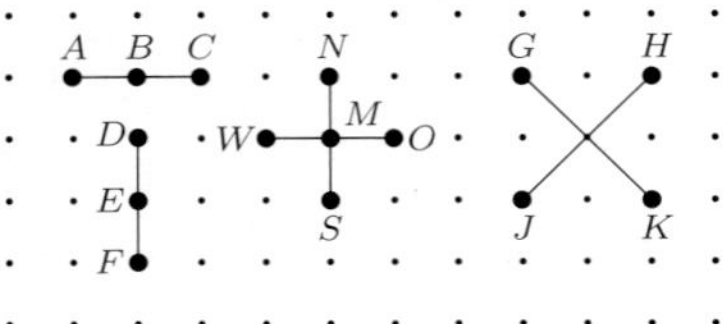


*Figure 7.4. Difference approximations.*

where $u_{kl} := u(kh, lh)$. Because of the five lattice points which are used, visualized as the points $M, N, S, O, W$ in Figure 7.4, equation (7.4) is also referred to as the *five-point formula.*

With the differences considered above, all partial derivatives of the form $u_x, u_y, u_{xx}, u_{yy}, \Delta u$ can be approximated. The mixed second derivative $u_{xy}$ can be approximated by the product $\partial^0_{h,x}\partial^0_{h,y}$:

$$\frac{1}{4h^2}(u_{k+1,l+1} + u_{k-1,l-1} - u_{k+1,l-1} - u_{k-1,l+1}) = u_{xy}(kh, lh) + O(h^2), \qquad h \to 0$$

(see the vertices $G, H, J, K$ in Figure 7.4). There is an abbreviating notation known as the star notation, which is described in [442]. The generalization of the difference approximations in the case of $d$ independent variables with a $d$-dimensional lattice is obvious. Thus, also higher derivatives than just the second can be approximated.

Although we have kept to equidistant lattices up to now, there is no difficulty in accommodating $d$-dimensional lattices with different step widths $h_i$ in the different $x_i$-directions. If the step width in one of the axis directions is not equidistant, then the derivatives in that directions can be approximated with the Newton divided differences method. A general, irregular lattice with no structure in the direction of the base coordinates is, however, not appropriate for the procedures to be discussed here, as one requires collinear lattice points for the calculation of difference approximations of second derivatives. This makes it quite clear that the rigidity of a geometric lattice structure causes the difference method to be inflexible, so that particulars like a local subdivision of the lattice are difficult to accommodate with this method.

#### 7.7.2.2 Ritz–Galerkin procedure

If $Lu = f$ is the differential equation and $(u, v) := \int\limits_D uv\,\mathrm{d}x$ is the $L^2$-scalar product on the domain of definition $D$, the solution $u$ necessarily also satisfies the equation

$$\boxed{(Lu, v) = (f, v)} \tag{7.5}$$

for all *test functions* $v$. In general one rewrites the left-hand side of (7.5) using integration by parts, leading to the so-called *weak formulation* (*variational formulation*) of the differential equation:

$$\boxed{a(u, v) = f(v).} \tag{7.6}$$

Here $a$ denotes a *bilinear form* and $f$ denotes a functional on appropriate function spaces $U$ and $V$, in which $u$ and $v$ vary. In what follows we restrict ourselves to the standard case $U = V$.

The *Ritz–Galerkin method* approximates, instead of the differential operator $L$, the entire space $V$, by replacing $V$ with a finite-dimensional function space $V_n$, $n = \dim V_n$, and reformulating the problem:

$$\boxed{\text{determine } u \in V_n \quad \text{with} \quad a(u, v) = f(v) \quad \text{for all} \quad v \in V_n.} \tag{7.7}$$

Boundary conditions of the Dirichlet type are so-called essential conditions and are part of the definition of $V_n$. Other boundary conditions, the so-called *natural boundary conditions*, which arise from the variational formulation of the problem, are not explicitly part of the formulation and are only approximately satisfied (see [442]).

For the concrete numerical calculation one chooses a basis $\{\varphi_1, \varphi_2, \dots, \varphi_n\}$ of $V_n$. The solution $u$ of (7.7) is to be determined in the form $\sum \xi_k \varphi_k$. The problem (7.7) is then equivalent to the linear system of equations

$$\boxed{Ax = b} \tag{7.8a}$$

where $x$ contains the coefficients $\xi_k$ which one wishes to determine and the so-called *stiffness matrix* $A$ and the right-hand side $b$ is defined by

$$A = (a_{ik}), \qquad a_{ik} = a(\varphi_k, \varphi_i), \qquad b = (b_i), \qquad b_i = f(\varphi_i). \tag{7.8b}$$

Since the residue $r = Lu - f$ in the weighting $(r, \varphi_i)$ vanishes (see 7.5), the method is also referred to as the 'procedure of the weighted residues'.

#### 7.7.2.3 Finite element methods (FEM)

The finite element method, abbreviated FEM, is the Ritz–Galerkin procedure with special function spaces, which are accordingly called the finite elements (FE). In general, the stiffness matrices of the Galerkin procedures can be fully populated. In order to reduce these to sparse matrices as in difference methods, one tries to use basis elements $\varphi_k$ with as small a support as possible. (The *support* of a function $\varphi$ is the closure of the set of $x$ with $\varphi(x) \neq 0$.) The effect of this is that the functions which occur in $a(\varphi_k, \varphi_i)$ have, with few exceptions, disjoint support and therefore $a_{ik} = 0$. The requirement is for function spaces with global polynomials or other globally defined functions not satisfied. Instead, one uses here functions which are defined piecewise. Their definition contains two aspects: (a) the geometric elements (a disjoint decomposition of the domain of definition), and (b) the analytic functions defined on these parts of the domain.

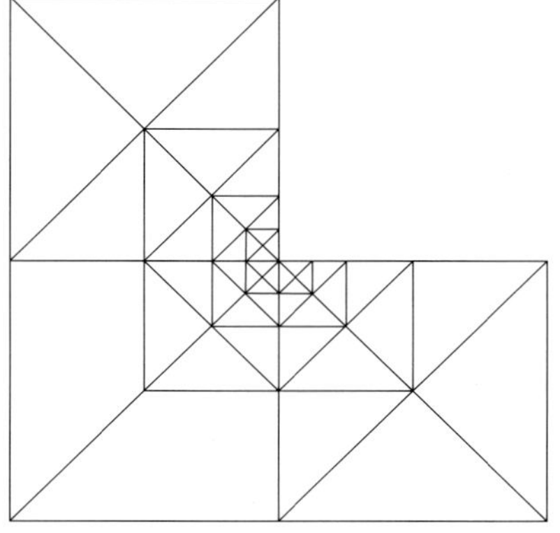

Figure 7.5. An irregular triangulation.

A typical example for the geometric elements is the decomposition of a two-dimensional domain of definition into triangles (*triangulation*). These triangle can have some regular structure, an example of which arises after dividing all the rectangles in Figure 7.3 in to two triangles each, but the structure can also be irregular as in Figure 7.5. A triangulation is said to be *admissible* if the intersection of two distinct triangles is either empty or a common vertex or edge. A triangulation is said to be *quasiuniform*, if the ratio of the size of the triangles (measured on the longest side) is bounded. A triangulation is said to be *regular*, if for all triangles the ratio of the outer to the inner radius is uniformly bounded.

Given a triangulation, different functions can be defined on the triangles. Examples are piecewise constant functions (in which case the dimension $n$ of $V_n$ is the number of triangles), piecewise linear functions (affine on each triangle, globally continuous; this forms a space of dimension equal to the number of vertices of the triangulation), or piecewise quadratic functions (in which the dimension is equal to the sum of the number of edges and vertices).

Instead of triangles, one can also take rectangles. Moreover there are analogs in three dimensions (with tetrahedra instead of triangles, cubes instead of squares, etc.). For more details on finite element methods we recommend [440].

Since the triangles (and squares, etc.) of the triangulation go into the finite element equations (7.8a) and (7.8b) as domains of integration, for the finite element methods the more important aspect of the lattice is the surfaces, rather than the vertices or edges.

An appropriate method for generating a triangulation consists in starting with a coarse triangulation and then making it finer, mentioned in sections 7.7.6.2 and 7.7.7.6 below.

#### 7.7.2.4 The Petrov–Galerkin procedure

If the functions $u, v$ in (7.6) are in *different* spaces $U$ (the function space) and $V$ (the space of test functions), then one gets a generalization of the Ritz–Galerkin procedure, called the Petrov–Galerkin procedure.

#### 7.7.2.5 Finite volume methods

The finite volume method (sometimes also referred to as the box-method) is a mixture of the difference method and the finite element method. As for difference procedures, one often uses square lattices as in Figure 7.3, and the interest is in the flows along the edges of that lattice.

To formulate the procedure mathematically, one chooses in (7.5) $v$ to be the characteristic function of an element $E$ (i.e., $v = 1$ on a square $E$, otherwise $v = 0$). The left-hand side in (7.5) is the integral $\int\limits_E Lu\,\mathrm{d}x$ over $E$. Integration by parts yields boundary integrals over the sides of the rectangles $\partial E$, which can be approximated in different ways.

If the differential operator $L$ has the form $Lu = \mathbf{div}\, Mu$ (for example $M = \mathbf{grad}$), then integration by parts yields

$$\int\limits_{\partial E} (Mu)\mathbf{n}\,\mathrm{d}F = \int\limits_E f\,\mathrm{d}x,$$

where $(Mu)\mathbf{n}$ is the scalar product with the outer normal unit vector $\mathbf{n}$. If the normal vector of a common side of two of the elements has the opposite sign in each element, then the sum of the all elements $E$ represents the *conservation law* on the domain of definition $D$:

$$\int\limits_{\partial D} (Mu)\mathbf{n}\,\mathrm{d}F = \int\limits_D f\,\mathrm{d}x.$$

This conservation property is often the decisive reason for the choice of the finite volume method. To justify the name, consider here the three-dimensional case with cubes as 'finite volumes'.

#### 7.7.2.6 Spectral methods and collocation

The finite element method uses the approximation by piecewise polynomials of fixed degree, while the size of the elements is decreased. With this procedure, only an approximation of *fixed* order can be achieved. Indeed, typically errors of the form $Ch^p$ arise, where $h$ denotes the size of the elements and $p$ is the maximal possible order. In the two-dimensional case $h$ and the dimension $n$ of the finite element space $V_n$ are connected by the relation $n = C/h^2$. Thus the error is as a function of dimension $O(n^{-p/2})$. Estimates by $O(\mathrm{e}^{-\alpha n^b})$, $\alpha > 0$, $b > 0$ describe, on the other hand, the *exponential* speed of convergence, which can be achieved with global polynomials or trigonometric functions in the approximation of smooth solutions.

The spectral method uses these global functions to an extent, combined with special geometries (for example rectangles), while the discrete equations are derived with *collocation.* This denotes the process in which the differential equation $Lu = f$ is only required to hold at certain collocation points instead of in the entire domain. Formally, the Petrov–Galerkin method can be interpreted in terms of distributions on the test space.

The disadvantage of the spectral method is the fact that the matrix is fully filled and that the domain of definition is required to have a special structure. In addition, the required smoothness of the solution is in general not true globally.

#### 7.7.2.7 *h*-, *p*- and *hp*-methods

The usual finite element method, in which the step width $h$ approaches zero, is also called the $h$-method. If, on the other hand, one fixes the size of the lattice, but lets the degree $p$ of the piecewise polynomial functions grow as in the spectral method, then one speaks of the $p$-method.

A combination of both of these methods is the so-called $hp$-method. In this case, the function space consists of finite element functions of degree $p$ on the geometric elements of size $h$. If one adapts both $h$ and $p$ to the problem at hand, then one obtains a very precise approximation. The way in which the quantities $h$ and $p$ are locally chosen is a typical topic of adaptive discretization, as it is sketched in 7.7.6. The test space here is taken to be identical with the function space, so that it is, in fact, a special case of the Ritz–Galerkin method.

## 7.7.3 Elliptic differential equations

### 7.7.3.1 Positive definite boundary value problems

Scalar differential equations in general lead to the problems we now take up. Systems of differential equations, however, can be of the type of a saddle point problem, which puts new requirements on the finite element discretization, which will be discussed in 7.7.3.2.

**7.7.3.1.1 Model cases (the Poisson and Helmholtz equations).** Let $\Omega$ be a bounded domain of $\mathbb{R}^2$ with boundary $\Gamma := \partial\Omega$. The Laplace operator is defined by $\Delta u := u_{xx} + u_{yy}$. The prototype of all differential equations of order two is given by the *Poisson equation*

$$\boxed{-\Delta u = f \text{ on } \Omega.} \tag{7.9a}$$

The function $f = f(x, y)$ is considered as given; this is the source. The problem is to determine the function $u = u(x, y)$. A boundary problem is formulated by extending the differential equation (7.9a) by a boundary condition, for example by a *Dirichlet condition*

$$\boxed{u = g \text{ on the boundary } \Gamma.} \tag{7.9b}$$

The solution of this boundary value problem is uniquely determined. If $f = 0$, then the equation is a *Laplace-* or *potential equation*, for which the following maximum principle holds: the solution $u$ takes on minimum and maximum on the boundary $\Gamma$.

For later applications it is convenient to restrict ourselves to homogenous boundary conditions, which means $g = 0$. Here one requires an arbitrary (smooth) extension $G$ of the inhomogeneous boundary data $g$ on $\Gamma$ to all of $\Omega$ (i.e., $G = g$ on $\Gamma$). One introduces an auxiliary function $\widetilde{u} := u - G$, which satisfies the homogeneous boundary condition $\widetilde{u} = 0$ on $\Gamma$, and the new differential equation $-\Delta\widetilde{u} = \widetilde{f}$ with $\widetilde{f} := \Delta G + f$.

As a second example, let us introduce the *Helmholtz equation*

$$\boxed{-\Delta u + u = f \text{ on } \Omega} \tag{7.10a}$$

for the function $u = u(x, y)$ with the *Neumann boundary condition*

$$\boxed{\frac{\partial u}{\partial n} = g \text{ on } \Gamma.} \tag{7.10b}$$

Here,

$$\frac{\partial u}{\partial n} := \mathbf{n}(\mathbf{grad}\, u)$$

denotes the outer normal derivative at a boundary point, i.e., $\mathbf{n}$ is the out normal unit vector.

Under appropriate assumptions on the behavior of the function $u$ at infinity, the Neumann boundary problem is uniquely solvable even for unbounded domains $\Omega$.

In $d$ dimensions the Poisson equation is $-\Delta u = f$, where

$$-\Delta u := -u_{x_1x_1} - \ldots - u_{x_dx_d} = -\mathbf{div}\,\mathbf{grad}\, u.$$

Moreover, we set $A = A(x_1, \ldots, x_d)$, a $(d \times d)$ matrix, $\mathbf{b} = \mathbf{b}(x_1, \ldots, x_d)$, a $d$-dimensional vector-valued function, and $c = c(x_1, \ldots, x_d)$, a scalar function. Then

$$-\mathbf{div}\,(A\,\mathbf{grad}\, u) + \mathbf{b}\,\mathbf{grad}\, u + cu = f \tag{7.11}$$

is the general linear differential equation of second order. It is said to be *elliptic*, if $A(x_1, \ldots, x_d)$ is positive definite. Here, $-\mathbf{div}\,(A\,\mathbf{grad}\, u)$ is a *diffusion* term, $\mathbf{b}\,\mathbf{grad}\, u$ is a *convection* term, and $cu$ is a *reaction* term. Both the Poisson equation and the Helmholtz equation are special cases of (7.11) when $A = I$ and $\mathbf{b} = 0$.

**7.7.3.1.2 Formulation as a variational problem.** Because of the Green formula

$$-\int_\Omega (\Delta u)v\,\mathrm{d}x = \int_\Omega \mathbf{grad}\, u\,\mathbf{grad}\, v\,\mathrm{d}x - \int_\Gamma \frac{\partial u}{\partial n} v\,\mathrm{d}F,$$

which corresponds to an integration by parts, we obtain from $-\Delta u = f$ the equation

$$\boxed{\int_\Omega \mathbf{grad}\, u\,\mathbf{grad}\, v\,\mathrm{d}x = \int_\Omega fv\,\mathrm{d}x + \int_\Gamma \frac{\partial u}{\partial n} v\,\mathrm{d}F}$$

with

$$\mathbf{grad}\, u\, \mathbf{grad}\, v := \sum_{i=1}^{d} u_{x_i} v_{x_i} \,,$$

where $u_x$ denotes the partial derivative of $u$ with respect to $x_i$. In the case of $\mathbb{R}^2$, we have $d = 2$.

According to 7.7.3.1.1, we may assume that we have homogeneous Dirichlet boundary conditions, i.e., $u = 0$ on $\Gamma$. We assume therefore that both $u$ and $v$ vanish at the boundary $\Gamma$.

*Example 1:* The classical homogeneous Dirichlet problem for the *Poisson equation* is:

$$-\Delta u = f \quad \text{on} \quad \Omega, \qquad u = 0 \quad \text{on} \quad \Gamma.$$

If we multiply this equation with an arbitrary smooth function $v$ which vanishes on the boundary $\Gamma$, then we obtain by virtue of the Green formula the so-called *weak formulation*:

$$\boxed{\begin{aligned} \int_\Omega \mathbf{grad}\, u\, \mathbf{grad}\, v \,\mathrm{d}x &= \int_\Omega f v \,\mathrm{d}x, \\ u &= 0 \text{ on } \Gamma. \end{aligned}}$$

Not that the boundary integral $\int_\Gamma (\partial u/\partial n) v \,\mathrm{d}F$ in Green's formula vanishes due to '$v = 0$ on $\Gamma$'.

In order to insure the existence of a solution, it is necessary to use Sobolev spaces. The final variational problem is to find a function

$$u \in H_0^1(\Omega),$$

so that

$$\boxed{a(u,v) = b(v) \text{ for all } v \in H_0^1(\Omega).} \tag{7.12}$$

Here we have set

$$a(u,v) := \int_\Omega \mathbf{grad}\, u\, \mathbf{grad}\, v \,\mathrm{d}x, \qquad b(v) = \int_\Omega f v \,\mathrm{d}x.$$

**Sobolev spaces:** In equation (7.12), $H_0^1(\Omega)$ denotes the so-called *Sobolev space.* Roughly speaking, the Sobolev space $H^1(\Omega)$ consists of functions $u$ which are quadratically integrable, together with their first partial derivatives. In other words, we have

$$\int_\Omega (u^2 + |\mathbf{grad}\, u|^2) \,\mathrm{d}x < \infty.$$

The space $H^1(\Omega)$ can be given the structure of a Hilbert space, by introducing a scalar product by the formula

$$(u,v) := \int_\Omega (uv + \mathbf{grad}\, u\, \mathbf{grad}\, v) \,\mathrm{d}x.$$

The Sobolev space $H_0^1(\Omega)$ consists of all functions in $H^1(\Omega)$ with $u = 0$ on $\Gamma$ (in the sense of so-called generalized boundary values). With respect to the scalar product

$$(u,v) := \int_\Omega \mathbf{grad}\, u\, \mathbf{grad}\, v\, \mathrm{d}x$$

this space also has the structure of Hilbert space.

The precise definitions used here are presented in [212]. Note that $H^1(\Omega)$ (resp. $H_0^1(\Omega)$) corresponds to the space $W_p^1(\Omega)$ (resp. $\overset{\circ}{W}{}_p^1(\Omega)$) with $p = 2$.

*Example 2:* We now consider the Helmholtz equation with the Neumann boundary conditions

$$-\Delta u + u = f \quad \text{on} \quad \Omega, \qquad \frac{\partial u}{\partial n} = g \quad \text{on} \quad \Gamma.$$

It is now important that the function $v$ is not restricted in any way along the boundary. If we multiply this equation with $v$, then we obtain in a manner similar to that of Example 1 the following formulation of the problem as a variational problem. The problem is to determine a function

$$u \in H^1(\Omega)$$

so that

$$\boxed{a(u,v) = b(v) \text{ for all } v \in H^1(\Omega).} \tag{7.13}$$

Here we have used the notations

$$a(u,v) := \int_\Omega (\mathbf{grad}\, u\, \mathbf{grad}\, v + uv)\, \mathrm{d}x,$$

$$b(v) := \int_\Omega fv\, \mathrm{d}x + \int_\Gamma gv\, \mathrm{d}F.$$

In Examples 1 and 2 the bilinear forms $a(.,.)$ occuring are *strongly positive*,[3] i.e., one has the decisive inequality

$$\boxed{a(u,u) \geq c\,\|u\|_V^2 \text{ for all } u \in V \text{ and some fixed } \ c > 0.} \tag{7.14}$$

In Example 1, one must choose $V$ to be $V = H_0^1(\Omega)$ with the normed square

$$\|u\|_V^2 = (u,u)_V = \int_\Omega \mathbf{grad}\, u\, \mathbf{grad}\, u\, \mathrm{d}x.$$

On the other hand, in Example 2 we should take $V = H^1(\Omega)$ with

$$\|u\|_V^2 = (u,u)_V = \int_\Omega \left(u^2 + |\mathbf{grad}\, u|^2\right) \mathrm{d}x.$$

We will discuss in [212] that the equation

$$a(u,v) = b(v) \text{ for all } v \in V \text{ and fixed } u \in V$$

[3]Instead of strongly positive, one also use the more precise terminology of *V-elliptic.*

is, under appropriate assumptions, equivalent to the quadratic variational problem

$$\frac{1}{2}a(u,u) - b(u) \overset{!}{=} \min., \qquad u \in V.$$

This legitimates the use of the name 'variational problem'.

The inequality (7.14) insures a unique solution of the variational problems (7.12) and (7.13).

**7.7.3.1.3 Application to the finite element method.** For a *conformal* finite element method, the space of functions $V_n$ must be a subset of the functional spaces $H_0^1(\Omega)$ and $H^1(\Omega)$ used in (7.12) and (7.13). For the piecewise defined functions discussed above, this means that the functions must be, in fact, globally continuous. The simplest choice are the piecewise linear functions on an admissible triangulation of $\Omega$. For simplicity's sake we assume that $\Omega$ is polygonal, so that an exact triangulation is possible.

First we consider the example given in (7.13). As a basis of the finite element space we choose the Lagrange functions $\{\varphi_P : P \in E\}$, where $E$ is the set of the vertices of the triangulation. The *Lagrange function* is uniquely defined as the piecewise continuous function with $\varphi_P(Q) = \delta_{PQ}$ ($P, Q \in E$; $\delta$ Kronecker symbol). The support consists of all triangles which have a common vertex with $P$. The finite element space $V_n \subset H^1(\Omega)$ is spanned by all the basis functions $\{\varphi_P : P \in E\}$. The Lagrange basis is also referred to as the *standard basis*. According to (7.8b) one must compute the coefficients $a_{ik} = a(\varphi_k, \varphi_i)$ of the *stiffness matrix* $A$, where the indices $i, k = \{1, \ldots, n\}$ are to be identified with the vertices $\{P_1, \ldots, P_n\} \in E$ of the triangulation.

For the Dirichlet problem (7.12) the space $V_n \subset H_0^1(\Omega)$ must satisfy in addition the zero boundary conditions. This is the case, if $v(Q) = 0$ for $v \in V_n$ at all vertices $Q \in E \cap \Gamma$. Therefore, $V_n$ is spanned by all Lagrange functions $\{\varphi_P : P \in E_0\}$, where the subset $E_0 \subset E$ consists of all the *interior vertices* of $E$, i.e., $E_0 := E \backslash \Gamma$.

**7.7.3.1.4 Representation of the finite element matrix.** The coefficients $a_{ik} = a(\varphi_k, \varphi_i)$ can only be different from zero, if the support of $\varphi_k, \varphi_i$ contains common interior points. This will be the case if the corresponding vertices $P_k, P_i$ coincide or are connected by an edge of the triangulation (see Figure 7.6(a)). The number of non-vanishing entries in the $k^{th}$ row of the matrix $A$ therefore is one plus the number of neighboring vertices of $P_k$. A *neighbor* $P_i$ of $P_k$ is defined by the condition that $P_k P_i$ is an edge of the triangulation. In representing this matrix, it is advantageous to use the data structure, which is used for saving the geometric information about the triangulation anyway. The vertex $P_k$ is associated to the matrix element $a_{kk}$, and the pointer of $P_k$ to $P_i$ is associated to the matrix element $a_{ki}$. Every matrix–vector product $Ax$ requires only a summation over the $a_{ki}x_i$ for all relevant $i$ (i.e., $i$ such that $a_{ki} \neq 0$). These are $i = k$ and all $i$, for which the pointer explained above exists.

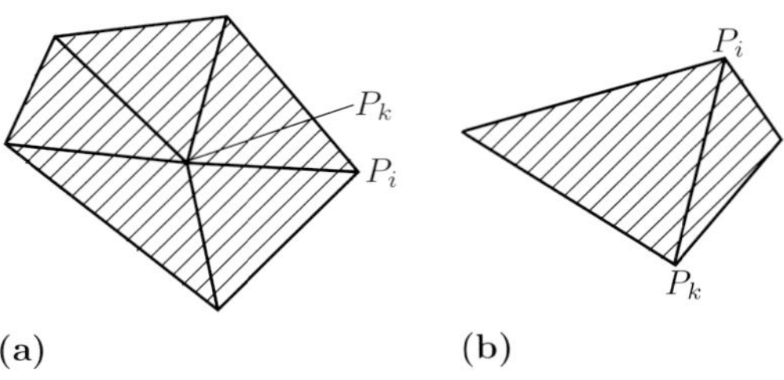


*Figure 7.6. Triangulations and finite element matrices.*

**7.7.3.1.5 Calculation of the finite element matrix.** The coefficients $a_{ik} = a(\varphi_k, \varphi_i)$ are in case of (7.12) equal to $\int_\Omega \mathbf{grad}\,\varphi_k\ \mathbf{grad}\,\varphi_i\,\mathrm{d}x$. The domain of integration $\Omega$ can be reduçed to the intersection of the supports of $\varphi_k$ and $\varphi_i$. For $i = k$, this is the union of all triangles having $P_k$ as vertex (see Figure 7.6(a)), while for $i \neq k$ it is the union of the two triangles which have $P_kP_i$ as a common edge. Thus, the integration problem is reduced to the calculation of $\int_\Delta \mathbf{grad}\,\varphi_k\ \mathbf{grad}\,\varphi_i\,\mathrm{d}x$ for a few triangles $\Delta$.

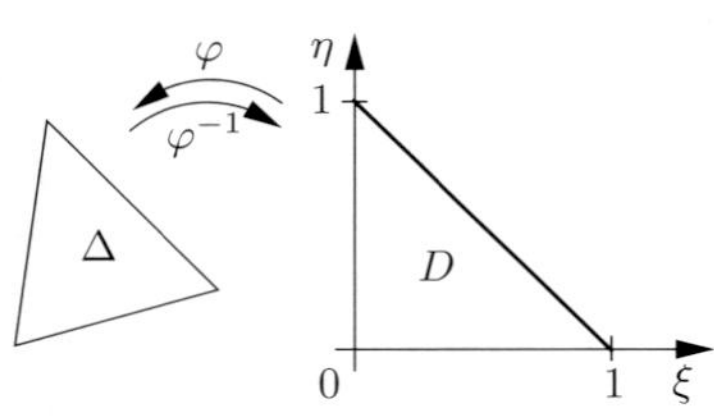


*Figure 7.7. Using the unit triangle D.*

Since the triangles of a triangulation can have different shapes, the calculation can be simplified by mapping all triangles to a unit triangle $D = \{(\xi, \eta) : \xi \geq 0,\ \eta \geq 0,\ \xi + \eta \leq 1\}$ by means of a linear map; this triangle is exhibited in Figure 7.7. For details, see [442], §8.3.2. The integration is thus reduced to a numerical quadrature over the unit triangle $D$. For piecewise linear functions the gradients in $\int_\Delta \mathbf{grad}\,\varphi_k\ \mathbf{grad}\,\varphi_i\,\mathrm{d}x$ are constant, and a one-point quadrature yields an exact result. For (7.13) there is the additional term $\varphi_k\varphi_i$ in $a_{ik} = a(\varphi_k, \varphi_i) = \int_\Omega (\mathbf{grad}\,\varphi_k\ \mathbf{grad}\,\varphi_i + \varphi_k\varphi_i)\,\mathrm{d}x$, which is quadratic, so that higher quadrature formulas need to be applied [432]. In more general cases like (7.11) with variable coefficients, a quadrature error is invariable and must be accounted for by the numerical analysis.

**7.7.3.1.6 Stability conditions.** Stability insures that the inverse of the stiffness matrix $A$ exists and remains bounded in an appropriate sense. Inequality (7.14) is a very strong stability condition. Under the assumption (7.14), the Ritz–Galerkin procedure (and in particular the finite element method) has a solution for every choice of $V_n \subset V$. If, in addition, the symmetry $a(u, v) = a(v, u)$ holds for the function $a$, the stiffness matrix is positive definite.

In general cases there is a necessary and sufficient stability criterion, given by the Babuška condition (infinimum–supremum condition):

$$\inf\big(\sup\{|a(u, v)| : v \in V_n,\ \|v\|_V = 1\} \,|\, u \in V_n,\ \|u\|_V = 1\big) := \varepsilon_n > 0$$

([442], §6.5). If we are given a family of finite element lattices with growing dimension $n = \dim V_n$, then the estimate $\inf_n \varepsilon_n > 0$ must be insured, as otherwise the convergence of the finite element method to the actual solution is problematic.

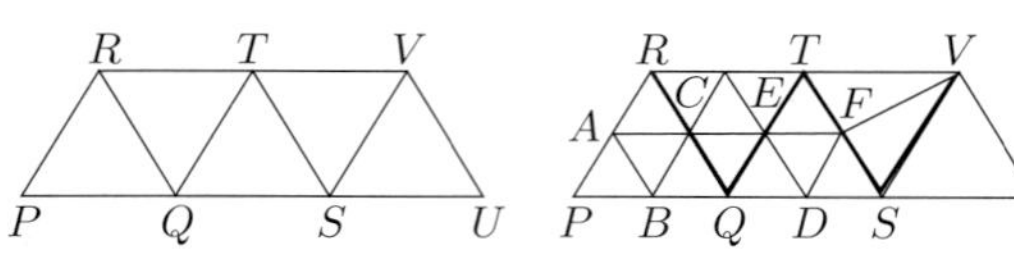


*Figure 7.8. Refinement of a finite element net.*

**7.7.3.1.7 Isoparametric elements and hierarchical bases.** The inverse of the map depicted in Figure 7.7 maps the unit triangle $D$ linearly to an arbitrary triangle $\Delta$. A new circle of problems arises by allowing maps $\Phi : D \to \Delta$ which are not linear (for example quadratic). If $\Omega$ is a non-polygonal domain, there always remain, after the triangulation, crooked pieces on the boundary, which can be approximated by $\Phi(D)$. On $\Phi(D)$ we use the function $v \circ \Phi^{-1}$, where $v$ is a linear function on $D$. The calculation of the matrix elements can then be reduced to an integration on $D$.

The finite element space $V_n$ belongs to a triangulation, which afterwards can be refined by subdividing its triangles (see Figure 7.8). The new finite element space $V_N$ contains the previous one $V_n$. Therefore, one can add to the basis of $V_n$ Lagrange functions in $V_N$, belonging to the newly added vertices, and obtain in this way an alternative basis of $V_N$. In the situation shown in Figure 7.8, the new basis contains the basis functions of the coarser triangulation to the vertices $P, Q, \ldots, T$ and the basis functions of the finer triangulation corresponding to the vertices $A, \ldots, F, V$. In particular, if the refinement of the triangulation is repeated many times, one speaks of a *hierarchical basis* ([442], §8.7.5 and [444], §11.6.4). It is used among other things to define iteration procedures (see 7.7.7.7).

**7.7.3.1.8 Difference procedures.** To obtain a solution of the Poisson problem (7.9a), one covers the domain with a lattice as shown in Figure 7.3. For every interior vertex of the lattice, we use the five-point formula (7.4) as a difference approximation to the Laplace operator $\Delta$. If one of the neighboring points lies on the boundary, then the known value from (7.9b) is used instead. In this way, a system of equations with a sparse $(n \times n)$ matrix $A$ arises, where $n$ is the number of interior vertices of the lattice. For each row of $A$ there are at most five non-vanishing entries. Because of this, the matrix product $Ax$ is quickly calculated. This fact is taken advantage of in the iteration procedure for solving the system of equations $Ax = b$.

For general domains $\Omega$, where the boundary does not coincide with edges of the lattice, there are differences on the boundary of non-equidistant points (see also 7.7.2.1). The *Shorley–Weller* procedure is made to deal with this; see [442], §4.8.1 for details.

In addition to the consistency (approximation of the differential operator $L$ by the difference formula), also the stability of the procedure is necessary; this can be expressed as the boundedness of the inverse matrix $A^{-1}$ (see [442], §4.8.1). Often the stability follows from the $M$-matrix property of $A$.

**7.7.3.1.9 *M*-matrices.** A matrix $A$ is said to be a *$M$-matrix*, if $a_{ii} \geq 0$ and $a_{ik} \leq 0$ for $i \neq k$ and moreover, all components of $A^{-1}$ are non-negative. The first condition on the sign is satisfied, for example, for the negative five-point formula (7.4). A sufficient condition for this requirement on $A^{-1}$ is the irreducible diagonal dominance, which is satisfied in the present situation. For details, see [442], §4.5.

**7.7.3.1.10 The convection–diffusion equation.** Even is the main term (diffusion term) $-\mathbf{div}\,(A\,\mathbf{grad}\,u)$ in (7.11) is responsible for the elliptic character of the differential equation, the convection term $\mathbf{b}\,\mathbf{grad}\,u$ can play a dominant role, as soon as $\|A\|$ is relatively small compared with $\|\mathbf{b}\|$. The problems which occur will be illustrated in the one-dimensional example

$$\boxed{-u'' + \beta u' = f \qquad \text{on} \quad [0,1]}$$

(note this is (7.11) with $d = 1$, $A = 1$, $\mathbf{b} = \beta$, $c = 0$). The combination of second differences (7.3a) for $-u''$ and the symmetric difference (7.2a) for $\beta u'$ yield for the approximation $u_k$ of $u(x_k)$ ($x_k := kh$, $h$ lattice width) the discretization $-\partial_h^2 u + \partial_h^0 u = f$:

$$\boxed{-\left(1 + \frac{h\beta}{2}\right) u_{k-1} + 2u_k - \left(1 - \frac{h\beta}{2}\right) u_{k+1} = h^2 f(x_k).}$$

For $|h\beta| \leq 2$ we have a $M$-matrix, so that in this case stability is insured. Since $\partial_h^2$ and $\partial_h^0$ are exact of second order (cf. (7.2b), (7.3b)), $u_k$ is exact up to $O(h^2)$. As soon as $|h\beta|$ becomes larger than 2, the sign conditions for the $M$-matrix property are no longer satisfied and the difference solution starts to get instable, leading to oscillations (see [442], §10.2.2). The solution $u_k$ which then is determined is in general useless. The condition $|h\beta| \leq 2$ tells us that $h$ is either sufficiently small or that the convection term is not dominant.

In case of $|h\beta| > 2$ one can replace $\partial_h^0$ by the forward or the backward difference in (7.1a,b), depending on the sign of $\beta$. For negative $\beta$, for example, we get

$$-(1 + h\beta)u_{k-1} + (2 - h\beta)u_k - u_{k+1} = h^2 f(x_k).$$

Here we again have a $M$-matrix. However, the approximation is, due to (7.1c), only exact to the first order.

As the usual finite element method becomes instable for large $\beta$, also finite elements requires some kind of stabilization.

#### 7.7.3.2 Saddle point problems

While systems like the Lamé equations again lead to bilinear forms that satisfy inequality (7.14), the Stokes equation, which we now discuss, leads to an *indefinite* bilinear form.

**7.7.3.2.1 Model: Stokes equation.** The *Navier–Stokes* equation, which is absolutely fundamental in hydrodynamics, is, for incompressible fluids:

$$\begin{aligned} -\eta\Delta\mathbf{v} + \varrho(\mathbf{v}\,\mathbf{grad})\mathbf{v} + \mathbf{grad}\,p &= \mathbf{f}, \\ \mathbf{div}\,\mathbf{v} &= 0. \end{aligned}$$

Here $\mathbf{v} = (v_1, \ldots, v_d)$ is a velocity vector, $p$ is the pressure, $f$ the density of the exterior force, $\eta$ is the viscosity constant and $\varrho$ is the density of the fluid.

If $\eta$ is very large compared with $\varrho$, then the term $\varrho(\mathbf{v}\,\mathbf{grad})\mathbf{v}$ can be approximately neglected. With the normalization $\eta = 1$ we then obtain the *Stokes equation*

$$-\Delta\mathbf{v} + \mathbf{grad}\,p = \mathbf{f}, \tag{7.15a}$$

$$-\mathbf{div}\,\mathbf{v} = 0. \tag{7.15b}$$

Equation (7.15a) has $d$ components of the form $-\Delta v_i + \partial p/\partial x_i = f_i$. Since the pressure is only determined up to a constant, one takes $\int p\,\mathrm{d}x = 0$ as normalization condition. One must in addition put boundary conditions on the velocity field $\mathbf{v}$, which we for simplicity ignore in the following discussion. For the pression $p$ there is no such natural boundary condition.

The problem formulated in (7.15a,b) is an example for a system of differential equations of the block form

$$\begin{bmatrix} A & B \\ B^* & 0 \end{bmatrix} \begin{bmatrix} \mathbf{v} \\ p \end{bmatrix} = \begin{bmatrix} \mathbf{f} \\ 0 \end{bmatrix}, \tag{7.16}$$

where $A = -\Delta$, $B = \mathbf{grad}$, $B^* = -\mathbf{div}$ ($B^*$ denotes the adjoint operator to $B$). If one replaces the derivative operators $\partial/\partial x_i$ by real numbers $\xi_i$, $i = 1, \dots, d$, then $A, B, B^*$ transform into $-|\xi|^2 I$ ($I$ is the $(3 \times 3)$ unit matrix), $\xi$ and $-\xi^{\mathsf{T}}$. The block differential operator $L = \begin{bmatrix} A & B \\ B^* & 0 \end{bmatrix}$ transforms under the substitution to the matrix $\widehat{L}(\xi)$ with $|\det L(\xi)| = |\xi|^{2d}$. The positivity for $\xi \neq 0$ classifies the Stokes equation as an *elliptic system*. More details on the Agmon–Douglas–Nirenberg definition of elliptic systems can be found, for example, in [442], §12.1.

If one collects the unknowns of the problem into the vector function $\varphi = \begin{bmatrix} \mathbf{v} \\ p \end{bmatrix}$, then (7.16) can be written as $L_\varphi = \begin{bmatrix} \mathbf{f} \\ 0 \end{bmatrix}$ as in (7.15a). Upon multiplication with $\psi = \begin{bmatrix} \mathbf{w} \\ q \end{bmatrix}$ followed by integration, we obtain the bilinear form

$$c(\varphi, \psi) := a(\mathbf{v}, \mathbf{w}) + b(p, \mathbf{w}) + b(q, \mathbf{v}). \tag{7.17a}$$

Here we have set

$$a(\mathbf{v}, \mathbf{w}) := \sum_{i=1}^{d} \int_\Omega \mathbf{grad}\, v_i \,\mathbf{grad}\, w_i \,\mathrm{d}x,$$

$$b(p, \mathbf{w}) := \int_\Omega (\mathbf{grad}\, p)\mathbf{w} \,\mathrm{d}x,$$

$$b^*(\mathbf{v}, q) := b(q, \mathbf{v}) = \int_\Omega q \,\mathbf{div}\,\mathbf{v} \,\mathrm{d}x.$$

The weak formulation is: determine $\varphi = \begin{bmatrix} \mathbf{v} \\ p \end{bmatrix}$ with

$$\boxed{c(\varphi, \psi) = \int_\Omega \mathbf{f}\mathbf{w} \,\mathrm{d}x, \quad \text{for all } \psi = \begin{bmatrix} \mathbf{w} \\ q \end{bmatrix}.} \tag{7.17b}$$

The variational representation of the problem, equivalent to (7.17b), and analogous to (7.15a,b), is

$$a(\mathbf{v}, \mathbf{w}) + b(p, \mathbf{w}) = \int_\Omega \mathbf{w}\mathbf{f} \,\mathrm{d}x \quad \text{for all } \mathbf{w} \in V, \tag{7.18a}$$

$$b^*(\mathbf{v}, q) = 0 \quad \text{for all } q \in W. \tag{7.18b}$$

Appropriate function spaces $V$ and $W$ in (7.18a,b) are in the case of the Stokes equations $V = [H^1(\Omega)]^d$ and $W = L^2(\Omega)/\mathbb{R}$, the latter being the quotient space with respect to constant functions.

The quadratic functional $F(\varphi) := c(\varphi, \varphi) - 2\int \mathbf{f}\mathbf{v} \mathrm{d}x$ is explicitly

$$F(\mathbf{v}, p) = a(\mathbf{v}, \mathbf{v}) + 2b(p, \mathbf{v}) - 2\int_\Omega \mathbf{f}\mathbf{v} \,\mathrm{d}x. \tag{7.19}$$

If the bilinear form $c(\varphi, \psi)$ were symmetric and positive definite, then the solution of (7.17b) would be the minimizing argument of $F(\mathbf{v}, p) \overset{!}{=} \min$.

But since, for example $c(\varphi, \varphi) = 0$ for $\varphi = \begin{bmatrix} 0 \\ p \end{bmatrix} \neq 0$, we do not have a positive definite bilinear form. Let $(\mathbf{v}^*, p^*)$ denote the solution of (7.17b) or (7.18a,b). This is a *saddle point* of $F$, i.e., we have

$$F(\mathbf{v}^*, p) \leq F(\mathbf{v}^*, p^*) \leq F(\mathbf{v}, p^*) \text{ for all } \mathbf{v}, p. \tag{7.20}$$

This set of inequalities describe that $F$ is minimal at $(\mathbf{v}^*, p^*)$ with respect to $\mathbf{v}$ and maximal with respect to $p$. Moreover, we have

$$F(\mathbf{v}^*, p^*) = \min_{\mathbf{v}} F(\mathbf{v}, p^*) = \max_{p} \min_{\mathbf{v}} F(\mathbf{v}, p). \tag{7.21}$$

We refer to [442], §12.2.2 for the equivalence of (7.18a,b), (7.20) and (7.21).

One particular interpretation of the saddle point problem arises when we introduce $V_0 \subset V$ as the set of functions $\mathbf{v}$ which satisfy the constraint $B^*\mathbf{v} = 0$ in (7.16). For the Stokes problem this is the set of divergence-free functions ($\mathbf{div}\,\mathbf{v} = 0$). The solutions $\mathbf{v}^*$ of (7.18)-(7.20) are obtained from the variational problem 'minimize $a(\mathbf{v}, \mathbf{v}) - 2\int_\Omega f\mathbf{v}\,\mathrm{d}x$ on $V_0$'. The pressure $p$ is in this formulation a Lagrange variable expressing the constraint $\mathbf{div}\,\mathbf{v} = 0$.

A sufficient and necessary condition for (7.18a,b) to have a solution are the following *Babuška–Brezzi conditions*, which we formulate here for the symmetric case $a(\mathbf{v}, \mathbf{w}) = a(\mathbf{w}, \mathbf{v})$:

$$\inf\big(\sup\{|a(\mathbf{u}, \mathbf{v})| : \mathbf{v} \in V_0,\ \|\mathbf{v}\|_V = 1\}\,\big|\,\mathbf{u} \in V_0,\ \|\mathbf{u}\|_V = 1\big) > 0, \tag{7.22a}$$

$$\inf\big(\sup\{|b(p, \mathbf{v})| : \mathbf{v} \in V,\ \|\mathbf{v}\|_V = 1\}\,\big|\,p \in W,\ \|p\|_W = 1\big) > 0. \tag{7.22b}$$

For the Stokes problem, (7.22a) is also correct for the stronger formulation with $V$ instead of $V_0$.

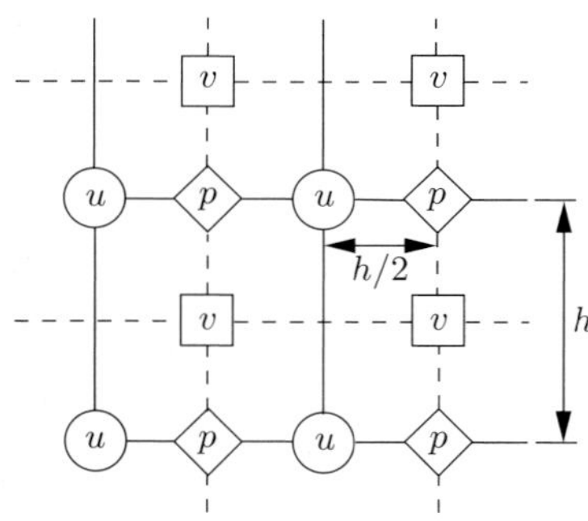


*Figure 7.9. Lattice of variables $u, v, p$.*

**7.7.3.2.2 Difference procedures.** We now consider the plane case $d = 2$. We write the two-dimensional velocity vector $\mathbf{v}$ as $(u, v)$. As opposed to the method described in paragraph 7.7.3.1.8 we do not use a square lattice here, but rather three different lattices for $u$, $v$ and $p$. As shown in Figure 7.9, the $u$-lattice (resp. $v$-lattice) is shifted with respect to the $p$-lattice half a step width in the $x$-direction (resp. $y$-direction). This insures that the vertices of the $u$-lattice satisfy not only the five-point formula for $\Delta_h$ of (7.4), but in addition allow the formation of the symmetric difference $\partial^0_{h/2,x}p(x, y) := [p(x + h/2) - p(x - h/2)]/h$, which compared with (7.2a) is defined with a half-step shift. In this way the first equation of (7.15a), $-\Delta u + \partial p/\partial x = f_1$, is approximated by the difference equation

$$\boxed{-\Delta_h u + \partial^0_{h/2,x} p = f_1 \quad \text{on the } u\text{-lattice}}$$

exactly to second order. Similarly,

$$\boxed{-\Delta_h v + \partial^0_{h/2,y} p = f_2 \quad \text{on the } v\text{-lattice.}}$$

*Figure* 7.9

The incompressibility condition (7.15b) is explicitly $\partial u/\partial x + \partial v/\partial y = 0$. In each lattice point $(x, y)$ of the $p$-lattice there are $u$-values at $(x \pm h/2, y)$ and $v$-values at $(x, y \pm h/2)$. Therefore, the difference equations

$$\boxed{\partial^0_{h/2,x} u + \partial^0_{h/2,y} v = 0 \quad \text{on the } p\text{-lattice}}$$

can be introduced and are again exact to order two.

**7.7.3.2.3 Mixed finite element methods.** The finite element discretization of the saddle point problem (7.18a,b) arises after the substitution of the infinite-dimensional spaces $V$ and $W$ by finite-dimensional spaces $V_h$ and $W_h$, which consist of appropriately chosen finite element functions. The index $h$ indicates the size of the triangles of the underlying triangulation. The finite element solutions $\mathbf{v}^h \in V_h$ and $p^h \in W_h$ must satisfy the variational problem

$$a(\mathbf{v}^h, \mathbf{w}) + b(p^h, \mathbf{w}) = \int_\Omega \mathbf{f}\mathbf{w}\, \mathrm{d}x \quad \text{for all} \quad \mathbf{w} \in V_h, \tag{7.23a}$$

$$b(q, \mathbf{v}^h) = 0 \qquad\qquad \text{for all} \quad q \in W_h. \tag{7.23b}$$

Let $V_{h,0}$ be the space of all functions $\mathbf{v}^h \in V_h$ which satisfy the constraint (7.23b). Equation (7.23b) is just an approximation of the original divergence condition (7.15b). Therefore the functions of $V_{h,0}$ are not contained in the subspace $V_0$ introduced in paragraph 7.7.3.2.1. This is the reason for the name '*mixed* finite element method'.

In $V_h$ and $W_h$ one chooses basis $\{\varphi_1^V, \ldots, \varphi_n^V\}$ and $\{\varphi_1^W, \ldots, \varphi_m^W\}$ where $n = \dim V_h$, $m = \dim W_h$. The system of equations which results has the same block structure as the operator in (7.16). The total matrix is $C := \begin{bmatrix} A & B \\ B^\mathsf{T} & 0 \end{bmatrix}$. The coefficients of the block matrices are $a_{ik} = a(\varphi_k^V, \varphi_i^V)$ and $b_{ik} = b(\varphi_k^W, \varphi_i^V)$.

In contrast with section 7.7.3.1, we here must be very careful about the choice of finite element spaces $V_h$ and $W_h$. A necessary condition for (7.23a,b) to have a solution is $n \geq m$ (i.e. $\dim V_h \geq \dim W_h$). Otherwise the matrix $C$ is singular! We have the paradoxical situation that an 'improvement' of the approximation of the pressure through the use of a higher-dimensional finite element space $W_h$ ruins the numerical solution. Necessary and sufficient conditions for the existence of a solution are the *Babuška–Brezzi conditions*

$$\inf\big(\sup\{|a(\mathbf{u}, \mathbf{v})| : \mathbf{v} \in V_{h,0},\ \|\mathbf{v}\|_V = 1\} \,\big|\, \mathbf{u} \in V_{h,0},\ \|\mathbf{u}\|_V = 1\big) =: \alpha_h > 0, \tag{7.24a}$$

$$\inf\big(\sup\{|b(p, \mathbf{v})| : \mathbf{v} \in V_h,\ \|\mathbf{v}\|_V = 1\} \,\big|\, p \in W_h,\ \|p\|_W = 1\big) =: \beta_h > 0. \tag{7.24b}$$

The indices on $\alpha_h > 0$, $\beta_h > 0$ in (7.24a,b) indicate that these quantities can change with the size $h$ of the triangulation. As in 7.7.3.1.6 we require the uniform boundedness by a positive number which is bounded below, as $h$ tends to 0: $\inf_h \alpha_h > 0$, $\inf_h \beta_h > 0$.

If, for example, we only have $\beta_h \geq \text{const} \cdot h > 0$, then the estimate of the error of the finite element solution can be worse by a factor of $h^{-1}$ than the best approximation in finite element spaces.

The verification of the stability condition (7.24a,b) for concretely given finite element spaces can be quite complicated. Since (7.24a,b) are necessary and sufficient, these conditions can not be replaced by simple ones (for example so-called patch tests).

The choice of finite element functions should be based on the same triangulation for all components $\mathbf{v}$ and $p$. The most obvious choice of piecewise linear elements for $\mathbf{v}$ and $p$ does not satisfy the stability condition (7.24a,b). According to the necessary condition $\dim V_h \geq \dim W_h$, it makes sense to extend the finite element space $V_h$ by further functions. For example the choice of piecewise quadratic elements for $\mathbf{v}$ and piecewise linear elements for $p$ does the trick. An interesting variant consists of the piecewise linear functions and in addition a 'bubble function'. The 'bubble function' is defined on the unit triangle $D$ of Figure 7.7 by $\xi\eta(1-\xi-\eta)$. It vanishes on the boundary of the triangle and is positive on the interior. Using the linear map from $D$ to an arbitrary triangle as discussed in that section can be used to define the bubble function on an arbitrary triangle. The function space $V_h$ obtained this way satisfies the Babuška–Brezzi conditions ([442], §12.3.3.2).

## 7.7.4 Parabolic differential equations

### 7.7.4.1 Model problems

The typical example of a parabolic differential equation is the *heat equation*

$$\boxed{u_t - \Delta u = f \quad \text{for } t > t_0 \text{ and } x \in \Omega,} \tag{7.25a}$$

where the function $u = u(t,x)$ to be determined depends on position variables $x = (x_1, \ldots, x_d) \in \Omega$ and the time $t$. Instead of $-\Delta$ there could be here a more general elliptic differential operator $L$. This operator only acts on the $x$-variables, but can have $t$-dependent coefficients. As in the case of elliptic differential equations appropriate boundary conditions for $u(t,.)$ are part of the problem, for example the Dirichlet data:

$$\boxed{u(t,x) = \varphi(t,x) \quad \text{for } t > t_0 \text{ and } x \in \Gamma.} \tag{7.25b}$$

In addition, there are initial values given at time $t_0$:

$$\boxed{u(t_0,x) = u_0(x) \quad \text{for } x \in \Omega.} \tag{7.25c}$$

The problem (7.25a-c) is referred to as a *initial boundary value problem.*

Note that the time direction plays a special role in these equations. The problem presented in (7.25a-c) has a certain asymmetry: it can only be solved in one direction (future) $(t > t_0)$, but not in the other (past) $(t < t_0)$.

Even if the initial and boundary data are not compatible (i.e., $u_0(x) = \varphi(t_0,x)$ does not hold for all $x \in \Gamma$), a solution still exists which is smooth for $t > t_0$ but discontinuous at the boundary for $t \to t_0 + 0$.

#### 7.7.4.2 Discretization in time and space

The discretization is done in the time direction and the position variables separately. In the case of a difference procedure, we cover the domain $\Omega$ as in 7.7.3.1.8 with a lattice $\Omega_h$ (where $h$ again denotes the size of the lattice). Correspondingly we replace the differential operator $-\Delta$ by a difference operator $-\Delta_h$. Independently of this, we replace the time derivative $u_t$ by an appropriate difference, for example by $u_t \approx [u(t + \delta t, .) - u(t, .)]/\delta t$ with the time step width $\delta t$. One possible discretization is the *explicit Euler procedure*

$$\frac{1}{\delta t}\big(u(t+\delta t, x) - u(t,x)\big) - \Delta_h u(t,x) = f(t,x) \qquad \text{for } x \in \Omega_h. \tag{7.26a}$$

At the boundary we substitute for $u(t,x)$ the boundary values (7.25b). Solving (7.26a) with respect to the time step $t + \delta t$ yields the algorithm

$$\boxed{u(t, +\delta t, x) = u(t,x) + \delta t \Delta_h u(t,x) + \delta t f(t,x) \qquad \text{for } x \in \Omega_h.} \tag{7.26a$'$}$$

Starting with the initial values (7.25c) we obtain from (7.26a′) an approximation for the time value $t_k = t_0 + k\delta t$.

If, on the other hand, we evaluate the position discretization at $t + \delta t$ instead of at $t$, we arrive at the *implicit Euler procedure*

$$\frac{u(t+\delta t, x) - u(t,x)}{\delta t} - \Delta_h u(t+\delta t, x) = f(t,x) \qquad \text{for } x \in \Omega_h. \tag{7.26b}$$

The new value $u(t + \delta t, .)$ is the solution of the following system of equations

$$\boxed{(I - \delta t \Delta_h) u(t+\delta t, x) = u(t,x) + \delta t f(t,x) \qquad \text{for } x \in \Omega_h.} \tag{7.26b$'$}$$

A discretization which is symmetric with respect to $t + \delta t/2$ is the *Crank–Nicholson scheme*

$$\boxed{\begin{aligned} &\frac{1}{\delta t}\big(u(t+\delta t, x) - u(t,x)\big) - \frac{1}{2}\Delta_h\big(u(t,x) + u(t+\delta t, x)\big) \\ &= \frac{1}{2}\big(f(t,x) + f(t+\delta t, x)\big) \qquad \text{for } x \in \Omega_h. \end{aligned}} \tag{7.26c}$$

#### 7.7.4.3 Stability of difference procedures

The explicit procedure (7.26′), which at first sight appears to be much simpler to apply, is often in applications *not* an appropriate algorithm, as $\delta t$ must satisfy a very restrictive stability condition. In the case of the five point operator $\Delta_h$ presented in equation (7.4), this stability condition is

$$\boxed{\lambda := \delta t/h^2 \le 1/4.} \tag{7.27}$$

Consistency of the approximation requires that the convergence of the discrete solution to the exact one in (7.25a-c) is equivalent to the stability condition, as discussed in the theorem of section 7.7.5.5 below. Without (7.27) one obtains only useless solutions. In the case of $\lambda > 1/4$ perturbations of initial values after $k = (t - t_0)/\delta t$ steps of the algorithm are amplified by a factor of $[1-8\lambda]^k$. The condition (7.27) couples $\delta t$ with the square $h^2$ of the lattice size of the position lattice, since second derivatives in $x$ but only first derivatives of $t$ occur. For applications, the condition $\delta t \le h^2/4$ means that, precisely when the high dimension of the space variables require a high computational cost per time step, the number of necessary steps of the approximation increases drastically.

For general differential operators $L$ instead of $\Delta$, it is essentially the same stability condition (7.27) which must be satisfied, with only the constant of $1/4$ being replaced by another constant.

The implicit Euler procedure (7.26b) as well as the Crank–Nicholson scheme (7.26c) are both what are called *absolutely stable* procedures, meaning that they are stable for every value of $\lambda = \delta t/h^2$.

#### 7.7.4.4 Semi-discretization

If one discretizes in (7.25a) only the space variable differential operator, by not the time derivative, one obtains

$$\boxed{u_t - \Delta_h u = f \quad \text{for } t > t_0 \text{ and } x \in \Omega_h.} \tag{7.28}$$

Since in this notation $u$ is the vector of the $n$ vertices of $\Omega_h$, we have here a system of ordinary differential equations, for which (7.25c) plays the role of the initial values at $t = t_0$. As the eigenvalues of the system matrix $\Delta_h$ are on the order of 1 to $h^{-2}$, (7.28) is a *stiff* system of differential equations (see 7.6.1.1). This explains why explicit procedures for solving (7.28) only work for sufficiently small time steps. The implicit trapeze rule is, according to 7.6.1.1, absolutely stable. If we apply it to (7.28), we end up with the Crank–Nicholson scheme (7.26c). The implicit Euler procedure is, in fact, even *strongly* absolutely stable, meaning that perturbations in the form of strong oscillations are actually damped by this algorithm, while this fact is not true for the Crank–Nicholson scheme.

#### 7.7.4.5 Step-size control

As usual for differential equations, it is not necessary that the time step width $\delta t$ remain constant throughout the algorithm; indeed, it can be advantageously adapted to the situation at hand. However, this requires that the underlying algorithm is implicit and absolutely stable, so that $\delta t$ is not bounded above by stability conditions like (7.27).

One will choose $\delta t$ to be larger as the difference of $u(t+\delta t, x)$ and $u(t, x)$ gets smaller and smaller. This is in particular the case, if (7.25a) is to be solved with time independent functions $f$ and $\varphi$ for $t \to \infty$. In this case $u$ tends to the solution of the stationary equation $-\Delta u = f$ with boundary condition (7.25b). We note at this point that this procedure is not appropriate if one is only interested in obtaining a stationary solution. In fact, although formulating the problem as a parabolic differential equation does lead to an iteration procedure which converges to the stationary solution, this procedure is in general not at all effective, as discussed in section 7.7.7 below.

Conversely, there are good reasons for choosing $\delta t$ to be small near the initial time ($t \approx t_0$). As mentioned toward the end of 7.7.4.1, $u$ can be discontinuous for $t = t_0$ and $x \in \Gamma$. To simulate a smoothing numerically, one needs small step widths and must avoid implicit procedures like (7.26c), which are not strongly stable. In such a case it makes sense to carry out several time steps with the explicit procedure (7.26a′) and $\delta t \le h^2/8$ instead of (7.27).

#### 7.7.4.6 Finite element solutions

The simplest way to obtain the finite element discretization for a parabolic differential equation is to first carry out the semi-discretization with the help of the finite element method. Let $V_n$ denote the finite element space. From equation (7.25a) we obtain the system of ordinary differential equations for a function $u(t,x)$ with $u(t,.) \in V_n$, which is formulated in the weaker form

$$\boxed{(u_t, v) + a\big(u(t), v\big) = b(v) \quad \text{for } v \in V_n, \quad t > t_0,} \tag{7.29}$$

where $a$ and $b$ are defined as in (7.12); we take $u$ to have the form

$$u(t) := \sum y_k(t)\varphi_k \qquad (\varphi_k \text{ basis functions of } V_n)$$

with time-dependent coefficients $y_k(t)$. In matrix form, this system can be written

$$My_t + Ay = b \quad \text{for } t > t_0,$$

where $A$ and $b$ are the quantities given in (7.8b). The 'mass matrix' $M$ has the components $M_{ij} = \int \varphi_i\varphi_j \,\mathrm{d}x$. If one now applies the Euler procedure for time discretization, one gets for example $M[y(t+\delta t) - y(t)]/\delta t + Ay(t) = b$, and from this the recursion formula

$$\boxed{y(t+\delta t) = y(t) + \delta t M^{-1}[b - Ay(t)] \qquad \text{for } t > t_0.} \tag{7.30}$$

To avoid having to solve this system of equations with matrix $M$, the matrix $M$ is often replaced by a diagonal matrix (for example, one with diagonal elements which are the row sums of $M$). This so-called *lumping* does not reduce the quality of the approximation (see [453]).

The initial values (7.25c) are carried over to the finite element solution $u(t,.) \in V_n$ with the aid of a $L^2$-projection:

$$\int_\Omega u(t_0, x)v(x)\,\mathrm{d}x = \int_\Omega u_0(x)v(x)\,\mathrm{d}x.$$

For the coefficients $y(t_0)$ of $u(t_0,x)$ this means we have the equation $My(t_0) = c$ with $c_i = \int u_0(x)\varphi_i(x)\,\mathrm{d}x$.

It is also possible to carry out finite element discretization in *both* the time and the space variables. In the simplest cases (for example functions which are piecewise continuous in the time variable on $[t, t+\delta t]$, in the space corresponding to $V_n$), this procedure, however, leads back to the Euler discretization of (7.29) and thus to (7.30).

### 7.7.5 Hyperbolic differential equations

#### 7.7.5.1 Initial value and initial boundary value problems

The simplest example of a hyperbolic equation is

$$\boxed{u_t(t,x) + a(t,x)u_x(t,x) = f(t,x) \qquad (a, f \text{ given, } u \text{ to be determined}).} \tag{7.31}$$

Every solution of the ordinary linear differential equation

$$x'(t) = a\big(t, x(t)\big) \tag{7.32}$$

is called a *characteristic* of (7.31). The family of all characteristics $x(t) = x(t; x_0)$ with initial values $x(0) = x_0 \in \mathbb{R}$ is said to be a *family of characteristics* for the differential equation. Along a characteristic, the function $U(t) := u\big(t, x(t)\big)$ satisfies the ordinary differential equation

$$U_t(t) = f\big(t, x(t)\big). \tag{7.33}$$

If we are dealing with a pure initial value problem, then the initial values are prescribed along a curve (for example the line $t = 0$) by

$$\boxed{u(0, x) = u_0(x), \qquad -\infty < x < \infty.} \tag{7.34}$$

This implies for the equation (7.33) along the characteristic $x(t; x_0)$ the initial values $U(0) = u_0(x_0)$. The combination of (7.31) and (7.34) is called a *initial value problem.*

If the initial values (7.34) are only given for a bounded interval $[x_\ell, x_r]$, then we also require boundary values for either $x = x_\ell$ (left boundary) or for $x = x_r$ (right boundary). Which boundary we should take here depends on the sign of $a(t, x)$: the characteristic is required to run from the outside to the inside when it intersects the boundary curve (given here by $x = \text{const}$). In case $a > 0$, the appropriate boundary condition is

$$u(t, x_\ell) = u_\ell(t), \qquad t \geq 0. \tag{7.35}$$

The combination of (7.31) and (7.34) on $[x_\ell, x_r]$ and (7.35) is what is called an *initial value–boundary problem.*

A typical property of hyperbolic differential equations is the conservation of *discontinuities.* If the initial value $u_0$ jumps at a point $x = x_0$, then this discontinuity is extended along the characteristic $x(t; x_0)$ into the interior (in case $f = 0$, the discontinuity is preserved). This property is in contrast to the property of solution of elliptic and parabolic differential equations, which become smoother in the interior.

#### 7.7.5.2 Hyperbolic systems

Let $u = u(t, x)$ be a vector-valued function: $u = (u_1, \ldots, u_n)$. The differential equation

$$\boxed{Au_x + Bu_t = f} \tag{7.36}$$

with $(n \times n)$-matrices $A$ and $B$ is hyperbolic, if the *generalized eigenvalue problem*

$$e^{\mathsf{T}}(B - \lambda A) = 0, \qquad e \neq 0$$

has $n$ linearly independent (left-)eigenvectors $e_i$ $(1 \leq i \leq n)$ with real eigenvalues $\lambda_i$. Instead of the family of characteristics in this case, one has $n$ such families, which are given by[4]

$$\frac{\mathrm{d}t}{\mathrm{d}x} = \lambda_i, \qquad 1 \leq i \leq n.$$

If one defines the derivative of the $i^{th}$ characteristic direction by $(\varphi)_i = \varphi_x + \lambda_i \varphi_t$, then one obtains instead of (7.33) the ordinary differential equation

$$\big(e_i^{\mathsf{T}} A\big)(u)_i = e_i^{\mathsf{T}} f, \qquad 1 \leq i \leq n. \tag{7.37}$$

[4]In case $1/\lambda_i = 0$ we choose the equation $\mathrm{d}x/\mathrm{d}t = 0$.

In the linear case, $A, B, e_i, \lambda_i$ and $(t, x)$ depend only on $(t, x)$, and in the general case they in addition depend on $u$.

The initial value condition for the system (7.36) is just as in (7.34), if we view $u$ and $u_0$ there to be vector-valued. With respect to the boundary value prescriptions, note that at $x_\ell$ there will be $k_\ell$ conditions, where $k_\ell$ is the number of eigenvalues with $\lambda_i(t, x_l) > 0$. Similarly, there will be $k_r$ boundary conditions at $x_r$. If we have $\lambda_i \neq 0$ for all eigenvalues, then the numbers $k_\ell$ and $k_r$ are constant and add up to $n$.

Hyperbolic system like (7.31) occur quite often after manipulations on a scalar equation of higher order.

#### 7.7.5.3 Characteristics as a tool

In the scalar case discussed in 7.7.5.1, we can reduce the solution of the partial differential equation to the (numerical approximation of the) ordinary differential equations (7.32) and (7.33). A corresponding procedure can also be carried out for $n = 2$ if the eigenvalues $\lambda_1$ and $\lambda_2$ are always distinct. To see this, suppose we are given the values $x, t, u$ at the points $P$ and $Q$ as shown in Figure 7.10. The characteristic of the first family of characteristics which passes through $P$ and the characteristic of the second family which passes through $Q$ meet at a point $R$. The differences $(e_1^\top A)(u_R - u_P)$ and $(e_2^\top A)(u_R - u_Q)$ approximate the left-hand side of (7.37) and yield the equations which determine $u$ at the point $R$. Repeatedly applying this procedure yields solutions at the vertices of a lattice, which follows both families of characteristics (this can be interpreted as an equidistant lattice with respect to the so-called characteristic coordinates).

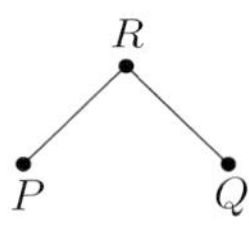


*Figure 7.10.*

Sometimes difference procedures are incorrectly referred to as characteristic procedures, if characteristics were used in their theoretical derivation.

#### 7.7.5.4 Difference procedures

In what follows we use an equidistant lattice with step width $\Delta x$ in the $x$-direction and $\Delta t$ in the $t$-direction. Let $u_\nu^m$ be the approximation of the solution $u$ at $t = t_m = m\Delta t$ and $x = x_\nu = \nu\Delta x$. The initial values define $u_\nu^m$ for $m = 0$:

$$u_\nu^0 = u_0(x_\nu), \qquad -\infty < \nu < \infty.$$

If one replaces $u_t$ in (7.31) by the forward difference $(u_\nu^{m+1} - u_\nu^m)/\Delta t$ and $u_x$ by the symmetric difference $(u_{\nu+1}^m - u_{\nu-1}^m)/(2\Delta x)$, one obtains the difference equation

$$u_\nu^{m+1} := u_\nu^m + \frac{a\lambda}{2}(u_{\nu+1}^m - u_{\nu-1}^m) + \Delta t f(t_m, x_\nu) \tag{7.38}$$

which, as we shall see, turns out to be *totally useless*. The parameter

$$\lambda := \Delta t/\Delta x \tag{7.39}$$

corresponds to the parameter $\lambda := \Delta t/\Delta x^2$ with the same name in (7.27) in the parabolic case.

If one uses instead of the symmetric difference the left or right-sided difference (7.1a,b), then one obtains the following difference equations, which are named after *Courant–Isaacson–Rees*:

$$u_\nu^{m+1} := (1 + a\lambda)u_\nu^m - a\lambda u_{\nu-1}^m + \Delta t f(t_m, x_\nu), \tag{7.40a}$$

$$u_\nu^{m+1} := (1 - a\lambda)u_\nu^m + a\lambda u_{\nu+1}^m + \Delta t f(t_m, x_\nu). \tag{7.40b}$$

The combination of the symmetric difference for the space variables and the unusual appearing time difference $(u_\nu^{m+1} - \frac{1}{2}[u_{\nu+1}^m + u_{\nu-1}^m])/\Delta t$ yields the *Friedrichs scheme*

$$u_\nu^{m+1} := (1 - a\lambda/2)u_{\nu-1}^m + (1 + a\lambda/2)u_{\nu+1}^m + \Delta t f(t_m, x_\nu). \tag{7.40c}$$

If $a$ does not depend on $t$, then the following *Lax–Wendorff procedure* defines a discretization procedure of *second order*:

$$u_\nu^{m+1} := \frac{1}{2}(\lambda^2 a^2 - \lambda a)u_{\nu-1}^m + (1 - \lambda^2 a^2)u_\nu^m + \frac{1}{2}(\lambda^2 a^2 + \lambda a)u_{\nu+1}^m + \Delta t f(t_m, x_\nu). \tag{7.40d}$$

Figure 7.11 gives a schematic picture how the new values $u_\nu^{m+1}$ depend on the values

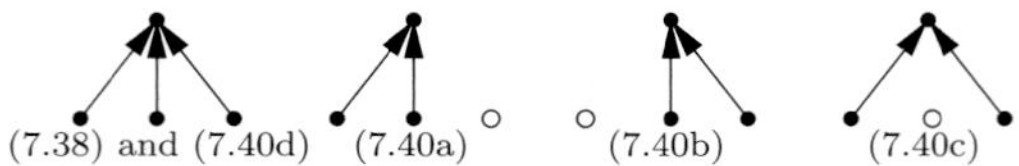


*Figure 7.11. Difference molecules.*

of the $m^{th}$ level. All examples are special *explicit difference procedures* of the form

$$u_\nu^{m+1} := \sum_{\ell=-\infty}^{\infty} c_\ell u_{\nu+\ell}^m + \Delta t g_\nu, \qquad -\infty < \nu < \infty,\ m \geq 0. \tag{7.41}$$

The coefficients $c_\ell$ depend on $t_m, x_n, \Delta x$ and $\Delta t$. In the case of a vector-valued function $u \in \mathbb{R}^n$, the $c_\ell$ are real $(n \times n)$-matrices. In general the sum in (7.41) contains only finitely many non-vanishing coefficients.

### 7.7.5.5 Consistency, stability and convergence

The theoretical analysis of these points is easier if we formulate the difference equation (7.41) not only on the vertices of the lattice, but on all of $\mathbb{R}$:

$$u^{m+1}(x) := \sum_{\ell=-\infty}^{\infty} c_\ell u^m(x + \ell\Delta x) + \Delta t g(x), \qquad -\infty < x < \infty,\ m \geq 0. \tag{7.41$'$}$$

The algorithm described by (7.41′) defines the action of the difference operator $C = C(\Delta t)$

$$u^{m+1}(x) := C(\Delta t)u^m + \Delta t g, \qquad m \geq 0. \tag{7.41$''$}$$

Now let $B$ be some appropriate Banach space[5] containing the function $u^m$. The standard choice is $B = L^2(\mathbb{R})$ with the norm $\|u\| = \left(\int_{\mathbb{R}} |u(x)|^2 \mathrm{d}x\right)^{\frac{1}{2}}$ or $B = L^\infty(\mathbb{R})$ with the norm

[5] Fundamental notions and result about Banach spaces can be found in [212].

$\|u\| = \text{ess sup}\left\{|u(x)| : x \in \mathbb{R}\right\}$. Let $B_0$ be a dense subset of $B$ and $u(t)$ the solution of (7.31) with $f = 0$ for an arbitrary starting value $u_0 \in B_0$. The difference operator $C(\Delta t)$ is said to be *consistent* (in the interval $[0, T]$ and with respect to $\|.\|$), if

$$\sup_{0 \le t \le T} \|C(\Delta t)u(t) - u(t + \Delta t)\| / \Delta t \to 0 \qquad \text{for } \Delta t \to 0.$$

The objective of the discretization is to approximate $u(t)$ by $u^m(m\Delta t \to t)$. Correspondingly, we call an algorithm *convergent* (in $[0, T]$ and with respect to $\|.\|$), if $\|u^m - u(t)\| \to 0$ for $\Delta t \to 0$ with $m\Delta t \to t \in [0, T]$. Here we have set $\lambda := \Delta t / \Delta x$, which is taken to be fixed, so that $\Delta t \to 0$ also implies $\Delta x \to 0$.

The consistency, which is in general easy to verify, is however definitely not sufficient to insure convergence. Instead, we have the following result.

**Equivalence theorem:** If we take consistency for granted, we have convergence if and only if the difference operator $C(\Delta t)$ is stable.

Here we define the notion of *stability* (in $[0, T]$ and with respect to $\|.\|$) of the operator $C(\Delta t)$ by the requiring the estimate

$$\|C(\Delta t)^m\| \le K \qquad \text{for all } m \text{ and } \Delta t \text{ with } 0 \le m\Delta t \le T \tag{7.42}$$

for some fixed $K$.

Formulated in the negative, this theorem tells us that instable difference operators can yield *ridiculous* results, where the instability usually expresses itself in rapid oscillations of the solution. Note that (7.41) is a single step procedure. In contrast with consistent single step procedures for ordinary differential equations, which converge in general as discussed in section 7.6.1.1, while for them we only meet with stability problems in the case of multiple step procedures, for explicit difference procedures a *conditional stability* is the best we can hope for, meaning there are restrictions on $\lambda$.

#### 7.7.5.6 Stability conditions

**7.7.5.6.1 CFL-conditions as necessary conditions for stability.** The stability conditions we now wish to discuss originate in work of Courant, Friedrichs and Lewy, and are usually just referred to as the *CFL-conditions.* This criterion, which is relatively easy to check, is a necessary condition for stability.

In the sum (7.41′), let $\ell_{\min}$ (resp. $\ell_{\max}$) be the smallest (resp. largest) index for which $c_\ell \neq 0$. In the scalar case ($u \in \mathbb{R}^1$), the CFL-conditions are

$$\ell_{\min} \le \lambda a(t, x) \le \ell_{\max} \qquad \text{for all } x \text{ and } t \tag{7.43}$$

where $a(t, x)$ is the coefficient from (7.31). In the case of higher dimensions ($u \in \mathbb{R}^n, n \ge 2$) the quantity $a(t, x)$ in (7.43) must be replaced by the set of all eigenvalues of the $(n \times n)$-matrix $a(t, x)$.

Note that the only property of $C(\Delta t)$ which is of importance in the CFL-condition is that the boundaries of the indices are $\ell_{\min}$ and $\ell_{\max}$. Even the particular choice of norm in (7.42) is irrelevant.

Aside from the trivial case $a = 0$, the CFL-condition shows that a value of $\lambda$ which is too large always leads to instability. On the other hand, we can force conditional stability with the help of an *implicit* difference procedure.. This can be written formally as (7.42), with but an infinite sum. Because $-\ell_{\min} = \ell_{\max} = \infty$, the CFL-condition is then automatically satisfied.

The CFL-conditions are in general not sufficient for stability. If a procedure happens to be stable under precisely the restriction (7.43) on $\lambda$, we refer to it as *optimally stable.* The stronger the restrictions on $\lambda$ and hence on $\Delta t = \lambda \Delta x$ are, the more steps we require for (7.41″) to arrive at $t = m\Delta t$.

**7.7.5.6.2 Sufficient stability conditions.** A sufficient condition for stability, which implies (7.42) with $K := \exp(TK')$, is

$$\|C(\Delta t)\| \leq 1 + \Delta t K'. \tag{7.44}$$

In the scalar case and for the choice of $B = L^\infty(\mathbb{R})$ we have $\|C(\Delta t)\| = \sum_l |c_\ell|$. This shows that the procedure (7.40a-c) is stable with respect to the supremum norm, provided $|\lambda a| \leq 1$; in addition we must require $a \leq 0$ or $a \geq 0$ for the Courant–Isaacson–Rees scheme (7.40a,b). By the equivalence theorem in 7.7.5.5, the approximations then converge uniformly to the exact solution.

The *Lax–Wendorff* procedure (7.40d), for which the sufficient condition (7.44) for $\|.\| = \|.\|_\infty$ is not satisfied, is really instable with respect to the supremum norm. The reason for this is that the procedure is of consistency order two, and as such, cannot be stable with respect to $L^\infty(\mathbb{R})$. We see here a conflict between the (higher) consistency requirements and the requirements for stability.

From now on we assume that the coefficients $c_\ell = c_\ell(\Delta t, \lambda)$ in (7.41) do not depend on $x$. The $L^2(\mathbb{R})$-stability can then be described with the help of the following *amplification matrix*

$$G = G(\Delta t, \xi, \lambda) := \sum_{\ell=-\infty}^{\infty} c_\ell(\Delta t, \lambda)\, \mathrm{e}^{\mathrm{i}\ell\xi}, \qquad \xi \in \mathbb{R}.$$

$G$ is a $2\pi$-periodic function, which in the case of several variables ($n > 1$) is matrix-valued. For example, in the case of the Lax–Wendorff procedure (7.40d), the amplification matrix is $G(\Delta t, \xi, \lambda) = 1 + \mathrm{i}\lambda a \sin(\xi) - \lambda^2 a^2(1 - \cos(\xi))$.

The $L^2(\mathbb{R})$-stability property (7.42) is equivalent to

$$|G(\Delta t, \xi, \lambda)^m| \leq K \text{ for all } |\xi| \leq \pi \text{ and } 0 \leq m\Delta t \leq T$$

with the same constant $K$ as in (7.42). Here $|.|$ denotes the spectral norm. From this we obtain a further stability criterion, the *von Neumann condition*: for all $|\xi| \leq \pi$ let the eigenvalues $\gamma_j = \gamma_j(\Delta t, \xi, \lambda)$ of the amplification matrix $G(\Delta t, \xi, \lambda)$ satisfy the inequality

$$|\gamma_j(\Delta t, \xi, \lambda)| \leq 1 + \Delta t K', \qquad 1 \leq j \leq n. \tag{7.45}$$

For $n = 1$, (7.45) is just the statement $|G(\Delta t, \xi, \lambda)| \leq 1 + \Delta t K'$. The von Neumann condition is in general only a necessary condition. It is, however, even *sufficient*, if one of the following assumptions is satisfied:

1) $n = 1$,

2) $G$ is a normal matrix,

3) there is a similarity transformation which is independent of $\Delta t$ and $\ell$ and brings all coefficients $c_\ell(\Delta t, \lambda)$ into diagonal form,

4) $|G(\Delta t, \xi, \lambda) - G(0, \xi, \lambda)| \leq L\Delta t$ and $G(0, \xi, \lambda)$ satisfies one of the above conditions.

It follows from the von Neumann condition that the examples (7.40a-d) are $L^2(\mathbb{R})$-stable for $|\lambda a| \leq 1$ (where as above we require $a \leq 0$ or $a \geq 0$ for (7.40a,b)). From the fact that the Lax–Wendorff procedure is $L^2(\mathbb{R})$-stable but not $L^\infty(\mathbb{R})$-stable, we can conclude from the equivalence theorem that the solutions converge in quadratic means but not uniformly to the exact solution. The difference procedure (7.38) leads to $G(\Delta t, \xi, \lambda)1 + \mathrm{i}\lambda a \sin(\xi)$ and is therefore instable with the trivial exception $a = 0$.

In the case of $x$-dependent coefficients $c_\ell$ one uses the technique of *frozen coefficients.* Let $C_{x_0}(\Delta t)$ be the difference operator which arises when we replace all coefficients $c_\ell(x, \Delta t, \lambda)$ (depending on $x$) with the coefficients $c_\ell(x_0, \Delta t, \lambda)$ (independent of $x$). The stability of $C(\Delta t)$ and of $C_{x_0}(\Delta t)$ for all $x_0 \in \mathbb{R}$ are almost equivalent. Under mild technical assumptions the stability of $C(\Delta t)$ in fact implies that of $C_{x_0}(\Delta t)$ for all $x_0 \in \mathbb{R}$. For the converse direction one needs the fact that $C(\Delta t)$ is dissipative. This notion is defined as follows. $C$ is *dissipative of order* $2r$ if the inequality $|\gamma_j(\Delta t, \xi, \lambda)| \leq 1 - \delta|\xi|^{2r}$ for $|\xi| \leq \pi$ holds with a fixed $\delta > 0$. For more details and further stability criteria we refer the reader to [452].

#### 7.7.5.7 Approximation of discontinuous solutions ('shock capturing')

In 7.7.5.1 we already mentioned that discontinuous initial values can lead to solutions which remain discontinuous along the characteristic. In the non-linear case, discontinuities (*shocks*) can occur, even if the initial values are arbitrarily smooth. For this reason one requires that hyperbolic discretizations – in contrast with elliptic or parabolic cases – also yield a good approximation of a discontinuous solution.

Two phenomena which one wants to avoid can occur during an approximation of a discontinuous point of $u_\nu^m$:

1) the discontinuity becomes smoother as $m$ grows larger, and

2) the approximation oscillates at the discontinuity.

The second case occurs in procedures of higher order. There are procedures, called *high resolution* procedures, which have a higher approximation order in smooth domains, but estimate the discontinuity very closely without breaking out in oscillations, can be constructed for example with *flux–limit* methods.

#### 7.7.5.8 Properties of the non-linear case, conservation form and entropy

Non-linear hyperbolic equations with discontinuous solutions lead to difficulties, which do not occur for linear hyperbolic equations or non-linear hyperbolic equations with smooth solutions.[6] The formulation of the equation in the *conservation form* is

$$\boxed{u_t(t,x) + f\big(u(t,x)\big)_x = 0} \tag{7.46}$$

with the *flux function* $f$. The equation is hyperbolic, if $f'(u)$ is not diagonalizable over the reals. Since the 'solution' of (7.46) is not necessarily differentiable, one seeks a 'generalized' or *weak solution* which satisfies the relation

$$\boxed{\int_0^\infty \int_{\mathbb{R}} [\varphi_t u + \varphi_x f(u)]\,\mathrm{d}x\,\mathrm{d}t = -\int_{\mathbb{R}} \varphi(0,x) u_0(x)\,\mathrm{d}x} \tag{7.47}$$

for all differentiable functions $\varphi = \varphi(x,t)$ with bounded support. The initial value condition (7.43) is already taken into account by (7.47).

[6]See also the detailed discussion in section 1.13.1.2.

The name 'conservation form' for the equations (7.46) and (7.47) comes from the fact that $\int_{\mathbb{R}} u(t,x)\,\mathrm{d}x$ remains constant for all $t$ (for example conservation of energy, momentum and mass in the case of the Euler equations).

The discontinuity of the function $\varphi = \varphi(t,x)$ with right- or left-sided limits $\varphi(t,x+0), \varphi(t,x-0)$ will be denoted by $[\varphi](t,x) := \varphi(t,x+0) - \varphi(t,x-0)$. If the weak solution $u(t,x)$ of (7.47) has a jump ('shock') along the curve $(t,x(t))$, then there is the following relation between the slope $\mathrm{d}x/\mathrm{d}t$ of the curve and the jumps:

$$\frac{\mathrm{d}x}{\mathrm{d}t}[u] = [f(u)] \qquad \textit{(Rankine–Hugoniot discontinuity condition).} \tag{7.48}$$

The importance of the weak formulation (7.47) can be illustrated with the following example. The equations $u_t - (u^2/2)_x = 0$ and $v_t - (2v^{3/2}/3)_x = 0$ are equivalent under the substitution $v = u^2$, provided the equations are classical, i.e., differentiable. But since the formulation uses all possible flux functions, (7.48) yields in the case of a shock different slopes $\mathrm{d}x/\mathrm{d}t$ and therefore also different solutions.

A weak solution is in general not uniquely determined. The physically sensible solution is characterized by the *entropy condition.* The simplest formulation of this condition is that we have $f'(u_\ell) > \mathrm{d}x/\mathrm{d}t > f'(u_r)$ along the shock wave given by $u_\ell := u(t,x(t)-0)$ and $u_r := u\big(t,x(t)+0\big)$. For generalizations and formulations with an entropy function, we refer to [447]. *Entropy solutions*, by which we mean solutions of (7.47) which satisfy the entropy condition, can also be obtained as the limit of $u_t + f(u)_x = \varepsilon u_{xx}$ $(\varepsilon > 0)$ as $\varepsilon \to 0$.

While smooth solutions of hyperbolic equations are reversible, this is not the case for non-continuous entropy solutions.

#### 7.7.5.9 Numerical properties of the non-linear case

For numerical approximations two new questions arise here: Suppose the discretization converges to a function $u$. Then 1) is $u$ a weak solution in the sense of (7.47), and 2) is $u$ an entropy solution?

In order to answer the first question, we formulate the difference procedure in a conservation form:

$$u_\nu^{m+1} := u_\nu^m + \lambda\big[F(u_{\nu-p}^m, u_{\nu-p+1}^m, \ldots, u_{\nu+q}^m) - F(u_{\nu-p-1}^m, u_{\nu-p}^m, \ldots, u_{\nu+q-1}^m)\big]$$

with $\lambda$ from (7.39). The function $F$ is aptly referred to as the *numerical flux.* Solutions of these equations have the discrete conservation property $\sum_\nu u_\nu^m = \text{const}$. The Friedrichs procedure (7.40c) can be written in its non-linear form with the numerical flux

$$F(U_\nu, U_{\nu+1}) := \frac{1}{2\lambda}(U_\nu - U_{\nu+1}) + \frac{1}{2}\big(f(U_\nu) + f(U_{\nu+1})\big) \tag{7.49}$$

(while the linear case $f(u) = au$, $a = \text{const}$ corresponds to (7.40c)). The consistency of the procedure can be expressed among others through the condition $F(u,u) = f(u)$. If consistent difference conditions in conservation form converge, then the solution is a weak solution of (7.47) but need not satisfy the entropy condition.

The procedure (7.49) is *monotonous*, i.e., two initial values $u^0$ and $v^0$ with $u^0 \le v^0$ yield $u^m \le v^m$. Procedures of order greater than one can not be monotonous. Procedures which are monotonous and consistent converge to entropy solutions.

From the monotonousness a further property, the *TVD property* (total variation diminishing) follows, which means that the total variation $TV(u^m) := \sum_{\nu} |u_\nu^m - u_{\nu+1}^m|$ is monotonously decreasing as $m$ grows. This property prevents for example the above mentioned oscillations near to the shock waves.

### 7.7.6 Adaptive discretization procedures

#### 7.7.6.1 Variable grid size

The discretization procedures for ordinary and partial differential equations generally use a grid or a triangulation or some similar kind of decomposition of the domain. In the simplest case the structure to be used is chosen to have equidistant grid size $h$. The error analysis is generally carried out for this case and yields error estimates of the form $c(u)h^\kappa$, where $\kappa$ is the consistency order and $c(u)$ is a quantity which is independent of $h$, but generally depends on bounds of (higher) derivatives of $u$. As long as the mentioned derivatives have roughly the same size, there is no problem with taking the grid in this form.

There are, however, very many causes for the situation that the derivatives have very different magnitudes at different locations, and can even have singularities, i.e., points at which their values are unbounded. In the case of a simple elliptic differential equation $Lu = f$ (cf. 7.7.3.1.1.), edges or corners of the boundary of the domain generally lead to solutions whose higher derivatives have singularities at these points. Also a special right-hand side of the equation (for example $f$ a point force) can make $u$ less smooth at an arbitrary point. Singular perturbations can lead to extended boundary layers (meaning that there are large gradients in the direction normal to the boundary). If the grid is equidistant of size $h$, then the preciseness of numerical approximations is compromised by the existence of singularities. In order to obtain the same precision as for a smooth solution, we would have to choose a much smaller size $h$ of the grid, which in applications soon leads to impracticability because of limited computational resources. Instead, the idea is to use the finer grid only near those points where it is necessary, leaving the coarser grid for the majority of the domain. This means that we require a grid with different size scales at different points, or a triangulation with triangles of different sizes, as depicted in Figure 7.5.

In what follows, we will use the simple problem of a numerical integration as an illustration of the discussion above. If the integral $\int_0^1 f(x)\,dx$ is approximated with the summed trapeze rule from 7.3.3.1 for a twice continuously differentiable function $f$, then the error is estimated by $h^2 f''(\zeta)/12$, where $h = 1/N$ denotes the equidistant step size, $N+1$ denotes the number of vertices of the grid and $\zeta$ denotes a mean value. The computational cost is essentially given by $N+1$ functional evaluations of $f$. The error can then be written in dependency on $N$ as $O(N^{-2})$. For the integrand $f(x) := x^{0.1}$, already the first derivative is unbounded at the left boundary point $x = 0$. For $N+1$ equidistant vertices of the grid $x_i = ih = i/N$ one finds an error of magnitude $O(N^{-1.1})$. If instead one chooses a variable step width with vertices $x_i = (i/N)^{3/1.1}$, which are distributed more frequently near the singularity $x = 0$, then we again have a quadrature error of magnitude $O(N^{-2})$ as in the smooth case. If one wants a quadrature result with for example an absolute error of $\leq 10^{-6}$, then one requires $N = 128\,600$ evaluations in the equidistant case, but only $N = 391$ in the case of the variable step size.

The choice of $x_i = (i/N)^{3/1.1}$ follows the general strategy of *even distribution of the error*, i.e., the local errors (here given by the trapeze rule on $[x_i, x_{i+1}]$) should be as close to each other as possible.

In the example just discussed, the behavior of the solution (location and exponent of the singularity) were known and the discretization could accordingly be optimally fit to the problem. For the following reasons, an *a priori* adaption of the grid size will be an exception.

(a) Whether there are any singularities at all, or, if they exist, where they are located, is generally not known from the onset (in particular not so for a non-specialist).

(b) Even if we know the behavior of singularities ahead of time, taking account of this in the algorithm requires the 'insider' knowledge of a numerical analyst in addition to requiring more work for its implementation.

#### 7.7.6.2 Self-adaptivity and error indicators

The natural alternative to making an *a priori* choice of local mesh sizes is to use the insights provided by a running numerical procedure to deduce the information necessary. A simple case is the method for changing the step width for ordinary differential equations discussed in section 7.6.1.1. There the length of the *next* step is optimally chosen with all the information provided by the procedure up to that point. For boundary value problems, however, one obtains no information without already constructing a preliminary solution by means of some chosen mesh. Therefore, the following steps have to be iterated several times.

(a) Solve the problem with a given mesh.

(b) Determine improved local mesh sizes by using the information of the solution obtained in (a).

(c) Construct a new mesh with the information provided by (a) and (b).

At this point the discretization and the approximation of the solution are blended in an inseparable way to a single unit. Since the adaption is now part of the algorithm, one speaks of self-adaptivity.

Questions which arise in connection with the steps (a)-(c) are:

(1) How can we obtain local mesh sizes in (b)?

(2) How are the improved meshs constructed?

(3) When is the mesh adaptive enough that we can stop the iteration of (a)-(c)?

Here are some answers, by no means exhaustive, but first hints.

**Ad (1):** Let for example a mesh have the form of a finite element triangulation $\tau$ and let $\tilde{u}$ be a corresponding solution. An *error indicator* is a function $\varphi$ of $\tilde{u}$ which associates to each $\Delta \in \tau$ a value $\varphi(\Delta)$. The idea is that $\varphi(\Delta)$ is closely connected with the error on $\tau$ or the part of the error due to $\tau$. For the adaption of the step size, there are two strategies.

($\alpha$) If we have an appropriate theory at our disposal, we can exhibit a function $H(\varphi)$ which suggests a new mesh size $h = H(\varphi(\Delta))$ on $\Delta$.

($\beta$) This strategy starts with the ideal of the *equidistribution*. Suppose $\varphi$ has the same magnitude at all $\Delta \in \tau$, requiring perhaps a uniform subdivision of the mesh for this magnitude to be acceptable. Until we have reached this state of affairs, we make the mesh finer at those points for which, for example $\varphi(\Delta)$ is greater than $0.5 \cdot \max\{\varphi(\Delta) : \Delta \in \tau\}$.

The error indicator can for example be defined via the residue. The *residue* $r = f - L\tilde{u}$ is obtained by inserting an approximation to the solution $\tilde{u}$ into the differential equation

$Lu = f$, which then has to be evaluated on $\Delta$ (see also (7.51)).

**Ad (2):** In strategy ($\alpha$) we have produced an everywhere defined optimal step size $h = h(x)$. There is an algorithm with which a triangulation can be constructed with triangles of just this size. Still this global adaptation is less appropriate for step (b), as the cost for the recalculation of the mesh is tremendous. Moreover, all quantities calculated up to that point (for example a finite element matrix) can no longer be used. The strategy ($\beta$) is more appropriate for a *local* adaption of the mesh. Only triangles which have been distinguished for subdivision are decomposed into smaller ones. (Note that this process may require a subdivision also of neighboring triangles, to obtain in sum an admissible triangulation, see the triangle STU in Figure 7.8 as well as [445], §3.8.2.) This has the advantage that only certain new finite element matrix coefficients need to be recalculated. Also, this provides us with a hierarchy of triangulations, which can be taken advantage of by, for example, the multiple step procedures.

**Ad (3):** Stopping as soon as the condition $\varphi(\Delta) \leq \varepsilon$ for all $\Delta \in \tau$ is satisfied seems to be quite natural. It would be ideal if this would really insure that the actual discretization error is less than $\varepsilon$. Error indicators $\varphi$ which are so closely related to the actual errors will be discussed in the next section.

### 7.7.6.3 Error estimators

Let $e(\tilde{u})$ be the error of a finite element solution $\tilde{u}$ with respect to an exact solution, measured in some appropriate norm. Furthermore, let $\varphi$ be the error indicator described above, which summed over all triangles of the mesh yields the quantity $\Phi(\tilde{u}) := \Big[\sum_{\Delta\in\tau} \varphi(\Delta)^2\Big]^{\frac{1}{2}}$. The error indicator $\varphi$ is said to be a *error estimator*, if we have the inequalities

$$A\Phi(\tilde{u}) \leq e(\tilde{u}) \leq B\Phi(\tilde{u}), \qquad 0 < A \leq B, \tag{7.50}$$

or at least asymptotic approximations of these. The second inequality is sufficient for insuring an error of $e(\tilde{u}) \leq \varepsilon$ upon stopping the algorithm with $\Phi(\tilde{u}) \leq \eta := \varepsilon/B$. A $\Phi$ which satisfies the second inequality is said to be *dependable*. If it satisfies in addition the first inequality it is said to be *efficient*, since the meshs which are too fine (and thus too costly) can be avoided. Indeed, as soon as the error estimate $e(\tilde{u}) \leq \varepsilon A/B$ is realized, the criterion $\Phi(\tilde{u}) < \eta$ for stopping the procedure takes effect. In the best possible case the error analyzer is asymptotically optimal, meaning that we have asymptotically $A, B \to 1$ in (7.50). Since from (7.50) we can determine the error from the calculation, one speaks in this case of an *a posteriori* estimate.

There are a series of suggestions for error estimators. However, take note of the fact that all error estimators $\varphi$ which require only finitely many evaluations never guarantee that the error can be estimated from above and below as in (7.50). Inequalities of the form of (7.50) can only be deduced from theoretical assumptions on the solutions. Note also that these theoretical assumptions are of a qualitative kind and as opposed to the situation discussed in 7.7.6.1 do not reenter the implementation themselves.

In the case of the Poisson equation (7.9a) with homogeneous Dirichlet boundary values (7.9b) (and $g = 0$) and a discretization in terms of piecewise linear finite elements on triangles, the Babuška–Rheinboldt error estimator on a triangle $\Delta \in \tau$ is given by

$$\varphi(\Delta) := \left(h_\Delta^2 \int_\Delta f(x)^2 \mathrm{d}x + \frac{1}{2}\sum_K h_K \int_K \left[\frac{\partial \tilde{u}}{\partial n}\right]^2 \mathrm{d}s\right)^{\frac{1}{2}}. \tag{7.51}$$

Here $h_\Delta$ denotes the diameter of $\Delta$, and the sum is over the three sides of the triangle;

$h_K$ is the length of the edge and $[\partial u/\partial n]$ denotes the jump of the normal derivative along the edge $K$.

## 7.7.7 Iterative solutions of systems of equations

### 7.7.7.1 Generalities

When systems of linear equations arise through the discretization of a differential equation, then on the one hand the dimension of these systems is quite high (typically of magnitude 10,000 to 10,000,000), on the other the matrix is generally sparse, i.e., it contains per row a small number of non-vanishing entries which is independent of the dimension; the latter has already been discussed in section 7.7.3.1.4. In the case of the discrete Poisson equation (7.4), for example, this number of non-vanishing entries is five. If one were to apply straightforward methods for the solution (Gauss elimination, Cholesky decomposition, Householder procedure), then during the process of the algorithm we would be creating non-vanishing elements at spots in the matrix where the entries previously had vanished. This in turn would lead to difficulties in storing the matrix entries as discussed in 7.7.3.1.4. Moreover, the computation cost would grow more than just linearly with the dimension, as we have seen in previous discussions. Compared with this situation, the method of matrix-vector multiplication requires only the non-vanishing elements of the matrix and the computational cost is in this case proportional to the dimension. Iteration procedures which are based on this operation (matrix-vector multiplication) have therefore a minor computational cost. If, in addition, the convergence to a solution is fast, then the iterative methods are the ideal procedures for solving large systems of equations.

**7.7.7.1.1 Richardson iteration.** In what follows we deal with the system of equations

$$\boxed{Ax = b.} \tag{7.52}$$

The only assumption we make at this point is that $A$ is non-singular, so that a solution of (7.52) is insured. The basic model for every iteration is the Richardson iteration, given by the algorithm

$$\boxed{x^{m+1} := x^m - (Ax^m - b),} \tag{7.53}$$

with an arbitrary initial vector $x^0$.

**7.7.7.1.2 General linear iteration.** The general algorithm given by linear iteration is

$$\boxed{x^{m+1} := Mx^m + Nb, \quad \text{(first normal form)}} \tag{7.54a}$$

with matrices $M$ and $N$ which are connected by the relation $M + NA = I$. If one eliminates the *iteration matrix* $M$ from (7.54a) with the aid of $M + NA = I$, one obtains

$$\boxed{x^{m+1} := x^m - N(Ax^m - b), \quad \text{(second normal form).}} \tag{7.54b}$$

Since a singular matrix $N$ generates divergences, we assume here further that $N$ is invertible with inverse matrix $N^{-1} = W$. An implicit formulation of (7.54b) is

$$\boxed{W(x^m - x^{m+1}) = Ax^m - b, \quad \text{(third normal form).}} \tag{7.54c}$$

**7.7.7.1.3 Convergence of iteration procedures.** The iteration procedure described by (7.54a-c) is said to be *convergent*, if the iterated $\{x^m\}$ converge to the same solution (which is then necessarily the solution of (7.52)) for all initial values $x^0$. The procedure (7.54a) is convergent if and only if the spectral radius satisfies the condition $\rho(M) < 1$, i.e., all the eigenvalues of $M$ are less than unity in absolute value.

The so-called *speed of convergence* is of particular interest. If we have

$$\boxed{\rho(M) = 1 - \eta < 1} \tag{7.55a}$$

for small $\eta$, then one requires only roughly $1/\eta$ iteration steps for the error to improve by a factor of $1/\mathrm{e}$ (where $\mathrm{e} = 2.71\ldots$). In fact, we would like

$$\rho(M) \leq \text{const} < 1 \tag{7.55b}$$

to be valid, where the constant does not depend on the dimension of the system of equations (for example, does not depend on the grid size of the discretization procedure which gave rise to the system of equations in the first place). In this case we can achieve a fixed precision (for example an estimate $\|x - x^m\| < \varepsilon$) using only a constant number $m$ of iterations.

**7.7.7.1.4 Generation of an iteration procedure.** There are two different methods which can be applied to obtain the rules for the iteration. The first of these is called the *splitting method.* The matrix $A$ is additively split as

$$A = W - R, \tag{7.56}$$

where $W$ is required not only to be invertible, but also have the property that systems of equations of the form $W\nu = d$ are relatively easy to solve. The idea behind this is that $W$ contains the essential information about $A$, and the 'rest' $R$ is thought of as being 'small'. Using $Wx = Rx + b$ one obtains the iteration $x^{m+1} := W^{-1}(Rx^m + b)$ which coincides with (7.54b) for the choice of $N = W^{-1}$.

If one chooses $W$ to be the diagonal of $A$, then we have produced the *Jacobi procedure* already discussed in 7.2.2.1. In the case of the *Gauss–Seidel procedure*, the rest $R$ in (7.56) consists of the upper right corner of the matrix $A$, i.e., $R_{ij} = A_{ij}$ for $j > i$ and $R_{ij} = 0$ otherwise.

A *regular decomposition* (7.56) is realized if we have $W^{-1} \geq 0$ and $W \geq A$ in the sense of element-wise inequalities. This turns out to automatically imply convergence (see [444], §6.5).

A different technique is to take the left transformation of equation (7.52) by a nonsingular matrix $N$ so that $NAx = Nb$. If one writes for this $A'x = b'$ ($A' = NA$, $b' = Nb$), and applies the Richardson iteration (7.53), then one obtains a transformed iteration $x^{m+1} := x^m - (A'x^m - b')$, which again can be written in the form

$$x^{m+1} := x^m - N(Ax^m - b)$$

and therefore coincides with the second normal form (7.54b).

Both of the described methods make it, at least in principle, possible to generate *every* iteration. Conversely, every iteration (7.54b) can be interpreted as a Richardson iteration (7.53) applied to $A'x = b'$ with $A' = NA$.

The matrices $N$ and $W$ need not be available in a component-wise stored form. It is only important that the matrix-vector multiplication $d \to Nd$ is easily carried out. In the case of an *incomplete block ILU decomposition* (described in more detail in [444], §8.5.3), $N$ has the form $N = (U' + D)^{-1}D(L' + D)^{-1}$ with strictly lower (resp. upper) triangular matrices $L'$ (resp. $U'$) and a block diagonal matrix $D$.

**7.7.7.1.5 Efficient iteration schemes.** An iteration procedure should be, on the one hand, fast (see (7.55a,b)), on the other it should have as small a computation cost per iteration as possible (for more information on determining the 'effective cost' see [444], §3.3.2). We have a fundamental dilemma in the fact that these two requirements work against each other. The fastest convergence can be achieved for $W = A$, as then $M = 0$, and the exact solution is obtained after a single step, but requires directly solving the systems of equations (7.54c) with matrix $A$. At the same time, a simple choice of $W$ as diagonal or lower-diagonal matrix, yielding to the Jacobi or Gauss–Seidel procedures, leads to speeds of convergence which for a discretized Poisson equation (7.4) with step width $h$ are of the form (7.55a) with $\eta = O(h^2)$. According to 7.7.7.1.3, we then require $O(h^{-2})$ iteration steps, a number which increases rapidly.

### 7.7.7.2 Positive definite matrices

The analysis we are discussing is greatly simplified when the matrices $A$ and $N$ (and consequently also $W$) are positive definite and symmetric. In what follows we shall thus make this assumption.

**7.7.7.2.1 Matrix condition and speed of convergence.** By assumption $A$ has only positive eigenvalues. Let $\lambda = \lambda_{\min}(A)$ be the smallest and $\Lambda = \lambda_{\max}(A)$ the largest of these eigenvalues. The condition number $\kappa(A)$, introduced in 7.2.1.7, (using the Euclidean norm as vector norm) then has the value $\kappa(A) = \Lambda/\lambda$. The condition number does not change when $A$ is multiplied by a constant factor, $\kappa(A) = \kappa(\Theta A)$. After an appropriate scalar multiplication we may assume that $\Lambda + \lambda = 2$. Under these assumptions the Richardson iteration (7.53) has the rate of convergence

$$\rho(M) = \rho(I - A) = (\kappa(A) - 1)/(\kappa(A) + 1) = 1 - 2/(\kappa(A) + 1) < 1,$$

in other words the value $\eta$ from (7.55a) is $\eta = 2/(\kappa(A) + 1)$. Matrices which have good condition (meaning that $\kappa(A) = O(1)$) therefore lead to satisfactory convergence, while matrices which arise during the discretization of boundary value problems have a condition of magnitude $O(h^{-2})$.

The scalar multiplication with $\Theta$ corresponds in general to the (optimal) *damped iteration*

$$x^{m+1} := x^m - \Theta N(Ax^m - b) \quad \text{with} \quad \Theta := 2/(\lambda_{\max}(NA) + \lambda_{\min}(NA)). \qquad (7.57)$$

The above considerations remain valid if $A$ is similar to a positive definite matrix, as the spectral quantities are invariant under similarity transformations.

**7.7.7.2.2 Preconditioning.** The transformation described in 7.7.7.1.4 leads to a Richardson iteration for a new matrix $A' = NA$ (this matrix is not necessarily positive definite, but it is similar to a positive definite matrix). If $A'$ has a smaller condition number than $A$, then the procedure (7.54b) arising from the transformation (Richardson with matrix $A'$) has a speed of convergence which is superior to that of (7.53). In this sense $N$ is called a *preconditioning matrix* and (7.54b) is called the preconditioning iteration. If this iteration is optimally damped as in (7.57), then its speed of convergence is

$$\rho(M) = \rho(I - \Theta NA) = (\kappa(NA) - 1)/(\kappa(NA) + 1) = 1 - 2/(\kappa(NA) + 1) < 1. \quad (7.58)$$

**7.7.7.2.3 Spectral equivalence.** In what follows we let the notation $A \leq B$ stand for the situation that $A$ and $B$ are symmetric and $B - A$ is positive semidefinite.[7] $A$ and $W$ are said to be *spectral equivalent* (with equivalence constant $c$), if

$$A \leq cW \quad \text{and} \quad W \leq cA. \quad (7.59)$$

The case in which $c$ does not depend on parameters like the dimension of the discretization matrices is particularly interesting. The spectral equivalence (7.59) insures the estimate $\kappa(NA) \leq c^2$ for the condition number, where $N = W^{-1}$. If one can find a matrix $W$ corresponding to $A$ which is easy to invert, then the iteration (7.54c) has (perhaps after including dampening) the speed of convergence $1 - 2/(c^2 + 1)$.

**7.7.7.2.4 Transformation utilizing a hierarchical basis.** Suppose the matrix $A$ arises from a finite element discretization with some standard grid. In the case of elliptic problems of second order, considered in 7.7.3, the condition number is $\kappa(A) = O(h^{-2})$. The transformation $x = Tx'$ between the coefficients $x$ of the vertices of the grid and the coefficients $x'$ of the hierarchical basis introduced in 7.7.3.1.7 can be implemented in such a way that the multiplications $T^{\mathsf{T}}$ and $T$ are easily carried out. Through the two way transformation given by (7.52) one obtains $T^{\mathsf{T}}ATx' = T^{\mathsf{T}}b$, in other words $A'x' = b'$ with the stiffness matrix $A' = T^{\mathsf{T}}AT$ with respect to the hierarchical basis. By expressing the $x$-quantities with the Richardson iteration $x'^{m+1} := x'^m - (A'x'^m - b')$, one obtains $x^{m+1} := x^m - TT^{\mathsf{T}}(Ax^m - b)$, i.e., (7.54b) with $N = TT^{\mathsf{T}}$. In the case of elliptic equations in two space variables one can show $\kappa(A') = O(|\log h|)$. Because of this, the transformed (hierarchical) iteration has with $N = TT^{\mathsf{T}}$ an almost optimal (only weakly dependent on $h$) speed of convergence, which is $\rho(M) = 1 - O(|\log h|)$.

### 7.7.7.3 Semi-iterative procedures

A *semi-iterative* procedure consists of the iteration (7.57), provided that the dampening parameter $\Theta$ is allowed to vary during the iteration:

$$\boxed{x^{m+1} := x^m - \Theta_m N(Ax^m - b).} \quad (7.60)$$

The essential property of semi-iterations are described by the polynomials $p_m$, given by

$$p_m(\zeta + 1) := (\Theta_0\zeta + 1)(\Theta_1\zeta + 1) \cdot \ldots \cdot (\Theta_m\zeta + 1).$$

If one knows the extremal eigenvalues $\Lambda = \lambda_{\max}(NA)$ and $\lambda = \lambda_{\min}(NA)$, then $p_m$ can be chosen to be a Chebychev polynomial (cf. 7.5.1.3), which is transformed from the

[7] This means that all eigenvalues of $B - A$ are non-negative.

interval $[-1, 1]$ to $[\lambda, \Lambda]$ and which is normalized according to $p_m(1) = 1$. The speed of convergence improves then from $(\kappa(NA) - 1)/(\kappa(NA) + 1)$ for a simple iteration (see (7.58) to the asymptotic speed of convergence

$$(\sqrt{\kappa(NA)} - 1)/(\sqrt{\kappa(NA)} + 1) \tag{7.61}$$

of the semi-iteration. In particular for slower iterations (i.e. for which $\kappa(NA) \gg 1$) is replacement of $\kappa(NA)$ by $\sqrt{\kappa(NA)}$ is essential. As an aid for the actual implementation, one uses instead of (7.60) the *three term relation*

$$\boxed{x^m := \sigma_m\{x^{m-1} - \Theta N(Ax^{m-1} - b)\} + (1 - \sigma_m)x^{m-2} \qquad (m \geq 2)} \tag{7.62}$$

with $\sigma_m := 4/\{4 - [(\kappa(NA) - 1)/(\kappa(NA) + 1)]^2\sigma_{m-1}\}, \sigma_1 = 2$ and $\Theta$ as in (7.57). For this initial term with $m = 2$ one uses $x^1$ from (7.57).

### 7.7.7.4 Gradient methods and conjugate gradients

The semi-iteration (7.60) is a method for accelerating the basic iteration (*basis iteration* (7.54b)). The iterates from (7.60) or (7.62) remain linearly independent from the initial value $x^0$. In contrast, the procedures we now describe are non-linear methods, i.e., $x^m$ does not depend linearly on $x^0$. Note that the gradient methods we describe do not replace iterations, but rather are combined with a basis iteration to improve the latter.

**7.7.7.4.1 Gradient procedure.** The gradient procedure, applied to the basis iteration (7.54b) with positive definite matrices $A$ and $N$, is given by

$$x^0 \text{ arbitrary}, \qquad r^0 := b - Ax^0, \qquad \text{(start)} \tag{7.63a}$$

$$q := Nr^m, \quad a := Aq, \quad \lambda := \langle q, r^m\rangle/\langle a, q\rangle, \qquad \text{(recursion)} \tag{7.63b}$$

$$x^{m+1} := x^m + \lambda q, \qquad r^{m+1} := r^m - \lambda a. \tag{7.63c}$$

Here not only the iterate $x^m$, but also the residue $r^m := b - Ax^m$ is recalculated in each step. The vectors $q$ and $a$ are only used to store intermediate results, as for each gradient step of the procedure only one matrix-vector multiplication is required. For a derivation of this procedure, see [444], §9.2.4.

The asymptotic speed of convergence is as in (7.58) $(\kappa(NA) - 1)/(\kappa(NA) + 1)$. Thus the gradient method is just as fast as the optimal damped iteration (7.57). In contrast with (7.57), the gradient method reaches this rate without any explicit knowledge of the extreme eigenvalues $\lambda_{\max}(NA)$ and $\lambda_{\min}(NA)$, which are necessary for (7.57).

**7.7.7.4.2 Conjugate gradients.** The procedure we now discuss, also known as the 'CG-method', can like the gradient method just discussed be applied to a basis iteration (7.54b) with positive definite matrices $A$ and $N$. It is given by

$$x^0 \text{ arbitrary}, \quad r^0 := b - Ax^0, \quad p^0 := Nr^0, \quad \rho_0 := \langle p^0, r^0\rangle, \qquad \text{(start)} \tag{7.64a}$$

$$a := Ap^m, \quad \lambda := \rho_m/\langle a, p^m\rangle, \qquad \text{(recursion)} \tag{7.64b}$$

$$x^{m+1} := x^m + \lambda p^m, \quad r^{m+1} := r^m - \lambda a, \tag{7.64c}$$

$$q^{m+1} := Nr^{m+1}, \quad \rho_{m+1} := \langle q^{m+1}, r^{m+1}\rangle, \quad p^{m+1} := q^{m+1} + (\rho_{m+1}/\rho_m)p^m. \tag{7.64d}$$

Here the 'search direction' $p^m$ itself is part of the recursion.

The asymptotic speed of convergence of this method is as in (7.61) at least $(\sqrt{\kappa(NA)}-1)/(\sqrt{\kappa(NA)}+1)$. In comparing this method with the semi-iteration (7.62), it is important that the CG-method requires no previous knowledge of the spectral data $\lambda_{\max}(NA)$ and $\lambda_{\min}(NA)$ and can be even faster than (7.61).

In case in (7.64b) a division by zero occurs, because of $\langle a, p^m\rangle = 0$, $x^m$ is already the exact solution of the problem!

The CG-method (7.64) is basically speaking a direct procedure, as after $n$ steps at the latest (where $n$ is the dimension of the system of equations) the exact solution is attained. However, this property is meaningless in applications, as for large system of equations the maximal number of iteration should be far below the actual dimension $n$.

The application of (7.64) to the Richardson iteration (7.53) as basis iteration (i.e. $N = I$) reduces the algorithm to the scheme presented in 7.2.2.2.

### 7.7.7.5 Multi-grid methods

**7.7.7.5.1 Generalities.** Multi-grid methods are iteration procedures which can be applied to the discretization of elliptic differential equations and have *optimal convergence.* By this we mean that the speed of convergence does not depend on the discretization step width and therefore is also independent of the dimension of the system of equations (see (7.55b)). In contrast with the CG-methods just discussed, it is not important for multi-grid methods that the matrix $A$ is positive definite or symmetric.

The multi-grid method has two complementary components, a *smoothing iteration* and a *coarse grid correction.* The smoothing iteration are classical iteration procedures which smooth the error (not the solution!). The coarse grid correction reduces the 'smooth' error produced by the smoothing iteration. It uses discretization on coarser grids as a tool, and this is what gives the method its name. However, the name does not mean that the procedure is restricted to discretization in regular grids. It can just as well be applied to general finite element methods, in which case it is useful if the finite element spaces form a hierarchy.

**7.7.7.5.2 An example of a smoothing iteration.** Simple examples for smoothing iterations are the Gauss–Seidel iteration described in 7.2.2.1 and the Jacobi procedure, damped with $\Theta = 1/2$:

$$\boxed{x^{m+1} := x^m - \frac{1}{2}D^{-1}(Ax^m - b).} \tag{7.65}$$

In the case of the five point formula (7.4), the vector $x^m$ consists of the components $u_{ik}^m$. Equation (7.65) can be read component-wise as

$$u_{ik}^{m+1} := \frac{1}{2}u_{ik}^m + \frac{1}{8}\left(u_{i-1,k}^m + u_{i+1,k}^m + u_{i,k-1}^m + u_{i,k+1}^m\right) + \frac{1}{8}h^2 f_{ik}.$$

Let $e^m := x^m - x$ be the error of the $m^{th}$ iteration step. In (7.65) this error satisfies the recursion formula

$$e_{ik}^{m+1} := \frac{1}{2}e_{ik}^m + \frac{1}{8}\left(e_{i-1,k}^m + e_{i+1,k}^m + e_{i,k-1}^m + e_{i,k+1}^m\right).$$

The right-hand side is an average value, which is formed with neighboring points. This makes it clear that oscillations are quickly damped and the error thus is indeed smoothed.

**7.7.7.5.3 Coarse grid correction.** Let $\tilde{x}$ be the result of several smoothing iteration steps as just described. The error $\tilde{e} := \tilde{x} - x$ is the solution of $A\tilde{e} = \tilde{d}$, where the *defect* is calculated from $\tilde{d} := A\tilde{x} - b$. Let $X_n$ be the $n$-dimensional space of vectors $x$. As $\tilde{e}$ is smooth, it can be approximated by coarser grids. So let $A'$ be the discretization matrix of a coarser step width (or a coarser finite element space), and let $x'$ be the corresponding coefficient vector in the lower dimensional space $X_{n'}$ $(n' < n)$.

Between $X_n$ and $X_{n'}$ we introduce two linear mappings: the *restriction* $r : X_n \to X_{n'}$ and the prolongation $p : X_{n'} \to X_n$.

In the case of a one-dimensional Poisson equation discretized by the difference (7.3a) on the grids of sizes $h$ and $h' := 2h$, one chooses for $r : X_n \to X_{n'}$ the weighted mean

$$d' = rd \in X_{n'} \qquad \text{with} \quad d'(\nu h') = d'(2\nu h) := \frac{1}{2}d(2\nu h) + \frac{1}{4}\big[d\big((2\nu+1)h\big) + d\big((2\nu-1)h\big)\big].$$

For $p : X_{n'} \to X_n$ one chooses the linear interpolation

$$u = pu' \in X_n \quad \text{with} \quad u(\nu h) := u'\left(\frac{\nu}{2}h'\right) \qquad \text{for even } \nu$$

$$u(\nu h) := \frac{1}{2}\left[u'\left(\frac{\nu-1}{2}h'\right) + u'\left(\frac{\nu+1}{2}h'\right)\right] \qquad \text{for odd } \nu$$

(see Figure 7.12). In more general cases, $r$ and $p$ can be chosen such that applying them implies a minimum of computational cost.

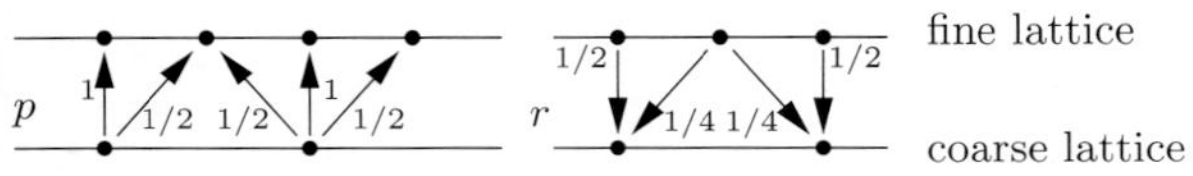


*Figure 7.12. Grid transfer p and r.*

The equation $A\tilde{e} = \tilde{d}$ for the error $\tilde{e} := \tilde{x} - x$ corresponds on the coarser grid to the so-called coarse grid equation

$$A'e' = d' \quad \text{with} \quad d' = rd.$$

Its solution yields $e'$ and the prolonged value $e := pe'$. By definition $x = \tilde{x} - \tilde{e}$ is the exact solution, so $\tilde{x} - pe'$ should be a good approximation. Correspondingly the *coarse grid correction* is

$$\tilde{x} \longmapsto \tilde{x} - pA'^{-1}r(A\tilde{x} - b). \tag{7.66}$$

**7.7.7.5.4 The two-grid procedure.** The two-grid method is the product of a smoothing iteration from 7.7.7.5.2 and the coarse grid correction (7.66). If $x \mapsto \mathscr{S}(x, b)$ denotes the smoothing iteration (for example (7.65)), then the two-grid algorithm is given by

$$x := x^m\,; \tag{7.67a}$$

$$\text{for} \quad i := 1 \quad \text{to} \quad \nu \quad \text{do} \quad x := \mathscr{S}(x, b)\,; \tag{7.67b}$$

$$d' := r(Ax - b)\,; \tag{7.67c}$$

$$\text{solve} \quad A'e' = d'\,; \tag{7.67d}$$

$$x^{m+1} := x - pe'\,; \tag{7.67e}$$

Here $\nu$ denotes the number of smoothing iterations. Usually $\nu$ is on the order of $2 \leq \nu \leq 4$. The two-grid method itself has little practical relevance, since (7.67d) requires the exact solution of the (lower-dimensional) system of equations.

**7.7.7.5.5 Multi-grid methods.** In order to approximatively solve (7.67d), the procedure is recursively applied. For this we require coarse discretizations. Altogether it is necessary to have a hierarchy of discretizations:

$$A_\ell x_\ell = b_\ell, \qquad \ell = 0, 1, \ldots, \ell_{\max}, \tag{7.68}$$

where for the maximal level $\ell = \ell_{\max}$ equation (7.68) coincides with the original equation $Ax = b$. For $\ell = \ell_{\max} - 1$, the system $A'$ used in (7.67d) arises. For $\ell = 0$ the dimension $n_0$ is assumed to be so small (for example $n_0 = 1$) that the exact solution of $A_0 x_0 = b_0$ can be directly calculated.

The multi-grid method for solving $A_\ell x_\ell = b_\ell$ is characterized by the following algorithm. The function $MGM(\ell, x, b)$ yields for $x = x_\ell^m$ and $b = b_\ell$ the next iterate of the algorithm, $x_\ell^{m+1}$:

function $MGM(\ell, x, b)$; (7.69a)

if $\ell = 0$ then $MGM := A_0^{-1} b$ else (7.69b)

begin for $i := 1$ to $\nu$ do $x := \mathscr{S}_\ell(x, b)$; (7.69c)

$d := r(A_\ell x - b)$; (7.69d)

$e := 0$; for $i := 1$ to $\gamma$ do $e := MGM(\ell - 1, e, d)$; (7.69e)

$MGM := x - pe$ (7.69f)

end;

Here $\gamma$ is the number of coarse lattice corrections; the only cases of interest are $\gamma = 1$ (*V-cycle*) and $\gamma = 2$ (*W-cycle*).

For details on implementing this algorithm and further numerical examples we refer to [445] and [444], §10.

**7.7.7.5.6 Numerical examples for discrete Poisson equations.** As system of equations we choose the discretization (7.4) of (7.9a) for $\Omega = (0,1) \times (0,1)$ with Dirichlet boundary values (7.9b). The step width is taken to be $h = 1/64$, so that $63^2 = 3\,969$ is the number of unknowns. The $m^{th}$ error $e^m := u^m - u$ is measured in the energy norm $\|e\| := \langle Ae, e\rangle^{1/2}$. The initial error $e^0$ has the norm 2.47E−1. The Gauss–Seidel iteration has, even after 300 iteration steps, only reduced the initial error by a factor of 10. To accelerate the procedure we use the so-called SOR-method with optimal over-relaxation parameter. Then the bound on the error of 1E − 6 is reached after 161 iterations.

| The multi-grid method requires for this only five steps. |
|---|

If one were to further reduce the step width $h$, then the speed of convergence for the first two procedures would get even worse, while the number of iterations for the multi-grid method (with a checkerboard–Gauss–Seidel smoothing before and after each step) does not increase, as the necessary number of such, $m$, is proportional to, in the three examples listed, $h^{-2}, h^{-1}$ and a constant, respectively. The procedure using conjugate gradients applied to the Richardson iteration shows a similar speed as the SOR-method listed in Table 7.5. If one wishes to accelerate the latter with the aid of the CG-method, then one must replace SOR by a symmetric version SSOR. For this, one obtains already after 22 steps an error which is below 1E–6.

*Table 7.5. The error of the $m^{th}$ iterate $u^m$, measured in the energy norm.*[8]

| *m* | *Gauss–Seidel* | *m* | *SOR* | *m* | *Multi-grid method* |
|---|---|---|---|---|---|
| 10 | 9.382E − 2 | 10 | 1.02931E − 1 | 1 | 1.711472796E − 2 |
| 20 | 7.324E − 2 | 20 | 5.43417E − 2 | 2 | 9.659697997E − 4 |
| 50 | 5.023E − 2 | 50 | 1.29191E − 2 | 3 | 5.501125568E − 5 |
| 100 | 3.575E − 2 | 100 | 8.51213E − 4 | 4 | 3.206732671E − 6 |
| 200 | 2.371E − 2 | 150 | 2.50194E − 6 | 5 | 1.891178440E − 7 |
| 300 | 1.755E − 2 | 161 | 9.94034E − 7 | 6 | 1.128940250E − 8 |

*Table 7.6. Iteration methods accelerated by the CG-method.*

| *m* | *Richardson* | *m* | *SSOR* | *m* | *Multi-grid method* |
|---|---|---|---|---|---|
| 10 | 6.6931E − 2 | 5 | 1.17912E − 2 | 1 | 1.135035786E − 2 |
| 20 | 4.0034E − 2 | 10 | 1.00844E − 3 | 2 | 7.254914612E − 4 |
| 50 | 1.2571E − 2 | 15 | 6.70161E − 5 | 3 | 4.298721850E − 5 |
| 100 | 2.5151E − 4 | 20 | 3.70379E − 6 | 4 | 2.274098344E − 6 |
| 120 | 1.6995E − 5 | 21 | 1.78763E − 6 | 5 | 1.313049259E − 7 |
| 142 | 9.0458E − 7 | 22 | 8.78341E − 7 | 6 | 7.171669050E − 9 |

Also the multi-grid method (symmetric smoothing with lexicographical Gauss–Seidel procedure) can in principle be accelerated with the aid of the CG-method. However, as the advantage produced by this acceleration is small for procedures which are fast anyway, it is generally not worth the trouble. The asymptotic number of iterations required for reaching a given precision is $O(h^{-1}), O(h^{-1/2})$ and $O(1)$, respectively, in the three cases discussed.

The examples have been calculated with the programs given in [444]. There one can also find more details on the above procedures.

### 7.7.7.6 Nested iterations

The iteration error $e^m = x^m - x$ of the $m^{th}$ iteration can be estimated by $\|e^m\| \leq \rho^m \|e^0\|$, where $\rho$ denotes the speed of convergence. In order to reduce the size of the error $e^m$, one should not only have a good speed of convergence, but also a small initial error $\|e^0\|$. This strategy is realized with the following algorithm ('nested iterations'), which just like the multi-grid methods uses different levels of discretizations $\ell = 0, \ldots$. To solve $A_\ell x_\ell = b_\ell$ for $\ell = \ell_{\max}$, also the coarser discretizations for $\ell < \ell_{\max}$ are solved. Since $x_{\ell-1}$ and $px_{\ell-1}$ should represent a good approximation to $x_\ell$, but because of the lower-dimensionality can be calculated at lower cost, it is more efficient to first approximate $x_{\ell-1}$ (approximation $\tilde{x}_{\ell-1}$) and then to use $p\tilde{x}_{\ell-1}$ as an initial value for the iteration of level $\ell$. In the algorithm we now present, we let $x^{m+1} := \Phi_\ell(x^m, b_\ell)$ denote an arbitrary

[8]The notation $9.382\text{E} - 2$ means here $9.382 \cdot 10^{-2}$

iteration for solving $A_\ell x_\ell = b_\ell$.

$\tilde{x}_0$ solution (or approximation) of $A_0 x_0 = b_0$ ;
for $\ell := 1$ to $\ell_{\max}$ do
begin $\tilde{x}_\ell := p\tilde{x}_{\ell-1}$ ; (* initial value; $p$ from (7.69f) *)
for $i := 1$ to $m_\ell$ do $\tilde{x}_\ell := \Phi_\ell(\tilde{x}_\ell, b_\ell)$ (* $m_\ell$ iterations *)
end;

In the cases in which $\Phi_\ell$ represents a multi-grid procedure, we may choose $m_\ell$ to be constant; in fact, it is often sufficient to take $m_\ell = 1$ in order to obtain an iteration error $\|\tilde{x}_\ell - x_\ell\|$ which is of the same magnitude as the discretization error .

### 7.7.7.7 Partial decomposition of the approximation space

Let $Ax = b$ be the system of equations we are dealing with, where $x$ is an element of a solution space $X$. We say there is an admissible decomposition of $X$ into subspaces $X^{(\nu)}$, $\nu = 0, \ldots, k$, if $\sum_\nu X^{(\nu)} = X$. Here we allow the subspaces to be overlapping, i.e., non-disjoint. The goal of the method we discuss here is to find an iteration

$$\boxed{x^{m+1} := x^m - \sum_\nu \delta^{(\nu)}}$$

for which the correction factors $\delta^{(\nu)}$ are elements of $X^{(\nu)}$. To represent a vector $x^{(\nu)} \in X^{(\nu)} \subseteq X$, we need a coefficient vector $x_\nu$ in a space $X_\nu = \mathbb{R}^{\dim(X^{(\nu)})}$. The unique correspondence between $X_\nu$ and $X^{(\nu)}$ is obtained with the help of a linear 'prolongation',

$$p_\nu : X_\nu \to X^{(\nu)} \subset X, \qquad \nu = 0, \ldots, k.$$

This means that $p_\nu X_\nu = X^{(\nu)}$. The 'restriction' $r_\nu := p_\nu^\mathsf{T} : X^{(\nu)} \to X_\nu$ is the transposed mapping. Then the basic version of the partial decomposition method (also called the *additive Schwarz iteration*) is given by

$$d := Ax^m - b\,; \tag{7.70a}$$

$$d_\nu := r_\nu d\,; \qquad \nu = 0, \ldots, k, \tag{7.70b}$$

$$\text{solve} \quad A_\nu \delta_\nu = d_\nu\,; \qquad \nu = 0, \ldots, k, \tag{7.70c}$$

$$x^{m+1} := x^m - \omega \sum_\nu p_\nu \delta_\nu. \tag{7.70d}$$

The matrix appearing in (7.70c) of dimension $n_\nu := \dim X^{(\nu)}$ is the product

$$A_\nu := r_\nu A p_\nu, \qquad \nu = 0, \ldots, k.$$

The dampening parameter $\omega$ in (7.70d) is included to improve the convergence (see (7.57)). If one inserts the iteration into a CG-method (discussed in 7.7.7.4), then the choice of $\omega \neq 0$ is irrelevant.

The local problems in (7.70c) can be solved independently of one another, so that they are interesting for algorithms for parallel computers. The exact solution of (7.70c) can be iteratively approximated (application of a secondary iteration).

The hierarchical iterations discussed in 7.7.7.2.4, as well as variants of the multi-grid method and the domain decomposition method to be discussed presently, fall into the abstract context of (7.70).

In the case of a hierarchical iteration, $X_0$ contains all vertices of the original triangulation of $\tau_0$, $X_1$ contains all vertices of the next triangulation $\tau_1$ which are not vertices of $\tau_0$, and so forth. The prolongation $p_1 : X_1 \to X$ is the piecewise linear interpolation (evaluation of the finite element functions at the new vertices).

The theory of the convergence for decomposition methods is more or less restricted to matrices $A$ which are positive definite. The same is true for the multiplicative Schwarz procedure, in which before each correction $x \mapsto x - p_\nu \delta_\nu$ the steps (7.70a-c) are repeated.

#### 7.7.7.8 Domain decomposition

The domain decomposition to be discussed now has two completely different interpretations. The first views the decomposition of the domain as a decomposition of data, the second is a special kind of iteration method which uses decompositions of the domain as a tool.

In the case of data decomposition, one divides the coefficient vector $x$ in blocks, as follows: $x = (x^0, \ldots, x^k)$. Every block $x^\nu$ contains the data of the vertices of the lattice in some partial domain $\Omega^\nu$ of the underlying domain $\Omega$ of the boundary value problem. If the matrix is sparse, then the basic operations (most important, matrix multiplication) require only information of the neighborhood, i.e., mostly from the same subdomain. If every block $x^\nu$ is assigned to a processor of a parallel computer, the procedure only requires communication along the boundaries of the subdomains. Since the number of vertices of the mesh contained in these boundaries must be at least one magnitude smaller than the total number of vertices, one could hope that the necessary communication is computationally of little cost compared with the main computational cost of the algorithm.

Next we consider the domain decomposition as a special kind of iteration method. Let $\overline{\Omega} = \bigcup \overline{\Omega^\nu}$ be a not necessarily disjoint decomposition of the underlying domain $\overline{\Omega}$ of the partial differential equation. The vertices which are in $\overline{\Omega^\nu}$ for the space $X^\nu$ were already discussed in 7.7.7.7. The prolongation $p_\nu : X_\nu \to X$ can for example be defined by a zero extension, i.e., for all vertices outside of $\overline{\Omega^\nu}$, it vanishes. Then this method is defined by (7.70).

Let $k$ be the number of subdomains $\overline{\Omega^\nu}$. Since $k$ could be the number of processors of a parallel computer, one would like to obtain a speed of convergence which depends not only on the number $n$ of dimensions of the problem, but also on $k$. This cannot be achieved with a pure decomposition, so that one adds to the spaces a coarse lattice space $X_0$. In this respect, this method is quite similar to the two-grid method described above.

#### 7.7.7.9 Non-linear systems of equations

In the case of non-linear systems of equations there are numerous new questions to be answered. In particular, solutions no longer need to be unique. Therefore, we shall assume that the system $F(x) = 0$ has, in a neighborhood of $x^*$, the unique solution $x^*$.

To solve $F(x) = 0$, two strategies offer themselves. The first is to apply a variant of the Newton iteration, which requires per Newton step solving a linear system of equations (again as secondary iteration), for which the methods discussed in section 7.7.7.1 to 7.7.7.8 can be used. Whether this can be carried out or not depends among other things on how computationally intensive it is to calculate the Jacobi matrix $F'$. A second strategy tries to extend the methods discussed above for linear systems directly to non-linear systems. For example, the non-linear analog of the Richardson iteration (7.53) for solving $F(x) = 0$ would be

$$\boxed{x^{m+1} := x^m - F(x^m).}$$

The multi-grid method can also be extended to the non-linear case. If one carries this out, then the asymptotic speed of convergence of the non-linear iteration coincides with the speed of the linear multi-grid method, which one applies to the linearized equation (this is $A = F'(x^*)$); see [445], §9 for details.

In the case of multiple solutions, iteration methods can have a *local* convergence at best. The computationally most intensive part of the method is often determining appropriate initial values $x^0$. For this, the nested iteration of 7.7.7.6 can be of some help. The choice of initial iterate is then basically restricted to the lower-dimensional system of level $\ell = 0$.

### 7.7.8 Boundary element methods

Replacing a differential equation with an integral equation is part of the integral equation method. The boundary element method proper arises when we then discretize this integral equation.

#### 7.7.8.1 The method of integral equations

A homogeneous differential equation $Lu = 0$ with constant coefficients has a fundamental solution $U_0$ (see [212]). Here we deal with the following problem: a boundary value problem on a domain $\Omega \subset \mathbb{R}^d$ with boundary values on $\Gamma := \partial\Omega$. One approach is to suppose $u$ has the form of a boundary or a surface integral

$$\boxed{u(x) := \int_\Gamma k(x,y)\varphi(y)\,\mathrm{d}F_y, \qquad x \in \Omega,} \tag{7.71}$$

with an arbitrary weight function $\varphi$. Then $u$ satisfies the equation $Lu = 0$ in $\Omega$ if the kernel function $k$ coincides with $U_0(x-y)$ or a derivative of this function. For $k(x,y) = U_0(x-y)$, $u$ as in (7.71) is a *single layer potential* (see Example 2 in 10.4.3). The normal derivative $k(x,y) = \partial U_0(x-y)/\partial n_y$ with respect to $y$ defines what is known as the *double layer potential.*

We now derive an integral equation for the function $\varphi$ occuring in (7.71), so that the solution $u$ of (7.71) satisfies the boundary condition. Since the single layer potential is continuous in $x \in \mathbb{R}^d$, the Dirichlet values $u = g$ on $\Gamma$ lead directly to

$$\boxed{g(x) = \int_\Gamma k(x,y)\varphi(y)\,\mathrm{d}F_y.} \tag{7.72}$$

This is a Fredholm integral equation of the first kind for determining $\varphi$. In the case of Neumann boundary conditions (7.10b) or Dirichlet conditions in connection with the double layer potential, also the discontinuities on the boundary must be taken into consideration. The integral equation for $\varphi$ which then arises is discussed in [212]. It has, generally speaking, the form

$$\lambda\varphi(x) = \int_{\Gamma} \kappa(x,y)\varphi(y)\,\mathrm{d}F_y + h(x), \tag{7.73}$$

in which $\kappa(x,y)$ is either the kernel $k(x,y)$ of (7.7.2) or the derivative $B_x k(x,y)$, where $B$ is arises from the boundary condition $Bu = g$ (for example, we could have $B = \partial/\partial n$).

Advantages of the integral equation method are the following.

(1) The domain of definition of the function to be determined is only $(d-1)$-dimensional, which following the discretization leads to an enormous reduction in the size of the system of equations.

(2) The method of integral equations is equally apt to apply to exterior and interior problems. In the case on an *exterior boundary problem* $\Omega$ is an unbounded domain which is the complement of the interior of the surface or curve $\Gamma$. For finite element methods, this leads to difficult problems, because of the unboundedness. For the exterior boundary problems there are additional boundary conditions at infinity $x = \infty$, which are automatically satisfied by integral equation methods.

(3) Finally, in many case the solution of the boundary value problem is not necessary in the entire domain, rather one requires only certain boundary values or data (for example the normal derivative if the boundary value is given). In that case, most of the values calculated by the finite element methods in the domain would be unnecessary.

As opposed to simpler integral equations, (7.73) does present the following difficulties.

(1) From theory we have for dipole integral operators compactness; however this property no longer holds for non-smooth boundaries.

(2) All integrals which appear are surface or curve integrals, so that for their solution we generally require concrete parameterizations.

(3) By definition, the kernel is singular. The strength of the singularity of the fundamental solution depends on the order of the differential equation. If $\kappa$ is obtained through further differentiations, then the singularity is intensified (gets 'worse'). Integral equations of type (7.73) which typically occur in applications have improper integrable kernels, strongly singular integrals of the type of Cauchy principal values, or hypersingular integrals, which are defined with the help of 'part-fini' integrals according to Hadamard. Contrary to what one would tend to believe, strong singularities turn out to be an advantage for a numerical treatment.

#### 7.7.8.2 Discretization through collocation

There is a discretization procedure due to Nyström, which we discuss in [212]. This method is however seldomly applied to boundary element methods. Instead, one uses one of two types of projection methods: either the collocation method, which is the projection onto lattice vertices, or the Galerkin procedure, which is an orthogonal projection on the space of approximating functions. In the case of collocation, one replaces in (7.73) the unknown function $\varphi$ by a function $\tilde{\varphi} = \sum c_i\varphi_i$. Here we could for example have finite element functions $\varphi_i$, which belong to the lattice vertices $x_i \in \Gamma$. In the two-dimensional case, when $\Gamma$ is a curve, we could also take a global point of view using trigonometric functions. The collocation equations arise when (7.73) is satisfied at all collocation points $x = x_i$. The matrix coefficients which arise in this manner are described by the integrals $\int_{\Gamma} \kappa(x_i,y)\varphi_j(y)\,\mathrm{d}F_y$.

#### 7.7.8.3 Galerkin method

According to 7.7.2.2 one obtains a Galerkin discretization after an additional integration with $\varphi_i$ acting as a test function. The matrix coefficients then contain the double integrals $\int_\Gamma\int_\Gamma \varphi_i(x)\kappa(x,y)\varphi_j(y)\,\mathrm{d}F_y\,\mathrm{d}F_x$, which in the case of surfaces $\Gamma$ are four-dimensional. Although this method is more complicated, it has better stability properties and higher precision in an appropriate norm.

Even if the name finite element method (FEM) would be legitimate here, the discretizations of integral equations like (7.73) are generally grouped together as 'boundary element methods', BEM.

#### 7.7.8.4 Numerical properties of boundary element methods

If one compares the discretizations of a boundary value problem in $\Omega \subseteq \mathbb{R}^d$ given by the finite element method with that given by boundary element methods, the properties can be summed up as follows.

For a lattice size $h$, one obtains systems of equations which have the magnitude of $O(h^{-d})$ for FEM, and magnitude $O(h^{1-d})$ for BEM. The condition numbers for the BEM matrixes are, compared with the FEM case, harmless.

One decisive disadvantage of the BEM is the fact that the matrices are not sparse, but rather fully populated. This leads to problems with computation time and storage. Therefore, there are different approaches to representing the matrices in a more compact form (for example panel clustering in [443] or matrix compression using wavelet basis). For a numerical quadrature of the singular integrals there are modern methods, which can also deal with double surface integrals $\int_\Gamma\int_\Gamma \varphi_i(x)\kappa(x,y)\varphi_j(y)\,\mathrm{d}F_y\,\mathrm{d}F_x$ quickly and sufficiently precisely, as described for example in [443], §9.4.

### 7.7.9 Harmonic analysis

#### 7.7.9.1 Discrete Fourier transformation and trigonometric interpolation

With the help of complex-valued coefficients $c_0, c_1, \ldots, c_{n-1}$ we define the *trigonometric polynomial*

$$\boxed{y(x) := \frac{1}{\sqrt{n}}\sum_{\nu=0}^{n-1} c_\nu \mathrm{e}^{\mathrm{i}\nu x}, \qquad x \in \mathbb{R}.} \tag{7.74}$$

It can be interpreted as a genuine polynomial $\sum c_\nu z^\nu$ by using the substitution $z = \mathrm{e}^{\mathrm{i}x}$, which restricts the argument $z$ to the unit circle, as $|z| = |\mathrm{e}^{\mathrm{i}x}| = 1$. If one evaluates the function $y$ from (7.74) at the equidistant interpolation points $x_\mu = 2\pi\mu/n$, where $\mu = 0, 1, \ldots, n-1$, then we obtain the values at these points

$$\boxed{y_\mu = \frac{1}{\sqrt{n}}\sum_{\nu=0}^{n-1} c_\nu \mathrm{e}^{2\pi\mathrm{i}\nu\mu/n}, \qquad \mu = 0, 1, \ldots, n-1.} \tag{7.75}$$

The *trigonometric interpolation problem* can then be formulated as follows. For given values $y_\mu$, determine the *Fourier coefficients* $c_\nu$ from equation (7.75). The solution can be described by means of the following *backward transformation*:

$$c_\nu = \frac{1}{\sqrt{n}} \sum_{\mu=0}^{n-1} y_\mu \mathrm{e}^{-2\pi \mathrm{i} \nu\mu/n}, \qquad \nu = 0, 1, \ldots, n-1. \tag{7.76}$$

To formulate this in matrix notation, we introduce the vectors $c = (c_0, \ldots, c_{n-1}) \in \mathbb{C}^n$ and $y = (y_0, \ldots, y_{n-1}) \in \mathbb{C}^n$. The mapping $c \mapsto y$ according to (7.75) is referred to as *discrete Fourier synthesis*, while the mapping $y \mapsto c$ in the other direction is referred to as *discrete Fourier analysis*. If we use the matrix $T$ which has coefficients $T_{\nu\mu} := n^{-1/2}\mathrm{e}^{2\pi\mathrm{i}\nu\mu/n}$, then we can write (7.75) and (7.76) as

$$y = Tc, \qquad c = T^* y. \tag{7.77}$$

Here $T^*$ denotes the adjoint matrix of $T : (T^*)_{\nu\mu} = \overline{T_{\mu\nu}}$. In the present case $T$ is unitary, i.e., $T^* = T^{-1}$. This property corresponds to the fact that (7.74) is an expansion with respect to the *orthonormal basis* $\{n^{-1/2}\mathrm{e}^{2\pi\mathrm{i}\nu\mu/n} : \mu = 0, 1, \ldots, n-1\}$.

In the sum (7.74) we can shift the index set $\nu = 0, 1, \ldots, n-1$. In this way, evaluation of (7.75) at the interpolation points $x_\mu = 2\pi\mu/n$ is not affected, as $\exp(\mathrm{i}\nu x_\mu) = \exp(\mathrm{i}(\nu \pm n)x_\mu)$, while at the same time the points in between are affected. For example, for even $n$ we may choose the index set $\{1 - n/2, \ldots, n/2 - 1\}$. Because of the relation

$$c_{-\nu}\mathrm{e}^{-\mathrm{i}\nu x} + c_{+\nu}\mathrm{e}^{+\mathrm{i}\nu x} = (c_{-\nu} + c_{+\nu})\cos \nu x + \mathrm{i}(c_{+\nu} - c_{-\nu})\sin \nu x$$

we then obtain a linear combination of the real trigonometric functions $\{\sin \nu x, \cos \nu x : 0 \le \nu \le n/2 - 1\}$.

### 7.7.9.2 Fast Fourier transformation (FFT)

In many practical applications the Fourier synthesis $c \to y$ and the Fourier analysis $y \to c$ of (7.77) play an important role, so that it is desirable to carry these out with as little computational cost as possible. Equation (7.76) can be written, up to a scalar factor of $n^{-1/2}$, which we for simplicity neglect here, in the form

$$c_\nu^{(n)} = \sum_{\mu=0}^{n-1} y_\mu^{(n)} \omega_n^{\nu\mu}, \qquad \nu = 0, 1, \ldots, n-1 \tag{7.78}$$

with the $n^{th}$ root of unity $\omega_n := \mathrm{e}^{-2\pi\mathrm{i}/n}$. The synthesis (7.75) also takes the form of (7.78), after exchanging the symbols $c$ and $y$ and using $\omega_n := \mathrm{e}^{2\pi\mathrm{i}/n}$. The index $n$ occurring in $y_\mu^{(n)}, c_\nu^{(n)}$ and $\omega_n$ is meant to indicate the dimension of the Fourier transformation we are working with.

If one evaluates (7.78) in the usual way, one requires $n$ (complex) multiplications and $n-1$ additions for each coefficient. To evaluate all components $c_\nu^{(n)}$ we therefore need $2n^2 + O(n)$ operations. Here we assume that the values of $\Omega = \{\omega_n^{\nu\mu} : 0 \le \nu, \mu \le n-1\}$ are known. Since $\omega_n^{\nu\mu}$ only depends on the residue class of $\nu\mu$ modulo $n$, $\Omega$ only contains $n$ distinct values and can be calculated with a cost of $O(n)$.

The cost $O(n^2)$ for evaluating (7.78) can be significantly reduced is $n$ happens to be a power of 2, $n = 2^p$ with $p \ge 0$. If $n$ is even, the coefficients to be determined can be written with a sum consisting only of $n/2$ summands:

$$c_{2\nu}^{(n)} = \sum_{\mu=0}^{n/2-1} \left[y_\mu^{(n)} + y_{\mu+n/2}^{(n)}\right]\omega_n^{2\nu\mu}, \qquad 0 \le 2\nu \le n-1, \tag{7.79a}$$

$$c_{2\nu+1}^{(n)} = \sum_{\mu=0}^{n/2-1} \left[\left(y_\mu^{(n)} - y_{\mu+n/2}^{(n)}\right)\omega^\mu\right]\omega_n^{2\nu\mu}, \qquad 0 \le 2\nu \le n-1. \tag{7.79b}$$

The $c$-coefficients in (7.79a,b) form vectors

$$c^{(n/2)} = \left(c_0^{(n)}, c_2^{(n)}, \ldots, c_{n-2}^{(n)}\right), \qquad d^{(n/2)} = \left(c_1^{(n)}, c_3^{(n)}, \ldots, c_{n-1}^{(n)}\right)$$

of $\mathbb{C}^{n/2}$. If we introduce further the coefficients

$$y_\mu^{(n/2)} := y_\mu^{(n)} + y_{\mu+n/2}^{(n)}, \qquad z_\mu^{(n/2)} := \left(y_\mu^{(n)} - y_{\mu+n/2}^{(n)}\right)\omega^\mu, \qquad 0 \le \mu \le n/2-1$$

and note that $(\omega_n)^2 = \omega_{n/2}$, then we obtain the new set of equations

$$c_\nu^{n/2} = \sum_{\mu=0}^{n/2-1} y_\mu^{(n/2)}\omega_{n/2}^{\mu\nu}, \qquad d_\nu^{n/2} = \sum_{\mu=0}^{n/2-1} y_\mu^{(n/2)}\omega_{n/2}^{\mu\nu}, \qquad 0 \le \nu \le n/2-1.$$

Both sums here have the form of (7.78) with $n$ replaced by $n/2$. This reduces the $n$-dimensional problem (7.78) to two $(n/2)$-dimensional problems. Because of $n = 2^p$, this process can be iterated $p$ times and yields in the end $n$ one-dimensional problems! (Note that in the one-dimensional case we have $y_0 = c_0$.) The algorithm we have obtained in this manner can be formulated in the following fashion:

```
procedure FFT(ω, p, y, c);  {y : input-, c : output-vector}          (7.80)
if p = 0 then c[0] := y[0] else
begin n2 := 2^(p−1);
      for μ := 0 to n2 − 1 do yy[μ] := y[μ] + y[μ + n2];             (7.80a)
      FFT(p − 1, ω², yy, cc);  for ν := 0 to n2 do c[2ν] := cc[ν];
      for μ := 0 to n2 − 1 do yy[μ] := (y[μ] − y[μ + n2]) * ω^μ;      (7.80b)
      FFT(p − 1, ω², yy, cc);  for ν := 0 to n2 do c[2ν + 1] := cc[ν]
end;
```

Since there have been $p$ steps of halving the dimension, in which $n$ evaluations of (7.80a,b) have to be carried out, the computational cost is in sum $p \cdot 3n = O(n \log n)$ operations.

#### 7.7.9.3 Applications to periodic Toeplitz matrices

A matrix $A$ is by definition a *Toeplitz matrix*, if the coefficients $a_{ij}$ only depend on the difference $i-j$ modulo $n$. In this case, $A$ has the form

$$A = \begin{bmatrix} c_0 & c_1 & c_2 & \cdots & c_{n-1} \\ c_{n-1} & c_0 & c_1 & \cdots & c_{n-2} \\ \vdots & \vdots & \vdots & \vdots & \vdots \\ c_2 & c_3 & c_4 & \cdots & c_1 \\ c_1 & c_2 & c_3 & \cdots & c_0 \end{bmatrix}. \tag{7.81}$$

Every periodic Toeplitz matrix can be diagonalized with the aid of the Fourier transformation, i.e., with the matrix $T$ from (7.77):

$$T^*AT = D := \operatorname{diag}\{d_1, d_2, \ldots, d_n\}, \qquad d_\mu := \sum_{\nu=0}^{n-1} c_\nu \mathrm{e}^{2\pi\mathrm{i}\nu\mu/n}. \tag{7.82}$$

A basic operation which is used particularly often is the matrix-vector multiplication of a periodic Toeplitz matrix with a vector $x$. If $A$ is fully filled, a standard multiplication would mean $O(n^2)$ multiplications. In contrast, multiplication with the diagonal matrix $D$ of (7.82) requires only $O(n)$ operations. The factorization $Ax = T(T^*AT)T^*x$ yields the following algorithm.

$$x \mapsto y := T^*x \qquad \text{(Fourier analysis)}, \tag{7.83a}$$
$$y \mapsto y' := Dy \qquad (D \text{ from (7.82)}), \tag{7.83b}$$
$$y' \mapsto Ax := Ty' \qquad \text{(Fourier synthesis)}. \tag{7.83c}$$

Under the assumption $n = 2^p$, the fast Fourier transformation (7.80) may be applied, so that the matrix-vector multiplication $x \to Ax$ can be carried out with a cost of $O(n \log n)$.

The solution of the system of equations $Ax = b$ with a periodic Toeplitz matrix $A$ is just as simple as the matrix-vector multiplication, as in (7.83b) we only have to replace $D$ by $D^{-1}$.

The inverse of $A$ from (7.81) is of the same form, only with $\zeta_\nu$ replacing $c_\nu$, where $\zeta_\nu$ arises from

$$1/d_\mu = n^{-1} \sum_{\nu=0}^{n-1} \zeta_\nu \mathrm{e}^{2\pi\mathrm{i}\nu\mu/n} \tag{7.84}$$

(see equation (7.75)). The interpolation problem (7.84) can again be solved with computational cost $O(n \log n)$.

Similarly, the product of periodic Toeplitz matrices, polynomials $P(A)$ or other functions of the matrix $A$ (for example the square root in case $d_\mu \geq 0$) can be computed with low cost.

#### 7.7.9.4 Fourier series

Let now $\ell^2$ denote the space of all sequences of coefficients $\{c_\nu : \nu \text{ integer}\}$ whose norm $\sum_{\nu=-\infty}^{\infty} |c_\nu|^2$ is finite (note that whereas earlier we used $\ell$ as an index, here the combination $\ell^2$ denotes a space, and not the square of an index). To every $c \in \ell^2$ we associate the $2\pi$-periodic function

$$\boxed{f(x) = \frac{1}{\sqrt{2\pi}} \sum_{\nu=-\infty}^{\infty} c_\nu \mathrm{e}^{\mathrm{i}\nu x}} \tag{7.85}$$

(again, *Fourier synthesis*). The sum converges in quadratic means and $f$ satisfies the *Parseval equation*

$$\int_{-\pi}^{\pi} |f(x)|^2 \, \mathrm{d}x = \sum_{\nu=-\infty}^{\infty} |c_\nu|^2.$$

The inverse transformation (again, *Fourier analysis*) is

$$c_\nu = \frac{1}{\sqrt{2\pi}} \int_{-\pi}^{\pi} f(x)\, \mathrm{e}^{-\mathrm{i}\nu x}\, \mathrm{d}x. \tag{7.86}$$

While the periodic function $f$ is often viewed as an initial quantity of kinds, for which Fourier coefficients are to be determined, one can turn this point of view around. Suppose we are given a lattice function $\varphi$ by means of the values $c_\nu = \varphi(\nu h)$, where $h$ denotes the size of the lattice and $\nu$ is an integer. For the purpose of the analysis, the function which is associated to it by (7.85) turns out to be quite convenient.

The condition $\sum\limits_{\nu=-\infty}^{\infty} |c_\nu|^2 < \infty$ for $c \in \ell^2$ can be weakened. Let $s \in \mathbb{R}$. Then the condition

$$\sum_{\nu=-\infty}^{\infty} (1+\nu^2)^{\frac{s}{2}}\, |c_\nu|^2 < \infty$$

strongly increases the decreasing property of the coefficients if $s > 0$, while if $s < 0$, the opposite is true: the coefficients may even increase instead of decreasing. For $s > 0$, (7.85) defines a function in the Sobolev space $H^s_{\text{periodic}}(-\pi,\pi)$, while for $s < 0$, (7.85) can be taken as a formal definition of a *distribution* on the space $H^s_{\text{periodic}}(-\pi,\pi)$.

#### 7.7.9.5 Wavelets

**7.7.9.5.1 Non-localness of the Fourier transformation.** The characterizing property of the Fourier transformation is the decomposition of the functions according to their *frequencies*. These are in the case of (7.85) and (7.86) discrete; in the case of the integral Fourier transformation

$$\widehat{f}(\xi) = \frac{1}{\sqrt{2\pi}} \int_{-\infty}^{\infty} f(x)\, \mathrm{e}^{-\mathrm{i}\xi x}\, \mathrm{d}x, \qquad f(x) = \frac{1}{\sqrt{2\pi}} \int_{-\infty}^{\infty} \widehat{f}(\xi)\, \mathrm{e}^{\mathrm{i}\xi x} \mathrm{d}\xi, \tag{7.87}$$

they are continuous. A decisive *disadvantage* of the Fourier transformation, on the other hand, is its inability to resolve details of the position. Depending on the application, we have to replace 'position' by 'time'.

As an example we consider the periodic function which is given on $[-\pi,+\pi]$ by $f(x) = \operatorname{sgn}(x)$ (sign of $x$). Its periodic extension to the real line has discontinuities at all integral multiples of $\pi$. The Fourier coefficients of $f$ are $c_\nu = C/\nu$ ($C = -2\mathrm{i}/\sqrt{2\pi}$) for odd $\nu$ and $c_\nu = 0$ otherwise. The small rate of decrease $c_\nu = O(1/\nu)$ of the coefficients yields a global statement about the non-smoothness of the function $f$. But while the series (7.85) shows a slow convergence for *all* $x$ (no absolute convergence), this is really only true for $x$ near the discontinuities.

The reason for the non-localness of the Fourier transformation lies in the fact that the functions $\mathrm{e}^{\mathrm{i}\nu x}$ in the series have no privileged position, but rather are characterized only in terms of the frequency $\nu$.

**7.7.9.5.2 Wavelets and the wavelet transformation.** In order to alleviate the problem just mentioned, one replaces the $\{\mathrm{e}^{\mathrm{i}\xi x} : \xi \in \mathbb{R}\}$ by functions which depend on, in addition to the frequency, also further coordinates which are related to position. Just as $\{e^{i\xi x} : \xi \in \mathbb{R}\}$ arises from the single function $e^{ix}$ by means of the dilation $x \to \xi x$, we can generate the family of functions required by doing the same to a function which we call a wavelet and denote by $\psi$.

A wavelet is not a uniquely defined object, rather all square integrable functions $f \in L^2(\mathbb{R})$ can be used, for which the corresponding Fourier transform $\widehat{\psi}$ according to (7.87) leads to a positive, finite integral $\int_{\mathbb{R}} |\widehat{\psi}(\xi)|^2/|\xi|\,d\xi$. Every wavelet has a vanishing mean, $\int_{\mathbb{R}} \psi(\xi)\,dx = 0$.

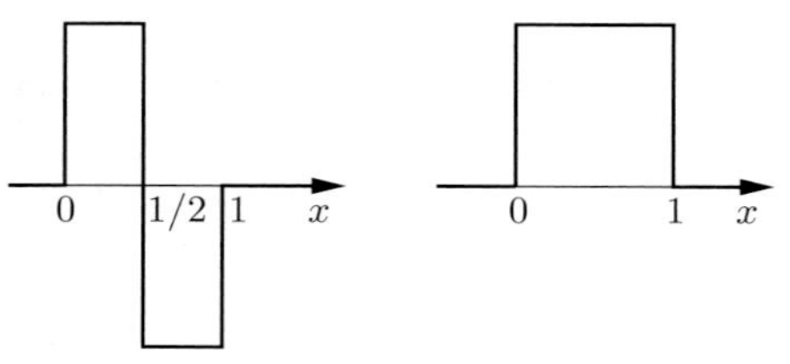


**(a)** Haar wavelet **(b)** scaling function $\chi_{[0,1]}$

*Figure 7.13. The simplest wavelet and the simplest scaling function.*

The simplest wavelet is the function first considered by Haar, shown in Figure 7.13(a), which for $0 \le x \le 1$ is the sign of $1 - 2x$ and vanishes elsewhere. Since all functions $\psi \neq 0$ in $L^2(\mathbb{R})$ with compact support and $\int_{\mathbb{R}} \psi(\xi)\,dx = 0$ are already wavelets, the Haar function is also a wavelet.

By dilating $\psi$ one obtains a family $\{\psi_a : a \neq 0\}$ with $\psi_a(x) := |a|^{-1/2}\psi(x/a)$. For $|a| > 1$, the function is stretched out, for $|a| < 1$ it is compressed. For $a < 0$ there is in addition a reflection involved. The factor $|a|^{-1/2}$ is only a scalar normalization. The parameter $a$ plays the role of the inverse frequency $1/\xi$ in the function $e^{i\xi x}$.

In contrast with the Fourier method, there is in addition to the dilation also a translation. The shift parameter $b$ characterizes the position (or the time). The family of functions generated is $\{\psi_{a,b} : a \neq 0,\ b \text{ real}\}$ with

$$\boxed{\psi_{a,b}(x) := \frac{1}{\sqrt{|a|}}\,\psi\Big(\frac{x-b}{a}\Big).} \tag{7.88}$$

The *wavelet transformation* $L_\psi f$ is a function of both position and frequency, where $a$ is the frequency and $b$ the position variable. Explicitly, we have

$$\boxed{L_\psi f(a,b) := c\int_{\mathbb{R}} f(x)\psi_{a,b}(x)\,dx = \frac{c}{\sqrt{|a|}}\int_{\mathbb{R}} f(x)\psi\Big(\frac{x-b}{a}\Big)\,dx} \tag{7.89a}$$

with $c = \left(2\pi \int_{\mathbb{R}} |\widehat{\psi}(\xi)|^2/|\xi|\,d\xi\right)^{-1/2}$. The *inverse transformation* if given by

$$\boxed{f(x) = c\int_{\mathbb{R}} L_\psi f(a,b)\psi_{a,b}(x)a^{-2}\,da\,db.} \tag{7.89b}$$

For $f \in L^2(\mathbb{R})$ the wavelet transformation $f \mapsto L_\psi f$ is bijective.

**7.7.9.5.3 Properties of wavelets.** The Haar wavelet (Figure 7.13a) has compact support (here $[0,1]$) but it not continuous. The so-called *Mexican hat*, a function with the opposite properties (it is in fact infinitely often differentiable), is given by $\psi(x) := (1-x^2)\exp(-x^2/2)$.

The $k^{th}$ moment of a wavelet $\psi$ is

$$\mu_k := \int_{\mathbb{R}} x^k \psi(x)\, \mathrm{d}x.$$

The *order* of a wavelet $\psi$ is the smallest natural number $N$ for which the $N^{th}$ moment is non-vanishing. Since the mean of $\psi$ is zero, we have $\mu_k = 0$ for all $0 \le k \le N-1$. If $\mu_k = 0$ for all $k$, then $\psi$ has infinite order. However, wavelets with compact support always have finite order (for example the Haar wavelet has $N = 1$, the Mexican hat has $N = 2$).

A wavelet of order $N$ is orthogonal to all polynomials of degree $\le N-1$. Thus $L_\psi f(a,b)$ for sufficiently smooth $f$ depends only on the $N^{th}$ remainder term of the Taylor series of $f$. Up to a scalar multiple, $L_\psi f(a,b)$ converges as $a \to 0$ to the $N^h$ derivative $f^{(N)}(b)$.

The Fourier transform $\widehat{f}(\xi)$ in (7.87) decreases more rapidly to zero as $|\xi| \to \infty$ when $f$ is smooth. In contrast, for wavelets as $|a| \to 0$, $L_\psi f(a,b)$ tends to $O(|a|^{k-1/2})$ uniformly with respect to $b$ only if $f$ has a bounded $k^{th}$ derivative and $k \le N$. In this case, the rate of convergence is bounded by the order.

#### 7.7.9.6 Multiresolution analysis

**7.7.9.6.1 Introduction.** The true importance of wavelets becomes clear in the concept of *multiple scale analysis* (also called *multiresolution analysis*, the term we shall use henceforth), which can initially be introduced without wavelets. The wavelet transformation (7.89a,b) is an analog of the Fourier integral transformation (7.87). For practical purposes it would in fact be better to have a discrete version, which then would correspond to the Fourier series (7.85). But while only $2\pi$-periodic functions can be represented by Fourier series, the multiresolution analysis can represent an arbitrary $f \in L^2(\mathbb{R})$.

The scale index $m$ of the subspace $V_m$ to be defined below therefore corresponds to a frequency range up to $O(2^m)$, as in $V_m$ all 'details' up to magnitude $O(2^{-m})$ can be described.

The notion of a *Riesz basis* is of importance in connection with the multiresolution analysis. Let $\varphi_k \in L^2(\mathbb{R})$ be a family of functions which is dense in a subspace $V$ of $L^2(\mathbb{R})$. Suppose there are constants $0 < A \le B < \infty$ with

$$A \sum_k |c_k|^2 \le \int_{\mathbb{R}} \Big| \sum_k c_k \varphi_k(x) \Big|^2 \mathrm{d}x \le B \sum_k |c_k|^2 \tag{7.90}$$

for all coefficients with finite square sum $\sum_k |c_k|^2$. The family $\{\varphi_k\}$ is called a Riesz basis in $V$ with Riesz bounds $A$, $B$.

**7.7.9.6.2 Scaling functions and multiresolution analysis.** A multiresolution analysis is generated by a single function $\varphi \in L^2(\mathbb{R})$, called the *scaling function.* Its name comes from the following *scaling equation*, which it is to satisfy for appropriately chosen coefficients $h_k$:

$$\boxed{\varphi(x) = \sqrt{2} \sum_{k=-\infty}^{\infty} h_k \varphi(2x - k) \quad \text{for all } x.} \tag{7.91}$$

Equation (7.91) is also called the *mask equation* or the *refinement equation*. With an eye toward practical applications it is desirable that the sum in (7.91) be finite and contain as few summands as possible. The simplest example is the characteristic function $\varphi = \chi_{[0,1]}$, i.e. $\varphi(x) = 1$ for $x \in [0,1]$ and $\varphi(x) = 0$ otherwise, depicted in Figure 7.13(b). Then $\varphi(2x) = \chi_{[0,1/2]}$ is the characteristic function of $[0,1/2]$ and $\varphi(2x-1)$ is that of $[1/2,1]$ so that $\varphi(x) = \varphi(2x) + \varphi(2x-1)$, i.e., in (7.91) we have $h_0 = h_1 = 1/\sqrt{2}$ and $h_k = 0$ otherwise.

The translate $x \to \varphi(x-k)$ of $\varphi$ generates the subspace $V_0$:

$$V_0 := \Big\{ f \in L^2(\mathbb{R}) \quad \text{with} \quad f(x) = \sum_{k=-\infty}^{\infty} a_k \varphi(x-k) \Big\}. \tag{7.92}$$

In the case of the example $\varphi = \chi_{[0,1]}$, $V_0$ contains the functions which are piecewise constant on every subinterval $]\ell, \ell+1[$ ($\ell$ integer).

If one carries out in addition a dilation with $a = 2^m$, then one obtains the family of functions

$$\boxed{\varphi_{m,k}(x) := 2^{m/2} \varphi(2^m x - k) \quad \text{for all integers } m, k}$$

(see (7.88)). For all scales $m$ we can construct a subspace $V_m$ in the same way as in (7.92) as the closure of span $\{\varphi_{m,k} : k \text{ integer}\}$. By definition, $V_m$ is only a stretched or compressed copy of $V_0$. In particular, we have

$$f(x) \in V_m \quad \text{if and only if} \quad f(2x) \in V_{m+1}. \tag{7.93}$$

The scaling equation (7.91) implies $\varphi_{0,k} \in V_1$ and thus the inclusion $V_0 \subseteq V_1$, which can be extended to inclusions $V_m \subseteq V_{m+1}$ at all scales. Conversely, $V_0 \subseteq V_1$ implies the representation (7.91). The chain of inclusions ('ladder') which ensues,

$$\ldots \subseteq V_{-2} \subseteq V_{-1} \subseteq V_0 \subseteq V_1 \subseteq V_2 \subseteq \ldots \subseteq L^2(\mathbb{R}) \tag{7.94}$$

suggests that the spaces $V_m$ become larger as $m \to \infty$ and in the end fill out $L^2(\mathbb{R})$ completely. This idea can be made precise by the conditions

$$\bigcup_{m=-\infty}^{\infty} V_m \quad \text{is dense in } L^2(\mathbb{R}), \qquad \bigcap_{m=-\infty}^{\infty} V_m = \{0\}. \tag{7.95}$$

The ladder (7.94) is a *multiresolution analysis* if (7.93) and (7.95) hold and there is a scaling function whose translates $\varphi_{0,k}$ form a Riesz basis of $V_0$.

The Riesz basis property just mentioned can be directly read off the Fourier transform $\widehat{\varphi}$ of the scaling function; indeed, (7.90) is equivalent to

$$0 < A \le 2\pi \sum_{k=-\infty}^{\infty} \left|\widehat{\varphi}(\xi + 2\pi k)\right|^2 \le B \qquad \text{for} \quad |\xi| \le \pi.$$

**7.7.9.6.3 Orthonormality and filter.** The translates $x \to \varphi(x-k)$ of $\varphi$ form an orthonormal basis of $V_0$ if and only if the Riesz bounds in (7.90) are $A = B = 1$. In this case we call $\varphi$ an *orthogonal scaling function.* For example, the function $\varphi = \chi_{[0,1]}$ is orthogonal. Given any (not necessarily orthogonal) $\varphi$, we can construct an associated orthogonal scaling function $\tilde{\varphi}$; therefore, in what follows we shall assume that $\varphi$ is orthogonal.

The coefficients $h_k$ of the scaling equation (7.91) form a sequence $\{h_k\}$ known as a *filter.* For orthogonal $\varphi$ we have the equations

$$h_k = \int_{\mathbb{R}} \varphi(x)\varphi(2x-k)\,\mathrm{d}x \quad \text{and} \quad \sum_{k=-\infty}^{\infty} h_k h_{k+\ell} = \delta_{0,\ell} \quad (\delta \text{ Kronecker symbol}).$$

The Fourier series formed from the filter coefficients

$$H(\xi) := \frac{1}{\sqrt{2}} \sum_{k=-\infty}^{\infty} h_k \mathrm{e}^{-\mathrm{i}k\xi} \tag{7.96}$$

is called a *Fourier filter.* It can directly calculated from the Fourier transform $\widehat{\varphi}$ by using the formula $\widehat{\varphi}(x) = H(\xi/2)\widehat{\varphi}(\xi/2)$.

**7.7.9.6.4 Wavelets and multiresolution analysis.** Because of the inclusion $V_0 \subseteq V_1$ we can write $V_1$ as a direct sum of $V_0$ and the orthogonal complement $W_0 := \{f \in V_1 : \int_{\mathbb{R}} fg\,\mathrm{d}x = 0 \text{ for all } g \in V_0\}$:

$$V_1 = V_0 \oplus W_0.$$

Similarly, $V_0$ can be decomposed into a sum $V_{-1} \oplus W_{-1}$. Recursively, we then obtain

$$V_m = V_\ell \oplus \bigoplus_{j=\ell}^{m-1} W_j, \qquad V_m = \bigoplus_{j=-\infty}^{m-1} W_j, \qquad L^2(\mathbb{R}) = \bigoplus_{j=-\infty}^{\infty} W_j. \tag{7.97}$$

Every function $f \in L^2(\mathbb{R})$ can, according to (7.97), be written as $f = \sum_j f_j,\ f_j \in W_j$, which is an orthogonal decomposition. $f_j$ contains the 'details' of level $j$, where the index $j$ here indicates the frequency. A further resolution of $f_j$ into the position components follows in the next step.

Just as the spaces $V_m$ can be generated by $\varphi_{m,k}$, we can generate the spaces $W_m$ by

$$\boxed{\psi_{m,k}(x) := 2^{m/2}\psi(2^m x - k), \qquad m, k \text{ integer},}$$

in which $\psi$ is now a wavelet. For every orthogonal scaling function $\varphi$ we can construct an appropriate wavelet as follows. We set

$$g_k = (-1)^{k-1} h_{1-k}, \qquad \psi(x) = \sqrt{2} \sum_{k=-\infty}^{\infty} g_k \varphi(2x-k). \tag{7.98}$$

If the scaling function is the function $\varphi = \chi_{[0,1]}$ in Figure 7.13b, then the corresponding wavelet is the Haar wavelet shown in Figure 7.13a.

The translates $\{\psi_{m,k} : k \text{ integer}\}$ of the functions $\psi_m$ at scale $m$ form not only an orthonormal basis of $V_m$, but moreover $\{\psi_{m,k} : m, k \text{ integer}\}$ is an orthonormal basis of the entire space $L^2(\mathbb{R})$. There is a relation between the Fourier transforms of $\varphi$ and $\psi$, namely

$$\widehat{\psi} = \exp(-\mathrm{i}\xi/2)\overline{H(\pi + \xi/2)}\widehat{\varphi}(\xi/2)$$

where $H$ is the Fourier filter of (7.96).

**7.7.9.6.5 Fast wavelet transformations.** Suppose that for a function $f \in V_0$, the coefficients in the representation

$$f = \sum_{k=-\infty}^{\infty} c_k^0 \varphi_{0,k} \tag{7.99}$$

are known. According to the orthogonal decomposition $V_0 = V_{-M} \oplus W_{-M} \oplus \ldots \oplus W_{-2} \oplus W_{-1}$ (see (7.97)) we would like to decompose $f$ in

$$f = \sum_{j=-M}^{0} f_j + F_{-M} \quad \text{with} \quad f_j \in W_j,\ F_{-M} \in V_{-M} \tag{7.100a}$$

and

$$f_j = \sum_k d_k^j \psi_{j,k}\,, \qquad F_{-M} = \sum_k c_k^{-M} \varphi_{-M,k}. \tag{7.100b}$$

The function $F_{-M}$ contains the 'greater part' of $f$. The details at scale $j$ are decomposed in (7.100b) into the local components $d_k^j \psi_{j,k}$.

The coefficients $\{c_k^{-M}, d_k^j : k \text{ integer}, -M \le j \le -1\}$ can in principle be calculated using the scalar products $\int_{\mathbb{R}} f\psi_{j,k}\,\mathrm{d}x$, etc. But even if we were to know the functions $\psi_{j,k}$, actually carrying out this computation would be a hopeless task. Instead we can use the scaling equation (7.91), which leads us to the *fast wavelet transformation*:

$$\begin{aligned}&\text{for } j = -1 \text{ down to } -M \text{ do for all integers } k \text{ do}\\ &\text{begin } c_k^j := \sum_\ell h_{\ell-2k} c_\ell^{j+1}; \qquad d_k^j := \sum_\ell g_{\ell-2k} c_\ell^{j+1} \text{ end.}\end{aligned} \tag{7.101}$$

Note that the wavelet $\psi$ does not explicitly enter the computation, but only its coefficients $g_k$ from (7.98). In putting this algorithm to practice one must of course assume that $f$ is given by a *finite* sum (7.99). If $k_{\min}$ and $k_{\max}$ are the smallest and largest indices $k$ with $c_k^0 \neq 0$, then this corresponds to a 'signal length' $n = k_{\max} - k_{\min}$. We furthermore assume that the filter $\{h_k\}$ is finite. Then the fast wavelet transformation (7.101) requires $O(n) + O(M)$ operations, where $M$ is the *decomposition depth.* If one assumes that $M \ll n$, then the fast wavelet transformation has cost which grows only linearly with the signal length and hence is much cheaper than the Fourier transformation.

If conversely one wishes to deduce the coefficients $c_k^0$ in (7.99) from the coefficients $\{c_k^{-M}, d_k^j : -M \le j \le -1\}$, then one applies the *fast inverse wavelet transformation*, given by

$$\text{for } j = -M \text{ to } -1 \text{ do for all } k \text{ do } c_k^{j+1} := \sum_\ell (h_{2\ell-k} c_\ell^j + g_{2\ell-k} d_\ell^j)\,;$$

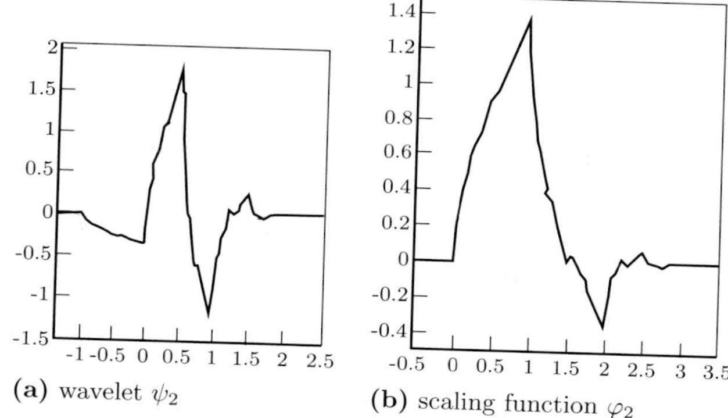


**(a)** wavelet $\psi_2$ **(b)** scaling function $\varphi_2$

*Figure 7.14. The Daubechies wavelet $\psi_2$ and the scaling function $\varphi_2$.*

**7.7.9.6.6 Daubechies wavelets.** The difficulties of the multiresolution analysis lie in the concrete scaling function $\varphi$, the wavelet $\psi$, and – even more importantly – the filter $\{h_k\}$. The Haar wavelet is the only wavelet with a simple description. Attempting to work with spline functions as in 7.3.1.6 leads to an infinite filter length.

Ingrid Daubechies made a significant breakthrough when she constructed a family $\{\psi_N : N > 0\}$ of orthogonal wavelets with the property that $\psi_N$ has order $N$, compact support and a filter of length $2N - 1$.

For $N = 2$ the non-vanishing filter coefficients are, for example

$$h_0 = (1+\sqrt{3})/(4\sqrt{2}), \quad h_1 = (3+\sqrt{3})/(4\sqrt{2}),$$
$$h_3 = (3-\sqrt{3})/(4\sqrt{2}), \quad h_4 = (1-\sqrt{3})(4\sqrt{2}).$$

The scaling function $\varphi = \varphi_2$ and the wavelet $\psi = \psi_2$ can, however, not be explicitly exhibited. Graphs of the functions are shown in Figure 7.14. The sharp corners testify to the fact that $\varphi_2$ and $\psi_2$ both are only Hölder-continuous with exponent 0.55. The smoothness of $\psi_N$ increases with increasing $N$. Starting with $N = 3$, the functions are even differentiable.

**7.7.9.6.7 Data compression and adaptivity.** The wavelet transformation has a variety of applications. One example is the data compression, which we sketch briefly here. The wavelet transformation maps a data package $c^0 = \{c_k^0\}$ belonging to a function $f$ to the 'smooth part' $c^{-M} = \{c_k^{-M}\}$ and the details $d^j = \{d_k^j\}$, which have scales $-M \leq j \leq -1$ (see (7.99)). This does not imply that the corresponding data package is $(M+1)$ times as large. For a finite filter and length $n$ of the initial sequence $c^0$, the sequences $c^j$ and $d^j$ which arise from these have asymptotically the length $2^j n$ $(j < 0)$. The sum of the lengths of $c^{-M}$ and $d^j$ for $-M \leq j \leq -1$ is, as before, just $O(n)$. For smooth $f$, the coefficients decrease for increasing $j$. If the function $f$ is only locally smooth – for example just in an interval $I$ – then the same holds for the $d_k^j$, which belong to the $\psi_{j,k}$ which have support in $I$. One can then replace coefficients which are sufficiently small by zero. Using this, one can generally find an approximation $\tilde{f}$ which is described by decisively fewer than $n$ amounts of data. The corresponding representation may be viewed as an *adaptive* approximation for $f$.

**7.7.9.6.8 Variants.** Since not all of the desirable properties (finite filter, orthogonality, high order and smoothness, explicit representation of $\varphi$ and $\psi$) can be satisfied at once, there are different variants which are adapted to the application one has in mind with the most important properties for that application. These differ from the multiresolution context described above.

One example is the notion of *pre-wavelets*, in which not all of the $\psi_{m,k}$ are orthonormal, but rather orthonormal to $\psi_{m,k}$ and $\psi_{j,\ell}$ with different scales $m \neq j$.

Another is the notion of *biorthogonal wavelets*, in which one applies two different multiresolution analysis spaces $\{V_m\}$ and $\{\tilde{V}_m\}$ with corresponding scaling functions $\varphi$ and $\tilde{\varphi}$ and wavelets $\psi$ and $\tilde{\psi}$, in which the latter form a biorthonormal system, i.e.,

$$\int_{\mathbb{R}} \psi_{m,k}\tilde{\psi}_{j,\ell}\,\mathrm{d}x = \delta_{mj}\delta_{k\ell}.$$

Generalizations of multiresolution analysis to several dimensions (for example for $L^2(\mathbb{R}^d)$) are possible (see [449]). It is more difficult to adapt multiresolution analysis to intervals or general domains in $\mathbb{R}^d$.

## 7.7.10 Inverse problems

### 7.7.10.1 Well-posed problems

If the problem

$$\boxed{Ax = b, \qquad x \in X}$$

with a given $b \in Y$ is to be solved numerically, then generally one requires that for every $b$ in some domain of definition $B \subseteq Y$ there is (at least locally) a unique solution $x \in U \subseteq X$, which depends on $b$ in a continuous manner. In this case, we refer to the problem $Ax = b$ as being *well-posed* or having good condition. Only under this assumption can we be assured that small perturbations in the 'data' $b$ (in the $Y$-topology) also lead to small perturbations of the solution $x$ (in the $X$-topology). Otherwise very small changes in the initial values of the problem or due to other causes (for example, non-exact arithmetic due to a implementation on a computer) lead to results which are total nonsense.

### 7.7.10.2 Ill-posed problems

If one of the assumptions just described is not satisfied, then the problem is said to be *ill-posed.* In the finite dimensional case the problem $Ax = b$ can be ill-posed because the matrix $A$ is singular. Infinite-dimensional problems are more interesting, leading to ill-posed problems when the operator $A$ has a trivial kernel and an unbounded inverse. Problems of this kind come up often if, for example, $A$ is an integral operator. An interesting and important example fore this is image reconstruction in tomography (see [450] and [448], §6).

If $A : X \to Y$ is compact, with $Y = X$, then $A$ has non-vanishing eigenvalues $\lambda_n \to 0$ $(n \geq 1)$. If $A$ is self-adjoint (the general case often reduces to this one if the singular value decomposition of section 7.2.4.3 is used), the corresponding eigenfunctions $\varphi_n$ form an orthonormal system. The solution of $Ax = b$ is then formally

$$x = \sum_{n=1}^{\infty} \frac{\alpha_n}{\lambda_n} \varphi_n \quad \text{with} \quad \alpha_n := \langle \varphi_n, b \rangle. \tag{7.102}$$

The solutions displayed belong to the solution space $X$ if $\sum_n (\alpha_n/\lambda_n)^2 < \infty$, where note must be taken of the fact that $\lambda_n \to 0$. In general one does not have $\sum_n (\alpha_n/\lambda_n)^2 < \infty$, as $b \in X$ generally only insures $\sum_n \alpha_n^2 < \infty$.

Even if one restricts the consideration to $b$ which belong to $x$ as in (7.102) with $x \in X$, the difficulties are not yet over. We have $b^{(n)} := b + \varepsilon\varphi_n$ for arbitrary $n$ is only perturbed by $\varepsilon$, i.e., $||b^{(n)} - b|| = \varepsilon$. But the unique (existing) solution of $Ax^{(n)} = b^{(n)}$ is $x^{(n)} = x + (\varepsilon/\lambda_n)\varphi_n$, and therefore we have an error of $||x^{(n)} - x|| = \varepsilon/\lambda_n$, which tends to infinity as $n$ grows beyond bounds. This shows again how arbitrarily small perturbations of $b$ lead to arbitrarily large changes in $x$.

The growth of the inverse eigenvalues $1/\lambda_n$ determines how ill-posed the problem is. If $1/\lambda_n$ grows like $O(n^{-\alpha})$ for some $\alpha > 0$, then $A$ is ill-posed to order $\alpha$. $A$ is said to be *exponentially ill-posed*, if $1/\lambda_n > \exp(\gamma n^{\delta})$ for $\gamma, \delta > 0$.

#### 7.7.10.3 Problems of ill-posed problems

In addition to what has been said already, solving the equation $Ax = b$ makes little sense even if a solution actually exists. Instead, it makes sense to change the questions about the problem, i.e., pose different problems with the ill-posed problem, which have a chance of being solvable.

The coefficient $\beta_n := \langle \varphi_n, x \rangle$ of a function $x$ describes in general its smoothness, in the following sense. The faster $\beta_n$ approaches 0 as $n \to \infty$, the smoother the function $x$. To quantify this vague statement, we define for real $\sigma$ the space

$$X_\sigma := \left\{ x = \sum_{n=1}^{\infty} \beta_n \varphi_n \quad \text{with} \quad ||x||_\sigma^2 := \sum_{n=1}^{\infty} (\beta_n/\lambda_n^\sigma)^2 < \infty \right\}.$$

For $\sigma = 0$ we have $X_0 = X$, while for $\sigma = 1$ we have $X_1 = \operatorname{Im} A$. The solution $x$ given by (7.102) belongs to $X_{-1}$ for $b \in X$.

A further essential assumption we can make is that the solution $x$ to be found belongs to some $X_\sigma$ for a positive $\sigma$, i.e., this solution has the property of greater smoothness. Assume the corresponding norm is bounded by $\rho$:

$$||x||_\sigma \le \rho. \tag{7.103a}$$

The ideal 'data' of the 'state' $x$ is $b := Ax$. We cannot expect that $b$ is given to us exactly. Instead, we assume that the known data $b$ are precise up to a factor of $\varepsilon$, as follows.

$$||b - \tilde{b}|| \le \varepsilon. \tag{7.103b}$$

In this case, the problem can be formulated as follows. Let $\tilde{b}$ be given. Find $x \in X_\sigma$ whose exact image $b = Ax$ approximates the data $\tilde{b}$ and satisfies, for example, the equation (7.103b). This problem definitely does not have a unique solution. If, however, $x'$ and $x''$ are two possible solutions, which satisfy (7.103a): $\|x'\|_\sigma \le \rho$, $\|x''\|_\sigma \le \rho$ and (7.103b): $\|b' - \tilde{b}\| \le \varepsilon$, $\|b'' - \tilde{b}\| \le \varepsilon$ for $b' := Ax'$, $b'' := Ax''$, then for the differences $\delta x := x' - x''$ and $\delta b := b' - b''$ we have by the triangle inequality

$$A\delta x = \delta b, \qquad \|\delta x\|_\sigma \le 2\rho, \qquad \|\delta b\| \le 2\varepsilon.$$

This yields the estimate

$$\boxed{\|\delta x\| \le 2\varepsilon^{\sigma/(\sigma+1)}\rho^{1/(\sigma+1)}.} \tag{7.104}$$

To interpret equation (7.104), one identifies the sought-for solution $x$ with $x'$ and views $x''$ as an approximate solution. Then equation (7.104) yields a bound on the remaining error. The bound $\rho$ in (7.103a) will have magnitude $O(1)$, so that $\rho^{1/(\sigma+1)}$ is constant. Only $\varepsilon$ can be assumed to be small. Because of $\sigma > 0$ the uncertainty $\|\delta x\|$ is then also small. However, the exponent will be worse when the order of smoothness of $\sigma$ is weak.

Independent of the numerical methods which are applied to determine $x$, (7.104) indicates the inevitable imprecision. Conversely, an approximation methods is said to be *optimal*, if the results of this method have at most the error given by (7.104).

#### 7.7.10.4 Regularization

Let $Ax = b$ be a given problem, and let $b^\varepsilon$ denote an approximation with $\|b^\varepsilon - b\| \le \varepsilon$. For positive $\gamma$ the map $T_\gamma$ is supposed to produce approximations $T_\gamma b^\varepsilon$ of $x$. If there is a *regularization parameter* $\gamma = \gamma(\varepsilon, b^\varepsilon)$ with the properties

$$\gamma(\varepsilon, b^\varepsilon) \to 0 \quad \text{and} \quad T_{\gamma(\varepsilon,b^\varepsilon)}b^\varepsilon \to x \qquad \text{for} \quad \varepsilon \to 0, \tag{7.105}$$

then one calls the family of mappings $\{T_\gamma : \gamma > 0\}$ a (linear) *regularization.*

A simple example is given by truncating the expansion in (7.102):

$$T_\gamma b^\varepsilon := \sum_{\lambda_n \ge \gamma} (\langle \varphi_n, b^\varepsilon\rangle/\lambda_n)\varphi_n.$$

In this case (7.105) is assured if we take $\gamma = \gamma(\varepsilon) = O(\varepsilon^\kappa)$ with $\kappa < 1$. In particular, for $\gamma(\varepsilon) = (\varepsilon/(\sigma\rho))^{1/(\sigma+1)}$ ($\varepsilon, \sigma, \rho$ as in (7.103a,b)), then this regularization yields optimal orders (see [448], §4.1).

A regularization which is often used is the *Tychonov–Phillips regularization.* In this case one seeks the minimizing element of the functional

$$\boxed{J_\gamma(x) := \|Ax - b\|^2 + \gamma\|x\|_\sigma^2 \qquad (\sigma \text{ as in } (7.103\text{a})).}$$

In this connection the quantity $\gamma > 0$ is referred to as the *penalty term.* For more on the choice of $\gamma$ and questions related to optimality, see [449].

Some regularizations are indirect. The usual discretization of infinite-dimensional problems $Ax = b$ can also represent a regularization. Moreover, $m = m(\gamma)$ steps of an iteration, for example of the Landweber iteration (this is (7.54b) with $N = \omega A^*$) can serve as a regularization.

# *Sketch of the history of mathematics*

*There is no such thing as patriotic art or patriotic science. Both of these belong, as does everything good, to the whole world and can only be promoted through a general, free interaction of all living people, taking into account what is known from the past.*

*Johann Wolfgang von Goethe* (1749–1832)

To exemplify historical correlations, we give some dates from the lives of several artists, scientists and philosophers, as well as of some important historical events, of course without attempting any kind of completeness. The dates given from ancient times are partly only approximately known; this will not be mentioned explicitly below.

The ensuing rough story of the formation of our universe and our planet should help to put the relatively short period of human endeavor into its proper perspective. This process of human culture and science has accelerated at a breathtaking rate in the twentieth century. In the coming millennium, mankind will have to learn to use its knowledge more responsibly than has been the case up to now.

| | |
|---|---|
| 14 billion years ago | The big bang starts the history of our universe.<br>Roughly three minutes after the big bang the universe has cooled down to about 900 million degrees, and the synthesis of helium began. In this way the most important source of energy of the young universe is formed.[1] |
| 13 billion years ago | Quasars form. |
| 10 billion years ago | Galaxies form. |
| 4.6 billion years ago | The formation of our solar system, including the earth, takes place. |
| 4 billion years ago | The first primitive forms of life form on the earth. |
| 2 billion years ago | The earth's crust forms. |
| 248 million years ago, and 213 million years ago | Two global ecological catastrophes occur and wipe out many of the existing life forms on the earth. The ancestors of mammals barely escape extinction. |
| 65 million years ago | The dinosaurs die out. |
| 5 million years ago | The ancestor of ours, *Australopithecus*, lives in eastern Africa and learns to walk upright.<br>In 1974 the 3.2 million year old skeleton "Lucy" was discovered; in 1995 a 1.2 meter in height, 4.4 million year old skeleton of an *Ardipithecus ramidus* (a ground-living ape which is an older ancestor of humanoids) was discovered in Ethiopia. |

[1]The early development of the universe after the big bang has been presented to a wide audience in an exciting form in his best seller *The first three minutes* by Steven Weinberg (1977). The theory of the development of the universe can be found in the monograph *The Early Universe* by G. Börner, Springer-Verlag, Berlin, 2003.

| | |
|---|---|
| 2.5 million years ago | *Homo habilis* (skilled man) lives in Africa and uses primitive stone-age tools. The brain starts an accelerated growth. |
| 1.6 million years | *Homo erectus* (standing man) lives in Africa and migrates into much of Asia and Europe. |
| 100,000 years ago | *Homo sapiens* (reasoning man) lives in Africa. |
| 30,000 years ago | The Neanderthals die out after an existence of 100,000 years in Europe and Asia, and are replaced by homo sapiens. |

## The beginnings of culture

13,000 BC — Cave paintings in France and Spain testify that ancient man has a keen sense for forms. During this time period also the oldest figures representing numbers in the form of slashes on cave walls and grooves on sticks have their origin (early stone age).

8000 BC — The glaciers covering Asia and Europe melt. This is a time of transition from hunting societies to farming communities; pottery displays geometric shapes (late stone age).

7000 BC — In Jarmo, Iraq, more than 1,000 solid balls dating from this time are found. Balls of this kind were presumably included in closed trading goods, to indicate to the recipient the number of items.

## Mathematics of prehistoric times

3200 BC — The Sumerians, whose origin is unknown to us, settle in the Tiger–Euphrates river valley (Mesopotamia, in what is now Iraq), and found city states like Ur. Even today, one can find traces of this Sumerian civilization in our modern culture.

3000 BC — The first written alphabets develop in Mesopotamia and Egypt.
Decorations on the club of the Egyptian king Narmer display written letters and a well-developed number system. (Ashmolean Museum in Oxford, England)

2600 BC — Construction of the great Pyramid of Cheops.

2000 BC — In Mesopotamia the Sumerians use a well-developed numeral system with base 60 (sexagesimal system). Around this time the culture of the Sumerians is replaced by that of the Babylonians.

1800 BC — Egyptian papyrus scrolls contain a developed arithmetic with fractions. In Egypt geometry is developed in connection with geodesy.

1800 BC — King Hammurabi reigns in the old Babylonian kingdom. Cuneiform writing shows the blossoming of Babylonian mathematics, which is in a position to solve linear, quadratic (and even some cubic and degree four) algebraic equations. The theorem of Pythagoras is well-known to the Babylonians. Their mathematics is strongly influenced by algebra (as compared with the later Greek mathematics, which had a strong geometric flavor).

575 BC — The Babylonian culture reaches its zenith under King Nebukadnezar. At this time the numeral 0 first appears as the place holder for an empty slot in the sexagesimal system.

## Mathematics of ancient times

| | |
|---|---|
| ca. 800 BC | Homer writes his famous works, the *Ilias* and the *Odyssey*. |
| 735 BC | The legendary founding of Rome by Romulus. |
| 624–547 BC | The Greek trader and natural philosopher Thales of Milet, who traveled extensively in Babylonia and Egypt and is considered to be the founder of Greek mathematics, lives. |
| 580–500 BC | The Greek Pythagoras of Samos learns the highly developed mathematics of the Babylonians and the Egyptians in Phoenicia (today the coastal part of Syria). He founds the school of the Pythagoreans. |
| 551–478 BC | Confucius (Master Kung), Chinese philosopher and statesman, lives. |
| 550–480 BC | The Indian prince Gautama Buddha founds Buddhism. |
| 500 BC | The Indian religious text "Súlvasũtras" contains instructions for the construction of squares and rectangles as well as the basics of the theorem of Pythagoras. |
| 500 BC | A Pythagorean discovers the existence of incommensurable lengths ($\sqrt{2}$ is an irrational number). This creates a crisis in the foundations of Greek mathematics. |
| 469–399 BC | Socrates teaches that mankind cannot recognize the essence of the world: he can only recognize himself. |
| 460–371 BC | Democritus of Abdera founds the theory of atoms. |
| 428–348 BC | Plato adopts the theory of the general notions as the essence of things from his teacher, Socrates.<br>However, he splits the general notions of the things from these and considers them to be eternal, absolute ideas, which exist in a world of their own. The ideas of Plato influenced Werner Heisenberg in an essential way in his development of the abstract ideas leading to the theory of quantum mechanics in 1924.<br>Plato believes in mathematics as a science in its own right, which should be studied for its own sake and not just for applications. |
| 408–355 BC | Eudoxs of Knidos creates a theory of proportions, in which also incommensurable lengths (irrational numbers) have a place. This saves Greek mathematics from the crisis in its foundations. |
| 300 BC to 400 AD | Alexandria (founded in 331 BC by Alexander the Great in the Nile delta) is the scientific and cultural center of the world of Greek culture, as well as of the Roman empire. The huge library, containing 700,000 papyrus scrolls, is destroyed in a battle with the Romans. After being conquered by the Arabians in 642 AD, the Arabian science is dominant. |
| 384–322 BC | Aristotle – the greatest mind of ancient times (a student of Plato) – creates formal logic and scientific classification. He summarizes the known status of the knowledge of his times in esthetics, astronomy, biology, ethics, history, metaphysics, philosophy, psychology and rhetoric, and develops them further.<br>The teachings of Aristotle are dominant for the next 2000 years in science, until Galileo Galilei founds modern physics, based on experimental evidence. |

365–300 BC — Euclid of Alexandria lives; his famous work *The elements* is a standard reference for geometry and a model for the axiomatic method in mathematics.

356–323 BC — Alexander the Great lives; he conquers Persia and Egypt and presses on to India. His goal, which he did not achieve, was to form a unified empire in all of the orient and occident, with a unified Greek culture.

287–212 BC — Archimedes of Syracuse lives, the most important mathematician of ancient times and the founder of mathematical physics. He determines the center of mass of simple surfaces and bodies and he derives formula for the workings of levers, equilibrium of floating bodies, determination of areas and volumes, which marked the birth of calculus. He was the last great mathematician of ancient times.

200 BC to 200 AD — China is ruled by the Han dynasty; the treatise *Nine books on the art of mathematics* appears, which treats practical problems of mathematics (for example the determination of square and cube roots). In connection with the solution of a quadratic equation, negative numbers appear for the first time.

180–125 BC — Hipparchus of Nicæa, an important astronomer of ancient times, lives.

100–44 BC — Gaius Julius Caesar – Roman field general and later emperor of Rome, lives (the German word for emperor, *Kaiser*, comes from the word Caesar). He writes *De bello Gallico* (On the Gallic war) about his experiences in war in northern Europe.

0–30 AD — Jesus founds Christianity.

85–169 AD[2] — Ptolemy of Alexandria, Greek mathematician and astronomer, lives. His main contribution to astronomy is *Almagest* (Big system), which applies studies of Hipparchus and furthers these. The work also contains elements of plane and spherical trigonometry in a geometric form. His picture of the world has the earth at the center.

ca. 100 — Heron of Alexandria, an engineer and applied mathematician, lives. His collected works summarize the practical knowledge of the times, and is complementary to Euclid's Elements. The books *Mechanica* (levers, inclined plane, pulleys), *Pneumatica* (presses), *Dioprica* (geodesy) and *Belopoika* (weaponry) are examples of works which flowed from his pen.

ca. 250 (?) — Diophantus of Alexandria, an important number theoretician which influenced even modern times, lives. His most important contribution to mathematics, *Arithmetica*, originally consists of 13 volumes, of which seven are preserved; dates about his life are very uncertain.

ca. 320 — Pappus of Alexandria, the last important mathematician of ancient times, lives. He takes first steps in the direction of projective geometry.

395 — The Roman empire is divided into an eastern empire, centered at Constantinople, and a western empire centered at Rome.

476 — The western Roman empire ceases to exist; the eastern part exists until 1453.

529 — Plato's Academy in Athens is closed by force by the Roman emperor Justitian, which signaled the decline of the mathematics of ancient times.

---

[2] Henceforth, all year numbers are AD.

570–632 The Islamic religion is founded by Mohammed.

ca. 800 Al–Khowarizmi of Choresm (near the Aral Sea in central Asia) lives. He is the first Islamic author to write about the solution of equations (linear and quadratic) in his book *Algebra*. A Latin translation of one of his books, which was not preserved in the original, translates his name as *Algoritmi*, which is the origin of the word 'algorithm'.

## Mathematics during the middle ages

1155–1227 Jinghis Khan, the founder of the Mongolian empire, lives. Under the reign of his sons this empire extends to Europe.

1160–1227 The famous minnesinger Walther von der Vogelweide lives.

1180–1250 Leonardo of Pisa (Fibonacci) lives; with his work in algebra and number theory, he revives the occidental art of mathematics after more than a millennium of decay and stagnation following the fall of the Roman empire. He presents the method of calculations with Arabic numbers in his book *Abacus* which appeared in 1202, and contributes to the dissemination of Indian and Islamic mathematics in Europe.

1199 The University of Bologna is established (the oldest University in the world); at the beginning of the 13th century, Universities are established in Paris, Oxford and Cambridge; a century later a wave of establishments of universities sweeps across Europe, including the universities in Prague (1348), Vienna (1365), Heidelberg (1386), Cologne (1388), Erfurt (1392), Leipzig (1409), Rostock (1419) and following this in several other cities.

1225–1274 Thomas Aquinas, an Italian theologian and philosopher whose theories influence thinking even today, lives.

1248 The construction of the cathedral in Cologne begins (completed in 1880).

1254–1324 Marco Polo, the famous Venician trader who traveled as far as China, lives.

1260 The Islamic mathematician at–Tusi puts trigonometry in its place as a separate discipline in mathematics in his main work, which collected the progress of the Islamic mathematicians over the preceding four centuries.

1265–1321 Dante Alighieri, the writer of the *Divina Comedia* (divine comedy) lives.

1339–1453 The Hundred Years' War between the English and the French Royal families devastates much of Europe.

1415 Jan Hus, a Czechoslovakian supporter of the Reformation and rector of the University of Prague, is burned at the stake.

1431 Jeanne d'Arc is banned.

## Mathematics during the Renaissance[3]

1436–1476 Regiomontanus (John Müller), the most important mathematician of the 15th century, lives.

His main work, the book *De triangulis omnimodis libri quinque* (Five books on the types of triangles) was not published until 1533; this book began the modern treatment of trigonometry.

[3] The Renaissance (rebirth) began in Florence in the 15th century.

| | |
|---|---|
| 1452–1519 | Leonardo da Vinci, the jack of all trades, painter, sculptor, architect and scientist, lives. |
| 1470 | The establishment of the Academy in Florence. |
| 1473–1543 | Nicolaus Copernicus, the creator of the modern heliocentric picture of our solar system, which places the sun at the center, lives.<br>In 1543 his main work appeared, under the title *De revolutionibus orbium coelestium.* |
| 1475–1564 | Michelangelo Buonarroti, painter, sculptor and architect, lives; he works in Florence under commission of the Medici and in Rome on the Cathedral of Saint Peter. |
| 1483–1520 | Raphael Santi, famous painter, lives; since 1515 he is the main architect working on the Cathedral of Saint Peter in Rome. |
| 1492 | Christopher Columbus discovers the Americas. |
| 1492–1559 | Adam Ries – who was chief of calculations in Annaberg, lives; he popularizes the art of doing arithmetic.<br>In 1524 his book *Coss* appears. |
| 1506 | Start of construction on the Cathedral of Saint Peter in Rome; the main architects were among others Bramante, Raphael and Michelangelo, |
| 1517 | Martin Luther (1483–1546), founder of the reformation, posts his thesis on the door of the Schlosskirche in Wittenberg, signaling the beginning of the reformation. |
| 1519 | Hernando Cortez (1485–1547) begins with the blood-thirsty conquest of Mexico, which totally destroys the blossoming culture there. |
| 1525 | Albrecht Dürer's (1471–1528) book *Unterweisung der Messung mit Zirkel und Richtscheit* (Directions for measurements with compass and level) appears, in which perspective drawing is described. This kind of technique goes back to the painter Leon Alberti (1404–1472) and Leonardo da Vinci. |
| 1531–1534 | Francisco Pizarro (1475–1541) conquers in another brutal war the thriving Inca kingdom (where the modern countries Chile and Peru are situated). |
| 1540–1603 | Francis Vieta (also known as Vieta) lives; he is responsible for the use of letters in mathematics as is customary today. |
| 1544 | Michael Stifel (1487–1567) publishes his three-volume *Arithmetica Integra* – a methodologically mature summary of mathematics of his time (addition, subtraction, multiplication, division as well as quadratic and cubic equations).[4] |
| 1545 | Geronimo Cardano (1501–1576), an Italian mathematician, publishes his book *Ars Magna* (Big Art), which contains a method of solving algebraic equations of third and fourth degrees. This is the first significant step which goes beyond the mathematics of ancient times. |
| 1550 | Rafael Bombielli introduces in his books *Geometry* (1550) and *Algebra* (1572) the imaginary unit $\sqrt{-1}$ and uses complex numbers systematically to solve algebraic equations of the third degree. |

[4] As a matter of curiosity, it is interesting to note that as an application of his calculations, Stifel predicted the end of the world on October 18, 1533 at 8.00 AM.

### Mathematics during the age of rationalism

1561–1626 Francis Bacon, philosopher, lives. He is the founder of empiricism, which as opposed to Aristotle tried to obtain his insights by means of experiments rather than by rational thinking.

1562–1598 War of the Huguenots in France.

1564–1642 Galileo Galilei, the discoverer of the law of falling bodies, lives. This marks the beginning of modern experimental physics. As opposed to Aristotle he observes that all bodies fall with the same velocity.

In 1609 he discovers four of the twelve moons of Jupiter with a self-made telescope.

In 1632 his thesis *Discorsi* (Discourses) appears.

In 1633 he is tried and convicted by a tribunal of the inquisition because of his heliocentric view of the world; he denies himself and remains a prisoner in Florence for the rest of his life.

1564–1616 William Shakespeare, the perhaps greatest play–writer of all times, lives.

1567–1643 Claudio Monteverdi, a composer of madrigals and a master of early European opera, lives.

1571–1630 Johannes Kepler, mathematician and astronomer (as well as astrologer) at the Royal court in Prague, lives. He is the discoverer of the three laws of planetary motion named after him.

In 1609 his work *Astronomia Nova* appears, in which these three laws are formulated, which are based on extensive observations of Tycho Brahe (1546–1601).

In 1627 he publishes his *Rudolphian tables* (tables of logarithmic values), which are indispensable for centuries for astronomy and navigation.

1587 Execution of Maria Stuart (Queen of Scotts).

1596–1650 René Descartes, mathematician, scientist and philosopher, lives. This marked the beginning of the age of modern mathematics. Together with Fermat he founded analytic geometry, in which geometry and algebra are synthesized.

In 1637 his book *Discours de la méthode* (Lecture on the method) appears, which contains his presentation of the foundations of analytic geometry.

1598–1647 Bonaventura Cavalieri lives; his principle of calculation of the volume of bodies is a precursor of the calculus developed later by Newton and Leibniz.

1600 Giordano Bruno (1548–1600), and Italian philosopher, is burned at the stake for his belief in the heliocentric nature of the solar system.

1601–1665 Pierre de Fermat, number theorist, lives. He is the founder (together with Descartes) of analytic geometry, and carries out in addition first investigations on probability (together with Pascal) as well as creating the methods for determining minima and maxima, which are closely related to differential calculus.

In 1629 his book *Maxima and Minima* appears. In this work, the fundamental Fermat principle in geometric optics is formulated: light moves in such a way as to require the smallest amount of time. This book in also a precursor of calculus.

In 1637 he writes in the margin of a book by Diophantus his famous claim; for centuries number theorists search for a proof, without success. In wasn't until a series of new, very abstract ideas were introduced that finally led Andrew Wiles in 1994 to his proof of *Fermat's last theorem*, as it has been known to be called.

1606–1669 Rembrandt van Rijn, famous Dutch painter of portraits, etc., lives.

1614 John Neper (1550–1617), a Scotch landowner, publishes his book *Mirifici logarithmorum canonis description*, in which the prototype of logarithms is introduced. Kepler is especially active in propagating the use of logarithms.

1618–1648 Thirty Year's War rampages in Europe.

1623 Wilhelm Schickgard (1552–1635), mathematician, theologian and orientalist, constructs for Kepler the first calculating machine in Tübingen; this machine can already do addition, subtraction, multiplication and division, applying the idea of logarithms due to Neper.

1623–1662 Blaise Pascal, lives. He works in geometry, hydrostatics and probability theory; he also constructs a machine which can add in 1652.

1623–1789 The famous scholarly family Bernoulli produces eight professors of mathematics, physics and other scientific disciplines.

1629–1695 Christian Huygens lives; he is the founder of wave optics, the inventor of the pendulum clock and uses continued fractions.

1632–1677 Baruch Spinoza, philosopher, lives. For him, God and nature are a united entity; he does not believe in free will.

1635 Establishment of the Académie Française in Paris.

1636 Harvard University in Boston is founded; this is the oldest university in the United States.

### Mathematics in the age of enlightenment

1643–1727 Isaac Newton, he creator of modern physics (mechanics) and mathematics (calculus), lives. Newton's life marks an abrupt change in the intellectual development of civilization.

In 1676 he communicates to Leibniz (in encoded form) his discovery of the basics of calculus.

In 1687 his main work *Philosophiae naturalis principia mathematica* appears.

1645 The English revolution.

1646–1716 Gottfried Wilhelm Leibniz, philosopher and universal scholar, lives. He is the co-founder of calculus, as well as the inventor of the convenient formal rules for expressing its laws.

In 1674 he gives the construction of a calculating machine in commission, based on his invention of stacked cylinders. This calculating machine can also do addition, subtraction, multiplication and division.

In 1677 he writes to Newton, explaining to him his *Calculo differentiali.*

In 1682 he founds the *Acta Eruditorim* (Scholarly journal) in Leipzig, and publishes in 1684 in it his first seminal paper *Nova methodus ...*

(the title in English: A new method for maxima and minima as well as for tangents, which is not impinged upon by rational and irrational quantities, and a specific way of calculating them.)

Leibniz is the first to realize the importance of well-designed symbols and notations for progress in mathematics (equality sign, multiplication point, division colon (still in use in Europe today), use of indices, symbols for differentiation and integration, functions, determinants.)

In 1700 he founds the Berlin Akademie and becomes its first president.

Leibniz teaches that the world consists of spiritual units, which he calls monads. This philosophical point of view influences his work in the development of calculus. For him the existing world is the best of all thinkable worlds. His main works in philosophy are *Theodizee* (1710) and *Monadologie* (1714).

1652 The founding of the Deutsche Akademie der Naturforscher Leopoldina in Halle (the oldest continually active such scientific academy); following this also the Royal Society in London is established in 1663, the Académie des sciences in Paris in 1666, the Berliner Akademie in 1700 and the Petersburg Academy in 1725.

1654–1705 Jakob Bernoulli works on calculus and probability theory.

In 1713, eight years after his death, his work *Ars Conjectandi* (The way of mathematical conclusion) appears. in this book the first limit theorem of probability theory is laid down; this is the Jakob Bernoulli law of large numbers.

1665 Robert Hooke (1635–1703) discovers under the microscope the existence of cells in plants.

1667–1748 Johann Bernoulli makes contributions to calculus.

In 1691 his work *Lectiones mathematicae de methodo integralium* appears; this is the first text book on calculus.

In 1697 he appeals in the *Acta Eruditorum* to the mathematicians of his time to find a solution of the Brachystochrons problem; this sparks the development of the calculus of variations.

1685–1731 Brook Taylor, the inventor of the Taylor series, produces a basic tool in the investigation of the local properties of functions.

1685–1750 Johann Sebastian Bach works as composer and organist in the Thomas church in Leipzig.

1694–1778 François–Marie Voltaire, philosopher and one of the main proponents of the enlightenment, lives.

1707–1783 Leonhard Euler, the most productive mathematician of all times, lives. His collected works fill 72 volumes. He works on all areas of mathematics, applications to hydrodynamics and to elasticity.

In 1744 his work *Methodus inveniendi lineas curvas maximi minive proprietate gaudentes, sive solution problematis isoperimetrici latissimo sensu accepti* appears, in which he develops the calculus of variations systematically.

1724–1804 Immanuel Kant, philosopher and natural scientist, lives. He is the author of *Kritik der reinen Vernunft* (Critique on pure reason), which appears in 1781. He differentiates between a-priori notions, which are intrinsic to human nature, and a-posteriori notions, which humans acquire through experience.

| | |
|---|---|
| 1735 | Karl von Linné (1707–1778) publishes his *Systema naturae* (Systems of nature), on which the modern classification of the plant and animal kingdom is based. |
| 1736–1813 | Joseph Louis Lagrange, who completes the building of Newtonian mechanics with his analytic mechanics, lives. He also wrote several fundamental works on celestial mechanics, algebra and number theory.<br>In 1762 he founds the calculus of variations of several variables, which is a basic up through modern times in formulating the basic laws of physics.<br>In 1788 he publishes his *Méchanique analytique*, on which the calculus of variations is based. |
| 1738 | Daniel Bernoulli's (1700–1782) thesis *Hydrodynamica sive de viribus et motibus fluidorom* appears. |
| 1738 | The Italian city of Herculeanum, which was destroyed by the eruption of Mount Vesuv in 79 AD, is discovered and excavated. |
| 1749–1832 | Johann Wolfgang von Goethe, poet, writer, painter and natural scientist, lives. |
| 1749–1827 | Pierre Simon Laplace, physicist and mathematician, lives. He makes important contributions to celestial mechanics, capillary theory as well as to probability theory.<br>In 1799–1825 his publishes his five-volume *Mécanique céleste* (Celestial mechanics).<br>In 1812, his work *Théorie analytique des probabilité* (Analytic theory of probability) appears, which is the first systematic presentation of probability theory. |
| 1756–1791 | Wolfgang Amadeus Mozart, child prodigy and one of the most famous composers of all times, lives. |
| 1759–1805 | Friedrich Schiller, poet and historian, lives. |
| 1764 | Johann Winckelmann (1717–1768) founds with his *Geschichte der Kunst des Altertums* (History of the Art of Ancient Times) the scientific investigation of ancient civilizations (archaeology). |
| 1769–1821 | Napoléon Bonaparte lives; he influenced in a profound way the following development of Europe.<br>In 1798–1799 scientists and artists accompany him at his invitation to an expedition to Egypt. Among these is Vivant Denon, who draws many of the Egyptian art treasures and in 1809–1813 publishes his 24-volume opus *Description de l'Égypte* (Description of Egypt). |
| 1770–1827 | Ludwig van Beethoven, perhaps the most famous composer of all times, lives. |
| 1776 | American Declaration of Independence. |
| 1777–1855 | Karl Friedrich Gauss, the *princeps mathematicorum* (prince of mathematicians, which is engraved on a current coin), lives. He makes fundamental contributions to all areas of pure and applied mathematics, conducts important astronomical observations, works in geodesy, on explaining electromagnetic phenomena, and influences all of mathematics and physics of the twentieth century. |

In 1796 he discovers in the context of the theory of cyclotomic fields which he has developed, that the 17-gon can be constructed with ruler and compass. At the same time he characterizes all regular polygons which can be constructed in this way. He thus solved a problem which had been open for 2000 years.

In 1799 he gives the first complete proof of the fundamental theorem of algebra in his dissertation.

In 1801 his *Disquisitiones arithmeticae* (Foundations of modern number theory) appears. In particular he proves the law of quadratic reciprocity and develops the theory of cyclotomic fields.

In 1807 he becomes professor in Göttingen.

In 1809 he publishes the book *Theoria motus corporum coelestium* (Theory of the motion of the planets), in which his methods break new ground in astronomy.

In 1827 he founds differential geometry with the appearance of his thesis *Disquisitiones generales circa superficies curvas* (General investigations on curved surfaces).

In 1839 he publishes his work on potential theory *Allgemeine Lehrsätze für die im verkehrten Verhältnis des Quadrates der Entfernung wirkenden Anziehungs- und Abstossungskräfte* (General theorems about the forces of attraction and repelling which depend on the square of the inverse of the distance).

In 1844/1847 his monographs *Untersuchungen über Gegenstände der höheren Geodäsie* (Investigations on the objects of advanced geodesy) appear.

1789–1794 The French Revolution takes place.

1789–1857 Augustin–Louis Cauchy, one of the most productive mathematicians of all times, publishing seven books and 800 papers, makes fundamental contributions to real and complex analysis, as well as to elasticity theory.

In 1821, his book *Cours d'anlyse de l'école polytechnique* (Course of analysis of the polytechnical University in Paris) appears. Cauchy tries to provide a rigorous foundation for analysis; the modern notion of limit is due to him.

1791–1867 Michael Faraday provides the experimental evidence for Maxwell's theory of electromagnetism. He causes a revolution in physics by breaking with Newton's idea of force acting at distances and introduces the notion of field (forces acting locally). The notion of field is basic for modern physical theories.

In 1831 he discovers electromagnetic induction and develops the idea of the existence of an electromagnetic field.

1794 The Jacobian government establishes the Ècole Polytechnique in Paris, which quickly develops to a center of mathematics and natural sciences. Among others, Lagrange, Laplace, Monge, Cauchy, Poncelet, Ampère, Gay–Lussac, Fresnel, Dulong and Petit are professors at this institution.

1798 Gaspard Monge (1746–1818) publishes his work *Géométrie descriptive* and makes his favorite area, representative geometry, a full-fledged mathematical discipline.

**Mathematics in the nineteenth century**

1802 In order to win a bet, Georg Friedrich Grotefend (1775–1853) manages to decipher the Cuneiform writing of the Assyrians and Babylonians, and opens the door to understanding these ancient cultures.

1804–1851 Carl Gustav Jacob Jacobi lives; he makes significant contributions to the theory of elliptic functions (theta functions), the calculus of variations, number theory (the law of reciprocity for cubic rests), the theory of quadratic forms and determinants as well as to the algebraic theory of elimination and celestial mechanics.

1805–1865 William Rowan Hamilton, the creator of Hamiltonian mechanics, lives.

In 1843 he discovers the quaternions and introduces vectors.

1805–1859 Peter Gustav Lejeune Dirichlet, the creator of analytic number theory, lives. Among his many students are the great mathematicians Eisenstein, Kronecker, Kummer and Riemann. In 1855 he succeeds Gauss in Göttingen.

1815–1897 Karl Weierstrass, who dedicates his life to making complex analysis rigorous, lives. Motivated by the Jacobian inversion problem, he makes significant contributions to the theory of elliptic functions and Abelian integrals, to algebra (the theory of elementary divisors for matrices) and to the calculus of variations (sufficient conditions for the existence of minima), as well as providing rigorous foundations in analysis.

In 1864 he becomes professor at the Berlin University. In his lectures there he gathers a large, gifted group of students, putting particular emphasis on mathematical rigor, which is therefore sometimes also referred to as Weierstrass rigor.

1817 Bernhard Bolzano (1781–1848), a Czechoslovakian pastor, also attempts to give rigorous foundations to analysis and provides in a long manuscript what in those days was a rigorous proof of the means value theorem for continuous functions (mean value theorem of Bolzano). This mean value theorem is later, in the twentieth century, extended to a fundamental topological existence principle (fixed point theorem of Brouwer and Schauder; notion of degree of a mapping).

1821–1894 Pafnuti Lvovitsch Chebychev, who gave fundamental contributions to probability theory, number theory and approximation theory, lives. He exerts great influence on the development of the Russian school of mathematics.

1822 Jean–François Champoillon (1790–1832) deciphers the Egyptian hieroglyphs, after he had learned more than a dozen languages as preparation for this attempt.

1822 Jean–Victor Poncelet (1788–1867) founds the discipline of projective geometry with his book *Traité des propriétés projectives des figures.*

1822 The book *Théorie anlaytique de la chaleur* (Analytic theory of heat) by Jean Baptiste Joseph de Fourier (1768–1830) appears. In this text, Fourier series and Fourier integrals are developed systematically for the solution of partial differential equations.

1826 Niels Henrik Abel (1802–1829) proves that the general equation of degree $\geq 5$ is not solvable by means of radicals. He also makes significant contributions to the theory of algebraic functions, of which the Abelian functions and Abelian integrals (as well as Abelian groups) are named after him.

1826–1866 Georg Friedrich Bernhard Riemann, one of the dominating mathematicians of the century, lives. He makes fundamental contributions to complex functions theory, differential geometry (Riemannian geometry), number theory (the famous Riemann hypothesis), topology (Riemannian surfaces) as well as to mathematical physics (for example gas dynamics). Riemann has influence much of the mathematics of the twentieth century with his idea of the Riemannian surface. He is among the most ingenious mathematicians of all times.

In 1851 he founds with his dissertation geometric number theory, centered around conformal mappings.

In 1854 he holds his famous Habilitations lecture *Über die Hypothese, welche der Geometrie zu Grunde liegen* (On the hypothesis on which geometry is based). In this lecture he develops a far-reaching program to describe the geometry of higher-dimensional curved spaces. On his way home he was accompanied by the very elderly Gauss, who was very impressed. These ideas of Riemann were decisive in the years 1907 to 1915 for Einstein in the formulation of his theory of general relativity.

In 1859, Riemann becomes Dirichlet's successor in Göttingen.

In that year, he also studies the $\zeta$-function in connection with the distribution of prime numbers and formulates the *Riemann hypothesis* on the distribution of zeros of that function.

1829 Nikolai Ivanovitch Lobatchevski (1793–1856) proves the existence of a hyperbolic non-Euclidean geometry. An independent proof of this fact is given in 1832 by Janos Bólyai (1802–1860).

1831 Évariste Galois (1811–1832) develops a general theory on the solution of equations, based on the theory of *Galois groups*. This makes him the founding father of modern structural theory.

1831–1879 James Clerk Maxwell, who makes fundamental contributions to the theory of electrodynamics and kinetic gas theory, lives.

In 1864 he formulates the basic equations for all electromagnetic phenomena, known as the *Maxwell equations*.

1842 Robert Meyer (1814–1878), a doctor in Heilbronn (Germany), publishes the law of conservation of energy, which he had discovered.

1842–1899 Sophus Lie, the creator of the theory of continuous groups (Lie groups and Lie algebras), lives. With this theory he provides physics with a fundamental instrument to describe symmetries which occur in nature in a mathematically precise manner.

1844 Joseph Liouville (1809–1882) gives a constructive proof of the existence of transcendental numbers.

1844 Hermann Grassmann's (1809–1877) book *Lineare Ausdehnungstheorie* (Theory of linear extensions) appears. This work contains much of the modern theory of linear and multilinear algebra, but is not understood by his contemporaries. Grassmann algebras are the algebraic core of the modern supersymmetric theories in the physics of elementary particles.

1844–1906 Ludwig Boltzmann, the creator of statistical physics, lives. He realized that both entropy and the second fundamental theorem of thermodynamics have a statistical character.

1845–1918 Georg Cantor creates with his set theory a new way of thinking and a new language in mathematics, which has allowed the mathematicians of the twentieth century to formulate abstract and deep ideas in a compact way.

In 1874 he publishes his first work on set theory and gives a non-constructive, purely set-theoretic proof of the existence of transcendental numbers with the help of a countability argument.

1846 Austen Layard (1817–1894) excavates the Assyrian–Babylonian site Ninive.

1847 Ernst Eduard Kummer (1810–1893) develops a theory of divisibility, using *ideal numbers*, and proves some special cases of Fermat's last theorem with these methods.

In 1855 he becomes the successor of Dirichlet in Berlin.

1847 George Boole (1815–1869) publishes his book *The Mathematical Analysis of Logic* and shows that one can do calculations not only with numbers, but also with sets (Boolean algebras). This paves the way for modern logic and computer science.

1849–1925 Felix Klein, one of the leading mathematicians in Germany in the second half of the nineteenth century, lives. He makes contributions to geometry, algebra (icosahedral group and equations of fifth degree), to complex function theory, and is practically the creator – in collaboration and in competition with Henri Poincaré – of the theory of automorphic functions. He reforms the form of mathematical instruction, both at the university as well as in schools in Germany, and wrote one of the most influential histories of mathematics of the nineteenth century.

In 1872 he formulates his *Erlanger Programm*, which expresses his view that geometry is essentially the invariant theory of symmetry groups of the space in question. This makes it possible to study different phenomena from a unified point of view.

1854–1912 Henri Poincaré, an extremely versatile and creative mathematician and mathematical physicist, lives. His contributions to complex function theory, to celestial mechanics, to partial differential equations, to number theory, to topology as well as to philosophical questions have influenced modern mathematic perhaps more than any those of any other mathematician of the nineteenth century. He is the creator of the theory of dynamical systems and algebraic topology. He is thus the founding father of the mathematics of qualitative behavior.

1858–1947 Max Planck, who revolutionized physics by creating quantum theory, lives. He also makes fundamental contributions to thermodynamics.

1859 Charles Darwin (1809–1882) publishes his seminal book *On the Origin of Species by Means of Natural Selection.*

1862–1943 David Hilbert, considered by many to be the last living mathematician familiar with all disciplines in mathematics, lives. His tremendously decisive contributions include the areas of algebra, analysis, geometry, foundations of mathematics, mathematical physics, number theory and philosophical aspects. He is the creator of the modern axiomatic method in mathematics.

In 1895 he goes to Göttingen and continues the tradition of Gauss and Riemann there.

1864–1909 Hermann Minkowski, the creator of the theory of geometry of numbers and the theory of convex bodies, lives. He also derived the geometrization of Einstein's theory of general relativity. His notion of convexity is fundamental in the modern theory of optimization, which arose between 1950 and 1960.

1869–1951 Élie Cartan, the creator of modern differential geometry with it emphasis on differential forms and symmetries by Lie groups, lives. His ideas on forming a theory of curvature in terms of principal fiber bundles is applied today in modern gauge theories in the physics of elementary particles.

1869 Dmitri Ivanovitch Mendeleyev (1834–1907) is the first to describe the modern periodic system of elements in chemistry.

1870–1955 Albert Einstein, one of the greatest geniuses of human kind, lives. He revolutionizes with his theory of relativity the understanding of space and time, and realizes that matter and energy are equivalent. This relation is the basis for the production of energy in stars by means of fusion.

1870 Heinrich Schliemann (1822–1890) believes that the *Ilias* and the *Odyssey* by Homer are not just legends, and sets out to find Troy; he succeeds after extensive excavations.

1871 Richard Dedekind (1831–1916) publishes his theory of ideals.

In 1872 his book *Stetigkeit und irrationale Zahlen* (Continuity and irrational numbers) appears. This book provides, for the first time, a completely rigorous construction of the real numbers, providing a sound foundation for analysis.

1873 Charles Hermite (1822–1901) proves the transcendence of the Euler number e.

1878 William Clifford (1845–1879) introduces certain algebras, which in modern times are referred to as Clifford algebras, which are essential in giving a mathematical description of fermions (elementary particles with half-integer spin).

1882 Ferdinand Lindemann (1852–1939) proves the transcendence of the number $\pi$ and demonstrates as a by-product that the problem of the quadrature of the circle, which had been open for over 2000 years, has no solution. Lindemann is Hilbert's teacher.

1882 Leopold Kronecker (1823–1891) publishes his thesis on the foundations of the theory of algebraic numbers and paves the way for the subsequent elegant theory of Hilbert describing class field theory.

1885–1955 Hermann Weyl, one of the most important mathematicians in the transition to modern mathematics, lives. He makes fundamental contributions to the theory of Riemann surfaces, the representation theory of Lie groups, invariant theory, spectral theory, differential geometry, the general theory of relativity and quantum theory. He is the founding father of modern harmonic analysis, and also made in-depth studies of philosophical questions.

1890 David Hilbert founds the Deutsche Mathematiker Vereinigung (union of German mathematicians).

1896 Jacques Hadamard (1865–1963) and Charles de la Vallée–Poussin (1866–1962) give independently proofs of the famous theorem on the distribution of prime numbers. This law had been discovered more than 100 years earlier by Gauss and Legendre through empirical studies, but had long resisted all attempts at proof.

1897 The first International Congress of Mathematicians takes place in Zürich; since 1900 these congresses take place every four years, with the exception years of the second world war.

1898 Marie Sklodovska–Curie (1867–1934) and Pierre Curie (1859–1906) discover the first radioactive elements Polonium and Radium.

1899 Hilbert's book *Grundlagen der Geometrie* (Foundations of Geometry) appears. This book introduces the modern axiomatic approach to mathematics, and is after 2000 years an advancement over Euclid's *Elements*.

1899 Robert Koldewey (1855–1925) begins the excavation of Babylon.

**Mathematics in the twentieth century**

1900 Planck makes the hypothesis that there is a smallest unit of energy and that the energy of the harmonic oscillator is quantized. From this he obtains the correct laws for radiation from stars. This is the moment of birth of a completely new kind of physics – quantum physics is born.

1900 Hilbert proves the Dirichlet principle and thus founds the development line of direct methods in the calculus of variations.

1900 Hilbert formulates his famous 23 problems at the International Congress of Mathematicians in Paris, which are to influence the development of all disciplines in mathematics throughout the twentieth century.

1903–1957 Janos (John) von Neumann, the most important and influential applied mathematician of the twentieth century, lives. He makes fundamental contributions to game theory, mathematical economics, mathematical foundations of quantum physics, spectral theory and the theory of operator algebras, ergodic theory, numerical analysis and computer science. He also makes important discoveries in pure mathematics, in the foundations of set theory, the solution of the fifth Hilbert problem for compact Lie groups, the theory of almost periodic functions on groups and in the theory of locally convex spaces and functional analysis.

1904 Hilbert begins with the publication of his general theory of integral equations, based on work of Fredholm (1866–1927) from 1900, and lays the foundations for functional analysis.

1905 Einstein publishes three fundamental papers on the electrodynamic of moving bodies (the special theory of relativity), on Brownian motion (the basis for the theory of stochastic processes) and on the photon theory of light (the basis for the ensuing theory of quantum electrodynamics).

1907 Henri Poincaré (1854–1912) and Paul Koebe (1882–1945) prove independently the uniformization theorem for Riemann surfaces, which completely determines the structure theory for Riemann surfaces.

1908 Minkowski shows that space and time form a geometric unit (this is the geometrization of Einstein's special theory of relativity).

1910–1913 Bertrand Russel (1872–1970) and Alfred Whitehead (1861–1947) publish the *Principia Mathematica*. The monumental, three volume opus contains
a completely rigorous development of formal logic and much of the then known mathematics.

1913 Niels Bohr (1885–1962) develops his model of the atom, which describes the spectrum of hydrogen.

1913 Hermann Weyl's book *Die Idee der Riemannischen Fläche* (The notion of a Riemann surface) appears. In this book, the ingenious ideas of Riemann are mixed with modern theory.

1914–1918 The First World War destroys much of Europe.

1915 Einstein publishes the basic equations of the general theory of relativity and provides the mathematical and physical foundation for modern cosmology. The gravitational force of Newton is replaced by the curvature of space and time.

1918 Hermann Weyl's book *Raum, Zeit, Materie* (Space, time, matter) appears.

1922 Howard Carter (1873–1939) discovers in the Valley of the Kings the totally preserved grave of Tut-Ench-Amun (from ca. 1340 BC).

1923 Norbert Wiener (1894–1964) publishes a paper on Brownian motion and lays the foundations for a mathematically rigorous theory of stochastic processes.

1924–1925 Werner Heisenberg (1901–1976) creates the mathematical basis for quantum mechanics by using the commutation relations of infinite matrices.

1926 Erwin Schrödinger (1887–1961) publishes the Schrödinger equation for calculating quantum processes in the context of wave mechanics, which is in turn soon shown to be equivalent (via the theory of abstract Hilbert spaces) to Heisenberg's theory.

1926 Max Born (1882–1970) formulates the statistical interpretation of the Schrödinger wave mechanics.

1926 John von Neumann lays the foundations for game theory with his dissertation.

1927 Heisenberg discovers the uncertainty principle in quantum mechanics; as opposed to classical mechanics, the position and velocity of a particle cannot both be determined with certainty. This leads to deep changes in the way of thinking in physics.

1928 Paul Dirac (1902–1984) derives the basic equation for the relativistic electron (the Dirac equation). He uses the Clifford algebras for this description, and predicts the existence of a positron, which is then discovered in 1932 by Anderson in cosmic radiation. This was the first discovery of an anti-particle.

1928 Edwin Hubble (1889–1953) discovers the Hubble effect, the red-shift of light from far galaxies, which is caused by the expansion of the universe following the big bang.

1930 Kurt Gödel (1906–1978) proves the completeness of predicate logic of the first class.

1930 Hermann Weyl becomes Hilbert's successor in Göttingen.

1930 The Institute for Advanced Studies in Princeton (New Jersey, USA) is established as an independent research institute.

1931 Kurt Gödel discovers the existence of non-decidable problems in mathematical theories.

1932 John von Neumann's monograph *Mathematische Grundlagen der Quantenmechanik* (Mathematical Foundations of Quantum Mechanics) appears.

1933 Andrei Nikolaievich Kolmogorov (1903–1987) publishes his book *Foundations of probability theory.* With this document, the axiomatic method is introduced into probability theory.

1933 Emigration of Emmy Noether (1882–1935), Paul Bernays (1888–1977), Otto Blumenthal (1876–1944), Max Born (1882–1970), Richard Courant (1888–1972), Albert Einstein (1879–1955), Hermann Weyl (1885–1955) and many other scientists from fascist Germany.

The famous number theoretician Edmund Landau (1877–1938) and the recipient of the Nobel Award for Physics, James Franck (1882–1964) are removed from their positions in Göttingen.

Albert Einstein, John von Neumann and Herrmann Weyl move to the newly founded Institute of Advanced Studies in Princeton and establish the reputation of the institute.

1935 Richard Courant founds the Department of Mathematics at the Graduate School of Arts and Sciences at the New York University (now famous as the Courant Institute).

1936 Alan Turing (1912–1954) establishes the modern theory of robotics and algorithms through the introduction of a theoretical, universally applicable machine, which is today known as a Turing machine.

1936 Konrad Zuse (1910–1995) builds the first mechanical computer $Z1$.

1938 Otto Hahn (1879–1968) and Fritz Strassmann (1902–1980) discover fission of uranium. An important contribution to this work was done by Lise Meitner (1878–1986). This physical process is the basis for the construction of the atomic bomb a few years later.

1939–1945 World War II brings infinite suffering to millions around the world.

1942 Robert Oppenheimer (1904–1967) becomes the director of the Manhattan project in Los Alamos (New Mexico, USA), and gathers the most brilliant scientist around him to work on the construction of the atomic bomb.

1944 The mathematical Research Institute in Oberwolfach (Black Forest) is established.

1944 John von Neumann and Oscar Morgenstern publish the book *Theory of Games and Economical Behaviour.*

1945 Two atomic bombs are dropped by US forces on the Japanese cities Hiroshima and Nagasaki.

1946 ENIAC – the first computer with computational power – is constructed. The predecessor MANIAC is used in the construction of the bomb.

1946–1949 Richard Feynman (1918–1988), Julian Schwinger (born 1918) and Sin-Itiro Tomonaga (1906–1979) lay independently of one another with different methods the foundations of quantum electrodynamics. Together they receive in 1965 the Nobel Award for Physics.

1948 Norbert Wiener publishes his book *Cybernetics*, which is the moment of birth of computer science.

1948 Claude Shannon (born 1916) establishes information theory.

1948 John Bardeen (born 1908), Walter Brattain (born 1902) and William Shockley (born 1910) develop in Bell Laboratories the transistor, which is based on the quantum mechanics of solid bodies. In 1956 they receive the Nobel award in Physics for this work. The transistor leads to a technical revolution.

1950–1960 Pure mathematics breaks new ground through the introduction of abstract theories like sheaf theory, fiber bundles, homological algebra and cohomology theory.

1950–1960 In applied mathematics the theory of optimization including the theory of optimal processes is born. The theory of partial differential equations and dynamical systems are intensively developed (for example through the KAM theory of Andrei Kolmogorov (1903–1987), Vladimir Arnol'd (born 1937) and Jürgen Moser (born 1928)).

1956 Tsung Dao Lee (born 1926) and Chen Ning Yang (born 1922) propose the theory that there is a fundamental asymmetry in the world of elementary particles: in processes with the weak force, one of the three fundamental symmetries (reflection) is broken. Both physicists get the Nobel Award for Physics one year later for this work.

1957 The *Sputnik* orbits as first unmanned satellite the earth, and shocks the western world by showing how advanced Soviet technology is.

1961 Yuri Gagarin is the first human to orbit the earth in outer space.

1962 Francis Crick (born 1916) and James Watson (born 1928) receive the Nobel Award in Biology for the description of the helix model of DNA (Deoxyribonucleic acid). The twisted double helix structure is of fundamental importance for the inheritance of genetic information.

1963 Paul Cohen (born 1934) proves the independence of the continuum hypothesis from the other axioms of set theory. This has the remarkable epistemological consequence that the structuring of infinity, which had been ingeniously described by Cantor at the end of the nineteenth century, is not unique, and there are more than one way to do this all of which are free of contradictions.

1963 The proof of the Atiyah–Singer index theorem by Michael Atiyah (born 1929) and Isadore Singer (born 1924) is published. This theorem, which combines analysis and topology in a unique and far-reaching way is the crowning completion of a long and fruitful line of development in mathematics, and belongs to the most profound results of this century. The index theorem is a far-reaching generalization of the deep theorem of Riemann–Roch–Hirzebruch, which was proved by Friedrich Hirzebruch (born 1927) in 1953.

1964 Murray Gell–Mann (born 1929) proposes the theory that protons are not elementary particles, but rather are composed of quarks; for this he receives the Nobel Award for Physics in 1969. The mathematical background for this theory is based on the interpretation of experimental data with the help of the Lie group $SU(3)$ and its Lie algebra.

1965 Arnold Penzias (born 1933) and Robert Wilson (born 1936) discover the background microwave radiation, a remnant of the big bang; they receive the Nobel Award in Physics in 1978 for this discovery.

1969 Neil Armstrong (born 1930) is the first man to walk on the moon on July 20 of this year.

1970–1980 The widespread use of the finite elements method revolutionizes numerical analysis.

1979 Abdus Salam (born 1926), Sheldon Glashow (born 1932) and Steven Weinberg (born 1933) receive the Nobel Award for Physics for their work on the standard model, uniting electromagnetism and the weak force in the context of a gauge field theory. The bosons which this theory predicts and are responsible for the weak force according to the theory (the $Z$ boson and the $W^{\pm}$ bosons) are identified in experiments at the accelerator at CERN near Geneva in 1983.

The standard model, which is today universally accepted, uses a gauge field theory with 6 quarks and 6 leptons (for example electron and neutrino) as basic building blocks. The interactions are described by 12 particles (8 gluons, the photon, the $Z$ boson and the $W^{\pm}$ bosons).

1980 Personal computers start conquering the world and revolutionize mathematics which, like physics, gains a branch devoted to experiments.

1983 Gerd Faltings (born 1954) proves the Mordell conjecture for Diophantine equations over number fields (rational points on algebraic curves).

1994 Andrew Wiles (born 1953) proves *Fermat's last theorem* in a tour de force using incredibly abstract and new mathematics.

1994 The sixth quark, which had long been searched for, was discovered in an experiment in the accelerator at Fermi lab near Chicago.

1995 The Hubble telescope, based in orbit around the earth, sends sharp photographs of the birth and death of stars and verifies in this way the predictions of astrophysicist, which are based on mathematical and physical theories.

**The recipients of the Fields Medal**

In the twentieth century many deep and interesting, as well as remarkable results have been obtained. To give a sketch of these, we list the winners of the Fields Medal, the highest award for mathematicians; it is awarded every four years at the International Congress of Mathematicians who are under 40, and is comparable for mathematicians to the Nobel Awards for physicists and other scientists.

1936 *Lars Ahlfors* (born 1907) (analysis; function theory and quasiconformal mappings);

*Jesse Douglas* (1897–1965) (analysis; proof of existence of minimal surfaces).

1950 *Laurent Schwartz* (born 1915) (analysis; theory of distributions – differential calculus and Fourier transformations for generalized functions);

*Atle Selberg* (born 1917) (number theory; elementary proof of the theorem on the distribution of prime numbers).

1954 *Kunihiko Kodaira* (born 1915) (analysis and differential geometry; harmonic integrals on manifolds and algebraic geometry; generalization of Riemann's ideas for algebraic functions and algebraic integrals);

*Jean–Pierre Serre* (born 1926) (algebra and topology; fiber bundles; proof of the finiteness of homotopy groups of spheres (up to trivial exceptions)).

1958 *Klaus Roth* (born 1925) (number theory; solution of the old problem on the approximation of algebraic numbers by rationals);

*René Thom* (born 1923) (differential topology; construction of cobordism theory as a tool in understanding deep structural properties of manifolds).

1962 *Lars Hörmander* (born 1931) (analysis; general theory of linear partial differential equations with constant coefficients);

*John Milnor* (born 1931) (topology; discovery of exotic spheres, which show that the topological structure and differential structure of manifolds need not coincide).

1966 *Michael Atiyah* (born 1929) (topology, differential geometry and partial differential equations; K-theory to describe deep structural properties of manifolds through the set of vector bundles which live on them; Atiyah–Singer index theorem);

*Paul Cohen* (born 1934) (foundations of mathematics; proof of the independence of the axiom of choice and the continuum hypothesis from the other axioms of set theory);

*Alexander Grothendieck* (born 1928) (analysis – theory of nuclear spaces; algebraic geometry – revolution of this mathematical discipline with the powerful apparatus of schemes);

*Stephen Smale* (born 1930) (topology and analysis; deep structural work in dynamical systems and their chaotic behavior).

1970 *Alan Baker* (born 1939) (number theory; theory of transcendental numbers);

*Heisuke Hironaka* (born 1931) (algebraic geometry; proof of the resolution of singularities);

*Sergei Novikov* (born 1938) (topology; important contributions to homology and homotopy theory);

*John Thompson* (born 1932) (algebra; important contributions to group theory).

1974 *Enrico Bombieri* (born 1940) (analytic number theory and geometry of numbers; algebraic surfaces);

*David Mumford* (born 1937) (algebraic geometry; structure of Abelian varieties, which are broad generalizations of elliptic curves and have developed historically from the theory of Abelian integrals).

1978 *Pierre Deligne* (born 1944) (algebraic geometry; proof of the (generalized) Riemann hypothesis for algebraic varieties over finite fields, as conjectured by André Weil);

*Charles Fefferman* (born 1949) (analysis; behavior of higher-dimensional Fourier series and singular integral operators; function theory of several complex variables);

*Grigori Margulis* (born 1946) (differential geometry; structure of discrete subgroups $\Gamma$ of Lie groups $G$ for which the volume $G/\Gamma$ is finite);

*Daniel Quillen* (born 1940) (algebra and topology; cohomology of groups; proof of the Serre conjecture that projective modules over a polynomial ring with coefficients in a field are free, i.e., these modules have the structure of linear spaces).

1982 *Alain Connes* (born 1947) (functional analysis; structure of von Neumann algebras of type III);

*Shing Tung Yau* (born 1949) (global analysis; differential equations on manifolds and general relativity; proof the theorem on the existence of a positive gravitational energy[5] and the proof of the Calabi conjecture for Kähler manifolds);

*William Thurston* (born 1946) (topology; structure of three-dimensional manifolds in connection with hyperbolic geometry).

1986 *Simon Donaldson* (born 1957) (global analysis and topology; the Yang–Mills equations, which come from gauge theory in theoretical physics, are used to get new insights into the smooth (differentiable) structure of four-manifolds);

*Gerd Faltings* (born 1954) (algebraic geometry and number theory; proof of the Mordell conjecture for Diophantine equations);

*Michael Freedman* (born 1951) (topology; proof of the Poincaré conjecture in dimension four: the four dimensional sphere is the only compact four-dimensional (topological) manifold whose homology coincides with the homology of the sphere; counterexample to the main conjecture in combinatorial topology via construction of homeomorphic four-manifolds which do not posses equivalent triangulations).

1990 *Vladimir Drinfeld* (born 1954) (algebraic geometry; applications to solutions of Yang–Mills equations (instantons); quantum groups; proof of the Langlands conjecture on Galois groups);

*Vaughan Jones* (born 1955) (functional analysis and topology; relation between von Neumann algebras and knot theory; applications to statistical physics (Yang–Baxter equations));

[5]This theorem, which is physically evident but mathematically difficult, says that the energy of gravitation of interacting matter in the general theory of relativity is always positive.

*Shigefumi Mori* (born 1951) (algebraic geometry; classification of three-dimensional algebraic varieties and proof of the minimal model conjecture in that dimension);

*Edward Witten* (born 1951) (ingenious combination of methods of quantum field theory (Feynman integral) and topology, differential geometry and algebraic geometry; actually a physicist, Witten has led to new deep insights in several branches of mathematics, for example the formal similarity between supersymmetry and Morse theory; the relations between quantum field theory and knot theory; proof of the theorem on positive gravitational energy with the help of the Dirac equation; the Seiberg–Witten theory, conceived by Witten and Nathan Seiberg (both physicists) in 1994 has led to a remarkable simplification of Donaldson theory for four-manifolds, by replacing the Yang–Mills equations by a physically *dual* set of equations, Dirac equations, which have a much simpler structure and lead to an easier and more complete analysis).

1994 *Jean Bourgain* (born 1954) (analysis; non-linear differential equations of mathematical physics; geometry of Banach spaces; ergodic theory; harmonic analysis; analytic number theory);

*Pierre–Louis Lions* (born 1956) (analysis and applied mathematics; new methods for solving non-linear partial differential equations; the viscosity method for the Hamilton–Jacobi equation and the Bellman equation in control theory; the method of concentration of energy; solution of the Hartree–Fock equation for treatment of atoms with many electrons; solution of the Boltzmann equation for gases with interacting particles; compressible liquids; construction of sharp computer images with the help of the anisotropic diffusion equation);

*Jean–Christophe Yoccoz* (born 1956) (analysis; stability of dynamical systems; the combination of extensive computer animations together with deep theoretical investigations);

*Efim Selmanov* (born 1955) (algebra; Lie algebras, Jordan algebras and the structure of finite groups).

1998 *Richard Borcherds* (born 1959) (algebra and geometry, proof of the moonshine conjecture which represents an unexpected relation between the monster group (a huge finite group) and elliptic functions, Kac–Moody algebras and automorphic forms);

*Maxim Kontsevich* (born 1964) (string theory and quantum field theory, mathematical equivalence of two models in quantum gravitation, Poisson structures and quantum deformations, knot invariants in topology);

*William Gowers* (born 1963) (sophisticated relations between functional analysis and combinatorics, development of new techniques for studying sophisticated geometric properties of Banach spaces including the problem of unconditional bases, construction of a Banach space which has almost no symmetry);

*Curtis McMullen* (born 1958) (geometry and the structure of chaotic dynamical systems, effective algorithms for constructing approximations of solutions, nonexistence proof for a universal generalized Newton method, deep structural result on the relation between hyperbolic dynamics, Julia sets, and Mandelbrot sets).

2002 *Laurent Lafforgue* (born 1966) (sophisticated relations between number theory, analysis, and group representation theory, proof of the global Langlands correspondence for function fields);

*Vladimir Voevodsky* (born 1966) (sophisticated relations between number theory and algebraic geometry, construction of a new powerful cohomology theory called motivic cohomology for algebraic varieties, proof of the Milnor conjecture in algebraic K-theory).

### The Nevanlinna Award

Since 1982, the Nevanlinna Award is given for exceptional work in the area of mathematical methods of computer science.

1982 *Robert Tarjan* (born 1948) (construction of particularly effective algorithms for computer calculations).

1986 *Leslie Valiant* (born 1949) (algebraic complexity theory; effective stochastically weighted algorithms; artificial intelligence).

1990 *Alexander Rasborow* (born 1960) (investigations on the complexity of networks).

1994 *Avi Widgerson* (born 1956) (verification of proofs for which details of the proof are unknown with stochastic criteria; application of this method in computer networks).

1998 *Peter Shor* (born 1959) (theory of quantum computing, coding theory, construction of a very fast algorithm for factorizing large prime numbers on quantum computers).

2002 *Madhu Sudan* (born 1951) (probabilistically checkable proofs, non-approximabality of optimization problems, error-correcting codes).

### The Abel Prize

This prize was founded by the Norwegian government in 2002; it is called the Nobel prize in mathematics.

2003 *Jean-Pierre Serre* (born 1926) (algebra and topology).

# *Bibliography*

## Encyclopaedias


[1] Eisenreich, G., Sube, R.: Mathematics (four-lingual dictionary in English, French, German and Russian), 3rd ed. Berlin: Verlag Technik 1985.

[2] Encyclopaedia Britannica: 32 Vols. New York: Encyclopaedia Britannica Corporation 1974-1987.

[3] Encyclopaedia of Mathematical Sciences, Vol. 1ff. Berlin: Springer-Verlag 1990ff. (transl. from Russian).

[4] Encyclopaedia of Mathematics, Vols. 1-10. Edited by M. Hazewinkel. Dordrecht: Kluwer 1987-1993. Revised Translation from the Russian.

[5] Encyclopedia of Mathematical Physics, Vols. 1-5, edited by J. Françoise, G. Naber, and T. Tsun. Amsterdam: Elsevier 2005 (to appear).

[6] Encyclopaedic Dictionary of Mathematics, Vols. 1,2, 2nd ed. Cambridge, Massachusetts: MIT Press 1993.

[7] Fiedler, B., Hasselblatt, B.: Handbook of Dynamical Systems. Vol. 1. Amsterdam: Elsevier 2002.

[8] Fiedler, B.: Handbook of Dynamical Systems, Vol. 2. Amsterdam: North Holland 2002.

[9] Gribbin, J.: Q is for Quantum.: Particle Physics from A-Z. London: Weidenfeld 1998.

[10] Sube, R., Eisenreich, G.: Physics (four-lingual dictionary in English, French, German and Russian), 3rd ed. Berlin: Verlag Technik 1985.


## Mathematics in the Real World


[11] Engquist, B., Schmid, W. (eds.), Mathematics Unlimited - 2001 and Beyond. With contributions written by 80 leading mathematicians. New York: Springer-Verlag 2001.

[12] Friedman, A., Littman, W.: Industrial Mathematics: A Course in Solving Real-World Problems. Philadelphia: SIAM 1994.


**How can you get informed about current developments in mathematics?**
The periodical *Mathematical Intellegencer*, published by Springer Verlag, contains interesting articles which are aimed at a general audience and contain contributions about current developments in mathematics as well as about the history of mathematics.
The periodical *Bulletin of the American Mathematical Society*, published by the largest scientific society in the United States specialized in mathematics, the American Mathematical Society (AMS), is the leading magazine containing survey articles about current developments in mathematics, here aimed at a 'general mathematical community', meaning people who have studied math and work with it to some extent. In addition there are many book reviews which are characterized by the fact that the general history of the book being reviewed as well as its relation to already existing literature is presented in some detail.
Survey articles on important new developments in physics can be found in the periodical *Reviews of Modern Physics*.
For presentations on all topics of the natural sciences aimed at a 'general scietific community', meaning those who have studied and work in some scientific or technical discipline, no periodical can compete with *Scientific American*, published by Scientific American Inc. More scientific presentations, at a weekly pace, are given in the 'Grandfather of scientific Journals', *Science*.

Every four years the *International Congress of Mathematics* takes place, which always publishes several volumes of proceedings. The invited lectures in those proceedings are an important source of presentations of new developments.
All mathematical journal articles and books are reviewed by *Mathematical Reviews*, published by the AMS, as well as *Zentralblatt der Mathematik*, published by Springer Verlag. In addition to these voluminous reviews there are yearly index volumes which collect lists of all publications, ordered by author and topic.
The journal *Current Mathematical Publications*, also published by the AMS, lists all new publications. Much of this information is also available through the internet.

## 0. Tables


[13] Abramowitz, M., Stegun, I. (ed.): Handbook of Mathematical Functions with Formulas, Graphs, and Mathematical Tables. Reprint of the 1972 edition. New York and Washington D.C.: Wiley and National Bureau of Standards 1984.

[14] Bateman, H. (ed.): Higher Transcendental Functions, Vols. 1-3. New York: McGraw-Hill 1953-1955.

[15] Beckenbach, E. and Bellman, R.: Inequalities. Berlin: Springer-Verlag 1983.

[16] Carlson, B.: Special Functions of Applied Mathematics. New York: Academic Press 1988.

[17] Fisz, M.: Wahrscheinlichkeitssrechnung und mathematische Statistik, translation from Polish, 5th ed. Berlin: Deutscher Verlag der Wissenschaften, 1970.

[18] Gradshteyn, I., Ryzhik, I.: Tables of Integrals, Series, and Products. New York: Academic Press 1980.

[19] Hardy, G., Littlewood, J., Pólya, G.: Inequalities. Cambridge, UK: Cambridge University Press 1978.

[20] Iwasaki, K. et al.: From Gauss to Painlevé. A Modern Theory of Special Functions. Wiesbaden: Vieweg 1991.

[21] Jahnke, E., Emde, F., Lösch, F.: Tafeln höherer Funktionen, 7th ed. Stuttgart: Teubner-Verlag 1966.

[22] Klein, Felix: Vorlesungen über das Ikosaeder and die Auflösung der Gelichungen vom fünften Grade. Edited by P. Slodowy, Basel, Stuttgart, Leipzig: Birkhäuser and Teubner-Verlag 1993.

[23] Luke, Y. Mathematical Functions and their Approximations. New York: Academic Press 1975.

[24] Magnus, W., Oberhettinger, F., Soni, R.: Formulas and Theorems for the Special Functions of Mathematical Physics, 3rd ed., Berlin: Springer-Verlag 1966.

[25] Owen, D.: Handbook of Statistical Tables. Reading, Massachusetts: Addison-Wesley 1965.

[26] Prudnikov, A., Brychkov, Yu., Manichev, O.: Integrals and Series, Vols. 1-5. New York: Gordon and Breach 1986-1990 (transl. from Russian).

[27] Smirnow, W.: Lehrgang der höhreren Mathematik, Volume III/2, 2nd ed. Berlin: Deutscher Verlag der Wissenschaften.

[28] Smoot, G., Davidson, K.: Wrinkels in Time, New York: Morrow 1993 (a history of modern cosmology).

[29] Spanier, J., Oldham, K.: An Atlas of Functions. Berlin: Springer-Verlag 1987.

[30] Szegö G.: Orthonormal Polynomials, 4th ed., New York: Amer. Math. Soc. Coll. 1975.

# 1. Analysis

## 1.1. Elementary Analysis

[31] Aigner, M., Ziegler, G.: Proofs from the Book, 2nd ed. Berlin: Springer-Verlag 2001 (a collection of elegant proofs for gems of mathematics).

[32] Conway, J., Guy, R.: The Book of Numbers. New York: Copernicus 1996.

[33] Courant, R., Robbin, H.: What is Mathematics? Oxford: Oxford University Press 1941 (a classic).

[34] Ebbinghaus, H. et al. (eds.): Numbers, 3rd ed. New York: Springer-Verlag (transl. from German).

[35] Hardy, G., Littlewood, J., Pólya, G.: Inequalities. Cambridge, UK: Cambridge University Press 1978.

[36] Koshy, T.: Fibonacci and Lukas Numbers with Applications. New York: Wiley 2001.

## 1.2. – 1.7. Limits, Differentiation and Integral Calculus

[37] Amann, H., Escher, J.: Analysis, Vols. 1-3. Basel: Birkhäuser 1998-2001 (English translation in preparation).

[38] Birkhoff, G.: A Source Book in Classical Analysis. Cambridge, MA: Harvard University Press 1973.

[39] Chaichian, M., Denichev, A.: Path Integrals in Physics, Vols. 1,2. Bristol, UK: Institute of Physics 2001.

[40] Choquet-Bruhat, Y., DeWitt-Morette, C., and Dillard-Bleick, M.: Analysis, Manifolds, and Physics. Vol. 1: Basics; Vol 2: 92 Applications. Amsterdam: Elsevier 1996 (standard textbook).

[41] Courant, R., John, F.: Introduction to Calculus and Analysis, Vols. 1, 2, 2nd ed. New York: Springer-Verlag 1988 (a classic).

[42] Fikhtengol'ts, G.: The Fundamentals of Mathematical Analysis. Oxford: Pergamon Press 1965 (transl. from Russian), (a classic).

[43] Grosche, C., Steiner, F.: Handbook of Feynman Path Integrals. New York: Springer-Verlag 1998.

[44] Jost, J.: Postmodern Analysis. Berlin: Springer-Verlag 1998.

[45] Hairer, E., Wanner, G.: Analysis by its History. New York: Springer-Verlag 1996.

[46] Hardy, G.: A Course of Pure Mathematics. Cambridge: Cambridge University Press 1992 (a classic).

[47] Harvard Calculus. New York: Wiley 1994.

[48] Lang, S.: Analysis I, II. Reading, MA: Addison Wesley 1969.

[49] Lang, S.: Real and Functional Analysis. New York: Springer-Verlag 1993.

[50] Lang, S.: Undergraduate Analysis, 2nd ed. New York: Springer-Verlag 1997.

[51] Lax, P., Burstein, S., and Lax, A.: Calculus with Applications and Computing. New York: Springer-Verlag 1976.

[52] Lieb, E., Loss, M.: Analysis. Providence, RI: American Mathematical Society 1997.

[53] Loeb, P., Wolff, M.: Nonstandard Analysis for the Working Mathematician. Boston: Kluwer Academic Publishers 2000.

[54] Marsden, J., Tromba, A., and Weinstein, A.: Basic Multivariable Calculus. New York: Springer-Verlag 1993.

[55] Marsden, J., Weinstein, A.: Calculus I, II, 2nd ed. New York: Springer-Verlag 1985.

[56] Maurin, K.: Analysis I, II. Boston, MA: Reidel 1976-80.

[57] Priestley, H.: Introduction to Integration. Oxford: Clarendon Press 1997.

[58] Prudnikov, A., Brychkov, J., and Marichev, O.: Integrals and Series, Vols. 1-5. New York: Gordon and Breach 1986-1992 (transl. from Russian).

[59] Royden, H.: Real Analysis. New York: McMillan 1989.

[60] Rudin, W.: Real and Complex Analysis, 3rd ed. New York: McGraw-Hill 1987.

[61] Steward, J.: Calculus, 3rd ed. Pacific Grove, CA: Brooks 2001.

[62] Whittaker, E., Watson, G.: A Course of Modern Analysis. Cambridge, UK: Cambridge University Press 1965.

### 1.8. – 1.9. Vector Calculus, Differential Forms, and Physical Fields

[63] Abraham, R., Marsden, J., and Ratiu, T.: Manifolds, Tensor Analysis, and Applications. New York: Springer-Verlag 1989.

[64] Agricola, I., Friedrich, T.: Global Analysis: Differential Forms in Analysis, Geometry, and Physics. Providence, RI: Amer. Math. Soc. 2002 (transl. from German).

[65] Guillemin, V., Pollack, A.: Differential Topology. Englewood Cliffs, NJ: Prentice-Hall 1974.

[66] Marsden, J., Tromba, A.: Vector Calculus. New York: Freeman 1996.

[67] Morse, P., Feshbach, H.: Methods of Theoretical Physics, Vols. 1, 2. New York: McGraw-Hill 1953 (a classic).

[68] Stein, E., Shakarchi, R.: Princeton Lectures in Analysis, Vol. 1: Fourier Analysis. Princeton, NJ: Princeton University Press 2003.

[69] von Westenholz, C.: Differential Forms in Mathematical Physics Amsterdam: North-Holland 1981.

### 1.10. Infinite Series

[70] Edwards, R.: Fourier Series. A Modern Introduction, Vols. 1, 2, 2nd ed. New York: Springer-Verlag 1979-1982.

[71] Erdélyi, A.: Asymptotic Expansions. Dover: New York 1965.

[72] Hardy, G.: Divergent Series. Oxford: Clarendon Press 1949.

[73] Hardy, G., Rogosinsky, W.: Fourier Series. New York: Cambridge University Press 1950.

[74] Jeffrey, A., Ryzhik, I.: Table of Integrals, Series and Products, 6th ed. San Diego, CA: Academic Press 2000.

[75] Knopp, K.: Theory and Applications of Infinite Series: New York, Dover: 1989 (transl. from German).

[76] Prudnikov, A., Brychkov, J., and Marichev, O.: Integrals and Series, Vols. 1-5. New York: Gordon and Breach 1986-1992 (transl. from Russian).

### 1.11. Integral Transforms

[77] Bracewell, R.: The Fourier Transform and its Applications, 4th ed. Boston, MA: McGraw-Hill 2000.

[78] Davies, B.: Integral Transforms and their Applications, 3rd ed. New York: Springer-Verlag 2002.

[79] Doetsch, G.: Theorie und Anwendungen der Laplace-Transformation, Vols. 1-3. Berlin: Springer-Verlag 1956 (the classic handbook).

[80] Mikusinski, J.: Operational Calculus. Oxford: Pergamon Press 1959 (transl. from Polish).

[81] Sneddon, I.: The Use of Integral Transforms. New York: McGraw-Hill 1972.

[82] Stein, E., Weiss, G.: Fourier Analysis on Euclidean Spaces. Princeton, NJ: Princeton University Press 1971.

[83] Titchmarsh, E.: Introduction to the Theory of Fourier Integrals. New York: Chelsea 1962.

[84] Widder, D.: The Laplace Transform. Princeton, NJ: Princeton University Press 1944.

### 1.12. Ordinary Differential Equations

[85] Abraham, R., Marsden, J.: Foundations of Mechanics, 2nd ed. Reading, MA: Addison Wesley 1985.

[86] Amann, H.: Ordinary Differential Equations: An Introduction to Nonlinear Analysis. Berlin: De Gruyter 1990 (transl. from German).

[87] Arnol'd, V.: Ordinary Differential Equations, 3rd ed. Berlin, New York: Springer-Verlag 1992.

[88] Arnol'd, V. (ed.): Dynamical Systems, Vols. 1-8. Encyclopaedia of the Mathematical Sciences. New York: Springer-Verlag 1988-1993 (transl. from Russian).

[89] Arnol'd, V.: Geometrical Methods in the Theory of Ordinary Differential Equations. New York: Springer-Verlag 1983 (transl. from Russian).

[90] Arnol'd, V.: Mathematical Methods of Classical Mechanics. Berlin: Springer-Verlag 1978 (transl. from Russian).

[91] Bhatia, N., Szegö, G.: Stability Theory of Dynamical Systems. Berlin: Springer-Verlag 2002.

[92] Boccaletti, D., Pucacco, G.: Theory of Orbits. Vol. 1: Integrable Systems and Non-Perturbative Methods; Vol. 2: Perturbative and Geometrical Methods. Berlin: Springer-Verlag 1996-1998.

[93] Chow, S., Hale, J.: Methods of Bifurcation Theory. New York: Springer-Verlag 1982.

[94] Coddington, E. Levinson, N.: Theory of Ordinary Differential Equations. New York: McGraw Hill 1955 (a classic).

[95] Goldstein, H., Poole, C., Safko, J.: Classical Mechanics, 3rd ed. San Francisco: Addison-Wesley 2001 (a classic).

[96] Gray, J.: Linear Differential Equations and Group Theory: From Riemann to Poincaré. Boston: Birkhäuser 2000.

[97] Hale, J., Koçak, H.: Dynamics and Bifurcations. New York: Springer-Verlag 1991.

[98] Hale, J., Verduyn, L.: Introduction to Functional Differential Equations. New York: Springer-Verlag 1993.

[99] Hartman, P.: Ordinary Differential Equations, 2nd ed. Basel: Birkhäuser 1982 (a classic).

[100] Hirsch, M., Smale, S.: Differential Equations, Dynamical Systems, and Linear Algebra. New York: Academic Press 1974.

[101] Ibragimov, N.: CRC Handbook of Lie Group Analysis of Differential Equations. Roca Baton, FL: CRC Press 1993.

[102] Katok, A., Hasselblatt, B.: Introduction to the Modern Theory of Dynamical Systems. Cambridge, UK: Cambridge University Press 1995.

[103] Lichtenberg, A., Lieberman, M.: Regular and Chaotic Dynamics. New York: Springer-Verlag 1992.

[104] Moser, J.: Lectures on Hamiltonian Systems. Providence, RI: Amer. Math. Soc. 1968.

[105] Moser, J.: Stable and Random Motion in Dynamical Systems. Princeton, NJ: Princeton University Press 1973.

[106] Murray, J.: Mathematical Biology. Berlin: Springer-Verlag 1989.

[107] Nayfeh, A., Balachandran, B.: Applied Nonlinear Dynamics; Analytical, Computational and Experimental Methods. New York: Wiley 1995.

[108] Nayfeh, A.: Perturbation Methods. New York: Wiley 1973.

[109] Nayfeh, A., Mook, D.: Nonlinear Oscillations. New York: Wiley 1979.

[110] Olver, P.: Applications of Lie Groups to Differential Equations. New York: Springer-Verlag 1993.

[111] Papastavridis, J.: Analytical Mechanics: a Comprehensive Treatise on the Dynamics of Constrained Systems for Engineers, Physicists and Mathematicians. Oxford-New York: Oxford University Press 2002.

[112] Peitgen, H., Richter, P.: The Beauty of Fractals. Berlin: Springer-Verlag 1986.

[113] Polianin, A., Zaitsev, V.: Handbook of Exact Solutions for Ordinary Differential Equations. Boca Raton, FL: CRC Press 1995.

[114] Scheck, F.: Mechanics from Newton's Laws to Deterministic Chaos, 4th ed. Berlin: Springer-Verlag 2000.

[115] Siegel, C., Moser, J.: Lectures on Celestial Mechanics. Berlin: Springer-Verlag 1971 (a classic).

[116] Titchmarsh, E.: Eigenfunction Expansions Associated with Second-Order Differential Equations, Vols. 1,2. Oxford: Clarendon Press.

[117] Walter, W.: Ordinary Differential Equations. New York: Springer-Verlag 1998 (transl. from German), (recommended as an introduction).

[118] Wasow, W.: Asymptotic Expansions for Ordinary Differential Equations. New York: Interscience 1965.

[119] Zwillinger, D.: Handbook of Differential Equations. New York: Academic Press 1992.

### 1.13. Partial Differential Equations

[120] Antman, S.: Nonlinear Problems of Elasticity. New York: Springer-Verlag 1995.

[121] Arnol'd, V. (Editor): Dynamical Systems, Volumes 1-8, in the *Encycolpaedia of the Mathematical Sciences*, translated from the Russian, New York: Springer-Verlag 1998-1993 (applications to celestial mechanics can be found in Volume 3).

[122] Arnol'd, V.: Mathematical Methods of Classical Mechanics. New York, Heidelberg, Berlin: Springer-Verlag 1978.

[123] Arseniev, A.: The Mathematical Theory of Kinetic Equations. Singapore: World Scientific 1999.

[124] Auber, G., Kornprobst, P.: Mathematical Problems in Image Processing. New York: Springer-Verlag 2002.

[125] Barton, G.: Elements of Green's Functions and Propagation: Potentials, Diffusion and Waves. Oxford: Clarendon Press 1989.

[126] Baym, G.: Lectures on Quantum Mechanics. Menlo Park, CA: Benjamin 1969.

[127] Berezin, F., Shubin, M.: The Schrödinger Equation. Dordrecht: Kluwer 1991 (transl. from Russian).

[128] Brokate, M., Sprekels, J.: Hysteresis and Phase Transitions. New York: Springer-Verlag 1996.

[129] Bryant, R., et al.: Exterior Differential Systems, New York: Springer-Verlag 1991.

[130] Cercigniani, C.: Theory and Applications of the Boltzmann Equation. Edinburgh: Scottish Academic Press 1975.

[131] Colton, D., Dress, R.: Inverse Acoustic and Electromagnetic Scattering Theory, 2nd ed. New York: Springer-Verlag 1997.

[132] Courant, R., Hilbert, D.: Methods of Mathematical Physics, Vols. 1, 2. New York: Wiley 1989 (transl. from German), (the classic textbook on partial differential equations).

[133] Courant, R.: Dirichlet's Principle, Conformal Mapping, and Minimal Surfaces. New York: Interscience 1950.

[134] Dautray, R., Lions, J.: Mathematical Analysis and Numerical Methods for Science and Technology, Vols 1-6. New York: Springer-Verlag 1988 (transl. from French), (comprehensive presentation of modern methods).

[135] Dierkes, U., Hildebrandt, S., Küster, A., and Wohlrab, O.: Minimal Surfaces, Vols. 1, 2. Berlin: Springer-Verlag 1992.

[136] Dirac, P.: The Principles of Quantum Mechanics, 4th ed. Oxford: Clarendon Press 1981 (the classical textbook on quantum mechanics).

[137] Dirac, P.: General Theory of Relativity. Princeton, NJ: Princeton University Press 1996.

[138] Dolzmann, G.: Variational Methods for Crystalline Microstructure - Analysis and Computation. New York: Springer-Verlag 2003.

[139] Egorov, Yu., Shubin, M.: Partial Differential Equations, Vols. 1-4, Encyclopaedia of Mathematical Sciences, New York: Springer-Verlag 1991 (transl. from Russian).

[140] Engquist, B., Schmid, W. (eds.): Mathematics Unlimited - 2001 and Beyond. New York: Springer-Verlag 2001 (collection of 80 survey articles).

[141] Evans, C.: Partial Differential Equations, Providence, RI: Amer. Math. Soc. 1998 (standard textbook on the modern theory of partial differential equations).

[142] Feynman, R., Leighton, R., and Sands, M.: The Feynman Lectures in Physics. Reading, MA: Addison-Wesley 1963 (a classic).

[143] Finn, R.: Equilibrium Capillary Surfaces. Berlin: Springer-Verlag 1985.

[144] Friedman, A.: Mathematics in Industrial Problems, Vols. 1-6. New York: Springer-Verlag 1988-1995.

[145] Friedman, A.: Variational Principles and Free-Boundary Value Problems. New York: Wiley 1983.

[146] Galdi, G.: An Introduction to the Mathematical Theory of the Navier-Stokes Equations, Vols. 1, 2. New York: Springer-Verlag 1994.

[147] Gelfand, I., Shilov, E.: Generalized Functions, Vols. 1-5. New York: Academic Press 1964.

[148] Giaquinta, M., Hildebrandt, S.: Calculus of Variations, Vols. 1, 2. Berlin: Springer-Verlag 1995.

[149] Gilbarg, D., Trudinger, N.: Elliptic Partial Differential Equations of Second Order, 2nd ed. New York: Springer-Verlag 1994.

[150] Gilkey, P.: Invariance Theory, the Heat Equation, and the Atiyah-Singer Index Theorem, 2nd ed. Boca Raton, FL: CRC Press 1995.

[151] Greiner, W. et al.: Course of Modern Theoretical Physics, Vols. 1-13. New York: Springer-Verlag 1996 (including many exercises with solutions) (transl. from German).

[152] Guillemin, V., Sternberg, S.: Geometric Asymptotics. Providence, RI: Amer. Math. Soc. 1989.

[153] Guillemin, V., Sternberg, S.: Symplectic Techniques in Physics. Cambridge, UK: Cambridge University Press 1990.

[154] Henry, D.: Geometric Theory of Semilinear Parabolic Equations. New York: Springer-Verlag 1981 (a classic).

[155] Hislop, P., Sigal, I.: Introduction to Spectral Theory With Applications to Schrödinger Operators, New York: Springer-Verlag 1996.

[156] Hofer, H., Zehnder, E.: Symplectic Invariants and Hamitonian Dynamics, Basel: Birkhäuser 1994.

[157] Hörmander, L.: The Analysis of Linear Partial Differential Equations, Vols. 1-4. New York: Springer-Verlag 1983 (standard textbook).

[158] Isakov, V.: Inverse Problems for Partial Differential Equations. New York: Springer-Verlag 1998.

[159] John, F.: Partial Differential Equations, 4th ed. New York: Springer-Verlag 1982.

[160] Jost, J., Li-Jost, X.: Calculus of Variations. Cambridge, UK: Cambridge University Press 1998.

[161] Jost, J.: Riemannian Geometry and Geometric Analysis, 3rd ed. Berlin: Springer-Verlag 2001.

[162] Jost, J.: Partial Defferential Equations. Berlin: Springer-Verlag 2002.

[163] Kichenassamy, S.: Nonlinear Waves. London: Pitman 1993.

[164] Lahiri, A., Pal, B.: A First Book of Quantum Field Theory. Pangbourne, India: Alpha Science International 2001.

[165] Landau, L., Lifshitz, E.: Course of Theoretical Physics, Vols. 1-10 (standard text in theoretical physics), 2nd English ed. Oxford: Butterworth-Heinemann 1987 (transl. from Russian).

[166] Lax, P., Phillips, R.: Scattering Theory. New York: Academic Press 1989.

[167] Leis, R.: Initial Boundary Value Problems in Mathematical Physics. Stuttgart: Teubner-Verlag 1986.

[168] López, G.: Partial Differential Equations of First Order and Applications in Physics. Singapore: World Scientific 1999.

[169] Marchioro, C., Pulvirenti, M.: Mathematical Theory of Incompressible Nonviscous Fluids. New York: Springer-Verlag 1994.

[170] Markowich, P.: Semiconductor Equations. Berlin: Springer-Verlag 1990.

[171] Marsden, J., Hughes, T.: Mathematical Foundations of Mechanics. Englewood Cliffs, NJ: Prentice Hall 1983.

[172] Martin, P., Rothen, F.: Many-Body Problems and Quantum Field Theory. Berlin: Springer-Verlag 2002.

[173] Milton, G.: Composite Materials. New York: Cambridge University Press 2001.

[174] Misner, C., Thorne, K., and Wheeler, A.: Gravitation. San Francisco, CA: Freeman 1973 (a classic).

[175] Müller, S.: Variational Models for Microstructure and Phase Transitions. Leipzig: Max-Planck Institute for Mathematics in the Sciences, www.mis.mpg.de/preprints/ln. Lecture Note Nr. 2 1998.

[176] Natterer, F.: The Mathematics of Computerized Tomography. Philadelphia: SIAM 2001.

[177] Øksendal, B.: Stochastic Differential Equations, 5th ed. New York: Springer-Verlag 1998.

[178] Peskin, M., Schröder, D.: An Introduction to Quantum Field Theory. Reading, MA: Addison-Wesley 1995.

[179] Pike, E., Sarkar, A.: The Quantum Theory of Radiation. Oxford: Clarendon Press 1995.

[180] Reed, M., Simon, B.: Methods of Modern Mathematical Physics, Vols. 1-4. New York: Academic Press 1972 (standard textbook).

[181] Renardy, M., Rogers, R.: Introduction to Partial Differential Equations. New York: Springer-Verlag 1993.

[182] Riemann, B.: Gesammelte mathematische Werke, wissenschaftlicher Nachlass und Nachträge, edited by R. Narisimhan, New York, Leipzig: Springer-Verlag and Teubner-Verlag 1990.

[183] Risken, H.: The Fokker–Planck Equation: Methods of Solutions and Applications, 2nd ed. New York: Springer-Verlag 1996.

[184] Roy, B.: Fundamentals of Classical and Statistical Thermodynamics. New York: Wiley 2002.

[185] Sachdev, P.: A Compendium on Nonlinear Partial Differential Equations. New York: Wiley 1997.

[186] Schiff. L.: Quantum Mechanics. New York: McGraw-Hill 1968 (a classic).

[187] Smoller, J.: Shock Waves and Reaction–Diffusion Equations. New York: Springer-Verlag 1983 (a classics).

[188] Sommerfeld, A.: Lectures on Theoretical Physics, Vols. 1-6, New York: Academic Press 1949 (transl. from German), (the classical textbook on theoretical physics).

[189] Stephani, H. et al.: Exact Solutions of Einstein's Field Equations, 2nd ed. Cambridge, UK: Cambridge University Press 2003 (transl. from German).

[190] Strauss, W.: Partial Differential Equations. New York: Wiley 1992.

[191] Stroke, H. (ed.): The Physical Review: The First Hundred Years - A Selection of Seminal Papers and Commentaries. New York: American Institute of Physics 1995.

[192] Struwe, M.: Variational Methods, 2nd ed. New York: Springer-Verlag 1996.

[193] Sulem, C., Sulem, P.: Nonlinear Schrödinger Equations: Self-Focusing and Wave Collapse. New York: Springer-Verlag 1999.

[194] Temam, R.: Infinite-Dimensional Dynamical Systems in Mechanics and Physics, 2nd ed. New York: Springer-Verlag 1997.

[195] Taylor, M.: Partial Differential Equations, Vols. 1-3. New York: Springer-Verlag 1996.

[196] Thaller, B.: The Dirac Equation. New York: Springer-Verlag 1992.

[197] Thirring, W.: A Course in Mathematical Physics, Vols 1-4. New York: Springer-Verlag 1981 (transl. from German).

[198] Thirring, W.: Classical Mathematical Physics: Dynamical Systems and Fields, New York: Springer-Verlag 1997.

[199] Thirring, W.: Quantum Mathematical Physics: Atoms, Molecules and Large Systems, New York: Springer-Verlag 2002.

[200] Toda, M.: Nonlinear Waves and Solitons. Dordrecht: Kluwer 1989.

[201] Triebel, H.: Higher Analysis. Leipzig: Barth 1992 (transl. from German).

[202] Triebel, H.: Analysis and Mathematical Physics. Dordrecht: Kluwer 1987 (transl. from German).

[203] Vishik, M., Babin, A.: Attractors of Evolution Equations. Amsterdam; New York: North-Holland 1992.

[204] Vishik, M., Fursikov, A.: Mathematical Problems of Statistical Hydromechanics. Boston: Kluwer 1988.

[205] Visintin, A.: Differential Models of Hysteresis. Berlin: New York: Springer-Verlag 1994.

[206] Visintin, A.: Models of Phase Transitions. Boston: Birkhäuser 1996.

[207] Vladimirov, V.: Equations of Mathematical Physics. New York: M. Dekker 1971 (transl. from Russian).

[208] Weinberg, S.: Gravitation and Cosmology: Principles and Applications of the General Theory of Relativity. New York: Wiley 1972.

[209] Weinberg, S.: Quantum Field Theory, Vols. 1-3. Cambridge, UK: Cambridge University Press 1995.

[210] Weyl, H.: Raum, Zeit, Materie. Berlin: Springer-Verlag 1918 (English translation: Space, Time, Matter, 4th ed. New York: Dover 1950).

[211] Yang, Y.: Solitons in Field Theory and Nonlinear Analysis. New York: Springer-Verlag 2001.

[212] Zeidler, E. (ed.): Teubner-Taschenbuch der Mathematik, Vol. 2. Leipzig-Stuttgart: Teubner-Verlag 1995 (English edition in preparation).

[213] Zeidler, E.: Nonlinear Functional Analysis and its Applications. Vol. I: Fixed-Point Theory (3rd ed. 1998), Vol. IIA: Linear Monotone Operators (2nd ed. 1997), Vol. IIB: Nonlinear Monotone Operators, Vol. III: Variational Methods and Optimization, Vol. IV: Applications to Mathematical Physics (2nd ed. 1995), New York: Springer-Verlag 1986 ff.

[214] Zeidler, E.: Applied Functional Analysis: Applications to Mathematical Physics, 2nd ed. New York: Springer-Verlag 1997.

[215] Zeidler, E.: Applied Functional Analysis: Main Principles and their Applications. New York: Springer-Verlag 1995.

### 1.14. Complex Function Theory

[216] Apostol, T.: Modular Functions and Dirichlet Series in Number Theory, 2nd ed. New York: Springer-Verlag 1990.

[217] Barrow-Green, J.: "Poincaré and the Three Body Problem", History of Mathematics, Volume 11, AMS and LMS, Providence RI, 1997.

[218] Farkas, M. and Kra, I: Riemann Surfaces, 2nd ed. New York : Springer-Verlag 1992.

[219] Ford, L.: Automorphic Functions. New York: McGraw-Hill 1931 (a classic).

[220] Forster, O.: Lectures on Riemann Surfaces. Berlin: Springer-Verlag 1981 (transl. from German).

[221] Hörmander, L.: An Introduction to Complex Analysis in Several Variables, Van Nostrand 1966.

[222] Hurwitz, A., Courant, R.: Vorlesungen über allgemeine Funktionentheorie und elliptische Funktionen. 4. Aufl. Berlin. Springer-Verlag 1964 (a classic).

[223] Iwasaki, K., et al.: From Gauss to Painlevé. A Modern Theory of Special Functions. Wiesbaden: Vieweg 1991.

[224] Jost, J.: Compact Riemann Surfaces: An Introduction to Contemporary Mathematics. Berlin: Springer-Verlag 1997.

[225] Lang, S.: Complex Analysis, 4th ed. New York: Springer-Verlag 1999.

[226] Lang. S.: Elliptic Functions. New York: Addison-Wesley 1973.

[227] Lang, S.: Introduction to Algebraic and Abelian Functions, 2nd ed. Berlin: Springer-Verlag 1995.

[228] Lang, S.: Introduction to Modular Forms, 2nd ed. Berlin: Springer-Verlag 1995.

[229] Magnus, W.: Formulas and Theorems for the Special Functions of Mathematical Physics. Berlin: Springer-Verlag 1966.

[230] Maurin, K.: Riemann's Legacy: Riemann's Ideas in Mathematics and Physics of the 20th Century. Dordrecht: Kluwer 1997.

[231] Milnor, J.: Dynamics in One Complex Variable: Introductory Lectures, 2nd ed. Wiesbaden: Vieweg 2000.

[232] Patterson, S.: An Introduction to the Theory of the Riemann Zeta Function. Cambridge, UK: Cambridge University Press 1995.

[233] Remmert, R.: Theory of Complex Functions, Vols. 1, 2. New York: Springer-Verlag 1991 (recommended as an introduction).

[234] Vladimirov, V.: Methods of the Theory of Many Complex Variables. Cambridge, MA: MIT Press 1966 (transl. from Russian).

[235] Wells, R.: Differential Analysis on Complex Manifolds. New York: Springer-Verlag 1980.

[236] Weyl, H.: Die Idee der Riemannschen Fläche (The Notion of Riemann Surface), Leipzig: Teubner-Verlag 1913. Reprinted Leipzig: Teubner-Verlag 1999.

## 2. Algebra

[237] Birkhoff, G., Bartee, T.: Modern Applied Algebra. New York: McGraw-Hill 1970.

[238] Cameron, P.: Introduction to Algebra. Oxford: Oxford University Press 1998.

[239] Eisenbud, D.: Commutative Algebra with a View Toward Algebraic Geometry. Berlin: Springer-Verlag 1994.

[240] Isham, C.: Lectures on Groups and Vector Spaces for Physicists. Singapore: World Scientific 1989.

[241] Kostrikin, A., Shafarevich, I. (eds.): Algebra, Vols. 1,2. Encyclopaedia of Mathematical Sciences. New York: Springer-Verlag 1990, 1991 (transl. from Russian).

[242] Lang, S.: Algebra, 3rd ed. Reading, MA: Addison-Wesley 1993.

[243] Spindler, K.: Abstract Algebra with Applications, Vols. 1,2. New York: Marcel Dekker 1994.

[244] Stillwell, J.: Elements of Algebra, Geometry, Numbers, Equations. Berlin: Springer-Verlag 1994.

[245] Springer, T.: Linear Algebraic Groups, Boston: Birkhäuser 1981.

[246] Tits, J.: Tabellen zu den einfachen Liegruppen und ihren Darstellungen. Berlin: Springer-Verlag 1967.

[247] Vinberg, E.: A Course in Algebra. Providence, RI: Amer. Math. Soc. 2003 (transl. from Russian).

[248] Waerden, B., van der: Modern Algebra, Vols. 1,2. New York: Frederyck Ungar 1975 (transl. from German).

[249] Waerden, B., van der: Group Theory and Quantum Mechanics. New York: Springer-Verlag 1974 (transl. from German).

### 2.2. Matrices

[250] Baker, A.: Matrix Groups: An Introduction to Lie Group Theory. New York: Springer-Verlag 2002.

[251] Bellman, R.: Introduction to Matrix Analysis, 2nd ed. Philadelphia: SIAM 1997.

[252] Curtis, M.: Matrix Groups. New York: Springer-Verlag 1987.

### 2.3. Linear Algebra

[253] Greub, W.: Linear Algebra, 4th ed. New York: Springer-Verlag 1975.

[254] Halmos, P.: Finite-Dimensional Vector Spaces. New York: Springer-Verlag 1974.

[255] Kostrikin, A., Manin, Y.: Linear Algebra and Geometry. New York: Gordon and Breach 1989 (transl. from Russian).

[256] Spindler, K.: Abstract Algebra with Applications, Volume I. New York, Basel, Hong Kong: Marcel Dekker 1994.

### 2.4. Multilinear Algebra

[257] Frappat, L., Sciarinno, A., Sorba, P.: Dictionary of Lie Algebras and Super Lie Algebras. New York: Academic Press 2000.

[258] Fuchs, J.: Affine Lie Algebras and Quantum Groups: An Introduction with Applications in Conformal Field Theory. Cambridge, UK: Cambridge University Press 1992.

[259] Fuchs, J., Schweigert, C.: Symmetries, Lie Algebras, and Representations: A Graduate Course for Physicists. Cambridge, UK: Cambridge University Press 1997.

[260] Greub, W.: Multilinear Algebra, 2nd ed. New York: Springer-Verlag 1978.

### 2.7. Number Theory

[261] Apostol, T.: Introduction to Analytic Number Theory, 3rd ed. New York: Springer-Verlag 1986.

[262] Apostol, T.: Modular Functions and Dirichlet Series in Number Theory, 2nd ed. New York: Springer-Verlag 1986.

[263] Berndt, B. (ed.): Ramanunjan's Notebook, Parts I-IV. New York: Springer-Verlag 1985-1994.

[264] Borel, A.: Linear Algebraic Groups, 2nd ed. New York: Springer-Verlag 1991.

[265] Borevich, Z., Shafarevich, I.: Number Theory. New York: Academic Press 1966 (transl. from Russian).

[266] Borwein, J., Borwein, P.: Ramanunjan, Modular Equations, and Approximations to p, or How to Compute One Billion Digits of p. *The American Monthly* **96**, 201-219 (1989).

[267] Cohen, H.: A Course in Computational Algebraic Number Theory. Berlin: Springer-Verlag 1993.

[268] Dunlap, R.: The Golden Ratio and Fibonacci Numbers. Singapore: World Scientific 1997.

[269] Ebbinghaus, H. et al. (eds.): Numbers, 3rd ed. New York: Springer-Verlag (transl. from German).

[270] Hardy, G., Wright, E.: An Introduction to the Theory of Numbers, 5th ed. New York: Oxford University Press 1996.

[271] Hellegouarch, Y.: Invitation to the Mathematics of Fermat-Wiles. New York: Academic Press 2002 (transl. from French).

[272] Hua, L., Wang, Y.: Applications of Number theory to Numerical Analysis. New York: Springer-Verlag 1981.

[273] Hua, L.: Introduction to Number Theory. Berlin: Springer-Verlag 1982 (transl. from the Chinese).

[274] Ireland, K., Rosen, M.: A Classical Introduction to Modern Number Theory, 2nd ed. New York: Springer-Verlag 1990.

[275] John, P.: Algebraic Numbers and Algebraic Functions. London: Chapman & Hall 1991.

[276] Kaku, M.: Strings, Conformal Fields, and Topology. New York: Springer-Verlag 1991.

[277] Koblitz, N.: p-adic Numbers, p-adic Analysis, and Zeta functions, 2nd ed. New York: Springer-Verlag1984.

[278] Koblitz, N.: A Course in Number Theory and Cryptography. New York: Springer-Verlag 1994.

[279] Lang, S.: Algebraic Number Theory. New York: Springer-Verlag 1986.

[280] Lang, S.: An Introduction to Diophantine Approximations. Berlin: Springer-Verlag 1995.

[281] Lang, S.: Introduction to Modular Forms. Berlin: Springer-Verlag 1995.

[282] Neukirch, J.: Class Field Theory, Berlin, Heidelberg, New York: Springer-Verlag 1986.

[283] Parshin, A., Shafarevich, I. (eds.): Number Theory, Vols. 1,2. Berlin: Springer-Verlag 1995 (transl. from Russian).

[284] Ribenboim, P.: The New Book of Prime Number Records, 3rd ed. New York: Springer-Verlag 1995.

[285] Scharlau, W., Opolka, H.: Von Fermat bis Minkowski. Berlin: Springer-Verlag 1980.

[286] Schroeder, M.: Number Theory in Science and Communication. With Applications in Cryptography, Physics, Biology, Digital Information, and Computing. Berlin: Springer-Verlag1986.

[287] Serre, J.-P.: Local Fields. New York, Heidelberg, Berlin: Springer-Verlag 1979.

[288] Waldschmidt, M. et al. (eds.): From Number Theory to Physics. Berlin: Springer-Verlag 1992.

[289] Weil, A.: Basic Number Theory, 3rd ed. New York: Springer-Verlag 1994.

[290] Zagier, D.: Zetafunktionen und quadratische Körper. Eine Einführung in die höhere Zahlentheorie. Berlin: Springer-Verlag 1981.

## 3. Geometry


[291] Berger, M.: Geometry, Vols. 1,2. Berlin: Springer-Verlag 1987.

[292] Chandrasekhar, S.: The Mathematical Theory of Black Holes. Oxford, UK: Clarendon Press 1983.

[293] Choquet-Bruhat, Y., DeWitt-Morette, C., Dillard-Bleick, M.: Analysis, Manifolds, and Physics, Vol. 1: Basics; Vol. 2: 92 Applications. Amsterdam: Elsevier 1996.

[294] Connes, A.: Noncommutative Geometry. New York: Academic Press (1994).

[295] Dubrovin, B., Fomenko, A., Novikov, S.: Modern Geometry, Vols. 1-3. New York: Springer-Verlag 1985-1995 (transl. from Russian).

[296] Gilbert, J., Murray, M.P: Clifford Algebras and Dirac Operators in Harmonic Analysis. Cambridge, England: Cambridge University Press 1991.

[297] Gilkey, P.: Invariance Theory, the Heat Equation, and the Atiyah-Singer Index Theorem, 2nd ed. Boca Raton, FL: CRC Press 1995.

[298] Gracia–Bondia, J., Vrilly, J., Figueroa, H.: Elements of Noncommutative Geometry. Boston: Birkhäuser 2000.

[299] Green, M., Schwarz, J., Witten, E.: Superstrings, Vols. 1,2. Cambridge, UK: Cambridge University Press 1987.

[300] Guillemin, V., Sternberg, S.: Symplectic Techniques in Physics. Cambridge, UK: Cambridge University Press 1990.

[301] Haag, R.: Local Quantum Physics. Fields, Particles, Algebras. Berlin: Springer-Verlag 1993.

[302] Hilbert, D.: Grundlagen der Geometrie, 13. Aufl., Stuttgart: Teubner-Verlag 1997 (The first edition of this classics was published in 1899).

[303] Hofer, H., Zehnder, E.: Symplectic Invariants and Hamiltonian Dynamics. Basel: Birkhäuser 1994.

[304] Isham, C.: Modern Differential Geometry for Physicists. Singapore: World Scientific 1993.

[305] Jost, J.: Compact Riemann Surfaces: an Introduction to Contemporary Mathematics. Berlin: Springer-Verlag 1997.

[306] Lüst, D., Theissen, S.: Lectures on String Theory. New York: Springer-Verlag 1989.

[307] Kirsten, K.: Spectral Functions in Mathematics and Physics. Boca Raton, FL: Chapman & Hall 2002.

[308] Madore, J.: An Introduction to Noncommutative Differential Geometry and its Applications. Cambridge, UK: Cambridge University Press 1995.

[309] Majid, M.: Foundations of Quantum Group Theory. Cambridge, UK: Cambridge University Press 1995.

[310] Marathe, K., Martucci, G.: The Mathematical Foundations of Gauge Theories, Amsterdam: North-Holland 1992.

[311] Misner, C., Thorne, K., Wheeler, A.: Gravitation. San Francisco, CA: Freeman 1973.

[312] Nakahara, M.: Geometry, Topology, and Physics. Bristol, UK: Adam Hilger 1990.

[313] Nikulin, V., Shafarevich, I.: Geometries and Groups. Berlin: Springer-Verlag 1987 (transl. from Russian).

[314] Sternberg, S.: Group Theory and Physics. Cambridge, UK: Cambridge University Press 1994.

[315] Weinberg, S.: Gravitation and Cosmology. New York: Wiley 1972.

[316] Wess, J., Bagger, J.: Supersymmetry and Supergravity, 2nd ed. Princeton, NJ: Princeton University Press 1991.

[317] Wigner, E.: Group Theory and its Applications to the Quantum Mechanics of Atomic Spectra. New York: Academic Press 1959.

[318] Woodhouse, N.: Geometric Quantization, 3rd ed. New York: Oxford University Press 1997.

**3.2. Elementary Geometry**

[319] Lang, S., Murrow, G.: A High School Course, 2nd ed. New York: Springer-Verlag 1991.

**3.5. Projective Geometry**

[320] Coxeter, H.: Projective Geometry, 2nd ed. New York: Springer-Verlag 1987.

**3.6. Differential Geometry**

[321] Berger, M., Gostiaux, B.: Differential Geometry. Berlin: Springer-Verlag 1988.

[322] Dubrovin, B., Fomenko, A., Novikov, S.: Modern Geometry, Vols. 1-3. New York: Springer-Verlag 1985-1995 (transl. from Russian).

[323] Guillemin, V., Pollack, A.: Differential Topology. Englewood Cliffs, New Jersey: Prentice Hall 1974.

[324] Isham, C.: Modern Differential Geometry for Physicists. Singapore: World Scientific 1993.

[325] Jost, J.: Differentialgeometrie und Minimalflächen. Berlin: Springer-Verlag 1994.

[326] Jost, J.: Riemannian Geometry and Geometric Analysis, 3rd ed. Berlin: Springer-Verlag 2002.

[327] Kobayashi, S., Nomizu, K.: Foundations of Differential Geometry, Vols, 1,2. New York: Wiley 1963, 1965.

[328] Stoker, J.: Differential Geometry. New York: Wiley 1989.

[329] Struik, D.: Lectures on Classical Differential Geometry, 2nd ed. New York: Dover 1988.

**3.8. Algebraic Geometry**

[330] Brieskorn, E., Knörrer, H.: Ebene algebraische Kurven. Basel: Birkhäuser 1981.

[331] Eisenbud, D.: Commutative Algebra with a View Toward Algebraic Geometry. Berlin: Springer-Verlag 1994.

[332] Griffiths, P., Harris, J.: Principles of Algebraic Geometry. New York: Wiley 1978.

[333] Hartshorne, R.: Algebraic Geometry, 3rd ed. New York: Springer-Verlag 1983.

[334] Hirzebruch, F.: Topological Methods in Algebraic Geometry. New York: Springer-Verlag 1995.

[335] Lang, S.: Introduction to Algebraic Geometry. Reading, MA: Addison Wesley 1972.

[336] Lang, S.: Abelian Varieties. New York: Springer-Verlag 1983.

[337] Shafarevich, I.: Basic Algebraic Geometry, Vol. 1: Varieties in Projective Space. Vol. 2: Schemes and Complex Manifolds, 2nd ed. Berlin: Springer-Verlag 1994 (transl. from Russian).

[338] Waerden, B., van der: Einführung in die algebraische Geometrie, 2. Aufl., Berlin: Springer-Verlag 1973.

### 3.9. Geometry in Modern Physics

[339] Abraham, R., Marsden, J.: Foundations of Mechanics. Reading, MA: Benjamin Company 1978.

[340] Aebischer, B. et al.: Symplectic Geometry. An Introduction. Basel: Birkhäuser 1994.

[341] Benn, I., Tucker, R.: An Introduction to Spinors and Geometry with Applications in Physics. Bristol, UK: Adam Hilger 1987.

[342] Bredon, G.: Topology and Geometry. New York: Springer-Verlag 1993.

[343] Dubrovin, B., Fomenko, A., Novikov, S.: Modern Geometry, Vols. 1-3. New York: Springer-Verlag 1985-1995 (transl. from Russian).

[344] Felsager, B.: Geometry, Particles, and Fields. New York: Springer-Verlag 1997.

[345] Frankel, T.: The Geometry of Physics. Cambridge, UK: Cambridge University Press 1999.

## 4. Foundations of Mathematics

[346] Ebbinghaus, H., Flum, J., Thomas, W.: Mathematical Logic, 2nd ed. New York: Springer-Verlag 1989.

[347] Halmos, P.: Naive Set Theory. New York: Springer-Verlag 1974.

[348] Hilbert, D., Bernays, P.: Grundlagen der Mathematik, Bd. 1,2. Berlin: Springer-Verlag 1934, 1939, 2. Aufl. 1968.

[349] Manin, Yu.: A Course in Mathematical Logic. New York: Springer-Verlag 1977 (transl. from Russian).

[350] Russell, B.: Introduction to Mathematical Philosophy. New York: Dover Publications 1993.

[351] Tarski, A.: Introduction to Logic and to the Methodology of the Deductive Sciences. New York: Oxford University Press 1994.

## 5. Calculus of Variations and Optimization

[352] Aubin, J.: Optima and Equilibria. New York: Springer-Verlag 1993.

[353] Bellman, A.: Dynamic Programming. Princeton, NJ: Princeton University Press 1957.

[354] Bronstein, I., Semendjajew, K.: Taschenbuch der Mathematik, 25th ed. Leipzig: Teubner Verlag 1991.

[355] Carathéodory, C.: Calculus of Variations and Differential Equations of First Order. New York: Chelsea 1982 (transl. from German).

[356] Dantzig, G.: Linear Programming and Extensions. Princeton, NJ: Princeton University Press 1963.

[357] Davis, D.: Foundations of Deterministic and Stochastic Control. Boston: Birkhäuser 2002.

[358] Dierkes, U., Hildebrandt, S., Küster, A., Wohlrab, O.: Minimal Surfaces, Vols. 1,2,. Berlin: Springer-Verlag 1992.

[359] Ekeland, I., Teman, R.: Convex Analysis and Variational Problems. Amsterdam: North-Holland 1976.

[360] Eschrig, H.: The Fundamentals of Density Functional Theory. Leipzig: Teubner-Verlag 1996.

[361] Finn, R.: Equilibrium Capillary Surfaces. New York: Springer-Verlag 1985.

[362] Friedman, A.: Variational Principles and Free Boundary Value Problems. New York: Wiley 1982.

[363] Funk, P.: Variationsrechnung und ihre Anwendung in Physik und Technik (in German), 2nd ed. Berlin: Springer-Verlag 1970.

[364] Gamkrelidze, R.,: Principles of Optimal Control Theory. New York: Plenum Press 1978 (transl. from Russian).

[365] Giaquinta, M., Hildebrandt, S.: Calculus of Variations, Vols. 1,2. Berlin: Springer-Verlag 1995.

[366] Grötschel, M., Lovsz, L., Schrijver, A.: Geometric Algorithms and Combinatorial Optimization, 2nd ed. New York: Springer-Verlag 1993.

[367] Hildebrandt, S., Tromba, A.: The Parsimonious Universe: Shape and Form in the Natural World. New York: Copernicus 1996.

[368] Hiriart-Urruty, J., Lemaréchal, C.: Convex Analysis and Minimization Algorithms, Vols. 1,2. New York: Springer-Verlag 1993.

[369] Jost, J., Li-Jost, X.: Calculus of Variations. Cambridge, UK: Cambridge University Press 1998.

[370] Lions, J.: Optmial Control of Systems Governed by Partial Differential Equations. Berlin: Springer-Verlag 1971 (transl. from French).

[371] Luenberger, D.: Optimization by Vector Space Methods. New York: Wiley 1969.

[372] Soper, D.: Classical Field Theory. New York: Wiley 1975.

[373] Struwe, M.: Variational Methods, 2nd ed. New York : Springer-Verlag 1996.

[374] Zabcyk, J.: Mathematical Control Theory. Basel: Birkhäuser 1992.

[375] Zeidler, E.: Nonlinear Functional Analysis and its Applications, Vol. 3: Variational Methods and Optimization. New York: Springer-Verlag 1984. See [213] for all volumes.

[376] Zeidler, E.: Nonlinear Functional Analysis and its Applications. Vol. 4: Applications to Mathematical Physics. New York: Springer-Verlag 1990. See [213] for all volumes.

[377] Zeidler, E.: Applied Functional Analysis. Applications to Mathematical Physics. Applied Mathematical Sciences, Vol. 108. New York: Springer-Verlag 1995.

[378] Zeidler, E.: Applied Functional Analysis. Main Principles and Their Applications. Applied Mathematical Sciences, Vol. 109. New York: Springer-Verlag 1995.

## 6. Stochastics – Mathematics of Chance

### 6.1 Elementary Probability

[379] Gnedenko, B., Khinchin, A.: An Elementary Introduction to the Theory of Probability. New York: Dover 1962 (transl. from Russian).

[380] Rozanov, Y.: Introductory Probability Theory. Englewood Cliffs, NJ: Prentice-Hall 1969.

### 6.2. Theory of Probability

[381] Bass, R.: Probabilistic Techniques in Analysis. New York: Springer-Verlag 1995.

[382] Bauer, H.: Probability Theory. Berlin: De Gruyter 1996.

[383] Bouwmeester, D., Ekert, A., Zeilinger, A.: The Physics of Quantum Information: Quantum Cryptography, Quantum Teleportation, Quantum Computation. New York: Springer-Verlag 2000.

[384] Cardy, J.: Scaling and Renormalization in Statistical Physics, Cambridge, UK: Cambridge University Press 1997.

[385] Cercigniani, C.: Theory and Applications of the Boltzmann Equations. Edinburgh: Scottish Academic Press 1975.

[386] Chaichian, M., Demichev, A.: Path Integrals in Physics, Vol. 1: Stochastic Processes and Quantum Mechanics; Vol. 2: Quantum Field Theory, Statistical Physics, and other Modern Applications. Bristol, UK: Institute of Physics Publishing 2001.

[387] Emch, G.: Liu, C.: The Logic of Thermostatistical Physics. New York Springer-Verlag 2002.

[388] Feller, W.: An Introduction to Probability Theory and its Applications, Vols. 1,2. New York: Wiley 1966/71.

[389] Gnedenko, B.: The Theory of Probability. New York: Chelsea 1963.

[390] Gut, A.: An Intermediate Course in Probability. New York: Springer-Verlag 1995.

[391] Jacod, H., Protter, P.: Probability Essentials. Berlin: Springer-Verlag 2000 (transl. from French).

[392] Khinchin, A.: Mathematical Foundations of Statistical Mechanics. New York: Dover 1949 (transl. from Russian).

[393] Khinchin, A.: Mathematical Foundations of Quantum Statistics. Mineola, NY: Dover 1998 (transl. from Russian).

[394] Khinchin, A.: Mathematical Foundations of Information Theory. New York: Dover 1998 (transl. from Russian).

[395] Kolmogorov, A.: Foundation of the Theory of Probability. New York: Chelsea 1956 (transl. from Russian).

[396] Minlos, R.: Introduction to Mathematical Statistical Physics. Providence, RI: Amer. Math. Soc. 2000.

[397] Stoyan, D., Kendall, W. Mecke, J.: Stochastic Geometry and its Applications. New York: Wiley 1987.

**6.3. Mathematical Statistics**

[398] Anderson, T.: An Introduction to Multivariate Statistical Analysis, 2nd ed. New York: Wiley 1984.

[399] Berger, J.: Statistical Decision Theory and Bayesian Analysis, 2nd ed. Berlin: Springer-Verlag 1985.

[400] Bickel, P., Doksum, K.: Mathematical Statistics. San Francisco, CA: Holden-Day 1977.

[401] Brockwell, P., Davis, R.: Time Series: Theory and Methods, 2nd ed. Berlin: Springer-Verlag 1991.

[402] Krickeberg, K., Ziesold, H.: Stochastische Methoden, 4th ed. Berlin: Springer-Verlag 1995.

[403] Pratt, J., Gibbons: Concepts of Nonparametric Theory. Berlin: Springer-Verlag 1981.

[404] Särnal, C., Swensson, B., Wretman: Model Assisted Survey Sampling. New York: Springer-Verlag 1992.

[405] Tuckey, J.: Exploratory Data Analysis. Reading, MA: Addison-Wesley 1977.

[406] Überla, K.: Faktoranalyse, 2nd ed. Berlin: Springer-Verlag 1977.

[407] Waerden, B., van der: Mathematical Statistics. Berlin: Springer-Verlag 1969 (transl. from German).

**6.4. Stochastic Processes**

[408] Asmussen, S.: Ruin Probabilities. Singapore: World Scientific.

[409] Chung, K., Zhao, Z.: From Brownian Motion to Schrödinger's Equation. Berlin: Springer-Verlag 1995.

[410] Doob, J.: Stochastic Processes. New York: Wiley 1953.

[411] Gerber, H.: Life Insurance Mathematics. Berlin: Springer-Verlag 1990.

[412] Karlin, S.: A First Course in Stochastic Processes, New York: Academic Press 1968.

[413] Karlin, S., Taylor, M.: A Second Course in Stochastic Processes. New York: Academic Press 1980.

[414] Kloeden, P., Platen, E., Schurz, H.: Numerical Solution of Stochastic Differential Equations through Computer Experiments. Berlin: Springer-Verlag 1994.

[415] Protter, P.: Stochastic Integration and Differential Equations, 2nd ed. New York: Springer-Verlag 1995.

[416] Resnick, S.: Adventures in Stochastic Processes, 2nd ed. Basel: Birkhäuser 1994.

[417] Rolski, T.: Schmidli, H., Schmidt, V., Teugels, J.: Stochastic Processes for Insurance and Finance. Chichester: Wiley 1999.

[418] Schuss, Z.: Theory and Applications of Stochastic Differential Equations. New York: Wiley 1980.

[419] Sharpe, M.: General Theory of Markov Processes. New York: Academic Press 1988.

[420] Todorovic, P.: An Introduction to Stochastic Processes and their Applications. Berlin: Springer-Verlag 1992.

[421] Williams, D.: Probability with Martingales. Cambridge, UK: Cambridge University Press 1991.

## 7. Numerical Mathematics and Scientific Computing

[422] Alefeld, G., Herzberger, J.: Introduction to Interval Computations. New York: Springer-Verlag 1983.

[423] Allgower, E., Georg, K.: Numerical Continuation Methods. New York: Springer-Verlag 1993.

[424] Atkinson, K.: An Introduction to Numerical Analysis, 2nd ed. New York: Wiley 1989.

[425] Axelsson, O., Kolotilina, L. (eds.): Preconditioned Conjugate Gradient Methods. Berlin: Springer-Verlag 1990.

[426] Ciarlet, P.: Handbook of Numerical Analysis, Vols. 1-9. Amsterdam: North Holland 1990 ff.

[427] Crandall, R.: Topics in Advanced Scientific Computation. Berlin: Springer-Verlag 1995.

[428] Deuflhard, P., Hohmann, A.: Numerical Analysis: A First Course in Scientific Computation. Berlin de Gruyter 1995 (transl. from German).

[429] Golub, G. Ortega, J.: Scientific Computing: An Introduction with Parallel Computing. Boston: Academic Press 1993.

[430] Kulisch, U., Miranker, W.: Computer Arithmetic in Theory and Practice. New York: Academic Press 1981.

[431] Press, W. et al.: Numerical Recipies. The Art of Scientific Computing. Cambridge, UK: Cambridge University Press 1989.

[432] Schwarz, H.: Numerische Mathematik, 3. Aufl., Stuttgart: Teubner-Verlag 1993.

[433] Stoer, J., Bulirsch, R.: Introduction to Numerical Analysis. New York: Springer-Verlag 1993 (transl. from German).

[434] Zeidler, E.: Nonlinear Functional Analysis and its Applications, Vol. 2A: Linear Monotone Operators, Vol. 2B: Nonlinear Monotone Operators, Vol. 3: Variational Methods and Optimization. New York: Springer-Verlag 1990 (numerical functional analysis).

### 7.6. Ordinary Differential Equations

[435] Deuflhard, P., Bornemann, F.: Numerical Mathematics II: Integration of Ordinary Differential Equations. New York: Springer-Verlag 1999 (transl. from German).

[436] Fehlberg, E.: Klassische Runge–Kutta-Formeln vierter und niedrigerer Ordnung mit Schrittweitenkontrolle und ihre Anwendung auf Wärmeleitungsprobleme. *Computing* **6**, 1970.

[437] Hairer, E., Nörsett, S., Wanner, G.: Solving Ordinary Differential Equations 1. Nonstiff Problems. Berlin: Springer-Verlag 1987.

[438] Hairer, E., Wanner, G.: Solving Ordinary Differential Equations 2. Stiff Problems. Berlin: Springer-Verlag 1991.

[439] Lambert, J.: Numerical Methods for Ordinary Differential Systems. The Initial-Value Problem. New York: Wiley 1991.

### 7.7. Partial Differential Equations and Scientific Computing

[440] Ciarlet, P., Lions, J.: Handbook of Numerical Analysis, Vol. 2. Finite Element Methods. Amsterdam: North-Holland 1991.

[441] Dautray, R., Lions, J.: Mathematical Analysis and Numerical Methods for Science and Technology, Vols. 1-6. New York: Springer-Verlag 1988-1992.

[442] Hackbusch, W.: Elliptic Differential Equations: Theory and Numerical Treatment. Berlin; New York: Springer-Verlag 1992.

[443] Hackbusch, W.: Integral Equations. Theory and Numerical Treatment. Basel: Birkhäuser 1995.

[444] Hackbusch, W.: Iterative Solution of Large Sparse Systems of Equations. New York: Springer-Verlag 1994.

[445] Hackbusch, W.: Multi-Grid Methods and Applications. Berlin: Springer-Verlag 1985.

[446] Knabner, P., Angermann, L.: Numerical Methods for Elliptic and Parabolic Partial Differential Equations. New York: Springer-Verlag 2003 (transl. from German).

[447] LeVeque, R.: Numerical Methods for Conservation Laws. Basel: Birkhäuser 1992.

[448] Louis, A.: Inverse und schlecht gestellte Probleme. Stuttgart: Teubner-Verlag 1989.

[449] Louis, A., Maass, P., Rieder, A.: Wavelets. Stuttgart: Teubner-Verlag 1994.

[450] Natterer, F.: The Mathematics of Computerized Tomography. New York and Stuttgart: Wiley and Teubner-Verlag 1986.

[451] Quarteroni, A., Valli, A.: Numerical Approximation of Partial Differential Equations. Berlin: Springer-Verlag 1994.

[452] Richtmyer, R., Morton, K.: Difference Methods for Initial-Value Problems, 2nd ed. New York: Interscience Publishers 1967.

[453] Thomée, V.: Galerkin Finite Element Methods for Parabolic Problems. Berlin: Springer-Verlag 1984.

## History of mathematics

[454] Albers, D., Alexanderson, G., Reid, C.: International Mathematical Congresses. An Illustrated History. New York: Springer-Verlag 1987.

[455] Albers, D., Alexanderson, G. (eds.): Mathematical People. Profiles and Interviews. Basel: Birkhäuser 1985.

[456] Albers, D., Alexanderson, G., Reid, C. (eds.): More Mathematical People. New York: Academic Press 1995.

[457] Alexander, D.: A History of Complex Dynamics. From Schröder to Fatou and Julia. Wiesbaden: Vieweg 1994.

[458] Arnold, V.: Huygens and Borrow, Newton and Hooke. Pioneers in Mathematical Analysis and Catastrophe Theory from Evolvents to Quasicrystals. Basel: Birkhäuser 1990 (transl. from Russian).

[459] Artin, M., Kraft, H., Remmert R.: Duration and Change. Fifty Years at Oberwolfach. Berlin: Springer-Verlag 1994.

[460] Atiyah, M., Iagolnitzer, D. (eds.): Fields Medalists' Lectures. Singapore: World Scientific 2000.

[461] Atiyah, M.: Mathematics in the 20th Century. *Bull. London Math. Soc.* **34** (2002), 1-15.

[462] Auglin, W.: Mathematics. A Concise History and Philosophy. Berlin: Springer-Verlag 1994.

[463] Bell, E.: Men of Mathematics: Biographies of the Greatest Mathematicians of All Times. New York: Simon 1986.

[464] Borel, A.: Twentyfive Years with Nicolas Bourbaki, 1949–1973. *Notices Amer. Math. Soc.* **45**, 3 (1998), 373-380.

[465] Born, M.: Physics in My Generation. New York: Springer-Verlag 1969.

[466] Born, M.: My Life: Recollections of a Nobel Laureat, Charles Sribner's Sons. New York 1977 (transl. from German).

[467] Browder, F.: Reflections on the Future of Mathematics. *Notices Amer. Math. Soc.* **49**, 6 (2002), 658-662.

[468] Bühler, W.: Gauss. A Bibliographical Study. Berlin: Springer-Verlag 1981.

[469] Cassidy, D.: Werner Heisenberg. Heidelberg: Spektrum 1995.

[470] Chandrasekhar, S.: Newton's Principia for the Common Reader. Oxford: Oxford University Press.

[471] Chern, S., Hirzebruch, F.: Wolf Prize in Mathematics. Singapore: World Scientific 2000.

[472] Cropper, W.: Great Physicists: The Lives and Times of Leading Physicists from Galileo to Hawking. Oxford: Oxford University Press 2001.

[473] Dieudonné, J. et al. (eds.): Abrégé d'histoire des mathématiques, 1700-1900, Vols. 1,2. Paris: Hermann: 1978 (in French).

[474] Dieudonné, J.: A History of Functional Analysis. Amsterdam: North-Holland 1981.

[475] Dieudonné, J.: History of Functional Analysis, 1900-1975. Amsterdam: North-Holland 1981.

[476] Dieudonné, J.: History of Algebraic Geometry, 400 BC-1985. New York: Chapman 1985.

[477] Dieudonné, J.: A History of Algebraic and Differential Topology, 1900-1960. Boston: Birkhäuser 1989.

[478] Feynman, R., Leighton, R.: Surely You're Joking Mr. Feynman: Adventures of a Curious Character. New York: Norton 1985.

[479] Feynman, R., Leighton, R.: What Do You Care What Other People Think? Further Adventures of a Curious Character. New York: Norton 1988.

[480] Fritzsch, H.: Quarks. London: Penguin 1983 (transl. from German).

[481] Gamov, G.: The Great Physicists from Galileo to Einstein. New York: Dover 1961.

[482] Gottwald, S. (ed.): Lexikon bedeutender Mathematiker. Leipzig: Bibliographisches Institut 1990.

[483] Gray, J.: The Hilbert Challenge: A Perspective on 20th Century Mathematics. Oxford: Oxford University Press 2000.

[484] Gribbin, J., White, M.: Stephen Hawking: A Life in Science. London: Penguin Books 1992.

[485] Gribbin, J.: In Search of the Double Helix. London: Penguin Books 1995.

[486] Gribbin, J.: Schrödinger's Kitten. London: Weidenfeld 1996.

[487] Gribbin, J.: Richard Feynman: A Life in Science. London: Viking 1997.

[488] Halmos, P.: I Want To Be A Mathematician. New York: Springer-Verlag 1983.

[489] Halmos, P.: I Have a Photographic Memory. Providence, RI: American Mathematical Society 1987.

[490] Hilbert, D.: Mathematical Problems. Lecture delivered before the Second International Congress of Mathematicians at Paris 1900. *Bull. Amer. Math. Soc.* **8** (1902), 473-479 (reprinted in F. Browder (ed.), (1976), Vol. 1, 1-34).

[491] Kanigel. R.: The Man Who Knew Infinity: A Life of the Genius Ramanujan (1887-1920). New York: Scribner's 1991.

[492] Klein, F.: Vorlesungen über die Entwicklung der Mathematik im 19. Jahrhundert, Volumes 1 and 2. Berlin: Springer-Verlag 1926, 1927, reprinting 1979.

[493] Klein. F.: Development of Mathematics in the 19th Century. New York: Math. Sci. Press 1979 (transl. from German).

[494] Kline, M.: Mathematical Thought from Ancient to Modern Times. New York: Oxford University Press 1972.

[495] Kolmogorov, A., Yushkevich, A.: Mathematics of the 19th Century: Mathematical Logic, Algebra, Number Theory, Probability Theory. Basel, Boston: Birkhäuser Verlag 1992 (transl. from Russian).

[496] Kolmogorov, A., Yushkevich, A.: Mathematics of the 19th Century: Function Theory According to Chebyshev, Ordinary Differential Equations, Calculus of Variations, Theory of Finite Differences. Basel: Birkhäuser Verlag 1998 (transl. from Russian).

[497] Kragh, H.: Quantum Generations: A History of Physics in the Twentieth Century. Princeton, NJ: Princeton University Press 2000.

[498] Lorentz, H., Einstein, A., Minkowski, H., Weyl, H.: The Principle of Relativity. New York: Dover 1952 (a collection of classical papers).

[499] Maurin, K.: Riemann's Legacy: Riemann's Ideas in Mathematics and Physics of the 20th Century. Dordrecht: Kluwer 1997.

[500] Mehra, J., Rechenberg, H.: The Historical Development of Quantum Mechanics. Vols. 1-6. New York: Springer-Verlag 2002.

[501] Mehra, J., Milton, K.: Climbing the Mountain: The Scientific Biography of Julian Schwinger. Oxford: Oxford University Press 2000.

[502] Monastirsky, M.: Riemann, Topology, and Physics. Basel: Birkhäuser 1987.

[503] Monastirsky, M.: Modern Mathematics in the Light of Fields Medals. Wellersley, MA: Peters 1997.

[504] Nasar, S.: A Beautiful Mind: A Biography of John Forbes Nash, Jr. New York: Simon & Schuster 1998.

[505] Nobel Prize Lectures Stockholm: Nobel Foundation 1954ff.

[506] Pais, A.: Subtle is the Lord: the Science and the Life of Albert Einstein. Oxford: Oxford University Press 1982.

[507] Pais, A.: Niels Bohr's Times. Oxford: Oxford University Press 1993.

[508] Pais, A.: The Genius of Science: A Portrait Gallery. Oxford: Oxford University Press 2000.

[509] Pier, J. (ed.): Development of Mathematics 1900-1950. Basel: Birkhäuser 1994.

[510] Regis, E.: Who Got Einstein's Office? Eccentricity and Genius at the Institute for Advanced Study in Princeton. Reading, MA: Addison-Wesley 1989.

[511] Reid, C.: Hilbert. New York: Springer-Verlag 1970.

[512] Reid, C.: Courant in Göttingen and New York. New York: Springer-Verlag 1976.

[513] Reid, C.: Courant: the Life of an Improbable Mathematician. New York: Springer-Verlag 1976.

[514] Rife, P.: Lise Meitner and the Dawn of the Nuclear Age. Basel: Birkhäuser 1995.

[515] Schweber, S.: QED (Quantum Electrodynamics) and the Men Who Made It: Dyson, Feynman, Schwinger, and Tomonaga. Princeton, NY: Princeton University Press 1994 (history of quantum electrodynamics).

[516] Singh, S.: Fermat's Last Theorem: The Story of a Riddle that Confounded the World's Greatest Minds for 358 Years. London: Fourth Estate 1997.

[517] Stillwell, J., Clayton, V.: Mathematics and its History, 2nd ed. New York: Springer-Verlag 1991.

[518] Stubhang, A.: The Mathematician Sophus Lie: It was the Audacity of My Thinking. New York: Springer-Verlag 2002 (transl. from Norwegian).

[519] Thorne, K.: Einstein's Outrageous Legacy. New York: Norton 1993.

[520] Tian Yu Cao (ed.): Conceptual Foundations of Quantum Field Theory. Cambridge, UK: Cambridge University Press 1998.

[521] Treiman, S.: The Odd Quantum. Princeton, NJ: Princeton University Press 1999.

[522] Waerden, B., van der: Sources of Quantum Mechanics. New York: Dover 1968.

[523] Waerden, B., van der: Geometry and Algebra in Ancient Civilizations. Berlin: Springer-Verlag 1983.

[524] Waerden, B., van der: A History of Algebra. From al-Khwarizhmi to Emmy Noether. Berlin: Springer-Verlag 1985.

[525] Weil, A.: The Apprenticeship of a Mathematician. Basel: Birkhäuser 1992.

[526] Weyl, H.: David Hilbert and His Mathematical Work. *Bull. Amer. Math. Soc.* **50** (1944), 612-654.

[527] Yandell, B.: The Honors Class: Hilbert's Problems and Their Solvers. Natick, MA: Peters Ltd. 2001.

[528] Yang, C.: Hermann Weyl's Contributions to Physics. In: Hermann Weyl (1885-1985). Berlin: Springer-Verlag 1985.

## Mathematics and Human Culture

[529] Adams, F., Laughlin, G.: The Five Ages of the Universe: Inside the Physics of Eternity. New York: Simon & Schuster 1999.

[530] Adams, C.: The Knot Book. Cambridge, UK: Cambridge University Press 1994.

[531] Aigner, M., Ziegler, G.: Proofs from the Book, 2nd ed. Berlin: Springer-Verlag 2001.

[532] Bochner, S.: The Role of Mathematics and the Rise of Science, 4th ed. Princeton, NJ: Princeton University Press 1984.

[533] Bodanis, D.: $E = mc^2$: A Biography of the World's Most Famous Equation. New York: Walker 2000. (The appendix of this book includes extensive hints to the literature on the history of modern physics.)

[534] Bovill, C.: Fractal Geometry in Architecture and Design. Basel: Birkhäuser 1995.

[535] Cascuberta, C., Castellet, M.: Mathematical Research Today and Tomorrow: Viewpoint of Seven Fields Medalists. New York: Springer-Verlag 1992.

[536] Casti, J.: Five More Golden Rules: Knots, Codes, Chaos and Other Great Theories of 20th Century Mathematics. New York: Wiley 2000.

[537] Connes, A., Lichnerowicz, A., Schützenberger, M.: Triangle of Thoughts. Providence, RI: Amererican Mathematical Society 2001.

[538] Cottingham, W., Greenwood, D.: An Introduction to the Standard Model of Particle Physics. Cambridge, UK: Cambridge University Press 1998.

[539] Courant, R., Robbin, H.: What is Mathematics? Oxford: Oxford University Press 1941 (a classic).

[540] Davies, P. (ed.): The New Physics. Cambridge, UK: Cambridge University Press 1990.

[541] Davis, D.: The Nature and Power of Mathematics. Princeton, NJ: Princeton University Press 1993.

[542] Dieudonné, J.: Mathematics – the Music of Reason. Berlin: Springer-Verlag 1992.

[543] Dirac, P.: Directions in Physics. New York: Wiley 1978.

[544] Dyson, F.: Disturbing the Universe. New York: Harper and Row 1979.

[545] Dyson, F.: Origins of Life. Cambridge, UK: Cambridge University Press 1999.

[546] Dyson, F.: The Sun, the Genome and the Internet: Tool of Scientific Revolution. New York: Oxford University Press 1999.

[547] Ebbinghaus, H. et al. (eds.): Numbers, 3rd ed. New York: Springer-Verlag 1995.

[548] Einstein, A.: Essays in Science. New York: Philosophical Library 1933.

[549] Einstein, A.: The Meaning of Relativity. Princeton, NJ: Princeton University Press 1955.

[550] Emch, G., Liu, C.: The Logic of Thermostatistical Physics. New York: Springer-Verlag 2002.

[551] Engquist, B., Schmid, W. (eds.): Mathematics Unlimited – 2001 and Beyond. New York: Springer-Verlag 2001 (80 articles on modern mathematics and its applications written by leading experts).

[552] Ferris, T.: The Red Limit: the Discovery of Quasars, Neutron Stars, and Black Holes. New York: Morrow 1977.

[553] Ferris, T.: The World Treasury of Physics, Astronomy, and Mathematics. Boston, MA: Brown 1991.

[554] Feynman, R.: The Character of Physical Law. Cambridge, MA: MIT Press 1966.

[555] Gardner, M.: Riddles of the Sphinx and other Mathematical Puzzle Tales. Washington, DC: Mathematical Association of America 1987.

[556] Gardner, M.: Mathematical Magic Show. Washington, DC: Mathematical Association of America 1990.

[557] Gell-Mann, M.: The Quark and the Jaguar. New York: Freeman 1994.

[558] Goldstine, H.: The Computer from Pascal to von Neumann. Prineton, NJ: Princeton University Press 1993.

[559] Golubitsky, M.: Stewart, I.: The Symmetry Perspective from Equilibrium to Chaos in Phase Space and Physical Space. Basel: Birkhäuser 2002.

[560] Green, B.: The Elegant Universe: Supersymmetric Strings, Hidden Dimensions and the Quest for the Ultimate Theory. New York: Norton 1999.

[561] Halmos, P.: Selecta: Expository Writings. New York: Springer-Verlag 1985.

[562] Hawking, S.: A Brief History of Time. New York: Bantam Books 1988.

[563] Hawking, S., Penrose, R.: The Nature of Space and Time. Princeton, NJ: Princeton University Press 1997.

[564] Hawking, S.: The Universe in a Nut-Shell. New York: Bantam Books 2001.

[565] Heisenberg, W.: Physics and Beyond: Encounters and Conversations. New York: Harper and Row 1970 (transl. from German).

[566] Hildebrandt, S., Tromba, A.: The Parsimonious Universe: Shape and Form in the Natural World. New York: Copernicus 1996.

[567] Hofstadter, D.: Gödel, Escher, Bach: An Eternal Golden Braid. New York: Basic Books 1979.

[568] Jaffe, A.: Ordering the Universe. The Role of Mathematics. *Notices Amer. Math. Soc.* **31** , 236 (1984), 589-608.

[569] Kähler, E.: Über die Beziehungen der Mathematik zu Astronomie und Physik. *Jahresbericht der Deutschen Mathematikervereinigung* **51** (1941), 52-63.

[570] Kline, M.: Mathematical Thought from Ancient to Modern Times. New York: Oxford University Press 1972.

[571] Manin, Yu.: Mathematics and Physics. Boston: Birkhäuser 1981 (transl. from Russian).

[572] Mathematics - the Unifying Thread in Science. *Notices Amer. Math. Soc.* **33** (1986), 716-733.

[573] Mazolla, G.: The Topos of Music. Basel: Birkhäuser 2002.

[574] Monastirsky, M.: Riemann, Topology, and Physics. Basel: Birkhäuser 1987.

[575] Peitgen, H., Richter, P.: The Beauty of Fractals. Berlin: Springer-Verlag 1986.

[576] Peitgen, H, Jürgens, H., Saupe, D.: Chaos and Fractals. New Frontiers of Science. New York: Springer-Verlag 1992.

[577] Penrose, R.: The Emperors New Mind: Concerning Computers, Minds, and the Laws of Physics. New York: Penguin Books 1991.

[578] Penrose, R.: Shadows of the Mind: a Search for the Missing Science of Consciousness. New York: Oxford University Press 1994.

[579] Penrose, R. et al.: The Large, the Small, and the Human Mind, ed. by M. Longair, Cambridge, UK: Cambridge University Press 1997.

[580] Ruelle, D.: Chance and Chaos. Princeton, N.J: Princeton University Press 1991.

[581] Smoot, G., Davidson, K.: Wrinkles in Time. New York: Morrow 1994. (This book reports on the COBE project; in 1990, this famous satellite experiment established the anisotropy of the 3K radiation which comes to us as a relict of the very early universe.)

[582] Stroke, H. (ed.): The Physical Review: The First Hundred Years - A Selection of Seminal Papers and Commentaries. New York: American Institute of Physics 1995 (14 survey articles on general developments, 200 fundamental articles, 800 additional articles on CD).

[583] Taylor, A.: Mathematics and Politics, Strategy, Voting, Power, and Proof. Berlin: Springer-Verlag 1995.

[584] t'Hooft, G.: In Search for the Ultimate Building Blocks. Cambridge, UK: Cambridge University Press 1996.

[585] Veltman, M.: Facts and Mysteries in Elementary Particle Physics. Singapore: World Scientific 2003.

[586] Weinberg, S.: Gravitation and Cosmology. New York: Wiley 1972.

[587] Weinberg, S.: The First Three Minutes: A Modern View of the Origin of the Universe. New York: Basic Books 1977.

[588] Weinberg, S.: Dreams of a Final Theory. New York: Pantheon Books 1992.

[589] Weyl, H.: Philosophy of Mathematics and Natural Sciences. Princeton, NJ: Princeton University Press 1949.

[590] Weyl, H.: Symmetry. Princeton, NJ: Princeton University Press 1952.

[591] Wigner, Philosophical Reflections and Syntheses, annotated by G. Emch. New York: Springer-Verlag 1995.

[592] Zeidler, E.: Mathematics: a Cosmic Eye of Humanity. Internet: `http://www.mis.mpg.de`.

# List of Names

# *Index*

Definitions of terms are indicated by **boldface** page numbers.

# *Mathematical Symbols*

The following list only contains those symbols used most often.

**Logical symbols**

$\mathscr{A} \longrightarrow \mathscr{B}$ — From statement $\mathscr{A}$ the statement $\mathscr{B}$ follows. An alternative way of saying this is: $\mathscr{A}$ is *sufficient* for $\mathscr{B}$, and $\mathscr{B}$ is *necessary* for $\mathscr{A}$.

$\mathscr{A} \longleftrightarrow \mathscr{B}$ — The statement $\mathscr{A}$ is *equivalent to* the statement $\mathscr{B}$. Alternatively, one also says: $\mathscr{A}$ holds if and only if $\mathscr{B}$ holds. Equivalently: $\mathscr{A}$ is *necessary and sufficient* for $\mathscr{B}$.

$\{x : \ldots\}$ or $\{x| \ldots\}$ — The set of all elements $x$, which have the property indicated by the ellipses ...

$\square$ — The end of a proof; a different way of indicating this is the abbreviation *q.e.d.* (quod erat demonstandum – which was to be shown).

$\mathscr{A} \vee \mathscr{B}$ — The statement $\mathscr{A}$ *or* the statement $\mathscr{B}$ is true.

$\mathscr{A} \wedge \mathscr{B}$ — Both statements $\mathscr{A}$ *and* $\mathscr{B}$ are true.

$\neg\mathscr{A}$ — The statement $\mathscr{A}$ does *not* hold (the negation of $\mathscr{A}$).

$\forall_x | \ldots$ — All elements $x$ have the property indicated by the ellipses ...

$\exists_x | \ldots$ — There is (exists) an element $x$ with the property indicated by the ellipses ...

$\exists!_x | \ldots$ — There exists precisely one element $x$ with the property ...

$a \sim b$ — $a$ is equivalent to $b$ (under an equivalence relation).

$X/\sim$ — The set of all equivalence classes in $X$ under an equivalence relation $\sim$.

$a = b$ — $a$ is equal to $b$.

$a \neq b$ — $a$ is not equal to $b$.

$f(x) := x^2$ — The function $f(x)$ is *defined* to be equal to $x^2$.

$f(x) \equiv 0$ — The function $f(x)$ is identically equal to 0, i.e., $f(x) = 0$ for all $x$.

$f = \text{const}$ — The function $f$ is constant, i.e., $f(x)$ attains the same value for all values of the argument $x$.

$\mathbb{N}$ — The set of natural numbers $0, 1, 2, \ldots$

$\mathbb{N}_+$ — The set of positive natural numbers $1, 2, \ldots$

$\mathbb{Z}$ — The set (ring) of integers.

$\mathbb{Q}$ — The set (field) of rational numbers.

$\mathbb{R}$ — The set (field) of real numbers.

$\mathbb{C}$ — The set (algebraically closed field) of complex numbers.

$\mathbb{K}$ — A general notation for a field, usually $\mathbb{R}$ or $\mathbb{C}$.

$\mathbb{R}^n$ — The set of all $n$-tuples $(x_1, \ldots, x_n)$, where all $x_j$ are elements of $\mathbb{R}$.

$\mathbf{xy}$ — The Euclidean scalar product $\sum_{j=1}^{n} x_j y_j$ for $\mathbf{x} = (x_1, \ldots, x_n)$, $\mathbf{y} = (y_1, \ldots, y_n) \in \mathbb{R}^n$, i.e., $\mathbf{x}$ and $\mathbf{y}$ $n$-dimensional Euclidean vectors.

$(\mathbf{x}, \mathbf{y})$ — The unitary scalar product $\sum_{j=1}^{n} x_j \overline{y}_j$ for $\mathbf{x} = (x_1, \ldots, x_n)$, $\mathbf{y} = (y_1, \ldots, y_n) \in \mathbb{C}^n$, i.e., $\mathbf{x}$ and $\mathbf{y}$ $n$-dimensional complex vectors.

$|\mathbf{x}|$ — The Euclidean norm (length), defined by $\sqrt{\mathbf{xx}}$.

$\pi$ — The Ludolf number $\pi$ (spoken 'pie'), $\pi = 3.14159\ldots$

$\mathrm{e}$ — The Euler number $\mathrm{e} = 2.7182818\ldots$

$C$ — The Euler constant $C = 0.5772\ldots$

$\mathrm{i}$ — The imaginary unit for which $\mathrm{i}^2 = -1$.

| | |
|---|---|
| $n!$ | Pronounced $n$-factorial, this is the product of the first $n$ postive natural numbers $1, 2, \ldots, n$. |
| $\binom{n}{m}$ | A binomial coefficient. |
| $\operatorname{Re} z,\ \operatorname{Im} z$ | The real (resp. imaginary) part $x$ (resp. $y$) of a complex number $z = x + \mathrm{i}y$. |
| $\overline{z}$ | The complex conjugate number to $z = x + \mathrm{i}y$, i.e., $\overline{z} = x - \mathrm{i}y$. |
| $\arg z$ | The argument (angle) of the complex number $z$. |
| $a \leq b$ | $a$ is less than or equal to $b$. |
| $a < b$ | $a$ is strictly less than $b$. |
| $a \equiv b \bmod p$ | $a$ is congruent to $b$ modulo $p$, i.e., the difference $b - a$ is divisible by $p$. |
| $[a, b]$ | The closed interval $\{x \in \mathbb{R} \mid a \leq x \leq b\}$. |
| $]a, b[$ | The open interval $\{x \in \mathbb{R} \mid a < x < b\}$, also denoted $(a, b)$. |
| $[a, b[$ | The half-open interval $\{x \in \mathbb{R} \mid a \leq x < b\}$, also denoted $[a, b)$. |
| $]a, b]$ | The half-open interval $\{x \in \mathbb{R} \mid a < x \leq b\}$, also denoted $(a, b]$. |
| $\operatorname{sgn} a$ | The sign of a number $a$ or of a permutation $a$. |
| $\sum_{j=1}^{n} a_j$ | The sum of the numbers $a_j$, $a_1 + \cdots + a_n$. |
| $\prod_{j=1}^{n} a_j$ | The product of the numbers $a_j$, $a_1 \cdot a_2 \cdots a_n$. |
| $\min\{a, b\}$ | The smaller of the two numbers $a$ and $b$. |
| $\max\{a, b\}$ | The larger of the two numbers $a$ and $b$. |

**Elementary functions**

| | |
|---|---|
| $\sqrt{x}$ | The positive square root of a positive real number $x$ (e.g., $\sqrt{4} = 2$) or a fixed square root of a complex number $x$. |
| $\sqrt[n]{x}$ | An $n^{th}$ root of a number $x$, e.g., $\sqrt[3]{8} = 2$ because $2^3 = 8$ and $\sqrt[3]{-8} = -2$ because $(-2)^3 = -8$; the other third roots of $-8$ are complex. |
| $\mathrm{e}^x$ | The value of the exponential function at the point $x$, or the exponential function (as a function of $x$). |
| $\ln x$ | The natural logarithm of $x$ (logarithmus naturalis). |
| $\log_a x$ | The logarithm of $x$ to the basis $a$. |
| $x^\alpha$ | The general power function (as a function of $x$) or the value of the $\alpha^{th}$-power of the number $x$ ($x^\alpha = \mathrm{e}^{\alpha \ln x}$). |
| $\sin x,\ \cos x$ | The sine and cosine functions. |
| $\tan x,\ \cot x$ | The tangent and cotangent functions. |
| $\arcsin x,\ \arccos x$ | The arc-sine and arc-cosine functions, the inverse functions of the sine and cosine functions. |
| $\sinh x,\ \cosh x$ | The hyperbolic sine and cosine functions. |
| $\tanh x,\ \coth x$ | The hyperbolic tangent and cotangent functions. |
| $\operatorname{arsinh} x,\ \operatorname{arcosh} x$ | The inverse functions of the hyperbolic sine and cosine functions. |
| artanh x, arcoth x | The inverse functions of the hyperbolic tangent and cotangent functions. |
| $\lim_{n\to\infty} x_n$ | The limit of the sequence $\{x_n\}$; another notation for this is $x_n \to L$ as $n \to \infty$, where $L$ is the limit value. |
| $\lim_{x\to a} f(x)$ | The limit of the function $f$ as $x$ tends to $a$. |
| $f'(x)$ | The derivative of the function $f$ at $x$. |
| $f'',\ f^{(2)}$ | The second derivative of the function $f$ at $x$. |

| | |
|---|---|
| $\dfrac{\partial f}{\partial x}$, $f_x$ | The partial derivative of the function $f$ with respect to $x$. |
| $\dfrac{\partial^2 f}{\partial x \partial y}$, $f_{xy}$ | The second partial derivative of the function $f$ with respect to the variable $x$ (first), then $y$. |
| $\partial_j f$ | The partial derivative of the function $f$ with respect to the variable $x_j$. |
| $\partial^\alpha f$ | An abbreviation for the partial derivative of the function $f$ given by $\partial_1^{\alpha_1}\partial_2^{\alpha_2}\cdots\partial_n^{\alpha_n} f$, which means more precisely $$\partial^\alpha f := \frac{\partial^{\lvert\alpha\rvert} f}{\partial^{\alpha_1} x_1 \cdots \partial^{\alpha_n} x_n},$$ where $\lvert\alpha\rvert := \alpha_1 + \cdots + \alpha_n$. |
| $\mathrm{d}f$ | The total differential of the function $f$. |
| $\mathrm{d}\omega$ | The Cartan derivative of a differential form $\omega$. |
| $\int_a^b f(x)\mathrm{d}x$ | The integral of the function $f$ on the interval $[a, b]$. |
| $\int_G f(x)\mathrm{d}x$ | The integral of the function $f$ on the set $G$. |
| $\int_G \omega$ | The integral of a differential form $\omega$ on the set $G$. |
| $\int_M f\mathrm{d}F$ | A surface integral. |
| $\mathbf{grad}\, T$ | The gradient of (for example) a temperature field $T$. |
| $\mathrm{Div}\, \mathbf{E}$ | The divergence of the electromagnetic field $\mathbf{E}$. |
| $\mathbf{curl}\, \mathbf{E}$ | The curl of the electromagnetic field $\mathbf{E}$. |
| $\Delta T$ | The Laplace operator of the temperature field $T$, i.e., $\Delta T := \mathrm{Div}\, \mathbf{grad}\, T$. |
| $\nabla$ | The nabla operator, defined as $\nabla := \dfrac{\partial}{\partial x}\mathbf{i} + \dfrac{\partial}{\partial y}\mathbf{j} + \dfrac{\partial}{\partial z}\mathbf{k}$. |
| $f = o(g),\ x \to a$ | The quotient $f(x)/g(x)$ tends to zero as $x \to a$. |
| $f = O(g),\ x \to a$ | The quotient $f(x)/g(x)$ is bounded in a neighborhood of $a$ (no statement about the behavior at $a$). |
| $f \cong g,\ x \to a$ | The quotient $f(x)/g(x)$ tends to unity as $x \to a$. |
| $\partial U$ | The boundary of a set $U$. |
| $\overline{U}$ | The closure of a set $U$, i.e., $\overline{U} = U \cup \partial U$. |
| $\mathrm{int}\, U$ | The interior of a set $U$. |
| $C^k(G)$ | The set of all functions $f : G \longrightarrow \mathbb{R}$, which have continuous partial derivatives of all orders up to and including $k$ on the open set $G$. |
| $C^k(\overline{G})$ | The set of all functions $f : G \longrightarrow \mathbb{R}$, which, together with their partial derivatives of all orders up to and including $k$, can be continously extended to the boundary $\overline{G}$. |
| $C^\infty(G)$ | The set of all functions $f : G \longrightarrow \mathbb{R}$, which have continuous partial derivatives of arbitrary order on the open set $G$. |
| $C_0^\infty(G)$ | The set of all functions $f \in C^\infty(G)$, which vanish outside a compact subset of $G$. |
| $L_2(G)$ | The set of all measurable functions $f$, i.e., for which $\int_G \lvert f(x)\rvert^2 \mathrm{d}x < \infty$; the integral is to be understood in the sense of Lebesgue, which includes the classical integral. |
| $A^\mathsf{T}$ | The transposed matrix of a given matrix $A$ (rows and columns are exchanged). |
| $A^*$ | The adjoint matrix of a given matrix $A$ (transposed and complex-conjugated). |

| | |
|---|---|
| $\mathrm{rank} A$ | The rank of a matrix $A$. |
| $\det A$ | The determinant of a quadratic matrix $A$. |
| $\mathrm{tr}\, A$ | The trace of a quadratic matrix $A$. |
| $\delta_{jk}$ | The Kronecker symbol defined by the conditions $\delta_{jk} = 0$ for $j \neq k$ and $\delta_{jk} = 1$ for $j = k$. |
| $E,\ I$ | A unit matrix. |
| $\mathbf{ab}$ | The scalar product of two vectors **a** and **b**. |
| $\mathbf{a} \times \mathbf{b}$ | The vector product of two vectors **a** and **b**. |
| $(\mathbf{abc})$ | The triple product defined by $(\mathbf{a} \times \mathbf{b})\mathbf{c}$. |
| $\mathbf{i},\ \mathbf{j},\ \mathbf{k}$ | The orthonormal standard basis vectors of a Cartesian coordinate system. |
| $X \bigoplus Y$ | The vector space direct sum of linear spaces $X$ and $Y$. |
| $X \bigotimes Y$ | The tensor product of two linear spaces $X$ and $Y$. |
| $X \bigwedge Y$ | The exterior product (Grassmann product) of two linear spaces $X$ and $Y$. |
| $X/Y$ | The factor space of a linear space $X$ with respect to a linear subspace $Y$. |
| $a \otimes b$ | The tensor product of two multilinear forms $a$ and $b$. |
| $a \wedge b$ | The exterior product of two alternating linear forms $a$ and $b$. |

**Sets and maps**

| | |
|---|---|
| $x \in M$ | $x$ is an element of a given set $M$. |
| $x \notin M$ | $x$ is not an element of a given set $M$. |
| $A \subseteq M,\ A \subset M$ | $A$ is a subset of a given set $M$, i.e., every element of $A$ also is an element of $M$. |
| $A \subsetneq M$ | $A$ is a proper subset of $M$, i.e., $A \subset M$ and $A \neq M$. |
| $A \cap B$ | The intersection set of two given sets $A$ and $B$. |
| $A \cup B$ | The union set of two given sets $A$ and $B$. |
| $A - B$ | The difference of two sets $A$ and $B$, i.e., the set of elements of $A$ not belonging to $B$. |
| $A \times B.$ | The product set of all ordered pairs $(a, b)$ with $a \in A$ and $b \in B$. |
| $2^A$ | The power set of $A$, i.e., the set whose elements are the subsets of $A$. |
| $\emptyset$ | The empty set. |
| $f : A \subseteq M \longrightarrow B$ | A function $f$ from a set $A$ to a set $B$. |
| $D(f),\ \mathrm{Dom}\, f$ | The domain of a mapping or function $f$. |
| $R(f),\ \mathrm{Im}\, f$ | The range of a mapping or function $f$. |
| $f(A)$ | The image of a set $A$ under a mapping or function $f$. |
| $f^{-1}(B)$ | The inverse image of a set $B$ under the mapping or function $f$, i.e., the set of all $x$ for which $f(x) \in B$. |
| $I,\ \mathrm{id}$ | The identity operator or unit operator on a set, i.e., $Ix = x$ for all $x$. |
| $\mathrm{span}\, L$ | The linear hull (span) of a set $L$. |
| $\mathrm{meas}\, M$ | The measure of a set $M$. |

# *Dimensions of important physical quantities*

| ***Basic quantities*** | | | |
|---|---|---|---|
| length | m | meter | |
| time | s | second | |
| mass | kg | kilogram[1] | |
| temperature | K° | degree Kelvin | |
| current | A | ampere | |
| amount of substance | mol | 1 mol $= L$ particles<br>($L$ is Avagadro's number, $L = 6.022 \cdot 10^{23}$) | |
| light intensity | cd | candela | |
| ***Derived quantities*** | | | |
| speed | m/s | meters per second<br>(distance per time) | |
| acceleration | m/s$^2$ | meters per second squared<br>(change in speed per time) | |
| (mass) density | kg/m$^3$ | kilograms per cubic meter<br>(mass per volume) | |
| force | N | Newton<br>(force time acceleration) | $\mathrm{N} = \mathrm{kg{\cdot}m/s^2}$ |
| pressure | Pa | Pascal<br>(force per surface area) | $\mathrm{Pa} = \mathrm{N/m^2}$ |
| (the mean air pressure at sea level is roughly $10^5$ Pascal) | | | |
| work | J | joule<br>(force time distance) | $\mathrm{J} = \mathrm{Nm}$ |
| power | W | watt<br>(performed work per time, energy per time) | $\mathrm{W} = \mathrm{J/s} = \mathrm{VA}$ |
| energy | J<br>eV | joule<br>electron volt<br>(performed work, mass time speed squared) | $\mathrm{J} = \mathrm{Nm} = \mathrm{kg{\cdot}m^2/s^2} = \mathrm{Ws}$<br>$1\ \mathrm{eV} = 1.6 \cdot 10^{-19}\ \mathrm{J}$ |
| action | Js | joule-second<br>(energy times time) | |
| heat | J | joule | $\mathrm{J} = \mathrm{Nm}$ |

[1]Modern physics requires the so-called *atomic unit* u. This is one-twelfth of the mass of a $^{12}$C carbon atom, for which $\mathrm{u} = 1.661 \cdot 10^{-27}$ kg.

| | | | |
|---|---|---|---|
| | cal | calorie (energy equivalent) | 1 cal = 4.1868 J |
| (heat) capacity | J/K | joules per degree Kelvin (absorbed heat per change in temperature) | |
| specific heat | J/(K·kg) | joules per degree Kelvin per kilogram (heat capacity per unit mass) | |
| entropy | J/K | joules per degree Kelvin (contributed heat per temperature) | |
| electric charge | C | Coulomb, (current strength times time) | C = As |
| voltage | V | volt (electric power per current) | V = W/A |
| electric field strength | V/m | volts per meter (force per charge, voltage difference per length) | |
| magnetic flux | Wb | Weber (induced voltage in a coil per time) | Wb = Vs |
| magnetic field strength | T<br>G | tesla<br>Gauss<br>(magnetic flux per surface area) | $T = Wb/m^2$<br>$1\ G = 10^{-4}\ T$ |
| (the mean magnetic field strength of earth's magnetic field is roughly 0.5 Gauss) | | | |
| electric resistance | Ω | ohm (voltage per current strength) | Ω = V/A |
| (electric) capacity | F | Farad (charge per voltage) | F = C/V |
| induction | H | Henry (magnetic flux per current strength) | H = Wb/A |
| frequency | Hz | Hertz (number of oscillations per second) | $Hz = s^{-1}$ |

# *Fundamental constants in physics*

The values for the constants presented here as well as the error estimates are taken from the list provided by the *Task Group on Fundamental Constants of the Committee on Data for Science and Technology* (CODATA) of the Internatinal Council of Scientific Unions (ISCU) which was recommended for general use in science and technology.[9]

The numbers in parenthesis following one of the values denotes the error in the last places of that constant.

*Example:* The value $h = 6.6260755(40)$ means $h = 6.6260755 \pm 0.0000040$.

The errors indicated are the standard deviation of the value (cf. CODATA Bulletin No. 63, November 1986 and E. Cohen, B. Taylor, *Review of Modern Physics* **59**, 4 (1987)).

[9] For mathematical constants, see section 0.1.1 of the book.

| *Name* | *Symbol and formula* | *Numerical value without power* | *Power of value and unit* | *Relative error* |
|---|---|---|---|---|
| velocity of light in a vacuum | $c_0, c$ | 2.997 924 58 | $10^8 \mathrm{ms}^{-1}$ | 0 |
| magnetic field constant | $\mu_0 = 1/\varepsilon_0 c_0^2$ | $4\pi$ | $10^{-7}\mathrm{N} \cdot \mathrm{A}^{-2}$ | 0 |
| | | $= 1.256\,637\,061\,4\ldots$ | $10^{-6}\mathrm{N} \cdot \mathrm{A}^{-2}$ | |
| electromagnetic field constant | $\varepsilon_0 = 1/\mu_0 c_0^2$ | 8.854 187 817... | $10^{-12}\mathrm{F} \cdot \mathrm{m}^{-1}$ | 0 |
| gravitational constant | $G$ | 6.672 59(85) | $10^{-11}\ \mathrm{m}^3 \cdot \mathrm{kg}^{-1} \cdot \mathrm{s}^{-2}$ | $128 \cdot 10^{-6}$ |
| acceleration due to the earth's gravity | $g$ | 9.806 65 | $\mathrm{m} \cdot \mathrm{s}^{-2}$ | 0 |
| Planck action quantum, Planck constant | h | 6.626 075 5(40) | $10^{-34}\ \mathrm{J} \cdot \mathrm{s}$ | $6.0 \cdot 10^{-7}$ |
| | | 4.135 669 2(12) | $10^{-15}\ \mathrm{eV} \cdot \mathrm{s}$ | $3.0 \cdot 10^{-7}$ |
| | $\hbar = \mathrm{h}/2\pi$ | 1.054 572 66(63) | $10^{-34}\ \mathrm{J} \cdot \mathrm{s}$ | $6.0 \cdot 10^{-7}$ |
| | | 6.582 122 0(20) | $10^{-16}\ \mathrm{eV} \cdot \mathrm{s}$ | $3.0 \cdot 10^{-7}$ |
| elementary charge | $e$ | 1.602 177 33(49) | $10^{-19}\mathrm{C}$ | $3.0 \cdot 10^{-7}$ |
| | $e/\mathrm{h}$ | 2.417 988 36(72) | $10^{14}\mathrm{A} \cdot \mathrm{J}^{-1}$ | $3.0 \cdot 10^{-7}$ |
| (magnetic) field flux quantum | $\Phi_0 = \mathrm{h}/2e$ | 2.067 834 61(61) | $10^{-15}\mathrm{Wb}$ | $3.0 \cdot 10^{-7}$ |
| Josephson constant | $2e/\mathrm{h}$ | 4.835 976 7(14) | $10^{14}\mathrm{Hz} \cdot \mathrm{V}^{-1}$ | $3.0 \cdot 10^{-7}$ |
| von Klitzing constant | $\mathrm{h}/e^2$ | 2.581 280 56(12) | $10^4\Omega$ | $4.5 \cdot 10^{-8}$ |
| | $e^2/\mathrm{h}$ | 3.874 046 14(17) | $10^{-5}\Omega^{-1}$ | $4.5 \cdot 10^{-8}$ |
| Bohr-magneton | $\mu_B = e\hbar/2\mathrm{m_e}$ | 9.274 015 4(31) | $10^{-24}\mathrm{J} \cdot \mathrm{T}^{-1}$ | $3.4 \cdot 10^{-7}$ |
| | | 5.788 382 63(52) | $10^{-5}\mathrm{eV} \cdot \mathrm{T}^{-1}$ | $8.9 \cdot 10^{-8}$ |
| nucleus magneton | $\mu_N = e\hbar/2\mathrm{m_p}$ | 5.050 786 6(17) | $10^{-27}\mathrm{J} \cdot \mathrm{T}^{-1}$ | $3.4 \cdot 10^{-7}$ |
| | | 3.152 451 66(28) | $10^{-8}\mathrm{eV} \cdot \mathrm{T}^{-1}$ | $8.9 \cdot 10^{-8}$ |
| Sommerfeld fine structure constant | $\alpha = \mu_0 c_0 e^2/2\mathrm{h}$ | 7.297 353 08(33) | $10^{-3}$ | $4.5 \cdot 10^{-8}$ |
| | $\alpha^{-1}$ | 1.370 359 895(61) | $10^2$ | $4.5 \cdot 10^{-8}$ |
| | $\alpha^2$ | 5.325 136 20(48) | $10^{-5}$ | $9.0 \cdot 10^{-8}$ |
| Rydberg constant | $R_\infty = m_e c_0 \alpha^2/2\mathrm{h}$ | 1.097 373 153 4(13) | $10^7\ \mathrm{m}^{-1}$ | $1.2 \cdot 10^{-9}$ |
| | $R_\infty \mathrm{h} c_0$ | 2.179 874 1(13) | $10^{-18}\mathrm{J}$ | $6.0 \cdot 10^{-7}$ |
| | | 1.360 569 81(40) | $10^1\mathrm{eV}$ | $3.0 \cdot 10^{-7}$ |
| Bohr radius | $a_0 = \alpha/4\pi R_\infty$ | 0.529 177 249(24) | $10^{-10}\mathrm{m}$ | $4.5 \cdot 10^{-8}$ |
| circulation quantum | $\mathrm{h}/2m_e$ | 3.636 948 07(33) | $10^{-4}\mathrm{m}^2 \cdot \mathrm{s}^{-1}$ | $8.9 \cdot 10^{-8}$ |
| rest mass of an electron | $m_e$ | 9.109 389 7(54) | $10^{-31}\mathrm{kg}$ | $5.9 \cdot 10^{-7}$ |
| | | 5.485 799 03(13) | $10^{-4}\mathrm{u}$ | $2.3 \cdot 10^{-8}$ |
| its energy equivalent in eV | | 0.510 999 06(15) | $10^6\mathrm{eV}$ | $3.0 \cdot 10^{-7}$ |
| specific electron charge | $-e/m_e$ | -1.758 819 62(53) | $10^{11}\mathrm{C} \cdot \mathrm{kg}^{-1}$ | $3.0 \cdot 10^{-7}$ |
| Compton wavelength of an electron | $\lambda_C = \mathrm{h}/m_e c_0$ | 2.426 310 58(22) | $10^{-12}\mathrm{m}$ | $8.9 \cdot 10^{-8}$ |
| (classical) electron radius | $r_e = \alpha^2 a_0$ | 2.817 940 92(38) | $10^{-15}\mathrm{m}$ | $1.3 \cdot 10^{-7}$ |

| *Name* | *Symbol and formula* | *Numerical value without power* | *Power of value and unit* | *Relative error* |
|---|---|---|---|---|
| magnetic moment of an electron | $\mu_e$ | 9.284 770 1(31) | $10^{-24}$J · T$^{-1}$ | $3.4 \cdot 10^{-7}$ |
| | $\mu_e/\mu_B$ | 1.001 159 652 193(10) | | $10^{-11}$ |
| | $\mu_e/\mu_N$ | 1.838 282 000(37) | $10^3$ | $2.0 \cdot 10^{-8}$ |
| g-factor of an electron | $g_e = 2\mu_e/\mu_B$ | 2.002 319 304 386(20) | | $10^{-11}$ |
| rest mass of a mion | $m_\mu$ | 1.883 532 7(11) | $10^{-28}$kg | $6.1 \cdot 10^{-7}$ |
| | | 0.113 428 913(17) | u | $1.5 \cdot 10^{-7}$ |
| its energy equivalent in eV | | 1.056 583 89(34) | $10^8$eV | $3.2 \cdot 10^{-7}$ |
| ratio rest mass mion to rest mass electron | $m_\mu/m_e$ | 2.067 682 62(30) | $10^2$ | $1.5 \cdot 10^{-7}$ |
| magnetic moment of a mion | $\mu_\mu$ | 4.490 451 4(15) | $10^{-26}$J · T$^{-1}$ | $3.3 \cdot 10^{-7}$ |
| | $\mu_\mu/\mu_B$ | 4.841 970 97(71) | $10^{-3}$ | $1.5 \cdot 10^{-7}$ |
| | $\mu_\mu/\mu_N$ | 8.890 598 1(13) | | $1.5 \cdot 10^{-7}$ |
| rest mass of a proton | $m_p$ | 1.672 623 1(10) | $10^{-27}$kg | $5.9 \cdot 10^{-7}$ |
| | | 1.007 276 470 (12) | u | $1.2 \cdot 10^{-8}$ |
| its energy equivalent in eV | | 9.382 723 1(28) | $10^8$eV | $3.0 \cdot 10^{-7}$ |
| ration rest mass proton to rest mass electron | $m_p/m_e$ | 1.836 152 701 (37) | $10^3$ | $2.0 \cdot 10^{-8}$ |
| ratio rest mass proton to rest mass mion | $m_p/m_\mu$ | 8.880 244 4(13) | | $1.5 \cdot 10^{-7}$ |
| specific proton charge | $e/m_p$ | 9.578 830 9(29) | $10^7$C · kg$^{-1}$ | $3.0 \cdot 10^{-7}$ |
| Compton wavelength of a proton | $\lambda_{C,p} = \mathrm{h}/m_p c_0$ | 1.321 410 02(12) | $10^{-15}$m | $8.9 \cdot 10^{-8}$ |
| magnetic moment of a proton | $\mu_p$ | 1.410 607 61(47) | $10^{-26}$J · T$^{-1}$ | $3.4 \cdot 10^{-7}$ |
| | $\mu_p/\mu_B$ | 1.521 032 202(15) | $10^{-3}$ | $1.0 \cdot 10^{-8}$ |
| | $\mu_p/\mu_N$ | 2.792 847 386(63) | | $2.3 \cdot 10^{-8}$ |
| gyromagnetic ratio of a proton | $\gamma_p$ | 2.675 221 28(81) | $10^8$s$^{-1}$ · T$^{-1}$ | $3.0 \cdot 10^{-7}$ |
| rest mass of a neutron | $m_n$ | 1.674 928 6(10) | $10^{-27}$kg | $5.9 \cdot 10^{-7}$ |
| | | 1.008 664 904 (14) | u | $1.4 \cdot 10^{-8}$ |
| its energy equivalent in eV | | 9.395 656 3(28) | $10^8$eV | $3.0 \cdot 10^{-7}$ |
| ratio rest mass neutron to rest mass electron | $m_n/m_e$ | 1.838 683 662(40) | $10^3$ | $2.2 \cdot 10^{-8}$ |
| ratio rest mast neutron to rest mass proton | $m_n/m_p$ | 1.001 378 404(9) | | $0.9 \cdot 10^{-8}$ |
| Compton wavelength of a neutron | $\lambda_{C,n} = \mathrm{h}/m_n c_0$ | 1.319 591 10(12) | $10^{-15}$m | $8.9 \cdot 10^{-8}$ |
| magnetic moment of a neutron | $\mu_n$ | 0.996 237 07(40) | $10^{-26}$J · T$^{-1}$ | $4.1 \cdot 10^{-7}$ |
| | $\mu_n/\mu_B$ | 1.041 875 63(25) | $10^{-3}$ | $2.4 \cdot 10^{-7}$ |
| | $\mu_n/\mu_N$ | 1.913 042 75(45) | | $2.4 \cdot 10^{-7}$ |

| *Name* | *Symbol and formula* | *Numerical value without power* | *Power of value and unit* | *Relative error* |
|---|---|---|---|---|
| rest mass of a deuteron | $m_d$ | 3.343 586 0(20) | $10^{-27}$kg | $5.9 \cdot 10^{-7}$ |
| | | 2.013 553 214 (24) | u | $1.2 \cdot 10^{-7}$ |
| its energy equivalent in eV | | 1.875 613 39(57) | $10^{9}$eV | $3.0 \cdot 10^{-7}$ |
| ratio rest mass deuteron to rest mass electron | $m_d/m_e$ | 3.670 483 014(75) | $10^{3}$ | $2.0 \cdot 10^{-8}$ |
| ratio rest mass deuteron to rest mass proton | $m_d/m_p$ | 1.999 007 496(6) | | $0.3 \cdot 10^{-8}$ |
| magnetic moment of a deuteron | $\mu_d$ | 0.433 073 75(15) | $10^{-26}\mathrm{J} \cdot \mathrm{T}^{-1}$ | $3.4 \cdot 10^{-7}$ |
| | $\mu_d/\mu_B$ | 0.466 975 447 9(91) | $10^{-3}$ | $1.9 \cdot 10^{-8}$ |
| | $\mu_d/\mu_N$ | 0.857 438 230 (24) | | $2.8 \cdot 10^{-8}$ |
| Avogadro's number | $N_A$ | 6.022 136 7(36) | $10^{23}\ \mathrm{mol}^{-1}$ | $5.9 \cdot 10^{-7}$ |
| atomic mass constant | $m_a = m(^{12}C)/12$ | 1.660 540 2(10) | $10^{-27}$kg | $5.9 \cdot 10^{-7}$ |
| | | 1 | u | |
| its energy equivalent in eV | | 9.314 943 2(28) | $10^{8}$eV | $3.0 \cdot 10^{-7}$ |
| Faraday constant | $F = N_A \cdot e$ | 9.648 530 9(29) | $10^{4}\mathrm{C} \cdot \mathrm{mol}^{-1}$ | $3.0 \cdot 10^{-7}$ |
| molar Planck constant | $N_A \cdot \mathrm{h}$ | 3.990 313 23(36) | $10^{-10}\mathrm{J} \cdot \mathrm{s} \cdot \mathrm{mol}^{-1}$ | $8.9 \cdot 10^{-8}$ |
| | $N_A \cdot \mathrm{h}c_0$ | 0.119 626 58(11) | $\mathrm{J} \cdot \mathrm{m} \cdot \mathrm{mol}^{-1}$ | $8.9 \cdot 10^{-8}$ |
| universal (molar) gas constant | $R$ | 8.314 510(70) | $\mathrm{J} \cdot \mathrm{mol}^{-1} \cdot \mathrm{K}^{-1}$ | $8.4 \cdot 10^{-6}$ |
| Boltzmann constant | $k = R/N_A$ | 1.380 658 (12) | $10^{-23}\mathrm{J} \cdot \mathrm{K}^{-1}$ | $8.5 \cdot 10^{-6}$ |
| | | 8.617 385(73) | $10^{-5}\mathrm{eV} \cdot \mathrm{K}^{-1}$ | $8.4 \cdot 10^{-6}$ |
| (molar) volume of ideal gas per normed volume | $RT/p$ | | | · |
| $T = 273.15K$, $p = 101\,325Pa$ | $V_m$, $V_0$ | 2.241 410(19) | $10^{-2}\mathrm{m}^3 \cdot \mathrm{mol}^{-1}$ | $8.4 \cdot 10^{-6}$ |
| $T = 273.15K$, $p = 100kPa$ | $V'_m$ | 2.271 108(19) | $10^{-2}\mathrm{m}^3 \cdot \mathrm{mol}^{-1}$ | $8.4 \cdot 10^{-6}$ |
| Loschmidt constant | $n_0 = N_A/V_m$ | 2.686 763(23) | $10^{25}\mathrm{m}^{-3}$ | $8.5 \cdot 10^{-6}$ |
| Stefan–Boltzmann constant | $\sigma = (\pi^2/60)k^4/\hbar^3 c_0^2$ | 5.670 51(19) | $10^{-8}\mathrm{W} \cdot \mathrm{m}^{-2} \cdot \mathrm{K}^{-4}$ | $3.4 \cdot 10^{-5}$ |
| first Planck radiation constant | $c_1 = 2\pi \mathrm{h} c_0^2$ | 3.741 774 9(22) | $10^{-16}\mathrm{W} \cdot \mathrm{m}^2$ | $6.0 \cdot 10^{-7}$ |
| second Planck radiation constant | $c_2 = \mathrm{h}c_0/k$ | 1.438 769(12) | $10^{-2}\mathrm{m} \cdot \mathrm{K}$ | $8.4 \cdot 10^{-6}$ |
| constant of the Wien dislocation law | $b = \lambda_{\max} T$ | | | · |
| | $= c_2/4.965\,114\,23...$ | 2.897 756(24) | $10^{-3}\mathrm{m} \cdot \mathrm{K}$ | $8.4 \cdot 10^{-6}$ |

## *Decimal powers*

| *Power notation* | *Decimal notation* | *Prefix* | *Abbreviation* |
|---|---:|---|---|
| $10^1$ | 10 | deca | da |
| $10^2$ | 100 | hecto | h |
| $10^3$ | 1,000 | kilo | k |
| $10^6$ | 1,000,000 | mega | M |
| $10^9$ | 1,000,000,000 | giga | G |
| $10^{12}$ | 1,000,000,000,000 | tera | T |
| $10^{15}$ | 1,000,000,000,000,000 | peta | P |
| $10^{-1}$ | 0.1 | deci | d |
| $10^{-2}$ | 0.01 | centi | c |
| $10^{-3}$ | 0.001 | milli | m |
| $10^{-6}$ | 0.000 001 | micro | $\mu$ |
| $10^{-9}$ | 0.000 000 001 | nano | n |
| $10^{-12}$ | 0.000 000 000 001 | pico | p |
| $10^{-15}$ | 0.000 000 000 000 001 | femto | f |

## *The Greek Alphabet*

| | | | | | | | | |
|---|---|---|---|---|---|---|---|---|
| $\alpha$ | $A$ | alpha | $\iota$ | $I$ | iota | $\varrho$ | $P$ | rho |
| $\beta$ | $B$ | beta | $\kappa$ | $K$ | kappa | $\sigma$ | $\Sigma$ | sigma |
| $\gamma$ | $\Gamma$ | gamma | $\lambda$ | $\Lambda$ | lambda | $\tau$ | $T$ | tau |
| $\delta$ | $\Delta$ | delta | $\mu$ | $M$ | mu | $\upsilon$ | $\Upsilon$ | upsilon |
| $\varepsilon$ | $E$ | epsilon | $\nu$ | $N$ | nu | $\varphi$ | $\Phi$ | phi |
| $\zeta$ | $Z$ | zeta | $\xi$ | $\Xi$ | xi | $\chi$ | $X$ | chi |
| $\eta$ | $H$ | eta | $o$ | $O$ | omicron | $\psi$ | $\Psi$ | psi |
| $\vartheta$ | $\Theta$ | theta | $\pi$ | $\Pi$ | pi | $\omega$ | $\Omega$ | omega |

3980023